LDA	lithium diisopropylamide
LDMAN	lithium 1-(dimethylamino)naphthalenide
LHMDS	= LiHMDS
LICA	lithium isopropylcyclohexylamide
LiHMDS	lithium hexamethyldisilazide
LiTMP	lithium 2,2,6,6-tetramethylpiperidide
LTMP	= LiTMP
LTA	lead tetraacetate
lut	lutidine
m-CPBA	*m*-chloroperbenzoic acid
MA	maleic anhydride
MAD	methylaluminum bis(2,6-di-*t*-butyl-4-methylphenoxide)
MAT	methylaluminum bis(2,4,6-tri-*t*-butylphenoxide)
Me	methyl
MEK	methyl ethyl ketone
MEM	(2-methoxyethoxy)methyl
MIC	methyl isocyanate
MMPP	magnesium monoperoxyphthalate
MOM	methoxymethyl
MoOPH	oxodiperoxomolybdenum(pyridine)-(hexamethylphosphoric triamide)
mp	melting point
MPM	= PMB
Ms	mesyl (methanesulfonyl)
MS	mass spectrometry; molecular sieves
MTEE	methyl *t*-butyl ether
MTM	methylthiomethyl
MVK	methyl vinyl ketone
n	refractive index
NaHDMS	sodium hexamethyldisilazide
Naph	naphthyl
NBA	*N*-bromoacetamide
nbd	norbornadiene (bicyclo[2.2.1]hepta-2,5-diene)
NBS	*N*-bromosuccinimide
NCS	*N*-chlorosuccinimide
NIS	*N*-iodosuccinimide
NMO	*N*-methylmorpholine *N*-oxide
NMP	*N*-methyl-2-pyrrolidinone
NMR	nuclear magnetic resonance
NORPHOS	bis(diphenylphosphino)bicyclo[2.2.1]-hept-5-ene
Np	= Naph
PCC	pyridinium chlorochromate
PDC	pyridinium dichromate
Pent	*n*-pentyl
Ph	phenyl
phen	1,10-phenanthroline
Phth	phthaloyl
Piv	pivaloyl
PMB	*p*-methoxybenzyl
PMDTA	*N,N,N′,N″,N″*-p[illegible] triamine
PPA	polyphosphoric [illegible]
PPE	polyphosphate ester
PPTS	pyridinium *p*-toluenesulfonate
Pr	*n*-propyl
PTC	phase transfer catalyst/catalysis
PTSA	*p*-toluenesulfonic acid
py	pyridine
RAMP	(*R*)-1-amino-2-(methoxymethyl)pyrrolidine
rt	room temperature
salen	bis(salicylidene)ethylenediamine
SAMP	(*S*)-1-amino-2-(methoxymethyl)pyrrolidine
SET	single electron transfer
Sia	siamyl (3-methyl-2-butyl)
TASF	tris(diethylamino)sulfonium difluorotrimethylsilicate
TBAB	tetrabutylammonium bromide
TBAF	tetrabutylammonium fluoride
TBAD	= DBAD
TBAI	tetrabutylammonium iodide
TBAP	tetrabutylammonium perruthenate
TBDMS	*t*-butyldimethylsilyl
TBDPS	*t*-butyldiphenylsilyl
TBHP	*t*-butyl hydroperoxide
TBS	= TBDMS
TCNE	tetracyanoethylene
TCNQ	7,7,8,8-tetracyanoquinodimethane
TEA	triethylamine
TEBA	triethylbenzylammonium chloride
TEBAC	= TEBA
TEMPO	2,2,6,6-tetramethylpiperidinoxyl
TES	triethylsilyl
Tf	triflyl (trifluoromethanesulfonyl)
TFA	trifluoroacetic acid
TFAA	trifluoroacetic anhydride
THF	tetrahydrofuran
THP	tetrahydropyran; tetrahydropyranyl
Thx	thexyl (2,3-dimethyl-2-butyl)
TIPS	triisopropylsilyl
TMANO	trimethylamine *N*-oxide
TMEDA	*N,N,N′,N′*-tetramethylethylenediamine
TMG	1,1,3,3-tetramethylguanidine
TMS	trimethylsilyl
Tol	*p*-tolyl
TPAP	tetrapropylammonium perruthenate
TBHP	*t*-butyl hydroperoxide
TPP	tetraphenylporphyrin
Tr	trityl (triphenylmethyl)
Ts	tosyl (*p*-toluenesulfonyl)
TTN	thallium(III) nitrate
UHP	urea–hydrogen peroxide complex
Z	= Cbz

D0847236

Handbook of Reagents for Organic Synthesis

Reagents, Auxiliaries, and Catalysts for C–C Bond Formation

OTHER TITLES IN THIS COLLECTION

Oxidizing and Reducing Agents
Edited by Steven D. Burke and Rick L. Danheiser
ISBN 0 471 97926 0

Acidic and Basic Reagents
Edited by Hans J. Reich and James H. Rigby
ISBN 0 471 97925 2

Activating Agents and Protecting Groups
Edited by Anthony J. Pearson and William R. Roush
ISBN 0 471 97927 9

Handbook of Reagents for Organic Synthesis

Reagents, Auxiliaries, and Catalysts for C–C Bond Formation

Edited by

Robert M. Coates
University of Illinois at Urbana-Champaign

and

Scott E. Denmark
University of Illinois at Urbana-Champaign

JOHN WILEY & SONS LTD

Chichester • New York • Weinheim • Brisbane • Toronto • Singapore

Copyright © 1999 John Wiley & Sons Ltd
Baffins Lane, Chichester
West Sussex PO19 1UD, UK

National 01243 779777
International (+44) 1243 779777
e-mail (for orders and customer service enquiries): cs-books@wiley.co.uk
Visit our Home Page on
http://www.wiley.co.uk
or http:/www.wiley.com

Reprinted February 2000, February 2001

All rights reserved. No part of this publication may be reproduced, stored in a retrieval system, or transmitted, in any form or by any means, electronic, mechanical, photocopying, recording or otherwise, except under the terms of the Copyright Designs and Patents Act 1988 or under the terms of a license issued by the Copyright Licensing Agency, 90 Tottenham Court Road, London, W1P 9HE, UK, without the permission in writing of the publisher.

Other Wiley Editorial Offices

John Wiley & Sons Inc., 605 Third Avenue,
New York, NY 10158-0012, USA

Wiley-VCH Verlag GmbH, Pappelallee 3,
D-69469 Weinheim, Germany

Jacaranda Wiley Ltd, 33 Park Road, Milton,
Queensland 4064, Australia

John Wiley & Sons (Asia) Pte Ltd, 2 Clementi Loop #02-01,
Jin Xing Distripark, Singapore 129809

John Wiley & Sons (Canada) Ltd, 22 Worcester Road,
Rexdale, Ontario M9W 1L1, Canada

Library of Congress Cataloguing-in-Publication Data

Handbook of reagents for organic synthesis.
p. cm.
Includes bibliographical references.
Contents: [1] Reagents, auxiliaries, and catalysts for C–C bond formation / edited by Robert M. Coates and Scott E. Denmark [2] Oxidising and reducing agents / edited by Steven D. Burke and Riek L. Danheiser [3] Acidic and basic reagents / edited by Hans J. Reich and James H. Rigby [4] Activating agents and protecting groups / edited by Anthony J. Pearson and William R. Roush
ISBN 0-471-97924-4 (v. 1). ISBN 0-471-97926-0 (v. 2)
ISBN 0-471-97925-2 (v. 3) ISBN 0-471-97927-9 (v. 4)
1. Chemical tests and reagents. 2. Organic compounds–Synthesis.
QD77.H37 1999
547′.2 dc 21 98-53088
CIP

British Library Cataloguing in publication Data

A catalogue record for this book is available from the British Library

ISBN 0 471 97924 4

Typset by Thomson Press (India) Ltd., New Delhi
Printed and bound in Great Britian by Antony Rowe, Chippenham, Wilts
This book is printed on acid-free paper responsibly manufactured from sustainable forestry, in which at least two trees are planted for each one used in paper production.

This volume is dedicated to the memory of William G. Dauben, an inspiring mentor and forthright colleague. Wc acknowledge the authors of the original articles in the *Encyclopedia of Reagents for Organic Synthesis* whose work forms the large body of this new edition. We would like to thank Professor Jeremiah P. Freeman and *Organic Syntheses* for providing the original graphics for the *Organic Syntheses* procedures presented in Section 3C.

We appreciate the assistance of Shirley Pierson and Linda Hirsch for the reformatting and compilation of much of the introductory material.

Robert M. Coates and **Scott E. Denmark**

Editorial Board

Editor-in-Chief

Leo A. Paquette
The Ohio State University, Columbus, OH, USA

Editors

Steven D. Burke
University of Wisconsin at Madison, WI, USA

Scott E. Denmark
University of Illinois at Urbana-Champaign, IL, USA

Dennis C. Liotta
Emory University Atlanta, CA, USA

Robert M. Coates
University of Illinois at Urbana-Champaign, IL USA

David J. Hart
The Ohio State University Columbus, OH, USA

Anthony J. Pearson
Case Western Reserve University, Cleveland, OH, USA

Rick L. Danheiser
Massachusetts Institute of Technology, Cambridge, MA, USA

Lanny S. Liebeskind
Emory University Atlanta, GA, USA

Hans J. Reich
University of Wisconsin at Madison, WI, USA

James H. Rigby
Wayne State University Detroit, MI, USA

William R. Roush
Indiana University Bloomington, IN, USA

Assistant Editors

James P. Edwards
Ligand Pharmaceuticals San Diego, CA, USA

Mark Volmer
Emory University Atlanta, GA, USA

International Advisory Board

Leon A. Ghosez
Université Catholique de Louvain, Belgium

Chun-Chen Liao
National Tsing Hua University, Hsinchu, Taiwan

Ryoji Noyori
Nagoya University, Japan

Pierre Potier
CNRS, Gif-sur-Yvette France

Hishashi Yamomoto
Nagoya University, Japan

Jean-Marie Lehn
Université Louis Pasteur Strasbourg, France

Lewis N. Mander
Australian National University, Canberra Australia

Gerald Pattenden
University of Nottingham UK

W. Nico Speckamp
Universiteit van Amsterdam The Netherlands

Steven V. Ley
University of Cambridge UK

Giorgio Modena
Università di Padua Italy

Edward Piers
University of British Columbia, Vancouver Canada

Ekkehard Winterfeldt
Universität Hannover Germany

Managing Editor

Colin J. Drayton
Woking, Surrey, UK

Contents

* Numbers in bold indicate the class number to within the Reagent belongs.

Preface

As stated in its Preface, the major motivation for our undertaking publication of the *Encyclopedia of Reagents for Organic Synthesis* was "to incorporate into a single work a genuinely authoritative and systematic description of the utility of all reagents used in organic chemistry." By all accounts, this reference compendium has succeeded admirably in attaining this objective. Experts from around the globe contributed many relevant facts that define the various uses characteristic of each reagent. The choice of a masthead format for providing relevant information about each entry, the highlighting of key transformations with illustrative equations, and the incorporation of detailed indexes serve in tandem to facilitate the retrieval of desired information.

Notwithstanding these accomplishments, the editors have since recognized that the large size of this eight-volume work and its cost of purchase have often detered the placement of copies of the *Encyclopedia* in or near laboratories where the need for this type of information is most critical. In an effort to meet this demand in a cost-effective manner, the decision was made to cull from the major work that information having the highest probability for repeated consultation and to incorporate same into a set of handbooks. The latter would also be purchasable on a single unit basis.

The ultimate result of these deliberations is the publication of the *Handbook of Reagents for Organic Synthesis* consisting of the following four volumes:

Reagents, Auxiliaries, and Catalysts for C–C Bond Formation
edited by Robert M. Coates and Scott E. Denmark

Oxidizing and Reducing Agents
edited by Steven D. Burke and Rick L. Danheiser

Acidic and Basic Reagents
edited by Hans J. Reich and James H. Rigby

Activating Agents and Protecting Groups
edited by Anthony J. Pearson and William R. Roush

Each of the volumes contains a selected compilation of those entries from the original *Encyclopedia* that bear on the specific topic. Ample listings can be found to functionally related reagents contained in the original work. For the sake of current awareness, references to recent reviews and monographs have been included, as have relevant new procedures from *Organic Syntheses*.

The end product of this effort by eight of the original editors of the *Encyclopedia* is an affordable, enlightening set of books that should find their way into the laboratories of all practicing synthetic chemists. Every attempt has been made to be of the broadest synthetic relevance and our expectation is that our colleagues will share this opinion.

Leo A. Paquette
Columbus, Ohio USA

Introduction

Every practicing organic chemist recognizes the critical importance of selecting the most appropriate reagent and reaction conditions for executing a given chemical transformation. While a working knowledge of the most common reagents employed in organic synthesis is essential for every synthetic chemist, even the most dedicated practitioner can hardly claim fluency with the overwhelming array of inorganic and organic reagents now at his or her disposal. Moreover, the development of new and more selective reagents has accelerated exponentially and still constitutes one of the most vigorous areas of organic chemistry research. In recognition of this challenge, the Editors of the *Encyclopedia of Reagents for Organic Synthesis (EROS)* have selected approximately 500 of the most important and useful reagents employed in organic synthesis. In keeping with our goal to provide a concise, desktop reference work, we culled the most important and commonly used reagents from the over 3000 entries in the original *EROS*. These selections comprise the *Handbook of Reagents for Organic Synthesis* and cover the full range of chemical types and transformations. The *Handbook* is divided into four volumes that contain reagents of similar type and/or function to assist the user in rapidly locating a reagent of choice: *Reagents, Auxiliaries, and Catalysts for C–C Bond Formation; Oxidizing and Reducing Agents; Acidic and Basic Reagents*; and *Activation Agents and Protection Groups*.

This volume entitled *Reagents, Auxiliaries, and Catalysts for C–C Bond Formation* contains the largest and most diverse group of reagents by virtue of the broad scope of chemical reactions included in this general category. Two major types of reagents selected from the parent *EROS* are essential carbon-nucleophiles and carbon-electrophiles widely used to form C–C bonds in synthesis. The C-nucleophile group includes numerous organometallic reagents, carbanions, enolates, ylides, and their precursors. Familiar alkylating, acylating, and cyclopropanating reagents together with Michael acceptors and other electron-deficient olefins comprise a fundamental group of C-electrophiles. An indispensable subset of reagents for cycloadditions includes acetylenes, allenes, dienes, dienophiles, and ketenes. However, many other reagents extensively represented in this volume are critical participants in C–C bond forming reactions, which are not themselves incorporated into the products, such as catalysts, chiral auxiliaries, and selected heteroatom electrophiles. Particular attention was paid to the selection of the most important chiral catalysts and auxiliaries in recognition of the growing importance of enantioselective and diastereoselective transformations. Numerous transition-metal catalysts and stoichiometric metalloid reagents vital for C–C coupling reactions and cyclopropanations were deemed appropriate. While efforts were made to minimize duplication, a number of entries found in this volume have multiple applications and therefore appear in other volumes, e.g. 1,4-benzoquinone (see also *Oxidizing and Reducing Agents*), cerium trichloride (see also *Acidic and Basic Reagents*), chromium(II) chloride (see also *Oxidizing and Reducing Agents*), diazomethane (see also *Activating Agents and Protecting Groups*) and thexylborane (see also *Oxidizing and Reducing Agents*). On the other hand, some obvious redox reagents such as nickel(II) chloride, palladium(II) acetate and chloride, and manganese(III) acetate were chosen exclusively for this volume on account of their primary function in C–C bond formation. Although chlorotrimethylsilane, tri-*n*-butylchlorostannane, (*R*)- and (*S*)-menthyl *p*-toluenesulfinates, *p*-toluenesulfonyl azide, and *p*-toluenesulfonylhydrazide are not themselves used for C–C bond formation, their crucial role in the synthesis of C–C bond forming reagents provided a compelling rationale for their inclusion in this volume. In contrast, reagents such as the isomeric butyllithiums and pyrrolidine, while obviously closely associated with the generation of C–C bond forming reagents, are assigned exclusively to *Acidic and Basic Reagents* in recognition of their primary function as bases.

To familiarize the user with the spectrum of reagents contained in this volume and to organize the pertinent reference material, we have subdivided the 203 featured reagents into 22 classes based on their chemical structures. Clearly many reagents are multi-functional and could be logically assigned to several different classes. In many cases the class assignment was based upon the functionality that would normally be retained in the product of their common synthetic applications. Some examples are ethyl bromozincacetate (Class 9: "Enolates, Homoenolates, and Dicarbonyl Compounds"), lithium bis[dimethyl(phenyl)silyl]cuprate (Class 18: "Silicon and Tin Reagents"), and trifluoromethyltrimethylsilane (Class 11: "Halo Compounds"). All but one (nitromethane) of the designated "C_1 Reagents" in Class 3 are carbonyl compounds commonly utilized as electrophilic reagents to introduce oxygen functionality in C–C bond forming reactions (e.g. formaldehyde and *N,N*-dimethylformamide). On the other hand, the common electrophilic C_1 reagents used to introduce nitrogen functionality are found in Class 12: "Imines, Iminium ions, and Amide Acetals" (e.g. formaldehyde-dimethlyamine and dimethylformamide acetal). These classifications should afford conceptual and organizational guidance for the users. It is worth while to remind readers that although headings designate a single compound, often related reagents (e.g. other esters, enantiomers of chiral reagents) are also covered. The reagent articles are arranged alphabetically in the volume, as they are in the original *EROS*. In addition, the Table of Contents also shows the class number to aid readers in finding reagents of similar structure and function.

1. Acetylene and Allenes
2. Aluminum and Boron Reagents
3. C_1 Reagents
4. Chiral Auxiliaries and Reagents
5. Copper Reagents
6. Cyano, Isocyano, and Isocyanato Reagents
7. Diazo, Hydrazido, and Azido Reagents
8. Dienes, Dienophiles, and Michael Acceptors
9. Enolates, Homoenolates, and Dicarbonyl Compounds
10. Epoxides
11. Halo Compounds
12. Imines, Iminium Ions, and Amide Acetals
13. Ketenes, Ketene Acetals, and Ortho Esters
14. Lithium and Magnesium Reagents
15. Nickel Reagents
16. Palladium Reagents
17. Phosphorus Reagents
18. Silicon and Tin Reagents
19. Sulfur Reagents
20. Titanium Reagents
21. Transition Metal and Lanthanide Reagents
22. Zinc Reagents

Each of the reagent entries appears almost verbatim from the original work (*EROS*). However, to make this new work as current as possible and to provide additional, useful information, the section Editors have compiled lists of recent reviews (1992- and 1998) in the general areas of each Volume. In addition, a collection of recent procedures from *Organic Syntheses* (*Vol. 70–75*) that feature preparations, reactions or applications pertinent to the content of this volume is provided in the front material. The reagent classes have been used to good advantage for the organization of this additional material that appears after the Table of Contents. The recent reviews and selected *Organic Syntheses* procedures are sorted and presented under the appropriate rubric of the 22 reagent classes. The Editors have taken a liberal view in the selection of recently published reviews, chapters, and monographs concerning important C–C bond forming methods related to each reagent class, whether or not the specific reagents appear in this volume. Furthermore, the Editors have culled out only the most relevant chemical transformations from multistep *Organic Syntheses* procedures. Finally, while the bulk of the entries are directly reproduced from *EROS*, they now all contain a **Related Reagents** section that guides the reader to reagents of similar structure or function that can be found either in the other volumes or in the original work.

The Editors of the *Handbook* are pleased with the content and organization of each of the four volumes in this set and we hope that your chemical enterprises will benefit from the efforts of the original researchers, authors and editors who have contributed to the creation of this valuable new resource.

Robert M. Coates
University of Illinois at Urbana-Champaign

Scott E. Denmark
University of Illinois at Urbana-Champaign

CLASS 1 Acetylenes and Allenes

A. Reagents

Acetylene
Bis(trimethylsilyl)acetylene
Ethoxyacetylene
Lithium Acetylide
1-Methyl-1-(trimethylsilyl)allene
Trimethylsilylacetylene

B. Reviews (1992–1998)

(1) Gleiter, R. and Kratz, D. Conjugated Enediynes-An Old Topic in a Different Light. *Angew. Chem., Int. Ed. Engl.* **1993**, *32*, 842–45.

(2) Hopf, H. and Witulski, B. Cyanoalkynes: Magic Wands for the Preparation of Novel Aromatic Compounds. *Pure Appl. Chem.* **1993**, *65*, 47–56.

(3) Kanematsu, K. Molecular Design and Syntheses of Biologically Active Compounds Via Intramolecular Allene Cycloaddition Reaction Strategy. *Rev. Heteroatom. Chem.* MYU K.K.; Tokyo; **1993**, Vol. 9.

(4) Zimmer, R. Alkoxyallenes-Building Blocks in Organic Synthesis. *Synthesis* **1993**, 165–78.

(5) Magnus, P. A General Strategy Using $\eta^2 - CO_2(CO)_6$ Acetylene Complexes For the Synthesis of the Enediyne Antitumor Agents Esperamicin, Calicheamicin, Dynemicin, and Neocarzinostatin. *Tetrahedron* **1994**, *50*, 1397–418.

(6) Masuda, T. and Tachimori, H. Design, Synthesis, and Properties of Substituted Polyacetylenes. *J. Macromol. Sci. Chem.* **1994**, *31*, 1675–90.

(7) Nicolaou, K. The Magic of Enediyne Chemistry. *Chem. Britain* **1994**, *30*, 33–6.

(8) Casson, S. and Kocienski, P. The Hydrometallation, Carbometallation, and Metallometallation of Heteroalkynes. *Contemp. Org. Synth.* **1995**, *2*, 19–34.

(9) Stang, P. J. and Diederich, F. *Modern Acetylene Chemistry*; VCH: Weinheim, 1995.

C. Organic Syntheses Procedures (Vols. 70–75)

Authors	Citation	Title
Peter J. Stang and Tsugio Kitamura	*Org. Synth.* **1992**, *70*, 215.	ALKYNYL(PHENYL)IODONIUM TOSYLATES: PREPARATION AND STEREOSPECIFIC COUPLING WITH VINYLCOPPER REAGENTS. FORMATION OF CONJUGATED ENYNES. 1-HEXYNYL(PHENYL)-IODONIUM TOSYLATE AND (*E*)-5-PHENYLDODEC-5-EN-7-YNE

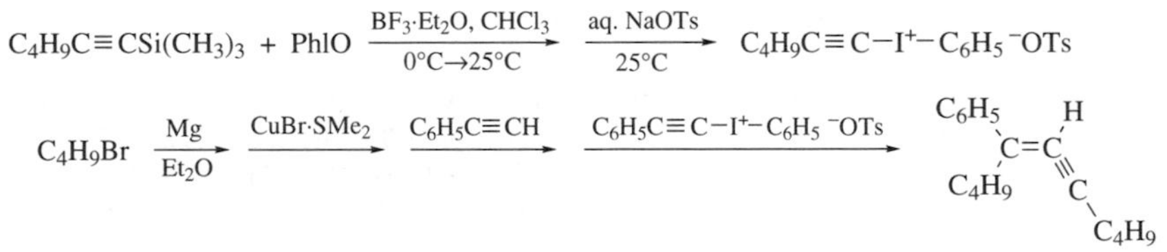

$$C_4H_9C{\equiv}CSi(CH_3)_3 + PhIO \xrightarrow[0°C\to25°C]{BF_3{\cdot}Et_2O,\ CHCl_3} \xrightarrow[25°C]{aq.\ NaOTs} C_4H_9C{\equiv}C{-}I^+{-}C_6H_5\ ^-OTs$$

$$C_4H_9Br \xrightarrow[Et_2O]{Mg} \xrightarrow{CuBr{\cdot}SMe_2} \xrightarrow{C_6H_5C{\equiv}CH} \xrightarrow{C_6H_5C{\equiv}C{-}I^+{-}C_6H_5\ ^-OTs} (C_6H_5)(C_4H_9)C{=}CH{-}C{\equiv}C{-}C_4H_9$$

CLASS 2 Aluminum and Boron Reagents

A. Reagents

Allenylboronic Acid
B-Allyldiisopinocampheylborane
9-Borabicyclo[3.3.1]nonane Dimer
Di-*n*-butylboryl Trifluoromethanesulfonate
Diisopinylcampheylboron Trifluoromethanesulfonate
Diisopropyl 2-Crotyl-1,3,2-dioxaborolane-4,5-dicarboxylate
Thexylborane
Triethyaluminum

Triethylborane
Trimethylaluminum

B. Reviews (1992–1998)

(1) Lohray, B. B. and Bhushan, V. Oxazaborolidines and Dioxaborolidines in Enantioselective Catalysis. *Angew. Chem., Int. Ed. Engl.* **1992**, *31*, 729–30.

(2) Deloux, L. and Srebnik, M. Asymmetric Boron-Catalyzed Reactions. *Chem. Rev.* **1993**, *93*, 763–84.

(3) Imamoto, T. Synthesis and Reactions of New Phosphine-Boranes. *Pure Appl. Chem.* **1993**, *65*, 655–60.

(4) Brown, H. C. and Ramachandran, P. V. Recent Advances in the Boron Route to Asymmetric Synthesis. *Pure Appl. Chem.* **1994**, *66*, 201–12.

(5) Kabalka, G. W. *Current Topics in the Chemistry of Boron*; Royal Society of Chemistry: Cambridge, U.K., 1994.

(6) Suzuki, A. New Synthetic Transformations Via Organoboron Compounds. *Pure Appl. Chem.* **1994**, *66*, 213–22.

(7) Brown, H. C. and Ramachandra, P. V. In *Advances in Asymmetic Synthesis*; JAI Press: Greenwich 1995; Vol. 1.

(8) Matteson, D. S. *Stereodirected Synthesis with Organoboranes*; Springer: Berlin, 1995.

(9) Suzuki, A. Haloboration of 1-Alkynes and Its Synthetic Application. *Rev. Heteroatom Chem.* **1997**, *17*, 271–314.

(10) Yamamoto, H. *Organoaluminum Compounds*; In *Organometallics in Synthesis – A Manual* Wiley: New York, 1994; Chapt. 7.

C. Organic Syntheses Procedures (Vols. 70–75)

Authors	Citation	Title
John A. Soderquist and Alvin Negron	*Org. Synth.* **1992**, *70*, 169.	9-BORABICYCLO[3.3.1]NONANE DIMER
Tatsuo Ishiyama, Norio Miyaura, and Akira Suzuki	*Org. Synth.* **1992**, *71*, 89.	PALLADIUM(0)-CATALYZED REACTION OF 9-ALKYL-9-BORABICYCLO[3.3.1]-NONANE WITH 1-BROMO-1-PHENYLTHIOETHENE: 4-(3-CYCLOHEXENYL)-2-PHENYLTHIO-1-BUTENE
T. Ooi, K. Maruoka, and H. Yamamoto	*Org. Synth.* **1993**, *72*, 95.	REARRANGEMENT OF *trans*-STILBENE OXIDE TO DIPHENYLACETALDEHYDE WITH CATALYTIC METHYLALUMINUM BIS(4-BROMO-2,6-DI-*tert*-BUTYLPHENOXIDE)
George W. Kabalka, John T. Maddox, Timothy Shoup, and Karla R. Bowers	*Org. Synth.* **1995**, *73*, 116.	A SIMPLE AND CONVENIENT METHOD FOR THE OXIDATION OF ORGANOBORANES USING SODIUM PERBORATE: (+)-ISOPINOCAMPHEOL

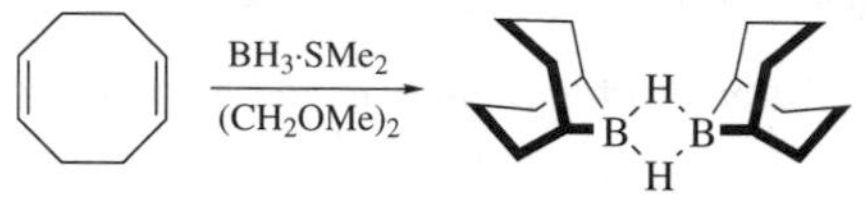

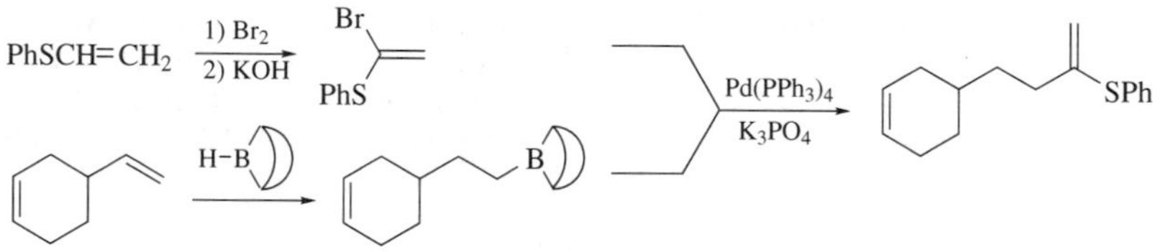

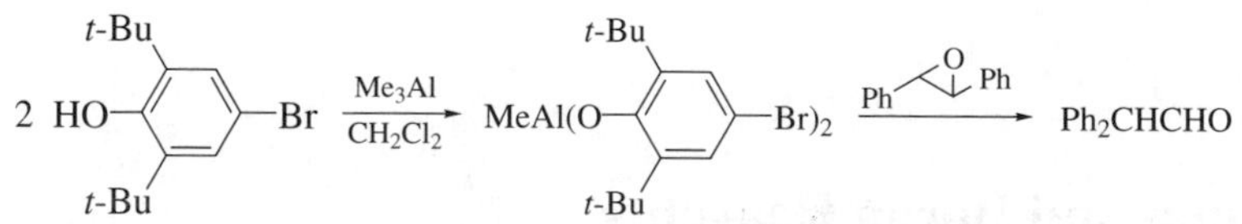

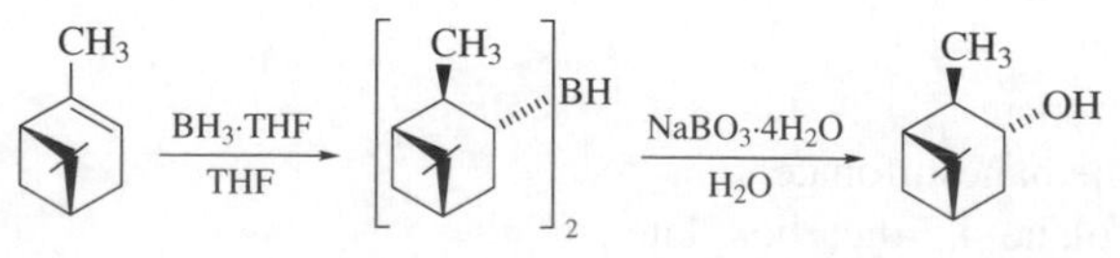

Shoji Hara and Akira Suzuki	*Org. Synth.* **1997**, *75*, 129	SYNTHESIS OF 4-(2-BROMO-2-PROPENYL)-4-METHYL-γ-BUTYROLACTONE BY THE REACTION OF ETHYL LEVULINATE WITH (2-BROMOALLYL) DIISOPROPOXYBORANE PREPARED BY HALOBORATION OF ALLENE

$H_2C=C=CH_2$ → (1) BBr_3; 2) i-Pr_2O) → Br, B(OPr-i)$_2$ → (CO_2Et) → Br, OH, COOEt → (CF_3COOH) → O, O, Br

CLASS 3 C_1 Reagents

A. *Reagents*

Carbon Monoxide
Diethyl Carbonate
Dimethoxycarbenium Tetrafluoroborate
N,*N*-Dimethylformamide
Formaldehyde
Methyl Cyanoformate
Methyl Formate
Nitromethane
Paraformaldehyde

B. *Reviews (1992–1998)*

(1) Sonoda, N. Selenium-Assisted Carbonylation With Carbon Monoxide. *Pure Appl. Chem.* **1993**, *65*, 699–706.

(2) Tsuda, T. Utilization of Carbon-Dioxide in Organic Synthesis and Polymer Synthesis by the Transition Metal Catalyzed Carbon-Dioxide Fixation into Unsaturated Hydrocarbons. *Gazz. Chim. Ital.* **1995**, *125*, 101–10.

(3) Ono, Y. Dimethyl Carbonate for Environmentally Benign Reactions. *Pure Appl. Chem.* **1996**, *68*, 367–76.

(4) Ryan, T. A. *Phosgene and Related Compounds*; Elsevier Science: New York, 1996.

(5) Ryu, I. and Sonoda, N. Free-Radical Carbonylations: Then and Now. *Angew. Chem., Int. Ed. Engl.* **1996**, *35*, 1050–66.

C. *Organic Syntheses Procedures (Vols. 70–75)*

Authors	Citation	Title
Simon R. Crabtree, Lewis N. Mander, and S. Paul Sethi	*Org. Synth.* **1992**, *70*, 256.	SYNTHESIS OF β-KETO ESTERS BY *C*-ACYLATION OF PREFORMED ENOLATES WITH METHYL CYANOFORMATE: PREPARATION OF METHYL (1α, 4aβ, 8aα)-2-OXO-DECAHYDRO-1-NAPHTHOATE

O → (Li, NH_3 (liq.); $(CH_3)_3COH$) → H, LiO, H → ($NCCO_2$ Me; Et_2O) → H, O, H, CO_2Me

X. Wang, S. O. deSilva, J. N. Reed, R. Billadeau, E. J. Griffen, A. Chan, and V. Snieckus	*Org. Synth.* **1993**, *72*, 163.	7-METHOXYPHTHALIDE

OMe, $CONEt_2$ → (1. s-BuLi/TMEDA; 2. DMF) → OMe, $CONEt_2$, CHO → (1. $NaBH_4$/MeOH; 2. 6M HCl) → OMe, O, O

CLASS 4 Chiral Auxiliaries and Reagents

A. *Reagents*

(*S*)-1-Amino-2-methoxymethylpyrrolidine
(*S*)-4-Benzyl-2-oxazolidinone
(*R*)- and (*S*)-2,2′-Bis(diphenylphosphino)-1,1′-binaphthyl
(*R*)-2-*t*-Butyl-6-methyl-4*H*-1,3-dioxin-4-one
(*R*,*R*)-2-*t*-Butyl-5-methyl-1,3-dioxolan-4-one
(*R*)-(+)-Butyl 2-(*p*-Tolylsulfinyl)acetate
10,2-Camphorsultam
Chloro(cyclopentadienyl)-
bis[3-*O*-(1,2;5,6-di-*O*-isopropylidene-α-D-glucofuransoyl)]titanium
10-Dicyclohexylsulfonamidoisoborneol
(*S*)-(+)-2,5-Dihydro-2-isopropyl-3,6-dimethylpyrazine
(*R**,*R**) − α-2,6-Diisopropylbenzyloxy-5-oxo-1,3,2-dioxaborolane-4-acetic Acid
(*S*)-*N*,*N*-Dimethyl-*N*′-(1-*t*-butoxy-3-methyl-2-butyl)formamidine
2,2′-(Dimethylmethylene)bis(4-*t*-butyl-2-oxazoline)
trans-2,5-Dimethylpyrrolidine
(*R*,*R*)-1,2 Diphenyl-1,2-diaminoethane, *N*,*N*′-Bis[3,5-bis(trifluoromethyl)benzenesulfonamide]
1*R*,2*S*-Ephedrine
(*R*)- and (*S*)-Ethyl 3-Hydroxybutyrate
(*S*)- and (*R*)-Ethyl lactate
3-Hydroxyisoborneol
(*S*)-(+)- and (*R*)-(−)-Mandelic Acid
p-(*S*)- and (*R*)-Menthyl Toluenesulfinate
(*S*)-2-Methoxymethylpyrrolidine
(1*R*,2*S*) − *N*-Methylephedrine
α-Methyltoluene-2,α-sultam
(*R*)-Pantolactone
(2*R*,4*R*)-2,4-Pentanediol
(−)-8-Phenylmenthol
1,1,2-Triphenyl-1,2-ethanediol

B. *Reviews (1992–1998)*

(1) Blaser, H.-U. The Chiral Pool as a Source of Enantioselective Catalysts and Auxiliaries. *Chem. Rev.* **1992**, *92*, 935–52.

(2) Braun, M. Recent Developments in Stereoselective Aldol Reactions; *Advances in Carbanion Chemistry*; JAI Press: Greenwich, CT, 1992, Vol. 1.

(3) Chelucci, G. Chiral Ligands Based on the Pyridine Framework: Synthesis and Application in Asymmetric Catalysis. *Gazz. Chim. Ital.* **1992**, *122*, 89–98.

(4) Faber, K. and Riva, S. Enzyme-Catalyzed Irreversible Acyl Transfer. *Synthesis* **1992**, 895–910.

(5) Halgas, J. *Biocatalysts in Organic Synthesis (Studies in Organic Chemistry)*; Elsevier: Amsterdam, 1992.

(6) Hayashi, T.; Kubo, A. and Ozawa, F. Catalytic Asymmetric Arylation of Olefins. *Pure Appl. Chem.* **1992**, *64*, 421–7.

(7) Inoue, Y. Asymmetric Photochemical Reactions in Solution. *Chem. Rev.* **1992**, *92*, 741–70.

(8) Knochel, P. Asymmetric Deprotonations as an Efficient Enantioselective Preparation of Functionalized Secondary Alcohols. *Angew. Chem., Int. Ed. Eng.* **1992**, *31*, 1459–61.

(9) Poppe, L. and Novak, L. *Selective Biocatalysis. A Synthetic Approach*; VCH: Weinheim, 1992.

(10) Rosini, C.; Franzini, L.; Raffaelli, A. and Salvadori, P. Synthesis and Applications of Binaphthylic C_2-Symmetry Derivatives as Chiral Auxiliaries in Enantioselective Reactions. *Synthesis* **1992**, 503–17.

(11) Santaniello, E.; Ferraboschi, P.; Grisenti, P. and Manzoocchi, A. The Biocatalytic Approach to the Preparation of Enantiomerically Pure Chiral Building Blocks. *Chem. Rev.* **1992**, *92*, 1071–140.

(12) Sawamura, M. and Ito, Y. Catalytic Asymmetric Synthesis by Means of Secondary Interaction Between Chiral Ligands and Substrates. *Chem. Rev.* **1992**, *92*, 857–781.

(13) Whitesell, J. K. Cyclohexyl-based Chiral Auxiliaries. *Chem. Rev.* **1992**, *92*, 953–64.

(14) Williams, R. M. Asymmetric Syntheses of α-Amino Acids. *Aldrichimica Acta* **1992**, *24*, 11–13 & 15–25.

(15) Banfi, L. and Guanti, G. Asymmetrized 2-Methyl-1,3-propanediol and its Equivalents: Preparation and Synthetic Applications. *Synthesis* **1993**, 1029–56.

(16) Brunner, H. and Zettlmeier, W. *Handbook of Enantioselective Catalysis*; VCH: Weinheim, 1993, Vols. 1 & 2.

(17) Danieli, B.; Lesma, G.; Passarella, D. and Riva, S. Chiral Synthons Via Enzyme-Mediated Asymmetrization of Meso-Compounds. *Advances in the Use of Synthons in Organic Chemistry*; JAI Press: Greenwich, 1993, Vol. 1.

(18) Fuji, K. Asymmetric Creation of Quaternary Carbon Centers. *Chem. Rev.* **1993**, *93*, 2037–66.

(19) Kim, B. H. and Curran, D. P. Asymmetric Thermal Reactions with Oppolzer's Camphor Sultam. *Tetrahedron* **1993**, *49*, 293–318.

(20) Kunz, H. and Rück, K. Carbohydrates as Chiral Auxiliaries in Stereoselective Synthesis. *Angew. Chem., Int. Ed. Engl.* **1993**, *32*, 336–58.

(21) Ojima, I. *Catalytic Asymmetric Synthesis*; VCH: New York, 1993.

(22) Pfaltz, A. Chiral Semicorrins and Related Nitrogen Heterocycles as Ligands in Asymmetric Catalysis. *Acc. Chem. Res.* **1993**, *26*, 339–45.

(23) Sakai, K. and Suemune, H. Application of Chiral Cyclic Diols to Asymmetric Synthesis. *Tetrahedron: Asymmetry* **1993**, *4*, 2109–18.

(24) Winterfeldt, E. Enantiomerically Pure Cyclopentadienes. *Chem. Rev.* **1993**, *93*, 827–43.

(25) Bach, T. Catalytic Enantioselective C–C Coupling – Allyl Transfer and Mukaiyama Aldol Reaction. *Angew. Chem., Int. Ed. Engl.* **1994**, *33*, 417–19.

(26) Brunner, H. Natural Products by Enantioselective Catalysis With Transition Metal Compounds. *Pure Appl. Chem.* **1994**, *66*, 2033–6.

(27) Cervinka, O. *Enantioselective Reactions in Organic Chemistry*; Horwood: London, 1994.

(28) Duthaler, R. O. Recent Developments in the Stereoselective Synthesis of α-Amino Acids. *Tetrahedron* **1994**, *50*, 1539–650.

(29) Franklin, A. S. and Paterson, I. Recent Developments in Asymmetric Aldol Methodology. *Contemp. Org. Synth.* **1994**, *1*, 317–38.

(30) Gant, T. G. and Meyers, A. I. The Chemistry of 2-Oxazolines (1985-present). *Tetrahedron* **1994**, *50*, 2297–360.

(31) Harada, T. and Oku, A. Enantiodifferentiating Transformation of Prochiral Polyols by Using Menthone as Chiral Template. *Synlett* **1994**, 95–104.

(32) Jones, J. B. Probing the Specificity of Synthetically Useful Enzymes. *Aldrichimica Acta* **1994**, *26*, 105–112.

(33) Nógrádi, M. *Stereoselective Synthesis: A Practical Approach, 2nd Ed.*; VCH: Weinheim, 1994.

(34) Noyori, R. *Asymmetric Catalysis in Organic Synthesis*; Wiley: New York, 1994.

(35) Atkinson, R. S. *An Introduction to Stereoselective Synthesis*; Wiley: New York, 1995.

(36) Kunz, H. Stereoselective Syntheses Using Carbohydrates as Chiral Auxiliaries. *Pure Appl. Chem.* **1995**, *67*, 1627–36.

(37) Waldmann, H. Amino Acid Esters: Versatile Chiral Auxiliary Groups for the Asymmetric Synthesis of Nitrogen Heterocycles. *Synlett* **1995**, 133–41.

(38) Williams, R. M. Asymmetric Synthesis of α-Amino Acids. *Advances in Asymmetic Synthesis*; JAI Press: Greenwich, 1995, Vol. 1.

(39) Ager, D. J. and East, M. B. *Asymmetric Synthetic Methodology*; CRC Press: Boca Raton, 1996.

(40) De Lucchi, O. High Symmetry Chiral Auxiliaries Containing Heteroatoms. *Pure Appl. Chem.* **1996**, *68*, 945–50.

(41) Enders, D. and Klatt, M. Asymmetric Synthesis with (*S*)-2-Methoxymethyl Pyrrolidine (SMP); A Pioneer Auxiliary. *Synthesis* **1996**, 1403–18.

(42) Gawley, R. E. and Aube, J. *Principles of Asymmetric Synthesis*; Elsevier: Amsterdam, 1996.

(43) Klabunovskii, E. I. Catalytic Asymmetric Synthesis of β-Hydroxyacids and Their Esters. *Russ. Chem. Rev.* **1996**, *65*, 329–44.

(44) Pfaltz, A. Design of Chiral Ligands for Asymmetric Catalysis: From C2-Symmetric Semicorrins and Bisoxazolines to Non-Symmetric Phosphinooxazolines. *Acta Chem. Scand.* **1996**, *50*, 189–94.

(45) Studer, A. Amino Acids and Their Derivatives as Stoichiometric Auxilaries in Asymmetric Synthesis. *Synthesis* **1996**, 793–815.

(46) Abiko, A. Isoxazolidine-Based Chiral Auxiliaries for Asymmetric Syntheses. *Rev. Heteroatom Chem.* **1997**, *17*, 51–72.

(47) Ager, D. J.; Prakash, I. and Schaad, D. R. Chiral Oxazolidinones in Asymmetric Synthesis. *Aldrichimica Acta* **1997**, *30*, 3–12.

(48) Bennani, Y. L. and Hanessian, S. *Trans*-1,2-Diaminocyclohexane Derivatives as Chiral Reagents, Scaffolds, and Ligands for Catalysis: Applications in Asymmetric Synthesis and Molecular Recognition. *Chem. Rev.* **1997**, *97*, 3161–95.

(49) Coppola, G. M. and Schuster, H. F. *Chiral α-Hydroxy Acids in Enantioselective Synthesis*; Wiley-VCH: Weinheim, 1997.

(50) Fessner, W. D. and Walter, C. Enzymatic C–C Bond Formation in Asymmetric Synthesis. *Top. Curr. Chem.* **1997**, *184*, 97–194.

(51) Hoppe, D. and Hense, T. Enantioselective Synthesis with Lithium/(−)-Sparteine Carbanion Pairs. *Angew. Chem., Int. Ed. Engl.* **1997**, *36*, 2282–316.

(52) Hultin, P. G.; Earle, M. A. and Sudharshan, M. Synthetic Studies with Carbohydrate-Derived Chiral Auxiliaries. *Tetrahedron* **1997**, *53*, 14823–70.

(53) Kiyooka, S. Development of a Chiral Lewis Acid-Promoted Asymmetric Aldol Reaction Using Oxazaborolidinone. *Rev. Heteroatom. Chem.* **1997**, *17*, 245–70.

(54) Seebach, D.; Sting, A. R. and Hoffmann, M. Self-Regeneration of Stereocenters (SRS); Applications, Limitations, and Abandonment of a Synthetic Principle. *Angew. Chem., Int. Ed. Engl.* **1997**, *35*, 2708–48.

(55) Ghosh, A. K.; Mathivanan, P. and Cappiello, J. C_2-Symmetric Chiral Bis(Oxazoline)-Metal Complexes in Catalytic Asymmetric Synthesis. *Tetrahedron: Asymmetry* **1998**, *9*, 1–45.

C. Organic Syntheses Procedures (Vols. 70–75)

Authors	Citation	Title
Philip Garner and Jung Min Park	*Org. Synth.* **1992**, *70*, 18	1,1,-DIMETHYLETHYL (*S*)- OR (*R*)-4-FORMYL-2,2-DIMETHYL-3-OXAZOLIDINECARBOXYLATE: A USEFUL SERINAL DERIVATIVE
S. Pikul and E. J. Corey	*Org. Synth.* **1992**, *71*, 22.	(1*R*,2*R*)-(+)- AND (1*S*,2*S*)-(−)-1,2-DIPHENYL-1,2-ETHYLENEDIAMINE
Dieter Seebach, Albert K. Beck, Richard Breitschuh, and Kurt Job	*Org. Synth.* **1992**, *71*, 39.	DIRECT DEGRADATION OF THE BIOPOLYMER POLY[(*R*)-3-HYDROXY-BUTYRIC ACID] TO (*R*)-3-HYDROXY-BUTANOIC ACID AND ITS METHYL ESTER
Larry E. Overman and Gilbert M. Rishton	*Org. Synth.* **1992**, *71*, 56.	3-(*S*)-[(*tert*-BUTYLDIPHENYLSILYL)-OXY]-2-BUTANONE

Larry E. Overman and Gilbert M. Rishton

Org. Synth. **1992**, *71*, 63.

STEREOCONTROLLED PREPARATION OF 3-ACYLTETRAHYDROFURANS FROM ACID-PROMOTED REARRANGEMENTS OF ALLYLIC KETALS: (2*S*,3*S*)-3-ACETYL-8-CARBOETHOXY-2,3-DIMETHYL-1-OXA-8-AZASPIRO[4.5]DECANE

C. Hubschwerlen and J.-L. Specklin

Org. Synth. **1993**, *72*, 1.

L-(*S*)-GLYCERALDEHYDE ACETONIDE

C. R. Schmid and J. D. Bryant

Org. Synth. **1993**, *72*, 6.

D-(*R*)-GLYCERALDEHYDE ACETONIDE

M. Braun, S. Gräf, and S. Herzog

Org. Synth. **1993**, *72*, 32.

(*R*)-(+)-2-HYDROXY-1,2,2-TRIPHENYLETHYL ACETATE

F. J. Lakner, K. S. Chu, G. R. Negrete, and J. P. Konopelski

Org. Synth. **1995**, *73*, 201.

SYNTHESIS OF ENANTIOMERICALLY PURE β-AMINO ACIDS FROM 2-*tert*-BUTYL-1-CARBOMETHOXY-2,3-DIHYDRO-4(1*H*)-PYRIMIDINONE: (*R*)-3-AMINO-3-(*p*-METHOXYPHENYL)-PROPIONIC ACID

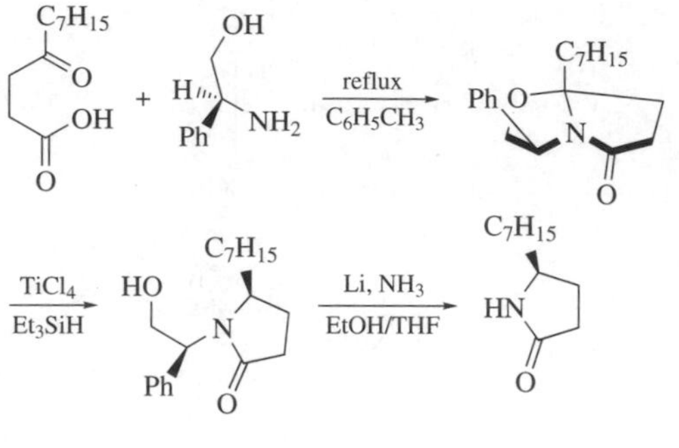

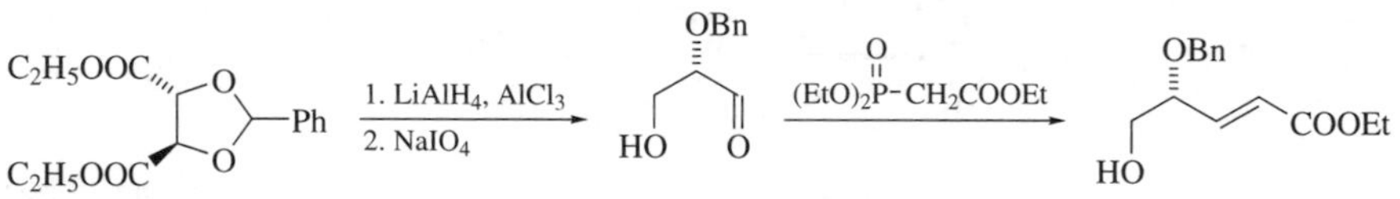

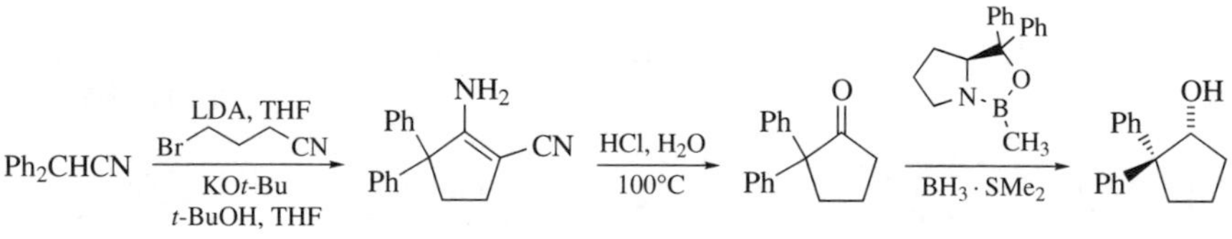

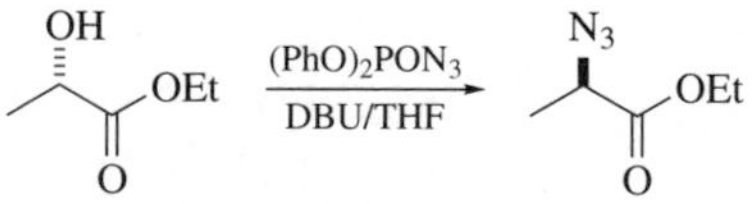

Authors	Citation	Title
M. A. Tschantz, L. E. Burgess, and A. I. Meyers	*Org. Synth.* **1995**, *73*, 221.	*S*-(−)-5-HEPTYL-2-PYRROLIDINONE and CHIRAL BICYCLIC LACTAMS AS TEMPLATES FOR PYRROLIDINES AND PYRROLIDINONES
B. Steuer, V. Wehner, A. Lieberknecht, and V. Jäger	*Org. Synth.* **1996**, *74*, 1.	(−)-2-*O*-BENZYL-L-GLYCERALDEHYDE AND ETHYL (*R*,*E*)-4-*O*-BENZYL-4,5-DIHYDROXY-2-PENTENOATE
Scott E. Denmark, Larry R. Marcin, Mark E. Schnute, and Atli Thorarensen	*Org. Synth.* **1996**, *74*, 33.	(*R*)-(−)-2,2-DIPHENYLCYCLOPENTANOL
Andrew S. Thompson, Frederick W. Hartner, Jr., and Edward J. J. Grabowski	*Org. Synth.* **1997**, *75*, 31	ETHYL (*R*)-2-AZIDOPROPIONATE
T. Akiba, O. Tamura, and S. Terashima	*Org. Synth.* **1997**, *75*, 45	(4*R*,5*S*)-4,5-DIPHENYL-3-VINYL-2-OXAZOLIDINONE

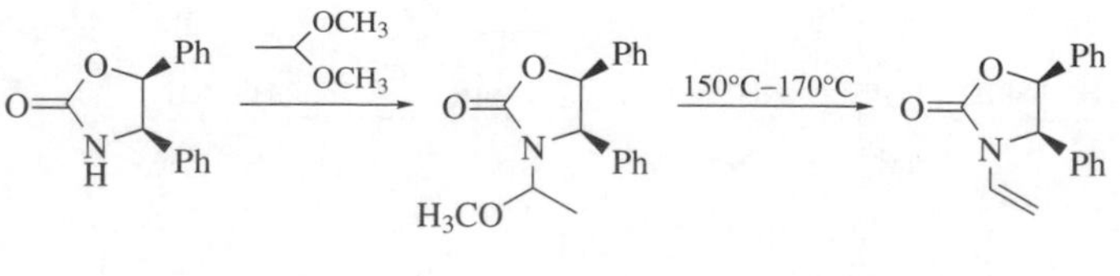

CLASS 5 Copper Reagents

A. Reagents

Copper(II) Acetylacetonate
Copper(I) Bromide
Copper(II) Bromide
Copper Bronze
Copper(I) Chloride
Copper(II) Chloride
Copper(I) Cyanide
Copper(I) Iodide
Dilithium Tetrachlorocuprate(II)
Lithium Bis(1-Ethoxyvinyl)cuprate
Lithium Dimethylcuprate
Lithium Dimethylcuprate–Boron Trifluoride
Lithium Divinylcuprate
Methylcopper–Boron Trifluoride Complex

B. Reviews (1992–1998)

(1) Ibuka, T. and Yamamoto, Y. New Aspects of Organocopper Reagents: 1,3- and 1,2-Chiral Induction and Reaction Mechanism. *Synlett.* **1992**, 769–77.

(2) Rossiter, B. E. and Swingle, N. M. Asymmetric Conjugate Addition. *Chem. Rev.* **1992**, *92*, 771–806.

(3) Negoro, T. and Oae, S. Ligand Coupling With Hypervalent Species II On Copper Atom. *Rev. Heteroatom Chem.* MYU K.K.; Tokyo, 1993, Vol. 9.

(4) Wipf, P. Transmetalation Reactions in Organocopper Chemistry. *Synthesis* **1993**, 537–57.

(5) Lipschutz, B. H.; Bhandari, A.; Lindsley, C.; Keil, R. and Wood, M. R. New Synthetic Methods Based on Organozirconium and Organocopper Chemistry. *Pure Appl. Chem.* **1994**, *66*, 1493–1500.

(6) Smith, R. A. J. and Vellekoop, A. S. 1,4-Addition Reactions of Organocuprates with α, β-Unsaturated Ketones. *Advances in Detailed Reaction Mechanisms*; JAI Press: Greenwich, 1994, Vol. 3.

(7) van Koten, G. Asymmetric Catalysis With Chiral Organocopper-Copper Arenethiolates. *Pure Appl. Chem.* **1994**, *66*, 1455–62.

(8) Lipshutz, B. H. Recent Progress in Higher Order Cyanocuprate Chemistry. *Advances in Metal-Organic Chemistry*; JAI Press: Greenwich, 1995, Vol. 4.

(9) Rossi, R.; Carpita, A. and Bellina, F. Palladium-and or Copper-Mediated Cross-Coupling Reactions Between 1-Alkynes and Vinyl, Aryl, 1-Alkynyl, 1,2-Propadienyl, Propargyl and Allylic Halides or Related Compounds. *Org. Prep. Proc. Int.* **1995**, *27*, 127–60.

(10) Taylor, R. J. K. *Organocopper Reagents: A Practical Approach*; Oxford University Press: Oxford, 1995.

(11) Lang, H. and Weinmann, M. Bis(Alkynyl) Titanocenes as Organometallic Chelating Ligands for the Stabilization of Monomeric Organo Copper(I) Compounds. *Synlett* **1996**, 1–10.

(12) Krause, N. and Gerold, A. Regioselective and Stereoselective Syntheses with Organocopper Reagents. *Angew. Chem., Int. Ed. Engl.* **1997**, *36*, 187–204.

C. Organic Syntheses Procedures (Vols. 70–75)

Authors	Citation	Title
Paul A. Wender, Alan W. White, and Frank E. McDonald	*Org. Synth.* **1992**, *70*, 204.	SPIROANNELATION VIA ORGANOBIS-(CUPRATES): 9,9-DIMETHYLSPIRO-[4.5]DECAN-7-ONE

Cl ~~~ Cl → (Li°, Et_2O, 0°C) Li ~~~ Li → (1. CuSPh, THF, −78°C to −15°C; 2. 3-chloro-5,5-dimethylcyclohex-2-enone, THF, −15°C) 9,9-dimethylspiro[4.5]decan-7-one

Authors	Citation	Title
G. Cahiez, S. Marquais, and M. Alami	*Org. Synth.* **1993**, *72*, 135.	MANGANESE-COPPER-CATALYZED CONJUGATE ADDITION OF ORGANO-MAGNESIUM REAGENTS TO α, β-ETHYLENIC KETONES: PREPARATION OF 2-(1,1-DIMETHYL-PENTYL)-5-METHYLCYCLOHEXANONE FROM PULEGONE

BuMgCl, 30%$MnCl_2$,3%CuCl; THF, 0°C

Authors	Citation	Title
Yoshiaki Horiguchi, Eiichi Nakamura, and Isao Kuwajima	*Org. Synth.* **1995**, *73*, 123.	DOUBLE HYDROXYLATION REACTION FOR CONSTRUCTION OF THE CORTICOID SIDE CHAIN: 16α-METHYLCORTEXOLONE

MeMgBr, cat. Cu^+, Me_3SiCl; HMPA, THF → Me_3SiO … 1. MCPBA (2.5 eq), $KHCO_3$ (6.4 eq); CH_2Cl_2; 2. H^+

CLASS 6 Cyano, Isocyano, and Isocyanato Reagents

A. *Reagents*

t-Butyl Isocyanide
Chlorosulfonyl Isocyanate
Cyanotrimethylsilane
Diethylaluminum Cyanide
Ethyl Isocyanoacetate
Hydrogen Cyanide
Lithium Cyanide
Potassium Cyanide
(*p*-Tolylsulfonyl)methyl Isocyanide

B. *Reviews (1992–1998)*

(1) Gorbatenko, V. I. Chemistry of Chlorocarbonyl Isocyanate. *Tetrahedron* **1993**, *49*, 3227–57.

(2) Marcaccini, S. and Torroba, T. The Use of Isocyanides in Heterocyclic Synthesis. *Org. Prep. Proc. Int.* **1993**, *25*, 141–208.

(3) North, M. Catalytic Asymmetric Cyanohydrin Synthesis. *Synlett.* **1993**, 807–20.

(4) Easton, C. J.; Hughes, C. M. M.; Savage, G. P. and Simpson, G. W. Cycloaddition Reactions of Nitrile Oxides With Alkenes. *Adv. Heterocycl. Chem.* **1994**, 261–328.

(5) Effenberger, F. Synthesis and Reactions of Optically Active Cyanohydrins. *Angew. Chem., Int. Ed. Engl.* **1994**, *33*, 1555–64.

(6) Lentz, D. Fluorinated Isocyanides – More Than Ligands With Unusual Properties. *Angew. Chem., Int. Ed. Engl.* **1994**, *33*, 1315–31.

(7) Ryu, I.; Sonoda, N. and Curran, D. P. Tandem Radical Reactions of Carbon Monoxide, Isonitriles, and Other Reagent Equivalents of the Geminal Radical Acceptor/Radical Precursor Synthon. *Chem. Rev.* **1996**, *96*, 177–94.

(8) Ulrich, H. *Chemistry and Technology of Isocyanates*; Wiley: Chichester, 1996.

C. Organic Syntheses Procedures (Vols. 70–75)

Authors	Citation	Title
Martine Bonin, David S. Grierson, Jacques Royer, and Henri-Philippe Husson	*Org. Synth.* **1992**, *70*, 54.	A STABLE CHIRAL 1,4-DIHYDROPYRIDINE EQUIVALENT FOR THE ASYMMETRIC SYNTHESIS OF SUBSTITUTED PIPERIDINES: 2-CYANO-6-PHENYLOXAZOLOPIPERIDINE

OH NH$_2$ + CHO CHO — 1) KCN, H_2O, pH 4; 1) CH_2Cl_2, $ZnBr_2$ → NC N O

Authors	Citation	Title
Jonathan L. Sessler, Azadeh Mozaffari, and Martin R. Johnson	*Org. Synth.* **1992**, *70*, 68.	3,4-DIETHYLPYRROLE AND 2,3,7,8,12, 13,17,18-OCTAETHYL-PORPHYRIN

O + O_2N — KF/*i*-PrOH → NO_2 OH — Ac_2O, H_2SO_4 → NO_2 OAc

— $C\equiv NCH_2CO_2Et$, DBU, *i*-PrOH → H N H CO_2Et — NaOH/$(CH_2OH)_2$, 190°C → H N H H

Authors	Citation	Title
Fen-Tair Luo, May-Wen Wang, and Ren-Tzong Wang	*Org. Synth.* **1997**, *75*, 146	PREPARATION OF CYANOALKYNES: 3-PHENYL-2-PROPYNENITRILE

$C\equiv CH$ — CuCN, NaI(cat.), DMSO/CH_3CN/H_2O, TMSCl, 50°C, 72 hr → $C\equiv C-CN$

CLASS 7 Diazo, Hydrazido, and Azido Reagents

A. Reagents

Benzenediazonium Tetrafluoroborate
Diazomethane
Diethyl (Diazomethyl)phosphonate
Ethyl Diazoacetate
p-Toluenesulfonyl Azide
p-Toluenesulfonylhydrazide
Trimethylsilyldiazomethane

B. Reviews (1992–1998)

(1) Bertrand, G. Bis(diisopropylamino)Phosphinyl-Diazomethane: A Building Block for the Synthesis of Stable Carbene and Nitrilimines. *Heteroatom Chem.* **1991**, *2*, 29–38.

(2) Khaskin, B. A.; Molodova, O. D. and Torgasheva, N. A. Reactions of Phosphorus-Containing Compounds With Diazo Compounds and Carbenes. *Russ. Chem. Rev.* **1992**, *61*, 306–334.

(3) Padwa, A. and Krumpe, K. E. Application of Intramolecular Carbenoid Reactions in Organic Synthesis. *Tetrahedron* **1992**, *48*, 5385–453.

(4) Zecchi, G. Chemistry of *C*-Azidohydrazones: Developments and Perspectives. *Synlett* **1992**, 858–60.

(5) Davies, H. M. L. Tandem Cyclopropanation/Cope Rearrangement: A General Method for the Construction of Seven-Membered Rings. *Tetrahedron* **1993**, *49*, 5203–23.

(6) Shapiro, E. A.; Dyatkin, A. B. and Nefedov, O. M. Carbene Reactions of Diazoesters With σ-Bonds as an Effective Method for the Alkoxycarbonylmethylenation of Organic Compounds. *Russ. Chem. Rev.* **1993**, *62*, 447.

(7) Shioiri, T. and Aoyama, T. Trimethylsilyldiazomethane. A Versatile Synthon for Organic Chemistry. *Advances in the Use of Synthons in Organic Chemistry*; JAI Press: Greenwich, 1993, Vol. 1.

(8) Sutton, D. Organometallic Diazo Compounds. *Chem. Rev.* **1993**, *93*, 995–1022.

(9) Tomilov, Y. V.; Dokichev, V. A.; Dzhemilev, U. M. and Nefedov, O. M. Catalytic Decomposition of Diazomethane as a General Method for the Methylenation of Chemical Compounds. *Russ. Chem. Rev.* **1993**, *62*, 803.

(10) Grishchuk, B. D.; Gorbovoi, P. M.; Ganushchak, N. I. and Dombrovskii, A. V. d. Reactions of Aromatic Diazonium Salts With Unsaturated Compounds in the Presence of Nucleophiles. *Russ. Chem. Rev.* **1994**, *63*, 257.

(11) Padwa, A. and Austin, D. J. Ligand Effects on the Chemoselectivity of Transition Metal-Catalyzed Reactions of α-Diazo Carbonyl Compounds. *Angew. Chem., Int. Ed. Engl.* **1994**, *33*, 1797–815.

(12) Ye, T. and McKervey, M. A. Organic Synthesis With α-Diazo Carbonyl Compounds. *Chem. Rev.* **1994**, *94*, 1091–160.

(13) Zollinger, H. D. *Diazo Chemistry I. Aromatic and Heteroaromatic Compounds*; VCH: Weinheim, 1994.

(14) Zollinger, H. D. *Diazo Chemistry II: Aliphatic, Inorganic and Organometallic Compounds*; VCH: Weinheim, 1995.

(15) Doyle, M. P. Chiral Dirhodium Carboxamidates: Catalysts for Highly Enantioselective Syntheses of Lactones and Lactams. *Aldrichimica Acta* **1996**, *29*, 3–11.

(16) Doyle, M. P.; McKervey, M. A. and Ye, T. *Modern Catalytic Methods for Organic Synthesis with Diazo Compounds*; Wiley-Interscience; New York, 1998.

C. Organic Syntheses Procedures (Vols. 70–75)

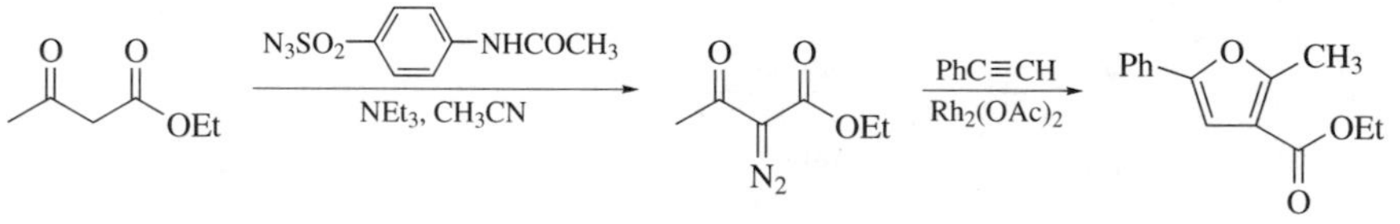

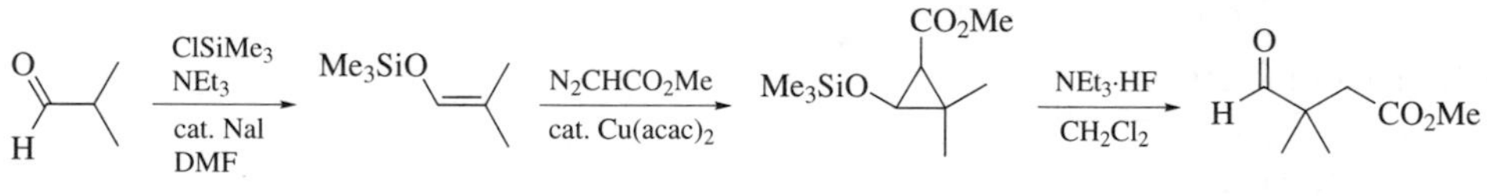

Authors	Citation	Title
Huw M. L. Davies, William R. Cantrell, Jr., Karen R. Romines, and Jonathan S. Baum	*Org. Synth.* **1992**, *70*, 93.	SYNTHESIS OF FURANS VIA RHODIUM(II) ACETATE-CATALYZED REACTION OF ACETYLENES WITH α-DIAZO CARBONYLS: ETHYL 2-METHYL-5-PHENYL-3-FURANCARBOXYLATE
Hans-Ulrich Reissig, Ingrid Reichelt, and Thomas Kunz	*Org. Synth.* **1992**, *71*, 189.	METHOXYCARBONYLMETHYLATION OF ALDEHYDES VIA SILOXYCYCLOPROPANES: METHYL 3,3-DIMETHYL-4-OXOBUTANOATE
Michael P. Doyle, William R. Winchester, Marina N. Protopopova, Amy P. Kazala, and Larry J. Westrum	*Org. Synth.* **1995**, *73*, 13.	(1*R*,5*S*)-(−)-6,6-DIMETHYL-3-OXABICYCLO[3.1.0]HEXAN-2-ONE. HIGHLY ENANTIOSELECTIVE INTRAMOLECULAR CYCLOPROPANATION CATALYZED BY DIRHODIUM(II) TETRAKIS[METHYL 2-PYRROLIDONE-5(*R*)-CARBOXYLATE]3

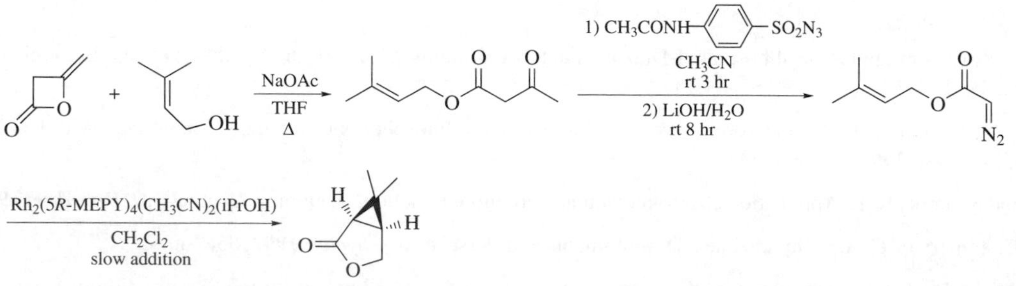

Rick L. Danheiser, Raymond F. Miller, and Ronald G. Brisbois — *Org. Synth.* **1995**, *73*, 134. — DETRIFLUOROACETYLATIVE DIAZO GROUP TRANSFER: (*E*)-1-DIAZO-4-PHENYL-3-BUTEN-2-ONE

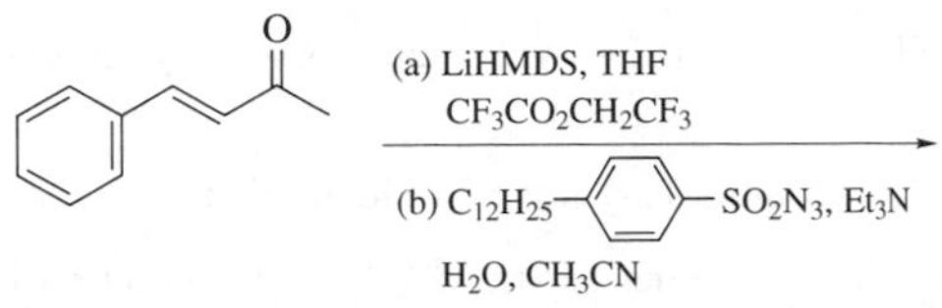

(a) LiHMDS, THF
$CF_3CO_2CH_2CF_3$
(b) $C_{12}H_{25}$–C_6H_4–SO_2N_3, Et_3N
H_2O, CH_3CN
N_2

G. G. Hazen, F. W. Bollinger, F. E. Roberts, W. K. Russ, J. J. Seman, and S. Staskiewicz — *Org. Synth.* **1995**, *73*, 144. — 4-DODECYLBENZENESULFONYL AZIDES

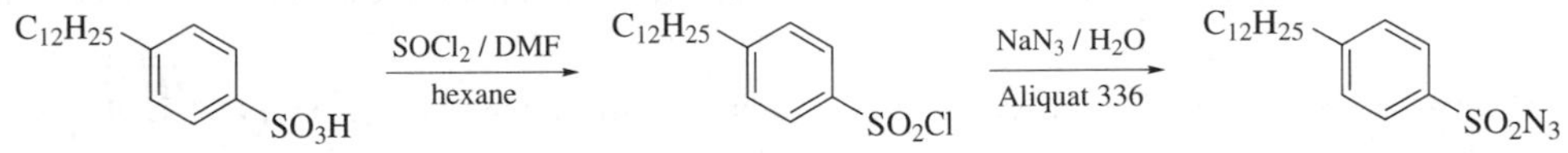

$C_{12}H_{25}$ SO_3H → $SOCl_2$ / DMF, hexane → $C_{12}H_{25}$ SO_2Cl → NaN_3 / H_2O, Aliquat 336 → $C_{12}H_{25}$ SO_2N_3

Jack R. Reid, Richard F. Dufresne, and John J. Chapman — *Org. Synth.* **1996**, *74*, 217 — MESITYLENESULFONYLHYDRAZINE, AND (1α, 2α, 6β)-2,6-DIMETHYLCYCLOHEXANECARBONITRILE AND (1α, 2β, 6α)-2,6-DIMETHYLCYCLOHEXANECARBONITRILE AS A RACEMIC MIXTURE

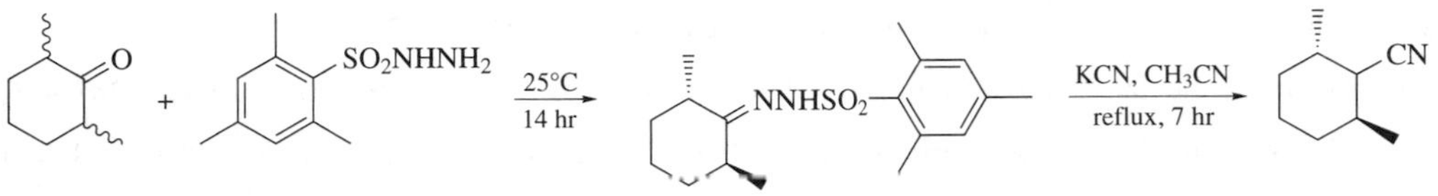

O + SO_2NHNH_2 → 25°C, 14 hr → $NNHSO_2$ → KCN, CH_3CN, reflux, 7 hr → CN

Joshua S. Tullis and Paul Helquist — *Org. Synth.* **1996**, *74*, 229 — RHODIUM-CATALYZED HETEROCYCLOADDITION OF A DIAZOMALONATE AND A NITRILE: 4-CARBOMETHOXY-5-METHOXY-2-PHENYL-1,3-OXAZOLE

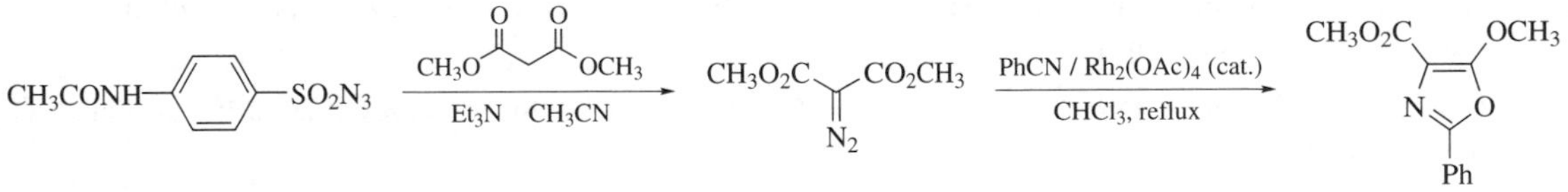

CH_3CONH–C_6H_4–SO_2N_3 → CH_3O–CO–CH_2–CO–OCH_3, Et_3N CH_3CN → CH_3O_2C–C(N_2)–CO_2CH_3 → PhCN / $Rh_2(OAc)_4$ (cat.), $CHCl_3$, reflux → CH_3O_2C, OCH_3, N, O, Ph

CLASS 8 Dienes, Dienophiles, and Michael Acceptors

A. Reagents

Acrylonitrile
1,4-Benzoquinone
1,3-Butadiene
2-Chloroacrylonitrile
Cyclopentadiene
Dimethyl Acetylenedicarboxylate
Maleic Anhydride
1-Methoxy-3-trimethylsilyloxy-1,3-butadiene
Methyl Acrylate
Methyl Vinyl Ketone
Nitroethylene
2,2,6-Trimethyl-4H-1,3-dioxin-4-one

1-Trimethylsilyloxy-1,3-butadiene
2-Trimethylsilyloxy-1,3-butadiene

B. Reviews (1992–1998)

(1) Afarinkie, K.; Vinader, V.; Nelson, T. D. and Posner, G. H. Diels-Alder Cycloadditions of 2-Pyrones and 2-Pyridones. *Tetrahedron* **1992**, *48*, 9111–71.

(2) Bryce, M. R.; Becher, J. and Fält-Hansen, B. Heterocyclic Synthesis Using New Heterodienophiles. *Adv. Heter. Cycl. Chem.* **1992**, *55*, 2.

(3) d'Angelo, J.; Desmaele, D; Dumas, F. and André, G. The Asymmetric Michael Addition Reaction Using Chiral Imines. *Tetrahedron: Asymmetry* **1992**, *3*, 456–505.

(4) Kagan, H. B. and Riant, O. Catalytic Asymmetric Diels-Alder Reactions. *Chem. Rev.* **1992**, *92*, 1007–19.

(5) Perlmutter, P. *Conjugate Addition Reactions in Organic Synthesis*; Pergamon Press: Oxford, 1992.

(6) Arai, Y. and Koizumi, T. Chiral Sulfinylethenes as Efficient Dienophiles for Asymmetric Diels-Alder Reactions. *Sulfur Reports* **1993**, *15*, 41–65.

(7) Barluenga, J. and Tomas, M. Synthesis of Heterocycles From Azadienes. *Adv. Heter. Cycl. Chem.* **1993**, *57*, 1–80.

(8) Pindur, U.; Lutz, G. and Otto, C. Acceleration and Selectivity Enhancement of Diels-Alder Reactions by Special and Catalytic Methods. *Chem. Rev.* **1993**, *93*, 741–61.

(9) Ripoll, J.-L. Synthetic Applications of the Retro-Ene Reaction. *Synthesis*. **1993**, 659–77.

(10) Ando, K. and Takayama, H. Heteroaromatic-Fused 3-Sulfolenes. *Heterocycles* **1994**, *37*, 1417–39.

(11) Coxon, J. A.; McDonald, D. Q. and Steel, P. J. Diastereofacial Selectivity in the Diels-Alder Reaction. *Advances in Detailed Reaction Mechanisms*; JAI Press: Greenwich, 1994, Vol. 3.

(12) Knight, D. W. Synthetic Approaches to Butenolides. *Contemp. Org. Synth.* **1994**, *1*, 287.

(13) Leonard, J. Control of Asymmetry Through Conjugate Addition Reactions. *Contemp. Org. Synth.* **1994**, *1*, 387–416.

(14) Oh, T. and Reilly, M. Reagent-Controlled Asymmetric Diels-Alder Reactions. *Org. Prep. Proc. Int.* **1994**, *26*, 129–58.

(15) Pearson, A. J. Cyclohexadienes. *Sec. Supp. 2nd Ed. Rodd's Chem. Carbon Compd.*; Elsevier: Amsterdam, 1994, 2.

(16) Perekalin, V. V.; Lipina, E. S.; Berestovitskaya, V. M. and Efremov, D. A. *Nitroalkenes: Conjugated Nitro Compounds*; Wiley: Chichester, 1994.

(17) Sainsbury, M. Cyclopentadiene *Sec. Supp. 2nd Ed. Rodd's Chem. Carbon Compd.*; Elsevier: Amsterdam, 1994.

(18) Waldmann, H. Asymmetric Hetero-Diels-Alder Reactions. *Synthesis* **1994**, 535–551.

(19) Bruce, J. M. Benzoquinones and Related Compounds. *Sec. Supp. 2nd Ed. Rodd's Chem. Carbon Compd. Vol.* $III^B/III^C/III^D$ *(partial): Aromatic Compounds*; Elsevier: Amsterdam, 1995.

(20) Casiraghi, G. and Rassu, G. Furan-, Pyrrole-, and Thiophene-Based Siloxydienes for Syntheses of Densely Functionalized Homochiral Compounds; *Synthesis* **1995**, 607–26.

(21) Gololobov, Y. G. and Krylova, T. O. 2-Cyanoacrylates as Reagents in Heteroatomic Synthesis. *Heteroatom. Chem.* **1995**, *6*, 271–80.

(22) Nielsen, A. T. *Nitrocarbons*; VCH: New York, 1995.

(23) Robertson, G. Nitro and Related Compounds. *Contemp. Org. Synth.* **1995**, *2*, 357–63.

(24) Basavaiah, D.; Rao, P. D. and Hyma, R. S. The Baylis-Hillman Reaction; A Novel Carbon-Carbon Bond-Forming Reaction. *Tetrahedron* **1996**, *52*, 8001–62.

(25) Boger, D. L. Cycloaddition Reactions of Azadienes, Cyclopropenone Ketals, and Related Systems: Scope and Applications. *CHEMTRACTS: Organic Chemistry* **1996**, *9*, 149–89.

(26) Collins, I. Saturated and Unsaturated Lactones. *Contemp. Org. Synth.* **1996**, *3*, 295–321.

(27) Denmark, S. E. and Thorarensen, A. Tandem [4 + 2]/[3 + 2] Cycloadditions of Nitroalkenes. *Chem. Rev.* **1996**, *96*, 137–65.

(28) Gallagher, P. T. The Synthesis of Quinones. *Contemp. Org. Synth.* **1996**, *3*, 433–46.

(29) Gilchrist, T. L., Gonsalves, A. M. d. A. R. and Pinho e Melo, T. M. V. D. The Use of 2-Azadienes in the Diels-Alder Reaction. *Pure Appl. Chem.* **1996**, *68*, 859–62.

(30) Padwa, A.; Ni, Z. and Watterson, S. H. Cycloaddition Reactions of Unsaturated Sulfones. *Pure Appl. Chem.* **1996**, *68*, 831–6.

(31) Winkler, J. D. Tandem Diels-Alder Cycloadditions in Organic Synthesis. *Chem. Rev.* **1996**, *96*, 167–76.

(32) Adams, J. P. and Box, D. S. Nitro and Related Compounds. *Contemp. Org. Synth.* **1997**, *4*, 415–34.

(33) Aversa, M. C.; Barattucci, A.; Bonaccorsi, P. and Giannetto, P. Chiral Sulfinyl-1,3-dienes. Synthesis and Use in Asymmetric Reactions. *Tetrahedron: Asymmetry* **1997**, *8*, 1339–67.

(34) Dias, L. C. Chiral Lewis Acid Catalysts in Diels-Alder Cycloadditions: Mechanistic Aspects and Synthetic Applications of Recent Systems. *Brazil. Chem. Soc.* **1997**, *8*, 289–332.

(35) Gololobov, Y. G. and Gruber, W. 2-Cyanoacrylates. Synthesis, Properties and Applications. *Russ. Chem. Rev.* **1997**, *66*, 953–62.

(36) Kaberdin, R. V., Potkin, V. I. and Zapol'skii, V. A. Nitrobutadienes and Their Halogen Derivatives: Synthesis and Reactions. *Russ. Chem. Rev.* **1997**, *66*, 827–42.

(37) Winterfeldt, E.; Borm, C. and Nerenz, F. Preparation and Application of Chiral Cyclopentadienes. *Advances in Asymmetric Synthesis*; JAI Press: Greenwich, 1997, Vol. 2.

C. Organic Syntheses Procedures (Vols. 70–75)

Authors	Citation	Title
G. Revial and M. Pfau	*Org. Synth.* **1992**, *70*, 35.	(*R*)-(−)-10-METHYL-1(9)-OCTAL-2-ONE

(±) + NH_2 Ph H → [N, Ph H] → $H_2O/AcOH$ → MeONa/MeOH →

Authors	Citation	Title
Dale L. Boger, James S. Panek, and Mona Patel	*Org. Synth.* **1992**, *70*, 79.	PREPARATION AND DIELS-ALDER REACTION OF A REACTIVE, ELECTRON-DEFICIENT HETEROCYCLIC AZADIENE: DIMETHYL 1,2,4,5-TETRAZINE-3,6-DICARBOXYLATE. 1,2-DIAZINE AND PYRROLE INTRODUCTION

CO_2CH_3 N N N N CO_2CH_3 + $OSiMe_3$ —dioxane, 25 °C→ Ph CO_2CH_3 N N CO_2CH_3 —Zn, CH_3COOH→ Ph CO_2CH_3 NH CO_2CH_3

Authors	Citation	Title
David C. Myles and Mathew H. Bigham	*Org. Synth.* **1992**, *70*, 231.	PREPARATION OF (*E*,*Z*-1-METHOXY-2-METHYL-3-(TRIMETHYLSILOXY)-1,3-PENTADIENE

O —1. HCO_2Et, NaH, benzene; 2. H_3O^+→ O, OH —MeOH, TsOH, PhH→ O, OMe —TMSOTf, Et_3N, ether→ OTMS, OMe

S. Pikul and E. J. Corey

Org. Synth. **1992**, *71*, 30.

ENANTIOSELECTIVE, CATALYTIC DIELS-ALDER REACTION: (1*S*)-*endo*-3-(BICYCLO[2.2.1]HEPT-5-EN-2-YLCARBONYL)-2-OXAZOLIDINONE

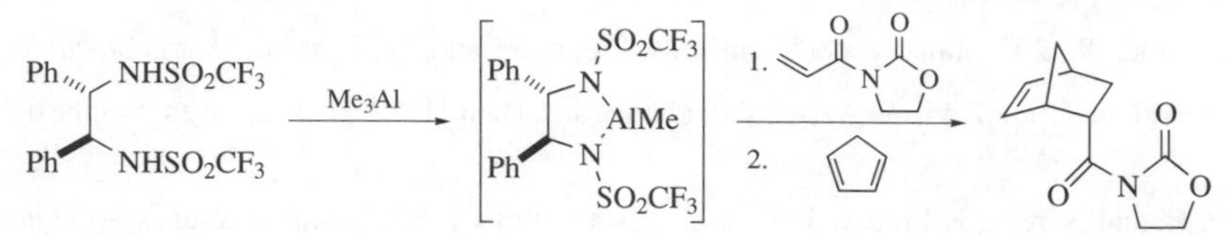

Ph, $NHSO_2CF_3$ / Ph $NHSO_2CF_3$ → (Me_3Al) [Ph, Ph, N(SO_2CF_3), AlMe, N(SO_2CF_3)] → 1. (acryloyl oxazolidinone) 2. (cyclopentadiene)

Karl R. Dahnke and Leo A. Paquette

Org. Synth. **1992**, *71*, 181.

INVERSE ELECTRON-DEMAND DIELS-ALDER CYCLOADDITION OF A KETENE DITHIOACETAL. COPPER HYDRIDE-PROMOTED REDUCTION OF A CONJUGATED ENONE. 9-DITHIOLANOBICYCLO[3.2.2]NON-6-EN-2-ONE FROM TROPONE

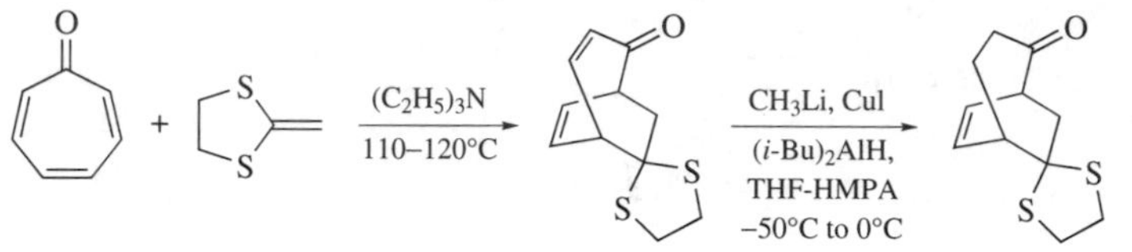

Tropone + ketene dithioacetal → ($(C_2H_5)_3N$, 110–120°C) → (CH_3Li, CuI, $(i\text{-Bu})_2AlH$, THF-HMPA, −50°C to 0°C)

Clarisse Mühlemann, Peter Hartmann, and Jean-Pierre Obrecht

Org. Synth. **1992**, *71*, 200.

TITANIUM-MEDIATED ADDITION OF SILYL DIENOL ETHERS TO ELECTROPHILIC GLYCINE: A SHORT SYNTHESIS OF 4-KETOPIPECOLIC ACID HYDROCHLORIDE

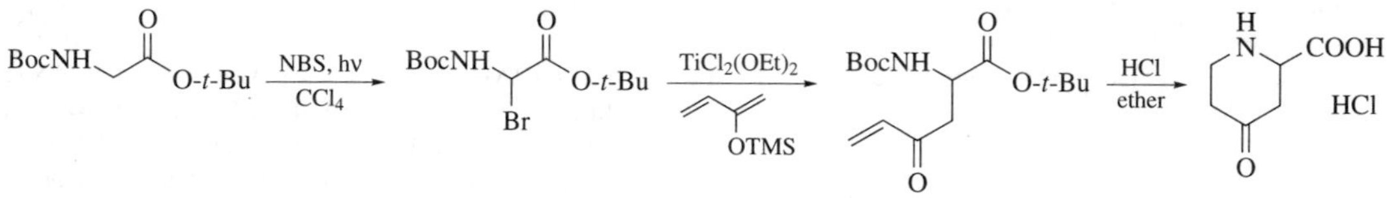

BocNH–CH$_2$–CO–O-*t*-Bu → (NBS, hν, CCl_4) → BocNH–CH(Br)–CO–O-*t*-Bu → ($TiCl_2(OEt)_2$, OTMS diene) → BocNH ... O-*t*-Bu → (HCl, ether) → 4-oxopiperidine-2-COOH · HCl

G. H. Posner, K. Afarinkia, and H. Dai

Org. Synth. **1995**, *73*, 231.

AN IMPROVED PREPARATION OF 3-BROMO-2(*H*)-PYRAN-2-ONE: AN AMBIPHILIC DIENE FOR DIELS-ALDER CYCLOADDITIONS

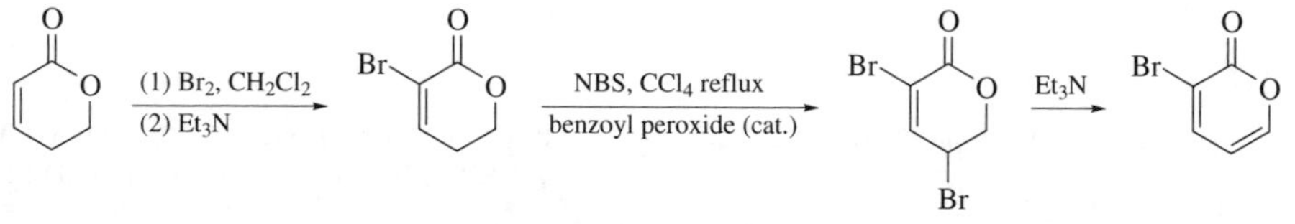

→ ((1) Br_2, CH_2Cl_2 (2) Et_3N) → Br ... → (NBS, CCl_4 reflux, benzoyl peroxide (cat.)) → Br, Br → (Et_3N) → Br

Masaji Oda, Takeshi Kawase, Tomoaki Okada, and Tetsuya Enomoto

Org. Synth. **1995**, *73*, 253.

2-CYCLOHEXENE-1,4-DIONE

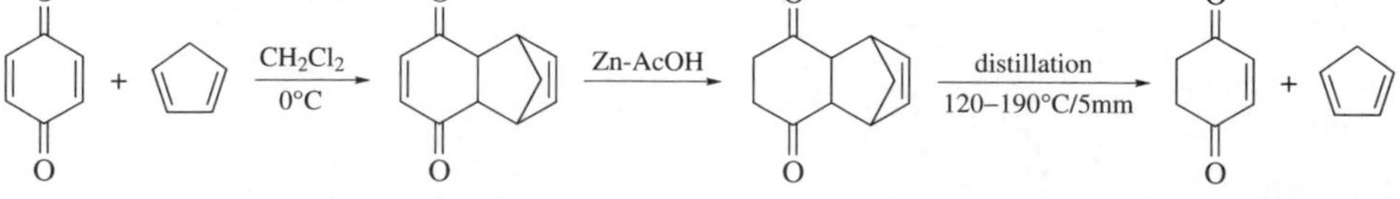

+ → (CH_2Cl_2, 0°C) → (Zn-AcOH) → (distillation, 120–190°C/5mm) → +

Carsten Behrens and Leo A. Paquette

Org. Synth. **1997**, *75*, 106

N-BENZYL-2,3-AZETIDINEDIONE

CH_3CHO + (methyl acrylate, COOMe) $\xrightarrow[CH_3OH]{DABCO}$ (OH, COOMe product)

Khalil Shahlai, Samuel Osafo Acquaah, and Harold Hart

Org. Synth. **1997**, *75*, 201

USE OF 1,2,4,5-TETRABROMOBENZENE AS A 1,4-BENZADIYNE EQUIVALENT: *anti*- AND *syn*-1,4,5,8-TETRAHYDROANTHRACENE 1,4:5,8-DIEPOXIDES

(1,2,4,5-tetrabromobenzene: Br, Br, Br, Br) + (furan, O) $\xrightarrow{BuLi}$ (O, O) + (O, O)

CLASS 9 Enolates, Homoenolates, and Dicarbonyl Compounds

A. Reagents

Diethyl Malonate
Dilithioacetate
Dimethyl 1,3-Acetonedicarboxylate
(3-Ethoxy-3-oxopropyl)iodozinc
1-Ethoxy-1(trimethylsilyloxy)cyclopropane
Ethyl Acetoacetate
Ethyl Bromozincacetate
Ethyl Cyanoacetate
Methyl Dilithioacetoacetate

B. Reviews (1992–1998)

(1) Fu, X. and Cook, J. M. The Synthesis of Polyquinanes and Polyquinenes via the Weiss Reaction. *Aldrichimica Acta* **1992**, *25*, 43–55.

(2) Hermez, I.; Kereszturi, G. and Vasvari-Debreczy, L. Aminomethylenemalonates and Their Use in Heterocyclic Synthesis; *Adv. Heterocycl. Chem.* **1992**, *54*, 1.

(3) Jackman, L. M. and Bortiatynski, J. Structures of Lithium Enolates and Phenolates in Solution; *Advances in Carbanion Chemistry*; JAI Press: Greenwich, 1992, Vol. 1.

(4) Paterson, I. New Methods and Strategies for the Stereocontrolled Synthesis of Polypropionate-Derived Natural Products. *Pure Appl. Chem.* **1992**, *64*, 1821–30.

(5) Williams, R. M. and Hendrix, J. A. Asymmetric Synthesis of Arylglycines. *Chem. Rev.* **1992**, *92*, 889–917.

(6) Crimmins, M. T. and Nantermet, P. G. Homoenolates and Other Functionalized Organometallics. *Org. Prep. Proc. Int.* **1993**, *25*, 41–81.

(7) Erian, A. W. The Chemistry of β-Enaminonitriles as Versatile Reagents in Heterocyclic Synthesis. *Chem. Rev.* **1993**, *93*, 1991–2005.

(8) Kovacs, L. Methods for the Synthesis of α-Keto Esters. *Rec. Trav. Chim. Pays-Bas* **1993**, *112*, 471–96.

(9) Mueller-Westerhoff, U. T. and Zhou, M. α-Diones From Cyclic Oxamides and Organolithium Reagents: A New, General and Environmentally Beneficial Synthetic Method. *Synlett* **1994**, 975–84.

(10) Mukaiyama, T. and Kobayash, S. Tin(II) Enolates in the Aldol, Michael, and Related Reactions. *Org. React.* **1994**, *46*, 1–104.

(11) O'Donnell, M. J.; Wu, S. D. and Huffman, J. C. A New Active Catalyst Species for Enantioselective Alkylation by Phase-Transfer Catalysis. *Tetrahedron* **1994**, *50*, 4507–18.

(12) Rappoport, Z. *The Chemistry of Enamines*; Wiley: Chichester, 1994.

(13) Schmittel, M. Umpolung of Ketones Via Enol Radical Cations. *Top. Curr. Chem.* **1994**, 183.

(14) Shibata, I. and Baba, A. Organotin Enolates in Organic Synthesis. *Org. Prep. Proc. Int.* **1994**, *26*, 85–100.

(15) Thompson, C. M. *Dianion Chemistry*; CRC: Boca Raton, 1994.

(16) Ward, R. S. *Bifunctional Compounds*; Oxford University Press: Oxford, 1994.

(17) Benetti, S.; Romagnoli, R.; De Risi, C.; Spalluto, G. and Zanirato, V. Mastering β-Keto Esters. *Chem. Rev.* **1995**, *95*, 1065–115.

(18) Bernardi, A. Stereoselective Conjugate Addition of Enolates to α, β-Unsaturated Carbonyl Compounds. *Gazz. Chim. Ital.* **1995**, *125*, 539.

(19) Moreno-Manas, M.; Marquet, J. and Vallribera, A. Transformations of β-Dicarbonyl Compounds by Reactions of Their Transition Metal Complexes with Carbon and Oxygen Electrophiles. *Tetrahedron* **1996**, *52*, 3377–401.

(20) Frederick, M. A. and Hulce, M. Ambiphilic Allenyl Enolates. *Tetrahedron* **1997**, *53*, 10197–226.

(21) Guingant, A. Asymmetric Syntheses of α, α-Disubstituted β-Diketones and β-Keto Esters; *Advances in Asymmetric Synthesis*; JAI Press: Greenwich; 1997, Vol. 2.

(22) Lue, P. and Greenhill, J. V. Enamines in Heterocyclic Synthesis. *Adv. Heterocycl. Chem.* **1997**, *67*, 207–343.

C. *Organic Syntheses Procedures (Vols. 70–75)*

Authors	Citation	Title
Philip G. Meister, Matthew R. Sivik, and Leo A. Paquette	*Org. Synth.* **1992**, *70*, 226.	2-METHYL-1,3-CYCLOPENTANEDIONE

O, OH, OH, O + C_2H_5COCl → ($AlCl_3$; CH_3NO_2; Δ) O, CH_3, O

Authors	Citation	Title
L. F. Tietze and U. Beifuss	*Org. Synth.* **1992**, *71*, 167.	DIASTEREOSELECTIVE FORMATION OF TRANS-1,2-DISUBSTITUTED CYCLOHEXANES FROM ALKYLIDENEMALONATES BY AN INTRAMOLECULAR ENE REACTION: DIMETHYL (1′*R*, 2′*R*, 5′*R*)-2-(2′-ISOPROPENYL-5′-METHYLCYCLOHEX-1′-YL)-PROPANE-1,3-DIOATE

O + O, OMe, O, OMe → (AcOH, piperidine; 3Å mol sieves) MeO_2C, CO_2Me

→ (0.1 eq. $FeCl_3/Al_2O_3$; CH_2Cl_2) MeO_2C, CO_2Me + MeO_2C, CO_2Me

Authors	Citation	Title
Lutz F. Tietze and Matthias Bratz	*Org. Synth.* **1992**, *71*, 214.	DIALKYL MESOXALATES BY OZONOLYSIS OF DIALKYL BENZALMALONATES: PREPARATION OF DIMETHYL MESOXALATE

COOR, COOR → (1) O_3, CH_2Cl_2; 2) Me_2S; (3)O_2) O, COOR, COOR

Authors	Citation	Title
R. Zibuck and J. Streiber	*Org. Synth.* **1992**, *71*, 236.	ETHYL 3-OXO-4-PENTENOATE (NAZAROV'S REAGENT)

O, OEt → (1) LDA, THF, −78°C; 2) CHO) OH, O, OEt → (H_2CrO_4; acetone) O, O, OEt

M. Braun and S. Gräf — *Org. Synth.* **1993**, *72*, 38.

STEREOSELECTIVE ALDOL REACTION OF DOUBLY DEPROTONATED 2-HYDROXY-1,2,2-TRIPHENYLETHYL ACETATE (HYTRA): (*R*)-3-HYDROXY-4-METHYLPENTANOIC ACID

2 LDA, THF / 0°C; CHO, −126°C; KOH / H_2O / MeOH, reflux

A. K. Beck, S. Blank, K. Job, D. Seebach, and Th. Sommerfeld — *Org. Synth.* **1993**, *72*, 62.

SYNTHESIS OF (*S*)-2-METHYLPROLINE: A GENERAL METHOD FOR THE PREPARATION OF α-BRANCHED AMINO ACIDS

t-BuCHO, cat. CF_3COOH; 1. (*i*-Pr)$_2$NLi, 2. MeI; 1. 3 N HCl, reflux, 2. Dowex® 50W × 8(H+)

S. P. Modi, R. C. Oglesby, and S. Archer — *Org. Synth.* **1993**, *72*, 125.

AN EFFICIENT SYNTHESIS OF INDOLE-2-ACETIC ACID METHYL ESTERS

1. EtOC(O)–C(O)OEt, NaOMe, 2. H_2O_2 – NaOH; 1. $SOCl_2$, 2. $CH_3COCH_2COOCH_3$, NaH, THF; NH_4OH, MeOH; 10% Pd-C, $HCOONH_4$

Rick L. Danheiser, James S. Nowick, Janette H. Lee, Raymond F. Miller, and Alexandre H. Huboux — *Org. Synth.* **1995**, *73*, 61.

SYNTHESIS OF β-LACTONES AND ALKENES via THIOL ESTERS: (*E*)-2,3-DIMETHYL-3-DODECENE

LDA, then $CH_3COCH(CH_3)_2$; SiO_2, cyclohexane, reflux

Michael J. Bradlee and Paul Helquist — *Org. Synth.* **1996**, *74*, 137

CYCLOPENTANONE ANNULATION VIA CYCLOPROPANONE DERIVATIVES: (3aβ,9bβ)-1,2,3a,4,5,9b-HEXAHYDRO-9b-HYDROXY-3a-METHYL-3*H*-BENZ[e]INDEN-3-ONE

$BrMgN(i\text{-}Pr)_2$, Et_2O; + [EtO OMgBr cyclopropane]; NaH, Et_2O

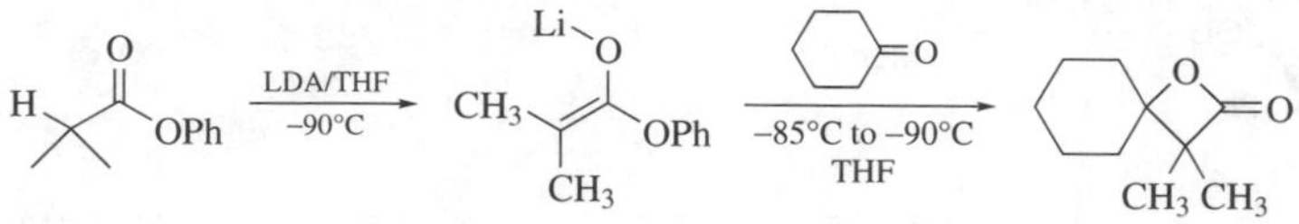

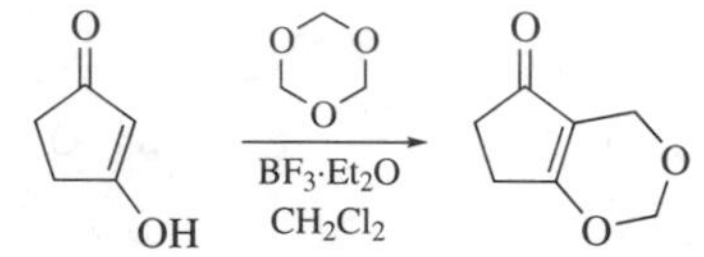

Authors	Citation	Title
Christine Wedler and Hans Schick	*Org. Synth.* **1997**, *75*, 116	SYNTHESIS OF β-LACTONES BY ALDOLIZATION OF KETONES WITH PHENYL ESTER ENOLATES: 3,3-DIMETHYL-1-OXASPIRO[3.5]NONAN-2-ONE

LDA/THF, −90°C; −85°C to −90°C, THF

Kwunmin Chen, Christopher S. Brook, and Amos B. Smith, III	*Org. Synth.* **1997**, *75*, 189	6,7-DIHYDROCYCLOPENTA-1,3-DIOXIN-5(4*H*)-ONE

$BF_3{\cdot}Et_2O$, CH_2Cl_2

Jean-Pierre Deprés and Andrew E. Greene	*Org. Synth.* **1997**, *75*, 195	3-CYCLOPENTENE-1-CARBOXYLIC ACID

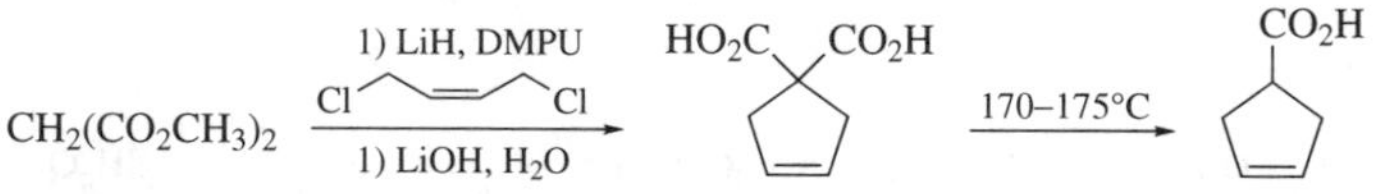

$CH_2(CO_2CH_3)_2$ → 1) LiH, DMPU, Cl–CH=CH–Cl; 1) LiOH, H_2O → HO_2C, CO_2H → 170–175°C → CO_2H

CLASS 10 Epoxides

A. Reagents

Epichlorohydrin
Ethylene Oxide
Glycidol

B. Reviews (1992–1998)

(1) Lohray, B. B. Cyclic Sulfites and Cyclic Sulfates: Epoxide-Like Synthons. *Synthesis* **1992**, 1035–52.

(2) Paterson, I. and Berrisford, D. J. Meso-Epoxides in Asymmetric Synthesis: Enantioselective Opening By Nucleophiles in the Presence of Chiral Lewis Acids. *Angew. Chem., Int. Ed. Engl.* **1992**, *31*, 1179–80.

(3) Schurig, V. and Betschinger, F. Metal-mediated Enantioselective Access to Unfunctionalized Aliphatic Oxiranes: Prochiral and Chiral Recognition. *Chem. Rev.* **1992**, *92*, 873–88.

(4) Taylor, S. K. Biosynthetic, Biomimetic, and Related Epoxide Cyclizations. *Org. Prep. Proc. Int.* **1992**, *24*, 245–84.

(5) Dale, J. The Contrasting Behavior of Oxirane and Oxetane in Cationic Cycloologomerization and Polymerization. *Tetrahedron* **1993**, *49*, 8707–25.

(6) deBont, J. A. M. Bioformation of Optically Pure Epoxides. *Tetrahedron: Asymmetry* **1993**, *4*, 1331–40.

(7) Besse, P. and Veschambre, H. Chemical and Biological Synthesis of Chiral Epoxides. *Tetrahedron* **1994**, *50*, 8885–927.

(8) Bonini, C. and Righi, G. Regio-and Chemoselective Synthesis of Halohydrins by Cleavage of Oxiranes With Metal Halides. *Synthesis* **1994**, 225–38.

(9) Jankowski, P.; Raubo, P. and Wicha, J. Tandem Transformations Initiated by the Migration of a Silyl Group. Some New Synthetic Applications of Silyloxiranes. *Synlett* **1994**, 985–92.

(10) Hodgson, D. M.; Gibbs, A. R. and Lee, G. P. Enantioselective Desymmetrisation of Achiral Epoxides. *Tetrahedron* **1996**, *52*, 14361–84.

(11) Satoh, T. Oxiranyl Anions and Aziridinyl Anions. *Chem. Rev.* **1996**, *96*, 3303–25.

(12) Archer, I. V. J. Epoxide Hydrolases as Asymmetric Catalytsts. *Tetrahedron* **1997**, *53*, 15617–62.

(13) Lohray, B. B. and Bhushan, V. 1,3,2-Dioxathiolane Oxides: Epoxide Equivalents and Versatile Synthons. *Adv. Heterocycl. Chem.* **1997**, *68*, 89–180.

(14) Sainsbury, M. Three-, Four-, and Five-membered Monoheterocyclic Compounds. *Rodd's Chemistry of Carbon Compounds, 2nd Ed.: Second Supplement to Volume IV, Heterocyclic Compounds; Part A*, Elsevier: Amsterdam, 1997.

C. Organic Syntheses Procedures (Vols. 70–75)

Authors	Citation	Title
Ryu Oi and K. Barry Sharpless	*Org. Synth.* **1995**, *73*, 1.	3-[(1*S*)-1,2-DIHYDROXYETHYL]-1,5 DIHYDRO-3*H*-2,4-BENZO-DIOXEPINE

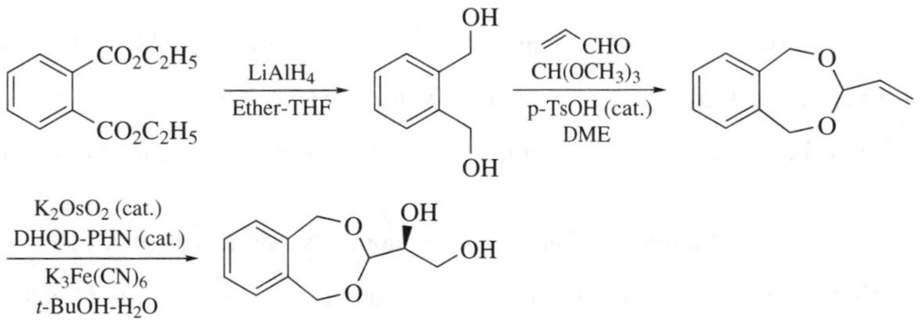

$CO_2C_2H_5$, $CO_2C_2H_5$ → LiAlH$_4$, Ether-THF → OH, OH → CHO, CH(OCH$_3$)$_3$, p-TsOH (cat.), DME → O, O

K$_2$OsO$_2$ (cat.), DHQD-PHN (cat.), K$_3$Fe(CN)$_6$, *t*-BuOH-H$_2$O → O, O, OH, OH

Authors	Citation	Title
Y. Petit and M. Larchevêque	*Org. Synth.* **1997**, *75*, 37	ETHYL GLYCIDATE FROM (*S*)-SERINE: ETHYL (*R*)-(+)-2,3-EPOXYPROPANOATE

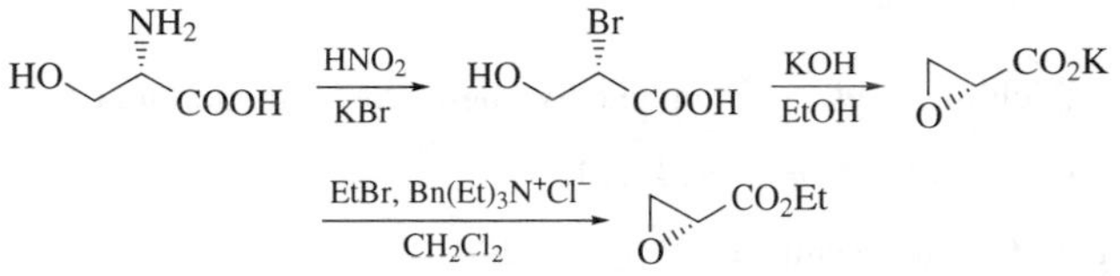

HO, NH$_2$, COOH → HNO$_2$, KBr → HO, Br, COOH → KOH, EtOH → O, CO_2K

EtBr, Bn(Et)$_3$N$^+$Cl$^-$, CH$_2$Cl$_2$ → O, CO_2Et

CLASS 11 Halo Compounds

A. Reagents

Acetyl Chloride
Allyl Bromide
2-(2-Bromoethyl)-1,3-dioxane
Bromoform
Carbon Tetrabromide
Diiodomethane
Ethyl 2-(Bromomethyl)acrylate
Iodoform
Iodomethane
Phenyl(trichloromethyl)mercury
Propargyl Chloride
Trichloroethylene
Trifluoromethyltrimethylsilane

B. Reviews (1992–1998)

(1) Yoshida, M. and Kamigata, N. Recent Progress in Perfluoroalkylation by Radical Species With Special Reference to the Use of Bis(perfluoroalkanoyl) Peroxides. *J. Fluorine Chem.* **1990**, *49*, 1–20.

(2) Gukasyan, A. O.; Galstyan, L. K. and Avetisyan, A. A. Synthesis, Structure, and Reactivity of α-(Trihalomethyl) carbinols. *Russ. Chem. Rev.* **1991**, *60*, 1318–30.

(3) Deev, L. E.; Nazarenko, T. I.; Pashkevich, K. I. and Ponomarev, V. G. Iodoperfluoroalkanes. *Russ. Chem. Rev.* **1992**, *61*, 40–54.

(4) McClinton, M. A. and McClinton, D. A. Trifluoromethylations and Related Reactions in Organic Chemistry. *Tetrahedron* **1992**, *48*, 6555–666.

(5) Sato, F. and Kobayashi, Y. Synthesis of Enantiomerically Pure Secondary γ-Halo Allylic Alcohols and Their Use in the Synthesis of Leukotrienes. *Synlett* **1992**, 849–57.

(6) Wakselman, C. Single Electron-Transfer Processes in Perfluoroalkyl Halide Reactions. *J. Fluorine Chem.* **1992**, *59*, 367–78.

(7) Yamazaki, T. and Kitazume, T. Stereoselective Preparation of Trifluoromethylated Organic Molecules. *Rev. Heterotom Chem.* Oae, S., Ed.; MYU K.K.: Tokyo, Japan, 1992, Vol. 7.

(8) Jackson, A. and Angoh, G. Small Molecule, Big Potential (Oxalyl Chloride). *Chem. Britain* **1993**, *29*, 1046–48.

(9) Maletina, I. I.; Mironova, A. A.; Orda, V. V. and Yagupolskii, L. M. Arylperfluoroalkyl- and Arylpolyfluoroalkyliodonium Salts. *Rev. Heteroatom Chem.* MYU K.K.: Tokyo, 1993, Vol. 8.

(10) Resnati, G. Synthesis of Chiral and Bioactive Fluoroorganic Compounds. *Tetrahedron* **1993**, *49*, 9385–445.

(11) Ritter, K. Synthetic Transformations of Vinyl and Aryl Triflates. *Synthesis* **1993**, 735–62.

(12) Sawada, H. Fluorinated Organic Peroxides—Their Decomposition Behavior and Applications. *Rev. Heteroatom Chem.* MYU K.K.: Tokyo, 1993, Vol. 8.

(13) Uno, H. and Suzuki, H. Nucleophilic Perfluoroalkylation Using Perfluoroalkyllithiums. *Synlett* **1993**, 91–6.

(14) Banks, R. E.; Smart, B. E. and Tatlow, J. C. *Organofluorine Chemistry: Principles and Commercial Applications*; Plenum: New York, 1994.

(15) Burton, D. J.; Yang, Z. Y. and Morken, P. A. Fluorinated Organometallics: Vinyl, Alkynyl, Allyl, Benzyl, Propargyl and Aryl Fluorinated Organometallic Reagents in Organic Synthesis. *Tetrahedron* **1994**, *50*, 2993–3063.

(16) Grushin, V. V. and Apler, H. Transformations of Chloroarenes, Catalyzed by Transition-Metal Complexes. *Chem. Rev.* **1994**, *94*, 1047–62.

(17) Hayashi, T. and Soloshonok, V. A. (Eds), Enantiocontrolled Synthesis of Fluoro-Organic Compounds. *Tetrahedron: Asymmetry* **1994**, *5*, 955–1538.

(18) Kaberdin, R. V. and Potkin, V. I. Trichloroethylene in Organic Synthesis. *Rev. Heteroatom Chem.* **1994**, *63*, 641–59.

(19) Spargo, P. L. Organic Halides. *Contemp. Org. Synth.* **1994**, *1*, 113.

(20) Benneche, T. α-Monohalo Ethers in Organic Syntheses. *Synthesis* **1995**, 1–27.

(21) Elguero, J.; Fruchier, A.; Jagerovic, N. and Werner, A. Trifluoromethyl and Perfluoroalkyl Derivatives of Azoles. A Review. *Org. Prep. Proc. Int.* **1995**, *27*, 33–74.

(22) Ohwada, T. Reactive Carbon Electrophiles in Friedel-Craft Reactions. *Rev. Heteroatom Chem.* **1995**, *12*, 179–209.

(23) Patai, S. and Rappoport, Z. *Chemistry of Halides, Pseudo-Halides and Azides, Part I.* Wiley: Chichester, 1995.

(24) Patai, S. and Rappoport, Z. *Chemistry of Halides, Pseudo-Halides and Azides, Part 2.* Wiley: Chichester, 1995.

(25) Kiselyov, A. S. and Strekowski, L. The Trifluoromethyl Group in Organic Synthesis. A Review. *Org. Prep. Proc. Int.* **1996**, *28*, 291–318.

(26) Marsden, S. P. Organic Halides. *Contemp. Org. Synth.* **1996**, *3*, 133–50.

(27) Burton, D. J. and Lu, L. Fluorinated Organometallic Compounds. *Top. Curr. Chem.* **1997**, *193*, 45–89.

(28) Marsden, S. P. Organic Halides. *Contemp. Org. Synth.* **1997**, *4*, 118–35.

(29) Prakash, G. K. S. and Yudin, A. K. Perfluoroalkylation with Organosilicon Reagents. *Chem. Rev.* **1997**, *97*, 757–86.

C. *Organic Syntheses Procedures (Vols. 70–75)*

Authors	Citation	Title
Rajarathnam E. Reddy and Conrad J. Kowalski	*Org. Synth.* **1992**, *71*, 146.	ETHYL 1-NAPHTHYLACETATE: ESTER HOMOLOGATION VIA YNOLATE ANIONS

CO_2Et + CH_2Br_2 → CH_2CO_2Et

1) LiTMP, −78°C
2) LiHMDS + LiOEt −78°C → −20°C
3) *sec*-BuLi, −78°C
4) BuLi, ~20°C
5) EtOH/AcCl

J. Gonzalez, M. J. Foti, and S. Elsheimer *Org. Synth.* **1993**, *72*, 225. (3,3-DIFLUOROALLYL)TRIMETHYLSILANE

Me$_3$Si–CH=CH$_2$ $\xrightarrow[\text{H}_2\text{NCH}_2\text{CH}_2\text{OH},\ (\text{CH}_3)_3\text{COH}]{\text{CF}_2\text{Br}_2,\ \text{CuCl}}$ Me$_3$SiCH$_2$CH(Br)CH$_2$CF$_2$Br $\xrightarrow[\text{DMSO}]{\text{NaBH}_4}$ Me$_3$SiCH$_2$CH$_2$CH$_2$CF$_2$Br $\xrightarrow{\text{DBU}}$ Me$_3$SiCH$_2$CH=CF$_2$

P. Ramaiah, R. Krishnamurti, and G. K. S. Prakash *Org. Synth.* **1993**, *72*, 232. 1-TRIFLUOROMETHYL-1-CYCLOHEXANOL

(CH$_3$)$_3$SiCF$_3$ + cyclohexanone $\xrightarrow[\text{THF}]{(\text{C}_4\text{H}_9)_4\text{N}^+\ \text{F}^-\cdot 3\text{H}_2\text{O}}$ [1-F$_3$C-1-OSi(CH$_3$)$_3$-cyclohexane] $\xrightarrow[\text{THF}]{\text{3N HCl}}$ 1-F$_3$C-1-OH-cyclohexane

Kathleen Mondanaro Lynch and William P. Dailey *Org. Synth.* **1997**, *75*, 89 3-CHLORO-2-(CHLOROMETHYL)-1-PROPENE

C(CH$_2$OH)$_4$ $\xrightarrow[\text{2. SOCl}_2]{\text{1. pyridine}}$ HOCH$_2$C(CH$_2$Cl)$_3$ $\xrightarrow{\text{HNO}_3}$ HOC(=O)C(CH$_2$Cl)$_3$ $\xrightarrow{200\text{–}215^\circ\text{C}}$ CH$_2$=C(CH$_2$Cl)$_2$

Kathleen Mondanaro Lynch and William P. Dailey *Org. Synth.* **1997**, *75*, 98 [1.1.1]PROPELLANE

CH$_2$=C(CH$_2$Cl)$_2$ $\xrightarrow[\text{18-crown-6}]{\text{CHBr}_3,\ \text{NaOH}}$ 1,1-dibromo-2,2-bis(chloromethyl)cyclopropane $\xrightarrow{\text{MeLi}}$ [1.1.1]propellane

CLASS 12 Imines, Iminium Ions, and Amide Acetals

A. Reagents

t-Butoxybis(dimethylamino)methane
N′-*t*-Butyl-*N*,*N*-dimethylformamidine
N,*N*-Dimethylacetamide Dimethyl Acetal
Dimethyl(chloromethylene)ammonium Chloride
Dimethylformamide Diethyl Acetal
Dimethyl(methylene)ammonium Iodide
Formaldehyde–Dimethylamine
N-Methyl-*N*-(2-pyridyl)formamide

B. Reviews (1992–1998)

(1) Meth–Cohn, O. The Synthesis of Pyridines, Quinolines and Other Related Systems by the Vilsmeier and the Reverse Vilsmeier Method. *Heterocycles* **1993**, *35*, 539–57.

(2) Bertrand, G. and Wentrup, C. Nitrile Imines: From Matrix Characterization to Stable Compounds. *Angew. Chem., Int. Ed. Engl.* **1994**, *33*, 527–45.

(3) Huang, Z. and Wang, M. Heterocyclic Ketene Aminals. *Heterocycles* **1994**, *37*, 1233–62.

(4) Rádl, S. Mono- and Diazaquinones. *Adv. Heterocycl. Chem.* **1994**, 142–208.

(5) Tramontini, M. and Angiolini, L. *Mannich Bases*; CRC Press: Boca Raton, 1994.

(6) Borzilleri, R. M. and Weinreb, S. M. Imino Ene Reactions in Organic Synthesis. *Synthesis* **1995**, 347–60.

(7) Alexakis, A.; Mangeney, P.; Lensen, N.; Tranchier, J.-P.; Gosmini, R. and Raussou, S. Chiral Aminals: Powerful Auxiliaries in Asymmetric Synthesis. *Pure Appl. Chem.* **1996**, *68*, 531–4.

(8) Adams, J. P. Imines, Enamines and Oximes. *Contemp. Org. Synth.* **1997**, *4*, 517–43.

(9) Adams, J. P. and Robertson, G. Imines, Enamines and Related Functional Groups. *Contemp. Org. Synth.* **1997**, *4*, 183–260.

(10) Jones, G. and Stanforth, S. P. The Vilsmeier Reaction of Fully Conjugated Carbocycles and Heterocycles. *Org. React.* **1997**, *49*, 1–330.

C. Organic Syntheses Procedures (Vols. 70–75)

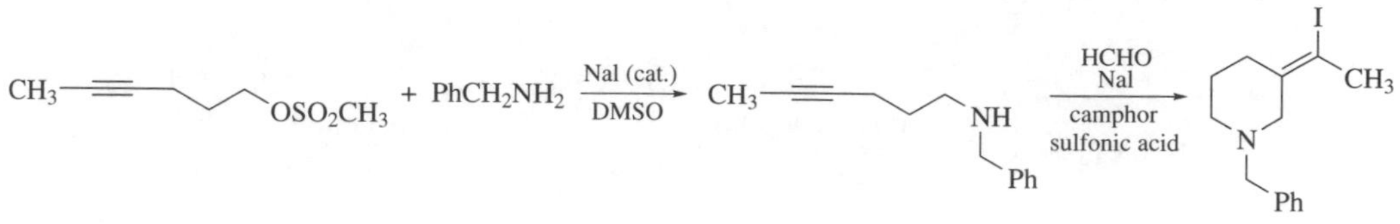

Authors	Citation	Title
H. Arnold, L. E. Overman, M. J. Sharp and M. C. Witschel	*Org. Synth.* **1992**, *70*, 111.	(*E*)-1-BENZYL-3-(1-IODOETHYLIDENE)PIPERIDINE: NUCLEOPHILE-PROMOTED ALKYNE-IMINIUM ION CYCLIZATIONS
Johann Polin and Herwig Schottenberger	*Org. Synth.* **1995**, *73*, 262.	CONVERSION OF METHYL KETONES INTO TERMINAL ACETYLENES: ETHYNYLFERROCENE

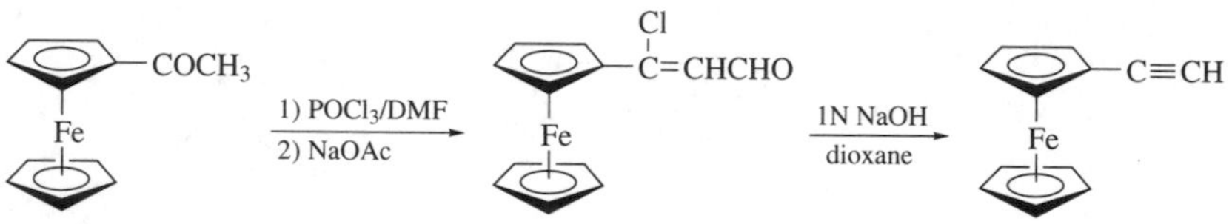

CLASS 13 Ketenes, Ketene Acetals, and Ortho Esters

A. Reagents

Dichloroketene
Diketene
Ketene
Ketene *t*-Butyldimethylsilyl Methyl Acetal
Ketene Diethyl Acetal
1-Methoxy-1-trimethylsilyloxy-1-propene
Triethyl Orthoacetate
Triethyl Orthoformate

B. Reviews (1992–1998)

(1) Moore, H. W. and Yerxa, B. R. Ring Expansions of Cyclobutenones-Synthetic Utility. *Chemtracts: Org. Chem.* **1992**, *5*, 273.

(2) Snider, B. B. Cycloadditions of Ketenes and Keteniminium Salts with Alkenes. *Advances in Strain in Organic Chemistry*; JAI Press: Greenwich, CT, 1992, Vol. 2.

(3) Brown, R. F. C. and Eastwood, F. W. Gas Phase Chemistry of Ketenes, Carbenes, Acetylenes and Arynes: Synthetic Consequences. *Synlett* **1993**, 9–19.

(4) Wentrup, C.; Heilmayer, W. and Kollenz, G. α-Oxoketenes-Preparation and Chemistry *Synthesis* **1994**, 1219–48.

(5) Allen, A. D; Ma, J; McAllister, M. A.; Tidwell, T. T. and Zhao, D. New Tricks From an Old Dog: Bisketenes After 90 Years. *Acc. Chem. Res.* **1995**, *28*, 265–72.

(6) Tidwell, T. T. *Ketenes*; Wiley: New York, 1995.

(7) Hyatt, J. A.; Raynolds, P.W. Ketene Cycloadditions; *Org. React.* **1994**, *45*, 159–646.

C. Organic Syntheses Procedures (Vols. 70–75)

Authors	Citation	Title
C. Hubschwerlen and J.-L. Specklin	*Org. Synth.* **1993**, *72*, 14.	(3*S*,4*S*)-3-AMINO-1-(3,4-DIMETHOXYBENZYL)-4-[(*R*)-2,2-DIMETHYL-1,3-DIOXOLAN-4-YL]-2-AZETIDINONE

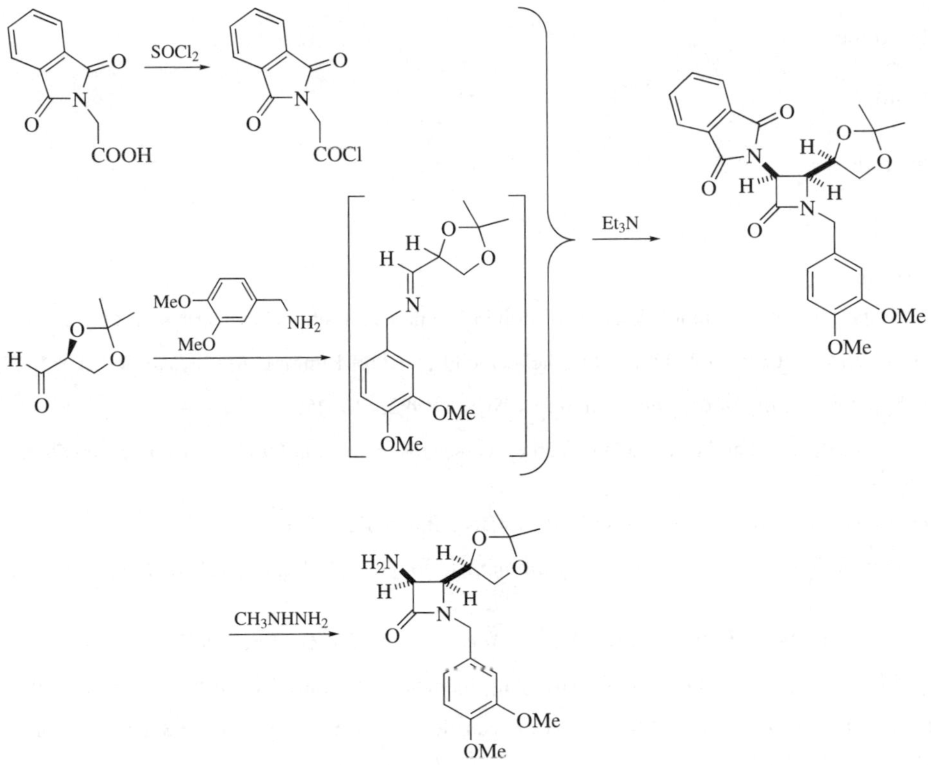

$SOCl_2$

COOH → COCl

MeO, MeO, NH_2

Et_3N

OMe, OMe

CH_3NHNH_2

H_2N

Authors	Citation	Title
P. Schiess, P. V. Barve, F. E. Dussy, and A. Pfiffner	*Org. Synth.* **1993**, *72*, 116.	BENZOCYCLOBUTENONE BY FLASH VACUUM PYROLYSIS

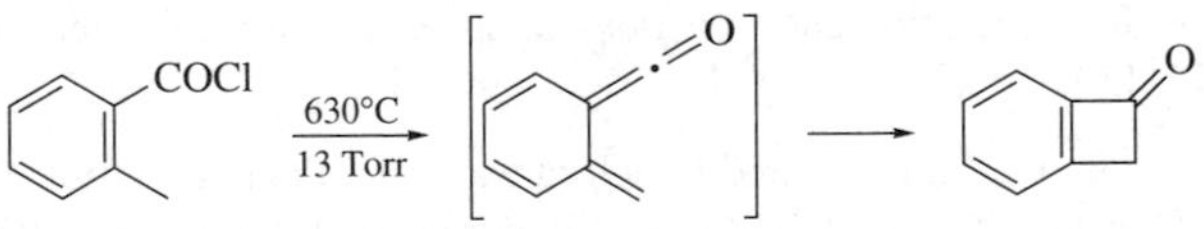

COCl —630°C, 13 Torr→

Authors	Citation	Title
V. Jäger and P. Poggendorf	*Org. Synth.* **1996**, *74*, 130.	NITROACETALDEHYDE DIETHYL ACETAL

$$O_2N{-}CH_3 + HC(OC_2H_5)_3 \xrightarrow[-90^\circ C]{ZnCl_2} O_2NCH_2CH(OC_2H_5)_2$$

Authors	Citation	Title
Andreas Rolfs and Jürgen Liebscher	*Org. Synth.* **1996**, *74*, 257	3-MORPHOLINO-2-PHENYLTHIOACRYLIC ACID MORPHOLIDE AND 5-(4-BROMOBENZOYL-2-(4-MORPHOLINO)-3-PHENYLTHIOPHENE

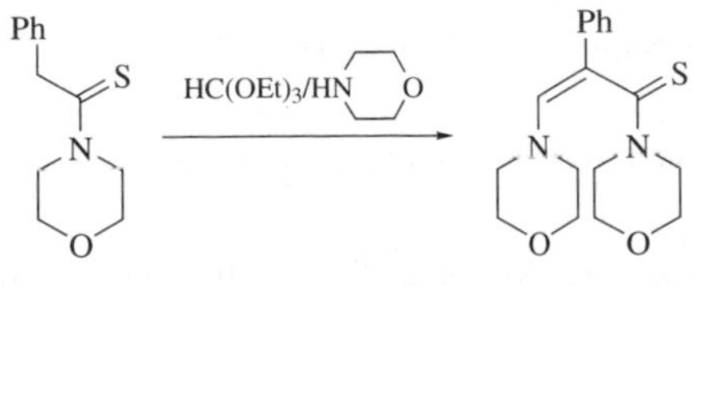

Ph, S → $HC(OEt)_3$/HN O → Ph, S

Avoid Skin Contact with All Reagents

CLASS 14 Lithium and Magnesium Reagents

A. Reagents

Allylmagnesium Bromide
1-(Benzyloxymethoxy)propyllithium
1-Ethoxyvinyllithium
Ethynylmagnesium Bromide
Methyllithium
Methylmagnesium Bromide
Phenyllithium
Phenylmagnesium Bromide
Vinyllithium
Vinylmagnesium Bromide

B. Reviews (1992–1998)

(1) Bickelhaupt, F. The Importance of (Intramolecular) Solvation in Organomagnesium Chemistry. *Acta Chem. Scand.* **1992**, *46*, 409–17.

(2) Collum, D. B. Is *N,N,N′,N′*-Tetramethylethylenediamine a Good Ligand for Lithium? *Acc. Chem. Res.* **1992**, *25*, 448–54.

(3) Comins, D. L. The Synthetic Utility of α-Amino Alkoxides. *Synlett* **1992**, 615–25.

(4) Blomberg, C. The Barbier Reaction and Related One-Step Processes *Reactivity and Structure. Concepts in Organic Chemistry*, Springer-Verlag: Berlin, 1993; Vol. 31.

(5) Cintas, P. *Activated Metals in Organic Synthesis*; CRC Press: Boca Raton, FL, 1993.

(6) Comasseto, J. V. Vinylic Tellurides: Precursors of Vinyllithium and Vinylcopper Reagents. *Rev. Heteroatom Chem.* MYU K.K.: Tokyo, 1993, Vol. 9.

(7) Fürstner, A. Chemistry of and With Highly Reactive Metals. *Angew. Chem., Int. Ed. Engl.* **1993**, *32*, 164–89.

(8) Hanusa, T. P. Ligand Influences on Structure and Reactivity in Organoalkaline Earth Chemistry. *Chem. Rev.* **1993**, *93*, 1023–36.

(9) Natale, N. R. and Mirzaei, Y. R. The Lateral Metalation of Isoxazoles. *Org. Prep. Proc. Int.* **1993**, *25*, 515–56.

(10) Perutz, R. N. Organometallic Intermediates: Ultimate Reagents. *Chem. Soc. Rev.* **1993**, *22*, 361–9.

(11) Reetz, M. T. Structural, Mechanistic, and Theoretical Aspects of Chelation-Controlled Carbonyl Addition Reactions. *Acc. Chem. Res.* **1993**, *26*, 462–8.

(12) Rewcatle, G. W. and Katritzky, A. R. Generation and Reactions of sp^2-Carbanionic Centers in the Vicinity of Heterocyclic Nitrogen Atoms. *Adv. Heterocycl. Chem.* **1993**, *56*, 157.

(13) Schlosser, M., Desponds, O., Lehmann, R., Moret, E. and Rauchschwalbe, G. Polar allyl Type Organometallics as Key Intermediates in Regio- and Stereocontrolled Reactions: Conformational Mobilities and Preferences. *Tetrahedron* **1993**, *49*, 10175–203.

(14) Wilson, S. R. Anion-Assisted Sigmatropic Rearrangements. *Org. React.* **1993**, *43*, 93–250.

(15) Aggarwal, V. K. Enantioselective Transformations and Racemization Studies of Heteroatom Substituted Organolithium Compounds. *Angew. Chem., Int. Ed. Engl.* **1994**, *33*, 175–7.

(16) Bailey, W. F. and Ovaska, T. V. Generation and Cyclization of Unsaturated Organolithiums; *Advances in Detailed Reaction Mechanisms*; JAI Press: Greenwich, 1994, Vol. 3.

(17) Blomberg, C. *The Barbier Reaction and Related One-Step Processes* Springer-Verlag: Heidelberg, 1994.

(18) Grimmett, M. R. and Iddon, B. Synthesis and Reactions of Lithiated Monocyclic Azoles Containing Two or More Hetero-Atoms, Part 3. *Heterocycles* **1994**, *37*, 2087–147.

(19) Hoppe, D.; Hintze, F.; Tebben, P.; Paetow, M.; Ahrens, H.; Schwerdtfeger, J.; Sommerfield, P.; Haller, J.; Guarnieri, W.; Kolczewski, S.; Hense, T. and Hoppe, I. Enantioselective Synthesis Via Sparteine Induced Asymmetric Deprotonation. *Pure Appl. Chem.* **1994**, *66*, 1479–92.

(20) Iddon, B. Synthesis and Reactions of Lithiated Monocyclic Azoles Containing Two or More Heteroatoms. *Heterocycles* **1994**, *37*, 1321–46.

(21) Iddon, B. and Ngochindo, R. I. Synthesis and Reactions of Lithiated Monocyclic Azoles Containing 2 or More Hetero-Atoms. *Heterocycles* **1994**, *38*, 2487–568.

(22) Katritzky, A. R.; Yang, Z. and Cundy, D. J. Benzotriazole-Stabilized Carbanions: Generation, Reactivity, and Synthetic Utility. *Adv. Org. Chem.* **1994**, *27*, 31–8.

(23) Katritzky, A. R.; Lan, X. and Fan, W. Q. Benzotriazole as a Synthetic Auxiliary: Benzotriazolylalkylations and Benzotriazole-Mediated Heteroalkylation. *Synthesis* **1994**, 445–56.

(24) Schlosser, M. *Organometallics in Synthesis*; Wiley: Chichester, 1994.

(25) Wills, M. Main Group Organometallics in Synthesis. *Contemp. Org. Synth.* **1994**, *1*, 339–66.

(26) Abel, E. W. *Organometallic Chemistry*; The Royal Society of Chemistry: Cambridge, 1995.

(27) Clark, R. D. and Jahangir, A. Lateral Lithiation Reactions Promoted by Heteroatomic Substituents. *Org. React.* **1995**, *47*, 1–314.

(28) Fürstner, A. *Active Metals*; VCH: Weinheim, 1995.

(29) Grimmett, M. R. and Iddon, B. Synthesis and Reactions of Lithiated Monocyclic Azoles Containing Two or More Hetero-Atoms. Part VI. Triazoles, Tetrazoles, Oxadiazoles, and Thiadiazoles. *Heterocycles* **1995**, *41*, 1525–74.

(30) Iddon, B. Synthesis and Reactions of Lithiated Monocyclic Azoles Containing Two or More Heteroatoms. *Heterocycles* **1995**, *41*, 533–93.

(31) Nájera, C. and Yus, M. Acyl Main Group Metal and Metalloid Derivatives in Organic Synthesis. A Review. *Org. Prep. Proc. Int.* **1995**, *27*, 383–456.

(32) Wakefield, B. J. *Organomagnesium Methods in Organic Chemistry*; Academic Press: San Diego, 1995.

(33) Beak, P.; Basu, A.; Gallagher, D. J.; Park, Y. S. and Thayumanavan, S. Regioselective, Diastereoselective, and Enantioselective Lithiation-Substitution Sequences; Reaction Pathways and Synthetic Applications. *Acc. Chem. Res.* **1996**, *29*, 552–60.

(34) Li, C. J. Aqueous Barbier-Grignard Type Reaction: Scope, Mechanism, and Synthetic Applications. *Tetrahedron* **1996**, *52*, 5643–68.

(35) Marek, I. and Normant, J.-F. Synthesis and Reactivity of sp^3-Geminal Organodimetallics. *Chem. Rev.* **1996**, *96*, 241–68.

(36) Silverman, G. S. and Rakita, P. E. *Handbook of Grignard Reagents*; Dekker: New York, 1996.

(37) Wills, M. Main Group Organometallics in Synthesis. *Contemp. Org. Synth.* **1996**, *3*, 201–28.

(38) Yus, M. Arene-catalysed Lithiation Reactions. *Chem. Soc. Rev.* **1996**, *25*, 155–62.

(39) Coldham, I. Main Group Organometallics in Synthesis. *Contemp. Org. Synth.* **1997**, *4*, 136–63.

(40) Hoppe, D. and Hense, T. Enantioselective Synthesis with Lithium/(−)-Sparteine Carbanion Pairs. *Angew. Chem., Int. Ed. Engl.* **1997**, *36*, 2282–316.

(41) Komiya, S. *Synthesis of Organometallic Compounds*; Wiley: New York, 1997.

(42) Yus, M. and Foubelo, F. Reductive Opening of Saturated Oxa-, Aza- and Thia-Cycles by Means of an Arene-Promoted Lithiation: Synthetic Applications. *Rev. Heteroatom Chem.* **1997**, *17*, 73–108.

C. Organic Syntheses Procedures (Vols. 70–75)

Authors	Citation	Title
Albert I. Meyers and Mark E. Flanagan	*Org. Synth.* **1992**, *71*, 107.	2,2′-DIMETHOXY-6-FORMYLBIPHENYL

OMe MgBr + (oxazoline) OMe OMe → (biaryl oxazoline, MeO OMe) —a) MeOTf b) $NaBH_4$ c) $(COOH)_2$→ MeO OMe OHC (biaryl aldehyde)

Authors	Citation	Title
Carl R. Johnson, John R. Medich, Rick L. Danheiser, Karen R. Romines, Hiroo Koyama, and Stephen K. Gee	*Org. Synth.* **1992**, *71*, 140.	PREPARATION AND USE OF (METH-OXYMETHOXY)METHYLLITHIUM: 1-HYDROXYMETHYL)CYCLOHEPTANOL

$Bu_3SnCH_2OCH_2OCH_3$ —1) BuLi, THF, −78°C 2) cycloheptanone→ $CH_3OCH_2OCH_2$ OH (cycloheptane) —HCl / MeOH→ HOCH$_2$ OH (cycloheptane)

Authors	Citation	Title
Marcus A. Tius and G. S. Kamali Kannangara	*Org. Synth.* **1992**, *71*, 158.	BENZOANNELATION OF KETONES: 3,4-CYCLODODECENO-1-METHYL-BENZENE

1. Me_3SiCl, Et_3N; 2. $CH_2{=}C(CH_3)CH_2MgCl$, Et_2O → OSiMe₃ intermediate; TsOH, $PhCH_3$, 80°C → CH_3 product

Authors	Citation	Title
T. F. Jamison, W. D. Lubell, J. M. Dener, M. J. Krisché, and H. Rapoport	Org. Synth., **1992**, *71*, 220.	9-BROMO-9-PHENYLFLUORENE

Br; 1. C_4H_9Li/ether; 2. (fluorenone)/THF; 3. H_2O → OH

Authors	Citation	Title
R. D. Rieke, T.-C. Wu, and L. I. Rieke	*Org. Synth.* **1993**, *72*, 147.	HIGHLY REACTIVE CALCIUM FOR THE PREPARATION OF ORGANO-CALCIUM REAGENTS: PREPARATION OF 1-ADAMANTYL CALCIUM HALIDES AND THEIR ADDITION TO KETONES: 1-(1-ADAMANTYL)CYCLOHEXANOL

$CaBr_2$ + Li^+ [biphenyl]$^{\dot{-}}$ —THF, RT→ Ca*

Br; Ca*, THF, −78°C; (cyclohexanone) −78°C→ −20°C → HO

Authors	Citation	Title
B. Mudryk and T. Cohen	*Org. Synth.* **1993**, *72*, 173	1,3-DIOLS FROM LITHIUM β-LITHIOALKOXIDES GENERATED BY THE REDUCTIVE LITHIATION OF EPOXIDES: 2,5-DIMETHYL-2,4-HEXANEDIOL

Li + (4,4'-di-tert-butylbiphenyl) —THF, 0°C, 4 hr→ [ ]$^{\dot{-}}$ Li^+

LDBB

O —2 LDBB, THF, −78°C→ Li OLi —1. *i*-PrCHO, −78°C; 2. H_2O, 0°C→ OH OH

Authors	Citation	Title
Masatomo Iwao and Tsukasa Kuraishi	*Org. Synth.* **1995**, *73*, 85	SYNTHESIS OF 7-SUBSTITUTED INDOLINES via DIRECTED LITHIATION OF 1-(*tert*-BUTOXYCARBONYL)INDOLINE: 7-INDOLINECARBOXALDEHYDE

N Boc —1. s-BuLi/TMEDA/ether/−78°C; 2. DMF/−78°C; 3. aq NH_4Cl→ OHC N Boc —1. conc. HCl; 2. aq NH_3→ OHC N H

Kohei Tamao, Yoshiki Nakagawa, and Yoshihiko Ito — *Org. Synth.* **1995**, *73*, 94. — REGIO- AND STEREOSELECTIVE INTRAMOLECULAR HYDROSILYLATION OF α-HYDROXY ENOL ETHERS: 2,3-*syn*-2-METHYOXYMETHOXY-1,3-NONANEDIOL

OCH$_2$OMe — 1) *sec*-BuLi, THF, −78→30°C; 2) C$_6$H$_{13}$CHO → C$_6$H$_{13}$ OCH$_2$OMe OH

M. A. Tschantz, L. E. Burgess, and A. I. Meyers — *Org. Synth.* **1995**, *73*, 215. — 4-KETOUNDECANOIC ACID

1. *t*-BuLi, −78°C→0°C; 2. CH$_3$(CH$_2$)$_6$I → C$_7$H$_{15}$ — H$_2$CrO$_4$, 0°C → C$_7$H$_{15}$ O OH O

Nikola A. Nikolic and Peter Beak — *Org. Synth.* **1996**, *74*, 23. — (*R*)-(+)-2-(DIPHENYLHYDROXYMETHYL)PYRROLIDINE

N Boc — *sec*-BuLi, (−)-sparteine, Et$_2$O, −78°C → [N Li *t*-BuO O] — Ph(C=O)Ph → N Boc Ph Ph OH — NaOH/EtOH, reflux → N H Ph Ph OH

Tina Morwick and Leo A. Paquette — *Org. Synth.* **1996**, *74*, 169 — PREPARATION OF POLYQUINANES BY DOUBLE ADDITION OF VINYL ANIONS TO SQUARATE ESTERS: 4,5,6,6a-TETRAHYDRO-3a-HYDROXY-2,3-DIISOPROPOXY-4,6a-DIMETHYL-1(3a*H*)-PENTALENONE

i-PrO *i*-PrO O O — H$_2$C=C(CH$_3$)Li, THF, −78°C→20°C; NH$_4$Cl, H$_2$O → *i*-PrO *i*-PrO O CH$_3$ OH CH$_3$ + *i*-PrO *i*-PrO CH$_3$ O OH CH$_3$ (33:1)

Akira Yanagisawa, Katsutaka Yasue, and Hisashi Yamamoto — *Org. Synth.* **1996**, *74*, 178 — REGIO- AND STEREOSELECTIVE CARBOXYLATION OF ALLYLIC BARIUM REAGENTS: (E)-4,8-DIMETHYL-3,7-NONADIENOIC ACID

BaI$_2$ + 2 Li$^+$[biphenyl]$^{\cdot-}$ — THF, rt, 30 min → Ba*

Cl — Ba*, THF, −78°C → CO$_2$, −78°C → CO$_2$H

Authors	Citation	Title
Ilane Marek, Christophe Meyer, and Jean-F. Normant	*Org. Synth.* **1996**, *74*, 194	A SIMPLE AND CONVENIENT METHOD FOR THE PREPARATION OF (*Z*)-β-IODO-ACROLEIN AND OF (*Z*)- OR (*E*)-γ-IODO ALLYLIC ALCOHOLS: (*Z*)- AND (*E*)-1-IODOHEPT-1-EN-3-OLS

H, H, I, CO_2Et —(i-Bu_2AlH, −78°C)→ [H, H, I, OEt, OAl-i-Bu_2] —BuMgBr→ H, H, I, Bu, OH (Z); —(1. 0°C, 15 min; 2. BuMgBr)→ I, H, H, Bu, OH (E)

Authors	Citation	Title
Mercedes Amat, Sabine Hadida, Swargam Sathyanarayana, and Joan Bosch	*Org. Synth.* **1996**, *74*, 248	REGIOSELECTIVE SYNTHESIS OF 3-SUBSTITUTED INDOLES: 3-ETHYLINDOLE

Br, N, $SiMe_2t$-Bu —t-BuLi/THF→ Li, N, $SiMe_2t$-Bu —EtI/THF→ CH_2CH_3, N, $SiMe_2t$-Bu —TBAF/THF→ CH_2CH_3, N, H

CLASS 15 Nickel Reagents

A. *Reagents*

Bis(1,5-cyclooctadiene)nickel(0)
Nickel(II) Acetylacetonate
Nickel(II) Chloride
Tetrakis(triphenylphosphine)nickel(0)

B. *Reviews (1992–1998)*

(1) Tamao, K.; Kobayashi, K. and Ito, Y. Nickel(0)-Mediated Intramolecular Cyclizations of Enynes, Dienynes, Bis-dienes, and Diynes. *Synlett* **1992**, 539–46.

(2) Campora, J.; Paneque, M.; Poveda, M. L. and Carmona, E. Organonickel Chemistry in Organic Synthesis, Some Applications of Alkyl and Metallacyclic Derivatives. *Synlett* **1994**, 465–70.

(3) Gai, Y. H.; Julia, M. and Verpeaux, J. N. Conversion of Nonactivated Alkenes into Cyclopropanes with Lithiated Sulfones Under Nickel Catalysis. *Bull. Soc. Chim. Fr.* **1996**, *133*, 817–29.

C. *Organic Syntheses Procedures (Vols. 70–75)*

Authors	Citation	Title
Zhi-Jie Ni and Tien-Yau Luh	*Org. Synth.* **1992**, *70*, 240.	NICKEL-CATALYZED SILYLOLEFINATION OF ALLYLIC DITHIOACETALS: (*E*,*E*)-TRIMETHYL(4-PHENYL-1,3-BUTADIENYL)SILANE

$$Me_3SiCH_2Cl + Mg \longrightarrow Me_3SiCH_2MgCl$$

Ph, S, S + Me_3SiCH_2MgCl —$NiCl_2(PPh_3)_2$→ Ph, $SiMe_3$

K. Takai, K. Sakogawa, Y. Kataoka, K. Oshima, and K. Utimoto

Org. Synth. **1993**, *72*, 180.

PREPARATION AND REACTIONS OF ALKENYLCHROMIUM REAGENTS: 2-HEXYL-5-PHENYL-1-PENTEN-3-OL

OTf + OHC Ph $\xrightarrow[\text{DMF, 25°C}]{\text{CrCl}_2\text{, cat. NiCl}_2}$ Ph OH

Tien-Min Yuan and Tien-Yau Luh

Org. Synth. **1996**, *74*, 187

NICKEL-CATALYZED, GEMINAL DIMETHYLATION OF ALLYLIC DITHIO ACETALS: (*E*)-1-PHENYL-3,3-DIMETHYL-1-BUTENE

S S Ph Me + MeMgI $\xrightarrow{\text{NiCl}_2\text{(dppp)}}$ Ph Me Me Me

CLASS 16 Palladium Reagents

A. Reagents

Bis(dibenzylideneacetone)palladium(0)
Dichloro[1,1′-bis(diphenylphosphine)ferrocene]palladium(II)
Palladium(II) Acetate
Palladium(II) Chloride
Tetrakis(triphenylphosphine)palladium(0)

B. Reviews (1992–1998)

(1) Frost, C.; Howarth, J. and Williams, J. M. J. Selectivity in Palladium-Catalyzed Allylic Substitution. *Tetrahedron: Asymmetry* **1992**, *3*, 1089–1122.

(2) Mitchell, T. N. Palladium-Catalyzed Reactions of Organotin Compounds. *Synthesis* **1992**, 803–15.

(3) Martin, A. R. and Yang, Y. Palladium-Catalyzed Cross-Coupling Reactions of Organoboronic Acids With Organic Electrophiles. *Acta Chem. Scand.* **1993**, *47*, 221–30.

(4) Reiser, O. Palladium-Catalyzed, Enantioselective Allylic Substitution. *Angew. Chem., Int. Ed. Engl.* **1993**, *32*, 547–49.

(5) de Meijere, A. and Meyer, F. E. Fine Feathers Make Fine Birds: The Heck Reaction in Modern Garb. *Angew. Chem., Int. Ed. Engl.* **1994**, *33*, 2379–411.

(6) Hiyama, T. and Hatanaka, Y. Palladium-Catalyzed Cross-Coupling Reaction of Organometalloids Through Activation With Fluoride Ion. *Pure Appl. Chem.* **1994**, *66*, 1471–8.

(7) Larock, R. C. Palladium-catalyzed Vinylic Substitution. *Advances in Metal-Organic Chemistry*; JAI Press: Greenwich, 1994, Vol. 3.

(8) Masuyama, Y. Palladium-Catalyzed Carbonyl Allylation Via π-Allylpalladium Complexes. *Advances in Metal-Organic Chemistry*; JAI Press: Greenwich, 1994, Vol. 3.

(9) Cabri, W. and Candiani, I. Recent Developments and New Perspectives in the Heck Reaction. *Acc. Chem. Res.* **1995**, *28*, 2–7.

(10) Heumann, A. and Reglier, M. The Stereochemistry of Palladium-Catalysed Cyclisation Reactions-B. Addition to Pi-Allyl Intermediates. *Tetrahedron* **1995**, *51*, 975–1016.

(11) Martin, A. R. and Zheng, Q. Palladium-Catalyzed Coupling Reactions of Indoles. *Advances in Nitrogen Heterocycles*; JAI Press: Greenwich, 1995, Vol 1.

(12) Miyaura, N. and Suzuki, A. Palladium-Catalyzed Cross-Coupling Reactions of Organoboron Compounds. *Chem. Rev.* **1995**, *95*, 2457–83.

(13) Rossi, R.; Carpita, A. and Bellina, F. Palladium-and/or Copper-Mediated Cross-Coupling Reactions Between 1-Alkynes and Vinyl, Aryl, 1-Alkynyl, 1,2-Propadienyl, Propargyl and Allylic Halides or Related Compounds. *Org. Prep. Proc. Int.* **1995**, *27*, 127–60.

(14) Tsuji, J. *Palladium Reagents and Catalysis: Innovations in Organic Synthesis*; Wiley: New York, 1995.

(15) Tsuji, J. and Mandai, T. Palladium-Catalyzed Reactions of Propargylic Compounds in Organic Synthesis. *Angew. Chem., Int. Ed. Engl.* **1995**, *34*, 2589–612.

(16) Undheim, K. and Benneche, T. Organometallics in Coupling Reactions in π-Deficient Azaheterocycles. *Ad. Heterocycl. Chem.* **1995**, *62*, 305–418.

(17) Andersson, P. G. and Bäckvall, J. E. Synthesis of Heterocyclic Natural Products Via Regio- and Stereocontrolled Palladium-Catalyzed Reactions. *Advances in Heterocyclic Natural Product Synthesis*; JAI Press: Greenwich, 1996, Vol. 3.

(18) Bäckvall, J. E. Enantiocontrol in Some Palladium- and Copper-Catalyzed Reactions. *Acta Chem. Scand.* **1996**, *50*, 661–5.

(19) Gibson, S. E. and Middleton, R. J. The Intramolecular Heck Reaction. *Contemp. Org. Synth.* **1996**, *3*, 447–72.

(20) Heumann, A. and Reglier, M. The Stereochemistry of Palladium-Catalyzed Cyclization Reactions. C. Cascade Reactions. *Tetrahedron* **1996**, *52*, 9289–346.

(21) Jeffery, T. Recent Improvements and Developments in Heck-Type Reactions and Their Potential in Organic Synthesis. *Advances in Metal-Organic Chemistry*;s JAI Press: Greenwich, 1996, Vol. 5.

(22) Moreno-Mañas, M. and Pleixats, R. Palladium(0)-Catalyzed Allylation of Ambident Nucleophilic Aromatic Heterocycles. *Adv. Heterocycl. Chem.* **1996**, *66*, 73–129.

(23) Negishi, E.; Copéret, C.; Ma, S.; Liou, S. and Liu, F. Cyclic Carbopalladation. A Versatile Synthetic Methodology for the Construction of Cyclic Organic Compounds. *Chem. Rev.* **1996**, *96*, 365–93.

(24) Yamamoto, Y. Palladium-Catalyzed Hydrocarbonation of Olefins. *Pure Appl. Chem.* **1996**, *68*, 9–14.

(25) Abad, J.-A. Synthesis and Reactivity of Organopalladium Complexes. *Gazz. Chim. Ital.* **1997**, *127*, 119–30.

(26) Shibasaki, M.; Boden, C. D. J. and Kojima, A. The Asymmetric Heck Reaction. *Tetrahedron* **1997**, *53*, 7371–93.

(27) Diederich, F. and Stang, P. J. *Metal-Catalyzed Cross-Coupling Reactions*; Wiley-VCH: Weinheim, 1998.

C. Organic Syntheses Procedures (Vols. 70–75)

Authors	Citation	Title
A. G. Myers and P. S. Dragovich	*Org. Synth.* **1993**, *72*, 104.	SYNTHESIS OF FUNCTIONALIZED ENYNES BY PALLADIUM/COPPER-CATALYZED COUPLING REACTIONS OF ACETYLENES WITH (*Z*)-2,3-DIBROMOPROPENOIC ACID ETHYL ESTER: (*Z*)-2- BROMO-5-(TRIMETHYLSILYL)-2-PENTEN-4-YNOIC ACID ETHYL ESTER

EtO, O, H; Br_2, CCl_4, 70°C; EtO, O, Br, H, Br; H—≡—TMS, (*i*-Pr)$_2$NEt, CuI, Pd(PPh$_3$)$_4$, DMF, 0°C; EtO, O, Br, H, TMS

Authors	Citation	Title
Bret E. Huff, Thomas M. Koenig, David Mitchell, and Michael A. Staszak	*Org. Synth.* **1997**, *75*, 53	SYNTHESIS OF UNSYMMETRICAL BIARYLS USING A MODIFIED SUZUKI CROSS COUPLING: 4-BIPHENYL-CARBOXALDEHYDE

B(OH)$_2$ + Br—C$_6$H$_4$—CHO; Pd(OAc)$_2$, PPh$_3$, Na_2CO_3, PrOH, H_2O; —CHO

Authors	Citation	Title
Felix E. Goodson, Thomas I. Wallow, and Bruce M. Novak	*Org. Synth.* **1997**, *75*, 61	ACCELERATED SUZUKI COUPLING VIA A LIGANDLESS PALLADIUM CATALYST: 4-METHOXY-2′-METHYLBIPHENYL

B(OH)$_2$, CH_3 + I—C$_6$H$_4$—OCH_3; Pd(OAc)$_2$, K_2CO_3, acetone, H_2O; —OCH_3, CH_3

Frederic S. Ruel, Matthew P. Braun, and Carl R. Johnson	*Org. Synth.* **1997**, *75*, 69	2-(4-METHOXYPHENYL)-2-CYCLOHEXEN-1-ONE: PREPARATION OF 2-IODO-2-CYCLOHEXEN-1-ONE AND SUZUKI COUPLING WITH 4-METHOXYPHENYL BORONIC ACID

$$\text{2-iodo-2-cyclohexen-1-one} + (HO)_2B\text{–}C_6H_4\text{–OMe} \xrightarrow[\text{THF, } H_2O\text{, 25°C}]{Ag_2O \\ Ph_3As\text{, } PdCl_2(PhCN)_2} \text{2-(4-methoxyphenyl)-2-cyclohexen-1-one}$$

CLASS 17 Phosphorus Reagents

A. Reagents

Ethoxycarbonylmethylenetriphenylphosphorane
Formylmethylenetriphenylphosphorane
Methoxymethylenetriphenylphosphorane
Methyl Bis(2,2,2-trifluoroethoxy)phosphinylacetate
Methylenetriphenylphosphorane
Triethyl Phosphonoacetate
Triphenylphosphine–Carbon Tetrabromide
Triphenylphosphine–Carbon Tetrachloride

B. Reviews (1992–1998)

(1) Trofimov, B. A.; Rakhmatulina, T. N.; Gusarova, N. K. and Malysheva, S. Elemental Phosphorus-strong Base as a System for the Synthesis of Organophosphorus Compounds. *Russ. Chem. Rev.* **1991**, *60*, 1360–67.

(2) Eguchi, S.; Matsushita, Y. and Yamashita, K. The Aza-Wittig Reaction in Heterocyclic Synthesis. *Org. Prep. Proc. Int.* **1992**, *24*, 209–243.

(3) Engel, R. *Handbook of Organophosphorus Chemistry*; Dekker: New York, 1992, Vol. 2.

(4) Hartley, F. R., Ed. *The Chemistry of Organophosphorus Compounds*; Wiley: Chichester, 1992, Vol. 2.

(5) Okuma, K. Synthesis and Reaction of Thioaldehydes from Wittig Reagents. *Rev. Heteratom Chem.* MYU K.K.: Tokyo, 1992, Vol. 7.

(6) Johnson, A. W. *Ylides and Imines of Phosphorus*; Wiley: New York, 1993.

(7) Power, P. P. Tantalum (V) Phosphinidene Complexes as Phospha-Wittig Reagents. *Angew. Chem., Int. Ed. Engl.* **1993**, *32*, 850–51.

(8) Molina, P. and Vilaplana, M. J. Iminophosphoranes: Useful Building Blocks for the Preparation of Nitrogen-Containing Heterocycles. *Synthesis* **1994**, 1197–218.

(9) Vedejs, E. and Peterson, M. J. Stereochemistry and Mechanism in the Wittig Reaction. *Top. Stereochem.* **1994**, *21*, 1–157.

(10) Allen, D. W. *Organophosphorus Chemistry*. The Royal Society of Chemistry: Cambridge, 1995.

(11) Allen, D. W. *Organophosphorus Chemistry*. The Royal Society of Chemistry: Cambridge, 1997, Vol. 28.

(12) Waschbusch, R.; Carran, J.; Marinetti, A. and Savignac, P. The Synthesis of Dialkyl α-Halogenated Methylphosphonates. *Synthesis* **1997**, 727–43.

(13) Wiemer, D. F. Synthesis of Nonracemic Phosphonates. *Tetrahedron* **1997**, *53*, 16609–44.

C. Organic Syntheses Procedures (Vols. 70–75)

Authors	Citation	Title
B. C. Hamper	*Org. Synth.* **1992**, *70*, 246.	α, β-ACETYLENIC ESTERS FROM α-ACYLMETHYLENEPHOSPHORANES: α-ETHYL 4,4,4-TRIFLUOROTETROLATE

$$[Ph_3P^+CH_2CO_2Et]\,Br^- \xrightarrow[(CF_3CO)_2O]{\text{2 equiv. } Et_3N} Ph_3P{=}C(CO_2Et)C(O)CF_3 \xrightarrow[\text{1–2 mmHg}]{150°\text{–}200°C} CF_3{-}C{\equiv}C{-}CO_2Et$$

Authors	Citation	Title
J. R. McCarthyl, D. P. Mathews, and J. P. Paolini	*Org. Synth.* **1993**, *72*, 216.	STEREOSELECTIVE SYNTHESIS OF 2,2-DISUBSTITUTED 1-FLUORO-ALKENES: (*E*)-[(FLUORO-(2-PHENYLCYCLOHEXYLIDENE)-METHYL]SULFONYL]BENZENE AND (*Z*)-[2-(FLUOROMETHYLENE)-CYCLOHEXYL]BENZENE

$[PhSO_2CH_2F + ClP(O)(OEt)_2]$ → LiHMDS → $PhSO_2CF(Li)P(O)(OEt)_2$; 2-phenylcyclohexanone → $PhSO_2C(F)$=cyclohexylidene(2-Ph) → (Bu_3SnH, AIBN) → $Bu_3SnC(F)$=cyclohexylidene(2-Ph)

Authors	Citation	Title
P. Savignac and C. Patois	*Org. Synth.* **1993**, *72*, 241.	DIETHYL 1-PROPYL-2-OXOETHYL-PHOSPHONATE

$(EtO)_2P(O)OEt$ → (2 *n*-BuLi, THF, −78°C → 0°C) → $(EtO)_2P(O)CHLi(CH_2CH_2CH_3)$ → (1. HCOOEt, THF, −78°C; 2. HCl, H_2O, THF, 0°C) → $(EtO)_2P(O)CH(CH_2CH_2CH_3)CHO$

Authors	Citation	Title
Carl Patois, Philippe Savignac, Elie About-Jaudet, and Noël Collignon	*Org. Synth.* **1995**, *73*, 152.	BIS(TRIFLUOROETHYL) (CARBO-ETHOXYMETHYL)PHOSPHONATE

$Cl_2P(O)CH_3$ → (2 CF_3CH_2OH, 2 NEt_3, THF, rt) → $(CF_3CH_2O)_2P(O)CH_3$ → (1. 2 LiHMDS, −78°C; 2. $ClCO_2Et$) → $(CF_3CH_2O)_2P(O)CH_2C(O)OC_2H_5$

Authors	Citation	Title
Angela Marinetti and Philippe Savignac	*Org. Synth.* **1996**, *74*, 108	DIETHYL (DICHLOROMETHYL) PHOSPHONATE. PREPARATION AND USE IN THE SYNTHESIS OF ALKYNES: (4-METHOXYPHENYL)ETHYNE

$(C_2H_5O)_2P(O)CHCl_2 + CH_3O-C_6H_4-CHO$ → (1. LDA/THF, −78°C → 0°C; 2. BuLi, −78°C → 0°C; 3. H_2O, HCl) → $CH_3O-C_6H_4-C{\equiv}CH$

Authors	Citation	Title
John Mann and Alexander C. Weymouth-Wilson	*Org. Synth.* **1997**, *75*, 139	PHOTOINDUCED-ADDITION OF METHANOL TO (5*S*)-(5-*O-tert*-BUTYL-DIMETHYLSILOXYMETHYL)FURAN-2(5*H*)-ONE: (4*R*,5*S*)-4-HYDROXY-METHYL-(5-*O-tert*-BUTYLDIMETHYL-SILOXYMETHYL)FURAN-2(5*H*)-ONE

Acetonide aldehyde (CHO) → ($Ph_3P{=}CHCO_2Me$, MeOH, 0°C) → acetonide $CH{=}CHCO_2Me$ → (1. H^+/MeOH; 2. TBDMSCl/imidazole) → RO-furanone, R = TBDMS → (MeOH/hv, benzophenone) → RO-lactone with HO-CH₂ substituent

Jean-Pierre Bégué,
Daniéle Bonnet-Delpon,
and Andrei Kornilov

Org. Synth. **1997**, *75*, 153

WITTIG OLEFINATION OF PERFLUOROALKYL CARBOXYLIC ESTERS; SYNTHESIS OF 1,1,1-TRIFLUORO-2-ETHOXY-5-PHENYLPENT-2-ENE AND 1-PERFLUOROALKYL EPOXY ETHERS: 1,1,1-TRIFLUORO-2-ETHOXY-2,3-EPOXY-5-PHENYLPENTANE

Ph_3P^+ Br^- Ph —(1. NaH, THF; 2. CF_3COOEt)→ CF_3 EtO H Ph —(*m*-CPBA)→ CF_3 EtO O H Ph

CLASS 18 Silicon and Tin Reagents

A. Reagents

Allyltributylstannane
Allyltrimethylsilane
(Chloromethyl)trimethylsilane
Chlorotrimethylsilane
Crotyltributylstannane
Hexabutyldistannane
Hexamethyldistannane
Lithium Bis[dimethyl(phenyl)silyl]cuprate
Tri-*n*-butylchlorostannane
Tributylstannyllithium
Trimethylsilylmethylmagnesium Chloride
Vinyltributylstannane

B. Reviews (1992–1998)

(1) Chan, T. H. and Wang, D. Chiral Organosilicon Compounds in Asymmetric Synthesis. *Chem. Rev.* **1992**, *92*, 995–1006.

(2) Cirillo, P. F. and Panek, J. S. Recent Progress in the Chemistry of Acylsilanes. *Org. Prep. Proc. Int.* **1992**, *24*, 553–82.

(3) Hayashi, T. and Uozumi, Y. Catalytic Asymmetric Synthesis of Optically Active Alcohols Via Hydrosilylation of Olefins With a Chiral Monophosphine-Palladium Catalyst. *Pure Appl. Chem.* **1992**, *64*, 1911–16.

(4) Hosomi, A. Structure and Reactivities of Pentacoordinate Organosilicon Compounds. Application to Highly Selective Organic Synthesis. *Rev. Heteroatom Chem.* MYU K.K.: Tokyo, 1992, Vol. 7.

(5) Stadnichuk, M. D. and Voropaeva, T. I. Silicon-Containing 1,3-Alkenynes and More Unsaturated Compounds. *Russ. Chem. Rev.* **1992**, *61*, 1091.

(6) Hudrlik, P. F. and Hudrlik, A. M. α,β-Epoxysilanes. *Advances in Silicon Chemistry*; JAI Press: Greenwich, 1993, Vol. 2.

(7) Jastrzebski, J. T. B. H. and Van Koten, G. Intramolecular Coordination in Organotin Chemistry. *Adv. Orgometal. Chem.* **1993**, *35*, 242.

(8) Luh, T.-Y. and Wong, K.-T. Silyl-Substituted Conjugated Dienes: Versatile Building Blocks of Organic Synthesis. *Synthesis* **1993**, 349–70.

(9) Nishigaichi, Y.; Takuwa, A.; Naruta, Y. and Maruyama, K. Versatile Roles of Lewis Acids in the Reactions of Allylic Tin Compounds. *Tetrahedron* **1993**, *49*, 7395–426.

(10) Yamamoto, Y. and Asao, N. Selective Reactions Using Allylic Metals. *Chem. Rev.* **1993**, *93*, 2207–93.

(11) Schubert, U. Formation and Breaking Si–E (E=C, Si) Bonds by Oxidative Addition and Reductive Elimination Reactions. *Angew. Chem., Int. Ed. Engl.* **1994**, *33*, 419–21.

(12) Sita, L. R. Heavy-Metal Organic Chemistry: Building With Tin. *Acc. Chem. Res.* **1994**, *27*, 191–7.

(13) Yamamoto, Y. and Shida, N. Stereochemistry and Mechanism of Allylic Tin-Aldehyde Condensation Reactions. *Advances in Detailed Reaction Mechanisms*; JAI Press: Greenwich, 1994, Vol. 3.

(14) Chan, T. H. and Wang, D. Silylallyl Anions in Organic Synthesis: A Study in Regio- and Stereoselectivity. *Chem. Rev.* **1995**, *95*, 1279–92.

(15) Horn, K. A. Regio- and Stereochemical Aspects of the Palladium-Catalyzed Reactions of Silanes. *Chem. Rev.* **1995**, *95*, 1317–50.

(16) Hwu, J. R. and Patel, H. V. Recent Development of Novel Organic Reactions Controlled by Silicon. *Synlett* **1995**, 989–96.

(17) Langkopf, E. S. and Schintzer, D. Uses of Silicon-Containing Compounds in the Synthesis of Natural Products. *Chem. Rev.* **1995**, *95*, 1375–408.

(18) Masse, C. E. and Panek, J. S. Diastereoselective Reactions of Chiral Allyl- and Allenylsilanes with Activated C=X π-Bonds. *Chem. Rev.* **1995**, *95*, 1293–316.

(19) Mukaiyama, T. and Kobayashi, S. Tin(II) Enolates in the Aldol, Michael, and Related Reactions. *Org. React.* **1995**, *46*, 1–104.

(20) Stadnichuk, M. D. and Voropaeva, T. I. Silicon-Containing Alka-1,3-Dienes and Their Functional Derivatives in Organic Synthesis. *Russ. Chem. Rev.* **1995**, *64*, 25–46.

(21) Tamao, K. and Kawachi, A. Silyl Anions. *Adv. Organometal. Chem.* **1995**, *38*, 1–58.

(22) Farina, V. New Perspectives in the Cross-Coupling Reactions of Organostannanes. *Pure Appl. Chem.* **1996**, *68*, 73–8.

(23) Farina, V. and Roth, G. P. Recent Advances in the Stille Reaction *Advances in Metal-Organic Chemistry*; JAI Press: Greenwich, 1996, Vol. 5.

(24) Jones, G. R. and Landais, Y. The Oxidation of the Carbon-Silicon Bond. *Tetrahedron* **1996**, *52*, 7599–662.

(25) Marshall, J. A. Chiral Allylic and Allenic Stannanes as Reagents for Asymmetric Synthesis. *Chem. Rev.* **1996**, *96*, 31–47.

(26) Cozzi, P. G., Tagliavini, E. and Umani-Ronchi, A. Enantioselective Addition of Allylic Silanes and Stannanes to Aldehydes Mediated by Chiral Lewis Acids. *Gazz. Chim. Ital.* **1997**, *127*, 247–54.

(27) Davies, A. G. *Organotin Chemistry*; Wiley-VCH: Weinheim, 1997.

(28) Farina, V.; Krishnamurthy, V. and Scott, W. J. The Stille Reaction. *Org. React.* **1997**, *50*, 1–652.

(29) Fleming, I.; Barbero, A. and Walter, D. Stereochemical Control in Organic Synthesis Using Silicon-Containing Compounds. *Chem. Rev.* **1997**, *97*, 2063–92.

C. *Organic Syntheses Procedures (Vols. 70–75)*

Authors	Citation	Title
Wha Chen, E. Kyle Stephenson, Michael P. Cava, and Yvette A. Jackson	*Org. Synth.* **1992**, *70*, 151	2-SUBSTITUTED PYRROLES FROM *N-tert*-BUTOXYCARBONYL-2-BROMOPYRROLE: *N-tert*-BUTOXY-CARBONYL-2-TRIMETHYLSILYL-PYRROLE

N-(CO_2CMe_3)-2-bromopyrrole →(BuLi, THF) N-(CO_2CMe_3)-2-lithiopyrrole →(Me_3SiCl) N-(CO_2CMe_3)-2-($SiMe_3$)pyrrole

Authors	Citation	Title
J. K. Stille, Antonio M. Echavarren, Robert M. Williams, and James A. Hendrix	*Org. Synth.* **1992**, *71*, 97.	4-METHOXY-4′-NITROBIPHENYL

p-$MgBrC_6H_4OCH_3$ + Bu_3SnCl →(THF) p-$Bu_3SnC_6H_4OCH_3$

p-$CF_3SO_2OC_6H_4NO_2$ + p-$Bu_3SnC_6H_4OCH_3$ →($PdCl_2(PPh_3)$, LiCl, DMF) O_2N–C_6H_4–C_6H_4–OCH_3

Authors	Citation	Title
Yoshinori Naruta, Yutaka Nishigaichi, and Kazuhiro Maruyama	*Org. Synth.* **1992**, *71*, 118.	TRIBUTYL(3-METHYL-2-BUTENYL)TIN

$$Bu_3SnCl + Me_2C{=}CHCH_2Cl \xrightarrow[\text{ultrasound}]{\text{Mg, THF}} Bu_3SnCH_2CH{=}CMe_2$$

Yoshinori Naruta and Kazuhiro Maruyama

Org. Synth. **1992**, *71*, 125.

UBIQUINONE-1

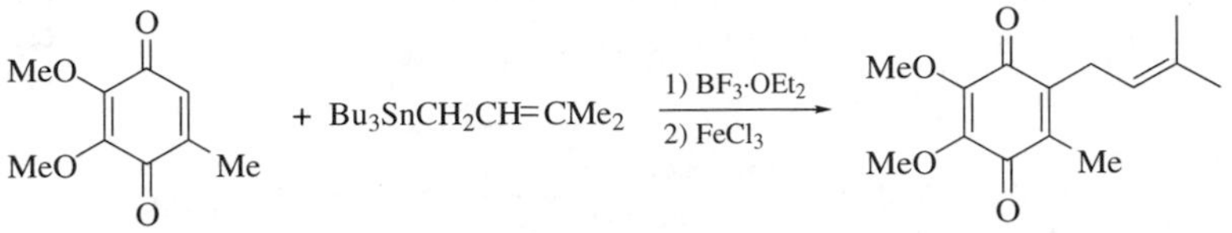

MeO, MeO, Me (quinone) + $Bu_3SnCH_2CH{=}CMe_2$ — 1) $BF_3{\cdot}OEt_2$; 2) $FeCl_3$ →

Rick L. Danheiser, Karen R. Romines, Hiroo Koyama, Stephen K. Gee, Carl R. Johnson, and John R. Medich

Org. Synth. **1992**, *71*, 133.

A HYDROXYMETHYL ANION EQUIVALENT: TRIBUTYL[(METHOXY-METHOXY)METHYL]STANNANE

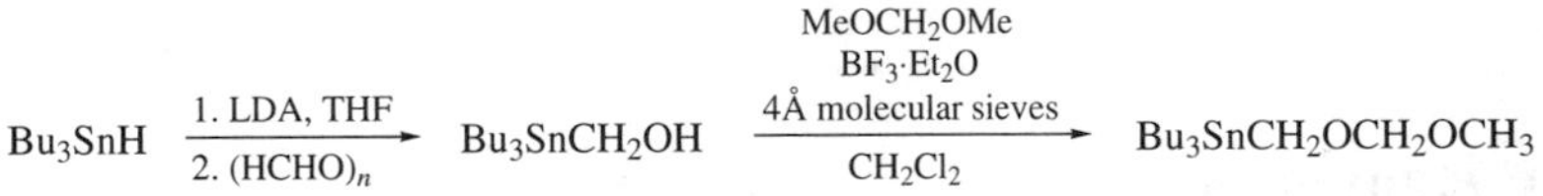

Bu_3SnH — 1. LDA, THF; 2. $(HCHO)_n$ → Bu_3SnCH_2OH — $MeOCH_2OMe$, $BF_3{\cdot}Et_2O$, 4Å molecular sieves, CH_2Cl_2 → $Bu_3SnCH_2OCH_2OCH_3$

T. V. Lee and J. R. Porter

Org. Synth. **1993**, *72*, 189

SPIROANNELATION OF ENOL SILANES: 2-OXO-5-METHOXY-SPIRO[5.4]DECANE

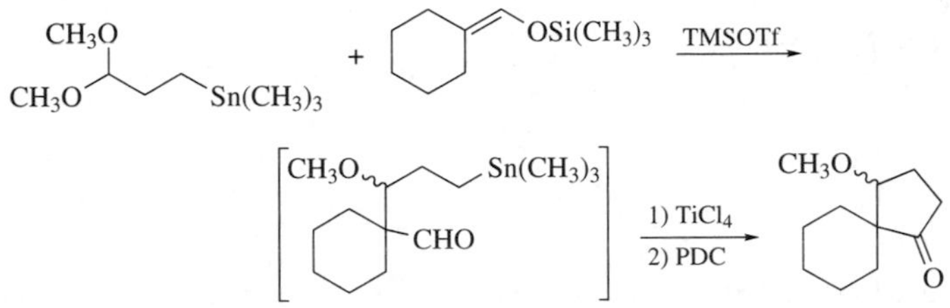

CH_3O, CH_3O, $Sn(CH_3)_3$ + $OSi(CH_3)_3$ — TMSOTf →

[CH_3O, $Sn(CH_3)_3$, CHO] — 1) $TiCl_4$; 2) PDC → CH_3O, O

Anthony G. M. Barrett, John A. Flygare, Jason M. Hill and Eli M. Wallace

Org. Synth. **1995**, *73*, 50

STEREOSELECTIVE ALKENE SYNTHESIS via 1-CHLORO-1-[(DIMEHYL)PHENYLSILYL]ALKANES] and α-(DIMETHYL)PHENYLSILYL KETONES: 6-METHYL-6-DODECENE

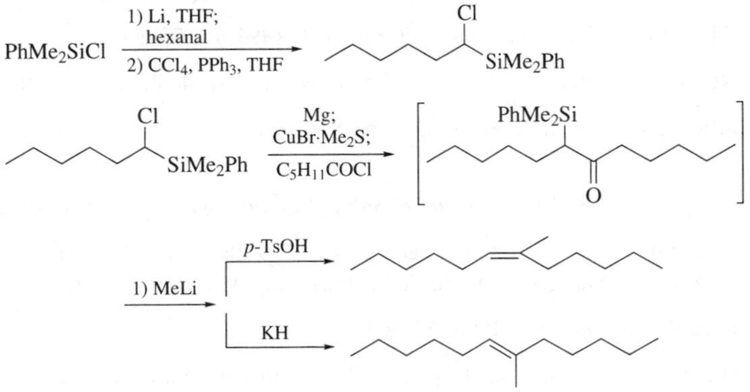

$PhMe_2SiCl$ — 1) Li, THF; hexanal; 2) CCl_4, PPh_3, THF → Cl, $SiMe_2Ph$

Cl, $SiMe_2Ph$ — Mg; $CuBr{\cdot}Me_2S$; $C_5H_{11}COCl$ → [$PhMe_2Si$, O]

— 1) MeLi → *p*-TsOH →; KH →

Gary E. Keck and Dhileepkumar Krishnamurthy

Org. Synth. **1997**, *75*, 12

CATALYTIC ASYMMETRIC ALLYLATION REACTIONS: (*S*)-1-PHENYLMETHOXY)-4-PENTEN-2-OL

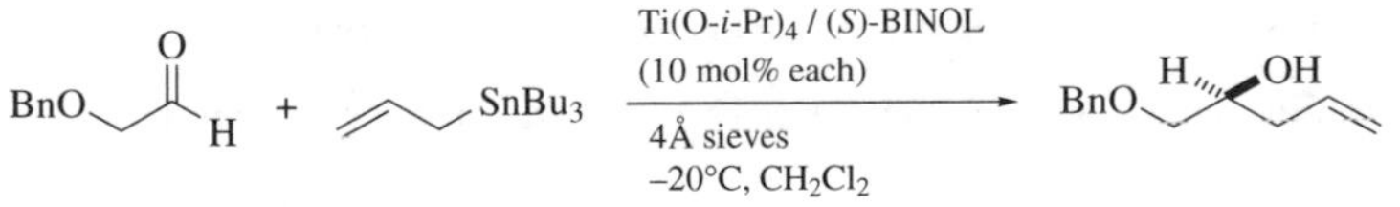

BnO, O, H + $SnBu_3$ — $Ti(O\text{-}i\text{-}Pr)_4$ / (*S*)-BINOL (10 mol% each), 4Å sieves, −20°C, CH_2Cl_2 → BnO, H, OH

Authors	Citation	Title
Richard T. Beresis, Jason S. Solomon, Michael G. Yang, Nareshkumar F. Jain, and James S. Panek	*Org. Synth.* **1997**, *75*, 78	SYNTHESIS OF CHIRAL (*E*)-CROTYLSILANES: [3*R*- AND 3*S*-]-(4*E*)-METHYL 3-(DIMETHYL-PHENYLSILYL)-4-HEXENOATE

(*S*)-**3a** ($SiPhMe_2$, HO) → propionic acid (cat.), $(MeO)_3CCH_3$(4 equiv), toluene, Δ → (*S*)-**4a** (CO_2Me, $PhSiMe_2$)

(*R*)-**3c** ($SiPhMe_2$, HO) → propionic acid (cat.), $(MeO)_3CCH_3$(4 equiv), toluene, Δ → (*R*)-**4b** (CO_2Me, $PhSiMe_2$)

CLASS 19 Sulfur Reagents

A. Reagents

Cyclopropyldiphenylsulfonium Tetrafluoroborate
Dimethylsulfonium Methylide
Dimethylsulfoxonium Methylide
Formaldehyde Dimethyl Thioacetal Monoxide
2-Lithio-1,3-dithiane
α-Phenylsulfonylmethyllithium
p-Tolylsulfinylmethyllithium

B. Reviews (1992–1998)

(1) Drabowicz, J.; Kielbasinski, P. and Lyzwa, P. Stereoselective Reactions at the α-Carbon Atom in Organosulfur Compounds. *Sulfur Reports* **1992**, *12*, 2163–96.

(2) Hua, D. H. Chiral Sulfinylallyl and α-Sulfinyl Ketimine Anions in Organic Synthesis. *Advances in Carbanion Chemistry*; JAI Press: Greenwich, 1992, Vol. 1.

(3) Ishibashi, H. and Ikeda, M. Recent Advances in Carbon-Carbon Bond Forming Reactions Using α-Thiocarbocations. *Rev. Heterocycl. Chem.* MYU K.K.: Tokyo, Japan, 1992, Vol. 7.

(4) Metzner, P. The Use of Thiocarbonyl Compounds in Carbon-Carbon Bond Forming Reactions. *Synthesis* **1992**, 1185–1199.

(5) Pyne, S. G. Diastereoselective Reactions of Sulfoximines. *Sulfur Reports* **1992**, *12*, 57–89.

(6) Satoh, T. and Yamakawa, K. Recent Developments in Organic Synthesis With 1-Haloalkyl Aryl Sulfoxides. *Synlett* **1992**, 455–68.

(7) Walker, A. J. Asymmetric Carbon-Carbon Bond Formation Using Sulfoxide-Stabilized Carbanions. *Tetrahedron: Asymmetry* **1992**, *3*, 961–998.

(8) Patai, S. and Rapport, Z., Eds. Supplement S. *The Chemistry of Sulfur-Containing Functional Groups*; Wiley: Chichester, 1993.

(9) Freeman, R.; Haynes, R. K.; Loughlin, W. A.; Mitchell, C. and Stokes, J. V. Synthetic Utilization of Highly Stereoselective Conjugate Addition Reactions of Phosphorus- and Sulfur-Stabilized Allylic Carbanions. *Pure Appl. Chem.* **1993**, *65*, 647–54.

(10) Koval', I. V. Thiols as Synthons. *Russ. Chem. Rev.* **1993**, *62*, 769.

(12) McGregor, W. M. and Sherrington, D. C. Some Recent Synthetic Routes to Thioketones and Thioaldehydes. *Chem. Soc. Rev.* **1993**, *22*, 199–204.

(13) Simpkins, N. S. *Sulphones in Organic Synthesis*; Pergamon: Oxford, 1993, Vol. 10.

(14) Bertrand, M. P. Recent Progress in the Use of Sulfonyl Radicals in Organic Synthesis. *Org. Prep. Proc. Int.* **1994**, *26*, 257–90.

(15) Metzner, P. and Thuillier, A. *Sulfur Reagents in Organic Synthesis*; Academic Press: London, 1994.

(16) Rayner, C. M. Thiols, Sulfides, Sulfoxides, and Sulfones. *Contemp. Org. Synth.* **1994**, *1*, 191.

(17) Solladie, G.; Antonio, A. and Carmen, D. Asymmetric Synthesis of Natural Products Monitored by Chiral Sulfoxides. *Pure Appl. Chem.* **1994**, *66*, 2159–62.

(18) Carreño, M. C. Applications of Sulfoxides to Asymmetric Synthesis of Biologically Active Compounds. *Chem. Rev.* **1995**, *95*, 1717–60.

(19) Dondoni, A. *New Formyl Anion and Cation Equivalents*; JAI Press: London, 1995, Vol. 2.

(20) Ohno, A. and Higaki, M. Stereochemistry of α-Sulfinyl Carbanions. *Rev. Heteroatom. Chem.* **1995**, *13*, 1–24.

(21) Page, P. *Organosulfur Chemistry: Synthetic Aspects*; Academic Press: London, 1995.

(22) Whitham, G. H. *Organosulfur Chemistry*; Oxford University Press: Oxford, 1995.

(23) Cremlyn, R. *Organosulfur Chemistry: An Introduction*; Wiley: New York, **1996**

(24) Hiroi, K. Transition Metal or Lewis Acid-Catalyzed Asymmetric Reactions with Chiral Organosulfur Functionality. *Rev. Heteroatom Chem.* **1996**, *14*, 21–58.

(25) Hua, D. H. and Chen, J. Asymmetric Syntheses of Alkaloids and Amino Acids Via Sulfur-Containing Compounds. *Advances in Heterocyclic Natural Product Synthesis*; JAI Press: Greenwich, 1996, Vol 3.

(26) Dondoni, A. and Perrone, D. Thiazole-Based Routes to Amino Hydroxy Aldehydes, and Their Use for the Synthesis of Biologically Active Compounds. *Aldrichimica Acta* **1997**, *30*, 35–46.

(27) Lee, A. W. M. and Chan, W. H. Chiral Acetylenic Sulfoxides and Related Compounds in Organic Synthesis. *Top. Curr. Chem.* **1997**, *190*, 103–29.

(28) Mikolajczk, M.; Drabowicz, J. and Kielbasinski, P. *Chiral Sulfur Reagents: Applications in Asymmetric and Stereoselective Synthesis*; CRC: Boca Raton, 1997.

(29) Weinreb, S. M. *N*-Sulfonyl Imines; Useful Synthons in Stereoselective Organic Synthesis. *Top. Curr. Chem.* **1997**, *190*, 131–84.

(30) Davies, F. A.; Zhou, P. and Chen, B.-C. Asymmetric Synthesis of Amino Acids Using Sulfinimines (Thiooxime *S*-Oxides). *Chem. Soc. Rev.* **1998**, *27*, 13–8.

(31) Smith, A. B., III; Condon, S. M. and McCauley, J. A. Total Synthesis of Immunosuppressants: Unified Strategies Exploiting Dithiane Couplings and σ-Bond Olefin Constructions. *Acc. Chem. Res.* **1998**, *31*, 35–46.

C. *Organic Syntheses Procedures (Vols. 70–75)*

Authors	Citation	Title
Dearg S. Brown and Steven V. Ley	*Org. Synth.* **1992**, *70*, 157.	SUBSTITUTION REACTIONS OF 2-BENZENESULFONYL CYCLIC ETHERS: TETRAHYDRO-2-(PHENYLETHYNYL)-2*H*-PYRAN

$PhC{\equiv}CH$, *i*-PrMgCl; $ZnBr_2$, THF; SO_2Ph; O; Ph

Authors	Citation	Title
Karl R. Dahnke and Leo A. Paquette	*Org. Synth.* **1992**, *71*, 175.	2-METHYLENE-1,3-DITHIOLANE

$ClCH_2CH(OMe)_2$ + HS⌒SH; HCl (conc.), 0°C to 25°C; S, S, H, Cl; MeLi·LiBr, Et_2O; S, S

Authors	Citation	Title
A. Dondoni and P. Merino	*Org. Synth.* **1993**, *72*, 21.	DIASTEREOSELECTIVE HOMOLOGATION OF D-(*R*)-GLYCERALDEHYDE ACETONIDE USING 2-(TRIMETHYLSILYL)-THIAZOLE: 2-*O*-BENZYL-3,4-ISOPROPYLIDENE-D-ERYTHROSE

O, O, CHO; N, S, $SiMe_3$; Bu_4NF; O, O, S, N, OH

$NaH, C_6H_5CH_2Br$; O, O, S, N, $OCH_2C_6H_5$; 1. MeI, 2. $NaBH_4$, 3. $HgCl_2$, H_2O; O, O, CHO, $OCH_2C_6H_5$

Authors	Citation	Title
P. A. Magriotis and J. T. Brown	*Org. Synth.* **1993**, *72*, 252.	PHENYLTHIOACETYLENE

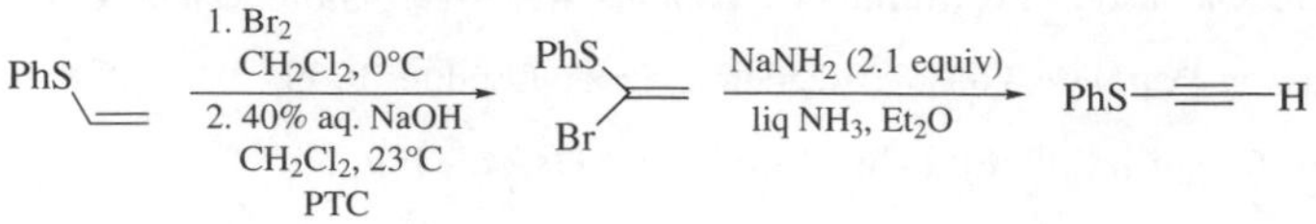

1. Br_2, CH_2Cl_2, 0°C; 2. 40% aq. NaOH, CH_2Cl_2, 23°C, PTC; $NaNH_2$ (2.1 equiv), liq NH_3, Et_2O; PhS—≡—H

Authors	Citation	Title
Scott H. Watterson, Zhijie Ni, Shaun S. Murphree, and Albert Padwa	*Org. Synth.* **1996**, *74*, 115.	2,3-DIBROMO-1-(PHENYLSULFONYL)-1-PROPENE AS A VERSATILE REAGENT FOR THE SYNTHESIS OF FURANS AND CYCLOPENTENONES: 2-METHYL-4-[(PHENYLSULFONYL)-METHYL]FURAN AND 2-METHYL-3-[(PHENYLSULFONYL)METHYL]-2-CYCLOPENTEN-1-ONE

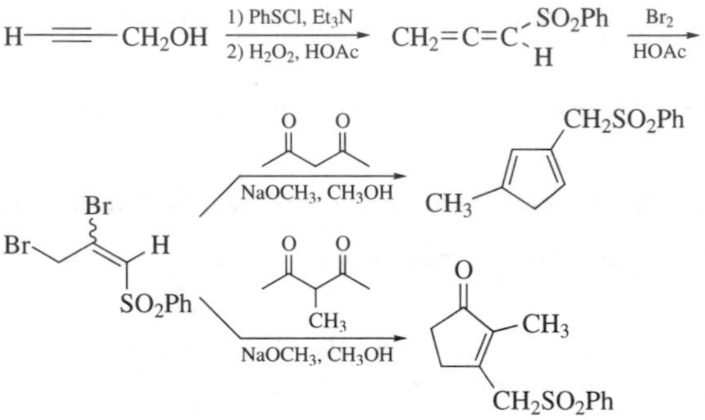

H—≡—CH_2OH → 1) PhSCl, Et_3N; 2) H_2O_2, HOAc → $CH_2{=}C{=}C(SO_2Ph)H$ → Br_2, HOAc

$NaOCH_3$, CH_3OH; CH_2SO_2Ph; CH_3; SO_2Ph; Br; H; CH_3; $NaOCH_3$, CH_3OH; O; CH_3; CH_2SO_2Ph

Authors	Citation	Title
Daniel S. Reno and Richard J. Pariza	*Org. Synth.* **1996**, *74*, 124.	PHENYL VINYL SULFIDE

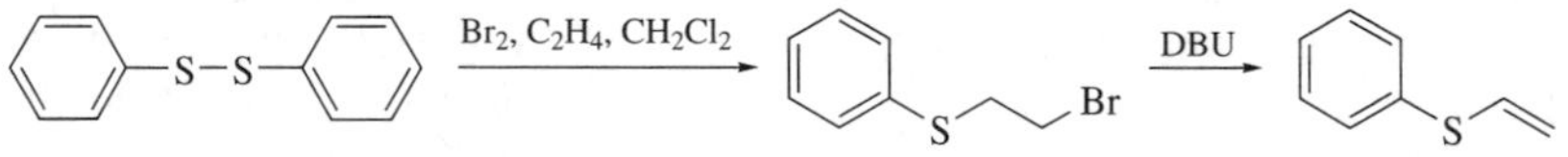

S–S; Br_2, C_2H_4, CH_2Cl_2; S; Br; DBU; S

Authors	Citation	Title
Albert Padwa, Scott H. Watterson, and Zhijie Ni	*Org. Synth.* **1996**, *74*, 147	[3 + 2]-ANIONIC ELECTROCYCLIZATION USING 2,3-BIS(PHENYL-SULFONYL)-1,3-BUTADIENE: *trans*-4,7,7-TRICARBOMETHOXY-2-PHENYL-SULFONYLBICYCLO[3.3.0]OCT-1-ENE

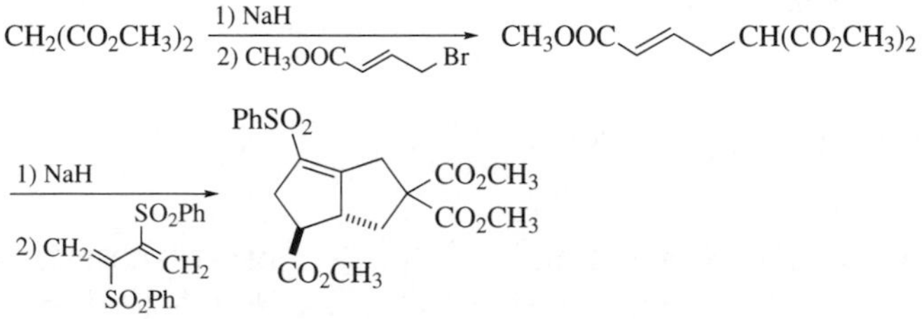

$CH_2(CO_2CH_3)_2$ → 1) NaH; 2) CH_3OOC–CH=CH–CH_2Br → CH_3OOC–CH=CH–CH_2–$CH(CO_2CH_3)_2$

1) NaH; 2) $CH_2{=}C(SO_2Ph){-}C(SO_2Ph){=}CH_2$; $PhSO_2$; CO_2CH_3; CO_2CH_3; CO_2CH_3

CLASS 20 Titanium Reagents

A. Reagents

1,1-Bis(cyclopentadienyl)-3,3-dimethyltitanocyclobutane
Bis(cyclopentadienyl)(dimethyl)Titanium
μ-Chlorobis(cyclopentadienyl)(dimethylaluminum)-μ-methylenetitanium
Diiodomethane–Zinc-Titanium(IV) Chloride
Methyltitanium Trichloride

B. Reviews (1992–1998)

(1) Duthaler, R. O. and Hafner, A. Chiral Titanium Complexes for Enantioselective Addition of Nucleophiles to Carbonyl Groups. *Chem. Rev.* **1992**, *92*, 807–32.

(2) Naraska, K. and Iwasawa, N. Asymmetric Reaction Promoted By Titanium Reagents. *Organic Synthesis-Theory and Applications*; JAI Press: Greenwich, 1993.

(3) Pine, S. H. Carbonyl Methylenation and Alkylidenation Using Titanium-Based Reagents. *Org. React.* **1993**, *43*, 1–92.

(4) Donohoe, T. J. Stoichiometric Applications of Organotransition Metal Complexes in Organic Synthesis. *Contemp. Org. Synth.* **1996**, *3*, 1–18.

(5) Fürstner, A. and Bogdanović, B. New Developments in the Chemistry of Low-Valent Titanium. *Angew. Chem., Int. Ed. Engl.* **1996**, *35*, 2442–69.

(6) Hoveyda, A. H. and Morken, J. P. Enantioselective C–C and C–H Bond Formation Mediated or Catalyzed by Chiral ebthi [ethylenebis(tetrahydroindenyl)] Complexes of Titanium and Zirconium. *Angew. Chem., Int. Ed. Engl.* **1996**, *35*, 1262–84.

(7) Ohff, A.; Pulst, S.; Lefeber, C.; Peulecke, N.; Arndt, P.; Burkalov, V. V. and Rosenthal, U. Unusual Reactions of Titanocene- and Zirconocene-Generating Complexes. *Advances in Heterocyclic Natural Product Synthesis*; JAI Press: Greenwich, 1996, Vol. 3.

C. Organic Syntheses Procedures (Vols. 70–75)

Authors	Citation	Title
Koichi Mikami, Masahiro Terada, Satoshi Narisawa, and Takeshi Nakai	*Org. Synth.* **1992**, *71*, 14.	ASYMMETRIC CATALYTIC GLYOXYLATE-ENE REACTION: METHYL (2*R*)-2-HYDROXY-4-PHENYL-4-PENTENOATE

A. $TiBr_4$ + $(i\text{-PrO})_4Ti$ —hexane→ 2 $(i\text{-PrO})_2TiBr_2$

B. Ph(C=CH$_2$)CH$_3$ + H(C=O)CO_2CH_3 —$(i\text{-PrO}_2TiBr_2/(R)\text{-}(+)\text{-}1,1'\text{-bi-2-naphthol}$ (0.5 mol% each), 4Å sieves, −30°C, CH_2Cl_2→ Ph(C=CH$_2$)CH$_2$CH(OH)CO_2CH_3

Authors	Citation	Title
Kazuhiko Takai, Yasutaka Kataoka, Jiro Miyai, Takashi Okazoe, Koichiro Oshima, and Kiitiro Utimoto	*Org. Synth.* **1995**, *73*, 73.	ALKYLIDENATION OF ESTER CARBONYL GROUPS: (*Z*)-1-ETHOXY-1-PHENYL-1-HEXENE

A. CH_2Br_2 —1) LDA, THF-ether-hexane, −90°C; 2) C_4H_9I, HMPA→ $C_4H_9CHBr_2$

B. Ph(C=O)OEt + $C_4H_9CHBr_2$ —$TiCl_4$, TMEDA, Zn, cat. $PbCl_2$, THF, 25°C→ Ph(OEt)C=CHC_4H_9

CLASS 21 Transition Metal and Lanthanide Reagents

A. Reagents

Cerium (III) Chloride
Chlorobis(cyclopentadienyl)hydridozirconium
Chlorobis(cyclopentadienyl)methylzirconium
Chromium(II) Chloride
Chromium(II) Chloride-Haloform
Chromium(II) Chloride-Nickel(II) Chloride
Dirhodium(II) Tetraacetate
Dirhodium(II) Tetrakis(methyl 2-pyrrolidone-5-(*S*)-carboxylate)
Manganese(III) Acetate
Trimethylaluminum-Dichlorobis(η^5-cyclopentadienyl)zirconium

B. *Reviews (1992–1998)*

(1) Brunner, H. Enantioselective Rhodium (II) Catalysts. *Angew. Chem., Int. Ed. Engl.* **1992**, *31*, 1183–85.

(2) Davies, S. G.; Donohoe, T. J. and Williams, J. M. J. Stereoselective Manipulation of Acetals Derived From *O*-Substituted Benzaldehyde Chromium Tri-Carbonyl Complexes. *Pure Appl. Chem.* **1992**, *64*, 379–86.

(3) Duthaler, R. O. Stereoselective Transformations Mediated by Chiral Monocyclopentadienyl Titanium, Zirconium, and Hafnium Complexes. *Pure Appl. Chem.* **1992**, *64*, 1897–910.

(4) Gerasimov, O. V. and Parmon, V. N. Photocatalysis by Transition Metal Complexes. *Russ. Chem. Rev.* **1992**, *61*, 154–67.

(5) Halterman, R. L. Synthesis and Applications of Chiral Cyclopentadienylmetal Complexes. *Chem. Rev.* **1992**, *92*, 965–94.

(6) Kasatkin, A. N.; Tsypyshev, O. Y.; Romanova, T. Y. and Tolstikov, G. A. Organic Derivatives of Manganese(II) in Organic Synthesis. *Russ. Chem. Rev.* **1992**, *61*, 537.

(7) Parshall, G. W. and Ittel, S. D. *Homogeneous Catalysis: The Applications and Chemistry of Catalysis by Soluble Transition Metal Complexes*; Wiley: New York, 1992.

(8) Annby, U.; Karlsson, S.; Gronowitz, S.; Hallberg, A.; Alvhall, J. and Svenson, R. Hydrozirconation-Isomerization Reactions of Terminally Functionalized Olefins With Zirconocene Hydrides and General Aspects. *Acta Chem. Scand.* **1993**, *47*, 425–33.

(9) Casey, C. P. Organorhenium Chemistry. *Science* **1993**, *259*, 1552–8.

(10) Davies, S. G. and Donohoe, T. J. Arene Chromium Tricarbonyl-Stabilized Benzylic Carbocations. *Synlett* **1993**, 323–32.

(11) Melikyan, G. G. Manganese(III) Mediated Reactions of Unsaturated Systems. *Synthesis* **1993**, 833–50.

(12) Rigby, J. H. Transition Metal-Promoted Higher-Order Cycloaddition Reactions in Organic Synthesis. *Acc. Chem. Res.* **1993**, *26*, 579–85.

(13) Sodeoka, M. and Shibasaki, M. Arene Chromium Tricarbonyl-Catalyzed Reactions in Organic Synthesis. *Synthesis* **1993**, 643–58.

(14) Blagg, J. Stoichiometric Applications of Organotransition Metal Complexes in Organic Synthesis. *Contemp. Org. Synth.* **1994**, *1*, 125.

(15) Dawson, G. J. and Williams, J. M. J. Catalytic Applications of Transition Metals in Organic Synthesis. *Contemp. Org. Synth.* **1994**, *1*, 77.

(16) Hegedus, L. S. *Transition Metals in the Synthesis of Complex Organic Molecules*; University Science Books: Mill Valley, 1994.

(17) Henderson, R. A. *The Mechanisms of Reactions at Transition Metal Sites*; Oxford University Press: New York, 1994.

(18) Imamoto, T. *Lanthanides in Organic Synthesis*; Academic Press: London, 1994.

(19) Negishi, E.-I. and Takahashi, T. Patterns of Stoichiometric and Catalytic Reactions of Organozirconium and Related Complexes of Synthetic Interest. *Acc. Chem. Res.* **1994**, *27*, 124–30.

(20) Pearson, A. J. *Iron Compounds in Organic Synthesis*; Academic Press: London, 1994.

(21) Schaverien, C. J. Organometallic Chemistry of the Lanthanides. *Adv. Org. Chem.* **1994**, *36*, 283.

(22) Schmalz, H.-G. Chromium Carbene Complexes in Organic Synthesis: Recent Developements and Perspectives. *Angew. Chem., Int. Ed. Engl.* **1994**, *33*, 303–5.

(23) Schrock, R. R. Recent Advances in the Chemistry and Applications of High Oxidation State Alkylidene Complexes. *Pure Appl. Chem.* **1994**, *66*, 1447–54.

(24) Agbossou, F.; Carpentier, J.-F. and Mortreux, A. Asymmetric Hydroformylation. *Chem. Rev.* **1995**, *95*, 2485–506.

(25) Blagg, J. Stoichiometric Organotransition Metal Complexes in Organic Synthesis. *Contemp. Org. Synth.* **1995**, *2*, 43–64.

(26) Frost, C. G. and Williams, J. M. J. Catalytic Applications of Transition Metals in Organic Synthesis. *Contemp. Org. Synth.* **1995**, *2*, 65–83.

(27) Grubbs, R. H.; Miller, S. J. and Fu, G. C. Ring-Closing Metathesis and Related Processes in Organic Synthesis. *Acc. Chem. Res.* **1995**, *28*, 446–52.

(28) Hidai, M. and Ishii, Y. Novel Carbonylation Reactions Catalyzed by Transition Metal Complexes. *Advances in Metal-Organic Chemistry*; JAI Press: Greenwich, 1995, Vol. 4.

(29) Kauffmann, T. High Selectivity Induced by Neighbouring-Group Effects in C–C Bond-Forming Reactions with Organotransition Metal Reagents. *Synthesis* **1995**, *7*, 745–55.

(30) Khumtaveeporn, K. and Alper, H. Transition Metal Mediated Carbonylative Ring Expansion of Heterocyclic Compounds. *Acc. Chem. Res.* **1995**, *28*, 414–22.

(31) Negishi, E.-I. Recent Advances in the Chemistry of Zirconocene and Related Compounds. *Tetrahedron* **1995**, *51*, 4255–570.

(32) Rigby, J. H. and Krueger, A. C. Mechanisms of Synthetically Useful Cycloaddition Reactions Mediated by Transition Metals. *Advances in Detailed Reaction Mechanisms. Vol 4: Synthetically Useful Reactions*; JAI Press: Greenwich, 1995.

(33) Roundhill, D. M. Organotransition Metal Chemistry and Homogeneous Catalysis in Aqueous Solution. *Adv. in Organomet. Chem.* **1995**, *38*, 155–88.

(34) Dawson, G. J.; Bower, J. F. and Williams, J. M. J. Catalytic Applications of Transition Metals in Organic Synthesis. *Contemp. Org. Synth.* **1996**, *3*, 277–93.

(35) Donohoe, T. J. Stoichiometric Applications of Organotransition Metal Complexes in Organic Synthesis. *Contemp. Org. Synth.* **1996**, *3*, 1–18.

(36) Lautens, M.; Klute, W. and Tam, W. Transition Metal-Mediated Cycloaddition Reactions. *Chem. Rev.* **1996**, *96*, 49–92.

(37) Ley, S. V.; Cox, L. R. and Meek, G. (π-Allyl)tricarbonyliron Lactone Complexes in Organic Synthesis: A Useful and Conceptually Unusual Route to Lactones and Lactams. *Chem. Rev.* **1996**, *96*, 423–42.

(38) Luh, T.-Y. Transition Metal-Catalyzed Cross Coupling Reactions of Unactivated Aliphatic C–X Bonds. *Rev. Heteroatom Chem.* **1996**, *15*, 61–82.

(39) Malacria, M. Selective Preparation of Complex Polycyclic Molecules from Acyclic Precursors via Radical Mediated- or Transition Metal-Catalyzed Cascade Reactions. *Chem. Rev.* **1996**, *96*, 289–30.

(40) Negishi, E.-I. and Kondakov, D. Y. An Odyssey from Stoichiometric Carbotitanation of Alkynes to Zirconium-Catalysed Enantioselective Carboalumination of Alkenes. *Chem. Soc. Rev.* **1996**, *25*, 417–26.

(41) Snider, B. B. Manganese(III)-Based Oxidative Free-Radical Cyclizations. *Chem. Rev.* **1996**, *96*, 339–63.

(42) Trost, B. M. and Van Vranken, D. L. Asymmetric Transition Metal-Catalyzed Allylic Alkylations. *Chem. Rev.* **1996**, *96*, 395–422.

(43) Wipf, P. and Jahn, H. Synthetic Applications of Organochlorozirconocene Complexes. *Tetrahedron* **1996**, *52*, 12853–910.

(44) Brandsma, L.; Vasilevsky, S. F. and Verkruijsse, H. D. *Application of Transition Metal Catalysts in Organic Synthesis*; Springer: Berlin, 1997.

(45) Davies, H. M. L. Asymmetric Synthesis Using Rhodium-Stabilized Vinylcarbenoid Intermediates. *Aldrichimica Acta* **1997**, *30*, 107–14.

(46) Donohoe, T. J.; Guyo, P. M.; Moore, P. R. and Stevenson, C. A. Applications of Stoichiometric Organotransition Metal Complexes in Organic Synthesis. *Contemp. Org. Synth.* **1997**, *4*, 22–39.

(47) Frühauf, H.-W. Metal-Assisted Cycloaddition Reactions in Organotransition Metal Chemistry. *Chem. Rev.* **1997**, *97*, 523–96.

(48) Hegedus, L. S. Chromium Carbene Complex Photochemistry in Organic Synthesis. *Tetrahedron* **1997**, *53*, 4105–4127.

(49) Iqbal, J.; Mukhopadhyay, M. and Mandal, A. D. Cobalt-Catalyzed Organic Transformations: Highly Versatile Protocols for Carbon-Carbon and Carbon-Heteroatom Bond Formation. *Synlett* **1997**, 876–86.

(50) Kauffmann, T. Organomolybdenum and Organotungsten Reagents. 7. Novel Reactions of Organomolybdenum and Organotungsten Compounds: Additive-Reductive Carbonyl Dimerization, Spontaneous Transformation of Methyl Ligands into μ-Methylene Ligands, and Selective Carbonylmethylenation. *Angew. Chem., Int. Ed. Engl.* **1997**, *36*, 1259–75.

(51) Melikyan, G. G. Carbon-Carbon Bond-Forming Reactions Promoted by Trivalent Manganese. *Org. React.* **1997**, *49*, 427–675.

(52) Rossi, R. and Bellina, F. Selective Transition Metal-Promoted Carbon-Carbon and Carbon-Heteroatom Bond Formation. A Review. *Org. Prep. Proc. Int.* **1997**, *29*, 139–76.

(53) Schuster, M. and Blechert, S. Olefin Metathesis in Organic Chemistry. *Angew. Chem., Int. Ed. Engl.* **1997**, *36*, 2037–56.

(54) Tonks, L. and Williams, J. M. J. Catalytic Applications of Transition Metals in Organic Synthesis. *Contemp. Org. Synth.* **1997**, *4*, 353–72.

(55) Wipf, P., Xu, W., Takahashi, H., Jahn, H. and Coish, P. D. G. Synthetic Applications of Organozirconocenes. *Pure Appl. Chem.* **1997**, *69*, 639–44.

(56) Beller, M. and Bolm, C. *Transition Metals for Organic Synthesis.* Wiley-VCH: Weinheim, 1998; Vol. 1 and 2.

(57) Doyle, M. P.; McKervey, M. A. and Ye. T. *Modern Catalytic Methods for Organic Synthesis with Diazo Compounds.* Wiley-Interscience: New York, 1998.

C. Organic Syntheses Procedures (Vols. 70–75)

Authors	Citation	Title
Matthew N. Mattson, Edward J. O'Connor, and Paul Helquist	*Org. Synth.* **1992**, *70*, 177.	CYCLOPROPANATION USING AN IRON-CONTAINING METHYLENE TRANSFER REAGENT: 1,1-DIPHENYLCYCLOPROPANE

$$\mathrm{Cp(OC)(CO)Fe{-}CH_2S^+(CH_3)_2\ BF_4^-} \xrightarrow[\text{dioxane}]{\mathrm{Ph_2C{=}CH_2}} \text{1,1-diphenylcyclopropane}$$

Authors	Citation	Title
Stephen L. Buchwald, Susan J. LaMaire, Ralph B. Nielsen, Brett T. Watson, and Susan M. King	*Org. Synth.* **1992**, *71*, 77.	SCHWARTZ'S REAGENT

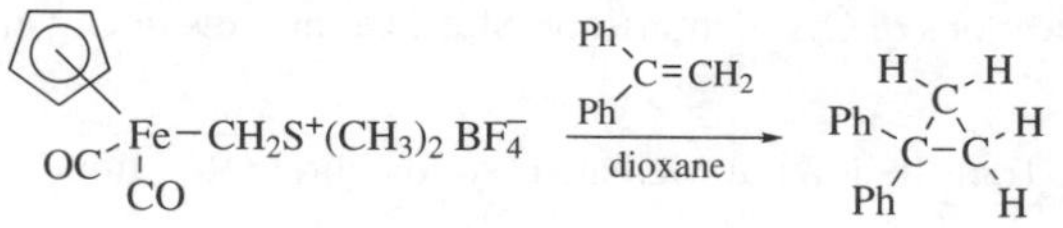

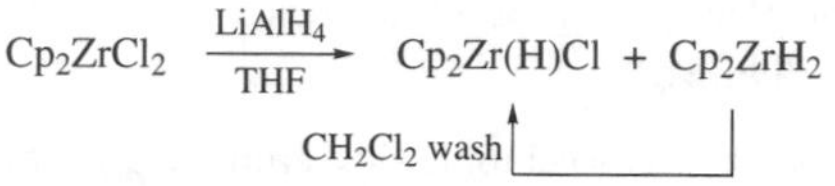

$$\mathrm{Cp_2ZrCl_2} \xrightarrow[\text{THF}]{\mathrm{LiAlH_4}} \mathrm{Cp_2Zr(H)Cl} + \mathrm{Cp_2ZrH_2}$$

$\mathrm{CH_2Cl_2}$ wash

Authors	Citation	Title
Ruen Chu Sun, Masami Okabe, David L. Coffen, and Jeffrey Schwartz	*Org. Synth.* **1992**, *71*, 83.	CONUGATE ADDITION OF A VINYLZIRCONIUM REAGENT: 3(1-OCTEN-1-YL)CYCLOPENTANONE

$$\mathrm{CP_2Zr(H)Cl} + \text{1-octyne} \xrightarrow{\text{THF}} [\mathrm{CP_2Zr(Cl)CH{=}CH(CH_2)_5CH_3}] \xrightarrow[\mathrm{Ni(acac)_2}]{\text{cyclopentenone}} \text{3-(1-octen-1-yl)cyclopentanone}$$

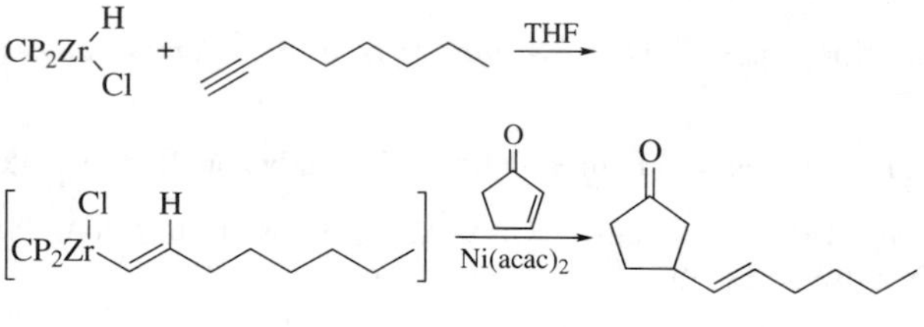

Authors	Citation	Title
Peter Wipf and Wenjing Xu	*Org. Synth.* **196**, *74*, 205	ALLYLIC ALCOHOLS BY ALKENE TRANSFER FROM ZIRCONIUM TO ZINC: 1-[(*tert*-BUTYLDIPHENYLSILYL)-OXY]-DEC-3-EN-5-OL

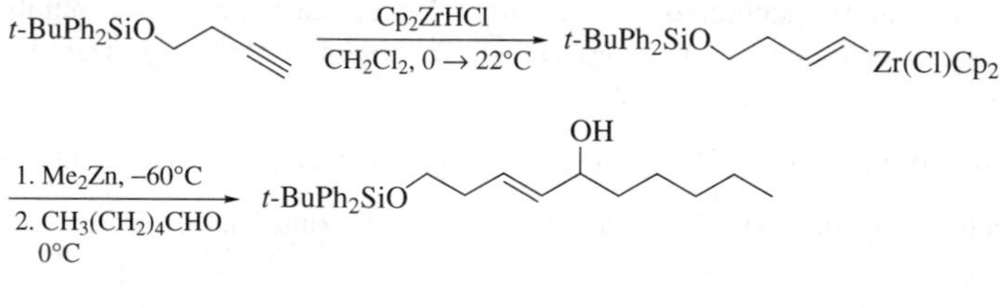

$$t\text{-}\mathrm{BuPh_2SiO(CH_2)_2C{\equiv}CH} \xrightarrow[\mathrm{CH_2Cl_2},\ 0 \rightarrow 22^\circ\mathrm{C}]{\mathrm{Cp_2ZrHCl}} t\text{-}\mathrm{BuPh_2SiO(CH_2)_2CH{=}CHZr(Cl)Cp_2}$$

$$\xrightarrow[\text{2. } \mathrm{CH_3(CH_2)_4CHO},\ 0^\circ\mathrm{C}]{\text{1. } \mathrm{Me_2Zn},\ -60^\circ\mathrm{C}} t\text{-}\mathrm{BuPh_2SiO(CH_2)_2CH{=}CHCH(OH)(CH_2)_4CH_3}$$

CLASS 22 Zinc Reagents

A. Reagents

Dibromomethane-Zinc-Copper Couple
Diethylzinc
Iodomethylzinc Iodide
Zinc-Copper Couple

B. Reviews (1992–1998)

(1) Erdik, E. Transition Metal Catalyzed Reactions of Organozinc Reagents. *Tetrahedron* **1992**, *48*, 9577–648.

(2) Normant, J. F.; Marek, I. and Lefrancois, J. M. Organobismetallic Zinc Reagents: Their Preparation and Use in Diastereoselective Reactions. *Pure Appl. Chem.* **1992**, *64*, 1857–64.

(3) Soai, K. and Niwa, S. Enantioselective Addition of Organozinc Reagents to Aldehydes. *Chem. Rev.* **1992**, *92*, 833–56.

(5) Knochel, P. and Singer, R. D. Preparation and Reactions of Polyfunctional Organozinc Reagents in Organic Synthesis. *Chem. Rev.* **1993**, *93*, 2117–88.

(6) Motherwell, W. B. and Nutley, C. J. The Role of Zinc Carbenoids in Organic Synthesis. *Contemp. Org. Synth.* **1994**, *1*, 219.

(7) Charette, A. B. and Marcoux, J.-F. The Asymmetric Cyclopropanation of Acyclic Allylic Alcohols: Efficient Stereocontrol with Iodomethylzinc Reagents. *Synlett* **1995**, *12*, 1197–207.

(8) Knochel, P. Stereoselective Reactions Mediated by Functionalized Diorganozincs. *Synlett* **1995**, *5*, 393–403.

(9) Knochel, P. New Preparation of Polyfunctional Dialkylzincs and Their Application in Asymmetric Synthesis *CHEMTRACTS*: *Org. Chem.* **1995**, *8*, 205–21.

(10) Tamaru, Y. Unique Reactivity of Functionalized Organozincs. *Advances in Detailed Reaction Mechanisms Volume 4: Synthetically Useful Reactions*; JAI Press: Greenwich, 1995, Vol. 4.

(11) van der Baan, J. L.; van der Heide, T. A. J.; van der Louw, J. and Klumpp, G. W. Preparation of Carbocyclic and Heterocyclic Compounds by the Use of Allylzinc and an Allylpalladium in Tandem. *Synlett* **1995**, 1–12.

(12) Erdik, E. *Organozinc Reagents in Organic Synthesis*; CRC Press: Boca Raton, 1996.

(13) Singh, V. K.; DattaGupta, A. and Sekar, G. Catalytic Enantioselective Cyclopropanation of Olefins Using Carbenoid Chemistry. *Synthesis* **1997**, 137–49.

C. Organic Syntheses Procedures (Vols. 70–75)

Authors	Citation	Title
Ming Chang P. Yeh, Huai Gu Chen, Paul Knochel	*Org. Synth.* **1992**, *70*, 195.	1,2-ADDITION OF A FUNCTIONALIZED ZINC-COPPER ORGANOMETALLIC [RCu(CN)ZnI] TO AN α, β-UNSATURATED ALDEHYDE: (*E*)-2-(4-HYDROXY-6-PHENYL-5-HEXENYL)-1*H*-ISOINDOLE-1,3(2*H*)-DIONE

1) Zn/THF
2) CuCN · 2 LiCl

Cu(CN)ZnI

1) $BF_3{\cdot}OEt_2$
2) Ph ... H (cinnamaldehyde)

OH Ph

Acetyl Chloride[1]

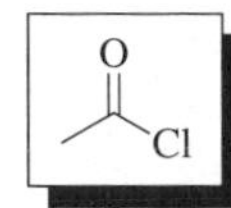

[75-36-5] C_2H_3ClO (MW 78.50)

(useful for electrophilic acetylation of arenes,[2] alkenes,[2a,3] alkynes,[4] saturated alkanes,[3a,5] organometallics, and enolates (on C or O);[6] for cleavage of ethers;[7] for esterification of sterically unhindered[8] or acid-sensitive[9] alcohols; for generation of solutions of anhydrous hydrogen chloride in methanol;[10] as a dehydrating agent; as a solvent for organometallic reactions;[11] for deoxygenation of sulfoxides;[12] as a scavenger for chlorine[13] and bromine;[14] as a source of ketene; and for nucleophilic acetylation[15])

Physical Data: bp 51.8 °C;[1a] mp −112.9 °C;[1a] *d* 1.1051 g cm^{-3};[1a] refractive index 1.38976.[1b] IR (neat) ν 1806.7 cm^{-1};[16] ^{1}H NMR ($CDCl_3$) δ 2.66 ppm; ^{13}C NMR ($CDCl_3$) δ 33.69 ppm (q) and 170.26 ppm (s); the bond angles (determined by electron diffraction[17]) are 127.5° (O–C–C), 120.3° (O–C–Cl), and 112.2° (Cl–C–C).

Analysis of Reagent Purity: a GC assay for potency has been described;[18] to check qualitatively for the presence of HCl, a common impurity, add a few drops of a solution of crystal violet in chloroform;[19] a green or yellow color indicates that HCl is present, while a purple color that persists for at least 10 min indicates that HCl is absent.[1b]

Preparative Methods: treatment of ***Acetic Acid*** or sodium acetate with the standard inorganic chlorodehydrating agents (PCl_3,[1b,23] SO_2Cl_2,[1a,24] or $SOCl_2$[1b,25]) generates material that may contain phosphorus- or sulfur-containing impurities.[1b,23a,26] Inorganic-free material can be prepared by treatment of HOAc with $Cl_2CHCOCl$ (Δ; 70%),[27] PhCOCl (Δ; 88%),[28] $PhCCl_3$ (cat. H_2SO_4, 90 °C; 92.5%),[29] or phosgene[30] (optionally catalyzed by DMF,[30e] magnesium or other metal salts,[30a,b,d] or activated carbon[30b,c]), or by addition of hydrogen chloride to acetic anhydride (85–90 °C; 'practically quantitative').[1a,31]

Purification: HCl-free material can be prepared either by distillation from dimethylaniline[11c,20] or by standard degassing procedures.[20c,21]

Handling, Storage, and Precautions: acetyl chloride should be handled only in a well-ventilated fume hood since it is volatile and toxic via inhalation.[22] It should be stored in a sealed container under an inert atmosphere. Spills should be cleaned up by covering with aq. sodium bicarbonate.[1a]

Friedel–Crafts Acetylation. Arenes undergo acetylation to afford aryl methyl ketones on treatment with acetyl chloride (AcCl) together with a Lewis acid, usually ***Aluminum Chloride***$_3$. This reaction, known as the Friedel–Crafts acetylation, is valuable as a preparative method because a single positional isomer is produced from arenes that possess multiple unsubstituted electron-rich positions in many instances.

For example, Friedel–Crafts acetylation of toluene ($AcCl/AlCl_3$, ethylene dichloride, rt) affords *p*-methylacetophenone predominantly (*p*:*m*:*o* = 97.6:1.3:1.2; eq 1).[32]

AcCl/$AlCl_3$, $ClCH_2CH_2Cl$, rt (1)

p:*m*:*o* = 97.6:1.3:1.2

Acetylation of chlorobenzene under the same conditions affords *p*-chloroacetophenone with even higher selectivity (*p*:*m* = 99.5:0.5).[33] Acetylation of bromobenzene[33] and fluorobenzene[33] afford the *para* isomers exclusively. The *para:meta*[34] and *para:ortho*[32,34] selectivities exhibited by $AcCl/AlCl_3$ are greater than those exhibited by most other Friedel–Crafts electrophiles.

Halogen substituents can be used to control regioselectivity. For example, by introduction of bromine *ortho* to methyl, it is possible to realize '*meta* acetylation of toluene' (eq 2).[35]

Br — AcCl/$AlCl_3$, CH_2Cl_2, rt — Br, O (2)

Regioselectivity is quite sensitive to reaction conditions (e.g. solvent, order of addition of the reactants, concentration, and temperature). For example, acetylation of naphthalene can be directed to produce either a 99:1 mixture of C-1:C-2 acetyl derivatives (by addition of a solution of arene and AcCl in CS_2 to a slurry of $AlCl_3$ in CS_2 at 0 °C) or a 7:93 mixture (by addition of the preformed $AcCl/AlCl_3$ complex in dichloroethane to a dilute solution of the arene in dichloroethane at rt).[36] Similarly, acetylation of 2-methoxynaphthalene can be directed to produce either a 98:2 mixture of C-1:C-6 acetyl derivatives (using the former conditions) or a 4:96 mixture (by addition of the arene to a solution of the preformed $AcCl/AlCl_3$ complex in nitrobenzene).[37] Also, acetylation of 1,2,3-mesitylene can be directed to produce either a 100:0 mixture of C-4:C-5 isomers or a 3:97 mixture.[36c]

Frequently, regioselectivity is compromised by side reactions catalyzed by the HCl byproduct. For example, acetylation of *p*-xylene by treatment with $AlCl_3$ followed by Ac_2O (CS_2, Δ, 1 h) produces a 69:31 mixture of 2,5-dimethylacetophenone and 2,4-dimethylacetophenone, formation of the latter being indicative of competitive acid-catalyzed isomerization of *p*-xylene to *m*-xylene.[38] Also, although acetylation of anthracene affords 9-acetylanthracene regioselectively, if the reaction mixture is allowed to stand for a prolonged time prior to work-up (rt, 20 h) isomerization to a mixture of C-1, C-2, and C-9 acetyl derivatives occurs.[39]

These side reactions can be minimized by proper choice of reaction conditions. Isomerization of the arene can be suppressed by adding the arene to the preformed $AcCl/AlCl_3$ complex. This order of mixing is known as the 'Perrier modification' of the Friedel–Crafts reaction.[40] Acetylation of *p*-xylene using this order of mixing affords 2,5-dimethylacetophenone exclusively.[38] Isomerization of the product aryl methyl ketone can be suppressed by crystallizing the product out of the reaction mixture as it is formed. For example, on acetylation of anthracene in benzene at 5–10 °C, 9-acetylanthracene crystallizes out of the reaction mixture (as its 1/1 $AlCl_3$ complex) in pure form.[39] Higher yields of purer products can also be obtained by substituting ***Zirconium(IV) Chloride***[41] or ***Tin(IV) Chloride***[42] for $AlCl_3$.

AcCl is not well suited for industrial scale Friedel–Crafts acetylations because it is not commercially available in bulk (only by the drum) and therefore must be prepared on site.[1] The combination of ***Acetic Anhydride*** and anhydrous ***Hydrogen Fluoride***, both of which are available by the tank car, is claimed to be more practical.[43] On laboratory scale, AcCl/$AlCl_3$ is more attractive than Ac_2O/HF or Ac_2O/$AlCl_3$. Whereas one equivalent of $AlCl_3$ is sufficient to activate AcCl, 1.5–2 equiv $AlCl_3$ (relative to arene) are required to activate Ac_2O.[36a,37b,38,44] Thus, with Ac_2O, greater amounts of solvent are required and temperature control during the quench is more difficult. Also, slightly lower isolated yields have been reported with Ac_2O than with AcCl in two cases.[36a,45] However, it should be noted that the two reagents generally afford similar ratios of regioisomers.[36a,38,46]

Acetylation of Alkenes. Alkenes, on treatment with AcCl/$AlCl_3$ under standard Friedel–Crafts conditions, are transformed into mixtures of β-chloroalkyl methyl ketones, allyl methyl ketones, and vinyl methyl ketones, but the reaction is not generally preparatively useful because both the products and the starting alkenes are unstable under the hyperacidic reaction conditions. Preparatively useful yields have been reported only with electron poor alkenes such as ethylene (dichloroethane, 5–10 °C; >80% yield of 4-chloro-2-butanone)[47] and ***Allyl Chloride*** (CCl_4, rt; 78% yield of 5-chloro-4-methoxy-2-pentanone after methanolysis),[48] which are relatively immune to the effects of acid.

The acetylated products derived from higher alkenes are susceptible to protonation or solvolysis which produces carbenium ions that undergo Wagner–Meerwein hydride migrations.[49] For example, on subjection of cyclohexene to standard Friedel–Crafts acetylation conditions (AcCl/$AlCl_3$, CS_2 −18 °C), products formed include not only 2-chlorocyclohexyl methyl ketone (in 40% yield)[50] but also 4-chlorocyclohexyl methyl ketone.[2a,51] If benzene is added to the crude acetylation mixture and the temperature is then increased to 40–45 °C for 3 h, 4-phenylcyclohexyl methyl ketone is formed in 45% yield (eq 3).[49a,b]

AcCl/$AlCl_3$, CS_2, −20 °C → [Cl-cyclohexyl methyl ketone, not isolated] → benzene, 40–45 °C, 3 h, 45% (3)

Wagner–Meerwein rearrangement also occurs during acetylation of methylcyclohexene, even though the rearrangement is anti-Markovnikov (β-tertiary → γ-secondary; eq 4).[52] Acetylation of *cis*-decalin[53] (see 'Acetylation of Saturated Alkanes' section below) also produces a β-tertiary carbenium ion that undergoes anti-Markovnikov rearrangement. The rearrangement is terminated by intramolecular *O*-alkylation of the acetyl group by the γ-carbenium ion to form a cyclic enol ether in two cases.[49c,53]

AcCl/$AlCl_3$, CS_2, −5 °C → rt; NaOH, MeOH, rt; *endo*:*exo* = 79:21 (4)

Higher alkenes themselves are also susceptible to protonation. The resulting carbenium ions decompose by assorted pathways including capture of chloride (with $SnCl_4$ as the catalyst),[51,54] addition to another alkene to form dimer or polymer,[5b,55] proton loss (resulting in *exo/endo* isomerization), or skeletal rearrangement.[56]

Higher alkenes can be acetylated in synthetically useful yield by treatment with AcCl together with various mild Lewis acids. One that deserves prominent mention is ***Ethylaluminum Dichloride*** (CH_2Cl_2, rt), which is useful for acetylation of all classes of alkenes (monosubstituted, 1,2-disubstituted, and trisubstituted).[57] For example, cyclohexene is converted into an 82/18 mixture of 3-acetylcyclohexene and 2-chlorocyclohexyl methyl ketone in 89% combined yield.

The following Lewis acids are also claimed to be superior to $AlCl_3$: Zn(Cu)/CH_2I_2 (AcCl, CH_2Cl_2, Δ), by which cyclohexene is converted into acetylcyclohexene in 68% yield (after treatment with KOH/MeOH);[58] $ZnCl_2$ (AcCl, Et_2O/CH_2Cl_2, −75 °C → −20 °C), by which 2-methyl-2-butene is converted into a 15:85 mixture of 3,4-dimethyl-4-penten-2-one and 4-chloro-3,4-dimethyl-2-pentanone in 'quantitative' combined yield;[59] and $SnCl_4$, by which cyclohexene (AcCl, CS_2, −5 °C → rt) is converted into acetylcyclohexene in 50% yield (after dehydrochlorination with $PhNEt_2$ at 180 °C),[60] methylcyclohexene (CS_2, rt) is converted into 1-acetyl-2-methylcyclohexene in 48% yield (after dehydrochlorination),[52] and camphene is converted into an acetylated derivative in ≈65% yield.[49c]

Conducting the acetylation in the presence of a non-nucleophilic base or polar solvent is reported to be advantageous. For example, methylenecyclohexane can be converted into 1-cyclohexenylacetone in 73% yield by treatment with $AcSbCl_6$ in the presence of Cy_2NEt (CH_2Cl_2, −50 °C → −25 °C, 1 h)[61] and cyclohexene can be converted into 3-acetylcyclohexene in 80% yield by treatment with $AcBF_4$ in $MeNO_2$ at −25 °C.[62]

Employment of Ac_2O instead of AcCl is also advantageous in some cases. For example, methylcyclohexene can be converted into 3-acetyl-2-methylcyclohexene in 90% yield by treatment with $ZnCl_2$ (neat Ac_2O, rt, 12 h).[63]

Finally, alkenes can be diacetylated to afford pyrylium salts by treatment with excess AcCl/$AlCl_3$,[55b,56,64] albeit in low yield (eq 5).[64a]

$AcCl/AlCl_3$, 25 °C, 20% (5) $^{-}AlCl_4$

Acetylation of Alkynes. Under Friedel–Crafts conditions ($AcCl/AlCl_3$, CCl_4, 0–5 °C), acetylene undergoes acetylation to afford β-chlorovinyl methyl ketone in 62% yield[4] and under similar conditions ($AcSbF_6$, $MeNO_2$, −25 °C) 5-decyne undergoes acetylation to afford 6-acetyl-5-decanone in 73% yield.[65]

Acetylation of Saturated Alkanes. Saturated alkanes, on treatment with a slight excess of $AcCl/AlCl_3$ at elevated temperature, undergo dehydrogenation (by hydride abstraction followed by deprotonation) to alkenes, which undergo acetylation to afford vinyl methyl ketones. The hydride-abstracting species is believed to be either the acetyl cation[66] or $HAlCl_4$,[67] with most evidence favoring the former. Perhaps because the alkenes are generated slowly and consumed rapidly, and therefore are never present in high enough concentration to dimerize, yields are typically higher than those of acetylation of the corresponding alkenes.[53b,68] A similar hypothesis has been offered to explain the phenomenon that the yield from acetylation of tertiary alkyl chlorides is typically higher than the yield from acetylation of the corresponding alkenes.[55b,64a] For example, methylcyclopentane on treatment with $AcCl/AlCl_3$ (CH_2Cl_2, Δ) undergoes acetylation to afford 1-acetyl-2-methylcyclopentene in an impressive 60% yield (eq 6).[53b,66a]

1.2 equiv $AcCl/AlCl_3$, CH_2Cl_2, Δ, 2 h; 60% (6)

If the reaction is carried out with excess alkane, a second hydride transfer occurs, resulting in reduction of the enone to the corresponding saturated alkyl methyl ketone.[69,70] For example, stirring $AcCl/AlCl_3$ in excess cyclohexane (30–35 °C, 2.5 h) affords 2-methyl-1-acetylcyclopentane in 50% yield (unpurified; based on AcCl)[55a,69,71] and stirring $AcCl/AlBr_3$ in excess cyclopentane (20 °C, 1 h) affords cyclopentyl methyl ketone in 60% yield (based on AcCl; eq 7).[55c]

$AcCl/AlBr_3$, 20 °C, 1 h, 60% (7)

If the reaction is carried out with a substoichiometric amount of alkane, the product is either a 2:1 adduct (if cyclic)[53b,66a] or pyrylium salt (if acyclic).[66b,68b]

Unbranched alkanes also undergo acetylation, but at higher temperature, so yields are generally lower. For example, acetylation of cyclohexane by $AcCl/AlCl_3$ requires refluxing in $CHCl_3$ and affords 1-acetyl-2-methylcyclopentene in only 36% yield.[55c,72]

Despite the modest to low yields, acetylation of alkanes provides a practical method for accessing simple methyl ketones because all the input raw materials are cheap.

Coupling with Organometallic Reagents. Coupling of organometallic reagents with AcCl is a valuable method for preparation of methyl ketones. Generally a catalyst (either a Lewis acid or transition metal salt) is required.

Due to the large number and varied characteristics of the organometallics, comprehensive coverage of the subject would require discussion of each organometallic reagent individually, which is far beyond the scope of this article. Information pertaining to catalyst and condition selection should therefore be accessed from the original literature; some seminal references are given in Table 1.

Table 1 Catalyst Selection Chart[i,ii]

Organometallic	R = Alkyl	Vinyl	Aryl	Alkynyl	Allyl[iii]
RLi	N[73 iv]				
R_2Mg	N[74 iv]				
RMgX	Fe[75] Cu[76] Mn[76b,77] D[78] N[79]				
$RCuL^-$	N[80]	N[81]	N[80,82]	N[80,83] I[83 a]	N[84]
RZnL	Pd[85,86] D[89] N[91]	Pd[85]	Pd[87] D[89]	Pd[85] N[90]	N[88 iv]
RCdL	N[91c,92]		N[91c,92b,d,e]		
RHgL	Al[93] Pd[96]	Al[94] Ti[94v]	Al[93] Pd[96,97]		Al[95]
RBL_3^-	N[98]		N[99,100 vi]	N[101 vi]	
$RAlL_2$	N[102] Cu[103]	N[85,102b] Pd[85]			
R_4Al^-	Fe[104] Cu[104]	Pd[85]			
$RTlL_2$	N[105]		N[105]		
$RSiL_3$	Al[106,107]	Al[107,108,109,vii]		Al[107,110]	Al[107,111,112 viii,113 ix] Ti[114 x]
$RGeL_3$	Al[106b]				
$RSnL_3$	Pd[115] Al[106b]	Pd[115,116] Ti[120 iv]	Pd[117] N[117c]	Pd[118]	Rh[119 x] N[121,122 iv]
$RBiL_2$			Pd[123]		
$RTiL_2$					N[124 iv]
$RZrL_3$	N[125]				
RVL_2			N[126]		
RMnL	Cu[127] N[128]	Cu[127] N[128]	Cu[127] N[128]	Cu[127] N[128]	Cu[127]
RRh			N[129]		
RNiL		N[130]			N[130,131]

[i] Codes (in headings): L = unspecified ligand and X = halogen; codes (in entries): N = no catalyst required, Fe = Fe^{III} salt, Cu = Cu^{I} salt, Mn = Mn^{I} salt, D = dipolar aprotic additive, I = LiI, Pd = Pd^0 or Pd^{II} salt, Al = $AlCl_3$, Ti = $TiCl_4$, and Rh = $ClRh(PPh_3)_3$. [ii] Coupling occurs at metal-bearing carbon with retention of configuration to afford RCOMe unless otherwise indicated. [iii] Coupling occurs at γ-carbon unless otherwise indicated. [iv] Product is tertiary alcohol. [v] Coupling occurs with inversion of configuration. [vi] Coupling occurs at β-carbon. [vii] Coupling occurs at δ-carbon. [viii] Substrate is allenic silane and product is furan. [ix] Substrate is propargylsilane. [x] Coupling occurs at α-carbon.

***C*-Acetylation of Enolates and Enolate Equivalents.** β-Diketones can be synthesized by treatment of metal enolates with

AcCl. *O*-Acetylation is often a significant side reaction, but the amount can be minimized by choosing a counterion that is bonded covalently to the enolate[6] such as copper[132] or zinc,[133] and by using AcCl rather than Ac_2O.[6a] Proton transfer from the product β-diketone to the starting enolate is another common side reaction.[134] Alternative procedures for effecting *C*-acetylation that avoid or minimize these side reactions include Lewis acid-catalyzed acetylation of the trimethylsilyl enol ether derivative (AcCl/cat. $ZnCl_2$, CH_2Cl_2 or CH_2Cl_2/Et_2O, rt)[135] and addition of ketene to the morpholine enamine (AcCl/Et_3N, $CHCl_3$, rt).[136]

Analogously, esters can be *C*-acetylated by conversion into the corresponding silyl ketene acetal followed by treatment with AcCl. Depending on the coupling conditions (neat AcCl[137] or AcCl/Et_3N[138]), either the *cis*-β-siloxycrotonate ester or the corresponding β,γ-isomer is produced (eqs 8 and 9). The third possible isomer (*trans*-β-siloxycrotonate) is accessible either by silylation of the acetoacetic ester (TMSCl, Et_3N, THF, Δ)[139] or by $HgBr_2/Et_3SiBr$-catalyzed equilibration of the *cis* isomer.[137]

OSiEt3 / OMe —AcCl (neat), rt→ [O, $\overset{+}{O}SiEt_3$, OMe, ⁻Cl] —67%→ Et3SiO, O, OMe (8)

OTBS / OEt —AcCl/Et3N, THF, rt→ [O⁻, $\overset{+}{O}TBS$, OEt] —86%→ TBSO, O, OEt (9)

The silyl ketene acetal strategy can also be used to effect γ-acetylation of α,β-unsaturated esters (AcCl/cat. $ZnBr_2$, CH_2Cl_2, rt)[140] and β-ketoesters (AcCl, Et_2O, $-78\,^{\circ}C$).[141]

Enol Acetylation. Enol acetylation of ketones can be effected by formation of a metal enolate in which the metal is relatively dissociated[6] (such as potassium[142] or magnesium[143]) followed by quenching with AcCl. Alternatively, enol acetates can be synthesized directly from the ketones. For example, 3-keto-$\Delta^{4,6}$-steroids can be converted into $\Delta^{2,4,6}$-trienol acetates by treatment with AcCl/$PhNMe_2$ or into $\Delta^{3,5,7}$-trienol acetates by treatment with AcCl/Ac_2O.[144]

Acetyl Bromide (AcBr) is apparently superior to AcCl as a catalyst for enol acetylation, based on a report that 17β-benzoyloxyestra-4,9(10)-dien-3-one is converted into estradiol 3-acetate-17-benzoate in higher yield at much lower temperature using AcBr rather than AcCl (87.5% yield with 1:2 AcBr:Ac_2O, CH_2Cl_2, rt, 1 h (eq 10) vs. 81.0% yield with 1:2 AcCl:Ac_2O, Δ, 4.5 h).[145]

OBz, H, H, O —1/2 AcBr/Ac2O, CH2Cl2, rt, 1 h, 87.5%→ OBz, H, H, H, AcO (10)

β-Keto esters can be converted into either *trans* or *cis* enol acetates. The *trans* isomer is accessible by treatment with AcCl/Et_3N (HMPA, rt)[146] or AcCl/DBU (MeCN, $5\,^{\circ}C \rightarrow$ rt; eq 11),[147] while the *cis* isomer is accessible by treatment with isopropenyl acetate/HOTs.[146] Each isomer couples with dialkylcuprates with retention of configuration to afford stereoisomerically enriched α,β-unsaturated esters.[146,148]

O, O, OMe —DBU, MeCN, 5 °C; then AcCl, rt, 3 h, 53%→ AcO, O, OMe (11)

cis:*trans* = 97:3

Attempted enol acetylation of β-keto esters by quenching the sodium enolate[146,147] or magnesium chelate[149] with AcCl afforded *C*-acetylated products.

Adducts with Aldehydes and Ketones. AcCl combines with aldehydes[150] (cat. $ZnCl_2$ or $AlCl_3$[150f]) to afford α-chloroalkyl acetates. The reaction is reversible,[151] but at equilibrium the ratio of adduct to aldehyde is usually quite high, and the reaction is otherwise clean (92% yield for acetaldehyde,[150e] 97% yield for benzaldehyde; eq 12[150f]).

O, H —AcCl/cat. ZnCl2, 0–10 °C, 1 h (neat), 97%→ AcO, Cl (12)

AcCl also adds to ketones,[150e,151,152] but the adducts are much less thermodynamically stable, so significant amounts of the starting materials are present at equilibrium.[151,152a,b] The equilibrium can be biased in favor of the adduct by employing high concentration, low temperature, a nonpolar solvent, excess AcCl, or AcBr or AcI instead of AcCl.[151,153] For example, the acetone/AcCl adduct can be obtained in good yield (85%) by treatment of acetone with excess (2 equiv) AcCl (cat. $ZnCl_2$, CCl_4, $-15\,^{\circ}C$).[152c]

Reduction of the aldehyde/AcBr adducts[151,154] with ***Zinc*** or ***Samarium(II) Iodide*** to α-acetoxyalkylzinc[154,155] and samarium[156] compounds, respectively, completes an umpolung of the reactivity of the aldehyde.

Cleavage of Ethers. THF can be opened by treatment with AcCl in combination with either ***Sodium Iodide*** (MeCN, rt, 21 h; 91% yield of 4-iodobutyl acetate)[157] or a Lewis acid such as $ZnCl_2$ (Δ, 1.5 h; 76% yield of 4-chlorobutyl acetate),[158] $SnCl_4$,[159] $CoCl_2$ (rt, MeCN; 90%),[160] $ClPdCH_2Ph(PPh_3)_2/Bu_3SnCl$ ($63\,^{\circ}C$, 48 h; 95%),[161] $Mo(CO)_6$ (hexane, Δ; 78%),[162] $KPtCl_3(H_2CCH_2)$, and $[ClRh(H_2CCH_2)_2]_2$ (rt; 75% and 83%, respectively).[163] Acyclic dialkyl ethers can also be cleaved efficiently and in many cases regioselectively.[159]

Many of these methods are applicable to deprotection of ether-type protecting groups. For example, benzyl and allyl ethers can be deprotected by treatment with AcCl/cat. $CoCl_2$[160] or AcCl/cat. $ClPdCH_2Ph(PPh_3)_2$/cat. Bu_3SnCl.[161] Dimethyl acetals can be cleaved selectively to aldehydes in the presence of ethylene acetals (AcCl/cat. $ZnCl_2$, Me_2S/THF, $0\,^{\circ}C$),[164] or to α-chloro ethers (AcCl/cat. $SOCl_2$, $55\,^{\circ}C$).[165] Tetrahydropyranyl (THP) ethers[166] and *t*-butyl ethers[167] can be deprotected by stirring in 1:10 AcCl:HOAc (40–50 °C).

Finally, *t*-alkyl esters can be cleaved to anhydrides and *t*-alkyl chlorides by treatment with AcCl ($MeNO_2$, 70 °C).[168]

Esterification. Although AcCl is intrinsically more reactive than Ac_2O, in combination with various acylation catalysts the reverse reactivity order is exhibited. For example, Ac_2O ***4-Dimethylaminopyridine*** (DMAP) acetylates ethynylcyclohexanol three times faster than AcCl/DMAP ($CDCl_3$, 27 °C).[169] Also, isopropanol does not react with AcCl/Bu_3P (CD_3CN, −8 °C; <5% conversion after 30 min), but after addition of sodium acetate reacts rapidly to form isopropyl acetate (complete in <10 min).[170] As a general rule, therefore, Ac_2O is preferable for acetylation of hindered alcohols while AcCl is preferable for selective monoacetylation of polyols.[171]

Examples of selective acetylations involving AcCl include: acetylation of primary alcohols in the presence of secondary alcohols by AcCl/2,4,6-collidine or *i*-$PrNEt_2$ (CH_2Cl_2, −78 °C);[8,172] acetylation of primary alcohols in the presence of secondary alcohols,[173] and secondary alcohols in the presence of tertiary alcohols,[174] by AcCl/pyridine (CH_2Cl_2, −78 °C); monoacetylation of a 2,4-dihydroxyglucopyranose by AcCl/pyridine/−15 °C (Ac_2O/pyridine/0 °C is less selective);[175] and acetylation of steroidal 5α-hydroxyls (not 5β) by AcCl/$PhNMe_2$ ($CHCl_3$, Δ).[176]

Although Ac_2O/DMAP[177] and Ac_2O/Bu_3P[170] are the preferred reagents for acetylation of most hindered alcohols, satisfactory results can be obtained with AcCl in combination with $PhNMe_2$ ($CHCl_3$, Δ),[178] $PhNEt_2$ ($CHCl_3$, Δ),[179] AgCN (benzene or HMPA, 80 °C),[180] magnesium powder (Et_2O, Δ; 45–55% yield of *t*-BuOAc),[181] and Na_2CO_3 (cat. $PhCH_2NEt_3Cl$, CH_2Cl_2, Δ; 79% yield of *t*-BuOAc).[182] Use of the combination of AcCl/DMAP is not recommended since unidentified byproducts may be generated.[169]

Although acetylations with AcCl/pyridine produce an acidic byproduct (pyridine hydrochloride), it is possible to acetylate highly acid-sensitive alcohols such as 2-(tributylstannylmethyl)allyl alcohol (eq 13)[9b] and 2-(trimethylsilylmethyl)allyl alcohol[9a] with AcCl/pyridine in >90% yield without competing protiodestannylation or protiodesilylation by selecting a solvent (CH_2Cl_2, 0 °C) in which the pyridine hydrochloride is insoluble.

n-Bu_3Sn … HO —[AcCl/pyridine; CH_2Cl_2, 90%]→ *n*-Bu_3Sn … AcO (13)

Alternatively, acid-sensitive alcohols may be acetylated by deprotonation with ***n-Butyllithium*** (THF, −78 °C)[183] or ***Ethylmagnesium Bromide*** (Et_2O, rt)[184] followed by quenching with AcCl.

Finally, by using a chiral tertiary amine as the base, it is possible to effect enantioselective acetylations. For example, racemic 1-phenethyl alcohol has been partially resolved by treatment with AcCl in combination with (*S*)-(−)-*N*,*N*-dimethyl-1-phenethylamine (CH_2Cl_2, −78 °C → rt; ee of acetate 52%, ee of alcohol 59.5%).[185]

Generation of Solutions of Anhydrous Hydrogen Chloride in Methanol. Esterification of alcohols by AcCl proceeds in the absence of HCl scavengers. For example, on addition of AcCl to methanol at rt, a solution of hydrogen chloride and methyl acetate in methanol forms rapidly.[10] This reaction provides a more practical method for access to solutions of HCl in methanol than the apparently simpler method of bubbling anhydrous HCl into methanol because of the difficulty of controlling the amount of anhydrous HCl delivered. Solutions of anhydrous HCl in acetic acid can presumably be prepared analogously by addition of AcCl and an equimolar amount of H_2O to HOAc.

Primary,[186] secondary,[187] and tertiary alcohols[178a,188] also react with AcCl, but the product is the alkyl chloride rather than the ester in most cases. Thus as a preparative esterification method this reaction has limited generality.

AcCl also reacts with anhydrous *p*-toluenesulfonic acid (3–4 equiv AcCl, Δ) to afford acetyl *p*-toluenesulfonate in 97.5% yield along with anhydrous HCl.[189] AcCl does not react with HOAc to generate HCl and Ac_2O, at least in appreciable amounts.[31]

Dehydrating Agent. AcCl reacts with H_2O to afford HCl and HOAc rapidly and quantitatively[31b] and thereby qualifies as a strong dehydrating agent. Examples of reactions in which AcCl functions as a dehydrating agent include: cyclization of dicarboxylic acids to cyclic anhydrides (neat AcCl, Δ);[190] cyclization of keto acids to enol lactones (neat AcCl, Δ);[191] dehydration of nitro compounds into nitrile oxides (by treatment with NaOMe followed by AcCl);[192] and conversion of allylic hydroperoxides into unsaturated ketones (AcCl/pyridine, $CHCl_3$, rt).[193] The dehydrating power of AcCl has been invoked as a possible explanation for its effectiveness for activation of zinc dust.[194]

In Situ Generation of High-Valent Metal Chlorides. Many high-valent metal chlorides are useful as reagents in organic synthesis but are difficult to handle due to their moisture sensitivity. AcCl can be used to generate such reagents in situ from the corresponding metal oxides[11] or acetates.[195] Examples include: α-chlorination of ketones by treatment with AcCl/***Manganese Dioxide*** (HOAc, rt);[196] *cis*-1,2-dichlorination of alkenes by treatment with AcCl/$(Bu_4N)_4Mo_8O_{26}$ (CH_2Cl_2, rt);[197] and dichlorination of alkenes by treatment with AcCl/MnO_2/$MnCl_2$ (DMF, rt).[198] Attempts to dichlorinate alkenes by treatment with AcCl/MnO_2 in THF, however, failed due to cleavage of THF to 4-chlorobutyl acetate.[196,199] A milder reagent that can be used to activate MnO_2 for dichlorination of alkenes in THF is ***Chlorotrimethylsilane***.[199]

Solvent for Organometallic Reactions. Because of its cheapness, volatility, and ability to form moisture-stable solutions of metal chlorides, AcCl is useful as a solvent for reactions involving hygroscopic metal salts.[11] For example, AcCl has been used as a co-solvent for 1,2-chloroacetoxylation of alkenes by ***Chromyl Chloride*** (1:2 AcCl:CH_2Cl_2, −78 °C → rt).[200]

Reaction with Heteroatom Oxides. The key step in a method for α-acetoxylation of aldehydes involves rearrangement of an

AcCl-nitrone adduct (eq 14).[201] Analogous methods for α-benzoylation and α-pivaloylation are higher yielding.

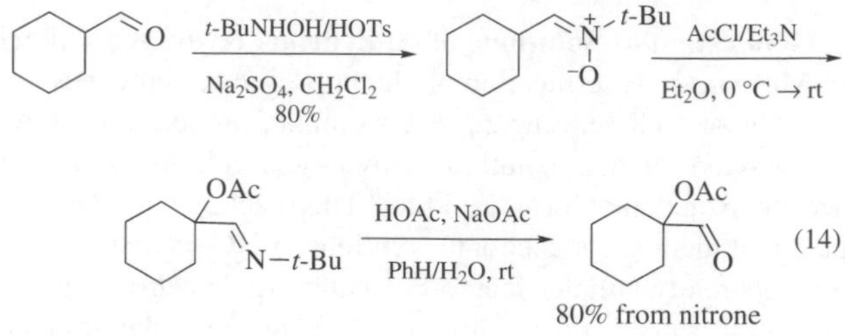

(14)

β-Nitrostyrenes cyclize to indolinones on treatment with AcCl ($FeCl_3$, CH_2Cl_2, 0 °C; eq 15).[202]

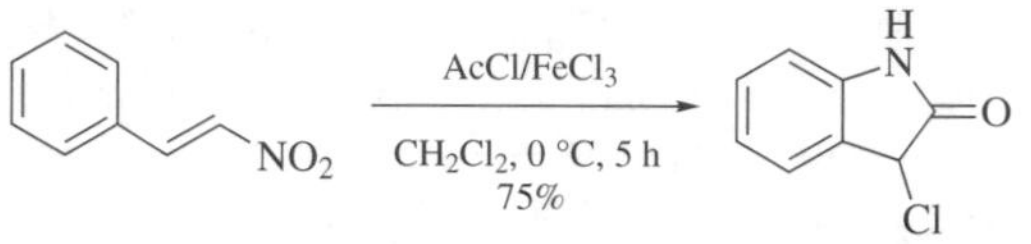

(15)

A high-yielding method for deoxygenation of sulfoxides to sulfides involves treatment with 1.1 equiv ***Tin(II) Chloride*** in the presence of a catalytic amount (0.4 equiv) of AcCl (MeCN/DMF, 0 °C → rt).[12] The mildness of this method is demonstrated by its usability for deoxygenation of a cephalosporin sulfoxide (eq 16).

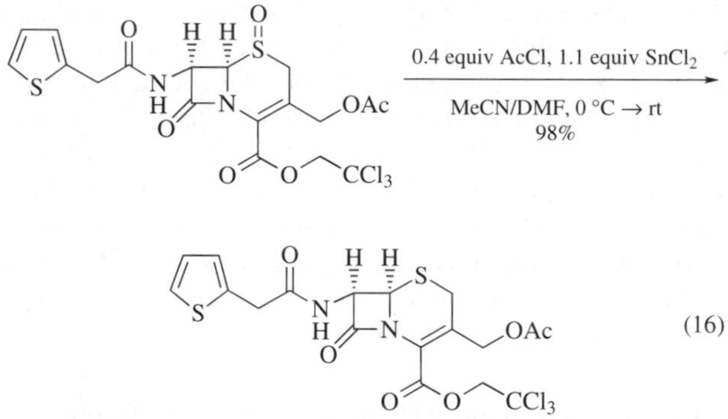

(16)

Another method for deoxygenation of sulfoxides involves treatment with two equiv AcCl (CH_2Cl_2, rt);[203] the oxidized by-product is claimed to be gaseous chlorine.[203]

Chlorine and Bromine Scavenger. AcCl (cat. H_2SO_4, 40–70 °C) scavenges Cl_2 efficiently (to afford chloroacetyl chloride in 87.1% yield).[13] AcCl also scavenges Br_2 efficiently at 35 °C.[14]

Source of Ketene. AcCl reacts with ***Triethylamine*** at low temperature (−20 °C) to afford acetyltriethylammonium chloride.[204] This salt functions as a source of ketene (or the functional equivalent). For example, it reacts with silyl ketene acetals (THF, rt) to afford silyl enol ethers of acetoacetic esters (eq 9),[138] with α-alkoxycarbonylalkylidenetriphenylphosphoranes (CH_2Cl_2, rt) to afford allenic esters,[205] with enamines (Et_2O, 0 °C) to afford cyclobutanones,[136a] with certain acyl imines (Et_2O, 0 °C) to form formal [4 + 2] ketene cycloadducts,[206] and with certain nonenolizable imines (Et_2O, rt) to afford formal [4 + 2] diketene cycloadducts in up to 55% yield.[207] Also, on refluxing in Et_2O in the absence of a trapping agent, diketene is formed in 50% yield.[208]

AcCl/$AlCl_3$ decomposes to acetylacetone on heating ($CHCl_3$, 54–61 °C, 6 h; 82.5% yield after aqueous work-up).[209] The mechanism presumably involves ketene as an intermediate. However, an attempt to trap the ketene was unsuccessful.[210]

***N*-Acetylation.** Primary and secondary amines can be *N*-acetylated to form acetamides by treatment with AcCl under Schotten–Baumann conditions (aq NaOH),[211] but hydrolysis of AcCl is a significant competing side reaction.[212] Use of Ac_2O (2.5 equiv; Δ, 10–15 min) is therefore recommended.[211]

Tertiary amines react with AcCl to afford acetylammonium salts. Ordinarily, these salts fragment to ketene on warming (see above). However, those that possess a labile alkyl group fragment by loss of the alkyl group (von Braun cleavage). For example, bis(dimethylamino)methane reacts with AcCl (Et_2O, rt) to afford chloromethyldimethylamine,[213] a useful Mannich reagent, and 1,3,5-trimethylhexahydro-*s*-triazine reacts with AcCl ($CHCl_3$, Δ, 1 h) to afford chloromethylmethyl acetamide,[214] a useful amidomethylation reagent.[215] Also, aziridines react with AcCl (PhH, 0 °C) to afford chloroethylacetamides.[216] Allylic amines react with in situ-generated AcI (AcCl/CuI, THF, rt) to afford acetamides.[217]

AcCl also activates pyridines toward nucleophilic addition. For example, phenylmagnesium chloride adds to pyridine in the presence of AcCl (cat. CuI, THF, −20 °C → rt) to afford, after catalytic hydrogenation, *N*-acetyl-4-phenylpiperidine in 65% yield.[218] Also, AcCl catalyzes the reaction between sodium iodide and 2-chloropyridine to afford 2-iodopyridine (MeCN, Δ, 24 h; 55%).[219]

N-Acetylation of enolizable imines to afford enamides can be accomplished by treatment with AcCl/$PhNEt_2$. For example, treatment of crotonaldehyde cyclohexylimine with AcCl followed by $PhNEt_2$ (toluene, rt) affords the enamide in 88% yield.[220]

Primary urethanes can be *N*-acetylated to afford imides by treatment with AcCl (100 °C, 1 h).[221] Alternatively, urethanes can be converted into acetamides by treatment with AcBr (120–130 °C)[221] or in situ-generated AcI[222] (MeCN, 60 °C).[223]

Finally, a convenient method for preparation of ***N-Trimethylsilylacetamide*** (MSA), a useful trimethylsilyl transfer reagent, involves treatment of ***Hexamethyldisilazane*** with AcCl (hexane, Δ; 88%).[224]

***S*-Acetylation.** Both aliphatic and aromatic thiols can be *S*-acetylated by treatment with AcCl (cat. $CoCl_2$, MeCN, rt).[225]

Nucleophilic Acetylation. AcCl together with SmI_2 (MeCN, rt) or $SmCp_2$ delivers the acetyl anion synthon to ketones to afford the corresponding acyloins (eq 17).[15]

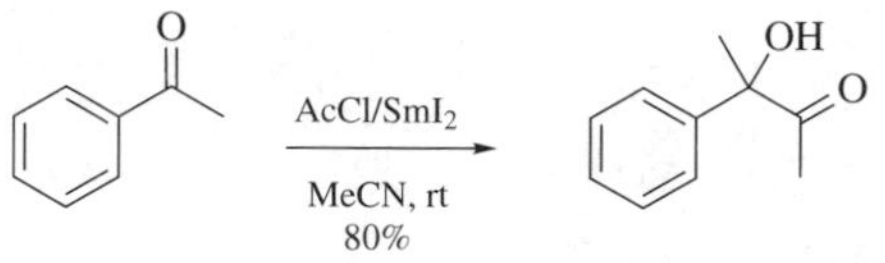

(17)

Related Reagents. Acetic Anhydride; Acetyl Bromide; Acetyl Fluoride.

---

1. (a) Moretti, T. A. *Kirk-Othmer Encyclopedia of Chemical Technology*, 3rd ed.; Wiley: New York, 1978; Vol. 1, p 162. (b) Wagner, F. S. Jr. *Kirk-Othmer Encyclopedia of Chemical Technology*, 4th ed.; Wiley: New York, 1991; Vol. 1, p 155.

2. (a) Baddeley, G. *QR* **1954**, *8*, 355. (b) Gore, P. H. *Friedel–Crafts and Related Reactions*; Wiley: New York, 1964; Vol. 3, p 1. (c) House, H. O. *Modern Synthetic Reactions*, 2nd ed; Benjamin: Menlo Park, 1972; p 797. (d) Olah, G. A. *Friedel–Crafts Chemistry*; Wiley: New York, 1973; pp 91, 191. (e) Heaney, H. *COC* **1979**, *1*, 241. (f) Olah, G. A.; Meidar, D. *Kirk-Othmer Encyclopedia of Chemical Technology*, 3rd ed.; Wiley: New York, 1980; Vol. 11, p 269. (g) Olah, G. A.; Prakash, G. K. S.; Sommer, J. *Superacids*; Wiley: New York, 1985; p 293.
3. (a) Nenitzescu, C. D.; Balaban, A. T. *Friedel–Crafts and Related Reactions*; Wiley: New York, 1964; Vol. 3, p 1033. (b) Groves, J. K. *CSR* **1972**, *1*, 73. (c) Olah, G. A. *Friedel–Crafts Chemistry*; Wiley: New York, 1973; pp 129, 200.
4. Price, C. C.; Pappalardo, J. A. *JACS* **1950**, *72*, 2613.
5. (a) Olah, G. A. *Friedel–Crafts Chemistry*; Wiley: New York, 1973; p 135. (b) Vol'pin, M.; Akhrem, I.; Orlinkov, A. *NJC* **1989**, *13*, 771.
6. (a) House, H. O.; Auerbach, R. A.; Gall, M.; Peet, N. P. *JOC* **1973**, *38*, 514. (b) Black, T. H. *OPP* **1989**, *21*, 179.
7. (a) Burwell, R. L. Jr. *CRV* **1954**, *54*, 615. (b) Johnson, F. *Friedel–Crafts and Related Reactions*; Wiley: New York, 1965; Vol. 4, p 1. (c) Bhatt, M. V.; Kulkarni, S. U. *S* **1983**, 249.
8. Ishihara, K.; Kurihara, H.; Yamamoto, H. *JOC* **1993**, *58*, 3791.
9. (a) Trost, B. M.; Chan, D. M. T. *JACS* **1983**, *105*, 2315. (b) Trost, B. M.; Bonk, P. J. *JACS* **1985**, *107*, 1778.
10. (a) Freudenberg, K.; Jacob, W. *CB* **1941**, *74*, 1001. (b) Riegel, B.; Moffett, R. B.; McIntosh, A. V. *OS* **1944**, *24*, 41. (c) Fraenkel-Conrat, H.; Olcott, H. S. *JBC* **1945**, *161*, 259. (d) Baker, B. R.; Schaub, R. E.; Querry, M. V.; Williams, J. H. *JOC* **1952**, *17*, 77. (e) De Lombaert, S.; Nemery, I.; Roekens, B.; Carretero, J. C.; Kimmel, T.; Ghosez, L. *TL* **1986**, *27*, 5099. (f) Nashed, E. M.; Glaudemans, C. P. J. *JOC* **1987**, *52*, 5255.
11. (a) Chretien, A.; Oechsel, G. *CR* **1938**, *206*, 254. (b) Paul, R. C.; Sandhu, S. S. *Proc. Chem. Soc.* **1957**, 262. (c) Paul, R. C.; Singh, D.; Sandhu, S. S. *JCS* **1959**, 315. (d) Paul, R. C.; Singh, D.; Sandhu, S. S. *JCS* **1959**, 319. (e) Maunaye, M.; Lang, J. *CR* **1965**, *261*, 3381, 3829.
12. Kaiser, G. V.; Cooper, R. D. G.; Koehler, R. E.; Murphy, C. F.; Webber, J. A.; Wright, I. G.; Van Heyningen, E. M. *JOC* **1970**, *35*, 2430.
13. Scheidmeir, W.; Bressel, U.; Hohenschutz, H. U.S. Patent 3 880 923, 1975.
14. Kharasch, M. S.; Hobbs, L. M. *JOC* **1941**, *6*, 705.
15. (a) Collin, J.; Namy, J.-L.; Dallemer, F.; Kagan, H. B. *JOC* **1991**, *56*, 3118. (b) Ruder, S. M. *TL* **1992**, *33*, 2621.
16. Pouchert, C. J. *The Aldrich Library of FT-IR Spectra*, ed. I; Aldrich: Milwaukee, 1985; Vol. 1, p 723A.
17. Tsuchiya, S.; Kimura, M. *BCJ* **1972**, *45*, 736.
18. *Reagent Chemicals*, 8th ed.; American Chemical Society: Washington, 1993; p 107.
19. Singh, J.; Paul, R. C.; Sandhu, S. S. *JCS* **1959**, 845.
20. (a) Whitmore, F. C. *RTC* **1938**, *57*, 562. (b) Cason, J.; Harman, R. E.; Goodwin, S.; Allen, C. F. *JOC* **1950**, *15*, 860. (c) Perrin, D. D.; Armarego, W. L. F. *Purification of Laboratory Chemicals*, 3rd ed.; Pergamon: Oxford, 1988; p 70.
21. Burton, H.; Praill, P. F. G. *JCS* **1952**, 2546.
22. Sax, N. I. *Dangerous Properties of Industrial Materials*, 6th ed.; Van Nostrand Reinhold: New York, 1984; p 106.
23. (a) Vogel, A. I. *A Text-book of Practical Organic Chemistry*, 3rd ed.; Wiley: New York, 1956; p 367. (b) Damjan, J.; Benczik, J.; Kolonics, Z.; Pelyva, J.; Laborczy, R.; Szabolcs, J.; Soptei, C.; Barcza, I.; Kayos, C. Br. Patent 2 213 144, 1989. (c) Valitova, L. A.; Popova, E. V.; Ibragimov, Sh. N.; Ivanov, B. E. *BAU* **1990**, *39*, 366.
24. Durrans, T. H. U.S. Patent 1 326 040, 1919.
25. Masters, C. L. U.S. Patent 1 819 613, 1931.
26. Montonna, R. E. *JACS* **1927**, *49*, 2114.
27. Mugdan, M.; Wimmer, J. Ger. Patent 549 725, 1931.
28. Brown, H. C. *JACS* **1938**, *60*, 1325.
29. Mills, L. E. U.S. Patent 1 921 767, 1933. U.S. Patent 1 965 556, 1934.
30. (a) Meder, G.; Eggert, E.; Grimm, A. U.S. Patent 2 013 988, 1935. (b) Meder, G.; Geissler, W.; Eggert, E. U.S. Patent 2 013 989, 1935. (c) Meder, G.; Bergheimer, E.; Geisler, W.; Eggert, E. Ger. Patent 638 306, 1936. (d) Eggert, E.; Grimm, A. Ger. Patent 655 683, 1938. (e) Christoph, F. J. Jr.; Parker, S. H.; Seagraves, R. L. U.S. Patent 3 318 950, 1967.
31. (a) Colson, A. *BSF* **1897**, *17*, 55. (b) Inoue, S.; Hayashi, K. *CA* **1954**, *48*, 8255g. (c) Satchell, D. P. N. *JCS* **1960**, 1752. (d) Satchell, D. P. N. *QR* **1963**, *17*, 160.
32. Brown, H. C.; Marino, G.; Stock, L. M. *JACS* **1959**, *81*, 3310.
33. Brown, H. C.; Marino, G. *JACS* **1962**, *84*, 1658.
34. (a) Brown, H. C.; Nelson, K. L. *JACS* **1953**, *75*, 6292. (b) Olah, G. A. *Friedel–Crafts Chemistry*; Wiley: New York, 1973; pp 448, 452.
35. (a) Elwood, T. A.; Flack, W. R.; Inman, K. J.; Rabideau, P. W. *T* **1974**, *30*, 535. (b) Todd, D.; Pickering, M. *J. Chem. Educ.* **1988**, *65*, 1100.
36. (a) Baddeley, G. *JCS* **1949**, S99. (b) Bassilios, H. F.; Makar, S. M.; Salem, A. Y. *BSF* **1954**, *21*, 72. (c) Friedman, L.; Honour, R. J. *JACS* **1969**, *91*, 6344.
37. (a) Girdler, R. B.; Gore, P. H.; Hoskins, J. A. *JCS(C)* **1966**, 181. (b) Arsenijevic, L.; Arsenijevic, V.; Horeau, A.; Jacques, J. *OSC* **1988**, *6*, 34. (c) Magni, A.; Visentin, G. U.S. Patent 4 868 338, 1989.
38. Friedman, L.; Koca, R. *JOC* **1968**, *33*, 1255.
39. (a) Merritt, C. Jr.; Braun, C. E. *OSC* **1963**, *4*, 8. (b) Gore, P. H.; Thadani, C. K. *JCS(C)* **1966**, 1729.
40. (a) Perrier, G. *CB* **1900**, *33*, 815. (b) Perrier, G. *BSF* **1904**, *31*, 859.
41. (a) Heine, H. W.; Cottle, D. L.; Van Mater, H. L. *JACS* **1946**, *68*, 524. (b) Gore, P. H.; Hoskins, J. A. *JCS* **1964**, 5666.
42. Johnson, J. R.; May, G. E. *OSC* **1943**, *2*, 8.
43. (a) Piccolo, O.; Visentin, G.; Blasina, P.; Spreafico, F. U.S. Patent 4 670 603, 1987. (b) Lindley, D. D.; Curtis, T. A.; Ryan, T. R.; de la Garza, E. M.; Hilton, C. B.; Kenesson, T. M. U.S. Patent 5 068 448, 1991.
44. Olah, G. A. *Friedel–Crafts Chemistry*; Wiley: New York, 1973; pp 106, 306.
45. Allen, C. F. H. *OSC* **1943**, *2*, 3.
46. Olah, G. A.; Moffatt, M. E.; Kuhn, S. J.; Hardie, B. A. *JACS* **1964**, *86*, 2198.
47. (a) Sondheimer, F.; Woodward, R. B. *JACS* **1953**, *75*, 5438. (b) Briner, P. H. U.S. Patent 5 124 486, 1992.
48. (a) Kulinkovich, O. G.; Tischenko, I. G.; Sorokin, V. L. *S* **1985**, 1058. (b) Kulinkovich, O. G.; Tischenko, I. G.; Sorokin, V. L. *JOU* **1985**, *21*, 1514. (c) Mamedov, E. I.; Ismailov, A. G.; Zyk, N. V.; Kutateladze, A. G.; Zefirov, N. S. *Sulfur Lett.* **1991**, *12*, 109.
49. (a) Nenitzescu, C. D.; Gavat, I. G. *LA* **1935**, *519*, 260. (b) Johnson, W. S.; Offenhauer, R. D. *JACS* **1945**, *67*, 1045. (c) Crosby, J. A.; Rasburn, J. W. *CI(L)* **1967**, 1365.
50. (a) Darzens, M. G. *CR* **1910**, *150*, 707. (b) Wieland, H.; Bettag, L. *CB* **1922**, *55*, 2246.
51. Royals, E. E.; Hendry, C. M. *JOC* **1950**, *15*, 1147.
52. (a) Turner, R. B.; Voitle, D. M. *JACS* **1951**, *73*, 1403. (b) Dufort, N.; Lafontaine, J. *CJC* **1968**, *46*, 1065.
53. (a) Baddeley, G.; Heaton, B. G.; Rasburn, J. W. *JCS* **1960**, 4713. (b) Morel-Fourrier, C.; Dulcere, J.-P.; Santelli, M. *JACS* **1991**, *113*, 8062.
54. Colonge, J.; Mostafavi, K. *BSF* **1939**, *6*, 335, 342.
55. (a) Nenitzescu, C. D.; Ionescu, C. N. *LA* **1931**, *491*, 189. (b) Baddeley, G.; Khayat, M. A. R. *Proc. Chem. Soc.* **1961**, 382. (c) Akhrem, I. S.; Orlinkov, A. V.; Mysov, E. I.; Vol'pin, M. E. *TL* **1981**, *22*, 3891.
56. Arnaud, M.; Roussel, C.; Metzger, J. *TL* **1979**, 1795.
57. Snider, B. B.; Jackson, A. C. *JOC* **1982**, *47*, 5393.
58. Shono, T.; Nishiguchi, I.; Sasaki, M.; Ikeda, H.; Kurita, M. *JOC* **1983**, *48*, 2503.

59. Baran, J.; Klein, H.; Schade, C.; Will, E.; Koschinsky, R.; Bauml, E.; Mayr, H. *T* **1988**, *44*, 2181.
60. Ruzicka, L.; Koolhaas, D. R.; Wind, A. H. *HCA* **1931**, *14*, 1151.
61. Hoffmann, H. M. R.; Tsushima, T. *JACS* **1977**, *99*, 6008.
62. (a) Smit, W. A.; Semenovsky, A. V.; Kucherov, V. F.; Chernova, T. N.; Krimer, M. Z.; Lubinskaya, O. V. *TL* **1971**, 3101. (b) Smit, V. A.; Semenovskii, A. V.; Lyubinskaya, O. V.; Kucherov, V. F. *DOK* **1972**, *203*, 272.
63. (a) Deno, N. C.; Chafetz, H. *JACS* **1952**, *74*, 3940. (b) Groves, J. K.; Jones, N. *JCS(C)* **1968**, 2215, 2898. (c) Beak, P.; Berger, K. R. *JACS* **1980**, *102*, 3848.
64. (a) Balaban, A. T.; Nenitzescu, C. D. *LA* **1959**, *625*, 74. (b) Balaban, A. T.; Schroth, W.; Fischer, G. *Adv. Heterocycl. Chem.* **1969**, *10*, 241. (c) Erre, C. H.; Roussel, C. *BSF(2)* **1984**, 454.
65. Roitburd, G. V.; Smit, W. A.; Semenovsky, A. V.; Shchegolev, A. A.; Kucherov, V. F.; Chizhov, O. S.; Kadentsev, V. I. *TL* **1972**, 4935.
66. (a) Tabushi, I.; Fujita, K.; Oda, R. *TL* **1968**, 5455. (b) Arnaud, M.; Pedra, A.; Roussel, C.; Metzger, J. *JOC* **1979**, *44*, 2972.
67. (a) Nenitzescu, C. D.; Dragan, A. *CB* **1933**, *66*, 1892. (b) Bloch, H. S.; Pines, H.; Schmerling, L. *JACS* **1946**, *68*, 153.
68. (a) Tabushi, I.; Fujita, K.; Oda, R.; Tsuboi, M. *TL* **1969**, 2581. (b) Arnaud, M.; Pedra, A.; Erre, C.; Roussel, C.; Metzger, J. *H* **1983**, *20*, 761.
69. Nenitzescu, C. D.; Cantuniari, J. P. *LA* **1934**, *510*, 269.
70. Nenitzescu, C. D.; Cioranescu, E. *CB* **1936**, *69*, 1820.
71. Hopff, H. *CB* **1932**, *65*, 482.
72. (a) Tabushi, I.; Fujita, K.; Oda, R. *TL* **1968**, 4247. (b) Harding, K. E.; Clement, K. S.; Gilbert, J. C.; Wiechman, B. *JOC* **1984**, *49*, 2049.
73. Gilman, H.; Van Ess, P. R. *JACS* **1933**, *55*, 1258.
74. Gilman, H.; Schulze, F. *JACS* **1927**, *49*, 2328.
75. (a) Percival, W. C.; Wagner, R. B.; Cook, N. C. *JACS* **1953**, *75*, 3731. (b) Fiandanese, V.; Marchese, G.; Martina, V.; Ronzini, L. *TL* **1984**, *25*, 4805. (c) Babudri, F.; D'Ettole, A.; Fiandanese, V.; Marchese, G.; Naso, F. *JOM* **1991**, *405*, 53.
76. (a) Hosomi, A.; Hayashida, H.; Tominaga, Y. *JOC* **1989**, *54*, 3254. (b) Sproesser, L.; Sperling, K.; Trautmann, W.; Smuda, H. Ger. Patent 3 744 619, 1989.
77. Cahiez, G.; Laboue, B. *TL* **1992**, *33*, 4439.
78. Fauvarque, J.; Ducom, J.; Fauvarque, J.-F. *CR(C)* **1972**, *275*, 511.
79. (a) Gilman, H.; Mayhue, M. L. *RTC* **1932**, *51*, 47. (b) Chan, T. H.; Chang, E.; Vinokur, E. *TL* **1970**, 1137. (c) Stowell, J. C. *JOC* **1976**, *41*, 560. (d) Sato, F.; Inoue, M.; Oguro, K.; Sato, M. *TL* **1979**, 4303.
80. Normant, J. F. *S* **1972**, 63.
81. (a) Marfat, A.; McGuirk, P. R.; Helquist, P. *TL* **1978**, 1363. (b) Fleming, I.; Newton, T. W.; Roessler, F. *JCS(P1)* **1981**, 2527. (c) Corriu, R. J. P.; Moreau, J. J. E.; Vernhet, C. *TL* **1987**, *28*, 2963.
82. (a) Jallabert, C.; Luong-Thi, N.-T.; Riviere, H. *BSF* **1970**, 797. (b) Ebert, G. W.; Rieke, R. D. *JOC* **1984**, *49*, 5280. (c) Ebert, G. W.; Rieke, R. D. *JOC* **1988**, *53*, 4482. (d) Rieke, R. D.; Wehmeyer, R. M.; Wu, T.-C.; Ebert, G. W. *T* **1989**, *45*, 443. (e) Zhu, L.; Wehmeyer, R. M.; Rieke, R. D. *JOC* **1991**, *56*, 1445. (f) Ebert, G. W.; Cheasty, J. W.; Tehrani, S. S.; Aouad, E. *OM* **1992**, *11*, 1560.
83. (a) Bourgain, M.; Normant, J.-F. *BSF* **1973**, 2137. (b) Logue, M. W.; Moore, G. L. *JOC* **1975**, *40*, 131.
84. (a) Corriu, R. J. P.; Guerin, C.; M'Boula, J. *TL* **1981**, *22*, 2985. (b) Fleming, I.; Pulido, F. J. *CC* **1986**, 1010.
85. Negishi, E.; Bagheri, V.; Chatterjee, S.; Luo, F.-T.; Miller, J. A.; Stoll, A. T. *TL* **1983**, *24*, 5181.
86. (a) Tamaru, Y.; Ochiai, H.; Nakamura, T.; Yoshida, Z. *AG(E)* **1987**, *26*, 1157. (b) Jackson, R. F. W.; James, K.; Wythes, M. J.; Wood, A. *CC* **1989**, 644. (c) Harada, T.; Kotani, Y.; Katsuhira, T.; Oku, A. *TL* **1991**, *32*, 1573. (d) Jackson, R. F. W.; Wishart, N.; Wood, A.; James, K.; Wythes, M. J. *JOC* **1992**, *57*, 3397.
87. Grey, R. A. *JOC* **1984**, *49*, 2288.
88. El Alami, N.; Belaud, C.; Villieras, J. *JOM* **1987**, *319*, 303.
89. Grondin, J.; Sebban, M.; Vottero, P.; Blancou, H.; Commeyras, A. *JOM* **1989**, *362*, 237.
90. Verkruijsse, H. D.; Heus-Kloos, Y. A.; Brandsma, L. *JOM* **1988**, *338*, 289.
91. (a) Blaise, E.; Koehler, A. *CR* **1909**, *148*, 489. (b) Jones, R. G. *JACS* **1947**, *69*, 2350. (c) Shirley, D. A. *OR* **1954**, *8*, 28.
92. (a) Gilman, H.; Nelson, J. F. *RTC* **1936**, *55*, 518. (b) Cason, J. *CRV* **1947**, *40*, 15. (c) Kollonitsch, J. *JCS(A)* **1966**, 453. (d) Jones, P. R.; Desio, P. J. *CRV* **1978**, *78*, 491. (e) Burkhardt, E. R.; Rieke, R. D. *JOC* **1985**, *50*, 416.
93. Kurts, A. L.; Beletskaya, I. P.; Savchenko, I. A.; Reutov, O. A. *JOM* **1969**, *17*, P21.
94. (a) Larock, R. C.; Bernhardt, J. C. *TL* **1976**, 3097. (b) Larock, R. C.; Bernhardt, J. C. *JOC* **1978**, *43*, 710.
95. (a) Bundel, Yu. G.; Rozenberg, V. I.; Kurts, A. L.; Antonova, N. D.; Reutov, O. A. *JOM* **1969**, *18*, 209. (b) Larock, R. C.; Lu, Y. *TL* **1988**, *29*, 6761.
96. For Pd-catalyzed coupling of Ph_2Hg and Et_2Hg with acyl bromides, see: Takagi, K.; Okamoto, T.; Sakakibara, Y.; Ohno, A.; Oka, S.; Hayama, N. *CL* **1975**, 951.
97. Bumagin, N. A.; Kalinovskii, I. O.; Beletskaya, I. P. *BAU* **1984**, *33*, 2144.
98. Negishi, E.; Chiu, K.-W.; Yosida, T. *JOC* **1975**, *40*, 1676.
99. Negishi, E.; Abramovitch, A.; Merrill, R. E. *CC* **1975**, 138.
100. Utimoto, K.; Okada, K.; Nozaki, H. *TL* **1975**, 4239.
101. (a) Paetzold, P. I.; Grundke, H. *S* **1973**, 635. (b) Naruse, M.; Tomita, T.; Utimoto, K.; Nozaki, H. *TL* **1973**, 795. (c) Naruse, M.; Tomita, T.; Utimoto, K.; Nozaki, H. *T* **1974**, *30*, 835.
102. (a) Adkins, H.; Scanley, C. *JACS* **1951**, *73*, 2854. (b) Carr, D. B.; Schwartz, J. *JACS* **1977**, *99*, 638. (c) Maruoka, K.; Sano, H.; Shinoda, K.; Nakai, S.; Yamamoto, H. *JACS* **1986**, *108*, 6036.
103. Takai, K.; Oshima, K.; Nozaki, H. *BCJ* **1981**, *54*, 1281.
104. Sato, F.; Kodama, H.; Tomuro, Y.; Sato, M. *CL* **1979**, 623.
105. Marko, I. E.; Southern, J. M. *JOC* **1990**, *55*, 3368.
106. (a) Frainnet, E.; Calas, R.; Gerval, P. *CR* **1965**, *261*, 1329. (b) Sakurai, H.; Tominaga, K.; Watanabe, T.; Kumada, M. *TL* **1966**, 5493. (c) Grignon-Dubois, M.; Dunogues, J.; Calas, R. *S* **1976**, 737. (d) Olah, G. A.; Ho, T.-L.; Prakash, G. K. S.; Gupta, B. G. B. *S* **1977**, 677. (e) Grignon-Dubois, M.; Dunogues, J.; Calas, R. *CJC* **1980**, *58*, 291. (f) Grignon-Dubois, M.; Dunogues, J. *JOM* **1986**, *309*, 35.
107. (a) Chan, T. H.; Fleming, I. *S* **1979**, 761. (b) Parnes, Z. N.; Bolestova, G. I. *S* **1984**, 991.
108. (a) Pillot, J.-P.; Dunogues, J.; Calas, R. *BSF* **1975**, 2143. (b) Fleming, I.; Pearce, A. *JCS(P1)* **1980**, 2485. (c) Babudri, F.; Fiandanese, V.; Marchese, G.; Naso, F. *CC* **1991**, 237.
109. Pillot, J.-P.; Dunogues, J.; Calas, R. *JCR(S)* **1977**, 268.
110. (a) Birkofer, L.; Ritter, A.; Uhlenbrauck, H. *CB* **1963**, *96*, 3280. (b) Walton, D. R. M.; Waugh, F. *JOM* **1972**, *37*, 45.
111. Pillot, J.-P.; Deleris, G.; Dunogues, J.; Calas, R. *JOC* **1979**, *44*, 3397.
112. Danheiser, R. L.; Stoner, E. J.; Koyama, H.; Yamashita, D. S.; Klade, C. A. *JACS* **1989**, *111*, 4407.
113. Pillot, J.-P.; Bennetau, B.; Dunogues, J.; Calas, R. *TL* **1981**, *22*, 3401.
114. (a) Franciotti, M.; Mordini, A.; Taddei, M. *SL* **1992**, 137. (b) Franciotti, M.; Mann, A.; Mordini, A.; Taddei, M. *TL* **1993**, *34*, 1355.
115. Stille, J. K. *AG(E)* **1986**, *25*, 508.
116. (a) Soderquist, J. A.; Leong, W. W.-H. *TL* **1983**, *24*, 2361. (b) Perez, M.; Castano, A. M.; Echavarren, A. M. *JOC* **1992**, *57*, 5047.
117. (a) Milstein, D.; Stille, J. K. *JACS* **1978**, *100*, 3636. (b) Milstein, D.; Stille, J. K. *JOC* **1979**, *44*, 1613. (c) Yamamoto, Y.; Yanagi, A. *H* **1982**, *19*, 41.
118. Logue, M. W.; Teng, K. *JOC* **1982**, *47*, 2549.

119. (a) Kosugi, M.; Shimizu, Y.; Migita, T. *JOM* **1977**, *129*, C36. (b) Andrianome, M.; Delmond, B. *JOC* **1988**, *53*, 542. (c) Andrianome, M.; Haberle, K.; Delmond, B. *T* **1989**, *45*, 1079.
120. Reetz, M. T.; Hois, P. *CC* **1989**, 1081.
121. Gambaro, A.; Peruzzo, V.; Marton, D. *JOM* **1983**, *258*, 291.
122. Yano, K.; Baba, A.; Matsuda, H. *CL* **1991**, 1181.
123. (a) Barton, D. H. R.; Ozbalik, N.; Ramesh, M. *T* **1988**, *44*, 5661. (b) Asthana, A.; Srivastava, R. C. *JOM* **1989**, *366*, 281.
124. Kasatkin, A. N.; Kulak, A. N.; Tolstikov, G. A. *JOM* **1988**, *346*, 23.
125. Hart, D. W.; Schwartz, J. *JACS* **1974**, *96*, 8115.
126. Hirao, T.; Misu, D.; Yao, K.; Agawa, T. *TL* **1986**, *27*, 929.
127. Cahiez, G.; Laboue, B. *TL* **1989**, *30*, 7369.
128. Normant, J.-F.; Cahiez, G. *Modern Synth. Methods* **1983**, *3*, 173.
129. (a) Hegedus, L. S.; Kendall, P. M.; Lo, S. M.; Sheats, J. R. *JACS* **1975**, *97*, 5448. (b) Pittman, C. U. Jr.; Hanes, R. M. *JOC* **1977**, *42*, 1194.
130. Inaba, S.; Rieke, R. D. *JOC* **1985**, *50*, 1373.
131. (a) Baker, R.; Blackett, B. N.; Cookson, R. C.; Cross, R. C.; Madden, D. P. *CC* **1972**, 343. (b) Inaba, S.; Rieke, R. D. *TL* **1983**, *24*, 2451.
132. (a) Tanaka, T.; Kurozumi, S.; Toru, T.; Kobayashi, M.; Miura, S.; Ishimoto, S. *TL* **1975**, 1535. (b) Kurozumi, S.; Toru, T.; Tanaka, T.; Miura, S.; Kobayashi, M.; Ishimoto, S. U.S. Patent 4 009 196, 1977. U.S. Patent 4 139 717, 1979. (c) Lee, S.-H.; Shih, M.-J.; Hulce, M. *TL* **1992**, *33*, 185.
133. Lapkin, I. I.; Saitkulova, F. G. *JOU* **1971**, *7*, 2586.
134. For examples of acetylations of enolates in which proton transfer is not competitive, see Refs. 132a,b, and Evans, D. A.; Ennis, M. D.; Le, T.; Mandel, N.; Mandel, G. *JACS* **1984**, *106*, 1154.
135. (a) Rasmussen, J. K. *S* **1977**, 91. (b) Tirpak, R. E.; Rathke, M. W. *JOC* **1982**, *47*, 5099.
136. (a) Hoch, H.; Hunig, S. *CB* **1972**, *105*, 2660. (b) Nilsson, L. *ACS(B)* **1979**, *33*, 710. (c) Zhang, P.; Li, L. *SC* **1986**, *16*, 957.
137. Burlachenko, G. S.; Mal'tsev, V. V.; Baukov, Yu. I.; Lutsenko, I. F. *JGU* **1973**, *43*, 1708.
138. Rathke, M. W.; Sullivan, D. F. *TL* **1973**, 1297.
139. Chiba, T.; Ishizawa, T.; Sakaki, J.; Kaneko, C. *CPB* **1987**, *35*, 4672.
140. Fleming, I.; Goldhill, J.; Paterson, I. *TL* **1979**, 3209.
141. Brownbridge, P.; Chan, T. H.; Brook, M. A.; Kang, G. J. *CJC* **1983**, *61*, 688.
142. Ladjama, D.; Riehl, J. J. *S* **1979**, 504.
143. (a) Heusler, K.; Kebrle, J.; Meystre, C.; Ueberwasser, H.; Wieland, P.; Anner, G.; Wettstein, A. *HCA* **1959**, *42*, 2043. (b) Ensley, H. E.; Parnell, C. A.; Corey, E. J. *JOC* **1978**, *43*, 1610.
144. Dauben, W. G.; Eastham, J. F.; Micheli, R. A. *JACS* **1951**, *73*, 4496.
145. Snozzi, C.; Goffinet, B.; Joly, R.; Jolly, J. U.S. Patent 3 117 142, 1964.
146. (a) Casey, C. P.; Marten, D. F. *TL* **1974**, 925. (b) Ouannes, C.; Langlois, Y. *TL* **1975**, 3461.
147. Ono, N.; Yoshimura, T.; Saito, T.; Tamura, R.; Tanikaga, R.; Kaji, A. *BCJ* **1979**, *52*, 1716.
148. Casey, C. P.; Marten, D. F.; Boggs, R. A. *TL* **1973**, 2071.
149. (a) Viscontini, M.; Merckling, N. *HCA* **1952**, *35*, 2280. (b) Rathke, M. W.; Cowan, P. J. *JOC* **1985**, *50*, 2622.
150. (a) Adams, R.; Vollweiler, E. H. *JACS* **1918**, *40*, 1732. (b) French, H. E.; Adams, R. *JACS* **1921**, *43*, 651. (c) Ulich, L. H.; Adams, R. *JACS* **1921**, *43*, 660. (d) Euranto, E. K.; Noponen, A.; Kujanpaa, T. *ACS* **1966**, *20*, 1273. (e) Kyburz, R.; Schaltegger, H.; Neuenschwander, M. *HCA* **1971**, *54*, 1037. (f) Neuenschwander, M.; Iseli, R. *HCA* **1977**, *60*, 1061. (g) Neuenschwander, M.; Vogeli, R.; Fahrni, H.-P.; Lehmann, H.; Ruder, J.-P. *HCA* **1977**, *60*, 1073. (h) Bigler, P.; Muhle, H.; Neuenschwander, M. *S* **1978**, 593.
151. Bigler, P.; Neuenschwander, M. *HCA* **1978**, *61*, 2165.
152. (a) Euranto, E.; Kujanpaa, T. *ACS* **1961**, *15*, 1209. (b) Euranto, E.; Leppanen, O. *ACS* **1963**, *17*, 2765. (c) Neuenschwander, M.; Bigler, P.; Christen, K.; Iseli, R.; Kyburz, R.; Muhle, H. *HCA* **1978**, *61*, 2047.
153. (a) Bigler, P.; Schonholzer, S.; Neuenschwander, M. *HCA* **1978**, *61*, 2059. (b) Bigler, P.; Neuenschwander, M. *HCA* **1978**, *61*, 2381.
154. (a) Chou, T.-S.; Knochel, P. *JOC* **1990**, *55*, 4791, 6232. (b) Knochel, P.; Chou, T.-S.; Jubert, C.; Rajagopal, D. *JOC* **1993**, *58*, 588.
155. Knochel, P.; Chou, T.-S.; Chen, H. G.; Yeh, M. C. P.; Rozema, M. J. *JOC* **1989**, *54*, 5202.
156. Enholm, E. J.; Satici, H. *TL* **1991**, *32*, 2433.
157. Oku, A.; Harada, T.; Kita, K. *TL* **1982**, *23*, 681.
158. Cloke, J. B.; Pilgrim, F. J. *JACS* **1939**, *61*, 2667.
159. Duboudin, J.-G.; Valade, J. *BSF* **1974**, 272.
160. (a) Ahmad, S.; Iqbal, J. *CL* **1987**, 953. (b) Iqbal, J.; Srivastava, R. R. *T* **1991**, *47*, 3155.
161. Pri-Bar, I.; Stille, J. K. *JOC* **1982**, *47*, 1215.
162. Alper, H.; Huang, C.-C. *JOC* **1973**, *38*, 64.
163. Fitch, J. W.; Payne, W. G.; Westmoreland, D. *JOC* **1983**, *48*, 751.
164. Chang, C.; Chu, K. C.; Yue, S. *SC* **1992**, *22*, 1217.
165. (a) Straus, F.; Heinze, H. *LA* **1932**, *493*, 191. (b) Quintard, J.-P.; Elissondo, B.; Pereyre, M. *JOM* **1981**, *212*, C31. (c) Quintard, J.-P.; Elissondo, B.; Pereyre, M. *JOC* **1983**, *48*, 1559.
166. (a) Bakos, T.; Vincze, I. *SC* **1989**, *19*, 523. (b) Sabharwal, A.; Vig, R.; Sharma, S.; Singh, J. *IJC(B)* **1990**, *29*, 890.
167. (a) Pop, L.; Oprean, I.; Barabas, A.; Hodosan, F. *JPR* **1986**, *328*, 867. (b) Oprean, I.; Ciupe, H.; Gansca, L.; Hodosan, F. *JPR* **1987**, *329*, 283.
168. Dutka, F.; Marton, A. F. *ZN(B)* **1969**, *24*, 1664.
169. Hofle, G.; Steglich, W.; Vorbruggen, H. *AG(E)* **1978**, *17*, 569.
170. (a) Vedejs, E.; Diver, S. T. *JACS* **1993**, *115*, 3358. (b) Vedejs, E.; Bennett, N. S.; Conn, L. M.; Diver, S. T.; Gingras, M.; Lin, S.; Oliver, P. A.; Peterson, M. J. *JOC* **1993**, *58*, 7286.
171. 1,2-Diols can be selectively monoacetylated by conversion into the cyclic dibutylstannylidene derivative followed by treatment with AcCl: (a) Anchisi, C.; Maccioni, A.; Maccioni, A. M.; Podda, G. *G* **1983**, *113*, 73. (b) Roelens, S. *JCS(P2)* **1988**, 2105. (c) Anderson, W. K.; Coburn, R. A.; Gopalsamy, A.; Howe, T. J. *TL* **1990**, *31*, 169. (d) Getman, D. P.; DeCrescenzo, G. A.; Heintz, R. M. *TL* **1991**, *32*, 5691.
172. γ-Picoline might be unsuitable as a pyridine replacement because it reacts with AcCl to form *N*-acetyl-4-(acetylmethylidene)-1,4-dihydropyridine under standard acetylation conditions (CH_2Cl_2, rt, 8–16 h): Ippolito, R. M.; Vigmond, S. U.S. Patent 4 681 944, 1987.
173. (a) Okamoto, K.; Kondo, T.; Goto, T. *T* **1987**, *43*, 5909. (b) McClure, K. F.; Danishefsky, S. J. *JACS* **1993**, *115*, 6094.
174. (a) Braun, M.; Devant, R. *TL* **1984**, *25*, 5031. (b) Devant, R.; Mahler, U.; Braun, M. *CB* **1988**, *121*, 397.
175. Capek, K.; Steffkova, J.; Jary, J. *CCC* **1966**, *31*, 1854.
176. Bladon, P.; Clayton, R. B.; Greenhalgh, C. W.; Henbest, H. B.; Jones, E. R. H.; Lovell, B. J.; Silverstone, G.; Wood, G. W.; Woods, G. F. *JCS* **1952**, 4883.
177. (a) Litvinenko, L. M.; Kirichenko, A. I. *DOK* **1967**, *176*, 763. (b) Steglich, W.; Hofle, G. *AG(E)* **1969**, *8*, 981. (c) Hofle, G.; Steglich, W. *S* **1972**, 619. (d) Scriven, E. F. V. *CSR* **1983**, *12*, 129.
178. (a) Norris, J. F.; Rigby, G. W. *JACS* **1932**, *54*, 2088. (b) Plattner, Pl. A.; Petrzilka, Th.; Lang, W. *HCA* **1944**, *27*, 513. (c) Hauser, C. R.; Hudson, B. E.; Abramovitch, B.; Shivers, J. C. *OSC* **1955**, *3*, 142. (d) Ohloff, G. *HCA* **1958**, *41*, 845.
179. (a) Plattner, Pl. A.; Furst, A.; Koller, F.; Lang, W. *HCA* **1948**, *31*, 1455. (b) Williams, K. I. H.; Rosenfeld, R. S.; Smulowitz, M.; Fukushima, D. K. *Steroids* **1963**, *1*, 377. (c) Kido, F.; Kitahara, H.; Yoshikoshi, A. *JOC* **1986**, *51*, 1478.
180. (a) Takimoto, S.; Inanaga, J.; Katsuki, T.; Yamaguchi, M. *BCJ* **1976**, *49*, 2335. (b) Amouroux, R.; Chan, T. H. *TL* **1978**, 4453.
181. Spassow, A. *OSC* **1955**, *3*, 144.
182. (a) Illi, V. O. *TL* **1979**, 2431. (b) Szeja, W. *S* **1980**, 402.
183. (a) Perriot, P.; Normant, J. F.; Villieras, J. *CR(C)* **1979**, *289*, 259. (b) Trost, B. M.; Tour, J. M. *JOC* **1989**, *54*, 484. (c) Corey, E. J.; Su, W. *TL* **1990**, *31*, 2089.

184. (a) Evans, D. D.; Evans, D. E.; Lewis, G. S.; Palmer, P. J.; Weyell, D. J. *JCS* **1963**, 3578. (b) Duboudin, J. G.; Ratier, M.; Trouve, B. *JOM* **1987**, *331*, 181.

185. Weidert, P. J.; Geyer, E.; Horner, L. *LA* **1989**, 533.

186. (a) Heyse, M. Ger. Patent 524 435, 1929. (b) Searles, S. Jr.; Pollart, K. A.; Block, F. *JACS* **1957**, *79*, 952. (c) Sharma, K. K.; Torssell, K. B. G. *T* **1984**, *40*, 1085.

187. Kotsuki, H.; Kataoka, M.; Nishizawa, H. *TL* **1993**, *34*, 4031.

188. Bryant, W. M. D.; Smith, D. M. *JACS* **1936**, *58*, 1014.

189. Karger, M. H.; Mazur, Y. *JOC* **1971**, *36*, 528.

190. (a) Lennon, J. J.; Perkin, W. H. Jr. *JCS* **1928**, 1513. (b) Zilkha, A.; Liwschitz, Y. *JCS* **1957**, 4397. (c) Bose, N. K.; Chaudhury, D. N. *T* **1964**, *20*, 49.

191. (a) Turner, R. B. *JACS* **1950**, *72*, 579. (b) Rosenmund, K. W.; Herzberg, H.; Schutt, H. *CB* **1954**, *87*, 1258. (c) Vignau, M.; Bucourt, R.; Tessier, J.; Costerousse, G.; Nedelec, L.; Gasc, J.-C.; Joly, R.; Warnant, J.; Goffinet, B. U.S. Patent 3 453 267, 1969.

192. (a) Harada, K.; Kaji, E.; Zen, S. *CPB* **1980**, *28*, 3296. (b) Fleming, I.; Moses, R. C.; Tercel, M.; Ziv, J. *JCS(P1)* **1991**, 617.

193. Farrissey, W. J. Jr. U.S. Patent 3 291 834, 1966.

194. Stirring zinc dust with AcCl and CuCl (Et_2O, rt → Δ) produces an active zinc couple capable of reacting with methylene bromide to form the Simmons–Smith reagent: Friedrich, E. C.; Lewis, E. J. *JOC* **1990**, *55*, 2491.

195. Watt, G. W.; Gentile, P. S.; Helvenston, E. P. *JACS* **1955**, *77*, 2752.

196. Bellesia, F.; Ghelfi, F.; Pagnoni, U. M.; Pinetti, A. *JCR(S)* **1990**, 188.

197. Nugent, W. A. *TL* **1978**, 3427.

198. Bellesia, F.; Ghelfi, F.; Pagnoni, U. M.; Pinetti, A. *SC* **1991**, *21*, 489.

199. Bellesia, F.; Ghelfi, F.; Pagnoni, U. M.; Pinetti, A. *JCR(S)* **1989**, 108.

200. Backvall, J. E.; Young, M. W.; Sharpless, K. B. *TL* **1977**, 3523.

201. Cummins, C. H.; Coates, R. M. *JOC* **1983**, *48*, 2070.

202. (a) Demerseman, P.; Guillaumel, J.; Clavel, J.-M.; Royer, R. *TL* **1978**, 2011. (b) Guillaumel, J.; Demerseman, P.; Clavel, J.-M.; Royer, R.; Platzer, N.; Brevard, C. *T* **1980**, *36*, 2459.

203. Numata, T.; Oae, S. *CI(L)* **1973**, 277.

204. (a) Adkins, H.; Thompson, Q. E. *JACS* **1949**, *71*, 2242. (b) Paukstelis, J. V.; Kim, M. *JOC* **1974**, *39*, 1503.

205. (a) Lang, R. W.; Hansen, H.-J. *HCA* **1980**, *63*, 438. (b) Lang, R. W.; Hansen, H.-J. *OS* **1984**, *62*, 202. (c) Abell, A. D.; Morris, K. B.; Litten, J. C. *JOC* **1990**, *55*, 5217.

206. (a) Burger, K.; Huber, E.; Sewald, N.; Partscht, H. *Chem.-Ztg.* **1986**, *110*, 83. (b) Sewald, N.; Riede, J.; Bissinger, P.; Burger, K. *JCS(P1)* **1992**, 267.

207. Maujean, A.; Chuche, J. *TL* **1976**, 2905.

208. Sauer, J. C. *JACS* **1947**, *69*, 2444.

209. Hunt, C. F. U.S. Patent 2 737 528, 1956.

210. Matoba, K.; Tachi, M.; Itooka, T.; Yamazaki, T. *CPB* **1986**, *34*, 2007.

211. Furniss, B. S.; Hannaford, A. J.; Smith, P. W. G.; Tatchell, A. R. *Vogel's Textbook of Practical Organic Chemistry*, 5th ed.; Longman/Wiley: New York, 1989; pp 916, 1273.

212. (a) Sonntag, N. O. V. *CRV* **1953**, *52*, 237. (b) To minimize the amount of hydrolysis, the acetylation should be run at pH = $(x + 13.25)/2$, where x is the pK_a of the protonated amine: King, J. F.; Rathore, R.; Lam, J. Y. L.; Guo, Z. R.; Klassen, D. F. *JACS* **1992**, *114*, 3028.

213. (a) Bohme, H.; Hartke, K. *CB* **1960**, *93*, 1305. (b) Kinast, G.; Tietze, L.-F. *AG(E)* **1976**, *15*, 239.

214. Kritzler, H.; Wanger, K.; Holtschmidt, H. U.S. Patent 3 242 202, 1966.

215. (a) Ikeda, K.; Morimoto, T.; Sekiya, M. *CPB* **1980**, *28*, 1178. (b) Ikeda, K.; Terao, Y.; Sekiya, M. *CPB* **1981**, *29*, 1156.

216. Okada, I.; Takahama, T.; Sudo, R. *BCJ* **1970**, *43*, 2591.

217. Caubere, P.; Madelmont, J.-C. *CR(C)* **1972**, *275*, 1305.

218. Comins, D. L.; Abdullah, A. H. *JOC* **1982**, *47*, 4315.

219. Corcoran, R. C.; Bang, S. H. *TL* **1990**, *31*, 6757.

220. (a) Oppolzer, W.; Bieber, L.; Francotte, E. *TL* **1979**, 981. (b) Ng, K. S.; Laycock, D. E.; Alper, H. *JOC* **1981**, *46*, 2899.

221. Ben-Ishai, D.; Katchalski, E. *JOC* **1951**, *16*, 1025.

222. Hoffmann, H. M. R.; Haase, K. *S* **1981**, 715.

223. Ihara, M.; Hirabayashi, A.; Taniguchi, N.; Fukumoto, K. *H* **1992**, *33*, 851.

224. (a) Pump, J.; Wannagat, U. *M* **1962**, *93*, 352. (b) Bowser, J. R.; Williams, P. J.; Kurz, K. *JOC* **1983**, *48*, 4111.

225. Ahmad, S.; Iqbal, J. *TL* **1986**, *27*, 3791.

Bruce A. Pearlman
The Upjohn Company, Kalamazoo, MI, USA

Acetylene

HC≡CH

[74-86-2] C_2H_2 (MW 26.04)

(ethynylation reagent,[1] hydrosilylation,[23] carbonylation,[24–26] cycloadditions,[31–33] cyclotrimerization[34])

Alternate Name: ethyne.
Physical Data: bp −83 °C.
Form Supplied in: widely available as compressed gas.
Drying: can be purified by passing through a trap at −80 °C followed by a column of sulfuric acid and then through a column of sodium hydroxide.[2]
Handling, Storage, and Precautions: flammable, colorless gas which possesses a garlic odor; use only with adequate ventilation; skin irritant. The low electrical conductivity of acetylene requires that care must be taken to eliminate any static charge buildup. Use in a fume hood.

Ethynylation. Due to the sp hybridization of acetylene's σ-bonds the methine proton has a pK_a of about 25 and can be removed with strong bases to form the acetylide anion.[3] Sodium acetylide is commonly prepared by deprotonation with sodium amide. ***Lithium Acetylide*** can be conveniently prepared using ***n-Butyllithium***, while the corresponding Grignard reagent is commonly generated by treatment with ***Ethylmagnesium Bromide***.[4] Alkyl cuprates undergo nucleophilic addition to acetylene.[5] The dianion of acetylene is available by using higher temperatures during deprotonation.[6] A variety of other counter cations of the acetylide ion have also been formulated, usually starting from the acetylides previously mentioned.[1a] The acetylide anion undergoes nucleophilic addition to a wide range of electrophiles such as halides,[7] alkyl halides,[8] alkyl sulfonates,[9] epoxides,[10] aldehydes,[11] ketones,[12] esters,[13] and imides.[14] In certain cases the yields of addition may be improved by additives such as ethylenediamine[15] or ***Copper(I) Iodide***.[16] When the electrophile is hindered, elimination may become the predominant pathway. This problem may be avoided with trialkylborane electrophiles (eq 1).[17] After treatment with iodine, the intermediate borate yields the substituted alkyne via migration of one of the alkyl groups. This sequence is not as sensitive to steric effects as simple substitutions with halides. Higher yields can be obtained if ***Lithium (Trimethylsilyl)acetylide*** is used instead (see also ***Trimethylsilylacetylene***).

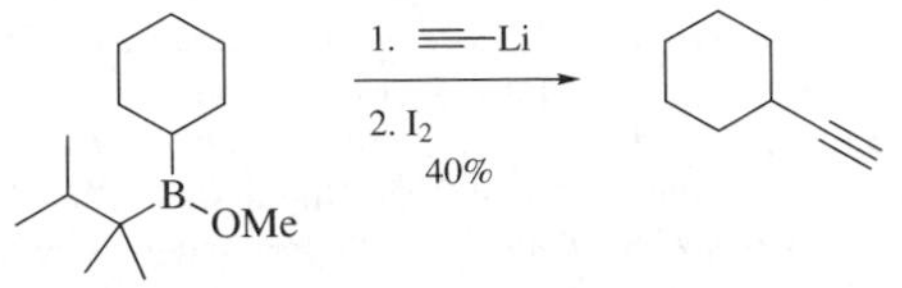

(1)

The nucleophilicity of the acetylide anion is such that discrimination between several electrophilic centers within a substrate is possible (eq 2).[18]

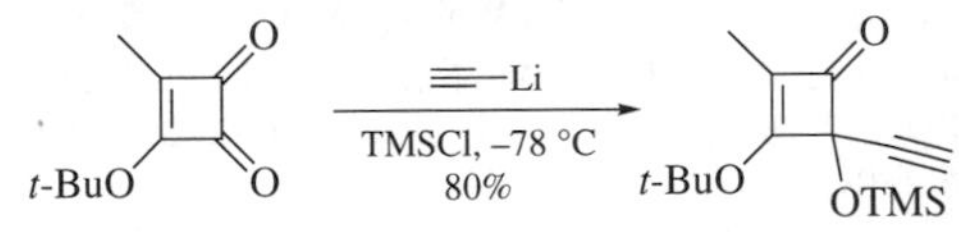

(2)

The regioselectivity of acetylide additions are sensitive to the counter cation. For example, ethynyltriisopropoxytitanium acetylide, prepared from ethynyllithium and ***Chlorotitanium Triisopropoxide***, gives a single addition product with pyrimidin-2(1*H*)-ones while the bromomagnesium and lithium acetylides give mixtures of regioisomers (eq 3).[19] There are also examples of diastereoselective additions of acetylide anions to ketones (eq 4).[20]

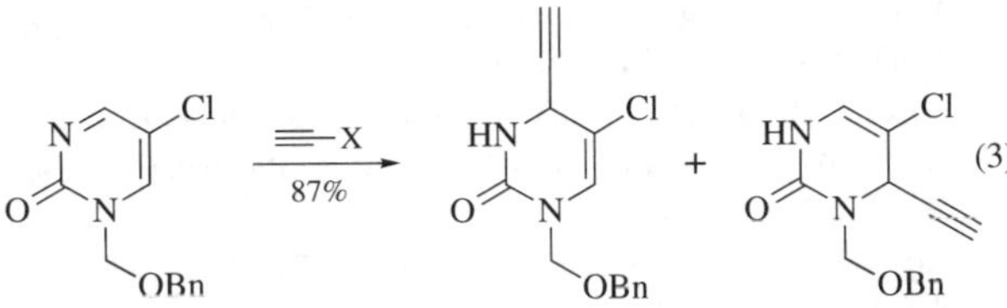

(3)

X =		
Li	20	80
MgBr	50	50
Ti(O-*i*-Pr)$_3$	0	100

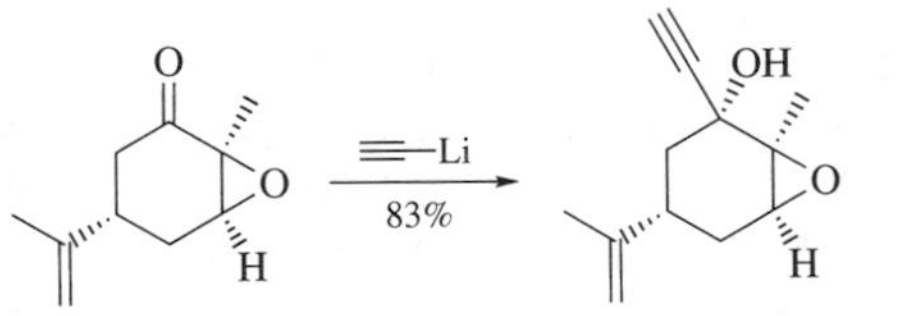

(4)

The acetylide anion is useful for the synthesis of enynes which have been used in palladium catalyzed cyclizations (eq 5).[21]

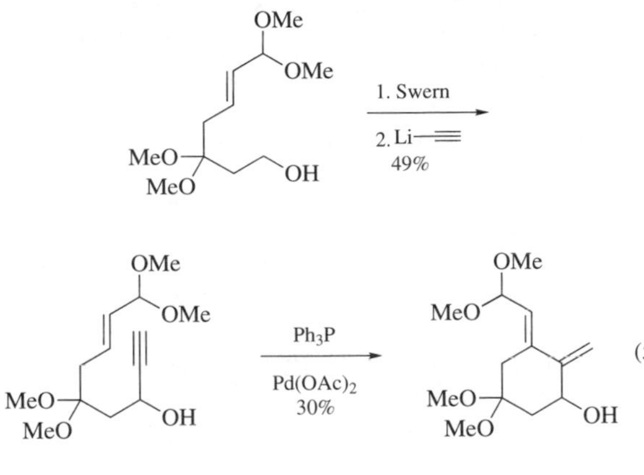

(5)

Ethynylation of aryl halides (Stephens–Castro coupling) with the copper acetylide is synthetically effective when one of the acetylenic methine protons is protected.[22]

Hydrosilylation. Acetylene can be converted[23] into a vinylsilane under appropriate hydrosilylation conditions using Pt, Rh, Ru, or Al catalysts, but the initial 1:1 product can react further to give a 1,2-disilylethane. Examples of these processes are shown in eq 6.

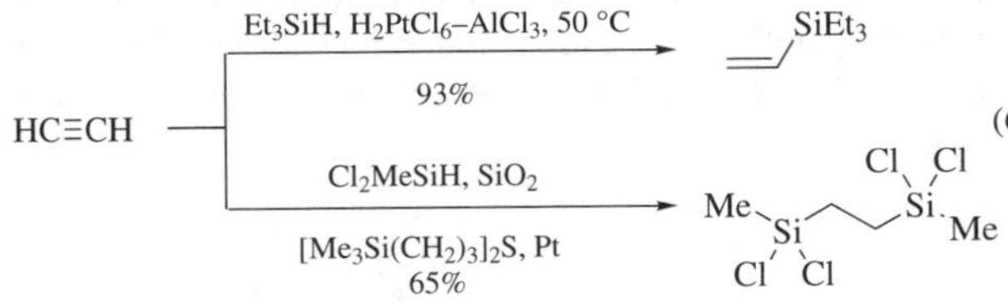

(6)

Reactions with CO and CO_2. Alkynes react with CO_2 and secondary amines in the presence of ruthenium complexes to afford vinyl carbamates (eq 7).[24] Mononuclear ruthenium complexes and $[RuCl_2(norbornadiene)]_n$ or ***Ruthenium(III) Chloride*** have been shown to be the best catalysts for monosubstituted alkynes and acetylene, respectively.

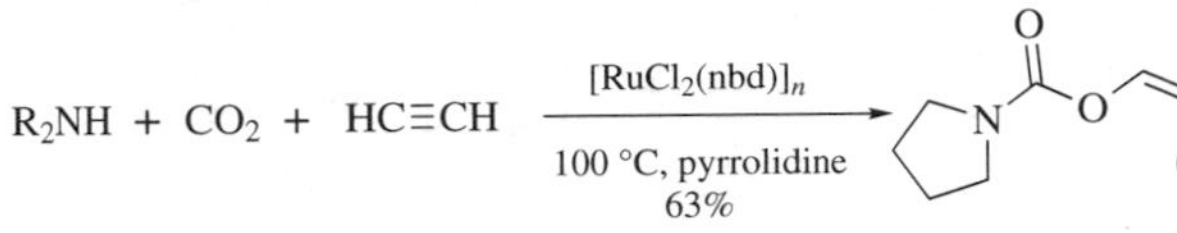

(7)

Acetylene reacts[25] with CO and O_2 in presence of ***Palladium(II) Chloride*** to give dimethyl maleate in 90% yield. Carboxylic acids have been formed by treating acetylene with CO and Ni or Pd catalysts.[25a] Acetylene reacts with CO in presence of $Co_2(CO)_8$ to give an (*E*)-bisbutenolide in aprotic solvents (eq 8), while the (*Z*)-isomer is obtained in the presence of tetramethylurea.[26]

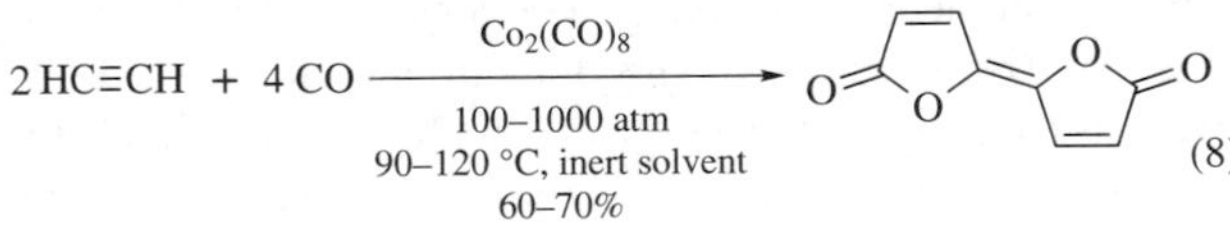

(8)

Addition. Acyl chlorides add across acetylene to give β-chloroenones in CCl_4 in the presence of ***Aluminum Chloride*** (eq 9).[27]

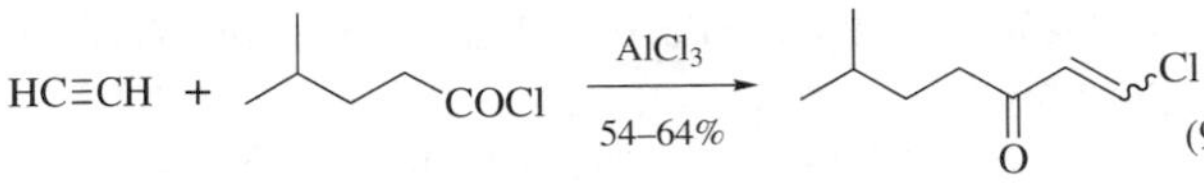

(9)

The acid catalyzed hydration of acetylene to acetaldehyde followed by condensation to give crotonaldehyde is a well known reaction first studied by Berthelot in 1862.[28] Although more efficient procedures have been developed since then, the mechanism of hydration of acetylene has not yet been unambiguously resolved.[29] Kinetic data based on NMR experiments suggest that a vinyl cation is probably an intermediate.[30]

Cycloadditions. Acetylene has low reactivity in Diels–Alder reactions and reacts with electron rich dienes only under severe conditions.[31] The reaction of dienes with vinyl bromide and subsequent elimination of HBr from the adduct leads to the same product as direct addition of acetylene. In addition a number of compounds have been suggested as acetylene surrogates, such as 2-phenyl- and ***2-Thiono-1,3-dioxol-4-ene***,[32] and phenyl vinyl sulfoxide.[33] Vinylene thioxocarbonate and 2-phenyl-1,3-dioxol-4-ene, on reaction[9] with anthracene in benzene at 170 °C for 16 h, gave the corresponding adducts in 60 and 65% yield, respectively. The [4 + 2] adducts yield the corresponding alkene upon treatment with trivalent phosphorus or *n*-butyllithium respectively. Similarly, heating anthracene and phenyl vinyl sulfoxide in chlorobenzene for 120 h afforded dibenzobarrelene in 83% isolated yield (eq 10).[9]

PhCl, Δ; 83% (10)

The following example (eq 11) involves in situ reductive extrusion of an oxygen atom. It has been suggested that the liberated PhSOH appears capable of deoxygenating isobenzofurans and related molecules.

toluene, Δ; SOPh

Cyclotrimerization. In addition to acetylene, many mono- or disubstituted alkynes undergo cyclotrimerization in the presence of transition metal complexes. Acetylene itself trimerizes to give benzene. Trimerization of unsymmetrically substituted alkynes gives rise mostly to benzenes in which the most sterically demanding substituents occupy positions 1, 2, and 4 around the ring and to a lesser extent a 1,3,5-substitution pattern is also obtained.[34] Although numerous mechanistic pathways have been postulated, depending on the metal involved, a common feature involves intermediacy of metallocyclopentadienes and complexation with a third alkyne which inserts to give a transient metallocycloheptatriene, leading finally to reductive elimination of the metal and the benzene product.

Selective intermolecular co-cycloaddition has been achieved, exploiting the relative unreactivity of phosphine nickel carbonyls towards trimerization of internal alkynes.[35,36] Thus two molecules of acetylene and one molecule of an internal alkyne give a mixture of the unsubstituted and substituted benzenes (eq 12).

THF, cat. $Ni(CO)_2(PPh_3)_2$, 5–15 atm, 80–160 °C, 3 h; 93%

46% + 54% (12)

Related Reagents. Bis(trimethylsilyl)acetylene; Phenylsulfonylacetylene; Phenylsulfinylethylene; Trimethylsilylacetylene.

1. (a) Garrat, P. J. *COS* **1991**, *3*, 271. (b) Friedrich, K. *Chemistry of Functional Groups, The Chemistry of Triple-Bonded Functional Groups*; Patai, S.; Rappoport, Z., Eds.; Wiley: Chichester, 1983; suppl. C, part 2, p 1380; (c) Brandsma, L. *Preparative Acetylene Chemistry*, 3rd ed.; Elsevier: Amsterdam, 1988. (d) Ben-Efraim, D. A. *The Chemistry of the Carbon–Carbon Triple Bond*; Patai, S., Ed.; Wiley: New York, 1978; pp 790–800.
2. Skattebøl, L.; Jones, E. R. H.; Whiting, M. C. *OSC* **1963**, *4*, 792.
3. (a) Cram, D. J. *Fundamentals of Carbanion Chemistry*; Academic Press: New York, 1965; pp 1–45. (b) Dessy, R. E.; Kitching, W.; Psarras, T.; Salinger, R.; Chen, A.; Chivers, T. *JACS* **1966**, *88*, 460.
4. Jones, E. R. H.; Skattebøl, L.; Whiting, M. C. *JACS* **1956**, 4765.
5. (a) Alexakis, A.; Barthel, A. M.; Normant, J. F.; Fugier, C.; Leroux, M. *SC* **1992**, *22*, 1839. (b) Furber, M; Taylor, R. J. K.; Buford, S. C. *JCS(P1)* **1986**, 1809.
6. Sudweeks, W. B.; Broadbent, H. S. *JOC* **1975**, *40*, 1131.
7. Brandsma, L.; Verkruijsse, H. D. *S* **1990**, 984.
8. Campbell, K. N.; Campbell, B. K. *OSC* **1963**, *4*, 117.
9. (a) Ireland, R. E.; Highsmith, T. K.; Gegnas, L. D.; Gleason, J. L. *JOC* **1992**, *57*, 5071. (b) Crombie, L.; Heavers, A. D. *JCS(P1)* **1992**, 1929.
10. Buist, P. H.; Adeney, R. A. *JOC* **1991**, *56*, 3449.
11. (a) Chiarino, D.; Fantucci, M. *JHC* **1991**, *28*, 1705. (b) Girard, S.; Deslongchamps, P. *CJC* **1992**, *70*, 1265.
12. Saunders, J. H. *OSC* **1955**, *3*, 416.
13. Bolitt, V.; Mioskowski, C.; Kollah, R. O.; Manna, S.; Rajapaksa, D.; Falck, J. R. *JACS* **1991**, *113*, 6320.
14. Omar, E. A.; Tu, C.; Wigal, C. T.; Braun, L. L. *JHC* **1992**, *29*, 947.
15. Smith, W. N.; Beumel, O. F. *S* **1974**, 441. (b) Beumel, O. F.; Harris, R. F. *JOC* **1963**, *28*, 2775.
16. (a) Bourgain, M.; Normant, J. F. *BSF(2)* **1973**, 1777; (b) Jeffery, T. *TL* **1989**, *30*, 2225.
17. (a) Brown, H. C.; Mahindroo, V. K.; Bhat, N. G.; Singaram, B. *JOC* **1991**, *56*, 1500. (b) Midland, M. M.; Sinclair, J. A.; Brown, H. C. *JOC* **1974**, *39*, 731.
18. Enhsen, A.; Karabelas, K.; Heerding, J. M.; Moore, H. W. *JOC* **1990**, *55*, 1177.
19. Gundersen, L. L.; Rise, F.; Undheim, K. *T* **1992**, *48*, 5647.
20. (a) Okamura, W. H.; Aurrecoechea, J. M.; Gibbs, R. A.; Norman, A. W. *JOC* **1989**, *54*, 4072. (b) Gordon, J.; Tabacchi, R. *JOC* **1992**, *57*, 4728.
21. Trost, B. M.; Dumas, J.; Villa, M. *JACS* **1992**, *114*, 9836.
22. (a) Stephens, R. D.; Castro, C. E. *JOC* **1963**, *28*, 3313. (b) Seiburth, S. M.; Chen, J. L. *JACS* **1991**, *113*, 8163.
23. Hiyama, T.; Kusumoto, T. *COS* **1991**, *8*, 769 and references cited therein.
24. Mahè, R.; Sasaki, Y.; Bruneau, C.; Dixneuf, P. H. *JOC* **1989**, *54*, 1518.
25. (a) Cassar, L.; Chiusoli, G. P.; Guerrieri, F. *S* **1973**, 509. (b) Chiusoli, G. P.; Venturello, C.; Merzoni, S. *Chim. Ind. (Milan)* **1968**, 977.
26. Sauer, J. C.; Cramer, R. D.; Engelhardt, V. A.; Ford, T. A.; Holmquist, H. E.; Howk, B. W. *JACS* **1959**, *81*, 3677.
27. Cooper, F. C.; Partridge, M. W. *OSC* **1963**, *4*, 769.
28. Berthelot, M. C. *CR* **1862**, *50*, 805.
29. Reviews: (a) Modena, G.; Tonellato, A. *CR* **1981**, *14*, 227. (b) Stang, P. J.; Rappoport, Z.; Hanak, M.; Subramanian, L. R. *Vinyl Cations*; Academic Press: New York, 1979. (c) Stang, P. J. *Prog. Phys. Org. Chem.* **1973**, *10*, 205. (d) Modena, G.; Tonellato, U. *Adv. Phys. Org. Chem.* **1971**, *9*, 185.
30. Lucchini, V.; Modena, G. *JOC* **1990**, *55*, 6291.
31. Sauer, J. *AG(E)* **1966**, *5*, 211.
32. Anderson, W. K.; Dewey, R. H. *JACS* **1973**, *95*, 7161.

33. Paquette, L. A.; Moerck, R. E.; Harirchian, B.; Magnus, P. D. *JACS* **1978**, *100*, 1597.
34. Schore, N. E. *COS* **1991**, *5*, 1144.
35. (a) Meriwether, L. S.; Colthup, E. C.; Kennerly, G. W.; Reusch, R. N. *JOC* **1961**, *26*, 5155. (b) Chalk, A. J.; Jerussi, R. A. *TL* **1972**, 61.
36. (a) Sauer, J. C.; Cairns, T. L. *JACS* **1957**, *79*, 2659. (b) Cope, A. C.; Handy, C. T. *CA* **1961**, *55*, 1527b. (c) Hübel, W.; Hoogzand, C. *CB* **1960**, *93*, 103. (d) Mills, O. S.; Robinson, G. *Proc. Chem. Soc. (London)* **1964**, 187.

Samit K. Bhattacharya, John E. Stelmach, & Jeffrey D. Winkler
University of Pennsylvania, Philadelphia, PA, USA

Acrylonitrile

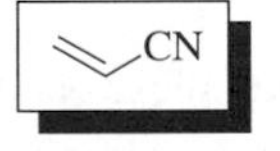

[107-13-1] C_3H_3N (MW 53.06)

(electrophile in 1,4-addition reactions; radical acceptor; dienophile; acceptor in cycloaddition reactions)

Physical Data: mp −83 °C; bp 77 °C; *d* 0.806 g cm^{-3}; n_D 1.3911.
Solubility: miscible with most organic solvents; 7.3 g of acrylonitrile dissolves in 100 g of water at 20 °C.
Form Supplied in: colorless liquid (inhibited with 35–45 ppm hydroquinone monomethyl ether); widely available.
Purification: the stabilizer can be removed prior to use by passing the liquid through a column of activated alumina or by washing with a 1% aqueous solution of NaOH (if traces of water are allowed in the final product) followed by distillation. For dry acrylonitrile, the following procedure is recommended. Wash with dilute H_2SO_4 or H_3PO_4, then with dilute aqueous Na_2CO_3 and water. Dry over Na_2SO_4, $CaCl_2$, or by shaking with molecular sieves. Finally, fractional distillation under nitrogen (boiling fraction of 75–75.5 °C) provides acrylonitrile which can be stabilized by adding 10 ppm *t*-butyl catechol or hydroquinone monomethyl ether. Pure acrylonitrile is distilled as required.[1a]
Handling, Storage, and Precautions: explosive, flammable, and toxic liquid. May polymerize spontaneously, particularly in the absence of oxygen or on exposure to visible light, if no inhibitor is present. Polymerizes violently in the presence of concentrated alkali. Highly toxic through cyanide effect. Use in a fume hood.

Deuterioacrylonitrile. Deuterium-labeled acrylonitrile can be obtained by reduction of propiolamide-d_3 with ***Lithium Aluminum Hydride***, followed by D_2O workup. The resulting acrylamide can then be dehydrated with P_2O_5.[1b]

Reactions of the Nitrile Group. Various functional group transformations have been carried out on the nitrile group in acrylonitrile. Hydration with concentrated ***Sulfuric Acid*** at 100 °C yields acrylamide after neutralization.[2] Secondary and tertiary alcohols produce *N*-substituted acrylamides under these conditions in excellent yield (Ritter reaction).[3] Heating in the presence of dilute sulfuric acid or with an aqueous basic solution yields acrylic acid.[4] Imido ethers have been prepared by reacting acrylonitrile with alcohols in the presence of anhydrous hydrogen halides.[5] Anhydrous ***Formaldehyde*** reacts with acrylonitrile in the presence of concentrated sulfuric acid to produce 1,3,5-triacrylylhexahydrotriazine.[6]

Reactions of the Alkene. Reduction with hydrogen in the presence of Cu,[7] Rh,[8] Ni,[9] or Pd[10] yields propionitrile. Acrylonitrile can be halogenated at low temperature to produce 2,3-dihalopropionitriles. For example, reaction with ***Bromine*** leads to dibromopropionitrile in 65% yield.[11] Also, treatment of acrylonitrile with an aqueous solution of ***Hypochlorous Acid***, gives 2-chloro-3-hydroxypropionitrile in 60% yield.[12] α-Oximation of acrylonitrile has been achieved using Co^{II} catalysts, ***n-Butyl Nitrite*** and phenylsilane.[13]

Nucleophilic Additions. A wide variety of nucleophiles react with acrylonitrile in 1,4-addition reactions. These Michael-type additions are often referred to as cyanoethylation reactions.[14] The following list illustrates the variety of substrates which will undergo cyanoethylation: ammonia, primary and secondary amines, hydroxylamine, enamines, amides, lactams, imides, hydrazine, water, various alcohols, phenols, oximes, sulfides, inorganic acids like HCN, HCl, HBr, chloroform, bromoform, aldehydes, and ketones bearing an α-hydrogen, malonic ester derivatives, and other diactivated methylene compounds.[15] Stabilized carbanions derived from ***Cyclopentadiene*** and fluorene and 1–5% of an alkaline catalyst also undergo cyanoethylation. The strongly basic quaternary ammonium hydroxides, such as ***Benzyltrimethylammonium Hydroxide*** (Triton B), are particularly effective at promoting cyanoethylation because of their solubility in organic media. Reaction temperatures vary from −20 °C for reactive substrates, to heating at 100 °C for weaker nucleophiles. The 1,4-addition of amines has recently been used in the synthesis of poly-(propyleneimine) dendrimers.[16]

Phosphine nucleophiles have been reported to promote nucleophilic polymerization of acrylonitrile.[17]

Addition of organometallic reagents to acrylonitrile is less efficient than to conjugated enones. Grignard reagents react with acrylonitrile by 1,2-addition and, after hydrolysis, give α,β-unsaturated ketones.[18] Lithium dialkylcuprate (R_2CuLi) addition in the presence of ***Chlorotrimethylsilane*** leads to double addition at the alkene and nitrile, giving a dialkyl ketone.[19] Yields of only 23–46% are obtained in the conjugate addition of *n*-$BuCu{\cdot}BF_3$ to acrylonitrile.[20] An enantioselective Michael reaction has been achieved with titanium enolates derived from *N*-propionyloxazolidone (eq 1).[21]

1. $(i\text{-}Pr)_2NEt$, $TiCl_3(O\text{-}i\text{-}Pr)$, 0 °C, 1 h
2. acrylonitrile, 0 °C, 6.5 h
93%
96% de (1)

Acrylonitrile fails to react with trialkylboranes in the absence of oxygen or other radical initiatiors. However, secondary trialkylboranes transfer alkyl groups in good yield when oxygen is slowly bubbled through the reaction mixture.[22] Primary and secondary alkyl groups can be added in excellent yields using copper(I) methyltrialkylborates.[23] Reaction of acrylonitrile with an organotetracarbonylferrate in a conjugate fashion provides 4-oxonitriles in moderate (25%) yields.[24]

Transition Metal-Catalyzed Additions. Palladium-catalyzed Heck arylation and alkenylation occurs readily with acrylonitrile (eq 2).[25] Double Heck arylation is observed in the Pd^{II}/montmorillonite-catalyzed reaction of aryl iodides with acrylonitrile.[26]

acrylonitrile, Et_3N; $Pd(OAc)_2$, Ph_3P; 86% (2)

Pd^{II} catalyzed oxidation of the double bond in acrylonitrile in the presence of an alcohol (Wacker-type reaction) produces an acetal in high yield.[27] When an enantiomerically pure diol such as ***(2R,4R)-2,4-Pentanediol*** is used, the corresponding chiral cyclic acetal is produced (eq 3).[28]

(*R,R*)-2,4-pentanediol; $PdCl_2$, CuCl, O_2, DME; 45% (3)

Hydrosilation[29a] of acrylonitrile with $MeCl_2SiH$ catalyzed by nickel gives the α-silyl adduct. The β-silyl adduct is obtained when copper(I) oxide is used.[29b] The regioselectivity of the cobalt catalyzed hydrocarboxylation to give either the 2- or 3-cyanopropionates can also be controlled by the choice of reaction conditions.[30] Hydroformylation of acrylonitrile has also been described.[31]

Cyclopropanation of the double bond has been achieved upon treatment with a Cu^I oxide/isocyanide or Cu^0/isocyanide complex. Although yields are low to moderate, functionalized cyclopropanes are obtained.[32,33] Photolysis of hydrazone derivatives of glucose in the presence of acrylonitrile provides the cyclopropanes in good yield, but with little stereoselectivity.[34] Chromium-based Fischer carbenes also react with electron deficient alkenes including acrylonitrile to give functionalized cyclopropanes (eq 4).[35]

$(CO)_5Cr{=}C(OMe)Ph$; acrylonitrile; 89% (4)

Radical Additions. Carbon-centered radicals add efficiently and regioselectively to the β-position of acrylonitrile, forming a new carbon–carbon bond.[36,37] Such radicals can be generated from an alkyl halide (using a catalytic amount of ***Tri-n-butyltin Hydride***, alcohol (via the thiocarbonyl/Bu_3SnH), tertiary nitro compound (using Bu_3SnH), or an organomercurial (using $NaBH_4$). The stereochemical course of the reaction has been examined in cyclohexanes and cyclopentanes bearing an α-stereocenter.[36] Cr^{II} complexes, vitamin B_{12}, and a Zn/Cu couple have been shown to mediate the intermolecular addition of primary, secondary, and tertiary alkyl halides to acrylonitrile.[38] Acyl radicals derived from phenyl selenoesters and Bu_3SnH also give addition products with acrylonitrile (eq 5).[39]

SePh; acrylonitrile, Bu_3SnH; 0.1 equiv AIBN, benzene, 80 °C; 46%; CN (5)

Radical additions with acrylonitrile have been used to prepare *C*-glycosides[36,37b] and in annulation procedures.[37c] Acrylonitrile has also been used in a [3 + 2] annulation based on sequential radical additions (eq 6).[40]

$h\nu$, cat. $(Bu_3Sn)_2$; benzene, 80 °C; 46% (6)

Alkyl and acyl Co^{III} complexes add to acrylonitrile and then undergo β-elimination to give a product corresponding to vinylic C–H substitution.[41] This methodology is complementary to the Heck reaction of aryl and vinyl halides, which fails for alkyl and acyl compounds.[25]

Radicals other than those based on carbon also add to acrylonitrile. Heating acrylonitrile and tributyltin hydride in a 2:3 molar ratio in the presence of a catalytic amount of ***Azobisisobutyronitrile*** yields exclusively the β-stannylated adduct in excellent yield.[42] Hydrostannylation in the presence of a Pd^0 catalyst gives only the α-adduct (eq 7).[42c]

CN: Et_3SnH, $Pd(PPh_3)_4$, 100% → Et_3Sn, CN; Bu_3SnH, AIBN, 80–90% → Bu_3Sn, CN (7)

Treatment of ethyl propiolate with Bu_3SnH in the presence of acrylonitrile results in addition of a tin radical to the β-site of the alkyne followed by addition to acrylonitrile. Use of excess acrylonitrile results in trapping of the radical followed by an annulation reaction, providing trisubstituted cyclohexenes.[43]

Thioselenation of the alkene using ***Diphenyl Disulfide***, ***Diphenyl Diselenide***, and photolysis gives the α-seleno-β-sulfide in 75% yield by a radical addition mechanism.[44] Similarly, ***Tris(trimethylsilyl)silane*** adds to acrylonitrile at 80–90 °C using AIBN to give the β-silyl adduct in 85% yield.[45]

Pericyclic Reactions. In the presence of a suitable alkene, the double bond in acrylonitrile undergoes a thermally induced ene reaction in low to moderate yield. For example, when (+)-limonene and acrylonitrile are heated in a sealed tube, the corresponding ene adduct is produced in 25% yield.[46]

The thermal [2 + 2] dimerization of acrylonitrile has been known for many years. Good regioselectivity is observed but the yield is low and a mixture of stereoisomers is produced.[47] *Cis*-1,2-dideuterioacrylonitrile was used in this reaction to study the stereochemical outcome of the cycloaddition. It was concluded that a diradical intermediate was involved.[1b]

Other [2 + 2] reactions have been reported. Regioselective cycloaddition between a silyl enol ether and acrylonitrile yields a cyclobutane in the presence of light and a triplet sensitizer.[48a] Reaction between acrylonitrile and a ketene silyl acetal in the presence of a Lewis acid gives either substituted cyclobutanes or γ-cyanoesters depending on the Lewis acid and solvent (eq 8).[48c]

Dihydropyridines undergo stereoselective cycloaddition with acrylonitrile under photolytic conditions.[48c] The combination of a Lewis acid (***Zinc Chloride***) and photolysis promotes cycloaddition between benzene and acrylonitrile.[48d] Allenyl sulfides undergo Lewis acid catalyzed [2 + 2] cycloaddition with electron deficient alkenes including acrylonitrile with good regioselectivity but little stereoselectivity (eq 9).[49]

Metal catalysts promote [3 + 2] cycloaddition reactions with acrylonitrile, leading to carbocyclic compounds. Reaction of acrylonitrile with a trimethylenemethane (TMM) precursor in the presence of Pd^0 provides an efficient route to methylenecyclopentanes in moderate yield (40%).[50] A similar yield is obtained when a Ni^0 or Pd^0 catalyzed cycloaddition is employed starting from methylenecyclopropane.[51] Moreover, a variety of substituted methylenecyclopropanes have also been used to furnish substituted methylenecyclopentanes (eq 10).[51b]

Five-membered heterocycles can be prepared from acrylonitrile by dipolar cycloadditions. Acrylonitrile undergoes efficient cycloaddition with 1,3-dipolar species[52] including nitrile oxides, nitrones, azomethine ylides, azides, and diazo compounds.[53] Cycloaddition of acrylonitrile with an oxopyrilium ylide generates stereoisomeric oxabicyclic compounds with excellent regioselectivity (eq 11).[54]

The dipolar cycloaddition of acrylonitrile with a hydroxypyridinium bromide is also highly regioselective.[55]

The [2 + 2 + 2] homo Diels–Alder cycloaddition between acrylonitrile and norbornadiene, substituted norbornadienes, or quadricyclane, has also been described under thermal and metal catalyzed conditions.[56] The effect of ligands and substituents on the stereo- and regioselectivity of the nickel catalyzed process has been investigated (eq 12).[56c,d]

Cobalt catalysts (***Octacarbonyldicobalt***) also promote the cycloaddition of 1,6-diynes with acrylonitrile, yielding cyclohexadienes which are readily aromatized.[57]

Diels–Alder reactions using acrylonitrile have been widely reported with many different dienes. These include alkyl, aryl, alkoxy, alkoxycarbonyl, amido, phenylseleno, phenylthio, and alkoxyboranato substituted butadienes.[58] Reactions between acrylonitrile and furans, thiophenes, and thiopyrans have been reported. In some instances, Lewis acids accelerate the reaction.[59] Heterodienes including 2-azabutadienes and the 4-(oxa, aza, and thio) derivatives also undergo cycloaddition. Reactive dienes such as *o*-quinodimethanes,[60] benzofurans,[61] and dimethylbenzodioxanes react efficiently with acrylonitrile (eq 13).[62]

Related Reagents. Diethyl Methylenemalonate; Ethyl Acrylate; Methyl Acrylate; Methyl Vinyl Ketone.

1. (a) Perrin, D. D.; Armarego, W. L. F. *Purification of Laboratory Chemicals*, 3rd ed.; Pergamon: Oxford, 1988. (b) von Doering, W. E.; Guyton, C. *JACS* **1978**, *100*, 3229.
2. Adams, R.; Jones, V. V. *JACS* **1947**, *69*, 1803.
3. Plaut, H.; Ritter, J. J. *JACS* **1951**, *73*, 4076.
4. Mamiya, Y. J. *Soc. Chem. Ind. Jpn.* **1941**, *44*, 860 (*CA* **1948**, *42*, 2108).
5. Price, C. C.; Zomlefer, J. *JOC* **1949**, *14*, 210.
6. Wegler, R.; Ballauf, A. *CB* **1948**, *81*, 527.
7. Reppe, W.; Hoffmann, U. U.S. Patent 1 891 055, 1932.
8. Hernandez, L. *Experienta* **1947**, *3*, 489.
9. Bruson, H. A. U.S. Patent 2 287 510, 1942.
10. Ali, H. M.; Naiini, A. A.; Brubaker Jr., C. H. *TL* **1991**, *32*, 5489.
11. Moureau, C.; Brown, R. L. *BSF(2)* **1920**, *27*, 901.
12. Tuerck, K. H. W.; Lichtenstein, H. J. U.S. Patent 2 394 644, 1946.
13. Kato, K.; Mukaiyama, T. *BCJ* **1991**, *64*, 2948.
14. (a) This reaction has been thoroughly reviewed, see: Bruson, H. A. *OR*, **1949**, *5*, 79. (b) *The Chemistry of Acrylonitrile*, 2nd ed.; American Cyanamid Co: 1959.
15. For some recent examples, see: (a) Thomas, A.; Manjunatha, S. G.; Rajappa, S. *HCA* **1992**, *75*, 715. (b) Fredriksen, S. B.; Dale, J. *ACS* **1992**, *46*, 574. (c) Nowick, J. S.; Powell, N. A.; Martinez, E. J.; Smith, E. M.; Noronha, G. *JOC* **1992**, *57*, 3763. (d) Genet, J. P.; Uziel, J.; Port, M.; Touzin, A. M.; Roland, S.; Thorimbert, S.; Tanier, S. *TL* **1992**, *33*, 77. (e) Kubota, Y.; Nemoto, H.; Yamamoto, Y. *JOC* **1991**, *56*, 7195.
16. (a) Buhleier, E.; Wehner, W.; Vögtle, F. *S* **1978**, 155. (b)Wörner, C.; Mülhaupt, R. *AG(E)* **1993**, *32*, 1306 (c) de Brabander-van den Berg, E. M. M.; Meijer, E. W. *AG(E)* **1993**, *32*, 1308.
17. Horner, L.; Jurgeleit, W.; Klüpfel, K. *LA* **1955**, *591*, 108.
18. (a) Kharash, M. S.; Reinmuth, O. *Grignard Reactions of Nonmetallic Substances*; Prentice Hall: New York, 1954; pp 782, 814. (b) Mukherjee, S. M. *JIC* **1948**, *25*, 155.

19. Alexakis, A.; Berlan, J.; Besace, Y. *TL* **1986**, *27*, 1047.
20. Yamamoto, Y.; Yamamoto, S.; Yatagai, H.; Ishihara, Y.; Maruyama, K. *JOC* **1982**, *47*, 119.
21. Evans, D. A.; Bilodeau, M. T.; Somers, T. C.; Clardy, J.; Cherry, D.; Kato, Y. *JOC* **1991**, *56*, 5750.
22. Brown, H. C.; Midland, M. M. *AG(E)* **1972**, *11*, 692.
23. Miyaura, N.; Itoh, M.; Suzuki, A. *TL* **1976**, 255.
24. Yamashita, M.; Tashika, H.; Uchida, M. *BCJ* **1992**, *65*, 1257.
25. (a) Heck, R. F. *Palladium Reagents in Organic Syntheses*; Academic Press: London, 1985 and references therein. (b) Bumagin, N. A.; More, P. G.; Beletskaya, I. P. *JOM* **1989**, *371*, 397.
26. Choudary, B. M.; Sarma, R. M.; Rao, K. K. *T* **1992**, *48*, 719.
27. Lloyd, W. G.; Luberoff, B. J. *JOC* **1969**, *34*, 3949.
28. Hosokawa, T.; Ohta, T.; Kanayama, S.; Murahashi, S.-I. *JOC* **1987**, *52*, 1758.
29. For reviews, see: (a) Speier, J. L. *Adv. Organomet. Chem.* **1979**, *17*, 407. (b) Ojima, I. *The Chemistry of Organic Silicon Compounds*; Patai, S.; Rappoport, Z. Eds.; Wiley: New York, 1989; Part 2, Chapter 25. For the specific examples described, see: (c) Boudjouk, P.; Han, B.-H.; Jacobsen, J. R.; Hauck, B. J. *CC* **1991**, 1424 and references therein. (d) Bank, H. M. *CA* **1992**, *116*, 255808a.
30. Pesa, F.; Haase, T. *J. Mol. Catal.* **1983**, *18*, 237.
31. Kollar, L.; Consiglio, G.; Pino, P. *C* **1986**, *40*, 428 and references therein.
32. (a) Saegusa, T.; Yonezawa, K.; Murase, I.; Konoike, T.; Tomita, S.; Ito, Y. *JOC* **1973**, *38*, 2319. (b) Saegusa, T.; Ito, Y. *S* **1975**, 291.
33. While the reactions of some copper complexes with substituted acrylonitriles give good yields, unsatisfactory yields were obtained using acrylonitrile; see: Saegusa, T.; Murase, I.; Ito, Y. *BCJ* **1972**, *45*, 830.
34. Somsak, L.; Praly, J.-P.; Descotes, G. *SL* **1992**, 119.
35. Wienand, A.; Reissig, H.-U. *OM* **1990**, *9*, 3133.
36. (a) Giese, B. *Radicals in Organic Synthesis: Formation of Carbon–Carbon Bonds*; Pergamon Press: Oxford, 1986. (b) Curran, D. P. *COS* **1991**, *4*, 715.
37. (a) Giese, B.; González-Gómez, J. A.; Witzel, T. *AG(E)* **1984**, *23*, 69 and references therein. (b) Dupuis, J.; Giese, B.; Hartung, J.; Leising, M.; Korth, H.-G, Sustmann, R. *JACS* **1985**, *107*, 4332. (c) Angoh, A. G.; Clive, D. L. J. *CC* **1985**, 980.
38. (a) For a recent example using Cr^{II}, see: Tashtoush, H. I.; Sustmann, R. *CB* **1992**, *125*, 287. (b) Scheffold, R.; Abrecht, S.; Orlinski, R.; Ruf, H.-R.; Stamouli, P.; Tinembart, O.; Walder, L.; Weymuth, C. *PAC* **1987**, *59*, 363. (c) Sarandeses, L. A.; Mourino, A.; Luche, J.-L. *CC* **1992**, 798. (d) Blanchard, P.; El Kortbi, M. S.; Fourrey, J.-L.; Robert-Gero, M. *TL* **1992**, *33*, 3319.
39. Boger, D. L.; Mathvink, R. J. *JOC* **1992**, *57*, 1429.
40. (a) Curran, D. P.; Chen, M.-H. *JACS* , **1987**, *109*, 6558. (b) For a recent example, see: Journet, M.; Malacria, M. *JOC* **1992**, *57*, 3085.
41. Pattenden, G. *CSR* **1988**, *17*, 361.
42. (a) Leusink, A. J.; Noltes, J. G. *TL* **1966**, 335. (b) Pereyre, M.; Colin, G.; Valade, J. *BSF(2)* **1968**, 3358. (c) Four, P.; Guibe, F. *TL* **1982**, *23*, 1825.
43. Lee, E.; Uk Hur, C. *TL* **1991**, *32*, 5101.
44. Ogawa, A.; Tanaka, H.; Yokoyama, H.; Obayashi, R.; Yakoyama, K.; Sonoda, N. *JOC* **1992**, *57*, 111.
45. Kopping, B.; Chatgilialoglu, C.; Zehnder, M.; Giese, B. *JOC* **1992**, *57*, 3994.
46. (a) Albisetti, C. J.; Fisher, N. G.; Hogsed, M. J.; Joyce, R. M. *JACS* **1956**, *78*, 2637. (b) Mehta, G.; Reddy, A. V. *TL* **1979**, 2625.
47. Coyner, E. C.; Hillman, W. S. *JACS* **1949**, *71*, 324.
48. (a) Mizuno, K.; Okamoto, H.; Pac, C.; Sakurai, H.; Murai, S.; Sonoda, N. *CL* **1975**, 237. (b) Adembri, G.; Donati, D.; Fusi, S.; Ponticelli, F. *JCS(P1)* **1992**, 2033. (c) Quendo, A.; Rousseau, G. *SC* **1989**, *19*, 1551. (d) Ohashi, M.; Yoshino, A.; Yamazaki, K.; Yonezawa, T. *TL* **1973**, 3395.
49. (a) Hayashi, Y.; Niihata, S.; Narasaka, K. *CL* **1990**, 2091. For other [2 + 2] cycloadditions of allenes, see: Pasto, D. J.; Sugi, K. D. *JOC* **1991**, *56*, 3795.
50. (a) Trost, B. M.; Chan, D. M. T. *JACS* **1983**, *105*, 2315. (b) Trost, B. M. *AG(E)* **1986**, *25*, 1.
51. (a) Noyori, R.; Odagi, T.; Takaya, H. *JACS* **1970**, *92*, 5780. (b) For a review, see: Binger, P.; Buch, H. M. *Top. Curr. Chem.* **1987**, *135*, 77.
52. For reviews, see: (a) Confalone, P. N.; Huie, E. M. *OR* **1988**, *36*, 1. (b) *Dipolar Cycloaddition Chemistry*; Padwa, A., Ed.; Wiley: New York, 1984; Vols. 1 and 2. (c) *Advances in Cycloaddition*, Curran, D. P., Ed.; JAI Press: Greenwich, CT, 1988-1993; Vols. 1–3.
53. Katritsky, A. R.; Hitchings, G. J.; Zhao, X. *S* **1991**, 863.
54. Wender, P. A.; Mascarenas, J. L. *TL* **1992**, *33*, 2115.
55. Jung, M. E.; Longmei, Z.; Tangsheng, P.; Huiyan, Z.; Yan, L.; Jingyu, S. *JOC* **1992**, *57*, 3528.
56. (a) Schrauzer, G. N.; Eichler, S. *CB* **1962**, *95*, 2764. (b) Yoshikawa, S.; Aoki, K.; Kiji, J.; Furukawa, J. *BCJ* **1975**, *48*, 3239. (c) Noyori, R.; Umeda, I.; Kawauchi, H.; Takaya, H. *JACS* **1975**, *97*, 812. (d) Lautens, M.; Edwards, L. E. *JOC* **1991**, *56*, 3761.
57. Zhou, Z.; Costa, M.; Chiusoli, G. P. *JCS(P1)* **1992**, 1399. For a review, see: Vollhardt, K. P. C. *AG(E)* **1984**, *23*, 539.
58. (a) Fringuelli, F.; Taticchi, A. *Dienes in the Diels–Alder Reaction*; Wiley: New York, 1990. (b) Ward, D. E.; Gai, Y.; Zoghaib, W. M. *CJC* **1991**, *69*, 1487 and references therein.
59. (a) Moore, J. A.; Partain, E. M. III *JOC* **1983**, *48*, 1105. (b) Brion, F. *TL* **1982**, *23*, 5299.
60. (a) Ito, Y.; Amino, Y.; Nakatsuka, M.; Saegusa, T. *JACS* **1983**, *105*, 1586. (b) For reactions of chromium complexed species, see: Kundig, E. P.; Bernardinelli, G.; Leresche, J. *CC* **1991**, 1713.
61. Rodrigo, R.; Knabe, S. M.; Taylor, N. J.; Rajapaksa, D.; Chernishenko, M. J. *JOC* **1986**, *51*, 3973 and references therein.
62. Ruiz, N.; Pujol, M. D.; Guillaumet, G.; Coudert, G. *TL* **1992**, *33*, 2965.

Mark Lautens & Patrick H. M. Delanghe
University of Toronto, Ontario, Canada

Allenylboronic Acid

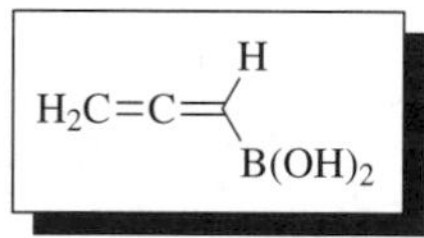

[83816-41-5] $C_3H_5BO_2$ (MW 83.88)

(propargylating reagent for carbonyl groups; chiral allenylboronic ester as a reagent for enantioselective alkylation;[1] 1,3-asymmetric induction in propargylation of β-hydroxy ketones[2])

Physical Data: mp 150 °C (dec).
Solubility: sol ether, alcohol.
Form Supplied in: white solid; widely available.
Analysis of Reagent Purity: ^1H NMR ($CDCl_3$–Me_2SO-d_6) δ 4.48 (d, J = 5.8 Hz, 1H), 4.51 (d, J = 7.2 Hz, 1H), 4.83 (dd, J = 5.8 and 7.2 Hz, 1H), 6.52 (br s, 2H).
Preparative Method: from ***Propargylmagnesium Bromide*** and ***Trimethyl Borate***.[1]
Handling, Storage, and Precautions: allenylboronic acid can be stored at −20 °C under argon. In air, it catches fire within several minutes. It should only be used in a fume hood.

Allenylboronic Acid.[1] Treatment of propargylmagnesium bromide with trimethyl borate followed by acidic workup gives, after recrystallization from hexane–ether, a single crystalline boronic acid in 40–60% yield (eq 1).[3] As far as can be described, a single compound results from this preparation.

$$HC{\equiv}C{-}CH_2Br \xrightarrow[\text{ether}]{\text{Mg (Hg)}} H_2C{=}C{=}CHMgBr \xrightarrow{B(OMe)_3} \xrightarrow{H_2O} H_2C{=}C{=}CHB(OH)_2 \quad (1)$$

Allenylboronic acid is an ambident nucleophile and its reactions with carbonyl compounds can be envisioned to occur either at the α or γ position. Treatment of cyclohexanecarbaldehyde with allenylboronic acid in toluene containing 4 equiv. of pentanol and molecular sieves produces exclusively the homopropargylic alcohol (**1**) in 40% yield. This result suggests that the reaction proceeds through the cyclic transition state (**2**).

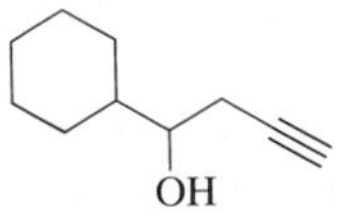

(1) 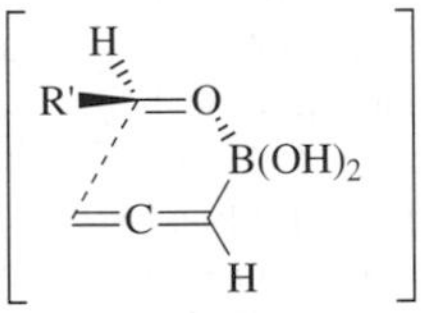 (2)

Chiral Allenylboronic Esters.[1] Condensations of aldehydes with chiral allenylboronic esters provide β-alkynic alcohols with an exceptionally high degree of enantioselectivity (eq 2). The reagents prepared using bulky tartrate esters like 2,4-dimethyl-3-pentyl as the chiral auxiliary are more enantioselective than reagents prepared from ethyl or isopropyl esters of tartaric acid. These results have been rationalized by the transition state shown in structure (**4**), in which the ester alkoxy group exerts a screening influence on the prochiral carbonyl compound such that R′ is positioned as far away from CO_2R as possible.

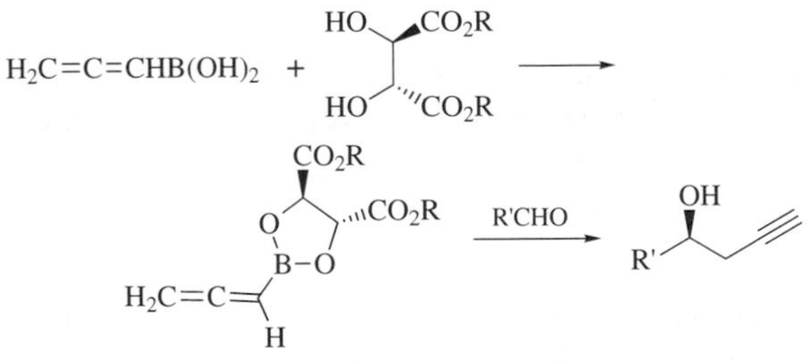

$H_2C{=}C{=}CHB(OH)_2$ + (HO, CO_2R tartrate) → (**3**) $\xrightarrow{R'CHO}$ (2)

R′ = cyclohexyl	R = Et	91% ee
	i-Pr	92% ee
	2,4-dimethyl-3-pentyl	98% ee

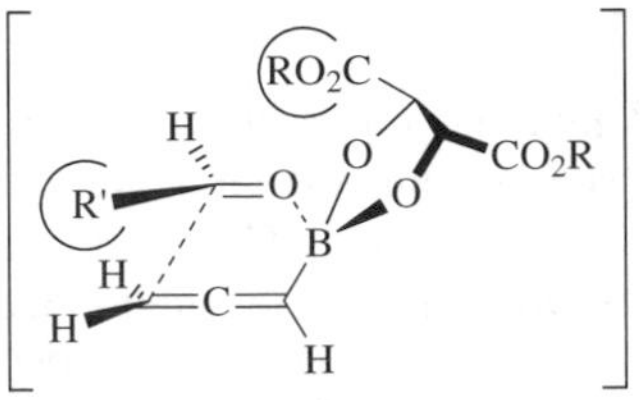

(4)

Condensation with aromatic aldehydes appears to be considerably less efficient than the reaction with saturated aldehydes. In addition, only low yields (<50%) of the homopropargylic alcohol are obtained by the reaction with a variety of unsaturated aldehydes with this reagent, indicating that the use of saturated aldehyde is crucial to the success of the asymmetric propargylation reaction. Additional results are summarized in Table 1. This method has been applied to the synthesis of (−)-ipsenol, as summarized in eq 3.[1b,4]

Table 1 Enantioselective Condensation of Allenyl Reagents (**3**) (R = 2,4-dimethyl-3-pentyl)

Aldehyde (equiv.)	Yield (%)	ee (%)
Hexanal (1.5)	81	94
Hexanal (5.0)	72	97
Cyclohexanecarbaldehyde (1.5)	88	98
Cyclohexanecarbaldehyde (3.0)	89	>99
Isovaleraldehyde (1.5)	78	99
Isovaleraldehyde (3.0)	74	>99
(*S*)-Citronellal (1.5)	74	92
(*S*)-Citronellal (5.0)	90	99
(*R*)-Citronellal (1.5)	67	98

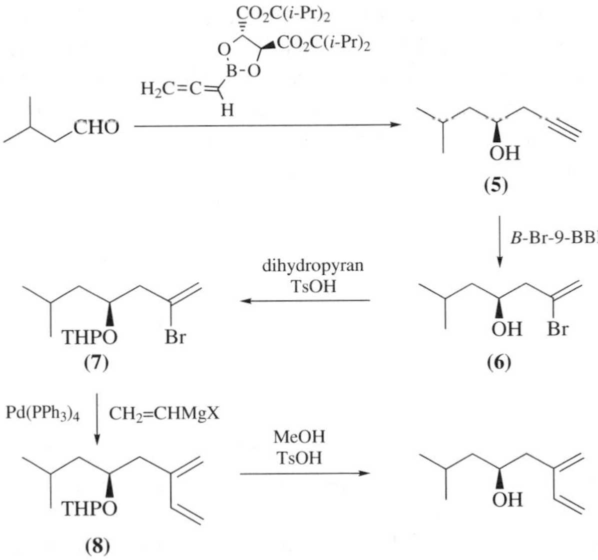

CHO → (**5**) $\xrightarrow{B\text{-Br-9-BBN}}$ (**6**) $\xrightarrow[\text{TsOH}]{\text{dihydropyran}}$ (**7**) $\xrightarrow{Pd(PPh_3)_4,\ CH_2{=}CHMgX}$ (**8**) $\xrightarrow[\text{TsOH}]{\text{MeOH}}$ (3)

1,3-Asymmetric Induction.[2] The reaction of allenylboronic acid with carbonyl compounds is slow relative to the reaction of allenylboronic esters.[1] However, the reaction of allenylboronic acid (1.2 equiv.) with β-hydroxy ketones in anhydrous ether at rt in the presence of molecular sieves (5Åfor 20 h), followed by treatment with basic ***Hydrogen Peroxide***, yields the *threo* diol (**10**) with excellent levels of 1,3-asymmetric induction (>99%, eq 4). Additional results are summarized in eqs 5–7. The yields in all cases are >90%. The reaction of propargylmagnesium bromide with (**9**) in ether affords a mixture of the *threo* and *erythro* diols, in agreement with related nonselective processes.[5]

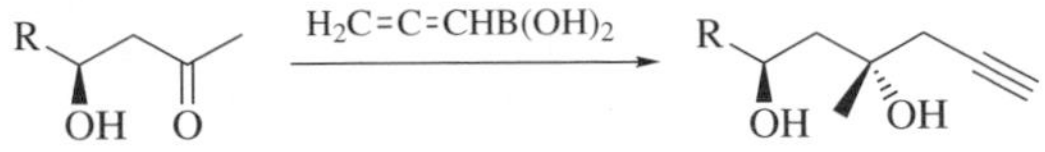

(**9**) $\xrightarrow{H_2C=C=CHB(OH)_2}$ (**10**) (4)

Ph… $\xrightarrow[>99\%\ de]{95\%}$ … $\xrightarrow{83\%}$ … (5)

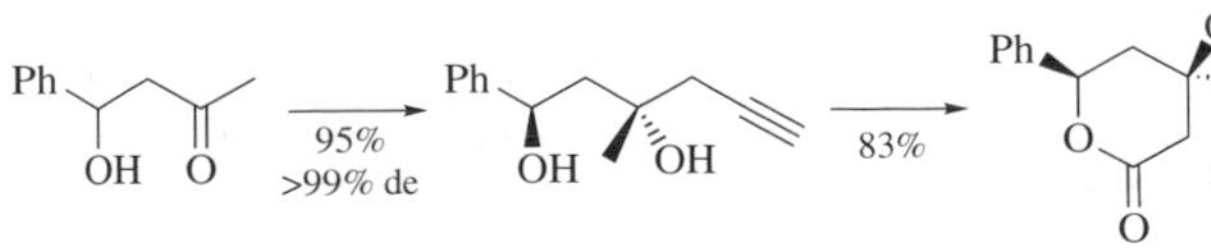

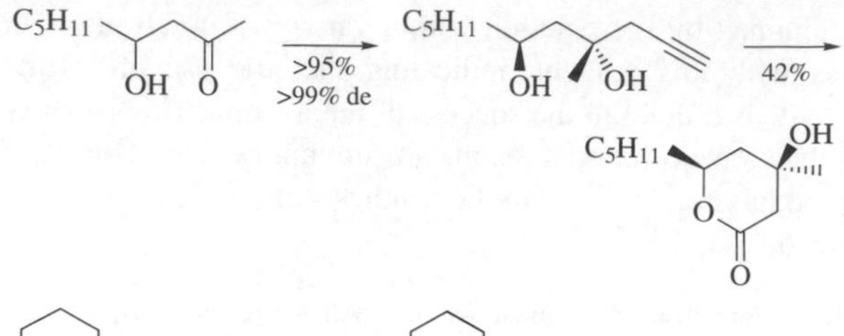

$$\text{C}_5\text{H}_{11}\text{-CH(OH)CH}_2\text{C(O)CH}_3 \xrightarrow[>99\%\ de]{>95\%} \xrightarrow{42\%} \quad (6)$$

$$\xrightarrow[>99\%\ de]{96\%} \xrightarrow{56\%} \quad (7)$$

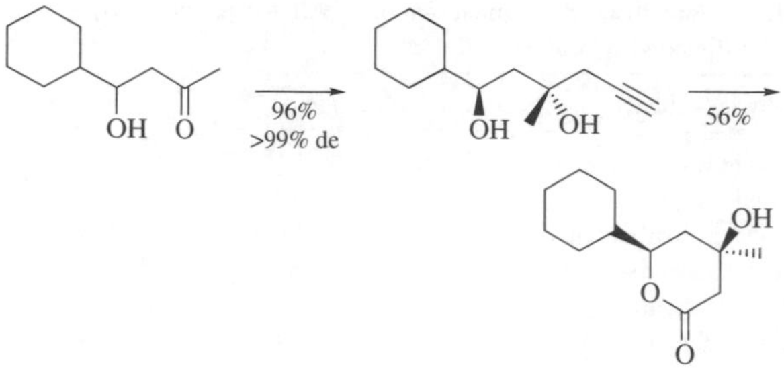

The very high levels of stereoselectivity observed in these reactions strongly suggest the existence of the covalently bonded organometallic species (**11**) as a reactive intermediate (eq 8). Indeed, treatment of a mixture of cyclohexanone and (*S*)-phenethyl alcohol with allenylboronic acid gave none of the expected homopropargylic alcohol under the above reaction conditions. Furthermore, treatment of a mixture of the ketone (**9**; R = cyclohexyl) and 2-hexanone with allenylboronic acid produced the diol (**10**) as a sole product.

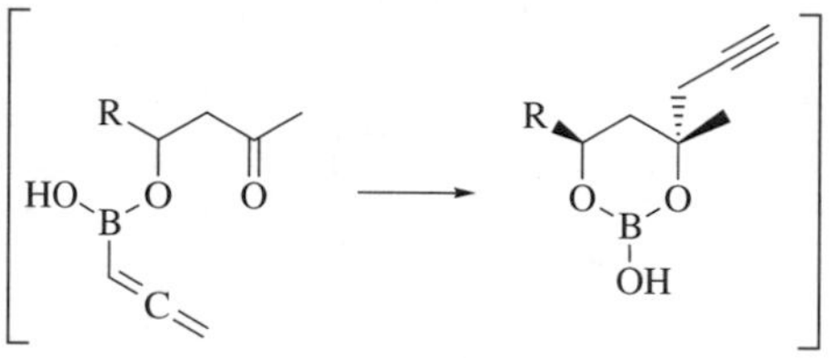

(8)

(**11**)

Treatment of *threo* diols (**10**) with ***Ruthenium(III) Chloride–Sodium Periodate***[6] gave the corresponding lactones in excellent yield (eqs 5–7).

The use of this method for the preparation of a diastereomer of the β-hydroxy-δ-lactone substituent of the HMG CoA reductase inhibitor mevinolin[7,8] is summarized in eq 9. Unfortunately, however, the product has the (4*S*,6*R*) configuration rather than the most active (4*R*,6*R*) configuration.[7]

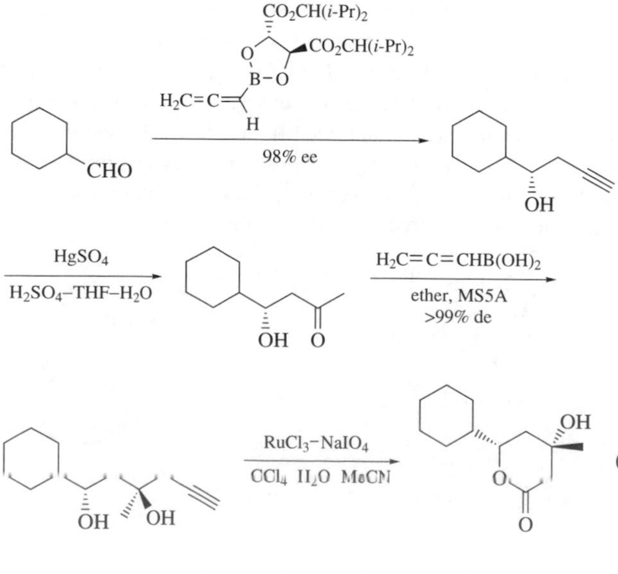

$CO_2CH(i\text{-}Pr)_2$, $CO_2CH(i\text{-}Pr)_2$, B–O, $H_2C{=}C{=}$, H, CHO, 98% ee, OH

$HgSO_4$ / H_2SO_4–THF–H_2O; $H_2C{=}C{=}CHB(OH)_2$ / ether, MS5A / >99% de

$RuCl_3$–$NaIO_4$ / CCl_4 H_2O MeCN (9)

Related Reagents. Allenyllithium; Propargylmagnesium Bromide; (Trimethylsilyl)allene.

1. (a) Haruta, R.; Ishiguro, M.; Ikeda, N.; Yamamoto, H. *JACS* **1982**, *104*, 7667. (b) Ikeda, N.; Arai, I.; Yamamoto, H. *JACS* **1986**, *108*, 483. Synthetic applications using chiral allenylboronic ester, see: (c) Tius, M. A.; Trehan, S. *JOC* **1986**, *51*, 765. (d) Fryhle, C. B.; Williard, P. G.; Rybak, C. M. *TL* **1992**, *33*, 2327.
2. Ikeda, N.; Omori, K.; Yamamoto, H. *TL* **1986**, *27*, 1175.
3. Favre, E.; Gaudemer, M. *CR(C)* **1966**, *262*, 1332.
4. Silverstein, R. M.; Rodin, J. O.; Wood, D. L. *Science* **1966**, *154*, 509.
5. For studies of propargyl Grignard reagent with 4-*t*-butyl-2-methoxycyclohexanone, see: Guillerm-Dron, D.; Capmau, M. L.; Chodkiewicz, W. *BSF(2)* **1973**, 1417.
6. Carlsen, H. J.; Katsuki, T.; Martin, V. S.; Sharpless, K. B. *JOC* **1981**, *46*, 3936. Ruthenium dioxide–periodate procedure is also effective for this oxidation: Corey, E. J.; Boaz, N. W. *TL* **1984**, *25*, 3059.
7. (a) Lee, T.; Holtz, W. J.; Smith, R. L. *JOC* **1982**, *47*, 4750. (b) Ferres, H.; Hatton, I. K.; Jennings, L. J. A.; Tyrrell, A. W. R.; Williams, D. J. *TL* **1983**, *24*, 3769.
8. (a) Prugh, J. D.; Rooney, C. S.; Deana, A. A.; Ramjit, H. G. *TL* **1985**, *26*, 2947. (b) Sletzinger, M.; Verhoeven, T. R.; Volante, R. P.; McNamara, J. M.; Corley, E. G.; Liu, T. M. H. *TL* **1985**, *26*, 2951 and references cited therein.

Kazuaki Ishihara & Hisashi Yamamoto
Nagoya University, Japan

Allyl Bromide

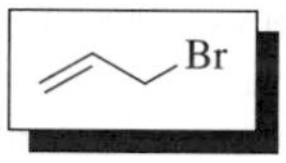

Br

[106-95-6] C_3H_5Br (MW 120.99)

(electrophilic allylating agent attacking C, N, O, S, Se, and Te nucleophiles; homoallylic alcohols obtained selectively from aldehydes by various organometallic intermediates; addition reactions provide further reagents of wide applicability)

Physical Data: mp −119.4 °C; bp 71.3 °C; *d* 1.398 g cm^{-3}.
Solubility: miscible with organic solvents; sparingly sol H_2O.
Form Supplied in: yellow to brown liquid.
Purification: wash with water and with aqueous $NaHCO_3$. Dry ($MgSO_4$ or Na_2SO_4) and fractionally distill.
Handling, Storage, and Precautions: highly toxic; cancer suspect agent. Protect from light in brown glass.

Allylating Agent. Carbon alkylation generally requires nucleophilic carbanions; thus the allylation of PhC≡CH is promoted with powdered ***Potassium Hydroxide*** alone or with ***Tetra-n-butylammonium Bromide*** in dioxane.[1] Dimeric side products (allyl ether, Ph_2C_4) and rearranged enynes (PhC≡CCH=CHMe; *E/Z*, 3:1) accompany PhC≡CCH$_2$CH=CH$_2$. Carbanions from acetoacetic esters,[2a,2b] ketones,[2c] malonates (K_2CO_3–Me_2CO or PhH),[3] acetonitrile,[4a] and cyanoacetates[4b] readily undergo allylation (eq 1); the necessary base may be generated electrochemically, as in the use of pyrrolidone anion to bring about allylation of dimethyl 2-(trifluoromethyl)malonate.[5] Perfluoro-2-methyl-2-pentyl carbanion is generated by the addition of F^- (KF or CsF) to $(CF_3)_2CFCF{=}CFCF_3$; upon allylation (RX; X = Cl, Br, I) the rearranged product $(CF_3)_2C(R)CF_2CF_2CF_3$ results.[6] Mn enolates of dialkyl ketones may be allylated (RBr; THF–sulfolane); thus Pr_2CO gives PrCOCH(R)Et in 98% yield.[7]

$$\text{X–CH}_2\text{–Y} + [\text{X–CH–Y}]^- \xrightarrow{\text{RBr}} \text{X–CHR–Y} \quad (1)$$
$$\text{X, Y} = \text{R'CO, CO}_2\text{R', CN, H}$$

The homoallylic alcohols $RCH(OH)CH_2CH{=}CH_2$ are formed from the reaction of RCHO and allyl bromide through organometallic intermediates, especially those involving allyl magnesium bromide. Ketones react similarly, but more slowly. The conventional Barbier–Grignard processes have been replaced by a reductive allylation. Thus Al brings about the reaction (i) in the presence of 'catalytic' amounts of $PbBr_2$ in DMF, aq THF, and/or aq MeOH[8a] or (ii) in the presence of $BiCl_3$ in aq THF.[8b] The process may be specific to aldehydes; Pb–Me_3SiCl–Bu_4NBr in DMF promotes allylation of aldehydes without significantly attacking ketones or α-hydroxycarboxylic esters, while esters, lactones, acid anhydrides, and acid chlorides are effectively inert.[9] Correspondingly, homoallylic alcohols are formed from aldehydes and allyl chloride, bromide, or iodide using Zn–$BiCl_3$ or Fe–$BiCl_3$, but ketones, esters, and benzoic acid are unaffected and do not interfere.[10] Fe or Al with $SbCl_3$ in aq DMF behaves similarly.[11] Electrochemical variants are reported in which Bi acts as the reductant towards the aldehyde.[12] Aldehydes and ketones react in the presence of Ph_3Bi to give homoallylic alcohols or their allylic ethers.[13] Zn alone in DMF gives 86% yields with allyl bromide and MeCH=CHCHO;[14a] Cd in DMF similarly shows only 1,2-addition with RCHO or RCOR′ (eq 2).[14b] A complex involving Ph_2CO and Yb provides Ph_2RCOH (R = allyl) with allyl bromide.[15]

$$\text{CH}_2{=}\text{CHCH}_2\text{Br} + \text{RCOR'} \xrightarrow{\text{M}} \text{CH}_2{=}\text{CHCH}_2\text{C(R')(OH)R} \quad (2)$$
$$\text{R} = \text{H, alkyl; M} = \text{Zn, Cd, Al etc.}$$

Asymmetric allylation occurs in a number of appropriately substituted systems. The Schiff base between (+)-camphor and $H_2NCHR'P(O)(OEt)_2$ is metalated (***n-Butyllithium***) and then reacts with allyl bromide to give the (1*S*,4*S*) analog (R = allyl) with >95% diastereomeric excess (eq 3). Sequential hydrolysis provides the (*S*)-ester and the (*S*)-phosphonic acid without appreciable racemization;[16a] (1*R*,2*R*,5*R*)-(+)- and (1*S*,2*S*,5*S*)-(−)-2-hydroxy-3-pinanone behave analogously.[16b] The bis(cyclohexylidene) acetal of D-galactodialdehyde similarly gives Schiff bases with α-alkylated glycine esters; these may be metalated (BuLi) and the anion quenched with allyl bromide with 76% diastereomeric excess.[17] ***Lithium Diisopropylamide*** metalation of a chiral lactam enolate and allylation similarly provides a considerable diastereomeric enhancement.[18] Formation of the enol from tetralone with (*R*)-$RCH_2CHPhN(CH_2CH_2OCH_2CH_2OMe)Li$ (R = piperidino) and allylation provides (*R*)-2-allyl-1-tetralone in 92% enantiomeric excess and 89% overall yield; ***Lithium Bromide*** is a necessary co-reagent (eq 4).[19]

(3) [Scheme: camphor-derived Schiff base bearing CH(R′)P(O)(OEt)₂ → (BuLi; CH₂=CHCH₂Br) → allylated product, C(R′)(R)P(O)(OEt)₂]

(4) [Scheme: cyclohexanone → (chiral lithium amide with Ph, piperidino, O, OMe; CH₂=CHCH₂Br) → 2-allylcyclohexanone]

Similarly, allylation of the lactam by allyl bromide–LDA[20] proceeds stereoselectively to give (**1**).

(**1**) [Structure: allyl-substituted β-lactam, CO₂H, N-TBDMS]

Allylation of Ni complexes of some Schiff bases with glycine are reported[21] to yield *S*-α-allylglycine. A three-step synthesis of α-amino acid HCl salts relies upon diastereoselective allylation of a glycine enolate synthon with >97.6% de and in 73–90% yield.[22] *N,N*-Dimethylhydrazones of α,β-unsaturated aldehydes[23a] and $RCH_2CH{=}CHCH{=}CHCHO$[23b] metalate (BuLi) and allylate with rearrangement, giving $RCH(R')(CH{=}CH)_nCHO$ analogs (R′ = allyl; $n = 1$ or 2). $Ph_2C{=}NCH_2CO_2$-*t*-Bu undergoes allylation (RBr, 50% aq NaOH, CH_2Cl_2, rt) in the presence of chiral PTC based upon cinchonine; the products show considerable ee (50–60%).[24]

Carbonyl insertion occurs when Me_3P-coordinated π-allyl Pd complexes are treated with CO in CH_2Cl_2 at rt, giving 3-butenoyl derivatives,[25] and allylation of a vinyl rhenium CO complex provides an allyl vinyl ketone complex.[26] Organotin species couple with allyl bromide (catalyzed by Pd complexes); while β-elimination may supervene, CHO, CO_2H, and OH groups do not interfere.[27a] Similar coupling occurs with $RSiMe_3$ or $RSiMe_2F$ and allyl bromide,[27b] but alkenylboranes in the presence of ***Tetrakis(triphenylphosphine)palladium(0)*** give alkenes by allyldeboronation.[27c] ***Tetracarbonylnickel*** with allyl bromide (RBr) gives π-allyl nickel bromide complexes. These can act as intermediates in the coupling of allylic systems either symmetrically[28a] or to give substituted alkenes by unsymmetrical coupling (eq 5).[28b]

$$2\ \text{RCH}{=}\text{CHCH}_2\text{Br} \longrightarrow \text{RCH}{=}\text{CHCH}_2\text{CH}_2\text{CH}{=}\text{CHR} \quad (5)$$

π-Allyl nickel bromide complexes allylate C-2 of benzoquinones.[28c] Oxygen-centered attack occurs readily; the K salt of L-ascorbic acid (**2**) gives (allyl bromide–Me_2CO) a lactone (**3**) which with ***Palladium(II) Chloride*** in aq DMF provides (40%) a bicyclic ketone in which the allylic side chain has become the acetonyl ($MeCOCH_2$) residue (eq 6).[29]

(6) [Scheme: (**2**) → (**3**)]

(**2**) (**3**)

Stannylation of monoalkylated oxiranes by ***Trimethylstannyllithium*** gives lithium alkoxides $XOCH(R)CH_2SnMe_3$ (X = Li) which react conventionally with allyl bromide to give the corresponding allyl ether (X = allyl).[30] Attack at oxygen may also be achieved using other displaced groups. Stannylene acetals of acyclic diols are monoallylated using F^- in a mild, selective, and high yield process.[31]

N-Allylation takes place easily; phthalimide reacts readily with allyl bromide (K_2CO_3–PEG 400; 90 °C).[32] Indoles[33a] and pyrazoles[33b] may be allylated on nitrogen using PTC such as Bu_4NBr (eq 7).

RBr (X = N, CH) (7)

Salts such as Na phenylsulfinate form allyl esters; Al_2O_3, ultrasound, and microwaves all influence the yield.[34] Correspondingly, sulfur,[35,36] selenium,[36] or tellurium[37] attack is preparatively useful (eq 8).

$$RBr + R'X^- \longrightarrow R'XR + Br^- \quad (X = O, S, Se, Te) \qquad (8)$$

Alkyl coupling reactions, mediated by Cu, are exemplified by the synthesis of $CF_3CH_2CH{=}CH_2$ using $(CF_3)_2Cu^{III}$(*N,N*-diethyldithiocarbamato),[38a] or using FO_2SCF_2I in DMF.[38b]

The allyl system is susceptible to further chemistry, notably epoxidation and other addition processes; the intermediates in such processes may show their own idiosyncratic chemistry[39] as in the cyclization of the thioallyl substituent (**4**) (eq 9), itself obtained by allylation of the thiophenoxide.

CO_2H (**4**) X_2 CO_2H X_3^- (9)

Related Reagents. Allyl Chloride; Allyl Iodide; Allyllithium; Allylmagnesium Bromide.

1. Paravyan, S. L.; Torosyan, G. O.; Babayan, A. T. *ZOR* **1986**, *22*, 706.
2. (a) Tsuji, J.; Yamada, T.; Minami, I.; Yuhara, M.; Nisar, M.; Shimizu, J. *JOC* **1987**, *52*, 2988. (b) Hughes, P.; De Virgilio, J.; Humber, L. G.; Chau Thuy; Weichman, B.; Neuman, G. *JMC* **1989**, *32*, 2134. (c) Vanderwerf, C. A.; Lemmerman, L. V. *OSC* **1955**, *3*, 44.
3. Liu, H.; Cheng, G. *Huaxue Shiji* **1991**, *13*, 248, 202 (*CA* **1991**, *115*, 255 598m).
4. (a) Tamaru, Y. *JACS* **1988**, *110*, 3994. (b) Abd el Samii, Z. K. M.; Al Ashmawy, M. I.; Mellor, J. M. *JCS(P1)* **1988**, 2523.
5. Fuchigami, T.; Nakagawa, Y. *JOC* **1987**, *52*, 5276.
6. Dmowski, W.; Wozniacki, R. *JFC* **1987**, *36*, 385.
7. Cahiez, G.; Figadere, B.; Tozzolino, P.; Clery, P. Eur. Patent 373 993, **1990** (*CA* **1991**, *114*, 61 550y).
8. (a) Tanaka, H.; Yamashita, S.; Hamatani, T.; Ikemoto, Y.; Torii, S. *SC* **1987**, *17*, 789. (b) Wada, M.; Ohki, H.; Akiba, K. *CC* **1987**, 708.
9. Tanaka, H.; Yamashita, S.; Hamatani, T.; Ikemoto, Y.; Torii, S. *CL* **1986**, 1611.
10. Wada, M.; Ohki, H.; Akiba, K. *TL* **1986**, *27*, 4771.
11. Wang, W.; Shi, L.; Huang, Y. *T* **1990**, *46*, 3315.
12. (a) Minato, M.; Tsuji, J. *CL* **1988**, 2049. (b) Tanaka, H.; Nakahara, T.; Dhimane, H.; Torii, S. *TL* **1989**, *30*, 4161.
13. Huang, Y.; Liao, Y. *HC* **1991**, *2*, 297 (*CA* **1991**, *115*, 91 330q).
14. (a) Shono, T.; Ishifune, M.; Kashimura, S. *CL* **1990**, 449. (b) Araki, S.; Ito, H.; Butsugan, Y. *JOM* **1988**, *347*, 5.
15. Takaki, K.; Tsubaki, Y.; Beppu, F.; Fujiwara, Y. *Chem. Express* **1991**, *6*, 57 (*CA* **1991**, *114*, 163 659h).
16. (a) Schöllkopf, U.; Schuetze, R. *LA* **1987**, 45. (b) Jacquier, R.; Ouazzani, F.; Roumestant, M. L.; Viallefont, P. *PS* **1988**, *36*, 73.
17. Schoellkopf, U.; Toelle, R.; Egert, E.; Nieger, M. *LA* **1987**, 399.
18. Wuensch, T.; Meyers, A. I. *JOC* **1990**, *55*, 4233.
19. Murakata, M.; Nakajima, M.; Koga, K. *CC* **1990**, 1657.
20. Baldwin, J. E.; Adlington, R. M.; Gollins, D. W.; Schofield, C. J. *T* **1990**, *46*, 4733.
21. (a) Belokon, Yu. N.; Chernoglazova, N. I.; Ivanova, E. V.; Popkov, A. N.; Saporovskaya, M. B.; Suvorov, N. N.; Belikov, V. M. *IZV* **1988**, 2818. (b) Belokon, Yu. N.; Maleev, V. I.; Saporovskaya, M. B.; Bakhmutov, V. I.; Timofeeva, T. V.; Batsanov, A. S.; Struchkov, Yu. T.; Belikov, V. M. *Koord. Khim.* **1988**, *14*, 1565 (*CA* **1989**, *111*, 16 646).
22. Dellaria, J. F.; Santarsiero, B. D. *JOC* **1989**, *54*, 3916.
23. (a) Yamashita, M.; Matsumiya, K.; Nakano, K.; Suemitsu, R. *CL* **1988**, 1215. (b) Matsumiya, K.; Nakano, K.; Suemitsu, R.; Yamashita, M. *CL* **1988**, 1837.
24. (a) O'Donnell, M. J.; Bennett, W. D.; Bruder, W. A.; Jacobsen, W. N.; Knuth, K.; Leclef, B.; Polt, R. L.; Bordwell, F. G.; Mrozack, S. R.; Cripe, T. A. *JACS* **1988**, *110*, 8520. (b) O'Donnell, M. J.; Bennett, W. D.; Wu, S. *JACS* **1989**, *111*, 2353. (c) O'Donnell, M. J.; Wu, S. *TA* **1992**, *3*, 591.
25. Ozawa, F.; Son, T.; Osakada, K.; Yamamoto, A. *CC* **1989**, 1067.
26. Casey, C. P.; Vosejpka, P. C.; Gavney, J. A. *JACS* **1990**, *112*, 4083.
27. (a) Stille, J. K. *AG(E)* **1986**, *25*, 508. (b) Hatanaka, Y.; Hiyama, T. *JOC* **1988**, *53*, 918. (c) Hatanaka, Y.; Hiyama, T. *JOC* **1989**, *54*, 268. (d) Matteson, D. S. *T* **1989**, *45*, 1859.
28. (a) Tamao, K.; Kumada, M. In *Chemistry of the Metal-Carbon Bond*; Hartley, F. R., Ed.; Wiley: New York, 1987; Vol. 4, pp 819–887. (b) Semmelhack, M. F. *OR* **1972**, *19*, 115. (c) Hegedus, L. S.; Waterman, E. L.; Catlin, J. E. *JACS* **1972**, *94*, 7155.
29. Poss, A. J.; Belter, R. K. *SC* **1988**, *18*, 417.
30. Mordini, A.; Taddei, M.; Seconi, G. *G* **1986**, *116*, 239.
31. (a) Nagashima, N.; Ohno, M. *CL* **1987**, 141. (b) Nagashima, N.; Ohno, M. *CPB* **1991**, *39*, 1972.
32. Vlassa, M.; Kezdi, M.; Fenesan, M. *Rev. Roum. Chim.* **1989**, *34*, 1607 (*CA* **1990**, *113*, 6079).
33. (a) Hlasta, D. J.; Luttinger, D.; Perrone, M. H.; Silbernagel, M. J.; Ward, S. J.; Haubrich, D. R. *JMC* **1987**, *30*, 1555. (b) Diez-Barra, E.; de la Hoz, A.; Sanchez-Migallon, A.; Tejeda, J. *SC* **1990**, *20*, 2849.
34. Villemin, D.; Ben Alloum, A. *SC* **1990**, *20*, 925.
35. Nishimura, H.; Ariga, T. Jpn. Patent 02 204 487, 1990 (*CA* **1990**, *114*, 6523s).
36. Barton, D. H. R.; Crich, D. *JCS(P1)* , **1986**, 1613.
37. (a) Higa, K. T.; Harris, D. C. *OM* **1989**, *8*, 1674. (b) Higa, K. T.; Harris, D. C. US Patent Appl. 66442, 1988 (*CA* **1988**, *109*, 190 579k).
38. (a) Willert-Porada, M. A.; Burton, D. J.; Baenziger, N. C. *CC* **1989**, 1633. (b) Chen, Q.; Wu, S. *JCS(P1)* **1989**, 2385.
39. Shestopalov, A. M.; Rodinovskaya, L. A.; Sharanin, Yu. A.; Litvinov, V. P. *Khim. Geterotsikl. Soedin* . **1990**, 256 (*CA* **1990**, *113*, 39 658x).

Roger Bolton
University of Surrey, Guildford, UK

B-Allyldiisopinocampheylborane[1]

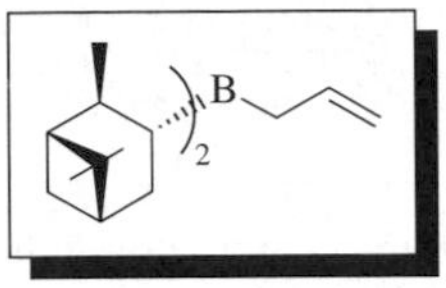

(+)-(Ipc_2BAll)
[85116-38-7] $C_{23}H_{39}B$ (MW 324.36)
(−)-(Ipc_2BAll)
[106356-53-0]

(reagent for the asymmetric allylboration of aldehydes to produce homoallylic alcohols[2,3])

Solubility: most often used as the crude preparation in Et_2O; however, reactions have also been conducted in CS_2, $CHCl_3$, CH_2Cl_2, toluene, and THF.

Analysis of Reagent Purity: ^{11}B NMR (δ +78, diethyl ether).

Preparative Methods: prepared in three steps from either (+)- or (−)-α-pinene (eq 1) (see also ***B-Methoxydiisopinocampheylborane***).

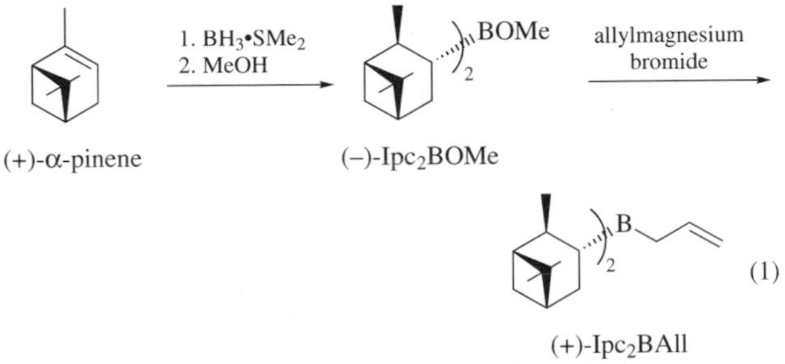

1. BH_3•SMe_2 2. MeOH; allylmagnesium bromide

(+)-α-pinene (−)-Ipc_2BOMe (+)-Ipc_2BAll (1)

Handling, Storage, and Precautions: removal of magnesium salts from the crude reagent preparation is accomplished by solvent exchange into pentane and filtration. Concentration in vacuo provides the neat allylborane as a colorless liquid that can be stored under argon atmosphere for an extended period of time.

Addition to Aldehydes. Condensation of *B*-allyldiisopinocampheylborane (Ipc_2BAll) with aldehydes occurs in good yield (50–90%) and with excellent enantioselection (83–96% ee) to provide secondary, homoallylic alcohols.[2] Improvements in the original procedure have produced a 'salt-free' reagent preparation that can increase the enantioselectivity of many additions to ≥99% ee (eqs 2 and 3).[3]

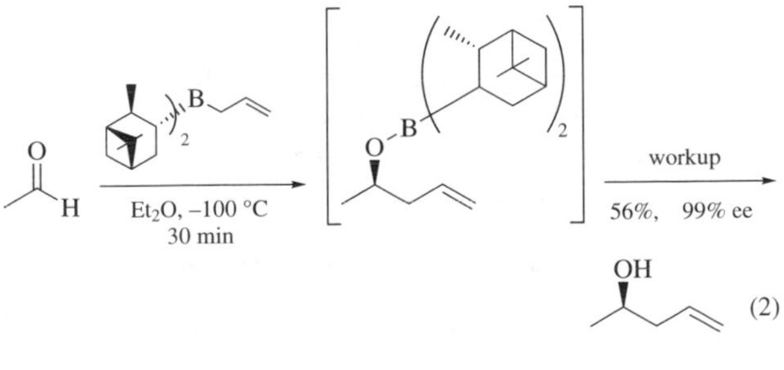

Et_2O, −100 °C 30 min; workup 56%, 99% ee (2)

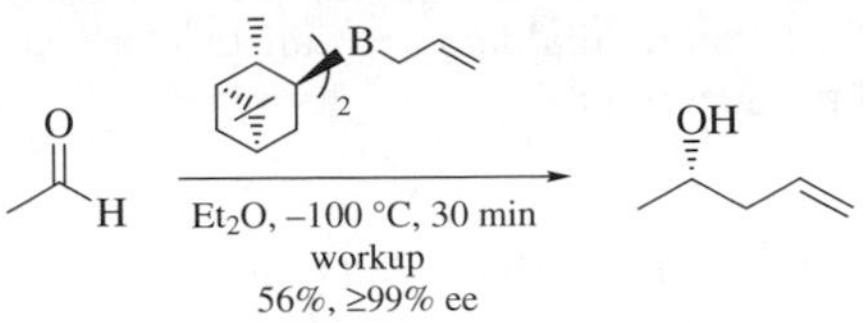

Et_2O, −100 °C, 30 min workup 56%, ≥99% ee (3)

Several methods have been employed in the reaction workup to decompose the initial borinic ester adduct (eq 4).[2,4] An oxidative workup procedure (H_2O_2, NaOH) delivers the desired alcohol product along with 2 equiv of isopinocampheol. The elimination workup (*i*-PrCHO, BF_3OEt_2) allows for recovery of the starting α-pinene, whereas the amino alcohol workup (ethanolamine or 8-hydroxyquinoline) precipitates the diisopinocampheylborinate–amino alcohol adducts which are removed by filtration, and which can also be recycled.

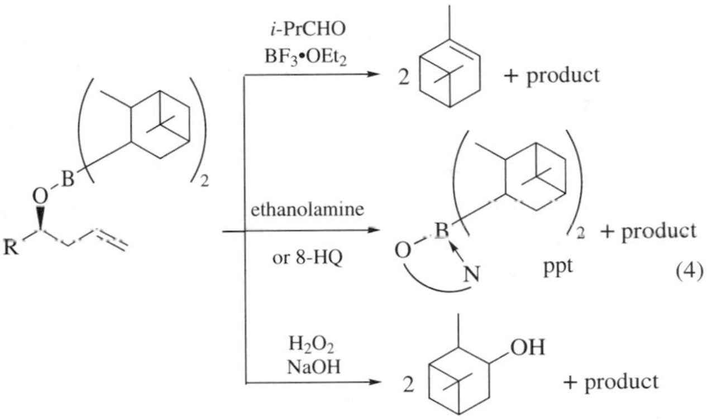

i-PrCHO BF_3•OEt_2 → 2 (α-pinene) + product
ethanolamine or 8-HQ → ppt + product
H_2O_2 NaOH → 2 (isopinocampheol) + product (4)

The (+)- and (−)-Ipc_2BAll reagents have been condensed with a variety of aldehyde structures[2,5] with no reported reversal of enantiospecificity. Because of this strict fidelity of addition, these reagents have been used in the structure determination of several natural products.[6] In the case of α-substituted aldehydes, however, an erosion of diastereofacial selectivity is sometimes observed for 'mismatched' reactant pairs (eqs 5–7).[7]

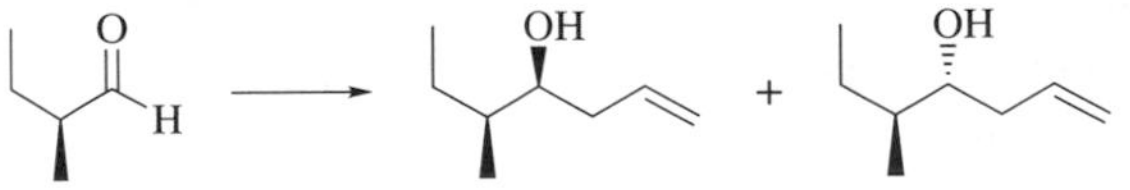

(5)

(+)-Ipc_2BAll	96:4
(−)-Ipc_2BAll	5:95

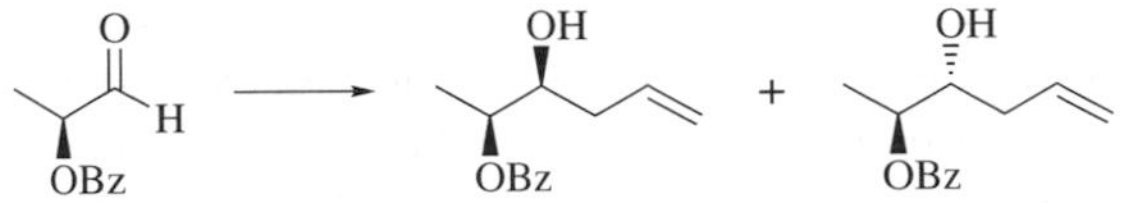

(6)

(+)-Ipc_2BAll	96:4
(−)-Ipc_2BAll	6:94

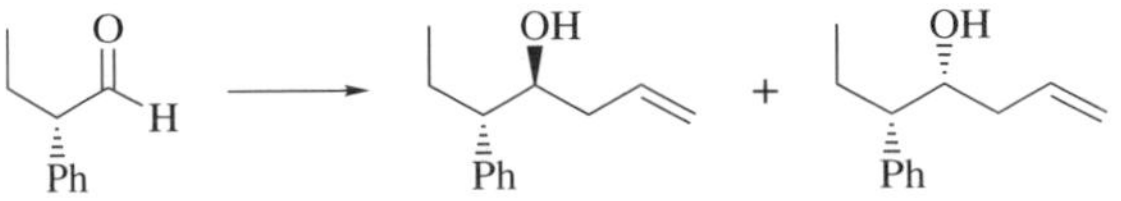

(7)

(+)-Ipc_2BAll 'matched'	97:3
(−)-Ipc_2BAll 'mismatched'	26:74

These reagents have been used for the conversion of C_s symmetrical chains into chiral, nonracemic products. A double condensation with *meso* dialdehydes can produce either optically active diol product (eq 8).[8]

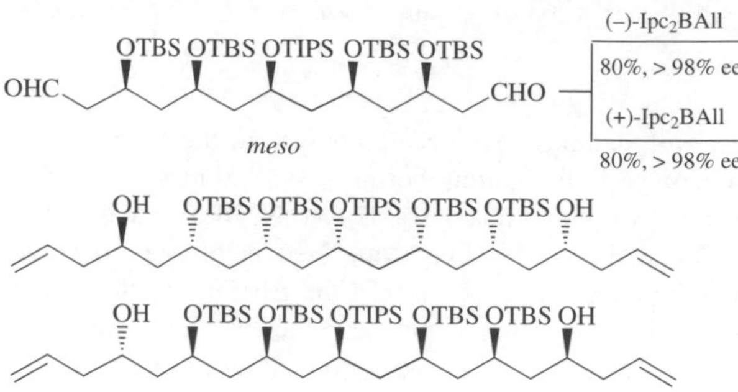

(8)

B-Allyldiisopinocampheylborane compares favorably in both yield and enantioselectivity to alternate methods of asymmetric allylboration[1] (see also ***B-Allyldiisocaranylborane***; ***Diisopropyl 2-Allyl-1,3,2-dioxaborolane-4,5-dicarboxylate***; ***(R,R)-2,5-Dimethylborolane***).

Addition to Ketones. Condensation of Ipc_2BAll with methyl ketones proceeds with modest to good enantioselectivity. Unlike with aldehydes, the degree of enantioselection is highly dependent upon ketone structure (eq 9).[2]

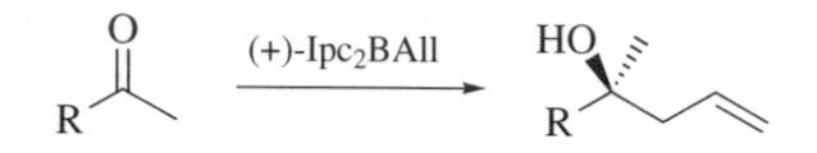

(9)

R =	yield (%) =	ee (%) =
ethyl	68	50
vinyl	79	35
phenyl	63	5
ethynyl	76	75

B*-Methallyldiisopinocampheylborane and Related Reagents.** Derivatives of the reagent containing substituted allyl groups have been used for the synthesis of more highly functionalized homoallylic alcohols (see also ***B-Crotyldiisopinocampheylborane and ***B-Allyldiisocaranylborane***). In general, these reagents are prepared in a manner analogous to Ipc_2BAll or by hydroboration of the corresponding diene with diisopinocampheylborane. For example, reagents containing alkyl substituents, such as *B*-methallyldiisopinocampheylborane and *B*-(3,3-dimethylallyl)diisopinocampheylborane, condense with aldehydes to give the corresponding substituted homoallylic alcohols (eqs 10 and 11).[9] An alkenyl-substituted derivative, *B*-2′-isoprenyldiisopinocampheylborane, has been used for the asymmetric isoprenylation of aldehydes (eq 12).[10]

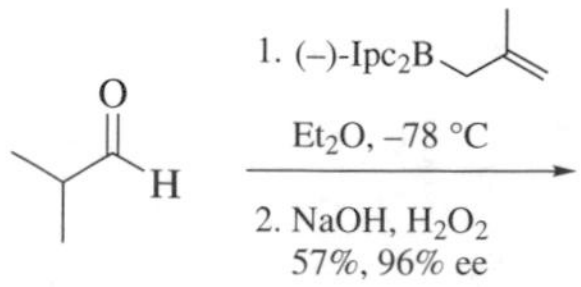

(10)

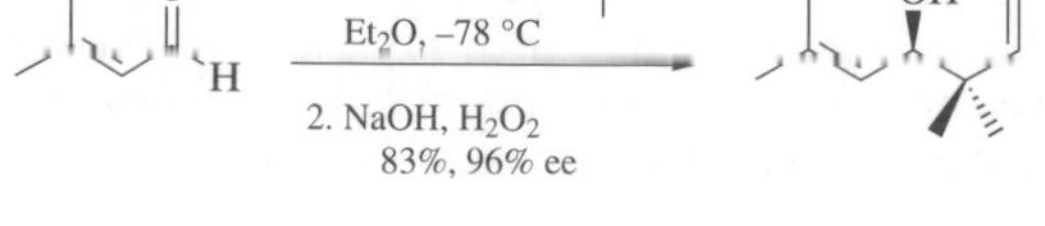

(11)

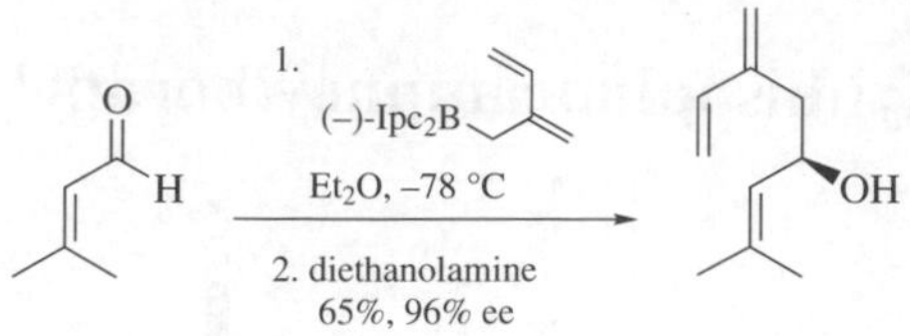

(12)

Derivatives containing heteroatoms have allowed for the preparation of allylic 1,2-diols and amino alcohols by this method. Aldehydes react with *B*-[(*Z*)-3-(methoxymethyloxy)-allyl]diisopinocampheylborane to give monoprotected diol adducts.[11] 1,2-Diols can be obtained directly by condensation with *B*-[(*E*)-3-((diisopropylamino)dimethylsilyl)allyl]diisopinocampheylborane followed by an oxidative reaction workup (eqs 13 and 14).[12]

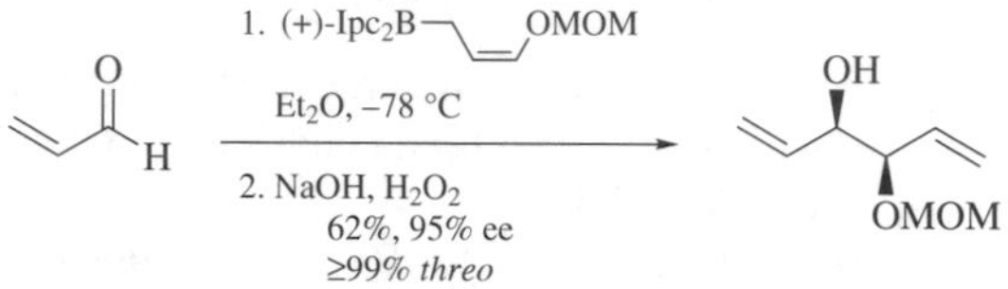

(13)

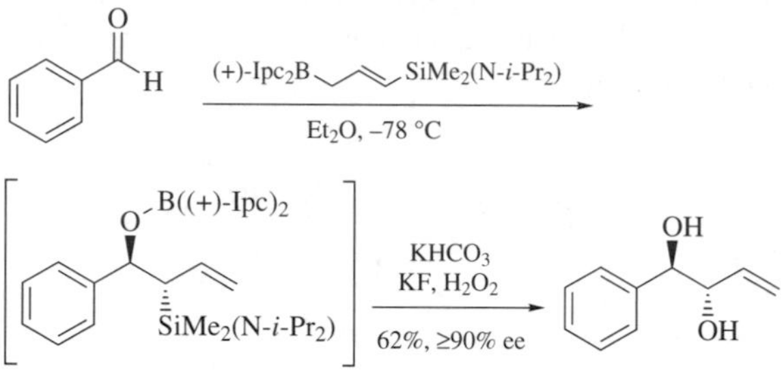

(14)

Amine-containing reagents, such as *B*-[(*E*)-3-(diphenylamino)allyl]diisopinocampheylborane, react in a similar manner to provide 1,2-amino alcohols (eq 15).[13] A further variation on this reagent is where the allylic group is contained in a ring. The cyclohexenyl reagent *B*-(cyclohex-2-enyl)diisopropylcampheylborane and the related cycloheptenyl and cyclooctenyl versions have been described.[14] At low temperatures ($< -25\,^\circ C$), these boranes are stable toward allylic isomerization and transfer with complete diastereoselectivity to provide *erythro* products (eq 16).

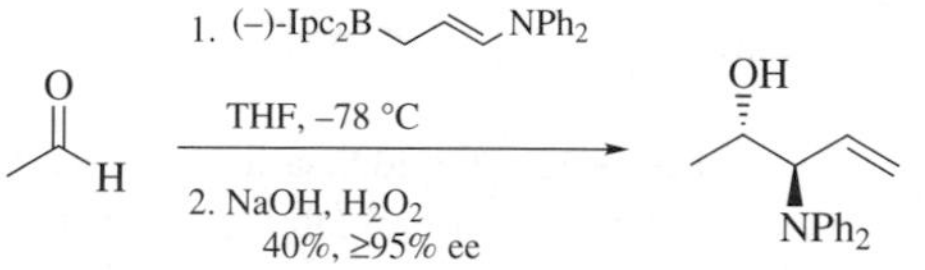

(15)

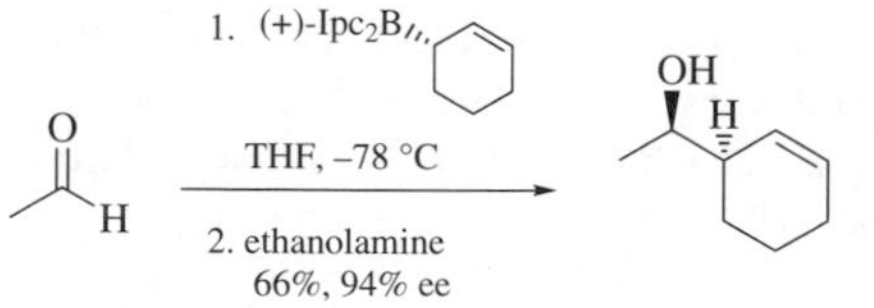

(16)

Related Reagents. *B*-Allyldiisocaranylborane; *B*-Crotyldiisopinocampheylborane; Diisopropyl 2-Allyl-1,3,2-dioxaborolane-4,5-dicarboxylate; Diisopropyl 2-Crotyl-1,3,2-dioxaborolane-4,5-dicarboxylate; (*R*,*R*)-2,5-Dimethylborolane.

1. Brown, H. C.; Ramachandran, P. V. *PAC* **1991**, *63*, 307.
2. (a) Brown, H. C.; Jadhav, P. K. *JACS* **1983**, *105*, 2092. (b) Jadhav, P. K.; Bhat, K. S.; Perumal, P. T.; Brown, H. C. *JOC* **1986**, *51*, 432.

3. Racherla, U. S.; Brown, H. C. *JOC* **1991**, *56*, 401.
4. Brown, H. C.; Racherla, U. S.; Liao, Y.; Khanna, V. V. *JOC* **1992**, *57*, 6608.
5. For a representation of functionality that is tolerated in this reaction see: (a) Stork, G.; Zhao, K. *JACS* **1990**, *112*, 5875. (b) Nicolaou, K. C.; Groneberg, R. D.; Stylianides, N. A.; Miyazaki, T. *CC* **1990**, 1275. (c) Rychnovsky, S. D.; Rodriguez, C. *JOC* **1992**, *57*, 4793. (d) Racherla, U. S.; Liao, Y.; Brown, H. C. *JOC* **1992**, *57*, 6614.
6. (a) Schreiber, S. L.; Goulet, M. T. *JACS* **1987**, *109*, 8120. (b) Nicolaou, K. C.; Ahn, K. H. *TL* **1989**, *30*, 1217. (c) Mori, Y.; Kohchi, Y.; Noguchi, H.; Suzuki, M.; Carmeli, S.; Moore, R. E.; Patterson, G. M. L. *T* **1991**, *47*, 4889.
7. (a) Brown, H. C.; Bhat, K. S.; Randad, R. S. *JOC* **1987**, *52*, 319. (b) Brown, H. C.; Bhat, K. S.; Randad, R. S. *JOC* **1989**, *54*, 1570.
8. Wang, Z.; Deschênes, D. *JACS* **1992**, *114*, 1090.
9. (a) Brown, H. C.; Jadhav, P. K. *TL* **1984**, *25*, 1215. (b) Brown, H. C.; Jadhav, P. K.; Perumal, P. T. *TL* **1984**, *25*, 5111. (c) Truesdale, L. K.; Swanson, D.; Sun, R. C. *TL* **1985**, *26*, 5009.
10. (a) Brown, H. C.; Randad, R. S. *TL* **1990**, *31*, 455. (b) Brown, H. C.; Randad, R. S. *T* **1990**, *46*, 4463.
11. (a) Brown, H. C.; Jadhav, P. K.; Bhat, K. S. *JACS* **1988**, *110*, 1535. (b) Burgess, K.; Chaplin, D. A.; Henderson, I.; Pan, Y. T.; Elbein, A. D. *JOC* **1992**, *57*, 1103. (c) Barrett, A. G. M.; Edmunds, J. J.; Horita, K.; Parkinson, C. J. *CC* **1992**, 1236.
12. (a) Barrett, A. G. M.; Malecha, J. W. *JOC* **1991**, *56*, 5243. (b) Barrett, A. G. M.; Edmunds, J. J.; Hendrix, J. A.; Malecha, J. W.; Parkinson, C. J. *CC* **1992**, 1240.
13. Barrett, A. G. M.; Seefeld, M. A. *CC* **1993**, 339.
14. (a) Brown, H. C.; Jadhav, P. K.; Bhat, K. S. *JACS* **1985**, *107*, 2564. (b) Brown, H. C.; Bhat, K. S.; Jadhav, P. K. *JCS(P1)* **1991**, 2633.

Mark T. Goulet
Merck Research Laboratories, Rahway, NJ, USA

Allylmagnesium Bromide[1]

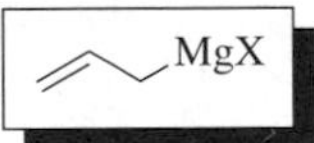

(X = Br)		
[1730-25-2]	C_3H_5BrMg	(MW 145.29)
(X = Cl)		
[2622-05-1]	C_3H_5ClMg	(MW 100.84)

(allylating agents capable of easy addition to carbonyl compounds,[15–23] nitro compounds[57] and nitriles,[58–61] imines[43–50] and aza aromatics,[51–56] unactivated alkenes,[30–42] and of displacement reactions on halo compounds[6,24–28,53,54])

Physical Data: Br: fp −40 °C; *d* 0.851 g cm^{-3} (commercial reagent). Cl: fp −17 °C; *d* 0.995 g mL^{-1} (commercial reagent).
Solubility: Br: very sol ether. Cl: slightly sol ether; sol THF.
Form Supplied in: Br: 1.0 M solution in Et_2O packaged under N_2. Cl: 2.0 M solution in THF packaged under N_2.
Preparative Methods: allylmagnesium bromide and chloride can be easily prepared by direct reaction of ***Magnesium*** with ***Allyl Bromide*** and ***Allyl Chloride***, respectively[2–5] in diethyl ether and/or tetrahydrofuran. An excess of Mg (preferably 10%) is recommended.[6,7] Prior amalgamation of *very pure* Mg[3,4] suppresses the formation of 1,5-hexadiene. An improved preparation[8] uses a THF slurry of Mg, obtained by co-condensation in a rotating solution reactor.[9] Preactivation of Mg by dry stirring in an inert atmosphere is also highly beneficial.[10] Allylpotassium can be readily converted into the magnesium analogs by simply adding a solution of ***Magnesium Bromide*** in THF.[11,12] Most of the work on allylic Grignards has been carried out with the bromide. Indeed, the bromide can be prepared in slightly higher yield than the chloride and shows a greater solubility in ether.[1]
Handling, Storage, and Precautions: both moisture-sensitive; highly inflammable; react violently with water; induce burns. Bottles of these reagents should be flushed with N_2, kept tightly sealed to preclude contact with oxygen, and stored far from sparks. In case of contact with eyes and skin, wash immediately with abundant water.

Structure of Allylmagnesium Halides. NMR data[3a] of several allylic Grignard reagents support the view that allylMgBr exists as a rapidly equilibrating mixture of the two classical structures (**1**) and (**2**) ($t_{1/2}$ <0.001 s) or as the bridged structure (**3**) (eq 1). This is supported by application of the Saunders isotopic perturbation technique[13] as well as by ^{25}Mg NMR studies.[14]

MgBr (**1**) ⇌ BrMg (**2**) (1)

H, H_2C, CH_2, Mg, Br (**3**)

Reactions with Carbonyls. Allylmagnesium halides react with carbonyl compounds to produce carbinols; 1,2-addition occurs with α,β-unsaturated carbonyls.[1,15d,16–18] Addition takes place also with severely hindered ketones, which with most other Grignards undergo either reduction or enolization. 'Allylmetallation' of carbonyls offers a complementary approach to the aldol reaction for acyclic stereocontrol.[15]

Complex (**4**), prepared by reacting the crystalline chiral complex $Cp(OR^*)_2TiCl$ with allylMgBr, adds to aldehydes to give homoallylic alcohols[19] with 85–94% ee (eq 2).

$Cp(OR^*)_2TiCl$ —(MgBr)→ R*O–Ti(Cp)–OR* (**4**) —(RCHO, ether, −30 °C)→ R–CH(OH)–allyl (2)

85–94% ee

R*OH =

Homoallylic alcohols have also been synthesized by ring opening of vinylic oxetanes[20] with copper-catalyzed allylmagnesium halides. C_2 homologation of aldehydes and ketones to 2-alkenals uses allylMgBr.[21]

AllylMgBr adds to the C=O of imides to give different products depending upon the imide.[22]

A new synthesis of propenyl ketones involves reaction of an ester with 2 equiv of allylMgCl, furnishing a tertiary bis-homoallylic alcohol that fragments at 80 °C to a mixture of β,γ- and α,β-unsaturated ketones.[23]

Displacement Reactions. The coupling of allylmagnesium halides with halogenated compounds provides a useful one-step method for introducing an alkenic bond directly into a molecule.[1,15d] Allylic Grignard halides couple with alkyl, allylic, and benzylic halides.[6,24] Sometimes, allylMgCl gives higher yields than the more commonly used allylMgBr.[24a] Alkylation in the presence of ***Copper(I) Iodide*** is much faster.[25]

AllylMgBr couples with fluoroalkenes in a rather unexpected manner.[26] Allylmagnesium halides can be hydroxymethylated in a two-step one-pot procedure by 1-chloro-2-(chloromethoxy)ethane.[27] A new synthesis of dienes, with the double bonds in predetermined positions, involves reaction of an α-chloro ketone with an allyl Grignard in ether at −60 °C followed by treatment with ***Lithium*** powder (eq 3).[29]

1. $H_2C{=}CHCH_2MgBr$; 2. Li ; 99% (3)

An efficient synthesis of 3-*t*-aminopropanols is based on the regioselective ring opening of tetrahydro-1,3-oxazines with allylMgCl.[28]

Allylmagnesium halides have been used to prepare different organometallics.[1,15d]

Addition to Alkenic Bonds. Allylmagnesium halides add smoothly to activated and some unactivated double bonds. Intermolecular and uncatalyzed carbomagnesiation of unfunctionalized alkenes, such as ***Ethylene*** and 1-octene, has been reported.[31–33] AllylMgCl[30] reacts with more than 90% regioselectivity with ***1,3-Butadiene*** (Mg at C-2) to afford 2,6-heptadienylmagnesium chloride (**5**) that, with a second carbomagnesiation, leads to 1:2 adduct (**6**), which undergoes a fast ring closure to (**7**) (eq 4).

MgCl — MgCl (**5**) — MgCl (**6**) — MgCl (**7**) (4)

The intramolecular carbomagnesiation involving reactive allylic Grignard reagents has been extensively studied.[34–38] The presence of a neighboring hydroxy, alkoxy, or amino group facilitates the addition of allylmagnesium halides.[39–42]

Addition to Azomethine Linkages. Allylmagnesium halides add to the azomethine linkage easier than the alkyl counterparts. Highly enantioselective syntheses of β- and γ-amino acids with good stereocontrol have been described;[43] the crucial step is the addition reaction of allylMgBr to chiral *N*-benzylidene-*p*-toluenesulfinamides followed by the elaboration of the allylic group (eq 5).

Tol, S, O, N, Ph — MgBr → Tol, S, O, N, H, Ph → H_3N^+, CO_2^-, Ph (5)

Resonance stabilization of the allyl anion results in a greater ionization of the carbon–metal bond and hence greater reactivity. This promotes addition to C=N linkages usually unaffected by nonresonance-stabilized organometallic reagents.[44] Thus the reaction of allylmagnesium halides with achiral aldimines affords homoallylamines.[45] The stereochemistry of the addition has been studied.[46]

The addition of allylmagnesium halides to the *O*-protected α-hydroxy-*N*-trimethylsilylimines (**8**), generated in situ, proceeds in a highly *anti* stereocontrolled manner to produce α-amino alcohols[47] that can be oxidized to α-amino acids (eq 6).[48]

OTBDMS, H, R, O — LiHMDS → OTBDMS, H, R, Me_3Si, N (**8**) — R'Li or R'MgBr(Cl) →

R = Me, Ph; R' = allyl — OTBDMS, R', R, NH_2 + OTBDMS, R', R, NH_2 (6)

anti > *syn*

AllylMgBr, complexed with ***Cerium(III) Chloride***, adds to chiral (*E*)-alkoxymethyl oxime ether (**9**) to give the corresponding (*S*)-amine derivative stereoselectively, while allyllithium gives mainly the opposite (*R*)-isomer (eq 7).[49]

BnO, N, H, O, OMe, Ph (**9**) — M; M = MgBr or MgBr–$CeCl_3$ → BnO, NH, O, OMe, Ph + BnO, NH, O, OMe, Ph (7)

Allylmagnesium halides also add to imines, iminium salts, *N*-hetero substituted imines (oximes, sulfenimines, etc.), *gem*-amino ethers, *gem*-amino nitriles, ethers, peroxides, disulfides, epoxides, acids and their derivatives, amides, orthoformates, *O*-alkylated oximes.[15d,50]

AllylMgBr adds to the C=N double bond of several aza aromatics to form allylated aza aromatics.[51]

The reaction of 2-hetero substituted benzothiazoles with allylMgBr affords 2-allylbenzothiazole or 2,2-diallylbenzothiazoline or *N*-triallylmethyl-*o*-aminobenzenethiol depending upon the experimental conditions;[52] a novel ring-opening reaction of the benzothiazole system with allylic Grignards has been reported.[53] 2-Alkylbenzoxazoles also react with allylic Grignards, undergoing ring opening to *N*-diallylalkyl-*o*-aminophenols, while 2-chlorobenzoxazole affords the cross-coupling product.[54] Mono and bis addition reactions of the allylic Grignard reagents to the C=N bond of quinoxalines afford high yields of allyldihydroquinoxalines and diallyltetrahydroquinoxalines; dehydrogenation with ***2,3-Dichloro-5,6-dicyano-1,4-benzoquinone*** leads to monoallyl- and diallylquinoxalines.[55] Similarly, reaction with pyrimidines furnishes excellent yields of allylpyrimidines.[56]

Reactions with Nitro Compounds and Nitriles. Allylmagnesium halides react with both aromatic and aliphatic nitro compounds via addition to the nitro group, leading to allylamines or to *N*-hydroxyallylamines (eq 8).[57]

PhNO$_2$ —(allyl-MgCl, THF, –70 °C)→ Ph–N$^+$(O$^-$)(OMgCl)–allyl —(1. LiAlH$_4$ 10% Pd/C; 2. NH$_4$Cl; 20 min / 0.5–48 h)→ Ph–N(OH)–allyl ; Ph–N(H)–allyl (8)

Allylmagnesium halides add to alkyl, aralkyl, and alkenyl cyanides in a preferred 2:1 ratio to yield trisubstituted primary amines.[58–60] The allyl groups on the carbinamine products are susceptible to catalytic hydrogenation to give *t*-alkyl primary amines, most of which are otherwise difficult to synthesize. Aryl nitriles react with allylmagnesium halides, leading to tetrahydropyridines.[61]

Related Reagents. Allyllithium; Allyltributylstannane; Allyltrimethylsilane; Crotylmagnesium Bromide; Methallylmagnesium Chloride; 3-Methyl-3-buten-1-ynylcopper(I).

1. Benkeser, R. A. *S* **1971**, 347.
2. (a) Young, W. G.; Lane, J. F.; Loshokoff, A.; Winstein, S. *JACS* **1937**, *59*, 2441. (b) Dewolfe, R. H.; Young, W. G. *CRV* **1956**, *56*, 753.
3. (a) Whitesides, G. M.; Nordlander, J. E.; Roberts, J. D. *Discuss. Faraday Soc.* **1962**, *34*, 185. (b) Grummit, O.; Budewitz, E. P.; Chudd, C. C. *OSC* **1963**, *IV*, 748.
4. Hwa, J. C. H.; Sims, H. *OS* **1961**, *41*, 49.
5. (a) Fieser, L. F.; Fieser M. *FF* **1967**, *1*, 415. (b) *OS* **1961**, *41*, 49.
6. Gilman, H.; Mc Glumphy, J. H. *BSF(4)* **1928**, *43*, 1322.
7. Kharasch, M. S.; Weinhouse, S. *JOC* **1936**, *1*, 209.
8. Oppolzer, W.; Kündig, E. P.; Bishop, P. M.; Perret, C. *TL* **1982**, *23*, 3901.
9. Klabunde, K. J.; Efner, H. F.; Satek, L.; Donley, W. *JOM* **1974**, *71*, 309.
10. Baker, K. V.; Brown, J. M.; Hughes, N.; Skarnulis, A. J.; Sexton, A. *JOC* **1991**, *56*, 698.
11. (a) Schlosser, M.; Hartmann, J. *AG* **1973**, *85*, 544. (b) *AG(E)* **1973**, *12*, 508. (c) *JACS* **1976**, *98*, 4674.
12. Desponds, O.; Schlosser, M. *JOM* **1991**, *409*, 93.
13. (a) Saunders, M.; Telkowski, L.; Kates, M. R. *JACS* **1977**, *99*, 8070. Saunders, M.; Kates, M. R. *JACS* **1977**, *99*, 8071. Saunders, M.; Kates, M. R.; Wiberg, K. B.; Pratt, W. *JACS* **1977**, *99*, 8072. (b) Schlosser, M.; Stähle, H. *AG(E)* **1980**, *19*, 487; Stähle, M.; Schlosser, M. *JOM* **1981**, *220*, 277.
14. Benn, R.; Lehmkuhl, H.; Mehler, K.; Rufińska, A. *AG(E)* **1984**, *23*, 534. *JOM* **1985**, *293*, 1.
15. (a) Yamamoto, Y. *ACR* **1987**, *20*, 243. (b) Hoffmann, R. W. *AG(E)* **1982**, *21*, 555. (c) Yamamoto, Y.; Maruyama, K. *H* **1982**, *18*, 357. (d) Courtois, G.; Miginiac, L. *JOM* **1974**, *69*, 1.
16. DeMeester, W. A.; Fuson, R. C. *JOC* **1965**, *30*, 4332.
17. Shtukin, G. I. *ZOB* **1940**, *10*, 77; *CA* **1940**, *34*, 4725.
18. Henze, H. R.; Allen, B. B.; Leslie, W. B. *JOC* **1942**, *7*, 326.
19. (a) Riediker, M.; Duthaler, R. O. *AG(E)* **1989**, *28*, 494. (b) Duthaler, R. O.; Herold, P.; Lottenbach, W.; Oertle, K.; Riediker, M. *AG(E)* **1989**, *28*, 495.
20. Larock, R. C.; Stolz-Dunn, S. K. *SL* **1990**, 341.
21. Chang, Y.-H.; Uang, B.-J.; Wu, C.-M.; Yu, T.-H. *S* **1990**, 1033.
22. (a) Lukes, R.; Cerny, M. *CCC* **1959**, *24*, 2722. (b) Flitsch, W. *LA* **1965**, *684*, 141.
23. Snowden, R. L.; Muller, B. L.; Schulte-Elte, K. H. *TL* **1982**, *23*, 335.
24. (a) Kharasch, M. S.; Fuchs, C. F. *JOC* **1944**, *9*, 359. (b) Colonge, J.; Poilane, G. *BSF(2)* **1955**, 953. (c) Martin, M. M.; Gleicher, G. J. *JACS* **1964**, *86*, 233.
25. Derguini-Boumechal, F.; Lorne, R.; Linstrumelle G. *TL* **1977**, 1181.
26. Tarrant, P.; Heyes, J. *JOC* **1965**, *30*, 1485.
27. Ogle, C. A.; Wilson, T. E.; Stowe, J. A. *S* **1990**, 495.
28. Alberola, A.; Alvarez, M. A.; Andrés, C.; Gonźalez, A.; Pedrosa, R. *SC* **1990**, *20*, 1149.
29. Barluenga, J.; Yus, M.; Bernad, P. *CC* **1978**, 847.
30. Lehmkuhl, H. *Organometallic in Organic Synthesis*; de Meijer, A.; tom Dieck, H., Eds.; Springer-Verlag; Berlin, 1987; p 185. Lehmkuhl, H.; Reinehr, D.; Mehler, K.; Schomburg, G.; Kötter, H.; Henneberg, D.; Schroth, G. *LA* **1978**, 1449.
31. Lehmkuhl, H. *BSF(2)* **1981**, 87.
32. Lehmkuhl, H.; Reinehr, D. *JOM* **1970**, *25*, C47.
33. (a) Lehmkuhl, H.; Reinehr, D.; Schomburg, G.; Henneberg, D.; Damen, H.; Schroth, G. *LA* **1975**, 103. (b) Lehmkuhl, H.; Mehler, K. *LA* **1978**, 1841.
34. (a) Hill, E. A. *Adv. Organomet. Chem.* **1977**, *16*, 131. (b) Hill, E. A. *JOM* **1975**, *91*, 123.
35. (a) Silver, M. S.; Shafer, P. R.; Nordlander, J. E.; Rüchardt, C.; Roberts, J. D. *JACS* **1960**, *82*, 2646. (b) Patel, D. J.; Hamilton, C. L.; Roberts, J. D. *JACS* **1965**, *87*, 5144. (c) Maercker, A.; Geuβ, R. *CB* **1973**, *106*, 773.
36. Oppolzer, C. W. *Selectivity–A Goal for Synthetic Efficiency*; Bartmann, W.; Trost, B. M., Eds.; Verlag Chemie: Weinheim, 1984; Vol. 14, p 137.
37. Kossa, W. C., Jr.; Rees, T. C.; Richey, H. G., Jr. *TL* **1971**, 3455.
38. Felkin, H.; Umpleby, J. D.; Hagaman, E.; Wenkert, E. *TL* **1972**, 2285.
39. (a) Eisch, J. J.; Husk, G. R. *JACS* **1965**, *87*, 4194. (b) Eisch, J. J.; Merkley, J. H. *JACS* **1979**, *101*, 1148. (c) Eisch, J. J.; Merkley, J. H.; Galle, J. E. *JOC* **1979**, *44*, 587.
40. (a) Chérest, M.; Felkin, H.; Frajerman, C.; Lion, C.; Roussi, G.; Swierczewski, G. *TL* **1966**, 875. (b) Felkin, H.; Kaeseberg, C. *TL* **1970**, 4587.
41. (a) Richey, H. G., Jr.; Wilkins, C. W., Jr. *JOC* **1980**, *45*, 5027. (b) Richey, H. G., Jr.; Bension, R. M. *JOC* **1980**, *45*, 5036. (c) Richey, H. G., Jr.; Moses, L. M.; Domalski, M. S.; Erickson, W. F.; Heyn, A. S. *JOC* **1981**, *46*, 3773. (d) Richey, H. G., Jr.; Domalski, M. S. *JOC* **1981**, *46*, 3780.
42. Lazzaroni, R.; Pini, D.; Bertozzi, S.; Fatti, G. *JOC* **1986**, *51*, 505.
43. Hua, D. H.; Miao S. W.; Chen J. S.; Iguchi, S. *JOC* **1991**, *56*, 4.
44. Gilman, H.; Eisch, J. *JACS* **1957**, *79*, 2150.

45. Barluenga, J.; Bayón, A. M.; Asensio, G. *CC* **1984**, 427.
46. Arous-Chtara, R.; Gaudemar, M.; Moreau, J. L. *CR(C)* **1976**, *282*, 687.
47. Cainelli, G.; Giacomini, D.; Mezzina, E.; Panunzio, M.; Zarantonello, P. *TL* **1991**, *32*, 2967.
48. Nikishin, G. I.; Elinson, N. M.; Makhova, I. R. *AG(E)* **1988**, *27*, 1716. Staunton, J.; Eisenbraun, E. J. *OS* **1962**, *42*, 4.
49. Ukaji, Y.; Kume, K.; Watai, T.; Fujisawa, T. *CL* **1991**, 173.
50. Kleinman, E. F.; Volkmann, R. A. *COS* **1991**, *2*, 975.
51. (a) Gilman, H.; Eisch, J.; Soddy, T. *JACS* **1957**, *79*, 1245. (b) Eisch, J. J.; Harrel, R. L., Jr. *JOM* **1970**, *21*, 21.
52. Florio, S.; Epifani, E.; Ingrosso, G. *T* **1984**, *40*, 4527.
53. Babudri, F.; Bartoli, G.; Ciminale, F.; Florio, S.; Ingrosso, G. *TL* **1984**, *25*, 2047.
54. Babudri, F.; Florio, S.; Ronzini, L. *T* **1986**, *42*, 3905.
55. Epifani, E.; Florio, S.; Ingrosso, G.; Sgarra, R.; Stasi, F. *T* **1987**, *43*, 2769.
56. Epifani, E.; Florio, S.; Ingrosso, G.; Babudri, F. *T* **1989**, *45*, 2075.
57. Bartoli, G.; Marcantoni, E.; Bosco, M.; Dalpozzo, R. *TL* **1988**, *29*, 2251.
58. Allen, B. B.; Henze, H. R. *JACS* **1939**, *61*, 1790.
59. Henze, H. R.; Allen, B. B.; Leslie, W. B. *JACS* **1943**, *65*, 87.
60. Henze, H. R.; Sutherland, G. L.; Edwards, G. D. *JACS* **1951**, *73*, 4915.
61. Grassberger, M. A.; Horvath, A.; Schulz, G. *TL* **1991**, *32*, 7393.

Saverio Florio & Vito Capriati
University of Bari, Italy

Allyltributylstannane[1]

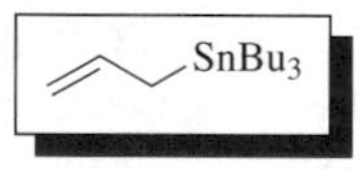

[24850-33-7] $C_{15}H_{32}Sn$ (MW 331.11)

(allylating reagent for many compounds, including alkyl halides, carbonyl compounds, imines, acetals, thioacetals, and sulfoximides)

Alternate Name: allyltributyltin.
Physical Data: bp 88–92 °C/0.2 mmHg; fp > 110 °C; *d* 1.068 g cm^{-3}.
Solubility: sol dichloromethane, diethyl ether, THF, toluene, benzene.
Form Supplied in: colorless liquid; widely available.
Purification: distillation.
Handling, Storage, and Precautions: all organotin compounds are highly toxic.

Allylstannanes are widely used as allyl anion equivalents.[1] They are less reactive than the corresponding magnesium or lithium reagents and, hence, can be classified as 'storable organometallic' reagents.[13b] This reduced activity increases the ease of handling in the laboratory; however, higher reaction temperatures or activation with Lewis acids are necessary. The relative reactivities of allyltriphenylsilane, -germane, and -stannane with diaryl carbenium ions are 1, 5.6, and 1600, respectively.[2] Allyltributylstannane is more reactive than ***Allyltriphenylstannane*** by three orders of magnitude. Allyl- and crotyltrialkyltin reagents undergo transmetalation reactions with strong Lewis acids through an S_E2' pathway.[3] Competing transmetalation processes can affect the mechanistic pathway and product distribution.[4] Radical processes can also be exploited in allylation reactions employing stannanes.[5] Because of the high toxicity of organotin reagents, allyltributyltin is more widely used than the more volatile allyltrimethyltin.

Additions to Aldehydes. Allyltrialkyltin reagents, such as allyltributyltin (**1**), react with carbonyl compounds[1] to form homoallylic alcohols under photolytic,[6] thermal,[7] high pressure,[8] or, more commonly, Lewis acidic conditions.[9] The order of reactivity is aldehydes > methyl ketones > internal ketones. A number of stereochemical issues are important when substituted allylic stannanes are utilized (see ***Crotyltributylstannane*** for details).

Selective conversion of protected α-hydroxy aldehydes (**2**) to monoprotected derivatives of *syn*- or *anti*-1,2-diols by reaction with allyltrialkylstannanes is realized with judicious choice of Lewis acid and protecting group.[10] ***Magnesium Bromide***, ***Titanium(IV) Chloride***, and ***Zinc Iodide*** favor *syn* products (**3**), especially with the benzyloxy derivative (**2a**), while use of ***Boron Trifluoride Etherate*** favors the *anti* products (**4**), particularly with the *t*-butyldimethylsilyl ether (**2b**) (eq 1).

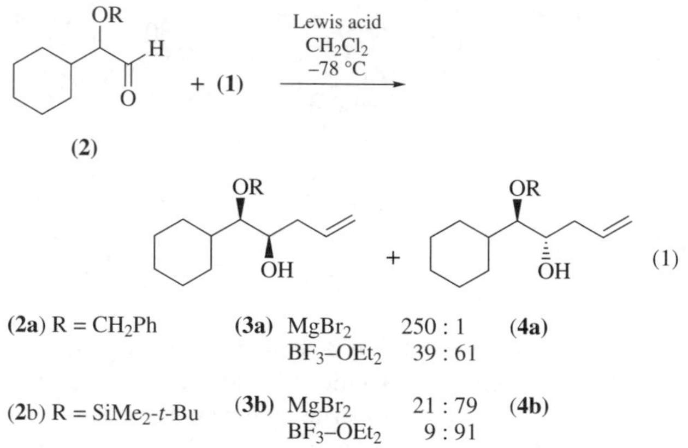

(**2**) + (**1**) —[Lewis acid, CH_2Cl_2, −78 °C]→ (**3**) + (**4**) (1)

(**2a**) R = CH_2Ph	(**3a**)	$MgBr_2$	250 : 1	(**4a**)
		BF_3–OEt_2	39 : 61	
(**2b**) R = $SiMe_2$-*t*-Bu	(**3b**)	$MgBr_2$	21 : 79	(**4b**)
		BF_3–OEt_2	9 : 91	

The ***Tin(IV) Chloride*** promoted allylation of α-methylthio aldehydes with allyltriphenyltin is highly selective for *anti* products, while the selectivity from $BF_3{\cdot}OEt_2$ mediation is variable.[11] A very mild promoter system, 5 M ***Lithium Perchlorate*** in diethyl ether, was used to allylate dialdose derivatives with high selectivity.[12] Acetals, ethers, and silyl ethers survive the allylation of aldehydes with this promoter.

β-Alkoxy aldehydes, with alkyl groups at C-2, readily form stable chelates with $TiCl_4$, $SnCl_4$, and $MgBr_2$ and consquently show high levels of *anti* selectivity in allylations with allyltributyltin.[13] High levels of diastereofacial selectivity in the Lewis acid mediated additions of allylstannanes to β-alkoxy aldehydes with substituents at C-3 are achieved when (a) the protecting group permits effective bidentate chelation between the aldehyde carbonyl and the ether oxygen and (b) the protecting group provides enough steric bulk to force C-3 substituents into an axial position in the six-membered chelate formed with the Lewis acid. $TiCl_4$ shows the highest *anti* selectivity when the protecting group is benzyl (R = *n*-hexyl; 96:1) and poor selectivity for methyl protection (R = *n*-hexyl; 3.8:1) (eq 2). $SnCl_4$ provides poor selectivities for all C-3 alkyl substituted β-alkoxy aldehydes. These results are consistent with predictions based upon ground state solution structures which show that the preferred conformation for $TiCl_4$ and $MgBr_2$ chelates has the alkyl group in a pseudoaxial position when the protecting group is ethyl or benzyl. Chelation is not involved in the reactions of α-or β-siloxy aldehydes.[14] In $TiCl_4$ promoted allyltriphenylstannane additions to β-alkoxyaldehydes with a methyl group at C-3, benzyl protection provides superior *anti* selectivity (29:1) to (methylthio)methyl (2:1) and (benzyloxy)methyl (9:1) groups.[15]

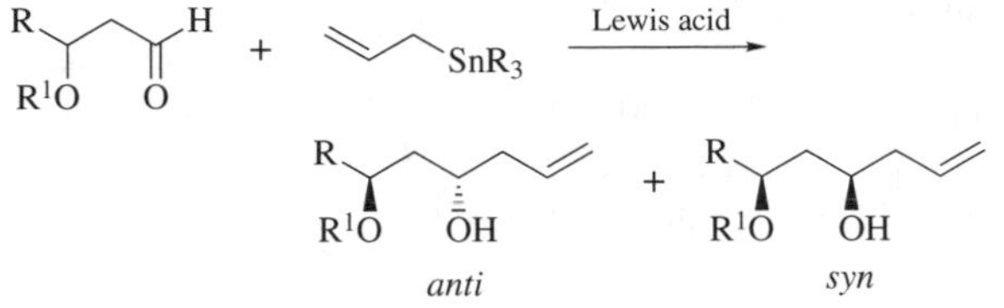

(2)

Allylation of Ketones. While irradiation of mixtures of aromatic ketones and allyltrialkylstannanes usually affords coupling products which are allylated at the carbonyl carbon,[6a,b] selective allylation at the α-carbon of aromatic α,β-epoxy ketones is observed (eq 3).[16] Yields of the α-allyl-β-hydroxy aryl ketones are highest when the *para* substituent is an electron-withdrawing group (CN) and lowest when it is an electron donor (MeO). Quinones undergo 1,4-monoallylation with allyltributyltin in the presence of BF_3 etherate (eq 4). However, 4-substituted 1,2-naphthoquinones and sterically hindered 3,5-di-*t*-butyl-*o*-quinones undergo 1,2-addition (eq 5).[17]

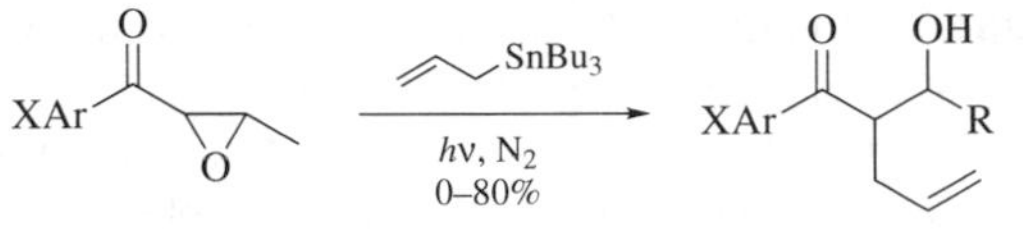

(3)

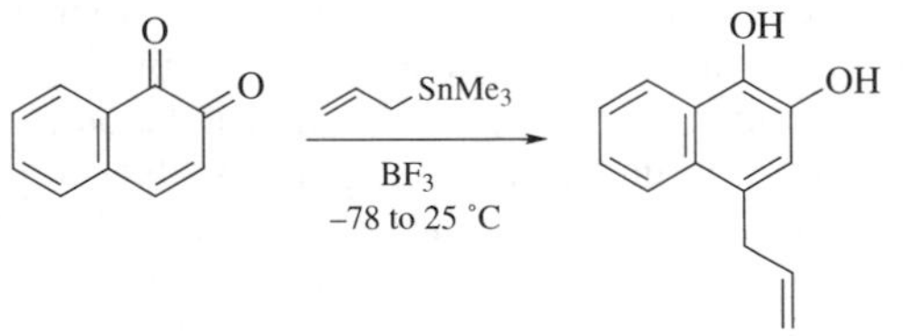

(4)

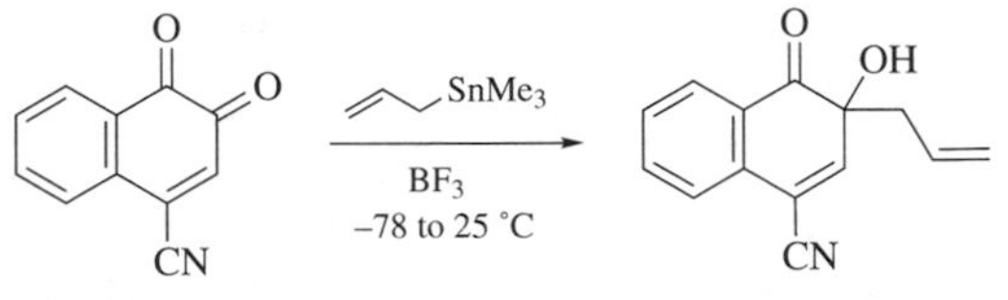

(5)

Allyltributylstannane in the presence of BF_3 etherate is a more efficient α-allylating reagent for quinones than the allylsilane–$TiCl_4$ reagent system. Eleutherin and isoleutherin were synthesized, in part, by this method.[18]

Unsymmetrical aryl alkyl α-diketones are regioselectively allylated by allylstannanes at the benzylic carbon under photolytic conditions and allylated at the acyl carbon in the presence of BF_3 etherate (eq 6).[19] Stannylated cyclopentanes are formed from the reaction of allyltributylstannane with aluminum trichloride-activated α,-β-unsatured acyliron complexes.[20] Stereochemistry about the alkene is preserved in this reaction.

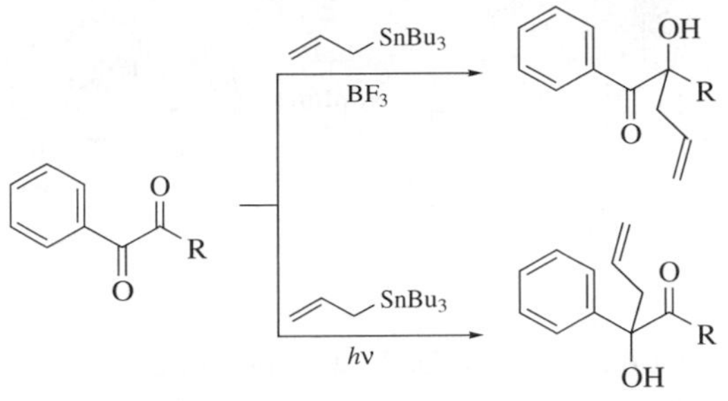

(6)

Allylation of Organohalides. Alkyl halides[21] and selenides[22] are allylated by allylstannes under thermal (with ***Azobisisobutyronitrile***), photochemical (with a tungsten lamp), or palladium-catalyzed conditions in high yield (eqs 7 and 8). Palladium catalyzes many reactions of allyltin reagents with various electrophiles, including allyl halides, aryl iodides and bromides, activated aryl chlorides, acid chlorides, vinyl halides, vinyl triflates, α-halo ketones and esters, and α-halo lactones.[23]

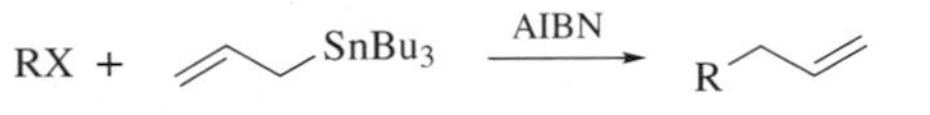

(7)

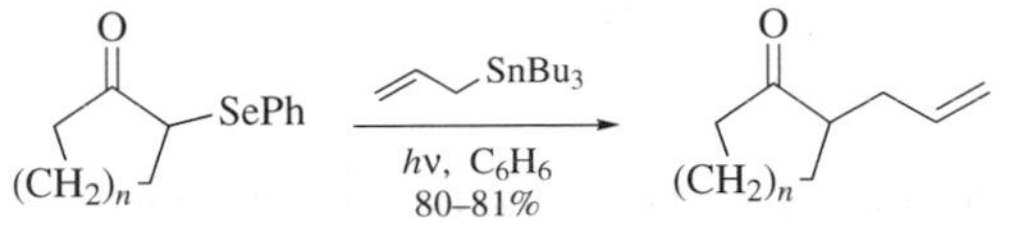

(8)

Aliphatic, aromatic, and heterocyclic acid chlorides react with allyltrialkylstannanes to give ketones in high yield (eq 9). Functional groups such as nitro, nitrile, haloaryl, methoxy, ester, and aldehyde are tolerated. An alternative palladium-catalyzed ketone synthesis involves the coupling of primary, secondary, or tertiary halides with carbon monoxide and allyltin (eq 10). Allyltributyltin adds to α-alkoxy-β-siloxy acylsilanes, with high *syn* selectivity (*syn*:*anti* = 91:9) in the presence of ***Zinc Chloride***.[24] A monoprotected *syn*-1,2,3 triol results from protiodesilylation. The palladium-catalyzed reaction of α-halo ketones with acetonyl- and allylstannanes produces oxiranes, oxetanes, and tetrahydrofurans in good yield.[25] Most allylic acetates do not react, although cinnamyl acetate and allyl acetate are exceptions.

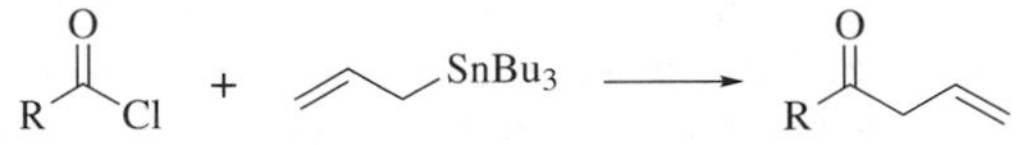

(9)

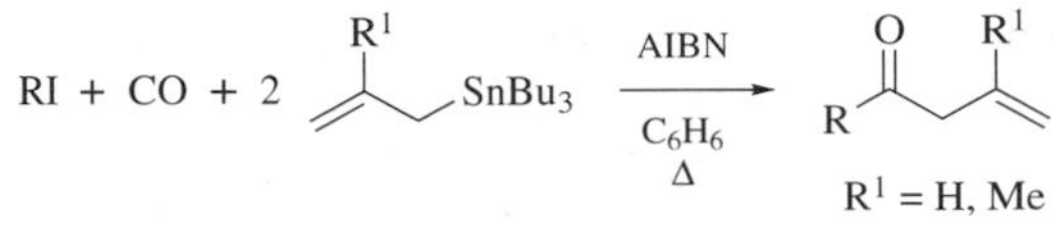

(10)

Stille Reaction. The reaction between phenyl triflates (a vinyl triflate) and allylic stannanes is useful for the synthesis of substituted aromatic compounds (eq 11).[26] The reaction works well with most highly substituted phenols except for hexasubstituted ones.[27] The reaction has been extended to the less expensive aryl fluorosulfonates[28] and aryl arenesulfonates.[29] These reactions proceed in good yield unless the aryl ring contains electron-donating substituents.

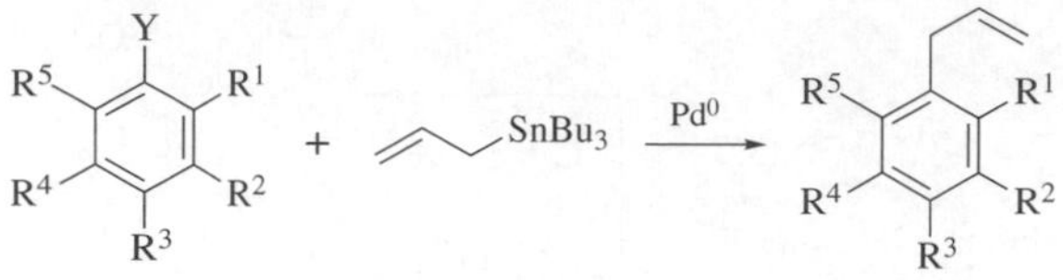

Y = OTf, SO_2 (11)

Allylation of Acetals. In the presence of a Lewis acid, 1,3-dioxolanes can be allylated with allyltributylstannanes or allylsilanes.[30] The Lewis acid promoted cleavage of chiral acetals with allylstannes affords chiral ethers with reported diastereoselectivites of >500:1 (eq 12).[31] Allylation of monothioacetals and dithioacetals occurs in a highly *syn* selective fashion to form homoallyl sulfides in good yield, particularly with $GaCl_3$ as the Lewis acid promoter.[32]

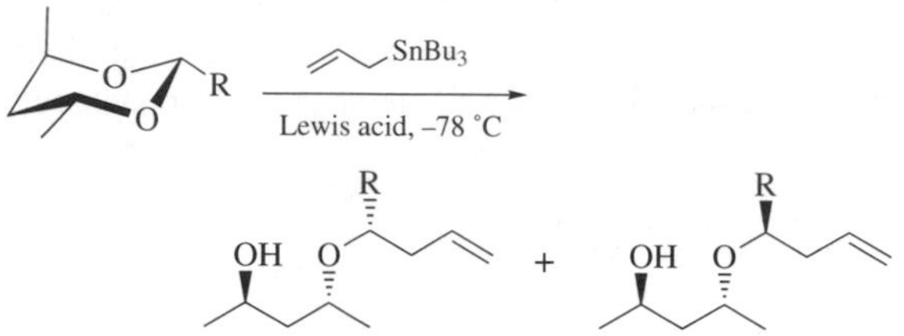

(12)

Allylation of Imines. Aldimines are converted to homoallylamines by allyltributyltin with Lewis acid promotion in moderate to high yield.[33] Likewise, β-methyl homoallylamines (predominantly *syn*) result from the reaction of ***Crotyltributylstannane*** with the $TiCl_4$ chelate of aldimines. In the $TiCl_4$-mediated allylstanne addition to (**5**), the Cram product is favored (92:8) (eq 13).[34]

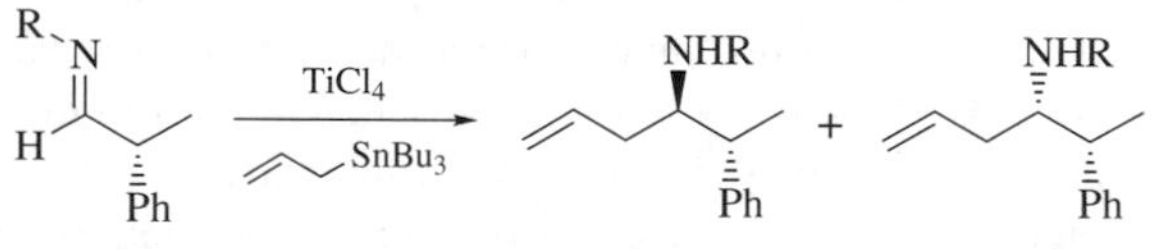

(**5**) Cram *anti*-Cram (13)

Allyltributyltin is also useful for α-allylations of *N*-acyl heterocycles, including pyridinium salts.[35] Isoquinolines (or dihydroisoquinolines) can be simultaneously acylated and allylated by the addition of α,β,γ,δ-unsaturated acyl chloride and allyltributyltin. The resulting adduct undergoes a Diels–Alder cyclization yielding an isoquinoline alkaloid precursor (eq 14).[36] Acylation and allylation of imidazoles is a particularly useful route to highly substituted 2-allylimidazolines (eq 15).[37]

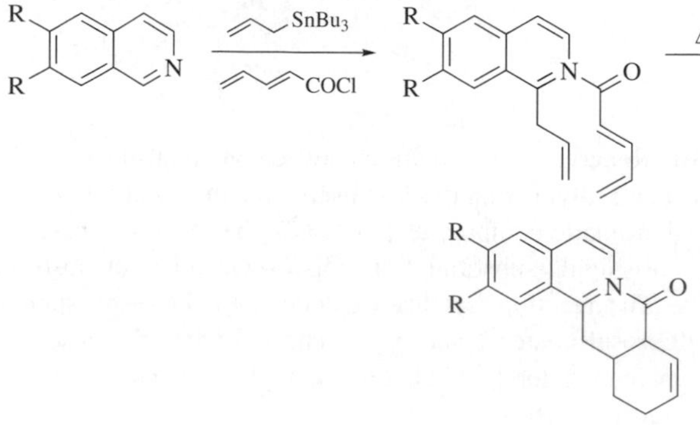

(14)

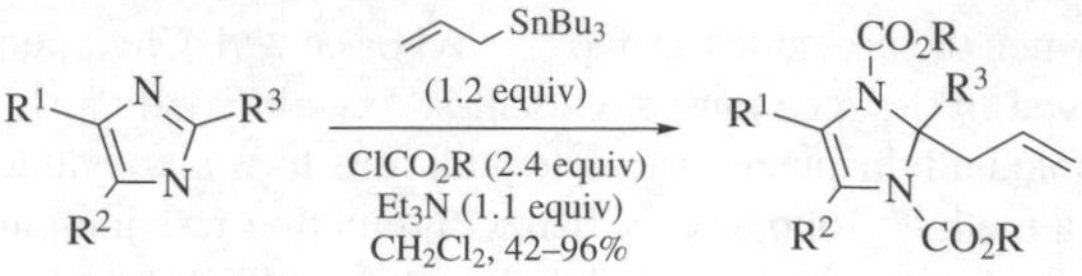

(15)

Acylimminium ions, formed by the reaction of α-alkoxy carbamates with Lewis acids, undergo allyl transfer from allylstannanes or silanes (eq 16).[38]

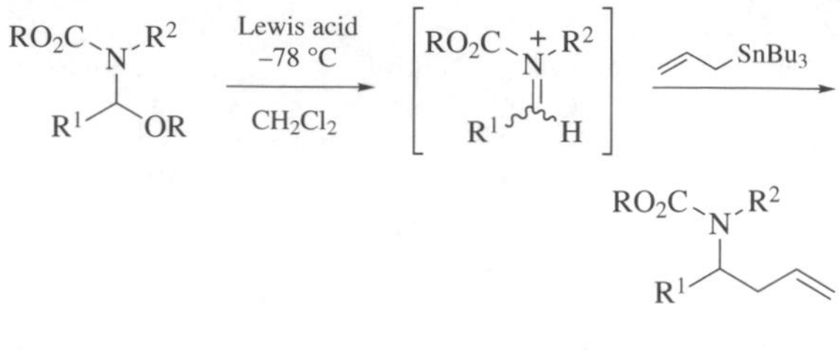

(16)

Allylation of Sulfoximidoyl Chlorides. A variety of *S*-allylsulfoximines can be synthesized in high yield by the allylation of sulfoximidoyl chlorides (eq 17).[39] Thiocarbonates are also allylated under photolytic conditions.[40]

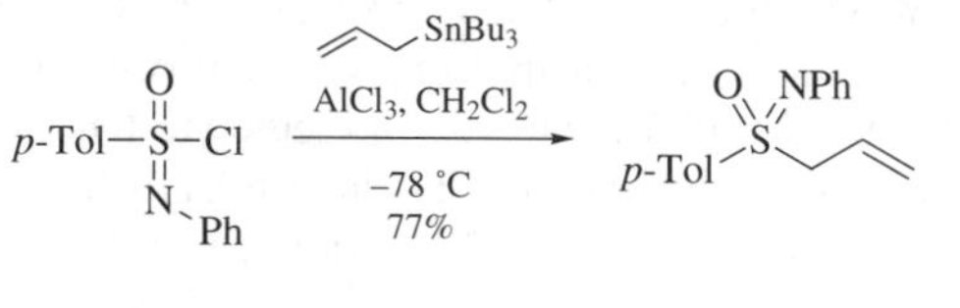

(17)

Radical Allylations. In addition to ionic pathways, radical processes can also be employed in allylations using stannanes.[5,21] The 1,2-asymmetric induction in radical allylations of α-alkoxycarbonyl radicals has been investigated. The observed selectivities, ranging from 1:1 to 99:1, are consistent with transition-state models which incorporate favorable stereoelectronic effects and the minimization of $A^{1,2}$, $A^{1,3}$, and torsional strain.[41] The camphorsultam derivative (**6**) undergoes thermal allylation (10% AIBN, 80 °C, benzene) with stannanes to give the allylated products in excellent yield with diastereoselectivities of 12:1 (eq 18).[42] Allyl transfer to quinones and α,β-epoxy ketones by single electron transfer pathways has also been investigated.[43]

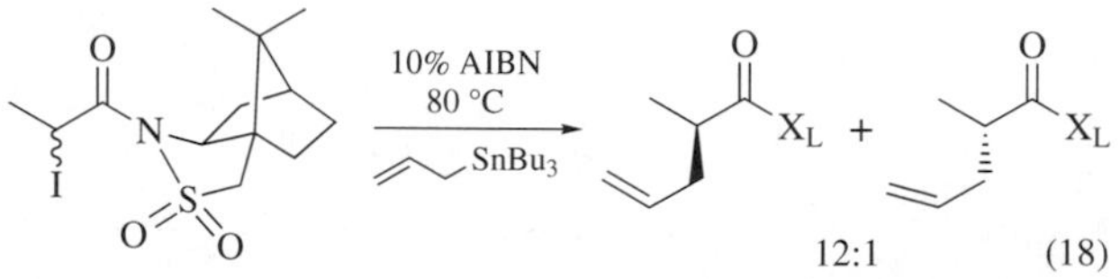

(**6**) 12:1 (18)
X_L = (+)-camphorsultam

Related Reagents. Allyltrimethylsilane; Allyltriphenylstannane; Crotyltributylstannane.

1. Reviews of allyl- and crotylmetal chemistry: (a) Hoffman, R. W. *AG(E)* **1982**, *21*, 555. (b) Yamamoto, Y.; Maruyama, K. *H* **1982**, *18*, 357. (c) Roush, W. R. *COS* **1991**, *2*, 1. (d) Yamamoto, Y. *ACR* **1987**, *20*, 243. (e) Yamamoto, Y. *Aldrichim. Acta* **1987**, *20*, 45. (f) Curran, D. P. *S*, **1988**, 489. (g) Yamamoto, Y. *Chemtracts–Org. Chem.* **1991**, 255. (h) Marshall, J. A. *Chemtracts–Org. Chem.* **1992**, 75. (i) Yamamoto, Y.; Asao, N. *CR* **1993**, *93*, 2207.
2. Hagen, G.; Mayr, H. *JACS* **1991**, *113*, 4954.

3. Naruta, Y.; Nishigaichi, Y; Maruyama, K. *T* **1989**, *45*, 1067.
4. (a) Keck, G. E.; Castellino, S.; Andrus, M. B. *Selectivities in Lewis Acid Promoted Reactions*, Schinzer, D., Ed.; Kluwer: Dordrecht, 1989; pp 73–105. (b) Keck, G. E.; Andrus, M. B.; Castellino, S. *JACS* **1989**, *111*, 8136. (c) Denmark, S. E.; Wilson, T.; Wilson, T. M. *JACS* **1988**, *110*, 984. (d) Boaretto, A.; Marton, D.; Tagliavini, G.; Ganis, P. *JOM* **1987**, *321*, 199. (e) Yamamoto, Y.; Maeda, N.; Maruyama, K. *CC* **1983**, 742. (f) Quintard, J. P.; Elissondo, B.; Pereyre, M. *JOC* **1983**, *48*, 1559.
5. (a) Pereyre, M.; Quintard, J. P.; Rahm, A. *Tin in Organic Synthesis*; Butterworths: London, 1987. (b) Yamamoto, Y. *T* , **1989**, *45*, 909. (c) Ref. 1f.
6. (a) Takuwa, A.; Tagawa, H.; Iwamoto, H.; Soga, O.; Maruyama, K. *CL* **1987**, 1091. (b) Takuwa, A.; Nishigaichi, Y.; Yamashita, K.; Iwamoto, H. *CL* **1990**, 639.
7. (a) Servens, C.; Pereyre, M. *JOM* **1972**, *35*, C20. (b) Abel, E. W.; Rowley, R. J. *JOM* **1975**, *84*, 199.
8. Yamamoto, Y.; Maruyama, K.; Matsumoto, K. *CC* **1983**, 489.
9. Naruta, Y.; Ushida, S.; Maruyama, K. *CL* **1979**, 919.
10. (a) Keck, G. E.; Boden, E. P. *TL* **1984**, *25*, 265. (b) Yamamoto, Y.; Komatsu, T.; Maruyama, K. *JOM* **1985**, *285*, 31.
11. Shimagaki, M.; Takubo, H.; Oishi, T. *TL* **1985**, *26*, 6235.
12. Henry, K. J. Jr.; Grieco, P. A.; Jagoe, C. T. *TL* **1992**, *33*, 1817.
13. (a) Keck, G. E.; Castellino, S. *JACS* **1986**, *108*, 3847. (b) Keck, G. E.; Castellino, S.; Wiley, M. R. *JOC* **1986**, *51*, 5478. (c) Keck, G. E.; Abbott, D. E. *TL* **1984**, *25*, 1883.
14. Keck, G. E.; Castellino, S. *TL* **1987**, *28*, 281.
15. Keck, G. E.; Murray, J. A. *JOC* **1991**, *56*, 6606.
16. Hasegawa, E.; Ishiyama, K.; Horaguchi, T.; Shimizu, T. *TL* **1991**, *32*, 2029.
17. (a) Maruyama, K.; Takuwa, A.; Naruta, Y.; Satao, K.; Soga, O. *CL* **1981**, 47. (b) Maruyama, K.; Naruta, Y. *CL* **1978**, 431. (c) Naruta, Y.; Maruyama, K. *CL* **1979**, 885. (d) Naruta, Y.; Maruyama, K. *CL* **1979**, 881.
18. Naruta, Y.; Uno, H.; Maruyama, K. *CC* **1981**, 1277.
19. Takuwa, A.; Nishigaichi, Y.; Yamashita, K.; Iwamoto, H. *CL* **1990**, 1761.
20. (a) Herndon, J. W.; Wu, C. *TL* **1989**, *30*, 5745.
21. (a) Kosugi, M.; Kurino, K.; Takayama, T.; Migata, T. *JOM* **1973**, *56*, C11. (b) Grignan, J.; Pereyre, M. *JOM* **1973**, *61*, C33. (c) Keck, G. E.; Yates, J. B. *JOC* **1982**, *47*, 3590. (d) Keck, G. E.; Yates, J. B. *JACS* **1982**, *104*, 5829.
22. Toru, T.; Okumara, T.; Ueno, Y. *JOC* **1990**, *55*, 1277.
23. For a review see: Stille, J. K. *AG(E)* **1986**, *25*, 508.
24. Cirillo, P. F.; Panek, J. S. *JOC* **1990**, *55*, 6071.
25. Pri-Bar, I.; Pearlman, P. S.; Stille, J. K. *JOC* **1983**, *48*, 4629.
26. (a) Stille, J. K. *PAC* **1985**, *57*, 1771. (b) Scott, W. J.; McMurray, J. E. *ACR* **1988**, *21*, 47. (c) Ref. 24.
27. Saá, J. M.; Martorell, G.; García-Raso, A. *JOC* **1992**, *57*, 678.
28. Roth, G. P.; Fuller, C. E. *JOC* **1991**, *56*, 3493.
29. Badone, D.; Cecchi, R.; Guzzi, U. *JOC* **1992**, *57*, 6321.
30. (a) Yamamoto, Y.; Abe, H.; Nishii, S.; Yamada, J. *JCS(P1)* **1991**, 3253. (b) Denmark, S. E.; Almstead, N. G. *JOC* , **1991**, *56*, 6458.
31. (a) Sammakia, T.; Smith, R. S. *JACS* **1992**, *114*, 10998. (b) Denmark, S. E.; Almstead, N. G. *JOC* , **1991**, *56*, 6485.
32. (a) Sato, T.; Otera, J.; Nozaki, H. *JOC* **1990**, *55*, 6116. (b) Saigo, K.; Hashimoto, Y.; Kihara, N. *CL* , **1990**, 1097.
33. Keck, G. E.; Enholm, E. J. *JOC* **1985**, *50*, 146.
34. (a) Yamamoto, Y.; Nishii, S.; Maruyama, K.; Komatsu, T.; Itoh, W. *JACS* **1986**, *108*, 7778. (b) Yamamoto, Y.; Komatsu, T.; Maruyama, K. *JACS* **1984**, *106*, 5031.
35. (a) Yamaguchi, R.; Moriyasu, M.; Yoshioka, M.; Kawanisi, M. *JOC* **1985**, *50*, 287. (b) Yamaguchi, R.; Moriyasu, M.; Yoshioka, M.; Kawanisi, M. *JOC* **1988**, *53*, 3507.
36. Yamaguchi, R.; Otsuji, A.; Utimoto, K. *JACS* **1988**, *110*, 2186.
37. Itoh, T.; Hasegawa, H.; Nagata, K.; Okada, M.; Ohsawa, A. *TL* **1992**, *33*, 5399.
38. (a) Yamamoto, T.; Schmid, M. *CC* **1989**, 1310. (b) Yamamoto, Y. Sato, H.; Yamada, J. *SL* **1991**, 339. (c) Wanner, K. T.; Wadenstorfer, E. Kärtner, A. *SL* **1991**, 797.
39. Harmata, M.; Claassen, R. J. II *TL* **1991**, *32*, 6497.
40. Kelly, M. J.; Roberts, S. M. *JCS(P1)* , **1991**, 787.
41. (a) Hart, D. J.; Krishnamurthy, R. *JOC* **1992**, *57*, 4457. (b) Hart, D. J.; Krishnamurthy, R. *SL* , **1991**, 412.
42. Curran, D. P.; Shen, W.; Zhang, J.; Heffner, T. A. *JACS* **1990**, *112*, 6738.
43. (a) Maruyama, K.; Imahori, H. *BCJ* **1989**, *62*, 816. (b) Hasegawa, E.; Ishiyama, K.; Horaguchi, T.; Shimizu, T. *TL* **1991**, *32*, 2029.

Stephen Castellino
Rhône-Poulenc, Research Triangle Park, NC, USA

David E. Volk
North Dakota State University, Fargo, ND, USA

Allyltrimethylsilane[1]

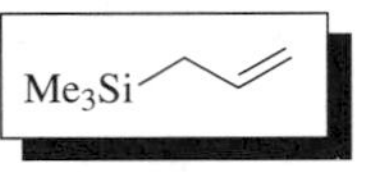

[762-72-1] $C_6H_{14}Si$ (MW 114.26)

(carbon nucleophile for the introduction of allyl groups by Lewis acid-catalyzed or fluoride ion-catalyzed reaction with acid chlorides, aldehydes, ketones, iminium ions, enones, and similar carbon electrophiles)

Physical Data: bp 85–86 °C, *d* 0.717 g cm^{-3}.
Solubility: freely sol all organic solvents.
Form Supplied in: colorless liquid. Methods for the synthesis of allylsilanes in general have been reviewed.[3]
Analysis of Reagent Purity: δ 5.74 (1H, ddt, *J* 16.9, 10.2 and 8), 4.81 (1H, dd, *J* 16.9 and 2), 4.79 (1H, dd, *J* 10.2 and 2), 1.49 (2H, d, *J* 8) and −0.003 (9H, s).[2]
Handling, Storage, and Precautions: inflammable; the vapor is irritating to the skin, eyes, and mucous membranes.

As a Carbon Nucleophile in Lewis Acid-Catalyzed Reactions. Allyltrimethylsilane is an alkene some 10^5 times more nucleophilic than propene, as judged by its reactions with diarylmethyl cations.[4] It reacts with a variety of cationic carbon electrophiles, usually prepared by coordination of a Lewis acid to a functional group, but also by chemical or electrochemical oxidation,[5] or by irradiation in the presence of 9,10-dicyanoanthracene.[6] The electrophile attacks the terminal alkenic carbon to give an intermediate cation, and the silyl group is lost to create a double bond at the other terminus. Among the more straightforward electrophiles are acid chlorides (eq 1),[7] aldehydes and ketones (eq 2),[8] their acetals (eq 3),[9] and the related alkoxyalkyl halides (eq 4),[10] iminium ions, and acyliminium ions (eqs 5 and 6),[11,12] and tertiary and allylic or benzylic alkyl halides (eqs 7 and 8).[13,14]

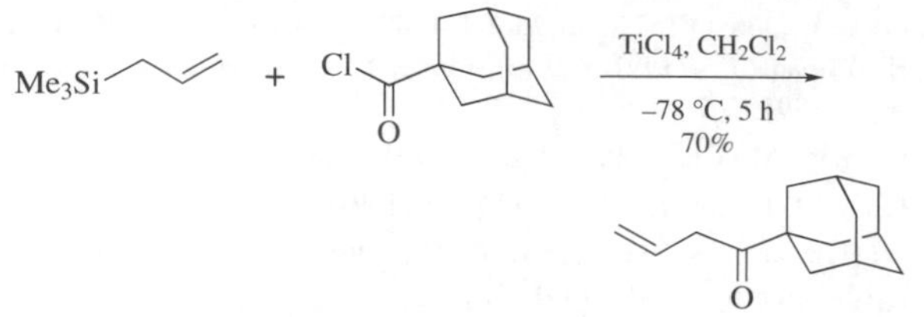

(1)

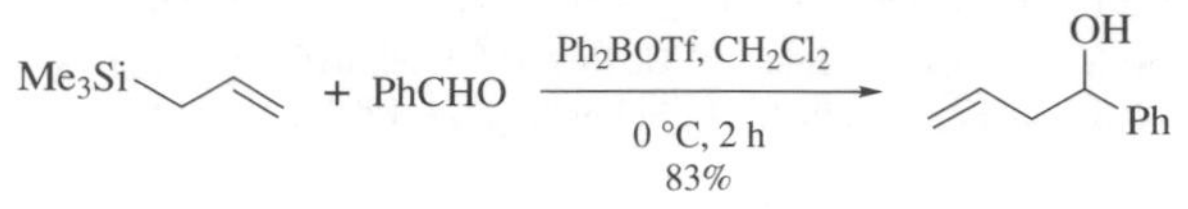

(2)

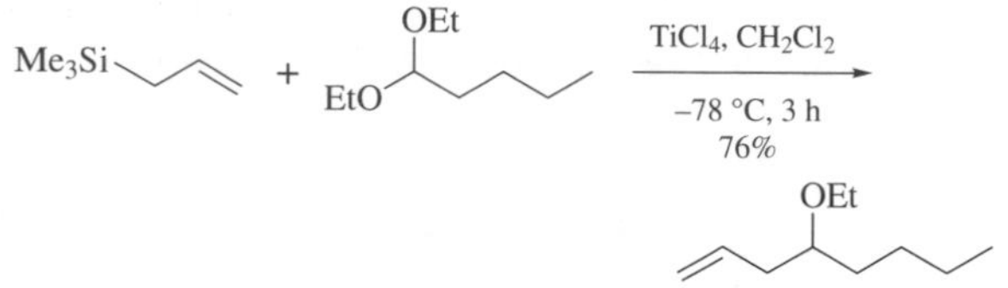

(3)

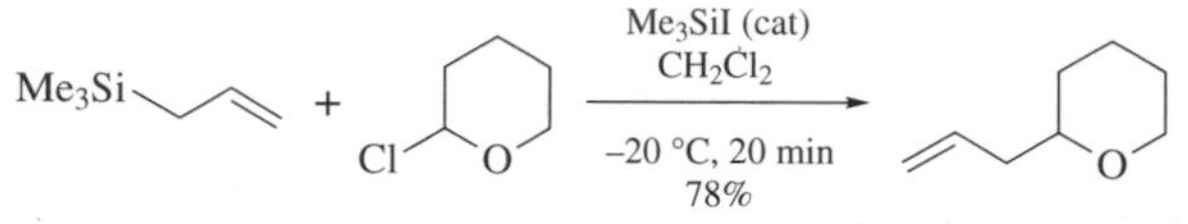

(4)

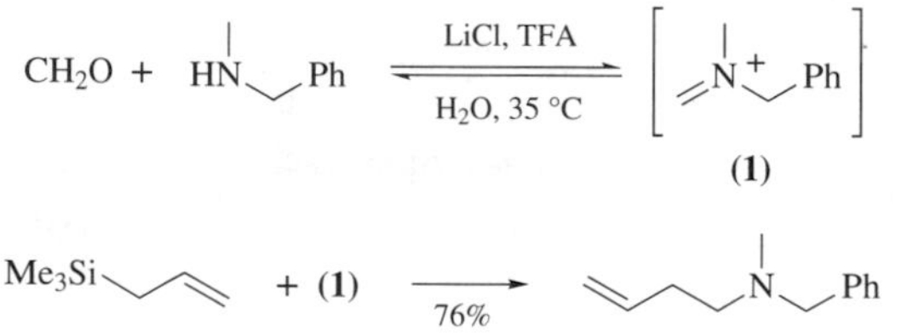

(5)

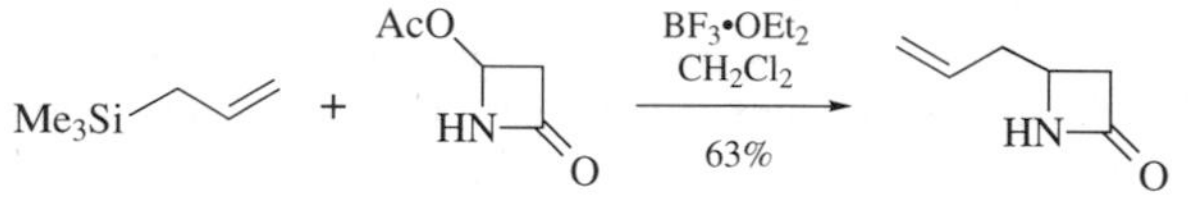

(6)

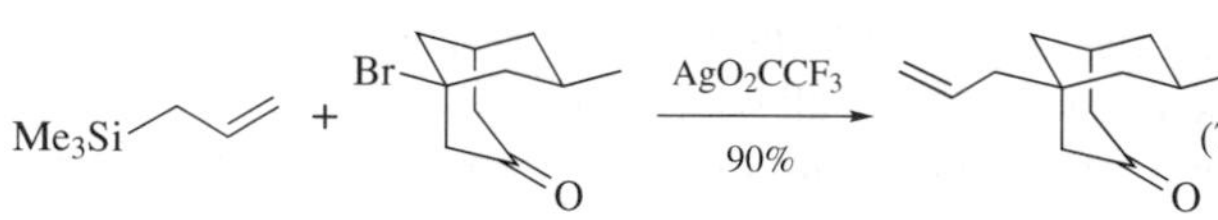

(7)

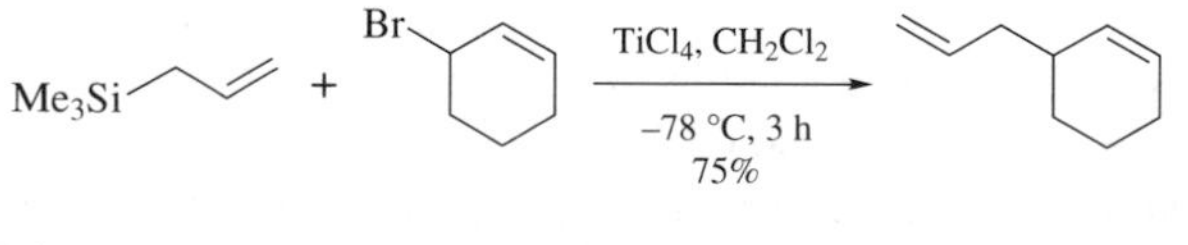

(8)

The reaction with acetals does not always need the acetal itself to be synthesized: it can be made in the same flask by mixing the allylsilane, the aldehyde, the silyl ether of the alcohol, and a catalytic amount of an acid such as ***Diphenylboryl Trifluoromethanesulfonate***,[15] ***Trimethylsilyl Trifluoromethanesulfonate***,[16] or ***Fluorosulfuric Acid***.[17]

The further reaction of the double bond in the first-formed product is an occasional complication, as in the formation of tetrahydropyrans from aldehydes when ***Aluminum Chloride*** is the Lewis acid (eq 9),[18] and of piperidines from primary amines and formaldehyde (eq 10).[11]

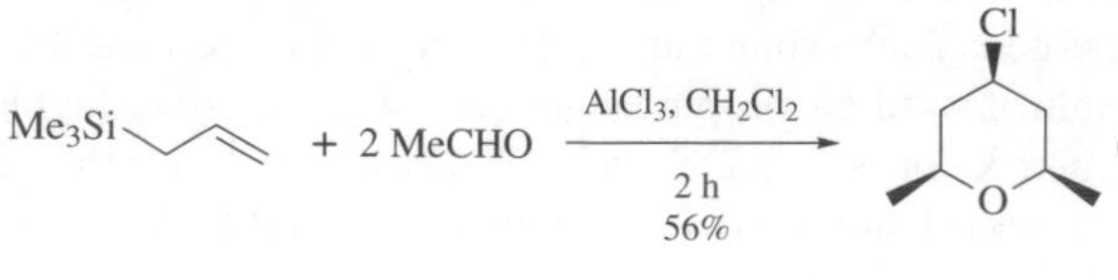

(9)

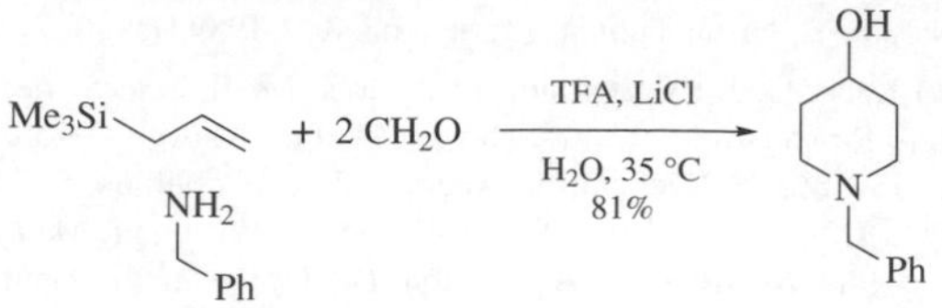

(10)

α,β-Unsaturated esters, aldehydes, ketones, and nitriles generally react in Michael fashion in what is called a Sakurai reaction (eq 11).[19] The intermediate silyl enol ether may be treated with a second electrophile to set up two C–C bonds in one operation (eq 12).[20] Occasionally the silyl group is not lost from the intermediate cation but migrates instead to give a cyclopentannulation byproduct (eq 13).[21]

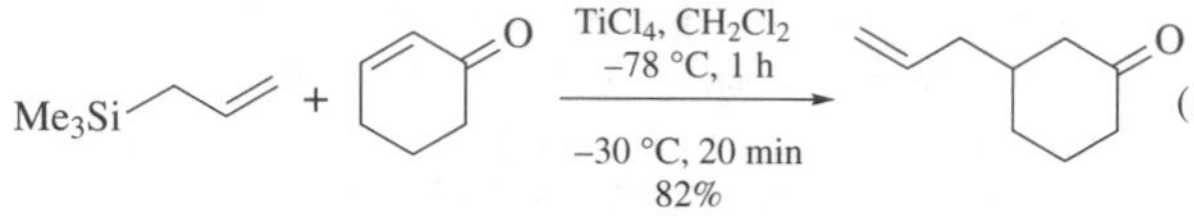

(11)

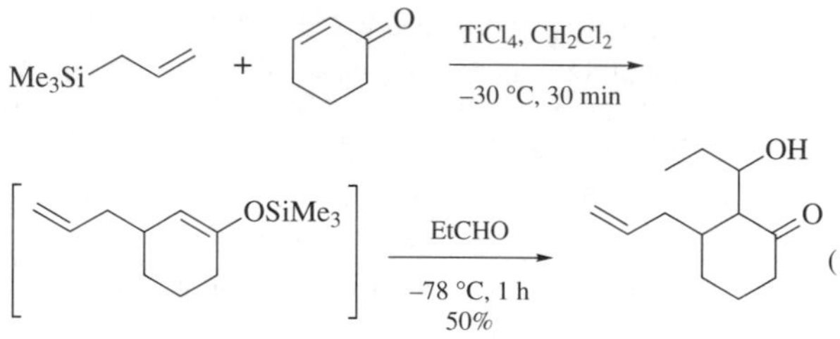

(12)

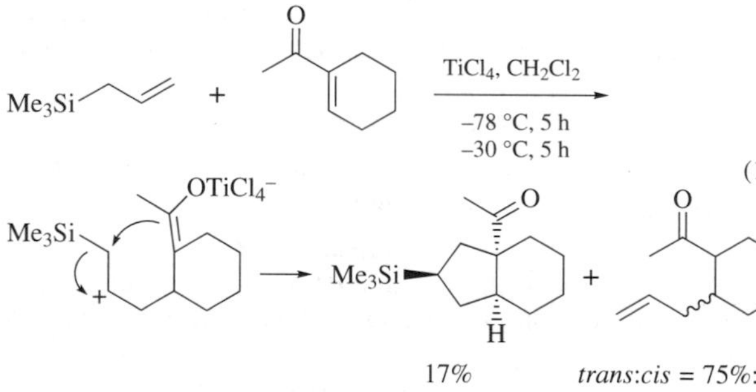

(13)

α,β-Unsaturated nitro compounds initially give nitronic acid derivatives, which can be reduced directly with ***Titanium(III) Chloride*** to give the corresponding ketone (eq 14).[22]

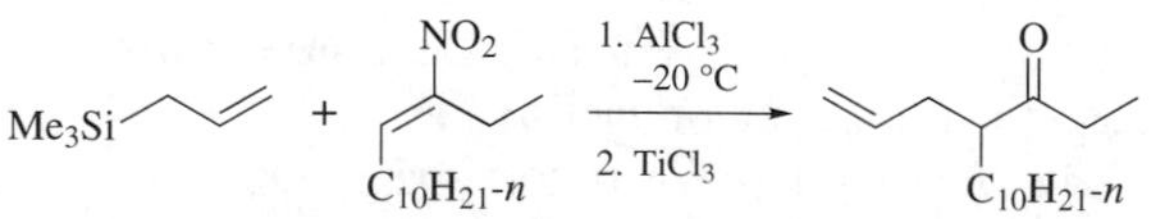

(14)

Some electrophiles require a separate activation step, as in the stereospecific reaction of alkenes with benzenesulfenyl chloride (eq 15).[23] The intermediate that actually reacts with the allylsilane is presumably the episulfonium ion. There is a corresponding reaction of epioxonium ions derived from 2-bromoethyl ethers.[24] Intramolecular hydrosilylation of an ester generates an acetal, which reacts with allyltrimethylsilane in the usual way and with high stereocontrol to make an *anti* 1,3-diol derivative (eq 16).[25]

The reactions with α- or β-oxygenated aldehydes can be made to give high levels of stereocontrol in either sense by choosing a chelating or nonchelating Lewis acid (eq 17).[26]

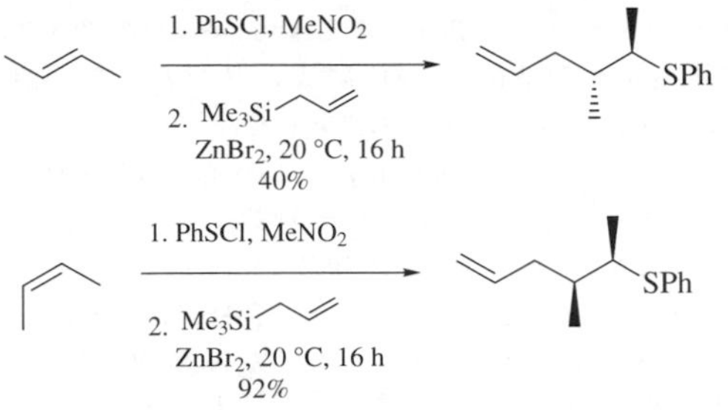

(15)

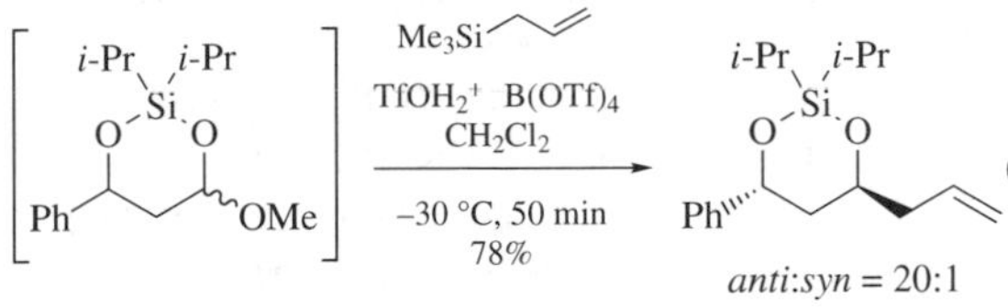

(16)

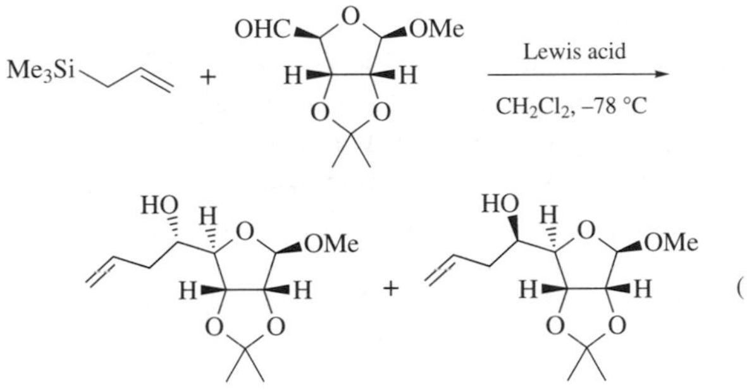

(17)

$TiCl_4$	20	:	1	89%
BF_3•OEt_2	1	:	90	80%

Sugar acetals (eq 18)[27] and glycal 3-acetates (eq 19)[28] and related compounds react stereoselectively in favor of axial attack at the anomeric carbon.

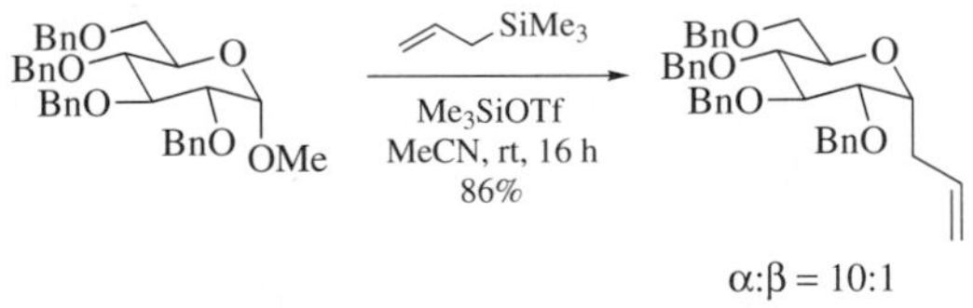

(18)

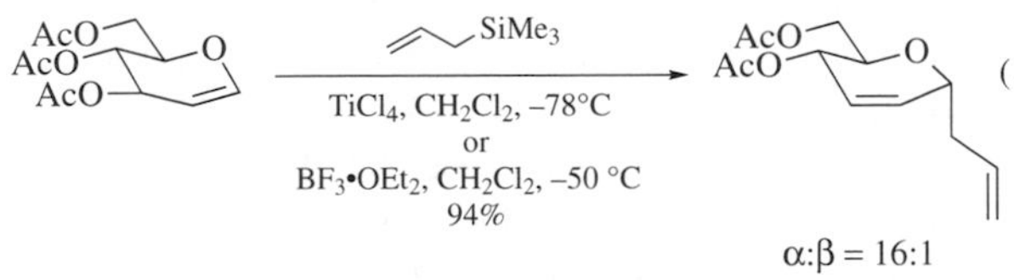

α:β = 10:1

(19)

α:β = 16:1

The choice of Lewis acid in all these reactions is often important in getting the best results, one Lewis acid being much the best in several cases. Most are used in molar amounts, but some, especially ***Iodotrimethylsilane***, ***Trimethylsilyl Trifluoromethanesulfonate***, and the more powerful trifluoromethanesulfoxonium tetrakis(trifluoromethanesulfonyl)boronate[29] $[TfOH_2{}^+B(OTf)_4{}^-]$, have the advantage that they can be used in catalytic quantities. In addition to the Lewis acids illustrated, the following less obvious Lewis acids have also been used with one or more of the electrophiles: ***Triphenylmethyl Perchlorate***,[30] titanocene ditriflate,[31] ***Lithium Perchlorate***,[32] ***Tin(IV) Chloride*** in the presence of either ***Tin(II) Trifluoromethanesulfonate***[33] or ***Zinc Chloride***,[34] ***Antimony(V) Chloride***,[35] ***Chlorotrimethylsilane*** combined with indium chloride,[36] ***Ethylaluminum Dichloride***,[37] and ***Diphenylboryl Trifluoromethanesulfonate***.[38] Among carbon electrophiles not illustrated are quinones,[39] cyclopropanedicarboxylic esters,[40] oxetanes,[41] nitriles,[42] dithioacetals,[43] diselenoacetals,[44] the intermediate sulfur-stabilized cation from a Pummerer rearrangement,[45] and propargyl ethers with[46] or without[47] octacarbonyldicobalt complexation.

As a Carbon Nucleophile in Uncatalyzed Reactions. Some electrophiles do not need Lewis acids, being already cationic and electrophilic enough to react with allyltrimethylsilane. Examples are the dithianyl cation (eq 20),[48] the tricarbonyl(cyclohexadienyl)iron cation (eq 21),[49] (π-allyl) tetracarbonyliron cations,[50] and ***Chlorosulfonyl Isocyanate*** (CSI).[51] Other reagents react directly by cycloaddition, but need further steps to achieve an overall electrophilic substitution, as in the reactions with nitrones (eq 22).[52]

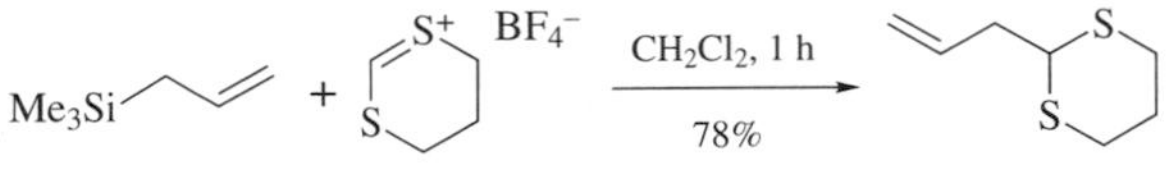

(20)

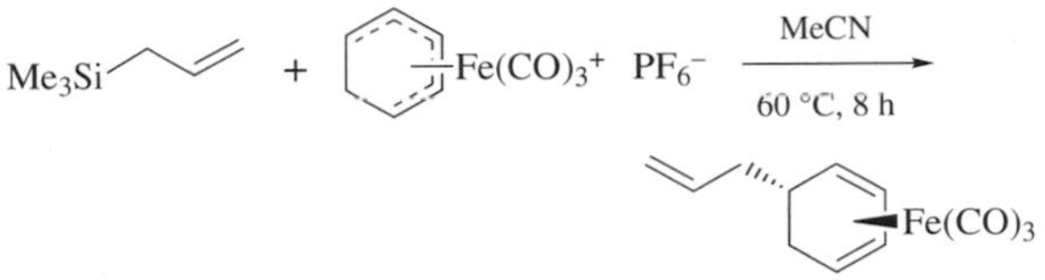

(21)

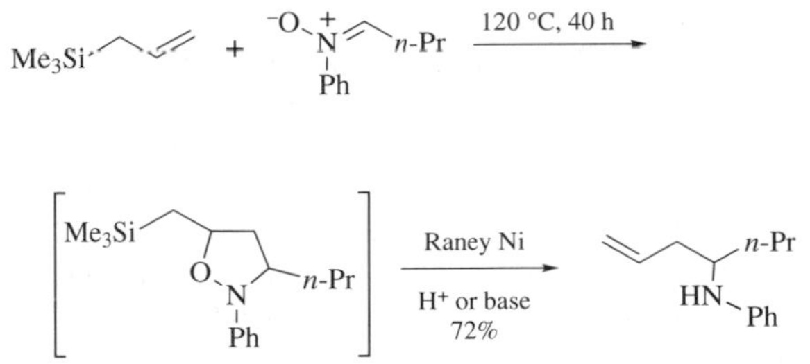

(22)

Although heteroatom (N-, P-, O-, S-, Se-, and halogen-based) electrophiles react with allylsilanes, the products with allyltrimethylsilane itself are usually too simple for this to be an important synthetic method. An exception perhaps is the reaction with ***Palladium(II) Chloride***, which gives the π-allylpalladium chloride cation.[53]

As a Carbon Nucleophile in Fluoride Ion-Catalyzed Reactions. The reactions with aldehydes, ketones (eq 23),[54] and α,β-unsaturated esters (eq 24)[55] can also be catalyzed by fluoride ion, usually introduced as ***Tetra-n-butylammonium Fluoride*** (TBAF), or other silicophilic ions such as alkoxide. These reactions produce silyl ether intermediates, which are usually hydrolyzed before workup. The stereochemistry of attack on chiral ketones can sometimes be different for the Lewis acid- and fluoride ion-catalyzed reactions.[56] In addition some electrophiles only react in the fluoride-catalyzed reactions, as with the addition to trinitrobenzene giving an allyl Meisenheimer 'complex'.[57]

Other Reactions. Allyltrimethylsilane reacts with some highly electrophilic alkenes, carbonyl compounds, azo compounds, and singlet oxygen to a greater or lesser extent in ene reactions that do not involve the loss of the silyl group, and hence give vinylsilanes in a solvent-dependent reaction (eq 25).[58]

Me_3Si + cyclohexanone — 1. TBAF, THF reflux, 16 h; 2. HCl, MeOH 60% → 1-allylcyclohexanol (OH) (23)

Me_3Si + 3-furyl-CH=CH-CO_2Et — TBAF, DMF, HMPA 83% → (24)

Me_3Si + N-phenyltriazolinedione (NPh) — solvent, 25 °C, 1 h → (25)

benzene	18	:	73	:	9
MeCN	6	:	28	:	67

Hydroalumination (and hydroboration) take place regioselectively to place the aluminum (or boron) atom at the terminus, creating a 3-trimethylsilylpropyl nucleophile.[59] Radicals attack allyltrimethylsilane at the terminus, and the intermediate radical reacts further without the trimethylsilyl group being expelled, as the corresponding germanium, tin, and lead groups are.[60] Treatment with strong bases gives trimethylsilylallyl–metal compounds (see ***Trimethylsilylallyllithium*** and ***Allyltrimethylsilylzinc Chloride***).

Related Reagents. Allyltributylstannane; Allyltrimethylsilylzinc Chloride; Crotyltributylstannane; Crotyltrichlorosilane; Crotyltrimethylsilane; Trimethylsilylallyllithium.

---

1. Fleming, I.; Dunoguès, J.; Smithers, R. *OR* **1989**, *37*, 57.
2. Delmulle, L.; van der Kelen, G. P. *J. Mol. Struct.* **1980**, *66*, 315.
3. Sarkar, T. K. *S* **1990**, 969 and 1101.
4. (a) Hagen, G.; Mayr, H. *JACS* **1991**, *113*, 4954. (b) Review: Mayr, H. *AG(E)* **1990**, *29*, 1371.
5. Yoshida, J.-I.; Murata, T.; Matsunaga, S.-I.; Tsuyoshi, M.; Shiozawa, S.; Isoe, S. *Rev. Heteroatom Chem.* **1991**, *5*, 193.
6. Pandey, G.; Rani, K. S.; Lakshmaiah, G. *TL* **1992**, *33*, 5107.
7. Sasaki, T.; Nakanishi, A.; Ohno, M. *JOC* **1982**, *47*, 3219.
8. Mukaiyama, T.; Nagaoka, H.; Murakami, M.; Oshima, M. *CL* **1985**, 977.
9. (a) Hosomi, A.; Endo, M.; Sakurai, H. *CL* **1976**, 941. (b) Review: Mukaiyama, T.; Murakami, M. *S* **1987**, 1043.
10. Sakurai, H.; Sakata, Y.; Hosomi, A. *CL* **1983**, 409.
11. Larsen, S. D.; Grieco, P. A.; Fobare, W. F. *JACS* **1986**, *108*, 3512.
12. Kraus, G. A.; Neuenschwander, K. *CC* **1982**, 134.
13. Kraus, G. A.; Hon, Y.-S. *JACS* **1985**, *107*, 4341.
14. Hosomi, A.; Imai, T.; Endo, M.; Sakurai, H. *JOM* **1985**, *285*, 95.
15. Mukaiyama, T.; Ohshima, M.; Miyoshi, N. *CL* **1987**, 1121.
16. Mekhalfia, A.; Marko, I. E. *TL* **1991**, *32*, 4779.
17. Lipshutz, B. H.; Burgess-Henry, J.; Roth, G. P. *TL* **1993**, *34*, 995.
18. Coppi, L. Ricci, A.; Taddei, M. *TL* **1987**, *28*, 973.
19. Hosomi, A.; Sakurai, H. *JACS* **1977**, *99*, 1673.
20. Hosomi, A.; Hashimoto, H.; Kobayashi, H.; Sakurai, H. *CL* **1979**, 245.
21. Pardo, R.; Zahra, J.-P.; Santelli, M. *TL* **1979**, *20*, 4557, with products reformulated in the light of Knölker, H. J.; Jones, P. G.; Pannek, J.-B. *SL* **1990**, 429 and Danheiser, R. L.; Dixon, B. R.; Gleason, R. W. *JOC* **1992**, *57*, 6094 and references therein.
22. Ochiai, M.; Arimoto, M.; Fujita, E. *TL* **1981**, *22*, 1115.
23. Alexander, R. P.; Paterson, I. *TL* **1983**, *24*, 5911.
24. Nishiyama, H.; Naritomi, T.; Sakuta, K.; Itoh, K. *JOC* **1983**, *48*, 1557.
25. Davis, A. P.; Hegarty, S. C. *JACS* **1992**, *114*, 2745.
26. Danishefsky, S. J.; DeNinno, M. P.; Phillips, G. B.; Zelle, R. L.; Lartey, P. A. *T* **1986**, *42*, 2809.
27. Hosomi, A.; Sakata, Y.; Sakurai, H. *TL* **1984**, *25*, 2383.
28. Danishefsky, S.; Kerwin, J. F. *JOC* **1982**, *47*, 3803.
29. Davis, A. P.; Jaspars, M. *JCS(P1)* **1992**, 2111.
30. Hayashi, M; Mukaiyama, T. *CL* **1987**, 289.
31. Hollis, T. K.; Robinson, N. P.; Whelan, J.; Bosnich, B. *TL* **1993**, *34*, 4309.
32. Pearson, W. H.; Schkeryantz, J. M. *JOC* **1992**, *57*, 2986.
33. Mukaiyama, T.; Shimpuku, T.; Takashima, T.; Kobayashi, S. *CL* **1989**, 145.
34. Hayashi, M.; Inubushi, A.; Mukaiyama, T. *BCJ* **1988**, *61*, 4037.
35. Mukaiyama, T.; Takenoshita, H.; Yamada, M.; Soga, T. *CL* **1990**, 1259.
36. Mukaiyama, T.; Ohno, T.; Nishimura, T.; Han, J. S.; Kobayashi, S. *CL* **1990**, 2239.
37. Simpkins, N. S. *T* **1991**, *47*, 323.
38. Mukaiyama, T.; Ohshima, M.; Miyoshi, N. *CL* **1987**, 1121.
39. (a) Hosomi, A.; Sakurai, H. *TL* **1977**, 4041. (b) Hosomi, A.; Sakurai, H. *TL* **1978**, 2589.
40. Bambal, R.; Kemmitt, R. D. W. *CC* **1988**, 734.
41. Carr, S. A.; Weber, W. P. *JOC* **1985**, *50*, 2782.
42. Hamana, H.; Sugasawa, T. *CL* **1985**, 921.
43. Mori, I.; Bartlett, P. A.; Heathcock, C. H. *JACS* **1987**, *109*, 7199.
44. Hermans, B.; Hevesi, L. *TL* **1990**, *31*, 4363.
45. Mori, I.; Bartlett, P. A.; Heathcock, C. H. *JOC* **1990**, *55*, 5966.
46. Schreiber, S. L.; Klimas, M. T.; Sammakia, T. *JACS* **1987**, *109*, 5749.
47. Hayashi, M.; Inubushi, A.; Mukaiyama, T. *CL* **1987**, 1975.
48. Hallberg, A.; Westerlund, C. *CL* **1982**, 1993.
49. Kelly, L. F.; Narula, A. S.; Birch, A. J. *TL* **1980**, *21*, 871.
50. Li, Z.; Nicholas, K. M. *JOM* **1991**, *402*, 105.
51. (a) Grignon-Dubois, M.; Pillot, J.-P.; Dunoguès, J.; Duffaut, N.; Calas, R.; Henner, B. *JOM* **1977**, *124*, 135. (b) Colvin, E. W.; Montieth, M. *CC* **1990**, 1230.
52. Hosomi, A.; Shoji, H.; Sakurai, H. *CL* **1985**, 1049.
53. (a) Kliegman, J. M. *JOM* **1971**, *29*, 73. (b) Yamamoto, K.; Shinohara, K.; Ohuchi, T.; Kumada, M. *TL* **1974**, 1153.
54. Hosomi, A., Shirahata, A.; Sakurai, H. *TL* **1978**, 3043.
55. Majetich, G.; Casares, A. M.; Chapman, D.; Behnke, M. *JOC* **1986**, *51*, 1745.
56. Taniguchi, M.; Oshima, K.; Utimoto, K. *CL* **1992**, 2135.
57. Artamkina, G. A.; Kovalenko, S. V.; Beletskaya, I. P.; Reutov, O. A. *JOM* **1987**, *329*, 139.
58. (a) Ohashi, S.; Ruch, W. E.; Butler, G. B. *JOC* **1981**, *46*, 614. (b) Review: Dubac, J.; Laporterie, A. *CRV* **1987**, *87*, 319.

59. Maruoka, K.; Sano, H.; Shinoda, K.; Nakai, S.; Yamamoto, H. *JACS* **1986**, *108*, 6036.
60. Light, J. P.; Ridenour, M.; Beard, L.; Hershberger, J. W. *JOM* **1987**, *326*, 17.

Ian Fleming
Cambridge University, UK

(*S*)-1-Amino-2-methoxymethylpyrrolidine[1]

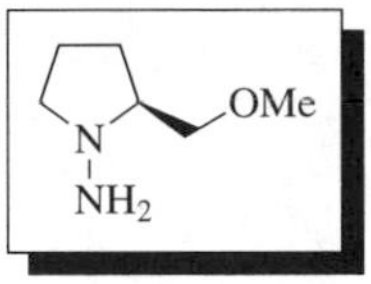

[59983-39-0] $C_6H_{14}N_2O$ (MW 130.19)

(chiral auxiliary; Enders' reagent; diastereo- and/or enantioselective alkylations,[2,3] aldol reactions,[4] Michael additions[5] and reductive or alkylative aminations,[6] resolutions,[7] ee determinations[8])

Alternate Name: SAMP.
Physical Data: bp 186–187 °C; *d* 0.977 g cm^{-3}; n_D^{20} 1.4650; α_D^{20} −80 to −82° (neat).
Solubility: sol H_2O, ether, dichloromethane.
Form Supplied in: colorless liquid or as crystalline colorless oxalate.
Handling, Storage, and Precautions: storage at 0–4 °C under argon atmosphere.

Since the pioneering times of the mid-1970s, (*S*)-1-amino-2-methoxymethylpyrrolidine (SAMP) and its enantiomer RAMP have been among the most powerful chiral auxiliaries in asymmetric synthesis, with a very broad range of applications. As a proline derivative it generally shows high stereoselectivities due to the rigidity of the five-membered ring and the ability to coordinate metal fragments[9] (see also ***(S)-2-Methoxymethylpyrrolidine***, SMP).

The procedure involves the transformation of carbonyl compounds to the corresponding SAMP or RAMP hydrazones, metalation, trapping of the intermediate azaenolates with various electrophiles, and either hydrazone cleavage (carbonyl compounds) or hydrazone reduction/N–N bond cleavage (amines).

The synthetic utility of the SAMP/RAMP hydrazone method is demonstrated in particular in the stereoselective alkylation of aldehyde[2] and ketone[3] SAMP/RAMP hydrazones. A great number of natural products have been synthesized using this method, like the principal alarm pheromone of the leaf cutting ant *Atta texana* (eq 1),[3a] the C(1)–C(15) segment of FK 506 (eq 2),[2b] the amino acid MeBMT (eq 3),[2c] and (−)-methyl kolavenate (eq 4).[3b]

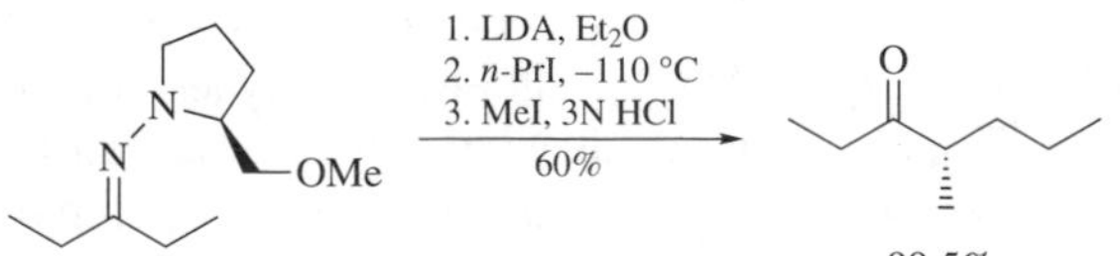

1. LDA, Et_2O
2. *n*-PrI, −110 °C
3. MeI, 3N HCl
60%

OMe

99.5% ee (1)

1. LDA, THF
2. MeI
3. cleavage
65%

OMe

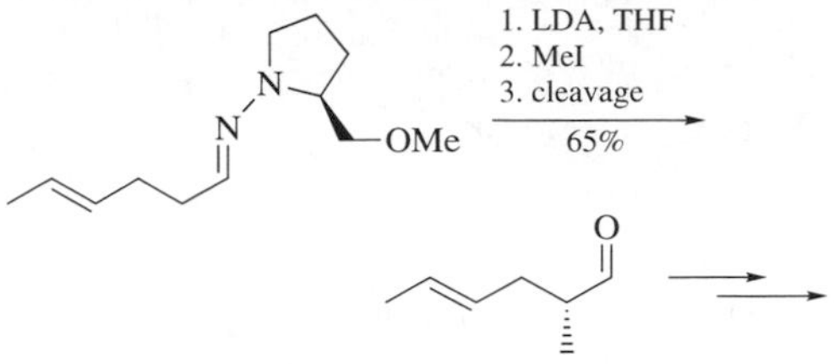

MeBMT (2)

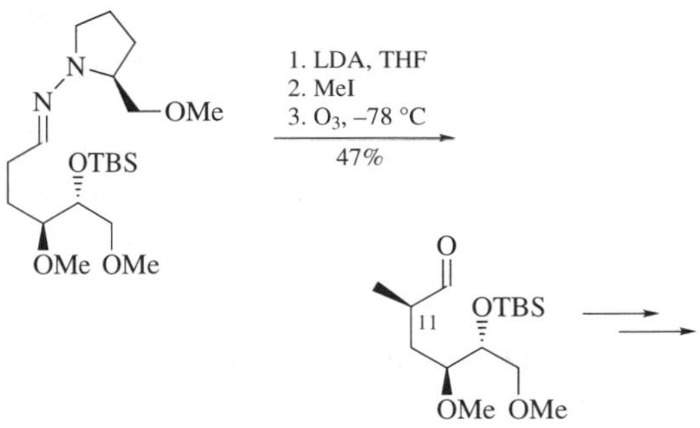

1. LDA, THF
2. MeI
3. O_3, −78 °C
47%

OMe OTBS OMe OMe

11 OTBS OMe OMe FK 506 (3)

1. LDA, THF
2. MeOTs
3. MeI, 2N HCl
no yield

OMe

ee >98%

CO_2Me

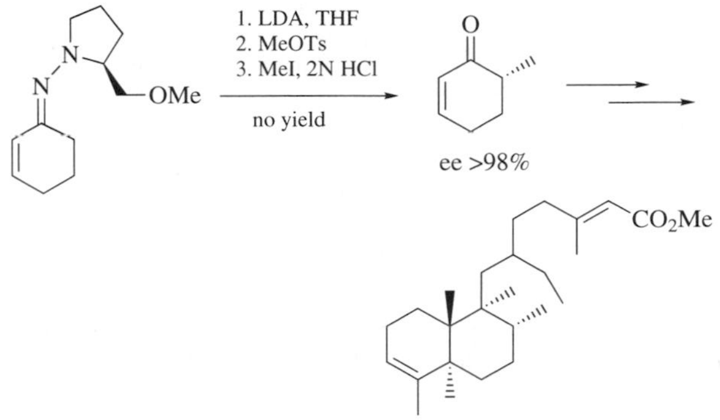

(4)

(−)-Methyl kolavenate

2,2-Dimethyl-1,3-dioxan-5-one SAMP/RAMP hydrazones[3f–j] were used as dihydroxyacetonephosphate equivalents in the synthesis of C_2 symmetric ketones (eq 5),[3g] aza sugars with novel substitution patterns,[3h] or C_5 to C_9 deoxy sugars.[3i] SAMP hydrazones of 2-oxo esters represent novel phosphoenolpyruvate (PEP) equivalents.[3k,l] α,α-Disubstituted spiroacetals are accessible via the alkylation of ketone SAMP/RAMP hydrazones.[3m]

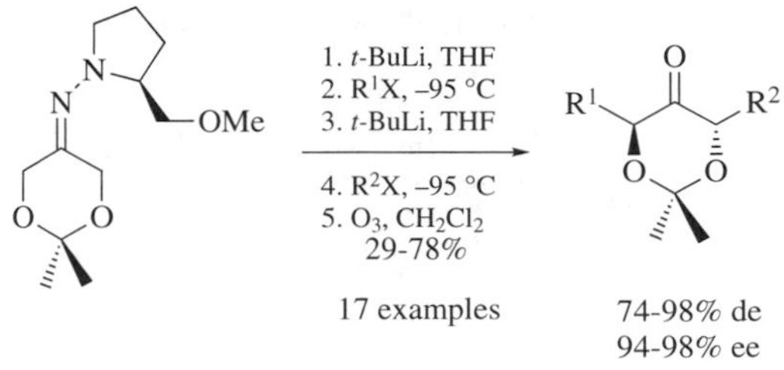

1. *t*-BuLi, THF
2. R^1X, −95 °C
3. *t*-BuLi, THF
4. R^2X, −95 °C
5. O_3, CH_2Cl_2
29-78%

17 examples

OMe

R^1 R^2

74-98% de
94-98% ee (5)

The aggregation pheromone of *Drosophila mulleri*, (*S*)-2-tridecanol acetate, is obtainable by alkylation of propiophenone SAMP hydrazone followed by a Baeyer–Villiger reaction of the ketone (eq 6).[3n]

The relative and absolute configuration of Stigmatellin A, one of the most potent inhibitors of the electron transport chain, was determined via alkylation of diethyl ketone SAMP hydrazone.[3o]

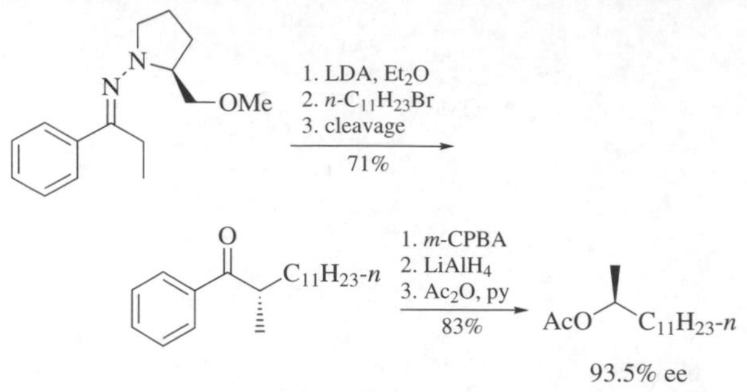

1. LDA, Et_2O
2. n-$C_{11}H_{23}Br$
3. cleavage
71%

1. m-CPBA
2. $LiAlH_4$
3. Ac_2O, py
83%

AcO $C_{11}H_{23}$-n (6)

93.5% ee

The aldol reaction is the preferred method for the stereoselective synthesis of 1,3-dioxygenated building blocks. In 1978, Enders et al.[4a,b] reported the first enantioselective aldol reaction in the difficult case of α-unsubstituted β-ketols using SAMP and RAMP. Diastereo- and enantiomerically pure *syn*-β-ketols are available by aldol reaction of SAMP/RAMP hydrazones (eq 7).[4c]

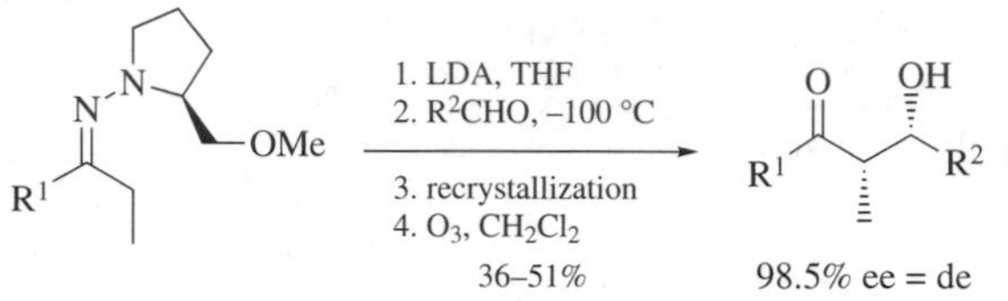

1. LDA, THF
2. R^2CHO, −100 °C
3. recrystallization
4. O_3, CH_2Cl_2
36–51%

(7)

98.5% ee = de

The aggregation pheromone of the rice and maize weevil was synthesized by aldol reaction of an enantiomerically pure α-silyl ketone, obtained by the SAMP/RAMP hydrazone method,[10a–d] with various aldehydes (eq 8).[4d]

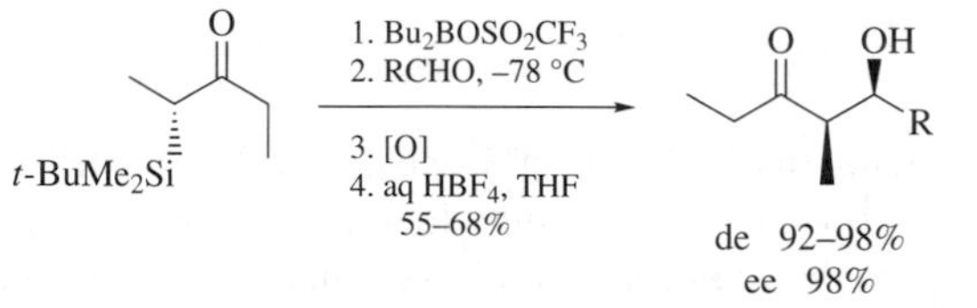

t-$BuMe_2Si$

1. $Bu_2BOSO_2CF_3$
2. RCHO, −78 °C
3. [O]
4. aq HBF_4, THF
55–68%

(8)

de 92–98%
ee 98%

The utility of the SAMP/RAMP hydrazone method in diastereo- and enantioselective Michael additions was demonstrated in the synthesis of 5-oxo esters[5a–e] (eq 9),[5b] δ-lactones (eq 10),[5c,e,f] oxo diesters and dinitriles,[5g] heterocyclic compounds (eq 11),[5h] MIRC (Michael initiated ring closure) reactions,[5i] and 2-substituted 4-oxo sulfones.[5i]

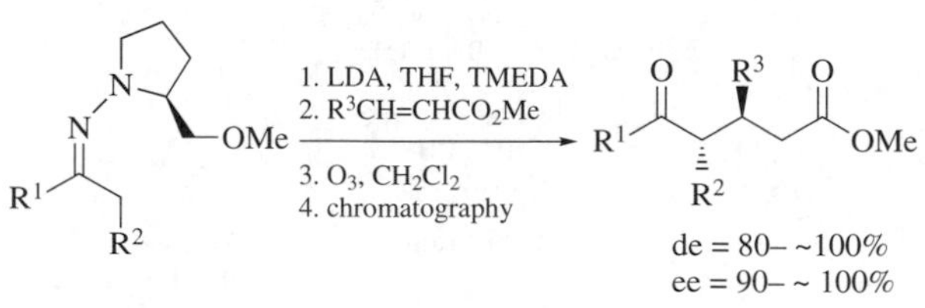

1. LDA, THF, TMEDA
2. $R^3CH{=}CHCO_2Me$
3. O_3, CH_2Cl_2
4. chromatography

(9)

de = 80– ~100%
ee = 90– ~ 100%

Organotin reagents can be added to cyclic α,β-unsaturated SAMP/RAMP hydrazones (eq 12).[5k]

Lithiated N-protected SAMP can be used as an ammonia equivalent in Michael and tandem Michael additions to α,β-unsaturated esters (eq 13).[5l] Furthermore, SAMP and RAMP can be employed for the asymmetric synthesis of α- and/or β-substituted primary amines with high regio-, diastereo-, and enantioselectivities.[6] This variant involves hydrazone reduction with a subsequent N–N bond cleavage and can be combined with a prior α-alkylation, as described above.

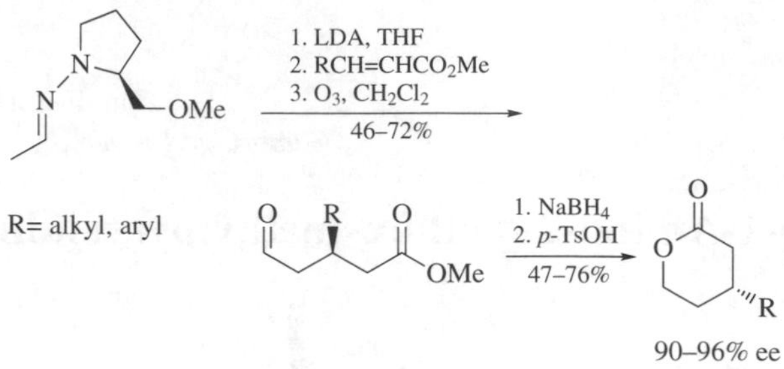

1. LDA, THF
2. $RCH{=}CHCO_2Me$
3. O_3, CH_2Cl_2
46–72%

R= alkyl, aryl

1. $NaBH_4$
2. p-TsOH
47–76%

(10)

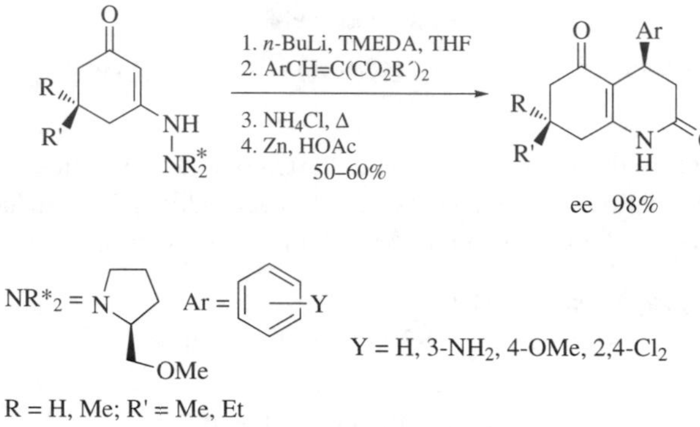

90–96% ee

1. n-BuLi, TMEDA, THF
2. $ArCH{=}C(CO_2R')_2$
3. NH_4Cl, Δ
4. Zn, HOAc
50–60%

(11)

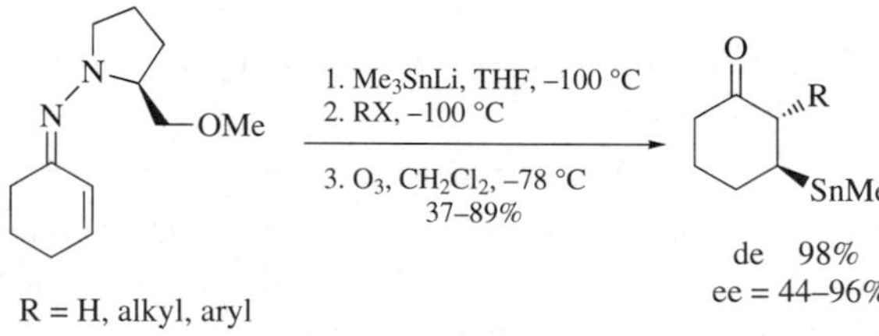

ee 98%

NR^*_2 = … Ar = … Y = H, 3-NH_2, 4-OMe, 2,4-Cl_2

R = H, Me; R' = Me, Et

1. Me_3SnLi, THF, −100 °C
2. RX, −100 °C
3. O_3, CH_2Cl_2, −78 °C
37–89%

(12)

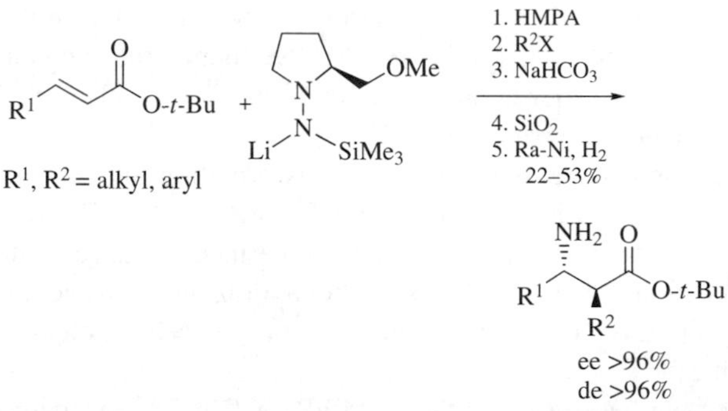

de 98%
ee = 44–96%

R = H, alkyl, aryl

1. HMPA
2. R^2X
3. $NaHCO_3$
4. SiO_2
5. Ra-Ni, H_2
22–53%

R^1, R^2 = alkyl, aryl

(13)

ee >96%
de >96%

This method was recently used in the synthesis of different natural products, like the ladybug defence alkaloid harmonine,[6d] α- and β-amino acetals and acids (eq 14),[6e,f] and both enantiomers of the hemlock alkaloid coniine,[6g] utilizing the nucleophilic 1,2-addition of organolithium and -lanthanoid reagents to SAMP/RAMP hydrazones.

The alkylation of SAMP/RAMP hydrazones with heteroelectrophiles leads to enantiomerically pure α-silyl aldehydes and ketones (eq 15),[10a–d] α-sulfenyl aldehydes and ketones (eq 16),[10e] and α-hydroxy aldehydes and ketones (eq 17).[10f]

These very interesting chiral building blocks are employed in aldol reactions,[4d] and in the synthesis of enantiomerically pure

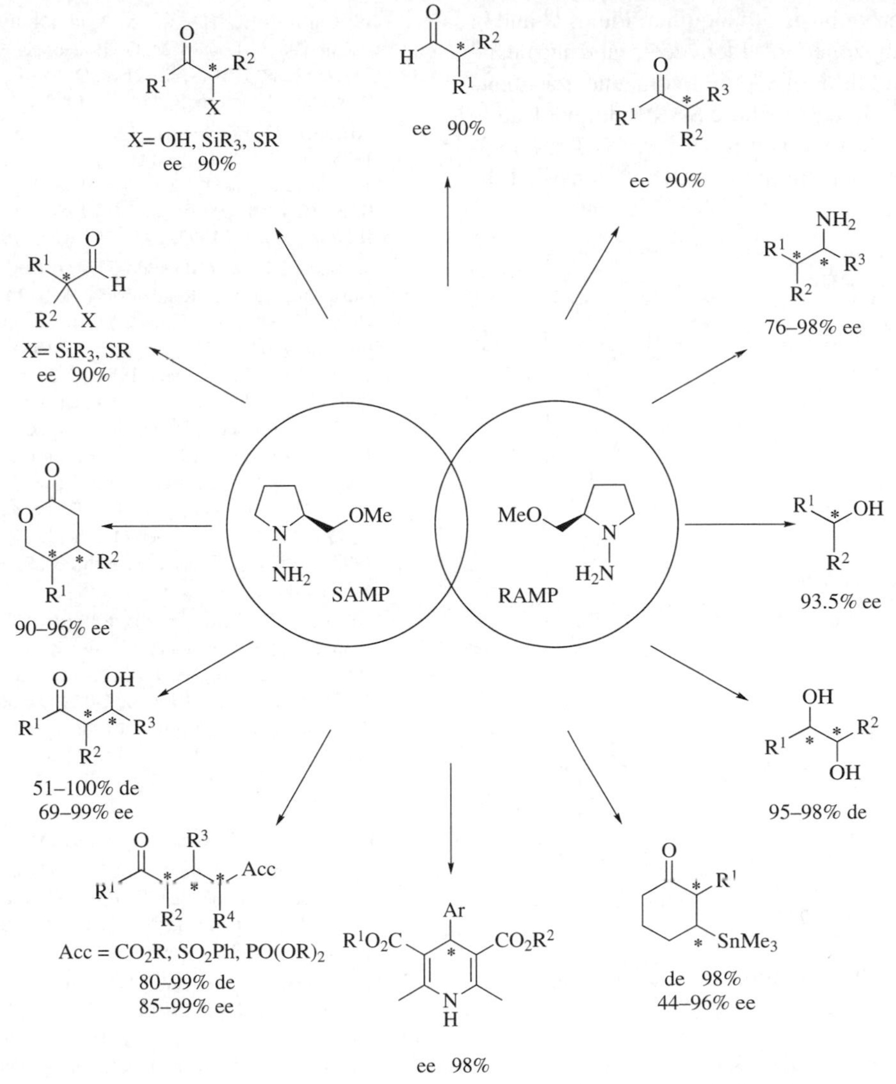

X= OH, SiR_3, SR
ee 90%

ee 90%

ee 90%

76–98% ee

X= SiR_3, SR
ee 90%

SAMP

RAMP

90–96% ee

93.5% ee

51–100% de
69–99% ee

95–98% de

Acc = CO_2R, SO_2Ph, $PO(OR)_2$
80–99% de
85–99% ee

ee 98%

de 98%
44–96% ee

Scheme 1

vicinal diols (eq 18)[10g,h] and 3-oxo esters and acids bearing quarternary stereogenic centers (eq 19).[10i]

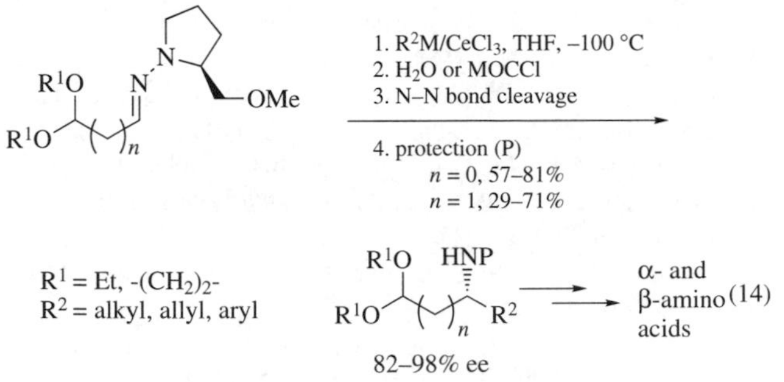

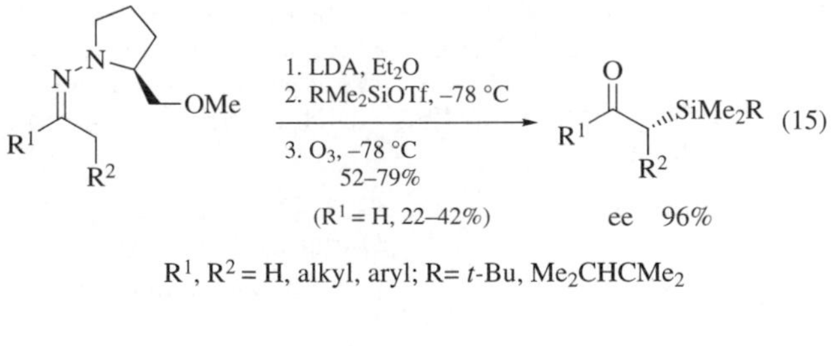

1. $R^2M/CeCl_3$, THF, −100 °C
2. H_2O or MOCCl
3. N–N bond cleavage
4. protection (P)
n = 0, 57–81%
n = 1, 29–71%

R^1 = Et, $-(CH_2)_2-$
R^2 = alkyl, allyl, aryl

82–98% ee

α- and β-amino acids (14)

1. LDA, Et_2O
2. RMe_2SiOTf, −78 °C
3. O_3, −78 °C
52–79%
(R^1 = H, 22–42%)

ee 96% (15)

R^1, R^2 = H, alkyl, aryl; R= t-Bu, Me_2CHCMe_2

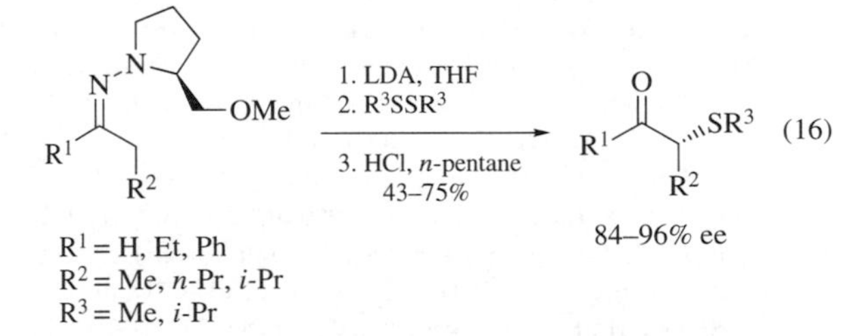

1. LDA, THF
2. R^3SSR^3
3. HCl, n-pentane
43–75%

(16)

R^1 = H, Et, Ph
R^2 = Me, n-Pr, i-Pr
R^3 = Me, i-Pr

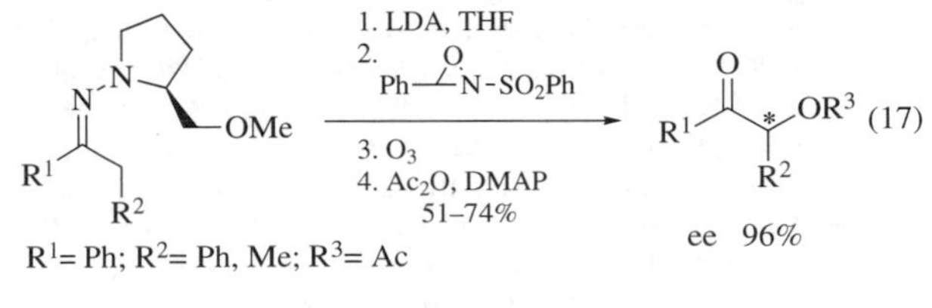

84–96% ee

1. LDA, THF
2. Ph–oxaziridine–SO_2Ph
3. O_3
4. Ac_2O, DMAP
51–74%

(17)

R^1= Ph; R^2= Ph, Me; R^3= Ac

ee 96%

Further applications can be mentioned briefly. SAMP was used in the resolution of 4-demethoxy-7-deoxydaunomycinone,[7] in ee determinations (Scheme 1),[8] as a chelate for tetracarbonylmolybdenum complexes,[11] in intramolecular Heck reactions,[12] as polysilylated hydrazine,[13] in the enantioselective synthesis of isoquinuclidines,[14] and in the conversion of hydrazones to aldehydes[15] and nitriles.[16] The structure of a chiral lithium SAMP hydrazone azaenolate has been determined.[17] In cases where SAMP did not lead to satisfactory inductions, a modified auxiliary, (*S*)-1-amino-2-dimethylmethoxymethylpyrrolidine (SADP),[18] enhanced the stereochemical control.

1. LDA, THF
2. $Me_2(O\text{-}i\text{-}Pr)SiCl$, –78 °C
3. O_3, –78 °C
4. L-Selectride
5. H_2O_2, KF, $KHCO_3$
14–48% (18)

1. LDA, THF
2. RI
72–99%

1. base, THF
2. MOCCl, –100 °C
3. cleavage
76–95% (19)

81–96% de

90–98% ee

X= SPh, R_3Si
R= alkyl

Related Reagents. (*S*)-2-Methoxymethylpyrrolidine.

1. Reviews (literature up to 1987): (a) Enders, D. In *Asymmetric Synthesis*; Morrison, J. D., Ed.; Academic Press: New York, 1984; Vol. 3, p 275. (b) Enders, D.; Fey, P.; Kipphardt, H. *OS* **1987**, *65*, 173, 183.
2. (a) Nicolaou, K. C.; Papahatjis, D. P.; Claremon, D. A.; Dolle, R. E. *JACS* **1981**, *103*, 6967. (b) Kocienski, P.; Stocks, M.; Donald, D.; Cooper, M.; Manners, A. *TL* **1988**, *29*, 4481. (c) Beulshausen, T.; Groth, U. M.; Schöllkopf, U. *JACS* **1994**, in press. (d) Enders, D.; Dyker, H. *LAC* **1990**, 1107. (e) Schmidt, U.; Siegel, W.; Mundinger, K. *TL* **1988**, *29*, 1269. (f) Hauck, R. S.; Wegner, C.; Blumtritt, P.; Fuhrhop, J. H.; Nau, H. *Life Sci.* **1990**, *46*, 513. (g) Kündig, P.; Liu, R.; Ripa, A. *HCA* **1992**, *75*, 2657.
3. (a) Enders, D.; Eichenauer, H. *AG* **1979**, *91*, 425; *AG(E)* **1979**, *18*, 397. (b) Hideo, I.; Mitsugu, M.; Kimikazu, O.; Tokoroyama, T. *CC* **1987**, 358. (c) Pennanen, S. I. *ACS* **1981**, *B35*, 555. (d) Mori, K.; Nomi, H.; Chuman, T.; Kohno, M.; Kato, K.; Noguchi, M. *T* **1982**, *38*, 3705. (e) Fischer, J.; Kilpert, C.; Klein, U.; Steglich, W. *T* **1986**, *42*, 2063. (f) Enders, D.; Bockstiegel, B. *S* **1989**, 493. (g) Enders, D.; Gatzweiler, W.; Jegelka, U. *S* **1991**, 1137. (h) Enders, D.; Jegelka, U. *SL* **1992**, 999. (i) Enders, D.; Jegelka, U.; Dücker, B. *AG* **1993**, *105*, 423; *AG(E)* **1993**, *32*, 423. (j) Enders, D.; Jegelka, U. *TL* **1993**, *34*, 2453. (k) Enders, D.; Dyker, H.; Raabe, G. *AG* **1992**, *104*, 649; *AG(E)* **1992**, *31*, 618. (l) Enders, D.; Dyker, H.; Raabe, G. *SL* **1992**, 901. (m) Enders, D.; Gatzweiler, W.; Dederichs, E. *T* **1990**, *46*, 4757. (n) Enders, D.; Plant, A. *LA* **1991**, 1241. (o) Enders, D.; Osborne, S. *CC* **1993**, 424. (p) Sainsbury, M.; Williams, C. S.; Naylor, A.; Scopes, D. I. C. *TL* **1990**, *31*, 2763. (q) Sainsbury, M.; Mahon, M. F.; Williams, C. S.; Naylor, A.; Scopes, D. I. C. *T* **1991**, *47*, 4195. (r) Ziegler, F. E.; Becker, M. R. *JOC* **1990**, *55*, 2800. (s) Warshawsky, A. M.; Meyers, A. I. *JACS* **1990**, *112*, 8090. (t) Andersen, M. W.; Hildebrandt, B.; Hoffmann, R. W. *AG* **1991**, *103*, 90; *AG(E)* **1991**, *30*, 90. (u) Clark, J. S.; Holmes, A. B. *TL* **1988**, *29*, 4333. (v) Carling, R. W.; Curtis, N. R.; Holmes, A. B. *TL* **1989**, *30*, 6081. (w) Curtis, N. R.; Holmes, A. B.; Looney, M. G.; Pearson, N. D.; Slim, G. C. *TL* **1991**, *32*, 537. (x) Hart, T. W.; Guillochon, D.; Perrier, G.; Sharp, B. W.; Toft, M. P.; Vacher, B.; Walsh, R. J. A. *TL* **1992**, *33*, 7211.
4. (a) Enders, D.; Friedrich, E.; Lutz, W.; Pieter, R. *AG* **1978**, *90*, 219; *AG(E)* **1978**, *17*, 206. (b) Enders, D.; Eichenauer, H.; Pieter, R. *CB* **1979**, *112*, 3703. (c) Enders, D. *Chem. Scr.* **1985**, *25*, 139. (d) Enders, D.; Lohray, B. B. *AG* **1988**, *100*, 594; *AG(E)* **1988**, *27*, 581. (e) Enders, D.; Dyker, H.; Raabe, G. *AG* **1993**, *105*, 420; *AG(E)* **1993**, *32*, 421.
5. (a) Enders, D.; Papadopoulos, K. *TL* **1983**, *24*, 4967. (b) Enders, D.; Papadopoulous, K.; Rendenbach, B. E. M.; Appel, R.; Knoch, F. *TL* **1986**, *27*, 3491. (c) Enders, D.; Rendenbach, B. E. M. *Pestic. Sci. Biotechnol., Proc. Int. Congr. Pestic. Chem., 6th* **1986**, 17. (d) Enders, D.; Rendenbach, B. E. M. *T* **1986**, *42*, 2235. (e) Enders, D.; Rendenbach, B. E. M. *CB* **1987**, *120*, 1223. (f) Tietze, L. F.; Schneider, C. *JOC* **1991**, *56*, 2476. (g) Enders, D.; Demir, A. S.; Rendenbach, B. E. M. *CB* **1987**, 1731. (h) Enders, D.; Demir, A. S.; Puff, H.; Franken, S. *TL* **1987**, *28*, 3795. (i) Enders, D.; Scherer, H. J.; Raabe, G. *AG* **1991**, *103*, 1676; *AG(E)* **1991**, *30*, 1664. (j) Enders, D., Papadopoulos, K.; Herdtweck, E. *T* **1993**, *49*, 1821. (k) Enders, D.; Heider, K. *AG* **1993**, *105*, 592; *AG(E)* **1993**, *32*, 598. (l) Enders, D.; Wahl, H.; Bettray, W. *AG* **1995**, *107*, 527; *AG(E)* **1995**, *34*, 455.
6. (a) Enders, D.; Schubert, H.; Nübling, C. *AG* **1986**, *98*, 1118; *AG(E)* **1986**, *25*, 1109. (b) Denmark, S. E.; Weber, T.; Piotrowski, D. W. *JACS* **1987**, *109*, 2224. (c) Weber, T.; Edwards, J. P.; Denmark, S. E. *SL* **1989**, 20. (d) Enders, D.; Bartzen, D. *LAC* **1991**, 569. (e) Enders, D.; Funk, R.; Klatt, M.; Raabe, G.; Hovestreydt, E. R. *AG* **1993**, *105*, 418; *AG(E)* **1993**, *32*, 418. (f) Enders, D.; Klatt, M.; Funk, R. *SL* **1993**, 226. (g) Enders, D.; Tiebes, J. *LAC* **1993**, 173. (h) Denmark, S. E.; Nicaise, O. *SL* **1993**, 359.
7. Dominguez, D.; Ardecky, R. J.; Cava, M. P. *JACS* **1983**, *105*, 1608.
8. (a) Günther, K.; Martens, J.; Messerschmidt, M. *J. Chromatogr.* **1984**, *288*, 203. (b) Effenberger, F.; Hopf, M.; Ziegler, T.; Hudelmeyer, J. *CB* **1991**, *124*, 1651. (c) Harden, R. C.; Rackham, D. M. *J. High Resolut. Chromatogr.* **1992**, *15*, 407.
9. (a) Enders, D.; Eichenauer, H. *AG* **1976**, *93*, 579. (b) Enders, D.; Fey, P.; Kipphardt, H. *OPP* **1985**, *17*, 1.
10. (a) Enders, D.; Bhushan, B. B. *AG* **1987**, *99*, 359; *AG(E)* **1987**, *26*, 351. (b) Enders, D.; Bhushan, B. B. *AG* **1988**, *100*, 594; *AG(E)* **1988**, *27*, 581. (c) Bhushan, B. B.; Enders, D. *HCA* **1989**, *72*, 980. (d) Bhushan, B. B.; Zimbinski, R. *TL* **1990**, *31*, 7273. (e) Enders, D.; Schäfer, T. publication in preparation. (f) Enders, D.; Bhushan, V. *TL* **1988**, *29*, 2437. (g) Enders, D.; Nakai, S. *HCA* **1990**, *73*, 1833. (h) Enders, D.; Nakai, S. *CB* **1991**, *124*, 219. (i) Enders, D.; Zamponi, A.; Raabe, G. *SL* **1992**, 897.
11. Ehlers, J.; Tom Dieck, H. *Z. Anorg. Allg. Chem.* **1988**, *560*, 80.
12. Grigg, R.; Dorrity, M. J. R.; Malone, J. F.; Mongkolaussavaratana, T.; Norbert, W. D. J. A.; Sridharan, V. *TL* **1990**, *31*, 3075.
13. Hwu, J. R.; Wang, N. *T* **1988**, *44*, 4181.
14. (a) Mehmandoust, M.; Marazano, C.; Singh, R.; Cesario, M.; Fourrey, J. L.; Das, B. C. *TL* **1988**, *29*, 4423. (b) Genisson, Y.; Marazano, C.; Mehmandoust, M.; Gnecco, D.; Das, B. C. *SL* **1992**, 431.
15. Enders, D.; Bhushan V. *Z. Naturforsch., Teil B* **1987**, *42B*, 1595; Enders, D.; Plant, A. *SL* **1990**, 725.
16. (a) Moore, J. S.; Stupp, S. I. *JOC* **1990**, *55*, 3374. (b) Enders, D.; Plant, A. *SL* **1994**, 1054.
17. Enders, D.; Bachstädter, G.; Kremer, K. A. M.; Marsch, M.; Harms, K.; Boche, G. *AG* **1988**, *100*, 1580; *AG(E)* **1988**, *27*, 1522.
18. Enders, D.; Kipphardt, H.; Gerdes, P.; Breña-Valle, J.; Bushan, V. *Bull. Soc. Chim. Belg.* **1988**, *97*, 691. Applications: (a) Enders, D.; Müller, S.; Demir, A. S. *TL* **1988**, *29*, 6437. (b) Enders, D.; Dyker, H.; Raabe, G. *AG* **1993**, *105*, 420; *AG(E)* **1993**, *32*, 421. (c) Enders, D.; Bhushan, V. *TL* **1988**, *29*, 2437.

Dieter Enders & Martin Klatt
RWTH Aachen, Germany

B

Benzenediazonium Tetrafluoroborate[1]

$Ph{-}\overset{+}{N}{\equiv}N\ BF_4^-$

[369-57-3] (MW 191.92)

(much higher stability than the corresponding chlorides;[2] shock-insensitive; often used when pure, isolated arenediazonium salts are needed;[1] reagent for introducing aryl, arylazo, arylhydrazono, or amino groups; forms fluoroarenes upon heating;[3] building block for heterocycles)

Alternate Name: phenyldiazonium tetrafluoroborate.

Physical Data: colorless solid after recrystallization; turns pink at 80 °C; decomposes at 114–116 °C.

Solubility: fairly sol polar solvents such as acetonitrile, acetone, pyridine, DMF, DMSO, and HMPA with decomposition; slightly sol water; insol hydrocarbons and Et_2O; solubilized in nonpolar media by crown ethers.

Form Supplied in: 4-nitrobenzenediazonium tetrafluoroborate is widely available (nearly colorless solid). A large variety of arenediazonium BF_4 salts are readily prepared from aromatic amines (see below).

Preparative Methods: [1a,1d,3,4] The most commonly used procedures consist in diazotization of an aromatic amine ($ArNH_2$) with $NaNO_2$ in aqueous HCl or H_2SO_4 followed by precipitation of the salt with added $NaBF_4$ or fluoroboric acid.[5] Alternatively, the diazotization can be carried out directly in 40–50% aqueous HBF_4.[2c] Aromatic amines that do not dissolve in aqueous mineral acids can be reacted with $NO^+BF_4^-$ in an anhydrous organic solvent or in liquid SO_2.[6] This method can also be applied to *N*-(trimethylsilyl)anilines[7] and for the preparation of those salts that are not easily isolated from water.[8] An improved, high-yielding one-pot procedure[9] employs $ArNH_2$, *t*-BuONO, and $BF_3{\cdot}Et_2O$ in an anhydrous organic solvent, typically CH_2Cl_2.

Handling, Storage, and Precautions: the dry parent salt can be stored for more than a month at rt or for a few years at −20 °C under N_2 in the dark, but decomposes when exposed to direct sunlight. Rapid recrystallization from warm water[2b] or from acetonitrile–Et_2O is possible without decomposition. Although arenediazonium tetrafluoroborates, in contrast to the chloride salts, are renowned in general for their enhanced thermal stability and shock-insensitivity, some care should nevertheless be taken. Some salts are known to decompose while drying, e.g. 3-methoxybenzenediazonium, 2-methylbenzenediazonium, and certain heteroarenediazonium tetrafluoroborates.[3,10] Avoid contact with metals.

General Reactivity Patterns. Similar to the chloride salts, arenediazonium tetrafluoroborates can be transformed into covalent azo compounds by addition of a nucleophile and into various functionalized arenes by displacement of the N_2^+ group. While the chloride salts are typically prepared in situ in an acidic aqueous or ethanolic solution, the tetrafluoroborates are usually isolated and can be employed as pure compounds in the solvent of choice or in suspension. Their low solubility in nonpolar organic solvents may be a problem, but it can be overcome by phase-transfer techniques.[1c,11]

Addition to Unsaturated Compounds with Retention of N_2. Diarylazo compounds ($Ar^1{-}N{=}N{-}Ar^2$) are obtained by coupling of arenediazonium salts with sufficiently electron-rich (hetero)aromatic substrates. The vast majority of these important azo coupling reactions have been carried out in the reaction media obtained by diazotization of aromatic amines with $NaNO_2$ in aqueous HCl or H_2SO_4 (see ***Benzenediazonium Chloride***). Comparisons of reactivity between in situ generated arenediazonium chlorides and tetrafluoroborates (used as isolated salts) are rare.[12] Thus 4-nitrobenzenediazonium chloride reacts only slowly with indole and its 1-, 2-, and 3-Me derivatives in aqueous neutral solution, whereas the corresponding BF_4 salt in H_2O–ethanol undergoes azo coupling rapidly and almost quantitatively (eq 1).[13a] Similarly, several arenediazonium tetrafluoroborates couple smoothly to 2-amino-4-(alkyl or aryl)-1,3-oxazoles in aqueous $NaHCO_3$ solution, whereas neutral or weakly acidic solutions of salts prepared by in situ diazotization give intractable product mixtures.[13b] In other cases, the use of pure, water-free $ArN_2^+BF_4^-$ salts in an organic solvent may be necessary or may simply give better yields. Examples include azo coupling of $PhN_2^+BF_4^-$ with 2-dimethylaminopyrazoles in MeCN[14] and of λ^5-phosphinines in MeOH–benzene (eq 2).[15] Uncommon substrates such as calixarene,[16] ditellurafulvene,[17] sesquifulvalene,[18] and coordinated cyclooctatetraene[19] are also subject to the reaction.

R^1 = H, Me; R^2 = H, Me (1)

2 $PhN_2^+BF_4^-$; NaOAc, MeOH, benzene; 77% (2)

Azo coupling is promoted by various phase-transfer catalysts such as crown ethers,[20,21] NaOAc,[22] Me_4NCl,[23] and Na[B(3,5-di-CF_3-C_6H_3)$_4$].[24]

Arenediazonium BF_4^- or PF_6^- salts undergo a facile, probably concerted, [2 + 4] cycloaddition to various acyclic methyl- or aryl-substituted 1,3-dienes, and 1,6-dihydropyridazines are isolated (eq 3).[25,26] In contrast, the products isolated with cyclopentadiene result from an initial azo coupling process.[27]

$$\mathrm{ArN_2^+BF_4^-} + \text{2,3-dimethylbutadiene} \xrightarrow[0\,^\circ C]{MeCN} [\text{cyclic adduct}] \xrightarrow{-HBF_4} \text{N-aryl dihydropyridazine} \quad (3)$$

Azomethine ylides and other 1,3-dipoles also afford cycloadducts with diazonium salts.[28] Ene-type reactions with unsaturated lactones have also been reported.[29]

Addition to Unsaturated Compounds with Loss of N_2 (Introduction of Aryl Groups). Arylation of alkenic compounds by diazonium salts, yielding saturated or unsaturated products, is known as the Meerwein reaction and is usually catalyzed by copper salts. The procedure can be efficiently applied to intramolecular cyclizations (eq 4).[30] In the presence of iodide ions the reaction takes place without copper salts.[31] Titanium(III) salts also facilitate the addition of alkenic compounds to give reduced products (eq 5).[32] Arylation of ferrocenyl-substituted alkynes and alkenes succeeds without such catalysts.[33]

$$\text{2-(2-methylallyloxy)benzenediazonium } BF_4^- \xrightarrow[\text{rt, 89\%}]{CuBr_2,\ DMSO} \text{3-(bromomethyl)-3-methyl-2,3-dihydrobenzofuran} \quad (4)$$

$$MeO_2C\text{-CH=CH-}CO_2Me \xrightarrow[\text{acetone, 0–5 °C, 72\%}]{PhN_2^+BF_4^-,\ TiCl_3,\ MeCO_2Na} MeO_2C\text{-CH(Ph)-CH}_2\text{-}CO_2Me \quad (5)$$

When the thermal decomposition of $ArN_2^+BF_4^-$ is conducted in the presence of pyridine, biaryls are formed in yields up to 30–40% by a radical mechanism, and no ArF is found.[34] Much higher yields of unsymmetrical biaryls can be obtained when arenediazonium BF_4^- or PF_6^- salts are allowed to react with an aromatic compound in the presence of KOAc and a phase-transfer catalyst[35] (eq 6). This procedure often gives better yields than the classical Gomberg–Bachmann–Hey reaction that employs arenediazonium chlorides in a two-phase system. The method also holds promise for the intramolecular version of aryl–aryl coupling (Pschorr cyclization).[11] Titanium(III) salts enhance *ortho* selectivity in the arylation of phenols.[36]

$$Cl\text{-}C_6H_4\text{-}N_2^+BF_4^- \xrightarrow[\text{KOAc, 18-crown-6, 80\%}]{C_6H_6} Cl\text{-}C_6H_4\text{-}C_6H_5 \quad (6)$$

Diazonium salts add to nitriles with loss of dinitrogen to form *N*-arylnitrilium ions which further react in situ with nucleophiles or aromatics. Hence, intramolecular cyclizations take place starting from *ortho*-substituted diazonium salts (eqs 7 and 8); substituents used are azido,[37] carboxyl,[38] alkylthio,[39] hydroxymethyl,[40] and aryl.[41] In this connection, intermolecular addition of carboxylates to nitrilium ions gives unsymmetrical *N*-arylimides.[42] In place of nitriles, isocyanides also undergo intramolecular cyclizations with *ortho*-substituted diazonium salts.[43] Reactions with carbon monoxide under high pressure[44] or with ***Tetracarbonylnickel***[45] result in the formation of carboxylic acid derivatives.

$$[\text{2-(azidomethyl)benzenediazonium}]BF_4^- \xrightarrow[\text{MeCN, 80 °C}]{-N_2} [\text{nitrilium ion}]BF_4^- \longrightarrow [\text{cyclized } N\text{-}N_2^+]BF_4^- \xrightarrow[70\%]{-HBF_4} \text{2-methylquinazoline} \quad (7)$$

$$\text{2-biphenylyldiazonium } BF_4^- \xrightarrow[\text{MeCN, 82 °C}]{-N_2} [\text{nitrilium ion}]BF_4^- \xrightarrow[86\%]{-HBF_4} \text{6-methylphenanthridine} \quad (8)$$

Coupling with Nucleophiles with Retention of N_2 (Introduction of Azo or Hydrazono Groups). The coupling of diazonium salts with active methylene compounds is known as the Japp–Klingemann reaction, a process applicable to nucleoside synthesis.[46] On the other hand, α-amination of simple esters is achieved through the hydrogenation of azo or hydrazono esters formed from ketene silyl acetals (eq 9).[47] The reaction with enamines likewise leads to α-iminio hydrazones which readily rearrange to heterocycles (eq 10).[48] The nucleophilic addition of ***p-Tolylsulfonylmethyl Isocyanide*** also results in cyclization, affording triazoles.[49] In addition, lithium enolates of α-substituted ketones,[50] Grignard reagents,[50,51] organozinc reagents,[52,53] and allylsilanes[54] produce corresponding azo or hydrazono compounds. ***Potassium Cyanide*** in the presence of ***18-Crown-6*** yields diazo cyanides that act as efficient dienophiles in the Diels–Alder reaction (eq 11).[55]

$$PhCH_2CO_2Me \longrightarrow PhCH=C(OMe)(OSiMe_3) \xrightarrow[\text{py, 0 °C, 83\%}]{PhN_2^+BF_4^-} PhC(=NNHPh)CO_2Me \xrightarrow[100\%]{H_2} PhCH(NH_2)CO_2Me \quad (9)$$

$$\text{piperidino-CH=CH-}CO_2Et + \text{3-MeO-}C_6H_4N_2^+BF_4^- \xrightarrow[\text{rt}]{MeCN} [\text{iminium hydrazone}]BF_4^- \longrightarrow \text{7-methoxycinnoline-3-}CO_2Et\ (72\%) \quad (10)$$

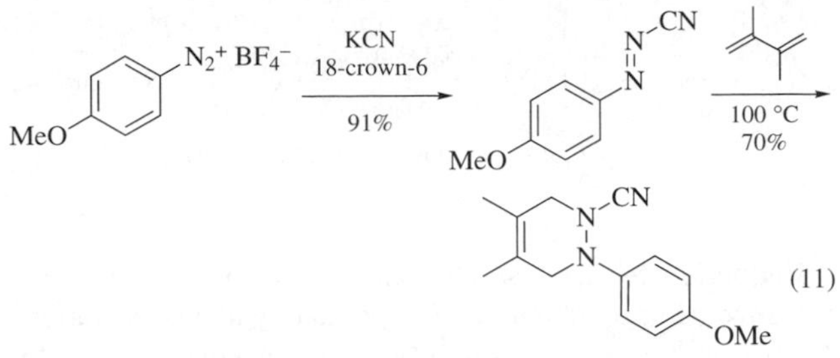

(11)

Amines also combine with the terminal nitrogen of diazonium salts to form triazenes (eq 12),[56] the utility of which is summarized in a review.[57] Couplings with ***Difluoramine*** or isopropyl fluorocarbamate yield azides through triazenes as intermediates.[58] Reaction with hydrazonomethanesulfonates or guanidine derivatives give five-membered azacycles.[59] The addition of sodium sulfinates produces arylazo sulfones.[60]

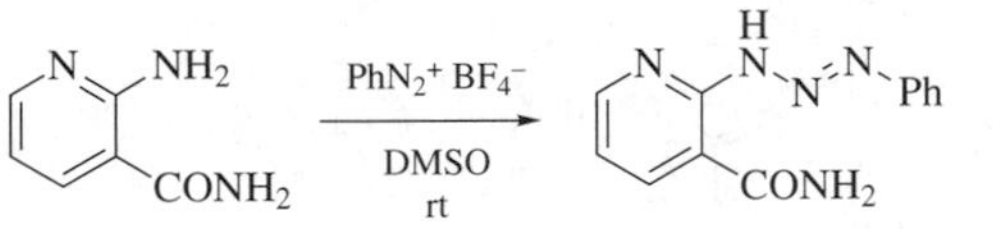

(12)

Coupling with Nucleophiles with Loss of N_2 (Introduction of Aryl Groups). Arenediazonium salts act as aryl cations toward various nucleophiles. Silyl enol ethers derived from aryl ketones react with diazonium tetrafluoroborates in pyridine to afford α-aryl ketones (eq 13).[61]

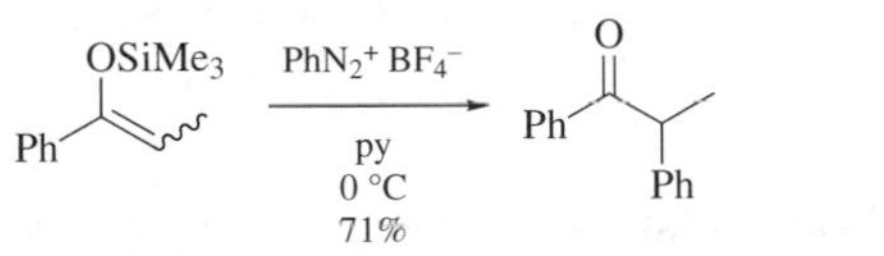

(13)

Solvolysis of diazonium tetrafluoroborates in ***Trifluoromethanesulfonic Acid*** results in the formation of aryl triflates (eq 14).[62] Similarly, decomposition in methanol[63] or ***Trifluoroacetic Acid***[64] leads to the corresponding ethers or esters; the latter is utilized for phenol synthesis. ***Pyridine N-Oxide*** reacts at the oxygen atom to give *N*-(aryloxy)pyridinium tetrafluoroborates.[65] Phenyl ethers and esters can be prepared via the reaction of trimethylsilyl ethers and esters with PhN_2BF_4.[66]

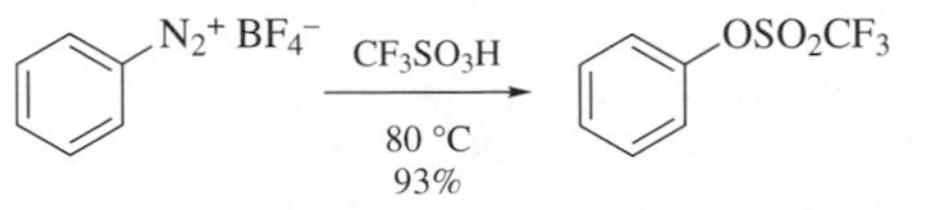

(14)

Unsymmetrical diaryl sulfides are obtained from diazonium fluoroborates and sodium aryl thiolates in DMSO (eq 15).[67] Similarly, aryl thiolesters are formed using thiocarboxylates.[68] Diaryl selenides and diaryl tellurides can be prepared by the reaction of arenediazonium salts with ***Sodium Selenide*** and ***Sodium Telluride*** (eq 16).[69]

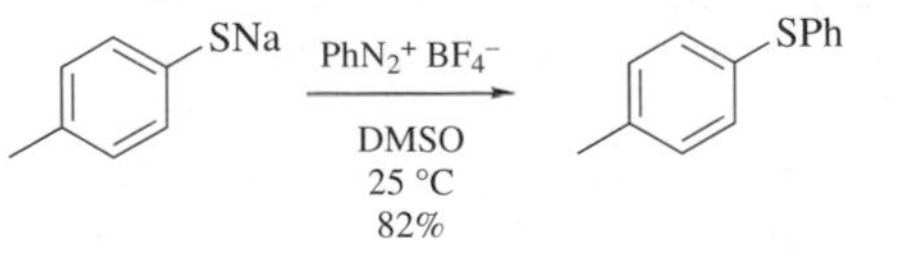

(15)

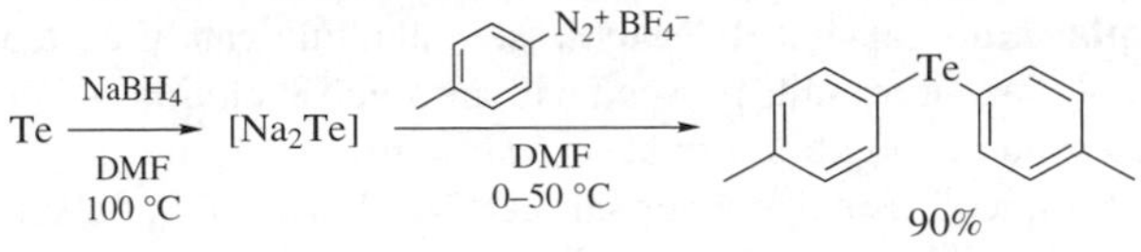

(16)

Controlled thermal decomposition of dry arenediazonium tetrafluoroborates affords fluoroarenes in normally good to high yield (Balz–Schiemann reaction[3]) (eq 17). Typically, the reaction is carried out with the solid salt,[5] but decomposition in suspension or in solution[2b,3,70] has also been reported. Nitroarenediazonium BF_4^- salts are usually mixed with sand, NaF, or $BaSO_4$ to avoid violent decomposition.[3] The BF_3 etherate complex may be useful for an efficient transformation.[71]

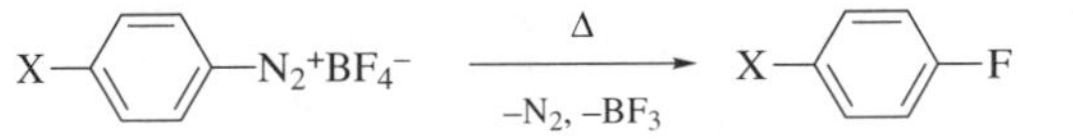

(17)

X = H, 51–57%
X = CO_2Et, 84–89%

The formation of fluoroarenes proceeds via an aryl cation intermediate that reacts with the BF_4^- ion in the ion pair,[72] and is accelerated by photolysis.[73] In some cases, the use of hexafluoroantimonates or hexafluorophosphates affords better yields.[74] On the other hand, aryl iodides are obtained via the reaction with ***Potassium Iodide/Iodine*** in DMSO.[75] The reactions with trimethylsilyl halides or ***Azidotrimethylsilane*** yield aryl halides or aryl azides, respectively (eq 18).[76] Organic halides such as $BrCCl_3$ and ***Iodomethane*** also act as halogen sources.[77]

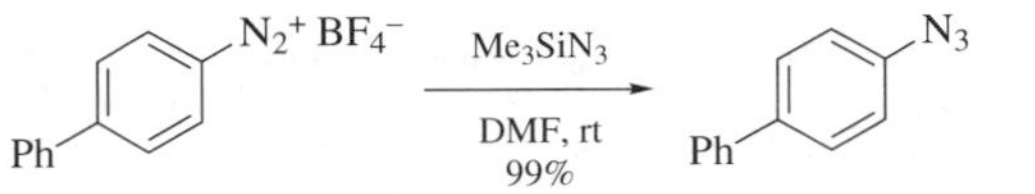

(18)

Diazonium fluoroborates transform iminophosphoranes into phosphonium salts.[78] Similarly, sulfones are converted to sulfoxonium salts which can be reduced in situ to sulfoxides with ***Sodium Borohydride***.[79]

Nucleophilic substitution of diazonium salts in the presence of Cu^I salts is known as the Sandmeyer reaction. Nitrodediazoniation of diazonium tetrafluoroborate with ***Sodium Nitrite*** in the presence of copper powder occurs smoothly (eq 19).[2c] An important access to arenephosphonic acids is provided by the Cu^I-catalyzed reaction of $ArN_2^+BF_4^-$ salts with ***Phosphorus(III) Chloride*** in an organic solvent, followed by hydrolysis (eq 20).[80] The method tolerates a wide range of substituents.[81] The thermal decomposition of arenediazonium salts in dilute aqueous solution in the presence of Cu_2O and a large excess of $Cu(NO_3)_2$ affords phenols in good yield; notably, this method does not require a strongly acidic medium.[82]

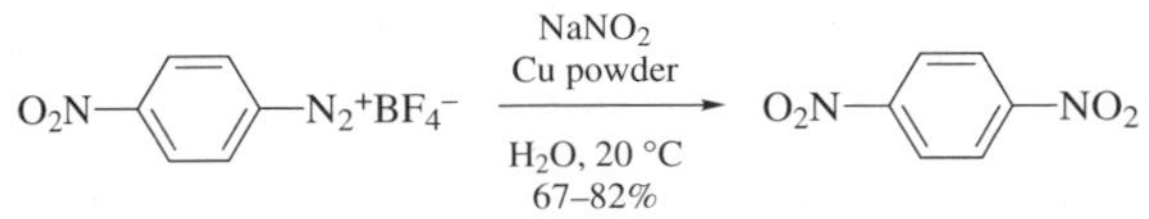

(19)

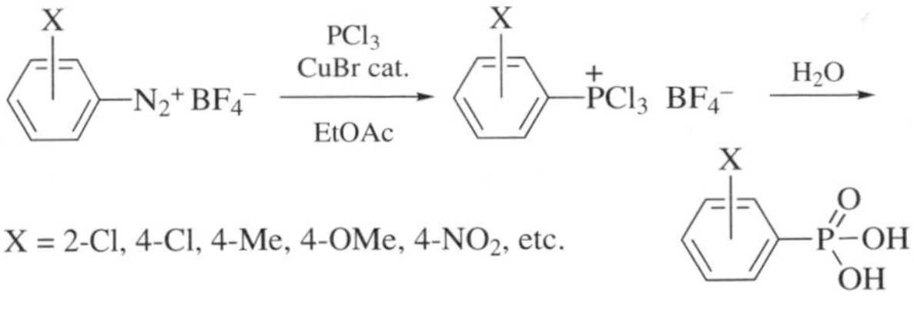

X = 2-Cl, 4-Cl, 4-Me, 4-OMe, 4-NO_2, etc. (20)

Palladium-Catalyzed Reactions. Palladium-catalyzed reactions of diazonium salts proceed with nitrogen evolution and provide a mild method to introduce aryl groups.

While the classical, copper salt-catalyzed arylation of alkenes (Meerwein arylation) requires alkene activation by an electron-withdrawing substituent[83] (see ***Benzenediazonium Chloride***), the Pd-catalyzed modification succeeds well not only with acrylic aldehydes and esters, but also with styrene, ethylene, and other non-activated acyclic and cyclic alkenes (eqs 21 and 22).[84] Coupling reactions with isolated $PhN_2^+BF_4^-$ in general give much better yields than with the in situ generated salt $PhN_2^+X^-$ (X = Cl or OAc). In an analogous reaction, silyl-, germyl-, or stannyl-substituted alkenes result in the loss of heteroatom substituents (eq 23).[85–87]

N_2^+ + $H_2C{=}CH_2$ — Pd(dba)$_2$, NaOAc; CH_2Cl_2, MeCN, 20 °C, 68% → (21)

$PhN_2^+BF_4^-$ + — Pd(dba)$_2$, NaOAc; CH_2Cl_2 or MeCN → Ph + Ph (22)

CH_2Cl_2: 10:90 (70%)
MeCN: 20:80 (67%)

$PhN_2^+BF_4^-$ + Ar / $SiMe_3$ — Pd(dba)$_2$ (5%); MeCN, 25 °C, 100% →

Ar = 4-MeC_6H_4

Ar / Ph + Ar / Ph (23)

74:26

The palladium-catalyzed carbonylation of an arenediazonium tetrafluoroborate with CO in the presence of Et_3SnH or a poly(methylhydrosiloxane) affords substituted benzaldehydes (eq 24).[88] Formation of ArH is sometimes a minor side-reaction, but predominates (81%) in the 2-nitro case. In related Pd-catalyzed reactions, diaryl ketones (eq 25),[89] aryl alkyl ketones,[89] substituted benzoic acids,[90] and mixed arylalkylcarboxylic anhydrides (eq 26)[91] can be obtained. Thermal disproportionation of the latter can be used to prepare homo arenecarboxylic anhydrides.[91]

X —$N_2^+BF_4^-$ + CO — Et_3SnH, Pd(OAc)$_2$ cat.; MeCN, Et_2O → X —CHO (24)

X = 2-Me, 2-Ph, 4-Me, 4-Cl, 4-NO_2, 4-COMe, 4-CO_2Et, etc

$Ar^1N_2^+BF_4^-$ + CO — Ar^2-$SnMe_3$, Pd(OAc)$_2$ cat.; MeCN, Et_2O →

Ar^1–CO–Ar^2 + Ar^2–CO–Ar^2 (25)
(minor)

X —$N_2^+BF_4^-$ + CO — $MeCO_2Na$, Pd(OAc)$_2$ cat.; MeCN → X —CO_2COMe (26)

X = 4-I (68%), 3-NO_2 (49%), 4-NO_2 (65%), etc.

Reduction to Arenes. The removal of an NH_2 function by reductive dediazoniation is an important synthetic operation on the aromatic nucleus. As with other benzenediazonium salts, a number of methods exist for the BF_4^- salts, but the radical-chain reaction with H_3PO_2 is a particularly good method (eq 27).[92] Hydro-dediazoniation with $NaBH_4$ in methanol[93] is also quite convenient, but does not always occur cleanly.[94] Other effective and easily available reducing agents are ***Thiophenol***,[95] hydrosilanes,[96] and hydrostannanes.[96] Decomposition in HMPA[97] or in the presence of crown ethers[98] also gives good results. Catalysis by rhodium phosphine complexes in DMF does not seem very effective.[99]

Br —$N_2^+BF_4^-$ — H_3PO_2, cat. Cu_2O; $CHCl_3$, 0 °C, 99% → Br— (27)

Benzeneselenol is a unique reagent for producing arylhydrazines from diazonium fluoroborates (eq 28); the reaction is applicable to the synthesis of indazolone from *ortho*-carbamoyl diazonium salts.[100]

$N_2^+BF_4^-$ — PhSeH; CH_2Cl_2, reflux, 80% → $NHNH_2$ (28)

Radical Initiation. Diazonium fluoroborates can be used as radical initiators. Applied examples are alkylation of heteroaromatics with alkyl iodides[101] and polymerization of vinylic compounds.[102]

Related Reagents. Benzenediazonium-2-carboxylate; Benzenediazonium Chloride.

---

1. (a) Pütter, R. *MOC* **1965**, *X/3*, 7. (b) Wulfman, D. S. In *The Chemistry of Diazonium and Diazo Groups, Part 2*; Patai, S., Ed.; Wiley: Chichester, 1978; pp 247–339. (c) Bartsch, R. A. In *The Chemistry of Triple-Bonded Functional Groups, Part 2*; Patai, S.; Rappoport, Z., Eds.; Wiley: Chichester, 1983; pp 889–915. (d) Engel, A. *MOC* **1990**, *E16a*, 1052. (e) Galli, C. *CRV* **1988**, *88*, 765.
2. (a) Wilke-Dörfurt, E.; Balz, G. *CB* **1927**, *60*, 115. (b) Balz, G.; Schiemann, G. *CB* **1927**, *60*, 1186. (c) Starkey, E. B. *OS* **1939**, *19*, 40; *OSC* **1943**, *2*, 225.
3. Roe, A. *OR* **1949**, *5*, 193.
4. Suschitzky, H. *Adv. Fluorine Chem.* **1965**, *4*, 1.
5. (a) Flood, D. T. *OSC* **1943**, *2*, 295. (b) Schiemann, G.; Winkelmüller, W. *OSC* **1943**, *2*, 188, 299.
6. Wannagat, U.; Hohlstein, G. *CB* **1955**, *88*, 1839.
7. Weiss, R.; Wagner, K.-G.; Hertel, M. *CB* **1984**, *117*, 1965.
8. Vonznesenskii, S. A.; Kurskii, P. P. *ZOB* **1938**, *8*, 524 (*CA* **1938**, *32*, 8379).
9. Doyle, M. P.; Bryker, W. J. *JOC* **1979**, *44*, 1572.
10. Doak, G. O.; Freedman, L. D. *Chem. Eng. News* **1967**, *45(53)*, 8.

11. Gokel, G. W.; Ahern, M. F.; Beadle, J. R.; Blum, L.; Korzeniowski, S. H.; Leopold, A.; Rosenberg, D. E. *Israel J. Chem.* **1985**, *26*, 270.
12. Szele, I.; Zollinger, H. *Top. Curr. Chem.* **1983**, *112*, 1.
13. (a) Jackson, A. H.; Lynch, P. P. *JCS(P2)* **1987**, 1483. See also: Jackson, A. H.; Prasitpan, N.; Shannon, P. V. P.; Tinker, A. C. *JCS(P1)* **1987**, 2543. (b) Crank, G.; Mekonnen, B. *JHC* **1992**, *29*, 1469.
14. Gompper, R.; Guggenberger, R.; Zentgraf, R. *AG* **1985**, *97*, 998; *AG(E)* **1985**, *24*, 984.
15. Märkl, G.; Liebl, R. *S* **1978**, 846.
16. Shinkai, S.; Araki, K.; Shibata, J.; Tsugawa, D.; Manabe, O. *JCS(P1)* **1990**, 3333.
17. Lakshmikantham, M. V.; Cava, M. P.; Albeck, M.; Engman, L.; Wudl, F.; Aharon-Shalom, E. *CC* **1981**, 828.
18. Araki, S.; Butsugan, Y. *TL* **1984**, *25*, 441.
19. Connelly, N. G.; Lucy, A. R.; Whiteley, M. W. *CC* **1979**, 985.
20. Hashida, Y.; Kubota, K.; Sekiguchi, S. *BCJ* **1988**, *61*, 905.
21. Butler, A. R.; Shepherd, P. T. *JCR(S)* **1978**, 339.
22. Anderson, Jr., A. G.; Grina, L. D.; Forkey, D. M. *JOC* **1978**, *43*, 664.
23. Korzeniowski, S. H.; Gokel, G. W. *TL* **1977**, 1637.
24. Kobayashi, H.; Sonoda, T.; Iwamoto, H. *CL* **1981**, 579.
25. Carlson, B. A.; Sheppard, W. A.; Webster, O. W. *JACS* **1975**, *97*, 5291.
26. Bronberger, F.; Huisgen, R. *TL* **1984**, *25*, 57.
27. Huisgen, R. Bronberger, F. *TL* **1984**, *25*, 61.
28. Bronberger, F.; Huisgen, R. *TL* **1984**, *25*, 65.
29. (a) Boyd, G. V.; Monteil, R. L.; Lindley, P. F.; Mahmoud, M. M. *JCS(P1)* **1978**, 1351. (b) Baydar, A. E.; Boyd, G. V. *JCS(P1)* **1978**, 1360.
30. Meijs, G. F.; Beckwith, A. L. J. *JACS* **1986**, *108*, 5890.
31. Beckwith, A. L. J.; Meijs, G. F. *JOC* **1987**, *52*, 1922.
32. Citterio, A.; Cominelli, A.; Bonavoglia, F. *S* **1986**, 308.
33. Nock, H.; Schottenberger, H. *JOC* **1993**, *58*, 7045.
34. (a) Abramovitch, R. A.; Saha, J. G. *T* **1965**, *21*, 3297. (b) Abramovitch, R. A.; Koleoso, O. A. *JCS(B)* **1968**, 1292.
35. Beadle, J. R.; Korzeniowski, S. H.; Rosenberg, D. E.; Garcia-Slanga, B. J.; Gokel, G. W. *JOC* **1984**, *49*, 1594; Korzeniowski, S. H.; Blum, L.; Gokel, G. W. *TL* **1977**, 1871; Rosenberg, D. E.; Beadle, J. R.; Korzeniowski, S. H.; Gokel, G. W. *TL* **1980**, *21*, 4141.
36. Caronna, T.; Ferrario, F.; Servi, S. *TL* **1979**, 657.
37. Kreher, R.; Bergmann, U. *TL* **1976**, 4259.
38. Schmidt, R. R.; Schneider, W. *TL* **1970**, 5095.
39. Lankin, D. C.; Petterson, R. C.; Velazquez, R. A. *JOC* **1974**, *39*, 2801.
40. Schmidt, R. R.; Schneider, W.; Karg, J.; Burkert, U. *CB* **1972**, *105*, 1634.
41. Petterson, R. C.; Bennett, J. T.; Lankin, D. C.; Lin, G. W.; Mykytka, J. P.; Troendle, T. G. *JOC* **1974**, *39*, 1841. See also Ref. 40.
42. Kikukawa, K.; Kono, K.; Wada, F.; Matsuda, T. *BCJ* **1982**, *55*, 3671.
43. Schmidt, R. R.; Vatter, H. *TL* **1971**, 1925.
44. Ravenscroft, M. D.; Skrabal, P.; Weiss, B.; Zollinger, H. *HCA* **1988**, *71*, 515.
45. Clark, J. C.; Cookson, R. C. *JCS* **1962**, 686.
46. Kozikowski, A. P.; Floyd, W. C. *TL* **1978**, 19.
47. (a) Sakakura, T.; Tanaka, M. *CC* **1985**, 1309. (b) As for amino cation equivalents, see also Erdik, E.; Ay, M. *CRV* **1989**, *89*, 1947.
48. (a) Kanner, C. B.; Pandit, U. K. *T* **1981**, *37*, 3513. (b) Manhas, M. S.; Brown, J. W.; Pandit, U. K.; Houdewind, *T* **1975**, *31*, 1325. (c) Katritzky, A. R.; Ürögdi, L.; Patel, R. C. *JCS(P1)* **1982**, 1349.
49. van Leusen, A. M.; Hoogenboom, B. E.; Houwing, H. A. *JOC* **1976**, *41*, 711.
50. Garst, M. E.; Lukton, D. *SC* **1980**, *10*, 155.
51. Stang, P. J.; Mangum, M. G. *JACS* **1977**, *99*, 2597.
52. Curtin, D. Y.; Tveten, J. L. *JOC* **1961**, *26*, 1764.
53. Examples: Enders, E. *MOC* **1965**, *X/3*, 477.
54. Mayr, H.; Grimm, K. *JOC* **1992**, *57*, 1057.
55. (a) Ahern, M. F.; Leopold, A.; Beadle, J. R.; Gokel, G. W. *JACS* **1982**, *104*, 548. (b) Gapinski, D. P.; Ahern, M. F. *TL* **1982**, *23*, 3875.
56. (a) Debeljak-Šuštar, M.; Stanovnik, B.; Tišler, M.; Zriměk, Z. *JOC* **1978**, *43*, 393. (b) Baldwin, J. E.; Harrison, P.; Murphy, J. A. *CC* **1982**, 818. Julliard, M.; Vernin, G.; Metzger, J. *S* **1980**, 116.
57. Vaughan, K.; Stevens, M. F. G. *CSR* **1978**, *7*, 377.
58. Baum, K. *JOC* **1968**, *33*, 4333.
59. (a) Hanley, R. N.; Ollis, W. D.; Ramsden, C. A. *JCS(P1)* **1979**, 736. (b) Baydar, A. E.; Boyd, G. V.; Lindley, P. F.; Walton, A. R. *JCS(P1)* **1985**, 415.
60. Ref. 55a and Kobayashi, M.; Gotoh, M.; Yoshida, M. *BCJ* **1987**, *60*, 295.
61. Sakakura, T.; Hara, M.; Tanaka, M. *CC* **1985**, 1545.
62. Yoneda, N.; Fukuhara, T.; Mizokami, T.; Suzuki, A. *CL* **1991**, 459.
63. Broxton, T. J.; Bunnett, J. F.; Paik, C. H. *JOC* **1977**, *42*, 643.
64. Horning, D. E.; Ross, D. A.; Muchowski, J. M. *CJC* **1973**, *51*, 2347.
65. (a) Abramovitch, R. A.; Alvernhe, G.; Bartnik, R.; Dassanayake, N. L.; Inbasekaran, M. N.; Kato, S. *JACS* **1981**, *103*, 4558. (b) Abramovitch, R. A.; Inbasekaran, M. N.; Kato, S.; Singer, G. M. *JOC* **1976**, *41*, 1717.
66. Olah, G. A.; Wu, A. *S* **1991**, 204.
67. Petrillo, G.; Novi, M.; Garbarino, G.; Dell'erba, C. *T* **1986**, *42*, 4007.
68. Petrillo, G.; Novi, M.; Garbarino, G.; Filiberti, M. *T* **1989**, *45*, 7411.
69. (a) Li, J.; Lue, P.; Zhou, X.-J. *S* **1992**, 281. (b) Chen, C.; Qiu, M.; Zhou, X. J. *SC* **1991**, *21*, 1729.
70. (a) Swain, C. G.; Rogers, R. J. *JACS* **1975**, *97*, 799. (b) Becker, H. G. O.; Israel, G. *JPR* **1979**, *321*, 579. (c) Abramovitch, R. A.; Saha, J. G. *CJC* **1965**, *43*, 3269.
71. Shinhama, K.; Aki, S.; Furuta, T.; Minamikawa, J. *SC* **1993**, *23*, 1577.
72. (a) Hegarty, A. F. In *The Chemistry of the Diazo and Diazonium Groups, Part 2*, Patai, S., Ed.; Wiley: Chichester, 1978; pp 511–591. (b) Zollinger, H. In *The Chemistry of Triple-Bonded Functional Groups, Part 1*, Patai, S.; Rappoport, Z., Eds.; Wiley: Chichester, 1983; pp 603–669.
73. (a) Kirk, K. L.; Nagai, W.; Cohen, L. A. *JACS* **1973**, *95*, 8389. (b) Petterson, R. C.; DiMaggio, III, A.; Hebert, A. L.; Haley, T. J.; Mykytka, J. P.; Sarkar, I. M. *JOC* **1971**, *36*, 631.
74. (a) Sellers, C.; Suschitzky, H. *JCS(C)* **1968**, 2317. (b) Rutherford, K. G.; Redmond, W.; Rigamonti, J. *JOC* **1961**, *26*, 5149.
75. Citterio, A.; Arnoldi, A. *SC* **1981**, *11*, 639.
76. Keumi, T.; Umeda, T.; Inoue, Y.; Kitajima, H. *BCJ* **1989**, *62*, 89.
77. Korzeniowski, S. H.; Gokel, G. W. *TL* **1977**, 3519.
78. Takeishi, M.; Shiozawa, N. *BCJ* **1989**, *62*, 4063.
79. Still, I. W. J.; Szilagyi, S. *SC* **1979**, *9*, 923.
80. Doak, G. O.; Freedman, L. D. *JACS* **1951**, *73*, 5658.
81. Examples: Sasse, K. *MOC* **1963**, *12/1*, 368.
82. Cohen, T.; Dietz, A. G., Jr.; Miser, J. R. *JOC* **1977**, *42*, 2053.
83. Rondestvedt, C. S. *OR* **1960**, *11*, 189; *OR* **1977**, *24*, 225.
84. (a) Kikukawa, K.; Nagira, K.; Terao, N.; Wada, F.; Matsuda, T. *BCJ* **1979**, *52*, 2609. (b) Kikukawa, K.; Nagira, K.; Wada, F.; Matsuda, T. *T* **1981**, *37*, 31; (c) Yong, W.; Yi, P.; Zhuangyu, Z.; Hongwen, H. *S* **1991**, 967.
85. Kikukawa, K.; Umekawa, H.; Matsuda, T. *JOM* **1986**, *311*, C44.
86. Ikenaga, K.; Kikukawa, K.; Matsuda, T. *JCS(P1)* **1986**, 1959.
87. Ikenaga, K.; Matsumoto, S.; Kikukawa, K.; Matsuda, T. *CL* **1990**, 185 and references cited therein.
88. Kikukawa, K.; Totoki, T.; Wada, F.; Matsuda, T. *JOM* **1984**, *270*, 283.
89. Kikukawa, K.; Idemoto, T.; Katayama, A.; Kono, K.; Wada, F.; Matsuda, T. *JCS(P1)* **1987**, 1511.
90. Nagira, K.; Kikukawa, K.; Wada, F.; Matsuda, T. *JOC* **1980**, *45*, 2365.
91. Kikukawa, K.; Kono, K.; Nagira, K.; Wada, F.; Matsuda, T. *JOC* **1981**, *46*, 4413.

92. Korzeniowski, S. H.; Blum, L.; Gokel, G. W. *JOC* **1977**, *42*, 1469.
93. Hendrickson, J. B. *JACS* **1961**, *83*, 1251.
94. Severin, T.; Schmitz, R.; Loske, J.; Hufnagel, J. *CB* **1969**, *102*, 4152.
95. Shono, T.; Matsumura, Y.; Tsubata, K. *CL* **1979**, 1051.
96. Nakayama, J.; Yoshida, M.; Simamura, O. *T* **1970**, *26*, 4609.
97. Tröndlin, F.; Rüchardt, C. *CB* **1977**, *110*, 2494.
98. Hartman, G. D.; Biffar, S. E. *JOC* **1977**, *42*, 1468.
99. Marx, G. S. *JOC* **1971**, *36*, 1725.
100. James, F. G.; Perkins, M. J.; Porta, O.; Smith, B. V. *CC* **1977**, 131.
101. Minisci, F.; Vismara, E.; Fontana, F.; Morini, G.; Serravalle, M. *JOC* **1986**, *51*, 4411.
102. Druliner, J. D. *Macromolecules* **1991**, *24*, 6079.

Gerhard Maas
University of Kaiserslautern, Germany

M. Tanaka & Toshiyasu Sakakura
National Institute of Materials & Chemical Research, Tsukuba, Japan

1,4-Benzoquinone

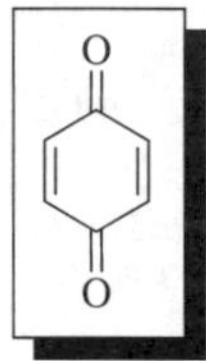

[106-51-4] $C_6H_4O_2$ (MW 108.10)

(useful as an oxidizing[1] or dehydrogenation agent;[2] can function as a dienophile[3] in the Diels–Alder reaction or as a dipolarophile to prepare 5-hydroxyindole derivatives[4])

Alternate Name: *p*-benzoquinone.
Physical Data: mp 115.7 °C; *d* 1.318 g cm^{-3}.
Solubility: slightly sol water; sol alcohol, ether, hot petroleum ether, and aqueous base.
Form Supplied in: yellowish powder; widely available.
Handling, Storage, and Precautions: the solid has an irritating odor and can cause conjunctivitis, corneal ulceration, and dermatitis. In severe cases, there can be necrotic changes in the skin. Use in a fume hood.

As Oxidizing or Dehydrogenation Agent. The ease of reduction of 1,4-benzoquinone to hydroquinone by various compounds renders it useful as an oxidizing or dehydrogenation agent.[2,5] The literature shows that 1,4-benzoquinone prefers to oxidize conjugated primary allylic alcohols over other alcohols (eq 1).[6] Kulkarni has demonstrated the selective oxidation of cinnamyl alcohols to cinnamaldehydes in the presence of secondary or benzylic alcohols. Primary alcohols have been oxidized to the corresponding aldehydes by using 1,4-benzoquinone as the hydrogen acceptor and hydrous zirconium(IV) oxide[7] as a catalyst (eqs 2 and 3).[1]

1,4-benzoquinone (1.5 equiv), diglyme or xylene, 120 °C, 80% (1)

1,4-benzoquinone (6 equiv), xylene, reflux, 86% (2)

1,4-benzoquinone (6 equiv), xylene, reflux, 46% (3)

Nitrogenous compounds can be oxidized by 1,4-benzoquinone in refluxing benzene.[8,9] Aurich reported the conversion of hydroxyamine (**1**) to nitrone (**2**) through the reaction with 1,4-benzoquinone (eq 4).[9] Wiberg used 1,4-benzoquinone to transform tetrazene (**3**) into molecular nitrogen (eq 5).[10] Rossazza reported the oxidation of leurosine at the carbon α to nitrogen (eq 6).[11]

1,4-benzoquinone, benzene, rt, 47% (4)

1,4-benzoquinone (2 equiv), benzene, 80 °C, ~100%; 2 TMSO–C₆H₄–OTMS + 2 N_2 (5)

1,4-benzoquinone, 0.2 M potassium phosphate buffer (pH = 7.0), 50% (6)

In the oxidation of alkenes with palladium(II) acetate, 1,4-benzoquinone serves as a cooxidant to reoxidize palladium(0) to palladium(II).[12,13] Davidson reported the oxidation of alkene (**4**) to a vinyl acetate with 0.1 equiv of palladium acetate and 1 equiv of 1,4-benzoquinone (eq 7).[14]

$Pd(OAc)_2$, 1,4-benzoquinone, AcOH, 25 °C, 50% (7)

Backvall reported that the reaction of 1,3-cyclohexadiene with ***Palladium(II) Acetate*** and 1,4-benzoquinone gave 1,4-diacetoxy-2-cyclohexene in high yield. The stereochemistry of the products was influenced by additives. In the presence of lithium acetate, the major *trans*-diacetate was obtained in 90% yield.[15] Without the addition of lithium acetate, a 1:1 mixture of *trans* and *cis* isomers was produced. However, the *cis*-diacetate was the major (>93%) product when both lithium acetate and lithium chloride were added (eq 8).[15] When the reaction was carried out in acetic acid containing trifluoroacetic acid and lithium trifluoroacetate, the *trans* isomer of 1-acetoxy-4-trifluoroacetoxy-2-cyclohexene was the major product in 67% yield (eq 9).[16] Similar oxidative 1,4-additions of other dienes, such as 1,3-butadiene, also can be accomplished in a regio- and stereoselective fashion (eq 10).[17] However, 1,4-benzoquinone has a high tendency to undergo Diels–Alder reactions when used as a cooxidant for $Pd(OAc)_2$ in such reactions. Many reactions of this type use only catalytic amounts of 1,4-benzoquinone and stoichiometric amounts of external oxidants, such as $Ce(SO_4)_2$, $Tl(OAc)_3$, and MnO_2.[18]

cat. $Pd(OAc)_2$; 1,4-benzoquinone; AcOH, LiOAc; 68% → OAc, OAc; >90% *trans* (8)

cat. $Pd(OAc)_2$; 1,4-benzoquinone; 74% → AcO, OAc; >93% *cis*

$Pd(OAc)_2$ (5 mol%); 1,4-benzoquinone; CF_3CO_2H, LiO_2CF_3; AcOH; 67% → CF_3CO_2, OAc (9)

$Pd(OAc)_2$; 1,4-benzoquinone; AcOH, LiOAc; LiCl; 78% → Cl, OAc (10)

The combination of 1,4-benzoquinone and a catalytic amount of palladium acetate can transform silyl enol ethers into conjugated enones.[19] This reaction is not only regiospecific (eq 11) but also stereospecific to give the more stable *trans* acyclic enone (eq 12).

OTMS; $Pd(OAc)_2$ (0.5 equiv); 1,4-benzoquinone (0.5 equiv); OTMS → O, 94%; O, 85% (11)

OTMS; *E/Z* mixture; $Pd(OAc)_2$; 1,4-benzoquinone; 97% → O; only *E* isomer (12)

Antonsson used ***Manganese Dioxide*** as the oxidant and 0.05 equiv of palladium acetate and 0.20 equiv of 1,4-benzoquinone as catalyst for oxidative ring closure of 1,5-hexadienes to give cyclopentane derivatives in good yield (eq 13).[20] With 1,4-benzoquinone and 10 mol% of ***Palladium(II) Chloride***, a homoallylic alcohol underwent oxidative ring closure to give a γ-butyrolactol (eq 14).[21]

H, H; $Pd(OAc)_2$ (0.05 equiv); 1,4-benzoquinone (0.2 equiv); MnO_2 (1 equiv), AcOH; 70% → H, OAc, H (13)

OH, C_6H_{11}; $PdCl_2$ (1 mol%); 1,4-benzoquinone (3 equiv); wet DMF; 64% → HO, O, C_6H_{13} (14)

The combination of palladium acetate and 1,4-benzoquinone was used in the ring opening of an α,β-epoxysilane to give 19% of oct-2-enal in the presence of oxygen gas (eq 15).[22] Similar reagents are effective in the oxidative coupling of carbon monoxide in methanol to give methyl oxalate (eq 16).[12]

O, TMS; $Pd(OAc)_2$; 1,4-benzoquinone; O_2, DMF; 19% → CHO (15)

CO + MeOH; $Pd(OAc)_2$ (1 mol%); 1,4-benzoquinone; PPh_3 (3 mol%); 83% → MeO, O, O, OMe (16)

As Dienophile. 1,4-Benzoquinone is an excellent dienophile in the Diels–Alder reaction toward electron rich dienes.[23,24] The bicyclic cycloadducts of these reactions are frequently used as the starting materials for the synthesis of natural products. A typical example is demonstrated by Mehta in a total synthesis of capnellene, in which 1,4-benzoquinone is allowed to react with 1-methyl-1,3-cyclopentadiene to give compound (**5**) (eq 17).[25] The cycloaddition of 1,4-benzoquinone and another activated diene is shown in eq 18.[26]

+ O, O; THF, rt; 70% → O, O (**5**) → → H, O (17)

O, O + OTMS; H_3O^+; 75~80% → O, H, O, H, OR (18)

Asymmetric Diels–Alder reactions of 1,4-benzoquinone have been reported. Several examples are shown in eqs 19–21.[3,27,28]

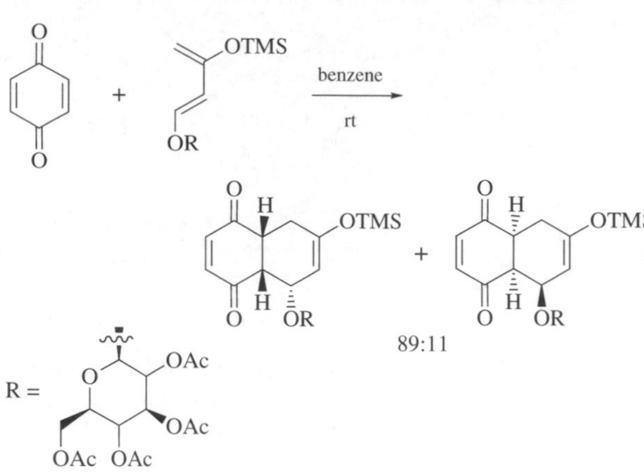

(19)

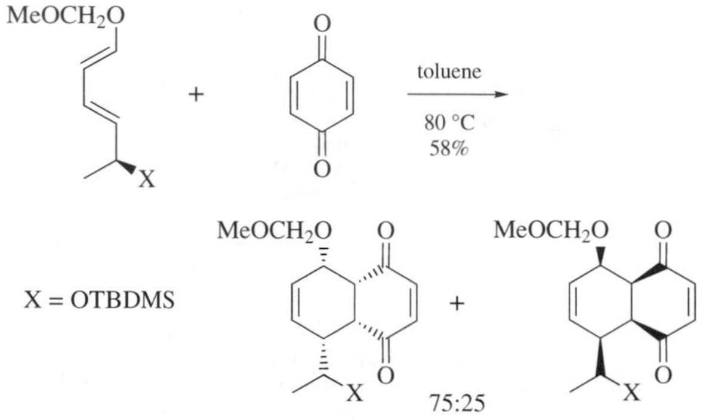

89:11

R =

(20)

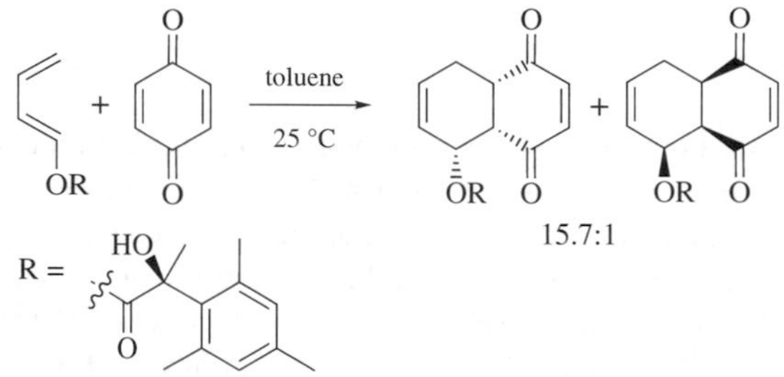

X = OTBDMS

75:25

(21)

15.7:1

R =

Under high pressure, even electron deficient dienes react with 1,4-benzoquinone. Dauben reported the asymmetric cycloaddition of a chiral dienic ester and 1,4-benzoquinone to produce chiral adducts with moderate enantioselectivity (eq 22).[29]

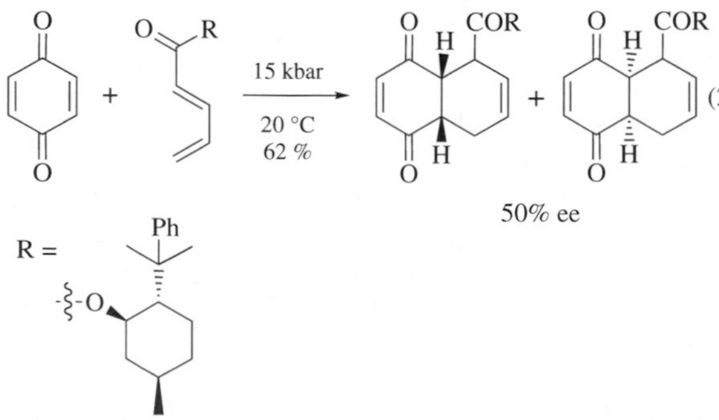

(22)

50% ee

R =

Preparation of 5-Hydroxyindoles. The reaction of 1,4-benzoquinone with certain enamines has been used to prepare 5-hydroxyindole derivatives.[4,30] The first example was reported by Nenitzescu, in which he successfully prepared ethyl 5-hydroxy-2-methylindole-3-carboxylate from 1,4-benzoquinone and ethyl 3-aminocrotonate (eq 23).[4] The ease of the reaction and the availability of the reactants have made it a popular method for indole synthesis. However, the yields of this type of reaction vary over a wide range (5–90%).

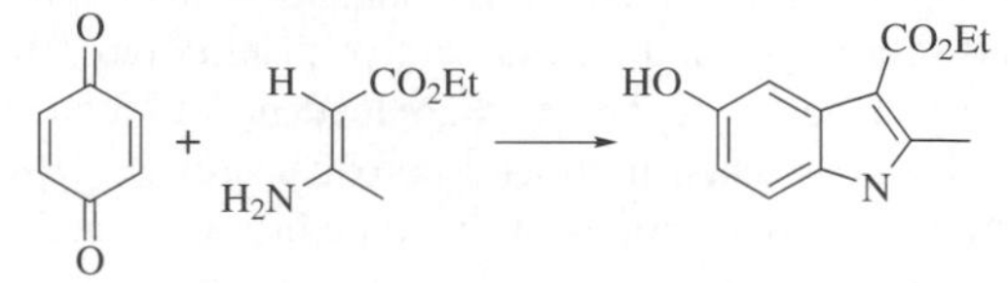

$+ H_2O$ (23)

As Reaction Promoter. In the hydrogenation of nitrobenzene with platinum catalyst in DMSO, addition of 1,4-benzoquinone accelerates the reaction rate.[31]

1. Kuno, H.; Shibagaki, M.; Takahashi, K.; Matsushita, H. *BCJ* **1991**, *64*, 312.
2. Walker, D.; Hiebert, J. D. *CRV* **1967**, *67*, 153.
3. Tripathy, R.; Carroll, P. J.; Thornton, E. R. *JACS* **1990**, *112*, 6743.
4. Nenitzescu, C. D. *Bull. Soc. Chim. Rom.* **1929**, *11*, 37 (*CA* **1930**, *24*, 110).
5. Turner, A. B.; Ringold, H. J. *JCS(C)* **1967**, 1720.
6. Kulkarni, M. G.; Mathew, T. S. *TL* **1990**, *31*, 4497.
7. Shibagaki, M.; Takahashi, K.; Matsushita, H. *BCJ* **1988**, *61*, 3283.
8. Fujita, S.; Sano, K. *JOC* **1979**, *44*, 2647.
9. Aurich, H. G.; Mobus, K. D. *T* **1989**, *45*, 5815.
10. Wiberg, N.; Bayer, H.; Vasisht, S. K.; Meyers, R. *CB* **1980**, *113*, 2916.
11. Goswami, A.; Schaumberg, J. P.; Duffel, M. W.; Rosazza, J. P. *JOC* **1987**, *52*, 1500.
12. Current, S. P. *JOC* **1983**, *48*, 1779.
13. Backvall, J.-E. *ACR* **1983**, *16*, 335.
14. Brown, R. G.; Chaudhari, R. V.; Davidson, J. M. *JCS(D)* **1977**, 183.
15. Backvall, J.-E.; Nordberg, R. E. *JACS* **1981**, *103*, 4959.
16. Backvall, J.-E.; Vagberg, J.; Nordberg, R. E. *TL* **1984**, *25*, 2717.
17. (a) Backvall, J.-E.; Nordberg, R. E.; Nyström, J. E. *TL* **1982**, *23*, 1617. (b) Backvall, J. E.; Nyström, J.-E.; Nordberg, R. E. *JACS* **1985**, *107*, 3676.
18. Backvall, J.-E.; Bystrom, S. E.; Nordberg, R. E. *JOC* **1984**, *49*, 4619.
19. Ito, Y.; Hirao, T.; Saegusa, T. *JOC* **1978**, *43*, 1011.
20. Antonsson, T.; Heumann, A.; Moberg, C. *CC* **1986**, 518.
21. Nokami, J.; Ogawa, H.; Miyamoto, S.; Mandai, T.; Wakabayashi, S.; Tsuji, J. *TL* **1988**, *29*, 5181.
22. Hirao, T.; Murakami, T.; Ohno, M.; Ohshiro, Y. *CL* **1991**, 299.
23. (a) Marchand, A. P.; Allen, R. W. *JOC* **1974**, *39*, 1596. (b) Hill, R. K.; Newton, M. G.; Pantaleo, N. S.; Collins, K. M. *JOC* **1980**, *45*, 1593. (c) Jurczak, J.; Kozluk, T.; Filipek, S.; Eugster, C. H. *HCA* **1983**, *66*, 222. (d) Kozikowski, A. P.; Hiraga, K.; Springer, J. P.; Wang, B. C.; Xu, Z.-B. *JACS* **1984**, *106*, 1845. (e) Burnell, D. J.; Valenta, Z. *CC* **1985**, 1247. (f) Pandey, B.; Zope, U. R.; Ayyangar, N. R. *SC* **1989**, *19*, 585.
24. Danishefsky, S.; Craig, T. A. *T* **1981**, *37*, 4081.
25. Mehta, G.; Reddy, D. S.; Murty, A. N. *CC* **1983**, 824.
26. Krohn, K. *TL* **1980**, *21*, 3557.
27. Gupta, R. C.; Raynor, C. M.; Stoodley, R. J.; Slawin, A. M. Z.; Williams, D. J. *JCS(P1)* **1988**, 1773.
28. McDougal, P. G.; Jump, J. M.; Rojas, C.; Rico, J. G. *TL* **1989**, *30*, 3897.
29. Dauben, W. G.; Bunce, R. A. *TL* **1982**, *23*, 4875.
30. Monti, S. A. *JOC* **1966**, *31*, 2669.

31. Kushch, S. D.; Izakovich, E. N.; Khidekel, M. L.; Strelets, V. V. *IZV* **1981**, *7*, 1500 (*CA* **1981**, *95*, 149 595s).

Teng-Kuei Yang & Chi-Yung Shen
National Chung-Hsing University, Taichung, Republic of China

(*S*)-4-Benzyl-2-oxazolidinone

(**1**; R^1 = H, R^2 = ◀ Bn) (*S*)
[90719-32-7] $C_{10}H_{11}NO_2$ (MW 177.20)
(Li salt)
[123731-35-1]
(**2**; R^1 = H, R^2 = ···Bn) (*R*)
[102029-44-7;] $C_{10}H_{11}NO_2$ (MW 177.20)
(Li salt)
[128677-61-2]
(**3**; R^1 = H, R^2 = ◀ *i*-Pr) (*S*)
[17016-83-0] $C_6H_{11}NO_2$ (MW 129.16)
(Li salt)
[96021-69-1]
(**4**; R^1 = ···Ph, R^2 = ···Me) (4*R*,5*S*)
[77943-39-6] $C_{10}H_{11}NO_2$ (MW 177.20)
(Li salt)
[92061-65-7]
(**5**; R^1 = ◀ Ph, R^2 = ◀ Me) (4*S*,5*R*)
[16251-45-9] $C_{10}H_{11}NO_2$ (MW 177.20)
(Li salt)
[127882-97-7]
(**6**; R^1 = H, R^2 = ◀ Ph) (*S*)
[99395-88-7] $C_9H_9NO_2$ (MW 163.18)
(**7**; R^1 = H, R^2 = ···Ph) (*R*)
[90319-52-1] $C_9H_9NO_2$ (MW 163.18)
(**8**; R^1 = H, R^2 = ◀ *t*-Bu) (*S*)
[54705-42-9] $C_7H_{13}NO_2$ (MW 143.19)

(chiral auxiliaries used in asymmetric alkylations,[1] acylations,[2] halogenations,[3] aminations,[4] hydroxylations,[5] aldol reactions,[6] conjugate additions,[7,8] Diels–Alder reactions,[9] acyl transfer,[10] and sulfinyl transfer[11])

Physical Data: (**1**) mp 87–89 °C; (**2**) mp 85–87 °C; (**3**) mp 71–72 °C; (**4**) mp 118–121 °C; (**5**) mp 118–121 °C; (**6**) mp 130–132 °C; (**7**) mp 130–132 °C; (**8**) mp 118–120 °C.
Solubility: sol most polar organic solvents.
Form Supplied in: white crystalline solid; commercially available.
Analysis of Reagent Purity: 99% purity attainable by GLC.
Handling, Storage, and Precautions: no special handling or storage precautions are necessary. There is no known toxicity. It may be harmful by inhalation, ingestion, or skin absorption and may cause skin or eye irritation.

Synthesis of the Chiral Oxazolidinone Auxiliaries. (*S*)-4-Benzyl- (**1**), (*R*)-4-benzyl- (**2**), (*S*)-4-*i*-propyl- (**3**), (4*R*,5*S*)-4-methyl-5-phenyl- (**4**), (*S*)-4-*t*-butyl- (**8**), and (*S*)-4-phenyl-2-oxazolidinones (**6**) are commercially available. Typical procedures to form these chiral auxiliaries involve the reduction of α-amino acids to the corresponding amino alcohols or the purchase of amino alcohols, followed by formation of the cyclic carbamate (eq 1). A number of high-yielding methods of reduction have been employed for this transformation, including ***Boron Trifluoride Etherate/Borane-Dimethyl Sulfide***,[12] ***Lithium Aluminum Hydride***,[1,6,13,14] ***Sodium Borohydride/Iodine***,[15] and ***Lithium Borohydride/Chlorotrimethylsilane***[1,2].[8] Selection among these methods is largely based upon cost of reagents and ease of performance. Reagents for effecting the second transformation include ***Diethyl Carbonate/Potassium Carbonate***[12] or ***Phosgene***,[16–18] with the former being preferable for large-scale production. Ureas,[19,20] dioxolanones,[21] chloroformates,[22] trichloroacetate esters,[22,23] ***N,N'-Carbonyldiimidazole***,[24] and ***Carbon Monoxide*** with catalytic elemental ***Sulfur***[25] or ***Selenium***[26,27] provide alternatives for the transformation of amino alcohols to the derived oxazolidinones.

(1)

Conversion of the appropriate α-amino acids to oxazolidinones may also be performed as a one-pot procedure, obviating the need to isolate the intermediate amino alcohols (eqs 2 and 3).[28,29] Overall isolated yields for these procedures are 70–80%.

(2) 1. $LiAlH_4$; 2. NaOH; 3. $(CCl_3O)_2CO$; 80%

(3) 1. $BF_3{\cdot}OEt_2$, $BH_3{\cdot}DMS$; 2. NaOH; 3. $(CCl_3O)ClCO$; 78%

Carbamate-protected amino alcohols also yield oxazolidinones upon treatment with base (eq 4)[30,31] or ***p-Toluenesulfonyl Chloride*** (eq 5).[32] The latter reaction requires the protection of the amino group as the *N*-methylated carbamate for selective inversion of the hydroxyl-bearing center.

(4) 1. $ClCO_2Ph$, pH 8.5, 100%; 2. $BH_3{\cdot}THF$, 94%; *t*-BuOK, 82%

(5) TsCl, 70–80%

Resolution of racemic oxazolidinones affords either enantiomer of the auxiliary and provides a versatile route to unusually substituted derivatives (eq 6).[33]

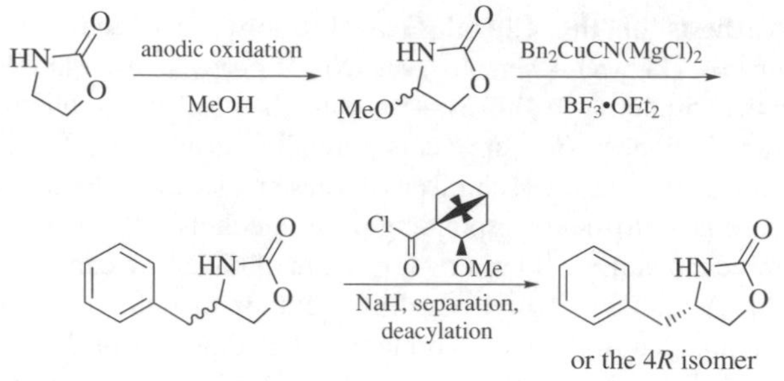

(6)

or the 4*R* isomer

Methods of *N*-Acylation. Lithiated oxazolidinones add to acid chlorides (eq 7)[6,34] and mixed anhydrides (eq 8)[35,36] in high yields to form the derived *N*-acyl imides. In the latter case the anhydride may be formed in situ with ***Trimethylacetyl Chloride***, and then condensed with the lithiated oxazolidinone selectively at the less hindered carbonyl moiety.

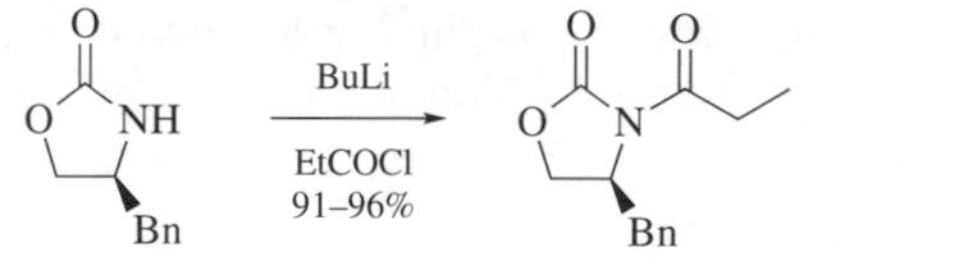

(7)

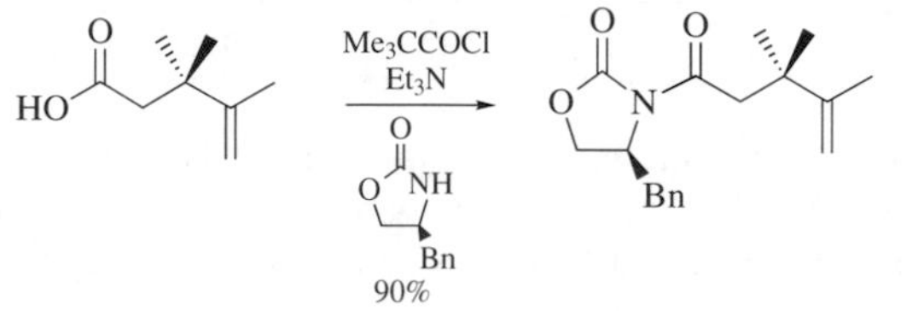

(8)

Acryloyl adducts cannot be formed through traditional acylation techniques due to their tendency to polymerize. These adducts may be obtained through reaction of acryloyl chloride with the bromomagnesium salt of the oxazolidinone auxiliary[9,37] or the *N*-trimethylsilyl derivative in the presence of ***Copper(II) Chloride*** and ***Copper*** powder.[38] These methods yield products in the range of 50–70%.

Methods of *N*-Alkylation. In analogy with acylation techniques, metalated oxazolidinones add to alkyl halides to afford the *N*-alkylated products in high yields.[39–42]

Enolization of *N*-Acyloxazolidinones. Various methods have been developed to effect the enolization of chiral *N*-acyloxazolidinones. In alkylation reactions, both ***Lithium Diisopropylamide*** and ***Sodium Hexamethyldisilazide*** deprotonate these imides to provide the (*Z*)-enolates in >100:1 selectivity.[1]

Di-n-butylboryl Trifluoromethanesulfonate with a tertiary amine also provides the (*Z*)-enolates of chiral acyl oxazolidinones in >100:1 selectivity for use in subsequent aldol additions.[6,14] With ***Triethylamine***, ***Diisopropylethylamine*** (Hünig's base), or ***2,6-Lutidine*** the order of addition is of no consequence to enolization.[43] Triethylamine has traditionally seen the greatest utilization in these reactions based upon cost considerations; however, with certain sensitive aldehyde substrates, lutidine provides milder reaction conditions.[44]

The (*Z*)-enolate is also accessed exclusively using titanium enolization procedures.[45,47] Irreversible complexation of ***Titanium(IV) Chloride*** with tertiary amine bases demands complexation of the substrate with the Lewis acid prior to treatment with either triethylamine or Hünig's base. Reactions using Hünig's base occasionally display higher diastereoselectivities, particularly in Michael additions.[7,45] Of the alkoxy titanium species employed in imide enolization, only $TiCl_3(O\text{-}i\text{-}Pr)$ is capable of quantitative enolate formation. In these reactions, order of addition of reagents is not significant. These enolates demonstrate enhanced nucleophilicity, albeit with somewhat diminished diastereoselectivity.

Other Lewis acids have been demonstrated to provide moderate levels of enolization, including ***Aluminum Chloride***, ***Magnesium Bromide***, and ***Tin(II) Trifluoromethanesulfonate***.[45,46] However, $SnCl_4$, Me_2AlCl, and $ZrCl_4$ failed to provide detectable enolization.[45]

Enolate Alkylation. Alkylation of chiral *N*-acyloxazolidinones by simple alkyl and allylic halides occurs through the chelated lithiated (*Z*)-imide enolates to afford products in greater than 93:7 diastereoselectivities (eqs 9 and 10).[1] For small electrophiles such as ***Iodomethane*** and ***Ethyl Iodide***, NaHMDS proved to be the enolization base of choice. On selected substrates, alkyl triflates also demonstrate promise as alkylating agents.[34]

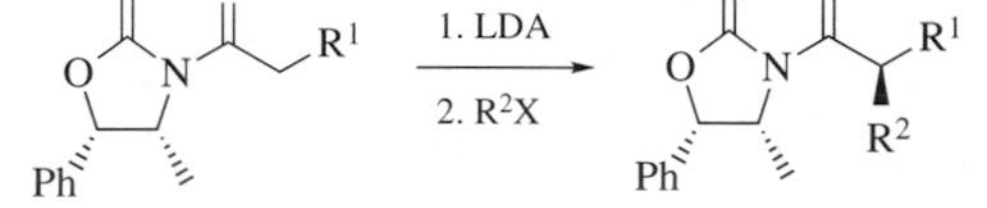

(9)

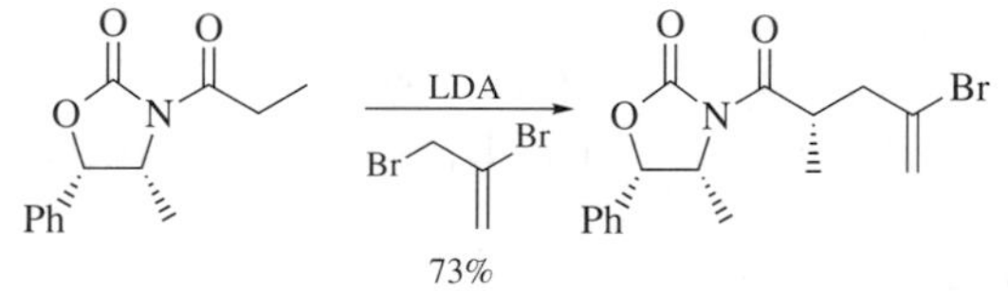

(10)

For benzyloxymethyl electrophiles, titanium enolates are superior to the corresponding lithium enolates in both yield and alkylation diastereoselectivity (eq 11). Unfortunately, the analogous *p*-methoxybenzyl-protected β-hydroxy adducts cannot be obtained by this method. In other cases the titanium methodology complements the corresponding reactions of the lithium and sodium enolates for S_N1-like electrophiles.[47] It is noteworthy that imides may be selectively enolized under all of the preceding conditions in the presence of esters (eq 12).

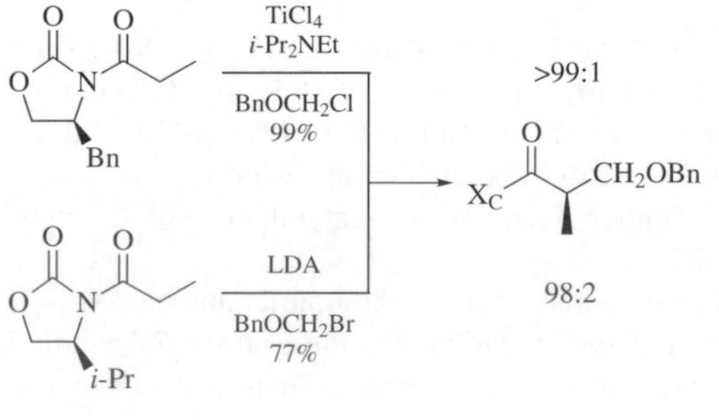

(11)

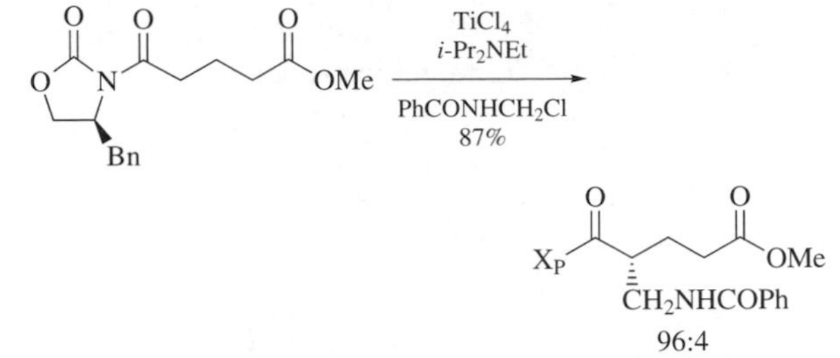

(12)

Treatment of the silyl enol ethers of *N*-acyloxazolidinones with selected electrophiles that do not require Lewis acid activation similarly results in high induction of the same enolate face (eq 13).[48] The facial bias of this conformationally mobile system improves with the steric bulk of the silyl group.

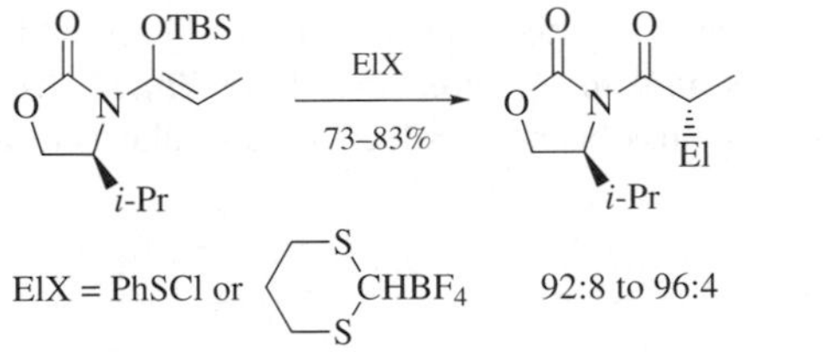

(13)

Chiral oxazolidinones have also been used to induce chirality in $TiCl_4$-mediated allylsilane addition reactions to α-keto imides (eq 14).[49]

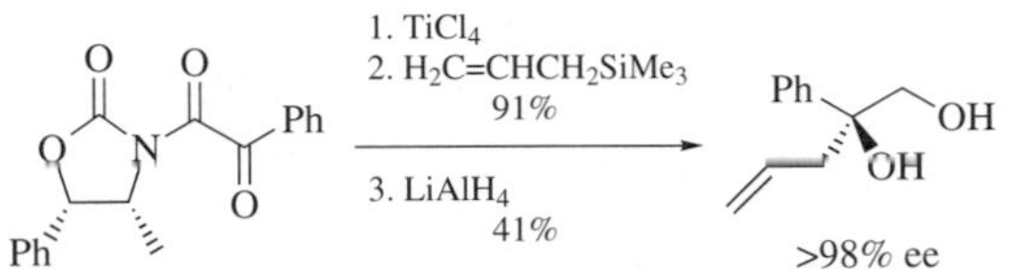

(14)

Enolate Alkylations with Transition Metal Coordinated Electrophiles. Coordination of various transition metals to dienes and aromatic compounds sufficiently activates these compounds to nucleophilic addition, resulting in high asymmetric induction at the α-center. However, the manganese complexes of various benzene derivatives couple with lithium enolates in low selectivity at the nascent stereogenic center on the ring (eq 15).[50]

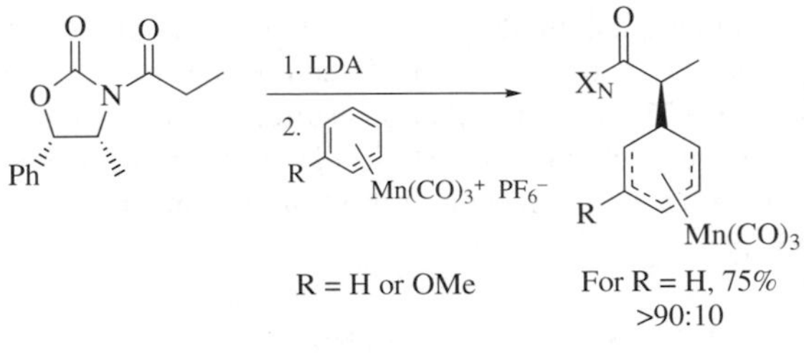

(15)

In contrast, molybdenum and iron diene complexes undergo the same type of reaction with chiral lithium imide enolates, with moderate to good induction at the β-position (eq 16).[51–53]

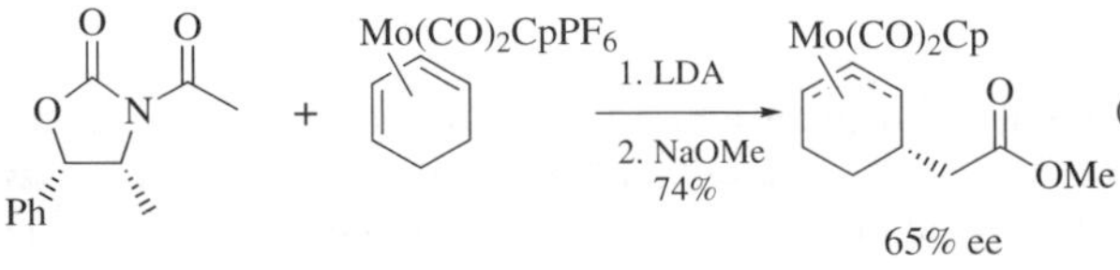

(16)

Dicobalt hexacarbonyl-coordinated propargyl ethers also combine with imide boron enolates through a kinetic resolution of the rapidly interconverting propargylic cation isomers to afford a 92:8 mixture of isomers at the β-center in 80% yield (eq 17). Stereocontrol of the α-center is 97:3.[54]

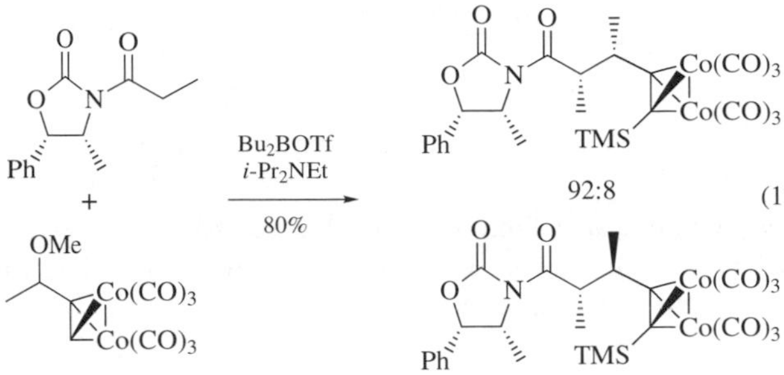

(17)

Enolate Acylation. Acylation of these enolates provides a direct route to β-dicarbonyl systems. Acylations generally proceed with >95% diastereoselection in 83–95% yields, with the valine-derived auxiliary providing slightly higher selectivity (eq 18).[2] The sense of induction is consistent with reaction through the chelated lithium (*Z*)-enolate, and the newly generated stereocenter is retained through routine manipulations.

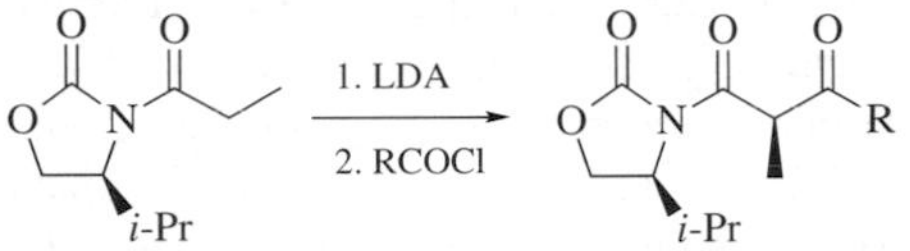

(18)

An alternate approach to these useful 1,3-dicarbonyl substrates may be achieved through enolate orthoester acylation. Titanium enolates have been employed to effect this transformation (eq 19).[45,47] Similarly, treatment of the titanium enolate of β-ketoimide with dioxolane orthoesters results in the formation of a masked tricarbonyl compound (eq 20). Trimethyl orthoacetate and ***Triethyl Orthoacetate*** are not appropriate partners in these coupling reactions.[45,47]

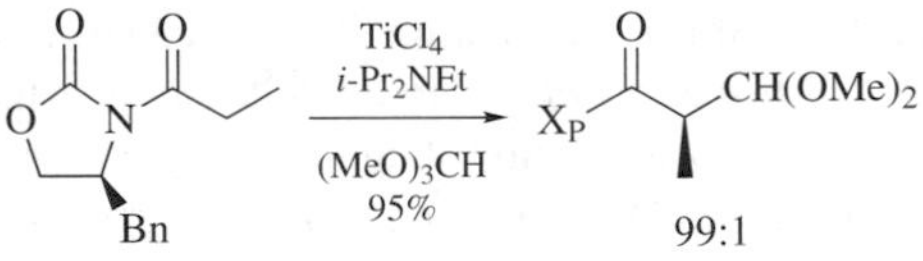

(19)

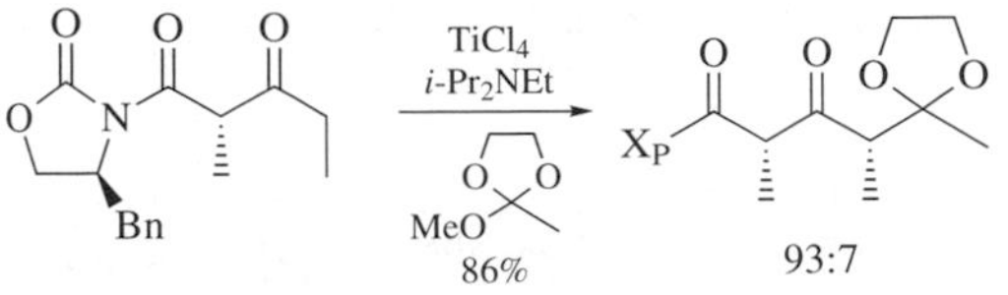

(20)

Michael Addition. Titanium imide enolates are excellent nucleophiles in Michael reactions. Michael acceptors such as ethyl vinyl ketone, ***Methyl Acrylate***, ***Acrylonitrile***, and *t*-butyl acrylate react with excellent diastereoselection (eq 21).[7,45] Enolate chirality transfer is predicted by inspection of the chelated (*Z*)-enolate. For the less reactive unsaturated esters and nitriles, enolates generated from $TiCl_3(O\text{-}i\text{-}Pr)$ afford superior yields, albeit with slightly lower selectivities. The scope of the reaction fails to encompass β-substituted, α,β-unsaturated ketones which demonstrate essentially no induction at the prochiral center. Furthermore, substituted unsaturated esters do not act as competent Michael acceptors at all under these conditions.

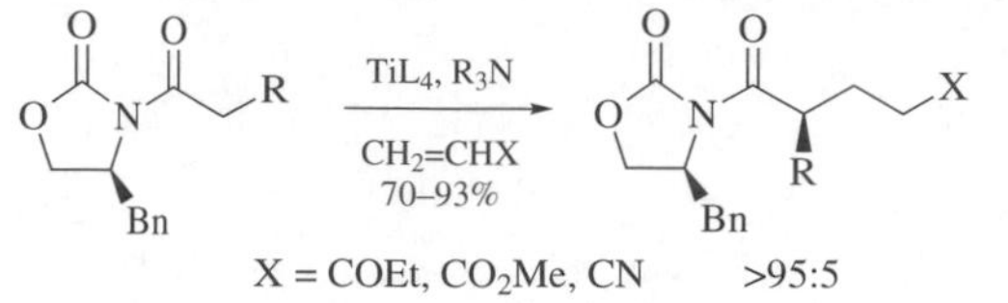

(21)

Various chelated lithium imide enolates have also served as nucleophiles in Michael additions to 3-trifluoromethyl acrylate, favoring the *anti* isomer (eq 22).[55]

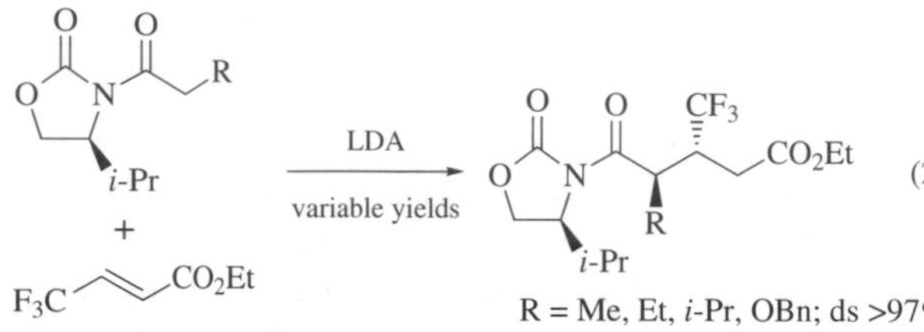

(22)

Enolate Hydroxylation. Treatment of the sodium enolates with the Davis oxaziridine reagent ***trans-2-(Phenylsulfonyl)-3-phenyloxaziridine*** affords the hydroxylated products with the same sense of induction as the alkylation products (eq 23).[5,35] Although high diastereoselectivity may be achieved with ***Oxodiperoxymolybdenum(pyridine)(hexamethylphosphoric triamide)*** (MoOPH), such reactions proceed in lower yields.

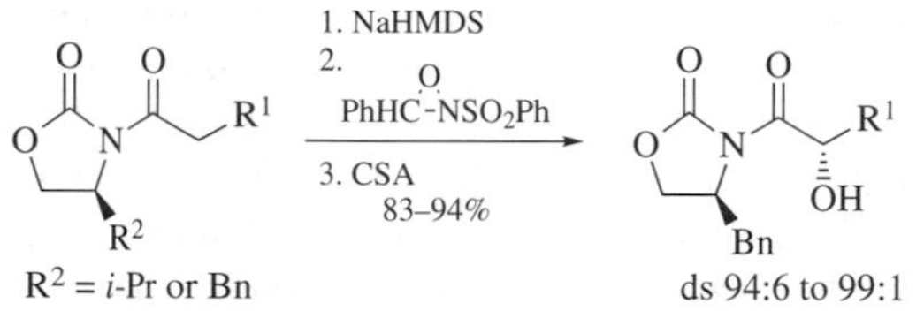

(23)

Enolate Amination. Amination likewise can be effected using ***Di-t-butyl Azodicarboxylate*** (DBAD).[4,56] Despite the excellent yields and diastereoselectivity obtained using this methodology (eq 24), the harsh conditions required for further transformation of the resultant hydrazide adducts (***Trifluoroacetic Acid*** and hydrogenation at 500 psi over ***Raney Nickel*** catalyst) limit its synthetic utility.

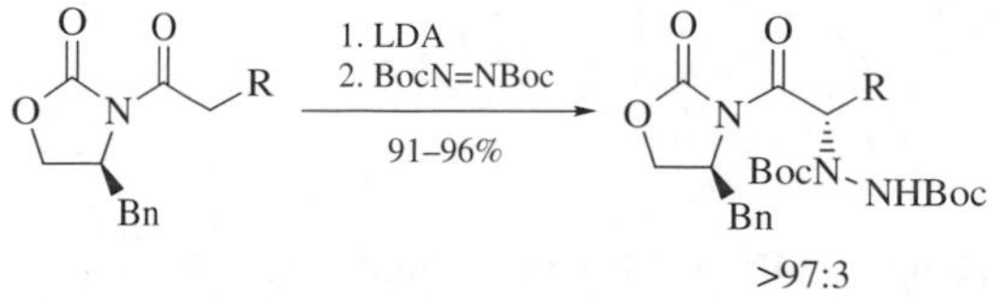

(24)

As a method for the synthesis of α-amino acids, the hydrazide methodology has now largely been supplanted by direct enolate azidation (eq 25).[4,57] These adducts are susceptible to mild chemical modification to afford *N*-protected α-amino acid derivatives. Under optimal conditions, yields range from 74–91% and selectivities from 91:9 to >99:1. Imide enolization can be carried out selectively in the presence of an enolizable *t*-butyl ester and suitably protected amino groups.

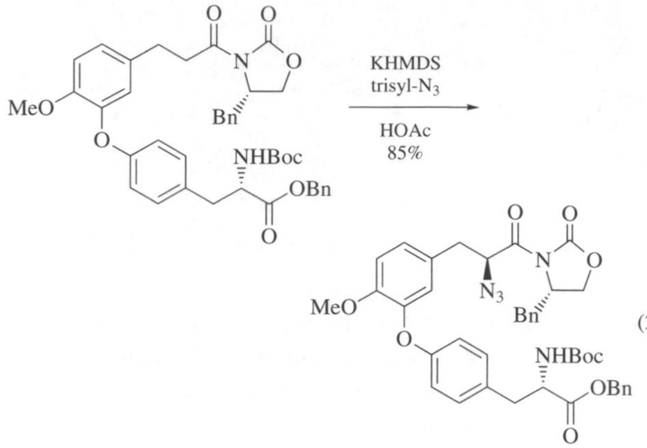

(25)

Hydrogenation of the azide moiety readily provides the amine using ***Palladium on Carbon*** and H_2 or ***Tin(II) Chloride***. This methodology has been extended to the synthesis of arylglycines (eq 26).[58]

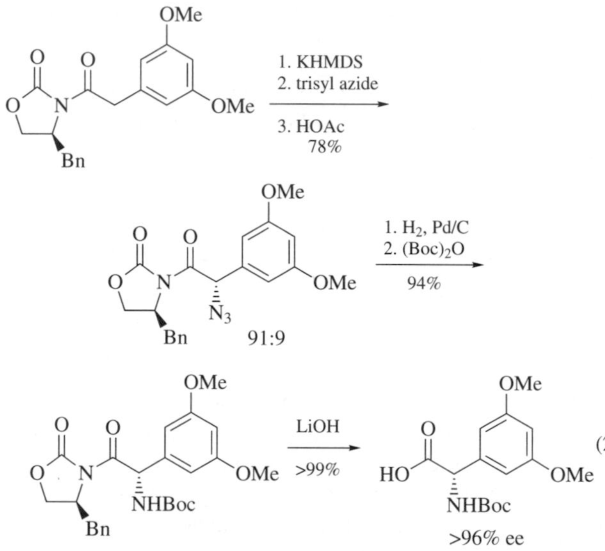

(26)

Failure to use ***2,4,6-Triisopropylbenzenesulfonyl Azide*** results in substantial diazo imide formation. However, optimization for the formation of the α-diazo imide compounds can be achieved with NaHMDS and *p*-nitrobenzenesulfonyl azide, followed by a neutral quench (eq 27).[4] These diazo compounds, however, have failed to demonstrate utility in asymmetric carbenoid chemistry.[59]

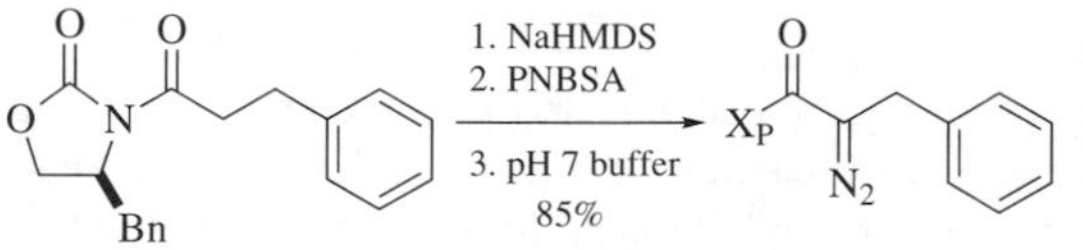

(27)

Enolate Halogenation. Enolate halogenation is achieved by reaction of the boryl enolate with ***N-Bromosuccinimide***, affording configurationally stable α-bromo imides in >94:6 diastereoselectivity in 80–98% yield (eq 28).[3,4] The sense of induction suggests halogenation of the chelated (*Z*)-enolate. Introduction of an α-fluoro substituent can be effected by the treatment of imide enolates or α,β-unsaturated enolates with *N*-fluoro-*o*-benzenedisulfonimide (eq 29).[60]

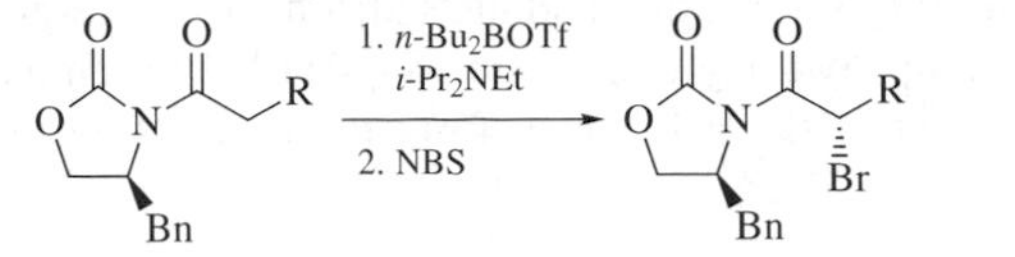

1. n-Bu_2BOTf, i-Pr_2NEt; 2. NBS (28)

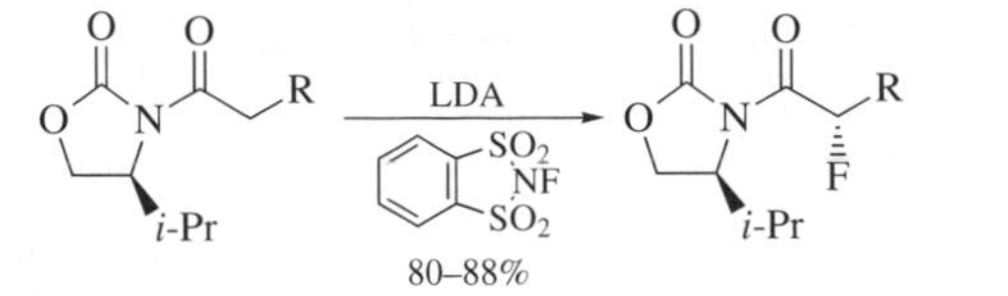

LDA, (29)

80–88%

R = *n*-Bu, *t*-Bu, Bn, Ph   86–97% de

Displacement of a halide at the α-position with tetramethylguanidinium azide (TMGA) introduces nitrogen functionality with inversion of the original halide configuration and <1% epimerization (eq 30).[3,4]

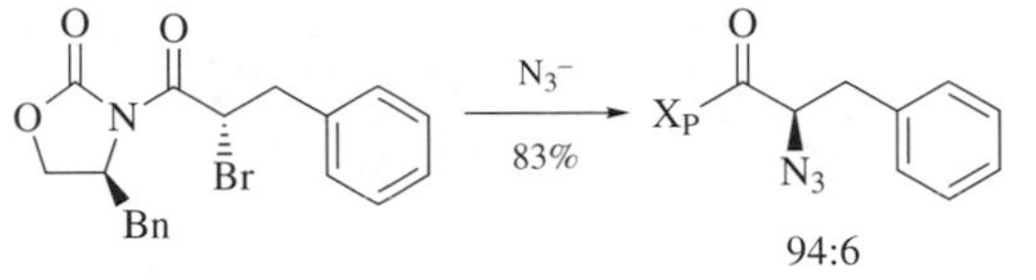

N_3^-, 83% (30)

94:6

In those transformations where other stereogenic centers reside proximal to the prochiral center, auxiliary control is dominant in most cases (eq 31).[61]

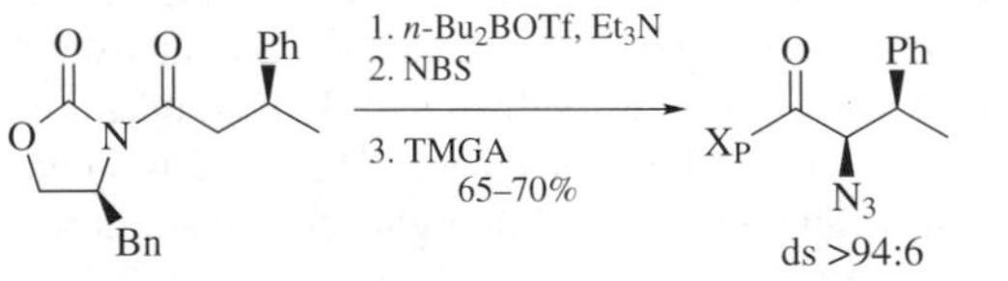

1. n-Bu_2BOTf, Et_3N; 2. NBS; 3. TMGA, 65–70% (31)

ds >94:6

Aldol Reactions. The dibutyl boryl enolates of chiral acyloxazolidinones react to afford the *syn*-aldol adducts with virtually complete stereocontrol (eq 32).[6,13,14,43,61–64] Notably, the sense of induction in these reactions is opposite to that predicted from the analogous alkylation reactions. This reaction is general for a wide range of aldehydes and imide enolates.[36,65–69] Enolate control overrides induction inherent to the aldehyde reaction partner.

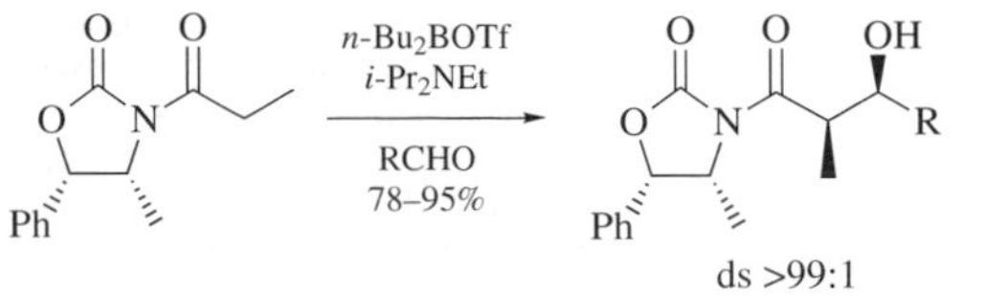

n-Bu_2BOTf, i-Pr_2NEt; RCHO, 78–95% (32)

ds >99:1

Titanium enolates of propionyloxazolidinones also undergo aldol reactions with the same sense of induction as the boryl counterparts, but require two or more equivalents of amine base to afford adducts in marginally higher yields but diminished selectivity (eq 33).[45]

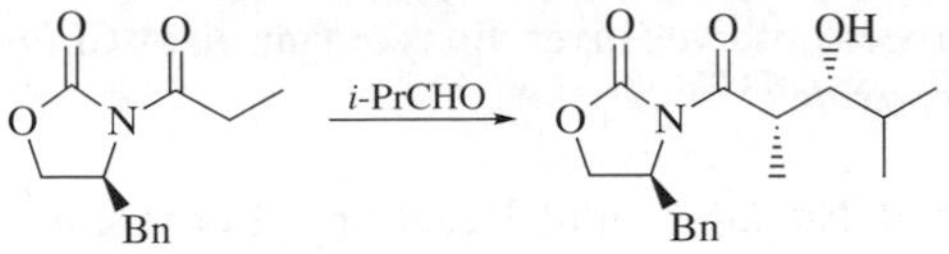

i-PrCHO (33)

Enolization	Yield	Stereoselection
$TiCl_4$, i-Pr_2NEt	87%	94:6
$TiCl_4$, TMEDA	84%	98:2
n-Bu_2BOTf, Et_3N	83%	>99:1

A second entry to dicarbonyl substrates utilizes the aldol reaction to establish the α-methyl center prior to oxidation of the β-hydroxyl moiety. Commonly, this oxidation is performed using the ***Sulfur Trioxide-Pyridine*** complex, which results in <1% epimerization of the methyl-bearing center (eq 34).[2] Interestingly, this procedure procures the opposite methyl stereochemistry from that obtained through enolate acylation of the same enantiomer of oxazolidinone.

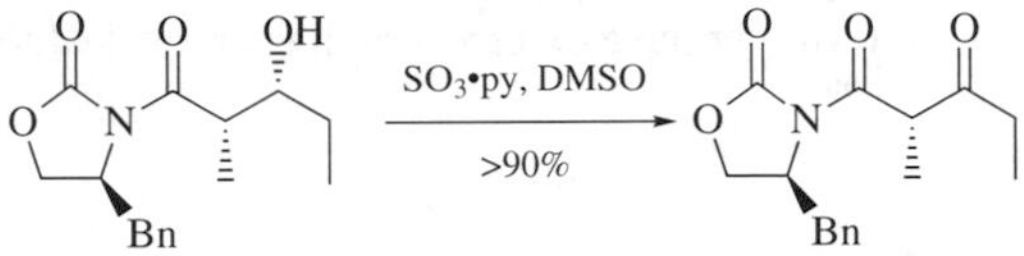

SO_3•py, DMSO, >90% (34)

Non-Evans Aldol Reactions. Either the *syn*- or *anti*-aldol adducts may be obtained from this family of imide-derived enolates, depending upon the specific conditions employed for the reaction. Although the illustrated boron enolate affords the illustrated *syn*-aldol adduct in high diastereoselectivity, the addition reactions between this enolate and Lewis acid-coordinated aldehydes afford different stereochemical outcomes depending on the Lewis acid employed (eq 35).[70] Open transition states have been proposed for the ***Diethylaluminum Chloride*** mediated, *anti*-selective reaction. These *anti*-aldol reactions have been used in kinetic resolutions of 2-phenylthio aldehydes.[71]

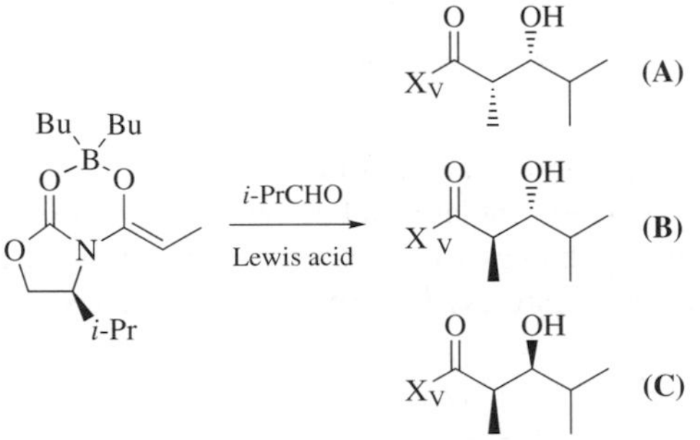

i-PrCHO, Lewis acid → (A), (B), (C) (35)

Lewis acid	Yield	A:B:C
$TiCl_4$ (2 equiv)	83%	0:16:84
$SnCl_4$ (2 equiv)	60%	0:13:87
Et_2AlCl (2 equiv)	63%	0:95:5

Enolates derived from α-haloimides also exhibit metal-dependent *syn*/*anti*-aldol diastereoselection. The derived Li, Sn^{IV}, and Zn enolates afford the *anti* isomer in reactions with aromatic aldehydes, while the corresponding B and Sn^{II} enolates lead to the conventional *syn* products.[72,73] The 'non-Evans' *syn* adducts have also been observed in reactions organized by ***Chlorotitanium Triisopropoxide***.[74,75]

Crotonyl Enolate Aldol Reactions. Boron enolates of the *N*-crotonyloxazolidinones have been shown to afford the expected *syn*-aldol adducts (eq 36).[76,77] The propensity for self-condensation during the enolization process is minimized by the use of triethylamine over less kinetically basic amines.

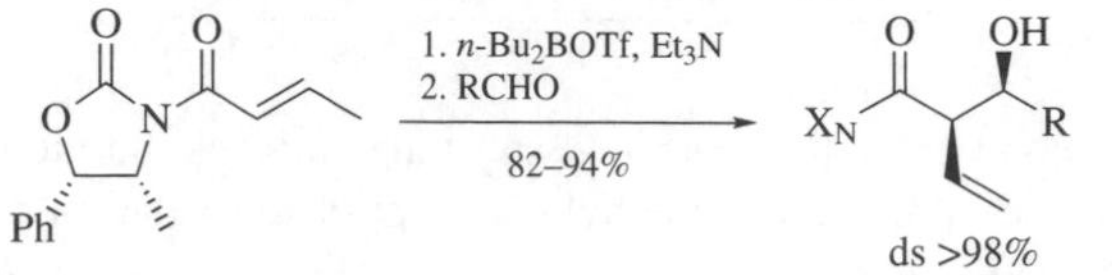

1. n-Bu_2BOTf, Et_3N; 2. RCHO; 82–94%; ds >98% (36)

α-Alkoxyacetate Aldol Reactions. The enolates derived from *N*-α-alkoxyacetyloxazolidinones also provide good yields of aldol adducts. Proper choice of reaction conditions leads to either the *syn* (eq 37)[78] or *anti* (eq 38)[46] adducts. In an application of this aldol reaction in the synthesis of cytovaricin, a complex chiral aldehyde was found to turnover the expected *syn* diastereoselectivity of the boron enolate.[66]

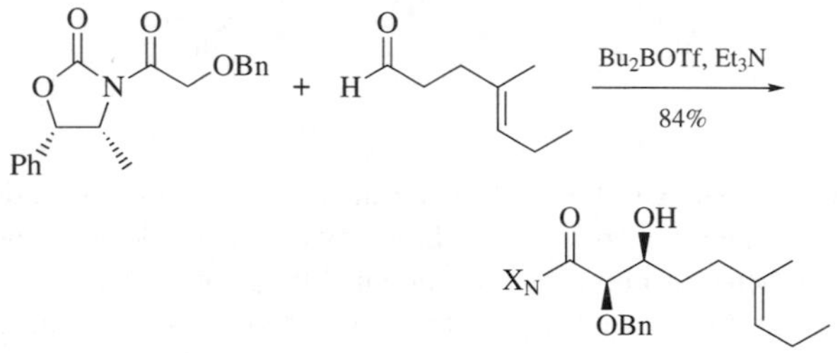

Bu_2BOTf, Et_3N; 84%; ds >95:5 (37)

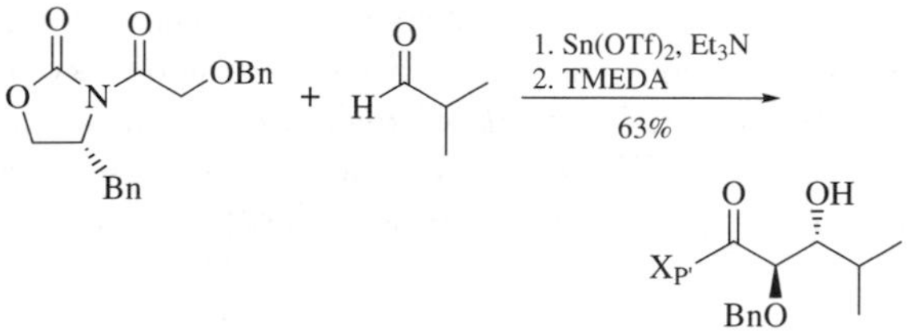

1. $Sn(OTf)_2$, Et_3N; 2. TMEDA; 63% (38)

ds of desired to the total amount of other isomers = 77:23

***N*-Isothiocyanoacetyl Aldol Reactions.** Auxiliary-controlled masked glycine enolate aldol reactions afford the chiral oxazolidine-2-thiones which can be cleaved to provide the *syn*-aldol adducts regardless of aldehyde stereochemistry (eq 39).[79]

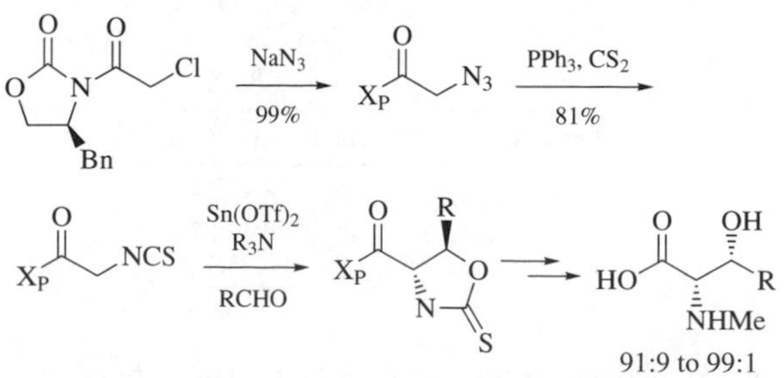

NaN_3, 99%; PPh_3, CS_2, 81%; $Sn(OTf)_2$, R_3N, RCHO; 91:9 to 99:1, 81–92% (39)

***N*-Haloacetyl Aldol Reactions.** *N*-Haloacetyloxazolidinones form suitable enolate partners in aldol reactions, although complete aldehyde conversion requires the use of a slight excess of imide (eq 40). The products can be chromatographed to diastereomeric purity.[4,80] Nucleophilic azide displacement of α-halo-β-hydroxy *syn* aldol adducts affords the corresponding *anti* α-amino-β-hydroxy compounds (eq 41).[4,80] Intramolecular displacement of the halogen to form the α-amino product is also possible (eq 42).[80]

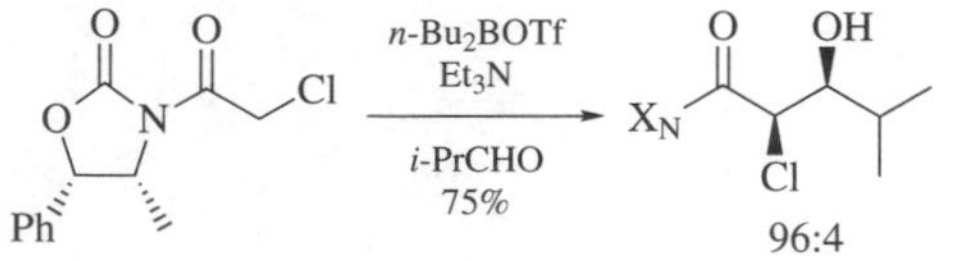

n-Bu_2BOTf, Et_3N; i-PrCHO; 75%; 96:4 (40)

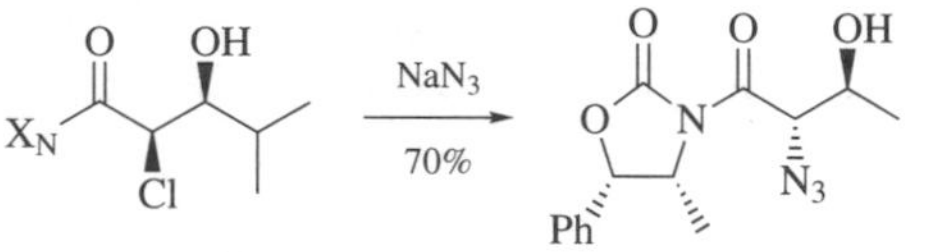

NaN_3; 70% (41)

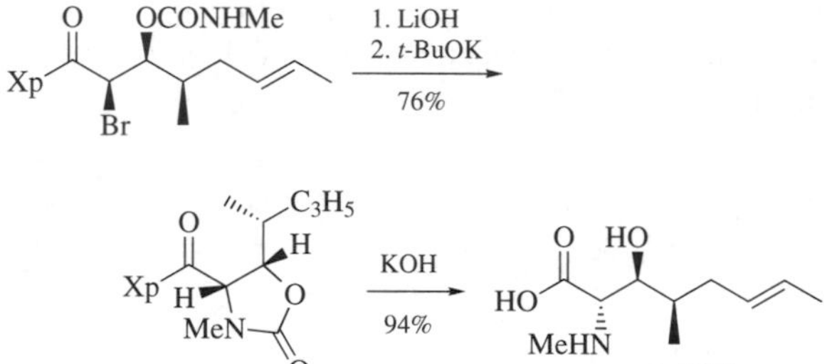

1. LiOH; 2. t-BuOK; 76%; KOH; 94%; ds 93:7 (42)

Acetate Aldol Equivalents. In contrast to the reliably excellent selectivities of α-substituted dibutylboryl imide enolates, boron enolates derived from *N*-acetyloxazolidinones lead to a statistical mixture of aldol adducts under the same reaction conditions. Acetate enolate equivalents may be obtained from these enolates bearing a removable α-substituent. To this end, thiomethyl- or thioethylacetyloxazolidinones (eq 43)[13] as well as haloacetyloxazolidinones can be submitted to highly selective boron-mediated aldol reactions. Products can be transformed to the acetate aldol products via desulfurization with either Raney Ni[81] or ***Tri-n-butyltin Hydride*** and ***Azobisisobutyronitrile***,[82] or via dehalogenation with ***Zinc–Acetic Acid*** (eq 44).[81] This latter procedure provides several advantages over the sulfur methodology, including ease of imide preparation and improved overall yields.

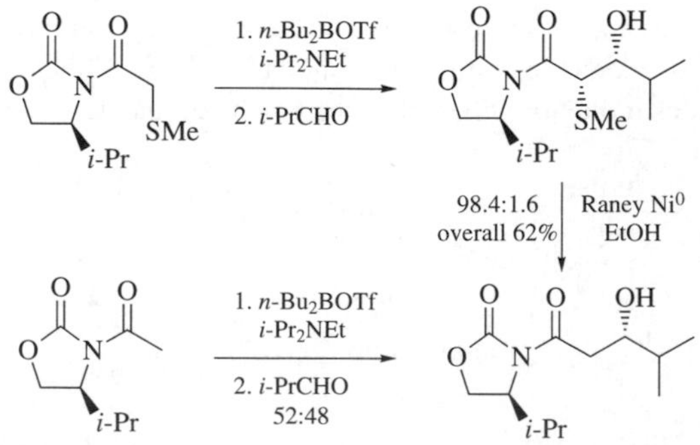

1. n-Bu_2BOTf, i-Pr_2NEt; 2. i-PrCHO; 98.4:1.6, overall 62%; Raney Ni^0, EtOH; 1. n-Bu_2BOTf, i-Pr_2NEt; 2. i-PrCHO; 52:48 (43)

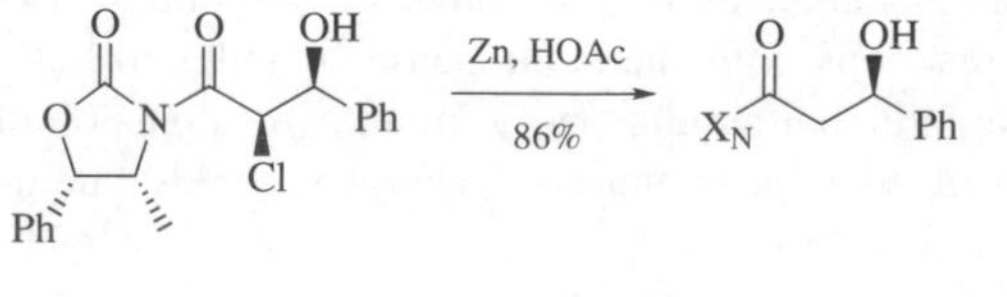

Zn, HOAc; 86% (44)

β-Ketoimide Aldol Reactions. As has been demonstrated, chiral oxazolidinones provide a gateway into asymmetric β-ketoimides via either an aldol–oxidation sequence, or enolate acylation. These substrates can then undergo an iterative aldol reaction, where chirality is induced by the methyl-bearing α-center. To date, three of the four diastereomeric aldol adducts may be selectively obtained with a variety of aldehydes (eq 45).[36,83–86]

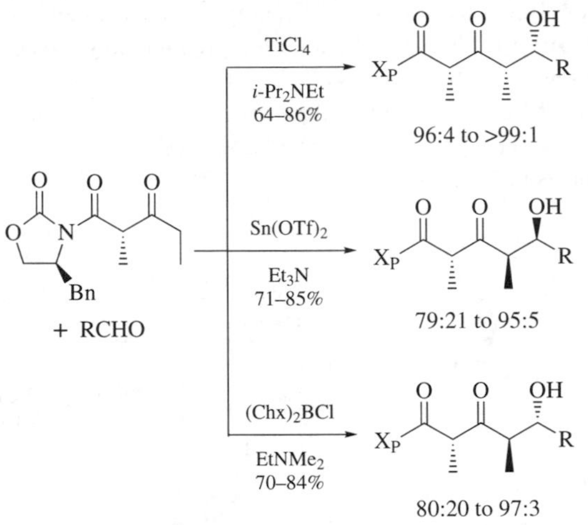

+ RCHO

$TiCl_4$, i-Pr_2NEt, 64–86%: 96:4 to >99:1

$Sn(OTf)_2$, Et_3N, 71–85%: 79:21 to 95:5 (45)

$(Chx)_2BCl$, $EtNMe_2$, 70–84%: 80:20 to 97:3

Reformatsky Reactions. The Reformatsky reaction of α-halooxazolidinones provides an alternative to the more conventional aldol reaction. Although the traditional zinc-mediated Reformatsky using valine-derived compounds proceeds nonselectively,[87,88] the Sn^{II} modification with 2-bromo-2-methylpropionyloxazolidinone proceeds well (eq 46).[89,90] In this particular case, however, the geminal dialkyl substituents favor the endocyclic carbonyl acyl transfer of the auxiliary by the aldolate oxygen.

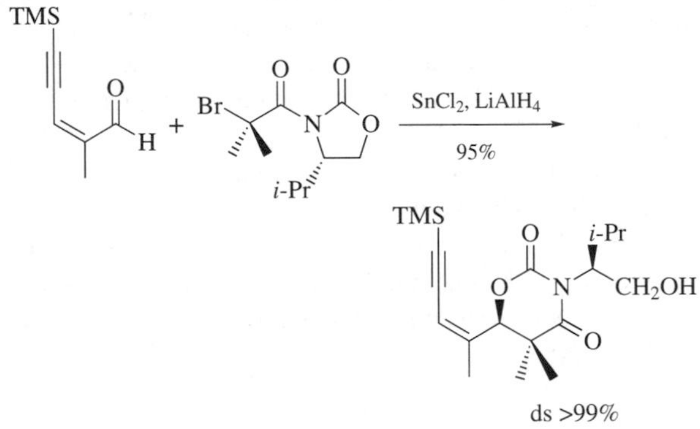

$SnCl_2$, $LiAlH_4$, 95% (46)

ds >99%

Acyl Transfer Reactions. (*S*)-*N*-benzoyloxazolidinones have been used as acyl transfer reagents to effect the kinetic resolution of racemic alcohols.[10] The bromomagnesium alkoxides formed from phenyl *n*-alkyl alcohols selectively attack the exocyclic benzoyl moiety to afford recovered auxiliary and the derived (*R*)-benzoates in >90% ee and >90% yield (eq 47). The scope of this reaction seems to be limited to this class of substrates as selectivity drops with increasing the steric bulk of the alkyl group.

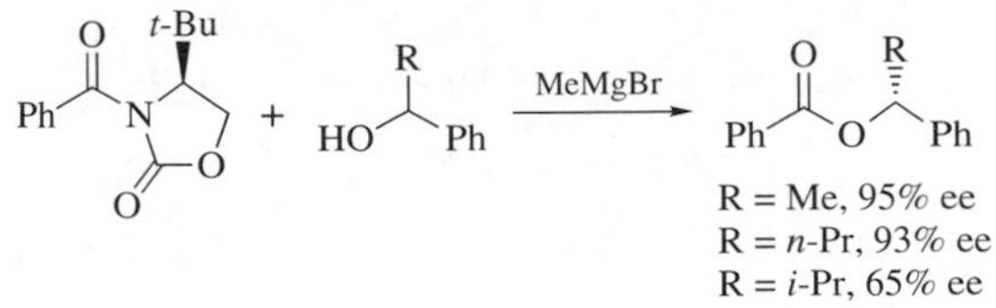

MeMgBr (47)

R = Me, 95% ee
R = *n*-Pr, 93% ee
R = *i*-Pr, 65% ee

Sulfinyl Transfer Reactions. Grignard reagents add to diastereomerically pure *N*-arylsulfinyloxazolidinones with inversion of configuration at sulfur to afford enantiopure dialkyl or aryl alkyl sulfoxides in excellent yields (eq 48).[11] Although broader in synthetic utility than the menthyl sulfinate esters,[91,92] this methodology is comparable to Kagan's chiral sulfite substrates as a strategy for constructing chiral sulfoxides.[93,94]

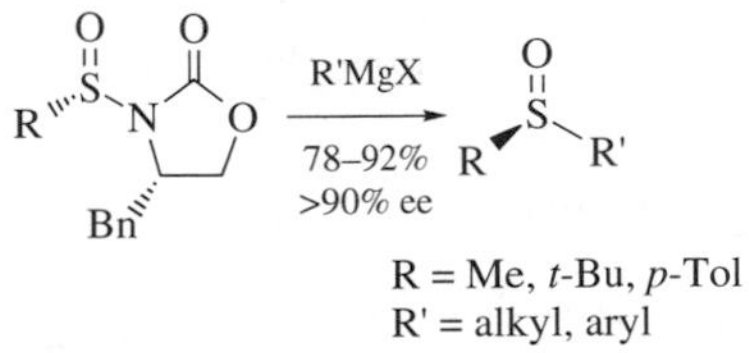

R'MgX, 78–92%, >90% ee (48)

R = Me, *t*-Bu, *p*-Tol
R' = alkyl, aryl

The *N*-arylsulfinyloxazolidinone methodology is readily extended to the formation of sulfinylacetates, sulfinates, and sulfinamides with >95% ee and high yields (eq 49).

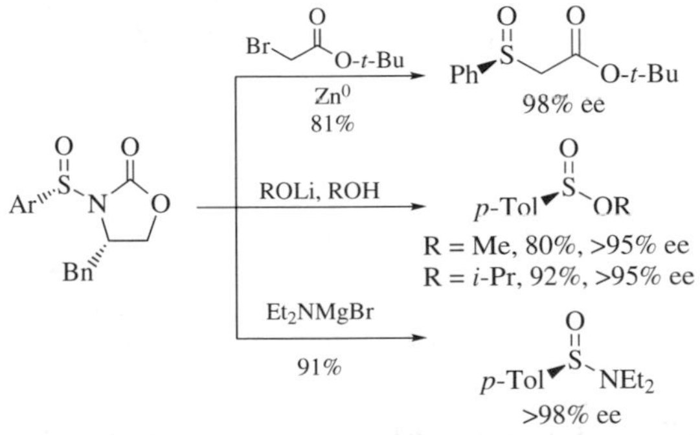

Zn^0, 81%: 98% ee

ROLi, ROH: R = Me, 80%, >95% ee; R = *i*-Pr, 92%, >95% ee (49)

Et_2NMgBr, 91%: >98% ee

Diels–Alder Reactions. Chiral α,β-unsaturated imides participate in Lewis acid-promoted Diels–Alder cycloaddition reactions to afford products in uniformly excellent *endo/exo* and *endo* diastereoselectivities (eqs 50 and 51).[9,37,95,96] Unfortunately, this reaction does not extend to certain dienophiles, including methacryloyl imides, β,β-dimethylacryloyl imides, or alkynic imides. Cycloadditions also occur with less reactive acyclic dienes with high diastereoselectivity (eq 52). Of the auxiliaries surveyed, the phenylalanine-derived oxazolidinones provided the highest diastereoselectivities. This methodology has been recently extended to complex intramolecular processes (eq 53).[68,95,97] In this case, use of the unsubstituted achiral oxazolidinone favored the undesired diastereomer.

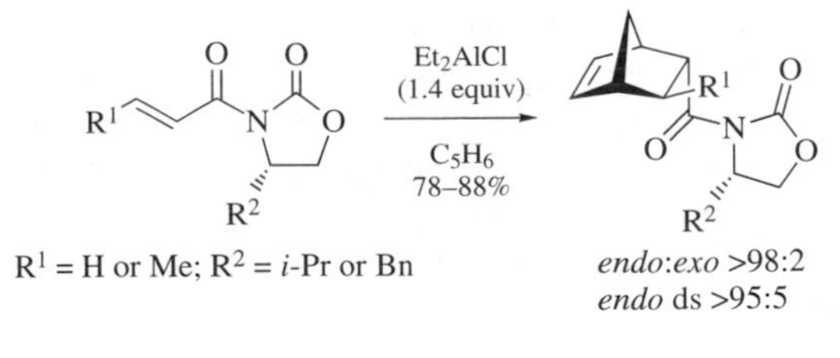

Et_2AlCl (1.4 equiv), C_5H_6, 78–88% (50)

R^1 = H or Me; R^2 = *i*-Pr or Bn

endo:exo >98:2
endo ds >95:5

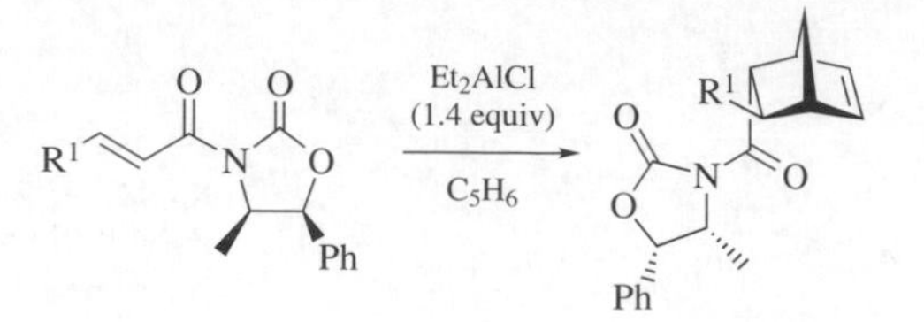

(51)

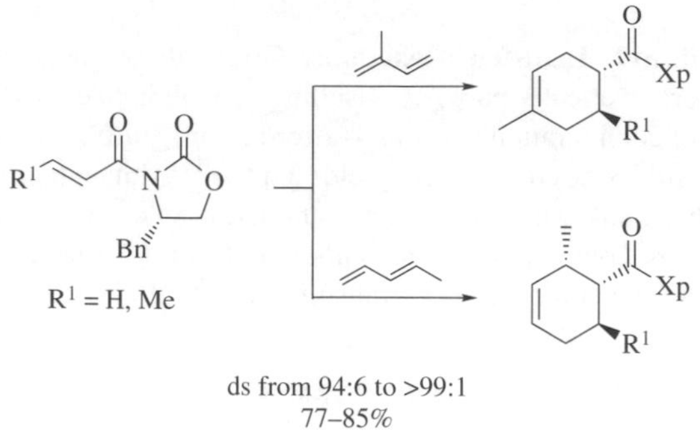

(52)

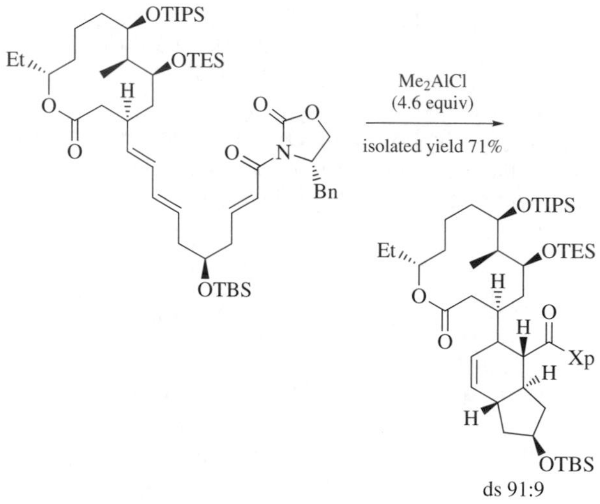

ds from 94:6 to >99:1
77–85%

(53)

ds 91:9

Staudinger Reactions. Chiral oxazolidinones have been employed as the chiral control element in the Staudinger reaction as well as the ultimate source of the α-amino group in the formation of β-lactams.[41] Cycloaddition of ketene derived from 4-(S)-phenyloxazolidylacetyl chloride with conjugated imines affords the corresponding β-lactams in 80–90% yields with excellent diastereoselectivity (eq 54). The auxiliary can then be reduced under Birch conditions to reveal the α-amino group.

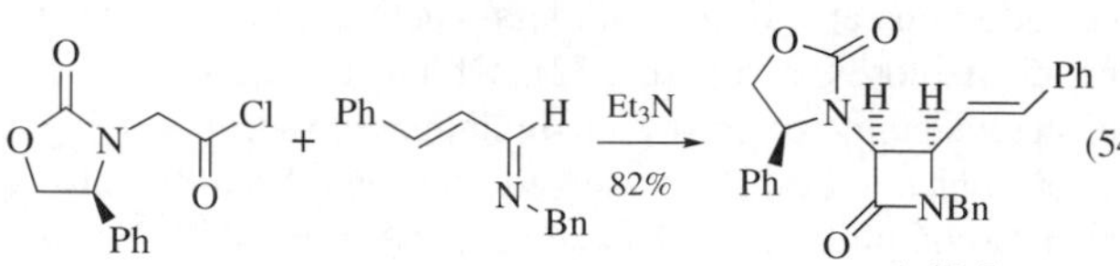

(54)

ds 95:5

Conjugate Addition Reactions. α,β-Unsaturated N-acyloxazolidinones have been implemented as Michael acceptors, inducing chirality in the same sense as in enolate alkylation reactions. Chiral α,β-unsaturated imides undergo 1,4-addition when treated with diethylaluminum chloride (eq 55). Photochemical initiation is required for the analogous reaction with ***Dimethylaluminum Chloride***.[96]

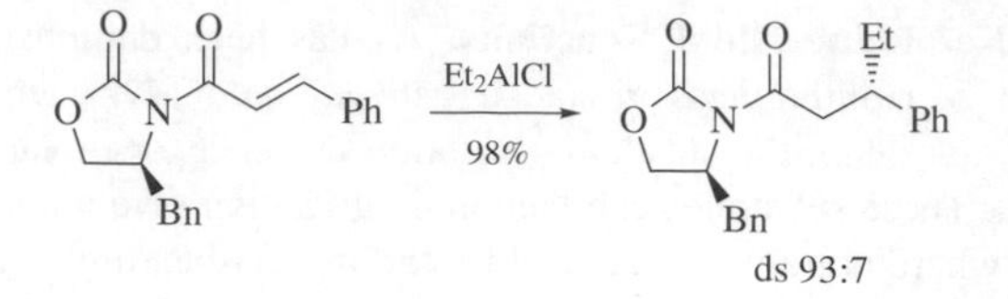

(55)

ds 93:7

Organocuprates also undergo conjugate addition with chiral α,β-unsaturated imides.[98] Treatment of the imides derived from 4-phenyl-2-oxazolidinone with methyl- or arylmagnesium halides and CuBr affords conjugate addition products in yields over 80% with few exceptions (eq 56).[8] Reaction diastereoselectivity appears to be contingent upon the use of the 4-phenyloxazolidinone auxiliary. The preceding methodology has been applied to the synthesis of β-methyltryptophan (eq 57).[99]

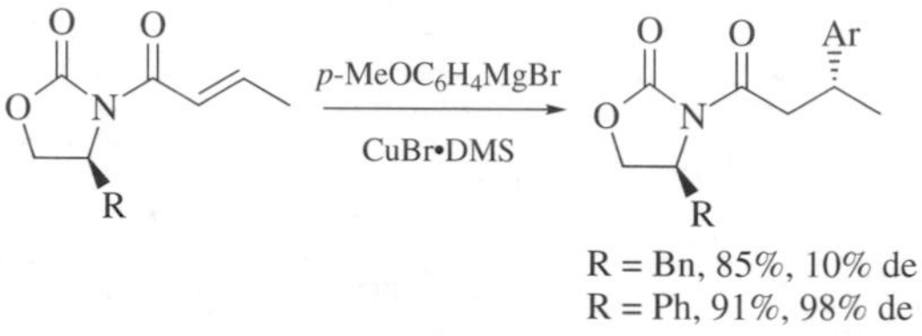

(56)

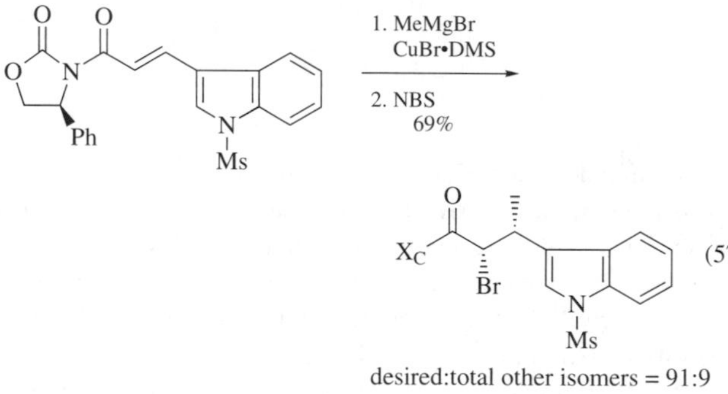

R = Bn, 85%, 10% de
R = Ph, 91%, 98% de

(57)

desired:total other isomers = 91:9

Similar chemistry using chiral sultam auxiliaries demonstrates superior yields and selectivities for specific cases of cuprate conjugate additions, but have not yet been extended to the more complex multistep transformation series illustrated above.[100,101] Moderate selectivities have been obtained in alkyl cuprate additions to γ-aminocrotonate equivalents where the nitrogen is derived from the oxazolidinone.[102]

The 4-phenyl-2-oxazolidinone auxiliary has also been employed in the $TiCl_4$-mediated conjugate additions of allylsilanes (eq 58).[103] Analogous reactions using the phenylalanine-derived auxiliary with dimethylaluminum chloride afforded lower selectivities.[104] In these reactions the oxazolidinones perform better than the sultams.

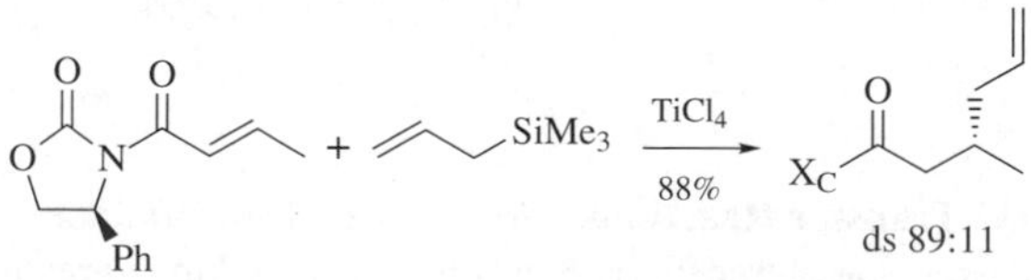

(58)

ds 89:11

Nucleophilic addition of thiophenol to chiral tiglic acid-derived imides proceeds in excellent yields and diastereoselectivities (eq 59).[105] Complete turnover of both the α- and β-centers results from the use of the (Z) rather than (E) isomer. Poor β-induction was found with the imides derived from cinnamic acid.

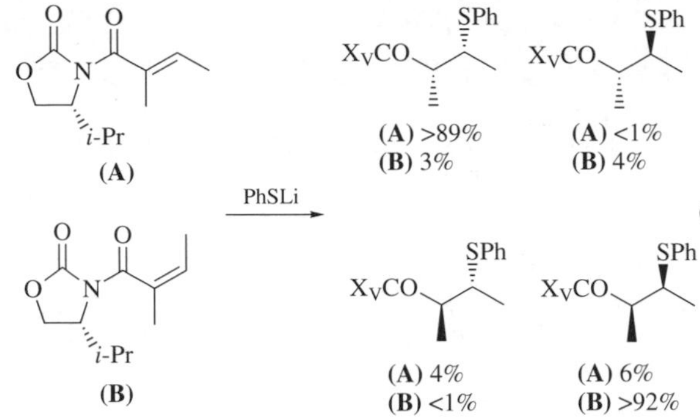

(59)

Dimethylaluminum chloride also catalyzes the ene reactions of chiral α,β-unsaturated imides with 1,1-disubstituted alkenes in moderate yields and selectivities.[104]

Oxazolidinone-Substituted Carbanions. Oxazolidinone-substituted organostannanes readily undergo transmetalation with alkyllithium reagents to the organolithium derivatives which then can undergo nucleophilic addition reactions. *N*-Substituted oxazolidinones can act in this capacity as both a nitrogen source and source of chirality (eq 60). Although the α-stereoselection in these reactions is excellent, a greater variety of reactant alkylstannanes are available using chiral imidazolidinones in place of oxazolidinones.[39,40,42]

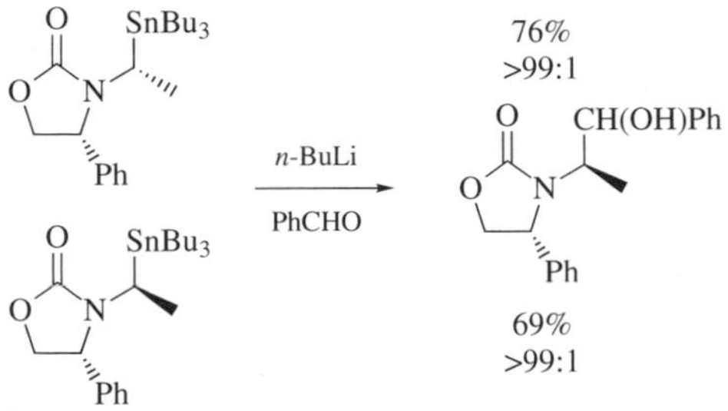

(60)

Synthesis of Cyclopropanes. Chiral imide enolates which contain γ-halide substituents undergo intramolecular displacement to form cyclopropanes.[106] Halogenation of γ,δ-unsaturated acyl imides occurs at the γ-position in 85% yield with modest stereoinduction. The (*Z*) sodium enolates of these compounds then cyclize through an intramolecular double stereodifferentiating reaction (eq 61).

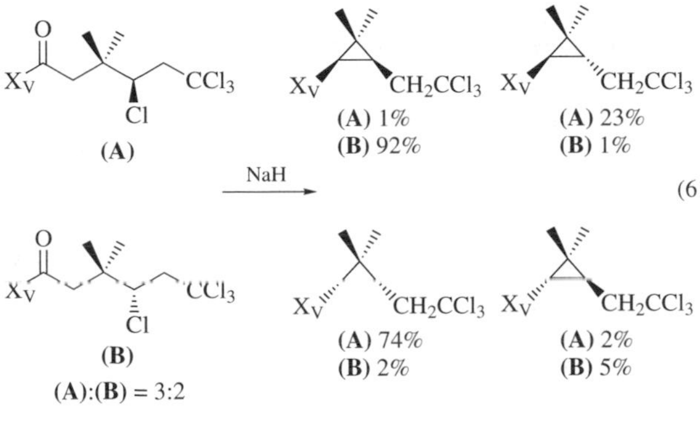

(61)

Stereoselective Cyclizations. Sultams have been demonstrated to be superior sources of chirality in selected cases of iodolactonizations,[107] oxidative 1,5-diene cyclizations,[108] and Claisen-type rearrangements of β-acetoxyl substrates.[109]

Chiral Ligands. Bidentate chelation of dirhodium(II) compounds by chiral oxazolidinones creates asymmetric sites on the metal, leading to induction in cyclopropanations and carbon–hydrogen insertion reactions. The oxazolidinones are less effective in this capacity than are the pyrrolidines.[110]

Removal of the Chiral Auxiliary. In each of the following transformations, the oxazolidinone auxiliary is recovered in high yields (eq 62).

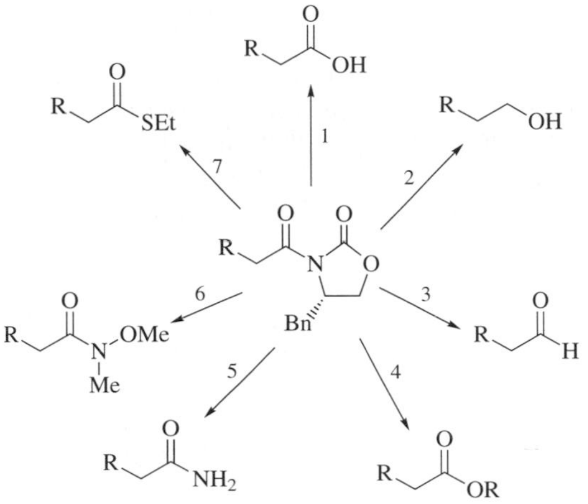

(62)

1a) KOH, MeOH; 1b) LiOH, MeOH or THF; 1c) LiOH, H_2O_2.
2a) $LiAlH_4$; 2b) $LiBH_4$; 2c) $LiBH_4$, H_2O; 2d) $LiBH_4$, MeOH; 2e) $LiAlH_4$, H_2, Lindlar cat., TFA; 2f) Bu_3B, HOAc, $LiBH_4$.
3a) i) $LiBH_4$, H_2O; ii) DMSO, $(COCl)_2$, Et_3N;
3b) i) $Me_2AlN(OMe)Me$; ii) DIBAL;
3c) i) LiSEt; ii) Et_3SiH, Pd/C.
4a) LiOBn; 4b) $Ti(OBn)_4$; 4c) ROMgBr; 4d) NaOMe; 4e) $Ti(OEt)_4$.
5) i) N_2H_4; ii) isopentyl nitrite, NH_4Cl.
6) $Me_2AlN(OMe)Me$.
7) LiSEt.

Conversion to the Acid. Hydroxide[6,111] and peroxide[112] agents saponify acyl imides in excellent yields; however, with sterically hindered acyl groups endocyclic cleavage may predominate upon treatment with ***Lithium Hydroxide***. ***Lithium Hydroperoxide***, however, is highly selective for the exocyclic carbonyl moiety.

Conversion to the Alcohol. Reduction of acyl imides to their corresponding alcohols is effected by a number of reagents, including ***Lithium Aluminum Hydride***,[1] ***Lithium Borohydride***,[1] $LiAlH_4/H_2$/Lindlar's cat./TFA,[113] $LiBH_4/H_2O/Et_2O$,[114] $LiBH_4$/MeOH/THF,[36] and Bu_3B/HOAc/$LiBH_4$.[77] Although the sole use of $LiAlH_4$ or $LiBH_4$ affords product often in low yields, the addition of an equivalent of H_2O or MeOH greatly enhances reaction efficiency. The MeOH/THF modification occasionally produces more consistent results. The last of the methods outlined above is effective in preventing retro-aldol cleavage in sensitive substrates such as crotyl or α-fluoro aldol adducts (eq 63).

Bu_3B, HOAc, $LiBH_4$, 89% (63)

Conversion to the Aldehyde. This transformation is accomplished through a two-step procedure. One such variant requires reduction to the alcohol (e.g. $LiAlH_4$, H_2O) and subsequent oxidation (e.g. Swern conditions).[36,85] Alternatively, Weinreb transamination[78,115–117] followed by ***Diisobutylaluminum Hydride***,[78] or conversion to the thioester (see below) and subsequent ***Triethylsilane*** reduction,[86] afford the desired aldehyde in excellent yields. Weinreb transamination proceeds with minimal endocyclic cleavage when there is a β-hydroxy moiety free for internal direction of the aluminum species.

Conversion to Esters. Ester formation is readily achieved by conventional alcoholysis with alkoxides such as LiOBn,[1] NaOMe,[6] or ROMgBr (eq 64).[76,118] In hindered cases, endocyclic cleavage becomes competitive. Various titanium(IV) alkoxides have also been employed to effect this transformation.[4,119]

PhMgBr, 79% (64)

Conversion to Amides. *N*-Acyloxazolidinones may be converted to the primary amide via the corresponding hydrazide.[120] Alternatively, trimethylaluminum/amine adducts form active transamination reagents,[36,117] providing amides of β-hydroxy acyloxazolidinones through intramolecular amine addition of the aminoaluminum species (eq 65).[105]

Me_3Al, 78% (65)

Conversion to Thioesters. The transformation of *N*-acyl imides into thioesters with lithium thiolate reagents proceeds with exceptional selectivity for the *exo* carbonyl moiety even in exceptionally hindered cases.[121] A recent application of this reaction in a complex setting has been reported (eq 66).[68,97] This transformation is significant in that the normally reliable peroxide hydrolysis procedure proved to be nonselective. The recently reported high yield reduction of thioesters to aldehydes[86] enhances the utility of these thioester intermediates.

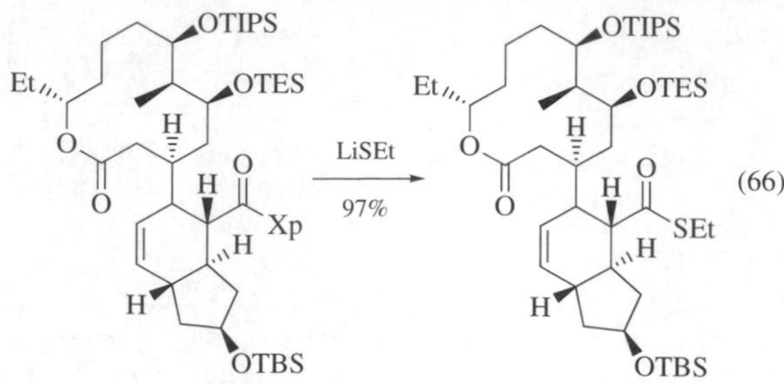

Related Reagents. 10-Dicyclohexylsulfonamidoisoborneol; (S)-Ethyl Lactate; 3-Hydroxyisoborneol; α-Methyltoluene-2,α-sultam.

1. Evans, D. A.; Ennis, M. D.; Mathre, D. J. *JACS* **1982**, *104*, 1737.
2. Evans, D. A.; Ennis, M. D.; Le, T.; Mandel, N.; Mandel, G. *JACS* **1984**, *106*, 1154.
3. Evans, D. A.; Ellman, J. A.; Dorow, R. L. *TL* **1987**, *28*, 1123.
4. Evans, D. A.; Britton, T. C.; Ellman, J. A.; Dorow, R. L. *JACS* **1990**, *112*, 4011.
5. Evans, D. A.; Morrissey, M. M.; Dorow, R. L. *JACS* **1985**, *107*, 4346.
6. Evans, D. A.; Bartroli, J.; Shih, T. L. *JACS* **1981**, *103*, 2127.
7. Evans, D. A.; Bilodeau, M. T.; Somers, T. C.; Clardy, J.; Cherry, D.; Kato, Y. *JOC* **1991**, *56*, 5750.
8. Nicolás, E.; Russell, K. C.; Hruby, V. J. *JOC* **1993**, *58*, 766.
9. Evans, D. A.; Chapman, K. T.; Bisaha, J. *JACS* **1988**, *110*, 1238.
10. Evans, D. A.; Anderson, J. C.; Taylor, M. K. *TL* **1993**, *34*, 5563.
11. Evans, D. A.; Faul, M. M.; Colombo, L.; Bisaha, J. J.; Clardy, J.; Cherry, D. *JACS* **1992**, *114*, 5977.
12. (a) Gage, J. R.; Evans, D. A. *OS* **1990**, *68*, 77. (b) Gage, J. R.; Evans, D. A. *OSC* **1993**, *8*, 528.
13. Evans, D. A.; Takacs, J. M.; McGee, L. R.; Ennis, M. D.; Mathre, D. J.; Bartroli, J. *PAC* **1981**, *53*, 1109.
14. Evans, D. A.; Nelson, J. V.; Taber, T. R. *Top. Stereochem.* **1982**, *13*, 1.
15. McKennon, M. J.; Meyers, A. I.; Drauz, K.; Schwarm, M. *JOC* **1993**, *58*, 3568.
16. Crowther, H. L.; McCombie, R. *JCS* **1913**, 27.
17. Newman, M. S.; Kutner, A. *JACS* **1951**, *73*, 4199.
18. Hyne, J. B. *JACS* **1959**, *81*, 6058.
19. Stratton, J. M.; Wilson, F. J. *JCS* **1932**, 1133.
20. Close, W. J. *JOC* **1950**, *15*, 1131.
21. Lynn, J. W. U.S. Patent 2 975 187 (*CA* **1955**, *49*, 16 568d).
22. Lesher, G. Y.; Surrey, A. R. *JACS* **1955**, *77*, 632.
23. Caccia, G.; Gladiali, S.; Vitali, R.; Gardi, R. *JOC* **1973**, *38*, 2264.
24. Saund, A. K.; Prashad, B.; Koul, A. K.; Bachhawat, J. M.; Mathur, N. K. *Int. J. Peptide Protein Res.* **1973**, *5*, 7.
25. Applegath, F. U.S. Patent 2 857 392 (*CA* **1953**, *47*, 5286d).
26. Koch, P.; Perrotti, E. *TL* **1974**, 2899.
27. Sonoda, N.; Yamamoto, G.; Natsukawa, K.; Kondo, K.; Murai, S. *TL* **1975**, 1969.
28. Correa, A.; Denis, J.-N.; Greene, A. E. *SC* **1991**, *21*, 1.
29. Pridgen, L. N.; Prol, J., Jr.; Alexander, B.; Gillyard, L. *JOC* **1989**, *54*, 3231.
30. Kano, S.; Yokomatsu, T.; Iwasawa, H.; Shibuya, S. *TL* **1987**, *28*, 6331.
31. Wuts, P. G. M.; Pruitt, L. E. *S* **1989**, 622.
32. Agami, C.; Couty, F.; Hamon, L.; Venier, O. *TL* **1993**, *34*, 4509.

33. Ishizuka, T.; Kimura, K.; Ishibuchi, S.; Kunieda, T. *CL* **1992**, 991.
34. Koch, S. S. C.; Chamberlin, A. R. *JOC* **1993**, *58*, 2725.
35. Evans, D. A.; Gage, J. R. *JOC* **1992**, *57*, 1958.
36. Evans, D. A.; Gage, J. R.; Leighton, J. L. *JACS* **1992**, *114*, 9434.
37. Evans, D. A.; Chapman, K. T.; Bisaha, J. *JACS* **1984**, *106*, 4261.
38. Thom, C.; Kocieński, P. *S* **1992**, 582.
39. Pearson, W. H.; Lindbeck, A. C. *JACS* **1991**, *113*, 8546.
40. Pearson, W. H.; Lindbeck, A. C.; Kampf, J. W. *JACS* **1993**, *115*, 2622.
41. Evans, D. A.; Sjogren, E. B. *TL* **1985**, *26*, 3783.
42. Pearson, W. H.; Lindbeck, A. C. *JOC* **1989**, *54*, 5651.
43. Evans, D. A.; Nelson, J. V.; Vogel, E.; Taber, T. R. *JACS* **1981**, *103*, 3099.
44. Carreira, E. M. Ph.D. Thesis, Harvard University, 1990.
45. Evans, D. A.; Bilodeau, M. Unpublished results.
46. Evans, D. A.; Gage, J. R.; Leighton, J. L.; Kim, A. S. *JOC* **1992**, *57*, 1961.
47. Evans, D. A.; Urpí, F.; Somers, T. C.; Clark, J. S.; Bilodeau, M. T. *JACS* **1990**, *112*, 8215.
48. Alexander, R. P.; Paterson, I. *TL* **1985**, *26*, 5339.
49. Soai, K.; Ishizaki, M.; Yokoyama, S. *CL* **1987**, 341.
50. Miles, W. H.; Smiley, P. M.; Brinkman, H. R. *CC* **1989**, 1897.
51. Green, M.; Greenfield, S., Kersting, M. *CC* **1985**, 18.
52. Pearson, A. J.; Khetani, V. D.; Roden, B. A. *JOC* **1989**, *54*, 5141.
53. Pearson, A. J.; Zhu, P. Y.; Youngs, W. J.; Bradshaw, J. D.; McConville, D. B. *JACS* **1993**, *115*, 10376.
54. Schreiber, S. L.; Klimas, M. T.; Sammakia, T. *JACS* **1987**, *109*, 5749.
55. Yamazaki, T.; Haga, J.; Kitazume, T. *CL* **1991**, 2175.
56. Evans, D. A.; Britton, T. C.; Dorow, R. L.; Dellaria, J. F. *JACS* **1986**, *108*, 6395.
57. Evans, D. A.; Britton, T. C. *JACS* **1987**, *109*, 6881.
58. Evans, D. A.; Evrard, D. A.; Rychnovsky, S. D.; Früh, T.; Whittingham, W. G.; DeVries, K. M. *TL* **1992**, *33*, 1189.
59. Doyle, M. P.; Dorow, R. L.; Terpstra, J. W.; Rodenhouse, R. A. *JOC* **1985**, *50*, 1663.
60. Davis, F. A.; Han, W. *TL* **1992**, *33*, 1153.
61. Dharanipragada, R.; Nicolas, E.; Toth, G.; Hruby, V. J. *TL* **1989**, *30*, 6841.
62. Evans, D. A.; Taber, T. R. *TL* **1980**, *21*, 4675.
63. Evans, D. A. *Aldrichim. Acta* **1982**, *15*, 23.
64. Hamada, Y.; Hayashi, K.; Shioiri, T. *TL* **1991**, *32*, 931.
65. Evans, D. A.; Bender, S. L. *TL* **1986**, *27*, 799.
66. Evans, D. A.; Kaldor, S. W.; Jones, T. K.; Clardy, J.; Stout, T. J. *JACS* **1990**, *112*, 7001.
67. Evans, D. A.; Miller, S. J.; Ennis, M. D.; Ornstein, P. L. *JOC* **1992**, *57*, 1067.
68. Evans, D. A.; Black, W. C. *JACS* **1993**, *115*, 4497.
69. Evans, D. A.; Dow, R. L. *TL* **1986**, *27*, 1007.
70. Walker, M. A.; Heathcock, C. H. *JOC* **1991**, *56*, 5747.
71. Chibale, K.; Warren, S. *TL* **1992**, *33*, 4369.
72. Abdel-Magid, A.; Pridgen, L. N.; Eggleston, D. S.; Lantos, I. *JACS* **1986**, *108*, 4595.
73. Pridgen, L. N.; Abdel-Magid, A.; Lantos, I. *TL* **1989**, *30*, 5539.
74. Nerz-Stormes, M.; Thornton, E. R. *TL* **1986**, *27*, 897.
75. Nerz-Stormes, M.; Thornton, E. R. *JOC* **1991**, *56*, 2489.
76. Evans, D. A.; Dow, R. L.; Shih, T. L.; Takacs, J. M.; Zahler, R. *JACS* **1990**, *112*, 5290.
77. Evans, D. A.; Sjogren, E. B.; Bartroli, J.; Dow, R. L. *TL* **1986**, *27*, 4957.
78. Evans, D. A.; Bender, S. L.; Morris, J. *JACS* **1988**, *110*, 2506.
79. Evans, D. A.; Weber, A. E. *JACS* **1986**, *108*, 6757.
80. Evans, D. A.; Sjogren, E. B.; Weber, A. E.; Conn, R. E. *TL* **1987**, *28*, 39.
81. Sjogren, E. B. Ph.D. Thesis, Harvard University, 1986.
82. Evans, D. A.; Shumsky, J. Unpublished results.
83. Evans, D. A.; Clark, J. S.; Metternich, R.; Novack, V. J.; Sheppard, G. S. *JACS* **1990**, *112*, 866.
84. Evans, D. A.; Ng, H. P.; Clark, J. S.; Rieger, D. L. *T* **1992**, *48*, 2127.
85. Evans, D. A.; Sheppard, G. S. *JOC* **1990**, *55*, 5192.
86. Evans, D. A.; Ng, H. P. *TL* **1993**, *34*, 2229.
87. Ito, Y.; Sasaki, A.; Tamoto, K.; Sunagawa, M.; Terashima, S. *T* **1991**, *47*, 2801.
88. Ito, Y.; Terashima, S. *T* **1991**, *47*, 2821.
89. Kende, A. S.; Kawamura, K.; Orwat, M. J. *TL* **1989**, *30*, 5821.
90. Kende, A. S.; Kawamura, K.; DeVita, R. J. *JACS* **1990**, *112*, 4070.
91. Andersen, K. K.; Gaffield, W.; Papnikolaou, N. E.; Foley, J. W.; Perkins, R. I. *JACS* **1964**, *86*, 5637.
92. Andersen, K. K. *TL* **1962**, 93.
93. Rebiere, F.; Samuel, O.; Ricard, L.; Kagan, H. B. *JOC* **1991**, *56*, 5991.
94. Rebiere, F.; Kagan, H. B. *TL* **1989**, *30*, 3659.
95. Evans, D. A.; Chapman, K. T.; Bisaha, J. *TL* **1984**, *25*, 4071.
96. Rück, K.; Kunz, H. *AG(E)* **1991**, *30*, 694.
97. Evans, D. A.; Black, W. C. *JACS* **1992**, *114*, 2260.
98. Pourcelot, G.; Aubouet, J.; Caspar, A.; Cresson, P. *JOM* **1987**, *328*, C43.
99. Boteju, L. W.; Wegner, K.; Hruby, V. J. *TL* **1992**, *33*, 7491.
100. Oppolzer, W.; Moretti, R.; Bernardinelli, G. *TL* **1986**, *27*, 4713.
101. Oppolzer, W.; Poli, G. *TL* **1986**, *27*, 4717.
102. Le Coz, S.; Mann, A. *SC* **1993**, *23*, 165.
103. Wu, M.-J.; Wu, C.-C.; Lee, P.-C. *TL* **1992**, *33*, 2547.
104. Snider, B.; Zhang, Q. *JOC* **1991**, *56*, 4908.
105. Miyata, O.; Shinada, T.; Ninomiya, I.; Naito, T. *TL* **1991**, *32*, 3519.
106. Kleschick, W. A.; Reed, M. W.; Bordner, J. *JOC* **1987**, *52*, 3168.
107. Yokomatsu, T.; Iwasawa, H.; Shibuya, S. *CC* **1992**, 728.
108. Walba, D. M.; Przybyla, C. A.; Walker, C. B., Jr. *JACS* **1990**, *112*, 5624.
109. Brandänge, S.; Leijonmarck, H. *TL* **1992**, *33*, 3025.
110. Doyle, M. P.; Winchester, W. R.; Hoorn, J. A. A.; Lynch, V.; Simonsen, S. H.; Ghosh, R. *JACS* **1993**, *115*, 9968.
111. Šavrda, J.; Descoins, C. *SC* **1987**, *17*, 1901.
112. Evans, D. A.; Britton, T. C.; Ellman, J. A. *TL* **1987**, *28*, 6141.
113. Tietze, L. F.; Schneider, C. *JOC* **1991**, *56*, 2476.
114. Penning, T. D.; Djuric, S. W.; Haack, R. A.; Kalish, V. J.; Miyashiro, J. M.; Rowell, B. W.; Yu, S. S. *SC* **1990**, *20*, 307.
115. Levin, J. L.; Turos, E.; Weinreb, S. M. *SC* **1982**, *12*, 989.
116. Basha, A.; Lipton, M.; Weinreb, S. M. *TL* **1977**, 4171.
117. Evans, D. A.; Sjogren, E. B. *TL* **1986**, *27*, 3119.
118. Evans, D. A.; Weber, A. E. *JACS* **1987**, *109*, 7151.
119. Harre, M.; Trabandt, J.; Westermann, J. *LA* **1989**, 1081.
120. Bock, M. G.; DiPardo, R. M.; Evans, B. E.; Rittle, K. E.; Boger, J. S.; Freidinger, R. M.; Veber, D. F. *CC* **1985**, 109.
121. Damon, R. E.; Coppola, G. M. *TL* **1990**, *31*, 2849.

David A. Evans & Annette S. Kim
Harvard University, Cambridge, MA, USA

1-(Benzyloxymethoxy)propyllithium

[83876-83-9] $C_{11}H_{15}LiO_2$ (MW 186.18)

(reagent for enantioselective nucleophilic α-hydroxyalkylation; example for several similar reagents)

Form Supplied in: in situ generation at low temperature in THF, ether, or hexane.

Preparative Method: the (preferentially enantioenriched) reagent is generated by tin–lithium exchange of [1-(benzyloxymethoxy)propyl]tributylstannane (eq 1)[1] which proceeds with stereoretention. The lithium carbanion is configurationally stable in THF below approx −40 °C. The nonracemic precursor is obtained by classic[1a] or enzymatic[2] resolution of the α-stannylalkanol, asymmetric reduction[3] of acylstannanes (eq 2), or from chiral (1-chloroalkyl)boronates.[4]

n-BuLi, THF; −78 °C; >90% (1)

BOM = $BnOCH_2$; (R)-(1) → >95% ee (S)-(2)

(S)-BINAL-H

$BnOCH_2Cl$; i-Pr_2NEt; 69%; 96% ee (2)

Handling, Storage, and Precautions: solution in THF must be kept under exclusion of air or moisture below −40 °C.

Reactions. Alkylation[1a] with ***Dimethyl Sulfate*** and carboxylation[5] proceed with complete stereoretention to yield BOM-protected (S)-2-butanol[1a] or (S)- or (R)-2-hydroxybutanoic acid (eq 3).[5] Analogously prepared (S)-1-(methoxymethoxy)propyllithium undergoes clean Michael addition to N,N′,N′-trimethylacryloylhydrazide, and the product can be converted to (S)-4-hexanolide (eq 4).[6] The acylation of the (R) enantiomer by an N,N-dimethylalkanamide leads to an α-hydroxy ketone, which was converted to (+)-*endo*-brevicomin (eq 5).[7]

(R)-(1) → (S)-(2); Me_2SO_4; 1. CO_2, 2. H_3O^+, 99%; >95% ee (3)

n-BuLi, THF, 10 min, −78 °C; 93% ee; 50%; $HClO_4$ in CH_2Cl_2, 87%; 92% ee (4)

1. n-BuLi, 2. ...; 98% ee; 50%; 1. $Zn(BH_4)_2$, 2. cat. $HClO_4$, 50%; 98% ee, 93% ds (5)

The addition of the lithium reagent to aldehydes gives monoprotected *syn*- and *anti-vic*-1,2-alkanediols with low diastereoselectivity (eq 6),[8] which is improved in favor of the *syn* diastereomers by lithium–magnesium exchange and by application of β-branched analogs.[8] Higher and opposite diastereoselectivity is achieved with the appropriate tributyltin or tributyllead reagents under Lewis acid catalysis; however, acetal O-protecting groups are not tolerated (eq 7).[9]

PhCHO (6)

Racemates
M = Li 53:47 75%
M = MgBr 63:37 77%

+ PhCHO; Lewis acid (LA); −78 to 0 °C (7)

Racemates
M = Sn, LA = $TiCl_4$ 99:1 90%
M = Pb, LA = $BF_3{\cdot}OEt_2$ 20:80 54%

The cuprate reagents, derived from 1-(methoxymethoxy)-alkyllithium, undergo 1,4-addition to enones[10] and to enals (in the presence of ***Chlorotrimethylsilane***) (eq 8).[11] Enantioenriched reagents racemize under these conditions.[12]

OMOM / C_5H_{11}–CH–Li → CuCN, THF → [C_5H_{11}–CH(OMOM)–]CuCNLi$_2$ → cyclohexenone → 3-[1-(OMOM)hexyl]cyclohexanone (8)

86%
diastereoisomers, 69:31

1-Lithioalkyl *N,N*-dialkylcarbamates are generally accessible by direct deprotonation of alkyl carbamates,[13] either in racemic form or highly enantiomerically enriched (see ***(−)-Sparteine***) (eq 9). Sterically blocked lithiated alkyl benzoates do not permit alkyl substitution at the carbanionic center.[14] Preexisting stereogenic centers in the precursor cause an efficient internal, kinetically controlled, chiral induction in the deprotonation step.[15] This was demonstrated by the synthesis of (2*S*,4*S*)-2-hydroxy-4-pentanolide (eq 10)[15a] from (*S*)-2,3-butanediol and of protected (2*S*,3*S*)-2-hydroxy-3-amino acids (eq 11)[15b] and the chain elongation of 2-amino alkanols.[15b]

OCby–CH$_2$R → s-BuLi, TMEDA, Et$_2$O, −78 °C → CbyO–CH(Li•TMEDA)–R → ElX, 70–90%

CbyO–CH(El)–R → 1. MeOH, MeSO$_3$H; 2. LiOH or Ba(OH)$_2$, 70–90% → HO–CH(El)–R (9)

R = alkyl, subst. alkyl
ElX = MeI, CO$_2$, R'COCl, R'$_3$SiCl, R$_3$SnCl, DOMe, R'CHO

CbyO⁀OCby → s-BuLi, TMEDA, Et$_2$O, −78 °C → [TMEDA•Li, CbyO ... OCby] → 1. CO$_2$; 2. 5N HCl, 36% → OH lactone (10)

CbyO–...(NBn$_2$) → 1. s-BuLi, TMEDA, Et$_2$O, −78 °C; 2. CO$_2$, 84% → CbyO–CH(CO$_2$H)–...(NBn$_2$) (11)

Racemic 1-methoxyalkyllithium[16] reagents are easily obtained from monothioacetals by reductive lithiation[17] with lithium salts of radical anions, such as ***Lithium 1-(Dimethylamino)naphthalenide*** (LDMAN).[18] The method is particularly useful for the preparation of substituted 2-lithiotetrahydropyrans[19] and 2-lithiotetrahydrofurans.[19] The axial lithium compound is preferentially formed from the radical intermediate under kinetic control, irrespective of the relative configuration of the precursor; it epimerizes above −30 °C to form the more stable equatorial epimer (eq 12).[19,20]

axial SPh + equatorial SPh, 77:23 → LDMAN, THF, −78 °C → radical (H) → axial Li → TMEDA, −30°C → equatorial Li

axial CH(OH)Ph, 95:5; equatorial CH(OH)Ph, 87:13 (12)

Protected phenylthio- and phenylsulfonyl-2-deoxypyranosides were also subjected to these conditions and substituted by electrophiles via the 1-lithiopyranosides.[21] Both α- and β-1-(3,4,6-tri-*O*-benzyl-2-deoxy-D-glucopyranosyl)cuprates, prepared by these methods, undergo highly stereoselective conjugate additions to enals and enones (eq 13).[22]

BnO, BnO, BnO pyranosyl–Li → 1. CuBrMe$_2$S; 2. methyl vinyl ketone, BF$_3$•OEt$_2$, 75% → BnO, BnO, BnO pyranosyl–CH$_2$CH$_2$C(O)CH$_3$ (H) (13)

Even (3,4,6-tri-*O*-benzyl-2-*O*-lithio-α-D-glucopyranosyl)lithium can be generated and added to aldehydes (eq 14).[23]

BnO, BnO, BnO, HO, Cl, H → 1. BuLi; 2. Li⁺ naphth⁻• → BnO, BnO, BnO, LiO, Li, H → 1. isobutyraldehyde; 2. H$_3$O⁺, 59% → BnO, BnO, BnO, HO, HO, H (14)

2 diastereomers, 1:1

4-Lithio-1,3-dioxanes[19c,24] derived from chiral 1,3-diols are similarly generated and proved to be versatile building blocks in the stereoselective synthesis of polyols, such as 1,3,5,7,9,10-decanehexaol (eq 15).

HO–CH₂CH(OH)–CH₂CO₂Me **(1)** → 1. TIPSOTf; 2. Me_3SiCl; 3. DIBAH; 4. $PhSSiMe_3$; 5. acetone, TMSOTf; 30% → **(2)** (TIPSO, SPh acetonide) → 2 LIDBB, THF, −78 °C → **(3)** (Li)

(3) → ethylene oxide, $BF_3•OEt_2$ → TIPSO…OH acetonide → 1. acetone, $CuSO_4$, PPTS; 2. Bu_4NF; 3. TsCl, NaOH; 53% → epoxide acetonide → **(3)**, $BF_3•OEt_2$; 62% → TIPSO bis-acetonide (OH) → 1. Bu_4NF; 2. MeOH, H^+; 100% → HO…OH OH OH OH OH (15)

LIDBB = lithium 4,4′-di-*t*-butylbiphenylide (Li^+)

The zinc–copper reagents, obtained from 1-(acetoxyalkyl)zinc bromides,[25] are subjected to efficient alkylations, alkynylations, acylations, and Michael addition reactions (eq 16).

1-bromo-2-methylpropyl acetate → 1. Zn, THF, DMSO, 0–10 °C; 2. CuCN•2LiCl → AcO–CH(*i*-Pr)–Cu(CN)ZnBr → $BrC{\equiv}CC_6H_{13}$, 76% → AcO–CH(*i*-Pr)–C≡C–C_6H_{13} (16)

Related Reagents. 1-Ethoxyvinyllithium[1].

1. (a) Still, W. C.; Sreekumar, C. *JACS* **1980**, *102*, 1201. (b) Sawyer, J. S.; Kucerovy, A.; Macdonald, T. L.; McGarvey, G. J. *JACS* **1988**, *110*, 842.
2. Chong, J. M.; Mar, E. K. *TL* **1991**, *32*, 5683.
3. (a) Chan, P. C.-M.; Chong, J. M. *JOC* **1988**, *53*, 5584. (b) Marshall, J. A.; Gung, W. Y. *T* **1989**, *45*, 1043. (c) Marshall, J. A. *Chemtracts, Org. Chem.* **1992**, 75.
4. Matteson, D. S.; Tripathy, P. B.; Sarkar, A.; Sadhu, K. M. *JACS* **1989**, *111*, 4399.
5. Chan, P. C.-M.; Chong, J. M. *TL* **1990**, *31*, 1985.
6. Chong, J. M.; Mar, E. K. *TL* **1990**, *31*, 1981.
7. Chong, J. M.; Mar, E. K. *T* **1989**, *45*, 7709.
8. McGarvey, G. J.; Kimura, M. *JOC* **1982**, *47*, 5420.
9. (a) Yamada, J.; Abe, H.; Yamamoto, Y. *JACS* **1990**, *112*, 6118. (b) Yamamoto, Y. *Chemtracts, Org. Chem.* **1991**, *4*, 255.
10. (a) Linderman, R. J.; Godfrey, A.; Horne, K. *T* **1989**, *45*, 495. (b) Linderman, R. J.; Godfrey, A.; Horne, K. *TL* **1987**, *28*, 3911. (c) Linderman, R. J.; Godfrey, A. *TL* **1986**, *27*, 4553.
11. Linderman, R. J.; McKenzie, J. R. *JOM* **1989**, *361*, 31.
12. (a) Linderman, R. J.; Griedel, B. D. *JOC* **1991**, *56*, 5491. (b) Linderman, R. J.; Griedel, B. D. *JOC* **1990**, *55*, 5428.
13. (a) Hoppe, D.; Hintze, F.; Tebben, P. *AG* **1990**, *102*, 1457; *AG(E)* **1990**, *29*, 1422. (b) Hintze, F.; Hoppe, D. *S* **1992**, 1216.
14. (a) Beak, P.; Carter, L. G. *JOC* **1981**, *46*, 2363. (b) Schlecker, R.; Seebach, D.; Lubosch, W. *HCA* **1978**, *61*, 512.
15. (a) Ahrens, H.; Paetow, M.; Hoppe, D. *TL* **1992**, *33*, 5327. (b) Schwerdtfeger, J.; Hoppe, D. *AG* **1992**, *104*, 1547; *AG(E)* **1992**, *31*, 1505. (c) Guarnieri, W.; Grehl, M.; Hoppe, D. *AG* **1994**, *106*, 1815; *AG(E)* **1994**, *33*, 1724.
16. Cohen, T.; Matz, J. R. *JACS* **1980**, *102*, 6900.
17. Cohen, T.; Bhupathy, M. *ACR* **1989**, *22*, 152.
18. Cohen, T.; Sherbine, J. P.; Matz, J. R.; Hutchins, R. R.; McHenry, B. M.; Willey, P. R. *JACS* **1984**, *106*, 3245.
19. Cohen, T.; Lin, M.-T. *JACS* **1984**, *106*, 1130.
20. (a) Verner, E. J.; Cohen, T. *JACS* **1992**, *114*, 375. (b) Verner, E. J.; Cohen, T. *JOC* **1992**, *57*, 1072. (c) Rychnovsky, S. D.; Mickus, D. E. *TL* **1989**, *30*, 3011.
21. (a) Beau, J.-M.; Sinay, P. *TL* **1985**, *26*, 6185. (b) Beau, J.-M.; Sinay, P. *TL* **1985**, *26*, 6189. (c) Beau, J.-M.; Sinay, P. *TL* **1985**, *26*, 6193. (d) Fernandez-Mayoralas, A.; Marra, A.; Trumtel, M.; Veyrières, A.; Sinay, P. *TL* **1989**, *30*, 2537.
22. Hutchinson, D. K.; Fuchs, P. L. *JACS* **1987**, *109*, 4930.
23. Wittmann, V.; Kessler, H. *AG* **1993**, *105*, 1138; *AG(E)* **1993**, *32*, 1091.
24. (a) Rychnovsky, S. D. *JOC* **1989**, *54*, 4982. (b) Rychnovsky, S. D.; Griesgraber, G. *JOC* **1992**, *57*, 1559.
25. Knochel, P.; Chou, T.-S.; Jubert, C.; Rajagopal, D. *JOC* **1993**, *58*, 588.

Dieter Hoppe
University of Münster, Germany

Bis(1,5-cyclooctadiene)nickel(0)[1]

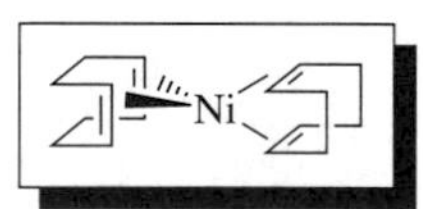

[1295-35-8] $C_{16}H_{24}Ni$ (MW 275.08)

(Ni(cod)$_2$ is a source of nickel(0) useful for the preparation of π-allylnickel halides,[2] for coupling of aryl and alkenyl halides,[3] and for the oligomerization and cycloaddition of strained alkenes,[4] of alkynes,[5] and of 1,3-dienes[6])

Physical Data: mp 60 °C dec (N_2).
Solubility: sol benzene, toluene, THF, ether, DMF, HMPA, *N*-methylpyrrolidinone.
Form Supplied in: yellow-orange crystals of 98+ % purity.
Analysis of Reagent Purity: ^{1}H NMR: δ 4.31 (br, =CH), 2.08 (br, CH_2).[1]
Preparative Methods: the standard preparation[2b] is a modification of the original procedure by Wilke and co-workers[7] and involves the reduction of ***Nickel(II) Acetylacetonate***, Ni(acac)$_2$, with ***Triethylaluminum*** in the presence of 1,5-cyclooctadiene

(cod) and 1,3-butadiene in toluene. A more convenient preparation utilizes ***Diisobutylaluminum Hydride*** (DIBAL) as the reducing agent.[8] In a typical reaction, 45.4 mL of a 1.0 M THF solution of DIBAL was added to a 250 mL Schlenk flask containing 4.67 g of $Ni(acac)_2$, and 7.93 g cod in THF solution under a nitrogen atmosphere at −78 °C. The resulting dark, reddish-brown solution was warmed to 0 °C and treated with diethyl ether to give a light yellow precipitate. Filtration under nitrogen gave a 72% yield of $Ni(cod)_2$ which was suitable for immediate use. Optional recrystallization from toluene under inert atmosphere gave bright yellow-orange needles with 40% recovery.

Handling, Storage, and Precautions: highly oxygen sensitive. Special inert-atmosphere techniques must be used.[9] Should be stored at 0 °C.

π-Allylnickel Halides. π-Allylnickel halide complexes are prepared by reaction of allylic halides with $Ni(cod)_2$ in a nonpolar solvent (eq 1).[10] The resulting dimeric species, isolated as a red solid, can be purified by crystallization, stored in the absence of air, and weighed out for reaction like any other moderately air-sensitive compound.[2b] $Ni(cod)_2$ is preferred over ***Tetracarbonylnickel*** for this transformation due to the mildness of conditions required, the extreme toxicity and thermal instability of $Ni(CO)_4$, and the absence of byproducts caused by CO insertion.

(1)

In polar, coordinating solvents such as DMF, HMPA, and *N*-methylpyrrolidinone, π-allylnickel halide complexes react with a wide range of alkyl, alkenyl, and aryl halides[11] to replace a halogen with the allyl group, as shown by the reactivity of π-(2-methallyl)nickel bromide (**1**) (eq 2).[12] These complexes also react with aldehydes and ketones to give homoallylic alcohols,[13] with quinones to produce allylquinones,[14] and with 2-pyridylcarboxylates to give β,γ-unsaturated ketones.[15] However, they do not react readily with esters, acid chlorides, amides, nitriles, or alcohols. Due to this significantly lower reactivity in these reactions, they offer a greater degree of selectivity over their allyllithium, -magnesium, and -zinc counterparts.[16] Mechanistic aspects of the coupling reaction between organic halides and π-allylnickel halides have been investigated and a radical-chain pathway has been proposed.[17]

(2)

The π-allylnickel halide formation reaction (eq 1) is tolerant to substitution on the allyl halide starting material. 1,1-Dimethylallyl,[13] 2-ethoxycarbonylallyl,[13] 2-methoxyallyl,[18] and 2-trimethylsilylmethyl[19] nickel halide complexes, to name a few, have all been synthesized and utilized in subsequent coupling reactions. Alkene-containing allyl halides have been also used to form dienes.[20] Allyl iodides are more reactive than allyl bromides for the generation of π-allylnickel halides and allyl chlorides are unreactive. Allyl mesylates and trifluoroacetates do not work as well as bromides due to extensive homocoupling of the substrates.[19] α,β-Unsaturated aldehydes and ketones react with $Ni(cod)_2$ in the presence of a trialkylsilyl chloride to give [1-[(trialkylsilyl)oxy]allyl]nickel chloride dimers (**2**) (eq 3).[21] These complexes couple to alkyl halides at the γ-position upon irradiation under a sun lamp to give silyl enol ethers (**3**), making the overall transformation a useful reversed-polarity complement to organocuprate conjugate addition chemistry. Chiral allyl acetals (**4**) have also been utilized to form chiral (*E*)-enol ethers (**5**) which serve as homoenolate equivalents of a new type (eq 4).[22]

(3)

(4)

Complexes having π-allylnickel halide-like reactivity may also be used as transient intermediates in polar media without the need for isolation. This type of reactivity is especially important for the intramolecular coupling of allyl groups (see ***Tetracarbonylnickel***). In these cases, allylic tosylates and carboxylates may also be used for the initial oxidative addition of the allyl substrate to nickel.[23] In one interesting group of reactions (eq 5),[24] a π-allyl intermediate undergoes an intramolecular alkene insertion to form a ring followed by β-hydride elimination to give diene (**6**). If carbon monoxide and methanol are included in the reaction mixture, CO insertion, followed by intramolecular insertion of the newly-formed double bond, followed by a final methoxycarbonylation gives rise to keto ester (**7**). The palladium catalyst ***Bis(dibenzylideneacetone)palladium(0)*** also catalyzes these reactions, but with poor diastereoselectivity. An allylic sulfonium ion (**8**) has shown a similar reactivity and has been utilized in the total synthesis of confertin.[25] Intramolecular coupling with an aldehyde followed by spontaneous lactonization yielded the tricyclic α-methylene-γ-lactone (**9**) (eq 6). Insertion of terminal alkynes into π-allylnickel complexes derived from allylic esters (**10**) has led to a catalytic synthesis of nonconjugated alkenynes (**11**) (eq 7).[26]

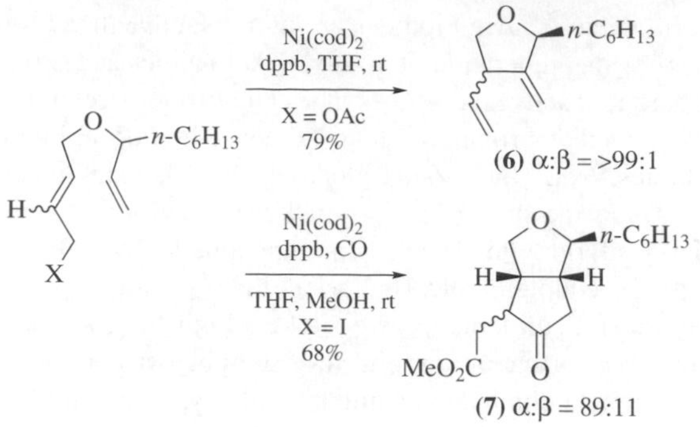

Coupling of Aryl Halides. $Ni(cod)_2$ reacts with a variety of aryl halides in DMF to give diaryl compounds in generally high yields (80–90%) (eq 8).[27]

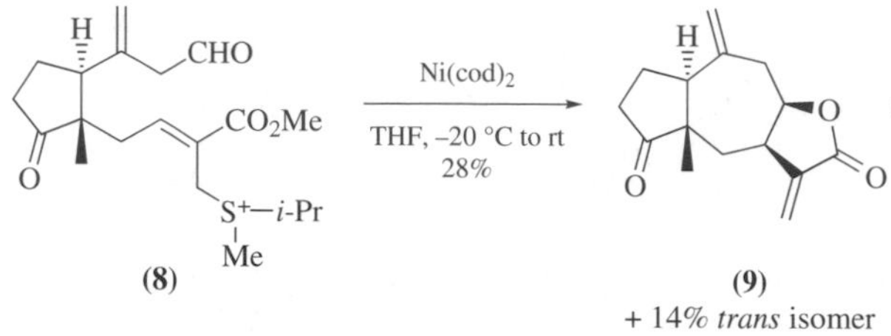

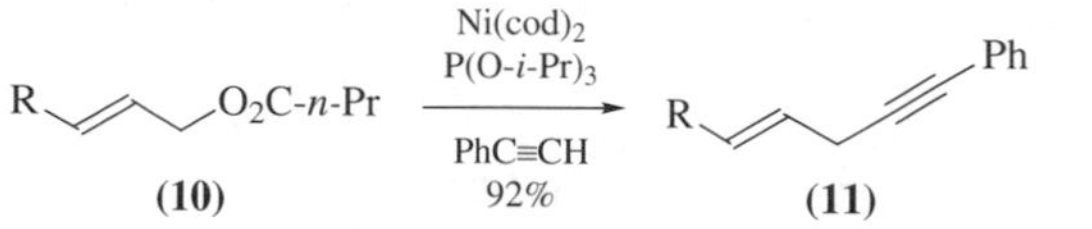

$$2\,ArX + Ni(cod)_2 \xrightarrow[25\text{–}40\ ^\circ C]{DMF} Ar_2 + NiX_2 + cod \quad (8)$$

Ortho substituents on the aryl halide drastically reduce the rate of the reaction and an increase in temperature leads only to decomposition of the nickel catalyst. The reaction is tolerant of most functional groups with both electron-donating and electron-withdrawing substituents allowing efficient coupling. Nitro groups, however, destroy the catalytic activity of the nickel complexes.[28] Acidic functionalities such as alcohols, phenols, or carboxylic acids cause reduction of the aryl halide in preference to coupling. The order of reactivity of substrates is I > Br > Cl with phenol *p*-toluenesulfonate esters being completely unreactive. This reaction may be contrasted with the copper-catalyzed Ullmann reaction,[29] which typically requires high temperatures (≥200 °C), or a two-step coupling procedure requiring intermediate arylmagnesium or -lithium reagents which are incompatible with many functional groups.

Attempted cross-coupling usually leads to mixtures of products due to extensive symmetrical coupling. Nevertheless, this strategy was utilized in the synthesis of a 2-benzazepine.[30] Intramolecular coupling reactions,[3,31] on the other hand, are generally quite efficient (eq 9)[32] and often benefit from added phosphine ligand (see also ***Tetrakis(triphenylphosphine)nickel(0)***).

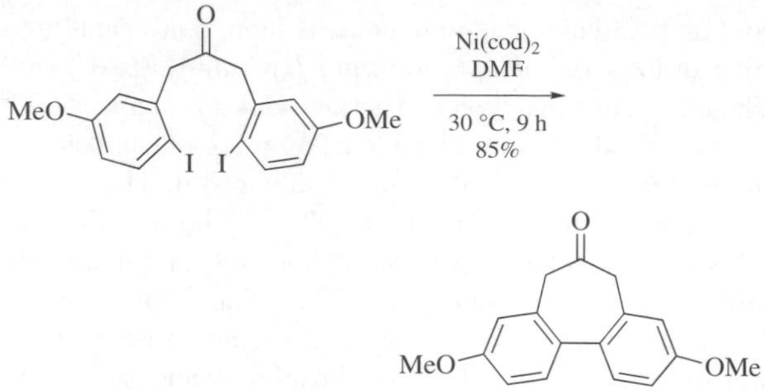

Coupling of Alkenyl Halides. $Ni(cod)_2$ reacts with alkenyl halides to produce symmetrical 1,3-dienes.[33] These reactions may be carried out in DMF, or in ether with an added ligand such as Ph_3P. Coupling of simple alkenyl halides gives only moderate yields (48–70%) with mixtures of geometrical isomers. Reactions of alkenyl halides bearing electron-withdrawing substituents such as α- and β-haloacrylates, however, are more efficient and highly stereoselective (eq 10).[3] The intramolecular coupling of a simple diiodide has also been shown to be effective (eq 11).[3] Finally, there is one example of the intramolecular coupling of alkyl halides in which α,ω-dihaloalkanes are cyclocoupled to give cycloalkanes by a $Ni(cod)_2$–bipyridyl complex.[34]

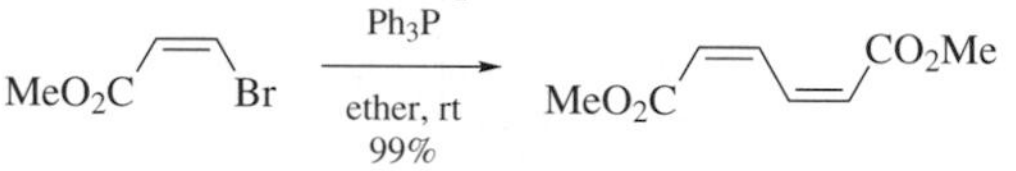

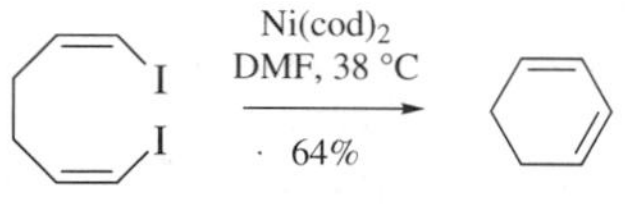

Oligomerization of Strained Alkenes and Alkanes. 3,3-Disubstituted cyclopropenes are known to react with electron-deficient alkenes in the presence of catalytic $Ni(cod)_2$ to give vinylcyclopropanes (eq 12).[35] The mechanism of the reaction begins with oxidative addition of Ni^0 into the C-1–C-3 bond, but there is some dispute as to whether a carbenoid intermediate is involved.[4] Methyl, phenyl, and methoxy groups have all been used as geminal substituents on the cyclopropene; however, the stereochemistry of the starting alkene is not preserved when methoxy is used.[36] Alkyl group substitution on the electron-deficient alkene disfavors this reaction pathway and leads to [2 + 2] cyclodimerization of the cyclopropene.[37] Addition of ligands has also been shown to affect the distribution of products. When the catalyst is modified by the bulky $P(i\text{-}Pr)_2(t\text{-}Bu)$, a [2 + 2 + 2] cycloaddition is observed (eq 13).[38]

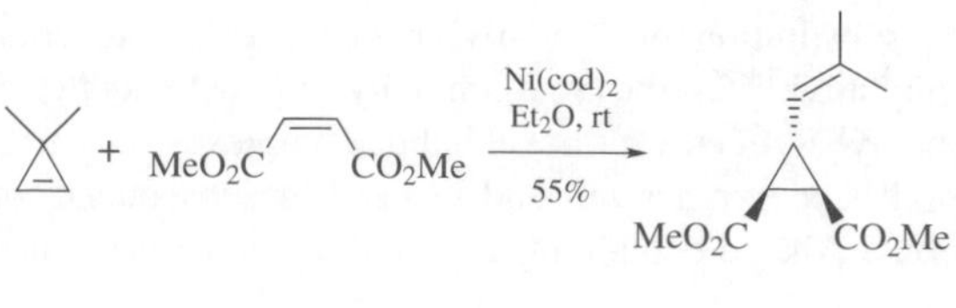

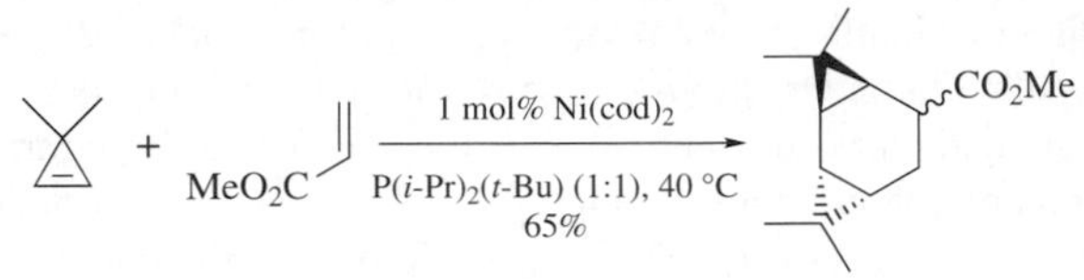

(13)

Cyclobutenones react with alkynes in the presence of $Ni(cod)_2$ to give substituted phenols. Many substituents are tolerated, but regioselectivity is poor (eq 14). The reaction proceeds by oxidative addition of nickel at the C-1–C-4 bond followed by insertion of the alkyne.[39] Cyclopropenones react to form benzoquinones.[40] This mechanism may involve a nickel-catalyzed [2 + 2] cyclodimerization followed by a thermal isomerization or an alkene methathesis.

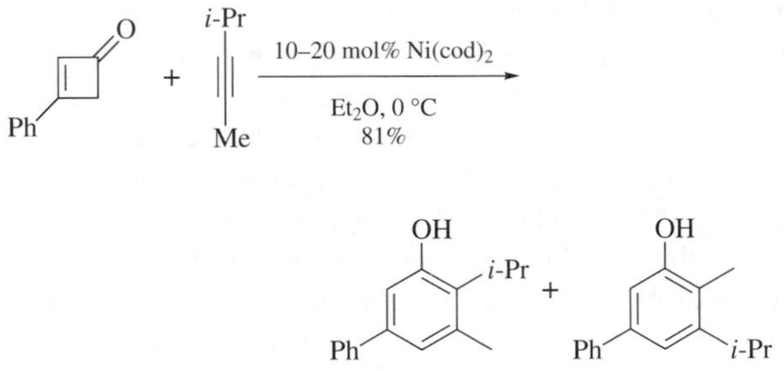

(14)

Norbornene derivatives[41] and cyclobutenes[42] have also been shown to react with alkenes under the influence of $Ni(cod)_2$ catalyst to give [2 + 2] cycloadducts. Norbornadiene substrates can react with electron-deficient alkenes in a [2 + 2 + 2] cycloaddition pathway to give homo-Diels–Alder products (eq 15) (see also ***Bis(acrylonitrile)nickel(0)***).[43] Norbornadiene also forms [2 + 2] cycloadducts with the exceptionally reactive methylene cyclopropane.[44] When this reaction was carried out in the presence of a chiral phosphine ligand, the product was obtained in an enantiomerically enriched form.

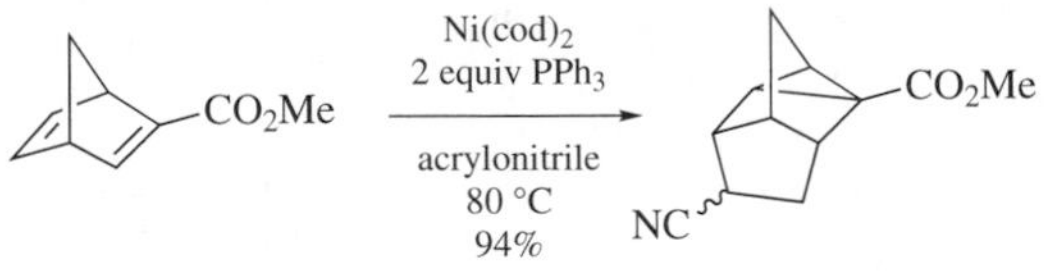

(15)

In addition to the ability to undergo [2 + 2] cycloadditions, methylenecyclopropanes have two additional and more useful [3 + 2] modes of reactivity available under nickel catalysis. As shown for the cycloaddition of methylenecyclopropane with an alkene (eq 16), distal ring opening by nickel leads to products of Type A, whereas Type B products are not a direct result of proximal ring opening, but are formed indirectly via reductive dimerization of two alkene units to give a metallacyclopentane followed by a cyclopropylmethyl/3-butenyl rearrangement. The course of the reaction is determined by many factors including the stoichiometry and physical properties of the ligands bonded to nickel, the number, type, and position of substituents on the methylenecyclopropane, and the nature of the substituents on the participating alkene. Because of these many contributing effects, it is difficult to predict which reaction pathway will be followed under any given condition; however, some general reactivity patterns can be deduced.[4]

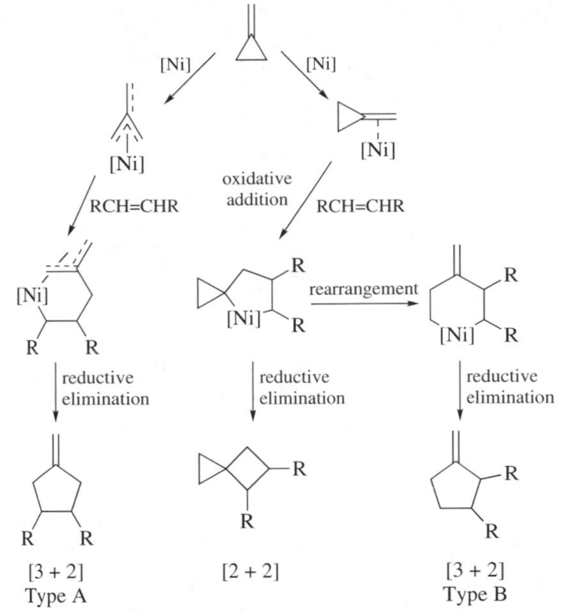

(16)

While the dimerization and trimerization of methylenecyclopropanes have been investigated,[45] these reactions are not useful from a synthetic standpoint. For this reason the present treatment will consider only the cycloadditions between methylenecyclopropanes and alkenes. The use of low valent nickel complexes, which have the ability to catalyze reactions of both Type A and Type B should be contrasted with the use of palladium(0) catalysts such as ***Bis(dibenzylideneacetone)palladium(0)***[46] or $Pd(\eta^3\text{-}C_3H_5)(\eta^5\text{-}C_5H_5)$[47] which lead exclusively to products of Type A.[4b]

Unsubstituted methylenecyclopropane can form 1:1 adducts with acrylates, crotonates, and maleates to give Type B cycloaddition products in high yields (eq 17)[48] (see also ***Bis(acrylonitrile)nickel(0)***). The use of enantiomerically pure alkyl acrylates in this type of reaction gave products with up to 64% de.[49] Addition of ligands requires higher reaction temperatures and results in decreased stereoselectivity.[50] Ligand addition also leads to Type A cycloadducts when highly electron-deficient alkenes such as dialkyl fumarates and maleates are used. It is clear that the outcome of these reactions is highly dependent upon the electronic properties of catalyst and substrate.

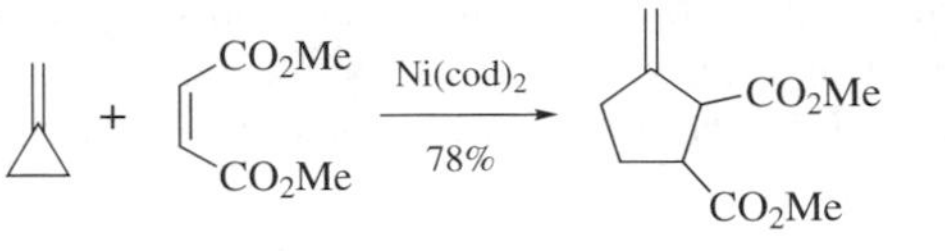

(17)

Substitution at the three-membered ring generally gives rise to cycloadducts of Type B.[4b] When (−)-camphorsulfamylacrylate was reacted with 2,2-dimethylmethylenecyclopropane, a methylenecyclopentane was obtained with a diastereomeric excess of 98% (eq 18).[51] When the substituents are phenyl groups, however, Type A cycloaddition results.[4]

Ni(cod)$_2$ 86% (18) 98% de

Substitution at the double bond almost always leads to cycloaddition of Type A. In these reactions there is an added complication of regioselectivity brought about by the intermediacy of a trimethylenemethane-like species. Thus ethyl crotonate reacts with isopropylidenecyclopropane to give a mixture of the alkylidenecyclopentane (**12**) and the 2-substituted methylenecyclopentane (**13**) (eq 19).[52] The use of bulky phosphite ligands and a high ligand:metal ratio favors the formation of 2-substituted methylenecyclopentanes. Alkenes without electron-withdrawing groups favor the formation of alkylidenecyclopentanes.[4,53] The use of triisopropylphosphine/palladium(0) catalysts also leads to this type of product.[4,54]Disubstituted alkynes can also be used in the cycloaddition to produce alkylidenecyclopentenes.[55] When ***Triethylborane*** was added along with the catalyst, systems that normally reacted along a Type A pathway were induced to react via Type B.[56] An interesting transannular cycloaddition was utilized to prepare a [3.3.3]propellane, although a palladium catalyst was found to be more suitable than nickel (eq 20).[57] Allene can also be oligomerized by Ni(cod)$_2$.[5a,58]

Ni(cod)$_2$ (19)

(**12**) 14% (**13**) 46%

catalyst (20)

20 mol% Ni(cod)$_2$/PPh$_3$, 110 °C 74%
10 mol% PdCl$_2$(PPh$_3$)$_2$/DIBAL, 130 °C 98%

Finally, some strained alkanes are known to undergo oligomerization in the presence of Ni^0 catalysts (see also ***Bis(acrylonitrile)nickel(0)***). Bicyclo[1.1.0]butanes react by suffering a geminal two-bond cleavage to form an allylcarbene intermediate which may be trapped stereoselectively by an electron-deficient alkene (eq 21).[59] Additionally, a nickel-catalyzed asymmetric vinylcyclopropane–cyclopentene rearrangement has been reported using Ni(cod)$_2$ with chiral phosphine ligands.[60]

Ni(cod)$_2$ 0 °C 92% (21)

cis:*trans* = 65:35

Oligomerization of Alkynes. The nickel catalyzed intermolecular oligomerizations of alkynes are some of the oldest and best-studied reactions in organometallic chemistry.[1,5a] Tetramerization and trimerization lead to cyclooctatetraenes and aromatic molecules, respectively. The cycloaddition of two equivalents of alkyne with one equivalent of alkene provides an interesting route to cyclohexadienes.[61] Addition of isocyanides leads to iminocyclopentadienes,[62] insertion of CO_2 gives pyrones,[63] and insertion of isocyanates yields 2-oxo-1,2-dihydropyridines.[64] Finally, hydroacylation of monoalkynes with aldehydes yields α,β-enones.[65]

The synthetic utility of these reactions is greatly increased when they are used intramolecularly with tethered alkynes since more than one ring can be formed and regiochemical problems are drastically reduced.[5b] Tethered diynes react with Ni(cod)$_2$ and CO_2 in the presence of trialkylphosphine ligands to give bicyclic α-pyrones (eq 22).[66] The reaction is catalytic in nickel and works well with three- and four-atom tethers, but the yield suffers when the tether length is raised to five. Tricyclohexylphosphine was found to be the best choice of ligand. When other ligands were used, dimerization of the starting material was observed. It is believed that an electron-donating ligand may be required for strong CO_2 coordination. The reaction is tolerant of many groups on the alkyne terminus including alkyl, hydrogen, and trimethylsilyl. The choice of these groups helps determine the regiochemistry of the CO_2 addition.[67] The use of aldehydes in the reaction allows for the catalytic generation of bicyclic α-pyrans (eq 23).[68] In these cases the structure of the added phosphine ligand does not play a crucial role; however, the length of the tether and the choice of alkyne substituents exert a great influence on the outcome of the reaction. Other products from this reaction such as oxoalkyl-substituted cyclopentenes are explained by various hydrogen transfer isomerizations of a strained 1,2-bis(alkylidene)cycloalkane intermediate.

Ni(cod)$_2$, CO_2, PCy$_3$, THF, rt (22)

n = 3 88%
n = 4 90%
n = 5 19%

Ni(cod)$_2$, PhCHO, PCy$_3$, THF, 120 °C (23)

R = Me 39%
R = Et 79%
R = n-Bu 78%

Diynes undergo cyclization with 2,6-dimethylphenyl isocyanide in the presence of a stoichiometric amount of Ni(cod)$_2$ to yield polycyclic iminocyclopentadienes.[69] In contrast to the previous cycloadditions, this reaction does not require a phosphine ligand, and seven-membered rings can be formed in moderate yields (47%). The products of these reactions could be hydrolyzed to the corresponding cyclopentadienones by ***10-Camphorsulfonic Acid***, used as diene moieties in Diels–Alder reactions, or stereoselectively substituted at the angular position by a 1,4-addition of an alkyllithium reagent.[5b] Enynes may also take part in this reaction to produce iminocyclopentenes with significant diastereoselectivity (eq 24).[70] No reaction occurs when carbon monoxide is used as the cyclization partner; however, there are titanium, zirconium, and cobalt catalysts that are able to carry out such a transformation directly.[71]

$Ni(cod)_2$, CNAr, $P(n\text{-}Bu)_3$, DMF, 100 °C, 83% (24)

Oligomerization of 1,3-Dienes. Extensive studies of the catalytic cyclooligomerization of butadiene have shown that many different products can be obtained from the reaction, depending upon the conditions chosen.[6] These products result from the variety of σ- and π-allylnickel intermediates that can be formed during the catalytic cycle (eq 25).[72] Low-temperature NMR studies of the reaction of butadiene with stoichiometric amounts of $Ni(cod)_2$ have shown that the η^1,η^3-octadienyl complex (**14**) is formed after reductive coupling of the two diene units.[73] This intermediate, which is probably important in the catalytic reaction as well, can isomerize to the bis-π-allyl complex (**15**). Reductive elimination to regenerate a nickel(0) species from these and other allylnickel(II) complexes can lead to 4-vinylcyclohexanes (VCH), 1,2-divinylcyclobutanes (DVCB), or 1,5-cyclooctadienes (cod). Insertion of alkenes or dienes into these intermediates leads to larger ring systems.[6a] These intermediates can also be induced to react with carbonyl compounds, but they are much less nucleophilic than π-allylhalide-like complexes.[74]

[Ni] (**14**) (**15**) VCH cod DVCB (25)

Due to the lack of general methods for the production of medium-sized rings, the synthetic value of these reactions lies especially in their ability to generate cyclooctane ring systems. Control of the cyclodimerization of 1,3-dienes to give cyclooctadiene products preferentially is contingent upon the substitution of the diene as well as the composition of the catalyst system, with both of these factors being mutually dependent. The choice of modifying ligand is extremely important in the selectivity of the reaction, with both electronic and steric parameters being important.[75] It has also been shown for some functionalized dienes that the method of generation of the Ni^0 catalyst is important to this selectivity. Catalysts generated from the reduction of nickel(II) salts such as $Ni(acac)_2$ with alkylaluminum derivatives favor formation of VCH derivatives, while aluminum-free reagents such as $Ni(cod)_2$ give preference to the cod products.[76] Several iron and cobalt catalysts have been reported to selectively catalyze cod production, while titanium, iron, and manganese catalysts have led to VCH products. Palladium catalysts, on the other hand, give DVCB products.[5a]

When [4+4] cycloadditions occur, simple alkyl substituents on the diene usually give 1,5-disubstituted cod products as a result of initial head-to-tail linking, but mixtures of products are inevitable.[1] Functional groups that can coordinate to nickel act as internal ligands to guide the reaction. Thus methyl 2,4-pentadienoate cyclodimerizes regio- and stereoselectively to give a *trans*-1,2-disubstituted cyclooctadiene (eq 26).[77] While ester and silyl ether substituents worked well in this reaction, amino and amido groups were not tolerated.

$Ni(cod)_2$, benzene, 60 °C (26)

$R = OSiMe_3$, 90%
$R = CO_2Me$, 83%

The synthetic utility of these reactions is significantly enhanced when the cycloadditions are carried out intramolecularly. When the diene units are connected by a tether, the production of byproducts can be essentially eliminated.[78] Terminally linked dienes undergo Type I [4+4] cycloadditions, giving rise to *cis*-fused bicycles when the tether length is three atoms, and *trans*-fused systems when the tether length is increased to four atoms. Diastereoselectivity is observed in these cycloadditions when substitution is made at an allylic position, and the degree of selectivity is related to the size of the substituent (eq 27).[79] A nickel catalyzed intramolecular [4+4] cycloaddition was used as the key step in the synthesis of (+)-asteriscanolide (eq 28).[80] Furthermore, when one of the diene fragments is connected at an internal position, bridged ring-systems can be assembled (eq 29).[81] This Type II class of cycloaddition was used in an approach to the taxane skeleton.[82] When trisubstituted dienes are used as substrates for the reaction, angularly substituted bicycles can be obtained.

$Ni(cod)_2$, PPh_3, toluene, 60 °C (27)

R	Yield	Ratio (α:β)
CN	66%	1.6:1
Me	88%	20:1
CH_2OAc	93%	22:1
CO_2Me	65%	65:1

$Ni(cod)_2$, PPh_3, toluene, 60 °C, 67% (28)

$Ni(cod)_2$, $P(O\text{-}o\text{-}BiPh)_3$, toluene, 85 °C, 74% (29)

β:α = 7:1

These reactions can also be tailored to give 1,4-cyclohexadiene products if an alkyne is tethered to the diene (eq 30).[83] These catalytic [4+2] cycloaddition reactions are an attractive alternative to the Diels–Alder reaction because they are very mild, tolerant of a wide array of functional groups, and are exempt from the restrictive electronic substitution requirements of the thermally activated reaction. If the alkyne portion of the molecule is replaced by an alkene, however, a rhodium catalyst must be used.[84] When an allene is used the regiochemistry of the cycloaddition can be controlled by the choice of catalyst (eq 31).[85]

OTBS; $Ni(cod)_2$, $P(O\text{-}o\text{-}BiPh)_3$, THF, rt, >99%; H, OTBS (30)

β:α = 2:1

$Ni(cod)_2$, $P(O\text{-}o\text{-}BiPh)_3$, THF, rt, 97%; H, OTBS; 2:1

OTBS (31)

$[Rh(cod)Cl]_2$, $P(O\text{-}o\text{-}BiPh)_3$, THF, 45 °C, 90%; H, OTBS

Linear dimerizations of 1,3-dienes may be induced by the addition of a hydrogen donor such as an amine, alcohol, or an aminophosphinate ligand to the catalyst.[5a,86] When a chiral aminophosphinate ligand was used, piperylene was dimerized to a 21:70 ratio of the two head-to-head 1,3,6-octatriene isomers having ~90% and 35% ee, respectively.[87]

Other Uses. Sulfur heterocycles can undergo hydrodesulfurization and ring contraction in the presence of nickel(0) complexes.[88] Symmetrical aromatic ketones may be prepared from aromatic carboxylic acids through the coupling of their *S*-(2-pyridyl) derivatives.[89] Alkenes can be formed from 1,2-diols by the stereospecific cleavage of thionocarbonates.[90] Alkane- and alkenecarboxylic acids can be generated from the reaction of alkenes with a nickel(0) catalyst and CO_2.[91] Geminal dihalides react with nickel(0) catalysts and electron deficient alkenes to give cyclopropanes.[92]

Related Reagents. Bis(acrylonitrile)nickel(0); Bis(allyl)-di-μ-bromodinickel; Tetrakis(triethyl phosphite)nickel(0); Tetrakis(triisopropyl phosphite)nickel(0); Tetrakis(triphenylphosphine)nickel(0); Tetrakis(triphenyl phosphite)nickel(0).

1. Jolly, P. W.; Wilke, G. *The Organic Chemistry of Nickel*; Academic: New York, 1974; Vols. I and II.
2. (a) Heimbach, P.; Jolly, P. W.; Wilke, G. *Adv. Organomet. Chem.* **1970**, *8*, 29. (b) Semmelhack, M. F. *OR* **1972**, *19*, 115. (c) Baker, R. *CRV* **1973**, *73*, 487. (d) Hegedus, L. S. *J. Organomet. Chem. Lib.* **1976**, *1*, 329. (e) Billington, D. C. *CSR* **1985**, *14*, 93. (f) Collman, J. P.; Hegedus, L. S. *Principles and Applications of Organotransition Metal Chemistry*; University Science Books: Mill Valley, CA, 1987.
3. Semmelhack, M. F.; Helquist, P.; Jones, L. D.; Keller, L.; Mendelson, L.; Ryono, L. S.; Smith, J. G.; Stauffer, R. D. *JACS* **1981**, *103*, 6460.
4. (a) Jolly, P. W. In *Comprehensive Organometallic Chemistry*; Wilkinson, G.; Stone, F. G. A.; Abel, E. W., Eds.; Pergamon: New York, 1982; Chapter 56.2, p 615. (b) Binger, P.; Büch, M. *Top. Curr. Chem.* **1987**, *135*, 77.
5. (a) Keim, W.; Behr, A.; Röper, M. In *Comprehensive Organometallic Chemistry*; Wilkinson, G.; Stone, F. G. A.; Abel, E. W., Eds.; Pergamon: New York, 1982; Chapter 52, p 371. (b) Tamao, K.; Kobayashi, K.; Ito, Y. *SL* **1992**, 539.
6. (a) Jolly, P. W. In *Comprehensive Organometallic Chemistry*; Wilkinson, G.; Stone, F. G. A.; Abel, E. W., Eds.; Pergamon: New York, 1982; Chapter 56.4, p 671. (b) Tolman, C. A.; Faller, J. W. In *Homogeneous Catalysis with Metal Phosphine Complexes*; Pignolet, L. H., Ed.; Plenum: New York, 1983, p 13.
7. Bogdonavich, B.; Kroner, M.; Wilke, G. *LA* **1966**, *699*, 1.
8. Krysan, D. J.; Mackenzie, P. B. *JOC* **1990**, *55*, 4229.
9. Shriver, D. F. *The Manipulation of Air-Sensitive Compounds*; McGraw-Hill: New York, 1969.
10. Wilke, G.; Bogdanovic, B.; Dardt, P.; Heimbach, P.; Keim, W.; Kroner, M.; Oberkirch, W.; Tanaka, K.; Walter, D. *AG(E)* **1966**, *5*, 151.
11. Kurosawa, H.; Ohnishi, H.; Emoto, M.; Kawasaki, Y.; Murai, S.; *JACS* **1988**, *110*, 6272.
12. Corey, E. J.; Semmelhack, M. F. *JACS* **1967**, *89*, 2755.
13. Hegedus, L. S.; Wagner, S. D.; Waterman, E. L.; Siirala-Hansen, K. *JOC* **1975**, *40*, 593.
14. Hegedus, L. S.; Waterman, E. L.; Catlin, J. *JACS* **1972**, *94*, 7155.
15. Onaka, M.; Goto, T.; Mukaiyama, T *CL* **1979**, 1483.
16. Katzenellenbogen, J. A.; Lenox, R. S. *JOC* **1973**, *38*, 326.
17. Hegedus, L. S.; Thompson, D. H. P. *JACS* **1985**, *107*, 5663.
18. Hegedus, L. S.; Stiverson, R. K. *JACS* **1974**, *96*, 3250.
19. Molander, G. A.; Shubert, D. C. *TL* **1986**, *27*, 787.
20. Hegedus, L. S.; Varaprath, S. *OM* , **1982**, *1*, 259.
21. Johnson, J. R.; Tully, P. S.; Mackenzie, P. B.; Sabat, M. *JACS* **1991**, *113*, 6172.
22. Krysan, D. J.; Mackenzie, P. B. *JACS* **1988**, *110*, 6273.
23. Yamamoto, T.; Ishizu, J.; Yamamoto, A. *JACS* **1981**, *103*, 6863.
24. (a) Oppolzer, W.; Bedoya-Zurita, M.; Switzer, C. Y. *TL* **1988**, *29*, 6433. (b) Oppolzer, W.; Keller, T. H.; Kuo, D. L.; Pachinger, W. *TL* **1990**, *31*, 1265.
25. Semmelhack, M. F.; Yamashita, A.; Tomesch, J. C.; Hirotsu, K. *JACS* **1978**, *100*, 5565.
26. Catellani, M.; Chiusoli, G. P.; Salerno, G.; Dallatomasina, F. *JOM* **1978**, *146*, C19.
27. Semmelhack, M. F.; Helquist, P. M.; Jones, L. D. *JACS* **1971**, *93*, 5907.
28. Negishi, E-i.; King, A. O.; Okukado, N. *JOC* **1977**, *42*, 1821.
29. (a) Ullmann, F.; Bielecki, J. *CB* **1901**, *34*, 2147. (b) Fanta, P. E.; *S* **1974**, 9. (c) Sainsbury, M. *T* **1980**, *36*, 3327.
30. Coffen, D. L.; Schaer, B.; Bizzarro, F. T.; Cheung, J. B. *JOC* **1984**, *49*, 296.
31. Kihara, M.; Itoh, J.; Iguchi, S.; Imakura, Y.; Kobayashi, S. *JCR(S)* **1988**, *8*.
32. Semmelhack, M. F.; Ryono, L. S. *JACS* **1975**, *97*, 3873.
33. Semmelhack, M. F.; Helquist, P. M.; Gorzynski, J. D. *JACS* **1972**, *94*, 9234.
34. Takahashi, S.; Suzuki, Y.; Hagihara, N. *CL* **1974**, 1363.
35. (a) Binger, P.; McMeeking, J. *AG* **1974**, *86*, 518. (b) Binger, P.; McMeeking, J. *AG(E)* **1974**, *13*, 466.
36. Binger, P.; Biedenbach, B. *CB* **1987**, *120*, 601.
37. Binger, P.; McMeeking, J.; Schäfer, H. *CB* **1984**, *117*, 1551.
38. Binger, P.; Brinkmann, A.; Wedemann, P. *CB* **1986**, *119*, 3089.
39. Huffman, M. A.; Liebeskind, L. S. *JACS* **1991**, *113*, 2771.
40. Noyori, R.; Umeda, I.; Takaya, H. *CL* **1972**, 1189.
41. (a) Takaya, H.; Yamakawa, M.; Noyori, R. *BCJ* **1982**, *55*, 852. (b) Voecks, G. E.; Jennings, P. W.; Smith, G. D.; Caughlan, C. N. *JOC* **1972**, *37*, 1460.

42. Kaufmann, D.; de Meijere, A. *CB* **1984**, *117*, 3134.
43. Lautens, M.; Edwards, L. G. *JOC* **1991**, *56*, 3761 and references therein.
44. Noyori, R.; Ishigami, T.; Hayashi, N.; Takaya, H. *JACS* **1973**, *95*, 1674.
45. (a) Binger, P.; *AG* **1972**, *84*, 352. (b) Binger, P. *S* **1973**, 427. (c) Binger, P.; McMeeking, J. *AG* **1973**, *85*, 1053. (d) Binger, P.; Brinkmann, A.; McMeeking, J. *LA* **1977**, 1065.
46. Ukai, T.; Kawazura, H.; Ishii, Y.; Bonnet, J. J.; Ibers, J. A. *JOM* **1974**, *65*, 253.
47. (a) Tatsuno, Y.; Yoshida, T.; Otsuka, S. *Inorg. Synth.* **1979**, *19*, 220. (b) Kühn, A.; Werner, H. *JOM* **1979**, *179*, 421.
48. (a) Noyori, R.; Odagi, T.; Takaya, H. *JACS* **1970**, *92*, 5780. (b) Noyori, R.; Kumagai, Y.; Umeda, I.; Takaya, H. *JACS* **1972**, *94*, 4018. (c) Noyori, R.; Yamakawa, M.; Takaya, H. *TL* **1978**, 4823. (d) Binger, P.; Brinkmann, A.; Wedemann, P. *CB* **1983**, *116*, 2920. (e) Buch, H. M.; Schroth, G.; Mynott, R.; Binger, P. *JOM* , **1983**, *247*, C63.
49. Binger, P.; Brinkmann, A.; Richter, W. J. *TL* **1983**, *24*, 3599.
50. Binger, P.; Wedemann, P. *TL* **1985**, *26*, 1045.
51. (a) Binger, P.; Schäfer, B. *TL* **1988**, *29*, 529. (b) Binger, P.; Brinkmann, A.; Roefke, P.; Schäfer, B. *LA* **1989**, 739.
52. Binger, P.; Wedemann, P. *TL* **1983**, *24*, 5847.
53. Binger, P.; Bentz, P. *JOM* **1981**, *221*, C33.
54. Binger, P.; Sternberg, E.; Wittig, U. *CB* **1987**, *120*, 1933.
55. (a) Binger, P.; Lü, Q-H; Wedemann, P. *AG* **1985**, *97*, 333. (b) Binger, P.; Lü, Q-H; Wedemann, P. *AG(E)* **1985**, *24*, 316.
56. Binger, P.; Schäfer, B. *TL* **1988**, *29*, 4539.
57. (a) Yamago, S.; Nakamura, E. *CC* **1988**, 1112. (b) Yamago, S.; Nakamura, E. *T* **1989**, *45*, 3081.
58. (a) Otsuka, S.; Nakamura, A.; Tani, K.; Ueda, S. *TL* **1969**, 297. (b) Otsuka, S.; Nakamura, A.; Yamagata, T.; Tani, K. *JACS* **1972**, *94*, 1037.
59. (a) Noyori, R.; Suzuki, T.; Kumagai, Y.; Takaya, H. *JACS* **1971**, *93*, 5894. (b) Noyori, R.; Kawauchi, H; Takaya, H. *TL* **1974**, 1749. (c) Takaya, H.; Suzuki, T.; Kumagai, Y.; Hosoya, M.; Kawauchi, H.; Noyori, R. *JOC* **1981**, *46*, 2854.
60. Hiroi, K.; Arinaga, Y.; Ogino, T. *CL* **1992**, 2329.
61. (a) Fahey, D. R. *JOC* **1972**, *37*, 4471. (b) Heimbach, P.; Ploner, K. J.; Thömel, F. *AG* **1971**, *83*, 285.
62. Eisch, J. J.; Aradi, A. A.; Han, K. I. *TL* **1983**, *24*, 2073.
63. Tsuda, T.; Hasegawa, N.; Saegusa, T. *CC* **1990**, 945.
64. Hoberg, H.; Oster, B. W. *S* **1982**, 324.
65. Tsuda, T.; Kiyoi, T.; Saegusa, T. *JOC* **1990**, *55*, 2554.
66. (a) Tsuda, T.; Sumiya, R.; Saegusa, T. *SC* **1987**, *17*, 147. (b) Tsuda, T.; Morikawa, S.; Sumiya, R.; Saegusa, T. *JOC* **1988**, *53*, 3140.
67. Tsuda, T.; Morikawa, S.; Hasegawa, N.; Saegusa, T. *JOC* **1990**, *55*, 2978.
68. Tsuda, T.; Kiyoi, T.; Miyane, T.; Saegusa, T. *JACS* **1988**, *110*, 8570.
69. Tamao, K.; Kobayashi, K.; Ito, Y. *JOC* **1989**, *54*, 3517.
70. Tamao, K.; Kobayashi, K.; Ito, Y. *JACS* **1988**, *110*, 1286.
71. See Refs. 5b and 57 and references cited therein.
72. Wilke, G. *AG(E)* **1988**, *27*, 186.
73. Benn, R.; Büssemeier, B.; Holle, S.; Jolly, P. W.; Mynott, R.; Tkatchenko, I.; Wilke, G. *JOM* **1985**, *79*, 63.
74. (a) Baker, R.; Nobbs, M. S.; Robinson, D. T. *JCS(P1)* **1978**, 543. (b) Baker, R.; Popplestone, R. J. *TL* **1978**, 3575. (c) Baker, R.; Crimmin, M. J. *JCS(P1)* **1979**, 1264. (d) Tsuda, T.; Chujo, Y.; Saegusa, T. *SC* **1979**, *9*, 427.
75. (a) Van Leeuwen, P. W. N. M.; Roobeek, C. F. *T* **1981**, *37*, 1973. (b) Heimbach, P.; Kluth, J.; Schenkluhn, H.; Weimann, B. *AG(E)* **1980**, *19*, 569, 570. (c) Bartik, T.; Heimbach, P.; Himmler, T. *JOM* **1984**, *276*, 399. (d) Tollman, C. A. *CRV* **1977**, *77*, 313.
76. Brun, P.; Tenaglia, A.; Waegell, B. *TL* **1985**, *26*, 5685.
77. (a) Brun, P.; Tenaglia, A.; Waegell, B. *TL* **1983**, *24*, 385. (b) Tenaglia, A.; Brun, P.; Waegell, B. *JOM* **1985**, *285*, 343.
78. Wender, P. A.; Ihle, N. C. *JACS* **1986**, *108*, 4678.
79. Ihle, N. C. Ph. D. Dissertation, Stanford University, 1988.
80. Wender, P. A.; Ihle, N. C.; Correia, C. R. D. *JACS* **1988**, *110*, 5904.
81. Wender, P. A.; Tebbe, M. J. *S* **1991**, 1089.
82. Wender, P. A.; Snapper, M. L. *TL* **1987**, 2221.
83. Wender, P. A.; Jenkins, T. E. *JACS* **1989**, *111*, 6432.
84. Jolly, R. S.; Luedtke, G.; Sheehan, D.; Livinghouse, T. *JACS* **1990**, *112*, 4965.
85. Jenkins, T. E. Ph. D. Dissertation, Stanford University, 1995.
86. (a) Denis, P.; Mortreux, A.; Petit, F.; Buono, G.; Peiffer, G. *JOC* **1984**, *49*, 5276. (b) Buono, G.; Siv, C.; Peiffer, G.; Triantaphylides, C.; Philippe, D.; Mortreux, A.; Petit, F. *JOC* **1985**, *50*, 1781. (c) Cros, P.; Triantaphylides, C.; Buono, G. *JOC* **1988**, *53*, 185. (d) Amrani, M. A.; Mortreux, A.; Petit, F. *TL* 1989, 6515. (e) Denis, P.; Jean, A.; Croizy, J. F.; Mortreux, A.; Petit, F. *JACS* **1990**, *112*, 1292.
87. Denis, P.; Croizy, J.-F.; Mortreux, A.; Petit, F. *J. Mol. Catal.* **1991**, *68*, 159.
88. (a) Eisch, J. J.; Im, K. R. *JOM* **1977**, *139*, C51. (b) Eisch, J. J.; Hallenbeck, L. E.; Han, K. I. *JOC* **1983**, *48*, 2963.
89. Goto, T.; Onaka, M.; Mukaiyama, T. *CL* **1980**, 51.
90. Semmelhack, M. F.; Stauffer, R. D. *TL* **1973**, 2667.
91. (a) Hoberg, H.; Ballesteros, A.; Sigan, A.; Jegat, C.; Milchereit, A. *S* **1991**, 395. (b) Hoberg, H.; Ballesteros, A.; Sigan, A.; Jegat, C.; Bärhausen, D.; Milchereit, A. *JOM* **1991**, *407*, C23.
92. (a) Furukawa, J.; Matsumura, A.; Matsuoka, Y.; Kiji, J. *BCJ* **1976**, *49*, 829. (b) Takahashi, S.; Suzuki, Y.; Sonogashira, K.; Hagihara, N. *CL* **1976**, 515. (c) Kanai, H.; Hiraki, N. *CL* **1979**, 761. (d) Kanai, H.; Hiraki, N.; Iida, S. *BCJ* **1983**, *56*, 1025.

Paul A. Wender & Thomas E. Smith
Stanford University, CA, USA

Bis(cyclopentadienyl)-3,3-dimethyltitanacyclobutane[1]

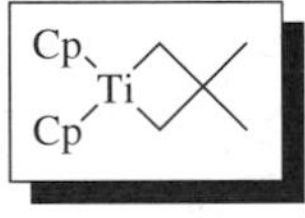

[80122-07-2] $C_{15}H_{20}Ti$ (MW 248.20)

(methylenating agent for alkenation of carbonyl compounds, particularly esters, ketones, and aldehydes;[1a,c] of low basicity and functions without epimerization at α-chiral centers; catalyst precursor in living ring-opening metathesis polymerization[1b])

Solubility: very sol dichloromethane, diethyl ether, and toluene; moderately sol pentane.[2]

Form Supplied in: red crystalline solid.

Analysis of Reagent Purity: by ^{1}H NMR (toluene-d_8, $-10\,^\circ$C)[3] or by bromination in ether at $-20\,^\circ$C followed by quantitation of the 1,3-dibromo-2,2-dimethylpropane product by GC.[4]

Preparative Methods: most often prepared from the Tebbe reagent (**1**) by reaction with isobutene and 4-dimethylaminopyridine in dichloromethane.[3] Alternatively, may be formed in 89% yield from Cp_2TiCl_2 and a 1,3-di-Grignard reagent, which is in turn obtained from 1,3-dibromo-2,2-dimethylpropane in 18% yield.[4,5]

Purification: may be recrystallized by slow cooling of a toluene solution from 0 °C to −78 °C under an inert atmosphere. Recrystallization from diethyl ether is also effective.

Handling, Storage, and Precautions: the title reagent (**2a**) may be stored indefinitely under nitrogen atmosphere, preferably in the dark. Although this titanacycle is moderately air- and water-sensitive, it may be handled in air for brief periods. In solution this complex is labile and will decompose to give the purple dimer $[Cp_2Ti(\mu\text{-}CH_2)]_2$,[6] as well as isobutene and other uncharacterized products. Solutions should be prepared at 0 °C and used immediately.[2]

Reactivity of Titanacyclobutanes. A number of titanacyclobutane complexes (e.g. (**2a–e**)) have been efficiently prepared from Tebbe reagent (**1**) (***μ-Chlorobis(cyclopentadienyl)-(dimethylaluminum)–μ-methylenetitanium***)[1c,2,7–9] and studied in significant detail (eq 1).[1] Typically pyridine bases are used; ether or tetrahydrofuran as bases give equilibrium concentrations of (**1**) and (**2**).[10] Certain titanacyclobutanes may be prepared in variable yields from 1,3-di-Grignard reagents;[4,5,11] a few examples are shown in eq 2. Very recently a route to titanacyclobutanes via cationic π-allyl species was demonstrated.[12] Also, titanacyclobutanes may be prepared by reaction of alkenes with less stable titanacyclobutanes.[3,8,9] These metallacycles cleave to afford '$Cp_2Ti{=}CH_2$' (as the free methylidene or, more likely, an alkene complex[13]) as shown in the ester to vinyl ether conversion of eq 3.[1a,1c,14]

$$Cp_2Ti(\mu\text{-}CH_2)(\mu\text{-}Cl)AlMe_2 \; (\mathbf{1}) + R^1R^2C{=}CHR^3 \xrightarrow[-\text{base}\cdot AlMe_2Cl]{\text{base}} \text{titanacyclobutane} \quad (1)$$

(**2a**) $R^1 = H$; $R^2, R^3 = Me$
(**2b**) $R^1 = H$; $R^2 = Me$; $R^3 = n\text{-Pr}$
(**2c**) $R^1, R^2 = H$; $R^3 = t\text{-Bu}$
(**2d**) $R^1, R^2 = H$; $R^3 = i\text{-Pr}$
(**2e**) $R^1, R^2 = (CH_2)_3$; $R^3 = H$

$$Cp_2TiCl_2 + (BrMgCH_2)_2CR^1R^2 \longrightarrow Cp_2Ti\text{-cyclobutane}(R^1)(R^2) \quad (2)$$

(**2a**) $R^1 = R^2 = Me$
(**2c**) $R^1 = H$; $R^2 = t\text{-Bu}$
(**2f**) $R^1 = R^2 = H$
(**2g**) $R^1 = H$; $R^2 = Me$

$$(\mathbf{2a}) \xrightarrow[\text{ca. 2 min}]{6\,°C} \left[\begin{array}{c} H_2C{=}CMe_2 \\ | \\ Cp_2Ti{=}CH_2 \end{array}\right] \xrightarrow[-[Cp_2TiO]_n,\; -Me_2C{=}CH_2]{R^1C(O)OR^2} R^1C({=}CH_2)OR^2 \quad (3)$$

The titanacyclobutanes may thus serve as aluminum-free alternatives to the Tebbe reagent for methylenation of a range of carbonyl compounds. Another application which involves 'retro 2 + 2' cleavage of metallacycles is the ring-opening metathesis polymerization (ROMP) of strained cycloalkenes via substituted methylidene intermediates (see below).[1b] As methylidene sources the titanacyclobutanes offer advantages over (**1**); primarily in avoidance of side reactions due to acidic aluminum species, ease of workup, and lesser air sensitivity. However, titanacyclobutanes have not been as widely employed because they are not commercially available and preparation requires Schlenk line and glovebox techniques.

The titanacyclobutanes undergo a number of potentially useful reactions in which the carbon skeleton of the ring remains intact in products such as alkanes, 1,3-dibromides, cyclopropanes, and acyloins. These products may be conveniently obtained from titanacycles formed in situ from (**1**).[1a]

Titanacyclobutanes as Methylidene Precursors. Most titanacyclobutanes, including (**2a–e**), thermally cleave by a 'retro 2 + 2' process to liberate alkene and afford the reactive methylidene species '$Cp_2Ti{=}CH_2$', or an alkene complex thereof, as shown in eq 3.[1f,13,14] Thus much of the chemistry of these metallacycles is qualitatively similar to that of the Tebbe reagent (**1**).[1c] For example, the titanium methylidene so produced affords vinyl ethers from a wide range of esters (eq 3), a conversion which is not possible using Wittig alkylidene phosphorane reagents.[15] Aldehydes and ketones are likewise methylenated and reaction takes place without epimerization at α-chiral centers,[1a] a frequent problem when using highly basic Wittig conditions.[16] Yields in these methylenations are generally good to excellent. The intermediate titanaoxacyclobutanes are not observed in these transformations, although stable analogs having β-alkylidene substitution have been characterized.[17] This methylenation chemistry has been the subject of a recent comprehensive review.[1c]

Different reactivity is observed for the titanacycles with acid chlorides as shown in eq 4.[18] The titanium enolates of methyl ketones which result do not undergo double-bond isomerization, and are useful in the aldol reaction. Although the Tebbe reagent shows similar reactivity, yields are much lower.[19] Reaction of the titanacyclobutanes with sterically hindered ketones also affords titanium enolate complexes; however, these are not active in the aldol reaction.[20] Reaction with anhydrides also gives enolate complexes which are not active in the aldol reaction. By contrast, unhindered imides may be methylenated in good yield with excellent selectivity in unsymmetrical cases.[21]

$$'Cp_2Ti{=}CH_2' + RC(O)Cl \longrightarrow Cp_2Ti\text{-oxatitanacyclobutane}(R)(Cl) \longrightarrow Cp_2Ti(Cl)OC(R){=}CH_2 \quad (4)$$

Reaction with certain alkyl halides occurs to give titanocene alkyl halides $[Cp_2Ti(Cl)CH_2R]$ via a radical mechanism.[22]

Although α-alkyl substituted titanacyclobutanes (e.g. **2h**, see eq 6) are thermally unstable,[10] α-alkylidene substitution is favorable[23–25] and leads to a versatile synthesis (via titanium vinylidene intermediates) of substituted allenes from ketones as shown in eq 5.[26]

$$(\mathbf{2a}) + H_2C{=}C{=}CMe_2 \xrightarrow{-H_2C{=}CMe_2} Cp_2Ti\text{-}(\alpha\text{-isopropylidene})\text{cyclobutane} \xrightarrow[-H_2C{=}CH_2,\; -[Cp_2TiO]_n,\; 80\%]{Ph_2C{=}O} Me_2C{=}C{=}CPh_2 \quad (5)$$

The titanacyclobutanes have also served as precursors in the synthesis of various interesting heterobimetallic μ-methylene complexes[27] and other organometallic species,[28] including titanium methylidene phosphine complexes.[29]

Choice of metallacycle is dictated by reaction conditions and ease of preparation. β-Substituted metallacycles such as (**2c** and **2d**) are especially stable and react at 50–60 °C, whereas β,β-disubstituted titanacycles such as (**2a** and **b**) react at about 0 °C.[1a,13] By contrast, the Tebbe reagent may be used at −40 °C. Titanacyclobutane (**2a**) is particularly convenient for NMR studies due to the simplicity of its spectrum, while (**2b**) is prepared by the same general method using a liquid alkene which is easier to handle than isobutene.

It should be noted that alternate methods for formation of active titanium methylidene species have been developed.[1c] A widely used system for methylenation of ketones[30] is the still undefined mixture formed from $Zn/CH_2X_2/TiCl_4$. Also, thermolysis of Cp_2TiMe_2 provides a clean aluminum-free source of $Cp_2Ti{=}CH_2$ and shows considerable synthetic promise.[31] Except in the case of titanacyclobutanes prepared from strained cycloalkenes (see below), or α-alkylidene substituted cases (see above), the titanacyclobutanes are efficient precursors only for the unsubstituted methylidene species; several approaches to substituted methylidene equivalents of titanium or zirconium have been devised,[1c,32] including a preparation of titanocene vinylcarbene complexes from cyclopropenes.[33]

Reactions Involving Substituted Alkylidene Intermediates: Ring-Opening Metathesis Polymerization (ROMP).[1b] The titanacyclobutanes were the first metallacycles to demonstrate the requisite reactivity for catalysis of alkene metathesis via alkylidene complex intermediates.[34] However, these species are not useful catalysts for productive metathesis of acyclic alkenes. This is due to the propensity of α-substituted titanacyclobutanes such as (**2h**) to cleave to the unsubstituted methylidene, as illustrated by eq 6.[10]

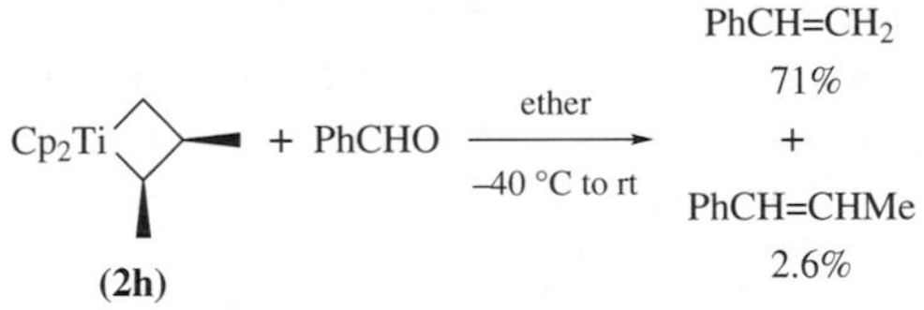

(6)

(2h)

However, titanacyclobutanes formed from strained cycloalkenes will ring open to afford substituted methylidenes. This useful mode of reactivity is illustrated for complex (**2i**), formed from 3,3-dimethylcyclopropene, which yields titanium alkylidene phosphine complex (**3**), as shown in eq 7.[9]

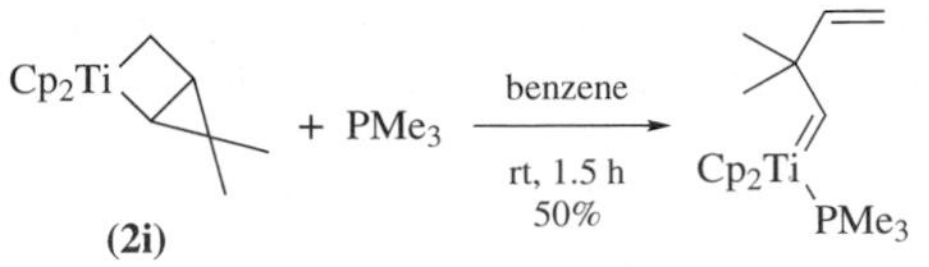

(7)

(2i) **(3)**

Likewise, (**2i**) reacts with benzophenone to afford 3,3-dimethyl-1,1-diphenyl-1,4-pentadiene.[9] Norbornene affords the extremely stable titanacyclobutane (**2j**) upon reaction with (**1**) or even the quite stable titanacycle (**2d**) under suitable conditions, and this species reacts further with norbornene (benzene solution, 65 °C) to produce ring-opened polynorbornene (*cis*:*trans* ratio 38:62) as illustrated in eq 8.[35] Because rates of chain transfer and chain termination are very slow relative to initiation and propagation in this system, a living polymerization process occurs. The low polydispersity (below 1.1) of the polynorbornene and the linear increase in molecular weight with time are consistent with this. Another entry into this chemistry involves thermolysis of dimethyltitanonocene.[31b] The living nature of this system allows for end capping of the polymer chains by Wittig-type reactions with, for example, benzophenone[36] or acetone,[37] and also, by sequential addition of different monomers, for the production of block copolymers.[37,38] Other monomer units employed include benzonorbornadiene, 6-methylbenzonorbornadiene, and *endo*- and *exo*-dicyclopentadiene, which have been used to synthesize diblock and triblock copolymers of low polydispersity as illustrated in eq 9.[37] It has been possible to end-cap ROMP polymers with an aldehyde group and to use aldol-group-transfer polymerization to make diblock polymers with poly(vinyl alcohol) or poly(silyl vinyl ether) segments.[38a] ROMP chemistry using titanium methylidene sources has also been used to produce the novel cross-conjugated poly(3,4-isopropylidenecyclobutene) shown in eq 10, which becomes conducting upon oxidative doping.[39]

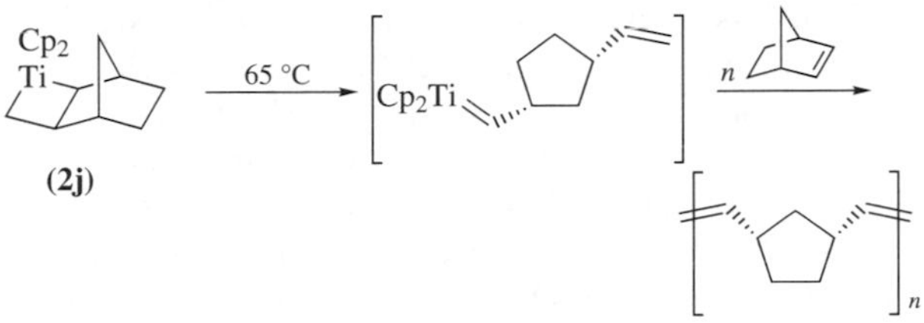

(8)

(2j)

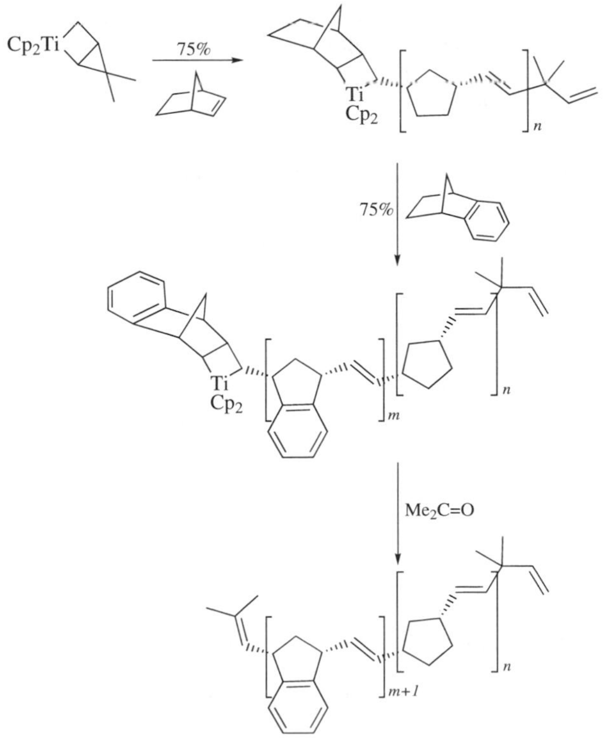

(9)

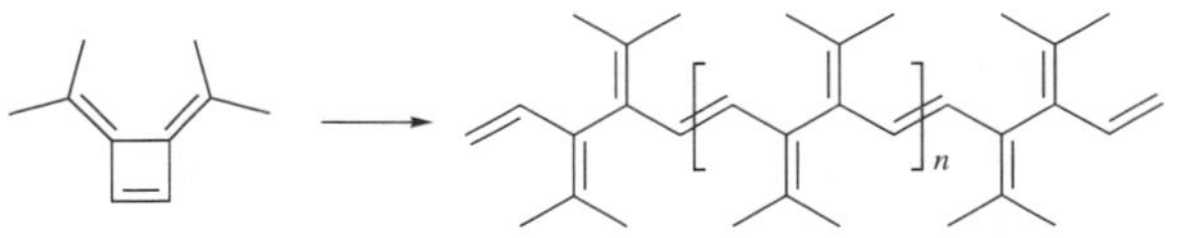

(10)

Numerous other ROMP catalysts have been developed,[1d,40] some of which are stable to air and water and are active at room temperature.[41]

It should be noted that in situ formation of a derivative of (**2j**), which cleaves to give a substituted methylidene, was employed in the synthesis of $\Delta^{9,12}$-capnellene.[42] Certain molybdenum alkylidenes show considerable promise for the stoichiometric[43] or catalytic[44] preparation of cycloalkenes, from unsaturated ketones or dienes, respectively, via metallacyclobutanes which cleave to afford substituted methylidenes.

Reactions Not Involving Methylidene Species. Alkanes are produced upon acidolysis;[34,45] for example (**2a**) affords neopentane.[1a] Titanacyclobutanes may be directly halogenated to afford 1,3-dibromides[1a] as shown in eq 11.[2a] This reaction is typically quantitative, and has been used to assay the titanacyclobutanes.[4]

$$\textbf{(2a)} \xrightarrow[\text{ether, }-20\ ^\circ\text{C; }100\%]{Br_2} (BrH_2C)_2CR^1R^2 \quad (11)$$

By contrast, iodination leads to cyclopropane formation in good yield,[1a] as shown in eq 12, a process which was demonstrated to proceed stepwise with retention of stereochemistry at one carbon and inversion at the other.[46] The titanacycles may be photolyzed, as in eq 13, to afford cyclopropanes via a 1,4-biradical intermediate.[47] Reaction with chemical oxidants such as tetrakis(trifluoromethyl)cyclopentadienone or ferricinium salts also affords cyclopropanes by reductive elimination.[48] Carbonylation leads to reductive coupling of two carbon monoxide units to afford a stable enediolate complex of titanocene which may be converted to a cyclic acyloin upon acidolysis; in the one-pot process of eq 14, metallacycle (**2c**) is generated in situ from the Tebbe reagent.[1a,2a]

$$\text{Cp}_2\text{Ti(CH}_2\text{CHPhCHD)} \xrightarrow[CH_2Cl_2;\ -Cp_2TiI_2]{I_2} \text{cyclopropane (Ph, D) } 30\% + \text{cyclopropane (Ph, D) } 30\% \quad (12)$$

trans-(**2k**-*d*)

$$\textbf{(2a)} \xrightarrow[PhC\equiv CPh;\ 95\%]{h\nu} \text{1,1-dimethylcyclopropane} + \text{Cp}_2\text{Ti(C}_4\text{Ph}_4) \quad (13)$$

$$\textbf{(1)} + t\text{-BuCHO} \xrightarrow[3.\ HCl;\ 54\%]{1.\ pyridine;\ 2.\ CO} \text{2-hydroxy-4-}t\text{-Bu-cyclopentanone} \quad (14)$$

Related Reagents. Bis(cyclopentadienyl)dimethyltitanium; μ-Chlorobis(cyclopentadienyl) (dimethylaluminum)-μ-methylenetitanium; Dibromomethane-Zinc-Titanium(IV) Chloride; Dichlorobis(cyclopentadienyl)zirconium-Zinc-Dibromomethane.

1. (a) Brown-Wensley, K. A.; Buchwald, S. L.; Cannizzo, L.; Clawson, L.; Ho, S.; Meinhardt, D.; Stille, J. R.; Straus, D. A.; Grubbs, R. H. *PAC* **1983**, *55*, 1733. (b) Grubbs, R. H.; Tumas, W. *Science* **1989**, *243*, 907. (c) Pine, S. H. *OR* **1993**, *43*, 1. (d) Feldman, J.; Schrock, R. R. *Prog. Inorg. Chem.* **1991**, *39*, 1. (e) Dang, Y.; Geise, H. J. *JOM* **1991**, *405*, 1. (f) Jørgensen, K. A.; Schiøtt, B. *CRV* **1990**, *90*, 1483. (g) Lenoir, D. *S* **1989**, 883. (h) Reetz, M. T. *Organotitanium Reagents in Organic Synthesis*; Springer: Berlin, 1986; pp 223–229. (i) Dötz, K. H. *AG(E)* **1984**, *23*, 587.
2. (a) Straus, D. A. Ph.D. Thesis, California Institute of Technology, 1983. (b) Straus, D. A.; Grubbs, R. H. In *Handbuch der Präparativen Anorganischen Chemie*; Herrmann, W. A., Ed.; Thieme: Munchen, 1994; in press.
3. Straus, D. A.; Grubbs, R. H. *OM* **1982**, *1*, 1658.
4. Bruin, J. W.; Schat, G.; Akkerman, O. S.; Bickelhaupt, F. *TL* **1983**, *24*, 3935.
5. Seetz, J. W. F. L.; Van de Heisteeg, B. J. J.; Schat, G.; Akkerman, O. S.; Bickelhaupt, F. *J. Mol. Catal.* **1985**, *28*, 71.
6. Ott, K. C.; Grubbs, R. H. *JACS* **1981**, *103*, 5922.
7. (a) Tebbe, F. N.; Parshall, G. W.; Reddy, G. S. *JACS* **1978**, *100*, 3611. (b) Klabunde, U.; Tebbe, F. N.; Parshall, G. W.; Harlow, R. L. *J. Mol. Catal.* **1980**, *8*, 37.
8. Lee, J. B.; Ott, K. C.; Grubbs, R. H. *JACS* **1982**, *104*, 7491.
9. Gilliom, L. R.; Grubbs, R. H. *OM* **1986**, *5*, 721.
10. Straus, D. A.; Grubbs, R. H. *J. Mol. Catal.* **1985**, *28*, 9.
11. (a) Seetz, J. W. F. L.; Schat, G.; Akkerman, O. S.; Bickelhaupt, F. *AG(E)* **1983**, *22*, 248. (b) Van de Heisteeg, B. J. J.; Schat, G.; Akkerman, O. S.; Bickelhaupt, F. *JOM* **1986**, *308*, 1.
12. Tjaden, E. B.; Casty, G. L.; Stryker, J. M. *JACS* **1993**, *115*, 9814.
13. Anslyn, E. V.; Grubbs, R. H. *JACS* **1987**, *109*, 4880.
14. Upton, T. H.; Rappé, A. K. *JACS* **1985**, *107*, 1206.
15. Pine, S. H.; Zahler, R.; Evans, D. A.; Grubbs, R. H. *JACS* **1980**, *102*, 3270.
16. Sowerby, R. L.; Coates, R. M. *JACS* **1972**, *94*, 4758.
17. Ho, S. C.; Hentges, S.; Grubbs, R. H. *OM* **1988**, *7*, 780.
18. Stille, J. R.; Grubbs, R. H. *JACS* **1983**, *105*, 1664.
19. Chou, T.-S.; Huang, S.-B. *TL* **1983**, *24*, 2169.
20. Clawson, L.; Buchwald, S. L.; Grubbs, R. H. *TL* **1984**, *25*, 5733.
21. Cannizzo, L. F.; Grubbs, R. H. *JOC* **1985**, *50*, 2316.
22. Buchwald, S. L.; Anslyn, E. V.; Grubbs, R. H. *JACS* **1985**, *107*, 1766.
23. Hawkins, J. M.; Grubbs, R. H. *JACS* **1988**, *110*, 2821.
24. Dennehy, R. D.; Whitby, R. J. *CC* **1990**, 1060.
25. Beckhaus, R.; Flatau, S.; Trojanov, S.; Hofmann, P. *CB* **1992**, *125*, 291.
26. Buchwald, S. L.; Grubbs, R. H. *JACS* **1983**, *105*, 5490.
27. (a) Park, J. W.; Henling, L. M.; Schaefer, W. P.; Grubbs, R. H. *OM* **1991**, *10*, 171. (b) Ozawa, F.; Park, J. W.; Mackenzie, P. B.; Schaefer, W. P.; Henling, L. M.; Grubbs, R. H. *JACS* **1989**, *111*, 1319. (c) Mackenzie, P. B.; Coots, R. J.; Grubbs, R. H. *OM* **1989**, *8*, 8. (d) Park, J. W.; Mackenzie, P. B.; Schaefer, W. P.; Grubbs, R. H. *JACS* **1986**, *108*, 6402. (e) Mackenzie, P. B.; Ott, K. C.; Grubbs, R. H. *PAC* **1984**, *56*, 59.
28. (a) Anslyn, E. V.; Santarsiero, B. D.; Grubbs, R. H. *OM* **1988**, *7*, 2137. (b) Paetzold, P.; Delpy, K.; Hughes, R. P.; Herrmann, W. A. *CB* **1985**, *118*, 1724. (c) Doxsee, K. M.; Farahi, J. B.; Hope, H. *JACS* **1991**, *113*, 8889.
29. (a) Meinhart, J. D.; Anslyn, E. V.; Grubbs, R. H. *OM* **1989**, *8*, 583. (b) Park, J. W.; Henling, L. M.; Schaefer, W. P.; Grubbs, R. H. *OM* **1990**, *9*, 1650.
30. For example: (a) Lombardo, L. *OS* **1987**, *65*, 81. (b) Eisch, J. J.; Piotrowski, A. *TL* **1983**, *24*, 2043.
31. (a) Petasis, N. A.; Bzowej, E. I. *JACS* **1990**, *112*, 6392. (b) Petasis, N. A.; Fu, D.-K. *JACS* **1993**, *115*, 7208.
32. (a) Tucker, C. E.; Knochel, P. *JACS* **1991**, *113*, 9888. (b) Okazoe, T.; Takai, K.; Oshima, K.; Utimoto, K. *JOC* **1987**, *52*, 4410. (c) Petasis, N. A.; Akritopoulou, I. *SL* **1992**, 665. (d) Petasis, N. A.; Bzowej, E. I. *TL* **1993**, *34*, 943. (e) Clift, S. M.; Schwartz, J. *JACS* **1984**, *106*, 8300.
33. Binger, P.; Müller, P.; Benn, R.; Mynott, R. *AG(E)* **1989**, *28*, 610.

34. Howard, T. R.; Lee, J. B.; Grubbs, R. H. *JACS* **1980**, *102*, 6876.
35. Gilliom, L. R.; Grubbs, R. H. *JACS* **1986**, *108*, 733.
36. Cannizzo, L. F.; Grubbs, R. H. *Macromolecules* **1987**, *20*, 1488.
37. Cannizzo, L. F.; Grubbs, R. H. *Macromolecules* **1988**, *21*, 1961.
38. (a) Risse, W.; Grubbs, R. H. *Macromolecules* **1989**, *22*, 1558. (b) Risse, W.; Grubbs, R. H. *Macromolecules* **1989**, *22*, 4462.
39. Swager, T. M.; Grubbs, R. H. *JACS* **1987**, *109*, 894.
40. (a) Schrock, R. R. *ACR* **1990**, *23*, 158. (b) McConville, D. H.; Wolf, J. R.; Schrock, R. R. *JACS* **1993**, *115*, 4413.
41. (a) Nguyen, S. T.; Grubbs, R. H.; Ziller, J. W. *JACS* **1993**, *115*, 9858. (b) Nguyen, S. T.; Johnson, L. K.; Grubbs, R. H.; Ziller, J. W. *JACS* **1992**, *114*, 3974.
42. (a) Stille, J. R.; Santarsiero, B. D.; Grubbs, R. H. *JOC* **1990**, *55*, 843. (b) Stille, J. R.; Grubbs, R. H. *JACS* **1986**, *108*, 855.
43. Fu, G. C.; Grubbs, R. H. *JACS* **1993**, *115*, 3800.
44. Fu, G. C.; Nguyen, S. T.; Grubbs, R. H. *JACS* **1993**, *115*, 9856.
45. Ikariya, T.; Ho, S. C. H.; Grubbs, R. H. *OM* **1985**, *4*, 199.
46. Ho, S. C. H.; Straus, D. A.; Grubbs, R. H. *JACS* **1984**, *106*, 1533.
47. Tumas, W.; Wheeler, D. R.; Grubbs, R. H. *JACS* **1987**, *109*, 6182.
48. (a) Burk, M. J.; Tumas, W.; Ward, M. D.; Wheeler, D. R.; *JACS* **1990**, *112*, 6133. (b) Burk, M. J.; Staley, D. L.; Tumas, W.; *CC* **1990**, 809.

Daniel A. Straus
San Jose State University, CA, USA

Bis(cyclopentadienyl)dimethyltitanium[1]

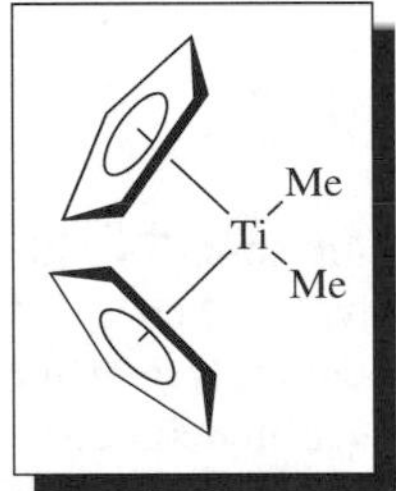

[1271-66-5] $C_{12}H_{16}Ti$ (MW 208.14)

(reagent of low basicity and acidity, useful for the methylenation of carbonyl compounds;[2] methylenates highly enolizable ketones,[2] esters[2] and lactones,[2] including aldonolactones[3] and acid-labile substrates;[4] homologs of the reagent perform carbonyl alkylidenations;[5–7] reacts with alkynes[8,9] and nitriles[8–10] and can initiate the ring-opening metathesis polymerization of strained cyclic alkenes;[11] it is a useful precursor for Ziegler–Natta polymerization catalysts,[12] a cocatalyst for alkene metathesis[13] and a catalyst for alkene hydrogenation[14] and the dehydrogenative coupling of silanes[15,16])

Alternate Name: dimethyltitanocene.
Physical Data: mp > 90 °C dec.
Solubility: sol most aprotic organic solvents, e.g. Et_2O, THF, CH_2Cl_2, toluene, hexanes.
Form Supplied in: orange crystals; not available commercially.
Analysis of Reagent Purity: ^1H NMR (250 MHz, C_6D_6) δ 0.04 (s, 6H), 5.69 (s, 10H).
Preparative Method: [17] a solution of ***Methyllithium*** (2.1 equiv) in Et_2O is added dropwise to a suspension of ***Dichlorobis(cyclopentadienyl)titanium*** in Et_2O (3–5 mL $mmol^{-1}$) cooled in an ice bath. After 1 h the reaction mixture is quenched with ice water, the ether layer is separated, dried over $MgSO_4$, filtered and evaporated. The resulting bright orange crystals of the reagent, obtained in 95% yield, are dissolved in THF or toluene and stored in the dark, in the refrigerator or freezer.
Handling, Storage, and Precautions: sensitive to oxygen and light, but can be briefly exposed to air and water. In the solid state the reagent is not stable for more than a few hours at rt, but it can be stored for months in the dark as a THF or toluene solution. Although it is stable in solution at rt, the reagent is best stored in a refrigerator.

Methylenation of Carbonyl Compounds[2]. Although the Wittig reagent[18] $CH_2{=}PPh_3$ (see ***Methylenetriphenylphosphorane***) and several other P-, Si-, or S-stabilized methylene anions[19] are commonly used for the methylenation of aldehydes and ketones, these reagents have several synthetic limitations due to their inherent basicity. Wittig-type alkenations generally fail with readily enolizable carbonyls or substrates that undergo a facile nucleophilic addition or elimination. Furthermore, sterically hindered substrates are often methylenated in low yields, while the alkenation of esters and lactones is usually not possible with these methods. Several Ti-based methylenation procedures[1c] have been developed that avoid most of the problems encountered with Wittig-type processes. The combination of ***Dibromomethane-Zinc-Titanium(IV) Chloride***,[20] particularly Lombardo's reagent,[21] is useful for the methylenation of sterically hindered and highly enolizable aldehydes and ketones. The similar reagent CH_2I_2–Zn–$TiCl_4$[22] is also used for this purpose, while two other variations, CH_2I_2–Zn–Ti(O-*i*-Pr)$_4$ and CH_2I_2–Zn–Me_3Al, allow the chemoselective methylenation of aldehydes in the presence of ketones.[23] Another widely used reagent is the Tebbe reagent (see ***μ-Chlorobis(cyclopentadienyl)(dimethylaluminum)-μ-methylenetitanium***[1]),[1b,1c,24,25] which is useful for the methylenation of a variety of carbonyls, including aldehydes, enolizable ketones, esters and lactones. This reagent, however, is not suitable for highly acid-labile substrates and requires stringent inert atmosphere techniques due to its high sensitivity to air and water.

Dimethyltitanocene is an experimentally convenient alternative to the Tebbe reagent.[2] It is easier to prepare and handle and is suitable for the methylenation of aldehydes, ketones, esters, and lactones, including highly enolizable and acid-labile substrates. Heating a solution of a carbonyl compound and Cp_2TiMe_2 (1–3 equiv) in the dark at 60–80 °C, in a sealed flask or under argon, results in a clean and efficient methylenation. Unlike the reactions of the Tebbe reagent, this procedure does not require aqueous workup. The product can be isolated by precipitation of the yellow titanium oxide byproduct, filtration and purification by chromatography or distillation. Although the yields for the methylenation of aldehydes are moderate (eq 1),[2] the reaction works well for alkyl and aryl ketones (eq 2),[2] as well as for cyclic ketones (eq 3).[26]

$$Me(CH_2)_8CH(R)CHO \xrightarrow[\text{THF, 60 °C}]{Cp_2TiMe_2} Me(CH_2)_8CH(R)C(=CH_2)H \quad (1)$$

R = H 48%
R = Me 62%

Cp_2TiMe_2, THF, 60 °C, 90% (2)

Cp_2TiMe_2, THF, 60 °C, 92% (3)

Aldehydes (eq 4)[27] and ketones (eq 5)[2] are methylenated faster than esters or lactones. Due to its suppressed basicity and similarity to the Tebbe reagent[28] and Takai's procedure,[22] Cp_2TiMe_2 also methylenates highly enolizable ketones (eq 6).[2]

Cp_2TiMe_2, THF, 65 °C, 8 h (4)
52% @ 62% completion

Cp_2TiMe_2, THF, 60 °C, 60% (5)

Cp_2TiMe_2, THF, 60 °C, 60% (6)

Esters (eq 7)[2] and lactones (eq 8),[29] including aldonolactones (eq 9),[3] are efficiently methylenated with this reagent. Since the resulting enol ether products undergo facile acid-catalyzed hydrolysis, their isolation and purification should be done under basic conditions. Highly acid-sensitive substrates, such as spiroketal lactones (eq 10)[4] which decompose with the Tebbe reagent, can be readily methylenated with Cp_2TiMe_2.

$Me(CH_2)_{11}O$ — Cp_2TiMe_2, THF, 60 °C, 65% — $Me(CH_2)_{11}O$ (7)

Cp_2TiMe_2, THF, 65 °C, 91% (8)

BnO, BnO, BnO, OBn — Cp_2TiMe_2, PhMe, 65 °C, 89% (9)

Cp_2TiMe_2 (10)

Alkylidenations of Carbonyl Compounds. While homologs of the Tebbe reagent are difficult to prepare, the corresponding dialkyltitanocene derivatives are readily prepared from Cp_2TiCl_2 and alkyllithium or Grignard reagents. With the exception of compounds that undergo facile β-hydride elimination, dialkyltitanocenes are adequately stable for isolation in a pure form. Thermolysis of such derivatives in the presence of aldehydes, ketones, esters, lactones, or amides, results in carbonyl alkylidenation. Thus dibenzyltitanocene gives benzylidenation products (eq 11),[5] bis(trimethylsilylmethyl)titanocene forms vinylsilanes (eq 12),[6] while bis(cyclopropyl)titanocene forms cyclopropylidenes (eq 13).[7]

$Cp_2Ti(CH_2Ph)_2$, PhMe, 45–55 °C (11)

		E:Z
$X = O(CH_2)_{11}Me$	80%	60:40
$X = NMe_2$	45%	71:29

$Cp_2Ti(CH_2SiMe_3)_2$, $(EtOCH_2)_2$, 110 °C, 80% (12)

$Cp_2Ti(c\text{-}C_3H_5)_2$, PhMe, 50 °C (13)
$X = CH_2$ 70%
$X = O$ 67%

Formation of Titanium Enolates with $(C_5Me_5)_2TiMe_2$. The pentamethyl-Cp derivative $(C_5Me_5)_2TiMe_2$ behaves differently from Cp_2TiMe_2. Upon thermolysis it forms a reactive titanium methylidene species which undergoes insertion into one of the Cp methyl groups.[30] Rather than carrying out methylenations, $(C_5Me_5)_2TiMe_2$ reacts with ketones to form titanium enolates with a high degree of regio- and stereocontrol (eq 14).[31] It also forms similar enolates from epoxides (eq 15).[32]

$(C_5Me_5)_2TiMe_2$, PhMe, 100 °C, 87% — $OTi(Me)(C_5Me_5)_2$ (14)
E:Z = 180:1

$(C_5Me_5)_2TiMe_2$, PhMe, 80 °C, 56% — $OTi(Me)(C_5Me_5)_2$ (15)

Reactions with Alkynes. Although addition–elimination pathways are also possible,[2,8,9] it is likely that the thermolysis of Cp_2TiMe_2 generates the highly reactive intermediate $Cp_2Ti{=}CH_2$, which can be trapped by various substrates. In the case of alkynes, clean quantitative conversions to the corresponding titanacyclobutenes take place.[8,9] Titanacyclobutenes[33] can also be prepared with the Tebbe reagent, and they react with aldehydes, ketones, nitriles, and other electrophiles to give a variety of titanium-free products (eq 16).[8] Interestingly, the reaction of Cp_2TiMe_2 with alkynes under photolytic conditions forms titanacyclopentadienyl derivatives,[34] presumably via a titanocene (Cp_2Ti) intermediate.

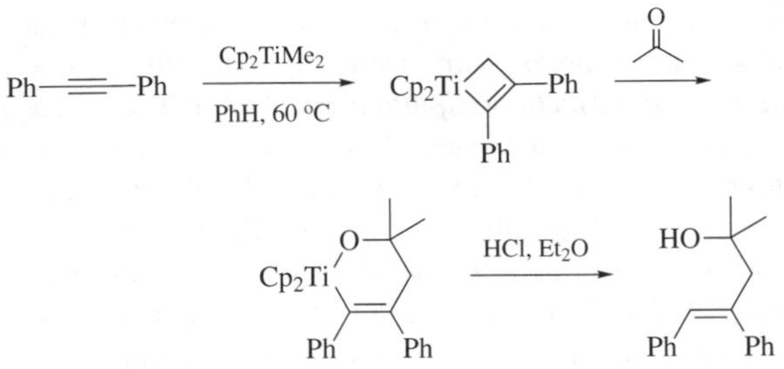

(16)

Reactions with Nitriles. Thermolysis of Cp_2TiMe_2 in the presence of nitriles (2 equiv) generates 1,3-diazatitanacycles,[8–10] which can be hydrolyzed to 4-amino-1-azadienes (eq 17).[10]

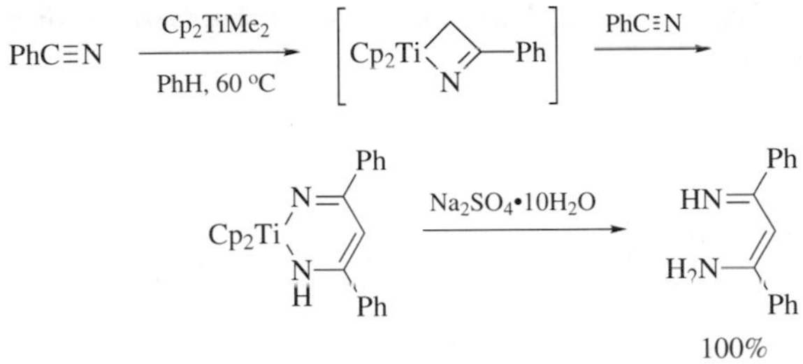

(17)

Ring-Opening Metathesis Polymerization. Norbornene readily undergoes ROMP with Cp_2TiMe_2 and other dialkyltitanocenes (eq 18).[11]

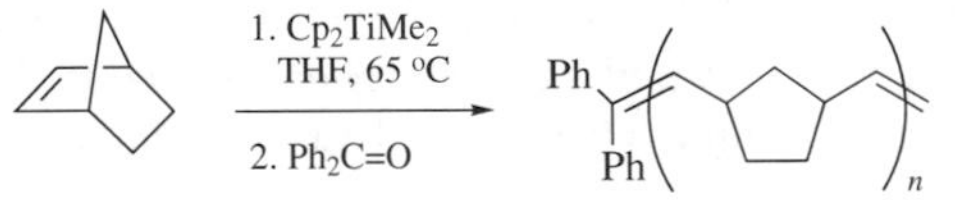

(18)

Alkene Metathesis. Among other related systems, the combination of Cp_2TiMe_2 with $WOCl_4$ shows the best catalytic activity for alkene metathesis (eq 19).[13]

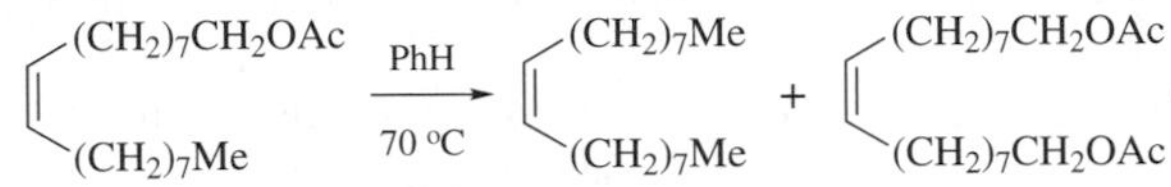

(19)

Cp_2TiMe_2, $WOCl_4$	43%	45%
$Cp_2TiClMe$, $WOCl_4$	41%	39%
Cp_2ZrMe_2, WCl_6	30%	36%
Cp_2TiMe_2, WCl_6	39%	16%

Ziegler–Natta Polymerization of Alkenes. In combination with Al compounds or other acids,[35] dialkyltitanocenes generate cationic titanocene complexes[12] which catalyze the Ziegler–Natta polymerization of alkenes (eq 20).[36–38]

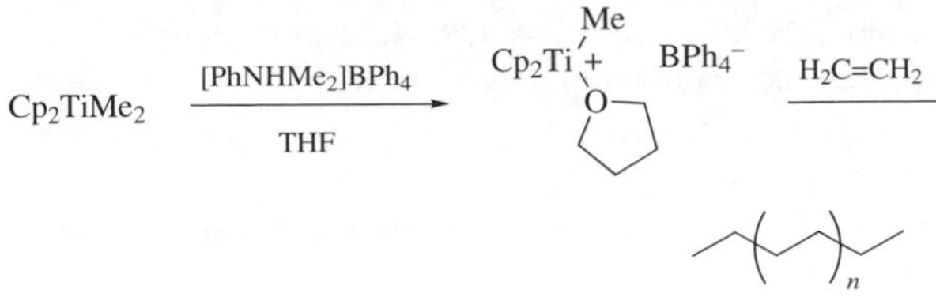

(20)

Hydrogenation of Alkenes. A variety of dicyclopentadienyl Ti derivatives, including Cp_2TiMe_2, can catalyze the hydrogenation of alkenes, presumably via a Ti–H species (eq 21).[14]

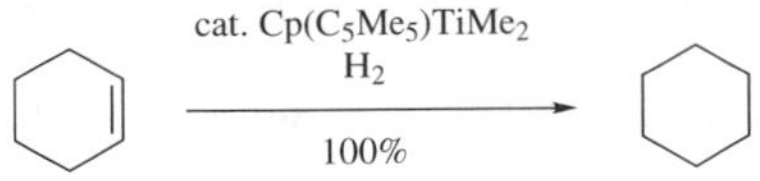

(21)

Reactions with Silanes. In the presence of catalytic amounts of Cp_2TiMe_2[15] or other similar derivatives,[16,39] hydrosilanes undergo dehydrogenative coupling to form oligomers and polymers (eq 22).[15] Alkene hydrogenation, isomerization, or hydrosilylation occurs when an alkene is present,[40,41] while in ammonia a different type of dehydrocoupling takes place (eq 23).[42]

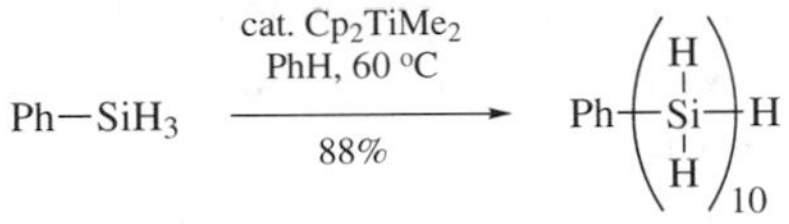

(22)

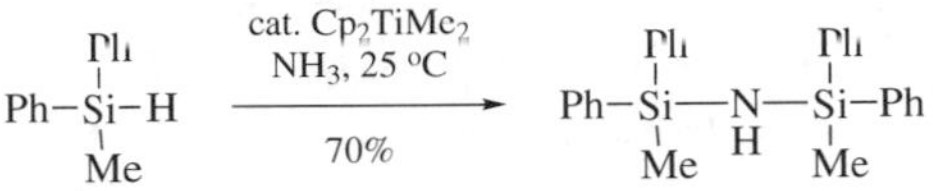

(23)

Related Reagents. Bis(cyclopentadienyl)-3,3-dimethyltitanacyclobutane; μ-Chlorobis(cyclopentadienyl)(dimethylaluminum)-μ-methylenetitanium; Dibromomethane-Zinc-Titanium(IV) Chloride; Dichlorobis(cyclopentadienyl)-zirconium-Zinc-Dibromomethane; Methylenetriphenylphosphorane.

1. (a) Bottrill, M.; Gavens, P. D.; Kelland, J. W.; McMeeking, J. In *Comprehensive Organometallic Chemistry*; Wilkinson, G., Ed.; Pergamon: Oxford, 1982; Vol. 3, pp 331–432. (b) Pine, S. H. *OR* **1993**, *43*, 1. (c) Stille, J. R. In *Comprehensive Organometallic Chemistry*, 2nd ed.; Wilkinson, G., Ed.; Pergamon: Oxford, 1995.
2. Petasis, N. A.; Bzowej, E. I. *JACS* **1990**, *112*, 6392.
3. Csuk, R.; Glänzer, B. I. *T* **1991**, *47*, 1655.
4. DeShong, P.; Rybczynski, P. J. *JOC* **1991**, *56*, 3207.
5. Petasis, N. A.; Bzowej, E. I. *JOC* **1992**, *57*, 1327.
6. Petasis, N. A.; Akritopoulou, I. *SL* **1992**, 665.
7. Petasis, N. A.; Bzowej, E. I. *TL* **1993**, *34*, 943.
8. Petasis, N. A.; Fu, D.-K. *OM* **1993**, *12*, 3776.
9. Doxsee, K. M.; Juliette, J. J. J.; Mouser, J. K. M.; Zientara, K. *OM* **1993**, *12*, 4682.
10. Barluenga, J.; Losada, C. P.; Olano, B. *TL* **1992**, *33*, 7579.
11. Petasis, N. A.; Fu, D.-K. *JACS* **1993**, *115*, 7208.
12. Jordan, R. F. *Adv. Organomet. Chem.* **1991**, *32*, 325.
13. Tsuji, J.; Hashiguchi, S. *JOM* **1981**, *218*, 69.
14. Cuenca, T.; Flores, J. C.; Royo, P. *JOM* **1993**, *462*, 191.
15. Aitken, C.; Harrod, J. F.; Gill, U. S. *CJC* **1987**, *65*, 1804.

16. Corey, J. Y. *Adv. Silicon Chem.* **1991**, *1*, 327.
17. Clauss, v. K.; Bestian, H. *LA* **1962**, *654*, 8.
18. Cadogan, J. I. G. *Organophosphorus Reagents in Organic Synthesis*; Academic: London, 1979; pp 1–15.
19. Larock, R. C. *Comprehensive Organic Transformations*; VCH: New York, 1989.
20. Takai, K.; Hotta, Y.; Oshima, K.; Nozaki, H. *BCJ* **1980**, *53*, 1698.
21. Lombardo, L. *OS* **1987**, *65*, 81.
22. Hibino, J.; Okazoe, T.; Takai, K.; Nozaki, H. *TL* **1985**, *26*, 5579.
23. Okazoe, T.; Hibino, J.; Takai, K.; Nozaki, H. *TL* **1985**, *26*, 5581.
24. Brown-Wensley, K. A.; Buchwald, S. L.; Cannizzo, L.; Clawson, L.; Ho, S.; Meinhardt, D.; Stille, J. R.; Straus, D.; Grubbs, R. H. *PAC* **1983**, *55*, 1733.
25. Pine, S. H.; Pettit, R. J.; Geib, G. D.; Cruz, S. G.; Gallego, C. H.; Tijerina, T.; Pine, R. D. *JOC* **1985**, *50*, 1212.
26. Petasis, N. A.; Patane, M. A. *TL* **1990**, *31*, 6799.
27. Colson, P.-J.; Hegedus, L. S. *JOC* **1993**, *58*, 5918.
28. Clawson, L.; Buchwald, S. L.; Grubbs, R. H. *TL* **1984**, *25*, 5733.
29. Petasis, N. A.; Bzowej, E. I. *TL* **1993**, *34*, 1721.
30. (a) Bercaw, J. E.; Marvich, R. H.; Bell, L. G.; Brintzinger, H. H. *JACS* **1972**, *94*, 1219. (b) McDade, C.; Green, J. C.; Bercaw, J. E. *OM* **1982**, *1*, 1629.
31. Bertz, S. H.; Dabbagh, G.; Gibson, C. P. *OM* **1988**, *7*, 563.
32. Gibson, C. P.; Dabbagh, G.; Bertz, S. H. *CC* **1988**, 603.
33. Doxsee, K. M.; Mouser, J. K. M.; Farahi, J. B. *SL* **1992**, 13.
34. Alt, H.; Rausch, M. D. *JACS* **1974**, *96*, 5936.
35. Soga, K.; Yanagihara, H. *Macromolecules* **1989**, *22*, 2875.
36. Bochmann, M.; Wilson, L. M. *CC* **1986**, 1610.
37. Taube, R.; Krukowka, L. *JOM* **1988**, *347*, C9.
38. Bochmann, M.; Jaggar, A. J.; Nicholls, J. C. *AG(E)* **1990**, *29*, 780.
39. Aitken, C.; Barry, J.-P.; Gauvin, F.; Harrod, J. F.; Malek, A.; Rousseau, D. *OM* **1989**, *8*, 1732.
40. Harrod, J. F.; Yun, S. S. *OM* **1987**, *6*, 1381.
41. Corey, J. Y.; Zhu, X.-H. *OM* **1992**, *11*, 672.
42. Liu, H. Q.; Harrod, J. F. *OM* **1992**, *11*, 822.

Nicos A. Petasis
University of Southern California, Los Angeles, CA, USA

Bis(dibenzylideneacetone)palladium(0)

$(PhCH{=}CHCOCH{=}CHPh)_2Pd$

[32005-36-0] $C_{34}H_{28}O_2Pd$ (MW 575.01)

(catalyst for allylation of stabilized anions,[1] cross coupling of allyl, alkenyl, and aryl halides with organostannanes,[2] cross coupling of vinyl halides with alkenyl zinc species,[3] cyclization reactions,[4] and carbonylation of alkenyl and aryl halides[5])

Physical Data: mp 135 °C (dec).
Form Supplied in: solid.
Preparative Method: prepared by adding sodium acetate to a hot methanolic solution of dibenzylideneacetone and $Na_2[Pd_2Cl_6]$, cooling, filtering, washing with MeOH, and air drying.[6]
Handling, Storage, and Precautions: moderately air stable in the solid state; slowly decomposes in solution to metallic palladium and dibenzylideneacetone.

Allylation of Stabilized Anions. $Pd(dba)_2$ is an effective catalyst for the coupling of electrophiles and nucleophiles, and has found extensive use in organic synthesis (for a similar complex with distinctive reactivities, see also ***Tris(dibenzylideneacetone)dipalladium***. Addition of a catalytic amount of $Pd(dba)_2$ activates allylic species, such as allylic acetates or carbonate derivatives, toward nucleophilic attack.[1] The intermediate organometallic complex, a π-allylpalladium species, is formed by backside displacement of the allylic leaving group, and stereochemical inversion of the original allylic position results. Subsequent nucleophilic attack on the external face of the allyl ligand displaces the palladium in this double inversion process to regenerate the original stereochemical orientation (eq 1).[7] The allylpalladium intermediate can also be generated from a variety of other substrates, such as allyl sulfones,[8] allenes,[9] vinyl epoxides,[10] or α-allenic phosphates.[11] In general, the efficiency of $Pd(dba)_2$ catalysis is optimized through the addition of either ***Triphenylphosphine*** or ***1,2-Bis(diphenylphosphino)ethane*** (dppe).

OAc; 1% $Pd(dba)_2$, dppe; $NaCH(CO_2Me)_2$; THF, rt, 48 h; 83%; CO_2Me; CO_2Me; >95% *cis* (1)

The anions of malonate esters,[12] cyclopentadiene,[12] β-keto esters,[13] ketones,[13,14] aldehydes,[14] α-nitroace tate esters,[15] Meldrum's acid,[15] diethylaminophosphonate Schiff bases,[16] β-diketones,[17] β-sulfonyl ketones and esters,[17] and polyketides[18,19] represent the wide variety of carbon nucleophiles effective in this reaction. Generation of the stabilized anions normally is accomplished by addition of ***Sodium Hydride***, ***Potassium Hydride***, or basic ***Alumina***.[15] However, when allyl substrates such as allylisoureas,[14] allyl oxime carbonates,[17] or allyl imidates[20] are used, the allylation reaction proceeds without added base. Nitrogen nucleophiles, such as azide[10] and nucleotide[21] anions, are useful as well.

The coupling reaction generally proceeds regioselectively with attack by the nucleophile at the least hindered terminus of the allyl moiety,[22] accompanied by retention of alkene geometry (eq 2). Even electron-rich enol ethers can be used as the allylic moiety when an allylic trifluoroacetyl leaving group is present.[23] When steric constraints of substrates are equivalent, attack will occur at the more electron rich site.[19] Although this reaction is usually performed in THF, higher yields and greater selectivity are observed for some systems with the use of DME, DMF, or DMSO.[14,16,20] Alternatively, $Pd(dba)_2$ can promote efficient substitution of allylic substrates in a two-phase aqueous–organic medium through the use of $P(C_6H_4\text{-}m\text{-}SO_3Na)_3$ as a phase transfer ligand.[24]

1% $Pd(dba)_2$, dppe, $NaCHE_2$; THF, rt, 48 h; E = CO_2Me (2)

(*E*) isomer	47%	95 (*E* only):5
(*Z*) isomer	62%	76 (*Z* only):23

Intramolecular reaction of a β-dicarbonyl functionality with a π-allyl species can selectively produce three-,[25] five-,[25] or six-membered[26] rings (eq 3).

NaH, $Pd(dba)_2$, PPh_3; THF, 65 °C, 10 h; 75% (3)

E = CO_2Me >98% *cis*

Asymmetric Allylation Reactions. Employing chiral bidentate phosphine ligands in conjunction with $Pd(dba)_2$ promotes allylation reactions with moderate to good enantioselectivities, which are dependent upon the solvent,[27] counterion,[28] and nature of the allylic leaving group.[27] Chiral phosphine ligands have been used for the asymmetric allylation of α-hydroxy acids (5–30% ee),[29] the preparation of optically active methylenecyclopropane derivatives (52% ee),[22] and chiral 3-alkylidenebicyclo[3.3.0]octane and 1-alkylidenecyclohexane systems (49–90% ee).[27] Allylation of a glycine derivative provides a route to optically active α-amino acid esters (eq 4).[28] The intramolecular reaction can produce up to 69% ee when vicinal stereocenters are generated during bond formation (eq 5).[30]

1. LDA, –78 °C, THF; 2. CH_2=$CHCH_2OAc$, 1.5% $Pd(dba)_2$, (+)-DIOP; 3. 10% HCl; 4. K_2CO_3; 60%, 57% ee (4)

7% $Pd(dba)_2$, (–)-CHIRAPHOS, K_2CO_3; THF, –51 °C, 3 h; 65%, 69% ee (5)

Cross-Coupling Reactions. Allylic halides,[5,31] aryl diazonium salts,[32] allylic acetates,[33] and vinyl epoxides[34] are excellent substrates for $Pd(dba)_2$ catalyzed selective cross-coupling reactions with alkenyl-, aryl-, and allylstannanes. The reaction of an allylic halide or acetate proceeds through a π-allyl intermediate with inversion of sp^3 stereochemistry, and transmetalation with the organostannane followed by reductive elimination results in coupling from the palladium face of the allyl ligand. Coupling produces overall inversion of allylic stereochemistry, a preference for reaction at the least substituted carbon of the allyl framework, and retention of allylic alkene geometry. In addition, the alkene geometry of alkenylstannane reagents is conserved (eq 6). Functional group compatibility is extensive, and includes the presence of CO_2Bn, OH, OR, CHO, OTHP, β-lactams, and CN functionality.

3% $Pd(dba)_2$, PPh_3; THF, 5 °C; 87% (6)

Similar methodology is used for the coupling of alkenyl halides and triflates with 1) alkenyl-, aryl-, or alkynylstannanes,[35] 2) alkenylzinc species,[3,36] or 3) arylboron species.[37] This methodology is applied in the synthesis of cephalosporin derivatives (eq 7),[35] and can be used for the introduction of acyl[3,36] and vinylogous acyl[3] equivalents (eq 8).

2% $Pd(dba)_2$, P(2-furyl)$_3$; Bu_3Sn-vinyl, $ZnCl_2$; 25 °C; 65% (7)

1. 5% $Pd(dba)_2$, PPh_3, THF; 2. HCl, H_2O; 68% (8)

Intramolecular Reaction with Alkenes. Palladium π-allyl complexes can undergo intramolecular insertion reactions with alkenes to produce five- and six-membered rings through a 'metallo-ene-type' cyclization.[4] This reaction produces good stereoselectivity when resident chirality is vicinal to a newly formed stereogenic center (eq 9), and can be used to form tricyclic and tetracyclic ring systems through tandem insertion reactions.[38] In the presence of $Pd(dba)_2$ and triisopropyl phosphate, α,β-alkynic esters and α,β-unsaturated enones undergo intramolecular [3 + 2] cycloaddition reactions when tethered to methylenecyclopropane to give a bicyclo[3.3.0]octane ring system (eq 10).[39]

7% $Pd(dba)_2$, PPh_3; AcOH, 80 °C, 6 h; 72%; *trans*:*cis* = 93:7 (9)

11% $Pd(dba)_2$, P(O-*i*-Pr)$_3$; toluene, 110 °C, 42 h; 47% (10)

Carbonylation Reactions. In the presence of CO and $Pd(dba)_2$, unsaturated carbonyl derivatives can also be prepared through carbonylative coupling reactions. Variations of this reaction include the initial coupling of allyl halides with carbon monoxide, followed by a second coupling with either alkenyl- or arylstannanes (eq 11).[5] This reaction proceeds with overall inversion of allylic sp^3 configurations and retains the alkene geometry of both the allyl species and the stannyl group. Similarly, aryl and alkenyl halides will undergo carbonylative coupling to generate intermediate acylpalladium complexes. Intermolecular reaction of these acyl complexes with $HSnBu_3$ produces aldehydes,[35,40] while reaction with MeOH or amines generates the corresponding carboxylic acid methyl ester[41] or amides, respectively.[42]

OMe
EtO_2C Br
+
Bu_3Sn
THPO
3% $Pd(dba)_2$
PPh_3
55 psi CO
THF, 50 °C
75%
OMe O
EtO_2C
THPO
(11)

Palladium acyl species can also undergo intramolecular acylpalladation with alkenes to form five- and six-membered ring γ-keto esters through exocyclic alkene insertion (eq 12).[43] The carbonylative coupling of *o*-iodoaryl alkenyl ketones is also promoted by $Pd(dba)_2$ to give bicyclic and polycyclic quinones through endocyclization followed by β-H elimination.[44] Sequential carbonylation and intramolecular insertion of propargylic and allylic alcohols provides a route to γ-butyrolactones (eq 13).[45]

I
600 psi CO
5% $Pd(dba)_2$
2 equiv MeOH
NEt_3, 36 h
73%
O
OMe
O
(12)

OH
20 atm CO
4% $Pd(dba)_2$
dppb
DME, 150 °C
48 h
80%
O O
(13)

Related Reagents. Tetrakis(triphenylphosphine)palladium(0); Tris(dibenzylideneacetone)dipalladium-Chloroform.

---

1. Trost, B. M. *AG(E)* **1989**, *28*, 1173.
2. Stille, J. K. *AG(E)* **1986**, *25*, 508.
3. (a) Rao, C. J.; Knochel, P. *JOC* **1991**, *56*, 4593. (b) Wass, J. R.; Sidduri, A.; Knochel, P. *TL* **1992**, *33*, 3717. (c) Knochel, P.; Rao. C. J. *T* **1993**, *49*, 29.
4. (a) Oppolzer, W.; Gaudin, J.-M. *HCA* **1987**, *70*, 1477. (b) Oppolzer, W.; Swenson, R. E.; Gaudin, J.-M. *TL* , **1988**, *29*, 5529. (c) Oppolzer, W.; Keller, T. H.; Kuo, D. L.; Pachinger, W. *TL* **1990**, *31*, 1265.
5. (a) Sheffy, F. K.; Stille, J. K. *JACS* **1983**, *105*, 7173. (b) Sheffy, F. K.; Godschalx, J. P.; Stille, J. K. *JACS* **1984**, *106*, 4833.
6. (a) Takahashi, Y.; Ito, T. S.; Sakai, S.; Ishii, Y. *CC* **1970**, 1065. (b) Rettig, M. F.; Maitlis, P. M. *Inorg. Synth.* **1977**, *17*, 134.
7. Fiaud, J.-C.; Legros, J.-Y. *JOC* **1987**, *52*, 1907.
8. Backväll, J.-E.; Juntunen, S. K. *JACS* **1987**, *109*, 6396.
9. (a) Ahmar, M.; Barieux, J.-J.; Cazes, B.; Goré, J. *T* **1987**, *43*, 513. (b) Chaptal, N.; Colovray-Gotteland, V.; Grandjean, C.; Cazes, B.; Goré, J. *TL* **1991**, *32*, 1795.
10. Tenaglia, A.; Waegell, B. *TL* **1988**, *29*, 4851.
11. Cazes, B.; Djahanbini, D.; Goré, J.; Genêt, J.-P.; Gaudin, J.-M. *S* **1988**, 983.
12. Fiaud, J. C.; Malleron, J. L. *TL* **1980**, *21*, 4437.
13. Fiaud, J.-C.; Malleron, J.-L. *CC* **1981**, 1159.
14. Inoue, Y.; Toyofuku, M.; Taguchi, M.; Okada, S.; Hashimoto, H. *BCJ* **1986**, *59*, 885.
15. Ferroud, D.; Genet, J. P.; Muzart, J. *TL* **1984**, *25*, 4379.
16. Genet, J. P.; Uziel, J.; Juge, S. *TL* **1988**, *29*, 4559.
17. Suzuki, O.; Hashiguchi, Y.; Inoue, S.; Sato, K. *CL* **1988**, 291.
18. Marquet, J.; Moreno-Mañas, M.; Prat, M. *TL* **1989**, *30*, 3105.
19. Prat, M.; Ribas, J.; Moreno-Mañas, M. *T* **1992**, *48*, 1695.
20. Suzuki, O.; Inoue, S.; Sato, K. *BCJ* **1989**, *62*, 239.
21. Liotta, F.; Unelius, R.; Kozak, J.; Norin, T. *ACS* **1992**, *46*, 686.
22. Stolle, A.; Ollivier, J.; Piras, P. P.; Salaün, J.; de Meijere, A. *JACS* **1992**, *114*, 4051.
23. RajanBabu, T. V. *JOC* **1985**, *50*, 3642.
24. Safi, M.; Sinou, D. *TL* **1991**, *32*, 2025.
25. (a) Ahmar, M.; Cazes, B.; Goré, J. *TL* **1985**, *26*, 3795. (b) Ahmar, M.; Cazes, B.; Goré, J. *T* **1987**, *43*, 3453. (c) Fournet, G.; Balme, G.; Barieux, J. J.; Goré, J. *T* **1988**, *44*, 5821. (d) Geng, L.; Lu, X. *JCS(P1)* **1992**, 17.
26. Bäckvall, J.-E.; Vågberg, J.-O.; Granberg, K. L. *TL* **1989**, *30*, 617.
27. Fiaud, J.-C.; Legros, J.-Y. *JOC* **1990**, *55*, 4840.
28. (a) Genet, J. P.; Ferroud, D.; Juge, S.; Montes, J. R. *TL* **1986**, *27*, 4573. (b) Genêt, J.-P.; Jugé, S.; Montès, J. R.; Gaudin, J.-M. *CC* **1988**, 718. (c) Genet, J.-P.; Juge, S.; Achi, S.; Mallart, S.; Montes, J. R.; Levif, G. *T* **1988**, *44*, 5263.
29. Moorlag, H.; de Vries, J. G.; Kaptein, B.; Schoemaker, H. E.; Kamphuis, J.; Kellogg, R. M. *RTC* **1992**, *111*, 129.
30. Genet, J. P.; Grisoni, S. *TL* **1988**, *29*, 4543.
31. Farina, V.; Baker, S. R.; Benigni, D. A.; Sapino, Jr., C. *TL* **1988**, *29*, 5739.
32. Kikukawa, K.; Kono, K.; Wada, F.; Matsuda, T. *JOC* **1983**, *48*, 1333.
33. Del Valle, L.; Stille, J. K.; Hegedus, L. S. *JOC* **1990**, *55*, 3019.
34. Tueting, D. R.; Echavarren, A. M.; Stille, J. K. *T* **1989**, *45*, 979.
35. Farina, V.; Baker, S. R.; Sapino, Jr., C. *TL* **1988**, *29*, 6043.
36. Russell, C. E.; Hegedus, L. S. *JACS* **1983**, *105*, 943.
37. (a) Legros, J.-Y.; Fiaud, J.-C. *TL* **1990**, *31*, 7453. (b) Tour, J. M.; Lamba, J. J. S. *JACS* **1993**, *115*, 4935.
38. Oppolzer, W.; DeVita, R. J. *JOC* **1991**, *56*, 6256.
39. Lewis, R. T.; Motherwell, W. B.; Shipman, M. *CC* **1988**, 948.
40. Baillargeon, V. P.; Stille, J. K. *JACS* **1986**, *108*, 452.
41. Takeuchi, R.; Suzuki, K.; Sato, N. *S* **1990**, 923.
42. Meyers, A. I.; Robichaud, A. J.; McKennon, M. J. *TL* **1992**, *33*, 1181.
43. Tour, J. M.; Negishi, E. *JACS* **1985**, *107*, 8289.
44. Negishi, E.; Tour, J. M. *TL* **1986**, *27*, 4869.
45. Ali, B. E.; Alper, H. *JOC* **1991**, *56*, 5357.

John R. Stille
Michigan State University, East Lansing, MI, USA

(R)- & (S)-2,2′-Bis(diphenylphosphino)-1,1′-binaphthyl[1]

PPh2 PPh2 (R) PPh2 PPh2 (S)

[76189-55-4] $C_{44}H_{32}P_2$ (MW 622.70)

(chiral diphosphine ligand for transition metals;[2] the complexes show high enantioselectivity and reactivity in a variety of organic reactions)

Alternate Name: BINAP.
Physical Data: mp 241–242 °C; $[\alpha]_D^{25}$ −229° (c = 0.312, benzene) for (*S*)-BINAP.[3]
Solubility: sol THF, benzene, dichloromethane; modestly sol ether, methanol, ethanol; insol water.
Form Supplied in: colorless solid.
Analysis of Reagent Purity: GLC analysis (OV-101, capillary column, 5 m, 200–280 °C) and TLC analysis (E. Merck Kieselgel 60 PF_{254}, 1:19 methanol–chloroform); R_f 0.42 (BINAPO, dioxide of BINAP), 0.67 (monoxide of BINAP), and 0.83 (BINAP). The optical purity of BINAP is analyzed after oxidizing to BINAPO by HPLC using a Pirkle column (Baker bond II) and a hexane/ethanol mixture as eluent.[3]
Preparative Methods: enantiomerically pure BINAP is obtained by resolution of the racemic dioxide, BINAPO, with camphorsulfonic acid or 2,3-di-*O*-benzoyltartaric acid followed by deoxygenation with ***Trichlorosilane*** in the presence of ***Triethylamine***.[3]
Handling, Storage, and Precautions: solid BINAP is substantially stable to air, but bottles of BINAP should be flushed with N_2 or Ar and kept tightly closed for prolonged storage. BINAP is slowly air oxidized to the monoxide in solution.

BINAP–Ru^{II} Catalyzed Asymmetric Reactions. Halogen-containing BINAP–Ru complexes are most simply prepared by reaction of $[RuCl_2(cod)]_n$ or $[RuX_2(arene)]_2$ (X = Cl, Br, or I) with BINAP.[4] Sequential treatment of $[RuCl_2(benzene)]_2$ with BINAP and sodium carboxylates affords $Ru(carboxylate)_2(BINAP)$ complexes. The dicarboxylate complexes, upon treatment with strong acid HX,[5] can be converted to a series of Ru complexes empirically formulated as $RuX_2(BINAP)$. These Ru^{II} complexes act as catalysts for asymmetric hydrogenation of various achiral and chiral unsaturated compounds.

α,β-Unsaturated carboxylic acids are hydrogenated in the presence of a small amount of $Ru(OAc)_2(BINAP)$ to give the corresponding optically active saturated products in quantitative yields.[6] The reaction is carried out in methanol at ambient temperature with a substrate:catalyst (S:C) ratio of 100–600:1. The sense and degree of the enantioface differentiation are profoundly affected by hydrogen pressure and the substitution pattern of the substrates. Tiglic acid is hydrogenated quantitatively with a high enantioselectivity under a low hydrogen pressure (eq 1), whereas naproxen, a commercial anti-inflammatory agent, is obtained in 97% ee under high pressure (eq 2).[6a]

CO2H → (4 atm H2, Ru(OAc)2[(R)-BINAP], MeOH) CO2H (1) 91% ee

MeO CO2H → (135 atm H2, Ru(OAc)2[(S)-BINAP], MeOH) MeO CO2H (2) 97% ee

Enantioselective hydrogenation of certain α- and β-(acylamino)acrylic acids or esters in alcohols under 1–4 atm H_2 affords the protected α- and β-amino acids, respectively (eqs 3 and 4).[2a,7] Reaction of *N*-acylated 1-alkylidene-1,2,3,4-tetrahydroisoquinolines provides the 1*R*- or 1*S*-alkylated products. This method allows a general asymmetric synthesis of isoquinoline alkaloids (eq 5).[8]

CO2Me NHAc → (100 atm H2, RuBr2[(R)-BINAP], MeOH) CO2Me NHAc (3) 85% ee

AcHN CO2Me → (1 atm H2, Ru(OAc)2[(R)-BINAP], MeOH) AcHN CO2Me (4) 96% ee

(R3O)n NCOR2 R1 → (1–4 atm H2, (R)-BINAP–Ru, EtOH–CH2Cl2) (R3O)n NCOR2 R1 (5) 92–100% ee

Geraniol or nerol can be converted to citronellol in 96–99% ee in quantitative yield without saturation of the C(6)–C(7) double bond (eq 6).[9] The S:C ratio approaches 50 000. The use of alcoholic solvents such as methanol or ethanol and initial H_2 pressure greater than 30 atm is required to obtain high enantioselectivity. Diastereoselective hydrogenation of the enantiomerically pure allylic alcohol with an azetidinone skeleton proceeds at atmospheric pressure in the presence of an (*R*)-BINAP–Ru complex to afford the β-methyl product, a precursor of 1β-methylcarbapenem antibiotics (eq 7).[10] Racemic allylic alcohols such as 3-methyl-2-cyclohexenol and 4-hydroxy-2-cyclopentenone can be effectively resolved by the BINAP–Ru-catalyzed hydrogenation (eq 8).[11]

100 atm H_2
$Ru(OAc)_2[(R)\text{-BINAP}]$
MeOH
OH → OH (6)
99% ee

TBDMSO, H H, OH, NH, O
1 atm H_2
$Ru(OAc)_2[(R)\text{-TolBINAP}]$
MeOH
TBDMSO, H H, OH, NH, O (7)
$\beta:\alpha = 99.9:0.1$

(±)- OH
4 atm H_2
$Ru(OAc)_2[(R)\text{-BINAP}]$
MeOH
OH (8)
46% recovery
>99% ee

Diketene is quantitatively hydrogenated to 3-methyl-3-propanolide in 92% ee (eq 9). Certain 4-methylene- and 2-alkylidene-4-butanolides as well as 2-alkylidenecyclopentanone are also hydrogenated with high enantioselectivity.[12]

100 atm H_2
$[RuCl[(S)\text{-BINAP}](C_6H_6)]Cl$, NEt_3
THF
(9)
92% ee

Hydrogenation with halogen-containing BINAP–Ru complexes can convert a wide range of functionalized prochiral ketones to stereo-defined secondary alcohols with high enantiomeric purity (eq 10).[13] 3-Oxocarboxylates are among the most appropriate substrates.[13a,4d] For example, the enantioselective hydrogenation of methyl 3-oxobutanoate proceeds quantitatively in methanol with an S:C ratio of 1000–10 000 to give the hydroxy ester product in nearly 100% ee (eq 11). Halogen-containing complexes RuX_2(BINAP) (X = Cl, Br, or I; polymeric form) or $[RuCl_2(BINAP)]_2NEt_3$ are used as the catalysts. Alcohols are the solvents of choice, but aprotic solvents such as dichloromethane can also be used. At room temperature the reaction requires an initial H_2 pressure of 20–100 atm, but at 80–100 °C the reaction proceeds smoothly at 4 atm H_2.[4c,4d]

$R^1C(O)R^2 \xrightarrow[(R)\text{-BINAP–Ru}]{H_2} R^1CH(OH)R^2$ (10)

R^1 = alkyl, aryl; R^2 = CH_2OH, CH_2NMe_2, CH_2CH_2OH, CH_2Ac, CH_2CO_2R, CH_2COSR, CH_2CONR_2, $CH_2CH_2CO_2R$, etc.

$RCOCH_2CO_2R' \xrightarrow[\text{MeOH or EtOH}]{100\text{ atm } H_2,\ (R)\text{-BINAP–Ru}} RCH(OH)CH_2CO_2R'$ (11)
98–100% ee
R = Me, Et, Bu, *i*-Pr; R′ = Me, Et, *i*-Pr, *t*-Bu

3-Oxocarboxylates possessing an additional functional group can also be hydrogenated with high enantioselectivity by choosing appropriate reaction conditions or by suitable functional group modification (eq 12).[13b,13c]

$ClCH_2COCH_2CO_2Et \xrightarrow[\text{EtOH, 100 °C, 5 min}]{100\text{ atm } H_2,\ RuCl_2[(S)\text{-BINAP}]} ClCH_2CH(OH)CH_2CO_2Et$ (12)
97% ee

The pre-existing stereogenic center in the chiral substrates profoundly affects the stereoselectivity. The (*R*)-BINAP–Ru-catalyzed reaction of (*S*)-4-(alkoxycarbonylamino)-3-oxocarboxylates give the statine series with (3*S*,4*S*) configuration almost exclusively (eq 13).[14]

$R^1CH(NHR^3)COCH_2CO_2R^2 \xrightarrow[\text{MeOH or EtOH}]{100\text{ atm } H_2,\ RuBr_2[(R)\text{-BINAP}]} R^1CH(NHR^3)CH(OH)CH_2CO_2R^2$ (13)
syn:*anti* = >99:1

Hydrogenation of certain racemic 2-substituted 3-oxocarboxylates occurs with high diastereo- and enantioselectivity via dynamic kinetic resolution involving in situ racemization of the substrates.[15] The (*R*)-BINAP–Ru-catalyzed reaction of 2-acylamino-3-oxocarboxylates in dichloromethane allows preparation of threonine and DOPS (anti-Parkinsonian agent) (eq 14).[16] In addition, a common intermediate for the synthesis of carbapenem antibiotics is prepared stereoselectively on an industrial scale from a 3-oxobutyric ester (**1**) with an acylaminomethyl substituent at the C(2) position.[16a] The second-order stereoselective hydrogenation of 2-ethoxycarbonylcycloalkanones gives predominantly the *trans* hydroxy esters (**2**) in high ee, whereas 2-acetyl-4-butanolide is hydrogenated to give the *syn* diastereomer (**3**).[17]

$RCOCH(NHCOR')CO_2Me \xrightarrow[CH_2Cl_2]{100\text{ atm } H_2,\ RuBr_2[(R)\text{-BINAP}]} RCH(OH)CH(NHCOR')CO_2Me$ (14)
syn:*anti* = 99:1
92–98% ee

(**1**) OH O, OMe, NHCOPh — *syn*:*anti* = 94:6, 98% ee

(**2**) OH O, OR′, R — R = CH_2, $(CH_2)_2$, $(CH_2)_3$; *trans*:*cis* = 93:7–99:1; 90–93% ee

(**3**) OH O, O, H — *syn*:*anti* = 98:2; 94% ee

Certain 1,2- and 1,3-diketones are doubly hydrogenated to give stereoisomeric diols. 2,4-Pentanedione, for instance, affords (*R,R*)- or (*S,S*)-2,4-pentanediol in nearly 100% ee accompanied by 1% of the *meso* diol.[13b]

A BINAP–Ru complex can hydrogenate a C=N double bond in a special cyclic sulfonimide to the sultam with > 99% ee.[18]

The asymmetric transfer hydrogenation of the unsaturated carboxylic acids using formic acid or alcohols as the hydrogen source is catalyzed by $Ru(acac\text{-}F_6)(\eta^3\text{-}C_3H_5)$(BINAP) or

$[RuH(BINAP)_2]PF_6$ to produce the saturated acids in up to 97% ee (eq 15).[19]

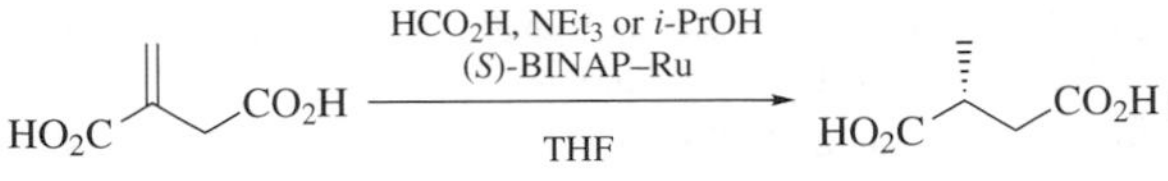

HO_2C … CO_2H — HCO_2H, NEt_3 or *i*-PrOH, (*S*)-BINAP–Ru, THF → HO_2C … CO_2H (15)

93–97% ee

BINAP–Ru complexes promote addition of arenesulfonyl chlorides to alkenes in 25–40% optical yield.[20]

BINAP–Rh^I Catalyzed Asymmetric Reactions. The rhodium(I) complexes $[Rh(BINAP)(cod)]ClO_4$, $[Rh(BINAP)(nbd)]ClO_4$, and $[Rh(BINAP)_2]ClO_4$, are prepared from $[RhCl(cod)]_2$ or ***Bis(bicyclo[2.2.1]hepta-2,5-diene)dichlorodirhodium*** and BINAP in the presence of $AgClO_4$.[21] $[Rh(BINAP)S_2]ClO_4$ is prepared by reaction of $[Rh(BINAP)(cod\ or\ nbd)]ClO_4$ with atmospheric pressure of hydrogen in an appropriate solvent, S.[21a] BINAP–Rh complexes catalyze a variety of asymmetric reactions.[2]

Prochiral α-(acylamino)acrylic acids or esters are hydrogenated under an initial hydrogen pressure of 3–4 atm to give the protected amino acids in up to 100% ee (eq 16).[21a] The BINAP–Rh catalyst was used for highly diastereoselective hydrogenation of a chiral homoallylic alcohol to give a fragment of the ionophore ionomycin.[22]

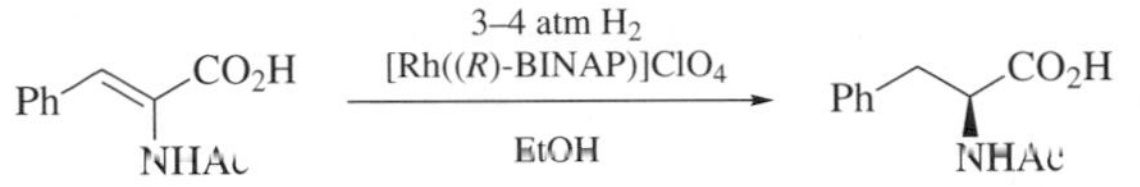

Ph, CO_2H, NHAc — 3–4 atm H_2, $[Rh((R)\text{-BINAP})]ClO_4$, EtOH → Ph, CO_2H, NHAc (16)

100% ee

The cationic BINAP–Rh complexes catalyze asymmetric 1,3-hydrogen shifts of certain alkenes. Diethylgeranylamine can be quantitatively isomerized in THF or acetone to citronellal diethylenamine in 96–99% ee (eq 17).[23] This process is the key step in the industrial production of (−)-menthol. In the presence of a cationic (*R*)-BINAP–Rh complex, (*S*)-4-hydroxy-2-cyclopentenone is isomerized five times faster than the (*R*) enantiomer, giving a chiral intermediate of prostaglandin synthesis.[24]

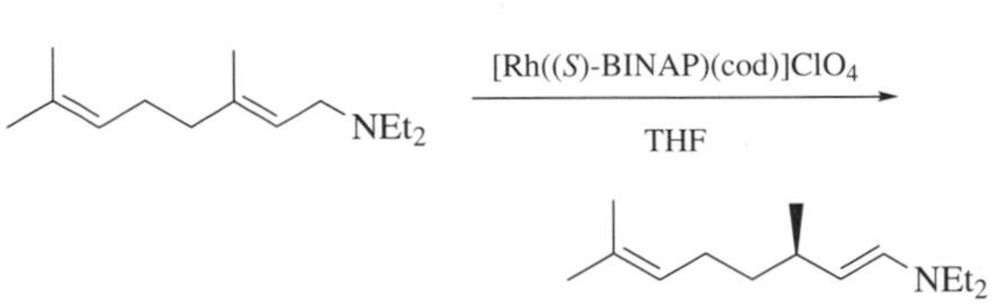

NEt_2 — $[Rh((S)\text{-BINAP})(cod)]ClO_4$, THF → NEt_2 (17)

99% ee

Enantioselective cyclization of 4-substituted 4-pentenals to 3-substituted cyclopentanones in >99% ee is achieved with a cationic BINAP–Rh complex (eq 18).[25]

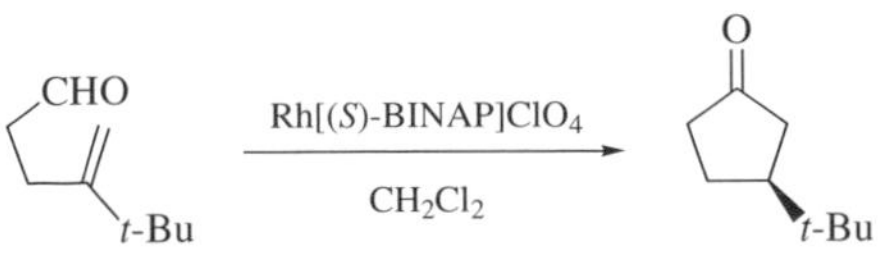

CHO, *t*-Bu — $Rh[(S)\text{-BINAP}]ClO_4$, CH_2Cl_2 → O, *t*-Bu (18)

>99% ee

Reaction of styrene and catecholborane in the presence of a BINAP–Rh complex at low temperature forms, after oxidative workup, 1-phenylethyl alcohol in 96% ee (eq 19).[26]

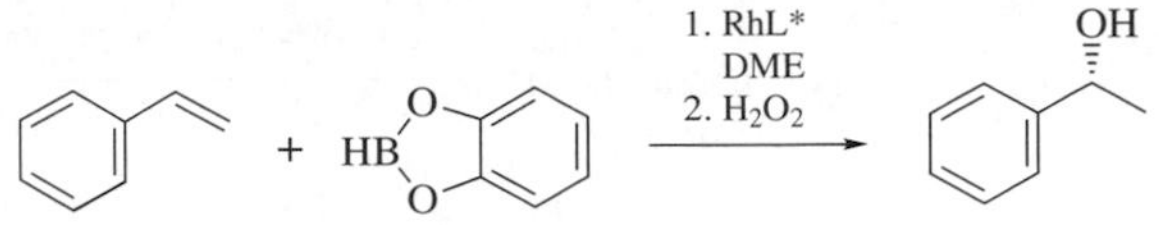

+ HB — 1. RhL*, DME; 2. H_2O_2 → OH (19)

RhL* = $[Rh(cod)_2]BF_4$ + (*R*)-BINAP

96% ee

Neutral BINAP–Rh complexes catalyze intramolecular hydrosilylation of alkenes. Subsequent ***Hydrogen Peroxide*** oxidation produces the optically active 1,3-diol in up to 97% ee (eq 20).[27]

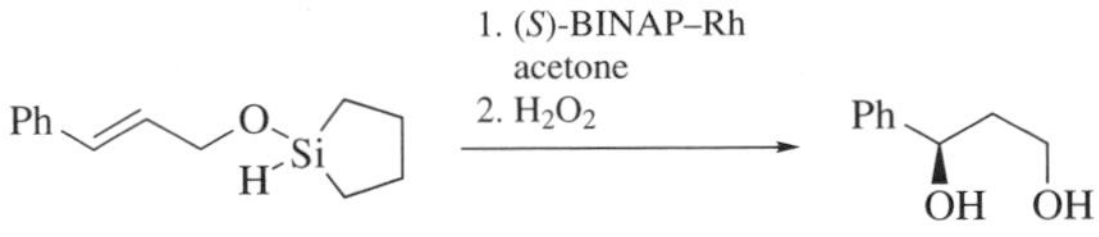

Ph, O, Si, H — 1. (*S*)-BINAP–Rh, acetone; 2. H_2O_2 → Ph, OH OH (20)

97% ee

BINAP–Pd Catalyzed Asymmetric Reactions. BINAP–Pd^0 complexes are prepared in situ from ***Bis(dibenzylideneacetone)palladium(0)*** or $Pd_2(dba)_3{\cdot}CHCl_3$ and BINAP.[28] BINAP–Pd^{II} complexes are formed from ***Bis(allyl)di-μ-chlorodipalladium***, ***Palladium(II) Acetate***, or $PdCl_2(MeCN)_2$ and BINAP.[29–31]

A BINAP–Pd complex brings about enantioselective 1,4-disilylation of α,β-unsaturated ketones with chlorinated disilanes, giving enol silyl ethers in 74–92% ee (eq 21).[29]

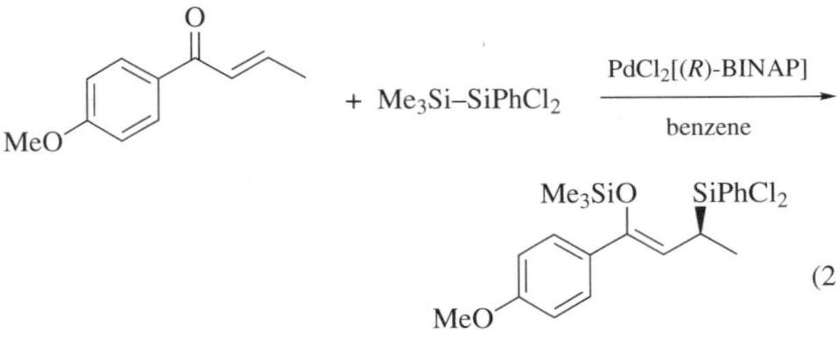

MeO, O + Me_3Si–$SiPhCl_2$ — $PdCl_2[(R)\text{-BINAP}]$, benzene → Me_3SiO, $SiPhCl_2$, MeO (21)

92% ee

A BINAP–Pd^{II} complex catalyzes a highly enantioselective C–C bond formation between an aryl triflate and 2,3-dihydrofuran (eq 22).[30] The intramolecular version of the reaction using an alkenyl iodide in the presence of $PdCl_2[(R)\text{-BINAP}]$ and ***Silver(I) Phosphate*** allows enantioselective formation of a bicyclic ring system (eq 23).[31]

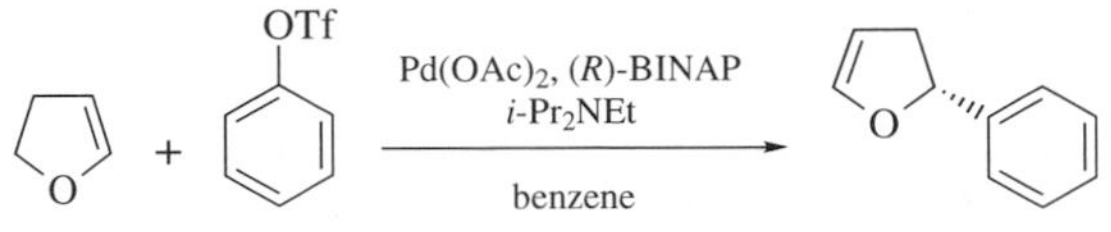

O + OTf — $Pd(OAc)_2$, (*R*)-BINAP, *i*-Pr_2NEt, benzene → O (22)

93% ee

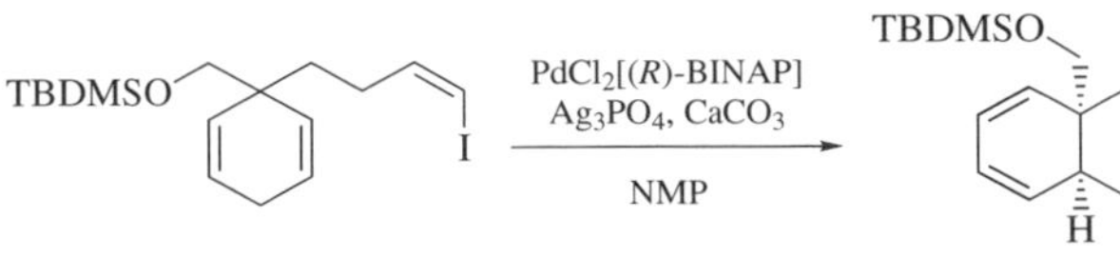

TBDMSO, I — $PdCl_2[(R)\text{-BINAP}]$, Ag_3PO_4, $CaCO_3$, NMP → TBDMSO, H (23)

80% ee

Enantioselective electrophilic allylation of 2-acetamidomalonate esters is effected by a BINAP–Pd^0 complex (eq 24).[32]

Ph(OAc)CH–CH=CH–Ph + $NaC(CO_2Me)_2NHAc$ $\xrightarrow[\text{THF}]{[PdCl(\eta^3\text{-}C_3H_5)]_2,\ (S)\text{-BINAP}}$ Ph–CH=CH–CH(Ph)$C(CO_2Me)_2NHAc$ (24)

94% ee

A BINAP–Pd^0 complex catalyzes hydrocyanation of norbornene to the *exo* nitrile with up to 40% ee.[28]

BINAP–Ir^I Catalyzed Asymmetric Reactions. $[Ir(BINAP)(cod)]BF_4$ is prepared from $[Ir(cod)(MeCN)_2]BF_4$ and BINAP in THF.[33]

A combined system of the BINAP–Ir complex and bis(*o*-dimethylaminophenyl)phenylphosphine or (*o*-dimethylaminophenyl)diphenylphosphine catalyzes hydrogenation of benzylideneacetone[33a] and cyclic aromatic ketones[33b] with modest to high enantioselectivities (eq 25).

1-tetralone $\xrightarrow[\substack{PPh(C_6H_4\text{-}o\text{-}NMe_2)_2 \\ \text{MeOH–dioxane}}]{\substack{\text{50–57 atm } H_2 \\ [Ir((R)\text{-BINAP})(cod)]BF_4}}$ 1-tetralol (25)

95% ee

Related Reagents. (+)-trans-(2*S*,3*S*)-Bis(diphenylphosphino)bicyclo[2.2.1]hept-5-ene; (2*R*,3*R*)-2,3-Bis(diphenylphosphino)butane; 1,3-Thiazolidine-2-thione.

---

1. (a) Miyashita, A.; Yasuda, A.; Takaya, H.; Toriumi, K.; Ito, T.; Souchi, T.; Noyori, R. *JACS* **1980**, *102*, 7932. (b) Noyori, R.; Takaya, H. *CS* **1985**, *25*, 83.
2. (a) Noyori, R.; Kitamura, M. In *Modern Synthetic Methods*; Scheffold, R., Ed.; Springer: Berlin, 1989; p 115. (b) Noyori, R. *Science* **1990**, *248*, 1194. (c) Noyori, R.; Takaya, H. *ACR* **1990**, *23*, 345. (d) Noyori, R. *Chemtech* **1992**, *22*, 360.
3. Takaya, H.; Akutagawa, S.; Noyori, R. *OS* **1988**, *67*, 20.
4. (a) Ikariya, T.; Ishii, Y.; Kawano, H.; Arai, T.; Saburi, M.; Yoshikawa, S.; Akutagawa, S. *CC* **1985**, 922. (b) Ohta, T.; Takaya, H.; Noyori, R. *IC* **1988**, *27*, 566. (c) Kitamura, M.; Tokunaga, M.; Ohkuma, T.; Noyori, R. *TL* **1991**, *32*, 4163. (d) Kitamura, M.; Tokunaga, M.; Ohkuma, T.; Noyori, R. *OS* **1992**, *71*, 1.
5. Kitamura, M.; Tokunaga, M.; Noyori, R. *JOC* **1992**, *57*, 4053.
6. (a) Ohta, T.; Takaya, H.; Kitamura, M.; Nagai, K.; Noyori, R. *JOC* **1987**, *52*, 3174. (b) Saburi, M.; Takeuchi, H.; Ogasawara, M.; Tsukahara, T.; Ishii, Y.; Ikariya, T.; Takahashi, T.; Uchida, Y. *JOM* **1992**, *428*, 155.
7. Lubell, W. D.; Kitamura, M.; Noyori, R. *TA* **1991**, *2*, 543.
8. (a) Noyori, R.; Ohta, M.; Hsiao, Y.; Kitamura, M.; Ohta, T.; Takaya, H. *JACS* **1986**, *108*, 7117. (b) Kitamura, M.; Hsiao, Y.; Noyori, R.; Takaya, H. *TL* **1987**, *28*, 4829.
9. (a) Takaya, H.; Ohta, T.; Sayo, N.; Kumobayashi, H.; Akutagawa, S.; Inoue, S.; Kasahara, I.; Noyori, R. *JACS* **1987**, *109*, 1596, 4129. (b) Takaya, H.; Ohta, T.; Inoue, S.; Tokunaga, M.; Kitamura, M.; Noyori, R. *OS* **1994**, *72*, 74.
10. Kitamura, M.; Nagai, K.; Hsiao, Y.; Noyori, R. *TL* **1990**, *31*, 549.
11. Kitamura, M.; Kasahara, I.; Manabe, K.; Noyori, R.; Takaya, H. *JOC* **1988**, *53*, 708.
12. Ohta, T.; Miyake, T.; Seido, N.; Kumobayashi, H.; Akutagawa, S.; Takaya, H. *TL* **1992**, *33*, 635.
13. (a) Noyori, R.; Ohkuma, T.; Kitamura, M.; Takaya, H.; Sayo, N.; Kumobayashi, H.; Akutagawa, S. *JACS* **1987**, *109*, 5856. (b) Kitamura, M.; Ohkuma, T.; Inoue, S.; Sayo, N.; Kumobayashi, H.; Akutagawa, S.; Ohta, T.; Takaya, H.; Noyori, R. *JACS* **1988**, *110*, 629. (c) Kitamura, M.; Ohkuma, T.; Takaya, H.; Noyori, R. *TL* **1988**, *29*, 1555. (d) Kawano, H.; Ishii, Y.; Saburi, M.; Uchida, Y. *CC* **1988**, 87. (e) Ohkuma, T.; Kitamura, M.; Noyori, R. *TL* **1990**, *31*, 5509.
14. Nishi, T.; Kitamura, M.; Ohkuma, T.; Noyori, R. *TL* **1988**, *29*, 6327.
15. (a) Kitamura, M.; Tokunaga, M.; Noyori, R. *JACS* **1993**, *115*, 144. (b) Kitamura, M.; Tokunaga, M.; Noyori, R. *T* **1993**, *49*, 1853.
16. (a) Noyori, R.; Ikeda, T.; Ohkuma, T.; Widhalm, M.; Kitamura, M.; Takaya, H.; Akutagawa, S.; Sayo, N.; Saito, T.; Taketomi, T.; Kumobayashi, H. *JACS* **1989**, *111*, 9134. (b) Genet, J. P.; Pinel, C.; Mallart, S.; Juge, S.; Thorimbert, S.; Laffitte, J. A. *TA* **1991**, *2*, 555. (c) Mashima, K.; Matsumura, Y.; Kusano, K.; Kumobayashi, H.; Sayo, N.; Hori, Y.; Ishizaki, T.; Akutagawa, S.; Takaya, H. *CC* **1991**, 609.
17. Kitamura, M.; Ohkuma, T.; Tokunaga, M.; Noyori, R. *TA* **1990**, *1*, 1.
18. Oppolzer, W.; Wills, M.; Starkemann, C.; Bernardinelli, G. *TL* **1990**, *31*, 4117.
19. (a) Brown, J. M.; Brunner, H.; Leitner, W.; Rose, M. *TA* **1991**, *2*, 331. (b) Saburi, M.; Ohnuki, M.; Ogasawara, M.; Takahashi, T.; Uchida, Y. *TL* **1992**, *33*, 5783.
20. Kameyama, M.; Kamigata, N.; Kobayashi, M. *JOC* **1987**, *52*, 3312.
21. (a) Miyashita, A.; Takaya, H.; Souchi, T.; Noyori, R. *T* **1984**, *40*, 1245. (b) Toriumi, K.; Ito, T.; Takaya, H.; Souchi, T.; Noyori, R. *Acta Crystallogr.* **1982**, *B38*, 807.
22. Evans, D. A.; Morrissey, M. M. *TL* **1984**, *25*, 4637.
23. (a) Tani, K.; Yamagata, T.; Otsuka, S.; Akutagawa, S.; Kumobayashi, H.; Taketomi, T.; Takaya, H.; Miyashita, A.; Noyori, R. *CC* **1982**, 600. (b) Inoue, S.; Takaya, H.; Tani, K.; Otsuka, S.; Sato, T.; Noyori, R. *JACS* **1990**, *112*, 4897. (c) Yamakawa, M.; Noyori, R. *OM* **1992**, *11*, 3167. (d) Tani, K.; Yamagata, T.; Tatsuno, Y.; Yamagata, Y.; Tomita, K.; Akutagawa, S.; Kumobayashi, H.; Otsuka, S. *AG(E)* **1985**, *24*, 217. (e) Otsuka, S.; Tani, K. *S* **1991**, 665.
24. Kitamura, M.; Manabe, K.; Noyori, R.; Takaya, H. *TL* **1987**, *28*, 4719.
25. Wu, X.-M.; Funakoshi, K.; Sakai, K. *TL* **1992**, *33*, 6331.
26. (a) Hayashi, T.; Matsumoto, Y.; Ito, Y. *JACS* **1989**, *111*, 3426. (b) Sato, M.; Miyaura, N.; Suzuki, A. *TL* **1990**, *31*, 231. (c) Zhang, J.; Lou, B.; Guo, G.; Dai, L. *JOC* **1991**, *56*, 1670.
27. Tamao, K.; Tohma, T.; Inui, N.; Nakayama, O.; Ito, Y. *TL* **1990**, *31*, 7333.
28. Hodgson, M.; Parker, D. *JOM* **1987**, *325*, C27.
29. Hayashi, T.; Matsumoto, Y.; Ito, Y. *JACS* **1988**, *110*, 5579.
30. Ozawa, F.; Hayashi, T. *JOM* **1992**, *428*, 267.
31. (a) Sato, Y.; Sodeoka, M.; Shibasaki, M. *CL* **1990**, 1953; Sato, Y.; Sodeoka, M.; Shibasaki, M. *JOC* **1989**, *54*, 4738. (b) Ashimori, A.; Overman, L. E. *JOC* **1992**, *57*, 4571.
32. Yamaguchi, M.; Shima, T.; Yamagishi, T.; Hida, M. *TL* **1990**, *31*, 5049.
33. (a) Mashima, K.; Akutagawa, T.; Zhang, X.; Takaya, H.; Taketomi, T.; Kumobayashi, H.; Akutagawa, S. *JOM* **1992**, *428*, 213. (b) Zhang, X.; Taketomi, T.; Yoshizumi, T.; Kumobayashi, H.; Akutagawa, S.; Mashima, K.; Takaya, H. *JACS* **1993**, *115*, 3318.

Masato Kitamura & Ryoji Noyori
Nagoya University, Japan

Bis(trimethylsilyl)acetylene

$Me_3Si—≡—SiMe_3$

[14630-40-1] $C_8H_{18}Si_2$ (MW 170.40)

(nucleophile in Friedel–Crafts type acylations and alkylations; lithium trimethylsilylacetylide precursor; cycloaddition substrate)

Alternate Name: BTMSA.
Physical Data: mp 26 °C; bp 136–137 °C; *d* 0.752 g cm^{-3}.
Solubility: sol all commonly used organic solvents.
Form Supplied in: clear liquid; widely available.
Preparative Method: from acetylene by treatment with ***n*-Butyllithium** followed by ***Chlorotrimethylsilane***.[1]
Handling, Storage, and Precautions: the liquid is flammable and an irritant. Use in a fume hood.

Friedel–Crafts Alkylation/Acylation and Related Reactions. Bis(trimethylsilyl)acetylene (BTMSA) can be added to a variety of Lewis acid-activated electrophiles. The most common variation which has been used in several total syntheses is the addition of BTMSA to acid chlorides in the presence of ***Aluminum Chloride*** to afford α,β-alkynic ketones in good yields (eq 1);[2,3] this complements the method by which monosubstituted alkynes are added to carboxylic acid anhydrides, esters and tertiary amides (see ***n-Butyllithium-Boron Trifluoride Etherate***). In addition to acid chlorides, BTMSA can be added to optically active acetals in the presence of ***Titanium(IV) Chloride*** to afford, after cleavage of the chiral template, α-hydroxyalkynes in excellent yield and high enantiopurity (eq 2).[4] BTMSA can also be added to the oxonium ion intermediate resulting from the treatment of tri-*O*-acetyl-D-glucal with ***Tin(IV) Chloride*** to give the α-alkynylated pyranose in excellent yield as a single stereoisomer (eq 3).[5] In a related example, it has been shown that BTMSA can be added to γ-lactols in the presence of ***Boron Trifluoride Etherate*** to afford highly substituted tetrahydrofuran-3-carboxylates in high stereoselectivity.[6]

BTMSA, $AlCl_3$, CH_2Cl_2, 0 °C, 60% (1)

BTMSA, $TiCl_4$, CH_2Cl_2, −78 °C, 98%; 93% de (2)

BTMSA, $SnCl_4$, CH_2Cl_2, −20 °C, 99% (3)

In addition, treatment of δ-methoxyproline derivatives[7] and γ-methoxy-γ-lactams[8] under the Lewis acid/BTMSA conditions gives δ-alkynic prolines and γ-alkynic γ-lactams, respectively. BTMSA will undergo a 1,4-addition to ethylenic acyl cyanides in the presence of $TiCl_4$ to give γ,δ-alkynic acyl cyanides in good yield (eq 4).[9] BTMSA will also add to tertiary[10] or activated halides in the presence of Lewis acids (eq 5).[11]

BTMSA, $TiCl_4$, CH_2Cl_2, −35 °C, 65% (4)

BTMSA, $AlCl_3$, CH_2Cl_2, rt (5)

Concerning noncarbon electrophiles, addition of BTMSA to arylsulfonyl chlorides in the presence of aluminum chloride affords ethynyl sulfones[12] which have been used as Michael acceptors and Diels–Alder dienophiles.[13] It is presumed that the Lewis acid catalyzed processes summarized above proceed via a silicon-stabilized vinyl cation, which then loses the trimethylsilyl moiety to afford the alkynic products.

Trimethylsilylacetylide Alkylation Reactions. While BTMSA adds to a number of electrophiles under Lewis acid-catalyzed conditions, this reagent can also be used as a trimethylsilylacetylide precursor. In this regard, it has been reported that treatment of BTMSA with ***Methyllithium***–LiBr complex in THF gives quantitative formation of lithium trimethylsilylacetylide (eq 6).[14] The utility of this reagent lies in its ability to provide access to a diverse group of protected, terminal alkynes. For example, this reagent, which is formed in situ, can be added to aldehydes to give secondary α-hydroxy alkynes,[14] ketones to afford tertiary α-hydroxy alkynes (eq 7),[15] epoxides to form β-hydroxy alkynes,[16] or alkyl halides to give monoalkylated alkynes.[17] Lithium trimethylsilylacetylide can be brominated with ***Bromine*** and coupled with copper acetylides to give 1,3-diynes[18] or iodinated with ***Iodine Monochloride***[19] and coupled with arylcopper reagents to afford aryl alkynes in good yields.[20] This lithium acetylide intermediate has also been transmetalated with ***Zinc Bromide*** and coupled with a vinyl iodide in the presence of Pd^0 to give a 1,3-enyne[21] (eq 8) or with ***Diethylaluminum Chloride*** and coupled with an enone in the presence of a nickel catalyst to afford a γ,δ-alkynic ketone (eq 9).[22]

$Me_3Si—≡—SiMe_3$ → (MeLi–LiBr, THF, rt) → $[Me_3Si—≡—Li]$ (6)

$SiMe_3—≡—Li$ followed by MsCl, THF, −78 °C (7)

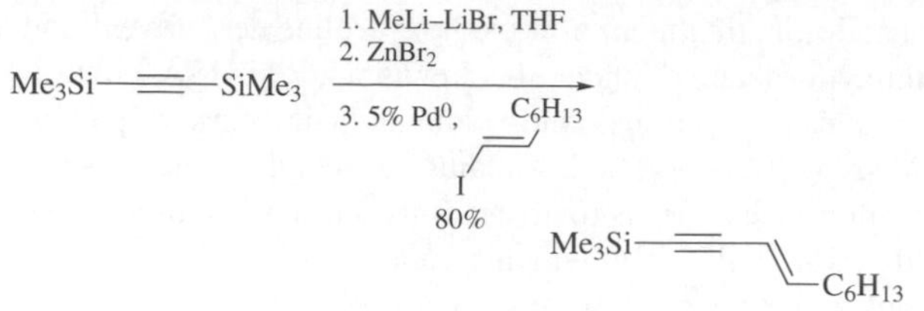

1. MeLi–LiBr, THF
2. $ZnBr_2$
3. 5% Pd^0, C_6H_{13}–CH=CH–I
80%

(8)

Me3Si–≡–CH=CH–C_6H_{13}

Me_3Si–≡–Li
Et_2AlCl
$Ni(acac)_2$, DIBAL
THF, Et_2O, 0 °C
83%

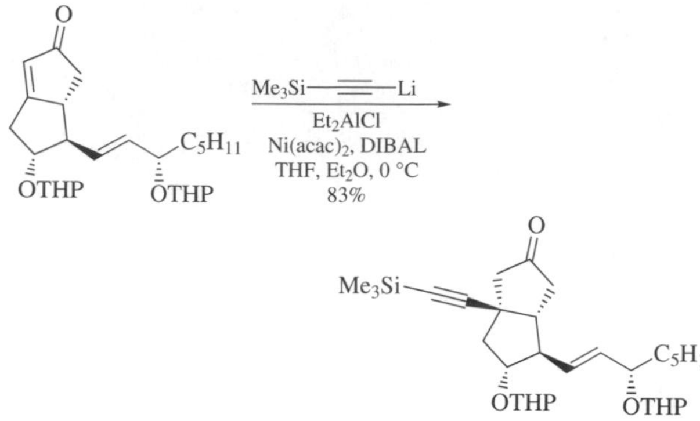

(9)

Cycloaddition Reactions. BTMSA is a powerful substrate in a variety of cycloaddition reactions. The most common use of this reagent in that sense involves the transition metal-catalyzed [2+2+2] cycloaddition.[23] BTMSA can be added to 1,5-hexadiyne in the presence of a cobalt catalyst to afford the highly strained, versatile synthetic intermediate 4,5-bis(trimethylsilyl)benzocyclobutene (eq 10).[24] When this reaction is run on a diyne with a properly tethered vinyl group the initial adduct isomerizes to the corresponding *o*-xylylene and undergoes an intramolecular [4+2] cycloaddition to afford complex polycycles in high yields (eq 11);[25] this method was used in a concise total synthesis of (±)-estrone.[26] Varying the BTMSA acceptor in this reaction gives a number of polycyclic heterocycles including 2*H*-pyrans,[27] 2,3-dihydro-5(1*H*)-indolizinones,[28] anthraquinones,[29] tetrahydroisoquinolines (eq 12)[30] and 1,3-diazabiphenylenes (eq 13).[31]

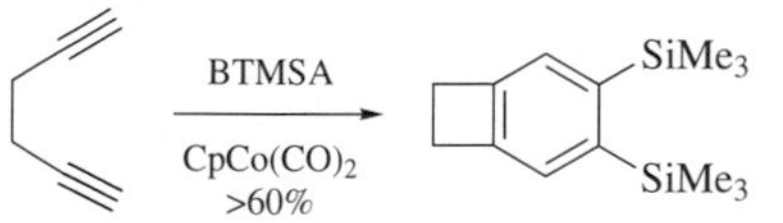

BTMSA
$CpCo(CO)_2$
>60%

(10)

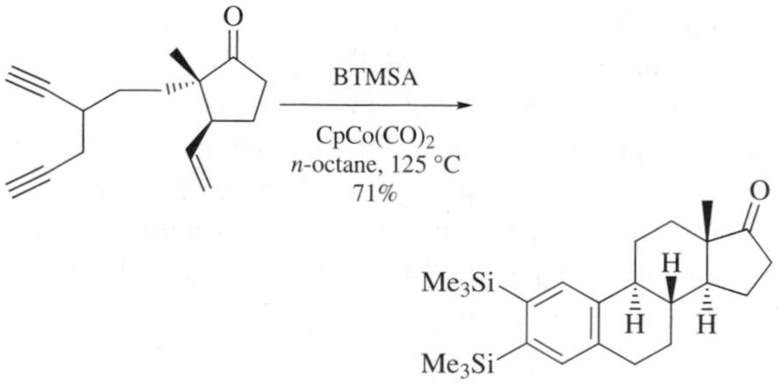

BTMSA
$CpCo(CO)_2$
n-octane, 125 °C
71%

(11)

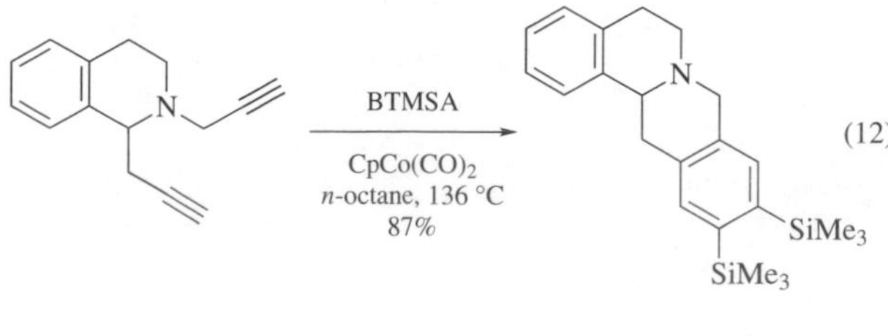

BTMSA
$CpCo(CO)_2$
n-octane, 136 °C
87%

(12)

BTMSA
$CpCo(CO)_2$, *hν*
xylenes, 136 °C
56%

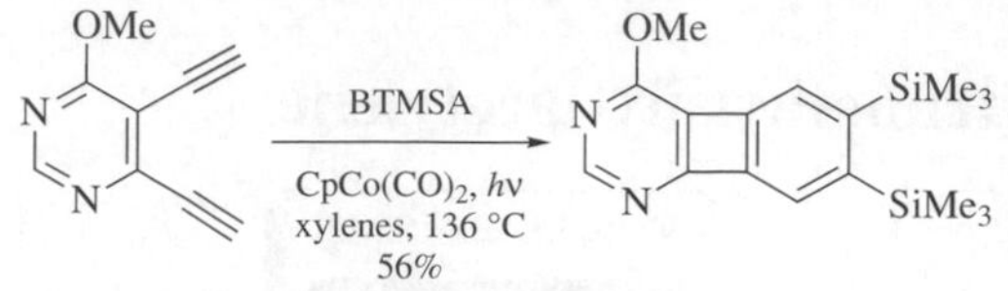

(13)

BTMSA can also be added to several dienes in Diels–Alder fashion. Pretreatment of hydrocarbon dienes such as 1,3-butadiene with catalytic $Et_2AlCl/TiCl_4$ followed by addition of BTMSA at 60 °C gives the corresponding bis(trimethylsilyl)cyclohexadienes in high yields.[32] BTMSA can be added to α-pyrone in decalin at 165 °C to give 1,2-bis(trimethylsilyl)benzene in good yield,[33] as well as to 6*H*-1,3-oxazin-6-ones in refluxing decalin to afford 3,4-bis(trimethylsilyl)pyridine in moderate yield.[34] In addition, BTMSA can be added to 4-phenyloxazole in the presence of catalytic triethylamine at 250 °C to give 3,4-bis(trimethylsilyl)furan in excellent yield.[35] BTMSA has also been used in other cycloadditions including cyclopropenation involving bis(methoxycarbonyl)carbene,[36] [2+2] cycloaddition with dichloroketene to give silylated cyclobutenedione,[37] 1,3-dipolar cycloaddition with pyridinium dicyanomethylides to afford silylated indolizines,[38] and with nitrile oxides to give isoxazoles.[39]

Miscellaneous Reactions. In addition to BTMSA being used as an acetylide synthon, it can also be converted to the corresponding alkynyl(phenyl)iodonium tosylate[40] and treated with nucleophiles such as 1,3-dicarbonyl compounds.[41] 1-Nitro-2-(trimethylsilyl)acetylene has been prepared from BTMSA and undergoes a series of cycloadditions analogous to BTMSA itself.[42] In addition, BTMSA will undergo hydroboration with dialkylborane to afford 1-substituted 1,2-bis(trimethylsilyl)ethenes[43] or with ***Borane-Dimethyl Sulfide*** to give an acylsilane.[44] Finally, it was reported that BTMSA could be treated with cyclopropylcarbene–chromium complex to give silylated 3-methoxy-2-cyclopentenone in good yield.[45]

Related Reagents. Trimethylsilylacetylene.

---

1. Walton, D. R. M.; Waugh, F. *JOM* **1972**, *37*, 45.
2. Treilhou, M.; Fauve, A.; Pougny, J.-R.; Prome, J.-C.; Veschambre, H. *JOC* **1992**, *57*, 3203.
3. For the conversion of the resulting α,β-alkynic ketones to α,β-unsaturated aldehydes, see: Newman, H. *JOC* **1973**, *38*, 2254.
4. Johnson, W. S.; Elliott, R.; Elliott, J. D. *JACS* **1983**, *105*, 2904. For a full procedural account of this method, see: Holmes, A. B.; Tabor, A. B.; Baker, R. *JCS(P1)* **1991**, 3301. For an example of this type of addition with an acetal resulting from an unusual one-pot Beckmann fragmentation/acetalization, see: Fujioka, H.; Kitagawa, H.; Yamanaka, T.; Kita, Y. *CPB* **1992**, *40*, 3118.
5. (a) Tsukiyama, T.; Isobe, M. *TL* **1992**, *33*, 7911. (b) Ichikawa, Y.; Isobe, M.; Konobe, M.; Goto, T. *Carbohydr. Res.* **1987**, *171*, 193.
6. Bruckner, C.; Holzinger, H.; Reissig, H.-U. *JOC* **1988**, *53*, 2450.
7. Manfre, F.; Kern, J.-M.; Biellmann, J.-F. *JOC* **1992**, *57*, 2060.
8. Lundkvist, J. R. M.; Ringdahl, B.; Hacksell, U. *JMC* **1989**, *32*, 863.
9. Jellal, A.; Zahra, J.-P.; Santelli, M. *TL* **1983**, *24*, 1395.
10. Capozzi, G.; Romeo, G.; Marcuzzi, F. *CC* **1982**, 959.
11. Casara, P.; Metcalf, B. W. *TL* **1978**, 1581.
12. Bhattacharya, S. N.; Josiah, B. M.; Walton, D. R. M. *Organomet. Org. Synth.* **1971**, *1*, 145.

13. For the preparation of ethynyl *p*-tolyl sulfone using this method along with references regarding its use, see: Waykole, L.; Paquette, L. A. *OS* **1988**, *67*, 149.
14. Holmes, A. B.; Jennings-White, C. L. D.; Schulthess, A. H.; Akinde, B.; Walton, D. R. M. *CC* **1979**, 840. For the preparation of dilithioacetylide from BTMSA, see: Ogawa, S.; Tajiri, Y.; Furukawa, N. *TL* **1993**, *34*, 839.
15. Takemoto, T.; Fukaya, C.; Yokoyama, K. *TL* **1989**, *30*, 723.
16. Negishi, E.; Boardman, L. D.; Sawada, H.; Bagheri, V.; Stoll, A. T.; Tour, J. M.; Rand, C. L. *JACS* **1988**, *110*, 5383.
17. Gorgen, G.; Boland, W.; Preiss, U.; Simon, H. *HCA* **1989**, *72*, 917.
18. Miller, J. A.; Zweifel, G. *S* **1983**, 128.
19. (a) Walton, D. R. M.; Webb, M. J. *JOM* **1972**, *37*, 41. (b) Al-Hassan, M. I. *JOM* **1989**, *372*, 183.
20. Oliver, R.; Walton, D. R. M. *TL* **1972**, 5209.
21. Alexakis, A.; Marek, I.; Mangeney, P.; Normant, J. F. *T* **1991**, *47*, 1677.
22. Aristoff, P. A.; Johnson, P. D.; Harrison, A. W. *JOC* **1983**, *48*, 5341.
23. For a review of this topic, see: Vollhardt, K. P. C. *ACR* **1977**, *10*, 1.
24. Aalbersberg, W. G. L.; Barkovich, A. J.; Funk, R. L.; Hillard III, R. L.; Vollhardt, K. P. C. *JACS* **1975**, *97*, 5600. For an example of this method in the synthesis of polycycles of theoretical interest, see: Schwager, H.; Spyroudis, S.; Vollhardt, K. P. C. *JOM* **1990**, *382*, 191.
25. Funk, R. L.; Vollhardt, K. P. C. *JACS* **1977**, *99*, 5483. For the full paper regarding this work, see: Funk, R. L.; Vollhardt, K. P. C. *JACS* **1980**, *102*, 5245.
26. Funk, R. L.; Vollhardt, K. P. C. *JACS* **1980**, *102*, 5253. For a synthesis of the phyllocladane diterpene skeleton using this methodology, see: Gotteland, J.-P.; Malacria, M. *TL* **1989**, *30*, 2541.
27. Harvey, D. F.; Johnson, B. M.; Ung, C. S.; Vollhardt, K. P. C. *SL* **1989**, 15.
28. Earl, R. A.; Vollhardt, K. P. C. *JOC* **1984**, *49*, 4786.
29. Hillard III, R. L.; Vollhardt, K. P. C. *JACS* **1977**, *99*, 4058.
30. Hillard III, R. L.; Parnell, C. A.; Vollhardt, K. P. C. *T* **1983**, *39*, 905.
31. Bakthavachalam, V.; d'Alarcao, M.; Leonard, N. J. *JOC* **1984**, *49*, 289.
32. Mach, K.; Antropiusova, H.; Petrusova, L.; Turecek, F.; Hanus, V.; Sedmera, P.; Schraml, J. *JOM* **1985**, *289*, 331. For an example of a [6+2] cycloaddition involving BTMSA under these conditions, see: Mach, K.; Antropiusova, H.; Petrusovă, L.; Hanuš, V.; Tureček, F.; Sedmera, P. *T* **1984**, *40*, 3295.
33. Jones, P. R.; Albanesi, T. E.; Gillespie, R. D.; Jones, P. C.; Ng, S. W. *Appl. Organomet. Chem.* **1987**, *1*, 521.
34. Yamamoto, Y.; Morita, Y. *H* **1990**, *30*, 771.
35. Song, Z. Z.; Zhou, Z. Y.; Mak, T. C. W.; Wong, H. N. C. *AG(E)* **1993**, *32*, 432.
36. Wheeler, T. N.; Ray, J. *JOC* **1987**, *52*, 4875. Also see: Garratt, P. J.; Tsotinis, A. *JOC* **1990**, *55*, 84.
37. Zhao, D.-C.; Tidwell, T. T. *JACS* **1992**, *114*, 10980.
38. Ikemi, Y.; Matsumoto, K.; Uchida, T. *H* **1983**, *20*, 1009. Also see: Matsumoto, K.; Uchida, T.; Ikemi, Y.; Tanaka, T.; Asahi, M.; Kato, T.; Konishi, H. *BCJ* **1987**, *60*, 3645.
39. Dondoni, A.; Fantin, G.; Fogagnolo, M.; Medici, A.; Pedrini, P. *S* **1987**, *11*, 998. Also see: Padwa, A.; MacDonald, J. G. *JOC* **1983**, *48*, 3189.
40. Stang, P. J.; Kitamura, T. *OS* **1991**, *70*, 215. For a review of this type of reagent, see: Stang, P. J. *AG(E)* **1992**, *31*, 274.
41. Bachi, M. D.; Bar-Ner, N.; Crittell, C. M.; Stang, P. J.; Williamson, B. L. *JOC* **1991**, *56*, 3912. Also see: Ochiai, M.; Ito, T.; Takaoka, Y.; Masaki, Y.; Kunishima, M.; Tani, S.; Nagao, Y. *CC* **1990**, 118.
42. Bottaro, J. C.; Schmitt, R. J.; Bedford, C. D.; Gilardi, R.; George, C. *JOC* **1990**, *55*, 1916.
43. Hoshi, M.; Masuda, Y.; Arase, A. *BCJ* **1993**, *66*, 914.
44. Miller, J. A.; Zweifel, G. *S* **1981**, 288.
45. Turner, S. U.; Herndon, J. W.; McMullen, L. A. *JACS* **1992**, *114*, 8394. For a related example, see: Xu, Y.-C.; Wulff, W. D. *JOC* **1987**, *52*, 3263.

Michael L. Curtin
Abbott Laboratories, Abbott Park, IL, USA

9-Borabicyclo[3.3.1]nonane Dimer[1]

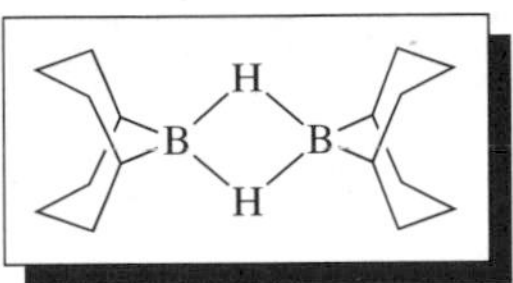

[70658-61-6] $C_{16}H_{30}B_2$ (MW 244.03)

(highly selective, stable hydroborating agent;[1,3] anti-Markovnikov hydration of alkenes and alkynes;[1d] effective ligation for alkyl-, aryl-, allyl-, allenyl-, alkenyl- and alkynylboranerts;[1a,4,5] forms stable dialkylboryl derivatives, borinate esters, and haloboranes;[1f] organoboranes from hydroboration and organometallic reagents;[1,5] precursor to boracycles;[1,11] can selectively reduce conjugated enones to allylic alcohols;[1a,30] organoborane derivatives for α-alkylation and arylation of α-halo ketones, nitriles, and esters;[1b] vinylation and alkynylation of carbonyl compounds;[1a,46] conjugate addition to enones;[1a,47] homologation; enantioselective reduction;[1,8] Diels–Alder reactions;[1a,18,50] enolboranes for crossed aldol condensations;[1a,20,52] Suzuki–Miyaura coupling[1a,54–57])

Alternate Name: 9-BBN-H.

Physical Data: mp 153–155 °C (sealed capillary); bp 195 °C/12 mmHg.[1,3]

Solubility: sparingly sol cyclohexane, dimethoxyethane, diglyme, dioxane (<0.1 M at 25 °C); sol THF, ether, hexane, benzene, toluene, CCl_4, $CHCl_3$, CH_2Cl_2, SMe_2 (ca. 0.2–0.6 M at 25 °C); reacts with alcohols, acetals, aldehydes, and ketones.[1,3]

Form Supplied in: colorless, stable crystalline solid; 0.5 M solution in THF or hexanes.

Analysis of Reagent Purity: the melting point of 9-BBN-H dimer is very sensitive to trace amounts of impurities. Recrystallization from dimethoxyethane is recommended for samples melting below 146 °C. The dimer exhibits a single ^{11}B NMR (C_6D_6) resonance at δ 28 ppm and ^{13}C NMR signals at 20.2 (br), 24.3 (t), and 33.6 (t) ppm.[1,3]

Handling, Storage, and Precautions: the crystalline 9-BBN-H dimer can be handled in the atmosphere for brief periods without significant decomposition. However, the reagent should be stored under an inert atmosphere, preferably below 0 °C. Under these conditions the reagent is indefinitely stable. In solution, 9-BBN-H is more susceptible both to hydrolysis and oxidation, and contact with the open atmosphere should be rigorously avoided. Many 9-BBN derivatives are pyrophoric and/or susceptible to hydrolysis so that individuals planning to use 9-BBN-H dimer should thoroughly familiarize themselves with the special techniques required for the safe handling of such reagents prior to their use.[1b] The reagent should be used in a well-ventilated hood.

Organoboranes from 9-BBN-H. First identified by Köster,[2] 9-BBN-H dimer is prepared from the cyclic hydroboration of 1,5-cyclooctadiene (eq 1).[3] As a dialkylborane, 9-BBN-H exhibits extraordinary steric- and electronic-based regioselectivities which distinguish these derivatives from the less useful polyhydridic reagents (Table 1).[4]

(1)

Table 1 Boron Atom Placement in the Hydroboration of Simple Alkenes

	Bu		Ph		i-Pr	
BH_3	94	6	81	19	43	57
$t\text{-HxBH}_2$	94	6	94	6	34	66
9-BBN-H	99.9	0.1	98.5	1.5	0.2	98.8

	$SiMe_3$		CH_2Cl		1-methylcyclohexene	
BH_3	40	60	60	40	50	50
Sia_2BH	95	5	95	5	52	48
9-BBN-H	100	0	98.9	1.1	80	20

However, in contrast to other dialkylborane reagents (e.g. ***Disiamylborane*** or ***Dicyclohexylborane***) which must be freshly prepared immediately prior to their use, 9-BBN-H dimer is a stable crystalline reagent[1,3] which is commercially available in high purity. This feature of the reagent facilitates the control of reaction stoichiometry at a level unattainable with most borane reagents. The remarkable thermal stability of 9-BBN derivatives permits hydroborations to be conducted over a broad range of temperatures (from 0 °C to above 100 °C) either neat or in a variety of solvents.[4] The *B*-alkyl-9-BBN products can frequently be isolated by distillation without decomposition and fully characterized spectroscopically.[1,4,5] The integrity of the 9-BBN ring is retained even at elevated temperatures (200 °C), but positional isomerization in the *B*-alkyl portion can take place at ca. 160 °C.[6]

Like most other dialkylboranes, 9-BBN-H exists as a dimer, but hydroborates as a monomer (eq 2).[7] In general, the rates of hydroboration follow the order $R_2C{=}CH_2 > RHC{=}CH_2 > cis\text{-}RHC{=}CHR > trans\text{-}RHC{=}CHR > RHC{=}CR_2 > R_2C{=}CR_2$.[1,4] For relatively unsubstituted alkenes the dissociation of the 9-BBN-H dimer is rate-limiting ($T_{1/2}$ at 25 °C ~20 min) so that the hydroborations of typical 1-alkenes are normally complete in less than 3 h at room temperature. Competitive rate studies have revealed that electron-donating groups enhance the rates within these groups, e.g. for $p\text{-XC}_6H_4CH{=}CH_2$, $k_{rel} = 1$ ($X = CF_3$), 5 (X = H), 70 (X = OMe).[4c]

$$[\text{9-BBN-H}]_2 \underset{k_{-1}}{\overset{k_1}{\rightleftharpoons}} 2\,[\text{9-BBN-H}] \qquad (2)$$

$$\text{9-BBN-H} \xrightarrow[k_2]{\text{alkene}} \text{R-9-BBN}$$

Hydroborations of more substituted alkenes such as α-pinene[8] or 2,3-dimethyl-2-butene[4b] with 9-BBN-H are slower (k_2 is rate-limiting) and require heating at reflux temperature in THF for 2 and 8 h, respectively, for complete reaction to occur. However, the enantioselective reducing agent[8a] Alpine-borane® (see ***B-3-Pinanyl-9-borabicyclo[3.3.1]nonane***) is formed quantitatively as a single enantiomer, the process taking place with complete Markovnikov regiochemistry, exclusively through *syn* addition from the least hindered face of the alkene (eq 3).

(3)

While the monohydroboration of symmetrical nonconjugated dienes with 9-BBN-H is thwarted by competing dihydroboration because these remote functionalities act as essentially independent entities, with nonequivalent sites the chemoselectivity of the reagent can be excellent (eq 4).[4] Also, whereas the monohydroboration of conjugated dienes is not always a useful process because of competitive dihydroboration, highly substituted dienes and 1,3-cyclohexadiene produce allylborane products efficiently. In contrast to the monohydroboration of allene itself, which gives a 1,3-diboryl adduct, 9-BBN-H is an effective reagent for the preparation of allylboranes from substituted allenes.[4g] For example, excellent selectivity has been observed for silylated allenes where hydroboration occurs *anti* to the silyl group on the allene and at the terminal position (eq 5).[9] It is also important to note that the diastereofacial selectivities of 9-BBN-H can be complementary to those obtained with Rh-catalyzed hydroborations (eq 6).[10]

(4)

(5)

(6)

9-BBN-H	11:89
CatBH, $ClRh(PPh_3)_3$	96:4

Medium-ring boracycles are efficiently prepared by the dihydroboration of α,ω-dienes with 9-BBN-H followed by exchange with borane.[11] In this process 9-BBN-H is particularly useful because it not only fixes the key 1,5-diboryl relationship, but also the 9-BBN ligands do not participate in the exchange process (eq 7).

$$\mathrm{Me_2Si(CH{=}CH_2)_2} \xrightarrow{\text{9-BBN-H}} \mathrm{Me_2Si(CH_2CH_2\text{-}9\text{-}BBN)_2} \xrightarrow{\mathrm{BH_3 \bullet SMe_2}}$$

$$\mathrm{Me_2Si\langle\ \rangle BH} \xrightarrow{\text{MeOH}} \mathrm{Me_2Si\langle\ \rangle BOMe} \quad (7)$$

Unlike most dialkylboranes, 9-BBN-H hydroborates alkenes faster than its does the corresponding alkynes, a feature which leads to the competitive formation of 1,1-diboryl adducts in the hydroboration of 1-alkynes with 9-BBN-H employing a 1:1 stoichiometry (eq 8).[1,12,13] In some cases, the (*E*)-1-alkenyl-9-BBN derivative can be efficiently prepared by employing either a large excess of the alkyne[12,13] or through the use of 1-trimethylsilyl derivatives.[6] However, these vinylboranes are now perhaps best prepared through the dehydroborylation of their 1,1-diboryl adducts with aromatic aldehydes (eq 8).[12]

$$\text{Bu—≡—H} \xrightarrow[\text{70\%}]{\text{9-BBN-H}} (E)\text{-BuCH=CH-9-BBN} \xrightarrow[\text{100\%}]{\text{9-BBN-H}} \text{BuCH}_2\text{CH(9-BBN)}_2 \quad (8)$$

(reverse step: 1-NaphCHO, 86%)

By contrast, 9-BBN-H effectively monohydroborates internal alkynes to produce the corresponding vinylboranes in >90% yields.[1,13] Compared to 1-alkynes, their 1-silyl counterparts also produce better yields of vinylboranes but, in contrast to normal internal alkynes, produce *vicinal* rather than *geminal* diboryl adducts with dihydroboration.[6,13,14] Larger silyl groups can effectively be used to redirect the boron to the internal position producing the 'silyl-Markovnikov' vinylborane, exclusively, without competitive dihydroboration.[6] For 1-halo-1-alkynes, hydroboration with 9-BBN is slow, but the (*Z*)-1-halovinylboranes (eq 9) are produced cleanly and these are protonolyzed to provide *cis*-vinyl halides.[15] The isomeric (*Z*)-2-bromovinyl-9-BBN derivatives are available from the bromoboration of 1-alkynes with *B*-Br-9-BBN[16] (see ***9-Bromo-9-borabicyclo[3.3.1]nonane***). It is important to point out that the preference of 9-BBN-H to hydroborate alkenes in the presence of alkynes can have useful synthetic applications (eq 10).[17]

$$\text{Bu—≡—Br} \xrightarrow{\text{9-BBN-H}} \text{Bu(Br)C=CH-9-BBN} \quad (9)$$

$$\text{Pr—≡—CH}_2\text{CH=CH}_2 \xrightarrow{\text{9-BBN-H}} \text{Pr—≡—CH}_2\text{CH}_2\text{CH}_2\text{-9-BBN} \quad (10)$$

Organometallic reagents can provide very useful entries to many *B*-substituted 9-BBN derivatives. These are particularly important for organoboranes which cannot be prepared by hydroboration.[5] Both *B*-alkoxy and *B*-halo derivatives serve as useful precursors to *B*-alkyl, -allyl, -aryl, -vinyl or -alkynyl-9-BBN derivatives (eqs 11–16). *B*-Halo-9-BBN derivatives are effectively vinylated with organotin reagents.[18] However, *B*-MeO-9-BBN is superior to its *B*-halo counterparts for secondary and tertiary alkyllithium reagents where the latter undergo some reduction to 9-BBN-H through β-hydride transfer from the organolithium.

$$\text{MeO-9-BBN} \xrightarrow[\text{–LiOMe, 82\%}]{t\text{-BuLi}} t\text{-Bu-9-BBN} \quad (11)$$

$$\text{MeO-9-BBN} \xrightarrow[\text{83\%}]{\mathrm{Al_2Br_3(CH_2CH{=}CH_2)_3}} \text{allyl-9-BBN} \quad (12)$$

$$\text{H-9-BBN} \xrightarrow[\text{2. MeI, 78\%}]{\text{1. LiPh}} \text{Ph-9-BBN} \quad (13)$$

$$\text{MeO-9-BBN} \xrightarrow[\text{2. } \mathrm{BF_3 \bullet Et_2O}]{\text{1. R—≡—M, THF, –78 °C}} \text{THF→9-BBN-C≡C-R} \quad (14)$$

M = Li, MgX

$$\text{9-BBN-Cl} + \mathrm{ViSnMe_3} \xrightarrow[\text{25 °C}]{\mathrm{CH_2Cl_2}} \text{9-BBN-CH=CH}_2 \quad (15)$$

$$\text{MeO-9-BBN} \xrightarrow[\text{2. } \mathrm{BF_3 \bullet Et_2O}]{\text{1. Bu-CH=CH-Li, } \mathrm{Et_2O}, \text{ –78 °C}} (Z)\text{-BuCH=CH-9-BBN} \quad (16)$$

Hydroboration of the byproduct alkene gives isomeric *B*-alkyl-9-BBN products.[5b] Generally, hydrocarbon solvents are preferable to ether or THF for this process because the greater stability of the intermediate methoxyborate complexes (i.e. Li[R(MeO)-9-BBN]) at −78 °C in these solvents prevents the product from being formed and competing with *B*-MeO-9-BBN for the alkyllithium reagent prior to its complete consumption.[5b] The complex is stable for alkenyl and alkynyl derivatives which require $BF_3 \cdot Et_2O$ to remove the methoxy moiety. The procedure has also been used for the preparation of *cis*-vinyl-9-BBN derivatives[19] since the normal route to such derivatives based upon the hydroboration of 1-haloalkynes, followed by hydride-induced rearrangement gives ring expansion products competitively with (*Z*)-1-halovinyl-9-BBNs.[20] Similar behavior has been observed for the reaction of α-methoxyvinyllithium with *B*-alkyl-9-BBNs (see ***1-Methoxyvinyllithium***).[21]

Derivatives of 9-BBN. Like other boron hydrides, a variety of proton sources (ROH, RCO_2H, RSO_3H, HX (X = Cl, Br, OH, SH, $O_2P(OH)_2$, NHR)) as well as boron halides can be effectively employed to prepare useful derivatives from 9-BBN-H.[1d,5b,22] The synthetic value of *B*-MeO-9-BBN lies principally in the preparation of *B*-alkyl derivatives through organometallic reagents as described above. As a byproduct in other processes, it is also easily converted to 9-BBN-H with BMS (eq 17).[22]

(17) [9-BBN-H ⇌ B-OMe-9-BBN; MeOH forward, BH_3•SMe_2 reverse]

Efficient procedures have been developed for the preparation of *B*-Cl-9-BBN from 9-BBN-H (HCl in Et_2O)[5b] and *B*-Br-9-BBN (BBr_3 in CH_2Cl_2),[23] the latter being a useful reagent for ether cleavage, the bromoboration of 1-alkynes, and for conjugative additions to enones (see ***9-Bromo-9-borabicyclo[3.3.1]nonane***). 9-BBN triflate is highly useful in formation of enolboranes for stereoselective crossed aldol reactions[24] (see ***9-Borabicyclononyl Trifluoromethanesulfonate***). The *B*-acyloxy-9-BBN derivatives have been employed in conjuction with borohydrides for the reduction of carboxylic acids to aldehydes[25] (see ***Lithium 9-boratabicyclo[3.3.1]nonane***). Amine complexes of 9-BBN-H and borohydride derivatives are easily prepared from the addition of amines or metal hydrides to 9-BBN-H.[26] *B*-Alkyl-9-BBNs and their borohydrides are very selective reducing agents and, with chiral terpenoid or sugar appendages, can also effectively function as enantioselective reagents (eq 18)[8,27] (see ***B-3-Pinanyl-9-borabicyclo[3.3.1]nonane*** and ***Potassium 9-Siamyl-9-boratabicyclo[3.3.1]nonane***). 9,9-Dialkylborate derivatives of 9-BBN are also highly selective reducing agents[28] (see ***Lithium 9, 9-Dibutyl-9-borabicyclo[3.3.1]nonanate***), transferring a bridgehead hydride with rearrangement to bicyclo[3.3.0]octylboranes. This process is best accomplished with ***Acetyl Chloride***[1] and provides a highly versatile entry to these organoboranes for subsequent conversions (eq 19).[29]

(18) [1. ^{d}Ipc-9-BBN; 2. [O]; 3. H_2 (cat)] R = n-C_8H_{17}, 97% ee

(19) [Li+ Me, Me B⁻ 9-BBN borate; AcCl → BMe_2 bicyclooctane, H]

Functional Group Conversions with 9-BBN-H. 9-BBN-H selectively reduces acid chlorides, aldehydes, ketones, lactones, and sulfoxides at 25 °C.[1e] Alcohol rather than amine products are produced as the major products from tertiary amides, while primary derivatives are not reduced effectively. Reduction is slow with esters, carboxylic acids, nitriles, and epoxides, and does not occur with nitro compounds, nor with alkyl or aryl halides. At 65 °C, carboxylic acids and esters are cleanly reduced to alcohols, the former being significantly slower (18 vs. 4 h). Moreover, 9-BBN-H is a highly selective reducing agent for the reduction of enones to allylic alcohols (eq 20).[30]

(20) [cyclopentenone → 9-BBN-H → cyclopentenol (OH)]

As noted earlier, *B*-substituted-9-BBN derivatives are available from a variety of sources and organoboranes serve as a versatile entry to other functionalities. Their oxidative conversion to alcohols with alkaline ***Hydrogen Peroxide*** or ***Sodium Perborate***[31] is quantitative and occurs with complete retention of configuration, making the process highly useful.[1] The 9-BBN moiety is oxidized to *cis*-1,5-cyclooctanediol (eq 21), a compound which can be removed from less polar products through extraction with water, selective crystallization, or by chromatography.[1] The monooxidation of 9-BBN derivatives with anhydrous ***Trimethylamine N-Oxide*** (TMANO) produces 9-oxa-10-borabicyclo[3.3.2]-decanes (eq 22),[14] many of which are air-stable and undergo useful coupling reactions.

(21) [R-9-BBN; H_2O_2, NaOH (aq) or $Na_2H_4B_2O_8$ → cis-1,5-cyclooctanediol (OH, HO) + ROH]

(22) [R-9-BBN; TMANO → R-B-O bicyclic]

Anomalous oxidation products are observed from the oxidation of tetraalkylborate salts (i.e. Li(R_2-9-BBN)), which produces bicyclo[3.3.0]octan-l-ol as a co-product through a skeletal rearrangement which occurs during the oxidation process.[32] Moreover, the alkaline hydrogen peroxide oxidation of 1,1-di-9-BBN derivatives gives primary alcohols rather than aldehydes because of their solvolysis prior to oxidation.[6,13]

While the protonolysis of *B*-alkyl-9-BBNs, like other trialkylboranes, with carboxylic acids takes place only at temperatures above 100 °C, *B*-vinyl derivatives are readily cleaved by HOAc at 0 °C with complete retention of configuration.[1,13] This can be combined with other 9-BBN processes (e.g. thermal isomerization or dehydroborylation) to give remarkable overall conversions (eq 23).[6] The hydrolysis of allylic and alkynic 9-BBN derivatives is more facile, occurring even with water or alcohols.[5c,32]

$$\text{—≡—SiEt}_3 \xrightarrow[170\ ^\circ\text{C, 2 h}]{\text{9-BBN-H}} \text{(9-BBN)}_2\text{CHCH}_2\text{CH}_2\text{SiEt}_3 \xrightarrow{\text{PhCHO}}$$

$$\text{9-BBN–CH=CHCH}_2\text{SiEt}_3 \xrightarrow{\text{DOAc}} \text{DCH=CHCH}_2\text{SiEt}_3 \quad (23)$$

In the absence of light, molecular bromine readily cleaves *B*-(*s*-alkyl)-9-BBN derivatives through a hydrogen abstraction process, to give excellent yields of the corresponding alkyl bromides, the 9-BBN moiety being converted to 9-Br-9-BBN (eq 24).[33] However, bicyclo[3.3.0]octylborinic and -boronic acids are produced from *B*-Me- and *B*-MeO-9-BBN through this radical bromination under hydrolytic conditions where the facile 1,2-alkyl migration of a ring B–C bond occurs (eq 25).[1b,34] This latter compound serves as a convenient source of 9-oxabicyclo[3.3.1]nonane through base-induced iodination via an S_E2-type inversion.[35]

$$\xrightarrow[90\%]{\text{Br}_2,\ \text{CH}_2\text{Cl}_2} \text{R–Br} + \text{9-Br-9-BBN} \quad (24)$$

$$\text{MeO–9-BBN} \xrightarrow[h\nu]{\text{Br}_2,\ \text{H}_2\text{O}} \text{bicyclo[3.3.0]octyl–B(OH)}_2 \quad (25)$$

Mechanistically similar to the oxidation of 9-BBN derivatives with TMANO, the amination of *B*-alkyl-9-BBN proceeds through ring B–C migration rather than through *B*-alkyl migration (eq 26).[36] Similar behavior is observed for the thermal reaction of organic azides with these derivatives. Dichloroboryl derivatives have proved to be more versatile and general for the synthesis of amines, including optically active derivatives.[37]

$$\text{Bu–9-BBN} \xrightarrow[\text{THF, reflux}]{\text{BuN}_3} \text{Bu–B–N(Bu) ring-expanded} \quad (26)$$

Carbon–Carbon Bond Formation via 9-BBN Derivatives. Consistent with the versatility of organoboranes in synthetically useful chemical transformations, most conversions with 9-BBN derivatives are very efficient and occur with strict stereochemical control.[1] Of particular importance are those which involve the formation of new carbon–carbon bonds because valuable R groups can often be selectively transferred to the substrates without competition from the 9-BBN ring. For example, whereas only one of the alkyl, vinyl or aryl groups can be transferred from BR_3 to the anions derived from α-halo ketones, esters and nitriles, these reactions are ideally suited to *B*-R-9-BBN derivatives which transfer the *B*-R group selectively (eq 27).[1a,38] The vinylogous γ-alkylation of γ-bromo-α,β-unsaturated esters efficiently leads to γ-substituted-β,γ-unsaturated esters, the double bond transposition being commonly observed in the kinetic protonation of enolates with extended conjugation. Both sulfur ylides and α-bromo sulfones undergo related alkylations.[1a]

$$\text{9-BBN–Ph} + \text{BrCH}_2\text{COMe} \xrightarrow{\text{2,6-di-}t\text{-Bu-C}_6\text{H}_3\text{OK}} \text{PhCH}_2\text{COMe} \quad (27)$$

Base-induced eliminations of γ-haloalkyl-9-BBN derivatives give cycloalkanes (C_3 to C_6)[1b] with inversion of configuration at both carbon centers.[1,39] 1,1-Diboryl adducts from the dihydroboration of 1-alkynes with 9-BBN-H serve as useful precursors to *B*-cyclopropyl-9-BBN derivatives by a similar process (eq 28).

$$\text{BrCH}_2\text{C≡CH} \xrightarrow{\text{9-BBN-H}} \text{BrCH}_2\text{CH}_2\text{CH(9-BBN)}_2 \xrightarrow{\text{LiMe}} \text{cyclopropyl–9-BBN} \quad (28)$$

The carbonylation of *B*-alkyl-9-BBNs at 70 atm, 150 °C in the presence of ethylene glycol produces intermediate boronate esters which are oxidized with alkaline hydrogen peroxide to give high yields of the corresponding carbinols (Scheme 1).[40] In the presence of hydride reducing agents (e.g. $LiHAl(OMe)_3$ or ***Potassium Triisopropoxyborohydride***), the carbonylation of *B*-R-9-BBN derivatives can be carried out at atmospheric pressures at 0 °C, producing an intermediate α-alkoxyalkylborane which can be further reduced with ***Lithium Aluminum Hydride***. This results in homologated organoboranes and, after oxidation, alcohols. Alternatively, the intermediate α-alkoxyalkylborane can be directly oxidized to produce aldehydes (Scheme 1).[1a,38c,41]

$$\text{cyclopentyl–9-BBN} \xrightarrow[\text{2. [O]}]{\text{1. CO, (CH}_2\text{OH)}_2\text{, 70 atm, 150 °C}} \text{9-cyclopentylbicyclo[3.3.1]nonan-9-ol}$$

$$\text{cyclopentyl–9-BBN} \xrightarrow{\text{CO, KHB(O-}i\text{-Pr)}_3\text{, 1 atm, 0 °C}} \left[\text{K}^+\ \text{9-BBN–CH(cyclopentyl)–O}\bar{\text{B}}\text{(O-}i\text{-Pr)}_3\right]$$

$$\xrightarrow{\text{H}_2\text{O}_2\text{, pH 7}} \text{cyclopentyl–CHO}$$

$$\xrightarrow[\text{2. [O]}]{\text{1. LiAlH}_4} \text{cyclopentyl–CH}_2\text{OH}$$

Scheme 1

As a useful alternative to carbonylation, the Brown dichloromethyl methyl ether (DCME) process has been effectively used for the synthesis of 9-alkylbicyclo[3.3.1]nonan-9-ols.[42] The ketone bicyclo[3.3.1]nonan-9-one (eq 29)[42b] has also been prepared from a hindered *B*-aryloxy-9-BBN derivative, with simple *B*-alkoxy-9-BBN derivatives failing to undergo this process. However, most borinate esters are smoothly converted to ketones through this process, including germa- and silaborinanes (eq 30).[11e,f] In these cases, 9-BBN-H provides the essential 1,5-diboryl relationship which allows the formation of borinane by the exchange reaction described earlier.

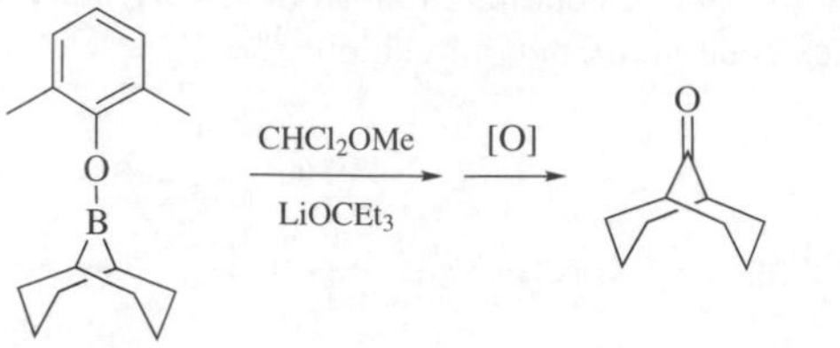

$CHCl_2OMe$, $LiOCEt_3$; [O] (29)

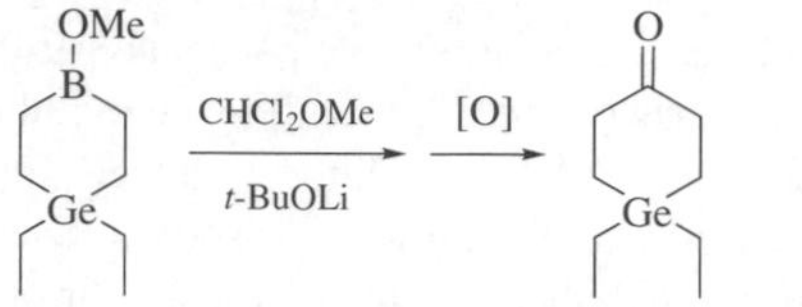

$CHCl_2OMe$, t-BuOLi; [O] (30)

Allylboration with 9-BBN derivatives (see ***B-Allyl-9-borabicyclo[3.3.1]nonane***) is an efficient process, resulting in the smooth formation of homoallylic alcohols (eq 31).[43] Alkynylboranes also undergo 1,2-addition to both aldehydes and ketones.[44] As with other reactions producing *B*-alkoxy-9-BBN byproducts, the conversion of these to alcohols with ***Ethanolamine*** also results in the formation of an alkane-insoluble 9-BBN complex which is conveniently removed, thereby greatly simplifying the workup procedure.

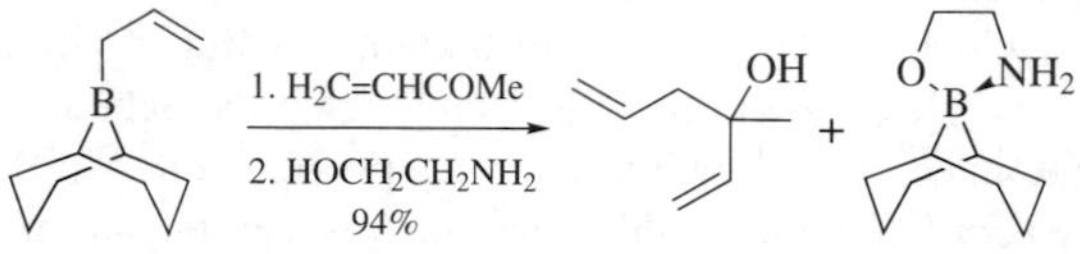

1. $H_2C{=}CHCOMe$; 2. $HOCH_2CH_2NH_2$; 94% (31)

Vinyl derivatives of 9-BBN uniquely undergo 'Grignard-like' 1,2-additions to aldehydes to produce stereodefined allylic alcohols (eq 32).[12,45] The thermal stability of the 9-BBN ring system, as well as its resistance to serve as a β-hydride source, facilitates this highly effective process. These vinylboranes are also the borane reagents of choice for the conjugate additions to enones which can adopt a *cisoid* conformation, providing a convenient entry to γ,δ-unsaturated ketones from enones (eq 33).[46] Alkynylboranes undergo a related addition–elimination process with β-methoxyenones, giving enynones (eq 34).[47]

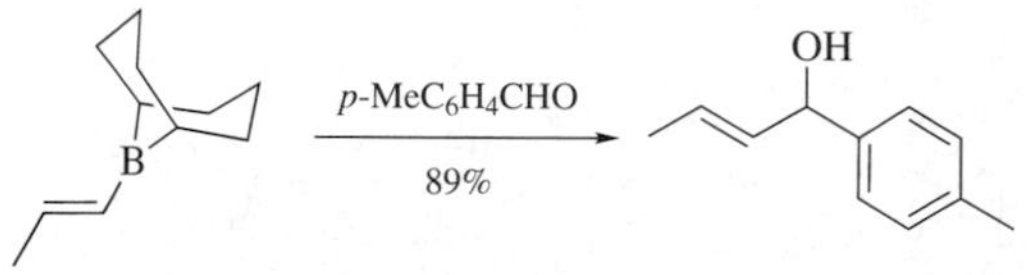

p-MeC_6H_4CHO; 89% (32)

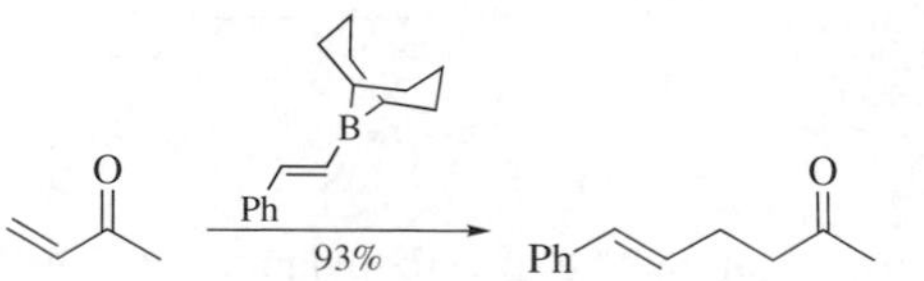

93% (33)

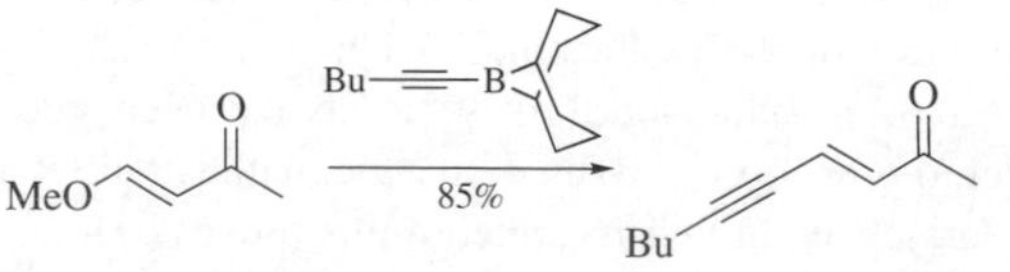

85% (34)

Vinyl(methoxy)-9-BBN 'ate' complexes undergo an unusual homocoupling reaction when treated with ***Zinc Chloride*** (0.5 equiv) (eq 35).[48] Related intermediates, formed through the Cu^I-catalyzed addition of stannylborate complexes to 1-alkynes, can be coupled either through catalytic palladium or stoichiometric copper chemistry to produce stereodefined vinylstannanes (eq 36).[49]

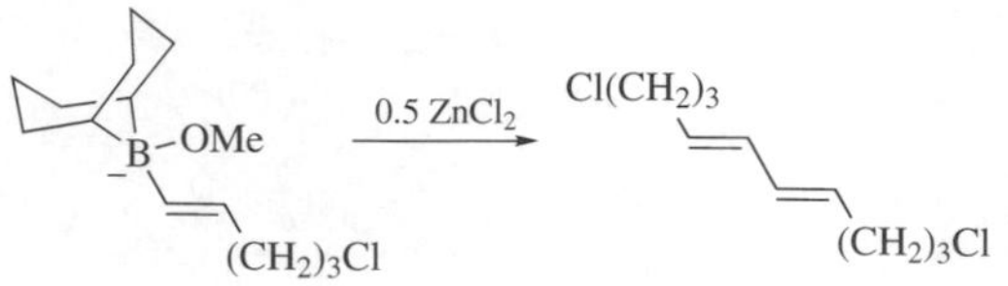

0.5 $ZnCl_2$ (35)

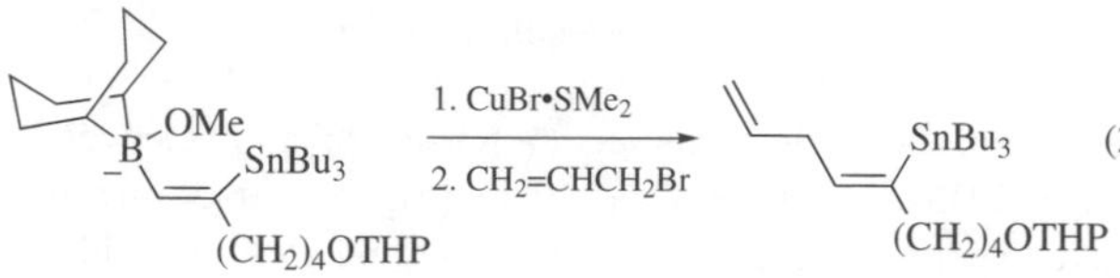

1. $CuBr{\bullet}SMe_2$; 2. $CH_2{=}CHCH_2Br$ (36)

Both vinyl- and alkynyl-9-BBN derivatives are effective dienophiles in Diels–Alder cycloadditions, leading to boron-functionalized cyclohexenes in a selective manner (eqs 37 and 38).[18,50] Silylated allenylboranes add selectively as allylboranes to aldehydes, a reaction which has been effectively used to prepare the steroid nucleus through a Hudrlik elimination followed by a Bergman rearrangement (eq 39).[51]

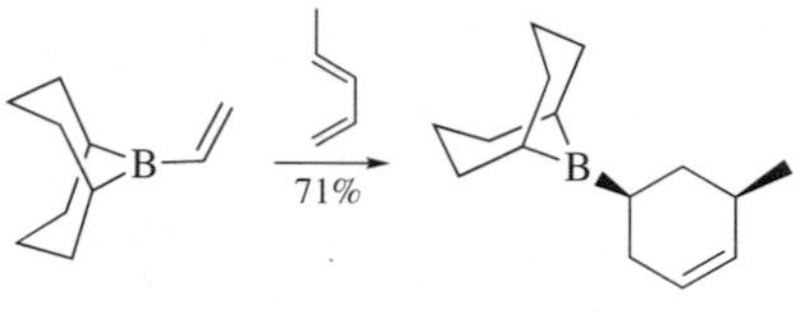

71% (37)

cis:*trans* = 92:8

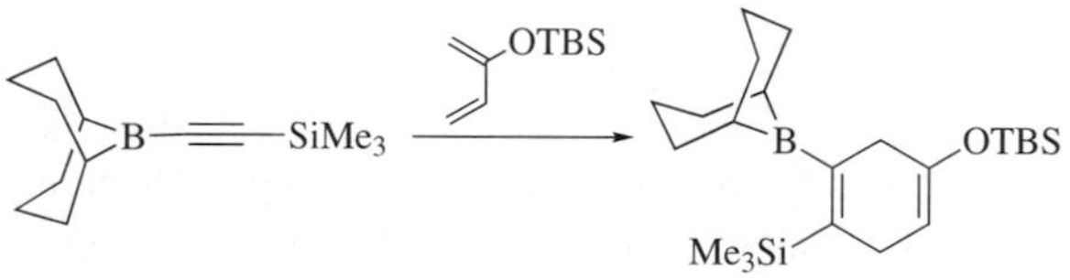

(38)

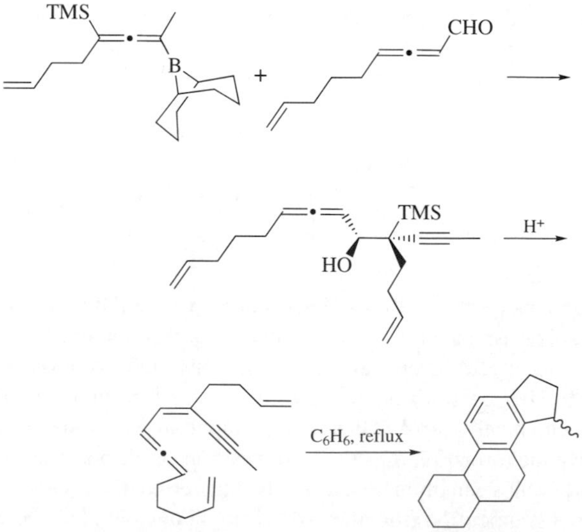

H^+; C_6H_6, reflux (39)

Stereodefined 9-BBN enolboranes which contain a directing chiral auxiliary undergo highly selective crossed aldol condensations as do other dialkylboryl systems (eq 40).[20,52] The conjugate addition of *B*-Br-9-BBN also produces enolboranes which condense with aldehydes to produce, after the elimination of the elements of *B*-HO-9-BBN, α-bromomethyl enones stereoselectively (eq 41).[53]

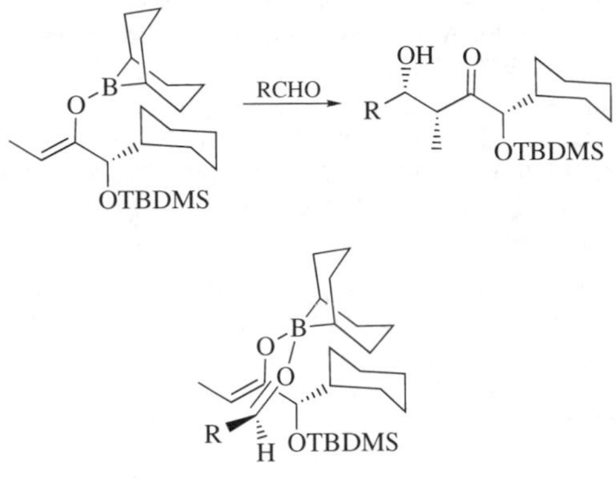

(40)

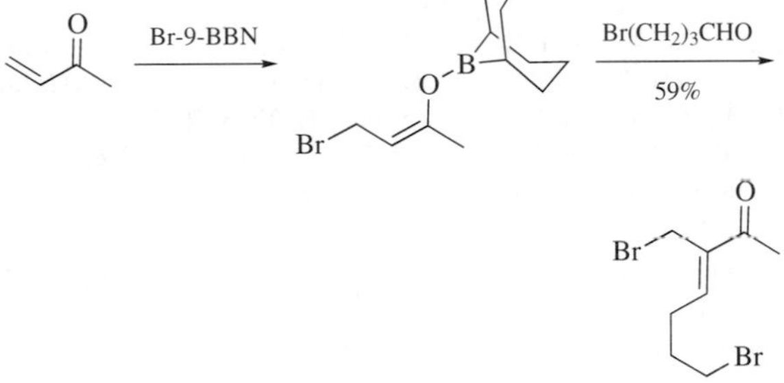

(41)

While catechol- and disiamylborane derivatives were originally employed in the Pd-catalyzed cross coupling of organoboranes to unsaturated organic halides under basic conditions (Suzuki–Miyaura coupling), 9-BBN has recently found an important place in this process. Initially, *B*-(primary alkyl)-9-BBNs, with added bases (NaOH, TlOH, NaOMe, or K_3PO_4), were found to undergo efficient coupling with iodobenzene using ***Dichloro[1, 1′-bis(diphenylphosphino)ferrocene]palladium(II)*** as the catalyst.[54] However, while secondary alkylboranes may require this catalyst, ***Tetrakis(triphenylphosphine)palladium(0)*** is perfectly satisfactory for the coupling of *n*- or *i*-alkyl-9-BBN derivatives to unsaturated bromides, iodides, or triflates under basic conditions (eqs 42–45).

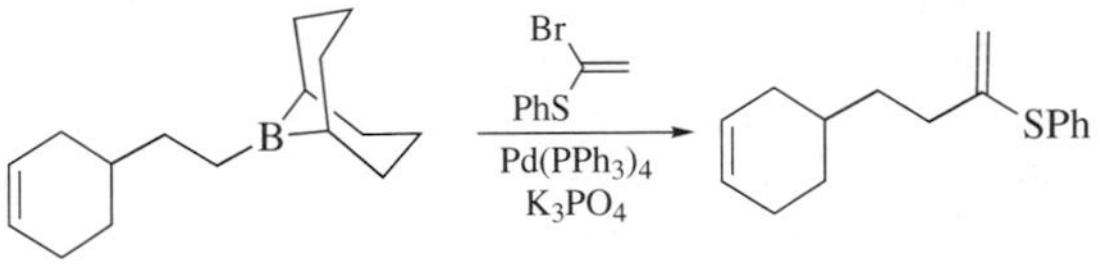

(42)

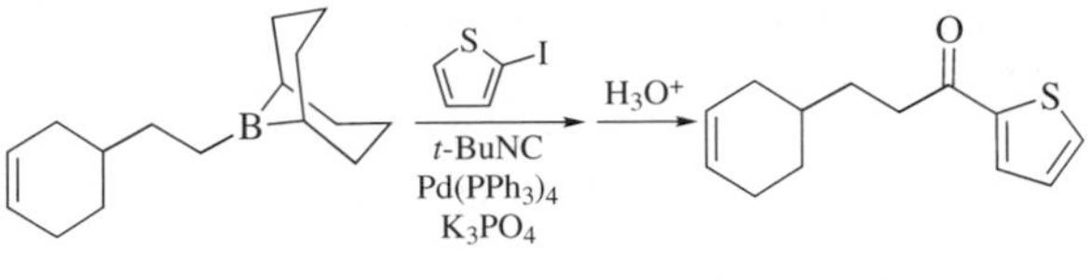

(43)

Several recent applications include the syntheses of pharmaceuticals, pheromones, and prostaglandins, with complete retention of configuration being observed with alkenyl substrates.[54b,55] Either carbon monoxide or ***t-Butyl Isocyanide*** can be used to prepare ketones through the sequential formation of two new carbon–carbon bonds with this reaction.[56] Moreover, vinyl-9-BBNs (eqs 46–48) are also smoothly cross-coupled with retention of configuration to these substrates and, with these now being readily available, their expanded use in this process should flourish.[12,57] It is important to mention that vinyl vs. primary alkyl group transfer is favored by oxygenated ligation on the borane.[52b,58]

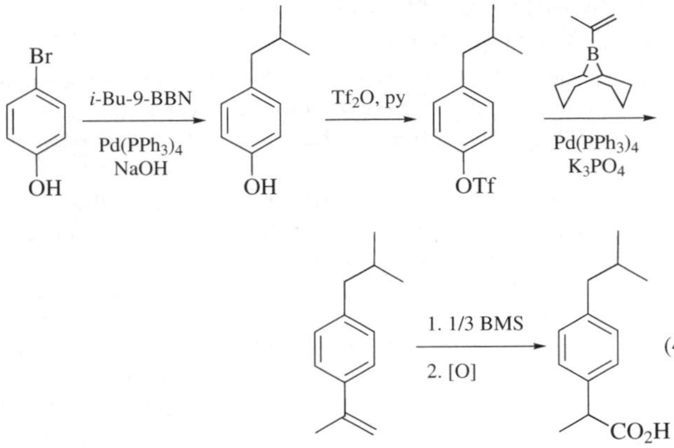

(44)

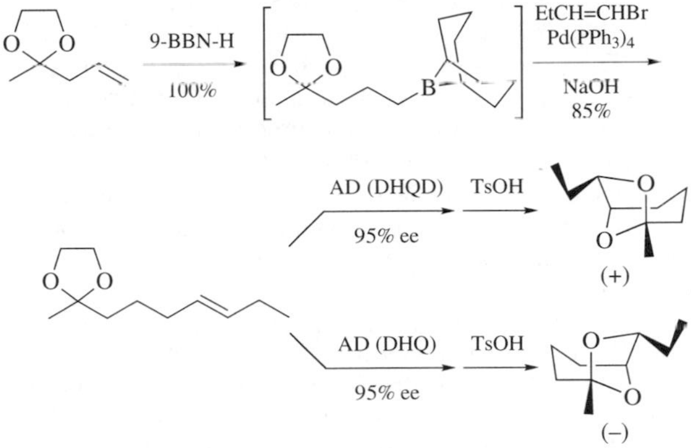

(45)

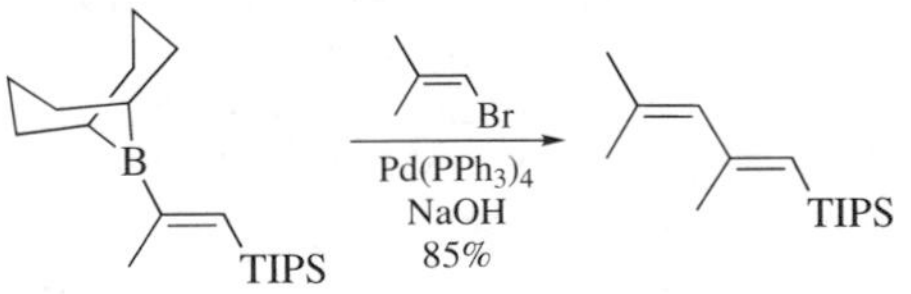

(46)

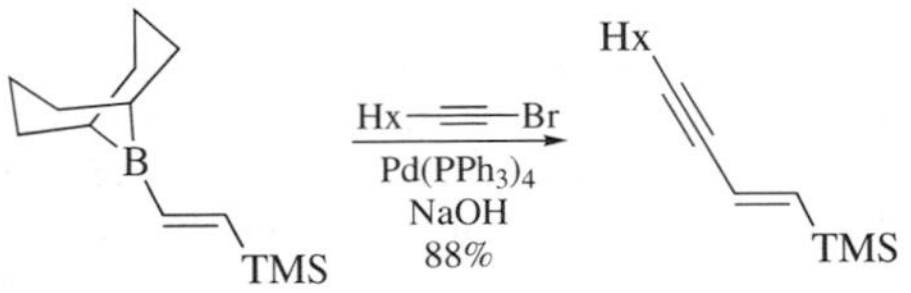

(47)

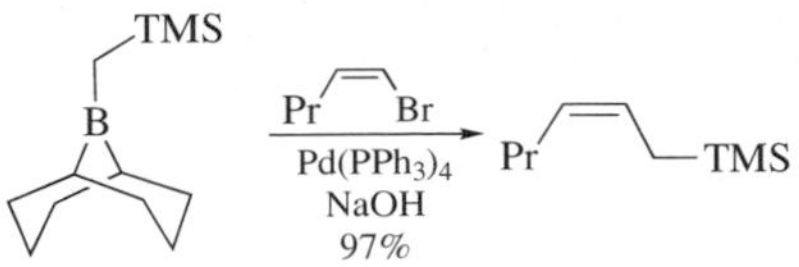

(48)

Related Reagents. B-Allyl-9-borabicyclo[3.3.1]nonane; Borane-Dimethyl Sulfide; Borane-Tetrahydrofuran; 9-Bromo-9-borabicyclo[3.3.1]nonane; Disiamylborane; Thexylborane.

---

1. (a) Pelter, A.; Smith, K.; Brown, H. C. *Borane Reagents*; Academic: London, 1988. (b) Brown, H. C.; Midland, M. M.; Levy, A. B.; Kramer, G. W. *Organic Synthesis via Boranes*; Wiley: New York, 1975. (c) Brown,

H. C.; Lane, C. F. *H* **1977**, *7*, 453. (d) Zaidlewicz, M. In *Comprehensive Organometallic Chemistry*; Wilkinson, G.; Stone, F. G. A.; Abel, E. W., Eds.; Pergamon: Oxford, 1982; Vol. 7 p 199. (e) Negishi, E.-I. In *Comprehensive Organometallic Chemistry*; Wilkinson, G.; Stone, F. G. A.; Abel, E. W., Eds.; Pergamon: Oxford, 1982; Vol 7, p 255. (f) Köster, R.; Yalpani, M. *PAC* **1991**, *63*, 387.

2. Köster, R. *AG* **1960**, *72*, 626.
3. (a) Knights, E. F.; Brown, H. C. *JACS* **1968**, *90*, 5281. (b) Soderquist, J. A.; Brown, H. C. *JOC* **1981**, *46*, 4599. (c) Soderquist, J. A.; Negron, A. *OS* **1991**, *70*, 169. (d) Brauer, D. J.; Kruger, C. *Acta Crystallogr. B*, **1973**, *29*, 1684.
4. (a) Scouten, C. G.; Brown, H. C. *JOC* **1973**, *38*, 4092. (b) Brown, H. C.; Knights, E. F. Scouten, C. G. *JACS* **1974**, *96*, 7765. (c) Brown, H. C.; Liotta, R.; Scouten, C. G. *JACS* **1976**, *98*, 5297. (d) Liotta, R.; Brown, H. C. *JOC* **1977**, *42*, 2836. (e) Brener, L.; Brown, H. C. *JOC* **1977**, *42*, 2702. (f) Brown, H. C.; Liotta, R.; Brener, L. *JACS* **1977**, *99*, 3427. (g) Brown, H. C.; Liotta, R.; Kramer, G. W. *JOC* **1978**, *43*, 1058. (h) Soderquist, J. A.; Hassner, A. *JOM* **1978**, *156*, C12. (i) Brown, H. C.; Liotta, R.; Kramer, G. W. *JACS* **1979**, *101*, 2966. (j) Brown, H. C.; Vara Prasad J. V. N.; Zee, S.-H. *JOC* **1985**, *50*, 1582. (k) Brown, H. C.; Vara Prasad J. V. N. *JOC* **1985**, *50*, 3002. (l) Brown, H. C.; Ramachandran, P. V.; Vara Prasad J. V. N. *JOC* **1985**, *50*, 5583. (m) Soderquist, J. A.; Anderson, C. L. *TL* **1986**, *27*, 3961. (n) Fleming, I. *PAC* **1988**, *60*, 71.
5. (a) Brown, H. C.; Rogić, M. M. *JACS* **1969**, *91*, 4304. (b) Kramer, G. W.; Brown, H. C. *JOM* **1974**, *73*, 1. (c) *ibid.*, *JOM* **1977**, *132*, 9. (d) Soderquist, J. A.; Brown, H. C. *JOC* **1980**, *45*, 3571. (e) Soderquist, J. A.; Rivera, I.; Negron, A. *JOC* **1989**, *54*, 4051.
6. (a) Soderquist, J. A.; Colberg, J. C.; Del Valle, L. *JACS* **1989**, *111*, 4873. See also: (b) Negishi, E.-I. In *Comprehensive Organometallic Chemistry*; Wilkinson, G.; Stone, F. G. A.; Abel, E. W., Eds.; Pergamon: Oxford, 1982; Vol 7, p 265.
7. (a) Brown, H. C.; Scouten, C. G.; Wang, K. K. *JOC* **1979**, *44*, 2589. (b) Brown, H. C.; Wang, K. K.; Scouten, C. G. *PNA* **1980**, *77*, 698. (c) Wang, K. K.; Brown, H. C. *JOC* **1980**, *45*, 5303. (d) Nelson, D. J.; Cooper, P. J. *TL* **1986**, *27*, 4693. (e) Brown, H. C.; Chandrasekharan, J.; Nelson, D. J. *JACS* **1984**, *106*, 3768. (f) Chandrasekharan, J.; Brown, H. C. *JOC* **1985**, *50*, 518.
8. (a) Midland, M. M. *CR* **1989**, *89*, 1553. (b) Brown, H. C.; Ramachandran, P. V. *PAC* **1991**, *63*, 307; *ibid.*, *ACR* **1992**, *25*, 16. (c) Srebnik, M.; Ramachandran, P. V. *Aldrichim. Acta* **1987**, *20*, 9. (d) Brown, H. C.; Srebnik, M.; Ramachandran, P. V. *JOC* **1989**, *54*, 1577.
9. Liu, C.; Wang, K. K. *JOC* **1986**, *51*, 4733.
10. (a) Evans, D. A.; Fu, G. C.; Hoveyda, A. H. *JACS* **1988**, *110*, 6917. (b) Burgess, K.; van der Donk, W. A.; Jarstfer, M. B.; Ohlmeyer, M. *JACS* **1991**, *113*, 6139.
11. (a) Negishi, E.-I.; Burke, P. L.; Brown, H. C. *JACS* **1972**, *94*, 7431. (b) Burke, P. L.; Negishi, E.-I.; Brown, H. C. *JACS* **1973**, *95*, 3654. (c) Brown, H. C.; Pai, G. G. *H* **1982**, *17*, 77. (d) *ibid.*, *JOM* **1983**, *250*, 13. (e) Soderquist, J. A.; Shiau, F.-Y.; Lemesh, R. A. *JOC* **1984**, *49*, 2565. (f) Soderquist, J. A.; Negron, A. *JOC* **1989**, *54*, 2462. However, for the unusual behavior of the 9-BBN systems with alkynyltins, see: (g) Bihlmayer, C.; Kerschl, S.; Wrackmeyer, B. *ZN(B)* **1987**, *42*, 715. (h) Wrackmeyer, B.; Abu-Orabi, S. T. *CB* **1987**, *120*, 1603. (i) Bihlmayer, C.; Abu-Orabi, S. T.; Wrackmeyer, B. *JOM* **1987**, *322*, 25.
12. Colberg, J. C.; Rane, A.; Vaquer, J.; Soderquist, J. A. *JACS* **1993**, *115*, 6065.
13. (a) Brown, H. C., Scouten, C. G.; Liotta, R. *JACS* **1979**, *101*, 96. (b) Wang, K. K.; Scouten, C. G.; Brown, H. C., *JACS* **1982**, *104*, 531. (c) Blue, C. D.; Nelson, D. J. *JOC* **1983**, *48*, 4538.
14. Soderquist, J. A.; Najafi, M. R. *JOC* **1986**, *51*, 1330.
15. Nelson, D. J.; Blue, C. D.; Brown, H. C. *JACS* **1982**, *104*, 4913.
16. Hara, S.; Dojo, H.; Takinami, S.; Suzuki, A. *TL* **1983**, *24*, 731.
17. Brown, C. A.; Coleman, R. A. *JOC* **1979**, *44*, 2328.
18. Singleton, D. A.; Martinez, J. P. *JACS* **1990**, *112*, 7423.
19. Brown, H. C.; Bhat, N. G.; Rajagopalan, S. *OM* **1986**, *5*, 816.
20. Campbell, Jr., J. B.; Molander, G. A. *JOM* **1978**, *156*, 71.
21. Soderquist, J. A.; Rivera, I. *TL* **1989**, *30*, 3919.
22. Soderquist, J. A.; Negron, A. *JOC* **1987**, *52*, 3441.
23. (a) Bhatt, M. V. *JOM* **1978**, *156*, 221. See also: (b) Köster, R.; Grassberger, M. A. *LA* **1968**, *719*, 169. (c) Brown, H. C.; Kulkarni, S. U. *JOC* **1979**, *44*, 281. (d) *ibid.*, *JOC* **1979**, *44*, 2422.
24. (a) Inoue, T.; Uchimaru, T.; Mukaiyama, T. *CL* **1977**, 153. (b) Inoue, T.; Mukaiyama, T. *BCJ* **1980**, *53*, 174. (c) Masamune, S.; Choi, W.; Kerdesky, F. A. J.; Imperiali, B. *JACS* **1981**, *103*, 1566. (d) Masamune, S.; Choi, W.; Peterson, S. S.; Sita, L. R. *AG(E)* **1985**, *24*, 1.
25. Cha, J. S.; Kim, J. E.; Oh, S. Y.; Kim, J. D. *TL* **1987**, *28*, 4575.
26. (a) Brown, H. C.; Kulkarni, S. U. *IC* **1977**, *16*, 3090. (b) *ibid.*, *JOC* **1977**, *42*, 4169. (c) Brown, H. C.; Soderquist, J. A. *JOC* **1980**, *45*, 846. (d) Brown, H. C.; Singaram, B.; Mathew, C. P. *JOC* **1981**, *46*, 2712. (e) Brown, H. C.; Mathew, C. P.; Pyun, C.; Son, J. C.; Yoon, Y. M. *JOC* **1984**, *49*, 3091. (f) Soderquist, J. A.; Rivera, I. *TL* **1988**, *29*, 3195. (g) Hubbard, J. L. *TL* **1988**, *29*, 3197.
27. (a) Brown, H. C.; Krishnamurthy, S. *T* **1979**, *35*, 567. (b) Brown, H. C.; Park, W. S.; Cho, B. T. *JOC* **1986**, *51*, 1934. (c) Brown, H. C.; Cho, B. T.; Park, W. S. *JOC* **1988**, *53*, 1231. (d) Narasimhan, S. *IJC(B)* **1986**, *25B*, 847 (e) Cha, J. S.; Yoon, M. S.; Kim, Y. S.; Lee, K. W. *TL* **1988**, *29*, 1069. (f) Soderquist, J. A.; Rivera, I. *TL* **1988**, *29*, 3195. (g) Cha, J. S.; Lee, K. W.; Yoon, M. S.; Lee, J. C.; Yoon, N. M. *H* **1988**, *27*, 1713. (h) Hutchins, R. O.; Abdel-Magid, A.; Stercho, Y. P.; Wambsgams, A. *JOC* **1987**, *52*, 702.
28. Toi, H.; Yamamoto, Y.; Sonoda, A.; Murahashi, S.-I. *T* **1981**, *37*, 2261.
29. (a) Kramer, G. W.; Brown, H. C. *JOM* **1975**, *90*, Cl. (b) *ibid.*, *JACS* , **1976**, *98*, 1964. (c) *ibid.*, *JOC* **1977**, *42*, 2832.
30. (a) Krishnamurthy, S.; Brown, H. C. *JOC* **1975**, *40*, 1864 (b) *ibid.*, *JOC* **1977**, *42*, 1197. (c) Molin, H.; Pring, B. G. *TL* **1985**, *26*, 677.
31. Kabalka, G. A.; Shoup, T. M.; Goudgaon, N. M. *JOC* **1989**, *54*, 5930.
32. Negishi, E.-I. In *Comprehensive Organometallic Chemistry*; Wilkinson, G.; Stone, F. G. A.; Abel, E. W., Eds.; Pergamon: Oxford, 1982; Vol 7, p 337.
33. (a) Lane, C. L.; Brown, H. C. *JOM* **1971**, *26*, C51. (b) Brown, H. C.; DeLue, N. R. *JACS* **1974**, *96*, 311.
34. (a) Yamamoto, Y.; Brown, H. C. *CC* **1973**, 801. (b) *ibid.*, *JOC* **1974**, *39*, 861.
35. Brown, H. C.; DeLue, N. R.; Kabalka, G. W.; Hedgecock, Jr., H. C. *JACS* **1976**, *98*, 1290.
36. Lane, C. L., PhD Thesis, Purdue University, 1972.
37. (a) Brown, H. C.; Midland, M. M.; Levy, A. B.; Suzuki, A.; Sano, S.; Itoh, M. *T* **1987**, *43*, 4079. (b) Brown, H. C.; Salunkhe, A. M.; Singaram, B. *JOC* **1991**, *56*, 1170. (c) For related behavior with α-diazo ketones, see: Hooz, J.; Gunn, D. M. *TL* **1969**, 3455.
38. (a) Brown, H. C.; Rogić, M. M. *JACS* **1969**, *91*, 2146. (b) Brown, H. C.; Rogić, M. M.; Nambu, H.; Rathke, M. W. *JACS* **1969**, *91*, 2147. (c) Brown, H. C.; Rogić, M. M.; Rathke, M. W.; Kabalka, G. W. *JACS* **1969**, *91*, 2150.
39. (a) Brown, H. C.; Rhodes, S. P. *JACS* **1969**, *91*, 2149. (b) *ibid.*, *JACS* **1969**, *91*, 4306.
40. Brown, H. C.; Knights, E. F. *JACS* **1968**, *90*, 5283.
41. (a) Brown, H. C.; Hubbard, J. L.; Smith, K. *S* **1979**, 701. (b) Hubbard, J. L.; Smith, K. *JOM* **1984**, *276*, C41. (c) Kabalka, G. W.; Delgado, M. C.; Kunda, U. S.; Kunda, S. A. *JOC* **1984**, *49*, 174.
42. (a) Carlson, B. A.; Katz, J.-J.; Brown, H. C. *JOC* **1973**, *38*, 3968. (b) Carlson, B. A.; Brown, H. C. *OSC* **1988**, *6*, 137.
43. (a) Kramer, G. W.; Brown, H. C. *JOC* **1977**, *42*, 2292. (b) Yamamoto, Y.; Yatagai, H.; Maruyama, K. *JACS* **1981**, *103*, 3229.
44. Brown, H. C.; Molander, G. A.; Singh, S. M.; Racherla, U. S. *JOC* **1985**, *50*, 1577.
45. (a) Jacob, III, P.; Brown, H. C. *JOC* **1977**, *42*, 579. (b) Soderquist, J. A.; Vaquer, J. *TL* **1990**, *31*, 4545.

46. (a) Jacob, III, P.; Brown, H. C. *JACS* **1976**, *98*, 7832. (b) Sinclair, J. A.; Molander, G. A.; Brown, H. C. *JACS* **1977**, *99*, 954.
47. Molander, G. A.; Brown, H. C. *JOC* **1977**, *42*, 3106.
48. Molander, G. A.; Zinke, P. W. *OM* **1986**, *5*, 2161.
49. (a) Sharma, S.; Oehlschlager, A. C. *TL* **1988**, *29*, 261. See also: (b) *ibid.*, *JOC* **1989**, *54*, 5064. (c) Hutzinger, M. W.; Singer, R. D.; Oehlschlager, A. C. *JACS* **1990**, *112*, 9397.
50. (a) Singleton, D. A.; Martinez, J. P. *TL* **1991**, *32*, 7365. (b) Singleton, D. A.; Leung, S.-W. *JOC* **1992**, *57*, 4796. (c) Singleton, D. A.; Redman, A. M. *TL* **1994**, *35*, 509.
51. Andemichael, Y. W.; Huang, Y.; Wang, K. K. *JOC* **1993**, *58*, 1651.
52. (a) Evans, D. A.; Bartroli, J.; Shih, T. L. *JACS* 1981, **103**, 2127. (b) Gage, J. R.; Evans, D. A. *OS* **1989**, *68, 77*, 83. (c) Evans, D. A.; Dow, R. L.; Shih, T. I.; Takacs, J. M.; Zahler, R. *JACS* **1990**, *112*, 5290. (d) Heathcock, C. H. *Aldrichim. Acta* **1990**, *23*, 99.
53. Shimizu, H.; Hara, S.; Suzuki, A. *SC* **1990**, *20*, 549.
54. (a) Miyaura, N.; Ishiyama, T.; Ishikawa, M.; Suzuki, A. *TL* **1986**, *27*, 6369. (b) Miyaura, N.; Ishiyama, T.; Sasaki, H.; Ishikawa, M.; Satoh, M.; Suzuki, A. *JACS* **1989**, *111*, 314.
55. (a) Hoshino, Y.; Ishiyama, T.; Miyaura, N.; Suzuki, A. *TL* **1988**, *29*, 3983. (b) Oh-e, T.; Miyaura, N.; Suzuki, A. *SL* **1990**, 221. (c) Soderquist, J. A.; Santiago, B.; Rivera, I. *TL* **1990**, *31*, 4981. (d) Soderquist, J. A.; Santiago, B. *TL* **1990**, *31*, 5113. (e) Suzuki, A. *PAC* **1991**, *63*, 419. (f) Ishiyama, T.; Miyaura, N.; Suzuki, A. *SL* **1991**, 687. (g) Ishiyama, T.; Miyaura, N.; Suzuki, A. *OS* **1992**, *71*, 89. (h) Miyaura, N.; Ishikawa, M.; Suzuki, A. *TL* **1992**, *33*, 2571. (i) Ishiyama, T.; Abe, S.; Miyaura, N.; Suzuki, A. *CL* **1992**, 691. (j) Santiago, B.; Soderquist, J. A. *JOC* **1992**, *57*, 5844. (k) Rivera, I.; Colberg, J. C.; Soderquist, J. A. *TL* **1992**, *33*, 6919. (l) Nomoto, Y.; Miyaura, N.; Suzuki, A. *SL* **1992**, 727. (m) Oh-e, T.; Miyaura, N.; Suzuki, A. *JOC* **1993**, *58*, 2201. (n) Soderquist, J. A.; Rane, A. M. *TL* **1993**, *34*, 5031. (o) Johnson, C. R.; Braun, M. P. *JACS* **1993**, *115*, 11014.
56. (a) Wakita, Y.; Yasunaga, T.; Akita, M.; Kojima, M. *JOM* **1986**, *301*, C17. (b) Ishiyama, T.; Miyaura, N.; Suzuki, A. *BCJ* **1991**, *64*, 1999. (c) *ibid.*, *TL* **1991**, *32*, 6923. (d) Ishiyama, T.; Oh-e, T.; Miyaura, N.; Suzuki, A. *TL* **1992**, *33*, 4465.
57. (a) Soderquist, J. A.; Colberg, J. C. *SL* **1989**, 25. (b) Soderquist, J. A.; Colberg, J. C. *TL* **1994**, *35*, 27.
58. Rivera, I.; Soderquist, J. A. *TL* **1991**, *32*, 2311.

John A. Soderquist
University of Puerto Rico, Rio Piedras, Puerto Rico

2-(2-Bromoethyl)-1,3-dioxane[1]

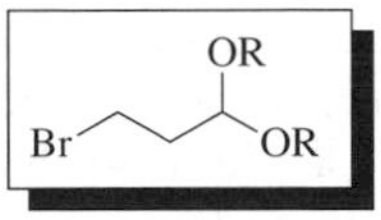

(**1**; R,R = $-(CH_2)_3-$)		
[33884-43-4]	$C_6H_{11}BrO_2$	(MW 195.06)
(**2**; R,R = $-(CH_2)_2-$)		
[18742-02-4]	$C_5H_9BrO_2$	(MW 181.03)
(**3**; R,R = Me,Me)		
[36255-44-4]	$C_5H_{11}BrO_2$	(MW 183.04)

(a group of bifunctional three-carbon reagents having the capacity to serve as electrophiles or as nucleophiles;[1] once the latent carbonyl is unmasked, annulation reactions of various types can be implemented; the application range of these reagents can be further enhanced by initial bromide displacement with nucleophiles possessing useful functionality)

Physical Data: (**1**) bp 67–70 °C/2.8 mmHg, *d* 1.431 g cm^{-3}; (**2**) bp 68–70 °C/8 mmHg, *d* 1.542 g cm^{-3}; (**3**) bp 58–60 °C/17 mmHg, *d* 1.341 g cm^{-3}.

Form Supplied in: all three acetals are colorless, commercially available liquids.

Preparative Methods: 3-bromopropionaldehyde acetals are prepared by bubbling anhydrous gaseous ***Hydrogen Bromide*** into a dichloromethane solution containing ***Acrolein*** and the alcohol or diol at or below rt.[2] More recently, the discovery has been made that ***Chlorotrimethylsilane***[1,2] can promote the conjugate addition of NaX (X = Br, I, SCN) to α,β-unsaturated acetals.[3]

Handling, Storage, and Precautions: these bromides are recognized to possess irritant properties and should therefore invariably be handled with gloved hands inside an efficient hood. It is also advisable that these reagents be refrigerated for long-term storage in order to minimize the potential for acid liberation.

As Electrophile in Alkylation Reactions. These three-carbon bifunctional reagents have proven to be very serviceable electrophiles in a variety of settings. Doubly activated anions such as those derived from β-diketones,[4] β-keto esters (eq 1),[5] malonate esters,[6] and phosphono esters[7] are readily homologated by this means. This level of activation is hardly necessary as demonstrated by the fact that anions derived from esters,[8] nitriles (eqs 2 and 3),[9] imines,[10] and isocyanides[11] behave comparably. Generally, the carbonyl group is subsequently unmasked and utilized in further chemical transformations. Where (**2**) and (**3**) are concerned, aqueous acid suffices to accomplish the hydrolysis. Dioxanes derived from (**1**) are much less reactive and often require prior conversion to the dimethyl acetal.[12] Sulfur-containing nucleophiles including α-sulfonyl,[13] thioallyl,[14] and 1,3-dithianyl anions[15] are also readily alkylated. In addition to (**4**) (eq 4), it has proven possible to prepare polyprenols stereoselectively,[14] to produce substituted 1,5-dienes, and to synthesize estrone in a modest number of steps.[15b]

Birch reduction of benzoic acid followed by direct exposure to (**2**) produces (**5**).[16] Bromide displacement by sodium cyclopentadienide, in situ hydrolysis, condensation with $Ph_3P{=}CHCO_2Me$, and ultimately warming to 115 °C to effect intramolecular Diels–Alder cycloaddition delivers (**6**).[17] The means for fusing a pyran ring to an existing oxygenated ring to arrive at a polyether backbone such as (**7**) begins by condensation of (**2**) with a metalated vinyl ether (eq 5).[18]

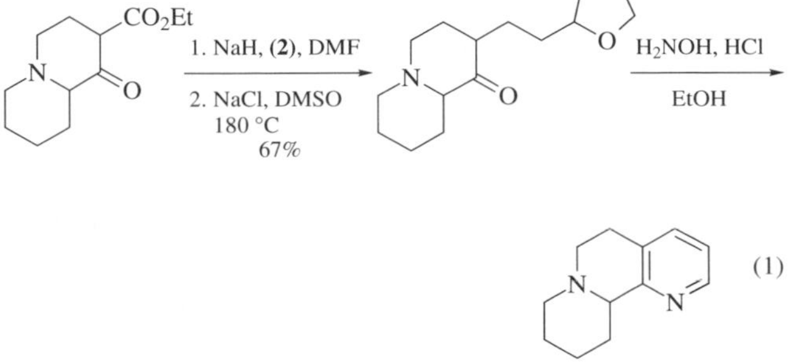

CO_2Et 1. NaH, (2), DMF 2. NaCl, DMSO 180 °C 67% H_2NOH, HCl EtOH (1)

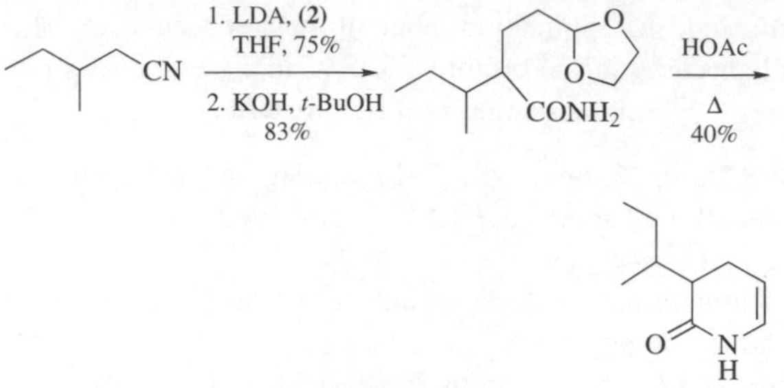

1. LDA, (2) THF, 75%; 2. KOH, t-BuOH 83%; HOAc Δ 40% (2)

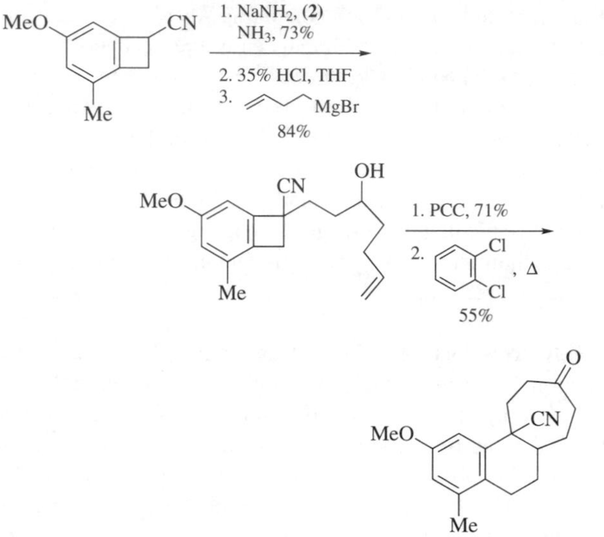

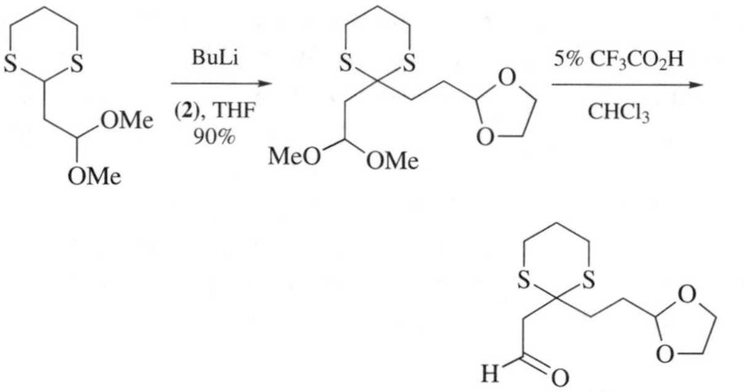

1. $NaNH_2$, (2) NH_3, 73%; 2. 35% HCl, THF; 3. MgBr 84%; 1. PCC, 71%; 2. Cl Cl, Δ 55% (3)

BuLi (2), THF 90%; 5% CF_3CO_2H $CHCl_3$ (4)

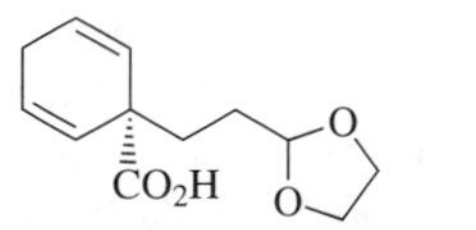

(5) 94%

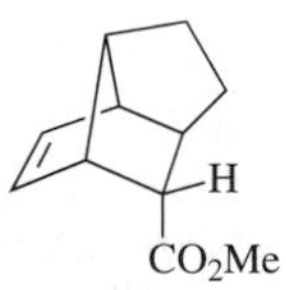

(6) alkylation step, 79%

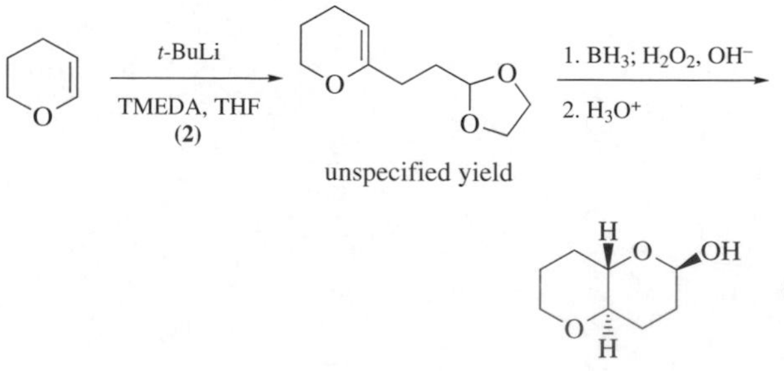

t-BuLi TMEDA, THF (2); unspecified yield; 1. BH_3; H_2O_2, OH^- 2. H_3O^+ (7) (5)

Alkynyl anions have frequently been combined with (**1**)–(**3**).[19] For example, the directness with which (**8**) is constructed underscores the abbreviated route that produces both diastereomers of (**9**) (eq 6).[20] Similarly, the generation of enediyne (**10**) is central to the successful realization of an intramolecular [2+2+2] cycloaddition that gives rise to diene (**11**) (eq 7).[21] Allenyllithiums have proven equally effective (eq 8).[22]

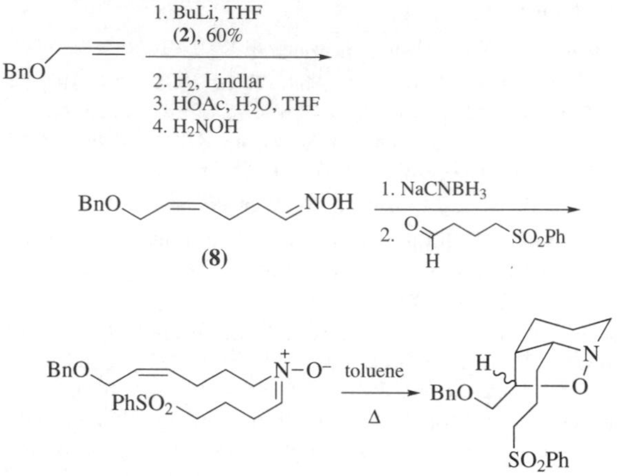

1. BuLi, THF (2), 60%; 2. H_2, Lindlar; 3. HOAc, H_2O, THF; 4. H_2NOH; (8); 1. $NaCNBH_3$ 2. SO_2Ph; toluene Δ (9) (6)

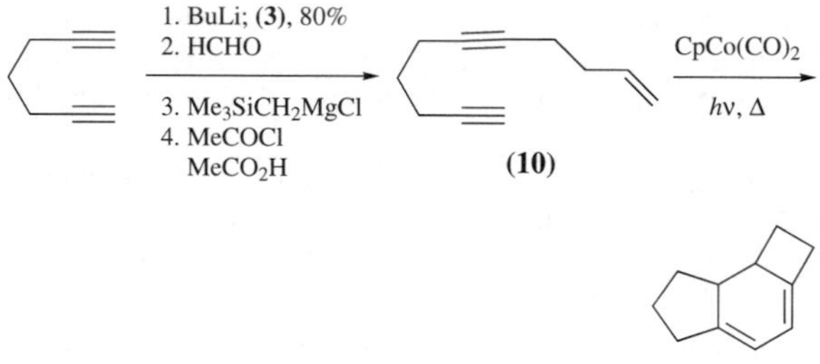

1. BuLi; (3), 80% 2. HCHO 3. Me_3SiCH_2MgCl 4. MeCOCl $MeCO_2H$; (10); $CpCo(CO)_2$ $h\nu$, Δ; (11) (7)

1. (2), THF 2. H_3O^+ 71% (8)

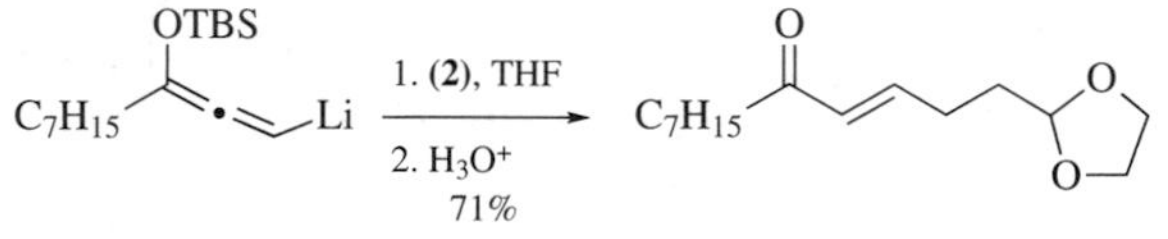

Since the lateral metalation of methyl-substituted aromatics can often be directly implemented, introduction of a functionalized sidechain is made possible by reaction of these anions with (**2**) and (**3**).[23] When dianions such as (**12**) and (**13**) are involved, chemoselectivity is manifested such that C–C bond formation is heavily favored kinetically (eqs 9 and 10).[24,25]

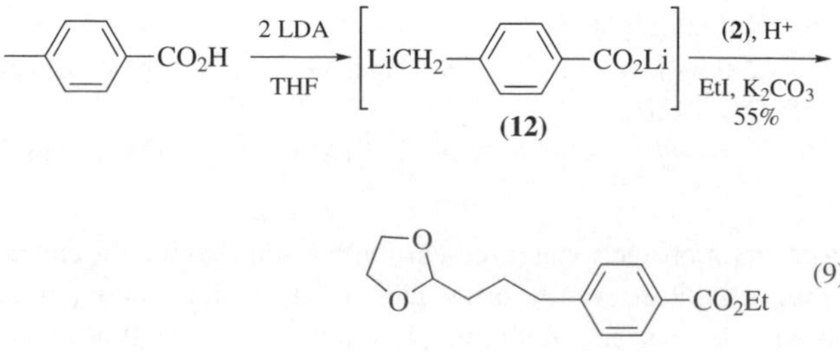

2 LDA THF; (12); (2), H^+ EtI, K_2CO_3 55% (9)

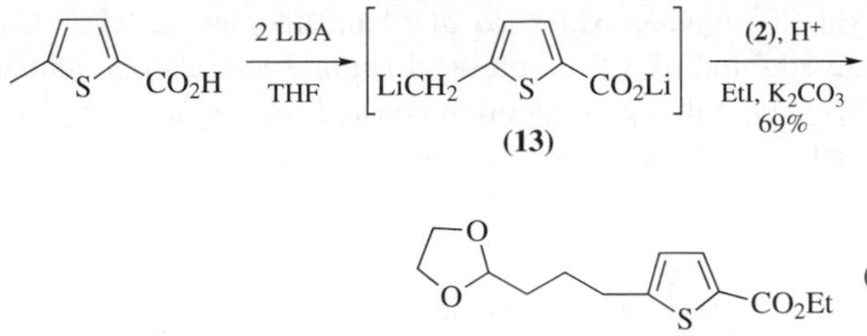

(10)

The (η^5-cyclohexadienylidene)$_2$[Cr(CO)$_3$]$_2$ dianion (**14**) reacts regioselectively with (**2**) to give intermediate (**15**); its further treatment, as shown in eq 11, leads ultimately to the doubly functionalized aromatic product (**16**).[26] The discovery has been made that the condensation of (**2**) with the anion of ethyl phenylsulfinylfluoroacetate leads to the α-fluoro acrylate (**17**) (eq 12).[27]

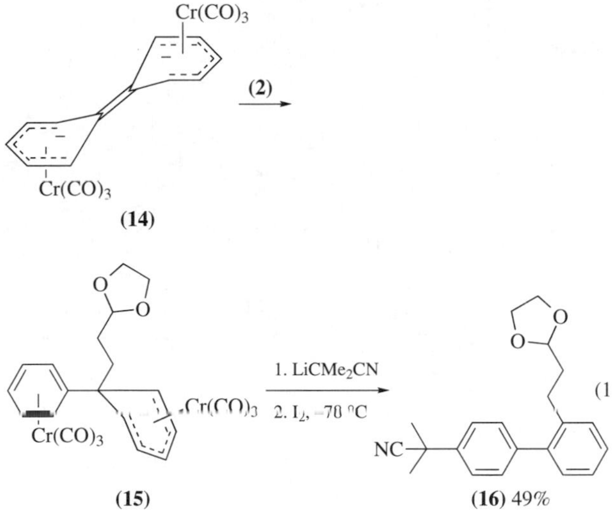

(11)

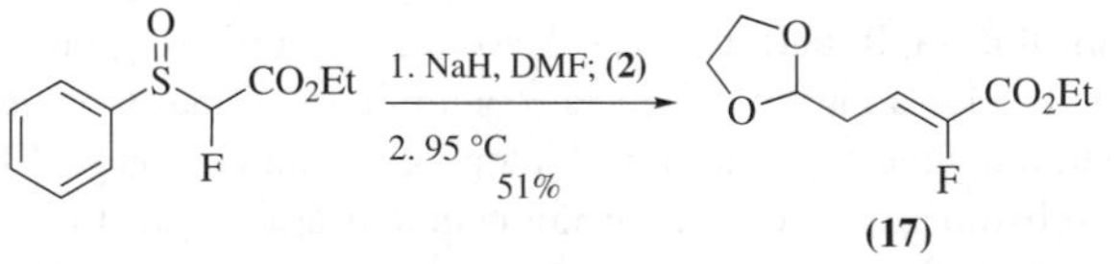

(12)

Alcohols and amines are also capable of condensing with these halo acetals. The conversion of a suitably protected hydroxyproline to (**18**) (eq 13)[28] and of the methylamino derivative to arecolone (**19**) are illustrative (eq 14).[29]

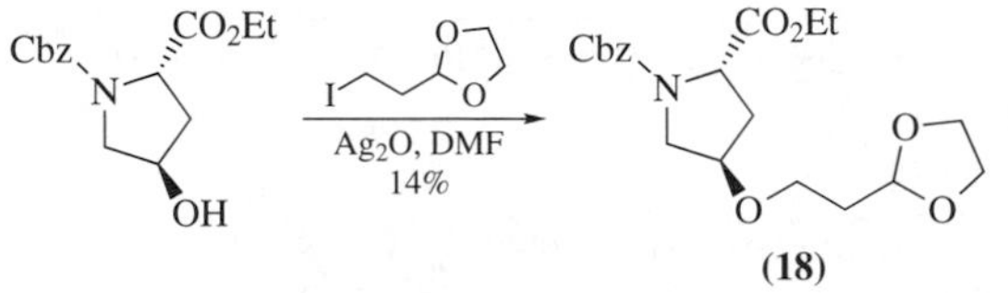

(13)

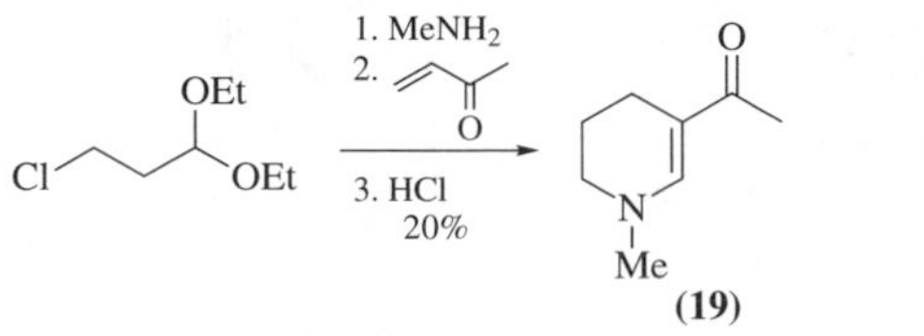

(14)

Use as Grignard Reagents. The Grignard reagents derived from (**1**)–(**3**) exhibit a stability order that parallels their hydrolytic behavior. That derived from (**1**) is thermally stable,[12] while the dimethyl acetal (**3**-MgBr) is too labile to be useful.[30] Since the magnesium reagent derived from (**2**) decomposes at rt and above,[30,31] it must be generated and used below 25 °C. As expected, examples of simple additions of (**1**-MgBr) and (**2**-MgBr) to aldehydes[32] and ketones[33] abound. In the first instance, this process has evolved into a facile means for preparing 5-substituted butyrolactones.[34] The conversion of ketone (**20**) to norbisabolide (**21**) shows that ketones can be analogously processed (eq 15).[35] Clever use has been made of this condensation reaction in a total synthesis of clavukerin A,[36] where Grob fragmentation of adduct (**22**) was utilized to obtain the functionalized 3-cycloheptenone (**23**) (eq 16).

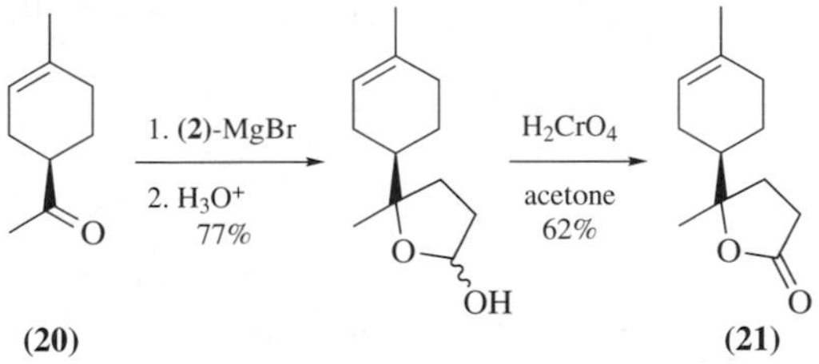

(15)

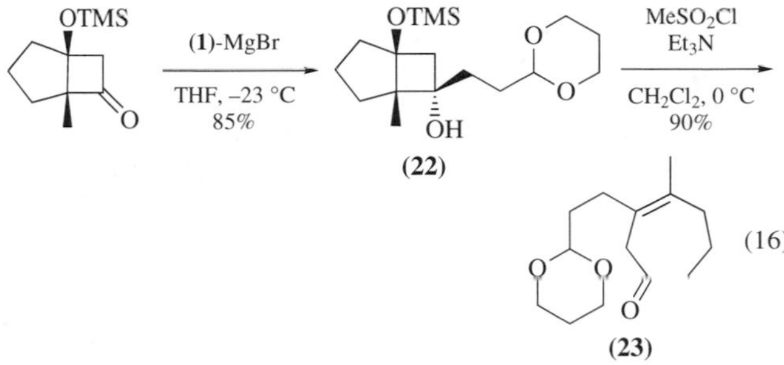

(16)

These Grignard reagents can be condensed directly with activated halides (eq 17).[37] Alternatively, related carbon–carbon bond formation can be realized by promoting the coupling with ***Dilithium Tetrachlorocuprate(II)***[1,2] (eq 18).[38]

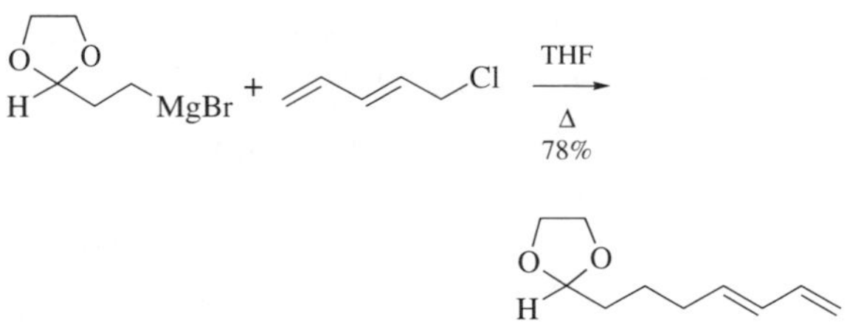

(17)

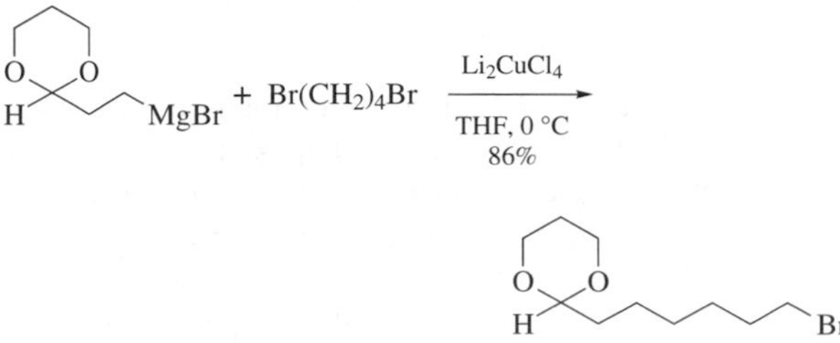

(18)

The presence of acetal oxygen atoms has been recognized to moderate the chemical reactivity of these Grignard reagents. As seen in eq 19, addition to cyclic enones at −78 °C affords principally the product of conjugate addition without the benefit of a Cu^I salt.[39] Additionally, these organometallics are widely known to react with acid chlorides to produce ketones without significant further conversion to tertiary alcohols.[12a,40] The substantial synthetic latitude provided by this process is reflected in eqs 20–22.[41–43]

(2)-MgBr, THF; at 0 °C 1:1; at −78 °C 1:20 73% (19)

(1)-MgBr, THF, −78 °C, 71%; 1. 10% HCl, dioxane, Δ, 82%; 2. $PhCH_2NH_2$, Et_2O, 83% (20)

1. (2)-MgBr; 2. H_3O^+, 81%; 1. $Ph_3P{=}CHCO_2Et$; 2. $Hg(OAc)_2$, NaCl; 3. $NaBH_4$ (21)

1. (2)-MgBr; 2. NaOH, 90%; 1. N-aminophthalimide, HCl; 2. H_2NNH_2 (22)

Other notable uses of (2-MgBr) can be found in syntheses of estafiatin (1,6-addition to tropone),[44] eupolauramine (displacement of an aryl oxazoline),[45] substituted eudistomins (capture by a nitrile group),[46] a tripeptide ACE inhibitor (addition to a thio ester),[47] the 1,7-dioxaspiro[5.5]undecane subunit of milbemycin β_3 (directed oxirane cleavage),[48] and the lipophilic sidechain of the cyclic hexadepsipeptide antibiotic L-156,602 (addition to an α-chloroboronate).[49]

Use as a Cuprate. As noted above (eq 18), (1-MgBr) and (2-MgBr) give coupling reactions with halides and tosylates under conditions of copper catalysis.[38,50] Cuprate formation also results in successful condensation with allylic acetates (eq 23),[51,52] epoxides,[53] vinyl epoxides (eq 24),[54] and alkynes (eq 25).[55] Conjugate addition to α,β-unsaturated ketones can also be realized in this way,[56] this process often constituting the first step of an annulation sequence.

(1)-MgBr, $CuBr{\bullet}Me_2S$, THF, −20 °C; high yield (23)

(2)-MgBr, CuCN, THF, 70% (24)

(3): 1. Mg; 2. $CuBr{\bullet}Me_2S$; 3. HC≡CH; 4. I_2; 42%; 50% HOAc, 91% (25)

Annulation Reactions. The keto acetals formed upon Cu^I-promoted 1,4-addition of acetal-containing Grignard reagents have found wide application in cyclopentene annulation.[57] When no α-substituent is present, the aldol condensation leads to a conjugated bicyclic enone (eq 26).[58] Dehydration is, of course, not spontaneous when an angular group is already in place; in such cases, dehydration must be implemented separately (eq 27).[59]

1. (2)-MgBr, $CuBr{\bullet}Me_2S$, THF, Me_2S, 74%; 2. HCl, aq THF, 77% (26)

(26): 1. (1)-MgBr, $CuBr{\bullet}Me_2S$, THF, 68%; 2. HCl, aq THF, 70%; 97%; 1. p-tolyl chlorothionoformate, py, THF; 2. 200 °C, 25 Torr, 84% (27)

When the site of connectivity is external to an aromatic ring, cyclodehydration results in the elaboration of a six-membered ring.[60] Depending upon the method employed and the degree of substitution, the annulated ring may be singly unsaturated (eq 28)[61] or fully aromatic (eq 29).[62]

1. LDA, (2) THF–HMPA, –78 °C 92%
2. TsOH, MeOH

TsOH C_6H_6, Δ (28) 93%

(2)-MgBr H_3O^+ 97% H_2SO_4 80% (29)

Pyridoannulations are also possible (eq 30).[63] Other nitrogen-containing rings have been elaborated either by radical means (eq 31)[64] or via the involvement of *N*-acyliminium ions (eq 32).[65] The potential for spirocyclic ring construction has been explored as well (eq 33).[66]

1. BuLi; (2), 66%
2. HCl, carbitol Δ, 80% (30)

1. KH, (2), DMF, 90%
2. HCl, aq THF
3. Bu_3SnH, AIBN C_6H_6, Δ
4. Ac_2O, py (31)

1. (2)-MgBr, 58%
2. HCO_2H

(32) 84%

1. LDA, (2), 94%
2. MeLi
3. citric acid, H_2O_2 CH_2Cl_2, 81%

1. $LiAlH_4$
2. H_2, Pd/C
3. BnBr, $NaHCO_3$ DMF
4. Swern; HCl, Δ (33) 60% overall

Allied Compounds. The lithium derivative of (**1**) has been satisfactorily generated by exposure of the iodide to ***t-Butyllithium*** in *n*-pentane–ether (3:2) at −78 °C.[67] Bromo acetals (**1**)–(**3**) are quite amenable to functional group exchange in the manner represented by (**24a–h**).[68–73] Of these, (**24b**)[69] has held particular fascination as a consequence of its ability to serve as a reaction partner in [3 + 2] cycloadditions (eq 34).[69d] The dithianyl acetals of type (**25**) are notably serviceable as cyclohexane annulating agents (eq 35).[72b,74] Ynamine acetal (**26**) has found similar application (eq 36).[75]

(**24a**) X = CO_2H (**24b**) X = NO_2 (**24c**) X = NH_2 (**24d**) X = CN
(**24e**) X = CH_2NH_2 (**24f**) X = $^+PPh_3Br^-$ (**24g**) X = SO_2Ph (**24h**) X = $SnMe_3$

(**25**)

(**24b**) MeNCO, Et_3N C_6H_6, 70 °C 68% (34)

1. (**25**), BuLi THF, HMPA, 56%
2. HCl, MeOH, 86% (35)

MeCN, Δ
1. NaOH, H_2O
2. 5% HCO_2H (36) 35% overall

Related Reagents. 2-(2-Bromoethyl)-2-methyl-1,3-dioxolane.

---

1. Stowell, J. C. *CRV* **1984**, *84*, 409.
2. Stowell, J. C.; Keith, D. R.; King, B. T. *OS* **1984**, *62*, 140 and references cited therein.
3. Feringa, B. L. *SC* **1985**, *15*, 87.
4. (a) Sliskovic, D. R.; Roth, B. D.; Wilson, M. W.; Hoefle, M. L.; Newton, R. S. *JMC* **1990**, *33*, 31. (b) Cheney, D. L.; Paquette, L. A. *JOC* **1989**, *54*, 3334.
5. (a) Das, T. K.; Dutta, P. C. *SC* **1976**, *6*, 253. (b) Huff, J. R., et al. *JMC* **1988**, *31*, 641.

6. Roush, W. R.; Myers, A. G. *JOC* **1981**, *46*, 1509.
7. Hammond, G. B.; Cox, M. B.; Wiemer, D. F. *JOC* **1990**, *55*, 128.
8. Grigg, R.; Markandu, J.; Perrior, T.; Surendrakumar, S.; Warnock, W. J. *T* **1992**, *48*, 6929.
9. (a) Stork, G.; Ozorio, A. A.; Leong, A. Y. W. *TL* **1978**, 5175. (b) Kametani, T.; Matsumoto, H.; Honda, T.; Fukumoto, K. *TL* **1980**, *21*, 4847. (c) Kametani, T.; Honda, T.; Shiratori, Y.; Matsumoto, H.; Fukumoto, K. *JCS(P1)* **1981**, 1386. (d) Shishido, K.; Hiroya, K.; Fukumoto, K.; Kametani, T. *TL* **1986**, *27*, 971. (e) Laredo, G. C.; Maldonado, L. A. *H* **1987**, *25*, 179. (f) Zhu, J.; Quirion, J.-C.; Husson, H.-P. *TL* **1989**, *30*, 6323. (g) Chelucci, G.; Giacomelli, G. *JHC* **1990**, *27*, 307.
10. (a) Chelucci, G.; Soccolini, F.; Botteghi, C. *SC* **1985**, *15*, 807. (b) Grigorieva, N. Ya.; Yudina, O. N.; Moiseenkov, A. M. *S* **1989**, 591.
11. Rama Rao, A. V.; Deshpande, V. H.; Pulla Reddy, S. *SC* **1984**, *14*, 469.
12. (a) Stowell, J. C. *JOC* **1976**, *41*, 560. (b) Stowell, J. C.; Keith, D. R. *S* **1979**, 132.
13. (a) Simpkins, N. S. *T* **1991**, *47*, 323. (b) Grigg, R.; Dorrity, M. J.; Heaney, F.; Malone, J. F.; Rajviroongit, S.; Sridharan, V.; Surendrakumar, S. *T* **1991**, *47*, 8297.
14. Grigorieva, N. Ya.; Avrutov, I. M.; Semenovsky, A. V. *TL* **1983**, *24*, 5531.
15. (a) Ellison, R. A.; Lukenbach, E. R.; Chiu, C.-w. *TL* **1975**, 499. (b) Jung, M. E.; Halweg, K. M. *TL* **1984**, *25*, 2121.
16. Chuang, C.-P.; Gallucci, J. C.; Hart, D. J. *JOC* **1988**, *53*, 3210.
17. Nickon, A.; Stern, A. G. *TL* **1985**, *26*, 5915.
18. Kozikowski, A. P.; Ghosh, A. K. *JOC* **1985**, *50*, 3017.
19. (a) Johnson, W. S.; Hughes, L. R.; Kloek, J. A.; Niem, T.; Shenvi, A. *JACS* **1979**, *101*, 1279. (b) Trost, B. M.; Runge T. A. *JACS* **1981**, *103*, 7559. (c) Yadav, J. S.; Chander, M. C.; Joshi, V. B. *TL* **1988**, *29*, 2737.
20. Holmes, A. B.; Hughes, A. B.; Smith, A. L.; Williams, S. F. *JCS(P1)* **1992**, 1089.
21. Dunach, E.; Halterman, R. L.; Vollhardt, K. P. C. *JACS* **1985**, *107*, 1664.
22. Matsuoka, R.; Horiguchi, Y.; Kuwajima, I. *TL* **1987**, *28*, 1299.
23. (a) Della Ciana, L.; Hamachi, I.; Meyer, T. J. *JOC* **1989**, *54*, 1731. (b) Bigge, C. F.; Drummond, J. T.; Johnson, G.; Malone, T.; Probert, Jr., A. W.; Marcoux, F. W.; Coughenour, L. L.; Brahce, L. J. *JMC* **1989**, *32*, 1580.
24. Bicking, J. B.; Bock, M. G.; Cragoe, Jr., E. J.; Di Pardo, R. M.; Gould, N. P.; Holtz, W. J.; Lee, T.-J.; Robb, C. M.; Smith, R. L.; Springer, J. P.; Blaine, E. H. *JMC* **1983**, *26*, 342.
25. Gould, N. P.; Lee, T.-J. *JOC* **1980**, *45*, 4528.
26. Yang, S. S.; Dawson, B. T.; Rieke, R. D. *TL* **1991**, *32*, 3341.
27. Allmendinger, T. *T* **1991**, *47*, 4905.
28. Mencel, J. J.; Regan, J. R.; Barton, J.; Menard, P. R.; Bruno, J. G.; Calvo, R. R.; Kornberg, B. E.; Schwab, A.; Neiss, E. S.; Suh, J. T. *JMC* **1990**, *33*, 1606.
29. (a) Ward, J. S.; Merritt, L. *JHC* **1990**, *27*, 1709. (b) Wohl, A.; Prill, A. *LA* **1924**, *440*, 139.
30. Feugeas, Cl. *BSF(2)* **1963**, 2568.
31. (a) Ponaras, A. A. *TL* **1976**, 3105. (b) Eaton, P. E.; Mueller, R. H.; Carlson, G. R.; Cullison, D. A.; Cooper, G. F.; Chou, T.-C.; Krebs, E.-P. *JACS* **1977**, *99*, 2751.
32. (a) Mann, J.; Shervington, L. A. *JCS(P1)* **1991**, 2961. (b) Hauser, F. M.; Hewawasam, P.; Rho, Y. S. *JOC* **1989**, *54*, 5110. (c) Paquette, L. A.; Wiedeman, P. E.; Bulman-Page, P. C. *JOC* **1988**, *53*, 1441. (d) Collins, P. W.; Kramer, S. W.; Gullikson, G. W. *JMC* **1987**, *30*, 1952. (e) Kelly, T. R.; Kaul, P. N. *JOC* **1983**, *48*, 2775. (f) Gradnig, G.; Berger, A.; Grassberger, V.; Stütz, A. E.; Legler, G. *TL* **1991**, *32*, 4889. (g) Eberle, M. K.; Weber, H.-P. *JOC* **1988**, *53*, 231. (h) Roush, W. R.; Gillis, H. R.; Ko, A. I. *JACS* **1982**, *104*, 2269. (i) Roush, W. R. *JACS* **1980**, *102*, 1390. (j) Pattenden, G.; Whybrow, D. *JCS(P1)* **1981**, 1046.
33. (a) Büchi, G.; Wüest, H. *JOC* **1969**, *34*, 1122. (b) Dittami, J. P.; Nie, X. Y.; Nie, H.; Ramanathan, H.; Breining, S.; Bordner, J.; Decosta, D. L.; Kiplinger, J.; Reiche, P.; Ware, R. *JOC* **1991**, *56*, 5572. (c) Fukumoto, K.; Suzuki, K.; Nemoto, H.; Kametani, T.; Furuyama, H. *T* **1982**, *38*, 3701. (d) Hatam, N. A. R.; Whiting, D. A. *TL* **1978**, 5145.
34. (a) Loozen, H. J. J.; Godefroi, E. F.; Besters, J. S. M. M. *JOC* **1975**, *40*, 892. (b) Kawashima, M.; Fujisawa, T. *BCJ* **1988**, *61*, 3377.
35. Feldstein, G.; Kocienski, P. J. *SC* **1977**, *7*, 27.
36. Kim, S. K.; Pak, C. S. *JOC* **1991**, *56*, 6829.
37. Gras, J.-L.; Bertrand, M. *TL* **1979**, 4549.
38. (a) Sato, T.; Naruse, K.; Fujisawa, T. *TL* **1982**, *23*, 3587. (b) Volkmann, R. A.; Davis, J. T.; Meltz, C. N. *JOC* **1983**, *48*, 1767.
39. Sworin, M.; Neumann, W. L. *TL* **1987**, *28*, 3217.
40. (a) Stowell, J. C.; Hauck, Jr., H. F. *JOC* **1981**, *46*, 2428. (b) Almquist, R. G.; Olsen, C. M.; Uyeno, E. T.; Toll, L. *JMC* **1984**, *27*, 115. (c) Boga, C.; Savoia, D.; Trombini, C.; Umani-Ronchi, A. *S* **1986**, 212. (d) Kruse, C. G.; Bouw, J. P.; van Hes, R.; van de Kuilen, A.; den Hartog, J. A. J. *H* **1987**, *26*, 3141. (e) Fiandanese, V.; Marchese, G.; Naso, F. *TL* **1988**, *29*, 3587. (f) Haynes, R. K.; Vonwiller, S. C.; Stokes, J. P.; Merlino, L. M. *AJC* **1988**, *41*, 881.
41. Kiyooka, S.-i.; Sekimura, Y.; Kawaguchi, K. *S* **1988**, 745.
42. Nishiyama, T.; Woodhall, J. F.; Lawson, E. N.; Kitching, W. *JOC* **1989**, *54*, 2183.
43. Bolós, J.; Pérez, Á.; Gubert, S.; Anglada, L.; Sacristán, A.; Ortiz, J. A. *JOC* **1992**, *57*, 3535.
44. Rigby, J. H.; Wilson, J. Z. *JACS* **1984**, *106*, 8217.
45. Levin, J. I.; Weinreb, S. M. *JOC* **1984**, *49*, 4325.
46. Rinehart, Jr., K. L.; Kobayashi, J.; Harbour, G. C.; Gilmore, J.; Mascal, M.; Holt, T. G.; Shield, L. S.; Lafargue, F. *JACS* **1987**, *109*, 3378.
47. Almquist, R. G.; Chao, W.-R.; Ellis, M. E.; Johnson, H. L. *JMC* **1980**, *23*, 1392.
48. Kocienski, P. J.; Street, S. D. A.; Yeates, C.; Campbell, S. F. *JCS(P1)* **1987**, 2171.
49. Caldwell, C. G.; Rupprecht, K. M.; Bondy, S. S.; Davis, A. A. *JOC* **1990**, *55*, 2355.
50. Stowell, J. C.; King, B. T. *S* **1984**, 278.
51. Sugahara, T.; Iwata, T.; Yamaoka, M.; Takano, S. *TL* **1989**, *30*, 1821.
52. Curran, D. P.; Chen, M.-H. *TL* **1985**, *26*, 4991.
53. Fujisawa, T.; Itoh, T.; Nakai, M.; Sato, T. *TL* **1985**, *26*, 771.
54. Liu, D.; Stuhmiller, L. M.; McMorris, T. C. *JCS(P1)* **1988**, 2161.
55. Cooke, Jr., M. P.; Widener, R. K. *JOC* **1987**, *52*, 1381.
56. Lipshutz, B. H.; Parker, D. A.; Nguyen, S. L.; McCarthy, K. E.; Barton, J. C.; Whitney, S. E.; Kotsuki, H. *T* **1986**, *42*, 2873.
57. (a) Marfat, A.; Helquist, P. *TL* **1978**, 4217. (b) Heathcock, C. H.; Tice, C. M.; Germroth, T. C. *JACS* **1982**, *104*, 6081. (c) Tsunoda, T.; Kodama, M.; Itô, S. *TL* **1983**, *24*, 83. (d) Paquette, L. A.; Leone-Bay, A. *JACS* **1983**, *105*, 7352. (e) Snider, B. B.; Faith, W. C. *JACS* **1984**, *106*, 1443. (f) Paquette, L. A.; Roberts, R. A.; Drtina, G. J. *JACS* **1984**, *106*, 6690. (g) Matlin, A. R.; Agosta, W. C. *JCS(P1)* **1987**, 365. (h) de Almeida Barbosa, L.-C.; Mann, J. *JCS(P1)* **1992**, 337. (i) Weyerstahl, P.; Brendel, J. *LA* **1992**, 669. (j) Franck-Neumann, M.; Miesch, M.; Lacroix, E.; Metz, B.; Kern, J.-M. *T* **1992**, *48*, 1911. (k) Wang, X.; Paquette, L. A. *TL* **1993**, *34*, 4579.
58. Bal, S. A.; Marfat, A.; Helquist, P. *JOC* **1982**, *47*, 5045.
59. Leone-Bay, A.; Paquette, L. A. *JOC* **1982**, *47*, 4173.
60. (a) Loozen, H. J. J.; Godefroi, E. F. *JOC* **1973**, *38*, 3495. (b) Loozen, H. J. J. *JOC* **1975**, *40*, 520. (c) Hatam, N. A. R.; Whiting, D. A. *JCS(P1)* **1982**, 461. (d) Teague, S. J.; Roth, G. P. *S* **1986**, 427. (e) Muratake, H.; Natsume, M. *H* **1989**, *29*, 783. (f) Rice, J. E.; He, Z.-M. *JOC* **1990**, *55*, 5490. (g) He, Z.-M.; Rice, J. E.; La Voie, E. J. *JOC* **1992**, *57*, 1784.
61. Honda, T.; Yamamoto, A.; Cui, Y.; Tsubuki, M. *JCS(P1)* **1992**, 531.
62. Muratake, H.; Natsume, M. *H* **1990**, *31*, 683.
63. (a) Chelucci, G.; Delogu, G.; Gladiali, S.; Soccolini, F. *JHC* **1986**, *23*, 1395. (b) Chelucci, G.; Gladiali, S.; Marchetti, M. *JHC* **1988**, *25*, 1761. (c) Chelucci, G.; Cossu, S.; Scano, G.; Soccolini, F. *H* **1990**, *31*, 1397.

64. Knapp, S.; Gibson, F. S. *JOC* **1992**, *57*, 4802.
65. Hart, D. J.; Yang, T.-K. *JOC* **1985**, *50*, 235.
66. Zhu, J.; Royer, J.; Quirion, J.-C.; Husson, H.-P. *TL* **1991**, *32*, 2485.
67. Bailey, W. F.; Punzalan, E. R. *JOC* **1990**, *55*, 5404.
68. Shea, K. J.; Wada, E. *JACS* **1982**, *104*, 5715.
69. (a) Crumbie, R. L.; Nimitz, J. S.; Mosher, H. S. *JOC* **1982**, *47*, 4040. (b) Rosini, G.; Ballini, R.; Petrini, M.; Sorrenti, P. *T* **1984**, *40*, 3809. (c) Kozikowski, A. P.; Li, C.-S. *JOC* **1985**, *50*, 778. (d) Schow, S. R.; Bloom, J. D.; Thompson, A. S.; Winzenberg, K. N.; Smith, III A. B. *JACS* **1986**, *108*, 2662. (e) Curran, D. P.; Heffner, T. A. *JOC* **1990**, *55*, 4585.
70. Gribble, G. W.; Switzer, F. L.; Soll, R. M. *JOC* **1988**, *53*, 3164.
71. Bazureau, J. P.; Person, D.; Le Corre, M. *TL* **1989**, *30*, 3065.
72. (a) Gaoni, Y.; Tomazic, A.; Potgieter, E. *JOC* **1985**, *50*, 2943. (b) St. Laurent, D. R.; Paquette, L. A. *JOC* **1986**, *51*, 3861. (c) LeBel, N. A.; Balasubramanian, N. *JACS* **1989**, *111*, 3363.
73. Lee, T. V.; Richardson, K. A.; Ellis, K. L.; Visani, N. *T* **1989**, *45*, 1167.
74. (a) Thomas, J. A.; Heathcock, C. H. *TL* **1980**, *21*, 3235. (b) Rosen, T.; Taschner, M. J.; Thomas, J. A.; Heathcock, C. H. *JOC* **1985**, *50*, 1190.
75. Ficini, J.; Guingant, A.; d'Angelo, J.; Stork, G. *TL* **1983**, *24*, 907.

Leo A. Paquette
The Ohio State University, Columbus, OH, USA

Bromoform

[75-25-2] $CHBr_3$ (MW 252.73)

(used in the synthesis of mono- and dibromocyclopropanes,[1] allenes[2] and cumulenes,[3] cyclopentadienes and fulvenes, α- and β-bromo-α,β-unsaturat ed carbonyl compounds, 1,1,3-tribromides, tribromomethyllithium carbonyl adducts, isocyanides,[4] α-hydroxy and α-aminoaryl acetic acids)

Alternate Name: tribromomethane.
Physical Data: mp 8.5 °C; bp 146–150 °C; *d* 2.827 g cm^{-3}.
Solubility: sol acetone, benzene, chloroform, ethanol, ether, petroleum ether.
Form Supplied in: commercially available with inhibitors present.
Analysis of Reagent Purity: technical (94%) and high purity (99+%).
Preparative Method: by reaction of acetone with sodium hypobromite.
Purification: wash with concentrated sulfuric acid and then water. The liquid is dried over $CaCl_2$ or K_2CO_3 then fractionally distilled.
Handling, Storage, and Precautions: store in closed containers away from light. Diphenylamine and ethanol are commonly used inhibitors. Do not breathe vapors as addiction may occur. Suspected cancer agent; lachrymator. Use in a fume hood.

Dibromocyclopropanes.[5] Bromoform is a routinely used precursor to dibromocarbene which can be generated by many different methods (eq 1).[6] Dibromocarbene adds stereospecifically to alkenes[7] and its reactivity towards alkyl substituted alkenes is analogous to that of dichlorocarbene: tetrasubstituted > trisubstituted > unsymmetrical disubstituted > symmetrical disubstituted > monosubstituted alkenes.[8] Dibromocarbene has been prepared by treatment of bromoform with ***Potassium t-Butoxide***.[9] The yields of dibromocyclopropanes obtained using this method are typically no more than 70% for good cases and are usually lower. For example, reaction of cyclooctene with bromoform and potassium *t*-butoxide gave a 52–65% yield of dibromide (eq 2).[10] Dibromocarbene has also been generated by adsorbing a mixture of alkene and bromoform on basic ***Alumina***.[11] At elevated temperatures, ***Potassium Carbonate*** can generate dibromocarbene from bromoform in the presence of ***18-Crown-6***. Cycloheptatriene at 140 °C with these reagents gave 1-bromobenzocyclobutane in 18–45% yield (eq 3).[12]

$$CHBr_3 + :B \rightleftharpoons {}^{-}CBr_3 + BH \qquad {}^{-}CBr_3 \rightleftharpoons :CBr_2 + Br^{-} \quad (1)$$

$$\text{cyclooctene} \xrightarrow[\text{KO-}t\text{-Bu, 52–65\%}]{CHBr_3} \text{9,9-dibromobicyclo[6.1.0]nonane} \quad (2)$$

$$\text{cycloheptatriene} \xrightarrow[\text{18-crown-6, 140 °C}]{CHBr_3,\ K_2CO_3} [\text{dibromocyclopropane adduct}] \longrightarrow \text{1-bromobenzocyclobutane (18–45\%)} \quad (3)$$

A significant improvement in the rate and yield of halocyclopropanes is realized if phase transfer catalysts[13] like triethylbenzylammonium chloride are used.[14] Yields for dibromocyclopropanes, however, while generally less than 50%, can be somewhat improved if a considerable excess of bromoform and 24–96 h reaction times are used.[15] Addition of a small amount of ethanol to the reaction mixture often results in 10–30% increase in the yield of dibromocyclopropanes. In a typical procedure, 50% aqueous NaOH is added over 10 min to a stirred mixture of bromoform, alkene, triethylbenzylammonium chloride (TEBA) and a small amount of ethanol at 40–45 °C. After 3 h at the same temperature the mixture is diluted with dichloromethane and poured into water. An extractive work-up and concentration of the organic phases provides the crude product which is typically purified by distillation. For this method a two- to three-fold molar excess of bromoform to alkene or the converse is ordinarily used.[16]

It has been observed that the nature of the phase transfer catalyst (PTC) can have a profound effect on the yield of product obtained.[17] For example, reaction of allylic bromides with bromoform can give either cyclopropanes or substitution products depending on the type of catalyst used (eq 4). When the phase transfer catalyst cetrimide $[(C_{16}H_{33})NMe_3Br]$ is used, the cyclopropane product predominates. The product ratio is reversed when tetraphenylarsonium chloride (Ph_4AsCl) is used as the catalyst and the substitution pathway is preferred.[18] When α,β-unsaturated carbonyls are used, the choice of phase transfer catalyst also impacts the product distribution obtained between the cyclopropyl product and that of Michael addition.[19] For systems where only ring formation is expected, the reaction selectivities between different alkenes are not affected by alteration of the phase transfer catalyst.[20] The use of an optically active phase transfer catalyst has led to unprecedented but low levels of asymmetric induction.[21]

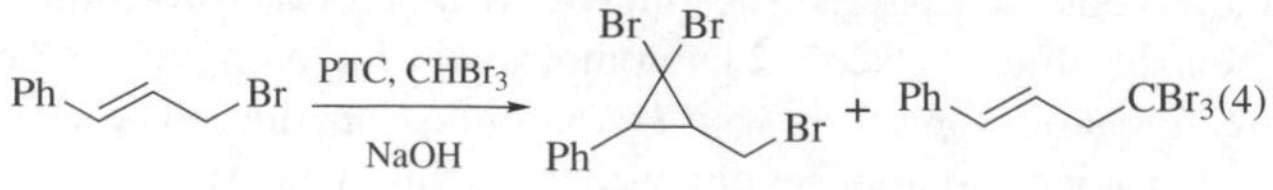

PTC	Product ratio cyclopropane:substitution
cetrimide	92:1
Ph_4AsCl	1:91

Dibromocarbene is also made via Seyferth's reagent.[22] The reagent, ***Phenyl(tribromomethyl)mercury***, is typically prepared from bromoform, a phenylmercury(II) halide, and potassium *t*-butoxide.[23] Another procedure for the synthesis of phenyl(tribromomethyl)mercury[24] using a phase transfer catalyst, ***Potassium Fluoride***, and ***Sodium Hydroxide*** has also been reported.[25] Using these types of methods, the carbene is released with elimination of phenylmercury(II) halide.[26] The ring expansion of 1-methylindenes into 3-bromo-1-methylnaphthalenes is accomplished in moderate yields (≈55%) by reaction of the dibromocarbene released from this type of mercury reagent (eq 5). The same reaction using bromoform/potassium *t*-butoxide gave a 2-bromo-1-methylnaphthalene in only 14% yield.[27] An alternative to other straightforward syntheses of monobromocyclopropanes[28] from alkenes (eq 6) uses ***Diethylzinc-Bromoform–Oxygen***.[29]

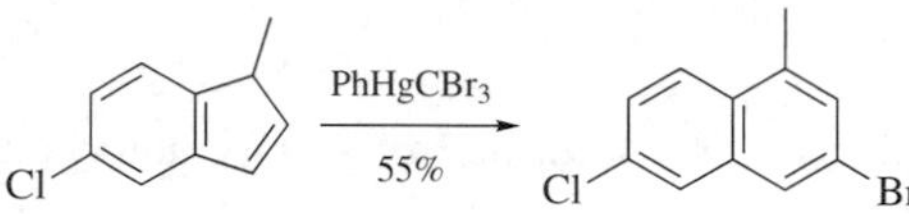

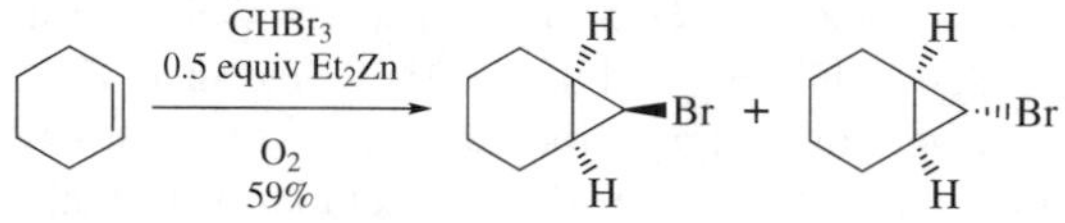

syn:*anti* = 62:38

Once having obtained the dibromocyclopropanes resulting from addition of dibromocarbene to different types of double bonds, chemists have been able to prepare a myriad of different structures (see also ***Chloroform*** and ***Iodoform***).

Allenes and Cumulenes. Allenes have been prepared by reaction of dibromocyclopropanes with ***Magnesium***, but in low yield.[30] The use of alkyllithiums leads to improved yields with ***Methyllithium*** being the reagent of choice.[31] 9,9-Dibromobicyclo[6.1.0]nonane is converted to 1,2-cyclononadiene with methyllithium (eq 7)[10,32,33] while the lower homolog, 8,8-dibromobicyclo[5.1.0]octane, leads to a complex mixture of products.[34] This is due, in large part, to intramolecular carbenoid insertions competing with allene formation. This property has been used to gain entry into the tricyclo[4.1.0.0^{2,7}]heptene ring system (eq 8).[35]

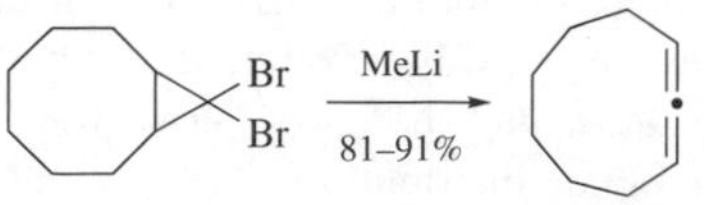

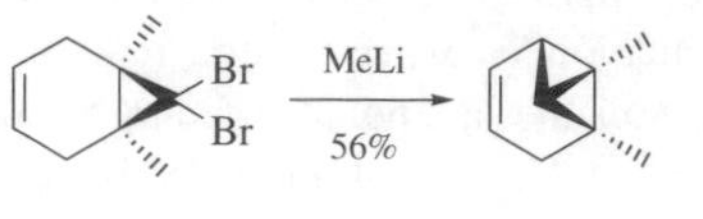

Reaction of dibromocarbene with allenes leads to cumulenes upon treatment of the intermediate dibromocyclopropane with methyllithium.[36,37] An example of the utility of this approach is shown in eq 9.[38,39]

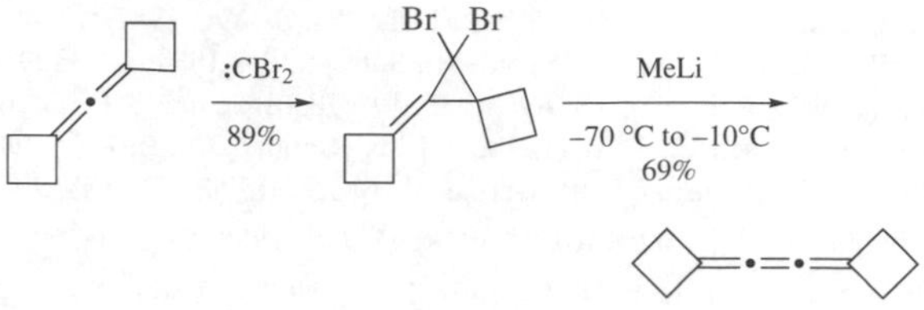

Cyclopentadienes and Fulvenes. The Skattebøl rearrangement occurs when conjugated dienes are converted into 1,1-dibromo-2-vinylcyclopropanes and then treated with MeLi at −78 °C. The major products are typically cyclopentadienes with small amounts of vinylallenes (eq 10). The ratio of products vary with the substitution on the starting vinylcyclopropane.[40] An example of the Skattebøl rearrangement in the synthesis of an optically active cyclopentadiene is shown in eq 11.[41,42]

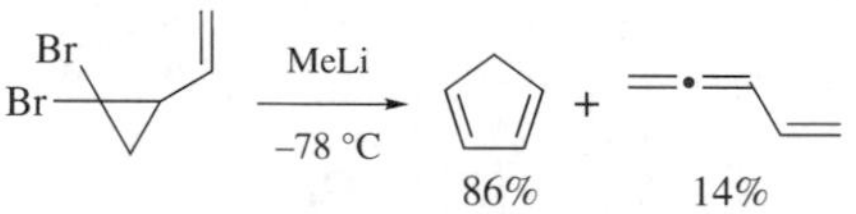

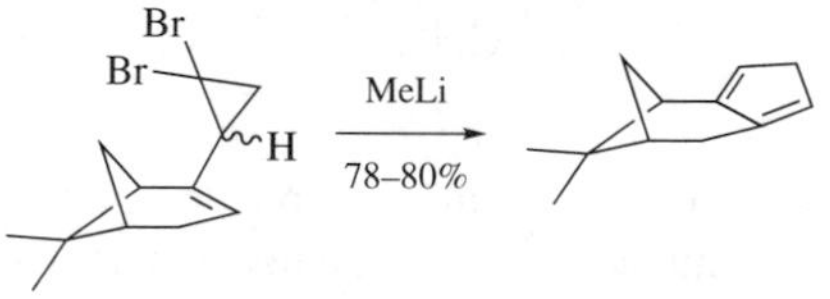

Similar treatment of 3,3,3′3′-tetramethyl-2,2,2′,2′-tetrabromobicycl opropyl with methyllithium yields a diallene as the major product (eq 12). When the dimethylbicyclopropyl analog is used, a fulvene is the major product with a diallene being a minor product (eq 13). This change in product distribution is believed to result from the difference in rates of allene formation versus that of rearrangement to a cyclopentylcarbene.[40,43]

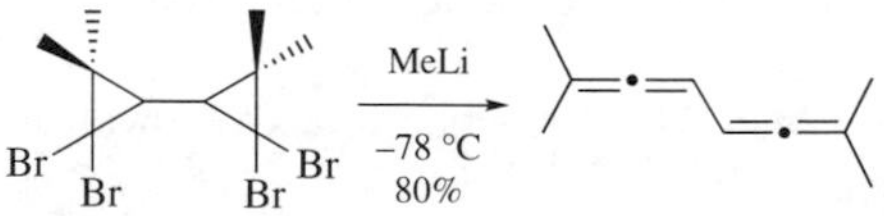

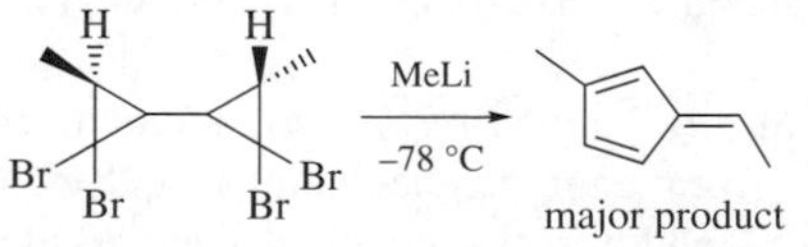

α-Bromo-α,β -Unsaturated Carbonyl Compounds. Addition of dibromocarbene to a silyl enol ether has been shown to generate α-bromo-α,β-unsatura ted carbonyl compounds after thermal or acid-catalyzed rearrangement of the intermediate dibromocyclopropane (eq 14).[44,45]

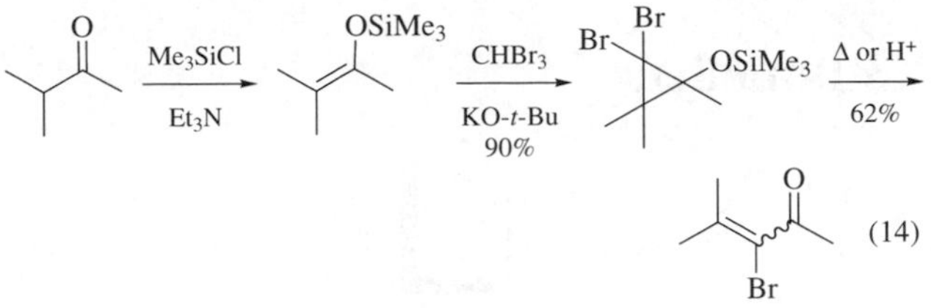

$$\text{(14)}$$

β-Bromo-α,β- Unsaturated Esters. Reaction of unsubstituted ketene trialkylsilyl acetals with bromoform and a catalytic amount of ***Triethylborane*** affords 3-bromoacrylate (eq 15). The same conditions using a substituted ketene acetal provides a 2-dibromomethyl ester (eq 16).[46]

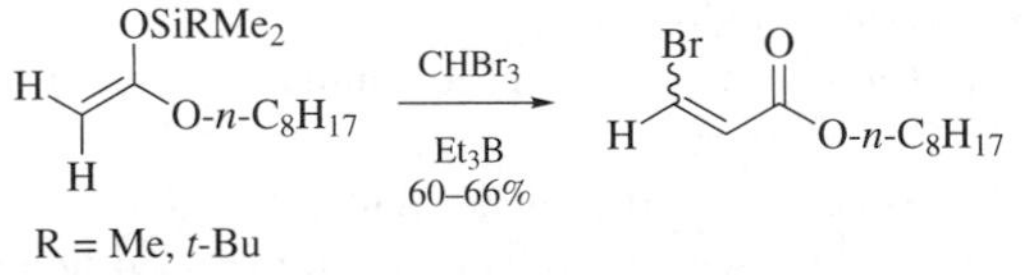

$$\text{(15)}$$

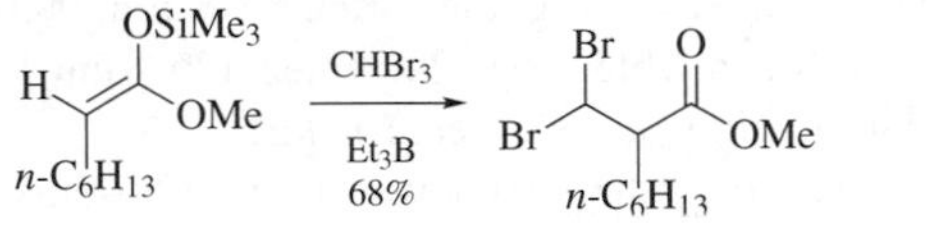

R = Me, *t*-Bu

$$\text{(16)}$$

1,1,3-Tribromides. Unlike chloroform, which adds under radical conditions across the double bond through the C–H bond,[47] bromoform typically reacts through one of the C–Br bonds (eq 17),[48] although mixtures of products are often obtained.[49]

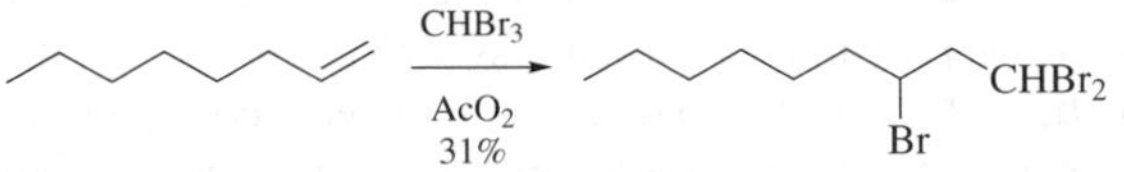

$$\text{(17)}$$

Tribromomethyllithium. Bromoform is converted in situ into ***Tribromomethyllithium*** upon treatment with lithium dicyclohexylamide. The anion reacts with cyclohexanone to afford the corresponding tribromomethyl alcohol (eq 18).[50]

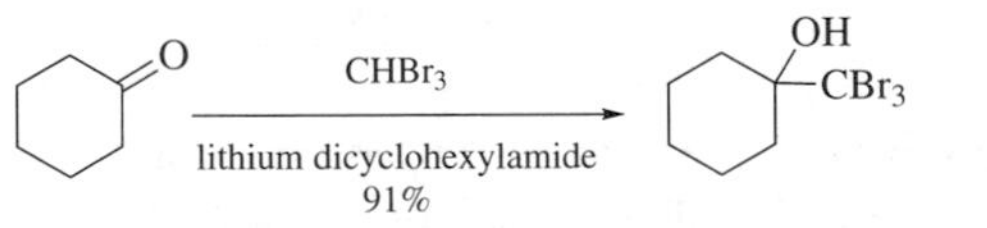

$$\text{(18)}$$

Isocyanides. The use of phase transfer catalysis in the Hofmann isocyanide synthesis[51] is reported to result in improved yields.[52] Either chloroform or bromoform can be used. Bromoform is often used in the preparation of low-boiling isocyanides since the products can be easily removed from the reaction mixture (eq 19).

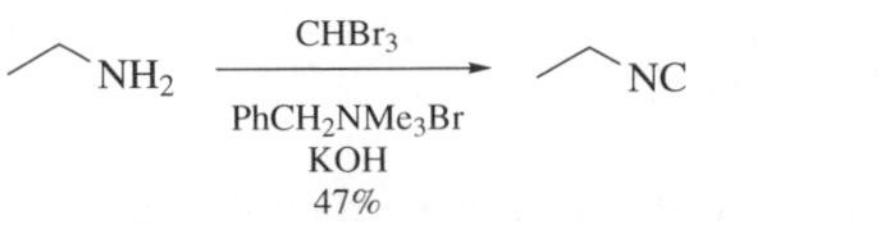

$$\text{(19)}$$

α-Hydroxyarylacetic and α-Aminoarylacetic Acids. A one-pot synthesis of α-hydroxyarylacetic acids from bromoform, an aryl aldehyde, and ***Potassium Hydroxide*** occurs with ***Lithium Chloride*** as catalyst (eq 20).[53] By a similar process, α-aminoarylacetic acids can be made using bromoform, an aryl aldehyde, and ***Ammonia*** with potassium hydroxide and ***Lithium Amide*** as catalyst. This method reportedly gives higher yields than other single-pot procedures.[54]

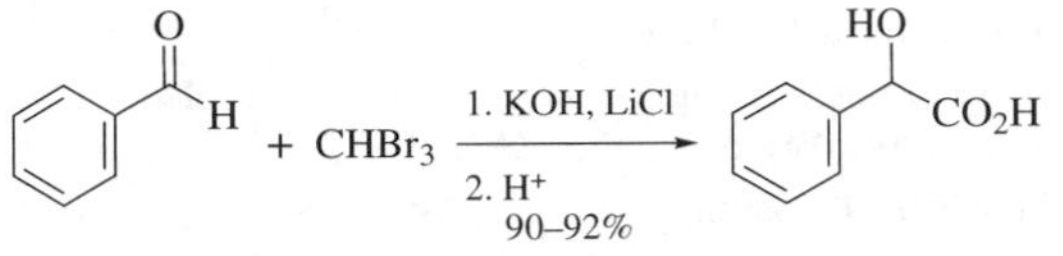

$$\text{(20)}$$

Related Reagents. Chloroform; Diethylzinc-Bromoform-Oxygen; Iodoform; Phenyl(tribromomethyl)mercury; Phenyl(trichloromethyl)mercury.

1. (a) Chinoporos, E. *CRV* **1963**, *63*, 235. (b) Parham, W. E.; Schweizer, E. E. *OR* **1963**, *13*, 55.
2. *The Chemistry of the Allenes*; Landor, S. R., Ed.; Academic: New York, 1983.
3. Brandsma, L.; Verkruijsse, H. D. *Synthesis of Acetylenes, Allenes and Cumulenes; A Laboratory Manual*; Studies in Organic Chemistry, Vol. 8; Elsevier: New York, 1981.
4. Ugi, I. *Isonitrile Chemistry*; Academic: New York, 1971.
5. For reviews of carbenes in general, see (a) Kirmse, W. E. *Carbene Chemistry*, 2nd ed.; Academic: New York, 1971. (b) Hine, J. *Divalent Carbon*; Ronald: New York, 1964.
6. Dehmlow, E. V. *MOC* **1989**, *E19b*, 1608.
7. (a) Skell, P. S.; Garner, A. Y. *JACS* **1956**, *78*, 3409. (b) Skell, P. S.; Woodworth, R. C. *JACS* **1956**, *78*, 4496.
8. Doering, W. von E.; Henderson Jr., W. A. *JACS* **1958**, *80*, 5274.
9. Doering, W. von E.; Hoffman, A. K. *JACS* **1954**, *76*, 6162.
10. Skattebøl, L.; Solomon, S. *OSC* **1973**, *5*, 306.
11. Serratosa, F. *J. Chem. Educ.* **1964**, *41*, 564.
12. DeCamp, M. R.; Viscogliosi, L. A. *JOC* **1981**, *46*, 3918.
13. For a survey see: Dehmlow, E. V.; Dehmlow, S. S. *Phase Transfer Catalysis*; VCH: Weinheim, 1993.
14. Makosza, M.; Wawrzyniewicz, M. *TL* **1969**, 4659.
15. Skattebøl, L.; Abskharoun, G. A.; Greibrokk, T. *TL* **1973**, 1367.
16. Makosza, M.; Fedorynski, M. *SC* **1969**, 4659.
17. (a) Fedorynski, M. *S* **1977**, 783. (b) Mandel'shtam, T. V.; Kharicheva, E. M.; Labeish, N. N.; Kostikov, R. R. *ZOR* **1980**, *16*, 2513 (*CA* **1981**, *94*, 120 501b) (c) Baird, M. S.; Baxter, A. G. W.; Devlin, B. R. J.; Searle, R. J. G. *CC* **1979**, 210.
18. Dehmlow, E. V.; Wilkenloh, J. *LA* **1990**, 125 (*CA* **1990**, *112*, 138 566b).
19. Dehmlow, E. V.; Wilkenloh, J. *CB* **1990**, *123*, 583 (*CA* **1990**, *112*, 157 708r).
20. Dehmlow, E. V.; Fastabend, U. *CC* **1993**, 1241.
21. Hiyama, T.; Sawada, H.; Tsukanaka, M.; Nozaki, H. *TL* **1975**, 3013.
22. (a) Seyferth, D.; Burlitch, J. M.; Minasz, R. J.; Mui, J. Y.-P.; Simmons Jr, H. D.; Treiber; A. J. H.; Dowd, S. R. *JACS* **1965**, *87*, 4259. (b) Seyferth, D.; Gordon, M. E.; Mui, J. Y.-P.; Burlitch, J. M. *JACS* **1967**, *89*, 959.
23. Seyferth, D. Burlitch, J. M. *JOM* **1965**, *4*, 127.
24. (a) Logan, T. J. *OSC* **1973**, *5*, 969. (b) Schweizer, E. E.; O'Neill, G. J. *JOC* **1963**, *28*, 851.
25. Fedorynski, M.; Makosza, M. *JOM* **1973**, *51*, 89.
26. (a) Seyferth, D.; Lambert Jr., R. L. *JOM* **1969**, *16*, 21. (b) Seyferth, D. *ACR* **1972**, *5*, 65.
27. Gillespie Jr., J. S.; Acharya, S. P.; Shamblee, D. A. *JOC* **1975**, *40*, 1838.
28. (a) Closs, G. L.; Coyle, J. J. *JACS* **1965**, *87*, 4270. (b) Seyferth, D.; Simmons Jr., H. D.; Singh, G. *JOM* **1965**, *3*, 337. (c) Nishimura, J.; Furukawa, J. *CC* **1971**, 1375. (d) Martel, B.; Hiriat, J. M. *S* **1972**, 201.

29. Miyano, S.; Matsumoto, Y.; Hashimoto, H. *CC* **1975**, 364.
30. (a) Doering, W. von E.; LaFlamme, P. M. *T* **1958**, *2*, 75. (b) Gardner, P. D.; Narayana, M. *JOC* **1961**, *26*, 3518.
31. (a) Moore, W. R.; Ward, H. R. *JOC* **1962**, *27*, 4179. (b) Skattebøl, L. *TL* **1961**, 167.
32. (a) Skattebøl, L. *ACS* **1963**, *17*, 1683. (b) Skattebøl, L. *JOC* **1964**, *29*, 2951. (c) Skattebøl, L. *JOC* **1966**, *31*, 2789.
33. For an example resulting in the synthesis of cyclononadienones, see Perez, G. H.; Weyerstahl, P. *S* **1985**, 174.
34. Marquis, E. T.; Gardner, P. D. *TL* **1966**, 2793.
35. Paquette, L. A.; Taylor, R. T. *JACS* **1977**, *99*, 5708.
36. (a) Skattebøl, L. *TL* **1965**, 2175. (b) Ball, W. J.; Landor, S. R.; Punja, N. *JCS(C)* **1967**, 194.
37. Dunkelblum, E.; Singer, B. *S* **1975**, 323.
38. Bee, L. K.; Beeby, J.; Everett, J. W.; Garratt, P. J. *JOC* **1975**, *40*, 2212.
39. (a) Garratt, P. J.; Nicolaou, K. C.; Sondheimer, F. *CC* **1971**, 1018. (b) Garratt, P. J.; Nicolaou, K. C.; Sondheimer, F. *JOC* **1973**, *38*, 2715.
40. Skattebøl, L. *T* **1967**, *23*, 1107.
41. Paquette, L. A.; McLaughlin, M. L. *OS* **1990**, *68*, 220.
42. For other examples see (a) Butler, D. N.; Gupta, I. *CJC* **1978**, *56*, 80. (b) Charumilind, P.; Paquette, L. A. *JACS* **1984**, *106*, 8225. (c) McLaughlin, M. L.; McKinney, J. A.; Paquette, L. A. *TL* **1986**, *27*, 5595. (d) Paquette, L. A.; Sivik, M. R. *OM* **1992**, *11*, 3503.
43. 6-Bromofulvene can be made directly from cyclopentadiene and bromoform: Washburn, W. N.; Zahler, R.; Chen, I. *JACS* **1978**, *100*, 5863.
44. Amice, P.; Blanco, L.; Conia, J. M. *S* **1976**, 196.
45. Hirao, T.; Hayashi, K.; Fujihara, Y.; Ohshiro, Y.; Agawa, T. *JOC* **1985**, *50*, 279.
46. Sugimoto, J.; Miura, K.; Oshima, K.; Utimoto, K. *CL* **1991**, 1319.
47. (a) Kharasch, M. S.; Jensen, E. V.; Urry, W. H. *JACS* **1946**, *68*, 154. (b) Kharasch, M. S.; Jensen, E. V.; Urry, W. H. *JACS* **1947**, *69*, 1100.
48. Meyers, A. I.; Babiak, K. A.; Campbell, A. L.; Comins, D. L.; Fleming, M. P.; Henning, R.; Heuschmann, M.; Hudspeth, J. P.; Kane, J. M.; Reider, P. J.; Roland, D. M.; Shimizu, K.; Tomioka, K.; Walkup, R. D. *JACS* **1983**, *105*, 5015.
49. Tamura, T.; Kunieda, T.; Takizawa, T. *JOC* **1974**, *39*, 38.
50. Taguchi, H., Yamamoto, H.; Nozaki, H. *JACS* **1974**, *96*, 3010.
51. Hofmann, A. W. *CB* **1870**, *3*, 761.
52. Weber, W. P.; Gokel, G. W.; Ugi, I. K. *AG(E)* **1972**, *11*, 530.
53. Compere, E. L., Jr. *JOC* **1968**, *33*, 2565.
54. Compere, E. L., Jr.; Weinstein, D. A. *S* **1977**, 852.

Matthew R. Sivik
The Lubrizol Corporation, Wickliffe, OH, USA

1,3-Butadiene

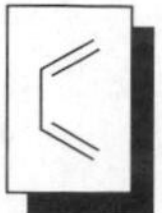

[106-99-0] C_4H_6 (MW 54.09)

(co-monomer in synthetic elastomers and polymers, polybutadiene rubber, chloroprene, and nylon intermediate;[1] 4π partner in Diels–Alder reactions[2])

Alternate Names: butadiene; α,γ-butadiene; bivinyl; divinyl; vinylethylene; biethylene.
Physical Data: bp −4.5 °C/760 mmHg; mp −109 °C; fp < −7 °C; ρ_4^{-6} 0.650.
Solubility: sol common organic solvents.
Form Supplied in: widely available in compressed liquid phase. Supplied in several grades ranging from 99.0–99.8% purity. Sizes available range incrementally from lecture bottles (100 g) to large gas cylinders (61 kg).
Preparative Methods: butadiene sulfone (2,5-dihydrothiophene 1,1-dioxide), a crystalline, nonhygroscopic, commercially available solid, can be used to generate 1,3-butadiene in situ: heating the sulfone at 100–130 °C induces loss of SO_2.[3b,3c]
Handling, Storage, and Precautions: extremely flammable liquid and gas. Avoid polymerization initiators. Relatively nontoxic. Suspected chronic carcinogen. Acutely irritating to respiratory tract. 1,3-Butadiene (**1**) is used in the laboratory by condensing the gas directly into a reaction vessel, or by forming a saturated solution of the gas in the reaction solvent. For reactions run below rt (e.g. Lewis acid-catalyzed reactions), normal reaction flasks can be used. For higher temperature reactions, sealed tubes are required to prevent escape of (**1**). A particularly convenient, reusable apparatus is available from Ace Glass, Inc., and consists of a heavy-walled glass tube sealed on one end and threaded on the other. The threaded end can be sealed with a heavy Teflon plug fitted with an O-ring.

Reactivity. 1,3-Butadiene (**1**) participates as the 4π partner in $[_{\pi}4_s + _{\pi}2_2]$ cycloaddition reactions with a variety of dienophiles. The majority of these cycloadditions are normal Diels–Alder reactions ($HOMO_{diene}$–$LUMO_{dienophile}$ controlled).[2,3a] Theoretical predictions on the mechanism of Diels–Alder reactions of butadiene have pointed towards a synchronous concerted mechanism.[4] 1,3-Butadiene (**1**) is typically less reactive than substituted butadienes, except when steric factors prevent the diene from attaining the reactive *s-cis* conformation. For example ***Tetracyanoethylene*** reacts more slowly with (**1**) than with (*E*)-1-methyl-1,3-butadiene (103×), 2-phenyl-1,3-butadiene (191×), or (*E*)-1-methoxy-1,3-butadiene (50 934×).

Lewis Acid-Catalyzed Cycloadditions. Friedel–Crafts type catalysts are reported to facilitate the cycloaddition reactions of (**1**) with α,β-unsaturated carbonyl compounds.[5] With methyl vinyl ketone (eq 1), cycloaddition was effected in the presence of less than 1 molar equivalent of ***Aluminum Chloride***, ***Tin(IV) Chlo-***

ride, ***Boron Trifluoride***, ***Iron(III) Chloride***, or ***Titanium(IV) Chloride*** for 1 h at 25 °C. For comparison, the uncatalyzed reaction required 8–10 h at 140 °C in a sealed tube (75%). ***Acrolein***, ***Methyl Vinyl Ketone***, and ***Acrylic Acid*** all underwent effective catalyzed Diels–Alder reactions with (**1**) in good yields. The principal limitation of this reaction is polymerization of the 1,3-diene. For example, 2,3-dimethylbutadiene and cyclopentadiene underwent polymerization and dimerization, respectively, under these conditions.

catalyst, benzene; 25 °C, 75–85% (1)

In the cycloaddition of (**1**) with 2,5-cyclohexadienones in an approach to *cis*-clerodanes, the facial selectivity depended on the length of the carboxylate side-chain (eq 2).[6] When the carboxylate was a significant distance from the ring ($n = 2$), cycloaddition occurred *syn* to the less bulky methyl group to afford the *anti* cycloadduct. Facial selectivity eroded with $n = 1$, and was reversed when the carboxylate was directly attached to the ring, to afford exclusively the *syn* diastereomer.

$ZnCl_2$; CH_2Cl_2, 25 °C (2)

	anti : *syn*	
$n = 2$	5:1	91%
$n = 1$	1.7:1	90%
$n = 0$	*syn*	90%

Lewis acids have also been used to catalyze the cycloaddition of (**1**) with alkynes (eq 3).[7] A mixture of $TiCl_4$ and ***Diethylaluminum Chloride*** promoted the [4 + 2] cycloaddition of phenyl(trimethylsilyl)acetylene and (**1**) to afford the dihydrobenzene adduct in 84% yield, which could be aromatized thermally to afford the *ortho*-disubstituted benzene in 92% yield. Cycloaddition of cycloalkenones with butadiene in the presence of $AlCl_3$ affords the expected *cis*-fused systems for $n = 2, 3, 4$ (eq 4).[8] Partial or total isomerization of the *cis* adduct to the more stable *trans* isomer was observed with R = H.

$TiCl_4$, Et_2AlCl; 60 °C, 1 h, benzene; then 260 °C, 1 h (3)

$AlCl_3$; R = H, Me (4)

A highly regioselective and moderately stereoselective cycloaddition of (**1**) with the γ,δ-double bond of a dienone was effected by $AlCl_3$ catalysis in 37% yield (eq 5).[9] The cycloadduct formally arising from approach of butadiene *syn* to the enone ring predominated initially, and was shown to isomerize to the more stable *anti* product under the reaction conditions. Contrasteric facial selectivity was observed in the Lewis acid-catalyzed cycloaddition of (**1**) with γ-alkoxycycloalkenones (eq 6).[10] When $n = 1$, cycloaddition at rt in the presence of catalytic $AlCl_3$ afforded predominately the *syn* cycloadduct (76%; 13:1 *syn/anti*). Under similar reaction conditions with $n = 2$, the *syn* cycloadduct was also the major product (10:1 *syn/anti*). The divergence of these results with those obtained with γ-alkylcycloalkenones was discussed, and a tentative theoretical model was proposed to account for the *syn* selectivity.

$AlCl_3$; *syn* : *anti* 1:2.3 (5)

OTBDMS; $AlCl_3$, toluene; *syn*, *anti* (6)

Montmorillonite K10 exchanged with Fe^{III} has been reported to catalyze the Diels–Alder reaction of butadiene with acrolein (eq 7).[11] Cycloaddition occurs at rt in high yield in the presence of 1.8 g catalyst per 15 mmol cycloaddition partners. Other dienes reacted with equal facility with acrolein under these conditions.

K10–Fe^{III}, CH_2Cl_2; 20 °C, 6 h; 83% (7)

Asymmetric Diels–Alder Reactions. Cycloaddition of (**1**) with a methacrylate dienophile using a camphor lactam imide in the presence of Lewis acid affords the cyclohexenecarboxylate cycloadduct in 61% yield with 85:15 π-facial selectivity.[12] Presumably the steric bulky of the coordinated metal forces the methacrylate to adopt an *s-trans* conformation as shown (eq 8), and cycloaddition occurs from the less crowded bottom face of the complex. A pyrrolidin-2-one auxiliary containing the bulky 5-(trityloxymethyl) group was effective at asymmetric induction in the Lewis acid-catalyzed cycloaddition of (**1**) with a fumarate dienophile (eq 9).[13] The cyclohexenedicarboxylate was obtained in 52% yield and 96% de. Other dienophiles and Lewis acids were examined and found to be equally effective, giving high *endo* selectivity (≥97:3) and excellent diastereoselectivity (≥94% de).

$MeAlCl_2$; X_c (8)

Ph_3CO; MeO_2C; $TiCl_4$, CH_2Cl_2; −45 °C; 52%; CO_2Me, X_c (9)

Carbohydrate-based chiral auxiliaries have proven effective in cycloadditions with (**1**).[14] The auxiliary derived from 3-

O-acryloyldihydro-D-glucal (eq 10; R = pivaloyl) afforded the (*S*)-cyclohexenecarboxylate (8:92 *R/S*), whereas the pseudo-enantiomeric system, 3-*O*-acryloyldihydro-L-rhamnal (eq 11; R = pivaloyl), afforded the (*R*)-cycloadduct (95:5 *R/S*). Complexation of the titanium catalyst to each of the ester carbonyl oxygens was postulated to rigidify the acryloyl group, locking it in the conformation depicted. In both cases, cycloaddition presumably occurs to the less hindered face of the acrylate away from the bulky 4-*O*-pivaloyl group.

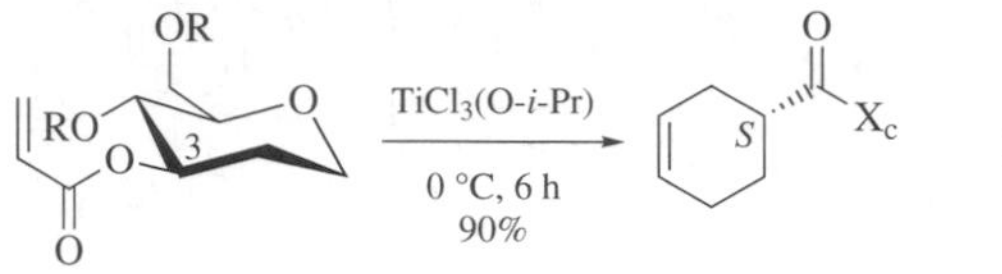

$TiCl_3(O$-*i*-$Pr)$, 0 °C, 6 h, 90% (10)

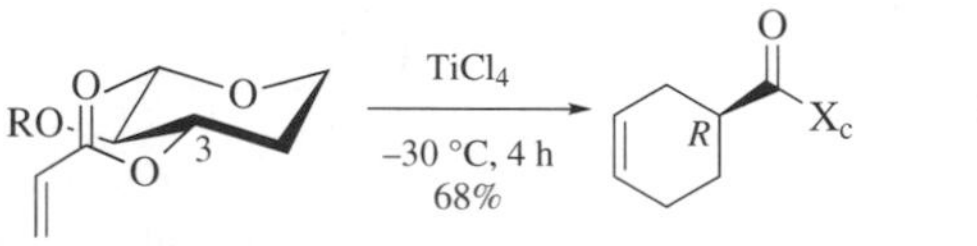

$TiCl_4$, −30 °C, 4 h, 68% (11)

Chiral titanium-based catalysts have been used in asymmetric Diels–Alder reactions of (**1**). A mixture of $TiCl_4$ and ***Titanium Tetraisopropoxide*** was used with a tartrate-derived auxiliary in the cycloaddition of (**1**) with a quinone (eq 12) to afford the expected bicyclic system in 88% yield and 63% ee.[15] Other dienes underwent stereoselective cycloadditions with better asymmetric induction. A similar catalyst system was used in the cycloaddition of (**1**) with fumarate[16] and acrylate[17] derivatives. The catalyst was prepared by alkoxy exchange with the tartrate-derived auxiliary and ***Dichlorotitanium Diisopropoxide***, and was treated with the cycloaddition partners at rt for 24 h to afford the cyclohexenecarboxylate in 81% yield and in 93% optical purity (eq 13). Other dienes underwent asymmetric cycloaddition with equal success. A model was proposed for the catalyst–dienophile complex.

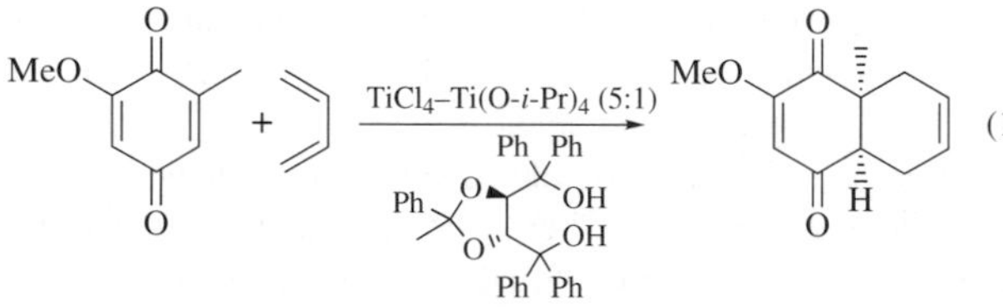

$TiCl_4$–$Ti(O$-*i*-$Pr)_4$ (5:1) (12)

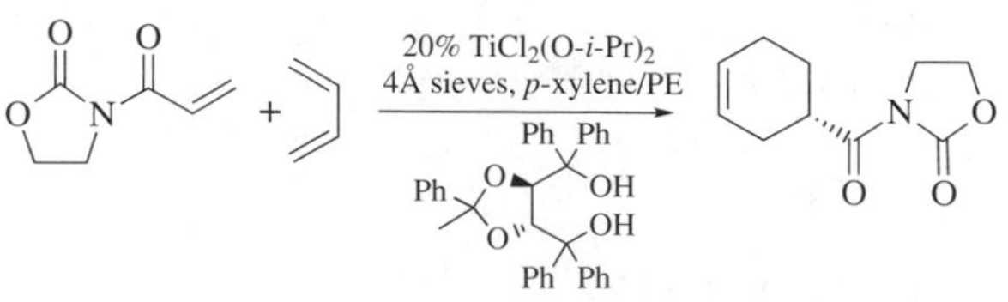

20% $TiCl_2(O$-*i*-$Pr)_2$, 4Å sieves, *p*-xylene/PE (13)

The C_2-symmetric hydrobenzoin complexed with $TiCl_4$ promoted the cycloaddition of (**1**) with ***Dimethyl Fumarate*** to afford the cyclohexenedicarboxylate in 78% yield with a modest 60% ee (eq 14).[18] Carboxylic ester dienophiles have typically participated ineffectively in asymmetric Diels–Alder reactions.

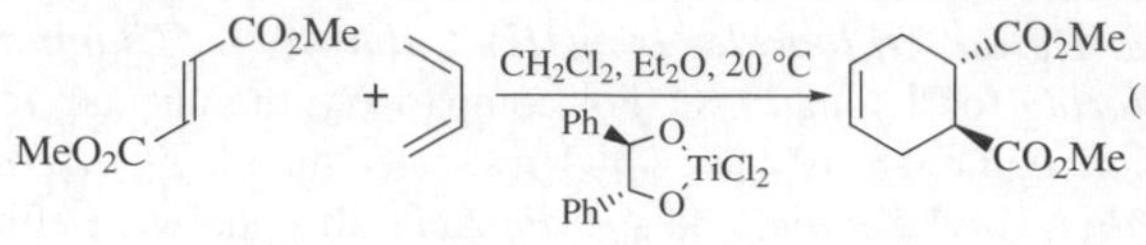

CH_2Cl_2, Et_2O, 20 °C (14)

Applications in Synthesis. Butadiene is a widely used reagent in organic synthesis. It is primarily used as the 4π participant in the Diels–Alder construction of carbocyclic and heterocyclic six-membered rings,[2] although numerous other uses have been recorded.

Butadiene cycloaddition was a key step in the synthesis of constrained phenanthrenamine derivatives that are active as noncompetitive NMDA antagonists.[19] Reaction of dihydronaphthalenecarboxylic acid with (**1**) in toluene at 110 °C in the presence of polymerization inhibitor afforded the tricyclic adduct in 60% yield (eq 15). Hydrogenation of the resulting double bond afforded the hydrophenanthrenecarboxylic acid, which was converted to a variety of tetracyclic structures where the carboxylate was converted to an amine by modified Curtius rearrangement.

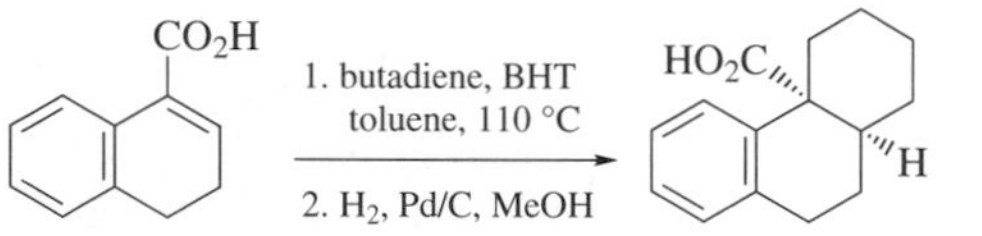

1. butadiene, BHT, toluene, 110 °C; 2. H_2, Pd/C, MeOH (15)

Three reports have described the use of butadiene in the asymmetric construction of the C-28–C-34 fragment of the immunosuppressant FK-506.[20–22] In the first report (eq 16), the acrylate ester of (*S*)-(+)-pantolactone was reacted with (**1**) in the presence of $TiCl_4$ to afford the cyclohexenecarboxylate in excellent yield and in 91% de.[20] In the second report, the same Diels–Alder reaction was performed with a chiral sultam auxiliary under $EtAlCl_2$ catalysis to give the cyclohexenecarboxylate in 86% yield in 93% ee (eq 17).[21] In the third report, an acryloyl-(*S*)-2-hydroxysuccinimide underwent cycloaddition with (**1**) under catalysis by $TiCl_4$ to afford the cyclohexenecarboxylate in 85% yield in 98% ee (eq 18).[22] The double bond of all three cycloadducts was subsequently converted to the *anti*-diol of FK-506. A cycloaddition reaction identical to that shown in eq 16 was used as a key step in the asymmetric synthesis of the cyclohexane ring of (+)-phyllanthocin.[23] In this instance the cycloaddition proceeded in 70% yield and in 97% ee.

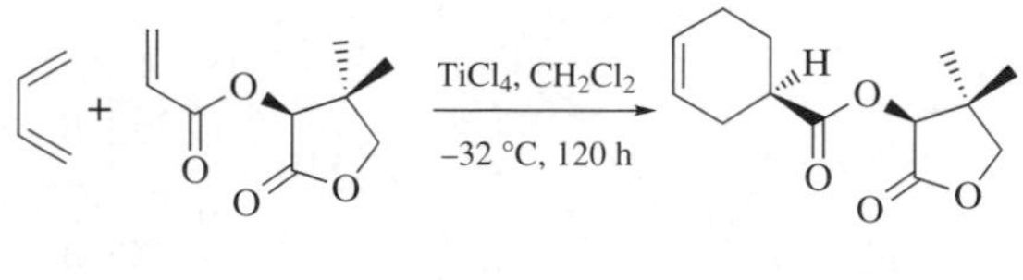

$TiCl_4$, CH_2Cl_2, −32 °C, 120 h (16)

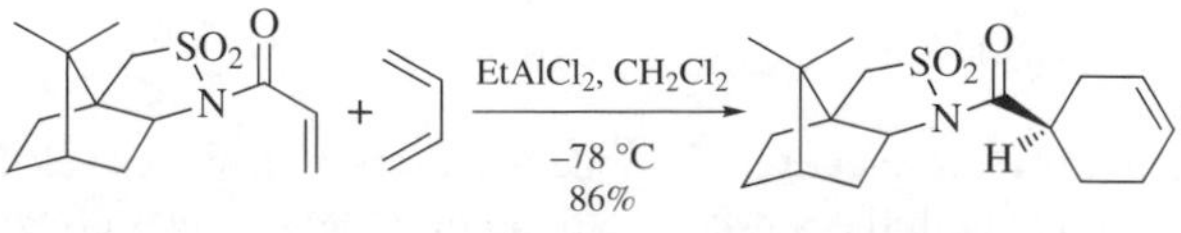

$EtAlCl_2$, CH_2Cl_2, −78 °C, 86% (17)

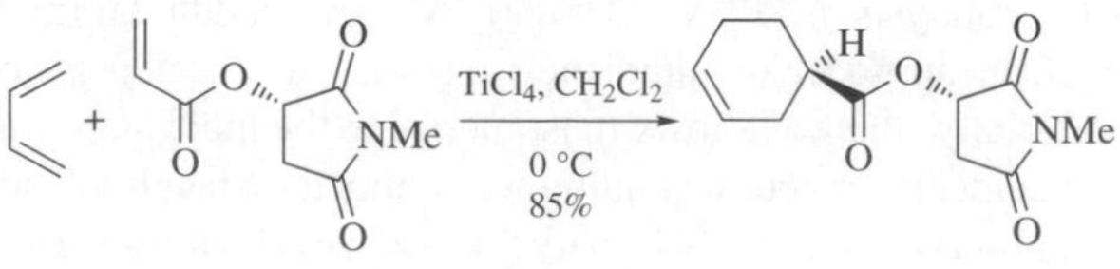

$TiCl_4$, CH_2Cl_2, 0 °C, 85% (18)

A β-pinene-derived dienophile was used as the 2π cycloaddition partner with (**1**) in an approach to the morphine skeleton (eq 19).[24] Thermal cycloaddition occurred stereoselectively from the face opposite the substituted bridge to afford the adduct in 82% yield. Cycloaddition was also effected in lower yield under Lewis acid-catalyzed conditions ($Cu(BF_4)_2$, CH_2Cl_2, 25 °C, 12 h, 34%). Further elaboration of the cycloadduct afforded the *cis*-Δ^6-1-octalone acetal possessing the required absolute configuration at three contiguous stereogenic centers for conversion to morphine-related compounds. In a synthetic approach to analogs of the potassium channel activator aprikalim, butadiene was used in a hetero Diels–Alder reaction for the construction of a thiopyran ring (eq 20).[25] Thus reaction of the thiocarbonyl group of an α-thioketo ester (generated in situ) with (**1**) afforded the desired thiopyran in 65% yield.

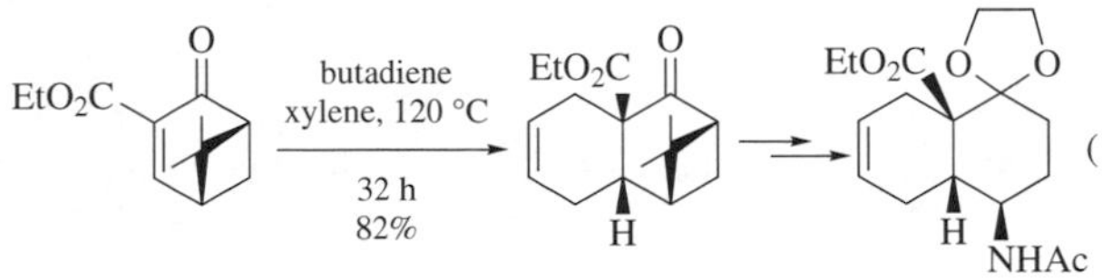

butadiene, xylene, 120 °C, 32 h, 82% (19)

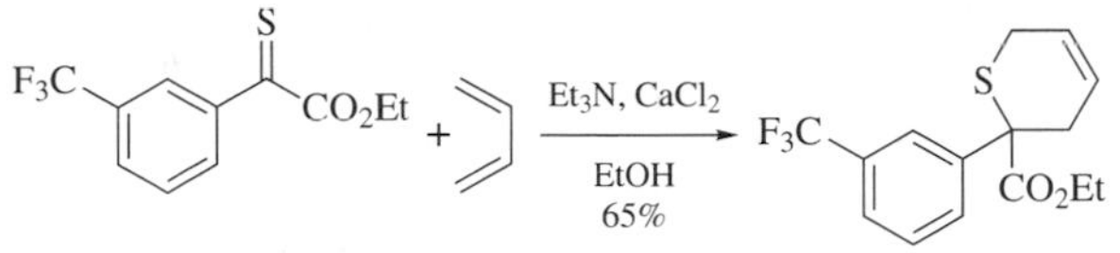

Et_3N, $CaCl_2$, EtOH, 65% (20)

An asymmetric Diels–Alder reaction of (**1**) with a phenylalaninol-derived auxiliary was used to construct the B-ring of (+)-compactin and (+)-mevinolin (eq 21).[26] Cycloaddition of the (*E*)-crotonic acid derivative with butadiene in the presence of catalytic Et_2AlCl afforded the desired cycloadduct in 56% yield and greater than 95% de. The auxiliary was removed and the desired *cis*-isomer of the 2-methyl-4-cyclohexenecarboxylate was obtained by base-promoted epimerization of the *trans*-isomer (65:35 *cis/trans* equilibrium mixture). Use of the corresponding (*Z*)-crotonoyl auxiliary was not examined as the (*Z*) configuration of the alkene is not preserved under the reaction conditions.[27]

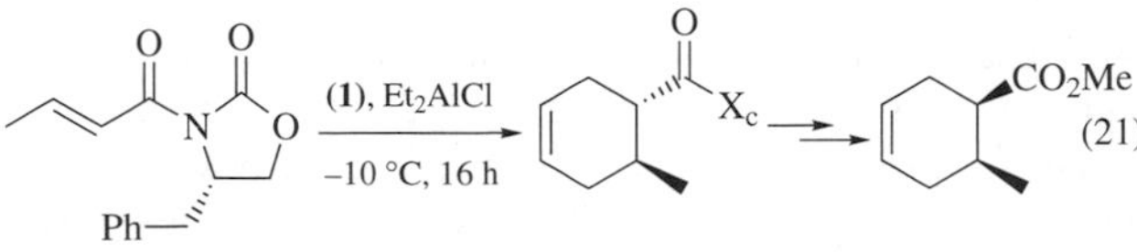

(**1**), Et_2AlCl, −10 °C, 16 h (21)

A rapid and stereospecific construction of the 2-azabicyclo[3.3.1]nonane ring system characteristic of *Strychnos* alkaloids was initiated with a Diels–Alder reaction of β-nitrostyrene and (**1**) (eq 22).[28] After reduction of the nitro group and *N*-acylation, the desired aminocyclohexene was obtained in 63% overall yield. Subsequent elaboration to the 2-azabicyclo[3.3.1]nonane system involved a high-yielding stereoselective Heck reaction. Butadiene sulfone was used for in situ generation of (**1**) in an approach to cephalotaxine analogs. In a key step, a β-nitrostyrene dienophile was treated with butadiene sulfone at 135 °C to afford the butadiene cycloadduct in 74% yield (eq 23).[29]

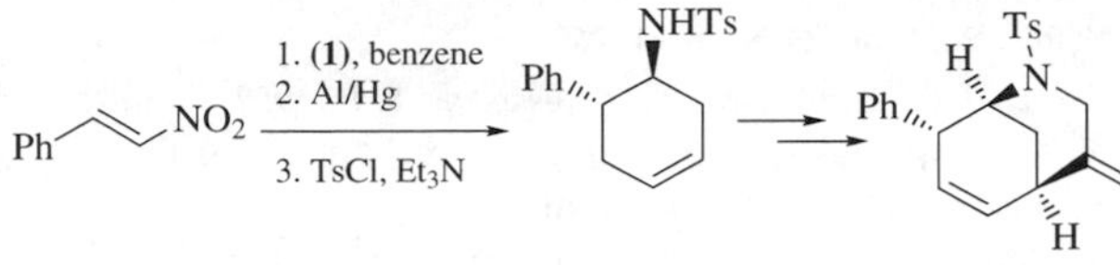

1. (**1**), benzene 2. Al/Hg 3. TsCl, Et_3N (22)

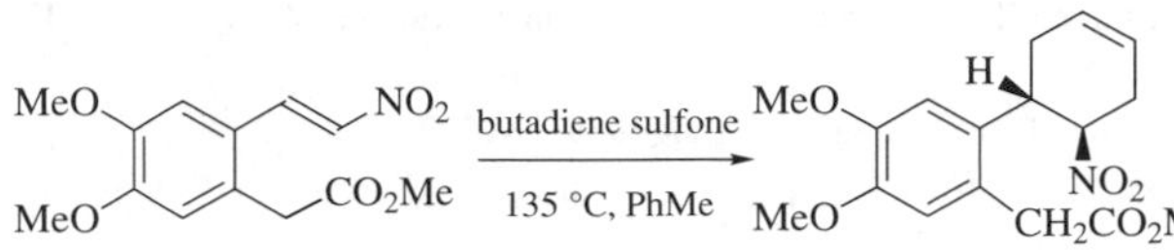

butadiene sulfone, 135 °C, PhMe (23)

Miscellaneous Reactions. The novel dienophile 2-trimethylsilylvinyl-9-BBN was shown to undergo facile Diels–Alder reaction with butadiene to afford the 2-trimethylsilylcyclohexenol in 85% yield after oxidation of the intermediate trialkylborane (eq 24).[30] The cycloadduct was shown to be a precursor to 5-trimethylsilyl-1,3-cyclohexadiene and -1,4-cyclohexadiene, thus making the dienophile an acetylene equivalent.

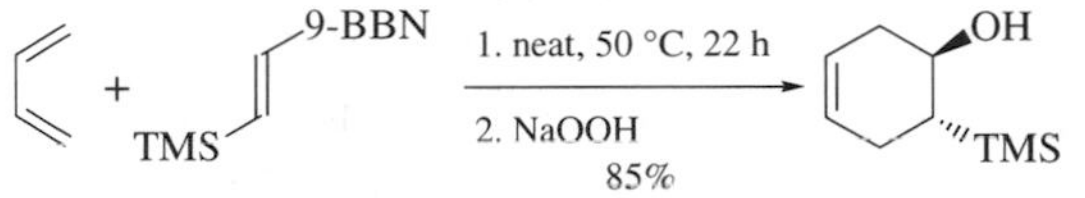

1. neat, 50 °C, 22 h 2. NaOOH, 85% (24)

Catalytic osmylation was used to convert 1,3-butadiene to a polyol with good control of relative stereochemistry (eq 25).[31] Under standard osmylation conditions, (**1**) was converted to the 2,3-*anti*-tetraol in 80% yield in a 5:1 *anti/syn* ratio. The stereoselectivity of this reaction is consistent with previous findings on the osmylation of allylic alcohols and ethers.

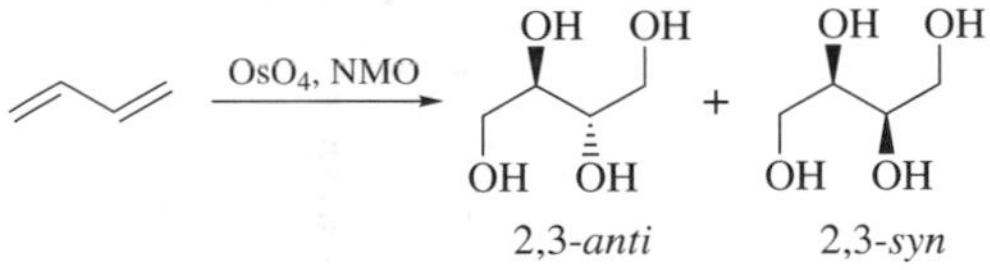

OsO_4, NMO; 2,3-*anti* + 2,3-*syn* (25)

Related Reagents. Isoprene; 1,3-Pentadiene; 3-Sulfolene.

1. Stevens, M. P. *Polymer Chemistry*; Oxford University Press: New York, 1990.
2. Fringuelli, F.; Taticchi, A. *Dienes in the Diels–Alder Reaction*; Wiley: New York, 1990.
3. (a) Sauer, J.; Sustmann, R. *AG(E)* **1980**, *19*, 779. (b) Backer, H. J.; Blaas, T. A. H. *RTC*, **1942**, *61*, 785. (c) Sample, T. E., Jr.; Hatch, L. F. *OS* **1970**, *50*, 43.
4. Houk, K. N.; Li, Y. *JACS* **1993**, *115*, 7478.
5. Fray, G. I.; Robinson, R. *JACS* **1961**, *83*, 249.
6. Liu, H.-J.; Han, Y. *TL* **1993**, *34*, 423.
7. Klein, R.; Sedmera, P.; Cejka, J.; Mach, K. *JOM* **1992**, *436*, 143.
8. Wenkert, E.; Fringuelli, F.; Pizzo, F.; Taticchi, A. *SC* **1979**, *9*, 391.
9. Minuti, L.; Selvaggi, R.; Taticchi, A.; Guo, M.; Wenkert, E. *CJC* **1992**, *70*, 1481.
10. Jeroncic, L. O.; Cabal, M.-P.; Danishefsky, S. J.; Shulte, G. M. *JOC* **1991**, *56*, 387.
11. Laszlo, P.; Moison, H. *CL* **1989**, 1031.
12. Boeckman, R. K., Jr.; Nelson, S. G.; Gaul, M. D. *JACS* **1992**, *114*, 2258.
13. Tomioka, K.; Hamada, N.; Suenaga, T.; Koga, K. *JCS(P1)* **1990**, *426*.

14. Stähle, W.; Kunz, H. *SL* **1991**, *260*.
15. Engler, T. A.; Letavic, M. A.; Takusagawa, F. *TL* **1992**, *33*, 6731.
16. Narasaka, K.; Iwasawa, N.; Inoue, M.; Yamada, T.; Nakashima, M.; Sugimori, J. *JACS* **1989**, *111*, 5340.
17. Narasaka, K.; Tanaka, H.; Kanai, F. *BCJ* **1991**, *64*, 387.
18. Devine, P. N.; Oh, T. *JOC* **1992**, *57*, 396.
19. Malone, T. C.; Ortwine, D. F.; Johnson, G.; Probert, A. W., Jr. *BML* **1993**, *3*, 49.
20. Corey, E. J.; Huang, H.-C. *TL* **1989**, *30*, 5235.
21. Smith, A. B., III; Hale, K. J.; Laakso, L. M.; Chen, K.; Riéra, A. *TL* **1989**, *30*, 6963.
22. Ireland, R. E.; Highsmith, T. K.; Gegnas, L. D.; Gleason, J. L. *JOC* **1992**, *57*, 5071.
23. Trost, B. M.; Kondo, Y. *TL* **1991**, *32*, 1613.
24. Boger, D. L.; Mullican, M. D.; Heilberg, M. R.; Patel, M. *JOC* **1985**, *50*, 1904.
25. Pinto, I. L.; Buckle, D. R.; Rami, H. K.; Smith, D. G. *TL* **1992**, *33*, 7597.
26. Clive, D. L. J.; Murthy, K. S. K.; Wee, A. G. H.; Prasad, J. S.; da Silva, G. V. J.; Majewski, M.; Anderson, P. C.; Evans, C. F.; Haugen, R. D.; Heerze, L. D.; Barrie, J. R. *JACS* **1990**, *112*, 3018.
27. Evans, D. A.; Chapman, K. T.; Bisaha, J. *JACS* **1988**, *110*, 1238.
28. Rawal, V. H.; Michoud, C. *TL* **1991**, *32*, 1695.
29. Bryce, M. R.; Gardiner, J. M. *CC* **1989**, 1162.
30. Singleton, D. A.; Martinez, J. P. *TL* **1991**, *32*, 7365.
31. Park, C. Y.; Kim, B. M.; Sharpless, K. B. *TL* **1991**, *32*, 1003.

Robert S. Coleman & Henry A. Alegria
University of South Carolina, Columbia, SC, USA

t-Butoxybis(dimethylamino)methane[1]

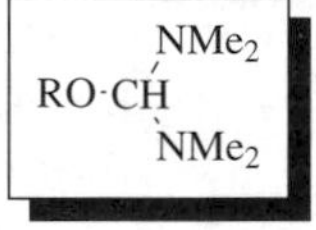

(**1**; R = *t*-Bu) (*t*-BAE)
[5815-08-7] $C_9H_{22}N_2O$ (MW 174.33)
(**2**; R = Me) (MAE)
[1186-70-5] $C_6H_{16}N_2O$ (MW 1328.24)

(aminal esters reactive as aminomethylenating reagents (formylating reagents) for CH_2- and NH_2-acidic compounds. The *t*-butyl analog is somewhat more reactive than the methyl derivative; bis(dimethylamino)carbene can be generated from both compounds)

Alternate Name: Bredereck's reagent.
Physical Data: (**1**) bp 50–52 °C/12 mmHg; $n_D^{20} = 1.4250$. (**2**) bp 128 °C/760 mmHg; 32–33 °C/25 mmHg; $n_D^{20} = 1.4158$.
Solubility: miscible with nonpolar aprotic waterfree solvents (benzene, toluene, cyclohexane, diethyl ether, etc.); react with protic solvents like water (hydrolysis to DMF) or alcohols (alcoholysis affords DMF acetals). Even common solvents like acetonitrile or acetone which are weakly CH-acidic react with the aminal esters on heating.
Form Supplied in: colorless to weak yellow, strong amine-smelling liquids.
Analysis of Reagent Purity: the best method is ^1H NMR spectroscopic examination of a neat sample.
Handling, Storage, and Precautions: both compounds should be handled in a fume hood, with strict exclusion of atmospheric moisture. They should be stored in tightly sealed containers in a refrigerator under an atmosphere of dry nitrogen or argon.

Introduction. Alkoxybis(dimethylamino)methanes can be obtained in good yields from *N,N,N′,N′*-tetraalkylformamidinium salts and alcohol-free alkoxides.[2] The reactions have to be conducted in inert, strictly anhydrous solvents like cyclohexane, hexane, ether, or THF, in which unfortunately both reagents are insoluble. As a consequence it is necessary to use relatively large amounts of the solvent and long reaction times (eq 1). In the patent literature it is claimed that by the use of a mixture of DMF and mesitylene as solvent, the yields can be improved to about 95%.[3]

$$H\text{-}C^{+}(NMe_2)_2\ X^- + NaOR \xrightarrow[-NaX]{\text{hexane, cyclohexane or ether, ~20 °C}} RO\text{·}CH(NMe_2)_2 \quad (1)$$

X	R	Solvent	Reaction time (h)	Yield (%)
$MeSO_4$	Me	Ether	1–2	62
$MeSO_4$	*t*-Bu	Cyclohexane	1–2	70
Cl	*t*-Bu	Ether	24	75

In another method,[4] bis(dimethylamino)acetonitrile is used as the starting compound. Alcohol-free alkoxides suspended in absolute ether substitute the cyano group to give the corresponding aminal ester (eq 2). In contrast to the aforementioned method, less solvent and shorter reaction times are needed because the nitrile is soluble in ether and the product yields are normally 5–20% higher.

$$NC\text{·}CH(NMe_2)_2 + NaOR \xrightarrow[\text{1 h, } -NaCN]{\text{ether, 15–20 °C}} RO\text{·}CH(NMe_2)_2 \quad (2)$$

R = Me, 82%
R = *t*-Bu, 84%

The purity and stability of the product are determined by the quality of the metal alkoxide. Traces of alcohol catalyze the disproportionation of the aminal ester to the corresponding DMF acetal and tris(dimethylamino)methane (eq 3),[5] which may complicate spectroscopic investigations. For preparative purposes the disproportionation reaction can be disregarded; at higher temperatures the orthoamides interconvert very rapidly to the aminal ester, and usually the condensation reactions of aminal esters are not affected by the equilibrium.

$$2\ RO\text{·}CH(NMe_2)_2 \rightleftharpoons RO\text{·}CH(OR)(NMe_2) + Me_2N\text{-}CH(NMe_2)_2 \quad (3)$$

Reactions. The reactivity of aminal esters is very similar to that of ***Tris(dimethylamino)methane***. In the first step of the reaction the compounds dissociate, forming strong basic alkoxide anions which set up an equilibrium with the XH_2-acidic compounds (eq 4). The HX^- anions thus formed combine with the formamidinium ions. β-Elimination of dimethylamine from the adduct is the final step of the condensation reaction.

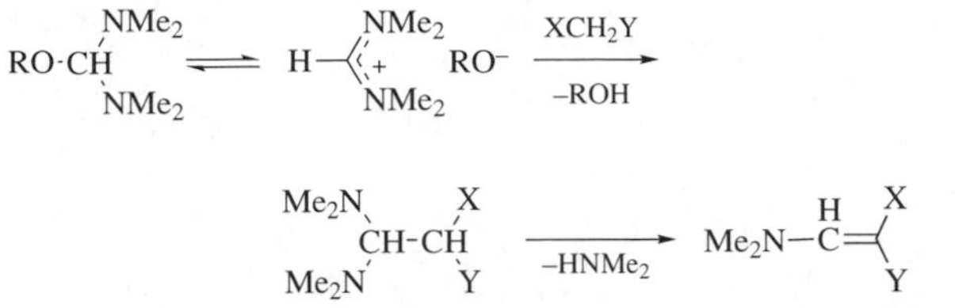

(4)

Because of their high reactivity with XH_2-acidic compounds, aminal esters have found widespread application as formylating reagents; especially with CH_2-acidic ketones such as simple dialkyl ketones,[6–9] aryl alkyl ketones,[6,10–16] heteroaryl alkyl ketones,[17–19] 1-alkenyl alkyl ketones,[20] α-dialkylamino ketones,[21] α-glycosyl alkyl ketones,[22] cycloalkanones,[10,11,16,20,23–28] 4-piperidinones,[29] and steroidal ketones.[30,31] If an additional acidic functional group is present (e.g. an NH_2 group), this may also be aminomethylenated by the reagent,[10] in some cases accompanied by ring closure.[32] The resulting difunctional compounds formed have been transformed into various types of product. Eqs 5–11 give an overview of these applications.[6–27,32]

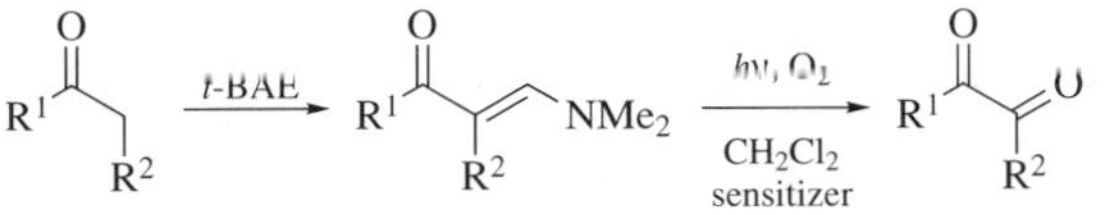

(5)

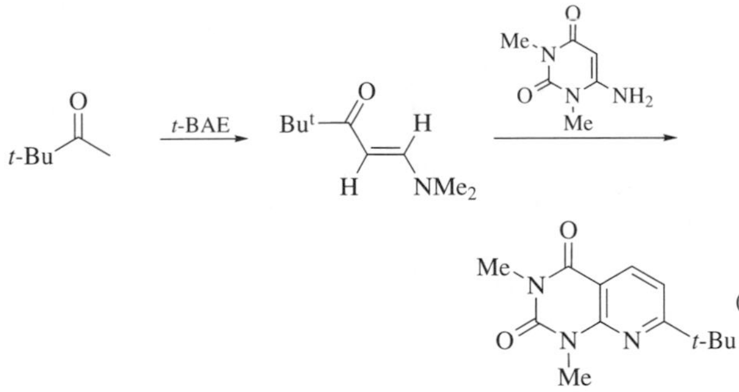

(6)

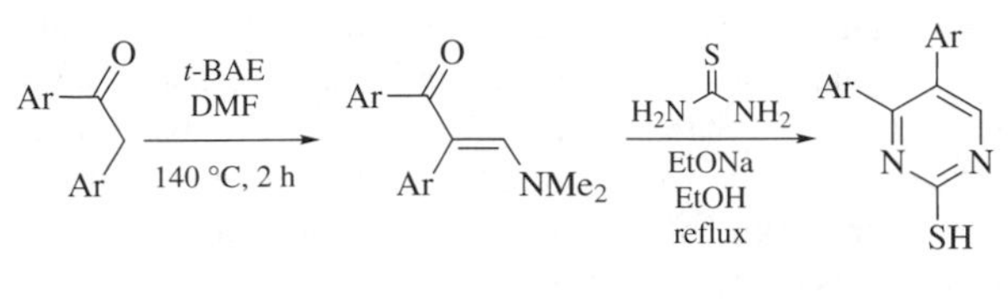

(7)

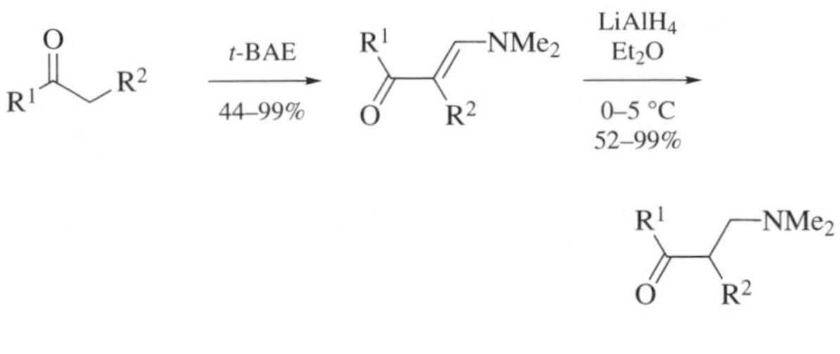

(8)

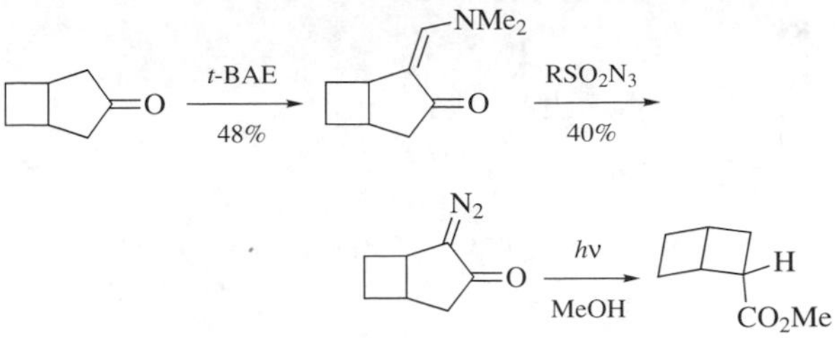

(9)

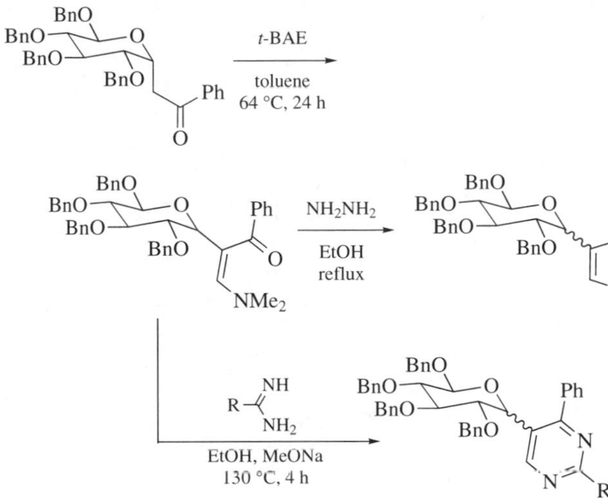

(10)

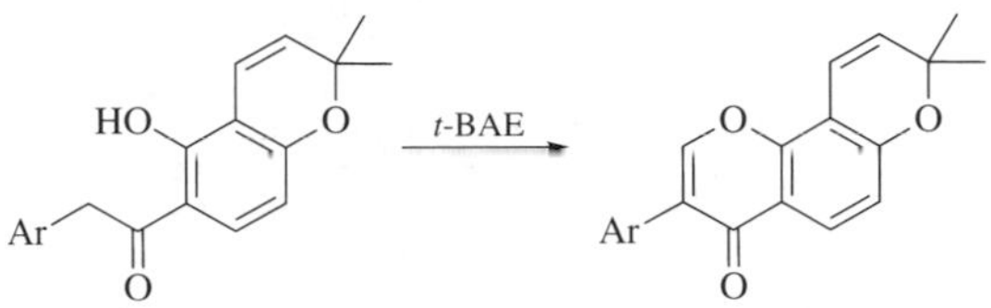

(11)

This type of condensation reaction of aminal esters may proceed with even weaker CH-acidic compounds (eq 12), such as esters[33–35] which have in the α-position, amino,[21,36–39] alkoxy[21,40] and heteroaryloxy,[39] thio,[21,39] phosphonyl,[41] and silyl groups.[41] Aminomethylenation reactions have also been reported for the following functionalities or compound classes: lactones,[33,42–51] α,β-unsaturated esters[38,52] and lactones,[53] thioesters,[33] imidates,[41] isocyanides[41] and nitriles,[6,21,41] amides[33] and lactams,[27,33,54–58] cyclic imides,[55,59] thioamides[33] and thiolactams,[33,60] amidines,[41] amidinium[41] and phosphonium salts,[61,62] phosphonates,[41,62] ketone hydrazones,[63,64] cyclopentadienes,[60–67] xanthene,[68] toluenes,[68–73] nitroarenes,[74] 2-methyl-1-pyrroline *N*-oxides,[75] 2-alkyl-5-nitroimidazole,[17] 4-methylpyridine,[68,73,76,77] methylpyrimidines,[68,77–79] 3-methylpyridazine,[78,80] 2-methylpyrazine,[78] 2-pyridones,[81] 1,3,4-thiadiazoles,[82] 1,4-dihydro-4-oxo-1,8-naphthyridine,[83] pyrimidine acids[84] and esters,[85] 3-arylidenebenzothiazines[86] and 3-methyl-1,2,4-triazines.[87]

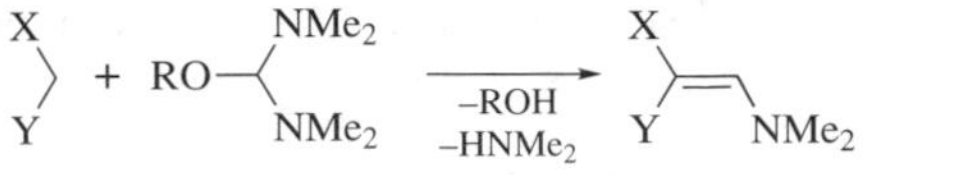

(12)

X = H, alkyl
Y = CO_2R, COSR, CS_2R, C(NR)OR, NC, CN, $CONR_2$, $CSNR_2$ $C(NR)NR_2$, $C(NMe_2)_2^+$, PPh_3^+, $PO(OR)_2$, C(NNHR')Ar aryl, heteroaryl

The condensation products thus available can be transformed into various types of compounds. Eqs 13–17 give some examples.[34,35,43,46,70,73,79,81]

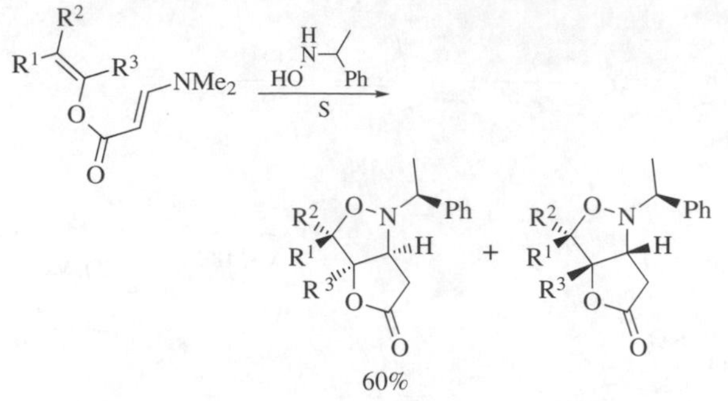

(13)

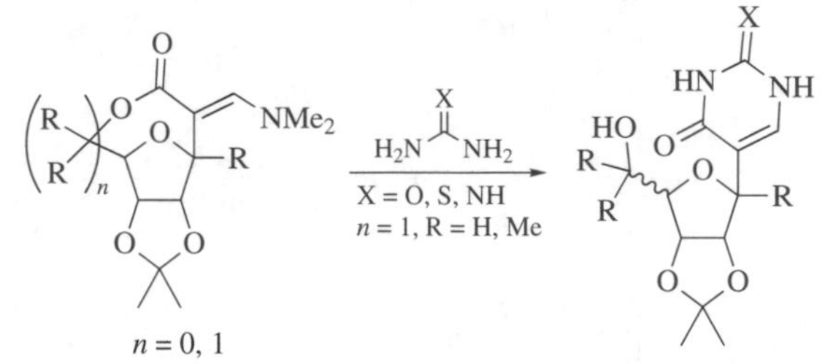

(14)

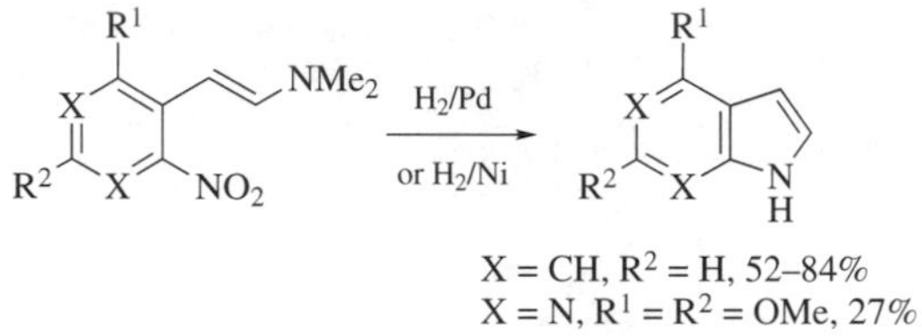

(15)

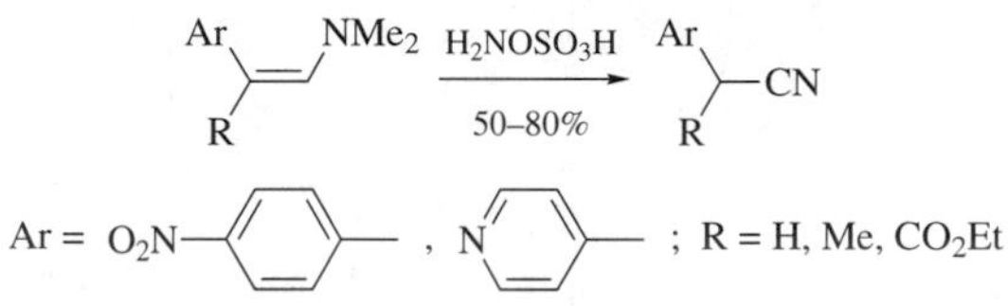

$X = CH, R^2 = H$, 52–84%
$X = N, R^1 = R^2 = OMe$, 27%

(16)

Ar = O_2N–C$_6$H$_4$–, pyridyl; R = H, Me, CO_2Et

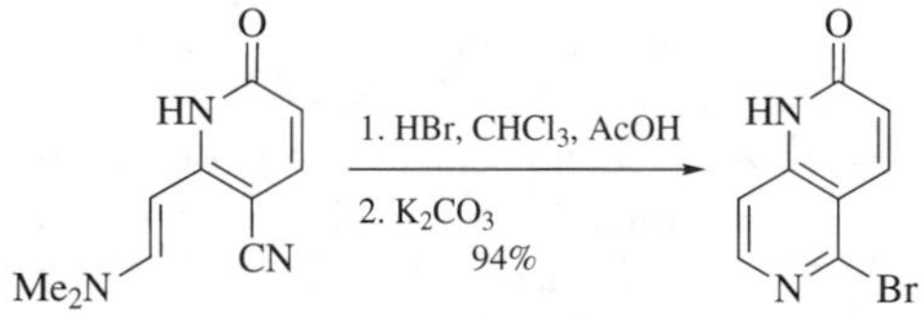

(17)

Compounds containing two acidic CH_2 groups can be formylated once or twice; examples of this type of reaction have been demonstrated with dinitriles,[59,88] 1,2-bis(phenylsulfonyl)ethane,[21] 2,2-dimethyl-1,3-cyclopentanedione,[59] 5,10-dihydroindeno[2,1-*a*]indene,[67] ethyl 4-oxoglutarate,[59] diethyl succinate,[59] diethyl adipate,[59] diethyl tetrathioadipate,[59] *N*-alkylsuccinimides,[59] and *N*-alkylglutarimides.[59] NH_2-Acidic compounds react with aminal esters analogously to CH_2-acidic compounds, i.e. the nitrogen atom is aminomethylenated. A few amides have been converted to the corresponding acylamidines (eq 18).[33,89] This condensation reaction is of less importance, because the same results can be achieved with the more accessible DMF acetals.

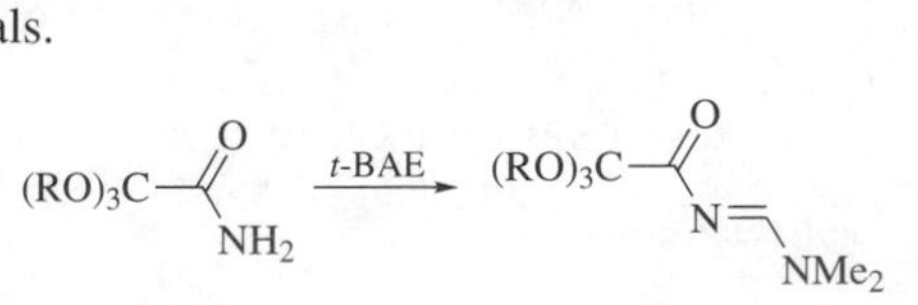

(18)

Since the primary condensation products from amides still contain an acidic methylene group, further formylation can occur (eq 19).[33] Similar results were obtained in the formylation of diethyl aminomethylphosphonate (eq 20).[90]

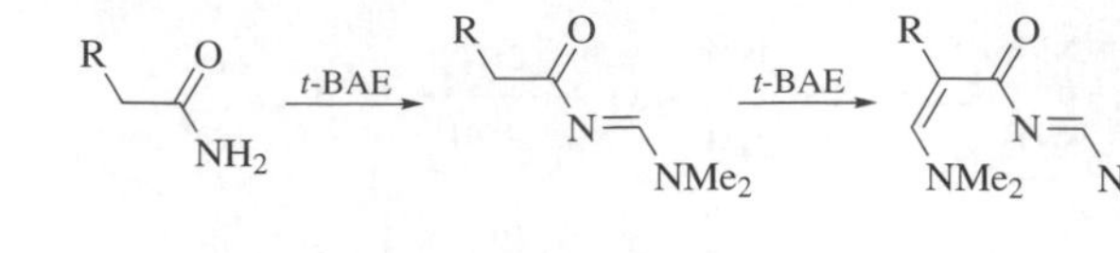

(19)

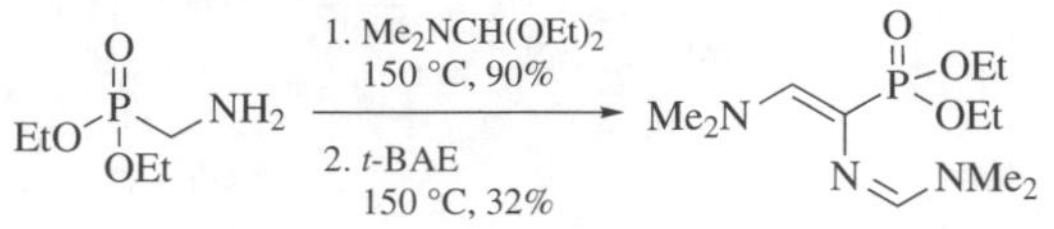

(20)

If two acidic functions of the type XH_2 and XH are present in suitable positions, the condensation reaction can be used for ring closure (eq 21).[91] In some cases the desired condensation reaction is accompanied by transformations of other functional groups present in the molecule (eqs 22 and 23).[77,92]

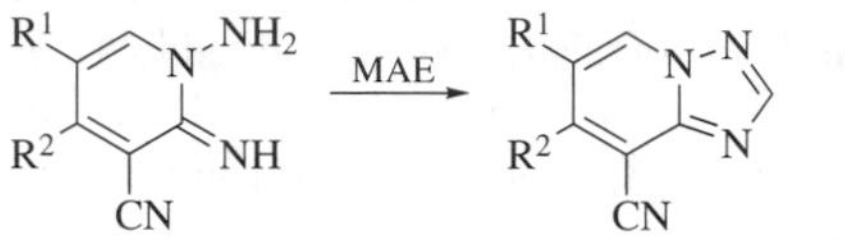

(21)

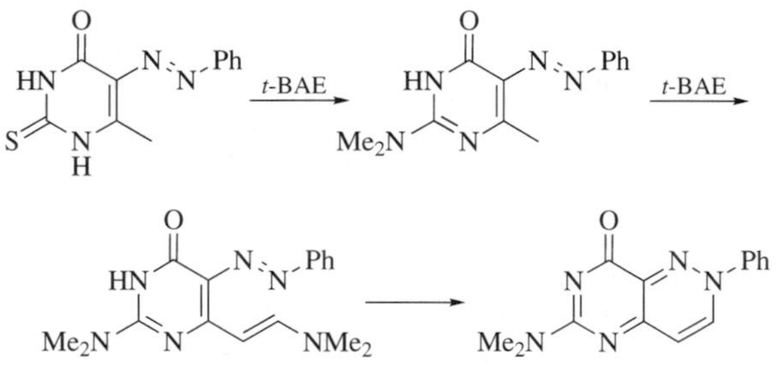

(22)

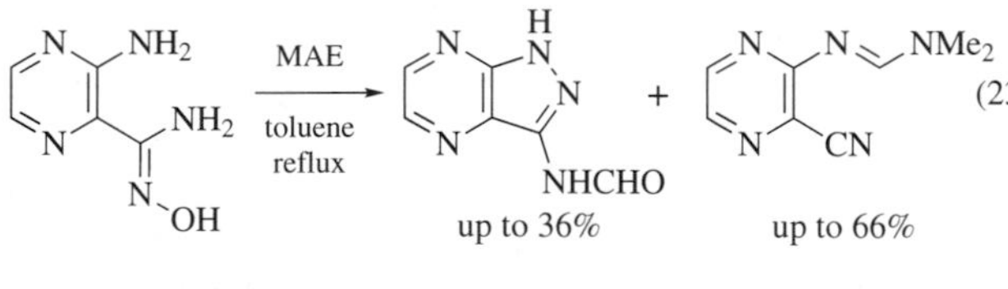

(23)

Weakly acidic heteronucleophiles substitute for the alkoxy group in aminal esters. Thus hydrocyanic acid,[83] terminal alkynes,[93] and thiols[94] afford cyano, alkynyl, and alkylthio aminals (eq 24). With secondary amines the substitution of the alkoxy group is accompanied by a transamination to give tris(dialkylamino)methanes (eq 25).[95] The alcoholysis of aminal esters proceeds very similarly, giving rise to the formation of DMF acetals,[4,84] whereas phenols are formylated in the 4-position. *N*-Monosubstituted formamides are converted by *t*-BAE to *N,N′,N′*-trisubstituted formamidines (eq 26).[38,43]

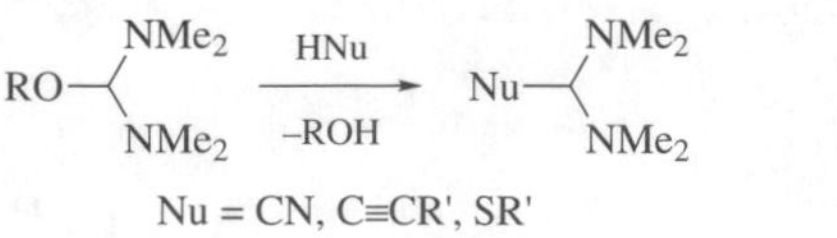

$$RO\text{–}CH(NMe_2)_2 \xrightarrow[-ROH]{HNu} Nu\text{–}CH(NMe_2)_2 \quad (24)$$

Nu = CN, C≡CR′, SR′

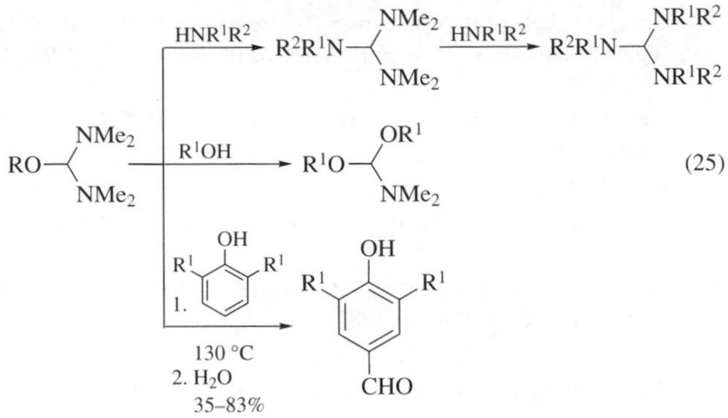

(25)

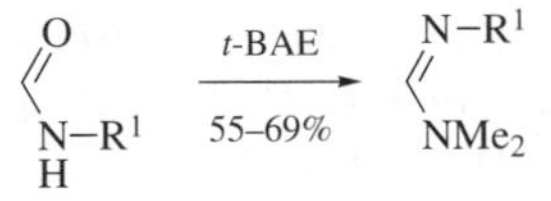

(26)

R^1 = Me, Et, *n*-Pr, $PhCH_2$, Ph

With stronger CH-, NH-, or OH-acidic compounds such as phenols bearing electron-attracting groups,[1c] imides of dicarboxylic acids,[59] tertiary CH-acidic acids,[93] and extremely strong CH_2-acidic compounds,[96] as well as mineral acids, the aminal esters act as a base to give formamidinium salts (eq 27).

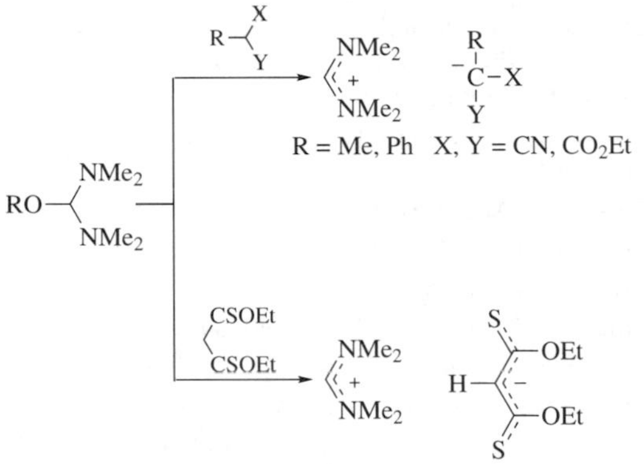

R = Me, Ph X, Y = CN, CO_2Et

(27)

The strong basicity of *t*-BAE is exploited in the synthesis of methylenedioxybenzenes from catechols and dichloromethane.[97] Electrophilic reagents such as acyl cyanides,[98] acyl azides,[99] CS_2,[1c] isothiocyanates,[100] and *N*-chloroamides[101] undergo a variety of unusual reactions with aminal esters. Although the mechanisms of these interesting reactions have been investigated thoroughly, they are – with a few exceptions – of less importance in the preparative sense because the resulting products are often obtainable with better yields from DMF acetals. Eq 28 gives three examples.

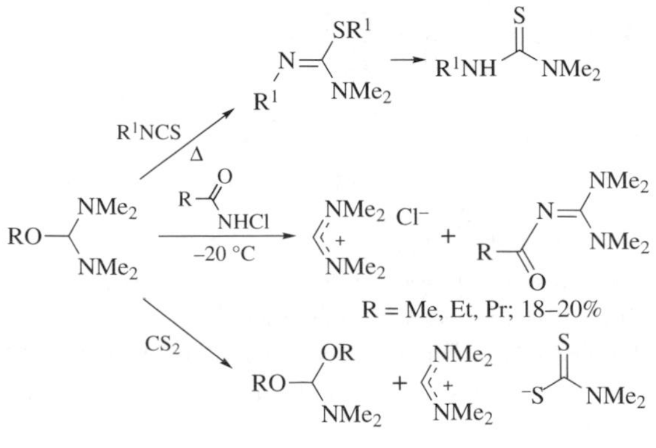

R = Me, Et, Pr; 18–20%

(28)

The reaction of *t*-BAE with Schiff bases of aromatic aldehydes, heteroaromatic aldehydes, and benzalazines at temperatures of about 160 °C represents a valuable synthesis of 1,1,2-triaminoethylenes and 1,1-diaminoethylenes (eq 29).[102] The reaction of *t*-BAE with aromatic aldehydes affords mixtures which are not easy to separate (eq 30).[103] 1,2-Diaminoethylene derivatives were isolated from reactions with *p*-tolyl- and anisaldehyde.[104]

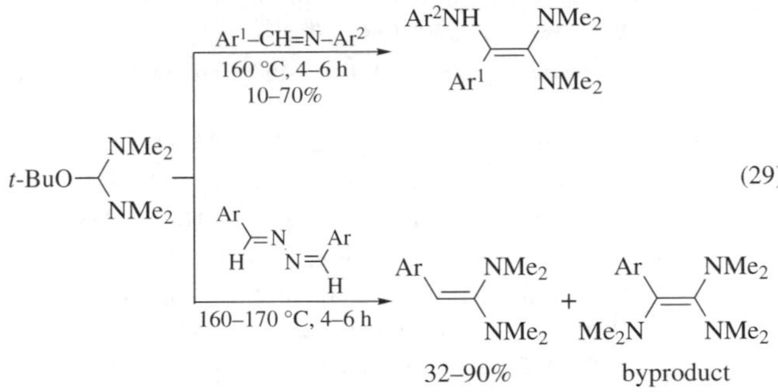

(29)

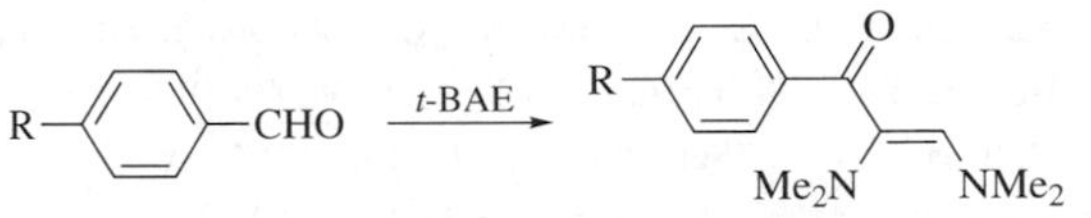

(30)

Heating selenium and *t*-BAE produces *N,N,N′,N′*-tetramethylselenourea.[105] Oxidation of MAE with peroxides affords *N,N,N′,N′*-tetramethylurea,[106,107] whereas with dibutyldiborane *N,N,N′,N′*-tetramethylformaldehydaminal is formed.[108] Formamidinium (dimethylcarbamoyltetracarbonyl)ferrate is obtained from ***Pentacarbonyliron*** and aminal esters (eq 31).[108]

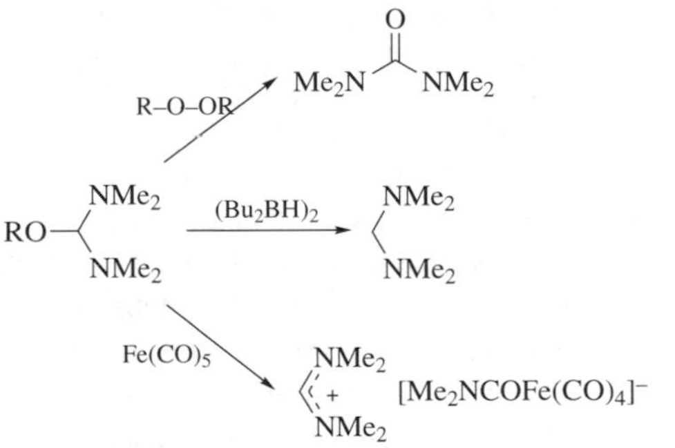

(31)

Related Reagents. Dimethylchloromethyleneammonium Chloride; *N,N*-Dimethylformamide[1]; *N,N*-Dimethylformamide Diethyl Acetal; Tris(dimethylamino)methane; Tris(formylamino)methane.

---

1. (a) De Wolfe, R. H. *Carboxylic Ortho Acid Derivatives*; Academic: New York, 1970. (b) Simchen, G. In *Methodicum Chimicum, C-N-Verbindungen*; Zymalkowski, F., Ed.; Thieme: Stuttgart, 1974; p 772. (c) Simchen, G. *Adv. Org. Chem.* **1979**, *9*, 393. (d) Kantlehner, W. in *The Chemistry of Functional Groups*; Patai, S., Ed.; Wiley: New York, 1979; Suppl. B, Part 1, p 533. (e) Effenberger, F. *AG* **1980**, *92*, 147; *AG(E)* **1980**, *19*, 151. (f) Stetter, H. *MOC* **1983**, *E3*, 8. (g) Simchen, G. *MOC* **1985**, *E5/1*, 1. (h) Kantlehner, W. *COS* **1991**, *6*, 485.
2. Bredereck, H.; Simchen, G.; Rebsdat, S.; Kantlehner, W.; Horn, P.; Wahl, R.; Hoffmann, H.; Grieshaber, P. *CB* **1968**, *101*, 41.
3. Blank, H. U.; Kraus, H.; Marzolph, G.; Müller, N.; Bayer, AG. Eur. Patent 525 536, 1993 (*CA* **1993**, *118*, 212 507a).
4. Kantlehner, W.; Speh, P. *CB* **1972**, *105*, 1340.

5. Simchen, G.; Hoffmann, H.; Bredereck, H. *CB* **1968**, *101*, 51.
6. Bredereck, H.; Effenberger, F.; Botsch, H. J. *CB* **1964**, *97*, 3397.
7. Bennett, G. B.; Simpson, W. R. J.; Mason, R. B.; Strohschein, R. J.; Mansukhani, R. *JOC* **1977**, *42*, 221.
8. Eiden, F.; Herdeis, C. *AP* **1978**, *311*, 287 (*CA* **1978**, *89*, 43 311).
9. Conrow, R.; Portoghese, P. S. *JOC* **1986**, *51*, 938.
10. Wasserman, H. H.; Ives, J. L. *JACS* **1976**, *98*, 7868.
11. Bennett, G. B.; Mason, R. B. *OPP* **1978**, *10*, 67.
12. Klose, W.; Schwarz, K. *JHC* **1982**, *19*, 1165.
13. Biere, H.; Boettcher, I.; Kapp, J. F. *AP* **1983**, *316*, 608 (*CA* **1983**, *99*, 88 106).
14. Takeuchi, N.; Okada, N.; Tobinaga, S. *CPB* **1983**, *31*, 4355.
15. Takeuchi, N.; Ochi, K.; Murase, M.; Tobinaga, S. *CPB* **1983**, *31*, 4360.
16. Schuda, P. F.; Ebner, C. B.; Morgan, T. M. *TL* **1986**, *27*, 2567.
17. Rufer, C.; Schwarz, K.; Winterfeldt, E. *LA* **1975**, 1465.
18. Dusza, J. P.; Tomcufcik, A. S.; Albright, J. D.; American Cyanamid Co. Eur. Patent 129 847, 1985 (*CA* **1985**, *102*, 220 889m).
19. Torley, L. W.; Johnson, B. B.; Dusza, J. P.; American Cyanamid Co. Eur. Patent 233 461, 1987 (*CA* **1988**, *108*, 112 478s).
20. Takeuchi, N. U.; Ochi, K.; Murase, M.; Tobinaga, S. *CC* **1980**, 593.
21. Bredereck, H.; Simchen, G.; Griebenow, W. *CB* **1974**, *107*, 1545.
22. Hoffmann, M. G.; Schmidt, R. R. *LA* **1985**, 2403.
23. Trost, B. M.; Preckel, M. *JACS* **1973**, *95*, 7862.
24. Trost, B. M.; Preckel, M.; Leichter, L. M. *JACS* **1975**, *97*, 2224.
25. Wiberg, K. B.; Furtek, B. L.; Olli, L. K. *JACS* **1979**, *101*, 7675.
26. Hutchinson, J. H.; Money, T. *TL* **1985**, *26*, 1819.
27. Wasserman, H. H.; Ives, J. L. *JOC* **1985**, *50*, 3573.
28. Fessner, W. D.; Sedelmeier, G.; Spurr, P. R.; Rihs, G.; Prinzbach, H. *JACS* **1987**, *109*, 4626.
29. Corbet, J. P.; Cotrel, C.; Farge, D.; Paris, J. M.; Rhône Poulenc Santé. Fr. Patent 2 549 064, 1985 (*CA* **1985**, *103*, 196 421a).
30. Neef, G.; Eder, U.; Haffer, G.; Sauer, G.; Schering AG. Ger. Patent 2 856 578, 1980 (*CA* **1980**, *93*, 186 672u).
31. Gonzales, F. B.; Neef, G.; Eder, U.; Wiechert, R.; Schillinger, E.; Nishino, Y. *Steroids* **1982**, *40*, 171 (*CA* **1983**, *98*, 179 733q).
32. Schuda, P. F.; Price, W. A. *JOC* **1987**, *52*, 1972.
33. Bredereck, H.; Simchen, G.; Funke, B. *CB* **1971**, *104*, 2709.
34. Wovkulich, P. M.; Uskokovic, M. R. *JACS* **1981**, *103*, 3956.
35. Wovkulich, P. M.; Uskokovic, M. R. *T* **1985**, *41*, 3455.
36. Chorvat, R. J.; Rorig, K. J. *JOC* **1988**, *53*, 5779.
37. Gayer, H.; Klausener, A.; Knueppel, P. C.; Maurer, F.; Bayer AG. Eur. Patent 471 262, 1992 (*CA* **1992**, *116*, 194 303h).
38. Klausener, A.; Knueppel, P. C.; Dehne, H. W.; Dutzmann, S.; Bayer, AG. Eur. Patent 471 261, 1992 (*CA* **1992**, *116*, 194 345y).
39. Knueppel, P. C.; Berg, D.; Dutzmann, S.; Dehne, H. W.; Haenssler, G.; Bayer, AG. Eur. Patent 503 436, 1992 (*CA* **1993**, *118*, 22 256v).
40. Schmidt, R. R.; Betz, R. *S* **1982**, 748.
41. Kantlehner, W.; Wagner, F.; Bredereck, H. *LA* **1980**, 344.
42. Noyori, R.; Sato, T.; Hayakawa, Y. *JACS* **1978**, *100*, 2561.
43. Sato, T.; Watanabe, M.; Noyori, R. *TL* **1978**, *19*, 4403.
44. Sato, T.; Kobayashi, H.; Noyori, R. *TL* **1980**, *21*, 1971.
45. Ziegler, F. E.; Fang, J.-M.; Tam, C. C. *JACS* **1982**, *104*, 7174.
46. Schmidt, R. R.; Beitzke, C.; Forrest, A. K. *CC* **1982**, 909.
47. Sato, T.; Watanabe, M.; Kobayashi, H.; Noyori, R. *BCJ* **1983**, *56*, 2680.
48. Schlessinger, R. H.; Schultz, J. A. *JOC* **1983**, *48*, 407.
49. Allan, R. D.; Johnston, G. A. R.; Kazlauskas, R.; Tran, H. *AJC* **1983**, *36*, 977.
50. Clarke, S. I.; Kasum, B.; Prager, R. H.; Ward, A. D. *AJC* **1983**, *36*, 2483.
51. Sato, T.; Hayakawa, Y.; Noyori, R. *BCJ* **1984**, *57*, 2515.
52. Farge, D.; LeRoy, P.; Moutonnier, C.; Peyronel, J. F.; Plau, B.; Rhône Poulenc Industries S. A. Eur. Patent 53 073, 1982 (*CA* **1982**, *97*, 198 035j).
53. Takeda, K.; Yano, S.-G.; Sato, M.-A.; Yoshii, E. *JOC* **1987**, *52*, 4135.
54. Danishefsky, S.; Berman, E.; Clizbe, L. A.; Hirama, M. *JACS* **1979**, *101*, 4385.
55. Tokmakov, G. P.; Zemlyanova, T. G.; Grandberg, I. I. *KGS* **1986**, 1662 (*CA* **1988**, *108*, 5877z).
56. Tokmakov, G. P.; Zemlyanova, T. G.; Grandberg, I. I. *Izv. Timiryazevks. S. Kh. Akad.* **1986**, 166 (*CA* **1987**, *107*, 217 536s).
57. Bowler, A. N.; Doyle, P. M.; Young, D. W. *CC* **1991**, 314.
58. Lessen, T. A.; Demko, D. M.; Weinreb, S. M. *TL* **1990**, *31*, 2105.
59. Bredereck, H.; Simchen, G.; Beck, G. *LA* **1972**, *762*, 62.
60. Duhamel, P.; Kotera, M.; Monteil, T. *BCJ* **1986**, *59*, 2353.
61. Bredereck, H.; Simchen, G.; Griebenow, W. *CB* **1973**, *106*, 3732.
62. Grassberger, A. M. *LA* **1974**, 1872.
63. Lieberman, D. F.; Albright, J. D. *JHC* **1988**, *25*, 827.
64. Kantlehner, W.; Speh, P.; Bräuner, H.-J.; Haug, E.; Mergen, W. W. *CZ* **1990**, *114*, 179.
65. Rufer, C.; Bahlmann, F.; Schröder, E.; Böttcher, I. *Eur. J. Med. Chem.-Chim. Ther.* **1982**, *17*, 173 (*CA* **1982**, *97*, 55 433b).
66. Alazard, J. P.; Brayer, J. L.; Tixidre, A.; Thal, C. *T* **1984**, *40*, 695.
67. Frank, W.; Gompper, R. *TL* **1987**, *28*, 3083.
68. Bredereck, H.; Simchen, G.; Wahl, R. *CB* **1968**, *101*, 4048.
69. Batcho, A. D.; Leimgruber, W.; Hofmann-La Roche. Ger. Offen. 2 057 840, 1971 (*CA* **1971**, *75*, 63 605v).
70. Kruse, L. I. *H* **1981**, *16*, 1119.
71. Schöllkopf, K.; Wachtel, H.; Schering AG. Eur. Patent 304 789, 1989 (*CA* **1989**, *111*, 97 237d).
72. Fischer, U.; Möhler, H.; Schneider, F.; Widmer, U. *HCA* **1990**, *73*, 763.
73. Biere, H.; Russe, R. *TL* **1979**, *20*, 1361.
74. Häfliger, W.; Knecht, H. *TL* **1984**, *25*, 285.
75. Black, D. C.; Strauch, R. J. *AJC* **1991**, *44*, 1217.
76. Bacon, E. R.; Daum, S. J. *JHC* **1991**, *28*, 1953.
77. Klein, R. S.; Lim, M.-I.; Tam, S. Y.-K.; Fox, J. J. *JOC* **1978**, *43*, 2536.
78. Copar, A.; Stanovnik, B.; Tisler, M. *BSB* **1991**, *100*, 533.
79. Cupps, T. L.; Wise, D. S.; Townsend, L. B. *JOC* **1983**, *48*, 1060.
80. Vors, J.-P. *JHC* **1991**, *28*, 1043.
81. Singh, B.; Lesher, G. Y. *JHC* **1990**, *27*, 2085.
82. Kantlehner, W.; Haug, E.; Hagen, H. *LA* **1982**, 298.
83. Domagala, J. M.; Peterson, P. *JHC* **1989**, *26*, 1147.
84. Rufer, C.; Schwarz, K. *Eur. J. Med. Chem.-Chim. Ther.* **1977**, *12*, 236 (*CA* **1977**, *87*, 135 244z).
85. Rufer, C.; Kessler, H. J.; Schwarz, K. *Eur. J. Med. Chem.-Chim. Ther* **1977**, *12*, 27 (*CA* **1977**, *87*, 68 197w).
86. Chorvat, R. J.; Radak, S. E.; Desai, B. N. *JOC* **1987**, *52*, 1366.
87. Reich, M. F.; Fabio, P. F.; Lee, V. J.; Kuck, N. A.; Testa, R. T. *JMC* **1989**, *32*, 2474.
88. Kantlehner, W.; Mergen, W.; Haug, E.; Speh, P.; Kapassakalidis, J. J.; Bräuner, H.-J. *LA* **1985**, 1804.
89. Kantlehner, W.; Kapassakalidis, J. J.; Maier, T. *LA* **1980**, 1448.
90. Biere, H.; Russe, R.; Seelen, W. *LA* **1986**, 1749.
91. Köckritz, P.; Liebscher, J.; Huebler, D. E. Ger. Patent 280 109, 1990 (*CA* **1991**, *114*, 102 014x).
92. Kocevar, M.; Tisler, M.; Stanovnik, B. *H* **1982**, *19*, 339.
93. Bredereck, H.; Simchen, G.; Horn, P. *CB* **1970**, *103*, 210.
94. Bredereck, H.; Simchen, G.; Hoffmann, H. *CB* **1973**, *106*, 3725.
95. Bredereck, H.; Simchen, G.; Schenck, H. U. *CB* **1968**, *101*, 3058.
96. Müller, H. G.; Hartke, K.; Kämpchen, T.; Massa, W.; Hahn, F.; *AP* **1988**, *321*, 873.

97. Leimgruber, W.; Wick, A. E.; Hoffmann-La Roche. Ger. Patent 2 214 498, 1972 (*CA* **1972**, *78*, 4233m).
98. Bredereck, H.; Simchen, G.; Kantlehner, W. *CB* **1971**, *104*, 924.
99. Bredereck, H.; Simchen, G.; Beck, G. *CB* **1971**, *104*, 3794.
100. Bredereck, H.; Simchen, G.; Rebsdat, S. *CB* **1968**, *101*, 1872.
101. Bredereck, H.; Simchen, G.; Porkert, H. *CB* **1970**, *103*, 245.
102. Bredereck, H.; Simchen, G.; Kapaun, G. *CB* **1971**, *104*, 792.
103. Bredereck, H.; Simchen, G.; Kapaun, G.; Wahl, R. *CB* **1970**, *103*, 2980.
104. Hauber, M.; Ph. D. Thesis, University of Stuttgart, 1989.
105. Kurbanov, D.; Pastushenko, E. V.; Zlotskii, S. S.; Rakhmankulov, D. L. *ZOR* **1985**, *21*, 899 (*CA* **1985**, *103*, 104 260w).
106. Kurbanov, D.; Khekimov, Yu. K.; Rol'nik, L. Z.; Zlotskii, S. S.; Rakhmankulov, D. L. *Otkrytiya Izobret* **1989**, 133 (*CA* **1990**, *112*, 20 694p).
107. Bagdasaryan, G. B.; Badalyan, K. S.; Sheiranyan, M. A.; Indzhikyan, M. G. *Arm. Khim. Zh.* **1982**, *35*, 379 (*CA* **1982**, *97*, 216 276v).
108. Daub, J.; Hasenhündl, A.; Krenkler, K. P.; Schmetzer, J. *LA* **1980**, 997.

Willi Kantlehner
Universität Stuttgart, Germany

N′-t-Butyl-*N,N*-dimethylformamidine[1]

[3717-82-6] $C_7H_{16}N_2$ (MW 128.22)

(lithiation of the title reagent generates an aminomethyl carbanion equivalent which reacts with electrophiles to give, after hydrolysis, homologated *N*-methylamines; exchange with a secondary amine allows for its subsequent lithiation and alkylation)

Physical Data: bp 133–134 °C.

Preparative Method: condensation of ***t*-Butylamine** with either *N,N*-dimethylformimidate salts or with dimethylformamide dimethyl acetal cleanly affords *N′-t*-butyl-*N,N*-dimethylformamidine (eq 1).[1,2a,b]

$Me_2N-CH(OMe)_2$ or $Me_2NCHO + Me_2SO_4$ —(t-BuNH$_2$, 80%)→ $Me_2N-CH=N-t$-Bu (1)

Handling, Storage, and Precautions: store under an inert atmosphere.

Synthetic Utility. Deprotonation of *N′-t*-butyl-*N,N*-dimethylformamidine produces a dipole-stabilized carbanion.[1,2c] Treatment of this species with electrophiles followed by hydrolysis of the formamidine affords the corresponding α-substituted amine (eq 2).[2a,d]

$Me_2N-CH=N-t$-Bu —(t-BuNH$_2$, PhCHO)→ $Me(N{=}CH{-}N\text{-}t\text{-Bu})CH_2CH(OH)Ph$ —(KOH, MeOH, H_2O, Δ, 76%)→ $MeNHCH_2CH(OH)Ph$ (2)

The related *N′-n*-butyl- and *N′*-cyclohexylformamidines (**1**) and (**2**) are also effective in the above procedure.[2a,3]

(**1**) $Me_2N-CH=N-n$-Bu (**2**) $Me_2N-CH=N$-cyclohexyl

When the title formamidine (or the *n*-butyl or cyclohexyl analog) is heated in the presence of a secondary amine, dimethylamine is displaced producing a new formamidine derivative of the secondary amine. Deprotonation and alkylation followed by formamidine hydrolysis allows entry into a host of 1-substituted amines.[1] For example, a variety of 1-substituted tetrahydroisoquinolines (eq 3) can be prepared by this method.

tetrahydroisoquinoline + $Me_2N-CH=N-t$-Bu —(PhMe, Δ, 89%)→ N-(t-BuN=CH)-tetrahydroisoquinoline —(1. s-BuLi, ethylene oxide; 2. KOH, Δ, MeOH, H_2O, 67%)→ 1-(2-hydroxyethyl)-tetrahydroisoquinoline (3)

In addition to the exchange method with the title reagent, the *t*-butylformamidines of secondary amines can be prepared by aminolysis of *O*-ethyl formimidate salts (eq 4).[4]

indoline —(1. HCO_2Et; 2. Et_3OBF_4; 3. t-BuNH$_2$; 87%)→ N-(t-BuN=CH)-indoline (4)

This route has been used to alkylate indolines (eq 5),[4a] tetrahydroquinolines,[4a] dihydroisoindoles,[4b] and *N*-methylanilines.[4a]

N-(t-BuN=CH)-indoline —(1. t-BuLi, MeI; 2. N_2H_4; 83%)→ 2-methylindoline (5)

Exchange of the title reagent with tetrahydro-β-carbolines followed by lithiation and alkylation gives access to a variety of indole alkaloids.[5] Piperidines,[2b,6,7] pyrrolidines,[2b,7] perhydroazepines (eq 6)[2b] and *cis*- and *trans*-decahydroquinolines,[8] as well as others,[7f] can also be alkylated via their formamidine derivatives.

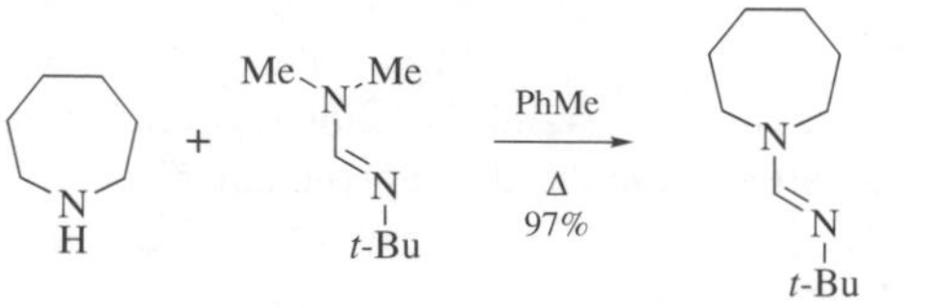

(6)

α-Arylpyrrolidines and -piperidines can be accessed by bridging dialkylations of *N*-benzyl-*N*-methylamines with 1,3- or 1,4-dihalides.[9] Other amine heterocycles, such as thiazolidine and 1,3-thiazine, may be alkylated at the carbon between the nitrogen and sulfur.[2b]

Unsaturated cyclic formamidines can be prepared by the sequence of formamidine exchange, metalation, selenation, and selenoxide elimination (eq 7).[10]

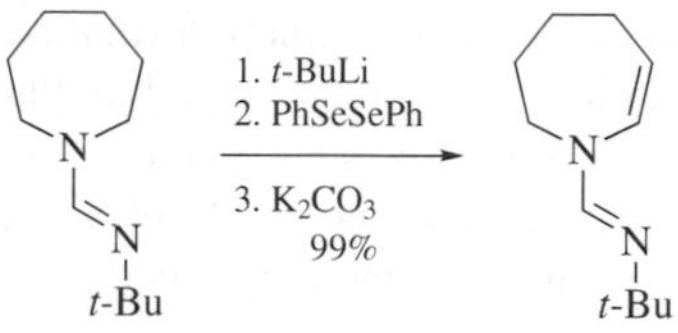

(7)

These unsaturated formamidines can be alkylated by the formamidine methodology. Metalation causes deprotonation of the alkenic proton adjacent to the ring nitrogen (eq 8).[10]

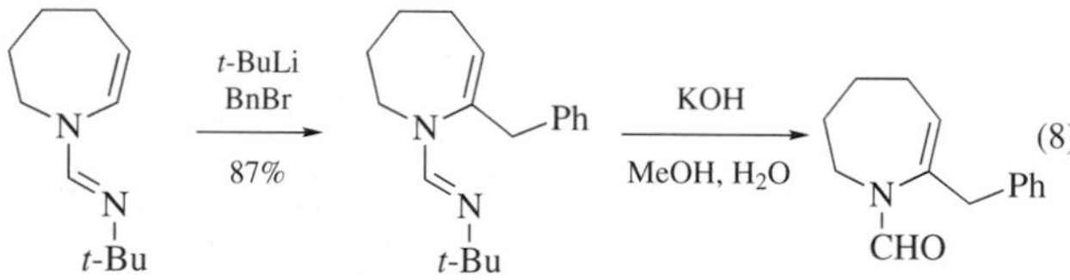

(8)

Nucleophilic additions adjacent to nitrogen have also been effected by using α′-methoxyformamidines. In this case the methoxyformamidines were not prepared from the title reagent and the formamidine is serving as a protecting and stabilizing group for the α-amino ether (eq 9).[11]

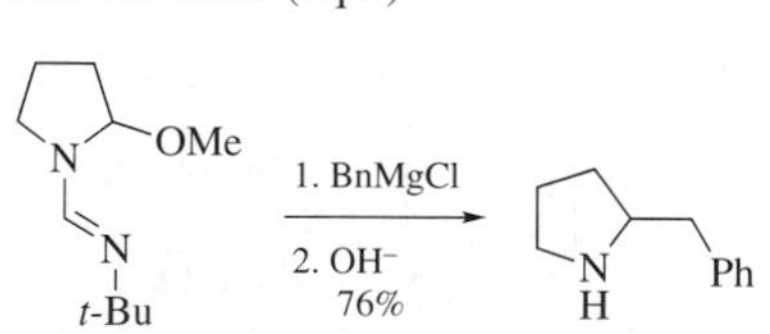

(9)

Various achiral formamidines that contain a chelating group have also been successfully utilized.[12] Thus by incorporation of a methoxy group on the formamidine, spiroisoquinolines (eq 10)[12a] and fused annulated isoquinolines can be prepared efficiently using the above methodology.

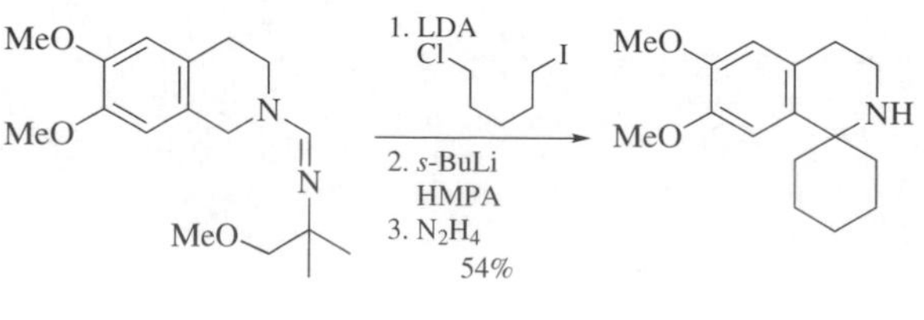

(10)

Conversion of the title reagent to ***N′-t-Butyl-N-methyl-N-trimethylsilylmethylformamidine*** (eq 11) allows for the one-carbon homologation of carbonyl compounds to nitriles, amines, aldehydes, and ketones.[13]

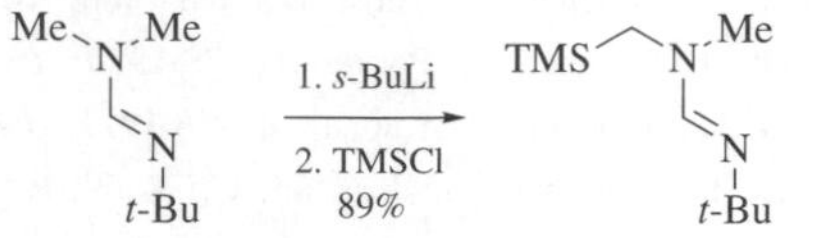

(11)

Formamidines form higher order cyanocuprates which add in a 1,4 fashion to Michael acceptors.[14] Incorporation of a chiral auxiliary in the formamidine results in high levels of asymmetric induction (see the chiral enantiopure ***(S)-N,N-Dimethyl-N′-(1-t-butoxy-3-methyl-2-butyl)formamidine***[1,2]).[1,15]

Related Reagents. *N′-t*-Butyl-*N*-methyl-*N*-trimethylsilylmethylformamidine; Dimethylaminomethyllithium; (*S*)-*N*,*N*-Dimethyl-*N′*-(1-*t*-butoxy-3-methyl-2-butyl)formamidine.

1. (a) Meyers, A. I. *Aldrichim. Acta* **1985**, *18*, 59. (b) Meyers, A. I.; Fuentes, L. M.; Bös, M.; Dickman, D. A. *CS* **1985**, *25*, NS25. Meyers, A. I. *H* **1984**, *21*, 360.
2. (a) Meyers, A. I.; Ten Hoeve, W. *JACS* **1980**, *102*, 7125. (b) Meyers, A. I.; Edwards, P. D.; Rieker, W. F.; Bailey, T. R. *JACS* **1984**, *106*, 3270. (c) Meyers, A. I.; Rieker, W. F.; Fuentes, L. M. *JACS* **1983**, *105*, 2082. (d) Solladié-Cavallo, A.; Bencheqroun, M. *TL* **1990**, *31*, 2157.
3. Meyers, A. I.; Hellring, S.; Ten Hoeve, W. *TL* **1981**, *22*, 5115.
4. (a) Meyers, A. I.; Hellring, S. *TL* **1981**, *22*, 5119. (b) Beeley, L. J.; Rockell, C. J. M. *TL* **1990**, *31*, 417.
5. (a) Meyers, A. I.; Miller, D. B.; White, F. H. *JACS* **1988**, *110*, 4778. (b) Meyers, A. I.; Loewe, M. F. *TL* **1984**, *25*, 2641. (c) Meyers, A. I.; Hellring, S. *JOC* **1982**, *47*, 2229.
6. Shawe, T. T.; Meyers, A. I. *JOC* **1991**, *56*, 2751.
7. (a) Edwards, P. D.; Meyers, A. I. *TL* **1984**, *25*, 939. (b) Thurkauf, A.; Mattson, M. V.; Richardson, S.; Mirsadeghi, S.; Ornstein, P. L.; Harrison, E. A.; Rice, K. C.; Jacobson, A. E.; Monn, J. A. *JMC* **1992**, *35*, 1323. (c) Heintzelman, G. R.; Parvez, M.; Weinreb, S. M. *SL* **1993**, 551. (d) Bolster, J. M.; Ten Hoeve, W.; Vaalburg, W.; Van Dijk, T. H.; Zijlstra, J. B.; Paans, A. M. J.; Wynberg, H.; Woldring, M. G. *Int. J. Appl. Radiat. Isot.* **1985**, *36*, 339. (e) Sanner, M. A. *TL* **1989**, *30*, 1909. (f) Monn, J. A.; Thurkauf, A.; Mattson, M. V.; Jacobson, A. E.; Rice, K. C. *JMC* **1990**, *33*, 1069.
8. Meyers, A. I.; Milot, G. *JACS* **1993**, *115*, 6652.
9. Meyers, A. I.; Marra, J. M. *TL* **1985**, *26*, 5863.
10. Meyers, A. I.; Edwards, P. D.; Bailey, T. R.; Jagdmann, G. E. *JOC* **1985**, *50*, 1019.
11. (a) Meyers, A. I.; Shawe, T. T.; Gottlieb, L. *TL* **1992**, *33*, 867. (b) Gottlieb, L.; Meyers, A. I. *TL* **1990**, *31*, 4723.
12. (a) Meyers, A. I.; Du, B.; Gonzalez, M. A. *JOC* **1990**, *55*, 4218. (b) Gonzalez, M. A.; Meyers, A. I. *TL* **1989**, *30*, 47. (c) Gonzalez, M. A.; Meyers, A. I. *TL* **1989**, *30*, 43.
13. For example, see: Santiago, B.; Meyers, A. I. *TL* **1993**, *34*, 5839.
14. Dieter, R. K.; Alexander, C. W. *TL* **1992**, *33*, 5693.
15. (a) Meyers, A. I. *T* **1992**, *48*, 2589. (b) Meyers, A. I.; Highsmith, T. K. In *Advances in Heterocyclic Natural Product Synthesis*; Pearson, W. H., Ed.; JAI: Greenwich, CT, 1990. (c) Dickman, D. A.; Boes, M.; Meyers, A. I. *OSC* **1993**, *8*, 204. (d) Meyers, A. I.; Boes, M.; Dickman, D. A. *OSC* **1993**, *8*, 573.

Todd D. Nelson & Albert I. Meyers
Colorado State University, Fort Collins, CO, USA

t-Butyl Isocyanide[1,2]

t-BuNC

[7188-38-7] C_5H_9N (MW 83.13)

(reactive amphiphilic reagent for 1,1-additions to form various heterocyclic *N*-*t*-butylimines; asymmetric amino acid synthesis; fragment condensation of peptides; mild carboxylic acid esterifications; metal-catalyzed cyanation of acetals, α,β-enones, and 1-alkenes; other applications involving various iminyl 1,1-addition reactions[1,2])

Physical Data: bp 91 °C/38 mmHg; *d* 0.74 g cm^{-3}.
Solubility: sol most organic solvents, including methanol, ethanol, ether, toluene, dichloromethane.
Form Supplied in: pure liquid; commercially available.
Analysis of Reagent Purity: by gas chromatography.
Preparative Methods: dehydration of ***N-t-Butylformamide*** with ***Phosphorus Oxychloride***[3] and by the Hofmann carbylamine reaction.[4]
Handling, Storage, and Precautions: awful smelling liquid: should be stored and used in a fume hood; contaminated equipment should be washed with 5% methanolic sulfuric acid.

Heterocyclic Synthesis. ***Silver(I) Cyanide*** catalyzes the formyl transfer from *t*-butyl isocyanide to 1,*n*-amino alcohols, diamines, and amino thiols.[5] An example is the conversion of γ-aminopropanol to 5,6-dihydro-4*H*-1,3-oxazine (eq 1). The reaction can be extended to *o*-aminophenol, *o*-phenylenediamine, and *o*-aminothiophenol, affording benzoxazole (54%), benzimidazole (64%), and benzothiazole (93%), respectively.

t-Bu–NC + [NH_2 OH] —AgCN, 66%→ [N O, H] + t-Bu–NH_2 (1)

Four-membered ring formation of 2,3-bis(alkylimino)oxetanes can be performed in high yields from ketones and 2 equiv of *t*-butyl isocyanide (eq 2).[6] Iminodioxolane derivatives are available in excellent yield from *t*-butyl cyanoketene (eq 3).[7]

(2): t-BuNC, pet. ether; BF_3•Et_2O, 88% → N-t-Bu / N-t-Bu

(3): t-BuNC, benzene, rt; 10 min, 97% → t-Bu, CN, NC, t-Bu, t-BuN

A spiro 1,4-dioxan-2-one is formed in a Lewis-acid catalyzed reaction in reasonable yields from cycloalkyl ethylene acetals, which are easily prepared from the ketone and ethylene glycol (eq 4).[8] [4 + 1] Cycloaddition of kinetically stabilized azetes with *t*-butyl isocyanide affords sterically hindered pyrrole imines (eq 5).[9] An intermediate ketenimine built from 3-aminoacrylamide reacts with *t*-butyl isocyanide, forming the diiminopyrroline system (eq 6).[10]

(4): 1. t-BuNC, $TiCl_4$, CH_2Cl_2, –70 °C to rt, 2–3 h; 2. $NaHCO_3$, 82%

(5): t-BuNC, CH_2Cl_2; rt, 24 h, 70% → N-t-Bu

(6): 1. PPh_3, NEt_3, CCl_4; 2. t-BuNC, 12 h, 70%; Ar = 2,6-$Me_2C_6H_4$

[4 + 1] Cycloaddition of 1,3-diaza-2-methylthiobutadienes with *t*-butyl isocyanide affords 4-imino-2-thioimidazolines (eq 7).[11] Isocyanides react with thiazolines, causing a ring transformation to imidazolines (eq 8).[12] In the presence of ***Boron Trifluoride Etherate***, *t*-butyl isocyanide dimerizes, forming pivalonitrile *N*-*t*-butylimine which reacts with *o*-phenylenediamine to afford 2-*t*-butylbenzimidazole (eq 9).[13]

(7): 2 equiv t-BuNC; $CHCl_3$, rt, 72 h, 71%

(8): 1. t-BuNC, Δ; 2. EtOH, reflux, 47%

(9): 2 t-BuNC —BF_3•Et_2O, pet. ether, rt, 5 d→ t-Bu–C(CN)=N-t-Bu + o-phenylenediamine —NaAc, HOAc, 120 °C, 30 min, 96%→ 2-t-butylbenzimidazole

A structurally interesting boron heterocycle, namely a borapyrrole derivative, can be made by [4 + 1] cycloaddition plus alkyl tautomerization (eq 10).[14]

(10): t-BuNC, 145 °C; 15 min, 79%

Asymmetric Synthesis of α-Amino Acids and Peptide Coupling. Enantioselective formation of α-amino acids can be performed by enantioselective Ugi reaction (eqs 11 and 12).[15,16]

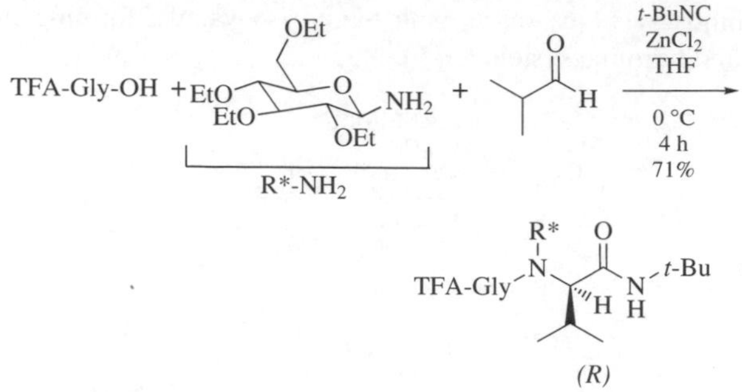

The chiral auxiliaries in eqs 11 and 12 are aminopyranoses with protected hydroxy functions, derived from glucose and galactose respectively. Another effective auxiliary is (*R*)-α-ferrocenylisobutylamine.[17] In eq 13 enantioselectively is quite high.

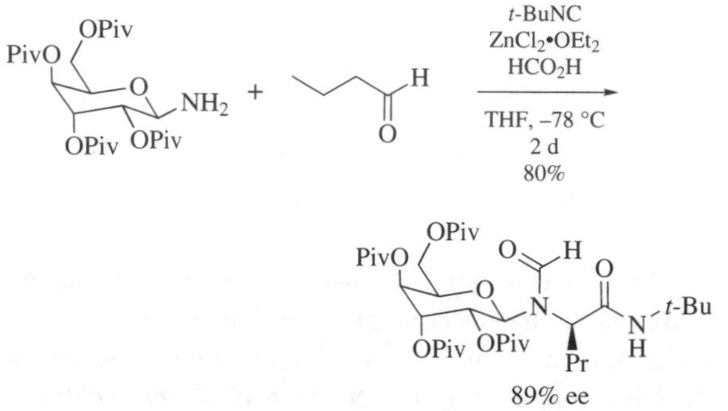

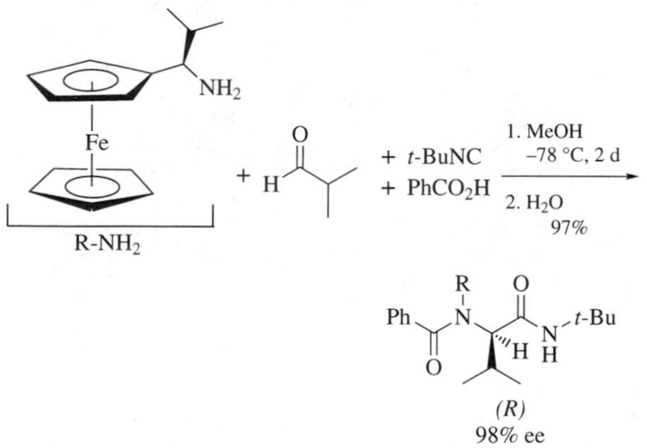

A method for fragment condensation of peptides involves using *t*-butyl isocyanide as coupling reagent (eq 14).[18]

Boc-Gly-Ala-OH + H-Phe-O-*i*-Pr —(*t*-BuNC, HOSu; MeOH, rt, 24 h, 45%)→ Boc-Gly-Ala-Phe-O-*i*-Pr (14)

Mild Esterification. In the chemistry of β-lactam antibiotics, especially of penicillins and cephalosporins, introduction of the ester group is often difficult. This problem can be solved by using the Passerini reaction with α-geminal polyhalogenated aldehydes and *t*-butyl isocyanide (eq 15).[19] The cleavage of β-haloalkyl esters can be performed under very mild conditions and regioselectively by cobalt(I) phthalocyanine anion without attack of the penicillin moiety.

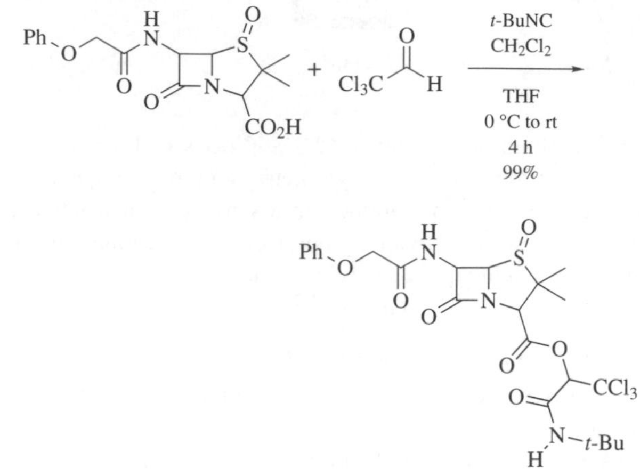

Esterification of carboxylic acids with absolute alcohols can be achieved under mild conditions using *t*-butyl isocyanide. Exceptionally high yields are achieved in forming dicarboxylic monoalkyl esters (eq 16).[20]

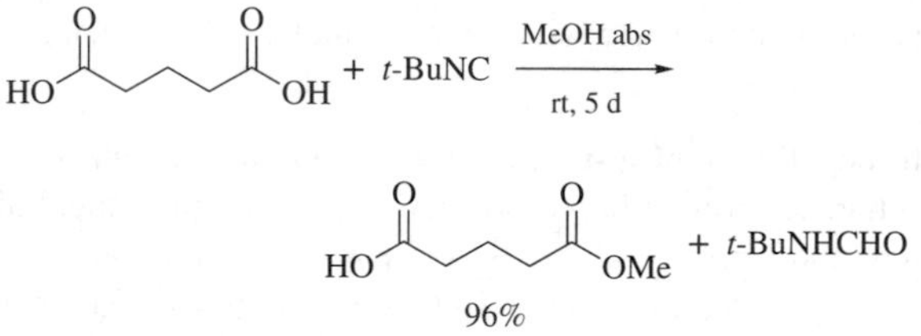

Cyanation Reactions. α-Imino nitriles are accessible in excellent yield from aldehyde acetals and *t*-butyl isocyanide (eq 17).[21] *t*-Butyl isocyanide serves as a cyanide source for cyanohydrin ether formation (eq 18).[8]

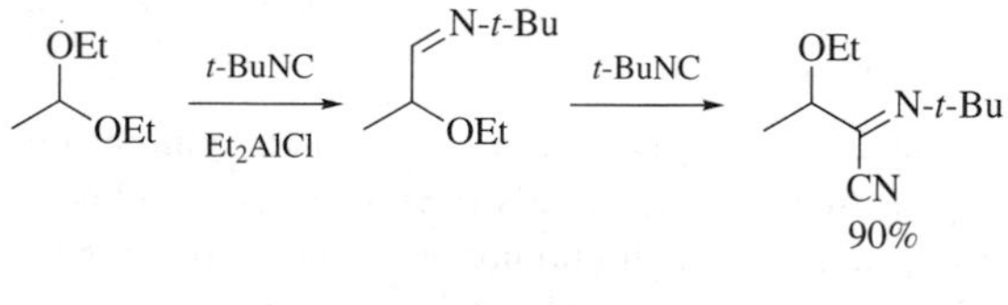

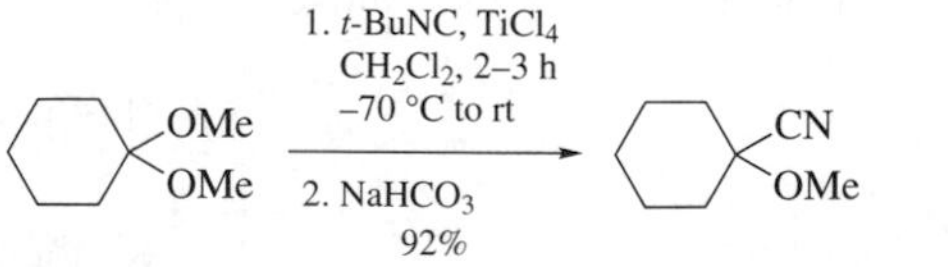

t-Butyl isocyanide undergoes formal [4 + 1] cycloaddition to α,β-enones in the presence of ***Diethylaluminum Chloride*** to give *N*-*t*-butylimines of β,γ-butenolides (∼65%). Catalytic hydrogenation and hydrolysis afford γ-butyrolactones.[22a] This tertiary isocyanide donates cyanide for conjugate hydrocyanation of α,β-enones, producing β-cyano ketones (eq 19).[22b]

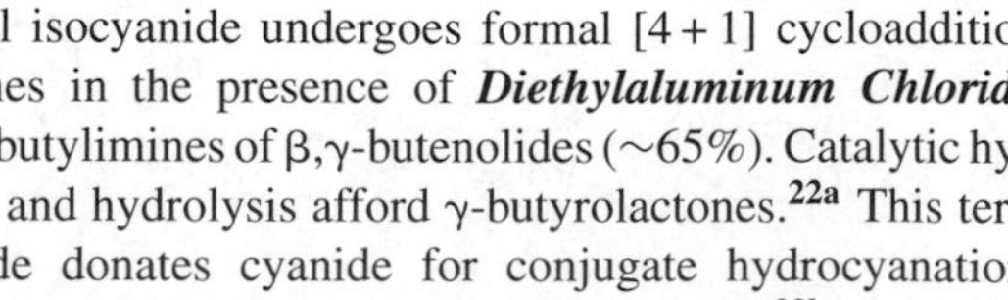

Regioselective hydrocyanation of terminal alkenes can be accomplished by hydrozirconation with Schwartz's reagent (***Chlorobis(cyclopentadienyl)hydridozirconium***), followed by imine insertion with *t*-butyl isocyanide and iodinolysis (eq 20).[23]

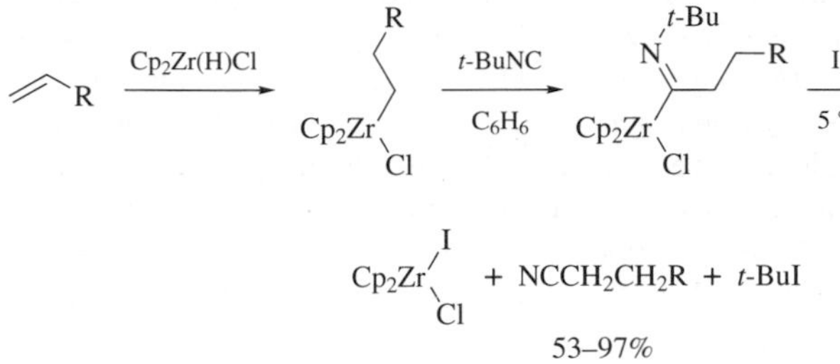

$Cp_2Zr(H)Cl$; *t*-BuNC, C_6H_6; I_2, 5 °C

Cp_2ZrICl + $NCCH_2CH_2R$ + *t*-BuI (20)

53–97%

Other Applications. Cyclohexenimines are accessible from ***Allyl Chloride***, ***Acrylonitrile***, and *t*-butyl isocyanide via *N*-*t*-butyl vinylketenimine in a [4+2] cycloaddition (eq 21).[24]

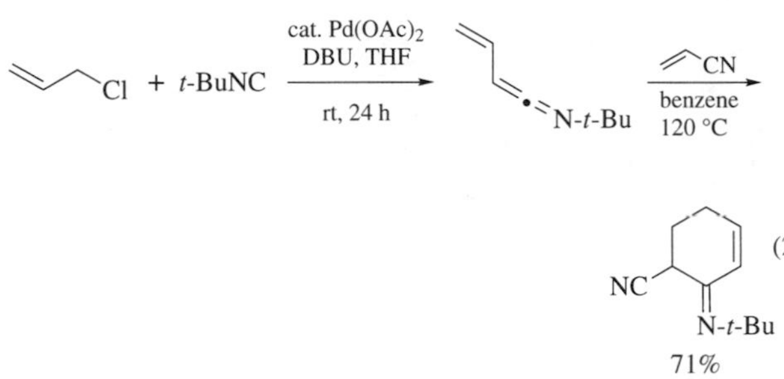

cat. $Pd(OAc)_2$, DBU, THF, rt, 24 h; CN, benzene, 120 °C (21)

71%

By a cross-coupling reaction of octyl-9-BBN, *t*-butyl isocyanide, and iodobenzene, an alkylarene ketimine is formed, which hydrolyzes by acid catalysis to the corresponding ketone (eq 22).[25] Unsymmetrical ketones may also be obtained by 1,1-addition of an alkyllithium reagent to *t*-butyl isocyanide, lithium–boron exchange with R_2BCl, thioglycolic acid-induced migration of one R group from boron to carbon, and alkaline oxidation (H_2O_2, OH^-).[26]

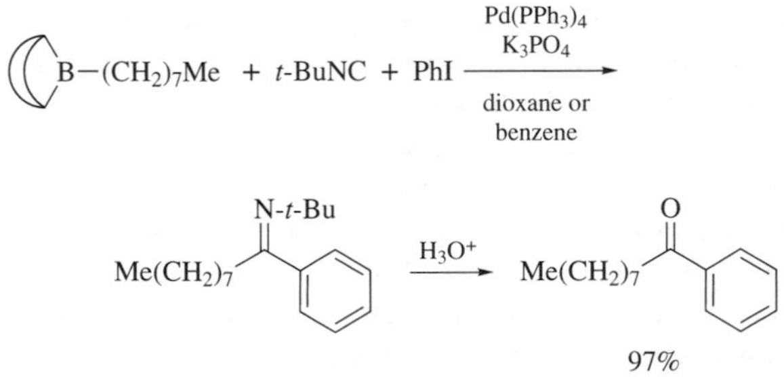

B–$(CH_2)_7Me$ + *t*-BuNC + PhI; $Pd(PPh_3)_4$, K_3PO_4, dioxane or benzene

$Me(CH_2)_7$C(=N-*t*-Bu)Ph; H_3O^+; $Me(CH_2)_7$C(=O)Ph (22)

97%

Arylimino esters are available by a palladium-catalyzed reaction of aryl halides with *t*-butyl isocyanide and ***Tri-n-butyl(methoxy)stannane***. The reaction conditions have been developed with great care (eq 23).[27] Cyclic imino esters are also formed by formal substitution of carbon monoxide with *t*-butyl isocyanide and 5-aryl-2,3-dihydro-1,3-furandiones via a ketene intermediate (eq 24).[28]

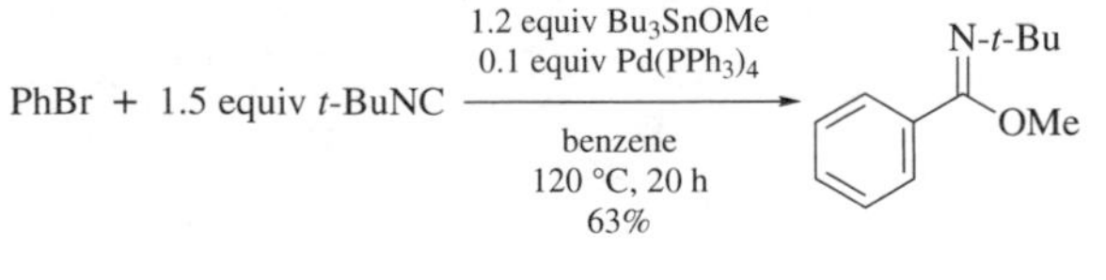

PhBr + 1.5 equiv *t*-BuNC; 1.2 equiv Bu_3SnOMe, 0.1 equiv $Pd(PPh_3)_4$, benzene, 120 °C, 20 h, 63%; N-*t*-Bu, OMe (23)

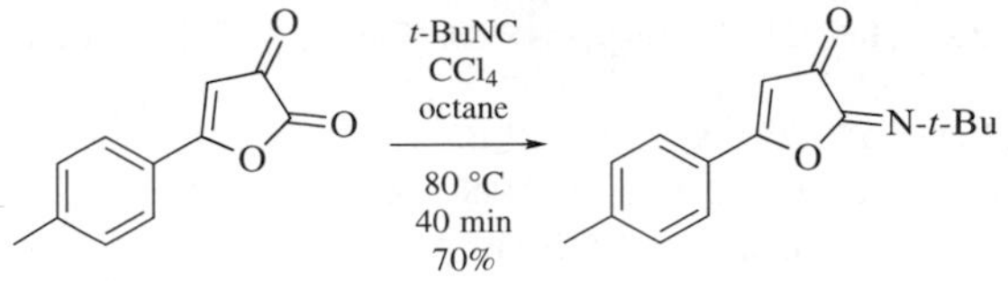

t-BuNC, CCl_4, octane, 80 °C, 40 min, 70%; N-*t*-Bu (24)

Radical carbon–carbon linkage of acrylonitrile and *t*-butyl isocyanide can be performed in acceptable yields by using ***Azobisisobutyronitrile*** as radical initiator and ***Tris(trimethylsilyl)-silane*** (eq 25).[29]

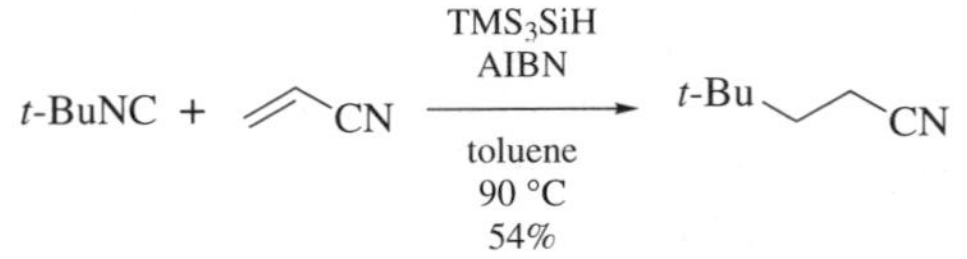

t-BuNC + CN; TMS_3SiH, AIBN, toluene, 90 °C, 54%; *t*-Bu CN (25)

Isocyanides can insert at a benzylic carbon between a carbon–sulfur and a methoxycarbonyl bond to form a ketene *S,N*-acetal containing the isocyanide carbon (eq 26).[30]

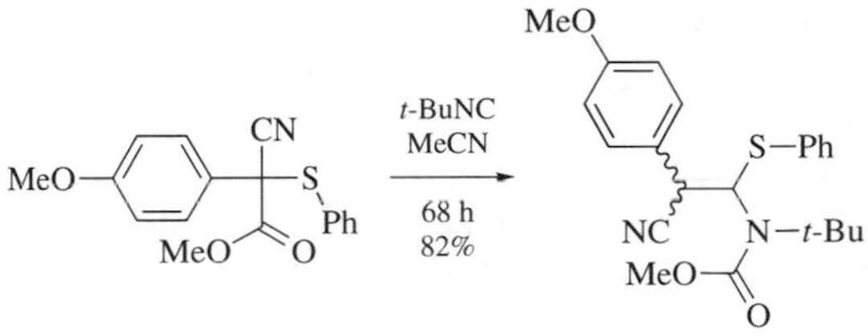

MeO, CN, S, Ph, MeO, O; *t*-BuNC, MeCN, 68 h, 82%; MeO, S–Ph, NC, N–*t*-Bu, MeO, O (26)

Polymerization of isocyanides under the action of Ni^{II} catalysis leads to helical polymethylenimines (eq 27).[31]

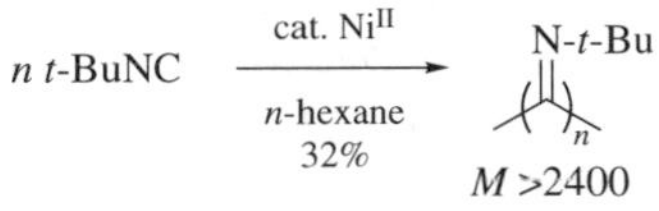

n t-BuNC; cat. Ni^{II}, *n*-hexane, 32%; N-*t*-Bu, *n*, *M* >2400 (27)

Oxidation of *t*-butyl isocyanide by elemental ***Sulfur*** with ***Tellurium*** as catalyst gives the corresponding isothiocyanate in reasonable yield (eq 28).[32] Direct oxidative addition to *t*-butyl isocyanide by elemental ***Selenium*** affords *t*-butyl isoselenocyanate.[33]

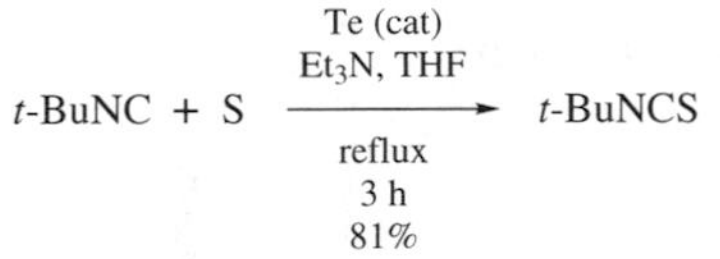

t-BuNC + S; Te (cat), Et_3N, THF, reflux, 3 h, 81%; *t*-BuNCS (28)

Related Reagents. Copper(I) Oxide-*t*-Butyl Isocyanide; Cyclohexyl Isocyanide; Ethyl Isocyanoacetate[1]; Methyl Isocyanide; Phenyl Isocyanide; 1,1,3,3-Tetramethylbutyl Isocyanide; *p*-Tolylthiomethyl Isocyanide; Triphenylmethyl Isocyanide.

1. For general reviews on isonitrile chemistry see (a) Ugi, I. *Isonitrile Chemistry*; Academic: New York, 1971. (b) Periasamy, M. P.; Walborsky, H. M. *OPP* **1979**, *11*, 295.
2. For applications of isonitriles in heterocyclic synthesis, see: Maraccini, S.; Torroba, T. *OPP* **1993**, *25*, 141.
3. Ugi, I.; Meyer, M.; Lipinski, M.; Bodensheim, F.; Rosendahl, F. *OS* **1961**, *41*, 13.
4. Ref. 1(a), Chapter II.

5. Ito, Y.; Inubushi, Y.; Zenbayashi, M.; Tomita, S.; Saegusa, T. *JACS* **1973**, *95*, 4447.
6. Kabbe, H. J. *CB* **1969**, *102*, 1404.
7. Moore, H. W.; Yu, C.-C. *JOC* **1981**, *46*, 4935.
8. Ito, Y.; Saegusa, T. *CL* **1984**, 937.
9. Hees, U.; Regitz, M. *S* **1990**, 834.
10. Capuano, L.; Dahm, B. *CB* **1988**, *121*, 271.
11. Morel, G.; Foucaud, A. *JOC* **1989**, *54*, 1185.
12. L'abbé, G.; Meutermanns, W. *BSB* **1986**, *95*, 1129.
13. Kabbe, H. J. *CB* **1969**, *102*, 1447.
14. Dorokhov, V.; Boldyreva, O. *H* **1982**, *18*, 87.
15. Goebel, M.; Ugi, I. *S* **1991**, 1095.
16. Kunz, H.; Pfrengle, W. *JACS* **1988**, *110*, 651.
17. Urban, R.; Ugi, I. *AG(E)* **1975**, *14*, 61.
18. Wackerle, L. *S* **1979**, 197.
19. (a) Eckert, H. *ZN(B)* **1990**, *45b*, 1715. (b) Eckert, H. *S* **1977**, 332.
20. Rehn, D.; Ugi, I. *JCR(S)* **1977**, 119.
21. (a) Pellissier, H.; Gil, G. *TL* **1989**, *30*, 171. (b) Ito, Y.; Kato, H.; Imai, H.; Saegusa, T. *JACS* **1982**, *104*, 6449.
22. (a) Ito, Y.; Kato, H.; Saegusa, T. *JOC* **1982**, *47*, 741. (b) Pellissier, H.; Gil, G. *T* **1989**, *45*, 3415.
23. Buchwald, S. L.; LaMaire, S. J. *TL* **1987**, *28*, 295.
24. (a) Ito, Y.; Saegusa, T. *SC* **1980**, *10*, 233. (b) Roesch, L.; Altnau, G. *AG(E)* **1979**, *18*, 60.
25. Ishiyama, T.; Oh-e, T. *TL* **1992**, *33*, 4465.
26. Yamamoto, Y.; Kondo, K.; Moritani, I. *TL* **1978**, 793.
27. Kosugi, M.; Migita, T. *CL* **1986**, 1197.
28. Andreichikov, Y.; Shurov, S. *JOU* **1986**, *22*, 766.
29. Chatgilialoglu, C.; Giese, B. *TL* **1990**, *31*, 6013.
30. Morel, G.; Foucaud, A. *T* **1984**, *40*, 1075.
31. Kamer, P. C.; Nolte, R. J. M. *JACS* **1988**, *110*, 6818.
32. Fujiwara, S.; Sonoda, N. *TL* **1992**, *33*, 7021.
33. Fujiwara, S.; Tsutomu, S. *TL* **1991**, *32*, 3503.

Heiner Eckert, Alfons Nestl & Ivar Ugi
Technische Universität München, Garching, Germany

(*R*)-2-*t*-Butyl-6-methyl-4*H*-1,3-dioxin-4-one[1]

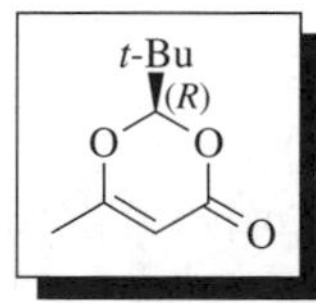

(*R*)
[107289-20-3] $C_9H_{14}O_3$ (MW 170.21)
(*S*)
[139973-88-9]

(enantiopure derivative of acetoacetic acid,[2–4] highly reactive Michael acceptor for Cu^I-doped Grignard and for Gilman reagents;[2,5] component for [2 + 2] photocycloadditions;[2] catalytic hydrogenation leads to the *cis*-disubstituted dioxanone;[2,5] the dienolate generated from the reagent can be used for chain elongations at the C(6)-Me carbon[6])

Physical Data: mp 59.8–60.2 °C; $[\alpha]_D^{rt} = -215°$ ($c = 1$, $CHCl_3$).

Solubility: sol most common organic solvents; poorly sol pentane at low temperature.

Preparative Methods: acid-catalyzed acetalization of ***Pivalaldehyde*** with (*R*)-3-hydroxybutanoic acid[8] gives the *cis*-1,3-dioxan-4-one (**2**) in 40% yield after recrystallization from ether/pentane. (Up to 60% yield can be obtained by using freshly loaded acidic ion-exchange resin and following the procedure of Seebach et al.[7b] Runs with up to 120 g hydroxy butanoic acid were performed in this way). Bromination with ***N-Bromosuccinimide*** leads to a mixture of brominated dioxinones which are debrominated hydrogenolytically to (**1**) (eq 1). The yield (**2**) → (**1**) is ~45% after recrystallization from pentane/ether (50:3) at −20 °C.[4,7] The enantiomer *ent*-(**1**) is of course equally readily available from (*S*)-3-hydroxybutanoic acid.[9] Both enantiomers of 3-hydroxybutanoic acid are commercially available.

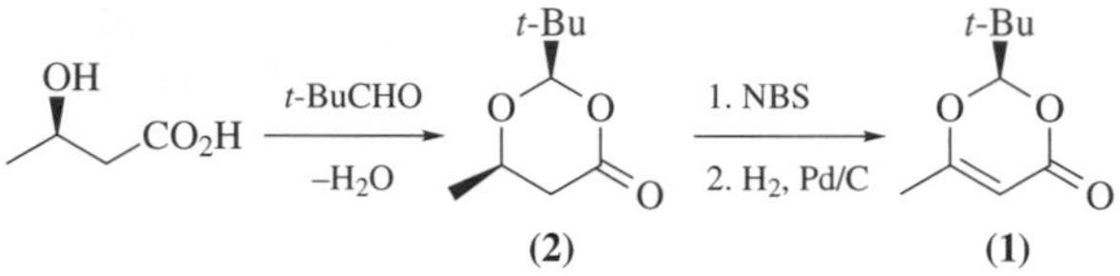

(1)

Handling, Storage, and Precautions: dioxinone (**1**) is commercially available and is indefinitely stable as a crystalline solid stored in a dark bottle at rt.

Reactions of 4*H*-1,3-Dioxin-4-one (1). The cuprate additions to (**1**) occur preferentially from the face *trans* to the *t*-Bu group. An example is the preparation of and correlation with mevalonolactone (**4**) in eq 2 by Michael addition of ***Lithium Diallylcuprate*** to give (**3**) and ozonolysis for degradation by one carbon.[5] Two examples of the use of the dienolate derived from dioxinone (**1**) are shown in eqs 3 and 4. The dienolate adds to aromatic aldehydes in a 1,2-fashion with reasonable diastereoselectivities at the exocyclic carbon atom. Oxidative degradation of the major diastereoisomer (**5**), obtained with benzaldehyde, leads to the β-hydroxy acid (**6**) of (*S*) configuration (eq 3).[6] With α,β-unsaturated aldehydes the exocyclic dienolate carbon reacts in a Michael addition. Thus the adduct (**7**) is isolated (53%) in a diastereomer ratio of 20:1 (eq 4).[6] Activation of the exocyclic methyl group in (**1**) is also realized by ***N-Bromosuccinimide*** bromination.[3,4,10] The resulting 6-bromomethyldioxinone has been employed in a vineomycinone B2 synthesis: see the intermediate (**8**) in eq 5.[11]

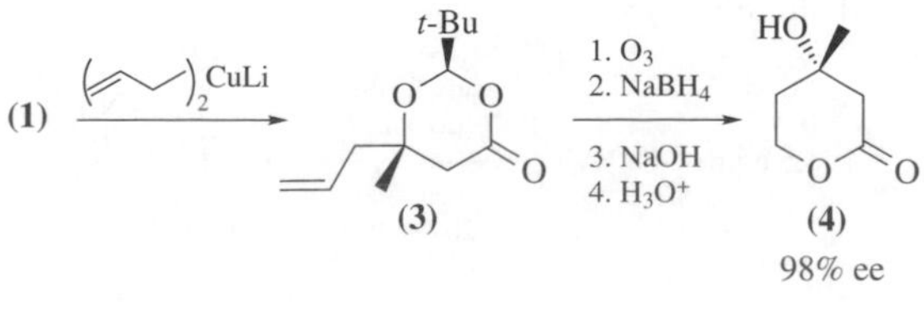

(2)

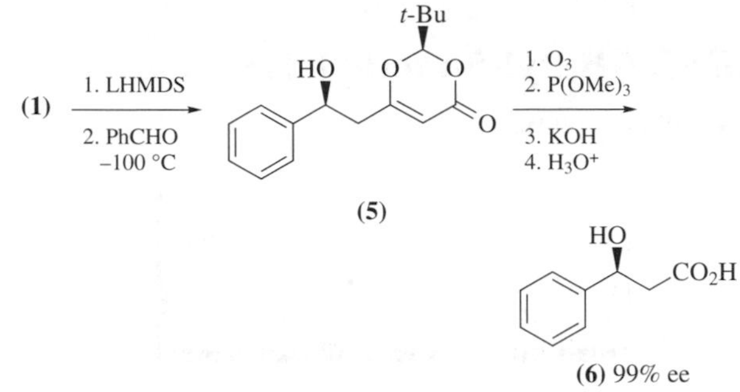

(3)

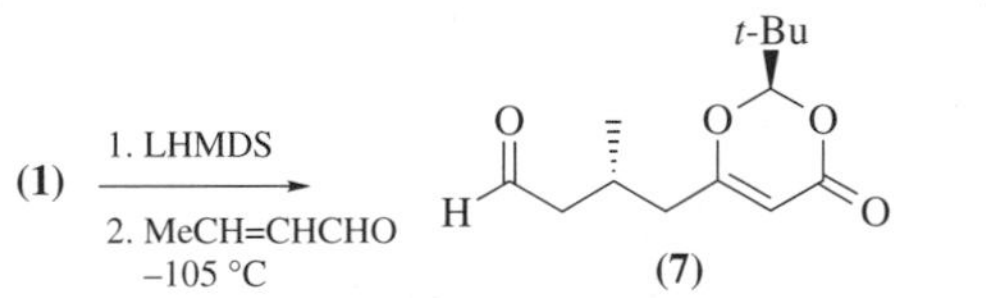

(6) 99% ee

(4)

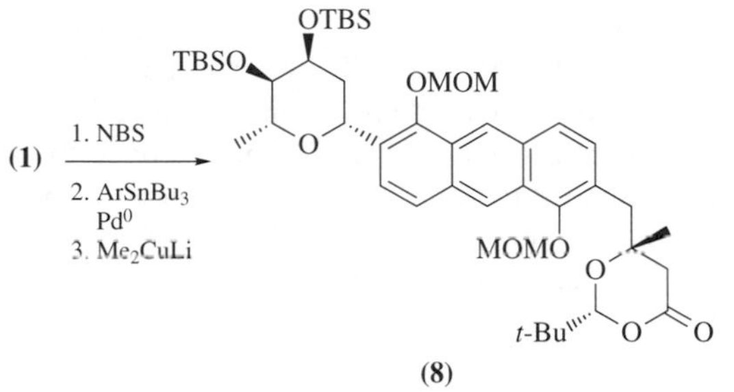

(5)

Other Enantiopure Dioxinones for Self-regeneration of the Stereogenic Center. The principle of preparing dioxinones from enantiopure β-hydroxy acids was also applied to 3-hydroxypentanoic acid,[3,6] 4,4,4-trifluoro-3-hydroxybutanoic acid,[12] 4,4,4-trichloro-3-hydroxybutanoic acid,[13] and (*S*)-serine;[3,14] aldehydes other than pivaldehyde and ketones[15] may be used for dioxinone preparation as well. Furthermore, numerous other dioxinones have been prepared from the parent compound (**1**) (eqs 3, 4, 6 and 7). The dibromide intermediate (**9**) in the preparation of (**1**) can be converted to an aldehyde, which after undergoing Wittig alkenation, followed by catalytic hydrogenation, leads to the 3-hydroxyadipic acid derivative (**10**) shown in eq 6, in an overall yield of 55%.[10] Aldol condensations of dioxanone (**2**) with aldehydes and shift of the double bond from the exo- to the endocyclic position produce 2,5,6-trisubstituted dioxinones such as (**11**), which can be used for the preparation of 2,3-disubstituted β-hydroxycarboxylic acids: see (**12**) in eq 7.[16,17] Such compounds are not accessible by current enantioselective aldol addition methodology. An example of the preparation of a CF_3-branched 3-hydroxycarboxylic acid derivative is shown in eq 8; trifluoromethyldioxinone (**13**) and ***Lithium Di-n-butylcuprate*** give a dioxanone which is solvolyzed in methanol to the hydroxy ester (**14**).[12b]

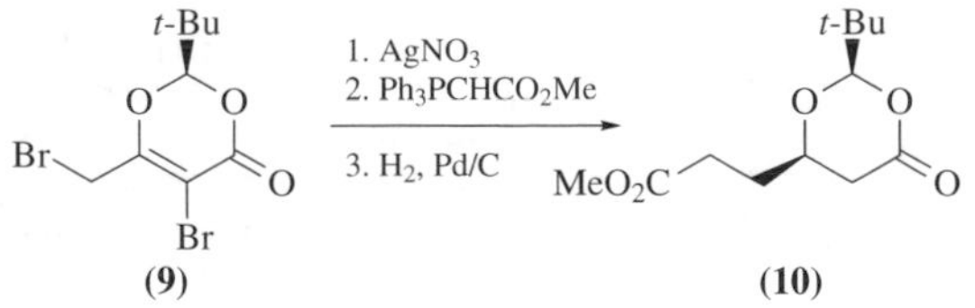

(6)

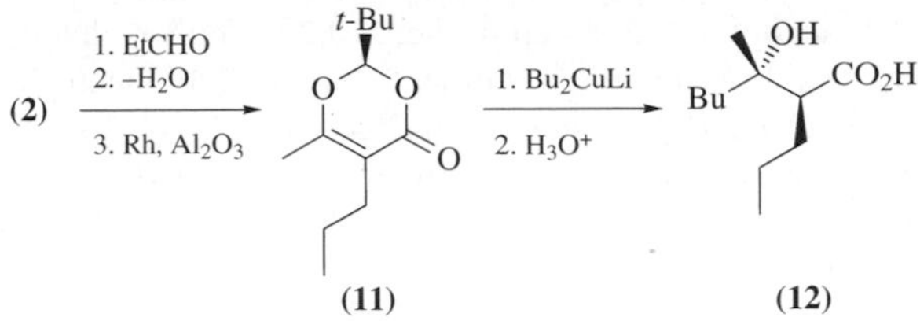

(7)

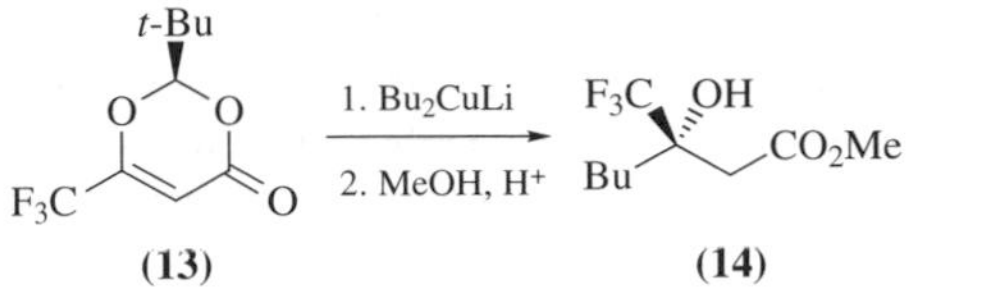

(8)

Dioxinones Obtained by Resolution or Prepared with a Chiral Auxiliary. 2-Phenyl-4*H*-1,3-dioxin-4-ones (**15**) derived from formylacetate or acetoacetate can be readily prepared in enantiopure form by preparative resolution[14,18] on cellulose triacetate.[18] These have been used for Michael additions and hydrolysis to long-chain β-hydroxycarboxylic acids, for example the tridecanoic acid (**16**) from (*R*)-(**15a**).[18] The cuprate adducts formed with the methylphenyldioxinone (*S*)-(**15b**) can be hydrogenolytically cleaved directly to β-branched β-hydroxy acids with benzyl protection of the hydroxy functional group; see (**17**) in eq 9.[18]

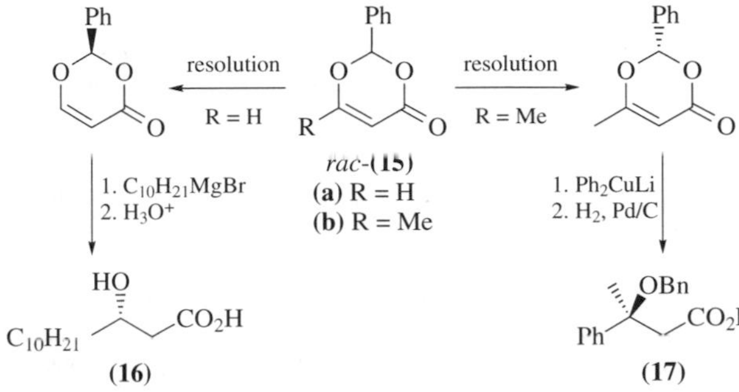

(9)

The chiral auxiliary approach involving dioxinones has been chosen by Demuth et al.[19] and, most extensively, by Kaneko and his collaborators.[20–22] They have used menthol esters (**18**) and (**19**) for typical diastereoselective reactions of dioxinones, with subsequent hydrolysis, for the preparation of various enantiopure products. For a review, also referring to the work of Winkler about photoreactions of *rac* or achiral dioxinones, see the articles by Kaneko.[23,24] For a table with enantiopure dioxinones as of mid-1991, see Kinkel et al.[14]

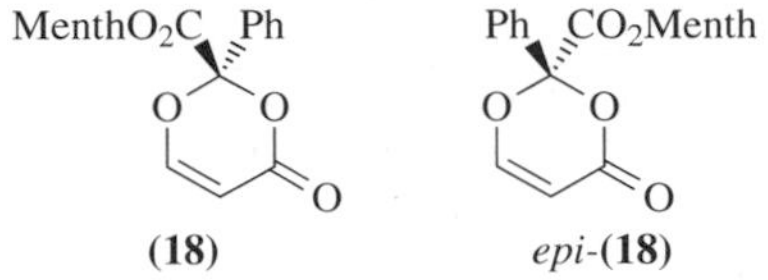

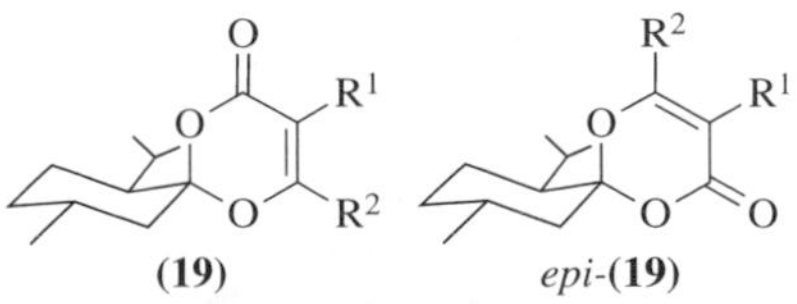

Related Reagents. (2*S*,4*S*)-3-Benzoyl-2-*t*-butyl-4-methyl-1,3-oxazolidin-5-one; (*S*)-4-Benzyl-2-oxazolidinone; (*R*,*R*)-2-*t*-

Butyl-5-methyl-1,3-dioxolan-4-one; 10,2-Camphorsultam; 10-Dicyclohexylsulfonamidoisoborneol; (*R,R*)-2,5-Dimethylborolane; (*R,R*)-1,2-Diphenyl-1,2-diaminoethane *N,N′*-Bis[3,5-bis-(trifluoromethyl)benzenesulfonamide]; Ethyl 3-Hydroxybutanoate; 2-Hydroxy-1,2,2-triphenylethyl Acetate; 2,2,6-Trimethyl-4*H*-1,3-dioxin-4-one.

1. Seebach, D.; Roggo, S.; Zimmermann, J. *Stereochemistry of Organic and Bioorganic Transformations*; Proceedings of the Seventeenth Workshop Conferences Hoechst; Bartmann, W., Sharpless, K. B., Eds; VCH: Weinheim, 1987; Vol. 17, pp 85–126.
2. Seebach, D.; Zimmermann, J. *HCA* **1986**, *69*, 1147.
3. Zimmermann, J.; Seebach, D. *HCA* **1987**, *70*, 1104.
4. Seebach, D.; Gysel, U.; Job, K.; Beck, A. K. *S* **1992**, 39.
5. Seebach, D.; Zimmermann, J.; Gysel, U.; Ziegler, R.; Ha, T.-K. *JACS* **1988**, *110*, 4763.
6. (a) Seebach, D.; Misslitz, U.; Uhlmann, P. *AG(E)* **1989**, *28*, 472. (b) Seebach, D.; Misslitz, U.; Uhlmann, P. *CB* **1991**, *124*, 1845 (*CA*, **1991**, *115*, 92 177g).
7. (a) Seebach, D.; Imwinkelried, R.; Stucky, G. *AG(E)* **1986**, *25*, 178. (b) Seebach, D.; Imwinkelried, R.; Stucky, G. *HCA* **1987**, *70*, 448 (*CA*, **1988**, *108*, 55 448f).
8. (a) Seebach, D.; Beck, A. K.; Breitschuh, R.; Job, K. *OS* **1992**, *71*, 39. (b) Kitamura, M.; Tokunaga, M.; Ohkuma, T.; Noyori, R. *OS* **1992**, *71*, 1.
9. (a) Seebach, D.; Sutter, M. A.; Weber, R. H.; Züger, M. F. *OS* **1985**, *63*, 1; *OSC* **1990**, *7*, 215. (b) Ehrler, J.; Giovannini, F.; Lamatsch, B.; Seebach, D. *C* **1986**, *40*, 172.
10. Noda, Y.; Seebach, D. *HCA* **1987**, *70*, 2137.
11. Tius, M. A.; Gomez-Galeno, J.; Gu, X.-q.; Zaidi, J. H. *JACS* **1991**, *113*, 5775.
12. (a) Acs, M.; von dem Bussche, C.; Seebach, D. *C* **1990**, *44*, 90. Beck, A. K.; Gautschi, M.; Seebach, D. *C* **1990**, *44*, 291 (*CA* **1991**, *114*, 101 862k). (b) Gautschi, M.; Seebach, D. *AC(E)* **1992**, *31*, 1083. Gautschi, M.; Schweizer, W. B.; Seebach, D. *CB* **1994**, *127*, 565 (*CA* **1994**, *121*, 107 565g).
13. Beck, A. K.; Brunner, A.; Montanari, V.; Seebach, D. *C* **1991**, *45*, 379 (*CA* **1992**, *116*, 174 083h).
14. Kinkel, J. N.; Gysel, U.; Blaser D.; Seebach D. *HCA* **1991**, *74*, 1622.
15. (a) Lange, G. L.; Organ, M. G. *TL* **1993**, *34*, 1425. (b) Organ, M. G.; Froese, R. D. J., Goddard, J. D., Taylor, N. J.; Lange, G. L. *JACS* **1994**, *116*, 3312.
16. Amberg, W.; Seebach, D. *CB* **1990**, *123*, 2429 (*CA* **1991**, *114*, 23 106a).
17. Pietzonka, T.; Seebach, D. *CB* **1991**, *124*, 1837.
18. Seebach, D.; Gysel, U.; Kinkel, J. N. *C* **1991**, *45*, 114 (*CA* **1991**, *115*, 136 015j).
19. Demuth, M.; Palomer, A.; Sluma, H.-D.; Dey, A. K.; Krüger, C.; Tsay, Y.-H. *AG(E)* , **1986**, *25*, 1117. Demuth, M.; Mikhail, G. *S* **1989**, 145.
20. Kaneko, C. *Organic Synthesis in Japan. Past, Present, and Future*, Noyori, R.; Ed., Tokyo Kagaku Dozin: Tokyo, 1992; pp 175–183.
21. Sato, M.; Murakami, M.; Kaneko, C.; Furuya, T. *T* **1993**, *49*, 8529.
22. See also: Jansen, U.; Runsink, J.; Mattay, J. *LA* **1991**, 283.
23. Kaneko, C.; Sato, M.; Sakaki, J.-i.; Abe, Y. *JHC* **1990**, *27*, 25.
24. (a) Cf. also: Takeshita, H.; Cui, Y.-S.; Kato, N.; Mori, A.; Nagano, Y. *BCJ* **1992**, *65*, 2940. (b) Winkler, J. D.; Shao, B. *TL* **1993**, *34*, 3355.

Albert K. Beck & Dieter Seebach
Eidgenössische Technische Hochschule, Zürich, Switzerland

(*R,R*)-2-*t*-Butyl-5-methyl-1,3-dioxolan-4-one[1]

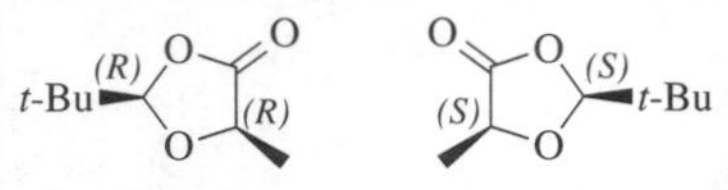

(*R,R*)-(**1**) (*S,S*)-(**1**)

(*R,R*)
[*104194-02-7*] $C_8H_{14}O_3$ (MW 158.20)
(*S,S*)
[*81037-06-1*]

(cyclic acetals from (*R*)- and (*S*)-lactic acid and pivaldehyde;[1b,2–6] reagents for the preparation of enantiopure α-hydroxy-α-methyl carboxylic acids by alkylation of the corresponding lithium enolate with alkyl,[1b,2–4] allyl,[1b,2,7,8] and benzyl[1b] halides, by hydroxyalkylation with aldehydes and ketones,[1b,3,9,10] and by Michael addition to nitroalkenes;[9,11] precursor to the 5-bromo derivative used for radical reactions;[12,13] precursor to 2-*t*-butyl-5-methylene-1,3-dioxolan-4-one;[13–15] an acceptor for radical additions;[15,16] and an ene component for Diels–Alder reactions leading to cyclic, heterocyclic, and bicyclic α-hydroxy carboxylic acids[14,17–19])

Physical Data: mp ca. 5 °C; bp 80 °C/20 mmHg; $[\alpha]^{rt}_D = +44.8°$ ($c = 1.83$, $CHCl_3$) for (*S,S*)-(**1**) containing 4% (2*R*,5*S*) epimer after two recrystallizations from ether/pentane at −75 °C.

Solubility: good to excellent in all common organic solvents.

Preparative Method: on a 0.5 mol scale reagent (*S,S*)-(**1**) is prepared by condensation of ***Pivalaldehyde*** and (*S*)-lactic acid under acid catalysis in pentane, with azeotropic removal of the water formed. The crude product is distilled in vacuo to give 93% of a 4:1 *cis*/*trans* mixture. Two recrystallizations from pentane/ether at −75 °C furnish 60% (*S,S*)-(**1**) (*cis*/*trans* = 96:4).

Handling, Storage, and Precautions: stable for many months under an inert atmosphere in a refrigerator.

Reactions of the Enolate of (1) with Electrophiles. Addition of the dioxolanones (**1**) to solutions of ***Lithium Diisopropylamide*** or ***Lithium Hexamethyldisilazide*** in THF at dry-ice temperature generates the corresponding enolates which react with alkyl halides,[1b,2–4,7,8] carbonyl compounds,[1b,3,9,10] and nitroalkenes[9,11] almost exclusively from the face remote from the *t*-Bu group to give products of type (**2**). These can be hydrolyzed to simple α-hydroxy-α-methyl carboxylic acids or further elaborated. Four examples are shown in (**3**)–(**6**) in which the part of the molecule originating from lactic acid is indicated in bold.

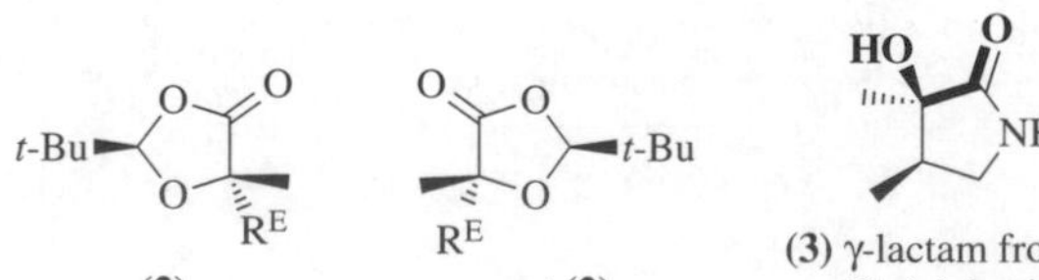

(**2**) *ent*-(**2**) (**3**) γ-lactam from (*S*)-(+)-lactic acid

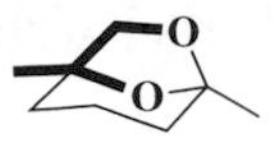

(4) (*S*)-(−)-Frontalin from (*R*)-(−)-lactic acid

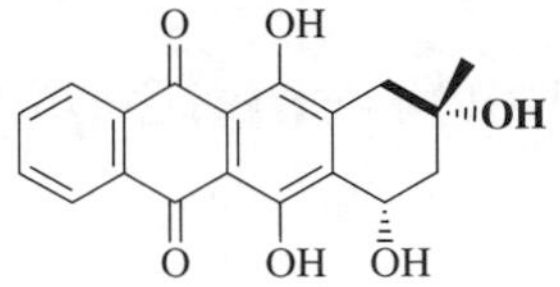

(5) 4-Demethoxyfeudomycinone C from (*S*)-(+)-lactic acid

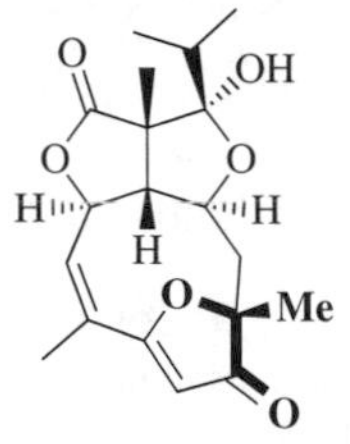

(6) (+)-Eremantholid A from (*S*)-(+)-lactic acid

Analogo us Transformations with other α-Hydroxy Carboxylic Acids. The conversion of lactic acid to products (**2**)–(**6**) is an example of the principle of self-regeneration of the stereogenic centers (SRSC) which is also applicable to β-hydroxy-, α- and β-amino acids (see Related Reagents). Many α-hydroxy carboxylic acids occur naturally, and most α-amino acids can be converted to hydroxy acids by diazotization, with retention of configuration,[20] producing a host of readily available starting materials for this kind of conversion, for example in the synthesis of the antitumor alkaloid (**7**).[21] Table 1 lists various dioxolanones (**8**) made from the corresponding hydroxy acids and aldehydes or ketones, together with information on reactions carried out with them. This method is also applicable to α-mercapto carboxylic acids.[1,22]

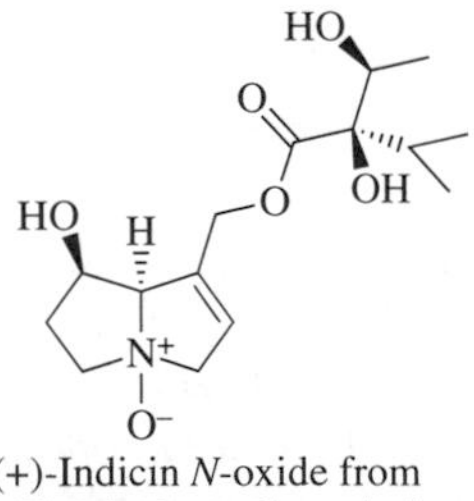

(7) (+)-Indicin *N*-oxide from (*S*)-2-hydroxy-3-methylbutanoic acid (from valine)

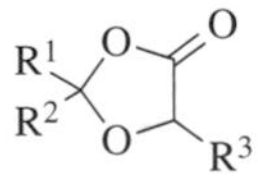

(8)

(*R*)- and (*S*)-*t*-Butyl-5-methylene-1,3-dioxolan-4-one, a Chiral α-Alkoxy Acrylate. It is also possible to introduce an exocyclic double bond onto the dioxolanone ring, as in compounds (**9**)–(**11**), derived from lactic[13,14,17–19,26] and malic[12,27] acids. These α,β-unsaturated carbonyl derivatives are acceptors for radical additions[12,15,16] and undergo cycloadditions with dienes[14,17–19,26,27] and heterodienes.[18] The Diels–Alder adduct (**12**) of *ent*-(**9**) with cyclopentadiene is formed[14,17a,19] with *exo* selectivity (96:4) and serves as a precursor to norbornenone (**13**).[14,19] Cycloadduct (**14**), obtained from methylenedioxolanone (**9**) and an open-chain triene, is also the result of an *exo* addition and is used in tetronolide synthesis.[17b]

Table 1 Various Dioxolanones (**8**) Derived from the Corresponding α-Hydroxy Acids

R^1	R^2	R^3	Reported reaction
Me	H	Me	–[2]
Ph	H	Me	–[6b]
Cy	H	Me	–[6b]
i-Pr	H	Me	–[6b]
Me	Bu	Me	–[23]
Me	Ph	Me	–[6a,23]
t-Bu	Me	Me	–[6a]
t-Bu	Ph	Me	–[6a]
t-Bu	Ph	Me	Aldol addition[24]
t-Bu	H	Et	Alkylation[7]
t-Bu	H	*i*-Pr	Aldol addition[21]
t-Bu	H	Bu	–[6b]
t-Bu	H	*s*-Bu	–[6b]
t-Bu	H	CH_2CO_2H	–[5]
t-Bu	H	CH_2CO_2H	Alkylation[25]
t-Bu	H	CH_2CO_2H	Barton reaction[15]
t-Bu	H	CH_2SPh	Oxidation/elimination[17a]
Cy	H	CH_2SPh	Oxidation/elimination[17a]
t-Bu	H	Bn	–[6b]
Mc	H	Ph	–[6b]
i-Pr	H	Ph	Alkylation[2]
t-Bu	H	Ph	–[5,6b]
t-Bu	H	Ph	Alkylation[2]
Me	Me	Ph	–[23]
C(Me)=CH_2	Me	Ph	–[23]
Bu	Me	Ph	–[23]
Ph	H	Ph	–[23]

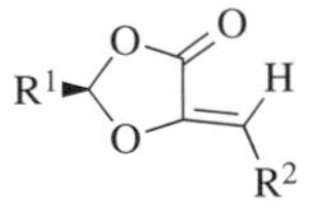

(9) R^1 = *t*-Bu, R^2 = H
(10) R^1 = C_6H_{11}, R^2 = H
(11) R^1 = *t*-Bu, R^2 = CO_2Me or CO_2Et

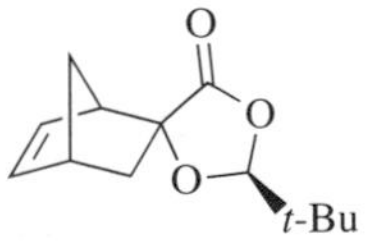

(12)

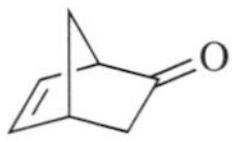

(13)

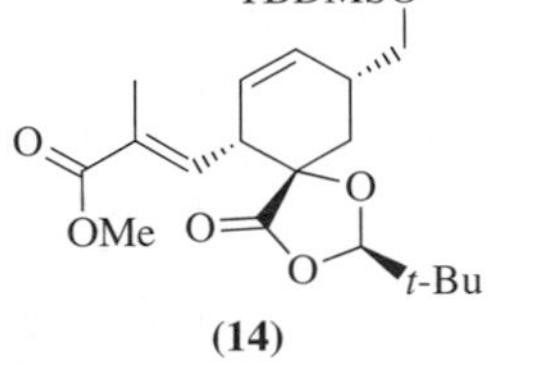

(14)

Menthone-Derived Dioxolanones: Chiral Glycolic Acid Derivatives[28]. Acetalization of menthone and phenylmenthone with glycolic acid leads to chromatographically separable mixtures of diastereoisomeric dioxolanones (**15**) and (**16**); they are precursors for chiral enolate derivatives of glycolic acid. Alkylations occur highly selectively, and the products can be solvolyzed with ethanol to give ethyl α-hydroxy carboxylates of either (*R*) or (*S*) configuration. Thus spiro compound (**16b**) gives the allylation product (**17**) (84%), from which pure ethyl (*R*)-2-hydroxypent-4-enoate is obtained.

(15) (a) R = H (b) R = Ph

(16) (a) R = H (b) R = Ph

(17) 123:1 from (16b)

Related Reagents. 1-Benzoyl-2-*t*-butyl-3,5-dimethyl-4-imidazolidinone; (2*S*,4*S*)-3-Benzoyl-2-*t*-butyl-4-methyl-1,3-oxazolidin-5-one; *t*-Butyl 2-*t*-Butyl-3-methyl-4-oxo-1-imidazolidinecarboxylate; (*R*)-2-*t*-Butyl-6-methyl-4*H*-1,3-dioxin-4-one; Ethyl 3-Hydroxybutanoate; Ethyl Mandelate; (*R*)-Methyl 2-*t*-Butyl-3(2*H*)-oxazolecarboxylate; Methyl *O*-Methyllactate; Phenoxyacetic Acid; (−)-8-Phenylmenthol.

1. (a) Seebach, D.; Imwinkelried, R.; Weber, T. In *Modern Synthetic Methods*; Springer: New York, 1986; Vol. 4, pp 125–259. (b) Seebach, D.; Naef, R.; Calderari, G. *T* **1984**, *40*, 1313.
2. Fráter, G.; Müller, U.; Günther, W. *TL* **1981**, *22*, 4221.
3. Seebach, D.; Naef, R. *HCA* **1981**, *64*, 2704.
4. Naef, R.; Seebach, D. *LA* **1983**, 1930.
5. Hoye, T. R.; Peterson, B. H.; Miller, J. D. *JOC* **1987**, *52*, 1351.
6. Greiner, A.; Ortholand, J.-Y. *TL* **1990**, *31*, 2135.
7. Krohn, K. *T* **1990**, *46*, 291.
8. Boeckman, R. K., Jr.; Yoon, S. K.; Heckendorn, D. K. *JACS* **1991**, *113*, 9682.
9. Suzuki, K.; Seebach, D. *LA* **1992**, 51.
10. Naef, R. Dissertation ETH Nr. 7442, **1983**.
11. Calderari, G.; Seebach, D. *HCA* **1985**, *68*, 1592 (*CA* **1986**, *105*, 133 326u).
12. Kneer, G.; Mattay, J. *TL* **1992**, *33*, 8051.
13. Zimmermann, J.; Seebach, D. *HCA* **1987**, *70*, 1104.
14. Mattay, J.; Mertes, J.; Maas, G. *CB* **1989**, *122*, 327.
15. Beckwith, A. L. J. Chai, C. L. L. *CC* **1990**, 1087.
16. Beckwith, A. L. J. *CSR* **1993**, *22*, 143.
17. (a) Roush, W. R.; Essenfeld, A. P.; Warmus, J. S.; Brown, B. B. *TL* **1989**, *30*, 7305. (b) Roush, W. R.; Koyama, K. *TL* **1992**, *33*, 6227.
18. Mattay, J.; Kneer, G.; Mertes, J. *SL* **1990**, 145.
19. Roush, W. R.; Brown, B. B. *JOC* **1992**, *57*, 3380.
20. Brewster, P.; Hiron, F.; Hughes, E. D.; Ingold, C. K.; Rao, P. A. D. S. *Nature* **1950**, *166*, 179.
21. Ogawa, T.; Niwa, H.; Yamada, K. *T* **1993**, *49*, 1571.
22. Strijtveen, B.; Kellogg, R. M. *T* **1987**, *43*, 5039.
23. Neveux, M.; Seiller, B.; Hagedorn, F.; Bruneau, C.; Dixneuf, P. H. *JOM* **1993**, *451*, 133.
24. Greiner, A.; Ortholand, J.-Y. *TL* **1992**, *33*, 1897.
25. Krohn, K.; Rieger, H. *LA* **1987**, 515 (*CA* **1987**, *107*, 58 713d).
26. Roush, W. R.; Brown, B. B. *TL* **1989**, *30*, 7309.
27. Kneer, G.; Mattay, J.; Raabe, G.; Krüger, C.; Lauterwein, J. *S* **1990**, 599.
28. Pearson, W. H.; Cheng, M.-C. *JOC* **1986**, *51*, 3746; **1987**, *52*, 1353, 3176; Pearson, W. H.; Hines, J. V. *JOC* **1989**, *54*, 4235.

Andrea Rolf Sting & Dieter Seebach
Eidgenössische Technische Hochschule Zürich, Switzerland

(*R*)-(+)-*t*-Butyl 2-(*p*-Tolylsulfinyl)acetate

[58059-08-8] $C_{13}H_{18}O_3S$ (MW 254.34)

(reagent for asymmetric aldol-type condensation;[1] used for the synthesis of sulfinyl dienophiles[13])

Physical Data: $[\alpha]_D^{20} = +149°$ (EtOH, $c = 2.25$)

Preparative Methods: conveniently prepared[2,3] by reaction of the magnesium enolate of *t*-butyl acetate (readily made with ***Bromomagnesium Diisopropylamide***) with ***(−)-(1R,2S,5R)-Menthyl (S)-p-Toluenesulfinate*** (eq 1). It was also made in 91% yield by reacting a solution of ***Lithium Diisopropylamide*** with (*R*)-(+)-methyl *p*-tolyl sulfoxide and *t*-butyl carbonate (eq 2).[4] It should be noted that asymmetric oxidation of *t*-butyl 2-(*p*-tolylsulfinyl)acetate with a modified Sharpless reagent gave a poor ee.[5]

(−)-(S) p-Tol S(O) OMenthyl → [$MeCO_2$-*t*-Bu, (*i*-Pr)$_2$NMgBr, Et_2O, −20 °C, 60%] p-Tol S(O) CO_2-*t*-Bu (+)-(R) (1)

O=C(O-*t*-Bu)$_2$ + p-Tol S(O) CH_2^- → (91%) p-Tol S(O) CO_2-*t*-Bu (+)-(R) (2)

Aldol-Type Addition. Aldol-type addition of the magnesium enolate of (*R*)-(+)-*t*-butyl 2-(*p*-tolylsulfinyl)acetate, prepared with *t*-butylmagnesium bromide, with aldehydes and ketones afforded, after desulfurization with ***Aluminum Amalgam***, β-hydroxy esters in very high diastereoselectivity (eq 3).[3,6,7] Two chiral centers are created in the first step with very high diastereoselectivity (mainly one diastereomer is formed). A model M based on the structure of the sulfinyl ester enolate (determined by ^{13}C NMR)[8] and on electrophilic assistance of magnesium to the carbonyl approach, was proposed to explain and predict the absolute configuration of the two created chiral centers.[3]

The first application of this aldol-type asymmetric synthesis was reported by Corey during the later stages of the total synthesis of maytansine.[9] This result (eq 4) showed that the *t*-butyl ester could be replaced by a phenyl ester as long as the same base, *t*-BuMgBr, is used for the condensation. The reaction of the α,β-unsaturated aldehyde gave, after desulfurization, the corresponding β-hydroxy ester in 80% yield and 86% de.

Enantiomerically enriched five- and six-membered lactones were also prepared by this aldol-type addition (eq 5).[10]

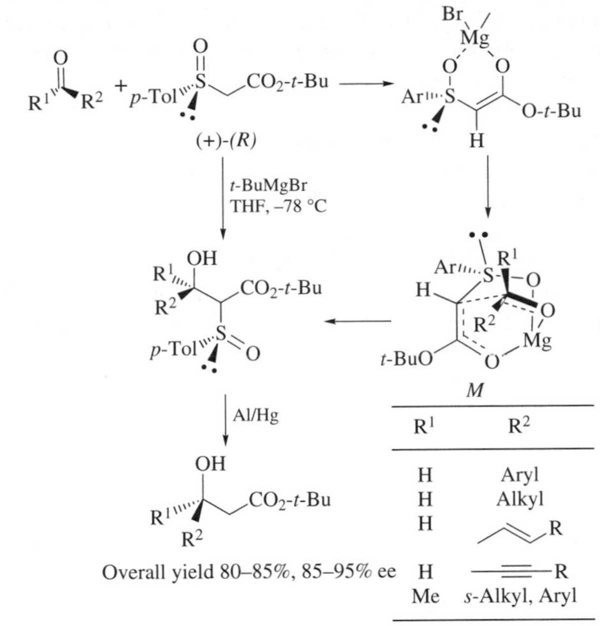

(3)

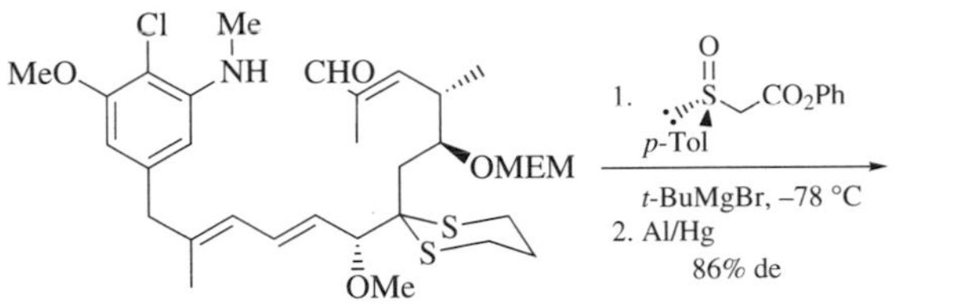

Overall yield 80–85%, 85–95% ee

R^1	R^2
H	Aryl
H	Alkyl
H	CH=CH–R
H	C≡C–R
Me	*s*-Alkyl, Aryl

1. *p*-Tol–S(O)–CH_2CO_2Ph, *t*-BuMgBr, –78 °C
2. Al/Hg

86% de (4)

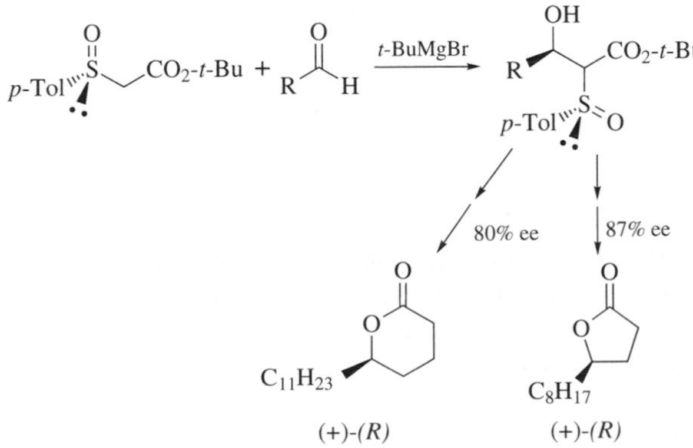

t-BuMgBr

80% ee; 87% ee (5)

(+)-(*R*) (+)-(*R*)

Propargylic aldehyde was also used to prepare, by condensation with (+)-(*R*)-*t*-butyl *p*-tolylsulfinylacetate, a precursor of the C-3–C-8 fragment of leukotriene B_4 (eq 6).[11]

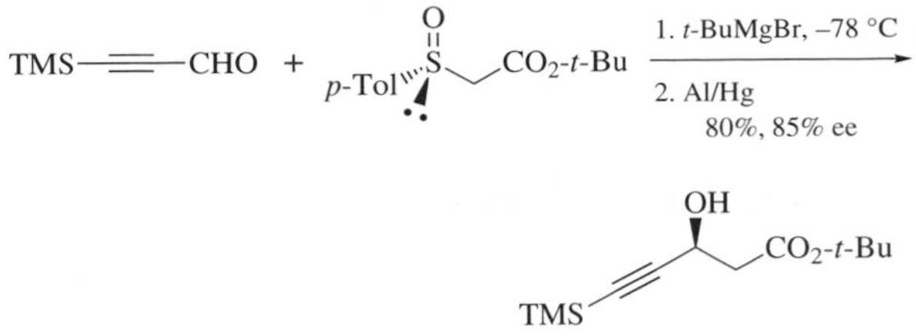

1. *t*-BuMgBr, –78 °C
2. Al/Hg
80%, 85% ee (6)

It should be noted that a poor ee was observed during the Michael addition of (+)-(*R*)-*t*-butyl *p*-tolylsulfinylacetate to an α,β-unsaturated ester.[12]

Preparation of Sulfinyl Dienophiles. This sulfinyl ester was also used to prepare optically active sulfinyl dienophiles by a Knoevenagel-type condensation of ***Glyoxylic Acid*** (eq 7).[13,14]

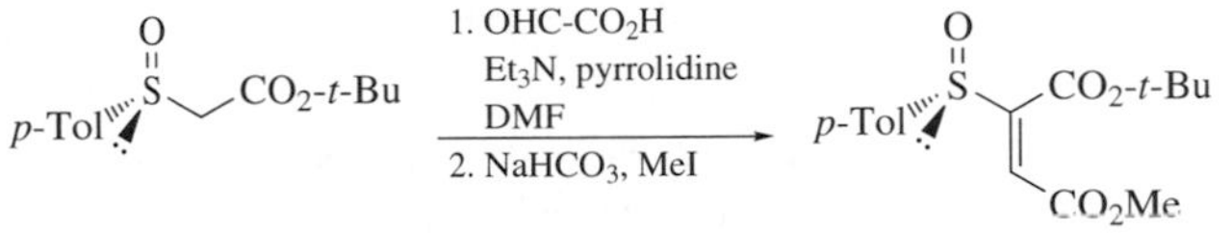

1. OHC-CO_2H, Et_3N, pyrrolidine, DMF
2. $NaHCO_3$, MeI (7)

Related Reagents. (*R*)-(+)-*t*-Butyl 2-(*p*-Tolylsulfinyl)-propionate; (*R*)-(+)-Methyl *p*-Tolyl Sulfoxide; (*R*)-(+)-Phenyl (*p*-Toluenesulfinyl)acetate.

1. Solladié, G. *S* **1981**, 185.
2. Solladié, G. In *Asymmetric Synthesis*; Morrison, J. D., Ed.; Academic: New York, 1983; Vol. 2, pp 157–198.
3. (a) Mioskowski, C.; Solladié, G. *TL* **1975**, 3341. (b) Mioskowski, C.; Solladié, G. *T* **1980**, *36*, 227.
4. Abushanab, E; Reed, D.; Suzuki, F.; Sih, C. J. *TL* **1978**, 3415.
5. Duñach, E.; Kagan, H. B. *NJC* **1985**, *9*, 1.
6. Mioskowski, C.; Solladié, G. *CC* **1977**, 162.
7. Solladié, G.; Fréchou, C.; Demailly, G. *NJC* **1985**, *9*, 21.
8. Solladié-Cavallo, A.; Mioskowski, C. *OMR* **1981**, *16*, 273.
9. Corey, E. J.; Weigel, L. O.; Chamberlin, A. R.; Cho, H.; Hua, D. H. *JACS* **1980**, *102*, 6613.
10. Solladié, G.; Matloubi-Moghadam, F. *JOC* **1982**, *47*, 91.
11. Solladié, G.; Hamdouchi C., *S* **1991**, 979.
12. Matloubi, F.; Solladié, G. *TL* **1979**, 2141.
13. Alonso, I.; Carretero, J. C.; García Ruano, J. L. *TL* **1991**, *32*, 947.
14. Alonso, I.; Cid, M. B.; Carretero, J. C.; García Ruano, J. L.; Hoyos, M. A. *TA* **1991**, *2*, 1193.

Guy Solladié & Françoise Colobert
University Louis Pasteur, Strasbourg, France

10,2-Camphorsultam[1]

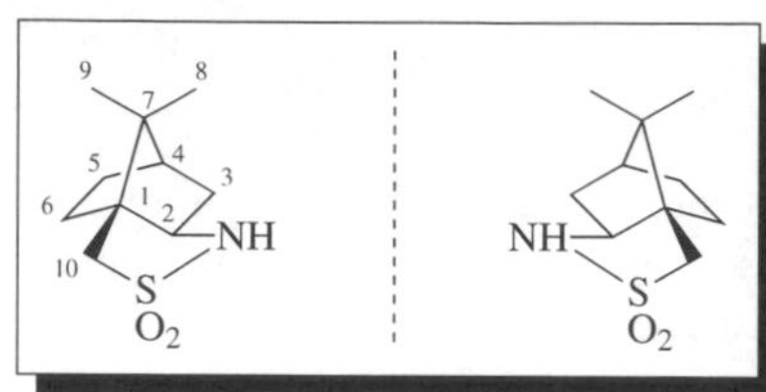

(−)-D-(2*R*)
[94594-90-8] $C_{10}H_{17}NO_2S$ (MW 215.31)
(+)-L-(2*S*)
[108448-77-7]

(versatile chiral auxiliary: *N*-enoyl derivatives undergo highly stereoselective [2 + 4] Diels–Alder[2] and [2 + 3][3] cycloadditions, cyclopropanations,[4] aziridinations,[5] dihydroxylations,[6] hydrogenations,[7] azido-iodinations[8] and conjugate hydride,[9] Grignard,[10] cuprate,[11] allylsilane[12] and thiolate[13] additions; radical additions[14] and S_N2' reactions[15] at the α-position also occur stereoselectively; enolates of *N*-acyl derivatives participate in highly stereoselective aldolizations,[16] alkylations,[17] halogenations, and 'aminations', the latter three types of reactivity being useful for α-amino acid preparation;[18] free radicals generated at the α-position of *N*-acyl derivatives participate in stereoselective intra- and intermolecular addition reactions;[19] the *N*-fluoro derivative functions as an enantioselective, electrophilic fluorinating reagent[20])

Alternate Name: bornane-10,2-sultam.
Physical Data: mp 183–185 °C (EtOH). (−)-D-(2*R*) enantiomer: $[\alpha]_D^{20}$ −31±1° ($CHCl_3$, *c* 2.3). (+)-L-(2*S*) enantiomer: $[\alpha]_D^{20}$ +34±1° (EtOH, *c* 1.00).
Form Supplied in: white crystalline solid; both enantiomers are commercially available (∼same price) or may be readily prepared (3 steps, >70% overall yield) from ***10-Camphorsulfonic Acid***.[21]
Handling, Storage, and Precautions: stable indefinitely at ambient temperature in a sealed container; mild irritant.

Introduction. Exploitation of chiral auxiliary controlled face discrimination in the reaction of a reactant with a prochiral molecule or functional group is a powerful strategy in asymmetric synthesis.[22] Clearly the choice of auxiliary for a desired chemical transformation is crucial for optimal synthetic efficiency. Hence, the ease with which the auxiliary can be introduced, the extent of stereoselection it imparts to the desired transformation, and the ease of its nondestructive removal are of critical importance. The 10,2-camphorsultam not only meets these criteria for a range of transformations, but also generally imparts crystallinity to all derived intermediates, thereby facilitating purification and isolation of enantiomerically pure products. Indeed, 10,2-camphorsultam derivatization alone allows for facile crystallographic determination of absolute configuration.[23]

Preparation of Derivatives. *N*-Acyl- and *N*-enoylsultam derivatives are routinely prepared in good yields using either sodium hydride–acid chloride[16a] or trimethylaluminum–methyl ester[18g] single-step protocols. A variant of the former method employing in situ stabilization of labile enoyl chlorides with CuCl/Cu has also been reported.[3k] A two-step procedure via the *N*-TMS derivative (**1**) is useful when a nonaqueous work-up is desirable and for synthesis of the *N*-acryloyl derivative.[24] *N*-Enoyl derivatives may also be prepared via the phosphonate derivative (**2**) by means of an Horner–Wadsworth–Emmons reaction (eq 1).[2c,2d]

Et3N, TMSCl
toluene, MeCN, rt
NH
N-TMS
(1)
1. BuLi
2. ClC(O)CH2Cl
3. P(OEt)3
NaH, R'COCl
toluene, rt
or
AlMe3, R'CO2Me
toluene, Δ
R'COCl
CuCl (cat.)
C6H6, 80 °C
(1)
P(OEt)3
(2)
R"CHO, LiCl, DBU
MeCN, rt
R = R', CH=CHR"

An *N*-acyl-β-keto derivative has been prepared by reaction with a diketene equivalent[17b] and the *trans*-N-cinnamoyl derivative by a Heck type coupling reaction.[4] The *N*-fluoro derivative (**3**) is prepared by direct fluorination (eq 2).[20]

10% F2/N2 (v/v), NaF
CHCl3, CFCl3, −40 °C
75% (cryst)
NH
NF
(3)
(2)

Reactions of *N*-Enoyl Derivatives.

***[4 + 2] Diels–Alder Cycloadditions (Alkene → Six-Membered Cycloadduct)*[2].** *N*-Enoylsultam derivatives were originally devised as 'activated chiral dienophiles' for stereoselective Diels–Alder reactions.[1,2a]

Thermal reactions of *N*-enoylsultams generally show only moderate *endo* and π-face selectivity, e.g. *N*-acryloyl- and *N*-crotonoyl-10,2-camphorsultams (**4**) and (**6**) with cyclopentadiene (eq 3, Table 1).[2g] The thermal hetero-Diels–Alder reaction of *N*-glyoxaloyl-10,2-camphorsultam with 1-methoxybuta-1,3-diene also proceeds with moderate *exo* and π-face selectivity (57% *exo*, 46% de).[2h] Thermal hetero-Diels–Alder reactions of *N*-acylnitroso-10,2-camphorsultam with cyclopentadiene and 1,3-cyclohexadiene, however, proceed with excellent selectivity (>98% ee, π-face selectivity not established).[2i]

Table 1 Intermolecular Diels–Alder Reactions of *N*-Enoylsultams (**4**)/(**5**) → (**6**) and (**4**) → (**7**)

Dienophile	Diene	Lewis acid[a]	Temp (°C)/time (h)	Adduct	Yield crude (cryst)[b] (%)	de crude (cryst) (%)
(**4**)	Cyclopentadiene	None	21 (72)	(**6**) $R^1 = H$	80[c]	66
(**5**)	Cyclopentadiene	None	21 (96)	(**6**) $R^1 = Me$	51[d]	52
(**4**)	Cyclopentadiene	$EtAlCl_2$	−130 (6)[e]	(**6**) $R^1 = H$	96 (83)	95 (99)
(**5**)	Cyclopentadiene	$TiCl_4$	−78 (1)	(**6**) $R^1 = Me$	98 (83)	93 (99)
(**4**)	1,3-Butadiene	$EtAlCl_2$	−78 (18)	(**7**) $R^2 = H$	93 (81)	94 (99)
(**4**)	Isoprene	$EtAlCl_2$	−94 (18)	(**7**) $R^2 = Me$	88 (68)	94 (99)

[a] $EtAlCl_2$ (1.5 equiv), $TiCl_4$ (0.5 equiv). [b] >98% *endo*. [c] 89% *endo*. [d] 79% *endo*. [e] EtCl as solvent.

(4) $R^1 = H$
(5) $R^1 = Me$

Lewis acid CH_2Cl_2 (3)

(6)

(7)

Lewis acid-mediated reactions of *N*-enoylsultams, on the other hand, occur under very mild conditions and with high levels of *endo* and π-face selectivity (eq 3, Table 1).[2b,2g] Dicoordinate $TiCl_4$, $EtAlCl_2$, and Me_2AlCl are particularly effective and their role in the stereodifferentiating process, which results in almost exclusive C(α)-*re* face dienophile attack, has been rationalized.[2g] Both inter- and intramolecular reactions proceed well even on a preparative scale (e.g. >100 g), often requiring just a single recrystallization to furnish isomerically pure products, valuable as synthetic intermediates (eq 4, the key step in a synthesis of (−)-pulo'upone).[2d] The hetero-Diels–Alder reaction of *N*-glyoxaloyl-10,2-camphorsultam with 1-methoxybuta-1,3-diene also proceeds efficiently and with high *endo* and π-face selectivity in the presence of 2% $Eu(fod)_3$ (90% *endo*, 88% de).[2h] These levels of asymmetric induction compare very favorably with those obtained using alternative auxiliaries (see Related Reagents below) for most substrates.

TBSO — OTBS

Me_2AlCl, CH_2Cl_2, −20 to −10 °C, 63% (cryst) (4)

all-*trans* isomer — ~100% *endo*, 93% de (crude)

[3+2] Cycloadditions (Alkene → Five-Membered Cycloadduct)[3]. The levels of selectivity found for 1,3-dipolar cycloaddition reactions are not as high as those obtained for Lewis acid-catalyzed Diels–Alder reactions. However, the 10,2-camphorsultam auxiliary can achieve synthetically useful levels of induction in these reactions, and this has been attributed to efficient enoyl conformational control by the sultam moiety leading to preferred C(α)-*re* face attack even in the absence of metal complexation.[1d]

The reactions of *N*-enoyl-10,2-camphorsultams with various nitrile oxides to give isoxazolines have been well studied.[3a–c] Indeed, the high regioselectivity and high π-face selectivity (62–90% de)[3a] observed in reactions with the *N*-acryloyl compound (**4**) have been exploited in synthesis (eq 5, the key step in a synthesis of (+)-hepialone[3b]), although related toluene-2,α-sultam auxiliaries provide still higher selectivity (see ***α-Methyltoluene-2,α-sultam***). Isoxazolines may also be obtained by regioselective and similarly π-face selective cycloadditions of silyl nitronates followed by acid catalyzed elimination of TMS alcohol[3d–f] (eq 6, the key step in a synthesis of (+)-methylnonactate).[3f] A cyclic, photochemically generated azomethine ylide also participates in *exo* and π-face selective 1,3-dipolar cycloaddition with (*ent*-**4**), a reaction for which alternative auxiliaries were significantly less effective (eq 7, the key step in a synthesis of (−)-quinocarcin).[3g–i]

Nickel catalyzed [3 + 2] cycloadditions of methylenecyclopropane and 2,2-dimethylmethylenecyclopropane with (**4**) afford 3-methylenecyclopentane derivatives with extremely high π-face selectivities (91% and 98% de respectively); five alternative auxiliaries were found to be less effective.[3j,3k] Palladium catalyzed [3 + 2] cycloaddition of 2-(TMS-methyl)-3-acetoxy-1-propene with an *N*-enoylsultam, however, proceeds with disappointing selectivity (4–26% de).[3l] A norephedrine derived auxiliary ((4*R*,5*S*)-4-methyl-5-phenyl-2-oxazolidinone) was similarly ineffective in this instance.[3l]

ent-(**4**) + [dioxolane nitro compound] — PhNCO, Et_3N, C_6H_6, 25 °C, 85% (FC) (5)

76% de (crude)

(**4**) + $EtNO_2$ — 1. TMSCl, Et_3N, toluene, rt 2. *p*-TsOH, ether, rt, 75% (cryst) (6)

86% de (crude)

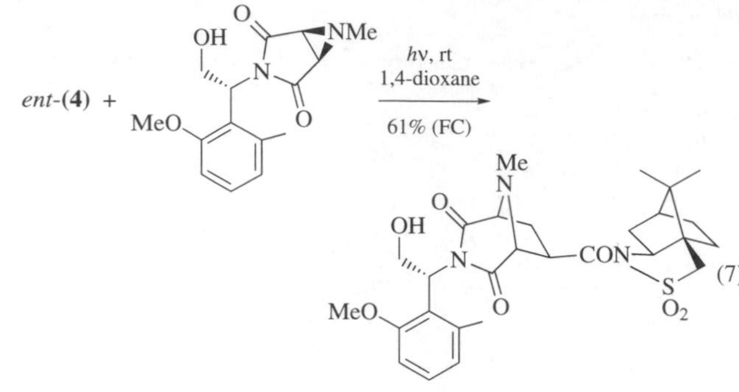

ent-(4) + [hν, rt, 1,4-dioxane, 61% (FC)] (7)

~100% *exo*, 92% de (crude)

Cyclopropanation and Aziridination (Alkene → Three-Membered Cycloadduct).[4,5] Cyclopropanation of various *trans*-*N*-enoyl derivatives using diazomethane with $Pd(OAc)_2$ as catalyst affords cyclopropyl products with good C(α)-*re* π-facial control (eq 8).[4] Similarly, aziridination with *N*-aminophthalimide–lead tetraacetate affords *N*-phthalimidoaziridines with variable but generally good π-face selectivity (33–95% de).[5]

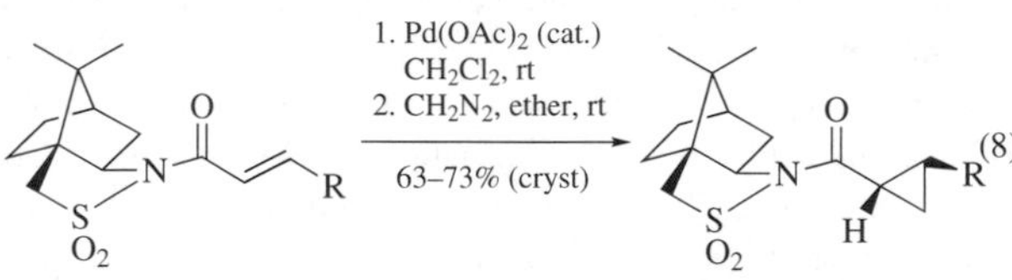

1. $Pd(OAc)_2$ (cat.) CH_2Cl_2, rt
2. CH_2N_2, ether, rt
63–73% (cryst) (8)

R = Me, Ph, *o*-$MeOC_6H_4$, *m*-$MeOC_6H_4$, 2-furyl
72.3–92.1% de (crude)

Dihydroxylation, Azido-Iodination and Hydrogenation (Alkene → α,β-A ddition Product).[6–8] *syn*-Dihydroxylation of β-substituted *N*-enoylsultams using *N*-methylmorpholine with a catalytic amount of OsO_4 affords vicinal diol products with good C(α)-*re* π-facial selectivity (80–90% de) (eq 9, the key step in a synthesis of (+)-LLP 880β).[6a] Similar levels of selectivity but lower chemical yields are obtained using $KMnO_4$ and *N*-dienoylsultams.[6b] Regioselective but poorly stereoselective *trans* addition of iodine azide to *N*-crotonoyl- and *N*-cinnamoyl-10,2-camphorsultams has also been reported (34% and 47% de, respectively). The sense of addition corresponds to iodonium ion formation from the C(α)-*re* face followed by S_N2 attack of azide at the β-position.[8] Heterogeneous *syn* hydrogenation of β,β-disubstituted enoylsultams over Pd/C using gaseous hydrogen (100 psi) affords reduced products, again with excellent C(α)-*re* topicity (90–96% de).[7]

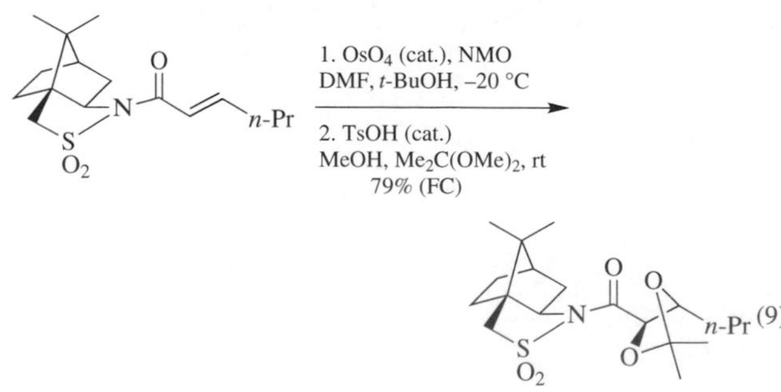

1. OsO_4 (cat.), NMO DMF, *t*-BuOH, –20 °C
2. TsOH (cat.) MeOH, $Me_2C(OMe)_2$, rt
79% (FC) (9)

83% de (crude)

1,4-Hydride, Grignard, Cuprate, Allylsilane, and Thiolate Addition (Alkene → β- or α,β-Functionalized Product).[9–13] β,β-Disubstituted enoylsultams undergo efficient reduction with L-Selectride®.[9a] The *syn* hydrogenated products obtained result from conjugate hydride delivery (and protonation) on the opposite π-face [i.e. C(α)-*si*] to that from hydrogenation (90–94% de) (eq 10).[9b] Similarly, simple alkylmagnesium chlorides also undergo 1,4-addition–protonation with *trans*-β-substituted enoylsultams from this face (72–89% de).[10] Use of α-substituted *N*-enoyl substrates,[9a] or trapping of the intermediate aluminum or magnesium enolates with other electrophiles, allows creation of two asymmetric centers in one synthetic operation.[9a,10] The observed topicity is that of *syn* addition from the C(α)-*si* face. As PBu_3 stabilized alkylcopper reagents,[11a,11b] Grignard reagents (in the presence of copper salts),[11c] and cuprates (Gilman reagents)[11d] participate in analogous reactions but show reversed π-face selectivity, an appropriate 1,4-addition–trapping protocol can be devised to generate products with any desired configuration at both the α- and β-positions (eq 11).[11c] This complementarity has been rationalized.[11] Phosphine stabilized alkyl- and alkenylcopper reagents also add to *N*-(β-silylenoyl)sultams (giving aldols after C–Si oxidative bond cleavage). In this case, either π-face selectivity can be achieved, depending on the promoting Lewis acid employed.[11b] Similar Lewis acid dependent selectivity is observed for addition of allyltrimethylsilane to *N*-enoylsultams.[12]

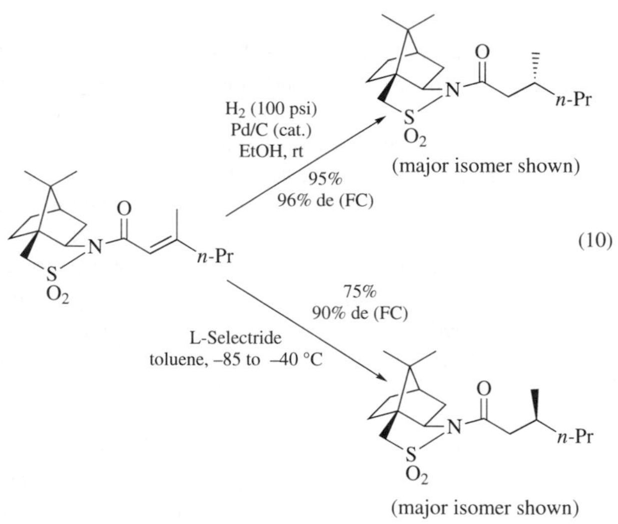

H_2 (100 psi) Pd/C (cat.) EtOH, rt; 95% 96% de (FC); (major isomer shown)
L-Selectride toluene, –85 to –40 °C; 75% 90% de (FC); (major isomer shown) (10)

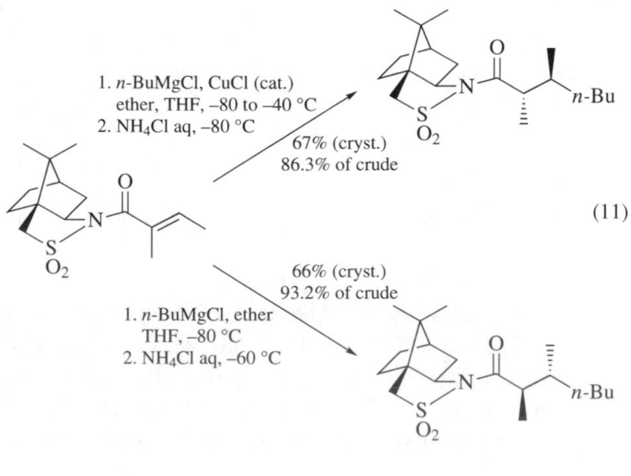

1. *n*-BuMgCl, CuCl (cat.) ether, THF, –80 to –40 °C
2. NH_4Cl aq, –80 °C
67% (cryst.) 86.3% of crude

1. *n*-BuMgCl, ether THF, –80 °C
2. NH_4Cl aq, –60 °C
66% (cryst.) 93.2% of crude (11)

Stereoselective *anti* addition of thiophenol to *N*-[β-(*n*-butyl)methacryloyl]-10,2-camphorsultam [the key step in a synthesis of (+)-*trans* whiskey lactone] has been explained by a sulfur-induced, stereoelectronically directed protonation following C(β)-*re* face conjugate addition.[13]

Radical Addition and S_N2' Displacement (Alkene → α-Functionalized Product).[14,15] Stereoselective radical additions to *N*-enoylsultams occur at the α-position, while additions to the β-position are essentially nonselective.[1d,14] The S_N2' displacement of γ-bromo-*N*-enoylsultams with higher order cyanocuprates occurs with good π-face selectivity (90–96% de).[15]

Reactions of *N*-Acyl Derivatives

Aldolization (Acyl Species → β-Hydroxyacyl Product).[16] Chiral oxazolidin-2-ones and 10,2-camphorsultams presently represent 'state of the art' aldol reaction mediators. Both auxiliaries have similarly high π-facial preferences (totally overwhelming any modest facial preference of most chiral aldehydes), allowing the predictable formation of essentially one (of four possible) diastereomeric aldol type products by judicious choice of auxiliary antipode and reaction conditions.[16a] Although sultam mediated aldolizations generally require a 2–3 fold excess of aldehyde to go to completion (cf. 1–2 equiv when oxazolidin-2-one mediated), which is clearly wasteful when employing a valuable aldehyde, the superior crystallinity and cleavage properties of the sultam adducts makes the choice of auxiliary for a given aldolization dependent on the specific substrate.

syn-Aldols with (*R*) configuration at the α-position are obtained from boryl enolates of (−)-10,2-camphorsultam derivatives (**8**) on condensation with aldehydes.[16b] The observed topicity is consistent with C(α)-*si*/C=O-*re* interaction of the 'nonchelated' (*Z*)-enolate and the aldehyde.[16b] *syn*-Aldols with (*S*) configuration at the α-position are obtained from lithium (BuLi–THF) or better tin(IV) enolates of the same derivatives (**8**), and this outcome is consistent with C(α)-*re*/C=O-*si* interaction of the 'chelated' (*Z*)-enolate and the aldehyde.[16b] *anti*-Aldols with (*S*) configuration at the α-position are obtained from in situ prepared *O*-silyl-*N,O*-ketene acetals of sultams (**8**) on condensation with aldehydes in the presence of $TiCl_4$[16c] (Mukaiyama aldolization). This topicity arises from C(α)-*re*/C=O-*re* interaction of the (*Z*)-*N,O*-ketene acetal and the Lewis acid coordinated aldehyde (eq 12).[16c] These same *anti*-aldols can also be obtained from sultams (**8**) with similarly excellent stereocontrol using boryl enolates in the presence of $TiCl_4$, and this unique procedure is the method of choice when using crotonaldehyde or methacrolein.[16d] *anti*-Aldols with (*R*) configuration at the α-position should be obtained from sultams (*ent*-**8**) using the above Mukaiyama conditions. Enantiocontrolled synthesis of α-unsubstituted β-hydroxy carbonyl compounds from the *N*-acetyl derivative is best accomplished using the Mukaiyama conditions (58–93% de).[16e] The synthesis of beetle sex pheromone (−)-serricorole serves to highlight the power of the above methods.[16f,16g]

Alkylation (Acyl Species → α-Alkylated Acyl Product).[17] An efficient procedure for the C(α)-*re* alkylation of lithium and sodium enolates of *N*-acylsultams with various (even nonactivated) primary halides in the presence of HMPA has been developed (88.7–99% de).[17a] 'Alkylation' with $ClCH_2NMeCO_2Bn$ enables a two-step β-lactam synthesis.[1c,17a] Michael-type alkylation of a β-keto derivative with arylidenemalononitriles in toluene containing piperidine has been reported to give 4*H*-pyrans (60–70% de).[17b]

(**8**)

1. R_2BOTf, *i*-Pr_2NEt CH_2Cl_2, −5 °C
2. R^2CHO, −78 °C
3. NH_4Cl aq
65–82% (cryst)
94.2–99% of crude

R = Et, Bu; R^1 = Me, Et
R^2 = Me, Et, *i*-Pr, Ph

1. *n*-BuLi, THF, −78 °C
2. $SnBu_3Cl$, −78 °C
3. R^2CHO
4. NH_4Cl aq
44–67% (cryst)
79.0–85.2% of crude

R^1 = Me, Et
R^2 = *n*-Pr, *i*-Pr, Ph

1. TBDMSOTf, Et_3N CH_2Cl_2, rt
2. $TiCl_4$, R^2CHO, −78 °C
3. NH_4Cl aq
4. MeCN
40% aq HF (10:1)
64–78% (cryst.)
98.9–>99% of crude

R^1 = Me;
R^2 = Me, Et, *i*-Pr, *i*-Bu, Ph

(12)

α-Amino Acid Preparation.[18] Three distinct strategies for the asymmetric preparation of α-amino acids using the 10,2-camphorsultam auxiliary have been developed. The first is a glycine anion strategy[18a] centered on alkylation, with excellent C(α)-*si* π-face stereocontrol, of lithium enolates of sultam derivative (**9a**) [mp 107–109 °C (EtOH)][18b,18c] to give adducts (**10**) (eq 13). α-Amino acids are obtained simply by Schiff base hydrolysis (0.5N HCl, rt) and auxiliary cleavage (LiOH, aq THF). Compound (**9b**) has also been reported to participate in analogous chemistry,[18d] but it is not crystalline and its derivatives require more vigorous hydrolysis. Commercially available (**9a**) is thus the preferred reagent, comparing favorably with other 'glycine anion' synthetic equivalents (see ***4-t-Butoxycarbonyl-5,6-diphenyl-2,3,5,6-tetrahydro-4H-oxazin-2-one***). Promising preliminary results of deprotonation–alkylation of (**9a**) under phase transfer catalysis have also been disclosed.[1c,18c,18j]

(**9a**) R = SMe
(**9b**) R = Ph

1. *n*-BuLi, THF, −78 °C
2. R^1I, HMPA
85–93% (cryst)

(**10**)

R^1 = Me, *i*-Pr, *i*-Bu, *n*-Bu, allyl, Bn
R^1 = *i*-Pr, 95% (FC)
94.7–97.7% de (crude)

(13)

The second strategy involves a bromination–azide displacement–hydrogenolysis protocol. Treatment of boryl enolates of *N*-acylsultams (**8**) with NBS provides the key α-bromo derivatives (**11**) with good C(α)-*re* topicity (eq 14).[1c,18b] Stereospecific substitution with tetramethylguanidinium azide $[(Me_2N)_2C{=}NH_2^+N_3^-]$, hydrogenolysis ($H_2$–Pd/C), and auxiliary cleavage provides α-amino acids in good overall yield.[1c,18b] As with the previous strategy, given that an appropriate derivative is crystallized to enantiomeric homogeneity, the enantiomeric purity of the product will reflect the extent of racemization during auxiliary hydrolysis (e.g. phenylglycine: 90.3% ee, isoleucine: >99% ee).[1c] This problem can be circumvented by the use of 'nonbasic' Ti(O-*i*-Pr)$_4$ assisted 'transesterification' with allyl alcohol then rhodium-catalyzed 'deprotection'. This allows for the preparation of either 'free' or *N*-Fmoc α-amino acids of excellent enantiomeric purity.[18e]

(**8**) 1. R_2BOTf, *i*-Pr_2NEt, CH_2Cl_2, –5 °C; 2. NBS, THF, –78 °C; 73–99% (cryst.); 85–>99% de (crude) → (**11**) (14)

R^1 = Me, Et, allyl, *i*-Bu, Ph

The third strategy involves electrophilic 'amination' of sodium enolates of *N*-acylsultams (**8**) using ***1-Chloro-1-nitrosocyclohexane*** as an $[NH_2^+]$ equivalent.[18f–i] The reaction proceeds via nitrone intermediates which are routinely hydrolyzed without isolation to give the key α-*N*-hydroxyamino derivatives (**13**) with outstanding C(α)-*re* π-facial control (eq 15). Nitrogen–oxygen bond hydrogenolysis (Zn, aq HCl, AcOH), then auxiliary cleavage, affords α-amino acids.[18f,18g] Omission of the hydrogenolysis step allows access to *N*-hydroxy-α-amino acids, which are extremely difficult to prepare by alternative means.[18f] The scope of the reaction has been extended to encompass the use of 1-chloro-1-nitrosocyclohexane as an electrophilic partner in conjugate addition–trapping reactions [allowing an expedient preparation of (2*S*,3*S*)-isoleucine],[18f,18g] *N*-alkyl-α-amino acid preparation,[18h] and enantiomerically pure α-substituted cyclic nitrone formation [giving a concise preparation of the piperidine alkaloid (−)-pinidine].[18i]

(**8**) 1. NaHMDS, THF, –78 °C; 2. 1-chloro-1-nitrosocyclohexane; 3. 1 N aq HCl, rt; 70–87% (cryst); >99% de (crude) → (**12**) (15)

R^1 = Me, *i*-Pr, allyl, *i*-Bu, Ph, *p*-$MeOC_6H_4$, Bn

α-Radical Addition (Acyl Species → α-Functionalized Acyl Product).[19] Radicals derived from α-iodo-*N*-acylsultams give high levels of asymmetric induction in intramolecular addition reactions with allyltributylstannanes (85–>94% de)[19a] and 5-*exo*-dig type cyclizations and annulations (eq 16).[19a] In addition, 'zipper' type manganese(III) promoted oxidative radical cyclization of *N*-(*trans*-4-methyl-4,9-nonadienoyl)-10,2-camphorsultam gives a *cis*-fused hydrindane derivative with modest (50% de) selectivity at the α-center.[19b] All these reactions proceed at or above room temperature, making the levels of induction remarkable. Furthermore, effective alternative auxiliaries are scarce.[1d,14]

1. $Bu_3SnSnBu_3$, C_6H_6, *hν*, 80 °C; 2. Bu_3SnH; 3. HI; 71% (FC); 80% de (crude); cryst. in unspecified yield (16)

Nondestructive Auxiliary Cleavage. One feature which makes the sultam chiral auxiliary, and to an even greater extent the related toluene-2,α-sultam auxiliaries (see ***α-Methyltoluene-2,α-sultam***), so versatile is the ease with which *N*-acyl bond fission occurs in derivatives. A great variety of extremely mild, bimolecular and intramolecular nondestructive cleavage protocols have been developed which tolerate a wide array of molecular functionality, simple extraction and crystallization usually providing almost quantitative auxiliary recovery without loss of enantiomeric purity.

Saponification with LiOH[6a] or H_2O_2–LiOH[16b] in aqueous THF is routinely employed for conversion of *N*-acylsultams to enantiomerically pure carboxylic acids. A variant conducted in aprotic media with phase transfer catalysis has also been reported.[18d] If base sensitive functionality is present, then the corresponding esters can be prepared by 'nonbasic' titanium mediated 'alcoholysis'. This can be accomplished with ethyl,[11b] benzyl,[4] or allyl[18e] alcohols, and in the latter two instances the carboxylic acids can be subsequently liberated by 'neutral' hydrogenolysis or $RhCl(PPh_3)_3$ catalyzed hydrolysis,[18e] respectively. Lactones and esters can also be formed by intra-[18j] and intermolecular[2d] sultam cleavage with lithium alkoxides and bromomagnesium alkoxides.[11d] β-Lactams can be prepared by intramolecular ring closure of metallated β-aminomethyl derivatives[1b,17a] and an aluminum 'thiobenzyloxy ate' complex has been used to obtain thioester derivatives.[13] Reductive cleavage of *N*-acylsultams using lithium aluminum hydride[2f] or L-Selectride®[3b,3f] in THF gives rise to the corresponding primary alcohols.

Auxiliary cleavage with concomitant carbon–carbon bond formation is a particularly attractive option, which has been demonstrated in a bimolecular sense using the dianion of methyl sulfone (giving a methyl ketone),[16f] and in an intramolecular sense using a Claisen-type condensation of a β-acetoxy enolate (giving a δ-lactone).[25] An interesting 'halolactonization' procedure has also been devised; for certain α-aryl-bis-(γ-unsaturated)-*N*-acyl derivatives this allows for highly efficient auxiliary cleavage and asymmetric formation of two stereocenters, one of which is quaternary (eq 17), the key step in a synthesis of (−)-mesembrine).[26]

I(collidine)$_2$ClO$_4$
CH_2Cl_2, MeOH, H_2O
−50 °C
[Ar = 3,4-(OMe)$_2$C$_6$H$_3$]
83% (FC)

(13) + **(14)** (17)

(13):**(14)** = 4.5:1
both diastereomers
88% ee (crude)

Enantioselective, Electrophilic Fluorination. (−)-*N*-Fluoro-10,2-camphorsultam (**3**) [mp 112–114 °C (CH_2Cl_2–pentane)] is an enantioselective, electrophilic fluorinating agent.[20] Fluorination of stabilized enolates occurs with highly variable yield (5–63%) and stereoselectivity (10–70% de).

Related Reagents. (*S*)-4-Benzyl-2-oxazolidinone; 10-Dicyclohexylsulfonamidoisoborneol; 2-Hydroxy-1,2,2-triphenylethyl Acetate; (4*S*,5*S*)-4-Methoxymethyl-2-methyl-5-phenyl-2-oxazoline; α-Methyltoluene-2,α-sultam.

1. (a) Oppolzer, W. *T* **1987**, *43*, 1969. (b) Oppolzer, W. *PAC* **1988**, *60*, 39. (c) Oppolzer, W. *PAC* **1990**, *62*, 1241. (d) Kim, B. H.; Curran, D. P. *T* **1993**, *49*, 293.
2. (a) Oppolzer, W. *AG(E)* **1984**, *23*, 876. (b) Oppolzer, W.; Chapuis, C.; Bernardinelli, G. *HCA* **1984**, *67*, 1397. (c) Oppolzer, W.; Dupuis, D. *TL* **1985**, *26*, 5437. (d) Oppolzer, W.; Dupuis, D.; Poli, G.; Raynham, T. M.; Bernardinelli, G. *TL* **1988**, *29*, 5885. (e) Smith III, A. B.; Hale, J. K.; Laahso, L. M.; Chen, K.; Riera, A. *TL* **1989**, *30*, 6963. (f) Vandewalle, M.; Van der Eycken, J.; Oppolzer, W.; Vullioud, C. *T* **1986**, *42*, 4035. (g) Oppolzer, W.; Rodriguez, I.; Blagg, J.; Bernardinelli, G. *HCA* **1989**, *72*, 123. (h) Bauer, T.; Chapuis, C.; Kozac, J.; Jurczak, J. *HCA* **1989**, *72*, 482. (i) Gouverneur, V.; Dive, G.; Ghosez, L. *TA* **1991**, *2*, 1173.
3. (a) Curran, D. P.; Kim, B. H.; Daugherty, H.; Heffner, T. A. *TL* **1988**, *29*, 3555. (b) Curran, D. P.; Heffner, T. A. *JOC* **1990**, *55*, 4585. (c) Kim, K. S.; Kim, B. H.; Park, W. M.; Cho, S. J.; Mhin, B. J. *JACS* **1993**, *115*, 7472. (d) Kim, B. H.; Lee, J. Y.; Kim, K.; Whang, D. *TA* **1991**, *2*, 27. (e) Kim, B. H.; Lee, J. Y. *TA* **1991**, *2*, 1359. (f) Kim, B. H.; Lee, J. Y. *TL* **1992**, *33*, 2557. (g) Garner, P.; Ho, W. B. *JOC* **1990**, *55*, 3973. (h) Garner, P.; Ho, W. B.; Grandhee, S. K.; Youngs, W. J.; Kennedy, V. O. *JOC* **1991**, *56*, 5893. (i) Garner, P.; Ho, W. B.; Shin, H. *JACS* **1992**, *114*, 2767. (j) Binger, P.; Schafer, B. *TL* **1988**, *29*, 529. (k) Binger, P.; Brinkmann, A.; Roefke, P.; Schafer, B. *LA* **1989**, 739. (l) Trost, B. M.; Yang, B.; Miller, M. L. *JACS* **1989**, *111*, 6482.
4. Vallgarda, J.; Hacksell, U. *TL* **1991**, *32*, 5625, and corrigendum *ibid. TL* **1991**, *32*, 7136.
5. Kapron, J. T.; Santarsiero, B. D.; Vederas, J. C. *CC* **1993**, 1074.
6. (a) Oppolzer, W.; Barras, J.-P. *HCA* **1987**, *70*, 1666. (b) Walba, D. M.; Przybyla, C. A.; Walker, Jr, C. B. *JACS* **1990**, *112*, 5624.
7. Oppolzer, W.; Mills, R. J.; Reglier M. *TL* **1986**, *27*, 183.
8. Lee, P.-C.; Wu, C.-C.; Cheng. M.-C.; Wang, Y.; Wu, M.-J. *J. Chinese Chem. Soc.* **1992**, *39*, 87.
9. (a) Oppolzer, W.; Poli, G. *TL* **1986**, *27*, 4717. (b) Oppolzer, W.; Poli, G.; Starkemann, C.; Bernardinelli, G. *TL* **1988**, *29*, 3559.
10. Oppolzer, W.; Poli, G.; Kingma, A. J.; Starkemann, C.; Bernardinelli, G. *HCA* **1987**, *70*, 2201.
11. (a) Oppolzer, W.; Mills, R. J.; Pachinger, W.; Stevenson, T. *HCA* **1986**, *69*, 1542. (b) Oppolzer, W.; Schneider, P. *HCA* **1986**, *69*, 1817. (c) Oppolzer, W.; Kingma, A. J. *HCA* 1989, *72*, 1337. (d) Oppolzer, W.; Kingma, A. J.; Poli, G. *T* **1989**, *45*, 479.
12. Wu, M.-J.; Wu, C.-C.; Lee, P.-C. *TL* **1992**, *33*, 2547.
13. Miyata, O.; Shinada, T.; Kawakami, N.; Taji, K.; Ninomiya, I.; Naito, T.; Date, T.; Okamura, K. *CPB* **1992**, *40*, 2579.
14. Porter, N. A.; Giese, B.; Curran, D. P. *ACR* **1991**, *24*, 296.
15. Girard, C.; Mandville, G.; Bloch, R. *TA* **1993**, *4*, 613.
16. (a) Heathcock, C. H. In *Modern Synthetic Methods*, VCH-VHCA: Basel, 1982, 1. (b) Oppolzer, W.; Blagg, J.; Rodriguez, I.; Walther, E. *JACS* **1990**, *112*, 2767. (c) Oppolzer, W.; Starkemann, C.; Rodriguez, I.; Bernardinelli, G. *TL* **1991**, *32*, 61. (d) Oppolzer, W.; Lienard, P. *TL* **1993**, *34*, 4321. (e) Oppolzer, W.; Starkemann, C. *TL* **1992**, *33*, 2439. (f) Oppolzer, W.; Rodriguez, I. *HCA* **1993**, *76*, 1275. (g) Oppolzer, W.; Rodriguez, I. *HCA* **1993**, *76*, 1282.
17. (a) Oppolzer, W.; Moretti, R.; Thomi, S. *TL* **1989**, *30*, 5603. (b) Martin, N.; Martinez-Grau, A.; Seoane, C.; Marco, J. L. *TL* **1993**, *34*, 5627.
18. (a) Williams, R. M. *Synthesis of Optically Active α-Amino Acids*, Pergamon: Oxford, 1989. (b) Oppolzer, W. *AP* **1990**, 190. (c) Oppolzer, W.; Moretti, R.; Thomi, S. *TL* **1989**, *30*, 6009. (d) Josien, H.; Martin, A.; Chassaing, G. *TL* **1991**, *32*, 6547. (e) Oppolzer, W.; Lienard, P. *HCA* **1992**, *75*, 2572. (f) Oppolzer, W.; Tamura, O. *TL* **1990**, *31*, 991. (g) Oppolzer, W.; Tamura, O.; Deerberg, J. *HCA* **1992**, *75*, 1965. (h) Oppolzer, W.; Cintas-Moreno, P.; Tamura, O. *HCA* **1993**, *76*, 187. (i) Oppolzer, W.; Merifield, E. *HCA* **1993**, *76*, 957. (j) Oppolzer, W.; Bienayme, H.; Genevois-Borella, A. *JACS* **1991**, *113*, 9660.
19. (a) Curran, D. P.; Shen, W.; Zhang, Z.; Heffner, T. A. *JACS* **1990**, *112*, 6738. (b) Zoretic, P. A.; Weng, X.; Biggers, C. K.; Biggers, M. S.; Caspar, M. L. *TL* **1992**, *33*, 2637.
20. Differding, E.; Lang, R. W. *TL* **1988**, *29*, 6087.
21. Weismiller, M. C.; Towson, J. C.; Davis, F. A. *OS* **1990**, *69*, 154.
22. Davies, S. G. *Chem. Br.* **1989**, *25*, 268.
23. Harada, N.; Soutome, T.; Nehira, T.; Uda, H. *JACS* **1993**, *115*, 7547.
24. Thom, C.; Kocienski, P. *S* **1992**, 582.
25. Brandange, S.; Leijonmarck, H. *TL* **1992**, *33*, 3025.
26. (a) Yokomatsu, T.; Iwasawa, H.; Shibuya, S. *CC* **1992**, 728. (b) Yokomatsu, T.; Iwasawa, H.; Shibuya, S. *TL* **1992**, *33*, 6999.

Alan C. Spivey
University of Cambridge, UK

Carbon Monoxide[1]

[630-08-0] CO (MW 28.01)

(carbonylation of various organic compounds[1])

Physical Data: mp −205.0 °C; bp −191.5 °C; *d* 1.250 g L^{-1} (d_4^0 at 760 mmHg).
Solubility: appreciably sol some organic solvents, such as EtOAc, $CHCl_3$, acetic acid.
Form Supplied in: cylinder types, valves, and pressure regulators.
Preparative Methods: may be generated by the dehydration of formic acid.
Purification: major impurities are CH_4 (∼0.2%), N_2 (∼0.5%), H_2, and O_2. O_2 can be removed from CO by passing the gas through a short bed of reduced Cu or MgO_2.

Handling, Storage, and Precautions: highly poisonous, odorless, colorless, and tasteless gas. All operations should be carried out in a fume hood.

Reactions with Carbocations. The reaction of carbocations with CO leading to acylium cations is a key step in the Koch reaction. The Koch reaction produces tertiary carboxylic acids by treating alcohols with CO in a strong acid. It can be applied to alkanes, alkenes, and other compounds equivalent to alcohols under acidic conditions. A variety of acid systems, including Brønsted acids such as H_2SO_4, HF, and H_3PO_4, as well as in combination with Lewis acids such as BF_3, $AlCl_3$, $SbCl_5$, and SbF_5, are effective. The use of HCO_2H as a CO source is a practical method for laboratory scale reactions and this method is often used as an in situ preparation of CO. The treatment of alcohols, including secondary ones, with HCO_2H in 96% H_2SO_4 gives tertiary carboxylic acids (eq 1).[2] Adamantane is also carbonylated to 1-adamantanecarboxylic acid (eq 2).[3] Besides the HCO_2H method, the Koch synthesis can also be conducted at normal pressure of CO in the presence of copper or silver salts. Cyclohexene, for example, yields only 1-methylcyclopentanecarboxylic acid via ring contraction of the first-formed carbocation under 1 atm of CO (eq 3).[4] The Gatterman–Koch reaction[5] is the formylation of aromatic hydrocarbons with CO and HCl in the presence of $AlCl_3$ (eq 4).[6] Other catalytic systems such as HF/BF_3, HF/SbF_5, $HF/CF_3SO_3H/BF_3$, and CF_3SO_3H have been investigated and found to be effective for this reaction. Formylation and sulfonation take place when aromatic compounds are exposed to an atmosphere of CO at 0 °C in a HSO_3F/SbF_5 system (eq 5).[7]

H_2SO_4, HCO_2H, 15–20 °C; CO_2H; 89–94% (1)

H_2SO_4, HCO_2H, t-BuOH, CCl_4, 17–25 °C, 56–61%; CO_2H (2)

Cu_2O, H_2SO_4, CO, 1 atm, 30 °C, 63%; CO_2H (3)

CuCl/$AlCl_3$, HCl, CO, 1 atm, 20 °C, 46–51%; CHO (4)

HSO_3F/SbF_5, CO, 1 atm, 0 °C, 93%; CHO; FO_2S (5)

Reactions with Carbanions. The reaction of organolithium reagents with CO gives carbonyl lithium reagents (acyllithium or aroyllithium), which have not been utilized in practical synthetic reactions until recently. Difficulties in controlling the reaction of carbonyl lithium reagents are attributed to their extremely high reactivity. Acyllithium can be trapped by ***Chlorotrimethylsilane*** under extremely careful reaction conditions (eq 6).[8] A controlled, slow-rate addition of alkyllithium to a solution of Me_3SiCl saturated with CO at −110 °C is recommended. Direct nucleophilic acylation of ketones by in situ generated acyllithium reagents leading to α-hydroxy ketones has been described (eq 7).[9] The use of aldehydes,[10] lactones,[11] CS_2,[12] isocyanate[13] disulfide,[14] and carbodiimide[15] as trapping electrophiles has also been reported. Acylcuprate reagents successfully undergo nucleophilic 1,4-addition to α,β-unsaturated ketones and aldehydes (eq 8).[16]

BuLi + CO —(THF, −110 °C)→ [Bu–C(=O)–Li] —(Me_3SiCl)→ Bu–C(=O)–$SiMe_3$ 77% (6)

BuLi + CO + (t-Bu ketone) —(THF, −110 °C; aq NH_4Cl, rt)→ Bu, t-Bu, OH 67–73% (7)

$Bu_2(CN)CuLi_2$ + CO + (cyclohexenone) —(THF, −110 °C; aq NH_4OH/NH_4Cl, rt, 86%)→ (3-(Bu–C(=O))cyclohexanone) (8)

The reaction of an α-silylalkyllithium with CO is a convenient access to acylsilane enolates, which can be trapped by electrophiles such as H^+, Me_3SiCl, and benzaldehyde (eq 9).[17] The formation of the enolates takes place in a highly stereoselective manner to give (*E*) enolates.

(Li, $SiMe_3$) + CO —(Et_2O, 15 °C)→ [(O, Li, $SiMe_3$) → (OLi, $SiMe_3$)]; Me_3SiCl, 88% → $OSiMe_3$, $SiMe_3$; PhCHO, 52% → Ph, OH, O, $SiMe_3$; *syn*:*anti* = 93:7 (9)

Reactions with Radicals. The reaction of alkyl halides with ***Tri-n-butyltin Hydride*** in the presence of ***Azobisisobutyronitrile*** (0.1–0.2 equiv) under CO pressure (65–80 atm) leading to aldehydes proceeds via the reaction of alkyl radicals with CO to give acyl radicals followed by abstraction of a hydrogen atom from $HSnBu_3$ (eq 10).[18] Aromatic halides can also be converted to aromatic aldehydes.[19] The use of ***Tris(trimethylsilyl)silane*** (TTMSS) in place of $HSnBu_3$ as a radical mediator permits the carbonylation under a lower pressure of CO (20 atm).[20] The radical carbonylation is applied to carbonylative cyclization of 4-alkenyl halides to cyclopentenes (eq 11).[21] The intramolecular trapping of acyl radicals by alkenes is utilized for the synthesis of functionalized unsymmetrical ketones.[22] The carbonylation of alkynes in the presence of thiols initiated by AIBN gives β-alkylthio-α,β-unsaturated aldehydes (eq 12).[23]

$$C_8H_{17}Br \xrightarrow[\text{AIBN, } C_6H_6,\ 65\text{–}80\text{ atm, } 80\,^\circ C]{CO,\ HSnBu_3} [C_8H_{17}\dot{C}{=}O] \longrightarrow C_8H_{17}CHO\ (61\%) \quad (10)$$

$$\text{(11) 6-methyl-5-heptenyl bromide} \xrightarrow[\text{AIBN, } C_6H_6,\ 75\text{ atm, } 80\,^\circ C,\ 65\%]{CO,\ HSnBu_3} \text{2-isopropylcyclopentanone}$$

$$C_8H_{11}C{\equiv}CH \xrightarrow[\text{AIBN, } C_6H_6,\ 80\text{ atm, } 100\,^\circ C,\ 70\%]{C_6H_{13}SH,\ CO} C_8H_{11}C(CHO){=}CH(SC_6H_{13}) \quad (12)$$

Transition Metal-Catalyzed Reactions. Ring-opening silosymethylation to 1,3-diol derivatives takes place in the reaction of cyclopentene oxide with $HSiEt_2Me$ and CO in the presence of $Co_2(CO)_8$ (eq 13).[24] The reagents $ML_n/HSiR_3/CO$ provide several important synthetic methods, which are covered under the reagent headings of ***Octacarbonyldicobalt–Diethyl(methyl)silane–Carbon Monoxide***, ***Dodecacarbonyltetrarhodium–Dimethyl(phenyl)silane–Carbon Monoxide***, and ***Tricarbonylchloroiridium–Diethyl-(methyl)silane–Carbon Monoxide***.

$$\text{cyclopentene oxide} \xrightarrow[C_6H_6,\ 1\text{ atm, } 25\,^\circ C,\ 88\%]{Co_2(CO)_8,\ HSiEt_2Me,\ CO} \text{trans-1-}OSiEt_2Me\text{-2-}CH_2OSiEt_2Me\text{-cyclopentane} \quad (13)$$

Hydroformylation has been an extremely important industrial process, and consequently has been the most extensively studied of all carbonylation reactions.[25] A wide variety of metals such as Pt, Co, Rh, Ir, and Ru exhibit catalytic activity for hydroformylation of alkenes. Among them, Rh has been preferred for laboratory use because of its higher activity (eq 14).[26] Various phosphine-modified catalysts such as $Co_2(CO)_8$/phosphine, (phosphine)$PtCl_2/SnCl_2$, and phosphine-modified Rh have been examined extensively to obtain high selectivity for the more desirable linear aldehydes. Practical, regioselective hydroformylation of functionalized terminal alkenes can be achieved by using a bis-organophosphite ligand (**1**).[27] Catalytic systems using diphosphines with natural bite angles near 120° increase regioselectivity for straight-chain aldehydes.[28] Highly selective formation of branched-chain aldehydes is attained by use of a zwitterionic Rh complex (eq 15).[29] Hydroformylation using $PtCl_2/SnCl_2$ in the presence of chiral ligands such as (−)-DIPHOS, (−)-DIOP, and chiraphos is reported to produce moderate enantioselectivity.[30] The highest level of enantioselective discrimination is realized with $PtCl_2/SnCl_2$ and (2*S*,4*S*)-4-(diphenylphosphino)-2-[(diphenylphosphino)methyl]pyrrolidine ((−)-BPPM) in the presence of ***Triethyl Orthoformate***, which converts the product aldehyde to its diethyl acetal, although the branched/linear ratios are low.[31]

$$\text{cyclohexene} \xrightarrow[C_6H_6,\ 150\text{ atm, } 100\,^\circ C,\ 82\text{–}84\%]{Rh_2O_3,\ H_2/CO = 1/1} \text{cyclohexane-CHO} \quad (14)$$

(1) [bis-organophosphite ligand: MeO, OMe, t-Bu, t-Bu substituted biphenyl bearing two O–P(O)O dibenzodioxaphosphepin units]

$$PhCH{=}CH_2 \xrightarrow[CHCl_3,\ 200\text{ psi, } 47\,^\circ C]{(\eta^6\text{-}C_6H_5BPh_3)Rh^+(cod),\ H_2/CO = 1/1} PhCH(CH_3)CHO + PhCH_2CH_2CHO\ \ (97.3{:}2.7) \quad (15)$$

Hydrocarboxylation or hydroesterification is also an important process of CO. The asymmetric synthesis of (+)-ibuprofen and (+)-naproxen is attained based on asymmetric hydrocarboxylation of vinylarenes using (*S*)-(+)-1,1′-binaphthyl-2,2′-diyl hydrogen phosphate ((*S*)-BNPPA) (eq 16).[32]

$$\text{4-}i\text{-Bu-}C_6H_4CH{=}CH_2 \xrightarrow[H_2O,\ HCl,\ 1\text{ atm, rt},\ 83\%\text{ ee}]{PdCl_2,\ CuCl_2,\ (S)\text{-BNPPA},\ CO,\ O_2} \text{4-}i\text{-Bu-}C_6H_4CH(CH_3)CO_2H \quad (16)$$

Terminal alkynes may be carbonylated to acetylenecarboxylic esters in the presence of $PdCl_2$ as catalyst and $CuCl_2$ as reoxidant in the presence of NaOAc (eq 17).[33] Hydroesterification of alkynes to α,β-unsaturated esters is catalyzed by $[P(p\text{-Tol})_3]_2PdCl_2/SnCl_2$,[34] $PdCl_2/CuCl/HCl/O_2$,[35] and $(PPh_3)_2PtCl_2/SnCl_2$.[36] Pd-catalyzed intramolecular carbonylation of alkynyl alcohols is an attractive route to α-methylene-γ-lactones (eq 18).[37]

$$PhC{\equiv}CH \xrightarrow[MeOH,\ 1\text{ atm, } 25\,^\circ C,\ 74\%]{PdCl_2,\ CuCl_2,\ CO,\ NaOAc} PhC{\equiv}CCO_2Me \quad (17)$$

$PdCl_2$, $SnCl_2$, PBu_3, CO, MeCN, 7.8 atm, 75 °C, 85% (18)

Ketones and aldehydes are formed using CO. Pd-catalyzed carbonylation of aryl or vinyl halides (or triflates) is an attractive method for a wide variety of carbonyl compounds. Carbonylation of aryl halides in the presence of hydrogen and a tertiary amine leads to the formation of aldehydes (eq 19).[38] Use of hydrogen donor such as HCO_2Na,[39] poly(methylhydrosiloxane) (PHMS),[40] $HSiEt_3$,[41] and $HSnBu_3$[42] also gives aldehydes. The Pd-catalyzed carbonylative coupling of aryl triflates with organostannanes is a potentially valuable route to unsymmetrical ketones (eq 20).[43] Organozinc reagents may be used in ketone syntheses.[44] Esters are obtained from aryl, vinyl, and heteroaryl halides (triflates) via the Pd-catalyzed carbonylation in the presence of alcohols.[45] The intramolecular version has been utilized for the preparation of α-methylene-γ-lactones from vinyl halides (eq 21).[46] The synthesis of lactams is also achieved in the presence of $Pd(OAc)_2/PPh_3$.[47] Aryl halides may be doubly carbonylated to α-keto amides in the Pd-catalyzed carbonylation of aryl halides with amines (eq 22).[48]

$(PPh_3)_2PdBr_2$, $H_2/CO = 1/1$, C_6H_6, Et_3N, 1510 psi, 150 °C, 76% (19)

$PdCl_2(dppf)$, LiCl, CO, Bu_3Sn–CH=CH–$SiMe_3$, DMF, 1 atm, 75 °C, 96% (20)

$Pd(PPh_3)_4$, CO, MeCN, 35 psi, 70 °C, 73% (21)

$(PMePh_2)_2PdCl_2$, CO, Et_2NH, 10 atm, 100 °C, 82% (22)

The catalytic carbonylation of benzyl chloride to arylacetic acid derivatives is shown to occur with Pd complexes as catalysts in the presence of a base under mild reaction conditions (eq 23).[49] $Co_2(CO)_8$-catalyzed carbonylation under phase transfer conditions (organic solvent/aq NaOH system containing catalytic amounts of quaternary ammonium salt) is also effective for the transformation of benzyl chlorides to arylacetic acids.[50]

$$PhCH_2Cl \xrightarrow[\text{200 psi, 80 °C; 68\%}]{(PPh_3)_2PdCl_2,\ NaOAc,\ CO,\ BuOH} PhCH_2CO_2Bu \quad (23)$$

Pd-catalyzed cyclocarbonylation of aryl- or heteroaryl-substituted allyl acetates with Ac_2O gives bicyclic aromatic systems (eq 24).[51] Allylic phosphonates are cleanly carbonylated to β,γ-unsaturated amides in the presence of a catalytic amount of $Rh_6(CO)_{16}/NBu_4Cl$ (eq 25).[52] Use of NH_4Cl/Et_3N in place of $PhCH_2NH_2$ is found to be effective for the preparation of primary amides. Under more harsh reaction conditions (50 atm, 110 °C), allylamines may be carbonylated directly to β,γ-unsaturated amides in the presence of $Pd(OAc)_2$/dppp.[53] The Pd-catalyzed carbonylation of propargyl carbonates in MeOH affords allenic esters (eq 26).[54]

$(PPh_3)_2PdCl_2$, CO, Ac_2O/Et_3N, C_6H_6, 70 atm, 170 °C, 85% (24)

$Rh_6(CO)_{16}$, Bu_4NCl, CO, $PhCH_2NH_2$, C_6H_6, 20 atm, 50 °C, 80% (25)

E:*Z* = 90:10

$Pd_2(dba)_3$, PPh_3, CO, MeOH, 30 atm, 40 °C, 71% (26)

Exposure of an azirine to CO in the presence of $Pd(PPh_3)_4$ affords bicyclic β-lactams (eq 27).[55] 2-Arylaziridines are carbonylated with retention of configuration in the presence of $[RhCl(CO)_2]_2$ to give monocyclic β-lactams (eq 28).[56] Similar carbonylative ring-expansion reactions are observed in the Rh-catalyzed reaction of azetidine-2,4-diones,[57] Co-catalyzed reaction of azetidines,[58] and Co/Ru-catalyzed reaction of oxetanes.[59]

$Pd(PPh_3)_4$, CO, C_6H_6, 1 bar, 40 °C, 63% (27)

$[RhCl(CO)_2]_2$, CO, C_6H_6, 20 atm, 90 °C, 93% (28)

Cyclic ethers, including oxiranes, oxetane, and tetrahydrofuran, react with CO and *N*-(trimethylsilyl)amine in the presence of $Co_2(CO)_8$ to form siloxy amides (eq 29).[60]

$Co_2(CO)_8$, CO, $BnNHSiMe_3$, C_6H_6, 60 kg cm^{-2}, 100 °C, 74% (29)

N-acyl-α-amino acids can be synthesized by amidocarbonylation of aldehydes with acetamide and synthesis gas (eq 30).[61] It is possible to achieve a direct, one-step synthesis of *N*-acyl-α-amino acids from precursors other than aldehydes themselves. For example, allyl alcohols can be converted to *N*-acyl-α-amino acids via isomerization of allyl alcohols to aldehydes followed by amidocarbonylation (eq 31).[62]

$$\text{BnCHO} \xrightarrow[\text{dioxane, 150 kg cm}^{-2}\text{, 140 °C, 54\%}]{Co_2(CO)_8,\ H_2/CO = 1/3,\ MeCONH_2} \text{BnCH(NHCOMe)CO}_2\text{H} \quad (30)$$

$$\text{CH}_2\text{=CHCH}_2\text{OH} \xrightarrow[\text{dioxane, 100 atm, 110 °C, 63\%}]{Co_2(CO)_8,\ HRh(CO)(PPh_3)_3,\ H_2/CO = 1/1,\ MeCONH_2} \text{CH}_3\text{CH}_2\text{CH(NHCOMe)CO}_2\text{H} \quad (31)$$

Heterocyclic rings can be constructed by oxidative carbonylation under mild reaction conditions. The formation of a pyran ring is achieved by oxidative carbonylation of hydroxy alkenes (eq 32).[63] Transformation of allenic amines to α-heterocyclic acrylic acid derivatives is also carried out under ambient conditions (eq 33).[64]

(32): hydroxy alkene $\xrightarrow[\text{MeOH, 1.1 atm, 25 °C, 74\%}]{PdCl_2,\ CuCl_2,\ CO}$ cis-tetrahydropyran-CH_2CO_2Me + trans isomer (CO_2Me), 20:1

(33): N-Bn allenic amine $\xrightarrow[\text{MeOH, 1 atm, rt, 67\%}]{PdCl_2,\ CuCl_2,\ CO}$ N-Bn pyrrolidinyl acrylate (CO_2Me)

CO acts as reducing agent in the water gas shift reaction (WGSR).[65] Application of WGSR in organic synthesis includes hydrogenation, related reductions, and carbonylation. Ru-catalyzed reduction of aromatic nitro compounds to aromatic amines under mild reaction conditions (at rt under 1 atm of CO) has been achieved (eq 34).[66] Montmorillonite–bipyridine–$Pd(OAc)_2/Ru_3(CO)_{12}$ is an active catalyst for the reductive carbonylation of aromatic nitro compounds to urethanes (eq 35).[67] The reaction is highly selective and no side products such as azo, azoxytoluene, or *N*-methylanilines are detected.

$$O_2N\text{-}C_6H_4\text{-}NO_2 \xrightarrow[\text{2-methoxyethanol, 1 atm, 25 °C, > 99\%}]{Rh_4(CO)_{12},\ \text{9,10-diaminophenanthrene},\ CO,\ NaOH} H_2N\text{-}C_6H_4\text{-}NH_2 \quad (34)$$

$$CH_3\text{-}C_6H_4\text{-}NO_2 \xrightarrow[\text{70 atm, 180 °C, 84\%}]{\text{Pd-cray},\ Ru_3(CO)_{12},\ CO,\ MeOH,\ bipy} CH_3\text{-}C_6H_4\text{-}NHCOMe \quad (35)$$

Miscellaneous. Selenium effectively catalyzes the carbonylation of various amines with CO and O_2 to give urea derivatives quantitatively (eq 36).[68] Turnover numbers of Se reach 10^4 by using 0.8 mg of Se (0.01 mmol) for the carbonylation of benzylamine (200 mmol). The reaction can be applied to the synthesis of cyclic ureas.[69]

$$\text{BnNH}_2 \xrightarrow[\text{120 °C}]{\text{Se (0.01 equiv)},\ CO\ (30\ atm),\ O_2\ (4\ atm)} \text{(BnNH)}_2\text{CO} \quad (36)$$

CO forms Lewis acid–base complexes with organoboranes. When the organoborane is heated with CO to 100–125 °C, a tertiary alcohol is obtained after workup by oxidation (eq 37).[70]

$$(s\text{-Bu})_3\text{B} \xrightarrow[\text{diglyme, 1 atm, 125 °C}]{CO} \xrightarrow{H_2O_2,\ OH^-} (s\text{-Bu})_3\text{COH}\ \ 87\% \quad (37)$$

Related Reagents. Dodecacarbonyltetrarhodium-Dimethyl(phenyl)silane-Carbon Monoxide; Octacarbonyldicobalt-Diethyl(methyl)silane-Carbon Monoxide; Tricarbonylchloroiridium-Diethyl(methyl)silane-Carbon Monoxide.

---

1. (a) Wender, I.; Pino, P. *Organic Syntheses via Metal Carbonyls*; Wiley: New York, 1977; Vol. 2. (b) Falbe, J. *New Syntheses with CO*; Springer: Berlin, 1980. (c) Thatchenko, I. In *Comprehensive Organometallic Chemistry*; Wilkinson, G.; Stone, F. G. A.; Abel, E. W., Eds.; Pergamon: Oxford, 1982; Vol. 8, p 101. (d) Sonoda, N.; Murai, S. *Yuki Gosei Kagaku Kyokai Shi* **1983**, *41*, 507. (e) Colquhoun, H. M.; Thompson, D. J.; Twigg, M. V. *Carbonylation. Direct Synthesis of Carbonyl Compounds*; Plenum: New York, 1991.
2. Haaf, W. *OSC* **1973**, *5*, 739.
3. Koch, H.; Haaf, W. *OSC* **1973**, *5*, 20.
4. (a) Souma, Y.; Sanao, H. *BCJ* **1973**, *46*, 3237. (b) Souma, Y.; Sano, H.; Iyoda, J. *JOC* **1973**, *38*, 2016.
5. Olah, G. A.; Ohannesian, L.; Arvanaghi, M. *CRV* **1987**, *87*, 671.
6. Coleman, G. H.; Craig, D. *OSC* **1943**, *2*, 583.
7. Tanaka, M.; Iyoda, J.; Souma, Y. *JOC* **1992**, *57*, 2677.
8. Seyferth, D.; Weinstein, R. M. *JACS* **1982**, *104*, 5534.
9. Hui, R. C.; Seyferth, D. *OS* **1990**, *69*, 114.
10. Seyferth, D.; Weinstein, R. M.; Wang, W.-L.; Hui, R. C. *TL* **1983**, *24*, 4907.
11. Weinstein, R. M.; Wang, W.-L.; Seyferth, D. *JOC* **1983**, *48*, 3367.
12. Seyferth, D.; Hui, R. C. *TL* **1984**, *25*, 2623.
13. Seyferth, D.; Hui, R. C. *TL* **1984**, *25*, 5251.
14. Seyferth, D.; Hui, R. C. *OM* **1984**, *3*, 327.
15. Seyferth, D.; Hui, R. C. *JOC* **1985**, *50*, 1985.
16. (a) Seyferth, D.; Hui, R. C. *JACS* **1985**, *107*, 4551. (b) Lipshutz, B. H.; Elworthy, T. R. *TL* **1990**, *31*, 477.
17. Murai, S.; Ryu, I.; Iriguchi, J.; Sonoda, N. *JACS* **1984**, *106*, 2440.
18. Ryu, I.; Kusano, K.; Ogawa, A.; Kambe, N.; Sonoda, N. *JACS* **1990**, *112*, 1295.
19. Ryu, I.; Kusano, K.; Masumi, N.; Yamazaki, H.; Ogawa, A.; Sonoda, N. *TL* **1990**, *31*, 6887.
20. Ryu, I.; Hasegawa, M.; Kurihara, A.; Ogawa, A.; Tsunoi, S.; Sonoda, N. *SL* **1993**, 143.
21. Ryu, I.; Kusano, K.; Hasegawa, M.; Kambe, N.; Sonoda, N. *CC* **1991**, 1018.
22. (a) Ryu, I.; Kusano, K.; Yamazaki, H.; Sonoda, N. *JOC* **1991**, *56*, 5003. (b) Ryu, I.; Yamazaki, H.; Kusano, K.; Ogawa, A.; Sonoda, N. *JACS* **1991**, *113*, 8558. (c) Ryu, I.; Yamazaki, H.; Ogawa, A.; Kambe, N.; Sonoda, N. *JACS* **1993**, *115*, 1187.
23. Nakatani, S.; Yoshida, J.-I.; Isoe, S. *CC* **1992**, 880.
24. Murai, T.; Yasui, E.; Kato, S.; Hatayama, Y.; Suzuki, S.; Yamasaki, Y.; Sonoda, N.; Kurosawa, H.; Kawasaki, Y.; Murai, S. *JACS* **1989**, *111*, 7938.

25. (a) Botteghi, C.; Ganzerla, R.; Lenard, M.; Moretti, G. *J. Mol. Catal.* **1987**, *40*, 129. (b) Kalck, P.; Peres, Y.; Jenck, J. *Adv. Organomet. Chem.* **1991**, *32*, 121.
26. Pino, P.; Botteghi, C. *OSC* **1988**, *6*, 338.
27. Cuny, G. D.; Buchwald, S. L. *JACS* **1993**, *115*, 2066.
28. (a) Casey, C. P.; Whiteker, G. T.; Melville, M. G.; Petrovich, L. M.; Gavney, J. A., Jr.; Powell, D. R. *JACS* **1992**, *114*, 5535. (b) Miyazawa, M.; Momose, S.; Yamamoto, K. *SL* **1990**, 711.
29. Amer, I.; Alper, H. *JACS* **1990**, *112*, 3674.
30. (a) Kagan, H. B. *BSF(2)* **1988**, 846. (b) Brunner, H. *S* **1988**, 645.
31. (a) Parrinello, G.; Stille, J. K. *JACS* **1987**, *109*, 7122. (b) Stille, J. K.; Su, H.; Brechot, P.; Parrinello, G.; Hegedus, L. S. *OM* **1991**, *10*, 1183.
32. Alper, H.; Hamel, N. *JACS* **1990**, *112*, 2803.
33. Tsuji, J.; Takahashi, M.; Takahashi, T. *TL* **1980**, *21*, 849.
34. Knifton, J. F. *J. Mol. Catal.* **1977**, *2*, 293.
35. Alper, H.; Despeyroux, B.; Woell, J. B. *TL* **1983**, *24*, 5691.
36. Tsuji, Y.; Kondo, T.; Watanabe, Y. *J. Mol. Catal.* **1987**, *40*, 295.
37. Murray, T. F.; Samsel, E. G.; Varma, V.; Norton, J. R. *JACS* **1981**, *103*, 7520.
38. Schoenberg, A.; Heck, R. F. *JACS* **1974**, *96*, 7761.
39. Ben-David, Y.; Portnoy, M.; Milstein, D. *CC* **1989**, 1816.
40. Pri-Bar, I.; Buchman, O. *JOC* **1984**, *49*, 4009.
41. Kikukawa, K.; Totoki, T.; Wada, F.; Matsuda, T. *JOM* **1984**, *270*, 283.
42. Baillargeon, V. P.; Stille, J. K. *JACS* **1986**, *108*, 452.
43. Echavarren, A. M.; Stille, J. K. *JACS* **1988**, *110*, 1557.
44. Tamaru, Y.; Ochiai, H.; Yamada, Y.; Yoshida, Z. *TL* **1983**, *24*, 3869.
45. (a) Schoenberg, A.; Bartoletti, I.; Heck, R. F. *JOC* **1974**, *39*, 3318. (b) Kobayashi, T.-A.; Abe, F.; Tanaka, M. *J. Mol. Catal.* **1988**, *45*, 91. (c) Adapa, S. R.; Prasad, C. S. N. *JCS(P1)* **1989**, 1706.
46. Martin, L. D.; Stille, J. K. *JOC* **1982**, *47*, 3630.
47. Mori, M.; Chiba, K.; Okita, M.; Kato, I.; Ban, Y. *T* **1985**, *41*, 375.
48. (a) Kobayashi, T.; Tanaka, M. *JOM* **1982**, *233*, C64. (b) Ozawa, F.; Soyama, H.; Yanagihara, H.; Aoyama, I.; Takino, H.; Izawa, K.; Yamamoto, T.; Yamamoto, A. *JACS* **1985**, *107*, 3235.
49. Stille, J. K.; Wong, P. K. *JOC* **1975**, *40*, 532.
50. Alper, H.; Des Abbayes, H. *JOM* **1977**, *134*, C11.
51. Ishii, Y.; Hidai, M. *JOM* **1992**, *428*, 279.
52. Imada, Y.; Shibata, O.; Murahashi, S.-I. *JOM* **1993**, *451*, 183.
53. Murahashi, S.-I.; Imada, Y.; Nishimura, K. *CC* **1988**, 1578.
54. Tsuji, J.; Sugiura, T.; Minami I. *TL* **1986**, *27*, 731.
55. Alper, H.; Mahatantila, C. P. *OM* **1982**, *1*, 70.
56. Calet, S.; Urso, F.; Alper, H. *JACS* **1989**, *111*, 931.
57. Roberto, D.; Alper, H. *OM* **1984**, *3*, 1767.
58. Roberto, D.; Alper, H. *JACS* **1989**, *111*, 7539.
59. Wang, M.-D.; Calet, S.; Alper, H. *JOC* **1989**, *54*, 20.
60. Tsuji, Y.; Kobayashi, M.; Okuda, F.; Watanabe, Y. *CC* **1989**, 1253.
61. Wakamatsu, H.; Uda, J.; Yamakami, N. *CC* **1971**, 1540.
62. Hirai, K.; Takahashi, Y.; Ojima, I. *TL* **1982**, *23*, 2491.
63. Semmelhack, M. F.; Bodurow, C. *JACS* **1984**, *106*, 1496.
64. Lathbury, D.; Vernon, P.; Gallagher, T. *TL* **1986**, *27*, 6009.
65. (a) Ford, P. C.; Rokicki, A. *Adv. Organomet. Chem.* **1988**, *28*, 139. (b) Laine, R. M.; Crawford, E. J. *J. Mol. Catal.* **1988**, *44*, 357.
66. Nomura, K.; Ishino, M.; Hazama, M. *BCJ* **1991**, *64*, 2624.
67. Valli, V. L. K.; Alper, H. *JACS* **1993**, *115*, 3778.
68. (a) Sonoda, N.; Yasuhara, T.; Kondo, K.; Ikeda, T.; Tsutsumi, S. *JACS* **1971**, *93*, 6344. (b) Sonoda, N. *PAC* **1993**, *65*, 699.
69. Yoshida, T.; Kambe, N.; Murai, S.; Sonoda, N. *BCJ* **1987**, *60*, 1793.
70. Brown, H. C.; Rathke, M. W. *JACS* **1967**, *89*, 2737.

Naoto Chatani & Shinji Murai
Osaka University, Japan

Carbon Tetrabromide

CBr_4

[558-13-4] CBr_4 (MW 331.65)

(brominating agent used in synthesis of α-acetoxycarboxylic acids[4] and allenes;[5–8] radical additions to alkenes[12–28]).

Alternate Name: tetrabromomethane.
Physical Data: shining plates, mp 88–90 °C; bp 190 °C (dec).
Solubility: insol in water, sol in organic solvents.
Form Supplied in: white solid; widely available.
Analysis of Reagent Purity: FT–IR data.[1]
Preparative Method: carbon tetrabromide is most conveniently prepared by the exhaustive bromination of acetone in the presence of alkali.[2]
Purification: can be sublimed in vacuo; bromide removal via reflux with dil aq Na_2CO_3, followed by steam-distillation and EtOH recrystallization.[29]
Handling, Storage, and Precautions: safety data are available.[3]

Carbon Tetrabromide–Tin(II) Fluoride. The reaction of aldehydes with CBr_4 and SnF_2 in DMSO at 25 °C gives 1-substituted 2,2,2-tribromoethanols in moderate to good yields. The acetate of the product can be hydrolyzed to an α-acetoxycarboxylic acid by $AgNO_3$ (eq 1).[4]

$$\text{PhCHO} \xrightarrow[78\%]{CBr_4,\ SnF_2,\ DMSO} \text{PhCH(OH)CBr}_3 \xrightarrow[40\%]{1.\ Ac_2O,\ py;\ 2.\ AgNO_3,\ H_2O} \text{PhCH(OAc)CO}_2\text{H} \quad (1)$$

2,3-Diacetyl-D-erythronolactone[4] has been prepared in a similar fashion (eq 2).

$$\text{(acetonide)CHO} \xrightarrow[61\%]{1)\ CBr_4,\ SnF_2;\ 2)\ Ac_2O,\ py} \text{(acetonide)CH(OAc)CBr}_3 \xrightarrow[57\%]{1.\ AgNO_3,\ H_2O;\ 2.\ Ac_2O,\ py} \text{lactone (AcO, OAc; major product)} + \text{lactone (AcO, OAc)} \quad (2)$$

Allene Synthesis. The system of carbon tetrabromide (1 equiv) with methyllithium (2 equiv) converts C_n alkenes into C_n

allenes.[5–8] The synthesis of 1,2,6-cyclodecatriene from *cis,cis*-1,5-cyclononadiene[7] serves as an example (eq 3).

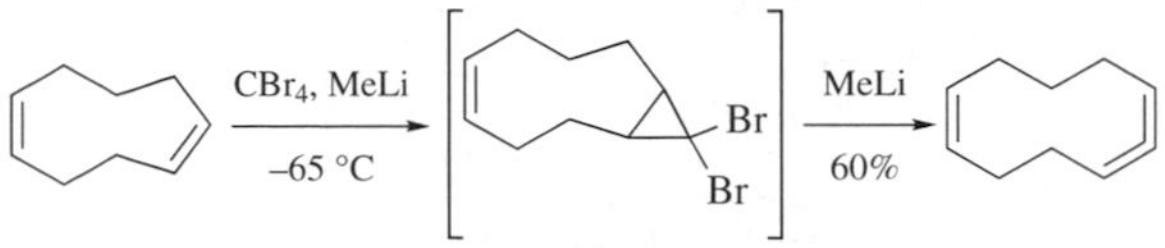

$$\xrightarrow[-65\,^\circ C]{CBr_4,\ MeLi} \left[\ \ \right] \xrightarrow[60\%]{MeLi} \quad (3)$$

When 1 equiv of MeLi is used, the intermediate dibromocyclopropane can be isolated (eqs 4 and 5). Bicyclobutanes are the sole products when the resulting allene would be highly strained (eq 4),[9] or they are significant byproducts (eq 5) when the allene possesses two bulky geminal groups.[10]

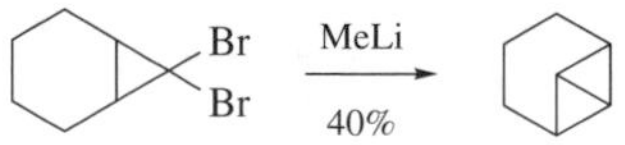

$$\xrightarrow[40\%]{MeLi} \quad (4)$$

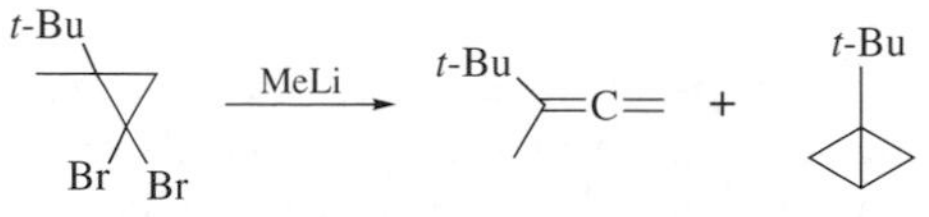

$$\xrightarrow{MeLi} \quad + \quad (5)$$

Dehalogenation of dibromocyclopropanes with an alkyllithium in the presence of (−)-sparteine gives optically active allenes of low optical purity.[11]

Radical Reactions. The addition of CBr_4, or other halogenocarbons, to alkenes is known as the Kharasch reaction. The reactions of terminal alkenes furnish the addition products in the highest yields (eq 6).

$$RCH{=}CH_2 + CBr_4 \longrightarrow RCHBrCH_2CBr_3 \quad (6)$$

The reaction can be initiated by photoirradiation,[12] radical initiators,[13] inorganic salts,[14,15] ruthenium complexes,[16–19] other transition metal complexes,[20–24] samarium diiodide,[25] or by a manganic salt generated electrochemically in situ.[26] The scope and limitations of this reaction have been reviewed in two monographs.[27,28]

Related Reagents. 1,2-Bis(diphenylphosphino)ethane; Tribromomethyllithium; Triphenylphosphine–Carbon Tetrabromide. Triphenylphosphine–Carbon Tetrabromide–Lithium Azide.

1. *The Aldrich Library of FT-IR Spectra*; Pouchert, C. J., Ed.; Aldrich: Milwaukee, 1989; Vol. 3, p 122.
2. Hunter, W. H.; Edgar, D. E. *JACS* **1932**, *54*, 2025.
3. *The Sigma-Aldrich Library of Chemical Safety Data*, 2nd ed.; Lenga, R. E, Ed.; Sigma-Aldrich: Milwaukee, 1988; Vol. 1, p 686.
4. Mukaiyama, T.; Yamaguchi, M.; Kato, J. *CL* **1981**, 1505.
5. Untch, K. G.; Martin, D. J.; Castellucci, N. T. *JOC* **1965**, *30*, 3572.
6. Moorthy, S. N.; Vaidyanathaswamy, R.; Devaprabhakara, D. *S* **1975**, 194.
7. Sharma, S. N.; Srivastava, R. K.; Devaprabhakara, D. *CJC* **1968**, *46*, 84.
8. (a) Moore, W. R.; Ozretich, T. M. *TL* **1967**, 3205; (b) Nozaki, H.; Kato, S.; Noyori, R. *CJC* **1966**, *44*, 1021.
9. Skattelbol, L. *TL* **1970**, 2361.
10. Brown, D. W.; Hendrick, M. E.; Jones, M. *TL* **1973**, 3951.
11. Nozaki, H.; Aratani, T.; Toroya, T.; Noyori, R. *T* **1971**, *27*, 905.
12. Kharasch, M. S.; Jensen, E. V.; Urry, W. H. *Science* **1945**, *102*, 128.
13. Kharasch, M. S.; Jensen, E. V.; Urry, W. H. *JACS* **1947**, *69*, 1100.
14. Asscher, M.; Vofsi, D. *JCS* **1963**, 1887.
15. Asscher, M.; Vofsi, D. *JCS* **1963**, 3921.
16. Matsumoto, H.; Nakano, T.; Nagai, Y. *TL* **1973**, 5147.
17. Kamigata, N.; Kameyama, M.; Kobayashi, M. *JOC* **1987**, *52*, 3312.
18. Kameyama, M.; Kamigata, N. *BCJ* **1987**, *60*, 3687.
19. Matsumoto, H.; Nikaido, T.; Nagai, Y. *TL* **1975**, 899.
20. Tsuji, J.; Sato, K.; Nagashima, H. *CL* **1981**, 1169.
21. Susuki, T.; Tsuji, J. *JOC* **1970**, *35*, 2982.
22. Shvekhgeimer, G. A.; Kobrakov, K. I.; Kartseva, O. I.; Balabanova, L. V. *KGS* **1991**, 369.
23. Davis, R.; Durrant, J. L. A.; Khazal, N. M. S.; Bitterwolf, T. *JOM* **1990**, *386*, 229.
24. (a) Davis, R.; Khazal, N. M. S.; Bitterwolf, T. E. *JOM* **1990**, *397*, 51; (b) Bland, W. J.; Davis, R.; Durrant, J. L. A. *JOM* **1985**, *280*, 95.
25. Ma, S.; Lu, X. *JCS(P1)* **1990**, 2031.
26. Nohair, K.; Lachaise, I.; Paugam, J.-P.; Nedelec, J.-Y. *TL* **1992**, *33*, 213.
27. Sosnovsky, G. *Free Radical Reactions in Preparative Organic Chemistry*; Macmillan: New York, 1964.
28. Giese, B. *Radicals in Organic Synthesis: Formation of Carbon-Carbon Bonds;* Pergamon: Oxford, 1986.
29. Perrin, D. D.; Armarego, W. L. F. *Purification of Laboratory Chemicals*, 3rd ed.; Pergamon: Oxford, 1988; p 116.

Lucjan Strekowski & Alexander S. Kiselyov
Georgia State University, Atlanta, GA, USA

Cerium(III) Chloride[1]

$CeCl_3{\cdot}7H_2O$

[7790-86-5]	$CeCl_3$(MW 246.47)	
($\cdot 7H_2O$)		
[18618-55-8]	$H_{14}CeCl_3O_7$	(MW 372.59)

(mild Lewis acid capable of selective acetalization;[2] organocerium reagents have increased oxo-[3] and azaphilicity[4] and greatly reduced basicity;[5] in combination with $NaBH_4$ is a selective 1,2-reducing agent[6])

Alternate Name: cerous chloride; cerium trichloride.
Physical Data: mp 848 °C; bp 1727 °C; *d* 3.92 g cm^{-3}.
Solubility: insol cold H_2O; sol alcohol and acetone; slightly sol THF.
Form Supplied in: white solid; widely available.
Drying: for some applications the cerium trichloride must be strictly anhydrous. The following procedure has proven most efficacious: a one-necked base-washed flask containing the heptahydrate and a magnetic stirring bar was evacuated to 0.1 Torr and heated *slowly* to 140 °C over a 2 h period. At 70 and 100 °C, considerable amounts of water are given off, and these critical temperature zones should not be passed through too quickly. The magnetically stirred white solid is heated overnight at 140 °C, cooled, blanketed with nitrogen, treated with dry THF (10 mL g^{-1}), and agitated at rt for 3 h. To guarantee the complete removal of water, *t*-butyllithium is added dropwise until an orange color persists.

Handling, Storage, and Precautions: anhydrous $CeCl_3$ should be used as prepared for best results. Cerium is reputed to be of low toxicity.

Organocerium Reagents. Organocerates, most conveniently prepared by the reaction of lithium compounds with anhydrous $CeCl_3$ in THF, are highly oxophilic and significantly less basic than their RLi and RMgBr counterparts. As a consequence, 1,2-addition reactions involving readily enolizable ketones are not adversely affected by competing enolization.[3] Well-studied examples of this phenomenon abound.[7] Although alkyl and vinyl cerates have most often been used,[3,5,7] Cl_2CeCH_2CN,[8] Cl_2CeCH_2COOR,[9] $Cl_2CeC{\equiv}CSiMe_3$,[10] and $Cl_2CeCH_2SiMe_3$[11] are known to be equally effective. The last reagent has been utilized for the methylenation of highly enolizable ketones (eq 1)[11a] and for the conversion of carboxylic acid halides, esters, and lactones into allylsilanes (eq 2).[11b–d] Amides and nitriles also condense with organocerates with equal suppression of enolization.[3,4] In general, high levels of steric hindrance in either reaction partner can be tolerated (eq 3). In select examples, the level of double stereodifferentiation can be impressively high.[5,7,12]

Me_3SiCH_2Li, $CeCl_3$, THF, –78 °C, 83%; HF, H_2O, MeCN, 87% (1)

Me_3SiCH_2MgBr, $CeCl_3$, THF, –70 °C → rt; SiO_2, CH_2Cl_2, 92% (2)

$BuCeCl_2$ (3 equiv), THF, –65 °C → rt, 90% (3)

The discovery has been made that allyl anions produced by the reductive metalation of allyl phenyl sulfides condense cleanly with α,β-unsaturated aldehydes in 1,2-fashion at the more substituted allyl terminus in the presence of ***Titanium Tetraisopropoxide***.[13] Use of the allylcerium reagent instead reverses the regioselectivity. Further, the stereochemical preference is *cis* at −78 °C (kinetic control) and *trans* when warmed to −40 °C (eq 4).[14] Grignard addition reactions to vinylogous esters are improved if promoted by $CeCl_3$ (eq 5).[15]

1. LDBB, –78 °C; 2. $CeCl_3$, –78 °C; 3. $H_2C{=}CHCHO$, –78 °C; 65%
1. LDBB, –78 °C; 2. $CeCl_3$, –78 °C → –40 °C; 3. $H_2C{=}CHCHO$, –78 °C; 48% (4)

$CeCl_3$, THF, 0 °C then H_3O^+, 75% (5)

Stereodivergent pathways relative to other organometallics have surfaced in the addition of organocerium(III) reagents to chiral α-keto amides (eq 6)[16] and 2-acyloxazolidines (eq 7).[17] Applications of this type hold considerable synthetic potential not only in additions to carbonyl compounds,[18] but also to oxime ethers,[19] oxiranes,[20] and oxazolidines.[21]

(6)
MeMgBr, Et_2O, –78 °C 76:24
MeMgBr, $CeCl_3$, THF, –78 °C 25:75

(7)
MeMgBr, Et_2O, –90 °C → –78 °C 94: 6
MeLi, $CeCl_3$, THF, –85 °C 16:84

The uniquely high reactivity of cerium reagents toward the often poorly electrophilic azomethine or C=N double bond of hydrazones has been extensively investigated. Smooth 1,2-addition occurs in good yield (67–81%) provided the air-sensitive hydrazones are acylated; when a chiral auxiliary is present, excellent diastereofacial control operates and enantiomerically enriched α-branched amines result (eq 8).[22] There appears to be a fundamental difference between 'warmed' reagents (−78 °C → 0 °C → −78 °C) and those generated and used at −78 °C. Sonication appears to facilitate conversion to the organocerate.[23] For SAMP hydrazones and related compounds the condensation is accommodating of a wide range of substitution both in the hydrazone and in the nucleophile. Subsequent reductive cleavage of the N–N bond with preservation of configuration can be performed by hydrogenolysis over W-2 ***Raney Nickel*** at 60 °C (free hydrazones; less preferred because of competing saturation of aromatic substituents if present) or by treatment with ***Lithium*** in liquid ammonia (acylated hydrazines).[22c]

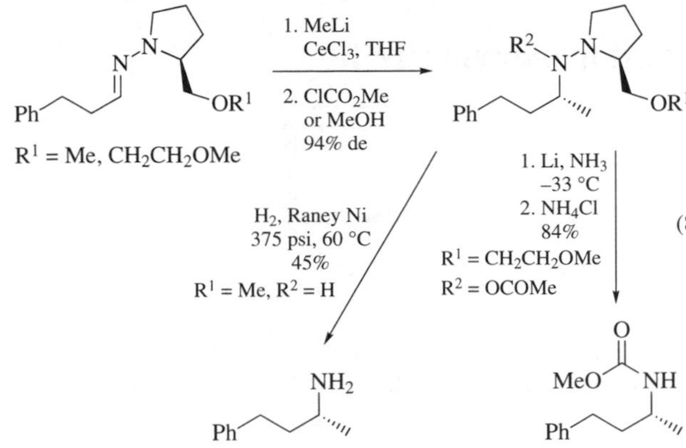

(8)

The preparation of (±)-1,3-diphenyl-1,3-propanediamine illustrates an alternative way in which this chemistry can be utilized (eq 9).[24] In all of the 1,2-addition reactions, the preferred reagent stoichiometry appears to be RLi:$CeCl_3$:hydrazone = 2:2:1. A detailed study of the consequences of varying the relative proportions of these constituents has indicated the optimal ratios of RLi:$CeCl_3$ to be 1:1, notwithstanding the fact that not all of the $CeCl_3$ is consumed in the transmetalation step.[25] The proposal has been made that the empirical formula of the cerate best approximates '$R_3CeCl_3Li_3$'.

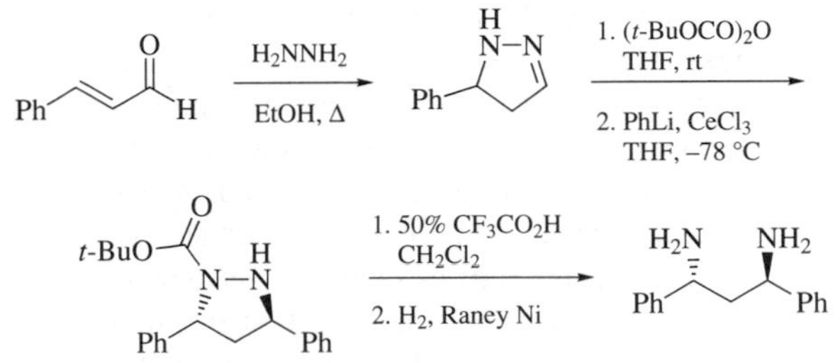

(9)

Cerium enolates, available by transmetalation of the lithium salts with $CeCl_3$ at −78 °C[26] or by reduction of α-bromo ketones with ***Cerium(III) Chloride–Tin(II) Chloride***,[27] give higher yields of crossed aldol products without altering the stereoselectivity. The $CeCl_3$–$SnCl_2$ reagent combination also acts on α,α′-dibromo ketones to give oxylallyl cations that can be captured in the usual way.[28]

Selective Reductions. Equimolar amounts of ***Sodium Borohydride***[4] and $CeCl_3{\cdot}7H_2O$ in methanol act on α,β-unsaturated ketones at rt or below to deliver allylic alcohols cleanly by 1,2-addition.[29] This widely used reducing agent[30] is not renowned for its diastereoselectivity[31] and asymmetric induction capabilities,[32] although exceptions are known.[33] Sometimes a stereofacial preference opposite that realized with other hydrides is encountered (eqs 10 and 11).[34]

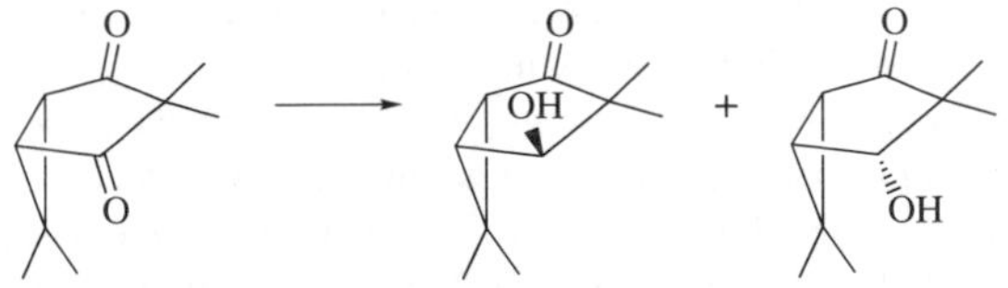

(10)

$NaBH_4$, MeOH, −78 °C 7:93
$NaBH_4$, $CeCl_3$, MeOH, −78 °C 96: 4

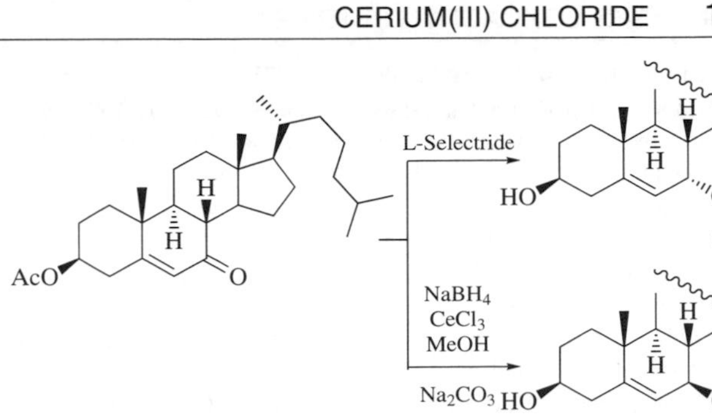

(11)

Allylic alcohol products that are especially sensitive are known to undergo ionization and solvent capture under these mildly acidic conditions (eq 12).[35] The Lewis acidic character of $CeCl_3$ has also been used to advantage in the selective acetalization[6] of saturated aldehydes under the Luche conditions, thereby preventing their reduction.[2] Since ketones and conjugated aldehydes are less responsive, these functionalities are not transiently protected and suffer reduction (eq 13).[36] $NaBH_4$–$CeCl_3$ in MeCN transforms cinnamoyl chlorides into cinnamyl alcohols,[37] while ***Lithium Aluminum Hydride–Cerium(III) Chloride*** in hot DME or THF can effectively reduce alkyl halides and phosphine oxides.[5]

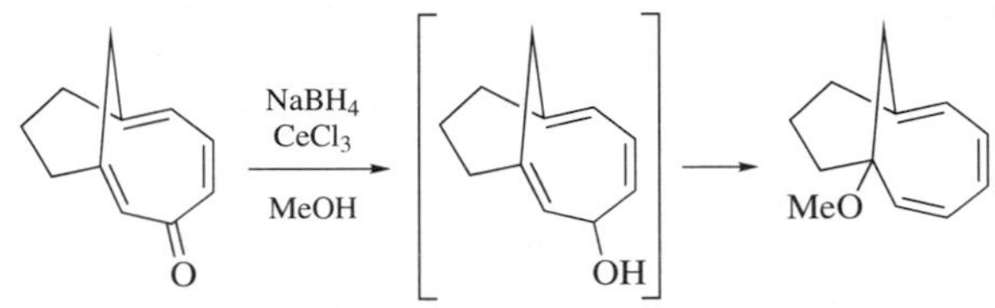

(12)

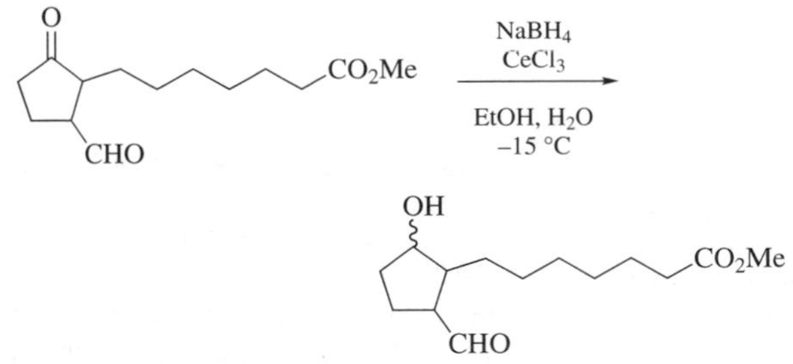

(13)

Related Reagents. Cerium(III) Iodide; Cerium(III) Methanesulfonate; Erbium(III) Chloride; Lanthanum(III) Chloride; Lanthanum(III) Triflate; Ytterbium(II) Chloride; Ytterbium(III) Trifluoromethanesulfonate.

1. (a) Kagan, H. B.; Namy, J. L. *T* **1986**, *42*, 6573. (b) Molander, G. A. *CRV* **1992**, *92*, 29.
2. Gemal, A. L.; Luche, J.-L. *JOC* **1979**, *44*, 4187.
3. Imamoto, T.; Takiyama, N.; Nakamura, K.; Hatajima, T.; Kamiya, Y. *JACS* **1989**, *111*, 4392.
4. Ciganek, E. *JOC* **1992**, *57*, 4521.
5. (a) Imamoto, T.; Takeyama, T.; Kusumoto, T. *CL* **1985**, 1491. (b) Paquette, L. A.; Learn, K. S.; Romine, J. L.; Lin, H.-S. *JACS* **1988**, *110*, 879.
6. Luche, J.-L.; Gemal, A. L. *CC* **1978**, 976.
7. (a) Paquette, L. A.; DeRussy, D. T.; Cottrell, C. E. *JACS* **1988**, *110*, 890. (b) Paquette, L. A.; He, W.; Rogers, R. D. *JOC* **1989**, *54*, 2291.

8. Liu, H.-J.; Al-said, N. H. *TL* **1991**, *32*, 5473.
9. (a) Nagasawa, K.; Kanbara, H.; Matsushita, K.; Ito, K. *TL* **1985**, *26*, 6477. (b) Imamoto, T.; Kusumoto, T.; Tawarayama, Y.; Sugiura, Y.; Mita, T.; Hatanaka, Y.; Yokoyama, M. *JOC* **1984**, *49*, 3904.
10. (a) Suzuki, M.; Kimura, Y.; Terashima, S. *CL* **1984**, 1543. (b) Tamura, Y.; Sasho, M.; Ohe, H.; Akai, S.; Kita, Y. *TL* **1985**, *26*, 1549.
11. (a) Johnson, C. R.; Tait, B. D. *JOC* **1987**, *52*, 281. (b) Anderson, M. B.; Fuchs, P. L. *SC* **1987**, *17*, 621. (c) Narayanan, B. A.; Bunnelle, W. H. *TL* **1987**, *28*, 6261. (d) Lee, T. V.; Channon, J. A.; Cregg, C.; Porter, J. R.; Roden, F. S.; Yeoh, H. T.-L. *T* **1989**, *45*, 5877.
12. (a) Paquette, L. A.; DeRussy, D. T.; Gallucci, J. C. *JOC* **1989**, *54*, 2278. (b) Paquette, L. A.; DeRussy, D. T.; Vandenheste, T.; Rogers, R. D. *JACS* **1990**, *112*, 5562.
13. Cohen, T.; Guo, B.-S. *T* **1986**, *42*, 2803.
14. Guo, B.-S.; Doubleday, W.; Cohen, T. *JACS* **1987**, *109*, 4710.
15. Crimmins, M. T.; Dedopoulou, D. *SC* **1992**, *22*, 1953.
16. Fujisawa, T.; Ukaji, Y.; Funabora, M.; Yamashita, M.; Sato, T. *BCJ* **1990**, *63*, 1894.
17. Ukaji, Y.; Yamamoto, K.; Fukui, M.; Fujisawa, T. *TL* **1991**, *32*, 2919.
18. Kawasaki, M.; Matsuda, F.; Terashima, S. *TL* **1985**, *26*, 2693.
19. Ukaji, Y.; Kume, K.; Watai, T.; Fujisawa, T. *CL* **1991**, 173.
20. (a) Vougioukas, A. E.; Kagan, H. B. *TL* **1987**, *28*, 6065. (b) Schaumann, E.; Kirschning, A. *TL* **1988**, *29*, 4281. (c) Cohen, T.; Jeong, I.-H.; Mudryk, B.; Bhupathy, M.; Awad, M. M. A. *JOC* **1990**, *55*, 1528. (d) Marczak, S.; Wicha, J. *SC* **1990**, *20*, 1511.
21. Pridgen, L. N.; Mokhallalati, M. K.; Wu, M.-J. *JOC* **1992**, *57*, 1237.
22. (a) Denmark, S. E.; Weber, T.; Piotrowski, D. W. *JACS* **1987**, *109*, 2224. (b) Weber, T.; Edwards, J. P.; Denmark, S. E. *SL* **1989**, 20. (c) Denmark, S. E.; Nicaise, O.; Edwards, J. P. *JOC* **1990**, *55*, 6219.
23. Greeves, N.; Lyford, L. *TL* **1992**, *33*, 4759.
24. Denmark, S. E.; Kim, J.-H. *S* **1992**, 229.
25. Denmark, S. E.; Edwards, J. P.; Nicaise, O. *JOC* **1993**, *58*, 569.
26. Imamoto, T.; Kusumoto, T.; Yokoyama, M. *TL* **1983**, *24*, 5233.
27. Fukuzawa, S.; Tsuruta, T.; Fujinami, T.; Sakai, S. *JCS(P1)* **1987**, 1473.
28. Fukuzawa, S.; Fukushima, M.; Fujinami, T.; Sakai, S. *BCJ* **1989**, *62*, 2348.
29. (a) Luche, J.-L. *JACS* **1978**, *100*, 2226. (b) Luche, J.-L.; Rodriquez-Hahn, L.; Crabbé, P. *CC* **1978**, 601.
30. Examples: Godleski, S. A.; Valpey, R. S. *JOC* **1982**, *47*, 381. Block, E.; Wall, A. *JOC* **1987**, *52*, 809. Marchand, A. P.; LaRoe, W. D.; Sharma, G. V. M.; Suri, S. C.; Reddy, D. S. *JOC* **1986**, *51*, 1622. Rubin, Y.; Knobler, C. B.; Diederich, F. *JACS* **1990**, *112*, 1607.
31. (a) Danishefsky, S. J.; DeNinno, M. P.; Chen, S. *JACS* **1988**, *110*, 3929. (b) DeShong, P.; Waltermire, R. E.; Ammon, H. L. *JACS* **1988**, *110*, 1901. (c) Abelman, M. M.; Overman, L. E.; Tran, V. D. *JACS* **1990**, *112*, 6959.
32. (a) Boutin, R. H.; Rapoport, H. *JOC* **1986**, *51*, 5320. (b) Paterson, I.; Laffan, D. D. P.; Rawson, D. J. *TL* **1988**, *29*, 1461. (c) Coxon, J. M.; van Eyk, S. J.; Steel, P. J. *T* **1989**, *45*, 1029.
33. (a) Nimkar, S.; Menaldino, D.; Merrill, A. H.; Liotta, D. *TL* **1988**, *29*, 3037. (b) Rücker, G.; Hörster, H.; Gajewski, W. *SC* **1980**, *10*, 623.
34. (a) Krief, A.; Surleraux, D. *SL* **1991**, 273. (b) Kumar, V.; Amann, A.; Ourisson, G.; Luu, B. *SC* **1987**, *17*, 1279.
35. Scott, L. T.; Hashemi, M. M. *T* **1986**, *42*, 1823.
36. Replacement of MeOH by DMSO reverses this chemoselectivity: Adams, C. *SC* **1984**, *14*, 1349.
37. Lakshmy, K. V.; Mehta, P. G.; Sheth, J. P.; Triverdi, G. K. *OPP* **1985**, *17*, 251.

Leo A. Paquette
The Ohio State University, Columbus, OH, USA

2-Chloroacrylonitrile

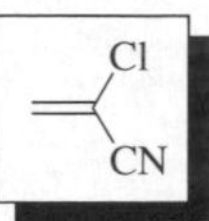

[920-37-6] C_3H_2ClN (MW 87.51)

(ketene equivalent for use in the Diels–Alder reaction[1] and other cycloaddition reactions; undergoes Michael additions and radical additions)

Physical Data: mp −65 °C; bp 88–89 °C; *d* 1.096 g cm^{-3}.
Solubility: sol most organic solvents other tn han pure hydrocarbons.
Form Supplied in: neat colorless liquid.
Analysis of Reagent Purity: ^{1}H NMR, ^{13}C NMR.
Handling, Storage, and Precautions: highly toxic; use only in a fume hood. Reagent can be absorbed through the skin. Always wear gloves when handling this reagent.

[4 + 2] Cycloaddition Reactions. 2-Chloroacrylonitrile is a reasonably good dienophile for the Diels–Alder reaction. It reacts thermally with a variety of cyclic dienes such as cyclopentadiene[2] and cyclohexadiene.[2,3] Most often the initial cycloadducts are hydrolyzed immediately to produce the corresponding ketones (eq 1). In this fashion, 2-chloroacrylonitrile, along with ***2-Acetoxyacrylonitrile***, is one of the most popular ketene equivalents for use in the Diels–Alder reaction.[1] In the case of substituted cyclohexadienes the starting diene is often an unconjugated 1,4-diene which under the reaction conditions isomerizes to the reactive 1,3-diene and produces the desired Diels–Alder adduct (eq 2).[4]

(1)

(2)

R = H, Me R = H, 50% R = Me, 75% R = H, 80% R = Me, 92%

Once again the product α-chloro nitrile is hydrolyzed to produce the bicyclo[2.2.2]octanone. The hydrolysis reaction is usually carried out with either ***Potassium Hydroxide*** in DMSO[2,5] or with ***Sodium Sulfide*** in refluxing ethanol.[6]

Copper salts have been used to catalyze the Diels–Alder reaction of 2-chloroacrylonitrile.[6] Most notable is the successful cycloaddition of the thermally sensitive substituted cyclopentadiene (**1**) (eq 3).[6a,b] Use of copper(II) fluoroborate allowed the reaction to take place at 0 °C. This ultimately produced ketone (**2**), an important intermediate in the early syntheses of prostaglandins.

Diels–Alder reactions between furans and 2-chloroacrylonitrile have also been achieved via copper catalysis (eq 4).[6d] Interestingly, the base-catalyzed hydrolysis of adduct (**3**) produced the amide (**4**) instead of the expected ketone. There have been other instances in which the intermediate α-chloronitrile yields products other than the hydrolyzed ketone.[7]

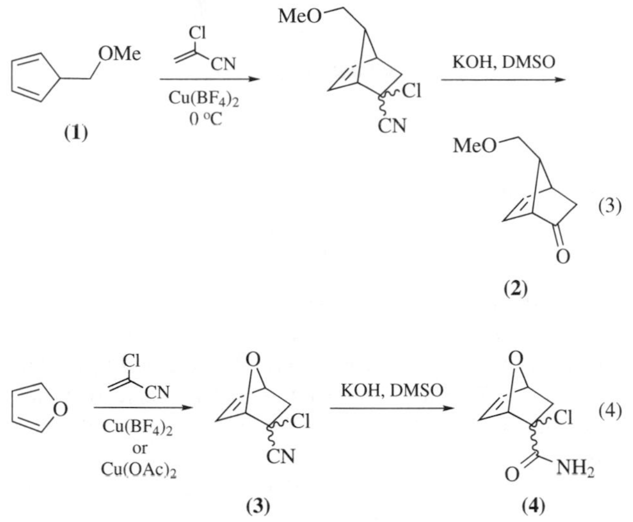

As seen from the above examples, cyclic dienes are usually used in conjunction with 2-chloroacrylonitrile.[8] Such dienes include substituted cyclohexadienes,[4,7a–d,9] substituted cyclopentadienes,[6a–c,7e] hydroxypyrones,[10] fulvenes,[11] furans,[6d,12] and isobenzofurans.[13] Other dienes which give [4 + 2] cycloadducts with 2-chloroacrylonitrile include vinyl heterocycles[14] and a few acyclic dienes.[15] It has been stated that thermal cycloadditions are limited to 140 °C due to polymerization of the chloroacrylonitrile above this temperature.[3] In addition to the aforementioned Cu^{2+} catalysis, ***Zinc Iodide***,[12b] triorganotin cations,[12e] and high pressure[12a,f] have been used to accelerate the cycloaddition reactions of 2-chloroacrylonitrile and furans.

In addition to 2-chloroacrylonitrile and ***2-Acetoxyacrylonitrile***, other ketene equivalents have been developed.[1] These include acrylonitrile,[2] 2-aminoacrylonitrile,[16a] 2-methylthioacrylonitrile,[16b] 2-chloroacryloyl chloride,[17a,b] 2-bromoacrolein,[17c] vinyl boronates,[4,18a] vinylboranes,[18b] nitroethylene,[19] and vinyl sulfoxides.[20] One comparative study found that 2-chloroacrylonitrile was both more reactive and more regioselective than either 2-acetoxyacrylonitrile or dibutyl vinyl boronate.[4] Chiral vinyl sulfoxides have been explored as possible chiral ketene equivalents.[17c,20]

Other Cycloaddition Reactions. In addition to [4 + 2] cycloadditions, 2-chloroacrylonitrile also undergoes [2 + 3][21,22] and [2 + 2][23] cycloaddition reactions. In particular, 2-chloroacrylonitrile is a good dipolarophile and has produced cycloadducts with nitrones[21] and a number of different azomethine ylides.[22] In the case of nitrones the cycloadducts have been hydrolyzed to isoxazolidinones; thus 2-chloroacrylonitrile once again functions as a ketene equivalent in these reactions (eq 5).[21c] Chiral nitrones have been found to undergo stereoselective cycloaddition reactions and have been used in the synthesis of chiral amino acids[21c] and carbapenem derivatives.[21a]

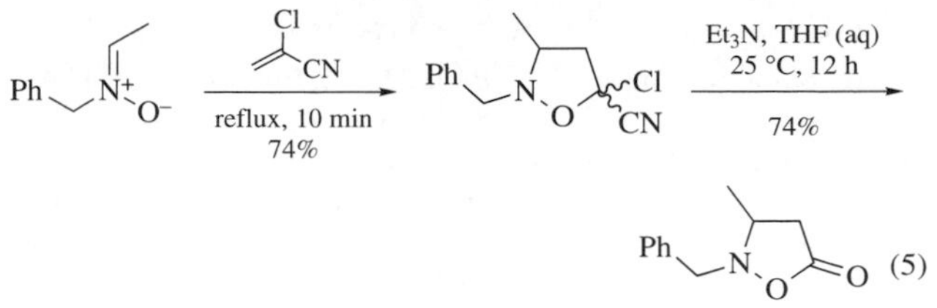

Michael Additions and Radical Additions. As with other acrylonitriles, 2-chloroacrylonitrile is an excellent acceptor of nucleophiles[23e,24] and nucleophilic radicals.[25] In a number of instances[24a,b,25b,e] the initial addition product can undergo a further cyclization, in which case 2-chloroacrylonitrile functions as a convenient two-carbon annulating agent. Such an annulation is illustrated for both a polar addition (eq 6)[24b] and a radical addition (eq 7).[25b] 2-Chloroacrylonitrile has been used extensively in the homologation of nucleophilic radicals[25a,c,f] and, in keeping with the captodative effect,[25d] has been shown to react 10–15 times faster than acrylonitrile.[25a,d]

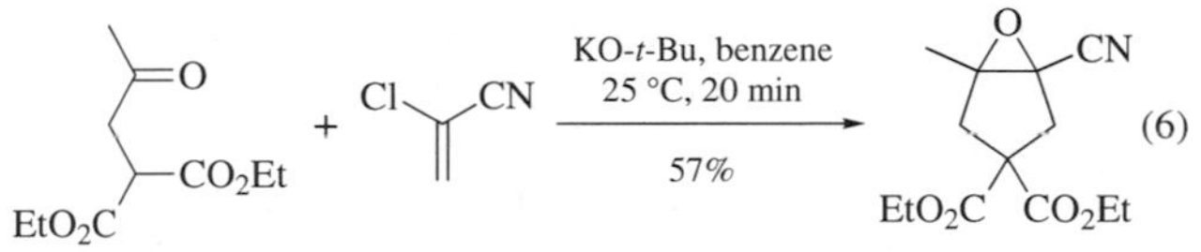

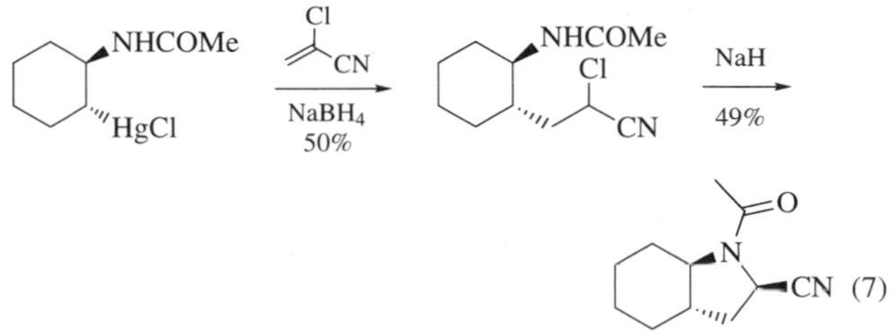

Related Reagents. 2-Acetoxyacrylonitrile; Phenylsulfinylethylene.

1. Ranganathan, S.; Ranganathan, D.; Mehrotra, A. K. *S* **1977**, 289.
2. Freeman, P. K.; Balls, D. M.; Brown, D. J. *JOC* **1968**, *33*, 2211.
3. Kreiger, H.: Nakajima, F. *Suom. Kemistil.* **1969**, *42*, 314 (*CA* **1969**, *71*, 112496p).
4. Evans, D. A.; Scott, W. L.; Truesdale, L. K. *TL* **1972**, 121.
5. Paasivirta, J.; Kreiger, H. *Suom. Kemistil.* **1965**, *B38*, 182 (*CA* **1966**, *64*, 4965g).
6. (a) Corey, E. J.; Weinshenker, N. M.; Schaaf, T. K.; Huber, W. *JACS* **1969**, *91*, 5675. (b) Corey, E. J.; Koelliker, U.; Neuffer, J. *JACS* **1971**, *93*, 1489. (c) Goering, H. L.; Chang, C.-S. *JOC* **1975**, *40*, 2565. (d) Vieira, E.; Vogel, P. *HCA* **1982**, *65*, 1700.
7. (a) Damiano, J.; Geribaldi, S.; Torri, G.; Azzaro, M. *TL* **1973**, 2301. (b) Yamada, Y.; Kimura, M.; Nagaoka, H.; Ohnishi, K. *TL* **1977**, 2379. (c) Clark, R. S. J.; Holmes, A. B.; Matassa, V. G. *TL* **1989**, *30*, 3223. (d) Clark, R. S. J.; Holmes, A. B.; Matassa, V. G. *JCS(P1)* **1990**, 1389. (e) Bull, J. R.; Grundler, C.; Niven, M. L. *CC* **1993**, 217.
8. Fringuelli, F.; Taticchi, A. *Dienes in the Diels–Alder Reaction*; Wiley: New York, 1990.
9. (a) Mirrington, R. N.; Greyson, R. P. *CC* **1973**, 598. (b) Munai, A.; Sato, S.; Masamune, T. *CL* **1981**, 429. (c) Oku, A.; Hasegawa, H.; Shimazu, H.; Nishimura, J.; Harada, T. *JOC* **1981**, *46*, 4152.

10. Corey, E. J.; Kozikowski, A. P. *TL* **1975**, 2389.
11. (a) Sakai, K.; Kobori, T. *TL* **1981**, *22*, 115. (b) Siegel, H. *S* **1985**, 798. (c) Nzabamwita, G.; Kolani, B.; Jousseaume, B. *TL* **1989**, *30*, 2207.
12. (a) Kotsuki, H.; Nishizawa, H. *H* **1981**, *16*, 1287. (b) Brion, F. *TL* **1982**, *23*, 5299. (c) Schuda, P. F.; Bennett, J. M. *TL* **1982**, *23*, 5525. (d) Moursoundis, J.; Wege, D. *AJC* **1983**, *36*, 2473. (e) Nugent, W. A.; McKinney, R. J.; Harlow, R. L. *OM* **1984**, *3*, 1315. (f) Kotsuki, H.; Mori, Y.; Ohtsuka, T.; Nishizawa, H.; Ochi, M.; Matsuoka, K. *H* **1987**, *26*, 2347.
13. Makhlouf, M. A.; Rickborn, B. *JOC* **1981**, *46*, 2734.
14. (a) Sasaki, T.; Ishibashi, Y.; Ohno, M. *JCR(S)* **1984**, 218. (b) Ohmura, H.; Motoki, S. *BCJ* **1984**, *57*, 1131. (c) Alexandre, C.; Rouessac, F.; Tabti, B. *TL* **1985**, *26*, 5453. (d) Pindur, U.; Eitel, M.; Abdoust-Houshang, E. *H* **1989**, *29*, 11.
15. (a) Kozikowski, A. P.; Hiraga, K.; Springer, J. P.; Wang, B. C.; Xu, Z. B. *JACS* **1984**, *106*, 1845. (b) Gordon, P. F. *CA* **1985**, *102*, 131717m. (c) Baldwin, J. E.; Otsuka, M.; Wallace, P. M. *T* **1986**, *42*, 3097.
16. (a) Boucher, O.-L.; Stella, L. *T* **1985**, *41*, 875. (b) Boucher, O.-L.; Stella, L. *T* **1986**, *42*, 3871.
17. (a) Corey, E. J.; Ravindranathan, T.; Terashima, S. *JACS* **1971**, *93*, 4326. (b) Van Tamelen, E. E.; Zawacky, S. R. *TL* **1985**, *26*, 2833. (c) Corey, E. J.; Loh, T.-P. *JACS* **1991**, *113*, 8966.
18. (a) Matteson, D. S.; Waldbillig, J. O. *JOC* **1963**, *28*, 366. (b) Singleton, D. A.; Martinez, J. P.; Watson, J. Y. *TL* **1992**, *33*, 1017.
19. (a) Bartlett, P. A.; Green, F. R.; Webb, T. R. *TL* **1977**, 33. (b) Ranganathan, D.; Rao, C. B.; Ranganathan, S.; Mehrotra, A. K.; Iyengar, R. *JOC* **1980**, *45*, 1185. (c) Mehta, G.; Subrahmanyam, D. *JCS(P1)* **1991**, 395.
20. Maignan, C.; Raphael, R. A. *T* **1983**, *39*, 3245. Lopez, R.; Carretero, J. C. *TA* **1991**, *2*, 93.
21. (a) Freer, A.; Overton, K.; Tomanek, R. *TL* **1990**, 1471. (b) Kurasawa, Y.; Kim, H. S.; Katoh, R.; Kawano, T.; Takada, A.; Okamoto, Y. *JHC* **1990**, *27*, 2209. (c) Keirs, D.; Moffat, D.; Overton, K.; Tomanek, R. *JCS(P1)* **1991**, 1041.
22. (a) Benages, I. A.; Albonica, S. M. *JOC* **1978**, *43*, 4273. (b) Pierini, A. B.; Cardozo, M. G.; Montiel, A. A.; Albonica, S. M.; Pizzorno, M. T. *JHC* **1989**, *26*, 1003. (c) Bonneau, R.; Liu, M. T. H.; Lapouyade, R. *JCS(P1)* **1989**, 1547. (d) Jones, R. C. F.; Nichols, J. R.; Cox, M. T. *TL* **1990**, *31*, 2333.
23. (a) Scheeren, H. W.; Frissen, A. E. *S* **1983**, 794. (b) De Cock, C.; Piettre, S.; Lahousse, F.; Janousek, Z.; Merenyi, R. Viehe, H. G. *T* **1985**, *41*, 4183. (c) Shimo, T.; Somekawa, K.; Wakikawa, Y.; Uemura, H.; Tsuge, O.; Imada, K.; Tanabe, K. *BCJ* **1987**, *60*, 621. (d) Schuster, D. I.; Heibel, G. E.; Brown, P.; Turro, N. J.; Kumar, C. V. *JACS* **1988**, *110*, 8261. (e) Quendo, A.; Rousseau, G. *SC* **1989**, *19*, 1551. (f) Narasaka, K.; Hayashi, Y.; Shimadzu, H.; Niihata, S. *JACS* **1992**, *114*, 8869.
24. (a) Bergmann, E. D.; Ginsburg, D.; Pappo, R. *OR* **1959**, *10*, 179. (b) White, D. R. *JCS(C)* **1975**, 95. (c) Joucla, M.; Fouchet, B.; Hamelin, J. *T* **1985**, *41*, 2707.
25. (a) Giese, B. *AG(E)* **1983**, *22*, 753. (b) Henning, R.; Urbach, H. *TL* **1983**, *24*, 5343. (c) Giese, B.; Horler, H. *T* **1985**, *41*, 4025. (d) Ito, O.; Arito, Y.; Matsuda, M. *JCS(P2)* **1988**, 869. (e) Srikrishna, A.; Hemamalini, P. *JCS(P1)* **1989**, 2511. (f) Barton, D. H. R.; Chern, C. Y.; Jaszberenyi, J. C. *TL* **1992**, *33*, 5017.

Patrick G. McDougal
Reed College, Portland, OR, USA

μ-Chlorobis(cyclopentadienyl)(dimethylaluminum)-μ-methylenetitanium[1]

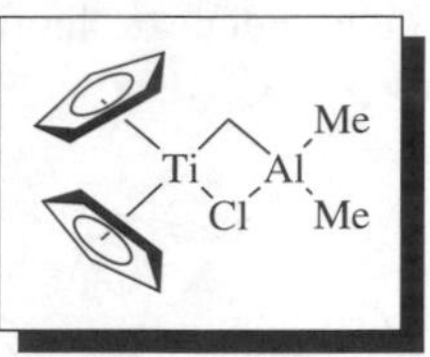

[67719-69-1] $C_{13}H_{18}AlClTi$ (MW 284.60)

(methylenating agent for alkenation of carbonyl compounds,[2–4] particularly esters, ketones, and aldehydes; of low basicity and functions without epimerization at α-chiral centers)

Alternate Name: Tebbe reagent.
Physical Data: red solid.
Solubility: highly sol toluene, benzene, or dichloromethane; will dissolve in THF at low temperature (but not stable for prolonged periods in ethereal solvents). Nearly insol saturated hydrocarbons.
Form Supplied in: may be purchased as the pure solid or prepared from titanocene dichloride and trimethylaluminum.
Handling, Storage, and Precautions: the dry solid is air-sensitive and may be pyrophoric in air, especially when impure; it must be handled under an atmosphere of nitrogen or argon. The reagent is most conveniently stored and handled as a toluene solution; in this form the reagent is somewhat pyrophoric in air and must be handled using syringe and cannula techniques as practiced when using Grignard or alkyllithium solutions.

Tebbe Reagent as a Source of '$Cp_2Ti{=}CH_2$'. The complex $Cp_2TiCH_2{\cdot}AlMe_2Cl$, commonly referred to as the Tebbe reagent,[5] is a source of the reactive titanium methylene species '$Cp_2Ti{=}CH_2$' as shown in eq 1. The methylene intermediate, formation of which is greatly accelerated by bases such as THF or pyridine, is useful for the methylenation of carbonyl compounds in a process similar to Wittig alkenation; the driving force is the high oxophilicity of titanium. The scope of reactivity is greater than with the corresponding phosphoranes;[6] thus esters and amides are converted to vinyl ethers and enamines by the Tebbe reagent.[2,7] The lower basicity of the Tebbe reagent offers further advantages over the Wittig procedure (see below). Although many examples of methylenation of unsaturated substrates are known, and carbon–carbon double bonds usually do not interfere, it should be noted that the Tebbe reagent will react with certain alkenes to form titanacyclobutanes.[8] These titanacycles, of importance as intermediates in degenerate metathesis of terminal alkenes and ring-opening metathesis polymerization of strained cyclic alkenes,[9] and as synthetic reagents themselves,[1b] have been studied in some detail. The most stable titanacyclobutanes are those derived from monosubstituted alkenes; even these will revert to free alkene and active methylene when heated to about 60 °C.[10] Therefore mild heating may be used to improve yields of irreversibly formed methylenation products for certain substrates.[3] Furthermore, use of excess Tebbe reagent will in certain cases lower the yield of alkene product.[3,11]

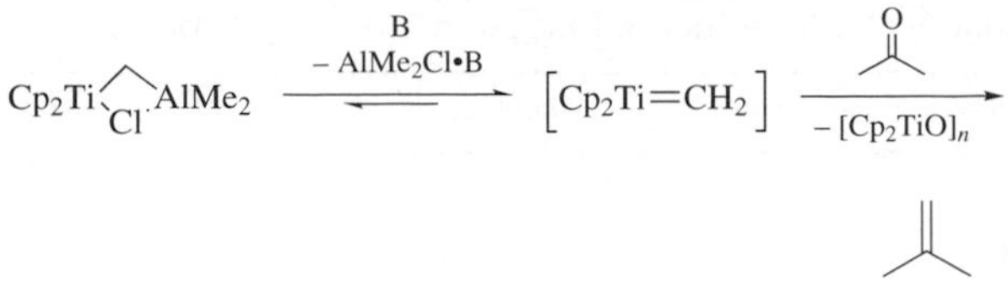

$$Cp_2Ti\langle Cl \rangle AlMe_2 \xrightarrow[\quad]{B,\ -AlMe_2Cl\bullet B} [Cp_2Ti{=}CH_2] \xrightarrow[-[Cp_2TiO]_n]{\text{acetone}} \text{isobutylene} \quad (1)$$

The reagent may be purchased as a pure solid, or synthesized as such using Schlenk line and glovebox techniques.[5a,12] Since isolation of the Tebbe reagent is time-consuming, in situ preparations of the complex have been developed.[3,13] It should be noted that in situ preparation produces a solution with one equiv of excess $AlMe_2Cl$ and often affords somewhat lower yields than the isolated reagent.[1a,13a] Typical conditions[2,13a] for use involve combining the reagent in toluene solution at low temperature with the substrate and a Lewis base such as THF and/or pyridine. After warming to room temperature, the solution is chilled and quenched with aqueous sodium hydroxide, diluted with ether, dried, filtered through Celite, and the product further purified, often by chromatography on neutral or basic alumina.

Tebbe methodology is specific for methylene transfer. However, approaches to analogous Group 4 metal reagents having substituted alkylidene units have been developed and show considerable promise.[1a,14] Several other titanium-based reagents for methylene transfer have been developed.[1a,15,16] A widely used system for methylenation of ketones[15a] is the still undefined mixture formed from $Zn/CH_2X_2/TiCl_4$. Also, thermolysis of Cp_2TiMe_2 provides a clean aluminum-free source of $Cp_2Ti{=}CH_2$ and shows considerable synthetic promise.[16] These various methylenation and alkylidenation processes have been comprehensively reviewed.[1a]

Methylenation of Aldehydes and Ketones. The Tebbe reagent accomplishes methylenation of aldehydes and ketones.[4,5a,11,17–23] Table 1 shows some representative conversions with yields for the Wittig reagent included where available; the titanium reagent affords consistently higher yields and is less sensitive to steric crowding than ***Methylenetriphenylphosphorane***.[4] It is particularly useful in preparation of exocyclic alkenes (e.g. entries 2, 4, 5 and 8); only in extremely hindered substrates such as fenchone (entry 3) is the reagent ineffective. Also, the titanium complex efficiently methylenates such readily enolizable ketones as β-tetralone (entry 4).

The Tebbe reagent will methylenate aldehydes and ketones without epimerization at α-stereogenic centers, as illustrated in eq (2).[22] The relatively low basicity of this reagent also permits conversions such as eq (3), which gives almost exclusively β-elimination across the C(1′)–C(2′) bond when attempted by the Wittig method.[11] When a 'large excess' of Tebbe reagent is used in eq (3), *gem*-dimethylation occurs via a titanacyclobutane. Ketones with α,α-disubstitution (e.g. eq 4) will enolize rather than methylenate;[23] these titanium enolates are not active in the aldol reaction. The Tebbe reagent shows synthetically useful selectivity for ketones in preference to esters, as in the conversion of entry 8 (where 15% methyl ketone byproduct was observed), which was unsuccessful using $Zn/CH_2X_2/TiCl_4$ due to cyclopropanation. However, the $Zn/CH_2X_2/TiCl_4$ mixture is much more widely used for ketone methylenation since in certain preparations it will not react with esters.[1a]

Table 1 Methylenation of Ketones with $Cp_2TiCH_2{\bullet}AlMe_2Cl$

Entry	Yield (%)	Substrate (X = O) and Product (X = CH_2)
1[4]	97; Wittig 46	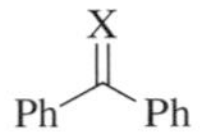 Ph(C=X)Ph
2[4]	96; Wittig 80	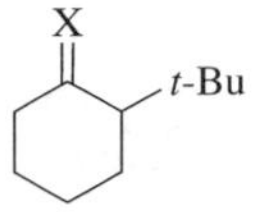 2-*t*-Bu-cyclohexan-X
3[4,23]	16; Wittig 5	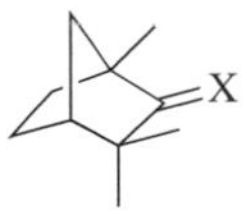 fenchone (=X)
4[23]	84	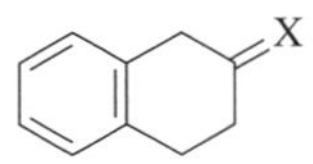 β-tetralone (=X)
5[19]	55	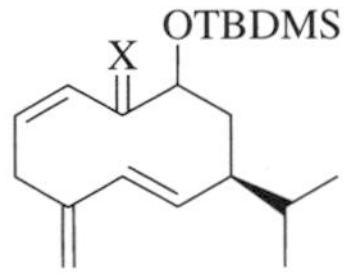 OTBDMS, X
6[4]	77; Wittig 4	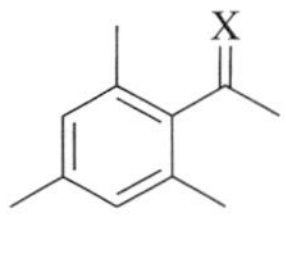 X
7[18]	93	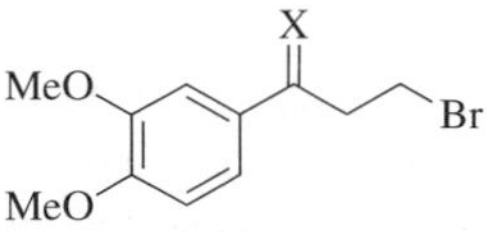 MeO, MeO, X, Br
8[21]	63	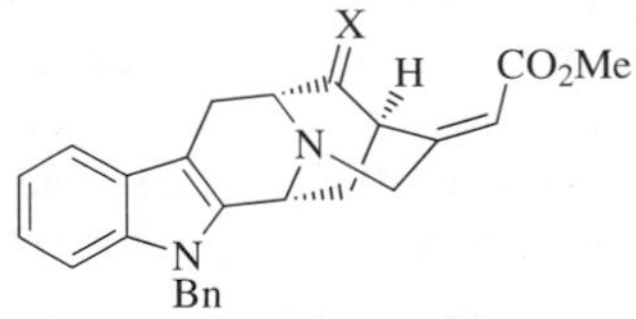 X, H, CO_2Me, N, N, Bn

$$\text{EtO}_2\text{C–CH(OTBS)–CH(OTBS)–CH(OBn)–CH(OBn)–CH(OTBS)–CH(OTBS)–CHO} \xrightarrow[\text{82\%}]{\text{Tebbe},\ -78 \rightarrow -10\ ^\circ\text{C},\ \text{toluene, THF, pyridine}} \text{EtO}_2\text{C–CH(OTBS)–CH(OTBS)–CH(OBn)–CH(OBn)–CH(OTBS)–CH(OTBS)–CH{=}CH}_2 \quad (2)$$

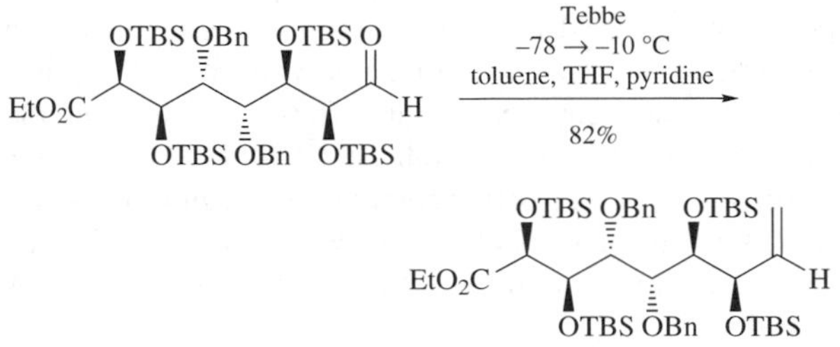

Methylenation of Esters and Lactones. The greatest advantage of the Tebbe reagent is that, unlike phosphorous ylides, it may be used to convert esters and lactones to versatile enol ethers.[2] The reaction is quite general and typically proceeds in good to excellent yield.[3,7,24–50] Several representative conversions are presented in Table 2.

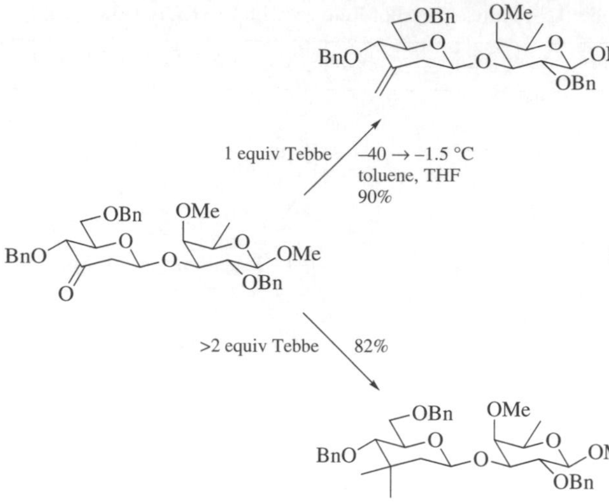

1 equiv Tebbe, −40 → −1.5 °C, toluene, THF, 90%

>2 equiv Tebbe, 82%

(3)

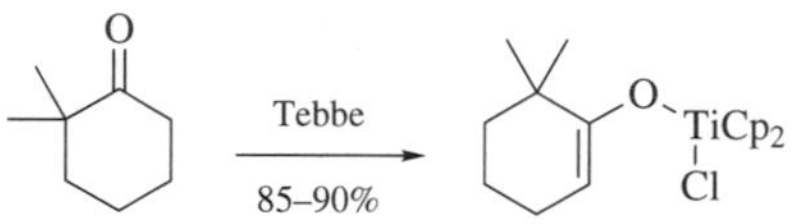

Tebbe, 85–90%

(4)

Table 2 Methylenation of Esters with $Cp_2TiCH_2 \cdot AlMe_2Cl$

Entry	Yield (%)	Substrate (X = O) and Product (X = CH_2)
1[7]	94	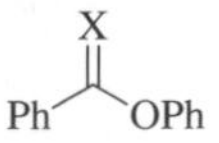
2[2]	87	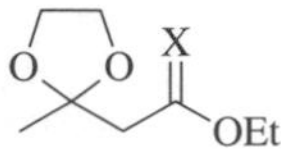
3[2]	97	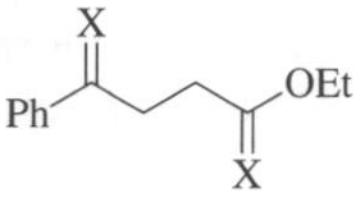
4[2]	96	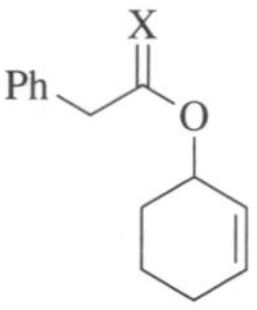
5[26]	85	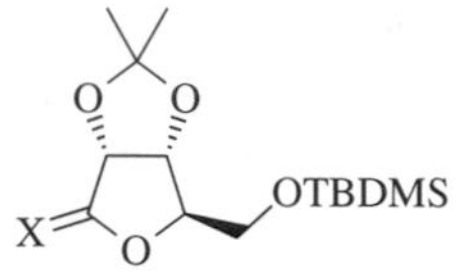
6[29]	R = Bn, 47 R = $CHPh_2$, 76 R = trityl, trace	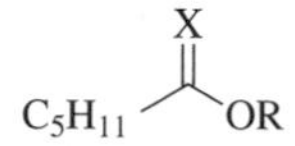
7[46]	76	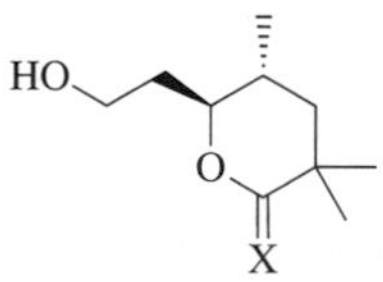
8[2]	79	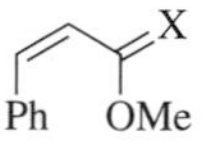
9[45]	–	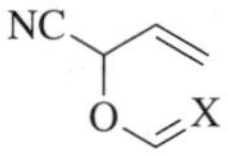
10[33]	R = Bn, 87–90	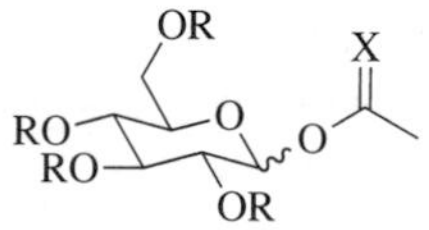
11[35]	R = Me, 95; R = Bn, 88	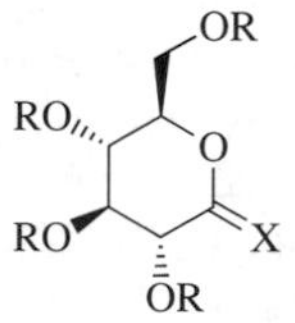
12[41]	ca. 50	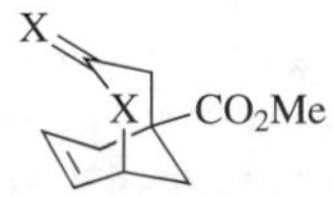

The procedure is tolerant of the acetal (entries 2 and 5) and cyano (entry 9) groups. Double bonds typically retain their configuration (entry 8) and position. In a few instances, double bond migration has been reported upon workup;[30,47] precautions to minimize such rearrangement are described.[13a] Other compatible functionalities include various siloxy groups,[36,46] halide,[44,48] and thioacetal.[24] Ketone and ester groups may be simultaneously methylenated (entry 3), or a keto group may be preferentially methylenated with 1 equiv of reagent.[7,21] Literature estimates of ketone versus ester selectivity are given as 4:1 in general[2] and 25–30:1 for acetophenone against methyl benzoate.[23] It has been observed that yields may differ depending on whether isolated Tebbe reagent is used or if the complex is generated in situ; for example, 94% in entry 1 compared with 68–70% by in situ methods.[13a] Nonetheless, yields are often excellent using the in situ reagent (e.g. entry 11 and others[3,35]), which is considerably less expensive than the isolated complex. Since enol ethers are not known to form titanacyclobutanes on reaction with the Tebbe complex, *gem*-dimethylation does not occur when excess reagent is employed.[3]

Methylenation of allyl esters (Table 2, entries 4, 9, 12, and 13) affords allyl vinyl ethers[2,24,25,27,32,42–45,49] which are useful substrates for the Claisen rearrangement (e.g. eq 5[25] and eq 6[24a]). Although the methylene group of the Ti complex is mildly basic, solutions of the reagent are Lewis acidic, particularly when an extra equivalent of $AlMe_2Cl$ is present from in situ reagent preparation. The acidic character of the reagent has been used to advantage in a mild one-pot synthesis of 1,5-dienes from allyl esters in which a Claisen rearrangement occurs under Lewis acid catalysis (eq 6 and entry 12).[24a]

The Tebbe complex has found frequent application in carbohydrate chemistry.[26,31b,33,35,36]

Table 2 Continued

Entry	Yield (%)	Substrate (X = O) and Product (X = CH_2)
13[43]	65	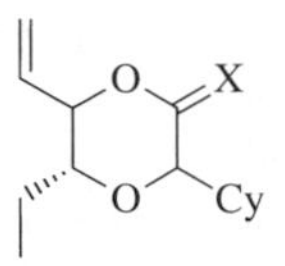
14[50]	94	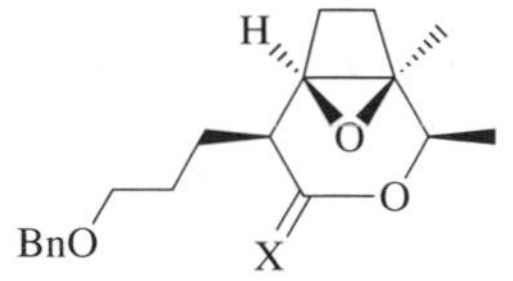

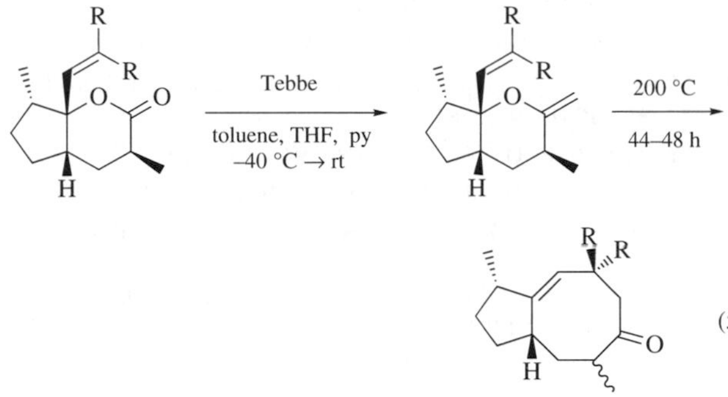

Tebbe; toluene, THF, py; –40 °C → rt; 200 °C; 44–48 h

R = H, 91%; R = Me, 87% (5)

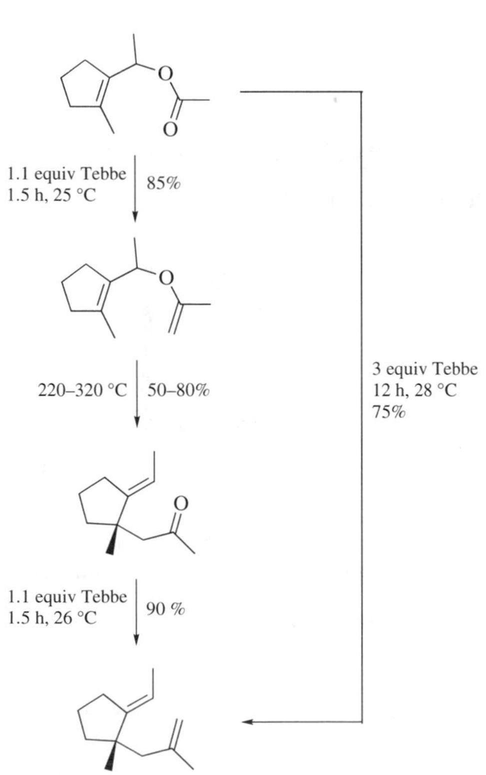

1.1 equiv Tebbe, 1.5 h, 25 °C, 85%; 220–320 °C, 50–80%; 1.1 equiv Tebbe, 1.5 h, 26 °C, 90 %; 3 equiv Tebbe, 12 h, 28 °C, 75% (6)

The titanium reagent allows transformations of esters without epimerization at α-chiral centers, as is illustrated in the methyl ketone preparation of eq 7.[28b] Cyclic carbonates may be methylenated as well (eq 8).[1b]

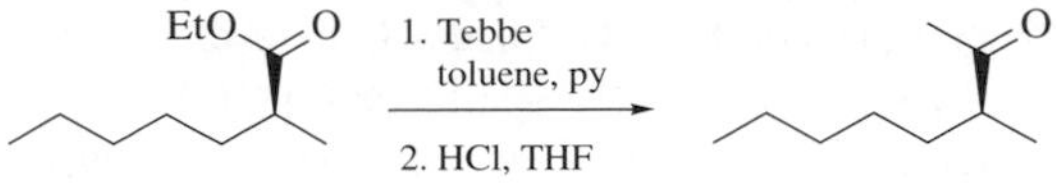

1. Tebbe, toluene, py; 2. HCl, THF (7)

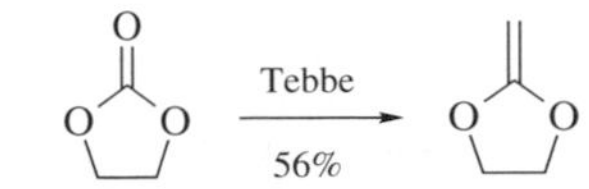

Tebbe, 56% (8)

In at least one case where the Tebbe procedure is ineffective, presumably for steric reasons, an ester has been efficiently methylenated with $Zn/CH_2X_2/TiCl_4$.[38] However, hindered esters may be methylenated with the Tebbe reagent (e.g. entry 7).

Methylenation of Amides. Amides have been converted to enamines of methyl ketones by the Tebbe reagent (eq 9).[1a,b,7,51] No added base is required. Isolation of enamines is possible, or in situ alkylation may be performed.[7] The reaction has been little developed.

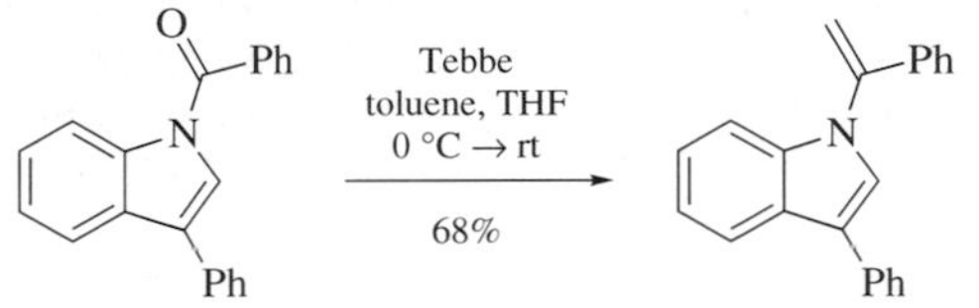

Tebbe, toluene, THF, 0 °C → rt, 68% (9)

Titanium Enolates from Acid Chlorides. Acid chlorides have been reported to react with the Tebbe reagent to afford titanium enolate complexes (eq 10),[52] although yields are much lower than those obtained using titanacyclobutane precursors.[53] These enolate complexes of methyl ketones are known to participate in aldol reactions.[53]

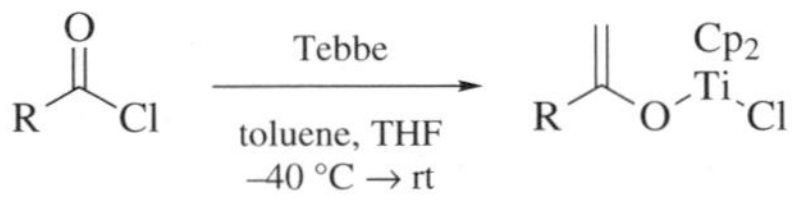

Tebbe; toluene, THF; –40 °C → rt (10)

Other Reactions of the Tebbe Reagent. Silyl esters and thioesters are methylenated by the Tebbe reagent.[1a] The complex is known to react with alkynes to give stable titanacyclobutenes.[54] Reaction with anhydrides gives enolate complexes which are not active in the aldol reaction.[55] By contrast, unhindered imides may be methylenated in good yield with excellent selectivity in unsymmetrical cases.[55] Nitriles react to afford vinylimidotitanium complexes, which have been studied in some detail.[56] The reagent has found application in the preparation of various organometallic complexes.[57]

Related Reagents. Bis(η^5-cyclopentadienyl)(diiodozinc)(μ-methylene)titanium; Bis(cyclopentadienyl)-3,3-dimethyltitanacyclobutane[1]; Bis(cyclopentadienyl)dimethyltitanium[1]; Dibromomethane–Zinc–Titanium(IV) Chloride; Diiodomethane.

1. (a) Pine, S. H. *OR* **1993**, *43*, 1. (b) Brown-Wensley, K. A.; Buchwald, S. L.; Cannizzo, L.; Clawson, L.; Ho, S.; Meinhart, D.; Stille, J. R.; Straus, D. A.; Grubbs, R. H. *PAC* **1983**, *55*, 1733. (c) Reetz, M. T. *Organotitanium Reagents in Organic Synthesis*; Springer: Berlin, 1986; pp 223–229.
2. Pine, S. H.; Zahler, R.; Evans, D. A.; Grubbs, R. H. *JACS* **1980**, *102*, 3270.

3. Cannizzo, L.; Grubbs, R. H. *JOC* **1985**, *50*, 2386.
4. Pine, S. H.; Shen, G. S.; Hoang, H. *S* **1991**, 165.
5. (a) Tebbe, F. N.; Parshall, G. W.; Reddy, G. S. *JACS* **1978**, *100*, 3611. (b) Klabunde, U.; Tebbe, F. N.; Parshall, G. W.; Harlow, R. L. *J. Mol. Catal.* **1980**, *8*, 37.
6. (a) Cadogan, J. I. G., Ed. *Organophosphorous Reagents in Organic Synthesis*; Academic: New York, 1979. (b) Johnson, W. A. *Ylid Chemistry*; Academic; New York, 1966; pp 132–192.
7. Pine, S. H.; Pettit, R. J.; Geib, G. D.; Cruz, S. G.; Gallego, C. H.; Tijerina, T.; Pine, R. D. *JOC* **1985**, *50*, 1212.
8. Howard, T. R.; Lee, J. B.; Grubbs, R. H. *JACS* **1980**, *102*, 6876.
9. (a) Gilliom, L. R.; Grubbs, R. H. *JACS* **1986**, *108*, 733. (b) Tumas, W.; Grubbs, R. H. *Science* **1989**, *243*, 907.
10. Straus, D. A.; Grubbs, R. H. *OM* **1982**, *1*, 1658.
11. Trumtel, M.; Tavecchia, P.; Veyrières, A.; Sinaÿ, P. *Carbohydr. Res.* **1990**, *202*, 257.
12. Lee, J. B.; Ott, K. C.; Grubbs, R. H. *JACS* **1982**, *104*, 7491.
13. (a) Pine, S. H.; Kim, G.; Lee, V. *OS* **1990**, *69*, 72. (b) Chou, T.-S.; Huang, S.-B.; Hsu, W.-H. *J. Chinese Chem. Soc.* **1983**, *30*, 277.
14. (a) Tucker, C. E.; Knochel, P. *JACS* **1991**, *113*, 9888. (b) Okazoe, T.; Takai, K.; Oshima, K.; Utimoto, K. *JOC* **1987**, *52*, 4410. (c) Petasis, N. A.; Akritopoulu, I. *SL* **1992**, 665. (d) Petasis, N. A.; Bzowej, E. I. *TL* **1993**, *34*, 943. (e) Clift, S. M.; Schwartz, J. *JACS* **1984**, *106*, 8300.
15. For example: (a) Lombardo, L. *OS* **1986**, *65*, 81. (b) Eisch, J. J.; Piotrowski, A. *TL* **1983**, *24*, 2043.
16. (a) Petasis, N. A.; Bzowej, E. I. *JACS* **1990**, *112*, 6392. (b) Petasis, N. A.; Fu, D.-K. *JACS* **1993**, *115*, 7208.
17. Bunelle, W. H.; Shangraw, W. R. *T* **1987**, *43*, 2005.
18. Winkler, J. D.; Muller, C. L.; Scott, R. D. *JACS* **1988**, *110*, 4831.
19. Hauptmann, H.; Mühlbauer, G.; Sass, H. *TL* **1986**, *27*, 6189.
20. McMurry, J. E.; Swenson, R. *TL* **1987**, *28*, 3209.
21. Magnus, P.; Mugrage, B.; DeLuca, M. R.; Cain, G. A. *JACS* **1990**, *112*, 5220.
22. Ikemoto, N.; Schreiber, S. L. *JACS* **1992**, *114*, 2524.
23. Clawson, L.; Buchwald, S. L.; Grubbs, R. H. *TL* **1984**, *25*, 5733.
24. (a) Stevenson, J. W. S.; Bryson, T. A. *TL* **1982**, *23*, 3143. (b) Stevenson, J. W. S.; Bryson, T. A. *CL* **1984**, 5.
25. Kinney, W. A.; Coghlan, M. J.; Paquette, L. A. *JACS* **1985**, *107*, 7352.
26. Wilcox, C. S.; Long, G. W.; Suh, H. *TL* **1984**, *25*, 395.
27. Ireland, R. E.; Varney, M. D. *JOC* **1983**, *48*, 1829.
28. (a) Hayashi, T.; Yamamoto, A.; Hagihara, T.; Ito, Y. *TL* **1986**, *27*, 191. (b) Hayashi, T.; Matsumoto, Y.; Ito, Y. *CL* **1987**, 2037.
29. Peterson, P. E.; Stepanian, M. *JOC* **1988**, *53*, 1903.
30. Adams, J.; Frenette, R. *TL* **1987**, *28*, 4773.
31. (a) Parshall, G. W.; Nugent, W. A.; Rajan-Babu, T. V. *CS* **1987**, *27*, 527. (b) Rajan-Babu, T. V.; Reddy, G. S. *JOC* **1986**, *51*, 5458.
32. Hayashi, T.; Yamamoto, A.; Ito, Y. *SC* **1989**, *19*, 2109.
33. Marra, A.; Esnault, J.; Veyrières, A.; Sinaÿ, P. *JACS* **1992**, *114*, 6354.
34. Maurya, R.; Pittol, C. A.; Pryce, R. J.; Roberts, S. M.; Thomas, R. J.; Williams, J. O. *JCS(P1)* **1992**, 1617.
35. Ali, M. H.; Collins, P. M.; Overend, W. G. *Carbohydr. Res.* **1990**, *205*, 428.
36. Barrett, A. G. M.; Bezuidenhoudt, B. C. B.; Gasiecki, A. F.; Howell, A. R.; Russell, M. A. *JACS* **1989**, *111*, 1392.
37. Wenkert, E.; Marsaioli, A. J.; Moeller, P. D. R. *J. Chromatog.* **1988**, *440*, 449.
38. Barrett, A. G. M.; Bezuidenhoudt, B. C. B.; Melcher, L. M *JOC* **1990**, *55*, 5196.
39. Ziegler, F. E.; Wester, R. T. *TL* **1984**, *25*, 617.
40. (a) Carling, R. W.; Curtis, N. R.; Holmes, A. B. *TL* **1989**, *30*, 6081. (b) Clark, J. S.; Holmes, A. B. *TL* **1988**, *29*, 4333. (c) Carling, R. W.; Holmes, A. B. *CC* **1986**, 565.
41. Bartlett, P. A.; Nakagawa, Y.; Johnson, C. R.; Reich, S. H.; Luis, A. *JOC* **1988**, *53*, 3195.
42. Daub, G. W.; McCoy, M. A.; Sanchez, M. G.; Carter, J. S. *JOC* **1983**, *48*, 3876.
43. Paquette, L. A.; Sweeney, T. J. *JOC* **1990**, *55*, 1703.
44. Gajewski, J. J.; Gee, K. R.; Jurayj, J. *JOC* **1990**, *55*, 1813.
45. Burrows, C. J.; Carpenter, B. K. *JACS* **1981**, *103*, 6983.
46. Ireland, R. E.; Thaisrivongs, S.; Dussalt, P. H. *JACS* **1988**, *110*, 5768.
47. Ziegler, F. E.; Kneisley, A.; Thottathil, J. K.; Wester, R. T. *JACS* **1988**, *110*, 5434.
48. Sørensen, P. E.; Pedersen, K. J.; Pedersen, P. R.; Kanagasabapathy, V. M.; McClelland, R. A. *JACS* **1988**, *110*, 5118.
49. Kang, H.-J.; Paquette, L. A. *JACS* **1990**, *112*, 3252.
50. Evans, D. A.; Dow, R. L.; Shih, T. L.; Takacs, J. M.; Zahler, R. *JACS* **1990**, *112*, 5290.
51. Wattanasin, S.; Kathwala, F. G. *SC* **1989**, *19*, 2659.
52. Chou, T.-S.; Huang, S.-B. *TL* **1983**, *24*, 2169.
53. Stille, J. R.; Grubbs, R. H. *JACS* **1983**, *105*, 1664.
54. McKinney, R. J.; Tulip, T. H.; Thorn, D. L.; Coolbaugh, T. S.; Tebbe, F. N. *JACS* **1981**, *103*, 5584.
55. Cannizzo, L. F.; Grubbs, R. H. *JOC* **1985**, *50*, 2316.
56. Doxsee, K. M.; Farahi, J. B.; Hope, H. *JACS* **1991**, *113*, 8889.
57. For example: (a) Ozawa, F.; Park, J. W.; Mackenzie, P. B.; Schaefer, W. P.; Henling, L. M.; Grubbs, R. H. *JACS* **1989**, *111*, 1319. (b) Awang, M. R.; Barr, R. D.; Green, M.; Howard, J. A. K.; Marder, T. B.; Stone, F. G. A. *JCS(D)* **1985**, 2009. (c) Herberich, G. E.; Englert, U.; Ganter, C.; Weseman, L. *CB* **1992**, *125*, 23.

Daniel A. Straus
San Jose State University, CA, USA

Chlorobis(cyclopentadienyl)-hydridozirconium[1]

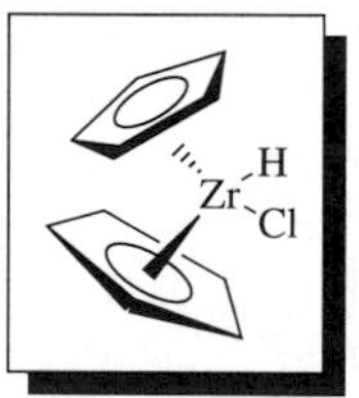

[37342-97-5] $C_{10}H_{11}ClZr$ (MW 257.87)

(hydrozirconation reagent; reducing reagent of unsaturated compounds; starting material for organozirconium complexes)

Alternate Name: Schwartz's reagent.
Solubility: insol; hydrolyzed in water; very slowly converted by CH_2Cl_2 to Cp_2ZrCl_2.
Form Supplied in: commercially available as white powder (95%).
Analysis of Reagent Purity: IR (nujol) 1390 cm^{-1} (metal–hydrogen bond). The purity of $Cp_2Zr(H)Cl$ is measured by reaction with acetone. A small sample is assayed in a 5 mm NMR tube in C_6D_6 by treatment with a known amount of excess acetone, and the relative areas of the signal for the mono- and diisopropoxides are determined by ^1H NMR.[2,3]

Preparative Methods: $Cp_2Zr(H)Cl$ is prepared by the reaction of ***Dichlorobis(cyclopentadienyl)zirconium*** with ***Lithium Tri-t-butoxyaluminum Hydride***,[4,5] ***Sodium Bis(2-methoxyethoxy)aluminum Hydride*** (Vitride or Red-Al),[3] or ***Lithium Aluminum Hydride***.[2] Zirconocene dichloride (100 g, 0.342 mol) is dissolved in dry THF (650 mL, heating required) in a 1 L Schlenk flask under argon. To this solution (at or slightly above rt) is added dropwise, over a 45 min period, a filtered solution of $LiAlH_4$ in diethyl ether (prepared from 3.6 g, 94.9 mmol of 95% $LiAlH_4$ and 100 mL dry ether, followed by filtration using a cannula fitted with a piece of glass fiber filter; a Schlenk filtered or commercial clear solution would work as well). The resulting suspension is allowed to stir at room temperature for 90 min. It is then Schlenk filtered under argon using a 'D' frit. The white solid is washed on the frit with THF (4 × 75 mL), CH_2Cl_2 (2 × 100 mL) with stirring or agitation of a stirbar immersed in the slurry and ether (4 × 50 mL). The resulting white solid is dried in vacuo to give a white powder: 65.5–81 g, 77–92% yield.[2]

Purification: over-reduction product Cp_2ZrH_2 reacts rapidly with methylene chloride to form $Cp_2Zr(H)Cl$.[2]

Handling, Storage, and Precautions: slowly develops a pink color when exposed to light and appears to be photosensitive. This compound should be handled in a fume hood.

Hydrozirconation of Unsaturated Compounds. Alkenes[6] react with $Cp_2Zr(H)Cl$ to give alkylzirconium complexes. Hydrozirconation of alkenes proceeds to place the zirconium moiety at the sterically least hindered position of the alkene as a whole. Formation of this product involves either the regiospecific addition of Zr–H to a terminal alkene or addition to an internal alkene followed by rapid rearrangement via Zr–H elimination and readdition to place the metal in the less hindered position of the alkyl chain (eq 1).[6,7] This is in sharp contrast to hydroboration or hydroalumination reactions. However, the tendency of zirconium to migrate towards the terminal position in the case of 1-arylalkenes is lower than simple alkenes.[8] Alkenes containing a heteroatom such as oxygen undergo C–heteroatom bond cleavage during the hydrozirconation reactions under some conditions (eq 2).[9] Alkylzirconium complexes obtained here are easily converted into alkanes,[10] alkyl halides (eq 3),[6] alcohols (eqs 4 and 5),[7] nitriles (eq 6),[11] aldehydes, carboxylic acids or esters (eq 7)[12] by hydrogenation (or hydrolysis), halogenation, oxidation, successive treatment with isocyanide and iodine, or carbonylation.

(1)

$Cp_2Zr(H)Cl$, benzene, rt; fast; 80% + $H_2C{=}CH_2$; $Cp_2Zr(H)Cl$; 20% (2)

Br_2, 96%, RBr; I_2, 91%, RI; $PhICl_2$, 65%, RCl; NBS, 95%, RBr; NCS, 67%, RCl; R = hexyl (3)

H_2O_2, H_2O, NaOH, 69%; t-BuOOH, 72%; m-CPBA, 45%; CrO_2Cl_2, 52%; octanol (4)

O_2, 91%; O_2, 77% (5)

$Cp_2Zr(H)Cl$; R'N≡C:; I_2 (6)

$$Cp_2Zr(Cl)R \xrightarrow{CO} Cp_2Zr(Cl)C(O)R \rightarrow \begin{cases} \xrightarrow{H_3O^+} RC(O)H \\ \xrightarrow{Br_2,\ MeOH} RC(O)OMe \\ \xrightarrow{H_2O_2} RC(O)OH \\ \xrightarrow{NBS} RC(O)Br \end{cases} \quad (7)$$

Hydrozirconation of alkynes gives alkenylzirconium complexes (eq 8).[13] Zr–H addition to the alkyne proceeds with *cis* stereochemistry. Hydrozirconation of various unsymmetrically disubstituted alkynes occurs readily to give a mixture of alkenylzirconium compounds. Over a period of several hours, this initial mixture of alkenyl species can be converted to one with higher regioselectivity, at room temperature, through catalysis with $Cp_2Zr(H)Cl$.[13] It is interesting to note that, whereas alkylzirconium complexes positionally rearrange rapidly, no such process occurs for purified alkenyl analogs. Alkenylzirconium complexes are also easily converted into alkenyl halides[13] and aldehydes (eq 9)[14] when they are treated with ***N-Bromosuccinimide*** and isocyanide/H^+, respectively.

$$Cp_2Zr(H)Cl \xrightarrow{R^2-\equiv-R^1} Cp_2Zr(Cl)C(R^2)=CHR^1 \rightleftharpoons Cp_2Zr(Cl)C(R^1)=CHR^2 \quad (8)$$

$$(R')H-\equiv-R' \xrightarrow{Cp_2Zr(H)Cl} Cp_2Zr(Cl)C((R')H)=CHR' \xrightarrow{RNC} ClCp_2Zr(\eta^2\text{-}C(=NR))C((R')H)=CHR' \xrightarrow{H_3O^+} OHC\text{-}C((R')H)=CHR' \quad (9)$$

In situ generation of $Cp_2Zr(H)Cl$ or in situ hydrozirconation is practically useful. There are several in situ procedures which consist of ***Dichlorobis(cyclopentadienyl)zirconium/ Sodium Bis(2-methoxyethoxy)aluminum Hydride*** (Red-Al), Cp_2ZrCl_2/ ***t-Butylmagnesium Chloride***[15] or Cp_2ZrCl_2/ ***Lithium Triethylborohydride***.[16] Comparison of hydrozirconation using $Cp_2Zr(H)Cl$ and these in situ procedures is shown in Table 1.[15,16]

Table 1 Comparison of Hydrozirconation using Schwartz's Reagent and Various in situ Procedures[15,16]

Alkyne	Schwartz's reagent	Cp_2ZrCl_2 + LiEt$_3$BH	Red-Al	*t*-BuMgCl
1-Octyne[a]	80–100	95	95	100[a]
$CH\equiv C(CH_2)_2OTHP$[b]	77	75	50	77[a]
1-Decyne[a]	80–100	88	80	95[a]

[a] Isolated yield (%) of H^+ quenched materials. [b] Isolated yields.

$(MeC_5H_4)_2Zr(H)Cl$ reacts about seven times faster with 1-hexene or acetophenone than $Cp_2Zr(H)Cl$.[17] Hydrozirconation of other various unsaturated bonds such as Schiff bases,[18] ketones,[3,4,19] diazoalkanes,[20] nitriles,[21] carbon dioxide (eq 10),[22] 1-ene-3-ynes,[23] dienes,[24] phosphaimines,[25] thioketones,[26] carbon oxide,[27] and isocyanide[28] has been investigated.

$$2\ Cp_2Zr(Cl)H + CO_2 \longrightarrow Cp_2Zr(Cl)\text{-}O\text{-}Zr(Cl)Cp_2 + H_2CO$$
$$Cp_2Zr(Cl)H + H_2CO \longrightarrow Cp_2Zr(Cl)OMe \quad (10)$$

Hydrozirconation of unsaturated bonds attached to metals also proceeds. Alkenylzinc or alkynylzinc compounds react with $Cp_2Zr(H)Cl$ to give unstable 1,1-bimetallic reagents which can be used for highly stereoselective alkenation of aldehydes (eq 11).[29] Similarly, addition of $Cp_2Zr(H)Cl$ to a solution of (neohexenyl)diisobutylaluminum gives a 1,1-bimetallic complex.[30] Its structural analysis by X-ray clearly shows the 1,1-bimetallic system.[31] Reaction of stannylacetylenes with $Cp_2Zr(H)Cl$ seems to give 1,1-bimetallic compounds, since the iodination product is 1-iodo-1-stannylalkenes.[32] The regioselectivity of hydrozirconation of unsaturated bonds attached to metals is noteworthy. Hydrozirconation of an alkynyl ligand attached to Ru[33] or Zr,[34] and an alkenyl ligand attached to Zr,[35] shows the opposite regioselectivity to those described above, which give 1,1-bimetallic compounds (eq 12). However, the reaction of $Cp_2Zr(H)Cl$ with $Cp(CO)_2Re{=}CHR$[36] gives a 1,1-bimetallic intermediate with Zr and Re. An acyl ligand on zirconium also reacts with $Cp_2Zr(H)Cl$.[37] When $Cp_2Zr(H)Cl$ is treated with potassium allyl alkoxide or homoallyl alkoxide, intramolecular hydrozirconation seems to give oxazirconacycles.[38]

$$R^1CH{=}C(R^2)ZnX \xrightarrow[25\,°C,\ 1\ min]{Cp_2Zr(H)Cl,\ CH_2Cl_2} \left[R^1CH_2C(R^2)(Zn)(Zr(Cl)Cp_2)\right] \xrightarrow{R^3C(O)R^4} R^1CH_2C(R^2){=}C(R^3)R^4 \quad (11)$$

$$CpRu(PMe_3)_2C\equiv CH \xrightarrow{Cp_2Zr(H)Cl} CpRu(PMe_3)_2CH{=}CH\text{-}Zr(H)(Cl)Cp_2 \xrightarrow{H_2O} CpRu(PMe_3)_2CH{=}CH_2 + [Cp_2ZrCl]_2O$$
$$\xrightarrow{Cp_2Zr(H)Cl} CpRu(PMe_3)_2CH_2CH_2\text{-}Zr(H)(Cl)Cp_2 \quad (12)$$

Carbon–Carbon Bond Forming Reactions of Hydrozirconation Products. Hydrozirconation products of unsaturated compounds can be used for further C–C bond forming reactions. Since alkyl or alkenyl carbon attached to zirconium is not nucleophilic like Grignard, lithium, or aluminum reagents, the direct C–C bond formation is very limited. Only carbonylation (eq 7) and acylation with ***Acetyl Chloride*** can be used. However, transmetalation of the alkyl or alkenyl moiety from zirconium to other metals changes the situation. It provides a variety of C–C bond formation reactions of hydrozirconation products.

The alkyl or alkenyl moiety of hydrozirconation products are transferred from zirconium to aluminum when they are treated with ***Aluminum Chloride***. These organoaluminum dichlorides obtained here are converted into ketones or α,β-unsaturated ketones in high yields (eq 13).[39] Interestingly, since hydrozirconation of alkenes gives the terminal alkylzirconium exclusively, this method can provide a single ketonic product from a mixture of isomeric alkenes. For alkenylzirconium, through this sequential use of two reactive organometallic species, the ketone is formed by overall *cis* addition of acyl–H to an alkyne.[39] Similar treatment of 1-(trimethylsilyl)-4-bromo-1-butyne with $Cp_2Zr(H)Cl$ followed by $AlCl_3$ affords 1-(trimethylsilyl)cyclobutene (eq 14).[40]

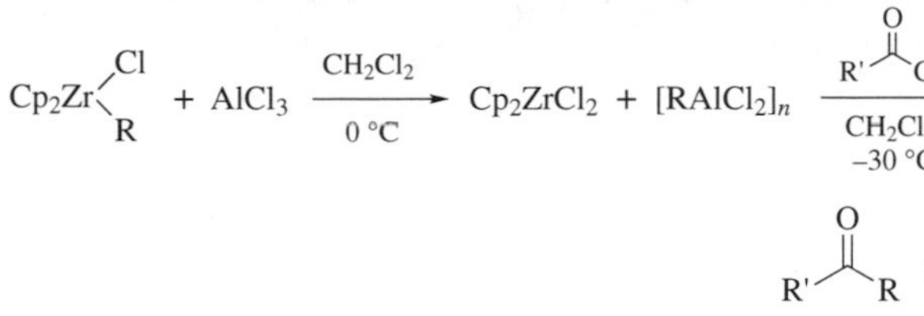

$Cp_2Zr(Cl)R$ + $AlCl_3$ $\xrightarrow[0\ °C]{CH_2Cl_2}$ Cp_2ZrCl_2 + $[RAlCl_2]_n$ $\xrightarrow[CH_2Cl_2,\ -30\ °C]{R'COCl}$ R'COR (13)

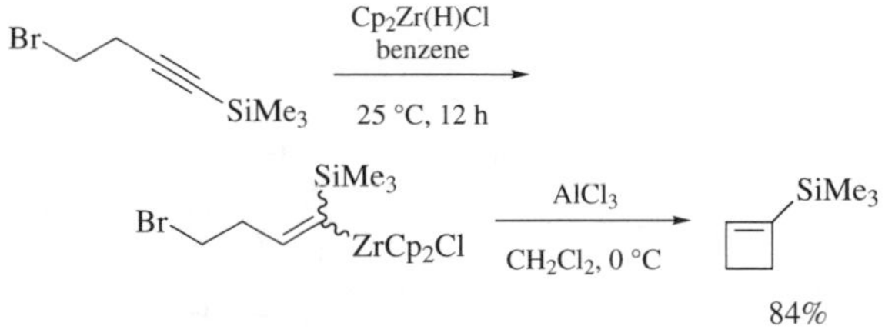

Br–CH_2CH_2–C≡C–$SiMe_3$ $\xrightarrow[25\ °C,\ 12\ h]{Cp_2Zr(H)Cl,\ benzene}$ Br–CH_2CH_2–CH=C($SiMe_3$)$ZrCp_2Cl$ $\xrightarrow[CH_2Cl_2,\ 0\ °C]{AlCl_3}$ 1-($SiMe_3$)cyclobutene, 84% (14)

Transmetalation from Zr to Cu produces the corresponding organocopper compounds which can be used for addition to α,β-unsaturated ketones (eq 15).[41] It is noteworthy that functionalized lithiocuprates, which contain –CN, $–CO_2Si(i\text{-}Pr)_3$, –OCOPh, or Cl groups, are easily prepared by this method.[42]

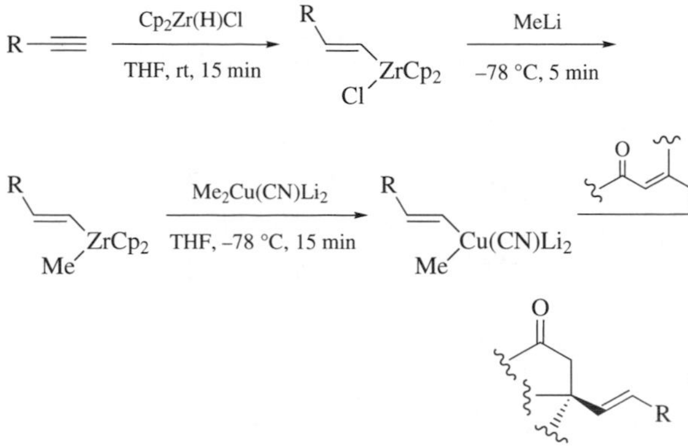

R–C≡CH $\xrightarrow[THF,\ rt,\ 15\ min]{Cp_2Zr(H)Cl}$ RCH=CH–$ZrCp_2Cl$ $\xrightarrow[-78\ °C,\ 5\ min]{MeLi}$ RCH=CH–$ZrCp_2Me$ $\xrightarrow[THF,\ -78\ °C,\ 15\ min]{Me_2Cu(CN)Li_2}$ RCH=CH–$Cu(CN)Li_2Me$ $\xrightarrow{\text{enone}}$ (15)

Migration of an alkenyl group from zirconium to boron (eq 16),[43] tin,[44] or selenium (eq 17)[44] also proceeds. Alkyltin compounds can be prepared from alkylzirconium and ***Tin(IV) Chloride***.[45] Imine transfer from zirconium to phosphorus or boron is also possible.[46]

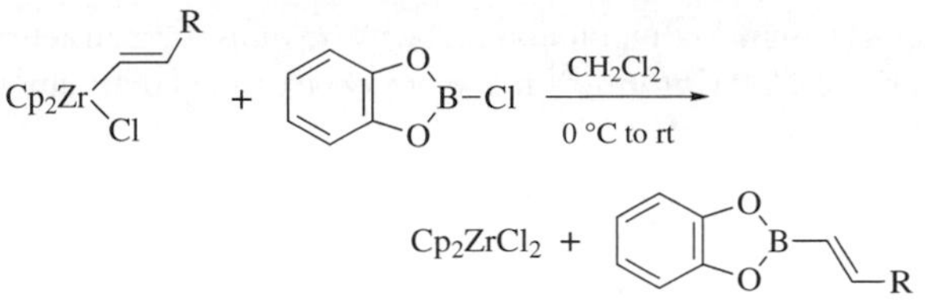

$Cp_2Zr(Cl)CH=CHR$ + catecholborane–Cl $\xrightarrow[0\ °C\ to\ rt]{CH_2Cl_2}$ Cp_2ZrCl_2 + catechol–B–CH=CHR (16)

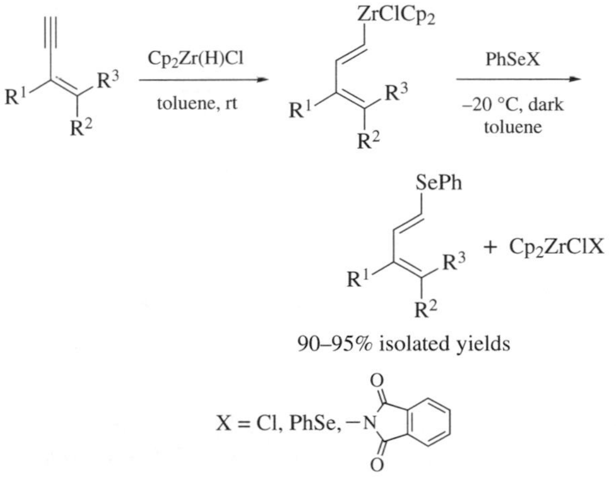

$\xrightarrow[toluene,\ rt]{Cp_2Zr(H)Cl}$ $\xrightarrow[-20\ °C,\ dark,\ toluene]{PhSeX}$ + Cp_2ZrClX (17)

90–95% isolated yields

X = Cl, PhSe, –N(phthalimide)

Catalytic reactions using transition metals have been investigated. Cross-coupling reaction of alkenylzirconium with aryl halides or alkenyl halides is catalyzed by Ni or Pd complexes (eq 18).[47] The effect of addition of metal salts such as ***Zinc Chloride*** and ***Cadmium Chloride*** is remarkable in this reaction.[48] Addition of alkenylzirconium to α,β-unsaturated ketones is also catalyzed by Ni (eqs 19 and 20).[49]

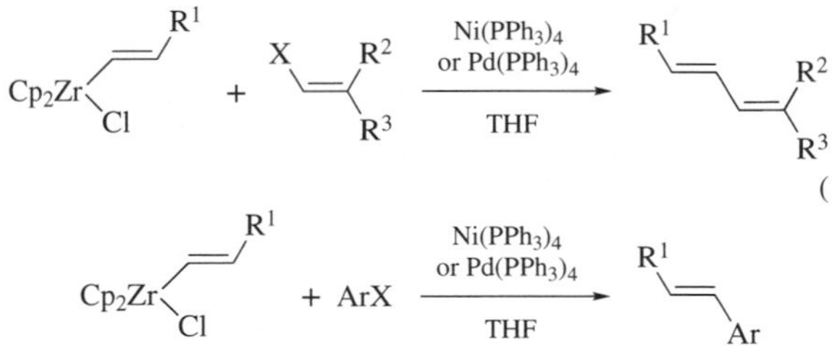

$\xrightarrow[THF]{Ni(PPh_3)_4\ or\ Pd(PPh_3)_4}$ (18)

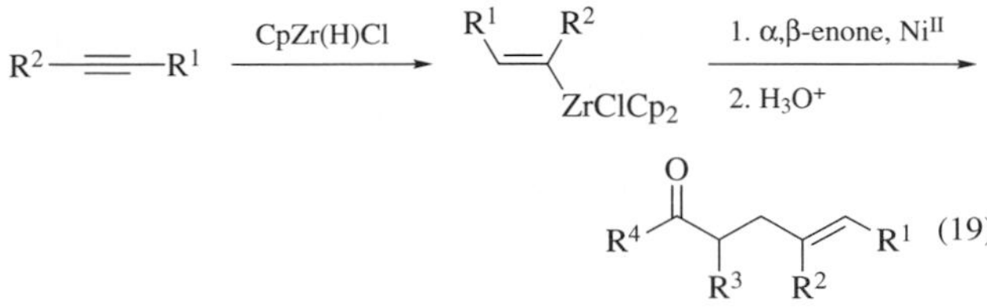

R^2–C≡C–R^1 $\xrightarrow{CpZr(H)Cl}$ $\xrightarrow[2.\ H_3O^+]{1.\ α,β\text{-enone},\ Ni^{II}}$ (19)

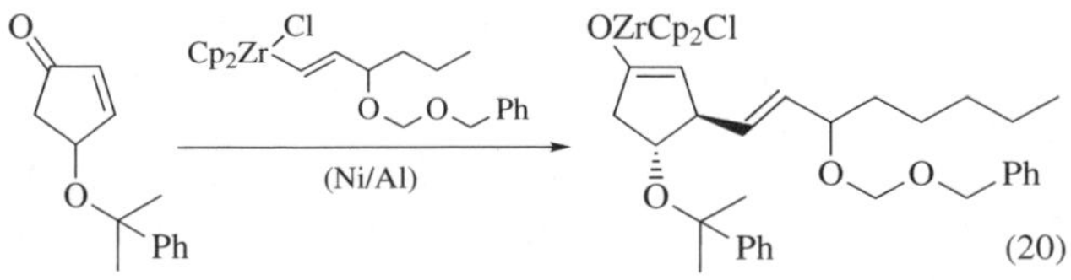

$\xrightarrow{(Ni/Al)}$ (20)

Alkylzirconium compounds react with α,β-unsaturated ketones (eq 21)[50] or ***Allyl Chloride***[51] in the presence of a catalytic amount of copper compound.

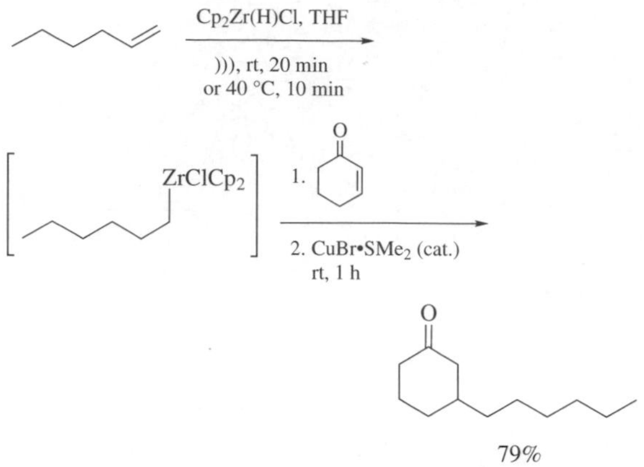

(21)

79%

Formation of cationic species from alkyl- or alkenylzirconium compounds with a catalytic amount of ***Silver(I) Perchlorate*** is one method to make direct C–C bonds with aldehydes (eq 22).[52] Both alkyl and alkenyl groups attack aldehydes to give desired products in high yields. When 3-trimethylsilyl-1-propyne is used as a starting alkyne, (*E*)-selective terminal 1,3-dienes are formed.[53] Alkynes, 1-ethoxyethyne, and (*Z*)-1-methoxy-1-buten-3-yne give two- and four-carbon homologation of aldehydes, respectively, after treatment with an acid (eq 23).[54] A direct C–C bond formation of alkenylzirconium compounds with epoxides also proceeds in the presence a catalytic amount of $AgClO_4$.[55]

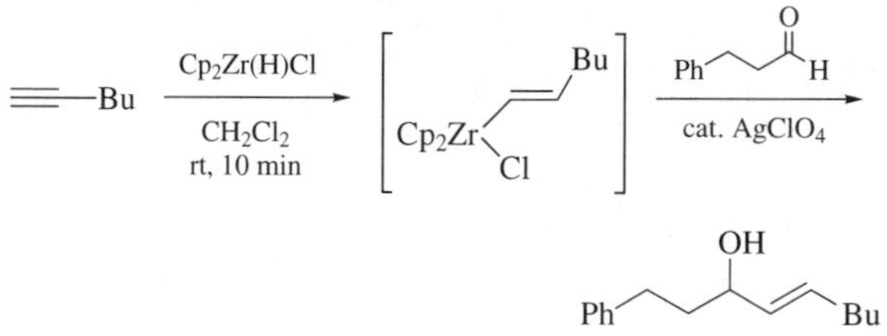

(22)

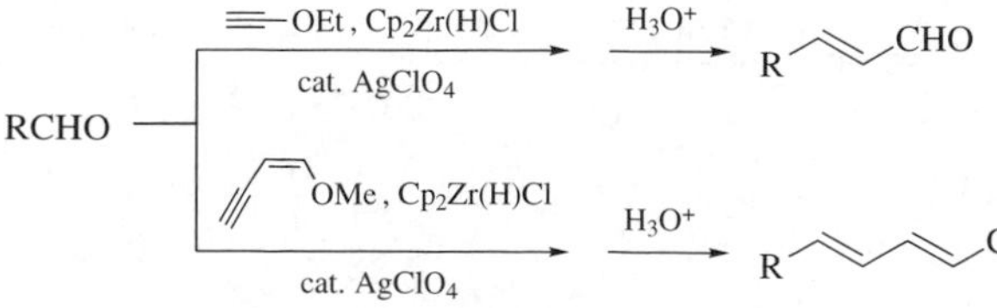

(23)

Organozirconium complexes, prepared by either hydrozirconation or the reaction of organolithium reagents with $Cp_2Zr(H)Cl$, can be converted into diene (eq 24),[56] alkyne (eq 25),[57] silanimine (eq 26),[58] and diazobenzene (or hydrazido(1−)) (eq 27)[59] Zr^{II} complexes which react with unsaturated compounds to afford various zirconacycles.

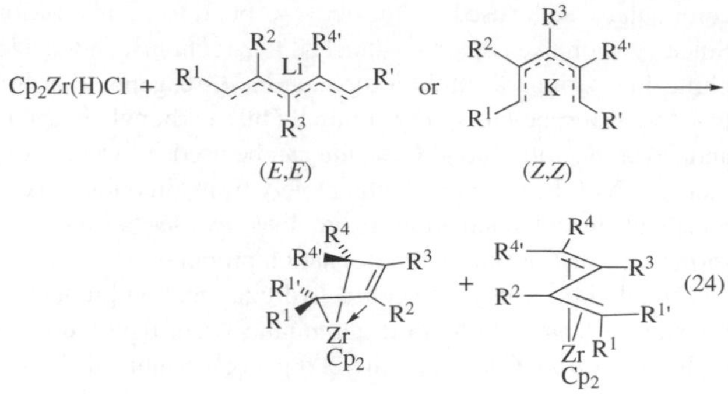

(24)

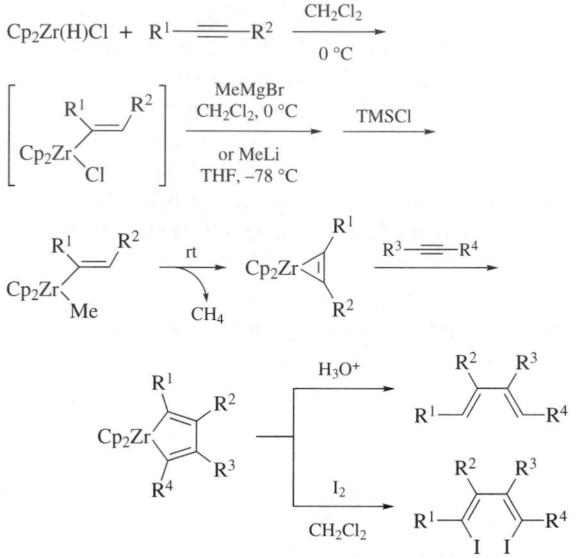

(25)

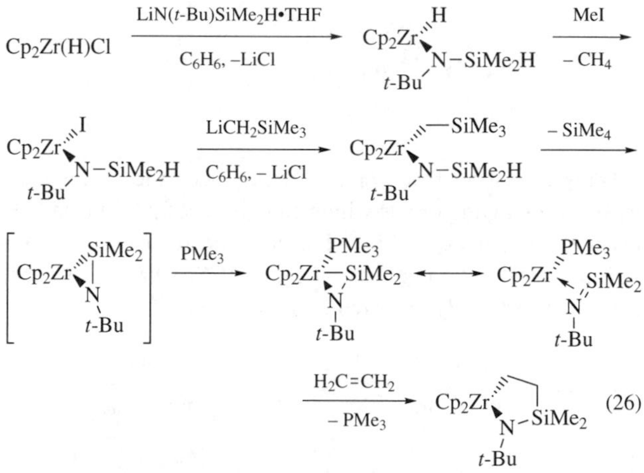

(26)

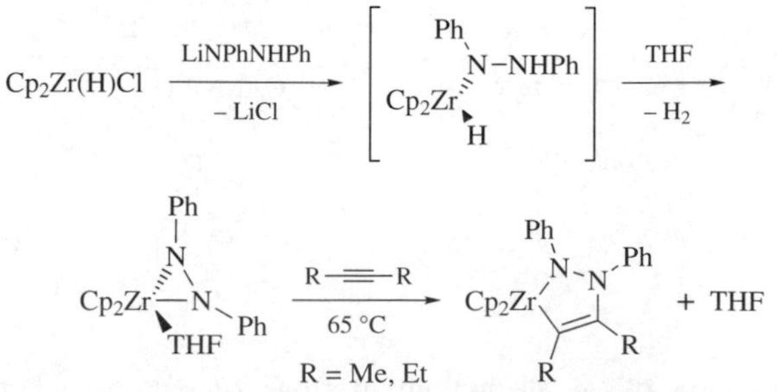

(27)

Related Reagents. Chlorobis(cyclopentadienyl)methylzirconium; Dichlorobis(cyclopentadienyl)zirconium.

1. (a) Schwartz, J. In *New Applications of Organometallic Reagents in Organic Synthesis*, D. Seyferth, Ed.; Elsevier: Amsterdam, 1976; p 46. (b) Schwartz, J.; Labinger, J. A. *AG(E)* **1976**, *15*, 333. (c) Cardin, D. J.; Lappert, M. F.; Raston, C. L.; Riley, P. I. In *Comprehensive Organometallic Chemistry*, Wilkinson, G.; Stone, F. G. A.; Abel, E. W. Eds.; Pergamon: Oxford, 1982; Vol. 3, pp 549–633. (d) Negishi, E.; Takahashi, T. *Aldrichim. Acta* **1985**, *18*, 31.
2. Buchwald, S. L.; LaMaire, S. J.; Nielsen, R. B.; Watson, B. T.; King, S. M. *TL* **1987**, *28*, 3895.
3. Carr, D. B.; Schwartz, J. *JACS* **1979**, *101*, 3521.
4. Wailes, P. C.; Weigold, H. *Inorg. Synth.* **1979**, *19*, 223.
5. Wailes, P. C.; Weigold, H. *JOM* **1970**, *24*, 405.
6. Hart, D. W.; Schwartz, J. *JACS* **1974**, *96*, 8115.
7. (a) Blackburn, T. F.; Labinger, J. A.; Schwartz, J. *TL* **1975**, 3041. (b) Gibson, T.; Tulich, L. *JOC* **1981**, *46*, 1821.
8. (a) Nelson, J. E.; Bercaw, J. E.; Labinger, J. A. *OM* **1989**, *8*, 2484. (b) Annby, U.; Gronowitz, S.; Hallberg, A. *JOM* **1989**, *365*, 233. (c) Annby, U.; Gronowitz, S.; Hallberg, A. *JOM* **1989**, *368*, 295.
9. (a) Buchwald, S. L.; Nielsen, R. B.; Dewan, J. C. *OM* **1988**, *7*, 2324. (b) Karlsson, S.; Hallberg, A.; Gronowitz, S. *JOM* **1991**, *403*, 133.
10. Gell, K. I.; Posin, B.; Schwartz, J.; Williams, G. M. *JACS* **1982**, *104*, 1846.
11. Buchwald, S. L.; LaMaire, S. J. *TL* **1987**, *28*, 295.
12. Bertelo, C. A.; Schwartz, J. *JACS* **1975**, *97*, 228.
13. Hart, D. W.; Blackburn, T. F.; Schwartz, J. *JACS* **1975**, *97*, 679.
14. Negishi, E.; Swanson, D. R.; Miller, S. R. *TL* **1988**, *29*, 1631.
15. (a) Negishi, E.; Miller, J. A.; Yoshida, T. *TL* **1984**, *25*, 3407. (b) Swanson, D. R.; Nguyen, T.; Noda, Y.; Negishi, E. *JOC* **1991**, *56*, 2590.
16. Lipshutz, B. H.; Keil, R.; Ellsworth, E. L. *TL* **1990**, *31*, 7257.
17. Erker, G.; Schlund, R.; Krüger, C. *OM* **1989**, *8*, 2349.
18. Ng, K. S.; Laycock, D. E.; Alper, H. *JOC* **1981**, *46*, 2899.
19. Cesarotti, E.; Chiesa, A.; Maffi, S.; Ugo, R. *ICA* **1982**, *64*, L207.
20. (a) Gambarotta, S.; Basso-Bert, M.; Floriani, C.; Guastini, C. *CC* **1982**, 374. (b) Gambarotta, S.; Floriani, C.; Chiesi-Villa, A.; Guastini, C. *IC* **1983**, *22*, 2029.
21. (a) Erker, G.; Frömberg, W.; Atwood, J. L.; Hunter, W. E. *AG(E)* **1984**, *23*, 68. (b) Frömberg, W.; Erker, G. *JOM* **1985**, *280*, 343.
22. Gambarotta, S.; Strologo, S.; Floriani, C.; Chiesi-Villa, A.; Guastini, C. *JACS* **1985**, *107*, 6278.
23. Fryzuk, M. D.; Bates, G. S.; Stone, C. *TL* **1986**, *27*, 1537.
24. Bertelo, C. A.; Schwartz, J. *JACS* **1976**, *98*, 262.
25. Dufour, N.; Caminade, A.-M.; Basso-Bert, M.; Igau, A.; Majoral, J.-P. *OM* **1992**, *11*, 1131.
26. Laycock, D. E.; Apler, H. *JOC* **1981**, *46*, 289.
27. Fachinetti, G.; Floriani, C.; Roselli, A.; Pucci, S. *CC* **1978**, 269.
28. Fromberg, W.; Erker, G. *JOM* **1985**, *280*, 355.
29. Tucker, C. E.; Knochel, P. *JACS* **1991**, *113*, 9888.
30. Hartner Jr., F. W.; Schwartz, J. *JACS* **1981**, *103*, 4979.
31. Hartner Jr., F. M.; Clift, S. M.; Schwartz, J.; Tulip, T. H. *OM* **1987**, *6*, 1346.
32. Lipshutz, B. H.; Keil, R.; Barton, J. C. *TL* **1992**, *33*, 5861.
33. (a) Bullock, R. M.; Lemke, F. R.; Szalda, D. J. *JACS* **1990**, *112*, 3244. (b) Lemke, F. R.; Szalda, D. J.; Bullock, R. M. *JACS* **1991**, *113*, 8466.
34. Erker, G.; Fromberg, W.; Angermund, K.; Schlund, R.; Krüger, C. *CC* **1986**, 372.
35. Erker, G.; Kropp, K.; Atwood, J. L.; Hunter, W. E. *OM* **1983**, *2*, 1555.
36. Casey, C. P.; Askham, F. R.; Petrovich, L. M. *JOM* **1990**, *387*, C31.
37. Erker, G.; Kropp, K.; Krüger, C.; Chiang, A.-P. *CB* **1982**, *115*, 2447.
38. (a) Takaya, H.; Yamakawa, M.; Mashima, K. *CC* **1983**, 1283. (b) Mashima, K.; Yamakawa, M.; Takaya, H. *JCS(D)* **1991**, *11*, 2851.
39. Carr, D. B.; Schwartz, J. *JACS* **1977**, *99*, 638.
40. Negishi, E.; Boardman, L. D.; Sawada, H.; Bagheri, V.; Stoll, A. T.; Tour, J. M.; Rand, C. L. *JACS* **1988**, *110*, 5383.
41. (a) Yoshifuji, M.; Loots, M. J.; Schwartz, J. *TL* **1977**, 1303. (b) Lipshutz, B. H.; Ellsworth, E. L. *JACS* **1990**, *112*, 7440. (c) Babiak, K. A.; Behling, J. R.; Dygos, J. H.; McLanghlin, K. T.; Ng, J. S.; Kalish, V. J.; Kramer, S. W.; Shone, R. L. *JACS* **1990**, *112*, 7441. (d) Lipshutz, B. H.; Kato, K. *TL* **1991**, *32*, 5647. (e) Lipshutz, B. H.; Fatheree, P.; Hagen, W.; Stevens, K. L. *TL* **1992**, *33*, 1041.
42. Lipshutz, B. H.; Keil, R. *JACS* **1992**, *114*, 7919.
43. (a) Fryzuk, M. D.; Bates, G. S.; Stone, C. *JOC* **1988**, *53*, 4425. (b) Cole, T. E.; Quintanilla, R.; Rodewald, S. *OM* **1991**, *10*, 3777. (c) Cole, T. E.; Quintanilla, R. *JOC* **1992**, *57*, 7366.
44. (a) Fryzuk, M. D.; Bates, G. S.; Stone, C. *TL* **1986**, *27*, 1537. (b) Fryzuk, M. D.; Bates, G. S.; Stone, C. *JOC* **1991**, *56*, 7201.
45. Das, V. G. K.; Chee, O. G. *JOM* **1987**, *321*, 335.
46. Boutonnet, F.; Dufour, N.; Straw, T.; Igau, A.; Majoral, J.-P. *OM* **1991**, *10*, 3939.
47. (a) Negishi, E.; Van Horn, D. E. *JACS* **1977**, *99*, 3168. (b) Negishi, E.; Takahashi, T.; Baba, S.; Van Horn, D. E.; Okukado, N. *JACS* **1987**, *109*, 2393. (c) Vincent, P.; Beaucourt, J.-P.; Pichat, L. *TL* **1982**, *23*, 63.
48. Negishi, E.; Okukado, N.; King, A. O.; Van Horn, D. E.; Spiegel, B. I. *JACS* **1978**, *100*, 2254.
49. (a) Loots, M. J.; Schwartz, J. *JACS* **1977**, *99*, 8045. (b) Loots, M. J.; Schwartz, J. *TL* **1978**, *45*, 4381. (c) Schwartz, J.; Hayasi, Y. *TL* **1980**, *21*, 1497. (d) Schwartz, J.; Loots, M. J.; Kosugi, H. *JACS* **1980**, *102*, 1333.
50. Wipf, P.; Smitrovich, J. H. *JOC* **1991**, *56*, 6494.
51. Venanzi, L. M.; Lehmann, R.; Keil, R.; Lipshutz, B. H. *TL* **1992**, *33*, 5857.
52. Maeta, H.; Hashimoto, T.; Hasegawa, T.; Suzuki, K. *TL* **1992**, *33*, 5965.
53. Maeta, H.; Suzuki, K. *TL* **1992**, *33*, 5969.
54. Meata, H.; Suzuki, K. *TL* **1993**, *34*, 341.
55. Wipf, P.; Xu, W. *JOC* **1993**, *58*, 825.
56. Yasuda, H.; Nagasuna, K.; Akita, M.; Lee, K.; Nakamura, A. *OM* **1984**, *3*, 1470.
57. (a) Buchwald, S. L.; Watson, B. T.; Huffman, J. C. *JACS* **1987**, *109*, 2544. (b) Buchwald, S. L.; Nielsen, R. B. *JACS* **1989**, *111*, 2870.
58. Procopio, L. J.; Carroll, P. J.; Berry, D. H. *JACS* **1991**, *113*, 1870.
59. Walsh, P. J.; Hollander, F. J.; Bergman, R. G. *JOM* **1992**, *428*, 13.

Tamotsu Takahashi & Noriyuki Suzuki
Institute for Molecular Science, Okazaki, Japan

Chlorobis(cyclopentadienyl)-methylzirconium[1]

[1291-48-8] $C_{11}H_{13}ClZr$ (MW 271.90)

(used to generate zirconocene complexes with stable and unstable organic molecules;[2] catalyzes the carboalumination of R_3Al to terminal alkynes[3])

Physical Data: pale yellow crystals.
Solubility: sol aromatic solvents.

Analysis of Reagent Purity: ^{1}H NMR (benzene-d_6) δ 5.73 (s, C_5H_5), 0.32 (s, CH_3) in a ratio of 10:3.

Preparative Methods: can be prepared by the reaction of Cp_2ZrMe_2 with ***Dichlorobis(cyclopentadienyl)zirconium***. A 0.5 M solution of each reagent in toluene is heated to 130 °C for 35 h. $Cp_2ZrMeCl$ is isolated in >90% yield by recrystallization from toluene/hexanes.[4] Alternatively, it can be prepared in pure form by the treatment of $(Cp_2ZrCl)_2O$ with ***Trimethylaluminum***.[5] Although $Cp_2ZrMeCl$ is available in a reasonable quantity as mentioned, it is often convenient to generate an analog of $Cp_2ZrMeCl$ in the reaction media and use it in situ. Thus treatment of Cp_2ZrCl_2 with ***t-Butyllithium*** generates $Cp_2Zr(i\text{-Bu})Cl$ through β-hydride elimination and insertion. $Cp_2Zr(i\text{-Bu})Cl$ shows a comparable reactivity to $Cp_2ZrMeCl$.[6]

Handling, Storage, and Precautions: the dry solid is highly sensitive to moisture, forming $(Cp_2ZrCl)_2O$, and should be kept tightly sealed to preclude moisture. Use of this reagent immediately following preparation is recommended.

Formation of Cp_2Zr Complex. $Cp_2ZrMeCl$ is a convenient reagent to produce mixed dialkylzirconocenes in the reaction with RLi (eq 1).[7] By the application of this method, zirconocene complexes of either unstable or stable organic molecules can be generated from $Cp_2ZrMeCl$.[8] The formation of zirconocene complexes is achieved by substituting the halogen atom of $Cp_2ZrMeCl$ with an alkyl or a vinyl organometallic reagent which possesses at least one β-hydrogen atom. Thermolysis then results in the concomitant generation of methane. This procedure is complementary to Cp_2Zr–butene (zirconocene equivalent) promoted zirconocene complex formation of unsaturated molecules,[9] since the zirconocene equivalent does not allow the isolation of Cp_2Zr complexes from very unstable or very reactive organic molecules. This β-hydrogen abstraction process is reported to involve an interaction of a β-C–H bond with the empty orbital of the zirconium valence shell.[10] The general scheme for the formation of complexes of unstable organic molecules and the organic precursor required for their syntheses is shown in eqs 2–8.

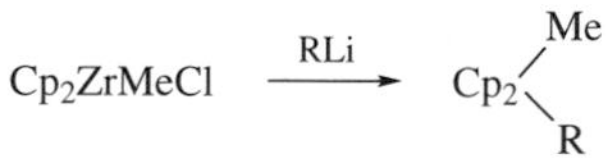

$Cp_2ZrMeCl \xrightarrow{RLi} Cp_2Zr(Me)(R)$ (1)

R = alkyl, alkenyl, aryl

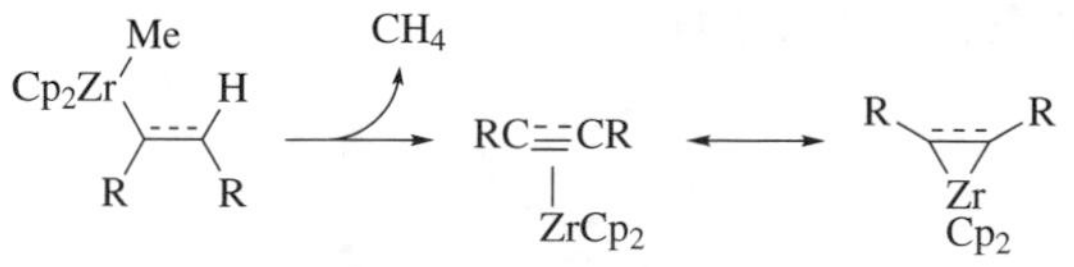

(2)

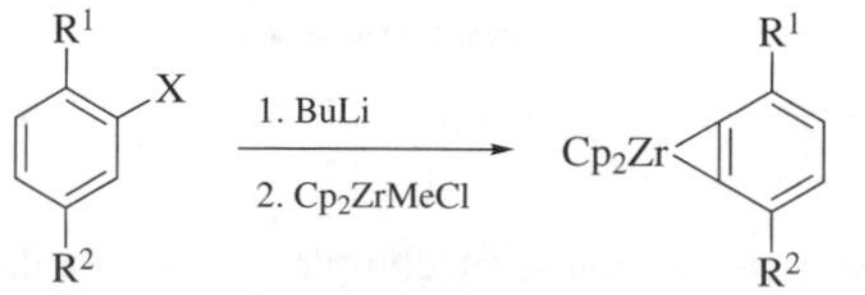

(3)

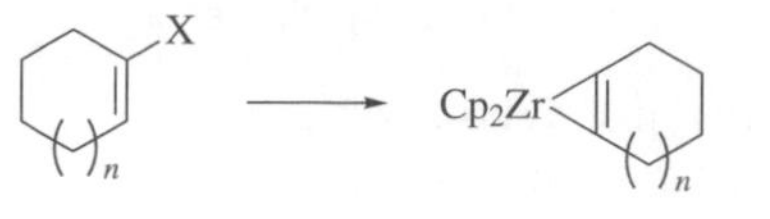

(4)

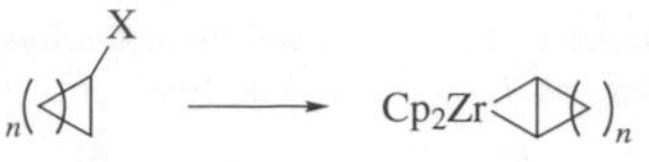

(5)

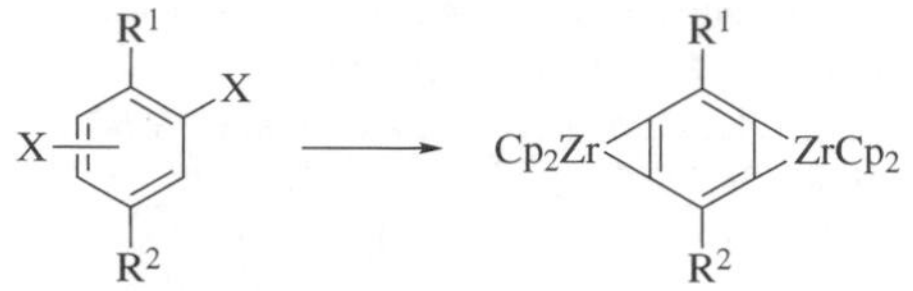

(6)

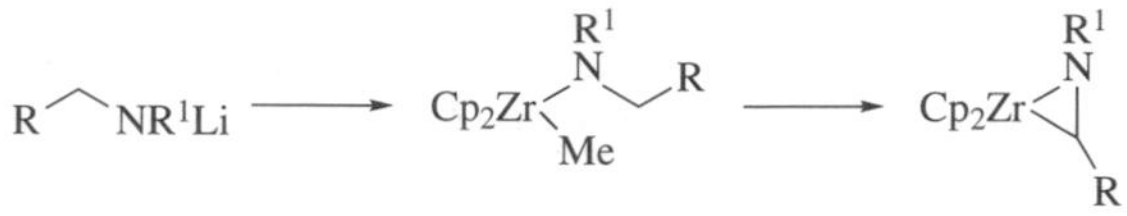

(7)

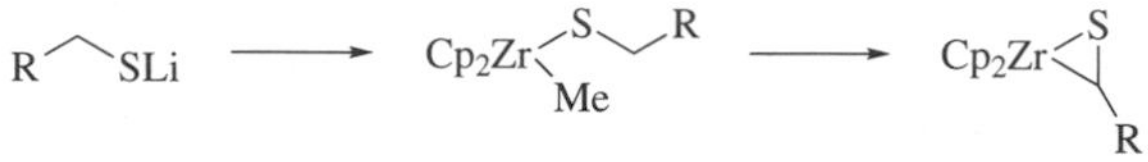

(8)

In each case, a metal–halogen exchange with an organolithium reagent followed by the addition of $Cp_2ZrMeCl$ produces the precursor for the zirconocene complex. Thermolysis of these precursors at temperatures ranging between 20–100 °C causes the evolution of methane to give the desired zirconocene complexes, which can be isolated as their trimethylphosphine adducts or trapped with an unsaturated compound. Through this procedure, Cp_2Zr complexes of arynes with a variety of substitution patterns (eq 3),[8a,11] cyclic alkynes containing five to eight membered rings (eq 4), cyclic alkenes (eq 5),[8b,c] benzdiynes (multiply unsaturated aromatic species) (eq 6),[12] imines (eq 7),[13] and thioaldehydes (eq 8)[14] have all been reported. Interestingly, methyl(2-norbornenyl)zirconocene is known to exist as a thermally stable complex (eq 9).[15] This stability is attributed to the strain and the bulkiness of the 2-norbornenyl ligand. Thermolysis of methyl(2-furyl)zirconocene, derived from $Cp_2ZrMeCl$ and 2-furyllithium, yields a ring-enlarged and methyl-migrated compound instead of a Cp_2Zr–heteroaryne complex (eq 10).[16]

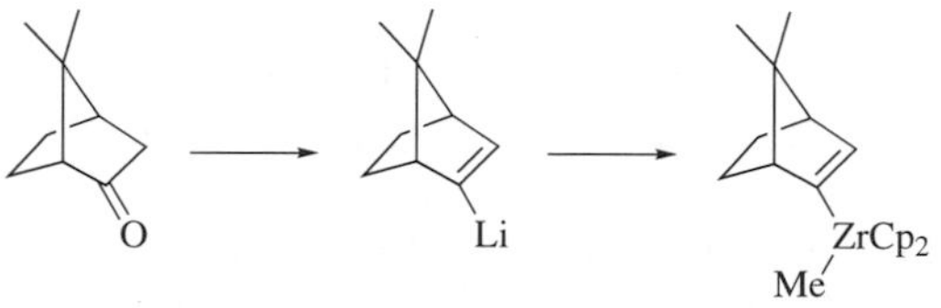

(9)

thermally stable

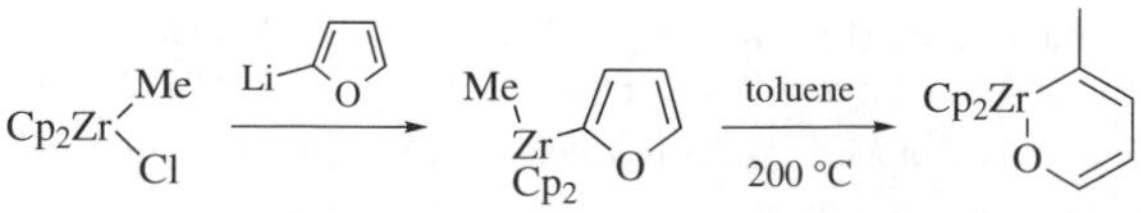

(10)

The reactions of Cp_2Zr complexes toward unsaturated molecules reveal their usefulness for preparing many functionalized compounds (eq 11). Thus the Cp_2Zr–benzyne complexes react with many unsaturated compounds to give benzo-fused compounds (eq 12).[8,17] A cyclopentyne–zirconocene complex has been used for the preparation of an optically active cyclopentenone derivative.[18] In this reaction the addition of an optically active allylic ether to the cyclopentyne complex proceeds with ~100% asymmetric induction (eq 13).

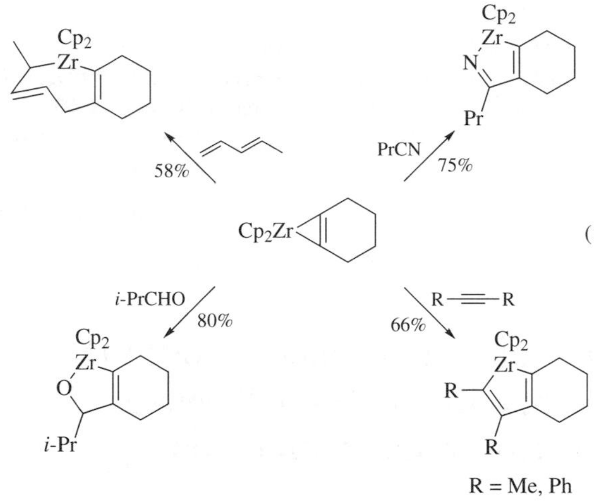

(11)

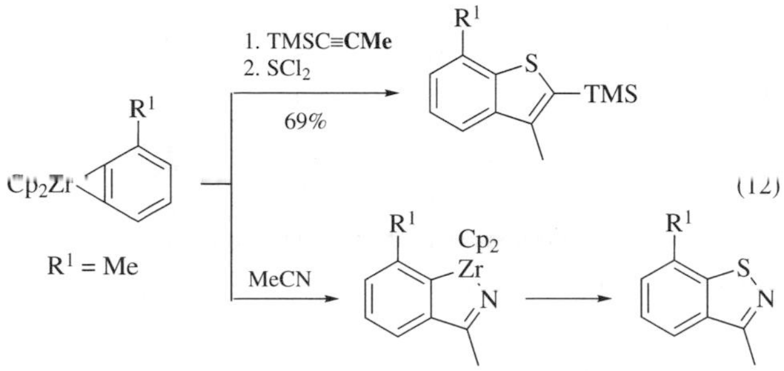

R = Me, Ph

(12)

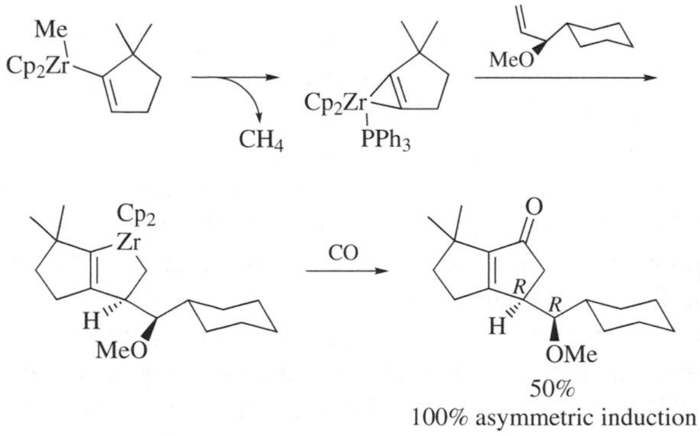

R^1 = Me

(13)

50%
100% asymmetric induction

Reaction of $Cp_2ZrMeCl$ with lithium amide derivatives and the subsequent extrusion of methane through β-hydrogen elimination is a very efficient way of generating imine–zirconocene complexes, zirconaaziridines.[13] Zirconaaziridines, which can be isolated as trimethylphosphine adducts or THF adducts, have proven to be valuable intermediates in organic synthesis. These complexes react with carbonyl compounds or alkynes to give amino alcohols or allylic amines. When the azazirconacyclopentene prepared from zirconaaziridine and alkyne is treated with ***Carbon Monoxide*** under high pressure, di- and trisubstituted pyrroles are obtained in synthetically viable yields (eq 14).[13b] In a similar manner, reaction of the lithium amide of an allylic amine with $Cp_2ZrMeCl$ when heated to 60 °C gives an azazirconacyclopentene, which is converted to a pyrrole when treated with an isocyanide.[13d] Notable and efficient application of the benzyne complex chemistry derived from $Cp_2ZrMeCl$ has been seen in the effective synthesis of the pharmacophore of potent antitumor antibiotics.[19]

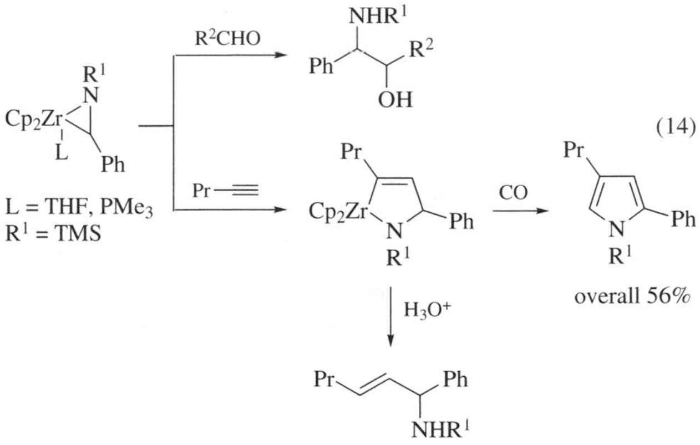

(14)

L = THF, PMe_3
R^1 = TMS

overall 56%

Hydroamination. Imidozirconocene complexes[20] can be generated in situ by treating $Cp_2ZrMeCl$ with a lithium amide (which does not have β hydrogen) followed by elimination of methane upon heating. This transient species reacts with an alkyne derivative to give an azazirconacyclobutene derivative which can be converted to a carbonyl compound via an enamine. Carbonyl compounds also react with imidozirconocene complexes giving imine compounds (eq 15).

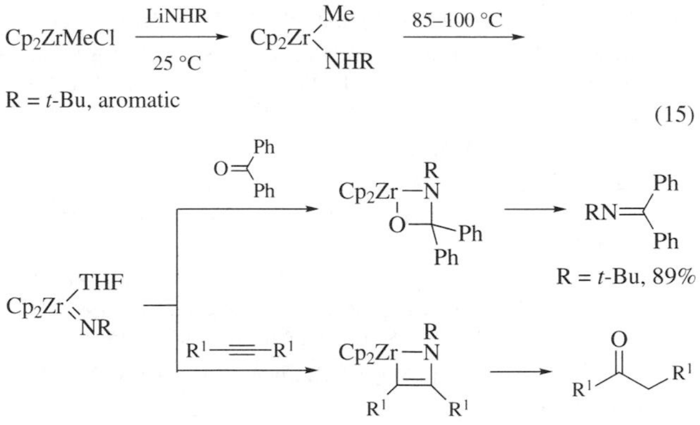

(15)

R = *t*-Bu, aromatic

R = *t*-Bu, 89%

Carbometalation. In the chemistry of carbometalation of alkylaluminum compounds mediated by zirconocene derivatives,[21] reaction of a stoichiometric amount of $Cp_2ZrMeCl$ with 1-pentynyldimethylalane is known to give an alkenylmetal species, which is converted to diiodo or 2-methyl-1-pentene in good yields upon treatment with excess iodine or water. This process is an Al-assisted direct addition of the Zr–carbon bond to the alkyne (eq 16). It is interesting to note that the $Cp_2ZrMeCl$-catalyzed R_3Al addition to alkyne is a Zr-assisted Al–carbon bond addition (eq 17).

Pr—≡—$AlMe_2$ —($Cp_2ZrMeCl$)→ [Pr—≡—$AlMe_2$ … Cl … Me—$ZrCp_2$] →

Pr, $AlMe_2$, Cl, $ZrCp_2$ —(I_2, 92%)→ Pr, I, I; —(3N HCl, 95%)→ Pr (16)

C_5H_{11}—≡ → [C_5H_{11}—≡ … Me—Al(Me)----X----$ZrCp_2ClX$] →

R, AlMeX —(I_2)→ R, I (17)

95%

Related Reagents. Chlorobis(cyclopentadienyl)hydridozirconium; Dichlorobis(tri-*o*-tolylphosphine)palladium(II).

---

1. (a) Buchwald, S. L.; Nielsen, R. B. *CRV* **1988**, *88*, 1047. (b) Buchwald, S. L.; Fisher, R. A. *CS* **1989**, *29*, 417. (c) Negishi, E.; Takahashi, T. *S* **1988**, 1.
2. Buchwald, S. L.; Watson, B. T.; Lum, R. T.; Nugent, W. A. *JACS* **1987**, *109*, 7137.
3. Negishi, E.; Van Horn, D. E.; Yoshida, T. *JACS* **1985**, *107*, 6639.
4. (a) Walsh, P. J.; Hollander, F. J.; Bergman, R. G. *JACS* **1988**, *110*, 8729. (b) Jordan, R. F. *JOM* **1985**, *294*, 321.
5. Wailes, P. C.; Weigold, H.; Bell, A. P. *JOM* **1971**, *33*, 181.
6. (a) Barr, K. J.; Watson, B. T.; Buchwald, S. L. *TL* **1991**, *32*, 5465. (b) Swanson, D. R.; Nguyen, T.; Noda, Y.; Negishi, E. *JOC* **1991**, *56*, 2590.
7. Surtees, J. R. *CC* **1965**, 567.
8. (a) Buchwald, S. L.; Nielsen, R. B. *CRV* **1988**, *88*, 1047. (b) Buchwald, S. L.; Fisher, R. A. *CS* **1989**, *29*, 417. (c) Negishi, E.; Takahashi, T. *S* **1988**, 1.
9. (a) Negishi, E. *CS* **1989**, *29*, 457. (b) Jensen, M.; Livinghouse, T. *JACS* **1989**, *111*, 4495. (c) Ito, H.; Taguchi, T.; Hanzawa, Y. *TL* **1992**, *33*, 4469. (d) Davis, J. M.; Whitby, R. J.; Jaxa-Chamiec, A. *CC* **1991**, 1743.
10. Negishi, E.; Nguyen, T.; Maye, J. P.; Choueiri, D.; Suzuki, N.; Takahashi, T. *CL* **1992**, 2367.
11. (a) Buchwald, S. L.; Watson, B. T.; Lum, R. T.; Nugent, W. A. *JACS* **1987**, *109*, 7137. (b) Buchwald, S. L.; Sayers, A.; Watson, B. T.; Dewan, J. C. *TL* **1987**, *28*, 3245. (c) Erker, G. *JOM* **1977**, *134*, 189. (d) Cuny, G. D.; Gutiérrez, A.; Buchwald, S. L. *OM* **1991**, *10*, 537.
12. Hsu, D. P.; Lucas, E. A.; Buchwald, S. L. *TL* **1990**, *31*, 5563.
13. (a) Buchwald, S. L.; Watson, B. T.; Wannamaker, M. W.; Dewan, J. C. *JACS* **1989**, *111*, 4486. (b) Buchwald, S. L.; Wannamaker, M. W.; Watson, B. T. *JACS* **1989**, *111*, 776. (c) Coles, N.; Whitby, R. J.; Blagg, J. *SL* **1990**, 271. (d) Davis, J. M.; Whitby, R. J.; Joxa-Chamiec, A. *CC* **1991**, 1743.
14. Buchwald, S. L.; Nielsen, R. B.; Dewan, J. C. *JACS* **1987**, *109*, 1590.
15. Erker, G.; Noe, R.; Albrecht, M. *JOM* **1993**, *450*, 137.
16. Erker, G.; Petrenz, R. *OM* **1992**, *11*, 1646.
17. (a) Cuny, G. D.; Gutierrez, A.; Buchwald, S. L. *OM* **1991**, *10*, 537. (b) Cuny, G. D.; Buchwald, S. L. *OM* **1991**, *10*, 363. (c) Buchwald, S. L.; King, S. M. *JACS* **1991**, *113*, 258. (d) Buchwald, S. L.; Fang, Q. *JOC* **1989**, *54*, 2793. (e) Buchwald, S. L.; Lucas, E. A.; Davis, W. M. *JACS* **1989**, *111*, 397.
18. Buchwald, S. L.; Lum, R. T.; Fisher, R. A.; Davis, W. M. *JACS* **1989**, *111*, 9113.
19. Tidwell, J. H.; Buchwald, S. L. *JOC* **1992**, *57*, 6380.
20. (a) Walsh, P. J.; Baranger, A. M.; Bergman, R. G. *JACS* **1992**, *114*, 1708. (b) Walsh, P. J.; Hollander, F. J.; Bergman, R. G. *OM* **1993**, *12*, 3705.
21. Negishi, E.; van Horn, D. E.; Yoshida, T. *JACS* **1985**, *107*, 6639.

Takeo Taguchi & Yuji Hanzawa
Tokyo College of Pharmacy, Japan

Chloro(cyclopentadienyl)bis[3-*O*-(1,2:5,6-di-*O*-isopropylidene-α-D-glucofuranosyl)]titanium

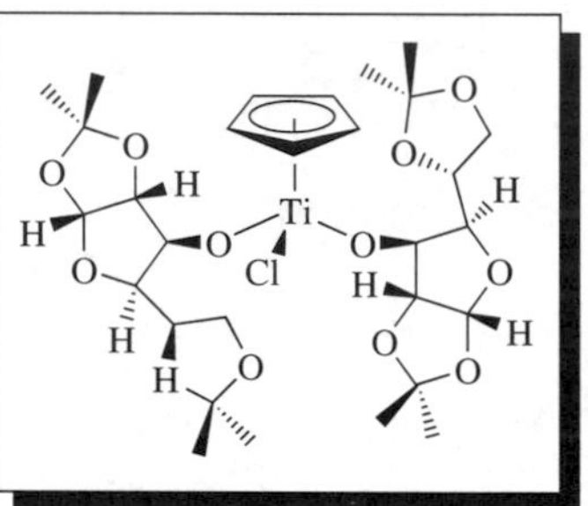

[119528-80-2] $C_{29}H_{43}ClO_{12}Ti$ (MW 667.05)

(highly enantio- and diastereoselective aldol reactions of acetic acid,[1] propionic acid,[2] and glycine[3] ester enolates with various aldehydes; stereoselective addition of allyl groups to aldehydes[4])

Physical Data: crystal structure; ^1H and ^{13}C NMR.[5]
Solubility: toluene (not determined, 0.155 M possible); Et_2O (not determined, 0.09 M possible).[4]
Analysis of Reagent Purity: ^1H NMR; test reaction.
Preparative Method: see ***Trichloro(cyclopentadienyl)titanium***.
Handling, Storage, and Precautions: best handled as stock solution either in Et_2O (ca. 0.1 M) or toluene (ca. 1.5 M), which must be protected from moisture and UV light. If handled under an inert atmosphere (argon), such solutions can be stored in a refrigerator (8 °C) for several months (possibly much longer) without deterioration. Reactions should be carried out in dry equipment and with absolute solvents under Ar or N_2.

Aldol Reactions. The titanium enolate (**2**) is obtained by addition of ca. 1.3 equiv of the title reagent (**1**) as a 0.1–0.15 M solution in toluene to the Li enolate of *t*-butyl acetate (**3**) generated at −78 °C with lithium dicyclohexylamide in Et_2O. This transmetalation takes about 24 h at −78 °C but is completed within 1 h at −30 °C (eq 1).[1a,b] The medium might also be important, as it has recently been reported that ***12-Crown-4*** has to be added for reproducible results in THF–Et_2O.[1c] The solution of (**2**) usually is recooled to −78 °C for the reaction with aldehydes, affording β-hydroxy esters (**4**) of high enantiomeric purity (90–90% ee) upon hydrolytic workup. Byproducts are insoluble cyclopentadienyltitanium oxides (**5**) and the ligand diacetone-glucose (DAGOH, **6**). The oxides (**5**) can be separated by filtration and may be recycled to $CpTiCl_3$. Ligand (**6**) and product are either separated by conventional methods (crystallization, distillation, chromatography), or glucose is extracted into the aqueous phase after acetonide cleavage in 0.1 N HCl (1.5 h at rt).[1a,b] In the case of isovaleraldehyde (R=*i*-Bu) it could be shown that the enantioselectivity (92–96% ee) is retained up to rt (27 °C).[1a,b]

1. $LiNCy_2 \cdot Et_2O$
2. (**1**), toluene −78 to −30 °C

DAGO–Ti(Cp)–ODAG

(**3**) (**2**)

1. RCHO, −78 °C
2. H_2O

$[CpTiO(OH)]_n$ + OH + (1)

(**5**) (**6**) DAGOH (**4**) 90–96% ee 51–87%

A clear drawback of this reagent is the availability of only one enantiomer, the one favoring the *re* attack to the aldehyde carbonyl, as only D-glucose is readily available. *si* Attack is observed with the analogous enolate prepared from ***Chloro(η⁵-cyclopentadienyl)[(4R,trans)-2,2-dimethyl-a,a,a′,a′-tetraphenyl-1, 3-dioxolane-4, 5-dimethanolato(2–)-Oᵃ, Oᵃᵃ]titanium***, but only with moderate enantioselectivity (78% ee).[1c,6] The reagent (**2**) is probably the most versatile chirally modified acetate enolate. Good results have also been obtained with the Mg enolate of 2-acetoxy-1,1,2-triphenylethanol[7] and with boron enolates derived from 2,4-dialkylborolanes.[8] Chiral Fe-acetyl complexes, which can be considered as acetate equivalents, give impressive stereocontrol upon enolization and aldol reaction.[9] Except for unsaturated residues R, β-hydroxy esters (**4**) of excellent optical purity can also be obtained by enantioselective hydrogenation of the corresponding β-keto esters catalyzed by $RuCl_2$(BINAP).[10]

For the propionate aldol reaction the Li enolate (**7**), generated by deprotonation of 2,6-dimethylphenyl propionate with ***Lithium Diisopropylamide*** in Et_2O,[11] was chosen.[2] Transmetalation with 1.25 equiv of an ethereal solution of (**1**) takes 24 h at −78 °C. The completion of this step is evident by the disappearance of racemic *anti*-aldol (**9**) in favor of optically active *syn*-isomer (**10**) (91–98% ee) upon reaction with an aldehyde (RCHO) and aqueous workup. At this point, 3–11% of *anti*-aldol (**9**) remaining in the reaction mixture is optically active as well (eq 2). This *anti*-isomer (**9**) (94–98% ee) becomes the major product if the reaction mixture, containing the putative (*E*)-titanium enolate derived from (**7**), is warmed for 4–5 h to −30 °C before reaction with an aldehyde (RCHO) again at −78 °C. Isomerization to the (*Z*)-titanium enolate is a possible explanation of this behavior. Some substrates, aromatic and unsaturated aldehydes, behave exceptionally, as a high proportion of *syn*-isomer (**10**) (19–77%) of lower optical purity (47–66% ee) is formed in addition to (**9**) (94–98% ee). After hydrolysis of the acetonide (**6**) the products (**9/10**) are isolated and separated by chromatography in 50–87% yield. The reactions of pivalaldehyde (R=*t*-Bu) are sluggish at −78 °C and have therefore been carried out at −50 to −30 °C.

LDA, Et_2O −78 °C

(**8**) (**7**)

(**1**). Et_2O 24 h, −78 °C

4–5 h, −30 °C (2)

−78 °C 1. RCHO 2. H_2O, NH_4Cl

−78 °C 1. RCHO 2. H_2O, NH_4Cl

(**6**) + (**5**) +

(**9**) 94–98% ee 23–89% ds

(**10**) 91–98% ee 89–97% ds

As above (eq 1), a major drawback of this reagent is the lack of a readily available enantiomer. There are many alternative methods for the enantioselective propionate aldol reaction. The most versatile chirally modified propionate enolates or equivalents are *N*-propionyl-2-oxazolidinones,[12] α-siloxy ketones,[13] boron enolates with chiral ligands,[14] as well as tin enolates.[15] Especially rewarding are new chiral Lewis acids for the asymmetric Mukaiyama reaction of *O*-silyl ketene acetals.[16] Most of these reactions afford *syn*-aldols; good methods for the *anti*-isomers have only become available recently.[8,17]

Transmetalation of the (*E*)-*O*-Li-enolate derived from the 'stabase'-protected glycine ethyl ester (**11**) with 1.1 equiv of (**1**) affords the chiral Ti enolate (**12**), which adds with high *re* selectivity to various aldehydes.[3,18] By mild acidic cleavage of the silyl protecting group, the primary product (**13**) can be transformed to various *N*-derivatives (**14**) of D-*threo*-α-amino-β-hydroxy acids in 45–66% yield and with excellent enantio- and *syn*(*threo*) selectivity (97–99%) (eq 3). An exception with lower enantioselectivity is glyoxylic ester (ethyl ester 78% ee; *t*-butyl ester 87% ee).

In this case the enantiomers are available by the analogous conversion of glycine *t*-butyl ester using ***Chloro-(η⁵-cyclopentadienyl)[(4R, trans)-2, 2-dimethyl-a, a, a′, a′-tetraphenyl-1,3-dioxolane-4,5-dimethanolato(2–)-Oᵃ, Oᵃ]-titanium***. An elegant alternative is the enantioselective addition of isocyanoacetate to aldehydes under the catalysis of a chiral Au^I complex.[19] Further methods, also for the *anti*(*erythro*) epimers, can be found in recent reviews of enantioselective α-amino acid synthesis.[20]

Allyltitanation of Aldehydes. The allyltitanium complex (**15**) is obtained by reaction of chloride (**1**) (1.1 equiv) with allylmagnesium chloride in Et_2O for 1 h at 0 °C.[4] The compound (**15**) has been characterized by ^{13}C NMR.[5] Reaction with various aldehydes (RCHO) at −78 °C and hydrolysis affords the homoallyl alcohols (**16**) (55–88%) of high optical purity (85–94% ee) (eq 4).[4] The isolation of the product is analogous to the aldol reactions (cf eq 1).

(11) → 1. LICA, THF; 2. (1), Et_2O, 17 h, –78 °C → (12) (3)

(12) → –78 °C, 1. RCHO, 2. H_2O → (13) + (5) + (6)

(13) → 1. AcOH, H_2O, THF; 2. *N*-protection → (14) R^1 = H, Boc, CHO, 97–99% ee and de

(15) → 1. RCHO, –78 °C; 2. H_2O → (16) 85–94% ee + (5) + (6) (4)

The enantiomers of (**16**) are obtained analogously by using ***Chloro(η⁵-cyclopentadienyl)[(4R,trans)-2,2-dimethyl-a,a′,a′,a′-tetraphenyl-1, 3-dioxolane-4, 5-dimethanolato(2–)-O^a, O^a]titanium***.[21] The stereoselectivity of this cyclic Ti complex in allyltitanations is better than the diacetone–glucose system (**1**). It is therefore advisable to use the (4*S*,*trans*) enantiomer instead of (**1**) for controlling the *re* addition to problematic substrates.

Related Reagents. Chloro(η^5-cyclopentadienyl)[(4*R*,trans)-2,2-dimethyl-*a,a′,a′*-tetraphenyl-1,3-dioxolane-4,5-dimethanolato(2–)-O^a,$O^{a'}$]titanium ; Trichloro(cyclopentadienyl)titanium.

1. (a) Duthaler, R. O.; Herold, P.; Lottenbach, W.; Oertle, K.; Riediker, M. *AG(E)*, **1989**, *28*, 495. (b) Oertle, K.; Beyeler, H.; Duthaler, R. O.; Lottenbach, W.; Riediker, M.; Steiner, E. *HCA* **1990**, *73*, 353. (c) Cambie, R. C.; Coddington, J. M.; Milbank, J. B. J.; Paulser, M. G.; Rustenhoven, J. J., Rutledge, P. S.; Shaw, G. L.; Sinkovich, P. I. *AJC* **1993**, *46*, 583.
2. Duthaler, R. O.; Herold, P.; Wyler-Helfer, S.; Riediker, M. *HCA* **1990**, *73*, 659.
3. Bold, G.; Duthaler, R. O.; Riediker, M. *AG(E)* **1989**, *28*, 497.
4. Riediker, M.; Duthaler, R. O. *AG(E)* **1989**, *28*, 494.
5. Riediker, M.; Hafner, A.; Piantini, U.; Rihs, G.; Togni, A. *AG(E)* **1989**, *28*, 499.
6. Duthaler, R. O.; Hafner, A.; Riediker, M. *PAC* **1990**, *62*, 631.
7. Braun, M. *AG(E)***1987**, *26*, 24.
8. (a) Masamune, S.; Sato, T.; Kim, B. M. Wollmann, T. A. *JACS* **1986**, *108*, 8279. (b) Reetz, M. T.; Rivadeneira, E.; Niemeyer, C. *TL* **1990**, *31*, 3863.
9. (a) Liebeskind, L. S.; Welker, M. E. *TL* **1984**, *25*, 4341. (b) Davies, S. G.; Dordor, I. M.; Warner, P. *CC* **1984**, 956. (c) Brunner, H. *AG* **1991**, *103*, A310.
10. Noyori, R.; Ohkuma, T.; Kitamura, M.; Takaya, H.; Sayo, N.; Kumobayashi, H.; Akutagawa, S. *JACS* **1987**, *109*, 5856.
11. Montgomery, S. H.; Pirrung, M. C.; Heathcock, C. H. *OS* **1985**, *63*, 99.
12. Evans, D. A.; Nelson, J. V.; Taber, T. R. *Top. Stereochem.* **1982**, *13*, 1.
13. (a) Masamune, S.; Choy, W.; Petersen, J. S.; Sita, L. R. *AG(E)* **1985**, *24*, 1. (b) Heathcock, C. H. *Aldrichim. Acta* **1990**, *23* 99.
14. (a) Corey, E. J.; Imwinkelried, R.; Pikul, S.; Xiang, Y. B. *JACS* **1989**, *111*, 5493. (b) Corey, E. J.; Kim, S. S. *JACS* **1990**, *112*, 4976.
15. Mukaiyama, T.; Kobayashi, S.; Sano, T. *T* **1990**, *46*, 4653.
16. (a) Kobayashi, S.; Uchiro, H.; Fujishita, Y.; Shiina, I.; Mukaiyama, T. *JACS* **1991**, *113*, 4247. (b) Kobayashi, S.; Uchiro, H.; Shiina, I.; Mukaiyama, T. *T* **1993**, *49*, 1761.
17. (a) Helmchen, G.; Leikauf, U.; Taufer-Knöpfel, I. *AG(E)* **1985**, *24*, 874. (b) Gennari, C.; Schimperna, G.; Venturini, I. *T* **1988**, *44*, 4221. (c) Oppolzer, W.; Starkemann, C.; Rodriguez, I.; Bernardinelli, G.; *TL* **1991**, *32*, 61. (d) Van Draanen, N. A.; Arseniyadis, S.; Crimmins, M. T.; Heathcock, C. H. *JOC* **1991**, *56*, 2499. (e) Myers, A. G.; Widdowson, K. L.; Kukkola, P. J. *JACS* **1992**, *114*, 2765. (f) Cardani, S.; De Toma, C.; Gennari, C.; Scolastico, C. *T* **1992**, *48*, 5557. (g) Oppolzer, W.; Lienard, P. *TL* **1993**, *34*, 4321.
18. Bold, G.; Allmendinger, T.; Herold, P.; Moesch, L.; Schär, H.-P.; Duthaler, R. O. *HCA* **1992**, *75*, 865.
19. Ito, Y.; Sawamura, M.; Hayashi, T. *JACS* **1986**, *108*, 6405.
20. (a) Williams, R. M. *Synthesis of Optically Active α-Amino Acids*; Pergamon: Oxford, 1989. (b) Duthaler, R. O. *T* **1994**, *50*, 1539.
21. Hafner, A.; Duthaler, R. O.; Marti, R.; Rihs, G.; Rothe-Streit, P.; Schwarzenbach, F. *JACS* **1992**, *114*, 2321.

Andreas Hafner
Ciba-Geigy, Marly, Switzerland

Rudolf O. Duthaler
Ciba-Geigy, Basel, Switzerland

(Chloromethyl)trimethylsilane[1]

Me_3SiCH_2Cl

[2344-80-1] $C_4H_{11}ClSi$ (MW 122.67)

(reagent for direct alkene synthesis;[2] electrophile for the formation of a variety of functionalized synthetic intermediates for alkene synthesis;[3] alkylation adducts are frequently a source of fluoride-induced reactive intermediates;[4] readily undergoes metal–halogen exchange to generate a reagent for Peterson methylenation[5])

Alternate Name: trimethylsilylmethyl chloride.
Physical Data: bp 97–98 °C; *d* 0.886 g cm^{-3}.
Form Supplied in: colorless liquid; widely available.
Handling, Storage, and Precautions: highly flammable liquid; corrosive; irritant; use in a fume hood.

Reactions with Carbonyl Compounds. Terminal alkenes are efficiently generated upon treatment of an aldehyde or ketone with Me_3SiCH_2Cl in the presence of ***Triphenylphosphine*** in a sealed tube (eq 1).[2] Reaction times are generally not more than 60 min, and the method is operationally simple, although use of this protocol for methylenation is limited to substrates which lack enolizable α-hydrogens, since in such cases studied, silyl enol ether formation is a competitive pathway.

$$\text{Ph}_2\text{C=O} \xrightarrow[90\%]{\text{Me}_3\text{SiCH}_2\text{Cl, PPh}_3,\ 120\ ^\circ\text{C, 10 min}} \text{Ph}_2\text{C=CH}_2 \quad (1)$$

Conversion of aldehydes and ketones to α,β-epoxy trimethylsilanes is readily achieved by treatment of Me_3SiCH_2Cl with the appropriate alkyllithium reagent at low temperature, and subsequent quenching with the desired carbonyl compound.[6] It is critical that ***s-Butyllithium*** be used as the lithiating reagent, as use of ***n-Butyllithium*** results in attack at silicon and use of ***t-Butyllithium*** results in metal–halogen exchange. The resultant α,β-epoxy trimethylsilanes undergo facile hydrolysis under relatively mild conditions to provide the corresponding aldehydes in good yields (eq 2). A variety of substrates have been successfully subjected to this sequence, including α,β-unsaturated carbonyl compounds and sterically congested aldehydes and ketones.

$$\text{Me}_3\text{SiCH}_2\text{Cl} \xrightarrow[-78\ ^\circ\text{C, TMEDA}]{s\text{-BuLi, THF}} \text{Me}_3\text{SiCH(Cl)Li} \xrightarrow[\text{THF}, -55\ ^\circ\text{C, 10 min}]{4\text{-}t\text{-Bu-cyclohexanone}}$$

$$\text{4-}t\text{-Bu-C}_6\text{H}_{10}\text{(OLi)CH(Cl)SiMe}_3 \longrightarrow \text{epoxide (SiMe}_3\text{), 88\%} \xrightarrow[72\%]{20\%\ \text{aq THF},\ 10\%\ \text{HClO}_4} \text{4-}t\text{-Bu-C}_6\text{H}_{10}\text{-CHO} \quad (2)$$

In the presence of ***Cesium Fluoride*** in DMF, Me_3SiCH_2Cl reacts with benzaldehyde to form phenyloxirane in 60% yield (eq 3);[7] this method has not yet proven to be a generally applicable route to epoxides.

$$\text{Me}_3\text{SiCH}_2\text{Cl} \xrightarrow[60\%]{\text{PhCHO (1.5 equiv), CsF, DMF, rt}} \text{phenyloxirane} \quad (3)$$

Electrophile for *O*-, *N*-, and *S*-Alkylations. A novel method for phenol homologation utilizes Me_3SiCH_2Cl to trap phenoxides, forming trimethylsilylmethyl phenyl ethers which, upon treatment with *s*-butyllithium, undergo 1,2-Wittig rearrangement of the intermediate lithiated trimethylsilylmethyl species to produce trimethylsilyl-substituted benzyl alcohols in good yields. Subjecting these phenylsilylmethanols to basic hydrolysis with methanolic potassium hydroxide produces the homologated phenols (eq 4).[8]

$$\text{Me}_3\text{SiCH}_2\text{Cl} \xrightarrow[87\%]{\text{PhOH, K}_2\text{CO}_3\text{, DMSO}} \text{Me}_3\text{SiCH}_2\text{OPh} \xrightarrow[\text{THF}, -78\ ^\circ\text{C}]{s\text{-BuLi}}$$

$$[\text{Me}_3\text{SiCH(Li)OPh}] \xrightarrow[28\ \text{h}]{25\ ^\circ\text{C}} \text{Me}_3\text{SiCH(Ph)OH} \xrightarrow[\text{MeOH}, \Delta]{\text{KOH}} \text{PhCH}_2\text{OH}\ \ 58\% \quad (4)$$

Much of the utility of Me_3SiCH_2Cl is derived from the lability of the methylene carbon–silicon bond in its alkylation products from nitrogen-containing nucleophiles. The subsequent, relatively mild conditions of fluoride treatment effect 1,3-dipole formation. This fluoride-promoted azomethine ylide generation methodology constitutes a highly stereoselective means for the synthesis of pyrroles by the trapping of the reactive intermediate with ***Dimethyl Fumarate*** or ***Dimethyl Maleate*** (eq 5).[4]

$$\text{PhCH}_2\text{NH}_2 \xrightarrow[50\%]{\text{1. Me}_3\text{SiCH}_2\text{Cl, 200 °C; 2. 1.0N HCl, THF, KCN, 37\% aq. HCHO}} \text{NCCH}_2\text{N(CH}_2\text{Ph)CH}_2\text{SiMe}_3$$

$$\xrightarrow[\text{rt, 10 h, 71\%}]{\text{MeO}_2\text{CCH=CHCO}_2\text{Me (fumarate), AgF, MeCN, dark}} \textit{trans}\text{-3,4-bis(CO}_2\text{Me)-1-benzylpyrrolidine}$$

$$\xrightarrow[\text{rt, 10 h, 45\%}]{\text{MeO}_2\text{CCH=CHCO}_2\text{Me (maleate), AgF, MeCN, dark}} \textit{cis}\text{-3,4-bis(CO}_2\text{Me)-1-benzylpyrrolidine} \quad (5)$$

Other dipolarophiles, including aldehydes, nitroalkenes, styrenes, and functionalized alkynes (eq 6), have also proven to be effective cycloaddition substrates for ylides generated in this fashion.[9]

$$\text{NCCH}_2\text{N(CH}_2\text{Ph)CH}_2\text{SiMe}_3 \xrightarrow[\text{AgF, MeCN, dark, rt, 10 h}]{\text{HC≡CCO}_2\text{Me}} [\text{1-benzyl-3-CO}_2\text{Me-3-pyrroline}] \xrightarrow{\text{DDQ}, \Delta} \text{1-benzyl-3-CO}_2\text{Me-pyrrole}\ \ 65\% \quad (6)$$

Intramolecular trapping of ylides derived from the Me_3SiCH_2Cl alkylation adducts of a variety of 3- and 4-substituted benzylamines proceeds through a Sommelet–Hauser pathway, providing the rearranged products in good yields (eq 7).[10]

$$\text{PhCH}_2\text{NHMe} \xrightarrow[\text{2. MeI, MeCN, 60 °C, 2 h; 38\%}]{\text{1. Me}_3\text{SiCH}_2\text{Cl, DMSO, 140 °C, 8 h}} \text{PhCH}_2\overset{+}{\text{N}}\text{Me}_2\text{CH}_2\text{SiMe}_3\ \text{I}^- \xrightarrow[\text{30 min, 10 °C}]{\text{CsF, HMPA}}$$

$$[\text{PhCH}_2\overset{+}{\text{N}}\text{Me}_2\text{CH}_2^-\ \text{I}^-] \longrightarrow \text{(6-methylenecyclohexadienyl)CH}_2\text{NMe}_2 \longrightarrow \text{2-MeC}_6\text{H}_4\text{CH}_2\text{NMe}_2\ \ 58\% \quad (7)$$

Azolylmethyl anions, generated by the fluoride-induced desilylation of Me_3SiCH_2Cl adducts of pyrroles, imidazoles, pyrazoles, triazoles, tetrazoles, and azinonylmethyl anions similarly derived from (trimethylsilylmethyl)azinones, have demonstrated efficacy in the addition reactions to carbonyl compounds.[11]

Sulfur alkylation products of Me_3SiCH_2Cl have also been used as a convenient source of reactive intermediates. Lithiated (2-benzothiazolylthio)(trimethylsilyl)methane (eq 8) functions as a synthetic equivalent for the mercaptomethyl anion by (a) reaction with an electrophile, (b) fluoride-promoted desilylation/reaction with a carbonyl compound, and (c) alkyllithium addition to the benzothiazole 2-position.[12]

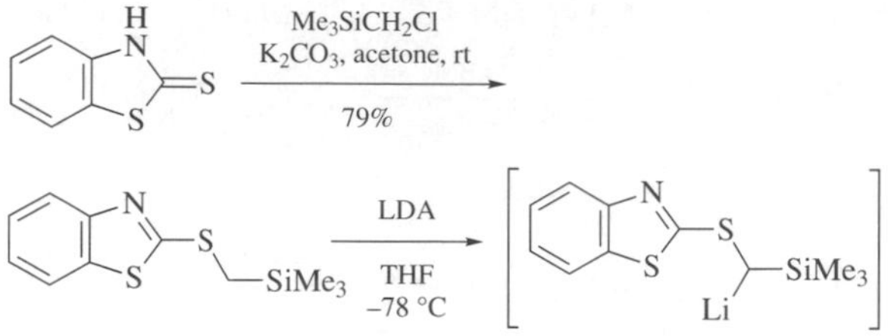

Me_3SiCH_2Cl, K_2CO_3, acetone, rt, 79%

LDA, THF, −78 °C (8)

Thiiranes have been prepared by the reaction of aldehydes with *S*-methyl *S*′-trimethylsilylmethyl *N*-(*p*-tolylsulfonyl)dithioiminocarbonate in the presence of cesium fluoride (eq 9).[13] This novel route to thiiranes utilizes the 1,3-dipolar cycloaddition of the iminothiocarbonyl ylide derived from a Me_3SiCH_2Cl alkylated thiol to aldehydes.

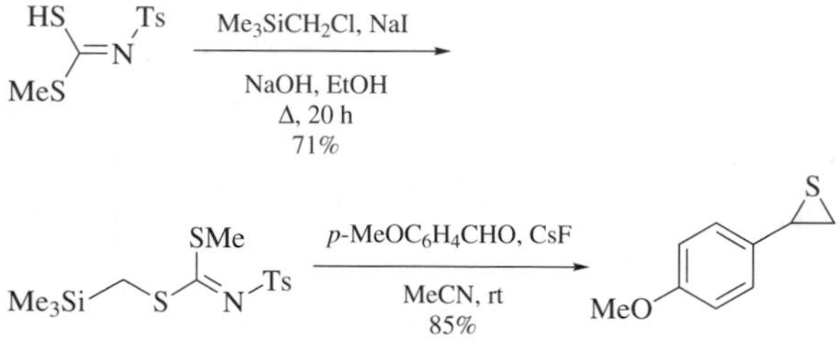

Me_3SiCH_2Cl, NaI, NaOH, EtOH, Δ, 20 h, 71%

p-$MeOC_6H_4CHO$, CsF, MeCN, rt, 85% (9)

Preparation of Other Reagents. One of the principal uses of this reagent has been for the formation of ***Trimethylsilylmethylmagnesium Chloride***, most often used for the methylenation of carbonyl compounds;[5] however, since both this reagent and the corresponding lithio reagent are now widely available commercially, its use in this capacity has diminished considerably. Similarly, chloromethyltrimethylsilane has been used to prepare many other now commercially available reagents, including trimethylsilylmethyl acetate, trimethylsilylmethyl isocyanide, ***Trimethylsilyldiazomethane*** and trimethylsilylmethyl trifluoromethanesulfonate, each useful for a variety of synthetic transformations.

Related Reagents. (Iodomethyl)trimethylsilane; Trimethylsilylmethyllithium; Trimethylsilylmethylmagnesium Chloride; Trimethylsilylmethylpotassium; Trimethylsilylmethyl Trifluoromethanesulfonate.

---

1. Anderson, R. *S* **1985**, 717.
2. Sekiguchi, A.; Ando, W. *JOC* **1979**, *44*, 413.
3. (a) Djahanbini, D.; Cazes, B.; Gore, J.; Gobert, F. *T* **1985**, *41*, 867. (b) Kawashima, T.; Ishii, T.; Inamoto, N. *BCJ* **1987**, *60*, 1831.
4. Padwa, A.; Chen, Y.-Y.; Dent, W.; Nimmesgern, H. *JOC* **1985**, *50*, 4006.
5. (a) Ager, D. J. *S* **1984**, 384. (b) Chan, T. H.; Chang, E. *JOC* **1974**, *39*, 3264. (c) Peterson, D. J. *JOC* **1968**, *33*, 780.
6. Burford, C.; Cooke, F.; Roy, G.; Magnus, P. *T* **1983**, *39*, 867.
7. (a) Kessar, S. V.; Singh, P.; Kaur, N. P.; Chawla, U.; Shukla, K.; Aggarwal, P.; Venugopal, D. *JOC* **1991**, *56*, 3908. (b) Kessar, S. V.; Singh, P.; Venugopal, D. *IJC(B)* **1987**, *26B*, 605.
8. Eisch, J. J.; Galle, J. E.; Piotrowski, A.; Tsai, M.-R. *JOC* **1982**, *47*, 5051.
9. (a) Padwa, A.; Chen, Y. Y.; Chiacchio, U.; Dent, W. *T* **1985**, *41*, 3529. (b) Padwa, A.; Dent, W. *OS* **1988**, *67*, 133. See also (c) Pandey, G.; Lakshmaiah, G.; Kumaraswamy, G. *CC* **1992**, 1313. (d) Anderson, W. K.; Kinder, F. R. *JHC* **1990**, *27*, 975.
10. Shirai, N.; Watanabe, Y.; Sato, Y. *JOC* **1990**, *55*, 2767.
11. (a) Shimizu, S.; Ogata, M. *JOC* **1988**, *53*, 5160. (b) Shimizu, S.; Ogata, M. *JOC* **1986**, *51*, 3897.
12. (a) Katritzky, A. R.; Kuzmierkiewicz, W.; Aurrecoechea, J. M. *JOC* **1987**, *52*, 844. See also (b) Terao, Y.; Aono, M.; Imai, N.; Achiwa, K. *CPB* **1987**, *35*, 1734.
13. (a) Tominaga, Y.; Ueda, H.; Ogata, K.; Kohra, S.; Hojo, M.; Ohkuma, M.; Tomita, K.; Hosomi, A. *TL* **1992**, *33*, 85. (b) Tominaga, Y.; Matsuoka, Y.; Kamio, C.; Hosomi, A. *CPB* **1989**, *37*, 3168.

Lawrence G. Hamann & Todd K. Jones
Ligand Pharmaceuticals, San Diego, CA, USA

Chlorosulfonyl Isocyanate[1]

$ClSO_2NCO$

[1189-71-5] $CClNO_3S$ (MW 141.53)

(most chemically reactive isocyanate;[1] CSI can undergo two types of nucleophilic addition, namely to the carbonyl carbon and to the sulfur of the sulfonyl chloride; the isocyanate portion can undergo formal cycloaddition)

Alternate Name: CSI.
Physical Data: mp −44 to −43 °C; bp 107–108 °C/760 mmHg, 38 °C/50 mmHg; $d_4^{20} = 1.626$ g cm^{-3}; $n_D^{27} = 1.4435$.
Solubility: sol most organic solvents; reacts violently with water.
Form Supplied in: colorless liquid; widely available.
Handling, Storage, and Precautions: reacts explosively with water to form sulfamic acid, hydrogen chloride, and carbon dioxide; fumes slightly in air, has a choking smell, and undergoes thermal decomposition only above about 300 °C. Glass stoppers of bottles used for storage of CSI stick fast after a short time, even when covered with silicone grease, so that polyethylene bottles with screw caps are best for longer storage times (a few weeks); can be kept indefinitely in sealed glass ampules. Suitable solvents include saturated aliphatic hydrocarbons (more or less limited solubility below 0 °C), aromatic hydrocarbons such as benzene and toluene, chlorinated hydrocarbons such as CH_2Cl_2, $CHCl_3$, CCl_4, and chlorobenzene, diethyl ether, diisopropyl ether, and acetonitrile. Liquid sulfur dioxide is particularly favorable, since it increases the reactivity of CSI still further. Solvents such as acetone and ethyl acetate can be used to a limited extent, but only at low temperatures.

Classification of Reactions. CSI is probably the most chemically reactive isocyanate known, and has been the subject of several reviews.[1] It can be synthesized by reaction of sulfur trioxide with cyanogen chloride.[2]

Reactions with CSI are classified according to the probable site of reaction. The CSI molecule has two electrophilic sites for attack by nucleophilic reagents, namely the carbonyl carbon (Class I) and the sulfur of the sulfonyl chloride group (Class III). The isocyanate portion can also undergo formal cycloaddition reactions (Class II), which makes CSI a very versatile reagent (eq 1).

$$\mathrm{Cl{-}S(=O)_2{-}N{=}C{=}O} \quad \text{(Nu: Class III at S; Nu: Class I at C; Cycloaddition Class II at N=C)} \tag{1}$$

Class I: Addition Involving Initial Attack on the Isocyanate Carbon. CSI undergoes nucleophilic additions by alcohols (thiols/phenols) and amines to yield *N*-chlorosulfonyl carbamates and urea derivatives, respectively, which can undergo further transformations with water/alcohol/amine (eq 2; X = OR, SH, or NR_2).[1] Primary alcohols can be selectively derivatized with CSI in the presence of other groups without affecting, in general, other sterocenters in complex molecules (eqs 3 and 4).[3a,3b]

$$\mathrm{Cl{-}S(=O)_2{-}N{=}C{=}O + H{-}X \longrightarrow ClSO_2NHCOX} \begin{cases} \xrightarrow{H_2O} NH_2COX \\ \xrightarrow{R^1OH} R^1OSO_2NHCOX \\ \xrightarrow{R^2R^3NH} R^2R^3NSO_2NHCOX \end{cases} \tag{2}$$

X = OR, SH or NR_2

$$\text{cephalosporin } \mathrm{CH_2OH} \xrightarrow[\text{2. acid hydrolysis}]{\text{1. CSI, MeCN}} \mathrm{CH_2OCONH_2} \tag{3}$$

R = 2-thienylacetyl

$$\mathrm{HOCH_2{-}R} \xrightarrow[\text{2. } H_2O]{\text{1. CSI}} \mathrm{H_2NOCOCH_2{-}R} \tag{4}$$

The reagent obtained by reaction of CSI with methanol followed by treatment of the intermediate with ***Triethylamine*** (eq 5) has been used to synthesize carbamates from primary alcohols and to dehydrate secondary alcohols.[4] The above intermediate itself has been used for the synthesis of various heterocycles.[5] Similar intermediates obtained from reaction of CSI and 2,4,6-trichlorophenol are found to be useful bactericides and fungicides,[6] and can be further transformed to the isocyanates by treatment with alcohols followed by pyrolysis of the resultant carbamates (eq 6).[7]

$$\mathrm{MeOH \xrightarrow{CSI} ClSO_2NHCO_2Me \xrightarrow[81\%]{\text{2 equiv } NEt_3,\ PhH,\ 25\,^\circ C} Et_3\overset{+}{N}SO_2\overset{-}{N}CO_2Me} \tag{5}$$

$$\mathrm{ClSO_2NHCO_2Ar \xrightarrow[-\,HCl]{ROH} ROSO_2NHCO_2Ar \xrightarrow[-\,ArOH]{\Delta} ROSO_2NCO} \tag{6}$$

R = Me, Pr, $MeOCH_2CH_2$

The reaction of CSI with hydroxy groups of α-hydroxy ketones, followed by thermal ring closure of the intermediate carbamate, results in the formation of oxazolones (eq 7).[8]

$$\mathrm{R^1COC(OH)R^2R^3 \xrightarrow[\text{2. } H_2O]{\text{1. CSI, PhH}} R^1COC(R^2R^3)OCONH_2 \xrightarrow{\Delta} oxazolone} \tag{7}$$

R^1 = Ph, *t*-Bu; R^2, R^3 = Me, Ph

Thiocarbamates are obtained from the reaction of CSI with thiols (eq 8), and have been used for the synthesis of potential herbicides.[9]

$$\mathrm{RSH \xrightarrow[PhH]{CSI} ClSO_2NHCOSR \xrightarrow{PhNH_2} PhNHSO_2NHCOSR} \quad 90\% \tag{8}$$

R = $C_{12}H_{25}$

CSI also undergoes reaction with amides, sulfonamides (eq 9), and certain phosphoramides.[10,1b] The addition products can be readily converted to the corresponding isocyanates by pyrolytic elimination of the sulfamoyl chloride. However, the dipolar intermediate postulated for the reaction of *N,N*-dialkylamides produces amidines through loss of carbon dioxide (eq 10).[11] The amidines were further derivatized to prepare insecticidal and acaricidal compounds.[12]

$$\mathrm{RSO_2NH_2 \xrightarrow{CSI} RSO_2NHCONHSO_2Cl \xrightarrow[-\,ClSO_2NH_2]{\Delta} RSO_2NCO} \quad 75\text{–}92\% \tag{9}$$

$$\mathrm{R^1CONR^2R^3 \overset{CSI}{\rightleftharpoons} \left[R^1\overset{+}{C}(NR^2R^3)OC(O)\overset{-}{N}SO_2Cl\right] \rightleftharpoons R^1C(NR^2R^3)(O)\text{-}C(=O)NSO_2Cl \longrightarrow R^2R^3NC(R^1)=NSO_2Cl} \tag{10}$$

The generality of the reaction of CSI with acetals of aliphatic aldehydes,[13] and similar reactivity of orthoesters,[1b] has been reviewed. The reaction of acetals has been used for the conversion of isopropylidine-protected sugars to the corresponding carbonates (eq 11).[14] A similar reaction sequence with carboxylic acids[15] sometimes provides intermediate *N*-chlorosulfonylcarboxamides which are stable enough to be isolated, or which can be treated with ***N,N-Dimethylformamide*** to provide a one-pot synthesis of nitriles (eq 12). The reported[16] anhydropenicillin rearrangement (eq 13) is believed to proceed by an intermediate similar to that described above. Interestingly, such an intermediate also has been used as a mild reagent for displacement of an *N*-trimethylsilyl group.[17]

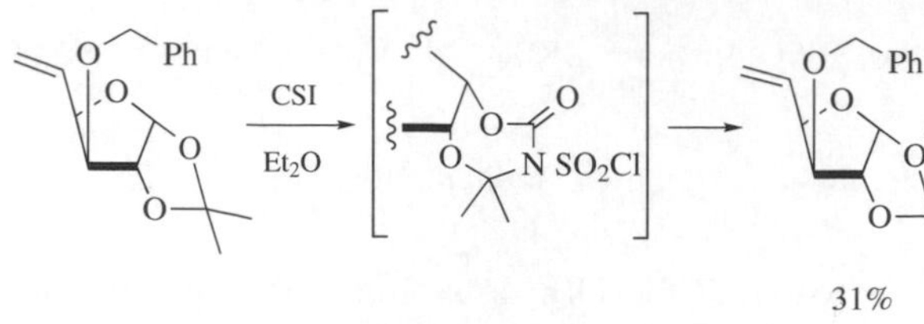

$$\text{(11)}\quad \xrightarrow[\text{Et}_2\text{O}]{\text{CSI}} \quad 31\%$$

$$RCO_2H \xrightarrow[-CO_2]{CSI,\ CH_2Cl_2} ClSO_2NHCOR \xrightarrow[-SO_3,\ -HCl]{DMF} RCN \quad 63\text{–}87\% \qquad (12)$$

R = *t*-Bu, Cy, ClCH$_2$CH$_2$, Bn, PhCH=CH, 1-naphthyl

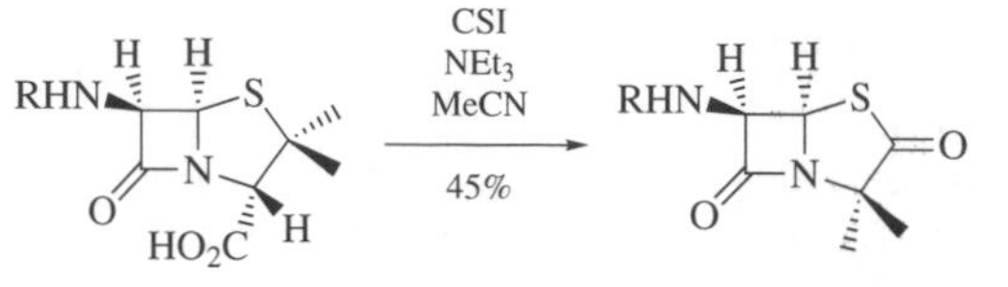

$$\xrightarrow[45\%]{CSI,\ NEt_3,\ MeCN} \qquad (13)$$

R = Ph$_3$C

Aromatic compounds undergo facile reaction with CSI. Treatment of the resultant *N*-chlorosulfonylcarboxamides with DMF in situ afforded the corresponding nitriles (eq 14).[18]

$$ArH \xrightarrow[CH_2Cl_2]{CSI} ClSO_2NHCOAr \xrightarrow[66\text{–}99\%]{DMF,\ -SO_3,\ -HCl} ArCN \qquad (14)$$

Nucleophilic attack by aldehydic carbonyl groups on the isocyanate carbon atom gives products as shown in eq 15, by way of 1,4-dipolar intermediates.[13] Reaction of arylmethylene malonaldehydes with CSI has provided bicyclic 1,3-oxazin-2-one derivatives (eq 16).[19]

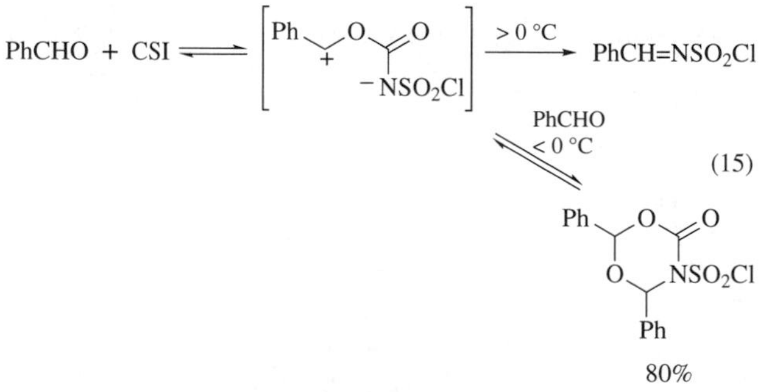

$$PhCHO + CSI \rightleftharpoons [\ ] \xrightarrow{>0\,°C} PhCH{=}NSO_2Cl \qquad (15)$$

PhCHO, < 0 °C: 80%

$$\xrightarrow{CSI} \qquad (16)$$

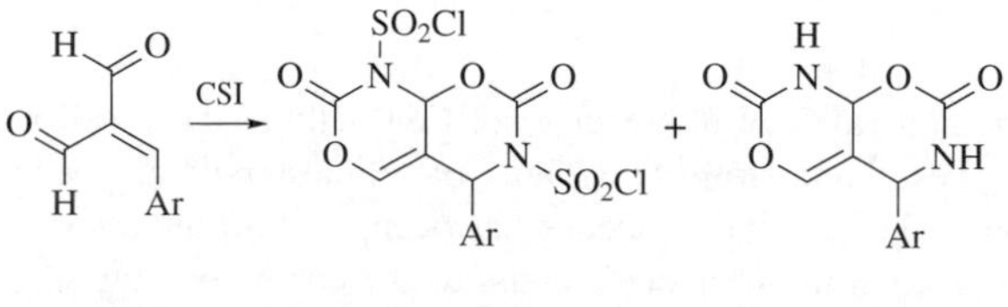

Enolizable ketones upon treatment with CSI produce *N*-chlorosulfonyl-β-ketocarboxamides, which have been used for the synthesis of β-ketonitriles by in situ treatment with DMF (eq 17).[20] However, in the presence of excess CSI, ketones with two α-hydrogens have been observed to undergo further transformations (eq 18).[21] The final product distribution depends on solvent, substituents, and concentration. Nonenolizable ketones, such as benzophenone and γ-pyrones,[22] produce azomethines (eq 19). The azomethine intermediate of benzophenone further cyclizes to provide benzoisothiazole (eq 20).[13] α,β-Unsaturated ketones are converted into 3,4-dihydro-1,3-oxazin-2-ones (eq 21).[13]

$$R^1COCHR^2R^3 \xrightarrow[2.\ DMF,\ 25\,°C]{1.\ CSI} R^1COCHR^2R^3CN \qquad (17)$$

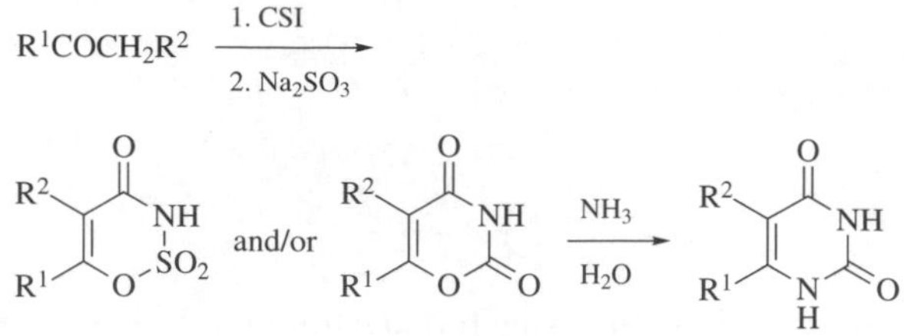

$$R^1COCH_2R^2 \xrightarrow[2.\ Na_2SO_3]{1.\ CSI} \text{ and/or } \xrightarrow[H_2O]{NH_3} \qquad (18)$$

$$\xrightarrow[72\%]{CSI,\ MeCN,\ -CO_2} \qquad (19)$$

$$\xrightarrow[MeCN]{CSI} \qquad (20)$$

$$+ CSI \longrightarrow \xrightarrow{hydrolysis} \quad 39\text{–}57\% \qquad (21)$$

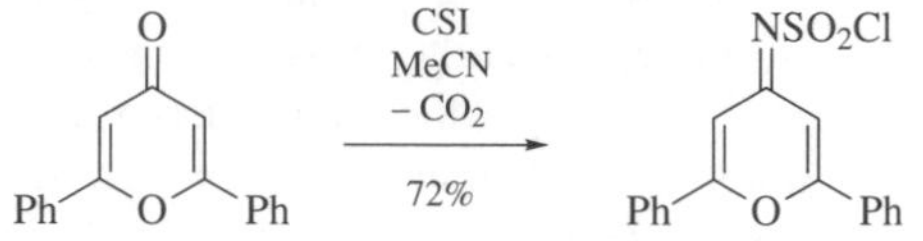

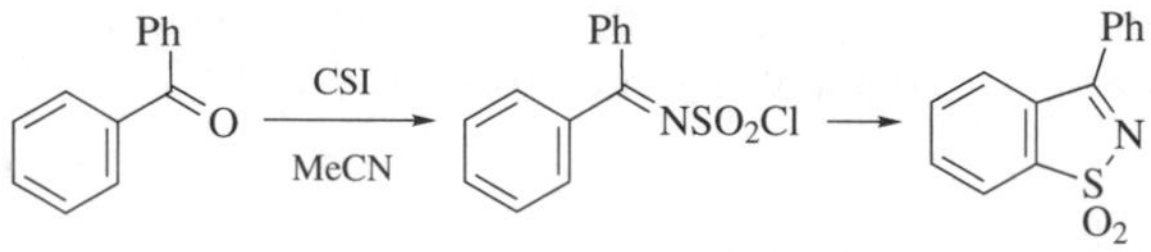

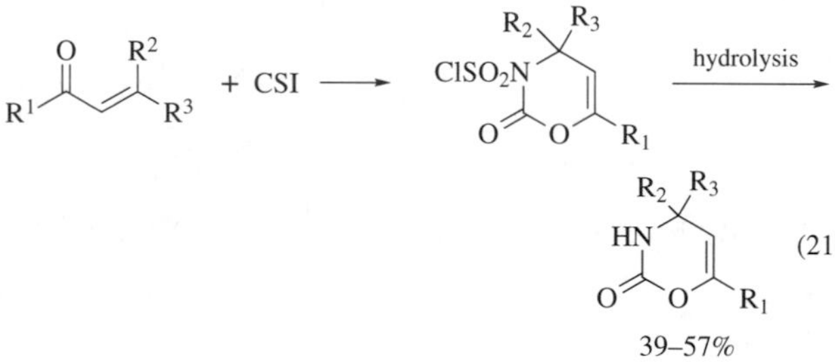

β-Diketones and β-keto esters react with CSI to produce amides (eqs 22 and 23).[13] Heterocycles have been obtained from the reaction of CSI with amino compounds having ester or carbonyl functions.[23] The reaction of CSI with ethyl 3-oxo-2-(arylhydrazono)butanoates gives thiazolotriazinediones (eq 24).[24]

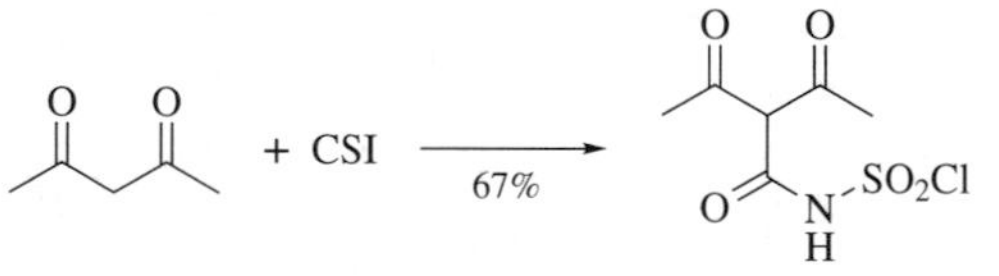

$$+ CSI \xrightarrow{67\%} \qquad (22)$$

$$\text{CO}_2\text{R} + CSI \xrightarrow{75\%} \qquad (23)$$

CSI 105–110 °C; CSI 0–5 °C (24)

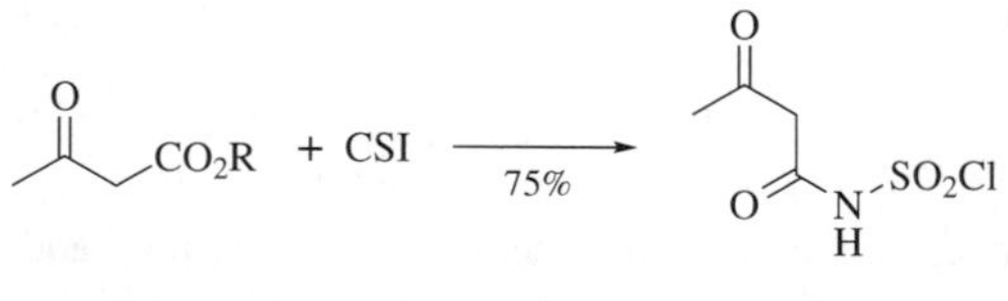

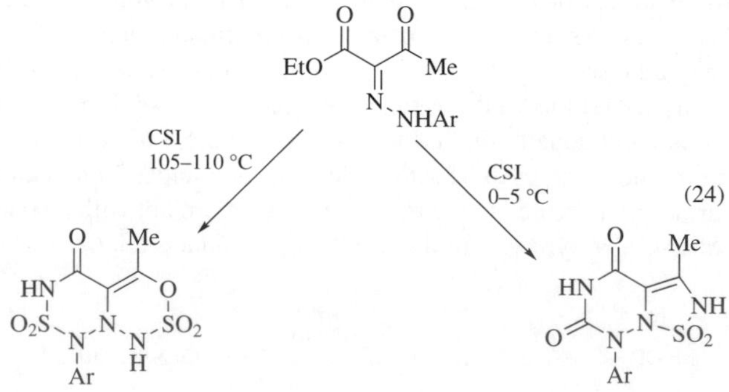

CSI has been used as a dipolar synthon of the type shown in eq 25 in its reaction with cyanohydrins. The final products of this one-pot reaction are 5,5-disubstituted 2,4-oxazolidinediones.[25]

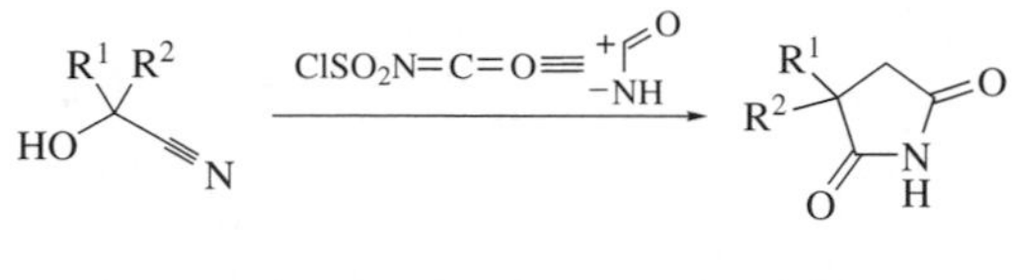

(25)

The reagent has been used to introduce the nitrile functionality in cyclic enamides (eq 26)[26] and in some other electron-rich alkenes (eqs 27 and 28).[27,28]

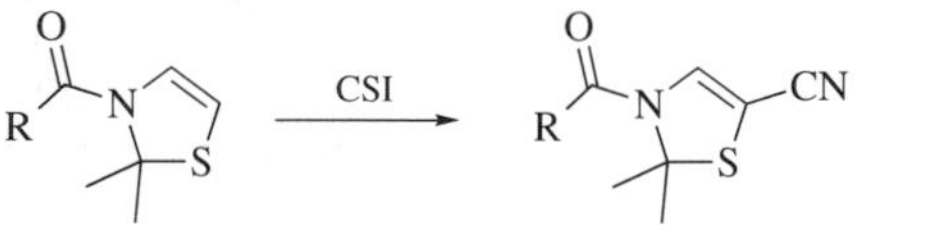

(26)

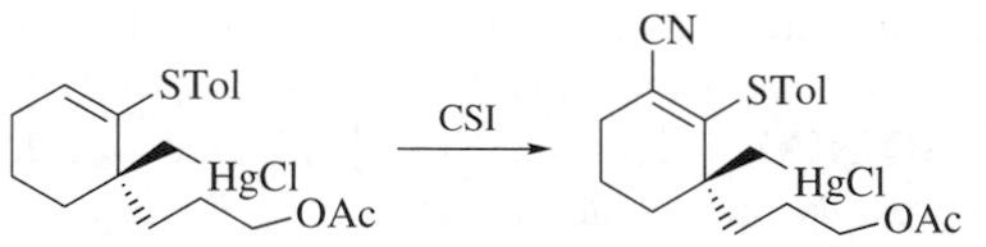

(27)

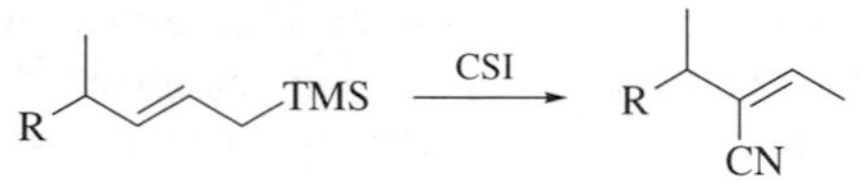

(28)

An attempt to use CSI as a dienophile in its reaction with 2-vinylindole was not successful (eq 29).[29] The product obtained was the indole-3-carboxamide, as reported earlier.[30] Similar electrophilic substitution on pyrroles[31] and thiophene[32] using CSI in the presence of DMF has furnished the respective cyanation products.

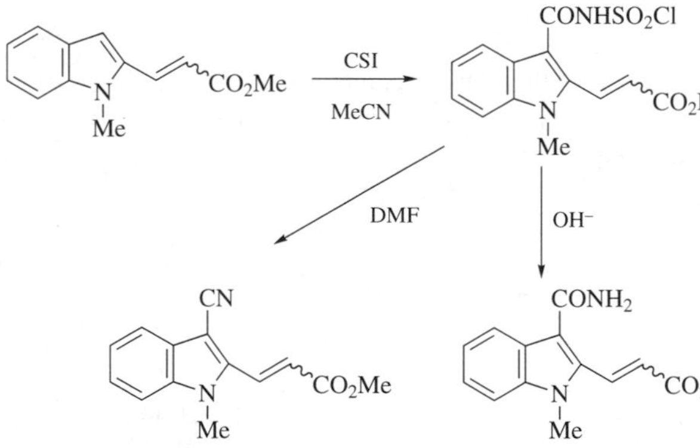

(29)

Brief treatment of CSI with epoxides at low temperature has provided the corresponding 1,3-dioxolan-2-ones (major products) and oxazolidine-2-one derivatives (eq 30).[33] This reaction is both regio- and stereospecific. Aziridines have also provided analogous products from their reaction with CSI.[33a] 1,2-Diols having substituents with good migratory aptitudes gave the rearrangement products, but unsubstituted or alkyl substituted 1,2-diols gave the corresponding carbamates as the major product (eq 31).[34]

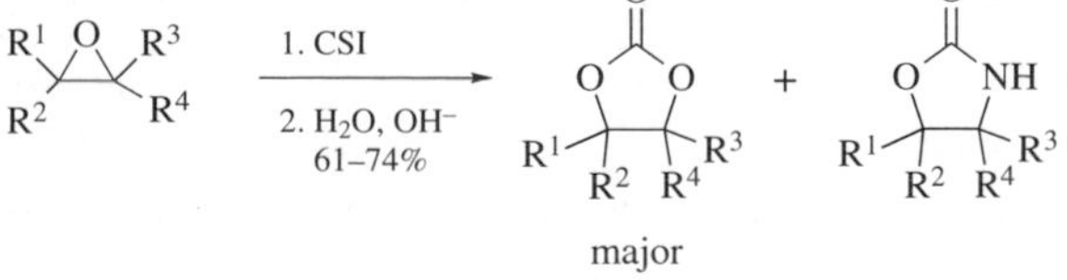

(30)

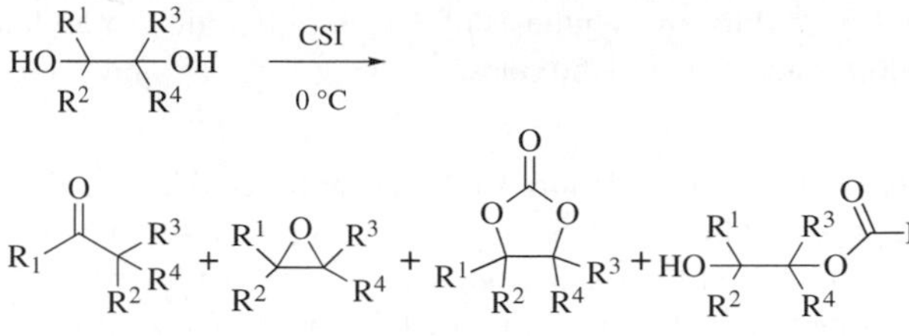

(31)

CSI also has been used as a mild chlorinating agent in its reaction with 6-aryl-3(2*H*)-pyridazinones to produce 6-aryl-3-chloropyridazines (eq 32).[35] 1-Trimethylsilylalkynes react with CSI to produce primary 2-alkynamides after hydrolysis (eq 33).[36] CSI reacts with nitrones derived from cyclic conjugated ketones to produce enamides (eq 34),[37] and with substituted nitrones to produce amides (eq 35).[38] Similar reactions with substituted 3,4-dihydro-2*H*-pyrrole 1-oxide produce 2*H*-pyrroles (eq 36).[39]

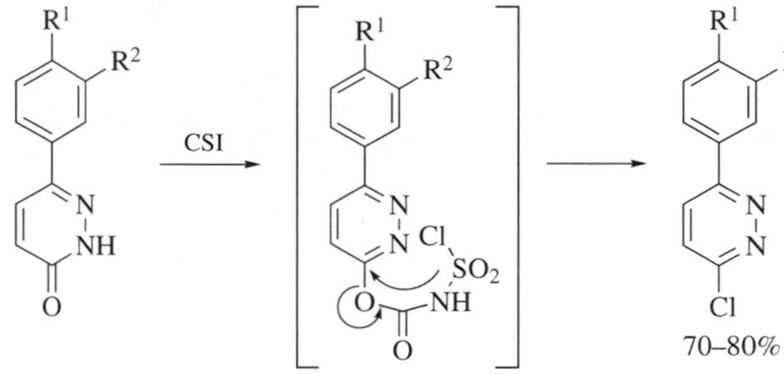

(32)

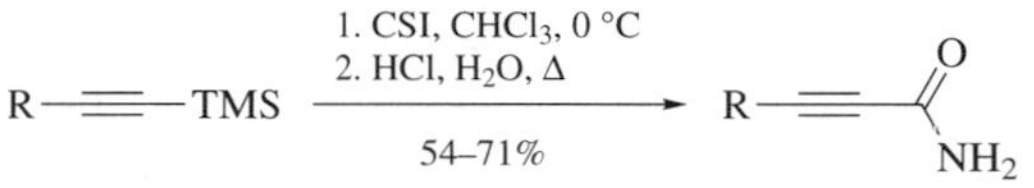

(33)

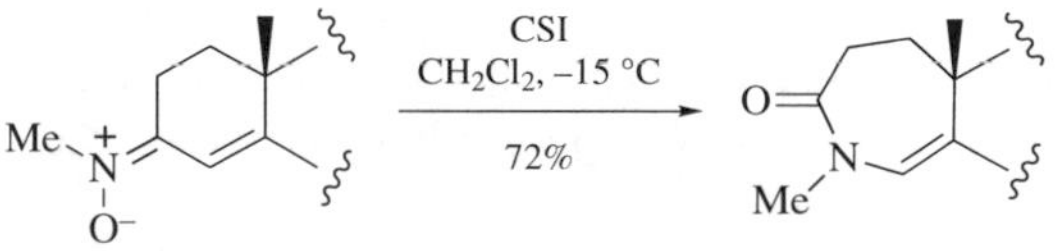

(34)

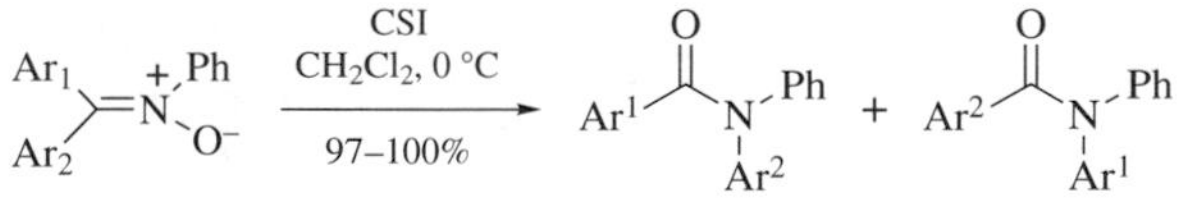

(35)

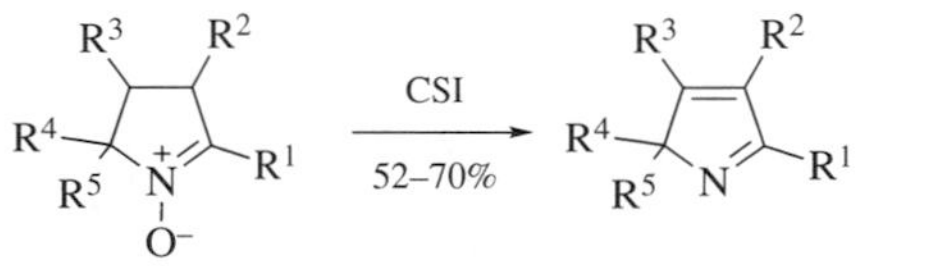

(36)

Recently, a facile denitrosation of *N*-nitrosoamines has been demonstrated by Dhar et al. (eq 37).[40] The reactivity of CSI towards sulfoxides[41] has been used for the reduction of sulfoxides to sulfides in the presence of ***Sodium Iodide*** (eq 38).[41c]

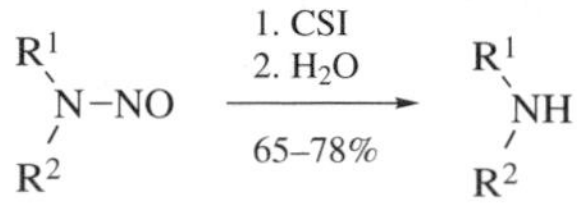

(37)

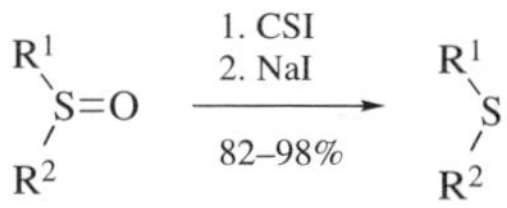

(38)

The new reagent *N*-carbo(trimethylsilyloxy)sulfamoyl chloride, obtained from CSI as a substitute for sulfamoyl chloride, has been used for the synthesis of 3-amino-4-*N*-alkyl-5-aryloxy-1,2,4,6-thiatriazine-1,1-dioxides.[42]

Class II: Cycloaddition to Isocyanate C=N. The formal [2 + 2] cycloaddition of CSI to a variety of alkenes to produce β-lactams has been the most thoroughly studied reaction (eq 39).[43] A competing elimination reaction (path b) forms an alkene byproduct. The ratio of β-lactam to alkene is determined by the pattern and type of substitution on the alkene. Both concerted and nonconcerted 1,4-dipolar mechanisms have been proposed for these reactions. Regiochemistry is dictated by the formation of the most stable carbocation, and the *cis* adduct is generally formed. Several examples are given in review articles,[1] which show the diversity of β-lactams available from this route. An improved procedure for β-lactam formation from volatile alkenes and CSI has been published.[44]

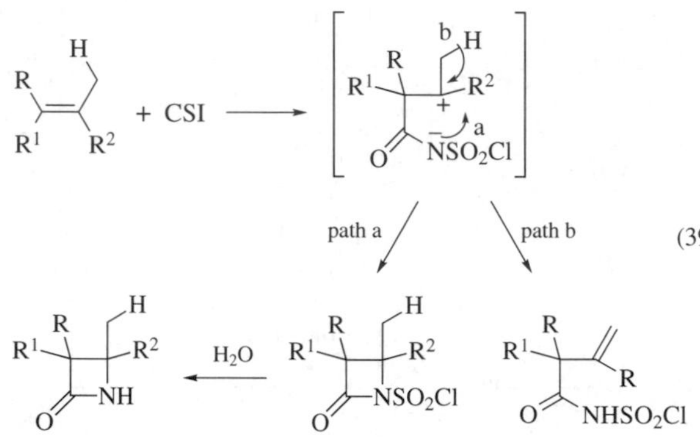

(39)

Reaction of CSI with conjugated or nonconjugated dienes also leads to 2-azetidinones after hydrolysis of the *N*-chlorosulfonyl group (eqs 40 and 41).[45]

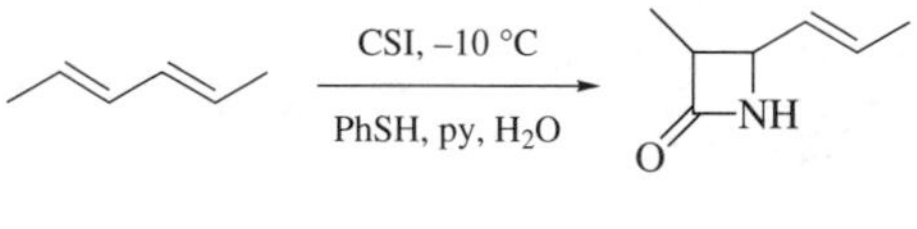

(40)

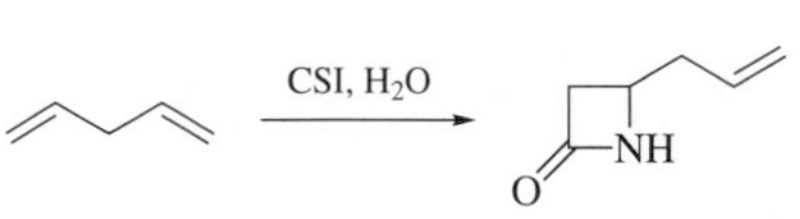

(41)

The reaction of CSI with vinyl acetates leads to 4-acetoxy-2-azetidinones upon hydrolysis (eq 42).[46] These β-lactams are ideal building blocks for synthesis of a wide variety of classes of antibiotics. The resulting β-lactams are also convenient precursors to *erythro*- and *threo*-amino acids.[47]

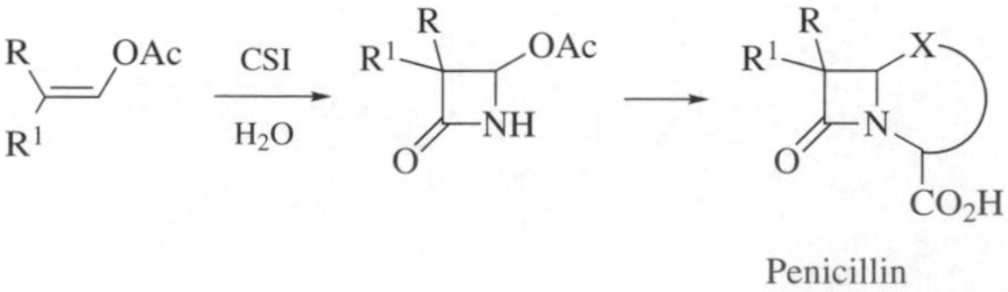

(42)

Penicillin
Cephalosporin
Oxacephem

CSI reacts with functionalized allenes to provide α-alkylidene β-lactams (eq 43).[48]

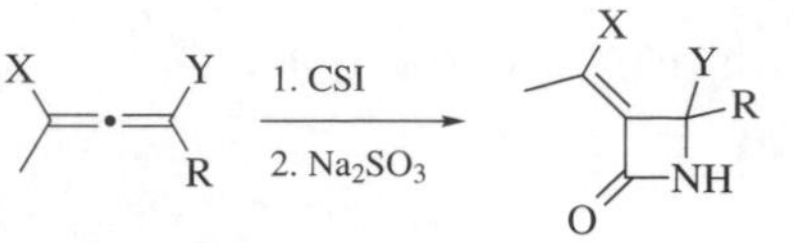

(43)

X	Y	R	Yield (%)
Me	SAr	Me	55[48d]
Me	SAr	TMS	87[48d]
Me	SAr	$(CH_2)OTBDMS$	66[48f]
Me	OAc	H	22[48e]
$Cl_2P(O)$	Me	Me	93[48g]

Alternative acyl and sulfonyl isocyanates that are preparatively useful for β-lactam synthesis include $Cl_3CH_2OSO_2NCO$, $Cl_3CH_2SO_2NCO$, and CF_3CONCO.[49] Methods for deprotection to the core β-lactam are described.

CSI has been useful as a mechanistic probe to study uniparticulate electrophilic addition to a number of fluxional molecules such as bullvalene, barrelene, and homobarrelene.[50]

CSI yields 2:1 adducts with Schiff's bases to form *s*-triazinediones (eq 44).[51]

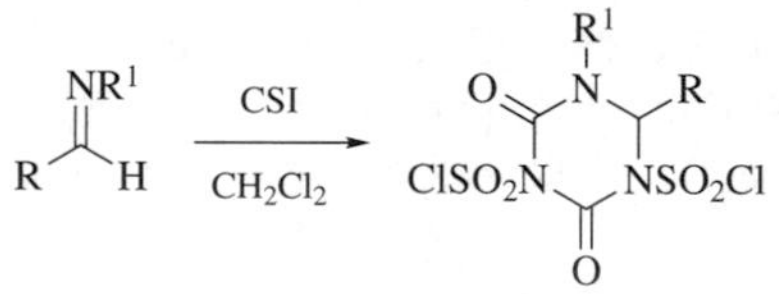

(44)

As a result of their large degree of p character, highly strained carbon–carbon single bonds can undergo formal cycloaddition with CSI (eq 45).[52]

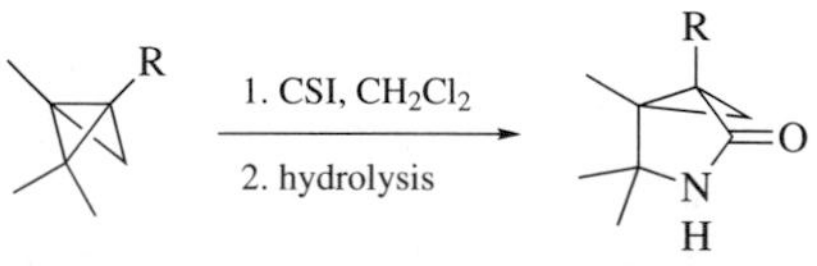

(45)

R = H, 43%
R = Me, 51%

CSI reacts with most alkynes to give six-membered heterocyclic ring structures. Hydrolysis affords ketones (eq 46), while methanolysis affords β-keto esters. Interestingly, only the alkynic portion of 1-octene-4-yne reacts with CSI.[53]

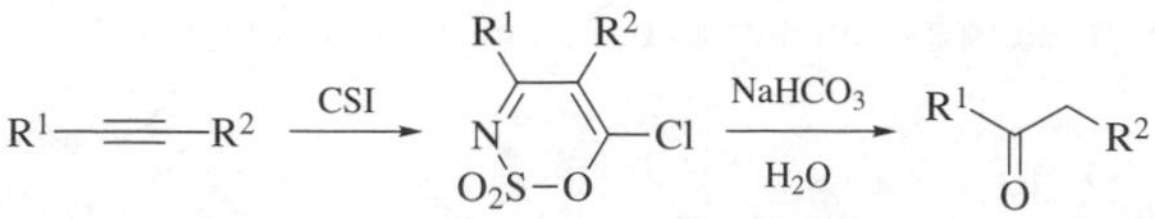

(46)

Azirines react with CSI to give three products derived from [2 + 2 + 2] cycloadducts (eq 47).[54]

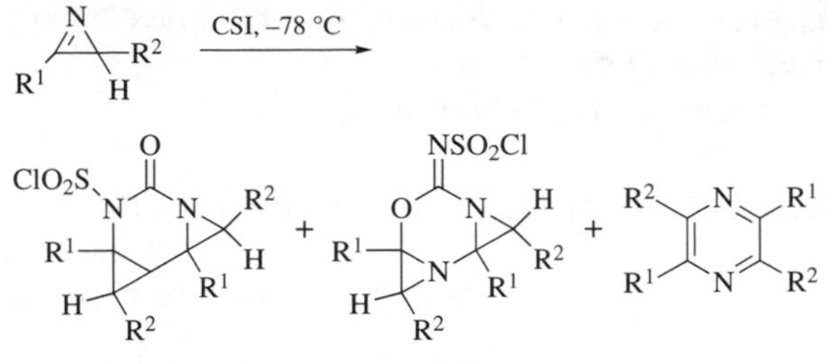

$R^1 = Ph, R^2 = H$ (47): 5 3.4%, 34.2%, 8.5%

Amidines react with CSI to form mesionic compounds with an interesting structure (eq 48).[55]

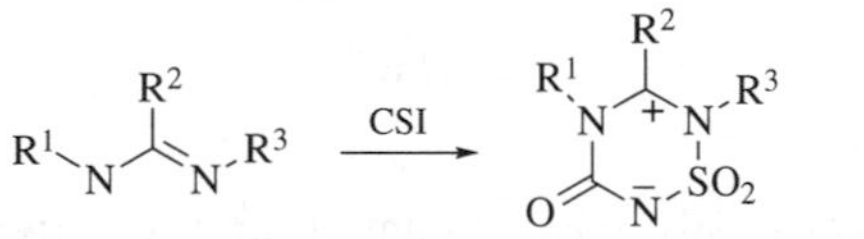

(48)

Regioselective 1,3 dipolar cycloaddition of nitrile oxides with CSI gives 3-aryl-substituted 1,2,4-oxadiazolin-5-ones after hydrolysis (eq 49).[56]

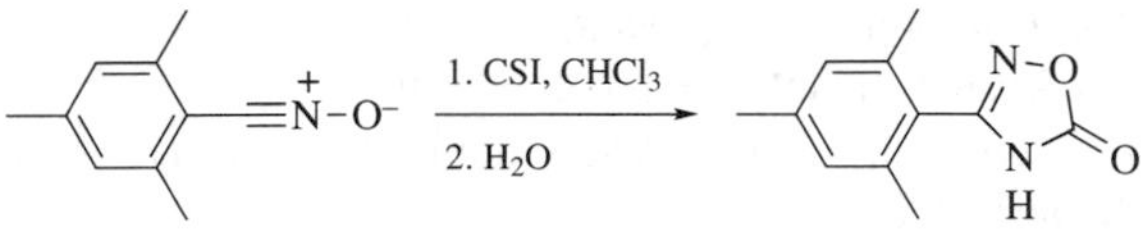

(49)

Transition metal 2-alkenyl and 2-alkynyl complexes undergo [3 + 2] cycloadditions with CSI, which lead to pyrrolidones and pyrrolinones respectively with migration of the metal atom (eqs 50 and 51).[57]

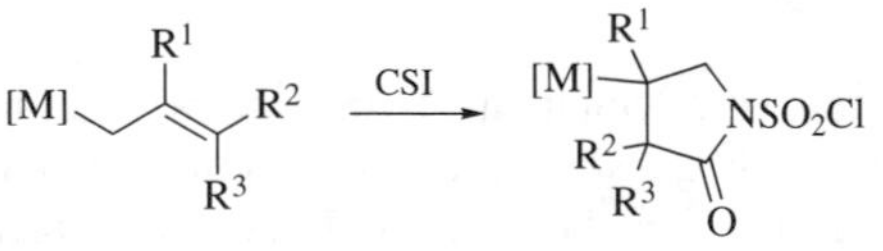

(50)

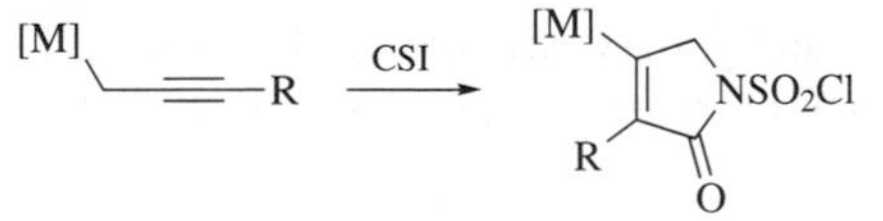

(51)

[M] = $CpFe(CO)_2$, $CpMo(CO)_3$, $Mn(CO)_5$, $Pt(PR_3)_2Cl$

Type III Reactions. Type III reactions involve compounds which react with the chlorosulfonyl portion of CSI. These include compounds which are unreactive towards the isocyanate group (eqs 52 and 53)[58,59] or which react with the chlorosulfonyl portion under special experimental conditions such as high temperature (eq 54)[60] or radical conditions (eq 55).[61]

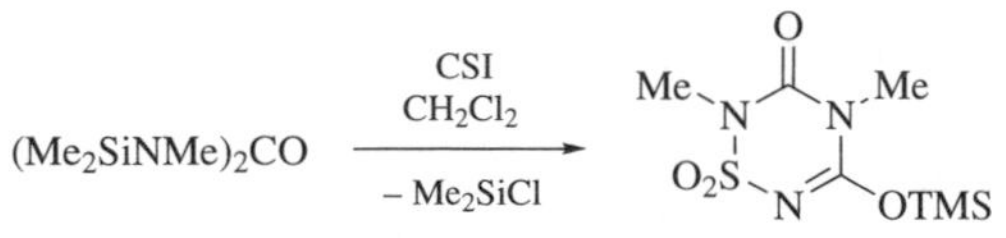

$$(Me_2SiNMe)_2CO \xrightarrow[-\,Me_2SiCl]{CSI,\ CH_2Cl_2} \quad (52)$$

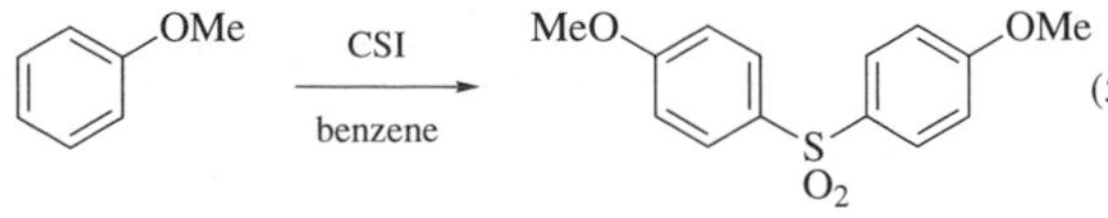

(53)

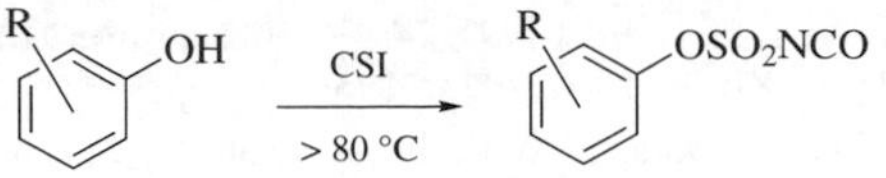

(54)

R = H, 3-Cl, 4-Cl, 4-Me, 4-MeO, 4-CN

$$RHC{=}CH_2 \xrightarrow[X\bullet \text{ or } h\nu]{CSI} ClCH\text{–}R\text{–}CH_2SO_2NCO \quad (55)$$

R = H, Me, Et, Bu, $ClCH_2$, Cl_2CHCH_2

By changing the reaction conditions in eq 55, a dramatically different product is obtained (eq 56).[1a]

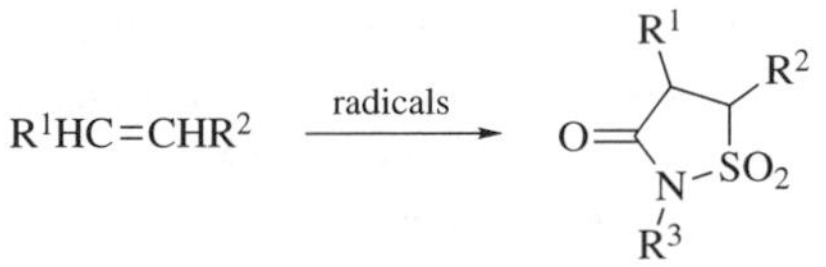

$$R^1HC{=}CHR^2 \xrightarrow{\text{radicals}} \quad (56)$$

Related Reagents. Phenyl Isothiocyanate.

1. (a) Rasmussen, J. K.; Hassner, A. *CRV* **1976**, *76*, 389. (b) Graf, R. *AG(E)* **1968**, *7*, 172. (c) Szabo, W. A. *Aldrichim. Acta* **1977**, *10*, 116. (d) Dhar, D. N.; Keshava Murthy, K. S. *S* **1986**, 437.
2. Graf, R. *OSC* **1973**, *5*, 226.
3. (a) Christensen, B. G.; Cama, L. D.; Kern, J. A. Ger. Offen. 2 264 651, 1974 (*CA* **1974**, *81*, 120 653j). (b) Tanino, H.; Nakata, T.; Kaneko, T.; Kishi, Y. *JACS* **1977**, *99*, 2818.
4. Burgess, E. M.; Penton, H. R. Jr.; Taylor, E. A. *JOC* **1973**, *38*, 26.
5. (a) *FF* **1974**, *4*, 331; (b) Burgess, E. M.; Williams, W. M. *JOC* **1973**, *38*, 1249.
6. Chiyomaru, I.; Ishihara, E.; Takita, K. Jpn. Kokai 75 94 129, 1975 (*CA* **1976**, *84*, 1223w).
7. Lattrell, R.; Lohaus, G. *CB* **1972**, *105*, 2800.
8. Hofmann, H.; Wagner, R.; Uhl, J. *CB* **1971**, *104*, 2134.
9. Sheers, E. H. US Patent 3 113 857, 1963 (*CA* **1964**, *60*, 5395b).
10. (a) Behrend, E.; Hass, M. *CZ* **1971**, *95*, 1009. (b) Appel, R.; Montenarh, M. *CB* **1974**, *107*, 706. (c) Roesky, H. W.; Tutkunkardes, S. *CB* **1974**, *107*, 508. (d) Roesky, W.; Janssen, E. *ZN(B)* **1974**, *29*, 174.
11. Graf, R.; Guenther, D.; Jensen, H.; Matterstock, K. Ger. Offen. 1 144 718, 1963 (*CA* **1963**, *59*, 6368c).
12. Beutel, P.; Adolphi, H.; Kiehs, K. Ger. Offen. 2 249 939, 1974 (*CA* **1974**, *81*, 13 108p).
13. Clauss, K.; Friedrich, H.-J.; Jensen, H. *LA* **1974**, 561.
14. Hall, R. H.; Jordaan, A.; Lourens, G. J. *JCS(P1)* **1973**, 38.
15. Lohaus, G. *CB* **1967**, *106*, 2719.
16. Faubl, H. *JOC* **1976**, *41*, 3048.
17. Fechtig, B.; Kocsis, K.; Bickel, H. Ger. Offen. 2 312 330, 1973 (*CA* **1974**, *80*, 3511e).
18. Lohaus, G. *OS* **1970**, *50*, 52.
19. Daniel, J.; Dhar, D. N. *T* **1992**, *22*, 4551.
20. Rasmussen, J. K.; Hassner, A. *S* **1973**, 682.
21. Hassner, A.; Rasmussen, J. K. *JACS* **1975**, *97*, 1451 and Ref. 6 cited therein.
22. Van Allan, J. A.; Chang, S. C.; Reynolds, G. A. *JHC* **1974**, *11*, 195.
23. (a) Kamal, A.; Sattur, P. B. *H* **1987**, *262*, 1057. (b) Kamal, A.; Rao, K. R.; Sattur, P. B. *S* **1985**, *10*, 729. (c) Heinrich, W.; Muntaz, E. *S* **1985**, 190.
24. Daniel, J.; Dhar, D. N. *H* **1991**, *32*, 1517.
25. (a) Mendez, J. C.; Villacampa, M.; Sollhuber, M. M. *H* **1991**, 469. (b) Garcia, M. V.; Mendez, J. C.; Villacampa, M.; Sollhuber, M. M. *H* **1991**, *9*, 697.

26. (a) Vorbruggen, H. *HCA* **1991**, *74*, 297. (b) Natsume, M.; Kumadaki, S.; Kanda, Y.; Kiuchi, K. *TL* **1973**, 2335.
27. Takemoto, Y.; Ohra, T.; Yonetoku, Y.; Imanishi, T.; Iwata, C. *CC* **1992**, 192.
28. Nativi, C.; Palio, G.; Taddeei, M. *TL* **1991**, 1583.
29. Pindur, U.; Kim, M.-H. *T* **1989**, 6427.
30. Mehta, C.; Dhar, D. N.; Suri, S. C. *S* **1978**, 374.
31. Loader, C. E.; Anderson, H. J. *CJC* **1981**, *59*, 2673.
32. Gronowitz, S.; Liljefors, S. *ACS* **1977**, *B31*, 771.
33. (a) Lorincz, T.; Erden, T. *SC* **1986**, *16*, 123. (b) Keshavamurthy, K. S.; Dhar, D. N. *SC* **1984**, *14*, 687.
34. Joseph, S. P.; Dhar, D. N. *SC* **1988**, *18*, 2295.
35. Srinivasan, T. N.; Rao, K. R.; Sattur, P. B. *SC* **1986**, *16*, 543.
36. Page, P. C. B.; Rosenthal, S.; Williams, R. V. *S* **1988**, 621.
37. Joseph, S. P.; Dhar, D. N. *T* **1988**, *44*, 5209.
38. Joseph, S. P.; Dhar, D. N. *T* **1986**, *42*, 5979.
39. Joseph, S. P.; Dhar, D. N. *SC* **1988**, *18*, 1743.
40. Dhar, D. N.; Bag, A. K. *IJC(B)* **1983**, *22*, 600.
41. (a) Graf, R. *AG(E)* **1968**, *7*, 172. (b) Olah, G. A.; Vankar, Y. D.; Arvanaghi, M. *S* **1980**, 141. (c) Keshava Murthy, K. S.; Vankar, Y. D.; Dhar, D. N. *IJC(B)* **1983**, *22*, 504.
42. Durham, P. J.; Galemmo, R. A. *TL* **1986**, *27*, 123.
43. (a) Ulrich, H. *Cycloaddition Reactions of Heterocumulenes*; Academic: New York, 1967, pp 135–141. (b) Clauss, K. *LA* **1969**, *722*, 110. (c) Isaacs, N. S. *CSR* **1976**, *5*, 181. (d) Graf, R. *LA* **1963**, *661*, 111. (e) Moriconi, E. J. *Mechanisms of Reactions of Sulfur Compounds*; Intra-Science Research Foundation: Santa Monica, CA, 1968; Vol. 3, p 131. (f) Moriconi, E. J.; Meyer, W. C. *JOC* **1971**, *36*, 2841. (g) Woodward, R. B.; Hoffmann, R. *AG(E)* **1969**, *8*, 781.
44. Hauser, F. M.; Ellenberger, S. R. *S* **1987**, 324.
45. (a) Moriconi, E. J.; Meyer, W. C. *JOC* **1971**, *36*, 2841. (b) Baxter, A. J. G.; Dickinson, K. H. *CC* **1979**, 236.
46. Mickel, S. J.; Hsiao, C. N.; Miller, M. J. *OS* **1987**, *65*, 135 and references therein.
47. Meyers, A. I. *Heterocycles in Organic Synthesis*; Wiley: New York, 1974; pp 285–286.
48. (a) Moriconi, E. J.; Kelly, J. F. *JOC* **1968**, *33*, 3036. (b) Gompper, R.; Lach, D. *TL* **1973**, 2683. (c) Poutsma, M. L.; Ibarbia, P. A. *JACS* **1971**, *93*, 440. (d) Buynak, J. D.; Rao, M. N.; Chandrasekaran, R. Y.; Haley, E. *TL* **1985**, *26*, 5001. (e) Buynak, J. D.; Rao, M. N.; Pajouhesh, H.; Chandrasekaran, R. Y.; Finn, K.; deMeester, P.; Chu, S. L. *JOC* **1985**, *50*, 4245. (f) Buynak, J. D.; Rao, M. N. *JOC* **1986**, *51*, 1571. (g) Mondeshka, D. Parashikov, V.; Angelov, C. *SC* **1989**, 3113.
49. Barrett, A. G. M.; Betts, M. J.; Fenwick, A. *JOC* **1985**, *50*, 169.
50. (a) Paquette, L. A.; Kirschner, S.; Malpass, J. R. *JACS* **1970**, *92*, 4330. (b) Paquette, L. A.; Volz, W. E. *JACS* **1976**, *98*, 2910. (c) Volz, W. E.; Paquette, L. A. *JOC* **1976**, *41*, 57.
51. Walrond, R. E.; Suschitzky, H. *CC* **1973**, 570.
52. (a) Volz, W. E.; Paquette, L. A.; Rogido, R. J.; Barton, T. J. *CI(L)* **1974**, 771. (b) Paquette, L. A.; Allen, G. R., Jr.; Broadhurst, M. J. *JACS* **1971**, *93*, 4503. (c) Paquette, L. A.; Lau, C. J.; Rogers, R. D. *JACS* **1988**, *110*, 2592.
53. Moriconi, E. J.; Shimakawa, Y. *JOC* **1972**, *37*, 196.
54. Daniel, J.; Dhar, D. N. *SC* **1991**, *21*, 1649.
55. Friedrichsen, W.; Mockel, G.; Debaerdemaecher, T. *H* **1984**, *22*, 63.
56. Rao, K. R.; Scrinivasan, T. N.; Sattur, P. B. *H* **1988**, *27*, 683.
57. (a) Yamamoto, Y.; Wojcicki, A. *Inorg. Nucl. Chem. Lett.* **1972**, *8*, 833. (b) *CC* **1972**, 1088. (c) Yamamoto, Y.; Wojcicki, A. *IC* **1973**, *12*, 1779. (d) Giering, W. P.; Raghu, S.; Rosenblum, M.; Cutler, A.; Ehntholt, D.; Fish, R. W. *JACS* **1972**, *94*, 8251. (e) Hu, Y.; Wojcicki, A.; Calligaris, M.; Nardin, G. *OM* **1987**, *6*, 1561.
58. Roesky, H. W.; Zamankhan, H. *CB* **1976**, *109*, 2107.
59. Effenberger, R.; Glitter, R.; Heider, L.; Neiss, R. *CB* **1968**, *101*, 502.
60. Lohaus, G. *CB* **1972**, *105*, 2791.
61. Gunther, P.; Slodan, F. *CB* **1970**, *103*, 663.

Marvin J. Miller, Manuka Ghosh, Peter R. Guzzo, Paul F. Vogt & Jingdan Hu

University of Notre Dame, IN, USA

Chlorotrimethylsilane[1,2]

$ClSiMe_3$

[75-77-4] C_3H_9ClSi (MW 108.64)

(protection of silyl ethers,[3] transients,[5–7] and silylalkynes;[8] synthesis of silyl esters,[4] silyl enol ethers,[9,10] vinylsilanes,[13] and silylvinylallenes;[15] Boc deprotection;[11] TMSI generation;[12] epoxide cleavage;[14] conjugate addition reactions catalyst[16–18])

Alternate Names: trimethylsilyl chloride; TMSCl.
Physical Data: bp 57 °C; *d* 0.856 g cm^{-3}.
Solubility: sol THF, DMF, CH_2Cl_2, HMPA.
Form Supplied in: clear, colorless liquid; 98% purity; commercially available.
Analysis of Reagent Purity: bp, NMR.
Purification: distillation over calcium hydride with exclusion of moisture.
Handling, Storage, and Precautions: moisture sensitive and corrosive; store under an inert atmosphere; use in a fume hood.

Protection of Alcohols as TMS Ethers. The most common method of forming a silyl ether involves the use of TMSCl and a base (eqs 1–3).[3,19–22] Mixtures of TMSCl and ***Hexamethyldisilazane*** (HMDS) have also been used to form TMS ethers. Primary, secondary, and tertiary alcohols can be silylated in this manner, depending on the relative amounts of TMS and HMDS (eqs 4–6).[23]

TMSCl, Et_3N, CH_2Cl_2, 88% (1)

TMSCl, Li_2S, MeCN, 90% (2)

TMSCl, K_2CO_3, Aliquat 336, 88% (3)

HMDS (1.2 equiv), TMSCl (0.1 equiv) (4)

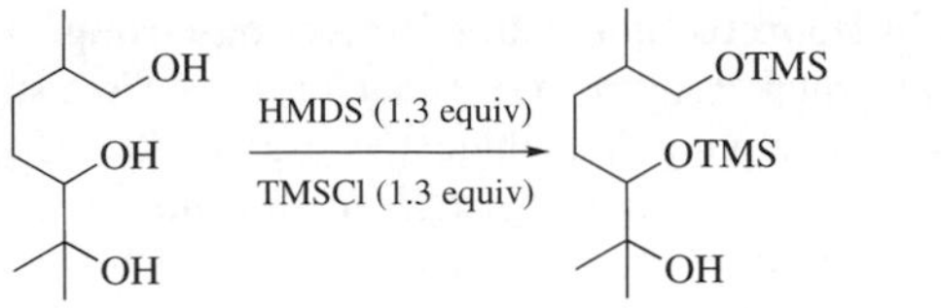

(5)

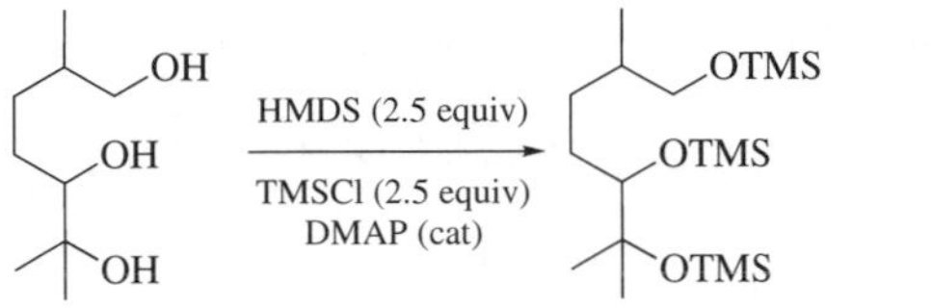

(6)

Trimethysilyl ethers can be easily removed under a variety of conditions,[19] including the use of ***Tetra-n-butylammonium Fluoride*** (TBAF) (eq 7),[20] citric acid (eq 8),[24] or ***Potassium Carbonate*** in methanol (eq 9).[25] Recently, resins (OH^- and H^+ form) have been used to remove phenolic or alcoholic TMS ethers selectively (eq 10).[26]

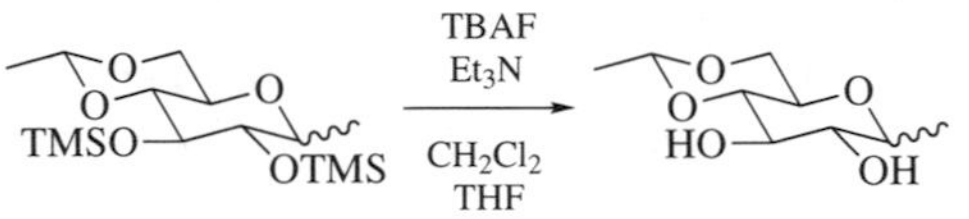

(7)

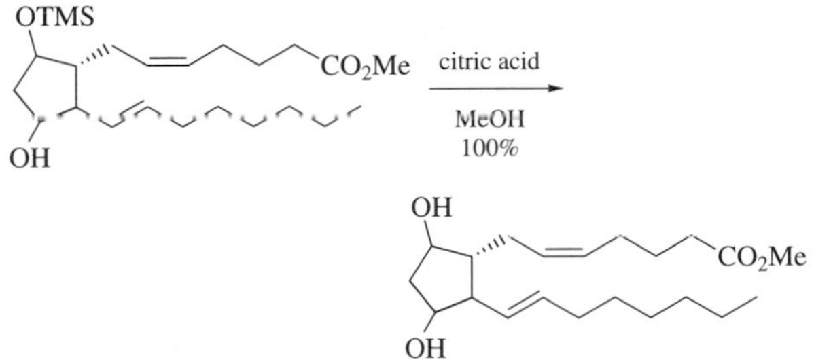

(8)

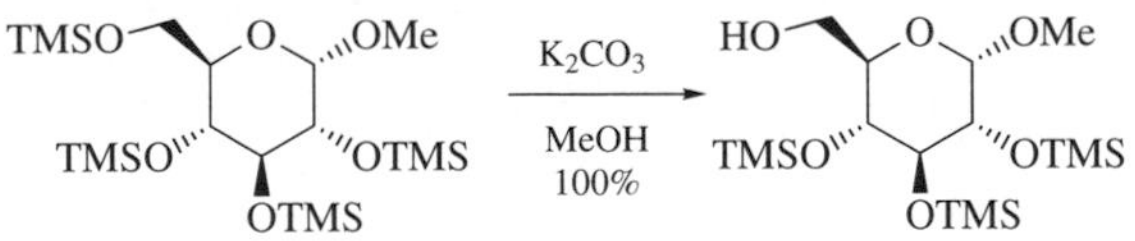

(9)

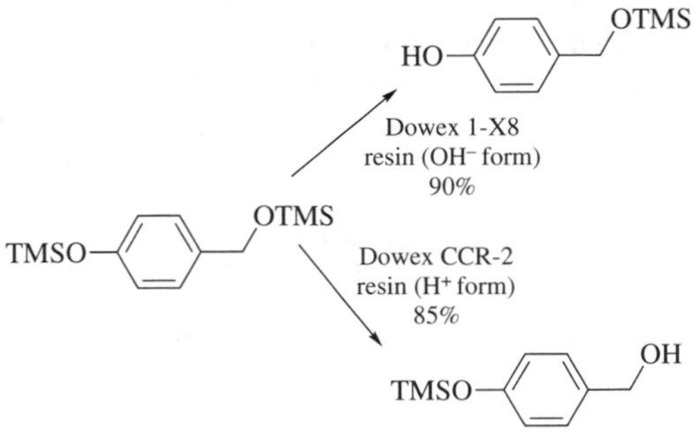

(10)

Transient Protection. Silyl ethers can be used for the transient protection of alcohols (eq 11).[27] In this example the hydroxyl groups were silylated to allow tritylation with concomitant desilylation during aqueous workup. The ease of introduction and removal of TMS groups make them well suited for temporary protection.

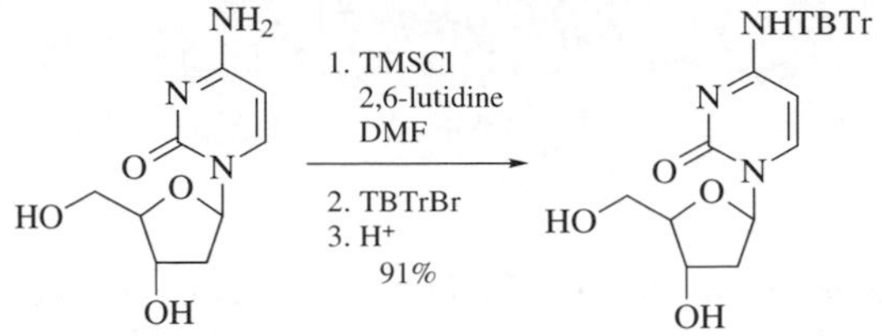

(11)

Trimethylsilyl derivatives of amino acids and peptides have been used to improve solubility, protect carboxyl groups, and improve acylation reactions. TMSCl has been used to prepare protected amino acids by forming the *O,N*-bis-trimethylsilylated amino acid, formed in situ, followed by addition of the acylating agent (eq 12).[5] This is a general method which obviates the production of oligomers normally formed using Schotten–Baumann conditions, and which can be applied to a variety of protecting groups.[5]

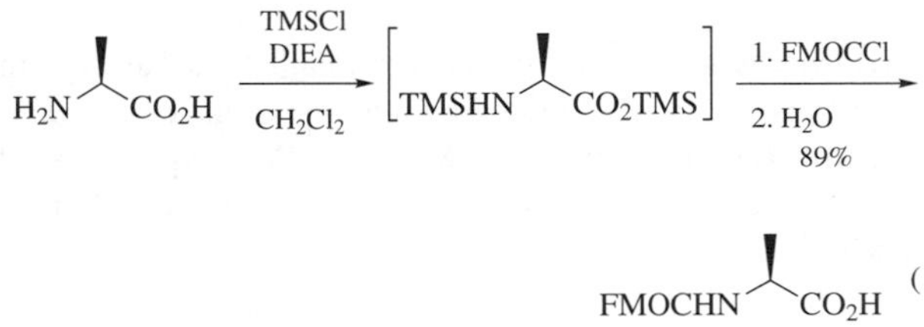

(12)

Transient hydroxylamine oxygen protection has been successfully used for the synthesis of *N*-hydroxamides.[6] Hydroxylamines can be silylated with TMSCl in pyridine to yield the *N*-substituted *O*-TMS derivative. Acylation with a mixed anhydride of a protected amino acid followed by workup affords the *N*-substituted hydroxamide (eq 13).[6]

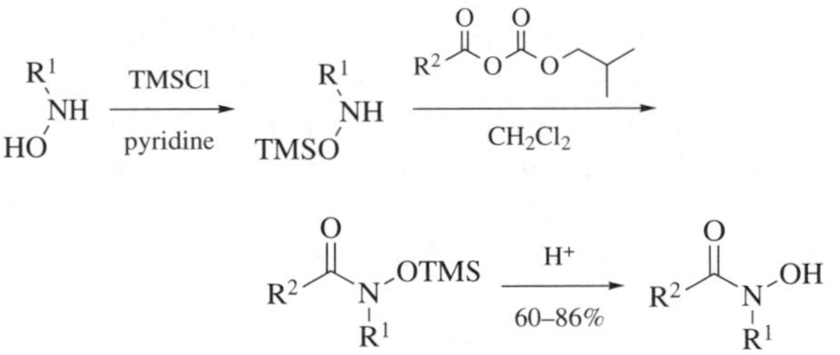

(13)

Formation of Silyl Esters. TMS esters can be prepared in good yields by reacting the carboxylic acid with TMSCl in 1,2-dichloroethane (eq 14).[4] This method of carboxyl group protection has been used during hydroboration reactions. The organoborane can be transformed into a variety of different carboxylic acid derivatives (eqs 15 and 16).[7] TMS esters can also be reduced with metal hydrides to form alcohols and aldehydes or hydrolyzed to the starting acid, depending on the reducing agent and reaction conditions.[28]

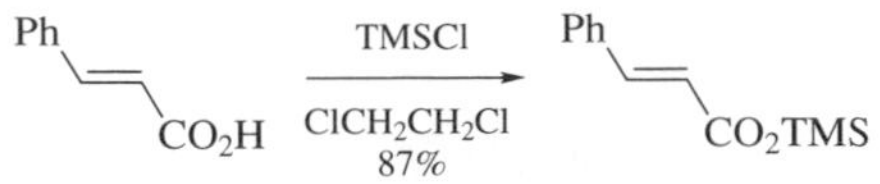

(14)

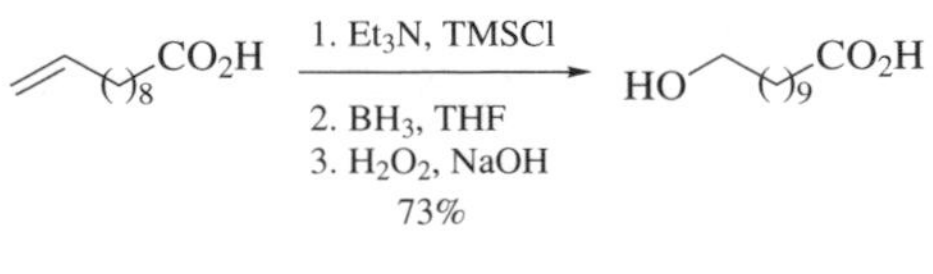

(15)

$$\text{CH}_2\text{=CH(CH}_2)_8\text{CO}_2\text{H} \xrightarrow[\text{2. BH}_3\text{, THF; 3. NaI; 79\%}]{\text{1. Et}_3\text{N, TMSCl}} \text{I(CH}_2)_9\text{CO}_2\text{H} \quad (16)$$

Protection of Terminal Alkynes. Terminal alkynes can be protected as TMS alkynes by reaction with ***n-Butyllithium*** in THF followed by TMSCl (eq 17).[8] A one-pot β-elimination–silylation process (eq 18) can also yield the protected alkyne.

$$\text{H–C≡C–O-}t\text{-Bu} \xrightarrow[\text{2. TMSCl; 80\%}]{\text{1. } n\text{-BuLi, THF}} \text{TMS–C≡C–O-}t\text{-Bu} \quad (17)$$

$$\text{BrHC=CH–O-}t\text{-Bu} \xrightarrow[\text{2. TMSCl; 74\%}]{\text{1. LDA, THF}} \text{TMS–C≡C–O-}t\text{-Bu} \quad (18)$$

Silyl Enol Ethers. TMS enol ethers of aldehydes and symmetrical ketones are usually formed by reaction of the carbonyl compound with ***Triethylamine*** and TMSCl in DMF (eq 19), but other bases have been used, including ***Sodium Hydride***[29] and ***Potassium Hydride***.[30]

$$\text{4-}t\text{-Bu-cyclohexanone} \xrightarrow[\text{DMF; 93\%}]{\text{TMSCl, Et}_3\text{N}} \text{4-}t\text{-Bu-1-OTMS-cyclohexene} \quad (19)$$

Under the conditions used for the generation of silyl enol ethers of symmetrical ketones, unsymmetrical ketones give mixtures of structurally isomeric enol ethers, with the predominant product being the more substituted enol ether (eq 20).[10] Highly hindered bases, such as ***Lithium Diisopropylamide*** (LDA),[31] favor formation of the kinetic, less substituted silyl enol ether, whereas ***Bromomagnesium Diisopropylamide*** (BMDA)[10] generates the more substituted, thermodynamic silyl enol ether. A combination of TMSCl/***Sodium Iodide*** has also been used to form silyl enol ethers of simple aldehydes and ketones[32] as well as from α,β-unsaturated aldehydes and ketones.[33] Additionally, treatment of α-halo ketones with ***Zinc***, TMSCl, and TMEDA in ether provides a regiospecific method for the preparation of the more substituted enol ether (eq 21).[34]

$$\text{2-methylcyclohexanone} \longrightarrow \text{OTMS enol ether (A)} + \text{OTMS enol ether (B)} \quad (20)$$

Reagents	Ratio (A):(B)
LDA, DME; TMSCl	1:99
NaH, DME; TMSCl	73:27
Et_3N, TMSCl, DMF	78:22
KH, THF; TMSCl	67:33
TMSCl, NaI, MeCN, Et_3N	90:10
BMDA, TMSCl, Et_3N	97:3

$$\text{2-chloro-2-methylcyclohexanone} \xrightarrow[\text{TMEDA, Et}_2\text{O; 85\%}]{\text{Zn, TMSCl}} \text{1-OTMS-2-methylcyclohexene} \quad (21)$$

Mild Deprotection of Boc Protecting Group. The Boc protecting group is used throughout peptide chemistry. Common ways of removing it include the use of 50% ***Trifluoroacetic Acid*** in CH_2Cl_2, ***Trimethylsilyl Perchlorate***, or ***Iodotrimethylsilane*** (TMSI).[19] A new method has been developed, using TMSCl–phenol, which enables removal of the Boc group in less than one hour (eq 22).[11] The selectivity between Boc and benzyl groups is high enough to allow for selective deprotection.

$$\text{Boc-Val-OCH}_2\text{-resin} \xrightarrow[\text{20 min; 100\%}]{\text{TMSCl, phenol}} \text{Val-OCH}_2\text{-resin} \quad (22)$$

In Situ Generation of Iodotrimethylsilane. Of the published methods used to form TMSI in situ, the most convenient involves the use of TMSCl with NaI in acetonitrile.[12] This method has been used for a variety of synthetic transformations, including cleavage of phosphonate esters (eq 23),[35] conversion of vicinal diols to alkenes (eq 24),[36] and reductive removal of epoxides (eq 25).[37]

$$\text{Bn–P(O)(OMe)}_2 \xrightarrow[\text{MeCN; 78\%}]{\text{TMSCl, NaI}} \text{Bn–P(O)(OH)}_2 \quad (23)$$

$$\text{vicinal diol} \xrightarrow[\text{MeCN; 90\%}]{\text{TMSCl, NaI}} \text{alkene} \quad (24)$$

$$\text{cyclohexene oxide} \xrightarrow[\text{MeCN; 94\%}]{\text{TMSCl, NaI}} \text{cyclohexene} \quad (25)$$

Conversion of Ketones to Vinylsilanes. Ketones can be transformed into vinylsilanes via intermediate trapping of the vinyl anion from a Shapiro reaction with TMSCl. Formation of either the tosylhydrazone[38] or benzenesulfonylhydrazone (eq 26)[13,39] followed by reaction with *n*-butyllithium in TMEDA and TMSCl gives the desired product.

$$\text{benzocycloheptanone} \xrightarrow[\text{2. } n\text{-BuLi, TMEDA; TMSCl; 88\%}]{\text{1. PhSO}_2\text{NHNH}_2\text{, } p\text{-TsOH, EtOH}} \text{vinyl–TMS} \quad (26)$$

Epoxide Cleavage. Epoxides open by reaction with TMSCl in the presence of ***Triphenylphosphine*** or tetra-*n*-butylammonium chloride to afford *O*-protected vicinal chlorohydrins (eq 27).[14]

$$\text{cyclohexene oxide} \xrightarrow[\text{CHCl}_3\text{; 99\%}]{\text{TMSCl, PPh}_3} \text{trans-2-chloro-1-OTMS-cyclohexane} \quad (27)$$

Formation of Silylvinylallenes. Enynes couple with TMSCl in the presence of Li/ether or Mg/***Hexamethylphosphoric Triamide*** to afford silyl-substituted vinylallenes. The vinylallene can be subsequently oxidized to give the silylated cyclopentanone (eq 28).[15]

1. Li, Et_2O; 2. TMSCl, 80%; [O] (28)

Conjugate Addition Reactions. In the presence of TMSCl, cuprates undergo 1,2-addition to aldehydes and ketones to afford silyl enol ethers (eq 29).[16] In the case of a chiral aldehyde, addition of TMSCl follows typical Cram diastereofacial selectivity (eq 30).[16,40]

Bu_2CuLi, TMSCl, THF, 75% (29)

Bu_2CuLi, TMSCl, THF, 96%; 9:1 (30)

Conjugate addition of organocuprates to α,β-unsaturated carbonyl compounds, including ketones, esters, and amides, are accelerated by addition of TMSCl to provide good yields of the 1,4-addition products (eq 31).[17,41,42] The effect of additives such as HMPA, DMAP, and TMEDA have also been examined.[18,43] The role of the TMSCl on 1,2- and 1,4-addition has been explored by several groups, and a recent report has been published by Lipshutz.[40] His results appear to provide evidence that there is an interaction between the cuprate and TMSCl which influences the stereochemical outcome of these reactions.

1. $(EtCH{=}CH)_2CuLi$, TMSCl, Et_2O; 2. H_2O, 86% (31)

The addition of TMSCl has made 1,4-conjugate addition reactions to α-(nitroalkyl)enones possible despite the presence of the acidic α-nitro protons (eq 32).[44] Copper-catalyzed conjugate addition of Grignard reagents proceeds in high yield in the presence of TMSCl and HMPA (eq 33).[45] In some instances the reaction gives dramatically improved ratios of 1,4-addition to 1,2-addition.

Bu_2CuLi, TMSCl, THF, 74% (32)

MgBr, 10% CuI, TMSCl, HMPA, THF, 82% (33)

Related Reagents. Bromotrimethylsilane; Hexamethyldisilazane; Iodotrimethylsilane; Trimethylsilyl Perchlorate; Trimethylsilyl Trifluoromethanesulfonate.

---

1. Colvin, E. *Silicon in Organic Synthesis*; Butterworths: Boston, 1981.
2. Weber, W. P., *Silicon Reagents for Organic Synthesis*; Springer: New York, 1983.
3. Langer, S. H.; Connell, S.; Wender, I. *JOC* **1958**, *23*, 50.
4. Hergott, H. H.; Simchen, G. *S* **1980**, 626.
5. Bolin, D. R.; Sytwu, I.-I; Humiec, F.; Meinenhofer, J. *Int. J. Peptide Protein Res.* **1989**, *33*, 353.
6. Nakonieczna, L.; Chimiak, A. *S* **1987**, 418.
7. Kabalka, G. W.; Bierer, D. E. *SC* **1989**, *19*, 2783.
8. Valenti, E.; Pericàs, M. A.; Serratosa, F. *JOC* **1990**, *55*, 395.
9. House, H. O.; Czuba, L. J.; Gall, M.; Olmstead, H. D. *JOC* **1969**, *34*, 2324.
10. Krafft, M. E.; Holton, R. A. *TL* **1983**, *24*, 1345.
11. Kaiser, E.; Tam, J. P.; Kubiak, T. M.; Merrifield, R. B. *TL* **1988**, *29*, 303.
12. Olah, G. A.; Narang, S. C.; Gupta, B. G. B.; Malhotra, R. *JOC* **1979**, *44*, 1247.
13. Paquette, L. A.; Fristad, W. E.; Dime, D. S.; Bailey, T. R. *JOC* **1980**, *45*, 3017.
14. Andrews, G. C.; Crawford, T. C.; Contillo, L. G. *TL* **1981**, *22*, 3803.
15. Dulcere, J.-P; Grimaldi, J.; Santelli, M. *TL* **1981**, *22*, 3179.
16. Matsuzawa, S.; Isaka, M.; Nakamura, E.; Kuwajima, I. *TL* **1989**, *30*, 1975.
17. Alexakis, A.; Berlan, J.; Besace, Y. *TL* **1986**, *27*, 1047.
18. Horiguchi, Y.; Matsuzawa, S.; Nakamura, E.; Kuwajima, I. *TL* **1986**, *27*, 4025.
19. Green, T. W.; Wuts, P. G. M., *Protective Groups in Organic Synthesis*; Wiley: New York, 1991.
20. Allevi, P.; Anastasia, M.; Ciufereda, P. *TL* **1993**, *34*, 7313.
21. Olah, G. A.; Gupta, B. G. B.; Narang, S. C.; Malhotra, R. *JOC* **1979**, *44*, 4272.
22. Lissel, M.; Weiffen, J. *SC* **1981**, *11*, 545.
23. Cossy, J.; Pale, P. *TL* **1987**, *28*, 6039.
24. Bundy, G. L.; Peterson, D. C. *TL* **1978**, 41.
25. Hurst, D. T.; McInnes, A. G. *CJC* **1965**, *43*, 2004.
26. Kawazoe, Y.; Nomura, M.; Kondo, Y.; Kohda, K. *TL* **1987**, *28*, 4307.
27. Sekine, M.; Masuda, N.; Hata, T. *T* **1985**, *41*, 5445.
28. Larson, G. L.; Ortiz, M.; Rodrigues de Roca, M. *SC* **1981**, 583.
29. Stork, G.; Hudrlik, P. F. *JACS* **1968**, *90*, 4462.
30. Negishi, E.; Chatterjee, S. *TL* **1983**, *24*, 1341.
31. Corey, E. J.; Gross, A. W. *TL* **1984**, *25*, 495.
32. Cazeau, P.; Duboudin, F.; Moulines, F.; Babot, O.; Dunogues, J. *T* **1987**, *43*, 2075.
33. Cazeau, P.; Duboudin, F.; Moulines, F.; Babot, O.; Dunogues, J. *T* **1987**, *43*, 2089.
34. Rubottom, G. M.; Mott, R. C.; Krueger, D. S. *SC* **1977**, *7*, 327.
35. Morita, T.; Okamoto, Y.; Sakurai, H. *TL* **1978**, *28*, 2523.
36. Barua, N. C.; Sharma, R. P. *TL* **1982**, *23*, 1365.
37. Caputo, R.; Mangoni, L.; Neri, O.; Palumbo, G. *TL* **1981**, *22*, 3551.
38. Taylor, R. T.; Degenhardt, C. R.; Melega, W. P.; Paquette, L. A. *TL* **1977**, 159.

39. Fristad, W. E.; Bailey, T. R.; Paquette, L. A. *JOC* **1980**, *45*, 3028.
40. Lipschutz, B. H.; Dimock, S. H.; James, B. *JACS* **1993**, *115*, 9283.
41. Nakamura, E.; Matsuzawa, S.; Horiguchi, Y.; Kuwajima, I. *TL* **1986**, *27*, 4029.
42. Corey, E. J.; Boaz, N. W. *TL* **1985**, *26*, 6015.
43. Johnson, C. R.; Marren, T. J. *TL* **1987**, *28*, 27.
44. Tamura, R.; Tamai, S.; Katayama, H.; Suzuki, H. *TL* **1989**, *30*, 3685.
45. Booker-Milburn, K. I.; Thompson, D. F. *TL* **1993**, *34*, 7291.

Ellen M. Leahy
Affymax Research Institute, Palo Alto, CA, USA

Chromium(II) Chloride

[10049-05-5] Cl_2Cr (MW 122.92)

(reducing agent for dehalogenation of organic halides, especially allylic and benzylic halides, and for transformation of carbon–carbon triple bonds leading to (*E*)-alkenes; conversion of dibromocyclopropanes to allenes; preparation and reaction of allylic chromium reagents; reduction of sulfur- or nitrogen-substituted alkyl halides to give hetero-substituted alkylchromium reagents)

Physical Data: mp 824 °C; d_4^{14} 2.751 g cm^{-3}.
Form Supplied in: off-white solid; commercially available.
Solubility: sol water, giving a blue solution; insol alcohol or ether.
Handling, Storage, and Precautions: very hygroscopic; oxidizes rapidly, especially under moist conditions; should be handled in a fume hood under an inert atmosphere (argon or nitrogen).

Reduction of Alkyl Halides.[1,2] Typically, the chromium(II) ion is prepared by reduction of chromium(III) salts with zinc and hydrochloric acid. Organochromium compounds produced in this way can subsequently be hydrolyzed to yield dehalogenated compounds (eq 1[3] and eq 2[4]). Anhydrous chromium(II) chloride is commercially available and can be used without further purification. The relative reactivities of various types of halide toward chromium(II) salts are shown in Scheme 1.

$$ClCH_2CO_2H \xrightarrow[HCl,\ H_2O]{CrCl_3,\ Zn} MeCO_2H \quad (1)$$

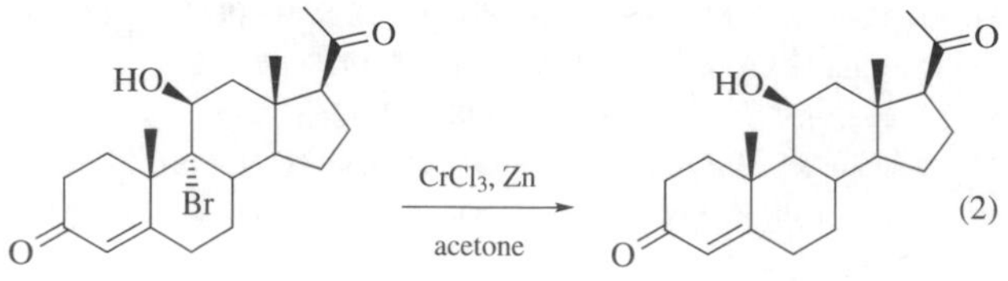

(2) [steroid bromide → dehalogenated steroid; $CrCl_3$, Zn, acetone]

Ph–CH₂X ≈ MeCOCH₂X > allyl–X > t-Bu–X > i-Pr–X > n-alkyl–X > vinyl–X ≈ Ar–X

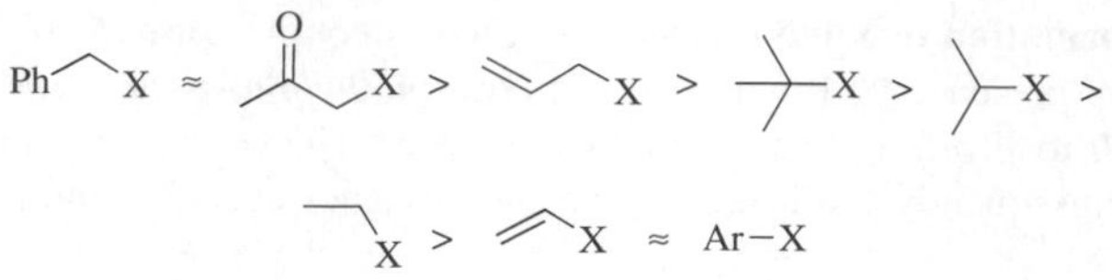

X = I > Br > Cl

Scheme 1

Conversion of Dihalocyclopropanes to Allenes. Reduction of geminal dihalides proceeds smoothly to give chromium carbenoids.[5] In the case of 1,1-dibromocyclopropanes, the intermediate carbenoids decompose instantaneously to give allenes (eq 3).[6,7]

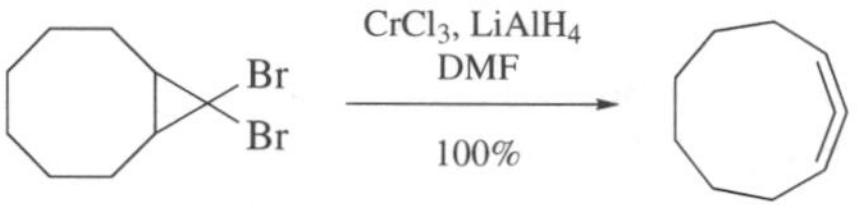

(3) [$CrCl_3$, $LiAlH_4$, DMF, 100%]

Reductive Coupling of Allylic and Benzylic Halides. Active halides, such as allyl and benzyl halides, are reduced with $CrCl_2$ smoothly to furnish homocoupling products. Allylic halides undergo coupling, forming mainly the head-to-head dimer (eq 4).[6,8]

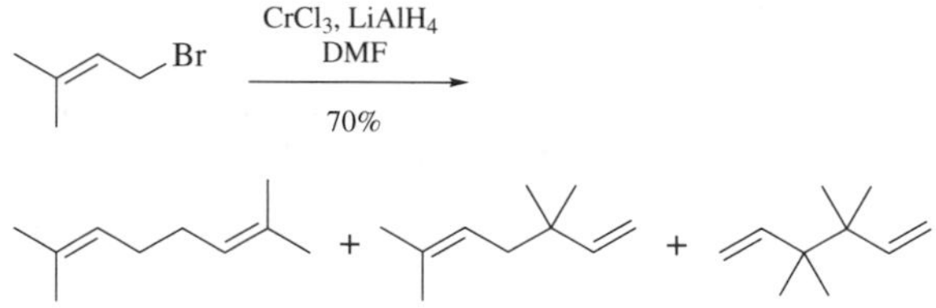

(4) [$CrCl_3$, $LiAlH_4$, DMF, 70%; 72:22:6]

Formation of *o*-Quinodimethanes. α,α′-Dibromo-*o*-xylenes are reduced with $CrCl_2$ in a mixed solvent of THF and HMPA to an *o*-quinodimethane, which can be trapped by a dienophile. The method has been applied to some anthracycline precursors (eq 5).[9]

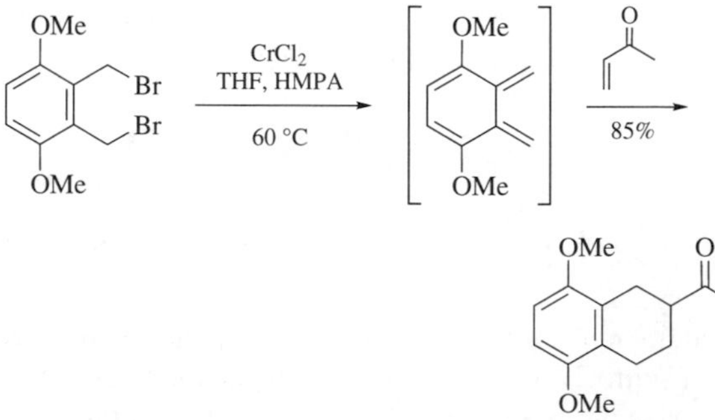

(5) [$CrCl_2$, THF, HMPA, 60 °C; then methyl vinyl ketone, 85%]

Reduction of Carbon–Carbon Unsaturated Bonds. The reduction of alkynes with chromium(II) salts in DMF leads to (*E*)-substituted alkenes.[10] The ease of reduction depends on the presence of an accessible coordination site in the molecule (eq 6). Chromium(II) chloride in THF/H_2O (2:1) (or ***Chromium(II) Sulfate*** in DMF/H_2O) is effective at reducing α-alkynic ketones to (*E*)-enones. Less than 2% of (*Z*)-enones are produced except in the case of highly substituted substrates, which also require longer reaction times.[11]

t-Bu–C≡C–C(O)CH$_3$ → ($CrCl_3$, Zn, THF, H_2O, 80%) → t-Bu–CH=CH–C(O)CH$_3$ (6)

Reduction of Other Functional Groups. Deoxygenation of α,β-epoxy ketones proceeds with acidic solutions of $CrCl_2$ to form α,β-unsaturated ketones.[12] Chromium(II) chloride has been regularly used in the deoxygenation of the limonoid group of triterpenes, in which ring D bears an α,β-epoxy-δ-lactone.[13] Treatment of nitrobenzene derivatives with $CrCl_2$ in methanol under reflux gives anilines (eq 7), while aliphatic nitro compounds afford aldehydes under the same reaction conditions.[14] Reduction of a nitroalkene with acidic solutions of $CrCl_2$ resulted in the formation of an α-hydroxy oxime.[15]

2-Nitrophenol (NO_2, OH) → ($CrCl_3$, Zn, MeOH, H_2O, reflux, 85%) → 2-aminophenol (NH_2, OH) (7)

Preparation of Allylic Chromium Reagents.[16] Allylic halides are reduced with low-valent chromium ($CrCl_3$–$LiAlH_4$) or $CrCl_2$ to give the corresponding allylic chromium reagents, which add to aldehydes and ketones in good to excellent yields.[17] The electronegativity of chromium is 1.6, almost the same as that of titanium (1.5). Therefore the nucleophilicity of organochromium reagents is not strong compared to the corresponding organolithium or -magnesium compounds. Chemoselective addition of allylic chromium reagents can be accomplished without affecting coexisting ketone and cyano groups (eq 8).[17] The reaction between crotylchromium reagents and aldehydes in THF proceeds with high diastereoselectivity (eq 9).[18,19] The *anti* (or *threo*) selectivity in the addition of acyclic allylic chromium reagents with aldehydes is explained by a chair-form six-membered transition state in which both R^1 and R^2 possess equatorial positions (**1** > **2**).

OHC–(CH$_2$)$_4$–C(O)C$_6$H$_{13}$ → (allyl bromide, $CrCl_3$, $LiAlH_4$ or $CrCl_2$, THF, 66%) → CH$_2$=CHCH$_2$CH(OH)–(CH$_2$)$_4$–C(O)C$_6$H$_{13}$ (8)

PhCHO + crotyl bromide → ($CrCl_2$ or $CrCl_3$, $LiAlH_4$, THF, 20 °C, 96–100%) → Ph–CH(OH)–CH(CH$_3$)–CH=CH$_2$ (9)

Transition states: (**1**) R equatorial, CrL_4; (**2**) R axial, CrL_4

Addition of crotylchromium reagents to aldehydes bearing a stereogenic center α to the carbonyl provides three of the four diastereomers (eq 10).[19] Excellent *anti* selectivity is observed with respect to the 1,2-positions, but the stereoselectivity with respect to the 2,3-positions (Cram/anti-Cram ratio) is poor.[20–22] High 1,2- and 2,3-diastereoselectivity is obtained with aldehydes having large substituents, especially a cyclic acetal group, on the α-carbon of the aldehyde (eq 11).[20,23] Reaction between chirally substituted acyclic allylic bromides and aldehydes proceeds with high stereocontrol (eq 12).[24]

OHC–CH(CH$_3$)CH$_2$CH$_3$ → (crotyl bromide, THF) → three diastereomeric homoallylic alcohols (OH) (10)

$CrCl_2$	90%, 62:31:7
$CrCl_3$, $LiAlH_4$	78%, 64:29:7

Acetonide aldehyde (OHC) + crotyl iodide → ($CrCl_3$, $LiAlH_4$, THF, rt, 86%) → homoallylic alcohol (OH, acetonide) + isomers (11)

>20:1

OBn-substituted acetonide allylic bromide (Br) + OHC–CH(OMOM)Bu → ($CrCl_3$, $LiAlH_4$, THF, 63%) → adduct (OBn, OMOM, Bu, OH) (12)

As with allylic halides, allylic diethylphosphates[25] and mesylates[26,27] are reduced with chromium(II) salts to give allylic chromium reagents which add to aldehydes regio- and stereoselectively. This transformation reveals conversion of the electronic nature of allylic phosphates (or mesylates) from electrophilic to nucleophilic by reduction with low-valent chromium.

The reaction of γ-disubstituted allylic phosphates with aldehydes mediated by $CrCl_2$ and a catalytic amount of LiI in DMPU is not stereoconvergent and proceeds with high stereoselectivity (eqs 13 and 14).[27] The presence of the two substituents at the γ-position slows down the process of equilibration between the intermediate allylic chromium reagents.

Geranyl–O–P(O)(OEt)$_2$ → (C_5H_{11}CHO, $CrCl_2$, LiI, DMPU, 93%) → homoallylic alcohol (OH, C_5H_{11}) (13)

94:6

$C_5H_{11}CHO$, $CrCl_2$, LiI, DMPU, 94% (14) 99:1

Because the coupling reaction between allylic halides and aldehydes proceeds under mild conditions, the reaction has been employed, in particular, in intramolecular cyclizations.[28–32] The intramolecular version also proceeds with high *anti* selectivity (eqs 15 and 16).

$CrCl_2$, THF, 25 °C, 64% (15) 4:1

$CrCl_2$, THF, 67–76% (16) + *trans* isomer 10–12%

Functionalized and Hetero-Substituted Allylic Chromium Reagents. When functionalized allylic halides are employed as precursors of allylic chromium reagents, an acyclic skeleton bearing a foothold for further construction is produced. Reaction of α-bromomethyl-α,β-un saturated esters with aldehydes mediated by $CrCl_2$ (or $CrCl_3$–$LiAlH_4$) affords homoallylic alcohols, which cyclize to yield α-methylene-γ-lactones in a stereoselective manner (eq 17).[33] Reaction between α-bromomethyl-α,β-un saturated sulfonates and aldehydes also proceeds with high stereocontrol.[34]

PhCHO + ... $CrCl_3$, $LiAlH_4$, THF, 25 °C, 86% (17)

The reaction of 3-alkyl-1,1-dichloro-2-propene with $CrCl_2$ results in α-chloroalkylchromium reagents, which react with aldehydes to produce a 2-substituted *anti*-(*Z*)-4-chloro-3-buten-1-ol in a regio- and stereoselective manner (eq 18).[35] Vinyl-substituted β-hydroxy carbanion synthons are produced by reduction of 1,3-diene monoepoxides with $CrCl_2$ in the presence of LiI, which react stereoselectively with aldehydes to give (R^*,R^*)-1,3-diols having a quaternary center at C-2.[36] Reduction in situ of acrolein dialkyl acetals with $CrCl_2$ in THF provides γ-alkoxy-substituted allylic chromium reagents which add to aldehydes at the same position of the alkoxy group to afford 3,4-butene-1,2-diol derivatives. The reaction rate and stereoselectivity are increased by addition of ***Iodotrimethylsilane*** (eq 19).[37]

$CrCl_2$, THF, DMF, 87% (18) mixture

$CrCl_2$, TMSI, THF, −30 °C, 99% (19) 88:12

Preparation of Propargylic Chromium Reagents. Propargyl halides react with aldehydes or ketones in the presence of $CrCl_2$ with HMPA as cosolvent to give allenes stereoselectively (eq 20).[38] The reaction was modified to include polyfunctional propargylic halides by using $CrCl_2$ and ***Lithium Iodide*** in DMA, and allenic alcohols accompanied only by small amounts of homopropargylic alcohols are produced.[39]

$CrCl_3$, $LiAlH_4$, THF, 25 °C, 80% (20)

Sulfur- and Nitrogen-Stabilized Alkylchromium Reagents. In combination with LiI, $CrCl_2$ reduces α-halo sulfides to (α-alkylthio)chromium compounds, which undergo selective 1,2-addition to aldehydes. Acetophenone is recovered unchanged under the reaction conditions. The (1-phenylthio)ethenylchromium reagents prepared in this way add to aldehydes under high stereocontrol in the presence of suitable ligands like 1,2-diphenylphosphinoethane (dppe) (eq 21).[40] The reaction of *N*-(chloromethyl)succinimide and -phthalimide with $CrCl_2$ provides the corresponding α-nitrogen-substituted organochromium reagents in the presence of LiI. These organochromium reagents react in situ with aldehydes, affording protected amino alcohols (eq 22).[41]

$$C_8H_{17}{-}CHO + Cl{-}CH(CH_3){-}SPh \xrightarrow[\text{53\%}]{CrCl_2,\ dppe,\ LiI,\ THF,\ 25\ °C} C_8H_{17}CH(OH)CH(SPh)CH_3\ (syn) + (anti) \quad >98{:}<2 \tag{21}$$

$$\text{N-(chloromethyl)phthalimide} + OHC{-}C_6H_{11} \xrightarrow[\text{95\%}]{CrCl_2,\ LiI,\ THF,\ 50\ °C} \text{PhthN-CH}_2\text{CH(OH)-}C_6H_{11} \tag{22}$$

Preparation of Alkylchromium Reagents.[42] With the assistance of catalytic amounts of ***Vitamin B$_{12}$*** or cobalt phthalocyanine (CoPc), $CrCl_2$ reduces alkyl halides, especially 1-iodoalkanes, to form alkylchromium reagents which add to aldehydes without affecting ketone or ester groups. The chemoselective preparation of organochromium reagents can be done by changing either the catalyst or the solvent. Alkenyl and alkyl halides remain unchanged under the conditions of the preparation of allylchromium reagents; on the other hand, alkenyl- and alkylchromium reagents are produced selectively under nickel and cobalt catalysis, respectively (eqs 23 and 24).

$$CH_2{=}CHCH_2Br + C_{12}H_{25}{-}Cl + C_9H_{19}C(Br){=}CH_2 \xrightarrow[\text{83\%}]{PhCHO,\ CrCl_2,\ THF,\ 25\ °C} CH_2{=}CHCH_2CH(OH)Ph + C_{12}H_{25}{-}Cl\ (\text{recov } 97\%) + C_9H_{19}C(Br){=}CH_2\ (\text{recov } 97\%) \tag{23}$$

$$C_{12}H_{25}{-}Cl + C_9H_{19}C(Br){=}CH_2 \xrightarrow[\text{DMF, 30 °C, 68\%}]{PhCHO,\ CrCl_2,\ cat\ B_{12}} C_{12}H_{25}CH(OH)Ph + C_9H_{19}C(Br){=}CH_2\ (\text{recov } 100\%)$$

$$C_{12}H_{25}{-}Cl + C_9H_{19}C(Br){=}CH_2 \xrightarrow[\text{DMF, 25 °C, 82\%}]{PhCHO,\ CrCl_2,\ cat\ NiCl_2} C_{12}H_{25}{-}Cl\ (\text{recov } 96\%) + C_9H_{19}C(=CH_2)CH(OH)Ph \tag{24}$$

Related Reagents. Chromium(II) Chloride–Haloform; Chromium(II) Chloride–Nickel(II) Chloride.

1. (a) Hanson, J. R.; Premuzic, E. *AG(E)* **1968**, *7*, 247; (b) Hanson, J. R. *S* **1974**, 1.
2. (a) Castro, C. E.; Kray, W. C., Jr. *JACS* **1963**, *85*, 2768. (b) Kray, W. C., Jr.; Castro, C. E. *JACS* **1964**, *86*, 4603. (c) Kochi, J. K.; Singleton, D. M.; Andrews, L. J. *T* **1968**, *24*, 3503.
3. Traube, W.; Lange, W. *CB* **1925**, *58*, 2773.
4. Beereboom, J. J.; Djerassi, C.; Ginsburg, D.; Fieser, L. F. *JACS* **1953**, *75*, 3500.
5. Castro, C. E.; Kray, W. C., Jr. *JACS* **1966**, *88*, 4447.
6. Okude, Y.; Hiyama, T.; Nozaki, H. *TL* **1977**, 3829.
7. Wolf, R.; Steckhan, E. *J. Electroanal. Chem.* **1981**, *130*, 367.
8. Sustmann, R.; Altevogt, R. *TL* **1981**, *22*, 5167.
9. Stephan, D.; Gorgues, A.; Le Coq, A. *TL* **1984**, *25*, 5649.
10. Castro, C. E.; Stephens, R. D. *JACS* **1964**, *86*, 4358.
11. Smith, A. B., III; Levenberg, P. A.; Suits, J. Z. *S* **1986**, 184.
12. Cole, W.; Julian, P. L. *JOC* **1954**, *19*, 131.
13. (a) Arigoni, D.; Barton, D. H. R.; Corey, E. J.; Jeger, O.; Caglioti, L.; Dev, S.; Ferrini, P. G.; Glazier, E. R.; Melera, A.; Pradhan, S. K.; Slhaffner, K.; Sternhell, S.; Templeton, J. F.; Tobinaga, S. *Experientia* **1960**, *16*, 41. (b) Akisanya, A.; Bevan, C. W. L.; Halsall, T. G.; Powell, J. W.; Taylor, D. A. H. *JCS* **1961**, 3705. (c) Ekong, D. E. U.; Olagbemi, O. E. *JCS(C)* **1966**, 944.
14. Akita, Y.; Inaba, M.; Uchida, H.; Ohta, A. *S* **1977**, 792.
15. (a) Hanson, J. R.; Premuzic, E. *TL* **1966**, 5441; (b) Rao, T. S.; Mathur, H. H.; Trivedi, G. K. *TL* **1984**, *25*, 5561.
16. Cintas, P. *S* **1992**, 248.
17. (a) Okude, Y.; Hirano, S.; Hiyama, T.; Nozaki, H. *JACS* **1977**, *99*, 3179. (b) Hiyama, T.; Okude, Y.; Kimura, K.; Nozaki, H. *BCJ* **1982**, *55*, 561.
18. Buse, C. T.; Heathcock, C. H. *TL* **1978**, 1685.
19. Hiyama, T.; Kimura, K.; Nozaki, H. *TL* **1981**, *22*, 1037.
20. (a) Nagaoka, H.; Kishi, Y. *T* **1981**, *37*, 3873. (b) Lewis, M. D.; Kishi, Y. *TL* **1982**, *23*, 2343.
21. Fronza, G.; Fganti, C.; Grasselli, P.; Pedrocchi-Fantoni, G.; Zirotti, C. *CL* **1984**, 335.
22. Evans, D. A.; Dow, R. L.; Shih, T. L.; Takacs, J. M.; Zahler, R. *JACS* **1990**, *112*, 5290.
23. Roush, W. R.; Palkowitz, A. D. *JOC* **1989**, *54*, 3009.
24. (a) Mulzer, J.; de Lasalle, P.; Freiler, A. *LA* **1986**, 1152. (b) Mulzer, J.; Kattner, L. *AG(E)* **1990**, *29*, 679. (c) Mulzer, J.; Kattner, L.; Strecker, A. R.; Schröder, C.; Buschmann, J.; Lehmann, C.; Luger, P. *JACS* **1991**, *113*, 4218.
25. Takai, K.; Utimoto, K. *J. Synth. Org. Chem. Jpn.* **1988**, *46*, 66.
26. Kato, N.; Tanaka, S.; Takeshita, H. *BCJ* **1988**, *61*, 3231.
27. Jubert, C.; Nowotny, S.; Kornemann, D.; Antes, I.; Tucker, C. E.; Knochel, P. *JOC* **1992**, *57*, 6384.
28. Still, W. C.; Mobilio, D. *JOC* **1983**, *48*, 4785.
29. Shibuya, H.; Ohashi, K.; Kawashima, K.; Hori, K.; Murakami, N.; Kitagawa, I. *CL* **1986**, 85.
30. Kato, N.; Tanaka, S.; Takeshita, H. *CL* **1986**, 1989.
31. Wender, P. A.; McKinney, J. A.; Mukai, C. *JACS* **1990**, *112*, 5369.
32. (a) Paquette, L. A.; Doherty, A. M.; Rayner, C. M. *JACS* **1992**, *114*, 3910. (b) Rayner, C. M.; Astles, P. C.; Paquette, L. A. *JACS* **1992**, *114*, 3926. (c) Paquette, L. A.; Astles, P. C. *JOC* **1993**, *58*, 165.
33. (a) Okuda, Y.; Nakatsukasa, S.; Oshima, Y.; Nozaki, H. *CL* **1985**, 481. (b) Drewes, S. E.; Hoole, R. F. A. *SC* **1985**, *15*, 1067.
34. (a) Auvray, P.; Knochel, P.; Normant, J. F. *TL* **1986**, *27*, 5091. (b) Auvray, P.; Knochel, P.; Vaissermann, J.; Normant, J. F. *BSF* **1990**, *127*, 813.
35. (a) Takai, K.; Kataoka, Y.; Utimoto, K. *TL* **1989**, *30*, 4389. (b) Wender, P. A.; Grissom, J. W.; Hoffmann, U.; Mah, R. *TL* **1990**, *31*, 6605. (c) Augé, J. *TL* **1988**, *29*, 6107.
36. Fujimura, O.; Takai, K.; Utimoto, K. *JOC* **1990**, *55*, 1705.
37. (a) Takai, K.; Nitta, K.; Utimoto, K. *TL* **1988**, *29*, 5263. (b) Roush, W. R.; Bannister, T. D. *TL* **1992**, *33*, 3587.
38. (a) Place, P.; Delbecq, F.; Gore, J. *TL* **1978**, 3801. (b) Place, P.; Venière, C.; Gore, J. *T* **1981**, *37*, 1359.
39. Belyk, K.; Rozema, M. J.; Knochel, P. *JOC* **1992**, *57*, 4070.

40. Nakatsukasa, S.; Takai, K.; Utimoto, K. *JOC* **1986**, *51*, 5045.
41. Knochel, P.; Chou, T.-S.; Jubert, C.; Rajagopal, D. *JOC* **1993**, *58*, 588.
42. Takai, K.; Nitta, K.; Fujimura, O.; Utimoto, K. *JOC* **1989**, *54*, 4732.

Kazuhiko Takai
Okayama University, Japan

Chromium(II) Chloride–Haloform

$CrCl_2$–CHX_3

($CrCl_2$) *[10049-05-5]*	Cl_2Cr	(MW 122.92)
(CHI_3) *[75-47-8]*	CHI_3	(MW 393.78)
($CHBr_3$) *[75-25-2]*	$CHBr_3$	(MW 252.77)
($CHCl_3$) *[67-66-3]*	$CHCl_3$	(MW 119.39)
($Me_3SiCHBr_2$) *[2612-42-2]*	$C_4H_{10}Br_2Si$	(MW 246.02)

(conversion of aldehydes to (*E*)-alkenyl halides or (*E*)-alkenylsilanes with one-carbon homologation; (*E*)-selective alkylidenation of aldehydes)

Physical Data: see entries for ***Chromium(II) Chloride***, ***Iodoform***, ***Bromoform***, ***Chloroform***, and ***(Dibromomethyl)-trimethylsilane***.

Preparative Method: prepared in presence of reactant aldehyde: to a suspension of anhyd $CrCl_2$ (0.74 g, 6.0 mmol) in THF (10 mL) under Ar, add a solution of aldehyde (1.0 mmol) and iodoform (0.79 g, 2.0 mmol) in THF (5 mL) dropwise at 0 °C, at 0 °C for 3 h, pour the reaction mixture into water (25 mL) and extract with ether; dry the combined extracts over Na_2SO_4 and concentrate; purification by column chromatography gives an alkenyl halide. Low-valent chromium derived by reduction of $CrCl_3$ (6.0 mmol) with ***Lithium Aluminum Hydride*** (3.0 mmol) can be used instead of $CrCl_2$.

Handling, Storage, and Precautions: use in a fume hood; chromium(II) chloride should be used under an inert atmosphere (argon or nitrogen).

Preparation of (*E*)-Alkenyl Halides.[1] The reagent prepared by treatment of CHI_3 or $CHCl_3$ with $CrCl_2$ gives iodo- and chloroalkenes, respectively (eq 1). When bromoform is employed, a mixture of bromo- and chloroalkenes is produced. This side-reaction is overcome by using a combination of $CrBr_3$ and $LiAlH_4$ instead of $CrCl_2$.[1] The rates of reaction of the haloforms are in the sequence I > Br > Cl. Reflux of a mixture of $CHCl_3$ and $CrCl_2$ in THF before addition of an aldehyde and sonication are reported to accelerate the reaction. An iodoalkene having a terminal ^{13}C can be prepared by using $^{13}CHI_3$ (eq 2).[2] An ene-reaction product is obtained as a byproduct in the case of an aldehyde having a suitable alkenic bond.[1,3]

CHX_3, $CrCl_3$, THF (1)

X = I, 0 °C, 82%, (*E*):(*Z*) = 83:17
X = Cl, 65 °C, 76%, (*E*):(*Z*) = 94:6

$^{13}CHI_3$, $CrCl_2$, THF, 71–88% (2)

(*E*)-Isomers of alkenyl halides are produced selectively. The (*E*):(*Z*) ratios of the alkenyl halides increase in the order I < Br < Cl. A mixed solvent of dioxane and THF is employed to improve the (*E*):(*Z*) ratios in certain cases.[4] Isomerization using NaOH in *t*-BuOH is also effective to obtain high (*E*):(*Z*) ratios (eq 3).[5]

1. CHI_3, $CrCl_2$, THF (*E*):(*Z*) = 4:1
2. NaOH, *t*-BuOH reflux (3)

(*E*):(*Z*) = >40:1

Although ketones are also converted into the corresponding alkenyl halides, they are less reactive than aldehydes. Selective conversion of an aldehyde into an (*E*)-iodoalkene can thus be accomplished without affecting the coexisting ketone group.[1] The following functional groups are also tolerated during the reaction: ester, lactone, amide, nitrile, 1,3-diene, alkyne, alkene, bromide, chloride, and acetal of ethylene glycol. A hydroxyl group can be protected by using the following groups: OMe, OCH_2Ph, $OSiMe_3$(*t*-Bu), OAc, OCOPh, OTHP, and OMPM. Because the basicity of the reagent is not strong, epimerization at the α-position of aldehydes does not generally take place. The alkenyl iodides (RCH=CHI) (**1**),[6] (**2**),[7] (**3**),[8] and (**4**)[9] are prepared from the corresponding aldehydes (RCHO).

THF, 0 °C, 75%, (*E*):(*Z*) = 4:1
(**1**)

THF, 25 °C, 70%
(**2**)

THF, 0 °C, 80%, (*E*):(*Z*) = 98:2
(**3**)

THF, 23 °C, 66%
(**4**)

Preparation of (*E*)-Alkenylsilanes.[10] When $Me_3SiCHBr_2$ is used instead of a haloform, (*E*)-alkenylsilanes are produced from aldehydes stereoselectively (eq 4). Because the reaction proceeds under mild conditions, selective transformation of an aldehyde to an (*E*)-alkenylsilane can be performed in the presence of ketone, cyano, ether, acetal, and ester groups (eq 5).[10–12]

Ph–CH₂CH₂–CHO —($TMSCHBr_2$, $CrCl_2$; THF; 86%)→ Ph–CH₂CH₂–CH=CH–TMS (4)

(5): $TMSCHBr_2$, $CrCl_2$; THF; 90%

R = $CH_2CH(OBn)CH_2CH_2OBn$

(*E*)-Selective Wittig-Type Alkylidenation of Aldehydes.[13] Treatment of aldehydes in THF with the reagent derived by reduction of 1,1-diiodoalkane with $CrCl_2$–DMF produces (*E*)-alkenes with high stereocontrol (eq 6).

(6): CHO —(*t*-$BuCHI_2$, $CrCl_2$; THF, DMF; 90%)→ *t*-Bu; (*E*):(*Z*) = 94:6

Related Reagents. Bromoform; Chloroform; Chromium(II) Chloride; 1,1-Diiodoethane; Diiodomethane; Iodoform.

1. Takai, K.; Nitta, K.; Utimoto, K. *JACS* **1986**, *108*, 7408.
2. Baker, K. V.; Brown, J. M.; Cooley, N. A. *J. Labelled Compd. Radiopharm.* **1988**, *25*, 1229.
3. Jung, M. E.; D'Amico, D. C.; Lew, W. *TL* **1993**, *34*, 923.
4. Evans, D. A.; Black, W. C. *JACS* **1993**, *115*, 4497.
5. (a) Wulff, W. D.; Powers, T. S. *JOC* **1993**, *58*, 2381. (b) Hayashi, T.; Konishi, M.; Okamoto, Y.; Kabeta, K.; Kumada, M. *JOC* **1986**, *51*, 3772.
6. Pontikis, R.; Randrianasolo, L. R.; Merrer, Y. L.; Nam, N. H.; Azerad, R.; Depezay, J.-C. *CJC* **1989**, *67*, 2240.
7. Kende, A. S.; Kawamura, K.; DeVita, R. J. *JACS* **1990**, *112*, 4070.
8. Roush, W. R.; Brown, B. B. *JACS* **1993**, *115*, 2268.
9. Kanda, Y.; Fukuyama, T. *JACS* **1993**, *115*, 8451.
10. Takai, K.; Kataoka, Y.; Okazoe, T.; Utimoto, K. *TL* **1987**, *28*, 1443.
11. Burke, S. D.; Takeuchi, K.; Murtiashaw, C. W.; Liang, D. W. M. *TL* **1989**, *30*, 6299.
12. Yoshida, J.; Maekawa, T.; Morita, Y.; Isoe, S. *JOC* **1992**, *57*, 1321.
13. Okazoe, T.; Takai, K.; Utimoto, K. *JACS* **1987**, *109*, 951.

Kazuhiko Takai
Okayama University, Japan

Chromium(II) Chloride–Nickel(II) Chloride

$CrCl_2$–$NiCl_2$

($CrCl_2$)		
[10049-05-5]	Cl_2Cr	(MW 122.92)
($NiCl_2$)		
[7718-54-9]	Cl_2Ni	(MW 129.61)

(catalyst for coupling reactions between haloalkenes (or alkenyl triflates) and aldehydes; active low-valent nickel catalyst)

Physical Data: see entries for ***Chromium(II) Chloride*** and ***Nickel(II) Chloride***.

Preparative Method: a mixture of anhyd $CrCl_2$ (0.47 g, 4.0 mmol) and a catalytic amount of $NiCl_2$ (2.6 mg, 0.020 mmol) in dry, oxygen-free DMF (10 mL) is stirred at 25 °C for 10 min under Ar; add successively a solution of an aldehyde (1.0 mmol) in DMF (5 mL) and a solution of a haloalkene (or an alkenyl triflate; 2.0 mmol) in DMF (5 mL) and stir at 25 °C for 1 h, dilute the reaction mixture with ether (20 mL), pour into water (20 mL), and extract with ether repeatedly; dry the combined extracts over Na_2SO_4 and concentrate; purification by column chromatography gives an allylic alcohol; DMSO or a DMF/THF mixture may be used as solvent (ether and THF are unsuitable).

Handling, Storage, and Precautions: chromium(II) chloride should be used in a fume hood under an inert atmosphere (argon or nitrogen).

Grignard-Type Carbonyl Addition of Haloalkenes to Aldehydes.[1–3] The carbonyl addition reaction mediated by $CrCl_2$ was first reported without any catalyst.[4] Later, a catalytic amount of nickel proved to be indispensable to promote the Grignard-type carbonyl addition of haloalkenes to aldehydes with good reproducibility (eq 1).[1,2] It is important to keep the content of $NiCl_2$ in $CrCl_2$ low (about 0.01–1% w/w) to avoid formation of dienes by homocoupling of the haloalkenes.[5] The Grignard-type reaction between alkenyl triflates and aldehydes proceeds under the same conditions.[1] In the case of α,β-unsaturated aldehydes, 1,2-addition products are the major products.

(1): CHO + X-cyclohexene —($CrCl_2$, cat. $NiCl_2$; DMF, 25 °C)→ OH

X = I, 93%; X = OTf, 83%

The process has many advantages. The basicity of the formed alkenylchromium reagent is not high, and epimerization at the α-position of aldehydes therefore generally does not take place. The regiochemistry of double bonds is maintained during the coupling reaction (eq 2).[6] Mild nucleophilicity of the alkenylchromium reagents permits addition to aldehydes in good to excellent yields without affecting coexisting ketone, ester, amide, acetal, silyl ether, cyano, and sulfinyl groups.[1–4,6–9] The system is especially effective for coupling reactions of highly oxygenated multifunctional substrates.[2,3,7] The configuration of *trans*- and *cis*-haloalkenes is retained, at least in the case of disubstituted haloalkenes and trisubstituted trans-haloalkenes (eqs 3 and 4).[1,2] Treatment of a trisubstituted *cis*-haloalkene (or an alkenyl triflate) with the $CrCl_2$–$NiCl_2$ system often results in a *cis–trans* isomerization–coupling reaction sequence or in recovery of the starting alkenyl halide. With respect to the newly introduced stereogenic center, the process produces a mixture of two possible diastereomers with a moderate to good preference for one stereoisomer. The major products produced from α-alkoxy- and α,β-bisalkoxyaldehydes have the stereochemistry opposite to cuprate or Grignard products.[2]

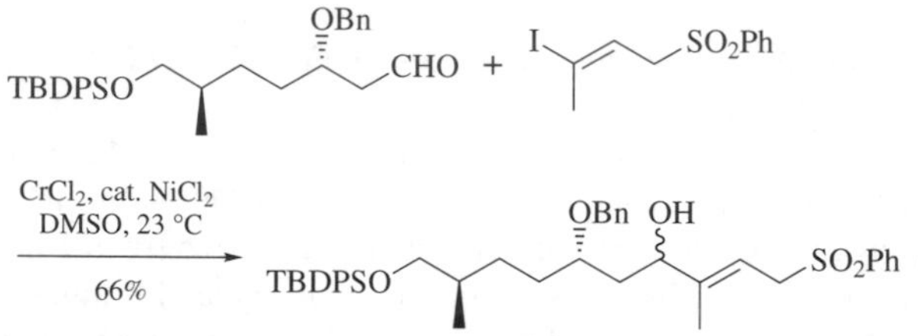

$CrCl_2$, cat. $NiCl_2$, DMSO, 23 °C, 66% (2)

1:1 mixture of epimers

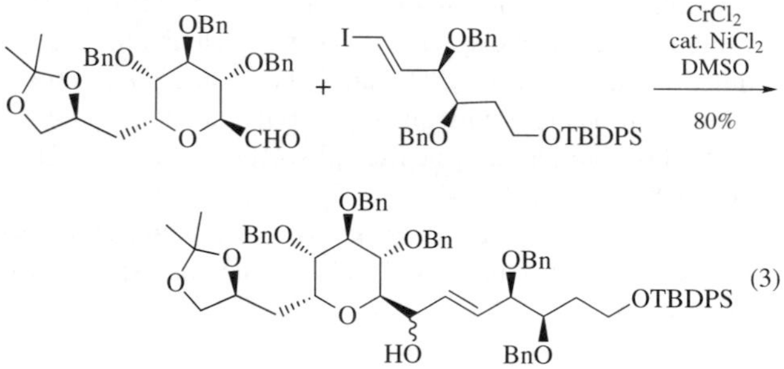

$CrCl_2$, cat. $NiCl_2$, DMSO, 80% (3)

1.3:1

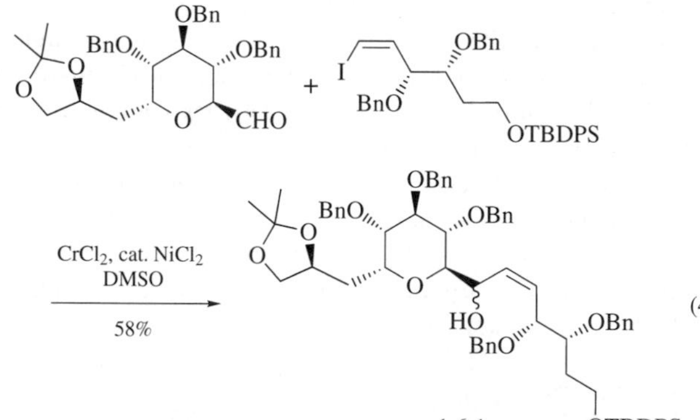

$CrCl_2$, cat. $NiCl_2$, DMSO, 58% (4)

1.6:1

Intramolecular cyclization using the reagent proceeds under mild conditions (eqs 5 and 6).[8,9]

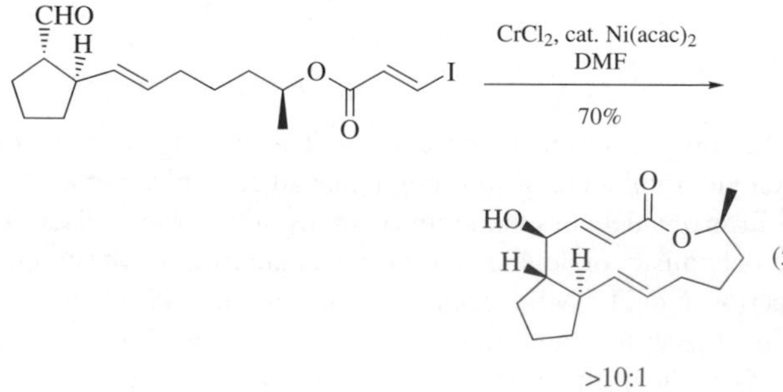

$CrCl_2$, cat. $Ni(acac)_2$, DMF, 70% (5)

>10:1

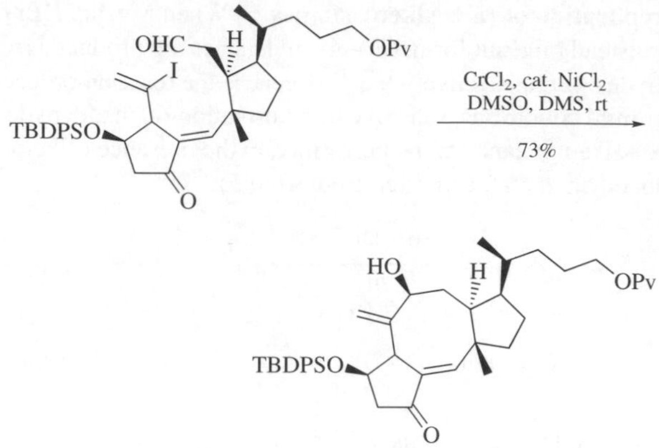

$CrCl_2$, cat. $NiCl_2$, DMSO, DMS, rt, 73% (6)

single diastereoisomer

Grignard-Type Carbonyl Addition of Haloalkynes to Aldehydes. Although reactions between simple haloalkynes and aldehydes proceed in the absence of $NiCl_2$, the $CrCl_2$–$NiCl_2$ system is employed in the case of highly-oxygenated substrates (eq 7)[10,11] and for intramolecular cyclizations (eq 8).[12,13] The process is exceptionally well suited for cases where the aldehyde is labile.

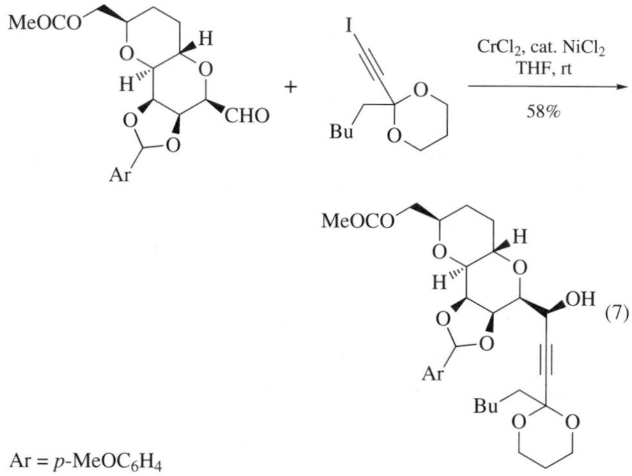

$CrCl_2$, cat. $NiCl_2$, THF, rt, 58% (7)

Ar = *p*-$MeOC_6H_4$

+ stereoisomer 7%

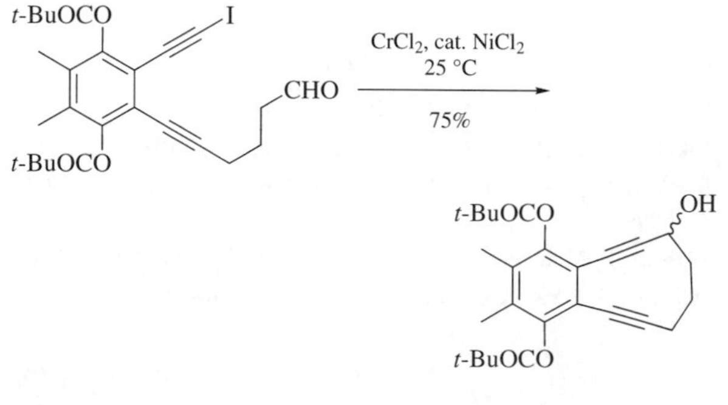

$CrCl_2$, cat. $NiCl_2$, 25 °C, 75% (8)

Active Low-Valent Nickel Catalyst. Intramolecular cyclization of enynes[14] and enallenes[15] proceeds with a catalytic amount of low-valent nickel derived by reduction of $NiCl_2$ with $CrCl_2$ in the presence of triphenylphosphine (eq 9). Decrease of catalytic activity of the nickel is suppressed by using a polymer-supported nickel catalyst.

$NiCl_2$(p-diphenylphosphinopolystyrene), $CrCl_2$, THF, EtOH, rt, 80% (9)

E = CO_2Me

Related Reagents. Chromium(II) Chloride.

1. Takai, K.; Tagashira, M.; Kuroda, T.; Oshima, K.; Utimoto, K.; Nozaki, H. *JACS* **1986**, *108*, 6048.
2. Jin, H.; Uenishi, J.; Christ, W. J.; Kishi, Y. *JACS* **1986**, *108*, 5644.
3. Kishi, Y. *PAC* **1992**, *64*, 343.
4. Takai, K.; Kimura, K.; Kuroda, T.; Hiyama, T.; Nozaki, H. *TL* **1983**, *24*, 5281.
5. Zembayashi, M.; Tamao, K.; Yoshida, J.; Kumada, M. *TL* **1977**, 4089.
6. Chen, S.-H.; Horvath, R. F.; Joglar, J.; Fisher, M. J.; Danishefsky, S. J. *JOC* **1991**, *56*, 5834.
7. (a) Aicher, T. D.; Buszek, K. R.; Fang, F. G.; Forsyth, C. J.; Jung, S. H.; Kishi, Y.; Matelich, M. C.; Scola, P. M.; Spero, D. M.; Yoon, S. K. *JACS* **1992**, *114*, 3162. (b) Aicher, T. D.; Buszek, K. R.; Fang, F. G.; Forsyth, C. J.; Jung, S. H.; Kishi, Y.; Scola, P. M. *TL* **1992**, *33*, 1549.
8. Schreiber, S. L.; Meyers, H. V. *JACS* **1988**, *110*, 5198.
9. (a) Rowley, M.; Kishi, Y. *TL* **1988**, *29*, 4909. (b) Rowley, M.; Tsukamoto, M.; Kishi, Y. *JACS* **1989**, *111*, 2735.
10. Aicher, T. D.; Kishi, Y. *TL* **1987**, *28*, 3463.
11. Wang, Y.; Babirad, S. A.; Kishi, Y. *JOC* **1992**, *57*, 468.
12. Crévisy, C.; Beau, J.-M. *TL* **1991**, *32*, 3171.
13. Nicolaou, K. C.; Liu, A.; Zeng, Z.; McComb, S. *JACS* **1992**, *114*, 9279.
14. Trost, B. M.; Tour, J. M. *JACS* **1987**, *109*, 5268.
15. Trost, B. M.; Tour, J. M. *JACS* **1988**, *110*, 5231.

Kazuhiko Takai
Okayama University, Japan

Copper(II) Acetylacetonate

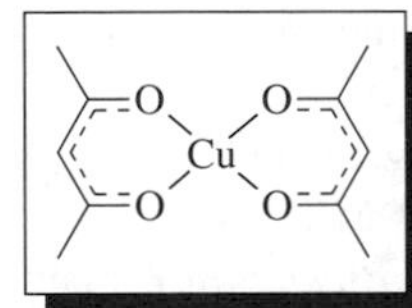

[13395-16-9; 46369-53-3] $C_{10}H_{14}CuO_4$ (MW 261.76)

(catalyst for decomposition of diazo compounds[1–4] and for coupling reactions of organometallics with organic halides and sulfones[9–11,13])

Alternate Names: cupric acetylacetonate; bis(2,4-pentanedionato-*O,O'*)copper.
Physical Data: crystalline solid, mp 284–288 °C (dec); structural studies have been reported.[14]
Solubility: sol $CHCl_3$; sl sol alcohols.
Form Supplied in: commercially available solid.
Preparative Methods: several available, e.g. addition of acetylacetone (***2,4-Pentanedione***) to an aq soln of $Cu(NH_3)_4{}^{2+}$ prepared from copper(II) nitrate trihydrate and conc aq NH_3.[15]
Purification: recrystallization from $CHCl_3$.[15a]
Handling, Storage, and Precautions: irritant.

Decomposition of Diazo Compounds. Copper(II) acetylacetonate has been used to catalyze the decomposition of diazo compounds in preparations of carbenoids. The influence of copper chelates on carbene reactions is due to coordination with the diazoalkane as a fifth ligand to give a complex which decomposes to a copper–carbene complex. ***Copper(I) Acetylacetonate*** was also reported to have similar functionality. The reaction mechanism has been studied, based on the decomposition of ***Diphenyldiazomethane*** (eqs 1 and 2).[1]

$$Ph_2CN_2 \xrightarrow{-N_2} [Ph_2C:] \xrightarrow[\text{medium}]{\text{aprotic}} Ph_2C{=}N{-}N{=}CPh_2$$
$$[Ph_2C:] \rightarrow Ph_2CH \xrightarrow[\text{medium}]{\text{protic}} Ph_2CHCHPh_2 \quad (1)$$

$$Ph_2CN_2 \xrightarrow[\text{cyclohexane}]{Cu(acac)_2} \underset{51\%}{Ph_2C{=}N{-}N{=}CPh_2} + \underset{43\%}{Ph_2C{=}CPh_2} \quad (2)$$

A variety of thermal cyclizations utilize copper(II) acetylacetonate-catalyzed carbene reactions (eqs 3–5).[2–6]

Cu(acac)₂, CuSO₄, benzene (3)

Cu(acac)₂, dioxane, reflux (4)

$$PhCOCHN_2 \xrightarrow{-N_2} [PhCOCH:]$$
Wolff rearrangement → PhCH=C=O; PhCN, $Cu(acac)_2$, 16% → 2,5-diphenyloxazole (5)

Copper(II) acetylacetonate catalyzes the reaction of 2-azido-1,4-quinones with conjugated alkadienyl side chains, giving dihydropyrroloindoloquinones or dihydropyrroloquinolinoquinones, which are not produced by uncatalyzed pyrolysis or photochemical reactions (eq 6).[7] In this example, $Cu(acac)_2$ showed greater activity then many other acetylacetonate complexes or copper powder.

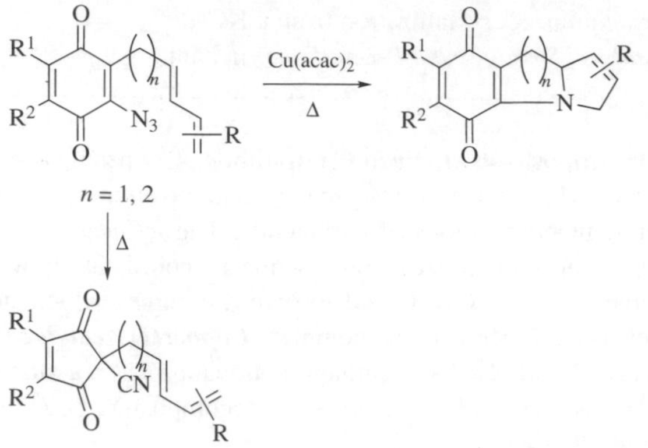

(6)

Copper(II) acetylacetonate has been used as a catalyst for the coupling of bis-diazo ketones to enediones (eq 7).[8]

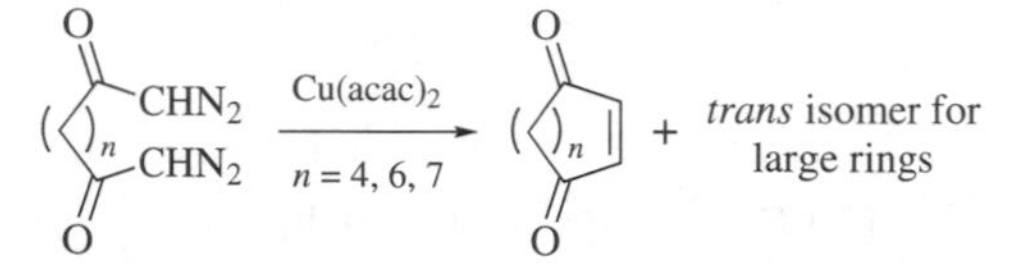

(7)

Coupling Reactions. The reactions of allylic sulfones and Grignard reagents take place readily under $Cu(acac)_2$ catalysis (eq 8).[9,10] The carbon–carbon bond-forming method is useful for introducing medium-length alkyl groups, e.g. butyl, hexyl, and octyl, and also succeeds with allylic chloride and acetate substrates at or below 20 °C.

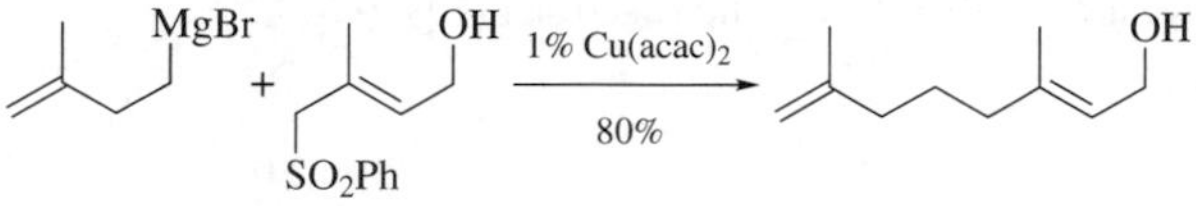

(8)

The reagent also catalyzes the stereospecific cross coupling of alkenyldicyclohexylboranes with allylic or alkynic halides (eq 9).[11] (*E*)-Enynes and 1,4-dienes are produced in good yields and high isomeric purity; ***Tetrakis(triphenylphosphine)palladium(0)*** gives comparative results in similar or better yields.[11]

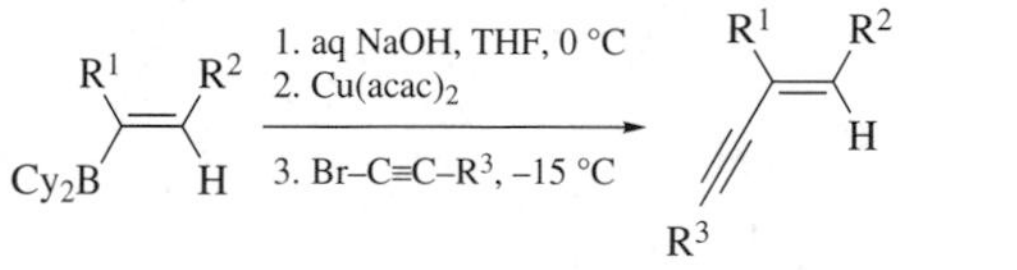

(9)

The coupling reaction of Reformatsky reagents with allylic halides can also be accomplished in high yield under the catalysis of copper(II) acetylacetonate.[13]

Reductive Reactions. It was reported that aromatic nitro compounds can be reduced to the corresponding amines by ***Sodium Borohydride***–$Cu(acac)_2$ in high yield.[12]

Related Reagents. Bis(acetylacetonato)zinc(II); Copper(I) Acetylacetonate; Manganese(III) Acetylacetonate; Nickel(II) Acetylacetonate[1]; Palladium(II) Acetylacetonate; 2,4-Pentanedione; Tris(acetylacetonato)indium; Tris(acetylacetonato)iron (III); Vanadyl Bis(acetylacetonate); .

1. Nozaki, H.; Takaya, H.; Moriuti, S.; Noyori, R. *T* **1968**, *24*, 3655.
2. Huisgen, R.; Binsch, G.; Ghosez, L. *CB* **1964**, *97*, 2628.
3. Fujita, M.; Hiyama, T.; Kondo, K. *TL* **1986**, *21*, 2139.
4. Hudlicky, T.; Koszyk, F. J.; Kutchan, T. M.; Sheth, J. P. *JOC* **1980**, *45*, 5020.
5. Hudlicky, T.; Short, R. P. *JOC* **1982**, *47*, 1522.
6. Hudlicky, T.; Natchus, M. G.; Sinai-Zingde, G. *JOC* **1987**, *52*, 4641.
7. Maruyama, K.; Nagai, N.; Naruta, Y. *CL* **1987**, 97.
8. Kulkowit, S.; McKervey, M. A. *JCS(C)* **1978**, 1069.
9. Julia, M.; Righin-Topie, A.; Verpeaux, J. *T* **1983**, *39*, 3283.
10. Julia, M.; Verpeaux, J. *T* **1983**, *39*, 3289.
11. Hoshi, M.; Masuda, Y.; Arase, A. *BCJ* **1983**, *56*, 2855.
12. Hanaya, K.; Muramatsu, T.; Kudo, H. *JCS(P1)* **1979**, 2409.
13. Gaudemar, M. *TL* **1983**, *24*, 2749.
14. For example: (a) Ferguson, J. *JCP* **1961**, *34*, 1609. (b) Thompson, D. W. *Struct. Bonding* **1971**, *9*, 27.
15. (a) Jones, M. M. *JACS* **1959**, *81*, 3188. (b) Mehrotra, R. C.; Gora, R.; Gaur, D. P. *The Chemistry of beta-Diketonates and Allied Derivatives*; Academic: London, 1978.

Edward J. Parish & Shengrong Li
Auburn University, AL, USA

Copper(I) Bromide[1]

CuBr

[7787-70-4]	BrCu	(MW 143.45)
($CuBr{\cdot}SMe_2$)		
[54678-23-8]	C_2H_6BrCuS	(MW 205.59)

(precursor for organocopper(I) reagents and organocuprates;[1a–f] catalyst for diazo chemistry)

Alternate Name: cuprous bromide.
Physical Data: mp 504 °C; the complex miwith dimethyl sulfide (DMS) decomposes at ca. 130 °C; d 4.720 g mL^{-1}.
Solubility: insol H_2O and most organic solvents; partially sol dimethyl sulfide.
Form Supplied in: light green or blue-tinged white solid. 99.999% grade available. The DMS complex is a white solid.
Handling, Storage, and Precautions: maintenance of a dry N_2 or Ar atmosphere is recommended. The DMS complex must be tightly sealed to prevent loss of DMS. Storage of this complex in a cold place is recommended.

Precursor for Organocopper(I) Reagents and Organocuprates. Although ***Phenylcopper*** was prepared from ***Copper(I) Iodide*** by Reich in 1923[2] and Gilman in 1936,[3] the material used for the modern characterization of this archetypal arylcopper(I) is prepared from CuBr,[4] which continues to be a favored precursor for new organocopper(I) compounds.[5–9] For example, Bertz discovered that halide-free organocopper compounds can be prepared from CuBr in ***Dimethyl Sulfide*** (DMS), owing to the precipitation of LiBr from this solvent.[5] Thus it was possible to prepare and structurally characterize the first bona fide 'higher order' cuprate.[5a,6] Weiss recently reported the second example, a higher order alkynyl cuprate,[9] prepared from CuBr·DMS. The chemistry of organocopper reagents in DMS has now become a flourishing subfield of organometallic chemistry.[5–9]

For the first decade of the modern era of organocopper reagents, CuI was used almost exclusively as the precursor to organocopper(I) and organocuprate reagents.[1f] In 1975, House introduced the DMS complex of CuBr, symbolized CuBr·SMe_2 or CuBr·DMS, as 'a convenient precursor for the generation of lithium organocuprates'.[10] Unlike the commercial CuBr, which is invariably contaminated with traces of colored Cu^{II} impurities, CuBr·DMS is a microcrystalline white solid. This material should be stored under a dry, inert atmosphere in a refrigerator in order to minimize the loss of DMS, which is quite volatile (bp 38 °C). It is not surprising that Lipshutz found that low quality material gave poor results.[11] This author has found that for ultraprecision work, where stoichiometry is of paramount importance, the ultrapure (99.999%) grade of CuBr is preferable.[5]

Nevertheless, in a side-by-side comparison of seven Cu^{I} salts (CuCN, CuI, CuBr, CuBr·DMS, CuCl, CuOTf, and CuSCN) as precursors of a typical alkyl and a typical aryl cuprate (***Lithium Di-n-butylcuprate*** and ***Lithium Diphenylcuprate***, respectively), CuBr·DMS and ***Copper(I) Cyanide*** were found to give the best results.[12] The comparison between ultrapure CuBr and CuBr·DMS is especially interesting, as it demonstrates a dramatic effect for just 1 equiv of DMS in THF and especially in ether. Another example of a significant difference between CuBr and CuBr·DMS is provided by Davis's study of 1,6 vs. 1,4 and 1,2-addition (eq 1).[13]

RMgCl / CuBr•DMS (1)

No Cu	4	1	3
CuBr	4	1	15
CuBr•DMS	3	1	30

Some of the most fundamental studies in organocopper chemistry have been carried out using CuBr or CuBr·DMS as starting materials. House showed that the chemoselectivity of ***Lithium Dimethylcuprate–Lithium Bromide*** could be completely controlled by the choice of solvent.[14] Thus a molecule with remote bromoalkane and α-enone functional groups gave only conjugate addition in ether–DMS and only displacement of the Br when HMPA (***Hexamethylphosphoric Triamide***) was present (eq 2). In a recent ^{13}C NMR study it was shown that phenylcuprates are dimeric in nonpolar solvents and monomeric in polar solvents.[15] It was further conjectured that the dimer is responsible for the conjugate addition reaction, and the monomer is responsible for the (much slower) S_N2-like displacement reaction.

The preparation of the first higher order cuprate, $Ph_5Cu_2Li_3 = [Ph_3CuLi_3][Ph_2Cu]$ or $Ph_3CuLi_2 + Ph_2CuLi$, from CuBr/DMS was mentioned above.[5] House first proposed higher order Ph_3CuLi_2 in solutions prepared from 3 equiv of PhLi and CuBr in ether to account for the higher reactivity observed for this mixture in certain coupling reactions.[16] However, ^{13}C NMR and ^{6}Li NMR studies did not detect any higher order phenylcuprate in ether or THF, only in DMS.[5] While the presence of a small amount of higher order cuprate acting as a catalytic intermediate cannot be ruled out, a more plausible explanation involves the attack of PhLi on a cuprate-complexed intermediate.

Me_2CuLi•LiBr, ether–DMS; Me_2CuLi•LiBr, ether–DMS–HMPA (2)

The first thermally stable phosphido- and amidocuprates were prepared from CuBr·SMe_2.[17] (It was also shown that LiBr has a beneficial effect on the reactions of organocuprates with typical substrates.[17c]) Chiral amidocuprates have been extensively studied because of their potential for asymmetric induction.[18–20] The chiral auxiliary has also been put on the substrate, e.g. cuprates have been added to chiral unsaturated imides.[21] A good recent review provides many more examples.[1b]

Whereas Grignard reagents and lithium reagents generally give thiophilic addition to dithioesters, the corresponding organocopper reagents give carbophilic addition.[22] The best yields were obtained with CuBr·DMS and ***Copper(I) Trifluoromethanesulfonate***; good results were also obtained with CuCN and CuI. This carbophilic addition has been applied to 1,3-thiazole-5(4*H*)-thiones.[23] It is interesting to note that CuBr has also been used in the preparation of the dithioesters (eq 3).[24]

R'MgX —(CS_2, CuBr (5–10%))→ —(MeI)→ R'C(=S)SMe —(R_2CuLi•LiBr)→ R'C(SH)R₂ (3)

Nakamura and Kuwajima have reported the CuBr·DMS catalyzed acylation and conjugate addition reactions of the Zn homoenolate from 1-alkoxy-1-siloxycyclopropanes and ***Zinc Chloride***.[25a] They have also reported the ***Chlorotrimethylsilane***/HMPA accelerated conjugate addition of stoichiometric organocopper reagents prepared from CuBr·DMS,[25b] and of catalytic copper reagents,[25c] to α,β-unsaturated ketones and aldehydes. This procedure appears to be more general than that based on putative cuprates of intermediate stoichiometry (see ***Copper(I) Iodide***). In a very significant observation, they report that 'reagents derived from cuprous iodide consistently gave lower yields'.[25b]

Wipf has used CuBr·DMS to catalyze the addition of alkyl and alkenylzirconocenes to acid chlorides to yield ketones,[26a] and also the 1,4-addition of alkylzirconocenes to α-enones.[26b] The hydrozirconation of alkynes followed by transmetalation to Cu^{I} was devised by Schwartz et al.,[27] who used CuI and ***Copper(I) Chloride***. Transmetalation from Al, B, Pb, Mn, Hg, Sm, Sn, Te, Ti, Zn, and Zr to Cu has been reviewed recently.[1h]

Carbocupration of alkynes by organocopper reagents is a very important area, as judged by the number of citations.[1a,i] An interesting example involves the use of organocopper reagents bearing protected α-hydroxy or α-thio functions.[28a] The preparation of γ-silylvinylcopper reagents via the addition of α-silylated organocopper reagents to alkynes has also been described.[28b] The carbocupration of alkynes is the key step in the synthesis of γ,γ-disubstituted allylboronates.[29] The stannylalumination of 1-alkynes is catalyzed by Cu^I and involves stannylcupration by an intermediate stannylcopper(I) reagent.[30]

The use of Grignard reagents in conjunction with Cu^I salts has been thoroughly reviewed.[1a,d] A very edifying example of the difference between organocopper reagents prepared from lithium reagents vs. Grignard reagents has been provided by Curran (eq 4).[31]

RLi + CuBr•DMS or R'MgBr + CuBr•DMS (4)

	CO_2H product 1	CO_2H product 2
$Me_2CuLi•LiBr$, THF	62	38
$Me_2CuLi•LiBr$, ether	54	46
$MeCu•MgBr_2$	98	2
MeCu•LiBr	86	14
MeCu•LiI	76	24
MeCu(CN)Li	75	25

Yields were 90–97%, except for the cuprate from CuCN: 60%

In chemistry that is clearly related to that of organocopper reagents, aryl bromides and aryl iodides undergo a Gabriel reaction with potassium phthalimide in the presence of CuBr (or CuI).[32] They also undergo coupling reactions with the sodium salts of active methylene compounds catalyzed by CuBr.[33] Copper-assisted nucleophilic substitution of aryl halogen has been reviewed.[1g] In a potentially far-reaching development, thermally stable, yet reactive formulations of organocuprates suitable for commercialization have been patented.[34]

Catalysts for Diazo Chemistry. CuBr has been used in other reactions besides those involving organocuprates. It is a popular catalyst for the activation of ***Diazomethane***, e.g. tropylium perchlorate is isolated in 85% yield starting from benzene.[35] CuBr has been used for the activation of diazoacetic esters,[1j] but not as often as CuCN, and especially CuCl. CuBr is the preferred catalyst for the Sandmeyer reaction of arenediazonium salts to afford bromoarenes,[36] and for the Meerwein reaction, the arylation of alkenes by diazonium salts.[37]

Related Reagents. Copper(I) Bromide-Lithium Trimethoxy aluminum Hydride; Copper(I) Bromide-Sodium Bis(2-methoxyethoxy)aluminum Hydride; Copper(I) Chloride; Copper(I) Chloride-Oxygen; Copper(I) Chloride-Sulfur Dioxide; Copper(I) Chloride-Tetrabutylammonium chloride; Copper(I) Cyanide; Copper(I) Iodide; Copper(I) Trifluoromethanesulfonate.

---

1. (a) Lipshutz, B. H.; Sengupta, S. *OR* **1992**, *41*, 135. (b) Rossiter, B. E.; Swingle, N. M. *CRV* **1992**, *92*, 771. (c) Chapdelaine, M. J.; Hulce, M. *OR* **1990**, *38*, 225. (d) Erdik, E. *T* **1984**, *40*, 641. (e) Posner, G. H. *An Introduction to Synthesis Using Organocopper Reagents*; Wiley: New York, 1980. (f) Posner, G. H. *OR* **1975**, *22*, 253; also see: Posner, G. H. *OR* **1972**, *19*, 1. (g) Lindley, J. *T* **1984**, *40*, 1433. (h) Wipf, P. *S* **1993**, 537. (i) Normant, J. F.; Alexakis, A. *S* **1981**, 841. (j) Dave, V.; Warnhoff, E. W. *OR* **1970**, *18*, 217.
2. Reich, M. R. *CR(C)* **1923**, *177*, 322.
3. Gilman, H.; Straley, J. M. *RTC* **1936**, *55*, 821.
4. Costa, G.; Camus, A.; Gatti, L.; Marsich, N. *JOM* **1966**, *5*, 568.
5. (a) Bertz, S. H.; Dabbagh, G. *JACS* **1988**, *110*, 3668. (b) Bertz, S. H.; Dabbagh, G. *T* **1989**, *45*, 425.
6. Olmstead, M. M.; Power, P. P. *JACS* **1990**, *112*, 8008.
7. Lenders, B.; Grove, D. M.; Smeets, W. J. J.; van der Sluis, P.; Spek, A. L.; van Koten, G. *OM* **1991**, *10*, 786.
8. Kapteijn, G. M.; Wehman-Ooyevaar, I. C. M.; Grove, D. M.; Smeets, W. J. J.; Spek, A. L.; van Koten, G. *AG(E)* **1993**, *32*, 72.
9. Olbrich, F.; Kopf, J.; Weiss, E. *AG(E)* **1993**, *32*, 1077.
10. House, H. O.; Chu, C.-Y.; Wilkins, J. M.; Umen, M. J. *JOC* **1975**, *40*, 1460.
11. Lipshutz, B. H.; Whitney, S.; Kozlowski, J. A.; Breneman, C. M. *TL* **1986**, *27*, 4273.
12. Bertz, S. H.; Gibson, C. P.; Dabbagh, G. *TL* **1987**, *28*, 4251.
13. Davis, B. R.; Johnson, S. J. *JCS(P1)* **1979**, 2840.
14. House, H. O.; Lee, T. V. *JOC* **1978**, *43*, 4369.
15. Bertz, S. H.; Dabbagh, G.; He, X.; Power, P. P. *JACS* **1993**, *115*, 11640.
16. House, H. O.; Koepsell, D. G.; Campbell, W. J. *JOC* **1972**, *37*, 1003.
17. (a) Bertz, S. H.; Dabbagh, G.; Villacorta, G. M. *JACS* **1982**, *104*, 5824. (b) Bertz, S. H.; Dabbagh, G. *CC* **1982**, 1030. (c) Bertz, S. H.; Dabbagh, G. *JOC* **1984**, *49*, 1119.
18. Bertz, S. H.; Dabbagh, G.; Sundararajan, G. *JOC* **1986**, *51*, 4953.
19. Rossiter, B. E.; Eguchi, M.; Miao, G.; Swingle, N. M.; Hernández, A. E.; Vickers, D.; Fluckiger, E.; Patterson, R. G.; Reddy, K. V. *T* **1993**, *49*, 965.
20. Dieter, R. K.; Lagu, B.; Deo, N.; Dieter, J. W. *TL* **1990**, *31*, 4105.
21. Melnyk, O.; Stephan, E.; Pourcelot, G.; Cresson, P. *T* **1992**, *48*, 841.
22. Bertz, S. H.; Dabbagh, G.; Williams, L. M. *JOC* **1985**, *50*, 4414.
23. Jenny, C.; Wipf, P.; Heimgartner, H. *HCA* **1986**, *69*, 1837.
24. Westmijze, H.; Kleijn, H.; Meijer, J.; Vermeer, P. *S* **1979**, 432.
25. (a) Nakamura, E.; Kuwajima, I. *JACS* **1984**, *106*, 3368. (b) Nakamura, E.; Matsuzawa, S.; Horiguchi, Y.; Kuwajima, I. *TL* **1986**, *27*, 4029. (c) Horiguchi, Y.; Matsuzawa, S.; Nakamura, E.; Kuwajima, I. *TL* **1986**, *27*, 4025.
26. (a) Wipf, P.; Xu, W. *SL* **1992**, 718. (b) Wipf, P.; Smitrovich, J. H. *JOC* **1991**, *56*, 6494.
27. Yoshifuji, M.; Loots, M. J.; Schwartz, J. *TL* **1977**, *18*, 1303.
28. (a) Gardette, M.; Alexakis, A.; Normant, J. F. *T* **1985**, *41*, 5887. (b) Foulon, J. P.; Bourgain-Commerçon, M.; Normant, J. F. *T* **1986**, *42*, 1389.
29. Hoffmann, R. W.; Schlapbach, A. *LA* **1990**, 1243.
30. Sharma, S.; Oehlschlager, A. C. *JOC* **1989**, *54*, 5064.
31. Curran, D. P.; Chen, M.-H.; Leszczweski, D.; Elliott, R. L.; Rakiewicz, D. M. *JOC* **1986**, *51*, 1612.
32. Bacon, R. G. R.; Karim, A. *CC* **1969**, 578.
33. Setsune, J.-i.; Matsukawa, K.; Kitao, T. *TL* **1982**, *23*, 663.
34. Hatch, H. B.; Wedinger, R. S. WO Patent Appl. 91 11494, 1991 (*CA* **1991**, *115*, 232514s).
35. Müller, E.; Fricke, H. *LA* **1963**, *661*, 38.
36. Buck, J. S.; Ide, W. S. *OSC* **1943**, *2*, 130.
37. Cleland, G. H. *JOC* **1961**, *26*, 3362.

Steven H. Bertz & Edward H. Fairchild
LONZA, Annandale, NJ, USA

Copper(II) Bromide

$CuBr_2$

[7789-45-9] Br_2Cu (MW 223.36)

(brominating agent; oxidizing agent; Lewis acid)

Alternate Name: cupric bromide.
Physical Data: mp 498 °C; *d* 4.770 g cm^{-3}.
Solubility: very sol water; sol acetone, ammonia, alcohol; practically insol benzene, Et_2O, conc H_2SO_4.
Form Supplied in: almost black solid crystals or crystalline powder; also supplied as reagent adsorbed on alumina (approx. 30 wt % $CuBr_2$ on alumina).
Purification: recryst from H_2O and dried in vacuo.[35]
Handling, Storage, and Precautions: anhydrous reagent is hygroscopic and should therefore be stored in the absence of moisture.

α-Bromination of Carbonyls. Copper(II) bromide is an efficient reagent for the selective bromination of methylenes adjacent to carbonyl functional groups.[1] Thus 2′-hydroxyacetophenone treated with a heterogeneous mixture of $CuBr_2$ in $CHCl_3$–EtOAc gives complete conversion to 2-bromo-2′-hydroxyacetophenone with no aromatic ring bromination (eq 1).[2]

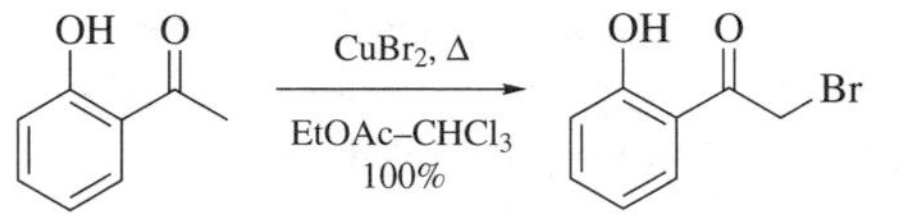

$CuBr_2$, Δ; EtOAc–$CHCl_3$; 100% (1)

Similar selectivity is obtained with a homogeneous solution of the reagent in dioxane.[3] A limitation of the reaction is observed with 2′-hydroxy-4′,6′-dimethoxyacetophenone, which undergoes aromatic nuclear bromination with $CuBr_2$.[4] Steroidal ketones have been selectively α-brominated with $CuBr_2$ in the presence of a double bond without bromination of the alkene (eq 2),[5] while γ-bromination occurs in other steroidal enones.[1]

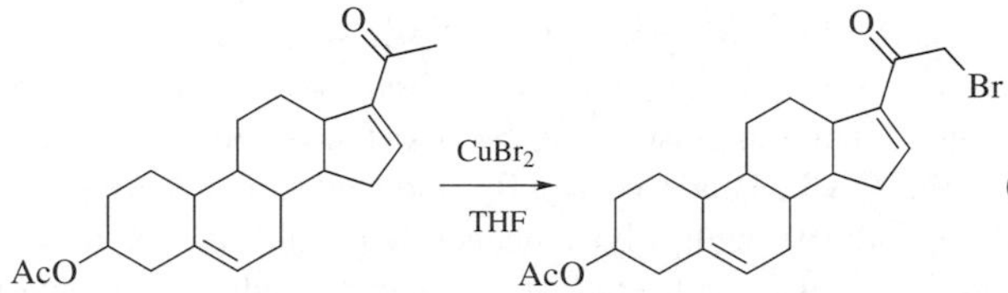

AcO; $CuBr_2$; THF; AcO; Br (2)

Copper(II) bromide has been used to α-brominate diketotetraquinanes[6] and to introduce a double bond into a prostanoid nucleus in a one-pot bromination–elimination procedure (eq 3).[7] 3,7-Dibromo-2*H*,6*H*-benzodithiophene-2,6-diones (eq 4)[8] and 5-bromo-4-oxo-4,5,6,7-tetrahydroindoles (eq 5)[9] are prepared by the selective α-bromination of their respective ketone starting materials without bromination of the aromatic or heterocyclic rings. 4-Carboxyoxazolines are converted to the corresponding oxazoles using a mixture of $CuBr_2$ and ***1,8-Diazabicyclo[5.4.0]undec-7-ene*** (eq 6).[10]

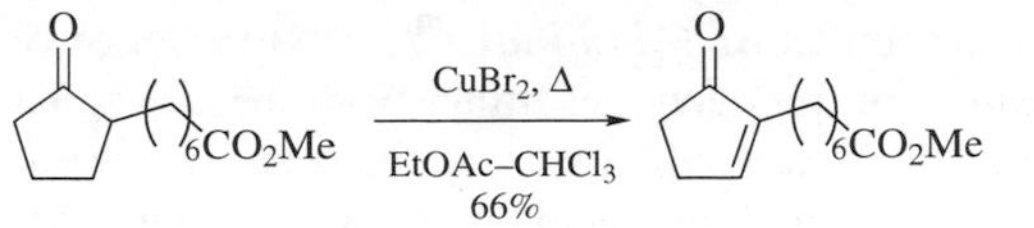

$(CH_2)_6CO_2Me$; $CuBr_2$, Δ; EtOAc–$CHCl_3$; 66% (3)

$CuBr_2$; MeOH; 83% (4)

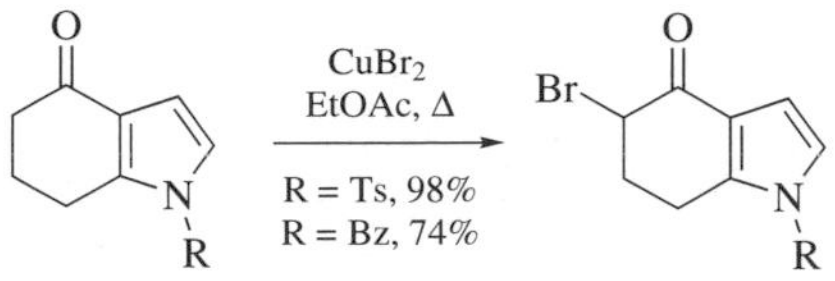

$CuBr_2$; EtOAc, Δ; R = Ts, 98%; R = Bz, 74% (5)

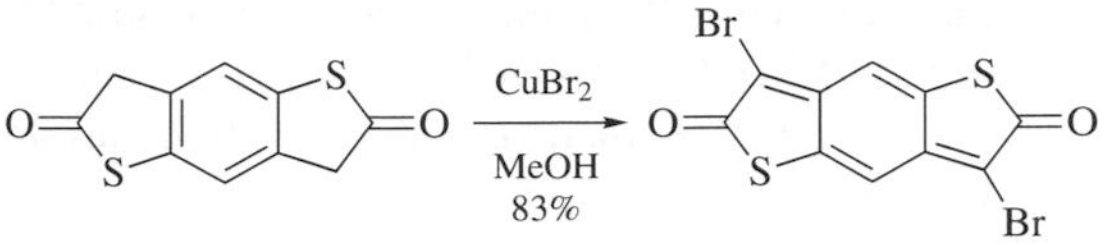

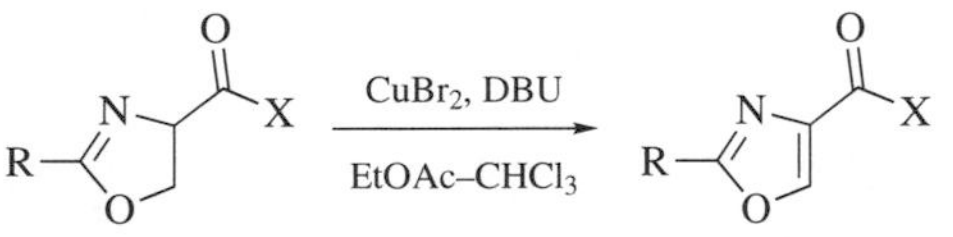

$CuBr_2$, DBU; EtOAc–$CHCl_3$; X = OR, NR_2 (6)

Bromination of Alkenes and Alkynes. Heating copper(II) bromide in methanol with compounds containing nonaromatic carbon–carbon multiple bonds leads to di- or tribromination.[11] For example, under these conditions allyl alcohol is converted to 1,2-dibromo-3-hydroxypropane in 99% yield (eq 7), while propargyl alcohol produces a mixture of *trans* di- and tribromoallyl alcohols (eq 8). 2′-Hydroxy-5′-methyl-4-methoxychalcone undergoes a bromination–ring-closure reaction, affording 3-bromo-6-methyl-5′-methoxyflavanone when heated with $CuBr_2$ in refluxing dioxane (eq 9).[12] The mechanism of the bromination of cyclohexene to 1,2-dibromocyclohexane with $CuBr_2$ has been studied.[13]

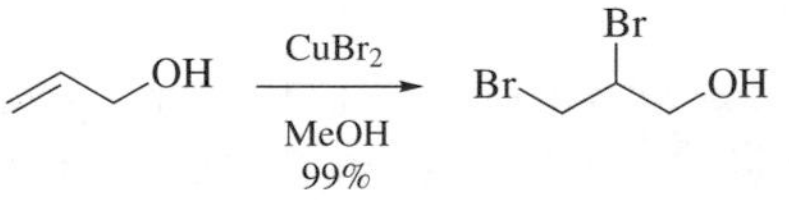

$CuBr_2$; MeOH; 99% (7)

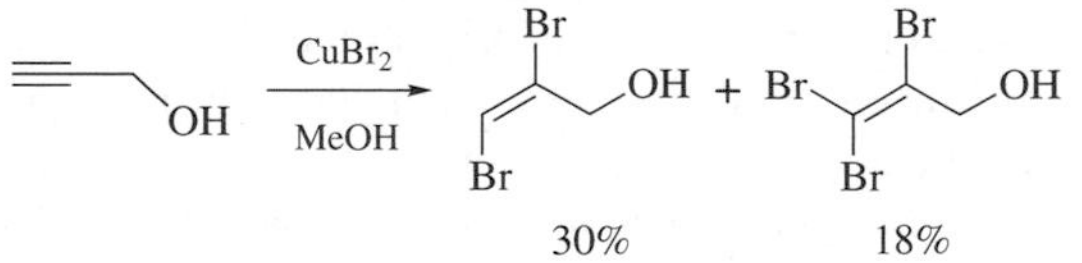

$CuBr_2$; MeOH; 30% + 18% (8)

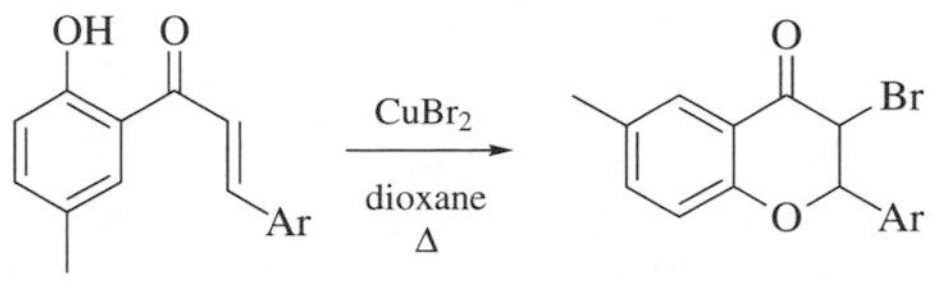

$CuBr_2$; dioxane; Δ (9)

Bromination of Aromatics. Aromatic systems are brominated by copper(II) bromide. For example, 9-bromoanthracene is prepared in high yield by heating anthracene and the reagent in carbon tetrachloride (eq 10).[14] When the 9-position is blocked by a halogen, alkyl, or aryl group, the corresponding 10-bromoanthracene is formed.[15] Under similar conditions, 9-acylanthracenes give 9-acyl-10-bromoanthracenes as the predominant products.[16] The aromatic nuclear bromination of monoalkylbenzenes has been shown to proceed cleanly under strictly anhydrous conditions (eq 11).[17a] Polymethylbenzenes are efficiently and selectively converted to the nuclear brominated derivatives by $CuBr_2$/***Alumina***[1].[17b] In the absence of alumina, a mixture of products resulting from benzylic halogenation is isolated. 3-Acetylpyrroles are nuclear monobrominated at the 4-position in high yield by $CuBr_2$ in acetonitrile at ambient temperature (eq 12).[18] The reaction also proceeds with ethyl 3-pyrrolecarboxylates to give 4-bromopyrrole derivatives,[19] while an excess of brominating agent at 60 °C affords 4,5-dibromopyrroles.[20]

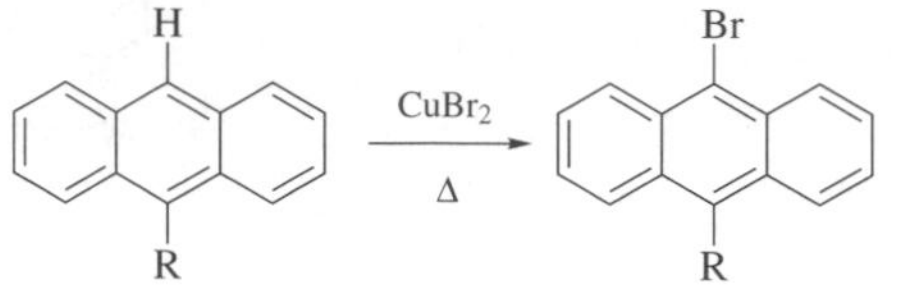

CuBr₂, Δ (10)

R = H, Me, Ph, Ac

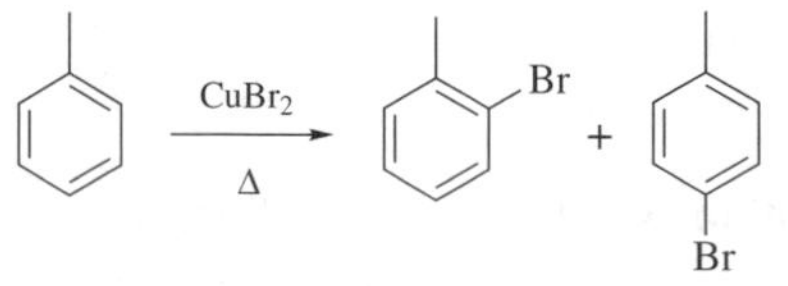

CuBr₂, Δ (11)

1:2

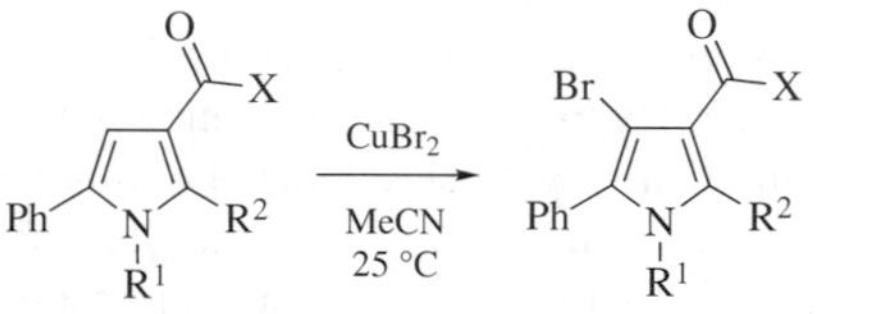

CuBr₂, MeCN, 25 °C (12)

R^1 = H, Me, Ph, Bn; R^2 = Me, Ph; X = Me, OH, OEt

Bromination of Allylic Alcohols. Silica gel-supported copper(II) bromide has been used for the regioselective bromination of methyl 3-hydroxy-2-methylenepropanoates and 3-hydroxy-2-methylenepropanenitriles (eq 13).[21a] In the absence of silica gel, no reaction occurs between $CuBr_2$ and these substrates, while adsorption onto Al_2O_3, MgO, or TiO_2 leads to side reactions rather than the clean allylic bromination observed with $CuBr_2/SiO_2$. The reaction is stereoselective with respect to formation of the (*Z*) isomer.

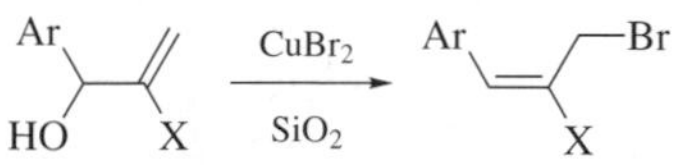

CuBr₂, SiO₂ (13)

X = CO_2Me, CN

Benzylic Bromination. Toluene and substituted methylbenzenes undergo benzylic bromination using $CuBr_2$ and ***t-Butyl Hydroperoxide*** in acetic acid or anhydride (eq 14).[21b] While the yields (43–95%) are not quite as high as those obtained using ***N-Bromosuccinimide***, the copper(II) bromide procedure allows the benzylic bromination of compounds which are insoluble in nonpolar solvents.

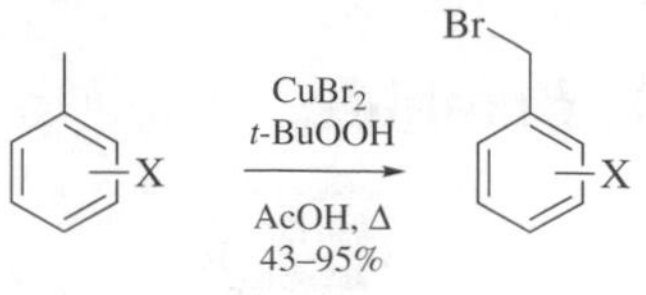

CuBr₂, *t*-BuOOH, AcOH, Δ, 43–95% (14)

X = H, hal, CO_2H

Esterification Catalyst. Highly sterically hindered esters are prepared by the reaction of *S*-2-pyridyl thioates and alcohols in acetonitrile with copper(II) bromide as the catalyst.[22] The reaction proceeds at ambient temperature under mild conditions and affords high yields of a range of sterically crowded esters such as *t*-butyl 1-adamantanecarboxylate (eq 15).

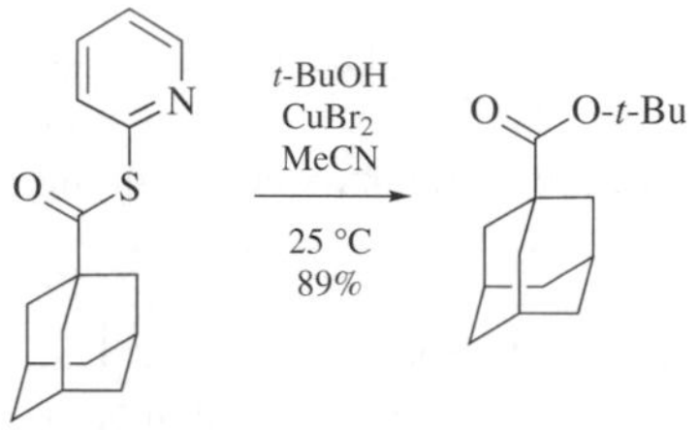

t-BuOH, CuBr₂, MeCN, 25 °C, 89% (15)

Conjugate Addition Catalyst. The 1,4-addition of Grignard reagents to α,β-unsaturated esters is promoted by catalytic $CuBr_2$ (1–5 mol%) with ***Chlorotrimethylsilane***/HMPA (eq 16).[23] Under these conditions the copper(II) species is not reduced by the Grignard reagent, resulting in high yields of the conjugate addition products.

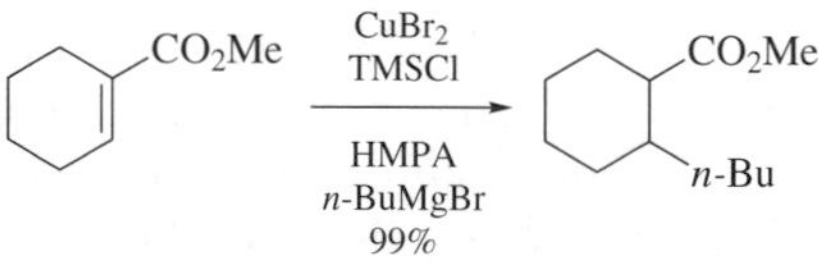

CuBr₂, TMSCl, HMPA, *n*-BuMgBr, 99% (16)

Oxidation of Stannanes and Alcohols. Allylstannanes have been oxidized with copper(II) bromide in the presence of various nucleophilic reagents (H_2O, ROH, AcONa, RNH_2) to afford the corresponding allylic alcohols, ethers, acetates, and amines.[24] This chemistry has been extended to trimethylsilyl enol ethers, which undergo a $CuBr_2$-induced carbon–carbon bond forming process with allylstannanes (eq 17).[25] Alkoxytributylstannanes may be converted to the corresponding aldehyde or ketone with two equivalents of $CuBr_2$/***Lithium Bromide*** in THF at ambient temperature (path a, eq 18).[26] A combination of copper(II) bromide/***Lithium t-Butoxide*** oxidizes alcohols to carbonyl compounds quite rapidly and in high yield (path b, eq 18).[27]

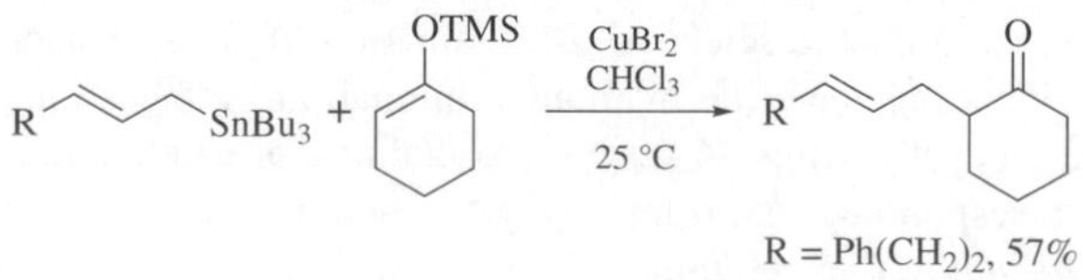

CuBr₂, CHCl₃, 25 °C (17)

R = $Ph(CH_2)_2$, 57%

(18)

(a) X = $SnBu_3$ **(a)** = $CuBr_2$, LiBr, Bu_3SnO-*t*-Bu
(b) X = H **(b)** = $CuBr_2$, LiO-*t*-Bu

Desilylbromination. β-Silyl ketones are desilylbrominated to α,β-unsaturated ketones with $CuBr_2$ in DMF.[28] This occurs spontaneously in cyclic ketones, while with open-chain ketones sodium bicarbonate is required to eliminate HBr from the β-bromo ketone thus formed. The carbon–silicon bond in organopentafluorosilicates prepared from alkenes and alkynes is cleaved with copper(II) bromide to give the corresponding alkyl and alkenyl bromides (eq 19).[29] The reaction is stereoselective; thus (*E*)-alkenyl bromides are obtained from (*E*)-alkenylsilicates.

1. $HSiCl_3$, H_2PtCl_6; 2. KF; $CuBr_2$ (19)

Reagent in the Sandmeyer and Meerwein Reactions. Diazonium salts of arylamines are converted to aryl halides (Sandmeyer reaction)[30] in the presence of copper(II) halides. Recent procedures have utilized *t*-butyl nitrite/$CuBr_2$[31] or *t*-butyl thionitrite/$CuBr_2$[32] combinations to afford aryl bromides from the corresponding arylamines in high yields (eq 20). The copper salt catalyzed haloarylation of alkenes with arenediazonium salts (Meerwein reaction) also proceeds with copper(II) halides. For example, treatment of *p*-aminoacetophenone with *t*-butyl nitrite/$CuBr_2$ in the presence of excess acrylic acid gives *p*-acetyl-α-bromohydrocinnamic acid (59% yield, eq 21).[33] The intramolecular version of this reaction, which affords halogenated dihydrobenzofurans, has been accomplished by reacting arenediazonium tetrafluoroborates with $CuBr_2$ in DMSO (eq 22).[34]

$CuBr_2$, *t*-BuONO, MeCN, Δ (20)

X = H, hal, CO_2R, NO_2, OMe, etc.

$CuBr_2$, *t*-BuONO, MeCN, 25 °C, 59% (21)

$CuBr_2$, DMSO, 25 °C, 82% (22)

Related Reagents. Bromine; N-Bromosuccinimide; Copper(I) Bromide.

1. (a) *FF* **1967**, *1*, 161. (b) Bauer, D. P.; Macomber, R. S. *JOC* **1975**, *40*, 1990.
2. King, L. C.; Ostrum, G. K. *JOC* **1964**, *29*, 3459.
3. Doifode, K. B.; Marathey, M. G. *JOC* **1964**, *29*, 2025.
4. Jemison, R. W. *AJC* **1968**, *21*, 217.
5. Glazier, E. R. *JOC* **1962**, *27*, 4397.
6. Paquette, L. A.; Branan, B. M.; Rogers, R. D. *T* **1992**, *48*, 297.
7. Miller, D. D.; Moorthy, K. B.; Hamada, A. *TL* **1983**, *24*, 555.
8. Nakatsuka, M.; Nakasuji, K.; Murata, I.; Watanabe, I.; Saito, G.; Enoki, T.; Inokuchi, H. *CL* **1983**, 905.
9. Matsumoto, M.; Ishida, Y.; Watanabe, N. *H* **1985**, *23*, 165.
10. Barrish, J. C.; Singh, J.; Spergel, S. H.; Han, W.-C.; Kissick, T. P.; Kronenthal, D. R.; Mueller, R. H. *JOC* **1993**, *58*, 4494.
11. Castro, C. E.; Gaughan, E. J.; Owsley, D. C. *JOC* **1965**, *30*, 587.
12. Doifode, K. B. *JOC* **1962**, *27*, 2665.
13. Koyano, T. *BCJ* **1971**, *44*, 1158.
14. (a) See Ref. 1a, p 162. (b) Mosnaim, D.; Nonhebel, D. C. *T* **1969**, *25*, 1591.
15. Mosnaim, D.; Nonhebel, D. C.; Russell, J. A. *T* **1969**, *25*, 3485.
16. Nonhebel, D. C.; Russell, J. A. *T* **1970**, *26*, 2781.
17. (a) Kovacic, P.; Davis, K. E. *JACS* **1964**, *86*, 427. (b) Kodomari, M.; Satoh, H.; Yoshitomi, S. *BCJ* **1988**, *61*, 4149.
18. Petruso, S.; Caronna, S.; Sprio, V. *JHC* **1990**, *27*, 1209.
19. Petruso, S.; Caronna, S.; Sferlazzo, M.; Sprio, V. *JHC* **1990**, *27*, 1277.
20. Petruso, S.; Caronna, S.; *JHC* **1992**, *29*, 355.
21. (a) Gruiec, A.; Foucaud, A.; Moinet, C. *NJC* **1991**, *15*, 943. (b) Chaintreau, A.; Adrian, G.; Couturier, D. *SC* **1981**, *11*, 669.
22. Kim, S.; Lee, J. I. *JOC* **1984**, *49*, 1712.
23. Sakata, H.; Aoki, Y.; Kuwajima, I. *TL* **1990**, *31*, 1161.
24. Takeda, T.; Inoue, T.; Fujiwara, T. *CL* **1988**, 985.
25. Takeda, T.; Ogawa, S.; Koyama, M.; Kato, T.; Fujiwara, T. *CL* **1989**, 1257.
26. Yamaguchi, J.; Takeda, T. *CL* **1992**, 423.
27. Yamaguchi, J.; Yamamoto, S.; Takeda, T. *CL* **1992**, 1185.
28. (a) *FF* **1989**, *14*, 100. (b) *FF* **1980**, *8*, 196.
29. Yoshida, J.; Tamao, K.; Kakui, T.; Kurita, A.; Murata, M.; Yamada, K.; Kumada, M. *OM* **1982**, *1*, 369.
30. Dickerman, S. C.; DeSouza, D. J.; Jacobson, N. *JOC* **1969**, *34*, 710.
31. Doyle, M. P.; Siegfried, B.; Dellaria, J. F. *JOC* **1977**, *42*, 2426.
32. Oae, S.; Shinhama, K.; Kim, Y. H. *BCJ* **1980**, *53*, 1065.
33. Doyle, M. P.; Siegfried, B.; Elliott, R. C.; Dellaria, J. F. *JOC* **1977**, *42*, 2431.
34. Meijs, G. F.; Beckwith, A. L. J. *JACS* **1986**, *108*, 5890.
35. Perrin, D. D.; Armarego, W. L. F. *Purification of Laboratory Chemicals*, 3rd ed.; Pergamon: New York, 1988; p 321.

Nicholas D. P. Cosford
SIBIA, La Jolla, CA, USA

Copper Bronze[1]

[7440-50-8] Cu (MW 63.54)

(Ullmann coupling;[6] thermal decomposition of diazo compounds;[9] vinyl ether synthesis;[10] Guerbet reaction[3])

Physical Data: mp 1083 °C; bp 2595 °C; *d* 8.95 g cm^{-3}.
Solubility: slowly sol in ammonia water.

Form Supplied in: copper-colored, finely ground powder; commercially available.
Preparative Methods: copper bronze can be activated either by the method of Vogel[1a] or by addition of a 2% solution of iodine in acetone.[2]
Handling, Storage, and Precautions: may lose its activity due to aging. Flammable when exposed to heat, sparks, and open flame. Irritant. May discolor on exposure to air and moisture. Incompatible with strong acids, strong oxidizing agents, acid chlorides, and halogens. Use of safety goggles and chemical-resistant gloves is strongly recommended. Avoid breathing the dust. Avoid contact in eyes, on skin, on clothing.

Guerbet Reaction. Preparation of ethylhexanol from butanol occurs within an autoclave at autogenous pressures of 50–60 atm and temperatures of about 300 °C in the presence of the hydrogenation–dehydrogenation catalyst, copper bronze. Butanol is first dehydrogenated to an aldehyde in the presence of sodium and copper bronze before undergoing aldol condensation and subsequent dehydration. The resulting α,β-unsaturated aldehyde is then hydrogenated, yielding 2-ethylhexanol.[3]

Ullmann Coupling. Under Ullmann conditions,[6] halogen-substituted aromatic compounds react with organic halides in the presence of activated copper bronze[1b,2,4] to give the corresponding coupled products in moderate yields. A recent example is the synthesis of perfluoroalkyl-substituted aromatic aldehydes via the coupling of two different substrates (i.e. perfluorooctyl iodide and *p*-iodobenzaldehyde) (eq 1).[5]

CHO, I + $CF_3(CF_2)_7I$ —(Cu, 112 °C; DMF)→ CHO, $CF_3(CF_2)_7$ (1)

Thermal Decomposition of Diazo Compounds. Copper bronze can promote the intramolecular cyclization of diazo ketones in boiling cyclohexane (eq 2); anhydrous ***Copper(II) Sulfate*** in boiling THF can also be used.[7] Other diazo compounds that have been shown to undergo copper-catalyzed thermal decomposition include ethyl diazopyruvates in the presence of enol ethers or alkynes[8] and diazoacetates (eq 3).[9]

MeO, O, N_2 —(Cu; cyclohexane; 55%)→ MeO, O (2)

OAc + N_2CHCO_2Me —(Cu; 48%)→ OAc, CO_2Me (3)

Vinyl Ether Synthesis. Rearrangement of β-alkoxycyclopropanecarboxylates in the presence of copper bronze catalyst produces vinyl ethers (eq 4). Vicinal alkoxy and ethoxycarbonyl groups activate the substrate as the reaction proceeds.[10]

R, MeO, CO_2Et —(Cu, reflux; 97%)→ R, MeO, CO_2Et (4)

Related Reagents. Copper.

---

1. *Vogel's Textbook of Practical Organic Chemistry*, 5th ed.; Wiley: New York, 1989; p 426. (b) Ref. 1(a), p 837. (c) *FF* **1967**, *1*, 155; **1972**, *3*, 62; **1974**, *4*, 100; **1984**, *11*, 49; **1986**, *12*, 140.
2. Fuson, R. C.; Cleveland, E. A. *OSC* **1955**, *3*, 339.
3. Weizmann, C.; Bergmann, E.; Sulzbacher, M. *JOC* **1950**, *15*, 54.
4. Weber, E.; Csöregh, I.; Stensland, B.; Czulger, M. *JACS* **1984**, *106*, 3297.
5. Paciorek, K. J. L.; Masuda, S. R.; Shih, J. G. *JFC* **1991**, *53*, 233.
6. Fanta, P. E. *S* **1974**, 9.
7. Chakrabortty, P. N.; Dasgupta, R.; Dasgupta, S. K.; Ghosh, S. R.; Ghatak, U. R. *T* **1972**, *28*, 4653.
8. Wenkert, E.; Alonso, M. E.; Buckwalter, B. L.; Sanchez, E. L. *JACS* **1983**, *105*, 2021.
9. Joshi, G. S.; Kulkani, G. H.; Shapiro, E. A. *CI(L)* **1989**, 424.
10. Doyle, M. P.; Van Leusen, D. *JACS* **1981**, *103*, 5917.

Edward J. Parish & Stephen A. Kizito
Auburn University, AL, USA

Copper(I) Chloride[1]

CuCl

[7758-89-6] ClCu (MW 99.00)

(catalyst for diazo and diazonium chemistry; for use with Grignard reagent[1])

Alternate Name: cuprous chloride.
Physical Data: mp 430 °C; *d* 4.140 g cm^{-3}.
Solubility: insol H_2O and most organic solvents; partially sol dimethyl sulfide (DMS).
Form Supplied in: light green-tinged white solid; 99.99% grade available commercially.
Handling, Storage, and Precautions: maintenance of a dry N_2 or Ar atmosphere is recommended.

General Discussion. Copper(I) chloride is the most popular Cu^I salt for use in many of the classic 'name reactions' of organic chemistry.[1a] Examples include the Sandmeyer reaction,[1b] the CuCl induced decomposition of arenediazonium salts to afford aryl chlorides,[2] and the related Bart reaction[1c] of arenediazonium salts with sodium metaarsenite to yield arylarsonic acids,[3] which is also catalyzed by CuCl. Kochi has elucidated the mechanisms of the Sandmeyer reaction and the Meerwein reaction[1d] of arenediazonium salts with alkenes and established that they proceed via a common radical intermediate which could be generated from other precursors.[4] Nagashima et al. took advantage of this fact when they cyclized an *N*-allyltrichloroacetamide derivative with CuCl,[5] in what is a significant extension and generalization of the Meerwein reaction (eq 1).

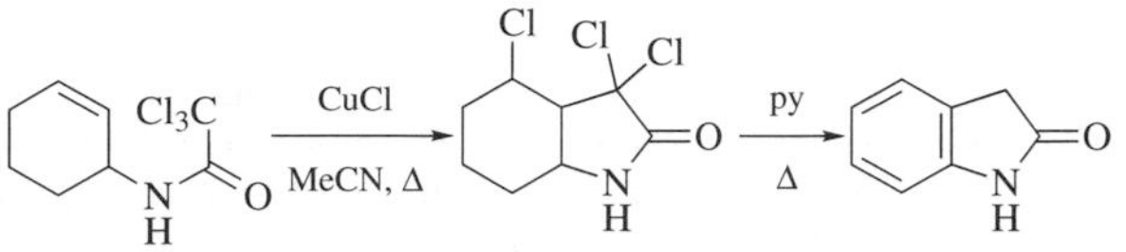

(1)

α-Diazo ketones with aryl substituents in the β′- or γ′-positions are decomposed by CuCl to give cyclized products. For example, Scott has reported an azulene synthesis starting from 3-phenylpropanoic acid,[6] and Iwata et al. have developed a spiroannulation method for phenolic α-diazo ketones,[7,8] as illustrated in eq 2 by the synthesis of α-chamigrene.[7] Many other examples[1e] of intramolecular reactions of diazo carbonyl compounds are known in an area that was pioneered by Stork,[9] who used ***Copper Bronze***. CuCl has also been used to catalyze cyclopropanations with ***Diazomethane***,[10] and with diazoacetic esters.[1f] It should be noted that CuI, CuBr, and CuCN have also been used to catalyze the reactions of diazo compounds, as have Cu^0 and Cu^{II}.[1e,f,9]

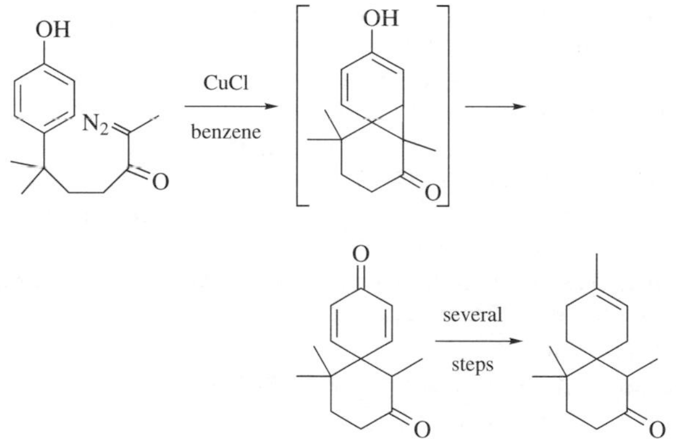

(2)

The Gattermann–Koch synthesis of aromatic aldehydes involves a Friedel–Crafts formylation by 'ClCHO' in the presence of ***Aluminum Chloride***[1] and CuCl catalysts.[1g,11] The reaction proceeds without Cu, but at much higher pressures.[1g]

The Glazer reaction couples terminal alkynes by using dioxygen and CuCl, usually in the presence of ammonia or another amine (pyridine, $EtNH_2$).[1h] Many examples are known; eq 3 shows an example in the overlapping areas of medium-sized ring formation and annulene chemistry.[12] A related reaction mixture which uses pyridine instead of ammonia oxidatively cleaves 1,2-diaminobenzene to *cis,cis*-muconontrile,[13a] and phenanthrenequinone to 2,2′-biphenyldicarboxylic acid.[14a] It also oxidizes the bishydrazones of α-diketones to alkynes and the monohydrazones to α-diazo ketones.[13b] 1,10-Phenanthroline has been used in conjunction with CuCl and dioxygen to oxidize alcohols to ketones or aldehydes.[14b]

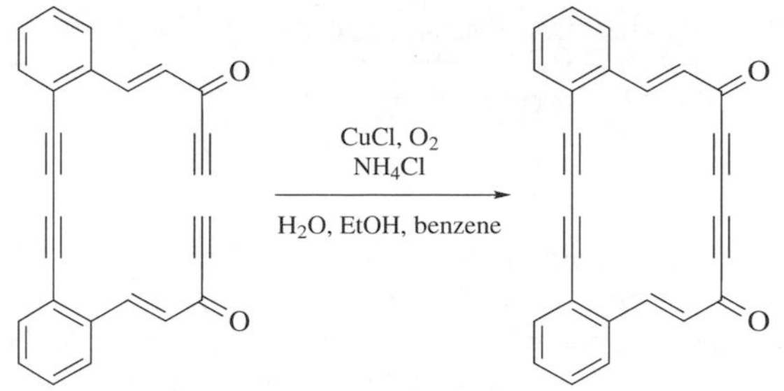

(3)

Marino has reported that cyclic α,β-unsaturated epoxides react with Cu carboxylates, generated in situ from Na carboxylates and CuCl, to give *syn*-1,4-diol derivatives.[15a] Goering has shown that allyl pivalates give α-alkylation with inversion of configuration with Grignard reagents catalyzed by CuCl, but (*anti*) γ-alkylation with ***Copper(I) Cyanide*** as catalyst.[15b]

Kharasch discovered that CuCl catalyzes the coupling of aryl Grignard reagents with aryl halides,[16a] and also the 1,4-addition of Grignard reagents to α,β-unsaturated ketones.[16b] While the former has been called 'the Kharasch reaction',[1i] today the latter is much more important synthetically,[1i–k] and it should be named for its inventor, who was one of the true pioneers in the synthetic applications of transition metals. One way to avoid confusion would be to call the reagent prepared from a Grignard reagent and a Cu^I salt a (catalytic or stoichiometric) 'Kharasch Reagent' by analogy with the commonly used 'Gilman Reagent', which is synonymous with cuprate prepared from 2 equiv of lithium reagent and a Cu^I salt.

While ***Copper(I) Iodide***, CuBr·DMS (see ***Copper(I) Bromide***[1]), and CuCN are now used more commonly than CuCl in the preparation of organocopper reagents, there are still interesting examples which utilize CuCl. For example, the addition of BuMgBr to sorbate esters gives a mixture of 1,2- and 1,4-addition products; however, upon the addition of CuCl, 1,6-addition predominates.[17] This example also illustrates one of the problems commonly encountered when trying to interpret the results of experiments using the Kharasch reagent. Experimenters frequently mix a Grignard reagent containing one halide with a Cu^I salt containing another. CuBr·DMS has also been shown to dramatically favor 1,6-addition of Grignard reagents (see ***Copper(I) Bromide***[1]).

Tandem organocuprate β-addition to α,β-unsaturated carbonyl compounds followed by α-alkylation, acylation or other functionalization of the regiospecifically generated α-enolate has become an important methodology.[1j,k] In what has been described as 'the first one-pot, three-component tandem vicinal difunctionalization reaction,'[1j] Stork used CuCl to catalyze the 1,4-addition of *m*-methoxyphenylmagnesium bromide to 5,5-dimethylcyclohex-2-en-1-one, followed by α-alkylation with allyl bromide,[18] which was the seminal step in the synthesis of lycopodine. Bunce and Harris have recently combined β-addition with *intramolecular* α-acylation in a tandem 'conjugate addition–Dieckmann condensation' reaction, which was used to prepare six-membered rings (eq 4).[19] It is interesting to note that the Stork conditions (RMgBr/CuCl) gave better results than the Gilman reagents prepared from 2 equiv of the corresponding Li reagent and either CuI or CuCN.[19]

$$\text{(diethyl 2-alkenylidene diester)} \xrightarrow[\substack{(6\text{ equiv each})\\ 96\%}]{\text{BuMgBr, CuCl}} \text{(2-ethoxycarbonyl-3-butylcyclohexanone)} \quad (4)$$

CuCl has been directly compared with CuBr, CuBr·DMS, CuI, CuCN, and ***Copper(I) Trifluoromethanesulfonate*** as a precursor for Gilman reagents,[20] and it appears to be inferior to the usual Cu^{I} salts for this important class of compounds. It was found early on that Cu^{II} salts catalyze many of the same organic reactions as the corresponding Cu^{I} salts,[1] e.g. a popular catalyst for coupling reactions with Grignard reagents is ***Dilithium Tetrachlorocuprate(II)*** = $CuCl_2 + 2LiCl$.[1i–l] It is generally assumed that the Cu^{II} salt is reduced to Cu^{I} under the reaction conditions; however, in most cases this has not been proven.

Along with CuBr and CuBr·DMS, CuCl continues to be a favored precursor for structural work and the preparation of new organocopper derivatives. For example, π-complexes between nonconjugated dienes and CuCl or CuBr have been characterized by X-ray crystallography.[21] Chaudret et al. have shown that $(C_5Me_5)Ru(PCy_3)H_3$ reacts smoothly with CuCl to give $(C_5Me_5)Ru(PCy_3)H_3CuCl$,[22] which was of interest in relation to the proposal that 'trihydrogen' ligands might exist. In these two cases the product still contains the Cl atom. Many interesting examples exist where it does not (see below). A study illustrating both Cl-containing and Cl-free products is provided by Zybill and Müller, who reacted hexaphenylcarbodiphosphorane with CuCl to form a 1:1 adduct, which they reacted further with $[C_5Me_5]^-$ to yield $(C_5Me_5)CuC(PPh_3)_2$.[23]

Both dinuclear and tetranuclear Cu alkyl complexes have been prepared from CuCl and 2-substituted pyridines containing bis(trimethylsilyl)methyl and trimethylsilylmethyl substituents, respectively.[24] $MeCu(PPh_3)_3$ has been prepared and studied by NMR and X-ray crystallography;[25] such phosphine-complexed organocopper reagents have proven useful for organic synthesis. Several copper(I) cupracarboranes have been synthesized, including an interesting trimer.[26] (Trimers are much less common than tetramers in Cu^{I} chemistry.[27]) In a most incredible communication, Fenske et al. characterize the brobdingnagian product of the reaction $146\,CuCl + 73\,Se(SiMe_3)_2 + 30\,PPh_3 \rightarrow Cu_{146}Se_{73}(PPh_3)_{30} + 146\,ClSiMe_3$.[28]

One of the most interesting arylcopper(I) compounds is mesitylcopper(I),[29] which is a pentamer in the solid state and a tetramer in the presence of S-ligands.[30] It has recently been shown to split dioxygen,[31] and is highlighted here because it is prepared from CuCl. Hexafluoroacetylacetonatocopper(I)–alkyne complexes (prepared from CuCl) have been shown to be useful in the chemical vapor deposition of Cu metal.[32] (*E*)-1-Trimethylsilyl-1-alkenes have been prepared in high yields by the CuCl-catalyzed decomposition of α-trimethylsilyldiazoalkanes.[33] Finally, Indian chemists have shown that 1,2-diketones are produced in good yield when $Na(RCO)Fe(CO)_4$ is treated with CuCl.[34]

Related Reagents. Copper(II) Chloride; Copper(II) Chloride–Copper(II) Oxide; Copper(I) Chloride–Oxygen; Copper(I) Chloride–Sulfur Dioxide; Copper(I) Chloride Tetrabutylammonium chloride; Iodine–Aluminum(III) Chloride–Copper(II) Chloride; Iodine–Copper(I) Chloride–Copper(II) Chloride; Iodine–Copper(II) Chloride; Methylmagnesium Iodide–Copper(I) Chloride; Palladium(II) Chloride–Copper(I) Chloride; Palladium(II) Chloride–Copper(II) Chloride; Phenyl Selenocyanate–Copper(II) Chloride; Vinylmagnesium Chloride–Copper(I) Chloride; Zinc–Copper(I) Chloride.

---

1. (a) *FF* **1975**, *5*, 164 and citations to previous volumes therein. (b) Mowry, D. T. *CRV* **1948**, *48*, 189. (c) Hamilton, C. S.; Morgan, J. F. *OR* **1944**, *2*, 415. (d) Rondestvedt, C. S., Jr., *OR* **1976**, *24*, 225. (e) Burke, S. D.; Grieco, P. A. *OR* **1979**, *26*, 361. (f) Dave, V.; Warnhoff, E. W. *OR* **1970**, *18*, 217. (g) Crounse, N. N. *OR* **1949**, *5*, 290. (h) Nakagawa, M. In *The Chemistry of the Carbon–Carbon Triple Bond*; Patai, S., Ed.; Wiley: New York, **1978**; Part 2, p 635. (i) Erdik, E. *T* **1984**, *40*, 641. (j) Chapdelaine, M. J.; Hulce, M. *OR* **1990**, *38*, 225. (k) Taylor, R. J. K. *S* **1985**, 364. (l) Lipshutz, B. H.; Sengupta, S. *OR* **1992**, *41*, 135.
2. (a) Marvel, C. S.; McElvain, S. M. *OSC* **1941**, *1*, 170. (b) Buck, J. S.; Ide, W. S. *OSC* **1943**, *2*, 130.
3. Ruddy, A. W.; Starkey, E. B. *OSC* **1955**, *3*, 665.
4. Kochi, J. K. *JACS* **1957**, *79*, 2942.
5. Nagashima, H.; Ara, K.-i.; Wakamatsu, H.; Itoh, K. *CC* **1985**, 518.
6. Scott, L. T. *CC* **1973**, 882.
7. Iwata, C.; Yamada, M.; Shinoo, Y. *CPB* **1979**, *27*, 274.
8. Iwata, C.; Yamada, M.; Shinoo, Y.; Kobayashi, K.; Okada, H. *CC* **1977**, 888.
9. Stork, G.; Ficini, J. *JACS* **1961**, *83*, 4678.
10. Doering, W. v. E.; Roth, W. R. *T* **1963**, *19*, 715.
11. Coleman, G. H.; Craig, D. *OSC* **1943**, *2*, 583.
12. Yamamoto, K.; Sondheimer, F. *AG(E)* **1973**, *12*, 68.
13. (a) Tsuji, J.; Takayanagi, H. *OSC* **1988**, *6*, 662. (b) Tsuji, J.; Takahashi, H.; Kajimoto, T. *TL* **1973**, *14*, 4573.
14. (a) Balogh-Hergovich, É.; Speier, G.; Tyeklár, Z. *S* **1982**, 731. (b) Jallabert, C.; Riviere, H. *TL* **1977**, *18*, 1215.
15. (a) Marino, J. P.; Jaén, J. C. *TL* **1983**, *24*, 441. (b) Tseng, C. C.; Yen, S.-J.; Goering, H. L. *JOC* **1986**, *51*, 2892.
16. (a) Kharasch, M. S.; Fields, E. K. *JACS* **1941**, *63*, 2316. (b) Kharasch, M. S.; Tawney, P. O. *JACS* **1941**, *63*, 2308.
17. Munch-Petersen, J.; Bretting, C.; Jørgensen, P. M.; Refn, S.; Andersen, V. K.; Jart, A. *ACS* **1961**, *15*, 277.
18. Stork, G. *PAC* **1968**, *17*, 383.
19. Bunce, R. A.; Harris, C. R. *JOC* **1992**, *57*, 6981.
20. Bertz, S. H.; Gibson, C. P.; Dabbagh, G. *TL* **1987**, *28*, 4251.
21. Håkansson, M.; Jagner, S.; Clot, E.; Eisenstein, O. *IC* **1992**, *31*, 5389.
22. Chaudret, B.; Commenges, G.; Jalon, F.; Otero, A. *CC* **1989**, 210.
23. Zybill, C.; Müller, G. *OM* **1987**, *6*, 2489.
24. Papasergio, R. I.; Raston, C. L.; White, A. H. *JCS(D)* **1987**, 3085.
25. Coan, P. S.; Folting, K.; Huffman, J. C.; Caulton, K. G. *OM* **1989**, *8*, 2724.
26. Kang, H. C.; Do, Y.; Knobler, C. B.; Hawthorne, M. F. *IC* **1988**, *27*, 1716.
27. Bertz, S. H.; Dabbagh, G.; He, X.; Power, P. P. *JACS* **1993**, *115*, 11640.
28. Krautscheid, H.; Fenske, D.; Baum, G.; Semmelmann, M. *AG(E)* **1993**, *32*, 1303.
29. (a) Tsuda, T.; Yazawa, T.; Watanabe, K.; Fujii, T.; Saegusa, T. *JOC* **1981**, *46*, 192. (b) Tsuda, T.; Watanabe, K.; Miyata, K.; Yamamoto, H.; Saegusa, T. *IC* **1981**, *20*, 2728.
30. (a) Gambarotta, S.; Floriani, C.; Chiesi-Villa, A.; Guastini, C. *CC* **1983**, 1156. (b) Meyer, E. M.; Gambarotta, S.; Floriani, C.; Chiesi-Villa, A.; Guastini, C. *OM* **1989**, *8*, 1067.
31. Håkansson, M.; Örtendahl, M.; Jagner, S.; Sigalas, M. P.; Eisenstein, O. *IC* **1993**, *32*, 2018.
32. Jain, A.; Chi, K.-M.; Kodas, T. T.; Hampden-Smith, M. J.; Farr, J. D.; Paffett, M. F. *Chem. Mater.* **1991**, *3*, 995.

33. Aoyama, T.; Shioiri, T. *TL* **1988**, *29*, 6295.
34. Periasamy, M.; Devasagayaraj, A.; Radhakrishnan, U. *OM* **1993**, *12*, 1424.

Steven H. Bertz & Edward H. Fairchild
LONZA, Annandale, NJ, USA

Copper(II) Chloride

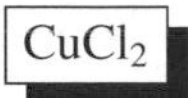

[7447-39-4]	Cl_2Cu	(MW 134.45)
($\cdot 2H_2O$)		
[10125-13-0]	$Cl_2CuH_4O_2$	(MW 170.48)

(chlorinating agent; oxidizing agent; Lewis acid)

Physical Data: anhydrous: *d* 3.386 g cm^{-3}; mp 620 °C (reported mp of 498 °C actually describes a mixture of $CuCl_2$ and CuCl); partially decomposes above 300 °C to CuCl and Cl_2; dihydrate *d* 2.51 g cm^{-3}; mp 100 °C.
Solubility: anhydrous: sol water, alcohol, and acetone; dihydrate: sol water, methanol, ethanol; mod sol acetone, ethyl acetate; sl sol Et_2O.
Form Supplied in: anhydrous: hygroscopic yellow to brown microcrystalline powder; dihydrate: green to blue powder or crystals; also supplied as reagent adsorbed on alumina (approx. 30 wt % $CuCl_2$ on alumina).
Analysis of Reagent Purity: by iodometric titration.[70]
Purification: cryst from hot dil aq HCl (0.6 mL g^{-1}) by cooling in a $CaCl_2$–ice bath.[71]
Handling, Storage, and Precautions: the anhydrous solid should be stored in the absence of moisture, since the dihydrate is formed in moist air. Irritating to skin and mucous membranes.

Chlorination of Carbonyls. Copper(II) chloride effects the α-chlorination of various carbonyl functional groups.[1] The reaction is usually performed in hot, polar solvents containing ***Lithium Chloride***, which enhances the reaction rate. For example, butyraldehyde is α-chlorinated in DMF (97% conversion, eq 1) while the same reaction in methanol leads to an 80% yield of the corresponding α-chloro dimethyl acetal (eq 2).[2]

$CuCl_2$; DMF, Δ; 97% (1)

$CuCl_2$; MeOH, Δ; 80% (2)

The process has been extended to carboxylic acids, anhydrides, and acid chlorides by using an inert solvent such as sulfolane.[3] 4-Oxo-4,5,6,7-tetrahydroindoles are selectively α-chlorinated, allowing facile transformation to 4-hydroxyindoles (eq 3).[4] The ability of the reaction to form α-chloro ketones selectively has been further improved by the use of trimethylsilyl enol ethers as substrates.[5] Recently, phase-transfer conditions have been employed in a particularly difficult synthesis of RCH(Cl)C(O)Me selectively from the parent ketones (eq 4).[6]

$CuCl_2$; aq AcOH, Δ; R = Ts, 86%; R = Bn, 81% (3)

$CuCl_2$, H_2O, $EtNH_3NO_3$; $C_5H_5N(CH_2)_{15}MeCl$; n = 4, 70% (4)
R = Me$(CH_2)_n$, n = 2–5, 8

Chlorination of Aromatics. Aromatic systems may be chlorinated by the reagent. For example, 9-chloroanthracene is prepared in high yield by heating anthracene and $CuCl_2$ in carbon tetrachloride (eq 5).[7] When the 9-position is blocked by a halogen, alkyl, or aryl group, the corresponding 10-chloroanthracenes are formed by heating the reactants in chlorobenzene.[8,9] Under similar conditions, 9-acylanthracenes give 9-acyl-10-chloroanthracenes as the predominant products.[10] Polymethylbenzenes are efficiently and selectively converted to the nuclear chlorinated derivatives by $CuCl_2$/***Alumina***[1] (eq 6).[11]

$CuCl_2$; Δ (5)
R = H, Me, Ph, Ac

$CuCl_2$, Al_2O_3; PhCl, Δ; 82% (6)

Reactions with Alkoxy and Hydroxy Aromatics. Hydroxy aromatics such as phenols and flavanones undergo aromatic nuclear chlorination with copper(II) chloride.[12] Thus heating 3,5-xylenol with a slight excess of the reagent in toluene at 90 °C gave a 93% yield of 4-chloro-3,5-xylenol (eq 7).[13] 2-Alkoxynaphthalenes are similarly halogenated at the 1-position.[14] Attempted reaction of $CuCl_2$ with anisole at 100 °C for 5 h gave no products; in contrast, it was found that alkoxybenzenes were almost exclusively *para*-chlorinated (92–95% *para*:0.5–3% *ortho*) using $CuCl_2/Al_2O_3$ (eq 8).[15] Anisole reacts with benzyl sulfides in the presence of equimolar $CuCl_2$ and ***Zinc Chloride*** to give anisyl(phenyl)methanes (*para*:*ortho* = 2:1, eq 9).[16,17]

$CuCl_2$; PhMe, Δ; 93% (7)

$$\text{PhOEt} \xrightarrow[\text{PhCl, }\Delta]{CuCl_2,\ Al_2O_3} \text{4-ClC}_6\text{H}_4\text{OEt (94\%)} + \text{2-ClC}_6\text{H}_4\text{OEt (<0.5\%)} \quad (8)$$

$$\text{PhOMe} \xrightarrow[\text{BnSMe, }\Delta,\ 53\%]{CuCl_2,\ ZnCl_2} \text{4-BnC}_6\text{H}_4\text{OMe} + \text{2-BnC}_6\text{H}_4\text{OMe} \ (2:1) \quad (9)$$

Reactions with Active Methylene-Containing Compounds. 9-Alkoxy(or acyloxy)-10-methylanthracenes react with $CuCl_2$ to give coupled products (eq 10), while the analogous 9-alkoxy(or acyloxy)-10-benzyl(or ethyl)anthracenes react at the alkoxy or acyloxy group to afford 10-benzylidene(or ethylidene)anthrones (eq 11).[18] The reactions are believed to proceed via a radical mechanism.

(10) R = Me, Ac; reagent $CuCl_2$

(11) R = Me, Ph; reagent $CuCl_2$

Under similar conditions, 9-alkyl(and aryl)-10-halogenoanthracenes give products resulting from replacement of the halogen, alkyl, or aryl groups with halogen from the $CuCl_2$.[19] Boiling toluene reacts with $CuCl_2$ to yield a mixture of phenyltolylmethanes.[20]

Lithium enolates of ketones[21] and esters[22] undergo a coupling reaction with copper(II) halides to afford the corresponding 1,4-dicarbonyl compounds. Thus treating a 3:1 mixture of *t*-butyl methyl ketone and acetophenone with ***Acetaldehyde 3-Bromopropyl Ethyl Acetal*** and $CuCl_2$ gives a 60% yield of the cross-coupled product (eq 12).

(12) 3:1; 1. LDA, 2. $CuCl_2$, 60%

The intramolecular variant of this reaction producing carbocyclic derivatives has been reported.[23] Copper(II) chloride catalyzes the Knoevenagel condensation of 2,4-pentanedione with aldehydes and tosylhydrazones (eq 13).[24] The reagent also catalyzes the reaction of various 1,3-dicarbonyls with dithianes such as benzaldehyde diethyl dithioacetal to give the corresponding condensation products (eq 14).[25]

(13) R = alkyl, aryl; $CuCl_2$, THF, 25 °C, 48–95%

(14) $CuCl_2$, THF, 25 °C; R = Me, 65%; R = OEt, 46%

Catalyst for Conjugate Additions. The catalytic effect of copper(II) chloride on the 1,4-addition of β-dicarbonyl compounds to (arylazo)alkenes[26,27] and aminocarbonylazoalkenes[28,29] has been studied in some detail. The reactions proceed at ambient temperature in THF and afford the corresponding pyrrole derivatives (eq 15). This mild method requires no other catalyst and succeeds with β-diketones, β-ketoesters, and β-ketoamides. Copper(II) chloride also catalyzes the addition of water, alcohols, phenol, and aromatic amines to arylazoalkenes (eq 16).[30]

(15) $CuCl_2$, THF, 25 °C; X = alkyl, aryl, OR, NHR; R^1 = Ar, ArNHCO; R^2, R^3 = alkyl, aryl, CO_2R

(16) $ArNH_2$ + $R^1N{=}NC(R^2){=}CHR^3$ → $R^1NHN{=}C(R^2)CH(R^3)NHAr$; $CuCl_2$, THF, 25 °C; R^1, R^2, R^3 = Ar

Oxidation and Coupling of Phenolic Derivatives. In the presence of oxygen, copper(II) chloride converts phenol derivatives to various oxidation products. Depending on the reaction conditions, quinones and/or coupled compounds are formed.[31] Several groups have examined different sets of conditions employing $CuCl_2$ to favor either of these products. Thus 2,3,6-trimethylphenol was selectively oxidized to trimethyl-*p*-benzoquinone with $CuCl_2$/amine/O_2 as the catalyst (eq 17),[32] while 2,4,6-trimethylphenol was converted to 3,5-dimethyl-4-hydroxybenzaldehyde using a catalytic system employing either acetone oxime or amine (eq 18).[33,34]

(17) $CuCl_2$, O_2, Et_2NH, ROH, 25 °C; 76.7% + 0.9% coupled product

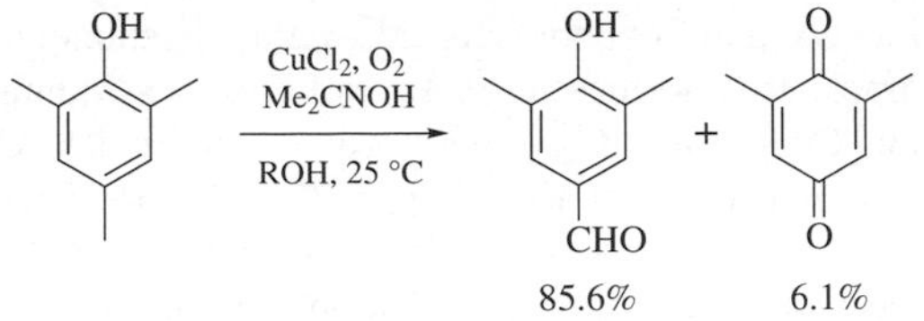

(18)

The oxidation of alkoxyphenols to the corresponding quinones has been studied,[35] and even benzoxazole derivatives are oxidized by a mixture of copper(II) chloride and ***Iron(III) Chloride*** (eq 19).[36] A $CuCl_2/O_2$/alcohol catalytic system has been used for the oxidative coupling of monophenols.[37]

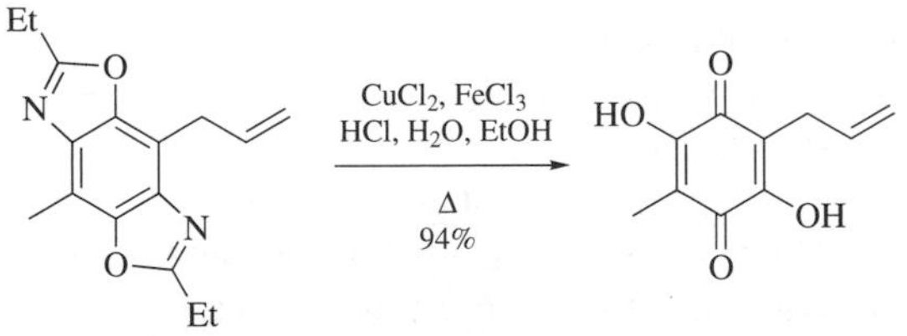

(19)

Copper(II) amine complexes are very effective catalysts for the oxidative coupling of 2-naphthols to give symmetrical 1,1′-binaphthalene-2,2′-diols.[38] Recent work has extended this methodology to the cross-coupling of various substituted 2-naphthols.[39,40] For example, 2-naphthol and 3-methoxycarbonyl-2-naphthol are coupled under strictly anaerobic conditions using $CuCl_2$/***t-Butylamine*** in methanol to give the unsymmetrical binaphthol in 86% yield (eq 20).

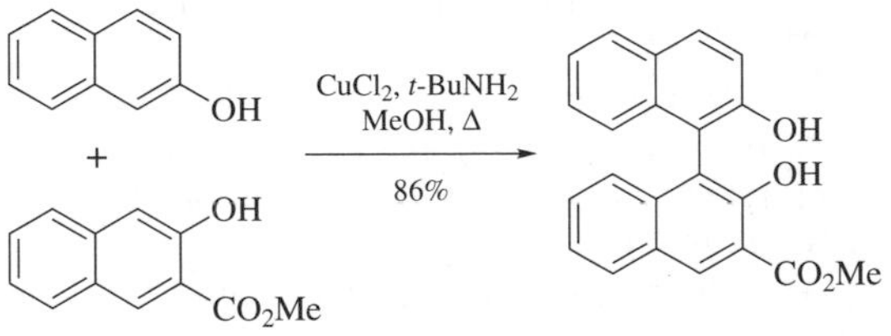

(20)

Other ligands such as methoxide are also effective; a mechanistic study indicates that the selectivity for cross- rather than homo-coupling is dependent upon the copper:ligand ratio.[41] A 1:1 mixture of 2-naphthol and 2-naphthylamine is cross-coupled with $CuCl_2$/benzylamine to give 2-amino-2′-hydroxy-1,1′-binaphthyl (68% yield, eq 21).[42] The cross-coupled products from these reactions are important in view of their use as chiral ligands for asymmetric synthesis.

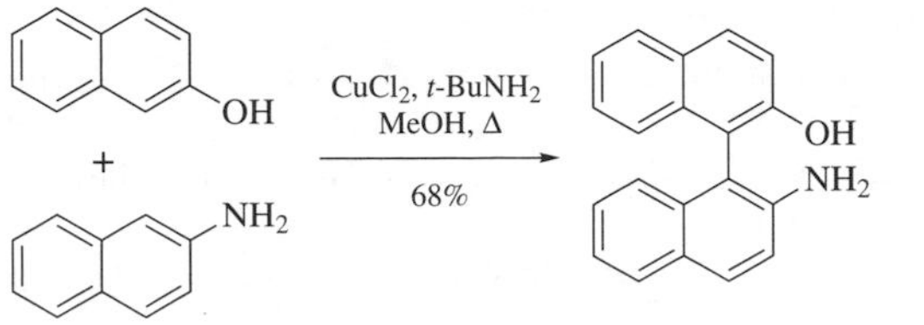

(21)

Dioxygenation of 1,2-Diones. 1,2-Cyclohexanedione derivatives have been converted to the corresponding 1,5-dicarbonyl compounds by oxidation with O_2 employing copper(II) chloride as the catalyst.[43] More recently, $CuCl_2$–***Hydrogen Peroxide*** has been used to prepare terminal dicarboxylic acids in high yield.[44] While 1,2-cyclohexanedione afforded α-chloroadipic acid in 85% yield, 1,2-cyclododecanedione was converted to 1,12-dodecanedioic acid in 47% yield under identical conditions (eq 22).

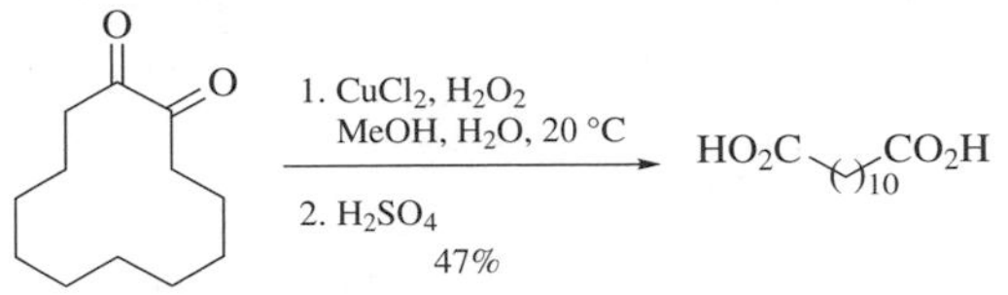

(22)

Addition of Sulfonyl Chlorides to Unsaturated Bonds. The addition of alkyl and aryl sulfonyl chlorides across double and triple bonds is catalyzed by copper(II) chloride.[45–51] The reaction appears to be quite general and proceeds via a radical chain mechanism. The 2-chloroethyl sulfones produced in the reaction with alkenes undergo base-induced elimination to give vinyl sulfones (eq 23).[45–48] 1,3-Dienes similarly react, yielding 1,4-addition products (eq 24) which may be dehydrohalogenated to 1,3-unsaturated sulfones.[45,49]

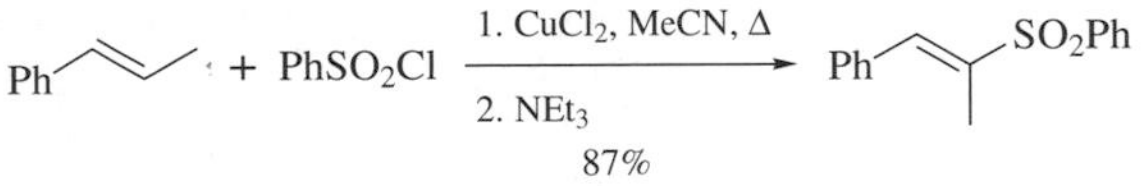

(23)

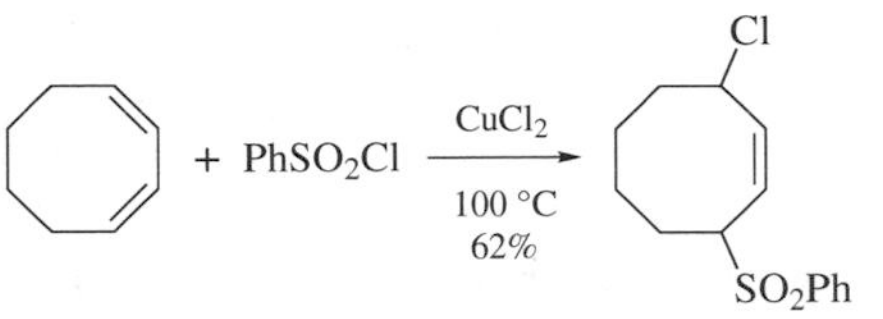

(24)

The stereoselectivity of the addition to alkynes can be controlled by varying the solvent or additive, and thus favoring either the *cis* or *trans* β-chlorovinyl sulfone.[50,51] For example (eq 25), when benzenesulfonyl chloride is reacted with phenylacetylene in acetonitrile with added triethylamine hydrochloride, the *trans*:*cis* ratio is 92:8, while the same reaction performed in CS_2 without additive favors the *cis* isomer (16:84).

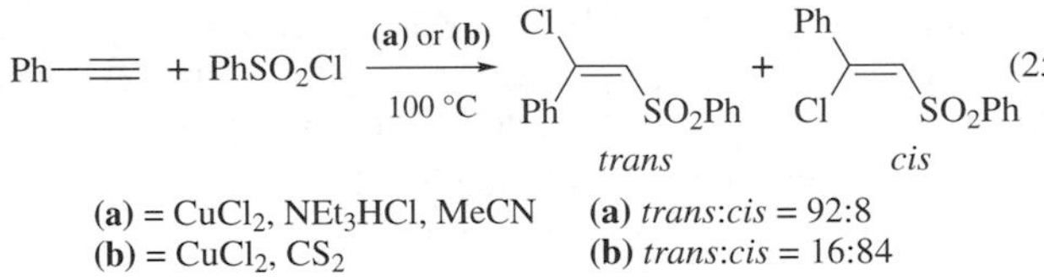

(25)

Acylation Catalyst. *N*-Trimethylsilyl derivatives of (+)-bornane-2,10-sultam (Oppolzer's chiral sultam) and chiral 2-oxazolidinones (the Evans chiral auxiliaries) are *N*-acylated with a number of acyl chlorides including acryloyl chloride in refluxing benzene in the presence of $CuCl_2$.[52] The *N*-acylated products were prepared in high yields; the method does not require an aqueous workup, making it advantageous for large-scale preparations.

Racemization Suppression in Peptide Couplings. A mixture of copper(II) chloride and ***Triethylamine*** catalyzes the formation of peptide bonds.[53] Furthermore, when used as an additive, $CuCl_2$ suppresses racemization in both the carbodiimide[54] and mixed anhydride[55] peptide coupling methods. Recently it was shown that a combination of ***1-Hydroxybenzotriazole*** and $CuCl_2$ gives improved yields of peptides while eliminating racemization.[56,57]

Reaction with Palladium Complexes. π-Allylpalladium complexes undergo oxidative cleavage with copper(II) chloride to form allyl chlorides with the concomitant release of $PdCl_2$ (eq 26).[58]

$$\text{(26)}$$

This methodology has been used in the dimerization of allenes to 2,3-bis(chloromethyl)butadienes.[59] 1,5-Bismethylenecyclooctane was transformed into the bridgehead-substituted bicyclo[3.3.1]nonane system using $CuCl_2$/HOAc/NaOAc, while the same substrate produced bicyclo[4.3.1]decane derivatives (eq 27) with a ***Palladium(II) Chloride***/$CuCl_2$ catalytic system.[60]

$$\text{(27)}$$

While reaction of a steroidal π-allylpalladium complex with AcOK yields the allyl acetate arising from *trans* attack, treatment of a steroidal alkene with $PdCl_2/CuCl_2$/AcOK/AcOH gave the allyl acetate arising from *cis* attack.[61]

Reoxidant in Catalytic Palladium Reactions. Copper(II) chloride has been used extensively in catalytic palladium chemistry for the regeneration of Pd^{II} in the catalytic cycle. In particular, the reagent has found widespread use in the carbonylation of alkenes,[62–64] alkynes,[65] and allenes[66,67] to give carboxylic acids and esters using $PdCl_2/CuCl_2$/CO/HCl/ROH, and in the oxidation of alkenes to ketones with a catalytic $PdCl_2/CuCl_2/O_2$ system (the Wacker reaction).[68] The $PdCl_2/CuCl_2$/CO/NaOAc catalytic system has been used in a mild method for the carbonylation of β-aminoethanols, diols, and diol amines (eq 28).[69]

$$\text{(28)}$$

Cyclopropanation with $CuCl_2$–$Cu(OAc)_2$ Catalyst. ***Ethyl Cyanoacetate*** reacts with alkenes under $CuCl_2$–***Copper(II) Acetate*** catalysis to give cyclopropanes.[72] Thus heating cyclohexene in DMF (110 °C, 5 h) with this reagent combination gives a 53% yield of the isomeric cyclopropanes. The reaction also proceeds with styrene, 1-decene, and isobutene. Byproducts formed from the addition to the alkene are removed with ***Potassium Permanganate***.

Related Reagents. Chlorine; N-Chlorosuccinimide; Copper(I) Chloride; Copper(II) Chloride–Copper(II) Oxide; Copper(I) Chloride–Oxygen Copper(I) chloride–tetrabutylammonium chloride Copper(I) Chloride–Sulfur Dioxide; Iodine–Aluminum(III) Chloride–Copper(II) Chloride; Iodine–Copper(II) Chloride; Iodine–Copper(I) Chloride–Copper(II) Chloride; Methylmagnesium Iodide–Copper(I) Chloride; Palladium(II) Chloride–Copper(I) Chloride; Palladium(II) Chloride–Copper(II) Chloride; Phenyl Selenocyanate–Copper(II) Chloride; Vinylmagnesium Chloride–Copper(I) Chloride; Zinc–Copper(I) Chloride.

1. *FF* **1969**, *2*, 84.
2. Castro, C. E.; Gaughan, E. J.; Owsley, D. C. *JOC* **1965**, *30*, 587.
3. Louw, R. *CC* **1966**, 544.
4. Matsumoto, M.; Ishida, Y.; Watanabe, N. *H* **1985**, *23*, 165.
5. *FF* **1982**, *10*, 106.
6. Atlamsani, A.; Brégeault, J.-M. *NJC* **1991**, *15*, 671.
7. (a) *FF* **1967**, *1*, 163. (b) Nonhebel, D. C. *OSC* **1973**, *5*, 206.
8. Mosnaim, D.; Nonhebel, D. C. *T* **1969**, *25*, 1591.
9. Mosnaim, D.; Nonhebel, D. C.; Russell, J. A. *T* **1969**, *25*, 3485.
10. Nonhebel, D. C.; Russell, J. A. *T* **1970**, *26*, 2781.
11. Kodomari, M.; Satoh, H.; Yoshitomi, S. *BCJ* **1988**, *61*, 4149.
12. *FF* **1980**, *8*, 120.
13. Crocker, H. P.; Walser, R. *JCS(C)* **1970**, 1982.
14. *FF* **1975**, *5*, 158.
15. Kodomari, M.; Takahashi, S.; Yoshitomi, S. *CL* **1987**, 1901.
16. Mukaiyama, T.; Narasaka, K.; Hokonoki, H. *JACS* **1969**, *91*, 4315.
17. Mukaiyama, T.; Maekawa, K.; Narasaka, K. *TL* **1970**, 4669.
18. (a) *FF* **1969**, *2*, 86. (b) Mosnaim, A. D.; Nonhebel, D. C.; Russell, J. A. *T* **1970**, *26*, 1123.
19. Mosnaim, D. A.; Nonhebel, D. C. *JCS(C)* **1970**, 942.
20. Cummings, C. A.; Milner, D. J. *JCS(C)* **1971**, 1571.
21. (a) *FF* **1977**, *6*, 139. (b) Ito, Y.; Konoike, T.; Harada, T.; Saegusa, T. *JACS* **1977**, *99*, 1487.
22. Rathke, M. W.; Lindert, A. *JACS* **1971**, *93*, 4605.
23. (a) *FF* **1981**, *9*, 123. (b) Babler, J. H.; Sarussi, S. J. *JOC* **1987**, *52*, 3462.
24. Attanasi, O.; Filippone, P.; Mei, A. *SC* **1983**, *13*, 1203.
25. Mukaiyama, T.; Narasaka, K.; Maekawa, K.; Hokonoki, H. *BCJ* **1970**, *43*, 2549.
26. Attanasi, O.; Santeusanio, S. *S* **1983**, 742.
27. Attanasi, O.; Bonifazi, P.; Foresti, E.; Pradella, G. *JOC* **1982**, *47*, 684.
28. Attanasi, O.; Filippone, P.; Mei, A.; Santeusanio, S.; Serra-Zanetti, F. *S* **1985**, 157.
29. Attanasi, O.; Filippone, P.; Mei, A.; Santeusanio, S. *S* **1984**, 671.
30. Attanasi, O.; Filippone, P. *S* **1984**, 422.
31. Hewitt, D. G. *JCS(C)* **1971**, 2967.
32. Shimizu, M.; Watanabe, Y.; Orita, H.; Hayakawa, T.; Takehira, K. *BCJ* **1992**, *65*, 1522.
33. Shimizu, M.; Watanabe, Y.; Orita, H.; Hayakawa, T.; Takehira, K. *BCJ* **1993**, *66*, 251.
34. Takehira, K.; Shimizu, M.; Watanabe, Y.; Orita, H.; Hayakawa, T. *TL* **1990**, *31*, 2607.
35. Matsumoto, M.; Kobayashi, H. *SC* **1985**, *15*, 515.
36. Hegedus, L. S.; Odle, R. R.; Winton, P. M.; Weider, P. R. *JOC* **1982**, *47*, 2607.
37. Takizawa, Y.; Munakata, T.; Iwasa, Y.; Suzuki, T.; Mitsuhashi, T. *JOC* **1985**, *50*, 4383.
38. Brussee, J.; Groenendijk, J. L. G.; Koppele, J. M.; Jansen, A. C. A. *T* **1985**, *41*, 3313.
39. Hovorka, M.; Günterová, J.; Závada, J. *TL* **1990**, *31*, 413.

40. Hovorka, M.; Ščigel, R.; Gunterová, J.; Tichý, M.; Závada, J. *T* **1992**, *48*, 9503.
41. Hovorka, M.; Závada, J. *T* **1992**, *48*, 9517.
42. Smrčina, M.; Lorenc, M.; Hanuš, V.; Kočovský, P. *SL* **1991**, 231.
43. Utaka, M.; Hojo, M.; Fujii, Y.; Takeda, A. *CL* **1984**, 635.
44. Starostin, E. K.; Mazurchik, A. A.; Ignatenko, A. V.; Nikishin, G. I. *S* **1992**, 917.
45. Asscher, M.; Vofsi, D. *JCS* **1964**, 4962.
46. *FF* **1975**, *5*, 158.
47. Truce, W. E.; Goralski, C. T. *JOC* **1971**, *36*, 2536.
48. Truce, W. E.; Goralski, C. T.; Christensen, L. W.; Bavry, R. H. *JOC* **1970**, *35*, 4217.
49. Truce, W. E.; Goralski, C. T. *JOC* **1970**, *35*, 4220.
50. Amiel, Y. *TL* **1971**, 661.
51. *FF* **1974**, *4*, 107.
52. Thom, C.; Kocieński, P. *S* **1992**, 582.
53. *FF* **1975**, *5*, 158.
54. Miyazawa, T.; Otomatsu, T.; Yamada, T.; Kuwata, S. *TL* **1984**, *25*, 771.
55. Miyazawa, T.; Donkai, T.; Yamada, T.; Kuwata, S. *CL* **1989**, 2125.
56. Miyazawa, T.; Otomatsu, T.; Fukui, Y.; Yamada, T.; Kuwata, S. *CC* **1988**, 419.
57. Miyazawa, T.; Otomatsu, T.; Fukui, Y.; Yamada, T.; Kuwata, S. *Int. J. Pept. Prot. Res.* **1992**, *39*, 308.
58. Castanet, Y.; Petit, F. *TL* **1979**, *34*, 3221.
59. Hegedus, L. S.; Kambe, N.; Ishii, Y.; Mori, A. *JOC* **1985**, *50*, 2240.
60. Heumann, A.; Réglier, M.; Waegell, B. *TL* **1983**, *24*, 1971.
61. Horiuchi, C. A.; Satoh, J. Y. *JCS(P1)* **1982**, 2595.
62. Alper, H.; Woell, J. B.; Despeyroux, B.; Smith, D. J. H. *CC* **1983**, 1270.
63. Inomata, K.; Toda, S.; Kinoshita, H. *CL* **1990**, 1567.
64. Toda, S.; Miyamoto, M.; Kinoshita, H.; Inomata, K. *BCJ* **1991**, *64*, 3600.
65. Alper, H.; Despeyroux, B.; Woell, J. B. *TL* **1983**, *24*, 5691.
66. Alper, H.; Hartstock, F. W.; Despeyroux, B. *CC* **1984**, 905.
67. Gallagher, T.; Davies, I. W.; Jones, S. W.; Lathbury, D.; Mahon, M. F.; Molloy, K. C.; Shaw, R. W.; Vernon, P. *JCS(P1)* **1992**, 433.
68. Januszkiewicz, K.; Alper, H. *TL* **1983**, *24*, 5159.
69. Tam, W. *JOC* **1986**, *51*, 2977.
70. *Reagent Chemicals: American Chemical Society Specifications*, 8th ed.; American Chemical Society: Washington, 1993; p 279.
71. Perrin, D. D.; Armarego, W. L. F. *Purification of Laboratory Chemicals*, 3rd ed.; Pergamon: New York, 1988; p 322.
72. Barreau, M.; Bost, M.; Julia, M.; Lallemand, J.-Y. *TL* **1975**, 3465.

Nicholas D. P. Cosford
SIBIA, La Jolla, CA, USA

Copper(I) Cyanide[1]

CuCN

[544-92-3]	CCuN	(MW 89.57)
(·NaCN)		
[13715-19-0]	C_2CuN_2Na	(MW 138.58)
(·LiCl)		
[59219-07-7]	CClCuLiN	(MW 131.96)
(·2LiCl)		
[121340-53-2]	CCl_2CuLi_2N	(MW 174.35)
(·2LiBr)		
[129126-28-9]	CBr_2CuLi_2N	(MW 263.25)

(useful precursor for organocopper(I) and organocuprate(I) reagents[1])

Alternate Name: cuprous cyanide.
Physical Data: mp 474 °C; *d* 2.920 g cm^{-3}.
Solubility: insol H_2O and most organic solvents. The lithium halide complexes are sol THF.
Form Supplied in: off-white solid.
Handling, Storage, and Precautions: highly toxic; must be handled with care.

General Discussion. The first investigation of organocopper chemistry was carried out by Kondyreva and Fomin,[2] who in 1915 studied the effect of heavy metal salts on organomagnesium compounds. They found that organocopper compounds, prepared from CuCl, CuBr, CuI, CuSCN, and CuCN, decomposed to give equimolar amounts of alkane and alkene when the organic group was *n*-alkyl, and biaryl when it was aryl. These observations were confirmed by Whitesides et al. a half century later.[3] The Russian authors also noted that the solutions prepared from Grignard reagents and copper salts no longer reacted with acetone to give *t*-butanol. While they did not find a synthetic application for their solutions, it should be noted that Gilman did not investigate the reaction chemistry of the solutions he prepared from 2 equiv of MeLi and copper(I) salts, either.[4] Nevertheless, organocuprates are now commonly called Gilman reagents.

The pioneering work on synthetic applications of organocopper reagents was done almost exclusively with ***Copper(I) Iodide***. Unfortunately, nearly two decades were to pass before it was found that Gilman reagents prepared from CuCN and 2 equiv of RLi were purer, more thermally stable,[5] and often gave higher yields of the desired product than Gilman reagents prepared from CuI.[1,7] Gilman reagents such as $Bu_2CuLi{\cdot}LiI$ appear to be more prone to electron transfer and radical reactions than the corresponding reagents prepared from CuCN.[6]

Based upon their apparent higher reactivity, it was originally claimed that the reagents prepared from 2 equiv of RLi and CuCN were 'higher order cyanocuprates', $R_2Cu(CN)Li_2$;[1a–c,7] however, NMR studies,[9–11] EXAFS investigations,[12] and theoretical calculations[13] have recently converged on the conclusion that the CN is not bonded to Cu. Lipshutz has recently conceded that the CN is not σ-bonded to Cu.[8] Thus these reagents should be represented as $R_2CuLi{\cdot}LiCN$ by analogy with $R_2CuLi{\cdot}LiI$ and $R_2CuLi{\cdot}LiBr$ for the cuprates prepared from CuI and ***Copper(I) Bromide***[1], respectively. It has been demonstrated by ^{13}C NMR and ^{6}Li NMR that LiI is not free in the Gilman reagent prepared from CuI and 2PhLi in DMS,[14] and that LiI and LiCN are not part of the cuprate bonding in THF.[9–11] It has recently been shown how the aggregation states of organocopper reagents in solution may be assigned by using ^{13}C NMR spectroscopy.[11]

In contrast, both NMR and EXAFS studies confirm that CN is bonded to Cu in the so-called 'lower order' cyanocuprates, $RCu(CN)Li$,[10,12] prepared from 1 equiv of RLi and CuCN. While they are relatively unreactive towards α-enones and alkyl halides,[15] they have been used successfully with more reactive substrates such as allylic carboxylates[16] and vinyloxiranes.[17] The introduction of additives such as ***Boron Trifluoride***[18,19] should give these reagents broader applicability. Trost used lower order cyanocuprates when he introduced the concept of 'chirality transfer' in acyclic systems,[20] which has been applied brilliantly by Ibuka and Yamamoto.[18]

Recently, several novel lower order cyanocuprates have found application. For example, Piers has converted 2-alkynoates into either (*Z*)- or (*E*)-3-trimethylstannyl-2-alkenoates by treatment with $Me_3SnCu(CN)Li$ under the appropriate conditions (eq 1).[21a] He also compared lower order phenyl cyano- and thiocuprates in developing an efficient methylenecyclopentane annulation (eq 2).[21b]

$R{-}C{\equiv}C{-}CO_2R'$ — $Me_3SnCu(CN)Li$, –48 °C, 2 h; 0 °C, 2 h → (*Z*)-$Me_3Sn(R)C{=}CH(CO_2R')$; $Me_3SnCu(CN)Li$, –78 °C, 4 h → (*E*)-$Me_3Sn(R)C{=}CH(CO_2R')$ (1)

Cu(X)Li (3-chloropropenyl), X = CN, SPh; KH, THF (2)

X = CN, 72%
X = SPh, 70%

In a most exciting development from the viewpoint of molecular complexity, Knochel has prepared highly functionalized cyanocuprates RCu(CN)ZnX, where X = I, Br, Cl, OMs, OTs, or $OP(O)(OR)_2$, from the corresponding alkylzinc reagents.[22] In a related development, Rieke has reported the 'direct' formation of highly functionalized allylic organocopper reagents via the reduction of CuCN·2LiBr, CuCN·LiCl, or CuCN·2LiCl with ***Lithium Naphthalenide*** to give a highly reactive form of Cu metal.[23] Linderman has prepared α-alkoxyorganocuprate reagents from the corresponding organotin compounds and reacted them with α,β-unsaturated carbonyl compounds (eq 3).[24]

cyclohexenone + $[\ldots]_2$CuLi•LiCN — TMSCl, THF → 3-substituted cyclohexanone (OR) (3)

R =	
MOM	96%
MEM	97%
SEM	62%
Me	76%

Corey has used $Bu_4NCu(CN)_2$ as a precursor to 'a new class of cuprate reagents', which he symbolizes as $RCu(CN)_2(NBu_4)Li$,[25] implying that they are 'higher order' cuprates with two CN ligands bonded to Cu, although this remains to be proven. The ^{13}C NMR spectroscopy of organocopper reagents has now been developed to the point where conjectures concerning the structures of new organocopper species should not be made without making sure that they are consistent with the NMR data.[11] At the time of this writing there is still a backlog of structures that need to be checked. Theory is also being developed to the point where it can shed considerable light on whether structures are feasible or not.[13]

As discussed above, when the Gilman reagents prepared from CuI were introduced, side-by-side comparisons were not made with the corresponding cuprates prepared from other Cu^I salts. (Or, if comparisons were made, they were not reported.) In other words, a depth-first search was done, rather than a breadth-first one.[26] Unfortunately, the same pattern was repeated when the cuprates prepared from CuCN were introduced: papers with tables of yields for typical substrates were published, but very few direct comparisons with other cuprates were made.[1b,c,7a] Thus the best advice we can give the synthetic chemist is to try several of the most successful versions of the Gilman reagent, viz, those prepared from CuI, CuBr·SMe_2 (see ***Copper(I) Bromide***[1]), and CuCN.[1a–c] It is also useful to compare organocopper reagents prepared from lithium reagents with those prepared from the corresponding Grignard reagents (See ***Copper(I) Chloride***).

In one of the few direct comparisons, Fleming has shown that the silylcupration of alkynes with $(Me_2PhSi)_2CuLi{\cdot}LiCN$ gives the opposite regiochemistry to $(Me_2PhSi)_2CuLi{\cdot}LiBr$, prepared from CuBr·DMS.[27] It is interesting to note that Fleming's work predates Lipshutz's, and that he uses the correct (lower order) formulation. Bertz has compared CuCN with CuI, CuBr, CuBr·DMS, CuCl, CuSCN and CuOTf as far as the formation of alkyl and aryl Gilman reagents is concerned,[28a] and also reacted a number of them, including those made from CuCN, with dithioesters, where they give carbophilic addition.[28b] For the preparation of stoichiometric cuprates from Grignard reagents, ***Copper(I) Trifluoromethanesulfonate*** (copper(I) triflate) was demonstrated to be superior to the other Cu^I salts, including CuCN.[28b]

While CuCN has been used in catalytic amounts with Grignard reagents (where the ate complex may be presumed to be the active species),[16b,29,30] CuCN has not been shown to be superior to the other Cu^I salts, and even some Cu^{II} salts appear to be equally effective.[1a] For example, in the addition of Grignard reagents to nitriles, CuCN, CuI, CuBr, CuBr·DMS, and CuCl all gave the same yield of product to within ±3%. Thus the yields of ketimine product from *t*-BuMgCl and 4-methoxybenzonitrile were 87–93% (i.e. 90 ± 3%) after 2 h in refluxing THF in the presence of 2 mol % of the Cu^I salt.[30] This is a very useful extension and generalization of Bertz's work on the activation of the carbon–nitrogen double bond by Cu^I.[31]

Lipshutz has described 'stoichiometric higher order, mixed lithio magnesio organocuprates' prepared from CuCN, 2-lithiothiophene, and Grignard reagent (1:1:1), although no evidence that a higher order species is present in such solutions has been adduced.[32a] In fact, the authors admit that these reagents 'are not discrete'.[32a] The corresponding organocuprates prepared from CuCN, 2-lithiothiophene, and a lithium reagent (or lithium 2-thienyl(cyano)cuprate(I) and a lithium reagent) have been assigned a 'higher order' structure,[1b] but are probably lower order R(2-thienyl)CuLi·LiCN reagents.[9–11] The 2-thienyl group was introduced as a nontransferred ligand for mixed cuprates by the Swedish group of Malmberg, Nilsson, and Ullenius.[32b] Whatever their structure may be, they have been applied to some interesting synthetic problems, such as the prostaglandin synthesis shown in eq 4.[33]

As mentioned above, $Bu_4NCu(CN)_2$ has been used as a cuprate precursor. The related alkali metal dicyanocuprates, especially $NaCu(CN)_2$, have been used in typical cuprate coupling reactions with aryl and vinyl halides.[34,35] In other typical organocopper reactions, CuCN itself reacts with acid chlorides to yield α-ketonitriles,[36] and with aryl iodides to afford aryl cyanides.[37]

It is noteworthy that the first sodium organocuprate, $Bu_2CuNa{\cdot}NaCN$, was prepared from BuNa and CuCN,[38] and Seyferth prepared the first acyl cuprates from $R_2CuLi{\cdot}LiCN$ and CO.[39] The latter development enables the synthesis of 1,4-diketones directly from α-enones via conjugate addition.

1. *t*-BuLi
2. (2-thienyl)Cu(CN)Li

Li•LiCN

EEO$(CH_2)_6$Cu(CN)MgCl

1. PPTS
2. CrO_3, H^+
3. HF, H_2O–MeCN

(4)

Whitesides et al. introduced the oxidative coupling of Cu^I ate complexes, which they prepared from CuI.[40] Bertz and Gibson demonstrated that the oxidation of organocuprates from CuCN gave significantly different results,[41] which Lipshutz et al. developed into a synthesis of unsymmetrical biaryls.[42] Oxidative couplings of heterocuprates usually proceed in poor yields;[1e,43] however, Snieckus et al. have developed useful synthetic methodology for the *N*-arylation of amines based upon the oxidation of aryl amidocuprates.[44]

In one of his many pioneering studies, House showed that the enolates formed upon the addition of $Me_2CuLi{\cdot}LiBr$ to α-enones were basically lithium enolates.[45] Nevertheless, Posner showed that the addition of 1 equiv of CuCN to a lithium enolate could dramatically decrease polyalkylation in cyclopentanone derivatives.[46]

Finally, CuCN has been used to catalyze the reactions of diazoacetic esters;[1f] however, ***Copper(I) Chloride*** is the Cu^I salt most frequently used for this purpose.

Related Reagents. Copper(I) Bromide, Copper(I) Chloride, Copper(I) Iodide (and their combination reagents); also Copper(I) Trifluoromethanesulfonate. For examples of cyanocuprates, see Lithium Butyl(cyano)cuprate, Lithium Cyano(methyl)cuprate, etc.

---

1. (a) Lipshutz, B. H.; Sengupta, S. *OR* **1992**, *41*, 135. (b) Lipshutz, B. H. *S* **1987**, 325. (c) Lipshutz, B. H.; Wilhelm, R. S.; Kozlowski, J. A. *T* **1984**, *40*, 5005. (d) Chapdelaine, M. J.; Hulce, M. *OR* **1990**, *38*, 225. (e) Kauffmann, T. *AG(E)* **1974**, *13*, 291. (f) Dave, V.; Warnhoff, E. W. *OR* **1970**, *18*, 217.
2. Kondyreva, N. V.; Fomin, D. A. *Zh. Russ. Fiz.-Khim. O-va, Chast Khim.* **1915**, *47*, 190 (*CA* **1915**, *9*, 1473).
3. Whitesides, G. M.; Stedronsky, E. R.; Casey, C. P.; San Filippo, Jr., J. *JACS* **1970**, *92*, 1426.
4. Gilman, H.; Jones, R. G.; Woods, L. A. *JOC* **1952**, *17*, 1630.
5. Bertz, S. H.; Dabbagh, G. *CC* **1982**, 1030.
6. Bertz, S. H.; Dabbagh, G.; Mujsce, A. M. *JACS* **1991**, *113*, 631.
7. (a) Lipshutz, B. H.; Wilhelm, R. S.; Floyd, D. M. *JACS* **1981**, *103*, 7672. (b) Lipshutz, B. H.; Sharma, S.; Ellsworth, E. L. *JACS* **1990**, *112*, 4032.
8. Lipshutz, B. H.; James, B. *JOC* **1994**, *59*, 7585.
9. Bertz, S. H. *JACS* **1990**, *112*, 4031.
10. Bertz, S. H. *JACS* **1991**, *113*, 5470.
11. Bertz, S. H.; Dabbagh, G.; He, X.; Power, P. P. *JACS* **1993**, *115*, 11 640.
12. Stemmler, T.; Penner-Hahn, J. E.; Knochel, P. *JACS* **1993**, *115*, 348.
13. Snyder, J. P.; Spangler, D. P.; Behling, J. R.; Rossiter, B. E. *JOC* **1994**, *59*, 2665.
14. Bertz, S. H.; Dabbagh, G. *JACS* **1988**, *110*, 3668.
15. Gorlier, J.-P.; Hamon, L.; Levisalles, J.; Wagnon, J. *CC* **1973**, 88.
16. (a) Underiner, T. L.; Goering, H. L. *JOC* **1988**, *53*, 1140. (b) Underiner, T. L.; Goering, H. L. *JOC* **1990**, *55*, 2757. (c) Tseng, C. C.; Paisley, S. D.; Goering, H. L. *JOC* **1986**, *51*, 2884.
17. Marino, J. P.; Jaén, J. C. *JACS* **1982**, *104*, 3165.
18. (a) Ibuka, T.; Tanaka, M.; Nishii, S.; Yamamoto, Y. *CC* **1987**, 1596. (b) Ibuka, T.; Tanaka, M.; Nishii, S.; Yamamoto, Y. *JACS* **1986**, *108*, 7420.
19. (a) Kang, S.-K.; Lee, D.-H.; Sim, H.-S.; Lim, J.-S. *TL* **1993**, *34*, 91. (b) Oppolzer, W.; Moretti, R.; Godel, T.; Meunier, A.; Löher, H. *TL* **1983**, *24*, 4971.
20. Trost, B. M.; Klun, T. P. *JOC* **1980**, *45*, 4256.
21. (a) Piers, E.; Wong, T.; Ellis, K. A. *CJC* **1992**, *70*, 2058. (b) Piers, E.; Karunaratne, V. *CC* **1983**, 935.
22. (a) Knochel, P.; Chou, T.-S.; Jubert, C.; Rajagopal, D. *JOC* **1993**, *58*, 588. (b) Jubert, C.; Knochel, P. *JOC* **1992**, *57*, 5425. (c) Achyutha Rao, S.; Knochel, P. *JOC* **1991**, *56*, 4591. (d) Knochel, P.; Yeh, M. C. P.; Berk, S. C.; Talbert, J. *JOC* **1988**, *53*, 2390.
23. Stack, D. E.; Dawson, B. T.; Rieke, R. D. *JACS* **1992**, *114*, 5110.
24. Linderman, R. J.; Godfrey, A.; Horne, K. *TL* **1987**, *28*, 3911.
25. Corey, E. J.; Kyler, K.; Raju, N. *TL* **1984**, *25*, 5115.
26. Cormen, T. H.; Leiserson, C. E.; Rivest, R. L. *Introduction to Algorithms*; MIT Press: Cambridge, MA, 1990; pp 469–485.
27. (a) Fleming, I.; Newton, T. W.; Roessler, F. *JCS(P1)* **1981**, 2527. (b) Fleming, I.; Roessler, F. *CC* **1980**, 276.
28. (a) Bertz, S. H.; Gibson, C. P.; Dabbagh, G. *TL* **1987**, *28*, 4251. (b) Bertz, S. H.; Dabbagh, G.; Williams, L. M. *JOC* **1985**, *50*, 4414.
29. Trost, B. M.; Merlic, C. A. *JACS* **1988**, *110*, 5216.
30. Weiberth, F. J.; Hall, S. S. *JOC* **1987**, *52*, 3901.
31. (a) Bertz, S. H. *JOC* **1979**, *44*, 4967. (b) Bertz, S. H. *TL* **1980**, *21*, 3151.
32. (a) Lipshutz, B. H.; Parker, D. A.; Nguyen, S. L.; McCarthy, K. E.; Barton, J. C.; Whitney, S. E.; Kotsuki, H. *T* **1986**, *42*, 2873. (b) Malmberg, H.; Nilsson, M.; Ullenius, C. *TL* **1982**, *23*, 3823.
33. Okamoto, S.; Kobayashi, Y.; Kato, H.; Hori, K.; Takahashi, T.; Tsuji, H.; Sato, F. *JOC* **1988**, *53*, 5590.
34. House, H. O.; Fisher, W. F., Jr. *JOC* **1969**, *34*, 3626.
35. Procházka, M.; Široký, M. *CCC* **1983**, *48*, 1765.
36. Normant, J. F.; Piechucki, C. *BSF(2)* **1972**, 2402.
37. Suzuki, H.; Hanafusa, T. *S* **1974**, 53.
38. Bertz, S. H.; Gibson, C. P.; Dabbagh, G. *OM* **1988**, *7*, 227.
39. (a) Seyferth, D.; Hui, R. C. *JACS* **1985**, *107*, 4551. (b) Seyferth, D.; Hui, R. C. *TL* **1986**, *27*, 1473.
40. Whitesides, G. M.; San Filippo, Jr., J.; Casey, C. P.; Panek, E. J. *JACS* **1967**, *89*, 5302.
41. Bertz, S. H.; Gibson, C. P. *JACS* **1986**, *108*, 8286.
42. Lipshutz, B. H.; Siegmann, K.; Garcia, E.; Kayser, F. *JACS* **1993**, *115*, 9276.
43. Åkermark, B.; Almemark, M.; Jutand, A. *ACS* **1982**, *B36*, 451.

44. Iwao, M.; Reed, J. N.; Snieckus, V. *JACS* **1982**, *104*, 5531.
45. House, H. O.; Wilkins, J. M. *JOC* **1976**, *41*, 4031.
46. Posner, G. H.; Lentz, C. M. *JACS* **1979**, *101*, 934.

Steven H. Bertz & Edward H. Fairchild
LONZA, Annandale, NJ, USA

Copper(I) Iodide[1]

[7681-65-4]	CuI	(MW 190.45)
($\cdot PBu_3$)		
[21591-31-1]	$C_{12}H_{27}CuIP$	(MW 392.81)
($\cdot(SBu_2)_2$)		
[35907-81-4]	$C_{16}H_{36}CuIS_2$	(MW 483.11)

(the classical precursor for organocopper(I) and organocuprate reagents[1])

Alternate Name: cuprous iodide.
Physical Data: mp 605 °C; *d* 5.620 g cm^{-3}.
Solubility: insol H_2O and most organic solvents; partially sol dimethyl sulfide (DMS).
Form Supplied in: off-white to grayish solid; 99.999% grade available.
Handling, Storage, and Precautions: maintenance of a dry N_2 or Ar atmosphere is recommended.

General Discussion. The first organocopper compound to be isolated, ***Phenylcopper***, was prepared in 1923 by Rene Reich from ***Phenylmagnesium Bromide*** and CuI.[2] In 1936, Gilman repeated the preparation of PhCu from CuI,[3a] and in 1952 he prepared ***Methylcopper*** and ***Lithium Dimethylcuprate*** from CuI and 1 and 2 equiv of ***Methyllithium***, respectively.[3b] An improved preparation of halide-free MeCu and Me_2CuLi from CuI has recently been reported,[4] and an appreciation of the role of the lithium salt from the preparation in the structure and reactivity of the reagent has developed in recent years.[5–10] Therefore it has been proposed that any salt from the metathesis reaction used to prepare the organocopper reagent should be explicitly shown in the formula, e.g. $Ph_2CuLi{\cdot}LiI$,[5] PhCu/LiI,[6] $Bu_2CuLi{\cdot}LiBr$,[7] $Bu_2CuNa{\cdot}NaCN$,[8a] etc. Unfortunately, many of the tables in the latest *Organic Reactions* chapter on organocopper reagents do not list the Cu^I salt from which the reagents were prepared.[1a]

Interest in organocopper reagents was rekindled in 1966 when House and Whitesides[9] showed that organocuprates were intermediates in what we propose to call the Kharasch reaction, the Cu-catalyzed 1,4-addition of Grignard reagents to enones.[11] They also reported 'the development of two organocopper reagents which have the stoichiometry $Li^+Me_2Cu^-$ and MeCuP(n-Bu)$_3$'.[9] Both of these classes of reagents are important today (see below). Corey and Posner also used CuI to prepare $Me_2CuLi{\cdot}LiI$[12] and $Bu_2CuLi{\cdot}LiI$,[13] which were coupled with alkyl halides and iodobenzene. These initial reports of the synthetic value of organocopper reagents for selective C–C bond formation led to an explosion of applications,[1a–g] which like the 'big bang' is still expanding today. The principal applications of organocopper reagents to C–C bond formation in the areas of conjugate addition to α,β-unsaturated carbonyl compounds, carbocupration of alkynes, and coupling reactions with oxiranes and alkyl, alkenyl, and aryl halides have been well reviewed.[1a–h] While CuI has been supplanted by CuBr·DMS (see ***Copper(I) Bromide***) and ***Copper(I) Cyanide*** for many purposes, it is still one of the main precursors that should be tried when optimizing an organocopper synthesis.[7]

In addition to the ate complexes prepared from 2 equiv of RLi, the organocopper(I) compounds RCu, prepared from 1 equiv of RLi and CuI, have found synthetic application in the presence of additives, which enhance the reactivity of these otherwise relatively unreactive compounds. Both Lewis bases and Lewis acids[1d] have been used for this purpose, and their utility has been extended to the organocuprates as well. Examples of the former are phosphines, such as ***Tri-n-butylphosphine***,[14] and sulfides, such as dibutyl sulfide[15] and ***Dimethyl Sulfide***;[6,16a] examples of the latter are ***Boron Trifluoride***[17,18] and ***Aluminum Chloride***.[1,17a,19] ***Chlorotrimethylsilane*** (TMSCl) is a useful additive,[20,21] especially in conjunction with HMPA.[6,22] In analogy with these major additives, triethylphosphine,[23a] triphenylphosphine,[23b,24] tricyclohexylphosphine,[24] dppe,[24,25] triethyl phosphite,[26] diisopropyl sulfide,[16b] triethylboron,[27] trimethylaluminum,[28] titanium tetrachloride,[17b,18a] TMSCl–DMAP,[22] TMSCl–TMEDA,[29] TMSI,[30,31] and TMSCN[31] have been used. One unique phosphine is polymer-supported $RPPh_2$,[32] where R is the polymer backbone.

The addition of BF_3 improves some of the usual organocopper reactions,[17] and it enables some unprecedented ones, e.g. the direct alkylation of allylic alcohols.[17c] It also favors 1,4- over 1,6-addition to methyl sorbate: 1,6-addition predominates with $Bu_2CuLi{\cdot}LiI$,[17b] and also with BuMgBr/CuCl,[11a] but 1,4-addition is observed for $BuCu{\cdot}BF_3$.[17b] One particularly interesting application of the BF_3 procedure is the conjugate addition of Cu^I aldimines to α,β-unsaturated carbonyl compounds (eq 1), which after hydrolysis gives 1,4-diketones.[18b]

R–N≡C: + R'Li ⟶ RN=C(Li)R' —CuI→ RN=C(Cu•LiI)R' —1. BF_3•ether; 2. α-enone→ 3-(RN=C(R'))cyclohexanone, 65–95% (1)

R = *t*-Bu
R' = Bu, *s*-Bu

α-enones:
2-cyclohexenone (shown)
3-penten-2-one
methyl vinyl ketone
benzalacetone
2-hexenal

In some cases the additives improve solubility; in other cases, entirely new reagents result. For example, PhCu is insoluble in ether, but it is soluble in DMS, where it has been shown to be an equilibrium mixture of $(PhCu)_4$ and $(PhCu)_3$.[25] When prepared from CuBr or CuCN in DMS, the product is Li-free, due to the precipitation of LiBr or LiCN from DMS.[6] This Li-free PhCu is relatively unreactive compared to the reagent prepared from CuI, PhCu/LiI, which still contains the LiI.[6] This LiI may be considered an activating 'additive'. The LiI present in $Me_2CuLi{\cdot}LiI$ also has an important effect in conjugate addition reactions, as the percentage of 1,2-addition can be significant for halide-free organocuprates.[4]

The LiI in PhCu/LiI is not incorporated into the organocopper(I) clusters;[6] however, in the case of $Ph_2CuLi{\cdot}LiI$, ^{13}C NMR shows that both I-containing (major) and I-free (minor) clusters are present.[5] House found that 'although the conjugate addition of lithium dimethylcuprate to α,β-unsaturated ketones appears to require no other species in the reaction solution, ***Trimethyl Phosphite*** and ***Tri-n-butylphosphine*** complexes of methylcopper will undergo conjugate addition only if various salts such as ***Lithium Iodide***, ***Lithium Bromide***, ***Magnesium Bromide***, or ***Lithium Cyanide*** are present in the reaction medium'.[10]

Organocopper reagents are assuming an increasing role in asymmetric synthesis and many of the procedures involve complex mixtures containing one or more of the additives discussed above.[15,17a,18a,c] This area has been reviewed recently,[1b] so we highlight one especially interesting example here, the enantiocontrolled synthesis of quaternary carbon centers via the asymmetric conjugate addition of organometallic reagents to enantiomerically pure 2-(arylsulfinyl)cycloalkenones.[33] Posner has noted that although $Me_2CuLi{\cdot}LiI$ and $Me_5Cu_3Li_2$ work well, $Me_2CuLi{\cdot}LiCN$ and $MeCu{\cdot}BF_3$ do not. He concludes that 'no one type of organocopper reagent will be universally preferred over others for all different kinds of carbon–carbon bond-forming reactions'.[33]

Tandem β-addition–α-functionalization reactions are an important strategy for rapidly building molecular complexity.[1c] A particularly impressive example (eq 2) is the 'three component coupling' used by Suzuki and Noyori to prepare prostaglandins,[14c,d] which owes its success to the use of derivatives of House's $RCu(PBu_3)$.[9,10,14a,b] An interesting intramolecular version involving conjugate addition–cycloacylation of alkynic diesters to give highly functionalized cyclopentenones has been developed by Crimmins.[34] Tandem organocopper addition–electrophilic functionalization of alkynes is well established,[1a,f] and Meyers has even performed such a sequence on a benzyne.[35]

1. 2 equiv t-BuLi; 2. CuI, 2.6 equiv PBu_3 → $(Bu_3P)_3Cu$…; 1. (RO-cyclopentenone); 2. $OHC(CH_2)_5CO_2Me$ → …; several steps → PGE_1 (2)

Useful applications of organocopper reagents to carbohydrate chemistry have been made by the Kocienski and Fraser-Reid groups. In the former case, an oxirane ring was opened regio- and stereoselectively.[36] In a contraintuitive finding, the latter group reported that the homogeneous system using the soluble $CuI{\cdot}PBu_3$ complex was less satisfactory than the heterogeneous system based upon CuI for the conjugate addition–alkylation of hex-2-enopyrano-4-ulosides.[37]

Some exciting recent developments extending the scope of organocopper chemistry involve CuI. Rieke has reported the preparation of $RCu{\cdot}PBu_3$ from a highly reactive copper intermediate prepared via the reduction of $CuI{\cdot}PBu_3$.[23,38] Ebert has used a similar procedure to prepare 'remote' ester and ketone functionalized organocopper reagents.[39]

Alkenes with perfluoroalkyl substituents have been prepared by the reaction of perfluoroalkyl iodides with terminal alkynes in the presence of ultrasonically dispersed ***Zinc*** and CuI.[40a] A general procedure for the formation of organocopper reagents from alkyl and aryl halides and ***Acetaldehyde*** metal in the presence of CuI or ***1-Pentynylcopper(I)*** under ultrasonic irradiation has also been described.[40b]

One of the extraordinary things about displacement reactions mediated by Cu^I is the fact that they occur with both facility at sp^3 and sp^2 centers, as eqs 3 and 4 illustrate. Eq 3 involves a classic S_N2-like reaction, but is noteworthy because of the complexity of the other functionality in the substrate, an α-amino acid derivative.[15a] Eq 4, the alkylation of alkenyl triflates,[41] adds a versatile alternative to the alkylation of alkenyl halides.[1a,g]

$R_2CuLi{\cdot}LiI$ (4 equiv), ether, –60 °C (3)

Y =	Bz	Z	Boc
R = Me	52	65	82
Et	70	82	87
Pr	75	84	87
Bu	74	85	90
vinyl	69	–	79
allyl	61	–	–
Ph	60	–	81

$R_2CuLi{\cdot}LiI$ or $R'_2CuMgI, MgBr_2$ (4)

R =	Me	75%
	Bu	100%
	Ph	75%
R' =	vinyl	62%
	cyclopropyl	86%
	other substrates	

Good reviews are available of stoichiometric cuprates prepared from lithium reagents,[1a] stoichiometric cuprates prepared from Grignard reagents,[1a] and the use of Grignard reagents with catalytic amounts of Cu^I salts.[1a,e] Sadly, direct comparisons have not been made among all the variations. When one adds further variables such as which Cu^I salt is used, what halide is present in the Grignard reagent, and whether the stoichiometry is 2:1 or 1:1, it is easy to see that the choice of conditions is a complex problem, even before all the possible additives are considered. The best advice we can give the synthetic chemist is to try several sets of conditions based upon similar examples in the literature. Fortunately, an exhaustive compilation has been published recently.[1a]

A species of intermediate stoichiometry between organocopper(I) and organocuprate(I) (RCu and R_2CuLi, respectively) has been described as $Me_5Cu_3Li_2 = MeCu + (Me_2CuLi)_2$. While the structure of the reagent has yet to be determined, it has proved to be useful for the conjugate addition of Me to α,β-unsaturated aldehydes.[42]

In recent work that is related to the organocuprate chemistry discussed above, Miura et al. have reported the CuI-catalyzed reaction of aryl and vinyl iodides with terminal alkynes,[43a] and of aryl iodides with active methylene compounds.[43b]

Finally, not all applications of CuI involve organocuprates or related reagents. For example, Corey used CuI to catalyze intramolecular diazoalkene cyclization reactions,[44] and Yates used CuI to catalyze the Wolff rearrangement of diazo ketones.[45] House recommended $CuI{\cdot}(SBu_2)_2$ for the intermolecular cyclopropanation of alkenes with α-diazo ketones. The intermediates are undoubtedly organocopper species of some kind, perhaps Cu–carbene complexes; however, it should be noted that Cu^0 and Cu^{II} are more commonly used as catalysts in conjunction with diazo compounds.[1i]

Related Reagents. See the copper(I) iodide combination reagents following this entry; also Copper(I) Bromide and Copper(I) Chloride (and their combination reagents), Copper(I) Cyanide, and Copper(I) Trifluoromethanesulfonate.

---

1. (a) Lipshutz, B. H.; Sengupta, S. *OR* **1992**, *41*, 135. (b) Rossiter, B. E.; Swingle, N. M. *CRV* **1992**, *92*, 771. (c) Chapdelaine, M. J.; Hulce, M. *OR* **1990**, *38*, 225. (d) Yamamoto, Y. *AG(E)* **1986**, *25*, 947. (e) Erdik, E. *T* **1984**, *40*, 641. (f) Normant, J. F.; Alexakis, A. *S* **1981**, 841. (g) Posner, G. H. *An Introduction to Synthesis Using Organocopper Reagents*; Wiley: New York, 1980. (h) Posner, G. H. *OR* **1975**, *22*, 253; **1972**, *19*, 1. (i) Burke, S. D.; Grieco, P. A. *OR* **1979**, *26*, 361.
2. Reich, M. R. *CR(C)* **1923**, *177*, 322 (*CA* **1924**, *18*, 383).
3. (a) Gilman, H.; Straley, J. M. *RTC* **1936**, *55*, 821. (b) Gilman, H.; Jones, R. G.; Woods, L. A. *JOC* **1952**, *17*, 1630.
4. (a) Bertz, S. H.; Smith, R. A. J. *JACS* **1989**, *111*, 8276. (b) Bertz, S. H.; Vellekoop, A. S.; Smith, R. A. J.; Snyder, J. P. *OM* **1995**, in press.
5. Bertz, S. H.; Dabbagh, G. *JACS* **1988**, *110*, 3668.
6. Bertz, S. H.; Dabbagh, G. *T* **1989**, *45*, 425.
7. Bertz, S. H.; Gibson, C. P.; Dabbagh, G. *TL* **1987**, *28*, 4251.
8. (a) Bertz, S. H.; Gibson, C. P.; Dabbagh, G. *OM* **1988**, *7*, 227. (b) Bertz, S. H.; Dabbagh, G.; Mujsce, A. M. *JACS* **1991**, *113*, 631.
9. House, H. O.; Respess, W. L.; Whitesides, G. M. *JOC* **1966**, *31*, 3128.
10. House, H. O.; Fischer, W. F. Jr. *JOC* **1968**, *33*, 949.
11. Kharasch, M. S.; Tawney, P. O. *JACS* **1941**, *63*, 2308.
12. Corey, E. J.; Posner, G. H. *JACS* **1967**, *89*, 3911.
13. Corey, E. J.; Posner, G. H. *JACS* **1968**, *90*, 5615.
14. (a) Suzuki, M.; Suzuki, T.; Kawagishi, T.; Morita, Y.; Noyori, R. *Isr. J. Chem.* **1984**, *24*, 118. (b) Suzuki, M.; Suzuki, T.; Kawagishi, T.; Noyori, R. *TL* **1980**, *21*, 1247. (c) Suzuki, M.; Kawagishi, T.; Suzuki, T.; Noyori, R. *TL* **1982**, *23*, 4057. (d) Suzuki, M.; Yanagisawa, A.; Noyori, R. *JACS* **1988**, *110*, 4718.
15. (a) Bajgrowicz, J. A.; el Hallaoui, A.; Jacquier, R.; Pigiere, C.; Viallefont, P. *T* **1985**, *41*, 1833. (b) Spescha, M.; Rihs, G. *HCA* **1993**, *76*, 1219.
16. (a) Clark, R. D.; Heathcock, C. H. *TL* **1974**, *15*, 1713. (b) Corey, E. J.; Carney, R. L. *JACS* **1971**, *93*, 7318.
17. (a) Ibuka, T.; Nakao, T.; Nishii, S.; Yamamoto, Y. *JACS* **1986**, *108*, 7420. (b) Yamamoto, Y.; Yamamoto, S.; Yatagai, H.; Ishihara, Y.; Maruyama, K. *JOC* **1982**, *47*, 119. (c) Yamamoto, Y.; Yamamoto, S.; Yatagai, H.; Maruyama, K. *JACS* **1980**, *102*, 2318.
18. (a) Ghribi, A.; Alexakis, A.; Normant, J. F. *TL* **1984**, *25*, 3083. (b) Ito, Y.; Imai, H.; Matsuura, T.; Saegusa, T. *TL* **1984**, *25*, 3083. (c) Oppolzer, W.; Löher, H. J. *HCA* **1981**, *64*, 2808.
19. (a) Ibuka, T.; Tabushi, E. *CC* **1982**, 703. (b) Ibuka, T.; Minakata, H.; Mitsui, Y.; Kinoshita, K.; Kawami, Y. *CC* **1980**, 1193. (c) Ibuka, T.; Minakata, H.; Mitsui, Y.; Kinoshita, K.; Kawami, Y.; Kimura, N. *TL* **1980**, *21*, 4073.
20. Corey, E. J.; Boaz, N. W. *TL* **1985**, *26*, 6019.
21. Alexakis, A.; Berlan, J.; Besace, Y. *TL* **1986**, *27*, 1047.
22. Nakamura, E.; Matsuzawa, S.; Horiguchi, Y.; Kuwajima, I. *TL* **1986**, *27*, 4029.
23. (a) Ebert, G. W.; Rieke, R. D. *JOC* **1984**, *49*, 5280. (b) Wehmeyer, R. M.; Rieke, R. D. *TL* **1988**, *29*, 4513.
24. Miyashita, A.; Yamamoto, A. *BCJ* **1977**, *50*, 1102.
25. Bertz, S. H.; Dabbagh, G.; He, X.; Power, P. P. *JACS* **1993**, *115*, 11640.
26. Normant, J. F.; Cahiez, G.; Bourgain, M.; Chuit, C.; Villieras, J. *BSF(2)* **1974**, 1656.
27. Yamamoto, Y.; Yatagai, H.; Maruyama, K. *JOC* **1979**, *44*, 1744.
28. Saddler, J. C.; Fuchs, P. L. *JACS* **1981**, *103*, 2112.
29. Johnson, C. R.; Marren, T. J. *TL* **1987**, *28*, 27.
30. Bergdahl, M.; Lindstedt, E.-L.; Nilsson, M.; Olsson, T. *T* **1989**, *45*, 535.
31. Bertz, S. H.; Smith, R. A. J. *T* **1990**, *46*, 4091.
32. Schwartz, R. H.; San Filippo, J., Jr. *JOC* **1979**, *44*, 2705.
33. Posner, G. H.; Kogan, T. P.; Hulce, M. *TL* **1984**, *25*, 383.
34. Crimmins, M. T.; Mascarella, S. W.; DeLoach, J. A. *JOC* **1984**, *49*, 3033.
35. Meyers, A. I.; Pansegrau, P. D. *CC* **1985**, 690.
36. Brockway, C.; Kocienski, P.; Pant, C. *JCS(P1)* **1984**, 875.
37. Yunker, M. B.; Plaumann, D. E.; Fraser-Reid, B. *CJC* **1977**, *55*, 4002.
38. Rieke, R. D.; Stack, D. E.; Dawson, B. T.; Wu, T.-C. *JOC* **1993**, *58*, 2483.
39. Ebert, G. W.; Klein, W. R. *JOC* **1991**, *56*, 4744.
40. (a) Kitazume, T.; Ishikawa, N. *CL* **1982**, 1453. (b) Luche, J. L.; Pétrier, C.; Gemal, A. L.; Zikra, N. *JOC* **1982**, *47*, 3805.
41. McMurry, J. E.; Scott, W. J. *TL* **1980**, *21*, 4313.
42. Clive, D. L. J.; Farina, V.; Beaulieu, P. L. *JOC* **1982**, *47*, 2572.
43. (a) Okuro, K.; Furuune, M.; Enna, M.; Miura, M.; Nomura, M. *JOC* **1993**, *58*, 4716. (b) Okuro, K.; Furuune, M.; Miura, M.; Nomura, M. *JOC* **1993**, *58*, 7606.
44. Corey, E. J.; Achiwa, K. *TL* **1970**, *11*, 2245.
45. Yates, P.; Fugger, J. *CI(L)* **1957**, 1511.
46. House, H. O.; Fischer, W. F., Jr.; Gall, M.; McLaughlin, T. E.; Peet, N. P. *JOC* **1971**, *36*, 3429.

Steven H. Bertz & Edward H. Fairchild
LONZA, Annandale, NJ, USA

Crotyltributylstannane[1]

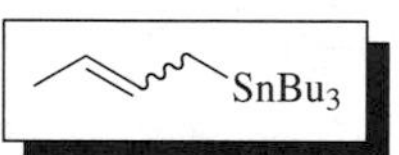

(E)		
[35998-93-7]	$C_{16}H_{34}Sn$	(MW 345.13)
(Z)		
[35998-94-8]		

(allylating reagent for many compounds, including alkyl halides, carbonyl compounds, imines, acetals, thioacetals, and sulfoximides)

Alternate Name: 2-butenyltributyltin.
Physical Data: bp 100–110 °C/1 mmHg;[2f] ^{1}H NMR (*E*) isomer δ 5.95–5.10 (m, 2 H), 2.00–0.70 (m, 32 H); ^{13}C NMR (*E*) isomer δ 130.5, 120.3, 29.3, 27.5, 27.4, 17.9, 13.8, 9.2;[2f] ^{1}H NMR (*Z*) isomer δ 6.84–5.96 (m, 2 H), 3.30–1.43 (m, 32 H); ^{13}C NMR (*Z*) isomer δ 129.6, 118.2, 29.3, 27.5, 27.4, 13.8, 12.5, 9.4.[2f]
Solubility: sol methylene chloride, diethyl ether, tetrahydrofuran, toluene, and benzene.
Form Supplied in: not commercially available.

Preparative Methods: several methods have been described.[2]
Purification: by distillation.
Handling, Storage, and Precautions: all organotin compounds are highly toxic. This reagent therefore must be used in a well-ventilated fume hood. Contact with the eyes, skin, and respiratory system should be avoided.

Introduction. Allylstannanes are widely used 'storable' allyl anion equivalents.[1] Since the chemistry associated with allystannanes and crotylstannanes is, in most cases, very similar, this section will focus on the reactions of crotyltrialkylstannane and crotyl derivatives (methallyltributylstannane, β-methylcrotylstannane, pent-3-en-2-yltrialkylstannanes, and cinnamylstannane. A more comprehensive review of allylation reactions employing allylic stannanes is presented in the article on ***Allyltributylstannane***. Related allylic stannane chemistry is also found under articles on α-alkoxyallylstannanes and ***Allyltriphenylstannane*** derivatives.

Additions to Aldehydes and Ketones. The addition of allylic reagents to carbonyl compounds has become a very attractive synthetic strategy in carbon–carbon bond formations. γ-Substituted allylmetal reagents are especially attractive in natural product synthesis since additions can provide two new adjacent stereocenters. Thus methods to control the stereochemical outcome of both diastereofacial and bond construction selectivity have received a lot of attention.

Crotyltrialkyltin reagents can undergo transmetalation reactions in the presence of Lewis acids.[3] Competing transmetalation process can affect the mechanistic pathway and product distribution (eq 1).[4] 1,3-Isomerization results in 'linear' (α) adducts, while 'branched' (γ) adducts occur through the Lewis acid–carbonyl complex or the transmetalated intermediate, a crotyltin halide. The course of the reaction depends upon the Lewis acid, temperature, substrate, stoichiometry, and order of addition.

R^1CH=CHCH$_2$SnBu$_3$ —(ML_n, $-R_3SnL$)→ R^1CH(ML_{n-1})CH=CH$_2$ —(R^2CHO)→ R^2CH(OH)CH$_2$CH=CHR^1 linear (*E*)/(*Z*) (1)

⇅ R^1CH=CHCH$_2$$ML_{n-1}$ —(R^2CHO)→ R^2CH(OH)CH(R^1)CH=CH$_2$ branched *syn*/*anti*

⇅ (*Z*)-R^1CH=CHCH$_2$$ML_{n-1}$ —(R^2CHO)→ R^2CH(OH)CH(R^1)CH=CH$_2$ branched *syn*/*anti*

In thermal reactions, the stereochemical outcome of crotyltin additions to aldehydes is dependent upon the geometry of the but-2-enyl unit. Thus the (*E*) isomer produces *anti* alcohols while the (*Z*) isomer produces the *syn* alcohols.[5] The allylic coupling to aldehydes also occurs at rt under high pressure.[6]

The Lewis acid-mediated reactions of crotylstannanes and silanes produces *syn* homoallylic alcohols predominantly, regardless of the geometry of the but-2-enyl unit (eq 2). The selectivity was rationalized with an extended acyclic transition state geometry. This is in contrast to other crotyl organometallic reagents in which the crotyl geometry is linked to the observed stereoselectivity.[7]

crotyl-SnBu$_3$ + RCHO —(BF_3•OEt_2, CH_2Cl_2)→ RCH(OH)CH(CH_3)CH=CH$_2$ (2)

>90% *syn* selectivity

(*E*)-Crotyltributylstannane has been shown to react faster with aldehyde–Lewis acid complexes than the corresponding (*Z*) isomer.[8] Thus even higher levels of *syn* selectivity (25:1; 94% yield) can be achieved when 2 equiv of crotylstannane, which consists of a mixture of the (*E*) and (*Z*) isomers, is employed in the ***Boron Trifluoride Etherate***-mediated addition to aldehydes (eq 3). A decrease in selectivity to 9:1 is observed when only 1 equiv of stannane is used. ***Titanium(IV) Chloride***-mediated reactions show *syn* or *anti* selectivity depending upon the order of addition. For example 'normal' addition (crotyltin added last) shows high *syn* selectivity (93:7) and only minor amounts of the linear alcohols (<3%). Under 'reverse' addition (stannane addition to $TiCl_4$, followed by aldehyde addition) *anti* selectivity is observed, again with only minor amounts of the linear alcohols (<5%). The use of chiral Lewis acid catalyst, as opposed to Lewis acid promoters, has also been successfully employed in the stereoselective allylations of aldehydes.[9]

c-C_6H_{11}CHO —(crotyl-SnBu$_3$, Lewis acid)→ *syn* branched + *anti* branched + (*Z*)-linear + (*E*)-linear (3)

Lewis acid	crotyltin equiv	*syn*	*anti*	(*Z*)-linear	(*E*)-linear
BF_3•OEt_2	2.1	96.1	3.9	–	–
$TiCl_4$ normal	1.05	90.5	7.0	2.1	0.5
$TiCl_4$ reverse	2.0	4.4	90.8	–	4.9

The effect of addition modes upon product distributions for (*E*)- and (*Z*)-crotyltributyltins with various Lewis acids and aldehydes has been investigated.[4d,10] The linear (α) homoallylic alcohol adduct with (*Z*) double bond geometry can be obtained selectively when (*Z*)-crotyltributylstannane is combined with an aldehyde followed by the addition of ***n-Butyltrichlorostannane*** (eq 4).[11] The reaction is believed to proceed through transmetalation to form the methallylbutyldichlorotin. The selectivity for the linear homoallylic products is reduced or lost when the order of addition of the aldehyde and $BuSnCl_3$ is reversed or when an (*E/Z*) mixture of crotylin is used. Similarly, the transmetalation of (*E*)-γ-(phenyl)allyltributyltin with $BuSnCl_3$ in the presence of an aldehyde produces the corresponding linear homoallylic alcohol adduct with (*Z*) double bond geometry.[12]

(*Z*)-crotyl-SnBu$_3$ + RCHO —($BuSnCl_3$)→ RCH(OH)CH$_2$CH=CHCH$_3$ (*Z*) (4)

	(*Z*) Linear:branched	yield (%)
R = Ph	96:4	96
R = *n*-Hex	95:5	94

The addition of crotyltributylstannane to aldehydes in the presence of ***Cobalt(II) Chloride*** also affords the linear homoallylic alcohols exclusively (yields: 54–74%) (eq 5).[13] The regiochemistry is not affected by the steric hinderance of the aldehyde or the order of addition.

$SnBu_3$ + RCHO —($CoCl_3$, MeCN, rt)→ (5)

In the Lewis acid-mediated additions of crotyltributylstannane to α-alkoxyaldehydes, $MgBr_2$ and $TiCl_4$ provide chelation control diastereofacial selectivity (≥99:1) (eq 6). The diastereofacial selectivity for all crotylin additions is higher than the corresponding allyl additions. Both diastereofacial and bond construction selectivity increase with the size of R^1 (cyclohexyl > n-butyl > Me). Within the chelation control manifold, *syn* selectivity for bond construction is ≥90:10; $MgBr_2$ is the Lewis acid of choice.[14]

Lewis acid (6)

Chelation control

Cram diastereofacial selectivity

The Lewis acid and hydroxy-protecting group also strongly influence stereoselectivity in the crotylstannane additions to 2-methyl-3-hydroxypropanal derivatives (β-alkoxyaldehydes). The promoter $BF_3{\cdot}OEt_2$ shows high *syn* diastereofacial selectivity (Cram) with the characteristic *syn* bond construction selectivity. The *syn,syn* isomer is formed in 92% yield with high diastereofacial selectivity (18:1) and essentially complete bond construction selectivity when the *t*-butyldimethylsilyl protecting group is present. Chelation control diastereofacial selectivity (*anti*) is high for all Lewis acids capable of chelation; however, bond construction selectivity is low. $MgBr_2$ provides the *anti, syn* isomer with 91:9 *anti* facial selectivity and 89:11 *syn* construction selectivity.[8] Using conditions of chelation control, the protected β-alkoxyaldehyde (**1**) was selectively allylated with crotyltriphenylstannane (as a 1:1 *cis*/*trans* mixture) to provide the desired stereoisomer in 60% yield and a 20% yield of a regiomeric crotyl addition product (eq 7).[15]

$SnPh_3$, $SnCl_4$, –78 °C, 60% (7)

(**1**)

High *syn* diastereofacial control and *anti* bond construction selectivity is observed, under conditions of chelation control, for the additions of β-methylcrotylstannanes to 2-(benzyloxy)-propanal (eq 8).[16]

$SnBu_3$ + H(C=O)CH(OBn)CH_3 —($MgBr_2$, CH_2Cl_2, –78 °C)→

(*E*):(*Z*) = 90:10

syn, anti 76:24 *syn, syn* (8)

The Lewis acid-mediated addition of crotyltributylstannane to pyruvates slightly favors the *anti* products when the ester group is small; however, *syn* products dominate as the size of the ester group is increased (eq 9).[17] The BF_3 etherate-mediated addition to glyoxylate esters is also influenced by the size of the ester group, providing a *syn*:*anti* ratio of 90:10 for the isopropyl ester and decreasing to 75:25 for the methyl ester.[18] Crotyltin addition to *meso*-dimethylglutaric hemialdehyde, also mediated by BF_3 etherate, shows chelation control *syn* selectivity. The selectivity is based upon the conformationally rigid substrate, which prefers a 'chelate-like' conformation.[19]

$SnBu_3$ + pyruvate —(Lewis acid, CH_2Cl_2, –78 °C)→ (9)

	yield (%)	*anti*:*syn*
R = Me	84	60:40
R = 2,6-dimethylphenyl	84	35:65

Imines and Imminium Ions. Crotyltributylstannane additions to aliphatic and furyl imines (R^1 = *i*-Pr, Cy, furyl; R^2 = Bn) provide addition products with very high *syn* selectivity (*syn*:*anti* > 20:1) (eq 10).[20] This high selectivity requires extended Lewis acid–imine complexation at −78 °C (2.5 h) prior to crotylstannane addition, otherwise a modest 4:1 *syn* selectivity is attained. Crotyltributylstannane additions to aromatic imines (R^1 = Ph; R^2 = Ph, *p*-tolyl) occur with much lower diastereoselectivity (*syn*:*anti* > 3:1).[21]

$SnBu_3$ + $HC(=NR^2)R^1$ —(Lewis acid, CH_2Cl_2, –78 °C)→ NHR^2 + NHR^2 (10)

Methallyl- and crotyltin reagents add to pyridine activated by ***Methyl Chloroformate*** (i.e. prior iminium salt formation) with poor regioselectivity (α:γ = 48:52 to 36:64) (eq 11).[22] Allylstannanes add in a highly α-selective fashion (86–99%) while prenyltributylstannane additions occur exclusively at the γ-carbon of pyridine. Crotyltin reagents add to larger aromatic N-heterocycles with excellent regioselectivity (>95%) (eq 12).

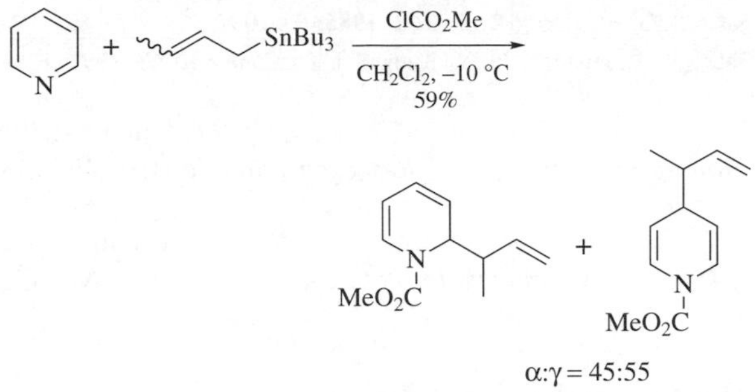

α:γ = 45:55 (11)

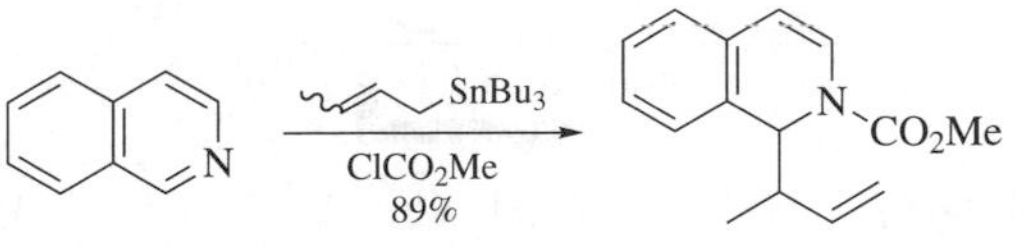

(12)

Conjugate Addition. Conjugate additions of crotyltributylstannanes to Michael acceptors occur with poor to modest *anti* selectivity in variable yields (eq 13).[23] ***Tin(IV) Chloride*** promotion provides the highest selectivity for the crotyltributylstannane addition to an enoate (90:10; 52%). Other Lewis acids provide slightly lower selectivities (>80:20) in higher yields (>94%). Conjugate additions to nitroalkenes occur with very low selectivity (*anti*:*syn* = 45:55 to 70:30). Yields are higher for the phenyl-substituted nitroalkene (53–92%) than for the methyl derivative (29–42%). The trialkylsilyl triflate-promoted conjugate addition of allylstannes to α,β-enones affords the corresponding β-alkylated silyl enol ethers in high yields (85–95%).[24]

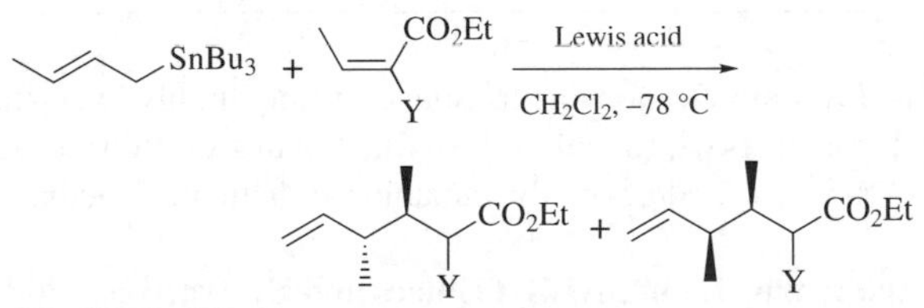

(13)

Y = CO_2Et, CN

Reactions with Orthoamides. The $BF_3 \cdot OEt_2$ promoted addition of crotyltributylstannane to orthoamides occurs with complete stereoselectivity (eq 14).[25] Interestingly, the $TiCl_4$-promoted reaction provides the other diastereomer exclusively.

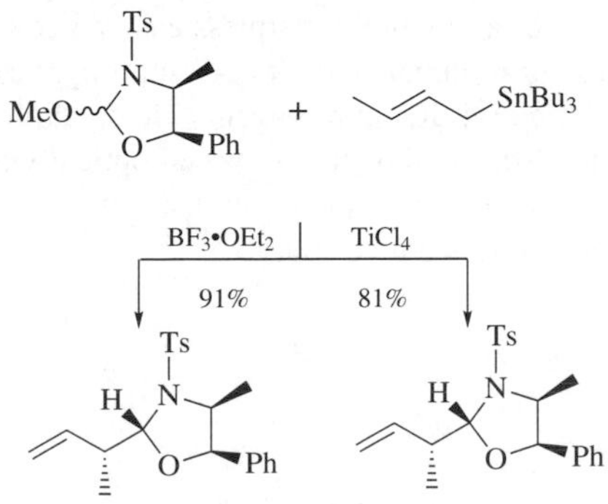

(14)

Intramolecular Cyclizations. Intramolecular condensation reactions between allylstannanes and a carbonyl group joined by a one-carbon tether afford silyloxyvinylcyclopropanes.[26] A variety of 1-butenyl- and 5-hexenyltributylstannanes undergo intramolecular cyclization in the presence of ***N-Phenylselenophthalimide*** and $SnCl_4$.[27]

Radical Reactions.[28] The radical allyl transfer reactions from allyl- or methallyltributylstannane to a variety of substrates under various initiation conditions have been demonstrated to be synthetically useful.[29] For example, thiophenyl glycosides are converted into C-glycosides by reaction with methallyltributylstannane. The stereoselectivity is dependent on the reaction conditions. Using photochemical initiation, the reaction of (**2**) with methallylbutylstannane gives an 87% yield of the C-glycoside with a 92:8 ratio of α:β anomers (eq 15). In contrast, the reaction in the presence of catalytic (0.2 equiv) tributylstannyl triflate provides a 95% yield preferentially of the β-isomer (99:1).[30]

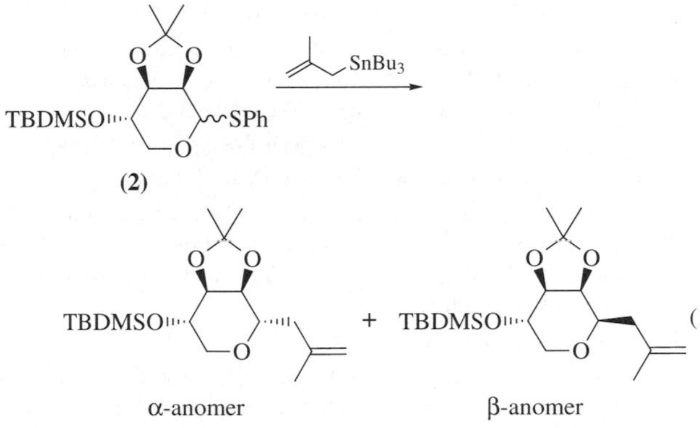

(15)

Crotylstannanes are less reactive than allyltributylstannane towards carbon-centered radicals. In contrast to allylstannanes, which readily react by radical pathways with alkyl halides, the use of crotylstannane affords only reduction products under a variety of initiation processes.[31] However, the successful allyl transfer reaction of crotylstannane has been reported for a bromine-substituted glycine (eq 16).[32] Thus a mixture of (*E*) and (*Z*) isomers (ca. 1:1) of crotyltributyltin reacted with (**3**) to produce a 57% yield of the allylated product as a 1:1 mixture of diastereomers. The reduction product and 5-methylallylglycine were not observed in the reaction mixture. Radical S_H2' substitution reaction using hetero-centered radicals (RS·, RSO_2·, and PhSe·) and crotylstannane has also been demonstrated.[33]

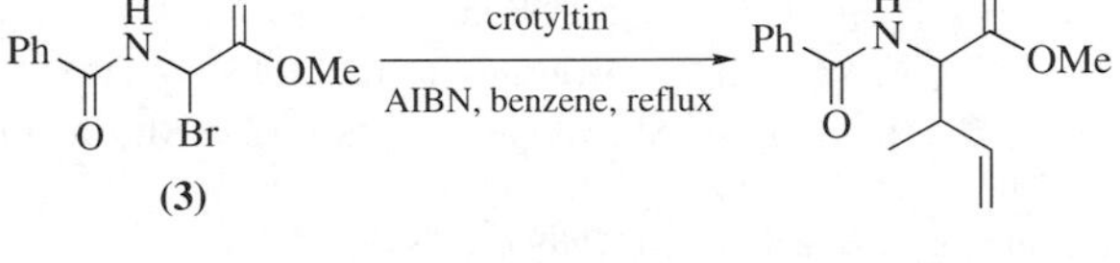

(16)

Irradiation of benzil in the presence of (*E*)- and (*Z*)-crotyl, -pent-2-enyl-, or -hex-2-enyltributylstannane affords predominantly the linear homoallylic alcohols in good yields with retention of the double bond configuration.[34]

Related Reagents. Allyltributylstannane; Allyltrimethylsilane; Allyltriphenylstannane; trans-Cinnamyltributylstannane; Crotyltrimethylsilane; Tri-*n*-butyl(1-methoxymethoxy-2-butenyl)stannane; Trimethyl(3-methyl-2-butenyl)stannane.

1. Reviews of allyl- and crotylmetal chemistry (a) Hoffman, R. W. *AG(E)* **1982**, *21*, 555. (b) Yamamoto, Y.; Maruyama, K. *H* **1982**, *18*, 357.

(c) Roush, W. R. *COS* **1990**, *2*, 1. (d) Yamamoto, Y. *ACR* **1987**, *20*, 243. (e) Yamamoto, Y. *Aldrichim. Acta* **1987**, *20*, 45. (f) Pereyre, M.; Quintard, J.-P.; Rahm, A. *Tin in Organic Synthesis*; Butterworths: London, 1987. (g) Curran, D. P. *S* **1988**, 489. (h) Yamamoto, Y. *Chemtracts–Org. Chem.* **1991**, 225. (i) Marshall, J. A. *Chemtracts–Org. Chem.* **1992**, 75. (j) Yamamoto, Y.; Asao, N. *CR* **1993**, *93*, 2207.
2. (a) Matarasso-Tchiroukhine, E.; Cadiot, P. *JOM* **1976**, *121*, 155. (b) *JOM* **1976**, *121*, 169. (c) Seyferth, D.; Weiner, M. A. *JOC* **1961**, *26*, 4797. (d) Jephcote, V. J.; Thomas, E. J. *JCS(P1)* **1991**, 429. (e) Jephcote, V. J.; Thomas, E. J. *TL* **1985**, *26*, 5327. (f) Aoki, S.; Mikami, K.; Terada, M.; Nakai, T. *T* **1993**, *49*, 1783.
3. Naruta, Y.; Nishigaichi, Y; Maruyama, K. *T* **1989**, *45*, 1067.
4. (a) Keck, G. E.; Castellino, S.; Andrus, M. B. *Selectivities in Lewis Acid Promoted Reactions*, Schinzer, D. Ed.; Kluwer: Dordrecht, 1989; pp 73–105. (b) Keck, G. E.; Andrus, M. B.; Castellino, S. *JACS* **1989**, *111*, 8136. (c) Denmark, S. E.; Wilson, T.; Wilson, T. M. *JACS* **1988**, *110*, 984. (d) Boaretto, A.; Marton, D.; Tagliavini, G.; Ganis, P. *JOM* **1987**, *321*, 199. (e) Yamamoto, Y.; Maeda, N.; Maruyama, K. *CC* **1983**, 742. (f) Quintard, J. P.; Elissondo, B.; Pereyre, M. *JOC* **1983**, *48*, 1559.
5. (a) Servens, C.; Pereyre, M. J. *JOM* **1972**, *35*, C20. (b) Abel, E. W.; Rowley, R. J. *JOM* **1975**, *84*, 199. (c) Pratt, A. J.; Thomas, E. J. *CC* **1982**, 1115. (d) Koreeda, M.; Tanaka, Y. *CL* **1982**, 1299.
6. (a) Yamamoto, Y.; Maruyama, K.; Matsumoto, K. *CC* **1983**, 489. (b) Isaacs, N. S.; Marshall, R. L.; Young, D. J. *TL* **1992**, *33*, 3023.
7. (a) Yamamoto, Y.; Yatagai, H.; Naruta, Y.; Maruyama, K. *JACS* **1980**, *102*, 7107. (b) Yamamoto, Y.; Yatagai, H.; Ishihara, Y.; Maeda, N.; Maruyama, K. *T* **1984**, *40*, 2239. (c) Yamamoto, Y.; Komatsu, T.; Maruyama, K. *JOM* **1985**, *285*, 31.
8. Keck, G. E.; Abbott, D. E. *TL* **1984**, *25*, 1883.
9. (a) Marshall, J. A.; Tang, Y. *SL* **1992**, 653. (b) Keck, G. E.; Tarbet, K. H.; Geraci, L. S. *JACS* **1993**, *115*, 8467. (c) Keck, G. E.; Geraci, L. S. *TL* **1993**, *34*, 7827.
10. Gambaro, A.; Marton, D.; Peruzzo, V.; Tagliavini, G. *JOM* **1982**, *226*, 149.
11. (a) Miyake, H.; Yamamura, K. *CL* **1992**, 1369. (b) Marshall, R. L.; Young, D. J. *TL* **1992**, *33*, 2369.
12. Miyake, H.; Yamamura, K. *CL* **1993**, 1473.
13. Iqbal, J.; Joseph, S. P. *TL* **1989**, *30*, 2421.
14. Keck, G. E., Boden, E. P. *TL* **1984**, *25*, 1879.
15. Jones, A. B.; Yamaguchi, M.; Patten, A.; Danishefsky, S. J.; Ragan, J. A.; Smith, D. B.; Schreiber, S. L. *JOC* **1989**, *54*, 17.
16. Mikami, K.; Kawamoto, K.; Loh, T.-P.; Nakai, T. J. *CC* **1990**, 1161.
17. Yamamoto, Y.; Komatsu, T.; Maruyama, K. *CC* **1983**, 191.
18. Yamamoto, Y.; Maeda, N.; Maruyama, K. *CC* **1983**, 774.
19. Yamamoto, Y.; Nemoto, H.; Kikuchi, R.; Komatsu, H.; Suzuki, I. *JACS* **1990**, *112*, 8598.
20. Keck, G. E.; Enholm, E. J. *JOC* **1985**, *50*, 146.
21. Yamamoto, Y.; Komatsu, T.; Maruyama, K. *JOC* **1985**, *50*, 3115.
22. Yamaguchi, R.; Moriyasu, M.; Yoshioka, M.; Kawanisi, M. *JOC* **1988**, *53*, 3507.
23. Yamamoto, Y.; Nishii, S. *JOC* **1988**, *53*, 3597.
24. Kim, S.; Lee, J. M. *SC* **1991**, *21*, 25.
25. Pasquarello, A.; Poli, G.; Potenza, D.; Scolastico, C. *TA* **1990**, *1*, 429.
26. Keck, G. E.; Tonnies, S. D. *TL* **1993**, *34*, 4607.
27. Herndon, J. W.; Harp, J. J. *TL* **1992**, *33*, 6243.
28. (a) Ref. 1f. (b) Ref. 1g (c) Yamamoto, Y. *T* **1989**, *45*, 909. (d) Giese, B. *Radicals in Organic Synthesis: Formation of C–C Bonds*; Pergamon: Oxford, 1986; pp 98–102.
29. Keck, G. E.; Enholm, E. J.; Yates, J. B.; Wiley, M. R. *T* **1985**, *41*, 4079.
30. Keck, G. E.; Enholm, E. J.; Kachensky, D. F. *TL* **1984**, *25*, 1867.
31. Keck, G. E.; Yates, J. B. *JOM* **1983**, *248*, C21.
32. (a) Easton, C. J.; Scharfbillig, I. M. *JOC* **1990**, *55*, 384. (b) Hamon, D. P. G.; Massy-Westropp, R. A.; Razzino, P. *CC* **1991**, 722.
33. Russell, G. A.; Herold, L. L. *JOC* **1985**, *50*, 1037.
34. Takuwa, A.; Nishigaichi, Y.; Yamaoka, T.; Iihama, K. *CC* **1991**, 1359.

Stephen Castellino
Rhône-Poulenc Ag. Co., Research Triangle Park, NC, USA

David E. Volk
North Dakota State University, Fargo, ND, USA

Cyanotrimethylsilane[1]

Me_3SiCN

[7677-24-9] C_4H_9NSi (MW 99.21)

(agent for the cyanosilylation of saturated and unsaturated aldehydes and ketones;[2] powerful silylating agent;[3] reagent for carbonyl umpolung[4])

Alternate Name: trimethylsilyl cyanide.
Physical Data: mp 11–12 °C; bp 118–119 °C; *d* 0.744 g cm^{-3}.
Solubility: sol organic solvents (CH_2Cl_2, $CHCl_3$); reacts rapidly with water and protic solvents.
Form Supplied in: colorless liquid; available commercially. Can also be prepared.[1e]
Handling, Storage, and Precautions: flammable liquid which must be stored in the absence of moisture and used in an inert atmosphere. Reagent is highly toxic: contact with water produces hydrogen cyanide.

Introduction. Cyanotrimethylsilane is a highly versatile reagent that reacts with a multitude of functional groups to yield an array of products and/or highly valuable synthetic intermediates.

Cyanohydrin Trimethylsilyl Ethers and Derivatives. Aldehydes and ketones are readily transformed into the corresponding cyanohydrin trimethylsilyl ethers when treated with cyanotrimethylsilane in the presence of Lewis acids (eq 1),[5,6] triethylamine,[7] or solid bases such as CaF_2 or hydroxyapatite.[8] The products can be readily hydrolyzed to the corresponding cyanohydrins. The cyanosilylation of aromatic aldehydes can be achieved with high enantioselectivity in the presence of catalytic amounts of a modified Sharpless catalyst consisting of ***Titanium Tetraisopropoxide*** and L-(+)-diisopropyl tartrate (eq 2).[9] Catalysis with chiral titanium reagents yields aliphatic and aromatic cyanohydrins in high chemical and optical yields (eq 3).[10] Cyanohydrins can be subsequently transformed into a variety of useful synthetic intermediates (eq 4).[11,12]

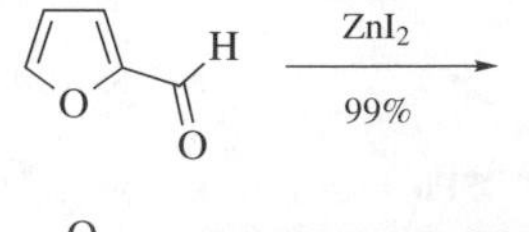

(1)

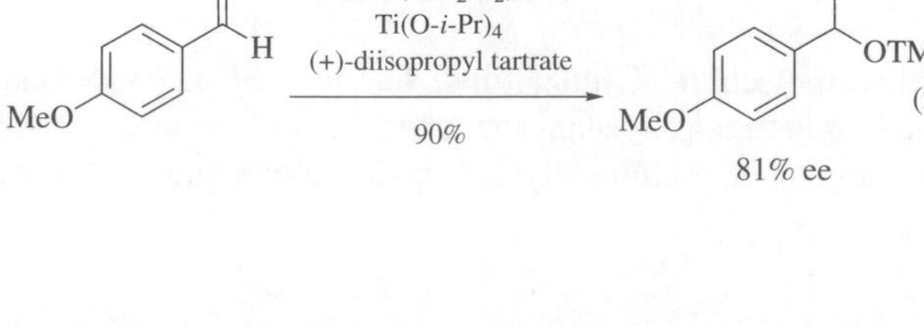

(2)

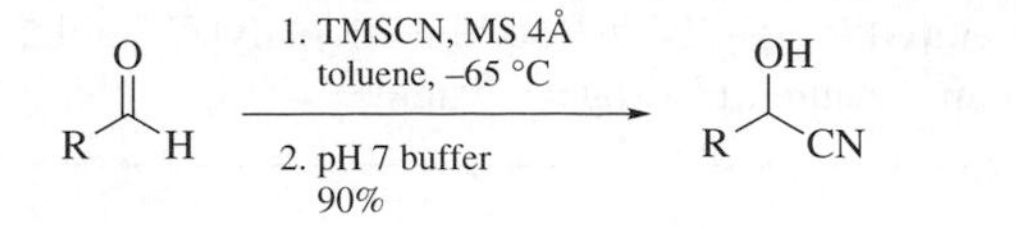

(3)

R = Ph, 79%, 96% ee
R = Bn, 66%, 77% ee
R = phenylethyl, 89%, 89% ee

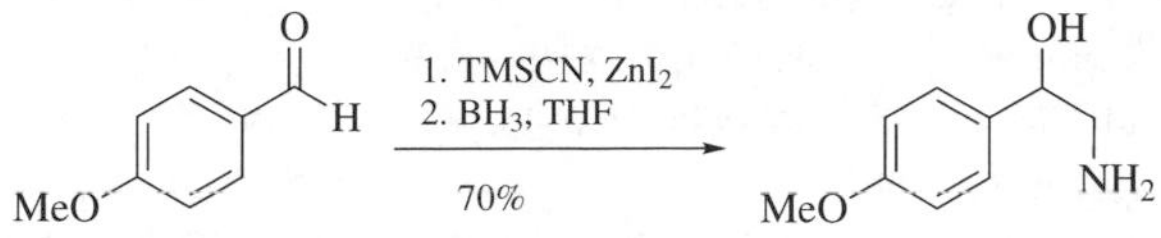

(4)

Conjugate Additions. Cyanotrimethylsilane reacts with α,β-unsaturated ketones in the presence of Lewis acids (***Aluminum Chloride***[1], ***Tin(II) Chloride***, ***Triethylaluminum***) to yield, upon hydrolysis, the corresponding 1,4-addition products (eqs 5 and 6).[13] This methodology is superior to other procedures.[14] By controlling the reaction conditions and the stoichiometry of the reaction, the kinetically controlled 1,2-addition products can also be obtained in high yields (eq 7).[13,15]

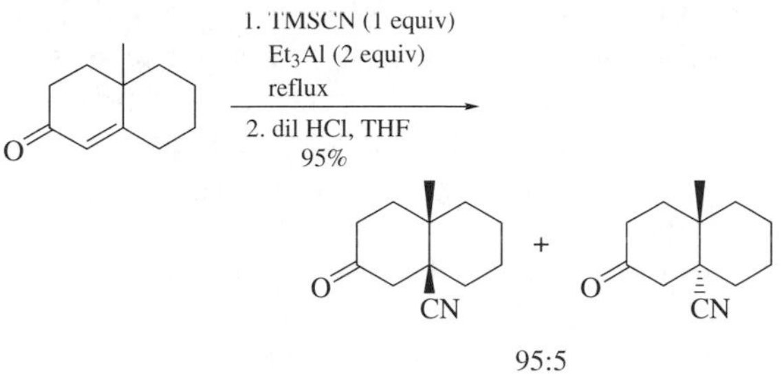

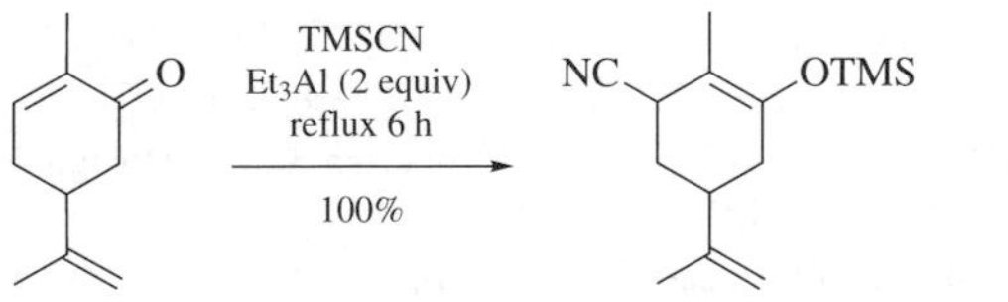

95:5

(6)

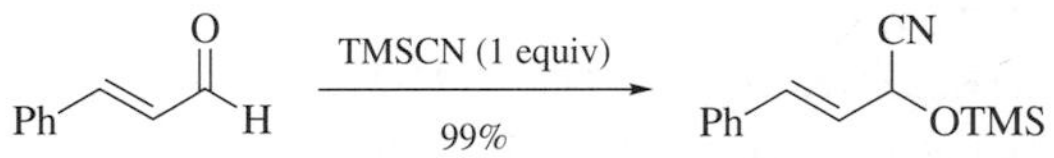

(7)

Regioselective cyanosilylation of unsaturated ketones can also be effected efficiently using a solid acid or solid base support.[16] 1,4-Adducts are obtained when a strong solid acid such as Fe^{3+}- or Sn^{4+}-montmorillonite is used (see ***Iron(III) Nitrate–K10 Montmorillonite Clay***), while 1,2-adducts are obtained in the presence of solid bases such as CaO and MgO.

Reactions with Oxiranes, Oxetanes, and Aziridines. Lewis acids, lanthanide salts, and titanium tetraisopropoxide or ***Aluminum Isopropoxide***[1] catalyze the reactions of cyanotrimethylsilane with oxiranes, oxetanes, and aziridines, yielding ring-opened products. The nature of the products and the regioselectivity of the reaction are primarily dependent on the nature of the Lewis acid, the substitution pattern in the substrate, and the reaction conditions. Monosubstituted oxiranes undergo regiospecific cleavage to form 3-(trimethylsiloxy)nitriles when refluxed with a slight excess of cyanotrimethylsilane in the presence of a catalytic amount of ***Potassium Cyanide–18-Crown-6*** complex (eqs 8–10).[17] The addition of cyanide occurs exclusively at the least-substituted carbon.

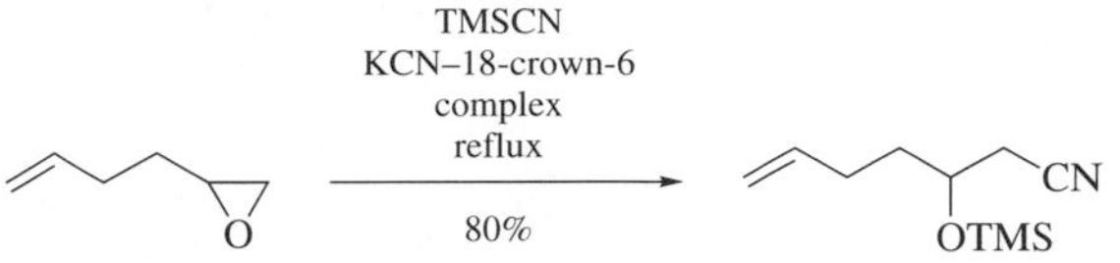

(8)

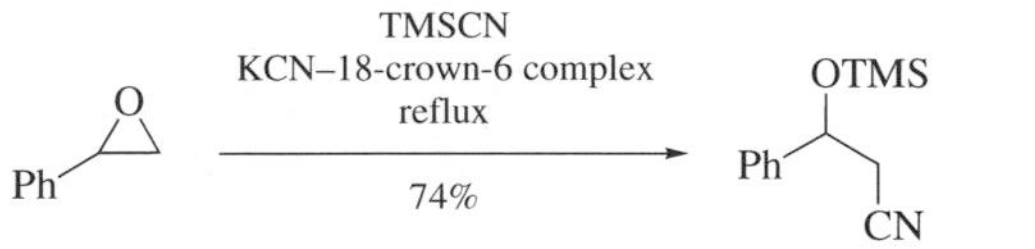

(9)

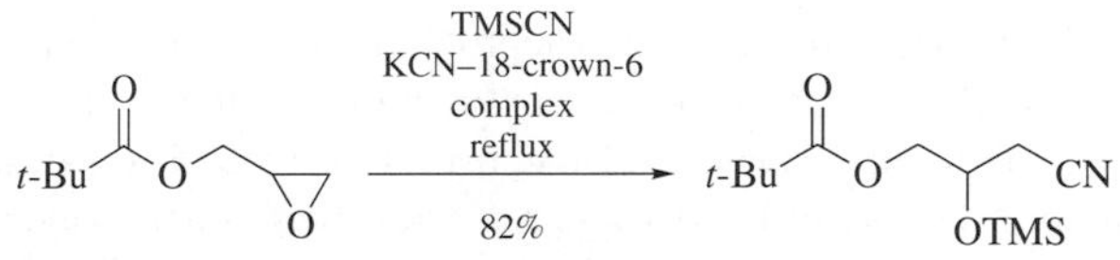

(10)

Good yields of 3-(trimethylsiloxy)nitriles are also obtained from the reactions of oxiranes with cyanotrimethylsilane in the presence of lanthanide salts,[18] or when the reaction is catalyzed by $AlCl_3$ or ***Diethylaluminum Chloride*** (eq 11).[19] Ring opening of chiral glycidyl derivatives by Me_3SiCN catalyzed by Ti(O-*i*-Pr)$_4$ or Al(O-*i*-Pr)$_3$ occurs in a regiospecific and highly stereoselective manner (eq 12).[20]

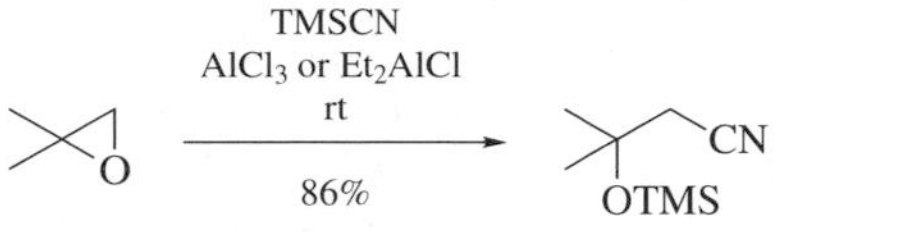

(11)

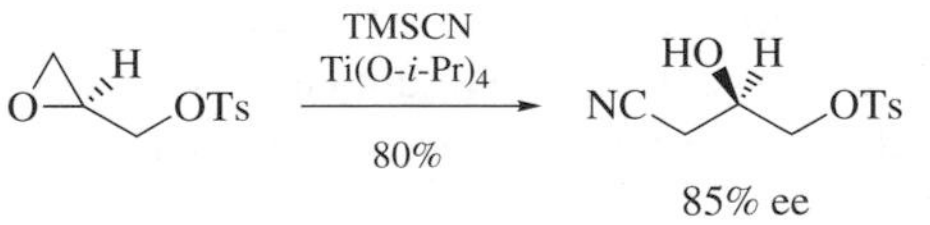

(12)

85% ee

Oxetanes give rise to 4-(trimethylsiloxy)propionitriles (eq 13).[19] Similar observations have been made in the reactions of *N*-tosylaziridines and cyanotrimethylsilane catalyzed by lanthanum salts (eq 14).[21]

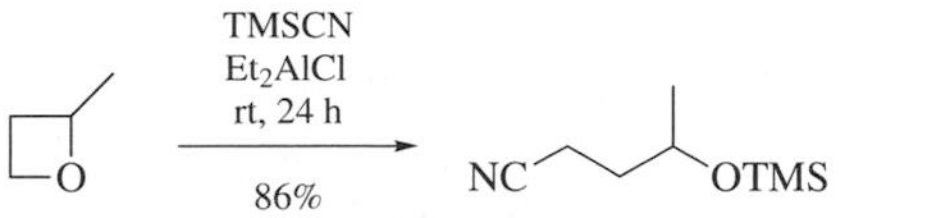

(13)

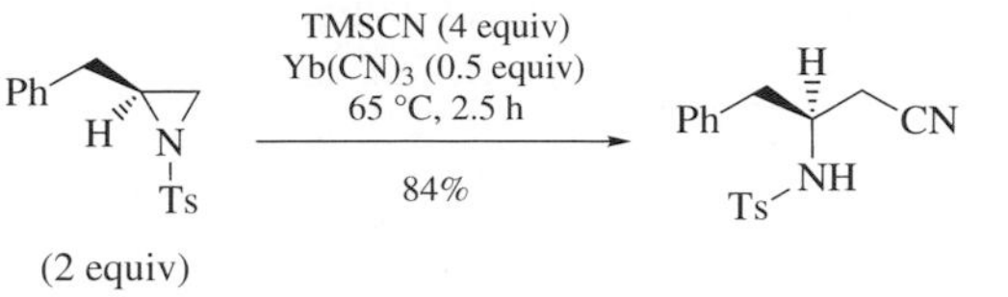

(14)

The ambident nature of cyanotrimethylsilane[22] can lead to the formation of nitriles or isocyanides, depending on the nature of the catalyst. For example, cyanotrimethylsilane reactions with epoxides and oxetanes catalyzed by soft Lewis acids give rise to 2-(trimethylsiloxy) isocyanides arising by attack on the more substituted carbon (eqs 15 and 16).[23–25] Milder reaction conditions and better yields of isocyanides can be realized when the reaction of cyanotrimethylsilane with oxiranes is carried out in the presence of $Pd(CN)_2$, $SnCl_2$, or Me_3Ga (eq 17).[26] Isocyanides are useful precursors for the synthesis of β-amino alcohols and oxazolines.

1. TMSCN, ZnI_2 2. KF; 77% (15)

TMSCN, ZnI_2; 91% (16)

TMSCN, $Pd(CN)_2$; 89% (17)

Carbonyl Umpolung. When deprotonated with a strong base, *O*-(trimethylsilyl) cyanohydrins can function as effective acyl anion equivalents that can be used to convert aldehydes to ketones,[27] and in the synthesis of 1,4-diketones,[28] tricyclic ketones,[29] and the highly sensitive α,β-epoxy ketone functionality (eq 18).[30]

1. TMSCN 2. LDA, DME 3. $MeCOCH_2Cl$ 4. CsF, MeCN (18)

Miscellaneous Transformations. Cyanotrimethylsilane effects the transformation of acyl chlorides to acyl cyanides,[31] α-chloro ethers and α-chloro thioethers to α-cyano ethers[32] and α-cyano thioethers (eq 19),[33] *t*-butyl chlorides to nitriles (eqs 20 and 21),[34] 1,3,5-trisubstituted hexahydro-1,3,5-triazines to aminoacetonitriles,[35] the cyanation of allylic carbonates and acetates (eqs 22 and 23),[36] and the formation of aryl thiocyanates from aryl sulfonyl chlorides and sulfinates.[37] The reagent has been used effectively in peptide synthesis[38] and in a range of other synthetic applications.[39–46]

TMSCN, $SnCl_4$, 0 °C, 1 h; 66% (19)

TMSCN, $SnCl_4$, CH_2Cl_2, rt, 24 h; 70% (20)

TMSCN, $SnCl_4$, CH_2Cl_2, rt, 24 h; 76% (21)

TMSCN, $Pd(CO)(PPh_3)_3$, THF, reflux; 98%; (*E*):(*Z*) = 99:1 (22)

TMSCN, $Pd(PPh_3)_3$, THF, reflux; 78%; *E*:*Z* = 80:20 (23)

Related Reagents. *t*-Butyldimethylsilyl Cyanide; Diethylaluminum Cyanide; Hydrogen Cyanide.

1. (a) Colvin, E. W. *Silicon in Organic Synthesis*; Butterworths: London, 1981. (b) Weber, W. P. *Silicon Reagents for Organic Synthesis*; Springer: Berlin, 1983. (c) *Advances in Silicon Chemistry 1*, Larson, G.; Ed.; JAI: Greenwich, CT, 1991. (d) Groutas, W. C.; Felker, D. *S* **1980**, 861. (e) Livinghouse, T. *OSC* **1990**, *7*, 517.
2. Gassman, P. G.; Talley, J. J. *TL* **1978**, 3773.
3. Mai, K.; Patil, S. *JOC* **1986**, *51*, 3545.
4. Hunig, S. *C* **1982**, *36*, 1.
5. Evans, D. A.; Carroll, G. L.; Truesdale, L. K. *JOC* **1974**, *39*, 914.
6. Lidy, W.; Sundermeyer, W. *CB* **1973**, *106*, 587.
7. Kobayashi, S.; Tsuchiya, Y.; Mukaiyama, T. *CL* **1991**, 537.
8. Onaka, M.; Higuchi, K.; Sugita, K.; Izumi, Y. *CL* **1993**, 1393.
9. Hayashi, M.; Matsuda, T.; Oguni, N. *JCS(P1)* **1992**, 3135.
10. Minamikawa, H.; Hayakawa, S.; Yamada, T.; Iwasawa, N.; Narasaka, K. *BCJ* **1988**, *61*, 4379.
11. Somanathan, R.; Aguilar, H. R.; Ventura, G. R. *SC* **1983**, *13*, 273.
12. Confalone, P. N.; Pizzolato, G. *JACS* **1981**, *103*, 4251.
13. Utimoto, K.; Obayashi, M.; Sishiyama, Y.; Inoue, M.; Nozaki, H. *TL* **1980**, *21*, 3389.
14. Nagata, W.; Yoshioka, M.; Hirai, S. *JACS* **1972**, *94*, 4635.
15. Evans, D. A.; Truesdale, L. K.; Carroll, G. L. *CC* **1973**, 55.
16. Higuchi, K.; Onaka, M.; Izumi, Y. *CC* **1991**, 1035.
17. Sassaman, M. B.; Prakash, G. K. S.; Olah, G. A. *JOC* **1990**, *55*, 2016.
18. Matsubara, S.; Onishi, H.; Utimoto, K. *TL* **1990**, *31*, 6209.
19. Mullis, J. C.; Weber, W. P. *JOC* **1982**, *47*, 2873.
20. Sutowardoyo, K. I.; Sinou, D. *TA* **1991**, *2*, 437.
21. Matsubara, S.; Kodama, T.; Utimoto, K. *TL* **1990**, *31*, 6379.
22. Seckar, J. A.; Thayer, J. S. *IC* **1976**, *15*, 501.
23. Gassman, P. G.; Guggenheim, T. L. *JACS* **1982**, *104*, 5849.
24. Gassman, P. G.; Haberman, L. M. *TL* **1985**, *26*, 4971.
25. Spessard, G. O.; Ritter, A. R.; Johnson, D. M.; Montgomery, A. M. *TL* **1983**, *24*, 655.
26. Imi, K.; Yanagihara, N.; Utimoto, K. *JOC* **1987**, *52*, 1013.
27. (a) Hunig, S.; Wehner, W. *S* **1975**, 180. (b) Deuchert, K.; Hertenstein, U.; Hunig, S. *S* **1973**, 777. (c) Hertenstein, U.; Hunig, S.; Oller, M. *S* **1976**, 416.
28. Hunig, S.; Wehner, G. *CB* **1980**, *113*, 324.
29. Fischer, K.; Hunig, S. *JOC* **1987**, *52*, 564.
30. Hunig, S.; Marschner, C. *CB* **1990**, *123*, 107.
31. Herrmann, K.; Simchen, G. *S* **1979**, 204.
32. Schwindeman, J. A.; Magnus, P. D. *TL* **1981**, *22*, 4925.
33. Fortes, C. C.; Okino, E. A. *SC* **1990**, *20*, 1943.
34. Reetz, M. T.; Chatziiosifidis, I.; Kunzer, H.; Muller-Starke, H. *T* **1983**, *39*, 961.
35. Ha, H.-J.; Nam, G.-S.; Park, K. P. *SC* **1991**, *21*, 155.
36. Tsuji, Y.; Yamada, J.; Tanaka, S. *JOC* **1993**, *58*, 16.
37. Kagabu, S.; Maehara, M.; Sawahara, K.; Saito, K. *CC* **1988**, 1485.
38. Anteunis, M. J. O.; Becu, C.; Becu, F. *BSB* **1987**, *96*, 133.
39. Molander, G. A.; Haar, J. P. *JACS* **1991**, *113*, 3608.
40. Linderman, R. J.; Chen, K. *TL* **1992**, *33*, 6767.
41. Utimoto, K.; Wakabayashi, Y.; Horiie, T.; Inoue, M.; Shishiyama, Y.; Obayashi, M.; Nozaki, H. *T* **1983**, *39*, 967.
42. Zimmer, H.; Reissig, H.-U.; Lindner, H. J. *LA* **1992**, 621.

43. Mukaiyama, T.; Soga, T.; Takenoshita, H. *CL* **1989**, 997.
44. Fuchiyami, T.; Ichikawa, S.; Konno, A. *CL* **1989**, 1987.
45. Krepski, L. R.; Lynch, L. E.; Heilmann, S. M.; Rasmussen, J. K. *TL* **1985**, *26*, 981.
46. Mai, K.; Patil, G. *TL* **1984**, *25*, 4583.

William C. Groutas
Wichita State University, KS, USA

Cyclopentadiene[1]

[542-92-7] C_5H_6 (MW 66.10)

(undergoes a variety of cycloaddition reactions;[5] can be deprotonated to form the anion which reacts with electrophiles[27] and forms stable transition metal complexes;[3] can be used to generate polymers[31])

Physical Data: bp 41–42 °C; *d* 0.805 g cm^{-3}.
Solubility: misc. EtOH, benzene.
Form Supplied in: obtained from dicyclopentadiene
Preparative Method: the volatile diene dimerizes readily and is prepared as required by depolymerization of technical dicyclopentadiene (widely available, bp 171–173 °C). The depolymerization is done by heating the dimer carefully under a fractionating column.[2–4]
Handling, Storage, and Precautions: toxic vapor; stench; highly flammable (flash point <25 °C); can dimerize explosively; forms unstable peroxides with oxygen; avoid skin contact. This reagent should be handled only in a fume hood.

Cycloadditions. Diels–Alder reactions of cyclopentadiene with dienophiles leads predominantly to the *endo* products. In the presence of Lewis acids the *endo* selectivity is greatly increased (eq 1).[5] The use of water as the reaction solvent,[6] or the presence of modified clays[7,8] or zeolites[9] in organic solvents, also leads to a pronounced increase in *endo* selectivity together with increases in reaction rates.

(1)
0 °C 82:18
0.1 equiv $AlCl_3{\cdot}OEt_2$, 0 °C 96:4

Cyclopentadiene has also been widely used in asymmetric Diels–Alder reactions (eq 2).[10] Chromium carbene complexes form cycloadducts with cyclopentadiene at increased reaction rates compared to the organic ester analogs (eq 3).[11–13]

(2)
5 mol%; CH_2Cl_2, −78 °C; 94%
(*R*):(*S*) = 200:1
exo:*endo* = 96:4

(3)
25 °C, 2 h; 85%
170 °C, 1 h; 22%

Cyclopentadiene reacts with benzyne to afford Diels–Alder adducts in high yields (eq 4).[14] This reaction is so efficient that it has been used as a diagnostic test for the formation of benzyne. Heteroatomic dienophiles such as imines,[15,16] thioaldehydes,[17] selenoaldehydes,[18] and sulfenes (eq 5)[19] also react with cyclopentadiene in a [4 + 2] manner.

(4)
Mg, THF; 67%

(5)
$Me_3SiCH_2SO_2Cl$ →(CsF) $[CH_2SO_2]$ →

Cycloaddition with cyclopentadiene has been used for protection of double bonds, which can be subsequently revealed upon pyrolysis (eq 6).[20a] This type of retrograde Diels–Alder reaction has been reviewed.[20b,c]

(6)
1. BuLi 2. RI; 600–650 °C and $-SO_2$

Cyclopentadiene undergoes a [4 + 3] cycloaddition upon reaction with 2-oxyallyl cations, providing bicyclo[3.2.1]oct-6-en-3-ones (eq 7).[21] These types of products have been employed as important precursors to natural products.[21]

(7)
Cu, NaI; MeCN; 82%
6.4:1

Cyclopentadiene reacts as a simple alkene with ketenes such as 2-carbonyl-1,3-dithiane (eq 8).[22] Likewise, cyclopentadiene reacts chemoselectively with alkynes in the Pauson–Khand cycloaddition. The chemical yields are generally excellent, but the regioselectivity is only moderate (eq 9).[23]

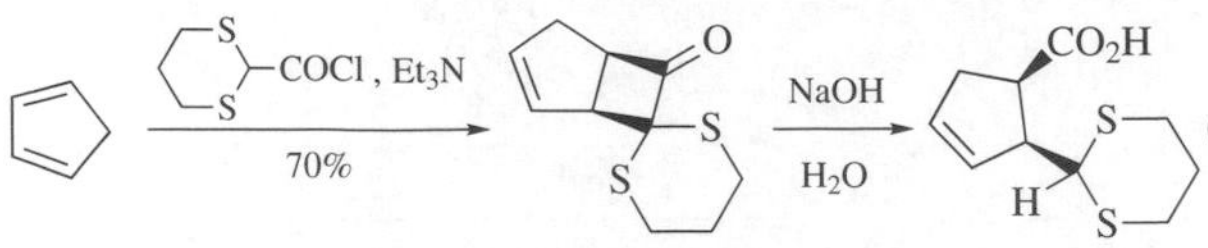

(8)

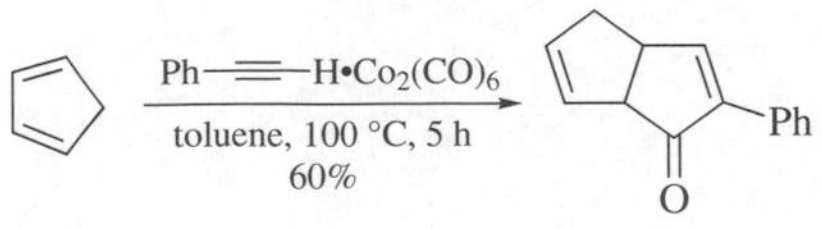

(9)

Deprotonation. Cyclopentadiene can be readily deprotonated by strong bases to give the stabilized (6π-electron) cyclopentadienyl anion. This anion is a ubiquitous ligand for transition metal complexes; examples are ruthenocene[24] and ferrocene (eq 10).[3]

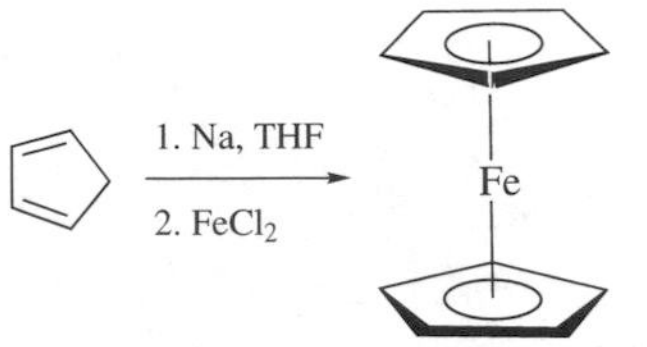

(10)

The cyclopentadienyl anion also reacts with ketones to generate fulvenes (eq 11).[25] Aldehydes only polymerize under these conditions; however, they may be used to form the corresponding fulvenes by employing a modified method.[26] Reaction of the anion with electrophiles often leads to mixtures of regioisomers due to facile ambient temperature 1,5-hydrogen shifts (eq 12).[27a] To circumvent these hydrogen shifts, the use of cyclopentadienylthallium at low temperature has been employed.[27b]

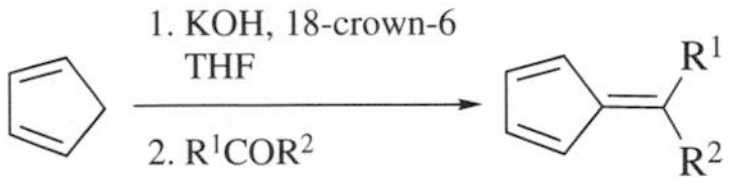

(11)

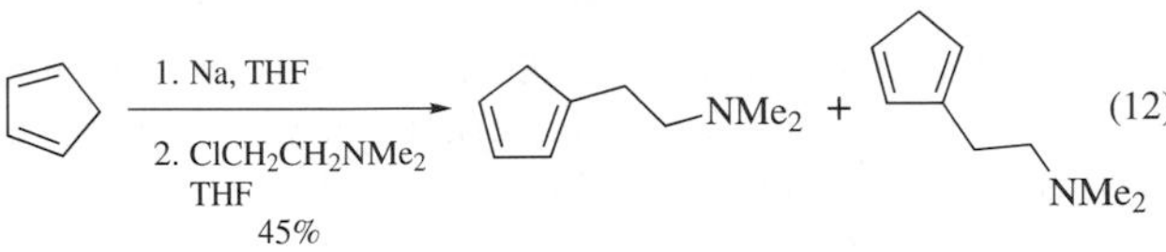

(12)

1,4-Addition Reactions. Cyclopentadiene reacts with ***Bromine*** to afford the *cis*-1,4-adduct in high yield (eq 13).[28] The *cis*-1,4-diol may be obtained by reaction with singlet oxygen and ***Thiourea*** (eq 14).[29] The diacetate of this diol can be desymmetrized with enzymes to afford the enantiomerically enriched alcohol, which is an important chiral building block (eq 15).[30]

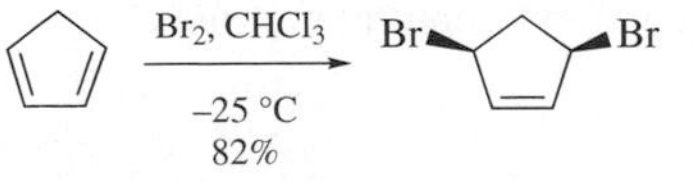

(13)

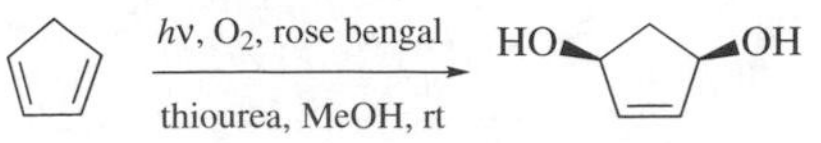

(14)

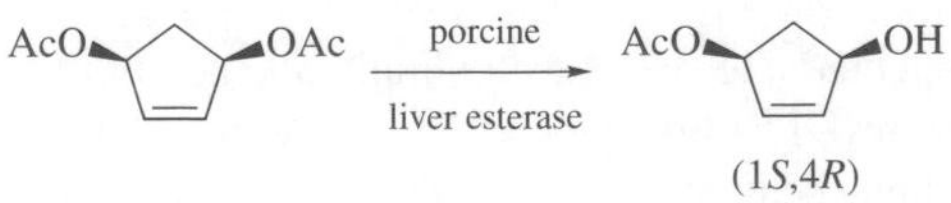

(15)

Polymerization. Cyclopentadiene undergoes polymerization in the presence of 0.1 mol % of a tungsten nitrosyl Lewis acid catalyst (eq 16).[31] The polymer is a mixture of 1,4- and 1,2-polycyclopentadiene.

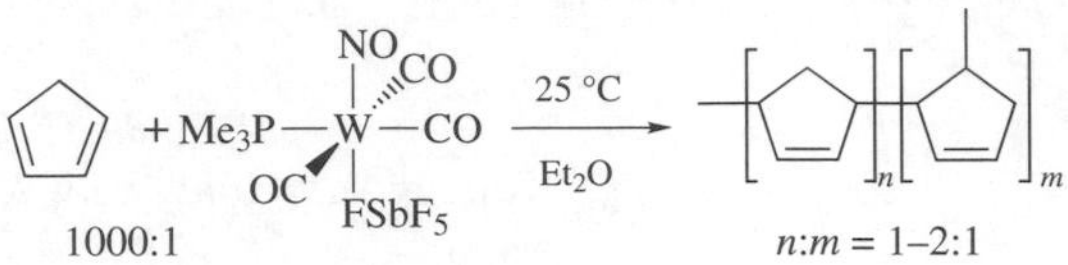

(16)

Related Reagents. 5-Bromo-1,3-cyclopentadiene; 5-(Methoxymethyl)-1,3-cyclopentadiene.

1. Wilson, P. J., Jr.; Wells, J. H. *CVR* **1944**, *34*, 1.
2. Moffett, R. B. *OSC* **1963**, *4*, 238.
3. Wilkinson, G. *OSC* **1963**, *4*, 473.
4. Korach, M.; Nielsen, D. R.; Rideout, W. H. *OSC* **1973**, *5*, 414.
5. Sauer, J.; Kredel, J. *TL* **1966**, 731.
6. (a) Rideout, D. C.; Breslow, R. *JACS* **1980**, *102*, 7816. (b) Breslow, R.; Guo, T. *JACS* **1988**, *110*, 5613.
7. Laszlo, P. *ACR* **1986**, *19*, 121.
8. Adams, J. M.; Martin, K.; McCabe, R. W. *Inclusion Phenom.* **1987**, *5*, 663.
9. Ipaktschi, J. *ZN(B)* **1986**, *41*, 496.
10. Corey, E. J.; Loh, T.-P. *JACS* **1991**, *113*, 8966.
11. Wulff, W. D.; Yang, D. C. *JACS* **1984**, *106*, 7565.
12. Dotz, K. H.; Kuhn, W. *JOM* **1985**, *286*, C23.
13. Dotz, K. H.; Kuhn, W.; Muller, G.; Huber, B.; Alt, H. G. *AG(E)* **1986**, *25*, 812.
14. Wittig, G.; Knauss, E. *CB* **1958**, *91*, 895.
15. (a) Krow, G.; Rodebaugh, R.; Marakowski, J.; Ramey, K. C. *TL* **1973**, 1899. (b) Krow, G. R.; Pyun, C.; Rodebaugh, R.; Marakowski, J. *T* **1974**, *30*, 2977.
16. Barco, A.; Benetti, S.; Baraldi, P. G.; Moroder, F.; Pollini, G. P.; Simoni, D. *LA* **1982**, 960.
17. Krafft, G. A.; Meinke, P. T. *TL* **1985**, *26*, 1947.
18. Krafft, G. A.; Meinke, P. T. *JACS* **1986**, *108*, 1314.
19. Block, E.; Aslam, M. *TL* **1982**, *23*, 4203.
20. (a) Bloch, R.; Abecassis, J. *TL* **1983**, *24*, 1247. (b) Sweger, R. W.; Czarnik, A. W. *COS* **1991**, *5*, 551. (c) Lasne, M.-C.; Ripoll, J.-L. *S* **1985**, 121.
21. Hosomi, A.; Tominaga, Y. *COS* **1991**, *5*, 593.
22. Cossement, E.; Biname, R.; Ghosez, L. *TL* **1974**, 997.
23. Khand, I. U.; Pauson, P. L.; Habib, J. A. *JCR(M)* **1978**, 4418.
24. Bublitz, D. E.; McEwen, W. E.; Kleinberg, J. *OSC* **1973**, *5*, 1001.
25. Alper, H.; Laycock, D. E. *S* **1980**, 799.
26. Stone, K. J.; Little, D. R. *JOC* **1984**, *49*, 1849.
27. (a) Rees, W. S., Jr.; Dippel, K. A. *OPP* **1992**, *24*, 527. (b) Corey, E. J.; Koelliker, U.; Neuffer, J. *JACS* **1971**, *93*, 1489.
28. Owen, L. N.; Smith, P. N. *JCS* **1952**, 4035.
29. Kaneko, C.; Sugimoto, A.; Tanaka, S. *S* **1974**, 877.
30. (a) Laumen, K.; Reimerdes, E. H.; Schneider, M. *TL* **1985**, *26*, 407. (b) Laumen, K.; Schneider, M. P. *CC* **1986**, 1298. (c) Johnson, C. R.; Bis, S. J. *TL* **1992**, *33*, 7287.
31. Honeychuck, R. V.; Bonnesen, P. V.; Farahi, J.; Hersh, W. H. *JOC* **1987**, *52*, 5293.

A. Chris Krueger
Wayne State University, Detroit, MI, USA

Cyclopropyldiphenylsulfonium Tetrafluoroborate[1]

$Ph_2\overset{+}{S}$–cyclopropyl BF_4^-

[33462-81-6] $C_{15}H_{15}BF_4S$ (MW 314.15)

(versatile C_3 building block; reagent for the spiroannulation of carbonyl compounds)

Physical Data: mp 136–138 °C.
Solubility: used as a suspension in THF, DME, DMSO, MeCN.
Form Supplied in: white solid; commercially available.
Preparative Methods: see eq 1.[2]

I(CH$_2$)$_3$Cl + Ph_2S —($AgBF_4$, $MeNO_2$, 87%)→ $Ph_2\overset{+}{S}$(CH$_2$)$_3$Cl BF_4^- —(NaH, THF, 79%)→ $Ph_2\overset{+}{S}$–cyclopropyl BF_4^- (1)

Introduction. Cyclopropyldiphenylsulfonium tetrafluoroborate is primarily used for the generation of the corresponding ylide which is synthetically useful. The ylide can be used in a variety of transformations including synthesis of cyclobutanones, γ-butyrolactones, cyclopentanones, and spiropentanes.

Generation of Diphenylsulfonium Cyclopropylide. The ylide can be formed by deprotonation of the sulfonium salt under reversible or irreversible conditions. Treatment with dimsylsodium in DME at −40 °C irreversibly generates the ylide.[3] Low temperatures are required in order to avoid decomposition of the ylide to cyclopropyl phenyl sulfide and benzyne. Generation of the ylide under reversible conditions with powdered ***Potassium Hydroxide*** in DMSO at room temperature[2b] minimizes thermal decomposition because the equilibrium between the sulfonium salt and ylide favors the salt.

Formation of Spiropentanes and Oxaspiropentanes. Conjugate addition of diphenylsulfonium cyclopropylide to α,β-unsaturated esters and ketones affords spiropentanes (eq 2).[4]

Analogous to other sulfur ylides (see ***Dimethylsulfonium Methylide***), diphenylsulfonium cyclopropylide reacts with saturated aldehydes and ketones to provide epoxide-type products known as oxaspiropentanes (eq 3).[5] The reaction is general in scope and works well even for readily enolizable ketones, though exceptions are known.[5,6] The good stereoselectivity in reactions with cyclic ketone partners, such as cyclohexanone, results from attack of the ylide from the less hindered equatorial direction. When epimerization of a diastereomeric mixture of ketones occurs under the reaction conditions, formation of one oxaspiropentane can result. This results because epimerization α to the ketone occurs faster than addition of the ylide to the ketone, and because one of the epimers reacts at a faster rate than the other.[1e] Isolated alkenes, esters, alcohols, ethers, and epoxides are some of the functional groups that can be tolerated under the reaction conditions. The ylide is chemoselective for saturated ketones over conjugated enones.[1e]

Methyl acrylate (CO_2Me) + $Ph_2\overset{+}{S}$–cyclopropyl BF_4^- —(KOH, DMSO, 83%)→ [MeO, O^-, Ph–$\overset{+}{S}$–Ph] → MeO(C=O)–spiropentane (2)

ketone (O) —(KOH, DMSO, 96%)→ oxaspiropentane (O, O) —(HBF_4, 96%)→ cyclobutanone (=O, O) (3)

The chemical versatility of these highly reactive oxaspiropentanes makes them useful building blocks. For example, such highly strained epoxides are very labile towards acid catalyzed rearrangements. This reactivity can be exploited to produce cyclobutanones (eq 3) or vinylcyclopropanol products (see below). In cases where the oxaspiropentanes possess substituents capable of stabilizing a carbonium ion, rearrangement to the cyclobutanone usually occurs under the reaction conditions.[5] In other cases the cyclobutanone can be obtained directly by acidic work-up of the ylide reaction. In general, the major diastereomer arises from inversion of configuration at the migration terminus, which is consistent with a concerted mechanism.[1e,7] If the cyclobutanone of opposite configuration is desired, an alternate procedure can be adopted. Opening of the oxaspiropentane with phenylselenide anion gives the hydroxyselenide, which is directly subjected to oxidation with ***m-Chloroperbenzoic Acid***. Under the reaction conditions, ionization of the selenoxide occurs, leading to rearrangement. The overall process involves two inversions, leading to the stereoreversed cyclobutanone.[8] Such stereoreversed cyclobutanones can be more efficiently prepared from the starting ketones using the alternative reagent 1-lithiocyclopropyl phenyl sulfide.[9] The ring strained spirocyclobutanones are useful for further structural elaborations.[10] Treatment with basic ***Hydrogen Peroxide*** affords γ-butyrolactones (via ring expansion with migration of the more substituted carbon),[11] and fused cyclopentenones by treatment of such γ-butyrolactones with acid (eq 4).[12] Direct transformation of vinylcyclobutanones to fused cyclopentenones can be effected by treatment with acid.[13] Cyclic enol esters can be formed via α-formylation and acid promoted fragmentation.[14] Synthetic elaboration of cyclobutanones via alkylation and subsequent cleavage to release ring strain results in the selective transformation of the original carbonyl group into a variety of substitution patterns. These include geminal dialkylation and reductive alkylation of the carbonyl group.[15–17]

1. Ph_2S–cyclopropyl BF_4^- 2. TsOH; NaOH H_2O_2

P_2O_5 $MeSO_3H$ (4)

Transformation of Oxaspiropentanes to Vinylcyclopropanol Derivatives. Treatment of oxaspiropentanes with strong base affords vinylcyclopropanols via a *cis,syn* elimination (eq 5). The regiochemistry of ring opening is dependent on the nature of the base, solvent, and factors determining whether the reaction is under thermodynamic or kinetic control.[18] If the oxaspiropentane does not possess a suitably placed hydrogen for base abstraction, treatment with sodium phenylselenide in a nonprotic solvent at rt can afford the β-hydroxy selenide which eliminates to the vinylcyclopropanol in situ (eq 6).[19] Thus the mechanism has switched from a *cis,syn* to a *trans,anti* elimination. Solvent plays an important role in this reaction; switching the solvent to ethanol affords the hydroxy selenide, an intermediate in the formation of stereoreversed cyclobutanones (see above). Attempts to effect a selenoxide elimination (oxidation of the selenide) often affords the cyclobutanone product.[8]

LDA hexane 30 h 99%; LiN(pyrrolidine) hexane 120 h 100%; C_6H_4-*p*-Me; OH (5)

PhSeNa 56%; OH; H (6)

Vinylcyclopropanol silyl ethers are obtained by quenching the alkoxide formed under strong base conditions with ***Chlorotrimethylsilane***. This composite functional group can undergo thermal rearrangements to cyclopentane trimethylsilyl enol ethers.[5,20] This rearrangement is accelerated by the presence of the siloxy group. The enol silyl ether produced can be further alkylated to provide substituted cyclopentanones. This sequence results in the regiospecific annulation of a cyclopentanone ring to a carbonyl group. In addition, vinylcyclopropanol silyl ethers can undergo ring expansion to cyclobutanones upon treatment with electrophiles.[17b,17c,21] Use of vinylcyclopropanol silyl ethers as cyclization terminators can result in good yields of six- to eight-membered rings (eq 7).[22] The cyclopropyl ring mediates the delocalization of the lone pair on the oxygen into the alkene, enhancing the alkene nucleophilicity, and the reaction is terminated effectively by rearrangement to the cyclobutanone.

OTMS; OMe; OMe; TMSOTf, py 0.01M CH_2Cl_2 –40 to –20 °C 85%; H; O; OMe; H (7)

Homologs of Cyclopropyldiphenylsulfonium Tetrafluoroborate. 2-Methylcyclopropyldiphenylsulfonium tetrafluoroborate has also been prepared and is formed as a mixture of *trans* (major) and *cis* isomers (eq 8). The stereochemical integrity of the ylide derived from this mixture of sulfonium salts is variable, depending on the conditions under which it is generated. The reactivity profile of the ylide is the same as the parent, but the presence of an additional chiral center often leads to a mixture of stereoisomers.[23]

$Ph_2\overset{+}{S}$ BF_4^- Cl; DMSO, THF *t*-BuOK; $Ph_2\overset{+}{S}$ BF_4^- (8)

Related Reagents. Dimethylsulfonium Methylide; Diphenylsulfonium Methylide.

---

1. (a) Salaün, J. R. Y. *Top. Curr. Chem.* **1988**, *144*, 1. (b) Reissig, H.-U. In *The Chemistry of the Cyclopropyl Group*, Rappoport, Z., Ed; Wiley: Chichester, 1987; Part I, pp 404–410. (c) Trost, B. M. *ACR* **1974**, *7*, 85. (d) Trost, B. M. *PAC* **1975**, *43*, 563. (e) Trost, B. M. *Top. Curr. Chem.* **1986**, *133*, 3.
2. (a) Trost, B. M.; Bogdanowicz, M. J. *JACS* **1971**, *93*, 3773. (b) Trost, B. M.; Bogdanowicz, M. J. *JACS* **1973**, *95*, 5298. (c) Badet, B.; Julia, M. *TL* **1979**, 1101. (d) Bogdanowicz, M. J.; Trost, B. M. *OS* **1974**, *54*, 27.
3. Corey, E. J.; Chaykovsky, M. *JACS* **1965**, *87*, 1345.
4. Trost, B. M.; Bogdanowicz, M. J. *JACS* **1973**, *95*, 5307.
5. Trost, B. M.; Bogdanowicz, M. J. *JACS* **1973**, *95*, 5311.
6. Trost, B. M.; Bogdanowicz, M. J. *TL* **1972**, 887.
7. Trost, B. M.; Rigby, J. H. *JOC* **1976**, *41*, 3217.
8. Trost, B. M.; Scudder, P. H. *JACS* **1977**, *99*, 7601.
9. (a) Trost, B. M.; Keeley, D. E.; Arndt, H. C.; Rigby, J. H.; Bogdanowicz, M. J. *JACS* **1977**, *99*, 3080. (b) Trost, B. M.; Keeley, D. E.; Arndt, H. C.; Bogdanowicz, M. J. *JACS* **1977**, *99*, 3088.
10. For reviews, see Conia, J. M.; Salaün, J. R. *ACR* **1972**, *5*, 33 and *Top. Curr. Chem.* **1986**, *133*, 85.
11. Trost, B. M.; Bogdanowicz, M. J. *JACS* **1973**, *95*, 5321.
12. (a) Baldwin, J. E.; Beckwith, P. L. M. *CC* **1983**, 279. (b) Paquette, L. A.; Wyvratt, M. J.; Schallner, O.; Muthard, J. L.; Begley, W. J.; Blankenship, R. M.; Balogh, D. *JOC* **1979**, *44*, 3616.
13. Matz, J. R.; Cohen, T. *TL* **1981**, *22*, 2459.
14. Trost, B. M.; Hiroi, K.; Holy, N. *JACS* **1975**, *97*, 5873.
15. Trost, B. M.; Bogdanowicz, M. J.; Frazee, W. J.; Salzmann, T. N. *JACS* **1978**, *100*, 5512.

16. (a) Trost, B. M.; Bogdanowicz, M. J.; Kern, J. *JACS* **1975**, *97*, 2218. (b) Trost, B. M.; Preckel. M.; Leichter, L. M. *JACS* **1975**, *97*, 2224.

17. For application of these methods in total syntheses, see (a) Trost, B. M.; Latimer, L. H. *JOC* **1978**, *43*, 1031. (b) Trost, B. M.; Mao, M. K.-T.; Balkovec, J. M.; Buhlmayer, P. *JACS* **1986**, *108*, 4965. (c) Trost, B. M.; Balkovec, J. M.; Mao, M. K.-T. *JACS* **1986**, *108*, 4974.

18. Trost, B. M.; Kurozumi, S. *TL* **1974**, *22*, 1929.

19. Trost, B. M.; Nishimura, Y.; Yamamoto, K.; McElvain, S. S. *JACS* **1979**, *101*, 1328.

20. Trost, B. M.; Bogdanowicz, M. J. *JACS* **1973**, *95*, 289.

21. Trost, B. M.; Brandi, A. *JACS* **1984**, *106*, 5041.

22. Trost, B. M.; Lee, D. C. *JACS* **1988**, *110*, 6556.

23. Trost, B. M.; Bogdanowicz, M. J. *JACS* **1973**, *95*, 5321.

Donna L. Romero
The Upjohn Company, Kalamazoo, MI, USA

William H. Pearson & P. Sivaramakrishnan Ramamoorthy
The University of Michigan, Ann Arbor, MI, USA

Diazomethane[1]

$H_2C=\overset{+}{N}=\overset{-}{N}$

[334-88-3] CH_2N_2 (MW 42.04)

(methylating agent for various functional groups including carboxylic acids, alcohols, phenols, and amides; reagent for the synthesis of α-diazo ketones from acid chlorides, and the cyclopropanation of alkenes[1])

Physical Data: mp −145 °C; bp −23 °C.

Solubility: diazomethane is most often used as prepared in ether, or in ether containing a small amount of ethanol. It is less frequently prepared and used in other solvents such as dichloromethane.

Analysis of Reagent Purity: diazomethane is titrated[2] by adding a known quantity of benzoic acid to an aliquot of the solution such that the solution is colorless and excess benzoic acid remains. Water is then added, and the amount of benzoic acid remaining is back-titrated with NaOH solution. The difference between the amount of acid added and the amount remaining reveals the amount of active diazomethane present in the aliquot.

Preparative Methods: diazomethane is usually prepared by the decomposition of various derivatives of *N*-methyl-*N*-nitrosoamines. Numerous methods of preparation have been described,[3] but the most common and most frequently employed are those which utilize ***N-Methyl-N-nitroso-p-toluenesulfonamide***(Diazald; **1**),[4] ***1-Methyl-3-nitro-1-nitrosoguanidine***(MNNG, **2**),[5] or *N*-methyl-*N*-nitrosourea (**3**).[2]

(1) (2) (3)

The various reagents each have their advantages and disadvantages, as discussed below. The original procedure[6] for the synthesis of diazomethane involved the use of *N*-methyl-*N*-nitrosourea, and similar procedures are still in use today. An advantage of using this reagent is that solutions of diazomethane can be prepared without distillation,[7] thus avoiding the most dangerous operation in other preparations of diazomethane. For small scale preparations (1 mmol or less) which do not contain any alcohol, a kit is available utilizing MNNG which produces distilled diazomethane in a closed environment. Furthermore, MNNG is a stable compound and has a shelf life of many years. For larger scale preparations, kits are available for the synthesis of up to 300 mmol of diazomethane using Diazald as the precursor. The shelf life of Diazald (about 1–2 years), however, is shorter than that of MNNG. Furthermore, the common procedure using Diazald produces an ethereal solution of diazomethane which contains ethanol; however, it can be modified to produce an alcohol-free solution. Typical preparations of diazomethane involve the slow addition of base to a heterogeneous aqueous ether mixture containing the precursor. The precursor reacts with the base to liberate diazomethane which partitions into the ether layer and is concomitantly distilled with the ether to provide an ethereal solution of diazomethane. Due to the potentially explosive nature of diazomethane, the chemist is advised to carefully follow the exact procedure given for a particular preparation. Furthermore, since diazomethane has been reported to explode upon contact with ground glass, apparatus which do not contain ground glass should be used. All of the kits previously mentioned avoid the use of ground glass.

Handling, Storage, and Precautions: diazomethane as well as the precursors for its synthesis can present several safety hazards, and must be used with great care.[8] The reagent itself is highly toxic and irritating. It is a sensitizer, and long term exposure can lead to symptoms similar to asthma. It can also detonate unexpectedly, especially when in contact with rough surfaces, or on crystallization. It is therefore essential that any glassware used in handling diazomethane be fire polished and not contain any scratches or ground glass joints. Furthermore, contact with certain metal ions can also cause explosions. Therefore metal salts such as calcium chloride, sodium sulfate, or magnesium sulfate must not be used to dry solutions of the reagent. The recommended drying agent is potassium hydroxide. Strong light is also known to initiate detonation. The reagent is usually generated immediately prior to use and is not stored for extended periods of time. Of course, the reagent must be prepared and used in a well-ventilated hood, preferably behind a blast shield. The precursors used to generate diazomethane are irritants and in some cases mutagens and suspected carcinogens, and care should be exercised in their handling as well.

Methylation of Heteroatoms. The most widely used feature of the chemistry of diazomethane is the methylation of carboxylic acids. Carboxylic acids are good substrates for reaction with diazomethane because the acid is capable of protonating the diazomethane on carbon to form a diazonium carboxylate. The carboxylate can then attack the diazonium salt in what is most likely an S_N2 reaction to provide the ester. Species which are not acidic enough to protonate diazomethane, such as alcohols, require an additional catalyst, such as ***Boron Trifluoride Etherate***, to increase their acidity and facilitate the reaction. The methylation reaction proceeds under mild conditions and is highly reliable and very selective for carboxylic acids. A typical procedure is to add a yellow solution of diazomethane to the carboxylic acid in portions. When the yellow color persists and no more gas is evolved, the reaction is deemed complete. Excess reagent can be destroyed by the addition of a few drops of acetic acid and the entire solution concentrated to provide the methyl ester.

Esterification of Carboxylic Acids and Other Acidic Functional Groups. A variety of functional groups will tolerate the esterification of acids with diazomethane. Thus α,β-unsaturated carboxylic acids and alcohols survive the reaction (eq 1),[9] as do ketones (eq 2),[10] isolated alkenes (eq 3),[11] and amines (eq 4).[12]

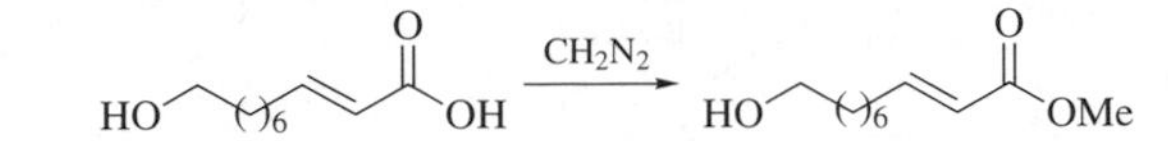

(1)

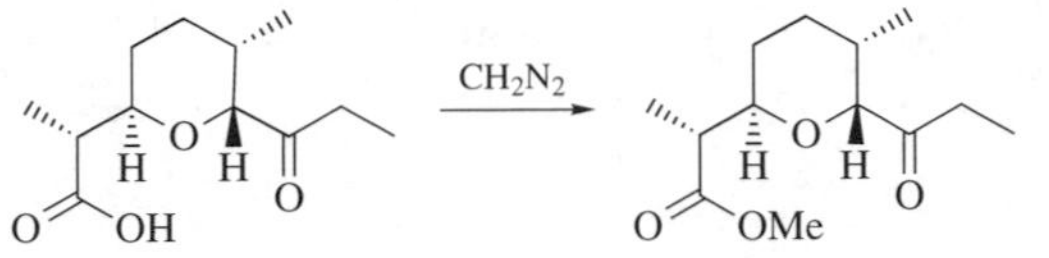

(2)

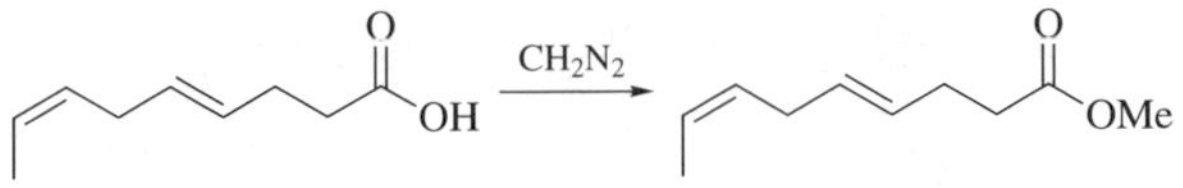

(3)

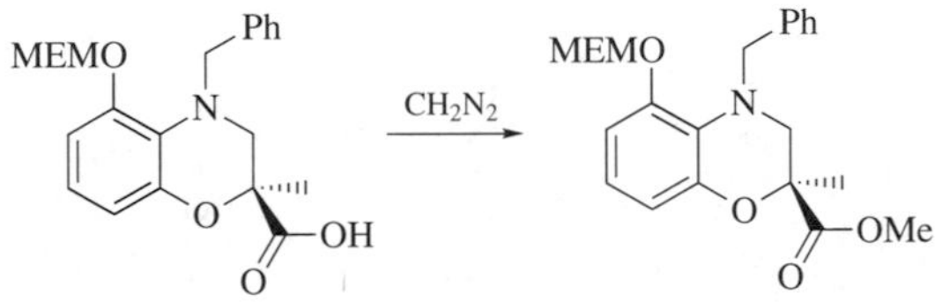

(4)

Other acidic functional groups will also undergo reaction with diazomethane. Thus phosphonic acids (eq 5)[13] and phenols (eq 6)[14] are methylated in high yields, as are hydroxytropolones (eq 7)[15] and vinylogous carboxylic acids (eq 8).[16] The origin of the selectivity in eq 6 is due to the greater acidity of the A-ring phenol.

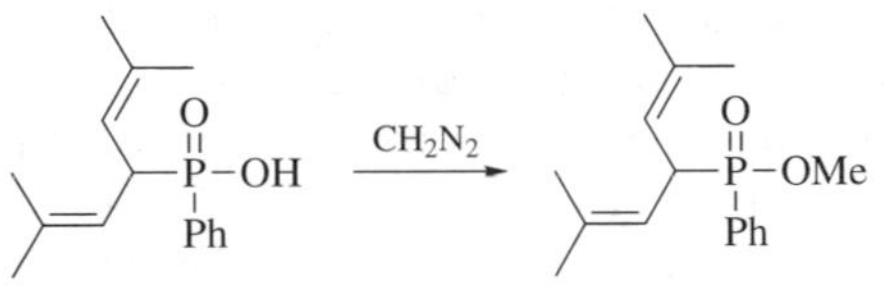

(5)

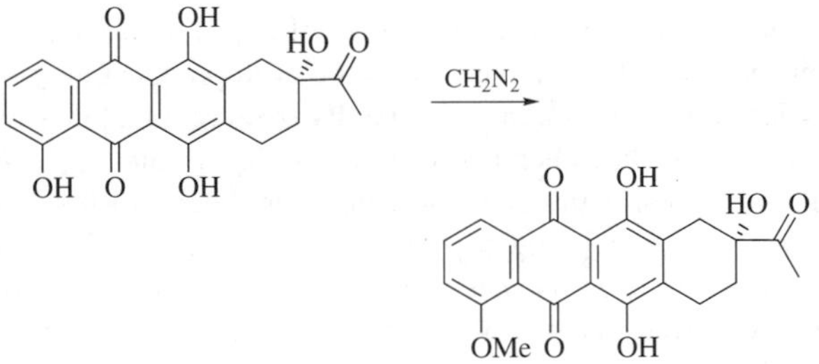

(6)

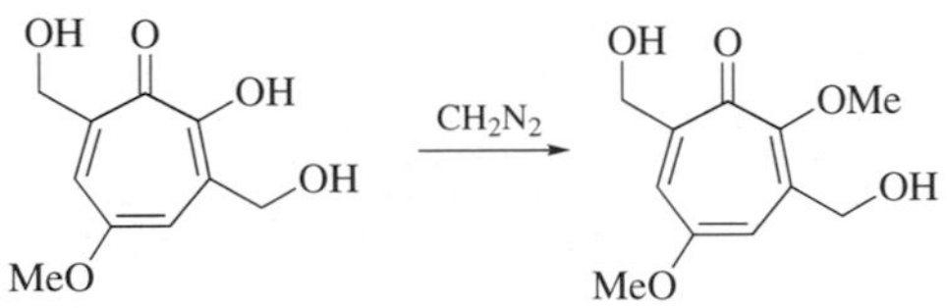

(7)

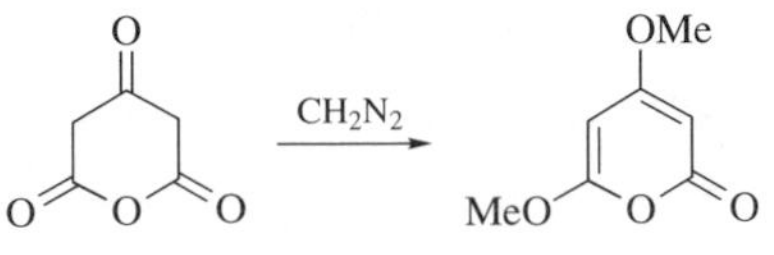

(8)

Selective monomethylation of dicarboxylic acids has been reported using ***Alumina*** as an additive (eq 9).[17] It is thought that one of the two carboxylic acid groups is bound to the surface of the alumina and is therefore not available for reaction. Carboxylic acids that are engaged as lactols will also undergo methylation with diazomethane to provide the methyl ester and aldehyde (eq 10).[18]

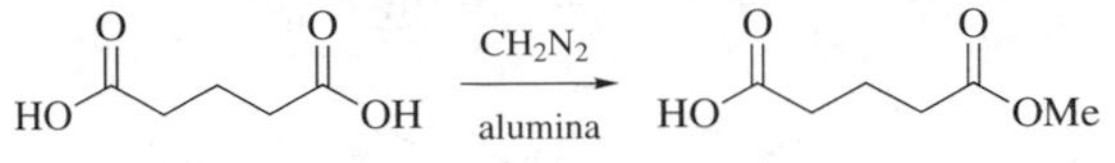

(9)

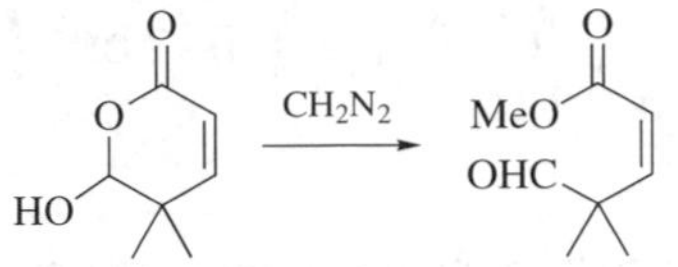

(10)

Methylation of Alcohols and Other Less Acidic Functional Groups. As previously mentioned, alcohols require the addition of a catalyst to react with diazomethane. The most commonly used is boron trifluoride etherate (eq 11),[19] but ***Tetrafluoroboric Acid*** has been used as well (eq 12).[20] Mineral acids are not effective since they rapidly react with diazomethane to provide the corresponding methyl halides. Acids as mild as silica gel are also effective (eq 13).[21] Monomethylation of 1,2-diols with diazomethane has been reported using various Lewis acids as promoters, the most effective of which is ***Tin(II) Chloride*** (eq 14).[22]

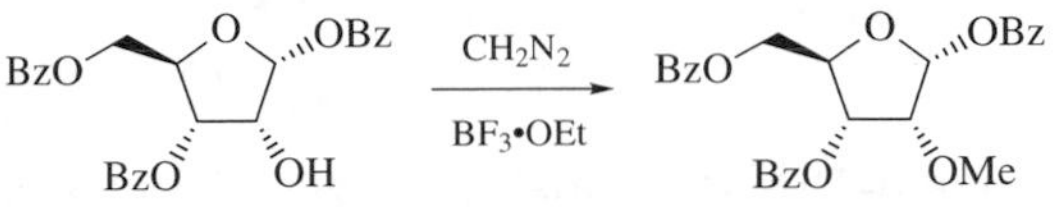

(11)

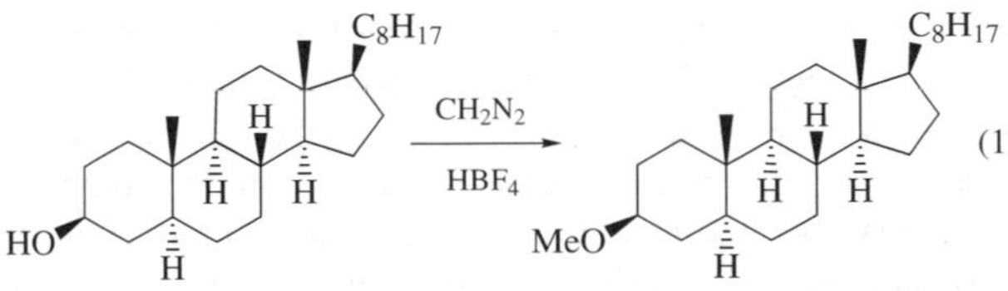

(12)

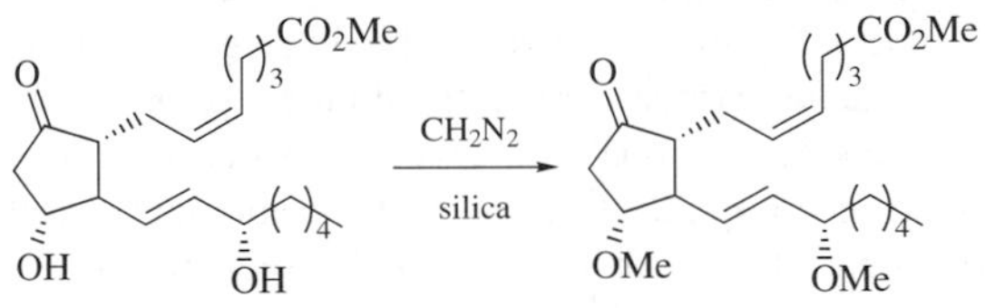

(13)

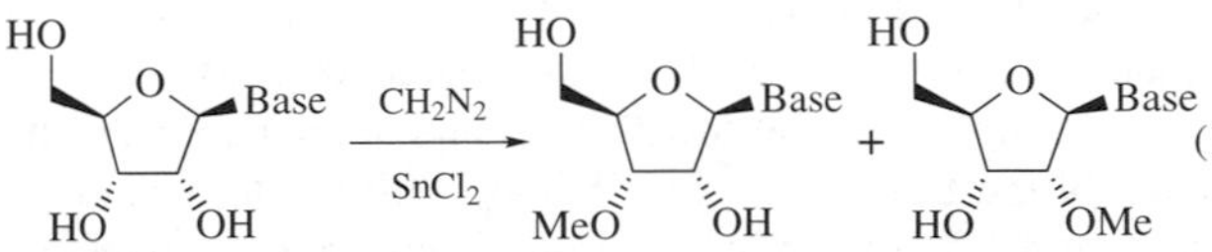

(14)

An interesting case of an alcohol reacting with diazomethane at a rate competitive with a carboxylic acid has been reported (eq 15).[23] In this case, the tertiary structure of the molecule is thought to place the alcohol and the carboxylic acid in proximity to each other. Protonation of the diazomethane by the carboxylic acid leads to a diazonium ion in proximity to the alcohol as well as the carboxylate. These species then attack the diazonium ion at competitive rates to provide the methyl ether and ester. No reaction is observed upon treatment of the corresponding hydroxy ester with diazomethane, indicating that the acid is required to activate the diazomethane.

$\xrightarrow{CH_2N_2}$ (15)

Amides can also be methylated with diazomethane in the presence of silica gel; however, the reaction requires a large excess of diazomethane (25–60 equiv, eq 16).[24] The reaction primarily provides *O*-methylated material; however, in one case a mixture of *O*- and *N*-methylation was reported. Thioamides are also effectively methylated with this procedure to provide *S*-methylated compounds. Finally, amines have been methylated with diazomethane in the presence of BF_3etherate, fluoroboric acid,[25] or copper(I) salts;[26] however, the yields are low to moderate, and the method is not widely used.

$\xrightarrow[\text{silica}]{CH_2N_2}$ (16)

The Arndt–Eistert Synthesis. Diazomethane is a useful reagent for the one-carbon homologation of acid chlorides via a sequence of reactions known as the Arndt–Eistert synthesis. The first step of this sequence takes advantage of the nucleophilicity of diazomethane in its addition to an active ester, typically an acid chloride,[27] to give an isolable α-diazo ketone and HCl. The HCl that is liberated from this step can react with diazomethane to produce methyl chloride and nitrogen, and therefore at least 2 equiv of diazomethane are typically used. The α-diazo ketone is then induced to undergo loss of the diazo group and insertion into the adjacent carbon–carbon bond of the ketone to provide a ketene. The ketene is finally attacked by water or an alcohol (or some other nucleophile) to provide the homologated carboxylic acid or ester. This insertion step of the sequence is known as the Wolff rearrangement[28] and can be accomplished either thermally (eq 17)[29] or, more commonly, by treatment with a metal ion (usually silver salts, eq 18),[30] or photochemically (eq 19).[31] It has been suggested that the photochemical method is the most efficient of the three.[32] As eqs 18 and 19 illustrate, retention of configuration is observed in the migrating group. The obvious limitations of this reaction are that there must not be functional groups present in the molecule which will react with diazomethane more rapidly than it will attack the acid chloride. Thus carboxylic acids will be methylated under these conditions. Furthermore, electron-deficient alkenes will undergo [2,3] dipolar cycloaddition with diazomethane more rapidly than addition to the acid chloride. Thus when the Arndt–Eistert synthesis is attempted on α,β-unsaturated acid chlorides, cycloaddition to the alkene is observed in the product. In order to prevent this, the alkene must first be protected by addition of HBr and then the reaction carried out in the normal way (eq 20).[33] Cycloaddition to isolated alkenes, however, is not competitive with addition to acid chlorides.

1. CH_2N_2; 2. 180 °C, collidine, BnOH (17)

CH_2N_2; $PhCO_2Ag$, MeOH (18)

1. CH_2N_2; 2. *hν*, MeOH (19)

1. HBr; 2. $(COCl)_2$; 3. CH_2N_2 (20)

Other Reactions of α-Diazo Ketones Derived from Diazomethane. Depending on the conditions employed, the Wolff rearrangement may proceed via a carbene or carbenoid intermediate, or it may proceed by a concerted mechanism where the insertion is concomitant with loss of N_2 and no intermediate is formed. In the case where a carbene or carbenoid is involved, other reactions which are characteristic of these species can occur, such as intramolecular cyclopropanation of alkenes. In fact, the reaction conditions can be adjusted to favor cyclopropanation or homologation depending on which is desired. Thus treatment of the dienoic acid chloride shown in eq 21 with diazomethane followed by decomposition of the α-diazo ketone with silver benzoate in the presence of methanol and base provides the homologated methyl ester. However, treatment of the same diazoketone intermediate with Cu^{II} salts provides the cyclopropanation products selectively.[34] This trend is generally observed; that is, silver salts as well as photochemical conditions (eqs 18 and 19) favor the homologation pathway while copper or rhodium salts favor cyclopropanation.[35] Using copper salts to decompose the diazo compounds, hindered alkenes as well as electron-rich aromatics can be cyclopropanated as illustrated in eqs 22 and 23,[36,37] respectively.

CH_2N_2; $PhCO_2Ag$, NEt_3, MeOH; $Cu(acac)_2$, benzene (21)

1. CH_2N_2; 2. Cu, Δ (22)

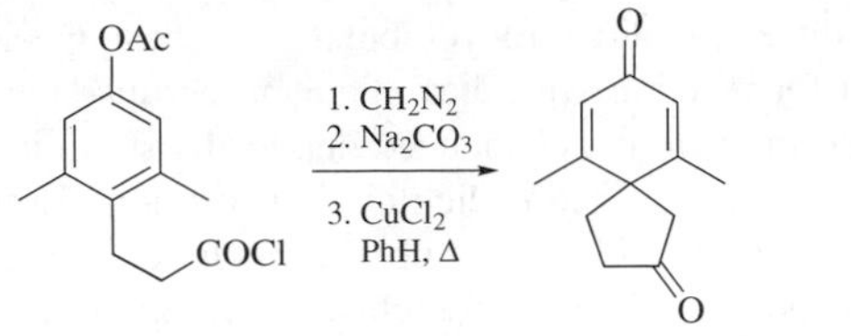

In addition to these reactions, α-diazo ketones will undergo protonation on carbon in the presence of protic acids[38] to provide the corresponding α-diazonium ketone. These species are highly electrophilic and can undergo nucleophilic attack. Thus if the proton source contains a nucleophile such as a halogen then the corresponding α-halo ketone is isolated (eq 24).[39] However, if the proton source does not contain a nucleophilic counterion then the diazonium species may react with other nucleophiles that are present in the molecule, such as alkenes (eq 25)[40] or aromatic rings (eq 26).[41] Note the similarity between the transformations in eqs 26 and 23 which occur using different catalysts and by different pathways. Also, eq 26 illustrates the fact that other active esters will undergo nucleophilic attack by diazomethane.

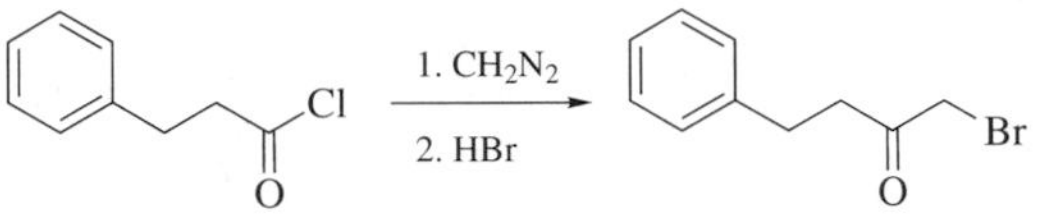

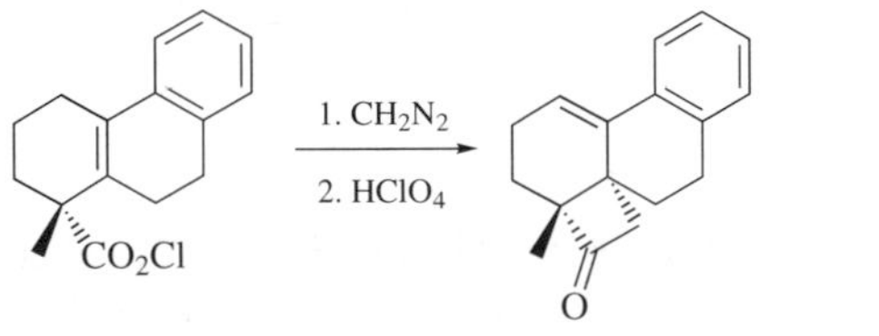

Lewis acids are also effective in activating α-diazo ketones towards intramolecular nucleophilic attack by alkenes and arenes.[42] The reaction has been used effectively for the synthesis of cyclopentenones (eq 27) starting with β,γ-unsaturated diazo ketones derived from the corresponding acid chloride and diazomethane. It has also been used to initiate polyalkene cyclizations (eq 28). Typically, boron trifluoride etherate is used as the Lewis acid, and electron-rich alkenes are most effective providing the best yields of annulation products.

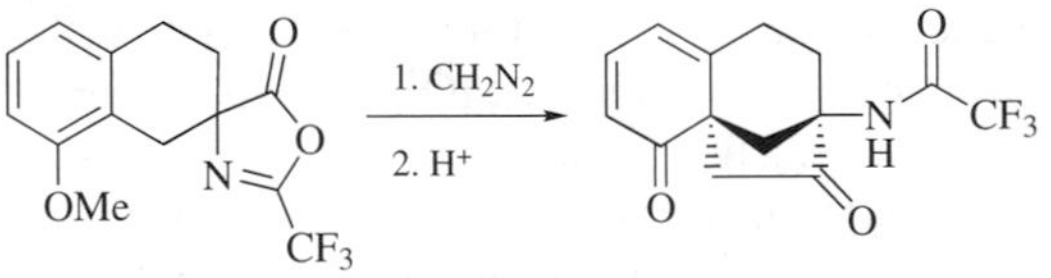

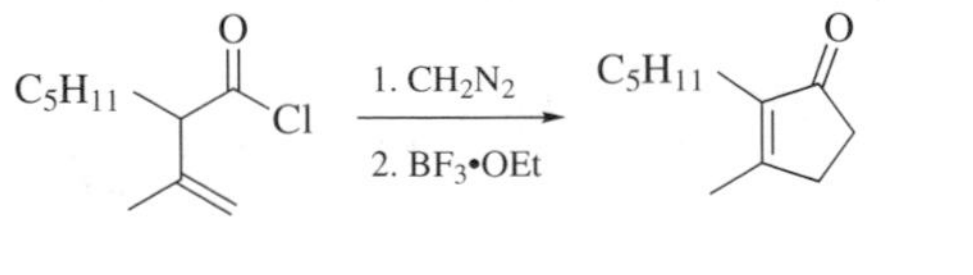

The Vinylogous Wolff Rearrangement. The vinylogous Wolff rearrangement[43] is a reaction that occurs when the Arndt–Eistert synthesis is attempted on β,γ-unsaturated acid chlorides using copper catalysis. Rather than the usual homologation products, the reaction proceeds to give what is formally the product of a [2,3]-sigmatropic shift, but is mechanistically not derived by this pathway.[44] The mechanism is thought to proceed by an initial cyclopropanation of the alkene by the α-diazo ketone to give a bicyclo[2.1.0]pentanone derivative. This compound then undergoes a fragmentation to a ketene alkene before being trapped by the solvent (eq 29). Inspection of the products reveals that they are identical with those derived from the Claisen rearrangement of the corresponding allylic alcohols, and as such this method can be thought of as an alternative to the Claisen procedure. However, the stereoselectivity of the alkene that is formed is not as high as is typically observed in the Claisen rearrangement (eq 30), and in some substrates the reaction proceeds with no selectivity (eq 31).

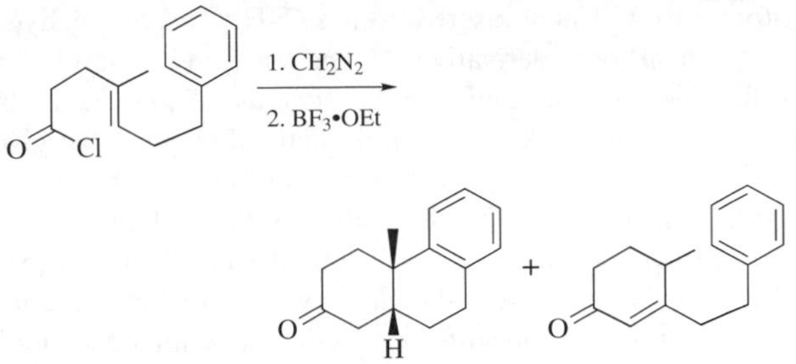

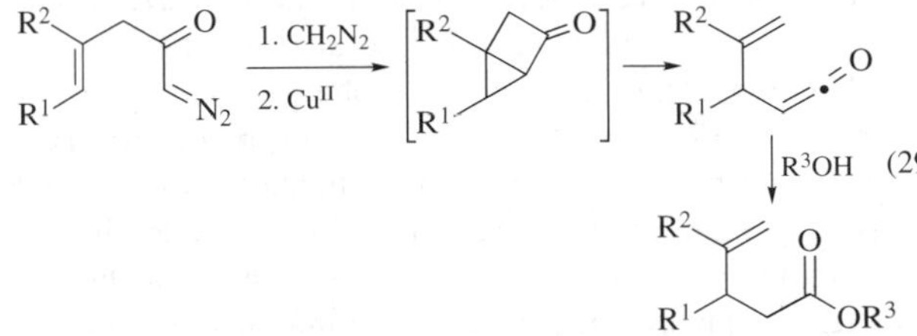

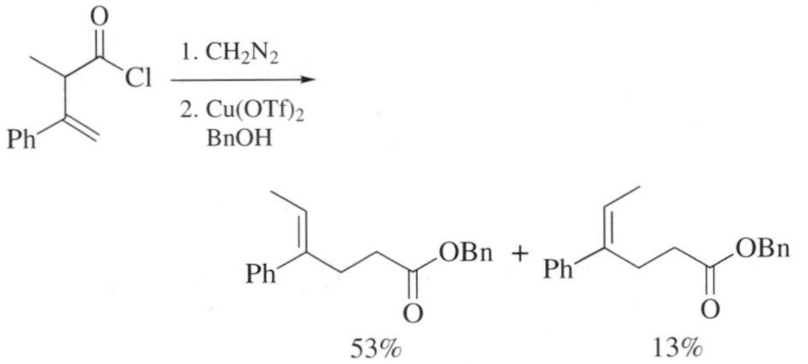

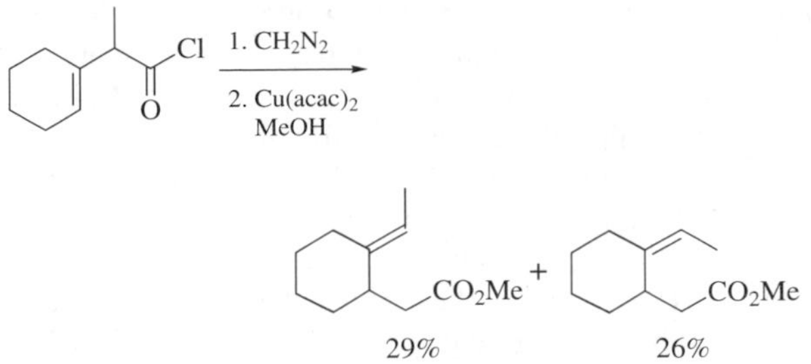

Insertions into Aldehyde C–H Bonds. The α-diazo ketones (and esters) derived from diazomethane and an acid chloride (or chloroformate) will also insert into the C–H bond of aldehydes to give 1,3-dicarbonyl derivatives.[45] The reaction is catalyzed by $SnCl_2$, but some simple Lewis acids, such as BF_3 etherate, also work. The reaction works well for aliphatic aldehydes, but gives variable results with aromatic aldehydes, at times giving none of the desired diketone (eq 32). Sterically hindered aldehydes will also participate in this reaction, as illustrated in eq 33 with the reaction of ethyl α-diazoacetate and pivaldehyde. In a related reaction, α-diazo phosphonates and sulfonates will react with aldehydes in the presence of $SnCl_2$ to give the corresponding β-keto phosphonates and sulfonates.[46] This reaction is a practical alternative to the Arbuzov reaction for the synthesis of these species.

(32)

R^1	R^2	Yield (%)
Ph	H	88
Ph	$PhCH_2$	90
$PhCH_2CH_2$	Ph	0

(33)

R^2	Yield (%)
t-Bu	65
Ph	50

Additions to Ketones. The addition of diazomethane to ketones[47] is also a preparatively useful method for one-carbon homologation. This reaction is a one-step alternative to the Tiffeneau–Demjanow rearrangement[48] and proceeds by the mechanism shown in eq 34. It can lead to either homologation or epoxidation depending on the substrate and reaction conditions. The addition of Lewis acids, such as BF_3 etherate, or alcoholic cosolvents tend to favor formation of the homologation products over epoxidation.

(34)

However, the reaction is limited by the poor regioselectivity observed in the insertion when the groups R^1 and R^2 in the starting ketone are different alkyl groups. What selectivity is observed tends to favor migration of the less substituted carbon,[49] a trend which is opposite to that typically observed in rearrangements of electron-deficient species such as in the Baeyer–Villiger reaction. Furthermore, the product of the reaction is a ketone and is therefore capable of undergoing further reaction with diazomethane. Thus, ideally, the product ketone should be less reactive than the starting ketone. Strained ketones tend to react more rapidly and are therefore good substrates for this reaction (eq 35).[50] This method has also found extensive use in cyclopentane annulation reactions starting with an alkene. The overall process begins with dichloroketene addition to the alkene to produce an α-dichlorocyclobutanone. These species are ideally suited for reaction with diazomethane because the reactivity of the starting ketone is enhanced due to the strain in the cyclobutanone as well as the α-dichloro substitution. Furthermore, the presence of the α-dichloro substituents hinders migration of that group and leads to almost exclusive migration of the methylene group. Thus treatment with diazomethane and methanol leads to a rapid evolution of nitrogen, and produces the corresponding α-dichlorocyclopentanone, which can be readily dehalogenated to the hydrocarbon (eq 36).[51] Aldehydes will also react with diazomethane, but in this case homologation is not observed. Rather, the corresponding methyl ketone derived from migration of the hydrogen is produced (eq 37).

(35)

(36)

(37)

Cycloadditions with Diazomethane. Diazomethane will undergo [3 + 2] dipolar cycloadditions with alkenes and alkynes to give pyrazolines and pyrazoles, respectively.[52] The reaction proceeds more rapidly with electron-deficient alkenes and strained alkenes and is controlled by FMO considerations with the HOMO of the diazomethane and the LUMO of the alkene serving as the predominant interaction.[53] In the case of additions to electron-deficient alkenes, the carbon atom of the diazomethane behaves as the negatively charged end of the dipole, and therefore the regiochemistry observed is as shown in eq 38. With conjugated alkenes, such as styrene, the terminal carbon has the larger lobe in the LUMO, and as such the reaction proceeds to give the product shown in eq 39. Pyrazolines are most often used as precursors to cyclopropanes by either thermal or photochemical extrusion of N_2. In both cases the reaction may proceed by a stepwise mechanism with loss of stereospecificity. As shown in eq 40, the thermal reaction provides an almost random product distribution, while the photochemical reaction provides variable results ranging from 20:1 to stereospecific extrusion of nitrogen.[54]

(38)

(39)

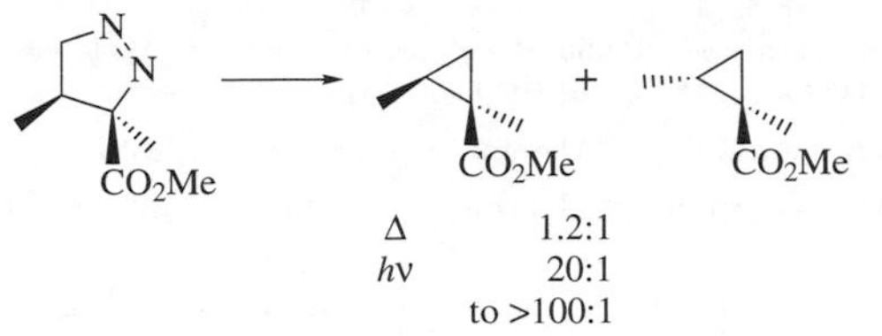

(40)

Cyclopropanes can also be directly synthesized from alkenes and diazomethane, either photochemically or by using transition metal salts, usually ***Copper(II) Chloride*** or ***Palladium(II) Acetate***, as promoters. The metal-mediated reactions are more commonly used than the photochemical ones, but they are not as popular as the Simmons–Smith procedure. However, they do occasionally offer advantages. Of the two processes, the Cu-catalyzed reaction produces a more active reagent[55] which will cyclopropanate a variety of alkenes, including enamines as shown in eq 41.[56] These products can then be converted to α-methyl ketones by thermolysis. The cyclopropanation of the norbornenol derivative shown in eq 42 was problematic using the Simmons–Smith procedure and provided low yields, but occurred smoothly using the $CuCl_2$/diazomethane method.[57]

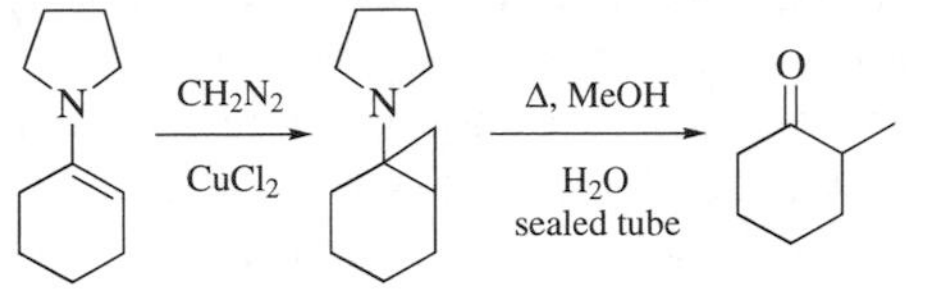

(41)

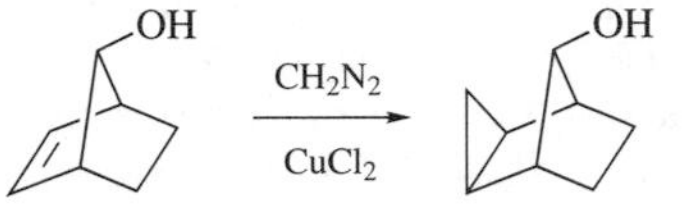

(42)

The $Pd(OAc)_2$-mediated reaction can be used to cyclopropanate electron-deficient alkenes as well as terminal alkenes. Thus selective reaction at a monosubstituted alkene in the presence of others is readily achieved using this method (eq 43).[58] The example shown in eq 44 is one in which the Simmons–Smith procedure failed to provide any of the desired product, whereas the current method provided a 92% yield of cyclopropane.[59]

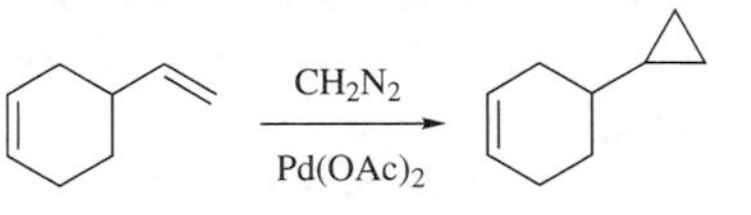

(43)

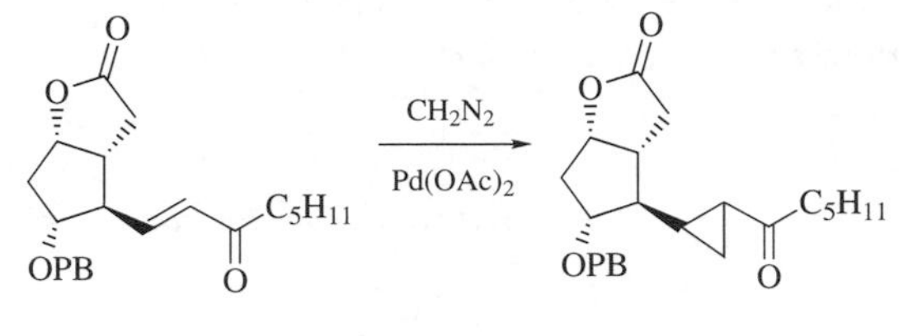

(44)

In the case of the photochemical reaction, irradiation of diazomethane in the presence of *cis*-2-butene provides *cis*-1,2-dimethylcyclopropane with no detectable amount of the *trans* isomer (eq 45).[60] This reaction is thought to proceed via a singlet carbene. However, if the same reaction is carried out via a triplet carbene, generated via triplet sensitization, then a 1.3:1 mixture of *trans* to *cis* dimethylcyclopropane is observed (eq 46).[61] The yields in the photochemical reaction are typically lower than the metal-mediated processes, and are usually accompanied by more side products.

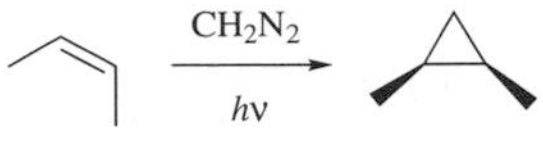

(45)

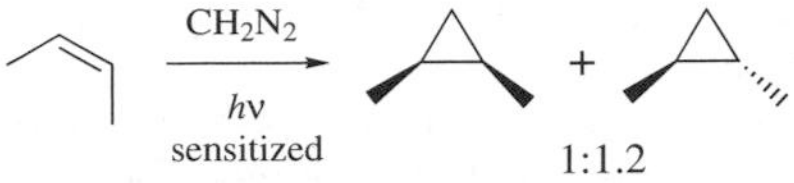

(46)

Additions to Electron-Deficient Species. Diazomethane will also add to highly electrophilic species such as sulfenes or iminium salts to give the corresponding three-membered ring heterocycles. When the reaction is performed on sulfenes, the products are episulfones which are intermediates in the Ramberg–Backlund rearrangement, and are therefore precursors for the synthesis of alkenes via chelotropic extrusion of SO_2. The sulfenes are typically prepared in situ by treatment of a sulfonyl chloride with a mild base, such as ***Triethylamine*** (eq 47).[62] Similarly, the addition of diazomethane to imminium salts has been used to methylenate carbonyls.[63] In this case, the intermediate aziridinium salt is treated with a strong base, such as ***n-Butyllithium***, to induce elimination (eq 48).

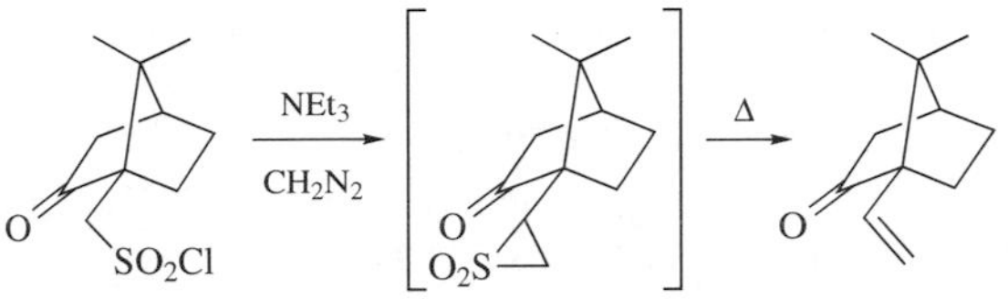

(47)

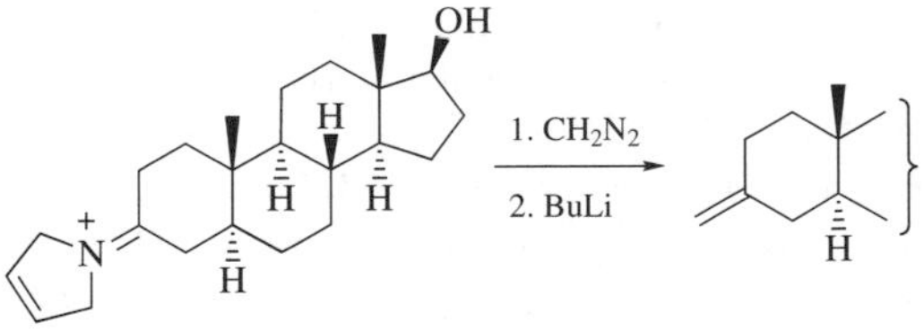

(48)

Miscellaneous Reactions. Diazomethane has been shown to react with vinylsilanes derived from α,β-unsaturated esters to provide the corresponding allylsilane by insertion of CH_2 into the C–Si bond (eq 49).[64] The reaction has been shown to be stereospecific, with *cis*-vinylsilane providing *cis*-allylsilanes; however, the mechanism of the reaction has not been defined. Diazomethane has also been used in the preparation of trimethyloxonium salts. Treatment of a solution of dimethyl ether and trinitrobenzenesulfonic acid with diazomethane provides trimethyloxonium trinitrobenzenesulfonate, which is more stable than the fluoroborate salt.[65]

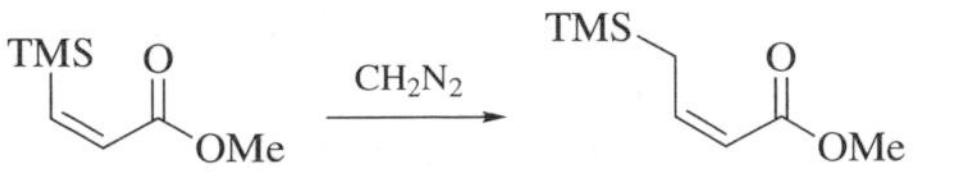

(49)

Related Reagents. 2-Diazopropane; 1-Diazo-2-propene; Diphenyldiazomethane; Ethyl Diazoacetate[1]; Phenyldiazomethane; Trimethylsilyldiazomethane.

1. (a) Regitz, M.; Maas, G. *Diazo Compounds, Properties and Synthesis*; Academic: Orlando, 1986. (b) Black, T. H. *Aldrichim. acta* **1983**, *16*, 3. (c) Pizey, J. S. *Synthetic Reagents*; Wiley: New York, 1974; Vol. 2, pp 65–142.
2. Arndt, F. *OSC* **1943**, *2*, 165.
3. Moore, J. A.; Reed, D. E. *OSC* **1973**, *5*, 351. Redemann, C. E.; Rice, F. O.; Roberts, R.; Ward, H. P. *OSC* **1955**, *3*, 244. McPhee, W. D.; Klingsberg, E. *OSC* **1955**, *3*, 119.
4. De Boer, Th. J.; Backer, H. J. *OSC* **1963**, *4*, 250. Hudlicky, M. *JOC* **1980**, *45*, 5377. See also Aldrich Chemical Company Technical Bulletins Number AL-121 and AL-131. Note that the preparation described in *FF* , **1967**, *1*, 191 is flawed and neglects to mention the addition of ethanol. Failure to add ethanol can result in a buildup of diazomethane and a subsequent explosion.
5. McKay, A. F. *JACS* **1948**, *70*, 1974. See also Aldrich Chemical Company Technical Bulletin Number AL-132.
6. von Pechman, A. *CB* **1894**, *27*, 1888.
7. Ref. 2, note 3.
8. For a description of the safety hazards associated with diazomethane, see: Gutsche, C. D. *OR* **1954**, *8*, 391.
9. Fujisawa, T.; Sato, T.; Itoh, T. *CL* **1982**, 219.
10. Nicolaou, K. C.; Paphatjis, D. P.; Claremon, D. A.; Dole, R. E. *JACS* **1981**, *103*, 6967.
11. Fujisawa, T.; Sato, T.; Kawashima, M.; Naruse, K.; Tamai, K. *TL* **1982**, *23*, 3583.
12. Kozikowski, A. P.; Sugiyama, K.; Springer, J. P. *JOC* **1981**, *46*, 2426.
13. De, B.; Corey, E. J. *TL* **1990**, *31*, 4831. Macomber, R. S. *SC* **1977**, *7*, 405.
14. Blade, R. J.; Hodge, P. *CC* **1979**, 85.
15. Kawamata, A.; Fukuzawa, Y.; Fujise, Y.; Ito, S. *TL* **1982**, *23*, 1083.
16. Ray, J. A.; Harris, T. M. *TL* **1982**, *23*, 1971.
17. Ogawa, H.; Chihara, T.; Taya, K. *JACS* **1985**, *107*, 1365.
18. Frimer, A. A.; Gilinsky-Sharon, P.; Aljadef, G. *TL* **1982**, *23*, 1301.
19. Chavis, C.; Dumont, F.; Wightman, R. H.; Ziegler, J. C.; Imbach, J. L. *JOC* **1982**, *47*, 202.
20. Neeman, M.; Johnson, W. S. *OSC* **1973**, *5*, 245.
21. Ohno, K.; Nishiyama, H.; Nagase, H. *TL* **1979**, *20*, 4405.
22. Robins, M. J.; Lee, A. S. K.; Norris, F. A. *Carbohydr. Res.* **1975**, *41*, 304.
23. Evans, D. S.; Bender, S. L.; Morris, J. *JACS* **1988**, *110*, 2506. For a similar example with the antibiotic lasalocid, see: Westly, J. W.; Oliveto, E. P.; Berger, J.; Evans, R. H.; Glass, R.; Stempel, A.; Toome, V.; Williams, T. *JMC* **1973**, *16*, 397.
24. Nishiyama, H.; Nagase, H.; Ohno, K. *TL* **1979**, *20*, 4671.
25. Muller, v. H.; Huber-Emden, H.; Rundel, W. *LA* **1959**, *623*, 34.
26. Seagusa, T.; Ito, Y.; Kobayashi, S.; Hirota, K.; Shimizu, T. *TL* **1966**, *7*, 6131.
27. In addition to acid chlorides, α-diazo ketones can be synthesized from carboxylic acid anhydrides; however, in this case one equivalent of the carboxylic acid is converted to the corresponding methyl ester. Furthermore, the anhydride can be formed in situ using DCC. See Hodson, D.; Holt, G.; Wall, D. K. *JCS(C)* **1970**, 971.
28. For a review of the Wolff rearrangement, see: Meier, H.; Zeller, K.-P. *AG(E)* **1975**, *14*, 32.
29. Bergmann, E. D.; Hoffmann, E. *JOC* **1961**, *26*, 3555.
30. Clark, R. D. *SC* **1979**, *9*, 325.
31. Smith, A. B.; Dorsey, M.; Visnick, M.; Maeda, T.; Malamas, M. S. *JACS* **1986**, *108*, 3110.
32. Smith, A. B.; Toder, B. H.; Branca, S. J.; Dieter, R. K. *JACS* **1981**, *103*, 1996.
33. Rosenquist, N. R.; Chapman, O. L. *JOC* **1976**, *41*, 3326.
34. Hudliky, T.; Sheth, J. P. *TL* **1979**, *20*, 2667.
35. For a review of intramolecular reactions of α-diazo ketones, see: Burke, S. D.; Grieco, P. A. *OR* **1979**, *26*, 361.
36. Murai, A.; Kato, K.; Masamune, T. *TL* **1982**, *23*, 2887.
37. Iwata, C.; Fusaka, T.; Fujiwara, T.; Tomita, K.; Yamada, M. *CC* **1981**, 463.
38. For a review on the reactions of α-diazo ketones with acid, see: Smith, A. B.; Dieter, R. K. *T* **1981**, *37*, 2407.
39. Ackeral, J.; Franco, F.; Greenhouse, R.; Guzman, A.; Muchowski, J. M. *JHC* **1980**, *17*, 1081.
40. Ghatak, U. R.; Sanyal, B.; Satyanarayana, G.; Ghosh, S. *JCS(P1)* **1981**, 1203.
41. Blair, I. A.; Mander, L. N. *AJC* **1979**, *32*, 1055.
42. Smith, A. B.; Toder, B. H.; Branca, S. J.; Dieter, R. K. *JACS* **1981**, *103*, 1996. Smith, A. B.; Dieter, K. *JACS* **1981**, *103*, 2009. Smith, A. B.; Dieter, K. *JACS* **1981**, *103*, 2017.
43. Smith, A. B.; Toder, B. H.; Branca, S. J. *JACS* **1984**, *106*, 3995.
44. Smith, A. B.; Toder, B. H.; Richmond, R. E.; Branca, S. J. *JACS* **1984**, *106*, 4001.
45. Holmquist, C. R.; Roskamp, E. J. *JOC* **1989**, *54*, 3258. Padwa, A.; Hornbuckle, S. F.; Zhang, Z. Z.; Zhi L. *JOC* **1990**, *55*, 5297.
46. Holmquist, C. R.; Roskamp, E. J. *TL* **1992**, *33*, 1131.
47. For a review of this reaction, see: Gutsche, C. D. *OR* **1954**, *8*, 364.
48. For a review, see: Smith, P. A. S.; Baer, D. R. *OR* **1960**, *11*, 157.
49. House, H. O.; Grubbs, E. J.; Gannon, W. F. *JACS* **1960**, *82*, 4099.
50. Majerski, Z.; Djigas, S.; Vinkovic, V. *JOC* **1979**, *44*, 4064.
51. Greene, A. E.; Depres, J-P. *JACS* **1979**, *101*, 4003.
52. For a review, see: Regitz, M.; Heydt, H. In *1,3-Dipolar Cycloadditions Chemistry*; Padwa, A.; Ed.; Wiley: New York, 1984; p 393.
53. For a discussion of the orbital interactions that control dipolar additions of diazomethane, see: Fleming, I. *Frontier Molecular Orbitals and Organic Chemical Reactions*; Wiley: New York, 1976; pp 148–161.
54. Van Auken, T. V.; Rienhart, K. L. *JACS* **1962**, *84*, 3736.
55. This is a very reactive reagent combination which will cyclopropanate benzene and other aromatic compounds. See: Vogel, E.; Wiedeman, W.; Kiefe, H.; Harrison, W. F. *TL* **1963**, *4*, 673. Muller, E.; Kessler, H.; Kricke, H.; Suhr, H. *TL* **1963**, *4*, 1047.
56. Kuehne, M. E.; King, J. C. *JOC* **1973**, *38*, 304.
57. Pincock, R. E.; Wells, J. I. *JOC* **1964**, *29*, 965.
58. Suda, M. *S* **1981**, 714.
59. Raduchel, B.; Mende, U.; Cleve, G.; Hoyer, G. A.; Vorbruggen, H. *TL* **1975**, *16*, 633.
60. Doering, W. von E.; LaFlamme, P. *JACS* **1956**, *78*, 5447.
61. Duncan, F. J.; Cvetanovic, R. J. *JACS* **1962**, *84*, 3593.
62. Fischer, N.; Opitz, G. *OSC* **1973**, *5*, 877.
63. Hata, Y.; Watanabe, M. *JACS* **1973**, *95*, 8450.
64. Cunico, R. F.; Lee, H. M.; Herbach, J. *JOM* **1973**, *52*, C7.
65. Helmkamp, G. K.; Pettit, D. J. *OSC* **1973**, *5*, 1099.

Tarek Sammakia
University of Colorado, Boulder, CO, USA

Dibromomethane–Zinc/Copper Couple[1]

CH_2Br_2–Zn/Cu

(CH_2Br_2)		
[74-95-3]	CH_2Br_2	(MW 173.83)
(Zn)		
[7440-66-6]	Zn	(MW 65.39)

(reagent combination for cyclopropanations)

Alternate Name: bromomethylzinc bromide.

Physical Data: CH_2Br_2: mp −52 °C; bp 96–98 °C; *d* 2.477 g cm^{-3}. Zn: mp 419.5 °C; *d* 7.140 g cm^{-3}.

Solubility: sol Et_2O, THF.

Preparative Methods: the original method for the preparation of a ***Zinc/Copper Couple*** that will react with ***Dibromomethane*** calls for the formal preparation of the Zn/Cu couple from 30-mesh granular ***Zinc*** instead of zinc dust.[2] In this procedure, 35 g of granular zinc (30 mesh) is added to a hot solution of 50 mL of acetic acid containing 0.5 g of CuOAc·H_2O (see ***Copper(I) Acetate***). The mixture is kept hot and shaken for 1–3 min. The acetic acid is carefully decanted from the couple. The couple is then washed with 50 mL of acetic acid and three 50 mL portions of Et_2O. The couple is shaken for 1 min with each washing. The Et_2O moistened couple is ready for use or can be dried with a stream of nitrogen. More recently a number of different methods for activation of the Zn/Cu couple have appeared. These newer modifications include the use of ultrasound[3] or catalytic amounts of ***Titanium(IV) Chloride***[4] or ***Acetyl Chloride***[1,5] They require no prior preparation of the couple. The couple is prepared in situ by reaction of zinc dust and ***Copper(I) Chloride***[1,6] The method of choice appears to be the acetyl chloride method.

Handling, Storage, and Precautions: CH_2Br_2 is toxic. Zn is a moisture-sensitive, flammable solid. When the reactions are complete, the reaction mixture is filtered from the zinc residue. It is recommended that the residue be wetted with water prior to disposal.

Cyclopropanations. The use of CH_2Br_2 in place of the more commonly employed ***Diiodomethane*** in the cyclopropanation of an alkene[1] has the advantages that it is less expensive and more stable. It has not been used as extensively because of the lower reactivity of the dibromide with the various forms of the Zn/Cu couple. The first reported form of a Zn/Cu couple that would react effectively with the dibromide was made from granular zinc.[2] It suffered from long reaction times and lower yields in comparison with the diiodomethane derived reagent. It was found that sonication could be used to effect cyclopropanations of alkenes with dibromomethane[3] and an in situ generated Zn/Cu couple prepared from Zn dust and CuCl.[6] The ultrasound method, however, suffered from irreproducibility. A more convenient method for activation involves the use of a catalytic amount (2 mol %) of $TiCl_4$. The use of more than 2% $TiCl_4$ results in unmanageable reactions. Activation of the couple with $TiCl_4$ is also effective when using diiodomethane. In either the dibromo- or diiodomethane reactions there is a competing formation of ethylene. This problem is greater for dibromomethane than for diiodomethane. The only drawback in the titanium mediated procedure is that it is not always tolerant of Lewis acid sensitive groups.[4] **Acetyl chloride** has been found to activate the Zn/Cu couple as well as $TiCl_4$ and has the advantage that Lewis acid sensitive groups are tolerated. This now appears to be the method of choice.[5] Table 1 shows the results[3–11] of some representative alkenes under some of the different cyclopropanation conditions. When the difference in yield between the diiodo and dibromo conditions are minimal and the alkene is not the limiting reagent, the dibromomethane conditions offer an inexpensive alternative. The somewhat lower yields obtained are offset by the cost savings.

Table 1 Comparison of Cyclopropanation Conditions for Various Alkenes

	Yield (%)					
	CH_2I_2 Zn/Cu	CH_2I_2 Zn/Cu $TiCl_4$[4]	CH_2Br_2 Zn/Cu	CH_2Br_2 Zn/Cu ultrasound[3]	CH_2Br_2 Zn/Cu $TiCl_4$[4]	CH_2Br_2 Zn/Cu AcCl[5]
	92[6]	69	61[2]	60	58	61
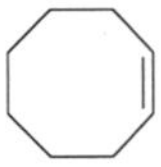	94[6]	90	56[2]	72	73	88
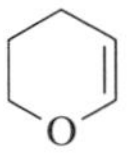	65[7]	<2	–	41	17	45
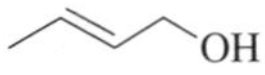	52[8]	42	68[11]	57	36	58
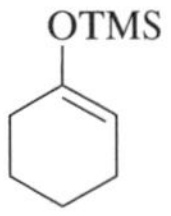	65[9]	–	–	–	–	60
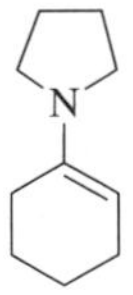	8[10]	–	–	–	–	22

The regioselectivity of the reagent in its reactions with limonene (eq 1) and 4-vinylcyclohexene (eq 2) has been studied. Using a ratio of diene:CH_2Br_2:Zn of 1:1:1.5, the major product of the reaction was recovered diene (47%). The monocyclopropanated products were isolated in ≈28% yield with a 2.5:1.0 ratio of exocyclic cyclopropane:ring cyclopropane. There was little stereoselectivity observed in ring cyclopropanated products, with a 1:1.1 ratio of *cis* and *trans* isomers being produced. A small (4%) amount of biscyclopropanation was obtained under these conditions. If the diene:CH_2Br_2:Zn ratio was changed to 1:2:3, the amount of recovered diene decreases but the amount of biscyclopropane increased. For 4-vinylcyclohexene with the 1:1:1.5 ratio, the ring cyclopropane was favored over the exocyclic cyclopropane by 1.75:1. The combined yield of these two products is again low (≈7%). The major product was recovered diene (53%).[12]

Miscellaneous. Reaction of ***Triethylsilane*** with dibromomethane and Zn/Cu couple results in an insertion of methylene into the Si–H bond. The reaction with triethylsilane (eq 3) produces triethylmethylsilane. The reaction with diiodomethane produces the insertion product in 64% yield.[13]

CH_2Br_2–Zn, CuCl, AcCl, Et_2O (1): 20%, 8%, 4%

CH_2Br_2–Zn, CuCl, AcCl, Et_2O (2): 2.4%, 4.2%

$$Et_3SiH \xrightarrow[55\%]{CH_2Br_2,\ Zn/Cu} Et_3SiMe \quad (3)$$

The ^1H and ^{13}C NMR data for bromomethylzinc bromide and bis(bromomethyl)zinc have been reported.[14]

Related Reagents. Diazomethane; Dibromomethane; Diiodomethane; Ethyliodomethylzinc; Tetraethylammonium Periodate.

---

1. Simmons, H. E.; Cairns, T. L.; Vladuchik, S. A.; Hoines, C. M. *OR* **1973**, *20*, 1.
2. LeGoff, E. *JOC* **1964**, *29*, 2048.
3. Friedrich, E. C.; Domek, J. M.; Pong, R. Y. *JOC* **1985**, *50*, 4640.
4. Friedrich, E. C.; Lunetta, S. E.; Lewis, E. J. *JOC* **1989**, *54*, 2388.
5. Friedrich, E. C.; Lewis, E. J. *JOC* **1990**, *55*, 2491.
6. Rawson, R. J.; Harrison, I. T. *JOC* **1970**, *35*, 2057.
7. Simmons, H. E.; Smith, R. D. *JACS* **1959**, *81*, 4256.
8. Bergman, R. G. *JACS* **1969**, *91*, 7405.
9. Murai, S.; Aya, T.; Renge, T.; Rhu, I.; Sonoda, N. *JOC* **1974**, *39*, 858.
10. Blanchard, E. P.; Simmons, H. E. *JOC* **1965**, *30*, 4321.
11. Friedrich, E. C.; Biresaw, G. *JOC* **1982**, *47*, 1615.
12. Friedrich, E. C.; Niyati-Shirkhodaee, F. *JOC* **1991**, *56*, 2202.
13. Seyferth, D.; Dertouzos, H.; Todd, L. J. *JOM* **1965**, *4*, 18.
14. Fabisch, B.; Mitchell, T. N. *JOM* **1984**, *269*, 219.

Michael J. Taschner
The University of Akron, OH, USA

Di-*n*-butylboryl Trifluoromethanesulfonate

n-Bu_2B–OSO_2CF_3

[60669-60-4] $C_9H_{18}BF_3O_3S$ (MW 274.11)

(Lewis acid for the preparation of vinyloxyboranes,[1–6] boryl azaenolates;[7,8] catalyst for macrolactonization[9])

Physical Data: bp 60 °C/2.0 mmHg.
Solubility: sol common inert organic solvents such as Et_2O, CH_2Cl_2, and hexane.
Form Supplied in: 1.0 M solution in CH_2Cl_2 and Et_2O.
Preparative Method: ***Trifluoromethanesulfonic Acid*** is added slowly to an equimolar amount of ***Tri-n-butylborane*** with gentle warming (50 °C) until butane evolution commences. The remaining acid is added at such a rate to maintain a reaction temperature of less than 50 °C. After the addition is complete, the mixture is stirred for an additional 30 min at 25 °C and then the product is isolated by vacuum distillation.[1c]
Handling, Storage, and Precautions: the neat liquid as well as solutions are moisture and air sensitive. Therefore the material should be stored and transferred under a dry inert atmosphere. Use in a fume hood.

Vinyloxyboranes (Boron Enolates). Dibutyl vinyloxyboranes are conveniently prepared from active methylene carbonyl containing compounds utilizing Bu_2BOTf in combination with a sterically hindered amine base. Typically the base, ***Diisopropylethylamine*** or ***2,6-Lutidine*** (2,6-lut), and Bu_2BOTf are premixed in Et_2O or CH_2Cl_2 at −78 °C and then the carbonyl component is added. Enolate generation is allowed to occur over the next 15–30 min at temperatures ranging from −78 °C to 0 °C, resulting in the formation of vinyloxyboranes which are suitable for reaction with aldehydes in a crossed aldol reaction. Other boryl triflate reagents have similar or complimentary utility in this process (see ***9-Borabicyclononyl Trifluoromethanesulfonate*** (9-BBNOTf) and ***Dicyclopentylboryl Trifluoromethanesulfonate*** ((c-$C_5H_9)_2$BOTf)).

With unsymmetrical methyl ketones, use of Bu_2BOTf and i-Pr_2NEt results in the regioselective formation of the vinyloxyborane at the least hindered carbon. However, if 9-BBNOTf and 2,6-lut are used, the regioselectivity of vinyloxyborane formation is reversed (eq 1).[10]

1. R_2BOTf, amine; 2. PhCHO (1)

Bu_2BOTf, i-Pr_2NEt	100:0
9-BBNOTf, 2,6-lut	12:88

The relative configuration of the two new chiral centers formed in the aldol product is a direct consequence of the vinyloxyborane enolate geometry with Z(O) enolates affording the 2,3-*syn* aldol products and the E(O) vinyloxyboranes leading to the 2,3-*anti* isomers. As a result, a number of studies have examined the effects of various boryl triflates and amines on the ratio of kinetic enolates formed. In general, it has been found that with a given base, more sterically hindered boryl triflates lead to increased amounts of the E(O) enolate. For example, when 3-pentanone is treated with Bu_2BOTf or (c-$C_5H_9)_2$BOTf in the presence of i-Pr_2NEt, the ratio of Z(O):E(O) enolates drops from 32:1 to 4:1 (eq 2).[1c]

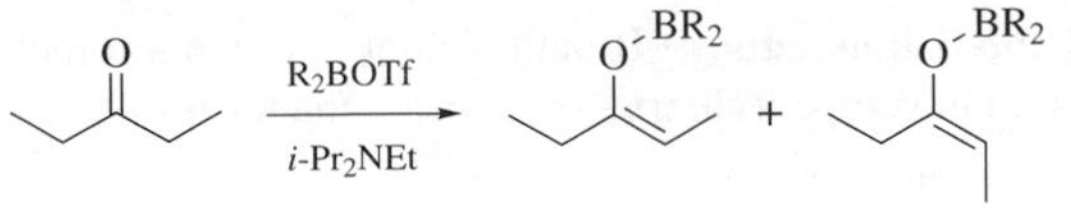

R_2	
Bu_2	>97:3
$(c\text{-}C_5H_9)_2$	82:18

With ketones in which no combination of boryl triflate and amine produces a vinyloxyborane of predominately one geometry, Bu_2BOTf has been utilized to catalyze the exchange with a trimethylsilyl enol ether of defined stereochemistry (eq 3). To maintain high stereoselectivity in the subsequent aldol reaction, it is necessary to remove the TMSOTf byproduct prior to the addition of the aldehyde (eq 3).[11]

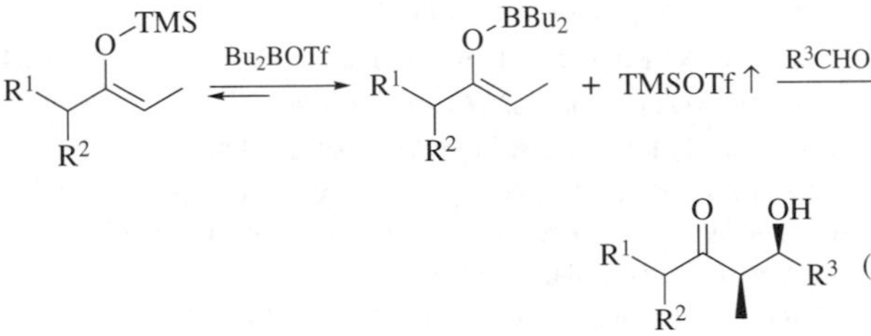

Probably the greatest utility of this reagent has been in the stereoselective formation of chiral vinyloxyboranes with defined enolate geometries from which the absolute configuration of the new chiral centers formed in the aldol process are controlled. A number of chiral masked propionate equivalents useful for the synthesis of polypropionate-like natural products have been developed for this purpose. Chiral α-hydroxy ethyl ketones derived from both enantiomers of hexahydromandelic acid have been shown to form the *Z*(O) vinyloxyborane stereospecifically with 9-BBNOTf, Bu_2BOTf, and $(c\text{-}C_5H_9)_2BOTf$. However, the size of the ligands attached to boron has a measurable effect upon the extent of the enantioselectivity achieved in the aldol process, as exemplified in eq 4.[1b]

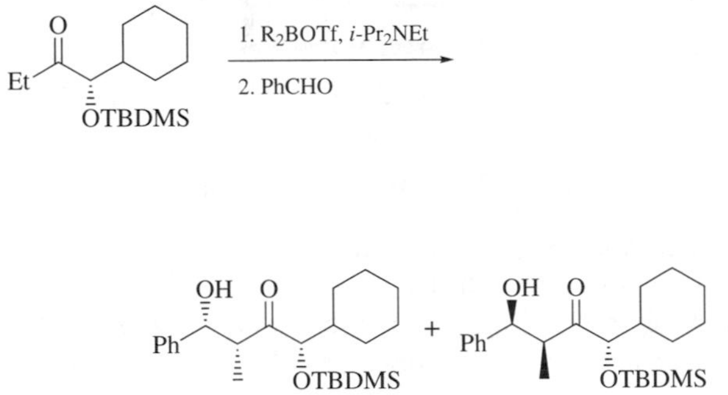

R_2	
9-BBN	14:1
Bu_2	40:1
$(c\text{-}C_5H_9)_2$	75:1

The other major class of chiral boron enolate reagents developed for this purpose employs a propionate unit attached to a chiral auxiliary. These are typified by the chiral oxazolidinones derived from α-amino alcohols initially developed by Evans (eq 5),[2a] with subsequent variants involving thiazolidinethiones, oxazolidinethiones,[3] sultams,[4] and camphor-derived oxazolidinones[2b] being reported more recently.

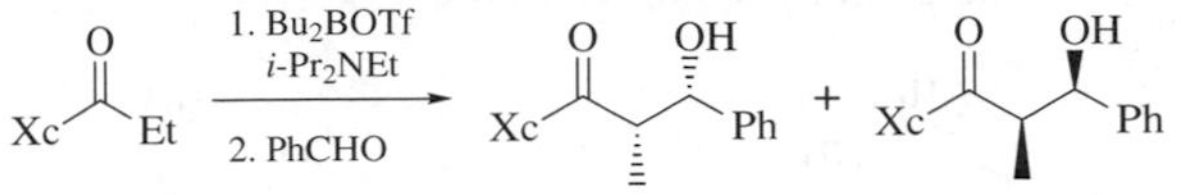

Xc		
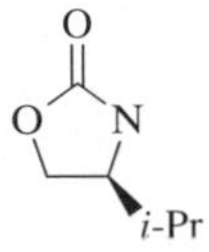 (4S)-isopropyloxazolidinone	>500:1	88%
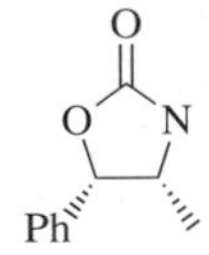 (4R,5S)-methylphenyloxazolidinone	<1:500	89%

Other functional groups which can be utilized for further synthetic manipulation are well tolerated on the chiral vinyloxyborane. These have included thioethers,[12a] selenoethers,[12b] carboxylic esters,[2a] and halogens. For example, the bromoacetyl oxazolidinone in eq 6 has been utilized in a Darzens-like condensation for the enantioselective synthesis of *cis*-substituted epoxides.[12c] This procedure is complemented by the synthesis of *trans*-substituted epoxide derivatives from the *E*(O) vinyloxyborane derived from α-bromo-*t*-butylthioacetate.

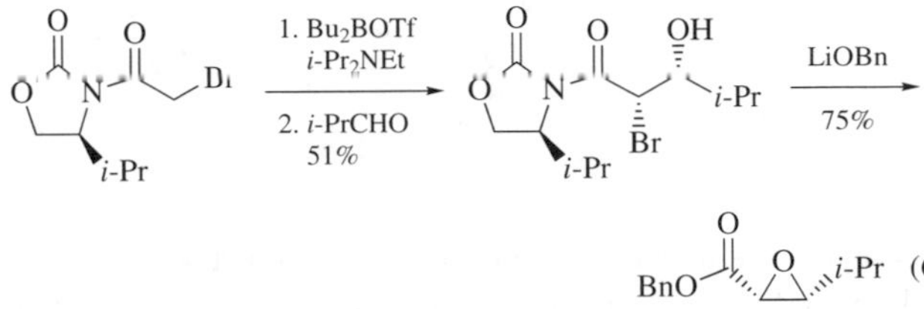

Vinyloxyboranes also react with electrophiles other than aldehydes. For example, the chiral vinyloxyborane generated from an *N*-propionyloxazolidinone, Bu_2BOTf, and $i\text{-}Pr_2NEt$ reacts with ***N*-Bromosuccinimide** to produce the chiral bromide shown in eq 7 as the principal adduct.[13] High enantioselectivity has also been achieved in sulfenylation and selenation reactions with chiral boron enolates, an example of which is shown in eq 8.[14]

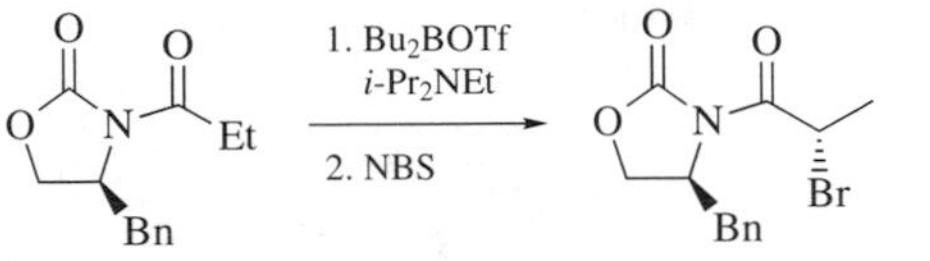

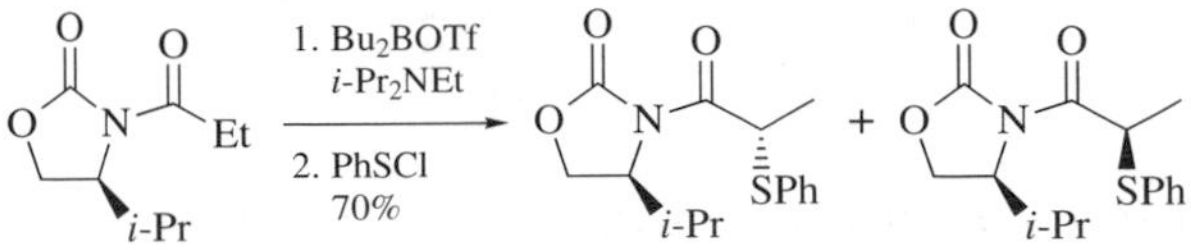

88:12

The boron enolate of *t*-butyl thioacetate has been employed in the addition with a variety of Schiff bases to produce β-amino acid derivatives. Reported yields range from 40–80% and the conditions are sufficiently mild to allow for the preparation of a highly functionalized intermediate used in the synthesis of bleomycin (eq 9).[5] The boron component of various vinyloxyboranes is also capable of activating ***Bis(dimethylamino)methane***, resulting in a convenient one-pot procedure for the preparation of β-dimethylamino carbonyl compounds (eq 10).[15]

t-BuS(C=CH$_2$)OBBu$_2$, 40% (9)

1. Bu_2BOTf, i-Pr_2NEt; 2. $Me_2NCH_2NMe_2$, 70% (10)

Simple carboxylic esters are not active enough to form enolates in the reaction with a boryl triflate and an amine; however, glycolate and thioglycolate esters are readily transformed into vinyloxyboranes which upon further reaction with aldehydes yield predominately the 2,3-*syn* aldol products (eq 11).[6]

R^1XCH$_2$COOR^2 (X = O, S); 1. Bu_2BOTf, i-Pr_2NEt; 2. R^3CHO → major + minor (11)

Boryl Azaenolates. Nitrile derivatives react with Bu_2BOTf and i-Pr_2NEt to produce enolates which condense with various substituted benzaldehydes to yield β-hydroxy nitrile products. Poor yields result when aliphatic aldehydes are utilized as reactants, thus limiting the scope of this reaction.[7] 2-Ethylpyridine yields predominately the 2,3-*syn* aldol product when the enolate is generated with Bu_2BOTf in the presence of ***Triethylamine*** (eq 12). No reaction occurs when i-Pr_2NEt is used.[8]

1. Bu_2BOTf, Et_3N; 2. PhCHO, 50%; 94:6 (12)

Macrolactonization. Trimethylsilyl ω-trimethylsilyloxycarboxylates cyclize to the corresponding macrolides in the presence of 1 equiv of a dialkylboryl triflate. Pr_2BOTf gives higher yields of product than either Et_2BOTf or Bu_2BOTf (eq 13).[9]

TMSO(CH$_2$)$_{11}$CO$_2$TMS → (R_2BOTf, toluene, Δ, 50–83%) → (CH$_2$)$_{11}$ lactone (13)

Related Reagents. 9-Borabicyclononyl Trifluoromethanesulfonate; Dicyclopentylboryl Trifluoromethanesulfonate.

1. (a) Mukaiyama, T.; Inoue, T. *CL* **1976**, 559. (b) Masamune, S.; Choy, W.; Kerdesky, F. A. J.; Imperiali, B. *JACS* **1981**, *103*, 1566. (c) Evans, D. A.; Nelson, J. V.; Vogel, E.; Taber, T. R. *JACS* **1981**, *103*, 3099.
2. (a) Evans, D. A.; Bartroli, J.; Shih, T. L. *JACS* **1981**, *103*, 2127. (b) Yan, T.-H.; Tan, C.-W.; Lee, H.-C.; Lo, H.-C. Huang, T.-Y. *JACS* **1993**, *115*, 2613.
3. Hsiao, C.-H.; Liu, L.; Miller, M. J. *JOC* **1987**, *52*, 2201.
4. Oppolzer, W.; Blagg, J.; Rodriguez, I.; Walther, E. *JACS* **1990**, *112*, 2767.
5. Otsuka, M.; Yoshida, M.; Kobayashi, S.; Ohno, M. *TL* **1981**, *22*, 2109.
6. (a) Sugano, Y.; Naruto, S. *CPB* **1988**, *36*, 4619. (b) Sugano, Y.; Naruto, S. *CPB* **1989**, *37*, 840.
7. Hamana, H.; Sugasawa, T. *CL* **1982**, 1401.
8. Hamana, H.; Sugasawa, T. *CL* **1984**, 1591.
9. Taniguchi, N.; Kinoshita, H.; Inomata, K.; Kotake, H. *CL* **1984**, 1347.
10. Inoue, T.; Mukaiyama, T. *BCJ* **1980**, *53*, 174.
11. Kuwajima, I.; Kato, M.; Mori, A. *TL* **1980**, *21*, 4291.
12. (a) Woo, P. W. K. *TL* **1985**, *26*, 2973. (b) Masamune, S.; Kaiho, T.; Garvey, D. S. *JACS* **1982**, *104*, 5521. (c) Abdel-Magid, A.; Lantos, I.; Pridgen, L. N. *TL* **1984**, *25*, 3273.
13. Evans, D. A.; Ellman, J. A.; Dorow, R. L. *TL* **1987**, *28*, 1123.
14. Paterson, I.; Osborne, S. *SL* **1991**, 145.
15. Nolen, E. G.; Aliocco, A.; Vitarius, J.; McSorley, K. *CC* **1990**, 1532.

David S. Garvey
Abbott Laboratories, Abbott Park, IL, USA

Dichloro[1,1′-bis(diphenylphosphino)ferrocene]palladium(II)

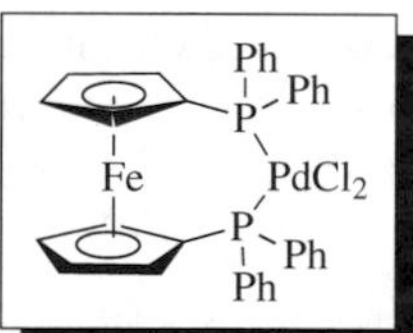

[72287-26-4] $C_{34}H_{28}Cl_2FeP_2Pd$ (MW 731.72)

(catalyst for cross-coupling reactions)

Physical Data: mp 265–268 °C (dec).
Solubility: sol ether, THF, benzene.
Form Supplied in: reddish brown solid; commercially available.
Purification: recrystallization from $CHCl_3$ under inert atmosphere.
Handling, Storage, and Precautions: the complex is an irritant; store under an inert atmosphere.

Cross Coupling. Like other Pd^{II} and Ni^{II} phosphine complexes, $PdCl_2$(dppf) is a very useful catalyst for the cross-coupling reactions of vinyl or aryl halides or triflates with Grignard reagents, leading to carbon–carbon bond formation.[1] $PdCl_2$(dppf)-catalyzed reaction of vinyl bromide with s-BuMgCl gives exclusively the desired coupling product (eq 1), while other

Pd catalysts also yield isomerized and reduced byproducts.[1a] (*E*)-1-Bromo-1-alkenes are more reactive than the (*Z*)-isomers.[1c] Selective monoalkylation of dichlorobenzene under refluxing conditions using $PdCl_2$(dppf) catalyst has been recorded.[2] Whereas palladium catalysts are usually unreactive in cross couplings of organosulfur compounds,[3] reactions of 2-methylthio-4,4-dimethyl-2-oxazoline with aryl Grignard reagents are catalyzed efficiently by $PdCl_2$(dppf) (eq 2).[4] Phototoxic terthiophenes have been prepared and the SMe group remains intact under the reaction conditions (eq 3).[5] Organozinc reagents behave similarly in these cross coupling reactions.[6] $PdCl_2$(dppf) has been found to catalyze iodine–zinc exchange reactions which lead to an interesting Pd-catalyzed intramolecular carbozincation of alkenes (eq 4).[7] It is noteworthy that various functional groups remain intact under these conditions.

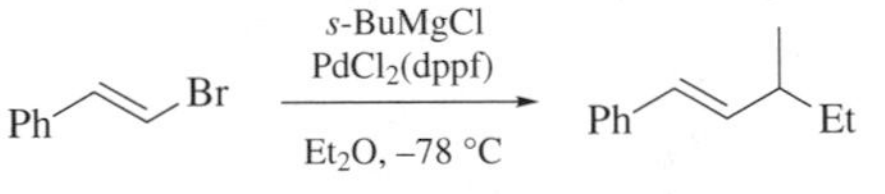

(1) Ph–CH=CH–Br → (s-BuMgCl, $PdCl_2$(dppf), Et_2O, −78 °C) → Ph–CH=CH–CH(Me)Et

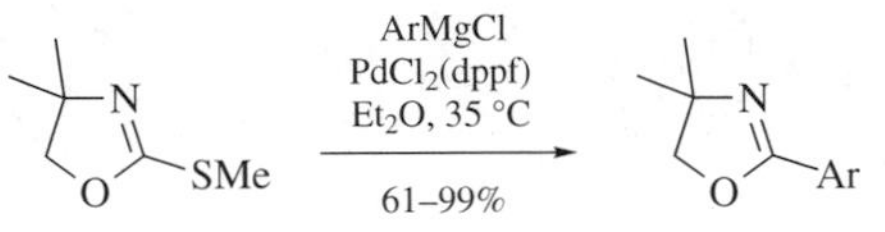

(2) SMe oxazoline → (ArMgCl, $PdCl_2$(dppf), Et_2O, 35 °C; 61–99%) → Ar oxazoline

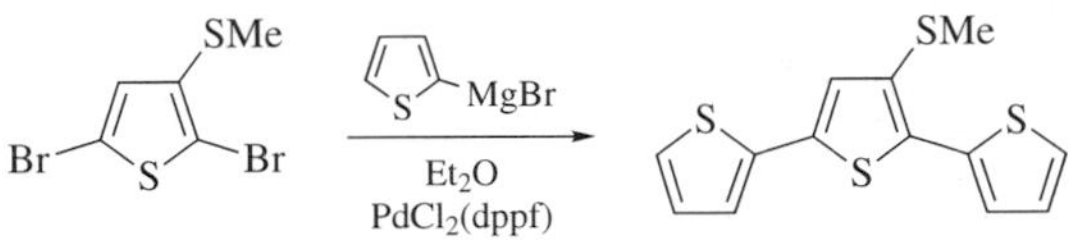

(3) dibromo(SMe)thiophene + 2-thienyl-MgBr → (Et_2O, $PdCl_2$(dppf)) → SMe terthiophene

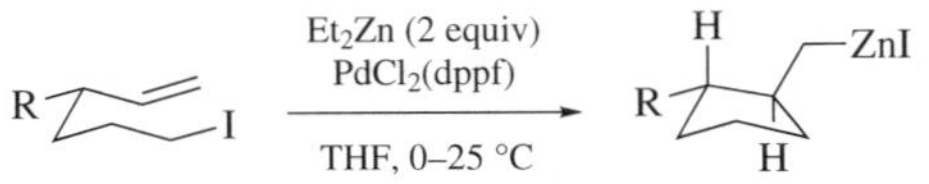

(4) → (Et_2Zn (2 equiv), $PdCl_2$(dppf), THF, 0–25 °C) → ZnI product

Aliphatic iodides are reduced upon treatment with alkyl Grignard reagents in the presence of $PdCl_2$(dppf) catalyst.[8] Aryl triflates are efficiently reduced by ***Sodium Borohydride*** or ***Ammonium Formate*** in the presence of $PdCl_2$(dppf).[9]

$PdCl_2$(dppf) also promotes the Stille reaction[10,11] as well as the Suzuki reaction.[12] Selective alkyl transfer in the Stille coupling reaction is effectively catalyzed by $PdCl_2$(dppf) (eq 5).[10a] Distannanes have been found to promote the $PdCl_2$(dppf) catalyzed allylation of aryl halides with allyl acetate.[10b] Intramolecular cyclization and cross coupling of alkynic aryl triflates with organotin reagents is promoted by $PdCl_2$(dppf) (eq 6).[10c] The $PdCl_2$(dppf)-catalyzed inter- and intramolecular coupling reactions of vinyl or aryl triflates with organostannanes in the presence of carbon monoxide and ***Lithium Chloride*** takes place under mild conditions to give good yields of vinyl or aryl ketones (eq 7).[11] Similar reactions have been performed with aryl triflates (eq 8).[13]

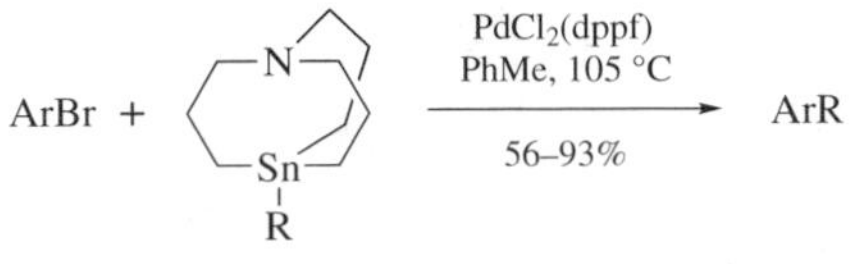

(5) ArBr + stannatrane (Sn–R) → ($PdCl_2$(dppf), PhMe, 105 °C; 56–93%) → ArR

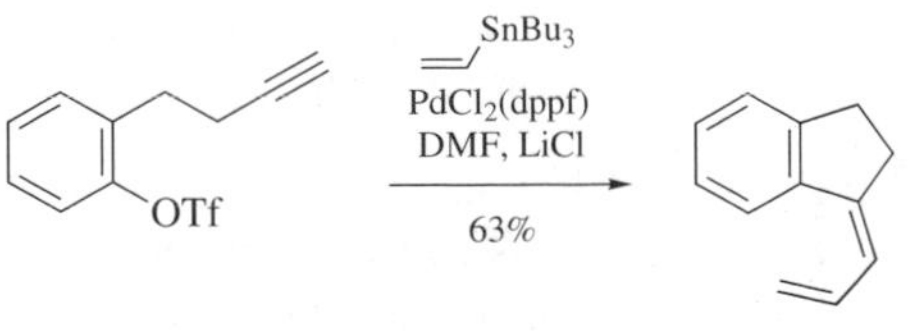

(6) aryl OTf alkyne + vinyl-$SnBu_3$ → ($PdCl_2$(dppf), DMF, LiCl; 63%) → indane diene

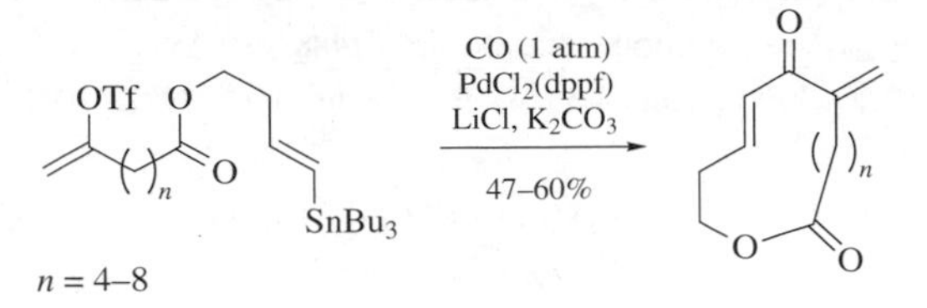

(7) OTf ester/$SnBu_3$, n = 4–8 → (CO (1 atm), $PdCl_2$(dppf), LiCl, K_2CO_3; 47–60%) → macrocyclic lactone ketone

(8) benzodioxole-OTf + $PhSnMe_3$ → ($PdCl_2$(dppf), CO (1 atm), 90 °C, 7 h; 88%) → aryl phenyl ketone

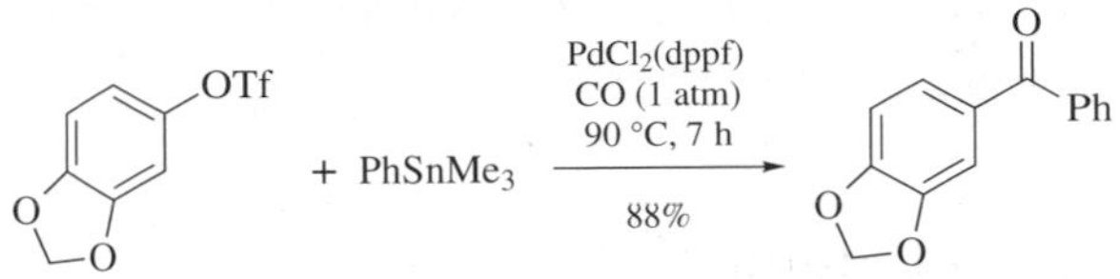

The title reagent has also been used to hydroesterify trimethylsilylalkynes, affording conjugated vinylsilanes in good yield (eq 9).[14]

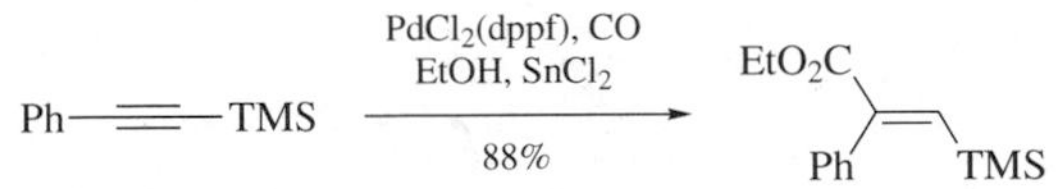

(9) Ph–C≡C–TMS → ($PdCl_2$(dppf), CO, EtOH, $SnCl_2$; 88%) → EtO_2C(Ph)C=CH–TMS

Related Reagents. Bis(acetonitrile)chloronitropalladium(II); Bis(acetonitrile)dinitropalladium(II); Bis(benzonitrile)dichloropalladium(II); Bis(dibenzylideneacetone)palladium(0); Dichloro[1,4-bis(diphenylphosphino)butane]palladium(II); Dichloro[1,2-bis(diphenylphosphino)ethane]palladium(II); Dichlorobis(methyldiphenylphosphine)palladium(II); Dichlorobis(tri-o-tolylphosphine)palladium(II); Dichlorobis(tri-p-tolylphosphine)palladium(II); Tetrakis(triphenylphosphine)palladium(0).

1. (a) Hayashi, T.; Konishi, M.; Kumada, M. *TL* **1979**, 1871. (b) Hayashi, T.; Konishi, M.; Yokota, K-I.; Kumada, M. *CL* **1980**, 767. (c) Rossi, R.; Carpita, A. *TL* **1986**, *27*, 2529. (d) Tamao, K.; Iwahara, T.; Kanatani, R.; Kumada, M. *TL* **1984**, *25*, 1909. (d) Hayashi, T.; Konishi, M.; Kobori, Y.; Kumada, M.; Higuchi, T.; Hirotsu, K. *JACS* **1984**, *106*, 158. (f) Hayashi, T.; Yamamoto, A.; Hagihara, T. *JOC* **1986**, *51*, 723.
2. Katayama, T.; Umeno, M. *CL* **1991**, 2073.
3. Okamura, H.; Miura, M.; Takei, H. *TL* **1979**, 43.
4. Pridgen, L. N.; Killmer, L. B. *JOC* **1981**, *46*, 5402.
5. (a) Rossi, R.; Carpita, A.; Ciofalo, M.; Lippolis, V. *T* **1991**, *47*, 8443. (b) Carpita, A.; Rossi, R.; Veracini, C. A. *T* **1985**, *41*, 1919.
6. (a) Asao, K.; Iio, H.; Tokoroyama, T. *TL* **1989**, *30*, 6401. (b) Campbell, J. B., Jr.; Firor, J. W.; Davenport, T. W. *SC* **1989**, *19*, 2265.
7. Stadtmüller, H.; Lentz, R.; Tucker, C. E.; Stüdemann, T.; Dörner, W.; Knochel, P. *JACS* **1993**, *115*, 7027.
8. (a) Yuan, K.; Scott, W. J. *TL* **1989**, *30*, 4779. (b) Yuan, K.; Scott, W. J. *JOC* **1990**, *55*, 6188.
9. Peterson, G. A.; Kunng, F.-A.; McCallum, J. S.; Wulff, W. D. *TL* **1987**, *28*, 1381.
10. (a) Vedejs, E.; Haight, A. R.; Moss, W. O. *JACS* **1992**, *114*, 6556. (b) Yokoyama, Y.; Ito, S.; Takahashi, Y.; Murakami, Y. *TL* **1985**, *26*, 6457. (c) Luo, F.-T.; Wang, R.-T. *TL* **1991**, *32*, 7703.
11. (a) Echavarren, A. M.; Stille, J. K. *JACS* **1988**, *110*, 1557. (b) Stille, J. K.; Su, H.; Hill, D. H.; Schneider, P.; Tanaka, M.; Morrison, D. L.; Hegedus, L. S. *OM* **1991**, *10*, 1993. (c) Torii, S.; Xu, L. H.; Okumoto, H. *SL* **1991**, 695.
12. Miyaura, N.; Ishiyama, T.; Ishikawa, M.; Suzuki, A. *TL* **1986**, *27*, 6369. Miyaura, N.; Ishiyama, T.; Sasaki, H.; Ishikawa, M.; Satoh, M.; Suzuki, A. *JACS* **1989**, *111*, 314. Ishiyama, T.; Miyaura, N.; Suzuki, A. *SL* **1991**, 687.

13. Echavarren, A. M.; Stille, J. K. *JACS* **1988**, *110*, 1557.
14. Takeuchi, R.; Sugiura, M.; Ishii, N.; Sato, N. *CC* **1992**, 1358.

Tien-Yau Luh, Lung-Lin Shiu & Sue-Min Yeh
National Taiwan University, Taipei, Taiwan

Timothy T. Wenzel
Union Carbide Corporation, South Charleston, WV, USA

Dichloroketene[1]

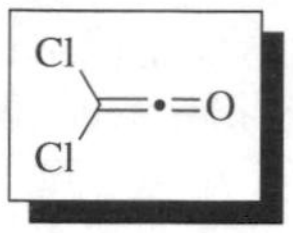

[4591-28-0] C_2Cl_2O (MW 110.93)

(dichloroacetylating agent;[2] undergoes cyclizations with numerous substrates[3])

Solubility: sol ether, pentane, hexane.

Form Supplied in: not commercially available.

Preparative Methods: typically prepared in situ by dehalogenation of trichloroacetyl halides or dehydrohalogenation of dichloroacetyl halides. Elimination of dichloroacetyl chloride with ***Triethylamine*** affords the desired ketene,[4] although material formed in this manner is of limited reactivity.[5] Reduction of ***Trichloroacetyl Chloride*** with ***Zinc/Copper Couple*** activated with either ***Phosphorus Oxychloride***[6] or ***1,1-Dimethoxyethane***[7] provides fully reactive dichloroketene. Recently, thermal activation of ***Zinc*** dust was used to generate dichloroketene.[8]

Handling, Storage, and Precautions: given the propensity of dichloroketene to dimerize or polymerize, the reagent must be made and used immediately. Polymeric material is obtained on heating or concentration of solutions of dichloroketene, and anhydrous conditions must be maintained.

Dichloroacetylation. In his original preparation of dichloroketene, Brady formed the reagent in the presence of aniline to form the corresponding acetanilide (eq 1).[2] Given the wealth of methods to perform this transformation, this is of limited synthetic utility.

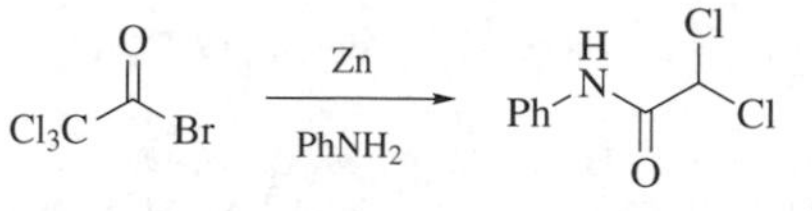

(1)

Cyclizations. The propensity of dichloroketene to undergo cycloaddition reactions with a variety of π-systems has rendered this reagent a powerful synthon.[1,3] Treatment of dichloroketene with numerous alkenes such as 1-pentene causes a [2 + 2] cycloaddition to occur (eq 2).[9] These cyclizations have been shown to occur in a *syn* fashion, as reaction of dichloroketene with *cis*- and *trans*-cyclooctene affords the corresponding *cis* and *trans* bicyclic products (eqs 3 and 4).[10] The cycloadditions of dichloroketene are often sensitive to the method of preparation. For example, conjugated dienes react with dichloroketene in a [2 + 2] fashion to form cyclobutanones, but the ketene must be generated by the dehydrohalogenation of dichloroacetyl chloride, as Lewis acids promote the polymerization of most dienes (eq 5).[11] In fact, the cyclization of acid sensitive alkenes such as vinyl ethers and styrene can only be realized with dichloroketene formed in this fashion.[3a,9] Cyclizations with tri- and tetrasubstituted alkenes are also possible, although in this case the dichloroketene must be generated from activated zinc (eq 6).[12] The cyclobutanones formed in this manner can undergo a vast array of transformations (see below).

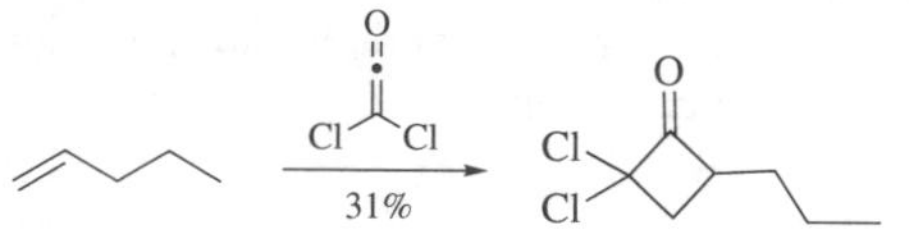

(2)

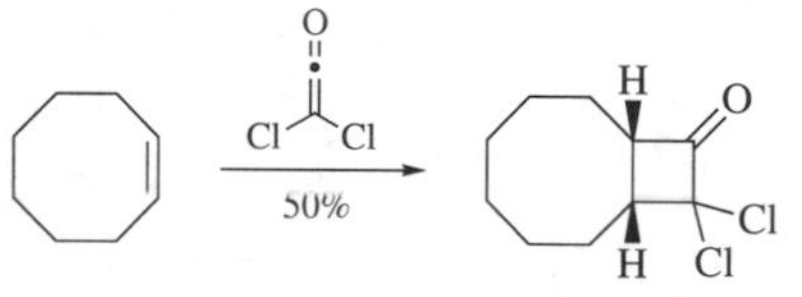

(3)

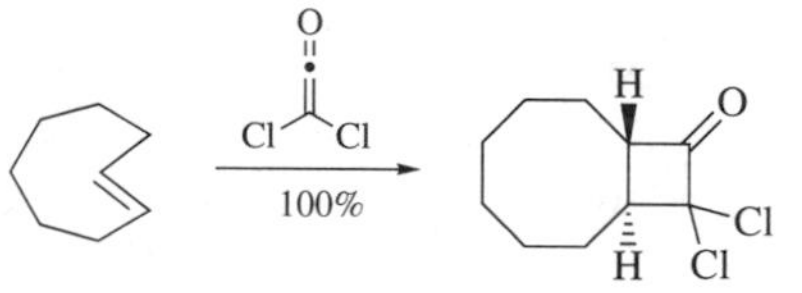

(4)

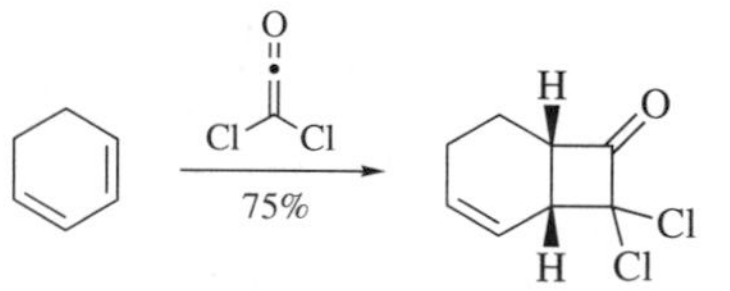

(5)

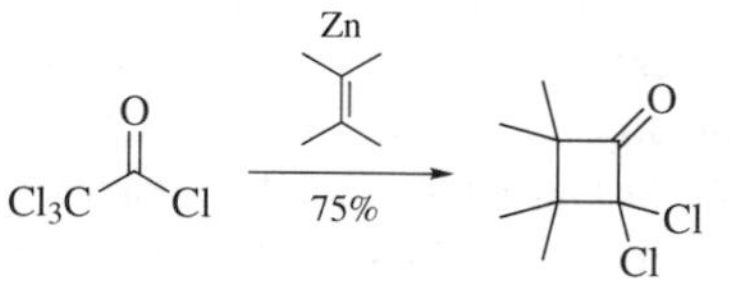

(6)

Cyclization of dichloroketene with unsymmetrical alkenes typically occurs in a regioselective fashion. For example, cyclization of 1-methylcyclohexene with dichloroketene gives a single product (eq 7).[12] Exocyclic methylenes undergo facile conversion to the corresponding cyclobutanones, allowing easy access to spiro compounds (eq 8).[13]

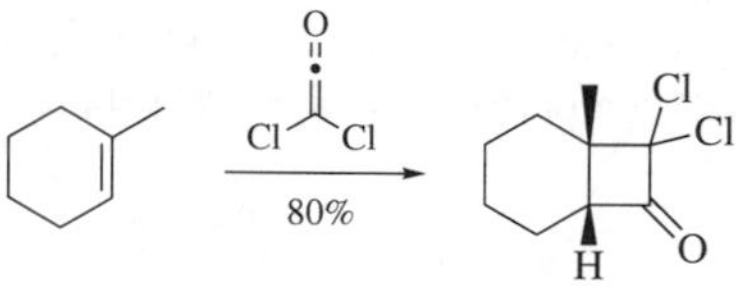

(7)

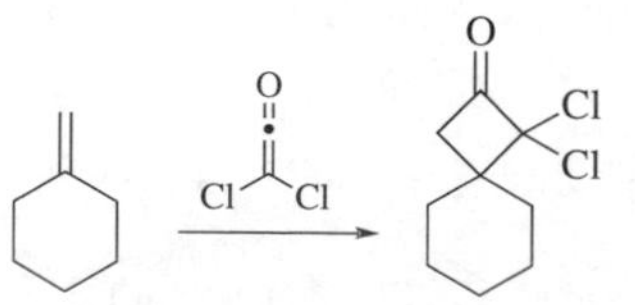

(8)

Dichloroketene also reacts with carbonyl groups to generate 2-oxetanones.[1,3] However, the method of preparation of the ketene is critical. Cyclohexanone, for example, is smoothly converted to the corresponding β-lactone on exposure to dichloroketene prepared via the activated zinc pathway (eq 9).[14] It has been proposed that the zinc present in the reaction activates the ketone carbonyl toward cycloaddition.[15] Dehydrohalogenation of dichloroacetyl chloride in the presence of acetone does not lead to useful product. However, dichloroketene generated in this fashion reacts readily with monosubstituted benzaldehydes (eq 10).[16] The 3,3-dichloro-2-oxetanones formed in this fashion can be made to undergo thermolysis under elevated temperatures, resulting in the generation of dichlorostyrenes (eq 11).[16] Cycloalkanones react with dichloroketene to form spirolactones (eq 12).[17] Thermolysis leads to exocyclic dichloroalkenes, and ***Sodium–Ammonia*** reduction affords the exocyclic methylene (eq 13).[17] This provides a useful complement to the methodology developed by Wittig and Tebbe. The chlorinated β-lactones formed via these cycloadditions are useful synthetic intermediates (see below).

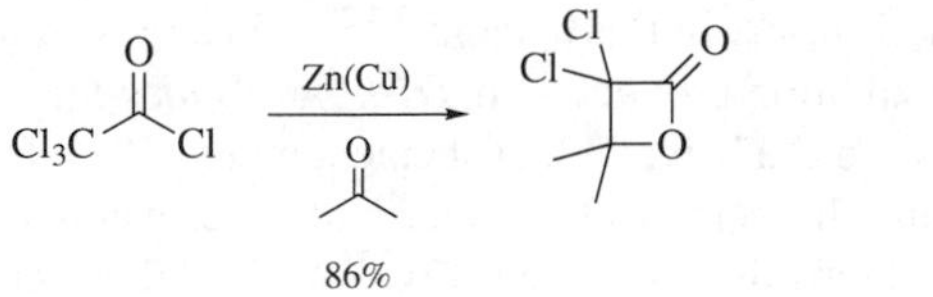

(9)

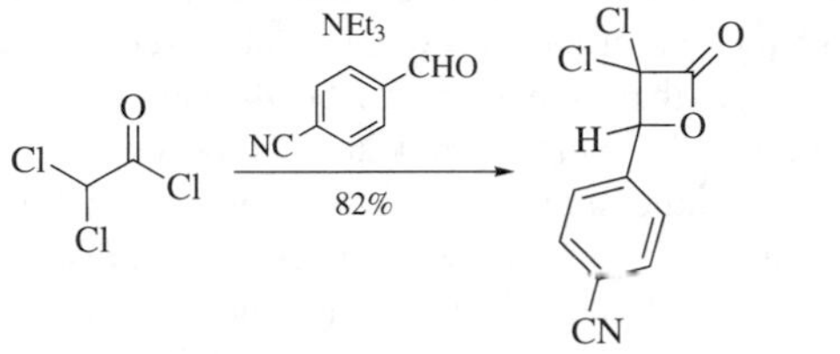

(10)

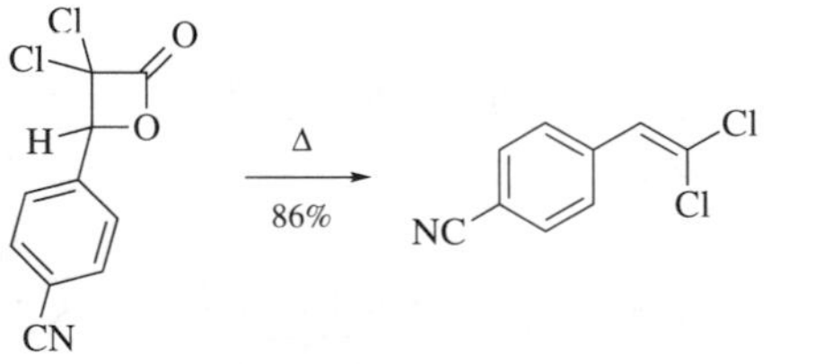

(11)

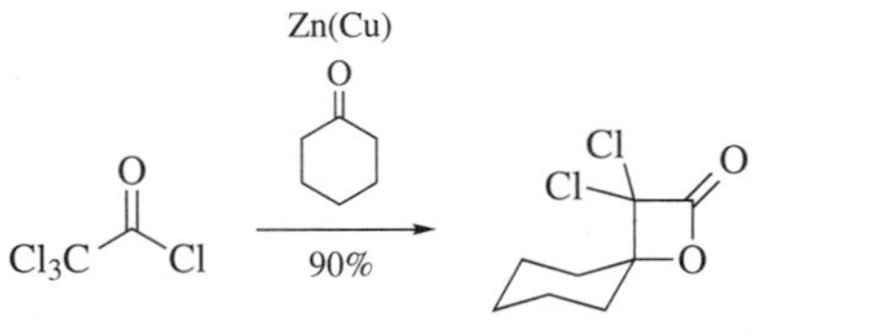

(12)

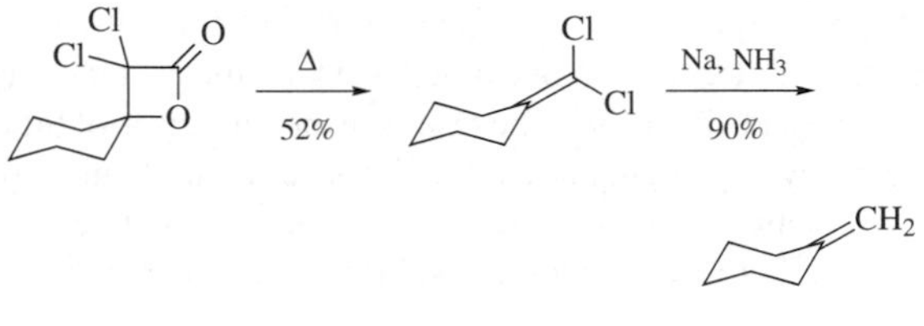

(13)

Dichloroketene undergoes [2 + 2] cyclizations with numerous other unsaturated compounds. Imines, for example, serve as excellent cycloaddition partners, providing a rapid route to the synthesis of β-lactams (eq 14).[18] Diimides also react readily with dichloroketene to form β-iminoazetidinones (eq 15).[19] Addition of trichloroacetyl chloride to alkynes in the presence of activated zinc yields 2-cyclobutenones (eq 16).[7] These reactions also occur in a regioselective manner. Ethoxyacetylene adds to dichloroketene, and subsequent hydrolysis gives the vinylogous acid (eq 17).[20] Cyclization of vinyl ethers with dichloroketene gives a single product, and chiral ethers have been used to generate enantiomerically pure cyclobutanones (eqs 18 and 19).[21,22] α-Alkylidenecyclobutanones can be formed on cyclization of dichloroketene with allenes (eq 20).[23]

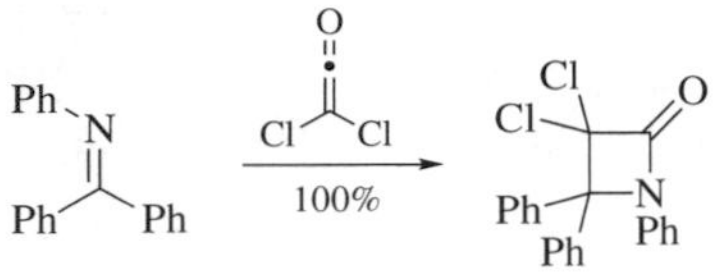

(14)

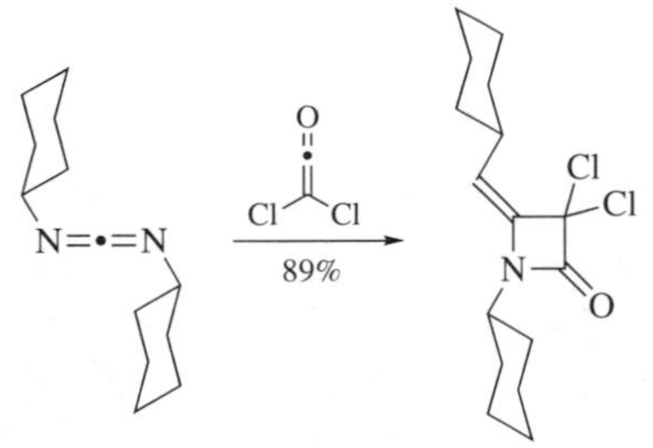

(15)

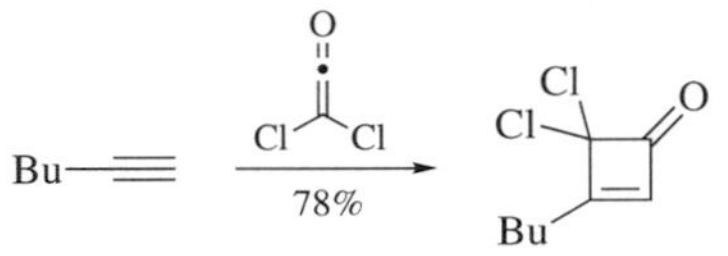

(16)

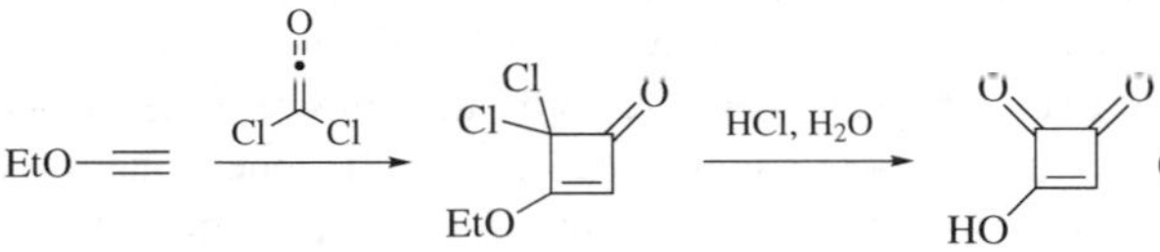

(17)

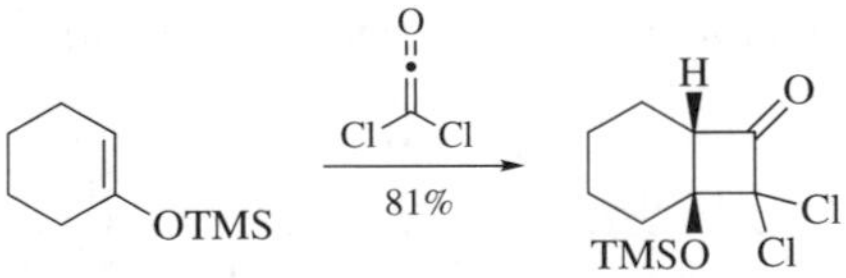

(18)

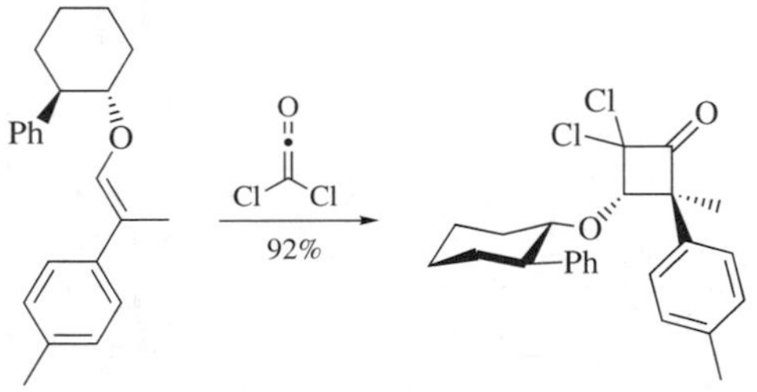

(19)

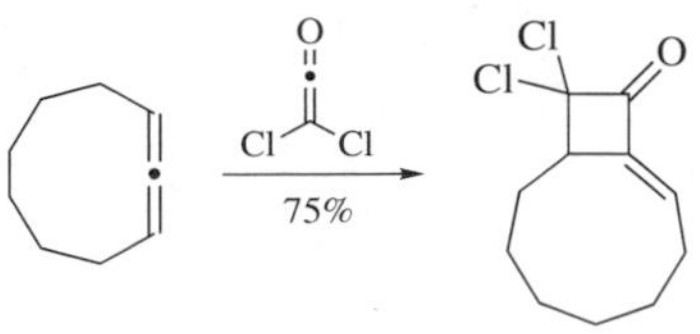

(20)

While [2 + 2] cyclizations are more common, dichloroketene does undergo [4 + 2] cycloaddition reactions in some cases. For example, β-dialkylamino-α,β-unsaturated enones add in an apparent [4 + 2] sense and, upon elimination of hydrogen chloride, furnish 2-pyrones (eq 21).[24] Some imines are also substrates for this reaction, and 2-pyridones can be prepared (eq 22).[25]

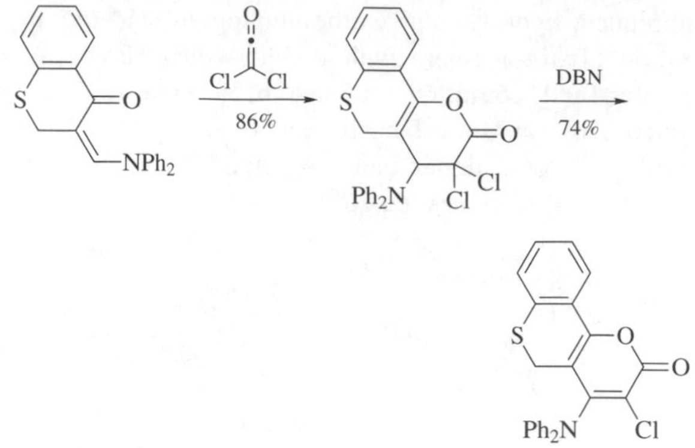

(21)

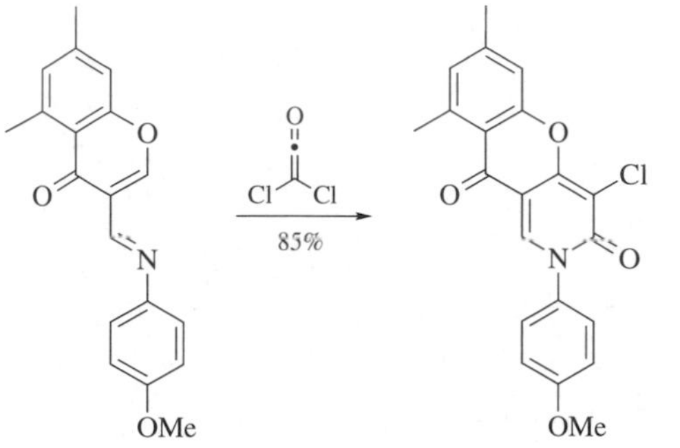

(22)

Reaction of vinyl sulfoxides with dichloroketene gives a novel synthetic route to butyrolactones (eq 23).[26] With the ready availability of enantiomerically pure sulfoxides, this provides an asymmetric source of γ-lactones (eq 24).[27] Allylic ethers and sulfides react with dichloroketene via a Claisen-type rearrangement to give esters, albeit in modest yields (eq 25).[28] This method can be used as a novel entry into macrolide synthesis (eq 26).[29]

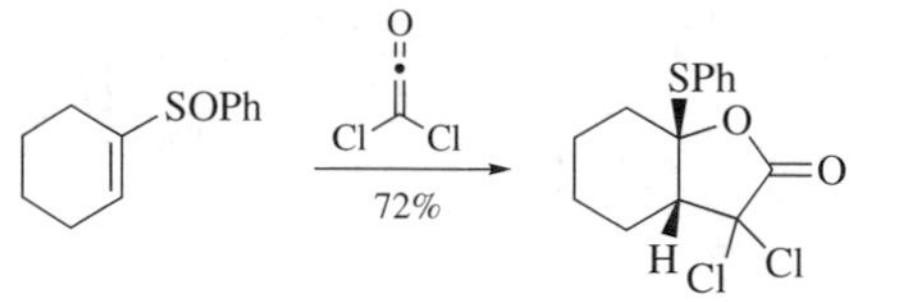

(23)

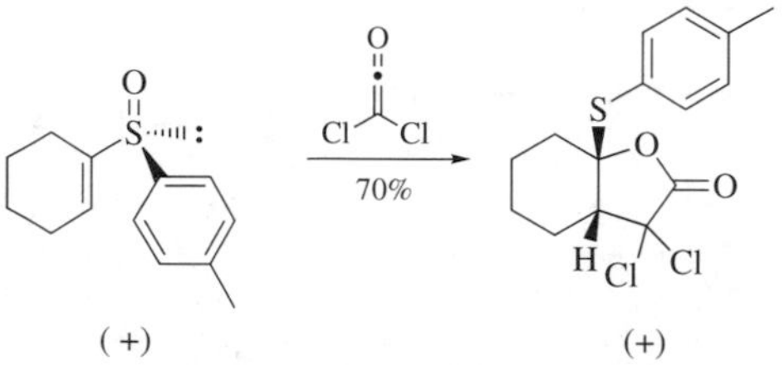

(24)

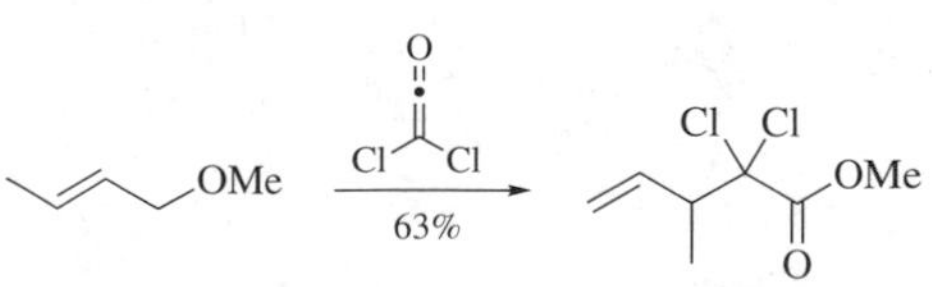

(25)

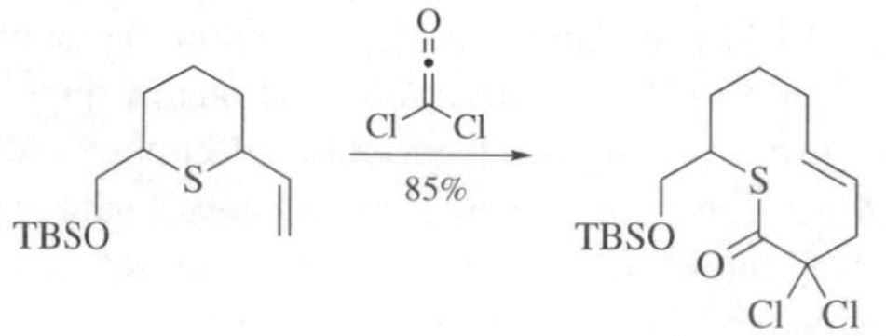

(26)

Reactions of Dichloroketene Adducts. The cycloadduct of cyclopentadiene with dichloroketene can be hydrolyzed to furnish tropolone (eq 27).[30] This is a fairly mild and general reaction, and a number of annulated tropolones can be formed in this fashion (eq 28).[3,31]

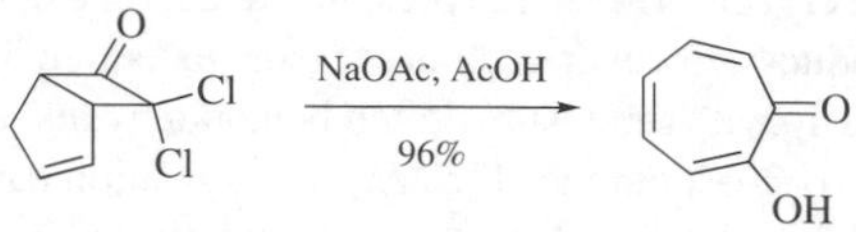

(27)

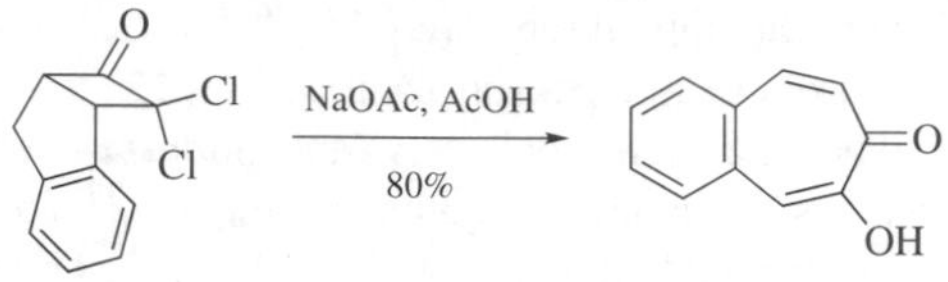

(28)

The reductive removal of one or both of the chlorine atoms of dichloroketene cycloadducts lends considerable flexibility to the synthetic utility of this reagent.[3,4,32] Reduction of the cyclopentadiene addition product with ***Tri-n-butylstannane*** or ***Zinc-Acetic Acid*** gave the bicyclo[3.2.0]heptenone (eq 29).[4] This compound can also be selectively reduced to give exclusively the *endo*-monochloro derivative (eq 30).[33,34] The facile removal of both chlorine substituents renders dichloroketene as a synthetic equivalent of the parent ketene. Since dichloroketene is considerably more reactive than ketene and can be made from inexpensive starting materials, it is superior to ketene in most cases.[1,3] In addition, many of the other adducts can also undergo dehalogenation, providing easy access to the β-lactones, β-lactams, and spiroalkanones (eq 31).[35] In fact, these compounds are substrates for many of the same reactions known for the cyclobutanones.[1,3,4]

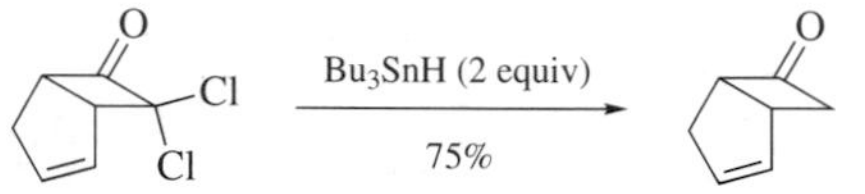

(29)

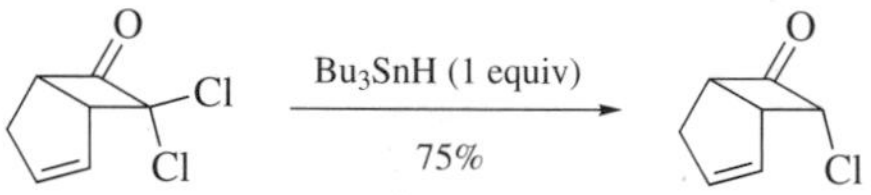

(30)

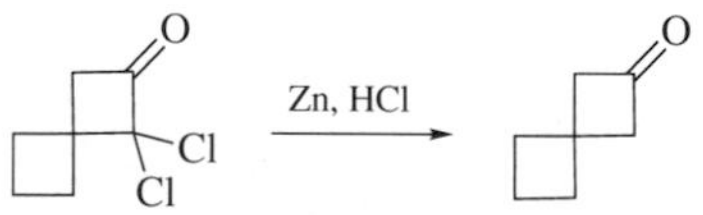

(31)

The dichlorocyclobutanones can easily be expanded or contracted, thereby providing a synthetic source of cyclopropanes or cyclopentanes. Reaction of diazomethane with the cycloadduct of cyclohexene and dichloroketene results in the formation of the corresponding cyclopentanone (eq 32).[36] These can undergo several reactions, including dehalogenation (eq 33),[36] reductive alkylation (eq 34),[37] and conversion to cyclopentenes (eq 35).[38,39]

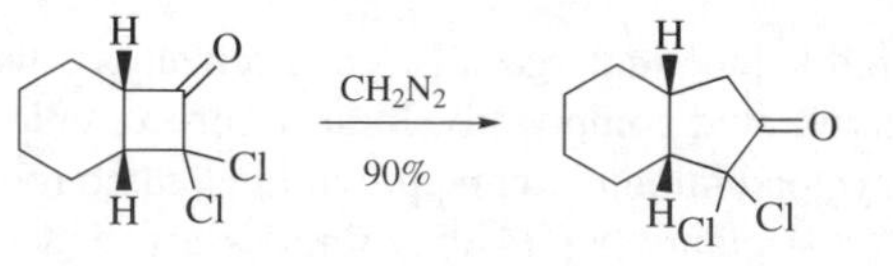

(32)

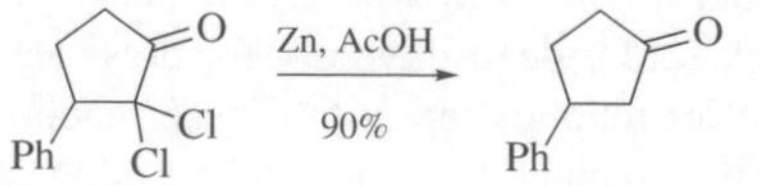

(33)

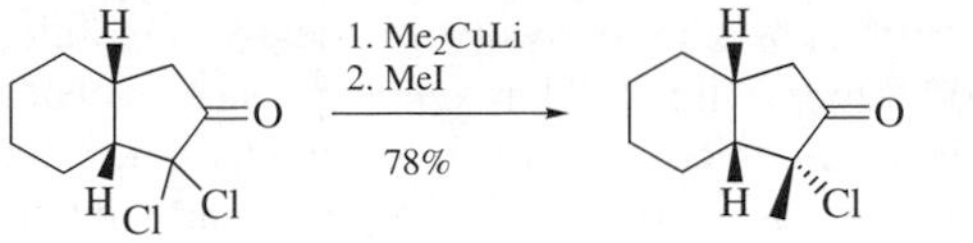

While these reactions are well precedented for the cyclopentanones, most are also amenable to the cyclobutanones.[1,3,4] Ring contractions of α-chlorocycloalkanones via Favorskii rearrangements are well known.[40] Direct exposure of dichloroketene cycloadducts to strong base, however, does not typically lead to ring contraction but rather to ring cleavage (eq 36).[41] One method that has been developed to circumvent this shortcoming is the reduction of the carbonyl of the cyclobutanone followed by treatment with base, which causes ring contraction to the corresponding cyclopropane carbaldehyde (eq 37).[42] Both *endo-* and *exo-*carbaldehydes can be made in this manner.[43]

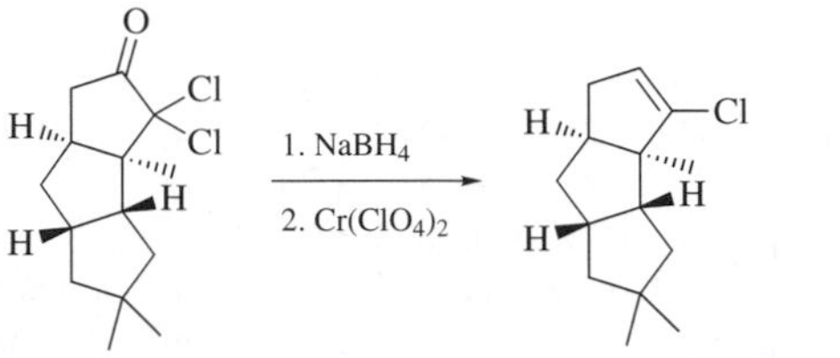

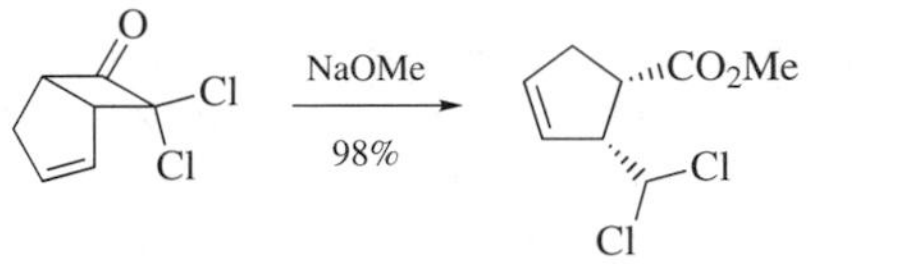

An alternative method of ring expansion is via oxidation. Baeyer–Villiger oxidation of the cyclobutanone adducts supplies a synthetic route to *cis*-bicyclic lactones (eq 38).[44] Ring expanded lactams can also be made, with migration of either of the alkyl substituents (eqs 39 and 40).[45]

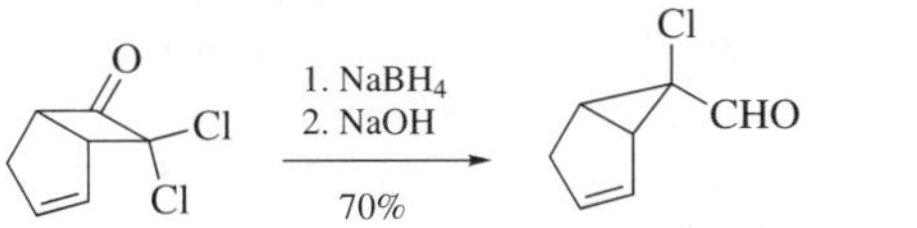

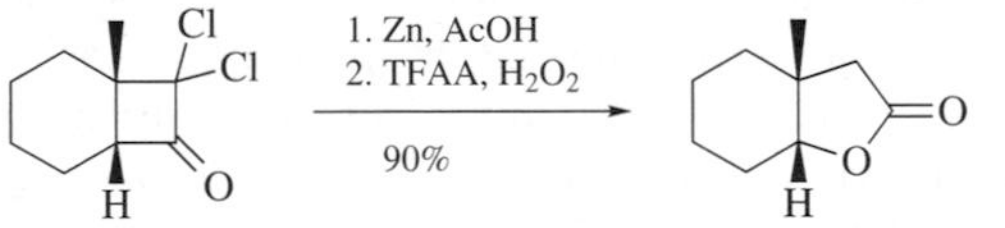

The dichlorocyclobutanones can be converted into several other useful materials. Following reductive removal of the chlorines, ***Selenium(IV) Oxide*** generates the 1,2-cyclobutanediones (eq 41).[46] Lithium–halogen exchange followed by acetylation of the enolate and ***Ruthenium(IV) Oxide*** oxidation affords a one-pot conversion to vicinal dicarboxylic acids (eq 42).[47] In tandem with the cycloaddition of dichloroketene, this provides a mild and general method for *syn*-vicinal dicarboxylations. In a similar fashion, the α-chloroenol acetates can be isolated (eq 43).[48] Thermolysis of these compounds results in a retrocyclization reaction to give exclusively the (*Z*)-α-chloroenone (eq 44).[48]

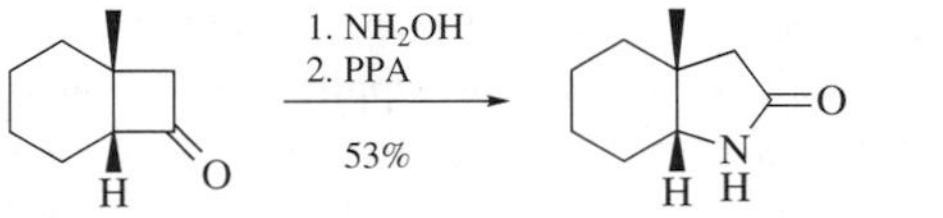

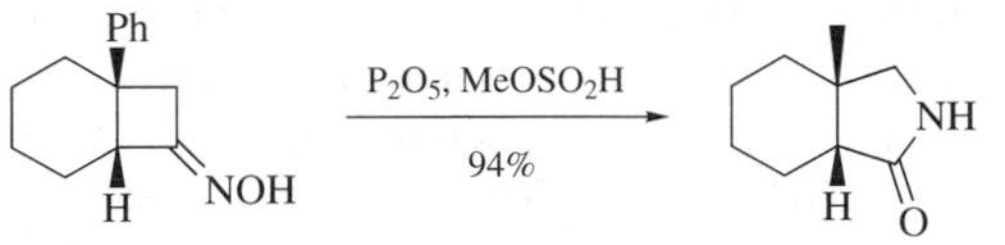

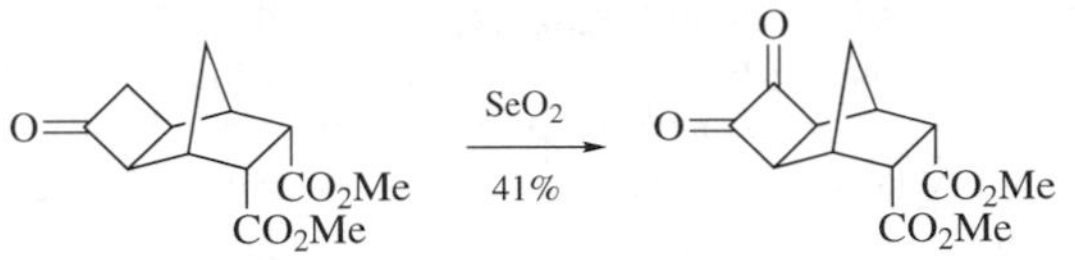

Related Reagents. 2-Chloroacrylonitrile; Chlorosulfonyl Isocyanate; Diiodomethane–Zinc–Titanium(IV) Chloride; Diketene; Ketene; Trichloroacetyl Chloride; Trimethylsilylketene; Zinc/Copper Couple.

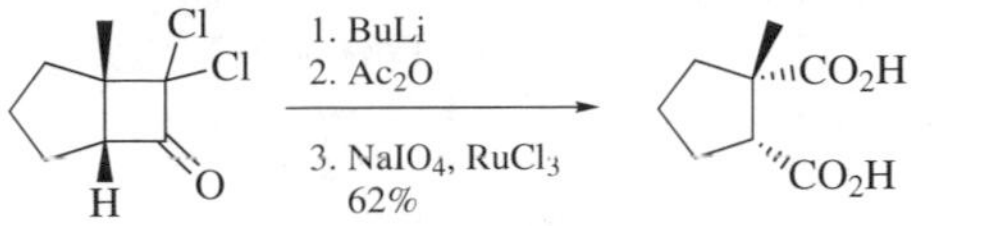

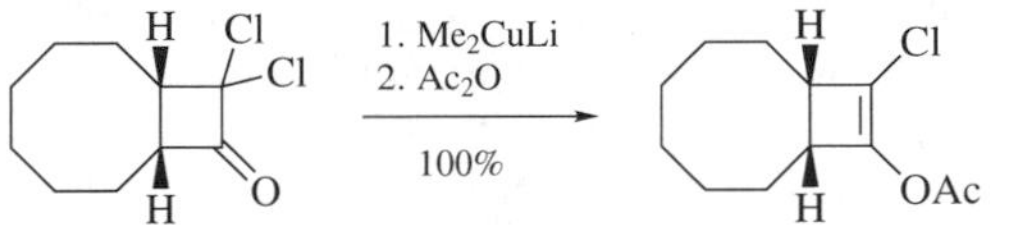

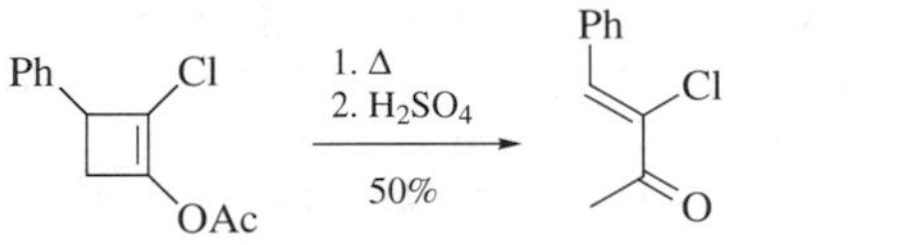

1. For a general review of the chemistry of ketenes, see *The Chemistry of Ketenes, Allenes and Related Compounds*; Patai, Ed.; Wiley: New York, 1980.
2. Brady, W. T.; Liddell, H. G.; Vaughn, W. L. *JOC* **1966**, *31*, 626.
3. For extensive reviews on synthetic uses of halogenated ketenes, see (a) Brady, W. T. *T* **1981**, *37*, 2949. (b) Brady, W. T. *S* **1971**, 415. For a recent review of ketenes, see Tidwell, T. T. *ACR* **1990**, *23*, 273.
4. Ghosez, L.; Montaigne, R.; Roussel, A.; Vanlierde, H.; Mollet, P. *T* **1971**, *27*, 615.
5. Hassner, A.; Fletcher, V. R.; Hamon, D. P. G. *JACS* **1971**, *93*, 264.
6. Krepski, L. R.; Hassner, A. *JOC* **1978**, *43*, 2879.
7. Danheiser, R. L.; Savariar, S.; Cha, D. D. *OSC* **1993**, *8*, 82.
8. Stenstrøm, Y. *SC* **1992**, *22*, 2801.
9. Brady, W. T.; Waters, O. H. *JOC* **1967**, *32*, 3703.
10. Montaigne, R.; Ghosez, L. *AG(E)* **1968**, *7*, 221.
11. Ghosez, L.; Montaigne, R.; Vanlierde, H.; Dumay, F. *AG(E)* **1968**, *7*, 643.
12. Bak, D. A.; Brady, W. T. *JOC* **1979**, *44*, 107.
13. Wiseman, J. R.; Chan, H.-F. *JACS* **1970**, *92*, 4749.
14. Brady, W. T.; Smith, L. *JOC* **1971**, *36*, 1637.
15. Brady, W. T.; Patel, A. D. *JHC* **1971**, *8*, 739.
16. Krabbenhoft, H. O. *JOC* **1978**, *43*, 1305.
17. Brady, W. T.; Patel, A. D. *S* **1972**, 565.
18. (a) Duran, F.; Ghosez, L. *TL* **1970**, 245. (b) Luttringer, J. P.; Streith, J. *TL* **1973**, 4163.
19. Hull R. *JCS(C)* **1967**, 1154.
20. Springer, J. P.; Clardy, J.; Cole, R. J.; Kirksey, J. W.; Hill, R. K.; Carlson, R. M.; Isidor, J. L. *JACS* **1974**, *96*, 2267.
21. Krepski, L. R.; Hassner, A. *JOC* **1978**, *43*, 3173.
22. Greene, A. E.; Charbonnier, F.; Luche, M.-J.; Moyano, A. *JACS* **1987**, *109*, 4752.

23. Brady, W. T.; Stockton, J. D.; Patel, A. D. *JOC* **1974**, *39*, 236.
24. (a) Mosti, L.; Schenone, P.; Menozzi, G. *JHC* **1978**, *15*, 181. (b) Bargagna, A.; Evangelisti, F.; Schenone, P. *JHC* **1979**, *16*, 93. (c) Mosti, L.; Schenone, P.; Menozzi, G. *JHC* **1979**, *16*, 913. (d) Bargagna, A.; Schenone, P.; Bondavalli, F.; Longobardi, M. *JHC* **1980**, *17*, 33. (e) Mosti, L.; Schenone, P.; Menozzi, G. *JHC* **1980**, *17*, 61.
25. Fitton, A. O.; Frost, J. R.; Houghton, P. G.; Suschitzky, H. *JCS(P1)* **1977**, 1450.
26. Marino, J. P.; Neisser, M. *JACS* **1981**, *103*, 7687.
27. Marino, J. P.; Perez, A. D. *JACS* **1984**, *106*, 7643.
28. Malherbe, R.; Belluš, D. *HCA* **1978**, *61*, 3096.
29. Vedejs, E.; Buchanan, R. A. *JOC* **1984**, *49*, 1840.
30. (a) Minns, R. A. *OSC* **1988**, *6*, 1037. (b) Stevens, H. C.; Reich, D. A.; Brandt, D. R.; Fountain, K. R.; Gaughan, E. J. *JACS* **1965**, *87*, 5257.
31. Turner, R. W.; Seden, T. *CC* **1966**, 399.
32. Ali, S. M.; Lee, T. V.; Roberts, S. M. *S* **1977**, 155.
33. Brady, W. T.; Hoff, E. F., Jr.; Roe, R., Jr.; Parry, F. H., III *JACS* **1969**, *91*, 5679.
34. Rey, M.; Huber, U. A.; Dreiding, A. S. *TL* **1968**, 3583.
35. (a) Brook, P. R.; Griffiths, J. G. *CC* **1970**, 1344. (b) Brady, W. T.; Patel, A. D. *JOC* **1972**, *37*, 3536.
36. Greene, A. E.; Deprés, J.-P. *JACS* **1979**, *101*, 4003.
37. (a) Deprés, J.-P.; Greene, A. E. *JOC* **1980**, *45*, 2036. (b) Greene, A. E.; Luche, M.-J.; Deprés, J.-P. *JACS* **1983**, *105*, 2435.
38. Greene, A. E.; *TL* **1980**, *21*, 3059.
39. Kochi, J. K.; Singleton, D. M. *JACS* **1968**, *90*, 1582.
40. For a review of the Favorskii rearrangement, see Kende, A. S. *OR* **1960**, *11*, 261.
41. (a) Brook, P. R.; Duke, A. J. *JCS(C)* **1971**, 1764. (b) Ghosez, L.; Montaigne, R.; Mollet, P. *TL* **1966**, 135.
42. Brook, P. R. *CC* **1968**, 565.
43. Brook, P. R.; Duke, A. J. *JCS(P1)* **1973**, 1013.
44. Jeffs, P. W.; Molina, G.; Cass, M. W.; Cortese, N. A. *JOC* **1982**, *47*, 3871.
45. Jeffs, P. W.; Molina, G.; Cortese, N. A.; Hauck, P. R.; Wolfram, J. *JOC* **1982**, *47*, 3876.
46. Ried, W.; Bellinger, O. *S* **1982**, 729.
47. Deprés, J.-P.; Greene, A. E. *OSC* **1993**, *8*, 377.
48. Deprés, J.-P.; Navarro, B.; Greene, A. E. *T* **1989**, *45*, 2989.

James W. Leahy
University of California, Berkeley, CA, USA

10-Dicyclohexylsulfonamidoisoborneol[1]

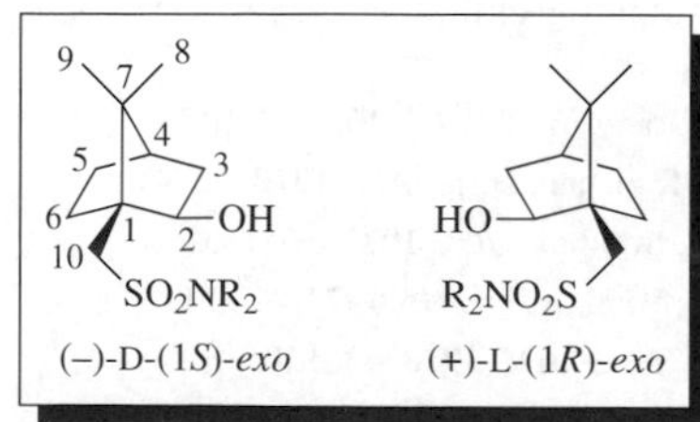

(−)-D-(1*S*)-*exo* (+)-L-(1*R*)-*exo*

(1*S*)-*exo*; R = cyclohexyl		
[96303-88-7]	$C_{22}H_{39}NO_3S$	(MW 397.69)
(1*S*)-*exo*; R = isopropyl		
[89156-11-6]	$C_{16}H_{31}NO_3S$	(MW 317.55)

(chiral auxiliary: enoate derivatives undergo stereoselective Diels–Alder[2,3] and 1,3-dipolar[3] cycloadditions and 1,4-cuprate additions;[4] enol ether derivatives undergo stereoselective [2 + 2] cycloadditions with dichloroketene;[5] ester enolate derivatives participate in stereoselective imine condensation,[6] alkylation,[4a] aldolization,[7] acetoxylation,[8] halogenation,[9] and 'amination'[10] reactions)

Physical Data: R = cyclohexyl: mp (from hexane) 163–164 °C; $[\alpha]_D^{21}$ −25.7 (c = 0.76, EtOH). R = isopropyl: mp (from hexane) 102–103 °C; $[\alpha]_D^{21}$ −34.4 (c = 4.74, EtOH).

Form Supplied in: white crystalline solids.

Preparative Methods: crystalline, enantiomerically pure 10-diisopropyl- and 10-dicyclohexylsulfonamidoisoborneol auxiliaries are readily prepared from the appropriate enantiomer of ***10-Camphorsulfonyl Chloride*** by successive amidation and *exo* selective reduction (eq 1).[2]

SO_2Cl → 1. HNR_2, isoquinoline, DMAP, DMF, 0 °C; 2. L-Selectride, THF, −78 °C → OH, SO_2NR_2 (1)

(**1a**) R = Cy, 55% (cryst)
(**1b**) R = *i*-Pr, 67% (cryst)

Simple acyl derivatives are prepared in good yields from carboxylic acids using Mukaiyama's ***2-Chloro-1-methylpyridinium Iodide*** coupling reagent[2] or from carboxylic acid chlorides using ***Silver(I) Cyanide***.[9a] The former method is also suitable for the preparation of enoyl derivatives, although a Horner–Wadsworth–Emmons reaction has also been employed for this purpose.[4b] The *cis*-propenyl enol ether derivative of 10-diisopropylsulfonamidoisoborneol was prepared by base-promoted isomerization of the corresponding allyl ether (the preparation of which was not described).[5]

Handling, Storage, and Precautions: these reagents are stable indefinitely at ambient temperature in sealed containers.

Introduction. The 10-dialkylsulfonamidoisoborneol auxiliaries exert a powerful topological bias over the π-facial reactivity of enoate, enol ether, and ester enolate derivatives in a wide range of asymmetric transformations. However, the subsequently developed ***10,2-Camphorsultam*** chiral auxiliary outperforms these auxiliaries both in terms of stereoinduction and ease of non-destructive cleavage for most applications. Consequently, only transformations for which the 10-dialkylsulfonamidoisoborneol auxiliaries are particularly advantageous, or for which the analogous transformations of the 10,2-camphorsultam have not been reported, are described here. It should be noted, however, that the origin of the stereoinduction provided by these two camphor-derived auxiliaries is fundamentally different;[1] hence key references for all transformations are provided above.

Reactions of Enoate, Enol Ether, and Acyl Derivatives.

1,4-Organocopper Addition (Alkene to β-Functionalized Product).[4] Tri-*n*-butylphosphine-stabilized organocopper reagents add in a conjugate fashion to *trans*-enoate derivatives of the 10-dicyclohexylsulfonamidoisoborneol auxiliary from the less hindered C(α)-*si* π-face with excellent selectivity (eq 2) (Table 1). This type of reaction has formed the basis of several natural product syntheses.[4]

(2) $\xrightarrow[\text{ether, THF, –78 °C to rt}]{R^2Cu\text{•}BF_3,\ Bu_3P}$ (3) (2)

Table 1 Conjugate Addition of Organocopper Reagents (**2**) → (**3**)

R^1	R^2	Yield (%)	de (%)
Bu	Me	93	97
Me	Bu	89	97
Pr	Me	89	94
Me	Pr	98	95
Me	CH_2=CH	80	98
Me	CH_2=CMe	84	94

[2+2] Dichloroketene Addition (Enol Ether to β-Alkoxy-α-dichlorocyclobutanone)[5]**.** Of six different chiral auxiliaries screened for their ability to control stereochemistry in the reaction of dichloroketene with derived *cis*-propenyl enol ethers, the 10-diisopropylsulfonamidoisoborneol auxiliary was the best. Thus, following ring expansion of the initially formed cyclobutanone (**4**) with ***Diazomethane–Chromium(II) Perchlorate***, α-chloro-γ-methylcyclopentenone was isolated in ~60% yield and 80% ee [C(α)-*si* face attack of the ketene] (eq 3). The auxiliary was also recovered in unspecified yield.

$SO_2N(i\text{-}Pr)_2$ $\xrightarrow[\text{ether, rt; >60\%}]{Cl_2C{=}C{=}O}$ (4) (3)

(**4**)
80% de
(major isomer shown)

Imine Condensation (Acyl Species to β-Lactam)[6]**.** Lithium enolates of acyl 10-diisopropylsulfonamidoisoborneols condense with *N*-aryl aldimines to give *cis*-disubstituted β-lactams with 56–92% ee, accompanied by 2.5–9% of their *trans* isomers (in undetermined ee) (eq 4) (the key step in a synthesis of the carbapenem antibiotic (+)-PS-5). Menthol was found to be a less efficient auxiliary for this application.[6]

1. LDA, THF, –70 °C; 2. Ph–CH=CH–CH=N–(C_6H_4-OMe); –70 °C to rt (4)

79% (FC), 91% ee
(+ recovered auxiliary 95%)

α-Acetoxylation and α-Halogenation (Acyl Species to α-Acetoxy or α-Halo Acyl Product)[8,9]**.** α-Acetoxylations of *O*-silyl enol ether derivatives of acyl 10-dicyclohexylsulfonamidoisoborneols with ***Lead(IV) Acetate*** proceed in high yield with excellent π-facial stereocontrol (95–100% de, with C(α)-*re* topicity).[8] Mechanistically related α-halogenations with *N*-halosuccinimides also proceed smoothly to afford α-halo acyl products in 76–96% de, but with C(α)-*si* topicity.[9] The observed topicities are consistent with initial attack of the electrophilic species from the less hindered C(α)-*si* face to give transient plumbonium/bromonium/chloronium ions. The plumbonium intermediates undergo S_N2-type attack by acetate at the β-position, whereas the bromonium/chloronium intermediates fragment with retention at C(β).[9a] The stereofacial influence of the auxiliary overrides any preexisting β-stereocenter. Hence, consecutive alkylcopper conjugate addition, then α-acetoxylation or α-bromination, allows the concise and stereocontrolled formation of two contiguous stereocenters. α-Acetoxy ester derivative (**6**) formed in this way is a precursor to a key intermediate for the synthesis of the elm bark beetle pheromone (eq 5), and α-bromo ester derivative (**7**) was converted via azide displacement, transesterification, and hydrogenolysis into L-*allo*-isoleucine (eq 6).[9b] α-Halo esters are also useful precursors of enantiomerically pure epoxides.[9a]

(**5**) 1. $PrCu\text{•}BF_3$, Bu_3P, ether, THF, –78 °C to rt; 2. H_2O; 3. LDA, TMSCl, THF, –78 °C; 4. $Pb(OAc)_4$, CH_2Cl_2, –15 °C; 5. Et_3NHF, rt → (**6**) (5)

(**6**)
66% (cryst)
97% de C(α), 92% de C(β) (crude)

ent-(**5**) 1. $EtCu\text{•}BF_3$, Bu_3P, ether, THF, –78 °C to rt; 2. H_2O; 3. LDA, TMSCl, THF, –78 °C; 4. NBS, THF, –78 °C → (**7**) (6)

(**7**)
60% (cryst)
99.3% de C(α), 97.8% de C(β)

α-'Amination' (Acyl Species to α-Amino Acyl Product).[10] Although asymmetric bromination and stereospecific azide displacement of *O*-silyl enol ethers of acyl 10-dicyclo hexylsulfonamidoisoborneols (as described above) is a generally applicable route to optically active α-amino acids,[9b] a complementary and more direct approach to this important class of compounds is via electrophilic 'amination' of these same compounds using ***Di-t-butyl Azodicarboxylate*** (DBAD). The initially formed α-(di-*N*-Boc-hydrazido)amino acid derivatives (**8**) (eq 7) may be efficiently converted to the corresponding α-amino acid hydrochlorides by successive deacylation, hydrogenolysis, transesterification, and hydrolysis. This reaction sequence has been shown to be efficient for the preparation of a wide range of α-amino acids in excellent enantiomeric purity[10] and compares favorably with closely related methods using alternative auxiliaries.[11]

1. LDA, TMSCl, THF, –78 °C
2. $TiCl_4$, $Ti(O\text{-}i\text{-}Pr)_4$, DBAD, CH_2Cl_2, –78 °C

SO_2NCy_2 ... NBocNHBoc (7)

(8)
R = Me, Et, Pr, *i*-Pr, Bu, *i*-Bu, Bn, Cy
69–84% (FC) [92.6–96.4% de (crude)]

Nondestructive Auxiliary Cleavage. The hindered ester linkage present in acyl derivatives of 10-dialkylsulfonamidoisoborneols is less readily cleaved than the corresponding sulfonamidic linkage of *N*-acyl-10,2-camphorsultam derivatives. However, it can be hydrolyzed and the auxiliary recovered intact under basic conditions using ***Potassium Hydroxide***[7] or ***Potassium Carbonate***[8] in MeOH, ***Sodium Hydroxide*** in aq EtOH,[4a] or ***Lithium Hydroxide*** in aq THF. Elevated temperatures are required to achieve acceptable reaction rates for all but the latter procedure which, although sluggish at ambient temperature, was employed for unmasking sensitive aldol products.[7] 'Nonbasic' transesterification using $Ti(OBn)_4$/BnOH affords benzyl esters which may be subject to hydrogenolysis to give the corresponding carboxylic acids.[9b] Alternatively, transesterification with $Ti(OEt)_4$/EtOH[10b] may be followed by hydrolysis under acidic conditions.[10]

Primary alcohols can be obtained by hydride reduction using either ***Lithium Aluminium Hydride*** in ether[2,8] or $Ca(BH_4)_2$ in THF,[9a] and this latter reagent is compatible with halogen functionality. A dimethyl tertiary alcohol was obtained by addition of 2 equiv of methyllithium in ether.[4b]

Related Reagents. 10,2-Camphorsultam; α-Methyltoluene-2,α-sultam.

1. Oppolzer, W. *T* **1987**, *43*, 1969.
2. Oppolzer, W.; Chapuis, C.; Bernardinelli, G. *TL* **1984**, *25*, 5885.
3. Curran, D. P.; Kim, B. H.; Piyasena, H. P.; Loncharich, R. J.; Houk, K. N. *JOC* **1987**, *52*, 2137.
4. (a) Oppolzer, W.; Dudfield, P.; Stevenson, T.; Godel, T. *HCA* **1985**, *68*, 212. (b) Oppolzer, W.; Moretti, R.; Bernardinelli, G. *TL* **1986**, *27*, 4713.
5. Greene, A. E.; Charbonnier, F. *TL* **1985**, *26*, 5525.
6. Hart, D. J.; Lee, C.-S.; Pirkle, W. H.; Hyon, M. H.; Tsipouras, A. *JACS* **1986**, *108*, 6054.
7. Oppolzer, W.; Marco-Contelles, J. *HCA* **1986**, *69*, 1699.
8. Oppolzer, W.; Dudfield, P. *HCA* **1985**, *68*, 216.
9. (a) Oppolzer, W.; Dudfield, P. *TL* **1985**, *26*, 5037. (b) Oppolzer, W.; Pedrosa, R.; Moretti, R. *TL* **1986**, *27*, 831.
10. (a) Oppolzer, W. In *Chirality in Drug Design and Synthesis*, Academic: New York, 1990. (b) Oppolzer, W.; Moretti, R. *HCA* **1986**, *69*, 1923. (c) Oppolzer, W.; Moretti, R. *T* **1988**, *44*, 5541.
11. (a) Gennari, C.; Colombo, L.; Bertolini, G. *JACS* **1986**, *108*, 6394. (b) Evans, D. A.; Britton, T. C.; Dorow, R. L.; Dellaria, J. F. *JACS* **1986**, *108*, 6395. (c) Evans, D. A.; Britton, T. C.; Dorow, R. L.; Dellaria, J. F. *T* **1988**, *44*, 5525. (d) Trimble, L. A.; Vederas, J. C. *JACS* **1986**, *108*, 6397.

Alan C. Spivey
University of Cambridge, UK

Diethylaluminum Cyanide[1]

Et_2AlCN

[5804-85-3] $C_5H_{10}AlN$ (MW 111.12)

(hydrocyanation of α,β-unsaturated ketones;[2] preparation of α-cyanohydrins from carbonyl compounds[3])

Physical Data: bp 162 °C/0.02 mmHg; colorless syrup; exists as a tetra- or pentamer in boiling benzene or diisopropyl ether and as a dimer in boiling THF.
Solubility: sol benzene, toluene, diisopropyl ether.
Form Supplied in: solution in toluene.
Preparative Method: by reaction of ***Triethylaluminum*** with a slight excess of ***Hydrogen Cyanide*** below rt.[1]
Handling, Storage, and Precautions: highly toxic. It can be stored as 1–2 M solutions in benzene, toluene, or diisopropyl ether. Use in a fume hood.

Hydrocyanation of α,β-Unsaturated Ketones. Diethylaluminum cyanide is a reagent for the conjugate hydrocyanation of α,β-unsaturated ketones. The reaction is carried out simply by adding a stock solution of the reagent to a solution of the substrate α-enone in an aprotic solvent with ice cooling, and then allowing the reaction mixture to stand at rt (eqs 1 and 2).[2,3] This hydrocyanation is reversible and therefore can be controlled both kinetically and thermodynamically; that is, the product is kinetically controlled in the early stage.

Et_2AlCN, benzene, 25 °C, 2 min (1)
trans 87%; *cis* 13%

OMe → CN: Et_2AlCN, benzene, 0–25 °C, 1 h, 55% (2)

Preparation of α-Cyanohydrins from Carbonyl Compounds. The preparation of cyanohydrins from ketones and aldehydes of low reactivity can be effected with Et_2AlCN (eq 3).[4]

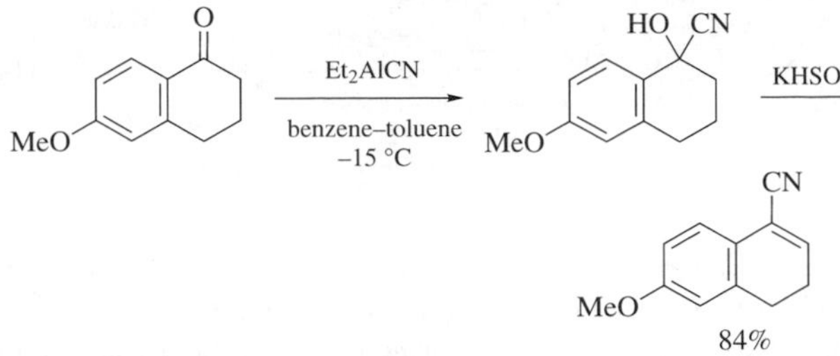

(3)

84%

Related Reagents. Cyanotributylstannane; Cyanotrimethylsilane; Hydrogen Cyanide.

1. Nagata, W.; Yoshioka, M. *OR* **1977**, *25*, 255.
2. Nagata, W.; Yoshioka, M.; Hirai, S. *JACS* **1972**, *94*, 4635.
3. Agosta, W. C.; Lowrance, W. W., Jr. *JOC* **1970**, *35*, 3851.
4. Nagata, W.; Yoshioka, M.; Murakami, M. *OS* **1972**, *52*, 96.

Takeshi Nakai & Katsuhiko Tomooka
Tokyo Institute of Technology, Japan

Diethyl Carbonate

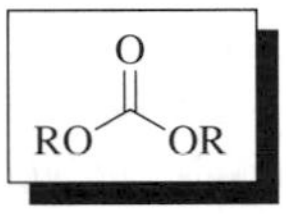

(R = Et)
[105-58-8] $C_5H_{10}O_3$ (MW 118.13)
(R = Me)
[616-38-6] $C_3H_6O_3$ (MW 90.08)

(*C*-alkoxycarbonylations of carbanions[2–12] and other nucleophiles;[13] synthesis of ketones[14] and tertiary alcohols;[15] allylations with allyl alkyl carbonates;[16] alkylations of nucleophilic substrates;[17–20] heterocyclic synthesis[21])

Physical Data: R = Et: bp 127 °C; *d* 0.975 g cm^{-3}. R = Me: bp 90.3 °C; *d* 1.070 g cm^{-3}.
Solubility: miscible organic alcohols, esters, ethers. R = Et: insol H_2O. R = Me: sol 13.9 g per 100 g H_2O at 20 °C.
Form Supplied in: readily available; inexpensive.
Handling, Storage, and Precautions: inhalation of the vapors of dimethyl carbonate may cause irritation.

Alkoxycarbonylations of Carbanions. Diethyl and dimethyl carbonate are valuable reagents for the *C*-alkoxycarbonylation of enolate anions derived from active methylene compounds.[1,2] The yields, in general, are comparable to those found using other electrophilic alkoxycarbonylating agents. Various bases such as ***Sodium Hydride***, NaH–KH mixtures, metal alkoxides, and non-nucleophilic species such as ***Acetaldehyde 3-Bromopropyl Ethyl Acetal*** at low temperature have been utilized to generate enolate anions. The latter method is particularly useful, since enolate formation is complete, and competitive condensations of the enolate anions with the substrate are avoided. Ketones[3] can readily be converted into β-keto esters and esters into malonate esters.[4] Selected examples for the conversion of a cyclic (eq 1)[3f] and an alicyclic ketone (eq 2)[3b] to the respective β-keto esters are illustrated. In the acyclic example, the kinetically formed enolate is *C*-ethoxycarbonylated.

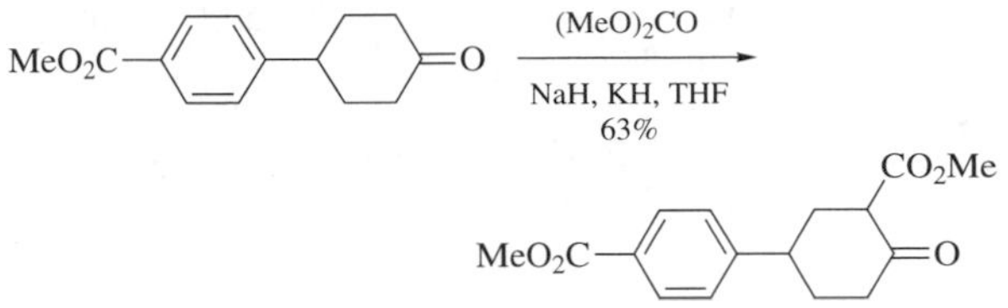

(1)

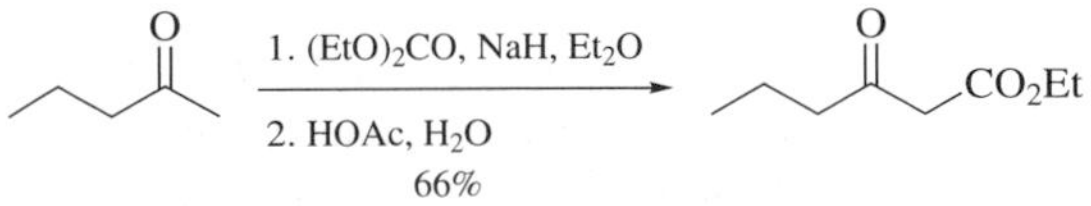

(2)

The conversion of *t*-butyl acetate to the *t*-butyl methyl malonate ester (eq 3)[4] and a *C*-methoxycarbonylation of γ-butyrolactone (eq 4) have been described.[5]

$$\mathrm{MeCO_2\text{-}}t\mathrm{\text{-}Bu} \xrightarrow[\text{2. (MeO)}_2\text{CO} \quad 73\%]{\text{1. LDA, THF, 0 °C}} \mathrm{MeO_2CCH_2CO_2\text{-}}t\mathrm{\text{-}Bu} \quad (3)$$

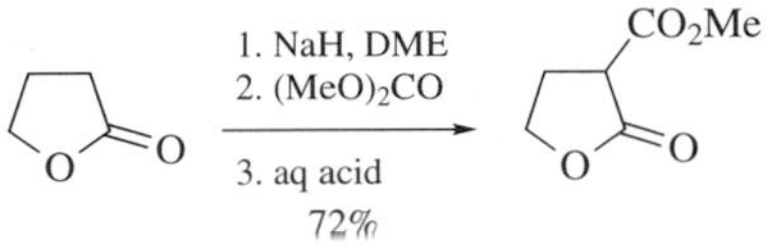

(4)

Aryl and alkyl cyanides are readily alkoxycarbonylated.[6] The α-ethoxycarbonylations of phenylacetonitrile (eq 5)[6c] and cyanopentanecarbonitrile (eq 6)[6a] are typical examples.

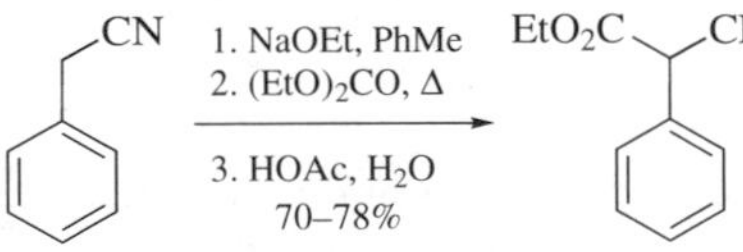

(5)

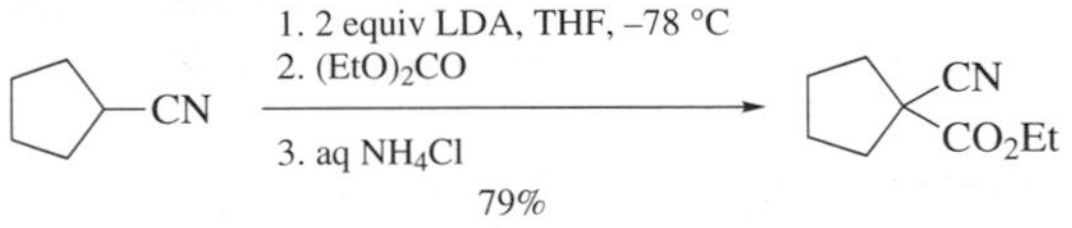

(6)

The transformations of primary nitro compounds (eq 7)[7] and isocyanides (eq 8)[8] into the α-ethoxycarbonylation products have been reported.

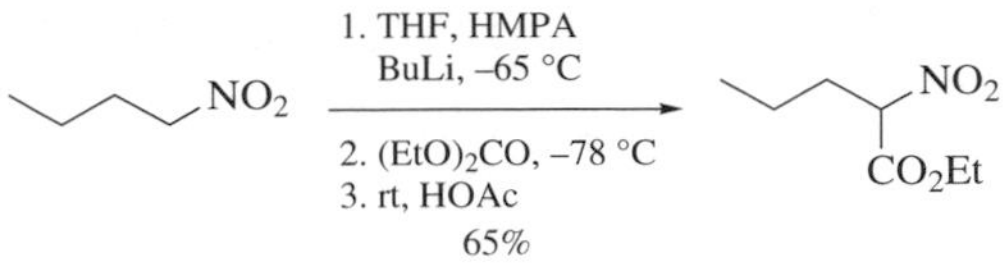

(7)

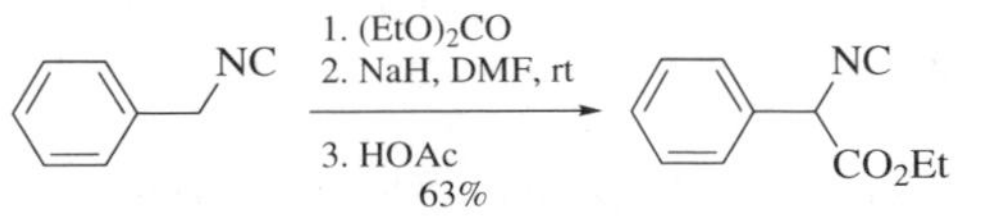

(8)

The conversion of 2,6-dimethylpyridine to ethyl 6-methylpyridine-2-acetate has been accomplished using ***Potassium Amide*** and diethyl carbonate (eq 9).[9]

1. KNH_2, NH_3 (liq); 2. $(EtO)_2CO$; 3. NH_4Cl; 59–75% → CO_2Et (9)

The dianions formed from cyclic or acyclic carboxylic acids on treatment with dimethyl or diethyl carbonate lead to monoesters of malonic acid (eq 10).[10]

CO_2H — 1. LDA, THF; 2. $(EtO)_2CO$; 3. aq HCl; 75% → EtO_2C CO_2H (10)

Carboxylic acid esters can be synthesized from aryl[11] or alkyl halides[11b] by treatment of the corresponding Grignard reagents with dimethyl or diethyl carbonate. For example, addition of ***Phenylmagnesium Bromide*** in THF to a THF solution of dimethyl carbonate (2.0 equiv) at 0–5 °C leads to methyl benzoate (87%).[11]

Carbanions formed via directed metalations undergo facile α-alkoxycarbonylations.[12] Lithiation of 1-(*t*-butoxycarbonyl)-1,2-dihydropyridine followed by addition of dimethyl carbonate leads to the α-methoxycarbonylated product (eq 11).[12a] The retention of configuration found in the alkoxycarbonylation of a benzyllithium analog is of interest (eq 12).[12b]

Cl, Bu, N, CO_2-*t*-Bu — 1. BuLi, THF, –42 °C; 2. $(MeO)_2CO$; 3. H_2O; 81% → Cl, Bu, N, CO_2Me, CO_2-*t*-Bu (11)

Ph, H, O, O, N(*i*-Pr)$_2$ — 1. *s*-BuLi, TMEDA, hexane, –78 °C; 2. $(MeO)_2CO$; 85% → Ph, MeO_2C, O, O, N(*i*-Pr)$_2$; 90% ee (12)

Other nucleophiles such as amines and hydrazines react with dimethyl or diethyl carbonate to yield the corresponding carbamates or carbazates. Treatment of hydrazine with diethyl carbonate leads to ethyl hydrazinecarboxylate (77–85%).[13]

Ketone Synthesis. The synthesis of ketones can be effected by treatment of organometallics with diethyl carbonate.[14] The addition of 2 equiv of 4-chloro-3-lithiopyridine to diethyl carbonate affords bis(4-chloro-3-pyridyl) ketone (eq 13).[14c]

Cl, N — 1. LDA, THF, –80 °C; 2. $(EtO)_2CO$; 3. H_2O; 65% → Cl, O, Cl, N, N (13)

Tertiary Alcohol Synthesis. Dimethyl or diethyl carbonate on treatment with organometallics lead to tertiary alcohols.[15]

Allylations and Alkylations. The allylations of β-keto esters, β-diketones, malonate esters, nitriles, and nitro analogs can be accomplished under neutral conditions by use of a variety of allylic carbonates in the presence of a palladium–phosphine catalyst.[16] A typical procedure is shown (eq 14).[16a]

O, CO_2Me — 1. OCO_2Me; 2. $Pd_2(dba)_3$•$CHCl_3$, Ph_3P, THF, 0.2 h, 30 °C; 92% → CO_2Me, O (14)

The selective mono-*N*-methylation of aromatic amines such as aniline[17] and mono-*C*-methylations of arylacetonitriles[18] by dimethyl carbonate can be performed by gas–liquid phase transfer catalysis. Dimethyl carbonate in the presence of potassium carbonate and ***18-Crown-6*** (or Aliquat 336) is useful for the *S*-methylation of alkyl or aryl thiols (eq 15), the *O*-methylation of phenols, and the *N*-methylation of imidazole.[19]

$C_6H_{13}SH$ — 1. K_2CO_3, $(MeO)_2CO$, 18-crown-6, 100 °C; 2. H_2O; 80% → $C_6H_{13}SMe$ (15)

The α-methylation and α-methoxycarbonylation of primary esters has been accomplished by addition of the ester in dimethyl carbonate to a mixture of NaH in dimethyl carbonate (eq 16).[20]

O, O, CO_2Me — 1. NaH; 2. $(MeO)_2CO$ (excess), 60 °C; 3. H_3O^+; 55% → O, CO_2Me, CO_2Me (16)

Synthesis of Heterocycles. Many substrates react with dimethyl or diethyl carbonate to form heterocyclic compounds. Treatment of 1,2-amino alcohols with dimethyl carbonate affords 2-oxazolidinones.[21] The preparation of (*S*)-4-(phenylmethyl)-2-oxazolidinone (eq 17) is illustrative.[21c]

H_2N, OH — K_2CO_3, $(EtO)_2CO$, 135 °C; 78% → HN, O, O (17)

Related Reagents. Benzyl Chloroformate; Carbon Dioxide; Carbon Oxysulfide; Di-*t*-butyl Dicarbonate; Diethyl Oxalate; Ethyl Chloroformate; Methyl Cyanoformate; Methyl Magnesium Carbonate.

1. (a) Mathieu, J.; Weill-Raynal, J. *Formation of C–C Bonds*; Thieme: Stuttgart, 1973; vol. 1, pp 325–345. (b) *Enichem Synthesis Technical Bulletin*; Enichem America: New York, 1992; a review of dimethyl carbonate with 577 references.
2. (a) House, H. O. *Modern Synthetic Reactions*, 2nd ed.; Benjamin: Menlo Park, CA, 1972; Chapter 11. (b) Stowell, J. C. *Carbanions in Organic Synthesis*; Wiley: New York, 1979; Chapters 5 and 6. (c) Carey, F. A.; Sundberg, R. J. *Advanced Organic Chemistry*, 3rd ed.; Plenum: New York, 1990; Part B, pp 55–94. (d) Davis, B. R.; Garratt, P. J. *COS* **1991**, *2*, 795.
3. (a) Krapcho, A. P.; Diamanti, J.; Cayen C.; Bingham, R. *OSC* **1973**, *5*, 198. (b) Crombie, L.; Jones, R. C. F.; Palmer, C. J. *JCS(P1)* **1987**, 317. (c) Kallury, K. R.; Krull, U. J.; Thompson, M. *JOC* **1988**, *53*, 1320. (d) Mori, K.; Kamada, A.; Mori, H.; *LA* **1989**, 303. (e) Alderdice, M.; Sum, F. W.; Weiler, L. *OSC* **1990**, *7*, 351. (f) DeGraw, J. I.; Christie, P. H.; Colwell, W. T.; Sirotnak, F. M. *JMC* **1992**, *35*, 320; (g) Ruest, L.; Blouin, G.; Deslongchamps, P. *SC* **1976**, *6*, 169.
4. Hill, J. E.; Harris, T. M. *SC* **1982**, *12*, 621.

5. (a) Fieser, L. F.; Fieser, M. *FF* **1975**, *5*, 214. (b) Quesada, M. L.; Schlessinger, R. H. *JOC* **1978**, *43*, 346.
6. (a) Albarella, J. P. *JOC* **1977**, *42*, 2009. (b) van den Berg, E. M. M.; Richardson, E. E.; Lugtenburg, J.; Jenneskens, L. W. *SC* **1987**, *17*, 1189. (c) Horning, E. C.; Finelli, A. F. *OSC* **1963**, *4*, 461. (d) Pakusch, J.; Beckhaus, H.-D.; Rüchardt, C. *CB* **1991**, *124*, 1191.
7. (a) Lehr, F.; Gonnermann, J.; Seebach, D. *HCA* **1979**, *62*, 2258. (b) Seebach, D.; Colvin, E. W.; Lehr, F.; Weller, T. *C* **1979**, *33*, 1.
8. Matsumoto, K.; Suzuki, M.; Miyoshi, M. *JOC* **1973**, *38*, 2094.
9. Kofron, W. G.; Baclawski, L. M. *OSC* , **1988**, *6*, 611.
10. Krapcho, A. P.; Jahngen, E. G. E., Jr.; Kashdan, D. S. *TL* **1974**, 2721.
11. (a) Whitmore, F. C.; Loder, D. J.; *OSC* **1943**, *2*, 282. (b) Bank, S.; Ehrlich, C. L.; Mazur, M.; Zubieta, J. A. *JOC* **1981**, *46*, 1243. (c) Satyanarayana, G.; Sivaram, S. *SC* **1990**, *20*, 3273.
12. (a) Comins, D. L.; Weglarz, M. A.; O'Connor, S. *TL* **1988**, *29*, 1751. (b) Hoppe, D.; Carstens, A.; Krämer, T. *AG(E)* **1990**, *29*, 1424. (c) Kawasaki, T.; Kodama, A.; Nishida, T.; Shimizu, K.; Somei, M. *H* **1991**, *32*, 221.
13. Cookson, R. C.; Gupte, S. S.; Stevens, I. D. R.; Watts, C. T. *OSC* **1988**, *6*, 936.
14. (a) Marsais, F.; Breant, P.; Ginguene, A.; Queguiner, G. *JOM* **1981**, *216*, 139. (b) Chen, L. S.; Chen, G. J.; Ryan, M. T.; Tamborski, C. *JFC* **1987**, *34*, 299. (c) Radinov, R.; Haimova, M.; Simova, E. *S* **1986**, 886. (d) Chen, G. J.; Chen, L. S. *JFC* **1991**, *55*, 119.
15. (a) Colle, T. H.; Lewis, E. S. *JACS* **1979**, *101*, 1810. (b) Syper, L. *Rocz. Chem.* **1973**, *47*, 433 (*CA* **1973**, *79*, 17 981z).
16. (a) Tsuji, J.; Shimizu, I.; Minami, I.; Ohashi, Y.; Sugiura, T.; Takahashi, K. *JOC* **1985**, *50*, 1523. (b) Tsuji, J.; Minami, I. *ACR* **1987**, *20*, 140.
17. Trotta, F.; Tundo, P.; Moraglio, G. *JOC* **1987**, *52*, 1300.
18. Tundo, P.; Trotta, F.; Moraglio, G. *JCS(P1)* **1989**, 1070.
19. Lissel, M.; Schmidt, S.; Neumann, B. *S* **1986**, 382.
20. Sengupta, D.; Venkateswaran, R. V. *JCR(S)* **1984**, 372.
21. (a) Schulz, K.-H.; Heine, H.-G.; Hartmann, W. *OSC* **1990**, *7*, 4. (b) Ito, Y.; Sasaki, A.; Tamoto, K.; Sunagawa, M.; Terashima, S. *T* **1991**, *47*, 2801. (c) Gage, J. R.; Evans, D. A. *OSC* **1993**, *8*, 528.

A. Paul Krapcho
The University of Vermont, Burlington, VT, USA

Diethyl Diazomethylphosphonate[1]

$N_2{=}CH{-}P(=O)(OR)_2$

(R = Et)
[25411-73-8] $C_5H_{11}N_2O_3P$ (MW 178.13)
(R = Me)
[27491-70-9] $C_3H_7N_2O_3P$ (MW 152.09)

(nucleophilic diazo species for the preparation of substituted diazophosphonates,[3,4] β-ketophosphonates,[5] alkynes,[6–8] electron-rich alkenes,[9] various cyclopropane derivatives,[10–12] and five-membered rings[13,14])

Alternate Name: DAMP.
Physical Data: yellow liquid, bp 51 °C/0.1 mmHg; 86–88 °C/0.2 mmHg.
Solubility: sol organic solvents.
Preparative Method: by diazotization of diethyl or dimethyl (aminomethyl)phosphonate using ***Sodium Nitrite*** under acidic conditions.[2]
Handling, Storage, and Precautions: should be stored under an inert atmosphere at 4 °C, and handled as any other diazo compound; *all distillations should be carried out behind a blast shield.*

Introduction. The corresponding dimethyl ester, dimethyl diazomethylphosphonate, is also widely used; this entry discusses both reagents. The acronym DAMP is used for both the ethyl and methyl derivatives.

Preparation of Diazophosphonates. Diethyl diazomethylphosphonate is readily lithiated (***n-Butyllithium***, THF, −100 °C) and the resulting lithio derivative can be acylated (***Phenacyl Bromide***, 33%) (eq 1) or silylated (poor yield).[3] The corresponding silver derivative, formed by treatment with ***Silver(I) Oxide***, can be alkylated with reactive allylic halides.[3] α-Diazo-β-hydroxyphosphonates result when the lithio derivative is treated with aryl aldehydes; alternatively, these can be prepared by direct reaction of DAMP with aldehydes in the presence of ***Triethylamine*** (eq 2).[4]

$$N_2{=}CH{-}PO(OEt)_2 \xrightarrow[\text{PhCOBr, } 33\%]{\text{BuLi, THF, } -100\,^\circ C} PhC(=O){-}C(=N_2){-}PO(OEt)_2 \quad (1)$$

$$N_2{=}CH{-}PO(OR^1)_2 \xrightarrow[\text{Et}_3\text{N, } 22\text{–}50\%]{R^2\text{CHO, ether, } 0\text{–}10\,^\circ C} R^2C(=O){-}C(=N_2){-}PO(OR^1)_2 \quad (2)$$

Preparation of β-Ketophosphonates. In the presence of ***Tin(II) Chloride***, DAMP adds to aldehydes to give β-ketophosphonates in reasonable yield (eq 3).[5]

$$N_2{=}CH{-}PO(OEt)_2 \xrightarrow[\text{CH}_2\text{Cl}_2]{\text{RCHO, SnCl}_2} RC(=O){-}CH_2{-}PO(OEt)_2 \quad (3)$$

R = CH_2CH_2Ph, 75%
R = CHMePh, 56%

Preparation of Alkynes. Diaryl ketones are converted into alkynes by reaction of the lithium (or potassium) salt of DAMP. The reaction presumably involves a Horner–Wadsworth–Emmons-like elimination to give the diazoalkene, followed by aryl migration with loss of nitrogen (eq 4)[6] (see also ***Diazo(trimethylsilyl)methyllithium***).

$$N_2{=}CH{-}PO(OEt)_2 \xrightarrow[\text{THF, } -78\,^\circ C]{Ar_2C{=}O,\ \text{BuLi or } t\text{-BuOK}} Ar_2C(O^-){-}C(=N_2){-}PO(OEt)_2 \longrightarrow Ar_2C{=}C{=}N_2 \longrightarrow Ar{-}C{\equiv}C{-}Ar \quad (4)$$

38–94%

Aldehydes react to give the homologous terminal alkyne by a similar mechanism involving hydrogen migration. The reaction appears to be quite general and has been applied to complex aldehydes for natural product synthesis (eqs 5 and 6).[7,8]

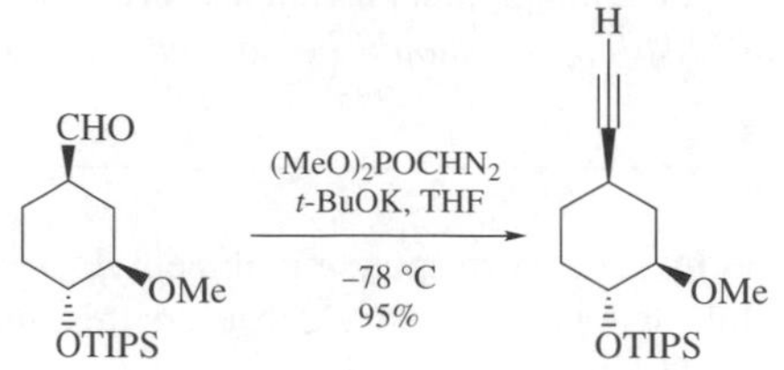

$(MeO)_2POCHN_2$, t-BuOK, THF, −78 °C, 95% (5)

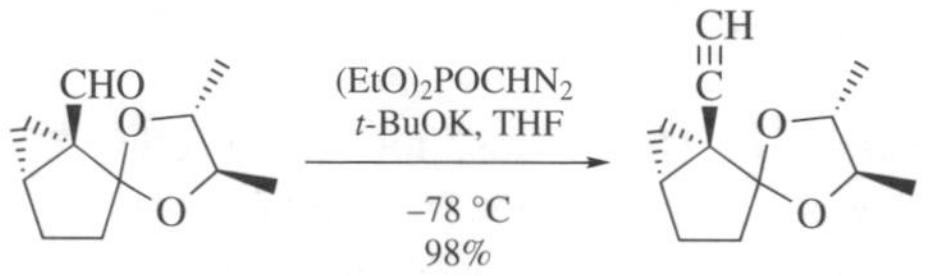

$(EtO)_2POCHN_2$, t-BuOK, THF, −78 °C, 98% (6)

Preparation of Electron-Rich Alkenes. Both acyclic and cyclic ketones are homologated to enol ethers or enamines by reaction with DAMP/***Potassium t-Butoxide*** followed by an alcohol or secondary amine (eqs 7 and 8).[9] The reaction presumably involves a diazoalkene intermediate, but, because alkyl groups migrate poorly, an alkyne is not formed; instead, reaction with the nucleophile occurs.

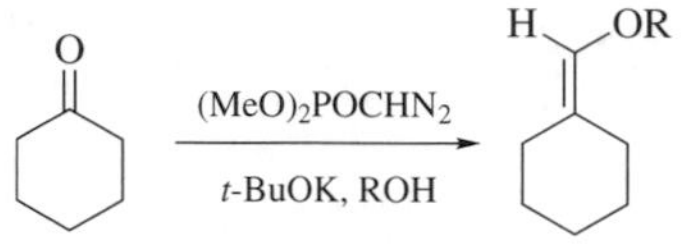

$(MeO)_2POCHN_2$, t-BuOK, ROH (7)

R = Me, 58%
R = t-Bu, 56%

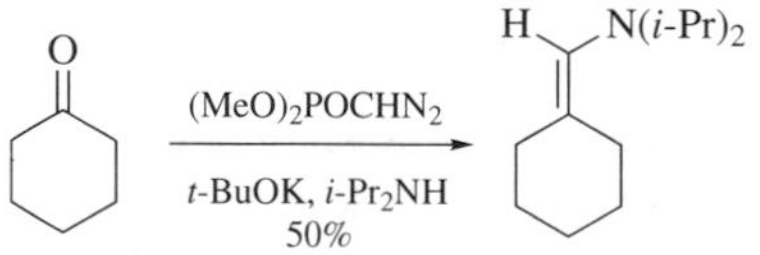

$(MeO)_2POCHN_2$, t-BuOK, i-Pr_2NH, 50% (8)

Preparation of Cyclopropanes. When the above reactions of ketones are carried out in the presence of alkenes, alkylidenecyclopropanes are formed in good yield by addition to the alkylidenecarbene (eq 9).[10]

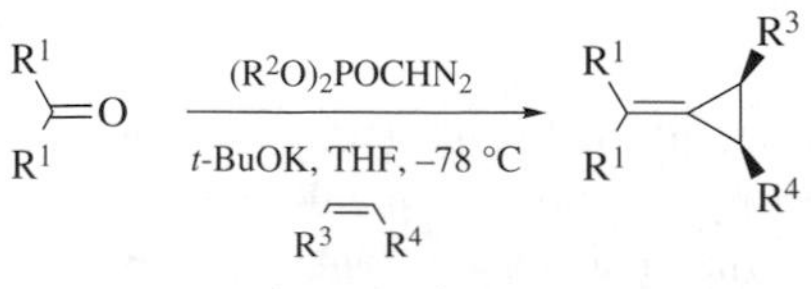

$(R^2O)_2POCHN_2$, t-BuOK, THF, −78 °C (9)

R^1 = Me, $R^3R^4 = (CH_2)_4$, 58%
$R^1 = R^3$ = Me, R^4 = i-Pr, 45%
$R^1R^1 = R^3R^4 = (CH_2)_4$, 70%

The reaction has been extended to α-phenylseleno ketones; in this case the resulting phenylseleno-substituted alkylidenecyclopropanes undergo rearrangement to vinylcyclopropanes.[2c,11]

An alternative route to alkylidenecyclopropanes involves ***Copper*** powder or ***Copper(I) Trifluoromethanesulfonate*** catalyzed reaction of the alkene with DAMP to give cyclopropylphosphonates, followed by Horner–Wadsworth–Emmons reaction with an aldehyde or ketone (eq 10).[2a,12]

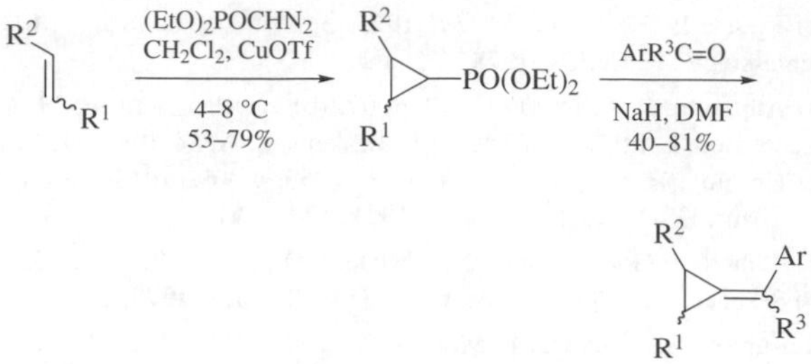

$(EtO)_2POCHN_2$, CH_2Cl_2, CuOTf, 4–8 °C, 53–79%; $ArR^3C=O$, NaH, DMF, 40–81% (10)

Preparation of Five-Membered Rings. The alkylidenecarbenes generated from diazoalkenes (resulting from reaction of ketones with DAMP) undergo intramolecular C–H insertion reactions to give cyclopentenes (eq 11)[13] and dihydrofurans, and hence furans (eq 12).[14]

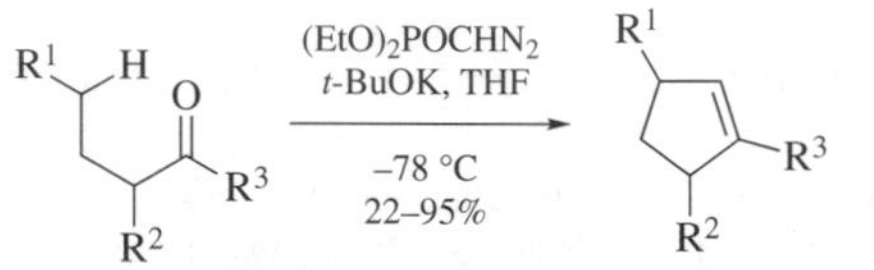

$(EtO)_2POCHN_2$, t-BuOK, THF, −78 °C, 22–95% (11)

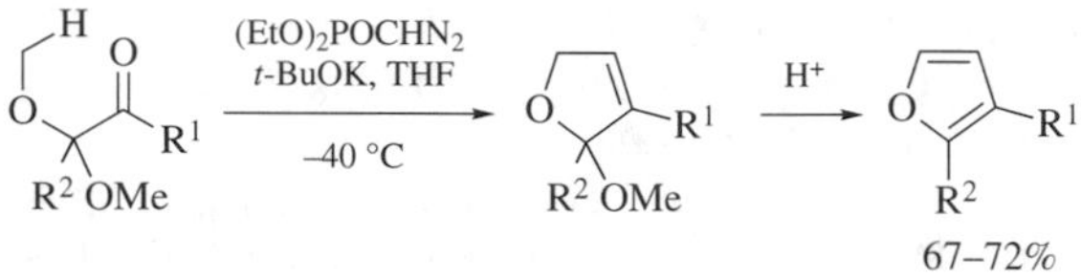

$(EtO)_2POCHN_2$, t-BuOK, THF, −40 °C; H^+, 67–72% (12)

Related Reagents. Diazo(trimethylsilyl)methyllithium; Ethyl Diazoacetate; Triphenylphosphine–Carbon Tetrabromide.

1. For general reviews, see: (a) Regitz, M. *AG(E)* **1975**, *14*, 222. (b) Regitz, M.; Maas, G. *Diazo Compounds. Properties and Synthesis*; Academic: Orlando, 1986.
2. (a) Seyferth, D.; Marmor, R. S.; Hilbert, P. *JOC* **1971**, *36*, 1379. (b) Regitz, M.; Liedhegener, A.; Eckstein, U.; Martin, M.; Anschütz, W. *LA* **1971**, *748*, 207. (c) Lewis, R. T.; Motherwell, W. B. *T* **1992**, *48*, 1465.
3. Regitz, M.; Weber, B.; Eckstein, U. *LA* **1979**, 1002.
4. Disteldorf, W.; Regitz, M. *CB* **1976**, *109*, 546.
5. Holmquist, C. R.; Roskamp, E. J. *TL* **1992**, *33*, 1131.
6. Colvin, E. W.; Hamill, B. J. *JCS(P1)* **1977**, 869.
7. Nakatsuka, M.; Ragan, J. A.; Sammakia, T.; Smith, D. B.; Uehling, D. E.; Schreiber, S. L. *JACS* **1990**, *112*, 5583.
8. Kabat, M.; Kiegiel, J.; Cohen, N.; Toth, K.; Wovkulich, P. M.; Uskoković, M. R. *TL* **1991**, *32*, 2343.
9. (a) Gilbert, J. C.; Weerasooriya, U. *TL* **1980**, *21*, 2041. (b) Gilbert, J. C.; Weerasooriya, U. *JOC* **1983**, *48*, 448. (c) Gilbert, J. C.; Weerasooriya, U.; Wiechman, B.; Ho, L. *TL* **1980**, *21*, 5003. (d) Gilbert, J. C.; Senaratne, K. P. A. *TL* **1984**, *25*, 2303.
10. Gilbert, J. C.; Weerasooriya, U.; Giamalva, D. *TL* **1979**, 4619.
11. Lewis, R. T.; Motherwell, W. B. *CC* **1988**, 751.
12. Lewis, R. T.; Motherwell, W. B. *TL* **1988**, *29*, 5033.
13. Gilbert, J. C.; Giamalva, D. H.; Weerasooriya, U. *JOC* **1983**, *48*, 5251.
14. Buxton, S. R.; Holm, K. H.; Skattebøl, L. *TL* **1987**, *28*, 2167.

Christopher J. Moody
Loughborough University of Technology, UK

Diethyl Malonate[1]

($\mathbf{1}$; $R^1 = R^2 = Et$)
[105-53-3] $C_7H_{12}O_4$ (MW 160.17)
($\mathbf{2}$; $R^1 = R^2 = Me$)
[108-59-8] $C_5H_8O_4$ (MW 132.12)
($\mathbf{3}$; $R^1 = R^2 = t\text{-Bu}$)
[541-16-2] $C_{11}H_{20}O_4$ (MW 216.28)
($\mathbf{4}$; $R^1 = t\text{-Bu}$, $R^2 = Me$)
[42726-73-8] $C_8H_{14}O_4$ (MW 174.20)
($\mathbf{5}$; $R^1 = t\text{-Bu}$, $R^2 = Et$)
[32864-38-3] $C_9H_{16}O_4$ (MW 188.22)
($\mathbf{6}$; $R^1 = R^2 = Bn$)
[15014-25-2] $C_{17}H_{16}O_4$ (MW 284.31)
($\mathbf{7}$; $R^1 = Bn$, $R^2 = Me$)
[52267-39-7] $C_{11}H_{12}O_4$ (MW 208.21)
($\mathbf{8}$; $R^1 = Bn$, $R^2 = Et$)
[42998-51-6] $C_{12}H_{14}O_4$ (MW 222.24)

(two- or three-carbon nucleophiles in enolate or enol alkylation, conjugate addition, and various condensation reactions[1])

Physical Data: (**1**) mp −50 °C; bp 199 °C; d^{20} 1.055 g cm^{-3}; n_D^{20} 1.4143; dipole moment 1.57 D. (**2**) mp −62 °C; bp 181 °C; d^{20} 1.154 g cm^{-3}; n_D^{20} 1.4140; dipole moment 2.39 D. (**3**) bp 112–115 °C/31 mmHg; d^{20} 0.965 g cm^{-3}. (**4**) bp 80 °C/11 mmHg; d^{20} 1.030 g cm^{-3}. (**5**) bp 83–85/8 mmHg; d^{20} 1.001 g cm^{-3}. (**6**) bp 188 °C/0.2 mmHg; d^{20} 1.158 g cm^{-3}. (**7**) bp 125 °C/0.5 mmHg; d^{20} 1.150 g cm^{-1}. (**8**) bp 138–139 °C; d^{20} 1.087 g cm^{-3}.

Solubility: (**1**) and (**2**) sparingly sol water; miscible in all proportions with ether and alcohol.

Form Supplied in: dimethyl and diethyl malonates are colorless liquids with a minimum assay of 99% (GC) which are widely available.

Preparative Methods: dibenzyl malonate is prepared by refluxing malonic acid with benzyl alcohol with catalytic amount of sulfuric acid.[2] Di-*t*-butyl malonate is usually obtained from malonic acid and isobutene with catalytic amount of sulfuric acid.[3] *t*-Butyl methyl and *t*-butyl ethyl malonates can be obtained through reaction of methyl and ethyl malonyl chlorides with *t*-butyl alcohol, e.g. over activated alumina as a catalyst.[4a] Unsymmetrical methyl, ethyl, and benzyl malonates may be prepared by reaction of the corresponding monoalkyl malonates with alkyl chloroformates.[4b]

Handling, Storage, and Precautions: no specific health hazard if handled with the usual precaution.

Introduction. The principal synthetic utility of the malonate esters may be summarized as follows:[1]

1) the acidity of the methylene group (pK_a ~13) is such that metal salts can be easily formed with the metal alkoxides. The resulting carbanions undergo acylation, alkylation,[5] aldol,[6] and Michael reactions;[7]
2) the possibility of hydrolyzing and subsequently decarboxylating only one of the ester functions;
3) the usual reactions of the ester functions.

In many syntheses reported in the literature, malonates react at both the methylene group and the ester functions, which make them very useful reagents, especially in the formation of heterocycles.

The main feature of di-*t*-butyl, dibenzyl, and the *t*-butyl and benzyl alkyl malonates is that the *t*-butyl ester and benzyl ester groups can easily be cleaved by hydrolysis/decarboxylation or by hydrogenolysis/decarboxylation, respectively. Selective removal of *t*-butyl and benzyl esters may also be accomplished with ***Iodotrimethylsilane***.[8] Cleavage and decarboxylation of dimethyl, diethyl, and unsymmetrical methyl and ethyl malonates can be effected by S_N2 attack on the methyl or ethyl ester carbon with strong nucleophilic anions such as chloride, iodide, cyanide, alkylthiolates, and phenylselenate.[9]

Acylation. Acylation of diethyl malonate followed by partial hydrolysis and decarboxylation is one of the methods used for synthesizing β-keto esters.[10] Nevertheless, further malonate derivatives such as Meldrum's acid (***2,2-Dimethyl-1,3-dioxane-4,6-dione***),[11a] and mixed malonates such as *t*-butyl ethyl malonate[11b] or potassium ethyl malonate,[11c] can also be advantageously used. The latter allows the synthesis of β-keto esters in high purity without the partial hydrolysis step so that they can be used without isolation in subsequent transformations to, for example, quinolone moieties (eq 1).[11c]

1. MeCN, $MgCl_2$, Et_3N, 10–25 °C; 2. HCl
1. $HC(OEt)_3$, Ac_2O; 2. cyclopropylamine, DMSO, 78%

(1)

Acylation of dibenzyl malonate[12a] or di-*t*-butyl malonate[12b] followed by hydrogenolysis/decarboxylation (eq 2) or hydrolysis/decarboxylation (eq 3) of the acylmalonates yields methyl ketones.

1. Na; 2. R^1COCl; hydrogenolysis (H_2, Pd); decarboxylation; + 2 CO_2 (2)

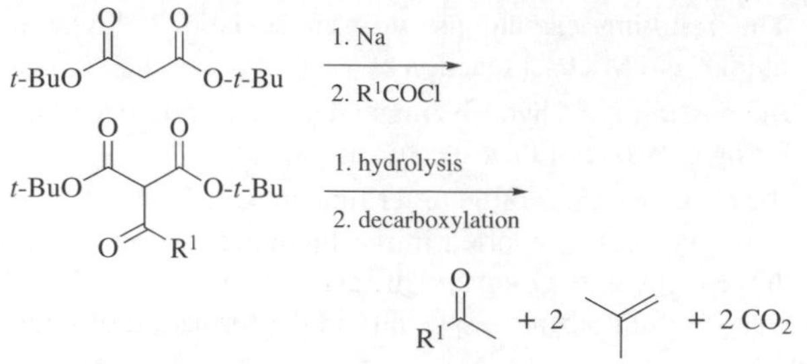

(3)

A similar synthesis of methyl 7-(2-hydroxy-5-oxo-1-cyclopentenyl)heptanoate has been reported (eq 4).[13] Condensation of orthoformates with malonates affords alkoxymethylene malonates which are widely used for heterocyclic synthesis.

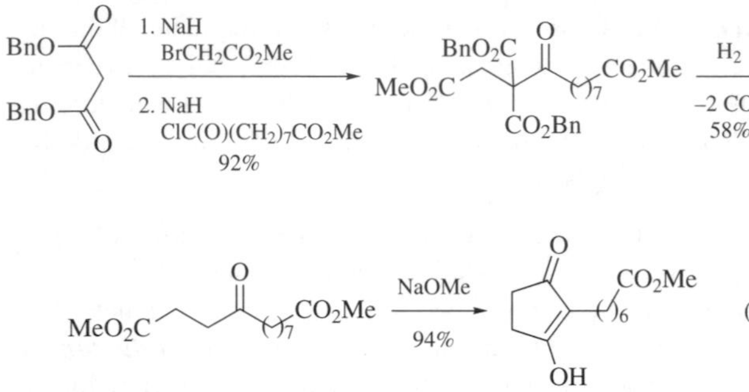

(4)

Alkylation.[1,5] Both α-hydrogens can be replaced by alkyl substituents (eq 5).[14]

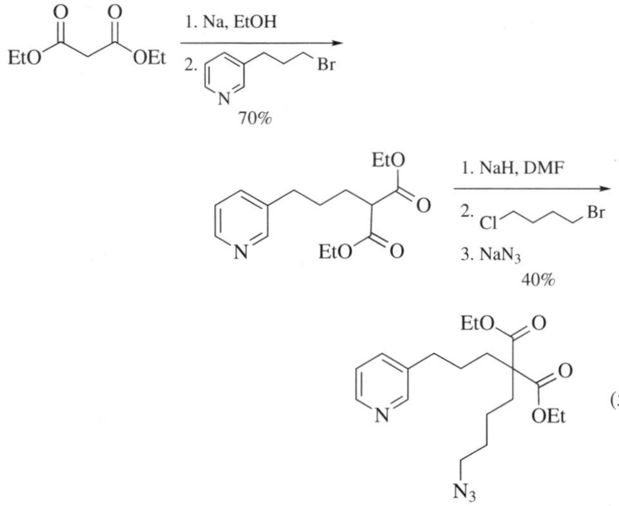

(5)

Mono- and dialkyl malonates can be further modified by partial hydrolysis and decarboxylation of an ester function,[9,14] or by reduction of both ester functions with ***Lithium Aluminum Hydride*** to give 2-substituted 1,3-propanediols (eq 6).[15] A useful synthesis of primary α-methylene alcohols involves $LiAlH_4$ reduction of the sodium salts formed from α-alkyl malonates (eq 7).[16] Cyclocondensation with *N*-ethylaniline affords a 4-hydroxy carbostyril (eq 8).[17]

α,α-Dialkylation of malonates with dihaloalkanes and intramolecular α-alkylation with α-(ω-haloalkyl)malonates provides effective approaches for construction of carbocyclic rings containing 3–21 carbon atoms.[18] Alkylation of sodio diethyl malonate with chiral 2,3-epoxybutane affords enantiomerically pure (>99% ee) *cis*-β,γ-dimethyl-γ- butyrolactone (eq 9).[19] Although α-alkylation of malonate carbanions with tertiary alkyl halides cannot usually be effected in practically useful yields owing to competing elimination, *t*-butylation of diethyl malonate could be accomplished in 56% yield under Lewis acidic conditions (BF_3, $CS_2/ClCH_2CH_2Cl$, Δ, 18 h), presumably via an S_N1-type mechanism involving the *t*-butyl carbenium ion (eq 10).[20]

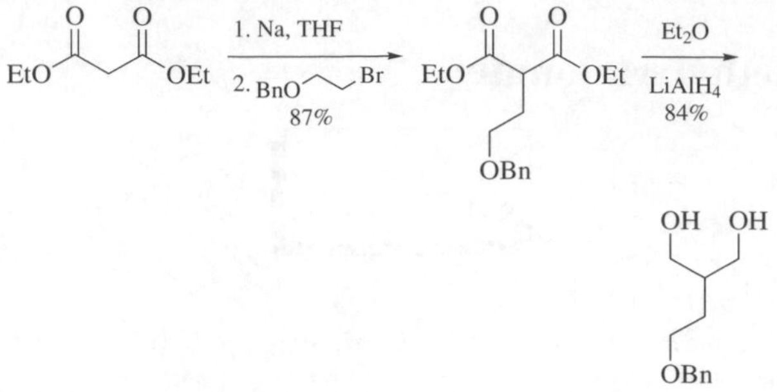

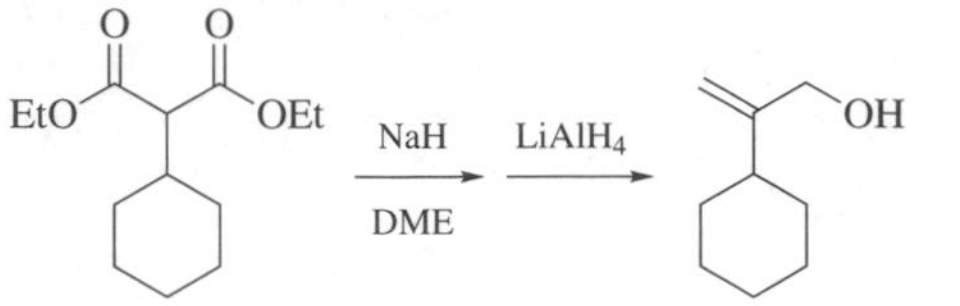

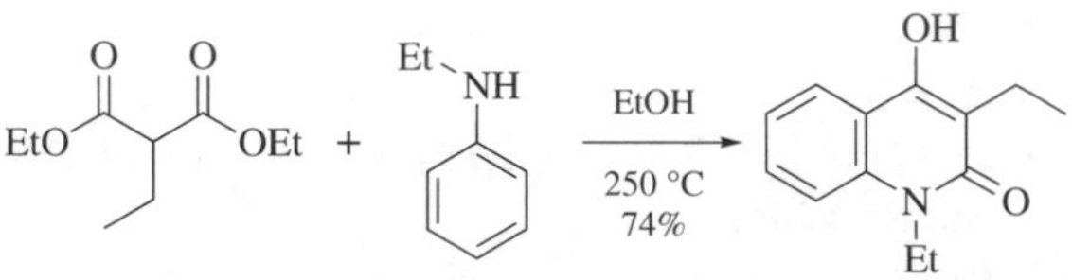

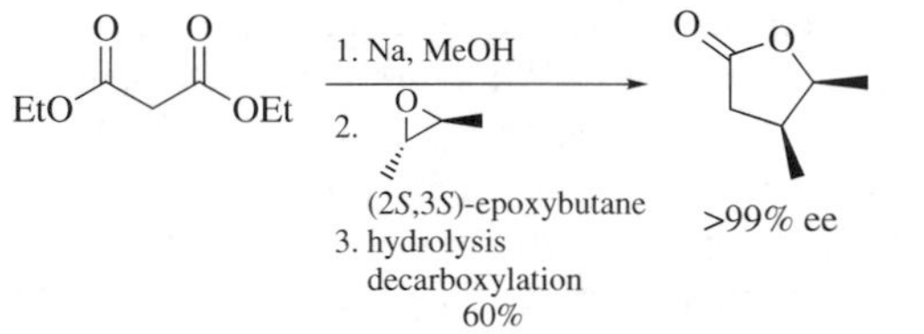

(6)

(7)

(8)

(9)

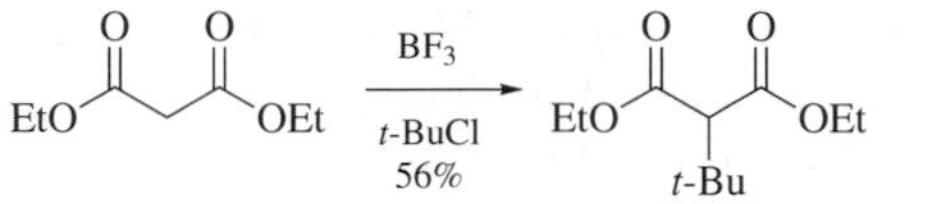

(10)

An important advance in the potential applications of malonate chemistry is the development of efficient procedures for α-allylation of malonate anions with allylic esters and lactones as well as vinyl epoxides (e.g. cyclopentadiene monoepoxide) by means of Pd^0 catalysts (eq 11).[21,22]

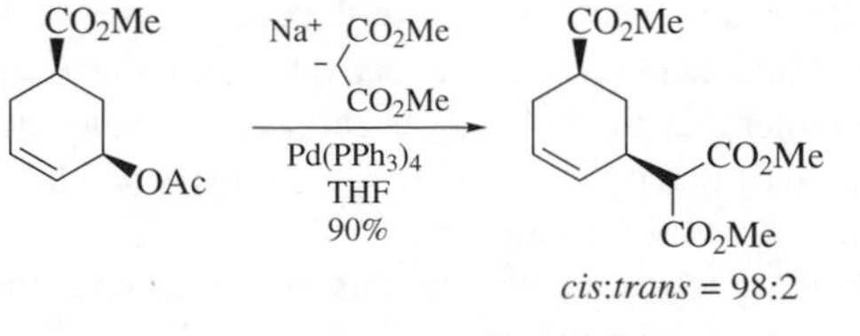

(11)

cis:*trans* = 98:2

Less economically attractive, but still quite useful, are the allyl coupling reactions that occur between π-allylpalladium complexes and malonate anions. The former Pd^0-catalyzed allylations usually occur with retention of the C–O bond configuration as a consequence of the two configurational inversions which occur during the formation of the π-allylpalladium intermediate and its subsequent displacement by the malonate group. It should be noted that mixtures of allylic isomers are often formed from unsymmetrical allylic esters owing to malonate attack at both C-1 and C-3 of the π-allylpalladium intermediate.

α-Alkylation of malonate esters by monosubstituted alkenes and isobutene occurs in the simultaneous presence of ***Manganese(III) Acetate*** (limiting stoichiometric oxidant) and catalytic amounts of ***Copper(II) Acetate***. The resulting α-alkenylmalonates (40–70% based on Mn^{III}) undergo hydrolysis and decarboxylation to γ,δ-unsaturated carboxylic acids.[23] α-Arylation of α-alkylmalonate diesters may be accomplished by means of coupling reactions with aryllead triacetates [$ArPb(OAc)_3$].[24] However, the unsubstituted parent malonates do not undergo arylation under the same conditions.

Knoevenagel Reactions. Such reactions[6] with aldehydes or ketones are frequently followed by cyclocondensation at one of the ester functions (eq 12).[25] When reacted with acetaldehyde, diethyl ethylidenemalonate is obtained which in turn can be ozonolyzed to diethyl oxomalonate (eq 13), a versatile reagent for organic synthesis.[26]

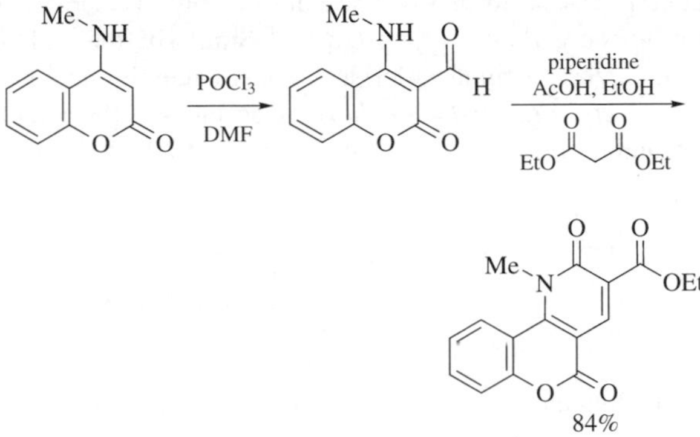

$POCl_3$, DMF; piperidine, AcOH, EtOH, EtO(O)CCH₂C(O)OEt; 84% (12)

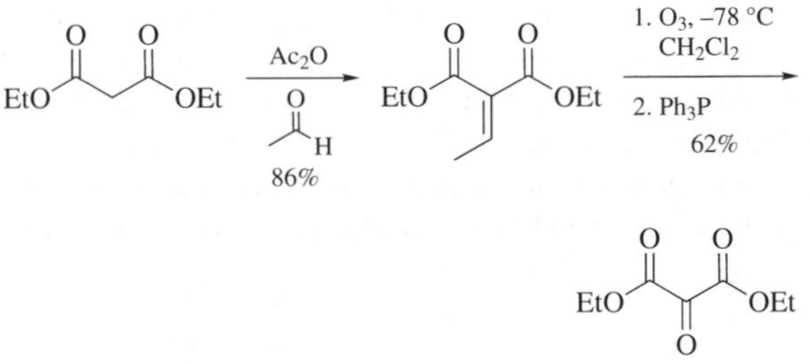

Ac_2O, CH_3CHO, 86%; 1. O_3, –78 °C, CH_2Cl_2; 2. Ph_3P, 62% (13)

Michael Reaction. The Michael addition[7] to activated double bonds such as those of α,β-unsaturated carbonyl compounds is frequently described in the literature. Followed by a ring-closing Claisen condensation and a subsequent decarboxylation, this sequence leads to the formation of 1,3-cyclohexanedione derivatives (eq 14).[27] They can further be aromatized to substituted resorcinols,[28] reduced to 5-substituted 2-cyclohexenones,[29] or involved in cyclocondensation reactions.[27]

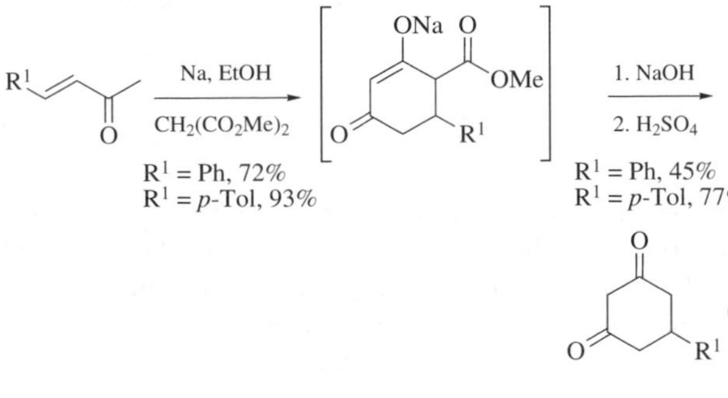

Na, EtOH, $CH_2(CO_2Me)_2$: R^1 = Ph, 72%; R^1 = *p*-Tol, 93%; 1. NaOH; 2. H_2SO_4: R^1 = Ph, 45%; R^1 = *p*-Tol, 77% (14)

In the case of β,β-bis(trifluoromethyl)acrylic esters as starting material, an anti-Michael addition followed by a fluoride elimination has been reported (eq 15).[30]

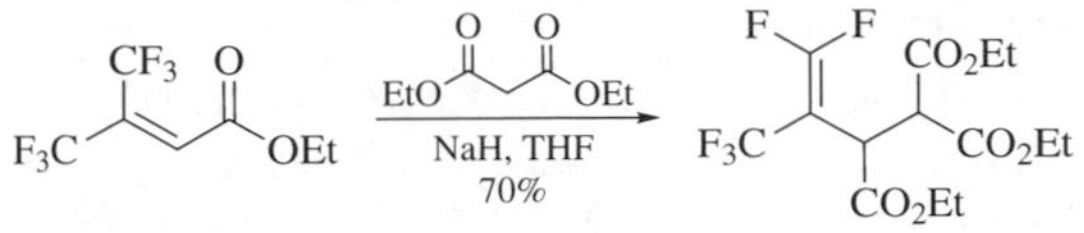

NaH, THF, 70% (15)

Other Reactions. A further reaction at the central methylene group is nitrosation with ***Sodium Nitrite***, giving oximinomalonates which are subsequently reduced to the corresponding 2-aminomalonates (eq 16).[31]

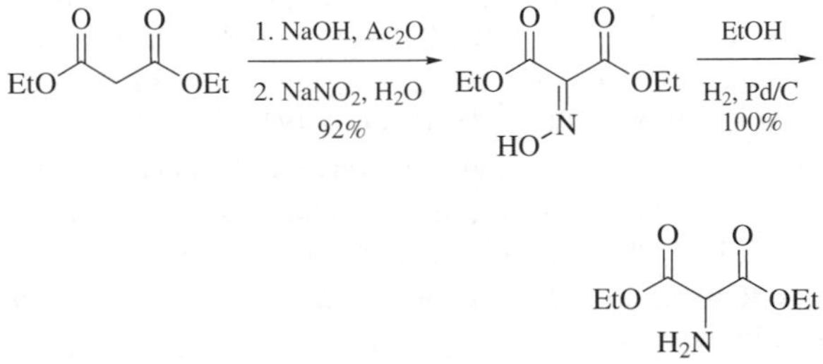

1. NaOH, Ac_2O; 2. $NaNO_2$, H_2O, 92%; EtOH, H_2, Pd/C, 100% (16)

Usual reactions at the ester functions include, for example, transesterification to mixed malonates like benzyl ethyl malonate[2] or base-catalyzed cyclocondensation with ureas, thioureas, guanidines, or amidines yielding the corresponding 2-substituted pyrimidines.[32]

Related Reagents. Bis(trimethylsilyl) Malonate; Diethyl Ethoxymethylenemalonate; Diethyl Ethylidenemalonate; Diethyl Ethoxymagnesiomalonate; Diethyl Methylenemalonate; Diethyl Oxomalonate; 2,2-Dimethyl-1,3-dioxane-4,6-dione; Ethyl Malonate; Ethyl Trimethylsilyl Malonate; Magnesium Ethyl Malonate; Manganese(III) Acetate; Tetrakis(triphenylphosphine)palladium(0).

1. (a) *Ullmann's Encyclopedia of Industrial Chemistry*, 5th ed.; VCH: Weinheim, 1990; Vol. A16, pp 63–75. (b) Hughes, D. W. In *Kirk-Othmer Encyclopedia of Chemical Technology*, 3rd ed.; Wiley: New York, 1976; Vol. 14, pp 794–804. (c) House, H. O. *Modern Synthetic Reactions*; Benjamin: Menlo Park, CA, 1972. (d) Stowell, J. C. *Carbanions in Organic Synthesis*; Wiley: New York, 1979; pp 192–197. (e) *FF* **1967**, *1*, 627, 887, 1069, 1251, 1268. (f) Henecka, H. *Chemie der Beta-Dicarbonyl Verbindungen*; Springer: Berlin, 1950.
2. Baker, B. R.; Schaub, R. E.; Querry, M. V.; Williams, J. A. *JOC* **1952**, *17*, 77.
3. MacCloskey, A. L.; Fonken, G. S.; Kluiber, R. W.; Johnson, W. S. *OSC* **1963**, *4*, 261.
4. (a) Nagasawa, K.; Yoshitake, S.; Amiya, T.; Ito, K. *SC* **1990**, *20*, 2033. (b) Gutman, A. L.; Boltanski, A. *TL* **1985**, *26*, 1573.
5. Cope, A. C.; Holmes, H. L.; House, H. O. *OR* **1957**, *9*, 107.
6. (a) Tietze, L. F.; Beifuss, U. *COS* **1991**, *2*, 341. (b) Tietze, L. F.; Beifuss, U. *OS* **1993**, *71*, 167. (c) Lehnert, W. *TL* **1970**, 4723.
7. (a) Jung, M. E. *COS* **1991**, *4*, 1. (b) Bergman, E. D.; Ginsburg, D.; Pappo, R. *OR* **1959**, *10*, 179.
8. (a) Groutas, W. C.; Felker, D. *S* **1980**, 861. (b) Jung, M. E. *JACS* **1977**, *99*, 968. (c) Ho, T.-L.; Olah, G. A. *AG(E)* **1976**, 749; *S* **1977**, 917.
9. (a) Krapcho, A. P. *S* **1982**, *805*, 893. (b) McMurry, J. *OR* **1976**, *24*, 187.
10. Pollet, P. L. *J. Chem. Educ.* **1983**, *60*, 244.

11. (a) Oikawa, Y.; Sugano, K.; Yonemitsu, O. *JOC* **1978**, *43*, 2087. (b) Pichat, L.; Beaucourt, J. P. *S* **1973**, 537. (c) Clay, R. J.; Collom, T. A.; Karrick, G. L.; Wemple, J. *S* **1993**, *3*, 290.
12. (a) Bowman, R. E. *JCS* **1950**, 325. (b) Fonken, G. S.; Johnson, W. S. *JACS* **1952**, *74*, 831.
13. Naora, H.; Ohnuki, T.; Nakamura, A. *BCJ* **1988**, *61*, 993.
14. Bhagwat, S. S.; Gude, C.; Boswell, C.; Contardo, N.; Cohen, D. S.; Dotson, R.; Mathis, J.; Lee, W.; Furness, P.; Zoganas, H. *JMC* **1992**, *35*, 4373.
15. Rastetter, W. H.; Phillion, D. P. *JOC* **1981**, *46*, 3204.
16. (a) Marshall, J. A.; Anderson, N. H.; Hochstetter, A. R. *JOC* **1967**, *32*, 113. (b) Corey, E. J.; Helquist, P. *TL* **1975**, 4091.
17. Stadlbauer, W.; Laschober, R.; Lutschounig, H.; Schindler, G.; Kappe, T. *M* **1992**, *123*, 617.
18. (a) Casadei, M. A.; Galli, C.; Mandolini, L. *JACS* **1984**, *106*, 1051. (b) Knipe, A. C.; Stirling, C. J. *JCS(B)* **1968**, 67. (c) Ref. 1c, pp 541–543.
19. Hedenström, E.; Högberg, H. E.; Wassgren, A. B.; Bergström, G.; Lötqvist, J.; Hansson, B.; Anderbrant, O. *T* **1992**, *48*, 3139.
20. Boldt, P.; Militzer, H.; Thielecke, W.; Schulz, L. *LA* **1968**, *718*, 101.
21. (a) Trost, B. M.; Verhoeven, T. R. *JACS* **1980**, *102*, 4730. (b) Trost, B. M. *JOM* **1986**, *300*, 263; *Chemtracts–Org. Chem.* **1988**, *1*, 415.
22. Heck, R. F. *Palladium Reagents in Organic Synthesis*; Academic: New York, 1985; pp 130–154.
23. Nikishin, G. I.; Vinogradov, M. G.; Fedorova, T. M. *CC* **1973**, 693.
24. Pinhey, J. T.; Rowe, B. A. *TL* **1980**, *21*, 965.
25. Ivanov, I. C.; Karagiosov, S. K.; Simeonov, M. F. *LA* **1992**, 203.
26. Jung, M. E.; Shishido, K.; Davis, L. H. *JOC* **1982**, *47*, 891.
27. Kesten, S. J.; Degnan, M. J.; Hung, J.; McNamara, D. J.; Ortwine, D. F.; Uhlendorf, S. E.; Werbel, L. M. *JMC* **1992**, *35*, 3429.
28. Kotnis, A. S. *TL* **1991**, *32*, 3441.
29. Hataba, H. M.; Sayed, M. A. *Egypt. J. Chem.* **1989**, *32*, 195.
30. Martin, V.; Molines, H.; Wackselman, C. *JOC* **1992**, *57*, 5530.
31. May, D. A., Jr.; Lash, T. D. *JOC* **1992**, *57*, 4820.
32. Brown, D. J. *The Chemistry of Heterocyclic Compounds: The Pyrimidines*, Weissberger, A., Ed.; Interscience: New York, 1962; Suppl. I, 1970; Suppl. II, 1985.

Gérard Romeder
Lonza, Basel, Switzerland

Diethylzinc[1]

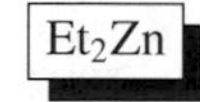

[557-20-0] $C_4H_{10}Zn$ (MW 123.50)

(useful organometallic reagent for organic synthesis;[1] polymerization catalyst)

Physical Data: mp −28 °C; bp 118 °C; bp 27 °C/30 mmHg; *d* 1.187 g cm^{-3}.

Solubility: sol most organic solvents; reacts with acidic hydrogens; pyrophoric; reacts violently with water.

Form Supplied in: colorless liquid; available as 1.0 M solution in hexanes and 1.1 M solution in toluene, or neat in a metal cylinder.

Analysis of Reagent Purity: can be titrated by iodometric methods[4] or by complexometry.[5]

Preparative Methods: prepared by the reaction of ***Zinc–Copper Couple*** with ***Ethyl Iodide*** followed by distillation (81–84%).[2] It can also be prepared via a B/Zn exchange reaction.[3]

Handling, Storage, and Precautions: highly flammable liquid; spontaneous ignition in air; reacts violently with water. Should be kept under an inert atmosphere.[6] Neat diethylzinc is best destroyed by first diluting with THF, followed by slow addition of ethanol at 0 °C.

Addition to Carbonyl Compounds. Diethylzinc, like other dialkylzincs, reacts only slowly with carbonyl compounds. ***Carbon Dioxide*** can be used as protecting gas for reactions performed with diethylzinc and the formation of a zinc propionate is observed only under a high pressure of CO_2 and at higher temperatures.[7] Whereas the reaction of ***Acetaldehyde*** with neat diethylzinc is complete within a few hours, the reaction with higher homologs requires several days.[8] Higher diorganozinc derivatives do not add to aldehydes in solution. The addition of magnesium salts to diethylzinc considerably accelerates the addition to the carbonyl group (eq 1).[9] Similarly, the addition of a tetraalkylammonium chloride or of a catalytic amount of ***N,N,N′,N′-Tetramethylethylenediamine*** accelerates the addition reaction and limits the formation of reduction product.[10]

$$Et_2Zn + MeCHO \xrightarrow{\text{ether}} \text{EtCH(OH)Me} \quad (1)$$

without $MgBr_2$, 5%
with $MgBr_2$, 60%

Neat diethylzinc does not add to ketones and its reaction with benzophenone produces only diphenylmethanol.[11] As in the case of aldehydes, the addition of diethylzinc is accelerated by the presence of magnesium salts.[9] With 1,2-diketones, diethylzinc undergoes an addition in the reverse way to the carbonyl group, producing an α-alkoxy ketone (eq 2).[12] A similar reaction with α-diimines leads to zincated enamines which can be converted to indolizidines after the addition of an aldehyde and heating (eq 3).[13]

$$\text{PhCOCOPh} \xrightarrow{Et_2Zn} \text{PhCH(OEt)COPh} \quad (2)$$

$$t\text{-Bu-N=CH-CH=N-}t\text{-Bu} \xrightarrow[\text{2. 2-pyridinecarboxaldehyde}]{\text{1. } Et_2Zn} \text{zinc chelate} \xrightarrow{\text{heat}} \text{indolizine} \quad (3)$$

70%

The presence of a heteroatom at the α-position to the carbonyl group accelerates the addition and diethylzinc adds to the aldehyde (**1**) via a chelate controlled pathway producing the alcohol (**2**), which was further converted to *exo*-brevicomin (eq 4).[14] The addition of diethylzinc to (−)-menthyl phenylglyoxylate gives, after saponification, an α-hydroxy acid with a good diastereoselectivity (eq 5).[15] The low reactivity of diethylzinc with aldehydes has been advantageously exploited by performing the addition reaction in the presence of a chiral catalyst.[16]

(4): (1) → (2); Et_2Zn, ether, 0–25 °C, 5 h, 76%

(5): 1. Et_2Zn; 2. H_3O^+, >80%; 71–75% de

Mukaiyama showed that the chiral proline derivative (**3**) catalyzes the addition of diethylzinc to benzaldehyde, unfortunately without any asymmetric induction.[17] Oguni found a few years later that (*S*)-leucinol (**4**) also catalyzes the reactions and the same addition proceeds with 48% ee.[18] Since this discovery, a variety of amino alcohols were found to catalyze the addition of diethylzinc to aldehydes with high enantioselectivity (up to 99% ee).[16,19–34] (−)-3-*Exo*-(dimethylamino)isoborneol, (−)-DAIB (**5**), has a particularly high catalytic activity. In the presence of 2 mol % of this catalyst, the addition of diethylzinc to benzaldehyde in toluene is complete within 6 h at 0 °C (98% ee).[20] This catalyst is effective for aromatic aldehydes but gives lower enantioselectivities with aliphatic aldehydes. A series of readily available pyrrolidinylmethanols (**6**), derived from proline, give excellent results both with aliphatic and aromatic aldehydes (eq 6).[21] Remarkably, chiral quaternary ammonium salts such as (**7**) catalyze in the solid state the addition of benzaldehyde to diethylzinc with a good enantioselectivity (eq 7).[22] Chiral metal complexes can also be used as catalysts, and the reaction of diethylzinc with aldehydes in the presence of a chiral oxazoborolidine produces ethylated secondary alcohols in 52–95% ee.[23]

(3) (4) (5) (6)

(6): Et_2Zn, 2 mol% (**6**), 100%; 98% ee

(7): PhCHO → Et_2Zn, hexane, 3 d; $Bn(Me)_2N^+$ X^- (**7**), 6 mol%, 90%; 74% ee

Very active titanium catalysts (**8**) and (**9**) have been developed respectively by Yoshioka and Ohno,[24] and Seebach.[25] Their high catalytic activity allows the extension to the enantioselective addition of higher dialkylzincs (eq 8)[25] and to both functionalized aldehydes and dialkylzincs (eqs 9–11).[26,27] The excellent functional group tolerance allows the use of many functionalized aldehydes.

(8) (9)

(8): $Ti(O\text{-}i\text{-}Pr)_4$, 10 mol% (**9**), 82%; 97% ee

(9): $(AcO(CH_2)_n)_2Zn$; $Ti(O\text{-}i\text{-}Pr)_4$, 8 mol% (**8**), 83%; 97% ee

(10): Et_2Zn, toluene; $Ti(O\text{-}i\text{-}Pr)_4$, 8 mol% (**8**), 90%; 95% ee

(11): $(Cl(CH_2)_4)_2Zn$, toluene; $Ti(O\text{-}i\text{-}Pr)_4$, 8 mol% (**8**), 69%; 95% ee

A new method for the enantioselective synthesis of lactones is based on the enantioselective addition of 3- or 4-formyl esters in the presence of catalytic amounts of the amino alcohol (**10**) (eq 12).[28] γ-Hydroxy ketones can be prepared enantioselectively by the addition of diethylzinc to the ketoaldehyde (**11**) (eq 13).[29] Optically active hydroxy aldehydes are obtained with high enantioselectivity by the addition of diethylzinc to monoprotected dialdehydes (eq 14).[30] A variety of other functionalized aldehydes have been ethylated under similar conditions (eqs 15 and 16).[31]

(12): 1. Et_2Zn, Bu_2N–OH cat. (**10**); 2. NaOH; 3. H^+; 82%; 88% ee

(13): (**11**), Et_2Zn, *ent*-(**10**) cat, 52%; 87% ee

(14): Et_2Zn, (**10**) cat; H_3O^+; 76%; 94% ee

(15): Et_2Zn, (**10**) cat; 74% ee

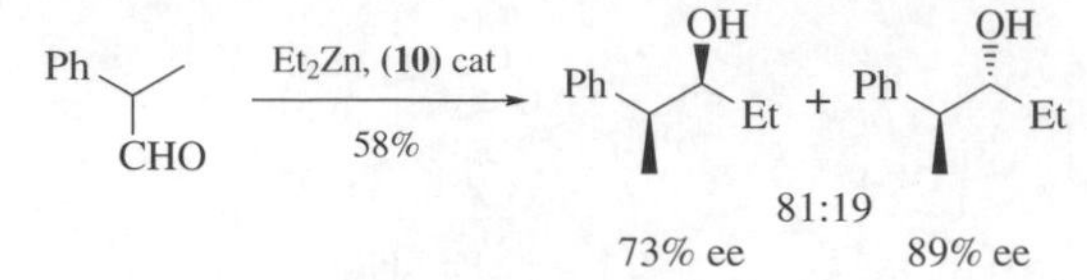

(eq 16)

The addition of diethylzinc to activated imines such as *N*-diphenylphosphinoylimines of type (**12**) proceeds in good yields and excellent enantioselectivity, allowing the synthesis of enantiomerically enriched secondary amines (eq 17).[32a] An enantioselective ethylation on *N*-(amidobenzyl)benzotriazoles catalyzed by chiral amino alcohols is also possible.[32b]

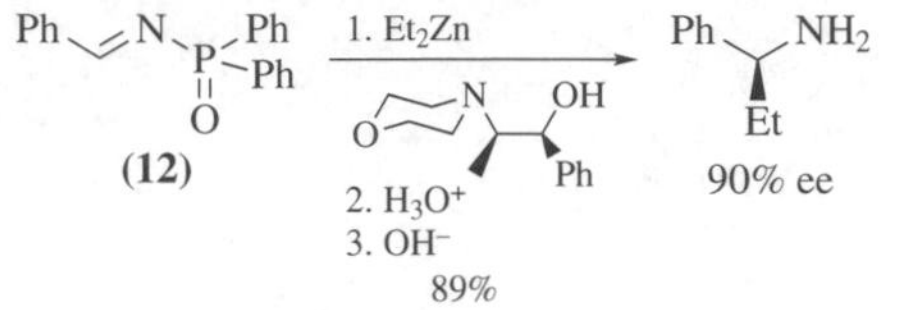

(eq 17)

Polymer-bound chiral catalysts[33] and chiral catalysts supported an silica gel and alumina[34] have been successfully employed for the enantioselective addition of diethylzinc to aldehydes. Little success was obtained in the addition of diethylzinc to esters or lactones. However, the addition to ***Diethyl Oxalate*** produces ethyl 2-ethyl-2-hydroxybutanoate in 75% yield.[9] The reaction with acid chlorides[1,35] or anhydrides[1,36] proceeds in some cases satisfactorily;[35,36] however, Pd^0 [37] or Cu^I [26,38] catalysis gives more predictable results and allows the extension of the reaction to higher functionalized homologs (eq 18).[26,38]

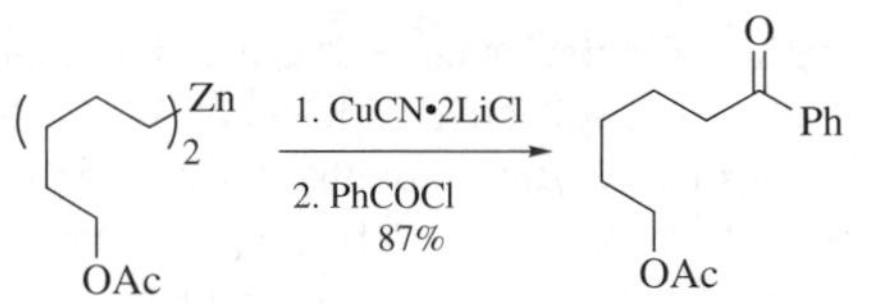

(eq 18)

Addition of Diethylzinc to Double or Triple Bonds. Diethylzinc, like higher homologs, is able to add to activated alkynes such as propiolic esters in the presence of a THF soluble copper salt such as $CuCN{\cdot}2LiCl$ (eq 19).[26] Less reactive alkynes such as 1-(methylthio)-1-hexyne or phenylacetylene also react with the reagent $EtCu(CN)ZnEt{\cdot}2LiCl$, leading stereospecifically to the *syn* adducts (eq 20).[39] Diethylzinc can be added to unfunctionalized alkynes in the presence of $ZrCp_2I_2$.[40] In the case of terminal alkynes, the addition shows a low regioselectivity, producing a mixture of alkenylzinc species (eq 21).[40] Remarkably, the addition to internal alkynes proceeds well, leading after iodolysis to tetrasubstituted alkenyl iodides (eq 22).[40] The addition to alkenes is more difficult. However, the intramolecular carbozincation of alkenes mediated by a palladium or nickel catalysis is possible (eq 23).[41] The role of diethylzinc in this reaction is to transmetalate the palladium(II) intermediate back to an organozinc reagent and to generate the Pd^0 catalyst. This Pd catalysis can also be used for an efficient generation of primary alkylzinc iodides (eq 24).[41] In the absence of palladium salts, the iodine-exchange reaction requires a temperature of 50 °C.[26,27b] This I/Zn exchange reaction constitutes a unique method for preparing functionalized dialkylzincs (eq 25).[26] The Michael addition of diethylzinc to α,β-unsaturated ketones proceeds well in the presence of a nickel catalyst. In the presence of a chiral nickel catalyst, good to excellent enantioselectivities are observed (eq 26).[42]

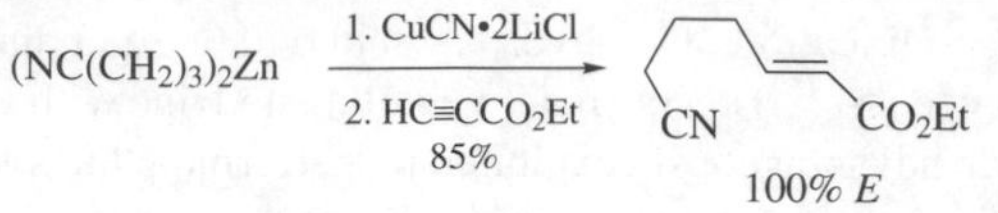

(eq 19)

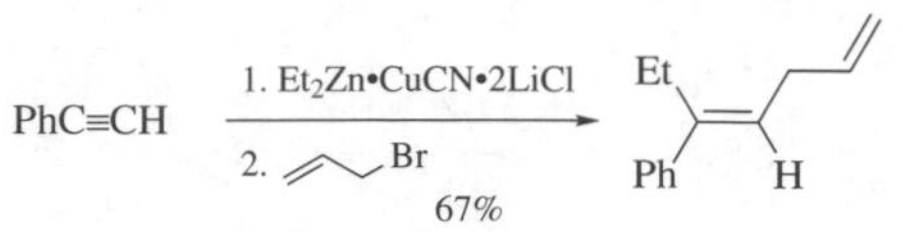

(eq 20)

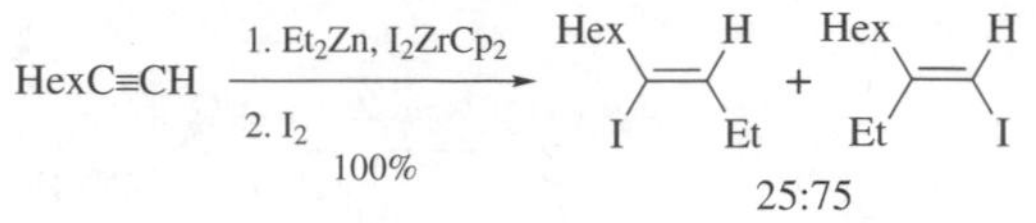

(eq 21)

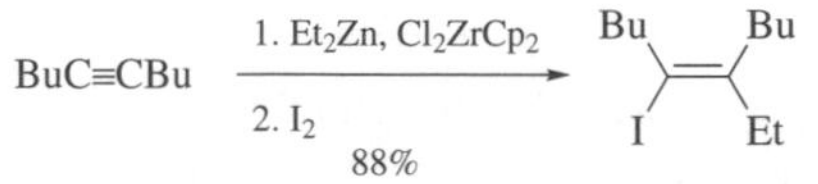

(eq 22)

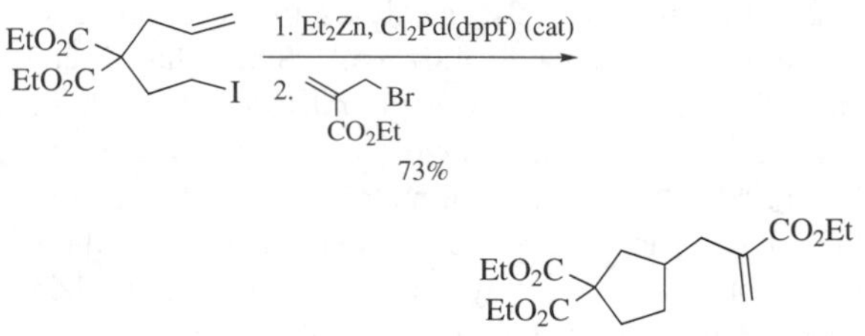

(eq 23)

(eq 24)

(eq 25)

(eq 26)

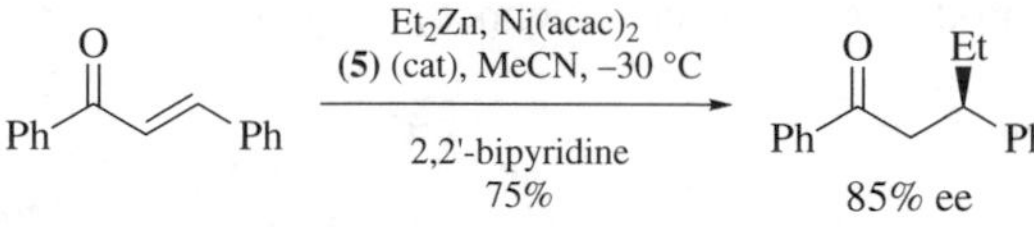

Cyclopropanation. The reaction of diethylzinc with ***Diiodomethane*** rapidly produces ICH_2ZnEt (see ***Ethyliodomethylzinc*** and ***Iodomethylzinc Iodide***).[1,43] A further exchange can occur, leading to $(ICH_2)_2Zn$.[43] Both of these iodomethylzinc derivatives are excellent cyclopropanation reagents, and their generation with diethylzinc (Furukawa method) allows convenient cyclopropanation reactions (eq 27).[44]

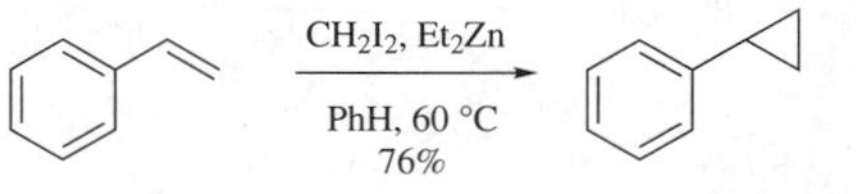

(eq 27)

The reaction gives especially good results with electron-rich alkenes (eq 28).[45] The cyclopropanation was originally performed under the strict absence of oxygen. It was later found that the presence of air accelerates the reaction and improves the yields (eq 29).[1,46] An extension of the cyclopropanation using higher 1,1-diiodoalkanes is possible, and leads preferentially to the *syn* products (eqs 30 and 31).[47] An *anti* stereoselectivity is observed with allylic alcohols (eq 32).[47] Benzene and polynuclear aromatic compounds are cyclopropanated by ***1,1-Diiodoethane*** and diethylzinc in moderate yields (eq 33).[48] Highly diastereoselective cyclopropanations can be performed with chiral allylic and homoallylic alcohols and acetal derivatives (eq 34).[49] The cyclopropanation of silylated ketene acetals using ***Bromoform*** and diethylzinc allows the generation of polyfunctional zinc carbenoids and the performance of intramolecular cyclopropanations (eq 35).[50]

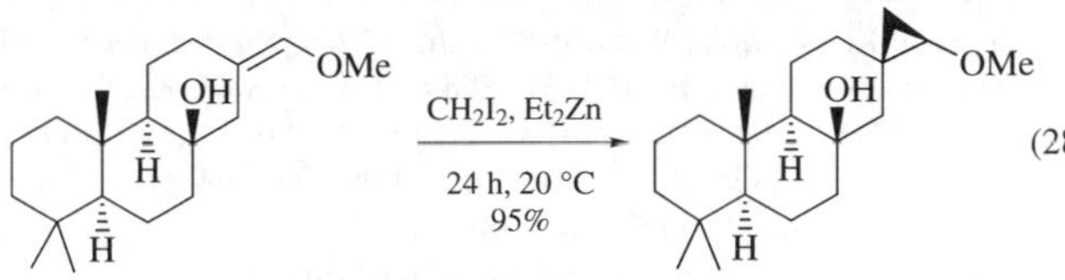
OMe
OH
H
H
CH_2I_2, Et_2Zn
24 h, 20 °C
95%

(28)

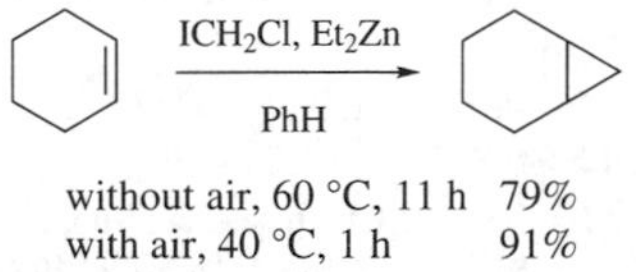
ICH_2Cl, Et_2Zn
PhH
without air, 60 °C, 11 h 79%
with air, 40 °C, 1 h 91%

(29)

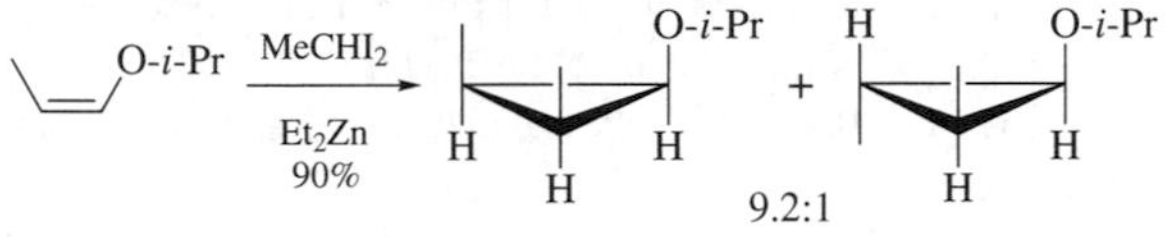
O-*i*-Pr
$MeCHI_2$
Et_2Zn
90%
9.2:1

(30)

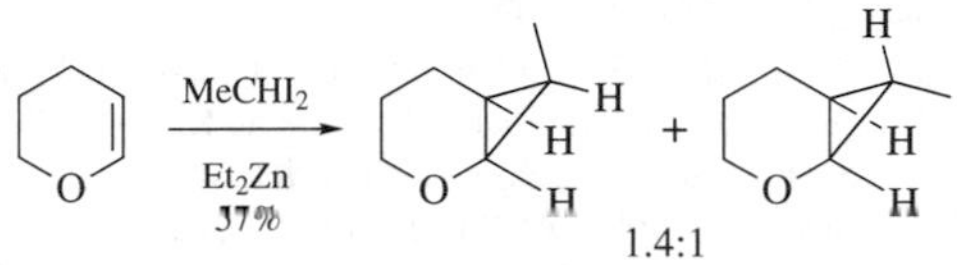
$MeCHI_2$
Et_2Zn
37%
1.4:1

(31)

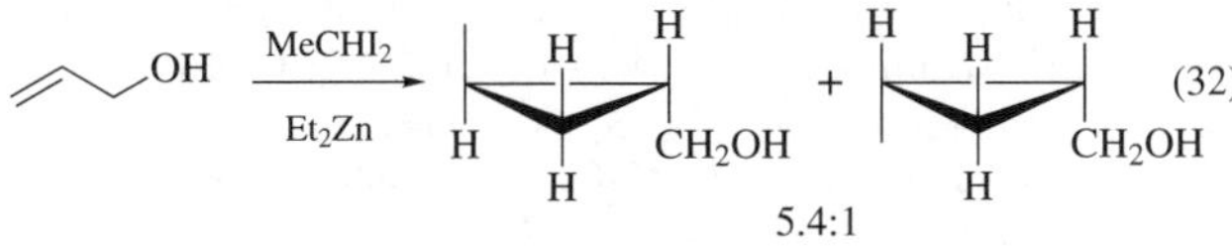
OH
$MeCHI_2$
Et_2Zn
CH_2OH
5.4:1

(32)

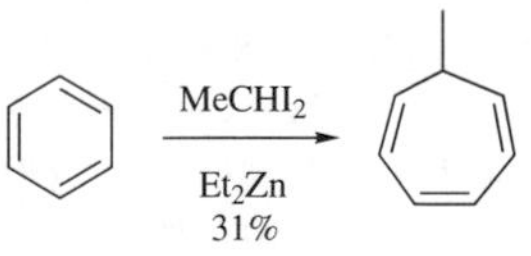
$MeCHI_2$
Et_2Zn
31%

(33)

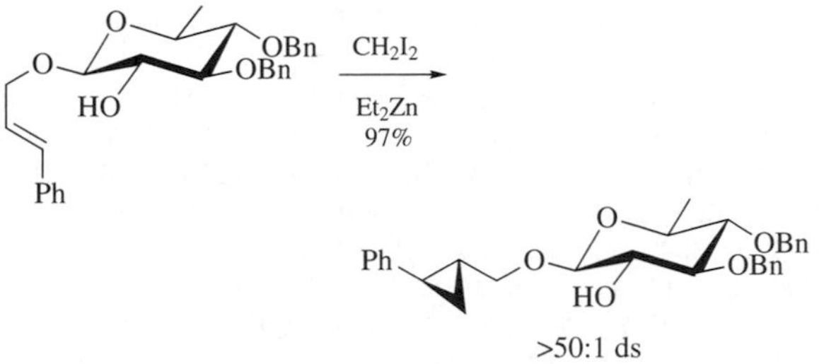
OBn
HO
Ph
CH_2I_2
Et_2Zn
97%
>50:1 ds

(34)

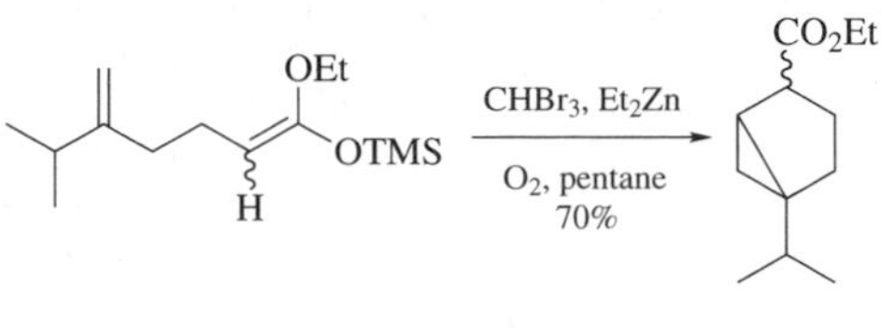
OEt
OTMS
H
$CHBr_3$, Et_2Zn
O_2, pentane
70%
CO_2Et

(35)

Miscellaneous Reactions. Diethylzinc undergoes a variety of substitution reactions with tertiary halides.[51] The elimination of HX is an important side reaction. By treating dithioacetals with ***Diiodomethane*** and diethylzinc, vinyl sulfides are obtained in satisfactory yields.[52] A direct treatment of diethylzinc with a dithioacetal produces the substitution product.[53] Ketocarbenoids have also been generated from α,α-dibromo ketones (eq 36).[54] Finally, diethylzinc has been used to prepare zinc enolates (eq 37)[55] or mixed lithium–zinc enolates (eq 38).[56]

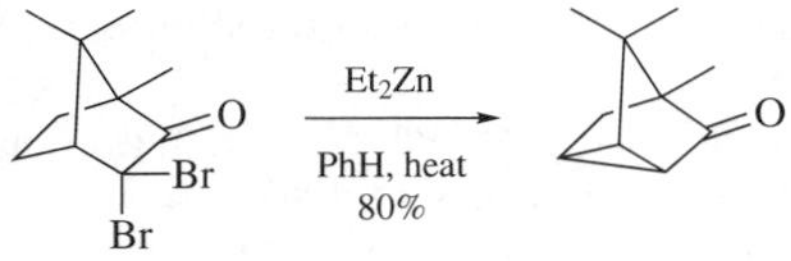
Br
Et_2Zn
PhH, heat
80%

(36)

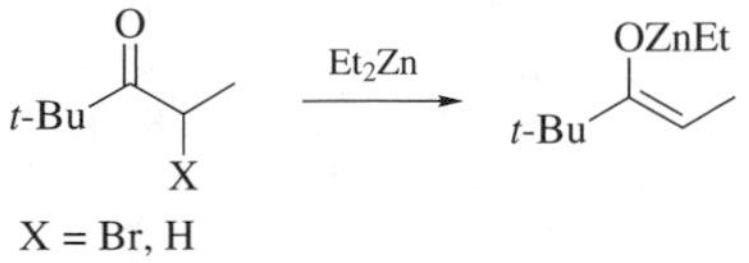
t-Bu
X
X = Br, H
Et_2Zn
OZnEt

(37)

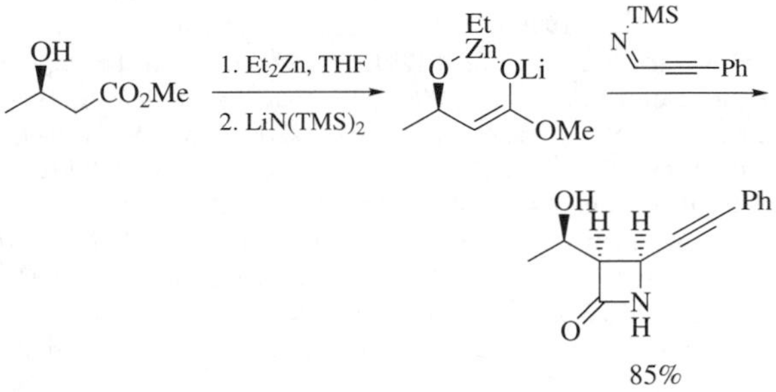
OH
CO_2Me
1. Et_2Zn, THF
2. $LiN(TMS)_2$
Et
Zn
OLi
OMe
TMS
Ph
85%

(38)

Related Reagents. 1,1-Dibromo-2-ethoxycyclopropane; 1,2-Dibromoethane; Diethylzinc–Bromoform–Oxygen; Diethylzinc–Iodoform; Ethyliodomethylzinc; Iodomethylzinc Iodide; Zinc/Copper Couple.

1. (a) Nützel, K. *MOC* **1973**, *13/2a*, 553. (b) Furukawa, J.; Kawabata, N. *Adv. Organomet. Chem.* **1974**, *12*, 83.
2. Noller, C. R. *OS* **1943**, *2*, 184.
3. (a) Zakharkin, L. I.; Okhlobystin, O. Y. *ZOB* **1960**, *30*, 2134 (*CA* **1961**, *55*, 9319). (b) Thiele, K.-H.; Zdunneck, P. *JOM* **1965**, *4*, 10. (c) Thiele, K.-H.; Engelhardt, G.; Köhler, J.; Arnstedt, M. *JOM* **1967**, *9*, 385.
4. (a) Job, A.; Reich, R. *BSF* **1923**, *33*, 1414. (b) Guffré, L.; Losio, E. *Chim. Ind. (Milan)* **1965**, *47*, 515.
5. Schwarzbach, G.; Biedermann, *C* **1948**, *2*, 56.
6. (a) Bretherick, L. *Handbook of Reactive Chemical Hazards*, 2nd ed.; Butterworths: London, 1979; p 508. (b) Sax, N. I. *Dangerous Properties of Industrial Materials*, 5th ed.; Van Nostrand-Reinhold: New York, 1979; p 1101. (c) *Hazards in the Chemical Laboratory*, 3rd ed.; Bretherick, L., Ed.; Royal Society of Chemistry: London, 1981; p 291.
7. Schmitt, R. *JPR* **1890**, *42*, 568.
8. (a) Wagner, G. *LA* **1876**, *181*, 261. (b) Gilman, H.; Marple, K. E. *RTC* **1936**, *55*, 131.
9. Marx, B.; Henry-Basch, E.; Frean, P. *CR(C)* **1967**, *264*, 527.
10. (a) Chastrette, M.; Amouroux, R. *TL* **1970**, 5165. (b) Soai, K.; Watanabe, M.; Koyano, M. *BCJ* **1989**, *62*, 2124.
11. Coates, G. E.; Ridley, D. *JCS(A)* **1966**, 1064.
12. (a) Eistert, B.; Klein, L. *CB* **1968**, *101*, 900. (b) Japp, F. R. *CB* **1880**, *13*, 761. (c) Goldschmidt, S.; Schmidt, W. *CB* **1922**, *55*, 3197.
13. (a) Klerks, J. M.; Jastrzebski, J. T. B. H.; van Koten, G; Vrieze, K. *JOM* **1982**, *224*, 107. (b) Kaupp, M.; Stoll, H.; Preuss, H.; Kain, W.; Stahl, T.; van Koten, G.; Wissing, E.; Havenith, R. W. A.; Boersma, J.; van Koten, G. *TL* **1992**, *33*, 7933.

14. Bhupathy, M.; Cohen, T. *TL* **1985**, *26*, 2619.
15. Boireau, G.; Deberly, A.; Abenhaïm, D. *TL* **1988**, *29*, 2175.
16. (a) Soai, K.; Niwa, S. *CR* **1992**, *92*, 833. (b) Evans, D. A. *Science* **1988**, *240*, 420. (c) Noyori, R.; Kitamura, M. *AG(E)* **1991**, *30*, 49. (d) Duthaler, R. O.; Hafner, A. *CR* **1992**, *92*, 807. (e) Knochel, P. *COS* **1991**, *1*, 211.
17. (a) Sato, T.; Soai, K.; Suzuki, K.; Mukaiyama, T *CL* **1978**, 601. (b) Mukaiyama, T.; Soai, K.; Sato, T.; Shimizu, H.; Suzuki, K. *JACS* **1979**, *101*, 1455.
18. (a) Oguni, N.; Omi, T. *TL* **1984**, *25*, 2823. (b) Oguni, N.; Omi, T.; Yamamotot, Y.; Nakamura, A. *CL* **1983**, 841.
19. (a) Smaardijk, A. A.; Wynberg, H. *JOC* **1987**, *52*, 135. (b) Muchow, G.; Vannoorenberghe, Y.; Buono, G. *TL* **1987**, *28*, 6163. (c) Chaloner, P. A.; Perera, S. A. R. *TL* **1987**, *28*, 3013. (d) Chaloner, P. A.; Langadianou, E. *TL* **1990**, *31*, 5185. (e) Corey, E. J.; Hannon, F. J. *TL* **1987**, *28*, 5233, 5237. (f) Soai, K.; Yokoyama, S.; Ebihara, K.; Hayasaka, T. *CC* **1987**, 1690. (g) Soai, K.; Yokoyama, S.; Hayasaka, T. *JOC* **1991**, *56*, 4264. (h) Oguni, N.; Matsuda, Y.; Kaneko, T. *JACS* **1988**, *110*, 7877. (i) Bohm, C.; Zehnder, M.; Bur, D. *AG(E)* **1990**, *29*, 205. (j) Tanaka, K.; Ushio, H.; Suzuki, H. *CC* **1989**, 1700. (k) Corey, E. J.; Yuen, P.-W.; Hannon F. J.; Wierda, D. A. *JOC* **1990**, *55*, 784. (l) Soai, K.; Watanabe, M.; Yamamoto, A.; Yamashita, T. *J. Mol. Catal.* **1991**, *64*, L27. (m) Watanabe, M.; Araki, S.; Butsugan, Y.; Uemura, M. *JOC* **1991**, *56*, 2218. (n) Soai, K.; Hyashi, H.; Haxgawa, H. *H* **1986**, *24*, 1287. (o) Soai, K.; Niwa, S.; Yamada, Y.; Inoue, H. *TL* **1987**, *28*, 4841. (p) Niwa, S.; Soai, K. *JCS(PI)* **1991**, 2717. (q) Rosini, C.; Franzini, L.; Pini, D.; Salvadori, P. *TA* **1990**, *1*, 587. (r) Kimura, K.; Sugiyama, E.; Ishizuka, T.; Kunieda, T. *TL* **1992**, *33*, 3147. (s) Watanabe, M.; Hashimoto, N.; Araki, S.; Butsugan, Y. *JOC* **1992**, *57*, 742. (t) Heatron, S. B.; Jones, G. B. *TL* **1992**, *33*, 1693. (u) Chelucci, G.; Soccolini, F. *TA* **1992**, *3*, 1235. (v) Kimura, K.; Sugiyama, E.; Ishizuka, T.; Kunieda, T. *TL* **1992**, *33*, 3147. (w) Mori, A.; Yu, D.; Inoue, S. *SL* **1992**, 427. (x) Shono, T.; Kise, N.; Fujimoto, T.; Tominaga, N.; Morita, H. *JOC* **1992**, *57*, 7175. (y) Conti, S.; Falorni, M.; Giacomelli, G.; Soccolini, F. *T* **1992**, *48*, 8993. (z) Andrés, C.; González, A.; Pedrosa, R.; Pérez-Encabo, A.; Garcia-Granda, S.; Salvadó, M. A.; Gómez-Beltrán, F. *TL* **1992**, *33*, 4743. (a.a) Stingl, Martens, J. *SC* **1992**, *22*, 2745.
20. (a) Kitamura, M.; Suga, S.; Kawai, K.; Noyou, R. *JACS* **1986**, *108*, 6071. (b) Kitamura, M.; Okada, S.; Suga, S.; Noyou, R. *JACS* **1989**, *111*, 4028. (c) Noyri, R.; Suga, S.; Kawai, K.; Okada, S.; Kitamura, M.; Oguni, N.; Hayashi, M.; Kaneko, T; Matsuda, Y. *JOM* **1990**, *382*, 19.
21. (a) Soai, K.; Ookawa, A. *CC* **1986**, 412. (b) Ookawa, A., Soai, K. *JCS(P1)* **1987**, 1465. (c) Soai, K.; Ookawa, A.; Ogawa, K., Kaba, T. *CC* **1987**, 467. (d) Soai, K.; Ookawa, A.; Kaba, T.; Ogawa, K. *JACS* **1987**, *109*, 7111. (e) Chelucci, G.; Falorni, M.; Giacomelli, G. *TA* **1990**, *1*, 843; (f) Asami, M.; Inoue, S. *CL* **1991**, 685.
22. Soai, K.; Watanabe, M. *CC* **1990**, 43.
23. Joshi, N. N.; Srebnik, M.; Brown, H. C. *TL* **1989**, *30*, 5551.
24. (a) Yoshioka, M.; Kawakita, T.; Ohno, M. *TL* **1989**, *30*, 1657. (b) Takahashi, H.; Kawakita, T.; Yoshioka, M.; Kobayashi, S.; Ohno, M. *TL* **1989**, *30*, 7095. (c) Takahashi, H.; Kawakita, T.; Ohno, M.; Yoshioka, M.; Kobayashi, S. *T* **1992**, *48*, 5691. (d) Alberts, A. H.; Wynberg, H. *JACS* **1989**, *111*, 7265.
25. (a) Schmidt, B.; Seebach, D. *AG(E)* **1991**, *30*, 99. (b) Seebach, D.; Behrendt, L.; Felix, D. *AG(E)* **1991**, *30*, 1008. (c) Beck, A. K.; Bastani, B.; Plattner, D. A.; Petter, W.; Seebach, D.; Braunschweiger, H.; Gysi, P.; La Vecchia, L. *C* **1991**, *45*, 238. (d) v. d. Bussche-Hünnefeld, J. L.; Seebach, D. *T* **1992**, *48*, 5719. (e) Seebach, D.; Plattner, D. A.; Beck, A. K.; Wang, Y. M.; Hunziker, D. *HCA* **1992**, *75*, 2171. (f) Schmidt, B.; Seebach, D. *AG(E)* **1991**, *30*, 1321.
26. Rozema, M. J.; AchyuthaRao, S.; Knochel, P. *JOC* **1992**, *57*, 1956.
27. (a) Brieden, W.; Ostwald, R.; Knochel, P. *AG(E)* **1993**, *32*, 582. (b) Rozema, M. J.; Eisenberg, C.; Lüttjens, H.; Ostwald, R.; Belyk, K.; Knochel, P. *TL* **1993**, *34*, 3115.
28. Soai, K.; Yokoyama, S.; Hayasaka, T.; Ebihara, K. *CL* **1988**, 843.
29. Soai, K.; Watanabe, M.; Koyano, M. *CC* **1989**, 534.
30. Soai, K.; Hori, H.; Kawahara, M. *TA* **1990**, *1*, 769.
31. (a) Soai, K.; Hori, H.; Niwa, S. *H* **1989**, *29*, 2065. (b) Soai, K.; Hori, H.; Kawahara, M. *CC* **1992**, 106. (c) Soai, K., Hori, H.; Kawahara, M. *TA* **1991**, *2*, 253. (d) Soai, K.; Niwa, S.; Hatanaka, T. *BCJ* **1990**, *63*, 2129. (e) Soai, K.; Niwa, S.; Hatanaka, T. *CC* **1990**, 709. (f) Niwa, S.; Hatanaka, T.; Soai, K. *JCS(P1)* **1991**, 2025. (g) Soai, K.; Niwa, S.; Hori, H. *CC* **1990**, 982. (h) Reetz, M. T.; Jung, A. *JACS* **1983**, *105*, 4833.
32. (a) Soai, K.; Hatanaka, T.; Miyazawa, T. *CC* **1992**, 1097. (b) Kateitzki, A. R.; Harris, P. A. *TA* **1992**, *3*, 437.
33. (a) Soai, K.; Niwa, S.; Watanabe, M. *JOC* **1988**, *53*, 927. (b) Soai, K.; Niwa, S.; Watanabe, M. *JCS(P1)* **1989**, 109. (c) Soai, K.; Watanabe, M. *TA* **1991**, *2*, 97. (d) Itsuno, S.; Sakurai, Y.; Ito, K.; Maruyana, T.; Nakahama, S.; Fréchet, J. M. J. *JOC* , **1990**, *55*, 304. (e) Itsuno, S.; Fréchet, J. M. J. *JOC* **1987**, *52*, 4140.
34. Soai, K.; Watanabe, M.; Yamamoto, A. *JOC* **1990**, *55*, 4832.
35. (a) Shirley, D. A. *OR* **1954**, *8*, 28. (b) Michel, J.; Henry-Basch, E.; Freon, P. *BSF* **1968**, 4898.
36. (a) Granichstädten, E.; Werner, F. *M* **1901**, *22*, 1. (b) Fieser, L. F.; Novello, F. C. *JACS* **1942**, *64*, 802.
37. (a) Negishi, E.; Bagheri, V.; Chatterjee, S.; Luo, F.-T.; Miller, J. A.; Stoll, A. T. *TL* **1983**, *24*, 5181. (b) Grey, R. A. *JOC* **1984**, *49*, 2288.
38. Knochel, P.; Yeh, M. C. P.; Berk, S. C.; Talbert, J. *JOC* **1988**, *53*, 2390.
39. AchyuthaRao, S.; Knochel, P. *JACS* **1991**, *113*, 5735.
40. Negishi, E.; Van Horn, D. E.; Yoshida, T.; Rand, C. L. *OM* **1983**, *2*, 563.
41. Stadtmüller, H.; Lentz, R.; Tucker, C. E.; Stüdemann, T.; Dörner, W.; Knochel, P. *JACS* **1993**, *115*, 7027.
42. (a) Soai, K.; Hayasaka, T.; Ugajin, S.; Yokoyama, S. *CL* **1988**, 1571. (b) Soai, K.; Yokoyama, S.; Hayasaka, T.; Ebihara, K. *JOC* **1988**, *53*, 4149. (c) Soai, K.; Hayasaka, T.; Ugajin, S. *CC* **1989**, 516. (d) Soai, K.; Okudo, M.; Okamoto, M. *TL* **1991**, *32*, 95. (e) Bolm, C.; Ewald, M. *TL* **1990**, *31*, 5011. (f) Bolm, C. *TA* **1991**, *2*, 701. (g) Jansen, J. F. G. A.; Feringa, B. L. *TA* **1992**, *3*, 581. (h) Bolm, C.; Felder, M.; Müller, J. *SL* **1992**, 439. (i) Uemura, M.; Miyake, R.; Nakayama, K.; Hayashi, Y. *TA* **1992**, *3*, 713.
43. (a) Furukawa, J.; Kawabata, N.; Nishimura, J. *TL* **1966**, 3353. (b) Furukawa, J.; Kawabata, N.; Nishimura, *T* **1968**, *24*, 53. (c) Furukawa, J.; Kawabata, N.; Nishimura, J. *TL* **1968**, 3495.
44. AchyuthaRao, S.; Rozema, M. J.; Knochel, P. *JOC* **1993**, *58*, 2694.
45. Due, D. K. M.; Fetizar, H.; Wenkert, E. *SC* **1973**, *3*, 277.
46. Miyano, S.; Yamashita, J.; Hashimoto, H. *BSJ* **1972**, *45*, 1946.
47. Nishimura, J.; Kawabata, N.; Furukawa, J. *T* **1969**, *25*, 2647.
48. (a) Nishimura, J.; Furukawa, J.; Kawabata, N.; Fujita, T. *T* **1970**, *26*, 2229. (b) Nishimura, J.; Furukawa, J.; Kawabata, N. *BSJ* **1970**, *43*, 2195. (c) Richardson, D. B.; Durrett, L. R.; Martin, J. M.; Putnam, W. E.; Slaymaker, S. C.; Dvoretzky, I. *JACS* **1965**, *87*, 2763.
49. (a) Charette, A. B.; Côté, B.; Marcoux, J.-F. *JACS* **1991**, *113*, 8166. (b) Mori, A.; Arai, I.; Yamamoto, H.; Nakai, H.; Arai, Y. *T* **1986**, *42*, 6447. (c) Mash, E. A.; Nelson, K. A. *JACS* **1985**, *107*, 8256. (d) Mash, E. A.; Nelson, K. A. *TL* **1986**, *27*, 1441. (e) Mash, E. A.; Fryling, J. A. *JOC* **1987**, *52*, 3000. (g) Mash, E. A. *JOC* **1987**, *52*, 4142. (h) Mash, E. A.; Math, S. K.; Flann, C. J. *TL* **1988**, *29*, 2147. (i) Mash, E. A.; Nelson, K. A.; Heidt, P. C. *TL* **1987**, *28*, 1865. (j) Mash, E. A.; Torok, D. S. *JOC* **1989**, *54*, 250. (k) Mash, E. A.; Nelson, K. A.; Van Deusen, S.; Hemperly, S. B. *OS* **1989**, *68*, 92. (l) Mash, E. A.; Hemperly, S. B.; Nelson, K. A.; Heidt, P. C.; Van Deusen, S. *JOC* **1990**, *55*, 2045. (m) Mash, E. A.; Hemperly, S. B. *JOC* **1990**, *55*, 2055. (n) Sugimura, T.; Futagawa, T.; Tai, A. *TL* **1988**, *29*, 5775. (o) Sugimura, T.; Futagawa, T.; Yoshikawa, M.; Tai, A. *TL* **1989**, *30*, 3807. (p) Imai, T.; Mineta, H.; Nishida, S. *JOC* **1990**, *55*, 4986. (q) Arai, I.; Mori, A.; Yamamoto, H. *JACS* **1985**, *107*, 8254.
50. (a) Rousseau, G.; Slougui, N. *JACS* **1984**, *106*, 7283. (b) Rousseau, G.; Slougui, N. *TL* **1983**, *24*, 1251.
51. (a) Kinney, C. R.; Spliethoff, W. L. *JOC* **1949**, *14*, 71. (b) Horton, A. W. *JACS* **1947**, *69*, 182. (c) Mathot, V. *BSB* **1950**, *59*, 111. (d) Morgan, G. T.; Carter, S. R.; Duck, A. E. *JCS* **1925**, 1252.

52. Rodriguez, A. D.; Nickon, A. *T* **1985**, *41*, 4443.
53. (a) Kozikowski, A. P.; Greco, M. N. In *Strategies and Tactics in Organic Synthesis*; Academic: London, 1989; vol. 2, p 263. (b) Kozikowski, A. P.; Konoike, T.; Ritter, A. *Carbohydr. Res.* **1989**, *171*, 109.
54. Scott, L. T.; Cotton, W. D. *JACS* **1973**, *95*, 2708.
55. Hansen, M. M.; Bartlett, P. A.; Heathcock, C. H. *OM* **1987**, *6*, 2069.
56. Oguni, N.; Ohkawa, Y. *CC* **1988**, 1376.

Paul Knochel
Philipps-Universität, Marburg, Germany

(2*S*)-(+)-2,5-Dihydro-2-isopropyl-3,6-dimethoxypyrazine[1]

[78342-42-4] $C_9H_{16}N_2O_2$ (MW 184.24)

(Schöllkopf–Hartwig bislactim ether reagent for asymmetric synthesis of amino acids by reaction of the metalated reagent with alkyl halides,[2] aldehydes,[3] ketones,[4] epoxides,[5] and enones,[6] and subsequent hydrolysis of the resulting bislactim ether adduct)

Physical Data: bp 103–104 °C/15 mmHg; *d* 1.03 g cm^{-3}; $[\alpha]_D^{20}$ −109 (*c* = 1, EtOH).
Solubility: sol ether, THF, *n*-hexane.
Form Supplied in: colorless liquid.
Analysis of Reagent Purity: NMR.[2]
Purification: distillation.
Handling, Storage, and Precautions: store refrigerated.

Bislactim Ether Method. The commercially available bislactim ethers of cyclo(L- and D-Val-Gly) and cyclo(L- and D-Val-Ala)[7] are very versatile reagents for the preparation of nonproteinogenic amino acids in high yields and with excellent enantioselectivities (typically >95%).[1]

Reactions of the Lithiated Bislactim Ether. The procedure involves metalation with ***n*-Butyllithium** in THF at −70 °C and reaction of the resulting azaenolate with alkyl halides (eq 1).[2] The latter enters with high diasteroselectivity *trans* to the isopropyl group. The alkylation products are hydrolyzed under mild acidic conditions to give the desired amino acid ester and the chiral auxiliary L-Val-OMe, which can be separated by distillation or chromatography (eq 1). As well as reacting with primary and secondary alkyl halides and sulfonates,[8] the lithiated bislactim ether reacts in good yields and with high *trans* diastereoselectivities (in general >95%) with a variety of other electrophiles such as ketones,[9] acyl chlorides,[10] thioketones,[11] epoxides,[5] α,β-unsaturated esters,[12] and arene–manganese tricarbonyl complexes.[13]

BuLi, THF, −70 °C; Br–CH₂–C≡C–Ph, −70 °C; 92%, >95% de; 0.25 N HCl, rt; L-Val-OMe; 86%, >95% ee (1)

Reaction of the Titanated Bislactim Ether. The titanium derivative of the bislactim ether of cyclo(L-Val-Gly) reacts with alkyl aldehydes,[3] aryl aldehydes,[14] and α,β-unsaturated aldehydes[15] highly diastereoselectively to give almost exclusively the *syn* addition products (eq 2). Hydrolysis with dilute ***Trifluoroacetic Acid***[3c] affords (2*R*, 3*S*)-β-hydroxy-α-amino acid methyl esters. α-Amino-γ-nitro amino acids can be obtained by 1,4-addition of the titanated bislactim ether to nitroalkenes and subsequent hydrolysis of the adduct.[16]

1. BuLi, THF, −70 °C; 2. ClTi(NEt₂)₃; 1. crotonaldehyde, −70 °C; 2. H⁺, H₂O; 91%, >95% de; 0.1 N CF₃CO₂H, rt, 63%; >95% ee (2)

Reactions of the Bislactim Ether Cuprate. The lithiated bislactim ether can be converted to an azaenolate cuprate by treatment with CuBr·SMe₂ (see ***Copper(I) Bromide***).[6] Conjugate addition of the cuprate to enones (eq 3)[6] and dienones,[17] or alkylation[6] with base labile electrophiles like ethyl 3-bromopropionate, proceeds with high *trans* diastereoselectivity. Hydrolysis of the Michael adducts and subsequent protection afford (2*R*,3*R*)-*N*-Boc-δ-oxo-α-amino acid methyl esters (eq 3).

1. BuLi THF, –70 °C
2. 0.5 equiv CuBr•SMe$_2$

CuLi –70 °C

1. 0.25 N HCl, rt
2. (Boc)$_2$O (3)

71%, (2*R*,1'*R*):(2*R*,1'*S*) = 98:2 71%

Reactions of the Bislactim Ether Carbene. Diazotization of the lithiated bislactim ether generates an electrophilic carbene species, which reacts in good yields and with high diastereomeric excess (<95%) with alkenes[18] and aryl alkynes[19] (eq 4) to give spirocyclopropanes and spirocyclopropenes, respectively. Hydrolysis of the latter affords the novel (*R*)-1-amino-2-arylcyclopropene-1-carboxylic acids.[19]

1. BuLi THF, –70 °C
2. PhSO$_2$N$_3$
3. BuLi

Ph—≡ 50–60 °C

1. 0.1 N HCl
2. (Boc)$_2$O, NEt$_3$
3. NaOH
4. 0.5 N HCl MeOH, 60 °C 58% (4)

50%, >95% de >95% ee

Related Reagents. 1-Benzoyl-2-*t*-butyl-3,5-dimethyl-4-imidazolidinone; (2*S*,4*S*)-3-Benzoyl-2-*t*-butyl-4-methyl-1,3-oxazolidin-5-one.

1. (a) Hartwig, W.; Schöllkopf, U. Ger. Patent 2934252, 1981. (b) Schöllkopf, U. In *Organic Synthesis: An Interdisciplinary Challenge*, Streith, H.; Prinzbach, G.; Schill, G., Eds.; Blackwell: Oxford, 1985; p 101. (c) Schöllkopf, U. *PAC* **1983**, *55*, 1799. (d) Schöllkopf, U. *CS* **1985**, *25*, 105. (e) Williams, R. M. *Synthesis of Optically Active α-Amino Acids*, Pergamon: Oxford, 1989, (f) Schöllkopf, U. *Top. Curr. Chem.* **1983**, *109*, 65.
2. (a) Schöllkopf, U.; Groth, U.; Deng, C. *AG(E)* **1981**, *20*, 798.
3. (a) Schöllkopf, U.; Nozulak, J.; Grauert, M. *S* **1985**, 55. (b) Grauert, M.; Schöllkopf, U. *LA* **1985**, 1817. (c) Beulshausen, T.; Groth, U.; Schöllkopf, U. *LA* **1991**, 1207.
4. (a) Schöllkopf, U.; Groth, U.; Gull, M.-R.; Nozulak, J. *LA* **1983**, 1133. (b) Neubauer, H.-J.; Balza, J.; Freer, J.; Schöllkopf, U. *LA* **1985**, 1508.
5. Gull, R.; Schöllkopf, U. *S* **1985**, 1052.
6. Schöllkopf, U.; Pettig, D.; Schulze, E.; Klinge, M.; Egert, E.; Benecke, B.; Noltemeyer, M. *AG(E)* **1988**, *27*, 1194.
7. Merck Suchardt D-6100 Hohenbrunn, Germany.
8. Baldwin, J. E.; Adlington, R. M.; Bebbington, D.; Russel, A. T. *CC* **1992**, 1249.
9. Schöllkopf, U.; Groth, U. *AG(E)* **1981**, *20*, 977.
10. Schöllkopf, U.; Westphalen, K.-O.; Schröder, J.; Horn, K. *LA* **1988**, 781.
11. Schöllkopf, U.; Nozulak, J.; Groth, U. *T* **1984**, *40*, 1409.
12. (a) Hartwig, W.; Born, L. *JOC* **1987**, *52*, 4352. (b) Schöllkopf, U.; Pettig, D.; Busse, U. *S* **1986**, 737.
13. (a) Pearson, A. J.; Bruhn, P. R.; Gouzoules, F.; Lee, S.-K. *CC* **1989**, *10*, 659. (b) Pearson, A. J.; Bruhn, P. R. *JOC* **1991**, *56*, 7092.
14. Schöllkopf, U.; Beulshausen, T. *LA* **1989**, 223.
15. Schöllkopf, U.; Bendenhaben, J. *LA* **1987**, 393.
16. (a) Schöllkopf, U.; Kühnle, W.; Egert, E.; Dyrbusch, M. *AG(E)* **1987**, *26*, 480. (b) Busch, K.; Groth, U.; Kühnle, W.; Schöllkopf, U. *T* **1992**, *27*, 5607.
17. Wild, H.; Born, L. *AG(E)* **1991**, *30*, 1685.
18. Schöllkopf, U.; Hauptreif, M.; Dippel, J.; Nieger, M.; Egert, E. *AG(E)* **1986**, *25*, 192.
19. Schöllkopf, U.; Hupfeld, B.; Küper, S.; Egert, E.; Dyrbusch, M. *AG(E)* **1988**, *27*, 433.

W. Hartwig & J. Mittendorf
Bayer, Wuppertal, Germany

Diiodomethane

CH_2I_2

[75-11-6] CH_2I_2 (MW 267.84)

(precursor to methylene transfer reagents: in combination with various metals or alkyl metals, generates carbenoids which cyclopropanate alkenes[1] or methylenate carbonyls;[2] precursor to ICH_2Met and I_2CHMet nucleophiles;[3] participates in radical-mediated couplings[4])

Alternate Name: methylene iodide.
Physical Data: mp 6 °C; bp 181 °C; *d* 3.325 g cm^{-3}.
Solubility: slightly sol H_2O; sol Et_2O, $CHCl_3$, EtOH, hexane, etc.
Form Supplied in: pale yellow liquid; crystallizes as pale yellow needles or plates; typically stabilized with copper powder.
Purification: fractional distillation from CaH_2 generally provides reagent of sufficient purity for typical uses.
Handling, Storage, and Precautions: store over copper powder to inhibit radical-induced decomposition; protect from light; incompatible with many metals (Al, Mg, Na, etc.) and strong bases; mildly corrosive; toxicity presumed to be on par with CH_2Cl_2, i.e. moderately toxic, mutagenic; use in a fume hood.

Methylenations. A wide variety of carbonyl methylenating reagents utilize CH_2I_2 as the carbon source. Several of these have proven to be useful alternatives to the Wittig reaction.[5] Cainelli and co-workers found that treatment of CH_2I_2 with 2 equiv of ***Magnesium Amalgam*** in the presence of ketones affords alkenes in good yields (eq 1).[6] The intermediate in these reactions is presumed to be $CH_2(MgI)_2$, and the reaction works well with ketones and aldehydes of variable structure. This reagent is particularly useful for the formation of dideuterated terminal alkenes (using CD_2I_2, eq 2),[7] since methylenation using deuterated Wittig reagents is often accompanied by various degrees of scrambling of the deuterium label.[8]

$$CH_2I_2 + Mg(Hg) \xrightarrow{Et_2O} [CH_2(MgI)_2] \xrightarrow[68\%]{\text{cyclohexanone}} \text{methylenecyclohexane} \quad (1)$$

$$\text{norbornenone} \xrightarrow[Et_2O,\ 61\%]{CD_2I_2,\ Mg(Hg)} \text{CD}_2\text{-norbornene} \quad (2)$$

Chromium(II) also mediates the alkenation of aldehydes by geminal diiodides (eq 3).[9] This transformation presumably proceeds via a 1,1-bis[chromium(III)] species, and works best with substituted diiodides. Generally, high levels of (*E*) selectivity are obtained.

$$\xrightarrow[THF]{2\text{ equiv } CrCl_2,\ DMF} \xrightarrow[85\%]{C_8H_{17}CHO} \quad C_8H_{17} \quad (3)$$

(E):(Z) = 94:6

Certain methylenating reagents utilize CH_2I_2 in combination with 2 equiv of ***Zinc*** and a Lewis acid.[2,10,11] A Lewis acid is not essential for this transformation,[11] but greatly accelerates the reaction and improves selectivity and yields. The most commonly used Lewis acids include ***Trimethylaluminum***,[2,10a] ***Titanium Tetraisopropoxide***,[10a] ***Titanium(IV) Chloride***,[10b,10c] ***Dichlorobis(cyclopentadienyl)zirconium***,[10d] and ***Dichlorobis(cyclopentadienyl)titanium***.[10e] Of a variety of Lewis acids and dihalomethanes examined, Takai and co-workers found that $CH_2I_2/Zn/Me_3Al$ and $CH_2Br_2/Zn/TiCl_4$ afforded the best results.[10f] Substrates for these reactions can vary from simple ketones such as acetophenone (eq 4)[2] to complex steroidal ketones.[10b] Chemoselective methylenation of an aldehyde in the presence of a ketone can also be achieved (eq 5).[10a] Other unsaturated functionalities also react with these reagents. For example, the Eisch reagent[10e] $[CH_2(ZnI)_2/Cp_2TiCl_2]$ has been shown to react with benzonitrile to afford, after hydrolysis, acetophenone (eq 6).

$$\text{PhCOCH}_3 \xrightarrow[67\%]{CH_2I_2,\ Zn,\ Me_3Al} \text{Ph-C(=CH}_2\text{)CH}_3 \quad (4)$$

$$\xrightarrow[THF,\ rt,\ 86\%]{CH_2I_2,\ Zn,\ Me_3Al} \quad (5)$$

$$\text{Zn (excess)} + CH_2I_2 \xrightarrow{THF,\ 45\ °C} \xrightarrow{Cp_2TiCl_2} \xrightarrow[2.\ D_2O,\ >80\%]{1.\ PhCN} \text{PhCOCH}_2\text{D} \quad (6)$$

Cyclopropanations. A variety of metallic species mediate the cyclopropanation of alkenes in the presence of CH_2I_2 (vide infra). A much simpler procedure is the photolysis of CH_2I_2 in the presence of alkenes.[12] Although this method was initially reported to give a mixture of products in poor yields,[12a] the addition of iodine scavengers and acid scavengers to the reaction medium results in clean, high-yielding reactions.[12b–d] Sterically hindered alkanes are cyclopropanated much more readily than with metal-based reagents (eq 7), and double bond geometry is generally retained (eq 8). Remarkably, products arising from C–H bond insertion are not observed, ruling out the formation of free carbene in these reactions. The iodomethyl cation (ICH_2^+) has been proposed as an intermediate in these reactions.[12c,12d]

$$t\text{-Bu-CH=CH}_2 \xrightarrow[CH_2Cl_2,\ aq\ Na_2S_2O_4,\ Na_2CO_3,\ 80\%]{CH_2I_2,\ h\nu} t\text{-Bu-cyclopropane} \quad (7)$$

$$\text{Et(H)C=C(R}^1\text{)(R}^2\text{)} \xrightarrow[CH_2Cl_2,\ aq\ Na_2S_2O_4,\ Na_2CO_3]{CH_2I_2,\ h\nu} \quad (8)$$

R^1 = Et, R^2 = H; 75%
R^1 = H, R^2 = Et; 76%

Perhaps the most common use of CH_2I_2 in organic synthesis is in metal-mediated cyclopropanations. Foremost among these are zinc-mediated cyclopropanations (Simmons–Smith reaction).[1,13] This is a widely utilized and versatile transformation. Treatment of CH_2I_2 and an alkene with ***Zinc–Copper Couple*** in refluxing Et_2O affords the corresponding cyclopropane, generally in good yield (eq 9).[14] The source of the zinc is crucial to the success of the reaction, and several reliable protocols exist.[15] The use of ***Diethylzinc*** place of the Zn/Cu couple generates a similar reagent;[16] this modification has several advantages, including the option of using noncoordinating solvents. In many cases, the use of ***Chloroiodomethane***/Et_2Zn in place of CH_2I_2/Et_2Zn is desirable, as the former is more reactive.[17] Regardless of the method of reagent generation, the stereochemical course of the reaction is strongly influenced by proximal oxygen substituents (eq 10), and several effective chiral auxiliaries have been developed.[18]

$$\xrightarrow[Et_2O,\ reflux,\ 87\%]{CH_2I_2,\ Zn/Cu\ couple} \quad (9)$$

$$\xrightarrow[DCE,\ 0\ °C,\ 99\%]{CH_2I_2,\ Et_2Zn} \quad (10)$$

OBn → OBn; >99% ee

Similarly, treatment of CH_2I_2 with a trialkylaluminum reagent (e.g. ***Triisobutylaluminum***) in the presence of alkenes also affords cyclopropanes.[19] This reagent system exhibits a reactivity pattern complementary to the zinc- and samarium-based systems, reacting preferentially with isolated alkenes rather than with allylic alcohols (eq 11).[19a]

$$\xrightarrow[CH_2Cl_2,\ rt]{i\text{-}Bu_3Al\text{–}CH_2I_2\ (2:1)} \quad (11)$$

OH → OH (64%) + OH (17%)

A very versatile reagent for the cyclopropanation of alkenes is derived from CH_2I_2 and Sm(Hg) or ***Samarium(II) Iodide***.[20] These reagents react well with allylic alcohols (eq 12)[20a,b] and enolates (eq 13),[20d] and are subject to the same hydroxy-directing effects[20a,b,f,g] as the zinc-based reagents (vide supra). In fact, an oxygen substituent is required for cyclopropanation to occur.[20b] A diastereoselective cyclopropanation utilizing a chiral acetal as a chiral auxiliary has also recently been reported (eq 14).[20g] The substitution of $ClCH_2I$ for CH_2I_2 in these reactions often results in higher yields.[20b] Interestingly, a comparison of the response of the CH_2I_2/Sm(Hg) and CH_2I_2/SmI_2 reagents towards various allylic alcohols showed little or no difference in either reactivity, chemoselectivity, or stereoselectivity.[20b]

excess Sm(Hg), $ClCH_2I$; THF, −78 °C to rt; 97% (12)

1. LDA, THF, −78 °C; 2. CH_2I_2, SmI_2, −78 °C to rt; 59% (13)

$ClCH_2I$, Sm(Hg); THF, −78 °C to rt; 94%; 90% ee (14)

Nucleophilic Additions of 'ICH_2'. Halomethyllithium reagents are generally unstable except at very low temperatures (<−90 °C),[21] but the use of additives such as ***Lithium Bromide*** results in greater stability, particularly for ***Chloromethyllithium*** and ***Bromomethyllithium***.[21c,22] The utility of ICH_2Li is still limited, however.[23] Fortunately, an alternative method for the generation of an iodomethyl nucleophile has been developed which utilizes ***Samarium(0)*** as the metal.[3,20c] Aldehydes, ketones, and enones all participate well in this reaction, and yields range from moderate to excellent (eq 15). In the illustrated example, the alkene is not cyclopropanated, consistent with the observation that tertiary allylic alcohols react sluggishly with samarium carbenoids.[20b]

excess CH_2I_2, Sm; THF, rt; 80% (15)

The replacement of an allylic alcohol oxygen by CH_2I has also been achieved by the use of aluminum reagents.[24] The combination of ***Triethylaluminum***, Et_2AlCl, and Et_2AlOEt mediates this reaction (eq 16).

Et_3Al, Et_2AlCl, Et_2AlOEt, CH_2I_2; toluene, rt; 71% (16)

Nucleophilic Additions of 'I_2CH'. Deprotonation of CH_2I_2 by base affords I_2CHMet derivatives which are more stable than the corresponding ICH_2Met species and react well with a variety of electrophiles.[25] Among the bases used successfully are Cy_2NLi,[25a] ***Sodium Hexamethyldisilazide*** (NaHMDS),[25b] ***Lithium Hexamethyldisilazide*** (LiHMDS),[25b] and ***Lithium Diisopropylamide*** (LDA).[25c] I_2CHLi reacts well with aldehydes (eq 17),[25a] and also with silacyclobutanes, affording 2-iodosilacyclopentanes after ring enlargement of the intermediate five-coordinate siliconate (eq 18).[25c] I_2CHNa has been shown to react with electrophiles such as ***Chlorotrimethylsilane***, ***Ethyl Iodide***, and ***Allyl Iodide***.[25b]

$CH_2I_2 + Cy_2NLi$ → (THF, −78 °C) → $C_8H_{17}CHO$, THF, −78 °C, 79% → $C_8H_{17}CH(OH)CHI_2$ (17)

CH_2I_2, LDA; THF, −78 °C; 83% (18)

A convenient synthesis of vinyl iodides using I_2CHLi has been reported by Julia and co-workers.[25b] Treatment of CH_2I_2 with LiHMDS followed by the addition of a lithiated sulfone affords vinyl iodides after aqueous workup. The selectivity is generally low, however (eq 19).

I_2CHLi; THF, −78 °C; 90%; (E):(Z) = 2:1 (19)

Radical Additions Utilizing CH_2I_2. The radical addition of the ICH_2 fragment to α,β-unsaturated ketones mediated by ***Triethylborane*** provides a route to γ-iodo ketones (eq 20).[4] The intermediate boron enolates can be either hydrolyzed or alkylated in some cases.[4] Vinylsilacyclobutanes are also alkylated by the putative ICH_2 radical, affording highly functionalized silylcyclobutane derivatives (eq 21).[26]

CH_2I_2, Et_3B; benzene; 69% (20)

CH_2I_2, Et_3B; hexane; 43% (21)

Alkylation Reactions. CH_2I_2 has seen limited use as an alkylating reagent, since heterodihalomethanes such as $ClCH_2I$ and $ClCH_2Br$ are preferred for these reactions.[27,28] CH_2I_2 has been used for alkylative cyclizations, however. A recent interesting application is the formation of 2-imino-1,3-dithiazetidines from thioureas (eq 22).[29a] Mixtures of isomers are obtained unless one of the urea nitrogen atoms is deactivated as a sulfonamide, amide, or carbamate. In addition, diamines are known to react with CH_2I_2: slow addition of the diamine to a solution of CH_2I_2 is necessary to obtain good yields (eq 23).[29b]

$$\text{TosHN-C(=S)-NHPh} \xrightarrow[\text{acetone, rt}\ 98\%]{CH_2I_2,\ Et_3N} \text{Tos-N=C(S-CH}_2\text{-)N-Ph (cyclic)} \quad (22)$$

$$\text{Me-N(bicyclic)N-Me} \xrightarrow[\text{EtOH, reflux}\ 76\%]{CH_2I_2} \text{Me-N}^{+}\text{-CH}_2\text{-N}^{+}\text{-Me (cage)}\ 2\,I^- \quad (23)$$

In a mechanistically distinct but related example, dithianes are produced by the Pt-catalyzed coupling of CH_2I_2 with thiols (eq 24).[30]

$$2\ HOCH_2CH_2SH \xrightarrow[K_2CO_3,\ \text{acetone, reflux}\ 73\%]{CH_2I_2,\ Pt(dppm)Cl_2\ (5\%)} (HOCH_2CH_2S)_2CH_2 \quad (24)$$

Other Uses. Several other transformations also utilize CH_2I_2. For example, heating CH_2I_2 with 2 equiv of ***Tin(II) Bromide*** and a catalytic amount of ***Triethylamine*** affords a di-tin compound[31a] which can be exhaustively methylated to afford bis(trimethylstannyl)methane (eq 25).[31b] In addition, a convenient procedure for the in situ preparation of the valuable reagent SmI_2 involves simply treating Sm powder with CH_2I_2 in THF.[32]

$$2\ SnBr_2 + CH_2I_2 \xrightarrow[140\ ^\circ C]{Et_3N\ (cat)} \xrightarrow[\text{reflux}\ 55\%]{\text{excess MeMgI}} (Me_3Sn)_2CH_2 \quad (25)$$

Related Reagents. Chloroiodomethane; Dibromomethane; 1,1-Diiodoethane; Diiodomethane–Zinc–Titanium(IV) Chloride; Ethyliodomethylzinc; Iodomethylzinc Iodide.

---

1. Simmons, H. E.; Cairns, T. L.; Vladuchick, S. A.; Hoiness, C. M. *OR* **1972**, *20*, 1.
2. Takai, K.; Hotta, Y.; Oshima, K.; Nozaki, H. *TL* **1978**, 2417.
3. (a) Imamoto, T.; Hatajima, T.; Takiyama, N.; Takeyama, T.; Kamiya, Y.; Yoshizawa, T. *JCS(P1)* **1991**, 3127. (b) Tabuchi, T.; Inanaga, J.; Yamaguchi, M. *TL* **1986**, *27*, 3891.
4. Nozaki, K.; Oshima, K.; Utimoto, K. *BCJ* **1991**, *64*, 403.
5. Maercker, A. *OR* **1965**, *14*, 270.
6. Cainelli, G.; Bertini, F.; Grasselli, P.; Zubiani, G. *TL* **1967**, 5153.
7. (a) Hasselman, D. *CB* **1974**, *107*, 3486. (b) Hoffman, R. W.; Riemann, A.; Mayer, B. *CB* **1985**, *118*, 2493.
8. Atkinson, J. G.; Fisher, M. H.; Horley, D.; Morse, A. T.; Stuart, R. S.; Synnes, E. *CJC* **1965**, *43*, 1614.
9. Okazoe, T.; Takai, K.; Utimoto, K. *JACS* **1987**, *109*, 951.
10. (a) Okazoe, T.; Hibino, J.-i.; Takai, K.; Nozaki, H. *TL* **1985**, *26*, 5581. (b) Lombardo, L. *TL* **1982**, *23*, 4293. (c) Hibino, J.-i.; Okazoe, T.; Takai, K.; Nozaki, H. *TL* **1985**, *26*, 5579. (d) Tour, J. M.; Bedworth, P. V.; Wu, R. *TL* **1989**, *30*, 3927. (e) Eisch, J. J.; Piotrowski, A. *TL* **1983**, *24*, 2043. (f) Takai, K.; Hotta, Y.; Oshima, K.; Nozaki, H. *BCJ* **1980**, *53*, 1698.
11. See, for example: (a) Miyano, S.; Hida, M.; Hashimoto, H. *JOM* **1968**, *12*, 263. (b) Turnbull, P.; Syhora, K.; Fried, J. H. *JACS* **1966**, *88*, 4764.
12. (a) Blomstrom, D. C.; Herbig, K.; Simmons, H. E. *JOC* **1965**, *30*, 959. (b) Pienta, N. J.; Kropp, P. J. *JACS* **1978**, *100*, 655. (c) Kropp, P. J.; Pienta, N. J.; Sawyer, J. A.; Polniaszek, R. P. *T* **1981**, *37*, 3229. (d) Kropp, P. J. *ACR* **1984**, *17*, 131.
13. (a) Simmons, H. E.; Smith, R. D. *JACS* **1958**, *80*, 5323. (b) Simmons, H. E.; Smith, R. D. *JACS* **1959**, *81*, 4256.
14. Koch, S. D.; Kliss, R. M.; Lopiekes, D. V.; Wineman, R. J. *JOC* **1961**, *26*, 3122.
15. (a) Shank, R. S.; Shechter, H. *JOC* **1959**, *24*, 1825. (b) LeGoff, E. *JOC* **1964**, *29*, 2048. (c) Rawson, R. J.; Harrison, I. T. *JOC* **1970**, *35*, 2057. (d) Denis, J. M.; Girard, C.; Conia, J. M. *S* **1972**, 549. (e) Friedrich, E. C.; Lewis, E. J. *JOC* **1990**, *55*, 2491.
16. (a) Furukawa, J.; Kawabata, N.; Nishimura, J. *TL* **1966**, 3353. (b) Furukawa, J.; Kawabata, N.; Nishimura, J. *T* **1968**, *24*, 53. (c) Nishimura, J.; Furukawa, J.; Kawabata, N.; Kitayama, M. *T* **1971**, *27*, 1799.
17. (a) Denmark, S. E.; Edwards, J. P. *JOC* **1991**, *56*, 6974. (b) Miyano, S.; Yamashita, J.; Hashimoto, H. *BCJ* **1972**, *45*, 1946.
18. See, for example: (a) Mash, E. A.; Hemperly, S. B. *JOC* **1990**, *55*, 2055, and references cited therein. (b) Mori, A.; Arai, I.; Yamamoto, H. *T* **1986**, *42*, 6447. (c) Charette, A. B.; Côté, B.; Marcoux, J.-F. *JACS* **1991**, *113*, 8166.
19. (a) Maruoka, K.; Fukutani, Y.; Yamamoto, H. *JOC* **1985**, *50*, 4412. (b) Fleming, I.; Lawrence, N. J.; Sarkar, A. K.; Thomas, A. P. *JCS(P1)* **1992**, 3303. (c) Miller, D. B. *TL* **1964**, 989. See also: (d) Hoberg, H. *LA* **1966**, *695*, 1. (e) Hoberg, H. *LA* **1962**, *656*, 1.
20. (a) Molander, G. A.; Etter, J. B. *JOC* **1987**, *52*, 3942. (b) Molander, G. A.; Harring, L. S. *JOC* **1989**, *54*, 3525. (c) Imamoto, T.; Takeyama, T.; Koto, H. *TL* **1986**, *27*, 3243. (d) Imamoto, T.; Takiyama, N. *TL* **1987**, *28*, 1307. (e) Imamoto, T.; Kamiya, Y.; Hatajima, T.; Takahashi, H. *TL* **1989**, *30*, 5149. (f) Lautens, M.; Delanghe, P. H. M. *JOC* **1992**, *57*, 798. (g) Kabat, M.; Kiegiel, J.; Cohen, N.; Toth, K.; Wovkulich, P. M.; Uskokovic, M. R. *TL* **1991**, *32*, 2343.
21. For a review, see: (a) Köbrich, G. *AG(E)* **1972**, *11*, 473. See also: (b) Köbrich, G.; Fischer, R. H. *T* **1968**, *24*, 4343. (c) Villieras, J.; Rambaud, M.; Kirschleger, B.; Tarhouni, R. *BSF(2)* **1985**, 837.
22. (a) Tarhouni, R.; Kirschleger, B.; Rambaud, M.; Villieras, J. *TL* **1984**, *25*, 835. (b) Barluenga, J.; Pedregal, B.; Concellón, J. M.; Yus, M. *TL* **1993**, *34*, 4563, and references cited therein.
23. For a recent use of ICH_2Li, see: Ambler, P. W.; Davies, S. G. *TL* **1988**, *29*, 6983.
24. Ukaji, Y.; Inomata, K. *CL* **1992**, 2353.
25. (a) Taguchi, H.; Yamamoto, H.; Nozaki, H. *JACS* **1974**, *96*, 3010. (b) Charreau, P.; Julia, M.; Verpeaux, J. N. *BSF(2)* **1990**, 275. (c) Matsumoto, K.; Oshima, K.; Utimoto, K. *TL* **1990**, *31*, 6055.
26. Matsumoto, K.; Miura, K.; Oshima, K.; Utimoto, K. *TL* **1991**, *32*, 6383.
27. For a recent discussion and references, see: Hanh, R. C. *JOC* **1988**, *53*, 1331.
28. For a recent attempt to use CH_2I_2 as a mono-alkylating reagent, see: Shatzmiller, S.; Lidor, R.; Bahar, E. *LA* **1991**, 381.
29. (a) Ried, W.; Mösinger, O. *CB* **1978**, *111*, 143. (b) Okajima, N.; Okada, Y. *JHC* **1991**, *28*, 177.
30. Page, P. C. B.; Klair, S. S.; Brown, M. P.; Smith, C. S.; Maginn, S. J.; Mulley, S. *T* **1992**, *48*, 5933.
31. (a) Bulten, E. J.; Gruter, H. F. M.; Martens, H. F. *JOM* **1976**, *117*, 329. (b) Sato, T.; Kikuchi, T.; Tsujita, H.; Kaetsu, A.; Sootome, N.; Nishida, K.-i.; Tachibana, K.; Murayama, E. *T* **1991**, *47*, 3281.
32. (a) Namy, J. L.; Girard, P.; Kagan, H. B.; Caro, P. E. *NJC* **1981**, *5*, 479. (b) Molander, G. A.; Kenny, C. *JACS* **1989**, *111*, 8236.

James P. Edwards
Ligand Pharmaceuticals, San Diego, CA, USA

Diiodomethane–Zinc–Titanium(IV) Chloride

CH_2I_2–Zn–$TiCl_4$

(CH_2I_2) *[75-11-6]*	CH_2I_2	(MW 267.84)
(Zn) *[7440-66-6]*	Zn	(MW 65.37)
($TiCl_4$) *[7550-45-0]*	Cl_4Ti	(MW 189.71)
(Ti(O-*i*-Pr)$_4$) *[546-68-9]*	$C_{12}H_{28}O_4Ti$	(MW 284.26)

(reagent combination for methylenation of aldehydes and ketones)

Physical Data: CH_2I_2: mp 6 °C; bp 181 °C; *d* 3.325 g cm^{-3}. Zn: mp 419.5 °C; *d* 7.140 g cm^{-3}. $TiCl_4$: mp −24 °C; bp 136.4 °C; *d* 1.730 g cm^{-3}. Ti(O-*i*-Pr)$_4$: mp 20 °C; bp 58 °C/1 mmHg; *d* 0.955 g cm^{-3}.
Solubility: sol THF.
Preparative Methods: prepared in situ.
Handling, Storage, and Precautions: CH_2I_2 is corrosive and is light sensitive; it is usually stabilized with a small amount of copper. $TiCl_4$ is highly toxic and extremely moisture sensitive. Ti(O-*i*-Pr)$_4$ is an irritant and is flammable. Zn is a moisture-sensitive, flammable solid.

General Discussion. The reagent prepared by the combination of ***Diiodomethane***, ***Zinc***, and ***Titanium(IV) Chloride*** is very effective at converting enolizable aldehydes and ketones into their methylene derivatives (eq 1).[1] The reaction many times provides improved yields over the conventional ***Methylenetriphenylphosphorane***. The β-acetoxy ketone was transformed into the alkene (eq 1) in 73% yield with minimal side reactions. The standard Wittig conditions produced only a 39% yield of the desired methylenated product along with a number of other side products derived from elimination reactions.[2]

Ph–CO–CH$_2$–CH(OAc)–C$_8$H$_{17}$ → (CH_2I_2, Zn / $TiCl_4$) → Ph–C(=CH$_2$)–CH$_2$–CH(OAc)–C$_8$H$_{17}$ (1)

73%
$Ph_3P=CH_2$ 39%

In the case of α-tetralone, the CH_2I_2 reagent combination was superior to the ***Dibromomethane–Zinc–Titanium(IV) Chloride*** reagent. The former conditions deliver the product in 88% yield, while the latter set produces only an 11% yield of alkene (eq 2).[2]

α-tetralone → (CH_2I_2, Zn / $TiCl_4$) → 1-methylenetetralin (2)

88%
CH_2Br_2 11%

By the appropriate choice of reaction conditions, chemoselective methylenations can be performed. One can chemoselectively convert an aldehyde into the methylene derivative in the presence of a ketone (eq 3). This requires that the reagent's reactivity be moderated slightly by using ***Titanium Tetraisopropoxide*** in place of $TiCl_4$. The reagent prepared from $TiCl_4$ is too reactive to offer any appreciable levels of selectivity. Alternatively, a ketone may be methylenated selectively in the presence of an aldehyde if the keto-aldehyde is first pretreated with ***Titanium Tetrakis(diethylamide)*** (eq 4). The ketones still require the $TiCl_4$ prepared reagent.[3]

keto-aldehyde (CHO) → (CH_2I_2, Zn / Ti(O-*i*-Pr)$_4$, 83%) → methyl ketone with terminal alkene (3)

keto-aldehyde (CHO) → (1. Ti(NEt$_2$)$_4$; 2. CH_2I_2, Zn, $TiCl_4$; 76%) → 1,1-disubstituted alkene aldehyde (CHO) (4)

A number of optically active α-amino ketones have been converted to allylic amines using this reagent (eq 5). The yields were moderate and the enantiomeric excesses generally were very high (>98%). The one exception was the α-phenylglycine-derived ketone which provided the alkene in only 48% ee.[4]

R–CH(NHTs)–CO–CH$_3$ → (CH_2I_2, Zn, $TiCl_4$; 35–61%) → R–CH(NHTs)–C(=CH$_2$)–CH$_3$ (5)

R = Bn, *i*-Pr, Ph

Related Reagents. Bis(η^5-cyclopentadienyl)(diiodozinc)-(μ-methylene)titanium; Bis(cyclopentadienyl)-3,3-dimethyltitanacyclobutane; Bis(cyclopentadienyl)dimethyltitanium; Dibromomethane-Zinc-Titanium(IV) Chloride; Diiodomethane.

---

1. Takai, K.; Hotta, Y.; Oshima, K.; Nozaki, H. *BCJ* **1980**, *53*, 1698.
2. Hibino, J.; Okazoe, T.; Takai, K.; Nozaki, H. *TL* **1985**, *26*, 5579.
3. Okazoe, T.; Hibino, J.; Takai, K.; Nozaki, H. *TL* **1985**, *26*, 5581.
4. Burgess, K.; Ohlmeyer, M. J. *JOC* **1991**, *56*, 1027.

Michael J. Taschner
The University of Akron, OH, USA

Diisopinocampheylboron Trifluoromethanesulfonate

$)_2B\text{-}OSO_2CF_3$

(+)
[108266-89-3] $C_{21}H_{34}BF_3O_3S$ (MW 434.36)
(−)
[108161-70-2]

(enolboration reagent for enantio- and diastereoselective aldol addition of oxazolines[1] and ketones;[2] also usoned for Ireland–Claisen rearrangement[3])

Alternate Name: diisopinocampheylboron triflate; Ipc_2BOTf.
Physical Data: colorless, viscous oil; bp ≤150 °C/0.01 mmHg ('bulb-to-bulb distillation'); $(-)Ipc_2BOTf$ $[\alpha]_D$ −43.5° (c = 30.4, hexane); ^{11}B NMR (hexane) broad singlet at $\delta = 60$ ppm (with reference to $BF_3 \cdot OEt_2$).
Solubility: highly sol in both polar and nonpolar aprotic solvents, e.g. diethyl ether, THF, dichloromethane, pentane, hexane, etc.
Preparative Methods: the (+) enantiomer of the reagent (first reported in 1981)[1] is prepared from commercially available (−)-α-pinene (~87% ee) by hydroboration with ***Borane–Dimethyl Sulfide*** in THF at 0 °C, which generates ***Diisopinocampheylborane***, (+)-Ipc_2BH, in more than 99% ee.[4] The crystalline Ipc_2BH is isolated and treated with ***Trifluoromethanesulfonic Acid*** at 0 °C either in dichloromethane[5a] or hexane.[5b] The reagent develops color in dichloromethane. However, it has been prepared in hexane as a clear and colorless solution which separates from an immiscible colored lower layer. In calculating the molarity of the reagent solution, a 60–70% conversion to triflate is assumed.[2] The reagent is usually prepared in situ from Ipc_2BH, and then its aldol reaction is carried out in the same flask by sequential addition of the required reagents[5,11] (procedure A).[6] Alternatively, the enantiomerically pure reagent can be conveniently prepared by treatment of commercially available (−)- or ***(+)-B-Chlorodiisopinocampheylborane*** (DIP-Cl™) with triflic acid at 0 °C in hexane (Scheme 1).[7] This method (procedure B) generates Ipc_2BOTf instantaneously in almost quantitative yield. The reagent generated by procedure B can be utilized for aldol reaction in the same manner as described for procedure A.

Handling, Storage, and Precautions: air sensitive; reacts instantaneously with protic solvents to liberate triflic acid; should be freshly prepared prior to use; the freshly prepared reagent turns from pale yellow to clear red upon standing. All transformations involving this reagent should be carried out under N_2 using standard techniques for air sensitive reagents; use in a fume hood.

Boron Azaenolates from Oxazolines. The reagent is useful for asymmetric aldol condensations of achiral oxazolines. Treatment of 2-ethyl-4-dimethyl-2-oxazoline with Ipc_2BOTf in the presence of a tertiary amine furnishes a boron azaenolate. Without isolation, treatment with an aldehyde in ether at −78 °C provides an alkylated oxazoline, which is hydrolyzed and converted to β-hydroxy ester via treatment with ***Diazomethane***. Although the yields for the four-step sequence are only moderate, the *anti* selectivities of the hydroxy acids are excellent with enantioselectivities of 77–85% ee (eq 1).[1]

1. $BH_3 \cdot SMe_2$, THF, 0 °C; 2. crystallization
87% ee (−)-α-pinene → 99% ee (+)-Ipc_2BH

CH_2Cl_2 or hexane, CF_3SO_3H, 0–20 °C

>99% ee (+)-Ipc_2BCl —CF_3SO_3H, hexane, 0 °C→ >99% ee (+)-Ipc_2BOTf

Scheme 1

Ipc_2BOTf, i-Pr_2NEt, ether, −78 to −20 °C; 1. RCHO; 2. H_2O_2

R = Et, Pr, pentyl, *i*-Pr
Cy, *i*-Bu

1. H^+; 2. CH_2N_2 (1)

anti:*syn* = 90:10 to 95:5
anti = 77–85% ee

Boron Enolates from Ketones. Boron enolates are highly versatile intermediates in organic synthesis.[8] Their high reactivity and stereoselectivity are often utilized for aldol condensation reactions.[9,10] The reagent has been elegantly applied for regio- and stereoselective enolboration of ketones and subsequent enantio- and diastereoselective aldol reactions with aldehydes.[11] For example, the aldol reaction between ethyl ketones and aldehydes using the (+) or (−) reagent in the presence of a tertiary amine in dichloromethane gives (via the desired (*Z*)-enolborinate) *syn*-α-methyl-β-hydroxy ketones in good enantiomeric excess (66–93% ee) and with high diastereoselectivity (≥95%) (eq 2). In contrast, the *anti* selectivity of the aldol product derived from diethyl ketone via formation of the (*E*)-enolate, derived from Ipc_2BCl (DIP-Cl™) with ***Methacrolein***, proceeds with negligible enantioselectivity.[11a] However, use of both the triflate and the chloride reagents in the aldol reaction of methyl ketones with aldehydes have been reported to give β-hydroxy ketones in moderate enantiomeric excess (53–78% ee) with a reversal in the enantioface selectivity of the aldehyde compared to the corresponding ethyl ketone *syn*-aldol. This variable selectivity is interpreted as evidence for the participation of competing chair and boat transition states.[11]

(–)-Ipc_2BOTf, i-Pr_2NEt, CH_2Cl_2, –78 °C or 0 °C; (Z) 97:3 (E); 1. R^2CHO 2. H_2O_2 45–99%

syn: 66–93% ee anti
95:5 to 98:2 syn:anti selectivity (2)

The aldol methodology mediated by Ipc_2BOTf was successfully applied to a macrolide antibiotic synthesis. Paterson reported a convenient asymmetric synthesis of a C_{19}–C_{27} segment of rifamycin S used in the Kishi synthesis, based on ethyl ketone aldol reactions mediated by optically pure reagent.[12a] He also reported the novel aldol approach to the synthesis of an enantiomerically pure C_7–C_{15} segment of tirandamycin A.[12b] This was prepared via enolboration of the (*R*)-ethyl ketone by (–)-Ipc_2BOTf in the presence of a tertiary amine. Addition of aldehyde to the corresponding enolborinate, followed by oxidative workup and chromatographic purification, led to the two separated *syn*-aldol isomers (8:1 ratio with 63% combined yield) with no *anti*-aldol product detected by HPLC. The major 1,2-*syn*-3,4-*syn* diastereomer is reported to be enantiomerically pure. Moreover, it was observed that enantiomeric excess of the major aldol isomer is significantly enhanced relative to the starting ketone. The corresponding aldol product is reduced to the 1,3-diol (eq 3) and subsequently converted to the enantiomerically pure C_7–C_{15} segment of tirandamycin A via pyranone synthesis.[12b]

(–)-Ipc_2BOTf, Et_3N, CH_2Cl_2, 0 °C, 2 h; 1. (furfural) CHO 2. H_2O_2 3. Me_4N•$HB(OAc)_3$; >95% ds (3)

Dihydropyrones are valuable intermediates for the synthesis of a variety of substituted tetrahydropyran rings. Recently, stereoselective aldol reactions of β-chlorovinyl ketones using the dienol boronate derivative derived from chiral Ipc_2BOTf was utilized for enantioselective formation of dihydropyrones. No detectable racemization was reported on the cyclization step (eq 4).[12c]

Ireland–Claisen Rearrangement. Oh et al. recently reported the Ireland–Claisen rearrangement of a variety of *O*-protected 2-butenyl glycolates via chelated boron and tin triflates to give, after esterification, methyl 2-methoxy/benzyloxy-3-methyl-4-pentenoates.[3] In reactions using Ipc_2BOTf, diastereoselection as high as 99.5% was reported. The diastereoselection obtained in reactions using ***Tin(II) Trifluoromethanesulfonate***, ***Zinc Trifluoromethanesulfonate***, and ***Di-n-butylboryl Trifluoromethanesulfonate*** was far lower than in reactions with Ipc_2BOTf. Moreover, the rate of rearrangement with boron enolates was found to be higher than the rates of rearrangement of the silyl ketene acetals or lithium enolate. As anticipated the *cis*-alkene gives better diastereoselectivity than the *trans* isomer (eq 5). With this high diastereoselection obtained using Ipc_2BOTf, it is surprising that the enantioselectivity of this reaction is only 0–10% ee.[3]

1. (–)-Ipc_2BOTf, i-Pr_2NEt, CH_2Cl_2, –78 °C, 16 h 2. PrCHO 3. H_2O_2 40%; 80% ee; Me_3SiOTf, i-Pr_2NEt 78%; 80% ee (4)

1. (–)-Ipc_2BOTf, i-Pr_2NEt, –78 to 25 °C, hexane 2. CH_2N_2 60–64%

1. *trans*, R = Me 1. 92:8
2. *cis*, R = Bn 2. <0.5:>99.5 (5)

Related Reagents. (+)-B-Chlorodiisopinocampheylborane; Diisopinocampheylborane; (*R*,*R*)-2,5-Dimethylborolane.

---

1. (a) Meyers, A. I.; Yamamoto, Y. *T* **1984**, *40*, 2309. (b) *JACS* **1981**, *103*, 4278.
2. Paterson, I.; Goodman, J. M.; Lister, M. A.; Schumann, R. C.; McClure, C. K.; Norcross, R. D. *T* **1990**, *46*, 4663.
3. Oh, T.; Wrobel, Z.; Devine, P. N. *SL* **1992**, 81.
4. (a) Brown, H. C.; Singaram, B. *JOC* **1984**, *49*, 945. (b) Brown, H. C.; Joshi, N. N. *JOC* **1988**, *53*, 4059.
5. (a) Paterson, I.; Lister, M. A.; McClure, C. K. *TL* **1986**, *27*, 4787. (b) Paterson, I.; Lister, M. A. *TL* **1988**, *29*, 585.
6. Purification of Ipc_2BOTf by distillation is unnecessary and probably inadvisable as optimum results are obtained with freshly prepared undistilled reagent.[2]
7. Dhar, R. K.; Brown, H. C.; unpublished results.
8. (a) Kim, B. M.; Williams, S. F.; Masamune, S. *COS* **1991**, *2*, Chapter 5. (b) Evans, D. A. In *Asymmetric Synthesis*; Morrison, J. D.; Ed.; Academic: New York, 1984; Vol. 3, Chapter 1. (c) Heathcock, C. H. In *Asymmetric Synthesis*; Morrison, J. D.; Ed.; Academic: New York, 1984; Vol. 3, Chapter 2. (d) Evans, D. A.; Nelson, J. V.; Taber, T. R. *Top. Stereochem.* **1982**, *13*, 1.
9. For an examination of the effect of leaving group (X) on the stereoselective enolboration of ketones with various R_2BX reagents (X = OTf, OMs, Cl, Br, I), see: (a) Brown, H. C.; Ganesan, K.; Dhar, R. K. *JOC* **1993**, *58*, 147. (b) Brown, H. C.; Dhar, R. K.; Bakshi, R. K.; Pandiarajan, P. K.; Singaram, B. *JACS* **1989**, *111*, 3441. (c) Goodman, J. M.; Paterson, I. *TL* **1992**, 7223.
10. Selective *trans* deprotonation of the ketone–L_2BOTf complex leads to formation of (*Z*)-enolborinate; see: Evans, D. A.; Nelson, J. V.; Vogel, E.; Taber, T. R. *JACS* **1981**, *103*, 3099. An alternative explanation for L_2BCl to (*E*)-enol borinate and L_2BOTf to (*Z*)-enol borinate has been proposed by Corey and Kim; see: Corey, E. J.; Kim, S. S. *JACS* **1990**, *112*, 4976.
11. (a) Paterson, I.; Goodman, J. M. *TL* **1989**, *30*, 997. (b) Paterson, I.; Goodman, J. M.; Isaka, M. *TL* **1989**, *30*, 7121. (c) Paterson, I.; McClure, C. K. *TL* **1987**, *28*, 1229.

12. (a) Paterson, I.; McClure, C. K.; Schumann, R. C. *TL* **1989**, *30*, 1293. (b) Paterson, I.; Lister, M. A.; Ryan, G. R. *TL* **1991**, *32*, 1749. (c) Paterson, I.; Osborne, S. *TL* **1990**, *31*, 2213.

Raj K. Dhar
Aldrich Chemical Company, Sheboygan Falls, WI, USA

(R^*,R^*)- α-(2,6-Diisopropoxybenzoyloxy)-5-oxo-1,3,2-dioxaborolane-4-acetic Acid[1]

(*R,R*)
[131703-55-4] $C_{17}H_{21}BO_9$ (MW 380.16)
(*S,S*)
[131703-56-5]

(chiral Lewis acid catalyst for Diels–Alder,[1,2] aldol-type,[3] allylation,[4] and hetero Diels–Alder[5] reactions)

Solubility: sol dichloromethane, propionitrile, THF.
Form Supplied in: the acyloxyborane·THF complex is available as a 0.1–0.2 M solution in dichloromethane or propionitrile.
Analysis of Reagent Purity: ^{1}H NMR (CD_2Cl_2, −95 °C, 500 MHz) δ 1.07–1.13 (m, 6H, $2CH_3$), 1.24 (br, 6H, $2CH_3$), 4.50 (br, 2H, $2(CH_3)_2CH$), 4.70–4.92 (m, 1H, $CHCO_2B$), 5.45–5.72 (m, 1H, $CHCO_2H$), 6.48 (br, 2H, 2*m*-H), 7.21 (br, 1H, *p*-H).
Preparative Methods: to a solution of (*R,R*)- or (*S,S*)-mono-(2,6-diisopropoxybenzoyl)tartaric acid (74 mg, 0.2 mmol) in dry dichloromethane or propionitrile (1 mL) is added BH_3·THF (0.189 mL of 1.06 M solution in THF, 0.2 mmol) at 0 °C under an argon atmosphere. The reaction mixture is stirred for 1 h at 0 °C to produce the chiral acyloxyborane. Only 2 equiv of hydrogen gas should evolve under these reaction conditions (0 °C). See also Furuta.[1]
Handling, Storage, and Precautions: the acyloxyborane solution should be flushed with Ar and stored tightly sealed (to preclude contact with oxygen and moisture) below 0 °C. Use in a fume hood.

Acyloxyborane as an Activating Device for Carboxylic Acids.[2a] The reduction of carboxylic acids by borane is an important procedure in organic synthesis. The remarkable reactivity of borane towards carboxylic acids over esters is characteristic of this reagent. Such selectivity is rarely seen with other hydride reagents.

The rapid reaction between carboxylic acids and borane is related to the electrophilicity of the latter. The carbonyl group of the initially formed acyloxyborane intermediate, which is essentially a mixed anhydride, is activated by the Lewis acidity of the trivalent boron atom. Addition of 1/3 equiv of the ***Borane-Tetrahydrofuran*** complex to acrylic acid in dichloromethane followed by addition of a diene at low temperature results in the formation of Diels–Alder adducts in good yield (eq 1). Further, the reaction is successful even with a catalytic amount of borane.

(1) CO_2H —[1/3 equiv BH_3; CH_2Cl_2, 0 °C]→ [(CO_2)$_3$B] —[−78 °C; cyclopentadiene]→ CO_2H adduct, 75%

Asymmetric Diels–Alder Reaction of Unsaturated Carboxylic Acids.[2a] A chiral acyloxyborane (CAB) complex (**1**) prepared from mono(2,6-dimethoxybenzoyl)tartaric acid and 1 equiv of borane is an excellent catalyst for the Diels–Alder reaction of α,β-unsaturated carboxylic acids and dienes. In the CAB-catalyzed Diels–Alder reaction, adducts are formed in a highly diastereo- and enantioselective manner under mild reaction conditions (eq 2). The reaction is catalytic: 10 mol % of catalyst is sufficient for efficient conversion, and the chiral auxiliary can be recovered and reused.

(2) CO_2H + cyclopentadiene —[(**1**) (10 mol%); CH_2Cl_2, −78 °C; 93%]→ CO_2H adduct, 78% ee (*R*), *endo*:*exo* = 96:4

Asymmetric Diels–Alder Reaction of Unsaturated Aldehydes.[1,2b–e] The boron atom of acyloxyborane is activated by the electron-withdrawing acyloxy groups, and consequently acyloxyborane derivatives are sufficiently Lewis acidic to catalyze certain reactions. Thus, asymmetric Diels–Alder reactions of α,β-enals with dienes using (**1**) as a Lewis acid catalyst have been developed. For example, the reaction of cyclopentadiene and methacrolein gives the adduct in 85% yield (*endo*:*exo* = 11:89) and 96% ee (major *exo* isomer) (eq 3). Some additional examples are listed in Figure 1. The α-substituent on the dienophile increases the enantioselectivity, while β-substitution dramatically decreases the selectivity. In the case of a substrate having substituents in both α- and β-positions, high enantioselectivity is observed; thus the α-substituent effect overcomes that of the β-substituent.

(3) CHO + cyclopentadiene —[10 mol% (**1**); CH_2Cl_2, −78 °C; 85%]→ CHO adduct, 96% ee (*R*), *endo*:*exo* = 11:89

The intramolecular Diels–Alder reaction of 2-methyl-(*E,E*)-2,7,9-decatrienal with CAB catalysis proceeds with high diastereo- and enantioselectivities.[2c]

Asymmetric Aldol-Type Reaction.[3] CAB complex (**2**) is an excellent catalyst for the Mukaiyama condensation of simple achiral enol silyl ethers of ketones with various aldehydes. The CAB-catalyzed aldol process allows the formation of adducts in a highly diastereo- and enantioselective manner (up to 96% ee) under mild reaction conditions (eqs 4 and 5). The reactions are catalytic: 20 mol % of catalyst is sufficient for efficient conversion, and the chiral auxiliary can be recovered and reused.

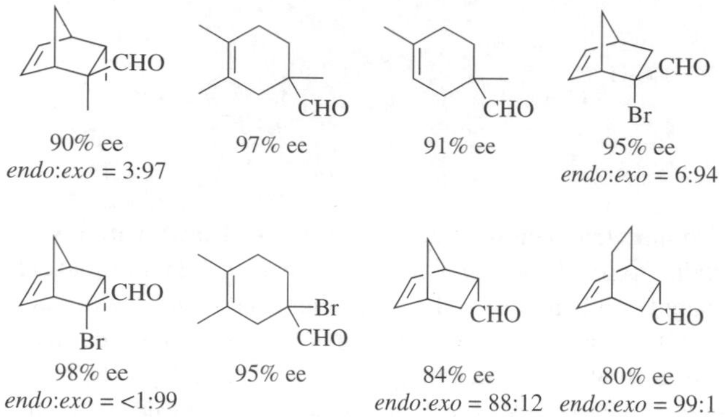

90% ee, *endo*:*exo* = 3:97; 97% ee; 91% ee; 95% ee, *endo*:*exo* = 6:94

98% ee, *endo*:*exo* = <1:99; 95% ee; 84% ee, *endo*:*exo* = 88:12; 80% ee, *endo*:*exo* = 99:1

Figure 1 Products of asymmetric Diels-Alder reactions

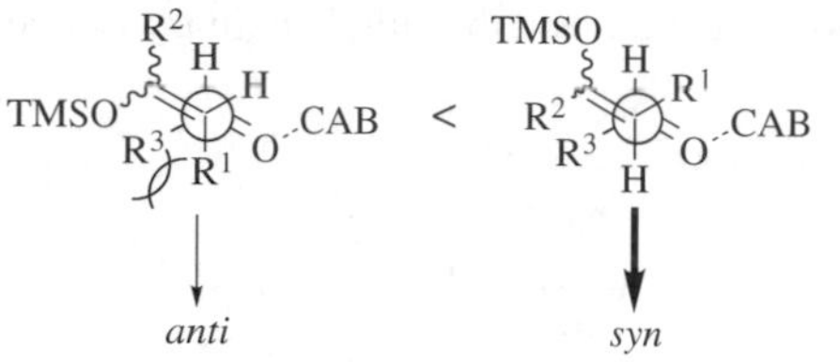

anti < syn

Figure 2 Extended transition state model

PhCHO + [Ph, OTMS enol ether] —(2) (20 mol%), EtCN, −78 °C, 98%→ TMSO…Ph (4) 85% ee

PhCHO + [OTMS, Et enol ether] —20 mol% (2), EtCN, −78 °C→ (5)

(*E*):(*Z*) = 4:1 96%, 96% ee, *syn*:*anti* = 94:6
(*E*):(*Z*) = 1:49 97%, 96% ee, *syn*:*anti* = 93:7

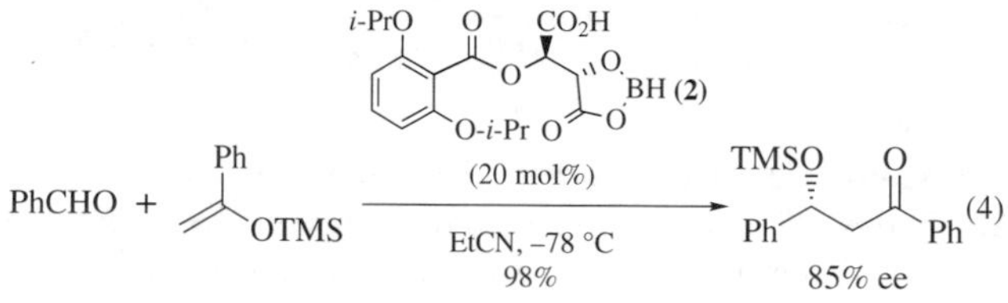

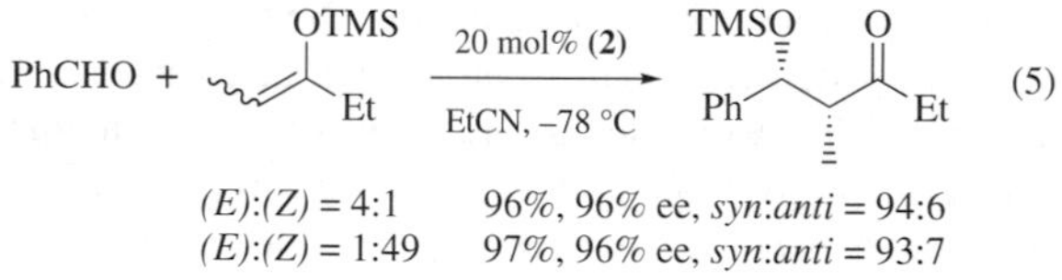

Almost perfect asymmetric induction is achieved in the *syn* adducts, reaching 96% ee, although a slight reduction in both the enantio- and diastereoselectivities is observed in the reactions with saturated aldehydes. It is noteworthy that, regardless of the stereochemistry of the starting enol silyl ethers, the CAB-catalyzed reaction is highly selective for *syn* adducts. The high *syn* selectivity and the independence of selectivity on the stereochemistry of silyl ethers in the CAB-catalyzed reactions are fully consistent with Noyori's ***Trimethylsilyl Trifluoromethanesulfonate***-catalyzed aldol reactions of acetals,[6] and thus may reflect the acyclic extended transition state mechanism postulated in the latter reactions (Figure 2). Judging from the product configurations, the CAB catalyst (from natural tartaric acid) should effectively cover the *si*-face of carbonyl following its coordination and the selective approach of nucleophiles from the *re*-face should result.

A catalytic asymmetric aldol-type reaction of ketene silyl acetals with achiral aldehydes also proceeds with the CAB catalyst (**2**), which can furnish *syn*-β-hydroxy esters with high enantioselectivity (eq 6).

This reaction is sensitive to the substituents of the starting ketene acetals. The use of ketene silyl acetals from phenyl esters leads to good diastereo- and enantioselectivities with excellent chemical yields.

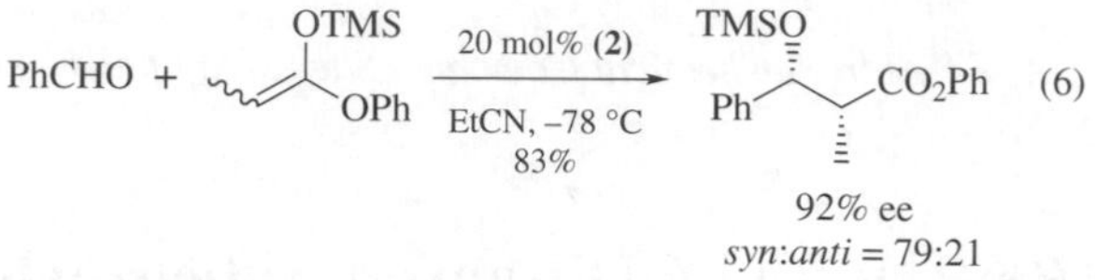

PhCHO + [OTMS, OPh ketene acetal] —20 mol% (2), EtCN, −78 °C, 83%→ TMSO, Ph, CO_2Ph (6)

92% ee, *syn*:*anti* = 79:21

Analogous with the previous results of enol silyl ethers of ketones, nonsubstituted ketene silyl acetals are found to exhibit lower levels of stereoregulation, while the propionate-derived ketene silyl acetals display a high level of asymmetric induction. The reactions with aliphatic aldehydes, however, resulted in a slight reduction in optical and chemical yields. With phenyl ester-derived ketene silyl acetals, *syn* adducts predominate, but the selectivities are moderate in most cases in comparison with the reactions of ketone-derived silyl enol ethers. Exceptions are α,β-unsaturated aldehydes, which revealed excellent diastereo- and enantioselectivities. The observed *syn* selectivity and *re*-face attack of nucleophiles on the carbonyl carbon of aldehydes are consistent with the aforementioned aldol reactions of ketone-derived enol silyl ethers.

Asymmetric Allylation (Sakurai–Hosomi Allylation).[4] Addition of allylsilanes to achiral aldehydes promoted by CAB catalyst (**2**) (20 mol %) at −78 °C in propionitrile produces homoallylic alcohols with excellent enantioselectivity (eq 7).

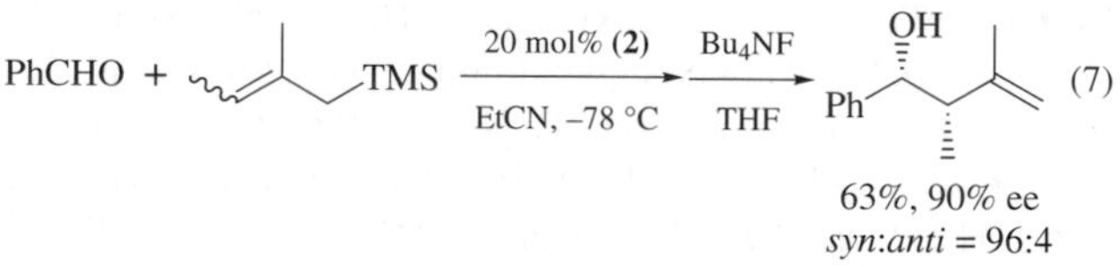

PhCHO + [allyl-TMS] —20 mol% (2), EtCN, −78 °C; Bu_4NF, THF→ OH, Ph (7)

63%, 90% ee, *syn*:*anti* = 96:4

Alkyl substitution at the alkene of the allylsilanes increases the reactivity, permitting lower reaction temperature and improved asymmetric induction. γ-Alkylated allylsilanes exhibit excellent diastereo- and enantioselectivities, affording *syn* homoallylic alcohols with high enantiomeric purity. The *syn* selectivity of these reactions is independent of the allylsilane configuration. Thus regardless of the geometry of the starting allylsilane, the predominant isomer in this reaction has *syn* configuration. The observed preference for relative and absolute configurations for the adduct alcohols derived from reaction catalyzed by the (2*R*,3*R*)-ligand–borane reagent can be rationalized on the basis of an extended transition state model similar to that for the CAB-catalyzed aldol reaction (see Figure 2).

Allystannanes are more nucleophilic than allylsilanes. Addition of achiral allylstannanes to achiral aldehydes in the presence of (**1**) (20 mol %) and ***Trifluoroacetic Anhydride*** (40 mol %) also affords homoallylic alcohols with high diastereo- and enantioselectivities (eq 8).

Asymmetric Hetero Diels–Alder Reaction.[5] In contrast to the CAB catalyst (**2**; R = H) which is stable and both air and moisture sensitive, the *B*-arylated CAB catalyst (**3**; R = Ph or alkyl) is stable and can be stored in a closed container at rt. A solution of the CAB (**3**; R = Ph) catalyzes Diels–Alder, aldol, and

Sakurai–Hosomi reactions. Although the asymmetric inductions achieved by these complexes are slightly less efficient than that of the corresponding hydride-type catalyst, the CAB catalyst (**3**; R = Ph) is shown to be an excellent system for hetero Diels–Alder reactions.

20 mol% (**1**), 40 mol% $(CF_3CO)_2O$, EtCN, –78 °C, 85%; 95% ee, *syn*:*anti* = 92:8 (8)

The *B*-arylated CAB catalyst (**3**) is easily prepared in situ by mixing a 1:1 molar ratio of tartaric acid derivative and phenylboronic acid in dry propionitrile at room temperature for 0.5 h. The hetero Diels–Alder reaction of aldehydes with Danishefsky dienes is promoted by 20 mol% of this catalyst solution at −78 °C for several hours to produce dihydropyranone derivatives of high enantiomeric purity (eq 9).

+ PhCHO, BAr (**3**) (20 mol%), EtCN, –78 °C (9)

R = H Ar = 2,4,6-*i*-$Pr_3C_6H_2$ 95% ee (*R*)
R = Me Ar = *o*-$MeOC_6H_4$ 97% ee (2*R*, 3*R*)

Related Reagents. (*R*)-1,1′-Bi-2,2′-naphthotitanium Dichloride; (*R*)-1,1′-Bi-2,2′-naphthotitanium Diisopropoxide; (*S*,*S*)-2,2′-(Dimethylmethylene)bis(4-*t*-butyl-2-oxazoline); (4*R*, 5*R*)-2,2-Dimethyl-4,5-bis(hydroxydiphenylmethyl)-1, 3-dioxolane-Titanium(IV) Chloride; 2,2-Dimethyl-α,α,α′,α′-tetraphenyl-1,3-dioxolane-4,5-dimethanolatotitanium Diisopropoxide.

1. Furuta, K.; Gao, Q.; Yamamoto, H. *OS* **1995**, *72*, 86.
2. (a) Furuta, K.; Miwa, Y.; Iwanaga, K.; Yamamoto, H. *JACS* **1988**, *110*, 6254. (b) Furuta, K.; Shimizu, S.; Miwa, Y.; Yamamoto, H. *JOC* **1989**, *54*, 1481. (c) Furuta, K.; Kanematsu, A.; Yamamoto, H. *TL* **1989**, *30*, 7231. (d) Ishihara, K.; Gao, Q.; Yamamoto, H. *JOC* **1993**, *58*, 6917. (e) Ishihara, K.; Gao, Q.; Yamamoto, H. *JACS* **1993**, *115*, 10412.
3. (a) Furuta, K.; Maruyama, T.; Yamamoto, H. *JACS* **1991**, *113*, 1041. (b) Furuta, K.; Maruyama, T.; Yamamoto, H. *SL* **1991**, 439. (c) Ishihara, K.; Maruyama, T.; Mouri, M.; Gao, Q.; Furuta, K.; Yamamoto, H. *BCJ* **1993**, *66*, 3483.
4. (a) Furuta, K.; Mouri, M.; Yamamoto, H. *SL* **1991**, 561. (b) Marshall, J. A.; Tang, Y. *SL* **1992**, 653. (c) Ishihara, K.; Mouri, M.; Gao, Q.; Maruyama, T.; Furuta, K.; Yamamoto, H. *JACS* **1993**, *115*, 11490.
5. (a) Gao, Q.; Maruyama, T.; Mouri, M.; Yamamoto, H. *JOC* **1992**, *57*, 1951. (b) Gao, Q.; Ishihara, K.; Maruyama, T.; Mouri, M.; Yamamoto, H. *T* **1994**, *50*, 979.
6. Noyori, R.; Murata, S.; Suzuki, M. *T* **1981**, *37*, 3899.

Kazuaki Ishihara & Hisashi Yamamoto
Nagoya University, Japan

Diisopropyl 2-Crotyl-1,3,2-dioxaborolane-4,5-dicarboxylate[1,2]

(**1**) L-(*R*,*R*)-tartrate (*E*)-crotyl
[99745-86-5] $C_{14}H_{23}BO_6$ (MW 298.14)
D-(*S*,*S*), (*Z*)
[99687-40-8]
L-(*R*,*R*), (*Z*)
[106357-20-4]
(**2**) D-(*S*,*S*), (*E*)
[106357-33-9]

(reagents for the asymmetric crotylboration of aldehydes to produce either *syn* or *anti* β-methylhomoallylic alcohols)[2]

Physical Data: bp 80 °C/0.1 mmHg.

Solubility: sol toluene, THF, ether, or CH_2Cl_2.

Analysis of Reagent Purity: ^{11}B NMR, data for (*E*)-crotyl: (δ 34.8, C_6D_6);[2b] the purity of the reagent is best determined by capillary GC;[2b] solutions of the reagent can be standardized using cyclohexanecarbaldehyde.[2b]

Preparative Method: prepared by treatment of (*E*)- or (*Z*)-crotylpotassium with ***Triisopropyl Borate*** followed by acidic extractive workup and direct esterification with diisopropyl tartrate (DIPT) (eqs 1 and 2).[2]

1. $(i\text{-PrO})_3B$, –78 °C 2. H_3O^+, Et_2O 3. DIPT, $MgSO_4$ (1)
(*R*,*R*)-(**1**) tartrate (*E*)-crotylboronate

1. $(i\text{-PrO})_3B$, –78 °C 2. H_3O^+, Et_2O 3. DIPT, $MgSO_4$ (2)
(*S*,*S*)-(**2**) tartrate (*E*)-crotylboronate

Handling, Storage, and Precautions: typically the reagents are handled as solutions in toluene (0.5–1M) and transferred by syringe under an inert atmosphere; stored neat or as a solution in toluene over 4Å molecular sieves under an argon atmosphere in a refrigerator (−20 °C), the reagent is stable for many months. In the presence of water, (**1**) rapidly hydrolyzes to achiral crotylboronic acid, the presence of which leads to reduced enantioselectivity in reactions with aldehydes.

Reactions with Achiral Aldehydes. The tartrate ester modified (*E*)- and (*Z*)-crotylboronates undergo rapid additions to aldehydes at −78 °C. The enantioselectivities obtained for aliphatic linear or α-monobranched aldehydes range from 72 to 91% ee.[2] When cyclohexanecarbaldehyde is treated with the (*E*)-crotylboronate reagent at −95 °C in toluene, the homoallylic alcohol is obtained in 98% yield and 91% ee (eq 3). The (*Z*)-crotylboronate reagent gives slightly lower selectivity (83% ee, eq 4). The *anti*:*syn*/*syn*:*anti* ratios obtained are also excellent for this reagent (typically greater than 98:2 and 2:98 for (**1**) and (**2**), respectively).

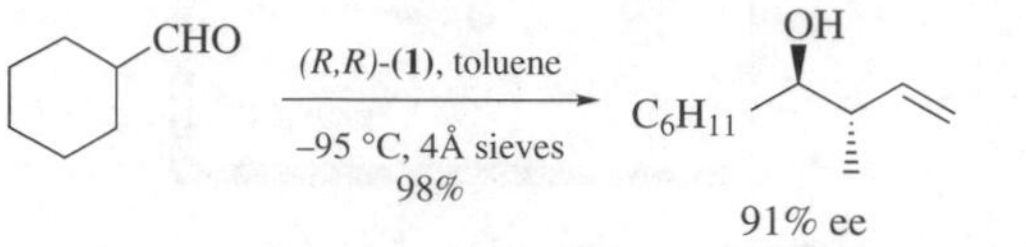

(3)

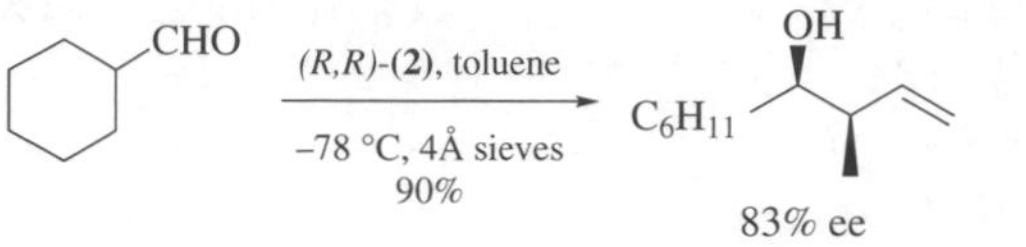

(4)

As with the corresponding allylboronate, the enantioselectivity of reactions with β-alkoxy and conjugated aldehydes are lower (55–74% ee). In the case of benzaldehyde (91%, 66% ee), selectivity can be improved by the use of the derived chromium tricarbonyl complex. The homoallylic alcohol is obtained after oxidative decomplexation in high yield and 92% enantiomeric purity (eq 5).[3]

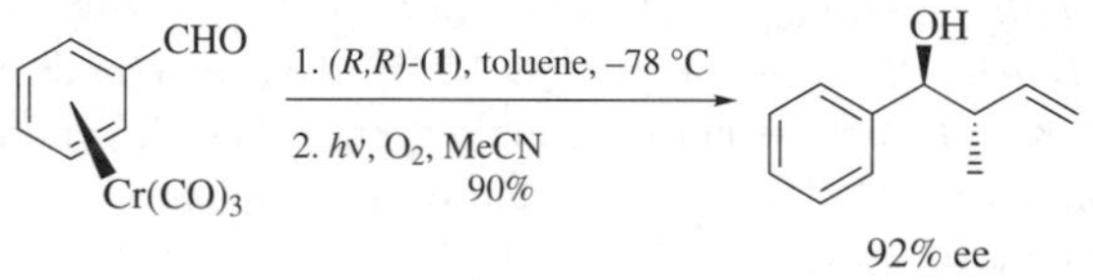

(5)

Reactions with Chiral Aldehydes. Addition of the (*E*)- or (*Z*)-crotylboronate reagent to optically active β-alkoxy-α-methylpropionaldehydes gives the corresponding polypropionate structures with good to excellent diastereoselection (eqs 6 and 7).[4] Three of the four stereochemical triads can be prepared in high yield with useful levels of selectivity. The all-*syn* stereoisomer of eq 7 is best prepared using other methods, such as the crotyltin methodology developed by Keck and co-workers.[5] The polypropionate structures with 1,3-*anti* relationships between branching methyl groups are prepared with excellent diastereoselection (via matched double asymmetric reactions). Those with a 1,3-*syn* relationship are more difficult to prepare. The relative diastereoselectivity of the reaction of α-methyl chiral aldehydes with (*E*)- and (*Z*)-crotylboronates can be predicted by use of the *gauche* pentane model.[6]

Both (*E*)- and (*Z*)-crotylboronates have been used in several applications in natural product synthesis.[4b,7] One application of both the allylboronate and (*E*)-crotylboronate reagents is found in the synthesis of the C(19)–C(29) segment of rifamycin S. The desired stereochemistry at C(25)–C(26) of the rifamycin ansa chain is set with excellent stereocontrol (>95:5) and high yield (87%) (eq 8).[4b,7a]

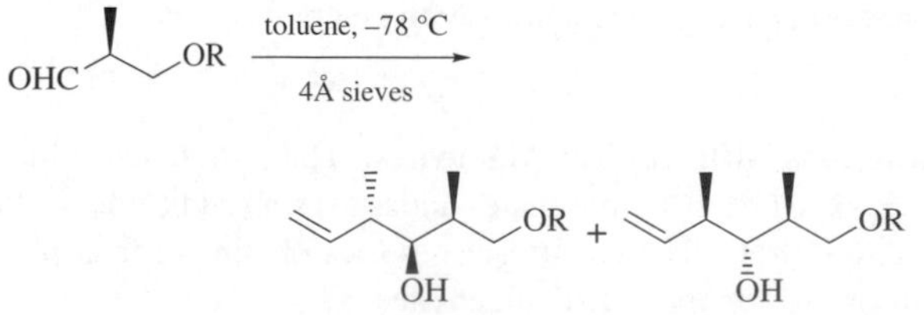

(6)

R = TBDMS, (*R,R*)-(**1**) 97:3 80%
R = TBDPS, (*S,S*)-(**1**) 10:90 77%

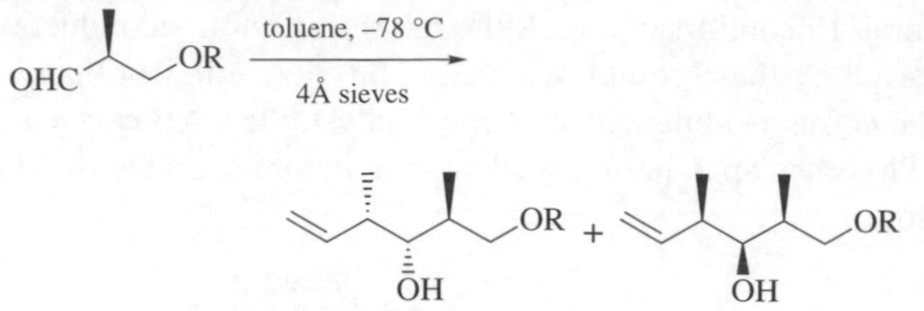

(7)

R = TBDMS, (*R,R*)-(**2**) 98:2 71%
R = TBDPS, (*S,S*)-(**2**) 27:73

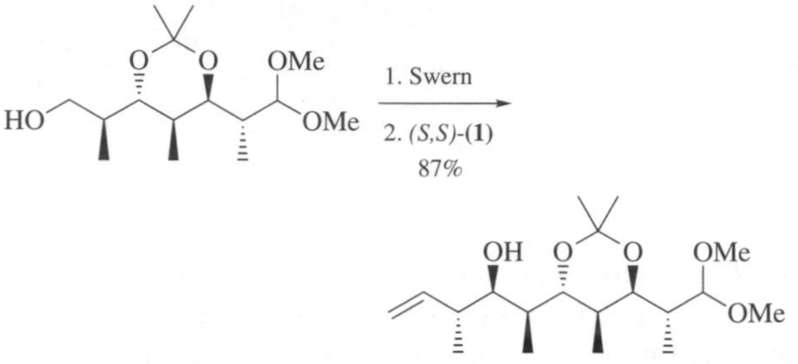

(8)

The (*E*)- and (*Z*)-crotylboronates provide selectivity in the best cases comparable to that obtained with other crotylboration procedures. Combining ease of preparation, stability, and selectivity the tartrate-modified (*E*)- and (*Z*)-crotylboronates are highly useful propionate enolate equivalents.

Related Reagents. *B*-Allyl-9-borabicyclo[3.3.1]-nonane; *B*-Allyldiisocaranylborane; *B*-Allyldiisopinocampheylborane; *B*-Crotyldiisopinocampheylborane; (*R,R*)-2,5-Dimethylborolane.

1. Roush, W. R. *COS* **1991**, *2*, 1.
2. (a) Roush, W. R.; Halterman, R. L. *JACS* **1986**, *108*, 294. (b) Roush, W. R.; Ando, K.; Powers, D. B.; Palkowitz, A. D.; Halterman, R. L. *JACS* **1990**, *112*, 6339; *JACS* **1991**, *114*, 5133.
3. Roush, W. R.; Park, J. C. *JOC* **1990**, *55*, 1143.
4. (a) Roush, W. R.; Palkowitz, A. D.; Palmer, M. A. J. *JOC* **1987**, *52*, 316. (b) Roush, W. R.; Palkowitz, A. D.; Ando, K. *JACS* **1990**, *112*, 6348.
5. Keck, G. E.; Abbott, D. E. *TL* **1984**, *25*, 1883.
6. (a) Evans, D. A.; Nelson, J. V.; Taber, T. R. *Top. Stereochem.* **1982**, *13*, 1. (b) Roush, W. R. *JOC* **1991**, *56*, 4151.
7. (a) Roush, W. R.; Palkowitz, A. D. *JACS* **1987**, *109*, 953. (b) Danishefsky, S. J.; Armistead, D. M.; Wincott, F. E.; Selnick, H. G.; Hungate, R. *JACS* **1987**, *109*, 8117. (c) Coe, J. W.; Roush, W. R. *JOC* **1989**, *54*, 915. (d) Roush, W. R.; Palkowitz, A. D. *JOC* **1989**, *54*, 3009. (e) Tatsuta, K.; Ishiyama, T.; Tajima, S.; Koguchi, Y.; Gunji, H. *TL* **1990**, *31*, 709. (f) Akita, H.; Yamada, H.; Matsukura, H.; Nakata, T.; Oishi, T. *TL* **1990**, *31*, 1735. (g) White, J. D.; Johnson, A. T. *JOC* **1990**, *55*, 5938. (h) Fisher, M. J.; Myers, C. D.; Joglar, J.; Chen, S.-H.; Danishefsky, S. J. *JOC* **1991**, *56*, 5826. (i) Roush, W. R.; Bannister, T. D. *TL* **1992**, *33*, 3587. (j) Roush, W. R.; Brown, B. B. *JACS* **1993**, *115*, 2268. (k) White, J. D.; Porter, W. J.; Tiller, T. *SL* **1993**, 535.

David J. Madar
Indiana University, Bloomington, IN, USA

Diketene[1]

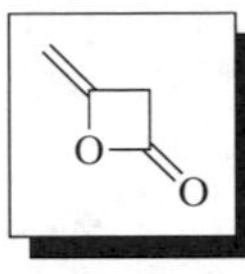

[674-82-8] $C_4H_4O_2$ (MW 84.07)

(acetoacetylation reagent; acetylketene equivalent for heterocyclic synthesis)

Alternate Name: 4-methyleneoxetan-2-one.
Physical Data: mp −7.5 °C; bp 69–70 °C/100 mmHg; *d* 1.090 g cm^{-3}.
Solubility: sol most organic solvents; immiscible H_2O, hexane.
Form Supplied in: neat liquid.
Analysis of Reagent Purity: IR (2150 cm^{-1}), NMR.
Preparative Method: via dimerization of ketene.[2]
Purification: vacuum distillation.
Handling, Storage, and Precautions: lachrymator; best stored as a solid (refrigerated) in plastic containers; avoid contamination; prepare for exothermic reactions. Handle in a fume hood.

Acetoacetylation Reagent. Diketene is most commonly used as a reagent for the acetoacetylation of nucleophiles, including alcohols, phenols, amines, anilines, thiols, and carbanions. ***4-Dimethylaminopyridine***[3] and tertiary amines are often the catalysts of choice, but sodium acetate, sulfuric acid, and a variety of other catalysts can also be used (eq 1). Amines can usually be acetoacetylated at 0–20 °C without any catalyst.

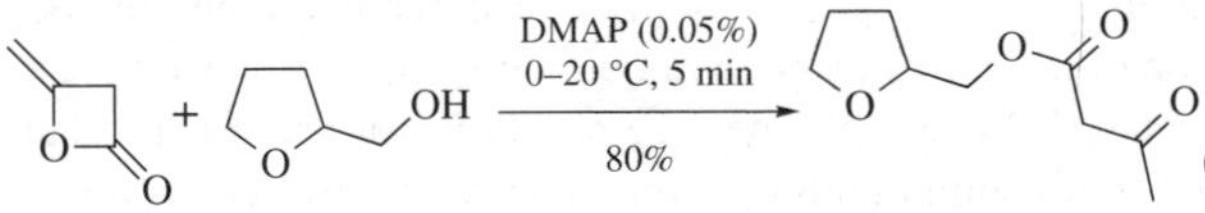

(1)

Alternate reagents for acetoacetylation include *t*-butyl acetoacetate[4] and ***2,2,6-Trimethyl-4H-1,3-dioxin-4-one*** (the diketene–acetone adduct). Both are prepared from diketene and are less reactive but easier to handle.

Acetoacetate esters are versatile synthetic intermediates; their chemistry is more thoroughly described elsewhere.[1a,b] Acetoacetate esters can be converted into γ,δ-unsaturated ketones (via Carroll rearrangement, eq 2),[5] into a variety of heterocycles via diazotization/cyclization (eq 3)[6] or the Hantzsch pyridine synthesis (eq 4),[7] or in the direct synthesis of other heterocycles, as discussed below.

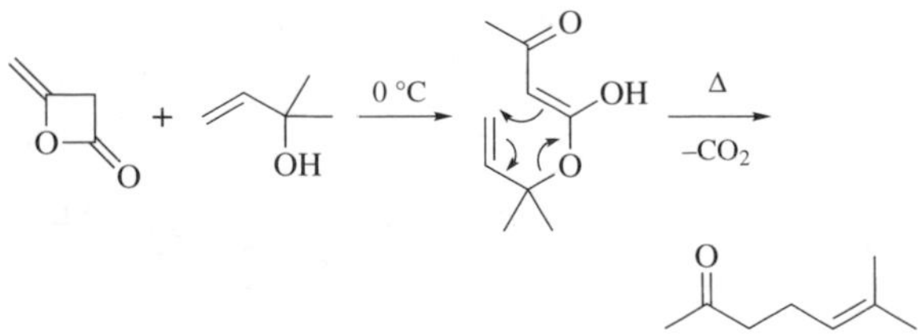

(2)

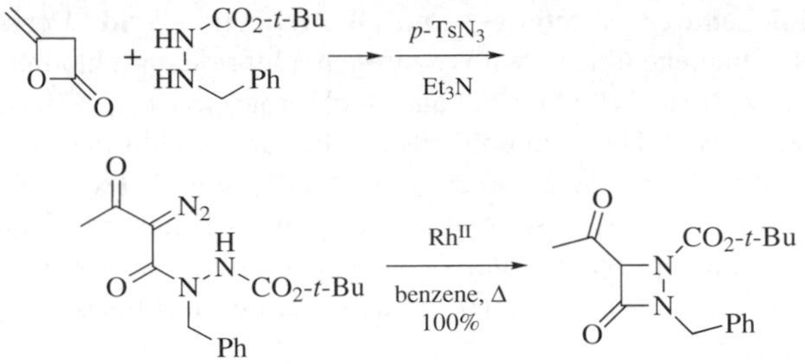

(3)

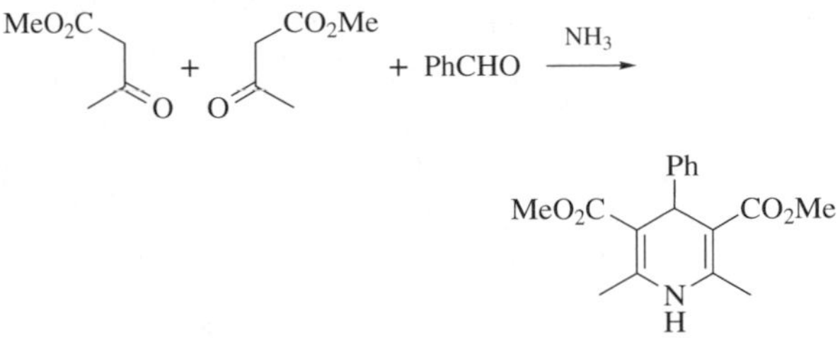

(4)

Preparation of Ketene. Diketene is a convenient laboratory source of ketene. The dimer can be 'cracked' at temperatures above 500 °C in a hot tube with a glowing filament (eq 5) to provide two molecules of ***Ketene***, free from the methylketene contaminant which is observed during the preparation of ketene by pyrolysis of acetone.[8]

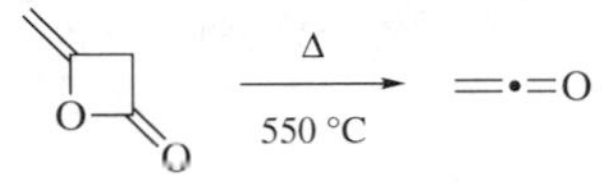

(5)

Malonic Anhydride. The ozonolysis of diketene provides malonic anhydride, which can be trapped with nucleophiles at low temperatures to afford malonate half-esters (eq 6).[9]

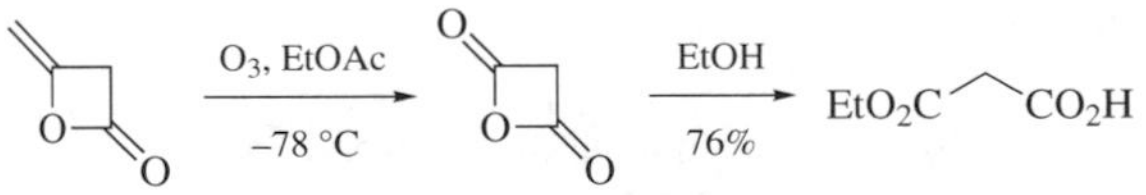

(6)

Preparation of 3-Methylenepropanoic Acids. Grignard reagents, in the presence of specific transition metal catalysts, react with diketene to afford 3-substituted 3-butenoic acids (eq 7).[10] The process has been optimized for large-scale synthesis of 3-aryl-3-butenoic acids.[11]

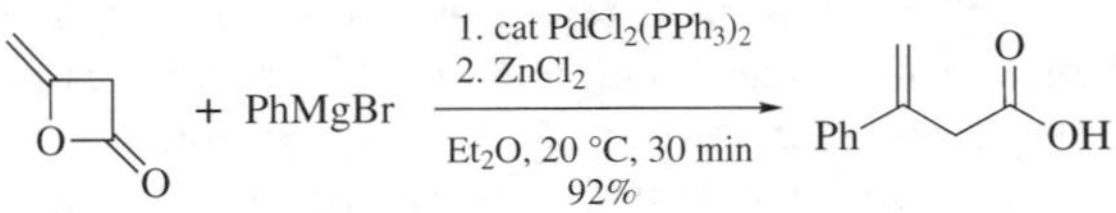

(7)

Acetoacetate Dianion Equivalent. In the presence of ***Titanium(IV) Chloride***, diketene reacts with aldehydes in a ring opening/condensation sequence to provide 4-substituted acetoacetate esters without the need for dianion formation (eq 8).[12] Reaction with ketones requires the use of the corresponding acetals.[13]

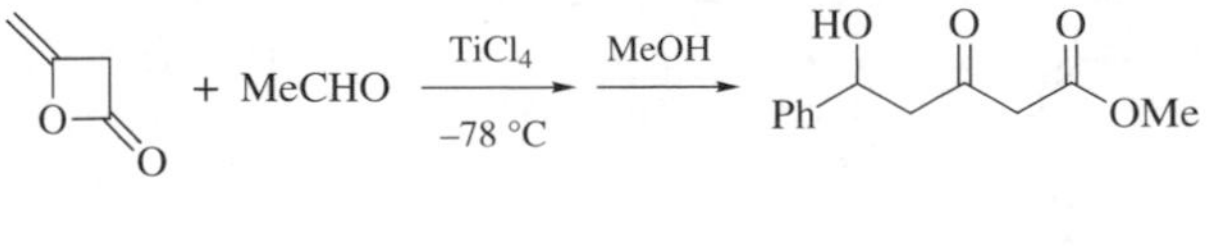

(8)

Halogenated Acetones and Acetoacetic Acid Derivatives. Diketene reacts with hydrogen chloride or chlorine to afford acetoacetyl chloride and 4-chloroacetoacetyl chloride, respectively.[14] These compounds can be further chlorinated, initially at C-2, to provide other chlorinated materials (eq 9). This work is thoroughly described in the patent literature.[1a] Workup with alcohols provides chloroacetoacetate esters, while hydrolysis and decarboxylation can be used to prepare substituted acetone derivatives.[15]

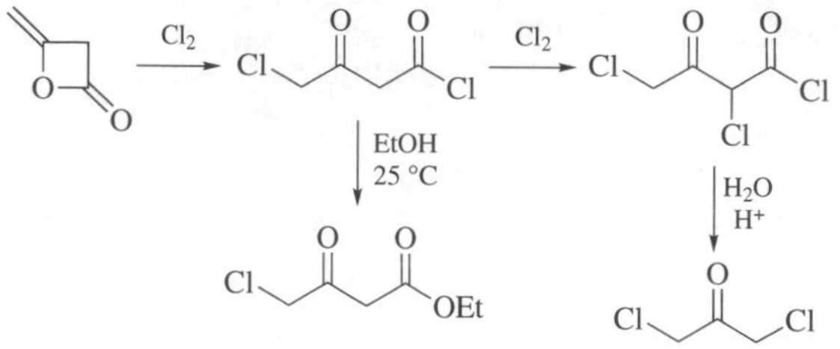

(9)

Acetone Enolate Equivalent. The careful hydrolysis of diketene with sodium hydroxide provides sodium acetoacetate which reacts with electrophiles at C-2 to form unstable β-keto acids; these decarboxylate on workup to provide substituted acetone derivatives (eq 10).[16]

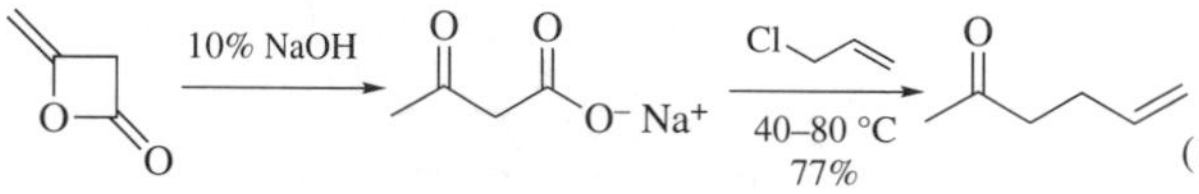

(10)

β-Butyrolactone. β-Butyrolactone, which is a valuable reagent for the preparation of β-hydroxybutyric acid derivatives and polymers, is easily prepared by reduction of diketene over ruthenium (eq 11).[17] Enantioselective reduction processes have been developed.[18] *Caution*: β-butyrolactone is a known carcinogen.

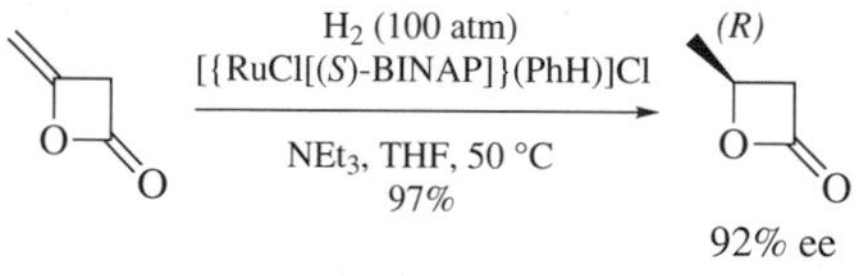

(11)

4-Substituted Acetoacetic Acid Derivatives: Radical, Carbene, and [2 + 2] Photocycloaddition. Radical and carbene additions, and photochemical [2 + 2] reactions of diketene, all involve the exocyclic methylene group of the diketene molecule. Carbene[19] and nitrene[20] additions (eq 12), as well as photocycloaddition (eq 13),[21] give spirocyclobutyrolactones. Ring opening these spirocyclic butyrolactones provides a synthesis of 4-substituted acetoacetates.

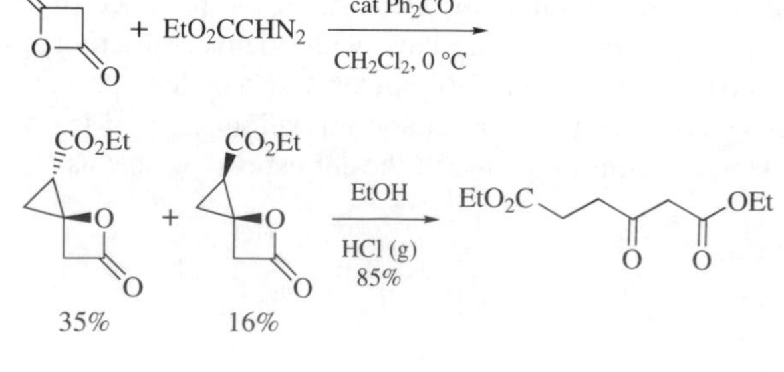

(12)

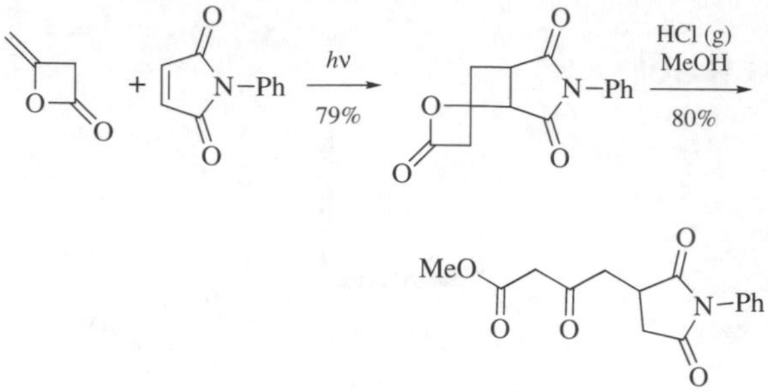

(13)

Likewise, radical addition to the double bond of diketene provides routes to substituted β-lactones and acetoacetic acid derivatives (eq 14).[22]

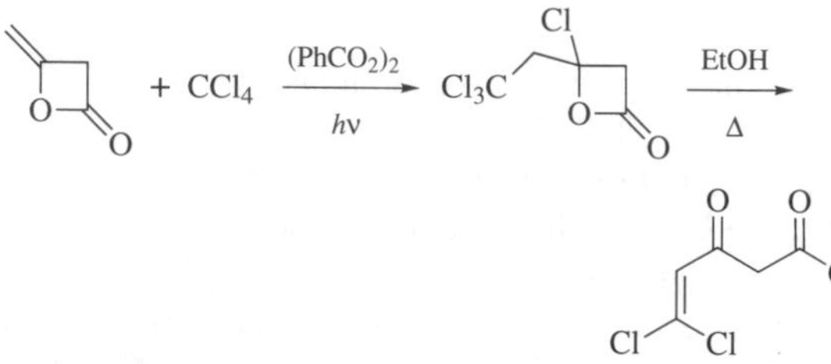

(14)

Preparation of Heterocycles. Diketene is used extensively for the preparation of heterocycles. The products can often be predicted by viewing diketene as an acetylketene equivalent. A leading example is provided for each of several types of heterocyclization; many other substrates react analogously. The reader is referred to more detailed reviews for further discussion.[1,23]

Pyrone and Pyridone Formation via Acetoacetylation of an Active Methylene Group: Dehydroacetic Acid. The fungicide dehydroacetic acid is prepared via the base-catalyzed dimerization of diketene (eq 15),[1a,24] which is based upon the acetoacetylation of the activated methylene group of a ring-opened diketene molecule. Other 4-pyrones are similarly formed when diketene reacts with compounds containing active methylene groups.[25]

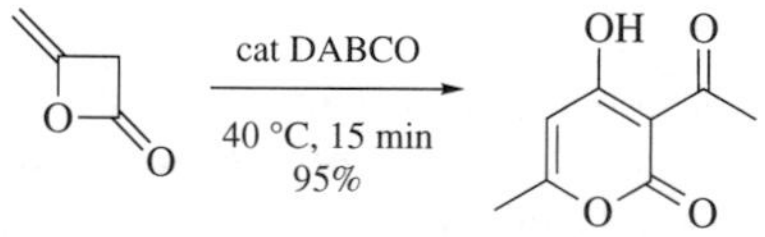

(15)

Likewise, diketene reacts with simple enamines to afford pyrones (eq 16).[26] Enamides, however, more commonly provide pyridones upon treatment with diketene (eq 17).[27] Because of the multiple functional groups normally present during heterocyclic synthesis with diketene, several different products can often be produced, depending on the reaction conditions.

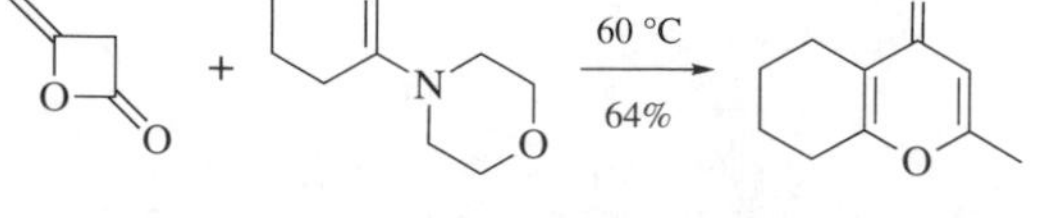

(16)

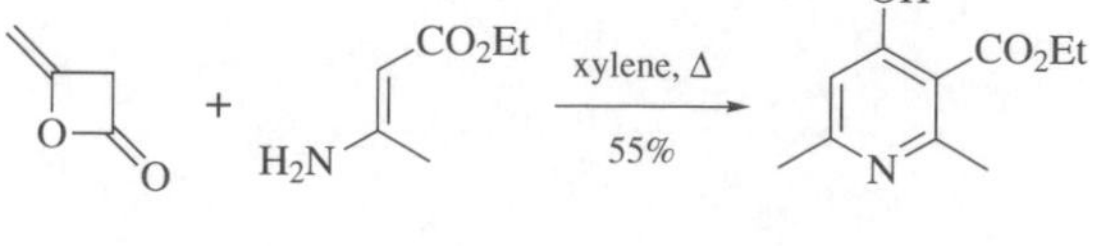

(17)

6-Methyluracil. The reaction of diketene with urea to afford 6-methyluracil (eq 18) is one of the classic heterocyclic syntheses which utilizes diketene;[28] many analogous heterocycles can be prepared by reaction of bisnucleophilic species with diketene.

Preparation of Furanones (Butenolides) and Pyrrolidinones. Diketene acetoacetylates α-hydroxy ketones[29] and α-hydroxy acids,[30] and α-amino ketones[31] and acids,[32] which undergo intramolecular condensation to afford five-membered heterocycles as shown (eqs 19 and 20).

β-Lactams. The imidazole-promoted reaction of imines with diketene can be used to prepare β-lactams (eq 21).[33]

Related Reagents. Acetoacetic Acid; Ethyl Acetoacetate; Ethyl 4-Chloroacetoaceta; Ketene; Methyl Dilithioacetoacetate; β-Methyl-β-propiolactone; 2,2,6-Trimethyl-4H-1,3-dioxin-4-one.

1. (a) Clemens, R. J. *CRV* **1986**, *86*, 241. (b) Clemens, R. J.; Witzeman, J. S. In *Acetic Acid and Derivatives*; Dekker: 1993; p 173. (c) Boese, A. B., Jr. *Ind. Eng. Chem.* **1940**, *32*, 16. (d) Bormann, D. *MOC* **1968**, *7*, 53.
2. William, J. W.; Krynitsky, J. A. *OSC* **1955**, *3*, 508.
3. Nudelman, A.; Kelner, R.; Broida, N.; Gottlieb, H. E. *S* **1989**, 387.
4. Lawesson, S. O.; Gronwall, S.; Sandberg, R. *OS* **1962**, *42*, 28.
5. (a) Caroll, M. F. *JCS* **1940**, 704; **1941**, 507. (b) Wilson, S. R.; Angelli, C. E. *OS* **1990**, *68*, 210.
6. (a) 1,2-Diazetidin-3-one synthesis; Lawton, G.; Moody, C. J.; Pearson, C. J. *JCS(P1)* **1987**, 899. (b) Furan synthesis; Padwa, A.; Kinder, F. R. *JOC* **1993**, *58*, 21.
7. Hantzsch, A. *LA* **1882**, *215*, 1.
8. Andreades, S.; Carlson, H. D. *OS* **1965**, *45*, 50.
9. (a) Perrin, C. L.; Arrhenius, T. *JACS* **1978**, *100*, 5249. (b) Hurd, C. D.; Blanchard, C. A. *JACS* **1950**, *72*, 1461.
10. (a) Fujisawa, T.; Sato, T.; Goto, Y.; Kawashima, M.; Kawara, T. *BCJ* **1982**, *55*, 3555. (b) Abe, Y.; Sato, M.; Goto, H.; Sugawara, R.; Takahashi, E., Kato, T. *CPB* **1983**, *31*, 4346.
11. Itoh, K.; Harada, T.; Nagashima, H. *BCJ* **1991**, *64*, 3746.
12. Izawa, T.; Mukaiyama, T. *CL* **1978**, 409; **1975**, 161.
13. Ishikawa, T.; Yamato, M. *CPB* **1982**, *30*, 1594.
14. Hurd, C. D.; Abernethy, J. L. *JACS* **1940**, *62*, 1147.
15. Nollett, A. J. H.; Ladage, J. W.; Mijs, W. J. *RTC* **1975**, *94*, 59.
16. (a) Kaku, T.; Katsuura, K.; Sawaki, M. (Nippon Soda Company, Ltd.) U.S. Patent 4 335 184, 1982 (*CA* **1981**, *94*, 120 871). (b) Moulin, F. (Lonza, AG) Swiss Patent 647 495, 1985 (*CA* **1985**, *102*, 203 605).
17. Sixt, J. (Wacker Chemie GmbH) U.S. Patent 2 763 664, 1956 (*CA* **1957**, *51*, 5115).
18. Ohta, T.; Miyake, T.; Takaya, H. *CC* **1992**, 1725.
19. Kato, T.; Katagiri, N. *CPB* **1973**, *21*, 729.
20. Kato, T.; Suzuki, Y.; Sato, M. *CPB* **1979**, *27*, 1181.
21. Kato, T.; Chiba, T.; Tsuchiya, S. *CPB* **1980**, *28*, 327; **1981**, *29*, 3715.
22. Dingwall, J. G.; Tuck, B. *JCS(P1)* **1986**, 2081.
23. (a) Kato, T. *ACR* **1974**, *7*, 265. (b) Kato, T. *Lect. Heterocycl. Chem.*, **1982**, *6*, 105.
24. Chick, F.; Wilsmore, N. T. M. *JSC* **1908**, 93, 946.
25. Kato, T.; Yamamoto, Y.; Hozumi, T. *CPB* **1973**, *21*, 1840.
26. (a) Hünig, S.; Benzing, E.; Hübner, K. *CB* **1961**, *94*, 486. (b) Eiden, F.; Wanner, K. T. *AP* **1984**, *317*, 958.
27. Hörlein, G.; Kübel, B.; Studeneer, A.; Salbeck, G. *LA* **1979**, 371.
28. (a) Gleason, A. H. (Standard Oil Development Corp.) U.S. Patent 2 174 239, 1939 (*CA* **1940**, *34*, 450). (b) Boese, A. B. (Carbide), U.S. Patent 2 138 756, 1938.
29. Lacey, R. N. *JCS* **1954**, 822.
30. Bloomer, J. L.; Kappler, F. E. *JOC* **1974**, *39*, 113.
31. Kato, T.; Sato, M.; Yoshida, T. *CPB* **1971**, *19*, 292.
32. Schmidlin, T.; Tamm, C. *HCA* **1980**, *63*, 121.
33. Sasaki, A.; Goda, K.; Enomoto, M.; Sunagawa, M. *CPB* **1992**, *40*, 1094.

Robert J. Clemens
Eastman Chemical Company, Kingsport, TN, USA

Dilithioacetate[1]

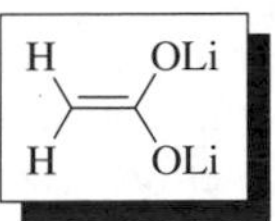

[31509-80-5] $C_2H_2Li_2O_2$ (MW 71.92)

(metalated species used to elaborate or functionally substitute the α-carbon of carboxylic acids;[3] useful for highly substituted examples;[4] possible alternatives for the malonic ester synthesis and the Haller–Bauer, Reformatsky,[23] and Wittig[28] reactions)

Physical Data: colorless solid.

Solubility: slightly sol THF; reacts with H_2O, air (O_2, CO_2), and protic solvents. Ethylene glycol-based solvents may react with lithium amides from which the dianions are formed. Dianions derived from disubstituted acetic acids are soluble in THF; less sol other ether solvents. Most carboxylate dianions are stable up to the boiling point of THF.

Formation and Reactivity. The metalation of carboxylic acids is a general phenomenon and many aliphatic, alicyclic, araliphatic, and functionally substituted acetic acids have been used in various applications.[1] The intermediate carboxylate dianions are commonly generated by one of four methods: (1) a carboxylic acid and 2 equiv of lithium (sodium) naphthalenide[2] or preferably (2) 2 equiv of LDA;[3] (3) a carboxylate salt and 1 equiv of LDA;[4] (4) a metathetical reaction of a dilithium carboxylate dianion and a metal halide. The more traditional method of reacting 2 equiv of a Grignard reagent with carboxylic acids (Ivanov reagents[1e]) is restricted to examples in which the carboxylate α-protons are relatively acidic, principally arylacetates ($pK_a \leq 22$).[5]

Accordingly, carboxylate dianions have been formed (method 3) from ***Acetaldehyde 3-Bromopropyl Ethyl Acetal*** and Li, Na, K, Ca, Mg, and Zn carboxylates. Metathesis of dilithium dianions with ***Copper(I) Iodide***, ***Zinc Chloride***, and several other metal halides produced carboxylate dianions with more selective reactivity.[6–8] Monosubstituted acetic acids yield dilithium or lithium–sodium dianions with limited solubilities, presumably due to polymeric aggregation of the metalated species.[9,10] HMPA has been recommended as cosolvent.[9] Monosubstituted acetic acids with bulky substituents (Ph, *t*-Bu) and disubstituted acetic acids form soluble (THF) dianions. Glycol ether solvents may react with LDA and THF is preferred.[11,12] LDA is preferred as base[1c] and, although more hindered bases[13] may be necessary for special applications, amines from less hindered lithium amides can react competitively with some electrophiles (TMSCl).[11] The lithium amide of hexamethyldisilazane is not sufficiently basic ($pK_a \geq 26$ (THF))[14] for general use (LDA, $pK_a \geq 36$ (THF)).[15] In examples where the carboxylate dianion is soluble, as little as 3 mol % of the amine can be used for metalation (eq 1).[1c] The latter is a useful variation for reactions with highly reactive electrophiles where diisopropylamine might compete for electrophiles (acylation).[16]

$$Me_2CHCO_2Na + LDA \xrightarrow[25\text{–}30\ ^\circ C]{} i\text{-}Pr_2NH + [Me_2CCO_2]^{2-}\ Li^+Na^+ \quad (1)$$

(BuLi recycles i-Pr_2NH to LDA)

Unsaturated carboxylic acids as their Cu dianions react selectively at the γ-rather than the α-position with allyl bromide and vinylic epoxides.[6,7] The reaction occurs by an S_N2' process and the yields and selectivity are impaired with substituted allyl bromides. Other electrophiles react poorly; however, unsaturated ketones react with the Cu dianion of acetic acid[8] predominantly by 1,2-addition.

LDA or LTBA metalation of Zn or Mg carboxylates offers considerable promise because they are soluble.[17,18] Small, aliphatic carboxylic acids, aromatic, and arylacetic acids form crystalline zinc carboxylates as their TMEDA complexes which are soluble in THF and hot toluene (eq 2). Butyric acid and larger aliphatic zinc carboxylates are soluble in hot toluene without TMEDA. The lithium amide displaces TMEDA as ligand on metalation and an additional equivalent abstracts the α-protons (eq 3). Reaction with carbonyl compounds provides adducts in yields equivalent to the Reformatsky reaction (eq 4). This metalated species is especially useful for acetic and propionic acids. Unlike alkali metal dianions, alkylation is not selective and disubstituted products are formed except when activating substituents are present (R = Ph). Like *t*-butylacetate anion,[19,20] alkylation of acetic acid dianion (**1**) (R = H) fails in the absence of dipolar solvents.

$$RCH_2CO_2H + ZnO + TMEDA \xrightarrow[90\text{–}95\%]{\Delta,\ toluene} [RCHCO_2]_2Zn\bullet TMEDA \quad (2)$$

$$[RCHCO_2]_2Zn\bullet TMEDA + 4\ LTBA \longrightarrow [(RCHCO_2)_2Zn\ (NH\text{-}t\text{-}Bu)_2]^{4-}\ 4Li^+ \quad (3)$$

R = Me, H (**1**)

$$(\mathbf{1}) + \text{cyclohexanone} \xrightarrow[78\%]{} \text{1-hydroxycyclohexyl–CH(R)CO}_2\text{H} \quad (4)$$

R = Me

Reactions with Electrophiles. Carboxylate dianions have been treated with many electrophiles,[1] but reactions with alkyl halides and carbonyl compounds have been used most widely. Most common carboxylate dianions have been reported to react successfully except acetic acid and cyclopropanecarboxylic acids where carboxyl derivatives must be used with few exceptions.[21] Carboxylate dianions are especially useful where forcing conditions are required, as in reactions with epoxides,[22] or in reactions with carbonyl substrates where hydrolysis of a carboxyl derivative causes product degradation.[23] Carbonyl addition is sensitive to steric effects, but steric hindrance must be severe to prevent adduct formation (eq 5).[23] The adduct may cyclize spontaneously on isolation, forming a β-lactone in highly constrained examples.[23b]

$$[C_6H_{10}\text{–}CO_2]^{2-}\ 2Li^+ + \text{dispirocyclopropyl cyclohexanone} \xrightarrow[61\%]{} \text{adduct (}HO_2C\text{, HO)} \quad (5)$$

A *C*-silylated analog, the trimethylsilylacetic acid dianion (**2**), forms carbonyl adducts which undergo spontaneous Peterson elimination (eq 6).[24] Mixtures of (*E*) and (*Z*) products are obtained. The limited accessibility of (**2**) has been overcome by use of a softer, *C*-silylating reagent (eq 7)[25]

$$\underset{(\mathbf{2})}{[Me_3SiCHCO_2]^{2-}\ 2Li^+} + \text{cyclopentanone} \xrightarrow[84\%]{} \text{cyclopentylidene–CHCO}_2\text{H} \quad (6)$$

$$[CH_2CO_2]^{2-}\ 2Li^+ \xrightarrow[60\%]{Ph_2MeSiCl} Ph_2MeSiCH_2CO_2H \quad (7)$$

Unsaturated carboxylate dianions have been used to demonstrate the reversibility of carbonyl addition. The α:γ ratio of adducts depends on the metal ion, temperature, solvent polarity, time, and steric effects (eqs 8 and 9).[26] Reaction with aldehydes gives *anti* products predominantly, and the *anti*:*syn* ratio can reach 49:1 if the substituents are sufficiently large and thermodynamic conditions are used (eq 10).[27] The major *anti* isomer (**5**) can produce either *trans* (**8**) or *cis* (**9**) alkene in many cases (eqs 11 and 12). Formation of β-lactone (**7**) is a general reaction[28] and *cis* elimination of carbon dioxide makes this sequence a useful alternative to the Wittig reaction.[28] Use of DMF acetals or DEAD effects an *anti* elimination through a charged intermediate and produces (*Z*)-alkene (**9**).[29] Lactones (**7**) can be alkylated stereoselectively (eq 13) if the lactone is sufficiently stable.[30,31] Combination of these procedures allows substantial control to be exercised over syntheses of di- and trisubstituted alkenes.

$[\ldots CO_2]^{2+}$ 2Li$^+$ + $Me_2C{=}O$ —(−78 °C, 5 min; 78%)→ (**3**) + (**4**) (8)

(**3**) 10:1 (**4**)

(**3**) —(2 LDA, 10 h; 25 °C)→ (**3**) + (**4**) 1:10 (9)

$[PhCHCO_2]^{2-}$ 2Li$^+$ + *i*-PrCHO —(77%)→ (**5**) (10)

anti (**5**):*syn* (**6**) = 14:1

(**5**) —($PhSO_2Cl$, py; 83%)→ (**7**) —(73%)→ (**8**) (11)

(**5**) —(DMF acetal; 92%)→ (**9**) (12)

(**7**) —(97%)→ (13)

Alkylation of carboxylate dianions in which one cation is lithium and the other an alkali metal is a simple synthetic operation.[1] Many primary and secondary halides (tosylates) have been used and success is independent of the leaving group and the order of introduction of the substituents.[1c] Further, successful reaction does not depend on obtaining a homogeneous solution of the dianion. Simple functionally substituted (ethers, double bonds, acetals, carboxylates) alkylating agents can be accommodated. Alkyl halides sensitive to elimination proceed normally unless steric hindrance in the dianion is extreme.[4] The stability of carboxylate dianions is important in uses with epoxides[22] or acylaziridines[32] where extended reaction times or forcing conditions may be necessary (eq 14). The scale and low temperatures are not important requirements. The production of the hypolipidemic agent gemfibrozil uses typical reaction conditions (eq 15).[17,33]

$[PhMeCCO_2]^{2-}$ 2Na$^+$ + NCOPh —(76%)→ HO_2C–C(Ph)(Me)–CH_2CH_2–NHCOPh (14)

$[Me_2CCO_2]^{2-}$ Li$^+$ Na$^+$ + ArO(CH$_2$)$_3$Cl —(85%)→ ArO(CH$_2$)$_3$C(Me)$_2$$CO_2H$ (15)

Chiral Syntheses. Asymmetric syntheses with carboxylate dianions requires a chiral amide that can also act as a chiral auxiliary by remaining associated with the dianion without being covalently bonded to it. Carbonyl addition gave as much as 85% ee in a carefully chosen model (eq 16).[34] Alkylation is much less stereospecific than carbonyl addition and it proved to be much less selective in the formation of a chiral product (eq 17).[35]

$PhCH_2CO_2H$ + LiHN(LiO, OLi, Ph amide) —(PhCHO; 80%)→ Ph–CH(OH)–CH(Ph)–CO_2H (16)

(2*S*,3*R*) 85% ee

$PhCH_2CO_2H$ + (*i*-Pr, O-*t*-Bu, LiN–*i*-Pr) —(EtI; 75%)→ PhCH(Et)C(O)OH (17)

(*S*) 20% ee

Related Reagents. *t*-Butyl α-Lithiobis(trimethylsilyl)acetate; *t*-Butyl α-Lithioisobutyrate; *t*-Butyl Trimethylsilylacetate; 2,6-Dimethylphenyl Propionate; Ethyl Bromozincacetate; Ethyl Lithioacetate; Ethyl Lithio(trimethylsilyl)acetate; Ethyl Trimethylsilylacetate; 2-Methyl-2-(trimethylsilyloxy)-3-pentanone; Trimethylsilylacetic Acid.

1. (a) Thompson, C. M.; Green, D. L. C. *T* **1991**, *47*, 4223. (b) Petragnani, N.; Yonashiro, M. *S* **1982**, 521. (c) Creger, P. L. *Annu. Rep. Med. Chem.* **1977**, *12*, 278. (d) Ivanov, D.; Vassilev, G.; Panayotov, I. *S* **1975**, 83. (e) Blagoev, B.; Ivanov, D. *S* **1970**, 615. (f) Ebel, H. F. in *MOC* **1970**, *13/1*, 445. (g) Morton, A. A. *Solid Organoalkali Metal Reagents*; Gordon & Breach: New York, 1964; p 47ff.
2. (a) Angelo, B. *CR(C)* **1973**, *276*, 293. (b) Angelo, B. *BSF(2)* **1970**, 1848.
3. Creger, P. L. *JACS* **1967**, *89*, 2500.
4. Creger, P. L. *OSC* **1988**, *6*, 517.
5. Gronert, S.; Streitweiser, A. *JACS* **1988**, *110*, 4418.
6. Savu, P. M.; Katzenellenbogen, J. A. *JOC* **1981**, *46*, 239.
7. Pitzele, B. S.; Baran, J. S.; Steinman, D. H. *T* **1976**, *32*, 1347.
8. Mulzer, J.; Brüntrup, G.; Hartz, G.; Kühl, U.; Blaschek, U.; Böhrer, G. *CB* **1981**, *14*, 3701.
9. Pfeffer, P. E.; Silbert, L. S. *JOC* **1971**, *36*, 3290.

10. Bauer, W.; Seebach, D. *HCA* **1984**, *67*, 1972.
11. House, H. O.; Liang, W. C.; Weeks, P. D. *JOC* **1974**, *39*, 3102.
12. Gall, M.; House, H. O. *OSC* **1988**, *6*, 121.
13. (a) Olofson, R. A.; Dougherty, C. M. *JACS* **1973**, *95*, 582. (b) Kopka, I. E.; Fataftah, Z. A.; Rathke, M. W. *JOC* **1987**, *52*, 448. (c) Prieto, J. A.; Suarez, J.; Larson, G. L. *SC* **1988**, *18*, 253.
14. Fraser, R. R.; Mansour, T. S.; Savard, S. *JOC* **1985**, *50*, 3232.
15. Fraser, R. R.; Bresse, M.; Mansour, T. S. *CC* **1983**, 620.
16. (a) Krapcho, A. P.; Kashdan, D. S.; Jahngen, Jr., E. G. E.; Lovey, A. J. *JOC* **1977**, *42*, 1189. (b) Krapcho, A. P.; Stephens, W. P. *JOC* **1980**, *45*, 1106.
17. Creger, P. L. U.S. Patent 3 674 836, 1972 (*CA* **1970**, *72*, 43 167p).
18. Creger, P. L. U.S. Patent 5 041 640, 1991 (*CA* **1991**, *115*, 255 632t).
19. Rathke, M. W.; Lindert, A. *JACS* **1971**, *93*, 2318.
20. Bos, W.; Pabon, H. J. J. *RTC* **1980**, *99*, 141.
21. (a) Häner, R.; Maertzke, T.; Seebach, D. *HCA* **1986**, *69*, 1655. (b) Jahngen, E. G. E.; Phillips, D.; Kobelski, R. J.; Demko, D. M. *JOC* **1983**, *48*, 2472. (c) Warner, P. M.; Le, D. *JOC* **1982**, *47*, 893.
22. Creger, P. L. *JOC* **1972**, *37*, 1907.
23. (a) Moersch, G. W.; Burkett, A. R. *JOC* **1971**, *36*, 1149. (b) Krapcho, A. P.; Jahngen, Jr., E. G. E. *JOC* **1974**, *39*, 1650.
24. Grieco, P. A.; Wang, C.-L. J.; Burke, S. D. *CC* **1975**, 537.
25. Larson, G. L.; Cruz de Maldonado, V.; Berrios, R. R. *SC* **1986**, *16*, 1347.
26. (a) Johnson, P. R.; White, J. D. *JOC* **1984**, *49*, 4424. (b) Ballester, P.; García-Raso, A.; Mestres, R. *S* **1985**, 802.
27. Mulzer, J.; Zippel, M.; Brüntrup, G.; Segner, J.; Finke, J. *LA* **1980**, 1108.
28. Adam, W.; Baeza, J.; Liu, J.-C. *JACS* **1972**, *94*, 2000.
29. (a) Rüttimann, A.; Wick, A.; Eschenmoser, A. *HCA* **1975**, *58*, 1450. (b) Mulzer, J.; Brüntrup, G. *CB* **1982**, *115*, 2057. (c) Hara, S.; Taguchi, H.; Yamamoto, H.; Nozaki, H. *TL* **1975**, 1545. (d) Mulzer, J.; Lammer, O. *AG(E)* **1983**, *22*, 628.
30. (a) Mulzer, J.; Zippel, M. *AG(E)* **1981**, *20*, 399. (b) Mulzer, J.; Kerkmann, T. *JACS* **1980**, *102*, 3620.
31. (a) Mulzer, J.; Brüntrup, G. *AG(E)* **1979**, *18*, 793. (b) Black, T. H.; Hall, J. A.; Shen, R. G. *JOC* **1988**, *53*, 2371. Not cited in text.
32. Stamm, H.; Weiss, R. *S* **1986**, 395.
33. Anderson, R. *CI(L)* **1984**, 205.
34. Mulzer, J.; deLasalle, P.; Chucholowski, A.; Blaschek, U.; Brüntrup, G.; Jibril, I.; Huttner, G. *T* **1984**, *40*, 2211.
35. Ando, A.; Shiori, T. *CC* **1987**, 656.

Paul L. Creger
Ann Arbor, MI, USA

Dilithium Tetrachlorocuprate(II)[1,2]

Li_2CuCl_4

[15489-27-7] Cl_4CuLi_2 (MW 219.24)

(organic soluble copper catalyst[1] used in catalytic quantities to improve the efficiency of halide[2] and allylic acetate[3] displacements with Grignard reagents; it is used in stoichiometric amounts to open epoxides[4] and aziridines[5a])

Physical Data: *d* 0.910 g cm^{-3}.
Solubility: sol up to concentrations of 0.5 M in THF.
Form Supplied in: commercially available as a 0.1 M solution in THF.[2a]
Preparative Methods: prepared in THF solution by reaction of ***Lithium Chloride*** (0.2 mol) and ***Copper(II) Chloride*** (0.1 mol).[5b]
Handling, Storage, and Precautions: moisture and oxygen sensitive.

Introduction. The addition of two or more equivalents of lithium chloride to copper(II) chloride will result in a dark red THF soluble solution of dilithium tetrachlorocuprate.[1,2] Dilithium tetrachlorocuprate has been used most frequently to enhance the efficiency of halide displacements with a Grignard reagent. The reagent is used catalytically in amounts less than 10 mol % and typically it will increase the displacement yield.

Halide Displacements with Grignard Reagents. The first report describing the use of dilithium tetrachlorocuprate was by Kochi and Tamura in 1971.[2] They described the coupling of *n*-butylmagnesium bromide with *n*-hexyl bromide using Li_2CuCl_4 to afford decane in 78% yield. This report was followed by the use of more complex substrates, including Grignard reagents containing protected ethers, unsaturation, and strained rings with homoallylic halides (eqs 1–3).[6–8] It is noteworthy that displacement predominated over dehydrohalogenation in these cases.

Br + $ClMg(CH_2)_6OTHP$ → (Li_2CuCl_4, THF, 65%) → $(CH_2)_8OTHP$ (1)

MgBr + I ~ OTHP → (Li_2CuCl_4, THF) → $(CH_2)_6OTHP$ (2)

MgCl + Cl ~ → (Li_2CuCl_4, THF, 57%) → (3)

Sterically hindered Grignard reagents were coupled with toluenesulfonates to afford sterols with modified side chains (eq 4)[9] and with benzylic halides resulting in neopentyl derivatives of naphthalenes (eq 5).[10] In this example the copper catalyst reduced homodimer formation (i.e. 1,2-bisnaphthylethane formation).

MeO ... OTs + RMgBr → (Li_2CuCl_4, THF) → R (4)

R = *i*-Pr (78%), Ph (85%), Cy (81%)

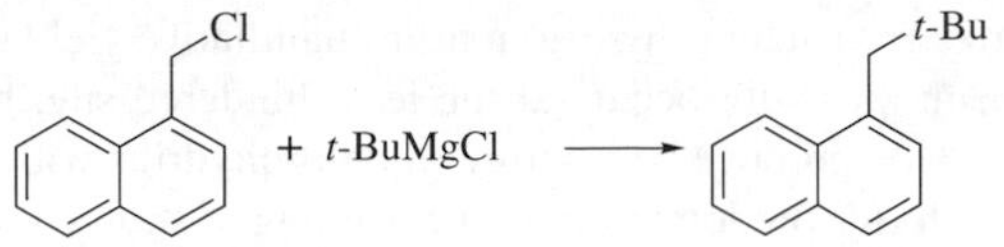

Cl + t-BuMgCl ⟶ t-Bu (5)

without Li_2CuCl_4 51%
with Li_2CuCl_4 86%

The benzylic type Grignard reagent derived from 3-chloromethylfuran would not displace an allylic halide unless dilithium tetrachlorocuprate was employed (eq 6).[11]

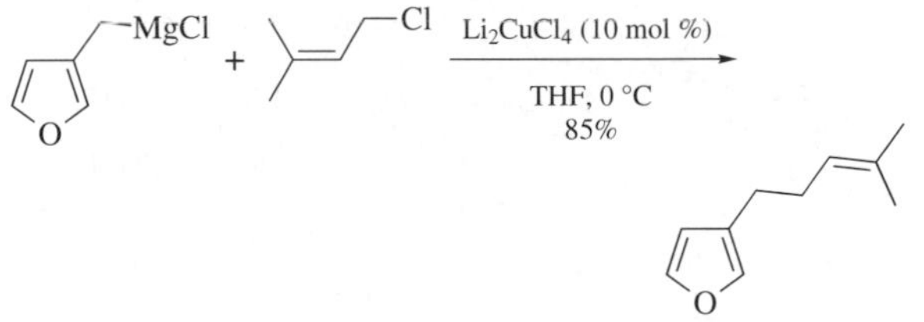

MgCl + Cl —Li_2CuCl_4 (10 mol %), THF, 0 °C, 85%⟶ (6)

no product without Li_2CuCl_4

Functionalized Grignard reagents were efficiently coupled with halides containing an ester (eq 7)[12] or carboxylic acid salt (eq 8).[13] In both cases there was little reaction at the carboxyl termini. The alkenic product from eq 8 was of interest for use in ultra thin-layer photoresists. High purity alkene in which double bond isomers were not present was obtained with Li_2CuCl_4 catalysis.

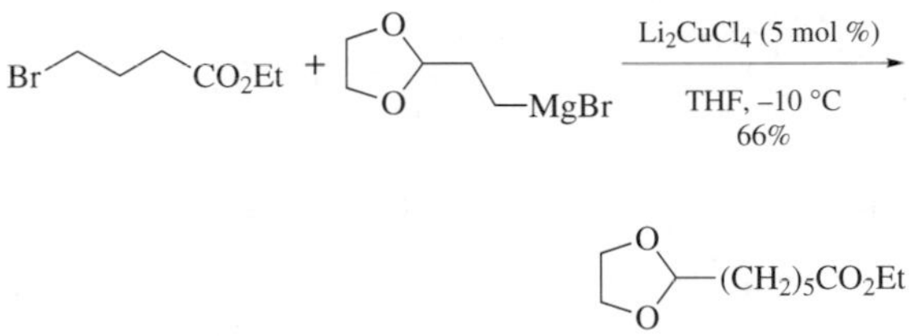

Br$\sim$$CO_2Et$ + MgBr —Li_2CuCl_4 (5 mol %), THF, −10 °C, 66%⟶ $(CH_2)_5CO_2Et$ (7)

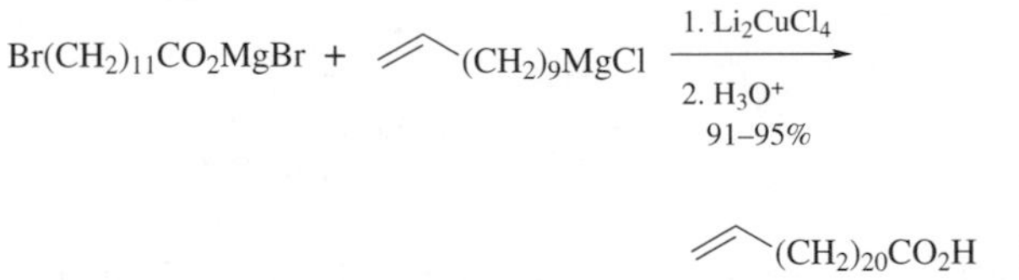

$Br(CH_2)_{11}CO_2MgBr$ + $(CH_2)_9MgCl$ —1. Li_2CuCl_4; 2. H_3O^+; 91–95%⟶ $(CH_2)_{20}CO_2H$ (8)

A vinylic type Grignard reagent was coupled with n-octyl iodide to afford 2-substituted 1,3-dienes in high yields. In this example, CuI was not an effective coupling catalyst (eq 9).[14]

MgCl + $C_8H_{17}I$ —Li_2CuCl_4 (2 mol %), THF, 20 °C, 95%⟶ C_8H_{17} (9)

A double displacement was reported by Schlosser using 1 mol % of the copper catalyst (eq 10).[15]

$TsO(CH_2)_{10}OTs$ + t-BuMgCl —Li_2CuCl_4 (1 mol %), THF, 85%⟶ t-Bu$(CH_2)_{10}$-t-Bu (10)

A cyclic α-halo ether was displaced with the aid of dilithium tetrachlorocuprate. The copper catalyst increased the yield and stereoselectivity for the displacement (eq 11).[16]

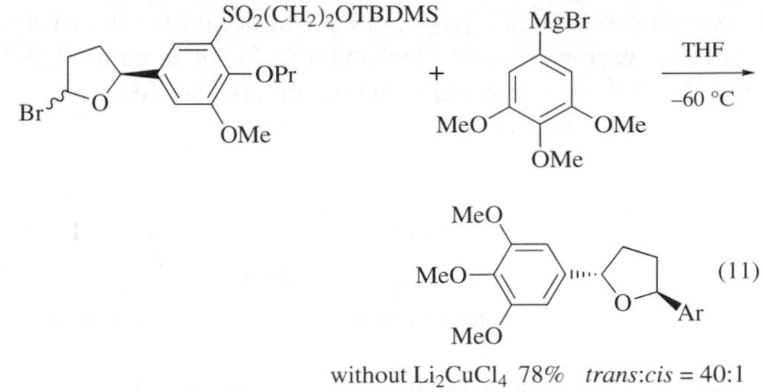

$SO_2(CH_2)_2OTBDMS$, OPr, OMe, Br + MgBr, MeO, OMe, OMe —THF, −60 °C⟶ MeO, MeO, MeO, O, Ar (11)

without Li_2CuCl_4 78% $trans$:cis = 40:1
with Li_2CuCl_4 88% $trans$:cis = 90:1

Allylic Acetate and Quaternary Ammonium Salt Displacements. The first report using dilithium tetrachlorocuprate in allylic acetate displacements was by Schlosser in which $trans$-crotyl acetate was converted into $trans$-2-octene (eq 12).[3] The reaction was reported to be regio- and stereoselective. In addition to the desired coupling product, the reaction also produced 10% of the carbonyl addition product 5-methyl-5-nonanol. The level of carbonyl addition product could be reduced using stoichiometric amounts of ***Copper(I) Iodide***. The coupling of a (Z)-trisubstituted allylic acetate also proceeded with 100% regio- and stereocontrol (eq 13).[17] In this latter example, nerol derivatives containing an allylic acetate were compared with those containing a bromide, sulfonium salt, ammonium salt, and a carbonate. The acetate was superior, both in terms of maintaining alkene geometry and regiocontrol (α- vs. γ-attack).

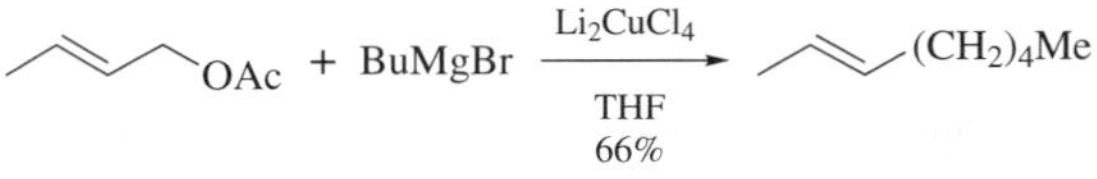

OAc + BuMgBr —Li_2CuCl_4, THF, 66%⟶ $(CH_2)_4Me$ (12)

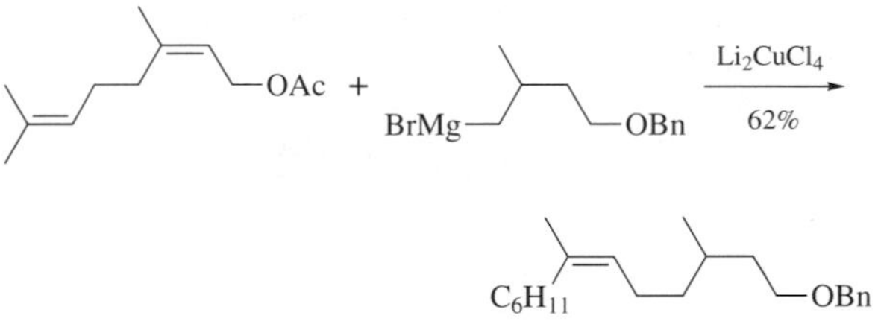

OAc + BrMg OBn —Li_2CuCl_4, 62%⟶ C_6H_{11} OBn (13)

100% (Z)

The regiocontrol was rationalized by Backvall, who observed drastic changes in regiochemistry with changes in reaction conditions. Rapid addition of the Grignard reagent resulted in α-attack (eq 14).[18] A slow Grignard addition in the presence of higher catalyst loadings (5 mol % vs. 2 mol %) completely reversed the regiochemistry of the displacement.

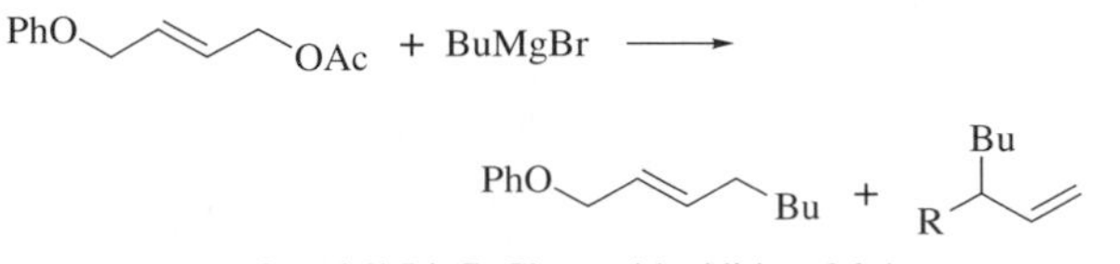

PhO OAc + BuMgBr ⟶ PhO Bu + R Bu (14)

2 mol % Li_2CuCl_4 rapid addition 96:4
5 mol % Li_2CuCl_4 slow addition 14:86

These results were explained by assuming the presence of a monoalkylcopper complex (**1**) and a dialkylcopper complex (**2**) (eq 15). With rapid addition, there is a buildup of the dialkylcopper species which favors the α-product. Support for this was provided by reacting preformed n-Bu_2CuLi with the acetate to afford a 96:4 mixture of regioisomers. A slow addition, on the other hand, will result in predominant reaction via the monoalkylcopper species, which favors γ-attack.

RMgX + CuX ⟶ X(R)Cu-MgX (1) favors γ-attack ⟶ (rapid addition resulting in an excess of RMgX) R(R)Cu-MgX (2) favors α-attack (15)

The displacement of allylic acetates was extended to diene systems. A study of all four stereoisomers of 2,4-heptadienyl acetate showed that only the (*E,E*) isomer was 100% stereospecific (eq 16).[19] The (2*Z*,4*E*) isomer was the least stereospecific, resulting in a 2:1 ratio of diene isomers (eq 17).

Et–diene–OAc + $ClMg(CH_2)_5OTHP$ (1. Li_2CuCl_4, –10 °C; 2. H_3O^+; 3. Ac_2O; 65%) ⟶ Et–diene–$(CH_2)_5OAc$ pure (*E,E*)-diene (16)

Et–diene–OAc + $ClMg(CH_2)_5OTHP$ (45%) ⟶ Et–diene–$(CH_2)_5$OAc + (*E,E*)-diene 2:1 (17)

The displacement reported by Roush in which an (*E,E*)-diene afforded a mixture of regioisomers may have been improved using the rapid addition technique or by using a preformed copper reagent (eq 18).[20]

diene–OAc + BrMg–$(CH_2)_3$–CH(OEt)OEt (Li_2CuCl_4, 45%) ⟶ diene–$(CH_2)_3$–$CH(OEt)_2$ + branched isomer–$(CH_2)_3$–$CH(OEt)_2$ 8:1 (18)

An allylic quaternary ammonium salt was also a good substrate for displacement under dilithium tetrachlorocuprate catalysis (eq 19).[21]

TMS-diene–N^+(Me)(Me)(Bn) Cl^- + BuMgBr (Li_2CuCl_4, THF, 92%) ⟶ TMS-substituted diene (19)

Epoxide and Aziridine Opening. When used in excess of stoichiometric quantities, the reagent will open epoxides[4] and activated aziridines.[5a] The epoxide of cyclohexene was opened at rt to afford a *trans*-chlorohydrin in near quantitative yield (eq 20).[4] The opening usually occurs at the least hindered site; however, styrene oxide produced a mixture of chlorohydrins with the secondary chloride predominating. The opening of strained rings was extended to aziridines which are activated as the toluenesulfonamide (eq 21).[5]

cyclohexene oxide (1.6 equiv Li_2CuCl_4, THF, 98%) ⟶ *trans*-2-chlorocyclohexanol (Cl, OH) (20)

bis(NTs-aziridinyl) acetonide (1.6 equiv Li_2CuCl_4, THF, 90%) ⟶ bis(Cl, NHTs) acetonide (21)

Cross-Coupling Reactions. There are only a few reports describing the use of dilithium tetrachlorocuprate as a catalyst for coupling Grignard reagents with vinylic and aryl iodides.[14,22–24] In one report, one of two iodides was replaced using Li_2CuCl_4 (eq 22).[23] An example using a 1,3-diene indicates that palladium catalysis may be more effective for these types of transformations (eq 23).[14]

diiodocycloheptatriene (RMgX, 3 mol % Li_2CuCl_4) ⟶ R, I-cycloheptatriene (22)

R = allyl (84%), Bu (51%) vinyl (88%)

PhI + CH_2=CH–C(=CH_2)MgCl ⟶ Ph–C(=CH_2)–CH=CH_2 (23)

with Li_2CuCl_4 45%
with $Pd(PPh_3)_4$ 75%

There is one report in which a carbonylative cyclization was observed, presumably by intramolecular trapping of an acylmetal derivative (eq 24).[24]

2-iodobenzyl malonate (CO_2Et, CO_2Et, I) (Li_2CuCl_4 (10 mol %), Et_3N, CO (600 psi), 100 °C, 91%) ⟶ indanone-2,2-dicarboxylate (CO_2Et, CO_2Et, O) (24)

Related Reagents. Copper(I) Acetylacetonate; Copper(II) Acetylacetonate; Copper(I) Bromide; Copper(II) Bromide; Copper(I) Chloride; Copper(II) Chloride; Copper(I) Iodide; Copper(I) Trifluoromethanesulfonate; Copper(II) Trifluoromethanesulfonate.

1. Eswein, R. P.; Howald, E. S.; Howald, R. A.; Keeton, D. P. *J. Inorg. Nucl. Chem.* **1967**, 437.
2. (a) Tamura, M.; Kochi, J. *S* **1971**, 303. (b) Tamura, M.; Kochi, J. K. *BCJ* **1971**, *44*, 3063. (c) Kochi, J. K. *Organometallic Mechanisms and Catalysis*; Academic: New York, 1978; p 381.

3. Fouquet, G.; Schlosser, M. *AG(E)* **1974**, *13*, 82.
4. Ciaccio, J. A.; Addess, K. J.; Bell, T. W. *TL* **1986**, *27*, 3697.
5. (a) Duréault, A.; Tranchepain, I.; Greck, C.; Depezay, J.-C. *TL* **1987**, *28*, 3341. (b) *FF* , **1990**, *4*, 163.
6. Corey, E. J.; Petrzilka, M. *TL* **1975**, 2537.
7. Rossi, R. *Chim. Ind. (Milan)* **1978**, *60*, 652 (*CA* **1978**, *89*, 196 934n).
8. Wey, H. G.; Betz, P.; Butenschön, H. *CB* **1991**, *124*, 465.
9. Herz, J. E.; Vazquez, E. *Steroids* **1976**, *27*, 133.
10. Bullpitt, M.; Kitching, W. *S* **1977**, 316.
11. Tanis, S. P. *TL* **1982**, *23*, 3115.
12. Volkmann, R. A.; Davis, J. T.; Meltz, C. N. *JOC* **1983**, *48*, 1767.
13. Mirviss, S. *JOC* **1989**, *54*, 1948.
14. Nunomoto, S.; Kawakami, Y.; Yamashita, Y. *JOC* **1983**, *48*, 1912.
15. Schlosser, M.; Bossert, H. *T* **1991**, *47*, 6287.
16. Thompson, A. S.; Tschaen, D. M.; Simpson, P.; McSwine, D. J.; Reamer, R. A.; Verhoeven, T. R.; Shinkai, I. S. *JOC* **1992**, *57*, 7044.
17. Suzuki, S.; Shiono, M.; Fujita, Y. *S* **1983**, 804.
18. Bäckvall, J.-E.; Sellén, M. *CC* **1987**, 827.
19. (a) Samain, D.; Descoins, C.; Commercon, A. *S* **1978**, 388. (b) Samain, D.; Descoins, C.; Langlois, Y. *NJC* **1978**, *2*, 249 (*CA* **1978**, *89*, 162 933w).
20. Roush, W. R.; Gillis, H. R. *JOC* **1982**, *47*, 4825.
21. Hosomi, A.; Hoashi, K.; Tominaga, Y.; Otaka, K.; Sakurai, H. *JOC* **1987**, *52*, 2947.
22. Commercon, A.; Normant, J. F.; Villieras, J. *JOM* **1977**, *128*, 1.
23. Okazaki, R.; O-oka, M.; Tokitoh, N.; Inamoto, N. *JOC* **1985**, *50*, 180.
24. Negishi, E.; Zhang, Y.; Shimoyama, I.; Wu, G. *JACS* **1989**, *111*, 8018.

Andrew S. Thompson
Merck & Co., Rahway, NJ, USA

Dimethoxycarbenium Tetrafluoroborate[1]

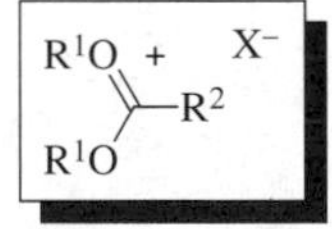

(**1**; $R^1 = Me$, $R^2 = H$, $X = BF_4$)
[18346-68-4] $C_3H_7BF_4O_2$ (MW 161.89)
(**2**; $R^1 = Me$, $R^2 = H$, $X = Br$)
[70146-62-2] $C_3H_7BrO_2$ (MW 154.99)
(**3**; $R^1 = Me$, $R^2 = H$, $X = SbCl_6$)
[27434-62-4] $C_3H_7Cl_6O_2Sb$ (MW 409.56)
(**4**; $R^1 = Me$, $R^2 = H$, $X = PF_6$)
[50318-32-6] $C_3H_7F_6O_2P$ (MW 220.05)
(**5**; $R^1 = Me$, $R^2 = H$, $X = SbF_6$)
[66226-85-5] $C_3H_7F_6O_2Sb$ (MW 310.83)
(**6**; $R^1 = Et$, $R^2 = H$, $X = BF_4$)
[1478-41-7] $C_5H_{11}BF_4O_2$ (MW 189.94)
(**7**; $R^1 = Me$, $R^2 = Et$, $X = BF_4$)
[64950-83-0] $C_5H_{11}BF_4O_2$ (MW 189.94)
(**8**; $R^1 = Et$, $R^2 = H$, $X = SbCl_6$)
[1663-59-8] $C_5H_{11}Cl_6O_2Sb$ (MW 437.61)
(**9**; $R^1 = Et$, $R^2 = H$, $X = PF_6$)
[97632-94-5] $C_5H_{11}F_6O_2P$ (MW 248.10)
(**10**; $R^1 = Et$, $R^2 = Me$, $X = BF_4$)
[21872-75-3] $C_6H_{13}BF_4O_2$ (MW 203.97)
(**11**; $R^1 = allyl$, $R^2 = H$, $X = BF_4$)
[18346-72-0] $C_7H_{11}BF_4O_2$ (MW 213.97)
(**12**; $R^1 = n\text{-Pr}$, $R^2 = H$, $X = BF_4$)
[18346-70-8] $C_7H_{15}BF_4O_2$ (MW 218.00)
(**13**; $R^1 = i\text{-Pr}$, $R^2 = H$, $X = BF_4$)
[18346-71-9] $C_7H_{15}BF_4O_2$ (MW 218.00)
(**14**; $R^1 = Me$, $R^2 = Ph$, $X = BF_4$)
[878-36-4] $C_9H_{11}BF_4O_2$ (MW 237.99)
(**15**; $R^1 = Et$, $R^2 = Ph$, $X = BF_4$)
[70861-63-1] $C_{11}H_{15}BF_4O_2$ (MW 266.04)
(**16**; $R^1 = Me$, $R^2 = 2,4,6\text{-}Me_3C_6H_2$, $X = BF_4$)
[15655-62-6] $C_{12}H_{17}BF_4O_2$ (MW 280.07)

(ambident electrophiles;[1,2] alkylation of S-,[3,4] N-,[5] and O-nucleophiles;[6,7] acylation of C-nucleophiles[8])

Physical Data: mp: (**1**) 20 °C; (**4**) 72 °C; (**6**) 20 °C; (**8**) 151–152 °C; (**10**) 55 °C; (**14**) 88–90 °C; (**15**) dec; (**16**) 89–91 °C.

Solubility: freely sol liquid sulfur dioxide and arsenic trifluoride.[9b,c] When prepared in situ, halogenated solvents such as dichloromethane or 1,1,2-trichlorotrifluoroethane are most commonly used.[2,5b,6a,b,8e,f] With these solvents, however, the reagents are only slightly soluble and usually exist as a slurry at −30 °C. Refer to Olah et al.,[9a] where solubilities follow the trend phosphates > antimonates > borates.

Form Supplied in: the reagents are prepared in situ. They exist as colorless solids when isolated.[6d] Stability of the carbenium ions varies considerably, especially in the substituted analogs.[6,8e,8f]

Preparative Methods: each reagent is conveniently prepared by introduction of the desired Lewis acid such as ***Boron Trifluoride Etherate***, ***Antimony(III) Chloride***, or phosphorus pentafluoride to a cooled solution containing the appropriate orthoformate (or orthoester, $R^2 \neq H$).[2] For dimethoxycarbenium bromide, the use of sulfur dioxide at −30 °C is required.[9b] For isolation techniques using a specially designed apparatus, see Pindur and Flo.[10]

Purification: since the reagents are prepared in situ, they are commonly used without further purification. When isolable, the reagents are purified by recrystallization.

Handling, Storage, and Precautions: dialkoxycarbenium salts are hygroscopic and should therefore be kept under an inert atmosphere (Ar or N_2). For optimal results, reaction vessels should be flame dried and purged with argon using anhydrous solvents in all instances.[6a] If isolable, storage in a dry box or a tightly sealed vessel under an inert atmosphere is recommended. Olah et al.[9a] describe the phosphate series to be stable under an inert atmosphere at rt without refrigeration. The borate and antimonate series are generally stable for weeks under an inert atmosphere at 0 °C. All dialkoxycarbenium salts are powerful acylating/alkylating agents; therefore skin and eyes should be protected when handling the solutions and especially the isolated salts. Use in a fume hood.

Introduction. Dialkoxycarbenium salts exhibit an interesting dichotomy in reactivity.[2g] These ambident electrophiles are powerful acylating and alkylating agents. The mode of reactivity can be altered depending upon the nucleophile, the reagent, the reaction temperature, the reaction time, and the solvent. Prepared in situ, these salts surpass in reactivity other common alkylating agents such as Meerwein's salt, alkyl triflates, and alkyl halides.[2,6,8] Each cationic component serves as a versatile electrophile which has proven to be an efficient formylating or alkylating reagent. All of the dialkoxycarbenium salts are accessible by judicious choice of orthoester and Lewis acid. Limitations include solubility and stability, which make the commercially available trialkyloxonium tetrafluoroborates attractive (see ***Triethyloxonium Tetrafluoroborate***, ***Trimethyloxonium Tetrafluoroborate***).[7a] However, dialkoxycarbenium salts provide a means for efficient acylation or alkylation of weak nucleophiles not accessible from their parent counterparts.

Alkylation of Sulfur Nucleophiles. Dimethoxycarbenium tetrafluoroborate has been extensively used in the conversion of numerous thiacyclophanes to polycyclic aromatic compounds.[3c] This method efficiently transforms sulfide linkages into carbon–carbon double bonds via a Stevens rearrangement followed by an elimination step. Using this sequence, dimethyldihydropyrene[3c,d] was synthesized in five steps, two of which involved the formation of a bis(sulfonium)cyclophane intermediate with dimethoxycarbenium tetrafluoroborate (eq 1).

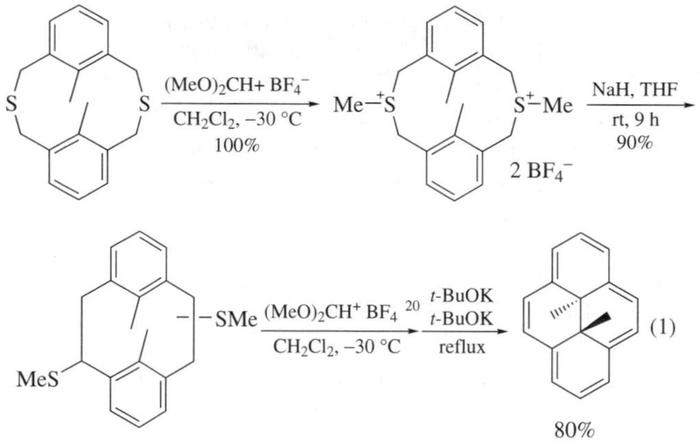

$(MeO)_2CH^+ BF_4^-$; CH_2Cl_2, –30 °C; 100% → 2 BF_4^-; NaH, THF; rt, 9 h; 90% → MeS ... –SMe; $(MeO)_2CH^+ BF_4^-$; CH_2Cl_2, –30 °C; *t*-BuOK, *t*-BuOK, reflux; 80% (1)

Formation of other bis(sulfonium) salts using dimethoxycarbenium tetrafluoroborate was successfully applied toward the synthesis of other pyrene derivatives such as [2.2]metaparacyclophene-1,9-diene,[3e] [7],[7.7] circulene,[3f] [2.2](2,6) pyridinophene-1,9-diene,[3g] and [2.2.2](1,3,5)cyclophane-1,9,17-triene.[3h] Another example of sulfur activation employing both dimethoxycarbenium tetrafluoroborate as well as hexafluorophosphate has been applied to the synthesis of various organic charge-transfer complexes.[4a,b] Critical for the success of these syntheses is the conversion of various thione derivatives to the corresponding selenone derivatives. Treatment of thieno[3,4-*d*]-1,3-dithiole-2-thione with dimethoxycarbenium hexafluorophosphate afforded the desired thionium hexafluorophosphate, which was subsequently transformed to the selenone analog in the presence of ***Hydrogen Selenide*** (eq 2).

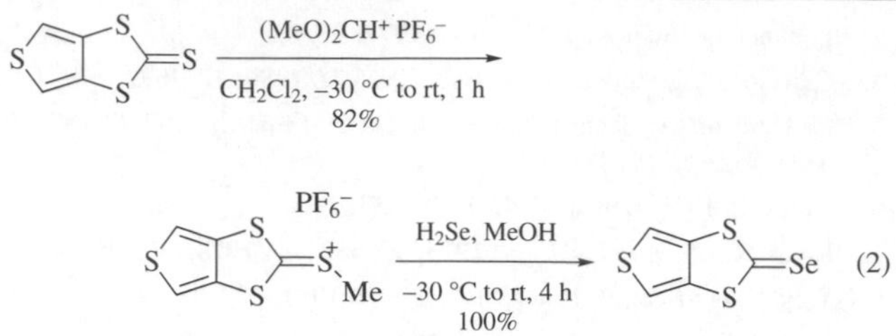

$(MeO)_2CH^+ PF_6^-$; CH_2Cl_2, –30 °C to rt, 1 h; 82% → PF_6^-, S–Me; H_2Se, MeOH; –30 °C to rt, 4 h; 100% → =Se (2)

Activation of the carbon–sulfur double bond with dialkoxycarbenium salts has led to the preparative synthesis of similar analogs. Thioimidates can be prepared by the selective alkylation of thioamides using either dimethoxy- or diethoxycarbenium tetrafluoroborate (eq 3).[4c,d] The products isolated are used as key intermediates in the synthesis of β-lactams. Meerwein's salt also carries out the transformation shown in eq 3. Other examples include the formation of hexachloroantimonate salts via ethylation of dithiolodithione dimers (eq 4).[4e] Lastly, heterolysis of disulfide bonds has been achieved by treatment of a diaryl disulfide with diethoxycarbenium tetrafluoroborate (eq 5).[4f]

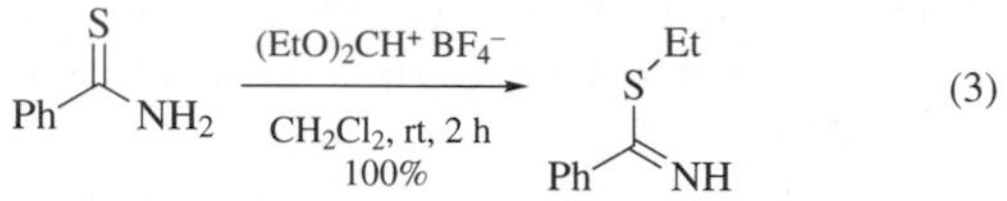

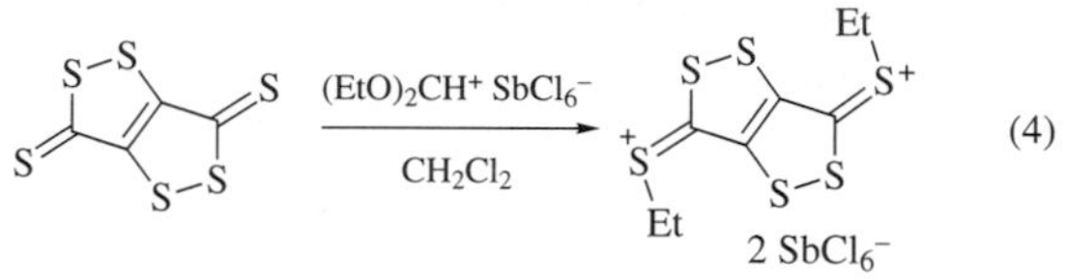

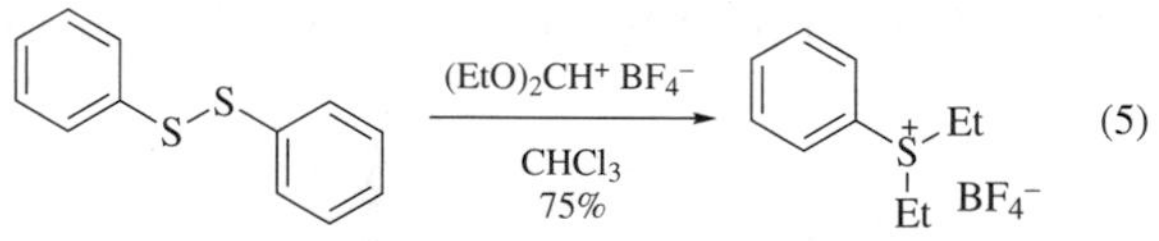

$PhC(=S)NH_2$ → $(EtO)_2CH^+ BF_4^-$; CH_2Cl_2, rt, 2 h; 100% → $PhC(SEt)=NH$ (3)

$(EtO)_2CH^+ SbCl_6^-$; CH_2Cl_2 → Et, S^+, 2 $SbCl_6^-$ (4)

$PhSSPh$ → $(EtO)_2CH^+ BF_4^-$; $CHCl_3$; 75% → PhS^+Et_2 BF_4^- (5)

Alkylation of Nitrogen Nucleophiles. Dialkoxycarbenium tetrafluoroborates provide an efficient means for the production of secondary amines through nitrilium salt intermediates.[5b] *N*-Alkylnitrilium salts other than *N*-methyl or *N*-ethyl are not readily accessible using the trialkyloxonium tetrafluoroborates. Treatment of a series of dialkoxycarbenium salts with benzonitrile furnished, after reductive workup, the desired substituted benzylamines (eq 6). Higher analogs such as *n*-propyl did not provide the desired *n*-propylbenzylamine but rather isopropylbenzylamine. This rearrangement does limit applications of other higher-order dialkoxycarbenium salts.

$$PhC{\equiv}N \xrightarrow[CH_2Cl_2,\ reflux]{(RO)_2CH^+ BF_4^-} [PhC{\equiv}N\text{-}R\ BF_4^-] \xrightarrow[2.\ NaBH_4,\ MeOH,\ 0\ °C,\ 1\ h]{1.\ EtOH,\ 0\ °C} PhCH_2NHR \quad (6)$$

R = Me, 79%; Et, 71%; *n*-Pr, 28%; *i*-Pr, 94%; allyl, 80%

Formation of nitrilium salts has also been applied to the field of organometallic chemistry.[5c] In these examples the reductive workup of the intermediate iminoester with sodium borohydride is eliminated. Diethoxycarbenium tetrafluoroborate is used to perform two nitrogen-based alkylations (eq 7). The resulting complex is isolated and purified by recrystallization.

$$[\text{NC}(\text{Ph}_3\text{P})\text{Pt}(\text{PPh}_3)\text{CN}] \xrightarrow[\text{CH}_2\text{Cl}_2,\ \text{rt},\ 18\ \text{h};\ 69\%]{(\text{EtO})_2\text{CH}^+\ \text{BF}_4^-} [\text{EtNC}(\text{Ph}_3\text{P})\text{Pt}(\text{PPh}_3)\text{CNEt}]^{2+}\ 2\text{BF}_4^- \quad (7)$$

Another example of activation/reduction is illustrated in the preparation of 1-alkyl-2,1-benzisoxazoline derivatives.[5d] Treatment of 2,1-benzisoxazoles with dialkoxycarbenium hexachloroantimonate affords the desired 1-alkyl-2,1-benzisoxazolium salts, which are subsequently treated with a variety of nucleophilic reagents (eq 8). The following transformation is catalyzed by in situ preparation of either dimethoxy- or diethoxycarbenium salts as well as the related trialkyloxonium tetrafluoroborates. Nucleophilic reagents employed include Grignards, cyanides, malonates, morpholines, and other amine derivatives. The end products were obtained in good yields without appreciable amounts of ring-opened byproducts.

$$\text{2,1-benzisoxazole (X, R)} \xrightarrow[\text{CH}_2\text{Cl}_2]{(\text{MeO})_2\text{CH}^+\ \text{SbCl}_6^-} \text{N-Me benzisoxazolium SbCl}_6^- \xrightarrow{\text{nucleophile}} \text{N-Me-3-Nuc-3-R-benzisoxazoline} \quad (8)$$

X = H; R = Me, 89%
X = Cl; R = Ph, 94%

A final example of nitrogen activation of weak organic bases is in the alkylation of *N,N*-disubstituted sulfonamides.[5e] The reagent of choice is dialkoxycarbenium hexachloroantimonate because sulfonamides are inert toward Meerwein's salt. Shown in eq 9 is the formation of two hexachloroantimonates from the corresponding *N,N*-disubstituted sulfonamides. Subsequent treatment of these salts with aqueous acid effectively cleaves the N–S bond into corresponding amine and sulfonic acid derivatives.

$$\text{Ts-NR}_2 \xrightarrow[\text{CH}_2\text{Cl}_2,\ \text{rt},\ 12\ \text{h}]{(\text{MeO})_2\text{CH}^+\ \text{SbCl}_6^-} \text{Ts-N}^+\text{R}_2\text{Me}\ \text{SbCl}_6^- \quad (9)$$

R = Me, 60%
R = -(CH$_2$)$_4$-, 80%

Alkylation of Oxygen Nucleophiles. Dialkoxycarbenium hexachloroantimonates are versatile reagents in the alkylation of a variety of oxygen-containing substrates.[6a] Even weak nucleophiles which are inert to trialkyloxonium salts are cleanly and rapidly alkylated. Treatment of diethoxycarbenium hexachloroantimonate with sulfoxides (eq 10),[6a] aldehydes, ketones, esters (eq 11),[6c] and amides (eq 12)[6a] yield alkylated products which can either be isolated as antimonate salts or their subsequent transformation products.

$$\text{Ph}_2\text{S=O} \xrightarrow[\text{CH}_2\text{Cl}_2,\ -30\ ^\circ\text{C};\ 92\%]{(\text{EtO})_2\text{CH}^+\ \text{SbCl}_6^-} \text{Ph}_2\text{S=O}^+\text{Et}\ \text{SbCl}_6^- \quad (10)$$

$$\text{CH}_2\text{=CHCO}_2\text{Et} \xrightarrow[\text{2. py, MeOH, }-78\ ^\circ\text{C};\ 95\%]{\text{1. }(\text{EtO})_2\text{CH}^+\ \text{SbCl}_6^-,\ \text{CH}_2\text{Cl}_2,\ -30\ ^\circ\text{C}} \text{CH}_2\text{=CHC(OMe)(OEt)}_2 \quad (11)$$

$$\text{PhC(O)NEt}_2 \xrightarrow[\text{CH}_2\text{Cl}_2,\ -30\ ^\circ\text{C};\ 95\%]{(\text{EtO})_2\text{CH}^+\ \text{SbCl}_6^-} \text{PhC(=O}^+\text{Et)NEt}_2\ \text{SbCl}_6^- \quad (12)$$

Dimethoxycarbenium tetrafluoroborates are stronger alkylating agents than their parent trialkyloxonium tetrafluoroborates. Interestingly, the preparation of methyl or ethyl Meerwein's salt can be accomplished by treatment of the appropriate dialkoxycarbenium salt with either dimethyl or diethyl ether (eq 13).[7] This procedure has been reported to yield the desired trialkyloxonium salt in very high purity.

$$\text{HC(OMe)}_3 \xrightarrow[\text{CH}_2\text{Cl}_2,\ -30\ ^\circ\text{C},\ 15\ \text{min}]{\text{Et}_2\text{O}\bullet\text{BF}_3} (\text{MeO})_2\text{CH}^+\ \text{BF}_4^- \xrightarrow[-15\ ^\circ\text{C},\ 3\ \text{h}]{\text{Me}_2\text{O},\ \text{CH}_2\text{Cl}_2} \text{Me}_3\text{O}^+\ \text{BF}_4^-\ \ 72\% \quad (13)$$

Aryl-substituted dimethoxycarbenium tetrafluoroborates, although sluggish in reactivity due to the steric conjestion about the carbenium center, will alkylate at oxygen.[6b] Treatment of 2,6-dimethyl-4-pyrone with an aryl-substituted tetrafluoroborate salt affords the desired 4-methoxy-2,6-dimethylpyrylium tetrafluoroborate in good yield (eq 14).

$$\text{2,6-dimethyl-4-pyrone} \xrightarrow[\text{CH}_2\text{Cl}_2,\ \text{rt};\ 60\%]{(\text{MeO})_2\text{CAr}^+\ \text{BF}_4^-} \text{4-methoxy-2,6-dimethylpyrylium BF}_4^- \quad (14)$$

Ar = Ph or 2,4,6-Me$_3$C$_6$H$_2$

Acylation of Carbon Nucleophiles. The foregoing examples illustrate the use of dialkoxycarbenium salts for the selective alkylation of weakly nucleophilic heteroatoms.[2g] In these examples, reaction conditions of thermodynamic control were generally employed. However, when dialkoxycarbenium salts are used under conditions of kinetic control a different type of product is obtained. Under the latter reaction conditions, carbocyclic or aromatic conjugated cyclic and acyclic ketones are converted into protected derivatives of formyl ketones.[8b] Employing diethoxycarbenium tetrafluoroborate prepared in situ with a variety of cyclic (eq 15) and acyclic (eq 16) ketones in the presence of ***Diisopropylethylamine*** at low temperatures in a methylene chloride solution affords the corresponding β-keto acetals.[8c] Other functional groups such as halide, nitrile, alkene, arene, and ester functionalities are compatible with this method. Regioselectivity is governed by steric effects; therefore reaction generally takes place at the less substituted α-position.

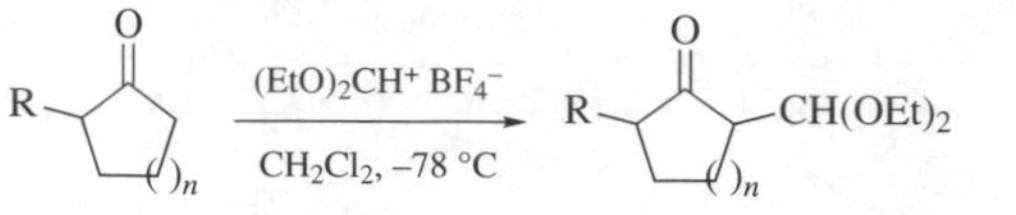

(15) [scheme: $(EtO)_2CH^+ BF_4^-$, CH_2Cl_2, −78 °C; product $CH(OEt)_2$]

R = H $n = 1$ 45%
R = H $n = 2$ 87%
R = Me $n = 2$ 11%

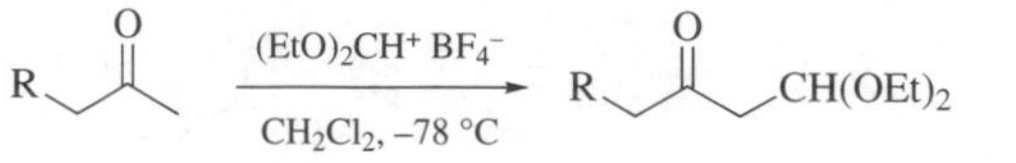

(16) [scheme: $(EtO)_2CH^+ BF_4^-$, CH_2Cl_2, −78 °C; product $CH(OEt)_2$]

R = Me 76%
R = $Me_2C{=}CH$ 58%
R = $EtO_2CCH_2CH_2$ 59%

This method has been extended to the preparation of α,β-unsaturated aldehydes via β-keto acetals (eq 17)[8c] and some aryl-substituted bicyclo[3.3.1]nonadienones through an acid-catalyzed cyclization of the α-(diethoxymethyl) ketones (eq 18).[8d]

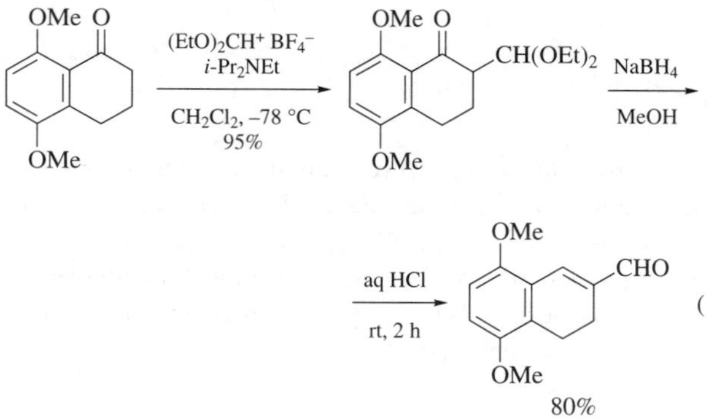

(17) [scheme: $(EtO)_2CH^+ BF_4^-$, i-Pr_2NEt, CH_2Cl_2, −78 °C, 95%; $NaBH_4$, MeOH; aq HCl, rt, 2 h; 80%]

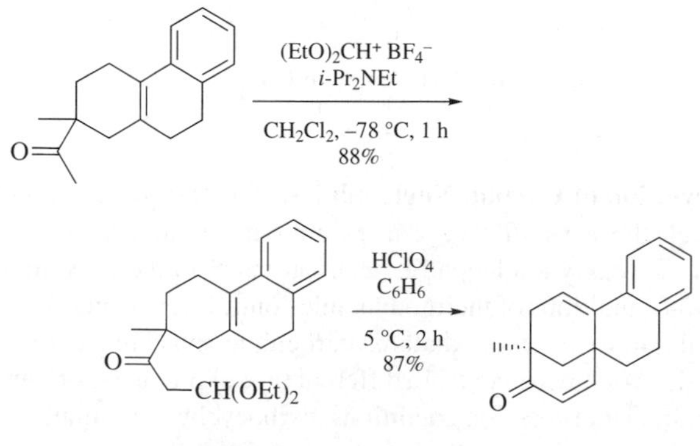

(18) [scheme: $(EtO)_2CH^+ BF_4^-$, i-Pr_2NEt, CH_2Cl_2, −78 °C, 1 h, 88%; $HClO_4$, C_6H_6, 5 °C, 2 h, 87%]

High levels of regioselectivity are observed when reactions between electron-rich heterocycles and dialkoxycarbenium tetrafluoroborates are carried out. From 2-methylindole, several 3-acylated indoles are produced without N-alkylation (eq 19).[8e] Conversion to the desired aldehyde or ketone after initial acylation is observed upon aqueous workup.

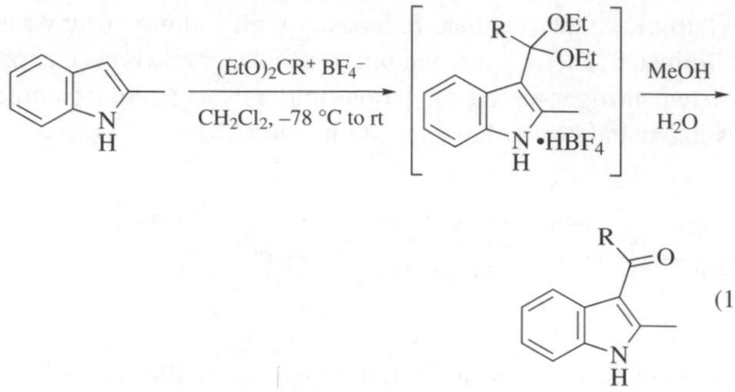

(19) [scheme: $(EtO)_2CR^+ BF_4^-$, CH_2Cl_2, −78 °C to rt; intermediate $N{\cdot}HBF_4$; MeOH, H_2O]

R = H 49%
R = Me 90%
R = Ph 25%

Treatment of 2,3-substituted indoles with diethoxycarbenium tetrafluoroborates results in selective acylation on the nitrogen.[8f] Interestingly, however, upon switching to the dimethoxy derivative, alkylation on nitrogen predominates even in the 2-methylindole series. As the steric size of the carbenium ion increases, for example when switching from dialkoxycarbenium salts to dialkoxyalkylcarbenium salts, one observes a shift from acylation to alkylation as well as a decrease in overall reactivity.

Another interesting variation in the electrophilic reactivity of these ambident reagents exists in the area of anionic acylations. Treatment of dimethoxycarbenium tetrafluoroborate, predominately an alkylating agent, with either lithium cyclononatetraenide[8g] or potassium cyclopentadienyl-irondicarbonyl[8h] results in the formation of the desired acylated derivatives.

Related Reagents. Dimethyliodonium Hexafluoroantimonate; O-Methyldibenzofuranium Tetrafluoroborate; Triethyl Orthoformate; Triethyloxonium Tetrafluoroborate; Trimethyloxonium Tetrafluoroborate.

1. (a) Pindur, U.; Müller, J.; Flo, C.; Witzel, H. *CSR* **1987**, *16*, 75. (b) Perst, H. *Oxonium Ions in Organic Chemistry*; Verlag Chemie: Weinheim, 1971; pp 145–149.
2. (a) Meerwein, H. *MOC* **1963**, *6/3*, 325. (b) Fieser, L. F.; Fieser, M. *FF* **1971**, *5*, 714. (c) Fieser, L. F.; Fieser, M. *FF* **1971**, *5*, 716. (d) Meerwein, H.; Hederich, V.; Morschel, H.; Wunderlich, K. *LA* **1960**, *635*, 1. (e) Meerwein, H.; Bodenbenner, K.; Borner, P.; Kunert, F.; Wunderlich, K. *LA* **1960**, *632*, 38. (f) Meerwein, H.; Borner, P.; Fuchs, O.; Sasse, H. J.; Schrodt, H.; Spille, J. *CB* **1956**, *89*, 2060. (g) Hünig, S. *AG(E)* **1964**, *3*, 548.
3. (a) Fieser, L. F.; Fieser, M. *FF* **1970**, *4*, 114. (b) Fieser, L. F.; Fieser, M. *FF* **1971**, *5*, 623. (c) Mitchell, R. H. *H* **1978**, *11*, 563. (d) Mitchell, R. H.; Boekelheide, V. *TL* **1970**, 1197. (e) Boekelheide, V.; Anderson, P. H. *TL* **1970**, 1207. (f) Yamamoto, K. *PAC* **1993**, *65*, 157. (g) Boekelheide, V.; Lawson, J. A. *CC* **1970**, 1558. (h) Boekelheide, V.; Hollins, R. A. *JACS* **1970**, *92*, 3512.
4. (a) Kobayashi, K. *CL* **1985**, 1423. (b) Chiang, L.-Y.; Shu, P.; Holt, D.; Cowan, D. *JOC* **1983**, *48*, 4713. (c) Casadei, M. A.; Rienzo, B. D.; Moracci, F. M. *SC* **1983**, *13*, 753. (d) Röhrich, J.; Müllen, K. *JOC* **1992**, *57*, 2374. (e) Richter, A. M.; Fanghänel, E. *TL* **1983**, *24*, 3577. (f) Miller, B.; Han, C.-H. *JOC* **1971**, *36*, 1513.
5. (a) Fieser, L. F.; Fieser, M. *FF* **1969**, *3*, 303. (b) Borch, R. F. *JOC* **1969**, *34*, 627. (c) Weigand, W.; Beck, W. *JOM* **1988**, *338*, 113. (d) Nakagawa, Y.; Aki, O.; Sirakawa, K. *CPB* **1972**, *20*, 2209. (e) Oishi, T.; Kamata, K.; Ban, Y. *CC* **1970**, 777.
6. (a) Kabuss, S. *AG(E)* **1966**, *5*, 675. (b) Dimroth, K.; Heinrich, P. *AG(E)* **1966**, *5*, 676. (c) Cook, G. A.; Butler, G. B. *J. Macromol. Sci., Chem.*

1985, *A22*, 1035 and references cited therein. (d) Pindur, U.; Flo, C. *SC* **1989**, *19*, 2307.

7. (a) Curphey, T. J. *OS* **1971**, *51*, 142. (b) Earle, M. J.; Fairhurst, R. A.; Giles, R. G.; Heaney, H. *SL* **1991**, 728. (c) Silverman, R. B.; Olofson, R. A. *CC* **1968**, 1313.
8. (a) Fieser, L. F.; Fieser, M. *FF* **1977**, *11*, 175. (b) Mock, W. L.; Tsou, H.-R. *JOC* **1981**, *46*, 2557. (c) Gupta, R. D.; Ranu, B. C.; Ghatak, U. R. *IJC(B)* **1983**, *22B*, 619. (d) Dasgupka, R.; Ghatak, U. R. *TL* **1985**, *26*, 1581. (e) Pindur, U.; Flo, C. *M* **1986**, *117*, 375. (f) Flo, C.; Pindur, U. *LA* **1987**, 509. (g) Sabbioni, G.; Neuenschwander, M. *HCA* **1985**, *68*, 623. (h) Theys, R. D.; Hossain, M. M. *TL* **1992**, *33*, 3447.
9. (a) Olah, G. A.; Olah, J. A.; Svoboda, J. J. *S* **1973**, 490. (b) Dusseau, Ch. H. V.; Schaafsma, S. E.; Steinberg, H.; de Boer, Th. J. *TL* **1969**, 467. (c) Ramsey, B. G.; Taft, R. W. *JACS* **1966**, *88*, 3058.
10. Pindur, U.; Flo, C. *SC* **1989**, *19*, 2307.

David C. Forbes
University of Illinois, Urbana-Champaign, IL, USA

N,N-Dimethylacetamide Dimethyl Acetal[1]

RO OR / NMe2 ⇌ OR / NMe2 + ROH

(1) R = Me (2)
(3) R = Et (4)

(**1**; R = Me)		
[18871-66-4]	$C_6H_{15}NO_2$	(MW 133.19)
(**2**; R = Me)		
[867-89-0]	$C_5H_{11}NO$	(MW 101.15)
(**3**; R = Et)		
[19429-85-7]	$C_8H_{19}NO_2$	(MW 161.24)
(**4**; R = Et)		
[816-65-9]	$C_6H_{13}NO$	(MW 115.17)

(Eschenmoser–Claisen rearrangements;[2] monoacylation of diols;[3] triazine synthesis[4])

Physical Data: mixture of (**1**) and (**2**) bp 105–110 °C; *d* 0.911 g cm^{-3}. Mixture of (**3**) and (**4**) bp 126–129 °C. (**2**) bp 104–105 °C. (**4**) bp 126 °C.
Solubility: sol benzene, toluene, xylene, DMF, chloroform.
Form Supplied in: amide acetal (**1**) is commercially available, usually as a mixture of (**1**) and (**2**). The commercial mixture is 85–95% pure; methanol (5–10%) may be present as a stabilizer.
Preparative Methods: O-methylation of *N,N*-dimethylacetamide with ***Dimethyl Sulfate*** (or diethyl sulfate), followed by reaction with ***Sodium Methoxide*** (or ethoxide). Fractional distillation affords mixtures of (**1**) and (**2**) (or **3** and **4**). Treatment with calcium metal and redistillation gives the ketene *O,N*-acetals (**2**) (58%) (or **4**; 74%).[5]

Introduction. Amide acetal reagent (**1**) is in equilibrium with its enamine (eq 1).[5] The stabilized carbenium ion, which is an intermediate in this equilibrium, adds to nucleophiles under neutral conditions.

MeO OMe / NMe2 (1) ⇌ OMe / + NMe2 MeO⁻ ⇌ OMe / NMe2 + MeOH (1)

Eschenmoser–Claisen Rearrangement. Condensation of (**1**) with allylic alcohols followed by loss of a second equivalent of methanol gives ketene *N,O*-acetals, which undergo 3,3-sigmatropic rearrangement to afford γ,δ-unsaturated amides (eq 2).

OH / R¹ R² —(1)→ [NMe2 / O / R¹ R²] → NMe2 / O / R¹ R² (2)

In a general procedure, the allylic alcohol and dimethylacetamide dimethyl acetal are heated at reflux in a high boiling solvent such as xylene or DMF with simultaneous distillation of methanol. Other suitable solvents include diglyme, dioxane, toluene, and benzene. Several papers indicate that the Eschenmoser–Claisen reaction proceeds at lower temperatures with higher yields and shorter reaction times than the classical Claisen and the orthoester Claisen rearrangements.[6] However, it suffers the disadvantage that the product is a disubstituted amide, a functional group which is not always easily converted to other groups.[7] Several examples of transformations of disubstituted amides to other functional groups are included in the discussion below.

Eschenmoser and co-workers demonstrated that the amide acetal rearrangement of cyclohexenols is stereospecific, i.e. the *cis*-4-substituted cyclohexenol gives a single stereoisomer of the cyclohexeneacetamide product (eq 3), whereas the *trans*-cyclohexenol affords its stereoisomer (eq 4).[8]

CO_2Me / OH —(1), xylene; 140 °C, 14 h; 70%→ Me_2N / O / CO_2Me (3)

CO_2Me / OH —(1), xylene; 140 °C, 14 h; 76%→ Me_2N / O / CO_2Me (4)

Benzyl alcohols rearrange to *o*-methylbenzeneacetamides (eq 5).

OH —(1), DMF; 20 h, 180 °C; 76%→ NMe_2 / O (5)

The amide acetal rearrangement, like the Claisen rearrangement, generally proceeds through a chairlike transition state with all substituents in equatorial positions, an arrangement which leads to *trans* geometry of the double bond in the product and chirality transfer (eq 6).[9]

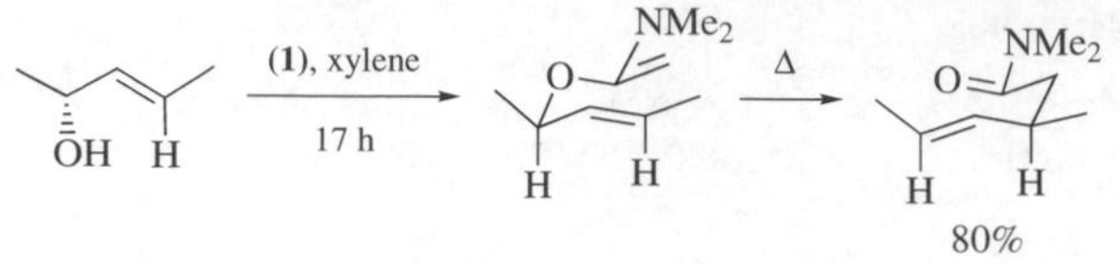

The products of the Eschenmoser–Claisen rearrangement with 3-(trimethylsilyl)allyl alcohols (eq 7) undergo protonolysis and desilylation with ***Hydrogen Fluoride*** to give β,γ-unsaturated amides which can be converted into the corresponding esters.[10]

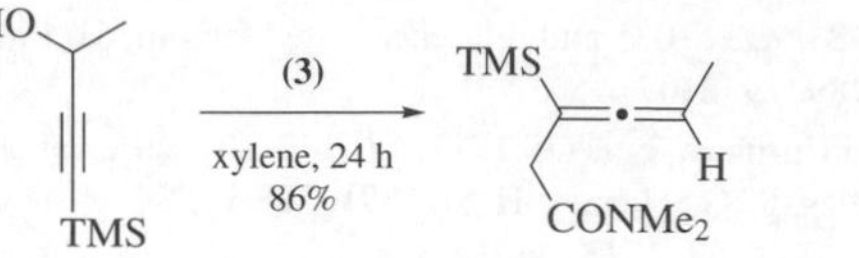

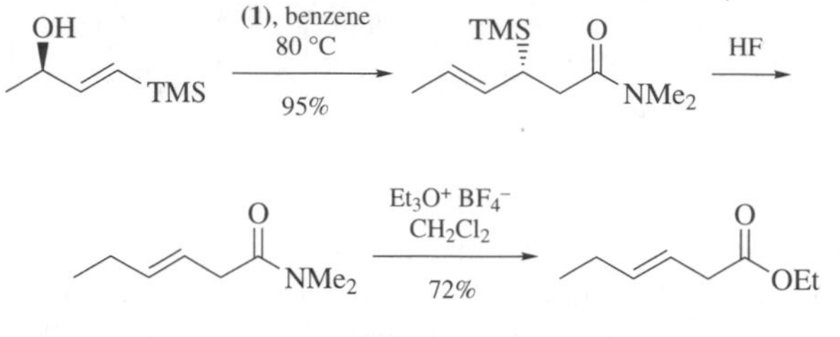

The Eschenmoser–Claisen method was used for the rearrangement of an allyl alcohol bearing two organometallic groups because of its nearly neutral conditions (eq 8). The product, obtained in 74% yield with 90% chirality transfer, was reduced with ***Lithium Triethylborohydride*** to the primary homoallylic alcohol.[11]

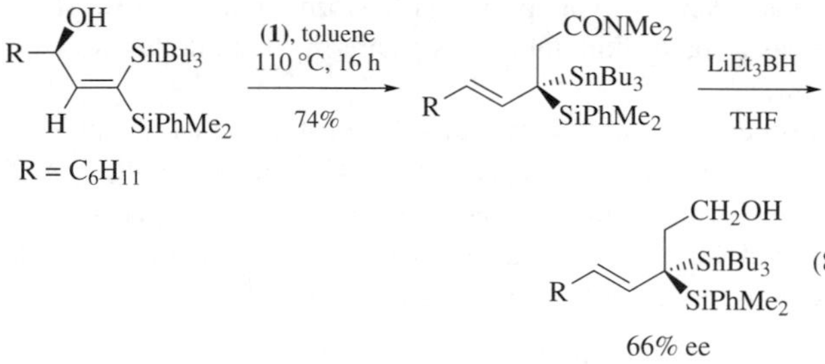

Parker and co-workers found that terminal propargyl alcohols condense with dimethylacetamide diethyl acetal (**3**) to form an intermediate which undergoes amine migration. This enamine subsequently rearranges through an orthoester Claisen rearrangement (eq 9).[12] A side product is the enol ether, which is formed from the enol derived from the enamine. Hydrolysis of the products gives a γ-keto ester.

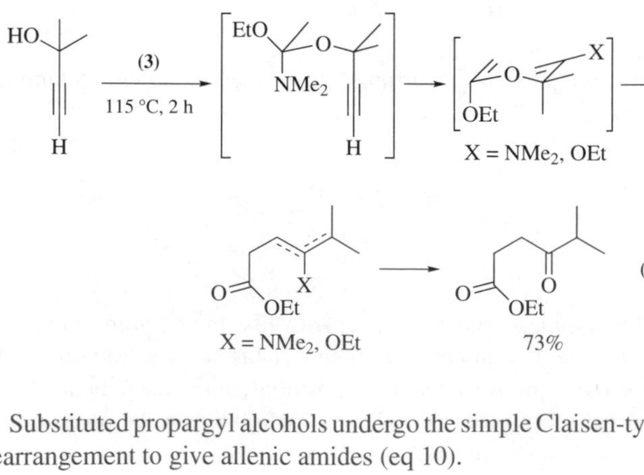

Substituted propargyl alcohols undergo the simple Claisen-type rearrangement to give allenic amides (eq 10).

Amide acetal Claisen rearrangements have been used in the synthesis of *C*-glycopyranosides,[6] pyranosides,[13] sugars,[14,15] and indole alkaloids.[7,16] In Corey's synthesis of thromboxane B_2, an amide product is transformed to a lactone by iodolactonization followed by deiodination with ***Tri-n-butyltin Hydride*** (eq 11).[14]

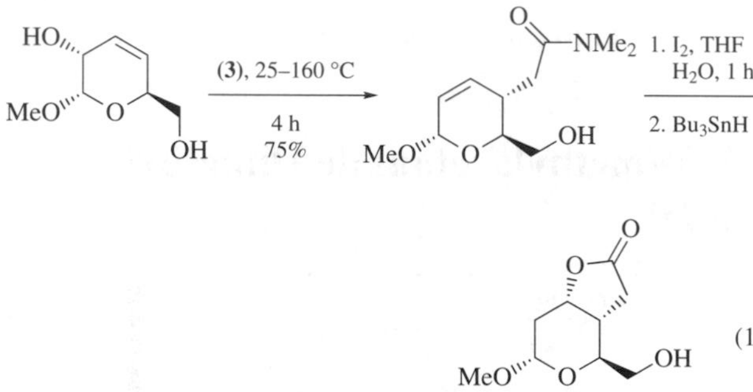

Monoacylation of Diols. Vicinal diols form (dimethylamino)ethylidene acetals, which can be cleaved with aqueous acetic acid to give monoacylated products.[3] In unsymmetrical diols, the less hindered hydroxy group is selectively protected. For example, with this method 1,2-*O*-isopropylidene-α-D-glucofuranose is selectively acylated at the C-6 hydroxy group.

Heterocycle Synthesis. *N,N*-Dimethylaminomethylenehydrazones are formed by addition of this reagent to α-ketohydrazones.[4] Subsequent heating with ammonium acetate in acetic acid causes cyclization to give 1,2,4-triazines (eq 12).

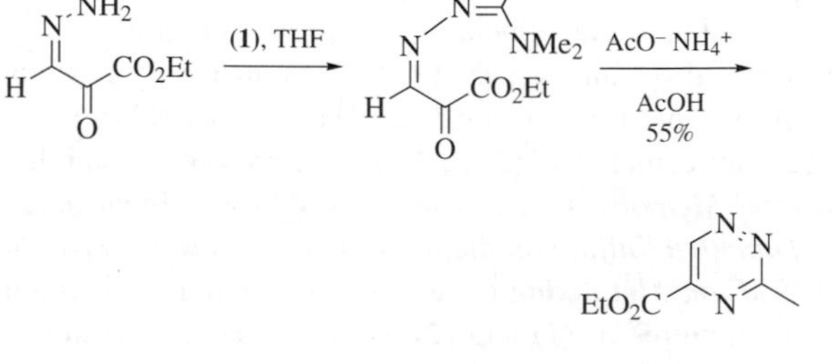

1,2,4-Triazoles and 1,2,4-oxadiazoles can be synthesized in high yields in a two-step procedure with this reagent. First the dimethyl acetal condenses with an amide to give an *N*-acylamidine. Subsequently, reaction with acetic acid and hydrazine gives triazoles, whereas addition of hydroxylamines gives oxadiazoles (eq 13).[17]

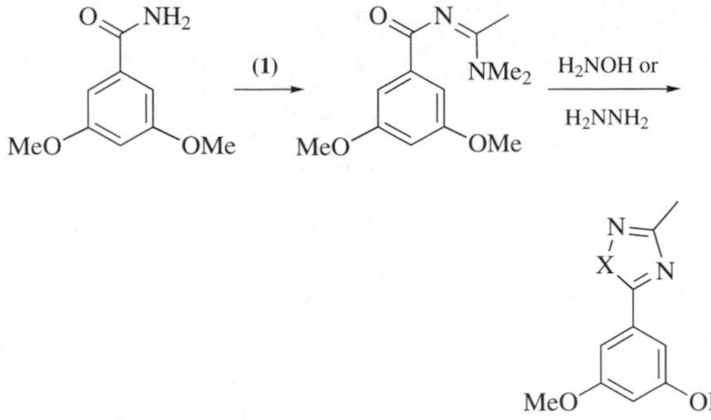

(13)

X = O, 95%; NH, 92%

Another method for the synthesis of 1,2,4-oxadiazoles is condensation of the amide acetal with an amidoxime (eq 14).[18] Fused amino pyridines have been formed through the cyclocondensation of an amide acetal with an amino ketone.[19]

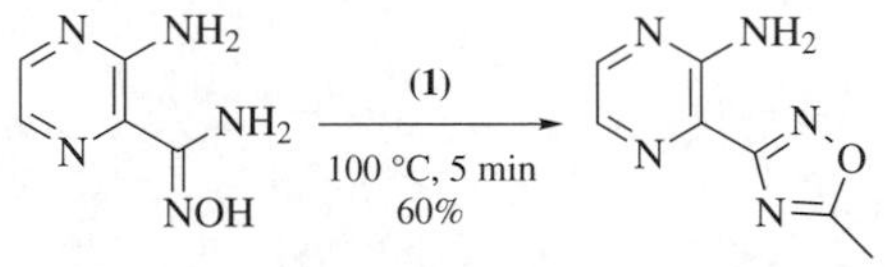

(14)

Related Reagents. Diketene; *N,N*-Dimethylformamide Diethyl Acetal; *N,N*-Dimethylpropionamide Dimethyl Acetal; Dimethyl Sulfate; Ethyl 3,3-Diethoxyacrylate; Ethyl Vinyl Ether; 2-Methoxy-1,3-butadiene; Triethyl Orthoacetate.

1. For reviews on amide acetal chemistry, see: (a) Pindur, U. In *The Chemistry of Acid Derivatives*; Patai, S., Ed.; Wiley: Chichester, 1992; Vol. 2, Suppl. B, p 1005. (b) Simchen, G. In *Iminium Salts in Organic Chemistry*; Böhme, H.; Viehe, H. G. Eds.; Wiley: New York, 1979, p 393. (c) Kantlehner, W. In *The Chemistry of Acid Derivatives*; Patai, S., Ed.; Wiley: Chichester, 1979; Part I, Suppl. B, p 533. (d) Dewolfe, R. H., *Carboxylic Ortho Acid Derivatives*; Academic: New York, 1970.
2. For recent reviews on Claisen rearrangements, see: (a) Wipf, P. *COS* **1991**, *5*, Chapter 7.2. (b) Kallmarten, J.; Wittman, M. D. *Stud. Nat. Prod. Chem.* , **1989**, *3*, 233. (c) Blechert, S. *S* **1989**, 71. (d) Ziegler, F. E. *CRV* **1988**, *88*, 1423. (e) Moody, C. J. *Adv. Heterocycl. Chem.* **1987**, *42*, 203.
3. (a) Hanessian, S.; Moralioglu, E. *CJC* **1972**, *50*, 233. (b) Hanessian, S.; Moralioglu, E. *TL* **1971**, *12*, 813.
4. (a) Ohsumi, T.; Neunoeffer, H. *T* **1992**, *48*, 5227. (b) Ohsumi, T.; Neunoeffer, H. *H* **1992**, *33*, 893.
5. (a) Meerwein, H.; Florian, W.; Schon, N.; Stopp, G. *LA* **1961**, *641*, 1. (b) Bredereck, H.; Effenberger, F.; Beyerlin, H. P. *CB* **1964**, *97*, 3081.
6. Moreau, P; Al Neirabeyeh, M.; Guillaumet, G.; Coudert, G. *TL* **1991**, *32*, 5525. (b) Tulshian, D. B.; Fraser-Reid, B. *JOC* **1984**, *49*, 518.
7. Ziegler, F. E.; Bennett, G. B. *JACS* **1973**, *95*, 7458.
8. (a) Wick, A. E.; Felix, D.; Steen, K.; Eschenmoser, A. *HCA* **1964**, *47*, 2425. (b) Felix, D; Gschwend-Steen, K.; Wick, A. E.; Eschenmoser, A. *HCA* **1969**, *52*, 1030.
9. Hill, R. K.; Soman, R.; Sawada, S. *JOC* **1972**, *37*, 3737.
10. Jenkins, P. R.; Gut, R.; Wetter, H.; Eschenmoser, A. *HCA* **1979**, *62*, 1922.
11. Lautens, M.; Huboux, A. H.; Chin, B.; Downer, J. *TL* **1990**, *31*, 5829.
12. (a) Parker, K. A.; Kosley, R. W. *TL* **1976**, *5*, 341. (b) Parker, K. A.; Petraitis, J. J.; Kosley, R. W.; Buchwald, S. L. *JOC* **1982**, *47*, 389.
13. Dickson, J. K.; Tsang, R.; Llera, J. M.; Fraser-Reid, B. *JOC* **1989**, *54*, 5350.
14. Corey, E. J.; Shibasaki, M; Knolle, J. *TL* **1977**, 1625.
15. Holzapfel, C. W.; Huyser, J. J.; Merwe, T. L.; Heerden, F. R. *H* **1991**, *32*, 1445.
16. Lounasmaa, M.; Jokela, R.; Tirkkonen, B.; Miettinen, J.; Halonen, M. *H* **1992**, *34*, 321.
17. Lin, Y.; Lang, S. A.; Lovell, M. F.; Perkinson, N. A. *JOC* **1979**, *44*, 4160.
18. Kocevar, M; Stanovnik, B.; Tisler, M. *JHC* **1982**, *19*, 1397.
19. Boamah, P. Y.; Haider, N.; Heinisch, G.; Moshuber, J. *JHC* **1988**, *25*, 879.

Kathlyn A. Parker & Lynn Resnick
Brown University, Providence, RI, USA

Dimethyl 1,3-Acetonedicarboxylate[1]

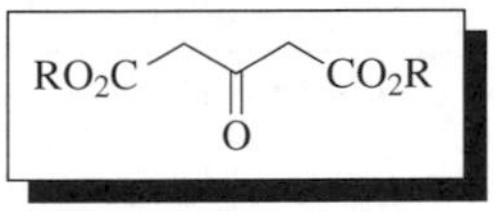

(**1**; R = Me)		
[1830-54-2]	$C_7H_{10}O_5$	(MW 174.15)
(**2**; R = Et)		
[105-50-0]	$C_9H_{14}O_5$	(MW 202.21)
(**3**; R = H)		
[542-05-2]	$C_5H_6O_5$	(MW 146.10)

(acidic β-keto esters, which undergo 2:1 condensations with α-dicarbonyl compounds to form bicyclo[3.3.0]octane-3,7-diones;[1–6] Robinson Schöpf condensation to form piperidones;[7] α,α- and α,α′-dialkylation with 1,*n*-dibromoalkanes; aldol and Claisen condensations with 1,3-dicarbonyl compounds;[13] various heterocyclizations with nitriles and aldehydes;[16] α,α′-aldol annulations with 1,2-enedials, producing tropone derivatives[20–22])

Physical Data: (**1**) bp 150 °C/25 mmHg; *d* 1.185 g cm^{-3}. (**2**) bp 250 °C/760 mmHg; bp 140 °C/12 mmHg; *d* 1.113 g cm^{-3}. (**3**) decomposes upon heating.
Solubility: sol ethanol, methanol.
Form Supplied in: colorless liquid.
Purification: distillation at reduced pressure.
Handling, Storage, and Precautions: can be stored in the refrigerator. Does not appear to be toxic.

Dimethyl and diethyl 1,3-acetonedicarboxylates (**1** and **2**) have been extensively used in the Weiss–Cook condensation reaction with various 1,2-dicarbonyl compounds, either at acidic or alkaline pH to form *cis*-bicyclo[3.3.0]octane-3,7-dione derivatives, as well as [*n*.3.3]propellanes in a single step (eqs 1–4).[1–6] These tetraesters are isolated in the *anti* bisenol form. They have been employed in the synthesis of several natural and nonnatural products.[1,2]

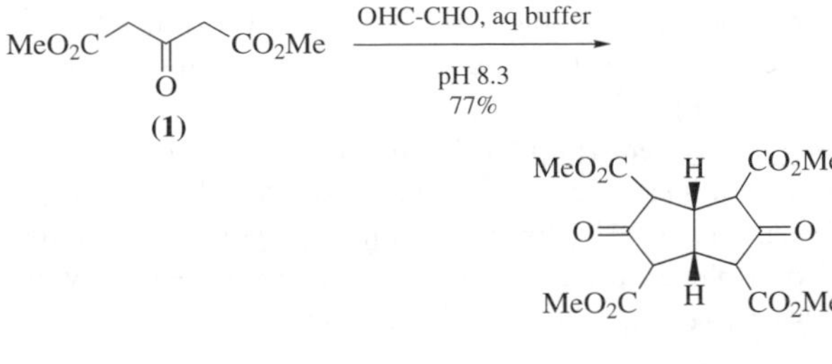

OHC-CHO, aq buffer, pH 8.3, 77%

(1)

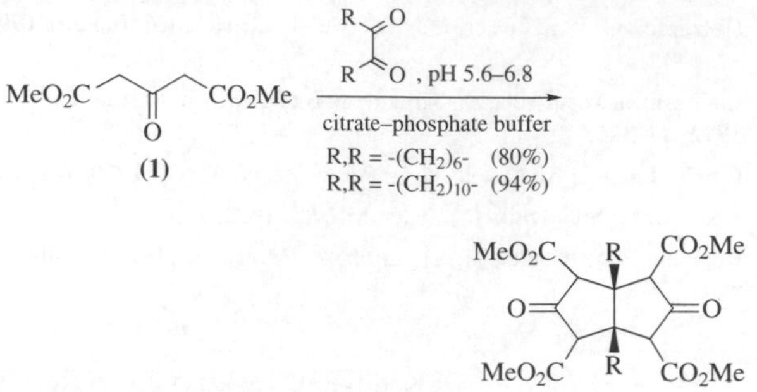

The diacid (**3**) undergoes the Robinson–Schöpf condensation with β-ethoxyglutaraldehyde and ammonium chloride to produce 7-ethoxy-9-azabicyclo[3.3.1]nonan-3-one (eq 5).[7]

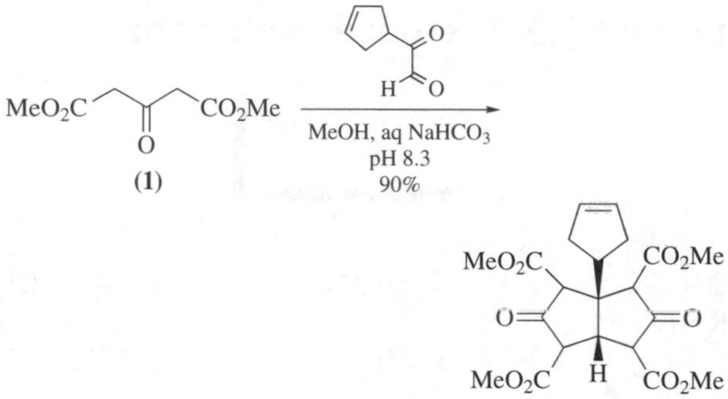

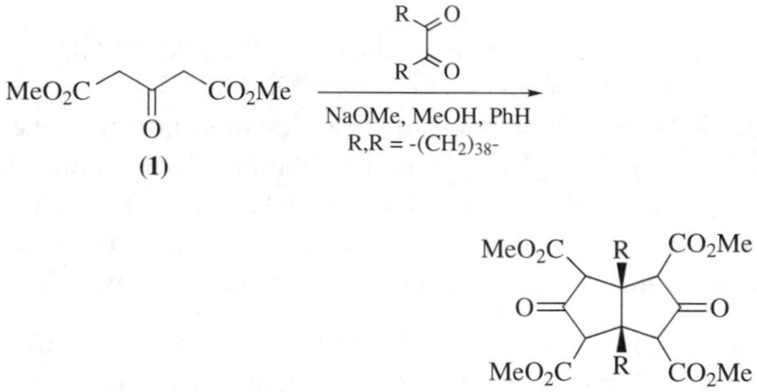

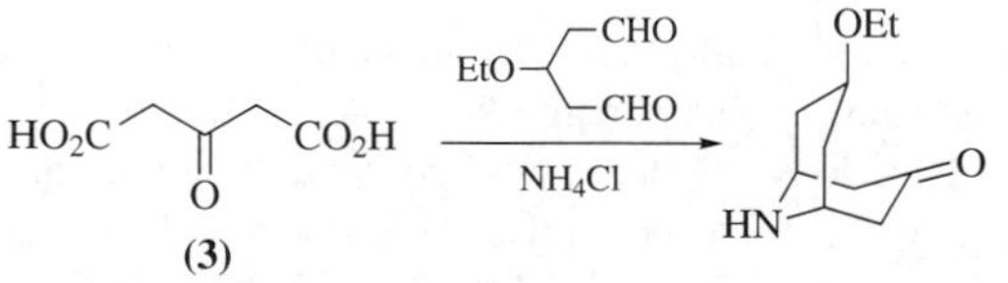

Dimethyl 1,3-acetonedicarboxylate (**1**), upon heating with acenaphthenequinone in the presence of glycine (catalyst) in norbornadiene as the solvent, affords dimethyl 7,10-fluoranthenedicarboxylate, a key intermediate in the synthesis of corannulene (eq 6).[8] Diethyl 1,3-acetonedicarboxylate (**2**) in the presence of ***Manganese(III) Acetate*** in acetic acid adds to the exocyclic enol lactone to provide an oxaspirolactone (eq 7).[9] Condensation of (**1**) (excess) with 1,3-cyclohexanedione in citrate phosphate buffer at pH 6.8 affords 5,6,7,8-tetrahydro-5-oxocoumarin-4-ylacetate (eq 8)[10] in good yield.

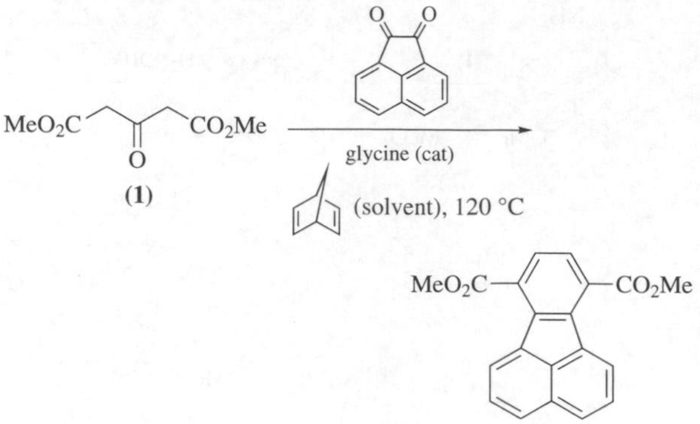

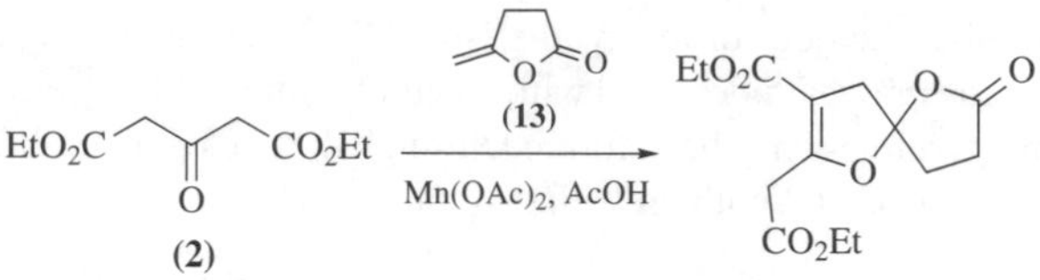

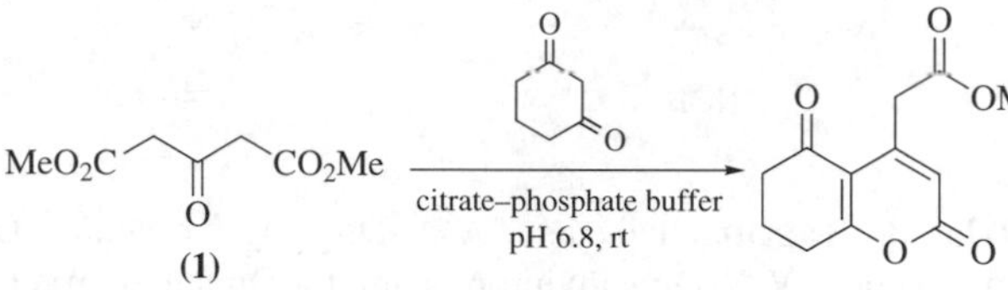

Alkylation of (**2**) with 1,3-dibromopropane and ***1,4-Dibromobutane***, individually, provides the cyclohexanone and cyclopentyl bis-esters, respectively (eqs 9 and 10).[11]

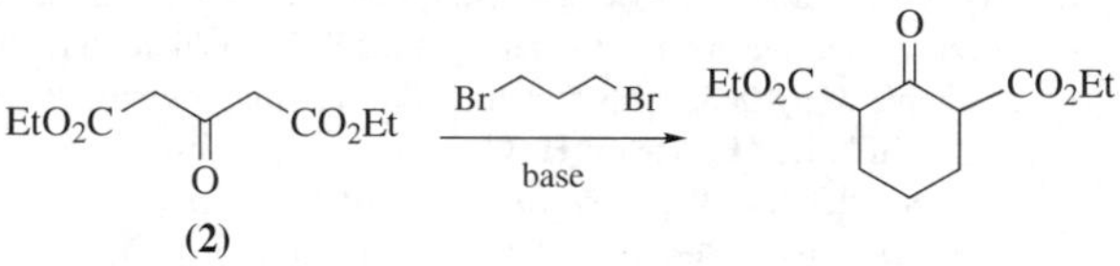

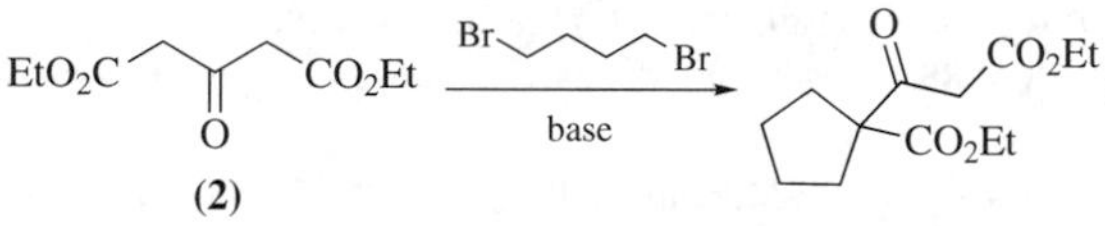

Reaction of (**1**) with ***α,α′-Dibromo-o-xylene*** under phase-transfer conditions yields bicyclo[4.4.1]undecan-11-one (eq 11).[12] The related naphthyl derivatives have also been prepared.

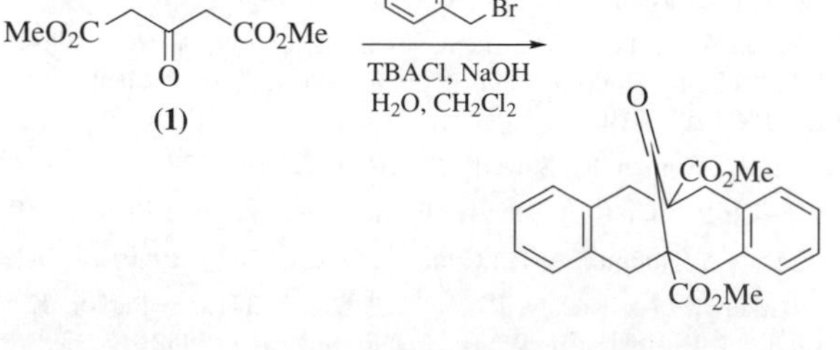

Condensation of (**2**) with ***Diethyl Malonate*** in the presence of ***Sodium Ethoxide*** gives diethyl 2,4,6-trihydroxybenzene-1,3-dicarboxylate (eq 12),[13] while condensation of either (**1**) or (**2**)

with an aldehyde and an amine in a 1:2:1 ratio provides the corresponding piperidones (eq 13).[14]

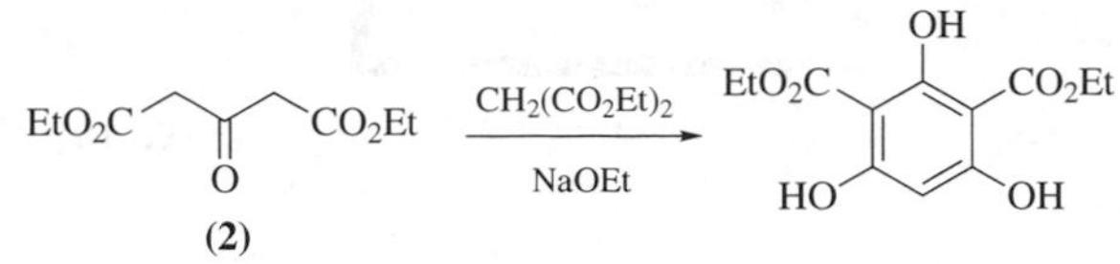

Condensation of (**3**) with 3,4-dihydro-β-carboline in aqueous methanol gives an amino acid intermediate which after esterification and treatment with aqueous ammonia cyclizes to the keto lactam illustrated in eq 14.[15]

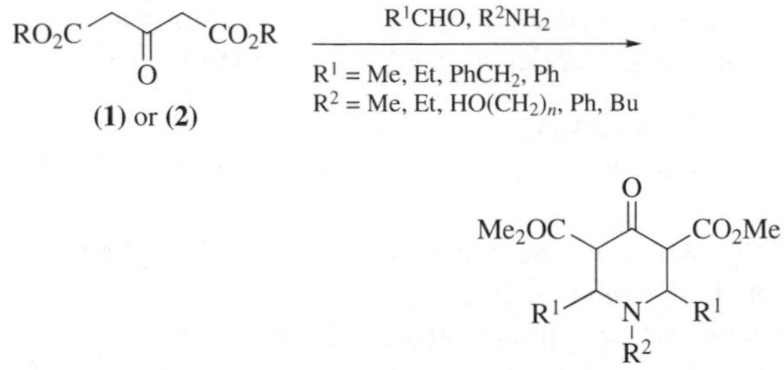

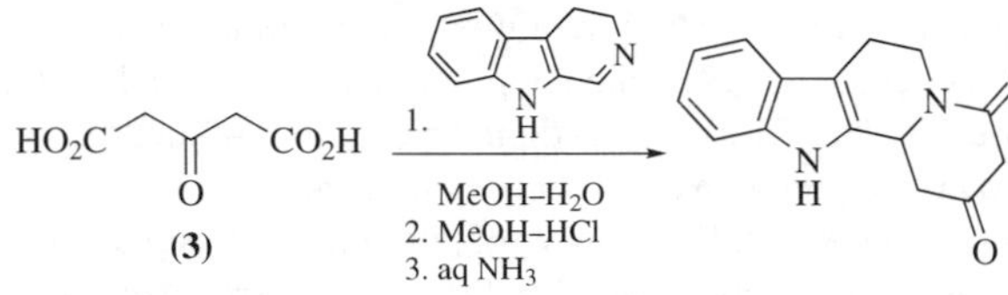

Reaction of (**1**) with trifluoroacetonitrile in the presence of ***Potassium t-Butoxide*** in THF produces methyl 4,6-dihydroxy-2-trifluoromethyl-3-pyridinecarboxylate (eq 15).[16]

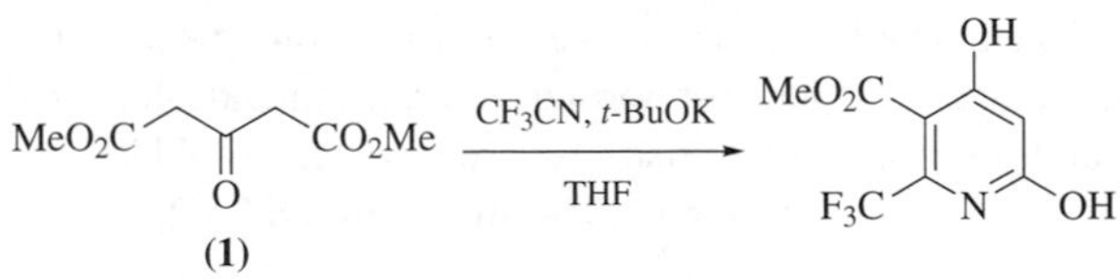

Reaction of (**2**) with ***Malononitrile*** and ***Sulfur*** in the presence of diethylamine as a catalyst affords ethyl 2-amino-3-cyano-5-(ethoxycarbonyl)thiophene-4-acetate (eq 16).[17] Diester (**1**) undergoes condensation with salicylaldehyde derivatives in the presence of piperidine to form the corresponding bis-coumarins represented in eq 17.[18] Condensation of (**2**) with *N*-methylaminoacetaldehyde at pH 9–10 produces ethyl 3-(ethoxycarbonyl)-1-methylpyrrole-2-acetate and 3-(ethoxycarbonyl)-1-methylpyrrole-2-acetic acid (eq 18).[19]

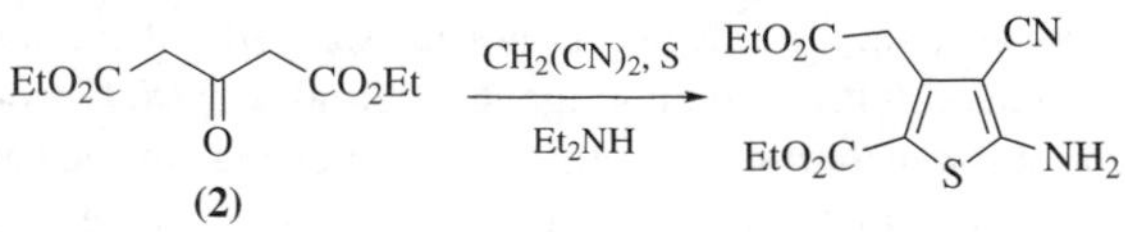

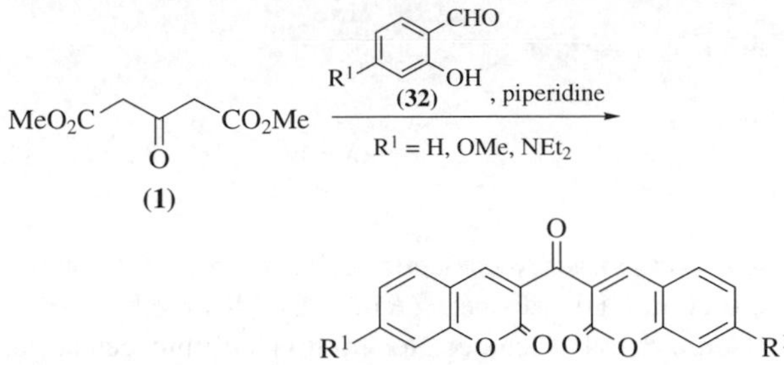

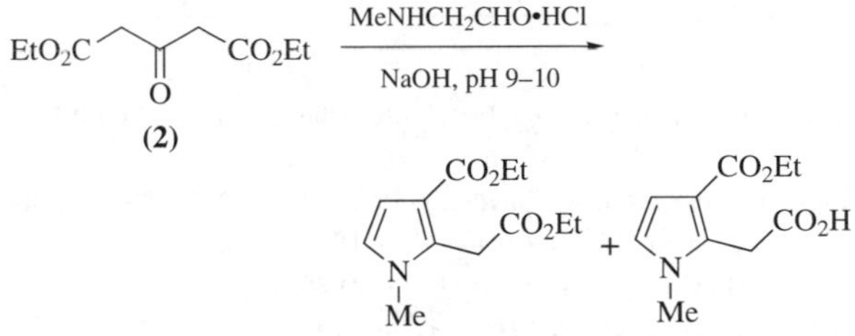

Diester (**1**) undergoes condensation with pyrrole-, *N*-methylindole- and 2,3-indenedicarbaldehydes to provide tropone derivatives (eqs 19–21).[20–22]

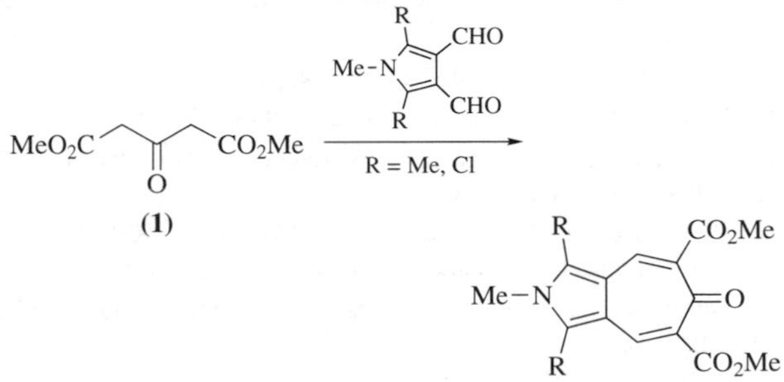

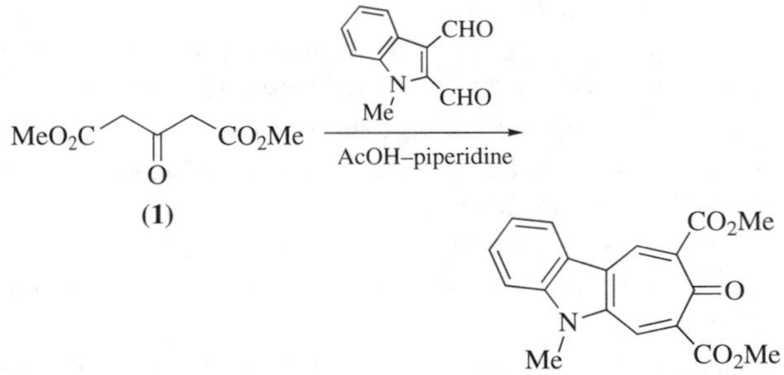

Reaction of (**1**) with activated ***Zinc*** in the presence of aniline in acetic acid affords the corresponding secondary amine depicted in eq 22.[23]

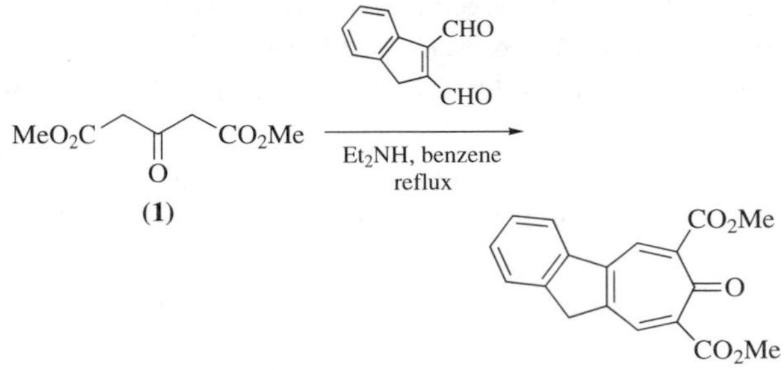

MeO_2C–CH_2–C(=O)–CH_2–CO_2Me (1) $\xrightarrow[60–70\ °C]{Zn,\ AcOH,\ ArNH_2}$ MeO_2C–CH_2–CH(NHAr)–CH_2–CO_2Me (22)

Related Reagents. Acetoacetic Acid; Acetone Cyclohexylimine; Acetone Hydrazone; (*R*)-2-*t*-Butyl-6-methyl-4*H*-1,3-dioxin-4-one; Ethyl Acetoacetate; Methyl Dilithioacetoacetate; 2,2,6-Trimethyl-4*H*-1,3-dioxin-4-one.

---

1. Ressig, H. U. *Nach. Chem. Tech. Lab.* **1986**, *34*, 162 (*CA* **1987**, *106*, 49 629h).
2. (a) Gupta, A. K.; Fu, X.; Snyder, J. P.; Cook, J. M. *T* **1991**, *47*, 3665. (b) Fu, X.; Cook, J. M. *Aldrichim. Acta* **1992**, *25*, 43.
3. Bertz, S. H.; Cook, J. M.; Weiss, U. *OS* **1986**, *64*, 27.
4. Yang, S.; Cook, J. M. *JOC* **1976**, *41*, 1903.
5. Mitschka, R.; Oehldrich, J.; Takahashi, K.; Cook, J. M.; Weiss, U.; Silverton, J. V. *T* **1981**, *37*, 4521.
6. (a) Ginsberg, D. Personal communication. (b) Ashkenazi, J.; Kettenring, J.; Migdal, S.; Gutman, A. L.; Ginsberg, D. *HCA* **1985**, *68*, 2033.
7. Momose, T.; Atarashi, S. *H* **1978**, *9*, 631.
8. Scott, L. T.; Hashemi, M. M.; Meyer, D. T.; Warren, H. B. *JACS* **1991**, *113*, 7082.
9. Mellor, J. M.; Mohammed, S. *TL* **1991**, *32*, 7107.
10. Oehldrich, J.; Cook, J. M. *JOC* **1977**, *42*, 889.
11. Naoshima, Y.; Yabuki, H.; Udea, H.; Nakamura, M. *ABC* **1987**, *51*, 1175 (*CA* **1988**, *108*, 55 473k).
12. Mataka, S.; Takahashi, K.; Hirota, T.; Takuma, K.; Kobayashi, H.; Tashiro, M.; Imada, K.; Kuniyoshi, M. *JOC* **1986**, *51*, 4618.
13. Chowdhri, B. L.; Haksar, C. N. *Labdev, Part A* **1972**, *10*, 40 (*CA* **1973**, *78*, 110 654j).
14. Zhelyazkov, L.; Bikova, N.; Krusteva, L.; Nikolova, M.; Vankov, M.; Nacheva, M.; Stefanova, D. *Tr. Nauchnoizsled. Khim.-Farm. Inst.* **1972**, *8*, 13 (*CA* **1973**, *78*, 147 746s).
15. (a) Akhrem, A. A.; Chernov, Y. G. *S* **1985**, 411. (b) Akhrem, A. A.; Chernov, Y. G.; *DOK* **1983**, *271*, 869 (*CA* **1984**, *100*, 51 479d).
16. Lee, L. F.; Normansell, J. E. *JOC* **1990**, *55*, 2964.
17. Soubnis, R. W.; Rangnekar, D. W. *J. Chem. Technol. Biotechnol.* **1990**, *47*, 39 (*CA* **1990**, *112*, 200 492e).
18. Specht, D. P.; Martic, P. A.; Farid, S. *T* **1982**, *38*, 1203.
19. Schneller, S. W., Luo, J.-K.; Hosmane, R. S.; Dürrfeld, R. H. *JHC* **1984**, *21*, 1153.
20. Vorkapic-Furac, J.; Suprina, M. *ZC* **1990**, *30*, 437 (*CA* **1991**, *114*, 206 940n).
21. Dupas, G.; Duflos, J.; Queguiner, J. D. G. *JHC* **1980**, *17*, 93.
22. Saito, M.; Morita, T.; Takase, K. *CL* **1974**, 955.
23. Mićović, I. V.; Ivanović, M. D.; Piatak, D. M.; Bojić, V. D. *S* **1991**, 1043.

Mundla S. Reddy & James M. Cook
University of Wisconsin-Milwaukee, WI, USA

Dimethyl Acetylenedicarboxylate

$MeO_2C—≡—CO_2Me$

[762-42-5] $C_6H_6O_4$ (MW 142.11)

(an electron deficient symmetrical alkynic diester useful as a dienophile and dipolarophile in cycloaddition reactions;[1] can undergo [2+2] cycloaddition;[2] Michael acceptor;[3] synthesis of phthalic acid derivatives[4] and heteroaromatics[5])

Alternate Name: DMAD.
Physical Data: bp 195–198 °C; bp 98 °C/20 mmHg; *d* 1.1564 g cm^{-3}.
Solubility: sol common organic solvents, e.g. diethyl ether, ethyl alcohol, carbon tetrachloride.
Form Supplied in: commercially available liquid.
Handling, Storage, and Precautions: lachrymator and vesicant; avoid inhalation and contact with skin or eyes. Use in a fume hood.

As a Dienophile in the Diels–Alder Cycloaddition. DMAD is a ubiquitous dienophile capable of facile cycloadditions with a wide variety of dienes. Much effort has been devoted to the development of dienes which retain additional functionality in the Diels–Alder adduct. New dienes are often reacted with DMAD in order to test their efficacy as a diene in the Diels–Alder reaction.[6] DMAD is a liquid with a high boiling point and can be used as the Diels–Alder solvent. The number of possible stereoisomers is limited. Due to the symmetry of DMAD there are no regio or *endo*/*exo* isomers with respect to the dienophile. For this reason, DMAD is not useful for determining any regioselective biases of a diene. The triple bond gives rise to a double bond in which the product methyl esters are necessarily *cis*. A maximum of two stereoisomers are possible only if the two faces of the diene are either electronically[7] or sterically[8] nonequivalent, otherwise one isomer is produced. The two esters make the triple bond electron poor, and therefore well suited for normal electron demand Diels–Alder cycloadditions. A comparison of relative Diels–Alder reaction rates of common dienophiles, including DMAD, has been performed.[9]

Diels–Alder adducts from DMAD are 1,4-cyclohexadienes, which are prone to aromatization. In fact, the Diels–Alder/aromatization sequence constitutes a powerful protocol for the synthesis of phthalate derivatives (see below). Aromatization is facile when loss of a small molecule is possible. In many cases a benzenoid compound, not the initial Diels–Alder adduct, is directly isolated. Loss of an alcohol is common (eq 1).[10]

OMe / TMSO diene $\xrightarrow[2.\ HCl,\ H_2O;\ 74\%]{1.\ DMAD,\ \Delta}$ HO–C_6H_3–$(CO_2Me)_2$ (1)

If loss of a small molecule is not possible, the diene adduct can be treated with oxidizing agents such as ***2,3-Dichloro-5,6-dicyano-1,4-benzoquinone***.[11] In the case of cyclic dienes, the initial adduct can undergo a retro-Diels–Alder reaction, extruding carbon dioxide,[12] carbon monoxide,[13] or ethylene[14] to produce aromatic products (eqs 2–4).

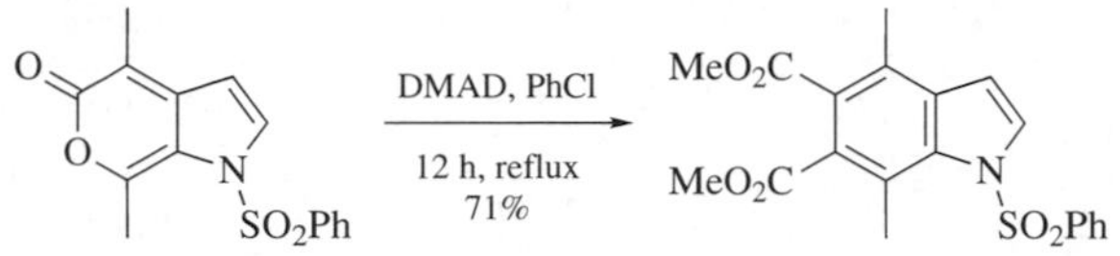

(2)

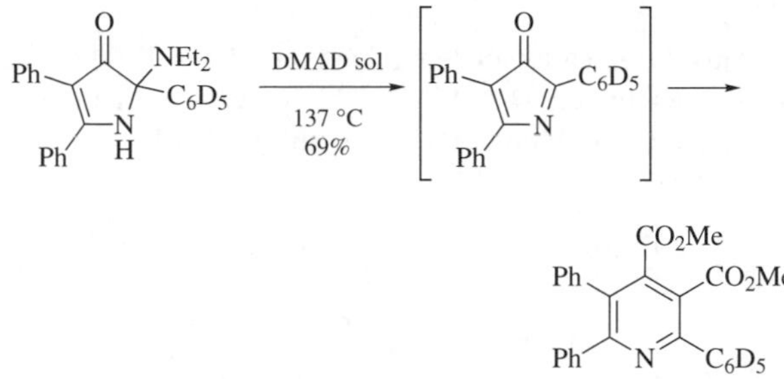

(3)

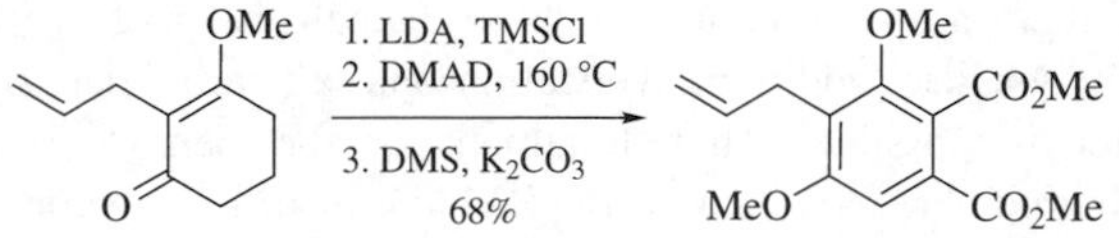

(4)

Adducts possessing a [2.2.1] ring system, obtained from five-membered cyclic dienes such as furans[15] and pyrroles,[16] may fragment to gain aromaticity. DMAD has been exceedingly useful for the synthesis of a number of heteroaromatics such as indoles[17] and furans.[18] Electron deficient alkynes are well suited for the synthesis of pyrimidines from 1,3-diazabutadienes (eq 5).[19]

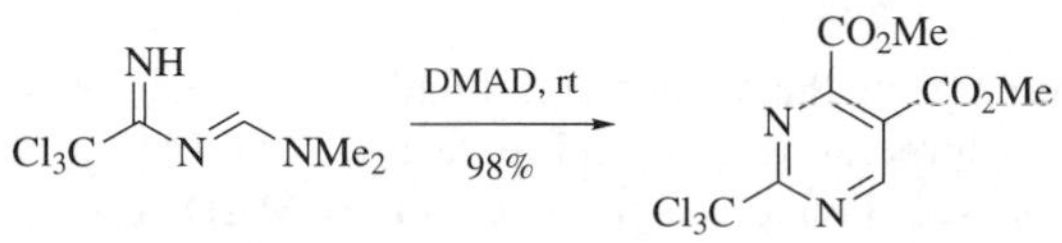

(5)

The cycloaddition of DMAD with anthracene[20] has been used for the synthesis of an interesting new iodate phosphorolytic agent.[21] DMAD has been used extensively in the development of *o*-quinodimethane equivalents (eq 6).[22]

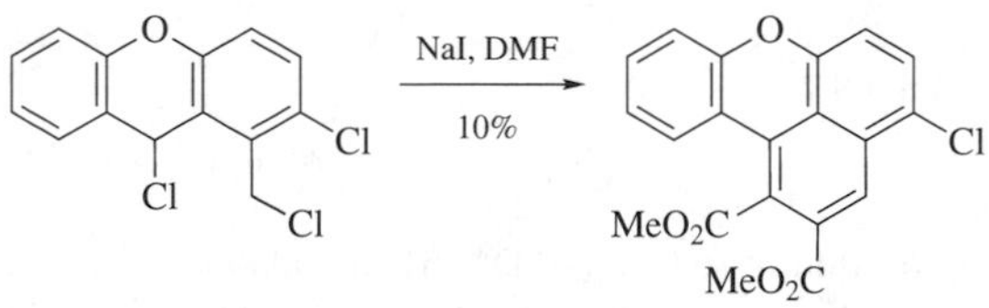

(6)

Examples of higher order cycloadditions with DMAD are abundant. Extended conjugated cycloalkenes such as heptafulvene,[23] fulvenes,[24] azulenes,[25] and indolizines[26] have furnished higher order cycloadducts on treatment with DMAD.

As a Dipolarophile. Not surprisingly, DMAD also undergoes cycloadditions with 1,3-dipoles. Cycloadditions with common 1,3-dipoles such as azides,[27] diazo alkanes,[28] nitrile imines,[29] nitrile ylides,[30] nitrile oxides,[31] azomethine imines,[32] azoxy compounds,[33] azomethine ylides,[34] and nitrones[35] have been reported. As in the case of the Diels–Alder reaction, the number of stereoisomers obtained is limited (see above). When the faces of the dipole are not equivalent, DMAD can add stereoselectively.[36] Two recent synthetically interesting examples of cycloaddition of DMAD with carbonyl ylides[37] and azomethine ylides[38] are illustrated (eqs 7 and 8).

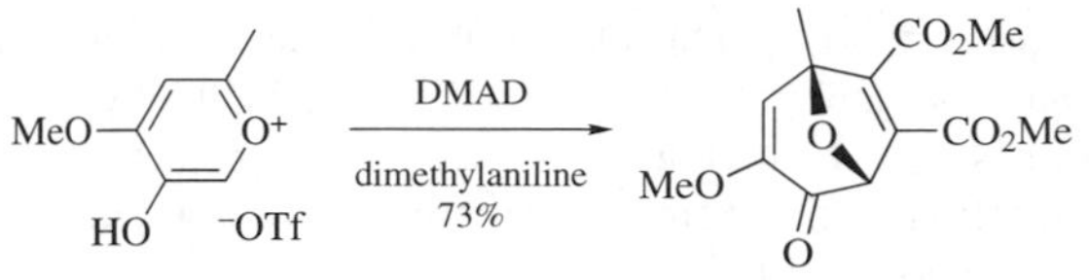

(7)

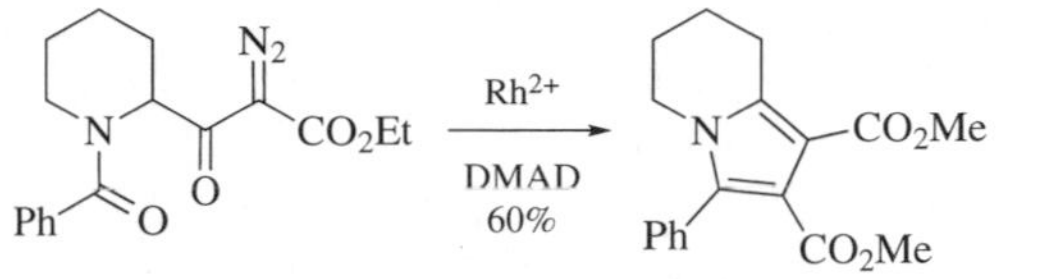

(8)

Due to the unsaturation remaining in the [3 + 2] adducts obtained with DMAD, aromatization and sigmatropic rearrangements are common. Adducts from α-diazo ketones tend to undergo 1,5-acyl sigmatropic rearrangements (eq 9).[39]

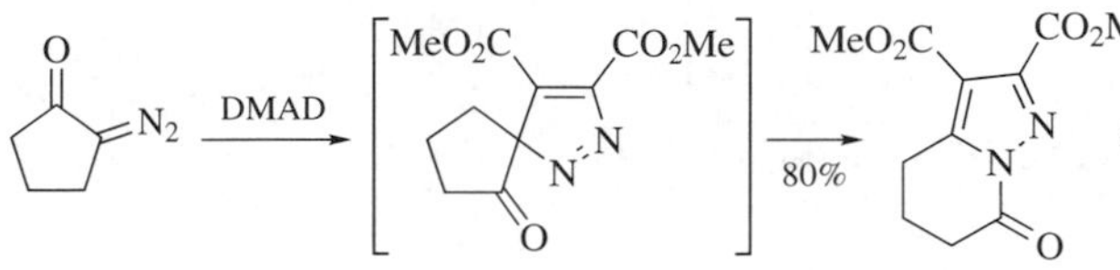

(9)

Simple fragmentations of the [3 + 2] adducts are also common. Reaction of DMAD with quinoxaline 4-oxide gives a pyrrole compound (eq 10).[40] The initial [3 + 2] adduct fragments and incorporates a second molecule of DMAD.

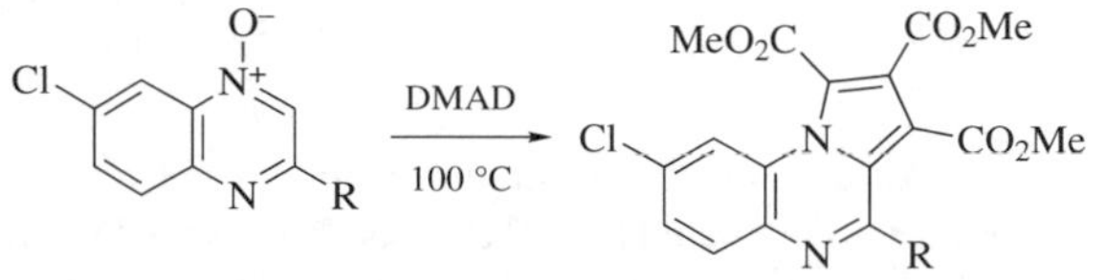

(10)

Cyclization with extended dipoles is known. Benzothiazole undergoes Michael addition with DMAD to form a 1,4-dipole intermediate which subsequently cyclizes with a second molecule of DMAD (eq 11).[41] 1,4-Thiazepines have been constructed via cyclization of DMAD with a 1,5-dipole (eq 12).[42]

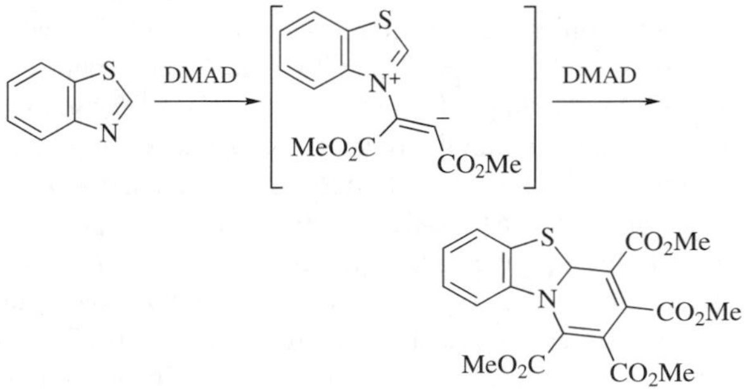

(11)

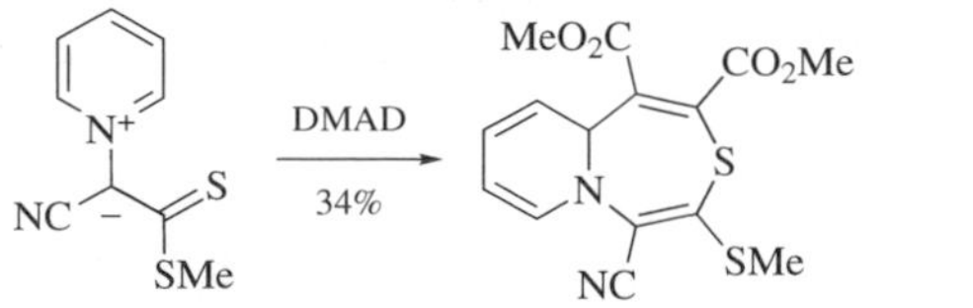

(12)

[2 + 2] Cycloaddition. DMAD readily undergoes [2 + 2] cycloadditions with enamines, often at ambient temperatures, to form cyclobutene compounds.[43] In many cases these compounds are not isolated and tend to fragment to form 1,3-dienamines.[44] When the enamine is cyclic the ring is enlarged by two carbon atoms (eq 13).[45] This chemistry has successfully been carried out on a range of 'enamine' substrates including indoles,[46] quinolines,[47] and vinylogous imides (enamino lactams).[48] The cycloaddition with enamines is compatible with a wide range of functionalities (eq 14).[49]

DMAD, 50% (13)

DMAD, HMPA, rt, 40% (14)

1,1-Dimethoxyalkenes,[50] enol ethers,[51] and silyl enol ethers[52] undergo Lewis acid mediated [2 + 2] cycloaddition with DMAD to afford cyclobutene compounds. In the case of silyl enol ethers the [2 + 2] adducts can generate fragmentation products analogous to the enamine substrates. The common isolation of simple Michael adducts and strong solvent dependency on yield is solid evidence for a stepwise mechanism involving dipolar intermediates.[53]

As a Michael Acceptor. A wide variety of nucleophiles undergo 1,4-addition with DMAD. There are examples where carbon,[54] oxygen,[55] nitrogen,[56] and halogen[57] nucleophiles undergo simple Michael addition, thus making DMAD useful for the synthesis of a great number of substituted alkenes. A predominance of the (*Z*)-isomer (methyl esters *trans*) is most common.[58] Frequently, the initial Michael adducts cyclize; this presents a facile route to heterocycles and especially heteroaromatics. With amine nucleophiles the resultant enamines (or vinylogous urethanes) often cyclize onto nearby electrophilic centers. The process is especially efficient when a five- or six-membered ring can be formed. Thus vinylogous amides yield pyridines on treatment with DMAD, whereas α-amino ketones give pyrroles (eqs 15 and 16).[59]

DMAD (15)

DMAD, 80% (16)

Alternatively, a nearby nucleophile in the Michael adduct may attack the ester moiety of DMAD. This process is facile when five- or six-membered rings can be formed (eq 17).[60]

$PhNHNH_2$ —DMAD→ ; HCl, 67 °C, 92% overall (17)

If there are no proton sources to quench the enolate formed by the Michael addition, cyclization onto a nearby electrophilic center is possible. Thus treatment of the benzyl anion of 5-pyrimidinecarboxylate with DMAD affords a quinazoline (eq 18).[61]

LDA, DMAD, 38% + 60% S.M. (18)

Alternatively, the enolate could undergo another Michael addition, incorporating a second molecule of DMAD which then cyclizes (see dipolar reactions, eq 11). DMAD reacts with *p*-toluenethiosulfonate at rt to form tetramethyl thiophenetetracarboxylate in good yield (eq 19).[62] Interestingly, phenylacetylene, diphenylacetylene, or 3-phenyl-2-propynoate did not give any thiophenes. Apparently only alkynes with two electron-withdrawing groups will work.

$ArSO_2SK$ + DMAD —rt, MeCN, 76%→ (19)

1,4-Addition of radicals to DMAD are also known.[63] Michael addition of the *t*-butoxy radical,[64] selenyl radicals,[65] and the isopropyl radical[66] (eq 20) have recently been studied.

DMAD, $(Bu_3Sn)_2$, $h\nu$, 81%; (*E*):(*Z*) = 4.6:1 (20)

DMAD has been used in a palladium-catalyzed annulation with aryl halides.[67]

Related Reagents. Methyl Propiolate.

---

1. (a) Carruthers, W. *Cycloaddition Reactions in Organic Synthesis*; Pergamon: Oxford, 1990. (b) Bastide, J.; Henri-Rousseau, O. *Chemistry*

of Functional Group, The Chemistry of Triple-Bonded Functional Groups; Patai, S.; Rappoport, Z., Eds; Wiley: Chichester, 1983; Suppl. C, Part 1, pp 447–522.

2. Cook, A. G. *Enamines: Synthesis, Structure and Reactions*, 2nd ed.; Cook, A. G., Ed.; Dekker: New York, 1988; pp 384–389.
3. Hendrickson, J. B.; Rees, R.; Templeton, J. F. *JACS* **1964**, *86*, 107.
4. Wenkert, E.; Johnston, D. B. R.; Dave, K. G. *JOC* **1964**, *29*, 2534.
5. (a) Muchowski, J. M.; Scheller, M. E. *TL* **1987**, *28*, 3453. (b) Belloir, P. F.; Laurent, A.; Mison, P. *S* **1986**, 683.
6. (a) Tada, M.; Schimizu, T. *BCJ* **1992**, *65*, 1252. (b) Murase, M.; Hosaka, T.; Yoshida, S.; Tobinaga, S. *CPB* **1992**, *40*, 1343. (c) Kanomata, N.; Kawaji, H.; Nitta, M. *JOC* **1992**, *57*, 618.
7. Halterman, R. L.; McCarthy, B. A.; McEvoy, M. A. *JOC* **1992**, *57*, 5585.
8. (a) Fessner, W. D.; Grund, C.; Prinzbach, H. *TL* **1991**, *32*, 5935. (b) Paquette, L. A.; Shen, C. C. *JACS* **1990**, *112*, 1159.
9. Sauer, J. *AG* **1961**, *73*, 545.
10. (a) Heffner, R. J.; Joullié, M. M. *SC* **1991**, *21*, 2231. (b) Danishefsky, S.; Kitahara, T. *JACS* **1974**, *96*, 7807.
11. Paquette, L. A.; Melega, W. P.; Kramer, J. D. *TL* **1976**, 4033.
12. Jackson, P. M.; Moody, C. J. *T* **1992**, *48*, 7447.
13. (a) Brown, R. F. C.; Choi, N.; Eastwood, F. W. *TL* **1992**, *35*, 3787. (b) Eicher, T.; Abdesaken, F.; Franke, G.; Weber, J. L. *TL* **1975**, 3915.
14. Rama Rao, A. V.; Reddy, R. G. *TL* **1992**, *28*, 4061.
15. (a) Krutôsiková, A.; Hanes, M. *CCC* **1992**, *57*, 1487. (b) Bloomer, J. L.; Lankin, M. E. *TL* **1992**, *33*, 2769.
16. Sha, C. K.; Liu, J. M.; Chiang, R. K.; Warg, S. L. *H* **1990**, *31*, 603.
17. Gribble, G. W.; Keavey, D. J.; Davis, D. A.; Saulnier, M. G.; Pelcman, B.; Barden, T. C.; Sibi, M. P.; Olson, E. R.; Belbruno, J. J. *JOC* **1992**, *57*, 5878.
18. Aken, K. V.; Hoornaert, G. *CC* **1992**, 895.
19. Guman, A.; Muchowski, J. M.; Romero, M.; Talamas, F. X. *TL* **1992**, *33*, 3449.
20. Paquette, L. A.; Bay, E. *JACS* **1984**, *106*, 6693.
21. Moss, R. A.; Zhang, H. *TL* **1992**, *33*, 4291.
22. (a) Greco, M. N.; Rasmussen, C. R. *JOC* **1992**, *57*, 5532. (b) Ando, K.; Hatano, C.; Akadegawa, N.; Shigihara, A.; Takayama, H. *CC* **1992**, 870. (c) Ruiz, N.; Pujol, M. D.; Guillaumet, G.; Coudert, G. *TL* **1992**, *33*, 2965. (d) Mertzanos, G. E.; Stephanidou-Stephanatou, J.; Tsolenidis, C. A.; Alexandrou, N. E. *TL* **1992**, *33*, 4499.
23. Doering, W. E.; Wiley, D. W. *T* **1960**, *11*, 183.
24. Liu, C. Y.; Ding, S. T. *JOC* **1992**, *57*, 4539.
25. Hafner, K.; Knaup, G. L.; Lindner, H. J. *BCJ* **1988**, *61*, 155.
26. (a) Blake, A. J.; Dick, J. W.; Leaver, D.; Strachan, P. *JCS(P1)* **1991**, 2991. (b) Matsumoto, K.; Uchida, T.; Yoshida, H.; Toda, M.; Kakehi, A. *JCS(P1)* **1992**, 2437.
27. Prabahar, J. K.; Shanmugasundaram, P.; Ananthanarayanan, C.; Ramakrishnan, V. T. *Indian J. Heterocycl. Chem.* **1992**, *1*, 157.
28. (a) Frampton, C. S.; Majchrzak, M. W.; Warkentin, J. *CJC* **1991**, *69*, 373. (b) Majchrzak, M. W.; Békhazi, M.; Tse-Sheepy, I.; Warkentin, J. *JOC* **1989**, *54*, 1842. (c) Birkhahn, M.; Hohlfeld, R.; Massa, W.; Schmidt, R.; Lorberth, J. *JOM* **1980**, *192*, 47.
29. (a) Baruah, A. K.; Prajapati, D.; Sandhu, J. S. *T* **1988**, *44*, 1241. (b) Tewari, R. S.; Dixit, P. D.; Parihar, P. *JHC* **1982**, *19*, 1573.
30. (a) Ibata, T.; Fukushima, K. *CL* **1992**, 2197. (b) Berée, F.; Marchand, E.; Morel, G. *TL* **1992**, *33*, 6155. (c) Bossio, R.; Marcaccini, S.; Paoli, P.; Pepino, R.; Polo, C. *H* **1990**, *31*, 1855. (d) Wipf, P.; Prewo, R.; Bieri, J. H.; Germain, G.; Heimgartner, H. *HCA* **1988**, *71*, 1177. (e) Frank-Neumann, M.; Buchecker, C. *TL* **1972**, 937.
31. (a) Yokoyama, M.; Sujino, K.; Irie, M.; Yamazaki, N.; Hiyama, T.; Yamada, N.; Togo, H. *JCS(P1)* **1991**, 2801. (b) Yokoyama, M.; Yamada, N. *TL* **1989**, *30*, 3675.
32. (a) Noguchi, M.; Kiriki, Y.; Tsuruoka, T.; Mizui, T.; Kajigaeshi, S. *BCJ* **1991**, *64*, 99. (b) Thesis, W.; Bethäuser, W.; Regitz, M. *CB* **1985**, *118*, 28. (c) Diez-Barra, E.; Pardo, C.; Elguero, J.; Arriau, J. *JCS(P2)* **1983**, 1317. (d) Burger, K.; Schickaneder, H.; Zettl, C.; Dengler, O. *LA* **1982**, 1730.
33. (a) Huisgen, R.; Gambra, F. P. *CB* **1982**, *115*, 2242. (b) Huisgen, R; Gombra, F. P. *TL* **1982**, *23*, 55.
34. (a) Padwa, A.; Austin, D. J.; Precedo, L.; Zhi, L. *JOC* **1993**, *58*, 1144. (b) Vedejs, E.; Piotrowski, D. W. *JOC* **1993**, 1341. (c) de Pablo, M. S.; Gandásegui, T.; Vaquero, J. L.; Navió, G.; Alvarez-Builla, J. *T* **1992**, *48*, 8793. (d) Pandey, G.; Lakshmaiah, G.; Kumaraswamy, G. *CC* **1992**, 1313. (e) Padwa, A.; Ku, H. *JOC* **1979**, *44*, 255.
35. (a) Purwono, B.; Smalley, R. K.; Porter, T. C. *SL* **1992**, 231. (b) Hippeli, C.; Reissig, H. O. *LA* **1990**, 475. (c) Bennett, G. A.; Mullen, G. B.; Georgiev, V. St. *HCA* **1989**, *72*, 1718.
36. Anslow, A. S.; Harwood, L. M.; Phillips, H.; Watkins, D.; Wong, L. F. *TA* **1991**, *2*, 1343.
37. Wender, P. A.; Mascarenas, J. L. *TL* **1992**, *33*, 2115.
38. Padwa, A.; Dean, D. C.; Hertzog, D. L.; Nadler, W. R.; Zhi, L. *T* **1992**, *48*, 7565.
39. (a) Jefferson, E.; Warkentin, J. *JACS* **1992**, *114*, 6318. (b) Katner, A. S. *JOC* **1973**, *38*, 825.
40. (a) Kurasaw, Y.; Katoh, R.; Takada, A.; Kim, H. S.; Okamoto, Y.; *JHC* **1992**, *29*, 1001. (b) Fukuchi, M.; Okamoto, M.; Takada, A.; Kim, H. S.; Okamoto, Y. *JHC* **1992**, *29*, 1009.
41. McKillop, A.; Sayer, T. S. B. *TL* **1975**, 3081.
42. Kakehi, A.; Ito, S.; Hakui, J. *CL* **1992**, 777.
43. Acheeson, R. M.; Wright, N. D.; Tasker, P. A. *JCS(P1)* **1972**, 2918.
44. (a) Brannock, K. C.; Burpitt, R. D.; Goodlett, V. W.; Thweatt, J. G. *JOC* **1963**, *28*, 1464. (b) Reinhoudt, D. N.; Kouwenhoven, C. G. *TL* **1973**, 3751. (c) Achcson, R. M.; Paglictti, G. *CC* **1973**, 665.
45. Huebner, C. F.; Dorfman, L.; Robinson, M. M.; Donoghue, E.; Pierson, W. G.; Strachan, P. *JOC* **1963**, *28*, 3134.
46. (a) Acheson, R. M.; Bridson, J. N.; Camcron, T. S. *JCS* **1972**, 968; (b) Plieninger, H.; Wild, D. *CB* **1966**, *99*, 3070.
47. Lehman, P. G. *TL* **1972**, 4863.
48. Heinicke, G. W.; Morella, A. M.; Orban, J.; Prager, R. H.; Ward, A. D. *AJC* **1985**, *38*, 1847.
49. Matsunaga, H.; Sonoda, M.; Tomioka, Y.; Yamazaki, M. *CPB* **1986**, *34*, 396.
50. Graziano, M. L.; Lesce, M. R.; Cermola, F.; Cimminiello, G. *JCS(P1)* **1992**, 1269.
51. Wang, Y. W.; Fang, J. M.; Wang, Y. K.; Wang, M. H.; Ko, T. Y.; Cherng, Y. J. *JCS(P1)* **1992**, 1209.
52. Clark, R. D.; Untch, K. G. *JOC* **1979**, *44*, 248.
53. Aue, H. A.; Thomas, D. *JOC* **1975**, *40*, 2360.
54. Parmer, R. H; Ghosal, S. C.; Kou, G. A. R. *JCS* **1961**, 1804.
55. (a) Ireland, R. E.; Obrecht, D. M. *HCA* **1986**, *69*, 1273. (b) Yokoyama, M.; Sujino, K.; Irie, M.; Togo, H. *TL* **1991**, *32*, 7269.
56. Heindel, N. D.; Bechara, I. S.; Kennewell, P. D.; Molnar, J.; Ohnmacht, C. J.; Lemke, S. M.; Lemke, T. F. *JMC* **1968**, *11*, 1218.
57. Johnson, A. W. *Chemistry of the Acetylenic Compounds*; Longmans, Green: New York, 1950; Vol. II, pp 199–266.
58. Hendrickson, J. B.; Rees, R. *JACS* **1961**, *83*, 1250.
59. James, D. S.; Fanta, P. E. *JOC* **1962**, *27*, 3346.
60. (a) Padwa, A.; MacDonald, J. G. *JHC* **1987**, *24*, 1225. (b) Reimlinger, H.; Jacquier, R.; Daunis, J. *CB* **1971**, *104*, 2702.
61. Wada, A.; Yamamoto, H.; Ohki, K.; Nagai, S.; Kanatomo, S. *JHC* **1992**, *29*, 911.
62. Kutateladze, T. G.; Kice, J. L. *JOC* **1992**, *57*, 5270.
63. Amiel, Y. *Chemistry of Functional Groups, The Chemistry of Triple-Bonded Functional Groups*; Patai, S.; Rappoport, Z., Eds; Wiley: Chichester, 1983; Suppl. C, Part 1, pp 341–382.

64. Bottle, S.; Bosfield, W. K.; Jenkins, I. D.; Skelton, B. W.; White, A. H.; Rizzardo, E.; Solomon, D. H. *JCS(P2)* **1991**, 1001.

65. (a) Kataoka, T.; Yoshimatsu, M.; Shimizu, H.; Hori, M. *TL* **1990**, *31*, 5927. (b) Kataoka, T.; Yoshimatsu, M.; Noda, Y.; Sato, T.; Shimizu, H.; Hori, M. *JCS(P1)* **1993**, 121.

66. Curran, D. P.; Dooseop, K. *T* **1991**, *47*, 6171.

67. (a) Dyker, G. *JOC* **1993**, *58*, 234. (b) Sakakibara, T.; Tanaka, Y.; Yamasaki, S. I. *CL* **1986**, 797.

John E. Stelmach & Jeffrey D. Winkler
University of Pennsylvania, Philadelphia, PA, USA

(*S*)-*N*,*N*-Dimethyl-*N*′-(1-*t*-butoxy-3-methyl-2-butyl)formamidine[1,2]

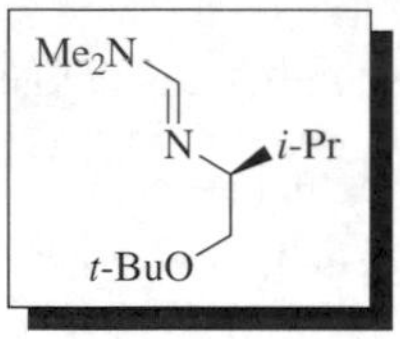

(*E*)-(*S*)
[114318-94-4] $C_{12}H_{26}N_2O$ (MW 214.40)
(*S*)
[66919-83-3]
(*R*)
[90482-06-7]

(chiral auxiliary for derivatization, directed lithiation, and asymmetric alkylation adjacent to nitrogen of benzylic or allylic secondary amines by formamidine exchange, metalation, alkylation, and hydrolysis[1,2])

Physical Data: bp 55–65 °C/0.05 mmHg; $[\alpha]_D$ −59.6° (*c* 3.5, EtOH).[2]

Preparative Methods: easily prepared from (*S*)-valinol in a high yielding four-step procedure (eq 1).[2] Other chiral formamidines can be prepared and used successfully in the methodology outlined here (eq 2).[2,3]

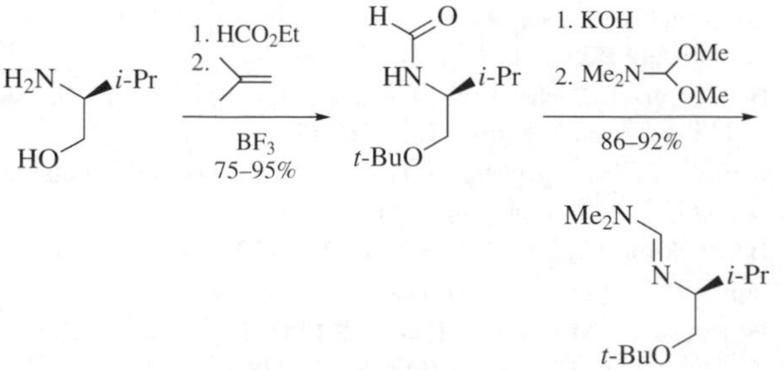

(1)

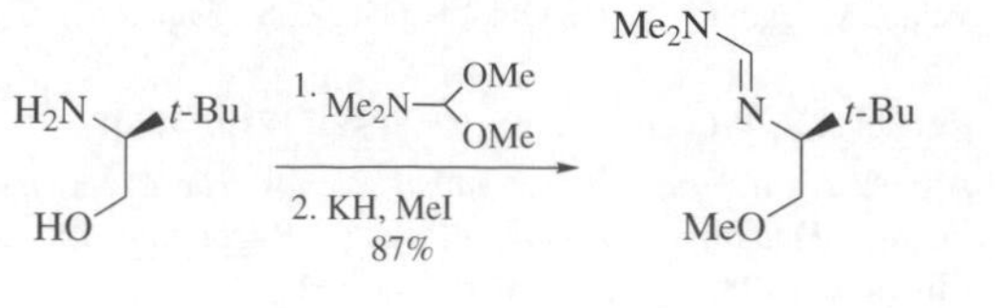

(2)

Handling, Storage, and Precautions: store under argon at rt; use in a fume hood.

Introduction. In the equations, 'VBE' will be used to depict the valinol *t*-butyl ether portion of the formamidine, while the *t*-leucinol methyl ether portion will be abbreviated 'LME'.

Upon heating the title formamidine with a secondary amine, dimethylamine is extruded, affording a chiral formamidine derivative of the original amine (eq 3).[4a]

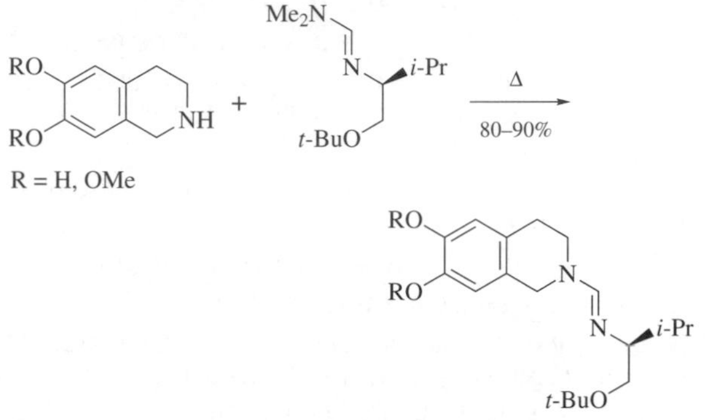

(3)

Deprotonation and alkylation followed by formamidine removal allows entry to a host of isoquinoline alkaloids (eq 4).[3b,4]

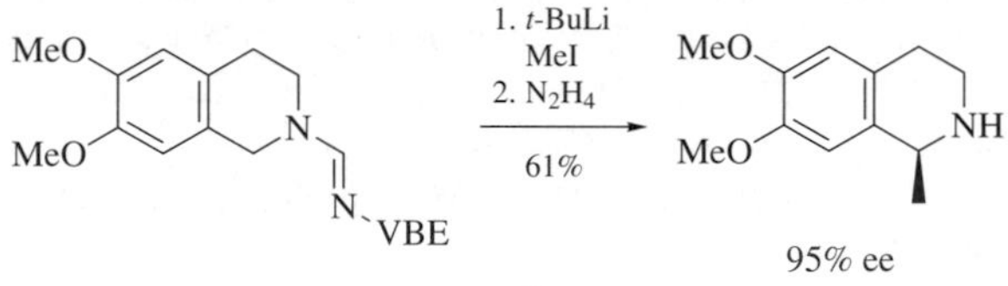

(4)

This protocol has also been an avenue to a variety of indole alkaloids (eq 5).[5]

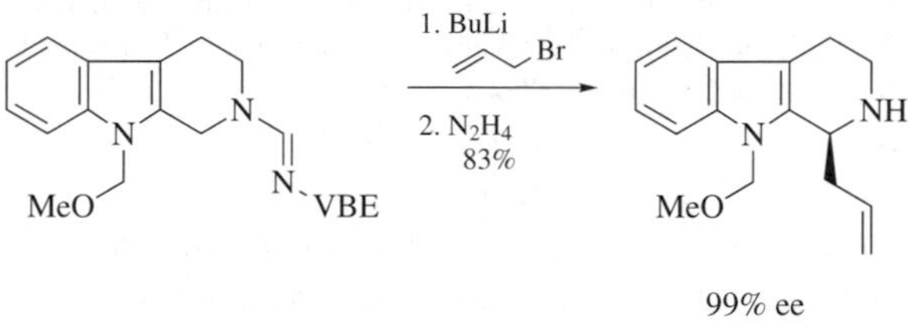

(5)

Chiral 1-alkyl-2-benzazepines can be formed by utilization of the same method (eq 6).[3a]

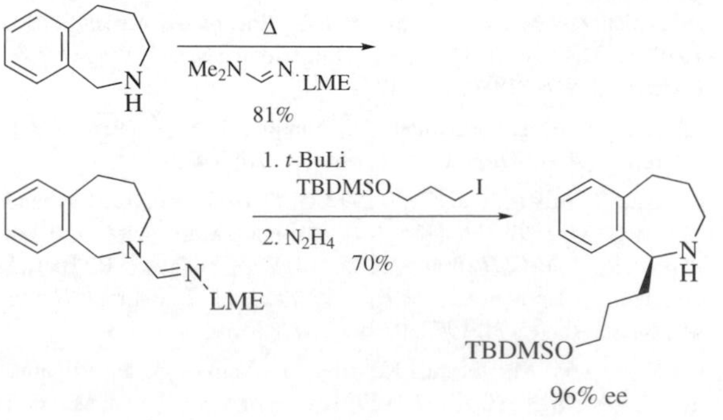

(6)

This strategy also works well for the asymmetric alkylation of 3-pyrrolines (eq 7)[6] and tetrahydropyridines (eq 8).[7]

1. t-BuLi; MeO–C6H4–CH2Cl; 2. N_2H_4; Δ, Me_2N–CH=N–VBE, 90–95%; 53%; 91% ee (7)

1. t-BuLi; MeO–C6H4–CH2Cl; 2. N_2H_4; Δ, Me_2N–CH=N–VBE, 71%; 44%; >98% ee (8)

The mechanistic pathway for these asymmetric alkylations and the configurational stability of the chiral lithioformamidines have been investigated.[8] A limitation of this strategy is that saturated, cyclic, secondary amines (e.g. pyrrolidines and piperidines) cannot be successfully alkylated in an asymmetric fashion.

Related Reagents. *N*-*t*-Butoxycarbonyl-*N*-methylaminomethyllithium; *N*′-*t*-Butyl-*N*,*N*-dimethylformamidine; *N*′-*t*-Butyl-*N*-methyl-*N*-trimethylsilylmethylformamidine; *N*,*N*-Dimethylformamide Diethyl Acetal; (*R*)-Methyl 2-*t*-Butyl-3(2*H*)-oxazolecarboxylate; *N*-Nitrosodimethylamine.

1. (a) Meyers, A. I. *T* **1992**, *48*, 2589. (b) Meyers, A. I.; Highsmith, T. K. In *Advances in Heterocyclic Natural Product Synthesis*; Pearson, W. H., Ed.; JAI Press: Greenwich, CT, 1990. (c) Meyers, A. I. *Aldrichim. Acta* **1985**, *18*, 59. (d) Meyers, A. I. *H* **1984**, *21*, 360.
2. (a) Dickman, D. A.; Boes, M.; Meyers, A. I. *OSC* **1993**, *8*, 204. (b) Meyers, A. I.; Boes, M.; Dickman, D. A. *OSC* **1993**, *8*, 573.
3. (a) Meyers, A. I.; Hutchings, R. H. *T* **1993**, *49*, 1807. (b) Meyers, A. I.; Elworthy, T. R. *JOC* **1992**, *57*, 4732.
4. (a) Meyers, A. I.; Dickman, D. A.; Boes, M. *T* **1987**, *43*, 5095. (b) Meyers, A. I.; Sielecki, T. M.; Crans, D. C.; Marshman, R. W.; Nguyen, T. H. *JACS* **1992**, *114*, 8483. (c) Sielecki, T. M.; Meyers, A. I. *JOC* **1992**, *57*, 3673. (d) Guiles, J. W.; Meyers, A. I. *JOC* **1991**, *56*, 6873. (e) Meyers, A. I.; Sielecki, T. M. *JACS* **1991**, *113*, 2789. (f) Gottlieb, L.; Meyers, A. I. *JOC* **1990**, *55*, 5659. (g) Meyers, A. I.; Guiles, J. *H* **1989**, *28*, 295.
5. (a) Meyers, A. I.; Highsmith, T. K.; Buonora, P. T. *JOC* **1991**, *56*, 2960. (b) Beard, R. L.; Meyers, A. I. *JOC* **1991**, *56*, 2091.
6. (a) Meyers, A. I.; Dupre, B. *H* **1987**, *25*, 113. (b) Warmus, J. S.; Dilley, G. J.; Meyers, A. I. *JOC* **1993**, *58*, 270.
7. Meyers, A. I.; Dickman, D. A.; Bailey, T. R. *JACS* **1985**, *107*, 7974.
8. (a) Castonguay, L. A.; Guiles, J. W.; Rappé, A. K.; Meyers, A. I. *JOC* **1992**, *57*, 3819. (b) Meyers, A. I.; Warmus, J. S.; Gonzalez, M. A.; Guiles, J.; Akahane, A. *TL* **1991**, *32*, 5509. (c) Meyers, A. I.; Guiles, J.; Warmus, J. S.; Gonzalez, M. A. *TL* **1991**, *32*, 5505. (d) Meyers, A. I.; Gonzalez, M. A.; Struzka, V.; Akahane, A.; Guiles, J.; Warmus, J. S. *TL* **1991**, *32*, 5501.

Todd D. Nelson & Albert I. Meyers
Colorado State University, Fort Collins, CO, USA

Dimethylchloromethyleneammonium Chloride

Me_2N^+=CHCl X^-

(**1**; X = Cl)		
[3724-43-4]	$C_3H_7Cl_2N$	(MW 128.01)
(**2**; X = PO_2Cl_2)		
[21382-90-1]	$C_3H_7Cl_3NO_2P$	(MW 226.43)

(formylation of arenes,[1–3] heterocycles,[1–3] and alkenes;[2,3] conversion of alcohols into chlorides,[4] and carboxylic acids into acid chlorides;[5] preparation of β-chlorovinyl aldehydes,[3,6] cyclic β-chloro-α,β-unsatura ted ketones,[7] and the conversion of acyclic precursors into ring systems[8])

Alternate Names: Vilsmeier reagent; Arnold's reagent.
Physical Data: (**1**) mp 132 °C (dec).
Solubility: fairly sol polar solvents.
Form Supplied in: the chloride (**1**) is a white hygroscopic powder and is commercially available. The dichlorophosphate (**2**) is made in situ.
Preparative Methods: ***Thionyl Chloride***,[4b] ***Oxalyl Chloride***,[5] or ***Phosgene*** is added slowly to ***N,N-Dimethylformamide***, with cooling, to give the chloride (**1**) (eq 1). A solvent (typically dichloromethane,[5] chloroform, or 1,1,2-trichloroethylene)[9] may be used, and is advisable for large scale preparations.

$$Me_2N\text{–CHO} + XOCl_2 \xrightarrow{-XO_2} Me_2N^+\text{=CHCl}\ Cl^- \quad (1)$$
X = C, S

The most commonly used Vilsmeier reagent is that formed by adding ***Phosphorus Oxychloride*** to DMF, to give dimethylchloromethyleneammonium dichlorophosphate (**2**).[9] Reagent (**2**) is more reactive than reagent (**1**). Recent reports suggest that DMF–***Pyrophosphoryl Chloride*** ($P_2O_3Cl_4$) gives an even more reactive iminium intermediate (eq 2) than either salt (**1**) or salt (**2**).[10]

$$Me_2N\text{–CHO} + Cl_2P(O)\text{–O–}P(O)Cl_2 \longrightarrow Me_2N^+\text{=CH–O–}P(O)Cl_2\ \ PO_2Cl_2^- \quad (2)$$

Handling, Storage, and Precautions: the chloride (**1**) is hygroscopic. DMF is an irritant and is harmful by skin absorption. The halogenating agents (thionyl chloride, phosgene, phosphorus oxychloride, etc.) are all corrosive and lachrymatory. Use in a fume hood.

Conversion of Alcohols into Chlorides. Reagent (**1**) has been used to convert alcohols into chlorides in moderate to high yields.[4] An alternative Vilsmeier reagent derived from *N,N*-diphenylbenzamide–oxalyl chloride has been used for the same transformation. Sensitive groups, such as acetals, are not affected by the reaction conditions (eq 3) and the yields are high.[11]

(3) [Ph(+)N=C(Cl)Ph Cl⁻; Et$_3$N; 78%]

Conversion of Carboxylic Acids to Acid Chlorides. Reagent (**1**) has been used to convert a wide variety of carboxylic acids into acid chlorides.[5] A catalytic quantity of DMF may be used; thus isobutyric acid is converted into isobutyroyl chloride (89%) upon treatment with DMF (3 mol %) and phosgene.[5c]

Formylation of Arenes and Heterocyclic Systems. Reagent (**2**) is the most commonly used Vilsmeier reagent for formylation of arenes and heterocycles. The subject has been extensively reviewed.[1–3] The formylation process is general for all activated aromatic systems (e.g. pyrene, anthracene, azulene, ferrocene, and *N,N*-dimethylaniline)[12a] and electron-rich heterocyclic systems (e.g. thiophenes, indoles,[12b] furans, and pyrroles (eq 4)).[12c] Formylations continue to be regularly reported in the literature.[13] However, simple aromatic hydrocarbons such as benzene and toluene do not react with Vilsmeier reagents, although the adduct formed from DMF–***Trifluoroacetic Anhydride*** has been used to formylate some unactivated aromatic hydrocarbons, e.g. naphthalene (50%) and 1,3,5-trimethylbenzene (60%).[14] DMF–$P_2O_3Cl_4$ has been reported to be a good formylating agent.[10] The time and temperature of the reaction, and the stoichiometry of the reagent, can be crucial to the outcome of the reaction.

(4) [DMF, POCl$_3$; 83%; CHO]

Formylation of Alkenes. Activated alkenes, e.g. styrene, undergo formylation by DMF–$POCl_3$, usually giving α,β-unsaturated aldehydes in high yields.[3,15] 1-Aryl-1,3-butadienes are similarly formylated.[16] Simple alkenes can give complex products as a result of a series of reactions. Thus isobutene affords a trimethinium salt after repeated iminoalkylation (eq 5).[17] However, by using *N*-formylmorpholine (NFM)–$POCl_3$, alkenes have been successfully monoformylated (eqs 6 and 7).[18]

(5) [1. DMF, POCl$_3$; 2. HClO$_4$; 73%; Me$_2$N, NMe$_2$; 3 ClO$_4^-$]

(6) [t-Bu; N-formylmorpholine, POCl$_3$; 80%; CHO]

(7) [N-formylmorpholine, POCl$_3$; 35%; CHO]

Depending upon the reaction conditions and the structure of the alkene, isomerization of the carbon–carbon double bond may occur prior to formylation (eq 8).[19]

(8) [DMF, POCl$_3$; 85 °C, 1 h; CHO; 20 °C, 24 h; DMF, POCl$_3$; 20 °C, 24 h; CHO]

Silyl enol ethers react with DMF–$POCl_3$ to give α-formyl carboxylic esters.[20]

Reaction of Carbonyl Compounds with DMF–$POCl_3$. This subject has been extensively reviewed.[21] Methyl and methylene ketones are the most studied classes of compound; they afford β-chlorovinyl aldehydes (β-CVAs) when allowed to react with DMF–$POCl_3$ (eq 9). Acyclic, cyclic, heterocyclic, and benzo-fused ketones all give β-CVAs, which have proved to be useful intermediates in organic synthesis.[3] The chlorine atom is readily displaced by nucleophiles and the aldehyde can then be transformed by means of condensations, reaction with Grignard reagents, or Wittig reagents. Condensations involving the aldehyde group, followed by base-catalyzed cyclization, allows the formation of a wide variety of heterocycles, e.g. thiophenes, isoxazoles (eq 10), pyrimidines (eq 10), and pyrazoles.[3,22] More than 50 different ring systems have been synthesized using Vilsmeier methodology.[22]

(9) [DMF, POCl$_3$; 67%; Cl; CHO]

(10) [R^1, R^2; H$_2$NCHO; 1. NH$_2$OH; 2. NaOH]

β-CVAs can be hydrogenated to give saturated aldehydes, allowing the introduction of an aldehyde β with respect to the original ketone function (eq 11).[23]

(11) [t-Bu; Cl; CHO; H$_2$, Pd/C]

β-CVAs have been used as dienophiles in Diels–Alder reactions.[24] Treatment of β-CVAs with strong aqueous base affords alkynes.[25] The chlorine atom can be removed using ***Zinc*** in ethanol,[26a] or hydrogen over ***Palladium on Carbon*** in the presence of ***Triethylamine***,[26b] giving α,β-unsaturated aldehydes.

Reaction of 1,3-Diketones. Salt (**1**) reacts with cyclic 1,3-diketones, giving β-chloro-α,β-unsaturated ketones.[7] Cyclic 1,3-diketones can give bis-β-CVAs upon treatment with reagent (**2**).[27] 2,4-Pentanedione gave 2,4-dichlorobenzaldehyde on reaction with DMF–$POCl_3$ (eq 12),[28] and 2,4-dichloro-1,3-benzene dicarbaldehyde on reaction with the adduct from NFM–$POCl_3$ (eq 13).[27b]

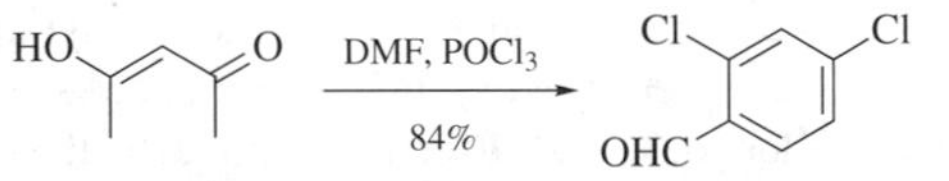
HO
O
DMF, POCl3
84%
Cl
Cl
OHC

(12)

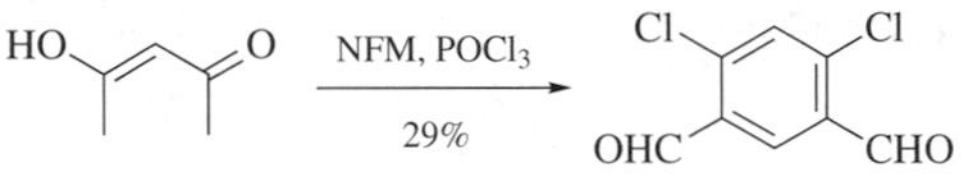
HO
O
NFM, POCl3
29%
Cl
Cl
OHC
CHO

(13)

Epoxidation of Alkenes. When reagent (**1**) is treated with ***Hydrogen Peroxide*** the reagent (**3**) is formed, which is capable of epoxidizing alkenes.[29] However, the selectivity of reagent (**3**) is poor, and chlorinated products are also obtained. The yields and selectivity are improved by reacting alkenes with the intermediate (**4**), obtained from *N*-methylpyrrolidine–$POCl_3$ and H_2O_2.[30] Little chlorination of the alkene was noted and no Baeyer–Villiger products were identified from ketone-containing substrates. Acetals are not cleaved under the reaction conditions. Epoxidation occurs at the most substituted carbon–carbon double bond (eq 14).[30]

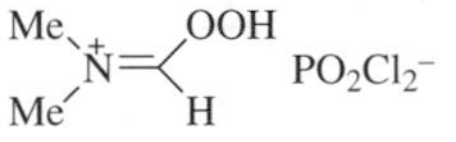
Me
Me
N
+
OOH
H
PO2Cl2−

(**3**)

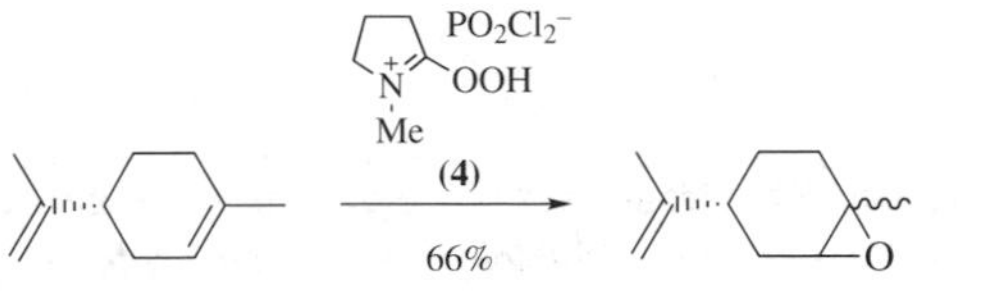
PO2Cl2−
N
+
OOH
Me
(4)
66%
O

(14)

Cyclizations under Vilsmeier Conditions. A large number of substrates undergo cyclization in the presence of DMF–$POCl_3$ and other Vilsmeier reagents.[8] The outcome of the reaction often depends upon the reaction time and temperature, and the nature of the Vilsmeier species. For example, benzylacetamides give quinolines and related systems; the nature of the product depends upon the conditions used (eq 15).[31]

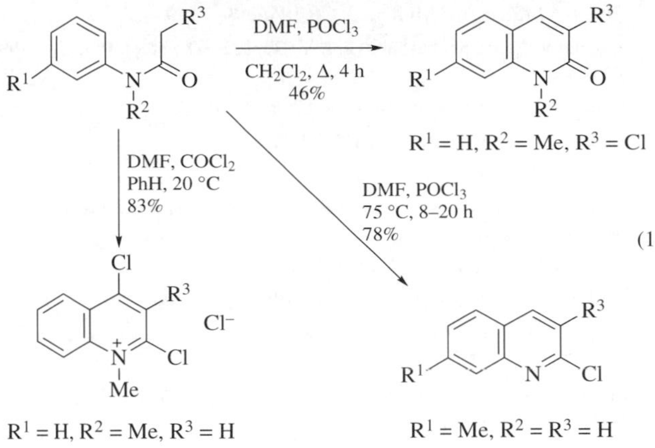
R3
R1
N
O
R2
DMF, POCl3
CH2Cl2, Δ, 4 h
46%
R1 = H, R2 = Me, R3 = Cl
DMF, COCl2
PhH, 20 °C
83%
DMF, POCl3
75 °C, 8–20 h
78%
Cl
Cl−
N
+
Me
R1 = H, R2 = Me, R3 = H
R1 = Me, R2 = R3 = H

(15)

2-Hydroxyacetophenones give 3-formylchromones.[32] Aryl benzyl ketones have been converted into 2-aryl-1,3-dichloroindenes.[33] Alkylidenemalononitriles give 2-chloro-3-cyanopyridines.[34] Unsaturated alcohols may undergo dehydration prior to formylation and cyclization (eq 16).[35]

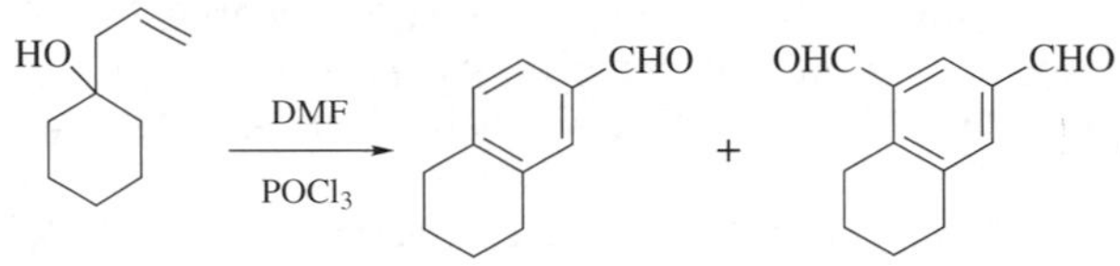
HO
DMF
POCl3
CHO
+
OHC
CHO

(16)

Related Reagents. Dimethylbromomethyleneammonium Bromide; *N*,*N*-Dimethylformamide; Formyl Chloride; *N*-Methyl-*N*-phenyl(chloromethylene)ammonium Phosphorochloridate.

1. Minkin, V. I.; Dorofeenko, G. N. *RCR* **1960**, *29*, 599.
2. de Maheas, M. *BSF* **1962**, 1989.
3. Jutz, C. *Adv. Org. Chem.* , **1976**, *9*, 225.
4. (a) Hepburn, D. R.; Hudson, H. R. *JCS(P1)* **1976**, 754. (b) Dods, R. K.; Roth, J. S. *TL* **1969**, 165; *JOC* **1969**, *34*, 1627. (c) Yoshihara, M; Eda, T.; Sakaki, K.; Maeshima, T. *S* **1980**, 746.
5. (a) Fujisawa, T.; Sato, T. *OS* **1988**, *66*, 121. (b) Bossard, H. H.; Zollinger, H. *AG* **1959**, *71*, 375. (c) Eilingsfeld, H.; Seefelder, M.; Weidinger, H. *AG* **1960**, *72*, 836.
6. Pulst, M.; Weissenfels, M. *ZC* **1976**, *16*, 337.
7. Mewshaw, R. E. *TL* **1989**, *30*, 3753.
8. Meth-Cohn, O.; Tarnowski, B. *Adv. Heterocycl. Chem.* **1982**, *31*, 207.
9. Paquette, L. A.; Johnson, B. A.; Hinga, F. M. *OSC* **1973**, *5*, 215.
10. (a) Cheung, G. K.; Downie, I. M.; Earle, M. J., Heaney, H., Matough, M. F. S; Shuhaibar, K. F.; Thomas, D. *SL* **1992**, 77. (b) Downie, I. M.; Earle, M. J., Heaney, H.; Shuhaibar, K. F. *T* **1993**, *49*, 4015.
11. Fujisawa, T.; Sato, T.; Iida, S. *CL* **1984**, 1173.
12. (a) Campaigne, E.; Archer, W. L. *OSC* **1963**, *4*, 331. (b) James, P. N.; Synder, H. R.; *OSC* **1963**, *4*, 539. (c) Silverstein, R. M.; Ryskiewicz, E. E.; Willard, C. *OSC* **1963**, *4*, 831.
13. (a) Yokoyama, Y.; Okuyama, N.; Iwadate, S.; Momoi, T.; Murakami, Y.; *JCS(P1)* **1990**, 1319. (b) Black, D. St. C.; Kumar, N.; Wong, L. C. H. *S* **1986**, 474.
14. Martinez, A. G.; Alvarez, R. M.; Barcina, J. O.; de la Moya Cerero, S.; Vilar, E. T.; Fraile, A. G.; Hanack, M.; Subramanian, L. R. *CC* **1990**, 1571.
15. Schmidle, C. J.; Barnett, P. G. *JACS* **1956**, *78*, 3209.
16. Jutz, C.; Heinicke, R. *CB* **1969**, *102*, 623.

17. Jutz, C.; Müller, W.; Müller, E. *CB* **1966**, *99*, 2479.
18. Katritzky, A. R.; Shcherbakova, I. V.; Tack, R. D.; Steel, P. J. *CJC* **1992**, *70*, 2040.
19. Grimwade, M. J.; Lester, M. G. *T* **1969**, *25*, 4535.
20. Reddy, C. P.; Tanimoto, S. *S* **1987**, 575.
21. Marson, C. M. *T* **1992**, *48*, 3659.
22. Marson, C. M.; Giles, P. R. *Synthesis using Vilsmeier Reagents*; CRC: New York, 1994.
23. Virgilio, J. A.; Heilweil, E. *OPP* **1982**, *14*, 9.
24. Willard, P. G.; de Laszlo, S. E. *JOC* **1985**, *50*, 3738.
25. (a) Bodendorf, K; Mayer, R. *CB* **1965**, *98*, 3554. (b) Alexander, C; Feast, W. J. *S* **1992**, 735.
26. (a) Hara, A.; Sekiya, M. *CPB* **1972**, *20*, 309. (b) Traas, P. C.; Takken, H. J.; Boelens, H. *TL* **1977**, 2027.
27. (a) Katritzky, A. R.; Marson, C. M. *TL* **1985**, 4715. (b) Katritzky, A. R.; Marson, C. M. *JOC* **1987**, *52*, 2726.
28. Holy, A.; Arnold, Z. *CCC* **1965**, *30*, 53.
29. Dulcere, J.-P.; Rodriguez, J. *TL* **1982**, 1887.
30. Rodriguez, J.; Dulcere, J.-P. *JOC* **1991**, *56*, 469.
31. (a) Ahlbrecht, H.; Vonderheid, C. *CB* **1975**, *108*, 2300. (b) Chupp, J. P.; Metz, S. *JHC* **1979**, *16*, 65. (c) Hayes, R.; Meth-Cohn, O.; Tarnowski, B. *JCR(S)* **1980**, 414. (d) Korodi, F.; Cziaky, Z. *OPP* **1990**, *22*, 579.
32. (a) Harnish, H. *LA* **1972**, *765*, 8. (b) Nohora, A; Umetani, T.; Sanno, Y. *TL* **1973**, 1995. (c) Nohora, A; Umetani, T.; Sanno, Y. *T* **1974**, *30*, 3553.
33. Elliot, I. W.; Evans, S. L.; Kennedy, L. T.; Parrish, A. E. *OPP* **1989**, *21*, 368.
34. Sreenivasulu, M.; Rao, G. S. K. *IJC(B)* **1989**, *28*, 584.
35. (a) Rao, M. S. C.; Rao, G. S. K. *IJC(B)* **1988**, *27*, 660. (b) Rao, M. S. C.; Rao, G. S. K. *IJC(B)* **1988**, *27*, 213.

Paul R. Giles & Charles M. Marson
University of Sheffield, UK

N,N-Dimethylformamide[1]

[68-12-2] C_3H_7NO (MW 73.11)

(formylating agent for certain organometallics[2–5]; used in combination with $POCl_3$ or acid halides to form the reactive Vilsmeier reagent[8,9])

Alternate Name: DMF.

Physical Data: mp −61 °C; bp 153 °C; *d* 0.944 g cm^{-3}; fp 57 °C.

Solubility: miscible with water and most organic solvents.

Form Supplied in: clear liquid; both d_7 and ^{13}C forms are available commercially; widely available.

Purification: decomposes slightly at its normal boiling point to give small quantities of dimethylamine and carbon monoxide. The decomposition is catalyzed by acidic or basic materials so that even at rt DMF is appreciably decomposed if allowed to stand for several hours with solid KOH, NaOH, or CaH_2. Drying agents suitable for DMF are $CaSO_4$, $MgSO_4$, silica gel, and 4Å molecular sieves. After drying, distillation under reduced pressure will give reasonably dry DMF adequate for most laboratory purposes. Anhydrous DMF can be prepared by using certain desiccants.[1b]

Handling, Storage, and Precautions: anhydrous DMF should be stored under a nitrogen atmosphere in sure seal bottles or metallic cylinders; vapors are harmful, irritating to skin, eyes, and mucous membranes; it is suspected to be a carcinogen; use in a fume hood.

Reaction with Organometallics. DMF is a good formylating agent for certain organometallic compounds. Even though formylation of some alkyllithium or Grignard reagents with DMF does not always give good yields of products due to side reactions,[1] sonication of a mixture of alkyl or aryl halide, lithium, and DMF substantially increases the rate of organometallic formation and yield of the formylated product.[2] Allenyl and vinyl organolithiums undergo formylation with DMF to produce good yields of unsaturated aldehydes.[3] DMF can be added to the dianion of cyclohexanone oxime to give an isoxazole in good yield (eq 1).[4] Many organometallic compounds have been formylated on reaction with DMF.[5]

1. 2 BuLi; 2. DMF; 3. H_2SO_4, H_2O; 87% (1)

Alkali metal reduction of arylalkenes in the presence of DMF results in α-hydroxy amides in high yield (eq 2).[6] Through a novel three-component alkene formation involving DMF, an organolithium reagent, and a Wittig reagent, alkenes can be synthesized in high yield (eq 3).[7]

+ 2e⁻; + 2 H⁺; 70–93% (2)

R = Ph, 2-pyridyl

$Ph_3\overset{+}{P}\text{-}CH_2R^1\ Br^- + R^2Li \xrightarrow[74-95\%]{DMF}$ (3)

$R^1 = CO_2Et$, Ph, $(CH_2)_6Me$, $(CH_2)_2CN$; R^2 = Bu, Ph

Vilsmeier Reaction. A Vilsmeier reagent is generated when DMF is treated with ***Phosphorus Oxychloride*** or other acid chlorides; this reagent has found extensive use as a formylating, halogenating, and dehydroxylating agent (see also ***Dimethylchloromethyleneammonium Chloride***).[8]

Formylation. Formylation can be achieved using the Vilsmeier reagent generated by treating DMF with an acid chloride. Indoles substituted at C-3 react with DMF–$POCl_3$ at 100 °C to give 2-formylindoles (eq 4).[9a] Similarly, 1,2-dihydropyridines[9b–d] and 1,4-dihydropyridines[9d] have been regioselectively formylated using DMF–$POCl_3$.

Alcohols undergo *O*-formylation when treated with DMF and benzoyl chloride, providing good yields of formates (eq 5).[10]

$$\text{R–OH} \xrightarrow[\text{2. } H_3O^+,\ 35\text{–}96\%]{\text{1. DMF, PhCOCl, }-10\text{ to }20\,°\text{C}} \text{R–OCHO} \quad (5)$$

A similar transformation can be achieved by treating an amine with ***Chlorotrimethylsilane*** and ***Imidazole*** in DMF.[11] In contrast, primary amines form the formamidine on treatment with DMF–$POCl_3$ (eq 6).[12] Phthalimides undergo a one-pot deprotection and *N*-formylation on reaction with DMF and hydrazine.[13]

Reaction with Alcohols. Alkyl halides can be easily synthesized on reaction of an alcohol with DMF and ***Phosgene***.[14] In an analogous manner, 1-triptycyl carbinols undergo a deoxygenative rearrangement and halogenation on treatment with ***Thionyl Chloride***–DMF.[15]

Reaction with Enolizable Ketones. Both acyclic and cyclic enolizable ketones react with DMF–$POCl_3$ to form the β-chloro-α,β-unsaturated aldehydes. The regioselectivity of this reaction in substituted cyclohexanones is dependent on steric effects (eq 7).[16]

Several unsaturated alkenones have been converted to chlorobenzene mono-, di-, and tricarbaldehydes.[8] An illustrative example is the conversion of 2-hexen-4-one to 5-methyl-4-chloro-1,3-benzenedicarbaldehyde using DMF–$POCl_3$ (eq 8).[17]

Benzofurans result on reaction of α-phenoxyacetophenones with DMF–$POCl_3$ (eq 9).[18]

$R^1 = R^2 = \text{OMe}$
$R^1 = \text{H}, R^2 = \text{OMe, OEt, NEt}_2$

Tetrahydro-4*H*-thiopyran-1-one and tetrahydro-4*H*-pyran-4-one are converted to the respective chlorovinyl aldehydes on treatment with DMF–$POCl_3$ at ambient temperature.[19] Thiachroman-4-one affords the β-chlorovinyl aldehyde on treatment with DMF–$POCl_3$ below 50 °C, but at 100 °C 3-formylthiachromone is formed in modest yield (eq 10).[19]

Certain ketones can be converted to chloroalkenes on reaction with the Vilsmeier reagent. 3-Halo-2-cyclopenten-1-one has been prepared in excellent yield by reaction of cyclopentene-1,3-dione with DMF and ***Oxalyl Chloride*** or ***Oxalyl Bromide***.[20] In contrast, cyclohexane-1,3-diones afford cross-conjugated dialdehydes on treatment with DMF–$POCl_3$ (eq 11).[21]

In a similar manner, 2,3-dihydro-4-pyridones have been converted to 4-chloro-1,2-dihydropyridines on treatment with DMF–$POCl_3$ (eq 12).[22]

Reaction with Amides. Quinoline-3-carbaldehydes can be prepared by reaction of acylanilides with DMF–$POCl_3$ (eq 13).[23]

In an analogous manner, dihydroisoquinolin-3-ones can be treated with DMF–$POCl_3$ followed by oxidation to give 3-chloro-4-formylisoquinolines.[24] Pyridines and related systems can be prepared using the reaction of the DMF–acid chloride adduct with carboxamides.[8]

Reaction with Acids. The adduct generated from DMF and oxalyl chloride easily converts a carboxylic acid to its acid chloride.[25] The same adduct in combination with a hydride reagent can be used to reduce acids to aldehydes.[26] DMF and $SOCl_2$ convert a carboxylic acid to an acyl azide in the presence of ***Sodium Azide***, pyridine, and tetrabutylammonium chloride as a catalyst.[27] Using the dehydrating ability of the DMF–$SOCl_2$ adduct, β-lactams can be synthesized from an imine and a carboxylic acid in the presence of ***Triethylamine*** (eq 14).[27] Generally a mixture of *cis*- and *trans*-azetidinones is formed.

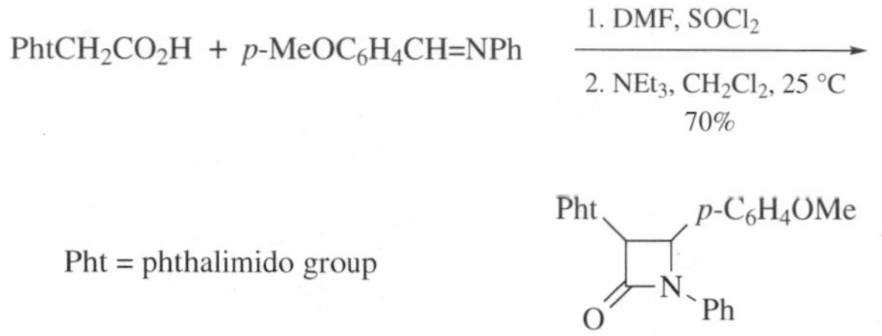

$PhtCH_2CO_2H$ + *p*-$MeOC_6H_4CH{=}NPh$ → (1. DMF, $SOCl_2$; 2. NEt_3, CH_2Cl_2, 25 °C; 70%)

Pht = phthalimido group (14)

1,4-Dihydrobenzoic acids react with DMF–$POCl_3$ to give benzene mono-, di-, and tricarbaldehydes. This procedure also affords naphthalenedicarbaldehyde (eq 15).[28]

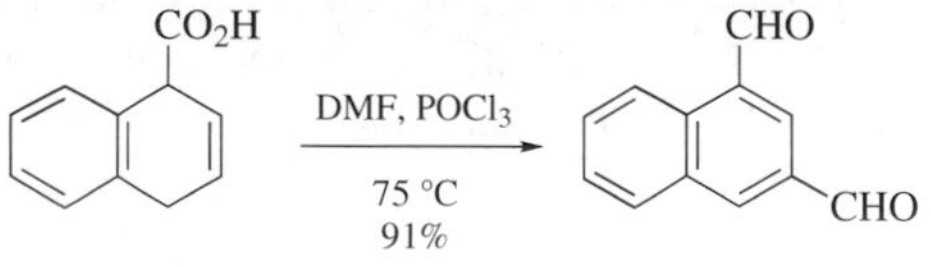

(DMF, $POCl_3$; 75 °C; 91%) (15)

Reaction with Esters and Lactonic Carbonyl Groups. Esters are normally inert to DMF–$POCl_3$ in the presence of other reactive groups.[29] The reaction of coumarone and DMF–$POCl_3$ gives a mixture of products.[30] 1,3-Oxazin-6-ones can be obtained by reaction of isoxazolin-5-ones with DMF–$POCl_3$ (eq 16).[31]

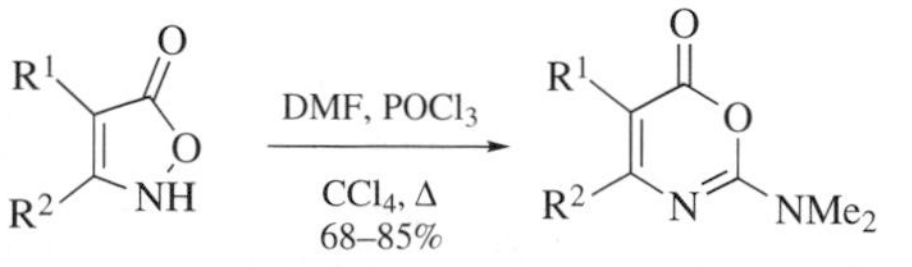

(DMF, $POCl_3$; CCl_4, Δ; 68–85%) (16)

In contrast, 3-phenyl-5-isoxazolinone reacts with DMF–$POCl_3$ to give different products depending on conditions (eq 17).[32]

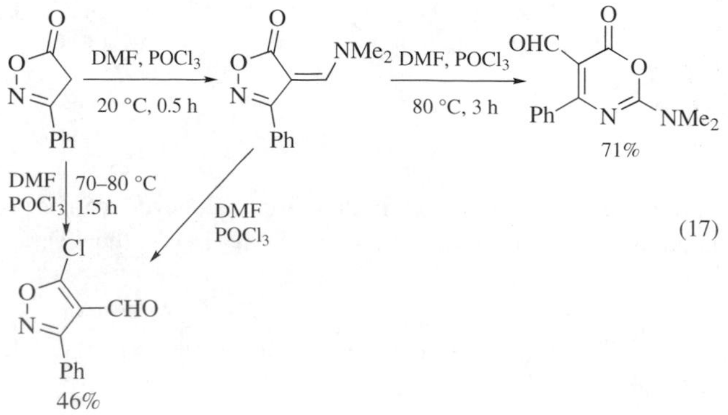

(DMF, $POCl_3$, 20 °C, 0.5 h; DMF, $POCl_3$, 80 °C, 3 h, 71%; DMF, $POCl_3$, 70–80 °C, 1.5 h, 46%) (17)

Miscellaneous Reactions. DMF reacts with α-bromopropionamides in presence of ***Sodium Hydride*** or ***Silver(I) Oxide*** to form oxazolidinones, where the carbonyl group of DMF becomes incorporated (eq 18).[33]

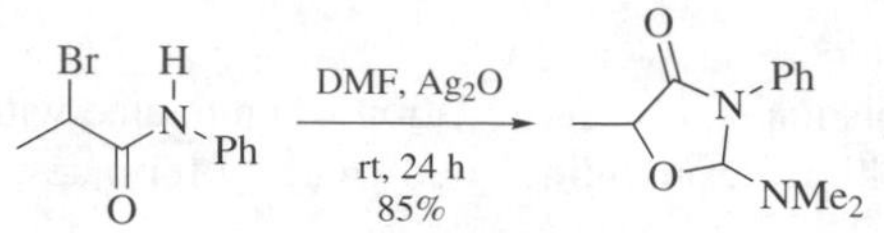

(DMF, Ag_2O; rt, 24 h; 85%) (18)

Formylation of secondary amines can be achieved on reaction with DMF and 2,3-dihydro-1,4-phthalazinedione.[34] The combination of DMF and $SOCl_2$ can dehydrate oximes to nitriles.[27] Trifluoromethylsulfonylalkynes react with DMF to form enamides.[35] *N,N*-Bis(trimethylsilyl)enamines give azabutadienes on reaction with ***Sodium Methoxide*** in DMF.[36] α,α-Dichlorination of ketones can be achieved with ***Chlorine*** using DMF as solvent or as catalyst at 80–100 °C.[37] Reaction of an α-bromo ketone with ***Lithium Bromide*** and ***Lithium Carbonate*** in DMF at 140 °C results in a cyclic ether (eq 19).[38]

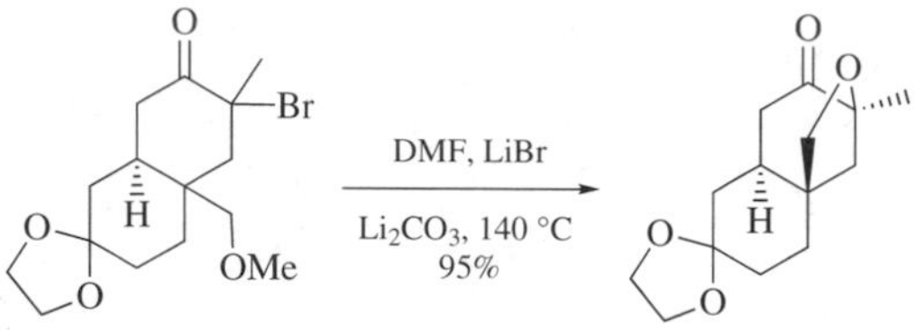

(DMF, LiBr; Li_2CO_3, 140 °C; 95%) (19)

Related Reagents. Acetic Formic Anhydride; Dimethylbromomethyleneammonium Bromide; Dimethylchloromethyleneammonium Chloride; Formyl Chloride; *N*-Formylpiperidine; *N*-Methylformanilide; Methyl Formate; *N*-Methyl-*N*-phenyl(chloromethylene)ammonium Phosphorochloridate; *N*-Methyl-*N*-(2-pyridyl)formamide.

1. (a) Evans, E. A. *JCS* **1956**, 4691. (b) Owsley, D. C.; Nelke, J. M.; Bloomfield, J. J. *JOC* **1973**, *38*, 901. (c) Burfield, D. R.; Smithers, R. H. *JOC* **1978**, *43*, 3966.
2. Petrier, C.; Gemal, A. L.; Luche, J.-L. *TL* **1982**, *23*, 3361.
3. Traas, P. C.; Boelens, H.; Takken, H. J. *TL* **1976**, 2287.
4. Barber, G. N.; Olofson, R. A. *JOC* **1978**, *43*, 3015.
5. (a) Marsais, F.; Cronnier, A.; Trecourt, F.; Queguiner, G. *JOC* **1987**, *52*, 1133. (b) Einhorn, J.; Luche, J. L. *TL* **1986**, *27*, 1793. (c) Wada, A.; Yamamoto, J.; Kanatomo, S. *H* **1987**, *26*, 585. (d) Mallet, M. *JOM* **1991**, *406*, 49. (e) Sawhney, I.; Wilson, J. R. H. *JCS(P1)* **1990**, 329. (f) Kawasaki, T.; Kodama, A.; Nishida, T.; Shimizu, K.; Somei, M. *H* **1991**, *32*, 221. (g) Kumagai, T.; Aga, M.; Okada, K.; Oda, M. *BCJ* **1991**, *64*, 1428. (h) Morita, H.; Shiotani, S. *JHC* **1986**, *23*, 1465. (i) Yang, Y.; Martin, A. P.; Nelson, D. L.; Regan, J. *H* **1992**, *34*, 1169. (j) Cohen, T.; Doubleday, M. D. *JOC* **1990**, *55*, 4784. (k) Bailey, W. F.; Khanolkar, A. D.; Gavaskar, K. V. *JACS* **1992**, *114*, 8053. (l) Ramon, D. J.; Yus, M. *JOC* **1991**, *56*, 3825.
6. Botteghi, C.; Gotta, S.; Marchetti, M.; Melloni, G. *TL* **1992**, *33*, 5601.
7. Turos, E.; Boy, K.; Ren, X.-F. *JOC* **1992**, *57*, 6667.
8. (a) Marson, C. M. *T* **1992**, *48*, 3659. (b) Meth-Cohn, O. *H* **1993**, *35*, 539.
9. (a) Walkup, R. E.; Linder, J. *TL* **1985**, *26*, 2155. (b) Comins, D. L.; Myoung, Y. C. *JOC* **1990**, *55*, 292. (c) Comins, D. L.; Mantlo, N. B. *JOC* **1986**, *51*, 5456. (d) Comins, D. L.; Herrick, J. J. *H* **1987**, *26*, 2159.
10. Barluenga, J.; Campos, P. J.; Gonzalez-Nunez, E.; Asensio, G. *S* **1985**, 426.
11. Berry, M. B.; Blagg, J.; Graig, D.; Willis, M. C. *SL* **1992**, 659.

12. Becalli, E. M.; Machesini, A.; Pilati, T. *S* **1991**, 127.
13. Iwata, M.; Kuzuhara, H. *CL* **1986**, 951.
14. Richter, R.; Tucker, B. *JOC* **1983**, *48*, 2625.
15. Crumrine, D. S.; Curtin, M. L.; Iwamura, H. *JOC* **1990**, *55*, 1076.
16. Kalsson, J. O.; Frejd, T. *JOC* **1983**, *48*, 1921.
17. Sreenivasulu, M.; Rao, G. S. K. *IJC(B)* **1989**, *28B*, 494.
18. Hirota, T.; Fujita, H.; Sasaki, K.; Namba, J.; Hayakawa, S. *H* **1986**, *24*, 771.
19. Giles, P. R.; Marson, C. M. *TL* **1990**, *31*, 5227.
20. Mewshaw, R. E. *TL* **1989**, *30*, 3753.
21. Katritzky, A. R.; Marson, C. M. *TL* **1985**, *26*, 4715.
22. Al-awar, R. S.; Joseph, S. P.; Comins, D. L. *TL* **1992**, *33*, 7635.
23. (a) Meth-Cohn, O.; Rhouati, S.; Tarnowski, B. *TL* **1979**, 4885. (b) Meth-Cohn, O.; Rhouati, S.; Tarnowski, B.; Robinson, A. *JCS(P1)* **1981**, 1537. (c) Meth-Cohn, O.; Narine, B.; Tarnowski, B. *JCS(P1)* **1981**, 1520. (d) Hayes, R.; Meth-Cohn, O.; Tarnowski, B. *JCR(S)* **1980**, 414.
24. Bartmann, W.; Konz, E.; Ruger, W. *S* **1988**, 680.
25. Burgstahler, A. W.; Wiegel, L. O.; Shaefer, G. G. *S* **1976**, 767.
26. (a) Fujisawa, T.; Mori, T.; Tsuge, S.; Sato, T. *TL* **1983**, *24*, 1543. (b) Johnstone, R. A. W.; Telford, R. P. *CC* **1978**, 354.
27. Arrieta, A.; Aizpurua, J. M.; Palomo, C. *TL* **1984**, *25*, 3365.
28. Raju, B.; Rao, G. S. K. *S* **1987**, 197.
29. Monge, A.; Palop, J. A.; Goni, T.; Martinez, A.; Fernandez-Alvarez, E. *JHC* **1985**, *22*, 1445.
30. Coppola, G. M. *JHC* **1981**, *18*, 845.
31. (a) Beccalli, E. M.; Marchesini, A.; Molinari, H. *TL* **1986**, *27*, 627. (b) Beccalli, E. M.; Marchesini, A.; Pilati, T. *T* **1989**, *45*, 7485.
32. Anderson, D. J. *JOC* **1986**, *51*, 945.
33. D'angeli, F.; Cavicchioni, G.; Catelani, G.; Marchetti, P.; Maran, F. *G* **1989**, *119*, 471.
34. Iwata, M.; Kuzuhara, H. *CL* **1989**, 2029.
35. Hanack, M.; Wilheim, B. *AG* **1989**, *101*, 1083.
36. Corriu, R. J. P.; Moreau, J. J. E.; Pataud-Sat, M. *JOC* **1990**, *55*, 2878.
37. De Kimpe, N.; De Bucyk, L.; Verhé, R.; Wychuyse, F.; Schamp, N. *SC* **1979**, *9*, 575.
38. Kanai, K.; Zelle, R. E.; Sham, H.-L.; Grieco, P. A.; Callant, P. *JOC* **1984**, *49*, 3867.

Daniel L. Comins & Sajan P. Joseph
North Carolina State University, Raleigh, NC, USA

N,N-Dimethylformamide Diethyl Acetal

R^1O, R^2 / R^1O, R^2 — $(R^1O)_2CH\text{–}NR^2_2$

(**1**; R^1 = Et, R^2 = Me)		
[1188-33-6]	$C_7H_{17}NO_2$	(MW 147.25)
(**2**; R^1 = Me, R^2 = Me)		
[4637-24-5]	$C_5H_{13}NO_2$	(MW 119.19)
(**3**; R^1 = $PhCH_2$, R^2 = Me)		
[2016-04-8]	$C_{17}H_{21}NO_2$	(MW 271.39)
(**4**; R^1 = Et, R^2 = Et)		
[22630-13-3]	$C_9H_{21}NO_2$	(MW 175.31)

(mild and selective reagents for alkylation, formylation, and aminomethylenation[1])

Alternate Name: DMF diethyl acetal.

Physical Data: (**1**) bp 134–136 °C/760 mmHg; pK_b 6.2.[5] (**2**) bp 102–104 °C/720 mmHg; pK_b 6.25.[5] (**3**) bp 138–140 °C/0.5 mmHg; pK_b 6.2.[5] (**4**) bp 57–58 °C/20 mmHg; pK_b 6.4.[5]

Solubility: sol a variety of inert solvents.

Form Supplied in: pure liquid.

Preparative Methods: (1) a solution of sodium alkoxide (1 mol) and the secondary amine (1.1 mol) in 200–300 mL of the respective alcohol is refluxed as chloroform (39.5 g, 0.33 mol) is added. The mixture is refluxed for 2 h and the filtrate obtained is distilled in vacuo. For example, *N,N*-diethylformamide diethyl acetal is obtained in 32% yield.[7] (2) A solution of *N,N*-dialkyl(chloromethylene)ammonium chloride (Vilsmeier reagent, 1 mol) in 640–500 mL of chloroform is stirred while sodium alkoxide (2.1 mol) in 1 L of the respective alcohol is added. After 1 h at 20 °C the sodium chloride is separated and the mixture is distilled in vacuo. For example, the yield of DMF dimethyl acetal is 55%.[2]

Handling, Storage, and Precautions: most of the known orthoamide derivatives are colorless, distillable liquids with an amine-like smell. If moisture and presence of acids and high temperature are avoided, most of the reagents can be stored almost indefinitely. No significant toxicity has been reported.

Reactions and Synthetic Applications: General. The formamide acetals enter into two main categories of reactions, namely alkylation and formylation, mostly via generation of aza-oxo stabilized carbenium ions (eq 1).[1–7] As alkylation reagents they have been used in the synthesis of esters from acids, in the synthesis of ethers and thioethers from phenols and aromatic and heterocyclic thiols, and in the alkylation of CH-active methines. As formylating agents, formamide acetals are useful in the synthesis of enamines from active methylene compounds and amidines from amines and amides, as well as in the formation and modification of many types of heterocyclic compounds. They can be used for the dehydrative decarboxylation of β-hydroxy carboxylic acids to alkenes, and for the cyclization of *trans* vicinal diols to epoxides.[1] From the numerous literature reports on applications of these reagents,[1–7] some representative reactions are discussed in the following sections.

$$(RO)_2CH\text{–}NMe_2 \rightleftharpoons RO(H)C{=}\overset{+}{N}Me_2 + {}^{-}OR \quad (1)$$

Alkylation Reactions. DMF dialkyl acetals undergo a variety of reactions with 1,2-diols.[1] For example, the reaction of *trans*-cyclohexane-1,2-diol with DMF dimethyl acetal leads to the formation of cyclohexane epoxide (eq 2)[8] with inversion of configuration. Similarly, *meso*-1,2-diphenyl-1,2-ethanediol gives *trans*-stilbene epoxide stereospecifically (eq 3).[8,9a] This method has also been applied in the synthesis of cholestane epoxide from vicinal diols.[8] If the intermediate 2-dimethylamino-1,3-dioxolane is treated with ***Acetic Anhydride***, reductive elimination to the alkene occurs with retention of stereochemistry (eq 4).[9b]

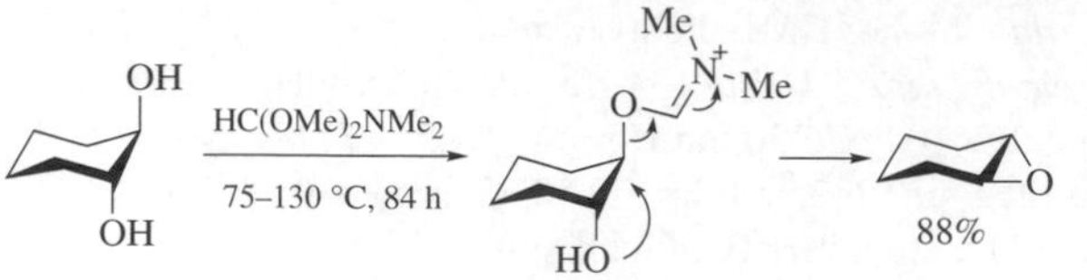

(2)

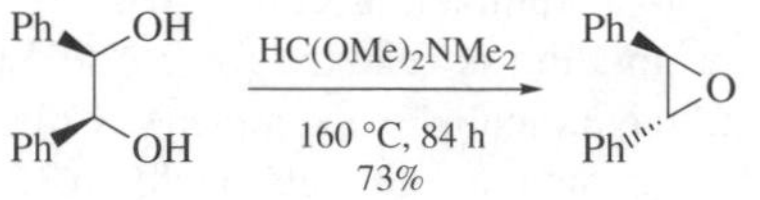

(3)

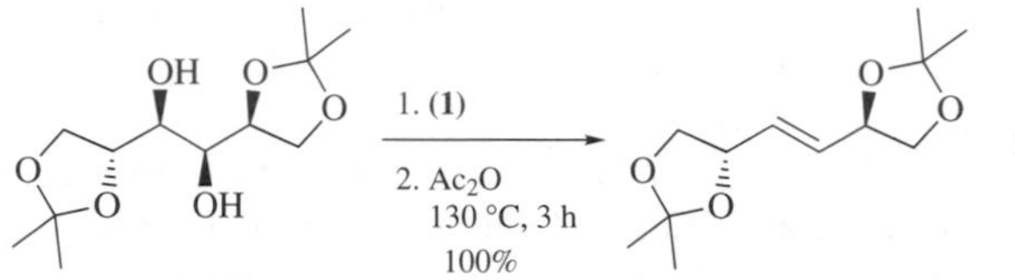

(4)

DMF dimethyl acetal is an effective methylating reagent. For example, heterocyclic thiols are transformed to *S*-methyl heterocycles in high yields (76–86%).[10] DMF dibenzyl acetal is an interesting reagent for selective protection of nucleosides. For example, uridine and guanosine are selectively blocked at the –CONH function (eq 5).[11]

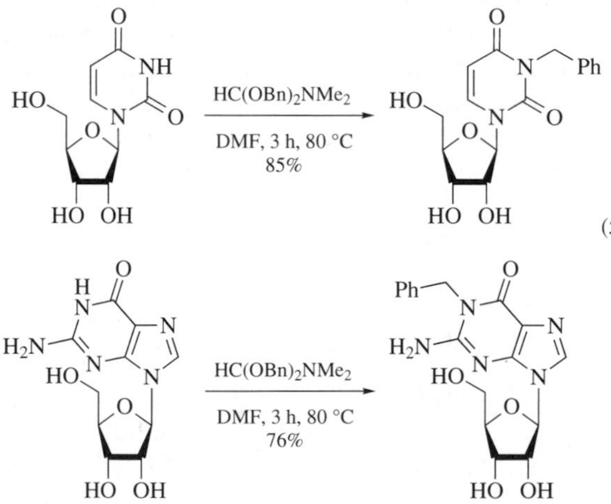

(5)

In a very simple procedure, carboxylic acids can be esterified under mild conditions with DMF dialkyl acetals. Some interesting uses are the conversion of carboxylic acids to ethyl and benzyl esters with DMF diethyl and dibenzyl acetals (yield 64–75%).[12] The dibenzyl acetal has been widely used as protecting group reagent for the carboxyl end group in peptides.[13] In several cases, DMF bis(4-dodecylbenzyl) acetal has also been used.[13,14] Some examples are given in Table 1.

Table 1 Esterification of Amino Acid and Peptide Derivatives with DMF Dibenzyl Acetal (**3**) and DMF Bis(4-dodecylbenzyl) Acetal (**5**)

RCO_2H^a	Acetal	Reaction conditions	Yield (%)
N-DOBC-L-Val	(**3**)	C_6H_6, 80 °C, 1.5 h	97[13,14]
N-DOBC-L-Phe	(**3**)	C_6H_6, 80 °C, 48 h	90[14]
N-DOBC-L-Try	(**3**)	C_6H_6, 80 °C, 2 h	78[13,14]
N-DOBC-Gly-L-Leu	(**3**)	C_6H_6, 80 °C, 1 h	73[13,14]
N-DOBC-Gly-L-Tyr	(**3**)	CH_2Cl_2, 25 °C, 48 h	80[14]
N-Boc-Gly-L-Phe	(**3**)	CH_2Cl_2, 25 °C, 90 h	80[14]
N-DOBC-L-Phe	(**5**)	CH_2Cl_2, 20 °C, 18 h	84[13]
N-DOBC-L-Try	(**5**)	CH_2Cl_2, 20 °C, 50 h	68[13]
N-DOBC-L-Phe-L-Ala	(**5**)	CH_2Cl_2, 20 °C, 120 h	75[13]

[a] DOBC = 4-decyloxybenzyloxycarbonyl; Boc = *t*-butoxycarbonyl.

DMF dineopentyl acetal has been used as a reagent for dehydrative decarboxylation in order to avoid the possibility of competing *O*-alkylation of the carboxyl group.[15] This conversion of β-hydroxy carboxylic acids to alkenes by reaction with a DMF acetal involves an *anti* elimination, and it is thus complementary to the *syn* elimination of these hydroxy acids via the β-lactone. These reactions have been used to obtain both (*E*)- and (*Z*)-1-alkoxy-1,3-butadienes (eq 6).[15] For additional examples of alkenes obtained from β-hydroxy acids with DMF acetals, see Scheeren et al.[7]

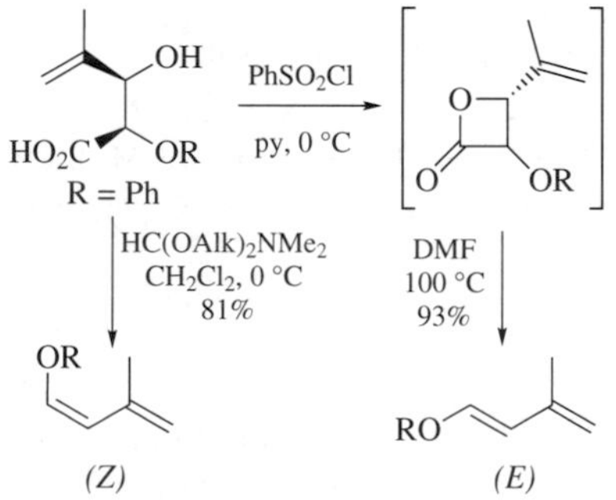

(6)

Formylation Reactions. DMF dialkyl acetals exhibit reactivity as a formyl cation synthon by introducing a C_1 unit at many nucleophilic centers (e.g. at N, S, O, or CH). For example, DMF dimethyl acetal can be used instead of formic acid for conversion of *o*-disubstituted aromatic systems into annulated heterocycles (eq 7).[16]

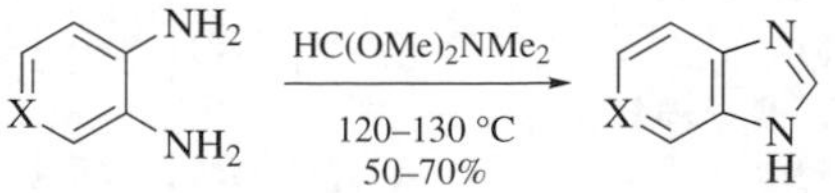

(7)

A nonacidic and regioselective route to Mannich bases from ketones and esters involves reaction with DMF acetals at a high temperature to form enamino ketones which are readily reduced by ***Acetaldehyde 3-Bromopropyl Ethyl Acetal*** to the Mannich bases (eq 8).[17]

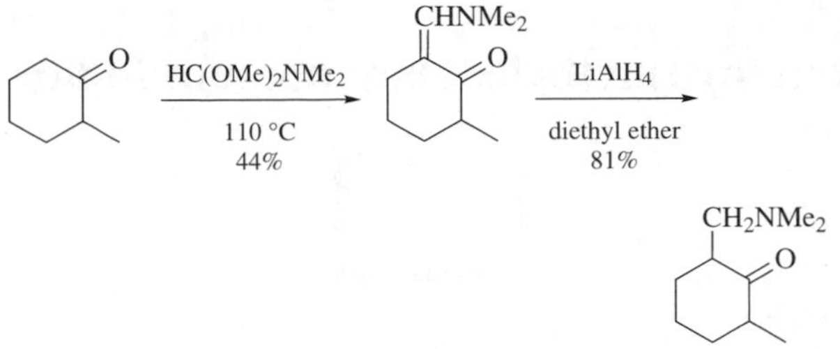

(8)

CH-acidic groups (including methyl) react with DMF acetals via carbon–carbon bond formation and subsequent elimination of the respective alkanols to form enamines (aminomethylenation). Thus 2,4,7-trimethyl-1,3-dithia-5,6-diazepine reacts with DMF dialkyl acetals to give mono- or bis-aminomethylenated products depending on the amount of reagent and the reaction conditions (eq 9).[18a] The same reagents convert the more nucleophilic 2,5-dimethyl-1,3,4-thiadiazolium salts to the monoenamines (eq 10).[18a] Similarly, 1,2-dimethyl-3-cyano-5-nitroindole condenses with DMF diethyl acetal to give the (*E*)-indol-2-yl enamine (eq 11).[18b]

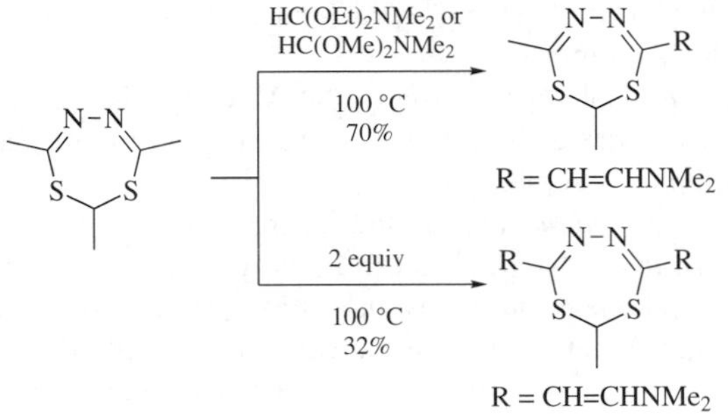

(9)

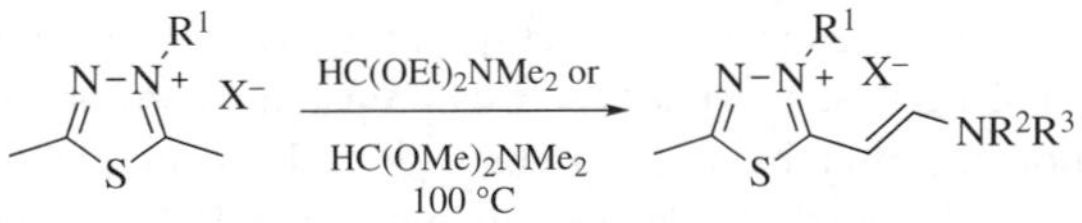

(10)

R^1	R^2	R^3	X	Yield (%)
Me	Me	Me	$MeSO_4$	93
Bz	Me	Me	Br	85
C_8H_{17}	Me	Me	I	85

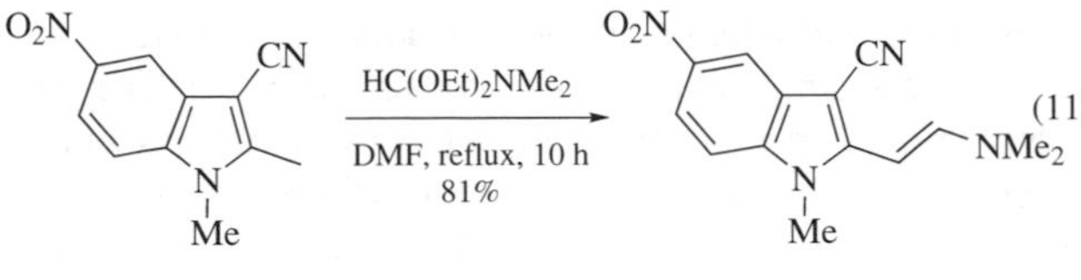

(11)

A general indole synthesis involves reaction of an *o*-nitrotoluene derivative with DMF dimethyl acetal in refluxing DMF (eq 12).[19,20] The initially formed *o*-nitroaryl-substituted (*E*)-*N,N*-dimethylenamine is submitted to catalytic hydrogenation to give the indole by spontaneous cyclization. According to a variation of this methodology,[20] 2-arylindoles are readily available by reaction of *o*-nitrotoluene with DMF diethyl acetal and *o*-halobenzoyl chloride. This reaction proceeds via benzoylation of the respective enamine.

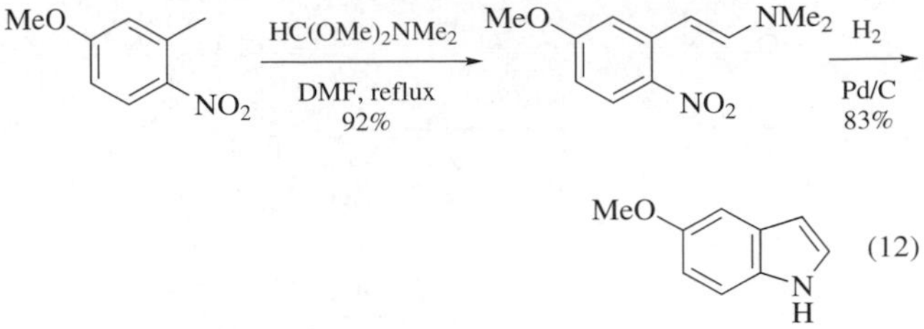

(12)

4-Dimethylamino-2-azabutadienes are readily accessible by the reaction of azomethines (imines) with DMF diethyl acetal (eq 13).[21] 1-Dimethylamino-1,3-butadienes can be synthesized in the same manner.[21] Reactions of 2-azavinamidinium salts with DMF diethyl acetal give rise to 2-aza- and 2,4-diazapentamethinium salts (eq 14).[22]

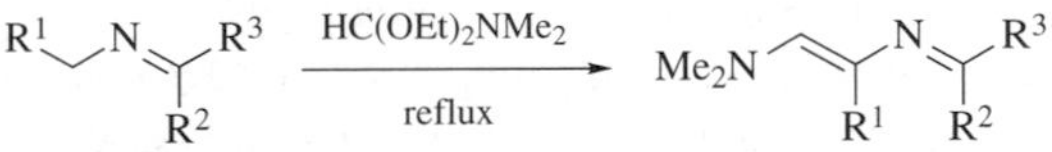

(13)

R^1	R^2	R^3	Yield (%)
CO_2Me	H	Ph	83
CO_2Me	H	p-ClC_6H_4	70
CO_2Me	H	p-Tol	82

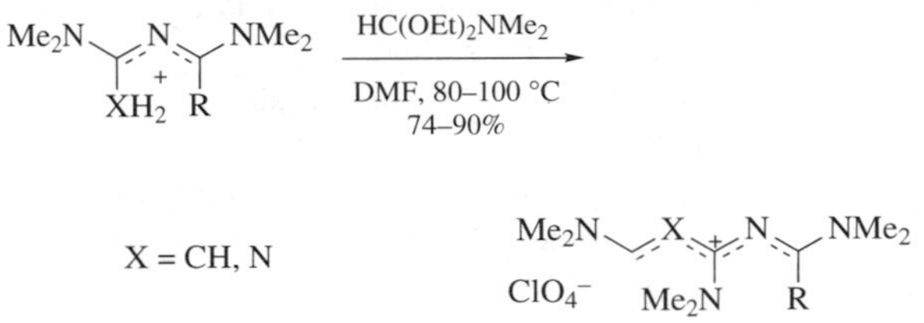

X = CH, N (14)

Miscellaneous Reactions. DMF acetals catalyze rearrangement reactions of allylic alcohols to β,γ-unsaturated amides.[23] This reaction, which involves a [2,3]-sigmatropic rearrangement, occurs with complete transfer of chirality. Thus the reaction of the (*R,Z*)-allylic alcohol (eq 15) with DMF dimethyl acetal gives the enantiomerically pure (*R,E*)-β,γ-unsaturated amide as the only product. The (*S,E*)-isomer also rearranges mainly to the (*R,E*)-amide, with only a trace of the (*S,Z*)-isomer. It has been suggested that both rearrangements proceed via a five-membered cyclic transition state with a carbene-like function.[23]

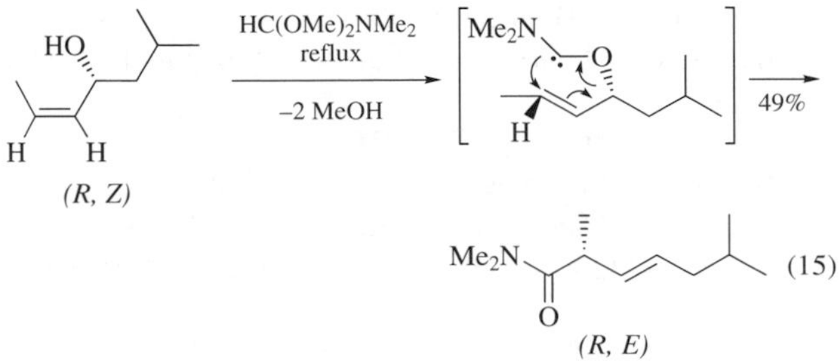

(15)

Enol acetates of oxo nucleosides are readily accessible by reaction of the nucleoside with DMF acetal and then with acetic anhydride (eq 16).[24] However, this method is limited to compounds with an oxo group in the 4′-position and with a free CO–NH group in the pyrimidine ring. This reaction is the first synthesis of oxo nucleoside enol acetates by direct enolization. Previous methods of enol formation fail with oxo nucleosides because of their instability in alkaline media.

1. $HC(OEt)_2NMe_2$
2. Ac_2O, py 60% (16)

Related Reagents. *t*-Butoxybis(dimethylamino)methane; *N*,*N*-Dimethylacetamide Dimethyl Acetal; Dimethylchloromethyleneammonium Chloride; *N*,*N*-Dimethylpropionamide Dimethyl Acetal; Triethyl Orthoformate; Tris(dimethylamino)-methane; Tris(formylamino)methane.

1. Abdulla, R. F.; Brinkmeyer, R. S. *T* **1979**, *35*, 1675.
2. Eilingsfeld, H.; Seefelder, M.; Weidinger, H. *CB* **1963**, *96*, 2671.
3. DeWolfe, R. H. *Carboxylic Ortho Acid Derivatives*, Academic: New York 1970.
4. Kantlehner, W. In *The Chemistry of Acid Derivatives*, Patai, S., Ed.; Wiley: Chichester 1979, Part 1, Suppl. B, p 533.
5. Simchen, G. In *Iminium Salts in Organic Chemistry*; Böhme, H.; Viehe, H. G., Eds.; Wiley: New York, 1979; p 393.
6. Pindur, U. In *The Chemistry of Acid Derivatives*, Patai, S. Ed.; Wiley: Chichester, 1992; Vol. 2, Suppl. B, p 1005.
7. Scheeren, J. W.; Nivard, R. J. F. *RTC* **1969**, *88*, 289.
8. Neumann, H. *C* **1969**, *23*, 267.
9. (a) Harvey, R. G.; Goh, S. H.; Cortez, C. *JACS* **1975**, *97*, 3468. (b) Eastwood, W.; Harrington, K. I.; Josan, J. S.; Pura, I. L. *TL* **1970**, 5223.
10. Holý, A. *TL* **1972**, 585.
11. Philips, K. D.; Horwitz, J. P. *JOC* **1975**, *40*, 1856.
12. (a) Vorbrüggen, H. *AG(E)* **1963**, *2*, 211. (b) Vorbrüggen, H. *LA* **1974**, 821.
13. (a) Brechbühler, H.; Büchi, H.; Hatz, E.; Schreiber, J.; Eschenmoser, A. *AG(E)* **1963**, *2*, 212; (b) Büchi, H.; Steen, K.; Eschenmoser, A. *AG(E)* **1964**, *3*, 62.
14. Brechbühler, H.; Büchi, H.; Hatz, E.; Schreiber, J., Eschenmoser, A. *HCA* **1965**, *48*, 1746.
15. (a) Luengo, J. L.; Koreeda, M. *TL* **1984**, *25*, 4881. (b) Koreeda, M.; Luengo, J. L. *JOC* **1984**, *49*, 2079.
16. Stanovnik, B.; Tisler, M. *S* **1974**, 120.
17. Schuda, P. F.; Ebner, C. B.; Morgan, T. M. *TL* **1986**, *27*, 2567.
18. (a) Kantlehner, W.; Haug, E.; Hagen, H. *LA* **1982**, 298. (b) Krichevski, E. S.; Granik, V. G. *KGS* **1992**, 502.
19. Batcho, A. D.; Leimgruber, W. U.S. Patent 3 976 639, 1973; 3 732 245, 1973.
20. Garcia, E. E.; Fryer, R. I. *JHC* **1974**, *11*, 219.
21. Gompper, R.; Heinemann, U. *AG(E)* **1981**, *20*, 296.
22. Gompper, R.; Heinemann, U. *AG(E)* **1981**, *20*, 297.
23. Yamamoto, H.; Kitatani, K.; Hiyama, T.; Nozaki, H. *JACS* **1977**, *99*, 5816.
24. Bessodes, M.; Ollapally, A.; Antonakis, K. *CC* **1979**, 835.

Ulf Pindur
University of Mainz, Germany

Dimethyl(methylene)ammonium Iodide[1]

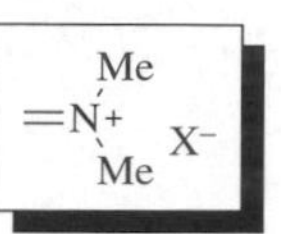

(**1**; X = I⁻)		
[36627-00-6]	C_3H_8IN	(MW 185.02)
(**2**; X = Cl⁻)		
[30438-74-5]	C_3H_8ClN	(MW 93.57)
(**3**; X = $CF_3CO_2^-$)		
[85413-84-9]	$C_5H_8F_3NO_2$	(MW 171.14)

(electrophilic aminomethylation (Mannich reaction) agent which condenses with active methylene compounds,[1] organometallic reagents,[2] and electron-rich aromatics and heteroaromatics[3] to produce tertiary amines; useful for the synthesis of α,β-unsaturated carbonyl compounds by elimination of the dimethylamino group[1])

Alternate Names: (**1**) Eschenmoser's salt; (**2**) Böhme's salt.
Physical Data: (**1**) mp 219 °C (dec). (**2**) mp 116 °C. (**3**) bp 100 °C/14 mmHg.
Solubility: sol DMF; partially sol[4] MeCN, CH_2Cl_2, THF (reactions in these solvents occur even though the reagent is not completely soluble).
Form Supplied in: salts (**1**) and (**2**) are widely available as solids.
Preparative Methods: salt (**1**) is prepared by thermolysis of (iodomethyl)trimethylammonium iodide[5] or cleavage of *N,N,N′,N′*-tetramethylmethanediamine by TMSI.[6] It is purified by recrystallization from tetrahydrothiophene dioxide or by sublimation at 120 °C/0.05 mmHg.[5] Salt (**2**) is prepared by cleavage of *N,N,N′,N′*-tetramethylmethanediamine by AcCl[7] or cleavage of methyl dimethylaminomethyl ether by TMSCl.[8] Salt (**3**), a liquid, is prepared by Polonovksi reaction of trimethylamine oxide with TFAA or cleavage of *N,N,N′,N′*-tetramethylmethanediamine with TFA. It is purified by distillation.[9]
Handling, Storage, and Precautions: salts (**1**)–(**3**) are very hygroscopic and must be used under anhydrous conditions to preserve their reactivity.

Dimethyl(methylene)ammonium Salts. These reagents comprise a special family of preformed iminium salts which have been widely utilized in condensation reactions with active methylene compounds and arenes in a variant of the Mannich reaction. They have their origins in the work of Böhme,[10] who prepared the first dimethyl(methylene)ammonium salt (**2**).[11] Higher concentrations of iminium salts are achieved using salts such as (**1**)–(**3**) than under the conditions of the classical Mannich reaction in which the iminium salt is generated reversibly; hence reactions are faster, proceed under milder conditions, and are more compatible with sensitive functionality. Preformed iminium salts are sufficiently soluble in many aprotic solvents, enabling the use of highly reactive nucleophiles which would ordinarily decompose under the protic conditions of the classical Mannich reaction. Moreover, salts (**1**)–(**3**) have special applications as one-carbon synthons for an *exo* methylene group, owing to the leaving group ability of the dimethylamino group on subsequent quaternization. In addition to the iodide salt (**1**), the chloride (**2**) and trifluoroacetate (**3**) have been increasingly utilized in recent years and are essentially equivalent Mannich reagents. In choosing a particular counterion form, factors such as solubility, ease of preparation and purification, and moisture sensitivity should be considered. A comparison study of the three reagents[12a] favors the trifluoroacetate (**3**) because it is the most soluble and can be transferred by syringe, although it is more tedious to prepare. Several other counterion forms have been prepared by anion exchange of the chloride but have had minimal synthetic applications.[10,13]

Reactions with Active Methylene Compounds and their Derivatives. Active methylene compounds with acidity comparable to or greater than that of a simple ketone or aldehyde condense directly and without prior activation with salts (**1**)–(**3**) in aprotic solvents to produce Mannich bases. Reaction conditions vary depending on the acidity of the substrate and may employ elevated temperatures and base additives to promote enolization. A comparison study finds that ketone Mannich bases are obtained in generally higher yields using salt (**2**) than under the conditions of the classical Mannich reaction, especially in the case of sterically crowded and α,β-unsaturated ketones.[7] For example, α-methylpropiophenone reacts with (**2**) to produce the corresponding Mannich base in 53% yield (eq 1), whereas under the classical conditions the yield is only 6%. Other ketone substrates reported to react with salts (**1**)–(**3**) in a superior manner to the classical Mannich reaction include chromanones,[14] thiochromanones,[14] and 1*H*-pyrido[3,2,1-*k*]phenothiazin-3(2*H*)-ones.[15] The regiochemistry of the reactions of unsymmetrical methyl ketones with salt (**3**) in TFA has been studied.[16] Under kinetic control, attack at the more substituted enol predominates in parallel to the classical Mannich reaction[17] to give the more substituted Mannich base, while under thermodynamic control reversibility favors the less substituted Mannich base (eq 2).[18] In addition to simple ketones and aldehydes, other classes of active methylene compounds known to react in this variant of the Mannich reaction include 2*H*-benzo[*b*]furan-3-ones,[19] 2*H*-benzo[*b*]thiophen-3-ones,[19] 1,3-dithian-5-ones,[20] α-methylene-β-diketones,[21] β-diketones,[22] malonates,[6,23–25] α-diazo ketones,[26] and *N,N*-diphenylglycinate esters.[27] These reactions have been used in the modification of natural products such as steroids,[9] hydrocodone,[28] oxycodone,[28] and spectinomycin.[26]

(**2**), MeCN, Δ, 53% → NMe_2•HCl (1)

(**3**), TFA, Δ → Me_2N … + … NMe_2 (2)

Temp		
72 °C	77%	16%
145 °C	–	55%

Preformed kinetic enolates undergo alkylation with salts (**1**)–(**3**), extending the scope of the Mannich reaction to weakly acidic active methylene components such as esters (eq 3),[12b,29,30] lactones,[12b,29,31–34] unsaturated lactones,[35] carboxylic acids,[12b] and nitriles.[36] Because of their enhanced nucleophilicity, enolates are able to add under extremely mild conditions (−78 °C to rt). Moreover, in the case of unsymmetrical ketones, predictable regiocontrol can be achieved if the enolate is prepared regiospecifically. Enolates used in these reactions have generally been prepared by deprotonation with strong bases such as ***Acetaldehyde 3-Bromopropyl Ethyl Acetal***[29–35] or ***Potassium Hydride***,[12,29] cleavage of silyl enol ethers[12b] or enol carbonates[37] with ***Methyllithium***, or by decomposition of α-diazo ketones in the presence of trialkylboranes to generate the corresponding boron enolate.[38] The latter method, which reliably generates ketone enolates regiospecifically, is excellent for the regiospecific synthesis of Mannich bases derived from simple unsymmetrical ketones (eq 4). Also, ester enolates generated by cleavage of cyclopropyl esters in the presence of ***Trimethylsilyl Trifluoromethanesulfonate*** add to salt (**2**).[39]

CO_2Me — 1. LDA, THF, −78 °C; 2. (**1**); 3. −78 °C to rt; 80% → CO_2Me, NMe_2 (3)

O, N_2 — Et_3B, THF → [$OBEt_2$] — (**1**), DMSO, 94% → O, NMe_2 (4)

Trimethylsilyl enol ethers derived from ketones and aldehydes add to salts (**1**)–(**3**) in a complementary fashion to enolates.[12a,32] The reaction proceeds via a silyloxonium ion which hydrolyzes on aqueous workup (eq 5). This method has been used for the synthesis of the two regioisomeric Mannich bases of 2-methylcyclohexanone starting from the individual silyl enol ethers[40] and for the synthesis of the *exo* Mannich base of (+)-camphor,[41] which does not undergo aminomethylation under the conditions of the classical Mannich reaction. Chromanone[42] and enone[40,43] trimethylsilyl enol ethers also react. In situ methods for the generation of salt (**1**), involving cleavage of *N,N,N′,N′*-tetramethylmethanediamine by ***Chloroiodomethane***[44] or cleavage of *n*-butyldimethylaminomethyl ether[45] by ***Iodotrimethylsilane***, are also compatible with these reactions and avoid handling the moisture-sensitive reagent (eq 6). In a minor variant of this reaction employing more stable *t*-butyldimethyl silyl enol ethers, the product retains the silyl enol ether following workup with retention or migration of the original position of the enol ether double bond, depending on the substrate (eq 7).[46]

OTMS — (**1**), CH_2Cl_2 → [$TMS\overset{+}{O}$ … NMe_2] — 1. H_3O^+; 2. OH^-; 65% → O, NMe_2 (5)

OTMS —[$(Me_2N)CH_2$, CH_2ClI, DMSO; 68%]→ (6)

OTBDMS —[(2), CH_2Cl_2; 89%]→ (7)

Reactions with Organometallic Reagents. In analogy to enolate anion additions, organometallic reagents such as alkyl bromomagnesium and lithium reagents,[2] vinyl and aromatic bromomagnesium reagents,[6] vinylcuprates,[47] and cyclopropyllithiums[48] add to salt (**1**) to produce *N,N*-dimethylamines (eq 8). Salt (**1**) is also capable of condensing directly with less nucleophilic organometallic reagents such as 1,4-bis(trimethylstannyl)-2-butyne to afford 2,3-bis(dimethylamino)methyl-1,3-butadiene.[49]

$$C_7H_{15}CH_2MgBr \xrightarrow[91\%]{(1),\ ether} C_7H_{15}CH_2CH_2NMe_2 \quad (8)$$

Synthesis of α,β-Unsaturated Carbonyl Compounds and Terminal Alkenes. Salts (**1**)–(**3**) serve as excellent one-carbon units for an α-methylene group, since the dimethylamino group of the derived Mannich base can be easily eliminated by quaternization with ***Iodomethane*** followed by treatment with base (eq 9). Active methylene compounds that have been converted to their α-methylene derivatives by this method include esters,[27,29,30,39] lactones,[29,32–34,50] and ketones.[37,51] Similarly, γ-methylene enones,[40] and γ-methylene butenolides[35] have been prepared from their respective Mannich bases. Quaternary ammonium salts from Mannich bases of α-alkoxymalonates[23,52] and α-alkoxy-α-methoxycarbonyl lactones[24] undergo base or thermal fragmentation to provide enolpyruvates for the synthesis of chorismic acid and related natural products (eq 10). Mannich base hydrochlorides derived from β-diketones and salt (**2**) undergo spontaneous elimination by addition of water to give enediones.[22] Tertiary amines derived from the addition of organometallic reagents to salts (**1**)–(**3**) can also be converted to terminal alkenes by thermolysis of the corresponding *N*-oxides.[2]

1, LDA, THF; 2. (**1**); 3. MeI; 4. DBU, dioxane; 65% (9)

1. (**1**), TEA, CH_2Cl_2; 2. MeI; 3. Δ; 78% (10)

Reactions with Aromatic Compounds. Several electron-rich aromatic and heteroaromatic systems have been shown to undergo condensation reactions with salts (**1**)–(**3**) with improved yields and regiocontrol over the conventional method. Monosubstituted phenols, which typically give mixtures of *ortho, para*, and diaminomethylated products under conventional conditions, react with salt (**1**) or (**2**) in the presence of ***Potassium Carbonate*** to afford the corresponding *ortho*-substituted Mannich base in high yield (eq 11).[53] An ion pair interaction between the phenoxide anion and the iminium salt is believed to be responsible for the directing effect. This reaction is also successful with deactivating groups in the *para* position. 2,3-Dimethylphenol,[54] 1-naphthol,[55] and 2-methyl-1-naphthol[56] also react in a superior manner with respect to yield and regiospecificity by simple treatment with salts (**1**) or (**2**) in aprotic solvents. In the case of 1-naphthol derivatives, 4-substituted Mannich bases mainly result. The reaction has been exploited for the preparation of aminomethylated derivatives of papulacandin A, a resorcinol-containing antibiotic,[57] and 4,4′- and 3,3′-dihydroxy-α,β-diethylstilbenes.[58] Among heterocyclic systems, salt (**2**) has been used to effect aminomethylation at the 2-position of furan[59] and thiophene,[60] the 3-position of 4-substituted indoles,[61–63] the 8-position of a 2,3-dihydro-5(1*H*)-indolizinone derivative,[64] and the 5-position of a 1,4-dihydropyridine derivative.[65] The cases of furan and thiophene are noteworthy, since they fail or condense poorly in the classical Mannich reaction. In situ methods for the generation of dimethyl(methylene)ammonium species involving cleavage of *N,N,N′,N′*-tetramethylmethanediamine by ***Sulfur Dioxide***[66] and by chlorosilanes[67] has been investigated in the Mannich reactions of *N*-methylpyrrole, indole, and *N*-methylindole (eq 12). Evidence suggests, however, that free iminium species are not involved in these reactions. An unusual out-of-ring aminomethylation reaction catalyzed by ***Acetyl Chloride*** occurs at the 1-methyl of a 1-methyl-6-phenyl-4*H*-*s*-triazolo[4,3-*a*][1,4]benzodiazepine derivative using salt (**2**), generated in situ by cleavage of *N,N,N′,N′*-tetramethylmethanediamine with excess AcCl.[68] In an interesting variant of the Mannich reaction, aryl- and heteroarylstannanes react with salt (**2**) to produce Mannich bases obtained by *ipso*-substitution of the stannyl group.[69] The directing effect of tin allows for the preparation of Mannich bases with substitution patterns not ordinarily obtained in the Mannich reaction, as in the case of 3-(*N,N*-dimethylaminomethyl)thiophene (eq 13).

(**1**), toluene, K_2CO_3; 85% (11)

$(Me_2N)_2CH_2$, TMSCl; 59% (12)

$SnMe_3$ —[(2), CH_2Cl_2; 66%]→ NMe_2 (13)

Heteroatom Addition. Salt (**2**) undergoes heteroatom addition to amides,[70] phosphoramides,[71] hydrazines,[72] oximes,[73] and secondary phosphines and arsines.[4]

Related Reagents. (Methylene)ammonium salts containing different N-alkyl substitution have been synthesized and undergo similar reactions to those of dimethyl(methylene)ammonium salts.[1,3,10]

1. Kleinman, E. F. *COS* **1991**, *2*, 899.
2. Roberts, J. L.; Borromeo, P. S.; Poulter, C. D. *TL* **1977**, 1299.
3. Heany, H. *COS* **1991**, *2*, 953.
4. Kellner, K.; Seidel, B.; Tzschach, A. *JOM* **1978**, *149*, 167.
5. Schreiber, J.; Maag, H.; Hashimoto, N.; Eschenmoser, A *AG(E)* **1971**, *10*, 330.
6. Bryson, T. A.; Bonitz, G. H.; Reichel, C. J.; Dardis, R. E. *JOC* **1980**, *45*, 524.
7. Kinast, G.; Tietze, L.-F. *AG(E)* **1976**, *15*, 239.
8. Rochin, C.; Babot, O.; Dunogues, J.; Duboudin, F. *S* **1986**, 228.
9. Ahond, A.; Cavé, A.; Kan-Fan, C.; Potier, P. *BSF(2)* **1970**, 2707.
10. For a comprehensive review of his work, see: Böhme, H.; Haake, M. In *Advances in Organic Chemistry: Methods and Results*; Taylor, E. C., Ed.; Wiley: New York, 1976; Vol. 9, Part 1, pp 107–213.
11. Böhme, H.; Mundlos, E.; Herboth, O.-E. *CB* **1957**, *90*, 2003.
12. (a) Holy, N.; Fowler, R.; Burnett, E.; Lorenz, R. *T* **1979**, *35*, 613. (b) Holy, N. L.; Wang, Y. F. *JACS* **1977**, *99*, 944.
13. Knoll, F.; Krumm, U. *CB* **1971**, *104*, 31.
14. Eiden, F.; Schmidt, M. *AP* **1987**, *320*, 1099.
15. Grol, C. J.; Dijkstra, D.; Schunselaar, W.; Westerink, B. H. C.; Martin, A. R. *JMC* **1982**, *25*, 5.
16. Jasor, Y.; Gaudry, M.; Luche, M. J.; Marquet, A. *T* **1977**, *33*, 295.
17. House, H. O.; Trost, B. M. *JOC* **1964**, *29*, 1339.
18. Gaudry, M.; Jasor, Y.; Bui Khac, T. *OSC* **1988**, *6*, 474.
19. Schaefer, M.; Weber, J.; Faller, P. *BSF(2)* **1978**, 241.
20. Mitsudera, H.; Uneme, H.; Okada, Y.; Numata, M.; Kato, A. *JHC* **1990**, *27*, 1361.
21. Dimmock, J. R.; Raghavan, S. K.; Logan, B. M.; Bigam, G. E. *Eur. J. Med. Chem.* **1983**, *18*, 248.
22. Möhrle, H.; Schaltenbrand, R. *Pharmazie* **1985**, *40*, 697.
23. Hoare, J. H.; Policastro, P. P.; Berchtold, G. A. *JACS* **1983**, *105*, 6264.
24. Ganem, B.; Ikota, N.; Muralidharan, V. B.; Wade, W. S.; Young, S. D.; Yukimoto, Y. *JACS* **1982**, *104*, 6787.
25. Landsbury, P. T.; Mojica, C. A. *TL* **1986**, *27*, 3967.
26. Thomas, R. C.; Fritzen, E. L. *J. Antibiotics* **1985**, *38*, 208.
27. Tarzia, G.; Balsamini, C.; Spadoni, G.; Duranti, E. *S* **1988**, 514.
28. Görlitzer, K.; Meyer, E. *AP* **1993**, *326*, 181.
29. Roberts, J. L.; Borromeo, P. S.; Poulter, C. D. *TL* **1977**, 1621.
30. Snider, B. B.; Phillips, G. P. *JOC* **1984**, *49*, 183.
31. Seebach, D.; Boes, M.; Naef, R.; Schweizer, W. B. *JACS* **1983**, *105*, 5390.
32. Danishefsky, S.; Kitahara, T.; McKee, R.; Schuda, P. F. *JACS* **1976**, *98*, 6715.
33. Rigby, J. H.; Wilson, J. Z. *JACS* **1984**, *106*, 8217.
34. Hanessian, S.; Liak, T. J.; Dixit, D. M. *Carbohydr. Res.* **1981**, *88*, C4.
35. Ley, S. V.; Trudell, M. L.; Wadsworth, D. J. *T* **1991**, *47*, 8285.
36. Böhme, H.; Hitzel, E. *AP* **1990**, *306*, 948.
37. Danishefsky, S.; Chackalamannil, S.; Harrison, P.; Silvestri, M.; Cole, P. *JACS* **1985**, *107*, 2474.
38. Hooz, J.; Bridson, J. N. *JACS* **1973**, *95*, 602.
39. Reissig, H. U.; Lorey, H. *LA* **1986**, 1914.
40. Danishefsky, S.; Prisbylla, M.; Lipisko, B. *TL* **1980**, *21*, 805.
41. McClure, N. L.; Dai, G.-Y.; Mosher, H. S. *JOC* **1988**, *53*, 2617.
42. Iwasaki, H.; Kume, T.; Yamamoto, Y.; Akiba, K. *TL* **1987**, *28*, 6355.
43. Trost, B. M.; Curran, D. P. *JACS* **1981**, *103*, 7380.
44. Miyano, S.; Hokari, H.; Hashimoto, H. *BCJ* **1982**, *55*, 534.
45. Hosomi, A.; Iijima, S.; Sakurai, H. *TL* **1982**, *23*, 547.
46. Wada, M.; Nishihara, Y.; Akiba, K. *TL* **1984**, *25*, 5405.
47. Germon, C.; Alexakis, A.; Normant, J. F. *BSF2* **1984**, 377.
48. Saimoto, H.; Nishio, K.; Yamamoto, H.; Shinoda, M.; Hiyama, T.; Nozaki, H. *BCJ* **1983**, *56*, 3093.
49. Reich, H. J.; Yelm, K. E.; Reich, I. L. *JOC* **1984**, *49*, 3440.
50. Marshall, J. A.; Flynn, G. A. *JOC* **1979**, *44*, 1391.
51. Greengrass, C. W.; Hughman, J. A.; Parsons, P. J. *CC* **1985**, 889.
52. Chouinard, P. M.; Bartlett, P. A. *JOC* **1986**, *51*, 75.
53. Pochini, A.; Puglia, G.; Ungaro, R. *S* **1983**, 906.
54. Möhrle, H.; Scharf, U. *AP* **1980**, *313*, 435.
55. Möhrle, H.; Tröster, K. *AP* **1982**, *315*, 397.
56. Möhrle, H.; Tröster, K. *AP* **1982**, *315*, 222.
57. Traxler, P.; Tosch, W.; Zak, O. *J. Antibiot.* **1987**, *XL*, 1146.
58. Schönenberger, H; Adam, D.; Alonso, G.; Adam, A. *AP* **1972**, *305*, 300.
59. Heaney, H.; Papegeorgiou, G.; Wilkins, R. F. *TL* **1988**, *29*, 2377.
60. Dowle, M. D.; Hayes, R.; Judd, D. B.; Williams, C. N. *S* **1983**, 73.
61. Kozikowski, A. P.; Ishida, H. *H* **1980**, *14*, 55.
62. Barrett, A. G. M.; Dauzonne, D.; O'Neil, I. A.; Renaud, A. *JOC* **1984**, *49*, 4409.
63. Matsumoto, M.; Kobayashi, H.; Watanabe, N. *H* **1987**, *26*, 1197.
64. Earl, R. A.; Vollhardt, K. P. C. *JOC* **1984**, *49*, 4786.
65. Bennasar, M.-L.; Vidal, B.; Bosch, J. *JACS* **1993**, *115*, 5340.
66. Eyley, S. C.; Heaney, H.; Papageorgiou, G.; Wilkins, R. F. *TL* **1988**, *29*, 2997.
67. Heany, H.; Papageorgiou, G.; Wilkins, R. F. *CC* **1988**, 1161.
68. Hester, J. B., Jr. *JOC* **1979**, *44*, 4165.
69. Cooper, M. S.; Fairhurst, R. A.; Heaney, H.; Papageorgiou, G.; Wilkins, R. F. *T* **1989**, *45*, 1155.
70. Abou-Gharbia, M.; Freed, M. E.; McCaully, R. J.; Silver, P. J.; Wendt, R. L. *JMC* **1984**, *27*, 1743.
71. Freeman, S.; Harger, M. J. P. *CC* **1985**, 241.
72. Böhme, H.; Martin, F. *CB* **1973**, *106*, 3540.
73. Unterhalt, B.; Koehler, H. *S* **1977**, 265.

Edward F. Kleinman
Pfizer Central Research, Groton, CT, USA

(*S*,*S*)-2,2′-(Dimethylmethylene)bis(4-*t*-butyl-2-oxazoline)[1]

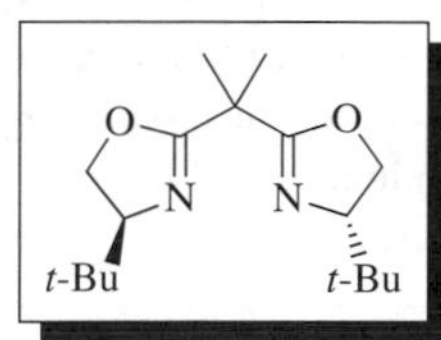

[131833-93-7] $C_{17}H_{30}N_2O_2$ (MW 294.49)

(versatile chiral ligands for enantiocontrol of metal-catalyzed reactions such as copper-catalyzed cyclopropanation[2–5] and aziridination of alkenes,[6] addition of cyanotrimethylsilane to aldehydes,[7] or Lewis acid-catalyzed Diels–Alder reactions[8,9])

Physical Data: mp 88–89 °C; $[\alpha]_D$ −108°; $[\alpha]_{365}$ −394° (*c* 0.97, CH_2Cl_2).

Solubility: insol H_2O; sol all common organic solvents.

Preparative Methods: ligand (**1**) and related C_2-symmetric bisoxazolines are readily prepared from chiral β-amino alcohols using standard methods for the synthesis of 2-oxazolines.[1] This is exemplified by the simple three-step procedure shown in eq 1, involving amide formation, conversion of the resulting bis(2-hydroxyalkyl)amide to the corresponding bis(2-chloroalkyl)amide, and subsequent base-induced cyclization.[3a,4,8a,10,11]

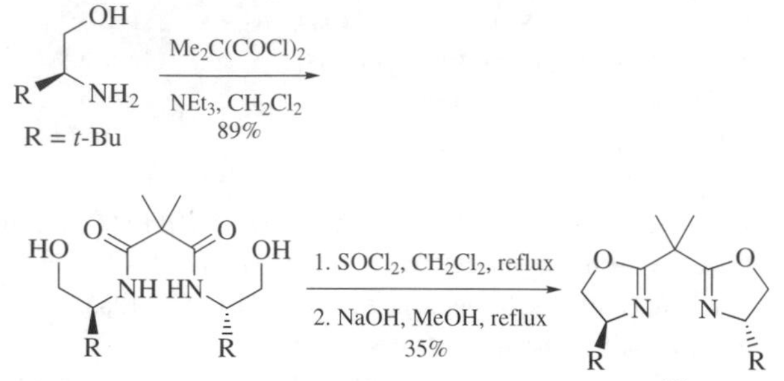

R = *t*-Bu; $Me_2C(COCl)_2$, NEt_3, CH_2Cl_2, 89%; 1. $SOCl_2$, CH_2Cl_2, reflux; 2. NaOH, MeOH, reflux, 35% (1)

(**1**)

There are several other convenient one- and two-step syntheses leading to enantiomerically pure bisoxazolines, e.g. condensation of amino alcohols with dicarboxylic acids,[2,7] dinitriles,[10a,12] or diimino esters[4,10,13] (cf. eq 2),[4,10a] or acid-catalyzed cyclization of (2-hydroxyalkyl)amides.[8b] By these methods, various types of differently substituted bisoxazoline ligands are readily available in both enantiomeric forms, often in high overall yield. [32]

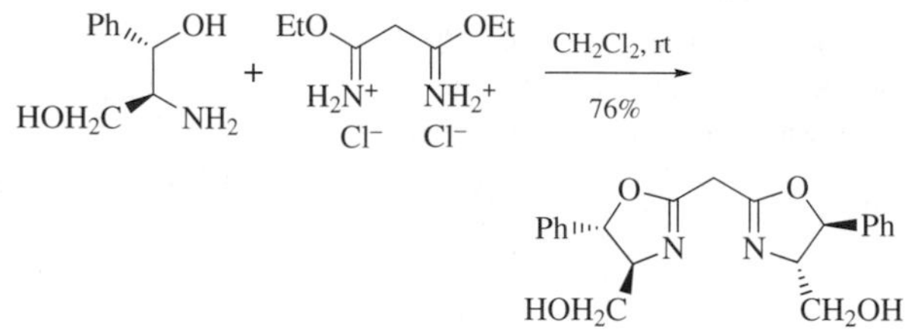

CH_2Cl_2, rt, 76% (2)

Purification: (**1**) can be purified by column chromatography (silica gel, hexane/EtOAc 7:3) and by recrystallization from pentane.

Handling, Storage, and Precautions: as a crystalline solid, (**1**) is stable at ambient temperature; for longer periods, storage at −20 °C is recommended.

C_2-Symmetric Bisoxazolines as Ligands in Asymmetric Catalysis. Methylenebis(oxazolines) such as (**1**), (**3**), and (**5**) are patterned after the semicorrins,[1] which have been successfully employed as ligands in enantioselective Cu-catalyzed cyclopropanations and other reactions (see ***(1S,9S)-1,9-Bis{[(t-butyl)dimethylsilyloxy]methyl}-5-cyanosemicorrin)***. The potential of bisoxazoline ligands of this type, which has been recognized independently by a number of research groups,[1–11,13–15] is demonstrated by a remarkable variety of different applications in asymmetric catalysis.

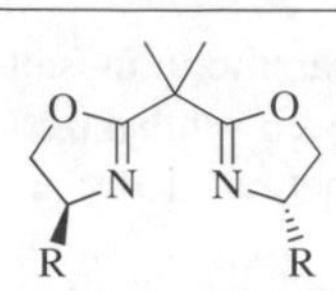

(**1**) R = *t*-Bu
(**2**) R = Ph
(**3**) 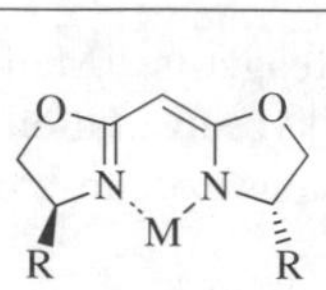
(**4**) 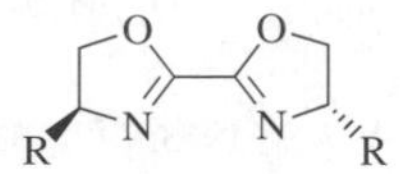
(**5**) 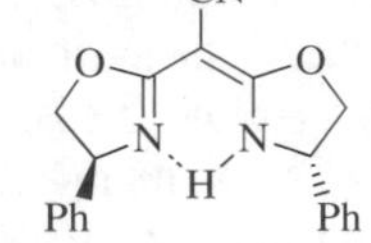

The short and simple syntheses of these compounds and the ease of modifying their structures make them ideal ligands for the stereocontrol of metal-catalyzed reactions. Using different amino alcohols and dicarboxylic acid derivatives as precursors, the steric and electronic properties, as well as the coordination geometry, can be adjusted to the specific requirements of a particular application. The neutral methylenebis(oxazoline) ligands (**1**) and (**2**), which form six-membered chelate rings, the bioxazolines (**4**), a class of neutral ligands with π-acceptor properties forming five-membered chelate rings, and the anionic methylenebis(oxazolines) of type (**3**) and (**5**) are representative examples.

Enantioselective Cyclopropanation of Alkenes. Cationic Cu^I complexes of methylenebis(oxazolines) such as (**1**), which have been developed by Evans and co-workers,[3] are remarkably efficient catalysts for the cyclopropanation of terminal alkenes with diazoacetates. The reaction of styrene with ethyl diazoacetate in the presence of 1 mol % of catalyst, generated in situ from ***Copper(I) Trifluoromethanesulfonate*** and ligand (**1**), affords the *trans*-2-phenylcyclopropanecarboxylate in good yield and with 99% ee (eq 3). As with other catalysts, only moderate *trans*/*cis* selectivity is observed. Higher *trans*/*cis* selectivities can be obtained with more bulky esters such as 2,6-di-*t*-butyl-4-methylphenyl[3] or dicyclohexylmethyl diazoacetate[5] (94:6 and 95:5, respectively). The efficiency of this catalyst system is illustrated by the cyclopropanation of isobutene, which has been carried out on a 0.3 molar scale using 0.1 mol % of catalyst derived from the (*R,R*)-enantiomer of ligand (**1**) (eq 4).[3] The remarkable selectivity of >99% ee exceeds that of Aratani's catalyst[16] which is used in this reaction on an industrial scale.

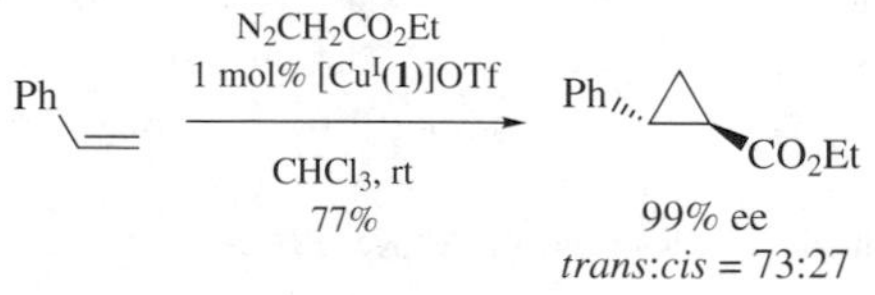

$N_2CH_2CO_2Et$, 1 mol% $[Cu^I(\mathbf{1})]OTf$, $CHCl_3$, rt, 77%; 99% ee, *trans*:*cis* = 73:27 (3)

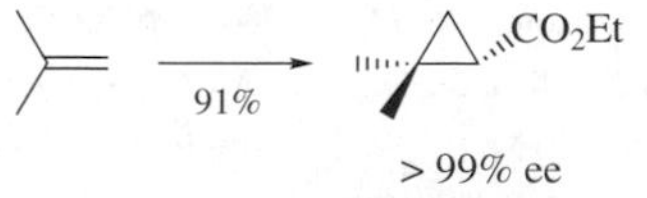

91%; > 99% ee (4)

For the cyclopropanation of terminal mono- and disubstituted alkenes, the cationic Cu^I complex derived from ligand (**1**) is clearly the most efficient catalyst available today, giving consistently higher enantiomeric excesses than related neutral semicorrin[1,17] or bisoxazoline Cu^I complexes of type (**3**),[1,2,4] which can induce enantiomeric excesses of up to 92% ee in the cyclopropanation of styrene with ethyl diazoacetate. High enantioselectivities, ranging between the selectivities of the Evans catalyst (eq 3) and complex (**3**) ($M = Cu^{I}$, $R = t$-Bu), have also been observed with cationic Cu^{I} complexes of azasemicorrins.[1,10a,18]

For analogous cyclopropanation reactions of trisubstituted and 1,2-disubstituted (*Z*)-alkenes, ligand (**1**) is less well suited. In these cases, better results have been obtained with the bisoxazoline ligand (**6**).[5] This is illustrated by the enantioselective cyclopropanation of 1,5-dimethyl-2,4-hexadiene, leading to chrysanthemates (eq 5).[5] The enantioselectivity in this reaction is comparable to the best results reported for Aratani's catalyst.[16] Ligand (**6**) has also been reported to induce high enantiomeric excesses in the cyclopropanation of (*Z*)-4,4-dimethyl-2-pentene, (*Z*)-1-phenylpropene, and 1,1-dichloro-4-methyl-1,3-pentadiene with (−)-menthyl diazoacetate (92–95% ee).[5] A mechanistic model rationalizing the stereoselectivity of Cu catalysts of this type has been published;[17] a comparison of different cyclopropanation catalysts is also available.[19]

(**6**)

N_2CHCO_2R, 1 mol% $[Cu^{I}(\mathbf{6})]ClO_4$, CH_2Cl_2, 0 °C, 78% (5)

CO_2R

R = dicyclohexylmethyl

94% ee

trans:*cis* = 95:5

Enantioselective Aziridination of Alkenes. Copper complexes with neutral methylenebis(oxazoline) ligands (**1**) and (**2**) have also been employed as enantioselective catalysts for the reaction of alkenes with (*N*-tosylimino)phenyliodinane, leading to *N*-tosylaziridines.[6] The best results have been reported for cinnamate esters as substrates, using 5 mol % of catalyst prepared from CuOTf and the phenyl-substituted ligand (**2**) (eq 6). The highest enantiomeric excesses are obtained in benzene, whereas in more polar and Lewis basic solvents, such as acetonitrile, the selectivities are markedly lower. The chemical yield can be substantially improved by addition of 4Å molecular sieves. Both Cu^{I}– and Cu^{II}–bisoxazoline complexes, prepared from Cu^{I} or Cu^{II} triflate, respectively, are active catalysts, giving similar results. In contrast to the Cu-catalyzed cyclopropanation reactions discussed above, in which only Cu^{I} complexes are catalytically active, here Cu^{II} complexes are postulated as the actual catalysts.[6]

PhI=NTs, 5 mol% CuOTf, 6 mol% (**2**), 4Å mol. sieves, C_6H_6, rt (6)

(**a**) R = Me — 63%; 94% ee

(**b**) R = Ph — 64%; 97% ee

Analogous naphthylacrylates also react with excellent enantioselectivity under these conditions. Styrene and (*E*)-β-methylstyrene afford the corresponding *N*-tosylaziridines with 63 and 70% ee, respectively. For these two substrates, the *t*-butyl-substituted bisoxazoline (**1**) rather than (**2**) proved to be the most effective ligand.

Similarly high enantioselectivities in aziridination reactions of this type have been reported for Cu catalysts with C_2-symmetric diimine ligands, derived from 1,2-diaminocyclohexane and aromatic aldehydes.[20] The best results in this case have been obtained with 7-cyano-2,2-dimethylchromene as substrate (>98% ee). At present, it is difficult to compare the diimine-based with the bisoxazoline-based catalysts because different substrates were examined in these studies, with the exception of styrene which gave very similar results with the two catalysts (66 and 63% ee, respectively).[20,6] Thus further work will be necessary to establish the full scope of these promising catalyst systems.

Enantioselective Diels–Alder Reactions. Methylenebis(oxazoline) complexes of Fe^{III}, Mg^{II}, and more recently also Cu^{II}, have been successfully employed as enantioselective Lewis acid catalysts in Diels–Alder reactions.[8,9] The most promising results have been obtained with Cu^{II} catalysts prepared from ligand (**1**) and ***Copper(II) Trifluoromethanesulfonate*** (eq 7).[9] In the presence of 10 mol % of catalyst in CH_2Cl_2 at −78 °C, acrylimide (**7a**) smoothly reacts with cyclopentadiene to afford the Diels–Alder product (**8a**) in 86% yield with excellent enantio and *endo*/*exo* selectivity. The crotonate derivative (**7b**) is less reactive, but at higher temperature also undergoes highly selective cycloaddition with cyclopentadiene. The fumarate (**7c**) gives similar results. In terms of selectivity and efficiency, this catalyst system can compete against the most effective chiral Lewis acid catalysts developed so far.[21]

(**7**) + cyclopentadiene → (**8**); 5–10 mol% $Cu(OTf)_2$, 5.5–11 mol% (**1**), CH_2Cl_2, 85–92% (7)

(**a**) R = H	−78 °C	>98% ee	*endo*:*exo* = 98:2
(**b**) R = Me	−15 °C	97% ee	*endo*:*exo* = 96:4
(**c**) R = CO_2Et	−55 °C	95% ee	*endo*:*exo* = 94:6

The thiazolidine-2-thione analogs of (**7b**) and (**7c**) are more reactive dienophiles and, therefore, the cycloaddition can be carried out at lower temperature. However, the selectivities and yields are similar as with (**7b**) and (**7c**).[9] The corresponding cinnamate derivative (**7**) (R = Ph), on the other hand, reacts with substantially lower enantioselectivity than the corresponding thiazolidine-2-thione analog (90% vs. 97% ee).

The stereochemical course of these reactions has been rationalized assuming a chelate complex between the (bisoxazoline)Cu catalyst and the dienophile as the reactive intermediate, with square planar coordination geometry of the Cu^{II} ion.[9]

Enantioselective Cyanohydrin Formation. Magnesium complexes formed with the anionic semicorrin-type ligand (**5**) catalyze the addition of ***Cyanotrimethylsilane*** to aldehydes, leading to optically active trimethylsilyl-protected cyanohydrins.[7] In the presence of 20 mol % of the chloromagnesium complex

(**9**), prepared from equimolar amounts of (**5**) and BuMgCl, cyclohexanecarbaldehyde is smoothly converted to the corresponding cyanohydrin derivative with 65% ee. Addition of 12 mol % of the bisoxazoline (**10**) results in a dramatic increase of enantioselectivity to 94% ee (eq 8). Replacement of (**10**) by its enantiomer reduces the selectivity to 38% ee. This remarkable effect has been proposed to arise from hydrogen-bond formation between the bisoxazoline (**10**) and HCN, which is generated in small amounts by hydrolysis of Me_3SiCN due to traces of water present in the reaction mixture. The chiral [(**10**)···HCN] aggregate is postulated as the reactive species undergoing nucleophilic addition to the aldehyde which, at the same time, is activated by coordination with the chiral magnesium complex (**9**).

(**9**) (**10**)

Reagents: Me_3SiCN, 20 mol% (**9**), 12 mol% (**10**); CH_2Cl_2, EtCN, –78 °C; 86–94% (8)

(**a**) R = C_6H_{13} 95% ee
(**b**) R = Et_2CH 91% ee
(**c**) R = Cy 94% ee

Heptanal, 2-ethylbutanal, and pivalaldehyde react with similarly high enantioselectivities, whereas benzaldehyde (52% ee) and certain α,β-unsaturated aldehydes such as geranial (63% ee) afford considerably lower enantiomeric excesses. Most other catalysts used for the addition of HCN or Me_3SiCN to aldehydes usually exhibit higher enantioselectivities with aromatic or α,β-unsaturated aldehydes than with alkyl carbaldehydes.[22]

Enantioselective Allylic Alkylation. Most ligands that have been employed in enantioselective Pd-catalyzed allylic substitutions are chiral diphosphines.[23] Recently, it has been found that chiral nitrogen ligands can also induce high enantioselectivities in such reactions.[1,18,24] The best results have been obtained with neutral azasemicorrin and methylenebis(oxazoline) ligands. In the presence of 1–2 mol % of catalyst, generated in situ from ***Bis(allyl)di-μ-chlorodipalladium*** and ligand (**11**), and a mixture of ***N,O-Bis(trimethylsilyl)acetamide*** (BSA) and catalytic amounts of KOAc in an apolar solvent like CH_2Cl_2 or toluene, racemic 1,3-diphenyl-2-propenyl acetate smoothly reacts with dimethyl malonate to afford the corresponding substitution product in high yield and with excellent enantioselectivity (eq 9).

(**11**)

Reagents: $CH_2(CO_2Me)_2$, [Pd(**11**)(C_3H_5)]Cl (2 mol%); BSA, KOAc, CH_2Cl_2, rt; 97% (9)

(*rac*) 97% ee

More recently, even higher selectivities of up to 99% ee have been achieved in this reaction with chiral phosphinooxazolines (see ***(S)-2-[2-(Diphenylphosphino)phenyl]-4-phenyloxazoline***).[24–26] The application range of (bisoxazoline)Pd catalysts is limited to relatively reactive substrates such as aryl-substituted allylic acetates.[24] Analogous reactions of 1,3-dialkyl-2-propenyl acetates, for example, are impracticably slow and unselective. In this case, phosphinooxazolines have proved to be the ligands of choice.[24,25]

The crystal structures of some (allyl)Pd^{II}–bisoxazoline complexes have been determined by X-ray analysis.[1] The structural data of these complexes provide some clues about how the chiral ligand controls the stereochemical course of eq 9.

Other Applications. In the reactions discussed so far, methylenebis(oxazolines) were found to be superior to bioxazolines of type (**4**). However, there are some enantioselective metal-catalyzed processes for which the bioxazolines (**4**) are better suited than neutral or anionic methylenebis(oxazolines). Two examples, the Ir-catalyzed transfer hydrogenation of aryl alkyl ketones[4] and the Rh-catalyzed hydrosilylation of acetophenone,[11] are given in eqs 10 and 11.

Reagents: 0.5 mol% $[Ir(cod)Cl]_2$, 2–4 mol% KOH, *i*-PrOH, 80 °C; (**12**) (1.3 mol%) (10)

(**a**) R = Me 58% ee
(**b**) R = Et 74% ee
(**c**) R = *i*-Pr 91% ee

Reagents: 1. Ph_2SiH_2, CCl_4, $[(cod)RhCl]_2$ (0.5 mol%); (**13**) (10 mol%); 2. H^+, H_2O; 59% (11)

84% ee

Using 1 mol % of catalyst generated in situ from ***Di-μ-chlorobis(1,5-cyclooctadiene)diiridium(I)*** and the bioxazoline (**12**) in refluxing isopropanol, various aryl alkyl ketones have been reduced in good yield with enantioselectivities ranging between 50–90% ee (eq 10).[10b] Dialkyl ketones are unreactive under these conditions. The highest enantiomeric excesses are obtained with phenyl isopropyl ketone (91% ee at 70% conversion, 88% ee at 93% conversion). Although these results compare favorably with the enantioselectivities reported for other Ir catalysts,[27] at present, (bioxazoline)Ir complexes cannot compete with the most efficient catalysts available for the enantioselective reduction of ketones[28] (see ***Tetrahydro-1-methyl-3,3-diphenyl-1H,3H-pyrrolo[1,2-c][1,3,2]oxazaborole***).Recently, high enantioselectivities in the transfer hydrogenation of certain aryl alkyl ketones have been achieved with chiral samarium catalysts.[29]

The dibenzylbioxazoline derivative (**13**) has been found to induce up to 84% ee in the Rh-catalyzed hydrosilylation of acetophenone with diphenylsilane (eq 11).[11] A large excess of ligand relative to [Rh] is necessary for optimal selectivity. Analogous bithiazoline derivatives were also investigated, but gave lower selectivities. In this case too, there are more selective catalysts available which afford high enantiomeric excesses in the hydrolsilylation of a wide range of ketones.[30]

Bioxazolines have also been employed in the enantioselective dihydroxylation of alkenes with ***Osmium Tetroxide***.[15] The best results have been obtained in the dihydroxylation of 1-phenylcyclohexene with a complex, formed between OsO_4 and the diisobutylbioxazoline (**4**) ($R = CH_2CHMe_2$), as a stoichiometric reagent (70% ee). Styrene and *trans*-stilbene afford enantioselectivities below 20% ee under these conditions (for highly enantioselective dihydroxylation catalysts,[31] see ***Dihydroquinine Acetate*** and ***Osmium Tetroxide***).

Related Reagents. (*R*)-1,1′-Bi-2,2′-naphthotitanium Dichloride; (*R*)-1,1′-Bi-2,2′-naphthotitanium Diisopropoxide; (1*S*, 9*S*)-1, 9-Bis[(*t*-butyl)dimethylsilyloxy]methyl-5-cyanosemicorrin; Dihydroquinine Acetate; 2,2-Dimethyl-α,α,α′,α′-tetraphenyl-1,3-dioxolane-4,5-dimethanolatotitanium Diisopropoxide.

1. Pfaltz, A. *ACR* **1993**, *26*, 339.
2. Lowenthal, R. E.; Abiko, A.; Masamune, S. *TL* **1990**, *31*, 6005.
3. (a) Evans, D. A.; Woerpel, K. A.; Hinman, M. M.; Faul, M. M. *JACS* **1991**, *113*, 726. (b) Evans, D. A.; Woerpel, K. A.; Scott, M. J. *AG* **1992**, *104*, 439; *AG(E)* **1992**, *31*, 430.
4. Müller, D.; Umbricht, G.; Weber, B.; Pfaltz, A. *HCA* **1991**, *74*, 232.
5. Lowenthal, R. E.; Masamune, S. *TL* **1991**, *32*, 7373.
6. Evans, D. A.; Faul, M. M.; Bilodeau, M. T.; Anderson, B. A.; Barnes, D. M. *JACS* **1993**, *115*, 5328.
7. Corey, E. J.; Wang, Z. *TL* **1993**, *34*, 4001.
8. (a) Corey, E. J.; Imai, N.; Zhang, H.-Y. *JACS* **1991**, *113*, 728. (b) Corey, E. J.; Ishihara, K. *TL* **1992**, *33*, 6807.
9. Evans, D. A.; Miller, S. J.; Lectka, T. *JACS* **1993**, *115*, 6460.
10. (a) Umbricht, G. Dissertation, University of Basel, 1993. (b) Müller, D. W. Dissertation, University of Basel, 1993.
11. Helmchen, G.; Krotz, A.; Ganz, K. T.; Hansen, D. *SL* **1991**, 257.
12. Witte, H.; Seeliger, W. *LA* **1974**, 996.
13. Hall, J.; Lehn, J.-M.; DeCian, A.; Fischer, J. *HCA* **1991**, *74*, 1.
14. Onishi, M.; Isagawa, K. *ICA* **1991**, *179*, 155.
15. Yang, R.; Chen, Y.; Dai, L. *Acta Chim. Sinica* **1991**, *49*, 1038 (*CA* **1992**, *116*, 41 342v).
16. Aratani, T. *PAC* **1985**, *57*, 1839.
17. (a) Fritschi, H.; Leutenegger, U.; Pfaltz, A. *HCA* **1988**, *71*, 1553. (b) Piqué, C. Dissertation, University of Basel, 1993.
18. Leutenegger, U.; Umbricht, G.; Fahrni, C.; von Matt, P.; Pfaltz, A. *T* **1992**, *48*, 2143.
19. (a) Doyle, M. P. In *Catalytic Asymmetric Synthesis*; Ojima, I., Ed.; VCH: New York, 1993; pp 63–99. (b) Doyle, M. P. *RTC* **1991**, *110*, 305.
20. Li, Z.; Conser, K. R.; Jacobsen, E. N. *JACS* **1993**, *115*, 5326.
21. Kagan, H. B.; Riant, O. *CRV* **1992**, *92*, 1007. Narasaka, K. *S* **1991**, 1.
22. (a) North, M. *SL* **1993**, 807. (b) Hayashi, M.; Miyamoto, Y.; Inoue, T.; Oguni, N. *JOC* **1993**, *58*, 1515.
23. (a) Consiglio, G.; Waymouth, R. M. *CRV* **1989**, *89*, 257. (b) Howarth, J.; Frost, C. G.; Williams, J. M. J. *TA* **1992**, *3*, 1089.
24. von Matt, P. Dissertation, University of Basel, 1993.
25. von Matt, P.; Pfaltz, A. *AG* **1993**, *105*, 614; *AG(E)* **1993**, *32*, 566.
26. (a) Sprinz, J.; Helmchen, G. *TL* **1993**, *34*, 1769. (b) Dawson, G. J.; Frost, C. G.; Williams, J. M. J.; Coote, S. J. *TL* **1993**, *34*, 3149.
27. Zassinovich, G.; Mestroni, G.; Gladali, S. *CRV* **1992**, *92*, 1051.
28. (a) Singh, V. K. *S* **1992**, 605. (b) Wallbaum, S.; Martens, J. *TA* **1992**, *3*, 1475.
29. Evans, D. A.; Nelson, S. G.; Gagné, M. R.; Muci, A. R. *JACS* **1993**, *115*, 9800.
30. Brunner, H.; Nishiyama, H.; Itoh, K. In *Catalytic Asymmetric Synthesis*; Ojima, I., Ed.; VCH: New York, 1993; pp 303–322.
31. (a) Johnson, R. A.; Sharpless, K. B. In *Catalytic Asymmetric Synthesis*; Ojima, I., Ed.; VCH: New York, 1993; pp 227–272. (b) Lohray, B. B. *TA* **1992**, *3*, 1317.
32. Evans, D. A.; Peterson, G. S.; Johnson, J. S.; Barnes, D. M.; Campos, K. R.; Woerpel, K. A. *JOC* **1998**, *63*, 4541.

Andreas Pfaltz
University of Basel, Switzerland

trans-2,5-Dimethylpyrrolidine

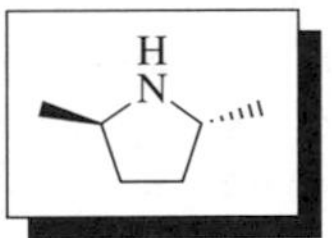

(racemate)		
[62617-69-0; 39713-72-9]	$C_6H_{13}N$	(MW 99.18)
(2*S*,5*S*)		
[117968-50-0]		
(2*R*,5*R*)		
[62617-70-3]		
(·HCl, racemate)	$C_6H_{14}ClN$	(MW 135.64)
[114143-75-8; 4832-49-9]		
(·HCl, 2*S*,5*S*)		
[138133-34-3]		
(·HCl, 2*R*,5*R*)		
[70144-18-2]		

(C_2 symmetric chiral pyrrolidine,[1] useful in optically active form as a chiral auxiliary in a variety of asymmetric reactions)

Physical Data: free amine: bp 102–103 °C; (2*S*,5*S*) $[\alpha]^{25}_D$+10.6° (*c* 1.0, EtOH);[2] (2*R*,5*R*) $[\alpha]^{25}_D$ −11.5° (*c* 1.0, EtOH).[2] Hydrochloride: racemate mp 187–189 °C;[3] (2*S*,5*S*) mp 200–201 °C,[4] $[\alpha]^{25}_D$ −5.63° (*c* 0.67, CH_2Cl_2);[4] (2*R*,5*R*) mp 200–203 °C,[5] $[\alpha]^{25}_D$ +5.57° (*c* 1.18, CH_2Cl_2).[5]

Form Supplied in: colorless oil; commercially available as a mixture of (±)-*trans* and *cis* isomers (the mixture is not easily separated).[6]

Purification: the free amine can be purified by fractional distillation; the hydrochloride salt can be recrystallized from absolute ethanol and diethyl ether.

Handling, Storage, and Precautions: irritant; flammable. Use in a fume hood.

Synthesis. Several routes are available for the synthesis of *trans*-2,5-dimethylpyrrolidine.[2–9,22] Discussed below are preparative scale procedures for the synthesis of the pure *trans* compound in racemic and enantiomerically pure form.

The racemic hydrochloride salt can be prepared in four steps and 70% overall yield (eq 1).[3] The synthesis is carried out on 2 mmol scale and starts with commercially available 5-hexen-2-one. The key step involves a mercury-catalyzed intramolecular amidomercuration to form the pyrrolidine ring. If desired, the racemate can be resolved via the salts of ***Mandelic Acid***.[2]

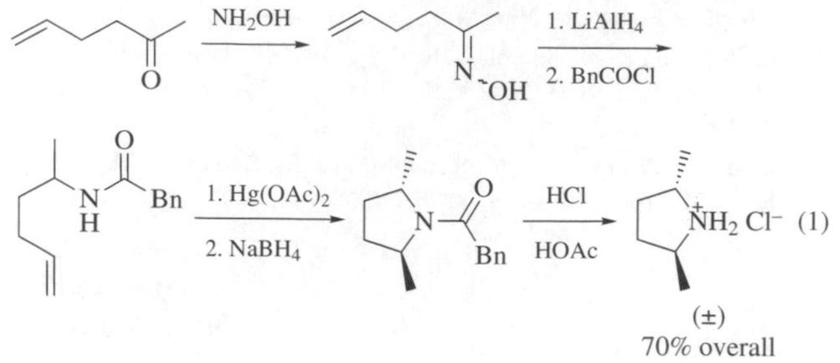

NH_2OH; 1. $LiAlH_4$ 2. BnCOCl; 1. $Hg(OAc)_2$ 2. $NaBH_4$; HCl HOAc; (±) 70% overall (1)

Alternatively, an efficient synthesis of either antipode starting from D- or L-alanine has been reported (eq 2).[9] The asymmetric synthesis conducted on 10 mmol scale involves a six-step sequence which incorporates the amidomercuration method.[3] The enantiomerically pure product is isolated as its hydrochloride salt in 44% overall yield. Furthermore, an optimization of the capricious cuprate reaction which improves both the yield and reproducibility has been described.[4]

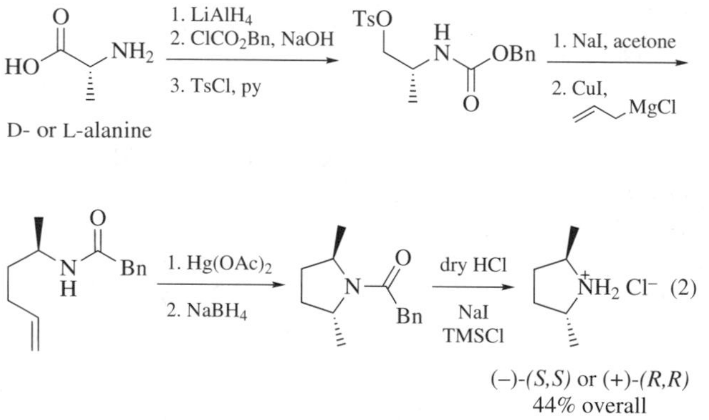

D- or L-alanine; 1. $LiAlH_4$ 2. $ClCO_2Bn$, NaOH 3. TsCl, py; 1. NaI, acetone 2. CuI, MgCl; 1. $Hg(OAc)_2$ 2. $NaBH_4$; dry HCl NaI TMSCl; (−)-(*S,S*) or (+)-(*R,R*) 44% overall (2)

More recently, a four-step synthetic sequence which provides expedient access to the (−)-(*R,R*)-enantiomer in 42% overall yield has been reported.[5] This route is convenient for large-scale preparation (0.2 mol scale), and is highlighted by an asymmetric ***Baker's Yeast*** reduction of 2,5-hexanedione. Subsequent mesylation, *N,N*-dialkylation, and deprotection provides the enantiomerically pure free pyrrolidine (eq 3). Alternatively, either enantiomer of the chiral pyrrolidine can be obtained in 15% overall yield from an isomeric mixture of 2,5-hexanediol, via a similar sequence in which (*S*)-α-methylbenzylamine is used as a chiral auxiliary.[22] Also, an enantioselective route to either (2*S*,5*S*)- or (2*R*,5*R*)-hexanediol has been reported.[23]

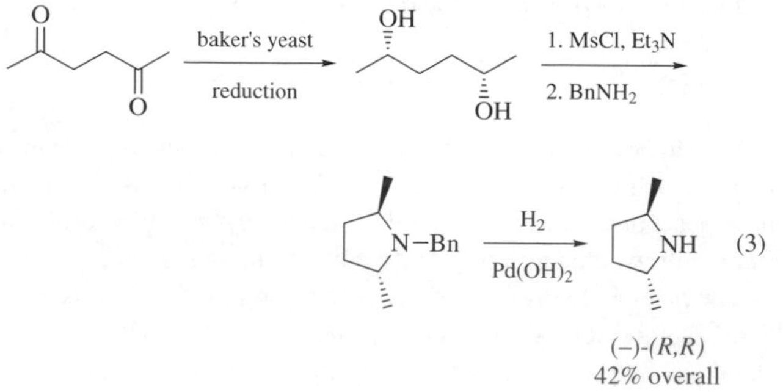

baker's yeast reduction; 1. MsCl, Et_3N 2. $BnNH_2$; H_2 $Pd(OH)_2$; (−)-(*R,R*) 42% overall (3)

Asymmetric Alkylations and Michael Additions. Asymmetric alkylation of the cyclohexanone enamine derived from (+)-*trans*-2,5-dimethylpyrrolidine has been studied (eq 4).[2] Alkylation with ***Iodomethane***, *n*-propyl bromide, and ***Allyl Bromide*** afforded the corresponding 2-*n*-alkylcyclohexanones in yields of 50–80% and with enantiomeric purities of 66, 86, and 64%, respectively.

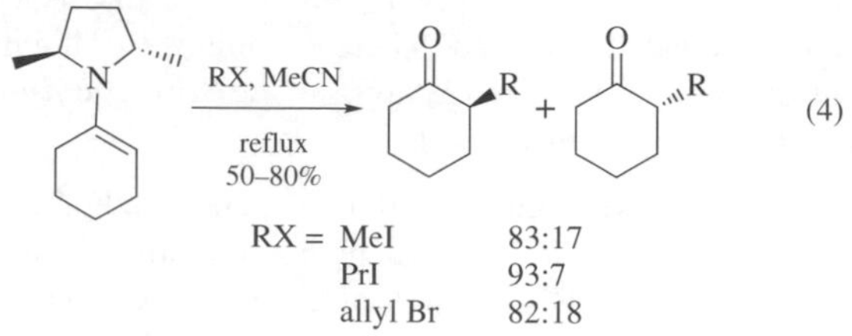

RX, MeCN reflux 50–80% (4)

RX =	
MeI	83:17
PrI	93:7
allyl Br	82:18

The lithium enolates from tetronic acid-derived vinylogous urethanes have been generated and their reactivity investigated with a variety of electrophiles (eq 5).[10,11] The reactions proceed with excellent regio- and diastereoselectivity and a variety of alkylating agents can be utilized.

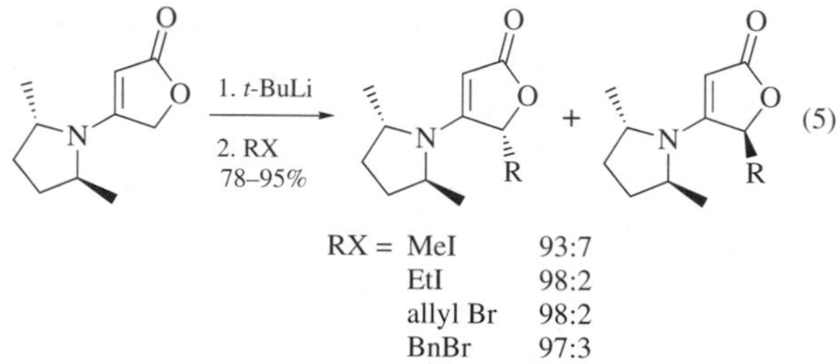

1. *t*-BuLi 2. RX 78–95% (5)

RX =	
MeI	93:7
EtI	98:2
allyl Br	98:2
BnBr	97:3

In the total synthesis of (−)-secodaphniphylline an asymmetric [1,4]-conjugate addition was used to establish relative and absolute stereocontrol.[12] The lithium enolate of a *trans*-2,5-dimethylpyrrolidine-derived amide adds in a Michael fashion to a cyclic α,β-unsaturated ester, with subsequent enolate trapping, to afford the desired product in 64% yield and 92:8 diastereoselection (eq 6).

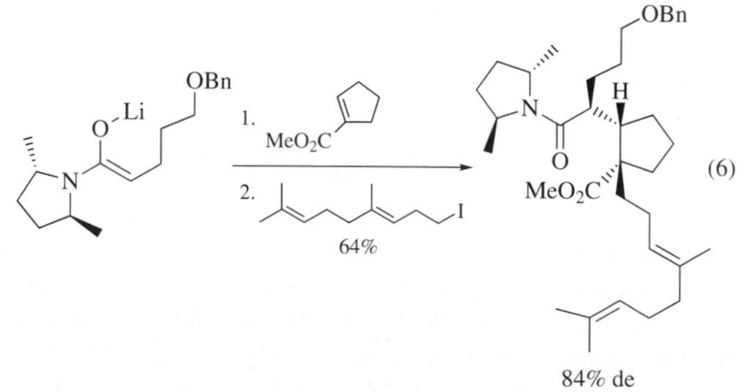

1. MeO_2C (cyclopentene) 2. (geranyl iodide) 64%; 84% de (6)

Asymmetric Radical Reactions. Several reports have documented the utility of nonracemic *trans*-2,5-dimethylpyrrolidine as a chiral auxiliary in asymmetric radical reactions.[13] For example, the addition of *n*-hexyl, cyclohexyl, and *t*-butyl radicals to the chiral acrylamide of 4-oxopentenoic acid provided four diastereomeric products resulting from α- and β-addition (eq 7).[14] The isomers resulting from β-addition were formed with no diastereoselectivity; however, the isomers resulting from α-addition were formed in ratios of 16:1, 24:1, and 49:1. Unfortunately, the application of this chemistry is limited due to the poor regioselectivity in the addition and difficulty in removal of the chiral auxiliary.

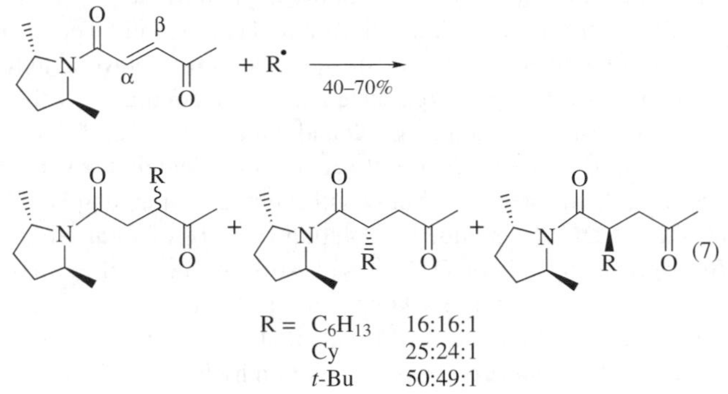

R =		
	C_6H_{13}	16:16:1
	Cy	25:24:1
	t-Bu	50:49:1

(7)

Similar results have been achieved in the addition of chiral amide radicals to activated alkenes.[13] For instance, a chiral amide radical, derived from (−)-*trans*-2,5-dimethylpyrrolidine, adds in a 1,4-fashion to ethyl acrylate in 35% yield and with 12:1 diastereoselectivity (eq 8).[15] Unfortunately, substantial amounts of higher oligomers are also formed. The radical telomerization of chiral acrylamides to afford nonracemic lower-order telomers ($n = 1$–5) has also been described.[16]

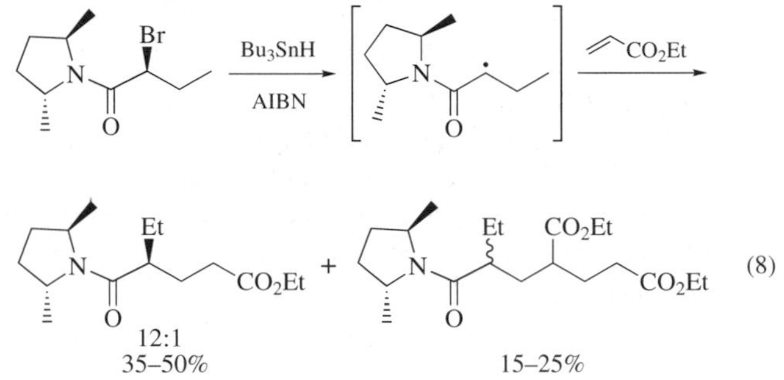

(8)

Asymmetric Pericyclic Reactions. Several reports illustrate the utility of *trans*-2,5-dimethylpyrrolidine as a chiral auxiliary in asymmetric Claisen-type rearrangements,[17] [4 + 2],[18,19] and [2 + 2] cycloaddition reactions.[20] The enantioselective Claisen-type rearrangement of *N,O*-ketene acetals derived from *trans*-2,5-dimethylpyrrolidine has been studied.[17] For example, the rearrangement of the *N,O*-ketene acetal, formed in situ by the reaction of *N*-propionyl-*trans*-(2*S*,5*S*)-dimethylpyrrolidine with (*E*)-crotyl alcohol, affords the [3,3]-rearrangement product in 50% yield and 10:1 diastereoselectivity (eq 9).

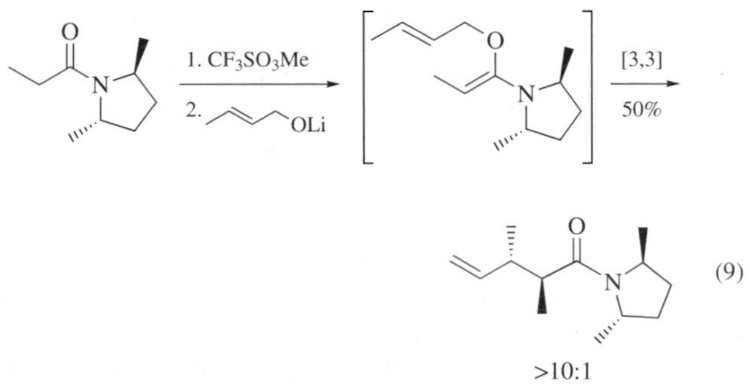

(9)

Carbamoyl nitroso dienophiles, derived from chiral pyrrolidines, have been generated and their reactivity with cyclohexadiene investigated.[18] Using (−)-*trans*-2,5-dimethylpyrrolidine as the auxiliary, the [4 + 2] cycloadduct is isolated in 82% yield and with 98% diastereomeric excess (eq 10). Similarly, chiral ynamine dienophiles have been utilized in asymmetric [4 + 2] cycloadditions with α,β-unsaturated nitroalkenes to afford cyclic nitronic esters.[19] The resulting esters subsequently undergo a rapid [1,3]-rearrangement to afford chiral cyclic nitrones in moderate yield and high diastereoselectivity (eq 11).

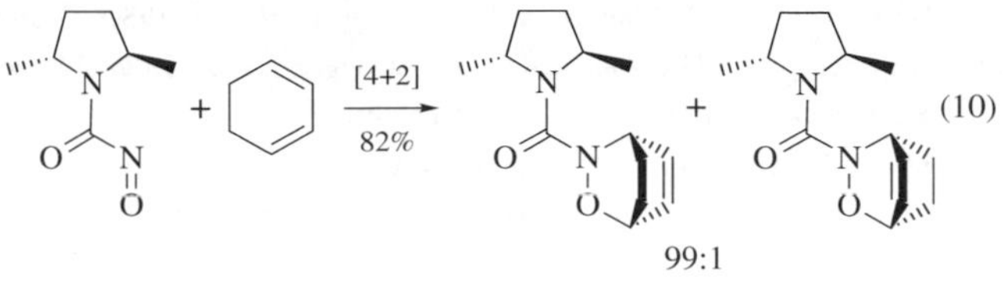

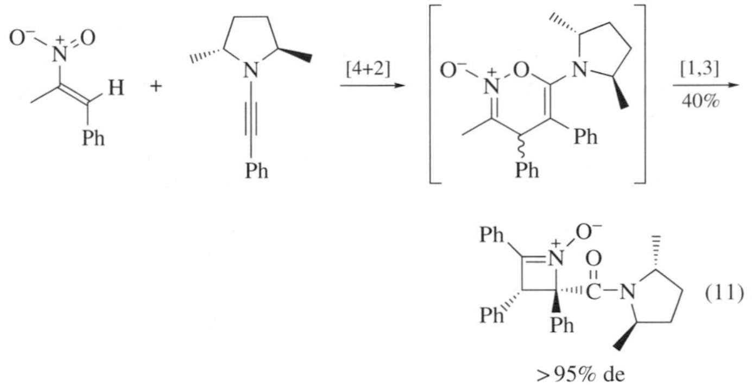

(10)

(11)

An asymmetric, thermal [2 + 2] cycloaddition of keteniminium salts derived from *trans*-2,5-dimethylpyrrolidine has been employed in the synthesis of prostaglandins.[20] An intramolecular [2 + 2] cycloaddition affords a *cis*-fused bicyclic system which is then further transformed into a common prostaglandin intermediate (eq 12).

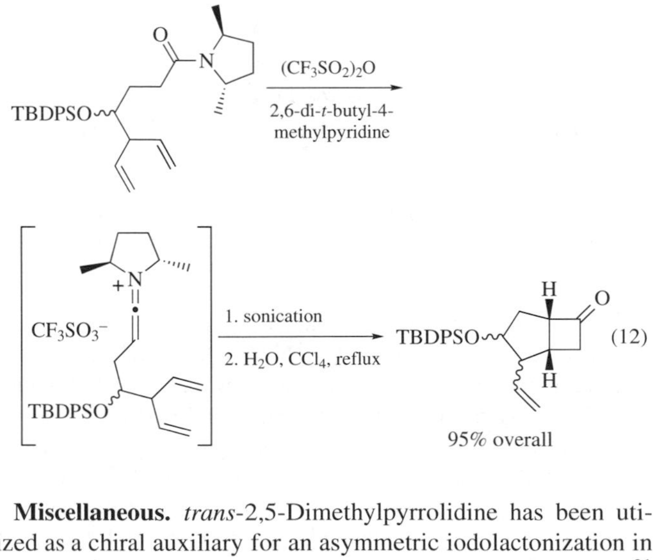

(12)

Miscellaneous. *trans*-2,5-Dimethylpyrrolidine has been utilized as a chiral auxiliary for an asymmetric iodolactonization in the total synthesis of (±)-pleurotin and (±)-dihydropleurotin.[21] The reaction affords the desired lactone in 47% yield and only 30% enantiomeric excess.

Related Reagents. trans-2,5-Bis(methoxymethyl)pyrrolidine.

1. Whitesell, J. K. *CRV* **1989**, *89*, 1581.
2. Whitesell, J. K.; Felman *JOC* **1977**, *42*, 1663.

3. Harding, K. E.; Burks, S. R. *JOC* **1981**, *46*, 3920.
4. Yamazaki, T.; Gimi, R.; Welch, J. T. *SL* **1991**, 573.
5. Short, R. P.; Kennedy, R. M.; Masamune, S. *JOC* **1989**, *54*, 1755.
6. House, H. O.; Lee, L. F. *JOC* **1976**, *41*, 863.
7. Dervan, P. B.; Uyehara, T. *JACS* **1976**, *98*, 2003.
8. Gagné, M. R.; Stern, C. L.; Marks, T. J. *JACS* **1992**, *114*, 275.
9. Schlessinger, R. H.; Iwanowicz, E. J. *TL* **1987**, *28*, 2083.
10. Schlessinger, R. H.; Iwanowicz, E. J.; Springer, J. P. *JOC* **1986**, *51*, 3070.
11. Schlessinger, R. H.; Iwanowicz, E. J.; Springer, J. P. *TL* **1988**, *29*, 1489.
12. Heathcock, C. H.; Stafford, J. A. *JOC* **1992**, *57*, 2566.
13. Porter, N. A.; Giese, B.; Curran, D. P. *ACR* **1991**, *24*, 296.
14. Porter, N. A.; Scott, D. M.; Rosenstein, I. J.; Giese, B.; Veit, V.; Zeitz, H. G. *JACS* **1991**, *113*, 1791.
15. Porter, N. A.; Swann, E.; Nally, J.; McPhail, A. T. *JACS* **1990**, *112*, 6740.
16. Porter, N. A.; Breyer, R.; Swann, E.; Nally, J.; Pradhan, J.; Allen, T.; McPhail, A. T. *JACS* **1991**, *113*, 7002.
17. Yamazaki, T.; Welch, J. T.; Plummer, J. S.; Gimi, R. H. *TL* **1991**, *32*, 4267.
18. Defoin, A.; Brouillard-Poichet, A.; Streith, J. *HCA* **1991**, *74*, 103.
19. Elburg, P. A.; Honig, G. W. N.; Reinhoudt, D. N. *TL* **1987**, *28*, 6397.
20. Chen, L.-Y.; Ghosez, L. *TA* **1991**, *2*, 1181.
21. Hart, D. J.; Huang, H.-C.; Krishnamurthy, R.; Schwartz, T. *JACS* **1989**, *111*, 7507.
22. Mariël, E. Z.; Meetsma, A.; Feringa, B. L. *TA* **1993**, *4*, 2163.
23. Burk, M. J.; Feaster, J. E.; Harlow, R. L. *TA* **1991**, *2*, 569.

Lawrence R. Marcin
University of Illinois at Urbana-Champaign, IL, USA

Dimethylsulfonium Methylide

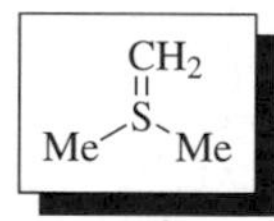

[6814-64-8] C_3H_8S (MW 76.18)

(exceedingly selective methylene transfer reagent capable of converting carbonyl compounds into oxiranes,[1] and imines into aziridines[2])

Form Supplied in: reactive intermediate generated in situ.

Preparative Methods: Method A.[1a,b] A solution of dimethylsulfonium methylide is readily prepared by addition of a solution of ***Trimethylsulfonium Iodide*** in ***Dimethyl Sulfoxide*** (800 mL per mol of sulfonium salt) with stirring (below 5 °C, because of the marked thermal instability of the methylide) to an equivalent molar amount of a 1.5–2 M solution of methylsulfinyl carbanion,[3] which is prepared by treatment of dry DMSO with ***Sodium Hydride*** under nitrogen followed by dilution with an equal amount of dry THF (to prevent solidification).

Method B.[1b] Alternatively, a solution of ***n-Butyllithium*** (11 mmol) in 8 mL of pentane may be added dropwise to a stirred suspension of 2.45 g (12 mmol) of powdered trimethylsulfonium iodide in 30 mL of dry THF under nitrogen at 0 °C.

Method C.[4] To a solution of ***Dimethyl Sulfate*** (189.2 g, 1.50 mol) in acetonitrile (700 mL) is added a solution of ***Dimethyl Sulfide*** (102.5 g, 1.65 mol) in acetonitrile (300 mL) with stirring at rt. After standing overnight, ***Sodium Methoxide*** (89.1 g, 1.65 mol) is added to the mixture at rt to give a solution of dimethylsulfonium methylide, which is successfully applied to the preparation of oxiranes from various kinds of carbonyl compound. In comparison with the former methods, this process is more favorable for preparing oxiranes on a large scale under mild conditions without using expensive alkylating agents and intractable solvents.

Method D.[1d] To a mixture of 0.1 mol of aldehyde (or ketone) and 22 g (0.14 mol) of trimethylsulfonium bromide dissolved in 60 mL of DMSO is added a solution of 14 g of ***Potassium t-Butoxide*** in 60 mL of DMSO with stirring and cooling under nitrogen over 30–45 min. After workup, the expected oxiranes can be separated and purified by distillation or recrystallization. In this method, ***Sodium Amide*** or sodium hydride can be used as a base instead of potassium *t*-butoxide, and DMF may be used as solvent. Benzalaniline may be converted by this method into the corresponding aziridine.

Method E.[5] Dimethylsulfonium methylide can also be obtained by phase-transfer catalysis. Thus trimethylsulfonium iodide does not react with benzaldehyde in the system CH_2Cl_2–NaOH, but on addition of 1–5 mol % of ***Tetra-n-butylammonium Iodide***, 2-phenyloxirane is obtained in >90% yield via the intermediate sulfonium methylide. Yields of oxiranes are rather low (20–35%) when ketones are used in this procedure.

Formation of Oxiranes. Dimethylsulfonium methylide reacts with a variety of aldehydes and ketones by net methylene transfer to form oxiranes. Thus several compounds, including benzophenone, benzaldehyde, and cycloheptanone were converted into the corresponding oxiranes by selective addition of methylene to the carbonyl group (eqs 1–3).[1b]

Ph(Ph)C=O —($Me_2S=CH_2$, 84%)→ 2,2-Ph,Ph-oxirane (1)

Ph(H)C=O —($Me_2S=CH_2$, 75%)→ 2-Ph-oxirane (2)

cycloheptanone —($Me_2S=CH_2$, 97%)→ spiro-oxirane (3)

It should be noted that dimethylsulfonium methylide prefers addition to the carbonyl group of α,β-unsaturated carbonyl compounds while ***Dimethylsulfoxonium Methylide***, which is a closely related reagent for methylene transfer to unsaturated compounds, tends to attack the α,β-double bond. For example, carvone was converted to the corresponding oxirane by treatment with the sulfonium ylide, whereas the sulfoxonium ylide reacts with carvone to give the cyclopropyl ketone (eq 4).[1b]

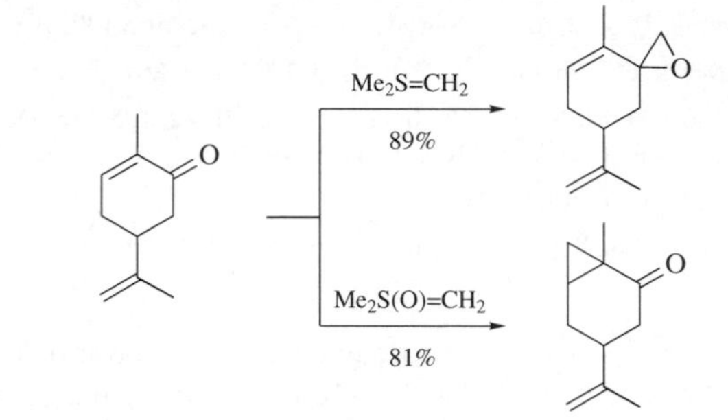

(4)

Similarly, selective oxirane formation was achieved in the cases of other α,β-unsaturated ketones such as eucarvone (93%), benzalacetone (87%), and pulegone (90%).[1b]

Epoxidation of β-ionone can also be achieved by the treatment of a mixture of trimethylsulfonium chloride (24 g), β-ionone (38.4 g), CH_2Cl_2 (100 mL), and ***Benzyltriethylammonium Chloride*** (1 g) with aqueous NaOH (18 M, 125 mL) at rt (eq 5).[6]

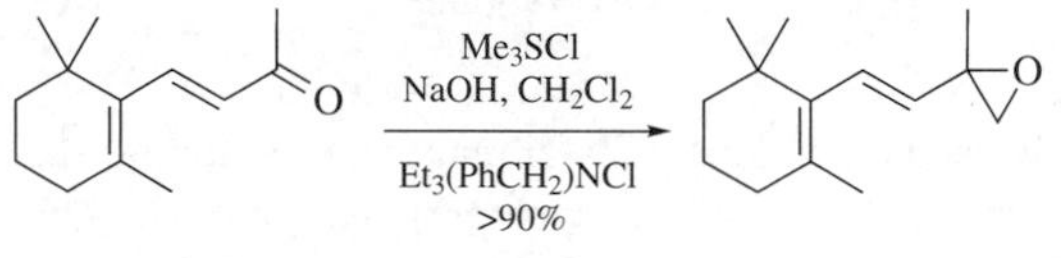

(5)

No epoxide is obtained, however, by reaction of β-ionone with dimethylsulfonium methylide generated by the conventional methods using other trimethylsulfonium halides[5,7] or trimethylsulfonium methylsulfate[4] as a precursor.

Dimethylsulfonium methylide undergoes stereoselective addition of two methylene units to 9,10-anthraquinone to afford the corresponding *trans*-bisepoxide (eq 6).[8]

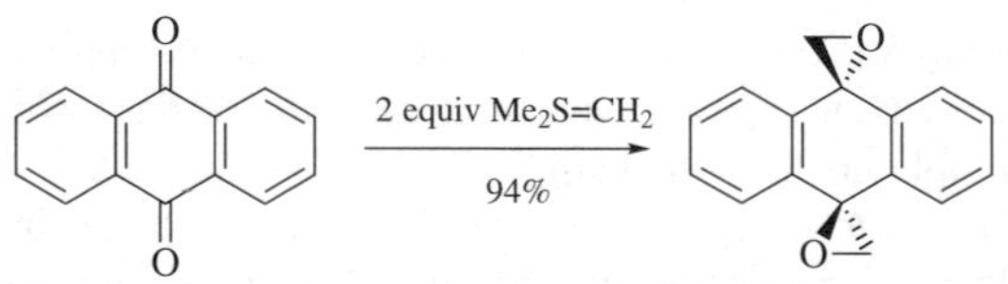

(6)

Asymmetric Oxirane Synthesis. A striking asymmetric synthesis of 2-phenyloxirane can be achieved by the reaction of benzaldehyde and dimethylsulfonium methylide generated from trimethylsulfonium iodide in 50% NaOH with the chiral phase-transfer catalyst (−)-*N*,*N*-dimethylephedrinium bromide (0.2 equiv) (eq 7).[9]

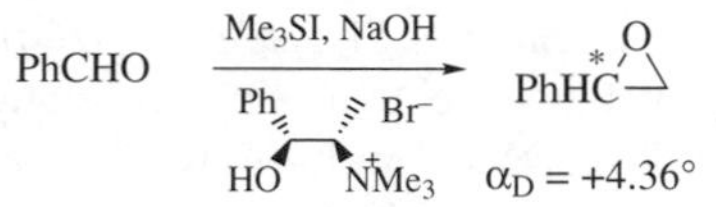

(7)

The enantiomeric excess of oxirane in eq 7 is 67%. If the asymmetric salt derived from ψ-ephedrine is used, the enantiomeric oxirane is formed preferentially. In this system the choice of the solvent is important. Little asymmetric induction is observed with THF or acetonitrile as solvent, while the use of benzene results in a high degree of induction.

Synthesis of Butenolides. The reaction of a cyclic α-keto ketene dithioacetal with dimethylsulfonium methylide followed by acid hydrolysis produced the corresponding butenolide in high yield (eq 8).[10]

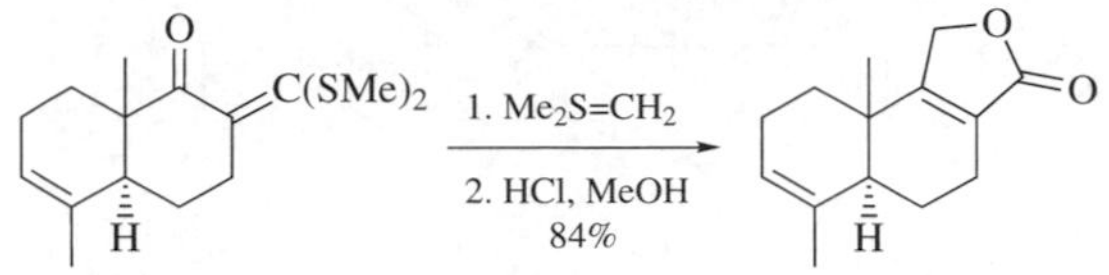

(8)

The reaction of dimethylsulfonium methylide with acyclic α-keto ketene dithioacetals also provides a simple synthesis of dihydrofurans, which can be converted into various furans and butenolides (eqs 9 and 10).[11]

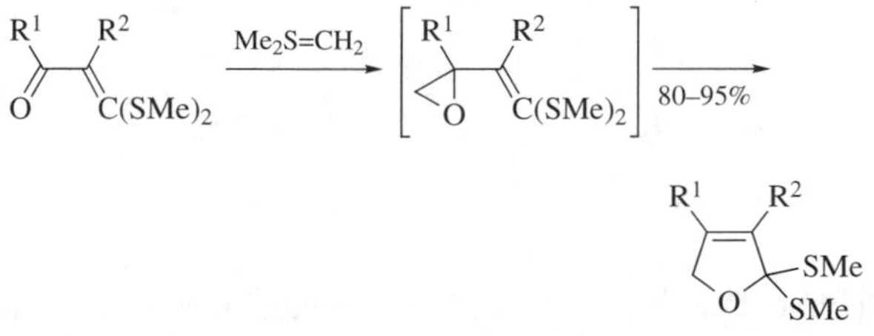

(9)

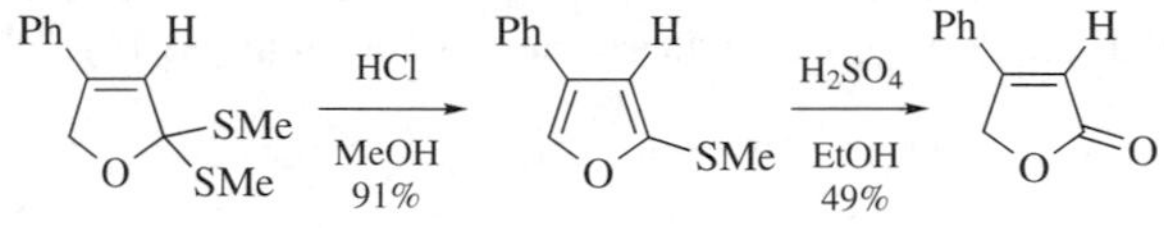

(10)

Cyclopropanation via Methylene Transfer to Carbon–Carbon Double Bonds. In contrast to α,β-unsaturated ketones, ethyl cinnamate, an α,β-unsaturated carboxylic acid ester, was transformed into 2-phenyl-1-cyclopropane carboxylate (eq 11).[1d]

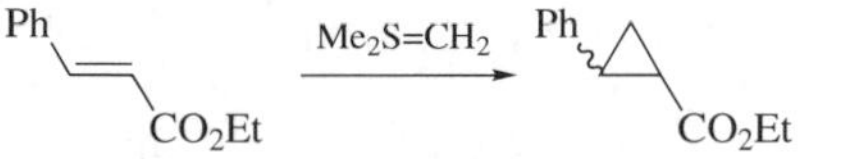

(11)

Since dimethylsulfonium methylide is a far more powerful methylene transfer reagent than dimethylsulfoxonium methylide, it can convert 1,1-diphenylethylene into 1,1-diphenylcyclopropane in 60% yield though it is necessary to use an excess amount of methylide (5 equiv) (eq 12).[1b] However, dimethylsulfonium methylide is inactive toward tolan or *trans*-stilbene, and unreactive toward nonconjugated carbon–carbon double bond compounds.

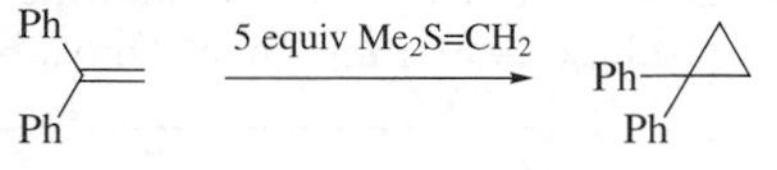

(12)

Methylation of Aromatic Rings. Dimethylsulfonium methylide reacts with acenaphthylene in DMSO–THF to give 3-methylacenaphthylene together with some 5-methylacenaphthylene (eq 13).[12] Similarly, fluoranthrene gives a mixture of 1- and 3-methylfluoranthrenes upon treatment with dimethylsulfonium methylide (eq 14).[12]

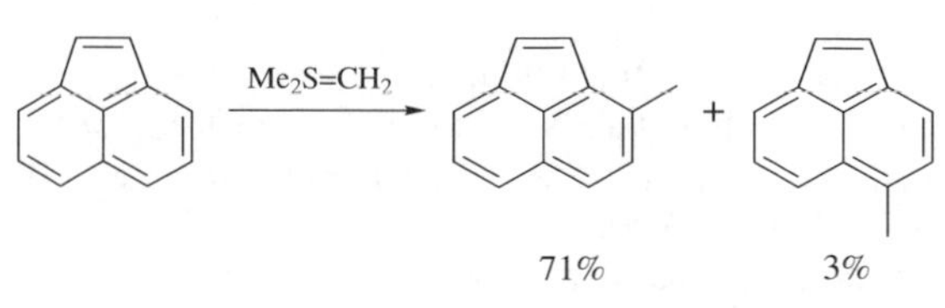

(13)

$Me_2S{=}CH_2$ 25% + 5% (14)

***N*-Methylation of Indoles.** Indoles can be selectively *N*-methylated by dimethylsulfonium methylide in THF at rt (eq 15).[13]

R^1, R^2, NH → $Me_2S{=}CH_2$ → R^1, R^2, N–Me (15)

Aziridine Formation. Dimethylsulfonium methylide reacts smoothly with imines. For example, benzalaniline can be converted to the corresponding aziridine in 91% yield (eq 16).[1b,d]

Ph, Ph, H, =N → $Me_2S{=}CH_2$, 91% → Ph, Ph, H, N (16)

Using this method the first known heterocyclic analog of bicyclobutane, 3-phenyl-1-azabicyclo[1.1.0]butane, has been obtained by the reaction of 3-phenyl-2*H*-azirine with dimethylsulfonium methylide in dry THF at −10 °C (eq 17).[2]

Ph, N → $Me_2S{=}CH_2$, 60% → Ph, N (17)

Alkene Formation. Dimethylsulfonium methylide reacts with benzyl chloride in the presence of excess amount of base to give styrene (eq 18).[1d]

$$PhCH_2Cl \xrightarrow[t\text{-BuOK}]{Me_2S=CH_2} PhCH{=}CH_2 \quad (18)$$

Related Reagents. Dimethyl Sulfide-Chlorine; Dimethyl Sulfoxide; Dimethylsulfoxonium Methylide; Thioanisole.

1. (a) Corey, E. J.; Chaykovsky, M. *JACS* **1962**, *84*, 3782. (b) Corey, E. J.; Chaykovsky, M. *JACS* **1965**, *87*, 1353. (c) Franzen, V.; Driesen, H. E. *TL* **1962**, 661. (d) Franzen, V.; Driesen, H. E. *CB* **1963**, *96*, 1881.
2. Hortmann, A. G.; Robertson, D. A. *JACS* **1967**, *89*, 5974. See also refs. 1b and 1d.
3. Corey, E. J.; Chaykovsky, M. *JACS* **1962**, *84*, 866.
4. Kutsuma, T.; Nagayama, I.; Okazaki, T.; Sakamoto, T.; Akaboshi, S. *H* **1977**, *8*, 397.
5. Merz, A.; Märkl, G. *AG(E)* **1973**, *12*, 845.
6. Rosenberger, M.; Jackson, W.; Saucy, G. *HCA* **1980**, *63*, 1665.
7. (a) Yoshimine, M.; Hatch, M. J. *JACS* **1967**, *89*, 5831. (b) Hatch, M. J. *JOC* **1969**, *34*, 2133. See also ref. 5.
8. McCarthy, T. J.; Connor, W. F.; Rosenfeld, S. M. *SC* **1978**, *8*, 379.
9. Hiyama, T.; Mishima, T.; Sawada, H.; Nozaki, H. *JACS* **1975**, *97*, 1626.
10. Garver, L. C.; van Tamelen, E. E. *JACS* **1982**, *104*, 867.
11. (a) Okazaki, R.; Negishi, Y.; Inamoto, N. *CC* **1982**, 1055. (b) Okazaki, R.; Negishi, Y.; Inamoto, N. *JOC* **1984**, *49*, 3819.
12. Trost, B. M. *TL* **1966**, 5761.
13. Bravo, P.; Gaudiano, G.; Umani-Ronchi, A. *G* **1970**, *100*, 652.

Renji Okazaki & Norihiro Tokitoh
The University of Tokyo, Japan

Dimethylsulfoxonium Methylide[1]

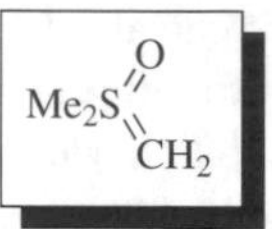

[6814-64-8] C_3H_8OS (MW 92.16)

(widely used methylene transfer reagent; used for the epoxidation of ketones and aldehydes,[1] cyclopropanation of α,β-unsaturated carbonyl compounds[1] and also as a methylation reagent[2])

Physical Data: pK_a of sulfoxonium ion ($Me_3\overset{+}{S}(O)$) at 25 °C in DMSO is 18.2;[3] mp 9–10 °C, bp 41–43 °C/0.1 mmHg.[4]
Solubility: sol THF, DMSO, 1,4-dioxane, DMF.
Preparative Method: usually generated in situ by the reaction of trimethylsulfoxonium iodide or chloride with strong base in a appropriate solvent without drying or further purification; not commercially available.
Analysis of Reagent Purity: determined by adding a small quantity of the reagent to water and titrating with standard acid using phenolphthalein as indicator.[5]
Handling, Storage, and Precautions: reactions using this nucleophilic reagent should be carried out under an inert atmosphere; it is more stable than ***Dimethylsulfonium Methylide*** and can be stored in solution under N_2 at 0 °C for months without appreciable decomposition.

Preparation. Dimethylsulfoxonium methylide is a stabilized ylide which is generally prepared and used at rt.[1,6,7] Typically, the reagent is obtained by the reaction of finely powdered ***Trimethylsulfoxonium Iodide*** (or trimethylsulfoxonium chloride) with a strong base such as ***Sodium Hydride*** or ***n-Butyllithium*** under nitrogen in a suitable solvent. A variety of anhydrous solvents such as DMSO, THF, DMF, and 1,4-dioxane are reported to work well for this reaction. When NaH is used as the base, after hydrogen evolution has completed, the resulting ylide is ready for further reaction.[8]

Recently, a less hazardous modification for large-scale production has been reported. This method involves the reaction of ***Potassium t-Butoxide*** with trimethylsulfoxonium iodide in DMSO. Use of potassium *t*-butoxide eliminates the hazards of handling sodium hydride on a large scale, and also eliminates problems with the isolation of products which may be contaminated with the mineral oil from sodium hydride dispersions.[9] Preparation of dimethylsulfoxonium methylide can also be achieved by using *t*-BuOK in *t*-butanol.[10] Alternatively, dimethylsulfoxonium methylide is accessible via a desilylation process.[11]

Preparations of sulfur ylides under phase-transfer catalyzed conditions have also been reported. One such system involved the use of CH_2Cl_2, ***Tetra-n-butylammonium Iodide*** (TBAI), NaOH, and trimethylsulfoxonium iodide.[12] *N*-Benzyl-*N,N,N*-trimethylammonium chloride has also been reported as an efficient catalyst for this reaction.[13] An alternative system using trimethylsulfoxonium iodide, cetrimide (cetyltrimethylammonium bromide), aqueous NaOH, and 1,1,1-trichloroethane has also been reported.[14] Usually, the phase-transfer catalyzed reactions are performed at elevated temperatures from 50 to 75 °C.

Epoxidation. Dimethylsulfoxonium methylide is commonly used for epoxide preparation by reaction with a variety of compounds which contain an aldehyde or a ketone group (eqs 1–5).[1,3,7,8,15–19] Normally very high yields and purities of the desired epoxides are obtained. Use of trimethylsulfoxonium iodide instead of trimethylsulfonium iodide eliminates the need for removal of the noxious byproduct Me_2S. The byproduct DMSO generated from dimethylsulfoxonium methylide preparation can be readily removed by using standard organic/aqueous extractions. The reaction involves initial nucleophilic addition (dimethylsulfoxonium methylide) to the carbonyl group and subsequent elimination of DMSO and epoxide formation. Enolization usually does not compete with nucleophilic addition to the carbonyl group.[3]

CHO; $Me_2S(O)=CH_2$; THF, 50 °C; 56% (1)

O; $Me_2S(O)=CH_2$; DMSO, 50 °C; 71%; O (2)

H; O; $Me_2S(O)=CH_2$; DMSO; 68%; O (3)

O; BnO OBn; $Me_2S(O)=CH_2$; DMSO, rt; 82%; O; BnO OBn (4)

O; N Bn; $Me_2S(O)=CH_2$; DMSO, 50 °C; 65%; O; N Bn (5)

Instead of using strong base in an anhydrous media, epoxides have also been prepared under phase-transfer catalysis conditions with a sulfur ylide.[12–14,20,21] Activated ***Barium Hydroxide*** was reported to catalyze the oxirane formation in interfacial solid–liquid conditions.[22,23] These reactions were performed in MeCN in the presence of $Ba(OH)_2$ with a small amount of H_2O.

The condensation of 1-tetralone with dimethylsulfoxonium methylide affords the corresponding oxirane in 75% yield.[24,25] However, if this parent ketone possesses an aromatic electron-releasing substituent such as a methoxy group, the resulting spiro-oxirane may not be stable and may readily rearrange to give ring-opened compounds.

Dimethylsulfoxonium methylide has also been employed in nucleoside preparations.[26] The reagent has been used to introduce the spiro epoxy group at the C-2′ position of 2′-keto-β-L-nucleosides. In this case, better yields of oxiranes are obtained with the use of trimethylsulfoxonium chloride instead of trimethylsulfoxonium iodide. Variation of the base used has a major effect on the course of the reaction. Only 2-L-galacto-spiro-epoxynucleoside is obtained if the dimethylsulfoxonium methylide is prepared from BuLi (eq 6).[27]

O; R; O; O; O; $Me_2S(O)=CH_2$; THF, rt; O; R; O; O; O; 40%; +; O; R; O; O; O; 40% (6)

R = thyminyl, theophyllinyl

In some examples, selective epoxidations were observed in the reaction of dimethylsulfoxonium methylide with dicarbonyl compounds. Selective condensation (83% yield) of dimethylsulfoxonium methylide with the six-membered-ring carbonyl group of *trans*-1,6-dimethylbicyclo[4.3.0]nonane-2,7-dione was observed, while the cyclopentanone remained unreacted (eq 7).[28]

O; O; $Me_2S(O)=CH_2$; DMSO, rt; 83%; O; O (7)

The further transformation of the product epoxides to other functionalities by treatment with nucleophiles or Lewis acids has been widely reported. Epoxidation of quadricyclanone and adamantanone with dimethylsulfoxonium methylide provided spiro-oxirane products in good yields.[29,30] In the case of adamantanone, the product oxirane was further converted to the rearranged acid by reaction with ***Boron Trifluoride Etherate*** followed by oxidation with Jones' reagent (eq 8). Similarly, reaction of dimethylsulfoxonium methylide with methylheptenone followed by treatment of $BF_3 \cdot OEt_2$ provided the rearranged 2-isopropyl-5-methylcyclopentanone (eq 9).[17]

O; $Me_2S(O)=CH_2$; DMSO, rt; 91%; O; 1. $BF_3 \cdot OEt_2$ benzene, rt; 2. CrO_3 aq acetone, rt; 79%; CO_2H; H (8)

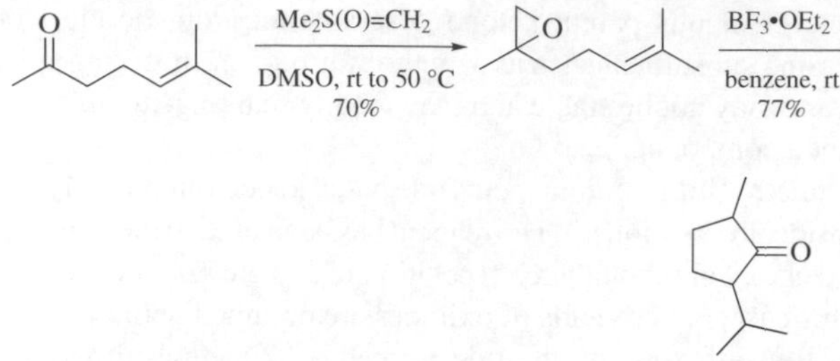

$Me_2S(O)=CH_2$, DMSO, rt to 50 °C, 70%; $BF_3•OEt_2$, benzene, rt, 77% (9)

Reactions of hydroxybenzopyranones, hydroxyphenylalkanones, and 2-(*o*-hydroxyphenyl)alkyl ketones with dimethylsulfoxonium methylide provide an efficient route to oxygen heterocycles (eq 10).[31–33] The mechanism involves an intramolecular ring opening of the oxirane intermediate.

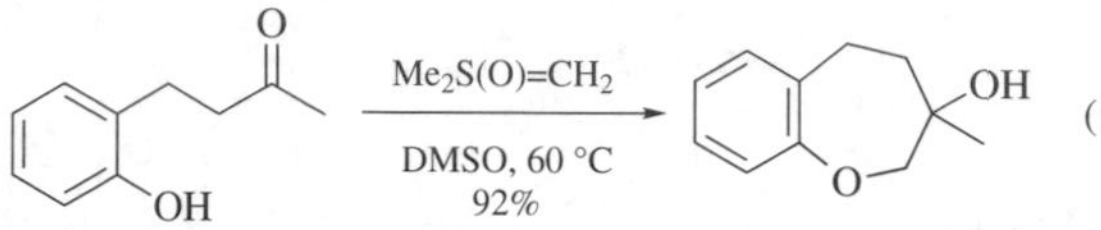

$Me_2S(O)=CH_2$, DMSO, 60 °C, 92% (10)

Stereoselective Epoxide Synthesis. Dimethylsulfoxonium methylide has been reported to show higher stereoselectivity in reactions with aldehydes and ketones than its sulfonium analog. Reaction of 4-*t*-butylcyclohexanone with dimethylsulfoxonium methylide gave exclusively the *cis*-oxirane, while reaction with ***Dimethylsulfonium Methylide*** gave products with a ratio of 17:83 of the *trans* and *cis* isomers (eq 11).[1,34] Presumably the sulfoxonium ylide provided the kinetically controlled product, while the smaller sulfonium ylide gave a mixture of the kinetically and thermodynamically favored products. Similarly, epoxidation of 4-protoadamantanone with dimethylsulfoxonium methylide gave almost exclusively the *exo* isomer (*exo*:*endo* = 15:1), while the dimethylsulfonium methylide provided a mixture of the *exo* and *endo* isomers in a 3:2 ratio (eq 12).[35]

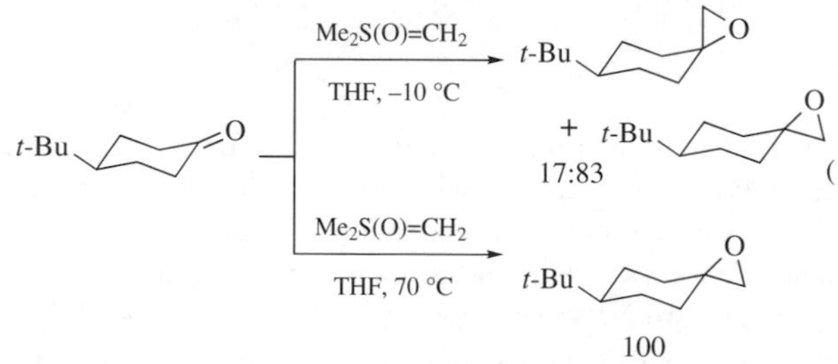

t-Bu; $Me_2S(O)=CH_2$, THF, –10 °C → *t*-Bu + *t*-Bu, 17:83; $Me_2S(O)=CH_2$, THF, 70 °C → *t*-Bu, 100 (11)

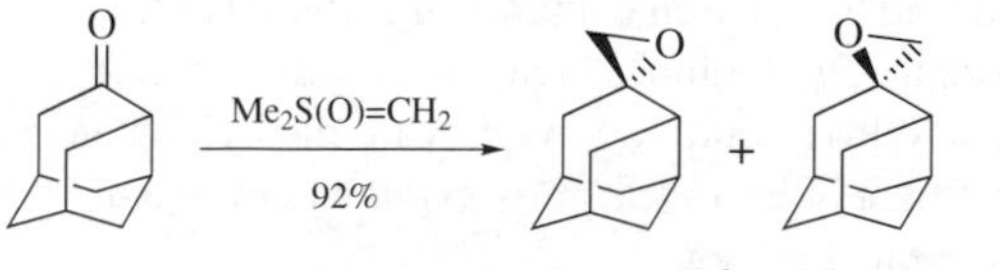

$Me_2S(O)=CH_2$, 92% (12)

exo:*endo* = 15:1

Several examples of conversion of chiral ketones and aldehydes to epoxides with high diastereomeric purity have been reported.[36–38] Stereoselective reaction of the nucleoside 2′-keto-3′,5′-*O*-(tetraisopropyldisiloxane-1,3-diyl)uridine with dimethylsulfoxonium methylide at 0 °C afforded only one isomer of the product epoxide in 63% yield.[39] Reaction of the heterocyclic carbonyl compounds with dimethylsulfoxonium methylide also provided the corresponding oxiranes, such as 1-oxa-6-heterospiro[2,5]octanes, in good yields and isomeric purities.[40,41]

One example showed that dimethylsulfoxonium methylide reacted chemoselectively with the five-membered ketone ring in the presence of the six-membered carbonyl function to give the monoepoxide in good yield (eq 13).

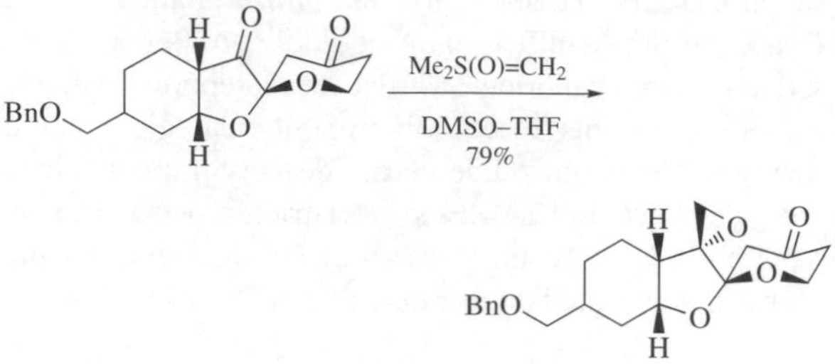

BnO; $Me_2S(O)=CH_2$, DMSO–THF, 79% → BnO (13)

In some cases, solvent appears to affect the stereoselectivity of the oxirane formation. Reaction of 3α,18-dihydroxy-17-noraphidicolan-16-one with dimethylsulfoxonium methylide in THF gives a 1:1 mixture of epoxides. When THF is replaced by DMSO as the solvent, the isomeric ratio is improved to 1:3. In the mixed solvent DMSO/DMI system, the ratio is further improved to 1:4.[42]

Generation of New Sulfur Ylides. Treatment of cyclic β-chloro-substituted enones and chloropyrimidines with dimethylsulfoxonium methylide results in the formation of new sulfoxonium ylides.[43–47] Dimethylsulfoxonium methylide also reacts with acid anhydride,[48] acid chloride,[49] isocyanate,[50] and other activated carbonyl groups[51,52] to generate new sulfur ylides for further reactions. This method has been used to prepare acylated stable ylides[53] which can be converted to other products or further reacted with a second mole of the α,β-unsaturated carbonyl compounds (eqs 14–16).

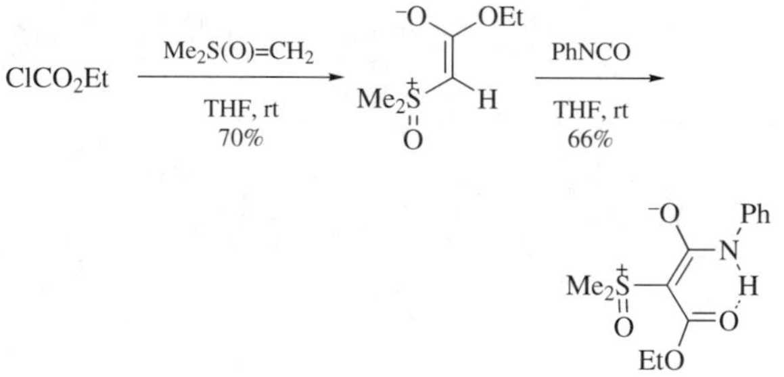

$ClCO_2Et$; $Me_2S(O)=CH_2$, THF, rt, 70%; PhNCO, THF, rt, 66% (14)

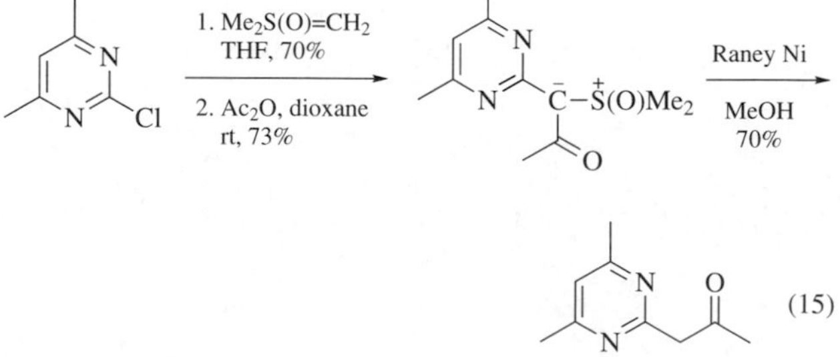

1. $Me_2S(O)=CH_2$, THF, 70%; 2. Ac_2O, dioxane, rt, 73%; Raney Ni, MeOH, 70% (15)

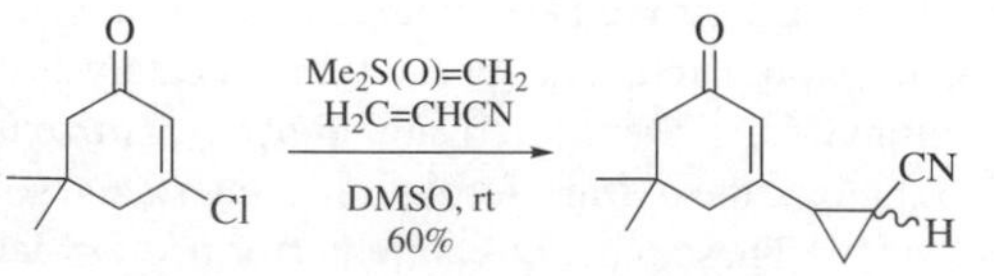

$Me_2S(O)=CH_2$, $H_2C=CHCN$, DMSO, rt, 60% (16)

Methylation. Dimethylsulfoxonium methylide is a useful methylation reagent for the conversion of acids,[54] oximes,[55]

N-heterocycles,[56] and aromatic hydrocarbons[57] to the corresponding methylated products. For example, nitrobenzene and 6-benzyladenine were methylated with dimethylsulfoxonium methylide in 67% and 63% yields (eqs 17 and 18).[2]

NO_2 — $Me_2S(O){=}CH_2$, DMSO, rt, 67% — (17)

Bn NH — $Me_2S(O){=}CH_2$, THF, rt, 63% — (18)

Preparation of Azetidine Derivatives. *N*-Arylsulfonyl-2-phenylazetidines can be prepared by reaction of dimethylsulfoxonium methylide with *N*-arylsulfonyl-2-phenylaziridines in 51–72% yield.[58] Stereospecific conversion of *N*-arylsulfonylaziridines to *N*-arylsulfonylazetidines has also been reported. Reactions with *cis*-aziridines give *trans*-azetidines, and reactions with *trans*-aziridines produce the *cis*-azetidines. Presumably an S_N2 1,4-elimination mechanism is involved (eq 19).[59] Although several 2- and 2,3-substituted *N*-arylsulfonylazetidines have also been synthesized in good yields from reactions with dimethylsulfoxonium methylide, the fused azetidines could not be prepared by this procedure.[60] Dimethylsulfoxonium methylide also converts *N*-arylsulfonyloxaziridines to azetidines via a methylene transfer reaction. Similar reactions with *N*-alkyloxaziridines give only the deoxygenated product.[61]

O_2S–*p*-Tol, Ph, N, Ph — $Me_2S(O){=}CH_2$, DMSO, rt, 78% — O_2S–*p*-Tol, N, Ph, Ph (19)

Preparation of Oxetanes. Dimethylsulfoxonium methylide is an efficient methylene transfer reagent in reactions with terminal epoxides to provide the corresponding oxetanes (eq 20).[10,62,63] The oxetanes can also be prepared from ketones or aldehydes with dimethylsulfoxonium methylide via double methylene transfer reactions.

Ph, Ph, O — $Me_2S(O){=}CH_2$, *t*-BuOH, 50 °C, 99% — Ph, Ph, O (20)

Cyclopropanation. Cyclopropanation using the Simmons–Smith conditions (see ***Iodomethylzinc Iodide***) does not work well for α,β-unsaturated carbonyl compounds. This difficulty has been overcome by the use of dimethylsulfoxonium methylide as the methylene transfer reagent. This method has been successfully applied to unsaturated ketones,[64–66] esters,[67–70] lactones,[71] amides,[72] nitriles,[73] and nitro compounds.[74,75] Dimethylsulfoxonium methylide appears to work better than the corresponding sulfonium analog in these reactions. Reaction of the sulfonium methylide with α,β-unsaturated carbonyl compounds gives the kinetically controlled 1,2-addition oxirane product,[1,66] while reaction with dimethylsulfoxonium methylide gives the 1,4-addition cyclopropane derivatives in synthetically useful yields (eqs 21–23).[64,70]

O — $Me_2S(O){=}CH_2$, DMSO, rt to 50 °C, 65% — O (21)

O, N(Me)OMe — $Me_2S(O){=}CH_2$, DMSO, rt to 50 °C, 88% — O, N(Me)OMe (22)

MEMO, CO_2Et, CO_2Et — $Me_2S(O){=}CH_2$, DMSO–THF, rt to 50 °C, 76% — MEMO, CO_2Et, CO_2Et (23)

The mechanism of cyclopropanation involves an initial reversible conjugate addition of dimethylsulfoxonium methylide to the α,β-unsaturated carbonyl compound, followed by an irreversible ring closure; if excess dimethylsulfoxonium methylide is used in the reaction, epoxidation of carbonyl group may follow the cyclopropanation (eq 24).[76] Usually, the reaction is carried out in DMSO over a wide range of temperatures under an inert atmosphere. Cyclopropanation using TBAI as phase-transfer catalyst in an aqueous system was also reported.[12] Reaction of dimethylsulfoxonium methylide with fluorine-containing α,β-unsaturated ketones can result in products from different reaction pathways.[38,77,78] Reaction of dimethylsulfoxonium methylide with α-fluoro-substituted enones gives the corresponding fluoroepoxides only.[79]

R^1, O, H, H, $C_6H_3R^2R^3$ — excess $Me_2S(O){=}CH_2$, DMSO, rt to reflux — R^1, O, H, H, $C_6H_3R^2R^3$ (24)

R^1 = Ph, C_6H_4Me, C_6H_4OMe, Me
R^2 = H, OMe
R^3 = H, OMe

The cyclopropanated carbonyl compounds can readily be further transformed to other functionalities.[72,80,81] For example, a substituted vinyl phenyl ketone was converted to the cyclopropane intermediate with dimethylsulfoxonium methylide and the product was further transformed to a 1-tetralone derivative by acid-catalyzed cyclization.[80] The cyclopropyl amide which is derived from reaction of an α,β-unsaturated amide with dimethylsulfoxonium methylide can be further converted to a ketone or an acid derivative by reaction with an organometallic nucleophile or an anhydrous base.[72]

Cyclopropanation with dimethylsulfoxonium methylide in multifunctional compounds can be chemoselective.[66,82,83] Dimethylsulfoxonium methylide reacts preferentially with the less sterically hindered α,β-unsaturated carbonyl function to give the corresponding cyclopropane (eq 25).[65,84,85] On the other hand, reaction of dimethylsulfoxonium methylide with bifunctional cyclic enediones affords the oxirane product instead of generating the cyclopropyl derivative (eq 26).[85]

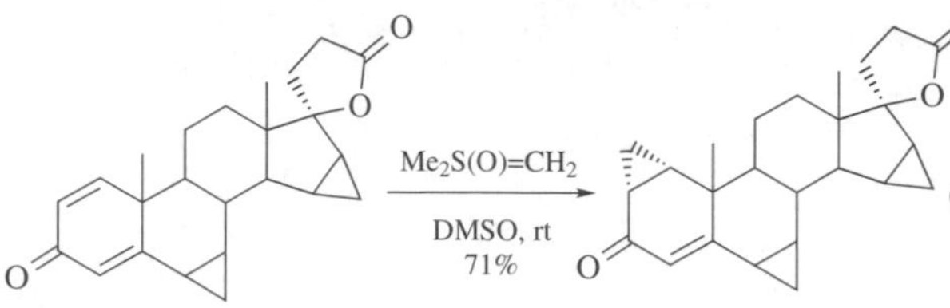

$Me_2S(O)=CH_2$; DMSO, rt; 71% (25)

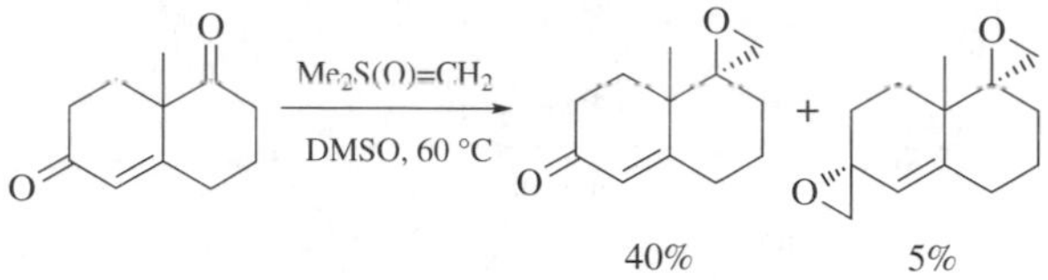

$Me_2S(O)=CH_2$; DMSO, 60 °C; 40% + 5% (26)

Chemoselective reaction of a methoxy-substituted dienone with dimethylsulfoxonium methylide results in cyclopropanation of the unsubstituted double bond.[83] Reaction of dimethylsulfoxonium methylide with the conjugated dienone 2,3-benzotropone gives a 1.1:1 mixture of the mono- and dicyclopropanated products.[86] One conjugated dienone, however, gave only the 1,6-addition product (eq 27).[82]

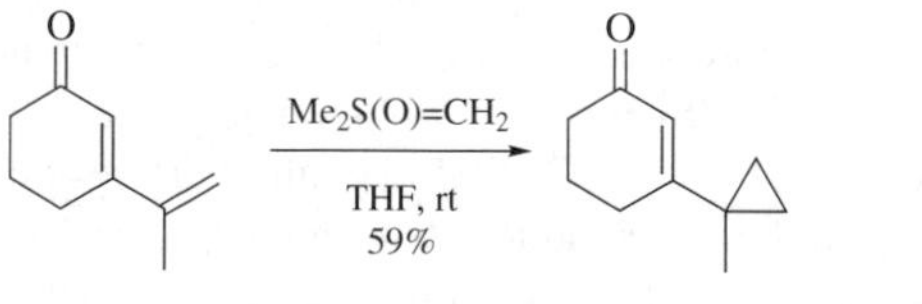

$Me_2S(O)=CH_2$; THF, rt; 59% (27)

Stereoselectivity of Cyclopropanation. Cyclopropanation of α,β-unsaturated carbonyl compounds with dimethylsulfoxonium methylide can be highly stereoselective. Reaction of methyl 2-(4-oxo-2-cyclohexenyl)acetate with dimethylsulfoxonium methylide gave the corresponding *anti*- and *syn*-cyclopropane derivatives in a 85:15 ratio; the favored *anti* isomer results from the nucleophilic addition of the dimethylsulfoxonium methylide to the less hindered side of the α,β-unsaturated ketone (eq 28).[87] This steric effect was also observed in the cyclopropanation of 3β-acetoxy-16α,17α-methylene-5-pregnen-20-one.[88] Stereoselective cyclopropanation of a ten-membered enone with dimethylsulfoxonium methylide gives the bicyclohumulenone in 90% yield, with none of the *cis* isomer observed. Peripheral addition of oxysulfurane proceeds through the lower-energy conformation which leads to the most likely enolate intermediate. Subsequent ring closure gives the *trans* cyclopropane.[89] A recently reported asymmetric cyclopropanation of a chiral vinyl sulfoxide with dimethylsulfoxonium methylide gave high yield and stereoselectivity ((*R,S*):(*S,S*) = 5.9:1) (eq 29).[90]

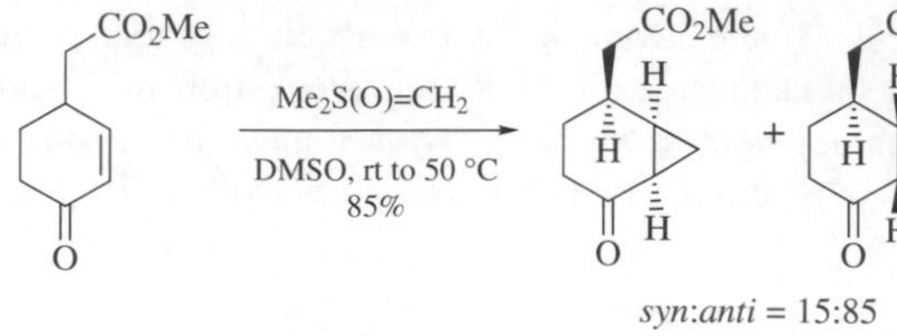

CO_2Me; $Me_2S(O)=CH_2$; DMSO, rt to 50 °C; 85%; CO_2Me + CO_2Me (28)

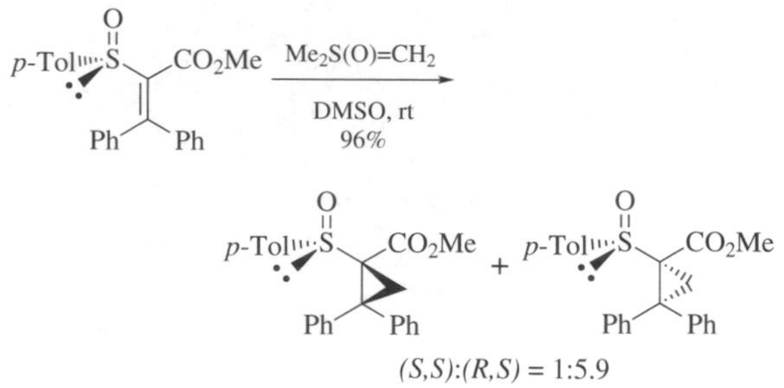

syn:*anti* = 15:85

p-Tol–S(O)–C(CO_2Me)=C(Ph)Ph; $Me_2S(O)=CH_2$; DMSO, rt; 96% (29)

(*S,S*):(*R,S*) = 1:5.9

Cyclopropanation of α,β-unsaturated bicyclic lactams with dimethylsulfoxonium methylide also proceeds with high diastereoselectivity. In these cases, a change of the angular substituent in the bicyclic lactam leads to a complete reversal in *endo–exo* selectivity (eq 30).[91]

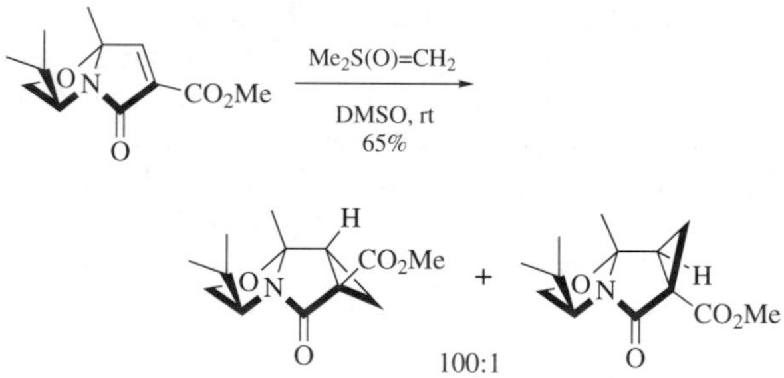

$Me_2S(O)=CH_2$; DMSO, rt; 65% (30)

100:1

Reaction with Other Functional Groups. Reaction of α-halo carbonyl compounds with dimethylsulfoxonium methylide provides the α-cyclopropyl carbonyl derivatives rather than the oxirane.[92] Presumably the mechanism involves a double methylene transfer with the dimethylsulfoxonium methylide. Dimethylsulfoxonium methylide has also been reported to react with 1,3-dipolar compounds for the preparation of heterocyclic products (eq 31).[93]

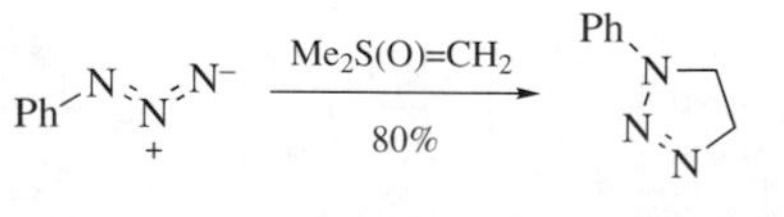

$Ph-N=N^+=N^-$; $Me_2S(O)=CH_2$; 80% (31)

Methylene transfer onto the triple bond of α,β-alkynyl ketones by reaction with dimethylsulfoxonium methylide has not been observed. Reaction of 1,3-diphenyl-2-propyn-1-one with dimethylsulfoxonium methylide gives only the Michael addition complex 1-methyl-3,5-diphenylthiabenzene 1-oxide.[94,95] Reaction of dimethylsulfoxonium methylide with thiobenzophenone gives 1,1-diphenylethylene sulfide in 71% yield.[96] Treatment of 4-dimethylamino-1-thia-3-azabutadienes with dimethylsulfoxonium methylide provided thiazol-2-ine derivatives. The mechanism involves regioselective ylide addition to the amidine group, followed by cyclization with elimination of DMSO (eq 32).[97]

$$\text{PhC(=S)N=CH–NMe}_2 \xrightarrow[\text{DMSO, rt, 70\%}]{\text{Me}_2\text{S(O)=CH}_2} \text{2-Ph-4-NMe}_2\text{-thiazoline} \qquad (32)$$

Methylene transfer of dimethylsulfoxonium methylide to imines results in aziridine derivatives.[98] A facile procedure for the preparation of 3-amino-2,3-dihydrobenzofuran from dimethylsulfoxonium methylide and *o*-hydroxybenzylideneaniline has been reported. The mechanism involves a ring opening of the initial aziridine intermediate (eq 33).[99,100]

$$o\text{-HOC}_6\text{H}_4\text{CH=N-C}_6\text{H}_4\text{Br-}p \xrightarrow[\text{DMSO, rt, 80\%}]{\text{Me}_2\text{S(O)=CH}_2} \text{3-(NH-C}_6\text{H}_4\text{Br-}p\text{)-2,3-dihydrobenzofuran} \qquad (33)$$

Reaction with Organometallics. Dimethylsulfoxonium methylide has been reported to react with transition metal organometallic compounds as simple terminal ligands, bridging groups, or chelating moieties.[101–106] For example, the reaction of two equivalents of dimethylsulfoxonium methylide with the metal-functionalized 1,3-diphospha-2-propanone $Fe_2(CO)_6[(i\text{-}Pr)_2N\text{–}PC(O)P\text{–}N(i\text{-}Pr)_2]$ affords the 1,5-diphospha-3-pentanone derivative, a product resulting from the insertion of a CH_2 group into the P–(CO) bond.[107]

Related Reagents. Dimethylsulfonium Methylide; Dimethyl Sulfoxide; Tetraethylammonium Periodate; Thioanisole.

---

1. Corey, E. J.; Chaykovsky, M. *JACS* **1965**, *87*, 1353.
2. Traynelis, V. J.; McSweeney, J. V. *JOC* **1966**, *31*, 243.
3. Johnson, C. R. In *Comprehensive Organic Chemistry*; Barton, D. H. R., Ed.; Pergamon: Oxford, 1979; Vol. 3, p 247.
4. Schmidbaur, H.; Tronich, W. *TL* **1968**, *51*, 5335.
5. *FF* **1968**, *1*, 315.
6. Gololobov, Y. G.; Nesmeyanov, A. N.; Lysenko, V. P.; Boldeskul, I. E. *T* **1987**, *43*, 2609.
7. Radunz, H. E.; Orth, D.; Baumgarth, M.; Schliep, H. J.; Enenkel, H. J. U.S. Patent 4 309 441, 1982 (*CA* **1982**, *96*, 199 394s).
8. Corey, E. J.; Chaykovsky, M. *OSC* **1973**, *5*, 755.
9. Ng, J. S. *SC* **1990**, *20*, 1193.
10. Wicks, D. A.; Tirrell, D. A. *J. Polym. Sci., Part A: Polym. Chem.* **1990**, *28*, 573.
11. Schmidbaur, H. *ACR* **1975**, *8*, 62.
12. Merz, A.; Märkl, G. *AG* **1973**, *85*, 867.
13. Abarca, B.; Ballesteros, R.; Jones, G. *JHC* **1984**, *21*, 1585.
14. Richardson, K.; Whittle, P. J. U.S. Patent 4 518 604, 1985 (*CA* **1985**, *103*, 178 265y).
15. Ogilvie, K. K.; Nguyen-ba, N.; Hamilton, R. G. *CJC* **1984**, *62*, 1622.
16. Popp, F. D.; Watts, R. F. *JHC* **1978**, *15*, 675.
17. Kulkarni, B. S.; Rao, A. S. *OPP* **1978**, *10*, 73.
18. Anderson, M. W.; Jones, R. C. F. *JCS(P1)* **1986**, *2*, 205.
19. Kongkathip, B.; Sookkho, R.; Kongkathip, N. *CL* **1985**, 1849.
20. Rosenberger, M.; Jackson, W.; Saucy, G. *HCA* **1980**, *63*, 1665.
21. Bouda, M.; Borredon, M. E.; Delmas, M.; Gaset, A. *SC* **1987**, *17*, 503.
22. Sinisterra, J. V.; Marinas, J. M.; Riquelme, F.; Arias, M. S. *T* **1988**, *44*, 1431.
23. Borredon, E.; Clavellinas, F.; Delmas, M.; Gaset, A.; Sinisterra, J. V. *JOC* **1990**, *55*, 501.
24. Crooks, P. A.; Sommerville, R. *CI(L)* **1984**, 350.
25. Moody, C. J.; Beck, A. L.; Coates, W. J. *TL* **1989**, *30*, 4017.
26. Matsuda, A.; Ueda, T. *CPB* **1986**, *34*, 1573.
27. Herscovici, J.; Egron, M. J.; Antonakis, K. *JCS(P1)* **1986**, 1297.
28. Martin, J. L.; Tou, J. S.; Reusch, W. *JOC* **1979**, *44*, 3666.
29. Bly, R. S.; Bly, R. K.; Shibata, T. *JOC* **1983**, *48*, 101.
30. Farcasiu, D. *S* **1972**, 615.
31. Bravo, P.; Ticozzi, C. *G* **1979**, *109*, 169.
32. Salimbeni, A.; Manghisi, E.; Arnone, A. *JHC* **1988**, *25*, 943.
33. Bravo, P.; Ticozzi, C.; Maggi, D. *CC* **1976**, 789.
34. Cambie, R. C.; Rutledge, P. S.; Strange, G. A.; Woodgate, P. D. *H* **1982**, *19*, 1501.
35. Abdel-Sayed, A. N.; Bauer, L. *T* **1988**, *44*, 1873.
36. Hoagland, S.; Morita, Y.; Bai, D. L.; Märki, H. P.; Kees, K.; Brown, L.; Heathcock, C. H. *JOC* **1988**, *53*, 4730.
37. Girijavallabham, V. M.; Ganguly, A. K.; Pinto, P. A.; Sarre, O. Z. *BML* **1991**, *1*, 349.
38. Koolpe, G. A.; Nelson, W. L.; Gioannini, T. L.; Angel, L.; Simon, E. J. *JMC* **1984**, *27*, 1718.
39. Sano, T.; Shuto, S.; Inoue, H. *CPB* **1985**, *33*, 3617.
40. Smith, A. B., III; Fukui, M. *JACS* **1987**, *109*, 1269.
41. Satyamurthy, N.; Berlin, K. D. *PS* **1984**, *19*, 113.
42. Ackland, M. J.; Gorden, J. F.; Hanson, J.; Ratcliffe, A. H. *JCS(P1)* **1988**, 2009.
43. Chalchat, J. C.; Garry, R.; Michet, A.; Vessiere, R. *CR(C)* **1978**, *286*, 329.
44. Bradbury, R. H.; Gilchrist, T. L.; Rees, C. W. *CC* **1979**, 528.
45. Niitsuma, S.; Sakamoto, T.; Yamanaka, H. *H* **1978**, *10*, 171.
46. Yamanaka, H.; Konno, S.; Sakamoto, T.; Niitsuma, S.; Noji, S. *CPB* **1981**, *29*, 2837.
47. Green, G. E.; Irving, E. Eur. Patent 35 969, 1981 (*CA* **1982**, *96*, 69 594w).
48. Johnson, C. R.; Rogers, P. E. *JOC* **1973**, *38*, 1798.
49. Truce, W. E.; Madding, G. D. *TL* **1966**, 3681.
50. König, H.; Metzger, H. *CB* **1965**, *98*, 3733.
51. Nagao, Y.; Inoue, T.; Fujita, E.; Terada, S.; Shiro, M. *T* **1984**, *40*, 1215.
52. Nozaki, H.; Tunemoto, D.; Matubara, S.; Kondo, K. *T* **1967**, *23*, 545.
53. Speakman, P. R. H.; Robson, P. *TL* **1969**, 1373.
54. Metzger, H.; König, H.; Seelert, K. *TL* **1964**, 867.
55. Bravo, P.; Gaudiano, G.; Ticozzi, C.; Umani-Ronchi, A. *CC* **1968**, 1311.
56. Kunieda, T.; Witkop, B. *JOC* **1970**, *35*, 3981.
57. Watson, C. R., Jr.; Pagni, R. M.; Dodd, J. R.; Bloor, J. E. *JACS* **1976**, *98*, 2551.
58. Nadir, U. K.; Koul, V. K. *CC* **1981**, 417.
59. Nadir, U. K.; Sharma, R. L.; Koul, V. K. *T* **1989**, *45*, 1851.
60. Nadir, U. K.; Sharma, R. L.; Koul, V. K. *JCS(P1)* **1991**, 2015.
61. Nadir, U. K.; Sharma, R. L.; Koul, V. K. *IJC(B)* **1989**, *28B*, 685.
62. Okuma, K.; Tanaka, Y.; Kaji, S.; Ohta, H. *JOC* **1983**, *48*, 5133.
63. Fitton, A. O.; Hill, J.; Jane, D. E.; Millar, R. *S* **1987**, 1140.
64. Taylor, K. G.; Hobbs, W. E.; Clark, M. S.; Chaney, J. *JOC* **1972**, *37*, 2436.
65. Tewari, R. S.; Nagpal, D. K. *JIC* **1979**, *56*, 911.
66. Johnson, C. R.; Mori, K.; Nakanishi, A. *JOC* **1979**, *44*, 2065.
67. Nichols, D. E.; Woodard, R.; Hathaway, B. A.; Lowy, M. T.; Yim, G. K. W. *JMC* **1979**, *22*, 458.
68. Baker, W. R.; Martin, S. F. World Patent 92 00 972, 1992 (*CA* **1992**, *116*, 256 057s).
69. Mapelli, C.; Turocy, G.; Switzer, F. L.; Stammer, C. H. *JOC* **1989**, *54*, 145.
70. Landor, S. R.; Punja, N. *JCS(C)* **1967**, 2495.

71. Minami, T.; Matsumoto, M.; Suganuma, H.; Agawa, T. *JOC* **1978**, *43*, 2149.
72. Rodriques, K. E. *TL* **1991**, *32*, 1275.
73. König, H.; Metzger, H.; Seelert, K. *CB* **1965**, *98*, 3712.
74. Sakakibara, T.; Sudoh, R.; Nakagawa, T. *BCJ* **1978**, *51*, 1189.
75. Radatus, B.; Williams, U.; Baer, H. H. *Carbohydr. Res.* **1986**, *157*, 242.
76. Donnelly, J. A.; O'Brien, S.; O'Grady, J. *JCS(P1)* **1974**, 1674.
77. Uno, H.; Yayama, A.; Suzuki, H. *T* **1992**, *48*, 8353.
78. Elkik, E.; Imbeaux-Oudotte, M. *BSF* **1987**, 861.
79. Elkik, E.; Imbeaux-Oudotte, M. *TL* **1985**, *26*, 3977.
80. Murphy, W. S.; Wattanasin, S. *JCS(P1)* **1981**, 2920.
81. Beal, R. B.; Dombroski, M. A.; Snider, B. B. *JOC* **1986**, *51*, 4391.
82. Piers, E.; Banville, J.; Lau, C. K.; Nagakura, I. *CJC* **1982**, *60*, 2965.
83. Evans, D. A.; Hart, D. J.; Koelsch, P. M. *JACS* **1978**, *100*, 4593.
84. Nickisch, K.; Bittler, D.; Casals-Stenzel, J.; Laurent, H.; Nickolson, R.; Nishino, Y.; Petzoldt, K.; Wiechert, R. *JMC* **1985**, *28*, 546.
85. Geetha, P.; Narasimhan, K.; Swaminathan, S. *TL* **1979**, 565.
86. Paquette, L. A.; Ewing, G. D.; Ley, S. V.; Berk, H. C.; Traynor, S. G. *JOC* **1978**, *43*, 4712.
87. Marshall, J. A.; Ellison, R. H. *JOC* **1975**, *40*, 2070.
88. Steinberg, N. G.; Rasmusson, G. H.; Reamer, R. A. *JOC* **1979**, *44*, 2294.
89. Takahashi, T.; Yamashita, Y.; Doi, T.; Tsuji, J. *JOC* **1989**, *54*, 4273.
90. Hamdouchi, C. *TL* **1992**, *33*, 1701.
91. Romo, D.; Romine, J. L.; Midura, W.; Meyers, A. I. *T* **1990**, *46*, 4951.
92. Bravo, P.; Gaudiano, G.; Ticozzi, C.; Umani-Ronchi, A. *TL* **1968**, 4481.
93. Gaudiano, G.; Umani-Ronchi, A.; Bravo, P.; Acampora, M. *TL* **1967**, 107.
94. Ide, J.; Kishida, Y. *TL* **1966**, 1787.
95. Kaiser, C.; Trost, B. M.; Beeson, J.; Weinstock, J. *JOC* **1965**, *30*, 3972.
96. Lecadet, D.; Paquer, D.; Thuillier, A. *CR* **1973**, *276*, 875.
97. Toure, S. A.; Voglozin, A.; Degny, E.; Danion-Bougot, R.; Danion, D.; Pradere, J. P.; Toupet, L.; N'Guessan, Y. T. *BSF* **1991**, *128*, 574.
98. Metzger, H.; Seelert, K. *ZN(B)* **1963**, *18B*, 335.
99. Nadir, U. K.; Chaurasia, B. P.; Sharma, R. L. *CL* **1989**, 2023.
100. Nadir, U. K.; Chaurasia, B. P. *IJC(B)* **1992**, *31B*, 189.
101. Seno, M.; Tsuchiya, S. *JCS(D)* **1977**, *8*, 751.
102. Weber, L. *ZN(B)* **1976**, *31B*, 780 (*CA* **1976**, *85*, 124 091e).
103. Weber, L.; Matzke, T.; Boese, R. *CB* **1990**, *123*, 739.
104. Porschke, K. R. *CB* **1987**, *120*, 425.
105. Yamamoto, Y. *BCJ* **1987**, *60*, 1189.
106. Lin, I. J. B.; Lai, H. Y. C.; Wu, S. C.; Hwan, L. *JOM* **1986**, *306*, C24.
107. Weber, L.; Lucke, E.; Boese, R. *CB* **1989**, *122*, 809.

John S. Ng & Chin Liu
Searle Research & Development, Skokie, IL, USA

(R,R)-1,2-Diphenyl-1,2-diaminoethane *N,N′*-Bis[3,5-bis(trifluoromethyl)-benzenesulfonamide]

$$\mathrm{R{-}SO_2{-}NH{-}CH(Ph){-}CH(Ph){-}NH{-}SO_2{-}R}$$

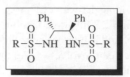

(**1**; R = 3,5-(CF_3)$_2C_6H_3$)
[127445-51-6] $C_{30}H_{20}F_{12}N_2O_4S_2$ (MW 764.66)
(**2**; R = CF_3)
[121788-73-6] $C_{16}H_{14}F_6N_2O_4S_2$ (MW 476.46)
(**3**; R = 4-MeC_6H_4)
[121758-19-8] $C_{28}H_{28}N_2O_4S_2$ (MW 520.72)
(**4**; R = 4-$NO_2C_6H_4$)
[121809-00-5] $C_{26}H_{22}N_4O_8S_2$ (MW 582.66)

(chiral controller group for enantioselective Diels–Alder reactions,[1] aldol additions,[2] Ireland–Claisen rearrangements,[3] ester-Mannich additions,[4] and carbonyl allylation[5] and propargylation[6])

Alternate Name: (*R,R*)-stilbenediamine *N,N′*-bis-3,5-bis(trifluoromethyl)benzenesulfonamide.

Physical Data: (**1**) mp 155–156 °C; α_D +83.7° (c = 1, $CHCl_3$). (**2**) mp 213–214 °C; α_D +6.6° (c = 1.4, $CHCl_3$). (**3**) mp 213–214 °C; α_D +43.9° (c = 1.74, $CHCl_3$). (**4**) mp 243 °C (dec); α_D 122° (c = 0.107, acetone).

Solubility: except for the nitro derivative, the sulfonamides are sol CH_2Cl_2.

Preparative Methods: the most convenient preparation of (*R,R*)-stilbenediamine is described in *Organic Syntheses*.[7] (For a new procedure see reference 10). Condensation of benzil and cyclohexanone in the presence of ammonium acetate and acetic acid (eq 1) produces a spirocyclic 2*H*-imidazole (mp 105–106 °C). Reduction with ***Lithium*** in THF/NH_3 followed by an ethanol quench and hydrolysis with aqueous HCl (eq 2) affords the racemic diamine as a pale yellow solid (mp 81–82 °C). Resolution is achieved by multiple recrystallizations of the tartaric acid salts from water/ethanol. The sulfonamides are prepared by reaction of the enantiomerically pure diamine with the appropriate anhydride[1b] or sulfonyl chloride[2a] in CH_2Cl_2 in the presence of ***Triethylamine*** and a catalytic amount of ***4-Dimethylaminopyridine*** (eq 3).

$$\text{PhCOCOPh} + \text{cyclohexanone} \xrightarrow[\text{AcOH}]{NH_4OAc} \text{2,3-diphenyl-1,4-diazaspiro[4.5]deca-1,3-diene} \quad (1)$$

$$\text{spiro-2}H\text{-imidazole} \xrightarrow[\text{2. ethanol}]{\text{1. Li, THF/}NH_3} \xrightarrow{H^+} \text{PhCH(NH}_2\text{)CH(NH}_2\text{)Ph} \ (\pm) \quad (2)$$

$$\text{H}_2\text{N–CH(Ph)–CH(Ph)–NH}_2 \xrightarrow[\text{Et}_3\text{N, DMAP, CH}_2\text{Cl}_2]{\text{RSO}_2\text{Cl or (RSO}_2)_2\text{O}} \text{R–SO}_2\text{–NH–CH(Ph)–CH(Ph)–NH–SO}_2\text{–R} \quad (3)$$

Handling, Storage, and Precautions: the sulfonamides are all stable, crystalline compounds that do not require any special precautions for storage or handling.

Diels–Alder Reactions. Reaction of the bis(triflamide) (**2**) with ***Diisobutylaluminum Hydride*** or ***Trimethylaluminum*** affords chiral Lewis acids that catalyze Diels–Alder reactions of acryloyl or crotonoyl derivatives with cyclopentadienes (eq 4).[1] The aluminum complex must be crystallized before use to remove traces of trimethylaluminum. High diastereo- and enantioselectivities are achieved with as little as 0.1 equiv of the Lewis acid, and the chiral sulfonamide is recoverable.

(2): CF_3-SO_2-N(Ph)…N-SO_2-CF_3, Al–Me; CH_2Cl_2, 94%

BnO … (4)

95% ee

Asymmetric Aldol Reactions. Reaction of (**1**) with ***Boron Tribromide*** in CH_2Cl_2 affords, after removal of solvent and HBr, a complex (**5**) useful for the preparation of chiral enolates (eq 5).[1a] Complex (**5**) is moisture sensitive and is generally prepared immediately before use. For propionate derivatives, either *syn* or, less selectively, *anti* aldol adducts may be obtained by selection of the appropriate ester derivative and conditions.[2a] Thus reaction of *t*-butyl propionate with (**5**) and triethylamine produces the corresponding *E*(O) enolate, leading to formation of *anti* aldol adducts upon addition to an aldehyde (eq 6). Selectivities may be enhanced by substitution of the *t*-butyl ester with the (+)-menthyl ester. Conversely, reaction of *S*-phenyl thiopropionate with (**5**) and ***Diisopropylethylamine*** affords the corresponding *Z*(O) enolates and *syn* aldol products (eq 7).[2a,c]

(**1**): 3,5-$(CF_3)_2C_6H_3$-SO_2-NH–CH(Ph)–CH(Ph)–NH-SO_2-3,5-$(CF_3)_2C_6H_3$ + BBr_3 ⟶ Ar-SO_2-N(Ph)…N-SO_2-Ar, B–Br (5)

(**5**)

O-*t*-Bu propionate + (**5**) + Et_3N —toluene–hexane→ OB^*L_2 enolate (O-*t*-Bu) —RCHO→ (6)

R = Ph	93%	98:2 ds	94% ee
R = Cy	82%	94:6 ds	75% ee

SPh thiopropionate + (**5**) + *i*-Pr_2NEt —CH_2Cl_2→ OB^*L_2 enolate (SPh) —RCHO→ (7)

R = Ph	93%	99:1 ds	97% ee
R = Cy	86%	98:2 ds	91% ee

Products with low enantiomeric purity are obtained by direct application of this chemistry to unsubstituted acetate esters. However, aldol reactions of *t*-butyl bromoacetate mediated by (**5**) afford synthetically useful bromohydrins (**6**) with high selectivities (eq 8).[2b] These may be reductively dehalogenated or converted to a variety of compounds by way of the derived epoxides.

Br-CH_2-CO-O-*t*-Bu + (**5**) —1. Et_3N, toluene; 2. RCHO→ R-CH(OH)-CH(Br)-CO-O-*t*-Bu (8)

(**6**)

R = Ph	94%	99:1 ds	98% ee
R = Cy	65%	98:2 ds	91% ee

Asymmetric Ireland–Claisen Rearrangements. Chiral enolates derived from the boron complex (**5**) and allyl esters rearrange with excellent selectivity upon warming to −20 °C for a period of 1–2 weeks (eqs 9 and 10).[3] As discussed above, the geometry of the intermediate enolate can be controlled by appropriate choice of base and solvent, thus allowing access to either *syn* or *anti* configuration in the product. The reaction can be completed in 2–4 days with little erosion in selectivity when run at 4 °C.

(**5**) + allyl ester (Et) —*i*-Pr_2NEt, CH_2Cl_2, −78 °C→ OB^*L_2 enolate —−20 °C→ HO acid (9)

75%, 99:1 ds, >97% ee

(**5**) + allyl ester (Et) —Et_3N, toluene–hexane, −78 °C→ OB^*L_2 enolate —−20 °C→ HO acid (10)

65%, 90:10 ds, 96% ee

Ester-Mannich Additions. The *E*(O) enolate (**7**) reacts with *N*-allyl or *N*-benzyl aldimines to afford chiral β-amino esters (eq 11).[4] As with the aldol reactions, best selectivities are achieved with imines derived from aromatic or unsaturated aldehydes. The method appears to have good potential for the synthesis of useful β-lactams if extended to other enolates.

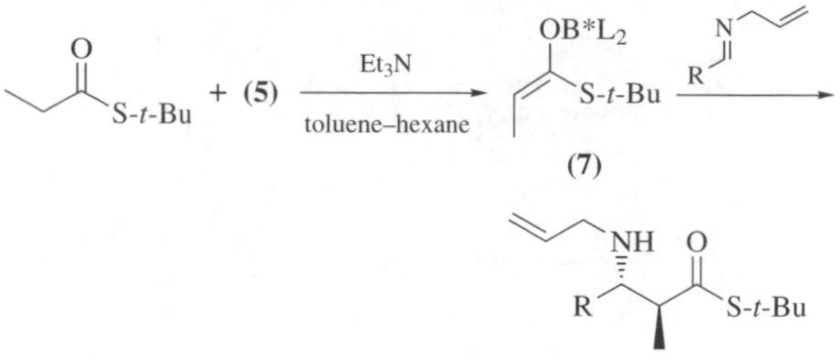

(11)

R = Ph 74% >99:1 ds 90% ee
R = hydrocinnamyl 67% 97:3 ds 70% ee

Carbonyl Allylation and Propargylation. Boron complex (**8**), derived from the bis(tosylamide) compound (**3**), transmetalates allylstannanes to form allylboranes (eq 12). The allylboranes can be combined without isolation with aldehydes at −78 °C to afford homoallylic alcohols with high enantioselectivity (eq 13).[5] On the basis of a single reported example, reagent control might be expected to overcome substrate control in additions to aldehydes containing an adjacent asymmetric center. The sulfonamide can be recovered by precipitation with diethyl ether during aqueous workup. Ease of preparation and recovery of the chiral controller makes this method one of the more useful available for allylation reactions.

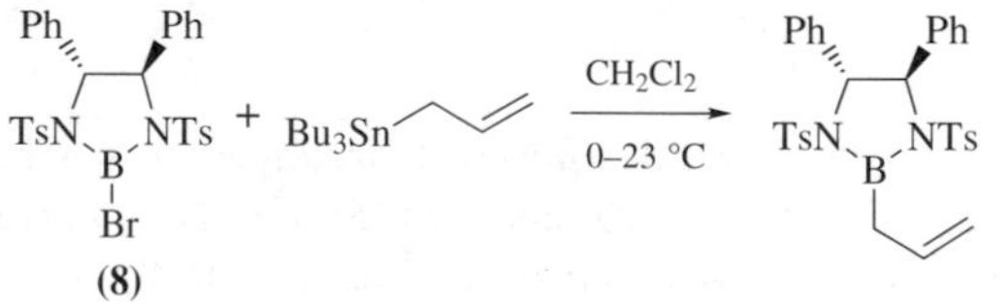

(12)

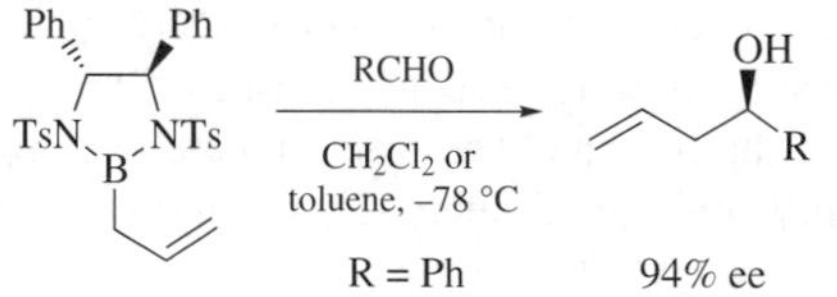

(13)

R = Ph 94% ee
R = cinnamyl 98% ee
R = cyclohexyl 93% ee
R = hexyl 90% ee

In the same way, reaction of (**8**) with allenyl- or propargylstannanes affords intermediate borane derivatives which, upon reaction with aldehydes, produce the expected adducts with high selectivities (eqs 14 and 15).[6]

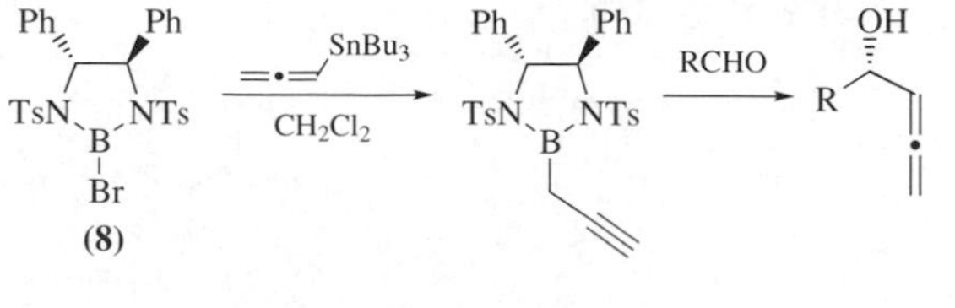

(14)

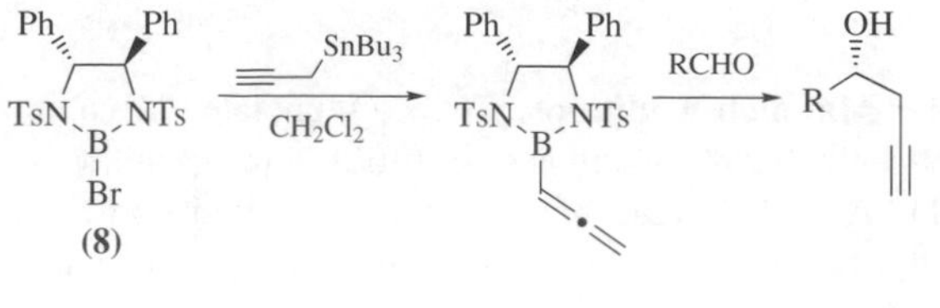

(15)

Other Applications. Other (*R,R*)-stilbenediamine derivatives have been used to direct the stereochemical course of alkene dihydroxylation[8] (with stoichiometric quantities of ***Osmium Tetroxide*** and epoxidation of simple alkenes with ***Sodium Hypochlorite*** and manganese(III) complexes.[9]

Related Reagents. *B*-Allyldiisopinocampheylborane; Chloro-(cyclopentadienyl)bis[3-*O*-(1, 2:5, 6-di-*O*-iso-propylidene-α-D-glucofuranosyl)]titanium; Chloro(η^5-cyclopentadienyl)[(4*R*, *trans*), 2, 2-dimethyl-α, α, α′,α′-tetraphenyl-1,3-dioxolane-4,5-dimethanolato(2−)-O^{α}, O^{α}]titanium; Diisopinocampheylboron Trifluoromethanesulfonate; Diisopropyl 2-Allyl-1,3,2-dioxaborolane-4,5-dicarboxylate; (4*R*,5*R*)-2,2-Dimethyl-4,5-bis(hydroxydiphenylmethyl)-1,3-dioxolane-Titanium(IV) Chloride; 2,2-Dimethyl-α,α,α′,α′-tetraphenyl-1,3-dioxolane-4,5-dimethanolatotitanium Diisopropoxide.

1. (a) Corey, E. J.; Imwinkelried, R.; Pikul, S.; Xiang, Y. B. *JACS* **1989**, *111*, 5493. (b) Pikul, S.; Corey, E. J. *OS* **1992**, *71*, 30.
2. (a) Corey, E. J.; Kim, S. S. *JACS* **1990**, *112*, 4976. (b) Corey, E. J.; Choi, S. *TL* **1991**, *32*, 2857. (c) Corey, E. J.; Lee, D.-H. *TL* **1993**, *34*, 1737.
3. Corey, E. J.; Lee, D.-H. *JACS* **1991**, *113*, 4026.
4. Corey, E. J.; Decicco, C. P.; Newbold, R. C. *TL* **1991**, *32*, 5287.
5. Corey, E. J.; Yu, C.-M.; Kim, S. S. *JACS* **1989**, *111*, 5495.
6. Corey, E. J.; Yu, C.-M.; Lee, D.-H. *JACS* **1990**, *112*, 878.
7. Pikul, S.; Corey, E. J. *OS* **1992**, *71*, 22.
8. Corey, E. J.; DaSilva Jardine, P.; Virgil, S.; Yuen, P.-W.; Connell, R. D. *JACS* **1989**, *111*, 9243.
9. Zhang, W.; Jacobsen, E. N. *JOC* **1991**, *56*, 2296.
10. Corey, E. J.; Kühnle, N. M. *TL* **1997**, *38*, 8631.

James R. Gage
The Upjohn Company, Kalamazoo, MI, USA

Dirhodium(II) Tetraacetate

$Rh_2(O_2CMe)_4$

[15956-28-2] $C_8H_{12}O_8Rh_2$ (MW 442.02)

(catalyst for carbenoid reactions of diazo compounds,[1] hydroboration of alkenes and alkynes,[2] hydrocarbon oxidation[3])

Physical Data: λ 590 nm, ϵ 210 (EtOH); λ 552 nm, ϵ 235 (MeCN).[4] IR $\bar{\nu}$ (CO_2) 1585 cm^{-1}.[5]
Solubility: sol MeOH, acetic acid, MeCN, acetone; slightly sol toluene; insol Et_2O, 1,2-dichloroethane, CH_2Cl_2.
Form Supplied in: emerald-green solid for anhydrous form.
Preparative Method: prepared from $RhCl_3{\cdot}xH_2O$.[6]
Handling, Storage, and Precautions: air stable, moderately hygroscopic; stored in desiccator. Removal of axially coordinated solvent can be achieved in vacuum oven at 60–80 °C.

Metal Carbene Transformations. When used in amounts as low as 0.05 mol %,[7] $Rh_2(OAc)_4$ serves as a highly effective and efficient catalyst for dinitrogen extrusion from diazo carbonyl compounds and subsequent metal carbene directed alkene[1,8] and alkyne[9] addition reactions, σ-bond insertion reactions,[1,10] ylide generation,[1,11] and carbene coupling processes,[12] among others. With diazoacetates and diazo ketones, catalytic reactions are generally performed in dichloromethane at rt. The diazo compound is added at such a rate so as to minimize its concentration in the reaction solution.[6] The less reactive α-diazo-β-carbonyl alkanoates, including diazomalonates and diazoacetoacetates, and related diazo dicarbonyl compounds require refluxing 1,2-dichloroethane or refluxing benzene for efficient catalyst turnovers. In laboratory scale reactions, $Rh_2(OAc)_4$ is removed from the reaction solution by filtering through a short column of silica or alumina, or the product(s) are distilled directly from the catalyst-containing mixture. (Rhodium acetate undergoes thermal decomposition at temperatures near 250 °C.)

Cyclopropanation/Cyclopropenation. Its catalytic uses for intermolecular cyclopropanation and cyclopropenation reactions, discovered only recently by Teyssié and co-workers,[9,13] have been the standard through which mechanistic[14] and synthetic (selectivity)[8,15] understanding of metal carbene addition reactions have been derived.[1] Intermolecular addition reactions of diazoacetates with alkenes occur at 25 °C in high yield. Their stereoselectivities and regioselectivities[16] are lower than those achieved with the use of ***Dirhodium(II) Tetraacetamide***,[8] but high selectivities are observed with vinylcarbenoid addition (eq 1)[17] and nitrocarbenoid addition[18] to alkenes. Cyclopropene formation occurs with a broad selection of alkynes (eq 2),[19a] but not phenylacetylene.[9,19] Neither cyclopropanation nor cyclopropenation occurs readily with α,β-unsaturated carbonyl compounds or nitriles.[20]

Ph–CH=CH$_2$ + EtO_2C–C(N_2)–CH=CH–CO_2Et —[$Rh_2(OAc)_4$, 95%]→ Ph-cyclopropane(CO_2Et)(CH=CH–CO_2Et) (1)

(E):*(Z)* = 92:8

N_2CHCO_2Et + PhC≡CMe —[$Rh_2(OAc)_4$, 93%]→ cyclopropene (Ph, Me, CO_2Et) (2)

Intramolecular analogs of these reactions suggest the overall viability of $Rh_2(OAc)_4$ catalysis (eqs 3 and 4),[21–23] although for alkyne addition the presumed cyclopropene intermediate is unstable and undergoes reactions characteristic of vinylcarbenoid species,[22,23] of which the most frequently encountered involve rearrangement to furans. The intramolecular Büchner reaction of aryl diazoketones (eq 5)[24] is, formally, cyclopropanation of an aromatic ring, for which $Rh_2(OAc)_4$ is especially suitable;[24] the diazoamide analog of this transformation is particularly facile.[25]

(dihydropyranyl)–$CH_2C(O)CHN_2$ —[$Rh_2(OAc)_4$, 61%]→ (3)

R–C≡C–CH_2CH_2–C(O)–C(N_2)–CO_2Me —[$Rh_2(OAc)_4$, 77%]→ [bicyclic cyclopropene intermediate (R, CO_2Me, O)] —[65%]→ MeO-furan(R)-fused cyclopentanone (4)

Ph–CH_2CH_2–C(O)–C(N_2)–Me —[$Rh_2(OAc)_4$, 67%]→ (5)

Carbon–Hydrogen Insertion. The utility of $Rh_2(OAc)_4$ as a catalyst for metal carbene transformations is most evident in its ability to effect insertion into unactivated carbon–hydrogen bonds.[1,10] Preference for the formation of five-membered rings is pronounced (eq 6),[26] and selectivity for insertion into tertiary C–H bonds is usually greater than for insertion into secondary C–H bonds, and primary C–H bonds are the least reactive.[27] Although these controlling influences are pervasive, an increasing number of examples suggest that the factors which control regioselectivity in these reactions are complex (e.g. eq 7).[28]

—[$Rh_2(OAc)_4$, 77%]→ (Ph, CO_2Me) (6)

—[$Rh_2(OAc)_4$, 87%]→ (7)

Heteroatom activation of adjacent C–H bonds[29] has made possible the construction of β-lactam derivatives from diazoacetoacetamides (eq 8)[30] and of β-lactones from selected diazomalonates.[31] With *N*-benzyl derivatives, the presence or absence of an acyl group uniquely defines the course of the catalytic reaction (eq 9; R = *t*-Bu; S = H, Me, OMe, Br).[32] Both electronic and conformational (steric) effects are responsible for the selectivity in these reactions.

—[$Rh_2(OAc)_4$, 96%]→ (EtO_2C, *t*-Bu) (8)

(9)

Intermolecular variants of carbon–hydrogen insertion reactions are generally of limited value because of lower reactivity and selectivity.[1] Insertion into vinylic or alkynic C–H bonds is not competitive with cyclopropanation or cyclopropenation under ordinary circumstances. The so-called 'allylic C–H insertion reaction', commonly observed in copper-catalyzed reactions of diazomalonates or β-diazo-α-keto esters,[33] is not common with $Rh_2(OAc)_4$ catalysis.[14b]

In contrast to the paucity of examples for C–H insertion into vinylic or alkynic C–H bonds, intramolecular 'insertion' into aromatic C–H bonds is well documented.[1] These reactions are most pronounced when the aromatic ring is activated for substitution by oxygen[34] or nitrogen[35] (e.g. eqs 10 and 11),[34,35a] and they are more suitably described as aromatic substitution reactions than as C–H insertion reactions. Aryl-substituted diazo ketones,[36] diazoacetates,[34] diazoacetoacetates,[34] and their corresponding amide derivatives[35] are all effective. The formation of a five-membered ring is preferred, but, with suitable structural demands in the diazo carbonyl reactant, six-membered ring formation occurs.[37]

(10)

(11)

Heteroatom–Hydrogen Insertion. One of the most important applications of metal carbene chemistry has been in the syntheses of penems, exemplified in eq 12, which is a key step in the total synthesis of the carbapenem thienamycin.[38] This reaction has become the method of choice for the synthesis of bicyclic β-lactams from 2-azetidinones substituted through the 4-position to a diazocarbonyl group.[39]

(12)

Oxygen–hydrogen insertion (with water) is a common undesirable side reaction in $Rh_2(OAc)_4$ catalyzed transformations, but its synthetic importance is evident in intramolecular processes. Cyclization via O–H insertion occurs readily and in high yield for five-, six-, and seven-membered ring formation (eq 13),[40] but C–H insertion becomes competitive in attempts to effect eight-membered ring formation. Intermolecular processes have also been examined,[41] but they have more limited usefulness.

(13)

Thiol insertion also occurs in both intermolecular and intramolecular transformations,[42,43] but these reactions are more difficult to perform because of the facile coordination of thiols and sulfides with $Rh_2(OAc)_4$. Intermolecular silicon–hydrogen insertion provides a convenient methodology for the synthesis of α-silyl esters and ketones (eq 14).[44] Overall, $Rh_2(OAc)_4$ is generally suitable, and often the catalyst of choice, for heteroatom–hydrogen insertion reactions of diazo esters and ketones.[43]

(14)

Ylide Generation. Metal carbenes produced by $Rh_2(OAc)_4$-catalyzed dinitrogen extrusion from diazo compounds are electrophilic.[1] Reactions of these reactive intermediates, which resemble metal-stabilized carbocations, with Lewis bases constitute a generally effective methodology for ylide generation (eq 15).[11] The relative reactivity of Lewis bases towards ylide generation follows the expected order of basicity for carbon–heteroatom compounds. Ylide products are further transformed by insertion, sigmatropic rearrangement, or dipolar addition reactions, dependent on the design of the ylide. Heteroatom–hydrogen insertion is reasonably regarded as an ylide transformation.

$$Rh_2(OAc)_4 + R_2C{=}N_2 \xrightarrow{-N_2}$$

$$[Rh_2(OAc)_4Rh{=}CR_2 \longleftrightarrow Rh(OAc)_4\bar{R}h{-}\overset{+}{C}R_2] \tag{15}$$

$$\Big\updownarrow \text{B:}$$

$$Rh_2(OAc)_4 + \overset{+}{B}{-}\bar{C}R_2 \rightleftharpoons Rh(OAc)_4\bar{R}h{-}CR_2{-}B^+$$

The relative reactivity for ylide generation, determined from subsequent [2,3]-sigmatropic rearrangement of the initially formed ylide in competition with cyclopropanation (eq 16),[45,46] shows that ylide formation increases in the order RI > RBr > RCl, $R_3N > R_2O$, and $R_2S > R_2O$. Allyl substituents facilitate ylide generation (eq 17),[46,47] and their influences on relative reactivities suggest that the formation of the metal-stabilized ylide and metal dissociation (eq 15) are equilibrium processes.

(16)

Z	Yield (%)	
NMe_2	95	100:0
SMe	96	100:0
I	98	100:0
Br	76	28:72
Cl	95	5:95
OEt	88	10:90

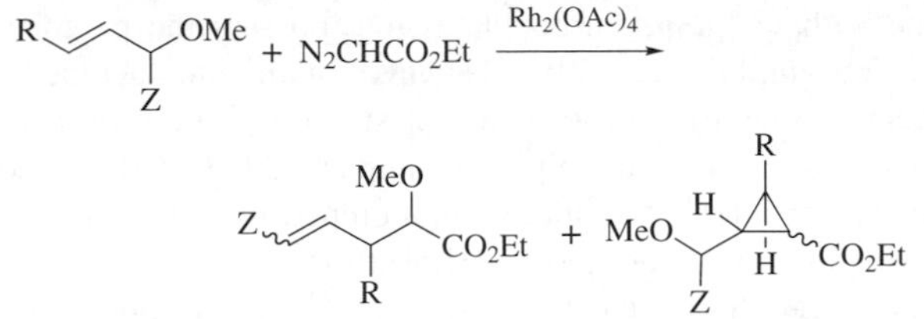

R	Z		Yield (%)
H	H	10:90	88
Me	H	92:8	86
Ph	H	73:27	95
TMS	H	94:6	68
H	OMe	77:23	75

Rhodium(II) acetate has two axial coordination sites at which reactions with diazo compounds occur. Strongly coordinating compounds, either reactants or solvents, occupy these sites and inhibit electrophilic addition to diazo compounds. Consequently, whereas reactions with diazo compounds that possess chloride, bromide, iodide, or ether functional groups take place at rt, those with sulfide or amine functional groups require higher temperatures for dinitrogen extrusion.

Stable sulfonium ylides have been produced by intermolecular reactions of thiophenes with dimethyl diazomalonate, catalyzed by $Rh_2(OAc)_4$ (eq 18),[48] as well as by intramolecular reactions (eq 19).[49] The formation of four- to seven-membered rings has been possible,[49,50] but C–H insertion is competitive with seven-membered ring ylide generation. The stability of the ylide is dependent on the sulfur substituent ($Ph > PhCH_2 >$ allyl) as well as on ring size. Stable sulfoxonium ylides have also been formed in $Rh_2(OAc)_4$-catalyzed reactions.[51]

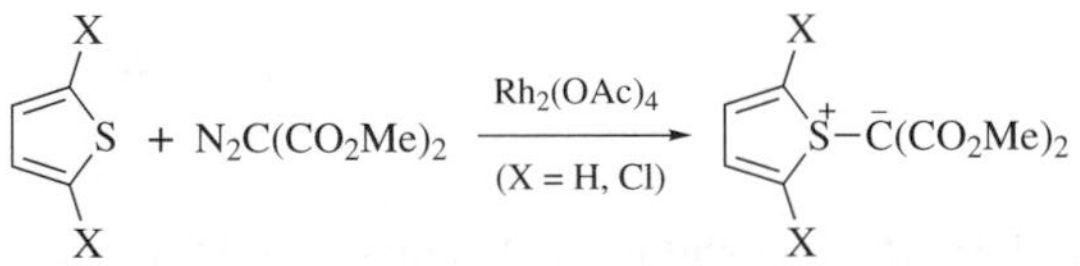

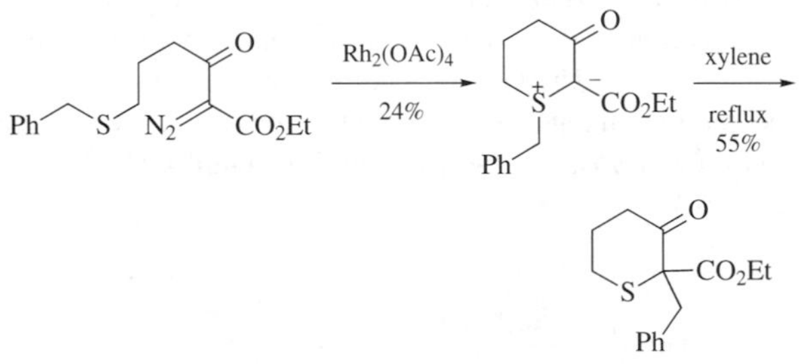

Use of the [2,3]-sigmatropic rearrangement of sulfonium ylides for ring enlargement has made possible the construction of medium ring compounds (eq 20).[47] β-Elimination is a competing process and becomes the favored transformation with the proper stereoelectronic arrangement in the reactant ylide (e.g. eq 21).[52] An intramolecular ring contraction methodology is also effectively promoted through sulfonium ylide generation with $Rh_2(OAc)_4$ catalysis (e.g. eq 22).[53] Sulfur ylides derived from $Rh_2(OAc)_4$-catalyzed reactions of diazo compounds have played important roles in β-lactam antibiotic syntheses.[54–56]

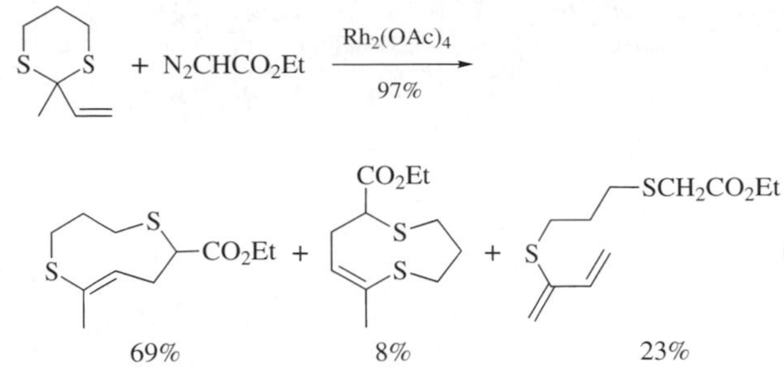

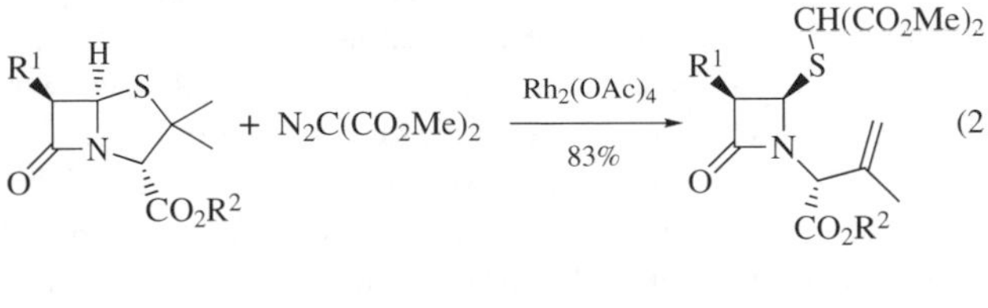

Nitrogen ylide formation is not reported as extensively as is sulfur ylide generation, but sufficient examples exist to suggest its versatility. The synthesis of allenes by [2,3]-sigmatropic rearrangement (eq 23)[57] of prop-2-yn-1-yl-dimethylammonium ylides exemplifies the potential diversity of its applications. A general methodology for oxazole synthesis has been developed through the use of nitrile ylides (eq 24).[58] Additional examples that suggest the advantages of the catalytic route to ylide generation are reported elsewhere.[11]

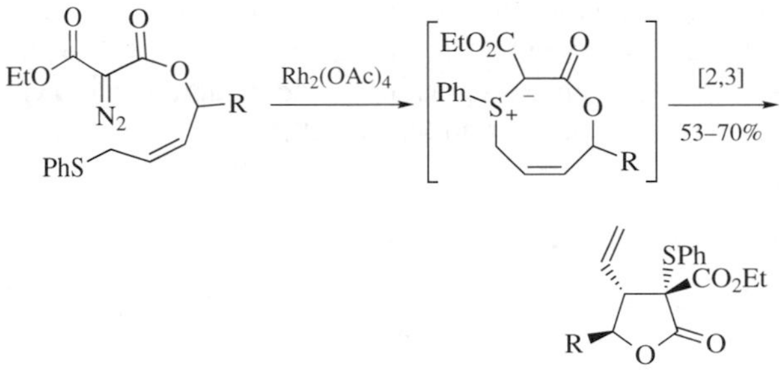

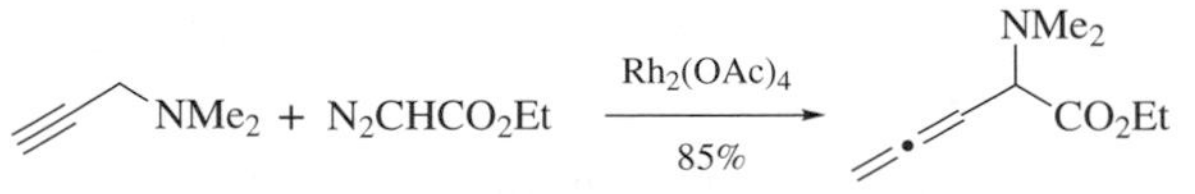

Relative to cyclopropanation (of alkenes) or cyclopropenation (of alkynes), oxonium ylide formation is generally disfavored in $Rh_2(OAc)_4$-catalyzed reactions of diazo compounds. Notable exceptions include those described in eq 17 and in intramolecular transformations. Cyclobutanone formation is the outcome of $Rh_2(OAc)_4$-catalyzed dinitrogen extrusion from 4-alkoxydiazo ketones (eq 25).[59] Allyl ethers offer a pathway to [2,3]-sigmatropic rearrangement products, including those leading to medium ring ethers (eq 26),[60,61] and tetrahydrofuran-3-ones have been prepared by a carbon–oxygen insertion methodology involving ylide intermediates,[62] but few other successful demonstrations of oxonium ylide generation have been reported.

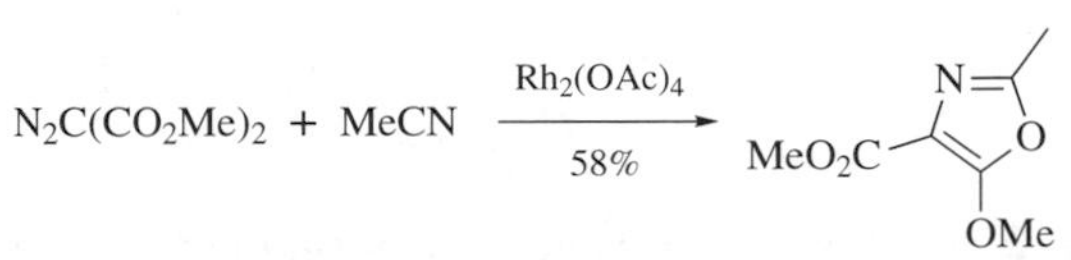

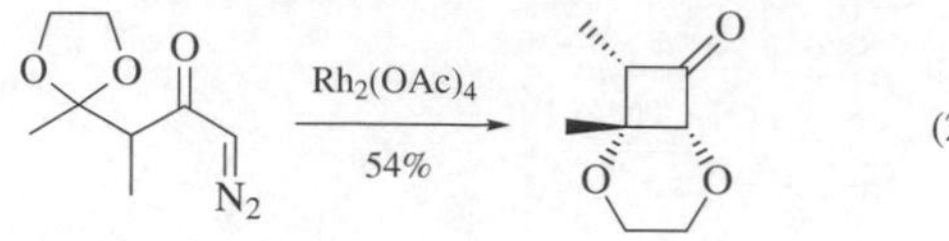

(25)

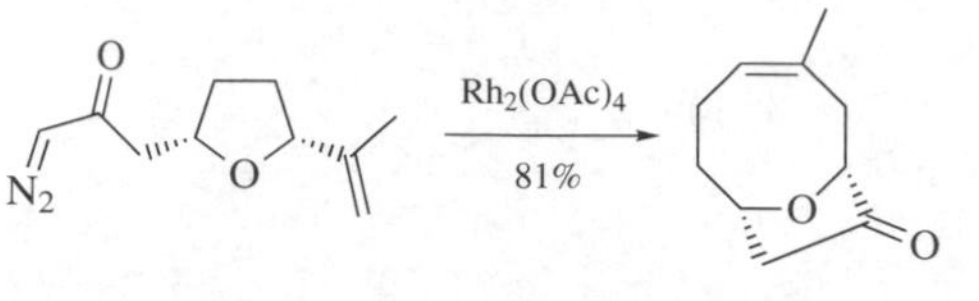

(26)

Dirhodium(II) tetraacetate has been shown to have superior capabilities for carbonyl ylide generation with diazo ketones,[11] especially in intramolecular reactions. The carbonyl ylide, when generated in the presence of selected dipolarophiles, readily undergoes 1,3-dipolar addition (e.g. eqs 27 and 28)[63,64] either intermolecularly (***Dimethyl Acetylenedicarboxylate***, ***N-Phenylmaleimide***, diethyl fumarate, aldehydes) or intramolecularly. Regioselectivity is predictable by frontier molecular orbital interactions.[64b] Carbonyl ylide generation with carboxylate esters is rare,[30b] but with amides, carbamates, and imides, carbonyl ylide formation is well documented (e.g. eq 29).[11,65]

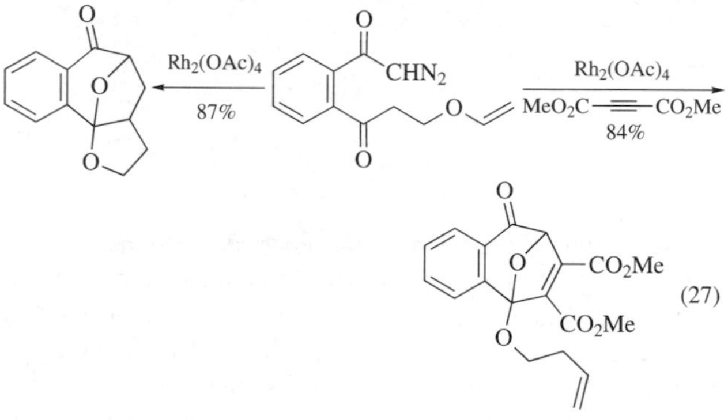

(27)

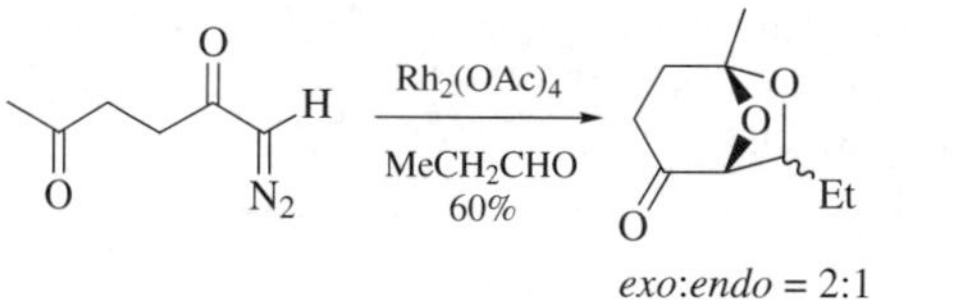

(28)

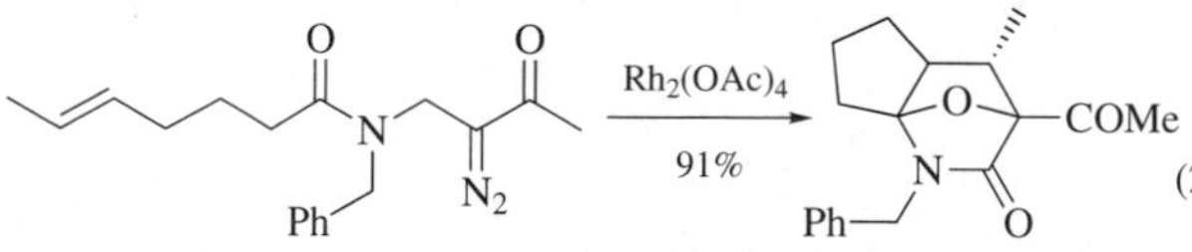

exo:*endo* = 2:1

(29)

Selectivity in Metal Carbene Transformations. The high reactivity of catalytically generated metal carbenes often limits their selectivity when more than one site for addition and/or insertion exists in a molecule. In this regard, $Rh_2(OAc)_4$ is often less selective for metal carbene transformations than is either ***Dirhodium(II) Tetrakis(perfluorobutyrate)*** or rhodium(II) carboxamides (either ***Dirhodium(II) Tetraacetamide*** or ***Dirhodium(II) Tetra(caprolactam)***).[8,28,66] In addition, use of $Rh_2(OAc)_4$ can lead to lower product yields than does use of alternative dirhodium(II) catalysts. Nevertheless, for most transformations where there is one preferred site for metal carbene reaction, $Rh_2(OAc)_4$ remains the catalyst of choice.

High diastereocontrol complements exceptional stereocontrol for geometrical isomer formation in the cyclopropanation of styrene with vinyldiazocarboxylates that possess the pantolactone chiral auxiliary (eq 30).[67] However, with pantolactone diazoacetate in the cyclopropanation of styrene, a 76:24 *trans*:*cis* product mixture forms in only 72% yield and 28–30% diastereomeric excess (de).[68] Similarly, in competitive reactions between cyclopropanation and C–H insertion, use of $Rh_2(OAc)_4$ provides very little selectivity (eq 31),[66] but dirhodium(II) tetrakis(perfluorobutyrate) promotes exclusive C–H insertion and use of dirhodium tetra(caprolactam) leads to exclusive cyclopropanation. A variety of examples of high and low selectivities has been reported for $Rh_2(OAc)_4$-catalyzed reactions,[1] and no simple explanation has emerged that will account for all of the results. Predictions are often empirical, but changing the ligands of the catalyst often provides a dirhodium(II) compound with the desired selectivity enhancement.

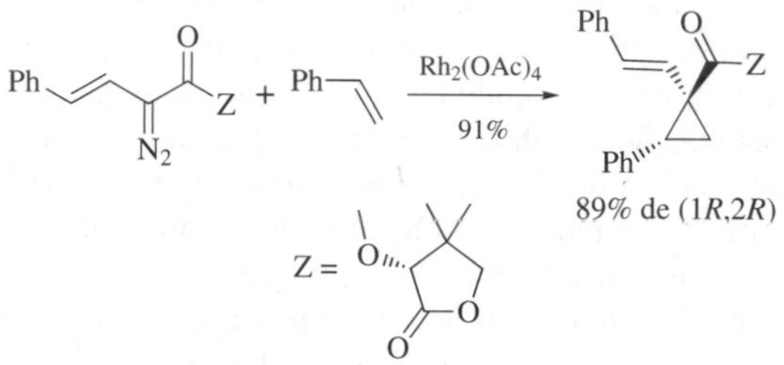

(30)

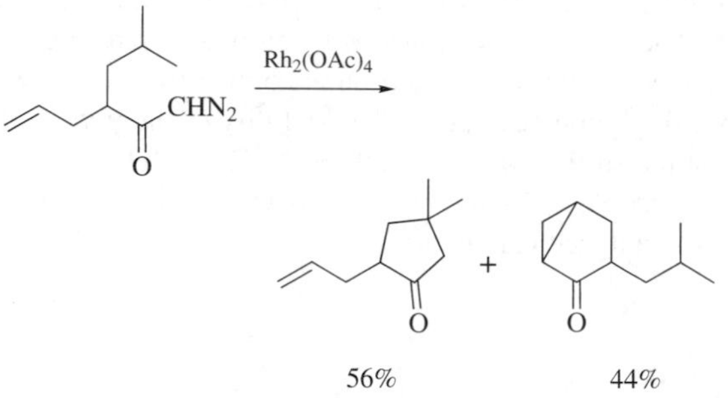

89% de (1*R*,2*R*)

(31)

Catalytic Hydroboration of Alkenes and Alkynes. The use of $Rh_2(OAc)_4$ to catalyze the hydroboration of alkenes and alkynes with ***Catecholborane*** represents a novel application of this transition metal compound.[2] In laboratory scale reactions, 0.5 mol % $Rh_2(OAc)_4$ is effective. Selectivity is dependent on the alkene (eq 32) and often differs from that found with ***Chlorotris(triphenylphosphine)rhodium(I)*** (Wilkinson's catalyst).[69]

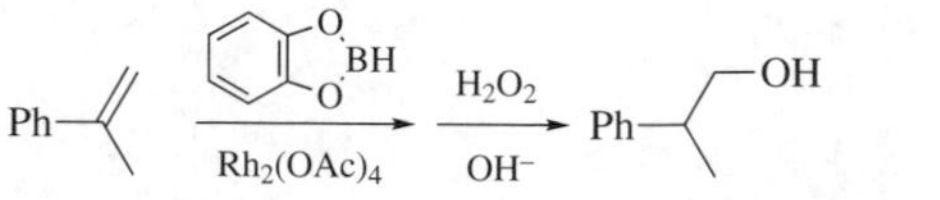

(32)

The hydroboration of alkynes is also promoted by rhodium acetate. However, since $Rh_2(OAc)_4$ is transformed during hydroboration, the nature of the active catalyst is at present unknown. The combination of catecholborane and $Rh_2(OAc)_4$, both in catalytic amounts, catalyzes isomerization of terminal alkenes.[2]

Autooxidation of Alkenes. In the presence of catalytic amounts of $Rh_2(OAc)_4$, cyclohexadienes undergo aromatization (eq 33), and dienes undergo oxidative cleavage (eq 34),[3a] at atmospheric pressure. *p*-Cymene is also reported to be oxidized to the tertiary alcohol. Hydroperoxide decomposition to alcohol and dioxygen, catalyzed by $Rh_2(OAc)_4$, has been demonstrated.

$$\xrightarrow[\text{O}_2,\ 93\%]{\text{Rh}_2(\text{OAc})_4} + H_2O \quad (33)$$

$$\text{Ph}\diagup\!\!\diagdown\!\!\diagup\!\!\diagdown \xrightarrow[\text{O}_2,\ 71\%]{\text{Rh}_2(\text{OAc})_4} \text{Ph}\diagup\!\!\diagdown\text{CHO} + \text{PhCHO} \quad (34)$$

Other Catalytic Reactions. Hydrogenation activity for $Rh_2(OAc)_4$ has been reported,[70] but these results could not be repeated.[71] Hydrosilylation of alkenes is catalyzed by $Rh_2(OAc)_4$, but reactions are much less efficient than those catalyzed by dirhodium(II) tetrakis(perfluorobutyrate).[71] The same is true of silylcarbonylation reactions with alkynes,[72] and $Rh_2(OAc)_4$ is not a hydroformylation catalyst.

Related Reagents. Dirhodium(II) Tetraacetamide; Dirhodium(II) Tetra(caprolactam); Dirhodium(II) Tetrakis-(methyl 2-pyrrolidone-5(S)-carboxylate); Dirhodium(II) Tetrakis(perfluorobutyrate); Dirhodium(II) Tetrakis(trifluoroacetate); Dirhodium(II) Tetraoctanoate.

---

1. (a) Doyle, M. P. *CRV* **1986**, *86*, 919. (b) Doyle, M. P. *ACR* **1986**, *19*, 348. (c) Maas, G. *Top. Curr. Chem.* **1987**, *137*, 75. (d) Padwa, A.; Krumpe, K. E. *T* **1992**, *48*, 5385.
2. Doyle, M. P.; Westrum, L. J.; Protopopova, M. N.; Eismont, M. Y.; Jarstfer, M. B. *Mendeleev Commun.* **1993**, 81.
3. (a) Doyle, M. P.; Terpstra, J. W.; Winter, C. H.; Griffin, J. H. *J. Mol. Catal.* **1984**, *26*, 259. (b) Noels, A. F.; Hubert, A. J.; Teyssié, Ph. *JOM* **1979**, *166*, 79.
4. Johnson, S. A.; Hunt, H. R.; Neumann, H. M. *IC* **1963**, *2*, 960.
5. Winkhaus, G.; Ziegler, P. *Z. Anorg. Allg. Chem.* **1967**, 350.
6. Rampel, G. A.; Legzdins, P.; Smith, H.; Wilkinson, G. *Inorg. Synth.* **1972**, *13*, 90.
7. Doyle, M. P.; van Leusen, D.; Tamblyn, W. H. *S* **1981**, 787.
8. Doyle, M. P.; Bagheri, V.; Wandless, T. J.; Harn, N. K.; Brinker, D. A.; Eagle, C. T.; Loh, K.-L. *JACS* **1990**, *112*, 1906.
9. Petiniot, N.; Anciaux, A. J.; Noels, A. F.; Hubert, A. J.; Teyssié, Ph. *TL* **1978**, 1239.
10. Adams, J.; Spero, D. M. *T* **1991**, *47*, 1765.
11. (a) Padwa, A.; Hornbuckle, S. F. *CRV* **1991**, *91*, 263. (b) Padwa, A. *ACR* **1991**, *24*, 22.
12. Shankar, B. K. R.; Shechter, H. *TL* **1982**, *23*, 2277.
13. Hubert, A. J.; Noels, A. F.; Anciaux, A. J.; Teyssié, P. *S* **1976**, 600.
14. (a) Doyle, M. P.; Dorow, R. L.; Buhro, W. E.; Griffin, J. H.; Tamblyn, W. H.; Trudell, M. L. *OM* **1984**, *3*, 44. (b) Doyle, M. P.; Griffin, J. H.; Bagheri, V.; Dorow, R. L. *OM* **1984**, *3*, 53. (c) Doyle, M. P.; Griffin, J. H.; da Conceicao, J. *CC* **1985**, 328.
15. (a) Anciaux, A. J.; Hubert, A. J.; Noels, A. F.; Petinoit, N.; Teyssié, P. *JOC* **1980**, *45*, 695. (b) Anciaux, A. J.; Demonceau, A.; Noels, A. F.; Warin, R.; Hubert, A. J.; Teyssié, P. *T* **1983**, *39*, 2169. (c) Demonceau, A.; Noels, A. F.; Hubert, A. J. *T* **1990**, *46*, 3889.
16. (a) Doyle, M. P.; Dorow, R. L.; Tamblyn, W. H.; Buhro, W. E. *TL* **1982**, *23*, 2261. (b) Doyle, M. P.; Wang, L. C.; Loh, K.-L. *TL* **1984**, *25*, 4087. (c) Doyle, M. P.; Dorow, R. L.; Terpstra, J. W.; Rodenhouse, R. A. *JOC* **1985**, *50*, 1663.
17. (a) Davies, H. M. L.; Clark, T. J.; Church, L. A. *TL* **1989**, *30*, 5057. (b) Davies, H. M. L.; Clark, T. J.; Smith, H. D. *JOC* **1991**, *56*, 3817. (c) Davies, H. M. L.; Hu, B. *JOC* **1992**, *57*, 3186.
18. O'Bannon, P. E.; Dailey, W. P. *T* **1990**, *46*, 7341.
19. (a) Cho, S. H.; Liebeskind, L. S. *JOC* **1987**, *52*, 2631. (b) Müller, P.; Pautex, N.; Doyle, M. P.; Bagheri, V. *HCA* **1990**, *73*, 1233.
20. Doyle, M. P.; Dorow, R. L.; Tamblyn, W. H. *JOC* **1982**, *47*, 4059.
21. Adams, J.; Frenette, R.; Belley, M.; Chibante, F.; Springer, J. P. *JACS* **1987**, *109*, 5432.
22. Hoye, T. R.; Dinsmore, C. J.; Johnson, D. S.; Korkovski, P. F. *JOC* **1990**, *55*, 4518.
23. (a) Padwa, A.; Krumpe, K. E.; Gareau, Y.; Chiacchio, U. *JOC* **1991**, *56*, 2523. (b) Padwa, A.; Kinder, F. R. *TL* **1990**, *31*, 6835.
24. Kennedy, M.; McKervey, M. A. *JCS(P1)* **1991**, 2565.
25. Doyle, M. P.; Shanklin, M. S.; Pho, H. Q. *TL* **1988**, *29*, 2639.
26. Taber, D. F.; Ruckle, R. E. *TL* **1985**, *26*, 3059.
27. Taber, D. F.; Ruckle, R. E. *JACS* **1986**, *108*, 7686.
28. Doyle, M. P.; Westrum, L. J.; Wolthuis, W. N. E.; See, M. M.; Boone, W. P.; Bagheri, V.; Pearson, M. M. *JACS* **1993**, *115*, 958.
29. (a) Adams, J.; Poupart, M.-A.; Grenier, L.; Schaller, C.; Quimet, N.; Frenette, R. *TL* **1989**, *30*, 1749. (b) Adams, J.; Poupart, M.-A.; Grenier, L. *TL* **1989**, *30*, 1753. (c) Spero, D. M.; Adams, J. *TL* **1992**, *33*, 1143.
30. (a) Doyle, M. P.; Taunton, J.; Pho, H. Q. *TL* **1989**, *30*, 5397. (b) Doyle, M. P.; Pieters, R. J.; Taunton, J.; Pho, H. Q.; Padwa, A.; Hertzog, D. L.; Precedo, L. *JOC* **1991**, *56*, 820.
31. Lee, E.; Jung, K. W.; Kim, Y. S. *TL* **1990**, *31*, 1023.
32. Doyle, M. P.; Shanklin, M. S.; Oon, S.-M.; Pho, H. Q.; van der Heide, F. R.; Veal, W. R. *JOC* **1988**, *53*, 3384.
33. (a) Wenkert, E. *ACR* **1980**, *13*, 27. (b) Wenkert, E. *H* **1980**, *14*, 1703.
34. Hrytsak, M.; Durst, T. *CC* **1987**, 1150.
35. (a) Doyle, M. P.; Shanklin, M. S.; Pho, H. Q.; Mahapatro, S. N. *JOC* **1988**, *53*, 1017. (b) Etkin, N.; Babu, S. D.; Fooks, C. J.; Durst, T. *JOC* **1990**, *55*, 1093.
36. Nakatani, K. *TL* **1987**, *28*, 165.
37. Taylor, E. C.; Davies, H. M. L. *TL* **1983**, *24*, 5453.
38. (a) Reider, P. J. Grabowski, E. J. J. *TL* **1982**, *23*, 2293. (b) Karady, S.; Amato, J. S.; Reamer, R. A.; Weinstock, L. M. *JACS* **1981**, *103*, 6765. (c) Sletzinger, M.; Liu, T.; Reamer, R. A.; Shinkai, I. *TL* **1980**, *21*, 4221. (d) Cama, L. D.; Christensen, B. G. *TL* **1978**, 4233.
39. (a) Sowin, T. J.; Meyers, A. I. *JOC* **1988**, *53*, 4154. (b) Fetter, J.; Lempert, K.; Gizur, T.; Nyitrai, J.; Kajtar-Peredy, M.; Simig, G.; Hornyak, G.; Doleschall, G. *JCS(P1)* **1986**, 221. (c) Mori, M.; Kagechika, K.; Sasai, H.; Shibasaki, M. *T* **1991**, *47*, 531.
40. (a) Heslin, J. C.; Moody, C. J. *JCS(P1)* **1988**, 1417. (b) Davies, M. J.; Moody, C. J.; Taylor, R. J. *JCS(P1)* **1991**, 1. (c) Davies, M. J.; Moody, C. J. *JCS(P1)* **1991**, 9. (d) Davies, M. J.; Moody, C. J.; Taylor, R. J. *SL* **1990**, 93. (e) Cox, G. G.; Moody, C. J.; Austin, D. J.; Padwa, A. *T* **1993**, *49*, 5109.
41. Hubert, A. J.; Noels, A. F.; Anciaux, A. J.; Teyssié, P. *S* **1976**, 600.
42. Moody, C. J.; Taylor, R. J. *TL* **1987**, *28*, 5351.
43. Moyer, M. P.; Feldman, P. L.; Rapoport, H. *JOC* **1985**, *50*, 5223.
44. Bagheri, V.; Doyle, M. P.; Taunton, J.; Claxton, E. E. *JOC* **1988**, *53*, 6158.
45. Doyle, M. P.; Tamblyn, W. H.; Bagheri, V. *JOC* **1981**, *46*, 5094.
46. Doyle, M. P.; Bagheri, V.; Harn, N. K. *TL* **1988**, *29*, 5119.
47. Doyle, M. P.; Griffin, J. H.; Chinn, M. S.; van Leusen, D. V. *JOC* **1984**, *49*, 1917.
48. (a) Gillespie, R. J.; Porter, A. E. A.; Willmott, W. E. *CC* **1978**, 85. (b) Gillespie, R. J.; Porter, A. E. A. *CC* **1979**, 50.
49. Moody, C. J.; Taylor, R. J. *TL* **1988**, *29*, 6005.
50. Davies, H. M. L.; Crisco, L. V. T. *TL* **1987**, *28*, 371.
51. Moody, C. J.; Slawin, A. M. Z.; Taylor, R. J.; Williams, D. J. *TL* **1988**, *29*, 6009.
52. Kametami, T.; Kanaya, N.; Mochizuki, T.; Honda, T. *H* **1983**, *20*, 455.

53. (a) Kido, F.; Sinha, S. C.; Abiko, T.; Yoshikoshi, A. *TL* **1989**, *30*, 1575. (b) Kido, F.; Sinha, S. C.; Abiko, T.; Watanabe, M.; Yoshikoshi, A. *CC* **1990**, 418.
54. (a) Prasad, K.; Kneussel, P.; Schulz, G.; Stütz, P. *TL* **1982**, *23*, 1247. (b) Kametani, T.; Kanaya, N.; Mochizuki, T.; Honda, T. *H* **1982**, *19*, 1023.
55. Kametami, T.; Kanaya, N.; Mochizuki, T.; Honda, T. *TL* **1983**, *20*, 221.
56. Chan, L.; Matlin, S. A. *TL* **1981**, *22*, 4025.
57. Doyle, M. P.; Bagheri, V.; Claxton, E. E. *CC* **1990**, 46.
58. Connell, R.; Scavo, F.; Helquist, P. *TL* **1986**, *27*, 5559.
59. Roskamp, E. J.; Johnson, C. R. *JACS* **1986**, *108*, 6062.
60. (a) Pirrung, M. C.; Werner, J. A. *JACS* **1986**, *108*, 6060. (b) Pirrung, M. C.; Brown, W. L.; Rege, S.; Laughton, P. *JACS* **1991**, *113*, 8561.
61. (a) Clark, J. S.; Krowiak, S. A.; Street, L. J. *TL* **1993**, *34*, 4385. (b) Clark, J. S. *TL* **1992**, *33*, 6193. (c) Kido, F.; Sinha, S. C.; Abiko, T.; Yoshikoshi, A. *TL* **1989**, *30*, 1575.
62. Eberlein, T. H.; West, F. G.; Tester, R. W. *JOC* **1992**, *57*, 3479.
63. Padwa, A.; Carter, S. P.; Nimmesgern, H.; Stull, P. D. *JACS* **1988**, *110*, 2894.
64. (a) Padwa, A.; Chinn, R. L.; Zhi, L. *TL* **1989**, *30*, 1491. (b) Padwa, A.; Fryxell, G. E.; Zhi, L. *JACS* **1990**, *112*, 3100.
65. Maier, M. E.; Evertz, K. *TL* **1988**, *29*, 1677.
66. Padwa, A.; Austin, D. J.; Hornbuckle, S. F.; Semones, M. A.; Doyle, M. P.; Protopopova, M. N. *JACS* **1992**, *114*, 1874.
67. Davies, H. M. L.; Cantrell Jr., W. R. *TL* **1991**, *32*, 6509.
68. Doyle, M. P.; Protopopova, M. N.; Brandes, B. D.; Davies, H. M. L.; Huby, N. J. S.; Whitesell, J. K. *SL* **1993**, 151.
69. Burgess, K.; Ohlmeyer, M. J. *CRV* **1991**, *91*, 1179.
70. Hui, B. C. Y.; Teo, W. K.; Rempel, G. L. *IC* **1973**, *12*, 757.
71. Doyle, M. P.; Devora, G. A.; Nefedov, A. O.; High, K. G. *OM* **1992**, *11*, 549.
72. Doyle, M. P.; Shanklin, M. S. *OM* **1993**, *12*, 11.

Michael P. Doyle
Trinity University, San Antonio, TX, USA

Dirhodium(II) Tetrakis(methyl 2-pyrrolidone-5(*S*)-carboxylate)

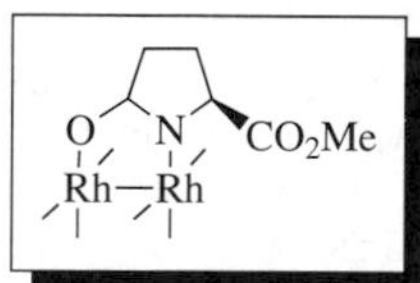

[131766-06-8] $C_{24}H_{36}N_4O_{12}Rh_2$ (MW 778.46)

(highly enantioselective catalyst for carbenoid reactions of diazo compounds)[1–3]

Physical Data: λ 615 nm, ε 211 ($ClCH_2CH_2Cl$). ^{1}H NMR ($CDCl_3$) of Rh_2(5*S*-MEPY)$_4$(MeCN)$_2$: δ 4.32 (dd, $J=8.8$, 3.0 Hz, 2H), 3.95 (dd, $J=8.6$, 2.1 Hz, 2H), 3.70 (s, 6H), 3.68 (s, 6H), 2.70–2.55 (m, 4H), 2.26 (s, 6H), 1.8–2.4 (m, 12H). $[\alpha]_D^{23} = -259.5°$ (MeCN, $c=0.098$).

Solubility: sol MeOH, MeCN, acetone; slightly sol CH_2Cl_2, $ClCH_2CH_2Cl$, toluene.

Form Supplied in: red crystals as the bis-acetonitrile complex; blue solid after removal of the axial nitrile ligands.

Preparative Method: from ***Dirhodium(II) Tetraacetate*** by ligand substitution with methyl 2-pyrrolidone-5(*S*)-carboxylate.[4,5]

Handling, Storage, and Precautions: air stable, weakly hygroscopic; stored in desiccator.

Introduction. The preparation of the title reagent, Rh_2(5*S*-MEPY)$_4$, is the same as that used for ***Dirhodium(II) Tetraacetamide***[6] or ***Dirhodium(II) Tetra(caprolactam)***.[7] Ligand exchange occurs in refluxing chlorobenzene, and the acetic acid that is liberated is trapped in a Soxhlet extraction apparatus by sodium carbonate. Purification occurs by chromatography on a CN-capped silica column; recrystallization from acetonitrile–2-propanol (1:1) provides Rh_2(5*S*-MEPY)$_4$(MeCN)$_2$(*i*-PrOH). Four 2-pyrrolidone-5(*S*)-carboxylate molecules ligate one dirhodium(II) nucleus; each rhodium is bound to two nitrogen and two oxygen donor atoms arranged in a *cis* configuration.[4] The methyl carboxylate substituents are positioned with a counterclockwise orientation on each rhodium face.

Metal Carbene Transformations. The effectiveness of Rh_2(5*S*-MEPY)$_4$ and its 5*R*-form, Rh_2(5*R*-MEPY)$_4$, is exceptional for highly enantioselective intramolecular cyclopropanation[8] and carbon–hydrogen insertion[9] reactions. Intermolecular cyclopropanation occurs with lower enantiomeric excesses[10] than with alternative chiral copper salicylaldimine[11] or C_2-symmetric semicorrin[12] or bis-oxazoline[13] copper catalysts, but intermolecular cyclopropenation exhibits higher enantiocontrol with Rh_2(MEPY)$_4$ catalysts.[14] The methyl carboxylate attachment of Rh_2(5*S*-MEPY)$_4$ is far more effective than sterically similar benzyl or isopropyl attachments for enantioselective metal carbene transformations.[4] The significant enhancement in enantiocontrol is believed to be due to carboxylate carbonyl stabilization of the intermediate metal carbene and/or to dipolar influences on substrate approach to the carbene center.

Enantioselective Intramolecular Cyclopropanation Reactions. The exceptional capabilities of the Rh_2(5*S*-MEPY)$_4$ and Rh_2(5*R*-MEPY)$_4$ catalysts for enantiocontrol are evident in results obtained with a series of allyl diazoacetates (eq 1).[5,8] Both high product yields and enantiomeric excess (ee's) are characteristic. Intramolecular cyclopropanation of (*Z*)-alkenes proceeds with a higher level of enantiocontrol than does intramolecular cyclopropanation of (*E*)-alkenes. In preparative scale reactions, less than 0.25 mol% of catalyst can be employed to achieve high yields of pure product.[5]

R^t, R^c, CHN_2, O, O — Rh_2(5*S*-MEPY)$_4$, CH_2Cl_2 → R^c, R^t, O, O (1)

R^t	R^c	% ee
H	H	95
Me	Me	98
H	Ph, Et, Bn, Bu_3Sn	≥94
Pr	H	83
Ph	H	65

Similar success in enantiocontrol has been achieved for intramolecular cyclopropanation of homoallyl diazoacetates (eq 2).[15] With these substrates the enantiomeric excesses do not extend beyond 90%, but they are virtually independent of double bond substituents.

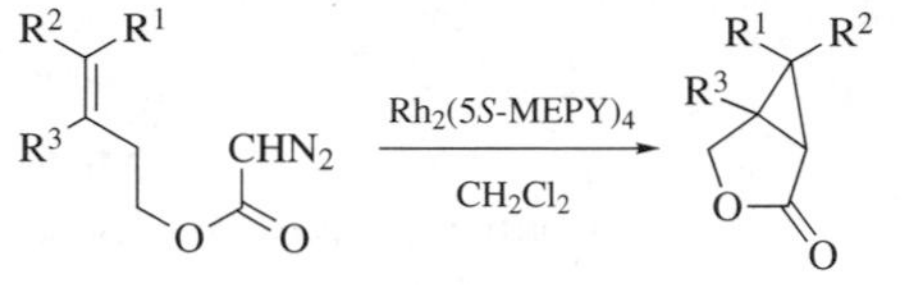

(2)

R^1	R^2	R^3	% ee
Me	Me	H	77
Et, Ph, Bn, TMS	H	H	80–90
H	Et, Ph	H	73–82
H	H	Me	79

Enantioselective Intermolecular Cyclopropenation Reactions. The use of $Rh_2(MEPY)_4$ catalysts for intermolecular cyclopropenation of 1-alkynes results in moderate to high selectivity. With propargyl methyl ether (or acetate), for example, reactions with (−)-menthyl [(+)-(1*R*,2*S*,5*R*)-2-isopropyl-5-methyl-1-cyclohexyl] diazoacetate catalyzed by $Rh_2(5S\text{-}MEPY)_4$ produces the corresponding cyclopropene product (eq 3) with 98% diastereomeric excess (de).[14,16]

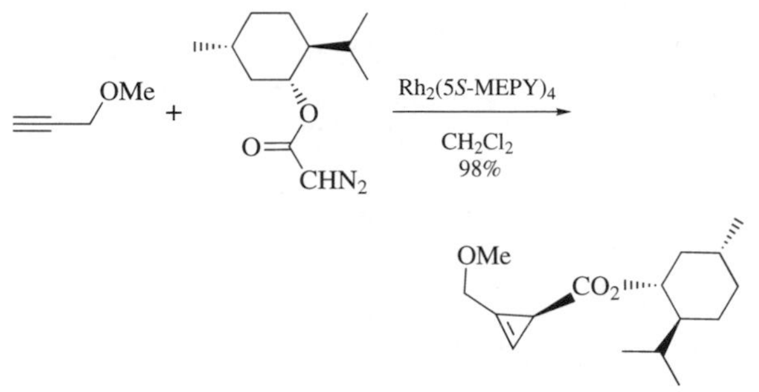

(3)

These reactions are subject to significant double diastereoselection with (+)- and (−)-menthyl diazoacetates. With ethyl diazoacetate, enantiomeric excesses are moderate (54–69% ee), but they increase up to 78% ee with *t*-butyl diazoacetate.[14] These are the first examples of enantioselective catalytic cyclopropenation reactions.

Enantioselective Intramolecular Carbon–Hydrogen Insertion Reactions. The suitability of $Rh_2(5S\text{-}MEPY)_4$ and $Rh_2(5R\text{-}MEPY)_4$ for enantioselective intramolecular C–H insertion reactions is evident in results with 2-alkoxyethyl diazoacetates (eq 4).[9] Both lactone enantiomers are available from a single diazo ester. Other examples have also been reported, especially those with highly branched diazo substrate structures.[9]

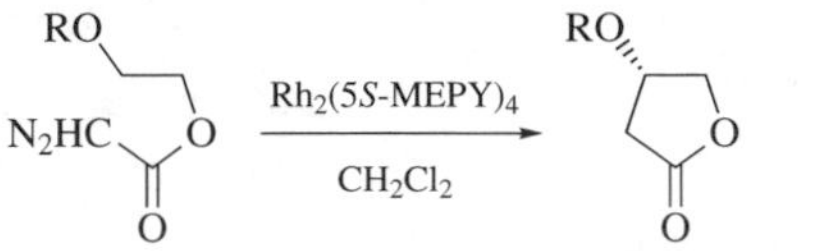

(4)

R = Me, 91% ee
R = Et, 89% ee
R = Bn, 87% ee

Diazoacetamides are robust diazo substrates, but they generally give lower enantioselection, and regioselectivity for γ-lactam formation is dependent on the substituents on carbon at which insertion occurs (e.g. eq 5).[17] With *N*-(*n*-butyl)-*N*-(*t*-butyl)diazoacetamide the ratio of γ:β-lactam is 88:12. A significant improvement in enantioselection (up to 78% ee) occurs with the use of the oxazolidinone analog of $Rh_2(5S\text{-}MEPY)_4$.[17]

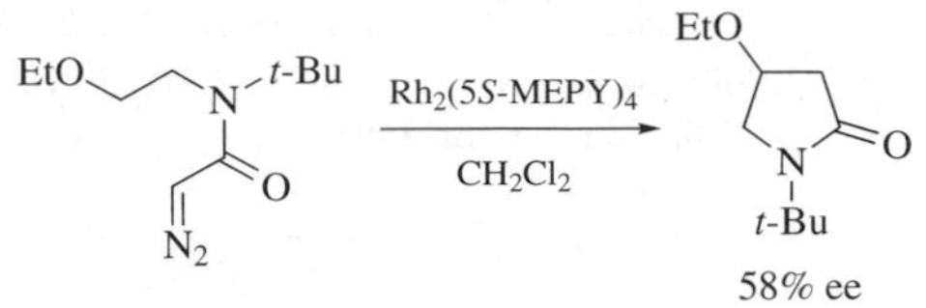

(5)

58% ee

Polyethylene-Bound, Soluble, Recoverable Dirhodium(II) 2-Pyrrolidone-5(*S*)-carboxylate. The homogeneous $Rh_2(5S\text{-}MEPY)_4$ catalyst has been attached to a polyethylene chain that is soluble in organic solvents at about 70 °C.[18] Ligand displacement of 2-pyrrolidone-5(*S*)-carboxylate from $Rh_2(5S\text{-}MEPY)_4$ by a soluble polyethylene-bound 2-pyrrolidone-5(*S*)-carboxylate produces a recoverable dirhodium(II) catalyst, PE–$Rh_2(5S\text{-}PYCA)_4$, in high yield. The effectiveness of this catalyst has been demonstrated by high enantioselection for intramolecular cyclopropanation of 3-methyl-2-buten-1-yl diazoacetate (see eq 1) in refluxing benzene solution (98% ee) and for intramolecular C–H insertion of 2-methoxyethyl diazoacetate (see eq 4) under the same conditions (72% ee). For both transformations, reactions catalyzed by $Rh_2(5S\text{-}MEPY)_4$ that occur at the same temperature give lower % ee values. Although diminished selectivity can occur with catalyst recovery and reuse under standard conditions, retention of catalyst effectiveness is achieved by using 2–3 mol % of the pyrrolidone ligand in up to seven subsequent runs with recovered, reused PE–$Rh_2(5S\text{-}PYCA)_4$.

Related Reagents. (S,S)-2,2'-(Dimethylmethylene)bis(4-t-butyl-2-oxazoline); Dirhodium(II) Tetraacetamide; Dirhodium(II) Tetraacetate; Dirhodium(II) Tetra(caprolactam); Dirhodium(II) Tetrakis(perfluorobutyrate); Dirhodium(II) Tetrakis(trifluoroacetate); Dirhodium(II) Tetraoctanoate.

1. Doyle, M. P. In *Selectivity in Catalysis*; Davis, M. E.; Suib, S. L., Eds.; American Chemical Society: Washington, 1993.
2. Doyle, M. P. In *Catalytic Asymmetric Synthesis*; Ojima, I., Ed.; VCH: New York, 1993.
3. Doyle, M. P. *RTC* **1991**, *110*, 305.
4. Doyle, M. P.; Winchester, W. R.; Hoorn, J. A. A.; Lynch, V.; Simonsen, S. H.; Ghosh, R. *JACS* **1993**, *115*, 9968.
5. Doyle, M. P.; Winchester, W. R.; Protopopova, M. N.; Kazala, A. P.; Westrum, L. J. *OS* **1994**, *73*, in press.
6. Doyle, M. P.; Bagheri, V.; Wandless, T. J.; Harn, N. K.; Brinker, D. A.; Eagle, C. T.; Loh, K.-L. *JACS* **1990**, *112*, 1906.
7. Doyle, M. P.; Westrum, L. J.; Wolthuis, W. N. E.; See, M. M.; Boone, W. P.; Bagheri, V.; Pearson, M. M. *JACS* **1993**, *115*, 958.
8. Doyle, M. P.; Pieters, R. J.; Martin, S. F.; Austin, R. E.; Oalmann, C. J.; Müller, P. *JACS* **1991**, *113*, 1423.
9. Doyle, M. P.; Van Oeveren, A.; Westrum, L. J.; Protopopova, M. N.; Clayton, T. W., Jr. *JACS* **1991**, *113*, 8982.
10. Doyle, M. P.; Brandes, B. D.; Kazala, A. P.; Pieters, R. J.; Jarstfer, M. B.; Watkins, L. M.; Eagle, C. T. *TL* **1990**, *31*, 6613.
11. Aratani, T. *PAC* **1985**, *57*, 1839.
12. Pfaltz, A. *ACR* **1993**, *26*, 339.

13. (a) Evans, D. A.; Woerpel, K. A.; Hinman, M. M.; Faul, M. M. *JACS* **1991**, *113*, 726. (b) Lowenthal, R. E.; Masamune, S. *TL* **1991**, *32*, 7373. (c) Müller, D.; Umbricht, G.; Weber, B.; Pfaltz, A. *HCA* **1991**, *74*, 232.
14. Protopopova, M. N.; Doyle, M. P.; Müller, P.; Ene, D. *JACS* **1992**, *114*, 2755.
15. Martin, S. F.; Oalmann, C. J.; Liras, S. *TL* **1992**, *33*, 6727.
16. Doyle, M. P.; Protopopova, M. N.; Brandes, B. D.; Davies, H. M. L.; Huby, N. J. S.; Whitesell, J. K. *SL* **1993**, 151.
17. Doyle, M. P.; Protopopova, M. N.; Winchester, W. R.; Daniel, K. L. *TL* **1992**, *33*, 7819.
18. Doyle, M. P.; Eismont, M. Y.; Bergbreiter, D. E.; Gray, H. N. *JOC* **1992**, *57*, 6103.

Michael P. Doyle
Trinity University, San Antonio, TX, USA

(1*R*,2*S*)-Ephedrine

[299-42-3] $C_{10}H_{15}NO$ (MW 165.23)

(chiral auxiliary for the following: diastereoselective alkylation and reduction of chiral hydrazones; diastereoselective alkylation of chiral amides; diastereoselective conjugate addition of organometallic reagents to unsaturated amides and imidazolidinones; diastereoselective alkylation and cyclopropanation of oxazepinediones and oxazolidines; diastereoselective homoaldol addition of *N*-allylimidazolidinone, and asymmetric coupling reaction of Grignard reagents; chiral ligand for enantioselective conjugate addition of organometallic reagents to enones; chiral ligand for enantioselective addition of dialkylzincs to aldehydes)

Alternate Name: [*R*-(*R**,*S**)]-α-[1-(Methylamino)ethyl]benzenemethanol.

Physical Data: mp 37–39 °C; bp 255 °C; $[\alpha]_D^{21}$ −41° (*c* 5, 1M HCl). Hydrochloride, mp 216–220 °C; $[\alpha]_D^{20}$ −34° (*c* 4, H_2O).

Solubility: sol alcohol, chloroform, ether, water.

Form Supplied in: waxy solid or crystals; also available as hydrochloride in either enantiomeric form.

General Features of Ephedrine. Ephedrine is a chiral β-amino alcohol which is available in either enantiomeric form. It is often utilized as a chiral auxiliary in asymmetric synthesis. By bond formation with the amino group of ephedrine, ephedrine can be derived into chiral hydrazones[2,3] and amides.[5,6] Highly diastereoselective asymmetric reactions are known using these chiral compounds. In reactions using organometallic reagents, the hydroxy groups of hydrazones and amides become metal alkoxides. Metal atoms of the alkoxide may chelate with nitrogen or oxygen atoms of chiral hydrazones and amides. This chelation may reduce the number of possible conformations of reactive species, and this may increase the diastereoselectivities.

On the other hand, by bond formation with amino and hydroxy groups of ephedrine, ephedrine can be converted into chiral ring systems such as imidazolidinones,[12,13] oxazepinediones,[14–18] and oxazolidines.[20,21] Diastereoselective reactions of derivatives of these chiral ring systems afford compounds with high de. The relatively rigid conformation of these ring systems is one of the reasons for high diastereoselectivities.

Ephedrine becomes a chiral ligand of metal atoms by the deprotonation of the hydroxy group and by the presence of the nitrogen atom.[22–26] Highly enantioselective asymmetric reactions are known using chiral ephedrine-type ligands.

In addition, ephedrine is a chiral base catalyst because of the presence of the amine group.[29–32] A highly enantioselective base-catalyzed reaction is known.

Diastereoselective Alkylation and Reduction of Chiral Hydrazones Derived from Ephedrine[2]. ***Methylmagnesium Bromide*** adds to the chiral hydrazone derived from *N*-aminoephedrine and benzaldehyde to afford the optically active chiral hydrazine in almost 100% de. Hydrogenolysis of the chiral hydrazine gives (*R*)-α-phenylethylamine with more than 97% ee (eq 1). Ephedrine is recovered in good yield and without any loss of enantiomeric purity.

MeMgBr; Pd/C, H_2; >97% ee (1)

On the other hand, the diastereoselective reduction of the chiral hydrazone derived from *N*-aminoephedrine and acetophenone and subsequent hydrogenolysis affords (*S*)-α-phenylethylamine with 30% ee.[3] Optically active α-phenylethylamine with high ee is obtained from the diastereoselective alkylation of chiral hydrazones derived from (*R*)- or (*S*)-1-amino-2-(methoxymethyl)pyrrolidine.[4a]

Diastereoselective Alkylation of Chiral Amides Derived from Ephedrine. Chiral amides derived from ephedrine are converted to the corresponding dianion. The subsequent diastereoselective alkylation with alkyl iodides affords chiral α-substituted amides with >90% de.[5] Acid hydrolysis affords optically active α-substituted acids with 78% ee as a result of racemization in the cleavage step (eq 2).

1. 2 equiv Mg, base; 2. R^2I, 90–95%; >90% de; H_3O^+, 55–68%; 77–78% ee (2)

On the other hand, treatment with ***Methyllithium*** affords optically active methyl ketone in 44–74% ee, also as a result of racemization. α-Chiral ketones with higher ee (99% ee) are obtained from the diastereoselective alkylation of chiral hydrazones derived from (*R*)- or (*S*)-1-amino-2-methoxymethylproline.[4b]

Diastereoselective Conjugate Addition of Organometallic Reagents to Chiral α,β-Unsaturated Amides and Imidazolidinones Derived from Ephedrine. Grignard reagents (2 equiv) add to chiral α,β-unsaturated amides derived from ephedrine in a 1,4-addition manner with high diastereoselectivities. Subsequent acidic hydrolysis affords optically active β-substituted carboxylic acids with 85–99% ee (eq 3).[6]

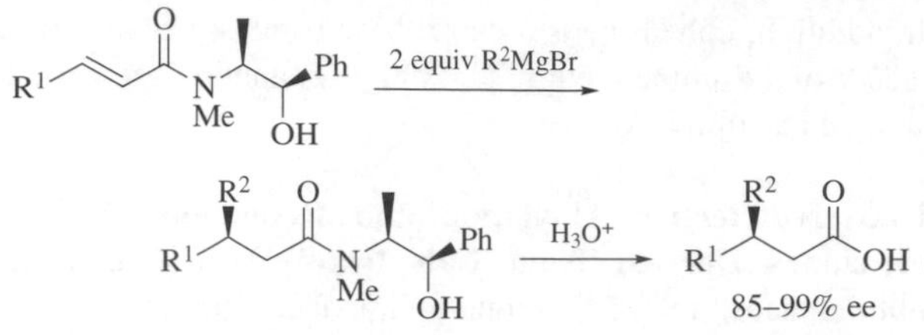

(3)

A seven-membered chelate intermediate is one of the reasons for the very high diastereoselectivities. The method is successfully applied to the asymmetric synthesis of malingolide.[7] Similar results are obtained in diastereoselective conjugate addition of Grignard reagents to unsaturated amides derived from (*S*)-2-(1-hydroxy-1-methylethyl)pyrrolidine. The presence of a tertiary amine (e.g. ***1,8-Diazabicyclo[5.4.0]undec-7-ene***) increases the diastereoselectivity, and subsequent hydrolysis affords β-substituted carboxylic acids with up to 100% ee.[8] Conjugate additions of alkyllithium or Grignard reagents to chiral *N*-crotonoylproline,[9] imides,[10] and *N*-enoyl sultams[11] also afford β-substituted carboxylic acids with 60% ee, 96% ee, and 96% ee, respectively.

The chiral imidazolidinone[12] derived from urea and ephedrine hydrochloride is utilized in a diastereoselective conjugate methylation.[13] Subsequent hydrolysis affords optically pure (−)-citronellic acid (eq 4).

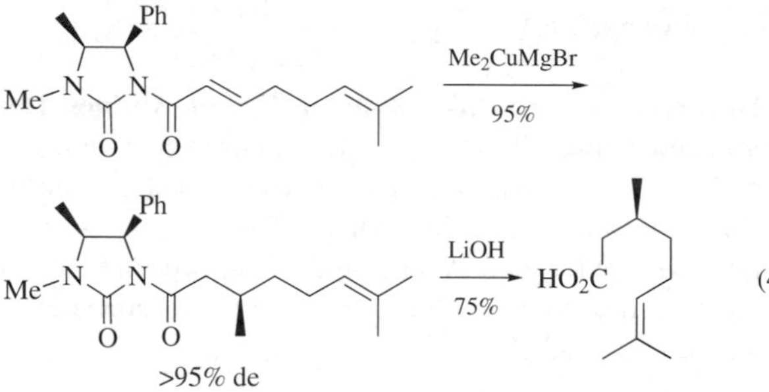

(4)

Diastereoselective Conjugate Additions to Chiral Oxazepinediones Derived from Ephedrine. Ephedrine can form a chiral seven-membered relatively rigid oxazepinedione ring by condensation with malonic acid monoester. Alkylidene derivatives of chiral oxazepinediones undergo highly diastereoselective additions with nucleophilic reagents. Grignard reagents in the presence of a catalytic amount of ***Nickel(II) Chloride*** add to chiral alkylideneoxazepinediones. Acid hydrolysis affords optically active β-substituted acids with up to >99% ee (eq 5).[14] The method is applied to the diastereoselective synthesis of (−)-indolmycin with 93% ee.[15]

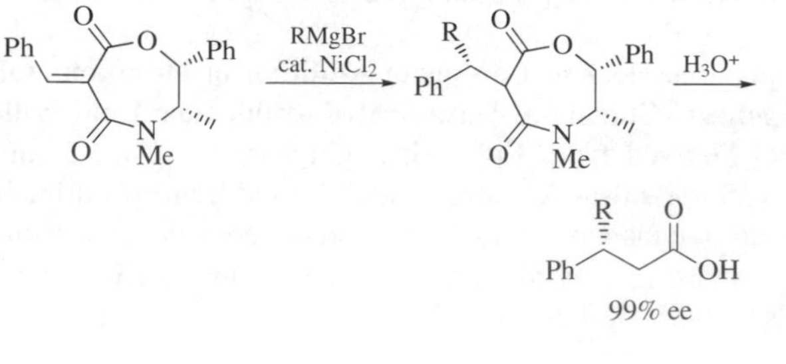

(5)

Diastereoselective addition of sulfoxonium ylides affords enantiomerically pure cyclopropanedicarboxylic acid diesters after removal of the chiral auxiliary (eq 6).[16]

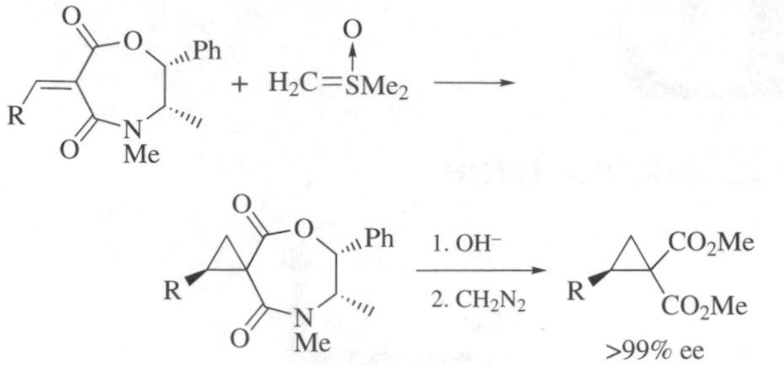

(6)

Diastereoselective addition of ***Phenylthiomethyllithium*** and subsequent treatment affords optically active lactones with >90% ee (eq 7).[17]

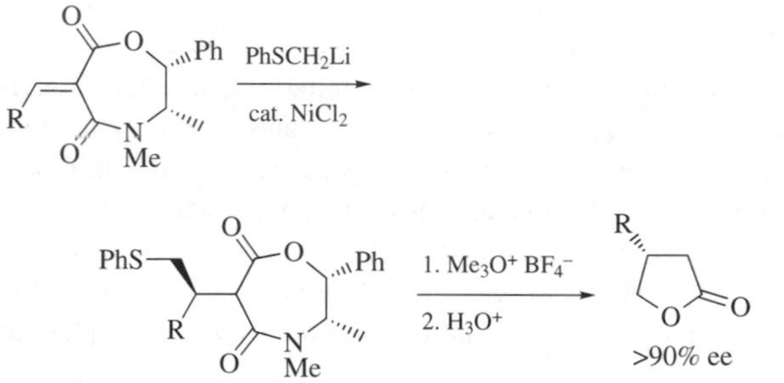

(7)

In addition, a chiral oxazepinedione plays the role of a nucleophile in the reaction with nitroalkenes in the presence of ***Potassium t-Butoxide*** and crown ether (eq 8).[18]

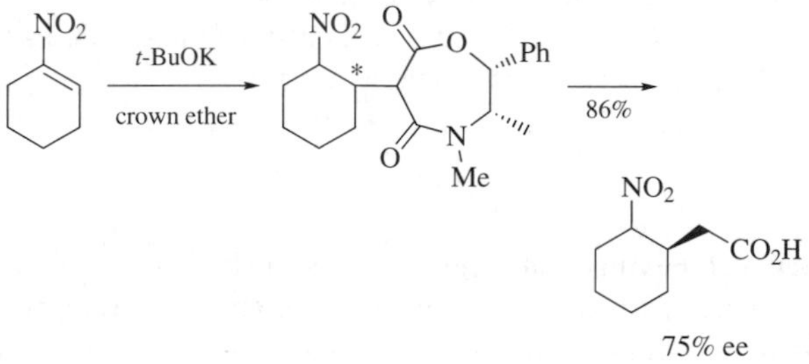

(8)

Homoaldol Addition with Chiral *N*-Allylimidazolidinone Derived from Ephedrine. The chiral allyltitanium compound derived from ephedrine reacts with carbonyl compounds with very high (>200:1) de. Subsequent hydrolysis and oxidation affords optically pure 4-substituted γ-lactones (eq 9).[12] 4-Substituted γ-lactones with 92% ee can also be synthesized by catalytic enantioselective alkylation of 3-formyl esters.[19]

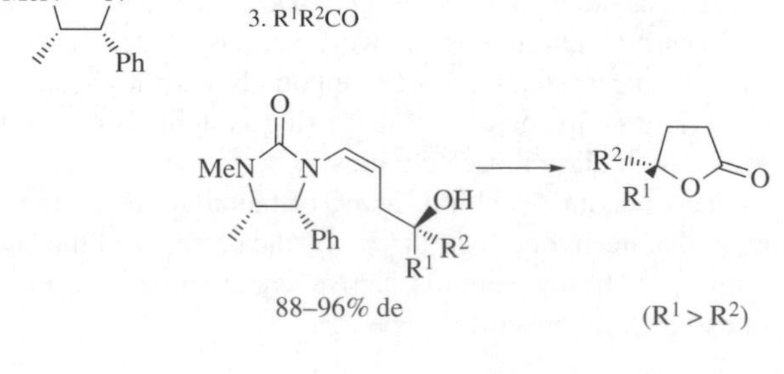

(9)

Diastereoselective Cyclopropanation and Alkylation of Chiral Oxazolidines Derived from Ephedrine. Ephedrine forms oxazolidines upon reaction with aldehydes. Chiral unsaturated oxazolidines derived from ephedrine and unsaturated aldehydes are treated with diazomethane in the presence of ***Palladium(II) Acetate***. Hydrolysis of the oxazolidine ring affords optically active formylcyclopropanes with >90% ee (eq 10).[20]

1. CH_2N_2, $Pd(OAc)_2$; 2. SiO_2, H_2O — 90% ee (10)

Diastereoselective addition of cuprate reagents to unsaturated oxazolidines and subsequent hydrolysis affords 3-substituted aldehydes with up to 81% ee (eq 11).[21]

1. RCu, LiX; 2. H_2O, KCN — 81% ee (11)

Asymmetric Coupling Reactions of Chiral Grignard Reagents Derived from Ephedrine Derivatives. Asymmetric coupling reactions of ***Allyl Bromide*** and chiral Grignard reagents derived from ephedrine methyl ether in the presence of ***Copper(I) Iodide*** (10 mol %) followed by oxidation affords optically active homoallyl alcohols with 60% ee (eq 12).[22]

1. allyl bromide, CuI; 2. HCl; 3. H_2O_2, KHF_2 — 60% ee (12)

Enantioselective Conjugate Addition to Prochiral Enones of Organometallic Reagents Modified with Ephedrine. Enantioselective conjugate addition to 2-cyclohexenone with chiral organo(alkoxo)cuprates [MCu(OR*)R] has been studied.[1a] When the cuprate is prepared from the lithium alkoxide of ephedrine, ***Phenyllithium***, and CuI, 3-phenylcyclohexanone with 50% ee is obtained.[23] The enantioselectivity reaches 92% ee in enantioselective ethylation when a chiral diamino alcohol derived from ephedrine is employed (eq 13).[24]

LiCu(OR*)R + 2-cyclohexenone — 52–90% — 77–92% ee (13)

On the other hand, enantioselective conjugate addition to 2-cyclohexenone with lithium dibutylcuprates (having a noncovalently bound chiral phosphorus ligand derived from ephedrine) affords 3-butylcyclohexanone with up to 76% ee (eq 14).[25]

BuLi, CuI, 4LiBr, 61% — 76% ee (14)

Isopropylmagnesium chloride adds to 2-cyclohexenone in 17% ee in the presence of a catalytic amount of chiral alkoxyzinc chloride derived from ephedrine and ***Zinc Chloride*** (eq 15).[26]

2-cyclohexenone + *i*-PrMgCl — chiral catalyst, 93% — 17% ee (15)

Concerning the catalytic enantioselective conjugate addition reaction, conjugate addition of dialkylzinc to chalcone in the presence of a catalytic amount of the chiral nickel complex derived from norephedrine affords β-substituted ketones with up to 90% ee (eq 16).[27]

chalcone + R_2Zn — chiral catalyst, $Ni(acac)_2$ — 90% ee (16)

Enantioselective Addition of Dialkylzincs to Aldehydes Using Chiral Amino Alcohols Derived from Ephedrine. Nucleophilic addition of dialkylzinc to aldehydes is usually very slow. Amino alcohols facilitate the addition of ***Diethylzinc*** to benzaldehyde to afford 1-phenylpropanol.[1b,28] When chiral amino alcohols possessing the appropriate structure are used as a precatalyst, optically active secondary alcohols are obtained.[1b] Highly enantioselective chiral catalysts derived from ephedrine are known. (1*R*,2*S*)-*N*-Isopropylephedrine functions as a precatalyst for the enantioselective addition of diethylzinc to benzaldehyde to afford (*R*)-1-phenylpropanol with 80% ee in 72% yield.[29] The use of an excess amount of diethylzinc increases the enantioselectivity up to 97% ee (eq 17).[30]

RCHO + Et_2Zn — chiral cat., 72–90% — 78–97% ee (17)

The lithium salt of (1*R*,2*S*)-*N*-[2-(dimethylamino)-ethyl]ephedrine acts as a precatalyst for the addition of diethylzinc to afford the alcohol with 90% ee (eq 18).[31]

RCHO + Et_2Zn — chiral cat. — 90% ee (18)

The dilithium salt of a chiral diaminodiol derived from ephedrine mediates the enantioselective addition of dialkylzinc to aldehydes to afford (*R*)-1-phenylethanol with 85% ee (eq 19).[32]

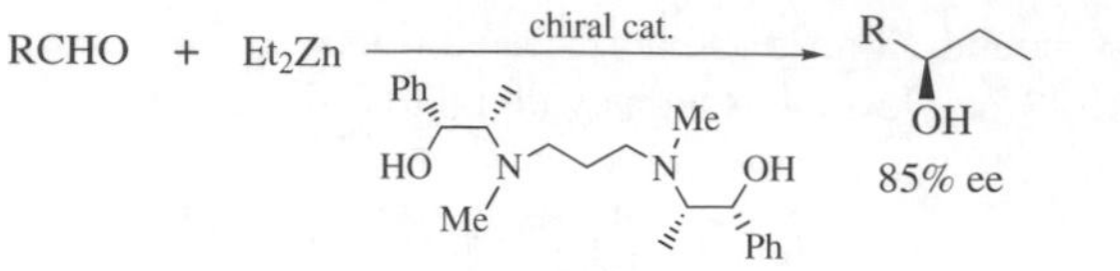

RCHO + Et_2Zn —chiral cat.→ R(OH) 85% ee (19)

Related Reagents. (*S*)-1-Amino-2-hydroxymethylindoline; (*S*)-1-Amino-2-methoxymethylpyrrolidine; (*S*)-4-Benzyl-2-oxazolidinone; Ephedrine-borane.

---

1. (a) Rossiter, B. E.; Swingle, N. M. *CRV* **1992**, *92*, 771. (b) Soai, K.; Niwa, S. *CRV* **1992**, *92*, 833.
2. Takahashi, H.; Tomita, K.; Noguchi, H. *CPB* **1981**, *29*, 3387.
3. Takahashi, H.; Tomita, K.; Otomasu, H. *CC* **1979**, 668.
4. (a) Enders, D.; Schubert, H.; Nübling, C. *AG(E)* **1986**, *25*, 1109. (b) Enders, D.; Eichenauer, H. *AG(E)* **1979**, *18*, 397.
5. Larcheveque, M.; Ignatova, E.; Cuvigny, T. *TL* **1978**, 3961.
6. Mukaiyama, T.; Iwasawa, N. *CL* **1981**, 913.
7. Kogure, T.; Eliel, E. L. *JOC* **1984**, *49*, 576.
8. (a) Soai, K.; Machida, H.; Yokota, N. *JCS(P1)* **1987**, 1909. (b) Soai, K.; Machida, H.; Ookawa, A. *CC* **1985**, 469.
9. Soai, K.; Ookawa, A. *JCS(P1)* **1986**, 759.
10. Tomioka, K.; Suenaga, T.; Koga, K. *TL* **1986**, *27*, 369.
11. (a) Oppolzer, W.; Mills, R. J.; Pachinger, W.; Stevenson, T. *HCA* **1986**, *69*, 1542. (b) Oppolzer, W.; Poli, G.; Kingma, A. J.; Starkemann, C.; Bernardinelli, G. *HCA* **1987**, *70*, 2201.
12. Roder, H.; Helmchen, G.; Peters, E. M.; Peters, K.; von Schnering, H. G. *AG(E)* **1984**, *23*, 898.
13. Stephan, E.; Pourcelot, G.; Cresson, P. *CI(L)* **1988**, 562.
14. (a) Mukaiyama, T.; Takeda, T.; Osaki, M. *CL* **1977**, 1165. (b) Mukaiyama, T.; Takeda, T.; Fujimoto, K. *BCJ* **1978**, *51*, 3368.
15. Takeda, T.; Mukaiyama, T. *CL* **1980**, 163.
16. Mukaiyama, T.; Fujimoto, K.; Takeda, T. *CL* **1979**, 1207.
17. Mukaiyama, T.; Fujimoto, K.; Hirose, T.; Takeda, T. *CL* **1980**, 635.
18. (a) Mukaiyama, T.; Hirako, Y.; Takeda, T. *CL* **1978**, 461. (b) Takeda, T.; Hoshiko, T.; Mukaiyama, T. *CL* **1981**, 797.
19. Soai, K.; Yokoyama, S.; Hayasaka, T.; Ebihara, K. *CL* **1988**, 843.
20. Abdallah, H.; Gree, R.; Carrie, R. *TL* **1982**, *23*, 503.
21. Berlan, J.; Besace, Y.; Pourcelot, G.; Cresson, P. *T* **1986**, *42*, 4757.
22. Tamao, K.; Kanatani, R.; Kumada, M. *TL* **1984**, *25*, 1913.
23. Bertz, S. H.; Dabbagh, G.; Sundararajan, G. *JOC* **1986**, *51*, 4953.
24. Corey, E. J.; Naef, R.; Hannon, F. J. *JACS* **1986**, *108*, 7114.
25. Alexakis, A.; Mutti, S.; Normant, J. F. *JACS* **1991**, *113*, 6332.
26. Jansen, J. F. G. A.; Feringa, B. L. *JOC* **1990**, *55*, 4168.
27. Soai, K.; Hayasaka, T.; Ugajin, S. *CC* **1989**, 516.
28. (a) Sato, T.; Soai, K.; Suzuki, K.; Mukaiyama, T. *CL* **1978**, 601. (b) Mukaiyama, T.; Soai, K.; Sato, T.; Shimizu, H.; Suzuki, K. *JACS* **1979**, *101*, 1455.
29. Chaloner, P. A.; Perera, S. A. R. *TL* **1987**, *28*, 3013.
30. (a) Chaloner, P. A.; Langadianou, E. *TL* **1990**, *31*, 5185. (b) Chaloner, P. A.; Langadianou, E.; Perera, S. A. R. *JCS(P1)* **1991**, 2731.
31. (a) Corey, E. J.; Hannon, F. J. *TL* **1987**, *28*, 5233. (b) Corey, E. J.; Hannon, F. J. *TL* **1987**, *28*, 5237.
32. Soai, K.; Nishi, M.; Ito, Y. *CL* **1987**, 2405.

Kenso Soai
Science University of Tokyo, Japan

Epichlorohydrin[1]

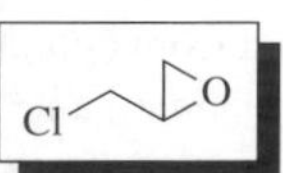

[106-89-8] C_3H_5ClO (MW 92.53)
(*R*)
[51594-55-9]
(*S*)
[67843-74-7]

(readily available three-carbon unit functionalized on every carbon; convenient HCl or HBr trap; linker for various polymers)

Alternate Name: chloromethyloxirane.
Physical Data: mp −57 °C; bp 115–117 °C; *d* 1.183 g cm^{-3}.
Solubility: 6.6 wt% in water; sol alcohol, acetone, THF, toluene, *n*-heptane.
Form Supplied in: neat liquid; both enantiomers available.
Handling, Storage, and Precautions: should only be handled in a well ventilated fume hood because of its low permissible exposure limit of 2 ppm and reports of allergic skin reactions and lung, liver, and kidney damage. MSDSs are available from the two principal manufacturers (Dow and Shell). The material is not moisture or air sensitive.

Introduction. Epichlorohydrin (**1**) is most widely used in polymer synthesis.[2] Other common uses include an in situ trapping agent for HCl, HBr,[3] or the alcohol generated during formation of Meerwein's reagent (eq 1).[4]

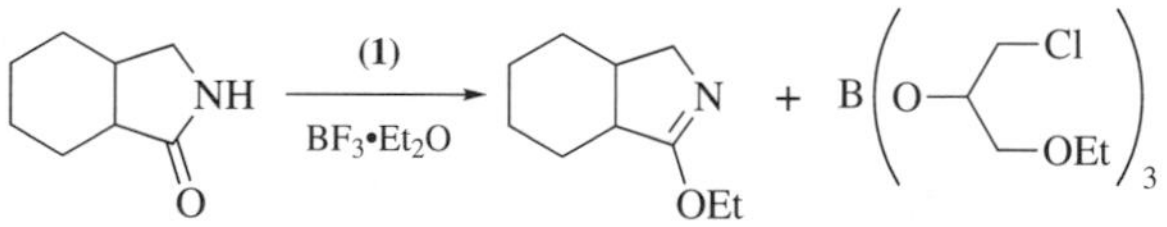

(lactam, NH, O) —(1), $BF_3 \cdot Et_2O$→ (imidate, N, OEt) + B(O–CH(CH$_2$Cl)(CH$_2$OEt))$_3$ (1)

Reactions with Nucleophiles. The epoxide is, by far, the more reactive site and a wide variety of nucleophiles have been used (eq 2) to open the ring at C-3 such as HCl (96%),[5] HOAc (>50%),[6] H_2S (65% as cyclized product 3-thietanol),[7] HCN (66%),[8] ethanol (90%),[9] *t*-butanol (86%),[10] phenyl or benzyl thiol (99% or 93%, respectively),[11] and phenyl selenide (generated in situ from the diselenide and sodium hydroxymethyl sulfite) (>55%).[12] If desired, the epoxide is easily formed from the chlorohydrin by treatment with excess KOH or Et_3N.

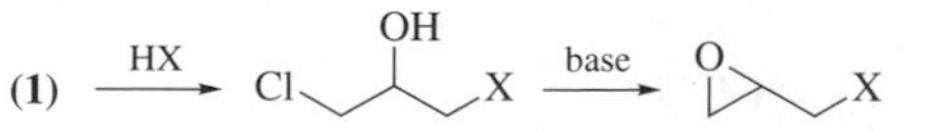

(**1**) —HX→ Cl–CH$_2$CH(OH)CH$_2$–X (**2**) —base→ (epoxide)–CH$_2$X (2)

The epoxide is also opened at C-3 by various electrophilic reagents that fit into the generalized scheme in eq 3. Examples include ***Chlorotrimethylsilane*** (TMSCl) (85%),[13] TMSCl/NaBr (X = Br) (85%),[14] ***Cyanotrimethylsilane*** (91%),[15] ***Azidotrimethylsilane*** (83%),[16] ***Thionyl Chloride*** (70%),[17] H_2NCOCl (96%),[18] and MeCH=CHCOCl (80%).[19]

The only report of unusual selectivity for opening the epoxide at C-2 was for ***Sulfuryl Chloride*** (eq 4).[20]

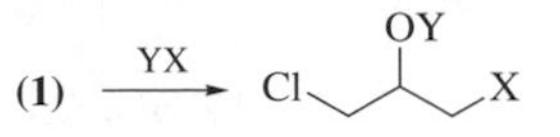

(1) —YX→ Cl–CH$_2$–CH(OY)–CH$_2$–X (3)

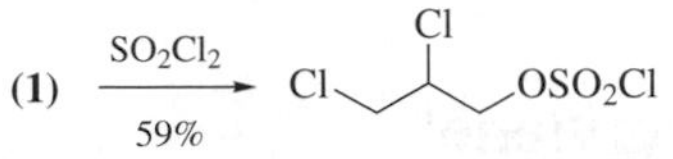

(1) —SO_2Cl_2, 59%→ Cl–CH$_2$–CH(Cl)–CH$_2$–OSO_2Cl (4)

A number of special catalysts have been developed to facilitate ring opening and improve the regioselectivity for reaction at C-3. For example, Sn^{II} halides are useful in preparations of (**2**) (X = Cl, 70%; X = Br, 63%; X = I, 90%).[21] An equimolar mixture of ***Lithium Bromide*** and ***Copper(II) Bromide*** gave (**2**) (X = Br, 93%).[22] The ring can be opened selectively by anilines in the presence of other amines when ***Cobalt(II) Chloride*** is the catalyst.[23] $MgSO_4$ was found to catalyze the addition of 2 mol of CN^- to (**1**) to afford 3-hydroxyglutaronitrile.[24] CaF_2 supported on KF was used in the conversion of (**1**) to epifluorohydrin.[25] A catalyst composed of a 1:2 mole ratio of ***Di-n-butyltin Oxide*** and tributyl phosphate was developed for ring opening by alcohols.[26] Other catalysts shown to be of value for the examples given above include $FeCl_3$,[6] $LiClO_4$,[11] Et_3N,[11] CAN,[9] DDQ,[10] Ti(O-*i*-Pr)$_4$,[15b] $CoCl_2$,[13,19] $YbCl_3$,[15a] and Al(O-*i*-Pr)$_3$.[16]

A variety of carbon nucleophiles react at C-3 with high regioselectivity. Examples include Grignard reagents,[27] aryllithium,[28] alkynyllithium,[29] and others (eqs 5–8).[30–33]

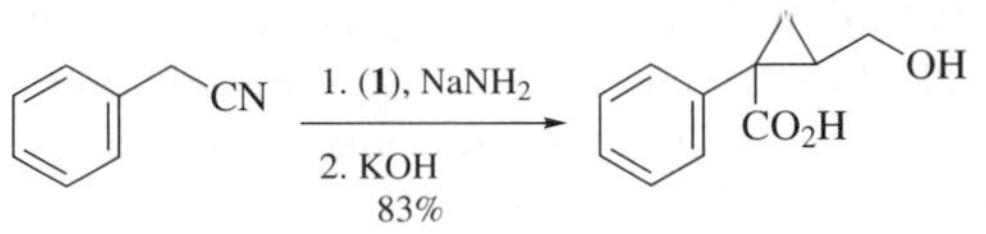

PhCH$_2$CN —1. (1), $NaNH_2$; 2. KOH; 83%→ (5)

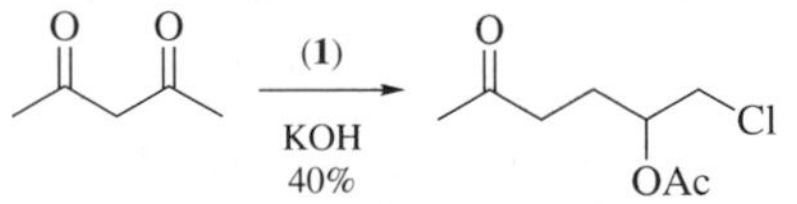

CO_2H, OH

acetylacetone —(1), KOH, 40%→ (6)

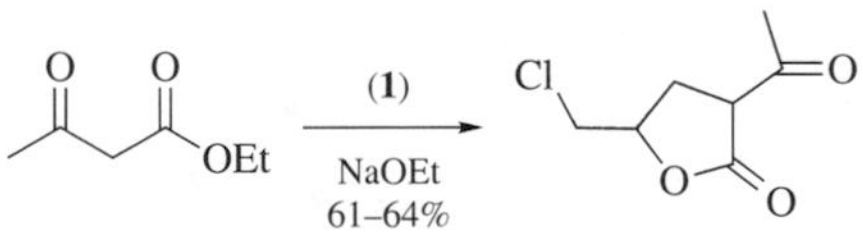

O, OAc, Cl

ethyl acetoacetate (OEt) —(1), NaOEt, 61–64%→ Cl (7)

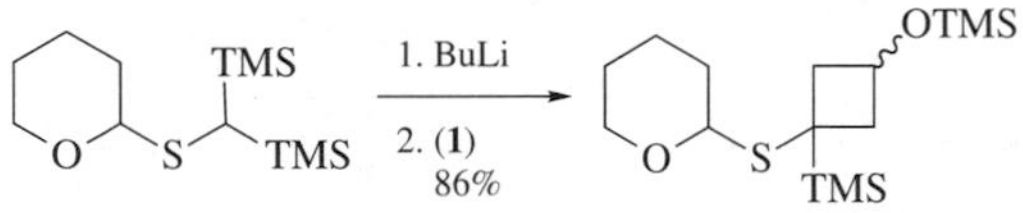

tetrahydropyranyl–S–CH(TMS)$_2$ —1. BuLi; 2. (1); 86%→ OTMS, TMS (8)

Compound (**1**) can be chain extended by one carbon by a Co-catalyzed CO insertion followed by reduction (eq 9).[34] A whole class of medicinally important compounds called β-blockers are prepared from (**1**) as illustrated in eq 10 for the synthesis of a propranolol analog.[35]

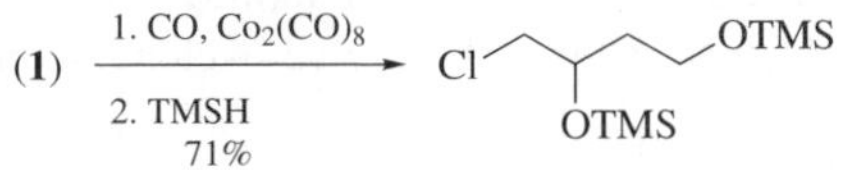

(1) —1. CO, $Co_2(CO)_8$; 2. TMSH; 71%→ Cl, OTMS, OTMS (9)

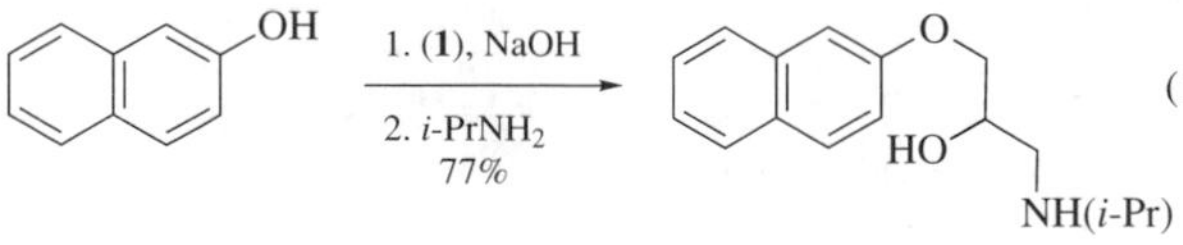

2-naphthol (OH) —1. (1), NaOH; 2. *i*-$PrNH_2$; 77%→ O, HO, NH(*i*-Pr) (10)

Preparation of Heterocycles. A wide array of heterocycles are available from (**1**). A few examples are shown in (eqs 11–16).[36–41]

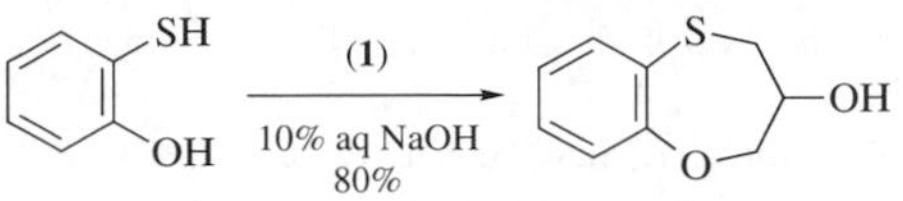

2-mercaptophenol (SH, OH) —(1), 10% aq NaOH, 80%→ S, OH, O (11)

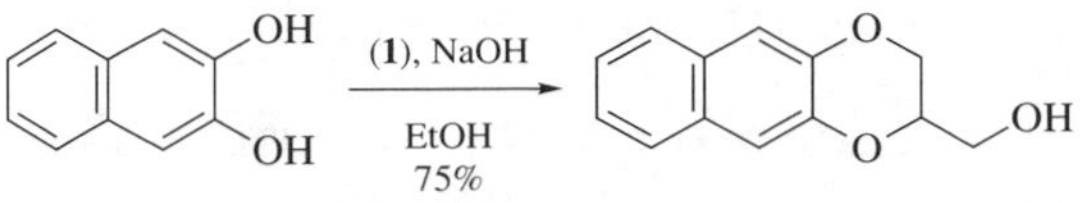

2,3-dihydroxynaphthalene (OH, OH) —(1), NaOH, EtOH, 75%→ O, O, OH (12)

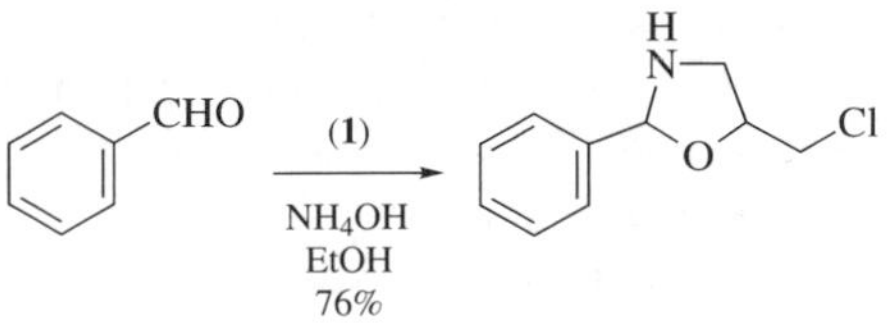

PhCHO —(1), NH_4OH, EtOH, 76%→ H, N, O, Cl (13)

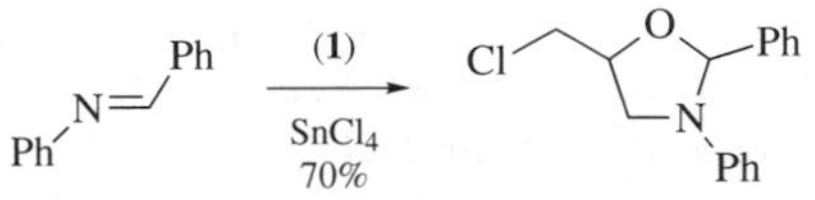

Ph–N=CH–Ph —(1), $SnCl_4$, 70%→ Cl, O, Ph, N, Ph (14)

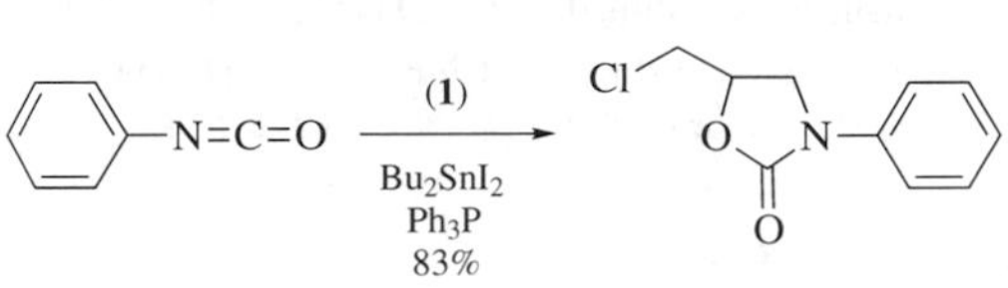

Ph–N=C=O —(1), Bu_2SnI_2, Ph_3P, 83%→ Cl, O, N, O (15)

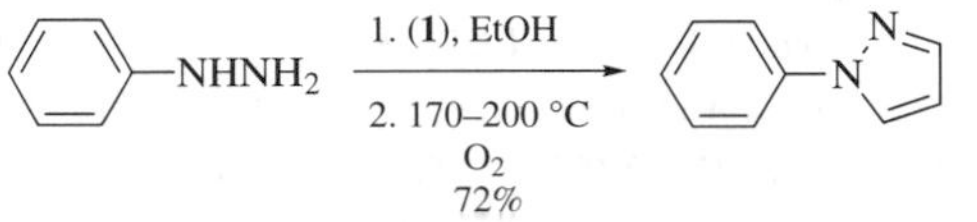

Ph–$NHNH_2$ —1. (1), EtOH; 2. 170–200 °C, O_2; 72%→ N, N (16)

In some cases the reaction rate and yield are dramatically improved if the product can be trapped as the TMS ether (eq 17).[42] Also, (**1**) lends itself well to nucleophilic opening under phase transfer conditions (eq 18).[43]

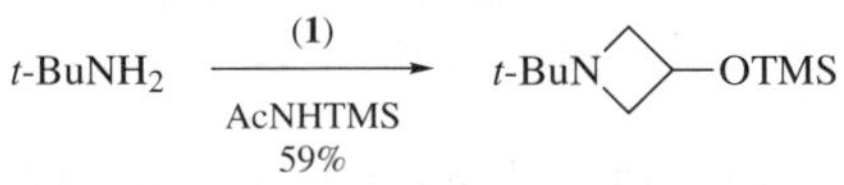

t-$BuNH_2$ —(1), AcNHTMS, 59%→ *t*-BuN(azetidine)–OTMS (17)

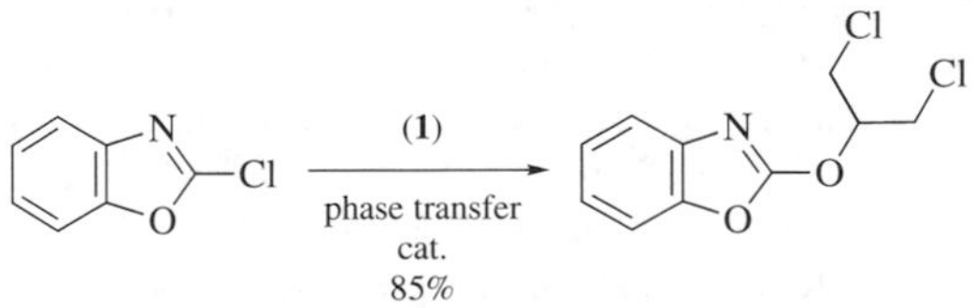

2-chlorobenzoxazole (N, O, Cl) —(1), phase transfer cat., 85%→ N, O, O, Cl, Cl (18)

The ready availability[44] of both enantiomers of (**1**) has greatly enhanced its value as a synthetic intermediate. The pheromone (*S*)-(−)-ipsenol (**2**), prepared[45] in 16% overall yield in four steps from (*R*)-(**1**), is just one of many examples of this utility. In practice, either isomer can sometimes be used by adjusting the order of addition of the groups at C-1 and C-3. The synthesis of (−)-anisomycin (**3**) illustrates this point.[46]

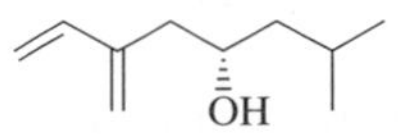

OH

(**2**) (*S*)-(−)-Ipsenol

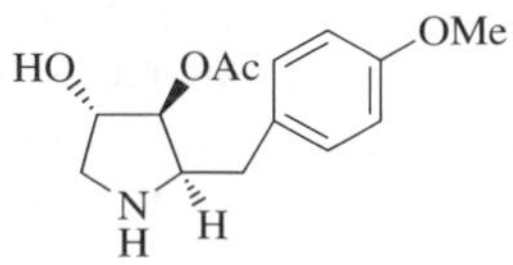

HO, OAc, OMe, N, H, H

(**3**) (−)-Anisomycin

Related Reagents. Glycidol.

1. *Encyclopedia of Chemical Technology*, 3rd ed.; Wiley: New York, 1978; Vol. 5, pp 858–864; 4th ed., 1991; Vol. 2, pp 146 and 156; 1991; Vol. 6, pp 140–155.
2. For example, *CA* 12th Coll. Index lists 133 pages of references to polymers.
3. Sato, K.; Kojima, Y.; Sato, H. *JOC* **1970**, *35*, 2374.
4. (a) Petersen, S.; Tietze, E. *LA* 1959, *623*, 166 (*CA* **1960**, *54*, 14 257i). (b) Meerwein, H. *OSC* **1973**, *5*, 1080. (c) Curphey, T. J. *OSC* **1988**, *6*, 1019.
5. Spadlo, M.; *Przem. Chem.* **1990**, *69*, 164 (*CA* **1990**, *113*, 190 697e).
6. Kozikowski, A. P.; Fauq, A. H. *SL* **1991**, 783.
7. Lamm, B.; Gustafsson, K. *ACS* **1974**, *B28*, 701.
8. Culvenor, C. C. J.; Davies, W.; Haley, F. G. *JCS* **1950**, 3123.
9. Iranpoor, N.; Baltork, I. M. *SC* **1990**, *20*, 2789.
10. Iranpoor, N.; Baltork, I. M. *TL* **1990**, *31*, 735.
11. Chini, M.; Crotti, P.; Giovani, E.; Macchia, F.; Pineschi, M. *SL* **1992**, 303.
12. Gasanov, F. G.; Aliev, A. Y.; Mamedov, E. G.; Akhmedov, I. M. *Azerb. Khim. Zh.* **1981**, *5*, 49 (*CA* **1982**, *96*, 217 607v).
13. Iqbal, J.; Khan, M. A. *CL* **1988**, 1157.
14. Iqbal, J.; Khan, M. A.; Ahmad, S. *SC* **1989**, *19*, 641.
15. (a) Matsubara, S.; Onishi, H.; Utimoto, K. *TL* **1990**, *31*, 6209. (b) Hayashi, M.; Tamura, M.; Oguni, N. *SL* **1992**, 663.
16. Emziane, M.; Lhoste, P.; Sinou, D. *S* **1988**, 541.
17. Etienne, A.; LeBerre, A.; Coquelin, J. *CR(C)* **1972**, *275*, 633 (*CA* **1973**, *78*, 123 928b).
18. Boberg, F.; Schultze, G. R. *CB* **1955**, *88*, 275 (*CA* **1956**, *50*, 1603e).
19. Iqbal, J.; Khan, M. A.; Srivastava, R. R. *TL* **1988**, *29*, 4985.
20. Malinovskii, M. S. *JGU* **1947**, *17*, 1559 (*CA* **1948**, *42*, 2229b).
21. Einhorn, C.; Luche, J. L. *CC* **1986**, 1368.
22. Ciaccio, J. A.; Heller, E.; Talbot, A. *SL* **1991**, 248.
23. Iqbal, J.; Pandey, A. *TL* **1990**, *31*, 575.
24. Johnson, F.; Panella, J. P. *OSC* **1973**, *5*, 614.
25. Ichihara, J.; Matsuo, T.; Hanafusa, T.; Ando, T. *CC* **1986**, 793.
26. Otera, J.; Yoshinaga, Y.; Hirakawa, K. *TL* **1985**, *26*, 3219.
27. DeCamp Schuda, A.; Mazzocchi, P. H.; Fritz, G.; Morgan, T. *S* **1986**, 309.
28. (a) Takano, S.; Yanase, M.; Sekiguchi, Y.; Ogasawara, K. *TL* **1987**, *28*, 1783. (b) Takano, S.; Yanase, M.; Ogasawara, K. *H* **1989**, *29*, 1825.
29. (a) South, M. S.; Liebeskind, L. S. *JACS* **1984**, *106*, 4181. (b) Hatakeyama, S.; Sugawara, K.; Kawamura, M.; Takano, S. *SL* **1990**, 691. (c) Russell, S. W.; Pabon, H. J. J. *JCS(P1)* **1982**, 545.
30. Mouzin, G.; Cousse, H.; Bonnaud, B. *S* **1978**, 304.
31. Sangwan, N. K.; Dhindsa, K. S. *OPP* **1989**, *21*, 241.
32. Zuidema, G. D.; vanTamelen, E.; VanZyl, G. *OSC* **1963**, *4*, 10.
33. Block, E.; Laffitte, J-A; Eswarakrishnan, V. *JOC* **1986**, *51*, 3428.
34. Murai, T.; Kato, S.; Murai, S.; Toki, T.; Suzuki, S.; Sonoda, N. *JACS* **1984**, *106*, 6093.
35. Farina, J. S.; Jackson, S. A.; Cummings, C. L. *OPP* **1989**, *21*, 173.
36. Cabiddu, S.; Melis, S.; Sotgiu, F. *PS* **1983**, *14*, 151.
37. Parekh, K. B.; Shelver, W. H.; Tsai, A.-Y. S.; Reopelle, R. *JPS* **1975**, *64*, 875.
38. Mazzetti, F.; Lemmon, R. M. *JOC* **1957**, *22*, 228.
39. Oda, R.; Okano, M.; Tokiura, S.; Miyasu, A. *BCJ* **1962**, *35*, 1216.
40. Baba, A.; Shibata, I.; Masuda, K.; Matsuda, H. *S* **1985**, 1144.
41. Finar, I. L.; Godfrey, K. E. *JCS* **1954**, 2293.
42. Higgins, R. H.; Watson, M. R.; Faircloth, W. J.; Eaton, Q. L.; Jenkins, H. *JHC* **1988**, *25*, 383.
43. Jin, R.-H.; Nishikubo, T. *S* **1993**, 28.
44. Baldwin, J. J.; Raab, A. W.; Mensler, K.; Arison, B. H.; McClure, D. E. *JOC* **1978**, *43*, 4876.
45. Imai, T.; Nishida, S. *JOC* **1990**, *55*, 4849.
46. Takano, S.; Iwabuchi, Y.; Ogasawara, K. *H* **1989**, *29*, 1861.

Joel E. Huber
The Upjohn Co., Kalamazoo, MI, USA

Ethoxyacetylene[1]

EtO—≡—H

[927-80-0] C_4H_6O (MW 79.09)

(preparation of alkenylcuprates,[2] anhydrides,[3] ketenes,[4,5] lactones;[6] reacts with allylboranes[7])

Alternate Name: ethyl ethynyl ether.
Physical Data: bp 48–50 °C; *d* 0.800 g cm^{-3} (0.745 g cm^{-3} in hexane).
Solubility: sol alcohol, ether, benzene, hexane.
Preparative Methods: from chloroacetaldehyde diethyl acetal (70%).[8,9]
Form Supplied in: 50 w/w % solution in hexane(s).
Handling, Storage, and Precautions: moisture sensitive; potential fire hazard with certain metal salts.[8] Use in a fume hood.

Cycloaddition Reactions. Ethoxyacetylene (**1**) participates in [2+2], [4+2], and 1,3-dipolar cycloaddition reactions. Diels–Alder reactions with oxazinones give substituted 3,5-disubstituted 2,6-dichloropyrimidines (eq 1).[10] 1,3-Dipolar cycloaddition reactions with aldoximes (eq 2)[11] and with nitrile oxides[12] give isoxazoles.

ethoxyacetylene; hexanes, 60 °C, 16 h, 90%; OEt; Cl N Cl; 2:1 (1)

NOH; R H; ethoxyacetylene; NaBrO$_2$, Bu$_3$SnCl (cat), CH$_2$Cl$_2$–H$_2$O; R; N O OEt; R = Me (72%), Ph (45%) (2)

Ethoxyacetylene and its derivatives thermally decompose to generate ketenes[4,5,13] which either add to themselves or undergo further cycloaddition reactions. In addition, other ketenes, such as dimethylketene, add to (**1**) to give cyclobutanone derivatives (eq 3).[14]

=•=O; EtO—≡—H; acetonitrile; rt to 45 °C, 3 h, 83%; O; EtO (3)

Ethoxyacetylene gives pyrazoles with ***Diazomethane***, triazoles with aryl and substituted alkyl azides, 2-quinolines with aryl isocyanates, and *o*-ethoxyphenylacetylene with benzyne.[1] Phospholes may be prepared in a [2 + 2 + 1] cycloaddition of a terminal phosphinidene complex with (**1**).[15] Organometallic reagents add to ethoxyacetylene to give 1,4-naphthoquinone–cobalt[16] and 2*H*-pyrrole–tungsten complexes,[17] and phenols.[18]

Nucleophilic Additions to Ethoxyacetylene. Vinyl anions can be generated by nucleophilic addition to ethoxyacetylene. Alkyl cuprates tend to transfer only one group while (*Z*)-alkenyl cuprates transfer both groups to give (*E*)-α,β-unsaturated ketones after hydrolysis of the initial enol ethers (eq 4).[19] These vinylcopper reagents have been coupled, condensed with alkyl and vinyl iodides, acid chlorides, vinylogous acid chlorides, and iodine to give a variety of derivatives.[2]

Cu–Li + 2 EtO—≡—H → (1. ether, THF, N_2, 2 h, –40 to –15 °C; 2. 20% HCl; 75%) (4)

Trimethylsilylmethylcopper adds to ethoxyacetylene with *cis* addition to give allylic silane (**2**) that, when quenched with electrophiles, gives (*E*)-3-substituted 2-alkoxy-2-alkenylsilanes (eq 5).[20]

H—≡—OEt → ($TMSCH_2MgCl$; $LiCuBr_2$, THF, –60 °C, 15 min, –30 °C, 0.5 h) → TMS/Cu/EtO/H (**2**) → (E^+, HMPA; –60 to 25 °C, 0.5 h) → TMS/E/EtO/H (*E*) only (5)

E	yield (%)
H	95
Cl	90
I	95
CH_2=$CHCH_2$	98
CO_2Me	70

Arylcopper reagents extend this chemistry. Thus, by coupling a Grignard reaction with a Wacker oxidation, 1,4-diketones can be prepared (eq 6).[21]

H—≡—OEt + PhMgBr → (allyl bromide; $LiCuBr_2$, THF, N_2, 4 h, –20 °C; 96%) → OEt → ($PdCl_2$, CuCl; O_2, DMF, H_2O, 3N HCl; 58%) (6)

Nucleophilic Reactions of Ethoxyacetylene. Anions derived from ethoxyacetylene, generated as the Grignard reagent,[22] or lithium[8,23] or aluminum[24] salts, add to alkyl halides,[25] aldehydes, ketones, and epoxides[22] to give precursors of α-hydroxy ketones,[26] aldehydes, acids, or α,β-unsaturated esters. Cyclohexene epoxide and ethoxyacetylene give a lactone upon treatment of the intermediate alcohol with acid. Conversion to the alane is necessary as no ring opening occurs with lithium ethoxyacetylide (eq 7). This methodology is superior to that which uses lithium *t*-butylacetate (Rathke's salt), which fails to react with hindered epoxides.[24]

(1. ethoxyacetylene; 2. BuLi, hexane; 3. diethylchloroalane; 4. cyclohexane epoxide, –40 °C to rt, 6 h; 80%) → EtO, HO → (TsOH, Δ; 88%) (7)

Ethoxyacetylene reacts with aldehydes and ketones in the presence of ***Boron Trifluoride*** to give α,β-unsaturated esters.[27]

Acylation Reactions. Amines can be acylated by ethoxyacetylene to give carboxylic acid derivatives in good yields.[13] 1-Ethoxyvinyl esters, which are useful as acylating agents, have been prepared from carboxylic acids and ethoxyacetylene in a Ru^{II}-catalyzed reaction.[28] Cyclic anhydrides have been formed by dehydrating diacids with (**1**) (eqs 8 and 9).[3,29] Weakly nucleophilic 2-fluoro-2,2-dinitroethanol yields bis(2-fluoro-2,2-dinitroethyl)ethoxy ethyl orthoester in a 95% yield in a Hg^{II}-catalyzed reaction with ethoxyacetylene.[30]

AcO, AcO, OH, HO → (ethoxyacetylene; CH_2Cl_2, 22 °C, 24 h; 91%) → AcO, AcO (8)

OH, OH → (ethoxyacetylene; 95%) (9)

Lactones have been prepared by thermolysis of adducts formed from lithium ethoxyacetylide and hydroxy ketones (eq 10). Higher reaction temperatures increase conversion and lower reaction times. Good yields are reported for five-, six-, and seven-membered rings. Larger macrocycles (9-, 10-, 15-membered) require Bu_3N as a catalyst to get respectable yields and to minimize oligomeric products (see also ***2-Chloro-1-methylpyridinium Iodide***).[6]

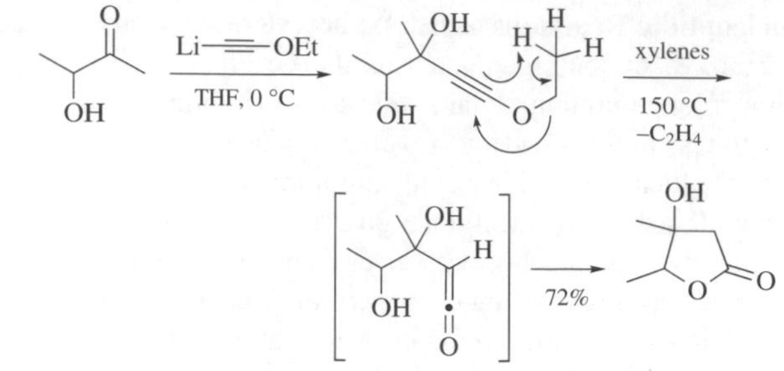

(10)

Insertion Reactions. Allyl(dialkyl)boranes react with ethoxyacetylene to give dialkyl(alkoxypentadienyl)boranes which may be converted to 2-alkoxy-1,4-pentadienes[7] or methyl allyl ketone[31] on protonolysis. In a one-pot reaction, triallylboranes and ethoxyacetylene give 1,4-pentenynes as shown in eq 11.[7] Similar chemistry is observed with tribenzylboranes.[31] The ethoxyacetylene/(iodo-9-BBN) adduct reacts with acyclic α,β-unsaturated ketones to give δ-keto esters (74–95%)[32] and with aryl aldehydes to give cinnamic esters (89% yield, 99% (*E*)).[33] Unlike Wittig chemistry, this methodology tolerates base-sensitive functional groups. In a ***Mercury(II) Acetate*** catalyzed reaction, ethoxyacetylene, diphenylboronic acid, and alkyl or aryl aldehydes give β-hydroxy and β-acetoxy esters.[34]

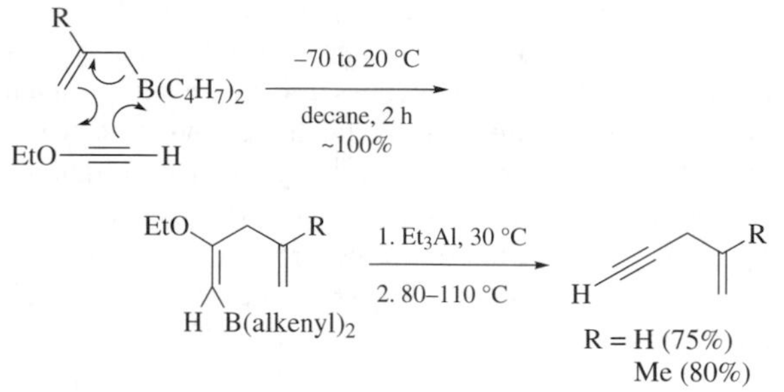

(11)

R = H (75%)
Me (80%)

Transition Metal Catalyzed Reactions. Nickel reagents catalyze a one-pot addition of ethoxyacetylene to aryl iodides or bromides to give (*E*)-β-ethoxyalkene derivatives (eq 12). A variety of functional groups are tolerated on the aryl halide (Cl, OMe, CN, CO_2Me).[35] This chemistry has been extended to indoles.[36] Ethoxyacetylene gives pyrones with CO_2 and ***Bis(1,5-cyclooctadiene)nickel(0)***,[37] and indenols with acetophenone and $PhCH_2Mn(CO)_5$.[38]

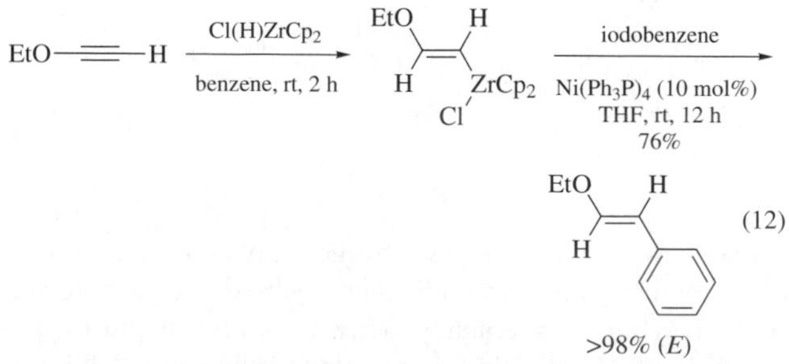

(12)

>98% (*E*)

Miscellaneous Synthetic Transformations. Ethoxyacetylene has been used to prepare substituted acylketenes,[39] (alkoxyethynyl)phosphonites,[40] and alkoxy metal carbenes.[41]

Related Reagents. Acetylene; Lithium Acetylide; Propynyllithium; Trimethylsilylacetylene.

1. (a) Brandsma, L.; Bos, H. J. T.; Arens, J. F. In *Chemistry of Acetylenes*; Viehe, H. C., Ed.; Dekker: New York, 1969; p 751. (b) Arens, J. F. *Advances in Organic Chemistry*; Interscience: New York, 1960; Vol. 2, p 117.
2. (a) Foulon, J. P.; Bourgain-Commerçon, M.; Normant, J. F. *T* **1986**, *42*, 1389. (b) Foulon, J. P.; Bourgain-Commerçon, M.; Normant, J. F. *T* **1986**, *42*, 1399.
3. Shealy, Y. F.; Clayton, J. D. *JACS* **1969**, *91*, 3075.
4. Rosebeek, B.; Arens, J. F. *RTC* **1962**, *81*, 549.
5. Ruden, R. A. *JOC* **1974**, *39*, 3607.
6. Liang, L.; Ramaseshan, M.; MaGee, D. I. *T* **1993**, *49*, 2159.
7. (a) Bubnov, Y. N.; Grigorian, M. S.; Tsyban, A. V.; Mikhailov, B. M. *S* **1980**, 902. (b) Bubnov, Y. N.; Tsyban, A. V.; Mikhailov, B. M. *S* **1980**, 904.
8. (a) Raucher, S.; Bray, B. L. *JOC* **1987**, *52*, 2332. (b) Jones, E. R. H.; Eglinton, G.; Whiting, M. C.; Shaw, B. L. *OSC* **1963**, *4*, 404.
9. Brandsma, L. *Preparative Acetylenic Chemistry*, 2nd ed.; Elsevier: New York, 1988; pp 171, 174.
10. Meerpoel, L.; Deroover, G.; Van Aken, K.; Lux, G.; Hoornaert, G. *S* **1991**, 765.
11. Moriya, O.; Nakamura, H.; Kageyama, T.; Urata, Y. *TL* **1989**, *30*, 3987.
12. Kozikowski, A. P.; Goldstein, S. *JOC* **1983**, *48*, 1139.
13. Nieuwenhuis, J.; Arens, J. F. *RTC* **1958**, *77*, 761.
14. Hasek, R. H.; Gott, P. G.; Martin, J. C. *JOC* **1964**, *29*, 2510.
15. Ngoc H. T. H.; Mathey, F. *PS* **1990**, *47*, 477.
16. Cho, S. H.; Wirtz, K. R.; Liebeskind, L. S. *OM* **1990**, *9*, 3067.
17. Aumann, R.; Hinterding, P. *CB* **1991**, *124*, 213.
18. Morris, K. G.; Saberi, S. P.; Thomas, S. E. *CC* **1993**, 209.
19. Alexakis, A.; Cahiez, G.; Normant, J. F. *T* **1980**, *36*, 1961.
20. Kleijn, H.; Vermeer, P. *JOC* **1985**, *50*, 5143.
21. Wijkens, P.; Vermeer, P. *JOM* **1986**, *301*, 247.
22. Vollema, G.; Arens, J. F. *RTC* **1963**, *82*, 305.
23. Smithers, R. H. *S* **1985**, 556.
24. Danishefsky, S.; Kitahara, T.; Tsai, M.; Dynak, J. *JOC* **1976**, *41*, 1669.
25. Nooi, J. R.; Arens, J. F. *RTC* **1962**, *81*, 517.
26. Sperna Weiland, J. H.; Arens, J. F. *RTC* **1960**, *79*, 1293.
27. Vieregge, H.; Schmidt, H. M.; Renema, J.; Bos, H. J. T.; Arens, J. F. *RTC* **1966**, *85*, 929.
28. Kita, Y.; Maeda, H.; Omori, K.; Okuno, T.; Tamura, Y. *SL* **1993**, 273.
29. Tamura, Y.; Akai, S.; Sasho, M.; Kita, Y. *TL* **1984**, *25*, 1167.
30. Cochoy, R. E.; McGuire, R. R.; Shackelford, S. A. *JOC* **1990**, *55*, 1401.
31. Mikhailov, B. M.; Bubnov, Y. N.; Korobeinikova, S. A.; Frolov, S. I. *JOM* **1971**, *27*, 165.
32. Kawamura, F.; Tayano, T.; Satoh, Y.; Hara, S.; Suzuki, A. *CL* **1989**, 1723.
33. Satoh, Y.; Tayano, T.; Hara, S.; Suzuki, A. *TL* **1989**, *30*, 5153.
34. Murakami, M.; Mukaiyama, T. *CL* **1982**, 241.
35. (a) Negishi, E.; Van Horn, D. E. *JACS* **1977**, *99*, 3168. (b) Negishi, E.; Takahashi, T.; Baba, S.; Van Horn, D. E.; Okukado, N. *JACS* **1987**, *109*, 2393.
36. Hegedus, L. S.; Toro, J. L.; Miles, W. H.; Harrington, P. J. *JOC* **1987**, *52*, 3319.
37. Tsuda, T.; Kunisada, K.; Nagahama, N.; Morikawa, S.; Saegusa, T. *SC* **1989**, *19*, 1575.
38. Liebeskind, L. S.; Gasdaska, J. R.; McCallum, J. S.; Tremont, S. J. *JOC* **1989**, *54*, 669.

39. Lukashev, N. V.; Fil'chikov, A. A.; Kazankova, M. A.; Beletskaya, I. P. *HC* **1993**, *4*, 403.
40. Lukashev, N. V.; Fil'chikov, A. A.; Kozlov, A. I.; Luzikov, Yu. N.; Kazankova, M. A. *ZOB* **1991**, *61*, 1739 (*CA* **1992**, *116*, 194 436d).
41. Dötz, K. H. *AG(E)* **1984**, *23*, 587.

Kenneth C. Caster
Union Carbide Corporation, South Charleston, WV, USA

(Ethoxycarbonylmethylene)triphenylphosphorane[1]

$Ph_3P{=}CHCO_2Et$

[1099-45-2] $C_{22}H_{21}O_2P$ (MW 348.38)

(stabilized phosphorane reagent useful for Wittig alkenation reactions to give α,β-unsaturated ethyl esters on reaction with carbonyl compounds;[1] heterocyclic ring construction;[2–4] synthesis of cyclopropanes from epoxides[5])

Physical Data: mp 128–130 °C.
Solubility: sol EtOH (42 g 100 mL^{-1}), THF (13 g 100 mL^{-1}), and $CHCl_3$ (29 g 100 mL^{-1}); insol water (<0.5 g 100 mL^{-1}).
Form Supplied in: white crystalline powder.
Handling, Storage, and Precautions: no special handling or storage requirements are necessary. The use of carefully dried reagents leads to a lower rate of Wittig alkenation reaction.[6] The compound is stable in water (>3 h), and >99% stable in 0.5 N NaOH in MeOH–water (1:1) for at least 3 h.

Weakly Basic Phosphorus Ylide. This phosphorane is readily prepared by deprotonation of ***((Ethoxycarbonylmethyl)-triphenylphosphonium Bromide*** (pK_a 8.95–9.2), typically with ***Sodium Ethoxide*** or ***Sodium Hydroxide***.[7] Deprotonation of the phosphonium bromide produces a phosphorane in which the negative charge on the α-carbon is stabilized by d_π-p_π bonding to the phosphorus and also by the adjacent ethoxycarbonyl group. Additionally, NMR studies have shown that this reagent exists as a ca. 3:1 mixture of *cis:trans* ylide/enolates as shown in eq 1, with a barrier to rotation estimated at ca. 10 kcal mol^{-1} for the methyl ester.[6,8] The high proportion of negative charge on the oxygen is reflected in *O*-ethylation with ***Triethyloxonium Tetrafluoroborate*** at –78 °C to give a 3:1 mixture of *cis:trans* ethyl vinyl ethers.[6] The presence of small amounts of water or other proton sources facilitate the reaction of the phosphorane, presumably by reversibly protonating the α-carbon and lowering the barrier to rotation, allowing for more reactive ylide forms with single bond character between the α- and β-carbons (viz. *cis*- and *trans*-ylide structures).[6] For this reason, it is counterproductive to be overly concerned with removing traces of moisture in standard reactions involving this phosphorane.

Ph_3P+ O^- / H OEt — *cis*-enolate; Ph_3P+ OEt / H O^- — *trans*-enolate; Ph_3P O / H OEt — *cis*-ylide; Ph_3P OEt / H O — *trans*-ylide (1)

Wittig Alkenation Reactions. This reagent is among those typically referred to as stabilized Wittig reagents, because of the extra negative charge stabilization by the ester functionality (see also ***(Methoxycarbonylmethylene)triphenylphosphorane***). As such, its basicity and reactivity are moderated relative to the non-stabilized alkylidenephosphoranes. If more reactive reagents are required, phosphonates such as $(EtO)_2P(O)CH_2CO_2Et$ can be employed in the Horner–Wadsworth–Emmons (HWE) alkenation procedure.[1] The phosphonate reagents have the added advantage that the phosphates obtained as side-products are easier to remove than triphenylphosphine oxide, the phosphorus-containing product of the phosphorane reactions. Although intermediates have not been observed in reactions of stabilized phosphoranes by spectroscopic techniques, the reaction mechanism would be expected to include a four-membered oxaphosphetane immediately prior to elimination to the alkene and triphenylphosphine oxide.[1a]

(Ethoxycarbonylmethylene)triphenylphosphorane reacts with the carbonyl groups of aldehydes,[9–26] ketones,[27–34] ketenes,[35–37] anhydrides,[39] imides,[40] and certain amides[41a,42] and esters[4,41] to insert the α,β-unsaturated ethyl ester (ethyl acrylate) functionality. The reaction has wide generality and has been extensively employed, especially en route to intermediates for further synthetic manipulation. Where the newly established double bond has stereochemistry, such as in reactions with aldehydes and unsymmetrical ketones, it is generally *trans* (*E*) in nature, although the (*E:Z*) ratio is influenced by a variety of factors, such as the nature of the aldehyde, solvent, and additives.

Reactions with Aldehydes. There are numerous instances in which this reaction has been carried out in organic synthesis, and only a few representative examples are given here. In the synthesis of rapamycin, the aldehyde group was converted to the expected (*E*) ester, without significant epimerization or elimination of the α-mesyloxy or β-benzyloxy substituents, respectively (eq 2).[10a] The relatively low basicity of the phosphorane allows for mild transformations of this type, without affecting base-sensitive groups.

R, OBn, MsO, CHO —$Ph_3P{=}CHCO_2Et$→ BnO, R, OMs, CO_2Et (2)

There are occasions when these mild conditions are preferred to the more basic conditions attendant to the HWE phosphonate reagent $(EtO)_2P(O)CHCO_2Et^-\ Na^+$. For example, in the preparation of a substrate for a quinolizine synthesis, the phosphorane produced the desired acrylate ester, whereas use of the HWE reagent resulted in considerable amounts of premature ring closure to give a piperidine side product (eq 3).[10b]

$Ph_3P{=}CHCO_2Et$

$(EtO)_2P(O)CHCO_2Et^-Na^+$

(3)

The alkenation reaction can be carried out by the in-situ generation of the phosphorane reagent by the use of ethylene oxide (eq 4).[11] The bromide counterion of the phosphonium salt opens the ethylene oxide, and the alkoxide which then forms deprotonates the α-carbon.

$BrCH_2CO_2Et$ — 1. Ph_3P, ethylene oxide; 2. PhCHO; 91% → Ph–CH=CH–CO_2Et (4)

When the phenyl ligands on the phosphorus atom are replaced with alkyl groups, the relative proportion of (*E*) isomer increases.[12] For example, in reactions with benzaldehyde under the same conditions, $Ph_3P{=}CHCO_2Et$ and $Bu_3P{=}CHCO_2Et$ gave (*E*:*Z*) ratios of 89:11 and 95:5, respectively.[12] The placement of ferrocenyl ligands on the phosphorus also resulted in the same degree of (*E*) stereoselectively,[13] but 2-furyl ligands produced more of the (*Z*) isomer in ethanol.[14] Rate studies on phosphoranes of type $(p\text{-}XC_6H_4)_3P{=}CHCO_2Et$ gave ρ values in the area of +2.5–3.0.[15] Bridged phosphorane species in which two of the phenyl ligands are attached via a divinyl spacer react slowly to give more (*Z*) isomer (24%) than use of the standard reagent (11%).[16]

Greater (*E*) selectivity is observed when the α-carbon is alkylated, such as in reactions of $Ph_3P{=}C(Me)CO_2Et$.[1,17] However, α-halogenation of the phosphorane and its use to prepare vinyl halides is less (*E*) selective (eq 5).[18]

(4)

R = H >90:<10
R = Br 80:20

Generally, the formation of the (*E*) isomer is favored by the use of aprotic solvents in the absence of lithium salts,[1a] and the influence of solvent can often be remarkable. For example, reaction of the phosphorane with a sugar-derived aldehyde gave the corresponding acrylates with (*E* : *Z*) mixtures that varied from 86:14 (DMF) to 8:92 (MeOH; eq 6).[19]

$Ph_3P{=}CHCO_2Et$ (6)

solvent	
DMF	86:14
PhH	80:20
acetone	53:47
CCl_4	47:53
MeOH	9:92

The high amounts of the (*Z*)-acrylate observed in eq 6 under certain conditions is characteristic of the reactions of α-alkoxy aldehydes, particularly in the presence of an additional β-alkoxy group and in an anhydrous alcoholic solvent.[1a] For example, treatment of isopylideneglyceraldehyde with the phosphorane reagent in methanol at 0 °C[20] or 25 °C[21] gave the expected acrylates with an (*E*:*Z*) ratio of ca. 10:90. This effect has been observed repeatedly in reactions with carbohydrate-based aldehydes, as they bear multiple oxygenated sites. In the reaction of a fucose-derived hemiacetal (reacting in the aldehyde form) a ca. 1:2 mixture of (*E*:*Z*) isomers formed (eq 7),[22] whereas reaction of 2-deoxyribose derivatives (lacking an α-alkoxy group) gave solely (*E*) acrylates.[23]

$Ph_3P{=}CHCO_2Et$ (7)

1:2

The reaction of sugar-derived hemiacetals with stabilized phosphoranes has provided an important starting point for the synthesis of key natural products such as the leukotrienes.[1c] Additionally, the hydroxyl group that remains from the hemiacetal can attack the β-carbon of the acrylate in a Michael fashion to form *C*-glycosides, useful as important synthetic intermediates, such as in the reaction of diisopropylideneallofuranose shown in eq 8.[24] The alkene product of the Wittig reaction can be isolated directly in some cases or the Michael closure can be so fast that only the *C*-glycoside products are detected. In certain circumstances, diene products formed by further elimination reactions are observed.[25] The nature of the ester group and the presence of free or masked hydroxyls on the sugar substrate are important determinants in whether cyclized or open chain products are isolated in these systems.[26]

R = H → $Ph_3P{=}CHCO_2Et$ → R = CH_2CO_2Et, ca. 3:1 β:α (8)

Reactions with Ketones. Ketones can be unreactive partners in condensations with stabilized phosphoranes. These reactions can be enhanced by reaction at 9 kbar.[27] For example, alkenation with benzophenone gave an 82% yield of product at 50 °C after 35 h, whereas refluxing in xylene after 1 d at 1 bar resulted in only 10% product. In addition, the reaction with ketones can be facilitated by prolonged heating or mixing the two reactants together at high temperature without the use of solvent.[28] In other cases, the addition of small amounts of an acid catalyst (e.g. benzoic acid) increases the yield of reactions with hindered carbonyls (eq 9).[29] The acid may be acting by increasing the electrophilicity of the carbonyl or by enhancing the rate of exchange of the various ylide forms as discussed above.

Keto sugars typically react very nicely with the phosphorane, often producing unexpected stereochemical outcomes. For example, the choice of a silyloxy or benzyl ether protecting group led to dramatically different mixtures of acrylates as shown in eq 10.[30] Alkoxy or hydroxy substituted ketones may actually react faster than their deoxy counterparts.[31]

$Ph_3P{=}CHCO_2Et$ + (methyl propyl ketone) —PhH, 80 °C; 83% with $PhCO_2H$; 12% without $PhCO_2H$→ (E)-acrylate + (Z)-acrylate, 66:34 (9)

$Ph_3P{=}CHCO_2Et$ (10)

R = *t*-BuMe$_2$Si 100:0
R = Bn 50:50

The use of a chiral carboxylic acid in reaction of the phosphorane and 4-methylcyclohexanone resulted in the expected ethyl acrylate with an enantiomeric excess (ee) of <10%.[32] However, the presence of a chiral host in the same sequence afforded the expected acrylate with an ee of 42% (eq 11).[33] There are only a few reported enantioselective Wittig alkenations, such as the use of bicyclic phosphonamide reagents, in which case as much as 90% ee is observed.[34]

$Ph_3P{=}CHCO_2Et$; (−) host, Ph_2COH, $C(OH)Ph_2$, 73% → (−)Me ... CO_2Et, H; 42% ee (11)

Reactions with Ketenes. Reaction of an acyl halide with the phosphorane in the presence of 1 mol equiv of ***Triethylamine*** affords allenic esters via in-situ generation of a ketene from the acid halide, followed by a standard Wittig reaction (eq 12).[35] In a similar fashion, carboxylic acids, activated by triethylamine and ***2-Chloro-1-methylpyridinium Iodide***, generate ketenes which can be reacted with the stabilized phosphoranes to afford allenic carboxylate esters.[36] The reaction of silicon- and germanium-substituted ketenes ($Me_3MCH{=}C{=}O$) with the phosphorane also gives allenic esters, which rearrange to the alkynes $Me_3MC{\equiv}CCH_2CO_2Et$ upon distillation.[37]

$Ph_3P{=}CHCO_2Et$ —EtCOCl, Et_3N, 70%→ allene (CO_2Et, H, H) (12)

Allenes are also formed by the use of $Ph_3P{=}C{=}C(OEt)_2$, produced by treatment of $Ph_3P{=}CHCO_2Et$ with ***Triethyloxonium Tetrafluoroborate*** followed by base treatment.[38] Reaction with fluorenone gave the expected allene product, which dimerized immediately to a spirofluorene adduct.[38]

Reactions with Anhydrides and Imides. Stabilized phosphoranes condense with anhydrides to yield enol lactones (see ***(Methoxycarbonylmethylene)triphenylphosphorane***).[39] Both the (*E*) and (*Z*) isomers can be obtained depending on the substrate, and also regioisomers are produced if the anhydride is unsymmetrical. There is predominant reaction at a site with weak steric hindrance (e.g. α-methyl) which changes to the less hindered carbonyl with added hindrance (α,α-dimethyl).[39] Imides also react to give *N*-vinyl amides in a similar manner.[40]

Reactions with Amides and Esters. Certain reactive or suitably disposed esters[4,41] also react in a Wittig fashion to yield vinyl ethers, and some amides[41a,42] such as those in penicillin and clavulanic acid derivatives afford vinyl amines.

Alkylations, Acylations, and Related Processes. Stabilized phosphoranes are sufficiently nucleophilic to be alkylated in order to construct more complicated reagents, particularly with reactive alkyl groups such as allyl, benzyl, and methyl.[1,43] Halogenation of the phosphorane leads to the α-halo derivatives, which give vinyl halides upon standard Wittig alkenations.[7c,44] The nucleophilic alkylation reaction of the phosphorane with a large variety of substrates has been investigated, such as in the ring opening of *N*-methoxypyridinium salts[45] and the β-alkylation of alkyl propynoates.[46] Treatment of the phosphorane in the absence of an added electrophile leads to phosphaketene $Ph_3P{=}C{=}C{=}O$.[47] The phosphorane coordinates with organometallics to form stable complexes, and can occasionally participate in subsequent reactions of the organometallic complex.[48]

Acylation of the phosphorane provides diacylphosphoranes, which are often so highly stabilized and unreactive that they do not perform Wittig alkenations well.[1] However, flash vacuum pyrolysis (FVP) can afford either terminal alkynes or propargylic esters, depending on the conditions (eq 13).[49]

(3-thienyl)COCl —$Ph_3P{=}CHCO_2Et$, Et_3N→ Ph_3P/EtO_2C acyl ylide —FVP, 750 °C, 57%→ 3-ethynylthiophene (13)

Heterocyclic Ring Formation. Reaction of the phosphorane with azides gives phosphazenes, whereas similar treatment of $Ph_3P=C(Me)CO_2Et$ results in triazole formation via 1,3-dipolar cycloaddition of the azide to the enolate resonance form of the phosphorane (eq 14).[50] A similar cyclocondensation occurs with hydrazonyl chlorides to give 5-alkoxy substituted pyrazoles.[2]

The combination of Wittig alkenation followed by further reaction of the α,β-unsaturated ester has resulted in the efficient preparation of several heterocyclic ring systems. In addition to the *C*-glycoside synthesis discussed earlier, pyrrolidines can be prepared in a similar 'Wittig/Michael' sequence starting from α-hydroxypyrrolidine carbamates.[51]

$$Ph_3P{=}CHCO_2Et \xrightarrow{RN_3} EtOC(=O)CHN_2 + RN{=}PPh_3 \quad 69\% \quad 60\%$$

$$Ph_3P{=}C(Me)CO_2Et \xrightarrow[73\%]{RN_3} \text{triazole (N-R, OEt)} \tag{14}$$

R = *m*-$NO_2C_6H_4C(O)$

Reaction of *o*-hydroxybenzaldehydes with the phosphorane is followed by condensation of the ester with the phenyl hydroxyl in a synthesis of coumarins (eq 15).[3,43,52]

$$\text{2-HOC}_6\text{H}_4\text{CHO} \xrightarrow[72\%]{Ph_3P=CHCO_2Et} \text{coumarin} \tag{15}$$

Nucleophilic Epoxide Opening. Under vigorous conditions, oxiranes react with the phosphorane to give ethoxycarbonylcyclopropanes (eq 16).[5] The mechanism proceeds via nucleophilic opening of the epoxide, followed by attack of the alkoxide on the phosphorus, cleavage of the alkyl-phosphorus bond, and backside displacement of triphenylphosphine oxide. Normal Wittig alkenations of aldehydes with the phosphorane are sufficiently mild so as to not react with the epoxide functionality.

$$\text{cyclohexene oxide} \xrightarrow[200\ ^\circ C,\ 56\%]{Ph_3P=CHCO_2Et} \text{bicyclo[4.1.0]heptane-}CO_2Et \tag{16}$$

Oxidation. Oxidation of the phosphorane yields symmetrical alkenes, or cyclic alkenes if bisylides are employed. This reaction was carried out originally using ***Triphenyl Phosphite Ozone***;[53] however, a newer procedure using oxaziridines is a more general method.[54]

Formation of Thio- and Selenoaldehydes. Treatment of the phosphorane with sulfur,[55] episulfides, [56] or selenium[57] results in formation of $Ph_3P{=}S$ or $Ph_3P{=}Se$ and $S{=}CHCO_2Et$ or $Se{=}CHCO_2Et$. These unstable thio- or selenoaldehydes can be trapped as their Diels-Alder adducts. In reaction with elemental sulfur or episulfides,[55,56] the symmetrical alkene ($EtO_2CCH{=}CHCO_2Et$) was formed by decomposition of the thioaldehyde. Treatment of thioaldehyde $S{=}CHCO_2Et$ with amines (e.g. ***Morpholine***) led to good yields of thioamides $R^1R^2NC({=}S)CO_2Et$.[55a]

Related Reagents. (Ethoxycarbonylmethylene)triphenylphosphorane; (Ethoxycarbonylmethyl)triphenylphosphonium Bromide; Ethyl Trimethylsilylacetate; Ethyl 4-(Triphenylphosphoranylidene)acetoacetate.

1. (a) Maryanoff, B. E.; Reitz, A. B. *CRV* **1989**, *89*, 863 (b) Gosney, I.; Rowley, A. G. *In Organophosphorus Reagents in Organic Synthesis*; Cadogan, J. I. G., Ed.; Academic: New York, 1979; p 17. (c) Schlosser, M. *Top Stereochem.* **1970**, *5*, 1. (d) Johnson, A. W. *Ylid Chemistry*; Academic: New York, 1966. (e) Hudson, R. F. *Structure and Mechanism in Organo-Phosphorus Chemistry*; Academic: New York, 1965. (f) Maercker, A. *OR* **1965**, *14*, 270. (g) Trippen, S. *QR* **1963**, *17*, 406.
2. Padwa, A.; MacDonald, J. G. *H* **1987**, *24*, 1225.
3. Mali, R. S.; Yadav, V. J. *S* **1977**, 464.
4. Hercouet, A.; Le Corre, M. *TL* **1979**, *20*, 2995.
5. Denney, D. B.; Vill, J. J.; Boskin, M. J. *JACS* **1962**, *84*, 3944.
6. (a) Kayser, M. M.; Hatt, K. L.; Hooper, D. L. *CJC* **1991**, *69*, 1929. (b) Kayser, M. M.; Hooper, D. L. *CJC* **1990**, *68*, 2123.
7. (a) Speziale, A. J.; Ratts, K. W. *JACS* **1963**, *85*, 2790. (b) Isler, O.; Gutmann, H.; Montavon, M.; Rüegg, R.; Ryser, G.; Zeller, P. *HCA* **1957**, *40*, 1242. (c) Denney, D. B.; Ross, S. T. *JOC* **1962**, *27*, 998.
8. Crews, P. *JACS* **1968**, *90*, 2961.
9. (a) Isler, O.; Gutmann, H.; Montavon, M.; Ruegg, R.; Ryser, G.; Zeller, P. *HCA* **1957**, *89*, 863. (b) Wittig, G.; Haag, W. *CB* **1955**, *88*, 1654.
10. (a) Chen, S.-H.; Horvath, R. F.; Joglar, J.; Fisher, M. J.; Danishefsky, S. J. *JOC* **1991**, *56*, 5834. (b) Ihara, M.; Kirihara, T.; Kawaguchi, A.; Tsuruta, M.; Fukumoto, K.; Kametani, T. *JCS(PI)* **1987**, 1719.
11. Buddrus, J. *CB* **1974**, *107*, 2050.
12. (a) Bissing, D. E. *JOC* **1965**, *30*, 1296. (b) Maryanoff, B. E.; Reitz, A. B.; Mutter, M. S.; Inners, R. R.; Almond, H. R., Jr.; Whittle, R. R.; Olofson, R. A. *JACS* **1986**, *108*, 7664.
13. McEwen, W. E.; Sullivan, C. E.; Day, R. O. *OM* **1983**, *2*, 420.
14. Allen, D. W.; Ward, H. *ZN(B)* **1980**, *35b*, 754.
15. (a) Giese, B.; Schoch, J.; Rüchardt, C. *CB* **1978**, *111*, 1395. (b) Ruchardt, C.; Panse, P.; Eichler, S. *CB* **1967**, *100*, 1144.
16. Hocking, M. B. *CJC* **1966**, *44*, 1581.
17. (a) Bernardi, A.; Cardani, S.; Scolastico, C.; Villa, R. *T* **1988**, *44*, 491. (b) Aparicio, F. J. L.; Cubero, I. I.; Olea, M. D. P. *Carbohydr. Res.* **1983**, *115*, 250. (c) Oikawa, Y.; Nishi, T.; Yonemitsu, O. *JCS(PI)* , **1985**, 7.
18. Elliott, M.; Janes, N. F.; Pulman, D. A. *JCS(PI)* **1974**, 2470.
19. Tronchet, J. M. J.; Gentile, B. *HCA* **1979**, *62*, 2091.
20. (a) Katsuki, T.; Lee, A. W. M.; Ma, P.; Martin, V. S.; Masamune, S.; Sharpless, K. B.; Tuddenham, D.; Walker F. J. *JOC* **1982**, *47*, 1373. (b) Häfele, B.; Jäger, V. *LA* **1987**, *85*. (c) Mann, J.; Partlett, N. K.; Thomas, A. *JCR(S)* **1987**, 369. (d) Kametani, T.; Suzuki, T.; Nishimura, M.; Sato, E.; Unno, K. *H* **1982**, *19*, 205.
21. Minami, N.; Ko, S. S.; Kishi, Y. *JACS* **1982**, *104*, 1109.
22. Franck, R. W.; Subramaniam, C. S.; John, T. V.; Blount, J. F. *TL* **1984**, *25*, 2439.
23. (a) Rokach, J.; Lau, C.-K.; Zamboni, R.; Guindon, Y. *TL* **1981**, *22*, 2763. (b) Leblanc, Y.; Fitzsimmons, B. J.; Zamboni, R.; Rokach, J. *JOC* **1988**, *53*, 265.
24. Ohrui H.; Jones, G. H.; Moffatt, J. G.; Maddox, M. L.; Christensen, A. T.; Byram, S. K. *JACS* **1975**, *97*, 4602.
25. (a) Nicotra, F.; Ronchetti, F.; Russo, G.; Toma, L. *TL* **1984**, *25*, 5697. (b) Nicotra, F.; Russo, G.; Ronchetti, F.; Toma, L. *Carbohydr. Res.* **1983**, *124*, C5.
26. (a) Collins, P. M.; Overend, W. G.; Shing, T. S. *CC* **1982**, 297. (b) Webb, T. H.; Thomasco, L. M.; Schlachter, S. T.; Gaudino, J. J.; Wilcox, C. S. *TL* **1988**, *29*, 6823.
27. Isaacs, N. S.; El-Din, G. N. *TL* **1987**, *28*, 2191.

28. (a) Roberts, D. L.; Heckman, R. A.; Hege, B. P.; Bellin, S. A. *JOC* **1968**, *33*, 3566. (b) Openshaw, H. T.; Whittaker, N. *Proc. Chem. Soc.* **1961**, 454.
29. (a) Ruchardt, C.; Panse, P.; Eichler, S. *CB* **1967**, *100*, 1144. (b) Rüchardt, C.; Eichler, S.; Panse, P. *AG(E)* **1963**, *2*, 619. (c) Bose, A. K.; Manhas, M. S.; Ramer, R. M. *JCS(C)* **1969**, 2728. (d) Mulzer, J.; Kappert, M. *AG(E)* **1983**, *22*, 63.
30. (a) Fraser-Reid, B.; Tsang, R.; Tulshian, D. B.; Sun, K. M. *JOC* **1981**, *46*, 3764. (b) Tulshian, D. B.; Tsang, R.; Fraser-Reid, B. *JOC* **1984**, *49*, 2347.
31. Garner, P.; Ramakanth, S. *JOC* **1987**, *52*, 2629.
32. Bestmann, H. J.; Lienert, J.; *CZ* **1970**, *94*, 487.
33. Toda, F.; Akai, H. *JOC* **1990**, *55*, 3446.
34. Rein, T.; Reiser, O. *ACS* **1996**, *50*, 369.
35. (a) Lang, R. W.; Hansen, H.-J. *HCA* **1980**, *63*, 438. (b) Lang, R. W.; Hansen, H. J. *OS* **1984**, *62*, 202.
36. Kohl-Mines, E.; Hansen, H.-J. *HCA* **1985**, *68*, 2244.
37. Orlov, V. Y.; Lebedev, S. A.; Ponomarev, S. V.; Lutsenko, I. F. *ZOB* **1975**, *45*, 708.
38. Bestmann, H. J.; Saalfrank, R. W.; Snyder, J. P. *CB* **1973**, *106*, 2601.
39. Kayser, M. M.; Breau, L. *CJC* **1989**, *67*, 1401; and references cited therein.
40. (a) Flitsch, W.; Peters, H. *TL* **1969**, 1161. (b) For reactions of the methyl ester with thioimides, see: Bishop, J. E.; O'Connell, J. F.; Rapaport, H. *JOC* **1991**, *56*, 5079.
41. (a) Murphy, P. J.; Brennan, J. *CSR* **1988**, *17*, 1. (b) Subramanyam, V.; Silver, E. H.; Soloway, A. H. *JOC* **1976**, *41*, 1272. (c) Le Corre, M.; Normant, H. *CR(C)* **1973**, *276*, 963.
42. Gilpin, M. L.; Harbridge, J. B.; Howarth, T. T. *JCS(PI)* **1987**, 1369 (methyl ester).
43. For allylation, see: Mali, R. S.; Tilve, S. G.; Yeola, S. N.; Manekar, A. R. *H* **1987**, *26*, 121.
44. Speziale, A. J.; Ratts, K. W. *JACS* **1962**, *84*, 854.
45. Schnekenburger, J.; Heber, D.; Heber-Brunschweiger, E. *T* **1977**, *33*, 457.
46. Barluenga, J.; Lopez, F.; Palacios, F.; Sanchez-Ferrando, F. *TL* **1988**, *29*, 381.
47. Appel, R.; Winkhaus, V.; Knoch, F. *CB* **1986**, *119*, 2466; and references cited therein.
48. Hegedus, L. S.; McGuire, M. A. *OM* **1982**, *1*, 1175.
49. Märkl, G. *CB* **1961**, *94*, 3005.
50. L'abbé, G.; Ykman, P.; Smets, G. *T* **1969**, *25*, 5421.
51. Nagasaka, T.; Yamamoto, H.; Hayashi, H.; Watanabe, M.; Hamaguchi, F. *H* **1989**, *29*, 155.
52. Brubaker, A. N.; DeRuiter, J.; Whitmer, W. L. *JMC* **1986**, *29*, 1094.
53. Bestmann, H. J.; Kisielowski, L.; Distler, W. *AG(E)* **1976**, *15* 298.
54. Davis, F. A.; Chen, B. *JOC* **1990**, *55*, 360.
55. (a) Okuma, K.; Komiya, Y.; Ohta, H. *BCJ* **1991**, *64*, 2402. (b) Sato, R.; Satoh, S.-I. *S* **1991**, 785. (c) Okuma, K.; Tachibana, Y.; Sakata, J.-I.; Komiya, T.; Kaneko, I.; Komiya, Y.; Yamasaki, Y.; Yamamoto, S.-I.; Ohta, H. *BCJ* **1988**, *61*, 4323.
56. Okuma, K.; Yamasaki, Y.; Komiya, T.; Kodera, Y.; Ohta, H. *CL* **1987**, 357.
57. (a) Okuma, K.; Kaneko, I.; Ohta, H.; Yokomori, Y. *H* **1990**, *31*, 2107, (b) Okuma, K.; Komiya, Y.; Kaneko, I.; Tachibana, Y.; Iwata, E.; Ohta, H. *BCJ* **1990**, *63*, 1653. (c) Okuma, K.; Sakata, J.-I.; Tachibana, Y.; Honda, T.; Ohta, H. *TL* **1987**, *28*, 6649.

Allen B. Reitz & Mark E. McDonnell
The R. W. Johnson Pharmaceutical Research Institute, Spring House, PA, USA

(3-Ethoxy-3-oxopropyl)iodozinc

I–Zn–$(CH_2)_n$–R

($n = 2$; R = CO_2Et)
[104089-16-9] $C_5H_9IO_2Zn$ (MW 293.43)

(organozinc reagents functionalized with ester, ketone, nitrile, amide, or halide functionalities, capable of C–C bond formation with a variety of organic electrophiles, e.g arylation, vinylation, ethynylation, acylation, allylation, Michael-type addition, and addition to carbonyl CO bonds)

Alternate Name: Tamaru's reagent.
Physical Data: dark powder; reagent prepared in situ (see below).
Analysis of Reagent Purity: aliquots of reactions in process can be checked by VPC or TLC after addition of 1 N HCl.[1c]
Preparative Methods: organozincs (**1**)–(**5**), possessing electrophilic functionality (ester, ketone, nitrile, amide, halide) in the same molecule, can be prepared in high reproducibility in mmol to mol scales by the direct reaction of the corresponding iodides and zinc metal. Four kinds of procedures have been reported. *Method A:*[1] reaction of ***Zinc/Copper Couple*** with an iodide in benzene containing 1.5–2 equiv of DMF or DMA at 60 °C for several hours. For the preparation of zincio ketone (**3**), the use of HMPA in place of DMF or DMA is essential.[2] For the preparation of β-zincio esters (**1a–c**) and β-zincio nitriles (**2a–c**), THF also may be used as solvent in place of benzene–DMF or DMA. *Method B:*[3] reaction of zinc/copper couple with an iodide in benzene–DMA (2.7 equiv) under sonication at 20–35 °C for 30 min. This method is applied to the preparation of (**1d**), characteristically possessing an acidic amide proton. *Method C:*[4] reaction of ***Zinc*** powder or foil, activated with ***1,2-Dibromoethane*** (4–5%) and ***Chlorotrimethylsilane*** (3%) in THF, with a primary iodide at 35–40 °C and with a secondary iodide at 25–30 °C in the same solvent (ca. 12 h). *Method D:*[5] reaction of zinc metal, prepared by a slow addition of ***Lithium Naphthalenide*** (2 equiv) to a THF solution of ***Zinc Chloride***, with an alkyl bromide at rt. The zinc powders produced by this method are very reactive and react even with vinyl bromides and aryl bromides (e.g. ethyl *o*- and *p*-bromobenzoates).

ZnI CO2Et (**1a**) ZnI CO2Et (**1b**) ZnI CO2Et (**1c**) NHBoc ZnI CO2Bn (**1d**)

ZnI CO2Et (**1e**) O ZnI N O CO2Bn (**1f**) ZnI CO2Et (**1g**)

ZnI CO2Et (**1h**)

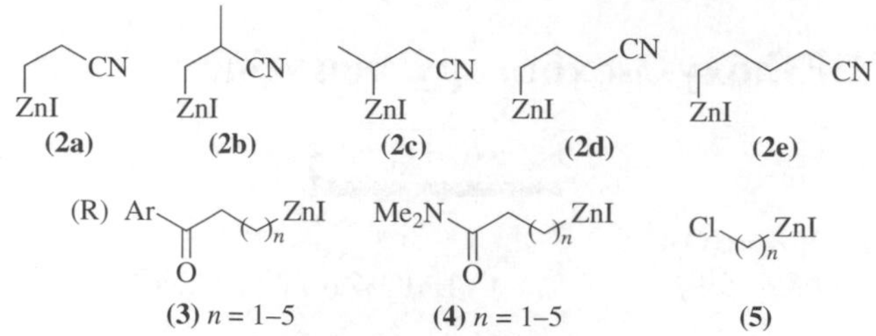

(2a) (2b) (2c) (2d) (2e)

(R) (3) n = 1–5 (4) n = 1–5 (5)

Handling, Storage, and Precautions: the iodozinc reagents can be prepared and handled like Grignard reagents, and may be stored at rt for ca. one week; however, freshly prepared samples are recommended. All reactions should be conducted in a well-ventilated fume hood.

Introduction. The title reagent is an example of a 'lower homolog' in a series of functionalized organozinc reagents with the general structure depicted above. See Table 1 for selected examples of reagents discussed in this entry.

Table 1 Selected Functionalized Organozinc Reagents I–Zn–$(CH_2)_n$–R

Compound	n	R	CAS number
(1a)	2	CO_2Et	*[104089-16-9]*
(1e)	3	CO_2Et	*[104089-17-0]*
(1g)	4	CO_2Et	*[109976-46-7]*
(1h)	5	CO_2Et	*[113274-32-1]*
(2a)	2	CN	*[121236-17-7]*
(2d)	3	CN	*[126761-11-3]*
(2e)	4	CN	*[142338-61-2]*
(3)	2	COPh	*[113274-38-7]*
	5	COPh	*[113274-37-6]*
(5)	4	Cl	*[112403-39-1]*

Reaction of Functionalized Organozincs. As evident from their high functional group compatibility, organozincs show a very low reactivity toward most organic electrophiles. However, as discussed here, they can undergo C–C bond forming reactions with many electrophiles in the presence of an appropriate transition-metal catalyst or a Lewis acid. Some reactions of functionalized organocopper reagents,[6] derived from organozincs by transmetalation with a stoichiometric amount of CuCN·2LiCl (see ***Copper(I) Cyanide***) are discussed for comparison. For the reactions of bis(β-alkoxycarbonylethyl)zinc reagents, see Nakamura et al.[7]

Arylation, Vinylation, and Ethynylation. Vinyl iodides, vinyl triflates, and aromatic iodides,[2,8] as well as aromatic bromides with electron-attracting substituents,[5] undergo coupling with organozincs in the presence of a catalytic amount of palladium (eqs 1–4). The vinylation (eq 3) proceeds with retention of configuration; the corresponding *trans*-iodide provides the *trans* coupling product exclusively in 71% yield.[8a] Arylation sometimes suffers from biaryl formation; for such cases, the use of the catalyst specified in eq 1 is recommended.[8a] The arylation tolerates an acidic amide proton (eq 2),[3,14c] and optically active phenylalanine derivatives are obtained according to this procedure.[3] Organocopper reagents directly react with ethynyl iodides and bromides under mild conditions (eq 5).[9]

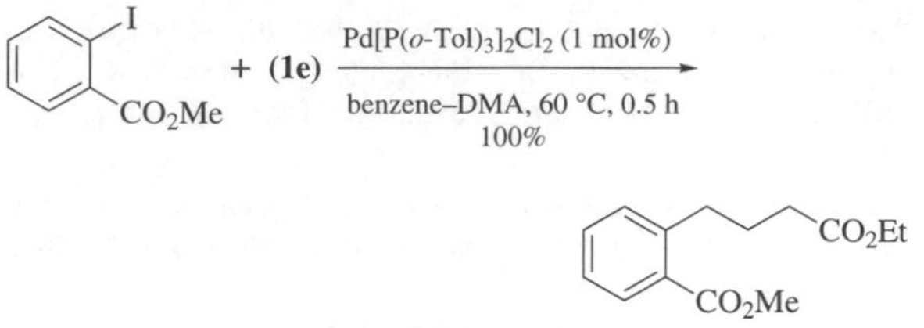

+ (1e) $\xrightarrow[\text{benzene–DMA, 60 °C, 0.5 h; 100\%}]{Pd[P(o\text{-}Tol)_3]_2Cl_2\ (1\ mol\%)}$ (1)

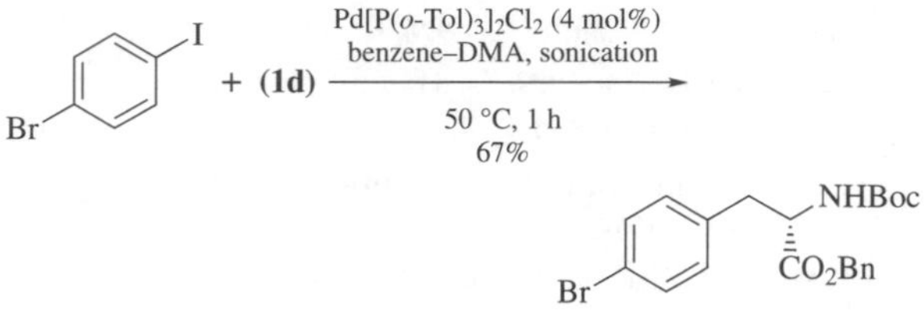

+ (1d) $\xrightarrow[\text{50 °C, 1 h; 67\%}]{Pd[P(o\text{-}Tol)_3]_2Cl_2\ (4\ mol\%),\ \text{benzene–DMA, sonication}}$ (2)

+ (1e) $\xrightarrow[\text{60 °C, 0.5 h; 89\%}]{Pd(PPh_3)_4\ (4\ mol\%),\ \text{benzene–DMA}}$ (3)

+ (3) $\xrightarrow[\text{40 °C, 1 h}]{Pd(PPh_3)_4\ (4\ mol\%),\ \text{benzene–HMPA}}$ (4)

n = 3, 67%; n = 6, 74%

+ (1e)•(CuCN) $\xrightarrow[\text{40 min}]{\text{THF, –55 °C}}$ (5)

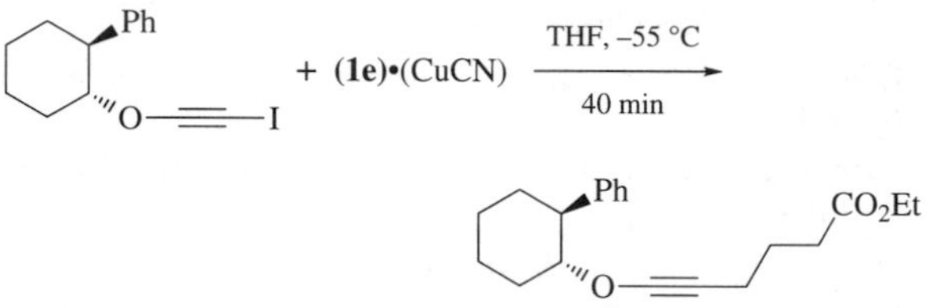

Acylation. Organozinc iodides are so unreactive that they do not react smoothly even with highly reactive acid chlorides, e.g. yielding butyrophenone in 20% yield by the reaction of n-propylzinc iodide with ***Benzoyl Chloride*** in benzene–DMF at 70 °C for 4 h.[10] The acylation is greatly accelerated by a palladium catalyst (eqs 6 and 7)[1,2] or a stoichiometric amount of CuCN·2LiCl (eq 8).[4a,5,11] The reactions in eqs 6 and 7, and the three-component connection reactions of allylic benzoates, ***Carbon Monoxide***, and organozincs (eq 9),[12] may proceed via acylpalladium intermediates.

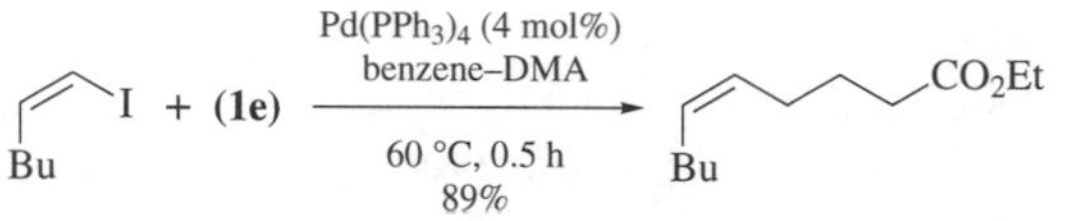

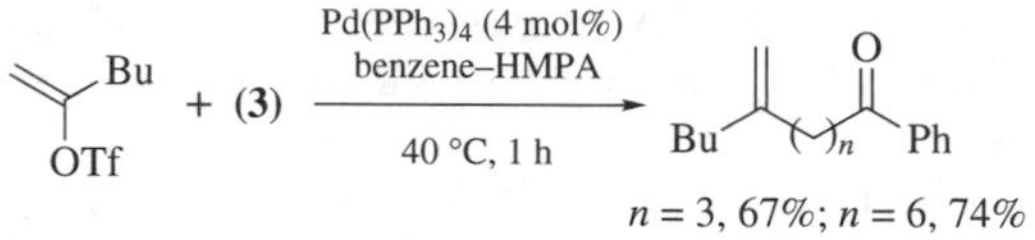

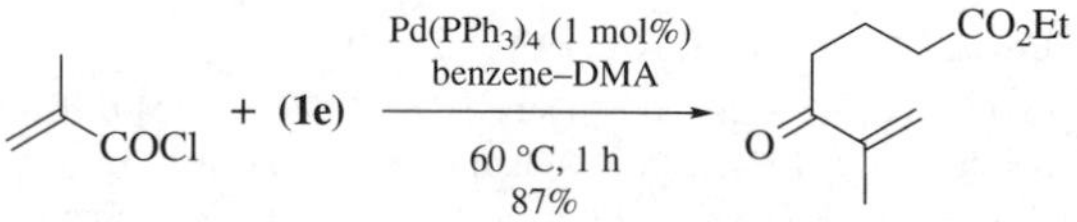

COCl + (1e) $\xrightarrow[\text{60 °C, 1 h; 87\%}]{Pd(PPh_3)_4\ (1\ mol\%),\ \text{benzene–DMA}}$ (6)

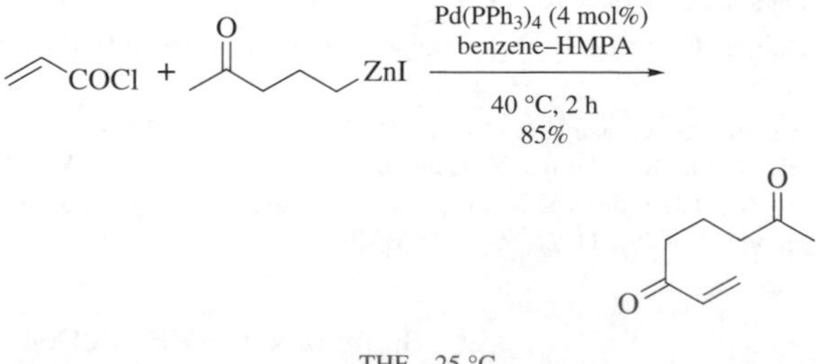

COCl + ZnI $\xrightarrow[\text{40 °C, 2 h; 85\%}]{Pd(PPh_3)_4\ (4\ mol\%),\ \text{benzene–HMPA}}$ (7)

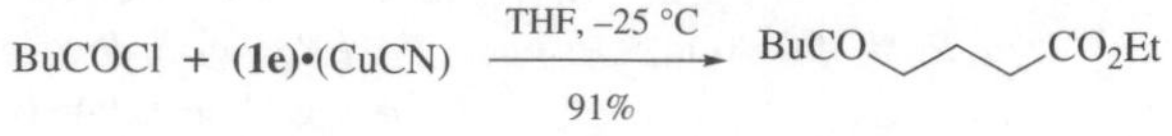

BuCOCl + (1e)•(CuCN) $\xrightarrow[\text{91\%}]{\text{THF, –25 °C}}$ BuCO$\sim\sim$$CO_2Et$ (8)

$$\text{(9)}$$

Pd(PPh$_3$)$_4$ (5 mol%), CO (1 atm), toluene–HMPA, rt, 85%

Addition to Activated Alkenes and Alkynes.

Addition to Activated Alkenes and Alkynes. Conjugate addition of organozincs to α,β-unsaturated aldehydes, ketones, esters, and nitro compounds takes place smoothly in the presence of ***Chlorotrimethylsilane*** (2 equiv) and either a catalytic[13] or a stoichiometric amount of ***Copper(I) Cyanide***.[14] The preservation of connectivity of the methyl substituent of (**1c**) (eq 11) is in contrast to the migration of the methyl group as observed for the reaction of (**1c**) with aldehydes (eq 18).[15] The reaction is not accompanied by racemization (eq 10)[13] and tolerates the acidic alkynic and amide protons (eq 12).[14c] Organozincs (e.g. **1a**, **1e**, **5**) nonselectively add to conjugated iminium salts in 1,2- and 1,4-fashion (eq 13),[16] while sorbaldehyde (2,4-hexadienal) undergoes a selective 1,4-addition reaction (vs. 1,2- and 1,6-additions) under the catalytic conditions shown in eq 10.[13]

CHO + (**1f**) — TMSCl (2 equiv), CuCN (0.35 equiv), dioxane–HMPA, 7 °C, 6 h, 70% → BnO$_2$C (10)

+ (**1b**) or (**1c**) — TMSCl (2 equiv), CuCN (0.35 equiv), dioxane–HMPA, 7 °C, 6 h → CO$_2$Et + CO$_2$Et (11)

(78% from (**1c**)) (70% from (**1b**))

H—≡—CONH$_2$ + (**1e**)•(CuCN) — THF, 0 °C, 19 h, 53% → CO$_2$Et, CONH$_2$ (12) 100% *E*

+ PhOCOCl → [N–CO$_2$Ph]$^+$ Cl$^-$ — (**5**), benzene–DMA, rt, 70% → N–CO$_2$Ph, Cl + N–CO$_2$Ph, Cl (13) 32:68

Allylation. Allylation of organozincs proceeds either catalytically (eqs 14 and 15)[17] or stoichiometrically (eqs 16 and 17)[14a] with respect to Cu^I salts. Generally, the stoichiometric reaction shows a higher regioselectivity (S$_N$2′/S$_N$2) and is applied to sequential double allylation reactions (eqs 16 and 17).[18,19]

Ph–CH=CH–CH$_2$Cl + (**1e**) — CuCN (20 mol%), THF–DMA, rt, 99% → Ph, CO$_2$Et (S$_N$2′ product) + Ph, CO$_2$Et (S$_N$2 product) 87:13 (14)

—≡—CH$_2$OTs + (**1e**) — CuCN (20 mol%), THF–DMA, 60 °C, 74% → CO$_2$Et (15)

ClCH$_2$—≡—CH$_2$Cl + (**2d**)•(CuCN) — THF, 84% → CN, CN (16)

SPh, Cl, Cl + (**1e**)•(CuCN) — THF, 25 °C, 89% → SPh, CO$_2$Et, CO$_2$Et (17)

Addition to Carbonyls. Functionalized organozincs react smoothly with aromatic aldehydes, but not with aliphatic aldehydes, in the presence of 2 equiv of chlorotrimethylsilane to give the alcohols in good yields (eqs 18 and 19).[15] Aliphatic aldehydes primarily provide the corresponding aldol products. Organocopper reagents may be sufficiently reactive toward both aromatic and aliphatic aldehydes in the presence of 2 equiv of ***Boron Trifluoride Etherate*** (eq 20).[20] Organotitanium reagents, prepared by the reaction of organozincs and ***Chlorotitanium Triisopropoxide***, are reactive toward aldehydes and ketones (eq 21).[21] Dialkylzincs (e.g. bis(γ-ethoxycarbonylpropyl)zinc) undergo an enantioselective addition to aldehydes in the presence of 2 equiv of ***Titanium Tetraisopropoxide*** and a catalytic amount of a chiral amide ligand.[22] In the presence of trialkylsilyl chloride and at elevated temperatures, β-zincio esters intramolecularly add to the ester carbonyl to provide allyloxysiloxycyclopropanes in moderate yields (eq 22).[23] These allyloxy derivatives are difficult to prepare according to Salaün's procedure.[24] The isomerization of (**1c**) to (**1b**) (eq 18) may involve cyclization similar to the one shown in eq 22.

PhCHO + (**1b**) or (**1c**) — TMSCl (2 equiv), toluene–DMA, 60 °C 17 h, 70% → Ph, OH, CO$_2$Et (18)

PhCHO + (**1g**) — TMSCl (2 equiv), toluene–DMA, rt, 18 h, 80% → Ph, OH, CO$_2$Et (19)

PhCHO + (**1e**)•(CuCN) —[BF_3•OEt_2 (2 equiv), THF, –30 °C, 4 h, 72%]→ PhCH(OH)CH$_2$CH$_2$CH$_2$CO$_2$Et (20)

BuCHO + (**3**)•[(*i*-PrO)$_3$TiCl] —[THF, 0 °C, 2 h, 75%]→ Bu(OH)…COPh (21)

ZnI ester (allyl) —[TBDMSCl (1.5 equiv), THF, 35 °C, overnight, 56%]→ O-allyl OTBDMS cyclopropane (22)

Related Reagents. 1-Ethoxy-1-(trimethylsilyloxy)cyclopropane; Ethyliodomethylzinc; Ethylzinc Iodide; 2-Iodopropane; Potassium Iodide-Zinc/Copper Couple; Zinc.

---

1. (a) Tamaru, Y.; Ochiai, H.; Nakamura, T.; Tsubaki, K.; Yoshida, Z. *TL* **1985**, *26*, 5559. (b) Tamaru, Y.; Ochiai, H.; Nakamura, T.; Yoshida, Z. *OS* **1989**, *67*, 98. (c) *OSC* **1993**, *8*, 274.
2. Tamaru, Y.; Ochiai, H.; Nakamura, T.; Yoshida, Z. *AG(E)* **1987**, *26*, 1157.
3. (a) Jackson, R. F. W.; Wythes, M. J.; Wood, A. *TL* **1989**, *30*, 5941. (b) Jackson, R. F. W.; James, K.; Wythes, M. J.; Wood, A. *CC* **1989**, 644. (c) Jackson, R. F. W.; Wishart, N.; Wood, A.; James, K.; Wythes, M. J. *JOC* **1992**, *57*, 3397; See also (d) Dunn, M. J.; Jackson, R. F. W.; Pietruszka, J.; Wishart, N.; Ellis, D.; Wythes, M. J. *SL* **1993**, 499. (e) Jackson, R. F. W.; Wishart, N.; Wythes, M. J. *SL* **1993**, 219.
4. (a) Knochel, P.; Yeh, M. C. P.; Berk, S. C.; Talbert, J. *JOC* **1988**, *53*, 2390. (b) Jubert, C.; Knochel, P. *JOC* **1992**, *57*, 5452.
5. Zhu, L.; Wehmeyer, R. M.; Rieke, R. D. *JOC* **1991**, *56*, 1445.
6. Knochel, P.; Rozema, M. J.; Tucker, C. E.; Retherford, C.; Furlong, M.; AchyuthaRao, S. *PAC* **1992**, *64*, 361.
7. Nakamura, E.; Aoki, S.; Sekiya, K.; Oshino, H.; Kuwajima, I. *JACS* **1987**, *109*, 8056.
8. (a) Tamaru, Y.; Ochiai, H.; Nakamura, T.; Yoshida, Z. *TL* **1986**, *27*, 955. (b) Sakamoto, T.; Nishimura, S.; Kondo, Y.; Yamanaka, H. *S* **1988**, 485. (c) For the palladium catalyzed coupling reaction of alkylzincs and vinyl iodides, see: Negishi, E.; Valente, L. F.; Kobayashi, M. *JACS* **1980**, *102*, 3298.
9. (a) Yeh, M. C. P.; Knochel, P. *TL* **1989**, *30*, 4799. (b) Sörensen, H.; Greene, A. E. *TL* **1990**, *31*, 7597.
10. Tamaru, Y.; Ochiai, H.; Sanda, F.; Yoshida, Z. *TL* **1985**, *26*, 5529.
11. Majid, T. N.; Yeh, M. C. P.; Knochel, P. *TL* **1989**, *30*, 5069.
12. Tamaru, Y.; Yasui, K.; Takanabe, H.; Tanaka, S.; Fugami, K. *AG(E)* **1992**, *31*, 645.
13. Tamaru, Y.; Tanigawa, H.; Yamamoto, T.; Yoshida, Z. *AG(E)* **1989**, *28*, 351.
14. (a) Yeh, M. C. P.; Knochel, P. *TL* **1988**, *29*, 2395. (b) Retherford, C.; Yeh, M. C. P.; Schipor, I.; Chen, H. G.; Knochel, P. *JOC* **1989**, *54*, 5200. (c) Knoess, H. P.; Furlong, M. T.; Rozema, M. J.; Knochel, P. *JOC* **1991**, *56*, 5974. (d) Retherford, C.; Knochel, P. *TL* **1991**, *32*, 441. (e) Sidduri, A.-R.; Knochel, P. *JACS* **1992**, *114*, 7579.
15. Tamaru, Y.; Nakamura, T.; Sakaguchi, M.; Ochiai, H.; Yoshida, Z. *CC* **1988**, 610.
16. (a) Comins, D. L.; O'Conner, S. *TL* **1987**, *28*, 1843. (b) Comins, D. L.; Foley, M. A. *TL* **1988**, *29*, 6711.
17. Ochiai, H.; Tamaru, Y.; Tsubaki, K.; Yoshida, Z. *JOC* **1987**, *52*, 4418.
18. Zhu, L.; Rieke, R. D. *TL* **1991**, *32*, 2865.
19. Chen, H. G.; Gage, J. L.; Barrett, S. D.; Knochel, P. *TL* **1990**, *31*, 1829.
20. Yeh, M. C. P.; Knochel, P.; Santa, L. E. *TL* **1988**, *29*, 3887.
21. Ochiai, H.; Nishihara, T.; Tamaru, Y.; Yoshida, Z. *JOC* **1988**, *53*, 1343.
22. (a) Rozema, M. J.; Sidduri, A.-R.; Knochel, P. *JOC* **1992**, *57*, 1956. (b) Yoshioka, M.; Kawakita, T.; Ohno, M. *TL* **1989**, *30*, 1657. (c) Takahashi, H.; Kawakita, T.; Yoshioka, M.; Kobayashi, S.; Ohno, M. *TL* **1989**, *30*, 7095.
23. Yasui, K.; Fugami, K.; Tanaka, S.; Ii, A.; Yoshida, Z.; Saidi, M. R.; Tamaru, Y. *TL* **1992**, *33*, 785.
24. (a) Salaün, J.; Marguerite, J. *OS* **1984**, *63*, 147; (b) *OSC* **1990**, *7*, 131.

Yoshinao Tamaru
Nagasaki University, Japan

1-Ethoxy-1-(trimethylsilyloxy)-cyclopropane

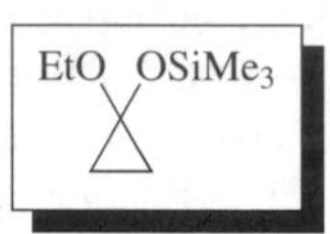

[27374-25-0] $C_8H_{18}O_2Si$ (MW 174.32)

(preparation of 3-metallopropionates;[1] metal homoenolate precursor[2])

Physical Data: bp 50–53 °C/22 mmHg.
Solubility: insol H_2O.
Form Supplied in: colorless liquid.
Analysis of Reagent Purity: GLC, NMR.
Preparative Methods: for the synthesis of the parent and the 2-monoalkyl-substituted compounds, reduction of ethyl 3-chloropropionate with ***Sodium-Potassium Alloy*** alloy in the presence of ***Chlorotrimethylsilane*** in ether.[3] A recent modification using ultrasound irradiation is much more convenient and more widely applicable.[4] Other substituted derivatives are prepared by cyclopropanation of alkyl silyl ketene acetals with the Furukawa reagent (***Diiodomethane/Diethylzinc***).[5]
Purification: distillation under reduced pressure.
Handling, Storage, and Precautions: moisture sensitive, yet, once purified by distillation, is stable for a long period of time in a tightly capped bottle at room temperature.

Stoichiometric Precursor of Metal Homoenolates. The reaction of 1-alkoxy-1-trimethylsilyloxycyclopropane with a variety of Lewis acidic metal chlorides affords the 3-metallated propionate esters in good to excellent yield (see below).[1,6,7] For instance, the reaction of 1-ethoxy-1-trimethylsilyloxycyclopropane with one equivalent of ***Tin(IV) Chloride*** gives a 3-stannylpropionate, which further reacts with another equivalent of the cyclopropane to give a dialkylated tin compound (eq 1).

The reaction of the siloxycyclopropane with ***Titanium(IV) Chloride*** produces the titanium homoenolate (3-titaniopropionate) in good yield; this, however, is relatively unreactive (eq 2).[8] Addition of one equivalent of $Ti(OR')_4$ generates a more reactive $RTiCl_2OR'$ species, which smoothly reacts with carbonyl compounds below room temperature.[9] The γ-hydroxy ester adducts are useful synthetic intermediates and serve as precursors to γ-lactones and cyclopropanecarboxylates.[10] A useful variation involves the use of the cyclopropanecarboxylate ester as a functionalized homoenolate precursor to obtain levulinic acid derivatives (eq 3).[11]

The zinc homoenolate prepared by the treatment of the siloxycyclopropane with ***Zinc Chloride*** is a versatile synthetic reagent (eq 4).[12] Reduction of 3-iodopropionate with activated ***Zinc*** also produces a zinc homoenolate species.[13]

Treatment of the silyloxycyclopropane with $ZnCl_2$ followed by addition of an enone, HMPA, THF, and a catalytic amount of a Cu^I salt results in quantitative formation of a conjugate adduct as an enol silyl ether (eq 4). The chlorosilane, a byproduct, is essential for the conjugate addition of the copper homoenolate.[14,15] ***Boron Trifluoride Etherate*** promotes the copper-catalyzed conjugate addition reaction with a different stereochemical outcome.[16] A useful application of the conjugate addition reaction is a [3 + 2] synthesis of cyclopentenones, wherein the homoenolate acts as a 1,3-dipole equivalent (eq 5).[17]

The zinc homoenolate undergoes copper-catalyzed allylation with allylic chlorides. The reaction is not only extremely S_N2' regioselective but stereoselective for δ-chiral allylic chlorides.[18] Arylation and vinylation of the zinc homoenolates proceed in the presence of a palladium–phosphine complex.[19] Similarly, palladium-catalyzed acylation reaction gives γ-keto esters (eq 6).

Catalytic Generation of Homoenolate Reactive Species. Homoaldol reaction between the siloxycyclopropane and an aldehyde with a catalytic amount of ***Zinc Iodide*** in methylene chloride affords a γ silyloxy ester (eq 7).[18,20] Arylation[21] and acylation[22,23] of the silyloxycyclopropanes in the presence of a palladium catalyst take place via direct attack of an aryl- or acylpalladium intermediate on the C–C bond of the cyclopropane (eqs 8 and 9). The reaction is applicable not only to ester synthesis but also to ketone and aldehyde synthesis. Heating a chloroform solution of the silyloxycyclopropane in the presence of a palladium–phosphine catalyst under 1 atm ***Carbon Monoxide*** produces a γ-keto pimelate (eq 10).[24]

R^1-cyclopropane(OTMS)(OR) + ArOTf $\xrightarrow{\text{cat. } PdCl_2(Ph_3P)_2}$ Ar–$CH_2CH(R^1)$–COR^2 (8)

R^2 = H, alkoxy, alkyl, aryl

R^1-cyclopropane(OTMS)(OR) + R^2COCl $\xrightarrow{\text{cat. } PdCl_2(Ph_3P)_2}$ R^2CO–$CH_2CH(R^1)$–COR (9)

R = alkoxy, alkyl, aryl

R^1-cyclopropane(OTMS)(OR) + CO $\xrightarrow[CHCl_3]{\text{cat. } PdCl_2}$ RO_2C–$CH(R^1)CH_2$–CO–$CH_2CH(R^1)$–CO_2R (10)

Precursor of Lithiocyclopropane. Bromination of the silyloxycyclopropane with ***Phosphorus(III) Bromide*** produces 1-bromo-1-ethoxycyclopropane. Successive treatment of the bromide with ***t-Butyllithium*** and an enal affords a cyclopropylcarbinol, which undergoes acid-catalyzed ring enlargement to give 2-vinylcyclobutanone (eq 11).[25]

cyclopropane(OTMS)(OEt) $\xrightarrow{PBr_3}$ cyclopropane(Br)(OEt) $\xrightarrow[\text{2. MeCH=CHCHO}]{\text{1. } t\text{-BuLi}}$ cyclopropane(OEt)–CH(OH)–CH=CHMe $\xrightarrow{HBF_4}$ 2-(propenyl)cyclobutanone (11)

Reactions with Azidoformates. Photolysis of an acetonitrile solution of the cyclopropane and ***Ethyl Azidoformate*** at rt gives a C–H insertion product (eq 12).[26] However, thermolysis of the same mixture in DMSO gives a 3-aminopropionate by insertion of nitrene into the cyclopropane ring (eq 13).[27]

cyclopropane(OTMS)(OEt) + N_3CO_2Et $\xrightarrow[MeCN]{h\nu}$ EtO_2CNH-cyclopropane(OTMS)(OEt) (12)

Me-cyclopropane(OTMS)(OMe) + N_3CO_2Et $\xrightarrow[DMSO]{120\ °C}$ EtO_2CNH–$CH(CH_3)$–CH_2CO_2Me (13)

Cyclopropanone Hemiacetals and Their Use. Mild alcoholysis of the silyloxycyclopropane gives a cyclopropanone hemiacetal. This compound serves as a stable equivalent of unstable cyclopropanones.[3] Treatment with two equivalents of alkynylmagnesium bromide gives a 1-ethynyl-1-hydroxycyclopropane (eq 14).[28]

cyclopropane(OTMS)(OEt) $\xrightarrow{EtOH}$ cyclopropane(OH)(OEt) $\xrightarrow[Et_2O]{R-C\equiv C-MgBr}$ cyclopropane(OH)(C≡C–R) (14)

The cyclopropanol also serves as a source of homoenolate radical species. Treatment of a mixture of the cyclopropanol and an enol silyl ether with manganese(III) 2-pyridinecarboxylate in DMF gives a 1,5-dicarbonyl compound (eq 15).[29]

cyclopropane(OH)(OEt) + CH_2=C(OTBDMS)Ph $\xrightarrow[\text{DMF, 0 °C}]{Mn(pic)_3}$ EtO–CO–$CH_2CH_2CH_2$–CO–Ph (15)

Strecker amino acid synthesis starting with the cyclopropanone hemiacetal provides a enantioselective route to a cyclopropane amino acid (eq 16).[30]

R_2-cyclopropane(OH)(OEt) $\xrightarrow[MeOH]{R^1NH_2\bullet HCl,\ NaCN,\ )))}$ R_2-cyclopropane(CN)(NHR^1) (16)

R = H, Me

Related Reagents. Cyclopropanone; (3-Ethoxy-3-oxopropyl)iodozinc; 1-Lithio-1-methoxycyclopropane.

1. Nakamura, E.;. Shimada, J.; Kuwajima, I. *OM* **1985**, *4*, 641.
2. Kuwajima, I.; Nakamura, E. *COS* **1991**, *2*, Chapter 1.14.
3. Salaün, J.; Marguerite, J. *OS* **1985**, *63*, 147.
4. Fadel, A.; Canet, J.-L.; Salaün, J. *SL* **1990**, 89.
5. Rousseau, G.; Slonghi, N. *TL* **1983**, *24*, 1251.
6. Murakami, M.; Inouye, M.; Suginome, M.; Ito, Y. *BCJ* **1988**, *51*, 3649.
7. Ryu, I.; Murai, S.; Sonoda, N. *JOC* **1986**, *51*, 2389.
8. Nakamura, E.; Kuwajima, I. *JACS* **1983**, *105*, 651.
9. Nakamura, E.; Oshino, H.; Kuwajima, I. *JACS* **1986**, *108*, 3745.
10. Nakamura, E.; Kuwajima, I. *JACS* **1985**, *107*, 2138.
11. Reissig, H.-U. *Top. Curr. Chem.* **1988**, *144*, 73.
12. Nakamura, E.; Aoki, S.; Sekiya, K.; Oshino, H.; Kuwajima, I. *JACS* **1987**, *109*, 8056.
13. Tamaru, Y.; Ochiai, H.; Nakamura, T.; Tsubaki, K.; Yoshida, Z.-i. *TL* **1985**, *26*, 5559. Tamaru, Y.; Ochiai, H.; Nakamura, T.; Yoshida, Z.-i. *TL* **1986**, *27*, 955. Yeh, M. C. P.; Knochel, P. *TL* **1988**, *29*, 2395. Tamaru, Y.; Ochiai, H.; Nakamura, T.; Yoshida, Z.-i. *AG(E)* **1987**, *26*, 1157.
14. Nakamura, E.; Kuwajima, I. *JACS* **1984**, *106*, 3368. Nakamura, E.; Kuwajima, I. *OS* **1987**, *66*, 43.
15. (a) Corey, E. J.; Boaz, N. W. *TL* **1985**, *26*, 6015, 6019–6021. (b) Alexakis, A.; Berlan, J.; Besace, Y. *TL* **1986**, *27*, 1047. (c) Horiguchi, Y.; Matsuzawa, S.; Nakamura, E.; Kuwajima, I. *TL* **1986**, *27*, 4025. (d) Nakamura, E.; Matsuzawa, S.; Horiguchi, Y.; Kuwajima, I. *TL* **1986**, *27*, 4029. (e) Matsuzawa, S.; Horiguchi, Y.; Nakamura, E.; Kuwajima, I. *T* **1989**, *45*, 349. (f) Nakamura, E. In *Organocopper Reagents*; Taylor, R. J. K., Ed.; Oxford: Oxford University Press, 1994; Chapter 6.
16. Horiguchi, Y.; Nakamura, E.; Kuwajiama, I. *JACS* **1989**, *111*, 6257.
17. Crimmins, M. T.; Nantermet, P. G.; Wesley, B.; Vallin, I. M.; Watson, P. S.; AcKerlie, L. A.; Reinhold, T. L.; Cheung, A. W.-H.; Stetson, K. A.; Dedopoulou, D.; Gray, J. L. *JOC* **1993**, *58*, 1038.
18. Nakamura, E.; Sekiya, K.; Arai, M.; Aoki, S. *JACS* **1989**, *111*, 3091.
19. Aoki, S.; Fujimura, T.; Nakamura, E.; Kuwajima, I. *TL* **1989**, *30*, 6541.
20. Gore, V. G.; Mahendra, D.; Chordia, D.; Narasimhan, S. *T* **1990**, *46*, 2483.
21. Aoki, S.; Fujimura, T.; Nakamura, E.; Kuwajima, I. *JACS* **1988**, *110*, 3296.
22. Fujimura, T.; Aoki, S.; Nakamura, E. *JOC* **1991**, *56*, 2810.
23. Aoki, S.; Nakamura, E. *T* **1991**, *47*, 3935.
24. Aoki, S.; Nakamura, E.; Kuwajima, I. *TL* **1988**, *29*, 1541.

25. Gadwood, R. C.; Rubino, M. R.; Nagarajan, S. C.; Michel, S. T. *JOC* **1985**, *50*, 3255.
26. Mitani, M.; Tachizawa, O.; Takeuchi, H.; Koyama, K. *CL* **1987**, 1029.
27. Mitani, M.; Tachizawa, O.; Takeuchi, H.; Koyama, K. *JOC* **1989**, *54*, 5397.
28. Salaün, J. *JOC* **1976**, *41*, 1237.
29. Iwasawa, N.; Hayakawa, S.; Isobe, K.; Narasaka, K. *CL* **1991**, 1193.
30. Fadel, A. *T* **1991**, *47*, 6265. Fadel, A. *SL* **1993**, 503.

Eiichi Nakamura
Tokyo Institute of Technology, Japan

1-Ethoxyvinyllithium[1]

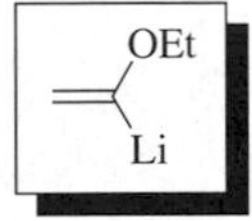

[40207-59-8] C_4H_7LiO (MW 78.04)

(reacts with aldehydes and ketones to form allylic alcohols;[1,2] adds to the carbon–nitrogen double bond of isoquinolinium salts;[3] reacts with chlorosilanes;[4] can add to conjugated sulfones[5])

Physical Data: ^{13}C NMR solution studies in THF have been reported.[6]
Solubility: sol THF.
Form Supplied in: prepared in situ and used directly.
Preparative Methods: prepared in situ under nitrogen by the reaction of ***Ethyl Vinyl Ether*** in anhydrous THF with ***t-Butyllithium***.
Handling, Storage, and Precautions: must be prepared and transferred under inert gas (Ar or N_2) to exclude oxygen and moisture.

Additions to Aldehydes and Ketones. 1-Ethoxyvinyllithium adds to benzaldehyde to yield an allylic alcohol (eq 1) which, upon acid hydrolysis, forms the α-hydroxy methyl ketone.[2] Reaction also occurs with ketones (eq 2).[7]

Ph–CHO → (OEt, Li reagent) → Ph–CH(OH)–C(=CH$_2$)OEt (1)

(2)

γ-Ethoxy dienones have been synthesized in good yields via the intermediate produced by the reaction of 1-ethoxyvinyllithium with 3-ethoxy-2-cyclohexenones.[8]

Addition of the reagent to a 2-dialkylaminocyclohexanone has been reported (eq 3).[9]

(3) (–78 °C)

The reagent adds to conjugated enones at the carbonyl group (eq 4).[10–12]

(4)

Addition of the reagent to an aldehyde in the total synthesis of nikkomycin B produced the vinyl ether, which was not isolated.[13] The reagent can also be used to prepare the corresponding cuprate, which adds 1,4 to enones.[14]

Addition to Carbon–Nitrogen Double Bonds. The reagent adds cleanly to isoquinolinium salts (eq 5) to yield, after acid hydrolysis, the methyl ketone.[3]

(5)

Reaction with Chlorosilanes. Chlorosilanes react with 1-ethoxyvinyllithium[4,15] to yield, after acid hydrolysis, acylsilanes with overall yields of 19–31% (eq 6).[4]

t-Bu–SiMe(Cl)–CH$_2$OR → 1. (OEt, Li reagent); 2. H_3O^+ → t-Bu–SiMe(COMe)–CH$_2$OR (6)

Addition to Conjugated Sulfones. The reagent adds to conjugated sulfones at the β-carbon in good yields (eq 7).[5]

PhO_2S–CH=C(SiMe)R → 1. (OEt, Li reagent); 2. KF → PhO_2S–CH$_2$–CH(R)–CO–OEt (7)

Related Reagents. 1,3-Dithiane; 5-Lithio-2,3-dihydrofuran; 6-Lithio-2,3-dihydro-4H-pyran; 2-Lithio-1,3-dithiane; 1-Methoxyvinyllithium.

---

1. (a) Lever, O. W., Jr. *T* **1976**, *32*, 1943. (b) Gschwend, H. W.; Rodriquez, H. R. *OR* **1979**, *26*, 1.
2. Schöllkopf, U.; Hänssle, P. *LA* **1972**, *763*, 208.
3. Tietze, L. F.; Brill, G. *LA* **1987**, 311.
4. Bienz, S.; Chapeaurouge, A. *HCA* **1991**, *74*, 1477.
5. Isobe, M.; Funabashi, Y.; Ichikawa, Y., Mio, S.; Goto, T. *TL* **1984**, *25*, 2021.
6. Oakes, F. T.; Sebastian, J. F. *JOC* **1980**, *45*, 4959.

7. Modi, S. P.; Michael, M. A.; Archer, S.; Carey, J. J. *T* **1991**, *47*, 6539.
8. Kraus, G. A.; Krolski, M. E. *SC* **1982**, *12*, 521.
9. Jacobsen, E. J.; Levin, J.; Overman, L. E. *JACS* **1988**, *110*, 4329.
10. Kanda, Y.; Saito, H.; Fukuyama, T. *TL* **1992**, *33*, 5701.
11. Braish, T. F.; Saddler, J. C.; Fuchs, P. L. *JOC* **1988**, *53*, 3647.
12. Kallmerten, J.; Plata, D. J. *H* **1987**, *25*, 145.
13. Barrett, A. G. M.; Lebold, S. A. *JOC* **1990**, *55*, 5818.
14. Boeckman, R. K., Jr.; Bruza, K. J. *JOC* **1979**, *44*, 4781.
15. Cunico, R. F.; Kuan, C.-P. *JOC* **1985**, *50*, 5410.

John F. Sebastian
Miami University, Oxford, OH, USA

Ethyl Acetoacetate[1]

(**1**; R^1 = Me, R^2 = Et)
[141-97-9] $C_6H_{10}O_3$ (MW 130.16)
(**2**; R^1 = R^2 = Me)
[105-45-3] $C_5H_8O_3$ (MW 116.13)
(**3**; R^1 = Me, R^2 = *t*-Bu)
[1694-31-1] $C_8H_{14}O_3$ (MW 158.22)
(**4**; R^1 = Ph, R^2 = Et)
[94-02-0] $C_{13}H_{16}O_3$ (MW 192.29)

(three- or four-carbon nucleophile in alkylations,[2] conjugate additions,[3] and various condensation reactions[4,5])

Physical Data: (**1**) bp 180–181 °C; *d* 1.025 g cm^{-3}. (**2**) bp 169–170 °C; *d* 1.076 g cm^{-3}. (**3**) bp 71–72 °C/11 mmHg; *d* 0.961 g cm^{-3}. (**4**) bp 266–270 °C; *d* 1.113 g cm^{-3}.

Preparative Methods: ethyl, methyl, and *t*-butyl acetoacetates and ethyl benzoyl acetate are commercially available. A wide variety of acetoacetic esters may be prepared via alcoholysis of diketene[6] and ester exchange reaction of *t*-butyl acetoacetate.[7]

Handling, Storage, and Precautions: use in a fume hood.

Alkylations. Alkylation at the α-carbon of ethyl acetoacetate via an alkali metal enolate and subsequent removal of the ester group by sequential hydrolysis and decarboxylation provides a classical synthesis of methyl ketones (eq 1).[8] The ambident reactivity of the acetoacetate anion often causes competing *C*- and *O*-alkylations.[9] The extent of *C* vs. *O* alkylation depends on the nature of the leaving group, metal counterion, and solvent.[10,11] In general, heterogeneous reaction conditions, less electropositive metal cations, e.g. lithium, nonpolar or protic solvents, and polarizable alkylating agents such as alkyl iodides and allylic, propargylic, and benzylic halides favor *C*-alkylation. Highly dipolar aprotic solvents, e.g. HMPA, and highly reactive alkylating agents such as alkyl sulfonates and sulfates tend to favor *O*-alkylation. Use of phase-transfer catalysis,[12] heating crystalline thallium(I) enolates with alkyl iodides,[13] and alkylation in the presence of ***Tetra-n-butylammonium Fluoride*** base (eq 2)[14] are among the methods that successfully avoid or minimize *O*-alkylation. While primary and unhindered secondary halides give good yields of *C*-alkylated products, tertiary halides undergo elimination under anionic alkylation conditions. In such cases, alkylation may be achieved via reaction of the enol with cationic species (eq 3).[15]

BuBr; NaOEt, EtOH, reflux, 69–72%; 1. NaOH, H_2O; 2. H_2SO_4, H_2O, reflux, 52–61% (1)

i-PrI; Et_4NF, DMF, rt, 94% (2)

t-BuBr; $AgClO_4$, $MeNO_2$, 68% (3)

Metal coordinated alkenes react with acetoacetate anion, affording an alternative type of alkylation.[16] π-Allyl palladium species, catalytically generated by reaction of Pd^0 complexes with allylic acetates or halides, provide efficient *C*-allylation of the acetoacetate anion.[17] Alkylations with vinyl epoxides, allylic carbonates, and allylic carbamates via Pd^0 catalysis proceed without the need for an external base since the π-allylpalladium species in these systems are sufficiently basic to deprotonate acetoacetate in situ.[17] The regio- and stereoselectivities in alkylations with cyclic vinyl epoxides (eq 4)[18] are opposite to those observed under standard carbanion conditions.[19] A highly efficient Pd^0-catalyzed addition of methyl acetoacetate to 1,3-dienes can be achieved using ***1,3-Bis(diphenylphosphino)propane*** ligand (eq 5).[20]

1.7% $Pd(PPh_3)_4$, 83% (4)

1% $[(\pi\text{-}C_3H_5)PdCl]_2$, 3% $Ph_2P(CH_2)_3PPh_2$, 3% NaOMe, 99% (5)

Manganese(III) Acetate promotes the radical addition of ethyl acetoacetate to enol ethers and subsequent oxidative cyclization to 1-alkoxy-1,2-dihydrofurans, which may be hydrolyzed to alkylated acetoacetate derivatives (eq 6).[21] The corresponding reaction with simple alkenes is less general.[22]

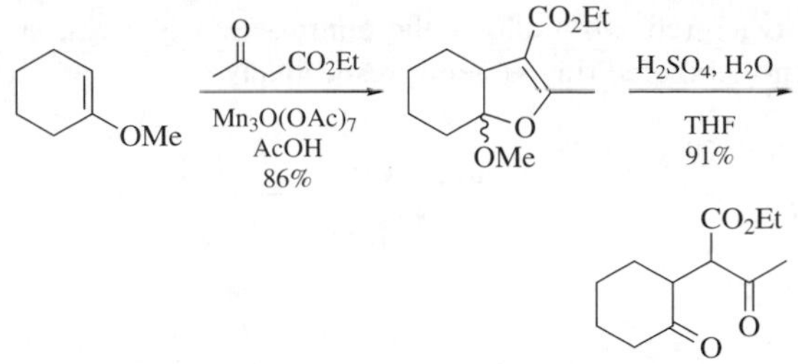

(6)

α,γ-Dianions of acetoacetates (see ***Methyl Dilithioacetoacetate***) can be generated by treatment with 1 equiv of ***Sodium Hydride***, followed by 1 equiv of ***n-Butyllithium***, or 2 equiv of ***Acetaldehyde 3-Bromopropyl Ethyl Acetal*** in THF.[23] Reaction of the dianion with a variety of alkylating agents including alkyl halides (eq 7)[23] and epoxides (eq 8)[24] results in regioselective γ-alkylation.

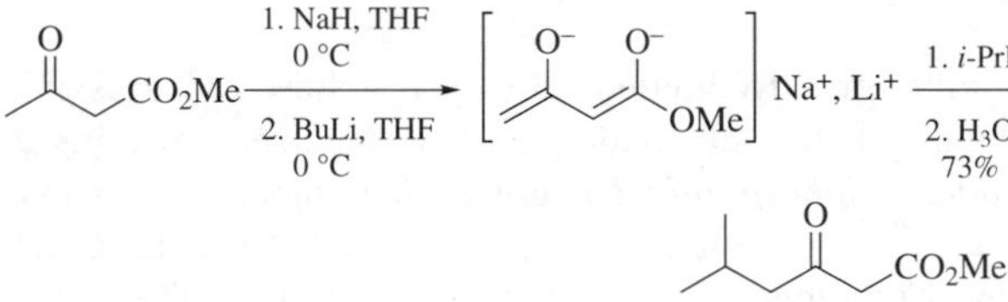

(7)

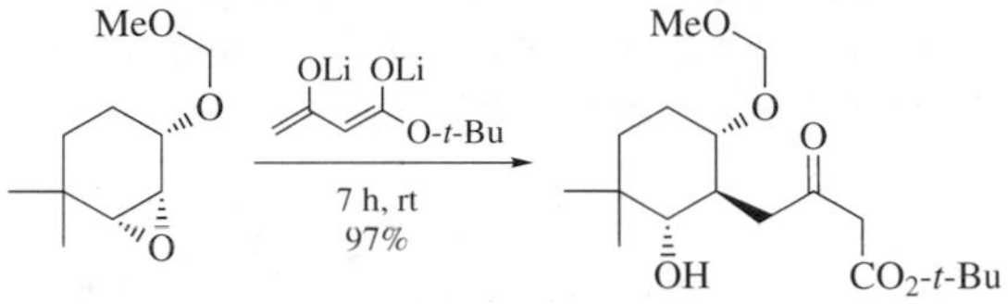

(8)

Acylations. *C*-acylation at the α-carbon of ethyl acetoacetate may be achieved by reaction of the magnesio or sodio derivative with an acid chloride.[25] Acylation may also be performed efficiently in the presence of $MgCl_2$ and pyridine (eq 9).[26]

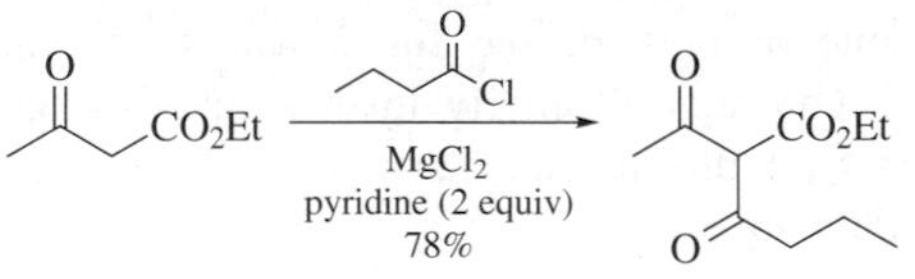

(9)

The dianions of acetoacetates undergo γ-*C*-acylation with esters.[27] Proton transfer occurs readily from product to the starting enolate. As a result, two equivalents of the dianion are needed to ensure complete conversion. This problem can be overcome by employing *N*-methoxy-*N*-methyl amides (see ***N-Methoxy-N-methylacetamide***) as acylating agents.[28] Addition of catalytic amounts of ***N,N,N′,N′-Tetramethylethylenediamine*** greatly improves the efficiency of acylation of dilithio ethyl acetoacetate (eq 10).[29] *N,N*-Dimethyl carboxamides can also be used as acylating agents in the presence of ***Boron Trifluoride Etherate***.[30]

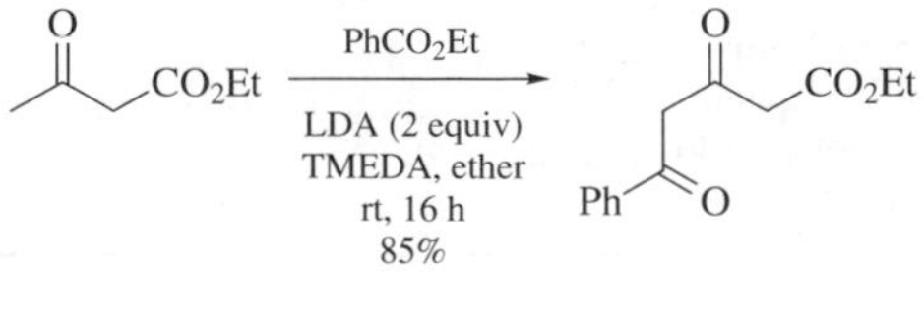

(10)

Carbonyl Additions and Condensations. Knoevenagel condensation of acetoacetates with carbonyl compounds may be carried out using a wide range of catalysts and reaction conditions.[4] A particularly mild and efficient procedure involves the use of ***Titanium(IV) Chloride*** and pyridine in THF at low temperatures, which allows the condensation of ethyl acetoacetate with aliphatic, aromatic, and heteroaromatic aldehydes to produce α-alkylideneacetoacetates (eq 11).[31] The condensation of *t*-butyl acetoacetate with aldehydes and subsequent cleavage of the *t*-butyl ester by heating with catalytic ***p-Toluenesulfonic Acid*** provides a convenient synthesis of α,β-unsaturated methyl ketones.[32] α-Alkyl-substituted acetoacetates can add to aldehydes; however, the adducts often undergo acyl cleavage under the reaction conditions to give α,β-unsaturated esters (eq 12).[33] The dianion of methyl acetoacetate reacts with aldehydes and ketones in the γ-position to give aldol adducts (eq 13).[34]

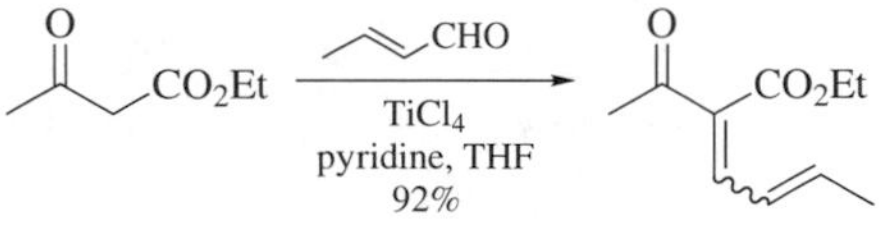

(11)

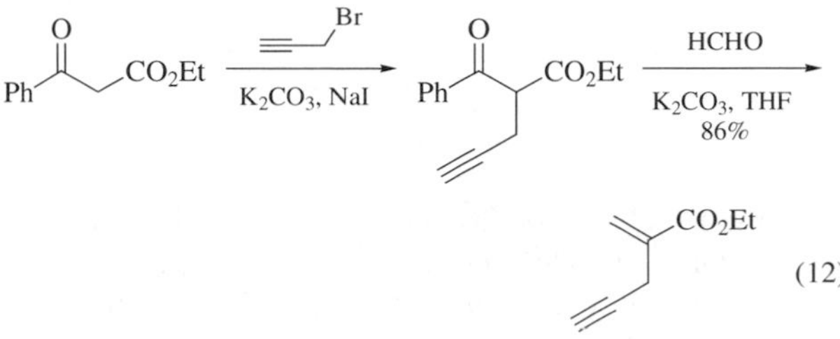

(12)

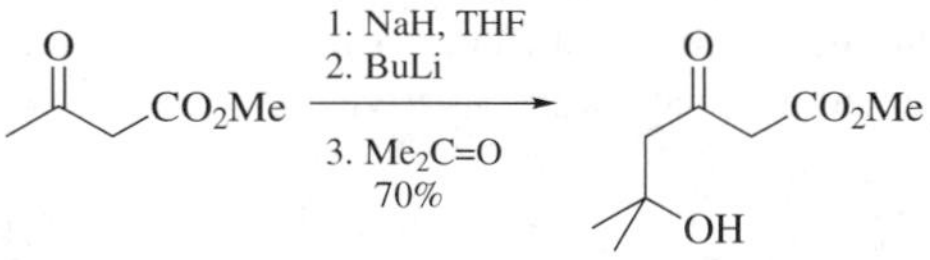

(13)

Conjugate Additions and Annulation Reactions. A wide variety of activated alkenes, alkynes, and allenes participate in conjugate additions or Michael reactions with acetoacetates. α,β-Unsaturated ketones, esters, nitriles, aldehydes, lactones, sulfoxides, sulfones, and nitro compounds are among the more commonly used alkene acceptors.[3] Although conventional procedures employ strongly basic catalysts, neutral reaction conditions (eq 14)[35,36] are often used to overcome side reactions such as bis additions and self condensations.

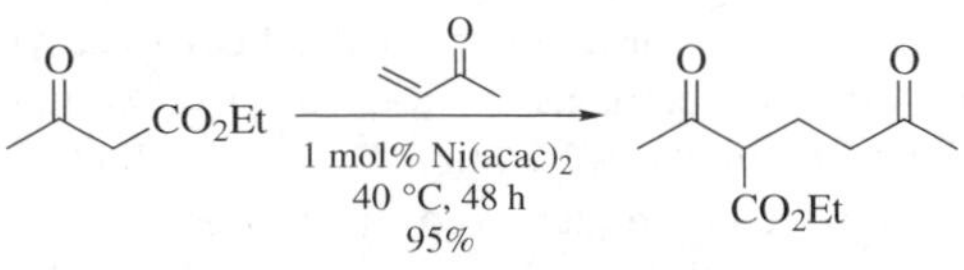

(14)

1,5-Diketones produced by the conjugate addition of acetoacetates to α,β-unsaturated ketones are useful in the synthesis of cyclohexenones via intramolecular aldol condensation.[37] A sequence of Knoevenagel condensation, Michael addition, aldol cyclization, and dealkoxycarbonylation may be carried in one pot using ethyl acetoacetate, an aliphatic aldehyde, and ***Piperidine*** catalyst (eq 15).[38] In general, however, annulation reactions of acetoacetates involve the Michael addition and the aldol cyclization as two separate steps (eq 16).[39]

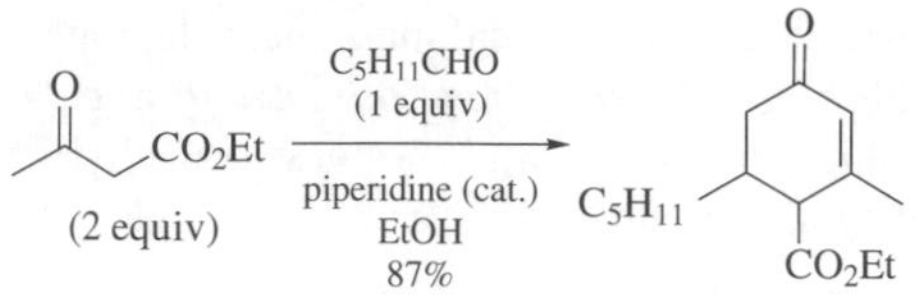

(2 equiv) — $C_5H_{11}CHO$ (1 equiv), piperidine (cat.), EtOH, 87% → CO_2Et (15)

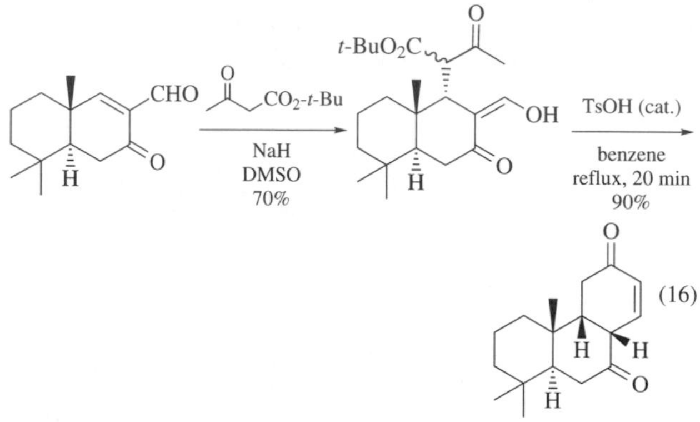

CHO; CO_2-*t*-Bu; NaH, DMSO, 70%; *t*-BuO_2C; OH; TsOH (cat.), benzene, reflux, 20 min, 90% (16)

Trimethylsilyl Enol Ethers. Successive silylation of methyl acetoacetate with ***Chlorotrimethylsilane*** gives the mono- and the bis(trimethylsilyl) enol ethers (eq 17),[40] which are useful as synthetic equivalents of acetoacetate in reactions with a variety of electrophiles. Lewis acid-catalyzed reaction of the mono(trimethylsilyl) enol ether with *N*-acylimines,[41] oxonium ions,[42] thianium ions[43] and vinyl iminium ions (eq 18)[44] gives α-*C*-alkylated acetoacetate derivatives.

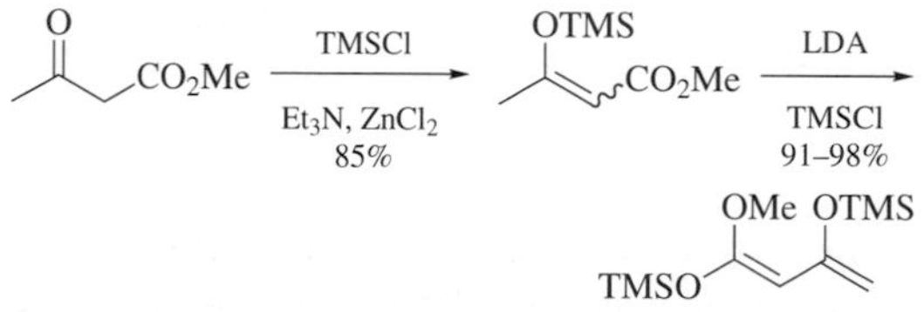

CO_2Me — TMSCl, Et_3N, $ZnCl_2$, 85% → OTMS CO_2Me — LDA, TMSCl, 91–98% → OMe OTMS, TMSO (17)

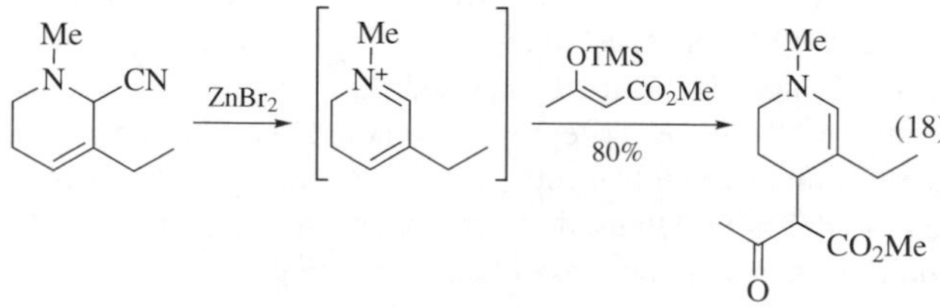

Me N CN — $ZnBr_2$ → [Me N^+] — OTMS CO_2Me, 80% → Me N, CO_2Me (18)

$TiCl_4$-catalyzed aldol addition of the bis(trimethylsilyl) enol ether of methyl acetoacetate to carbonyl compounds occurs exclusively in the γ-position.[45] Highly diastereoselective aldol additions can be achieved in reactions with α-alkoxy aldehydes (eq 19).[46] The bis(trimethylsilyl) enol ether may also be used as a 1,3-dianionic synthon for annulation with dicarbonyl electrophiles. Lewis acid-catalyzed annulation with α,α-dialkoxy ketones,[47] β,β-dialkoxy ketones,[40] and 1,4-keto aldehydes (eq 20)[48] allows the construction of five-, six-, and seven-membered ring systems, respectively.

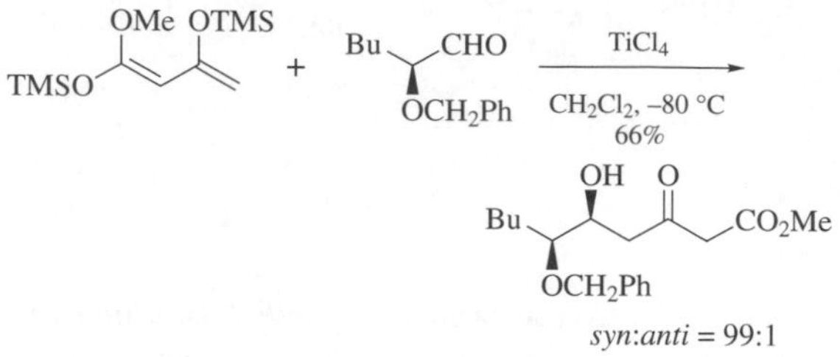

OMe OTMS, TMSO + Bu CHO, OCH_2Ph — $TiCl_4$, CH_2Cl_2, –80 °C, 66% → OH O, Bu, CO_2Me, OCH_2Ph (19)

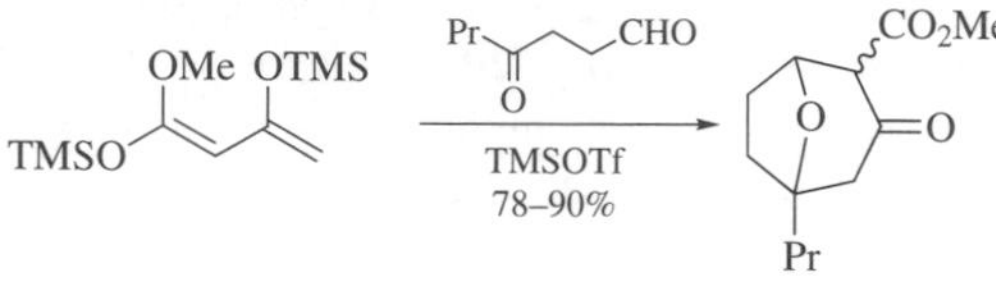

syn:*anti* = 99:1

OMe OTMS, TMSO — Pr CHO O, TMSOTf, 78–90% → CO_2Me, O, =O, Pr (20)

Asymmetric Reduction. Highly enantioselective asymmetric hydrogenation of acetoacetates can be achieved using ***2,2′-Bis(diphenylphosphino)-1,1′-binaphthyl***–ruthenium complexes (eq 21).[49] The resulting 3-hydroxybutanoates (see ***Ethyl 3-Hydroxybutanoate***) are versatile auxiliaries and building blocks for the synthesis of enantiomerically pure products. Racemic α-substituted acetoacetates may also be hydrogenated in high stereoselectivity via dynamic kinetic resolution.[50]

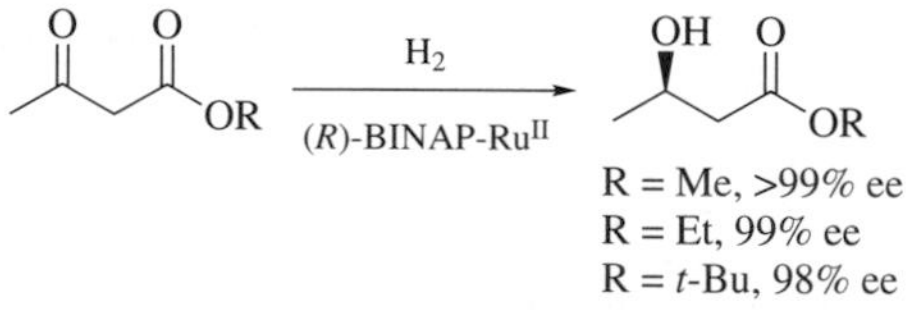

O O OR — H_2, (*R*)-BINAP-Ru^{II} → OH O OR (21)

R = Me, >99% ee
R = Et, 99% ee
R = *t*-Bu, 98% ee

Heterocyclic Synthesis. A wide variety of five- and six-membered heterocyclic systems may be derived by the cyclocondensation of acetoacetates with 1,2- and 1,3-bifunctional reagents wherein one or both of the functionalities are heteroatomic (Table 1). The regioselectivity in these cyclizations can often be altered by replacing acetoacetates with their enol ether or enamine derivatives.

Table 1 Heterocyclic Synthesis via Cyclocondensation of Ethyl Acetoacetate (and Analogs) with 1,2- and 1,3-Bifunctional Reagents

Bifunctional reagent	Heterocyclic products
Hydrazines[51]	pyrazolones
Hydroxylamines[52]	isoxazolones
Hydroxamoyl chlorides[53]	isoxazoles
Hydrazidoyl chlorides[53]	pyrazoles
α-Amino ketones[54]	pyrroles
α-Halo ketones[55]	furans
α-Hydroxy carbonyl compounds[4]	furanones
β-Hydroxy carbonyl compounds[4]	pyranones
Salicylaldehydes[4]	coumarins
o-Aminobenzaldehydes and *o*-aminoacetophenones[4,56]	quinolines
Ureas, thioureas, amidines and guanidines[57]	pyrimidines

The classical synthesis of coumarins via the von Pechmann reaction[58] involves the condensation of acetoacetates with phenols under acid catalysis (eq 22).[59] Chromones are produced when the phenol contains a deactivating group or the acetoacetates are α-substituted. The condensation of acetoacetates with anilines can give either 2-hydroxy- or 4-hydroxyquinolines, depending upon the reaction temperature.[60] In the Hantzch synthesis of symmetrical dihydropyridines, two equivalents of acetoacetates are condensed with aldehydes in the presence of ***Ammonia*** (eq 23).[61] A variation of the Hantzch synthesis for the preparation of unsymmetrical dihydropyridines involves the condensation of α-(alkylidene)acetoacetates with 3-aminocrotonate.[62]

HO, OH; O, CO_2Et; H_2SO_4, 82–90%; HO, O, O (22)

NO_2, CHO, (1 equiv); O, CO_2Me, (2 equiv); NH_3, 50%; NO_2, MeO_2C, CO_2Me, N, H (23)

Related Reagents. Acetoacetic Acid; Acetone; Acetone Cyclohexylimine; Acetone Hydrazone; N,O-Dimethylhydroxylamine; Ethyl 4-Chloroacetoacetate; Ethyl Cyanoacetate; Ethyl 4-(Triphenylphosphoranylidene)acetoacetate; 1-Methoxy-1,3-bis(trimethylsilyloxy)-1,3-butadiene; *N*-Methoxy-*N*-methylacetamide; Methyl Dilithioacetoacetate.

1. (a) Black, D. St. C.; Blackburn, G. M.; Johnston, G. A. R. In *Rodd's Chemistry of Carbon Compounds*; Coffey, S., Ed.; Elsevier: Amsterdam, 1965; Vol. 1, Part D, Chapter 16, pp 238–257. (b) Ames, D. E. In *Rodd's Chemistry of Carbon Compounds*; Ansell, M. F., Ed.; Elsevier: Amsterdam, 1973; Vol 1, Suppl., Part D, Chapter 16, pp 357–361. (c) Clemens, R. J.; Witzeman, J. S. In *Acetic Acid and Derivatives* ; Dekker: New York, 1993; p 173.
2. House, H. O. *Modern Synthetic Reactions*, 2nd ed.; Benjamin: Menlo Park, CA, 1972; pp 510–546.
3. (a) Bergmann, E. D.; Ginsburg, D.; Pappo, R. *OR* **1959**, *10*, 179. (b) Bruson, H. A. *OR* **1949**, *5*, 79. (c) Ref. 2, pp 595–623. (d) Jung, M. E. *COS* **1991**, *4*, Chapter 1.1.
4. Jones, G. *OR* **1967**, *15*, 204.
5. Tietze, L. F.; Beifuss, U. *COS* **1991**, *2*, Chapter 1.11.
6. Clemens, R. J. *CRV* **1986**, *86*, 241.
7. Witzeman, J. S.; Nottingham W. D. *JOC* **1991**, *56*, 1713.
8. (a) Marvel, C. S.; Hager, F. D. *OSC* **1944**, *1*, 248. (b) Johnson, J. R.; Hager, F. D. *OSC* **1944**, *1*, 351. (c) For a review of dealkoxycarbonylation methods, see: Krapcho, A. *S* **1982**, 893.
9. Jackman, L. M.; Lange, B. C. *T* **1977**, *33*, 2737.
10. Guibe, F.; Sarthou, P.; Bram, G. *T* **1974**, *30*, 3139.
11. (a) Brieger, G.; Pelletier, W. M. *TL* **1965**, 3555. (b) Hara, Y.; Matsuda, M. *BCJ* **1976**, *49*, 1126.
12. Durst, H. D.; Liebeskind, L. *JOC* **1974**, *39*, 3271.
13. Taylor, E. C.; Hawks III, G. H.; McKillop, A. *JACS* **1968**, *90*, 2421.
14. Clark, J. H.; Miller, J. M. *JCS(P1)* **1977**, 1743.
15. (a) Boldt, P.; Ludwig, A.; Militzer, H. *CB* **1970**, *103*, 1312. (b) Boldt, P.; Militzer, H.; Thielecke, W.; Schulz, L. *LA* **1968**, *718*, 101.
16. Tsuji, J. *ACR* **1969**, *2*, 144.
17. (a) Trost, B. M. *ACR* **1980**, *13*, 385. (b) Tsuji, J. *T* **1986**, *42*, 4361. (c) Tsuji, J. *JOM* **1986**, *300*, 281. (d) Trost, B. M. *AG(E)* **1986**, *25*, 1. (e) Trost, B. M. *JOM* **1986**, *300*, 263.
18. Larock, R. C.; Lee, N. H. *TL* **1991**, *32*, 5911.
19. Marino, J. P.; Fernández de la Pradilla, R.; Laborde, E. *JOC* **1987**, *52*, 4898.
20. Trost, B. M.; Zhi, L. *TL* **1992**, *33*, 1831.
21. (a) Corey, E. J.; Ghosh, A. K. *TL* **1987**, *28*, 175. (b) Corey, E. J.; Ghosh, A. K. *CL* **1987**, 223.
22. Dessau, R. M.; Heiba, E. I. *JOC* **1974**, *39*, 3456.
23. (a) Huckin, S. N.; Weiler, L. *JACS* **1974**, *96*, 1082. (b) Sum, P. E.; Weiler, L. *CC* **1977**, 91.
24. Kieczykowski, G. R.; Roberts, M. R.; Schlessinger, R. H. *JOC* **1978**, *43*, 788.
25. (a) Spassow, A. *OSC* **1955**, *3*, 390. (b) Shriner, R. L.; Schmidt, A. G.; Roll, L. J. *OSC* **1943**, *2*, 266. (c) Straley, J. M.; Adams, A. C. *OSC* **1963**, *4*, 415.
26. Rathke, M. W.; Cowan, M. W. *JOC* **1985**, *50*, 2622.
27. Huckin, S. N.; Weiler, L. *CJC* **1974**, *52*, 1343.
28. Gilbreath, S. G.; Harris, C. M.; Harris, T. M. *JACS* **1988**, *110*, 6172.
29. Narasimhan, N. S.; Ammanamanchi, R. *JOC* **1983**, *48*, 3945.
30. Yamaguchi, M.; Shibato, K.; Hirao, I. *CL* **1985**, 1145.
31. Lehnert, W. *T* **1972**, *28*, 663.
32. Lawesson, S.-O.; Larsen, E. H.; Sundström, G.; Jakobsen, H. J. *ACS* **1963**, *17*, 2216.
33. Queignec, R.; Kirschleger, B.; Lambert, F.; Aboutaj, M. *SC* **1988**, *18*, 1213.
34. (a) Huckin, S. N.; Weiler, L. *TL* **1971**, 4835. (b) Huckin, S. N.; Weiler, L. *CJC* **1974**, *52*, 2157.
35. Nelson, J. H.; Howells, P. N.; DeLullo, G. C.; Landen, G. L.; Henry, R. A. *JOC* **1980**, *45*, 1246.
36. (a) Antonioletti, R.; Bonadies, F.; Monteagudo, E. S.; Scettri, A. *TL* **1991**, *32*, 5373. (b) Corsico Coda, A.; Desimoni, G.; Righetti, P.; Tacconi, G. *G* **1984**, *114*, 417. (c) Watanabe, K.; Miyazu, K.; Irie, K. *BCJ* **1982**, *55*, 3212. (d) Boyer, J.; Corrin, R. J. P.; Perz, R.; Reye, C. *CC* **1981**, 122.
37. (a) Jung, M. E. *T* **1976**, *32*, 3. (b) Gawley, R. E. *S* **1976**, 777.
38. (a) McCurry, Jr., P. M.; Singh, R. K. *SC* **1976**, *6*, 75. (b) Horning, E. C.; Denekas, M. O.; Field, R. E. *OSC* **1955**, *3*, 317.
39. (a) Meyer, W. L.; Sigel, C. W.; Hoff, R. J.; Goodwin, T. E.; Manning, R. A.; Schroeder, P. G. *JOC* **1977**, *42*, 4131. (b) Meyer, W. L.; Manning, R. A.; Schroeder, P. G.; Shew, D. C. *JOC* **1977**, *42*, 2754.
40. Chan, T. H.; Brownbridge, P. *JACS* **1980**, *102*, 3534.
41. Bretschneider, T.; Miltz, W.; Münster, P.; Steglich, W. *T* **1988**, *44*, 5403.
42. Brown, D. S.; Bruno, M.; Davenport, R. J.; Ley, S. V. *T* **1989**, *45*, 4293.
43. Chikashita, H.; Motozawa, T.; Itoh, K. *SC* **1989**, *19*, 1119.
44. (a) Koskinen, A.; Lounasmaa, M. *TL* **1983**, *24*, 1951. (b) Cainelli, G.; Contento, M.; Drusiani, A.; Panunzio, M.; Plessi, L. *CC* **1985**, 240.
45. (a) Chan, T. H.; Brownbridge, P. *CC* **1979**, 578. (b) Brownbridge, P.; Chan, T. H.; Brook, M. A.; Kang, G. J. *CJC* **1983**, *61*, 688.
46. Hagiwara, H.; Kimura, K.; Uda, H. *CC* **1986**, 860.
47. Chan, T. H.; Brook, M. A. *TL* **1985**, *26*, 2943.
48. Molander, G. A.; Cameron, K. O. *JOC* **1991**, *56*, 2617.
49. Kitamura, M.; Tokunaga, M.; Ohkuma, T.; Noyori, R. *OS* **1993**, *71*, 1.
50. Kitamura, M.; Ohkuma, T.; Tokunaga, M.; Noyori, R. *TA* **1990**, *1*, 1.
51. Dorn, H. *Chem. Heterocycl. Compd.* **1980**, *16*, 1.
52. Krogsgaard-Larsen, P.; Christensen, S. B.; Hjeds, H. *ACS* **1973**, *27*, 2802.
53. Ulrich, H. *The Chemistry of Imidoyl Halides*; Plenum: New York, 1968; Chapter 6, pp 157–172; Chapter 7, pp 173–192.
54. Sundberg, R. J. In *Comprehensive Heterocyclic Chemistry*; Katritzky, A. R.; Rees, C. W., Eds.; Pergamon: Oxford, 1984; Vol. 4, Chapter 3.06, p 331.

55. Dean, F. M. *Adv. Heterocycl. Chem.* **1982**, *30*, 167.
56. Cheng, C. C.; Yan, S.-J. *OR* **1982**, *28*, 37.
57. Kenner, G. W.; Todd, A. In *Heterocyclic Compounds*; Elderfield, R. C., Ed.; Wiley: New York, 1957; Vol. 6.
58. Sethna, S.; Phadke, R. *OR* **1953**, *7*, 1.
59. Russell, A.; Frye, J. R. *OSC* **1955**, *3*, 281.
60. Jones, G. *Quinolines* ; Wiley; New York, 1977; Vol. 1.
61. (a) Sausins, A.; Duburs, G. *H* **1988**, *27*, 269. (b) Loev, B.; Goodman, M. M.; Snader, K. M.; Tedeschi, R.; Macko, E. *JMC* **1974**, *17*, 956.
62. Bossert, F.; Meyer, H.; Wehinger, E. *AG(E)* **1981**, *20*, 762.

Shridhar G. Hegde
Monsanto Company, St. Louis, MO, USA

Ethyl 2-(Bromomethyl)acrylate

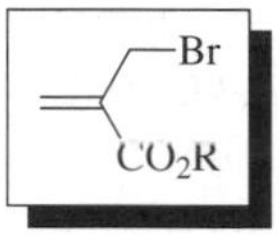

(R = Et)
[17435-72-2] $C_6H_9BrO_2$ (MW 193.04)
(R = *t*-Bu)
[53913-96-5] $C_8H_{13}BrO_2$ (MW 221.09)

(powerful allylic alkylating reagent[1–3] employed as its organozinc derivative to prepare α-methylene lactones[4] and lactams;[5] important source of functional acrylic monomers for polymerization[6,7])

Alternate Name: ethyl 2-(bromomethyl)propenoate.
Physical Data: R = Et, bp 40–41 °C/0.1 mmHg; R = *t*-Bu, bp 60–61 °C/0.1 mmHg.
Solubility: insol H_2O; sol ether, acetone, alcohol, THF, chloroform.
Form Supplied in: colorless liquid.
Preparative Methods: many preparations of these compounds have been described,[8] for example dehydrobromination of β,β′-dibromoisobutyric esters[9–11] or acid, or bromination of *t*-butyl or ***Ethyl α-(Hydroxymethyl)acrylate*** with ***Phosphorus(III) Bromide***.[12]
Handling, Storage, and Precautions: reputed to be lachrymatory, vesicatory, and toxic. Usual precautions of storage in darkness between 0 and 5 °C and handling with gloves under a well ventilated hood are recommended.

Electrophilic Alkylating Reagent. The electrophilic reactivity of the allylic bromide moiety is increased by the presence of the carboxylate group (Michael acceptor) and the reagent can thus act as a powerful allylic electrophile. It can be coupled with various organometallic reagents, *gem*-dimetallic reagents[2,3] or O-,[13,14], S-,[15–17], N-, and P-nucleophiles (eq 1).[18,19]

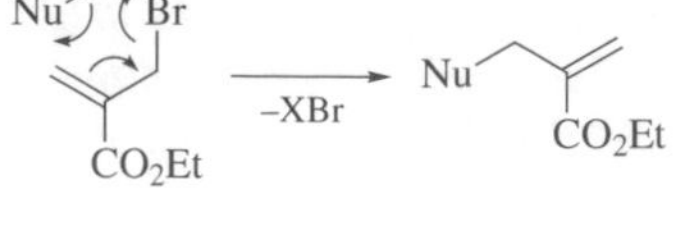

(1)

The alkylation with ***Cyanomethylcopper***[1] gave 89% yield of the corresponding 1-(3-cyanoethyl)acrylic ester, while π-2-(methoxyallyl)nickel bromide was tested as a source of 5-ethoxycarbonyl-5-hexen-2-one.[20] Palladium-catalyzed cross-coupling with vinyl- and aryltin[21] and vinylzirconium reagents[22] allowed the preparation of the corresponding 2-allyl- and 2-benzylacrylates. Coupling with functional zinc or zinc–copper organometallic derivatives[23–25] gives rise to various polyfunctional acrylic derivatives (eq 2).[26]

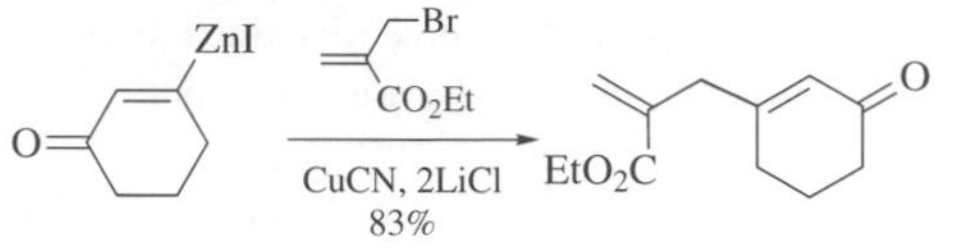

(2)

The allylation of other nucleophiles such as the enolates of cyclic 1,3-diketones[27] (or their enol ethers[28]), β-keto esters,[29] β-dimethylaminopropionates,[30] and bicyclic enolates[31] has been reported. Tandem alkylation–Michael addition with dienolates gives rise to cyclic or bicyclic compounds,[32] while tandem alkylation–Claisen condensations leads to the formation of functional α-methylenecyclopentanones (eq 3).[33]

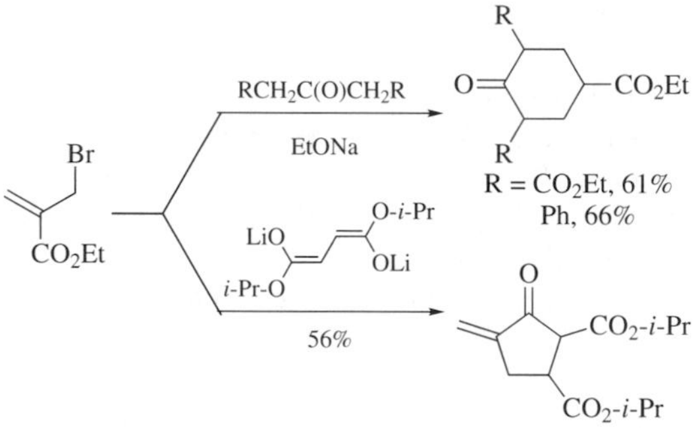

(3)

Alkylation of enamines[34] has been proposed as the first step in the synthesis of potential antitumor α-methylene δ-valerolactones.

Unsymmetrical organic sulfides containing the methacrylic ester group are produced under nonbasic reaction conditions.[35] Thio-Claisen rearrangement[36] took place via *S*-allylation with ethyl 2-(bromomethyl)acrylate (eq 4).

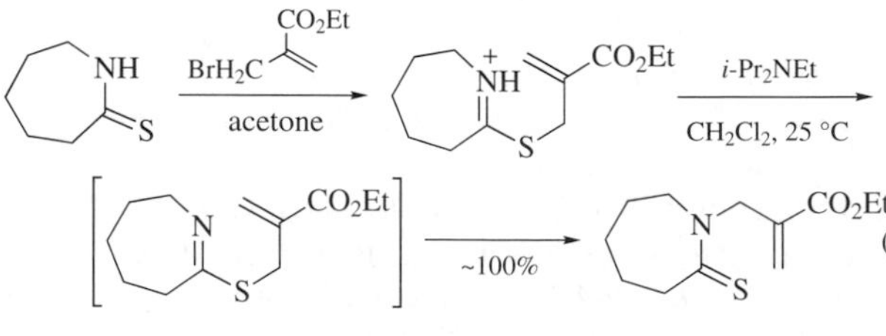

(4)

N-Allylation by ethyl 2-(bromomethyl)acrylate has been developed as key steps in the synthesis of bioactive molecules such as α-methylene-β-alanine,[37] or analogs of Iboga alkaloids.[38]

Synthesis of α-Methylene Lactones and Lactams. One of the most important applications of ethyl 2-(bromomethyl)acrylate was found in a pseudo-Reformatsky reaction. First described by Ohler et al.,[4] this reaction leads to the formation of potential antitumor α-methylene-γ-lactones,[39,40] via an organozinc intermediate (eq 5).

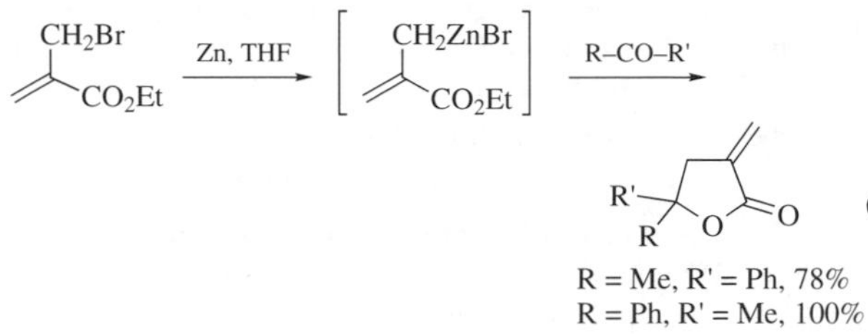

(5)

R = Me, R' = Ph, 78%
R = Ph, R' = Me, 100%

This reaction was extensively applied to various carbonyl compounds. 5-Mono- or 5,5-disubstituted α-methylene-γ-lactones bearing a steroidal moiety,[41] nucleic acid bases,[42] furanose systems,[43] six-membered[44] or five-membered[45] heterocyclic groups, adamantyl groups,[46] and bornyl groups[47] have been prepared and evaluated as antineoplasic and allergenic[48,49] compounds. Bis-α-methylene-γ-lactones have been produced from dicarbonyl compounds.[50]

The Reformatsky reaction with imines leads to the corresponding less toxic α-methylene-γ-lactams.[5] Bis addition of the organozinc intermediate to acyl chlorides or nitriles before cyclization leads to the formation of new α-methylene-γ-lactones and lactams (eq 6).[51]

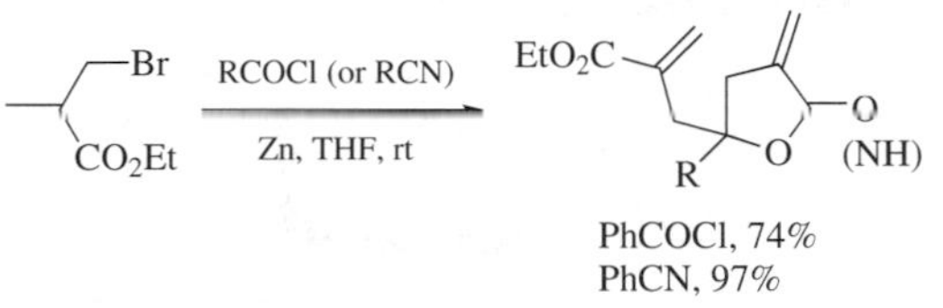

(6)

PhCOCl, 74%
PhCN, 97%

Attempts to improve the yield of the Reformatsky reaction included the use of ***Tin***,[52] tin–aluminum alloy,[53] ***Zinc*** dust in aqueous ammonium chloride solution,[54] or electrogenerated zinc.[55] Low valent metal halides (***Lithium Aluminum Hydride-Chromium(III) Chloride***)[56] and ***Tin(II) Chloride***[57] were found to undergo insertion into the carbon–bromine bond of 2-(bromomethyl)acrylates. Asymmetric induction from complex reagents made with $SnCl_2$, tartaric ester, and allylic bromides has been described.[58]

The preparation of intermediate organozinc derivatives of 2-(bromomethyl)acrylates[59] by direct action of zinc dust in THF (or DME) at controlled temperature allowed regio- and stereoselective condensations with polyfunctional electrophilic compounds without side reactions (eq 7).[60]

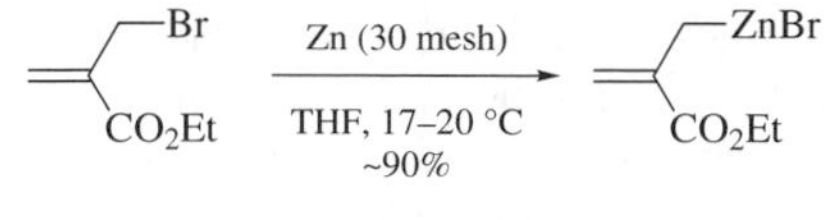

(7)

Their coupling with chiral imines, derived from α-amino esters[61] or β-amino alcohols,[62] leads to the formation of the corresponding α-methylene-γ-lactams (*R,R*) or (*S,S*) via a diastereoselective set of reactions (de > 95%).

Other organometallic reagents derived from ethyl 2-(bromomethyl)acrylate are known, such as organotin,[63] which allows palladium-promoted arylation, and organomercury,[64] for acylation reactions. Ethyl 2-(bromomethyl)acrylate/π-(allyl)nickel in DMF is a powerful selective nucleophilic reagent for haloaromatics[65] and iodoalkanes[66] (eq 8).

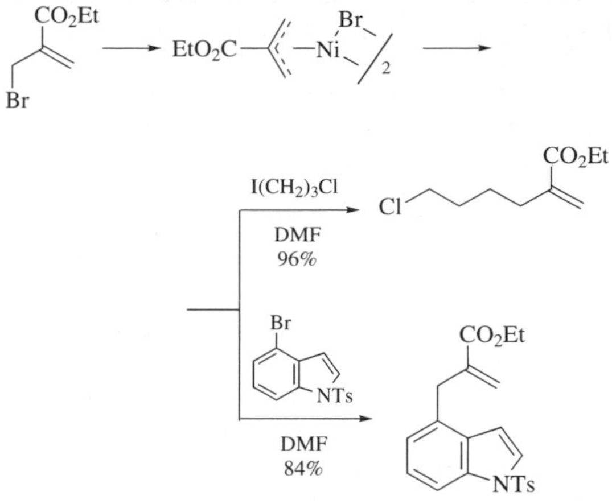

(8)

Polymer Chemistry. Ethyl 2-(bromomethyl)acrylate also has an important place in polymer chemistry as a monomer[6] or as a source of functional monomers such as sulfur-containing acrylic compounds[67] or 6-*O*-methylallylgalactose intermediates for the preparation of poly(vinyl)saccharides.[7] It also has been employed as a chain-transfer agent.[68]

Related Reagents. Ethyl α-(Hydroxymethyl)acrylate; Methyl 4-Bromocrotonate; Methyl (Z)-(2)-(Bromomethyl)-2-butenoate.

1. Corey, E. J.; Kuwajima, I. *TL* **1972**, 487.
2. Knochel, P. *JACS* **1990**, 112, 7431. Waas, J. R.; Sidduri, A.; Knochel, P. *TL* **1992**, *33*, 3717.
3. Knochel, P.; Normant, J. F. *TL* **1986**, *27*, 4431; *TL* **1986**, 27, 4427; *TL* **1984**, *25*, 1475.
4. Öhler, E.; Reininger, K.; Schmidt, U. *AG(E)* **1970**, *9*, 457.
5. Belaud, C.; Roussakis, C.; Letourneux, Y.; El Alami, N.; Villiéras, J. *SC* **1985**, *15*, 1233.
6. Yamada, B.; Otsu T. *Makromol. Chem., Rapid Commun.* **1990**, *11*, 513.
7. Klein, J.; Blumenberg, K. *Makromol. Chem.* **1988**, *189*, 805.
8. Ferris, A. F. *JOC* **1955**, *20*, 780.
9. Cassady, J. M.; Howie, G. A.; Robinson, J. M.; Stamos, I. K. *OS* **1983**, *61*, 77. Holm, A.; Scheuer, P. J. *TL* **1980**, *21*, 1125.
10. Charlton, J. L.; Sayeed, V. A.; Lypka, G. N. *SC* **1981**, *11*, 931.
11. Anzeveno, P. B.; Campbell, J. A.; White, W. L. *SC* **1986**, *16*, 387.
12. Villiéras, J.; Rambaud, M. *S* **1982**, 924. Villiéras, J.; Rambaud, M. *OS* **1988**, *66*, 220.
13. Golec, J. M. C.; Hedgecock, C. J. R.; Murdoch, R.; Tully, W. R. *TL* **1992**, *33*, 551.
14. Barton, D. H. R.; Fekih, A.; Lusinchi, X. *BSF(2)* **1988**, 681. Reisch, J.; Kamal, G. M.; Gunaherath, B. *JCS(P1)* **1989**, 1047.
15. Barton, D. H. R.; Gero, S. D.; Quiclet-Sire B.; Samadi, M. *JCS(P1)* **1991**, 981.
16. Colombani, D.; Navarro, C.; Degueil-Castaing, M.; Maillard, B. *SC* **1991**, *21*, 1481.
17. Haynes, R. K.; Katsifis, A.; Vonwiller, S. C. *AJC* **1984**, *37*, 1571.
18. Majewski, P. *PS* **1989**, *45*, 151.
19. Sanyal, U.; Chatterjee, R. S.; Das, S. K.; Chakraborti, S. K. *Neoplasma* **1984**, *31*, 149.

20. Hegedus, L. S.; Stiverson, R. K. *JACS* **1974**, *96*, 3250.
21. Sheffy, F. K.; Godschalx, J. P.; Stille, J. K. *JACS* **1984**, *106*, 4833.
22. Hayasi, Y.; Riediker, M.; Temple, J. S.; Schwartz, J. *TL* **1981**, *22*, 2629.
23. Knoess, H. P.; Furlong, M. T.; Rozema, M. J.; Knochel P. *JOC* **1991**, *56*, 5974.
24. Rozema, M. J.; Sidduri, A.; Knochel, P. *JOC* **1992**, *57*, 1956.
25. Majid, T. N.; Yeh, M. C. P.; Knochel, P. *TL* **1989**, *30*, 5069. Yeh, M. C. P.; Knochel, P. *TL* **1988**, *29*, 2395.
26. Rao, C. J.; Knochel, P. *JOC* **1991**, *56*, 4593.
27. De Groot, A.; Jansen, B. J. M. *RTC* **1974**, *93*, 153. Groutas, W. C.; Felker, D.; Magnin, D.; Meitzner, G.; Gaynor, T. *SC* **1980**, *10*, 1.
28. De Groot, A.; Jansen, B. J. M. *RTC* **1976**, *95*, 81.
29. Ameer, F.; Drewes, S. E.; Houston-McMillan, M. W.; Kaye, P. T. *S. Afr. J. Chem.* **1986**, *39*, 57.
30. Grigg, R.; Dorrity, M. J.; Heaney, F.; Malone, J. F.; Rajviroongit, S.; Sridharan, V.; Surendrakumar, S. *T* **1991**, *47*, 8297.
31. Herradón, B.; Seebach, D. *HCA* **1989**, *72*, 690.
32. Stetter, H.; Elfert, K. *S* **1974**, 36.
33. Furuta, K.; Misumi, A.; Mori, A.; Ikeda, N.; Yamamoto, H. *TL* **1984**, *25*, 669.
34. Marschall, H.; Vogel, F.; Weyerstahl, P. *CB* **1974**, *107*, 2852.
35. Labuschagne, A. J. H.; Malherbe, J. S.; Meyer, C. J.; Schneider, D. F. *JCS(P1)* **1978**, 955.
36. Gompper, R.; Kohl, B. *TL* **1980**, *21*, 907; *TL* **1980**, *21*, 917. Takahata, H.; Banba, Y.; Mozumi, M.; Yamazaki, T. *H* **1986**, *24*, 3347; *H* **1986**, *24*, 947.
37. Holm, A.; Scheuer, P. J. *TL* **1980**, *21*, 1125.
38. Sundberg, R. J.; Cherney, R. J. *JOC* **1990**, *55*, 6028.
39. Grieco, P. A. *S* **1975**, 67.
40. Rosowsky, A.; Papathanasopoulos N.; Lazarus, H.; Foley, G. E.; Modest, E. J. *JMC* **1974**, *17*, 672.
41. Lee, K.-H.; Ibuka, T.; Kim S-H.; Vestal, B. R.; Hall, I. H.; Huang, E.-S. *JMC* **1975**, *18*, 812.
42. Lee, K.-H.; Wu, Y-S.; Hall, I. H. *JMC* **1977**, *20*, 911. Sanyal, U.; Mitra, S.; Pal, P.; Chakraborti, S. K. *JMC* **1986**, *29*, 595. Lee., K.-H.; Rice, G. K.; Hall, I. H.; Amarnath, V. *JMC* **1987**, *30*, 586.
43. Csuk, R.; Fuerstner, A.; Sterk, H.; Weidmann, H. *J. Carbohydr. Chem.* **1986**, *5*, 459. Rauter, A. P.; Figueiredo, J. A.; Ismael, I.; Pais, M. S.; Gonzalez, A. G.; Diaz, J.; Barrera, J. B. *J. Carbohydr. Chem.* **1987**, *6*, 259.
44. Pantaleo, N. S.; Satyamurthy, N.; Ramarajan, K.; O'Donnell, D. J.; Berlin, K. D.; Van der Helm, D. *JOC* **1981**, *46*, 4284. Venkatapathy, S. V.; Ikramuddeen, T. M.; Chandrasekara, N.; Ramarajan, K.; Nanjappan, P. *IJC(B)* **1989**, *28*, 863.
45. Heindel, N. D.; Minatelli, J. A. *JPS* **1981**, *70*, 84.
46. Hickmott, P. W.; Wood, S. *S. Afr J. Chem.* **1985**, *38*, 61.
47. Ruecker, G.; Gajewski, W. *Eur. J. Med. Chem.* **1985**, *20*, 87.
48. Schlewer, G.; Stampf, J.-L.; Benezra, C. *JMC* **1980**, *23*, 1031.
49. Mattes, H.; Benezra, C. *JMC* **1987**, *30*, 165.
50. Stamos, I. K.; Howie, G. A.; Manni, P. E.; Haws, W. J.; Byrn, S. R.; Cassady, J. M. *JOC* **1977**, *42*, 1703.
51. El Alami, N.; Belaud, C.; Villiéras, J. *JOM* **1987**, *319*, 303.
52. Zhou, J.; Lu, G.; Wu, S. *SC* **1992**, *22*, 481.
53. Nokami, J.; Tamaoka, T.; Ogawa, H.; Wakabayashi, H. *CL* **1986**, 541.
54. Mattes, H.; Benezra, C. *TL* **1985**, *26*, 5697.
55. Rollin, Y.; Derien, S.; Dunach, E.; Gebehenne, C.; Perichon, J. *T* **1993**, *49*, 7723.
56. Okuda, Y.; Nakatsukasa, S.; Oshima, K.; Nozaki, H. *CL* **1985**, 481. Mattes, H.; Benezra, C. *JOC* **1988**, *53*, 2732.
57. Tomimori, K.; Tanaka, T.; Kurozumi, S. Jpn. Patent 62 209 095, 1987 (*CA* **1988**, *109*, 211 269b). Talaga, P.; Schaeffer, M.; Benezra, C.; Stampf, J.-L. *S* **1990**, 530.
58. Boldrini, G. P.; Lodi, L.; Tagliavini, E.; Tarasco, C.; Trombini, C.; Umani-Ronchi, A. *JOC* **1987**, *52*, 5447.
59. El Alami, N.; Belaud, C.; Villiéras, J. *JOM* **1988**, *348*, 1.
60. El Alami, N.; Belaud, C.; Villiéras, J. *JOM* **1988**, *353*, 157. El Alami, N.; Belaud, C.; Villiéras, J. *SC* **1988**, *18*, 2073. El Alami, N.; Belaud, C.; Villiéras, J. *TL* **1987**, *28*, 59.
61. Dembélé, Y. A.; Belaud, C.; Hitchcock, P.; Villiéras, J. *TA* **1992**, *3*, 351.
62. Dembélé, Y. A.; Belaud, C.; Villiéras, J. *TA* **1992**, *3*, 511.
63. Baldwin, J. E.; Adlington, R. M.; Birch, D. J.; Crawford, J. A.; Sweeney, J. B. *CC* **1986**, 1339.
64. Larock, R. C.; Lu, Y. D. *TL* **1988**, *29*, 6761.
65. Semmelhack, M. F. *OR* **1972**, *19*, 115. Hegedus, L. S.; Sestrick, M. R.; Michaelson, E. T.; Harrington, P. J. *JOC* **1989**, *54*, 4141.
66. Corey, E. J.; Semmelhack, M. F. *JACS* **1967**, *89*, 2755.
67. Cerf, M; Mieloszynski, J L; Paquer, D. Eur. Patent 463 947 (*CA* **1992**, *116*, 152 603a).
68. Yamada, B.; Kobatake, S.; Otsu, T. *Polym. J. (Tokyo)* **1992**, *24*, 281. Vertommen, L. L. T.; Meijer, J.; Maillard, B. J. Patent Int. Appl. 91 07 387 (*CA* **1991**, *115*, 160 039r). Meijs, G. F.; Rizzardo, E.; Thang, S. H. *Polym. Bull. (Berlin)* **1990**, *24*, 501.

Jean Villiéras & Monique Villiéras
Université de Nantes, France

Ethyl Bromozincacetate[1]

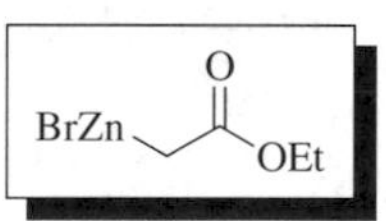

[5764-82-9] $C_4H_7BrO_2Zn$ (MW 232.37)

(chemoselective ester enolate reagent[1])

Alternate Name: Reformatsky reagent.
Preparative Methods: see below.
Handling, Storage, and Precautions: α-halo esters and α-halo ketones used for the preparation of Reformatsky reagents are lacrymators and should be used in a well ventilated hood; solutions of the reagent in ethereal solvents are stable at low temperatures for a few days. The reagent is prone to hydrolysis and must be handled in anhydrous solvents under an inert atmosphere.

Introduction. The Reformatsky reagent shown in the above title is the historically first[5] and most widely used zinc ester enolate prepared by zinc insertion into an α-halo ester. However, the Reformatsky reaction can be taken as subsuming all enolate formations by oxidative addition of a metal or a low-valent metal salt into an activated carbon–halogen bond.[1] It is this mode of enolate formation which distinguishes the Reformatsky reaction from other fields of metal–enolate chemistry.

The reagent is dimeric in solution except in the most polar solvents;[2] X-ray analysis of $BrZnCH_2CO_2$-*t*-Bu·THF shows a dimeric unit in the solid state with the metal being bound to an sp^3 carbon atom.[3] In contrast, zinc enolates of ketones are *O*-metalated species.[4]

Performance and Reagent Preparation. Originally,[5] the reaction was performed in a one-step fashion by adding a mixture of the halo ester and the electrophile (usually a carbonyl compound) to a ***Zinc*** suspension in an ethereal or hydrocarbon solvent. However, enolate formation and its reaction with electrophiles can be done successively. This two-step procedure has contributed to the advancement of the Reformatsky reaction, as electrophiles which would quarternize upon mixing with α-halo esters (e.g. azomethines) can be used under these conditions without problems. Moreover, the zinc enolates may be transmetalated prior to use in order to adapt their reactivity to electrophiles beyond the scope of classical Reformatsky reactions.[1]

Difficulties in initiating the reaction are avoided by the use of activated zinc samples.[1a,6] As they promote the reaction even at very low temperatures (Table 1), their use results in generally high yields, the suppression of undesirable side reactions, and an increase in diastereoselectivity. Less reactive donors such as α-chloro esters are equally suited when highly reactive metals are employed.[7] The improved reliability recommends metal-activation techniques for applications to natural product synthesis. Among the zinc samples described so far, finely dispersed and readily prepared zinc/silver on graphite[7] (see also ***Potassium-Graphite***) and zinc obtained by reduction of ***Zinc Chloride*** with ***Potassium Naphthalenide***, ***Lithium Naphthalenide***, or alkali naphthalenides are most effective (Table 1).[8] For large-scale experiments, the application of ultrasound[9] or of electrochemical support[10] for Reformatsky reactions is recommended.

Table 1 Comparison of the Reactivity of Different Zinc Samples for the Preparation of Ethyl Bromozincacetate followed by reaction with Benzaldehyde (A) or Cyclohexanone (B)

Reagent	Carbonyl compound	Solvent	T(°C)	t(min)	Yield (%)
Zinc dust	A	Benzene	+80	720	61
Zinc dust	B	Benzene	+80	–[a]	56
Zn/Cu couple	A	THF	+66	60	82
Zn/Cu couple	B	THF	+66	60	82
Rieke Zn	A	Et_2O	+25	60	98
Rieke Zn	B	Et_2O	+25	60	95
Zn–ultrasound	A	1,4-Dioxane	+25	5–300	98
$ZnCl_2$/Li[b]	A	Et_2O	0	30	95
$ZnCl_2$/Li[b]	A	Et_2O	–78	30	56
Zn/Ag–graphite	B	THF	–78	20	92

[a] Reaction time not reported. [b] Zinc sample obtained by reduction of $ZnCl_2$ with commercially available lithium dispersion in Et_2O.[8e]

In contrast to other metal enolates, Reformatsky reagents are reasonably stable over a wide temperature range (from −78 °C to above 80 °C for short periods of time) and can be prepared in solvents greatly differing in polarity (most commonly employed are THF, DME, Et_2O, 1,4-dioxane, benzene, toluene, dimethoxymethane, DMF, or mixtures thereof; scattered reports using hexane, acetonitrile, CH_2Cl_2, $B(OMe)_3$, DMSO, and HMPA may be found).[1] The major path for their decomposition is loss of EtOZnBr with formation of ketene which immediately acylates an intact zinc enolate, thus leading to β-keto esters.[11] Reformatsky reagents are therefore usually freshly prepared and used without delay, although solutions of $BrZnCH_2CO_2$-*t*-Bu in a number of solvents were reported to exhibit virtually unchanged reactivity after 4–6 days.[2]

Reformatsky Donors. Ethyl bromoacetate is the most widely used halogen compound for zinc insertion but other short-chain 2-bromo esters work equally well under standard conditions.[1] The alcohol part of the ester plays only a minor role, and by its proper choice (e.g. *t*-butyl, TMS, tetrahydropyranyl bromo esters)[12] β-hydroxy acids can be readily obtained by zinc-induced reaction of these donors with carbonyl compounds followed by hydrolysis of the respective β-hydroxy esters initially formed. Allyl bromo esters on treatment with zinc undergo Claisen rearrangements to 4-alkenoic acids (eq 1).[13] As the chain length of the 2-bromo ester increases, or upon switching to the less reactive 2-chloro ester, the use of highly activated zinc samples and/or more polar solvent systems becomes obligatory to accomplish zinc enolate formation.[1]

1. Zn, TMSCl, MeCN, 100 °C, 20 h; 2. H_2O; 78% (1)

Bromo(chloro)difluoro esters and ketones show no peculiarities in their behavior and have found widespread applications in the synthesis of fluorinated natural product analogs (eq 2).[14] In contrast, dibromo-, dichloro-, and trichloroacetates tend to polymerize on reaction with zinc dust at reflux temperatures;[15] however, they can be selectively transformed with more appropriate reagent systems. Depending on the conditions used, they either afford glycidates (with Zn/Ag–graphite at −78 °C)[16] or 2-alkenoates (with ***Zinc/Diethylaluminum Chloride*** at 0 °C),[17] respectively, on reaction with carbonyl compounds (eq 3).

BrF_2CCO_2Et, Zn, THF, 22 °C; 81%; 1.38:1 (2)

Cl_2CHCO_2R: 1. Zn/Ag–graphite, THF, –20 °C, R = Et; 2. K_2CO_3; 78% / Zn, Et_2AlCl, THF, 25 °C, R = *t*-Bu; 67% (3)

Haloacetonitriles and α-halo acetamides[1,18] are well suited as donors in Reformatsky reactions (eq 4). When *N*-methoxy-*N*-methylbromoacetamide is used (eq 5), the resulting products may be converted into β-hydroxy ketones by the Weinreb procedure.[19]

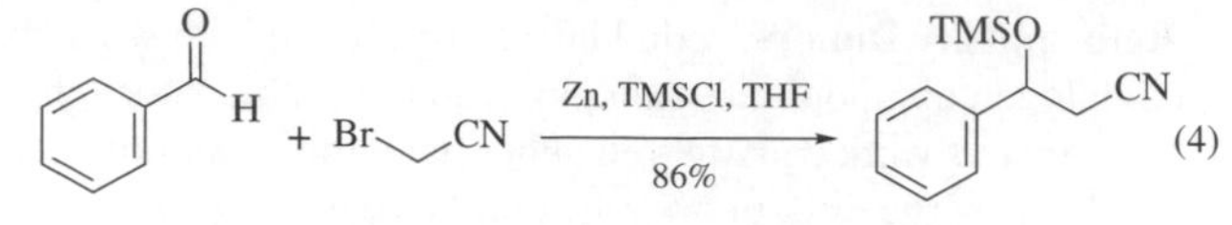

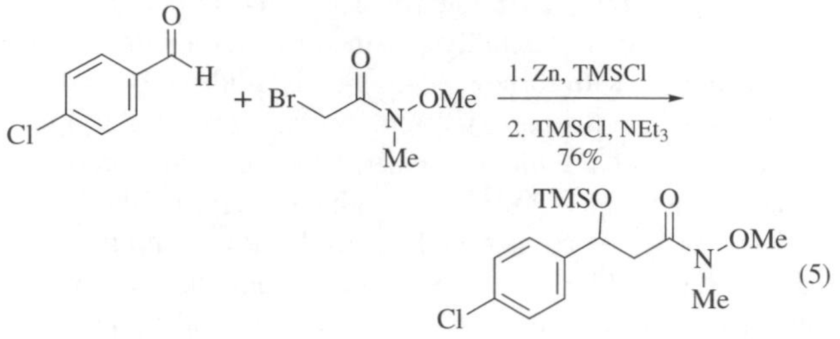

4-Bromo-2-butenoates as vinylogous bromo esters form ambident carbon nucleophiles upon treatment with zinc, which may either lead to α- or γ-substitution products on reaction with electrophiles. A set of conditions has been worked out that allows control of these pathways, with the polarity of the reaction medium and the temperature being the crucial parameters (eq 6).[20] The data suggest that kinetic control leads to α-products, whereas thermodynamic control affords the γ-substitution products.

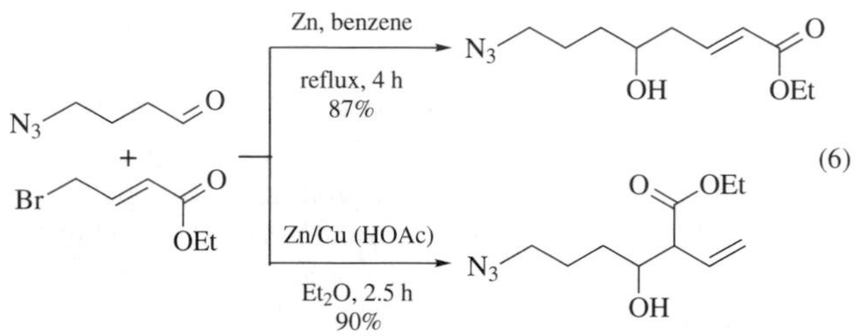

2-Bromomethyl-2-alkenoates react readily under allylic rearrangement with carbonyl compounds in the presence of zinc to form α-methylene-γ-lactones.[21] Because of the high biological relevance of that structural unit, this reaction has found widespread use in natural product synthesis,[22] both in an inter- (eq 7)[23] and intramolecular fashion (eq 8).[24] With diastereotopic ketones the stereoselectivity of the C–C bond-forming step may be significantly enhanced by using highly reactive Zn/Ag–graphite as promotor at very low temperatures.[25]

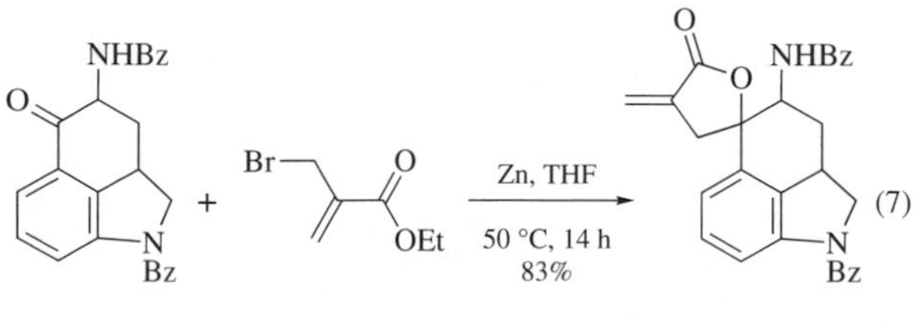

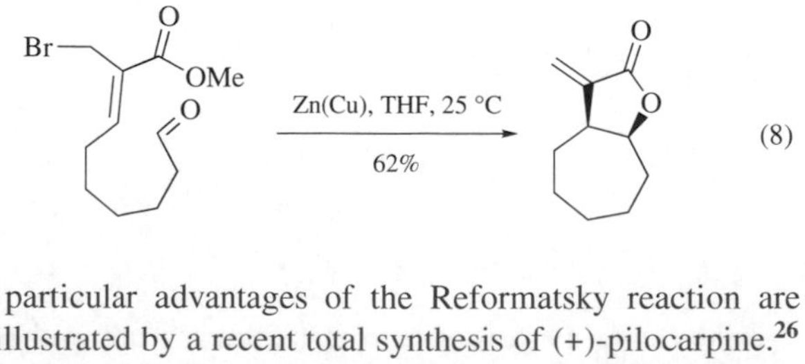

The particular advantages of the Reformatsky reaction are nicely illustrated by a recent total synthesis of (+)-pilocarpine.[26] Zinc insertion provides a reliable and regioselective access to monoenolates of succinic acid diesters, a difficult task with other methods of enolate formation (eq 9). Although some optimization was necessary to avoid undesirable side reactions (eliminations), dimethyl (2*S*,3*S*)-2-bromo-3-ethyl-1,4-butandioate could be transformed into the key intermediate of the pilocarpine synthesis in a highly selective manner and in excellent yield.

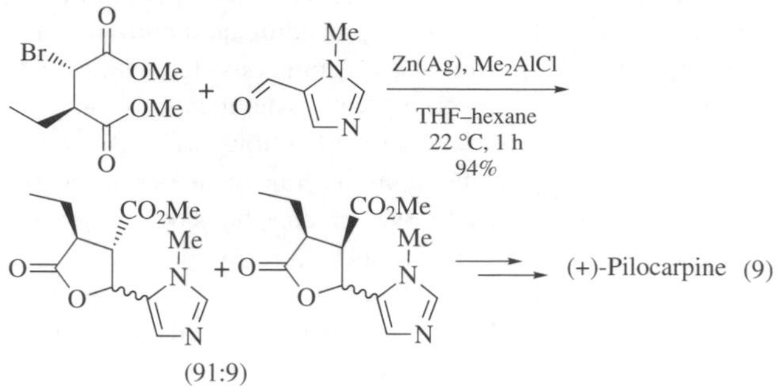

Electrophiles. Zinc enolates of esters or ketones show moderate reactivity compared to the respective of alkali metal enolates. Hence, they exhibit higher degrees of chemoselectivity upon treatment with different electrophiles. For a long time, aldehydes and ketones have been the only relevant group of substrates for these reagents and they are still widely used.[1] Examples of Reformatsky reactions with carbonyl compounds of almost any class of natural products can be found in the literature. Highly hindered ketones as well as readily enolizable carbonyl compounds and even acylsilanes (eq 10)[27] are prone to nucleophilic attack. α,β-Unsaturated carbonyl compounds react regioselectively in a 1,2-addition manner with only very few exceptions to this rule. The latter can easily be explained by steric hindrance (eq 11)[4b,28] or peculiar electronic properties of the acceptor molecule, as in aroyl ketene *S*,*N*-acetals (eq 12)[29] or 1,3-diaza-1,3-butadiene[30] derivatives.

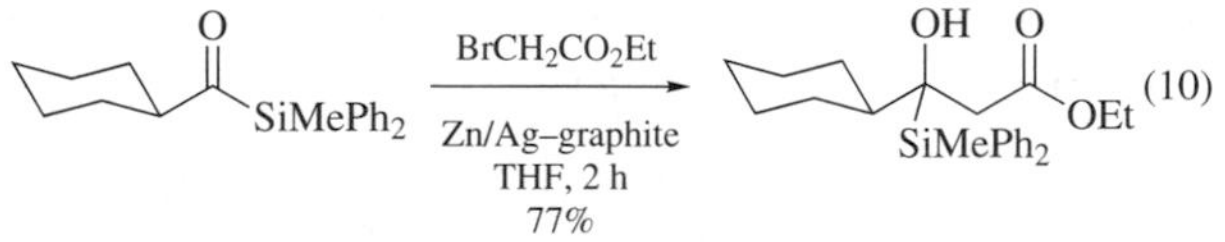

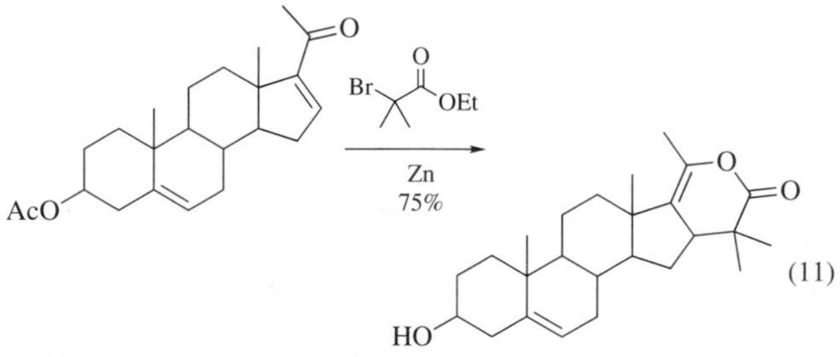

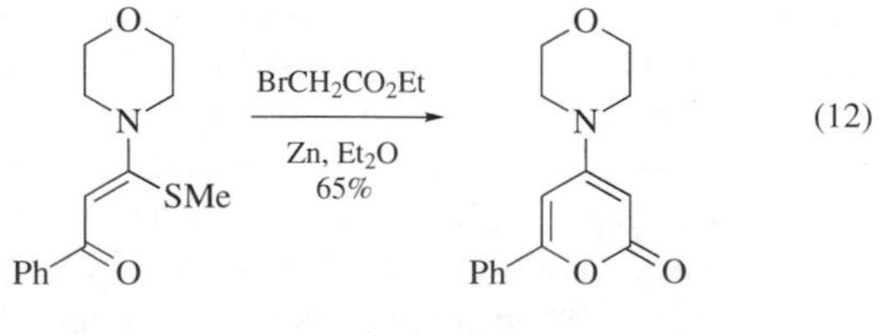

However, other kinds of electrophiles are also good acceptors for Reformatsky reagents, such as nitriles (sometimes called the Blaise reaction) (eq 13),[31,4b] azomethines (preferably via the two-step procedure giving rise to β-lactams) (eq 14),[32] and carboxylic acid chlorides (with assistance of catalytic amounts of Pd^0 complexes) (eq 15).[31c,33] Alkylation of zinc ester enolates, however, is troublesome and restricted to short-chain alkyl iodides, allyl and benzyl halides in aprotic dipolar solvents.[2,34] Under transition metal catalysis they also react with aryl and alkenyl halides or triflates.[35] *C*-Silylation of the zinc enolates by ***Chlorotrimethylsilane*** is essentially confined to bromoacetate (eq 16)[36] and haloacetonitrile.[37]

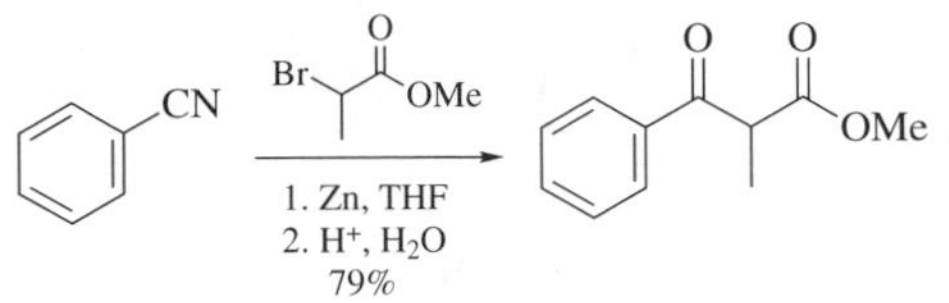

(13)

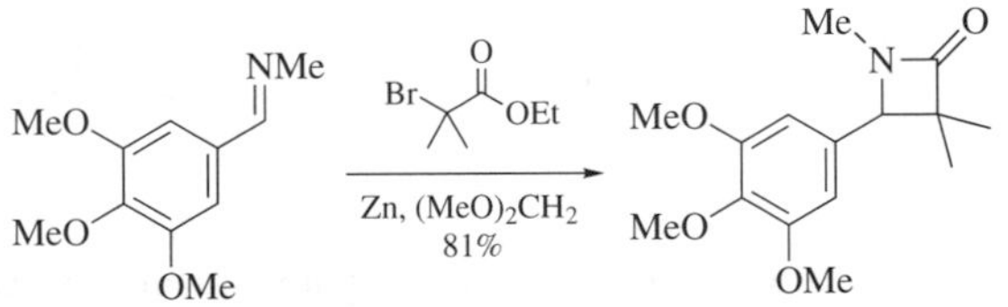

(14)

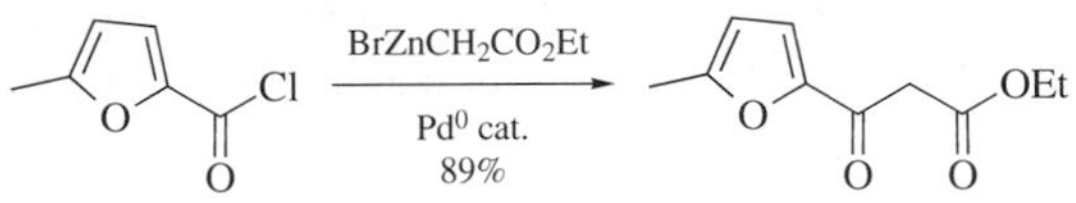

(15)

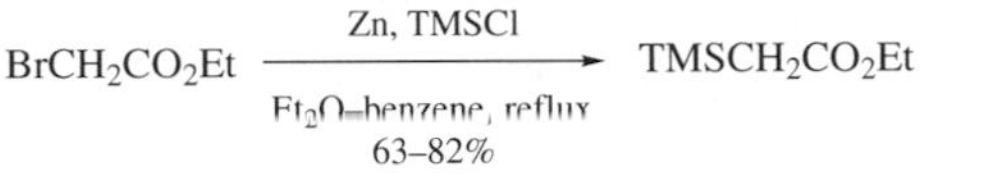

$$BrCH_2CO_2Et \xrightarrow[\text{Et}_2\text{O–benzene, reflux; 63–82\%}]{\text{Zn, TMSCl}} TMSCH_2CO_2Et \quad (16)$$

The slightly different reactivity of zinc enolates towards the aforementioned types of electrophiles allows chemoselective transformations of di- or polyfunctional substrates. Although in unsymmetrical diketones no differentiation among the carbonyl groups was reported, keto nitriles, keto amides, keto esters, or halogenated ketones may be attacked exclusively at the carbonyl group.[1]

Recent investigations show that 3-acyloxazolidin-2-ones or 3-acylthiazolidine-2-thiones constitute a promising group of acyl donors for zinc enolates (eq 17).[38] Different kinds of anhydrides (eq 18),[39] activated esters, and lactones,[40] including aldonolactones,[41] react smoothly with bromo esters under Reformatsky conditions. Particular emphasis is laid on the high yield of β-amino esters by reaction of zinc enolates with different kinds of *N,O*-acetals and aminals (eq 19).[42] Oxocarbenium cations obtained by in situ activation of acetals with Lewis acids are equally suited as electrophiles in Reformatsky reactions (eqs 20 and 21).[43] A summary of reactions with electrophiles other than carbonyl compounds is given in the literature.[1]

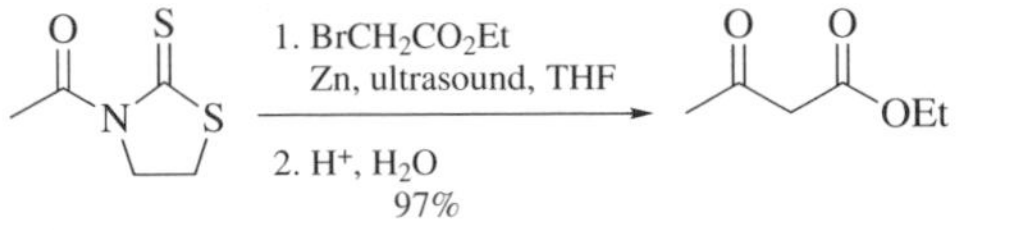

(17)

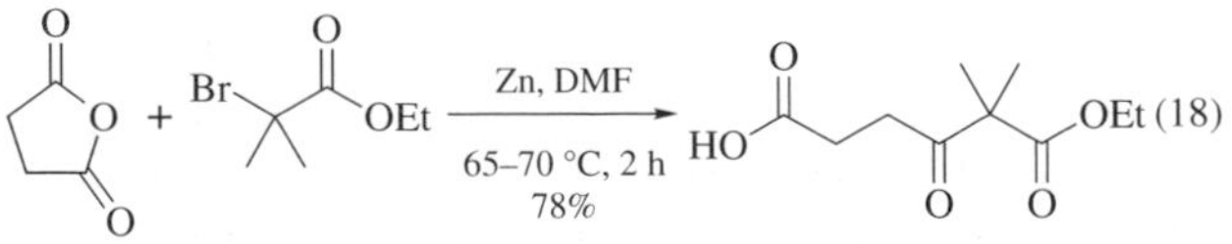

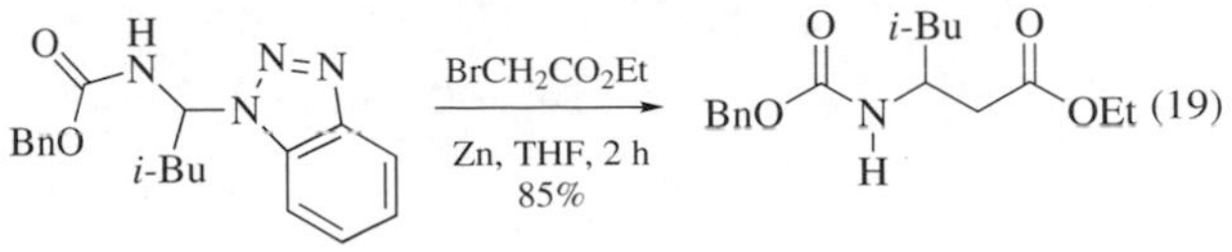

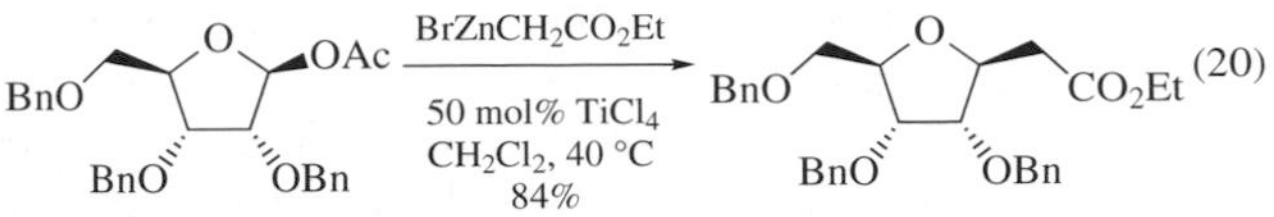

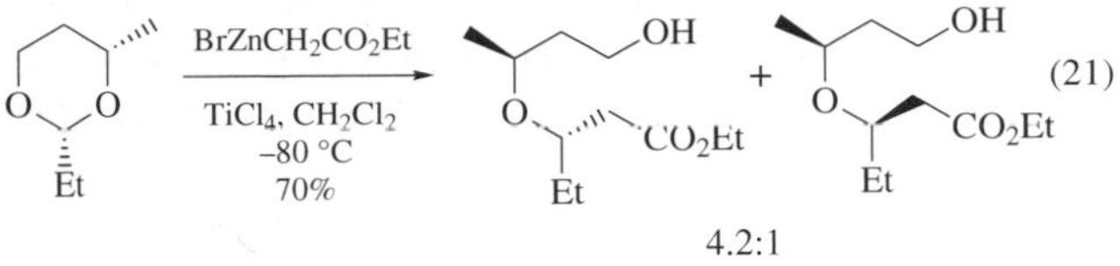

(18)

(19)

(20)

(21)

Intramolecular Reformatsky Reactions. As the site of enolate formation is determined by the halogen moiety, Reformatsky reactions are well suited for intramolecular aldolizations. This is a major advantage since regioselective enolate formation by proton abstraction in polycarbonyl systems is a rather difficult task. Thus a homologous series of ω-(α-bromoacetoxy)aldehydes has been cyclized to β-hydroxy lactones of ring size 13–16,[44] but smaller rings can also be formed in moderate to good yields.[45] This technique has been used in natural product synthesis (eq 22),[46] with the formation of the 11-membered ring of cyctochalasan being the most impressive example (eq 23).[47] Less conventional electrophiles such as nitriles[48] or imides[49] are equally well suited acceptors in entropically favored intramolecular reactions (eq 24).

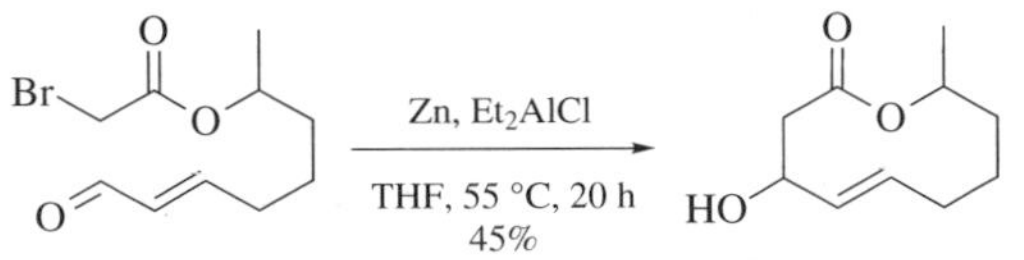

(22)

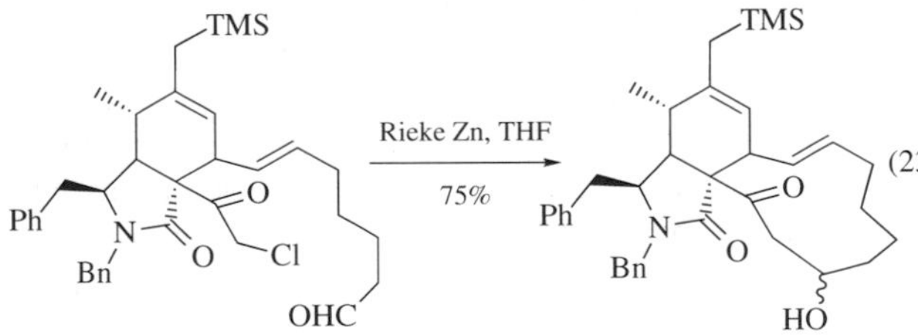

(23)

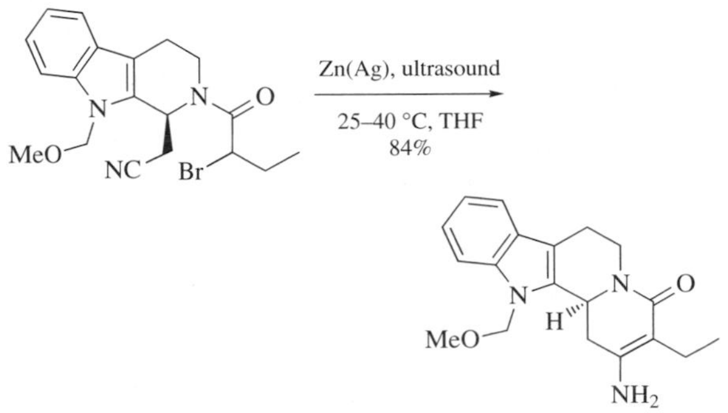

(24)

Samarium(II) Iodide as substitute for metallic zinc turned out to be highly advantageous in promoting such cyclizations with formation of normal-, medium-, and large-sized ring systems (eq 25).[50,51] Furthermore, high degrees of diastereocontrol may be exercised via the formation of rigid transition states with the strongly chelating Sm^{3+} counterion.[51]

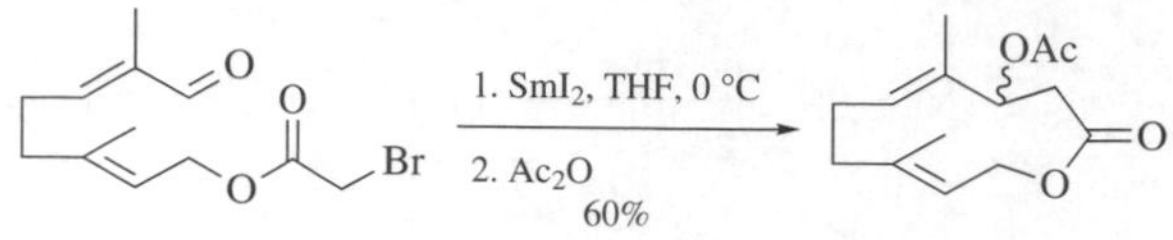

(25)

Tandem Reaction. Due to their high tolerance towards different functional groups, zinc enolates are predisposed for reaction sequences. The selective formation of either a carbocyclic or a heterocyclic ring from nitriles bearing an additional halo or sulfonyloxy group within the molecule is an illustrative example of how to impose control on such tandem reactions (eq 26).[52] In this specific case, the hard–soft acid–base (HSAB) principle determines whether the intermediate zinc enamides are *N*-alkylated (X = OSO_2Me) or *C*-alkylated (X = Br). Moreover, *O*-TMS cyanohydrins serve for lactone syntheses in a one-pot procedure.[31c]

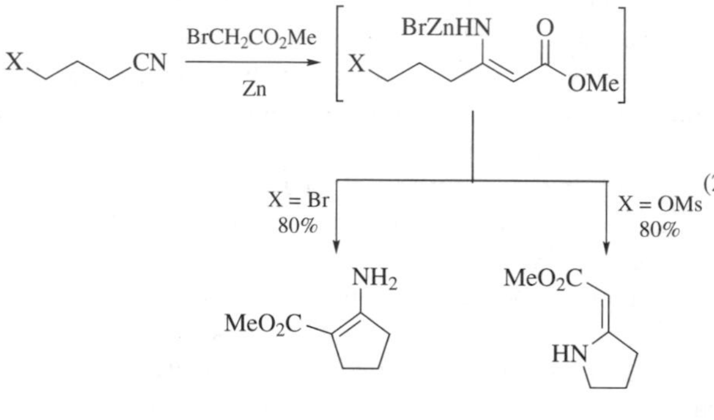

(26)

Under the rather drastic conditions of conventional Reformatsky reactions (reflux in ethereal or hydrocarbon solvents), the β-hydroxy esters formed may suffer subsequent dehydration in an unselective way. The use of α-silylated Reformatsky donors or of α-silyl ketones shows how to control the regioselectivity of the elimination step.[16,27a] Thus sequences of Reformatsky reactions followed by (in part spontaneous) Peterson eliminations determine the regiochemistry of the newly formed double bond (eqs 27 and 28). When bromo(trimethylsilyl)acetonitrile is used as donor, high to complete (*Z*) selectivity for the α,β-unsaturated nitriles obtained was reported.[53]

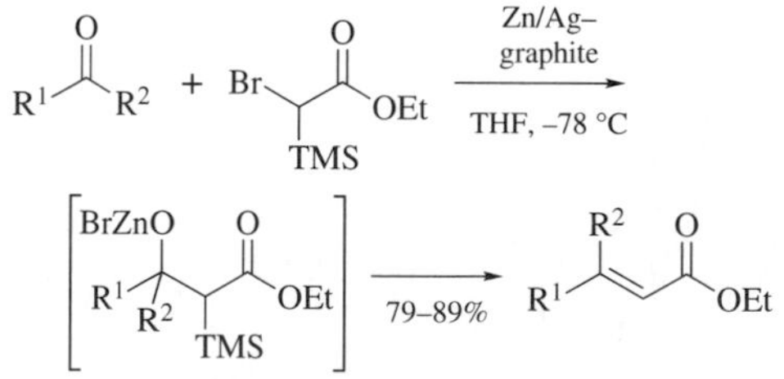

(27)

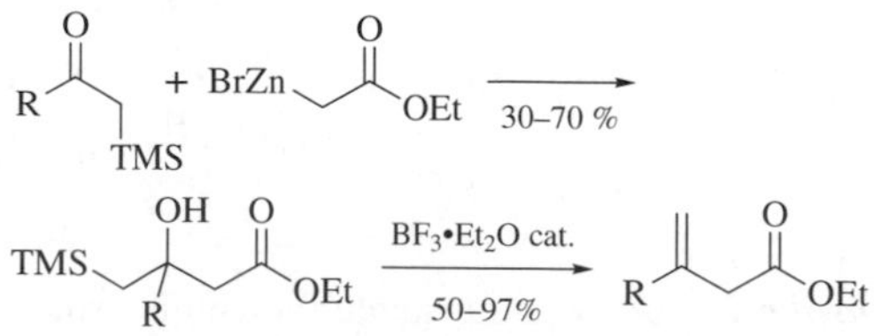

(28)

A sequence of an intramolecular Reformatsky reaction with a 2,5-dibromopentanoate donor, followed by etherification of the intermediate zinc aldolate with the terminal bromo group in the presence of HMPA, was ingeniously used to build-up the tricyclic skeleton of daphnilactone and related molecules starting from rather simple precursors (eq 29).[54]

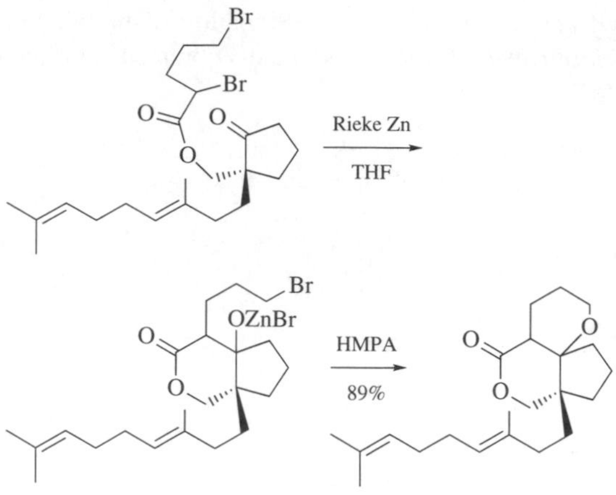

(29)

2-Oxoglycosyl bromides upon treatment with zinc give rise to carbohydrate intermediates with an 'umpoled' anomeric center; the aldolate initially formed on reaction with excess ***Formaldehyde***, together with the residual ketone group on the sugar, trap a second aldehyde molecule with formation of a stable lactol ring (eq 30).[55]

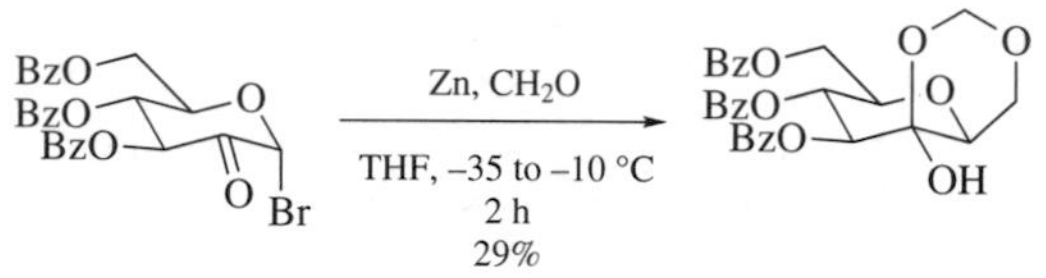

(30)

Stereoselectivity. A general method for highly diastereo- and enantioselective Reformatsky reactions is still missing. Some of the approaches described so far take advantage of the known propensity of zinc(II) to bind to nitrogen donor atoms which may be present in either one of the substrates or in a ligand added to the reaction mixture. With amino carbonyl compounds as electrophiles, for example, highly stereoselective additions are usually observed, and a direct relationship between the complexing ability of the amino group (location, basicity of the N atom) and the de values for the products obtained has been established.[56] A nice illustration of how uncommon electrophiles can be used to prepare enantiomerically enriched products is the ring-opening of enantiomerically pure oxazolidines by zinc ester enolates to β-amino esters, proceeding with inversion of the configuration at the *N,O*-acetal carbon atom (eq 31).[57]

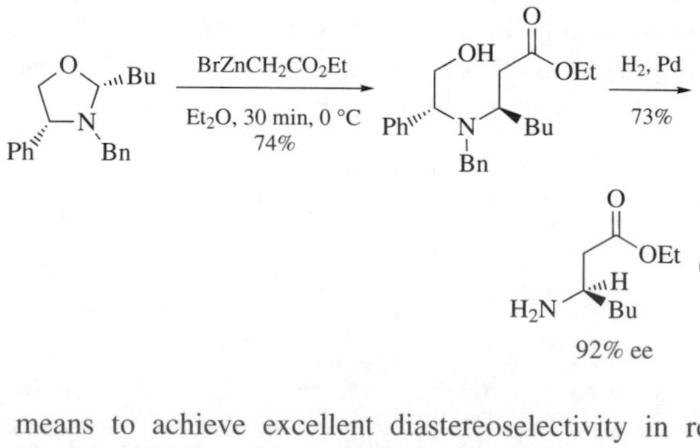

(31)

A means to achieve excellent diastereoselectivity in reactions of *ortho*-substituted benzaldehyde derivatives is the π-face selective attack of a zinc ester enolate on the corresponding tricarbonylchromium complexes (eq 32).[58]

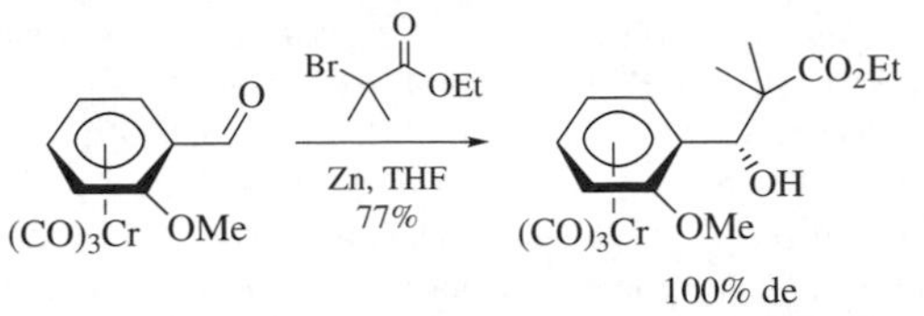

(32)

100% de

Preparatively useful degrees of diastereoselectivity have been observed with *N*-(α-bromoacyl)oxazolidinones[59] and/or with metals other than zinc exhibiting higher chelating abilities. For intramolecular reactions, SmI_2 served this purpose very well, because of the rigidity of a Sm^{3+}-chelated bicyclic transition state (eq 33).[51] An Evans-type auxiliary together with activated tin as promotor for the Reformatsky reaction have been employed in a diastereoselective approach to neooxazolomycin (eq 34).[60] With a few exceptions only, induction by chiral alcohol components in α-bromo esters is rather low[61] and until now hardly competitive with today's state of the art in aldol reactions.

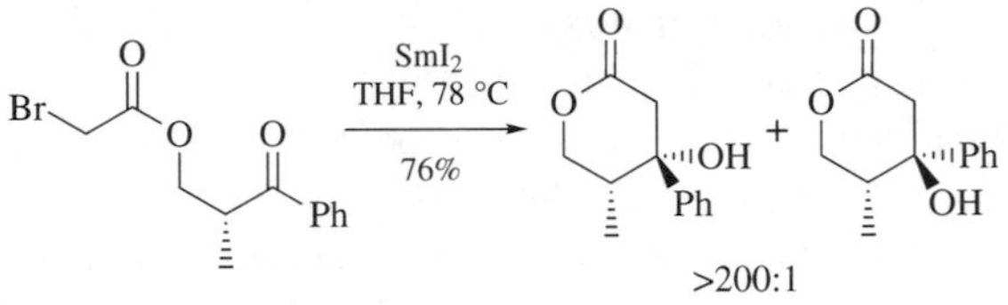

(33)

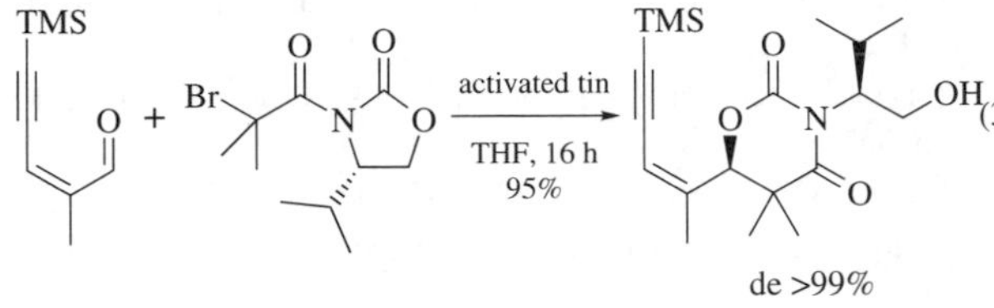

>200:1

(34)

de >99%

In addition to some early reports with (−)-sparteine as chiral ligand,[62] promising results of enantioselective zinc- or indium-induced reactions of haloacetates with carbonyl compounds in the presence of chiral amines or amino alcohols as ligands to the metal have been published recently (eq 35).[63]

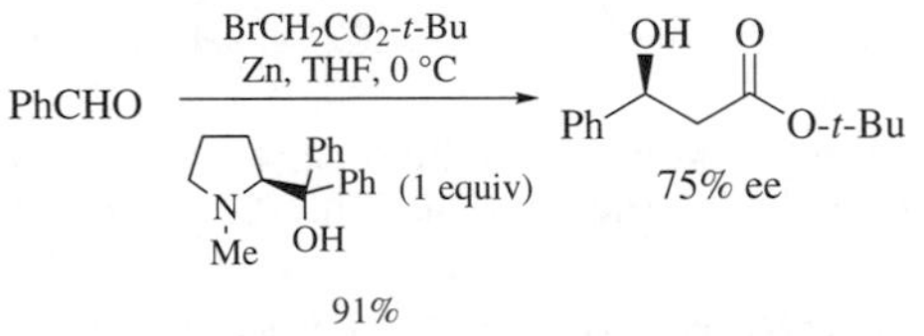

(35)

75% ee

91%

Substitution of Zinc by Other Metals. Oxidative addition into an activated carbon–halogen bond with formation of enolates is by no means restricted to metallic zinc.[1] Low-valent metal salts of adequate reduction potentials such as ***Samarium(II) Iodide***[41d,64,50,51] or ***Chromium(II) Chloride***[65] are equally suited. A great deal of work has been carried out with more or less activated forms of ***Magnesium***,[66] ***Tin***,[67] ***Nickel***,[68] ***Cerium***,[69] ***Indium***,[70] ***manganese***,[71] and ***cadmium***[72] as promotors for Reformatsky-type reactions. Although in specific cases advantages such as increased diastereoselectivity could be drawn from their use, only SmI_2 and to some extent cerium and indium (both are effective with the more reactive α-iodo esters) show reasonable scope. Some representative examples are compiled in Table 2.

Table 2 Reformatsky Reactions with Substitutes for Zinc

Donor	Electrophile	Metal	T(°C)	t(h)	Yield (%)
$BrCH_2CO_2Et$[72]	Cyclohexanone	Cd[a]	35	24	100
ICH_2CO_2Et[70a]	Cyclohexanone	In	20	1.5	65
$BrCH_2CO_2Et$[69]	Benzaldehyde	Ce/Hg	20	17	49
ICH_2CO_2Et[69]	Benzaldehyde	Ce/Hg	0	3.5	81
$BrCH_2CO_2Et$[69]	Acetophenone	Ce/Hg	20	46	60
$BrCH_2CO_2Et$[67]	Benzaldehyde	Sn[b]	25	2	84
$BrCH_2COPh$[67]	Benzaldehyde	Sn	−23	2	63
$BrCH_2CN$[68]	Benzaldehyde	Ni[c]	85	0.7	84
$BrCH(Me)CO_2$-Et[64a]	Cyclohexanone	SmI_2	–[d]	–[d]	90
$BrCH_2COPh$[64b]	Benzaldehyde	SmI_2	20	1	75

[a] Rieke Cd. [b] Activated Sn obtained by reduction of $SnCl_2$; only 6% yield with commercial Sn dust in DMF. [c] Rieke Ni. [d] Not reported.

Related Reagents. *t*-Butyl Chloroacetate; *t*-Butyl α-Lithioisobutyrate; Dilithioacetate; Ethyl 2-(Bromomethyl)-acrylate; Ethyl Lithioacetate; Methyl Bromoacetate; Methyl 4-Bromocrotonate; Methyl Chloroacetate; Potassium–Graphite Laminate; Zinc; Zinc/Copper Couple; Zinc–Graphite; Zinc/Silver Couple; Zinc–Zinc Chloride.

1. (a) Fürstner, A. *S* **1989**, 571. (b) Rathke, M. W.; Weipert, P. D. *COS* **1991**, *2*, 277. (c) Rathke, M. W. *OR* **1975**, *22*, 423. (d) Shriner, R. L. *OR* **1942**, *1*, 1. (e) Gaudemar, M. *Organomet. Chem. Rev. A* **1972**, *8*, 183. (f) Nützel, K. *MOC* **1973**, *XIII/2a*, 805.
2. Orsini, F.; Pelizzoni, F.; Ricca, G. *T* **1984**, *40*, 2781.
3. Dekker, J.; Budzelaar, P. H. M.; Boersma, J; van der Kerk, G. J. M.; Spek, A. L. *OM* **1984**, *3*, 1403.
4. (a) Dekker, J.; Schouten, A.; Budzelaar, P. H. M.; Boersma, J.; van der Kerk, G. J. M.; Spek, A. L.; Duisenberg, A. J. M. *JOM* **1987**, *320*, 1. (b) Hansen, M. M.; Bartlett, P. A.; Heathcock, C. H. *OM* **1987**, 6, 2069. (c) Kuroboshi, M.; Ishihara, T. *BCJ* **1990**, *63*, 428.
5. Reformatsky, S. *CB* **1887**, *20*, 1210.
6. Erdik, E. *T* **1987**, *43*, 2203.
7. (a) Csuk, R.; Fürstner, A.; Weidmann, H. *CC* **1986**, 775. (b) Fürstner, A. *AG(E)* **1993**, *32*, 164.
8. (a) Rieke, R. D.; Uhm, S. J. *S* **1975**, 452. (b) Rieke, R. D.; Li, P. T. J.; Burns, T. P.; Uhm, S. J. *JOC* **1981**, *46*, 4323. (c) Boldrini, G. P.; Savoia, D.; Tagliavini, E.; Trombini, C.; Umani-Ronchi, A. *JOC* **1983**, *48*, 4108. (d) Arnold, R. T., Kulenovic, S. T. *SC* **1977**, *7*, 223. (e) Bouhlel, E.; Rathke, M. W. *SC* **1991**, *21*, 133.
9. (a) Han, B. H.; Boudjouk, P. *JOC* **1982**, *47*, 5030. (b) Boudjouk, P.; Thompson, D. P.; Ohrbom, W. H.; Han, B. H. *OM* **1986**, *5*, 1257.
10. (a) Conan, A.; Sibille, S.; Périchon, J. *JOC* **1991**, *56*, 2018. (b) Schwarz, K. H.; Kleiner, K.; Ludwig, R.; Schick, H. *JOC* **1992**, *57*, 4013. (c) Zylber, N.; Zylber, J.; Rollin, Y.; Duñach, E.; Perichon, J. *JOM* **1993**, *444*, 1. (d) Rollin, Y.; Gebehenne, C.; Derien, S.; Duñach, E.; Perichon, J. *JOM* **1993**, *461*, 9. (e) Schick, H.; Ludwig, R.; Schwarz, K. H.; Kleiner, K.; Kunarth, A. *JOC* **1994**, *59*, 3161.
11. Vaughan, W. R.; Knoess, H. P. *JOC* **1970**, *35*, 2394.
12. (a) Gaudemar-Bardone, F.; Gaudemar, M.; Mladenova, M. *S* **1987**, 1130. (b) Bogavac, M.; Arsenijević, L.; Arsenijević, V. *BSF(2)* **1980**, 145. (c) Picotin, G.; Miginiac, P. *JOC* **1987**, *52*, 4796. (d) Liu, W. S.; Glover, G. I. *JOC* **1978**, *43*, 754. (e) Horeau, A. *TL* **1971**, 3227. (f) Bellassoued, M.; Dardoize, F.; Frangin, Y.; Gaudemar, M. *JOM* **1981**, *219*, C1.

13. (a) Baldwin, J. E.; Walker, J. A. *CC* **1973**, 117. (b) Greuter, H.; Lang, R. W.; Romann, A. J. *TL* **1988**, *29*, 3291.
14. Leading references: (a) Witkowski, S.; Rao, Y. K.; Premchandran, R. H.; Halushka, P. V.; Fried, J. *JACS* **1992**, *114*, 8464. (b) Kitazume, T. *S* **1986**, 855. (c) Lang, R. W.; Schaub, B. *TL* **1988**, *29*, 2943. (d) Fried, J.; Hallinan, E. A.; Szwedo, Jr. M. J. *JACS* **1984**, *106*, 3871. (e) Kitagawa, O.; Taguchi, T.; Kobayashi, Y. *TL* **1988**, *29*, 1803. (f) Curran, T. T. *JOC* **1993**, *58*, 6360.
15. Originally described by: (a) Darzens, G. *CR(C)* **1910**, *151*, 883. Failure of conventional conditions: (b) Miller, R. E.; Nord, F. F. *JOC* **1951**, *16*, 728.
16. Fürstner, A. *JOM* **1987**, *336*, C33.
17. (a) Takai, K.; Hotta, Y.; Oshima, K.; Nozaki, H. *BCJ* **1980**, *53*, 1698. (b) Ishihara, T.; Matsuda, T.; Imura, K.; Matsui, H.; Yamanaka, H. *CL* **1994**, 2167.
18. Palomo, C.; Aizpurua, J. M.; López, M. C.; Aurrekoetxea, N. *TL* **1990**, *31*, 2205.
19. Palomo, C.; Aizpurua, J. M.; Aurrekoetxea, N.; López, M. C. *TL* **1991**, *32*, 2525.
20. (a) Rice, L. E.; Boston, M. C.; Finklea, H. O.; Suder, B. J.; Frazier, J. O.; Hudlicky, T. *JOC* **1984**, *49*, 1845. (b) Bellassoued, M.; Gaudemar, M.; El Borgi, A.; Baccar, B. *JOM* **1985**, *280*, 165. (c) Bortolussi, M.; Seyden-Penne, J. *SC* **1989**, *19*, 2355. (d) Hudlicky, T.; Natchus, M. G.; Kwart, L. D.; Colwell, B. L. *JOC* **1985**, *50*, 4300.
21. Öhler, E.; Reininger, K.; Schmidt, U. *AG(E)* **1970**, *9*, 457.
22. (a) Grieco, P. A. *S* **1975**, 67. (b) Petragnani, N.; Ferraz, H. M. C.; Silva, G. V. *S* **1986**, 157.
23. Rebek, Jr. J.; Tai, D. F.; Shue, Y. K. *JACS* **1984**, *106*, 1813.
24. Semmelhack, M. F.; Wu, E. S. C. *JACS* **1976**, *98*, 3384.
25. (a) Csuk, R.; Fürstner, A.; Sterk, H.; Weidmann, H. *J. Carbohydr. Chem.* **1986**, *5*, 459. (b) Csuk, R.; Hugener, M.; Vasella, A. *HCA* **1988**, *71*, 609.
26. Dener, J. M.; Zhang, L. H.; Rapoport, H. *JOC* **1993**, *58*, 1159.
27. (a) Fürstner, A.; Kollegger, G.; Weidmann, H. *JOM* **1991**, *414*, 295. (b) Narasaka, K.; Saito, N.; Hayashi, Y.; Ichida, H. *CL* **1990**, 1411.
28. Gandolfi, C.; Doria, G.; Amendola, M.; Dradi, E. *TL* **1970**, 3923.
29. Datta, A.; Ila, H.; Junjappa, H. *S* **1988**, 248.
30. Mazumdar, S. N.; Mahajan, M. P. *TL* **1990**, *31*, 4215.
31. (a) Blaise, E. E. *Hebd. Sceances Acad. Sci.* **1901**, *132*, 478. For leading references, see: (b) El Alami, N.; Belaud, C.; Villieras, J. *JOM* **1987**, *319*, 303. (c) Krepski, L. R.; Lynch, L. E.; Heilman, S. M.; Rasmussen, J. K. *TL* **1985**, *26*, 981.
32. Originally described by: (a) Gilman, H.; Speeter, M. *JACS* **1943**, *65*, 2255. Leading references: (b) Luche, J. L.; Kagan, H. B. *BSF(2)* **1969**, 3500. (c) Bosch, J.; Domingo, A.; Lopez, F.; Rubiralta, M. *JHC* **1980**, *17*, 241. (d) Bose, A. K.; Gupta, K.; Manhas, M. S. *CC* **1984**, 86. (e) Palomo, C.; Cossío, F. P.; Arrieta, A.; Odriozola, J. M.; Oiarbide, M.; Ontaria, J. M. *JOC* **1989**, *54*, 5736. (f) Cossio, F. P.; Odriozola, J. M.; Oiarbide, M.; Palomo, C. *CC* **1989**, 74.
33. Sato, T.; Itoh, T.; Fujisawa, T. *CL* **1982**, 1559.
34. Orsini, F.; Pelizzoni, F. *SC* **1984**, *14*, 805. For recent advances, see: Bott, K. *TL* **1994**, *35*, 555.
35. (a) Fauvarque, J. F.; Jutand, A. *JOM* **1979**, *177*, 273. (b) Fauvarque, J. F.; Jutand, A. *JOM* **1977**, *132*, C17. (c) Fauvarque, J. F.; Jutand, A. *JOM* **1981**, *209*, 109. (d) Orsini, F.; Pelizzoni, F.; Vallarino, L. M. *JOM* **1989**, *367*, 375.
36. (a) Fessenden, R. J.; Fessenden, J. S. *JOC* **1967**, *32*, 3535. (b) Kuwajima, I.; Nakamura, E.; Hashimoto, K. *OS* **1983**, *61*, 122. (c) Nietzschmann, E.; Böge, O.; Tzschach, A. *JPOC* **1991**, *333*, 281.
37. Matsuda, I.; Murata, S.; Ishii, Y. *JCS(P1)* **1979**, 26.
38. Kashima, C.; Huang, X. C.; Harada, Y.; Hosomi, A. *JOC* **1993**, *58*, 793.
39. (a) Schick, H.; Ludwig, R. *S* **1992**, 369. (b) Gedge, D. R.; Pattenden, G.; Smith A. G. *JCS(P1)* **1986**, 2127.
40. (a) Hauser, F. M.; Rhee, R. P. *JOC* **1977**, *42*, 4155. (b) Warnhoff, E. W.; Wong, M. Y. H.; Raman, P. S. *CJC* **1981**, *59*, 688. (c) Gawronski, J. K. *TL* **1984**, *25*, 2605.
41. (a) Csuk, R.; Glänzer, B. I. *J. Carbohydr. Chem.* **1990**, *9*, 797. (b) Shrivastava, V. K.; Lerner, L. M. *JOC* **1979**, *44*, 3368. (c) Graberger, V.; Berger, A.; Dax, K.; Fechter, M.; Gradnig, G.; Stütz, A. E. *LA* **1993**, 379. (d) Hanessian, S.; Girard, C. *SL* **1994**, 865.
42. (a) Katritzky, A. R.; Yannakopoulou, K. *S* **1989**, 747. (b) Alberola, A.; Alvarez, M. A.; Andrés, C.; González, A.; Pedrosa, R. *S* **1990**, 1057. (c) Kise, N.; Yamazaki, H.; Mabuchi, T.; Shono, T. *TL* **1994**, *35*, 1561. (d) Nishiyama, T.; Kishi, H.; Kitano, K.; Yamada, F. *BCJ* **1994**, *67*, 1765.
43. (a) Hayashi, M.; Sugiyama, M.; Toba, T.; Oguni, N. *CC* **1990**, 767. (b) Basile, T.; Tagliavini, E.; Trombini, C.; Umani-Ronchi, A. *CC* **1989**, 596.
44. (a) Maruoka, K.; Hashimoto, S.; Kitagawa, Y.; Yamamoto, H.; Nozaki, H. *BCJ* **1980**, *53*, 3301. (b) Maruoka, K.; Hashimoto, S.; Kitagawa, Y.; Yamamoto, H.; Nozaki, H. *JACS* **1977**, *99*, 7705.
45. (a) Sato, A.; Ogiso, A.; Noguchi, H.; Mitsui, S.; Kaneko, I; Shimada, Y. *CPB* **1980**, *28*, 1509. (b) Stokker, G. E.; Hoffmann, W. F.; Alberts, A. W.; Cragoe, Jr. E. J.; Peana, A. A.; Gilfillan, J. L.; Huff, J. W.; Novello, F. C.; Prugh, J. D.; Smith, R. L.; Willard, A. K. *JMC* **1985**, *28*, 347.
46. Tsuji, J.; Mandai, T. *TL* **1978**, 1817.
47. Vedejs, E.; Ahmed, S. *TL* **1988**, *29*, 2291.
48. Beard, R. L.; Meyers, A. I. *JOC* **1991**, *56*, 2091.
49. Flitsch, W.; Rukamp, P. *LA* **1985**, 1398.
50. (a) Tabuchi, T.; Kawamura, K.; Inanaga, J.; Yamaguchi, M. *TL* **1986**, *27*, 3889. (b) Moriya, T.; Handa, Y.; Inananga, J.; Yamaguchi, M. *TL* **1988**, *29*, 6947. (c) Inanaga, J.; Yokoyama, Y.; Handa, Y.; Yamaguchi, M. *TL* **1991**, *32*, 6371.
51. Molander, G. A.; Etter, J. B.; Harring, L. S.; Thorel, P. J. *JACS* **1991**, *113*, 8036.
52. Hannick, S. M.; Kishi, Y. *JOC* **1983**, *48*, 3833.
53. Palomo, C.; Aizpurua, J. M.; Aurrekoetxea, N. *TL* **1990**, *31*, 2209.
54. (a) Heathcock, C. H.; Ruggeri, R. B.; McClure, K. F. *JOC* **1992**, *57*, 2585. (b) Ruggeri, R. B.; Heathcock, C. H. *JOC* **1987**, *52*, 5745.
55. Lichtenthaler, F. W.; Schwidetzky, S.; Nakamura, K. *TL* **1990**, *31*, 71.
56. (a) Lucas, M.; Guetté, J. P. *T* **1978**, *34*, 1681 and 1685. (b) Adlington, R. M.; Baldwin, J. E.; Jones, R. H.; Murphy, J. A.; Parisi, M. F. *CC* **1983**, 1479.
57. Andrés, C.; González, A.; Pedrosa, R.; Pérez-Encabo, A. *TL* **1992**, *33*, 2895.
58. (a) Brocard, J.; Mahmoudi, M.; Pelinski, L.; Maciejewski, L. *T* **1990**, *46*, 6995. (b) Brocard, J.; Pelinski, L.; Lebibi, J. *JOM* **1987**, *336*, C47.
59. Ito, Y.; Terashima, S. *TL* **1987**, *28*, 6625 and 6629.
60. Kende, A. S.; Kawamura, K.; DeVita, R. J. *JACS* **1990**, *112*, 4070.
61. (a) Palmer, M. H.; Reid, J. A. *JCS* **1962**, 1762. (b) Furukawa, M.; Okawara, T.; Noguchi, Y.; Terawaki, Y. *CPB* **1978**, *26*, 260. (c) Basavaiah, D.; Bharathi, T. K. *SC* **1989**, *19*, 2035. (d) Basavaiah, D.; Bharathi, T. *TL* **1991**, *32*, 3417.
62. Guette, M.; Capillon, J.; Guetté, J. P. *T* **1973**, *29*, 3659.
63. (a) Soai, K.; Kawase, Y. *TA* **1991**, *2*, 781. (b) Johar, P. S.; Araki, S.; Butsugan, Y. *JCS(P1)* **1992**, 711.
64. (a) Kagan, H. B.; Namy, J. L.; Girard, P. *T* **1981**, *37*, Suppl. 1, 175. (b) Zhang, Y.; Liu, T.; Lin, R. *SC* **1988**, *18*, 2003.
65. Dubois, J. E.; Axiotis, G.; Bertounesque, E. *TL* **1985**, *26*, 4371.
66. Moriwake, T. *JOC* **1966**, *31*, 983.
67. (a) Harada, T.; Mukaiyama, T. *CL* **1982**, 161. (b) Harada, T.; Mukaiyama, T. *CL* **1982**, 467.

68. Inaba, S. I.; Rieke, R. D. *TL* **1985**, *26*, 155.
69. Imamoto, T.; Kusumoto, T.; Tawarayama, Y.; Sugiura, Y.; Mita, T.; Hatanaka, Y.; Yokoyama, M. *JOC* **1984**, *49*, 3904.
70. (a) Chao, L. C.; Rieke, R. D. *JOC* **1975**, *40*, 2253. (b) Araki, S.; Ito, H.; Butsugan, Y. *SC* **1988**, *18*, 453. (c) Araki, S.; Katsumura, N.; Kawasaki, K. I.; Butsugan, Y. *JCS(P1)* **1991**, 499.
71. Cahiez, G.; Chavant, P. Y. *TL* **1989**, *30*, 7373.
72. Burkhardt, E. R.; Rieke, R. D. *JOC* **1985**, *50*, 416.

Alois Fürstner
Max-Planck-Institut für Kohlenforschung, Mülheim, Germany

Ethyl Cyanoacetate

(R = Et)
[105-56-6] $C_5H_7NO_2$ (MW 113.13)
(R = Me)
[105-34-0] $C_4H_5NO_2$ (MW 99.10)
(R = *t*-Bu)
[1116-98-9] $C_7H_{11}NO_2$ (MW 141.19)

(reagent with a doubly activated methylene group for free radical cyclizations, for preparing heterocycles, for nucleophilic substitution reactions, for preparing α-alkylidene derivatives, and for carbocyclic ring formation)

Physical Data: R = Me, bp 204–207 °C; *d* 1.123 g cm^{-3}; R = Et, bp 208–210 °C; *d* 1.063 g cm^{-3}; R = *t*-Bu, bp 90 °C/10 mmHg.
Solubility: sol EtOH, ether.
Form Supplied in: liquid; widely available.
Preparative Method: ethyl and methyl cyanoacetate are conveniently prepared from cyanoacetic acid and the corresponding ***Triethyloxonium Tetrafluoroborate*** and trimethyloxonium fluoroborate.[1] *t*-Butyl cyanoacetate is prepared from cyanoacetyl chloride and *t*-butyl alcohol.[2]
Analysis of Reagent Purity: gas chromatography.
Handling, Storage, and Precautions: ethyl cyanoacetate is a lachrymator irritant. Methyl cyanoacetate is described as an irritant. Keep containers well closed. Inhalation should be avoided. Hydrogen cyanide or other low molecular weight cyano compounds may be liberated in event of overheating. Handle in a fume hood.

Condensation Reactions With Carbonyl Groups. Ethyl cyanoacetate undergoes the Knoevenagel condensation[3] (or the Knoevenagel–Doebner modification) with aldehydes and ketones in the presence of β-alanine,[4] glycine,[5] ammonium acetate,[6,7] ***Piperidine***,[8] piperidine acetate,[8] bismuth trichloride,[9] ***Dihydridotetrakis(triphenylphosphine)ruthenium (II)***,[10] or weak bases.[11] The ammonium acetate-catalyzed reaction of acetophenone and ethyl cyanoacetate affords ethyl (1-phenylethylidene)cyanoacetate (eq 1).[7] Condensation of ethyl cyanoacetate with propanal followed by reduction with hydrogen and palladium yielded ethyl butylcyanoacetate.[8] Ethyl cyanoacetate and other active hydrogen compounds can also be alkylated by alcohols in modest yields using the combination of ***Diethyl Azodicarboxylate*** and ***Triphenylphosphine***.[12]

(1) [Ph(Me)C=O + NC–CH2–CO2Et, NH4OAc, 55% → Ph(Me)C=C(CN)CO2Et]

Reactions with α,β-Unsaturated Carbonyl Compounds. Activated nitriles such as ethyl cyanoacetate react with α,β-unsaturated carbonyl compounds in the presence of dihydridotetrakis(triphenylphosphine)ruthenium to afford Michael adducts which can undergo an aldol cyclization stereoselectively to afford cyclohexanes (eq 2).[10] Ethyl cyanoacetate reacts with *p*-benzoquinone to afford diethyl α,α′-dicyano-2,5-dihydroxy-*p*-benzenediacetate which is hydrolyzed to 2,5-dihydroxy-*p*-benzenediacetic acid (eq 3).[13] Methacrylamides react with ethyl cyanoacetate in the presence of ***Cesium Fluoride***–tetramethoxysilane to form dihydropyridinones (eq 4).[14]

(2) $MeHC{=}C(CO_2Et)_2$ + NC–CH2–CO2Et, Ru^{II}, 90% → $(EtO_2C)_2CH–CH(Me)–CH(CN)CO_2Et$

(3) *p*-benzoquinone + NC–CH2–CO2Et, NH_4OAc, 49% → 2,5-dihydroxy-1,4-bis[CH(CN)CO2Et]benzene

(4) $CH_2{=}C(Me)CONH_2$ + NC–CH2–CO2Et, CsF, $Si(OR)_4$, 84% → dihydropyridinone (HN, C=O, CN)

Synthesis of Cyclopropanes and Cyclopentanes. 1-Cyanocyclopropanecarboxylic acid (eq 5)[15] or 1-cyanocyclopentane-1-carboxylate[16] can be prepared by cyclocondensation (double alkylation) of ethyl cyanoacetate and ***1,2-Dibromoethane*** or ***1,4-Dibromobutane*** under phase transfer conditions. Ethyl cyanoacetate reacts (2 equiv) with alkenes (1 equiv) in the presence of ***Copper(II) Chloride*** and ***Copper(II) Acetate*** in DMF to afford *endo*- and *exo*-cyclopropane derivatives (yields based on the alkene) (eq 6).[17] Cyclopropanes were obtained under similar conditions from 1-decene and styrene. Use of dimethyl malonate in place of ethyl cyanoacetate gave lower yields of cyclopropanes.

(5) $BrCH_2CH_2Br$ + NC–CH2–CO2Et, TEBA, NaOH, H_2O, 86% → 1-cyanocyclopropane-1-CO_2H

$$\text{cyclohexene} + NC\text{-}CH_2\text{-}CO_2Et \text{ (excess)} \xrightarrow[\text{DMF, 110 °C; 53\%}]{CuCl_2} \text{7-cyano-7-ethoxycarbonylbicyclo[4.1.0]heptane} + \text{isomer} \quad (6)$$

85:15

Alkylation Reactions. Ethyl cyanoacetate undergoes nucleophilic aromatic substitution with hexafluorobenzene in DMF/***Potassium Carbonate*** to yield ethyl cyano(pentafluorophenyl)acetate which is converted to (pentafluorophenyl) acetonitrile.[18] Ethyl cyanoacetate reacts with *o*,α-dichlorotoluene to afford ethyl 2-(*o*-chlorobenzyl)cyanoacetic acid which is an intermediate in the synthesis of 1-cyanobenzocyclobutene (bicyclo[4.2.0]octa-1,3,5-triene-7-carbonitrile).[19] The nucleophilic substitution reaction of ethyl and *t*-butyl cyanoacetate with 3-bromochromone (eq 7) or 6-bromofurochromone affords the respective 3-substituted derivative,[20] presumably by an addition–elimination mechanism.

$$\text{3-bromochromone} + NC\text{-}CH_2\text{-}CO_2Et \xrightarrow[\text{MeCN}]{\text{DBU}} \text{3-(CH(CN)CO}_2\text{R)chromone} \quad (7)$$

R = Et, *t*-Bu

Radical Cyclizations. Ethyl cyanoacetate reacts with the tosylate of (*E*)-4-hexen-1-ol to afford ethyl (*E*)-2-cyano-6-octenoate which yields 1-cyano-2-methylcyclohexanecarboxylate on treatment with ***Dibenzoyl Peroxide*** (eq 8).[21] Similarly, oxidative free radical cyclization of ethyl (4*E*,8-nonadien-1-yl)cyanoacetate in the presence of the co-oxidants Mn^{III} and Cu^{II} acetate afforded ethyl 1-cyano-2-[1-(1,4-pentadienyl)]cyclopentanecarboxylate (35%).[22] In the absence of Cu^{II} acetate or with benzoyl peroxide alone in cyclohexane, a mixture of the stereoisomers of methylhydrindane is formed (eq 9). The Mitsunobo reactions of 1,*n*-diols with ethyl cyanoacetate in the presence of triphenylphosphine and diethyl azodicarboxylate afford 1-cyano-1-cycloalkanecarboxylates (eq 10).[23]

$$\text{(E)-hex-4-enyl-OTs} + NC\text{-}CH_2\text{-}CO_2Et \xrightarrow[\text{DMF; 53\%}]{\text{NaH}} \text{ethyl (E)-2-cyano-6-octenoate} \xrightarrow[\text{cyclohexane; 75\%}]{PhC(O)OOC(O)Ph} \text{ethyl 1-cyano-2-methylcyclohexanecarboxylate} \quad (8)$$

$$\text{ethyl 2-cyano-nonadienyl acetate} \xrightarrow[35\%]{Mn^{III}, \text{ EtOH}} \text{hydrindane (NC, CO}_2\text{Et)} \quad (9)$$

$$HO(CH_2)_nOH + NC\text{-}CH_2\text{-}CO_2Et \xrightarrow[-Ph_3PO;\ 20\text{–}60\%]{EtO_2C\text{-}N{=}N\text{-}CO_2Et,\ PPh_3} \text{cyclo-}C(CN)(CO_2Et)(CH_2)_n \quad (10)$$

$n = 4, 5, 6$

Synthesis of Heterocycles. Reaction of 2-arylazirines and ***Hexacarbonylmolybdenum*** with ethyl cyanoacetate furnishes the *trans* disubstituted succinimides (eq 11).[24] Ethyl cyanoacetate condenses with butanone in the presence of ammonium acetate and ammonia to yield α,α′-dicyano-β-ethyl-β-methylglutarimide (70%), which can be converted to β-ethyl-β-methylglutaric acid.[25] Ethyl cyanoacetate reacts with aromatic and heterocyclic hydrazines to form 1-aryl-3-amino-5-pyrazolones (eq 12).[26]

$$\text{Ph-azirine} + NC\text{-}CH_2\text{-}CO_2Et + Mo(CO)_6 \longrightarrow \text{trans-3-Ph-4-CN-succinimide} \quad (11)$$

$$PhNHNH_2 + NC\text{-}CH_2\text{-}CO_2Et \xrightarrow[\text{2. HOAc; 47\%}]{\text{1. NaOEt}} \text{1-phenyl-3-amino-5-pyrazolone} \quad (12)$$

Ethyl cyanoacetate condenses with urea to form 2,6-dihydroxy-5-aminouracil (eq 13), which is the first step in a synthesis of 5,6-diaminouracil hydrochloride.[27] Similarly, thiourea and ethyl ethoxymethylenecyanoacetate afford 2-mercapto-4-amino-5-ethoxycarbonylpyrimidine and 2-mercapto-4-hydroxy-5-cyanopyrimidine.[28] Guanidine condenses with ethyl cyanoacetate to produce 2,4-diamino-6-hydroxypyrimidine.[29]

$$H_2NC(O)NH_2 + NC\text{-}CH_2\text{-}CO_2Et \xrightarrow[\text{EtOH}]{\text{NaOEt}} \text{4-amino-2,6-dihydroxypyrimidine} \quad (13)$$

Other Applications. Novel carbon–carbon bond formation of nonactivated alkenes with ethyl cyanoacetate has been developed by anodic oxidation using Mn^{II} and Cu^{II} diacetates to give selectively either ethyl cyanoalkanecarboxylates or cyanoalkenecarboxylates (eq 14).[30,31] Preparation of cyanomethyl ketone derivatives of *N*-acetylphenylalanine and *N*-acetylleucylphenylalanine is accomplished by condensation of the corresponding activated carboxylic acids and the carbanion of *t*-butyl cyanoacetate.[32] ***Malononitrile***, benzoylacetonitrile, and nitroacetonitrile are also reactive. Reaction of ethyl cyanoacetate and butanone in the presence of ***Potassium Cyanide*** and ***Acetic Acid***, followed by acidic hydrolysis and heat, affords α-ethyl-α-methylsuccinic acid.[33]

$$NC\text{-}CH_2\text{-}CO_2Et + R\text{-}CH{=}CH\text{-}R' \begin{cases} \xrightarrow[51\text{–}64\%]{Mn(OAc)_2} \text{saturated adduct (R, R', CH(CN)CO}_2\text{Et)} \\ \xrightarrow[Cu(OAc)_2;\ 44\text{–}58\%]{Mn(OAc)_2} \text{unsaturated adduct (R, R', CH(CN)CO}_2\text{Et)} \end{cases} \quad (14)$$

Related Reagents. Acetonitrile; Cyanoacetic Acid; 3-Ethoxyacrylonitrile; Ethyl Ethoxymethylenecyanoacetate; Lithioacetonitrile; Malononitrile.

1. Raber, D. J.; Gariano, Jr., P.; Brod, A. O.; Gariano, A. L.; Guida, W. C. *OSC* **1988**, *6*, 576.
2. Ireland, R. E.; Chaykovsky, M. *OSC* **1973**, *5*, 171.
3. For a review, see: Tietze, L. F.; Beifuss, U. *COS* **1991**, *2*, 341.
4. (a) Prout, F. S.; Hartman, R. J.; Huang, E. P.-Y.; Korpics, C. J.; Tichelaar, G. R. *OSC* **1963**, *4*, 93. (b) Egawa, Y.; Suzuki, M.; Okuda, T. *CPB* **1963**, *11*, 589.
5. Bastus, J. B. *TL* **1963**, 955.
6. Cope, A. C.; Hancock, E. M. *OSC* **1955**, *3*, 399.
7. McElvain, S. M.; Clemens, D. H. *OSC* **1963**, *4*, 463.
8. Alexander, E. R.; Cope, A. C. *OSC* **1955**, *3*, 385.
9. Prajapati, D.; Sandhu, J. S. *CL* **1992**, 1945.
10. Naota, T.; Taki, H.; Mizuno, M.; Murahashi, S.-I. *JACS* **1989**, *111*, 5954.
11. Jones, C. *OR* **1967**, *15*, 204.
12. Wada, M.; Mitsunobu, O. *TL* **1972**, 1279.
13. Wood, J. H.; Cox, L. *OSC* **1955**, *3*, 286.
14. Chuit, C.; Corriu, R. J. P.; Perz, R.; Reye, C. *T* **1986**, *42*, 2293.
15. Singh, R. K.; Danishefsky, S. *JOC* **1975**, *40*, 2969.
16. Lin, Q.; Liu, Z. *Huaxue Shiji* **1992**, *14*, 310 (*CA* **1993**, *118*, 101 540).
17. Barreau, M.; Bost, M.; Julia, M.; Lallemand, J.-Y. *TL* **1975**, 3465.
18. Filler, R.; Woods, S. M. *OSC* **1988**, *6*, 873.
19. Skorcz, J. A.; Kaminski, F. E. *OSC* **1973**, *5*, 263.
20. Gammill, R. B.; Nash, S. A.; Bell, L. T.; Watt, W. *TL* **1992**, *33*, 997.
21. Julia, M.; Maumy, M. *OSC* **1988**, *6*, 586.
22. Snider, B. B.; Armanetti, L.; Baggio, R. *TL* **1993**, *34*, 1701.
23. Kurihara, T.; Nakajima, Y.; Mitsunobu, O. *TL* **1976**, 2455.
24. Alper, H.; Mahatantila, C. P.; Einstein, F. W. B.; Willis, A. C. *JACS* **1984**, *106*, 2708.
25. Farmer, H. H.; Rabjohn, N. *OSC* **1963**, *4*, 441.
26. Porter, H. D.; Weissberger, A. *OSC* **1955**, *3*, 708.
27. Sherman, W. R.; Taylor, Jr., E. C. *OSC* **1963**, *4*, 247.
28. Ulbricht, T. L. V.; Okuda, T.; Price, C. C. *OSC* **1963**, *4*, 566.
29. VanAllan, J. A. *OSC* **1963**, *4*, 245.
30. Shundo, R.; Nishiguchi, I.; Matsubara, Y.; Hirashima, T. *CL* **1990**, 2285.
31. Shundo, R.; Nishiguchi, I.; Matsubara, Y.; Hirashima, T. *T* **1991**, *47*, 831.
32. Brillon, D.; Sauve, G. *JOC* **1992**, *57*, 1838.
33. Prout, F. S.; Aguilar, V. N.; Girard, F. H.; Lee, D. D.; Shoffner, J. P. *OSC* **1973**, *5*, 572.

Fillmore Freeman
University of California, Irvine, CA, USA

Ethyl Diazoacetate[1]

[623-73-4] $C_4H_6N_2O_2$ (MW 114.10)

(for synthesis of cyclopropane- and cyclopropenecarboxylic acid derivatives,[1–3] β-keto esters,[4] and 2-substituted 2-diazoacetic esters,[5,6] for homologation of ketones[7–9])

Alternate Name: EDA.

Physical Data: mp −22 °C; bp 42 °C/5 mmHg 73 °C/80 mmHg, 141 °C/720 mmHg; $d_4^{17.6}$ 1.0852, d_4^{24} 1.083 g cm^{-3}; $n_D^{17.6}$ 1.4588, n_D^{25} 1.4616; volatile with steam, ether and benzene vapors.

Solubility: sol alcohol, acetone, benzene, ether, ligroin; slightly sol water.

Form Supplied in: yellow oil; commercially available from many suppliers; typical impurities include ethyl chloroacetate and solvents (CH_2Cl_2, ether).

Preparative Method: by the reaction of ***Sodium Nitrite*** with glycine ethyl ester hydrochloride.[5,6]

Handling, Storage, and Precautions: explodes when heated, or on contact with concentrated H_2SO_4 or HCl; decomposes under irradiation. The reagent has to be stored in a dark cold place. It is toxic and reactions involving the formation or reaction of this substance should be performed in well-ventilated fume hood and behind a safety shield. The chemist should wear a suitable face shield.

Cyclopropanation of Unsaturated Compounds. The most useful method of preparation of many cyclopropanecarboxylic acids is catalytic decomposition of ethyl diazoacetate in the presence of alkenes (eq 1).[1–3,10–12]

(1)

Addition of the transient ethoxycarbonyl carbenoid formed under catalytic decomposition of diazoacetate is *cis* stereospecific (eq 2).[1,13] The less sterically hindered cyclopropanecarboxylate usually predominates (eq 3).[1–3,11]

(2)

(3)

Decomposition of EDA with heterogeneous copper catalysts requires relatively high temperatures and gives moderate yields of cyclopropanecarboxylates.[1,2,14] Utilization of homogeneous catalysts such as ***Copper(I) Acetylacetonate*** increases the yields of cyclopropanes (eq 4). Rhodium(II) carboxylates (e.g. ***Dirhodium(II) Tetraacetate***) have been found to be even more effective, allowing the reaction to be carried out at room temperature or at even lower temperatures, with low concentrations of catalyst (0.5 mol %).[3,11] Rh^{II} complexes with chiral ligands have been used in stereocontrolled synthesis of chrysantemic and related acids.[2,15]

$$\text{(eq 4): } N_2CHCO_2Et,\ RhL^*_2,\ 64\%;\ \text{products } 38:62,\ 62\%\ ee,\ 68\%\ ee$$

A wide variety of unsaturated compounds react with EDA in the presence of catalysts to give cyclopropanecarboxylates, for example polynuclear aromatic or heteroaromatic compounds,[1,2] enol ethers,[12] and ketene dialkyl acetals.[16] Catalytic cyclopropanation with EDA has also been used in the synthesis of bicyclobutanes,[1] triangulanes,[17] and other polycyclic systems.[10,12]

Cyclopropene-3-carboxylate esters usually are prepared by rhodium-catalyzed decomposition of EDA and its analogs in the presence of alkynes. Target compounds can be obtained in a high yield even at room temperature from both terminal and disubstituted alkynes (eq 5).[1,18]

$$\text{(eq 5): } Rh_2(OAc)_4,\ 30\text{–}78\%;\quad R = Me, Ar, TMS, AcOCH_2$$

Copper catalysis requires a temperature of 90–140 °C and gives satisfactory yields of cyclopropene esters only in the case of disubstituted alkynes.[18] When double and triple bonds are both present in the same molecule, reaction takes place at both of them, a preference being shown for addition to the double bond.[1,6,19] Decomposition of EDA with chiral rhodium(II) compounds allows one to achieve high enantioselectivity in cyclopropene synthesis.[20]

Homologation of Ketones. EDA is the most useful diazoalkane for homologation of cyclic and acyclic ketones without the formation of epoxides as byproducts. This reaction is catalyzed by ***Boron Trifluoride Etherate***,[7–9] ***Tin(IV) Chloride***,[21] or ***Triethyloxonium Tetrafluoroborate***.[2] The homologation is not regiospecific and usually gives a mixture of isomeric 2-keto esters, but insertion occurs preferentially at the less substituted side of the carbonyl group (eq 6).[8]

$$\text{(eq 6): } BF_3\bullet Et_2O$$

$R^1 = Me, R^2 = H$	71%	14%
$R^1 = R^2 = Me$	96%	0%

A procedure for regiospecific homologation of ketones was developed by using α-halo-substituted ketones. Treatment of the halo ketone with EDA is followed by removal of the halogen using ***Zinc*** followed by decarboxylation of the resulting β-keto ester (eq 7).[7,22]

$$\text{(eq 7): 1. } N_2CHCO_2Et,\ BF_3\cdot Et_2O;\ \text{2. Zn; 3. } H_2O;\ 78\%$$

α-Diketones can be converted to β-diketones by reaction with EDA in the presence of ***Zinc Chloride*** followed by hydrolysis (eq 8).[23]

$$\text{(eq 8): 1. } N_2CHCO_2Et,\ ZnCl_2;\ \text{2. KOH; 3. HCl};\ 38\text{–}64\%$$

Synthesis of 2-Substituted Acetic Esters. EDA reacts with trialkylboranes,[24] dialkylchloroboranes,[25] or alkyl(aryl)-dichloroboranes[26] with loss of nitrogen to give, after hydrolysis, the homologated ethyl ester in a moderate to high yield. β,γ-Unsaturated carboxylic esters have been obtained using this procedure (eq 9).[25]

$$\text{(eq 9): 1. THF, –20 °C; 2. } H_2O;\ 80\text{–}90\%$$

Homologated ethyl esters have also been prepared by insertion of ethoxycarbonylcarbene generated by catalytic decomposition of EDA in C–H,[12,27] C–O and C–S,[6] or Si–H bonds.[28] Ethyl 2-alkoxyacetates have been obtained in good yield by Rh^{II}-catalyzed decomposition of EDA in the presence of various alcohols.[29]

Ethyl Diazolithioacetate. This reagent, which is stable only below −50 °C, can be prepared by treating EDA with ***n-Butyllithium*** in THF at −115 to −78 °C.[5] Reaction of ***Ethyl Diazolithioacetate*** with alkyl-, acyl-, and trialkylsilyl halides results in the formation of 2-substituted ethyl diazoesters (eq 10).[5,6,30]

$$\text{(eq 10): 1. BuLi, THF, –110 °C; 2. RX}$$

R = Alk, X = I	30–55%
R = RCO, X = Cl	45%
R = TMS, X = Cl	47%

This reagent also reacts with carbonyl compounds, forming 2-diazo-3-hydroxy esters (eq 11).[5,6]

$$\text{(eq 11): THF},\ 62\%$$

Transmetalation of ethyl diazolithioacetate gives access to silver, mercury, and tin derivatives of EDA.[5,6]

Cycloheptatrienes. Thermal or photochemical decomposition of EDA in benzene and its derivatives leads to the formation of cycloheptatriene esters in moderate yields.[1,2] ***Dirhodium(II)***

Tetrakis(trifluoroacetate) has been recently reported to be a very effective catalyst for this reaction and increased the yields of cycloheptatrienes (eq 12).[31]

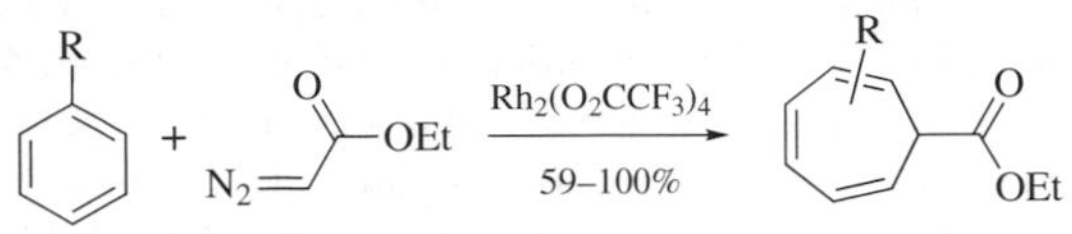

(12)

β-Keto Esters from Aldehydes. In addition to methods of β-keto ester synthesis described above, these compounds can be prepared by the addition of EDA to aldehydes in the presence of ***Tin(II) Chloride*** (eq 13).[4]

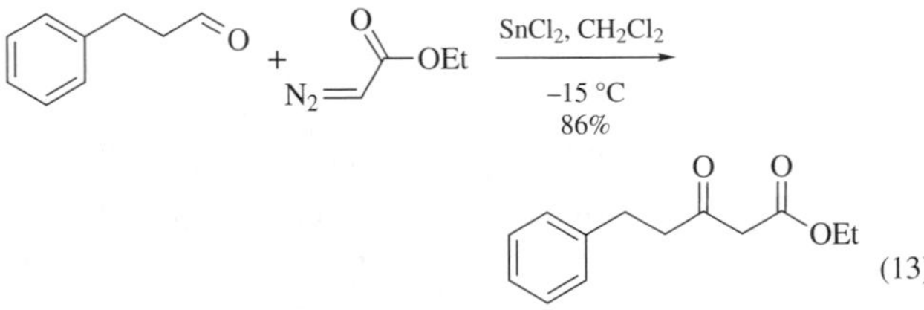

(13)

Ylide Rearrangements. Catalytic decomposition of EDA in the presence of sulfides, tertiary amines, or iodides leads to formation of intermediate ylides which undergo Stevens rearrangement (eq 14).[2,3,6] Ylides bearing an allylic double bond give the products of a [2,3]-sigmatropic rearrangement (eq 15).[3,32]

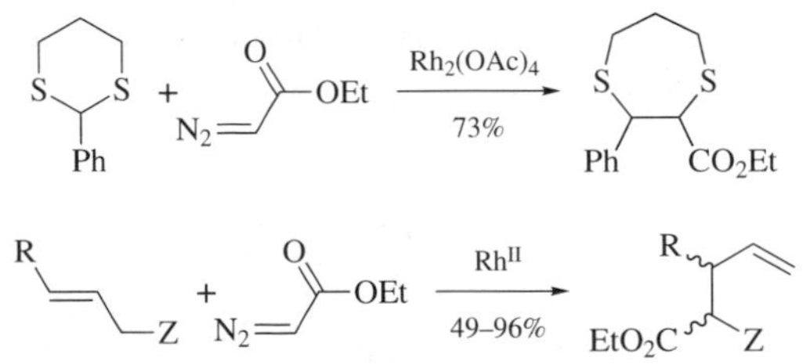

(14)

(15)

Dirhodium(II) Tetraacetate has been shown to be the most useful catalyst for this transformation.[3]

Related Reagents. Diazoacetaldehyde; Diazoacetone; α-Diazoacetophenone; Diazomethane; Dimethyl Diazomalonate; Dirhodium(II) Tetraacetate; Dirhodium(II) Tetrakis(trifluoroacetate); Ethyl Diazolithioacetate; Trimethylsilyldiazomethane.

1. Dave, V.; Warnhoff, E. V. *OR* **1970**, *18*, 218.
2. Wulfman, D. S.; Linstrumelle, G.; Cooper, C. F. In *The Chemistry of Diazonium and Diazo Groups*; Patai, S., Ed.; Wiley: New York, 1978; Part 2, p 823.
3. Doyle, M. P. *ACR* **1986**, *19*, 348.
4. Holmquist, C. R.; Roskamp, E. J. *JOC* **1989**, *54*, 3258.
5. Regitz, M.; Maas, G. *Diazo Compounds*; Academic: New York, 1986.
6. Regitz, M.; Korobizina, I. K.; Rodina, L. L. In *Methodicum Chimicum*; Korte, F., Ed.; Thieme: Stuttgart, 1975; Vol. 6, p 205.
7. Fieser, M. *FF* **1986**, *12*, 223.
8. Liu, H. J.; Majumdar, S. P. *SC* **1975**, *5*, 125.
9. Krow, G. R. *T* **1987**, *43*, 3.
10. Mandel'shtam, T. V. *RCR* **1989**, *58*, 1250.
11. Doyle, M. P.; van Leusen, D.; Tamblyn, W. H. *S* **1981**, 787.
12. Kulinkovich, O. G. *RCR* **1989**, *58*, 1233.
13. D'yakonov, I. A.; Komendantov, M. I.; Korichev, G. L. *JGU* **1962**, *32*, 917.
14. Doyle, M. P. In *Catalysis of Organic Reactions*; Augustine, R. L., Ed.; Dekker: New York, 1985; Chapter 4.
15. (a) Aratani, R.; Yoneyoshi, Y.; Nagase, T. *T* **1982**, *38*, 685. (b) Laidler, D. A.; Milner, D. J. *JOM* **1984**, *270*, 121. (c) Demonceau, A.; Noels, A. F.; Anciaux, A. J.; Hubert, A. J.; Teyssie, P. *BSB* **1984**, *93*, 949.
16. Graziano, M. L.; Iesce, M. R. *S* **1985**, 762.
17. (a) De Meijere, A.; Kozhushkov, S. I.; Spaeth, T.; Zefirov, N. S. *JOC* **1993**, *58*, 502. (b) Lukin, K. A.; Kozhushkov, S. I.; Andrievskii, A. A.; Ugrak, B. I.; Zefirov, N. S. *Mendeleev Commun.* **1992**, 51.
18. Protopopova, M. N.; Shapiro, E. A. *RCR* **1989**, *58*, 1145.
19. Danilkina, L. P.; D'yakonov, I. A., *JOU* **1966**, *2*, 1.
20. Protopopova, M. N.; Doyle, M. P.; Mueller, P.; Ene, D. *JACS* **1992**, *114*, 2755.
21. Dodd, J. H.; Schwender, C. F.; Gray-Nunez, Y. *JHC* **1990**, *27*, 1453.
22. Dave, V.; Warnhoff, E. W. *JOC* **1983**, *48*, 2590.
23. Korobitsyna, I. K.; Studzinskiii, O. P. *JOU* **1969**, *5*, 1246.
24. Hooz, J.; Linke, S. *JACS* **1968**, *90*, 6891.
25. (a) Fieser, M.; Fieser, L. F. *FF* **1974**, *4*, 229. (b) Brown, H. C.; Salunkhe, A. M.; *SL* **1991**, 684.
26. Fieser, M.; Fieser, L. F. *FF* **1975**, *5*, 295.
27. Demonceau, A.; Noels, A. F.; Anciaux, A. J.; Hubert, A. J.; Teyssie, P. *BSB* **1984**, *93*, 945.
28. Baghery, V.; Doyle, M. P.; Taunton, J.; Claxton, E. E. *JOC* **1988**, *53*, 6158.
29. Noels, A. F.; Demonceau, A.; Petiniot, N.; Hubert, A. J.; Teyssie, P. *T* **1982**, *38*, 2733.
30. Hoffman, K-L.; Maas, G.; Regitz, M. *JOC* **1987**, *52*, 3851.
31. Anciaux, A. J.; Demonceau, A.; Noels, A. F.; Hubert, A. J.; Warin, R.; Teyssie, P. *JOC* **1981**, *46*, 873.
32. (a) Doyle, M. P.; Tamblyn, W. H.; Bagheri, V. *JOC* **1981**, *46*, 5094. (b) Yoshimoto, M.; Ishihara, S.; Nakayama, E.; Soma, N. *TL* **1972**, 2923.

Vladimir V. Popik
St. Petersburg State University, Russia

Ethylene Oxide

[75-21-8] C_2H_4O (MW 44.05)

(electrophile; 2-hydroxyethyl equivalent)

Alternate Name: oxirane.
Physical Data: bp 13.5 °C/746 mmHg; *d* 0.8824 g cm^{-3}.
Solubility: sol water and most organic solvents.
Form Supplied in: liquid, in 100-mL sealed tubes; gas, in 100-mL cylinders.
Handling, Storage, and Precautions: extremely flammable; vapors can detonate; possible carcinogen; use in a fume hood.

Nucleophilic Opening. Ethylene oxide and epoxides in general have been the subject of extensive reviews.[1] Much of the chemistry involved with ethylene oxide revolves around the opening of the strained ring to form a 2-ethanol substituted moiety. The reagents employed have ranged from carbon nucleophiles, to alcohols and water, to amines.

Reaction with Organolithium Reagents. The carbon nucleophiles can be classified by the type of organometallic reagent involved in the opening reaction of the epoxide. The reagent of choice for this transformation is an organolithium reagent.[2] Treatment of an ester lithium enolate with an excess of ethylene oxide affords a spirolactone (eq 1).[2a]

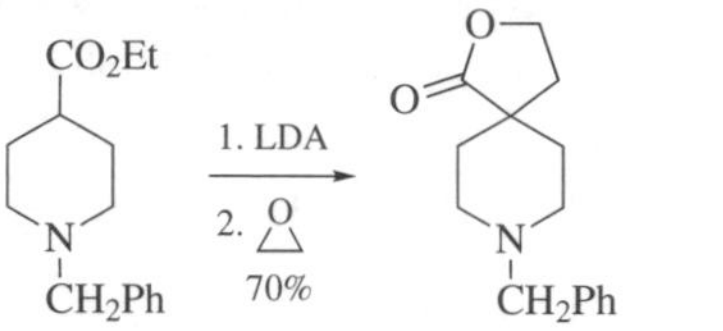

(1)

A lithio acetylide reacts with ethylene oxide in the presence of ***Boron Trifluoride Etherate*** (eq 2).[2c] While $BF_3{\cdot}Et_2O$ has been used to increase the yields of epoxide openings with organolithium reagents,[3] the reaction solvent also plays an important role in obtaining high yields of ring-opened material with a lithio acetylide (eq 3).[2b] While ***t-Butyllithium*** readily metalates some epoxides, it only opens ethylene oxide.[4]

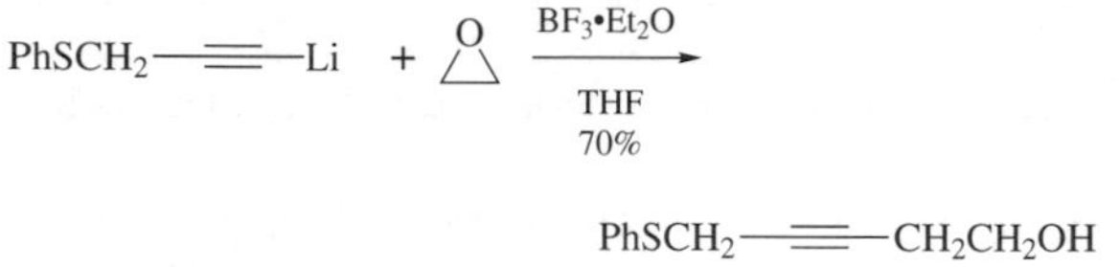

$PhSCH_2C{\equiv}CLi$ + ethylene oxide → ($BF_3{\cdot}Et_2O$, THF, 70%) → $PhSCH_2C{\equiv}CCH_2CH_2OH$ (2)

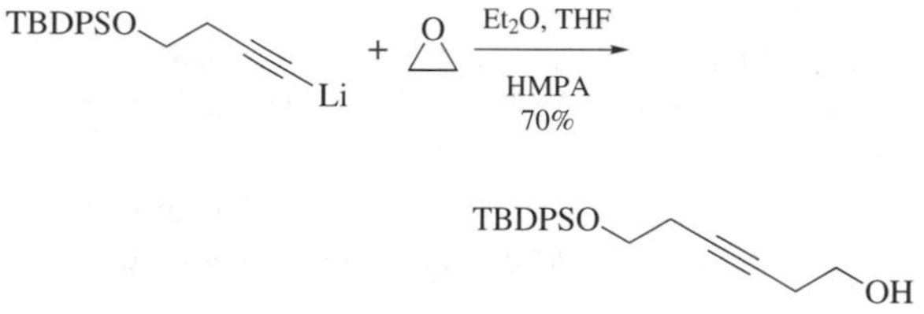

TBDPSO–alkynyllithium + ethylene oxide → (Et_2O, THF, HMPA, 70%) → TBDPSO–alkynyl–OH (3)

Reaction with Allylpotassium Reagents. Isoprene has been metalated with ***Potassium Diisopropylamide*** (KDA). Employing this nonnucleophilic, strong base lessened polymerization and permitted the reaction of the allylpotassium reagent with ethylene oxide to form the homologated alcohol (eq 4).[5]

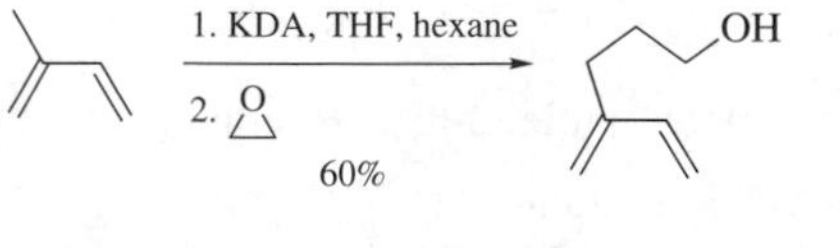

(1. KDA, THF, hexane; 2. ethylene oxide; 60%) (4)

Reaction with Grignard Reagents. Grignard reagents add with equal facility as organolithium reagents to ethylene oxide (eq 5).[6]

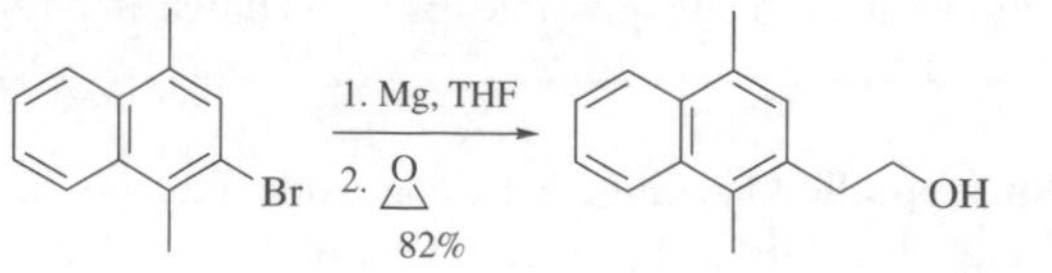

(1. Mg, THF; 2. ethylene oxide; 82%) (5)

Reaction with Lithium Dialkylcuprates. Lithium dialkylcuprates have long been known to react selectively and in high yield with oxiranes.[7] A number of more complex, modified cuprate reagents recently have been developed.[8] A vinylstannane was treated with the higher-order cuprate derived from ***Methyllithium*** and ***Copper(I) Cyanide*** and subsequent reaction with ethylene oxide afforded the homologated alcohol in 48% yield. The vinylstannane was also metalated with ***n-Butyllithium*** and reaction with 2-thienyl(cyano)copperlithium afforded the mixed higher-order vinyl cuprate that reacted with ethylene oxide in 38% yield. However, the authors found the easiest method for the formation of the homologated alcohol was to produce the mixed cuprate and treat it with ethylene oxide (eq 6).[8a]

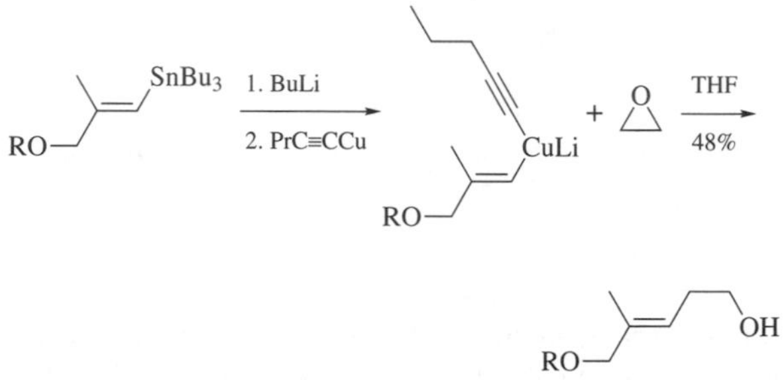

RO–vinyl–$SnBu_3$ → (1. BuLi; 2. $PrC{\equiv}CCu$) → mixed cuprate (CuLi) + ethylene oxide → (THF, 48%) → RO–vinyl–OH (6)

Reaction with Vinylaluminum Reagents. Vinylaluminum reagents react with ethylene oxide, but lithium alanates react in even higher yield (eq 7).[9]

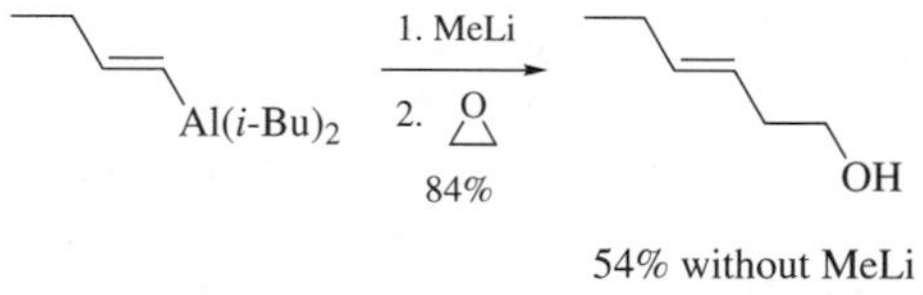

vinyl–$Al(i\text{-}Bu)_2$ → (1. MeLi; 2. ethylene oxide; 84%) → homoallylic alcohol (OH) (7)

54% without MeLi

Reaction of Ylides. A new β-oxidobenzyl ylide was formed by reaction of lithium diphenylphosphide with ethylene oxide followed by alkylation with ***Benzyl Bromide***. The reaction of these ylides with aldehydes afforded (*E*)-alkenes with good stereoselectivity (eq 8).[10] Another ylide formed from ethylene oxide undergoes a Wittig reaction with an aldehyde derived from (*R*)-limonene. The resulting alkene retains the ethanol moiety derived from ethylene oxide (eq 9).[11]

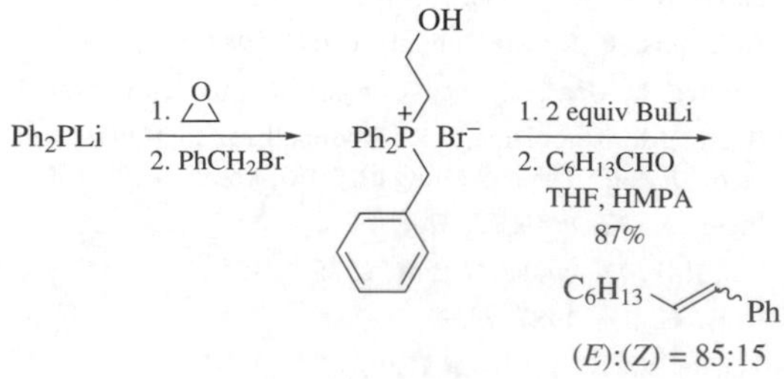

Ph_2PLi → (1. ethylene oxide; 2. $PhCH_2Br$) → $Ph_2P^+(CH_2CH_2OH)(CH_2Ph)\ Br^-$ → (1. 2 equiv BuLi; 2. $C_6H_{13}CHO$, THF, HMPA, 87%) → $C_6H_{13}CH{=}CHPh$ (8)

(*E*):(*Z*) = 85:15

$Ph_3\overset{+}{P}$ Br^- —1. $NaN(TMS)_2$ 2. ethylene oxide→ $Ph_3\overset{+}{P}$ ... O^- —1. $NaN(TMS)_2$ 2. RCHO 75%→ (9)

(Z):(E) = 67:33

Reaction with Amines. Heteroatom-containing moieties add to ethylene oxide to form 2-substituted ethanol derivatives.[1] Amines add in this fashion to form 1,2-ethanolamines.[12] For instance, bubbling ethylene oxide through a solution of dimethylethylenediamine in methanol at 35 °C affords the corresponding alcohol (eq 10).[13]

MeNH–CH2CH2–NHMe + ethylene oxide —MeOH, 68%→ MeNH–CH2CH2–N(Me)–CH2CH2–OH (10)

Reaction with Me_3SiI. By reacting ethylene oxide with ***Iodotrimethylsilane***, a simple derivative is formed that is also useful as a two-carbon synthon in the preparation of pheromone components (eq 11).[14]

ethylene oxide —TMSI, −10 °C, hexane, 91%→ TMSO–CH2CH2–I (11)

Reaction with Aldehydes. A very facile method for the formation of acetals under essentially neutral conditions involves the reaction of ethylene oxide with aldehydes in an autoclave at 110–220 °C (eq 12).[15]

Pr(Me)CH–CHO + ethylene oxide —$Et_4N^+Br^-$, 81%→ 2-substituted 1,3-dioxolane (12)

Related Reagents. 1,3-Butadiene Monoxide; Glycidol; Isoprene Epoxide; Propylene Oxide.

1. (a) Smith, J. G. *S* **1984**, 629. (b) Larock, R. C. *Comprehensive Organic Transformations*; VCH: New York, 1989; pp 508–520. (c) Bartock, M.; Lang, K. L. In *The Chemistry of Ethers, Crown Ethers, Hydroxyl Groups and Their Sulfur Analogs, Part II*; Patai, S., Ed.; Wiley: New York, 1980; Chapter 14.
2. (a) King, F. D.; Hadley, M. S.; Joiner, K. T.; Martin, R. T.; Sanger, G. J.; Smith, D. M.; Smith, G. E.; Smith, P.; Turner, D. H.; Watts, E. A. *JMC* **1993**, *36*, 683. (b) Viala, J.; Sandri, J. *TL* **1992**, *33*, 4897. (c) Kido, F.; Abiko, T.; Kato, M. *BCJ* **1992**, *65*, 2471. (d) Girard, S.; Deslongchamps, P. *CJC* **1992**, *70*, 1265. (e) Rao, A. V. R.; Sharma, G. V. M.; Bhanu, M. N. *TL* **1992**, *33*, 3907. (f) Bartoli, G.; Bosco, M.; Cimarelli, C.; Dalpozzo, R.; Palmieri, G. *JCS(P1)* **1992**, 2095. (g) Krief, A.; Hobe, M. *SL* **1992**, 317. (h) Lattuada, L.; Licandro, E.; Maiorana, S.; Papagni, A.; Zanotti-Geroso, A. *SL* **1992**, 315. (i) Narasaka, K.; Hayashi, Y.; Shimada, S.; Yamada, J. *Isr. J. Chem.* **1991**, *31*, 261. (j) Noe, C. R.; Knollmuller, M.; Dungler, K.; Gartner, P. *M* **1991**, *122*, 185. (k) Enders, D.; Dahman, W.; Dederichs, E.; Gatzweiler, W.; Weuster, P. *S* **1990**, 1013. (l) Al-Dulayyami, J. R.; Baird, M. S. *T* **1990**, *46*, 5703. (m) Jung, M. E.; Miller, S. J. *H* **1990**, *30*, 839. (n) Lee, T. V.; Leigh, A. J.; Chapleo, C. B. *T* **1990**, *46*, 921.
3. Eis, M. J.; Wrobel, J. E.; Ganem, B. *JACS* **1984**, *106*, 3693.
4. Eisch, J. J.; Galle, J. E. *JOC* **1990**, *55*, 4835.
5. Klusener, P. A. A.; Tip, L.; Brandsma, L. *T* **1991**, *47*, 2041.
6. (a) Sharma, P. K. *SC* **1993**, *23*, 389. (b) Dreger, E. E. *OSC* **1941**, *1*, 306.
7. Johnson, C. R.; Herr, R. W.; Wieland, D. M. *JOC* **1973**, *38*, 4263.
8. (a) Booker-Milburn, K. I.; Heffernan, G. D.; Parsons, P. J. *CC* **1992**, 350. (b) Barbero, A.; Cuadrado, P.; Gonzalez, A. M.; Pulido, F. J.; Fleming, I. *JCS(P1)* **1991**, 2811. (c) Dawson, I. M.; Gregory, J. A.; Herbert, R. B.; Sammes, P. G. *JCS(P1)* **1988**, 2585. (d) Marfat, A.; McGuirk, P. R.; Helquist, P. *JOC* **1979**, *44*, 3888. (e) Lipshutz, B. H.; Koerner, M.; Parker, D. A. *TL* **1987**, *28*, 945.
9. Warwel, S.; Schmitt, G.; Ahlfaenger, B. *S* **1975**, 632.
10. Boubia, B.; Mann, A.; Bellamy, F. D.; Mioskowski, C. *AG(E)* **1990**, *29*, 1454.
11. Becker, D.; Sahali, Y. *T* **1988**, *44*, 4541.
12. Alder, R. W.; Mowlam, R. W.; Vachon, D. J.; Weisman, G. R. *CC* **1992**, 507.
13. Pastor, S. D.; Togni, A. *HCA* **1991**, *74*, 905.
14. Poleschner, H.; Heydenreich, M.; Martin, D. *S* **1991**, 1231.
15. Nerdel, F.; Buddrus, J.; Scherowsky, G.; Klamann, D.; Fligge, M. *LA* **1967**, *710*, 85.

Edward W. Thomas
The Upjohn Company, Kalamazoo, MI, USA

Ethyl 3-Hydroxybutanoate[1,2]

(R) and (S) structures: $CH_3CH(OH)CH_2CO_2Et$

(*R*) [*24915-95-5*] (*S*) [*56816-01-4*]	$C_6H_{12}O_3$	(MW 132.18)

(chiral building block; stereoselective substitution of the dianion at C-2 by electrophiles[3])

Alternate Name: ethyl 3-hydroxybutyrate.
Physical Data: bp 71–73 °C/12 mmHg;[2] $[\alpha]_D^{20}$ +41.3° (c 1, $CHCl_3$) (97% ee);[4] $[\alpha]_D^{24}$ −43.6° (c 1.2, $CHCl_3$);[1] $[\alpha]_D^{24}$ −19.05° (neat) (100% ee).[1]
Solubility: sol common organic solvents such as ether, THF, *n*-hexane, CH_2Cl_2, EtOH, and also in water.
Form Supplied in: colorless liquid.
Analysis of Reagent Purity: NMR and optical rotation ($[\alpha]_D$).
Preparative Methods: reduction of ***Ethyl Acetoacetate*** with ***Baker's Yeast***[2,4–6,] yields the (*S*)-(+) enantiomer; reduction with *Geotrichum candidum*[4,7] yields the (*R*)-(−) enantiomer. The (*R*)-(−) enantiomer was also prepared[1] by depolymerization of poly-(*R*)-3-hydroxybutyric acid (PHB). Both enantiomers can be prepared by chemical reduction using

a ruthenium catalyst complexed with (*S*)- and (*R*)-BINAP, respectively.[8,9] The yeast reduction of ethyl acetoacetate has been varied: in petroleum ether (58%, 94% ee),[10] in the presence of ethyl chloroacetate (75%, 99% ee),[10,11] immobilized by magnesium alginate and under high concentration of Mg^{2+} ion (65%, 99% ee),[13] immobilized by calcium alginate in organic–water solvent system.[14]

Chemical reductions with ***Sodium Borohydride***–L-tartaric acid[15] ((*R*)-(−), 65%, 81% ee) and with a chiral oxazaphospholidine–borane complex[16] ((*R*)-(−), 80%, 99% ee) have been published.

Purification: the enrichment of the (*S*)-(+) enantiomer to 100% enantiomeric purity by means of the 3,5-dinitrobenzoate has been described.[2,17]

Handling, Storage, and Precautions: should be stored in a refrigerator because of possible transesterification–oligomerization at room temperature.[2]

Dianion Alkylation. β-Hydroxy carboxylates can be converted to their corresponding dianions through deprotonation of the hydroxy group and the α-carbon atom.[18,19] The so-formed double charged enolates are relatively stable species and can efficiently react with electrophiles in the α-position. The dianions of (3*R*)- and (3*S*)-ethyl 3-hydroxybutanoate were prepared[3,20] by lithiation with 2 equiv of ***Lithium Diisopropylamide*** in THF at −50 to −70 °C and 1 equiv of an alkyl halide was added, often in the presence of HMPA at −75 to 0 °C. The usual yield is 70–90%, and the stereoselectivity is generally better than 95:5 in favor of the *anti* isomer (eq 1).

OH CO2Et —LDA→ [Li+ O− O− Li+ OEt] —1. RX 2. H2O→ OH CO2Et R *anti* + OH CO2Et R *syn* (1)

Alkylations with ***Iodomethane***,[3,21–24] ***Allyl Bromide***,[3,20] ***Benzyl Bromide***,[3] *n*-butyl bromide,[20] dimethylallyl bromide,[25] ***Benzyl Chloromethyl Ether***,[26,27] (*S*)-2-methylbutyl iodide,[29] and (benzene)$Mn(CO)_3{}^+PF_6^-$[29] have been reported.

The mechanistic explanation for the ≈20:1 stereoselection is the formation of a chelate as depicted in (**1**), where the approach of the electrophile is more favorable *anti* to the sterically more demanding Me (or alkyl) group.

EtO O Li O X–R H

(**1**)

This method of dianion alkylation is an important and reliable preparative way to control the vicinal stereochemistry; the method is especially valuable, because both (*R*) and (*S*) starting materials are commercially available, although quite expensive, or may be easily prepared.

Dianion Amination. The electrophilic amination has been carried out with the synthetic equivalent of $[NH_2{}^+]$, i.e. ***Di-t-butyl Azodicarboxylate*** (TBAD) (eq 2).[30,31] The initially formed 94:6 mixture of the α-hydrazino-β-hydroxy ester was transformed into the α-amino-β-hydroxy derivative,[30] and was also used in the total synthesis of an amino sugar.[32]

OH CO2Et —1. 4.2 equiv LDA 2. 2.5 equiv TBAD −25 °C, 60 min 56%→ OH CO2Et N Boc BocHN + OH CO2Et N Boc BocHN (2) 94:6

Reaction of the Dianion with Imines: Formation of β-Lactams. The dianions of (*R*)- and (*S*)-ethyl 3-hydroxybutyrate have been added to different imines, yielding in one step the β-lactam, a precursor for, e.g. thienamycin (eq 3).[33–37]

OH CO2Et (*R*) —LDA→ Li O Li O OEt —R1–N=CH–R2→ OH H H R2 1' 3 4 N O 2 1 R1 (3)

A very clean and highly stereoselective reaction has been achieved with the Zn/Li dianion and *N*-trimethylsilylphenylpropargylidene imine and the corresponding cinnamylidene imine (eq 4). Only one isomer was observed out of the four possible in yields of 85% and 78%, respectively.[38]

OH CO2R (*R*) R = Me, *i*-Pr ⟶ Et Zn O O Li OR ⟶ OH H H R1 NH O (4)

R1 = C≡C–Ph, CH=CHPh (*E*)

Other Esters and General Remarks. The dianion chemistry of 3-hydroxy esters is of a very broad utility and synthetic importance.[39] Beside the cases above, a large variety of alkylations of 3-hydroxy ester dianions has been reported, e.g. (*S*)-dialkyl 2-hydroxysuccinate (malate)[40–42] and other examples.[3,43–45] It is of major importance to realize that a second alkylation allows the enantioselective preparation of a quaternary carbon center.[3,20,46–48] This is obviously also the case when cyclic analogs, such as (1*R*,2*S*)-ethyl 2-hydroxycyclohexanecarboxylates (eq 5, 95:5 de),[49] are alkylated by means of the dianion.[49–51]

OH CO2Et (1*R*,2*S*) —1. 2 equiv LDA 2. Br 72%→ OH CO2Et (1*S*,2*S*) (5)

Related Reagents. (*R*)-2-*t*-Butyl-6-methyl-4*H*-1,3-dioxin-4-one; Di-*t*-butyl Azodicarboxylate; Ethyl Acetoacetate.

---

1. Seebach, D.; Züger, M. *HCA* **1982**, *65*, 495.
2. Seebach, D.; Suter, M. A.; Weber, R. H.; Züger, M. F. *OSC* **1984**, *7*, 215.

3. Fráter, G.; Müller, U.; Günther, W. *T* **1984**, *40*, 1269.
4. Wipf, B.; Kupfer, E.; Bertazzi, R.; Leuenberger, H. G. W. *HCA* **1983**, *66*, 485.
5. Deol, B. S.; Ridley, D. D.; Simpson, G. W. *AJC* **1976**, *29*, 2459.
6. Fráter, G. *HCA* **1979**, *62*, 2825.
7. Buisson, D.; Azerad, R.; Sanner, C.; Larcheveque, M. *Biocatalysis* **1992**, *5*, 249.
8. (a) Noyori, R.; Ohkuma, T.; Kitamura, M.; Takaya, H.; Sayo, N.; Kumobayashi, H.; Akutagawa, S. *JACS* **1987**, *109*, 5586. (b) Kitamura, M.; Tokunaga, M. Ohkuma, T.; Noyori, R. *OS* **1993**, *71*, 1.
9. Keck, G. E.; Murry, J. A. *JOC* **1991**, *56*, 6606.
10. Jayasingh, L. Y.; Smallridge, A. J.; Trewhella, M. A. *TL* **1993**, *34*, 3949.
11. Nakamura, K.; Kawai, Y.; Ohno, A. *TL* **1990**, *31*, 267.
12. Nakamura, K.; Kawai, Y.; Nakajima, N.; Ohno, A. *JOC* **1991**, *56*, 4778.
13. Nakamura, K.; Kawai, Y.; Oka, S.; Ohno, A. *TL* **1989**, *30*, 2245.
14. Naoshima, Y.; Nishiyama, T.; Munakata, Y. *CL* **1989**, 1517.
15. Yatagai, M.; Ohnuki, T. *JCS(P1)* **1990**, 1826.
16. Brunel, J.-M.; Pardigon, O.; Faure, B.; Buono, G. *CC* **1992**, 287.
17. Hungerbühler, E.; Seebach, D.; Wasmuth, D. *HCA* **1981**, *64*, 1467.
18. Herrmann, J. L.; Schlessinger, R. H. *TL* **1973**, 2429.
19. Kraus, G. A.; Taschner, M. J. *TL* **1977**, 4575.
20. Fráter, G. *HCA* **1979**, *62*, 2825.
21. Suter, M. A.; Seebach, D. *LA* **1983**, 939.
22. Mori, K.; Ebata, T. *T* **1986**, *42*, 4413; Mori, K.; Takikawa, H. *T* **1990**, *46*, 4473.
23. Brooks, D. W.; Kellog, R. P. *TL* **1982**, *23*, 4991.
24. Keck, G. E.; Kachensky, D. F.; Enholm, E. J. *JOC* **1985**, *50*, 4317.
25. Kramer, A.; Pfander, H. *HCA* **1982**, *65*, 293.
26. Hatakeyama, S.; Ochi, N.; Numata, H.; Takano, S. *CC* **1988**, 1202.
27. Ireland, R. E.; Wardle, R. B. *JOC* **1987**, *52*, 1780.
28. Caldwell, C. G.; Rupprecht, K. M.; Bondy, S. S.; Davis, A. A. *JOC* **1990**, *55*, 2355.
29. Miles, W.; Smiley, P. M.; Brinkman, H. R. *CC* **1989**, 1987.
30. Guanti, G.; Banfi, L.; Narisano, E. *T* **1988**, *44*, 5553.
31. Genet, J. P.; Juge, S.; Mallart, S. *TL* **1988**, *29*, 6765.
32. Guanti, G.; Banfi, L.; Narisano, E.; Riva, R. *TL* **1992**, *33*, 2221.
33. (a) Georg, G. I. *TL* **1984**, *25*, 3779. (b) Georg, G. I.; Gill, H. S.; Gerhardt, C. *TL* **1985**, *26*, 3903.
34. (a) Ha, D.-C.; Hart, D. J.; Yang, T.-K. *JACS* **1984**, 106, 4819. (b) Hart, D. J.; Ha, D.-C. *TL* **1985**, *26*, 5493.
35. (a) Chiba, T.; Nagatsuma, M.; Nakai, T. *CL* **1984**, 1927. (b) Chiba, T.; Nakai, T. *CL* **1985**, 651.
36. Hatanaka, M.; Nitta, H. *TL* **1987**, *28*, 69.
37. Cainelli, G.; Panunzio, M.; Basile, T.; Bongini, A.; Giacomini, D.; Martelli, G. *JCS(P1)* **1987**, 2637.
38. Oguni, N.; Ohkawa Y. *CC* **1988**, 1376.
39. Seebach, D.; Roggo, S.; Zimmermann, J. In *Stereochemistry of Organic and Bioorganic Transformations*; Bartmann, W.; Sharpless, K. B. Eds.; Verlag Chemie: Weinheim, 1987; Vol. 17, p 85–126.
40. Seebach, D.; Wasmuth, D. *HCA* **1980**, *63*, 197.
41. Grossen, P.; Herold, P.; Mohr, P.; Tammer, C. *HCA* **1984**, *67*, 1625.
42. Miller, M. J., Bajwa, J. S.; Mattingly, P. G.; Peterson, K. *JOC* **1982**, *42*, 4928.
43. Still, W. C.; Romero, A. G. *JACS* **1986**, *108*, 2105.
44. Jones, A. B.; Yamaguchi, M.; Patten, A.; Danishefsky, S. J.; Ragan, J. A.; Smith, D. B.; Schreiber, S. L. *JOC* **1989**, *54*, 17.
45. Katsuki, T.; Hanomoto, T.; Yamaguchi, M. *CL* **1989**, 117.
46. Fráter, G. *TL* **1981**, *22*, 425.
47. Uno, T.; Watanabe H.; Mori, K. *T* **1990**, *46*, 5563.
48. Wasmuth, D.; Arigoni, D.; Seebach, D. *HCA* **1982**, *65*, 344.
49. Fráter, G. *HCA* **1980**, *63*, 1383.
50. Fráter, G.; Günther, W.; Müller, U. *HCA* **1989**, *72*, 1846.
51. Kitahara, T.; Touhara, K.; Watanabe, H.; Mori, K. *T* **1989**, *45*, 6387.

Georg Fráter
Givaudan-Roure Research Ltd, Duebendorf, Switzerland

Ethyl Isocyanoacetate[1]

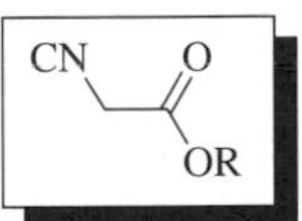

(R = Et)		
[2999-46-4]	$C_5H_8NO_2$	(MW 113.14)
(R = Me)		
[39687-95-1]	$C_4H_6NO_2$	(MW 99.11)

(heterocycle formation accompanied by cyclization onto the isocyano group;[1] forms stabilized carbanion for alkylations in amino acid synthesis;[1] α,α-additions to isocyano group and Passerini reactions[2])

Physical Data: ethyl ester, bp 89–91 °C/11 mmHg; methyl ester, bp 59–60 °C/3 mmHg.
Solubility: slightly sol H_2O; sol organic solvents.
Form Supplied in: yellow oils.
Preparative Methods: α-isocyanoacetate esters are most conveniently prepared by dehydration of the corresponding *N*-formylamino acid esters using ***Phosphorus Oxychloride*** in the presence of ***Triethylamine*** in CH_2Cl_2.[3]
Purification: by distillation.
Handling, Storage, and Precautions: isocyano compounds are generally quite sensitive to acids and immediately decompose to amine or formamide derivatives. They must be stored under N_2 at 0 to −20 °C. All operations should be performed in a well-ventilated fume hood because of the strong, disagreeable odor, and the equipment should be washed with acid after use.

Heterocycle Formation by Electrophiles. Isocyanoacetates having hydrogen at the carbon flanked with two activating groups can easily form the carbanion. α-Metalated (anionic) isocyanoacetates, which are obtained by means of the usual bases in carbanion chemistry such as NaH, *t*-BuOK, *n*-BuLi, DBU, and Et_3N, permit attack of various electrophiles. The ambivalent nature of the isocyanide carbon atom allows nucleophilic cyclization to form heterocycles after addition to polar multiple bonds such as carbonyl, imino, thioxo, nitrile, active alkene, etc. (eq 1).

$$:C{=}N{-}\overset{-}{C}H{-}CO_2R\ B^+ + Y{=}X{-}R' \longrightarrow {}^{-}Y{-}X(R'){-}CH(CO_2R){-}N{=}C:\ B^+ \longrightarrow \text{ring } (Y, X, R', N, CO_2R) \quad (1)$$

B = base

Reaction of α-anions from isocyanoacetate with acylating agents gives oxazole-4-carboxylates via cyclization of the enol form onto the isocyano group (eq 2).[4] 3-Amino-4-

hydroxycoumarins and 2(1*H*)-quinolinones are synthesized by reactions with acetyl or salicyloyl chlorides[5] and benzooxazinones[6] followed by acid hydrolysis in a straightforward reaction. The addition of an arylsulfenyl chloride to the carbenoid carbon of ethyl isocyanoacetate followed by treatment with ***Triethylamine*** affords 2-arylthio-5-ethoxyoxazoles in situ (eq 3).[7] Reaction with nitriles in the presence of ***Potassium Hydride*** in diglyme affords 5-substituted imidazole-4-carboxylates[8] and reactions with Schiff bases (imino compounds) form imidazolines.[9]

$$\mathrm{CNCH_2CO_2Me + RCOCl \xrightarrow[\text{THF}]{Et_3N}} \text{4-R-oxazole-5-}\mathrm{CO_2Me} \quad (2)$$

$$\mathrm{CNCH_2CO_2Et \xrightarrow[CH_2Cl_2]{PhSCl} Cl(PhS)C{=}NCH_2CO_2Et \xrightarrow[100\%]{Et_3N} \text{2-PhS-5-OEt-oxazole}} \quad (3)$$

Ethyl isocyanoacetate reacts with ***Carbon Disulfide*** to give thiazolethiolates in the presence of ***Potassium t-Butoxide***, which provide 5-(methylthio)thiazole-4-carboxylates with ***Iodomethane*** (eq 4).[10]

$$\mathrm{CNCH_2CO_2Et \xrightarrow[\text{t-BuOK, THF}]{CS_2} \text{thiazole(4-}EtO_2C\text{, 5-}S^-\text{)}\,K^+ \xrightarrow[77\%]{MeI} \text{thiazole(4-}EtO_2C\text{, 5-SMe)}} \quad (4)$$

The coupling reactions with arenediazonium ions give 1-aryl-1*H*-1,2,4-triazole-3-carboxylates via cyclization of hydrazone compounds in the presence of sodium acetate in aq MeOH.[11] Condensation with heteroallenes such as isocyanates and isothiocyanates generates oxazole and thiazole skeletons, but the reactions give complicated products under various reaction conditions.[12] Michael addition proceeds easily with C–C bond formation, while α,β-unsaturated carbonyl compounds on heating to 70–80 °C provide pyrrolines accompanying further cyclization by insertion of the anion into the isocyano group.[13] When a nitroalkene is the Michael acceptor, 3,4-disubstituted pyrrole-2-carboxylates are obtained by cyclization and elimination of the nitro group (eq 5).[14]

$$\mathrm{R^1CH{=}C(R^2)NO_2 \xrightarrow[\text{DBU, THF}]{CNCH_2CO_2Et} \text{3-}R^2\text{-4-}R^1\text{-pyrrole-2-}CO_2Et\ (NH)} \quad (5)$$

Reactions with Aldehydes and Ketones. Reactions with carbonyl compounds as electrophiles afford many valuable products. For example, formylaminoacrylates are obtained in the presence of *t*-BuOK or NaH via isomerization of the oxazolinyl anion by proton shift from C-4 to C-2 followed by elimination (eq 6).[15] In the presence of NaCN in EtOH, *trans*-2-oxazoline-4-carboxylates are produced by protonation of the oxazolinyl anion.[16] The use of metal catalysts such as NiCl and Cu_2O also gives the oxazoline products. Especially, in the presence of a gold(I) complex (**1**) coordinated to a chiral ferrocenylphosphine ligand (**2**), oxazoline-4-carboxylates are formed with high enantio- and diastereoselectivity (eq 7).[17]

$$\mathrm{CNCH_2CO_2R \xrightarrow[\text{t-BuOK, THF}]{R^1R^2CO} \left[\text{oxazolinyl anion } M^+ \rightarrow \text{oxazolinyl anion } M^+\right] \rightarrow R^1R^2C{=}C(NHCHO)CO_2R} \quad (6)$$

$$\mathrm{MeCHO + CNCH_2CO_2Me \xrightarrow[CH_2Cl_2,\ 100\%]{(1),\ (2)} \text{trans-oxazoline (72\% ee)} + \text{cis-oxazoline (44\% ee)}} \quad (7)$$

trans:*cis* = 84:16

$[Au(CyNC)_2]^+ BF_4^-$ (**1**)

Ferrocenyl-PPh_2 ligand with $CH_2CH_2NEt_2$ side chain (**2**)

Formamidine derivatives are also formed using pyrrolidine or piperidine as bases in MeOH by insertion of the bases into the isocyano group of α-isocyanoacrylate, which is proposed as an intermediate (eq 8).[18] The reaction with aldehydes in the presence of ***1,8-Diazabicyclo[5.4.0]undec-7-ene*** in THF affords pyrrole-2,4-dicarboxylates via Michael addition by another equivalent of methyl isocyanoacetate to the proposed intermediate, i.e. methyl isocyanoacrylates (eq 8).[19] Reaction with acetaldehyde in the presence of Et_3N in benzene affords methyl α-isocyano-β-hydroxybutyrate (eq 8).[20]

$$\mathrm{RCHO + CNCH_2CO_2Me \longrightarrow [RCH{=}C(NC)CO_2Me]} \quad (8)$$

R = Me, Et_3N: $\mathrm{CH_3CH(OH)CH(NC)CO_2Me}$; pyrrolidine (HN), MeOH: $\mathrm{RCH{=}C(CO_2Me)N{=}CH\text{-}N(pyrrolidinyl)}$; DBU, THF, $CNCH_2CO_2Me$: 3-R-pyrrole-2,4-dicarboxylate (MeO_2C, CO_2Me, NH)

α-Activating Group for Amino Acids Synthesis. Carbon–carbon bond formation at the α-carbon of α-isocyanoacetates by alkylation with carbon electrophiles provides a useful preparative method for the synthesis of various amino acid derivatives, since the isocyano group is easily converted to the amino group by acidification. Hence, isocyanoacetates are regarded as synthetic equivalents for α-activated amino acids (eq 9).[1]

R^1; H_2N CO_2H $\xrightarrow{+E^+}$ E; H_2N—CO_2H; R^1 (9)

R^1; CN CO_2R^2 $\xrightarrow{base}$ [B^+; CN CO_2R^2; R^1] $\xrightarrow{+E^+}$ E; CN—CO_2R^2; R^1

Reactions of α-metalated α-isocyanopropionate with alkyl halides and Michael acceptors followed by hydrolysis give α-methyldopa,[21] α-methylhistidine,[21] and α-methylglutamic acid.[22] Palladium-catalyzed allylation of ethyl α-isocyanopropionate with allyl acetate in the presence of ***Tetrakis(triphenylphosphine)palladium(0)*** affords α-allylalanine ethyl ester after acidification.[23] Furthermore, 2-oxazoline-4-carboxylates, which are accessed readily by reaction with aldehydes and ketones as described in the above section, are easily converted to β-hydroxy-α-amino acid derivatives by acid hydrolysis.[1] In the same way, oxazole-4-carboxylates obtained by reaction with acylating agents are hydrolyzed to give *C*-acylamino acid esters.[4]

Insertion into the Isocyano Group. α,α-Additions to the isocyano group are well known reactions in catalytic chemistry.[24] Furthermore, the Passerini reaction, which provides α-acyloxycarboxamide from isocyano compounds, ketones, and carboxylic acids,[25] and the Ugi reaction (four-component condensation: 4CC)[26] are unique peptide synthetic methods. As an application of the 4CC, the reaction of 3-aryloxy-4,5-dihydroxypyrrole, which is an intramolecular condensation product of the aldehyde and amine components, with ethyl isocyanoacetate and benzoic acid affords *N*-benzoyl-β-aryloxyprolylglycinate (eq 10).[27]

OAr; N + $PhCO_2H$ $\xrightarrow[\text{DBU, MeOH}]{CNCH_2CO_2Et}$ OAr; N; Ph; O; H; N; O; CO_2Et (10)

Ar = *p*-C_6H_4CN *cis*:*trans* = 63:37

Related Reagents. (2*S*,4*S*)-3-Benzoyl-2-*t*-butyl-4-methyl-1,3-oxazolidin-5-one; *t*-Butyl Isocyanide; Diethyl Isocyanomethylphosphonate; Ethyl *N*-(Diphenylmethylene)glycinate; Phenyl Isocyanide; 1,1,3,3-Tetramethylbutyl Isocyanide; *p*-Tolylsulfonylmethyl Isocyanide; *p*-Tolylthiomethyl Isocyanide.

1. (a) Hoppe, D. *AG(E)* **1974**, *13*, 789. (b) Schöllkopf, U. *AG(E)* **1977**, *16*, 339. (c) Matsumoto, K.; Moriya, T.; Suzuki, M. *J. Synth. Org. Chem. Jpn.* **1985**, *43*, 764.
2. Ugi, I. *Isonitrile Chemistry*; Academic: New York, 1971.
3. Haytman, G. D.; Weinstock, L. M. *OSC* **1988**, *6*, 620.
4. Suzuki, M.; Iwasaki, T.; Miyoshi, M.; Okumura, K.; Matsumoto, K. *JOC* **1973**, *38*, 3571.
5. Matsumoto, K.; Suzuki, M.; Miyoshi, M.; Okumura, K. *S* **1974**, 500.
6. (a) Matsumoto, K.; Suzuki, M.; Yoneda, N.; Miyoshi, M. *S* **1976**, 805. (b) Suzuki, M.; Matsumoto, K.; Miyoshi, M.; Yoneda, N.; Ishida, R. *CPB* **1977**, *25*, 2602.
7. Bossio, R.; Marcaccini, S.; Pepino, R. *H* **1986**, *24*, 2003.
8. Murakami, T.; Otsuka, M.; Ohno, M. *TL* **1982**, *23*, 4729.
9. Meyer, R.; Schöllkopf, U.; Böhme, P. *LA* **1977**, 1183.
10. (a) Schöllkopf, U.; Porsch, P. H.; Blume, E. *LA* **1976**, 2122. (b) Fleischmann, K.; Scheunemann, K. H.; Schorlemmer, H. U.; Dickneite, G.; Blumbach, J.; Fischer, G.; Dürckheimer, W.; Sedlacek, H. H. *AF* **1989**, *39*, 743.
11. Matsumoto, K.; Suzuki, M.; Tomie, M.; Yoneda, N.; Miyoshi, M. *S* **1975**, 609.
12. (a) Schröder, R.; Schöllkopf, U.; Blume, E.; Hoppe, I. *LA* **1975**, 533. (b) Suzuki, M.; Moriya, T.; Matsumoto, K.; Miyoshi, M. *S* **1982**, 874. (c) Solomon, D. M.; Rizvi, R. K.; Kaminski, J. J. *H* **1987**, *26*, 651.
13. (a) Schöllkopf, U.; Hantke, K. *LA* **1973**, 1571. (b) Saegusa, T.; Ito, Y.; Kinoshita, H.; Tomita, S. *JOC* **1971**, *36*, 3316.
14. (a) Barton, D. H. R.; Zard, S. Z. *CC* **1985**, 1098. (b) Ono, N.; Kawamura, H.; Bougauchi, M.; Maruyama, K. *T* **1990**, *46*, 7483.
15. Schöllkopf, U.; Gerhart, F.; Schroder, R.; Hoppe, D. *LA* **1972**, *766*, 116.
16. Hoppe, D.; Schöllkopf, U. *LA* **1972**, *763*, 1.
17. (a) Ito, Y.; Sawamura, M.; Hayashi, T. *JACS* **1986**, *108*, 6405. (b) Ito, Y.; Sawahara, M.; Shirakawa, E.; Hayashizaki, K.; Hayashi, T. *T* **1988**, *44*, 5253.
18. Suzuki, M.; Nunami, K.; Moriya, T.; Matsumoto, K.; Yoneda, N. *JOC* **1978**, *43*, 4933.
19. Suzuki, M.; Miyoshi, M.; Matsumoto, K. *JOC* **1974**, *39*, 1980.
20. Matsumoto, K.; Ozaki, Y.; Suzuki, M.; Miyoshi, M. *ABC* **1976**, *40*, 2045.
21. Suzuki, M.; Miyahara, T.; Yoshioka, R.; Miyoshi, M.; Matsumoto, K. *ABC* **1974**, *38*, 1709.
22. Schöllkopf, U.; Hantke, K. *LA* **1973**, 1571.
23. Ito, Y.; Sawamura, M.; Matsuoka, M.; Matsumoto, Y.; Hayashi, T. *TL* **1987**, 28, 4849.
24. Saegusa, T.; Ito, Y. *S* **1975**, 291.
25. Passerini, M. *G* **1971**, *51*, 126.
26. Ugi, I. *AG(E)* **1982**, *21*, 810.
27. Bowers, M. M.; Carroll, P.; Joullie, M. M. *JCS(P1)* **1989**, 857.

Kazuo Matsumoto & Mamoru Suzuki
Tanabe Seiyaku Co., Osaka, Japan

(*S*)-Ethyl Lactate

OH; CO_2R

(*S*)-(R = Et)		
[687-47-8]	$C_5H_{10}O_3$	(MW 118.15)
(*R*)-(R = Et)		
[97-64-3]	$C_5H_{10}O_3$	(MW 118.15)
(*S*)-(R = Me)		
[17392-83-5]	$C_4H_8O_3$	(MW 104.12)
(*R*)-(R = Me)		
[27871-49-4]	$C_4H_8O_3$	(MW 104.12)
(*R*)-(R = Bu)		
[34451-18-8]	$C_7H_{14}O_3$	(MW 146.21)

(chiral pool reagent for synthesis; occasionally used as a chiral auxiliary)

Alternate Name: ethyl L-(−)-lactate.

Physical Data: (*S*)-(R = Et) bp 154 °C, 69 °C/36 mmHg, *d* 1.031 g cm^{-3};[20] (*R*)-(R = Et) bp 58 °C/20 mmHg, *d* 1.032 g cm^{-3};[20] (*S*)-(R = Me) bp 40 °C/11 mmHg, *d* 1.086 g cm^{-3};[25] (*R*)-(R = Me) bp 58 °C/19 mmHg, *d* 1.091 g cm^{-3};[20] (*R*)-(R = Bu) bp 77 °C/10 mmHg, *d* 0.974 g cm^{-3}.[27]

Solubility: sol water, alcohols, ethers, THF, and common organic solvents.

Form Supplied in: liquid; commercially available.

Use as a Chiral Pool Reagent. (*S*)-Ethyl lactate has been extensively used as a chiral pool reagent, often via transformation into a diverse array of simple, enantiomerically pure analogs. Principal among these are a variety of *O*-protected (*S*)-2-hydroxypropanals (eq 1).

OH, CO_2Et ⟶ OP, CHO (1)

These have been prepared by various combinations of straightforward steps including ester to amide conversion, alcohol protection, direct reduction of the ester or amide to the aldehyde group, and reduction of the ester to the alcohol followed by reoxidation to the aldehyde. The sensitivity of the (*S*)-propanals to epimerization has been of paramount concern. One of the best procedures which avoids racemization and has been run on a preparative scale is noted (eq 2).[1]

OH, CO_2Et —Me_2NH, 70 °C, 93%→ OH, $CONMe_2$ —NaH, BnCl, DMF, 88%→
OBn, $CONMe_2$ —$Na(MeOCH_2CH_2O)_2AlH_2$, THF, toluene, 90%→ OBn, CHO —$NaBH_4$, EtOH, 0 °C, 89%→ OBn, OH (2)

Subsequent reduction also affords (*S*)-2-benzyloxypropanol in 89% yield. NMR assay of the (*R*)- and (*S*)-Mosher esters indicated no racemization over the sequence. Synthesis via oxidation of (*S*)-benzyloxypropanol (eq 3) provides the benzyloxypropanal with <8% racemization.[2]

OH, CO_2Et —BnBr, Ag_2O, Et_2O→ OBn, CO_2Et —$LiAlH_4$, Et_2O→
OBn, OH (85% overall, 100% ee) —DMSO, $(COCl)_2$, Et_3N, 83%→ OBn, CHO (3)

Attempted formation of the benzyl ether of (*S*)-ethyl lactate with NaH/BnBr results in considerable racemization (50–75% ee). This racemization is obviated by use of the amide analog noted in eq 2. ***Diisobutylaluminum Hydride*** has been used to convert the ester directly to the aldehyde employing the methoxymethyl,[3,4] benzyl,[2,5] 2,6-dichlorobenzyl,[6] *t*-butyldiphenylsilyl,[7] benzyloxymethyl,[8] THP,[9] trityl[10] and TBDMS[11] protecting groups. Protected (*S*)-2-hydroxypropanals have been used in synthetic studies relating to sugars,[12–15] amino sugars,[16] thiotetronic acids,[17] antimycin-A_3,[18] rhodinose,[19] aplysiatoxin via the (*R*)-lactate,[20] (−)-sarracenin,[21] and for preparation of enantiomerically pure 1-methyl-2-alkenyl-*N,N*-diisopropylcarbamates from the (*R*)- and (*S*)-lactates.[22]

(*S*)-Ethyl lactate has also been used as a ready source of (*S*)-propane-1,2-diol and (*S*)-methyloxirane (eq 4).[23,24] These compounds have been used for preparation of numerous natural products including nonactin,[25] sulcatol,[26] recifeiolide,[27] methyl-1,6-dioxaspiro[4,5]decanes, the pheromone components of *Paravespula vulgaris*,[28] and the rhynchosporosides.[29] The (*S*)-oxirane has also been used in the synthesis of chiral macrocyclic poly(ether diester)ligands.[30] A convenient procedure for preparation of the (*R*)-methyloxirane via mesylate activation, reduction, and internal inversion has been reported.[31]

OH, CO_2Et —LAH→ OH, OH —HBr, HOAc→ OAc, Br —$C_5H_{11}OK$, $C_5H_{11}OH$→ O (4)

A variety of inverted analogs of (*S*)-ethyl lactate have been prepared by standard activation displacement procedures. Included are the (*R*)-propionyloxypropionate (mesylation/$EtCO_2Cs$–DMF);[32] azide (Mitsunobu conditions);[33,34] aryloxy ethers (mesylation/aryl oxide);[35] chloride ($SOCl_2$–DMF);[36] bromide (sulfonation/$MgBr_2$);[37] mercapto analogs (Mitsonobu conditions);[38,39] amino analogs (triflate/amine);[40] hydroxylamines;[41,42] and selenides.[43]

Protected (*S*)-ethyl lactate cleanly acylates methyllithium to afford the 2-butanone with essentially complete enantiomeric fidelity and in nearly quantitative yield. Various diastereoselective constructions were achieved by nucleophilic addition to the ketone (eq 5).[44] For example, addition of vinyllithiums, followed by acetal formation and Lewis acid-mediated rearrangement, provided a ready entry into the indicated 3-acyltetrahydrofurans.

OTBDPS, CO_2Et —MeLi, THF, −100 °C→ OTBDPS, O ⟶ O, R^1, O, R^2 (5)

(*S*)-Ethyl lactate has been used to prepare (*S*)-2-methyloxetane in modest yield with <0.5% racemization by a series of standard transformations (eq 6).[45]

OH, CO_2Et —1. OEt 2. $LiAlH_4$ 3. TsCl 4. NaCN 5. EtOH, HCl→ OH, CO_2Et —1. OEt 2. $LiAlH_4$ 3. TsCl, HCl 4. NaCN, KOH→ O (6)

Other small chiral molecules have also been prepared by straightforward transformations (eq 7).[46]

(7)

(S)-Ethyl lactate has also been used as a chiral fragment for numerous other studies. Included are synthetic efforts relating to salenomycin,[47] (−)-biopterin,[48] (+)-polyoxamic acid,[49] jaspamide,[50] the enantiomeric 2-pentanols,[51] pumilitoxin B,[52,53] D-ristosamine,[54] protomycinolide IV,[55] and tirandamycin.[56]

Use as a Chiral Auxiliary. (S)-Ethyl lactate has been used as a chiral auxiliary in a variety of simple Diels–Alder reactions.[57–60] As the fumaric acid diester, the de employing cyclopentadiene can almost be completely reversed by addition of ***Titanium(IV) Chloride*** (eq 8).[61] In general, superior de values are achieved using ***(R)-Pantolactone*** in this context, and also for base-mediated addition to ketenes.[62]

(8)

hexane, CCl_4	98:2
CH_2Cl_2, >1.5 equiv $TiCl_4$	5:95

(S)-Ethyl lactate was used for diastereocontrol and asymmetric transmission in a sequential 2,3-Wittig–oxy-Cope rearrangement, affording product in 91% ee (eq 9).[63,64] Excellent asymmetric induction has also been noted in the Lewis acid-mediated ene reaction of (S)-ethyl lactate-derived intermediates (eq 10).[65]

(9)

(a) MeC≡CCH$_2$OC(=NH)CCl$_3$, H$^+$, 85%; DIBAL, 82%; PrPPh$_3$Br, BuLi, −78 °C, 75%
(b) BuLi, −78 °C, 75%; H$_2$, P-2 Ni, 95%
(c) KH, 18-crown-6, 25 °C, 75%

(10)

R = Et, Ph
R^1 = C$_6$H$_{13}$, aryl, styryl

86–97% ee

(S)-Ethyl lactate has been used to enantioselectively protonate the indicated enolate at −100 °C to afford the (R)-ketone in 73% yield and 73% ee (eq 11).[66]

(11)

Applications to Products of Commercial Interest. (S)-Ethyl lactate has been incorporated in chiral syntheses of (S)-2-arylpropionic acids, an important class of nonsteroidal anti-inflammatory agents, including ibuprofen and naproxen (eq 12).[67–69] These syntheses, though elegant in concept, are unlikely to compete with existing industrial methods for production of the (S) enantiomers of these drugs.

(12)

97.3% ee

(S)-Ethyl lactate has also been used to synthesize the important 4-acetoxyazetidinone intermediate, crucial to numerous carbapenem syntheses. The key step in its use was the diketene addition to the (S)-lactaldehyde imine, which in the best case proceeded in 67% yield with a 10:1 ratio of diastereomers (eq 13).[70,71]

(13)

Other applications to β-lactam syntheses have been reported.[72–74]

Related Reagents. Dimethyl L-Tartrate; Ethyl 3-Hydroxybutyrate; Ethyl Mandelate; Mandelic Acid; Methyl O-Methyllactate; (R)-Pantolactone.

1. Kobayashi, Y.; Takase, M.; Ito, Y.; Terashima, S. *BCJ* **1989**, *62*, 3038.
2. Takai, K.; Heathcock, C. H. *JOC* **1985**, *50*, 3247.
3. Wasserman, H. H.; Gambale, R. J. *T* **1992**, *48*, 7059.
4. Iida, H.; Yamazaki, N.; Kibayashi, C. *CC* **1987**, 746.
5. De Amici, M.; Dallanoce, C. de M. C.; Grana, E.; Dondi, G.; Ladinsky, H.; Schiavi, G.; Zonta, F. *Chirality* **1992**, *4*, 230.

6. Chan, T. H.; Li, C. J. *CJC* **1992**, *70*, 2726.
7. Braun, M.; Moritz, J. *SL* **1991**, 750.
8. Brown, P. A.; Bonnert, R. V.; Jenkins, P. R.; Lawrence, N. J.; Selim, M. R. *JCS(P1)* **1991**, 1893.
9. Kang, S.-K.; Lee, D.-H. *SL* **1991**, 175.
10. Mori, K.; Kikuchi, H. *LA* **1989**, 963.
11. Hirama, M.; Shigemoto, T.; Ito, S. *JOC* **1987**, *52*, 3342.
12. Hiyama, T.; Nishide, K.; Kobayashi, K. *TL* **1984**, *25*, 569.
13. Guanti, G.; Banfi, L.; Narisano, E. *G* **1987**, *117*, 681.
14. Yamamoto, Y.; Komatsu, T.; Maruyama, K. *JOM* **1985**, *285*, 31.
15. Guanti, G.; Banfi, L.; Guaragna, A.; Narisano, E. *JCS(P1)* **1988**, 2369.
16. Hiyama, T.; Kobayashi, K.; Nishide, K. *BCJ* **1987**, *60*, 2127.
17. Chambers, M. S.; Thomas, E. J.; Williams, D. J. *CC* **1987**, 1228.
18. Wasserman, H. H.; Gambale, R. J. *JACS* **1985**, *107*, 1423.
19. Kelly, T. R.; Kaul, P. N. *JOC* **1983**, *48*, 2775.
20. Ireland, R. E.; Thaisrivongs, S.; Dussault, P. H. *JACS* **1988**, *110*, 5768.
21. Baldwin, S. W.; Crimmins, M. T. *JACS* **1982**, *104*, 1132.
22. Schwark, J-R.; Hoppe, D. *S* **1990**, 291.
23. Ellis, M. K.; Golding, B. T. *OS* **1985**, *63*, 140.
24. Mori, K.; Senda, S. *T* **1985**, *41*, 541.
25. Schmidt, U.; Gombos, J.; Haslinger, E.; Zak, H. *CB* **1976**, *109*, 2628.
26. Johnston, B. D.; Slessor, K. N. *CJC* **1979**, *57*, 233.
27. Utimoto, K.; Uchida, K.; Yamaya, M.; Nozaki, H. *TL* **1977**, 3641.
28. Hintzer, K.; Weber, R.; Schurig, V. *TL* **1981**, *22*, 55.
29. Nicolaou, K. C.; Randall, J. L.; Furst, G. T. *JACS* **1985**, *107*, 5556.
30. Jones, B. A.; Bradshaw, J. S.; Izatt, R. M. *JHC* **1982**, *19*, 551.
31. Hillis, L. R.; Ronald, R. C. *JOC* **1981**, *46*, 3348.
32. Kruizinga, W. H.; Strijtveen, B.; Kellogg, R. M. *JOC* **1981**, *46*, 4321.
33. Viaud, M. C.; Rollin, P. *S* **1990**, 130.
34. Fabiano, E.; Golding, B. T.; Sadeghi, M. M. *S* **1987**, 190.
35. Burkard, U.; Effenberger, F. *CB* **1986**, *119*, 1594.
36. Biedermann, J.; Leon-Lomeli, A.; Borbe, H. O.; Prop, G. *JMC* **1986**, *29*, 1183.
37. Hanessian, S.; Kagotani, M.; Komaglou, K. *H* **1989**, *28*, 1115.
38. Rollin, P. *TL* **1986**, *27*, 4169.
39. Strijtveen, B.; Kellogg, R. M. *JOC* **1986**, *51*, 3664.
40. Effenberger, F.; Burkard, U.; Willfahrt, J. *LA* **1986**, 314.
41. Feenstra, R. W.; Stokkingreef, E. H. M.; Nivard, R. J. F. Ottenheijm, H. C. J. *T* **1988**, *44*, 5583.
42. Feenstra, R. W.; Stokkingreef, E. H. M.; Nivard, R. J. F.; Ottenheijm, H. C. J. *TL* **1987**, *28*, 1215.
43. Fitzner, J. N.; Shea, R. G.; Fankhauser, J. E.; Hopkins, P. B. *JOC* **1985**, *50*, 417.
44. Hopkins, M. H.; Overman, L. E.; Rishton, G. M. *JACS* **1991**, *113*, 5354.
45. Hintzer, K.; Koppenhoefer, B.; Schurig, V. *JOC* **1982**, *47*, 3850.
46. Berens, U.; Scharf, H. D. *S* **1991**, 832.
47. Horita, K.; Nagato, S.; Oikawa, Y.; Yonemitsu, O. *CPB* **1989**, *37*, 1705.
48. Kikuchi, H.; Mori, K. *ABC* **1989**, *53*, 2095.
49. Savage, I.; Thomas, E. *CC* **1989**, 717.
50. Chiarello, J.; Joullie, M. M. *SC* **1989**, *19*, 3379.
51. Cheskis, B.; Shpiro, N. A.; Moiseenkov, A. M. *ZOR* **1990**, *26*, 1864.
52. Overman, L. E.; McCready, R. J. *TL* **1982**, *23*, 2355.
53. Overman, L. E.; Bell, K. L.; Ito, F. *JACS* **1984**, *106*, 4192.
54. Hamada, Y.; Kawai, A.; Shiori, T. *CPB* **1985**, *33*, 5601.
55. Suzuki, K.; Tomooka, K.; Katayama, E.; Matsumoto, T.; Tsuchihashi, G. *JACS* **1986**, *108*, 5221.
56. Kelly, T. R.; Chandrakumar, N. S.; Cutting, J. D.; Goehring, R. R.; Weibel, F. R. *TL* **1985**, *26*, 2173.
57. Avenoza, A.; Cativiela, C.; Mayoral, J. A.; Peregrina, J. M.; Sinou, D. *TA* **1990**, *1*, 765.
58. Rebiere, F.; Riant, O.; Kagan, H. B. *TA* **1990**, *1*, 199.
59. Avenoza, A.; Cativiela, C.; Mayoral, J. A.; Peregrina, J. M. *TA* **1992**, *3*, 913.
60. Cativiela, C.; Mayoral, J.; Avenoza, A.; Peregrina, J. M.; Lahoz, F. J.; Gimeno, S. *JOC* **1992**, *57*, 4664.
61. Helmchen, G.; Abdel Hady, A. F.; Hartmann, H.; Karge, R.; Krotz, A.; Sartor, K.; Urmann, M. *PAC* **1989**, *61*, 409.
62. Larsen, R. D.; Corley, E. G.; Davis, P.; Reider, P. J.; Grabowski, E. J. J. *JACS* **1989**, *111*, 7650.
63. Wei, S. Y.; Tomooka, K.; Nakai, T. *JOC* **1991**, *56*, 5973.
64. Wei, S. Y.; Tomooka, K.; Nakai, T. *T* **1993**, *49*, 1025.
65. Tanino, K.; Shoda, H.; Nakamura, T.; Kuwajima, I. *TL* **1992**, *33*, 1337.
66. Matsumoto, K.; Ohta, H. *TL* **1991**, *32*, 4729.
67. Yamauchi, T.; Hattori, K.; Nakao, K.; Tamaki, K. *BCJ* **1987**, *60*, 4015.
68. Honda, Y.; Ori, A.; Tsuchihashi, G. *BCJ* **1987**, *60*, 1027.
69. Brown, J. D. *TA* **1992**, *3*, 1551.
70. Ito, Y.; Kawabata, T.; Terashima, S. *TL* **1986**, *27*, 5751.
71. Ito, Y.; Kobayashi, Y.; Kawabata, T.; Takase, M.; Terashima, S. *T* **1989**, *45*, 5767.
72. Okonogi, T.; Shibahara, S.; Murai, Y.; Inouye, S.; Kondo, S. *H* **1990**, *31*, 791.
73. Okonogi, T.; Shibahara, S.; Murai, Y.; Yoshida, T.; Inouye, S.; Kondo, S.; Christensen, B. G. *J. Antibiot.* **1990**, *43*, 357.
74. Pfaendler, H. R. In *Recent Advances in the Chemistry of β-Lactam Antibiotics;* Gregory, G. I., Ed.; Royal Society of Chemistry: London, 1981, p 368.

Edward J. J. Grabowski
Merck & Co., Rahway, NJ, USA

Ethynylmagnesium Bromide

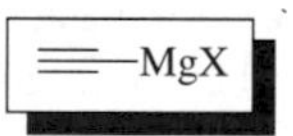

(X = Br)		
[4301-14-8]	C_2HBrMg	(MW 129.24)
(X = Cl)		
[65032-27-1]	C_2HClMg	(MW 84.79)

(addition of acetylene to ketones;[1] preparation of TMSC≡CH[2])

Preparative Methods: an *Organic Synthesis* preparation of ethynylmagnesium bromide[1] and chloride[2] from ***Acetylene*** and ***Ethylmagnesium Bromide*** or butylmagnesium chloride, respectively, is available. The use of butylmagnesium chloride instead of ethylmagnesium bromide is recommended[2] because of the greater solubility of the chloride.

Handling, Storage, and Precautions: ethynylmagnesium halides are air and moisture sensitive. They are usually used immediately after preparation.

Addition to Ketones and Aldehydes[1]. Ethynylmagnesium bromide adds to ketones and aldehydes to give ethynyl carbinols (eq 1). The reagent is less prone to disproportionation to XMgC≡CMgX than is monolithium acetylide. ***Lithium Acetylide***

provides the above product in 96% yield.[3] Ethynylmagnesium chloride also adds to aldehydes (eq 2).[4]

$$\text{PhCH=CHCHO} \xrightarrow[58\text{–}69\%]{\text{HC}\equiv\text{C–MgBr}} \text{PhCH=CHCH(OH)C}\equiv\text{CH} \quad (1)$$

$$\text{(MeO)}_2\text{CHCH=CHCHO} \xrightarrow[87\%]{\text{HC}\equiv\text{C–MgCl}} \text{(MeO)}_2\text{CHCH=CHCH(OH)C}\equiv\text{CH} \quad (2)$$

Preparation of TMS-acetylene[2]. Treatment of ethynylmagnesium chloride with ***Chlorotrimethylsilane*** provides a good route to ***Trimethylsilylacetylene*** (eq 3).

$$\text{HC}\equiv\text{C–MgCl} + \text{TMSCl} \xrightarrow[75\%]{} \text{HC}\equiv\text{C–TMS} \quad (3)$$

Related Reagents. Lithium Acetylide; Lithium Chloroacetylide; Lithium (Trimethylsilyl)acetylide.

1. Skattebol, L.; Jones, E. R. H.; Whiting, M. S. *OSC* **1963**, *4*, 792.
2. Holmes, A. B.; Sporikou, C. N. *OSC* **1993**, *8*, 606.
3. Midland, M. M. *JOC* **1975**, *40*, 2250.
4. Descoins, C.; Henrick, C. A.; Siddall, J. B. *TL* **1972**, 3777.

M. Mark Midland
University of California, Riverside, CA, USA

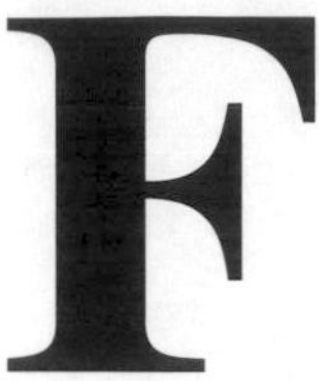

Formaldehyde

H₂C=O

[50-00-0]	CH_2O	(MW 30.03)
$((CH_2O)_n)$		
[30525-89-4]		
$(3CH_2O)$		
[110-88-3]		

(ene reaction;[1,2] Prins reaction;[1,3] enolate trapping;[4] alkylation; Mannich reaction; acetalization; condensation; reduction)

Alternate Names: paraformaldehyde; *s*-trioxane.

Physical Data: monomeric formaldehyde: mp −92 °C; bp −19.5 °C; *d* 1.067 g cm^{-3}. Paraformaldehyde: mp 163–165 °C (dec). *s*-Trioxane: mp 64 °C, bp 114.5 °C, *d* 1.17 g cm^{-3} (65 °C); sublimes readily.

Solubility: formaldehyde gas: v sol water, up to 55%; sol alcohol, ether. Paraformaldehyde: slowly sol cold water; readily sol hot water with evolution of formaldehyde; insol alcohol, ether. *s*-Trioxane: readily sol water (21.1 g/100 mL at 25 °C), alcohols, ether, acetone, chlorocarbons, aromatics; sl sol pentane.

Form Supplied in: 37% aqueous solution (with 10 to 15% methanol to prevent polymerization) known as formalin; colorless and pungent liquid. Paraformaldehyde: polymer as a white crystalline powder. *s*-Trioxane: crystalline solid.

Preparative Methods: dry formaldehyde may be generated by heating solid paraformaldehyde or by decomposing with barium peroxide; ethereal solutions of monomeric formaldehyde can be obtained by pyrolysis of paraformaldehyde at 150 °C in a special apparatus; can be generated in situ from trioxane by ***Methylaluminum Bis(2,6-diphenylphenoxide)*** (MAPH).[15]

Handling, Storage, and Precautions: reported to be storable for an extended period at −78 °C under argon (unchecked *OS* procedure);[5] irritating to mucous membranes; carcinogen; should be used only in a well-ventilated fume hood.

Ene Reaction. Thermal ene reactions of paraformaldehyde with reactive 1,1-di- and trisubstituted alkenes occur at 180–220 °C.[6,7] β-Pinene, for example, is transformed into nopol upon reaction with formaldehyde at 180 °C (eq 1).[6] Extension of this reaction has been reported using acetic acid–acetic anhydride as solvent.[8]

CH_2O; 180 °C; 95% (1)

Lewis acids such as ***Tin(IV) Chloride***[9] and ***Boron Trifluoride Etherate***[10] can facilitate addition of formaldehyde to alkenes to afford ene adducts. Thus reaction with 2,6-dimethyl-2,5-heptadiene in the presence of $SnCl_4$ as catalyst provides lavanduol (eq 2).[11] The $BF_3{\cdot}Et_2O$-promoted reaction with the ethylidene steroid shown in eq 3 takes place exclusively from the less hindered α-face to produce the ene adduct with the natural steroid configuration at C-20.[12]

CH_2O; $SnCl_4$; 55% (2)

CH_2O; $BF_3{\bullet}OEt_2$; 84% (3)

Organoaluminum reagents can also be utilized as promoters for ene reactions to give homoallylic alcohol ene adducts in high yield and selectivity, although γ-chloro alcohols are formed in some cases as byproducts.[13] ***Diethylaluminum Chloride***-induced ene reactions of formaldehyde with 1,4-dienes occur selectivity at the less deactivated terminal bond, and subsequent quasi-intramolecular Diels–Alder reaction with formaldehyde as dienophile leads to the formation of a dihydropyran (eq 4).[14] The formaldehyde–dimethylaluminum chloride complex acts on terminal alkynes to afford α-allenic alcohols and (*Z*)-3-chloroallylic alcohols as byproducts in a 2:3 ratio (eq 5).[13b] The aluminum reagent MAPH complexed with formaldehyde, generated from trioxane, also reacts with various alkenes to furnish ene adducts with excellent regioselectivities (see ***Methylaluminum Bis(2,6-diphenylphenoxide)***).[15]

CH_2O, $EtAlCl_2$; $MeNO_2$, CH_2Cl_2; 25 °C; 35–40% (4)

$C_5H_{11}CH_2C{\equiv}CH$ → $CH_2O{\bullet}Me_2AlCl$; 25 °C; 55–70% → allenic alcohol + (Z)-3-chloroallylic alcohol, 2:3 (5)

Prins Reaction. The condensation of aldehydes and ketones with alkenes in the presence of Brønsted acids is usually called the Prins reaction. The main products of this alkene–aldehyde condensation, depending on experimental conditions, may be a 1,3-dioxane, a 1,3-diol, a homoallylic alcohol, or an α-chloro ether (see ***Formaldehyde-Hydrogen Chloride***).[3]

Various catalysts have been used in the Prins reaction, although in the synthesis of 1,3-dioxane, sulfuric acid seems to be the most efficient. Thus stirring a mixture of formalin, sulfuric acid, and styrene under gentle reflux gives 4-phenyl-*m*-dioxane (eq 6).[16]

$$PhCH{=}CH_2 + 2\,CH_2O \xrightarrow[72\text{–}85\%]{H_2SO_4} \text{4-phenyl-}m\text{-dioxane} \quad (6)$$

Prins reactions involving formaldehyde have been used to prepare prostaglandin intermediates. An example is the acid-catalyzed addition of formaldehyde to the unsaturated lactone in acetic acid at 60–80 °C to give a prostaglandin intermediate (eq 7).[17]

(7) Reagents: CH_2O, H_2SO_4; AcOH; 75–85%. Product substituents: AcO, CH_2OAc.

Reaction of a 1,4-diaryl-1-butene with paraformaldehyde catalyzed by a 1:1 mixture of ***Methylaluminum Dichloride*** and ***Dimethylaluminum Chloride*** resulted in cylization to form a 1-aryltetralin (eq 8). The mixed catalyst was found to be more effective than Me_2AlCl alone.[18]

(8) Reagents: CH_2O, $MeAlCl_2$, Me_2AlCl (1:1); 88%. Starting material: OMe, OMe; *(E)*:*(Z)* = 67:33. Product: CH_2OH, OMe, OMe; 92% *trans*.

Enolate Trapping Reactions and Formation of α-Methylene Products. Lactones when treated with ***Lithium Diisopropylamide*** in THF at −78 °C form enolate anions that react with gaseous formaldehyde to give the α-hydroxymethyl derivative. The product can be easily transformed to an α-methylene lactone by conversion to mesylate followed by treatment with refluxing pyridine (eq 9).[19] This method was used for bis-α-methylenation in the total synthesis of deoxyvernolepin, which contains a dilactone structure (eq 10).[20]

(9) Step 1: 1. LDA, THF, −78 °C; 2. CH_2O, −20 °C; 3. 10% HCl. Step 2: 1. $MeSO_2Cl$, py; 2. py, Δ; 80%.

(10) 1. LDA, CH_2O, THF, HMPA; 2. $MeSO_2Cl$, py; 3. py, Δ; 32%.

α-Thiobutyrolactones,[21] carboxylic acids,[22] sulfonylhydrazones,[23] enol phosphinites,[24] and α,β-enones[25] on treatment with lithium reagents and subsequent reaction with formaldehyde undergo the same reaction, providing aldoladducts.

α-Methylenation products can also result from reaction of the carbanion of α-alkyl β-keto esters or lactones with paraformaldehyde in THF, initially at −78 °C and followed by reflux (eq 11).[26]

(11) Starting material: CO_2Et, CH_2Ph; 1. LDA, THF; 2. $(CH_2O)_n$; 96%; product: PhH_2C, CO_2Et.

Δ^1-5α-Androsten-3-one in DMSO containing paraformaldehyde and $BF_3{\cdot}Et_2O$ when heated resulted in formation of the 4-methylene ketone (eq 12).[27]

$$+ (CH_2O)_n + Me_2SO \xrightarrow[160\,°C,\ 4.5\,h;\ 75\%]{BF_3{\cdot}Et_2O} \quad (12)$$

Reductive Methylation. The classical method of dimethylation of primary amines or monomethylation of primary and secondary amines involves reaction with formaldehyde and ***Formic Acid*** (eq 13).[28]

$$PhCH_2CH_2NH_2 \xrightarrow[\Delta]{2\text{ equiv }CH_2O,\ 2\text{ equiv }HCO_2H} PhCH_2CH_2NMe_2 + 2\,CO_2 + 2\,H_2O \quad (13)$$

The use of either ***Sodium Borohydride***[29] or ***Sodium Cyanoborohydride***[30] in place of formic acid leads to the same reaction. However, $NaCNBH_3$ is usually the preferred reducing agent since the methylated amines can be formed without reduction of carbonyl compounds in the substrate.[31]

Potassium hydridotetracarbonylferrate prepared from ***Potassium Hydroxide*** and ***Pentacarbonyliron*** is likewise effective for the dimethylation of primary amines in ethanol under carbon monoxide at 20 °C (eq 14).[32] The use of a 1:1 molar ratio of amine and formaldehyde gives the monomethylated product.

$$PhNH_2 + 2\,CH_2O \xrightarrow[CO,\ EtOH;\ \sim100\%]{KHFe(CO)_4} PhNMe_2 \quad (14)$$

Carbonyl compounds having α-methyl or α-methylene groups also undergo methylation with formaldehyde in the presence of potassium or ***Sodium Tetracarbonylhydridoferrate*** (eqs 15 and 16).[33]

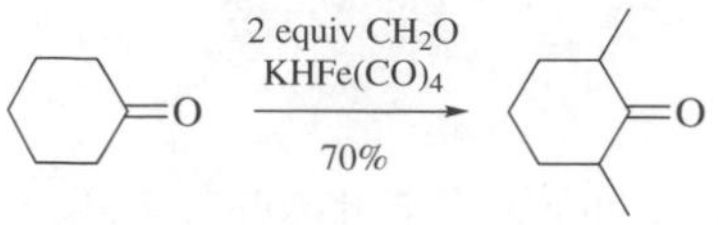

(15)

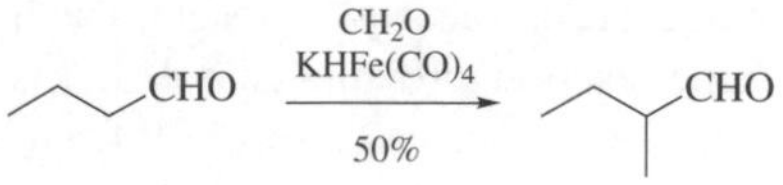

(16)

Ethyl *p*-nitrophenylacetate undergoes reductive methylation upon reaction with aqueous formaldehyde by catalytic hydrogenation with ***Palladium on Carbon*** (eq 17).[34]

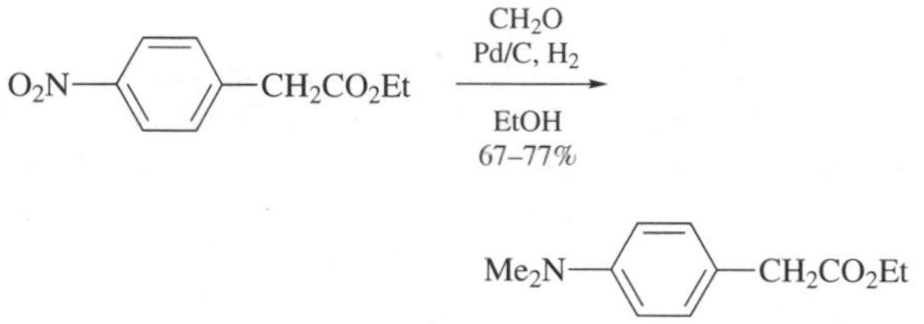

(17)

Hydroxymethylation. Reaction of carbonyl compounds with formaldehyde in the presence of base results in the formation of hydroxymethylated products. For example, allyl β-keto esters react with formaldehyde and ***Potassium Carbonate*** to give the α-hydroxymethyl ketone in quantitative yield (eq 18).[35] Sugar aldehydes undergo the same reaction with K_2CO_3.[36]

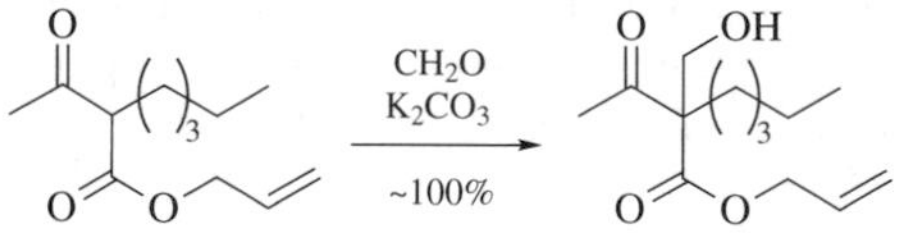

(18)

With a 2:1 ratio of formaldehyde and ***Diethyl Malonate*** and potassium bicarbonate as catalyst, the product is diethyl bis(hydroxymethyl)malonate (eq 19).[37]

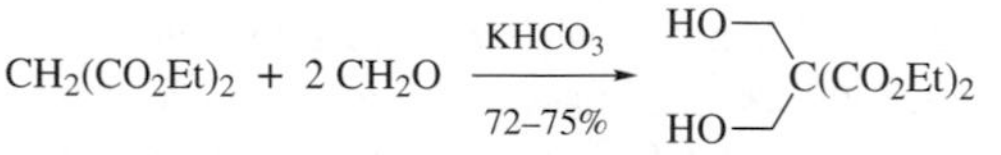

$$CH_2(CO_2Et)_2 + 2\ CH_2O \xrightarrow[72\text{–}75\%]{KHCO_3} (HOCH_2)_2C(CO_2Et)_2 \quad (19)$$

Nitriles possessing at least one α-aryl group can be hydroxymethylated using paraformaldehyde and Triton B (***Benzyltrimethylammonium Hydroxide***) in pyridine (eq 20).[38]

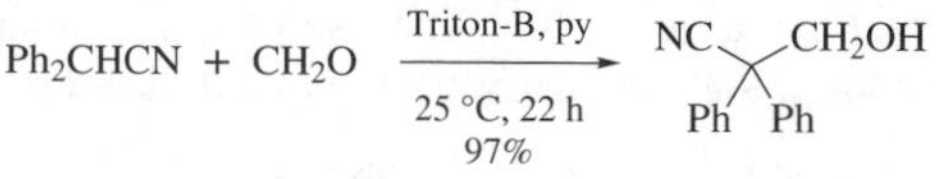

$$Ph_2CHCN + CH_2O \xrightarrow[25\ °C,\ 22\ h;\ 97\%]{\text{Triton-B, py}} NC\text{–}C(Ph)_2\text{–}CH_2OH \quad (20)$$

Chromium tricarbonyl complexes of methoxyalkylbenzenes in the presence of ***Potassium t-Butoxide*** and formaldehyde undergo regiospecific alkylation only at the *meta*-position with respect to the methoxy group (eq 21).[39]

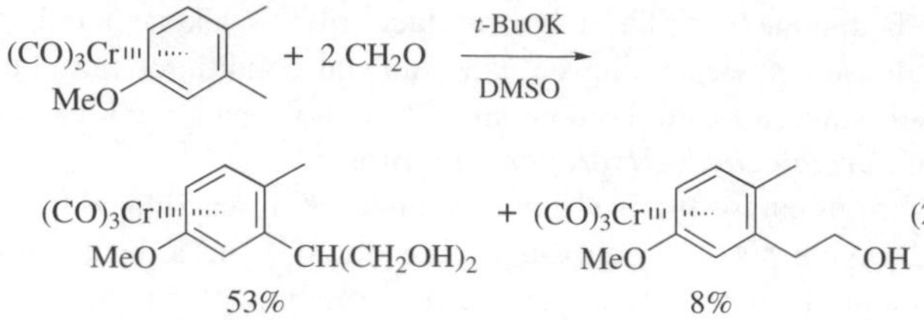

(21)

Reaction with Organometallic and Other Reagents. Hydroxymethylation of Grignard reagents with formaldehyde can be represented by the synthesis of cyclohexymethanol using cyclohexylmagnesium chloride.[40] ***Methylmagnesium Bromide***, in the presence of a nickel–aluminum catalyst, acts on 1-trimethylsilyl-1-octyne to provide a mixture of addition products which react with formaldehyde to give di- and trisubstituted vinylsilanes (eq 22).[41]

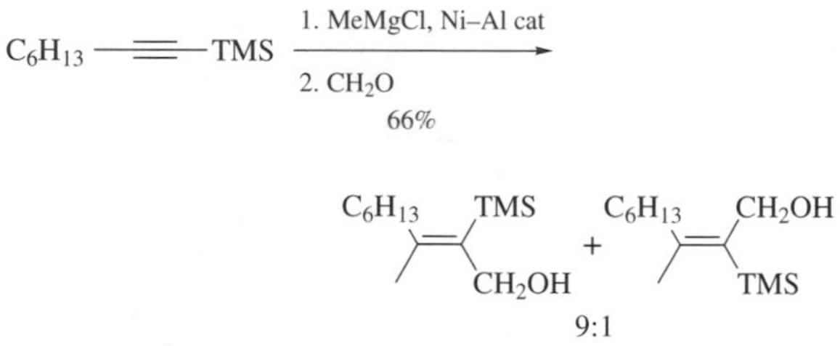

(22)

Boric Acid is a useful catalyst for the cyclocondensation of phenol with formaldehyde to give benzodioxaborin.[42] Phenylboronic acid, in combination with propionic acid, is found to be a more efficient catalyst.[43] Thus a high yield of dioxaborin is obtained in the reaction with a slight excess of phenylboronic acid, and the free hydroxymethylated phenol is liberated by an exchange reaction or by oxidation with ***Hydrogen Peroxide*** (eq 23).

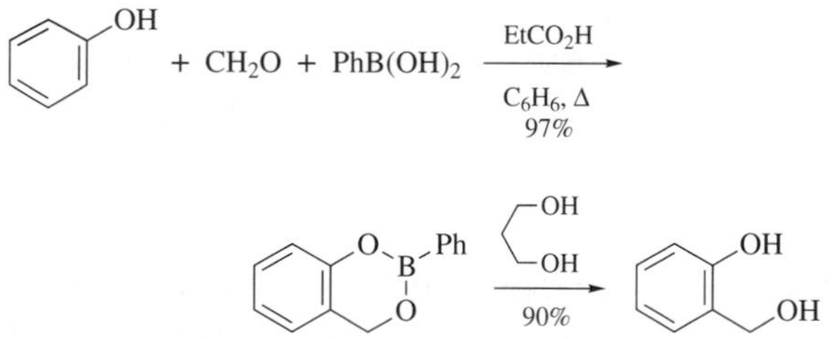

(23)

Silyl enol ethers of aldehydes and ketones react with trioxane in the presence of ***Titanium(IV) Chloride*** to give cross-aldol adducts in good yield (eq 24).[44]

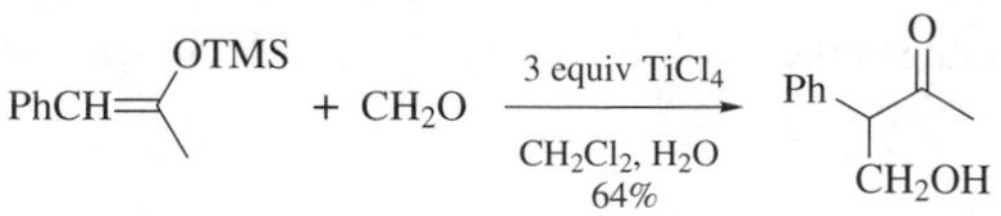

(24)

Ytterbium(III) Trifluoromethanesulfonate, a Lewis acid stable in aqueous media, is found to be effective in catalytic hydroxymethylation of silyl enol ethers with formalin (eq 25).[45]

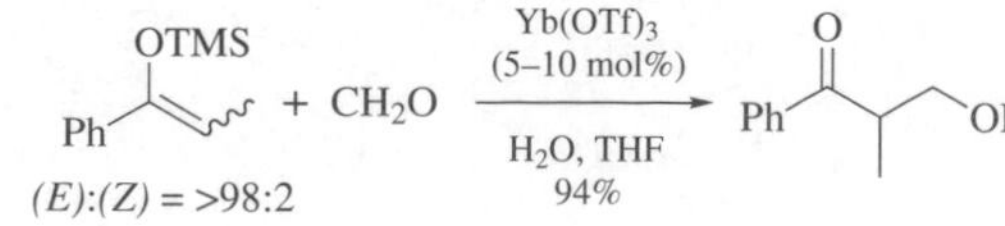

(25)

Synthesis of Heterocyclic Compounds. Condensation of ***Ethyl Acetoacetate***, formaldehyde, and ***Ammonia*** in 2:1:1 molar ratio furnishes a 1,4-dihydropyridine which is the first step in the synthesis of 2,6-lutidine (eq 26).[46]

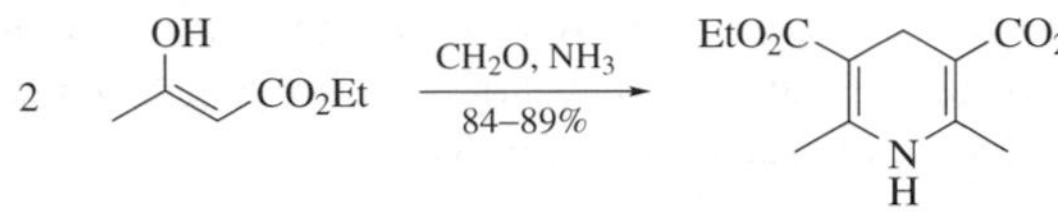

(26)

Symmetric trimers such as *sym*-trithiane[47] and hexahydro-1,3,5-tripropionyl-*sym*-triazine[48] are obtained from formaldehyde reactions. The former is synthesized by passing ***Hydrogen Sulfide*** into a mixture of formalin and concentrated HCl, while the latter is prepared by addition of trioxane in propionitrile to a mixture of propionitrile and concentrated H_2SO_4 (eqs 27 and 28).

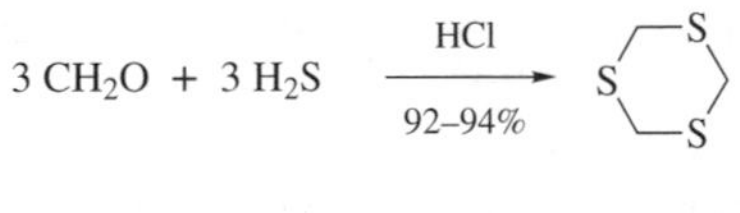

$$3\ CH_2O + 3\ H_2S \xrightarrow[92\text{–}94\%]{HCl} \text{trithiane} \quad (27)$$

$$3\ EtCN + 3\ CH_2O \xrightarrow[62\text{–}68\%]{H_2SO_4} \text{triazine} \quad (28)$$

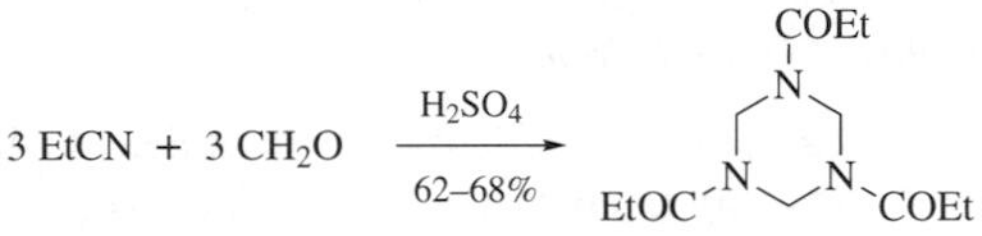

A general synthesis of piperidines involves the reaction of the acid salt of primary amines with allylsilanes and 2 equiv formaldehyde in H_2O. The iminium ion generated from the amine and formaldehyde reacts with the allylsilane to give a homoallyl amine, which then can form the second iminium ion and H_2O is captured to deliver the piperidinol (eq 29).[49]

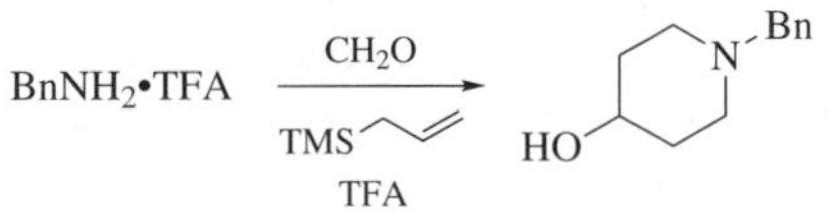

(29)

Cyclization of the alkynylamine *N*-benzyl-4-hexyn-1-amine with formaldehyde, ***Sodium Iodide***, and ***10-Camphorsulfonic Acid*** gives the iodomethylenepiperidine (eq 30).[50] In the absence of the iodide nucleophile, no cyclization is observed.

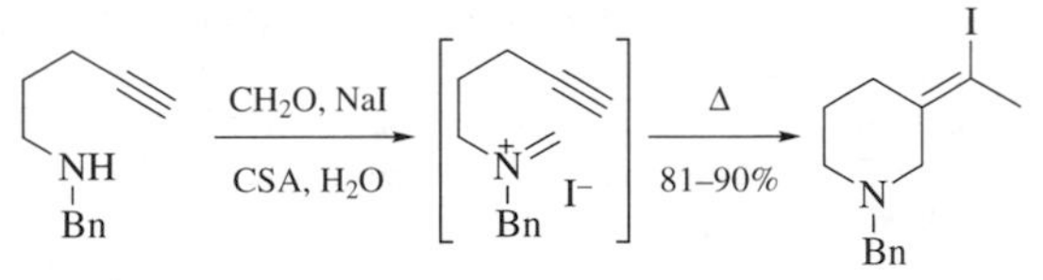

(30)

Reactions of (*E*)- or (*Z*)-vinylsilanes with an excess of paraformaldehyde and camphorsulfonic acid give the cyclized product with >98% retention of the double bond configuration of the starting reactants (eq 31).[51]

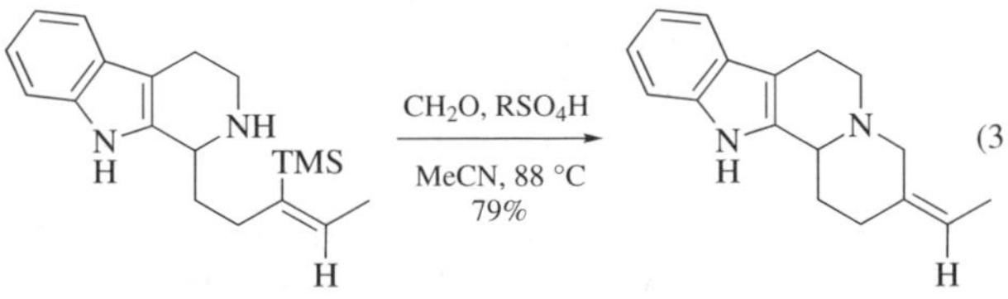

(31)

A series of imine condensations and chain cleavage occur when D-fructose is heated with ammonium hydroxide, basic copper(II) carbonate, and formaldehyde, which ultimately gives 4-hydroxymethylimidazole (eq 32).[52]

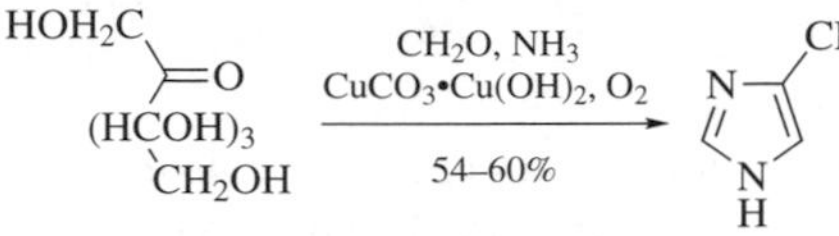

(32)

Condensation. For the preparation of aurin tricarboxylic acid, solid ***Sodium Nitrite*** was added to concentrated sulfuric acid to produce nitrogen oxides, then salicyclic acid and formalin were added dropwise with vigorous cooling (eq 33).[53]

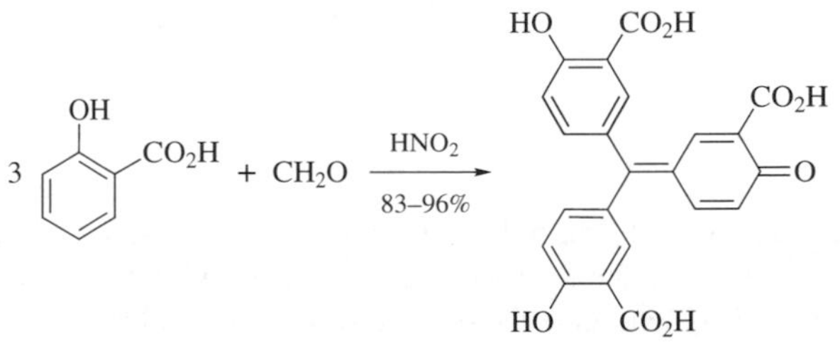

(33)

3,4-Diethylpyrrole can be converted to octaethylporphyrin by condensation with aqueous formaldehyde in benzene in the presence of ***p-Toluenesulfonic Acid***; the water is removed and the methylenepyrrole tetramer is oxidized with oxygen (eq 34).[54]

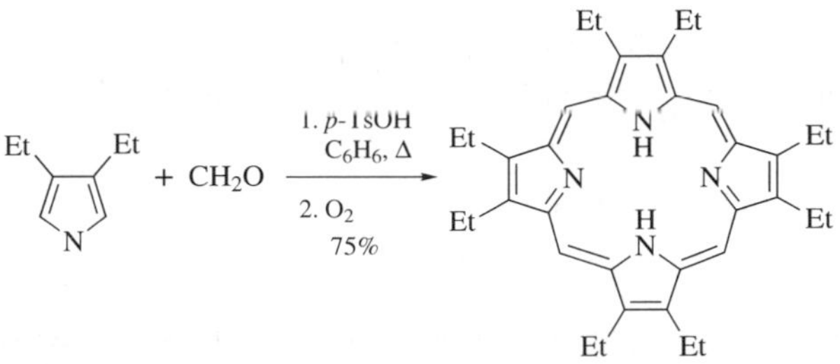

(34)

The synthesis of *p*-*t*-butylcalix[4]arene is a three-step process which involves initial reactions of *p*-*t*-butylphenol, formalin, and a base, followed by pyrolysis at a later stage (eq 35).[55] The procedure is limited to *p*-alkylphenols in which the *p*-alkyl group is highly branched at the position adjacent to the phenyl ring. A single-step procedure is also described for *p*-*t*-butylcalix[6]arene[56] and *p*-*t*-butylcalix[8]arene,[57] which requires a larger amount of base for the preparation of the hexamer.

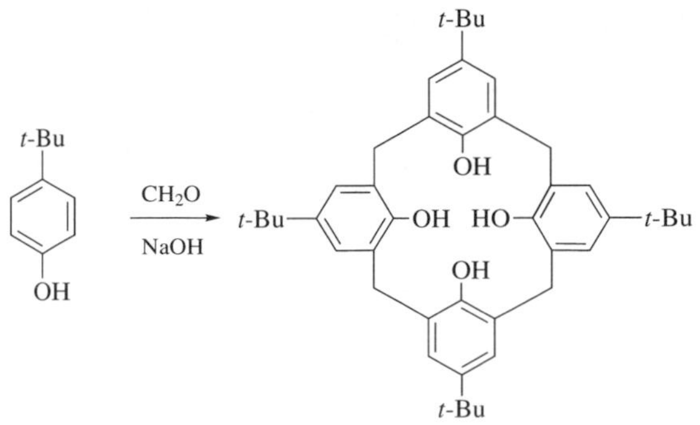

(35)

Acetalization. 1,3-Diols react with formaldehyde in an excess of alcohol under acid catalysis, such as *p*-TsOH or ion exchange resin, to form cyclic acetals (eq 36).[58]

$$\mathrm{(EtO_2C)_2C(CH_2OH)_2 \xrightarrow[\text{ROH, 91\%}]{CH_2O,\ p\text{-TsOH}} (EtO_2C)_2C(CH_2O)_2CH_2} \quad (36)$$

Cortisone, a steroid containing an α,α′-dihydroxyacetone side chain, is converted into the bismethylenedioxy derivative by stirring with chloroform, formalin, and concentrated HCl (eq 37).[59]

$$\text{Cortisone} \xrightarrow[\mathrm{CCl_4},\ 70\%]{\text{2 equiv } \mathrm{CH_2O, HCl}} \text{bismethylenedioxy derivative} \quad (37)$$

Cannizzaro and Related Reductions. In the presence of calcium oxide and water, cyclohexanone reacts with 5 moles of paraformaldehyde to give the pentaol shown in eq 38.[60]

$$\text{cyclohexanone} \xrightarrow[\mathrm{H_2O},\ 73\text{–}85\%]{\mathrm{CH_2O, CaO}} \text{2,2,6,6-tetrakis(hydroxymethyl)cyclohexanol} \quad (38)$$

p-Tolylmethanol is formed in a Cannizzaro reaction involving *p*-tolualdehyde and formalin in aqueous KOH (eq 39).[61]

$$\mathrm{p\text{-}CH_3C_6H_4CHO \xrightarrow[80\%]{CH_2O,\ KOH} p\text{-}CH_3C_6H_4CH_2OH} \quad (39)$$

An interesting preparation of *p*-dimethylaminobenzaldehyde from *p*-nitrosodimethylaniline is accomplished by reaction with dimethylaniline and 2 moles of formaldehyde in HCl solution. In the first step, one mole of formaldehyde is reduced and the other is oxidized to the formate level, furnishing the benzylidene derivative which is then hydrolyzed by exchange with formaldehyde (eq 40).[62]

$$\mathrm{Me_2N\text{-}C_6H_4\text{-}NO + C_6H_5NMe_2 \xrightarrow[HCl]{2\ equiv\ CH_2O} Me_2N\text{-}C_6H_4\text{-}N{=}CH\text{-}C_6H_4\text{-}NMe_2 \xrightarrow[AcOH]{CH_2O} Me_2N\text{-}C_6H_4\text{-}N{=}CH_2 + OHC\text{-}C_6H_4\text{-}NMe_2} \quad (40)$$

Other Reactions with Nitrogen Compounds. A reaction involving a mixture of formalin, ammonium chloride, and aqueous ***Sodium Cyanide*** provides *N*-(methylene)aminoacetonitrile (eq 41).[63]

$$\mathrm{2\ CH_2O \xrightarrow[0\ ^\circ C,\ 6\ h]{NaCN,\ NH_4Cl} H_2C{=}N\text{-}CH_2CN + NaCl} \quad \text{61–71\%} \quad (41)$$

Diethylaminoacetonitrile is prepared by adding formalin to a solution of sodium bisulfate, thereby forming the addition compound which is stirred in diethylamine after cooling, and the aqueous solution of sodium cyanide is added (eq 42).[64]

$$\mathrm{CH_2O \xrightarrow[Et_2NH]{NaCN,\ NaHSO_3} Et_2NCH_2CN + Na_2SO_3} \quad \text{88–90\%} \quad (42)$$

For the preparation of glycolonitrile, formalin is added to a stirred solution of ***Potassium Cyanide*** in water at 0–10 °C (eq 43).[65]

$$\mathrm{CH_2O \xrightarrow[0\text{–}10\ ^\circ C]{KCN,\ H_2O} HOCH_2CN + KOH} \quad \text{76–80\%} \quad (43)$$

For Mannich-type reactions involving formaldehyde, see ***Formaldehyde–Dimethylamine***.

Related Reagents. Acetaldehyde; Formaldehyde–Dimethylamine; Formaldehyde-Hydrogen Bromide; Formaldehyde-Hydrogen Chloride; Methylaluminum Bis(2,6-diphenylphenoxide); Paraformaldehyde; Sodium Cyanoborohydride.

1. Snider, B. B. *COS* **1991**, *2*, 527.
2. Mikami, K.; Shimizu, M. *CRV* **1992**, *92*, 1021.
3. Adams, D. R.; Bhatnagar, S. P. *S* **1977**, 661.
4. Chapdelaine, M. J.; Hulce, M. *OR* **1990**, *38*, 225.
5. Fieser, M. *FF* **1986**, *12*, 233; unchecked procedure submitted to *OS* **1983**.
6. Bain, J. P. *JACS* **1946**, *68*, 638.
7. Arnold, R. T.; Dowdall, J. F. *JACS* **1948**, *70*, 2590.
8. (a) Blomquist, A. T.; Verdol, J. A. *JACS* **1955**, *77*, 78. (b) Blomquist, A. T.; Passer, M.; Schollenberger, C. S.; Wolinsky, J. *JACS* **1957**, *79*, 4972. (c) Blomquist, A. T.; Verdol, J. A.; Adami, C. L.; Wolinsky, J.; Phillips, D. D. *JACS* **1957**, *79*, 4976.
9. (a) Addy, L. E.; Baker, J. W. *JCS* **1953**, 4111. (b) Yang, N. C.; Yang, D.-D. H.; Ross, C. B. *JACS* **1959**, *81*, 133. (c) Klimova, E. I.; Arbuzov, Y. A. *DOK* **1966**, *167*, 1060 (*CA* **1966**, *65*, 3736h). (d) Klimova, E. I.; Arbuzov, Y. A. *DOK* **1967**, *173*, 1332 (*CA* **1967**, *76*, 108 156c). (e) Klimova, E. I.; Treshckova, E. G.; Arbuzov, Y. A. *DOK* **1968**, *180*, 865 (*CA* **1968**, *69*, 67 173b).
10. (a) Blomquist, A. T.; Himics, R. J. *TL* **1967**, 3947. (b) Blomquist, A. T.; Himics, R. J. *JOC* **1968**, *33*, 1156. (c) Blomquist, A. T.; Himics, R. J.; Meador, J. D. *JOC* **1968**, *33*, 2462.
11. Cookson, R. C.; Mirza, N. A. *SC* **1981**, *11*, 299.
12. Batcho, A. D.; Berger, D. E.; Davoust, S. G.; Wovkulich, P. M.; Uskokovic, M. R. *HCA* **1981**, *64*, 1682.
13. (a) Snider, B. B.; Rodini, D. J. *TL* **1980**, *21*, 1815. (b) Rodini, D. J.; Snider, B. B. *TL* **1980**, *21*, 3857. (c) Snider, B. B.; Rodini, D. J.; Kirk, T. C.; Cordova, R. *JACS* **1982**, *104*, 555.
14. Snider, B. B.; Phillips, G. B. *JACS* **1982**, *104*, 1113.
15. (a) Maruoka, K.; Concepcion, A. B.; Hirayama, N.; Yamamoto, H. *JACS* **1990**, *112*, 7422. (b) Maruoka, K.; Concepcion, A. B.; Murase, N.; Oishi, M.; Hirayama, N.; Yamamoto, H. *JACS* **1993**, *115*, 3943.
16. Shriner, R. L.; Ruby, P. R. *OSC* **1963**, *4*, 786.
17. Tömösközi, I.; Gruber, L.; Kovács, G.; Székely, I.; Simonidesz, V. *TL* **1976**, 4639.

18. Snider, B. B.; Jackson, A. C. *JOC* **1983**, *48*, 1471.
19. (a) Grieco, P. A.; Hiroi, K. *CC* **1972**, 1317. (b) Grieco, P. A.; Hiroi, K. *TL* **1973**, 1831. (c) Grieco, P. A.; Hiroi, K. *TL* **1974**, 3467.
20. Grieco, P. A.; Noguez, J. A.; Masaki, Y. *TL* **1975**, 4213.
21. Lucast, D. H.; Wemple, J. *S* **1976**, 724.
22. (a) Pfeffer, P. E.; Silbert, L. S. *JOC* **1970**, *35*, 262. (b) Pfeffer, P. E.; Kinsel, E.; Silbert, L. S. *JOC* **1972**, *37*, 1256. (c) Pfeffer, P. E.; Silbert, L. S.; Chirinko, J. M. Jr. *JOC* **1972**, *37*, 451.
23. Chamberlain, A. R.; Bond, F. T. *S* **1979**, 44.
24. (a) Stork, G.; Isobe, M. *JACS* **1975**, *97*, 4745. (b) Stork, G.; Isobe, M. *JACS* **1975**, *97*, 6260.
25. Stork, G.; d'Angelo, J. *JACS* **1974**, *96*, 7114.
26. Ueno, Y.; Setoi, H.; Okawara, M. *TL* **1978**, 3753.
27. Lunn, W. H. W. *JOC* **1965**, *30*, 2925.
28. (a) Eschweiler, W. *CB* **1905**, *38*, 880. (b) Clarke, H. T.; Gillespie, H. B.; Weisshaus, S. Z. *JACS* **1933**, *55*, 4571. (c) Icke, R. N.; Wisegarver, B. B. *OSC* **1955**, *3*, 723.
29. (a) Sondengam, B. L.; Hémo, J. H.; Charles, G. *TL* **1973**, 261. (b) Barluenga, J.; Bayón, A. M.; Asensio, G. *CC* **1984**, 1334.
30. (a) Borch, R. F.; Hassid, A. I. *JOC* **1972**, *37*, 1673. (b) Nelsen, S. F.; Weisman, G. R. *TL* **1973**, 2321.
31. Kapnang, H.; Charles, G.; Sondengam, B. L.; Hemo, J. H. *TL* **1977**, 3469.
32. Watanabe, Y.; Mitsudo, T.; Yamashita, M.; Shim, S. C.; Takegami, Y. *CL* **1974**, 1265.
33. (a) Cainelli, G.; Panunzio, M.; Umani-Ronchi, A. *TL* **1973**, 2491. (b) Cainelli, G.; Panunzio, M.; Umani-Ronchi, A. *JCS(P1)* **1975**, 1273.
34. Romanelli, M. G.; Becker, E. I. *OS* **1967**, *47*, 69.
35. Tsuji, J.; Nisar, M.; Minami, I. *TL* **1986**, *27*, 2483.
36. Ho, P. T. *TL* **1978**, 1623.
37. Block, Jr. P. *OS* **1960**, *40*, 27.
38. Avramoff, M.; Sprinzak, Y. *JOC* **1961**, *26*, 1284.
39. (a) Jaouen, G.; Top, S.; Laconi, A.; Couturier, D.; Brocard, J. *JACS* **1984**, *106*, 2207. (b) Top, S.; Vessieres, A.; Abjean, J.-P.; Jaouen, G. *CC* **1984**, 428.
40. Gilman, H.; Catlin, W. E. *OSC* **1941**, *1*, 188.
41. Snider, B. B.; Karras, M.; Conn, R. S. E. *JACS* **1978**, *100*, 4626.
42. Peer, H. G. *RTC* **1960**, *79*, 825.
43. Nagata, W.; Okada, K.; Aoki, T. *S* **1979**, 365.
44. Mukaiyama, T.; Banno, K.; Narasaka, K. *JACS* **1974**, *96*, 7503.
45. Kobayashi, S. *CL* **1991**, 2187.
46. Singer, A.; McElvain, S. M. *OSC* **1943**, *2*, 214.
47. Bost, R. W.; Constable, E. W. *OSC* **1943**, *2*, 610.
48. Teeters, W. O.; Gradsten, M. A. *OSC* **1963**, *4*, 518.
49. (a) Larsen, S. D.; Grieco, P. A.; Fobare, W. F. *JACS* **1986**, *108*, 3512. (b) Grieco, P. A.; Bahsas, A. *JOC* **1987**, *52*, 1378.
50. Arnold, H.; Overman, L. E.; Sharp, M. J.; Witschel. M. C. *OS* **1992**, *70*, 111.
51. Overman, L. E.; Malone, T. C. *JOC* **1982**, *47*, 5297.
52. Totter, J. R.; Darby, W. I. *OSC* **1955**, *3*, 460.
53. Heisig, G. B.; Lauer, W. M. *OSC* **1941**, *1*, 54.
54. Sessler, J. L.; Mozaffari, A.; Johnson, M. R. *OS* **1992**, *70*, 68.
55. Gutsche, C. D.; Iqbal, M. *OSC* **1993**, *8*, 75.
56. Gutsche, C. D.; Dhawan, B.; Leonis, M.; Stewart, D. *OSC* **1993**, *8*, 77.
57. Munch, J. H.; Gutsche, C. D. *OSC* **1993**, *8*, 80.
58. Anteunis, M.; Becu, C. *S* **1974**, 23.
59. (a) Beyler, R. E.; Moriarity, R. M.; Hoffman, F.; Sarret, L. H. *JACS* **1958**, *80*, 1517. (b) Beyler, R. E.; Hoffman, F.; Moriariry, R. M.; Sarret, L. H. *JOC* **1961**, *26*, 242.
60. Wittcoff, H. *OSC* **1963**, *4*, 907.
61. Davidson, D.; Weiss, M. *OSC* **1943**, *2*, 590.
62. Adams, R.; Coleman, G. H. *OSC* **1941**, *1*, 214.
63. Adams, R.; Langley, W. D. *OSC* **1941**, *1*, 355.
64. Allen, C. F. H.; VanAllan, J. A. *OSC* **1955**, *3*, 275.
65. Gaurdy, R. *OSC* **1955**, *3*, 436.

Arnel B. Concepcion & Hisashi Yamamoto
Nagoya University, Japan

Formaldehyde–Dimethylamine[1]

HCHO–R^1R^2NH

(HCHO)		
[50-00-0]	CH_2O	(MW 30.03)
($(CH_2O)_n$)		
[30525-89-4]		
(R^1 = Me, R^2 = Me)		
[124-40-3]	C_2H_7N	(MW 45.10)
(R^1 = Me, R^2 = Me; HCl salt)		
[506-59-2]	C_2H_8ClN	(MW 81.56)
(R^1 = Bn, R^2 = H)		
[100-46-9]	C_7H_9N	(MW 107.17)
(R^1 = Bn, R^2 = H; HCl salt)		
[3287-99-8]	$C_7H_{10}ClN$	(MW 143.63)

(reagent for aminomethylation of active methylenes,[1a–c] aromatic compounds,[1d] and nucleophilic alkenes,[1e,f] reagent for generation of methyleneiminium species capable of participating in [4 + 2][2] and [3 + 2][3] cycloadditions)

Physical Data: Me_2NH, bp 7 °C (also supplied as a 40 wt % solution in water); $Me_2NH·HCl$, mp 170–173 °C; $BnNH_2$, bp 184–185 °C, *d* 0.981 g cm^{-3}; $BnNH_2·HCl$, mp 262 °C.

Form Supplied in: amine HCl salts are widely available as white solids; HCHO is available as 37 wt % soln in water; paraformaldehyde $(CH_2O)_n$ is available as a white solid.

Handling, Storage, and Precautions: amines are irritants and corrosive; HCHO is highly toxic and a cancer suspect agent; paraformaldehyde is a moisture-sensitive irritant which can be expected to give off HCHO. Use in a fume hood.

Mannich Aminomethylation. The generation of reactive methyleneiminium salts (eq 1) from formaldehyde and primary or secondary amines (usually as HCl salts) defines the classical Mannich conditions under which a wide variety of nucleophilic species can be aminomethylated. Historically, the most common nucleophiles have been active methylene compounds[1a–c] with a pK_a of 20 or less (e.g. ketones, aldehydes, and malonates) and nucleophilic aromatic compounds.[1d] Recent examples of each include the aminomethylation of a ketone in the synthesis of the alkaloid (±)-glaucine (eq 2)[4] and the aminomethylation of a phenol, which occurs regioselectively *ortho* to the hydroxyl group (eq 3).[5]

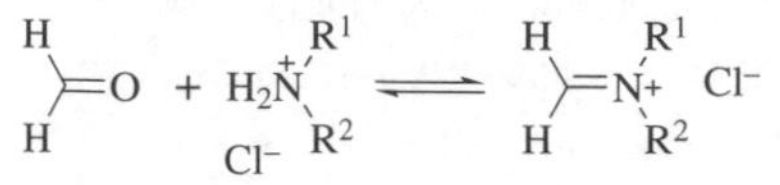

(1)

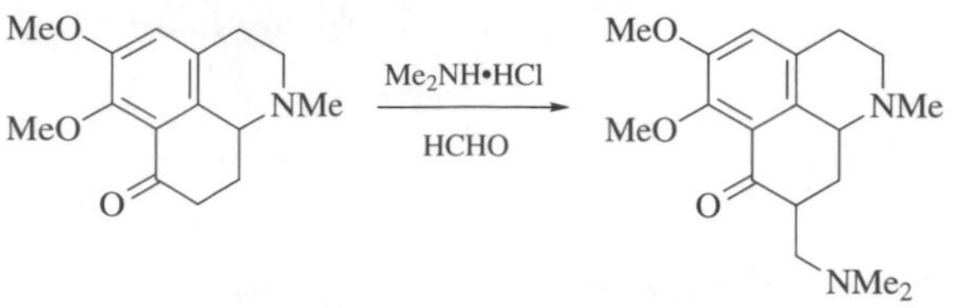

(2)

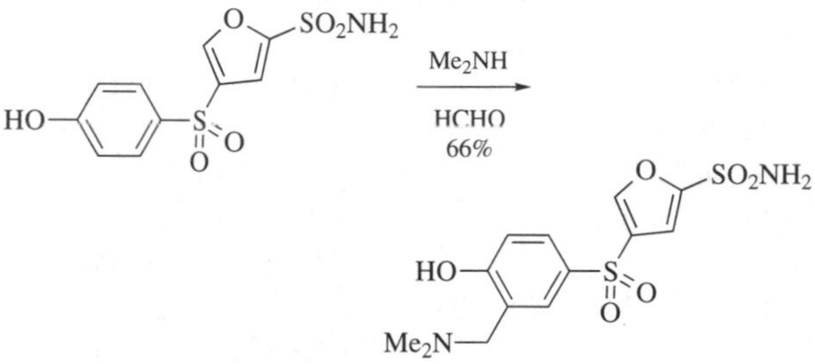

(3)

Aminomethylation of active methylene compounds is often accompanied by elimination of the amine, comprising one of the most useful syntheses of α-methylene carbonyl compounds.[1a–c] This methylenation procedure was recently incorporated into a cephalosporin synthesis (eq 4).[6] Formation of the double bond is also often accomplished separately by quaternization and base-induced elimination.

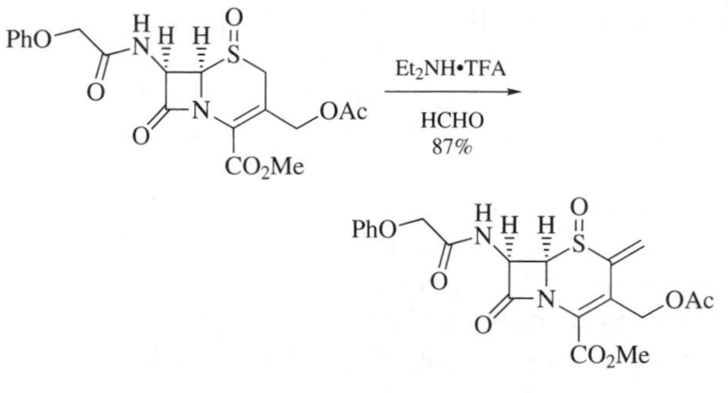

(4)

Methyleneiminium species are sufficiently reactive to be attacked by moderately nucleophilic alkenes.[1f,7–10] In most cases, this reaction is terminated by addition of an external nucleophile, such as halide or water, to the opposite terminus of the alkene.[9,10] However, the addition of an alkene to the methyleneiminium ion can also occur in the context of a pericyclic reaction. The elegant stereoselective pyrrolidine synthesis reported by Overman proceeds via a tandem aza-Cope rearrangement/Mannich condensation and has been incorporated into a number of alkaloid total syntheses.[1f] The method is exemplified in eq 5.[11] An unprecedented aza–ene reaction was also recently observed during an intramolecular reaction of an alkene with a methyleneiminium ion (eq 6).[12]

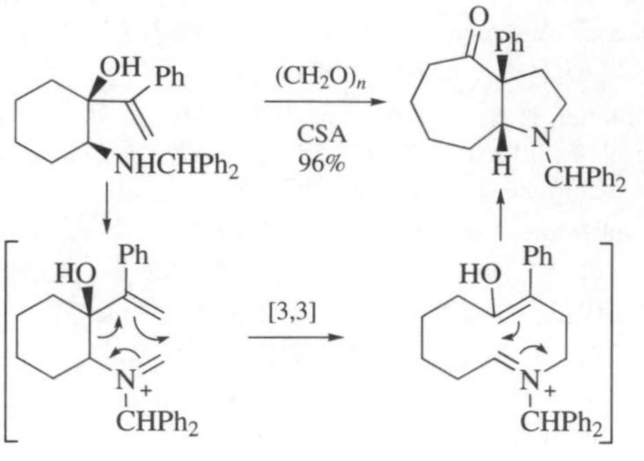

(5)

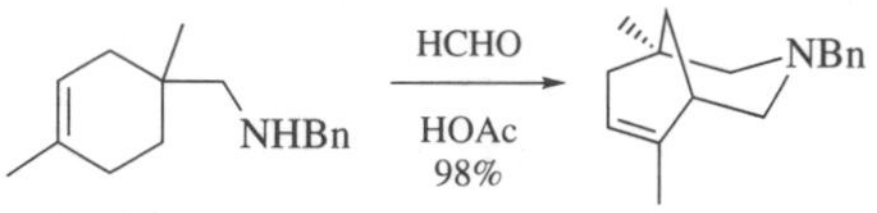

(6)

Terminal alkynes will condense with secondary amines and HCHO in the presence of Cu^+ or Zn^{2+} catalysts to afford aminomethylalkynes[13] or allenes.[14] Overman has exploited the intramolecular trapping of methyleneiminium ions with alkynes coupled with nucleophilic addition to the incipient vinyl cations to effect the syntheses of pyrrolidines and piperidines possessing exocyclic substituted methylenes of defined stereochemistry (eq 7).[15] Apparently the success of this cyclization depends on the presence of an external nucleophile.

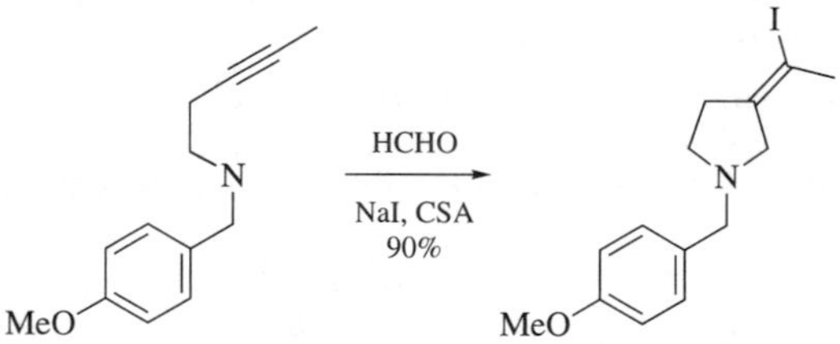

(7)

The enhanced nucleophilicity of allyl and vinyl silanes relative to alkenes renders them particularly susceptible to reaction with methyleneiminium ions. Grieco has demonstrated that allylsilanes and -stannanes react readily, both inter- and intramolecularly.[16] The resulting homoallyl amines may react further with formaldehyde, affording 4-hydroxypiperidines via the subsequent intramolecular addition of the terminal alkene to the methyleneiminium ion (eq 8).[16a] Vinylsilanes are particularly valuable nucleophiles in that they add intramolecularly in a stereospecific fashion to methyleneiminium ions (eq 9).[1f,17]

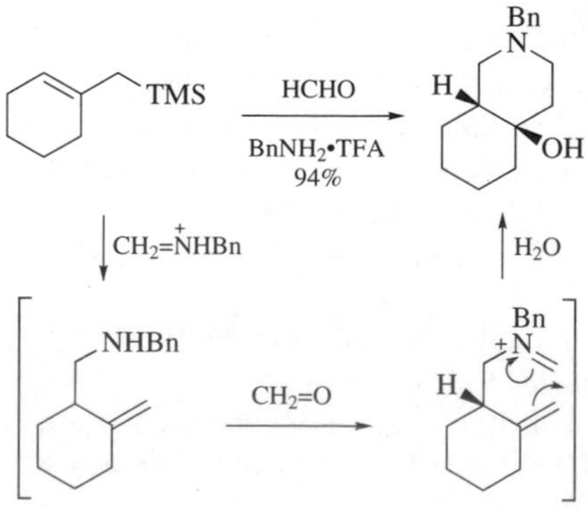

(8)

(9)

Cycloadditions. Böhme first demonstrated that dimethylmethyleneiminium ion is reactive enough to participate as a dienophile in Diels–Alder reactions with 1,3-dienes.[18] More recently it has been reported that methyleneiminium ions generated in situ from primary amines and HCHO undergo this reaction as well (eq 10).[2] Diastereoselectivity has been noted when amines bearing asymmetrical groups are employed (eq 11).[2a,19]

(10)

(11)

When HCHO is combined with a primary amine bearing an electron-withdrawing group (e.g. ester or nitrile) in the α-position, the resultant methyleneiminium ion can deprotonate at the α-carbon, generating an azomethine ylide capable of undergoing [3 + 2] cycloadditions with alkenic dipolarophiles.[3,20] A typical example is presented in eq 12.[20] This method suffers from a lack of regio- and stereoselectivity.

(12)

If the electron-withdrawing group is carboxyl, spontaneous decarboxylation occurs, comprising a remarkably mild and convenient generation of nonstabilized azomethine ylides.[21] Cycloaddition with various alkenes is stereospecific, providing 2,5-unsubstituted pyrrolidines (eq 13).[21b]

(13)

Related Reagents. Dimethyl(methylene)ammonium Iodide; Dimethyl(phenylthiomethyl)amine; Formaldehyde; *N*-(Methylthiomethyl)piperidine; Paraformaldehyde.

1. (a) Tramontini, M. *S* **1973**, 703. (b) Tramontini, M.; Angiolini, L. *T* **1990**, *46*, 1791. (c) Kleinman, E. F. *COS* **1991**, *2*, 893. (d) Heaney, H. *COS* **1991**, *2*, 953. (e) Kleinman, E. F.; Volkmann, R. A. *COS* **1991**, *2* 975. (f) Overman, L. E.; Ricca, D. J. *COS* **1991**, *2*, 1007.
2. (a) Larsen, S. D.; Grieco, P. A. *JACS* **1985**, *107*, 1768. (b) Grieco, P. A.; Larsen, S. D. *OS* **1990**, *68*, 206; *OSC* **1993**, *8*, 31. (c) Cortes, D. A. U.S. Patent 4 946 993, 1990. (d) Skvarchenko, V. R.; Lapteva, V. L.; Gorbunova, M. A. *JOU* **1990**, *26*, 2244.
3. Tsuge, O.; Kanemasa, S. *Adv. Heterocycl. Chem.* **1989**, *45*, 231.
4. Ozaki, Y.; Kim, S.-W. *CPB* **1991**, *39*, 1349.
5. Hartman, G. D.; Halczenko, W. *JHC* **1990**, *27*, 127.
6. Gunda, T. E. *LA* **1990**, 311.
7. Barnett, C. J.; Copley-Merriman, C. R.; Maki, J. *JOC* **1989**, *54*, 4795.
8. Tolstikov, G. A.; Shul'ts, É. É.; Baikova, I. P.; Spirikhin, L. V. *JOU* **1991**, *27*, 357.
9. Schmidle, C. J.; Mansfield, R. C. *JACS* **1956**, *78*, 1702.
10. (a) Grewe, R.; Hamann, R.; Jacobsen, G.; Nolte, E.; Riecke, K. *LA* **1953**, *581*, 85. (b) Grob, C. A.; Wohl, R. A. *HCA* **1966**, *49*, 2175. (c) Cope, A. C.; Burrows, W. D. *JOC* **1966**, *31*, 3099.
11. Overman, L. E.; Jacobsen, E. J.; Doedens, R. J. *JOC* **1983**, *48*, 3393.
12. Shishido, K.; Hiroya, K.; Fukumoto, K.; Kametani, T. *TL* **1986**, *27*, 1167.
13. (a) Amstutz, R.; Ringdahl, B.; Karlén, B.; Roch, M.; Jenden, D. J. *JMC* **1985**, *28*, 1760. (b) Stütz, A.; Georgopoulos, A.; Granitzer, W.; Petranyi, G.; Berney, D. *JMC* **1986**, *29*, 112.
14. (a) Crabbe, P.; Fillion, H.; André, D.; Luche, J.-L. *CC* **1979**, 859. (b) Yasukouchi, T.; Kanematsu, K. *TL* **1989**, *30*, 6559.
15. (a) Overman, L. E.; Sharp, M. J. *JACS* **1988**, *110*, 612. (b) Overman, L. E.; Sharp, M. J. *TL* **1988**, *29*, 901. (c) Arnold, H.; Overman, L. E.; Sharp, M. J.; Witschel, M. C. *OS* **1992**, *70*, 111.
16. (a) Larsen, S. D.; Grieco, P. A.; Fobare, W. F. *JACS* **1986**, *108*, 3512. (b) Grieco, P. A.; Fobare, W. F. *TL* **1986**, *27*, 5067. (c) Grieco, P. A.; Fobare, W. F. *CC* **1987**, 185. (d) Grieco, P. A.; Bahsas, A. *JOC* **1987**, *52*, 1378.
17. (a) Overman, L. E.; Malone, T. C. *JOC* **1982**, *47*, 5297. (b) Overman, L. E.; Bell, K. L.; Ito, F. *JACS* **1984**, *106*, 4192. (c) Daly, J. W.; McNeal, E. T.; Overman, L. E.; Ellison, D. H. *JMC* **1985**, *28*, 482.
18. Böhme, H.; Harke, K.; Müller, A. *CB* **1963**, *96*, 607.
19. (a) Grieco, P. A.; Bahsas, A. *JOC* **1987**, *52*, 5746. (b) Waldman, H. *AG(E)* **1988**, *27*, 274. (c) Waldman, H. *LA* **1989**, 231.
20. Tsuge, O.; Kanemasa, S.; Ohe, M.; Yorozu, K.; Takenaka, S.; Ueno, K. *BCJ* **1987**, *60*, 4067.
21. (a) Joucla, M.; Mortier, J. *CC* **1985**, 1566. (b) Tsuge, O.; Kanemasa, S.; Ohe, M.; Takenaka, S. *CL* **1986**, 973.

Scott D. Larsen
The Upjohn Co., Kalamazoo, MI, USA

Formaldehyde Dimethyl Thioacetal Monoxide[1]

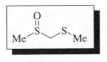

[33577-16-1] $C_3H_8OS_2$ (MW 124.25)

(versatile synthetic reagent used for the production of aldehydes, ketones, α-keto carboxylates, α-amino acid derivatives, and arylacetic acid derivatives)

Alternate Names: methyl (methylthio)methyl sulfoxide (MMSO); formaldehyde dimethyl dithioacetal *S*-oxide; methyl methylsulfinylmethyl sulfide; (methylsulfinyl)methylthiomethane; bis(methylthio)methane *S*-oxide.

Physical Data: bp 222–226 °C; *d* 1.22 g cm^{-3} (20 °C); viscosity 11.78 cP (20 °C); refractive index (20 °C) 1.5524; specific heat (20 °C) 336 cal g^{-1} $°C^{-1}$; surface tension 37.04 dyn cm^{-1} (20 °C); dielectric constant (15 °C) 3.2; pK_a (DMSO) 29.0.[2]

Solubility: sol acetic acid, alcohol, THF, CS_2, CCl_4, $CHCl_3$, acetone, benzene, DMSO; slightly sol cyclohexane, hexane.

Form Supplied in: available as a colorless, clear liquid; commercially available.

Preparative Methods: substitution of chloromethyl methyl sulfoxide with sodium methanethiolate[3] or oxidation of formaldehyde dimethyl dithioacetal with hydrogen peroxide.[4]

Purification: distillation under a reduced pressure at a temperature below 100 °C.

Analysis of Reagent Purity: GC.

Handling, Storage, and Precautions: reagent decomposes slightly when heated under acidic conditions. It is stable in alkaline and neutral media. It is desirable to keep the reagent in a tightly sealed container over 4 Å molecular sieves.

Aldehyde and Ketone Synthesis. The abstraction of a proton from MMSO with ***Sodium Hydride***, ***Lithium Diisopropylamide***, or ***Potassium Hydride*** affords a carbanion that readily reacts with a variety of electrophiles. Monoalkylation[5] employing alkyl halides or terminal epoxides,[6] followed by hydrolysis, yields the corresponding homologated aldehyde (eq 1). Dialkylation of this anion provides a pathway to symmetrical ketones.[7] The synthesis of cyclic ketones using this procedure has been effective in the production of both natural products (eq 2)[8] and theoretically interesting molecules.[9] It should be noted, however, that this method is not suitable for the synthesis of acyclic, unsymmetrical ketones as a protocol involving sequential alkylation with nonequivalent electrophiles failed.[10]

MeS(O)CH₂SMe → (1. NaH, THF; 2. BuI; 3. 1N H_2SO_4; 60%) BuCHO (1)

MeO_2SO, OSO_2Me, THPO, OR cyclopentane → (1. $MeS(O)CH_2SMe$, BuLi; 2. NBS, $CaCO_3$; 68%) bicyclic ketone, THPO, OR (2)

The addition of a Grignard reagent to MMSO, followed by hydrolysis of the intermediate dithioacetal, affords the corresponding aldehyde in moderate yield (eq 3).[11] This procedure is significant when compared to aldehyde formation through the use of ***1,3-Dithiane***. Since 1,3-dithiane requires an electrophilic coupling partner, the production of aromatic aldehydes is prohibited.

MeS(O)CH₂SMe → (1. naphthyl-MgBr; 2. HCl; 3. $SOCl_2$, SiO_2–H_2O; 61%) naphthyl-CHO (3)

The anion of MMSO also reacts at activated sp^2 carbon centers. Michael addition proceeds efficiently when lithio MMSO is added to α,β-unsaturated carbonyl compounds.[12] The sodium anion of MMSO adds to 2-bromopyridine, displacing the bromide substituent (eq 4).[13] In both cases, hydrolysis of the coupled intermediate affords the corresponding aldehyde. Moreover, the lithium derivative of MMSO condenses efficiently with ketones to provide, after hydrolysis, the corresponding α-hydroxy aldehydes (eq 5).[14] The mild nature of the addition/hydrolysis sequence lends itself to the production of sensitive compounds. For instance, the aldehyde generated in eq 5 is difficult to prepare using certain other methods due to its lability toward both acid and heat.

2-bromopyridine → (1. $MeS(O)CH_2SMe$, NaH, DME; 2. 150–200 °C; 74%) pyridine-2-CHO (4)

$Ph_2C{=}O$ + MeS(O)CH₂SMe → (1. BuLi; 2. H_2O; 3. HCl, THF; 40%) $Ph_2C(OH)CHO$ (5)

Synthesis of α-Keto Acid and α-Amino Acid Derivatives. Nitriles react with MMSO in the presence of NaH to give enamino sulfoxides. These versatile intermediates can serve as precursors for both α-keto acids (eq 6)[15,16] and α-amino acids (eq 7).[15,17] Treatment of the enamino sulfoxide with ***Copper(II) Chloride*** in ethanol provides an α-keto ethyl ester. Acetylation of the intermediate enamino sulfoxide, followed by methanolysis and reductive desulfurization, affords protected α-amino esters.

$PhCH_2CH_2CN$ → (MMSO, NaH, THF, rt) $H_2N(PhCH_2CH_2)C{=}C(SOMe)SMe$ → ($CuCl_2$, EtOH, rt) $PhCH_2CH_2COCO_2Et$ 77% (6)

PhCN → (MMSO, NaH, THF, rt) $H_2N(Ph)C{=}C(SOMe)SMe$ → (1. Ac_2O; 2. Et_3N, MeOH; 3. Raney Ni) $AcHNCH(Ph)CO_2Me$ 77% (7)

Synthesis of Arylacetic Acid Derivatives. When a mixture of MMSO and an aromatic aldehyde is heated in the presence of a base, a Knoevenagel-type condensation results (eq 8). Reaction of the condensation products with HCl in an alcohol affords arylacetic esters.[18] In the case of electron-rich aromatic aldehydes, hydrolysis of the condensation product often generates the corresponding α-(methylthio)aryl acetate (eq 9). This undesired substrate can, however, be converted to the aryl acetate upon treatment with ***Raney Nickel***.[19] For acid-labile arylacetic acids, a two-step hydrolysis involving base can be employed.[20]

MMSO, Triton B / THF, reflux; HCl / EtOH, rt; CH_2CO_2Et 65% (8)

HCl, EtOH / –5 to 5 °C; Raney Ni / EtOH, reflux; 60% (9)

Friedel–Crafts Alkylation. The α-thioalkylation of aromatic compounds can be carried out by subjecting the aromatic substrate to MMSO and a Lewis acid (eq 10).[21] Similar attempts employing chloromethyl methyl sulfide as the electrophile result in poor yields of the desired alkylated aromatics (35% as compared to the 90% in eq 10). ***Aluminum Chloride*** is the Lewis acid of choice, providing greater reproducibility than other catalysts that were examined. This method can be extended to heteroaromatic substrates. For example, thiophene and *N*-methylpyrrole are alkylated in 55% and 50%, respectively, when subjected to thioalkylation.

MMSO / $AlCl_3$, CH_2Cl_2 90% (10)

Heterocycle Addition. The exposure of quinoxaline to the dianion of MMSO results in the production of an appended third ring (eq 11).[22] ***Dimethyl Sulfoxide***, participating in a similar process, affords annulated substrate in a slightly lower yield (51%). This reaction course is in contrast to the addition of a typical organometallic, such as a Grignard reagent, to quinoxaline. For this case, the 2-alkylquinoxaline is usually generated.

1. LiMMSO / 2. H_2O 54% (11)

Ethyl Analog of MMSO. Formaldehyde diethyl dithioacetal monoxide can be used in analogous synthetic applications,[10,23] although it is not commercially available.

Related Reagents. *N,N*-Diethylaminoacetonitrile; *N,N*-Dimethyldithiocarbamoylacetonitrile; (4a*R*)-(4aα,7α,8aβ)-Hexahydro-4,4,7-trimethyl-4*H*-1,3-benzoxathiin; 2-Lithio-1,3-dithiane; Methylthiomethyl *p*-Tolyl Sulfone; Nitromethane; 1,1,3,3-Tetramethylbutyl Isocyanide; *p*-Tolylsulfonylmethyl Isocyanide; 2-(Trimethylsilyl)thiazole.

1. (a) Ogura, K. *PAC* **1987**, *59*, 1033. (b) Ogura, K. In *Studies in Natural Product Chemistry*; Atta-ur-Rahman, Ed.; Elsevier: Amsterdam, 1990; Vol. 6, p 307.
2. Bordwell, F. G.; Drucker, G. E.; Andersen, N. H.; Deniston, A. D. *JACS* **1986**, *108*, 7310.
3. Ogura, K.; Tsuchihashi, G. *CC* **1970**, 1689.
4. Ogura, K.; Tsuchihashi, G. *BCJ* **1972**, *45*, 2203.
5. Ogura, K.; Tsuchihashi, G. *TL* **1971**, 3151.
6. Torii, S.; Uneyama, K.; Ishihara, M. *JOC* **1974**, *39*, 3645.
7. Schill, G.; Jones, P. R. *S* **1974**, 117.
8. (a) Torisawa, Y.; Okabe, H.; Ikegami, S. *CC* **1984**, 1602. (b) For other syntheses of cycloalkenones see: Ogura, K.; Yamashita, M.; Suzuki, M.; Furukawa, S.; Tsuchihashi, G. *BCJ* **1984**, *57*, 1637. Ogura, K.; Yamashita, M.; Tsuchihashi, G. *TL* **1976**, 759. Ogura, K.; Yamashita, M.; Suzuki, M.; Tsuchihashi, G. *TL* **1974**, 3653.
9. Dowd, P.; Schappert, R.; Garner, P.; Go, C. L. *JOC* **1985**, *50*, 44.
10. Richman, J. E.; Herrmann, J. L.; Schlessinger, R. H. *TL* **1973**, 3267.
11. Hojo, M.; Masuda, R.; Saeki, T.; Fujimori, K.; Tsutsumi, S. *TL* **1977**, 3883.
12. (a) Ogura, K.; Yamashita, M; Tshuchihashi, G. *TL* **1978**, 1303. (b) Breukelman, S. P.; Meakins, G. D.; Roe, A. M. *JCS(P1)* **1985**, 1627. (c) Tanaka. K.; Kanemasa, S.; Ninomiya, Y.; Tsuge, O. *BCJ* **1990**, *63*, 466.
13. Newkome, G. R.; Robinson, J. M.; Sauer, J. D. *CC* **1974**, 410.
14. Ogura, K.; Tsuchihashi, G. *TL* **1972**, 2681.
15. Ogura, K.; Tsuchihashi, G. *JACS* **1974**, *96*, 1960.
16. Ogura, K.; Katoh, N.; Yoshimura, I.; Tsuchihashi, G. *TL* **1978**, 375.
17. Ogura, K.; Yoshimura, I.; Katoh, N.; Tsuchihashi, G. *CL* **1975**, 803.
18. (a) Ogura, K.; Tsuchihashi, G. *TL* **1972**, 1383. (b) Ogura, K.; Ito, Y.; Tsuchihashi, G. *BCJ* **1979**, *52*, 2013. (c) Artico, M.; Corelli, F.; Massa, S.; Stefancich, G. *JHC* **1982**, 1493. (d) Katagiri, N.; Kato, T.; Nakano, J. *CPB* **1982**, *30*, 2440. (e) Schuda, P. F.; Price, W. A. *JOC* **1987**, *52*, 1972.
19. (a) Cannon, J. R.; Lolanapiwatna, V.; Raston, C. L.; Sinchai, W.; White, A. H. *AJC* **1980**, *33*, 1073. (b) Rizzacasa, M. A.; Sargent, M. V. *JCS(P1)* **1987**, 2017.
20. (c) Ogura, K.; Ito, Y.; Tsuchihashi, G. *S* **1980**, 736.
21. Torisawa, Y.; Atsushi, S.; Ikegami, S. *TL* **1988**, *29*, 1729.
22. Vierfond, J.-M.; Legendre, L.; Mahuteau, J.; Miocque, M. *H* **1989**, *29*, 141.
23. (a) Herrmann, J. L.; Richman, J. E.; Schlessinger, R. H. *TL* **1973**, 3271. (b) Herrmann, J. L.; Richman, J. E.; Schlessinger, R. H. *TL* **1973**, 3275.

Katsuyuki Ogura
Chiba University, Japan

Jeffrey A. McKinney
Zeneca Pharmaceuticals, Wilmington, DE, USA

Formylmethylenetriphenylphosphorane[1]

$Ph_3P{=}CHCHO$

[2136-75-6] $C_{20}H_{17}OP$ (MW 304.33)

('stabilized' Wittig reagent used for the synthesis of α,β-unsaturated aldehydes from aldehydes; reagent for two-carbon chain homologation)

Alternate Name: (triphenylphosphoranylidene)acetaldehyde.
Physical Data: mp t 187–188 °C (dec).
Form Supplied in: white solid, 97%; commercially available.
Preparative Methods: reaction of ***Chloroacetaldehyde*** and ***Triphenylphosphine*** affords formylmethyltriphenylphosphonium chloride, which furnishes the phosphorane reagent on treatment with ***Triethylamine*** in ethanol.[2]
Handling, Storage, and Precautions: unstable to moisture and acids. Keep dry.

(*E*)-α,β-Unsaturated Aldehydes. This 'stabilized' phosphonium ylide reacts with a wide variety of aliphatic and aromatic aldehydes to give α,β-unsaturated aldehydes, with a high predominance of the (*E*)-alkene.[2–5] This reagent has been applied in many areas of organic synthesis including natural products, carbohydrates, and heterocycles. Thus in the synthesis of swainsonine from D-mannose the chain-extended (*E*)-alkene (**2**) was prepared from the aldehyde (**1**) in good yield (eq 1).[6]

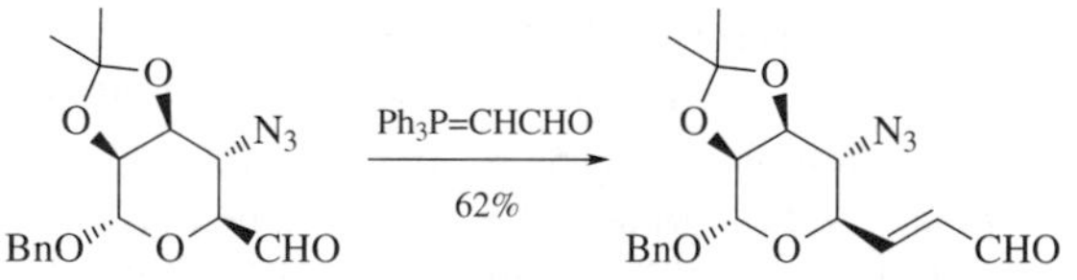

(1) → (2): $Ph_3P{=}CHCHO$, 62% (1)

Tandem addition of the ylide has been successfully applied to the synthesis of (*E,E*)-dienyl aldehydes. For example, in the synthesis of natural leukotrienes, Rokach et al. converted epoxy aldehyde (**3**) to intermediate (**4**) en route to LTA_4 (eq 2).[7]

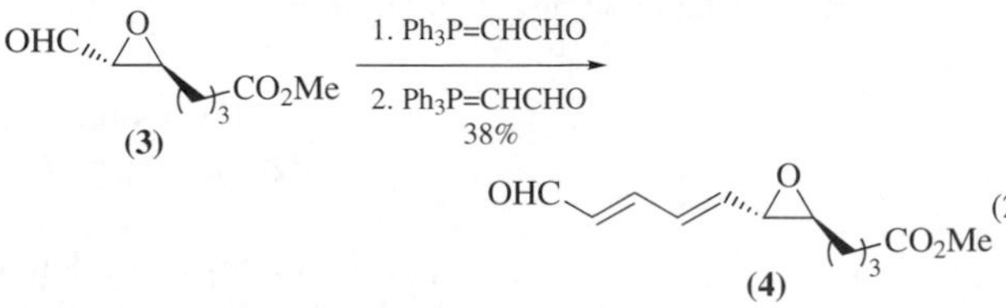

(3) → (4): 1. $Ph_3P{=}CHCHO$; 2. $Ph_3P{=}CHCHO$, 38% (2)

Reaction of the ylide with ketones is generally unsuccessful;[2] however, it has been reported to react with trifluoroacetone to give the expected trisubstituted alkene in low yield (eq 3).[8]

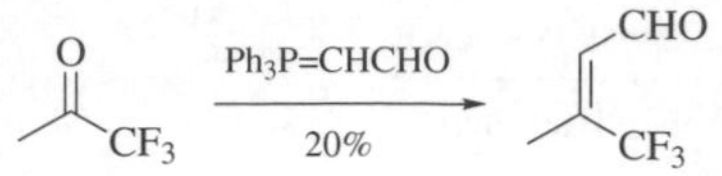

$Ph_3P{=}CHCHO$, 20% (3)

Alkenation with this ylide can be effected in good yield when a free hydroxyl group is present in the aldehyde substrate,[9,10] although a neighboring hydroxyl appears to detract from the normally high (*E*) stereoselectivity (eq 4).[10]

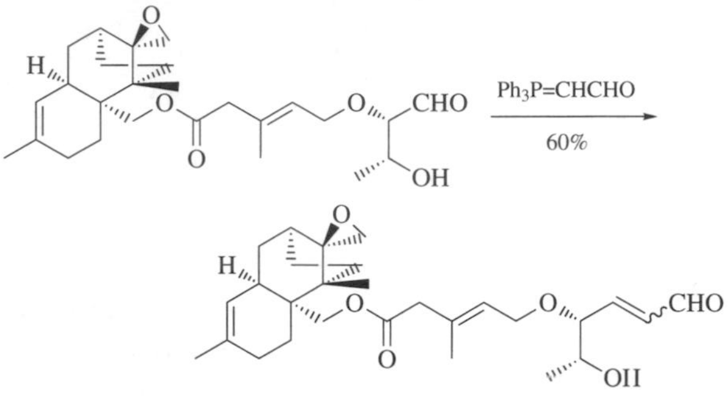

$Ph_3P{=}CHCHO$, 60% (4)

(*Z*):(*E*) = 1:4

(*Z*)-α,β-Unsaturated Aldehydes. In a complementary manner, Bestmann's group has shown that conversion of this ylide to a diethyl acetal in situ affords a new phosphorane that transforms aldehydes into α,β-unsaturated aldehydes with predominantly (*Z*) stereochemistry (86–97%).[11,12] The procedure involves treatment of the ylide with EtBr, followed by NaOEt, then addition of the aldehyde; the intermediate unsaturated diethyl acetals are hydrolyzed to α,β-unsaturated aldehydes (eq 5). Higher yields of unsaturated diethyl acetals were usually realized by an alternative reaction sequence that employed a different base system: 1) EtBr; 2) sodium amide; and 3) ethanol.

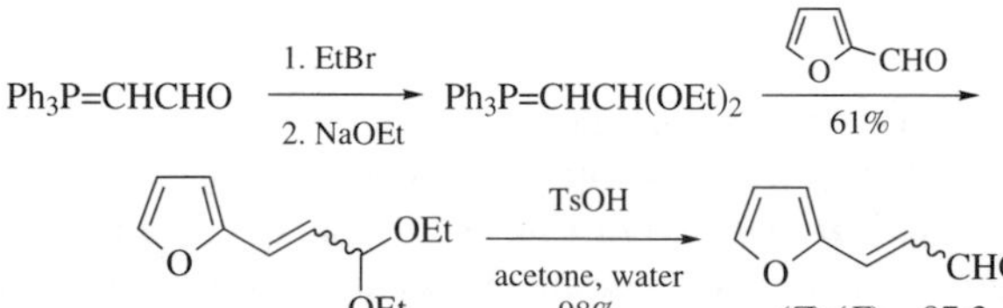

$Ph_3P{=}CHCHO$ → (1. EtBr; 2. NaOEt) → $Ph_3P{=}CHCH(OEt)_2$ → (furfural, 61%) → TsOH, acetone, water, 98% (5)

(*Z*):(*E*) = 97:3

Related Reagents. Acetylmethylenetriphenylphosphorane; (Ethoxycarbonylmethylene)triphenylphosphorane; (Methoxycarbonylmethylene)triphenylphosphorane; Methyl Bis(trifluoroethoxy)phosphinylacetate; Triethyl Phosphonoacetate; Trimethyl Phosphonoacetate.

---

1. For relevant reviews on the Wittig reaction: (a) Maryanoff, B. E.; Reitz, A. B. *CRV* **1989**, *89*, 863. (b) *Organophosphorus Reagents in Organic Synthesis*; Cadogan, J. I. G., Ed.; Academic: New York, 1979. (c) Bestmann, H. J.; Vostrowsky, O. *Top. Curr. Chem.* **1983**, *109*, 85.
2. Trippett, S.; Walker, D. M. *JCS* **1961**, 1266.
3. Olstein, R.; Stephenson, E. F. M. *AJC* **1979**, *32*, 681.

4. Venkataraman, H.; Cha, J. K. *TL* **1987**, *28*, 2455.
5. Ku, T. W.; McCarthy, M. E.; Weichman, B. M.; Gleason, J. G. *JMC* **1985**, *28*, 1847.
6. Fleet, G. W. J.; Gough, M. J.; Smith, P. W. *TL* **1984**, *25*, 1853.
7. Rokach, J.; Young, R. N.; Kakushima, M.; Lau, C.-K.; Seguin, R.; Frenette, R.; Guindon, Y. *TL* **1981**, *22*, 979.
8. Nägele, U. M.; Hanack, M. *LA* **1989**, 847.
9. Nishiyama, S.; Toshima, H.; Kanai, H.; Yamamura, S. *TL* **1986**, *27*, 3643.
10. Still, W. C.; Gennari, C.; Noguez, J. A.; Pearson, D. A. *JACS* **1984**, *106*, 260.
11. Bestmann, H. J.; Roth, K.; Ettlinger, M. *AG(E)* **1979**, *18*, 687.
12. Bestmann, H. J.; Roth, K.; Ettlinger, M. *CB* **1982**, *115*, 161.

David F. McComsey & Bruce E. Maryanoff
R. W. Johnson Pharmaceutical Research Institute, Spring House, PA, USA

Glycidol[1]

OH

[556-52-5]	$C_3H_6O_2$	(MW 74.08)
(*R*)		
[57044-25-4]		
(*S*)		
[60456-23-7]		
(±)		
[61915-27-3]		

(versatile bifunctional, three-carbon synthon)

Alternate Name: oxiranemethanol.

Physical Data: mp −53 °C; bp 161–163 °C (dec), 30 °C/1 mmHg, 54 °C/8 mmHg, 114 °C/114 mmHg; *d* 1.115 g cm^{-3}; $[\alpha]_D$ +15° (neat, L-(+)-glycidol).

Solubility: insol aliphatic hydrocarbons; sol H_2O, acetone, THF, toluene, most other organic solvents.

Form Supplied in: racemic, (*R*), and (*S*) forms; all as colorless, neat liquids. Solid derivatives: phenyl isocyanate, mp 60 °C; α-naphthyl isocyanate, mp 102 °C.

Analysis of Reagent Purity: ^{1}H NMR.

Handling, Storage, and Precautions: neat samples of glycidol should be stored in the freezer to slow the process of self-condensation; when stored neat, glycidol should be checked for purity before use and will usually require purification, which can be achieved by distillation under reduced pressure; self-condensation is greatly reduced by storage of glycidol in solutions, e.g. 50–70% in toluene or dichloromethane; distillation of glycidol should be done behind a safety shield; care should be taken when using glycidol under acidic conditions (e.g. acetic acid) since acid catalyzes self-condensation; use in a fume hood.

Glycidol and Glycidol Derivatives. Two excellent reviews of glycidol and glycidol derivatives are available. The first is a very thorough review of the properties and reactions of glycidol written by Kleemann and Wagner.[1a] In the second, the use of glycidol and glycidol derivatives as synthons, with a strong emphasis on nonracemic glycidol, is the subject of a superb review by Hanson.[1b]

Glycidol is a versatile three-carbon synthetic building block and its value is greatly expanded through derivatization of the hydroxyl group. The use in synthesis of derivatives such as *O*-aryl and *O*-arylmethyl (e.g. *O*-benzyl) ethers, sulfonates, carboxylates, and silyl ethers is integrated with those of glycidol for this review. In the following discussion, glycidol and derivatives are occasionally referred to collectively as glycidols. Also note that reactions of nonracemic glycidol are illustrated only with one enantiomer, but apply equally to use of both.

Glycidol, like all 2,3-epoxy alcohols, is susceptible to the Payne rearrangement when exposed to base. Payne rearrangement of (*R*)- or (*S*)-glycidol is degenerate; consequently racemization does not occur.

(*R*) and (*S*) are empirical designations of absolute configurations and in comparing glycidol and an *O*-substituted glycidol derivative having the same absolute configuration, the designation changes (see eq 1). For further discussion of this point, see Hanson's review.[1b]

Preparations. Racemic glycidol, (*R*) and (*S*)-glycidol, and a number of derivatives of each are commercially available. Preparations of these materials are described in the literature and a selected listing follows: (*S*)-glycidol via asymmetric epoxidation[2] and enzymatic kinetic resolution;[3] *O*-benzyl glycidol,[4] *O*-trityl glycidol;[5] (*R*)-(−)-***Glycidyl Tosylate***;[6] (*R*)-(−)-glycidyl 3-nitrobenzenesulfonate (a derivative whose optical purity is enhanced by recrystallization);[6] (*R*)-(−)-glycidyl *p*-nitrobenzoate (see eq 1).[2,7]

Ti(O-*i*-Pr)$_4$, (−)-DET, TBHP; (*S*)-glycidol → (*R*)-glycidol derivative (1)

R = Bn, Tr, PNB, Ts, Ns

Reactions at C-1 of Glycidol. A number of *O*-derivatives of glycidol are described in the preceding section and may be prepared directly from racemic or (*R*)- or (*S*)-glycidol. Alternatively, if carrying out the laboratory preparation of (*R*)- or (*S*)-glycidol, convenient in situ methods for derivatization have been developed.[2,7,8] Derivatization as *O*-sulfonate esters (e.g. tosylates) activates the C-1 position and permits displacement by nucleophiles. An example is displacement by phenolates to generate *O*-aryl glycidol ethers (see eq 2),[6] which find extensive use as intermediates in the synthesis of a variety of pharmacologically active agents (see additions of nitrogen at C-3, below).

(*S*)-glycidol tosylate → (O^-Na^+ naphthoxide, 89%) (2)

O-Aryl glycidol ethers can be prepared from glycidol by the Mitsunobu reaction with phenols (see eq 3)[9a] and are also made from direct displacement by glycidol on activated haloaryls.[9b]

MeO–C_6H_4–OH, PPh_3, $EtO_2CN{=}NCO_2Et$ (3)

Addition of Hydrogen at C-3. Both catalytic reduction of glycidol over Pd/C[10] and reaction with $MeLi/CuBr(PBu_3)_2$[11]

give propane-1,2-diol as a consequence of addition of hydrogen at C-3. (*S*)-Glycidyl tosylate is reduced to (*S*)-propane-1,2-diol 1-monotosylate with ***Borane-Tetrahydrofuran*** and a catalytic amount of ***Sodium Borohydride***, as shown in eq 4.[6]

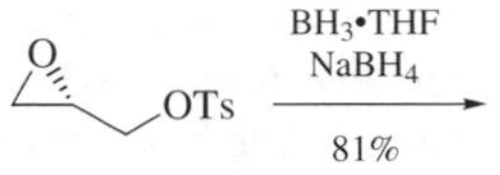

(4) [Scheme: glycidyl tosylate, BH_3•THF, $NaBH_4$, 81%]

Nucleophilic Additions of Carbon at C-3. One of the few reported additions of a carbon nucleophile to underivatized glycidol is that of diethyl sodiomalonate. The initial addition at C-3 is followed by lactonization between the C-2 hydroxyl group and one of the malonate carboxylic esters (eq 5).[12] Far more numerous are the additions of carbon nucleophiles to glycidol derivatives such as *O*-benzyl, *O*-phenyl, or *O*-tosyl glycidol. In addition to the examples included below, many others may be found in Hanson's review.[1b]

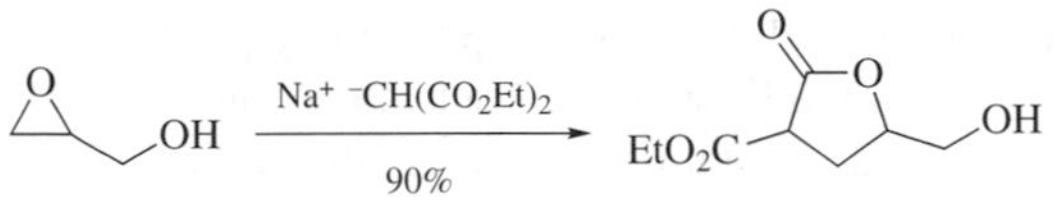

(5) [Scheme: glycidol, $Na^+\ ^-CH(CO_2Et)_2$, 90%, product EtO_2C lactone]

Single carbons can be added as cyanide using ***Acetone Cyanohydrin***,[6,7] diethylaluminum cyanide (eq 6),[6,7] or ***Lithium Cyanide***[13] or as methyl groups using an organocuprate (eq 7).[14] A single carbon may be added with dithiane salts and an example of addition of a substituted ***1,3-Dithiane*** to *O*-benzyl glycidol is shown in eq 8.[15]

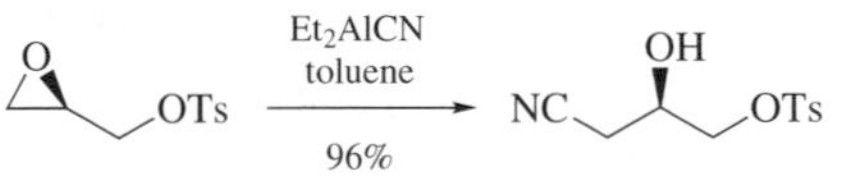

(6) [Scheme: Et_2AlCN, toluene, 96%]

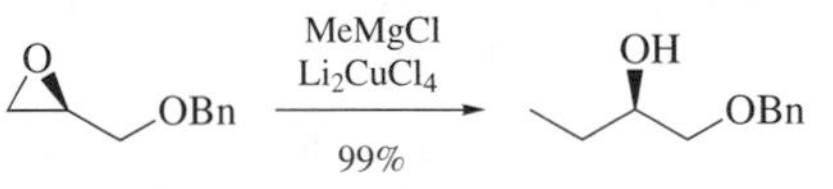

(7) [Scheme: MeMgCl, Li_2CuCl_4, 99%]

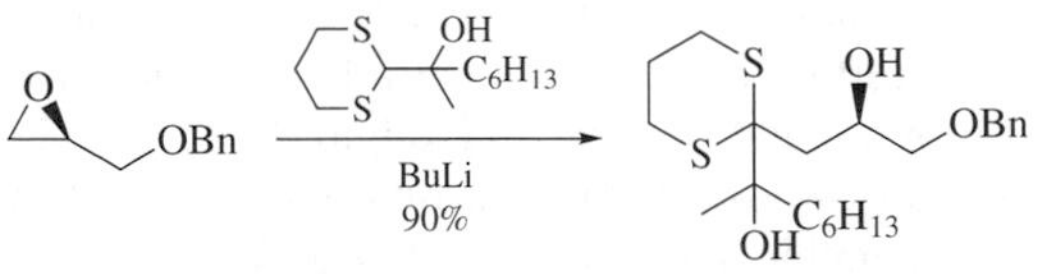

(8) [Scheme: dithiane, BuLi, 90%]

Other alkyl groups, alkenyl groups (eq 9), and aryl groups have been added to glycidol via organometallic reagents. The reactions with organometallic reagents often are sensitive to conditions and frequently are improved by the addition of Cu^I or Cu^{II} to the medium.[16] Alkynic salts add to glycidols, giving 3-alkynyl derivatives in yields which are generally good but which may be enhanced in some cases by the addition of a Lewis acid such as ***Boron Trifluoride Etherate*** to the reaction (eq 10).[17]

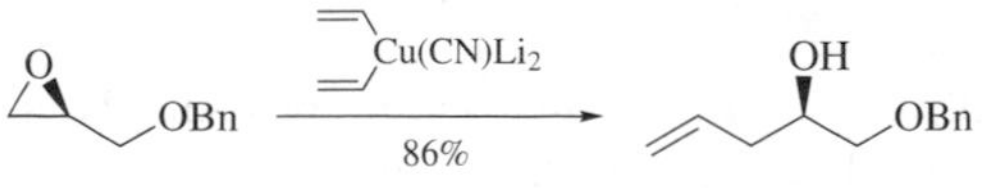

(9) [Scheme: vinyl cuprate $Cu(CN)Li_2$, 86%]

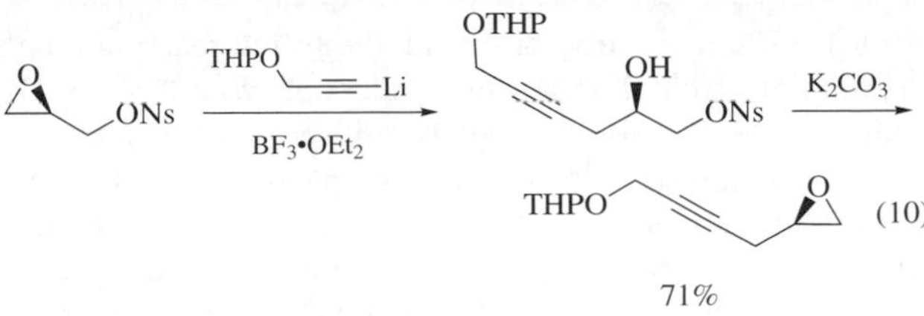

(10) [Scheme: THPO-propynyllithium, BF_3•OEt_2; K_2CO_3; 71%]

Opening at C-3 of glycidol sulfonates generates a 1,2-diol monosulfonate array which is ideally situated for closure under mildly alkaline conditions to a new epoxide group, as shown in eq 10 and also, below, in eq 18. Either the intermediate monosulfonate or the new epoxide present an activated electrophilic site for further synthetic transformations.

Carbon nucleophiles such as ester enolates and α-carboxylic acid anions add to glycidols by opening the oxirane ring and forming an intermediate C-2 alcohol. As shown above in eq 5, the intermediate can cyclize to a five-membered lactone via further reaction with the newly introduced carboxylic acid or ester.[18] Variations on the theme of intramolecular transformations following the initial addition to glycidol have been described. These include seven-membered lactone formation following addition of a sulfone-stabilized anion to *O*-benzylglycidol (eq 11),[19] and oxetane formation following addition of ***Dimethylsulfoxonium Methylide*** to glycidol (eq 12).[20]

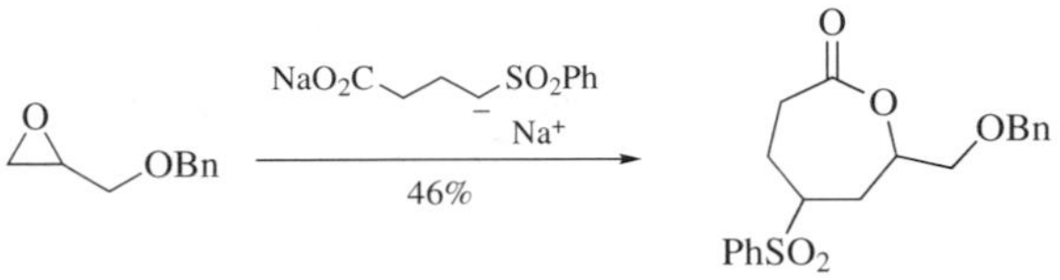

(11) [Scheme: NaO_2C sulfone anion Na^+, 46%, product with $PhSO_2$]

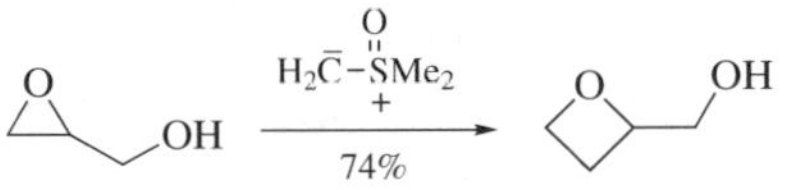

(12) [Scheme: $H_2\bar{C}$–$\overset{+}{S}(O)Me_2$, 74%]

Other examples of carbon nucleophiles which have been added to a glycidol include the lithium salt of 1-trimethylsilyl-3-phenylthioprop-1-yne (eq 13),[21] the lithium salt of 1-phenylsulfonyl-2-trimethylsilylethane,[22] the lithium salt of pentacarbonyl(methoxymethylcarbene)chromium,[23] the dimsyl anion,[24] and the lithium salt of acetone dimethylhydrazone.[25]

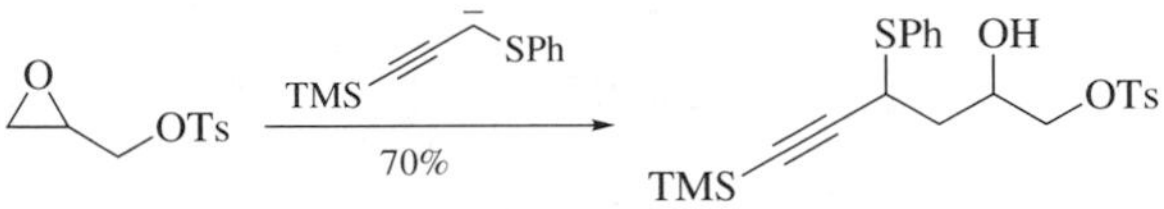

(13) [Scheme: TMS alkyne SPh anion, 70%]

Nucleophilic Addition of Oxygen at C-3. Addition of water to glycidol or a glycidol derivative produces glycerol or a substituted glycerol, respectively. The oxygen nucleophiles used most frequently for addition to glycidols are alcohols, phenols, and carboxylic acids and their close relatives. For glycidol itself, Kleemann and Wagner summarize extensive studies of additions of these classes of compounds.[1a] Very good yields of products are achieved with all three classes when acid or, preferentially, basic catalysts are added to the reactions. Careful analyses of the reaction products reveal that in addition to the primary opening of the oxirane at C-3, most reactions include small (2–10%) amounts of product derived from opening at C-2. Other byproducts can result from self-reaction of glycidol with the reaction products. Opening of glycidol with primary alcohols with 0.5% NaOH as catalyst yields 70% of the 1-*O*-alkylglycerol together with 3% of the 2-*O*-alkylglycerol.[1a] Opening with phenols and 0.03% NaOH gives 70–80% yields of 1-*O*- and 2-*O*-arylglycerols in ratios of 90–95:5–10.[1a] Glycidol generated in situ from hydrolysis of the *p*-nitrobenzoate ester with $MeOH/H_2SO_4$ reacts further at C-3 with the MeOH to give 1-*O*-methoxyglycerol (eq 14).[7]

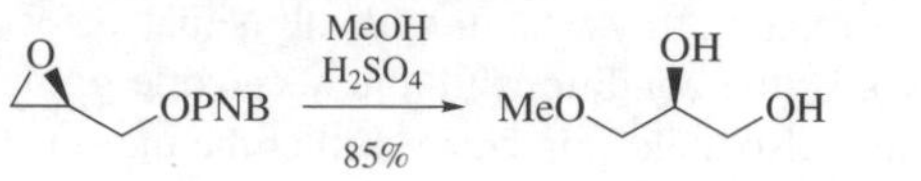

(14)

Lewis acid catalysis of additions to 2,3-epoxy alcohols often improves the regioselectivity of the ring-opening process.[26] $Ti(OR)_4$ catalyzed reaction of glycidol with primary alcohols gives 1-*O*-alkylglycerols in yields of 45–59%.[27] The addition of primary alcohols to (*R*)-glycidyl sulfonate esters give 1-*O*-alkylglycerol 3-sulfonates in yields of 73–89% when catalyzed with $BF_3{\cdot}OEt_2$ (eq 15).[28] Non-racemic glycidol, generated by catalytic asymmetric epoxidation of allyl alcohol with $Ti(O\text{-}i\text{-}Pr)_4$ and a (+)- or (−)-dialkyl tartrate, undergoes ***Titanium Tetraisopropoxide*** assisted reaction in situ with sodium phenolates to generate 1-*O*-arylglycerols (eq 16).[8,29] The $BF_3{\cdot}OEt_2$ addition of stearic anhydride to (*R*)-glycidyl tosylate gives (*R*)-1,2-distearoylglyceryl tosylate in 76% yield.[30]

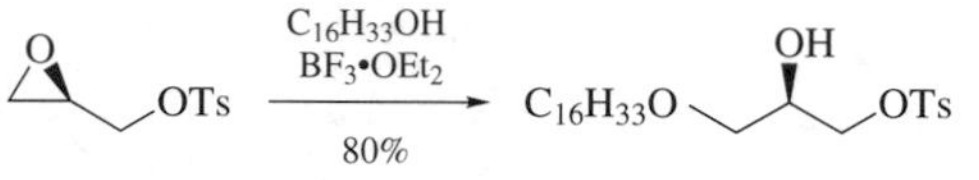

(15)

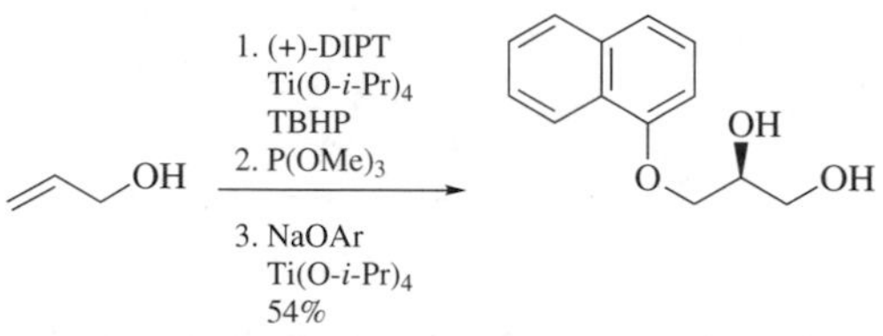

(16)

Examples of other oxygen nucleophiles that have been added at C-3 include phosphorylcholine (eq 17)[31] and ethyl *N*-hydroxyacetimidate (eq 18).[32]

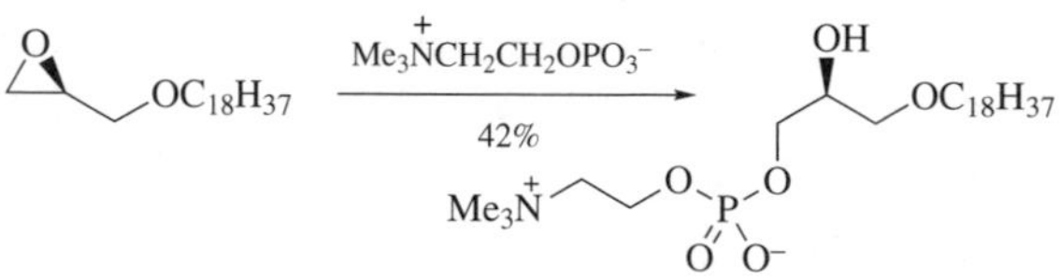

(17)

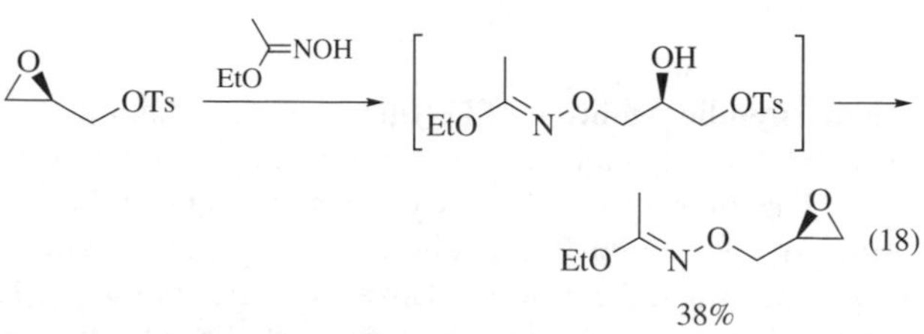

(18)

Nucleophilic Additions of Nitrogen at C-3. *Ammonia* and amines add readily to glycidol and glycidol derivatives, giving the 1-aminopropane-2,3-diols (eq 19).[33] With ammonia and primary amines, an excess of the amine often is used to reduce the amount of addition by a second glycidol to the 1-aminopropane-2,3-diol. Secondary amines are used with glycidols in an equimolar ratio. Azide ion also opens glycidols at C-3. $Ti(O\text{-}i\text{-}Pr)_4$ or ***Aluminum Isopropoxide*** assisted openings with ***Azidotrimethylsilane*** have been examined with glycidol and a variety of derivatives[34] and give excellent yields of 3-azido-2-hydroxypropane 1-*O*-derivatives (eq 20). ***Sodium Azide*** has also been used as a source of azide when combined with either ***Pyridinium p-Toluenesulfonate***,[7] NH_4Cl,[9,35] or ***Lithium Perchlorate***[36] to react with various glycidols.

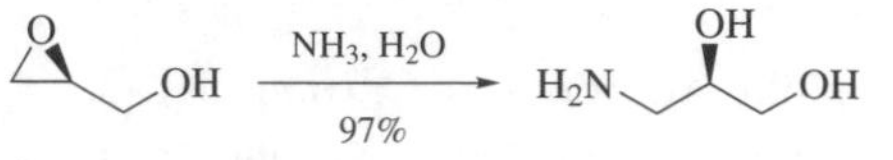

(19)

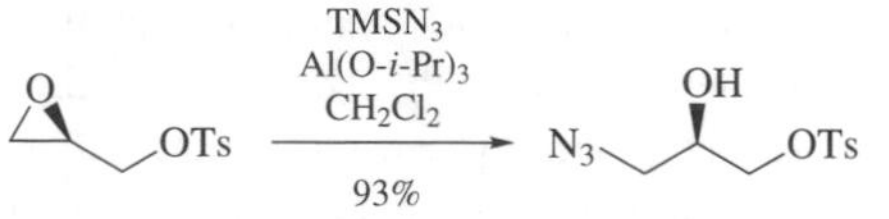

(20)

The opening of glycidols, especially of *O*-aryl glycidol ethers, at C-3 with amines has found extensive application in pharmaceutical research.[1b] A typical example is in the opening of *O*-(1-naphthyl)glycidol at C-3 with isopropylamine to generate the β-adrenergic blocking agent propranolol (eq 21).[29] A similar application is the addition of the 4-substituted piperazine to glycidol shown in eq 22.[37] With (*R*)- and (*S*)-glycidol now readily available, the synthesis of individual enantiomers or diastereoisomers of a pharmacological agent by methods such as those shown in eqs 21 and 22 becomes an attractive goal.

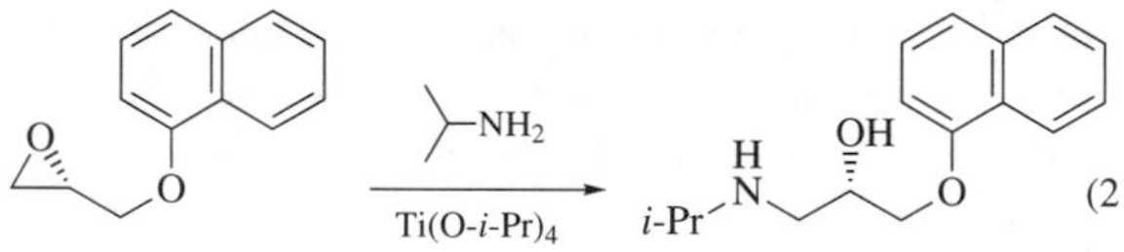

(21)

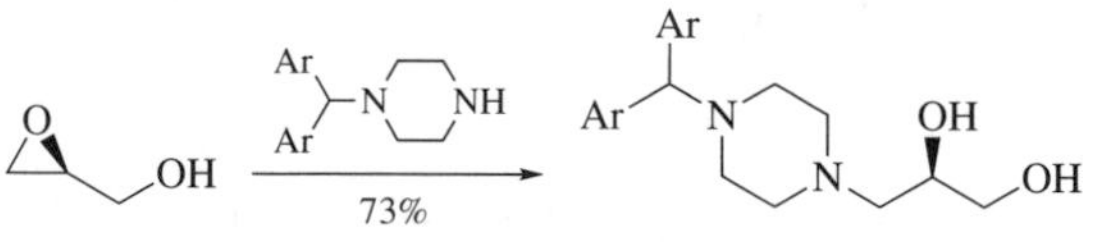

(22)

Other nitrogen nucleophiles added to glycidol include several heterocycles, an example of which is the addition of ***Imidazole***.[38] The iminodioxolane shown in eq 23 adds to glycidol and then undergoes further cyclization to a cyclic urethane.[39] ***Acetonitrile*** in a $BF_3{\cdot}OEt_2$- catalyzed reaction adds to glycidyl tosylate to form 2-methyl-4-(tosyloxy)methyloxazoline (eq 24).[40] Dibenzylamine adds via an amidocuprate at C-3 of *O*-phenylglycidol to give 3-dibenzylaminopropane-1,2-diol 1-*O*-phenyl ether in 94% yield.[41]

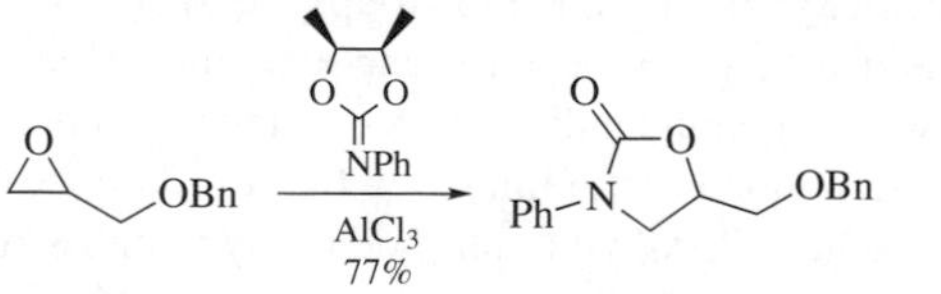

(23)

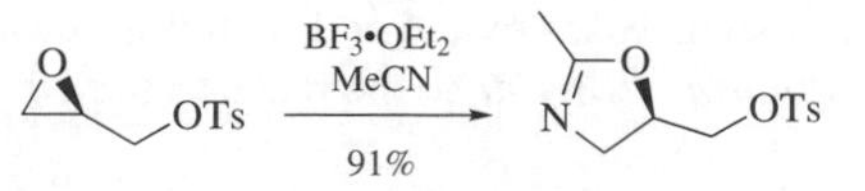

(24)

Additions of Other Nucleophiles at C-3. The halogens (F, Cl, Br, and I) and sulfur are the other elements most frequently found in C-3 additions to glycidols. Fluoride has been added to both glycidol and various glycidol (see eq 25) derivatives using tetrabutylammonium dihydrogentrifluoride.[42] Several methods have been used for the other three halogens, including reaction with the lithium salts in THF[43] or with the ammonium salts and $LiClO_4$ in AcCN.[44] Chlorine has also been added via HCl[45] or with ***Benzoyl Chloride/Cobalt(II) Chloride***;[46] the latter reaction also adds the benzoyl group to give the 2-*O*-benzoate derivative (eq 26). Bromine has been added with dimethylboron bromide[47] and iodine has been added with ***Sodium Iodide*** in a $NaOAc/HOAc/EtCO_2H$ system.[7]

(25) Glycidyl phenyl ether → F–CH2CH(OH)CH2OPh; reagents: $Bu_4NH_2F_3$, KHF_2; 90%

(26) Glycidyl phenyl ether → Cl–CH2CH(OCOPh)CH2OPh; reagents: PhCOCl, $CoCl_2$; 85%

Most additions of sulfur to glycidols have been of arylthiolates and are performed under either acidic [Ti(O-*i*-Pr)$_4$ (eq 27),[8] $BF_3{\cdot}OEt_2$[48]] or alkaline conditions.[7,49] $LiClO_4$[50] or $CoCl_2$[51] have also been used as catalysts for addition of aryl thiols. The additions of lithium alkylthiolates and of thiobenzoic acid to *O*-trityl glycidol have been reported.[5a]

(27) Glycidol → PhS–CH2CH(OH)CH2OH; reagents: PhSH, Ti(O-*i*-Pr)$_4$; 88%

Oxidation of Glycidol. Glycidol is oxidized to glycidic acid with ***Ruthenium(IV) Oxide***.[52] Glycidaldehyde is a mutagenic compound that has been prepared in racemic form by epoxidation of ***Acrolein***[53] and in nonracemic forms by the degradation of mannitol.[54] Alternately, (*R*)- and (*S*)-glycidaldehyde may be prepared and handled more conveniently via asymmetric dihydroxylation of acrolein benzene-1,2-dimethanol acetal followed by conversion of the diol to an epoxide (see eq 28).[55]

(28) Acrolein benzene-1,2-dimethanol acetal → diol; reagents: $K_2OsO_2(OH)_4$, DHQD-PHN, $K_3Fe(CN)_6$, K_2CO_3, *t*-BuOH, H_2O; 60%. Diol → epoxide; reagents: 1. TsCl, py 2. NaOMe; 85%

Miscellaneous. Glycidol reacts with dinitrogen pentoxide (N_2O_5) in CH_2Cl_2 in the presence of $AlCl_3$, giving trinitroglycerine (73%).[56] A useful review describing numerous synthetic transformations of 2,3-*O*-isopropylideneglyceraldehyde, a three-carbon synthon related to glycidol, has been published.[57]

Related Reagents. Epichlorohydrin; Ethylene Oxide; Glycidyl Tosylate; Propylene Oxide.

1. (a) Kleemann, A.; Wagner, R. M. *Glycidol*; Hüthig: New York, 1981. (b) Hanson, R. M. *CRV* **1991**, *91*, 437.
2. Gao, Y.; Hanson, R. M.; Klunder, J. M.; Ko. S. Y.; Masamune, H.; Sharpless, K. B. *JACS* **1987**, *109*, 5765.
3. (a) Ladner, W. E.; Whitesides, G. M. *JACS* **1984**, *106*, 7250. (b) Fu, H.; Newcomb, M.; Wong, C.-H. *JACS* **1991**, *113*, 5878.
4. Lipshutz, B. H.; Moretti, R.; Crow, R. *OS* **1990**, *69*, 80.
5. (a) Hendrickson, H. S.; Hendrickson, E. K. *Chem. Phys. Lipids* **1990**, *53*, 115. (b) Kim, M.-J.; Choi, Y. K. *JOC* **1992**, *57*, 1605.
6. Klunder, J. M.; Onami, T.; Sharpless, K. B. *JOC* **1989**, *54*, 1295.
7. Ko, S. Y.; Masamune, H.; Sharpless, K. B. *JOC* **1987**, *52*, 667.
8. Ko, S. Y.; Sharpless, K. B. *JOC* **1986**, *51*, 5413.
9. (a) Swindell, C. S.; Krauss, N. E.; Horwitz, S. B.; Ringel, I. *JMC* **1991**, *34*, 1176. (b) McClure, D. E.; Engelhardt, E. L.; Mensler, K.; King, S.; Saari, W. S.; Huff, J. R.; Baldwin, J. J. *JOC* **1979**, *44*, 1826.
10. Kötz, A.; Richter, K. *JPR[2]* **1925**, *111*, 373.
11. Mitani, M.; Matsumoto, H.; Gouda, N.; Koyama, K. *JACS* **1990**, *112*, 1286.
12. Michael, A.; Weiner, N. *JACS* **1936**, *58*, 999.
13. Ciaccio, J. A.; Stanescu, C.; Bontemps, J. *TL* **1992**, *33*, 1431.
14. Abushanab, E.; Sarma, M. S. P. *JMC* **1989**, *32*, 76.
15. Lipshutz, B. H.; Garcia, E. *TL* **1990**, *31*, 7261.
16. Lipshutz, B. H.; Kozlowski, J. A. *JOC* **1984**, *49*, 1147.
17. Burgos, C. E.; Nidy, E. G.; Johnson, R. A. *TL* **1989**, *30*, 5081.
18. Kraus, G. A.; Frazier, K. *JOC* **1980**, *45*, 4820.
19. Williams, K.; Thompson, C. M. *SC* **1992**, *22*, 239.
20. Fitton, A. O.; Hill, J.; Jane, D. E.; Millar, R. *S* **1987**, 1140.
21. Narjes, F.; Schaumann, E. *S* **1991**, 1168.
22. Lai, M.-t.; Oh, E.; Shih, Y.; Liu, H.-w. *JOC* **1992**, *57*, 2471.
23. Lattuada, L.; Licandro, E.; Maiorana, S.; Molinari, H.; Papagni, A. *OM* **1991**, *10*, 807.
24. Takano, S.; Tomita, S.; Iwabuchi, Y.; Ogasawara, K. *S* **1988**, 610.
25. Takano, S.; Shimazaki, Y.; Takahashi, M.; Ogasawara, K. *CC* **1988**, 1004.
26. Caron, M.; Sharpless, K. B. *JOC* **1985**, *50*, 1557.
27. Johnson, R. A.; Burgos, C. E.; Nidy, E. G. *Chem. Phys. Lipids* **1989**, *50*, 119.
28. (a) Guivisdalsky, P. N.; Bittman, R. *JOC* **1989**, *54*, 4637. (b) Guivisdalsky, P. N.; Bittman, R. *JOC* **1989**, *54*, 4643. (c) Kazi, A. B.; Hajdu, J. *TL* **1992**, *33*, 2291. (d) Liu, Y.-j.; Chu, T.-y.; Engel, R. *SC* **1992**, *22*, 2367.
29. Klunder, J. M.; Ko, S. Y.; Sharpless, K. B. *JOC* **1986**, *51*, 3710.
30. Ali, S.; Bittman, R. *JOC* **1988**, *53*, 5547.
31. Cimetiere, B.; Jacob, L.; Julia, M. *TL* **1986**, *27*, 6329.
32. Stanek, J.; Frei, J.; Mett, H.; Schneider, P.; Regenass, U. *JMC* **1992**, *35*, 1339.
33. (a) Deveer, A. M. Th. J.; Dijkman, R.; Leuveling-Tjeenk, M.; van den Berg, L.; Ransac, S.; Batenburg, M.; Egmond, M.; Verheij, H. M.; de Haas, G. H. *B* **1991**, *30*, 10034. (b) Sowden, J. C.; Fischer, O. L. *JACS* **1942**, *64*, 1291.
34. (a) Sutowardoyo, K. I.; Emziane, M.; Lhoste, P.; Sinou, D. *T* **1991**, *47*, 1435. (b) Sutowardoyo, K. I.; Sinou, D. *TA* **1991**, *2*, 437.
35. (a) Trinh, M.-C.; Florent, J.-C.; Grierson, D. S.; Monneret, C. *TL* **1991**, *32*, 1447. (b) Konosu, T.; Oida, S. *CPB* **1992**, *40*, 609.
36. Chini, M.; Crotti, P.; Macchia, F. *TL* **1990**, *31*, 5641.
37. Press, J. B.; Falotico, R.; Hajos, Z. G.; Sawyers, R. A.; Kanojia, R. M.; Williams, L.; Haertlein, B.; Kauffman, J. A.; Lakas–Weiss, C.; Salata, J. J. *JMC* **1992**, *35*, 4509.
38. Banfi, A.; Benedini, F.; Sala, A. *JHC* **1991**, *28*, 401.
39. Baba, A.; Seki, K.; Matsuda, H. *JOC* **1991**, *56*, 2684.
40. Delgado, A.; Leclerc, G.; Cinta Lobato, M.; Mauleon, D. *TL* **1988**, *29*, 3671.

41. Yamamoto, Y.; Asao, N.; Meguro, M.; Tsukada, N.; Nemoto, H.; Sadayori, N.; Wilson, J. G.; Nakamura, H. *CC* **1993**, 1201.
42. Landini, D.; Albanese, D.; Penso, M. *T* **1992**, *48*, 4163.
43. Bajwa, J. S.; Anderson R. C. *TL* **1991**, *32*, 3021.
44. Chini, M.; Crotti, P.; Gardelli, C.; Macchia, F. *T* **1992**, *48*, 3805.
45. Baldwin, J. J.; Raab, A. W.; Mensler, K.; Arison, B. H.; McClure, D. E. *JOC* **1978**, *43*, 4876.
46. Iqbal, J.; Srivastava, R. R. *T* **1991**, *47*, 3155.
47. Guindon, Y.; Therien, M.; Girard, Y.; Yoakim, C. *JOC* **1987**, *52*, 1680.
48. Guivisdalsky, P. N.; Bittman, R. *JACS* **1989**, *111*, 3077.
49. Takano, S.; Akiyama, M.; Ogasawara, K. *JCS(P1)* **1985**, 2447.
50. Chini, M.; Crotti, P.; Giovani, E.; Macchia, F.; Pineschi, M. *SL* **1992**, 303.
51. Iqbal, J.; Pandey A.; Shukla, A.; Srivastava, R. R.; Tripathi, S. *T* **1990**, *46*, 6423.
52. Pons, D.; Savignac, M.; Genet, J. P. *TL* **1990**, *31*, 5023.
53. Payne, G. B. *JACS* **1959**, *81*, 4901.
54. Schray, K. J.; O'Connell, E. L.; Rose, I. A. *JBC* **1973**, *248*, 2214.
55. Oi, R.; Sharpless, K. B. *TL* **1992**, *33*, 2095.
56. Golding, P.; Millar, R. W.; Paul, N. C.; Richards, D. H. *T* **1993**, *49*, 7037.
57. Jurczak, J.; Pikul, S.; Bauer, T. *T* **1986**, *42*, 447.

Roy A. Johnson
The Upjohn Company, Kalamazoo, MI, USA

Carmen E. Burgos-Lepley
Cortech, Denver, CO, USA

Hexabutyldistannane[1]

$Bu_3Sn\text{–}SnBu_3$

[813-19-4] $C_{24}H_{54}Sn_2$ (MW 580.20)

(source of tributylstannyl radicals; used in palladium-catalyzed tin–carbon bond formation; used for deoxygenation and desulfurization reactions)

Alternate Name: hexabutylditin.
Physical Data: bp 147–150 °C/0.2 mmHg, 198 °C/10 mmHg; *d* 1.1520 g cm^{-3}; n_D^{20} = 1.5120.
Solubility: sol most organic solvents.
Form Supplied in: colorless oil; readily available and not expensive.
Analysis of Reagent Purity: ^{119}Sn NMR recommended (δ −83 ppm, $^1J(^{119}Sn\text{–}^{119}Sn)$ 2748 Hz).[9]
Preparative Methods: various methods are available. The 'classical' method uses ***Tri-n-butylstannane*** and either ***Bis(tri-n-butyltin) Oxide*** (0.5 equiv), tributyldiethylaminotin, or ***Tri-n-butyl(methoxy)stannane*** in yields of 90–98%.[2] Since hexabutyldistannoxane is commercially available and not air-sensitive, its use is preferable. A second method involving the stannoxane involves its treatment with metals (Mg, Na, K, Ti/K); the yields lie between 70 and 80%.[3]

It can also be obtained from ***Tri-n-butylchlorostannane*** and either lithium, sodium,[4] or magnesium,[5] from tributyltin chloride and ***Tri-n-butylstannyllithium***[7] (the latter from tributyltin hydride and ***Lithium Diisopropylamide***[6]), or by catalytic elimination of hydrogen from tributyltin hydride. This can be effected either by various bases[8a] or by palladium catalysts such as ***Tetrakis(triphenylphosphine)palladium(0)***.[8b]
Purification: reversed-phase flash chromatography using C-18.[10]
Handling, Storage, and Precautions: must be stored in the absence of oxygen, moisture, and light (preferably under argon). Highly toxic. Use in a fume hood.

Introduction. General aspects of the preparation and chemistry of hexaalkyldistannanes are reviewed in the Houben–Weyl volume on organotin compounds (literature coverage up to 1977).[1]

The chemistry of hexabutyldistannane is determined by the weakness of the tin–tin bond and dominated by three aspects:

1) it dissociates on heating to give tributylstannyl radicals, which are of considerable importance in organic synthetic transformations;

2) under the influence of palladium catalysts it can be used for tin–carbon bond formation in the sense of either substitution reactions or addition reactions to multiply bonded systems;

3) it can readily take up oxygen or sulfur and can thus be used for deoxygenation and desulfurization reactions.

It can also be used for the preparation of other organometallic reagents via transmetalation.

Use as a Source of Tributylstannyl Radicals. Thermolysis or photolysis of hexabutyldistannane generates tributylstannyl radicals; these can in turn be used to generate other synthetically useful radicals, for example carbon radicals from reactions with organic halides or sulfides.[11] This reaction has the advantage over generation of stannyl radicals from triorganotin hydrides in that the latter can act as powerful hydrogen donors and thus strongly influence subsequent reactions.

The distannane can also be used in less than stoichiometric amounts, the reactants being irradiated in the presence of a small amount of the distannane (generally 0.1–0.3 equiv). This so-called 'atom transfer' method has been developed by Curran and also applied by other authors. Examples include the radical cyclization of α-iodo ketones (or esters with a suitably placed double bond) (eq 1),[12] annulation reactions of iodomalonates (eq 2),[13] cyclizations of unsaturated α-iodocarbonyls,[14] and radical cyclization of α-fluoro-α-iodo and α-iodo esters and amides (eq 3).[15] A novel cyclization of an iodo epoxide to a cyclopentanol possessing the hydrindane skeleton has also been reported (eq 4).[16] This type of reaction may be subject to a temperature effect; thus the cyclization of allylic α-iodo esters and amines proceeds much more efficiently at 80 °C than at 25 °C.[17]

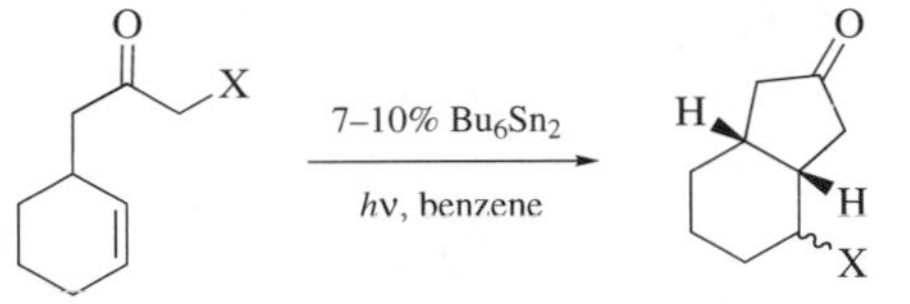

7–10% Bu_6Sn_2, *hν*, benzene; X = Br, I (1)

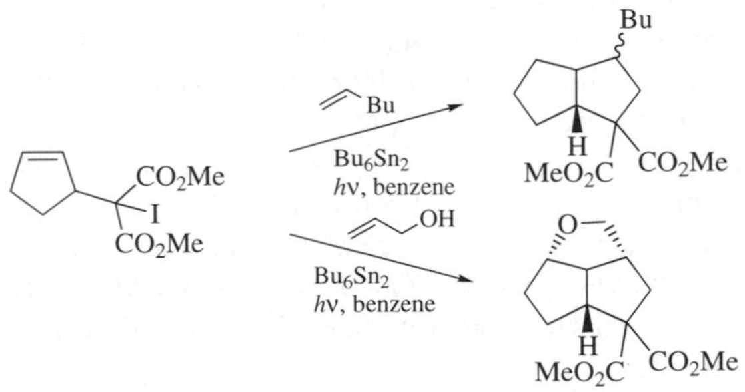

Bu_6Sn_2, *hν*, benzene (with 1-hexene or allyl alcohol) (2)

Bu_6Sn_2, *hν*, benzene (3)

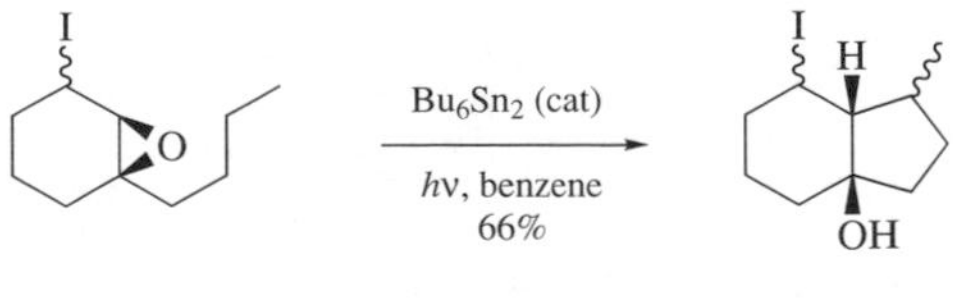

Bu_6Sn_2 (cat), *hν*, benzene, 66% (4)

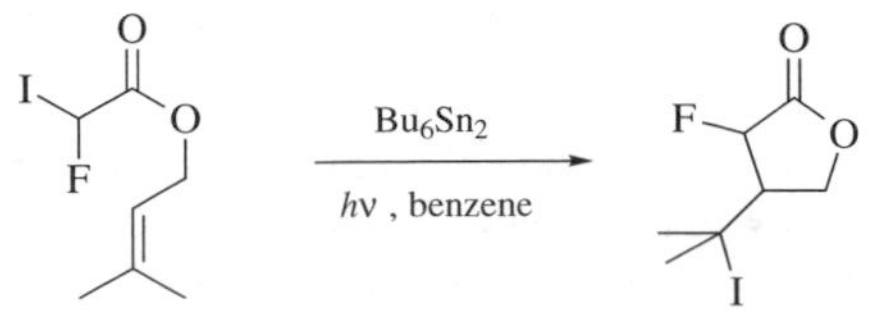

An in situ generation of nitrile oxides via photolysis of hexabutyldistannane has been reported (eq 5).[18]

$$R^1C(Cl)=NOH + HCR^2{=}CHR^3 \xrightarrow[h\nu,\ \text{benzene},\ 39\text{–}92\%]{Bu_6Sn_2} \text{isoxazoline} \quad (5)$$

R^1 = Ph, Et, Bu

Palladium-Catalyzed Reactions. These can be of two types, either substitution of a (generally halide) ligand by a tributylstannyl group or addition to multiple bonds. These reactions have been reviewed by Stille[19] and Mitchell.[20]

The substitution reactions provide a useful alternative to the conventional use of ***Tri-n-butylstannyllithium***, which is a very strong base. The halides used have mainly been aryl or heteroaryl halides, while ***Tetrakis(triphenylphosphine)palladium(0)***, dichlorobis(triphenylphosphine)palladium(II), and ***Bis(allyl)di-μ-chlorodipalladium/Tetra-n-butylammonium Fluoride*** have been employed as catalysts.

The synthesis of symmetrical biaryls has been described.[21] This principle has recently been applied in an intramolecular manner and extended to include more complex cyclization reactions. Thus as well as two (symmetrical or mixed) aryl and benzyl halide moieties, a combination of two aryl iodide moieties with a carbon–carbon double or triple bond can be used (eq 6).[22]

(6) 2 equiv Bu_6Sn_2, Pd^0, 48%

Unfortunately the reaction of Bu_6Sn_2 with acyl halides is not suitable for the preparation of tributylacylstannanes, in contrast to the corresponding reaction of ***Hexamethyldistannane***.[23]

A further interesting development is the use of three-component systems: a mixture of hexabutyldistannane, an allyl halide or carbonate, and a heteroaryl bromide leads to allylation of the heteroaromatic moiety; the catalyst used was the somewhat 'exotic' ***Dichloro[1,1′-bis(diphenylphosphino)ferrocene]palladium (II)***.[24] This method avoids the prior preparation of organotin reagents (eq 7).

$$\text{Ar–Br} + CH_2{=}CHCH_2OAc \xrightarrow[PdCl_2(dppf)]{Bu_6Sn_2} \text{Ar}CH_2CH{=}CH_2 \quad (7)$$

In the case of the addition reactions, hexabutyldistannane adds readily to a variety of allenes; it is often possible to distinguish a kinetic and a thermodynamic product (eq 8).[25]

$$Bu_6Sn_2 + RCH{=}C{=}CH_2 \xrightarrow{[Pd]}$$

$$Bu_3SnCH(R)C(SnBu_3){=}CH_2 \xrightarrow{\Delta} Bu_3SnCH_2C(SnBu_3){=}CHR \quad (8)$$

On the other hand, the addition to alkynes does not proceed in a quantitative manner at atmospheric pressure (in contrast to the behavior of hexamethyldistannane), though it can be forced to do so by the application of high pressure.[26]

Deoxygenation and Desulfurization Reactions. The deoxygenation of amine oxides has been described,[27] as has the photo-desulfurization of 1,3-dithiole-2-thiones to give tetrathiofulvalenes.[28]

Use as a Source of Other Tributylstannylmetal Compounds. Hexabutyldistannane can be cleaved by lithium metal[29] or a lithium alkyl (e.g. MeLi)[30] to give tributylstannyllithium. These methods cannot be recommended: Bu_3SnLi can be better obtained either in a one-step process from Bu_3SnCl and Li (via Bu_6Sn_2 which is not isolated) or from the reaction between Bu_3SnH and LDA[6] (see ***Tri-n-butylstannyllithium*** for details). Hexabutyldistannane can also serve as a source of stannylcuprates ($Bu_3SnCu(CN)Li$, $(Bu_3Sn)_2Cu(CN)Li_2$, $Bu_3Sn(R)Cu(CN)Li_2$).[31]

Use in Electron-Transfer Reactions. It has been shown in a very recent development that Bu_6Sn_2 (and in addition Bu_4Sn) can take part in novel substitution reactions when allowed to react with pyridine derivatives.[32] Thus treatment of 4-cyanopyridine with Bu_6Sn_2 leads to a new type of *ipso* substitution accompanied by a substitution of the type observed when lepidine reacts with Bu_6Sn_2 and *t*-butyl bromide (eqs 9 and 10).

$$\text{4-cyanopyridine} + Bu_6Sn_2 \xrightarrow{h\nu} \text{4-butylpyridine (54\%)} + \text{2-butyl-4-cyanopyridine (30\%)} \quad (9)$$

$$\text{lepidinium} + Bu_6Sn_2 + t\text{-BuBr} \xrightarrow{h\nu} \text{2-}t\text{-butyllepidine (50\%)} + \text{2-butyllepidine (50\%)} \quad (10)$$

Related Reagents. Allyltributylstannane; Chlorotrimethylstannane; Hexamethyldistannane; Tri-*n*-butylchlorostannane; Tri-*n*-butylstannane; Tri-*n*-butylstannyllithium; Trimethylstannane.

1. Bähr, G.; Pawlenko, S. *MOC* **1978**, *13/6*, 401.
2. Neumann, W. P.; Schneider, B.; Sommer, R. *LA* **1966**, *692*, 1.
3. Jousseaume, B.; Chanson, E.; Pereyre, M. *OM* **1986**, *5*, 1271.
4. Zimmer, H.; Homberg, O. A.; Jayawant, M. *JOC* **1966**, *31*, 3857.
5. Shirai, H.; Sato, Y.; Niwa, N. *YZ* **1970**, *90*, 59 (*CA* **1970**, *72*, 90 593).
6. Still, W. C. *JACS* **1978**, *100*, 1481.
7. Wittig, G.; Meyer, F. J.; Lange, G. *LA* **1951**, *571*, 167.
8. (a) Neumann, W. P. *AG* **1961**, *73*, 542. (b) Bumagin, N. A.; Gulevich, Yu. V.; Beletskaya, I. P. *IZV* **1984**, 1137; Mitchell, T. N.; Amamria, A.; Killing, H.; Rutschow, D. *JOM* **1986**, *304*, 257.

9. Mitchell, T. N.; Walter, G. *JCS(P2)* **1977**, 1842.
10. Farina, V. *JOC* **1991**, *56*, 4985.
11. Baldwin, J. E.; Kelly, D. R.; Ziegler, C. B. *CC* **1984**, 133.
12. Curran, D. P.; Chang, C.-T. *TL* **1987**, *28*, 2477; Curran D. P.; Chang, C.-T. *JOC* **1989**, *54*, 3140.
13. Curran, D. P.; Chen, M.-H.; Spletzer, E.; Seong, C. M.; Chang, C.-T. *JACS* **1989**, *111*, 8872.
14. Curran, D. P.; Chang, C.-T. *TL* **1990**, *31*, 933.
15. Barth, F.; Yang, C.-O. *TL* **1990**, *31*, 1121.
16. Rawal, V. H.; Iwasa, S. *TL* **1992**, *33*, 4687.
17. Curran, D. P.; Tamine, J. *JOC* **1991**, *56*, 2746.
18. Kim, B. H. *SC* **1987**, *17*, 1199.
19. Stille, J. K. *AG* **1986**, *98*, 504; *AG(E)* **1986**, *25*, 508.
20. Mitchell, T. N. *S* **1992**, 803.
21. Gulevich, Yu. V.; Beletskaya, I. P. *Metalloorg. Khim.* **1988**, *1*, 704.
22. Grigg, R.; Teasdale, A.; Sridharan, V. *TL* **1991**, *32*, 3859.
23. Mitchell T. N.; Kwetkat, K. *JOM* **1992**, *439*, 127.
24. Yokoyama, Y.; Ikeda, M.; Saito, M.; Yoda, T.; Suzuki, H.; Murakami, Y. *H* **1990**, *31*, 1505.
25. Mitchell T. N.; Schneider, U. *JOM* **1991**, *407*, 319.
26. Mitchell, T. N.; Dornseifer, N. M.; Rahm, A. *J. High Pressure Res.* **1991**, *7*, 165.
27. Jousseaume, B.; Chanson, E. *S* **1987**, 55.
28. Ueno, Y.; Nakayama, A.; Okawara, M. *JACS* **1976**, *98*, 7440.
29. Tamborski, C.; Ford, F. E.; Soloski, E. J. *JOC* **1963**, *28*, 237.
30. Still, W. C. *JACS* **1977**, *99*, 4836.
31. Singer, R. D.; Hutzinger, M. W.; Oehlschlager, A. C. *JOC* **1991**, *56*, 4933.
32. Minisci, F.; Fontana, F.; Caronna, T.; Zhao, L. *TL* **1992**, *33*, 3201.

Terence N. Mitchell
University of Dortmund, Germany

Hexamethyldistannane[1]

$Me_3Sn{-}SnMe_3$

[661-69-8] $C_6H_{18}Sn_2$ (MW 327.66)

(palladium-catalyzed substitution and addition reactions)

Physical Data: colorless oil, bp 62–63 °C/12 mmHg, 182 °C/756 mmHg; mp 23 °C; $n_D^{20} = 1.5321$.
Solubility: sol most organic solvents.
Form Supplied in: colorless oil; readily available but highly expensive.
Analysis of Reagent Purity: ^{119}Sn NMR recommended (δ −109 ppm, $^1J(^{119}Sn{-}^{119}Sn)$ 4404 Hz).[8]
Preparative Methods: although the preparation from ***Trimethylstannane*** and trimethyl(diethylamino)stannane affords an excellent yield (95%), it has the disadvantage of requiring two extremely air-sensitive starting materials.[2] Better methods involve the use of ***Chlorotrimethylstannane*** with either ***Acetaldehyde*** or ***Sodium***,[3] or ***Trimethylstannyllithium*** (the latter from trimethylstannane and ***Lithium Diisopropylamide***[4]).[5] Catalytic elimination of hydrogen from trimethylstannane also affords hexamethyldistannane and can be effected either by various bases[6a] or by palladium catalysts such as ***Tetrakis(triphenylphosphine)palladium(0)***.[6b] This latter method can be recommended. The preparation of hexamethyldistannane starting from bis(trimethylstannyl) sulfide has also been described.[7]
Handling, Storage, and Precautions: the compound must be stored in the absence of oxygen, moisture, and light (preferably under argon). Very highly toxic. Use in a fume hood.

Introduction. General aspects of the preparation and chemistry of hexaalkyldistannanes are reviewed in the Houben–Weyl volume on organotin compounds (literature coverage up to 1977).[1]

The chemistry of hexamethyldistannane is determined by the weakness of the tin–tin bond. The main reaction of synthetic importance is its use in palladium-catalyzed substitution reactions or addition reactions to multiply bonded systems.

While, in analogy to ***Hexabutyldistannane***, Me_6Sn_2 dissociates on heating to give trimethylstannyl radicals, these are generally no longer used in organic synthetic transformations because of the toxicity of methyltin compounds and their cost. Me_6Sn_2 can also be used in transmetalation reactions, leading to the formation of other trimethylstannylmetal reagents.

Palladium-Catalyzed Reactions. These can be of two types, either substitution of a ligand (generally halide) by a trimethylstannyl group or addition to multiple bonds. These reactions have been reviewed by Stille[9] and Mitchell.[10]

The substitution reactions provide an extremely useful alternative to the conventional use of ***Trimethylstannyllithium***, which is a very strong base. The halides used have mainly been aryl or heteroaryl halides, while ***Tetrakis(triphenylphosphine)palladium(0)***, dichlorobis(triphenylphospine)palladium(II), and ***Bis(allyl)di-μ-chlorodipalladium–Tetra-n-butylammonium Fluoride*** have generally been employed as catalysts. In more recent developments, couplings involving enol triflates[11] and aryl triflates[12] have been described.

The first example of a *cine* substitution was observed when a vinylstannane derived from a camphor triflate and hexamethyldistannane was allowed to react with bromobenzene (eq 1).[13]

The synthesis of symmetrical biaryls (using Bu_6Sn_2) has been described.[14] This principle has recently been applied in an intramolecular manner and extended to include more complex cyclization reactions; hexamethyldistannane can be used as an alternative to hexabutyldistannane. Thus as well as two (symmetrical or mixed) aryl and benzyl halide moieties, a combination of two aryl iodide moieties with a carbon–carbon double or triple bond can be used (eq 2).[15]

Me_6Sn_2, [Pd] (2)

The reaction of Me_6Sn_2 with acyl halides provides an excellent method for the preparation of trimethylacylstannanes (eq 3).[16]

$$Me_6Sn_2 + RCOCl \xrightarrow{[Pd]} Me_3SnCOR + Me_3SnCl \quad (3)$$

In the case of the addition reactions, hexamethyldistannane adds readily to a variety of allenes to give 2,3-distannyl-1-propenes; it is often possible to distinguish a kinetic and a thermodynamic product (eq 4).[17]

$$Me_6Sn_2 + RCH{=}C{=}CH_2 \xrightarrow{[Pd]}$$

(4)

The addition to 1-alkynes proceeds in a quantitative manner at atmospheric pressure to give (*Z*)-1,2-distannylalkenes, which can in some cases undergo photochemical isomerization to afford the corresponding (*E*)-products in quantitative yields (eq 5).[18]

$$RC{\equiv}CH + Me_6Sn_2 \xrightarrow{[Pd]}$$

(5)

The presence of various functional groups is tolerated; thus the preparation of trimethylstannyl-substituted allylglycine derivatives[19] and (*Z*)-4-trimethylstannyl-1,3-butadienes[20] (via the enones) have been described (eqs 6 and 7). The first tris(trimethylstannyl)ethylenes were obtained by using a more active precatalyst system.[21]

$+ Me_6Sn_2$, $Pd(PPh_3)_4$, 60% (6)

$+ Me_6Sn_2$, $Pd(PPh_3)_4$; $Ph_3P{=}CH_2$ (7)

In spite of the development of new catalyst systems, additions to nonterminal alkynes remain a virtually unsolved problem. Two groups have described the addition of hexamethyldistannane to 1,3-dienes; depending on the reaction conditions, the products are either (*Z*)-1,4-bis(trimethylstannyl)-2-butenes[22] or the products of addition/dimerization (eqs 8 and 9).[23]

$+ Me_6Sn_2$, $Pd(dba)_2$, $P(OEt)_3$ (8)

R = H, Me, Ph

$+ Me_6Sn_2$, $Pd(dba)_2$ (9)

R = H, Me, OTMS

Use as a Source of Trimethylstannyl Radicals. Thermolysis or photolysis of hexamethyldistannane generates trimethylstannyl radicals, which can in turn be used to generate other synthetically useful radicals. On this basis, Curran has developed the so-called 'atom transfer' method for radical cyclization, in which about 0.1 equiv. of the distannane is used. While in his original work[24] on the cyclization of α-iodo esters, ketones, and malonates Curran used both hexamethyl- and hexabutyldistannane, later papers by his and other groups have reported only the use of Bu_6Sn_2 (for reasons of cost and toxicity). Me_6Sn_2 has been used to effect a selective one-electron radical chain reduction of the 10-methylacridinium ion to 10,10′-dimethyl-9,9′-biacridine.[25] In combination with ***1,1-Di-t-butyl Peroxide*** (as the radical initiator) it can generate sulfonyl radicals from sulfonate esters.[26]

Use as a Source of Other Trimethylstannylmetal Reagents. Hexamethyldistannane can be cleaved by lithium metal[27] or a lithium alkyl[28] (e.g. MeLi) to give trimethylstannyllithium. These methods cannot be recommended: Me_3SnLi can be better obtained either in a one-step process from Me_3SnCl and Li (via Me_6Sn_2 which is not isolated) or from the reaction between Me_3SnH and LDA (see ***Trimethylstannyllithium*** for details). Hexamethyldistannane can also serve as a source of stannylcuprates ($Me_3SnCu(CN)Li$, $(Me_3Sn)_2Cu(CN)Li_2$,[29] $Me_3Sn(2\text{-thienyl})Cu(CN)Li_2$).[30]

Related Reagents. Chlorotrimethylstannane; Hexabutyldistannane; Trimethylstannane; Trimethylstannylcopper-Dimethyl Sulfide; Trimethylstannyllithium.

1. Bähr, G.; Pawlenko, S. *MOC* **1978**, *4*, 401.
2. Neumann, W. P.; Schneider, B.; Sommer, R. *LA* **1966**, *692*, 1.
3. Zimmer, H.; Homberg, O. A.; Jayawant, M. *JOC* **1966**, *31*, 3857.
4. Still, W. C. *JACS* **1978**, *100*, 1481.
5. Wittig, G.; Meyer, F. J.; Lange, G. *LA* **1951**, *571*, 167.
6. (a) Neumann, W. P. *AG* **1961**, *73*, 542. (b) Bumagin, N. A.; Gulevich, Yu. V.; Beletskaya, I. P. *IZV* **1984**, 1137. Mitchell, T. N.; Amamria, A.; Killing, H.; Rutschow, D. *JOM* **1986**, *304*, 257.
7. Capozzi, G.; Menichetti, S.; Ricci, A.; Taddei, M. *JOM* **1988**, *344*, 285.
8. Mitchell, T. N.; Walter, G. *JCS(P2)* **1977**, 1842.
9. Stille, J. K. *AG* **1986**, *98*, 504; *AGE* **1986**, *25*, 508.
10. Mitchell, T. N. *S* **1992**, 803.
11. Barber, C.; Jarowicki, K.; Kocienski, P. *SL* **1991**, 197.
12. Echavarren, A. M.; Stille, J. K. *JACS* **1987**, *109*, 5478.
13. Stork, G.; Isaacs, R. C. A. *JACS* **1990**, *112*, 7399.

14. Gulevich, Yu. V.; Beletskaya, I. P. *Metalloorg. Khim.* **1988**, *1*, 704.
15. Grigg, R.; Teasdale, A.; Sridharan, V. *TL* **1991**, *32*, 3859.
16. Mitchell, T. N.; Kwetkat, K. *JOM* **1992**, *439*, 127.
17. Mitchell, T. N.; Schneider, U. *JOM* **1991**, *407*, 319. Killing, H.; Mitchell, T. N. *OM* **1984**, *3*, 1318.
18. Mitchell, T. N.; Amamria, A.; Killing, H.; Rutschow, D. *JOM* **1986**, *304*, 257.
19. Crisp, G. T.; Glink, P. T. *TL* **1992**, *33*, 4649.
20. Piers, E.; Tillyer, R. D. *JCS(P1)* **1989**, 2124.
21. Mitchell, T. N.; Kowall, B. *JOM* **1992**, *437*, 127.
22. Mitchell, T. N.; Kowall, B.; Killing, H.; Nettelbeck, C. *JOM* **1992**, *439*, 101.
23. Tsuji, Y.; Kakehi, T. *CC* **1992**, 1000.
24. Curran, D. P.; Chang, C.-T. *TL* **1987**, *28*, 2477. Curran, D. P.; Chang, C.-T. *JOC* **1989**, *54*, 3140.
25. Fukuzumi, S.; Kitano, T.; Mochida, K. *JACS* **1990**, *112*, 3246.
26. Culshaw, P. N.; Walton, J. C. *JCS(P2)* **1991**, 1201.
27. Tamborski, C.; Ford, F. E.; Soloski, E. J. *JOC* **1963**, *28*, 237.
28. Still, W. C. *JACS* **1977**, *99*, 4836.
29. Singer, R. D.; Hutzinger, M. W.; Oehlschlager, A. C. *JOC* **1991**, *56*, 4933.
30. Piers, E.; Tillyer, R. D. *JOC* **1988**, *53*, 5366.

Terence N. Mitchell
University of Dortmund, Germany

Hydrogen Cyanide[1]

HCN

[74-90-8] CHN (MW 27.03)

(useful C_1 synthon)

Alternate Names: hydrocyanic acid; prussic acid.
Physical Data: mp −13.4 °C; bp 25.6 °C; *d* (gas) 0.941 g cm^{-3}; *d* (liquid) 0.687 g cm^{-3}.
Solubility: slightly sol ether; miscible with water and alcohol.
Form Supplied in: colorless gas at rt, or bluish-white when liquified, with a characteristic odor resembling bitter almonds.
Preparative Methods: anhydrous HCN is prepared on a large scale by the catalytic oxidation of ammonia–methane mixtures (Andrussow process).[1c] In the laboratory it can conveniently be prepared by acidifying NaCN or potassium hexacyanoferrate, $K_4[Fe(CN)_6]$.[3]

HCN oligomers:[4] the HCN dimer (HN=CHCN) is too reactive to be isolated. Under acidic conditions the reaction proceeds to the trimer *s*-triazine[5] and under basic conditions to the tetramer ***Diaminomaleonitrile***[6] via the trimer intermediate 2-aminomalononitrile.

Handling, Storage, and Precautions: **Intensely poisonous. Avoid skin contact and inhalation. Must be handled by specially trained experts**. Use in a fume hood. Flash point −17.8 °C (closed cup); explosion limits: upper 40%, lower 5.6% by vol in air; when not pure or stabilized, can polymerize explosively. The principal routes of occupational HCN exposure are inhalation and absorption through the skin. *Inhalation of large amounts of cyanide causes immediate unconsciousness, convulsions, and death from respiratory arrest within 1–15 min.* TLV ceiling of 10 ppm has been recommended since 1980.[2]

Electrophilic Aromatic Substitution. The preparation of aldehydes from phenols or phenol ethers by treatment of the aromatic substrate with HCN and ***Hydrogen Chloride*** in the presence of Lewis acid catalysts is known as the Gattermann aldehyde synthesis (eq 1).[7] The first stage consists of protonation of the nitrile to give a nitrilium ion or nitrile–Lewis acid complex, regarded as the electrophilic species attacking the substrates.[8]

PhOR —(HCN, HCl; $AlCl_3$ or $ZnCl_2$)→ [4-RO-C_6H_4-CH=NH•HCl] —(H_2O)→ 4-RO-C_6H_4-CHO + NH_4Cl (1)

R = Ph, 80%

Nucleophilic Substitution. ***Chlorotrimethylsilane*** is converted into ***Cyanotrimethylsilane*** (eq 2),[9] which is also a versatile cyanidation agent,[10] whereas carbonocyanidic acid methyl ester *[17640-15-2]* and ethyl ester *[623-49-4]*, introduced by Mander as selective *C*-acylation agents,[11] are made from the corresponding chloroformates (eq 3).[12]

Me_3SiCl —(HCN, TEA, −5 to 10 °C, Et_2O; 70%)→ Me_3SiCN (2)

$ClCO_2Et$ —(aromatic hydrocarbons, HCN, TEA, −14 °C; 82%)→ $NCCO_2Et$ (3)

Epoxide Opening. Addition of HCN to ***Epichlorohydrin*** gives an intermediate, which on quaternization with ***Trimethylamine*** and subsequent hydrolysis of the nitrile gives carnitine (eq 4).[13]

epichlorohydrin —(HCN)→ $ClCH_2CH(OH)CH_2CN$ —(TMA; H_2O)→ $^-O_2CCH_2CH(OH)CH_2\overset{+}{N}Me_3$ (4)

Additions to Alkenes, Dienes, and Alkynes. These occur in the presence of Ni^0, Pd^0, or cobalt carbonyl complexes.[14,15] The use of Lewis acids as cocatalysts generally improves the activity and lifetime of the catalyst.[16] Ethylene is hydrocyanated to propionitrile in the presence of ***Octacarbonyldicobalt*** or $[M(P(OR)_3)_4]$ (M = Ni, Pd; R = alkyl, aryl); with unsymmetrical higher alkenes, the cobalt catalyst gives exclusively the product from Markovnikov addition (e.g. propene gives 2-cyanopropane in 75% yield),[14] while the nickel and palladium phosphite systems give product mixtures.[17]

In the case of conjugated dienes, nickel catalysts are preferred; the mechanism involves a π-allyl intermediate. Both 1,4- and 1,2-addition occur, although the regiochemistry seems to be controlled by the thermodynamic stability of the intermediate allyl species, e.g. 1,3-pentadiene gives exclusively the branched nitrile (eqs 5 and 6).[18]

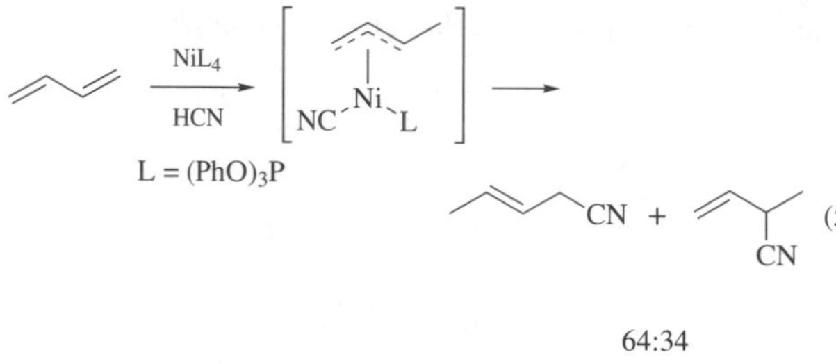

NiL_4, HCN; L = $(PhO)_3P$ (5)

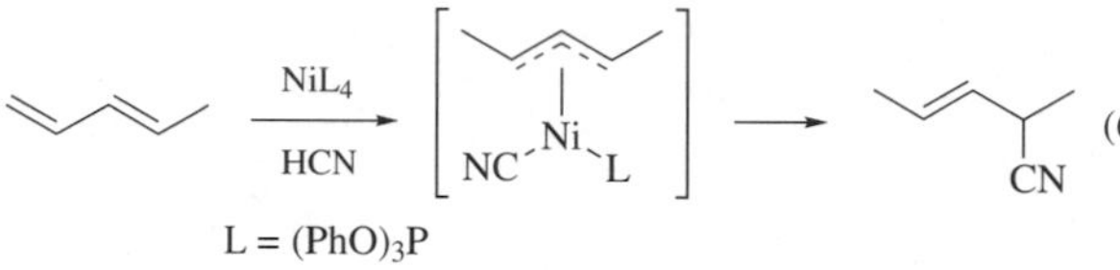

64:34

NiL_4, HCN; L = $(PhO)_3P$ (6)

In the presence of these catalysts, unconjugated dienes rearrange to conjugated dienes prior to the addition of HCN.

Addition to alkynes is best catalyzed by nickel complexes with *syn* stereochemistry. The regioselectivity is influenced by both steric and electronic effects.[19] The reactions with alkynols show evidence for some chelation between the oxygen and nickel in the transition state (eq 7).[20]

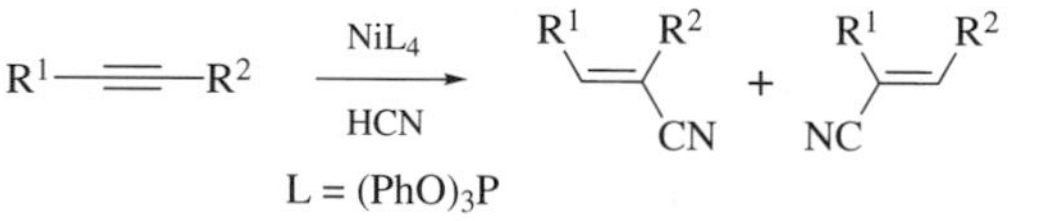

R^1—≡—R^2 $\xrightarrow[\text{HCN}]{NiL_4}$ L = $(PhO)_3P$ (7)

R^1 = Pr, R^2 = H	60%	12:88
R^1 = *t*-Bu, R^2 = Me	78%	90:10
R^1 = $Me_2C(OH)$, R^2 = H	30%	20:80

Hydrolysis and cyclization of some cyanoalkenes containing methoxymethyl ethers gives α-alkylidene-γ-lactones (eq 8).[21]

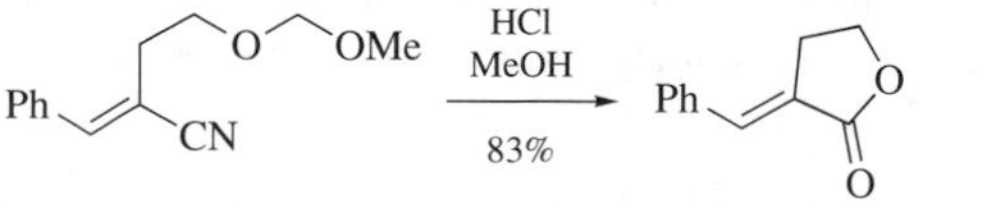

HCl, MeOH, 83% (8)

The asymmetric Markovnikov addition of HCN to vinylarenes in the presence of Ni^0 complexes of 1,2-diolphosphinites derived from sugars gives 2-aryl-2-propionitriles in high yield and enantioselectivity (eq 9).[22]

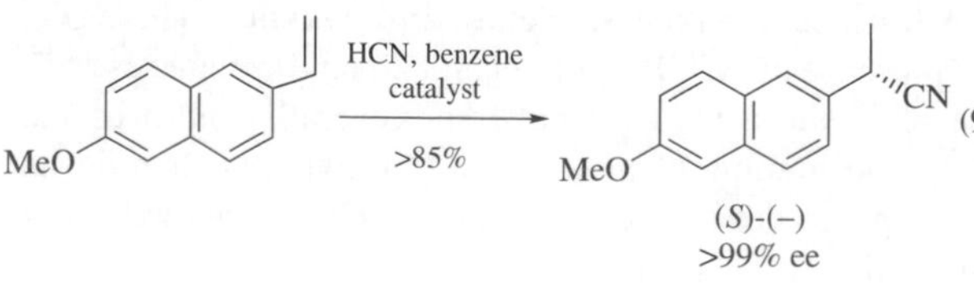

HCN, benzene, catalyst, >85% (9)

(*S*)-(−) >99% ee

Asymmetric Hydrocyanation. Asymmetric hydrocyanation of aldehydes to chiral cyanohydrins is a convenient route to chiral α-hydroxy carboxylic acids and chiral 1-amino-2-alkanols; catalysts are often used, e.g. the cyclic dipeptide cyclo((*S*)-phenylalanyl-(*S*)-histidyl) (eq 10)[23] or the enzyme (*R*)-oxynitrilase (eq 11).[24] This reaction proceeds normally in aqueous solutions with NaCN; however, the noncatalyzed reaction leading to racemic product can be efficiently suppressed by the use of anhydrous HCN in organic solvents.

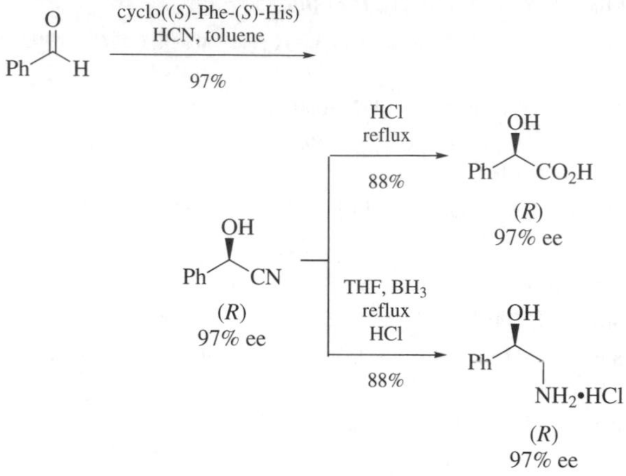

cyclo((*S*)-Phe-(*S*)-His), HCN, toluene, 97%; (*R*) 97% ee; HCl reflux 88% → (*R*) 97% ee; THF, BH_3 reflux, HCl 88% → NH_2•HCl (*R*) 97% ee (10)

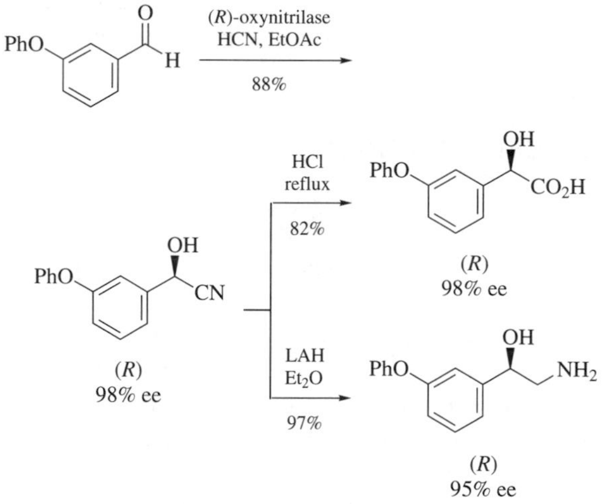

(*R*)-oxynitrilase, HCN, EtOAc, 88%; (*R*) 98% ee; HCl reflux 82% → (*R*) 98% ee; LAH Et_2O 97% → (*R*) 95% ee (11)

Enantioselective addition of HCN to ketones in organic solvents catalyzed by (*R*)-oxynitrilase proceeds similarly.[25]

Related Reagents. Acetone Cyanohydrin; Cyanotrimethylsilane; Diethylaluminum Cyanide; Lithium Cyanide; Potassium Cyanide; Sodium Cyanide; Zinc Cyanide.

---

1. (a) Jenks, W. R. In *Kirk-Othmer Encyclopedia of Chemical Technology*, 3rd ed.; Wiley: New York, 1979; Vol. 7, pp 307–319. (b) Klink, H.; Griffiths, A.; Huthmacher, K.; Ittzel, H.; Knorre, H.; Voigt, C.; Weitburg, O. In *Ullmann's Encyclopedia of Industrial Chemistry*, 5th ed.; VCH: Weinheim, 1978; Vol. A8, pp 159–165. (c) *The Merck Index*, 11th ed.; Budavari, S., Ed.; Merck: Rahway, NJ, 1989; p 760.
2. *Documentation of the Threshold Limit Values and Biological Exposure Indices*, 6th ed.; American Conference of Governmental Industrial Hygienists: Cincinnati, 1991; Vol. III, p 775.
3. Glemser, O. In *Handbook of Preparative Inorganic Chemistry*, 2nd ed.; Brauer, G., Ed.; Academic: New York, 1963; Vol. 1, pp 658–660.

4. Donald, D. S.; Webster, O. W. In *Advances in Heterocyclic Chemistry*; Academic: New York, 1987; Vol. 41, pp 1–36.
5. Grundmann, C. *AG(E)* **1963**, *2*, 309.
6. Okada, T.; Asai, N. Ger. Patent, 2 022 243, 1970 (*CA* **1971**, *74*, 22 456h).
7. Baltazzi, E.; Krimen, L. I. *CRV* **1963**, *63*, 526.
8. Amer, M. I.; Booth, B. L.; Noori, G. F. M.; Proenca, M. F. *JCS(P1)* **1983**, 1075.
9. Uznanski, B.; Stec, W. J. *S* **1978**, 154.
10. Rasmussen, J. K.; Heilmann, S. M.; Krepski, L. R. In *Advances in Silicon Chemistry*; Larson, G. L., Ed.; JAI: Greenwich CT, 1991; Vol. 1, pp 65–187.
11. Mander, L. N.; Sethi, S. P. *TL* **1983**, *24*, 5425.
12. Lonza A. G. Swiss Patent 675 875, 1990 (*CA* **1991**, *115*, 279 444j).
13. Yamaguchi, H. Jpn. Patent 139 559, 1989 (*CA* **1990**, *112*, 7049n).
14. Arthur, P.; England, D. C.; Pratt, B. C.; Whitman, G. M. *JACS* **1954**, *76*, 5364.
15. Jackson, W. R.; Lovel, C. G. *TL* **1982**, *23*, 1621.
16. Druliner, J. D. *OM* **1984**, 205.
17. Brown, E. S. In *Organic Syntheses via Metal Carbonyls*; Wender, I.; Pino, P., Eds.; Wiley: New York, 1977; Vol. 2, pp 655–672.
18. Keim, W.; Behr, A.; Lühr, H. O.; Weisser, J. *J. Catal.* **1982**, *78*, 209.
19. Jackson, W. R.; Lovel, C. G. *AJC* **1983**, *36*, 1975.
20. Jackson, W. R.; Lovel, C. G. *AJC* **1988**, *41*, 1099.
21. Jackson, W. R.; Perlmutter, P.; Smallridge, A. J. *AJC* **1988**, *41*, 251.
22. RajanBabu, T. V.; Casalnuovo, A. L. *JACS* **1992**, *114*, 6265.
23. Tanaka, K.; Mori, A.; Inoue, S. *JOC* **1990**, *55*, 181.
24. Ziegler, T.; Hörsch, B.; Effenberger, F. *S* **1990**, 575 (*CA* **1990**, *113*, 190 847d).
25. Effenberger, F.; Hörsch, B.; Weingart, F.; Ziegler, T.; Kühner, S. *TL* **1991**, *32*, 2605.

Gérard Romeder
Lonza, Basel, Switzerland

3-Hydroxyisoborneol[1]

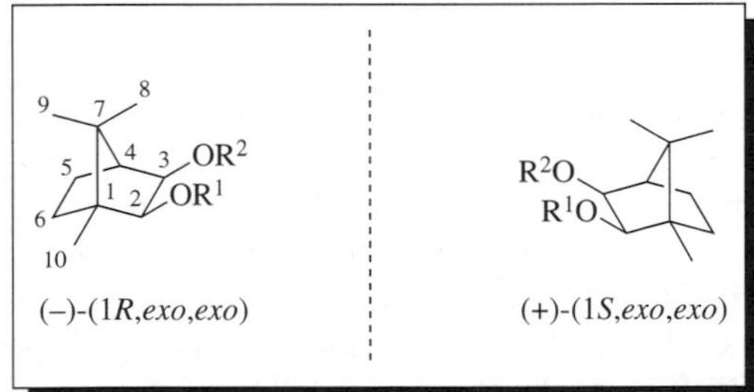

(**1**; R^1 = neopentyl, R^2 = H) (1*R*,*exo*,*exo*)
[85695-96-1] $C_{15}H_{28}O_2$ (MW 240.43)
(**2**; R^1 = neopentyl, R^2 = H) (1*S*,*exo*,*exo*)
[85718-76-9] $C_{15}H_{28}O_2$ (MW 240.43)
(**3**; R^1 = H, R^2 = neopentyl) (1*R*,*exo*,*exo*)
[85695-92-7] $C_{15}H_{28}O_2$ (MW 240.43)
(**4**; R^1 = benzyl, R^2 = H) (1*R*,*exo*,*exo*)
[104154-98-5] $C_{17}H_{24}O_2$ (MW 260.41)
(**5**; R^1 = H, R^2 = benzyl) (1*R*,*exo*,*exo*)
[73440-88-7] $C_{17}H_{24}O_2$ (MW 260.41)
(**6**; R^1 = Ph_2CH, R^2 = H) (1*R*,*exo*,*exo*)
[85695-93-8] $C_{23}H_{28}O_2$ (MW 336.51)
(**7**; R^1 = 1-naphthylmethyl, R^2 = H) (1*R*,*exo*,*exo*)
[85695-95-0] $C_{21}H_{26}O_2$ (MW 310.47)
(**8**; R^1 = 2-naphthylmethyl, R^2 = H) (1*R*,*exo*,*exo*)
[85695-94-9] $C_{21}H_{26}O_2$ (MW 310.47)

(chiral auxiliary; acrylate[2] and acyl nitroso[3] derivatives undergo stereoselective [4 + 2] cycloadditions; enoate derivatives undergo stereoselective 1,4-conjugate additions of organocopper reagents;[4] enol ether derivatives undergo stereoselective Pauson–Khand cyclizations,[5] [4 + 2][6] and [2 + 2][7] cycloadditions; alkynyl ether derivatives undergo stereoselective Pauson–Khand cyclizations[5])

Alternate Name: 1,7,7-trimethylbicyclo[2.2.1]heptane-2,3-diol.

Physical Data: (**1**) mp 4–5 °C; $[\alpha]_D^{25}$ (EtOH) −42.4° (c = 1.40). (**2**) mp 4–5 °C; $[\alpha]_D^{25}$ (EtOH) +42.6° (c = 2.52). (**3**) oil; $[\alpha]_D^{25}$ (EtOH) −18.8° (c = 1.08). (**4**) oil; bp 130 °C/0.05 mmHg; $[\alpha]_D^{25}$ ($CHCl_3$) −36.1° (c = 1.44). (**5**) mp 43 °C; $[\alpha]_D^{25}$ (EtOH) +0.4° (c = 4.99). (**6**) mp 57 °C; $[\alpha]_D^{25}$ (EtOH) −107.6° (c = 1.70). (**7**) mp 69–70 °C; $[\alpha]_D^{25}$ (EtOH) −79.2° (c = 0.90). (**8**) mp 70–71 °C; $[\alpha]_D^{25}$ (EtOH) −61.5° (c = 0.57).

Preparative Methods: the 3-hydroxyisoborneol derivatives are readily prepared from (+)- or (−)-camphor. The preparation of the 2-neopentyl ether derivative (**1**) is representative (eq 1).[2b] Analogous alkylation with different electrophiles allows for easy variation of the shielding moiety.[2c,8] The corresponding 3-substituted derivatives have been prepared from either 3-hydroxyisoborneol after separation of the regioisomeric mixture or from 3-*exo*-hydroxycamphor regioselectively after reduction with ***Lithium Tri-s-butylborohydride*** (L-Selectride).[2c,9]

SeO_2, 74%; 1. 1,2-ethylene glycol, TsOH; 2. L-Selectride, 68%; 1. NaH, NMP, neopentyl bromide; 2. 50% H_2SO_4, H_2O, 88%; $NaBH_4$, MeOH, 83% (1)

Handling, Storage, and Precautions: these auxiliaries vary from oils to white crystalline solids depending on the ether substituent and are stable indefinitely at ambient temperatures in sealed containers.

Introduction. The abundance of (+)-camphor in the chiral pool provided Oppolzer with an excellent framework to develop a chiral auxiliary which provides high levels of stereoselectivity in a wide range of reaction classes. The 3-hydroxyisoborneol skeleton provides two derivatizable positions at C-2 and C-3 of the molecule which are in close proximity to each other. By appending a reactive functionality to one and a sterically shielding appendage to the other, high stereodirecting ability can be envisioned. Likewise by reversing the roles of C-2 and C-3 it is possible to tune the auxiliary to fit the reaction parameters and desired product configuration. These characteristics have provided a means for π-facial differentiation to acrylates, enol ethers, and alkynyl ethers.

Preparation of Derivatives. Enoate derivatives are prepared from the corresponding chiral alcohol by treatment with acryloyl chloride in the presence of ***Triethylamine*** and catalytic ***4-Dimethylaminopyridine*** or the appropriate carboxylic acid chloride and ***Silver(I) Cyanide***.[2b] Alkynyl ethers are readily available from the potassium alkoxide by treating with ***Trichloroethylene***, in situ dechlorination with ***n-Butyllithium***, and electrophilic trapping.[10] Trapping the intermediate anion with a proton source or ***Iodomethane*** followed by Lindlar reduction of the alkynyl ether affords the corresponding vinyl and 1-(*Z*)-propenyl ether, respectively, while reduction of the alkynyl ether with ***Acetaldehyde 3-Bromopropyl Ethyl Acetal*** affords the 1-(*E*)-propenyl ether.

[4 + 2] Cycloadditions of Acrylate Derivatives.[2] Acrylate derivatives undergo highly stereoselective Diels–Alder cycloadditions with 1,3-dienes when promoted by a Lewis acid, ***Dichlorotitanium Diisopropoxide*** or ***Titanium(IV) Chloride*** (eq 2). With the latter, care must be taken to avoid acid-mediated cleavage of the auxiliary ether linkage. Generally, 2-substituted auxiliaries (**10**) show higher facial and *endo* selectivity than the corresponding 3-substituted analogs. This has been rationalized by a buttressing effect caused by the C-10 methyl forcing the ether side chain into close proximity to the acrylate. Of the range of shielding moieties examined, the neopentyl ether was shown to provide the highest selectivity. The stereochemical outcome can be explained by assuming that the acrylate adopts an *s-trans* conformation on coordination of the Lewis acid[11] and that the diene approaches from the face opposite the neopentyl ether. It should be noted that the analogous cycloadditions with crotonate derivatives give very poor yields (<7%).[2b] Similar highly stereoselective Diels–Alder cycloadditions have also been reported for fumarate and allenic ester derivatives.[12]

[4 + 2] Cycloadditions of Enol Ether Derivatives.[6] Asymmetric, inverse electron demand Diels–Alder reactions between nitroalkenes and alcohol (**1**)-derived vinyl and 1-(*E*)- and 1-(*Z*)-propenyl ethers have been reported to proceed with high stereoselectivity (eq 3). The resulting cycloadducts undergo an intramolecular [3 + 2] cycloaddition at rt to afford nitroso acetals which, after hydrogenolytic cleavage, provide tricyclic α-hydroxy lactams in high enantiomeric excess. The auxiliary alcohol can be recovered in 86–92% yield. The overall sense of asymmetric induction is dependent on the Lewis acid promoter employed, either $Ti(O\text{-}i\text{-}Pr)_2Cl_2$ or ***Methylaluminum Bis(2,6-diphenylphenoxide)*** (MAPh).[6b] This has been rationalized by a switch from a highly *endo* selective cycloaddition with $Ti(O\text{-}i\text{-}Pr)_2Cl_2$ to a highly *exo* selective cyclization with MAPh. When promoted by $Ti(O\text{-}i\text{-}Pr)_2Cl_2$ the corresponding 1-(*E*)-propenyl ether shows exclusive *endo* selectivity and 99% facial selectivity; however, facial selectivity for the 1-(*Z*)-propenyl ether is only 50%.

(**9**) (**10**) $Ti(O\text{-}i\text{-}Pr)_2Cl_2$ (2)

Acrylate	R	Temp (°C)	Yield (%)	*endo*:*exo*	*endo* adduct % de	*endo* adduct config.
(**9**)	Benzyl	0	91	86:14	46	(2*S*)
(**9**)	Diphenylmethyl	−20	94	86:14	64	(2*S*)
(**9**)	1-Naphthylmethyl	0	97	85:15	54	(2*S*)
(**9**)	2-Naphthylmethyl	−20	98	90:10	69	(2*S*)
(**9**)	Neopentyl	−20	95	96: 4	97	(2*S*)
(**10**)	Diphenylmethyl	−20	74	95: 5	91	(2*R*)
(**10**)	1-Naphthylmethyl	0	97	93: 7	88	(2*R*)
(**10**)	2-Naphthylmethyl	−20	98	95: 5	92	(2*R*)
(**10**)	Neopentyl	−20	96	96: 4	99	(2*R*)

MeO_2C; Lewis acid, CH_2Cl_2, −78 °C, 82–83%; H_2, Raney Ni, 86–88% (3)

98% ee (1*S*) $Ti(O\text{-}i\text{-}Pr)_2Cl_2$
99% ee (1*R*) MAPh

[4 + 2] Cycloadditions of Acyl Nitroso Derivatives.[3] In situ formation of the acyl nitroso derivative by oxidation of the hydroxy carbamic acid under Swern–Moffat conditions in the presence of a functionalized diene affords the corresponding cycloadduct in 94% yield and 96% diastereomeric excess (eq 4). The resulting cycloadduct can be further elaborated to prepare optically active functionalized amino alcohols.

1,4-Conjugate Additions of Enoate Derivatives.[4] Boron trifluoride-mediated conjugate additions of organocopper reagents to (*E*)-enoates derived from auxiliary (**1**) proceed with high stereoselectivity, affording optically active carboxylic acids after saponification (eq 5). The organocopper reagent is formed by addition of an alkyllithium reagent to ***Copper(I) Iodide***, ***Tri-n-butylphosphine***, and ***Boron Trifluoride Etherate*** in equimolar amounts, where Bu_3P is believed to stabilize the reagent. The overall sense of asymmetric induction can be controlled by changing either the order of substituent introduction (R^1 and R^2) or the configuration of the auxiliary. The stereochemical course of the reaction has been rationalized by assuming that the enoate exists in an *s-trans* conformation and the organocopper reagent approaches from the face opposite to the neopentyl ether. Conjugate additions of this type have been applied to the total synthesis of several natural products.

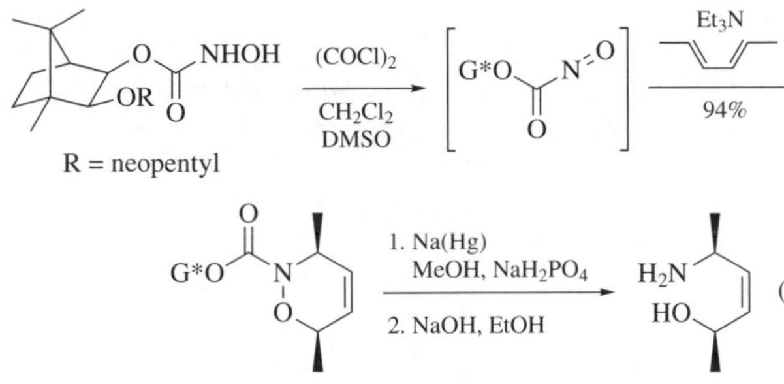

(4)

R = neopentyl

96% de

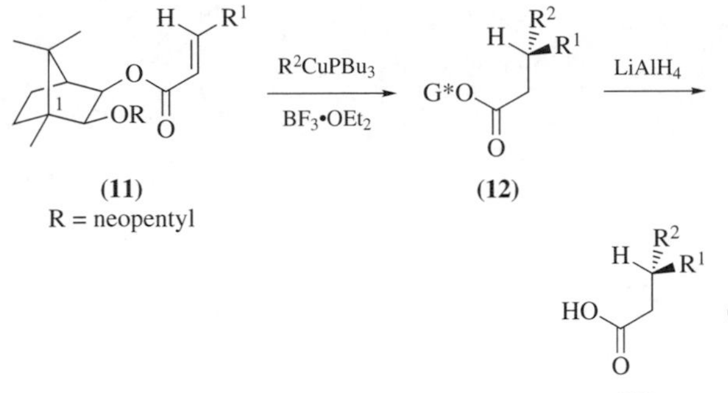

(5)

(11) R = neopentyl

(12)

(13)

(11) config.	R^1	R^2	(12) yield (%)	(13) % ee	config.
(1*R*)	Bu	Me	82	94	(*R*)
(1*R*)	Et	Me	85	92	(*R*)
(1*R*)	$Me_2C{=}CH(CH_2)_2$	Me	90	92	(*S*)
(1*S*)	Me	$Me_2C{=}CH(CH_2)_2$	81	98	(*S*)
(1*R*)	Me	$H_2C{=}CH$	85	94	(*R*)

Pauson–Khand Bicyclization.[5] Alkynyl and enol ether derivatives have been studied in the cobalt-mediated intramolecular Pauson–Khand reaction and found to provide high diastereoselectivity, superior to previous work with the auxiliary 2-phenylcyclohexanol.[13] The 3-substituted auxiliary alcohol (**3**) provides higher selectivity than the 2-substituted analog. Also, the alkynyl ether derivatives exhibit higher reactivity and selectivity than the corresponding enol ether derivatives (eq 6).

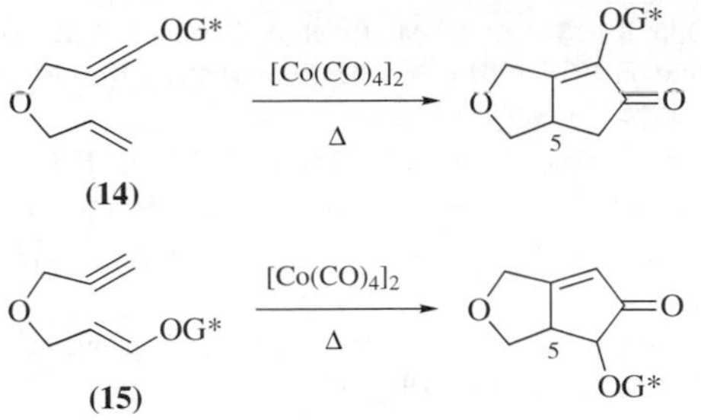

Photochemical [2+2] Cycloadditions.[7] Photochemical [2+2] cycloadditions between alkenes and chiral phenylglyoxylate derivatives of 3-hydroxyisoborneol show minimal diastereoselectivity (16% de).[14] Better results are obtained in [2+2] cycloadditions between chiral enol ethers and ***Dichloroketene*** (eq 7). After ring expansion and expulsion of the auxiliary (***Diazomethane***, ***Chromium(II) Perchlorate***), chiral α-chloro cyclopentenones are obtained in 60% yield. The observed diastereoselectivity is believed to arise from the enol ether *s-trans* conformation and approach of the ketene to the face opposite to the neopentyl ether.

(14)

(15)

(6)

G* =

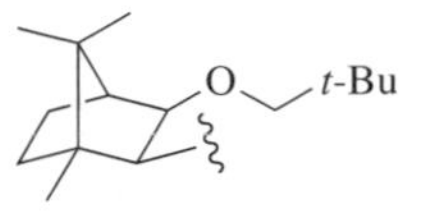

(3)

Enyne	Conditions	Yield (%)	Diastereomer ratio	Config.
(**14**)	18 °C, 2 h; 25 °C, 2 h; N_2	54	94: 6	(5*R*)
(**15**)	20 °C, 2 h; 50 °C, 12 h; CO	53	90:10	(5*S*)

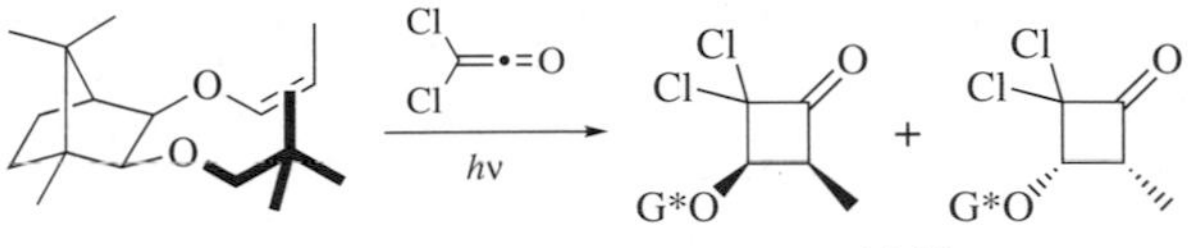

(7)

10:90

Non-destructive Auxiliary Cleavage. The high stability of the ether linkage to the shielding moiety generally allows for a very high recovery of the auxiliary alcohol. For acyl derivatives, primary alcohols can be obtained by $LiAlH_4$[2b] or AlH_3[15] reduction. Hydrolysis of the auxiliary under basic conditions providing the carboxylic acid has been accomplished with NaOH in aq. ethanol,[3] NaOH in methanol,[4b] or KOH in ethanol.[16] Intramolecular transesterification has been applied using KO-*t*-Bu in THF.[17] Enol ethers derived from Pauson–Khand cyclizations of alkynyl ether derivatives can be readily cleaved to the corresponding ketone and recovered auxiliary by catalytic HCl in methanol.[5]

Related Reagents. 10,2-Camphorsultam; 10-Dicyclohexylsulfonamidoisoborneol; (*S*)-Ethyl Lactate; α-Methyl-toluene-2, α-sultam; (*R*)-Pantolactone.

1. Oppolzer, W. *T* **1987**, *43*, 1969.
2. (a) Oppolzer, W. *AG(E)* **1984**, *23*, 876. (b) Oppolzer, W.; Chapuis, C.; Dupuis, D.; Guo, M. *HCA* **1985**, *68*, 2100. (c) Oppolzer, W.; Chapuis, C.; Dao, G. M.; Reichlin, D.; Godel, T. *TL* **1982**, *23*, 4781.
3. Martin, S. F.; Hartmann, M.; Josey, J. A. *TL* **1992**, *33*, 3583.
4. (a) Rossiter, B. E.; Swingle, N. M. *CRV* **1992**, *92*, 771. (b) Oppolzer, W.; Moretti, R.; Godel, T.; Meunier, A.; Löher, H. *TL* **1983**, *24*, 4971.
5. Verdaguer, X.; Moyano, A.; Pericàs, M. A.; Riera, A.; Greene, A. E.; Piniella, J. F.; Alvarez-Larena, A. *JOM* **1992**, *433*, 305.
6. (a) Denmark, S. E.; Senanayake, C. B. W.; Ho G.-H. *T* **1990**, *46*, 4857. (b) Denmark, S. E.; Schnute, M. E.; Senanayake, C. B. W. *JOC* **1993**, *58*, 1859.
7. Greene, A. E.; Charbonnier, F. *TL* **1985**, *26*, 5525.
8. Herzog, H.; Scharf, H.-D. *S* **1986**, 788.

9. (a) Oppolzer, W.; Kurth, M.; Reichlin, D.; Chapuis, C.; Mohnhaupt, M.; Moffatt, F. *HCA* **1981**, *64*, 2802. (b) Sasaki, S.; Kawasaki, M.; Koga, K. *CPB* **1985**, *33*, 4247.
10. Moyano, A.; Charbonnier, F.; Greene, A. E. *JOC* **1987**, *52*, 2919.
11. Loncharich, R. J.; Schwartz, T. R.; Houk, K. N. *JACS* **1987**, *109*, 14.
12. (a) Helmchen, G.; Schmieres, R. *AG(E)* **1981**, *20*, 205. (b) Oppolzer, W.; Chapuis, C. *TL* **1985**, *24*, 4665.
13. Castro, J.; Sörensen, H.; Riera, A.; Morin, C.; Moyano, A.; Pericàs, M. A.; Greene, A. E. *JACS* **1990**, *112*, 9388.
14. Herzog, H.; Koch, H.; Scharf, H.-D.; Runsink, J. *T* **1986**, *42*, 3547.
15. Oppolzer, W.; Pitteloud, R.; Bernardinelli, G.; Baettig, K. *TL* **1983**, *24*, 4975.
16. Cativiela, C.; López, P.; Mayoral, J. A. *TA* **1991**, *2*, 449.
17. Remiszewski, S. W.; Yang, J.; Weinreb, S. M. *TL* **1986**, *27*, 1853.

Mark E. Schnute
Stanford University, CA, USA

Iodoform

[75-47-8] CHI_3 (MW 393.72)

(precursor to both mono- and diiodocarbene for addition to alkenes and arenes; adds to alkenes by a radical chain mechanism when correctly initiated; iodine atom donor for quenching carbon radicals; used in conjunction with chromium(II) chloride for the homologation of aldehydes to (*E*)-iodoalkenes)

Alternate Name: triiodomethane.
Physical Data: mp 120–123 °C; *d* 4.008 g cm^{-3}.
Solubility: sol ethanol, acetone, ether, benzene, carbon disulfide, dichloromethane, THF; slightly sol petroleum ether.
Purification: crystallization from ethanol.
Handling, Storage, and Precautions: yellow crystalline solid with a disagreeable odor. Decomposes at high temp with evolution of iodine.

Diiodocarbenoid Precursor. Diiodocarbene, formed by base-induced decomposition of iodoform, with or without phase-transfer catalysis, adds to alkenes giving diiodocyclopropanes which may be rearranged with Ag^I to iodoalkenes (eq 1).[1] Under phase-transfer conditions, reaction with adamantane gives 1-(diiodomethyl)adamantane.[2]

CHI_3, *t*-BuOK, 58%; $AgClO_4$, MeOH, 90% (1)

The triiodomethyl anion, formed from iodoform and 50% ***Potassium Hydroxide***, can be trapped by addition to pyridinium and isoquinolinium salts (eq 2).[3]

CHI_3, 50% KOH, 90% (2)

DCB = 2,6-dichlorobenzyl

Monoiodocyclopropanes may be prepared by reaction of alkenes with iodoform in the presence of copper, albeit in only low yield.[4] The iodocarbenoid formed from iodoform and ***Diethylzinc*** adds to alkenes in moderate yield with clean retention of configuration and with high preference for the *syn*-isomer (eq 3).[5]

Et_2Zn, CHI_3, 30% (3)

syn only

Photolysis of iodoform in the presence of alkenes also leads to the formation of monoiodocyclopropanes, also with retention of configuration (eq 4). With *cis*-2-butene the *syn*:*anti* ratio was 1.6:1 (eq 5).[5c,6] Interestingly, with cyclic *cis*-alkenes such as cyclooctene only the *endo*-iodocyclopropane was formed both under photolytic conditions and with diethylzinc.[5c] Flash photolysis experiments indicated that the photochemical reaction proceeds via homolysis of iodoform to iodine atoms and diiodomethyl radicals.[7] Accordingly, photolysis in the presence of a more reactive alkene, norbornene, gave a mixture of radical and carbene addition products.[8]

Et_2Zn, CHI_3, 41% (4)

CHI_3, Na_2SO_3, *hν*; CH_2Cl_2, H_2O, 50% (5)

1.6:1

Arenes also react with the diethylzinc/iodoform couple to give a cyclopropane adduct that suffers further reaction with diethylzinc and rearrangement to cycloheptatrienes (eq 6).[9]

Et_2Zn, CHI_3; 35–60% (6)

R = H, Me, Et, *i*-Pr, *t*-Bu

mixture of isomers

Radical Reactions. ***Triethylborane*** initiates the radical addition of iodoform to ketene silyl acetals, resulting in the isolation of β-iodo-α,β-unsaturated esters (eq 7).[10] With tetraiodomethane the corresponding β,β-diiodide was obtained. Radical addition of iodoform to methyl acrylate is also reported to be initiated with ***Pentacarbonyliron*** (eq 8).[11]

OTBDMS, OC_8H_{17}; CHI_3, Et_3B, hexane, 37%; $CO_2C_8H_{17}$ (7)

CO_2Me; CHI_3, DMF, $Fe(CO)_5$, 33%; I_2HC, CO_2Me (8)

Addition to limonene (eq 9) serves to illustrate that terminal alkenes are more readily attacked than internal alkenes and that the reaction is not limited to electron-deficient substrates.[12]

CHI_3, AcOOAc, Δ, 35% (9)

Iodoform acts as an iodine atom donor in the Barton version of the Hunsdiecker reaction. Aliphatic, and to some extent aromatic, carboxylic acids undergo this oxidative decarboxylation and optimum yields were obtained when cyclohexene was used as solvent in place of benzene.[13] The reaction, which may be initiated thermally,[13,16] photolytically,[14] or with ultrasound,[15] tolerates many functional groups and has been applied to cyclopropanecarboxylic acids (eq 10)[16] and notably to electron-rich aromatic acids (eq 11).[13]

CHI_3, Δ, c-C_6H_{12}, 56% (10)

CHI_3, AIBN, toluene, Δ, 31% (11)

Alkenation Reactions. Takai has described a very simple procedure for the homologation of aldehydes to iodoalkenes which involves treatment with iodoform and ***Chromium(II) Chloride*** and which proceeds with high selectivity for the (*E*)-product (eqs 12 and 13).[17]

$CrCl_2$, CHI_3, THF, 84% (12)

(*E*):(*Z*) = 82:18

$CrCl_2$, CHI_3, THF, 70% (13)

Although ketones are less reactive, as demonstrated by competition reactions, good yields of iodoalkenes are nonetheless obtained (eq 14).[17]

$CrCl_2$, CHI_3, THF, 75% (14)

In the course of a synthesis of Macbecin I, Baker and Castro devised a synthesis of β-iodomethacrylic acid that involved alkylation of diethyl methylmalonate with iodoform followed by saponification and decarboxylation (eq 15).[18]

NaH, CHI_3, Et_2O, Δ, 65%; KOH, Δ, EtOH, H_2O, 89% (15)

Other Applications. Deuterated or tritiated methylene diiodide may be prepared from iodoform as reported by Saljoughian (eq 16), providing a convenient method for the incorporation of deuterium or tritium into cyclopropane rings via the Simmons–Smith reaction.[19]

$$CHI_3 + Na_3AsO_3 + NaOT \xrightarrow[\text{HTO, 74\%}]{60\,^\circ C} TCHI_2 + NaI + Na_3AsSO_4 \quad (16)$$

The Finkelstein reaction of iodoform with ***Silver(I) Fluoride*** or ***Mercury(II) Fluoride*** provides mixtures of diiodofluoromethane and difluoroiodomethane (eq 17).[20]

$$CHI_3 \xrightarrow[\text{or } HgF_2]{AgF} \underset{35\%}{CHFI_2} + \underset{15\%}{CHF_2I} \quad (17)$$

Related Reagents. Bromoform; Chromium(II) Chloride-Iodoform; Diethylzinc-Iodoform; Diiodomethane; Triphenylphosphine-Iodoform-Imidazole.

1. (a) Baird, M. S.; Gerrard, M. E. *JCR(S)* **1986**, 114. (b) Mathias, R.; Weyerstahl, P. *AG* **1974**, *86*, 42. (c) Baird, M. S. *CC* **1974**, 196.
2. Slobodin, Ya. M.; Ashkinazi, L. A.; Klimchuk, G. N. *ZOR* **1984**, *20*, 1238.
3. Duchardt, K. H.; Kröehnke, F. *CB* **1977**, *110*, 2669.
4. Kawabata, N.; Tanimoto, M.; Fujiwara, S. *T* **1979**, *35*, 1919.
5. (a) Nishimura, J.; Furukawa, J. *JCS(D)* **1971**, 1375. (b) Miyano, S.; Hashimoto, H. *BCJ* **1974**, *47*, 1500. (c) Dehmlow, E. V.; Stütten, J. *TL* **1991**, *32*, 6105.
6. (a) Yang, N. C.; Marolewski, T. A. *JACS* **1968**, *90*, 5644. (b) Marolewski, T. A.; Yang, N. C. *OS* **1972**, *52*, 132.
7. Cossham, J. A.; Logan, S. R. *JPP* **1988**, *A42*, 127.
8. Wang, C.-B.; Hsu, Y.-G.; Lin. L. C. *J. Chin. Chem. Soc. (Taipei)* **1977**, *24*, 53 (*CA* **1977**, *87*, 134 005k).
9. (a) Miyano, S.; Hashimoto, H. *CC* **1973**, 216. (b) Miyano, S.; Hashimoto, H. *BCJ* **1973**, *46*, 3257. (c) Miyano, S.; Minagawa, M.; Matsumoto, Y.; Hashimoto, H. *NKK* **1976**, 1255. (d) Miyano, S.; Higuchi, T.; Sato, F.; Hashimoto, H. *NKK* **1976**, 256.
10. Sugimoto, J.; Miura, K.; Oshima, K.; Utimoto, K. *CL* **1991**, 1319.
11. (a) Freidlina, R. K; Amriev, R. A.; Velichko, F. K.; Baibuz, O. P.; Rilo, R. P. *IZV* **1983**, 1456. (b) Vasil'eva, T. T.; Velichko, F. K.; Kochetkova, V. A.; Bondarenko, O. P. *IZV* **1987**, 1904.
12. Weizmann, M.; Israelashvili, S.; Halevy, A.; Bergmann, F. *JACS* **1947**, *69*, 2569.
13. (a) Barton, D. H. R.; Crich, D.; Motherwell, W. B. *T* **1985**, *41*, 3901. (b) Barton, D. H. R.; Lacher, B.; Zard, S. Z. *T* **1987**, *43*, 4321.
14. Dauben, W. G.; Kowalczyk, B. A.; Bridon, D. P. *TL* **1989**, *30*, 2461.
15. Dauben, W. G.; Bridon, D. P.; Kowalczyk, B. A. *JOC* **1989**, *54*, 6101.

16. Gawronska, K.; Gawronski, J.; Walborsky, H. M. *JOC* **1991**, *56*, 2193.
17. (a) Takai, K.; Nitta, K.; Utimoto, K. *JACS* **1986**, *108*, 7408. (b) Pontikis, R.; Musci, A.; Le Merrer, Y.; Depezay, J. C. *BSF(2)* **1991**, 968.
18. Baker, R.; Castro, J. L. *JCS(P1)* **1990**, 47.
19. Saljoughian, M.; Morimoto, H.; Williams, P. G.; DeMello, N. *CC* **1990**, 1652.
20. (a) Hine, J.; Butterworth, R.; Langford, P. B. *JACS* **1958**, *80*, 819. (b) Weyerstahl, P.; Mathias, R.; Blume, G. *TL* **1973**, 611.

David Crich, Milan Bruncko & Jarmila Brunckova
University of Illinois at Chicago, IL, USA

Iodomethane

MeI

[74-88-4] CH_3I (MW 141.94)

(methylating agent for carbon, oxygen, nitrogen, sulfur, and trivalent phosphorus)

Alternate Name: methyl iodide.
Physical Data: bp 41–43 °C; *d* 2.28 g cm^{-3}.
Solubility: sol ether, alcohol, benzene, acetone; moderately sol H_2O.
Form Supplied in: colorless liquid; stabilized by addition of silver wire or copper beads; widely available.
Purification: percolate through silica gel or activated alumina then distill; wash with dilute aqueous $Na_2S_2O_3$, then wash with water, dilute aqueous Na_2CO_3, and water, dry with $CaCl_2$ then distill.
Handling, Storage, and Precautions: toxic, corrosive and a possible carcinogen. Liquid should be stored in brown bottles to prevent liberation of I_2 upon exposure to light. Keep in a cool, dark place. Use only in well ventilated areas.

***C*-Methylation.** Methyl iodide is an 'active' alkylating agent employed in the *C*-methylation of carbanions derived from ketones, esters, carboxylic acids, amides, nitriles, nitroalkanes, sulfones, sulfoxides, imines, and hydrazones.[1] The quantity of methyl iodide utilized in methylations varies from a slight (1.1 equiv) to a large excess (used as solvent).

The monomethylation of carbanions derived from 1,3-cyclohexanedione and acetylacetone has been described (eqs 1 and 2).[2] Selective monomethylation of β-diketones is dependent upon the base employed; variable amounts of *O*-methylation, dimethylation, and carbon–carbon bond cleavage may occur. The tetraethylammonium enolate of β-diketones reportedly provides higher yields of *C*-methylation without competing side reactions (eq 3).[3] Dimethylation is sometimes a desired reaction pathway. In this case, a large excess of both methyl iodide and base favors the dimethylated product (eq 4).[4] Recently, the combination of a potassium base and a catalytic amount of ***18-Crown-6*** (eq 5) has been described to provide a higher yield of dimethylation.[5,6]

MeI; dioxane (aq); 54–56% (1)

MeI; K_2CO_3, acetone; 75–77% (2)

MeI; DMF; Et_4N^+; 92% (3)

t-BuOK, *t*-BuOH; MeI (xs); 57% (4)

MeI (2.2 equiv); *t*-BuOK, THF; 18-crown-6; 83% (5)

Methylation of kinctically derived enolates is most readily accomplished via the corresponding silyl enol ether.[7] Lithium enolates, generated by treatment of the silyl enol ether with ***Methyllithium***, may then be alkylated with methyl iodide (eq 6).[8] Alternatively, quarternary ammonium enolates are produced by treatment of the silyl enol ether with the corresponding fluoride salt. For example, monomethylation of a ketone is cleanly effected by treatment of an anhydrous mixture of the trimethylsilyl enol ether of the ketone in methyl iodide with benzyltrimethylammonium fluoride (eq 7).[9] Kinetic enolates produced from the conjugate addition of an organocuprate to an unsaturated ketone or a dissolving metal reduction of an enone may be methylated directly.[7c] However, the choice of solvent is crucial for the success of these reactions. Ether, the solvent typically used in organocopper conjugate additions, is a poor solvent for alkylation reactions.[10] ***N,N,N′,N′-Tetramethylethylenediamine*** (TMEDA), ***Hexamethylphosphoric Triamide*** (HMPA) and liquid ***Ammonia*** have been used as additives to increase the efficiency of alkylation. Alternatively, the solvent used in the conjugate addition can be removed in vacuo and replaced with a more effective medium for alkylation. For example the rate of methylation of the enolate produced in the conjugate addition of ***Lithium Dimethylcuprate*** to an enone is approximately 10^5 times faster in DME than in ether (eq 8).[11]

TMSO; MeLi; MeI; 80% (6)

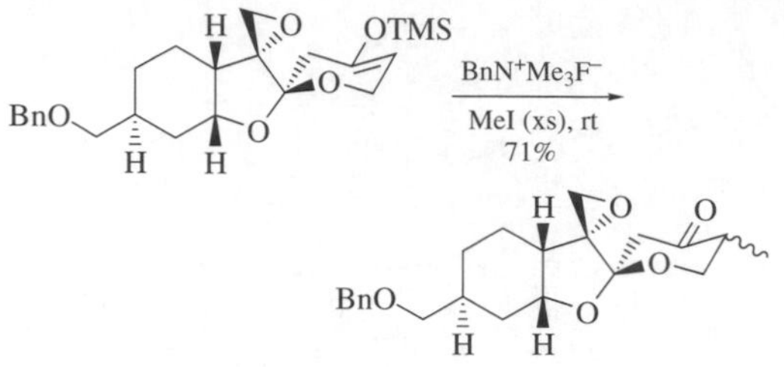

The stereoselectivity of the methylation of ketone enolates is determined by the structure of the substrate.[12] Stereoselective methylation of cyclic ketone enolates has been examined in detail and current models reliably predict the stereochemical outcome (eqs 9–11).[13–15] Diastereoselective methylation of acyclic ketone and ester enolates has been accomplished employing a variety of chiral auxiliaries (eq 12).[12,16] Efficient catalytic enantioselective methylation of 6,7-dichloro-5-methoxy-1-indanone has been accomplished via a chiral phase-transfer catalyst (eq 13).[17] An enantiomeric excess of 92% was observed when employing ***Chloromethane*** as the methylating agent, whereas methyl iodide provided a product of only 36% enantiomeric excess.

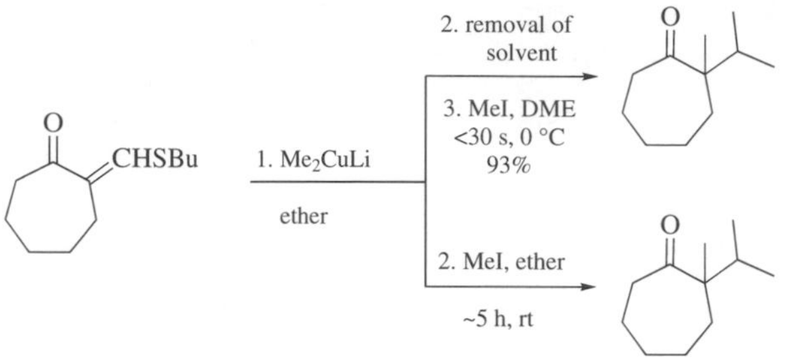

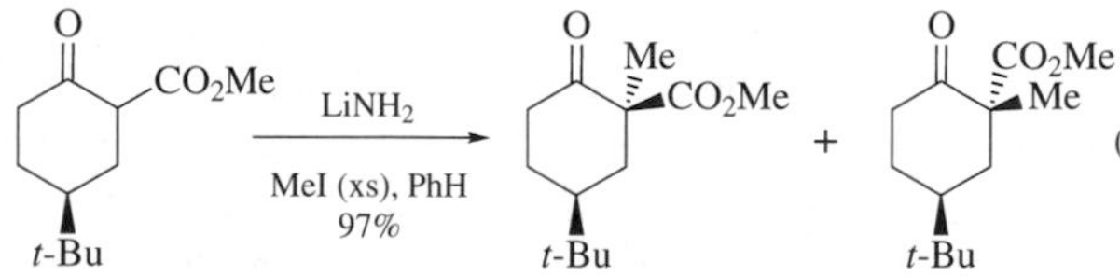

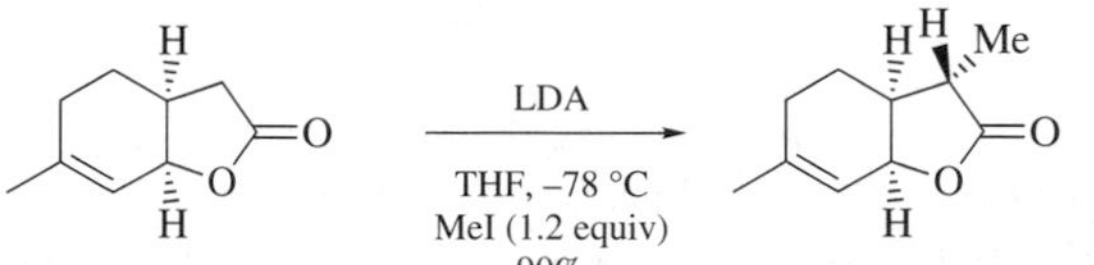

O*-Methylation.** Carboxylic acids can be converted to the corresponding methyl ester by stirring a mixture of the carboxylic acid in methanol with an excess of methyl iodide and ***Potassium Carbonate.[18] A recent report describes the esterification of carboxylic acids using ***Cesium Fluoride*** and methyl iodide in DMF (eq 14).[19] ***Dimethyl Sulfate*** has also been advantageously utilized to effect *O*-methylation of carboxylic acids as well as alcohols (eq 15).[18c,20] These methods often serve as useful alternatives to ***Diazomethane*** for preparative scale esterification of carboxylic acids.

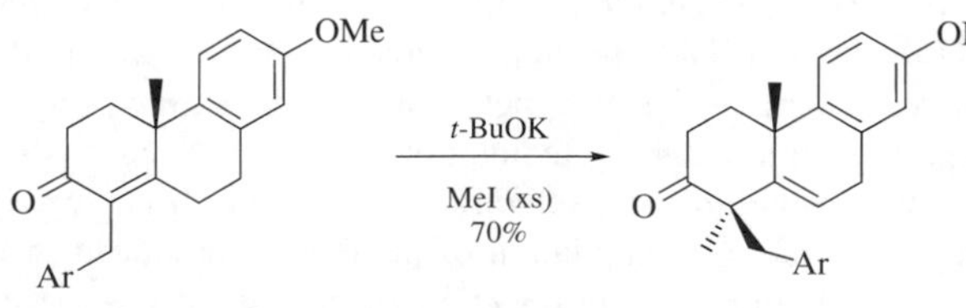

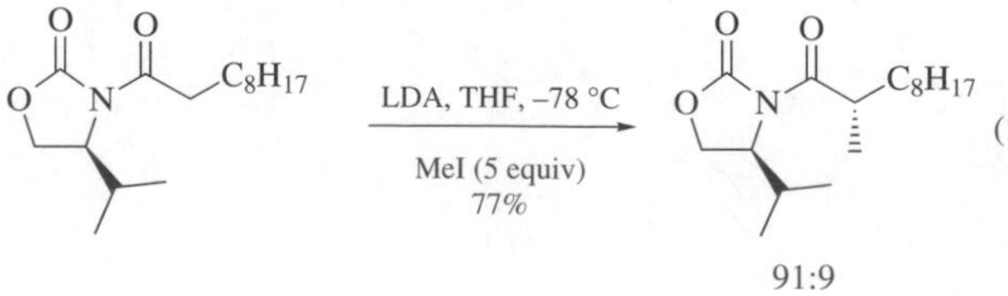

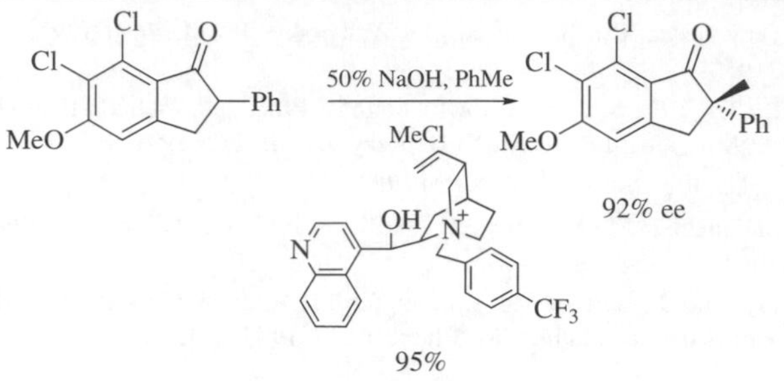

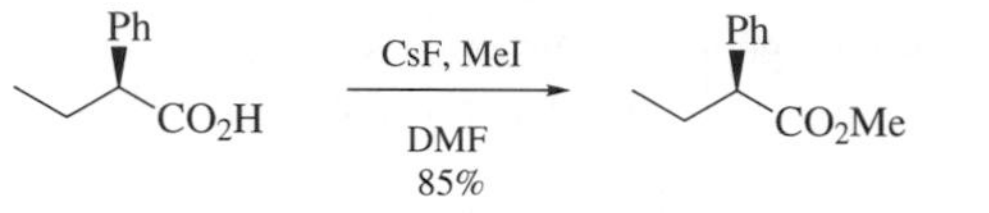

Phenolic hydroxyls are readily methylated by methyl iodide under basic conditions. The most common conditions are methyl iodide and potassium carbonate in acetone (eq 16).[18a,21] The use of ***Lithium Carbonate*** as the base allows for the selective protection of phenols with a $pK_a < 8$ (eq 17).[18c] The *peri*-hydroxy group of an anthraquinone is methylated using methyl iodide and ***Silver(I) Oxide*** in chloroform (eq 18).[18a,22]

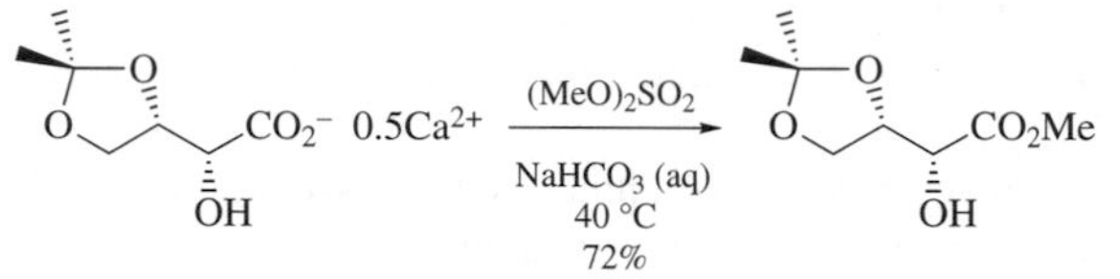

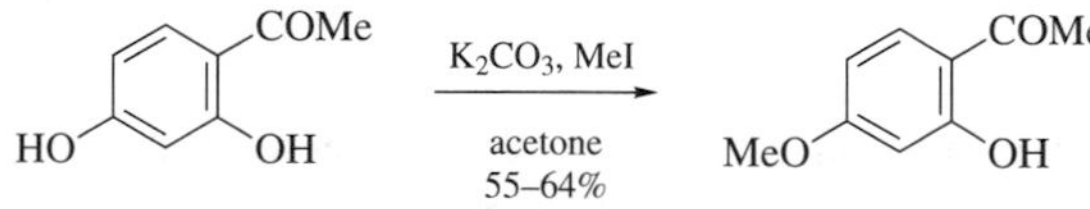

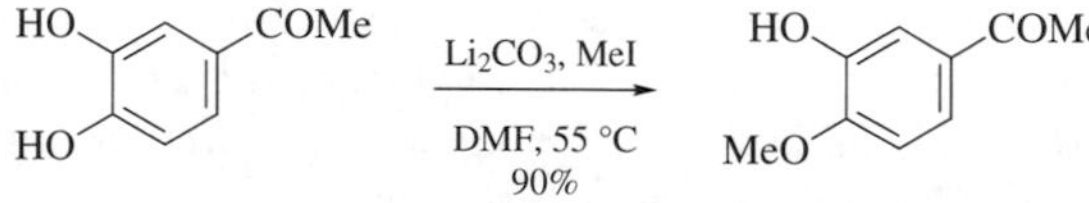

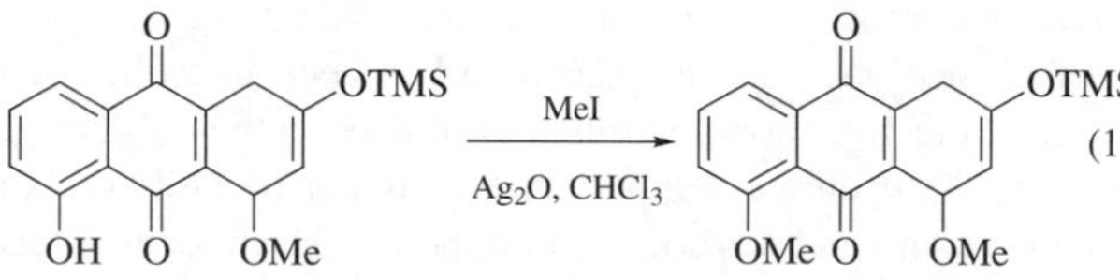

Aliphatic alcohols are also methylated by methyl iodide under basic conditions in dipolar aprotic solvents (eqs 19 and 20).[23] Typical conditions employ ***Sodium Hydride*** as a base, DMF as the solvent and an excess of methyl iodide.[24] Alternatively, dimethyl sulfate or ***Methyl Trifluoromethanesulfonate*** may be used as the methylating agent. Under acidic conditions, diazomethane will also methylate aliphatic hydroxy groups. Finally, methylation of a hydroxy group may be achieved under essentially neutral conditions using silver(1) oxide (eq 21).[23,25]

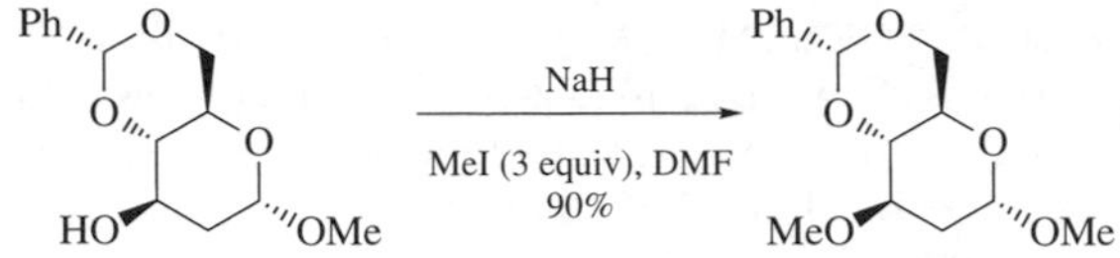
(19)

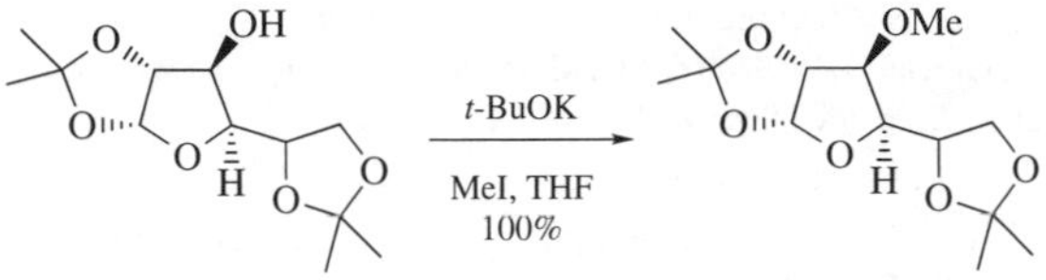
(20)

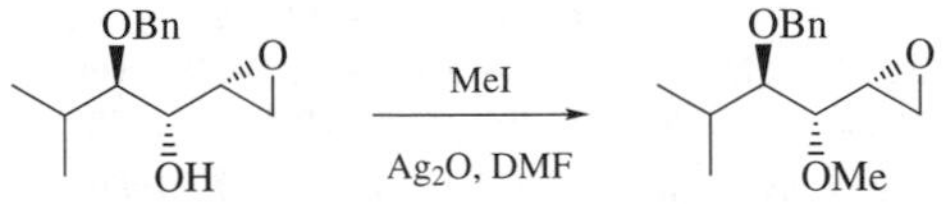
(21)

S*-Alkylation.** Methyl iodide alkylates thiolates and sulfides to produce sulfides and sulfonium ions, respectively. For example, thiolates produced from thiocarbonyl compounds are methylated with methyl iodide to generate the corresponding methyl thioether (eq 22).[26] Sulfonium halides, derived from the reaction of methyl iodide with an alkyl sulfide, are sometimes labile in solution and may undergo further reaction (eq 23).[27,28] ***Dimethyl Sulfoxide when refluxed with an excess of methyl iodide produces trimethyloxosulfonium iodide, which is collected as a white solid and recrystallized from water. Similarly, methylation of ***Dimethyl Sulfide*** produces trimethylsulfonium iodide.[29] Treatment of trimethyloxosulfonium and trimethylsulfonium salts with a base yields the corresponding ylides, which serve as useful methylene transfer reagents. ***Silver(I) Perchlorate*** promotes the methylation of less reactive sulfides (eq 24).[30]

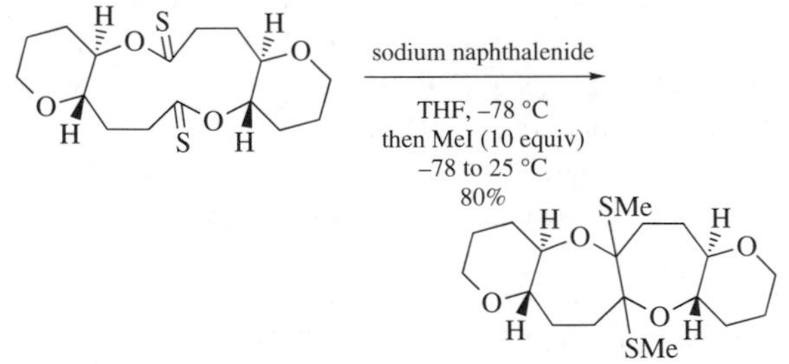
(22)

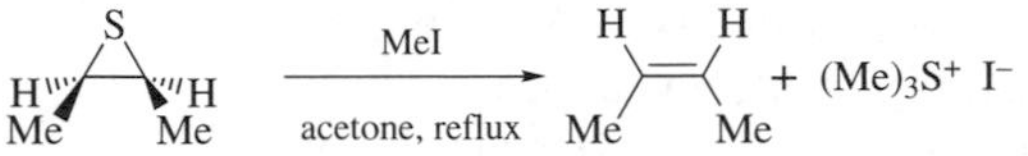
(23)

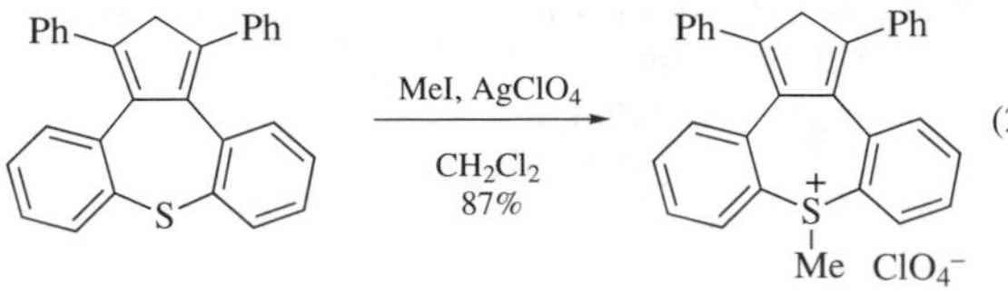
(24)

Hydrolysis of sulfonium salts serves as a useful protocol for removal of a protecting group or hydrolysis of a carboxylic acid derivative. For example, thioamides are converted into the corresponding methyl esters by methylation with methanolic methyl iodide followed by treatment with aqueous potassium carbonate (eq 25).[31] Methyl thiomethyl ethers are readily hydrolyzed using an excess of methyl iodide in aqueous acetone (eq 26).[32] Under similar reaction conditions, thioacetals are hydrolyzed to the corresponding carbonyl compounds (eq 27).[33]

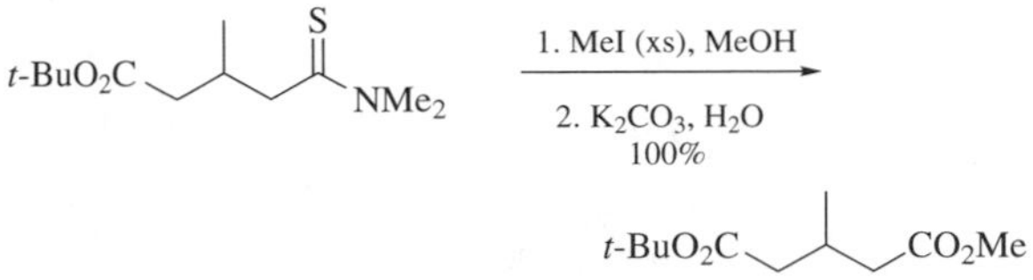
(25)

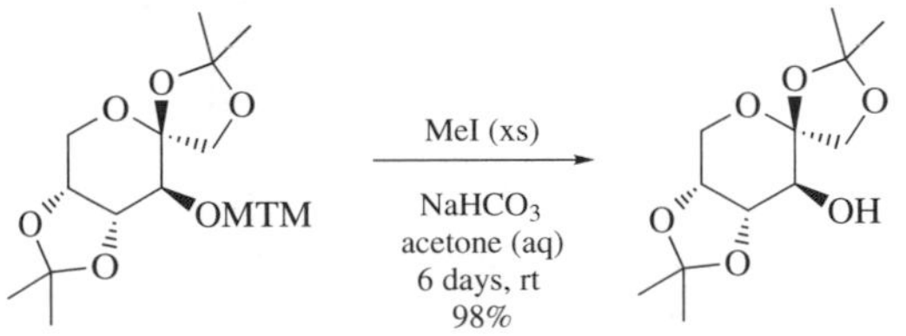
(26)

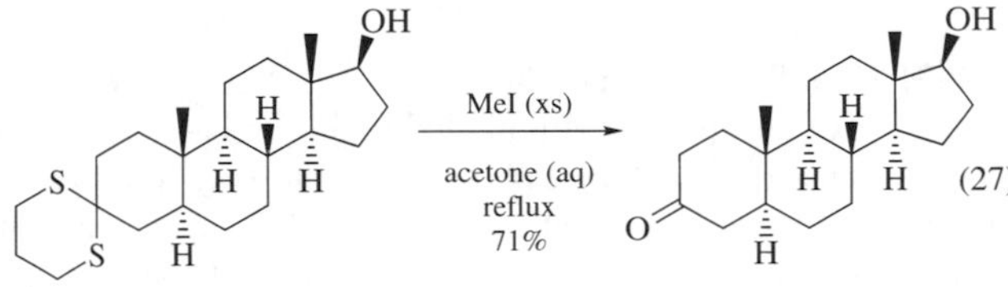
(27)

N*-Methylation.** The direct monomethylation of ammonia or a primary amine with methyl iodide is usually not a feasible method for the preparation of primary or secondary amines since further methylation occurs. However, methylation of secondary and tertiary amines leading to the production of tertiary amines and quaternary ammonium salts, respectively, is a useful method. Secondary *N*-methylalkylamines can be prepared from primary amines by a multi-step sequence involving first methylation of the benzylidene of the primary amine followed by hydrolytic removal of the benzylidene group (eq 28).[34] An alternative to the methylation procedure using methyl iodide is the employment of Eschweiler–Clarke conditions.[35] Exhaustive *N*-methylation of amines results in the production of a quaternized amine. The use of ***2,6-Lutidine as base is beneficial to carry out quaternization of amines due to the slow rate of methylation of 2,6-lutidine (eq 29).[36] Quaternized ammonium salts are employed in the Hofmann elimination (eqs 30 and 31).[37,38] As in the case of alcohol methylation, silver(I) salts may be used to facilitate the methylation process (eq 32).[39] Finally, conditions for the methylation of indole have been reported (eq 33).[40]

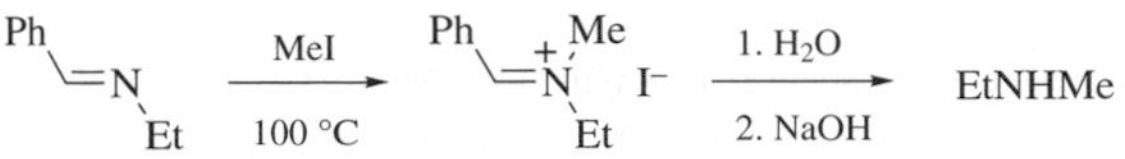
(28)

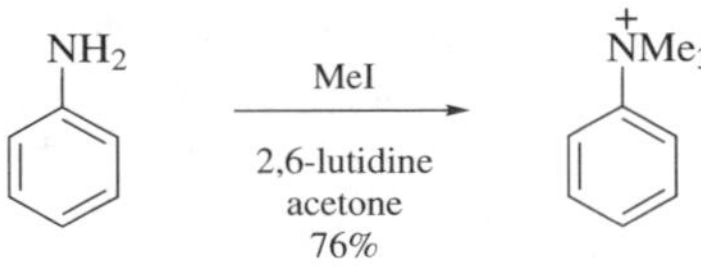
(29)

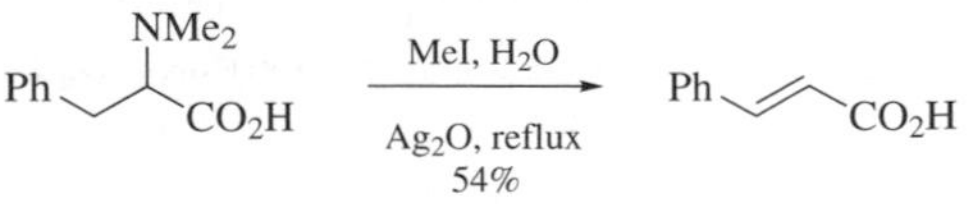
(30)

$$\text{cyclooctyl-NMe}_2 \xrightarrow[\text{then Ag}_2\text{O}]{\text{MeI, MeOH}} \text{cyclooctyl-}^{+}\text{NMe}_3\ ^{-}\text{OH} \xrightarrow[40\%]{120\ ^\circ\text{C}} \text{cyclooctene} \quad (31)$$

$$\xrightarrow[\text{AgCl}]{\text{MeI}} \quad (32)$$

$$\text{indole} \xrightarrow[\substack{\text{NaNH}_2,\ \text{NH}_3(\text{l}) \\ 85\text{–}95\%}]{\text{MeI}} \text{N-methylindole} \quad (33)$$

P*-Methylation.** Phosphonium salts are prepared by the quaternization of phosphines with methyl iodide.[41] The displacement reaction is usually conducted in polar solvents such as acetonitrile or DMF. Dialkyl phosphonates are prepared from the reaction of trialkyl phosphites with alkyl halides, commonly known as the Arbuzov reaction.[42] For example, diisopropyl methylphosphonate is prepared by heating a mixture of methyl iodide and ***Triisopropyl Phosphite (eq 34).[43]

$$(i\text{-PrO})_3\text{P} + \text{MeI} \xrightarrow[85\text{–}95\%]{\text{heat}} (i\text{-PrO})_2\text{P(=O)Me} \quad (34)$$

Related Reagents. Bromomethane; Chloromethane; Dimethyl Sulfate; Methyl Fluoride-Antimony(V) Fluoride; Methyl Trifluoromethysulfonate.

---

1. (a) Stowell, J. C. *Carbanions in Organic Synthesis*; Wiley: New York, 1979. (b) Caine, D. *COS* **1991**, *3*, Chapter 1. (c) House, H. O. *Modern Synthetic Organic Reactions*, 2nd ed.; Benjamin: Menlo Park, 1972; Chapter 9.
2. Mekler, A. B.; Ramachandran, S.; Swaminathan, S.; Newman, M. S. *OSC* **1973**, *5*, 742. (b) Johnson, A. W.; Markham, E.; Price, R. *OSC* **1973**, *5*, 785.
3. Shono, T.; Kashiura, S.; Sawamura, M.; Soejima, T. *JOC* **1988**, *53*, 907.
4. Nedelec, L.; Gasc, J. C.; Bucourt, R. *T* **1974**, *30*, 3263.
5. Prasad, G.; Hanna, P. E.; Noland, W. E.; Venkatraman, S. *JOC* **1991**, *56*, 7188.
6. Rubina, K.; Goldverg, Y.; Shymanska, M. *SC* **1989**, *19*, 2489.
7. (a) Rasmussen, J. K. *S* **1979**, 91. (b) Brownbridge, P. *S* **1983**, 1. (c) d'Angelo, J. *T* **1976**, *32*, 2979.
8. Stork, G.; Hudrlik, P. F. *JACS* **1968**, *90*, 4462, 4464.
9. (a) Kuwajima, I.; Nakamura, E. *ACR* **1985**, *18*, 181. (b) Kuwajima, I.; Nakamura, E. *JACS* **1975**, *97*, 3257. (c) Smith, A. B., III; Fukui, M. *JACS* **1987**, *109*, 1269.
10. Taylor, R. J. K. *S* **1985**, 364.
11. Coates, R. M.; Sandfur, L. O. *JOC* **1974**, *39*, 275.
12. Evans, D. A. In *Asymmetric Synthesis*; Morrison, J. D., Ed.; Academic: New York, 1984; Vol. 3, p 1.
13. Kuehne, M. E. *JOC* **1970**, *35*, 171.
14. Bartlett, P. A.; Pizzo, C. F. *JOC* **1981**, *46*, 3896.
15. Ireland, R. E.; Evans, D. A.; Glover, D.; Rubottom, G. M.; Young, H. *JOC* **1969**, *34*, 3717.
16. Evans, D. A.; Ennis, M. D.; Mathre, D. J. *JACS* **1982**, *104*, 1737.
17. (a) Dolling, U.-H.; Davis, P.; Grabowski, E. J. J. *JACS* **1984**, *106*, 446. (b) Hughes, D. L.; Dolling, U.-H.; Ryan, K. M.; Schoenewaldt, E. F.; Grabowski, E. J. J. *JOC* **1987**, *52*, 4745.
18. (a) *FF* **1967**, *1*, 682. (b) Haslam, E. In *Protective Groups in Organic Chemistry*; McOmie, J. F. W., Ed.; Plenum: New York, 1973; Chapter 5. (c) Greene, T. W.; Wuts, P. G. M. *Protective Groups in Organic Synthesis*, 2nd ed.; Wiley: New York, 1991; Chapter 5.
19. Sato, T.; Otera, J.; Nozaki, H. *JOC* **1992**, *57*, 2166.
20. Chung, C. W.; De Bernardo, S.; Tengi, J. P.; Borgese, J.; Weigele, M. *JOC* **1985**, *50*, 3462.
21. (a) Greene, T. W.; Wuts, P. G. M. *Protective Groups in Organic Synthesis*, 2nd ed.; Wiley: New York, 1991; Chapter 3. (b) Wymann, W. E.; Davis, R.; Patterson, Jr., J. W.; Pfister, J. R. *SC* **1988**, *18*, 1379.
22. Manning, W. B.; Kelly, T. R.; Muschik, G. M. *TL* **1980**, *21*, 2629.
23. Greene, T. W.; Wuts, P. G. M. *Protective Groups in Organic Synthesis*, 2nd ed.; Wiley: New York, 1991; Chapter 2.
24. Fisher, M. J.; Myers, C. D.; Joglar, J.; Chen, S.-H.; Danishefsky, S. J. *JOC* **1991**, *56*, 5826.
25. (a) Greene, A. E.; Le Drina, C.; Crabbe, P. *JACS* **1980**, *102*, 7583. (b) Finch, N.; Fitt, J. J.; Hsu, I. H. S. *JOC* **1975**, *40*, 206. (c) Ichikawa, Y.; Tsuboi, K.; Naganawa, A.; Isobe, M. *SL* **1993**, 907.
26. Nicolaou, K. C.; Hwang, C.-K.; Marron, B. E.; DeFrees, S. A.; Coulandouros, E. A.; Abe, Y.; Carroll, P. J.; Snyder, J. P. *JACS* **1990**, *112*, 3040. (b) Nicolaou, K. C.; McGarry, D. G.; Somers, P. K.; Kim, B. H.; Ogilvie, W. W.; Yiannikouros, G.; Prasad, C. V. C.; Veale, C. A.; Hark, R. R. *JACS* **1990**, *112*, 6263.
27. Barrett, G. C. In *Comprehensive Organic Chemistry*; Barton, D. H. R., Ed.; Pergamon: Oxford, 1979; Vol. 3, pp 105–120.
28. Helmkamp, G. K.; Pettitt, D. J. *JOC* **1960**, *25*, 1754.
29. (a) Corey, E. J.; Chaykovsky, M. *JACS* **1965**, *87*, 1353. (b) Kuhn, R.; Trischmann, H. *LA* **1958**, *611*, 117. (c) Emeleus, H. J.; Heal, H. G. *JCS* **1946**, 1126.
30. Hori, M.; Katakoka, T.; Shimizu, H.; Okitsu, M. *TL* **1980**, 4287.
31. Tamaru, Y.; Harada, T.; Yoshida, Z.-I. *JACS* **1979**, *101*, 1316.
32. Pojer, P. M.; Angyal, S. J. *TL* **1976**, 3067.
33. Fetizon, M.; Jurion, M. *CC* **1972**, 382.
34. Wawzonek, S.; McKillip, W.; Peterson, C. J. *OSC* **1973**, *5*, 785.
35. March, J. *Advanced Organic Chemistry*, 3rd. ed.; Wiley: New York, 1985; p 799.
36. Sommer, H. Z.; Jackson, L. L. *JOC* **1970**, *35*, 1558.
37. Cope, A. C.; Trumbull, E. R. *OR* **1960**, *11*, 317.
38. (a) Cope, A. C; Bach, R. D. *OSC* **1973**, *5*, 315. (b) Manitto, P.; Monti, D.; Gramatica, P.; Sabbioni, E. *CC* **1973**, 563.
39. Horwell, D. C. *T* **1980**, *36*, 3123.
40. Potts, K. T.; Saxton, J. E. *OSC* **1973**, *5*, 769.
41. Smith, D. J. H. In *Comprehensive Organic Chemistry*; Barton, D. H. R., Ed.; Pergamon: Oxford, 1979; Vol. 2, p 1160.
42. Arbuzov, B. A. *PAC* **1964**, *9*, 307.
43. Fieser, L. F.; Fieser, M. *FF* **1967**, *1*, 685.

Gary A. Sulikowski & Michelle M. Sulikowski
Texas A&M University, College Station, TX, USA

Iodomethylzinc Iodide[1]

ICH_2ZnI

(**1**; ICH_2ZnI)		
[4109-94-8]	CH_2I_2Zn	(MW 333.22)
(**2**; $BrCH_2ZnBr$)		
[4109-95-9]	CH_2Br_2Zn	(MW 239.22)
(**3**; $(ICH_2)_2Zn$)		
[14399-53-2]	$C_2H_4I_2Zn$	(MW 347.25)
(**4**; $(ICH_2)_2Zn{\cdot}DME$)		
[131457-21-1]	$C_6H_{14}I_2O_2Zn$	(MW 437.39)
(**5**; $(BrCH_2)_2Zn$)		
[92601-82-6]	$C_2H_4Br_2Zn$	(MW 253.25)
(**6**; $(ClCH_2)_2Zn{\cdot}DME$)		
[131457-22-2]	$C_6H_{14}Cl_2O_2Zn$	(MW 254.49)

(methylene transfer reagent: cyclopropanates alkenes,[1] a^1/d^1 multicoupling reagent,[2] transmetalation with various metal halides affords other iodomethylmetal compounds[3])

Alternate Name: Simmons–Smith reagent.

Physical Data: an X-ray crystal structure of $(ICH_2)_2Zn$ complexed to a glycol bis-ether is known;[4] DME complexes of $(ICH_2)_2Zn$ and $(ClCH_2)_2Zn$ and an acetone complex of $(ICH_2)_2Zn/ZnI_2$ have been characterized by NMR spectroscopy;[4] ^1H NMR spectra attributed to THF complexes of $BrCH_2ZnBr$ and $(BrCH_2)_2Zn$ have been reported.[5]

Solubility: ICH_2ZnI generated from either CH_2I_2/Zn–Cu couple or $EtZnI/CH_2I_2$ is generally prepared in ethereal solvents (Et_2O, DME). The Et_2Zn/CH_2I_2 method of reagent generation can utilize noncoordinating solvents (CH_2Cl_2, $ClCH_2CH_2Cl$, toluene, etc.).

Preparative Methods: the two most widely used methods of preparing halomethylzinc reagents are the Simmons–Smith and Furukawa procedures, utilizing ***Diiodomethane/Zinc/Copper Couple*** and CH_2I_2/***Diethylzinc*** (or ***Chloroiodomethane***/Et_2Zn),[6] respectively. The reagent is often prepared in the presence of the substrate (usually an alkene). Various methods of reagent preparation are discussed below. The precursors are widely available.

Reagent Preparation. There are a number of protocols for generating iodomethylzinc reagents, which can be categorized into three general classes: type 1, the oxidative addition of a dihalomethane to zinc metal, as typified by the original Simmons–Smith procedure;[7,8] type 2, the reaction of a zinc(II) salt with a diazoalkane, first reported by Wittig and co-workers;[9] and type 3, an alkyl exchange reaction between an alkyl zinc and a 1,1-dihaloalkane, often referred to as the Furukawa procedure.[10]

Type 1 reagent generation has been used most often in synthetic contexts due to the ease with which the reagent precursors can be handled. Although the initial method of preparation of the Zn–Cu couple was difficult and not easily reproducible,[7] several simpler and highly reproducible methods soon followed.[11] Treatment of the Zn–Cu couple with CH_2I_2 and a crystal of ***Iodine*** in Et_2O followed by heating to reflux generates the active reagent. Other modifications include the use of CH_2I_2/Zn/CuCl,[12a] CH_2I_2/Zn–Ag couple,[12b] $CH_2Br_2/Zn/TiCl_4$,[12c] and CH_2Br_2/Zn/AcCl/CuCl.[12d]

Type 2 reagent generation has been utilized much less frequently. The method consists of the treatment of an ethereal suspension of a zinc(II) salt ($ZnCl_2$, $ZnBr_2$, ZnI_2, or $Zn(OBz)_2$) with CH_2N_2 or an aryldiazomethane.[9a]

Type 3 halomethylzinc generation (originally reported in 1966)[10a] involves treatment of a solution (Et_2O, hexane, toluene, etc.) of Et_2Zn with CH_2I_2 to generate the reagent. The use of a 2:1 ratio of CH_2I_2 to Et_2Zn generates $(ICH_2)_2Zn$,[4] while a 1:1 ratio presumably generates $EtZnCH_2I$.[10] The reaction is accelerated by the presence of trace amounts of oxygen.[6b] Treatment of Et_2Zn with substituted diiodides, such as benzylidene and ethylidene iodide, also gives rise to active cyclopropanating reagents.[13] Recently, the substitution of $ClCH_2I$ for CH_2I_2 and the use of $ClCH_2CH_2Cl$ (DCE) as the reaction solvent has been demonstrated to provide a more reactive reagent for certain applications.[6a] In addition, the combination of EtZnI and CH_2I_2 has also been shown to provide ICH_2ZnI, thus avoiding the need for the highly pyrophoric Et_2Zn.[14]

Cyclopropanations. The cyclopropanation of alkenes utilizing halomethylzinc reagents (ICH_2ZnI being the prototypical reagent), known as the Simmons–Smith reaction,[7] has proven to be an extremely versatile and general reaction. Typical examples of alkenes that have been successfully cyclopropanated are provided in eqs 1–5. A variety of isolated alkenes have been cyclopropanated with the Simmons–Smith reagent (e.g. eq 1),[1a,12b] and ICH_2ZnI provides for a unique preparation of numerous spiro derivatives (eq 2).[15] Electron-rich alkenes such as enol ethers (eq 3)[16a–c] and enamines (eq 4)[16d,e] also have been found to be good substrates under the proper conditions, as have certain steroidal enones (eq 5).[16f,g] Simmons–Smith reagents thus have been demonstrated to cyclopropanate alkenes ranging from electron-rich to electron-deficient. This contrasts with the analogous reagents generated from CH_2I_2/R_3Al[17] and $ClCH_2I$/Sm(Hg):[18] the former reacts preferentially with isolated alkenes, while the latter cyclopropanates allylic alcohols almost exclusively. Certain vinyl metal species (Al, Si, Ge, Sn, B) can also be cyclopropanated with some success with the Simmons–Smith reagent.[19] For example, vinylalanes produced in situ from alkynes and ***Diisobutylaluminum Hydride*** react readily with CH_2Br_2/Zn–Cu couple; the intermediate cyclopropylalanes react with bromine to produce cyclopropyl bromides (eq 6).[19b] Generally, the reaction is most successful with electron-rich alkenes, indicative of the electrophilic nature of halomethylzinc reagents.[1a]

CH_2I_2, Zn–Cu couple; Et_2O, reflux; 92% (1)

CH_2I_2, Zn–Cu couple; Et_2O, reflux; 100% (2)

TMSO OTMS → CH_2I_2, Zn–Ag couple; Et_2O, reflux; 78% → TMSO OTMS (3)

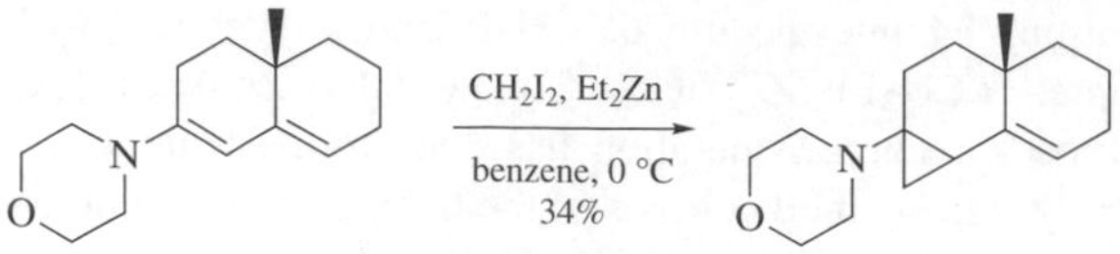

The reaction is not limited to unsubstituted methylene transfers.[13] The combination of $MeCHI_2$[13a–c] or $PhCHI_2$[13a,d] with Et_2Zn also provides active cyclopropanation reagents. The diastereoselectivity is highly substrate dependent, but good diastereoselectivity can be achieved in certain cases (eq 7), particularly with cyclic alkenes. The stereoselectivity is solvent dependent, with ethereal solvents affording the higher levels of selectivity.[13d] Halogen-substituted carbenoids can also be prepared from various $XCHI_2$ (X = I, Br, F) or X_2CHI (X = Br, Cl) and Et_2Zn (eq 8).[13e–g]

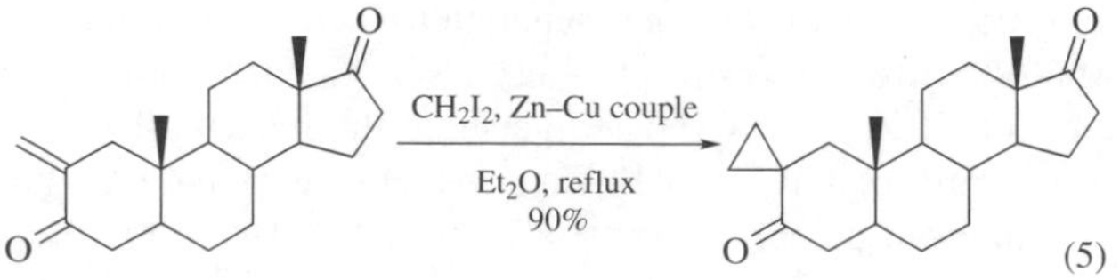

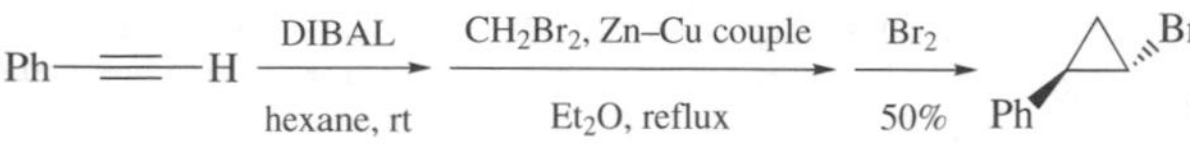

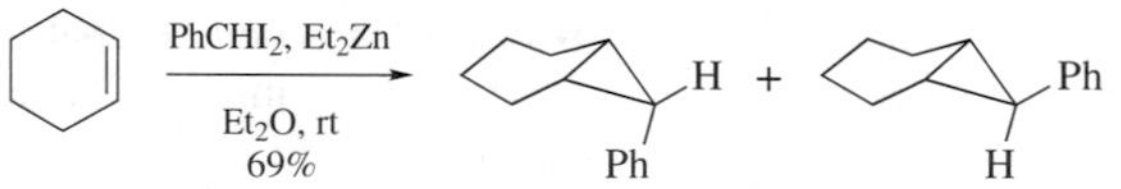

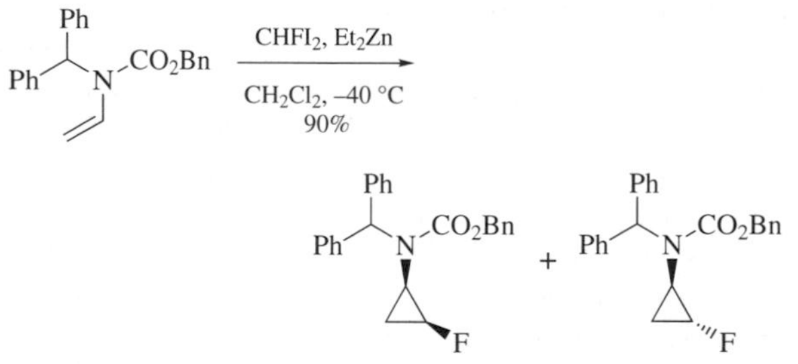

Perhaps the most intriguing aspect of the Simmons–Smith reaction is the strong accelerating and stereodirecting effect of oxygen functions proximal to the alkene. First discovered in 1959,[20] this reaction has been often utilized in synthetic efforts[21] and the reaction itself has been the subject of several investigations.[22] For example, cyclopropanation of 2-cyclohexen-1-ol provides the *syn*-cyclopropane almost exclusively.[22a] A study of various cyclic allylic alcohols demonstrates the generality of the effect (eq 9):[22c] the larger rings afford *trans* adducts due to conformational effects. The diastereoselectivity of the cyclopropanation of acyclic secondary allylic alcohols depends upon the configuration of the alkene. *cis*-Alkenes react with diastereoselectivities of >99:1 (eq 10), while *trans*-alkenes react with much less selectivity (<2:1).[22f] Homoallylic alcohols also show a similar directing effect in certain cases (eq 11).[20,21e]

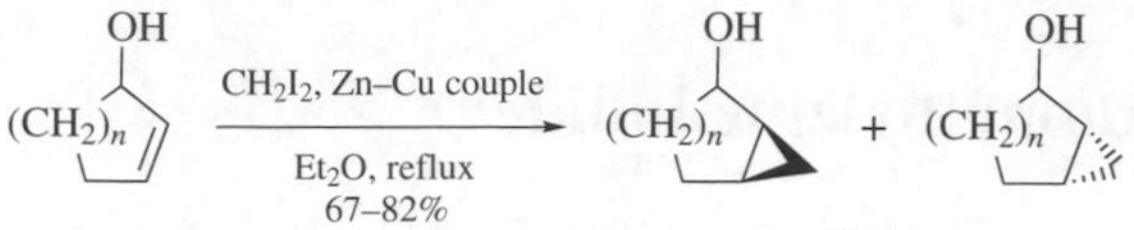

Chiral auxiliary mediated cyclopropanations which exploit this oxygen-directing effect have recently been developed. The first Simmons–Smith reactions exhibiting effective diastereofacial control by chiral auxiliaries were reported simultaneously by two groups in 1985.[23,24] Chiral acetals derived from cyclic enones undergo highly diastereoselective cyclopropanations upon treatment with CH_2I_2/Zn–Cu couple (eq 12). Acyclic enones are cyclopropanated with greatly attenuated diastereoselectivity.

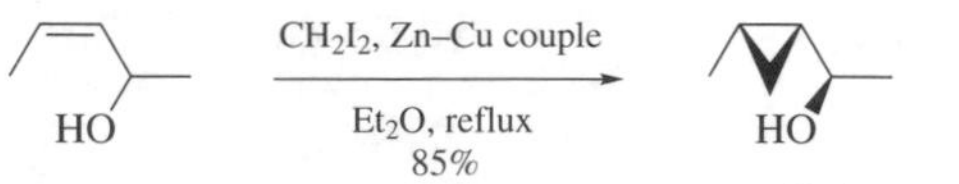

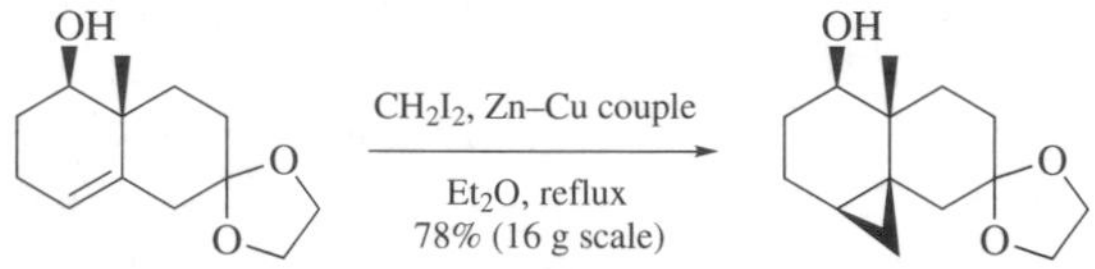

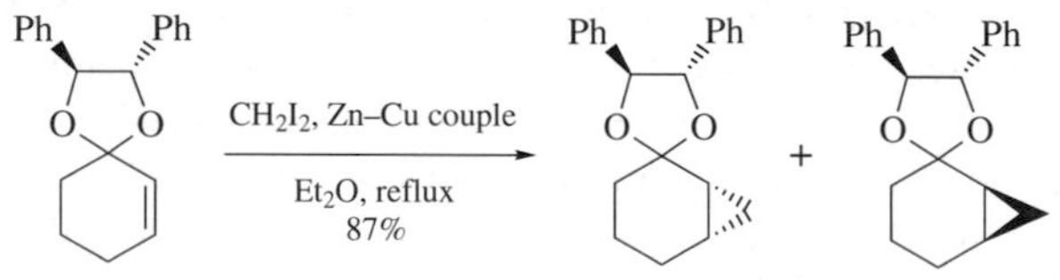

Similarly, chiral acetals[24] derived from α,β-unsaturated aldehydes and diisopropyl tartrate are cyclopropanated in a highly diastereoselective manner by CH_2I_2/Et_2Zn (eq 13). Diastereoselectivities are uniformly high for dioxolane acetals derived from *trans*-disubstituted α,β-unsaturated aldehydes, but acetals derived from α,β-unsaturated ketones react less selectively, as do 2-alkenyl-1,3-dioxane acetals.

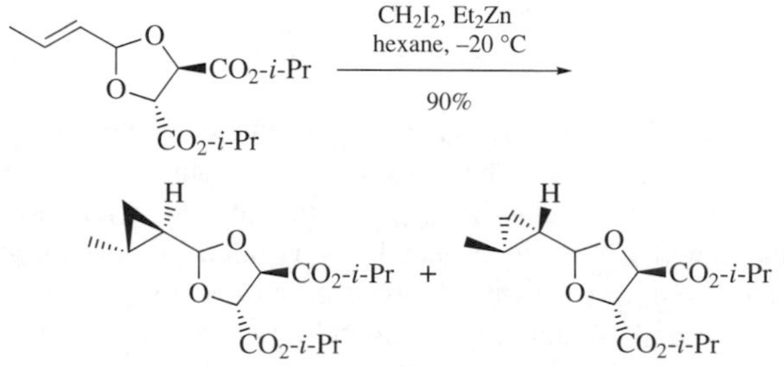

A related oxygen-directed cyclopropanation has also been reported.[25] Vinyl boronates derived from tartaric esters or amides were shown to undergo highly diastereoselective cyclopropanations upon treatment with CH_2I_2/Zn–Cu couple. These adducts were conveniently converted to enantiomerically enriched cyclopropanols. The carbohydrate 2-hydroxy-3,4,6-tri-*O*-benzyl-β-D-glucopyranose appended to an allylic alcohol also functions as an effective chiral auxiliary, affording cyclopropanes with extremely high levels of diastereoselectivity (eq 14).[26] Other chiral auxiliaries have also been shown to direct halomethylzinc cyclopropanations with good to excellent stereocontrol.[13g,27]

CH_2I_2, Et_2Zn; Et_2O, rt; 95% (14)

98% de

Although the potential for preparing enantioselective halomethylzinc reagents was recognized early on,[28] only since 1992 have encouraging levels of enantioselectivity been observed.[29] The best results reported to date utilize chiral C_2-symmetric sulfonamides in substoichiometric amounts as the source of chirality (eq 15).[29a] A zinc complex of this ligand is proposed to act as a chiral Lewis acid catalyst in this reaction. All of the enantioselective halomethylzinc cyclopropanations reported to date utilize allylic alcohols as substrates, and the free hydroxy group appears to play an essential role.[29]

CH_2I_2, Et_2Zn (3:2), 0.12 equiv L*; CH_2Cl_2–hexane (9:1), –23 °C; quant. (15)

82% ee

L* = (cyclohexane-1,2-diyl)bis($NHSO_2R$); R = p-$NO_2C_6H_4$

Methylene Homologation Reactions. The carbon bound iodine atom of ICH_2ZnI can be easily displaced by nucleophiles to generate new organozinc reagents.[9b] For example, various copper nucleophiles displace the carbon bound iodine from ICH_2ZnI or $(ICH_2)_2Zn$, generating new organometallic reagents that react with allyl halides.[2,30] Copper nucleophiles such as CuCN/LiCl, $NCCH_2Cu$, copper amides, vinylcoppers, and heteroarylcopper compounds all participate in this reaction (eq 16). This reaction has proven to be especially useful for the conversion of alkenylcoppers into allylic copper–zinc reagents which react with aldehydes affording homoallylic alcohols (eq 17). An expedient route to α-methylene-γ-butyrolactones that exploits this behavior has also been developed (eq 18).[30e]

$ICH_2ZnI + ZnI_2 +$ 2-thienyl-Cu; 1. THF, –60 °C to –40 °C; 2. 2-(bromomethyl)hexene, –40 to 0 °C; 96% (16)

ICH_2ZnI; $Me(CH_2)_4CHO$; 80% (17)

1. H—≡—CO_2Et; 2. PhCHO, $(ICH_2)_2Zn$, THF, –60 to 0 °C; 76% (18)

Transmetalation Reactions. Like other alkylzinc reagents,[31] halomethylzinc reagents have also been shown to participate in transmetalation reactions.[3,14a] This methodology provides an expedient route to iodomethylmercury and iodomethyltin compounds. For example, treatment of Me_3SnCl with ICH_2ZnI derived from EtZnI and CH_2I_2 provides Me_3SnCH_2I in 78% yield.[14a] Bu_3SnCH_2I may be prepared similarly in 96% yield.[32] Substituted diiodides also provide zinc reagents that participate well in this reaction.[14a]

[2,3]-Rearrangements. A method for the generation of sulfur ylides from allylic phenyl sulfides and CH_2I_2/Et_2Zn has been described.[33] The intermediate sulfur ylides undergo a sigmatropic [2,3]-rearrangement affording homoallylic sulfides (eq 19). The reaction gives (E)-alkenes selectively.

Et_2Zn, CH_2I_2; benzene, 55 °C; 75% (19)

Related Reagents. Dibromomethane–Zinc/Copper Couple; Diethylzinc; Diethylzinc-Bromoform-Oxygen; Diethylzinc-Iodoform; 1,1-Diiodoethane; Diiodomethane; Ethyliodomethylzinc; Ethylzinc Iodide; Zinc/Copper Couple.

1. (a) Simmons, H. E.; Cairns, T. L.; Vladuchick, S. A.; Hoiness, C. M. *OR* **1972**, *20*, 1. (b) Furukawa, J.; Kawabata, N. *Adv. Organomet. Chem.* **1974**, *12*, 83. (c) Zeller, K.-P.; Gugel, H. *MOC* **1989**, *E19b*, 1279. (d) Helquist, P. *COS* **1991**, *4*, Chapter 4.6.
2. Knochel, P.; Jeong, N.; Rozema, M. J.; Yeh, M. C. P. *JACS* **1989**, *111*, 6473.
3. Seyferth, D.; Andrews, S. B. *JOM* **1971**, *30*, 151.
4. (a) Denmark, S. E.; Edwards, J. P.; Wilson, S. R. *JACS* **1991**, *113*, 723. (b) Denmark, S. E.; Edwards, J. P.; Wilson, S. R. *JACS* **1992**, *114*, 2592.
5. Fabisch, B.; Mitchell, T. N. *JOM* **1984**, *269*, 219.
6. (a) Denmark, S. E.; Edwards, J. P. *JOC* **1991**, *56*, 6974. (b) Miyano, S.; Yamashita, J.; Hashimoto, H. *BCJ* **1972**, *45*, 1946.
7. (a) Simmons, H. E.; Smith, R. D. *JACS* **1958**, *80*, 5323. (b) Simmons, H. E.; Smith, R. D. *JACS* **1959**, *81*, 4256.
8. This reagent was first prepared by Emschwiller: Emschwiller, G. *CR(C)* **1929**, *188*, 1555.
9. (a) Wittig, G.; Schwarzenbach, K. *AG* **1959**, *71*, 652. (b) Wittig, G.; Schwarzenbach, K. *LA* **1962**, *650*, 1. (c) Wittig, G.; Wingler, F. *LA* **1962**, *656*, 18. (d) Wittig, G.; Wingler, F. *CB* **1964**, *97*, 2146. (e) Wittig, G.; Jautelat, M. *LA* **1967**, *702*, 24.
10. (a) Furukawa, J.; Kawabata, N.; Nishimura, J. *TL* **1966**, 3353. (b) Furukawa, J.; Kawabata, N.; Nishimura, J. *T* **1968**, *24*, 53. (c) Nishimura, J.; Furukawa, J.; Kawabata, N.; Kitayama, M. *T* **1971**, *27*, 1799.

11. (a) Shank, R. S.; Shechter, H. *JOC* **1959**, *24*, 1825. (b) LeGoff, E. *JOC* **1964**, *29*, 2048.

12. (a) Rawson, R. J.; Harrison, I. T. *JOC* **1970**, *35*, 2057. (b) Denis, J. M.; Girard, C.; Conia, J. M. *S* **1972**, 549. (c) Friedrich, E. C.; Lunetta, S. E.; Lewis, E. J. *JOC* **1989**, *54*, 2388. (d) Friedrich, E. C.; Lewis, E. J. *JOC* **1990**, *55*, 2491.

13. (a) Furukawa, J.; Kawabata, N.; Nishimura, J. *TL* **1968**, 3495. (b) Nishimura, J.; Kawabata, N.; Furukawa, J. *T* **1969**, *25*, 2647. (c) Nishimura, J.; Furukawa, J.; Kawabata, N. *BCJ* **1970**, *43*, 2195. (d) Nishimura, J.; Furukawa, J.; Kawabata, N.; Koyama, H. *BCJ* **1971**, *44*, 1127. (e) Nishimura, J.; Furukawa, J. *CC* **1971**, 1375. (f) Miyano, S.; Hashimoto, H. *BCJ* **1974**, *47*, 1500. (g) Tamura, O.; Hashimoto, M.; Kobayashi, Y.; Katoh, T.; Nakatani, K.; Kamada, M.; Hayakawa, I.; Akiba, T.; Terashima, S. *TL* **1992**, *33*, 3483; 3487.

14. (a) Seyferth, D.; Andrews, S. B.; Lambert, R. L. *JOM* **1972**, *37*, 69. (b) Sawada, S.; Inouye, Y. *BCJ* **1969**, *42*, 2669.

15. (a) Krapcho, A. P. *S* **1978**, 77. (b) Bee, L. K.; Beeby, J.; Everett, J. W.; Garratt, P. J. *JOC* **1975**, *40*, 2212. (c) Fitjer, L. *CB* **1982**, *115*, 1047. (d) Erden, I. *SC* **1986**, *16*, 117.

16. (a) Denis, J. M.; Conia, J. M. *TL* **1972**, 4593. (b) Ryu, I.; Murai, S.; Sonoda, N. *TL* **1977**, 4611. (c) Rubottom, G. M.; Lopez, M. I. *JOC* **1973**, *38*, 2097. (d) Kuehne, M. E.; King, J. C. *JOC* **1973**, *38*, 304. (e) Kuehne, M. E.; DiVencenzo, G. *JOC* **1972**, *37*, 1023. (f) Desai, U. R.; Trivedi, G. K. *LA* **1990**, 711. (g) Limasset, J.-C.; Amice, P.; Conia, J.-M. *BSF(2)* **1969**, 3981.

17. (a) Maruoka, K.; Fukutani, Y.; Yamamoto, H. *JOC* **1985**, *50*, 4412. (b) Miller, D. B. *TL* **1964**, 989.

18. (a) Molander, G. A.; Etter, J. B. *JOC* **1987**, *52*, 3942. (b) Molander, G. A.; Harring, L. S. *JOC* **1989**, *54*, 3525.

19. (a) Seyferth, D.; Cohen, H. M. *IC* **1962**, *1*, 913. (b) Zweifel, G.; Clark, G. M.; Whitney, C. C. *JACS* **1971**, *93*, 1305.

20. Winstein, S.; Sonnenberg, J.; deVries, L. *JACS* **1959**, *81*, 6523.

21. For some recent examples, see: (a) Corey, E. J.; Virgil, S. C. *JACS* **1990**, *112*, 6429. (b) Oppolzer, W.; Radinov, R. N. *JACS* **1993**, *115*, 1593. (c) Johnson, C. R.; Barbachyn, M. R. *JACS* **1982**, *104*, 4290. (d) Neef, G.; Cleve, G.; Otow, E.; Seeger, A.; Wiechert, R. *JOC* **1987**, *52*, 4143. (e) Grieco, P. A.; Collins, J. L.; Moher, E. D.; Fleck, T. J.; Gross, R. S. *JACS* **1993**, *115*, 6078.

22. (a) Dauben, W. G.; Berezin, G. H. *JACS* **1963**, *85*, 468. (b) Chan, J. H.-H.; Rickborn, B. *JACS* **1968**, *90*, 6406. (c) Poulter, C. D.; Friedrich, E. C.; Winstein, S. *JACS* **1969**, *91*, 6892. (d) Staroscik, J. A.; Rickborn, B. *JOC* **1972**, *37*, 738. (e) Kawabata, N.; Nakagawa, T.; Nakao, T.; Yamashita, S. *JOC* **1977**, *42*, 3031. (f) Ratier, M.; Castaing, M.; Godet, J.-Y.; Pereyere, M. *JCR(S)* **1978**, 179.

23. (a) Mash, E. A.; Nelson, K. A. *JACS* **1985**, *107*, 8256. (b) Mash, E. A.; Hemperly, S. B. *JOC* **1990**, *55*, 2055, and references cited therein.

24. (a) Arai, I.; Mori, A.; Yamamoto, H. *JACS* **1985**, *107*, 8254. (b) Mori, A.; Arai, I.; Yamamoto, H. *T* **1986**, *42*, 6447.

25. Imai, T.; Mineta, H.; Nishida, S. *JOC* **1990**, *55*, 4986.

26. (a) Charette, A. B.; Côté, B.; Marcoux, J.-F. *JACS* **1991**, *113*, 8166. (b) Charette, A. B.; Côté, B. *JOC* **1993**, *58*, 933.

27. (a) Sugimura, T.; Katagiri, K.; Tai, A. *TL* **1992**, *33*, 367, and references cited therein. (b) Seebach, D.; Stucky, G. *AG(E)* **1988**, *27*, 1351. (c) Fukuyama, Y.; Hirono, M.; Kodama, M. *CL* **1992**, 167. (d) Morikawa, T.; Sasaka, H.; Mori, K.; Shiro, M.; Taguchi, T. *CPB* **1992**, *40*, 3189. (e) de Frutos, M. P.; Fernandez, M. D.; Fernandez-Alvarez, E.; Bernabe, M. *TL* **1991**, *32*, 541.

28. (a) Sawada, S.; Oda, J.; Inouye, Y. *JOC* **1968**, *33*, 2141. (b) Furukawa, J.; Kawabata, N.; Nishimura, J. *TL* **1968**, 3495.

29. (a) Takahashi, H.; Yoshioka, M.; Ohno, M. Kobayshi, S. *TL* **1992**, *33*, 2757. (b) Ukakji, Y.; Nishimura, M.; Fujisawa, T. *CL* **1992**, 61. (c) Denmark, S. E.; Edwards, J. P. *SL* **1992**, 229. (d) Denmark, S. E.; Christenson, B. L.; Coe, D. M.; O'Connor, S. P. *TL* **1995**, *36*, 2215. (e) Denmark, S. E.; Christenson, B. L.; O'Connor, S. P. *TL* **1995**, *36*, 2219.

30. (a) Knochel, P.; Chou, T.-S.; Chen, H. G.; Yeh, M. C. P.; Rozema, M. J. *JOC* **1989**, *54*, 5202. (b) Knochel, P.; Jeong, N.; Rozema, M. J.; Yeh, M. C. P. *JACS* **1989**, *111*, 6474. (c) Knochel, P.; Rao, S. A. *JACS* **1990**, *112*, 6146. (d) Rozema, M. J.; Knochel, P. *TL* **1991**, *32*, 1855. (e) Knochel, P.; Rozema, M. J.; Tucker, C. E.; Retherford, C.; Furlong, M.; Sidduri, A. R. *PAC* **1992**, *64*, 361. (f) Sidduri, A. R.; Knochel, P. *JACS* **1992**, *114*, 7579.

31. Boersma, J. In *Comprehensive Organometallic Chemistry*; Wilkinson, G., Ed.; Pergamon: Oxford, 1984; Vol. 2, Chapter 16.

32. Still, W. C. *JACS* **1978**, *100*, 1481.

33. Kosarych, Z.; Cohen, T. *TL* **1982**, *23*, 3019.

James P. Edwards
Ligand Pharmaceuticals, San Diego, CA, USA

Paul Knochel
Philipps-Universität, Marburg, Germany

Ketene[1]

$H_2C{=}C{=}O$

[463-51-4] C_2H_2O (MW 42.04)

(mild acetylating reagent;[1] performs [2+2] cycloadditions with alkenes and alkynes;[1c] performs nucleophilic addition to acetyl chlorides;[2] capable of certain organometallic transformations[3])

Physical Data: mp −150 °C; bp −56 °C.

Solubility: generally introduced as a gas but is soluble in a wide variety of solvents.

Preparative Methods: by the thermal decomposition of acetone,[4] diketene,[5] acetic anhydride,[6] alkoxy[7] or siloxy acetylenes,[8,9] and carboxylic acid over zeolites.[10] Acetic anhydride gives the purest stream of this gas. The use of acetone, diketene, or acetic acid may leave impurities such as acetone and carbon monoxide.

Handling, Storage, and Precautions: poisonous gas with a toxicity approximately eight times greater than phosgene.[11] All operations utilizing ketene should be carried out in an appropriate isolated apparatus in a well-ventilated fume hood.[5,11]

Acetylating Reagent. The electrophilic nature of the sp carbon in ketene makes it a powerful yet mild acetylating reagent.[1,12,13] Unlike other reagents such as ***Acetic Anhydride***, ketene has no leaving group, the opening of the double bond being its equivalent.[14] This feature ensures fewer byproducts in the reaction mixture, allowing for easier purification of the product. Ketene will acetylate a wide variety of functional groups, including carboxylic acids, alcohols, amines, amides, thiols and 1,3-dione systems.[1a,14–16] One of the more recent uses of ketene in this area is the acetylation of β-diketone Cu^{II} intermediates (eq 1).[12] This reaction proceeds quickly at room temperature, without the need of a catalyst, the sole product arising from *C*-acetylation. Other attempts to introduce substitution into this system using alternative reagents not only required higher temperatures and the use of catalysts, but have resulted in the formation of both *C*- and *O*-acetylated products.[17,18] These reactions also resulted in the loss of the Cu–O bond, which the ketene reaction does not affect.

$H_2C{=}{\bullet}{=}O$; Ac_2O, 20 °C, 5–7 min; then HCl (1)

Ketene has also been used in acetylating cyclic diones to form the corresponding 1,3-cycloalkadienes (eq 2).[15,16] The formation of the 1,3-cycloalkadienes was previously accomplished by a Birch reduction of the corresponding aromatic derivatives.[15] For rings larger than six atoms in size, the use of ketene will result in *C*-acetylation as well as *O*-acetylation, and ***Isopropenyl Acetate*** has been shown to be more effective in this case.

$H_2C{=}{\bullet}{=}O$, Bu_2O, TsOH, 125 °C, 33–93%; $n = 1, 2$ (2)

Ketene is also a good acetylating reagent in reactions where the starting material has been absorbed onto a solid medium. The acetylation of phenol usually requires the use of high temperatures and a catalyst.[19] Even under these conditions, the reaction proves to be inconsistent. If phenol is first adsorbed onto silica, however, acetylation may take place in good yield by passing a stream of ketene through the medium as a gas at room temperature. Phenol derivatives adsorbed on alumina have also been shown to acetylate quantitatively at 0 °C when exposed to ketene gas.[19] This reaction has been shown to be quite versatile and may accommodate a wide variety of alcohols on many adsorbants including celite, magnesium oxide, and zinc oxide.[20] The ease of acetylation using ketene may also be shown by the following example (eq 3). The thermal fragmentation of 1-alkoxyalkynes in chloroform allows for the easy formation of ketene in situ. When this process is carried out in the presence of an amine, the ketene is trapped upon formation, giving the resulting amide.[7] As shown in eq 3, this reaction offers a mild process for acetylation which should be suitable for a wide variety of nucleophiles.[7]

$\equiv$—O—*t*-Bu + R^1R^2N–H → ($CHCl_3$, reflux 4 h, 82–99%) R^1R^2N–C(O)CH₃ (3)

R^1 = H, *i*-Pr
R^2 = C_6H_{13}, *t*-Bu, $HOCH_2CH_2$, Bn, Ph, *i*-Pr, Cy

Cycloadditions. Cycloaddition involving ketene and its derivatives is probably the most characteristic reaction involving this reagent. Examples of 1,2-cycloaddition to form four-membered rings are known to occur between ketene and C≡C, C=C, C=O, C=N, C=S, N=N, N=O, N=S, and P=N groups.[1c,1a] These reactions are generally considered to proceed via a $_{\pi}2_s + {}_{\pi}2_a$ thermally allowed cycloaddition following the Woodward–Hoffmann rules.[1c] The cycloaddition proves to be highly selective,[13] the most nucleophilic alkenic carbon becoming bonded to the sp carbon of ketene in the product. Addition occurs with retention of configuration of the alkene.[1a] In many cases, ketene proves to be too unreactive for the task at hand and ***Dichloroketene*** has become the reagent of choice in recent years. The relative reactivities of some ketenes in reference to cycloadditions are shown in Scheme 1.[1a]

$Cl_2C{=}{\bullet}{=}O > Ph_2C{=}{\bullet}{=}O > Me_2C{=}{\bullet}{=}O > H_2C{=}{\bullet}{=}O$

Scheme 1

Alkene Addition. There are few examples of alkene addition to ketene itself. The alkene generally must be 'activated' either through the use of an EDG or conjugation, such as the case with cyclopentadiene (eq 4)[21,22] and 1,3-cyclohexadiene (eq 5).[22] Linear conjugated dienes will sometimes react with ketene but yields are generally quite low and the reaction is not synthetically useful. Again, dichloroketene is generally used in this case owing to its greater reactivity (eq 6).[1c,23]

(4) 17% or 34%

(5) 5%

(6) 75%

One example of the cylcloaddition of ketene with an activated alkene involves vinyl ethers.[24,25] Staudinger had shown in 1920 that ***Diphenylketene*** adds to ethyl vinyl ether to form the 1,2-cycloaddition product.[1c] This reaction was repeated by others,[26,27] and in 1960 Hurd was able to show that ketene itself would also add to a vinyl ether.[24] Seija built upon this methodology later when he used a variety of vinyl ethers along with ketene to gain access to the synthetically interesting bicyclo[1.1.0]butanes, as shown in eq 7.[25] The reaction is regioselective in that the methylene group in the product is bonded to the carbon attached to the ether oxygen.[24,25]

(7) 100 °C, 4 h; 35–70%

Alkyne Addition. The alkyne bond is sufficiently high in energy to undergo 1,2-cycloadditions with ketene in the formation of cyclobutenones.[28] Wasserman[29] and Pericás[30] utilized this reaction in the formation of 1,3-cyclobutanedione. The latter synthesis used the alkyne as a source of ketene as well as the cycloaddition partner (eq 8). More recently, siloxyalkynes have been shown to undergo the same type of reaction utilizing a wide variety of R groups (eq 9); a multitude of cyclobutenone systems can thereby be constructed.[31]

(8) EtO, Δ, 31%; t-BuO, CH_2Cl_2, 30 °C, 70%

(9) TIPSO, 0 °C, 80–100%

C=X Cycloaddition. The cycloaddition reaction of ketene to a C=O bond is a very synthetically useful reaction which affords 2-oxetanones (β-lactone).[1] The rate of reaction is controlled somewhat by the polarization of the carbonyl bond. The greater the positive charge of the carbonyl carbon, the greater the rate of reaction.[32] The cycloaddition of ketene with the carbonyl bond requires a catalyst for reaction to occur.[1a] Ketene will readily couple with ***Chloral*** in the presence of a catalyst to form β-(trichloromethyl)-β-propiolactone.[33] Using chiral cinchona alkaloid catalysts, it is possible to form the chiral 4-(trichloromethyl)-2-oxetanone beginning with ketene and chloral. Acid hydrolysis of the cyclic product gives optically pure (*S*)- or (*R*)-malic acid in 79% yield (eq 10).[32,34]

(10) chiral cat., −50 °C, toluene; (*R*) or (*S*) → (*S*) or (*R*)

This compares quite well with more involved routes which begin with chiral tartaric acids[35] or the use of enzymatic catalysts, which allow for synthesis of the (*S*)-isomer but not the unnatural (*R*)-isomer.[36] Chiral polymeric cinchona alkaloids have also been employed in the reaction.[37] The use of polymeric catalysts allows greater ease of recycling and in some cases, such as poly(cinchona alkaloidacrylate), the enantioselectivity and yields were comparable to their monomeric counterparts.[37]

The cycloaddition of ketene to a C=N bond was first discovered by Staudinger in 1907 and is often referred to by that name.[38] The reaction proceeds regioselectively via a nonconcerted mechanism to give the β-lactam (eq 11).[39–41] Ketene itself is quite unreactive towards the imine group;[1c,42] however, reactions with the disubstituted ketenes proceed well.[42] Among the more popular ketenes utilized for this reaction are diphenyl-, chlorophenyl-, and dibromoketene.[42]

(11)

Diazomethane Cycloaddition. The reaction of ketene with ***Diazomethane*** offers a simple, high yielding synthesis of cyclopropanone, a molecule of both theoretical and synthetic interest.[43] In the presence of methanol and excess ketene, cyclopropanone may be further acetylated, as shown in eq 12. Reaction with excess diazomethane results in the formation of cyclobutanone.[43,44] Earlier methods to prepare cyclopropanone from cyclobutanediones met with little success.[45]

(12) (1) + CH_2N_2 (2); −78 °C, CH_2Cl_2, 90%; (1), MeOH, 99%; (2), 90%

1,3- and 1,4-Cycloaddition. Cyclizations of this type are quite rare using ketene itself, although ketene derivatives have been known to undergo 1,3- and 1,4-cyclizations.[1]

Allene Formation. The reaction of ketene with an ylide proves to be a convenient method for the formation of allenes. In the reaction with the ylide shown in eq 13, the α-vinylidene-γ-butyrolactone, which holds biological interest, may be formed in high yield.[46]

(13) [scheme: ylide lactone with $=PPh_3$ + $=\bullet=O$, CH_2Cl_2, 90% → α-vinylidene lactone]

R = H, Me

The same methodology has been employed in more complex systems, such as the formation of 4,5-dihydropyrazolo[1,5-*a*]pyridine beginning with azine phosphoranes (eq 14).[47]

(14) [scheme: azine phosphorane (R^1, R^2, Ph, PPh_3) + ketene → allene intermediate → pyrazolopyridine, 36%]

If a ketene derivative is utilized (phenylacetoxyketene or monophenyoxyketene), the product will spontaneously eliminate acetic acid or phenol to aromatize, giving the pyridine ring.[47] The ability to use either ketene or its derivatives offers good flexibility in the synthesis of these molecules.

Organometallic Compounds. Although not as well known as the organic chemistry of ketene, the organometallic chemistry of ketene has been studied.[3] An example is the insertion of ketene into metal hydrides, as shown in eq 15.[3,48]

$$HMn(CO)_5 + {=}{\bullet}{=}O \xrightarrow{-30\,^\circ C} CH_3C(O)Mn(CO)_5 \quad (15)$$

Ketene will also insert into metal–alkyl bonds.[3] The mechanism of these reactions is unknown, but it should be noted that the products follow the same type of pattern as the acetylation reactions. Some more recent work in the organometallic area is the formation of disilver ketenide (eq 16).[49] This complex was formed via reaction of silver acetate with ketene gas formed from pyrolysis of acetone or via in situ formation of ketene using acetic anhydride. The latter method could be run at temperatures as low as −28 °C in the presence of an amine base for a catalyst. The reaction was also run using enol acetates as the ketene source.[49] The formation of this complex shows the potential carbon acidity of the hydrogen atoms attached to ketene. The silver ketenide complex is quite stable but undergoes the transformations shown in eqs 17–19.

$$2\,AgOAc + {=}{\bullet}{=}O \xrightarrow[90\%]{py,\ rt} Ag_2C{=}{\bullet}{=}O \quad (16)$$

$$Ag_2C{=}{\bullet}{=}O + Br_2 \xrightarrow[98\%]{CCl_4,\ rt} 2\,AgBr + Br_2C{=}{\bullet}{=}O \quad (17)$$

$$Ag_2C{=}{\bullet}{=}O + 2\,HCl \xrightarrow[96\%]{H_2O} 2\,AgCl + MeCO_2H \quad (18)$$

$$Ag_2C{=}{\bullet}{=}O + 2\,HCl \xrightarrow[>98\%]{} 2\,AgCl + {=}{\bullet}{=}O \quad (19)$$

Although silver ketenide is relatively unreactive, it has the potential of being a useful synthetic reagent for the preparation of disubstituted ketenes or regeneration of ketene itself.[49]

Ketene as a Nucleophile. In acetylation reactions, ketene behaves as an electrophilic agent, but in the presence of hemiacetal chlorides,[2] acid chlorides, acetals,[2] etc., ketene behaves as a nucleophile. In a study by Hurd and Kimbrough, ketene was mixed with the hemiacetal chloride in the presence of ***Zinc Chloride*** to give, after workup, the corresponding acid.[50] Hurd extended the study to acetals as well as the chloro derivatives.[2] More recently, Goure reacted ***Trifluoroacetyl Chloride*** with ketene followed by addition of ethyl 3-amino-4,4,4-trifluoro-2-butenoate. Subsequent rearrangement of this intermediate led to the pyridinecarboxylate in 86% yield.[51] The use of ketene in this reaction allows a two-step, one-pot synthesis of the previously unknown 2-hydroxy-4,6-bis(trifluoromethyl)pyridine-5-carboxylate (eq 20). Other methods utilized to form this compound lack the ability to incorporate a functional group in the 5 position of the pyridine ring.[51,52]

(20) [scheme: F_3CCOCl + $=\bullet=O$, Et_2O, −23 °C → $[F_3CCOCH_2COCl]$ + $F_3CC(NH_2){=}CHCO_2R$ → 2-hydroxy-4,6-bis(CF_3)pyridine-5-CO_2R]

Miscellaneous. Recently, ketene has become a tool for mechanistic probes and the formation of high energy molecules. The transformation of methanol to hydrocarbons proceeds through a ketene intermediate at some point. The thermal decomposition of ketene is being studied on metal surfaces in an effort to elucidate the mechanism.[53,54]

Ketene is also used to form singlet methylene ($^1CH_2$) in an effort to study the insertion reactions of this short-lived intermediate.[55] Its use is also noted in the formation of distonic ions: ions that have a separated charge and radical site. The use of ketene in this instance offers an advantage over the known methods in that it allows greater ease of formation of the distonic ions and greater flexibility over the type of ion formed.[56]

Related Reagents. Acetic Anhydride; Chlorocyanoketene; Chloro(trimethylsilylmethyl)ketene; Dichloroketene; Diketene; Diphenylketene; Trimethylsilylketene.

1. (a) Patai, S. *The Chemistry of Ketenes, Allenes, and Related Compounds*; Wiley: New York, 1980; Part 1, Chapters 7 and 8. (b) Tidwell, T. T.; Seikaly, H. R. *T* **1986**, *42*, 2587. (c) Ulrich, H. *Cycloaddition Reactions*

of Heterocumulenes; Academic: New York, 1967; Chapter 2, pp 38–103, and references therein.
2. Hurd, C. D.; Kimbrough, R. D., Jr. *JACS* **1961**, *83*, 236.
3. Geoffroy, G. L.; Bassner, S. L. *Adv. Organomet. Chem.* **1988**, *28*, 1.
4. Hurd, C.; Williams, J. W. *JOC* **1940**, *5*, 122.
5. Carlson, H. D.; Andreades, S. *OS* **1965**, *45*, 50.
6. Fisher, G. J.; MacLean, A. F.; Schnizer, A. W. *JOC* **1953**, *18*, 1055.
7. Mana, D.; Serratosa, F.; Pericas, M. A.; Valenti, E. *JCR(S)* **1990**, 118.
8. Taylor, R.; Chapman, S. E. *JCS(P2)* **1991**, 1119.
9. Taylor, R. *CC* **1987**, 741.
10. Miller, I. J.; Bibby, D. M.; Parker, L. M. *J. Catal.* **1991**, *129*, 438.
11. *Chemistry of Acyl Halides*, Patai, S., Ed.; Interscience: New York, 1972; p 330.
12. Hrnciar, P.; Bohác, A.; Matare, G. J. *S* **1994**, 381.
13. Holder R. W. *J. Chem. Educ.* **1976**, 81.
14. Satchell, R. S.; Satchell, D. P. N. *CSR* **1975**, *4*, 231.
15. Hrnciar, P.; Bohác, A. *S* **1991**, 881.
16. Hrnciar, P.; Bohác, A. *CCC* **1991**, *56*, 2879.
17. Murdoch, H. D.; Nonhebel, D. C. *JCS(C)* **1968**, 2298.
18. Zagorevskii, V. A.; Akhrem, S. I. *JGU* **1959**, *21*, 615.
19. Ogawa, H.; Teratini, S.; Chihara, T. *CC* **1981**, 1120.
20. Ogawa, H.; Teratani, S.; Takagi, Y.; Chihara, T. *BCJ* **1982**, *55*, 1451.
21. Wilbert, G.; Brooks, B. T. *JACS* **1941**, *63*, 870.
22. Kwiatek, J.; Blomquist, A. T. *JACS* **1951**, *73*, 2098.
23. Stevens, H. C.; Reich, D. A.; Brandt, D. R.; Fountain, K. R.; Gaugan, E. J. *JACS* **1965**, *87*, 5257.
24. Hurd, C. D.; Kimbrough, R. D. *JACS* **1960**, *82*, 1373.
25. Sieja, J. B. *JACS* **1971**, *93*, 130.
26. Otto, P.; Feiler, L. A.; Huisgen, R. *TL* **1968**, *43*, 4485.
27. O'Neal, H. R.; Brady, W. T. *JOC* **1967**, *32*, 612.
28. Brandsma, L.; Bos, H. J. T.; Arens, J. F. *Chemistry of Acetylenes*, Viehe, H. G., Ed., Dekker: New York, 1969; pp 805–807.
29. Wasserman, H. H.; Dehmlow, E. V. *JACS* **1962**, *84*, 3786.
30. Valenti, E.; Serratosa, F.; Pericas, M. A. *S* **1985**, 1118.
31. Kowalski, C. J.; Lal, G. S. *JACS* **1988**, *110*, 3693.
32. Wynberg, H.; Staring, E. G. J. *JOC* **1985**, *50*, 1977.
33. (a) Bormann, D.; Wegler, R. *CB* **1966**, *99*, 1245. (b) Bormann, D.; Wegler, R. *CB* **1967**, *100*, 1575.
34. Wynberg, H.; Staring, E. G. J. *JACS* **1982**, *104*, 166.
35. Seebach, D.; Wasmuth, D.; Hungerbuler, E. *AG(E)* **1979**, *12*, 958.
36. Chibata, I. *PAC* **1978**, *50*, 667.
37. Song, C. E.; Ryu, T. H.; Roh, E. J.; Kim, I. O. *TA* **1994**, *5*, 1215.
38. Staudinger, H. *LA* **1907**, *356*, 61.
39. Gonzalez, J.; Sordo, J. A.; Sordo, T. L.; Lopez, R. *JOC* **1993**, *58*, 7036.
40. Pacansky, J.; Chong, J. B.; Brown, D. W.; Schwarz, W. *JOC* **1982**, *47*, 2233.
41. Paquette, L. A. In *Asymmetric Synthesis*; Morrison, J. D., Ed.; Academic: New York, 1984; Vol. 3, Chapter 7, p 484.
42. Brady, W. T.; Dorsey, E. D.; Parry, F. H., III. *JOC* **1969**, *34*, 2846.
43. Hammond, W. B.; Turro, N. J. *JACS* **1966**, *88*, 3672.
44. Hammond, W. B.; Turro, N. J. *JACS* **1966**, *88*, 2880.
45. Turro, N. J.; Hammond, W. B.; Leermakers, P. A. *JACS* **1965**, *87*, 2774.
46. Buono, G.; Archavlis, A.; Fotiadu, F. *TL* **1990**, *31*, 4859.
47. Schweizer, E. E.; Hayes, J. E.; Hirwe, S. N.; Rheingold, A. L. *JOC* **1987**, *52*, 1319.
48. Ungvary, F. *CC* **1984**, 824.
49. Blues, E. T.; Bryce-Smith, D.; Shaoul, R.; Hirsch, H.; Simons, M. J. *JCS(P2)* **1993**, 1631.
50. Hurd, C. D.; Kimbrough, R. D., Jr. *JACS* **1960**, *82*, 1373.
51. Balicki, R.; Nantka-Namirski, P. *Pol. J. Chem.* **1982**, *56*, 1125.
52. Portnoy, S. *JOC* **1965**, *30*, 3377.
53. Jackson, J. E.; Bertsch, F. M. *JACS* **1990**, *112*, 9085.
54. (a) Mitchell, G. E.; Radloff, P. L.; Greenlief, C. M.; Henderson, M. A.; White, J. M. *Surf. Sci.* **1987**, *183*, 402. (b) Mitchell, G. E.; Radloff, P. L.; Greenlief, C. M.; White, G. M. *Surf. Sci.* **1987**, *183*, 377.
55. Frey, H. M.; Walsh, R.; Watts, I. M. *CC* **1989**, *5*, 284.
56. Smith, R. L.; Franklin, R. L.; Stirk, K. M.; Kentämaa, H. I. *JACS* **1993**, *115*, 10 398.

Thomas M. Mitzel
The Ohio State University, Columbus, OH, USA

Ketene *t*-Butyldimethylsilyl Methyl Acetal[1]

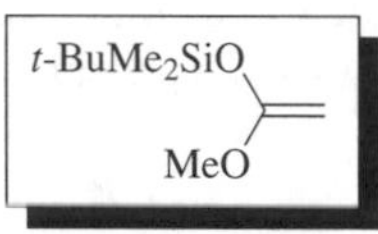

[91390-62-4] $C_9H_{20}O_2Si$ (MW 188.34)

(silylation of a variety of substrates;[3] promotion of Pummerer reaction of sulfoxides;[4–8] Lewis acid mediated aldol-type[8–22] and Michael additions,[31–38] under either stoichiometric or catalytic conditions)

Physical Data: colorless liquid; bp 76–76.5 °C/24 mmHg.
Solubility: sol *n*-pentane, diethyl ether, dichloromethane, etc.
Analysis of Reagent Purity: (NMR) ^{1}H 0.14 (s, 6H), 0.93 (s, 9H), 2.95 (d, 1H, J = 3.0 Hz), 3.10 (d, 1H, J = 3.0 Hz), 3.49 (s, 3H) ppm.
Preparative Method: obtained by reaction of methyl acetate with ***Lithium Diisopropylamide*** in THF/HMPA and subsequent trapping with ***t-Butyldimethylchlorosilane*** (72% yield).[2]
Handling, Storage, and Precautions: should be stored in the absence of moisture at −15 °C.

Silylation. The reagent silylates a variety of substrates (alcohols, acids, thiols, phenols, imides) under mild conditions (a catalytic amount of ***p-Toluenesulfonic Acid*** is occasionally added) with excellent yields (91–100%).[3]

Pummerer Reaction. The reagent transforms sulfoxides in the presence of catalytic amounts of ***Zinc Iodide*** into the corresponding α-silyloxy sulfides (eq 1).[4] Vinyl sulfoxides undergo a Michael–Pummerer type reaction to give γ-silyloxy-γ-phenylthio esters (eq 2).[5]

R–CH2–SOPh → [MeO/OTBDMS ketene acetal; MeCN, (ZnI_2); 42–79%] → R–CH(OTBDMS)–SPh (1)

R–CH=CH–SOPh → [MeO/OTBDMS ketene acetal; MeCN, (ZnI_2); 40–65%] → R–CH(CH_2CO_2Me)–CH(OTBDMS)–SPh (2)

As an extention of this reaction, the reagent promotes the intramolecular Pummerer-type rearrangement of ω-carbamoyl sulfoxides to give α-thio lactams.[6] The reaction proves particularly useful in the field of β-lactam synthesis (eq 3).[7,8a]

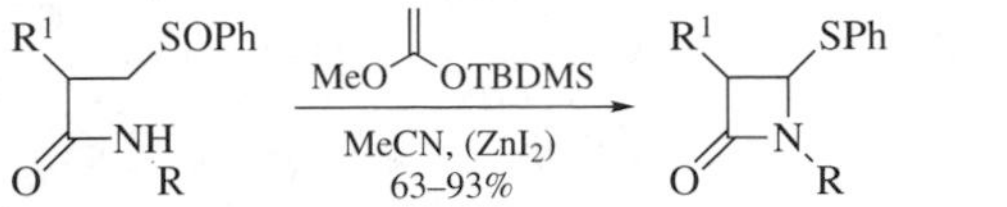

(3)

Addition to C=X Double Bonds (X = N, O).[1b] The reagent undergoes Lewis acid-catalyzed Mukaiyama-type additions[1b] to azetinones, or their corresponding iminium ions, to give *trans*-azetidin-2-one esters with good yields and excellent stereoselectivity (eq 4).[7b,8,9]

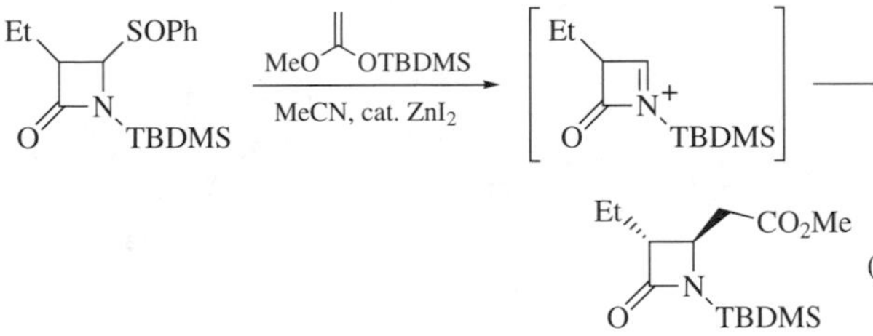

(4)

79%
racemate, *trans*:*cis* = 94:6

In the presence of a diphosphonium salt (7 mol %),[10a] ***Trimethylsilyl Trifluoromethanesulfonate*** (10 mol %),[10b] or ***Iron(III) Nitrate-K10 Montmorillonite Clay***,[10c] the reagent adds to imines or *N*-tosyliminium ions to give the corresponding β-amino esters in good yield.

ZnI_2-catalyzed additions of the reagent to chiral α,β-dialkoxy nitrones (eq 5; $R^1 = H$) proceed with good yield (86–100%) and high diastereoselectivity (ca. 90:10) in favor of the *syn* isomer ($R^1 = H$, $R^2 = CH_2Ph$, $R^3 = Me$). The *anti* isomer is obtained (ca. 90:10) by increasing the steric hindrance of R^2 and R^3 ($R^2 = CHPh_2$, $R^3 = t$-Bu).[11a,b] Addition to a different nitrone (eq 5; $R^1 = Me$) gives the *anti* isomer ($R^1 = Me$, $R^2 = CHMePh$, $R^3 = Me$) in quantitative yield and 100% diastereofacial selectivity. This material has been further elaborated to *N*-benzoyl-L-daunosamine.[11c,d]

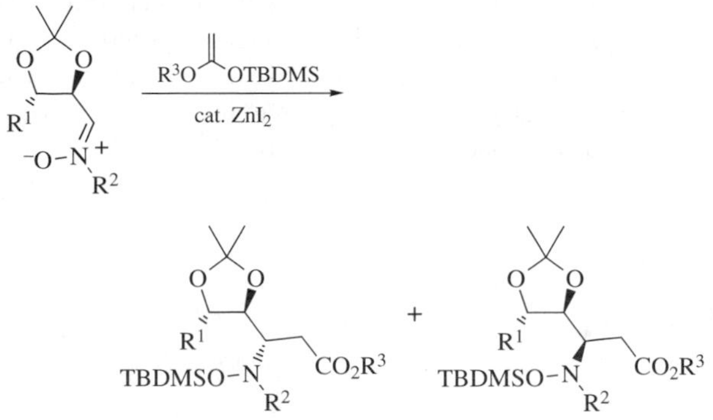

(5)

(*anti*) (*syn*)

New catalysts have been recently developed for promoting the aldol-type addition of acetate-derived silyl ketene acetals with high efficiency: 10-methylacridinium perchlorate (5 mol %),[12a] cationic mono- and dinuclear iron complexes (5 mol %),[12b] *t*-butyldimethylsilyl chloride–indium(III) chloride (10 mol %),[12c] [1,2-benzenediolato(2−)-*O,O′*]oxotitanium (20 mol %),[12d] phosphonium salts (7 mol %),[12e] and trityl salts (5–20 mol %).[12f–i] The reagent undergoes Lewis acid-promoted Mukaiyama-type additions[1b] to chiral aldehydes with moderate to good stereocontrol (eq 6).[13] It is remarkable that high 'chelation control' can be obtained by using a catalytic amount (3 mol %) of ***Lithium Perchlorate***.[13e,f]

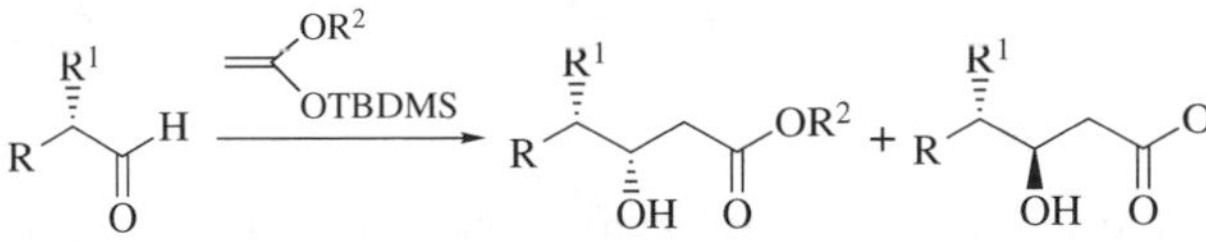

(6)

R	R^1	Promoter	R^2	Ratio	Yield (%)
Ph	Me	$BF_3•OEt_2$	*t*-Bu	97: 3	81[13a]
Ph	Me	10 kbar, 50 °C	Me	71:29	73[13b]
Me	OBn	$SnCl_4$	*t*-Bu	65:35	65[13c,d]
Me	OBn	cat. $LiClO_4$	Me	92: 8	84[13e,f]
$TBDMSOCH_2$	OBn	$SnCl_4$	Me	>98:<2	90[13g,h]
$CyCH_2$	$NHCO_2$-*i*-Pr	$TiCl_4$	Me	96: 4	95[13k,l]

With other substrates and under different conditions (eqs 7 and 8), 'non-chelated' products are obtained with excellent selectivity.[13f,14]

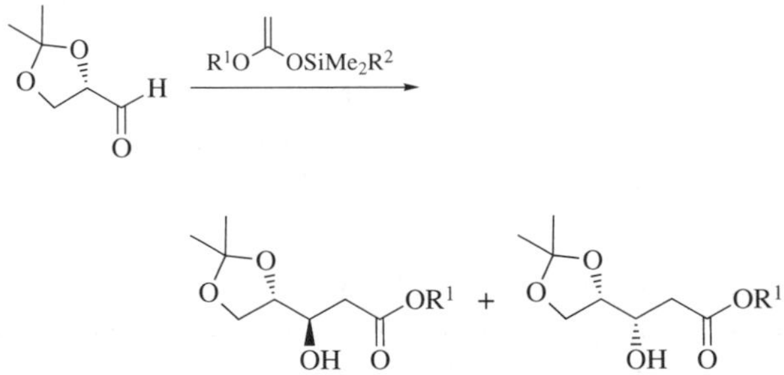

(7)

$R^1 = Me$, $R^2 = t$-Bu cat. ZnI_2 96:4 67%[14a,b]
$R^1 = Et$, $R^2 = Me$ cat. $Eu(dppm)_3$ 95:5 100%[14c]

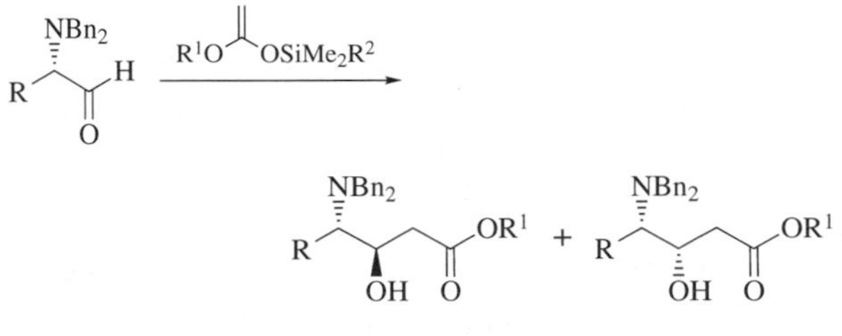

(8)

$R^1 = Bn$, $R^2 = Me$ $EtAlCl_2$ >99:<1 94%[14d]
$R^1 = Me$, $R^2 = t$-Bu cat. $LiClO_4$ in CH_2Cl_2 >98:<2 58%[13f]

In the field of *C*-glycoside synthesis, selective β-glycosylation is realized via neighboring group participation of a 2α-acyl group.[15a] In the case of 2-deoxy sugars the neighboring participation of a group at the 3α-position is exploited for selective formation of the β-anomer (β:α = 91:9) (eq 9).[15b]

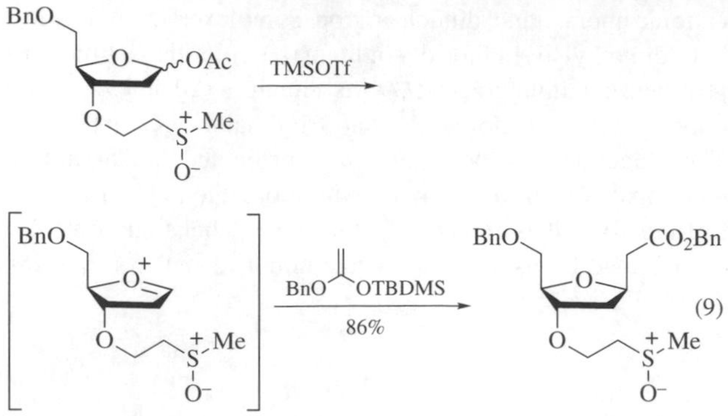

(9)

Enantioselective Aldol-Type Additions. Highly enantioselective aldol-type reactions are successfully carried out by the combined use of a chiral diamine-coordinated tin(II) triflate and tributyltin fluoride (eq 10).[16,17] A catalytic amount of chiral bis(sulfonamido)zinc(II), easily prepared from ***Diethylzinc*** and chiral sulfonamides, promotes the aldol addition in high yield and good enantiomeric excess (72–93%) only with chloral and bromal (CX_3CHO).[18]

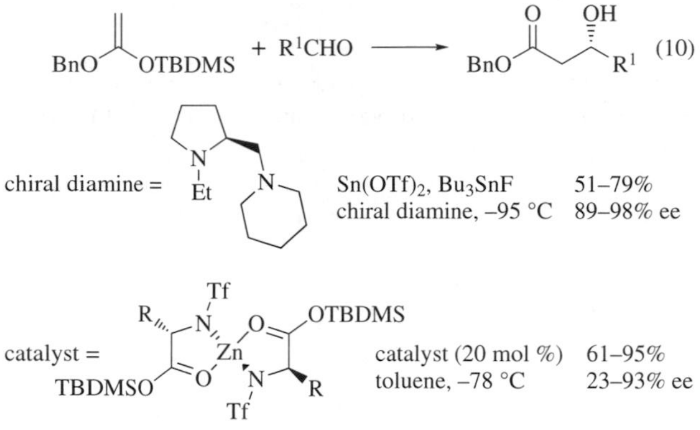

(10)

Chiral borane complexes (20 mol %) catalyze the aldol-type addition to achiral aldehydes in good to excellent yield and enantiomeric excess (eq 11).[19–21]

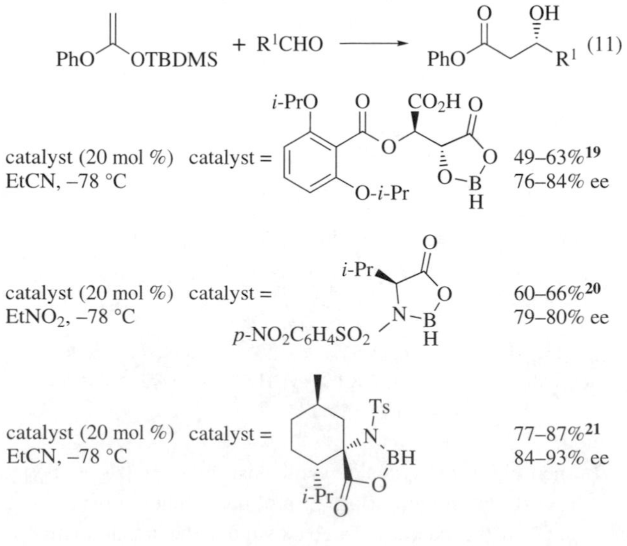

(11)

A chiral boron reagent, derived from equimolar amounts of (*R*)- or (*S*)-binaphthol and triphenyl borate, promotes the condensation of chiral imines with *t*-butyl acetate silyl ketene acetal in high diastereomeric excess (eq 12).[22]

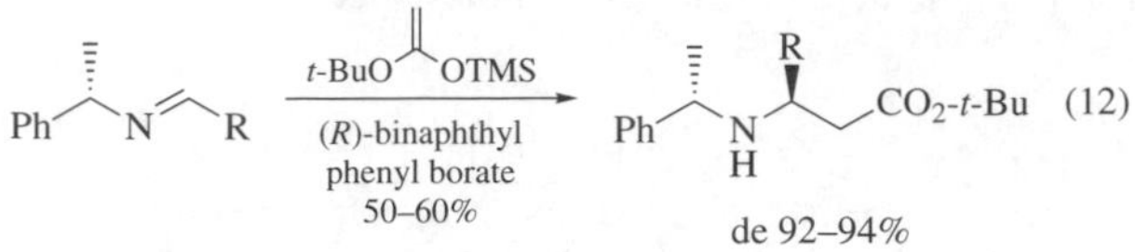

(12)

Addition to Various Electrophiles. Various Lewis acids promote the addition of the reagent to an allylic acetate, following a carbon-Ferrier rearrangement pathway.[23] ***Titanium(IV) Chloride*** promotes the addition of the reagent to 2,2-dialkoxycyclopropanecarboxylic esters to give 3-alkoxy-2-cyclopentenones (eq 13).[24]

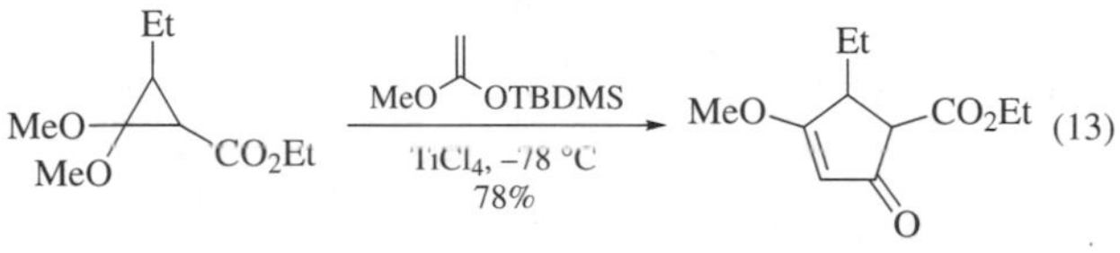

(13)

1,3-Dioxolan-2-ylium cations, derived from aldehyde ethylene acetals and trityl cation, react with the reagent to give the corresponding β-keto esters.[25] ***Montmorillonite K10*** catalyzes the addition of the reagent to pyridine derivatives with electron withdrawing groups to give *N*-silyldihydropyridines.[26] The ketene silyl acetal of ethyl acetate reacts with a chiral bromide to give the corresponding *syn*-lactone in 64% yield via a direct S_N2-type displacement and inversion of stereochemistry (eq 14).[27]

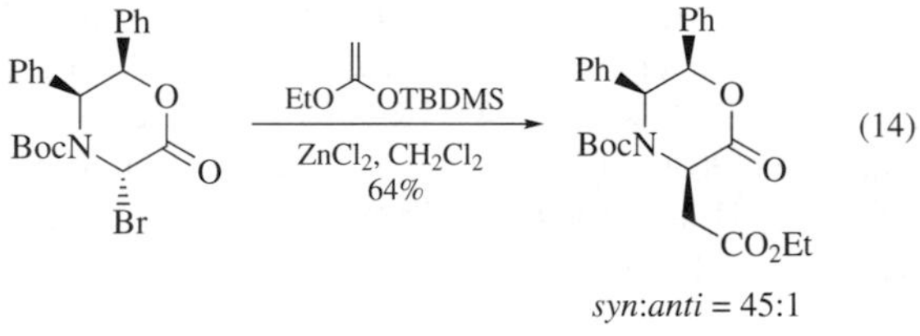

(14)

Six-membered chiral acetals, derived from aliphatic aldehydes, undergo aldol-type coupling reactions with silyl ketene acetals in the presence of $TiCl_4$ with high diastereoselectivity (eq 15).[28] This procedure, in combination with oxidative destructive elimination of the chiral auxiliary, has been applied to the preparation of (*R*)-(+)-α-lipoic acid[28a] and mevinolin analogs.[28b]

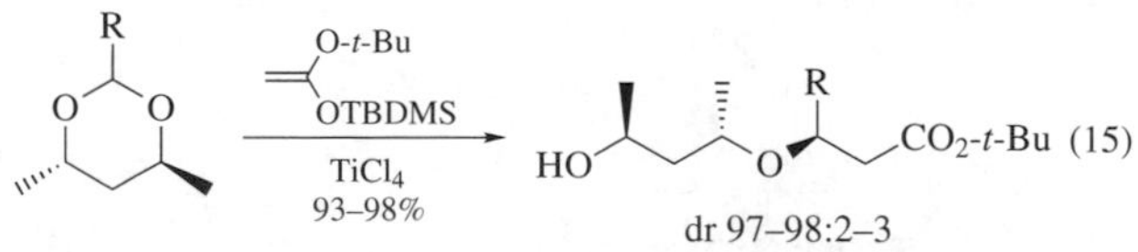

(15)

Addition to chiral, bicyclic acetals has been exploited in an approach to the synthesis of the tetrahydropyran subunit of the polyether nigerin.[29] The particular acetal generated by the ***Diisobutylaluminum Hydride*** reduction of aliphatic esters undergoes aldol addition in good yields (eq 16).[30]

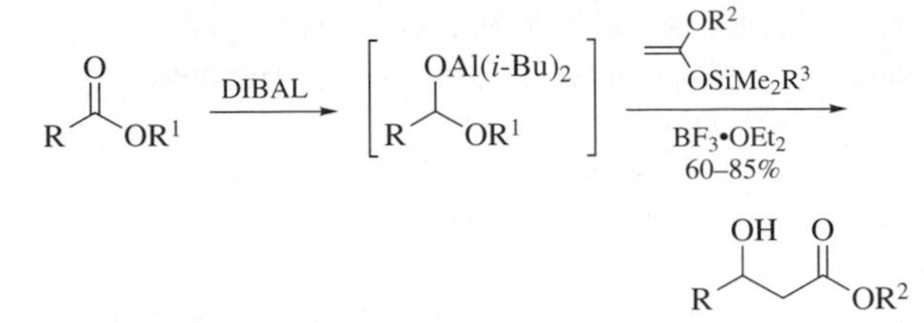

(16)

Michael Addition. The reagent undergoes Michael addition to α,β-enones in acetonitrile in the absence of a Lewis acid to afford the corresponding *O*-silylated Michael adducts in high yield. These silyl enol ethers undergo site-specific reaction with a variety of electrophiles (eq 17).[31a,b] Inability to repeat this procedure led to the discovery that the 'noncatalyzed' Michael reaction is due to traces of phosphorus compounds introduced by drying acetonitrile with P_4O_{10}. The new catalyst system, formed from P_4O_{10} in acetonitrile, was found to be highly effective with a variety of substrates.[31c]

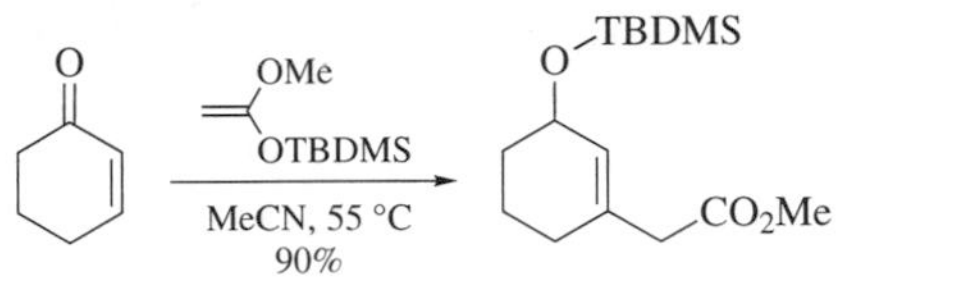

(17)

In those instances where the thermal Michael reaction is sluggish due to sterically hindered substrates, the use of high pressure (15 kbar, 20 °C),[32a] or of $LiClO_4$ (3 mol % in CH_2Cl_2[13f] or 1.0–2.5 M in Et_2O[32b]) prove extremely advantageous (eq 18). The lithium perchlorate-catalyzed Michael reaction can be carried out on α,β-unsaturated δ-lactones and on sterically demanding β,β-disubstituted unsaturated carbonyl systems in high yield and under mild conditions.[32b]

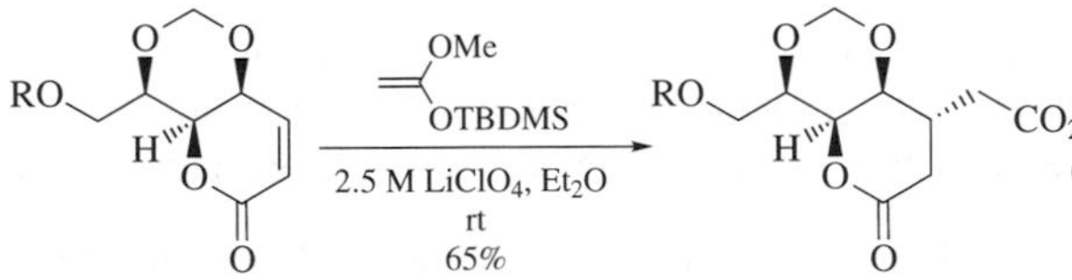

(18)

Michael addition of the reagent to enoates and enones occurs at low temperature (−50 to −78 °C) in the presence of catalytic amounts of various Lewis acids.[33] A catalytic amount of ***Triphenylmethyl Perchlorate*** (5 mol %) effectively catalyzes the tandem Michael reaction of ethyl acetate-derived silyl ketene acetal to α,β-unsaturated ketones and the sequential aldol addition to aldehydes with high stereoselectivity.[34] HgI_2 mediates the Michael addition to chiral enones, followed by Lewis acid-mediated addition to aldehydes. The Michael-aldol protocol has been used for the stereoselective synthesis of key intermediates on the way to prostaglandins, compactin, and ML-236A (eq 19).[35]

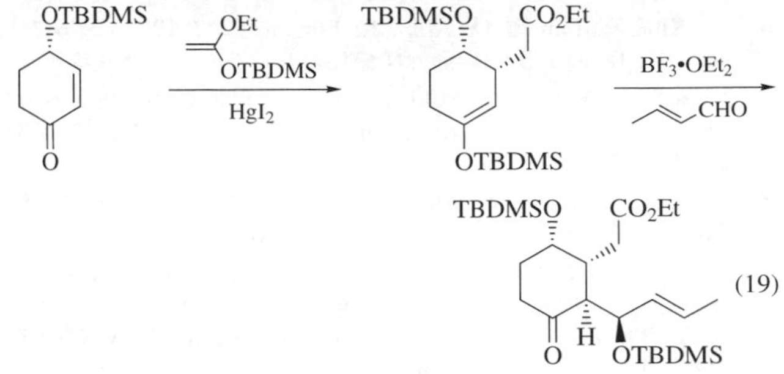

(19)

(single diastereomer)

The mechanism of the $TiCl_4$-mediated Michael addition of silyl ketene acetals has been investigated, and criteria for suppressing the electron transfer process have been devised.[36] Chiral enones show good to excellent diastereofacial preference in $TiCl_4$-mediated reactions with silyl ketene acetals (eq 20).[37]

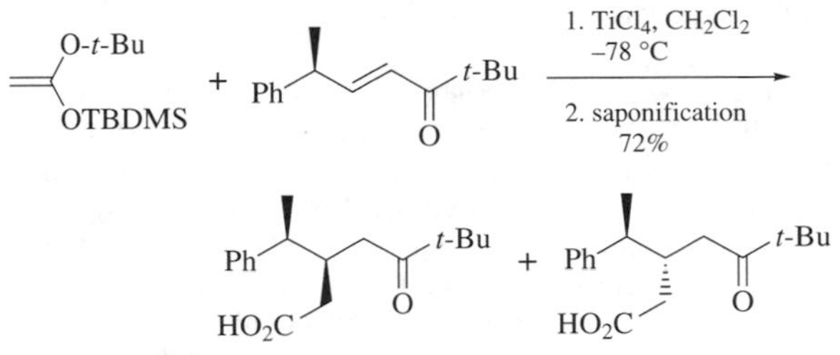

(20)

97:3

ZnI_2-mediated multiple Michael additions to bis-enoates proceed in good yield and with modest stereocontrol (eq 21).[38]

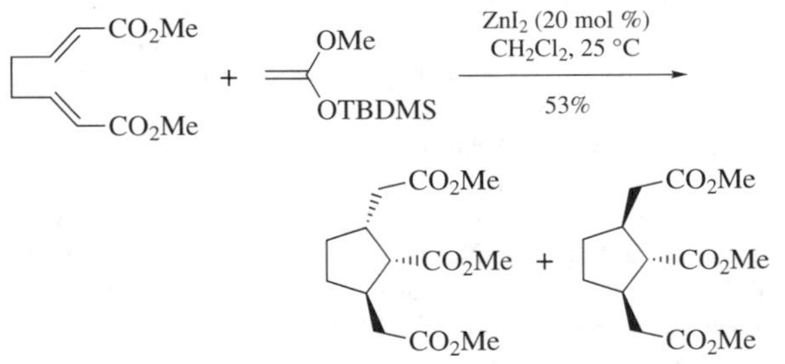

(21)

67:33

Related Reagents. 1-*t*-Butylthio-1-*t*-butyldimethylsilyloxyethylene; 1-*t*-Butylthio-1-*t*-butyldimethylsilyloxypropene; Ketene Bis(trimethylsilyl) Acetal; Ketene Diethyl Acetal; Methylketene Bis(trimethylsilyl) Acetal; Methylketene Dimethyl Acetal; 1-Methoxy-2-methyl-1-(trimethylsilyloxy)propene; 1-Methoxy-1-(trimethylsilyloxy)-1-propene.

1. (a) Fieser, L. F.; Fieser, M. *FF* **1984**, *11*, 279. (b) Gennari, C. *COS* **1991**, *2*, 629.
2. (a) Ainsworth, C.; Chen, F.; Kuo, Y.-N. *JOM* **1972**, *46*, 59. (b) Kita, Y.; Segawa, J.; Haruta, J.; Fujii, T.; Tamura, Y. *TL* **1980**, *21*, 3779.
3. Kita, Y.; Haruta, J.; Fujii, T.; Segawa, J.; Tamura, Y. *S* **1981**, 451.
4. (a) Kita, Y.; Yasuda, H.; Tamura, O.; Itoh, F.; Tamura, Y. *TL* **1984**, *25*, 4681. (b) Kita, Y.; Tamura, O.; Yasuda, H.; Itoh, F.; Tamura, Y. *CPB* **1985**, *33*, 4235.
5. Kita, Y.; Tamura, O.; Itoh, F.; Yasuda, H.; Miki, T.; Tamura, Y. *CPB* **1987**, *35*, 562.

6. (a) Kita, Y.; Tamura, O.; Miki, T.; Tamura, Y. *TL* **1987**, *28*, 6479. (b) Kita, Y.; Tamura, O.; Shibata, N.; Miki, T. *CPB* **1990**, *38*, 1473.
7. (a) Kita, Y.; Tamura, O.; Miki, T.; Tono, H.; Shibata, N.; Tamura, Y. *TL* **1989**, *30*, 729. (b) Kita, Y.; Tamura, O.; Shibata, N.; Miki, T. *JCS(P1)* **1989**, 1862.
8. (a) Kita, Y.; Shibata, N.; Miki, T.; Takemura, Y.; Tamura, O. *CC* **1990**, 727. (b) Kita, Y.; Shibata, N.; Tamura, O.; Miki, T. *CPB* **1991**, *39*, 2225.
9. (a) Chiba, T.; Nakai, T. *CL* **1987**, 2187. (b) Chiba, T.; Nagatsuma, M.; Nakai, T. *CL* **1985**, 1343. (c) Yoshida, A.; Tajima, Y.; Takeda, N.; Oida, S. *TL* **1984**, *25*, 2793. (d) Tajima, Y.; Yoshida, A.; Takeda, N.; Oida, S. *TL* **1985**, *26*, 673. (e) Murahashi, S.-I.; Saito, T.; Naota, T.; Kumobayashi, H.; Akutagawa, S. *TL* **1991**, *32*, 5991.
10. (a) Mukaiyama, T.; Kashiwagi, K.; Matsui, S. *CL* **1989**, 1397. (b) Åhman, J.; Somfai, P. *T* **1992**, *43*, 9537. (c) Onaka, M.; Ohno, R.; Yanagiya, N.; Izumi, Y. *SL* **1993**, 141.
11. (a) Kita, Y.; Tamura, O.; Itoh, F.; Kishino, H.; Miki, T.; Kohno, M.; Tamura, Y. *CC* **1988**, 761. (b) Kita, Y.; Tamura, O.; Itoh, F.; Kishino, H.; Miki, T.; Kohno, M,; Tamura, Y. *CPB* **1989**, *37*, 2002. (c) Kita, Y.; Itoh, F.; Tamura, O.; Yan Ke Y. Y.; Tamura, Y. *TL* **1987**, *28*, 1431. (d) Kita, Y.; Itoh, F.; Tamura, O.; Yan Ke Y. Y.; Miki, T.; Tamura, Y. *CPB* **1989**, *37*, 1446.
12. (a) Otera, J.; Wakahara, Y.; Kamei, H.; Sato, T.; Nozaki, H.; Fukuzumi, S. *TL* **1991**, *32*, 2405. (b) Bach, T.; Fox, D. N. A.; Reetz, M. T. *CC* **1992**, 1634. (c) Mukaiyama, T.; Ohno, T.; Han, J. S.; Kobayashi, S. *CL* **1991**, 949. (d) Hara, R.; Mukaiyama, T. *CL* **1989**, 1909. (e) Mukaiyama, T.; Matsui, S.; Kashiwagi, K. *CL* **1989**, 993. (f) Kobayashi, S.; Matsui, S.; Mukaiyama, T. *CL* **1988**, 1491. (g) Mukaiyama, T.; Leon, P.; Kobayashi, S. *CL* **1988**, 1495. (h) Homma, K.; Mukaiyama, T. *CL* **1990**, 161. (i) Homma, K.; Takenoshita, H.; Mukaiyama, T. *BCJ* **1990**, *63*, 1898.
13. (a) Heathcock, C. H.; Flippin, L. A. *JACS* **1983**, *105*, 1667. (b) Yamamoto, Y.; Maruyama, K.; Matsumoto, K. *TL* **1984**, *25*, 1075. (c) Heathcock, C. H.; Davidsen, S. K.; Hug, K. T.; Flippin, L. A. *JOC* **1986**, *51*, 3027. (d) Heathcock, C. H.; Hug, K. T.; Flippin, L. A. *TL* **1984**, *25*, 5973. (e) Reetz, M. T.; Raguse, B.; Marth, C. F.; Hügel, H. M.; Bach, T.; Fox, D. N. A. *T* **1992**, *48*, 5731. (f) Reetz, M. T.; Fox, D. N. A. *TL* **1993**, *34*, 1119. (g) Reetz, M. T.; Kesseler, K. *JOC* **1985**, *50*, 5434. (h) Reetz, M. T. *PAC* **1985**, *57*, 1781. (i) Rama Rao, A. V.; Chakraborty, T. K.; Purandare, A. V. *TL* **1990**, *31*, 1443. (j) Barrett, A. G. M.; Raynham, T. M. *TL* **1987**, *28*, 5615. (k) Takemoto, Y.; Matsumoto, T.; Ito, Y.; Terashima, S. *TL* **1990**, *31*, 217. (l) Takemoto, Y.; Matsumoto, T.; Ito, Y.; Terashima, S. *CPB* **1991**, *39*, 2425. (m) Gennari, C.; Cozzi, P. G. *T* **1988**, *44*, 5965. (n) Shirai, F.; Nakai, T. *CL* **1989**, 445. (o) Yamazaki, T.; Yamamoto, T.; Kitazume, T. *JOC* **1989**, *54*, 83.
14. (a) Kita, Y.; Yasuda, H.; Tamura, O.; Itoh, F.; Yuan Ke, Y.; Tamura, Y. *TL* **1985**, *26*, 5777. (b) Kita, Y.; Tamura, O.; Itoh, F.; Yasuda, H.; Kishino, H.; Yuan Ke, Y.; Tamura, Y. *JOC* **1988**, *53*, 554. (c) Mikami, K.; Terada, M.; Nakai, T. *TA* **1991**, *2*, 993. (d) Mikami, K.; Kaneko, M.; Loh, T.-P.; Terada, M.; Nakai, T. *TL* **1990**, *31*, 3909. (e) Reetz, M. T.; Schmitz, A.; Holdgrün, X. *TL* **1989**, *30*, 5421. (f) Annunziata, R.; Cinquini, M.; Cozzi, F.; Cozzi, P. G.; Consolandi, E. *JOC* **1992**, *57*, 456.
15. (a) Yokoyama, Y. S.; Elmoghayar, M. R. H.; Kuwajima, I. *TL* **1982**, *23*, 2673. (b) Narasaka, K.; Ichikawa, Y.; Kubota, H. *CL* **1987**, 2139.
16. Mukaiyama, T.; Kobayashi, S.; Sano, T. *T* **1990**, *46*, 4653.
17. Kobayashi, S.; Sano, T.; Mukaiyama, T. *CL* **1989**, 1319.
18. Mukaiyama, T.; Takashima, T.; Kusaka, H.; Shimpuku, T. *CL* **1990**, 1777.
19. Furuta, K.; Maruyama, T.; Yamamoto, H. *SL* **1991**, 439.
20. (a) Kiyooka, S.; Kaneko, Y.; Komura, M.; Matsuo, H.; Nakano, M. *JOC* **1991**, *56*, 2276. (b) Kiyooka, S.; Kaneko, Y.; Kume, K. *TL* **1992**, *33*, 4927.
21. Parmee, E. R.; Hong, Y.; Tempkin, O.; Masamune, S. *TL* **1992**, *33*, 1729.
22. Hattori, K.; Miyata, M.; Yamamoto, H. *JACS* **1993**, *115*, 1151.
23. (a) Paterson, I.; Smith, J. D. *JOC* **1992**, *57*, 3261. (b) Kozikowski, A. P.; Park, P. *JOC* **1990**, *55*, 4668.
24. Saigo, K.; Shimada, S.; Shibasaki, T.; Hasegawa, M. *CL* **1990**, 1093.
25. Hayashi, Y.; Wariishi, K.; Mukaiyama, T. *CL* **1987**, 1243.
26. Onaka, M.; Ohno, R.; Izumi, Y. *TL* **1989**, *30*, 747.
27. Williams, R. M.; Sinclair, P. J.; Zhai, D.; Chen, D. *JACS* **1988**, *110*, 1547.
28. (a) Elliott, J. D.; Steele, J.; Johnson, W. S. *TL* **1985**, *26*, 2535. (b) Johnson, W. S.; Kelson, A. B.; Elliott, J. D. *TL* **1988**, *29*, 3757.
29. Holmes, C. P.; Bartlett, P. A. *JOC* **1989**, *54*, 98.
30. Kiyooka, S.; Shirouchi, M. *JOC* **1992**, *57*, 1.
31. (a) Kita, Y.; Segawa, J.; Haruta, J.; Fujii, T.; Tamura, Y. *TL* **1980**, *21*, 3779. (b) Kita, Y.; Segawa, J.; Haruta, J.; Yasuda, H.; Tamura, Y. *JCS(P1)* **1982**, 1099. (c) Berl, V.; Helmchen, G.; Preston, S. *TL* **1994**, *35*, 233.
32. (a) Bunce, R. A.; Schlecht, M. F.; Dauben, W. G.; Heathcock, C. H. *TL* **1983**, *24*, 4943. (b) Grieco, P. A.; Cooke, R. J.; Henry, K. J.; VanderRoest, J. M. *TL* **1991**, *32*, 4665.
33. (a) Kawai, M.; Onaka, M.; Izumi, Y. *BCJ* **1988**, *61*, 2157. (b) Onaka, M.; Mimura, T.; Ohno, R.; Izumi, Y. *TL* **1989**, *30*, 6341. (c) Mukaiyama, T.; Hara, R. *CL* **1989**, 1171. (d) Minowa, N.; Mukaiyama, T. *CL* **1987**, 1719. (e) Hashimoto, Y.; Sugumi, H.; Okauchi, T.; Mukaiyama, T. *CL* **1987**, 1691.
34. Kobayashi, S.; Mukaiyama, T. *CL* **1986**, 1805; *H* **1987**, *25*, 205.
35. (a) Danishefsky, S. J.; Cabal, M. P.; Chow, K. *JACS* **1989**, *111*, 3456. (b) Danishefsky, S. J.; Simoneau, B. *JACS* **1989**, *111*, 2599. (c) Danishefsky, S. J.; Simoneau, B. *PAC* **1988**, *60*, 1555. (d) Chow, K.; Danishefsky, S. J. *JOC* **1989**, *54*, 6016. (e) Audia, J. E.; Boisvert, L.; Patten, A. D.; Villalobos, A.; Danishefsky, S. J. *JOC* **1989**, *54*, 3738.
36. (a) Sato, T.; Wakahara, Y.; Otera, J.; Nozaki, H.; Fukuzumi, S. *JACS* **1991**, *113*, 4028. (b) Otera, J.; Fujita, Y.; Sato, T.; Nozaki, H.; Fukuzumi, S.; Fujita, M. *JOC* **1992**, *57*, 5054.
37. Heathcock, C. H.; Uehling, D. E. *JOC* **1986**, *51*, 279.
38. Klimko, P. G.; Singleton, D. A. *JOC* **1992**, *57*, 1733.

Cesare Gennari
Università di Milano, Italy

Ketene Diethyl Acetal

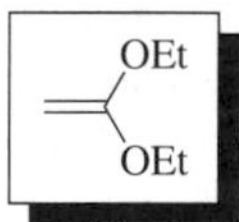

[2678-54-8] $C_6H_{12}O_2$ (MW 116.18)

(inverse electron demand Diels–Alder reactions; thermal and photochemical [2+2] cycloadditions; naphthoquinone and anthraquinone synthesis; Paterno–Büchi photocycloadditions)

Physical Data: bp 68 °C/100 mmHg, 84–86 °C/200 mmHg.
Solubility: sol THF, ether, benzene; decomposes in water, ethanol.
Form Supplied in: liquid, not commercially available.
Preparative Method: reaction of 2-bromo-1,1-diethoxyethane with potassium *t*-butoxide in *t*-butanol.[1]
Handling, Storage, and Precautions: storage in an alkaline glass bottle dusted with potassium *t*-butoxide is recommended.

Inverse Electron Demand Diels–Alder Reactions. Ketene diethyl acetal undergoes Diels–Alder reactions with various electron deficient dienes especially heterodienes such as α,β-unsaturated aldehydes and ketones (eq 1),[2] acylketenes (eq 2),[3] triazines,[4] acyliminium ions (eq 3),[5] and azodicarboxylates.[6] The Diels–Alder adducts often undergo further reaction, as in the case of cycloaddition of ketene diethyl acetal to 2-pyrones which undergo Diels–Alder reaction followed by retro-Diels–Alder reaction and elimination to allow rearomatization to produce good yields of substituted aromatics (eq 4).[7]

125 °C, 14 h; 70% (1)

PhMe, 110 °C; 70% (2)

C_6H_6, 80 °C, 15 h; H_2O; 100% (3)

PhMe, 140 °C, 21 h; 75% (4)

Thermal [2 + 2] Cycloadditions. Diethyl ketene acetal readily reacts with fumarate[8] and acrylate esters (eq 5)[9] in highly regioselective thermal [2 + 2] cycloadditions to give cyclobutanone acetals, and with propiolate esters[10] to provide cyclobutenone acetals (eq 6), in good yield. Reaction with aryl isocyanates can provide substituted β-lactams (eq 7),[11] but isocyanates with stabilizing groups on the nitrogen generally give ring opened products.[12] Dichloroketene undergoes thermal [2 + 2] cycloaddition with the ketene acetal followed by ring opening (eq 8).[13]

MeCN, heat; 60% (5)

CH_2Cl_2, 40 °C, 29 h; 65% (6)

4–85 °C, 3–72 h; 70–90% (7)

Et_3N, 25 °C; 51% (8)

Photochemical [2 + 2] Cycloadditions with Enones. One of the most common uses of ketene diethyl acetal has been in [2 + 2] photocycloadditions with enones.[14] This is a highly regioselective process which provides access to cyclobutanone acetals in high yields (eqs 9–11).[15–17] These intermediates can serve as important synthetic building blocks. One example which displays some asymmetric induction has been reported (eq 12).[18]

hν, C_6H_6, pyrex; 90% (9)

hν, pyrex; 90% (10)

hν, MeCN, acetone; 84% (11)

hν, PhMe; 75%; 56% ee (12)

Naphthoquinone and Anthraquinone Synthesis. The reaction of diethyl ketene acetal with benzo- and naphthoquinones to prepare naphtho- and anthraquinones respectively has been widely studied (eqs 13 and 14).[19–21] The yields vary depending on the substitution of the quinone, and benzofurans are sometimes byproducts.[22] A similar reaction has been carried out on pyridinium salts.[23]

AcOH, heat; 62–76% (13)

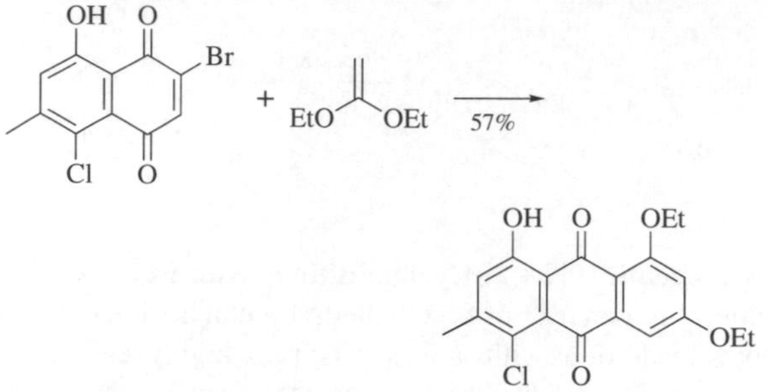

(14)

Paterno–Büchi [2 + 2] Photocycloadditions. Ketene diethyl acetal reacts with various ketones in the Paterno–Büchi reaction to produce substituted oxetanes in modest yields (eqs 15 and 16).[24–26] More electrophilic carbonyls tend to give higher yields.

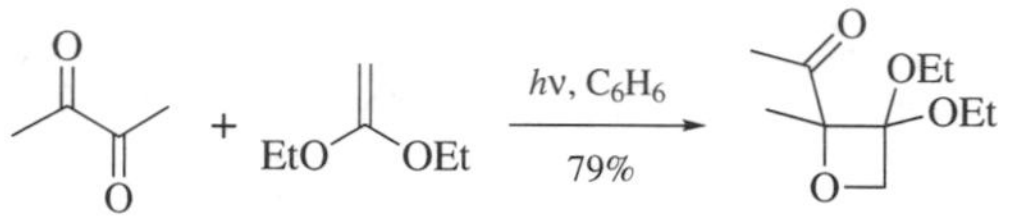

(15)

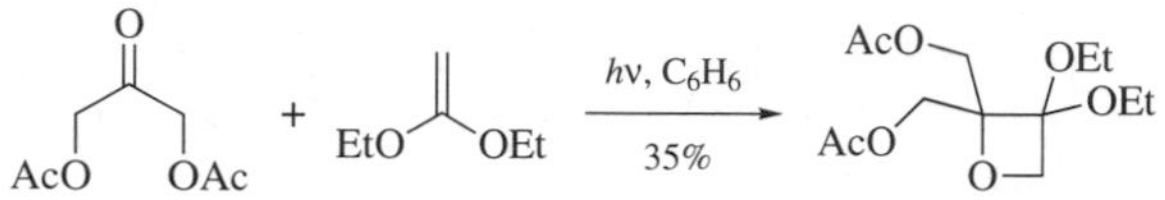

(16)

Miscellaneous Reactions. Ketene diethyl acetal has been used in the palladium-catalyzed orthoester Claisen rearrangement of allylic alcohols (eq 17)[27] and in reaction with phenyl selenocyanate (eq 18).[28]

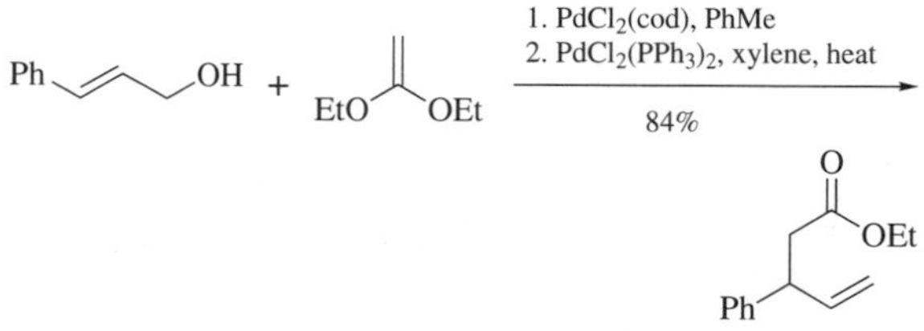

(17)

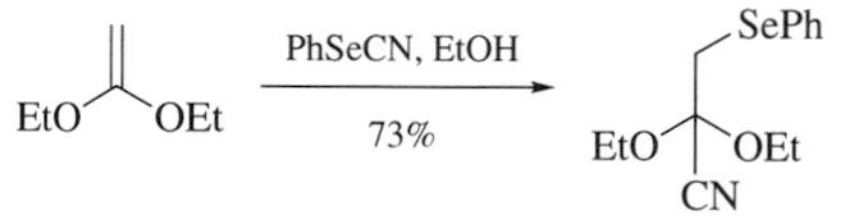

(18)

Related Reagents. Ethyl 3,3-Diethoxyacrylate; Ketene Bis(trimethylsilyl) Acetal; Ketene *t*-Butyldimethylsilyl Methyl Acetal; Methylketene Dimethyl Acetal.

1. McElvain, S. M.; Kundiger, D. *OSC* **1973**, *3*, 506.
2. Dauben, W. G.; Krabbenhoft, H. O. *JOC* **1977**, *42*, 282. Desimoni, G.; Tacconi, G. *CRV* **1975**, *75*, 651.
3. Sato, M.; Ogasawara, H.; Kato, K.; Sakai, M.; Kato, T. *CPB* **1983**, *31*, 4300. Stetter, H.; Schutte, M. *CB* **1975**, *108*, 3314.
4. Itoh, T.; Ohsawa, A.; Okada, M.; Kaihoh, T.; Igeta, H. *CPB* **1985**, *33*, 3030. Burg, B.; Dittmar, W.; Reim, H.; Steigel, A.; Sauer, J. *TL* **1975**, 2897.
5. Akiyama, T.; Urasato, N.; Imagawa, T.; Kawanisi, M. *BCJ* **1976**, *49*, 1105.
6. Hall, J.; Wojciechowska, M. *JOC* **1978**, *43*, 3348.
7. Boger, D. L.; Mullican, T. *TL* **1982**, *23*, 4551, 4555. Jung, M. E.; Hagenah, J. A. *H* **1987**, *25*, 117.
8. Slusarchyk, W. A.; Young, M. G.; Bisacchi, G. S.; Hockstein, D. R.; Zahler, R. *JMC* **1991**, *34*, 1415.
9. Amici, P.; Conia, J. M. *BSF(2)* **1974**, 1015; *TL* **1974**, 479.
10. Semmelhack, M. F.; Tomoda, S.; Nagaoka, H.; Boettger, S. D. Hurst, K. M. *JACS* **1982**, *104*, 747; *JACS* **1980**, *102*, 7567.
11. Graziano, M. L.; Cimminiello, G. *S* **1989**, 54.
12. Chitwood, J. L.; Gott, P. G.; Martin, J. C. *JOC* **1971**, *36*, 2228.
13. Scharf, H. D.; Sporrer, E. *S* **1975**, 733.
14. Crimmins, M. T.; Reinhold, T. L. *OR* **1993**, *44*, 297.
15. Liu, H. J.; Kulkarni, M. G. *TL* **1985**, *26*, 4847.
16. Smith, A. B., III; Richmond, R. E. *JACS* **1983**, *105*, 575.
17. Swenton, J. S.; Hyatt, J. A.; Lisy, J. M.; Clardy, J. *JACS* **1974**, *96*, 4885.
18. Herzog, H.; Koch, H.; Scharf, H. D.; Runsink, J. *T* **1986**, *42*, 3547.
19. Cameron, D. W.; Crossley, M. J.; Feutrill, G. I.; Griffiths, P. G. *AJC* **1978**, *31*, 1335, 1353; *CC* **1977**, 297.
20. Grandmaison, J.-L.; Brassard, P. *T* **1977**, *33*, 2047.
21. Banville, J.; Grandmaison, J.-L., Lang, G.; Brassard, P. *CJC* **1974**, *52*, 80.
22. McElvain, S. M.; Engelhardt, E. L. *JACS* **1944**, *66*, 1077.
23. Scherowsky, G.; Pickardt, J. *CB* **1983**, *116*, 186.
24. Mattay, J.; Buchkremer, K. *H* **1988**, *27*, 2153.
25. Araki, Y.; Nagasawa, J.; Ishido, Y. *JCS(P1)* **1981**, 12.
26. Rao, V. B.; Schroder, C.; Margaretha, P.; Wolff, S.; Agosta, W. C. *JOC* **1985**, *50*, 3881.
27. Oshima, M.; Murakami, M.; Mukaiyama, T. *CL* **1984**, 1535.
28. Tomoda, S.; Takeuchi, Y.; Nomura, Y. *CL* **1982**, 1733.

Michael T. Crimmins
University of North Carolina at Chapel Hill, NC, USA

2-Lithio-1,3-dithiane[1]

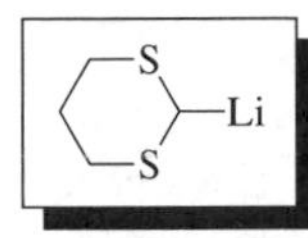

[36049-90-8] $C_4H_7LiS_2$ (MW 126.19)

(umpolung; C–C bond formation)

Physical Data: pK_a values for several 1,3-dithianes;[1e,2] measurements/calculations on relative acidity of axial vs. equatorial C-2 hydrogens;[3] crystal structure for 2-methyl- and 2-phenyl-2-lithio-1,3-dithiane;[4] low-temperature ^{13}C NMR spectra of ^{6}Li- and ^{13}C-labeled reagent.[5]

Solubility: sol ether, THF; slightly sol pentane.

Analysis of Reagent Purity: solutions of the reagent are colorless, with the exception of solutions of the 2-vinyl- and 2-phenyl-substituted analogs, which are yellow; solutions of the reagent should be titrated before use by one of the standard methods, e.g. MeOD quenching studies (NMR).

Preparative Methods: the parent compound is most conveniently prepared from commercially available ***1,3-Dithiane*** by metalation with one equivalent of ***n-Butyllithium*** at −20 °C in THF.[6] 2-Lithio-1,3-dithianes with a substituent at C-2 can be prepared similarly; the reaction time for metalation varies.[6] Alternatively, Sn/Li transmetalation (LDA or MeLi) of 2-trimethylstannyl- or 2,2-bis(trimethylstannyl)-1,3-dithiane[7] is a much faster process than the direct metalation and occurs within a minute at −78 °C. This allows preparation of substituted 2-lithio-1,3-dithianes, in which the substituent has electrophilic sites, which otherwise would not survive the metalation process.[7] Some sodium[8] and potassium[9] analogs have been reported.[1e] A comprehensive review on the preparation of dithianes is covered under ***1,3-Dithiane***.

Handling, Storage, and Precautions: must be prepared and transferred under inert gas (Ar or N_2) to exclude oxygen or moisture; solutions in THF are stable for weeks at −20 °C; at room temperature, 2-lithio-1,3-dithiane does not decompose via a carbenoid pathway, but abstracts a proton from the solvent.[1f] Handle in a fume hood.

Alkylation. On treatment with primary, secondary, or allylic alkyl halides, or primary sulfonates, the corresponding substituted dithianes are obtained, generally in good yield (eq 1).[1bf,10,11]

Formally, these transformations give access to ketones from aldehydes (or aldehydes from ***Formaldehyde***, R = H),[12] i.e. the dithiane moiety here functions in a reactivity umpolung mode,[13] as acyl anion **A** or acyl dianion equivalent **B**.

Allylic alcohols, after appropriate in situ activation, can be used as alternatives to the alkyl halides (eq 2).[14]

O=C⁻ **A**

(1) 50–80%; RX, BuLi, R^1X

O=C²⁻ **B**

(2) $Ph_3\overset{+}{P}O$; $[R_3CuNMePhLi_2]^-$; 78%

Considerable chemoselectivity is observed in reactions with multifunctionalized electrophiles,[1b–f] e.g. epoxide vs. primary halide (eq 3),[15] or allylic vs. nonallylic halide,[16] or iodide vs. allylic mesylate (eq 6).[17]

(3) 85%

Cycloalkylation products can be obtained from reactions with bifunctional electrophiles, an approach which has found use in the preparation of a variety of macrocycles (acyl dianion equivalent **B**, eq 4).[1b–f] Alternatively, dialkylation with the appropriate electrophiles can also be used for the straightforward preparation of large-ring compounds (eq 5)[1e,f,18] or to set the stage for the synthesis of polycyclic structures (eq 6).[17,19] More recent examples of alkylation of 2-lithio-1,3-dithianes include the preparation of backbone-rigid diamides,[20] the synthesis of (−)-ε-cadinene,[21] methyl ketones which undergo asymmetric reduction in the presence of the dithiane moiety,[22] aminoalkynyldithianes,[23] and dioxaspiro[5.5]undecane ring systems.[24]

(4) BuLi

(5) 1. BuLi; 2. $Br(CH_2)_3Cl$

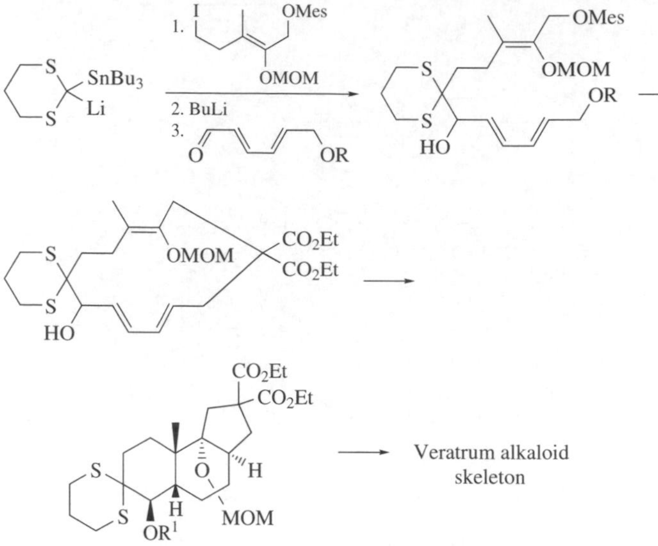

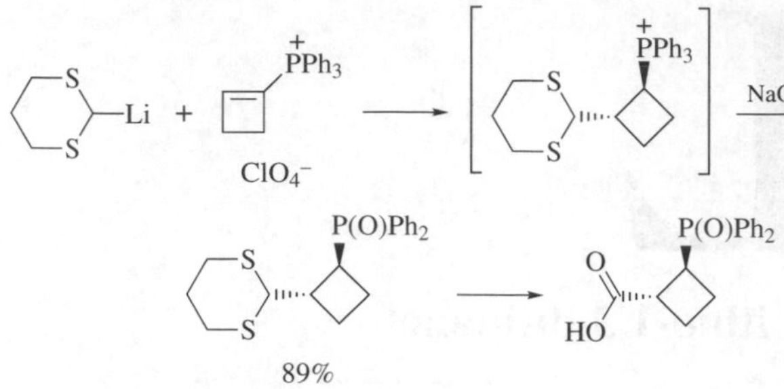

2-Lithio-1,3-dithiane reacts with arene–metal complexes to give the corresponding arene substituted derivatives (eq 7).[25] Some regiochemical control can be achieved; side products often limit the utility of these reactions.[26]

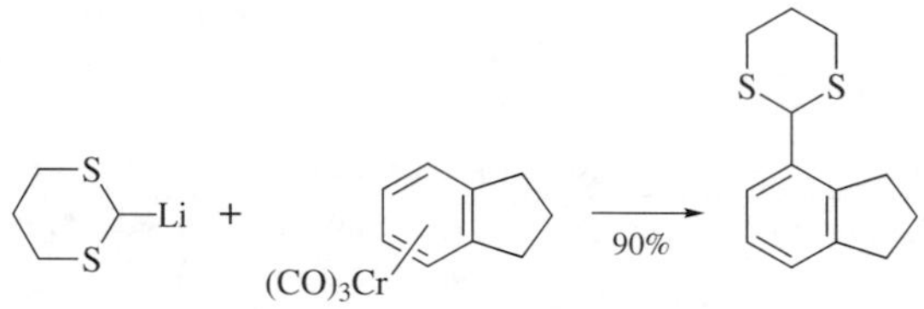

Nitroalkenes (eq 8) and nitroarenes react with 2-lithio-1,3-dithianes by conjugate addition.[27] Vinyl sulfones, as described for an oxygenated cyclopentene (eq 9)[28] and an α,β-(phenylseleno)vinyl selenoxide[29] example, display a similar reactivity pattern.

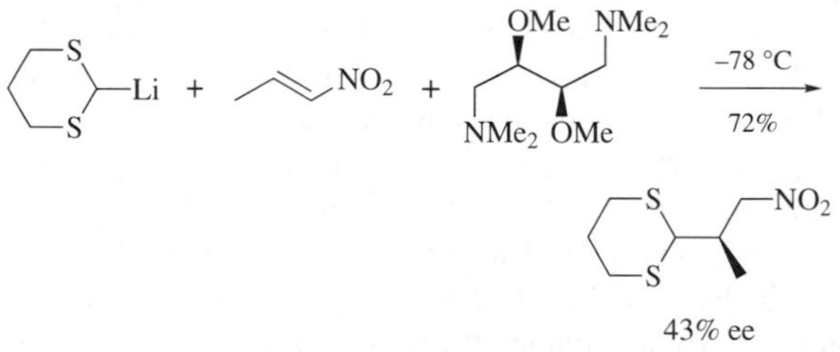

Vinylphosphonium salts react with 2-lithio-1,3-dithiane to form the β-phosphonio substituted derivative. The inherent properties of the dithiane moiety then allow for the preparation of β-phosphinyl carboxylic acids (eq 10)[30] or other functionalized phosphorus compounds,[31] valuable building blocks in organic synthesis.

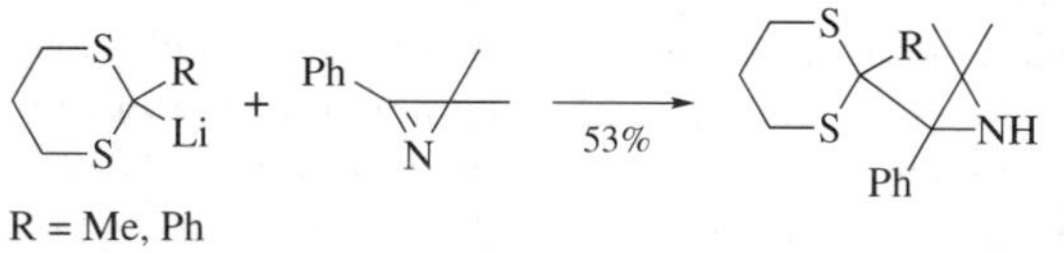

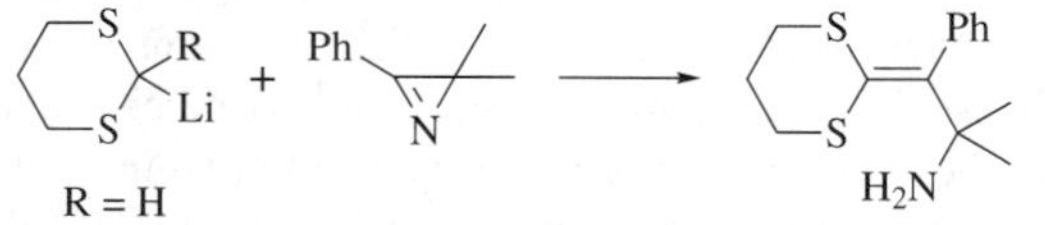

α/β-Aminoalkylation. The reagent reacts with activated imines to afford the corresponding α-aminoalkylated derivatives (eq 11). Typical reaction partners are phenylazirines (eq 12)[32] or iminium salts.[33] In the case of the former, only the C-2 substituted reagent gives the α-aminoalkylated compound; the parent reagent will afford β-aminoketene dithioacetals instead (eq 13).[32a] Recently, the reaction with *N*-tosyl aziridines has been reported, which leads to the β-amino substituted ketone derivatives.[34] 1-Methyl-4-quinolone as reaction partner gives the α-aminoalkylated product, which in this case derives from a 1,4-addition reaction (see also α-Hydroxyalkylation/γ-Ketoalkylation below).[35]

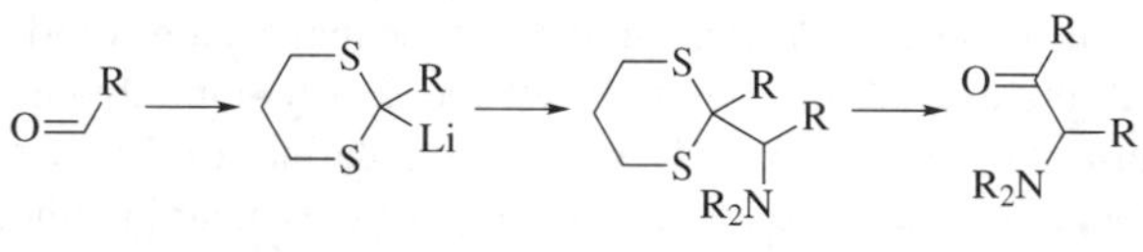

β-Hydroxyalkylation. Epoxide ring opening with 2-lithio-1,3-dithianes proceeds in general with good regio- and stereochemical control, following the rules for S_N2-type processes, and affording the β-hydroxyalkylated products in good to excellent yield, routinely exceeding 80% (eq 14).[1b–f]

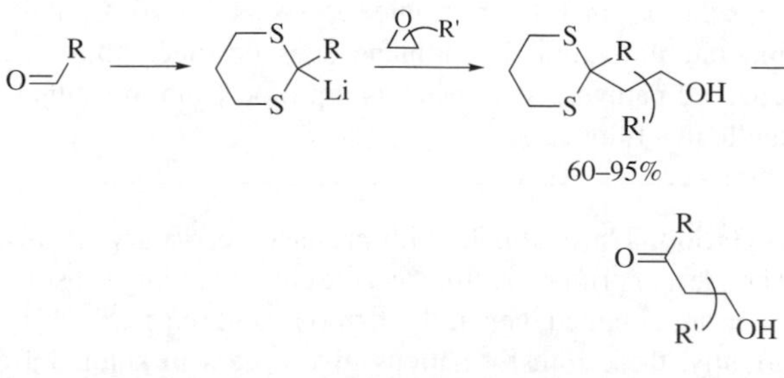

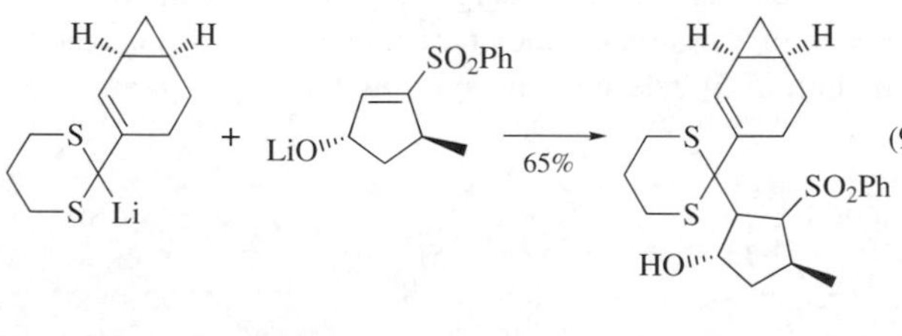

More recent examples include stereoselective epoxide opening to access enantiomerically pure hydroxylated compounds,[36] e.g. propanal-type aldol products.[37] This reaction sequence also proved to be useful for the synthesis of C-3 formyl sugar derivatives,[38] and chiral 4-hydroxycyclopentenones from glucose,[39] and it was employed as a key feature in the construction of the Kishi intermediate[40] for aplysiatoxin (eq 15).[41] Noteworthy is the high chemoselectivity displayed by the reagent in these reactions.

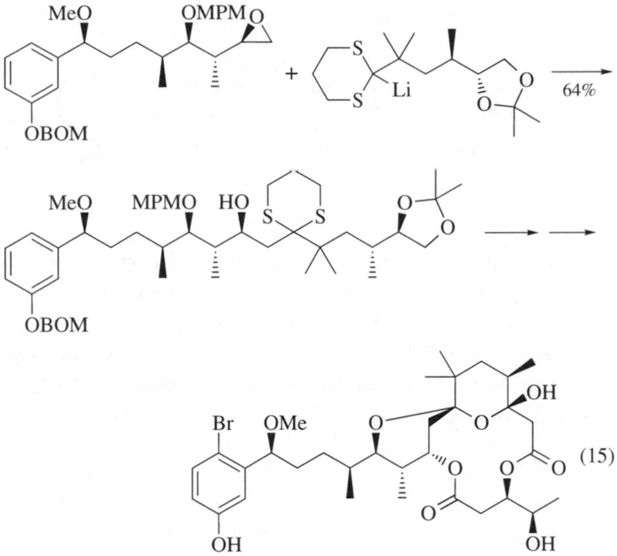

(15)

Aplysiatoxin

An unusual example, in which the reagent reacts preferentially with a fulvene functionality over epoxide opening, has been reported (eq 16).[42]

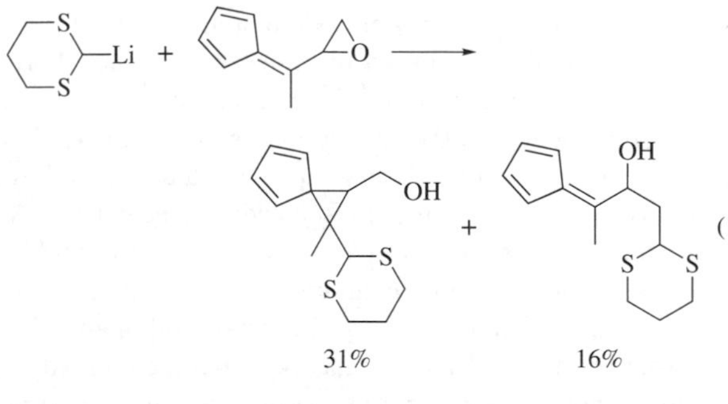

(16)

α-Hydroxyalkylation/γ-Ketoalkylation. Both types of product, obtained from the addition of 2-lithio-1,3-dithianes to aldehydes and ketones, or to the conjugated unsaturated analogs, represent one of the unique features available from the reactivity pattern of the reagent. By virtue of its role as acyl anion equivalent **A**, this reaction allows establishment of a 1,2-arrangement of oxygen functional groups, not so readily achieved by using classical reagents (eq 17). It is this feature of umpolung[1d] of the normal carbonyl reactivity pattern, which earned 2-lithio-1,3-dithianes so much attention in synthetic organic chemistry, and accounts for a significant proportion of all reactions involving the use of the reagent.[1b–f] The yields observed for these addition reactions are in general very good; product formation in the 70–90% range is not uncommon.

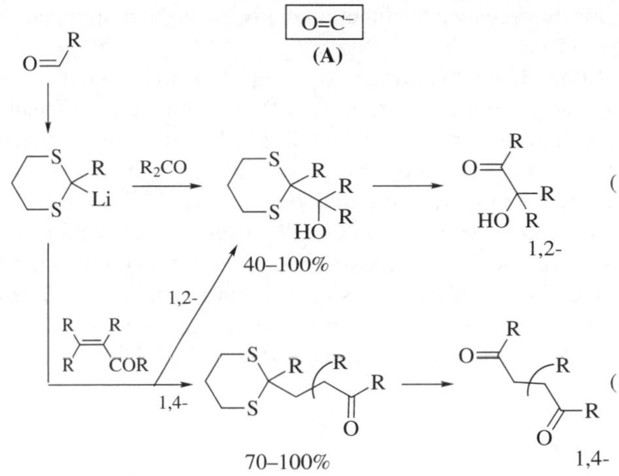

$O{=}C^-$ **(A)**

(17)

(18)

The scope of the reagent was broadened after a thorough investigation established the factors which influence the regioselectivity of the addition of the reagent to α,β-unsaturated aldehydes and ketones.[1d] As a complement to the 'standard' 1,2-addition mode described above, 2-lithio-1,3-dithianes can also be used deliberately to establish a 1,4-arrangement of oxygen functional groups (eq 18), again by functioning as an acyl anion equivalent **A**.[1b–f,35,43] The outcome of the addition to enones can be influenced not only by the choice of solvent,[1d] but also with the help of alkene-complexing agents, as has been demonstrated for the tricarbonyl(tropone)iron complex.[44]

Recent literature reports the addition reaction of 2-lithio-1,3-dithiane to aldehydes for assembling the C(10)–C(19) moiety of FK506,[45] and the possibility of achieving asymmetric induction in this reaction sequence has been investigated (eq 19).[46]

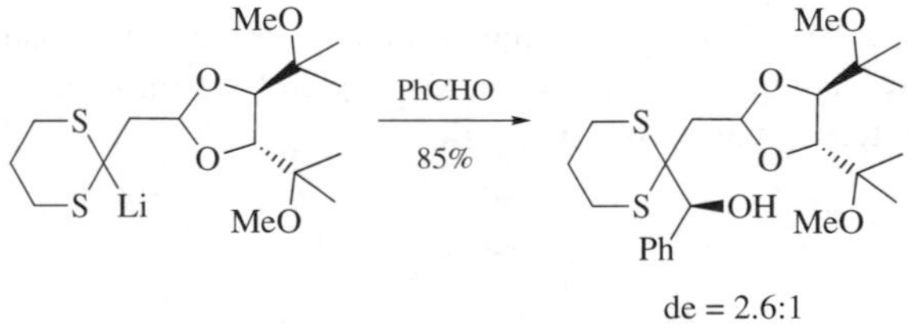

(19)

Cyclic hemiacetals can be used in place of the free carbonyl compound with equal success.[47] The stereoselectivity of the reaction can be influenced by neighboring polar groups, as illustrated in the example with (*R*)-pantolactone (eq 20).[47]

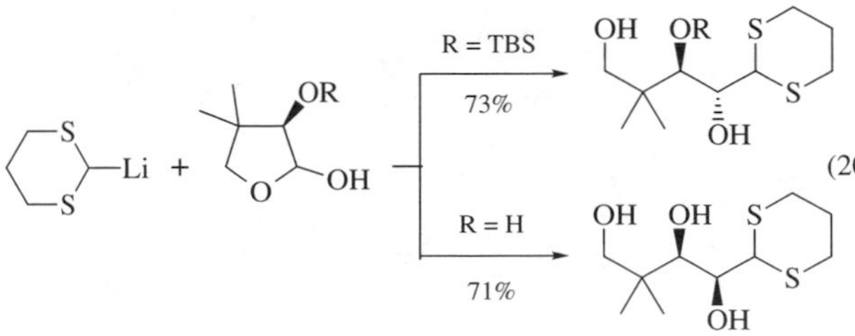

(20)

Acylation. The characteristic features of the reagent, as indicated in the foregoing section on addition reactions to carbonyl derivatives, also dominate the reactivity pattern observed for the reaction with carboxyl derivatives. The role of 2-lithio-1,3-dithianes as acyl anion equivalent **A**, establishing a 1,2-

relationship between the functional groups, is the dominating feature (eq 21).

2-Lithio-1,3-dithiane reacts with acyl derivatives, such as esters or lactones, amides or aziridinones,[48] to afford the corresponding acyl dithiane in variable yield (eq 21).[1b–f] Little investigation has yet been carried out into the regioselectivity for the reaction with the corresponding α,β-unsaturated derivatives.[49,50] With an electron releasing substituent on C-2 the reagent reacts with butenolides in a 1,4-conjugated fashion (eq 22), a scheme specifically exploited in several syntheses of steganin lignans.[49] Also, unsaturated amides, thioamides, and nitriles are reported to react with the reagent in the 1,4-mode indicated in eq 22.[50c–f]

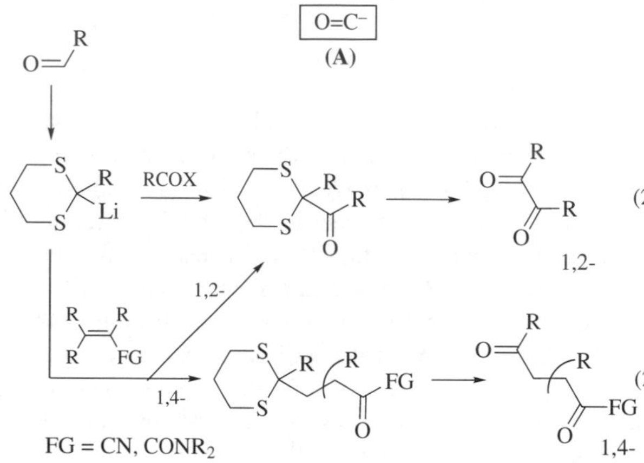

O=C⁻ (A)

RCOX (21)

1,2- 1,4- (22)

FG = CN, $CONR_2$

Acid chlorides will give the double addition carbinol product (eq 25),[51] nitriles afford either amino ketene dithioacetals (eq 24)[52,53] or ketones (eq 23),[54] nitrile oxides lead to oxime derivatives (eq 26), activated imides can be opened regioselectively (eq 29),[55] activated ketene acetals give the corresponding enol-type compound (eq 30),[56] and isocyanides afford the cyano substituted dithiane (eq 27).[57]

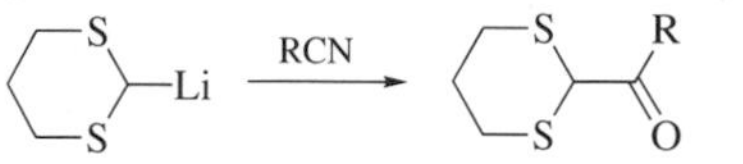

RCN (23)

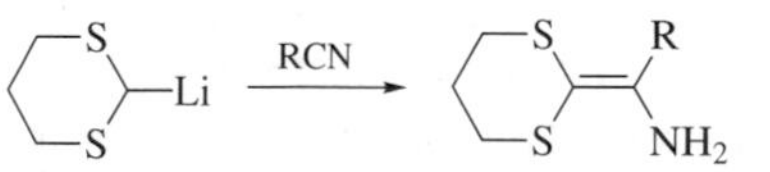

RCN (24)

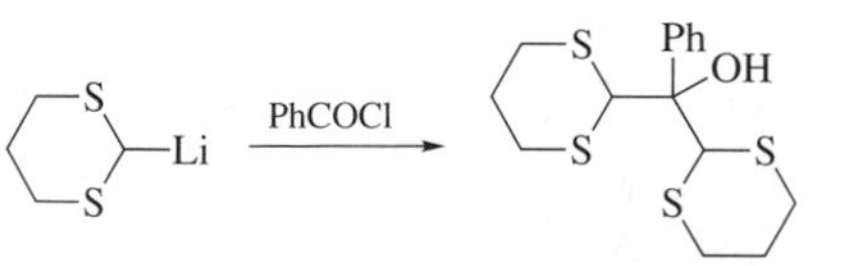

PhCOCl (25)

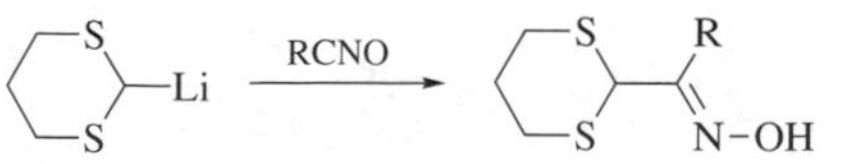

RCNO (26)

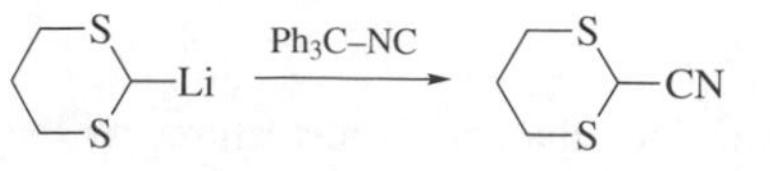

Ph_3C–NC (27)

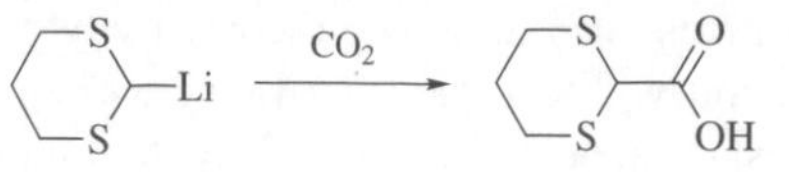

CO_2 (28)

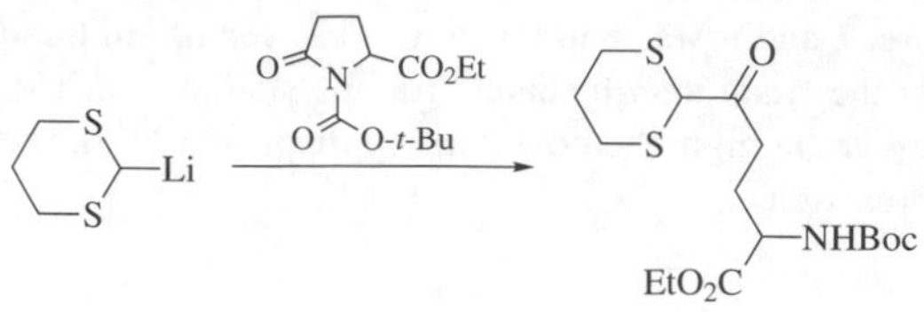

(29)

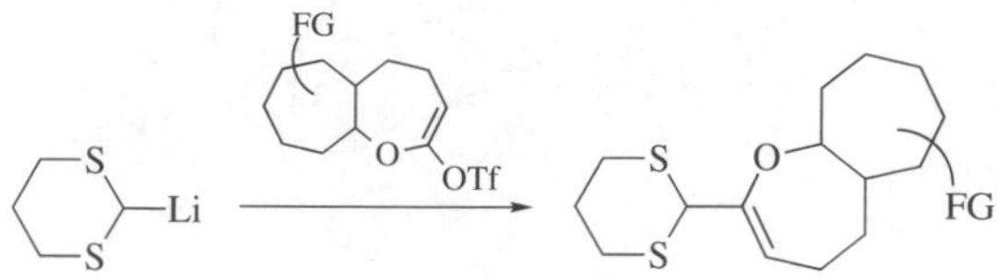

(30)

Carboxylation by reaction with ***Carbon Dioxide*** proceeds in excellent yield (eq 28).[1d,e,58] The reaction products find use as intermediates in a notable lactone synthesis.[59]

The reaction of 2-lithio-1,3-dithiane with carboxyl derivatives can be reversed. This result allows a synthetically useful C–C bond cleavage (eq 31).[1e]

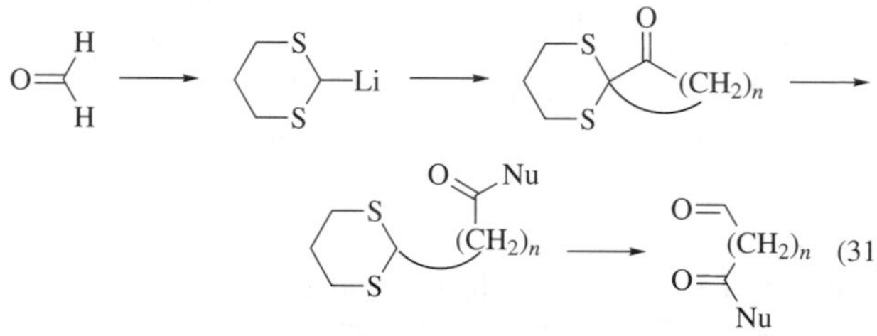

(31)

2-P, Sn, Si, Ge, and Ti Derivatives. The versatility of 2-lithio-1,3-dithiane as an acyl anion **A** or acyl dianion equivalent **B** is further amplified by the possibility of preparing the phosphorus,[60] tin,[1f,7] silicon,[60,61] germanium,[1f,62] or titanium[63] analogs, e.g. on simple treatment with the corresponding chlorides. The phosphorus and silicon analogs deserve special attention, as they allow a convenient entry to ketene dithioacetals (see below). Also, the silicon analog was used in the preparation of formyl-[61] and acylsilanes.[11] The tin and germanium, as well as silicon, derivatives react with acid chlorides to give 2-acyl-1,3-dithianes.[62] The special role of the tin derivative has been outlined already (see Preparation above);[7] the titanium analog was found to be more stable than the reagent, and as a consequence more selective.[63]

Ketene Dithioacetals. The lithio derivative obtained from deprotonation of ***2-Trimethylsilyl-1,3-dithiane*** deserves special mention. On reaction with carbonyl compounds, ketene dithioacetals are obtained, compounds which are responsive to both electrophilic as well as nucleophilic attack. Other methods for the synthesis of ketene dithioacetals have been reviewed.[60] In addition to their normal reactivity pattern, these derivatives offer an exceptionally broad umpolung-type reactivity pattern, i.e. acyl anion **A**, enolate carbanion **C**, homoenolate anion **D**, and chemistry of the homolog **E** becomes available (eq 32). The richness offered by this class of compounds in terms of their potential as strategical and useful intermediates in organic synthesis is documented amply, and reference to review articles have to suffice in this context.[1a,60]

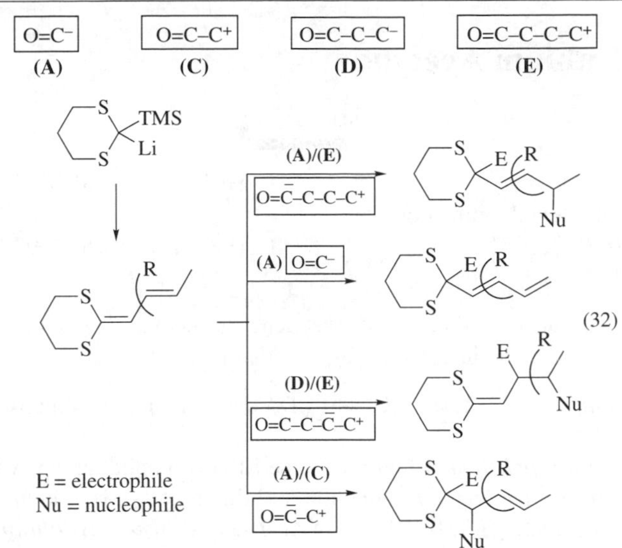

Related Reagents. Bis(methylthio)(trimethylsilyl)methane; Bis(phenylthio)methane; 2-Chloro-1,3-dithiane; *N*,*N*-Diethylaminoacetonitrile; 1,3-Dithiane; 1-Ethoxyvinyllithium; Formaldehyde Dimethyl Thioacetal Monoxide; 2-Lithio-2-trimethylsilyl-1,3-dithiane; Methoxy(phenylthio)methane; 1-Methoxyvinyllithium; 2-Trimethylsilyl-1,3-dithiane.

---

1. (a) Krief, A. *COS* **1991**, *3*, 134. (b) Krief, A. *COS* **1991**, *3*, 124, 131. (c) Larock, R. C. *Comprehensive Organic Transformations*; VCH: New York, 1989; p 721. (d) Page, P. C. B.; van Niel, M. B.; Prodger, J. C. *T* **1989**, *45*, 7643 and references therein. (e) Gröbel, B.-T.; Seebach, D. *S* **1977**, 357. (f) Seebach, D. *S* **1969**, 17.
2. (a) Xie, L.; Bors, D. A.; Streitwieser, A. *JOC* **1992**, *57*, 4986. (b) Bordwell, F. G.; Drucker, G. E.; Andersen, N. H.; Denniston, A. D. *JACS* **1986**, *108*, 7310. (c) Fraser, R. R.; Bresse, M.; Mansour, T. S. *CC* **1983**, 620. (d) Streitwieser, A., Jr.; Guibé, F. *JACS* **1978**, *100*,4532. (e) Streitwieser, A., Jr; Ewing, S. P. *JACS* **1975**, *97*, 190.
3. (a) Wolfe, S.; LaJohn, L. A.; Bernardi, F.; Mangini, A.; Tonachini, G. *TL* **1983**, *24*, 3789. (b) Wolfe, S.; Stolow, A.; LaJohn, L. A. *TL* **1983**, *24*, 4071. (c) Abatjoglou, A. G.; Eliel, E. L.; Kuyper, L. F. *JACS* **1977**, *99*, 8262. (a) Bernardi, F.; Csizmadia, I. G.; Mangini, A.; Schlegel, H. B.; Whangbo, M.-H.; Wolfe, S. *JACS* **1975**, *97*, 2209. (e) Eliel, E. L. *T* **1974**, *30*, 1503. (f) Eliel, E. L.; Hartman, A. A.; Abatjoglou, A. G. *JACS* **1974**, *96*, 1807. (g) Eliel, E. L.; Abatjoglou, A.; Hartmann, A. A. *JACS* **1972**, *94*, 4786. (h) Eliel, E. L. *AG* **1972**, *84*, 779; *AG(E)* **1972**, *11*, 739. (i) Eliel, E. L.; Hutchins, R. O. *JACS* **1969**, *91*, 2703.
4. (a) Amstutz, R.; Laube, T.; Schweizer, W. B.; Seebach, D.; Dunitz, J. D. *HCA* **1984**, *67*, 224. (b) Amstutz, R.; Dunitz, J. D.; Seebach, D. *AG* **1981**, *93*, 487; *AG(E)* **1981**, *20*, 465. (c) Amstutz, R.; Seebach, D.; Seiler, P.; Schweizer, B.; Dunitz, J. D. *AG* **1980**, *92*, 59; *AG(E)* **1980**, *19*, 53.
5. (a) Reich, H. J.; Borst, J. P. *JACS* **1991**, *113*, 1835. (b) Seebach, D.; Gabriel, J.; Hässig, R. *HCA* **1984**, *67*, 1083.
6. Seebach, D.; Corey, E. J. *JOC* **1975**, *40*, 231.
7. Seebach, D.; Willert, I.; Beck, A. K.; Gröbel, B.-T. *HCA* **1978**, *61*, 2510.
8. (a) Carre, M. C.; Ndebeka, G.; Riondel, A.; Bourgasser, P.; Caubere, P. *TL* **1984**, *25*, 1551. (b) Seebach, D.; Leitz, H. F.; Ehrig, V. *CB* **1975**, *108*, 1924. (c) Eliel, E. L.; Hartmann, A. A. *JOC* **1972**, *37*, 505.
9. Weil, R.; Collignon, N. *BSF(2)* **1974**, 253.
10. For some mechanistic investigation see: Juaristi, E.; Jimenez-Vazquez, H. A. *JOC* **1991**, *56*, 1623.
11. Reich, H. J.; Holtan, R. C.; Bolm, C. *JACS* **1990**, *112*, 5609.
12. Brook, A. G.; Duff, J. M.; Jones, P. F.; Davis, N. R. *JACS* **1967**, *89*, 431.
13. Seebach, D. *AG* **1979**, *91*, 259; *AG(E)* **1979**, *18*, 239.
14. (a) Tanigawa, Y.; Ohta, H.; Sonoda, A.; Murahashi, S.-I. *JACS* **1978**, *100*, 4610. (b) Tanigawa, Y.; Kanamaru, H.; Sonoda, A.; Murahashi, S.-I. *JACS* **1977**, *99*, 2361.
15. Hungerbühler, E.; Naef, R.; Wasmuth, D.; Seebach, D.; Loosli, H.-R.; Wehrli, A. *HCA* **1980**, *63*, 1960.
16. Orsini, F.; Pelizzoni, F. *JOC* **1980**, *45*, 4726.
17. Quimpere, M.; Ruest, L.; Deslongchamps, P. *CJC* **1992**, *70*, 2335.
18. (a) Spracklin, D. K.; Weiler, L. *CC* **1992**, 1347. (b) Finch, N.; Gemenden, C. W. *JOC* **1979**, *44*, 2804.
19. (a) Grigg, R.; Markandu, J.; Surendrakumar, S.; Thornton-Pett, M.; Warnock, W. J. *T* **1992**, *48*, 10399. (b) See also: Narasaka, K.; Saitou, M.; Iwasawa, N. *TA* **1991**, *2*, 1305. (c) Hammond, G. B.; Plevey, R. G. *OPP* **1991**, *23*, 735.
20. Liang, G.-B.; Desper, J. M.; Gellman, S. H. *JACS* **1993**, *115*, 925.
21. Narasaka, K.; Hayashi, Y.; Shimada, S.; Yamada, J. *Isr. J. Chem.* **1991**, *31*, 261.
22. Inoue, Y.; Tanimoto, S.; Nakamura, K.; Ohno, A. *Bull. Inst. Chem. Res., Kyoto Univ.* **1992**, *69*, 520.
23. Adams, T. C.; Dupont, A. C.; Carter, J. P.; Kachur, J. F.; Guzewska, M. E.; Rzeszotarski, W. J.; Farmer, S. G.; Noronha-Blob, L.; Kaiser, C. *JMC* **1991**, *34*, 1585.
24. Krohn, S.; Fletcher, M. T.; Kitching, W.; Drew, R. A. A.; Moore, C. J.; Francke, W. *J. Chem. Ecol.* **1991**, *17*, 485.
25. (a) Roell, Jr., B. C.; McDaniel, K. F.; Vaughan, W. S.; Macy, T. S. *OM* **1993**, *12*, 224. (b) Kündig, E. P.; Inage, M.; Bernardinelli, G. *OM* **1991**, *10*, 2921. (c) Kündig, E. P.; Grivet, C.; Wenger, E.; Bernardinelli, G.; Williams, A. F. *HCA* **1991**, *74*, 2009. (d) Uemura, M.; Minami, T.; Shinoda, Y.; Nishimura, H.; Shiro, M.; Hayashi, Y. *JOM* **1991**, *406*, 371. (e) Mandon, D., Astruc, D. *OM* **1990**, *9*, 341. (f) Cambie, R. C., Clark, G. R.; Gallagher, S. R.; Rutledge, P. S.; Stone, M. J.; Woodgate, P. D. *JOM* **1988**, *342*, 315. (g) Kündig, E. P.; Simmons, D. P. *CC* **1983**, 1320. (h) Semmelhack, M. F. *PAC* **1981**, *53*, 2379. (i) Kozikowski, A. P.; Isobe, K. *CC* **1978**, 1076. (j) Semmelhack, M. F.; Clark, G. *JACS* **1977**, *99*, 1675. (k) Raubenheimer, H. G.; Lotz, S. *CC* **1976**, 732.
26. See also: Roell, B. C., Jr.; McDaniel, K. F. *JACS* **1990**, *112*, 9004.
27. (a) Bartoli, G.; Dalpozzo, R.; Grossi, L.; Todesco, P. E. *T* **1986**, *42*, 2563. (b) Funabashi, M.; Wakai, H.; Sato, K.; Yoshimura, J. *JCS(P1)* **1980**, 14. (c) Funabashi, M.; Kobayashi, K.; Yoshimura, J. *JOC* **1979**, *44*, 1618. (d) Funabashi, M.; Yoshimura, J. *JCS(P1)* **1979**, 1425. (e) Seebach, D.; Langer, W. *HCA* **1979**, *62*, 1701. (f) Langer, W.; Seebach, D. *HCA* **1979**, *62*, 1710.
28. Saddler, J. C.; Fuchs, P. L. *JACS* **1981**, *103*, 2112.
29. Back, T. G.; Krishna, M. V. *JOC* **1987**, *52*, 4265.
30. (a) Okada, Y.; Minami, T.; Umezu, Y.; Nishikawa, S.; Mori, R.; Nakayama, Y. *TA* **1991**, *2*, 667. (b) Okada, Y.; Minami, T.; Sasaki, Y.; Umezu, Y.; Yamaguchi, M. *TL* **1990**, *31*, 3905.
31. Cristau, H. J.; El Hamad, K.; Torreilles, E. *PS* **1992**, *66*, 47.
32. (a) Ben Cheikh, R.; Bouzouita, N.; Ghabi, H.; Chaabouni, R. *T* **1990**, *46*, 5155. (b) Padwa, A.; Dharan, M.; Smolanoff, J.; Wetmore, Jr., S. I. *JACS* **1973**, *95*, 1954.
33. (a) Seebach, D.; Ehrig, V.; Leitz, H. F.; Henning, R. *CB* **1975**, *108*, 1946. (b) Duhamel, L.; Duhamel, P.; Mancelle, N. *BSF(2)* **1974**, 331.
34. Osborn, H. M. I.; Sweeney, J. B.; Howson, B. *SL* **1993**, 675.
35. Griera, R.; Rigat, L.; Alvarez, M.; Joule, J. A. *JCS(P1)* **1992**, 1223.
36. (a) Takano, S.; Setoh, M.; Takahashi, M.; Ogasawara, K. *TL* **1992**, *33*, 5365. (b) Dumortier, L.; Van der Eycken, J.; Vandewalle, M. *SL* **1992**, 245. (c) De Brabander, J.; Vanhessche, K.; Vandewalle, M. *TL* **1991**, *32*, 2821.
37. Pasquarello, A.; Poli, G.; Scolastico, C. *SL* **1992**, 93.
38. Benefice-Malouet, S.; Coe, P. L.; Walker, R. T. *Carbohydr. Res.* **1992**, *229*, 293.
39. Achab, S.; Das, B. C. *JCS(P1)* **1990**, 2863.

40. Park, P.; Broka, C. A.; Johnson, B. F.; Kishi, Y. *JACS* **1987**, *109*, 6205.
41. Okamura, H.; Kuroda, S.; Tomita, K.; Ikegami, S.; Sugimoto, Y.; Sakaguchi, S.; Katsuki, T.; Yamaguchi, M. *TL* **1991**, *32*, 5137.
42. Antczak, K.; Kingston, J. F.; Fallis, A. G. *TL* **1984**, *25*, 2077.
43. (a) For recent examples see: 1,2-Addition: Nishikawa, T.; Isobe, M.; Goto, T. *SL* **1991**, 393. (a) Gordon, P. M.; Siegel, C.; Razdan, R. K. *CC* **1991**, *692*. (c) 1,4-Addition: ref 35;
44. Rigby, J. H.; Ogbu, C. O. *TL* **1990**, *31*, 3385.
45. Gu, R.-L.; Sih, C. J. *TL* **1990**, *31*, 3283.
46. (a) Chikashita, H.; Yuasa, T.; Itoh, K. *CL* **1992**, 1457. (b) Jenkins, P. R.; Selim, M. M. R. *JCR(S)* **1992**, 85.
47. Roy, R.; Rey, A. W. *CJC* **1991**, *69*, 62.
48. Talaty, E. R.; Clague, A. R.; Behrens, J. M.; Agho, M. O.; Burger, D. H.; Hendrixson, T. L.; Korst, K. M.; Khanh, T. T.; Kell, R. A.; Dibaji, N. *SC* **1981**, *11*, 455.
49. (a) Tomioka, K.; Ishiguro, T.; Iitaka, Y.; Koga, K. *T* **1984**, *40*, 1303. (b) Tomioka, K.; Ishiguro, T.; Koga, K. *TL* **1980**, *21*, 2973. (c) Ziegler, F. E.; Schwartz, J. A. *JOC* **1978**, *43*, 985.
50. (a) For reaction with α,β-unsaturated acids: Cooke, M. P., Jr. *JOC* **1987**, *52*, 5729. (b) Majewski, M.; Snieckus, V. *JOC* **1984**, *49*, 2682. (c) For reaction with α,β-unsaturated amides and thioamides: Mpango, G. B.; Mahalanabis, K. K.; Mahdavi-Damghani, Z.; Snieckus, V. *TL* **1980**, *21*, 4823. (d) Mpango, G. B.; Snieckus, V. *TL* **1980**, *21*, 4827. (e) Tamaru, Y.; Harada, T.; Iwamoto, H.; Yoshida, Z. *JACS* **1978**, *100*, 5221. (f) For reaction with α,β-unsaturated nitriles: Basha, F. Z.; DeBernardis, J. F.; Spanton, S. *JOC* **1985**, *50*, 4160.
51. Kita, Y.; Sekihachi, J.; Hayashi, Y.; Da, Y. Z.; Yamamoto, M.; Akai, S. *JOC* **1990**, *55*, 1108.
52. Page, P. C. B.; van Niel, M. B.; Westwood, D. *JCS(P1)* **1988**, 269. Page, P. C. B.; van Niel, M. B.; Williams, P. H. *CC* **1985**, 742.
53. The ambident nucleophilicity of amino ketene dithioacetals make them useful in the synthesis of nitrogen heterocycles (e.g. Page, P. C. B.; van Niel, M. B.; Westwood, D. *CC* **1987**, 775), as well as reaction partners towards α,β-unsaturated ketones to give γ-diketones (homoenolate anion equivalent, e.g. Page, P. C. B.; Harkin, S. A.; Marchington, A. P.; van Niel, M. B. *T* **1989**, *45*, 3819. Page, P. C. B.; van Niel, M. B. *CC* **1987**, 43).
54. (a) Fuji, K.; Ueda, M.; Sumi, K.; Kajiwara, K.; Fujita, E.; Iwashita, T.; Miura, I. *JOC* **1985**, *50*, 657. (b) Fuji, K.; Ueda, M.; Sumi, K.; Fujita, E. *JOC* **1985**, *50*, 662. (c) Fuji, K.; Ueda, M.; Fujita, E. *CC* **1983**, 49. (d) Kawamoto, I.; Muramatsu, S.; Yura, Y. *TL* **1974**, 4223.
55. Ezquerra, J.; de Mendoza, J.; Pedregal, C.; Ramirez, C. *TL* **1992**, *33*, 5589.
56. Feng, F.; Murai, A. *CL* **1992**, 1587.
57. Khatri, H. N.; Walborsky, H. M. *JOC* **1978**, *43*, 734.
58. Knight, D. W.; Pattenden, G. *JCS(P1)* **1979**, 84.
59. Nicolaou, K. C.; Seitz, S. P.; Sipio, W. J.; Blount, J. F. *JACS* **1979**, *101*, 3884.
60. (a) Kolb, M. *S* **1990**, 171. (b) Kolb, M. In *The Chemistry of Ketenes, Allenes, and Related Compounds*; Patai, S., Ed.; Wiley: Chichester, 1980; Part 2, p 669. (c) Barrett, G. C. In *Comprehensive Organic Chemistry*; Barton, D. H. R.; Ollis, W. D., Eds.; Pergamon: Oxford, 1979; Vol. 3, Part 11.4.
61. (a) Soderquist, J. A.; Miranda, E. I. *JACS* **1992**, *114*, 10078. (b) See also: Hwu, J. R.; Lee, T.; Gilbert, B. A. *JCS(P1)* **1992**, 3219. (c) Silverman, R. B.; Lu, X.; Banik, G. M. *JOC* **1992**, *57*, 6617.
62. Jutzi, P.; Lorey, O. *PS* **1979**, *7*, 203.
63. Weidmann, B.; Widler, L.; Olivero, A. G.; Maycock, C. D.; Seebach, D. *HCA* **1981**, *64*, 357.

Michael Kolb

Marion Merrell Dow, Cincinnati, OH, USA

Lithium Acetylide

[1111-64-4] C_2HLi (MW 31.97)
(ethylenediamine complex)
[6867-30-7] $C_4H_9LiN_2$ (MW 92.09)

(reagent for addition of acetylene to ketones,[1] opening of epoxides,[2] and ethynylation of alkyl halides[3])

Solubility: sol THF; LiC≡CH·EDA sol alkyl amines, slightly sol THF.

Form Supplied in: lithium acetylide·EDA is supplied as a powder.

Preparative Method: monolithium lithium acetylide is prepared by treating a THF solution of ***Acetylene*** with ***n-Butyllithium*** at 78 °C (eq 1).[1]

$$\text{H—≡—H} + \text{BuLi} \xrightarrow[-78\,^\circ\text{C}]{\text{THF}} \text{H—≡—Li} \qquad (1)$$

It is critical to keep the temperature at −78 °C. The concentration should be 0.5 M or less. At higher concentrations and temperatures, a white precipitate forms (presumably ***Dilithium Acetylide***) and yields of addition product fall appreciably. Properly formed, lithium acetylide in THF should be a clear, colorless solution.

Handling, Storage, and Precautions: lithium acetylide/THF solution must be used immediately after preparation. It decomposes above −78 °C. A saturated solution of THF/acetylene should not be used during the preparation of lithium acetylide since this may lead to an acetylene atmosphere above the THF. *n*-Butyllithium dropping through the acetylene atmosphere may react violently. Lithium acetylide is air and moisture sensitive.

Addition to Ketones. Lithium acetylide rapidly adds to a variety of aldehydes and ketones (eq 2).[1]

(2) [ketone → alcohol; H—≡—Li, 90%; O; OH]

Hindered ketones such as di-*t*-butyl ketone give best results if an excess of reagent is used. The reagent freshly prepared in THF generally gives better results than lithium acetylide·EDA[4] or lithium acetylide in ammonia.[5] In the case of easily enolizable ketones, the cerium reagent works well (eq 3).[6]

(3) [ketone → alcohol; H—≡—$CeCl_2$, 89%; O; OH]

The TMS cerium acetylide reportedly gives better results than the unsubstituted cerium regent (eq 4).[7]

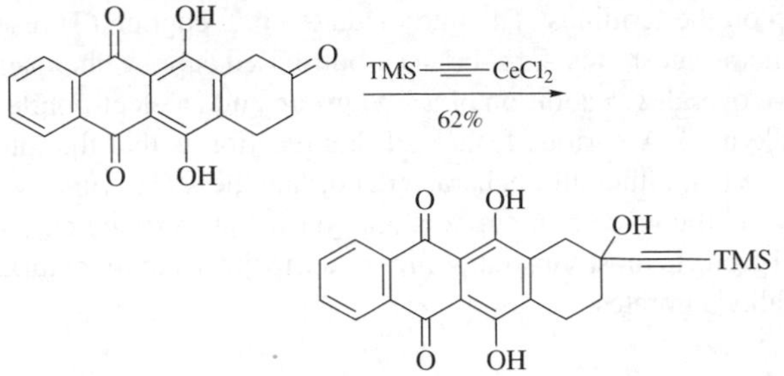

(4)

Reaction with Epoxides. Lithium acetylide in THF decomposes before it can react with epoxides. However, in the presence of ***Boron Trifluoride Etherate*** epoxides are readily opened (eq 5).[2]

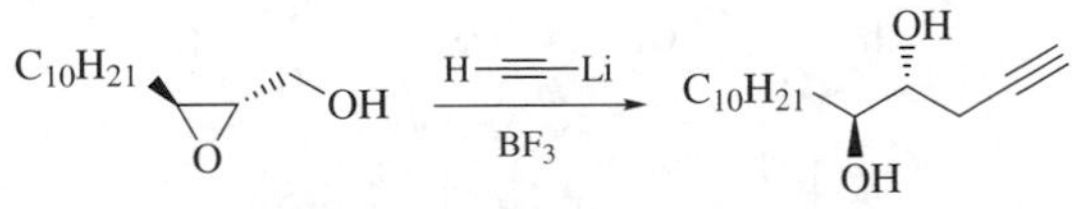

(5)

Lithium acetylide·EDA in THF/HMPA[8] or DMSO[9] may also be used to open epoxides (eq 6).

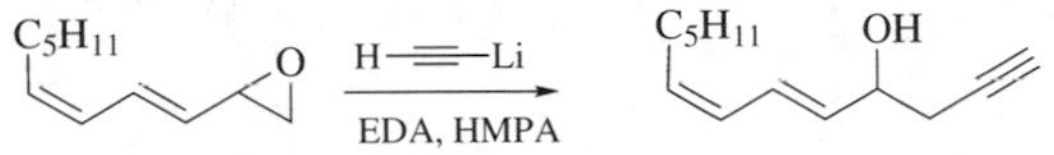

(6)

Ethynylation of Alkyl Halides. Lithium acetylide·EDA in DMSO can be used to convert alkyl halides and sulfates into terminal alkynes (eq 7).[3]

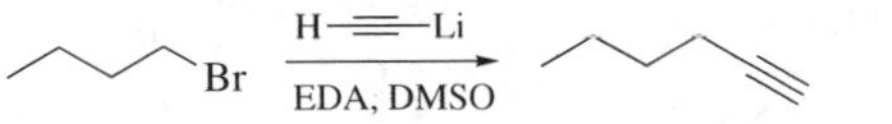

(7)

Monolithium acetylide in THF in the presence of HMPA is stable at 0 °C. This reagent may also be used to ethynylate alkyl halides (eq 8).[10]

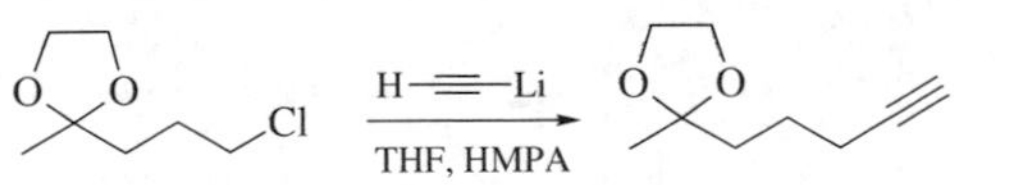

(8)

An alternative route to terminal alkynes involves iodination of lithium ethynyl trialkylborates (eq 9).[11] These are prepared by treating a trialkylborane with lithium acetylide·EDA. This method retains the stereochemistry of the boron–carbon bond.

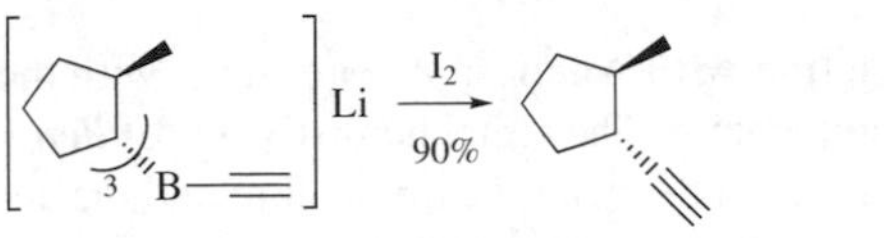

(9)

Related Reagents. Acetylene; Bis(trimethylsilyl)acetylene; Dilithium Acetylide; Ethynylmagnesium Bromide; Lithium Chloroacetylide; Lithium (Trimethylsilyl)acetylide; Propynyllithium; Trimethylsilylacetylene.

---

1. (a) Midland, M. M. *JOC* **1975**, *40*, 2250. (b) Midland, M. M.; McLoughlin, J. I., Werley, R. T. *OS* **1989**, *68*, 14. (c) Midland, M. M.; McLoughlin, J. I., Werley, R. T. *OSC* **1993**, *8*, 391.
2. Yamaguchi, M.; Hirao, I. *CC* **1984**, 202.
3. Smith, W. N.; Beumel, O. F. *S* **1974**, 441.
4. Beumel, O. F., Jr.; Harris, R. F. *JOC* **1964**, *29*, 1872.
5. Oroshnik, W.; Mebane, A. D. *JACS* **1949**, *71*, 2062.
6. Imamota, T.; Sugiura, Y.; Takiyama, N. *TL* **1984**, *25*, 4233.
7. Suzuki, M.; Kimura, Y.; Terashima, S. *CPB* **1986**, *34*, 1531.
8. Corey, E. J.; Kang, J. *JACS* **1981**, *103*, 4618.
9. Hanack, M.; Kunzmann, E.; Schumacher, W. *S* **1978**, 26.
10. (a) Johnston, B. D.; Oehlschlager, A. C. *JOC* **1982**, *47*, 5384. (b) Beckmann, W.; Doerjer, G.; Logemann, E.; Merkel, C.; Schill, C.; Zürcher, C. *S* **1975**, 423.
11. Midland, M. M.; Sinclair, J. A.; Brown, H. C. *JOC* **1974**, *39*, 731.

M. Mark Midland
University of California, Riverside, CA, USA

Lithium Bis[dimethyl(phenyl)silyl]cuprate[1]

$(PhMe_2Si)_2CuLi$

[75583-57-2] $C_{16}H_{22}CuLiSi_2$ (MW 341.01)

(dimethyl(phenyl)silyl nucleophile for making Si–C bonds by reaction with α,β-unsaturated carbonyl compounds, alkynes, allenes, and allylic acetates)

Physical Data: typically a 0.6M, dark, reddish-brown solution in THF.

Analysis of Reagent Purity: the silyllithium solution can be double-titrated for active reagent using ***Allyl Bromide***, but the cuprate is usually used without further checks; NMR (of the cuprate made with CuCN in THF-d_8): ^{1}H δ 0.09; ^{13}C δ 5.1; ^{29}Si δ −24.4; ^{7}Li δ −3.33.[2]

Preparative Methods: the silyl cuprate is prepared[1] from the corresponding silyllithium reagent (***Dimethylphenylsilyllithium***); commercially available ***Chlorodimethylphenylsilane*** is stirred with ***Lithium*** shot, wire, or powder under Ar or N_2 in THF at 0 °C for 4–12 h; the silyllithium solution may also be prepared, free of halide ion, by cleaving tetramethyldiphenyldisilane with lithium and ultrasound irradiation;[2] the silyllithium solution (2 equiv), after assay, is transferred by syringe on to anhyd ***Copper(I) Iodide***, ***Copper(I) Bromide***, or ***Copper(I) Cyanide*** (1 equiv), kept under argon or nitrogen at 0 °C, stirred at this temperature for 20 min, and then used immediately.

Handling, Storage, and Precautions: must be kept free of O_2 and H_2O; while somewhat more stable thermally than alkyl cuprates, surviving for a few hours at 0 °C, it is best used immediately after its preparation; the copper salts, and especially copper(I) cyanide, are toxic; the solutions should therefore be handled in a fume hood wearing impermeable gloves, and the aqueous washings disposed of appropriately, immediately after use.

Introduction. Because dimethyl(phenyl)silyllithium is much easier to prepare than ***Trimethylsilyllithium***, the most commonly used silyl cuprate reagent is derived from this silyl group. The reagent can be prepared using CuI, CuBr·SMe_2, or CuCN. The three reagents appear to be very similar in their reactivity, except for the higher regioselectivity of the cyanide-derived reagent with terminal alkynes. The dimethyl(phenyl)silyl group in products like allyl- and vinylsilanes appears to impart very similar reactivity to that imparted by the trimethylsilyl group, and it has an advantage over the trimethylsilyl group in that the presence of the phenyl group allows the dimethyl(phenyl)silyl group to be converted into a hydroxyl group with retention of configuration at carbon (eq 1). This transformation requires first a reaction with an electrophile, such as a proton,[3] bromine, or the mercury(II) cation,[4] to remove the phenyl ring and place a nucleofugal group X on the silicon atom. This step is followed by treatment either with peracid or with ***Hydrogen Peroxide*** and a base. The two steps may be combined in one pot;[4] bromine itself does not have to be used, since the ***Peracetic Acid*** oxidizes bromide ion to bromine in situ.

Ph–$SiMe_2$ —(H^+, Br_2 or Hg^{2+})→ X–$SiMe_2$ —(AcO_2H or H_2O_2)→ OH (1)

This capacity of the dimethyl(phenyl)silyl group cannot be drawn upon, however, when there is a C=C double bond in the molecule; no matter which electrophile is used, it attacks the double bond more rapidly than it removes the phenyl ring from the silyl group. This limitation has been overcome using the corresponding diethylamino(diphenyl)silyl-[5] and 2-methylbut-2-enyl(diphenyl)silylcuprate[6] reagents.

A mixed cuprate, [dimethyl(phenyl)silyl]methyl(or-butyl)cuprate,[7] containing one silyl and one alkyl group, has some advantages. Only the silyl group is transferred to the substrate, and hence only one silyl group is needed, and the byproduct of the silyl-cupration step, methane or butane, is volatile. Yields in silyl-cupration reactions carried out with only 1:1 stoichiometry are apt not to be quite so good, however. Other silyl copper reagents and cuprates with more specific or limited applications are (*t*-butyldimethylsilyl)butylcuprate,[8] triphenylsilylcopper,[9] the bis(*t*-butyldiphenylsilyl)cuprate,[10] and the bis[tris-(trimethylsilyl)silyl]cuprate.[11]

Reaction with α,β-Unsaturated Carbonyl Compounds. Although silyllithium reagents add kinetically at the β-position of α,β-unsaturated ketones,[12] the reactions are better with the cuprate when the enone is hindered.[13] The cuprate, unlike the lithium reagent, also reacts with α,β-unsaturated aldehydes, esters (eq 2), amides, and nitriles[13] and with vinyl sulfoxides.[14] With esters, the intermediate enolates, which have the (*E*) geometry (**1**), may be used directly in highly stereoselective reactions with alkyl halides and aldehydes.[15,16] The β-hydroxy esters, such as (**2**), can be used in the synthesis of allylsilanes.[17]

α,β-Unsaturated carbonyl compounds attached to a homochiral auxiliary, such as Koga's lactam or Oppolzer's sultam, give products with a stereogenic center carrying a silyl group having high levels of enantiomeric purity.[18,19]

Reaction with Alkynes. The silyl cuprate reacts with alkynes by *syn* stereospecific metallo-metallation (eq 3). Provided that the cuprate is derived from copper cyanide, the regioselectivity with terminal alkynes is highly in favor of the isomer with the silyl group on the terminus. The intermediate vinyl cuprate (**3**) reacts with many substrates, familiar in carbon-based cuprate chemistry, to give overall *syn* addition of a silyl group and an electrophile to the alkyne.[20] A curious feature of this reaction is that the intermediate (**3**), although uncharacterized, has the stoichiometry of a mixed silicon–carbon cuprate, and yet it transfers the carbon-based group to most substrates, in contrast to the behavior of mixed silyl alkyl cuprates.

Ph–CH=CH–CO_2R —$(PhMe_2Si)_2CuCNLi_2$→ [$PhMe_2Si$, OLi, Ph, OR] (**1**)
(**1**) —MeI, R = Me→ $PhMe_2Si$, Ph, CO_2Me 84%
(**1**) —MeCHO, R = Bn→ $PhMe_2Si$, Ph, CO_2Bn, OH (**2**) 70%
(**1**) —NH_4Cl, R = Et→ $PhMe_2Si$, Ph, CO_2Et 89% (2)

$(PhMe_2Si)_2CuCNLi_2$ + 1-hexyne —THF, –50 °C→ [$PhMe_2SiCu^-$, $PhMe_2Si$] (**3**) —E^+→ $PhMe_2Si$–CH=C(E)–Bu (3)

E = H (94%), I (88%), CO_2H (69%), Ac (72%), Me (71%), CH(OH)Me (64%), 3-oxocyclohexyl (54%)

Disubstituted alkynes also react, and, if the two substituents are well differentiated sterically, as with a methyl group on one side and a branched chain on the other, the regioselectivity is highly in favor of the silyl group appearing at the less hindered end.

Reaction with Allenes. Allenes react with the silyl cuprate at low temperature. The regiochemistry with ***Allene*** itself places the silyl group on the central carbon atom and the added electrophile at the terminus (eq 4).[21] One surprising exception to this rule is with ***Iodine*** as the electrophile, when the product (**5**) has the opposite regiochemistry even from that of the reaction with chlorine. Since the iodide (**5**) can be converted into a lithium reagent [and a cuprate that is not identical with (**4**)], it is possible to achieve overall either regiochemistry in additions to allene. Monosubstituted allenes give mixtures of regioisomers,[21,22] and disubstituted and trisubstituted allenes give largely allylsilanes whatever electrophile is used. The metallo-metallation step is stereospecifically *syn*.[23]

$(PhMe_2Si)_2CuCNLi_2$ + =•= —THF, −78 °C→ [$PhMe_2SiCu^-$ ... $PhMe_2Si$] (4) —E+, −78 to 0 °C→ $PhMe_2Si$ ... E (4)

E = H (99%), Cl (61%), Ac (73%), Me (75%), OH (60%), (83%) HO

(4) —I_2→ $PhMe_2Si$ ... I

(5) 90%

Reaction with Allylic Acetates. Silyl cuprates react with allylic acetates to give allylsilanes directly (eqs 5–7). Allylic acetates that are secondary at both ends are apt to give both regioisomers. Some control of the regiochemistry is possible, however, by using a *cis* double bond; this encourages reaction with allylic shift, especially when the silyl group is delivered to the less-hindered end of the allylic system (eq 5).[24] An alternative protocol, assembling a mixed silyl cuprate on a carbamate group, is even better in controlling the regioselectivity, usually giving complete allylic shift (eq 6).[24,25] Tertiary acetates, on the other hand, are very well behaved regiochemically, giving only the product with the silyl group at the less-substituted end of the allylic system (eq 7).[25,26] Allylsilanes have many uses as carbon nucleophiles in organic synthesis.[27]

OAc —$(PhMe_3Si)_2CuCNLi_2$, THF, Et_2O, 0 °C, 2 h, 51%→ $PhMe_2Si$ + $PhMe_2Si$ (5)

82:18

$PhNHCO_2$ —1. BuLi, THF, −78 °C; 2. CuI, 0 °C; 3. $PhMe_2SiLi$, 0 °C; 68%→ $PhMe_2Si$ (6)

OAc —$(PhMe_2Si)_2CuLi$, THF, 0 °C, 15 min, 85%→ $SiMe_2Ph$ (7)

The acetate reaction (eq 5) and the carbamate alternative (eq 6) are complementary in their stereochemistry, the former taking place stereospecifically *anti* and the latter stereospecifically *syn*.[24,25]

Other Reactions. Other substrates that have been found to react with silyl-copper reagents and with silyl cuprates are allyl chlorides,[28] an alkyl bromide,[29] epoxides,[8,22] acid chlorides,[9,30] propargyl acetates[31] and sulfides,[32] a vinyl iodide,[9] an aminomethyl acetate,[33] ethyl tetrolate,[34] and some strained allylic ethers.[35]

Related Reagents. Dilithium Cyanobis(dimethylphenylsilyl)cuprate; Lithium Bis(3-trimethylsilyl-1-propen-2-yl)cuprate; Lithium Cyano(dimethylphenylsilyl)cuprate; Trimethylsilylcopper; Trimethylsilyllithium; Trimethylsilylmethylcopper; Trimethylsilylpotassium.

1. Fleming, I. In *Organocopper Reagents*; R. J. K. Taylor, Ed.; OUP: Oxford, 1995; p. 257.
2. Sharma, S.; Oehlschlager, A. C. *T* **1989**, *45*, 557; *JOC* **1991**, *56*, 770.
3. Fleming, I.; Henning, R.; Plaut, H. *CC* **1984**, 29.
4. Fleming, I.; Sanderson, P. E. J. *TL* **1987**, *28*, 4229.
5. Tamao, K.; Kawachi, A.; Ito, Y. *JACS* **1992**, *114*, 3989.
6. Fleming, I.; Winter, S. B. D. *TL* **1993**, *34*, 7287.
7. Fleming, I.; Newton, T. W. *JCS(P1)* **1984**, 1805.
8. Lipshutz, B. H.; Reuter, D. C.; Ellsworth, E. L. *JOC* **1989**, *54*, 4975.
9. Duffaut, N.; Dunoguès, J.; Biran, C.; Calas, R.; Gerval, J. *JOM* **1978**, *161*, C23.
10. Cuadrado, P.; Gonzalez, A. M.; Gonzalez, B.; Pulido, F. J. *SC* **1989**, *19*, 275.
11. Chen, H.-M.; Oliver, J. P. *JOM* **1986**, *316*, 255.
12. (a) Still, W. C. *JOC* **1976**, *41*, 3063. (b) Still, W. C.; Mitra, A. *TL* **1978**, 2659.
13. Ager, D. J.; Fleming, I.; Patel, S. K. *JCS(P1)* **1981**, 2520.
14. Takaki, K.; Maeda, T.; Ishikawa, M. *JOC* **1989**, *54*, 58.
15. Crump, R. A.; Fleming, I.; Hill, J. H. M.; Parker, D.; Reddy, N. L.; Waterson, D. *JCS(P1)* **1992**, 3277.
16. Fleming, I.; Kilburn, J. D. *JCS(P1)* **1992**, 3295.
17. Fleming I.; Gil, S.; Sarkar, A. K.; Schmidlin, T. *JCS(P1)* **1992**, 3351.
18. Fleming, I.; Kindon, N. D. *CC* **1987**, 1177.
19. Oppolzer, W.; Mills, R. J.; Pachinger, W.; Stevenson, T. *HCA* **1986**, *69*, 1542.
20. Fleming, I.; Newton, T. W.; Roessler, F. *JCS(P1)* **1981**, 2527.
21. (a) Fleming, I.; Rowley, M.; Cuadrado, P.; González-Nogal, A. M.; Pulido, F. J. *T* **1989**, *45*, 413. (b) Morizawa, Y.; Oda, H.; Oshima, K.; Nozaki, H. *TL* **1984**, *25*, 1163.
22. Singh, S. M.; Oehlschlager, A. C. *CJC* **1991**, *69*, 1872.
23. Fleming, I.; Landais, Y.; Raithby, P. R. *JCS(P1)* **1991**, 715.
24. Fleming, I.; Higgins, D.; Lawrence, N. J.; Thomas, A. P. *JCS(P1)* **1992**, 3331.
25. Fleming, I.; Terrett, N. K. *JOM* **1984**, *264*, 99.
26. Fleming, I.; Marchi, D. *S* **1981**, 560.
27. Fleming, I.; Dunoguès, J.; Smithers, R. *OR* **1989**, *37*, 575.
28. Smith, J. G.; Drozda, S. E.; Petraglia, S. P.; Quinn, N. R.; Rice, E. M.; Taylor, B. S.; Viswanathan, M. *JOC* **1984**, *49*, 4112.
29. (a) Singer, R. D.; Oehlschlager, A. C. *JOC* **1991**, *56*, 3510. (b) Fürstner, A.; Weidmann, H. *JOM* **1988**, *354*, 15.
30. Brook, A. G.; Harris, J. W.; Lennon, J.; Sheikh, M. E. *JACS* **1979**, *101*, 83.
31. Fleming, I.; Takaki, K.; Thomas, A. *JCS(P1)* **1987**, 2269.
32. Casarini, A.; Jousseaume, B.; Lazzari, D.; Porciatti, E.; Reginato, G.; Ricci, A.; Seconi, G. *SL* **1992**, 981.
33. Nativi, C.; Ricci, A.; Taddei, M. *TL* **1990**, *31*, 2637.

34. Audia, J. E.; Marshall, J. A. *SC* **1983**, *13*, 531.

35. (a) Lautens, M.; Belter, R. K.; Lough, A. J. *JOC* **1992**, *57*, 422. (b) Lautens, M.; Ma, S.; Belter, R. K.; Chiu, P.; Leschziner, A. *JOC* **1992**, *57*, 4065.

Ian Fleming
Cambridge University, UK

Lithium Bis(1-ethoxyvinyl)cuprate[1,2]

[123206-02-0] $C_8H_{14}CuLiO_2$ (MW 212.71)

(vinylcuprate reagent and acyl anion equivalent which undergoes 1,4-addition reactions,[3] 1,2-addition reactions,[4] and substitution reactions[3])

Physical Data: homogenous, red solution displaying some temperature sensitivity; t>dec. estimated between −35 and −25 °C.[3]

Solubility: sol THF.

Preparative Methods: prepared under an inert atmosphere of argon by the rapid addition of 2 equiv of ***1-Ethoxyvinyllithium***[5] in THF at −78 °C to a suspension of purified ***Copper(I) Iodide*** at −78 °C, which is stirred at −30 °C for 1 h to afford a deep red solution.[3]

Handling, Storage, and Precautions: air- and moisture-sensitive reagent prepared in situ. Use in a fume hood.

Addition Reactions. Lithium bis(1-ethoxyvinyl)cuprate effects transfer of the 1-ethoxyvinyl ligand, an acyl anion equivalent,[6] via conjugate addition to α,β-unsaturated ketones[7] (25–91% yield)[3] (eqs 1 and 2). This reaction is sensitive to steric hindrance resulting from substitution in the α,β-enone, especially at the β-position, resulting in slow 1,4-addition. Use of excess reagent can sometimes overcome this limitation. The *n*-propynyl mixed cuprate reagent can be used effectively to give conjugate transfer of the 1-ethoxyvinyl ligand to 2-cyclohexen-1-one (83% yield).[3]

$[H_2C{=}C(OEt)]_2CuLi$; THF, −78 to 0 °C; 67% (1)

$[H_2C{=}C(OEt)]_2CuLi$; 82% (2)

Employment of ***Boron Trifluoride Etherate*** in conjunction with lithium bis(1-ethoxyvinyl)cuprate gives an enhancement of reagent efficiency as well as improved control over the regioselectivity in the reaction, preventing 1,2-addition (eq 3).[8]

MOMO, H, TBDMSO; $[H_2C{=}C(OEt)]_2CuLi$; $BF_3{\cdot}OEt_2$, THF; 76%; MOMO, H, OEt, TBDMSO (3)

A lithium bis(1-ethoxyvinyl)cyanocuprate reagent can be prepared by the transmetalation of the vinylstannane with lithium dimethylcyanocuprate to give effective conjugate addition of the 1-ethoxyvinyl ligand to an α,β-enone.[9] The resulting vinyl ether 1,4-adducts can be readily hydrolyzed under mildly acidic conditions (0.1 N HCl/ether, 0.1 N oxalic acid/MeOH, or SiO_2 in wet benzene) to afford 1,4-diketones in 57–95% yield.[3]

Chemoselective 1,2-addition of lithium bis(1-ethoxyvinyl)cuprate to an aldehyde can be achieved in moderate yield (eq 4).[4]

CHO, H, SO_2Ph, N H; 1. $[H_2C{=}C(OEt)]_2CuLi$, THF, ether, −78 °C; 2. $ClCO_2Me$, py, 0 °C; 42%; OCO_2Me, OEt, H, SO_2Ph, N H (4)

Substitution Reactions. Coupling of lithium bis(1-ethoxyvinyl)cuprate with allylic bromides occurs in good yield (eq 5).[3,10] However, this reagent fails to give coupling products with saturated primary and secondary alkyl bromides, affording instead cuprate reagent self-coupling products; attempted opening of oxiranes instead was also unsuccessful.[3]

Br; 5 equiv $[H_2C{=}C(OEt)]_2CuLi$; THF, −78 to 0 °C; 74%; OEt (5)

Related Reagents. 1-Ethoxyvinyllithium; Lithium (Z)-Bis(2-ethoxyvinyl)cuprate; Lithium Bis(1-methoxyvinyl)cuprate; Lithium Divinylcuprate; 1-Methoxyvinyllithium.

---

1. Lipshutz, B. H.; Sengupta, S. *OR* **1992**, *41*, 135.

2. *FF* **1977**, *6*, 204.

3. (a) Boeckman, R. K., Jr.; Bruza, K. J.; Baldwin, J. E.; Lever, O. W., Jr. *CC* **1975**, 519. (b) Boeckman, R. K., Jr.; Bruza, K. J. *JOC* **1979**, *44*, 4781.

4. Trost, B. M.; Ohmori, M.; Boyd, S. A.; Okawara, H.; Brickner, S. J. *JACS* **1989**, *111*, 8281.

5. (a) Schöllkopf, U.; Hänle, P. *LA* **1972**, *763*, 208. (b) Baldwin, J. E.; Höfle, G. A.; Lever, O. W., Jr. *JACS* **1974**, *96*, 7125. (c) Gschwend, H. W.; Rodriguez, H. R. *OR* **1979**, *26*, 1.

6. Lever, O. W., Jr. *T* **1976**, *32*, 1943.
7. Kotick, M. P.; Leland, D. L.; Polazzi, J. O.; Schut, R. N. *JMC* **1980**, *23*, 166.
8. Fujiwara, S.; Smith, A. B., III *TL* **1992**, *33*, 1185.
9. Behling, J. R.; Babiak, K. A.; Ng, J. S.; Campbell, A. L.; Moretti, R.; Koerner, M.; Lipshutz, B. H. *JACS* **1988**, *110*, 2641.
10. Canonne, P.; Boulanger, R.; Angers, P. *TL* **1991**, *32*, 5861.

Christopher W. Alexander
Emory University, Atlanta, GA, USA

Robert K. Boeckman, Jr.
University of Rochester, NY, USA

Lithium Cyanide

[2408-36-8] CLiN (MW 32.96)

(synthesis of nitriles from halides, alcohols, aldehydes, ketones; synthesis of cyanohydrins; deoxygenation of ketones)

Physical Data: mp 160 °C, *d* 1.025 g cm^{-3}; hygroscopic solid.
Solubility: sol H_2O, DMF, THF.
Form Supplied in: solid or 0.5 M solution in DMF.
Preparative Methods: from ***Lithium*** metal and ***Silver(I) Cyanide***,[2] from lithium hydride and ***Hydrogen Cyanide***,[3] or, more conveniently, from ***Acetone Cyanohydrin*** and lithium hydride.[4]
Handling, Storage, and Precautions: moisture-sensitive; like ***Sodium Cyanide*** and ***Potassium Cyanide***, LiCN is highly toxic; use in a fume hood.

Nitriles (from Halides). Nitriles can be prepared from halides (Br or I) under nonaqueous conditions using LiCN in refluxing THF for 1 h (seven examples; >90% yields).[5] A simple alkyl chloride gave only a 50% yield after 18 h (one example).

Nitriles (from Alcohols). The quantitative conversion of a tosylate to a nitrile (one example) in refluxing THF has been reported.[5]

Manna et al. have reported the conversion of alcohols to nitriles using LiCN under Mitsunobu conditions (***Diethyl Azodicarboxylate*** and ***Triphenylphosphine***; two examples, 50% yields).[6]

β-Hydroxynitriles (from Epoxides). Ciaccio et al. have reported the preparation of β-hydroxynitriles from epoxides using LiCN in 60–88% yield (10 examples; eq 1).[7] Cycloalkene oxides gave exclusively the *trans*-hydroxynitriles. Monosubstituted epoxides were reported to give exclusive attack at the less substituted carbon. Acetals, alkenes, and *t*-butyldimethylsilyloxy groups were shown to be stable to the reaction conditions. After an aqueous extractive workup or passage through a plug of Florisil, the crude compounds are reported to be pure by TLC, IR, and ^1H and ^{13}C NMR analysis.

1. 3–4 equiv LiCN, THF, reflux; 2. H_2O; 60–88% (1)

Previously, in a synthesis of a dideoxynucleoside, Matsuda et al. reported the opening of an epoxide with LiCN in refluxing THF; however, LiCN in DMF gave intractable mixtures.[8]

Alkylsilyl Cyanides. LiCN is reported to be the method of choice for the preparation of alkylsilyl cyanides from the corresponding chlorides.[4,9] Both dialkylsilyl[9] and trialkylsilyl[4,9] cyanides have been reported.

Cyanohydrins. Yoneda et al. have reported the preparation of trialkylsilyl cyanohydrins and acyl cyanohydrins from aldehydes and ketones.[10,11] Addition of LiCN to a THF solution of the carbonyl compound and ***Benzoyl Chloride*** (nine examples),[10] ***Acetyl Chloride*** (five examples),[10] ***Chlorotrimethylsilane*** (six examples),[10] or ***t-Butyldimethylchlorosilane*** (eight examples),[11] followed by stirring at rt, results in the appropriate cyanohydrin (eq 2). Yields ranged from 47 to 100% (generally >80%; distillation or column chromatography).

$R^1COR^2 + R_3Cl$ —LiCN, THF, rt→ $R^1R^2C(CN)OR^3$ (2)

R^3 = TMS, TBDMS, PhCO, MeCO

Conjugate Additions. Treatment of 4-cholesten-3-one with lithium cyanide (THF, 16 h, reflux) gives 5-cyanocholestan-3-one (80%, 5α:5β-cyano = 3:1).[5]

Cyanophosphates. Aldehydes and ketones react rapidly with LiCN and ***Diethyl Phosphorocyanidate*** (DEPC) in THF to form cyanophosphates (eq 3).[12] These cyanophosphates can serve as intermediates in the synthesis of structurally diverse nitriles. These syntheses are discussed in the appropriate sections below.

R^1COR^2 —LiCN, $(EtO)_2P(O)CN$→ $(EtO)_2P(O)–O–C(CN)R^1R^2$ (3)

α,β-Unsaturated Nitriles. Several procedures have appeared for the preparation of α,β-unsaturated nitriles using LiCN as a reagent (e.g. eq 4; six examples; 61–94%; column chromatography).[12] A cyanophosphate when treated with ***Boron Trifluoride Etherate*** will undergo elimination to form the α,β-unsaturated nitrile. However, in the compounds examined, this elimination only occurs if R^1 is an aromatic group. If R^1 is an alkyl group, the cyanophosphate does not eliminate. This reaction has been used in the synthesis of lysergic acid and related compounds.[13,14]

$BF_3{\bullet}Et_2O$, rt, 2 h, benzene (4)

A second synthesis of α,β-unsaturated nitriles, also from ketones, has been reported (eq 5).[15] Reaction of a ketone to form the vinyl trifluoromethanesulfonate is followed by treatment of the triflate with LiCN, ***Tetrakis(triphenylphosphine)palladium(0)***, and ***12-Crown-4*** to give the α,β-unsaturated nitrile. This procedure may provide an alternative to the procedure of Harusawa, since an aromatic ketone is not required. Eight examples (all cycloalkyl ketones) were given, including one β-keto ester and two α,β-unsaturated ketones. Isolated yields (chromatography or distillation) were 59–87%.

LiCN, $Pd(PPh_3)_4$, 12-crown-4, benzene, rt (5)

A similar transformation on an *O*-triflyl imidate has been reported to occur without the need for transition metal catalysis (eq 6).[16]

$(TfO)_2O$, *i*-Pr_2NEt, CH_2Cl_2; LiCN, 12-crown-4, THF (6)

A synthesis of α-methylenenitriles has been reported (eq 7).[17] Treatment of an *N*-(*p*-tolylsulfonyl)vinylsulfoximine with LiCN in DMF for 1 h at rt gave 63–81% yields of the α-methylenenitriles (three examples).

LiCN, DMF, rt (7)

β,γ-Unsaturated Nitriles. Aldehydes and ketones can be transformed into β,γ-unsaturated nitriles.[18] Treatment of the cyanophosphate derived from the carbonyl compound with ***Samarium(II) Iodide*** in THF gives the β,γ-unsaturated nitrile in 60–94% yield (eq 8; column chromatography; 14 examples).

SmI_2 (8)

Nitriles (from Carbonyl Compounds). Cyanophosphates can be used to prepare nitriles by treating them with either lithium in liquid ammonia (seven examples; 72–91% yields; column chromatography)[19] or samarium iodide (12 examples; 61–95%; column chromatography) (eq 9).[18] The Li/NH_3 reduction should be carried out at −78 °C; when the reaction is done in refluxing NH_3, reduction of the cyanophosphate to a methylene group can be the major product (see below).[20]

SmI_2 or Li, NH_3 (9)

If one of the R groups is aromatic, then hydrogenation with a Pd catalyst will also give nitriles (two examples; 75, 81%).[21]

Deoxygenation of Ketones. Cyanophosphates have also been used for the deoxygenation of α,β-unsaturated ketones and aromatic ketones (eq 10; five examples; 80–99%; no chromatography) by treatment with lithium in liquid ammonia.[20] As noted above, this reaction should be carried out in refluxing NH_3; at lower temperatures, reduction to the nitrile predominates. The reduction to the alkene is reported to be free of double bond isomerization or migration.

Li, liq NH_3 (10)

Allenic Nitriles. Treatment of a cyanophosphate (derived from an α,β-alkynyl ketone) with a higher-order cuprate gives trisubstituted allenes (eq 11; three ketone and one aldehyde examples; five cuprate examples; 32–87%; column chromatography).[22]

cuprate (11)

Related Reagents. Acetone Cyanohydrin; Cyanotrimethylsilane; Diethylaluminum Cyanide; Hydrogen Cyanide; Potassium Cyanide; Sodium Cyanide.

1. (a) Wells, A. F. *Structural Inorganic Chemistry*, 4th ed.; Clarendon: Oxford, 1975; pp 749–751. (b) Holliday, A. K.; Hughes, G.; Walker, S. M. In *Comprehensive Inorganic Chemistry*; Trotman-Dickenson, A. F., Ed.; Pergamon: New York, 1973; Vol. 1, p 1245.
2. Rossmanith, K. *M* **1965**, *96*, 1690.
3. Evans, D. A.; Carroll, G. L.; Truesdale, L. K. *JOC* **1974**, *39*, 914.
4. Livinghouse, T. *OSC* **1990**, *7*, 517.
5. Harusawa, S.; Yoneda, R.; Omori, Y.; Kurihara, T. *TL* **1987**, *28*, 4189.
6. Manna, S.; Falck, J. R.; Mioskowski, C. *SC* **1985**, *15*, 663.
7. Ciaccio, J. A.; Stanescu, C.; Bontemps, J. *TL* **1992**, *33*, 1431.
8. Matsuda, A.; Satoh, M.; Nakashima, H.; Yamamoto, N.; Ueda, T. *H* **1988**, *27*, 2545.
9. Mai, K.; Patil, G. *JOC* **1986**, *51*, 3545.
10. Yoneda, R.; Santo, K.; Harusawa, S.; Kurihara, T. *S* **1986**, 1054.
11. Yoneda, R.; Hisakawa, H.; Harusawa, S.; Kurihara, T. *CPB* **1987**, *35*, 3850.
12. Harusawa, S.; Yoneda, R.; Kurihara, T.; Hamada, Y.; Shioiri, T. *TL* **1984**, *25*, 427.
13. Yoneda, R.; Terada, T.; Harusawa, S.; Kurihara, T. *H* **1985**, *23*, 557.
14. Kurihara, T.; Terada, T.; Harusawa, S.; Yoneda, R. *CPB* **1987**, *35*, 4793.

15. Piers, E.; Fleming, F. F. *CC* **1989**, 756.
16. Sisti, N. J.; Fowler, F. W.; Grierson, D. S. *SL* **1991**, 816.
17. Bailey, P. L.; Jackson, R. F. W. *TL* **1991**, *32*, 3119.
18. Yoneda, R.; Harusawa, S.; Kurihara, T. *JOC* **1991**, *56*, 1827.
19. Yoneda, R.; Osaki, T.; Harusawa, S.; Kurihara, T. *JCS(P1)* **1990**, 607.
20. Yoneda, R.; Osaki, H.; Harusawa, S.; Kurihara, T. *CPB* **1989**, *37*, 2817.
21. Harusawa, S.; Nakamura, S.; Yagi, S.; Kurihara, T.; Hamada, Y.; Shioiri, T. *SC* **1984**, *14*, 1365.
22. Yoneda, R.; Inagaki, N.; Harusawa, S.; Kurihara, T. *CPB* **1992**, *40*, 21.

Ronald H. Erickson
Scios Nova, Baltimore, MD, USA

Lithium Dimethylcuprate[1]

Me_2CuLi

[15681-48-8] C_2H_6CuLi (MW 100.57)

(methylating reagent; undergoes conjugate addition reactions,[1a,b,e,f] 1,2-addition reactions,[2] substitution reactions with alkyl, vinyl, and allyl substrates,[1a,c,e,f] carbocupration of alkynes,[1a] and reduction of carbon–hetf eroatom bonds[3])

Physical Data: colorless solution in THF or Et_2O; ^1H NMR (Et_2O) δ −1.46 to −1.55 ppm (at 25 to −70 °C);[4,5] ^7Li NMR δ 0.75 to −0.38 ppm (at 25 to −70 °C for Me_2CuLi) and δ 0.495 to −0.38 ppm (at 25 to −70 °C for $Me_2CuLi + LiI$);[5] X-ray scattering measurements;[4] molecular association data.[4,6]

Solubility: sol THF, Et_2O.

Preparative Methods: prepared in situ from Cu^I salts (***Copper(I) Iodide***, ***Copper(I) Bromide***, ***Copper(I) Trifluoromethanesulfonate***, ***Copper(I) Chloride***) and ***Methyllithium*** under N_2 or argon.[1a] Impurities present in CuI or CuBr can promote decomposition of the reagent. Pure, light tan CuI is obtained by dissolving crude CuI in boiling NaI(aq) followed by cooling, precipitation of CuI by addition of H_2O, filtration, sequentially washing the solid with H_2O, EtOH, EtOAc, Et_2O, and pentane, and then drying in vacuo for 24 h.[7] CuBr can be purified in a similar manner[8] with NaBr (aq) or by dissolving in 48% HBr, precipitating with H_2O, sequentially washing the solid with H_2O, EtOH, and Et_2O, and drying in vacuo.[9] CuI has also been purified by continuous extraction with THF in a Soxhlet extractor (12 h) followed by drying in vacuo.[10] Me_2CuLi free of LiI can be prepared from isolated ***Methylcopper***,[8] and the reagent can also be prepared in the presence of complexing agents from $CuI\cdot SBu_2$,[8] ***Copper(I) Iodide-Tributylphosphine***,[8] or $CuBr\cdot SMe_2$.[11] Polymer-supported Me_2CuLi is also available.[12]

Handling, Storage, and Precautions: air- and moisture-sensitive; handle under inert atmosphere in a fume hood.

Organocuprate Reagents. Organocuprate reactivity and thermal stability are functions of Cu^I precursor,[1a,11b,13] solvent,[1,14] temperature,[1b] cuprate composition,[5,11b] cation (e.g. Li,[1a] Mg,[1a] Zn,[15a] Mn[15b]), and alkyl[1b] ligand. Experimental evidence for ligand and metal lability has been interpreted in terms of a dynamic equilibrium between Me_2CuLi, MeLi, and Me_3Cu_2Li.[5] Me_2CuLi is formulated as a dimer, and since cuprate reactivity is a function of composition (e.g. Me_3Cu_2Li, $Me_5Cu_3Li_2$[16]), stoichiometric considerations are important in preparing the reagent. For any given application the reactivity profile of Me_2CuLi should be compared with a range of mixed homocuprates (e.g. $MeCuC{\equiv}CRLi$, ***Lithium Cyano(methyl)cuprate***, and $Me_2CuCNLi_2$ or $Me_2CuCN\cdot LiCN$),[17] mixed heteroatom alkylcuprates (i.e. amido,[10,18] phosphido,[18] thioxy,[10] and alkoxy[10]), and methylcopper in the presence of additives (e.g. TMEDA/TMSCl,[19] TMSI,[20] and TMSCl + LiI[20]).

Conjugate Addition Reactions. Me_2CuLi transfers the methyl ligand in a 1,4-fashion to a wide range of α,β-unsaturated substrates which include enones,[1,21,22b] enals,[22] enoates,[1,14d,23] ynones,[24a] ynoates,[24b] and α,β-unsaturated lactones,[25] imides,[26] and phosphonates.[27] α,β-Alkenyl sulfoxides,[28] sulfones,[29] and phosphine oxides,[30] α,β-allenyl phosphine oxides,[31a] sulfoxides, sulfones,[31b] and ketones,[31c,d] ketoketenimines,[31e] and α,β-alkynyl sulfones[32] also undergo 1,4-addition of methyl upon reaction with Me_2CuLi. These reactions proceed most readily in nonpolar solvents (Et_2O, PhMe, CH_2Cl_2, pentane), since good donor solvents (e.g. THF, DME, HMPA) diminish cuprate reactivity toward conjugate addition.[1,14] Less reactive substrates can often be activated toward conjugate addition by use of Lewis acid additives[1] and by solvent modifications (eq 1[14d]).[14] Utilization of low temperatures and nonpolar solvents[22] gives, with the more reactive enals,[22] better ratios of 1,4-/1,2-addition products, and $Me_5Cu_3Li_2$[16] is generally more effective than Me_2CuLi (see ***Dilithium Pentamethyltricuprate***).

OMe, O, NMe_2 — 2 equiv Me_2CuLi, 0 °C, 60–90 min → OMe, O, NMe_2 (1)

Solvent	% Yield	% de
THF	<1	–
Diethyl ether	18	68
Diethyl ether, TMSI	83	0
Toluene	91	82
Dichloromethane	94	69–79

Conformational (eq 2)[33] and steric factors often control diastereoselectivity, although electronic[34] effects have been observed and stereodivergent pathways (eq 3)[35a] can sometimes be achieved by use of Lewis acid additives. Stereocontrol in conjugate addition to γ-heteroatom substituted α,β-unsaturated substrates is a function of heteroatom identity and substitution patterns (eq 4).[35b]

O, O — Me_2CuLi, Et_2O, 0 °C, 55% → O, O (2)

cis:*trans* = 80:20

Me_2CuLi, THF (3)

–78 to 0 °C	61%	8:92
TMSCl, –78 °C	70%	99:1

Bn_2N ... CO_2Et, R; Me_2CuLi, –78 °C, THF, 20–24 h → (4)

R			
R = H	TMSCl	80%	94:6
R = CO_2Et	–	68%	21:79

The conjugate addition reaction is compatible with alkyl bromide[36] and nitroalkane (eq 5)[37] functionality, although substitution of mesylates is competitive[36] with conjugate addition. Substrates with extended conjugation generally afford products of 1,6- or 1,8-addition.[38] Reaction of Me_2CuLi with ynenoates gives nonconjugated allenyl esters upon careful protonation of the enolate (eq 6).[38c] Bis-activation (e.g. alkylidenemalonates (eq 4), perhydro-1,4-oxazepine-5,7-diones,[39a] malononitriles,[39b] α-keto sulfoxides,[39c] and 2-acyl- or 2-alkoxycarbonyl-4-chromones[39d]) of the alkene facilitates conjugate addition, allowing participation of normally unreactive functionality (e.g. α,β-unsaturated nitriles and 4-chromones).

Me_2CuLi (1.1 equiv), TMSCl (2 equiv), –78 °C, 10 min, 90% (5)

1. Me_2CuLi, Et_2O, –20 °C, 1 h; 2. 2 N H_2SO_4, 25 °C, 75% (6)

Transition metal–alkene complexes undergo conjugate addition (eq 7),[40a] and Me_2CuLi and MeLi can effect different[40b] regioselectivities. Reaction at metal carbonyls can be a side reaction, and the efficiency of cuprate conjugate addition is dependent upon the ligands attached to the metal.[40b]

Me_2CuLi (2 equiv), THF, –78 °C, 91%; > 99% de (7)

The enolate resulting from conjugate addition can be trapped with a variety of electrophiles (e.g. alkyl halides, halosilanes (eq 5), acid chlorides, formaldehyde, sulfenyl halides, selenenyl halides, aldehydes, Mannich salts, triflating agents, CO_2, and chlorophosphates), although solvent modifications and use of additives (e.g. HMPA, ***Zinc Chloride***) are sometimes required.[41] The wide range and variability of reaction conditions required for successful tandem conjugate addition–enolate trapping reflect the complexity of these reaction mixtures and the generally diminished reactivity of the enolate, although enolates generated by conjugate addition of Me_2CuLi to γ-methylenebutenolides[42a] and simple butenolides[42b] participate in conjugate addition reactions with the starting lactone. Although acylation of these enolates is often problematic,[43a] *O*- vs. *C*-acylation can be controlled by use of ***Methyl Cyanoformate***[43b] in either Et_2O or THF (eq 8). A powerful application of cuprate 1,4-addition lies in the generation of regiospecific enolates which can undergo subsequent chemistry such as Dieckmann cyclization (eq 9)[23a] or Claisen rearrangement (eq 10).[44]

1. Me_2CuLi; 2. $NCCO_2Me$ (8)

THF	89%	95:5
Et_2O	84%	7:93

MeO_2C ... CO_2Me; Me_2CuLi, –25 °C, 30 min, 76% (9)

Me_2CuLi, THF, –78 to 0 °C, 100% (10)

α,β-Unsaturated substrates containing a good leaving group (e.g. halide, alkoxycarbonyl, sulfonate esters, phosphonate esters, alkyl- or arylthioxy, alkoxy) on the β-carbon atom undergo substitution reactions,[45] and tandem addition–elimination–addition pathways[46] can be designed (eq 11)[46b] for substrates with an α-CH_2L substituent; ***Chlorotrimethylsilane*** is necessary to prevent elimination of the nitro group in the reaction of Me_2CuLi with an α-nitromethylenone[37] (eq 5).

1. $ZnBr_2$ (1.2 equiv), THF, 25 °C, 10 min; 2. Me_2CuLi (2 equiv), hexane, –78 to 0 °C, 1 h; 3. aq NH_4Cl; 84%; 90% ee (11)

Several scalemic substrates[26,37,39a,47] (eqs 11 and 12[47c]) afford good to excellent asymmetric induction in the conjugate addition reaction, while more modest ee's have been achieved with chiral coordinating[47a] ligands.

Me_2CuLi, Et_2O, 0 °C, 85%; 94% ee (12)

Although simple cyclopropyl ketones do not react with Me_2CuLi, 1,1-bis-activated cyclopropanes (e.g. cyclopropylmalonates,[48a] β-keto esters,[48a] β-keto phosphonates[48b]) undergo a homoconjugate addition reaction (eq 13),[48a] and methylenecyclopropyl ketones[48c] participate in 1,5-additions.

Me_2CuLi (1.5 equiv), −30 to 0 °C (13)

X = O 90%
X = CH_2 –

Carbonyl Additions. 1,2-Addition of Me_2CuLi to ketones is generally limited to alkyl aryl[49a] and diaryl[49b] ketones, although an α,α′-dialkoxy ketone (eq 14)[49c] gave good yields of the 1,2-addition product. Although enals preferentially react in a 1,4-fashion, 1,2-addition is often a serious side reaction and is favored by addition of TMSCl (which also facilitates 1,2-addition to ketones[2]); this additive also increases Cram stereoselectivity.[2,50]

Me_2CuLi, Et_2O, −78 °C, 1 h, 80% (14)

76% de

Carbocupration. Carbocupration of alkynes with Me_2CuLi appears limited to propynal diethyl acetal;[51a] the reaction fails with a higher homolog. Carbocuprations of 1-methoxyallene[51a] and of cyclopropene derivatives (eq 15)[51b] have been reported.

Me_2CuLi, THF, DME, −70 °C, 89%; 92% de; PhCOCl, >76% (15)

Substitution Reactions. Me_2CuLi participates in S_N2-type substitution reactions[1a,c,e] with alkyl halides[8,52] (e.g. α-halo ketones[52a] and β-alkoxy-β-OLi intermediates[52b]), sulfonate esters,[53] oxiranes (epoxides),[54] tosylated aziridines,[55a] oxetanes,[55b] cyclic sulfinates,[56a] trifluoromethyl sulfonimides,[56b] β-lactones,[57] and with vinyl halides,[58a] triflates,[58b,c] and selenones.[58d] 1,1-Dibromocyclopropanes[59] can undergo a double substitution reaction with Me_2CuLi. Proximate heteroatoms facilitate the reaction of secondary tosylates.[53b] Alkyl substrates generally react with inversion of configuration, while substitution on alkenyl substrates (eq 16)[58a] proceeds with retention of configuration. Although Me_2CuLi reacts with iodobenzene, it is unreactive towards aryl triflates and bromides and thus the reaction[8] is unsatisfactory for effecting aryl substitution.

Me_2CuLi, diglyme, −30 to 25 °C, 72% (16)

Regioselective cleavage of oxiranes is a powerful synthetic method.[1a,c,e] Me_2CuLi cleaves epoxides at the least substituted center; 1,2-disubstituted epoxides afford mixtures of regioisomers. Regiocontrol is possible in 2,3-epoxy acids,[54a] esters,[54b,c] and 4-alkyl-substituted 2,3-epoxy alcohols,[54d] and regiocontrol is enhanced by solvent effects with 2,3-epoxy alcohols[54e] lacking a C-4 substituent.

Allylic substrates react with (S_N2') or without (S_N2) rearrangement via a π-allyl copper complex, with the *anti* S_N2' pathway predominating. Regio- and stereochemistry is generally governed by steric factors, although stereoelectronic effects may also operate.[1a] Allylic carbamates display high *syn* S_N2' stereo- and regioselectivity (eq 17[60a]).[60] Propargyl substrates uniformly give S_N2' substitution.[1a] Vinyloxetanes,[61] vinyloxiranes (eq 18),[62a] and allylic sulfoxides and sulfones[62b] undergo S_N2' substitutions, and the vinyloxiranes react with good diastereoselectivity.

Me_2CuLi (2 equiv), Et_2O, 0 °C, 80–90% (17)

96% de

Me_2CuLi, THF, Et_2O, 0 °C, 12 h, 88% (18)

12:88

Reaction of Me_2CuLi with acyl halides[1c] or thiol esters[63] results in nucleophilic acyl substitution, providing a useful route to ketones. Substitution reactions also occur between Me_2CuLi and sulfonium salts,[64] and via benzyne[65] intermediates.

Miscellaneous Reactions. The reagent adds to 1-acylpyridinium salts[66] and to transition metal carbene complexes.[67] Me_2CuLi can reduce α-halo[3a] and α-acetoxy ketones[3a] and α,β-epoxy[3b] ketones. Reduction of α,β-epoxy sulfoxides affords enolates.[68] Me_2CuLi reacts with α,α′-dibromo ketones to afford α-methyl ketones via a cyclopropanone intermediate (eq 19).[36a,68] Reaction of Me_2CuLi with α′-nucleofuge α,β-enones leads to mixtures of reduction and substitution products.[70] Chiral allenes have been prepared by reaction of Me_2CuLi with allylic sulfinyl mesylates.[71]

Me_2CuLi (5 equiv), Et_2O, −78 °C, 0.5 h, 70% (19)

Related Reagents. Dilithium Pentamethyltricuprate; Dilithium Trimethylcuprate; Lithium Cyano(methyl)cuprate; Lithium Diallylcuprate; Lithium Di-*n*-butylcuprate; Lithium Diethylcuprate; Lithium Diethylcuprate-Tributylphosphine; Lithium Dimethylcuprate–Boron Trifluoride; Lithium Diphenylcuprate; Lithium Di-*n*-propylcuprate; Lithium Divinylcuprate; Lithium Methyl(phenylthio)cuprate; Lithium Trimethylmanganate; Lithium Trimethylzincate; Lithium (3,3-Dimethyl-1-butynyl)methylcuprate; Methylcopper; Methylcopper–Boron Trifluoride Etherate; Methylcopper-Tributylphosphine.

---

1. (a) Lipshutz, B. H.; Sengupta, S. *OR* **1992**, *41*, 135. (b) Posner, G. H. *OR* **1972**, *19*, 1. (c) Posner, G. H. *OR* **1975**, *22*, 253. (d) Posner, G. H. *An Introduction to Synthesis Using Organocopper Reagents*; Wiley: New York, 1980. (e) Faust, J.; Froböse, R. In *Gmelin Handbook of Inorganic Chemistry*; Springer: Berlin, 1983; Copper, Part 2. (f) Perlmutter, P. *Conjugate Addition Reactions in Organic Synthesis*; Pergamon: Oxford, 1992.
2. Matsuzawa, S.; Isaka, M.; Nakamura, E.; Kuwajima, I. *TL* **1989**, *30*, 1975.
3. (a) Bull, J. R.; Tuinman, A. *TL* **1973**, 4349. (b) Bull, J. R.; Lachmann, H. H. *TL* **1973**, 3055.
4. Pearson, R. G.; Gregory, C. D. *JACS* **1976**, *98*, 4098.
5. Lipshutz, B. H.; Kozlowski, J. A.; Breneman, C. M. *JACS* **1985**, *107*, 3197.
6. Ashby, E. C.; Watkins, J. J. *JACS* **1977**, *99*, 5312.
7. Kauffman, G. B.; Teter, L. A. *Inorg. Synth.* **1963**, *7*, 9.
8. Whitesides, G. M.; Fischer, W. F., Jr.; San Filippo, J., Jr.; Bashe, R. W.; House, H. O. *JACS* **1969**, *91*, 4871.
9. Osterlof, J. *ACS* **1950**, *4*, 374.
10. Posner, G. H.; Whitten, C. E.; Sterling, J. J. *JACS* **1973**, *95*, 7788.
11. (a) House, H. O.; Chu, C.-Y.; Wilkins, J. M.; Umen, M. J. *JOC* **1975**, *40*, 1460. (b) Lipshutz, B. H.; Whitney, S.; Kozlowski, J. A.; Breneman, C. M. *TL* **1986**, *27*, 4273. (c) Wuts, P. G. M. *SC* **1981**, *11*, 139. (d) Theis, A. B.; Townsend, C. A. *SC* **1981**, *11*, 157.
12. Schwartz, R. H.; San Filippo, J., Jr. *JOC* **1979**, *44*, 2705.
13. Bertz, S. H.; Gibson, C. P.; Dabbagh, G. *TL* **1987**, *28*, 4251.
14. (a) House, H. O.; Wilkins, J. M. *JOC* **1978**, *43*, 2443. (b) Hallnemo, G.; Ullenius, C. *T* **1983**, *39*, 1621. (c) Christenson, B.; Ullenius, C.; Hakansson, M.; Jagner, S. *T* **1992**, *48*, 3623. (d) Christenson, B.; Hallnemo, G.; Ullenius, C. *T* **1991**, *47*, 4739.
15. (a) Knochel, P.; Yeh, M. C. P.; Berk, S. C.; Talbert, J. *JOC* **1988**, *53*, 2390. (b) Cahiez, G.; Alami, M. *T* **1989**, *45*, 4163.
16. Clive, D. L. J.; Farina, V.; Beaulieu, P. L. *JOC* **1982**, *47*, 2572.
17. (a) House, H. O.; Umen, M. J. *JOC* **1973**, *38*, 3893. (b) Corey, E. J.; Beames, D. J. *JACS* **1972**, *94*, 7210. (c) Corey, E. J.; Floyd, D.; Lipshutz, B. H. *JOC* **1978**, *43*, 3418. (d) Hamon, L.; Levisalles, J. *JOM* **1983**, *251*, 133.
18. (a) Bertz, S. H.; Dabbagh, G.; Villacorta, G. M. *JACS* **1982**, *104*, 5824. (b) Bertz, S. H.; Dabbagh, G. *CC* **1982**, 1030.
19. Johnson, C. R.; Marren, T. J. *TL* **1987**, *28*, 27.
20. Bergdahl, M.; Lindstedt, E.-L.; Nilsson, M.; Olsson, T. *T* **1988**, *44*, 2055.
21. Kraus, G. A.; Yi, P. *SC* **1988**, *18*, 473.
22. (a) Chuit, C.; Foulon, J. P.; Normant, J. F. *T* **1980**, *36*, 2305. (b) Kowalski, C. J.; Weber, A. E.; Fields, K. W. *JOC* **1982**, *47*, 5088.
23. (a) Nugent, W. A.; Hobbs, F. W., Jr. *JOC* **1983**, *48*, 5364. (b) Gustafsson, B.; Nilsson, M.; Ullenius, C. *ACS* **1977**, *31B*, 667. (c) Clark, R. D. *SC*, **1979**, *9*, 325.
24. (a) Degl'Innocenti, A.; Stucchi, E.; Capperucci, A.; Mordini, A.; Reginato, G.; Ricci, A. *SL* **1992**, 329. (b) Caine, D.; Smith, T. L., Jr. *SC* **1980**, *10*, 751.
25. (a) Mattes, H.; Hamada, K.; Benezra, C. *JMC* **1987**, *30*, 1948. (b) Roberts, R. A.; Schull, V.; Paquette, L. A. *JOC* **1983**, *48*, 2076. (c) Wurster, J. A.; Wilson, L. J.; Morin, G. T.; Liotta, D. *TL* **1992**, *33*, 5689.
26. Melnyk, O.; Stephan, E.; Pourcelot, G.; Cresson, P. *T* **1992**, *48*, 841.
27. Macomber, R. S.; Constantinides, I.; Garrett, G. *JOC* **1985**, *50*, 4711.
28. Sugihara, H.; Tanikaga, R.; Tanaka, K.; Kaji, A. *BCJ* **1978**, *51*, 655.
29. Posner, G. H.; Brunelle, D. J. *JOC* **1973**, *38*, 2747.
30. Pietrusiewicz, K. M.; Zablocka, M.; Monkiewicz, J. *JOC* **1984**, *49*, 1522.
31. (a) Berlan, J.; Koosha, K. *JOM* **1978**, *153*, 99. (b) Berlan, J.; Koosha, K. *JOM* **1978**, *153*, 107. (c) Berlan, J.; Battioni, J.-P.; Koosha, K. *JOM* **1978**, *152*, 359. (d) Berlan, J.; Battioni, J.-P.; Koosha, K. *BSF(2)* **1979**, 183. (e) de la Cal, M. T.; Cristobal, B. I.; Cuadrado, P.; Gonzalez, A. M.; Pulido, F. J. *SC* **1989**, *19*, 1039.
32. (a) Fiandanese, V.; Marchese, G.; Naso, F. *TL* **1978**, 5131. (b) Eisch, J. J.; Behrooz, M.; Dua, S. K. *JOM* **1985**, *285*, 121.
33. Still, W. C.; Galynker, I. *T* **1981**, *37*, 3981.
34. (a) Smith, A. B., III; Dunlap, N. K.; Sulikowski, G. A. *TL* **1988**, *29*, 439. (b) Smith, A. B., III; Trumper, P. K. *TL* **1988**, *29*, 443.
35. (a) Corey, E. J.; Boaz, N. W. *TL* **1985**, *26*, 6015. (b) Reetz, M. T.; Rohrig, D. *AG(E)* **1989**, *28*, 1706. (c) Hanessian, S.; Sumi, K. *S* **1991**, 1083.
36. (a) Posner, G. H.; Sterling, J. J.; Whitten, C. E.; Lentz, C. M.; Brunelle, D. J. *JACS* **1975**, *97*, 107. (b) Wenkert, E.; Wovkulich, P. M.; Pellicciari, R.; Ceccherelli, P. *JOC* **1977**, *42*, 1105.
37. Tamura, R.; Tamai, S.; Katayama, H.; Suzuki, H. *TL* **1989**, *30*, 3685.
38. (a) Marshall, J. A.; Ruden, R. A.; Hirsch, L. K.; Phillippe, M. *TL* **1971**, 3795. (b) Barbot, F.; Kadib-Elban, A.; Miginiac, Ph. *JOM* **1983**, *255*, 1. (c) Krause, N. *CB* **1990**, *123*, 2173. (d) Cheng, M.; Hulce, M. *JOC* **1990**, *55*, 964. (e) Schuster, D. I.; Wang, L.; van der Veen, J. M. *JACS* **1985**, *107*, 7045. (f) Kato, M.; Ouchi, A.; Yoshikoshi, A. *CL* **1983**, 1511.
39. (a) Mukaiyama, T.; Takeda, T.; Fujimoto, K. *BCJ* **1978**, *51*, 3368. (b) Baraldi, P. G.; Barco, A.; Benetti, S.; Pollini, G. P.; Polo, E.; Simoni, D. *JOC* **1985**, *50*, 23. (c) Posner, G. H. In *Asymmetric Synthesis*; Academic: New York, 1983; Vol. 2, pp 225–241. (d) Wallace, T. W. *TL* **1984**, *25*, 4299.
40. (a) Peng, T.-S.; Gladysz, J. A. *TL* **1990**, *31*, 4417. (b) Pearson, A. J.; Kole, S. L.; Ray, T. *JACS* **1984**, *106*, 6060. (c) Stephenson, G. R.; Howard, P. W.; Owen, D. A.; Whitehead, A. J. *CC* **1991**, 641.
41. (a) Taylor, R. J. K. *S* **1985**, 364. (b) Chapdelaine, M. J.; Hulce, M. *OR* **1990**, *38*, 225.
42. (a) Bigorra, J.; Font, J.; Jaime, C.; Ortuño, R. M.; Sanchez–Ferrando, F. *T* **1985**, *41*, 5577. (b) Ortuño, R. M.; Merce, R.; Font, J. *T* **1987**, *43*, 4497.
43. (a) Bernasconi, S.; Jommi, G.; Montanari, S.; Sisti, M. *G* **1987**, *117*, 125. (b) Crabtree, S. R.; Alex Chu, W. L.; Mander, L. N. *SL* **1990**, 169.
44. Koreeda, M.; Luengo, J. I. *JACS* **1985**, *107*, 5572.
45. (a) Dieter, R. K.; Silks, L. A., III *JOC* **1986**, *51*, 4687. (b) Sum, F.-W.; Weiler, L. *CJC* **1979**, *57*, 1431.
46. (a) Smith, A. B., III; Wexler, B. A.; Slade, J. S. *TL* **1980**, *21*, 3237. (b) Tamura, R.; Watabe, K.-i.; Katayama, H.; Suzuki, H.; Yamamoto, Y. *JOC* **1990**, *55*, 408.
47. (a) Rossiter, B. E.; Swingle, N. M. *CRV* **1992**, *92*, 771. (b) Fuji, K.; Tanaka, K.; Mizuchi, M.; Hosoi, S. *TL* **1991**, *32*, 7277. (c) Alexakis, A.; Sedrani, R.; Mangeney, P.; Normant, J. F. *TL* **1988**, *29*, 4411.
48. (a) Clark, R. D.; Heathcock, C. H. *TL* **1975**, 529. (b) Callant, P.; D'Haenens, L.; Vandewalle, M. *SC* **1984**, *14*, 155. (c) Thomas, E. W.; Szmuszkovicz, J. R. *JOC* **1990**, *55*, 6054.
49. (a) House, H. O.; Prabhu, A. V.; Wilkins, J. M.; Lee, L. F. *JOC* **1976**, *41*, 3067. (b) House, H. O.; Chu, C.-Y. *JOC* **1976**, *41*, 3083. (c) Carda, M.; Gonzalez, F.; Rodriguez, S.; Marco, J. A. *TA* **1992**, *3*, 1511.
50. Arai, M.; Nemoto, T.; Ohashi, Y.; Nakamura, E. *SL* **1992**, 309.
51. (a) Alexakis, A.; Normant, J. F. *S* **1985**, 72. (b) Isaka, M.; Nakamura, E. *JACS* **1990**, *112*, 7428.
52. (a) Dubois, J. E.; Lion, C.; Moulineau, C. *TL* **1971**, 177. (b) Barluenga, J.; Llavona, L.; Yus, M.; Concellon, J. M. *JCS(P1)* **1991**, 2890.

53. (a) Johnson, C. R.; Dutra, G. A. *JACS* **1973**, *95*, 7777. (b) Hanessian, S.; Thavonekham, B.; DeHoff, B. *JOC* **1989**, *54*, 5831.
54. (a) Chong, J. M.; Sharpless, K. B. *TL* **1985**, *26*, 4683. (b) Johnson, C. R.; Herr, R. W.; Wieland, D. M. *JOC* **1973**, *38*, 4263. (c) Hartman, B. C.; Livinghouse, T.; Rickborn, B. *JOC* **1973**, *38*, 4346. (d) Johnson, M. R.; Nakata, T.; Kishi, Y. *TL* **1979**, 4343. (e) Chong, J. M.; Cyr, D. R.; Mar, E. K. *TL* **1987**, *28*, 5009.
55. (a) Tanner, D.; Birgersson, C.; Dhaliwal, H. K. *TL* **1990**, *31*, 1903. (b) Welch, S. C.; Rao, A. S. C. P.; Lyon, J. T.; Assercq, J.-M. *JACS* **1983**, *105*, 252.
56. (a) Harpp, D. N.; Vines, S. M.; Montillier, J. P.; Chan, T. H. *JOC* **1976**, *41*, 3987. (b) Muller, P.; Phuong, N. T. M. *TL* **1978**, 4727.
57. Kawashima, M.; Sato, T.; Fujisawa, T. *T* **1989**, *45*, 403.
58. (a) Miller, J. A.; Leong, W.; Zweifel, G. *JOC* **1988**, *53*, 1839. (b) McMurry, J. E.; Scott, W. J. *TL* **1980**, *21*, 4313. (c) Bacigaluppo, J. A.; Colombo, M. I.; Zinczuk, J.; Ruveda, E. A. *SC* **1992**, *22*, 1973. (d) Shimizu, M.; Ando, R.; Kuwajima, I. *JOC* **1984**, *49*, 1230.
59. Kitatani, K.; Hiyama, T.; Nozaki, H. *BCJ* **1977**, *50*, 1600.
60. (a) Gallina, C.; Ciattini, P. G. *JACS* **1979**, *101*, 1035. (b) Goering, H. L.; Kanter, S. S.; Tseng, C. C. *JOC* **1983**, *48*, 715.
61. Larock, R. C.; Stolz-Dunn, S. K. *SL* **1990**, 341.
62. (a) Marshall, J. A.; Trometer, J. D.; Blough, B. E.; Crute, T. D. *JOC* **1988**, *53*, 4274. (b) Masaki, Y.; Sakuma, K.; Kaji, K. *CC* **1980**, 434.
63. Anderson, R. J.; Henrick, C. A.; Rosenblum, L. D. *JACS* **1974**, *96*, 3654.
64. Akiba, K.-Y.; Takee, K.; Shimizu, Y.; Ohkata, K. *JACS* **1986**, *108*, 6320.
65. Meyers, A. I.; Pansegrau, P. D. *CC* **1985**, 690.
66. Gosmini, R.; Mangeney, P.; Alexakis, A.; Commercon, M.; Normant, J.-F. *SL* **1991**, 111.
67. Casey, C. P.; Miles, W. H.; Tukada, H.; O'Connor, J. M. *JACS* **1982**, *104*, 3761.
68. Satoh, T.; Sugimoto, A.; Itoh, M.; Yamakawa, K. *BCJ* **1989**, *62*, 2942.
69. Posner, G. H.; Sterling, J. J. *JACS* **1973**, *95*, 3076.
70. Barbee, T. R.; Albizati, K. F. *JOC* **1991**, *56*, 6764.
71. de la Pradilla, R. F.; Rubio, M. B.; Marino, J. P.; Viso, A. *TL* **1992**, *33*, 4985.

R. Karl Dieter
Clemson University, SC, USA

Lithium Dimethylcuprate–Boron Trifluoride[1]

$Me_2CuLi–BF_3$

(Me_2CuLi)		
[15681-48-8]	C_2H_6CuLi	(MW 100.57)
(BF_3)		
[7637-07-2]	BF_3	(MW 67.81)
$(BF_3{\cdot}Et_2O)$		
[109-63-7]	$C_4H_{10}BF_3O$	(MW 141.95)

(methylating reagent that undergoes accelerated conjugate addition reactions,[1] 1,2-addition reactions,[2] substitution reactions with alkyl[1,3] and allylic substrates,[1,4] and reduction reactions[4])

Physical Data: clear solution in THF; ^1H NMR data for $Me_2CuLi + 2\ BF_3{\cdot}Et_2O$ at $-80\,°C$ is suggestive of a four-component structure: Me_2CuLi (δ −1.57 ppm) + Me_3Cu_2Li (δ −1.31 and −0.35 ppm) + $MeLi{\cdot}BF_3$ (δ 0.16 ppm) + BF_3.[1b]

Solubility: sol THF, Et_2O.

Preparative Methods: addition of 2 equiv of ***Boron Trifluoride Etherate*** to ***Lithium Dimethylcuprate*** prepared from Cu^I salts and ***Methyllithium*** at $-78\,°C$ gives a clear solution in THF; warming the solution from −78 to −60 °C over 10 min generates a solution of $Me_3Cu_2Li + BF_3 + MeLi{\cdot}BF_3$.[1b] The reagent can also be prepared by addition of ***Boron Trifluoride*** to Me_2CuLi, although use of $BF_3{\cdot}Et_2O$ is more convenient and affords identical results.[3b]

Handling, Storage, and Precautions: air- and moisture-sensitive; handle under inert atmosphere in a fume hood.

Addition Reactions. Addition of $BF_3{\cdot}Et_2O$ to solutions of Me_2CuLi affords a reagent and reaction conditions in which conjugate transfer of the methyl ligand to α,β-alkenyl ketones (eq 1[5b]),[5] enoates,[6] lactones[5b,7] and alkylidenemalonates[6a,8] is accelerated relative to Me_2CuLi. The reagents $MeCu{\cdot}BF_3$ and $Me_2Cu(CN)Li_2/BF_3$ can also be used, and the relative effectiveness of each reagent is generally substrate dependent. γ-Silyloxy- (eq 2)[6a] and γ-amino-α,β-enoates[6b] as well as γ-alkoxyalkylidenemalonates[6a] undergo conjugate addition reactions with these reagents with varying degrees of stereoselectivity and yield. The reagent has often been used on substrates with reduced reactivity resulting from substitution and steric factors. This reagent combination has been effectively used in a homoconjugate addition reaction with a cyclopropyl ketone (eq 3).[9]

Me_2CuLi, Et_2O; $BF_3{\cdot}Et_2O$; 93% — 99% *trans* (1)

$BF_3{\cdot}Et_2O$, Et_2O, −78 to −30 °C, 2 h (2)

MeCu	68:32	6%
Me_2CuLi	73:27	26%
$Me_2Cu(CN)Li_2$	92:8	92%

Me_2CuLi, Et_2O; $BF_3{\cdot}Et_2O$; −78 to 0 °C, 75 min; 45% (3)

1,2-Addition of $Me_2CuLi/BF_3{\cdot}Et_2O$ to chiral imines provides a synthetic route to chiral amines that is stereodivergent relative to the 1,2-addition reactions of organolithium and organocerium reagents (eq 4).[2]

(4)

Me_2CuLi, BF_3 (from CuI)	14:86	52%
Me_2CuLi, BF_3 (from $CuBr{\bullet}SMe_2$)	10:90	56%
MeLi	95:5	66%
$MeCeCl_2$	97:3	65%

R = $C_{11}H_{23}$ $(CH_2)_3$- (dioxolane)

Substitution Reactions. Addition of $BF_3{\cdot}Et_2O$ to solutions of Me_2CuLi provides a reagent and reaction conditions that can effect substitution reactions on substrates, such as aziridines (eq 5)[10] and acetals,[3] that are normally unreactive toward Me_2CuLi alone. Although ***Titanium(IV) Chloride*** can also be employed, superior yields are generally obtained with $BF_3{\cdot}Et_2O$ (eq 6).[3b] A β-bromobutenolide undergoes a substitution reaction with varying combinations of Me_2CuLi, MeLi, and $BF_3{\cdot}Et_2O$ (eq 7).[11]

Me_2CuLi (6 equiv), $BF_3{\bullet}Et_2O$ (3 equiv), THF → $RNHCH_2CH_2Me$ (5)

R = $PhCH_2$	80%
R = $(4\text{-}MeOC_6H_4)_2CH$	97%

3 equiv reagent, Et_2O, –78 to –50 °C, 30 min (6)

Me_2CuLi, BF_3	95%	67% de
$Me_2CuLi, TiCl_4$	48%	88% de
$MeCu, BF_3$	50%	72% de

1. THF, cuprate; 2. 3 N HCl, 0 °C (7)

Me_2CuLi (1.5 equiv), 0 °C, 2 h	26–29%
Me_2CuLi, $MeBF_3Li$ (1.5 equiv), 0 °C, 2 h	57%
$MeBF_3Li$ (2 equiv), then Me_2CuLi (2 equiv) 0 °C, 30 min	78%

Allylic acetals[12] and oxazolidines[13] undergo reaction with Me_2CuLi/BF_3 to afford mixtures of rearranged (S_N2') and unrearranged (S_N2) substitution products, while allylic acetals[12] afford, exclusively, substitution products via S_N2' pathways. Chiral allylic oxazolidines give substitution products with very modest asymmetric induction.[13] Good asymmetric induction has been achieved with chiral γ-sulfonyloxy-α,β-en oates, which undergo preferential allylic substitution (eq 8).[4]

THF, Et_2O, –78 °C, 30 min (8)

Me_2CuLi, BF_3	61%	>99% de	25%
$Me_2Cu(CN)Li_2$	78%	>99% de	11%
$Me_2Cu(CN)Li_2, BF_3$	96%	>99% de	-

Reduction Reactions. γ-Acyloxy-α,β-enoates[4] undergo reduction upon treatment with Me_2CuLi/BF_3, in contrast to the γ-alkoxy derivatives,[6a] which undergo conjugate addition, and the γ-mesyloxy derivatives,[4] which undergo allylic substitution. $BF_3{\cdot}Et_2O$ as an additive attenuates the electron transfer properties of Me_2CuLi.[14]

Related Reagents. Boron Trifluoride Etherate; Lithium Cyano(methyl)cuprate; Lithium Dimethylcuprate; Methylcopper; Methylcopper–Boron Trifluoride Etherate; Titanium(IV) Chloride.

1. (a) Lipshutz, B. H.; Sengupta, S. *OR* **1992**, *41*, 135. (b) Lipshutz, B. H.; Ellsworth, E. L.; Siahaan, T. J. *JACS* **1989**, *111*, 1351.
2. Ukaji, Y.; Watai, T.; Sumi, T.; Fujisawa, T. *CL* **1991**, 1555.
3. (a) Normant, J. F.; Alexakis, A.; Ghribi, A.; Mangeney, P. *T* **1989**, *45*, 507. (b) Ghribi, A.; Alexakis, A.; Normant, J. F. *TL* **1984**, *25*, 3083.
4. Ibuka, T.; Nakao, T.; Nishii, S.; Yamamoto, Y. *JACS* **1986**, *108*, 7420.
5. (a) Smith, A. B., III; Jerris, P. J. *JOC* **1982**, *47*, 1845. (b) Still, W. C.; Galynker, I. *T* **1981**, *37*, 3981. (c) Mehta, G.; Murthy, A. N.; Reddy, D. S.; Reddy, A. V. *JACS* **1986**, *108*, 3443. (d) Banik, B. K.; Chakraborti, A. K.; Ghatak, U. R. *JCR(S)* **1986**, 406. (e) Banik, B. K.; Ghosh, S.; Ghatak, U. R. *T* **1988**, *44*, 6947. (f) Cha, J. K.; Lewis, S. C. *TL* **1984**, *25*, 5263.
6. (a) Yamamoto, Y.; Chounan, Y.; Nishii, S.; Ibuka, T.; Kitahara, H. *JACS* **1992**, *114*, 7652. (b) Reetz, M. T.; Röhrig, D. *AG(E)* **1989**, *28*, 1706.
7. Still, W. C.; Galynker, I. *JACS* **1982**, *104*, 1774.
8. Ibuka, T.; Aoyagi, T.; Yamamoto, Y. *CPB* **1986**, *34*, 2417.
9. Adams, J.; Belley, M. *TL* **1986**, *27*, 2075.
10. Eis, M. J.; Ganem, B. *TL* **1985**, *26*, 1153.
11. Olsen, R. K.; Hennen, W. J.; Wardle, R. B. *JOC* **1982**, *47*, 4605.
12. Ghribi, A.; Alexakis, A.; Normant, J. F. *TL* **1984**, *25*, 3079.
13. Berlan, J.; Besace, Y.; Prat, D.; Pourcelot, G. *JOM* **1984**, *264*, 399.
14. Smith, R. A. J.; Vellekoop, A. S. *T* **1989**, *45*, 517.

R. Karl Dieter
Clemson University, SC, USA

Lithium Divinylcuprate[1]

$(CH_2{=}CH)_2CuLi$

[22903-99-7] C_4H_6CuLi (MW 124.59)

(vinylating reagent; undergoes conjugate addition reactions[1a,b,d] and substitution reactions[1a,c,d])

Physical Data: clear yellow (prep. from CuI·PBu_3[2a]) or light gray solution (prep. from CuBr·SMe_2)[2b]

Solubility: sol THF, Et_2O.

Preparative Methods: prepared in situ from Cu^I salts (***Copper(I) Iodide***,[2b] ***Copper(I) Bromide***·SMe_2[3,2b] ***Copper(I) Iodide-Tributylphosphine***[2a]) and ***Vinyllithium*** under N_2 or argon atmosphere (see ***Lithium Dimethylcuprate***).

Handling, Storage, and Precautions: air- and moisture-sensitive. Use in a fume hood.

Introduction. Lithium divinylcuprate displays the characteristic reactivity patterns of lithium diorganocuprates (see ***Lithium Di-n-butylcuprate***, ***Lithium Dimethylcuprate***, ***Lithium Diphenylcuprate***).

Addition Reactions. Lithium divinylcuprate reacts with α,β-alkenyl ketones,[1] esters (eq 1),[4] lactones,[5] nitro compounds,[6] and aldehydes[7] with conjugate transfer of the vinyl ligand (see ***Vinylcopper***). Addition of ***Chlorotrimethylsilane*** can accelerate the conjugate addition reaction and promote increased stereoselectivity (eq 2).[8]

2 equiv $[H_2C{=}CH]_2CuLi$, Et_2O, –35 °C (1)

(*E*) 78%
(*Z*) 65%

$[H_2C{=}CH]_2CuLi$, THF, –78 °C (2)

TMSCl	99:1
–	56:44

The reagent adds stereoselectively (*cis* addition) to α,β-alkynoates, affording 2,4-dienoates.[9] Dienyl esters exclusively afford 1,6-addition products.[10] Tandem conjugate addition–enolate trapping is a powerful synthetic method (eq 3).[11] The reagent adds to vinyltriphenylphosphonium bromides to provide phosphoranes which can generally be exploited in subsequent chemistry.[12] 2-Siloxypyrylium salts undergo 1,4-addition, leading to 4-substituted 4*H*-pyrans.[13] 1,1-Diactivated cyclopropanes undergo a homoconjugate addition reaction with $(CH_2{=}CH)_2CuLi$.[14]

1. $[H_2C{=}CH]_2CuLi$, Et_2O, THF; 2. TMS-substituted methyl vinyl ketone (3)

$n = 2$, 75%; $n = 1$, 50%

Substitution Reactions. Lithium divinylcuprate participates in substitution reactions with alkyl halides[1a,1c,15] as well as propargyl and allyl carboxylates.[3,16] Vinyl triflates derived from β-dicarbonyl compounds afford substitution products that may result from addition–elimination reaction pathways (eq 4).[17] *gem*-Dibromocyclopropanes give mono- and disubstituted products, depending upon the reaction temperature.[18] Allylic ammonium salts participate in *syn*-S_N2' pathways with good stereo- and regioselectivity.[19]

$[H_2C{=}CH]_2CuLi$, $BF_3{\cdot}OEt_2$, THF, –78 °C, 35% (4)

Miscellaneous Reactions. Alkylation of a tricarbonylcyclohexadienyliron salt with lithium divinylcuprate gives moderate yields.[20] Reaction of this reagent with oxiranes (eq 5; iodide from CuI),[21] 2-alkenyl oxetanes (eq 6),[22] activated cyclopropanes,[23] and aziridines[24] affords ring-opened products. Carbocuprations of cyclopropene derivatives have been reported.[25] Arylmercurials couple with this reagent, giving fair to good yields of cross-coupled products.[26]

THF, 2 h, –78 °C (5)

Reagent	Yield	Ratio
$[H_2C{=}CH]_2CuLi + 2\ BF_3$	93%	75:18
$[H_2C{=}CH]_3Cu_2Li + 2\ BF_3$	97%	95:2
$[H_2C{=}CH]_2CuLi$	28%	0:100
$[H_2C{=}CH]_3Cu_2Li$	28%	0:100
$[H_2C{=}CH]Cu + BF_3$	100%	72:28

$[H_2C{=}CH]_2CuLi$, Et_2O, –50 °C, 1 h, 48% (6)

(*E*):(*Z*) = 76:24

Related Reagents. Lithium Bis(1-ethoxyvinyl)cuprate; Lithium Bis(1-methoxyvinyl)cuprate; Lithium Diallylcuprate; Lithium Diisopropenylcuprate; Lithium Di-(*E*)-1-propenylcuprate; Lithium Divinylcuprate-Tributylphosphine; Vinylcopper; Vinyllithium; Vinylmagnesium Bromide; Vinylmagnesium Bromide-Copper(I) Iodide; Vinylmagnesium Bromide-Methylcopper; Vinylmagnesium Chloride–Copper(I) Chloride.

---

1. (a) Lipshutz, B. H.; Sengupta, S. *OR* **1992**, *41*, 135. (b) Posner, G. H. *OR* **1972**, *19*, 1. (c) Posner, G. H. *OR* **1975**, *22*, 253. (d) Faust, J.; Froböse, R. *Gmelin Handbook of Inorganic Chemistry*; Springer: Berlin, 1983; Copper, Part 2.

2. (a) Harmon, C. A.; Streitwieser, A., Jr. *JOC* **1973**, *38*, 549. (b) House, H. O.; Chu, C.-Y.; Wilkins, J. M.; Umen, M. J. *JOC* **1975**, *40*, 1460.
3. Brinkmeyer, R. S.; Macdonald, T. L. *CC* **1978**, 876.
4. Roush, W. R.; Lesur, B. M. *TL* **1983**, *24*, 2231.
5. (a) Wurster, J. A.; Wilson, L. J.; Morin, G. T.; Liotta, D. *TL* **1992**, *33*, 5689. (b) Iwai, K.; Kosugi, H.; Uda, H.; Kawai, M. *BCJ* **1977**, *50*, 242.
6. Baer, H. H.; Hanna, Z. *Carbohydr. Res.* **1980**, *85*, 136.
7. Roush, W. R.; Michaelides, M. R.; Tai, D. F.; Chong, W. K. M. *JACS* **1987**, *109*, 7575.
8. Corey, E. J.; Boaz, N. W. *TL* **1985**, *26*, 6015.
9. (a) Li, T.-t.; Wu, Y. L.; Walsgrove, T. C. *T* **1984**, *40*, 4701. (b) Li, T.-t.; Wu, Y. L. *JACS* **1981**, *103*, 7007. (c) Keck, G. E.; Nickell, D. G. *JACS* **1980**, *102*, 3632.
10. Corey, E. J.; Chen, R. H. K. *TL* **1973**, 1611.
11. (a) Zoretic, P. A.; Biggers, M. S.; Biggers, C. K.; Caspar, M. L. *SC* **1991**, *21*, 31. (b) Boeckman, R. K., Jr. *T* **1983**, *39*, 925.
12. Just, G.; O'Connor, B. *TL* **1985**, *26*, 1799.
13. Kume, T.; Iwasaki, H.; Yamamoto, Y.; Akiba, K.-y. *TL* **1987**, *28*, 6305.
14. Corey, E. J.; Fuchs, P. L. *JACS* **1972**, *94*, 4014.
15. (a) Erickson, S. D.; Still, W. C. *TL* **1990**, *31*, 4253. (b) Bajgrowicz, J. A.; El Hallaoui, A.; Jacquier, R.; Pigiere, C.; Viallefont, P. *TL* **1984**, *25*, 2759. (c) Bajgrowicz, J. A.; El Hallaoui, A.; Jacquier, R.; Pigiere, C.; Viallefont, P. *TL* **1984**, *25*, 2231. (d) Lipshutz, B. H.; Wilhelm, R. S. *JACS* **1981**, *103*, 7672. (e) Mathias, R.; Weyerstahl, P. *CB* **1979**, *112*, 3041.
16. Trost, B. M.; Tanigawa, Y. *JACS* **1979**, *101*, 4413.
17. Kant, J.; Sapino, C., Jr.; Baker, S. R. *TL* **1990**, *31*, 3389.
18. Kitatani, K.; Hiyama, T.; Nozaki, H. *BCJ* **1977**, *50*, 1600.
19. Hutchinson, D. K.; Fuchs, P. L. *JACS* **1985**, *107*, 6137.
20. Pearson, A. J. *AJC* **1977**, *30*, 345.
21. (a) Lipshutz, B. H.; Ellsworth, E. L.; Siahaan, T. J. *JACS* **1989**, *111*, 1351. (b) Corey, E. J.; Nicolaou, K. C.; Beames, D. J. *TL* **1974**, 2439.
22. Larock, R. C.; Stolz-Dunn, S. K. *SL* **1990**, 341.
23. Taber, D. F.; Amedio, J. C., Jr.; Raman, K. *JOC* **1988**, *53*, 2984.
24. Dureault, A.; Tranchepain, I.; Greck, C.; Depezay, J.-C. *TL* **1987**, *28*, 3341.
25. Nakamura, E.; Isaka, M.; Matsuzawa, S. *JACS* **1988**, *110*, 1297.
26. Larock, R. C.; Leach, D. R. *OM* **1982**, *1*, 74.

Shou-Yuan Lin & R. Karl Dieter
Clemson University, SC, USA

Maleic Anhydride[1]

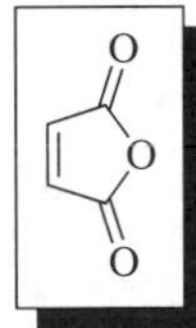

[108-31-6] $C_4H_2O_3$ (MW 98.06)

(reactive dienophile and dipolarophile; undergoes thermal and photochemical [2+2] cycloadditions; alkylating and acylating agent)

Alternate Name: MA.
Physical Data: mp 54–56 °C; bp 200 °C; *d* 1.314 g cm^{-3}.
Solubility: sol ether, acetone, chloroform, ethyl acetate, benzene, toluene, carbon tetrachloride, petroleum ether, dioxane.
Form Supplied in: white solid; widely available.
Drying: at 100 °C.
Purification: recrystallization from acetone/petroleum ether or from hot water. Alternatively sublimed.
Handling, Storage, and Precautions: toxic, an irritant, and corrosive. Avoid inhalation. Store in a tightly sealed brown bottle; avoid exposure to moisture.

Cycloadditions. Maleic anhydride has served as an excellent dienophile in a variety of Diels–Alder cycloadditions.[2] The use of purified maleic anhydride in the Diels–Alder reaction is recommended in order to remove trace amounts of maleic acid, which may lead to polymerization or diene isomerization. Of historical significance, the first [4+2] cycloaddition reported by Otto Diels and Kurt Alder employed maleic anhydride as a dienophile and butadiene as the diene component. Since the initial report, many variations on the cycloaddition process have been published. Reaction conditions vary from below rt to greater than 250 °C. As a representative example, the Diels–Alder reaction of ***Cyclopentadiene*** and MA proceeds at rt to provide the *endo* product in quantitative yield (eq 1).[3] Heterosubstituted dienes typically undergo *endo* selective cycloaddition reactions with maleic anhydride (eqs 2–4).[4] Maleic anhydride also reacts stereoselectively with *ortho*-quinodimethanes. For example, photoenolization of *o*-tolualdehyde stereoselectively generates an (*E*)-dienol, which is trapped by MA (eq 5).[5]

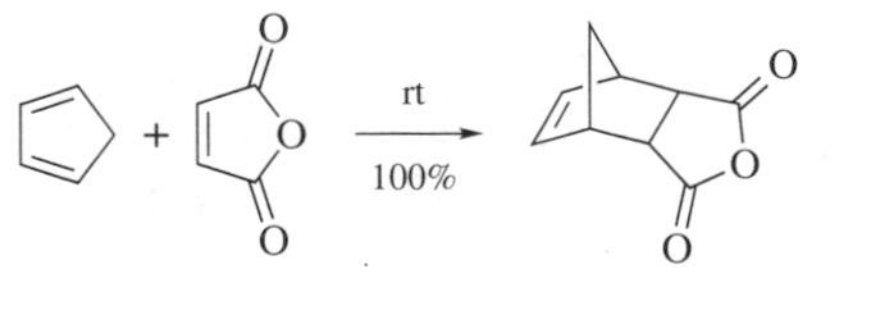

(1) rt, 100%

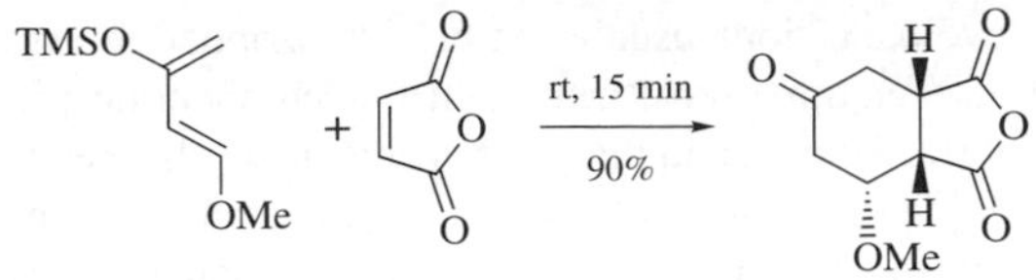

(2) rt, 15 min, 90%

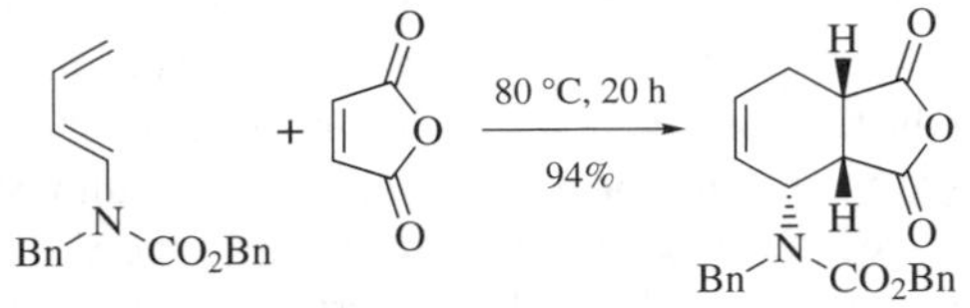

(3) 80 °C, 20 h, 94%

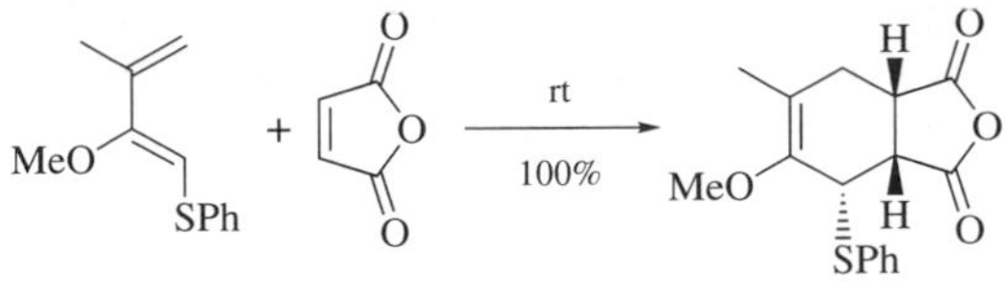

(4) rt, 100%

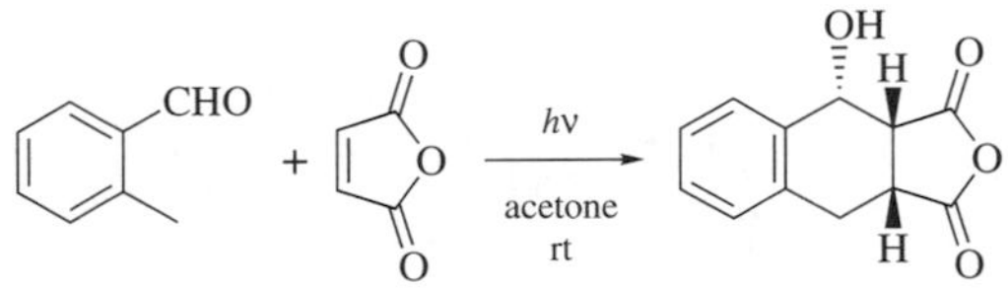

(5) *hν*, acetone, rt

A number of maleic anhydride derivatives have been examined in the Diels–Alder reaction. For example, the conjugate addition of thiophenol to maleic anhydride followed by chlorination and finally dehydrochlorination affords 2-(phenylthio)maleic anhydride in good overall yield (eq 6). 2-(Phenylthio)maleic anhydride reacts with cyclopentadiene at rt to provide the corresponding Diels–Alder adduct in 83% yield.[6] Maleic anhydride readily undergoes Diels–Alder reactions with furan. However, the Diels–Alder reaction of furan with dimethylmaleic anhydride fails due to the rapid reversion of the Diels–Alder adduct to starting material. On the other hand, the dihydrothiophene derivative of maleic anhydride reacts with furan at high pressure to produce a 20:80 mixture of *endo* and *exo* Diels–Alder adducts in good yield (eq 7).[7] Subsequent ***Raney Nickel*** desulfurization produces the dihydro version of the Diels–Alder adduct corresponding to the dimethylmaleic anhydride product.

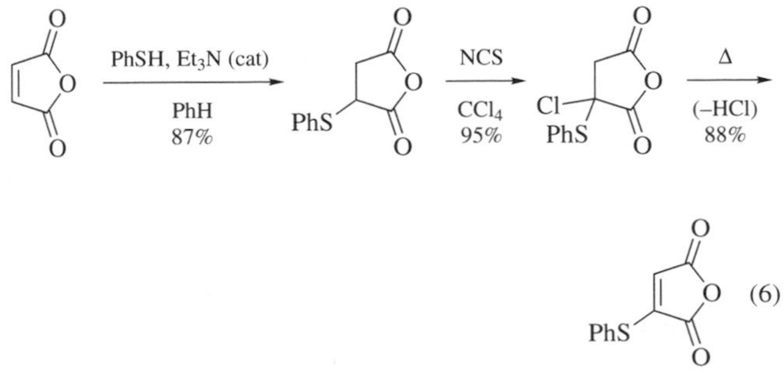

(6) PhSH, Et_3N (cat), PhH, 87%; NCS, CCl_4, 95%; Δ (–HCl), 88%

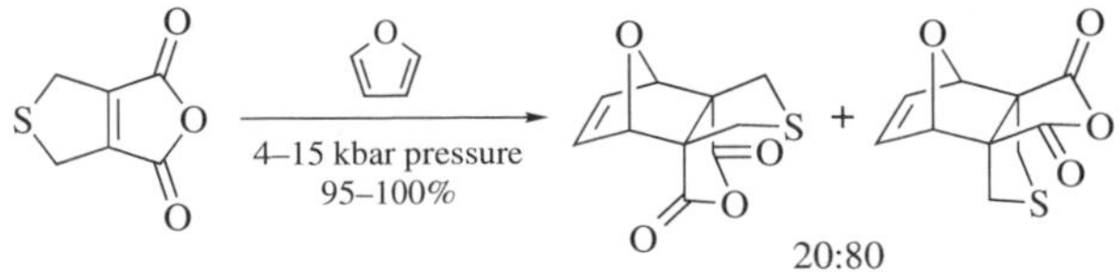

(7) furan, 4–15 kbar pressure, 95–100%; 20:80

Maleic anhydride undergoes [2+2] photocycloadditions with a variety of substrates. Aromatic species (eq 8),[8] isolated alkenes (eq 9),[9] and alkynes (eq 10)[10] readily undergo sensitized photochemical cycloadditions with maleic anhydride. Photochemical

[2+2] cycloadditions usually employ benzophenone as a sensitizer; however, other sensitizers such as acetophenone, propiophenone, benzaldehyde, and diacetylbenzene have also been utilized. For example, irradiation of a solution of benzene and maleic anhydride in the presence of benzophenone produces a 2:1 maleic anhydride/benzene adduct, which presumably arises from a photochemical [2+2] photocycloaddition followed by a Diels–Alder reaction of the resulting diene with a second equivalent of maleic anhydride (eq 8). Irradiation of a solution of maleic anhydride in acetone produces the Paterno–Büchi product in good yield (eq 11).[11] Maleic anhydride can also undergo thermal [2+2] cycloadditions. For example, thermolysis of a mixture of allene and maleic anhydride produces the corresponding cyclobutane in 22–26% yield (eq 12) (also see ***Dimethyl Maleate***).[12]

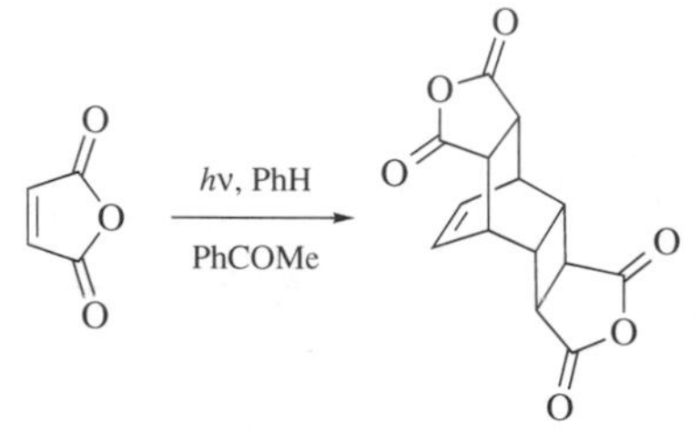

hν, PhH; PhCOMe (8)

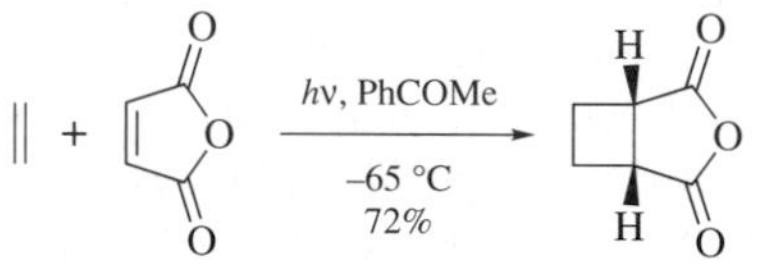

hν, PhCOMe; –65 °C; 72% (9)

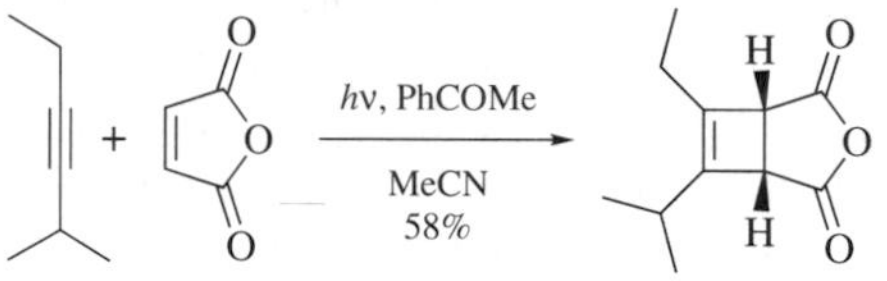

hν, PhCOMe; MeCN; 58% (10)

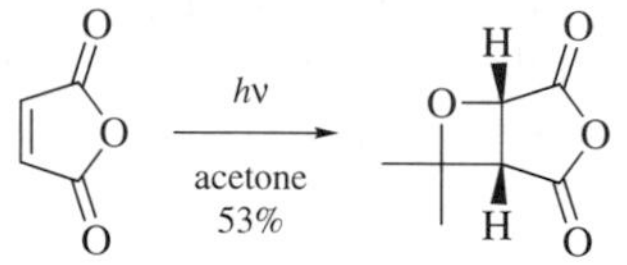

hν; acetone; 53% (11)

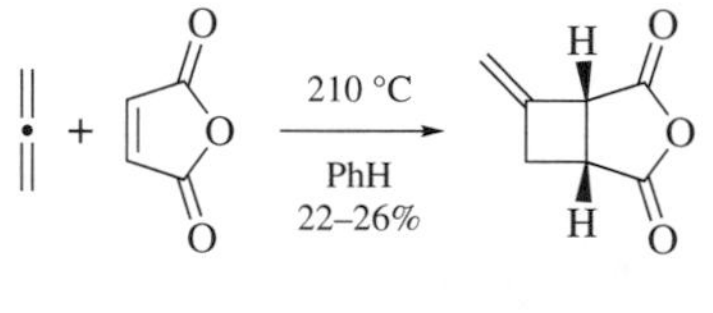

210 °C; PhH; 22–26% (12)

Maleic anhydride reacts with a variety of 1,3-dipoles to produce the corresponding cycloadducts in good to excellent chemical yield. For example, cycloadditions with nitrones,[13] diazo compounds,[14] nitrile oxide,[15] and azomethine ylides[16] have been reported (eqs 13–16).

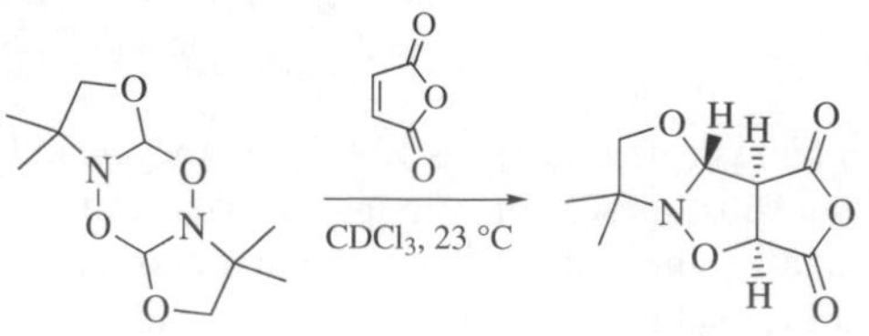

$CDCl_3$, 23 °C (13)

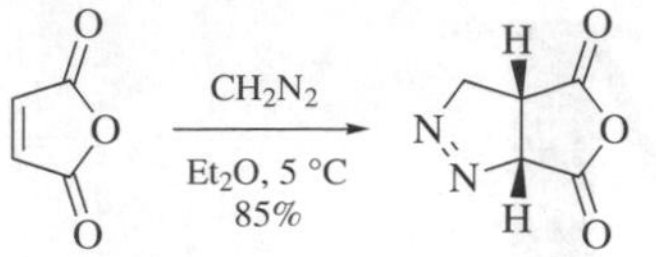

CH_2N_2; Et_2O, 5 °C; 85% (14)

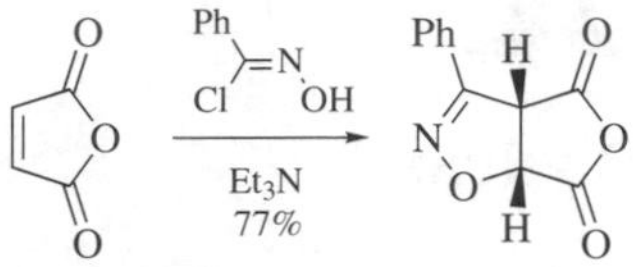

Et$_3$N; 77% (15)

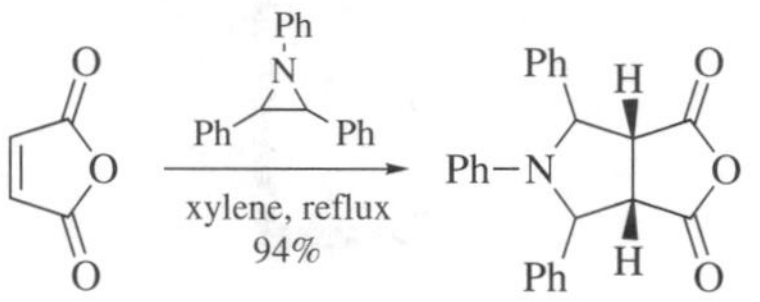

xylene, reflux; 94% (16)

Acylations and Alkylations. Maleic anhydride acylates aromatic ring systems under Friedel–Crafts acylation conditions.[17] A typical example is the acylation of benzene (eq 17) to provide β-benzoylacrylic acid in 80–85% yield.[18] The useful quinone naphthazarin can be prepared from dihydroxybenzene and MA under rigorous reaction conditions (eq 18).[19] Substituted aryl-1,3-dioxocarboxylic acids have been prepared by the generation of enolates from aryl ketones followed by acylation with MA (eq 19).[20]

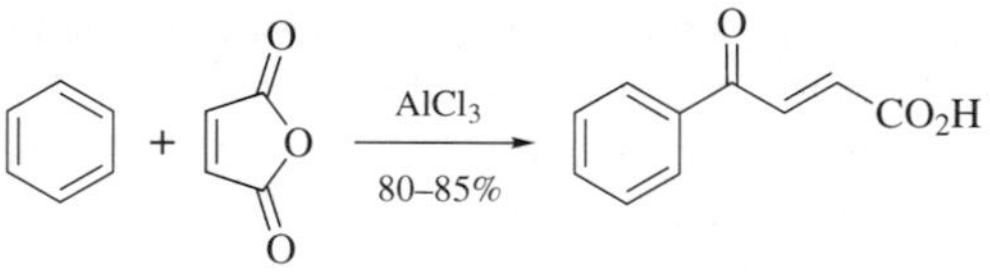

$AlCl_3$; 80–85% (17)

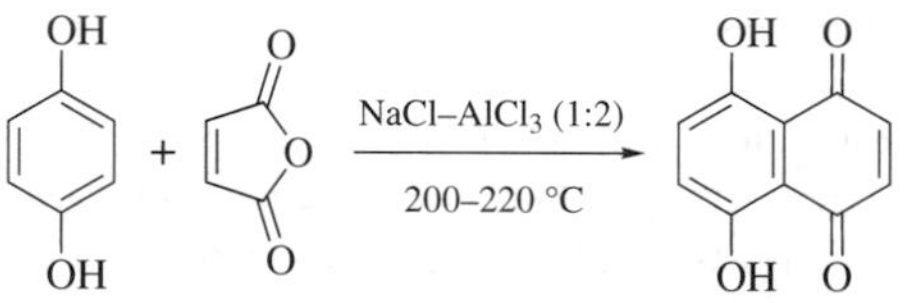

NaCl–$AlCl_3$ (1:2); 200–220 °C (18)

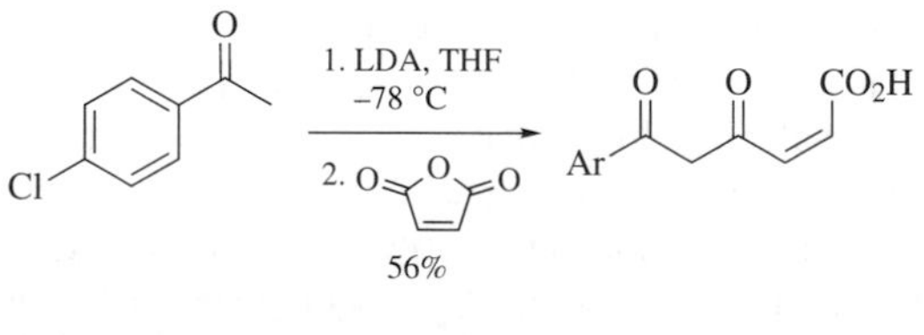

1. LDA, THF, –78 °C; 2. maleic anhydride; 56% (19)

Maleic anhydride can also partake in Michael-type reactions. As previously described, thiophenol adds in a conjugate fashion to maleic anhydride (eq 6).[6] On the other hand, amines undergo acylation with maleic anhydride to afford the corresponding amides (eqs 20 and 21).[21] It should be mentioned that maleic anhydride is a poor radical acceptor relative to other commonly used radical traps (see ***Dimethyl Fumarate*** and ***Dimethyl Maleate***) due to competing polymerization. For example, benzyl radicals generated by heating a mixture of dibenzylmercury and MA at 100 °C produce only 22% of benzylsuccinic anhydride (eq 22).[22]

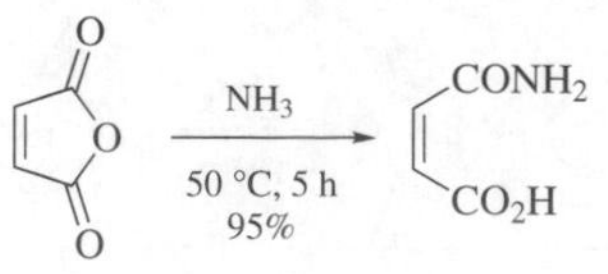

NH_3; 50 °C, 5 h; 95% (20)

(21)

(22)

Miscellaneous. Under thermal conditions, alkenes bearing an allylic hydrogen will undergo an ene reaction with maleic anhydride. Historically, the first reported ene reaction employed maleic anhydride as the enophile component. This reaction has proven to be quite general and typically requires elevated temperatures to occur (eqs 23 and 24). As an alternative to harsh thermal conditions, high pressure has been reported to induce the ene reaction. For example, maleic anhydride reacts with β-pinene at elevated temperatures to afford the corresponding ene adduct (eq 23), while the same reaction occurs at ambient temperature utilizing 4000 MPa pressure to afford the same adduct in 74% yield.[23]

(23)

(24)

Maleic anhydride has been utilized as a ligand in a coupling reaction between an alkenylzirconium(IV) complex and a (η^3-allylic)palladium chloride dimer to produce a 1,4-diene (eq 25).[24] In contrast, the overall rate of the coupling reaction was retarded by the addition of a phosphine ligand.

(25)

Related Reagents. Dimethyl Fumarate; Dimethyl Maleate; (Maleic Anhydride)bis(triphenylphosphine)palladium; Methylthiomaleic Anhydride; *N*-Phenylmaleimide; Succinic Anhydride.

1. (a) Trivedi, B. C.; Culbertson, B. M. *Maleic Anhydride*; Plenum: New York, 1982.
2. Kloetzel, M. C. *OR* **1948**, *4*, 1.
3. Alder, K.; In *Neuere Methoden der Praparativen Organischen Chemie*; Foerst, W., Ed.; Verlag Chemie: Berlin, 1943; Part I, p 251. (b) Alder, K.; In *Neuere Methoden der Praparativen Organischen Chemie*; Foerst, W., Ed.; Verlag Chemie: Berlin, 1953; Part II, p 125.
4. (a) Danishefsky, S.; Kitahara, T.; Schuda, P. F.; Golob, D.; Dynak, J.; Stevens, R. V. *OS* **1983**, *61*, 147 (b) Oppolzer, W.; Bieber, L.; Francotte, E. *TL* **1979**, 4537. (c) Kozikowski, A. P.; Huie, E.; Springer, J. P. *JACS* **1982**, *104*, 2059.
5. Sammes, P. G. *T* **1976**, *32*, 405.
6. Kaydos, J. A.; Smith, D. L. *JOC* **1983**, *48*, 1096.
7. (a) Dauben, W. G.; Kessel, C. R.; Takemura, K. H. *JACS* **1980**, *102*, 6893. (b) Dauben, W. G.; Gerdes, J. M.; Smith, D. B. *JOC* **1985**, *50*, 2576.
8. Grovenstein, Jr., E.; Rao, D. V.; Taylor, J. W. *JACS* **1961**, *83*, 1705.
9. Owsley, D. C.; Bloomfield, J. J. *JOC* **1971**, *36*, 3768.
10. Smith, III, A. B.; Boscelli, D. *JOC* **1983**, *48*, 1217.
11. Turro, N. J.; Wreide, P. A. *JOC* **1969**, *34*, 3562.
12. Stevenson, H. B.; Cripps, H. N.; Williams, J. K. *OS* **1963**, *43*, 27.
13. (a) Hendrickson, J. B.; Pearson, D. A. *TL* **1983**, *24*, 4657. (b) Huisgen, R.; Grashey, R.; Seidl, H. *CB* **1969**, *102*, 736. (c) Joucla, M.; Hamelin, D. G. J. *T* **1973**, *29*, 2315.
14. Hampel, W. *ZC* **1970**, *10*, 225.
15. Quilico, Q.; D'Aleontres, G. S.; Grunanger, P. *G* **1950**, *80*, 479.
16. Heine, H. W.; Peavy, R. *TL* **1965**, 3123.
17. (a) Berliner, E. *OR* **1949**, *5*, 229. (b) Pets, A. G. In *Friedel–Crafts and Related Reactions*; Olah, G., Ed.; Wiley: New York, 1964; Vol. 3.
18. Grummitt, O.; Becker, E. I.; Miesse, C. *OSC* **1955**, *3*, 109.
19. (a) Zahn, K. v.; Ochwat, P. *LA* **1928**, *462*, 72. (b) *FF* **1967**, *1*, 1027.
20. Murray, W. V.; Wachter, M. P. *JOC* **1990**, *55*, 3424.
21. (a) Anshutz, R. *LA* **1891**, *259*, 137. (b) Takaya, T.; Ono, T.; Okuda, Y. *CA* **1974**, *81*, 119 908. (c) Jacobi, P.; Blum C. A.; DeSimone, R. W.; Udodong, U. E. S. *JACS* **1991**, *113*, 5384.
22. Bass, K.; Nababsingh, P. *CI(L)* **1965**, 1599.
23. (a) Arnold, R. T.; Showell, J. S. *JACS* **1957**, *79*, 419. (b) Gladysz, J. A.; Yu, N. *CC* **1978**, 599.
24. Temple, J. S.; Riediker, M.; Schwartz, J. *JACS* **1982**, *104*, 1310.

Gary A. Sulikowski & Michelle M. Sulikowski
Texas A&M University, College Station, TX, USA

Mandelic Acid

(*S*)-(+)		
[17199-29-0]	$C_8H_8O_3$	(MW 152.16)
(*R*)-(−)		
[611-71-2]		

(useful reagent for the resolution of enantiomeric amines[1] and alcohols;[5] serves as a chiral nonracemic template for asymmetric reductions,[11–14] aldol condensations,[15–17] and Diels–Alder reactions;[18,19] chiral nonracemic starting material[22])

Alternate Names: phenylglycolic acid; α-hydroxyphenylacetic acid.

Physical Data: (*S*)-(+): mp 134–135 °C; $[\alpha]_D^{20}$ + 156.6° (H_2O, c = 2.9). (*R*)-(−): mp 133–135 °C; $[\alpha]_D^{20}$ − 158.0° (H_2O, c = 2.5). (±): mp 121–123 °C; d 1.341 g cm^{-3}; K_a 4.3 × 10^{-4} (25 °C).

Solubility: sol water (1 g/6.3 mL), ethanol (1 g/mL), acetic acid, chloroform; very sol ether.

Form Supplied in: white crystalline solid; widely available.

Handling, Storage, and Precautions: darkens and decomposes upon prolonged exposure to light.

Resolving Reagent. Due to the ready availability of both enantiomers of this compound in high enantiomeric purity, mandelic acid is widely used as a reagent for enantiomeric resolutions.[1] It is used in resolving racemic mixtures of amines[1] or diamines[2] as the diastereomeric ammonium salts. Amino esters[3] or amino lactams[4] are resolved by formation of the amides or ammonium salts,[4b] and alcohols[5] are resolved by formation of the corresponding diastereomeric esters[5a,b] or ethers.[5c] Generally, the derivatives are crystalline solids easily purified by recrystallization. In a related application, enantiomeric purity determinations of chiral nonracemic amines by ^{1}H NMR analysis are obtained using mandelic acid as a solvating agent.[6] As well, analysis absolute configuration determinations of enantiomers can be undertaken using either isomer of mandelic acid in conjunction with CD–ORD,[7] X-ray,[8] and mass spectral analyses.[9] Mandelic acid can also be utilized in enantiomeric chromatography[10] as a chiral nonracemic mobile phase additive[10a] or as a solid support component.[10b]

Asymmetric Reductions. Acyloxy-alkoxy borohydrides, produced from the reaction of mandelic acid enantiomers and ***Sodium Borohydride***,[11] and aluminum hydride reagents modified with ligands derived from mandelic acid,[12] will reduce ketones with poor stereoselectivity. Reactions of nitriles with a mixture of $NaBH_4$ and mandelic acid followed by alkylation of the intermediate *N*-boryl imine with organometallic reagents provide the corresponding primary amines in good yield but with low stereoselectivity (eq 1).[13] However, catalytic hydrogenations with chiral nonracemic phosphino ligands (**1**),[14b–d] easily derived from mandelic acid (eq 2), with rhodium have given the corresponding products (e.g. amino acids) with high enantiomeric purity (eq 3).[14]

PhCN —[$NaBH_4$, (*S*)-mandelic acid, THF, 70 °C]→ [PhCH=N–BH_3]$^-$ Na^+ —[BuLi, −80 °C]→ PhCH(NH_2)Bu, 25% ee (1)

(*S*)-mandelic acid —[1. LAH, THF; 2. MsCl, pyridine; 3. $LiPPh_2$, THF]→ PhCH(PPh_2)CH_2PPh_2 (**1**) (2)

Ph(H)C=C(NHAc)CO_2Me —[Rh$^+$(diene)$_2$ BF_4^-, (**1**), MeOH, H_2]→ 88% ee *N*-Ac-L-phenylalanine Me ester (3)

Aldol Condensations. Mandelic acid and a variety of easily prepared derivatives serve as excellent chiral nonracemic auxiliaries for aldol condensations, giving products in high diastereoselectivity. Lewis acid mediated condensation of silyl enol ethers or allylsilanes with 1,3-dioxolan-4-ones (**2**), produced from the reaction of mandelic acid with various aldehydes and ketones, gives the corresponding products in up to 86% de (eq 4).[15] The diastereomers are easily separated and the chiral nonracemic auxiliaries are readily removed with ***Lead(IV) Acetate*** without racemization, giving enantiomerically pure aldols or homoallylic alcohols.

(**2**) (from *S*-mandelic acid) —[Ph(C=CH_2)OTMS, CH_2Cl_2, BF_3•OEt_2, −80 to −30 °C, 98%]→ *SR*:*SS* = 94:6 (4)

Chiral nonracemic silyl ketene acetals produced from mandelic acid derived amino alcohols successfully undergo asymmetric Mukaiyama aldol condensations,[16] and the magnesium enolates of acetoxy-1,1,2-triphenylethanols (**3**) derived from mandelic acid (eq 5) condense with aldehydes with high stereoselectivity (eq 6).[17]

(*R*)-mandelic acid —[1. MeOH, H^+; 2. 3.5 equiv PhMgBr, THF; 3. AcCl, pyridine, CH_2Cl_2]→ (**3**) (5)

(**3**) —[1. LDA, THF, −78 °C; 2. $MgBr_2$; 3. PhCHO; 97%]→ *RR*:*RS* = 97:3 (6)

Diels–Alder Reactions. Hydroxamic acids of the mandelic acid enantiomers serve as precursors to chiral nonracemic acylnitroso dienophiles (**4**) (eq 7).[18] In most examples the stereoselectivity of the cycloaddition is relatively low. However, in some cases (with double asymmetric induction),[19] significant diastereoselectivities can be achieved (eq 8).

(R)-mandelic acid $\xrightarrow[CH_2Cl_2]{Pr_4NIO_4}$ [(4)] (7)

(4) $\longrightarrow$ SR:RR = 73:27 (8)

Asymmetric Organometallic Reagents. Amino derivatives (**5**) of mandelic acid serve as ligands for copper reagents, facilitating conjugate additions to enones with high enantioselectivity (eqs 9 and 10).[20] Lithio benzylmandelate enantioselectively cleaves tricyclic anhydrides to give enantiomers of bicyclic dicarboxylic acids.[21]

(R)-mandelic acid — 1. AcCl; 2. $SOCl_2$; 3. $MeHN(CH_2)_2NMe_2$; 4. LAH → (5) (9)

(5) — 1. BuLi, MeI; 2. CuI, DMS; 3. BuLi, MeI; 4. 2-cyclohexenone; 60% → 90% ee (10)

Chiral Nonracemic Starting Material. Both (S)-(+)- and (R)-(−)-mandelic acid are used extensively for enantiomeric syntheses.[22] It is a convenient starting material for enantiomerically pure phenylethanediol and styrene oxide (eq 11).[23] These methodologies are useful in preparing enantiomerically enriched deuterated compounds.[24] Enantiomerically pure benzoins produced from mandelic acid serve as templates for preparing enantiomerically enriched [^{16}O, ^{17}O, ^{18}O] phosphate esters[25] and sulfate esters.[26] The alcohol, acid, or aromatic ring functional groups can be interconverted in a variety of ways, so mandelic acid serves as an excellent chiral nonracemic starting material for numerous categories of compounds, including such complex molecules as macrolide antibiotics.[27]

(S)-mandelic acid — 1. LAH; 2. TsCl, pyridine; 3. KOH, MeOH → (11)

Related Reagents. (−)-(1S,4R) Camphanic Acid; Dimethyl L-Tartrate; Ethyl 3-Hydroxybutyrate; (S)-Ethyl Lactate; Ethyl Mandelate; α-Methoxy-α-(Trifluoromethyl)phenylacetic Acid; (R)-Pantolactone.

1. (a) Wilen, S. H. *Tables of Resolving Agents and Optical Resolutions*; Eliel, E. L., Ed.; University of Notre Dame: Notre Dame, 1972. (b) Newman, P. *Optical Resolution Procedures for Chemical Compounds*; Optical Resolution Information Center: Riverdale, New York, 1978; Vol. 1.
2. (a) Saigo, K.; Kubota, N.; Takebayashi, S.; Hasegawa, M. *BCJ* **1986**, *59*, 931. (b) Saigo, K.; Tanaka, J.; Nohira, H. *BCJ* **1982**, *55*, 2299.
3. Baldwin, J. E.; Adlington, R. M.; Rawlings, B. J.; Jones, R. H. *TL* **1985**, *26*, 485.
4. (a) Colon, D. F.; Pickard, S. T.; Smith, H. E. *JOC* **1991**, *56*, 2322. (b) Fitzi, R.; Seeback, D. *T* **1988**, *44*, 5277.
5. (a) Schmidlin, T.; Wallach, D.; Tamm, C. *HCA* **1984**, *67*, 1998. (b) Whitesell, J. K.; Reynolds, D. *JOC* **1983**, *48*, 3548. (c) Bihovsky, R.; Bodepudi, V. *T* **1990**, *46*, 7667.
6. Benson, S. C.; Cai, P.; Colon, M.; Haiza, M. A.; Tokles, M.; Snyder, J. K. *JOC* **1988**, *53*, 5335.
7. (a) Barth, G.; Voelter, W.; Mosher, H. S.; Bunnenberg, E.; Djerassi, C. *JACS* **1970**, *92*, 875. (b) Whitman, C. P.; Craig, J. C.; Kenyon, G. L. *T* **1985**, *41*, 1183.
8. (a) Patil, A. O.; Pennington, W. T.; Paul, I. C.; Curtin, D. Y.; Dykstra, C. E. *JACS* **1987**, *109*, 1529. (b) Lamm, B.; Simonsson, R.; Sundell, S. *TL* **1989**, *30*, 6423.
9. (a) Yang, H. J.; Chen, Y. Z. *Org. Mass Spectrom.* **1992**, *27*, 736. (b) Chen, Y. Z.; Li, H.; Yang, H. J.; Hua, S. M.; Li, H. Q.; Zhao, F. Z.; Chen, N. Y. *Org. Mass Spectrom.* **1988**, *23*, 821.
10. (a) Duprat, F.; Coyard, V. *Chromatographia* **1992**, *34*, 31. (b) Choulis, N. H. *JPS* **1972**, *61*, 1325.
11. (a) Nasipuri, D.; Sarkar, A.; Konar, S. K.; Ghosh, A. *IJC(B)* **1982**, *21*, 212. (b) Polyak, F. D.; Solodin, I. V.; Dorofeeva, T. V. *SC* **1991**, *21*, 1137.
12. Garry, S. W.; Neilson, D. G. *JCS(P1)* **1987**, 601.
13. Itsuno, S.; Hachisuka, C.; Ushijima, Y.; Ito, K. *SC* **1992**, *22*, 3229.
14. (a) Harada, T. *BCJ* **1975**, *48*, 3236. (b) Brown, J. M.; Murrer, B. A. *JCS(P2)* **1982**, 489. (c) Riley, D. P.; Shumate, R. E. *JOC* **1980**, *45*, 5187. (d) King, R. B.; Bakos, J.; Hoff, C. D.; Marko, L. *JOC* **1979**, *44*, 1729.
15. Mashraqui, S. H.; Kellogg, R. M. *JOC* **1984**, *49*, 2513.
16. Gennari, C.; Molinari, F.; Cozzi, P. G.; Oliva, A. *TL* **1989**, *30*, 5163.
17. (a) Devant, R.; Mahler, U.; Braun, M. *CB* **1988**, *121*, 397. (b) Braun, M.; Devant, R. *TL* **1984**, *25*, 5031. (c) Millar, A.; Mulder, L. W.; Mennen, K. E.; Palmer, C. W. *OPP* **1991**, *23*, 173.
18. (a) Defoin, A.; Brouillard-Poichet, A.; Streith, J. *HCA* **1992**, *75*, 109. (b) Kirby, G. W.; Nazeer, M. *TL* **1988**, *29*, 6173. (c) Snider, B. B.; Phillips, G. B.; Cordova, R. *JOC* **1983**, *48*, 3003.
19. Defoin, A.; Pires, J.; Tissot, I.; Tschamber, T.; Bur, D.; Zehnder, M.; Streith, J. *TA* **1991**, *2*, 1209.
20. Corey, E. J.; Naef, R.; Hannon, F. J. *JACS* **1986**, *108*, 7114.
21. (a) Ohtani, M.; Matsuura, T.; Watanabe, F.; Narisada, M. *JOC* **1991**, *56*, 2122. (b) Ohtani, M.; Matsuura, T.; Watanabe, F.; Narisada, M. *JOC* **1991**, *56*, 4120.
22. Szabo, W. A.; Lee, H. T. *Aldrichim. Acta* **1980**, *13*, 13.

23. Eliel, E. L.; Delmonte, D. W. *JOC* **1956**, *21*, 596.
24. (a) Elsenbaumer, R. L.; Mosher, H. S. *JOC* **1979**, *44*, 600. (b) Sankawa, U.; Sato, T. *TL* **1978**, 981. (c) Hill, R. K.; Prakash, S. R.; Zydowsky, T. M. *JOC* **1984**, *49*, 1666. (d) Morrison, J. D.; Tomaszewski, J. E.; Mosher, H. S.; Dale, J.; Miller, D.; Elsenbaumer, R. L. *JACS* **1977**, *99*, 3167.
25. Cullis, P. M.; Lowe, G. *JCS(P1)* **1981**, 2317.
26. Lowe, G.; Salamone, S. J. *CC* **1984**, 466.
27. Masamune, S.; Choy, W.; Kerdesky, F. A. J.; Imperiali, B. *JACS* **1981**, *103*, 1566.

Bruce D. Harris
The R. W. Johnson Pharmaceutical Research Institute, Spring House, PA, USA

Manganese(III) Acetate[1]

$Mn(OAc)_3$

($Mn(OAc)_3$)		
[993-02-2]	$C_6H_9MnO_6$	(MW 232.09)
($Mn(OAc)_3 \cdot 2H_2O$)		
[19513-05-4]	$C_6H_{13}MnO_8$	(MW 268.13)

(one-electron oxidant used to oxidize acetic acid and β-dicarbonyl compounds to the corresponding radical and for α′-acetoxylation of enones)

Physical Data: the commercially available dihydrate is cinnamon brown. The anhydrous form is dark brown. The crystal structure of the anhydrous form indicates an oxo-centered trimer with bridging acetates.[2]

Solubility: sol acetic acid, ethanol, and a variety of other organic solvents; disproportionates in water.

Preparative Methods: commercially available dihydrate is easily prepared by the reaction of manganese(II) acetate with potassium permanganate in acetic acid at reflux.[1] The anhydrous form is prepared in acetic acid and acetic anhydride.[1]

$Mn(OAc)_3$ has been extensively used for the oxidative addition of acetic acid to alkenes to give γ-butyrolactones.[3,4] The addition of ***Acetic Acid*** to alkenes in acetic acid at reflux is quite general.[1] ***Acetic Anhydride*** and sodium acetate are often used as additives that modify the rate of reaction and ratio of products. With simple carboxylic acids, this reaction is limited to acetic acid and other acids which can be used as solvent. Mechanistic studies[5a] have shown that the rate determining step is formation of a Mn^{III} enolate which rapidly transfers an electron to give the carboxymethyl radical and Mn^{II}. This radical adds to the alkene to give a γ-carboxyalkyl radical that is oxidized by a second equivalent of $Mn(OAc)_3$ to give mixtures of lactone, alkene, and γ-acetoxycarboxylic acid (eq 1), depending on the exact reaction conditions and structure of the alkene.[6]

(1) [Scheme: alkene → $Mn(OAc)_3$, AcOH, reflux → lactone 55% + unsaturated acid 22%]

The lactonization reaction is much more facile with ***Cyanoacetic Acid*** and ***Malonic Acid***, which undergo the rate-determining enolization more rapidly in acetic acid at rt (eq 2).[1,5,7] Malonic acid gives only bis-lactone 2:1 adducts resulting from the addition to two molecules of alkene (eq 3).[8]

(2) [Scheme: cyclohexene + $NCCH_2CO_2H$ → $Mn(OAc)_3$, AcOH, 25 °C, 51% → cyano lactone]

(3) [Scheme: α-methylstyrene + $CH_2(CO_2H)_2$ → $Mn(OAc)_3$, AcOH, 25 °C → spiro bis-lactone]

$Mn(OAc)_3$ oxidizes ketones and aldehydes to α-carbonyl radicals which add to alkenes to give γ-oxo radicals. If these radicals are secondary, they are not oxidized unless ***Copper(II) Acetate*** is used as a co-oxidant (eqs 4 and 5).[9,10] Instead they undergo chain-transfer reactions to give saturated carbonyl compounds. Tertiary radicals give mixtures of saturated and unsaturated carbonyl compounds.

(4) [Scheme: C_7H_{15}-alkene + propanal → $Mn(OAc)_3$, AcOH, 20 °C, 74% → C_7H_{15}…CHO]

(5) [Scheme: C_5H_{11}-alkene + acetone → $Mn(OAc)_3$, acetone, 65 °C, 68% → C_5H_{11}… ketone]

These reactions proceed in good yield based on oxidant consumed but must be carried out with a large excess of ketone or aldehyde, which is sometimes used as the solvent, since the products are oxidized further to give radicals at about the same rate as the starting carbonyl compound is oxidized.

A wide variety of β-dicarbonyl compounds can be oxidized by $Mn(OAc)_3$ to radicals in the presence of alkenes.[11] Addition of the radical to styrene and other electron-rich alkenes affords dihydrofurans. Addition to enol ethers occurs readily to give 1-alkoxy-1,2-dihydrofurans, which can be hydrolyzed to yield 1,4-diketones or dehydrated to form furans.[12]

Intramolecular versions of these reactions provide an efficient route to cyclic compounds. Oxidation of unsaturated β-keto acids affords cyclopentanones fused to γ-lactones (eq 6).[13]

(6) [Scheme: unsaturated β-keto acid → $Mn(OAc)_3$, AcOH, 20 °C, 80% → bicyclic keto lactone]

Oxidation of unsaturated β-keto esters affords α-keto radicals which add to alkenes to give alkyl radicals which are oxidized to alkenes in the presence of $Cu(OAc)_2$ (see ***Manganese(III) Acetate–Copper(II) Acetate***). If the alkyl radical is a 4-phenylbutyl radical, cyclization onto the aromatic ring results in the formation of a tetralin. This reaction occurs stereospecifically, as shown in the synthesis of *O*-methylpodocarpic acid (eq 7).[14]

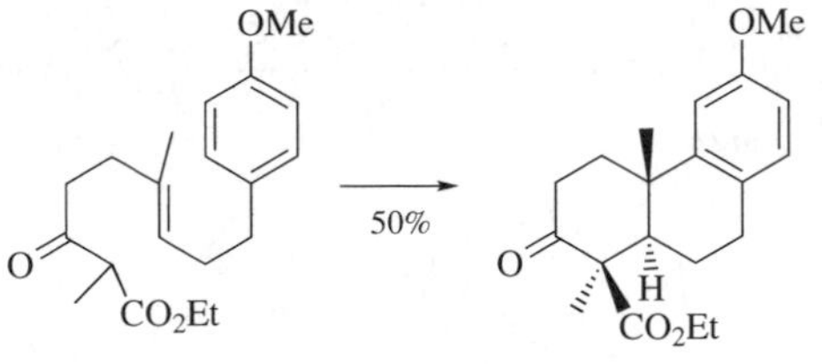

(7)

Haloalkenes are compatible with radical generation by oxidation of β-dicarbonyl compounds with $Mn(OAc)_3$ and can be used to control the regiochemistry of the cyclization (eq 8).[15] Loss of HCl from the intermediate affords the naphthol in 79% yield in one pot. This reaction is compatible with allylic oxygen functionality and has been used for the synthesis of okicenone (eq 9).[16]

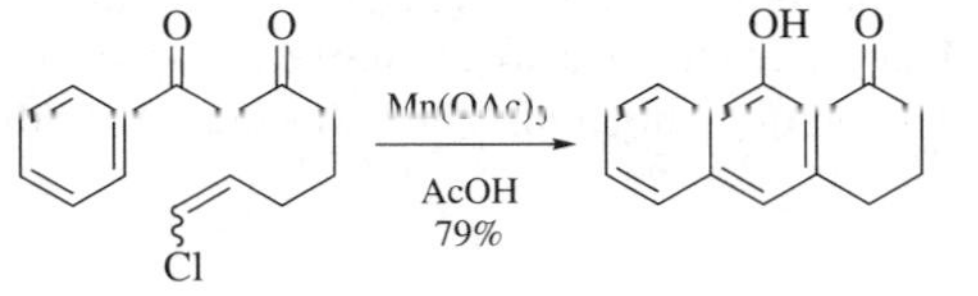

(8)

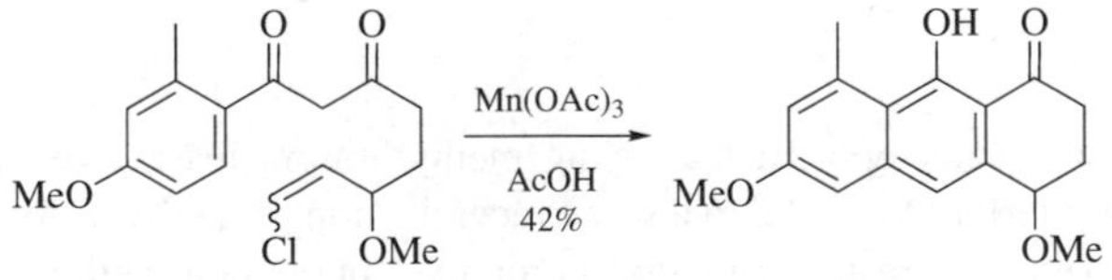

(9)

Oxidation of allylic β-keto amides with $Mn(OAc)_3$ in EtOH affords γ-lactams. The primary radical formed in the cyclization abstracts a hydrogen from the solvent (eq 10).[17] Similarly, saturated γ-lactones can be prepared by oxidation of allylic malonates and acetoacetates.[18]

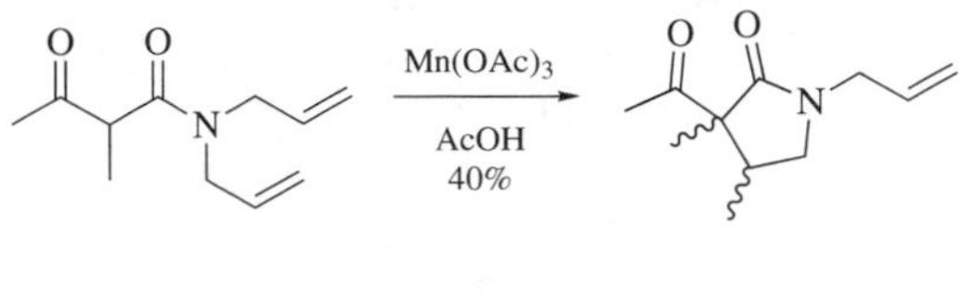

(10)

The radicals formed by oxidation of 1,3-dicarbonyl compounds will add to electron-rich naphthalenes (eq 11).[19] Tetralins and dihydronaphthalenes can be formed by oxidation of diethyl α-benzylmalonate in the presence of an alkene or alkyne (eq 12).[20–22]

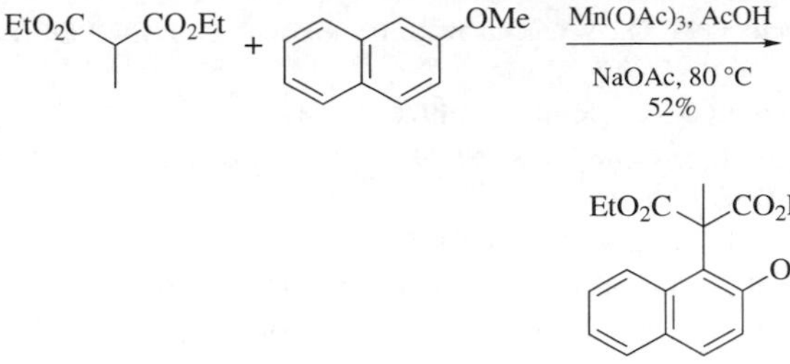

(11)

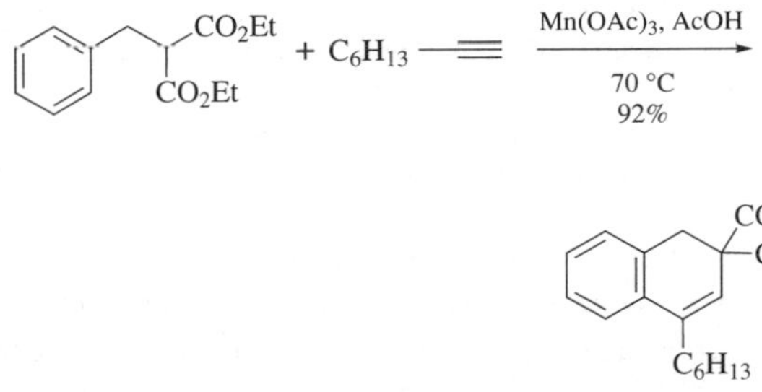

(12)

α′-Acetoxylation of conjugated enones with $Mn(OAc)_3$ in acetic acid takes place in modest yield.[23] The reaction proceeds in much better yield with well-dried oxidant in refluxing benzene (eq 13).[24] A wide variety of α-acyloxy groups can be introduced by exchange of carboxylic acids with $Mn(OAc)_3$ in benzene at reflux prior to the addition of the enone.[25] This procedure is applicable to α- and β-alkoxy enones[26] and is suitable for α-acetoxylation of aryl alkyl ketones.[27]

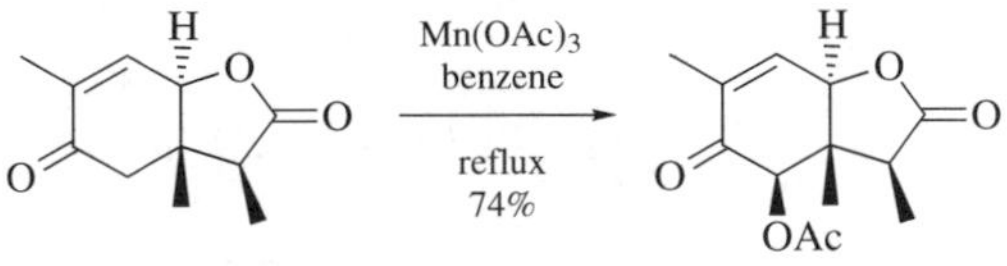

(13)

Related Reagents. Acetic Acid; Copper(II) Acetate; Cyanoacetic Acid; Malonic Acid; Manganese(III) Acetate-Copper(II) Acetate; Manganese(III) Acetylacetonate.

1. (a) de Klein, W. J. In *Organic Syntheses by Oxidation with Metal Compounds*; Mijs, W. J.; de Jonge, C. R. H. I., Eds.; Plenum: New York, 1986; Chapter 4. (b) Badanyan, Sh. O.; Melikyan, G. G.; Mkrtchyan, D. A. *RCR* **1989**, *58*, 286. (c) Snider, B. B. *Chemtracts–Org. Chem.* **1991**, *4*, 403. (d) Demir, A. S.; Jeganathan, A. *S* **1992**, 235. (e) Melikyan, G. G. *S* **1993**, 833.
2. Hessel, L. W.; Romers, C. *RTC* **1969**, *88*, 545.
3. Bush, J. B., Jr.; Finkbeiner, H. *JACS* **1968**, *90*, 5903.
4. Heiba, E. I.; Dessau, R. M.; Koehl, W. J., Jr. *JACS* **1968**, *90*, 5905.
5. (a) Fristad, W. E.; Peterson, J. R.; Ernst, A. B. *T* **1986**, *42*, 3429. (b) Snider, B. B.; Patricia, J. J.; Kates, S. A. *JOC* **1988**, *53*, 2137.
6. Okano, M. *BCJ* **1976**, *49*, 1041.
7. Corey, E. J.; Gross, A. W. *TL* **1985**, *26*, 4291.
8. (a) Fristad, W. E.; Hershberger, S. S. *JOC* **1985**, *50*, 3143. (b) Ito, N.; Nishino, H. Kurosawa, K. *BCJ* **1983**, *56*, 3527.
9. Nikishin, G. I.; Vinogradov, M. G.; Verenchikov, S. P.; Kosyukov, I. N.; Kereselidze, R. V. *ZOR* **1972**, *8*, 539.

10. Vinogradov, M. G.; Verenchikov, S. P.; Nikishin, G. I. *ZOR* **1972**, *8*, 2467.
11. Heiba, E. I.; Dessau, R. M. *JOC* **1974**, *39*, 3456.
12. (a) Corey, E. J.; Ghosh, A. K. *CL* **1987**, 223. (b) Corey, E. J.; Ghosh, A. K. *TL* **1987**, *28*, 175.
13. Corey, E. J.; Kang, M. *JACS* **1984**, *106*, 5384.
14. Snider, B. B.; Mohan, R.; Kates, S. A. *JOC* **1985**, *50*, 3659.
15. Snider, B. B.; Zhang, Q.; Dombroski, M. A. *JOC* **1992**, *57*, 4195.
16. Snider, B. B.; Zhang, Q. *JOC* **1993**, *58*, 3185.
17. (a) Cossy, J.; Leblanc, C. *TL* **1989**, *30*, 4531. (b) Cossy, J.; Bouzide, A.; Leblanc, C. *SL* **1993**, 202.
18. Bertrand, M. P.; Surzur, J. M.; Oumar-Mahamat, H.; Moustrou, C. *JOC* **1991**, *56*, 3089.
19. (a) Citterio A.; Santi, R.; Fiorani, T.; Strologo, S. *JOC* **1989**, *54*, 2703. (b) Citterio, A.; Fancelli, D.; Finzi, C.; Pesce, L.; Santi, R. *JOC* **1989**, *54*, 2713.
20. Snider, B. B.; Buckman, B. O. *T* **1989**, *45*, 6969.
21. Citterio, A.; Sebastiano, R.; Marion, A.; Santi, R. *JOC* **1991**, *56*, 5328.
22. Santi, R.; Bergamini, F.; Citterio, A.; Sebastiano, R.; Nicolini, M. *JOC* **1992**, *57*, 4250.
23. Williams, G. J.; Hunter, N. R. *CJC* **1976**, *54*, 3830.
24. (a) Dunlop, N. K.; Sabol, M. R.; Watt, D. S. *TL* **1984**, *25*, 5839. (b) Jeganathan, A.; Richardson, S. K.; Watt, D. S. *SC* **1989**, *19*, 1091. (c) Gross, R. S.; Kawada, K.; Kim, M.; Watt, D. S. *SC* **1990**, *19*, 1127.
25. (a) Demir, A. S.; Jeganathan, A.; Watt, D. S. *JOC* **1989**, *54*, 4020. (b) Demir, A. S.; Akgün, H.; Tanyeli, C.; Sayrac, T.; Watt, D. S. *S* **1991**, 719.
26. (a) Demir, A. S.; Sayrac, T.; Watt, D. S. *S* **1990**, 1119. (b) Demir, A. S.; Saatcioglu, A. *SC* **1993**, *23*, 571.
27. Demir, A. S.; Camkerten, N.; Akgun, H.; Tanyeli, C.; Mahasneh, A. S.; Watt, D. S. *SC* **1990**, *20*, 2279.

Barry B. Snider
Brandeis University, Waltham, MA, USA

(−)-(1*R*,2*S*,5*R*)-Menthyl (*S*)-*p*-Toluenesulfinate

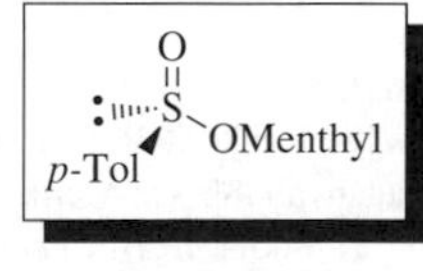

[1517-82-4] $C_{17}H_{26}O_2S$ (MW 294.50)

(agent used for the synthesis of chiral sulfoxides[1,3])

Physical Data: $[\alpha]_D = -202°$ (acetone, $c = 2.0$).

Preparative Method: obtained by reaction of (−)-menthol with *p*-toluenesulfinyl chloride. This esterification showed no particular stereoselectivity, giving an equal amount of the two sulfinate diastereomers.[1] Chromatographic separation may be avoided by epimerizing these sulfinate esters in acidic medium and displacing the resulting equilibrium towards the less soluble isomer, (−)-menthyl (*S*)-*p*-toluenesulfinate, in 80% yield (eq 1).[2] This procedure was later developed into a large scale preparation.[3]

p-Tol–S(O)–ONa
1. $SOCl_2$, benzene, 0 °C
2. (−)-menthol, pyridine ether, rt
(+)-(*R*) ⟶ crystallization, acetone, HCl, −20 °C, 80% ⟶ (−)-(*S*) (1)
HCl / H_2O via p-Tol–SCl_2–OMenthyl

The absolute configuration of (−)-menthyl (*S*)-*p*-toluenesulfinate was established by correlation with (−)-menthyl *p*-iodobenzenesulfinate, known from X-ray diffraction analysis.[4]

Synthesis of Chiral Sulfoxides.

Alkyl Sulfoxides. Grignard reagents react with (−)-menthyl (*S*)-*p*-toluenesulfinate and displace the menthoxy group with complete inversion of configuration at sulfur (eq 2; R = Me,[3,5] Et,[5,6] *n*-C_6H_{13}[7]).

p-Tol–S(O)–OMenthyl —RMgX→ p-Tol–S(O)–R (2)

It was also reported that using methyllithium instead of the methyl Grignard could give some racemization of methyl *p*-tolyl sulfoxide as a result of methyl group exchange via a methylene sulfine intermediate.[8]

(*R*)-4-Substituted cyclohexylmethyl *p*-tolyl sulfoxide (**1**)[9] as well as (*R*)-4-hydroxybutyl *p*-tolyl sulfoxide (**2**)[10] and (*R*)-3-butenyl *p*-tolyl sulfoxide (**3**)[11] were also obtained by reaction of (−)-menthyl (*S*)-*p*-toluenesulfinate and the corresponding Grignard reagent.

(**1**) R = Me, $MeOCH_2$, $ClCH_2$

(**2**)

(**3**)

Vinyl Sulfoxides. A stereocontrolled preparation of (*E*)-1-alkenyl *p*-tolyl sulfoxide from (−)-menthyl-(*S*) *p*-toluenesulfinate was reported (eq 3).[12a]

(3)

One example was also reported showing the formation of an (*E*)-alkenyl sulfoxide in the reaction of a vinylic lithium compound on menthyl sulfinate (eq 4).[12b]

(4)

(+)-(*S*)-2-(*p*-Tolylsulfinyl)-2-cyclopentenone was also prepared by reaction of a vinyllithium derivative and menthyl sulfinate (eq 5).[13]

(5)

The preparation of optically pure (*E*)- and (*Z*)-1-alkenyl *p*-tolyl sulfoxides was described via stereoselective reduction of 1-alkynyl *p*-tolyl sulfoxides (eq 6).[14]

(6)

Alkynic sulfoxides have been made from trimethylsilylethynylmagnesium bromide and the resulting alkyne desilylated on silica gel (eq 7).[15]

(7)

Chiral vinyl sulfoxides can also be prepared by Horner–Emmons reactions of carbonyl compounds with α-phosphoryl sulfoxides which are obtained from lithiated dimethyl methylphosphonate and (−)-menthyl (*S*)-*p*-toluenesulfinate (eq 8).[16] However, this reaction applied to carbonyl compounds often gives a mixture of the (*E*) and (*Z*) isomers of the vinylic sulfoxide.

(8)

The reaction of α-phosphoryl sulfoxide with the dimethyl acetal of pyruvic aldehyde allowed the preparation of the corresponding vinylic sulfoxide as a 1:1 mixture of (*E*) and (*Z*) isomers which could be isomerized with ***Lithium Diisopropylamide*** to the lithiated (*E*) isomer, used for the asymmetric synthesis of α-tocopherol (eq 9).[17]

(9)

The Wittig reaction of an optically active sulfinylphosphonium ylide was reported to yield only the (*E*)-vinylic sulfoxides (eq 10).[18]

(10)

Diaryl Sulfoxides. Optically active diaryl sulfoxides are prepared by reaction of an aryl Grignard with (−)-menthyl (*S*)-*p*-toluenesulfinate: 2,5-dimethoxyphenyl *p*-tolyl sulfoxide (**4**), a precursor of sulfinyl quinones,[19] and 3-pyridyl *p*-tolyl sulfoxide (**5**), a precursor of sulfinyl dihydropyridines (studied as NADH model compounds)[20] are two typical examples.

(**4**) (**5**)

Sulfinyl Esters and Derivatives. (*R*)-(+)-*t*-Butyl 2-(*p*-tolylsulfinyl)acetate is conveniently prepared by reaction of the magnesium enolate of *t*-butyl acetate (readily made with ***Bromomagnesium Diisopropylamide***) with (−)-menthyl (*S*)-*p*-toluenesulfinate (eq 11).[21]

(11)

Substituted sulfinyl esters (**6**) have also been prepared by this reaction using the same base[22a] or lithium cyclohexyl-(isopropyl)amide,[22b] which gives higher yields.

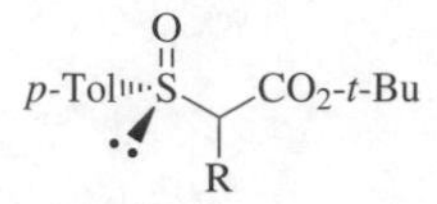

(**6**) R = Me, Et, $C_{14}H_{29}$

Lithioacetonitrile also reacts with (−)-menthyl (*S*)-*p*-toluenesulfinate to give the corresponding β-sulfinylacetonitrile (eq 12).[23]

MeCN — 1. LDA; 2. (−)-(*S*)-*p*-$TolSO_2$Menthyl → *p*-Tol-S(O)-CH_2CN (12)

Similarly, *exo*-metalation with LDA of the racemic 3-methyl-4,5-dihydroisoxazole and reaction with (−)-menthyl (*S*)-*p*-toluenesulfinate afforded the sulfinyl-4,5-dihydroisoxazole as a diastereomeric mixture;[24] lithiated *N,N*-dimethylthioacetamide leads to the sulfinyl *N,N*-dimethylthioacetamide,[25] and lithiated ethyl *N*-methoxyacetimidate leads to *p*-tolylsulfinylethyl-*N*-methoxyacetimidate (eq 13).[26]

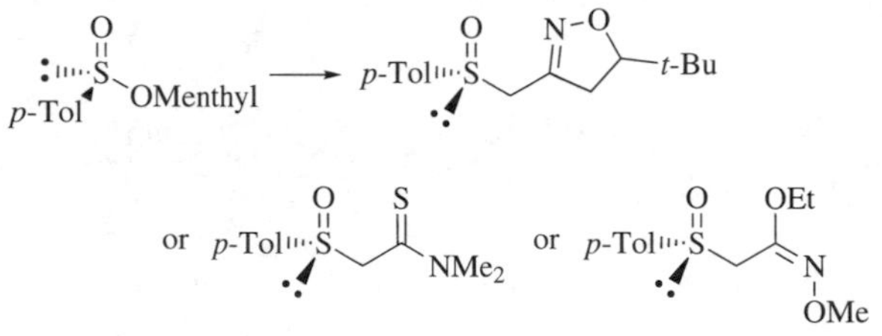

(13)

β-Keto Sulfoxides. Cyclic β-keto sulfoxides are readily obtained from the magnesium enolate of the ketone and (−)-menthyl (*S*)-*p*-toluenesulfinate[27] as a mixture of diastereomers in which the major epimer has the sulfoxide group in the equatorial orientation (eq 14).

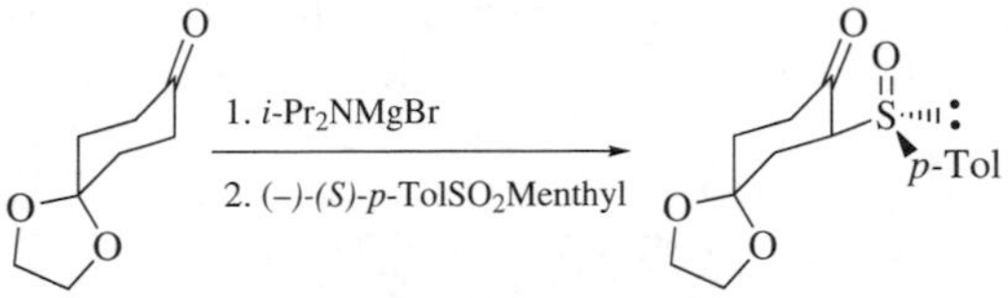

1. *i*-Pr_2NMgBr; 2. (−)-(*S*)-*p*-$TolSO_2$Menthyl (14)

66% de

By condensation of the dianion of *t*-butyl acetoacetate and (−)-menthyl (*S*)-*p*-toluenesulfinate, the corresponding β-keto sulfoxide was obtained in high yield (eq 15) and shown to be an efficient precursor of both enantiomers of β-hydroxybutyric acid via selective reduction of the ketone carbonyl group.[28] β,δ-Diketo sulfoxides were prepared in a similar way from diketone dianions (eq 16).[29]

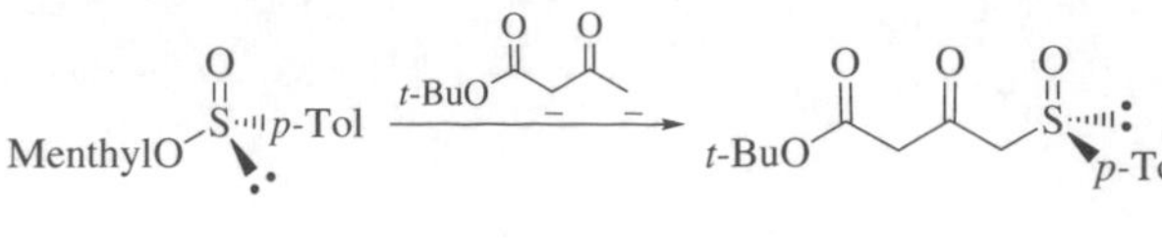

(15)

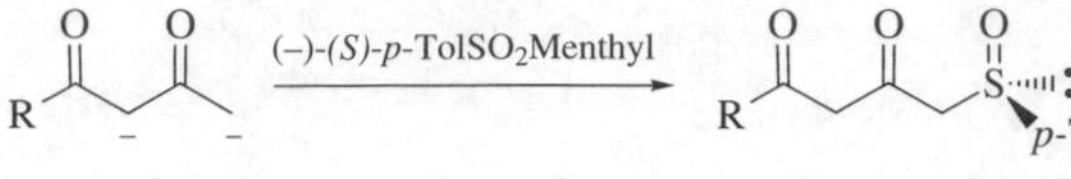

(−)-(*S*)-*p*-$TolSO_2$Menthyl (16)

R = Me, Ph

Imino Sulfoxides. Metalated imines reacted with (−)-menthyl (*S*)-*p*-toluenesulfinate to yield the corresponding sulfinylimines as a diastereoisomeric mixture (eq 17).[30]

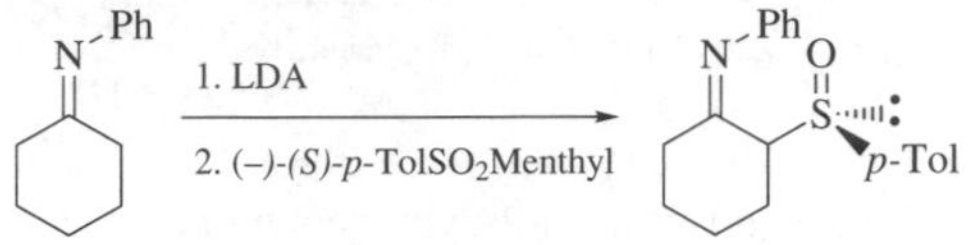

1. LDA; 2. (−)-(*S*)-*p*-$TolSO_2$Menthyl (17)

Similarly, *exo*-metalated cyclic imines afforded the sulfinylimines as an alkaloid precursor (eq 18).[31]

1. LDA; 2. (−)-(*S*)-*p*-$TolSO_2$Menthyl (18)

Miscellaneous. (*S,S*)-Bis(*p*-tolylsulfinyl)methane (**7**) is readily prepared from (−)-menthyl (*S*)-*p*-toluenesulfinate and ***(R)*-methyl *p*-tolyl sulfoxide**.[32] (+) (*S*)-*p*-Tolylsulfinylmethyl *t*-butyl sulfone (**8**) was made from the *t*-butyl methyl sulfone anion and (−)-menthyl (*S*)-*p*-toluenesulfinate.[33]

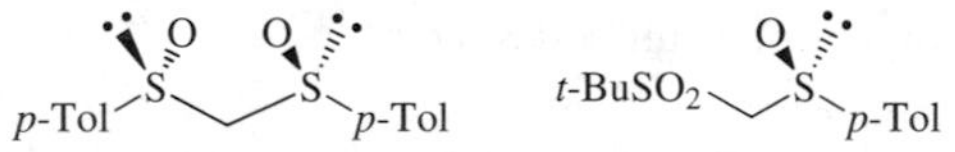

(**7**) (**8**)

Chiral *N*-benzylidene *p*-toluenesulfinamides were prepared by reaction of benzonitrile with an alkyllithium reagent followed by addition of (−)-menthyl (*S*)-*p*-toluenesulfinate and converted into optically active amines and amino acids (eq 19).[34]

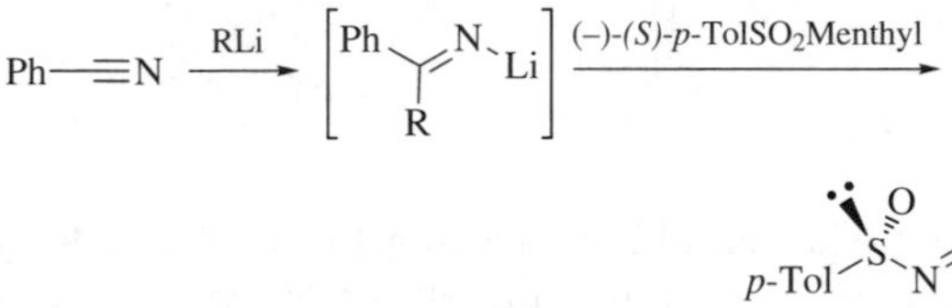

Ph—≡N — RLi → [Ph(R)C=N–Li] — (−)-(*S*)-*p*-$TolSO_2$Menthyl → (19)

R = Me (50%), Bu (75%)

Related Reagents. Benzenesulfinyl Chloride; (*R*)-(+)-*t*-Butyl 2-(*p*-Tolylsulfinyl)acetate; (*R*)-(+)-Methyl *p*-Tolyl Sulfoxide; Phenylmenthol; Phenyl (*p*-Tolylsulfinyl)acetate; *p*-Toluenesulfinic Acid; (R)-(+)-*p*-Tolylsulfinylacetic Acid.

1. Solladié, G. *S* **1981**, 185.
2. Mioskowski, C.; Solladié, G. *T* **1980**, *36*, 227.
3. Solladié, G.; Hutt, J.; Girardin, A. *S* **1987**, 173.
4. Mislow, K.; Green, M. M.; Laur, P.; Melillo, J. T.; Simmons, T.; Ternay, A. L. *JACS* **1965**, *87*, 1958.
5. Andersen, K. K. *JOC* **1964**, *29*, 1953.
6. Arai, Y.; Kuwayama, S.; Takeuchi, Y.; Koizumi, T. *TL* **1985**, *26*, 6205.
7. Bravo, P.; Resnati, G.; Viani, F.; Arnone, A. *T* **1987**, *43*, 4635.
8. Jacobus, J.; Mislow, K. *JACS* **1967**, *89*, 5228.

9. Solladié, G.; Zimmermann, R.; Bartsch, R. *S* **1985**, 662.
10. Iwata, C.; Fujita, M.; Hattori, K.; Uchida, S.; Imanishi, T. *TL* **1985**, *26*, 2221.
11. Arnone, A.; Bravo, P.; Cavicchio, G.; Frigerio, M.; Viani, F. *T* **1992**, *48*, 8523.
12. (a) Posner, G. H.; Tang, P. W. *JOC* **1978**, *43*, 4131. (b) Marino, J. P.; Fernández de la Pradilla, R.; Laborde, E. *S* **1987**, 1088.
13. Hulce, M.; Mallamo, J. P.; Frye, L. L.; Kogan, T. P.; Posner, G. H. *OS* **1986**, *64*, 196.
14. Kosugi, H.; Kitaoka, M.; Tagami, K.; Takahashi, A.; Uda, H. *JOC* **1987**, *52*, 1078.
15. Lee, A. W. M.; Chan, W. H.; Lee, Y. K. *TL* **1991**, *32*, 6861.
16. (a) Mikolajczyk, M.; Grzejszczak, S.; Zatorski, A. *JOC* **1975**, *40*, 1979. (b) Mikolajczyk, M.; Midura, W.; Grzejszczak, S.; Zatorski, A.; Chefczyńska, A. *JOC* **1978**, *43*, 473.
17. Moine, G.; Solladié, G. *JACS* **1984**, *106*, 6097.
18. Mikolajczyk, M.; Perlikowska, W.; Omelańczuk, J.; Cristau, H. J.; Perraud-Darcy, A. *SL* **1991**, 913.
19. (a) Carreño, C. M.; García Ruano, J. L.; Urbano, A. *TL* **1989**, *30*, 4003. (b) Carreño, C. M.; García Ruano, J. L.; Mata, J. M.; Urbano, A. *T* **1991**, *47*, 605.
20. Imanishi, T.; Hamano, Y.; Yoshikawa, H.; Iwata, C. *CC* **1988**, 473.
21. Mioskowski, C.; Solladié, G. *T* **1980**, *36*, 227.
22. (a) Solladié, G.; Matloubi-Moghadam, F.; Luttmann, C.; Mioskowski, C. *HCA* **1982**, *65*, 1602. (b) Nokami, J.; Ohtsuki, H.; Sokamoto, Y.; Mitsuoka, M.; Kunieda, N. *CL* **1992**, 1647.
23. Nokami, J.; Mandai, T.; Nishimura, A.; Takeda, T.; Wakabayashi, S.; Kunieda, N. *TL* **1986**, *27*, 5109.
24. Annunziata, R.; Cinquini, M.; Cozzi, F.; Gilardi, A.; Restelli, A. *JCS(P1)* **1985**, 2289.
25. Cinquini, M.; Manfredi, A.; Molinari, H.; Restelli, A. *T* **1985**, *41*, 4929.
26. Bernardi, A.; Colombo, L.; Gennari, C.; Prati, L. *T* **1984**, *40*, 3769.
27. (a) Carreño, M. C.; García Ruano, J. L.; Rubio, A. *TL* **1987**, *28*, 4861. (b) Carreño, M. C.; García Ruano, J. L.; Pedregal, C.; Rubio, A. *JCS(P1)* **1989**, 1335. (c) Carreño, M. C.; García Ruano, J. L.; Garrido, M.; Ruiz, M. P.; Solladié, G. *TL* **1990**, *31*, 6653.
28. (a) Schneider, F.; Simon, R. *S* **1986**, 582. (b) Solladié, G.; Almario, A. *TL* **1992**, *33*, 2477.
29. Solladié, G.; Ghiatou, N. *TA* **1992**, *3*, 33.
30. Carreño, M. C.; García Ruano, J. L.; Dominguez, E.; Pedregal, C.; Rodríguez, J. *T* **1991**, *47*, 10035.
31. (a) Hua, D. H.; Miao, S. W.; Bravo, A. A.; Takemoto, D. J. *S* **1991**, 970. (b) Hua, D. H.; Bharathi, S. N.; Panangadan, J. A. K.; Tsujimoto, A. *JOC* **1991**, *56*, 6998.
32. (a) Kunieda, N.; Nokami, J.; Kinoshita, M. *BCJ* **1976**, *49*, 256. (b) Solladié, G.; Colobert, F.; Ruiz, P.; Hamdouchi, C.; Carreño, C. M.; García Ruano, J. L. *TL* **1991**, *32*, 3695.
33. López, R.; Carretero, J. C. *TA* **1991**, *2*, 93.
34. Hua, D. H.; Miao, S. W.; Chen, J. S.; Iguchi, S. *JOC* **1991**, *56*, 4.

Guy Solladié & Françoise Colobert
University Louis Pasteur, Strasbourg, France

Methoxymethylenetriphenylphosphorane

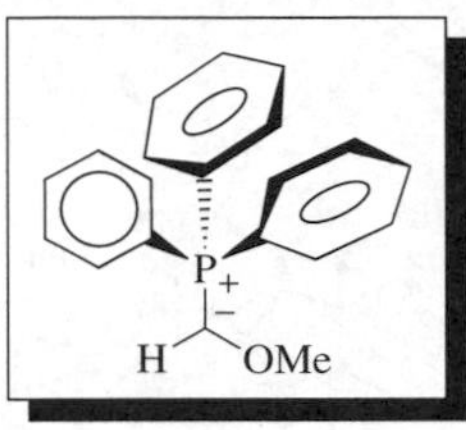

[20763-19-3] $C_{20}H_{19}OP$ (MW 306.36)

(Wittig reagent for carbonyl homologation; vinyl ether synthesis[1–8])

Physical Data: deep red in solution; unstable at 25 °C.[1–3]
Solubility: sol organic solvents; decomposes in protic solvents, aldehydes, and ketones.[1]
Analysis of Reagent Purity: unstable at 25 °C, decomposing completely in <24 h producing PPh_3 (70%; ^{31}P NMR δ −5.1 ppm) and other unidentified species.[2] The ylide gives a ^{31}P NMR signal at δ 7.9 ppm (−90 °C) which is shifted upfield to 5.8 ppm ($^2J_{P-H}$ = 48 Hz; $^3J_{P-H}$ = 12 Hz) at ambient temperatures.
Preparative Methods: by the deprotonation of methoxysymethyl(triphenyl)phosphonium chloride (from ***Triphenylphosphine/Chloromethyl Methyl Ether***)[4a,b] with a number of bases (alkyllithiums, alkoxides, dimsylsodium, amides).[4–8]
Handling, Storage, and Precautions: should be generated in situ at low temperature (e.g. −90 °C) under a nitrogen atmosphere and used immediately.

Vinyl Ether Homologs from Carbonyl Compounds. The Wittig alkenation of carbonyl compounds has evolved to occupy a preeminent position in the hierarchy of carbon–carbon bond-forming reactions.[1] With α-alkoxy substitution, Wittig reagents convert carbonyl derivatives into their vinyl ether homologs, and methoxymethyltriphenylphosphorane (**1**) has enjoyed widespread use in this regard since its generation in 1958 by Levine (eq 1).[4a] In a more extensive study of α-substituted Wittig reagents (e.g. Ph_3PCHX, X = OH, Cl, OMe, SMe, and OAr), it was determined that ***Phenyllithium*** is superior to ***n-Butyllithium*** for the deprotonation of the phosphonium salt and that X = OC_6H_4Me-*p* gives better yields of vinyl ether products than (**1**) in several cases.[4b,c] Moreover, the thermal instability of (**1**) has resulted in the reagent being commonly used in stoichiometric excess (2–4 equiv).[2–8]

While butylidene products have been observed on numerous occasions when *n*-BuLi has been used to generate (**1**),[4b,o,s] this base continues to be used for its preparation. Recently, it was discovered that *n*-BuLi, but not ***t-Butyllithium***, displaces PhLi from (Ph_3PCH_2OMe)Cl, ultimately producing $Ph_2(MeOCH_2)PCHPr$, which accounts for the observed butylidene products when employing this base.[2] Therefore, either PhLi[4a,b] or *t*-BuLi[4s] is superior to *n*-BuLi for the generation of (**1**). Alternatively, amide bases (e.g. ***Lithium Diisopropylamide***),[6] alkoxides (e.g. ***Potassium t-Butoxide***)[7] and ***Sodium Methylsulfinylmethylide***[8] are effective bases for the preparation of (**1**). As an example, (**1**) prepared from the *t*-BuOK method has been used for the effective conversion of conjugated enones to 1,3-dienyl ethers which function as useful Diels–Alder substrates (eq 2).[7d]

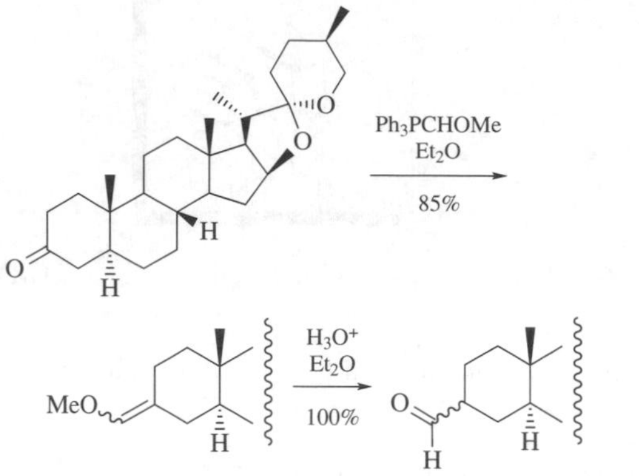

Ph$_3$PCHOMe
Et$_2$O
85%

H_3O^+
Et$_2$O
100%

(1)

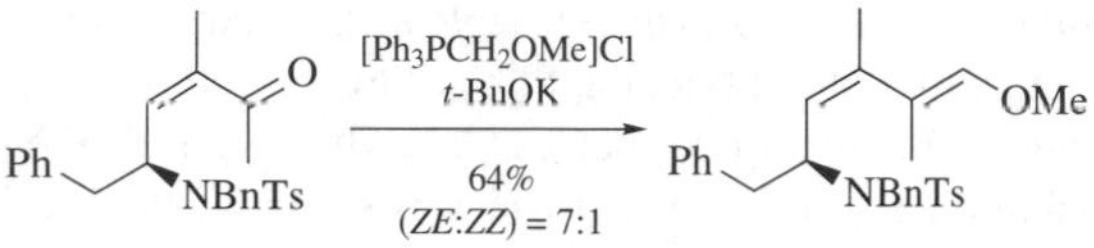

[Ph$_3$PCH$_2$OMe]Cl
t-BuOK
64%
(*ZE*:*ZZ*) = 7:1

(2)

Insufficient comparative data exist for a definitive judgement on the best method to generate (**1**). However, the fact that (**1**) decomposes at 25 °C suggests that its generation and reaction with carbonyl substrates should be conducted at low temperatures (i.e. −70 to −90 °C). A survey of representative procedures suggests that *t*-BuLi, which can be used to generate (**1**) at −90 °C, may eliminate the need for using a large excess of (**1**) because this method avoids its exposure to higher temperatures which result in its partial decomposition.[2] Moreover, even relatively hindered ketones are converted to the corresponding oxaphosphetanes at this temperature. Aliphatic carbonyl derivatives are likely to form these intermediates irreversibly, whereas their conjugated counterparts may well undergo complete or partial equilibration, ultimately giving rise to (*E*)-alkenes as the major vinyl ether products.[1a] With slow warming to 25 °C, the decomposition of the intermediate oxaphosphetane produces the corresponding vinyl ether very cleanly, at least in the systems examined (eq 3).[2] The excellent yields which have been achieved for many substrates add support to this generality.

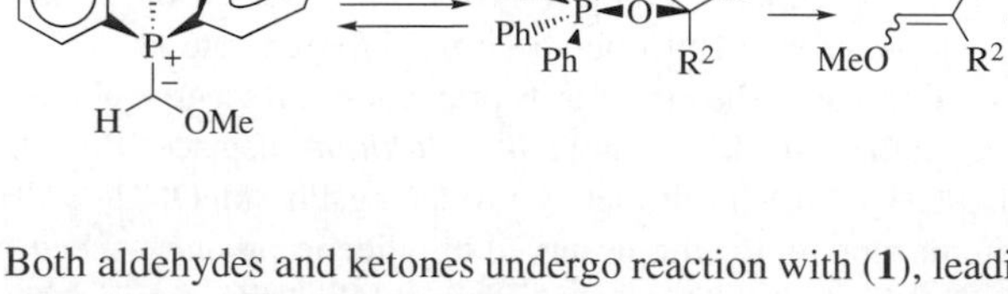

R^1R^2CO

(3)

Both aldehydes and ketones undergo reaction with (**1**), leading to vinyl ethers. While the product stereochemistry is not always reported, sufficient data are currently available for representative systems to indicate that unsymmetrical carbonyl compounds normally lead to (*Z*/*E*) product mixtures under Li$^+$-catalyzed conditions. A recent study employing the Levine–Wittig–Schlosser method (PhLi, Et$_2$O) provides useful stereochemical information for aromatic aldehydes which give roughly equal amounts of both *cis* and *trans* vinyl ethers under these conditions (e.g. PhCHO, 40:60; *p*-MeOC$_6$H$_4$CHO, 54:46; *o*-CF$_3$C$_6$H$_4$CHO, 45:55).[5g] With aliphatic derivatives, as mentioned above, kinetic control is operative and the process is more (*Z*) selective, as can be observed below for octanal[5g] and pentanoyltrimethylsilane[2] where the silyl group reverses the stereochemical relationship of the alkyl group with respect to the methoxy group in the vinyl ether products (**2**) and (**3**).

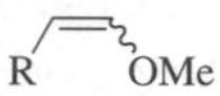

(**2**)
R = C_7H_{15} (54:46)

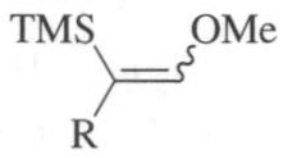

(**3**)
R = Bu (79:21)

Numerous synthetic applications have been found for (**1**) and several representative examples are shown in (**4**)–(**7**), which include the base and solvent employed for the generation of (**1**) as well as the yield and isomeric (*Z*/*E*) distribution of the vinyl ether products derived from their carbonyl precursors.[4d,4s,7c,7e]

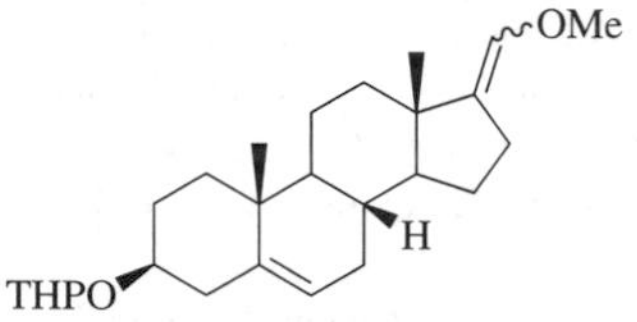

(**4**)
KOCMe$_2$Et, PhMe
100% (1:4)

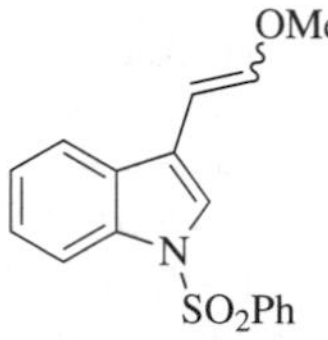

(**5**)
BuLi, THF
80% (1:3)

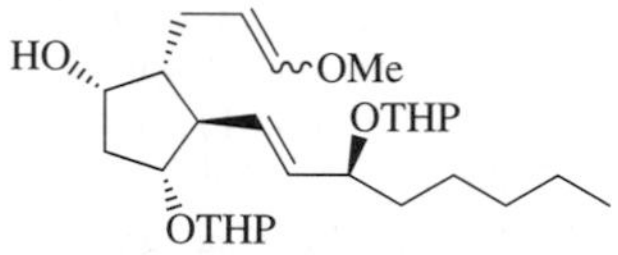

(**6**)
t-BuOK, THF
87% (1:2)

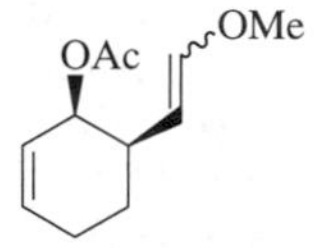

(**7**)
t-BuLi, THF
88% (2:3)

It can be noted from the above that the (*E*) isomer is isolated as the major product in each case, a result which may be attributable to oxaphosphetane equilibration either through its formation (i.e. highly conjugated systems) or through the participation of proximate functionality (e.g. alkoxy).[1a] In certain cases, these substrate features can result in a highly (*E*)-selective process, e.g. (**8**) and (**9**).[6g]

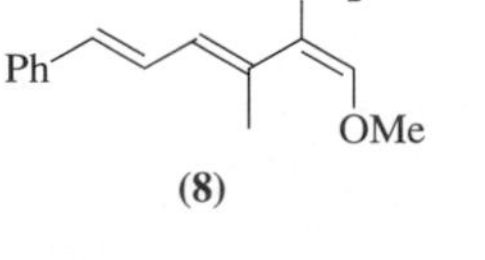

(**8**)

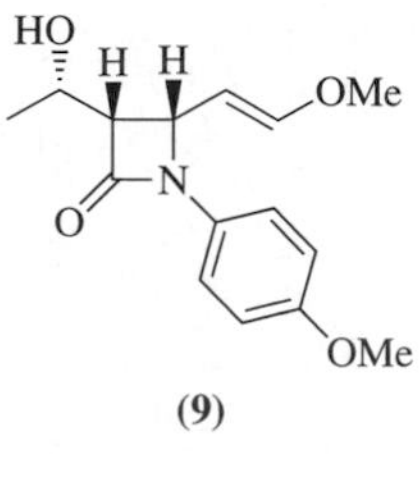

(**9**)

Homologation of Carbonyl Compounds. The vast majority of the synthetic applications of (**1**) involve the homologation of carbonyl compounds. Often the intermediate vinyl ether is converted directly to the aldehyde (or acetal) which itself is a required intermediate for the total synthesis of a target molecule which can be a natural product (e.g. (+)-pleuromutilin (eq 4))[6e] or compounds of theoretical interest, such as 2a,8a,8b,8c-tetrahydropentaleno[6,1,2-*aji*]azulene (eq 5).[4p] The intermediate vinyl ethers can also be oxidized to provide α,β-unsaturated aldehydes as an important extension of the methodology (eq 6).[5i]

(4)

(5)

(6)

The hydrolysis of the vinyl ethers normally occurs smoothly with dilute aqueous acid at 25 °C, mild conditions which make these derivatives especially attractive precursors to the desired aldehydes, particularly in systems where the aldehyde must be generated selectively in the presence of other protected carbonyl functionalities (eq 7).[5f] For certain applications, other alkoxymethylene ylides (Ph_3PCHOR, R = $(CH_2)_2SiMe_3$,[7d] THP,[5j] or $Ph_2P(O)CHLi(OMe)$,[5k] or $Ph_2PCHLi(OMe)$)[9] provide alternative or superior choices to (**1**) for the alkenation process or other subsequent conversions. However, (**1**) is a proven reagent for the effective homologation of aldehydes, both saturated and unsaturated, as well as their ketone counterparts.

(7)

Related Reagents. Bromomethylenetriphenylphosphorane; *N′*-*t*-Butyl-*N*-methyl-*N*-(trimethylsilyl)-methylformamidine; Diethyl Methoxymethylphosphonate; Diethyl Morpholinomethylphosphonate; Methoxymethyl(diphenyl)phosphine Oxide; (Methoxymethyl)trimethylsilane; Methylenetriphenylphosphorane; *N*-Morpholinomethyl(diphenyl)phosphine Oxide; (Phenylthiomethyl)trimethylsilane.

1. (a) Maryanoff, B. E.; Reitz, A. B. *CRV* **1989**, *89*, 863. (b) Murphy, P. J.; Brennen, J. *CSR* **1988**, *17*, 1. (c) Maryanoff, B. E.; Reitz, A. B. *PS* **1986**, *27*, 167. (d) Bestmann, H. J.; Vostrowsky, O. *Top. Curr. Chem.* **1983**, *109*, 85. (e) Schlosser, M. *Top. Stereochem.* **1970**, *5*, 13. (f) Maercker, A. *OR* **1965**, *14*, 270.
2. Anderson, C. L.; Soderquist, J. A.; Kabalka, G. W. *TL* **1992**, *33*, 6915.
3. Yamamoto, Y.; Kanda, Z. *BCJ* **1980**, *53*, 3436.
4. Alkyllithiums: (a) Levine, S. G. *JACS* **1958**, *80*, 6150. (b) Wittig, G.; Schlosser, M. *CB* **1961**, *94*, 1373. (c) Wittig, G.; Böll, W.; Krück, K.-H. *CB* **1962**, *95*, 2514. (d) Brewer, J. D.; Elix, J. A. *AJC* **1972**, *25*, 545. (e) Schlude, H. *T* **1975**, *31*, 89. (f) Field, D. J.; Jones, D. W.; Kneen, G. *JCS(P1)* **1978**, 1050. (g) Bishop, R.; Parker, W.; Stevenson, J. R. *JCS(P1)* **1981**, 565. (h) Oppolzer, W.; Grayson, J. I.; Wegmann, H.; Urrea, M. *T* **1983**, *39*, 3695. (i) Johnson, W. S.; Chen, Y-Q.; Kellogg, M. S. *JACS* **1983**, *105*, 6653. (j) Gibson, K. J.; d'Alarcao, M.; Leonard, N. J. *JOC* **1985**, *50*, 2462. (k) Brillon, D. *SC* **1986**, *16*, 291. (l) Shizuri, Y.; Okuno, Y.; Shigemori, H.; Yamamura, S. *TL* **1987**, *28*, 6661. (m) Johnson, W. S.; Telfer, S. J.; Cheng, S.; Schubert, U. *JACS* **1987**, *109*, 2517. (n) Nakamura, N.; Fujisaka, T.; Nojima, M.; Kusabayashi, S.; McCullough, K. J. *JACS* **1989**, *111*, 1799. (o) Pettit, G. R.; Green, B.; Dunn, G. L.; Sunder-Plassmann, P. *JOC* **1970**, *35*, 1385. (p) Trost, B. M.; Herde, W. B. *JACS* **1976**, *98*, 1988. (q) Weber, G. F.; Hall, S. S. *JOC* **1979**, *44*, 364. (r) Kozikowski, A. P.; Ishida, H.; Chen, Y-Y. *JOC* **1980**, *45*, 3350. (s) Trost, B. M.; Verhoeven, T. R. *JACS* **1980**, *102*, 4743. (t) Kano, S.; Sugino, E.; Shibuya, S.; Hibino, S. *JOC* **1981**, *46*, 3856. (u) Miyashita, M.; Makino, N.; Singh, M.; Yoshikoshi, A. *JCS(P1)* **1982**, 1303. (v) Kumar, K.; Wang, S-S.; Sukenik, C. N. *JOC* **1984**, *49*, 665. (x) Abarca, B.; Ballestros, R.; Jones, G. J. *JHC* **1984**, *21*, 1585. (y) Thompson, A.; Canella, K. A.; Lever, J. R.; Miura, K.; Posner, G. H.; Seliger, H. H. *JACS* **1986**, *108*, 4498. (z) Beautement, K.; Clough, J. M. *TL* **1987**, *28*, 475.
5. Alkyllithiums (continued): (a) Nagaoka, H.; Kobayashi, K.; Matsui, T.; Yamada, Y. *TL* **1987**, *28*, 2021. (b) Parkes, K. E. B.; Pattenden, G. *JCS(P1)* **1988**, 1119. (c) Soderquist, J. A.; Anderson, C. L. *TL* **1988**, *29*, 2425. (d) Rigby, J. H.; Kierkus, P. C. *JACS* **1989**, *111*, 4125. (e) Nakamura, N.; Fujisaka, T.; Nojima, M.; Kusabayashi, S.; McCullough, K. J. *JACS* **1989**, *111*, 1799. (f) Taber, D. F.; Mack, J. F.; Rheingold, A. L.; Geib, S. J. *JOC* **1989**, *54*, 3831. (g) Griesbaum, K.; Kim, W.-S.; Nakamura, N.; Mori, M.; Nojima, M.; Kusabayashi, S. *JOC* **1990**, *55*, 6153. (h) Schreck, V. A.; Serelis, A. K.; Solomon, D. H. *AJC* **1989**, *42*, 375. (i) Takayama, H.; Koike, T.; Aimi, N.; Sakai, S.-I. *JOC* **1992**, *57*, 2173. (j) Boger, D. L.; Palanki, M. S. S. *JACS* **1992**, *114*, 9318. (k) Srikrishna, A.; Krishnan, K. *JCS(P1)* **1993**, 667. (l) Harrison, P. J. *TL* **1989**, *30*, 7125.
6. Amides: (a) Mandai, T.; Osaka, K.; Wada, T.; Kawada, M.; Otera, J. *TL* **1983**, *24*, 1171. (b) Brillon, D. *SC* **1986**, *16*, 291. (c) Evans, E. H.; Hewson, A. T.; March, L. A.; Nowell, I. W.; Wadsworth, A. H. *JCS(P1)* **1987**, 137. (c) Paquette, L. A.; Schaefer, A. G.; Springer, J.

P. *T* **1987**, *43*, 5567. (d) Larock, R. C.; Hsu, M. H.; Narayanan, K. *T* **1987**, *43*, 2891. (e) Paquette, L. A.; Bulman-Page, P. C.; Pansegrau, P. D.; Wiedeman, P. E. *JOC* **1988**, *53*, 1450. (f) Majetich, G.; Defwauw, J. *T* **1988**, *44*, 3833. (g) Gallucci, J. C.; Ha, D.-C.; Hart, D. J. *T* **1989**, *45*, 1283.

7. Alkoxides: (a) Ireland, R. E.; Schiess, P. W. *JOC* **1963**, *28*, 6. (b) Casagrande, C.; Canonica, L.; Severini-Ricca, G. *JCS(P1)* **1975**, 1652. (c) Schow, S. R.; McMorris, T. C. *JOC* **1979**, *44*, 3760. (d) Pyne, S. G.; Hensel, M. J.; Fuchs, P. L. *JACS* **1982**, *104*, 5719. (e) Newton, R. F.; Wadsworth, A. H. *JCS(P1)* **1982**, 823. (f) Hamada, Y.; Kawai, A.; Shioiri, T. *TL* **1984**, *25*, 5409. (g) Dharanipragada, R.; Fodor, G. *JCS(P1)* **1986**, 545. (h) Kawai, A.; Hara, O.; Hamada, Y.; Shioiri, T. *TL* **1988**, *29*, 6331. (i) Hutchings, M. G.; Chippendale, A. M.; Ferguson, I. *T* **1988**, *44*, 3727. (j) Mehta, G.; Reddy, K. R. *TL* **1988**, *29*, 3607. (k) Kawai, A.; Hara, O.; Hamada, Y.; Shioiri, T. *TL* **1984**, *25*, 5409.
8. Dimsyl: (a) Hayakawa, Y.; Yokoyama, K.; Noyori, R. *JACS* **1978**, *100*, 1799. (b) Danishefsky, S.; Harvey, D. F. *JACS* **1985**, *107*, 6647.
9. Burke, S. D.; Cobb, J. E.; Takeuchi, K. *JOC* **1990**, *55*, 2138.

John A. Soderquist & Jorge Ramos-Veguilla
University of Puerto Rico, Rio Piedras, Puerto Rico

(*S*)-2-Methoxymethylpyrrolidine[1]

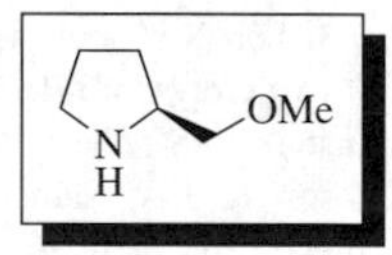

[63126-47-6] $C_6H_{13}NO$ (MW 115.20)

(chiral auxiliary; asymmetric syntheses with SMP enamines[2] and SMP amides;[3] asymmetric Birch reductions;[4] asymmetric Diels–Alder reactions[5])

Alternate Name: SMP.
Physical Data: bp 75 °C/40 mmHg; *d* 0.930 g cm$^{11-3}$; n_D^{20} 1.4467; α_D^{20} −3 to −4° (neat).
Solubility: sol H_2O, ether, dichloromethane.
Form Supplied in: colorless liquid.
Handling, Storage, and Precautions: store at 0–4 °C under an argon atmosphere.

General Considerations. Since the pioneering times of the mid-1970s, (*S*)-2-methoxymethylpyrrolidine has been one of the most generally useful chiral auxiliaries in asymmetric synthesis, with a very broad range of applications. As a proline derivative, it generally shows high stereoselectivities due to the rigidity of the five-membered ring and the ability to coordinate metal fragments[6,7] (see also ***(S)-1-Amino-2-methoxymethylpyrrolidine***, SAMP).

Lithiated SMP formamides and thioformamides have been used as acylanion equivalents (d^1 synthons) in the synthesis of enantiomerically pure α-hydroxy ketones and vicinal diols.[7] Metalated SMP aminonitriles have been used in nucleophilic acylation reactions to give α-hydroxy ketones.[8]

SMP enamines have a very broad range of applications as d^2 synthons. Cyclohexanone SMP enamine can be used for efficient Michael additions to nitroalkenes, Knoevenagel acceptors,[2a,b] and a nitroallylic ester in a [3 + 3] carbocyclization[2c] with excellent stereoselectivities (eq 1). The synthesis of γ-oxo-α-amino acids using SMP enamines has been developed (eq 2).[2d]

1. Ar–CH=C(X)–Y; 2. hydrolysis; 35–81%; X, Y = CO_2Me; X = H, Y = NO_2; 76–>90% de; 75–99% de (1)

1. *t*-$BuCO_2$–CH$_2$–C(NO_2)=CHPh; 2. HCl, reflux; $R^1 = R^2 = H$; 34%; >95% ds; >95% ee

R^1, R^2 = H, [benzo]

1. Ph–C(O)–N=CH–C(O)–O-menthyl, THF; 2. H_3O^+; 84–92%; X = CH_2, S; ≥ 98% de/ee (2)

SMP amide enolates have been employed by several research groups. Alkylation of SMP amide enolates gives α-substituted acids (eq 3).[3a,b] Excellent yields and stereoselectivities are observed in the Birch reduction of aromatic SMP amides with subsequent alkylation (eq 4).[4]

t-BuLi, THF, 25 °C

R^1 = alkyl; R^2 = alkyl, allyl, benzyl

(*Z*); 1. R^2X, −100 °C; 2. hydrolysis; 76–96%; 52–94% ee (3)

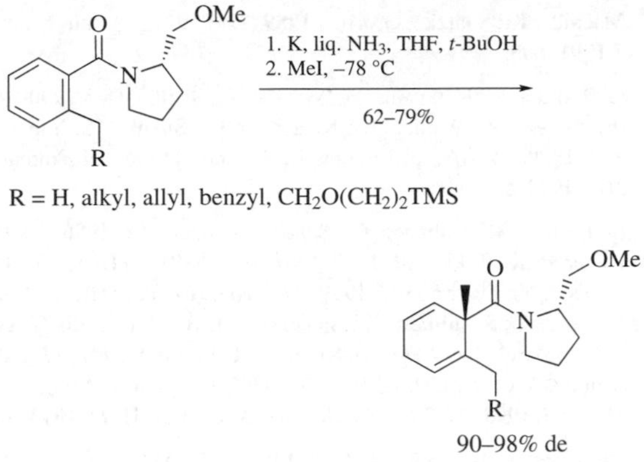

R = H, alkyl, allyl, benzyl, $CH_2O(CH_2)_2TMS$

1. K, liq. NH_3, THF, t-BuOH
2. MeI, −78 °C

62–79%

90–98% de (4)

SMP amides have been used in vanadium(II)-promoted pinacol cross-coupling[3c,d] and in asymmetric oxidations with chiral oxaziridines.[3e] The diastereoselective addition of thiocarboxylic acids to 1-(2-methylacryloyl) SMP amides[3f] and the stereocontrolled addition of various organometallics to α-keto SMP amides[3g] have been studied.

Metalated SMP allylamines or enamines have been used as the first chiral homoenolate equivalents (d^3 synthons; eq 5).[9]

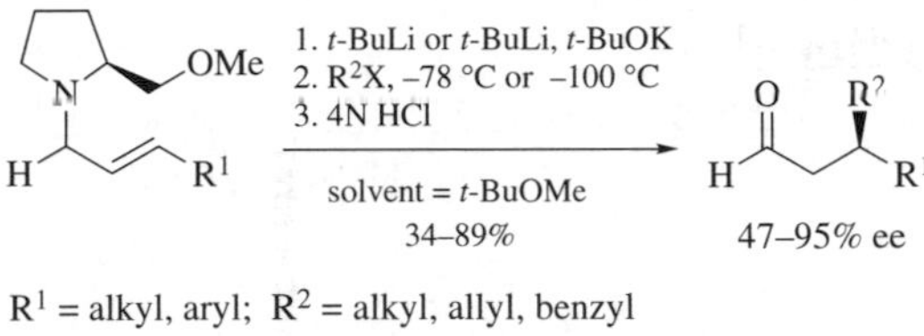

1. t-BuLi or t-BuLi, t-BuOK
2. R^2X, −78 °C or −100 °C
3. 4N HCl

solvent = t-BuOMe
34–89%

47–95% ee (5)

R^1 = alkyl, aryl; R^2 = alkyl, allyl, benzyl

SMP is a useful chiral auxiliary in various cycloaddition reactions. Chiral 2-amino-1,3-dienes have been used in the Diels–Alder reaction with 2-aryl-1-nitroethylenes,[5a,b] and 5-aryl-2-methyl-substituted 4-nitrocyclohexanones were obtained in excellent enantiomeric purities (ee = 95–99%) and diastereoselectivities (ds = 75–95%; eq 6). The photo-Diels–Alder reaction of SMP acrylonitrile with 1-acetylnaphthalene has been carried out.[5b] After hydrolysis of the adduct, the 1,4-diketone was obtained in excellent enantiomeric purity (ee ≥ 97%; eq 7).

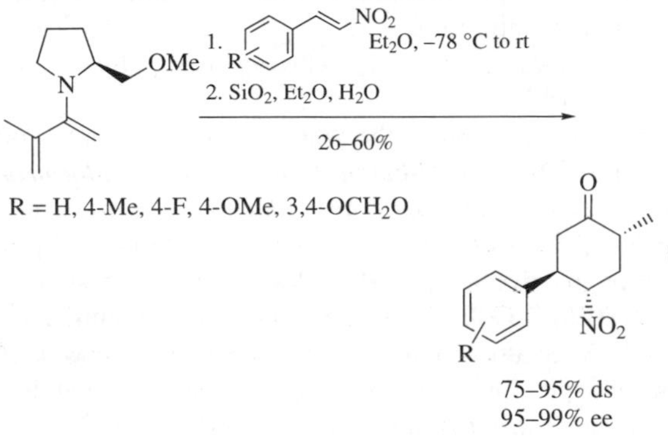

1. (R-substituted nitrostyrene), Et_2O, −78 °C to rt
2. SiO_2, Et_2O, H_2O

26–60%

R = H, 4-Me, 4-F, 4-OMe, 3,4-OCH_2O

(6)

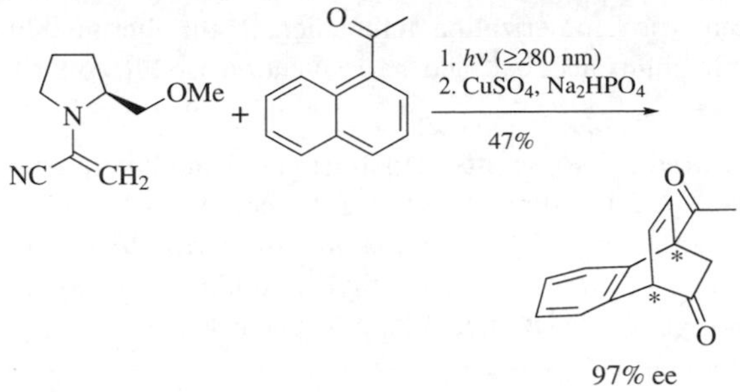

75–95% ds
95–99% ee

1. hν (≥280 nm)
2. $CuSO_4$, Na_2HPO_4

47%

(7)

97% ee

Stereoselective Diels–Alder reactions have been performed variously, using chirally modified sulfines as dienophiles,[5c] chiral ynamines,[5d] SMP enamines,[5e] SMP acrylamides,[5f] and the in situ preparation of SMP N-acylnitroso dienophiles.[9g,h,i] The [2 + 2] cycloaddition reactions of chiral keteniminium salts obtained from SMP amides with alkenes have been studied.[10]

Various metalated chiral organosilicon compounds bearing the SMP moiety have been alkylated to synthesize chiral alcohols.[11] Excellent regio- and stereoselectivities have been observed in the alkylation of chiral silylpropargyl anions (eq 8).[11f]

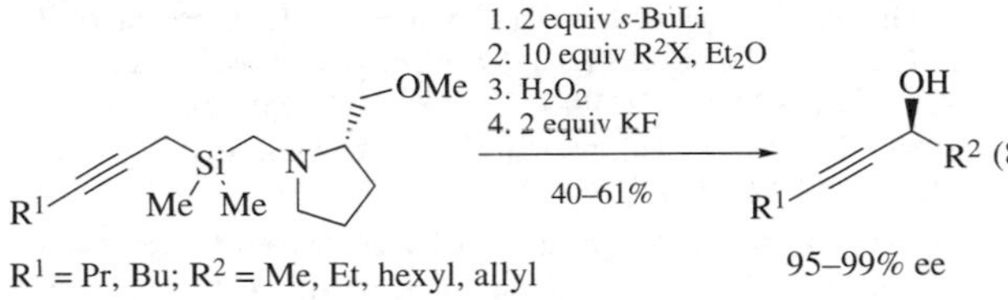

1. 2 equiv s-BuLi
2. 10 equiv R^2X, Et_2O
3. H_2O_2
4. 2 equiv KF

40–61%

(8)

R^1 = Pr, Bu; R^2 = Me, Et, hexyl, allyl

95–99% ee

The elegant application of SMP as a chiral leaving group has been studied,[13] using chiral nitroalkenes in the reaction with zinc enolates.[12] The coupling products were obtained in very good yields and enantiomeric purities (eq 9). SMP methyl-2-cycloalken-1-ones undergo conjugate addition with lithium diorganocuprates followed by elimination of the chiral auxiliary to form optically active α-methylene cycloalkanones (eq 10).

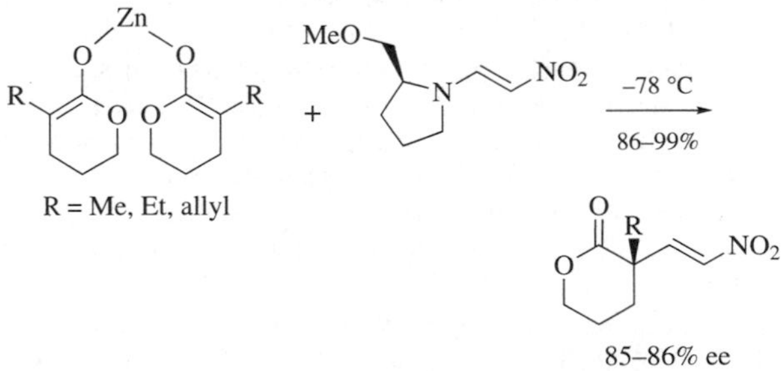

R = Me, Et, allyl

−78 °C

86–99%

(9)

85–86% ee

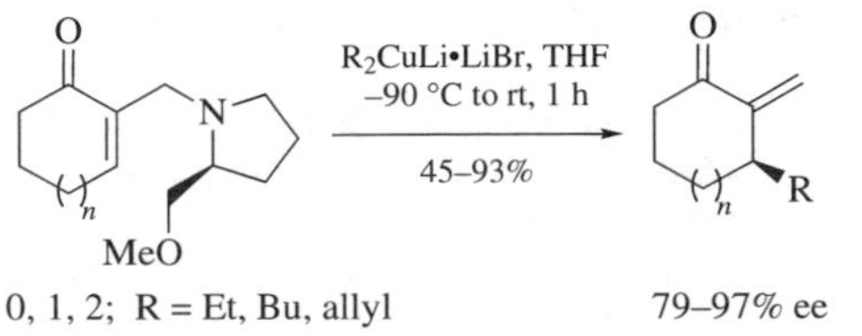

$R_2CuLi{\bullet}LiBr$, THF
−90 °C to rt, 1 h

45–93%

(10)

n = 0, 1, 2; R = Et, Bu, allyl

79–97% ee

Other applications are conjugate addition,[14] the ultrasound-promoted perfluoralkylation of SMP enamines,[15] the enantioselective fluorodehydroxylation of SMP 1-yl-sulfur trifluoride,[16] asymmetric telomerization of butadiene,[17] the chiral modification of ruthenium clusters,[18] and the application of SMP amide bases.[19]

Related Reagents. (*S*)-1-Amino-2-methoxymethylpyrrolidine; (*S*)-2-(Anilinomethyl)pyrrolidine; (*S*)-4-Benzyl-2-oxazolidinone; *trans*-2,5-Bis(methoxymethyl)pyrrolidine; 10,2-Camphorsultam; *trans*-2,5-Dimethylpyrrolidine; (−)-(*S*,*S*)-α,α′-Dimethyldibenzylamine; (1*R*,2*S*)-Ephedrine.

1. Review (literature up to 1985): Enders, D.; Kipphardt, H. *Nachr. Chem. Tech. Lab.* **1985**, *33*, 882.
2. (a) Blarer, S. J.; Seebach, D. *CB* **1983**, *116*, 2250 and 3086. (b) Blarer, S. J.; Schweizer, W. B.; Seebach, D. *HCA* **1982**, *65*, 1693. (c) Seebach, D.; Missbach, M.; Calderari, G.; Eberle, M. *JACS* **1990**, *112*, 7625. (d) Kober, R.; Papadopoulos, K.; Miltz, W.; Enders, D.; Steglich, W.; Reuter, H.; Puff, H. *T* **1985**, *41*, 1637. (e) Risch, N.; Esser, A. *LA* **1992**, 233. (f) Hodgson, A.; Marshall, J.; Hallett, P.; Gallagher, T. *JCS(P1)* **1992**, 2169. (g) Renaud, P.; Schubert, S. *SL* **1990**, 624.
3. (a) Sonnet, P. E.; Heath, R. R. *JOC* **1980**, *45*, 3137. (b) Evans, D. A.; Takacs, J. M. *TL* **1980**, *21*, 4233. (c) Annunziata, R.; Cinquini, M.; Cozzi, F.; Giaroni, P. *TA* **1990**, *1*, 355. (d) Annunziata, R.; Cinquini, M.; Cozzi, F.; Giaroni, P.; Benaglia, M. *T* **1991**, *47*, 5737. (e) Davis, F. A.; Ulatowski, T. G.; Haque, M. S. *JOC* **1987**, *52*, 5288. (f) Effenberger, F.; Isak, H. *CB* **1989**, *122*, 553. (g) Fujisawa, T.; Ukaji, Y.; Funabora, M.; Yamashita, M.; Sato, T. *BCJ* **1990**, *63*, 1894.
4. (a) Schultz, A. G.; Macielag, M.; Sundararaman, P.; Taveras, A. G.; Welch, M. *JACS* **1988**, *110*, 7828. (b) Schultz, A. G.; Green, N. J. *JACS* **1991**, *113*, 4931. (c) Schultz, A. G.; Taylor, R. E. *JACS* **1992**, *114*, 3937. (d) Schultz, A. G.; Hoglen, D. K.; Holoboski, M. A. *TL* **1992**, *33*, 6611.
5. (a) Enders, D.; Meyer, O.; Raabe, G. *S* **1992**, 1242. (b) Barluenga, J.; Aznar, F.; Valdes, C.; Martin, A.; Garcia-Granda, S.; Martin, E. *JACS* **1993**, *115*, 4403. (c) Döpp, D.; Pies, M. *CC* **1987**, 1734. (d) Van den Broek, L. A. G. M.; Posskamp, P. A. T. W.; Haltiwanger, R. C.; Zwanenburg, B. *JOC* **1984**, *49*, 1691. (e) Van Elburg, P. A.; Honig, G. W. N.; Reinhoudt, D. N. *TL* **1987**, *28*, 6397. (f) Bäckvall, J.-E., Rise, F. *TL* **1989**, *30*, 5347. (g) Lamy-Schelkens, H.; Ghosez, L. *TL* **1989**, *30*, 5891. (h) Brouillard-Poichet, A.; Defoin, A.; Streith, J. *TL* **1989**, *30*, 7061. (i) Defoin, A.; Pires, J.; Tissot, I.; Tschamber, T.; Bur, D.; Zehnder, M.; Streith, J. *TA* **1991**, *2*, 1209. (k) Defoin, A.; Brouillard-Poichet, A.; Streith, J. *HCA* **1992**, *75*, 109.
6. (a) Enders, D.; Eichenauer, H. *AG* **1976**, *93*, 579. (b) Seebach, D.; Kalinowski, H.-O.; Bastani, B.; Crass, G.; Daum, H.; Dörr, H.; Du Preez, N. P.; Ehrig, V.; Langer, W.; Nüssler, C.; Oei, H. A.; Schmidt, M. *HCA* **1977**, *60*, 301. (c) Enders, D.; Fey, P.; Kipphardt, H. *OPP* **1985**, *17*, 1.
7. (a) Enders, D.; Lotter, H. *AG* **1981**, *93*, 831. (b) Enders, D. In *Current Trends in Organic Synthesis*; Nozaki, H., Ed.; Pergamon: Oxford, 1983; p 151.
8. Enders, D.; Lotter, H.; Maigrot, N.; Mazaleyrat, J.-P.; Welvart, Z. *NJC* **1984**, *8*, 747. (b) Maigrot, Mazaleyrat, J.-P.; Welvart, Z. *CC* **1984**, 40.
9. (a) Ahlbrecht, H.; Bonnet, G.; Enders, D.; Zimmermann, G. *TL* **1980**, *21*, 3175. (b) Ahlbrecht, H.; Enders, D.; Santowski, L.; Zimmermann, G. *CB* **1989**, *122*, 1995. (c) Ahlbrecht, H.; Sommer, H. *CB* **1990**, *123*, 829.
10. (a) Saimoto, H.; Houge, C.; Hesbain-Frisque, A. M.; Mockel, A.; Ghosez, L. *TL* **1983**, *24*, 2251. (b) Houge, C.; Frisque-Hesbain, A. M.; Ghosez, L. *JACS* **1984**, *104*, 2920.
11. (a) Chan, T. H.; Pellon, P. *JACS* **1989**, *111*, 8737. (b) Chan, T. H.; Wang, D. *TL* **1989**, *30*, 3041. (c) Lamothe, S.; Chan, T. H. *TL* **1991**, *32*, 1847. (d) Chan, T. H.; Nwe, K. T. *JOC* **1992**, *57*, 6107. (e) Lamothe, S.; Cook, K. L.; Chan, T. H.; *CJC* **1992**, *70*, 1733. (f) Hartley, R. C.; Lamothe, S.; Chan, T. H. *TL* **1993**, *34*, 1449.
12. (a) Fuji, K.; Node, M.; Nagasawa, H.; Naniwa, Y.; Terada, S. *JACS* **1986**, *108*, 3855. (b) Fuji, K.; Node, M.; Nagasawa, H.; Naniwa, Y.; Taga, T.; Machida, K.; Snatzke, G. *JACS* **1989**, *111*, 7921. (c) Fuji, K.; Node, M. *S* **1991**, 603.
13. (a) Tamura, R.; Katayama, H.; Watabe, K.; Suzuki, H. *T* **1990**, *46*, 7557. (b) Tamura, R.; Watabe, K.; Katayama, H.; Suzuki, H.; Yamamoto, Y. *JOC* **1990**, *55*, 408. (c) Tamura, R.; Watabe, K.; Ono, N.; Yamamoto, Y. *JOC* **1992**, *57*, 4895.
14. (a) Bertz, S. H.; Dabbagh, G.; Sundararajan, G. *JOC* **1986**, *51*, 4953. (b) Dieter, R. K.; Tokles, M. *JACS* **1987**, *109*, 2040. (c) Dieter, R. K.; Lagu, B.; Deo, N.; Dieter, J. *TL* **1990**, *31*, 4105. (d) Quinkert, G.; Müller, T.; Königer, A.; Schultheis, O.; Sickenberger, B.; Dürner, G. *TL* **1992**, *33*, 3469. (e) Schultz, A. G.; Harrington, R. E. *JACS* **1991**, *113*, 4926. (f) Schultz, A. G.; Lee, H. *TL* **1992**, *33*, 4397. (g) Schultz, A. G.; Holoboski, M. A. *TL* **1993**, *34*, 3021. (h) Schultz, A. G.; Lee, H. *TL* **1993**, *34*, 4397.
15. Kitazume, T.; Ishikawa, N. *JACS* **1985**, *107*, 5186.
16. Hann, G. L.; Sampson, P. *CC* **1989**, 1650.
17. Keim, W.; Köhnes, A.; Roethel, T.; Enders, D. *JOM* **1990**, *382*, 295.
18. Süss-Fink, G.; Jenke, T.; Heitz, H.; Pellinghelli, M. A.; Tiripicchio, A. *JOM* **1989**, *379*, 311.
19. Hendrie, S. K.; Leonard, J. *T* **1987**, *43*, 3289.

Dieter Enders & Martin Klatt

RWTH Aachen, Germany

1-Methoxy-3-trimethylsilyloxy-1,3-butadiene[1]

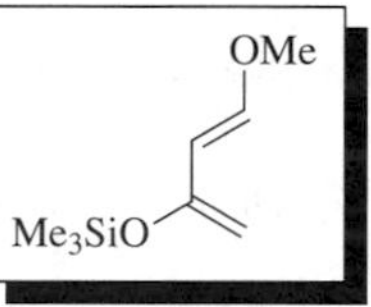

[59414-23-2] $C_8H_{16}O_2Si$ (MW 172.33)

(E)

[54125-02-9]

(highly reactive diene for Diels–Alder reactions[1a,b] and hetero-Diels–Alder reactions[1c])

Alternate Name: Danishefsky's diene.

Physical Data: bp 68–69 °C/14 mmHg; n_D^{20} 1.4540; *d* 0.885 g cm^{-3}.

Solubility: insol H_2O; sol most organic solvents.

Form Supplied in: colorless neat liquid; may contain 3–5% of 4-methoxy-3-buten-2-one. This level of purity is sufficient for most preparative purposes.

Analysis of Reagent Purity: checked easily by its ^1H NMR spectrum.[2,4]

Preparative Methods: the diene was originally prepared from ***4-Methoxy-3-buten-2-one*** with ***Chlorotrimethylsilane–Triethylamine–Zinc Chloride*** in benzene as shown below (68% yield, eq 1).[2a] A detailed experimental procedure for preparative scale work is described by Danishevsky et al. (45–50%).[3] TMSCl–***Sodium Iodide*** in acetonitrile is also a mild procedure for the preparation of this reagent (60%).[4] At the present time, the most efficient method is to use TMSCl–***Lithium Bromide***–Et_3N in THF (91%).[5]

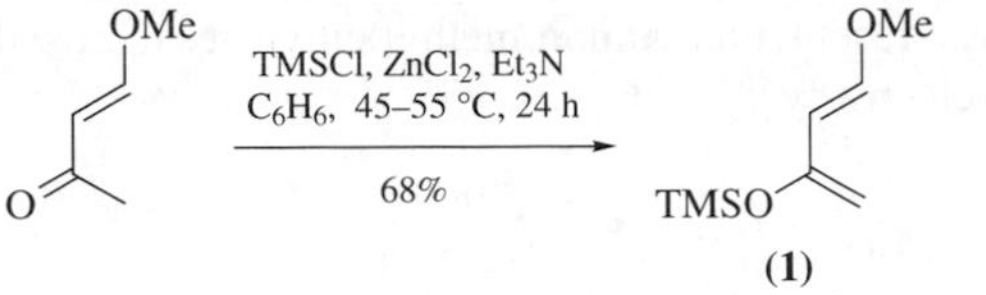

TMSCl, $ZnCl_2$, Et_3N, C_6H_6, 45–55 °C, 24 h; 68% (1)

Handling, Storage, and Precautions: this diene can be kept in a stoppered container without appreciable decomposition, but it is extremely moisture sensitive.

Diels–Alder Reactions.

General Aspects. 1-Methoxy-3-trimethylsilyloxy-1,3-butadiene (**1**) is one of the most reactive species for use in the Diels–Alder reaction and was the first example of a multioxygenated polar 1,3-diene for preparative use. More recently, analogous siloxy dienes with two or more oxygen functions have been prepared and their reactivities have been studied widely.[1a,b] The extremely high reactivity of this diene is illustrated by its reaction with ***Maleic Anhydride***, in which instantaneous exothermic reaction leads to the formation of the adduct via *cis-endo* addition. The stereochemical course of the reaction is unambiguously determined by the isolation of the initial adduct before acidic workup (eq 2).[2,3]

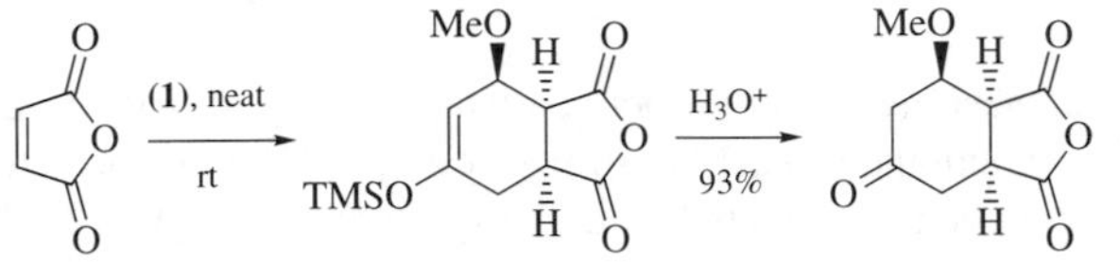

(2)

The utility of this diene for the preparation of functionalized aromatic compounds is demonstrated by the facile conversion of ***Dimethyl Acetylenedicarboxylate*** to a hydroxyphthalate (eq 3) and ***1,4-Benzoquinone*** to an oxygenated naphthalene (eq 4).[2]

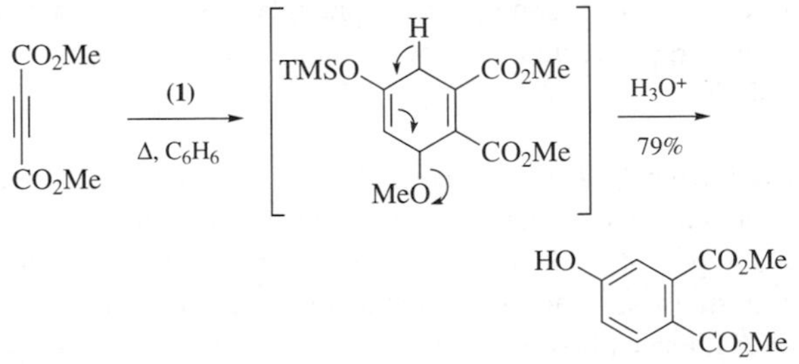

(3)

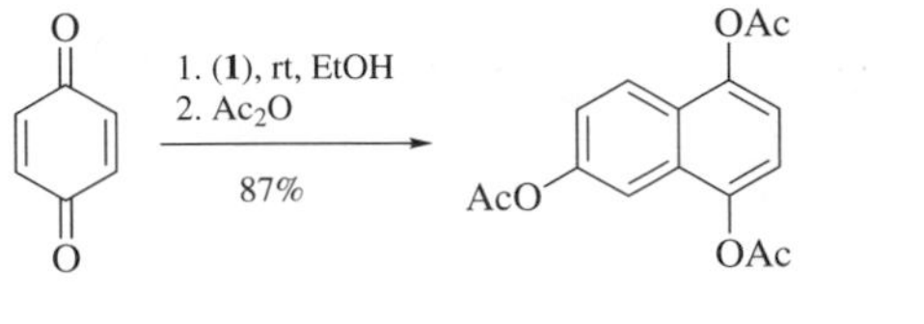

(4)

The second pathway leads to alicyclic compounds. In most cases, α,β-unsaturated cyclohexenones are produced. For example, treatment with a reactive dienophile such as ***Methacrolein*** or ***Methyl Acrylate*** gives the adducts shown (eqs 5 and 6).[2,6] The important feature is that only one regioisomer is obtained by high orientational selectivity of electron-donating substituents at the 1,3-positions of the diene.

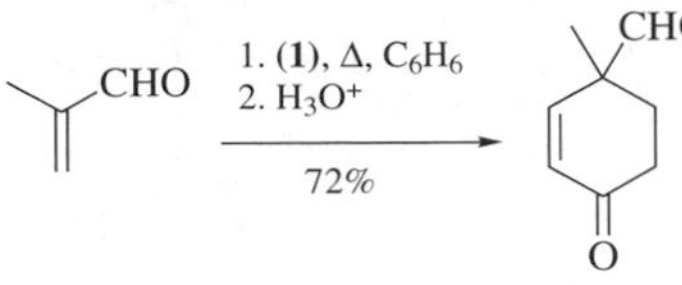

(5)

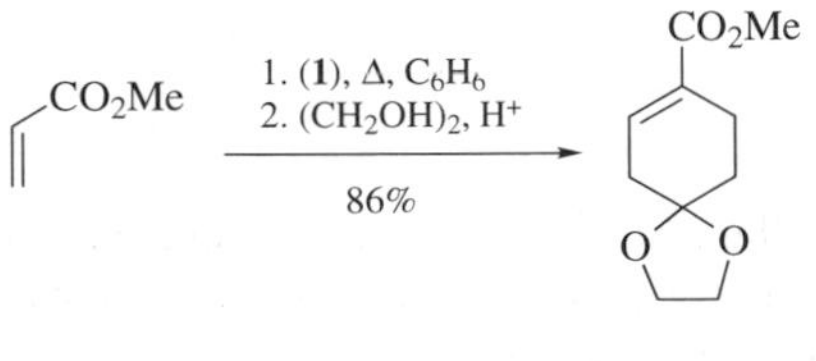

(6)

The remarkable reactivity of this siloxy diene is exemplified by successful reactions with poor dienophiles such as methyl 1-cyclohexenecarboxylate, which does not give Diels–Alder adducts with ordinary dienes like butadiene. It requires more severe conditions, but the reaction gives a Δ^1-*cis*-3-octalone carboxylate in good yield (eq 7); this is not very easy to access via other methods. Several other examples show the generality of this process.[6]

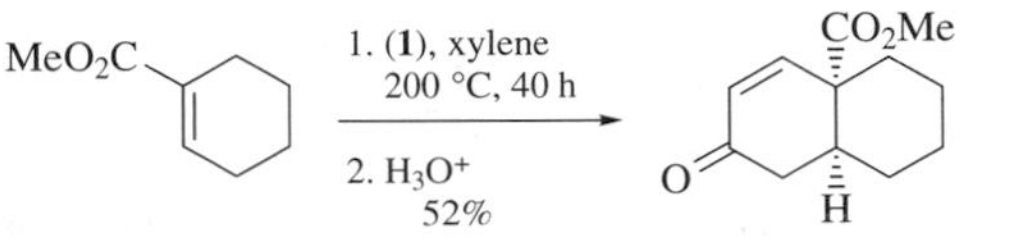

(7)

The excellent reactivity and regio- and stereoselectivity of this diene have been demonstrated, and there have been enormous numbers of publications on applications to the synthesis of various types of carbocyclic compounds.

Condensed Aromatic Compounds. Completely opposite regioselectivity is seen when juglone (eq 8) and its methyl ether (eq 9) react with this diene. Directing effects, with or without chelation by the hydroxy or methoxy group, clearly explain the selectivity.[7] The more polar carbonyl group, marked with an asterisk, controls the regiochemistry.

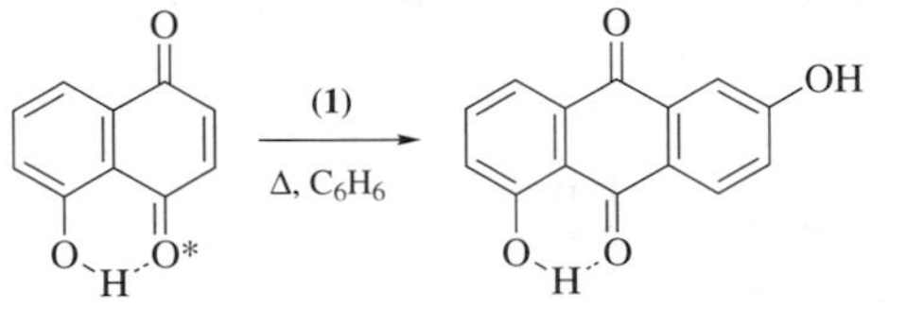

(8)

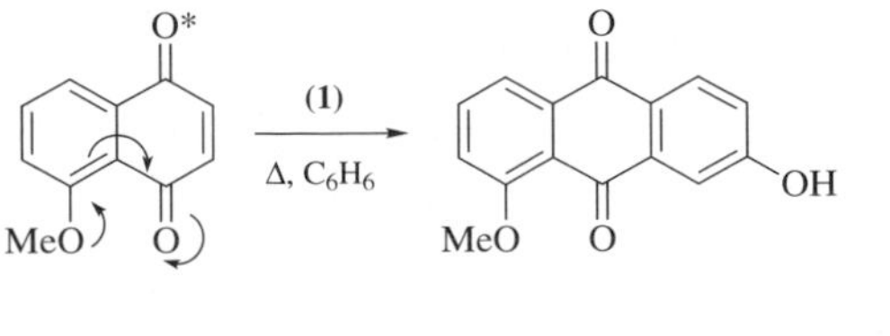

(9)

Similar regioselectivity is observed with 2,3-dichloro-5-hydroxy-1,4-naphthoquinone (eq 10),[8] 6,8-dimethoxy-1,4-naphthoquinone,[9] and anthraquinone derivatives.[10,11] These methodologies have been widely applied to the synthesis of alizarin analogs,[8] daumnomycinone,[12] and related anthracyclinones and their derivatives.[10,11,13]

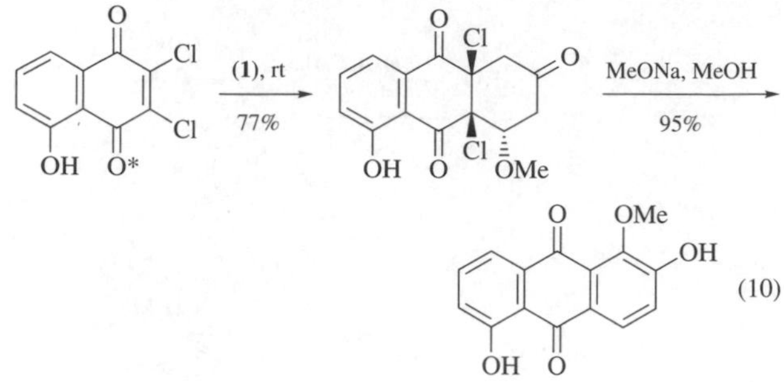

(10)

Alkyl or alkoxy quinones[14] also give single isomers regioselectively (eq 11), while a chloroquinone[15] gives the opposite type of adduct as the sole product (eq 12).

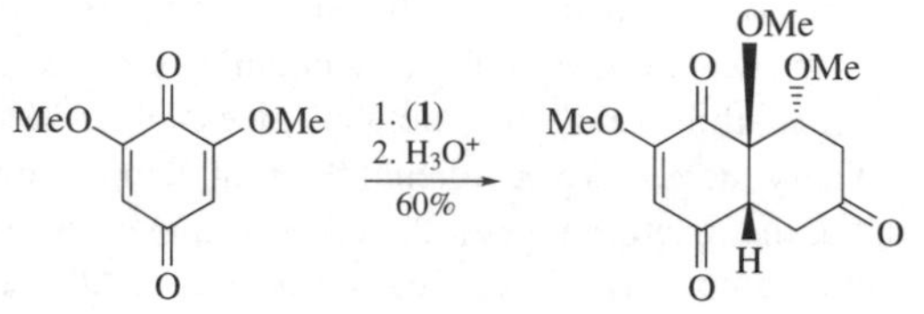

(11)

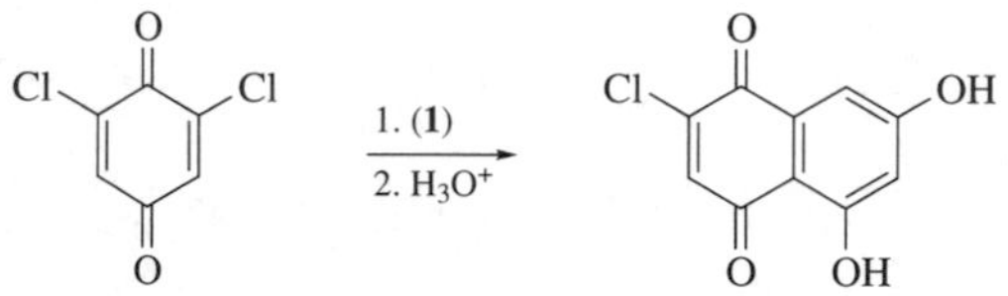

(12)

Stereoselective cycloaddition from the less hindered side of an epoxyanthraquinone derivative gives a single adduct, which has been converted to a daunomycinone analog (eq 13).[16]

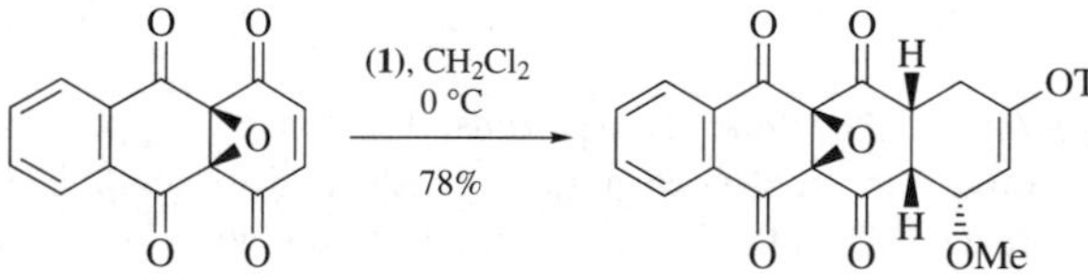

(13)

Reaction with an unsymmetrically substituted spiro-ene-dione dienophile also gives primarily the adduct from attack at the less hindered face (eq 14); the product is convertible to the fredericamycin A skeleton.[17]

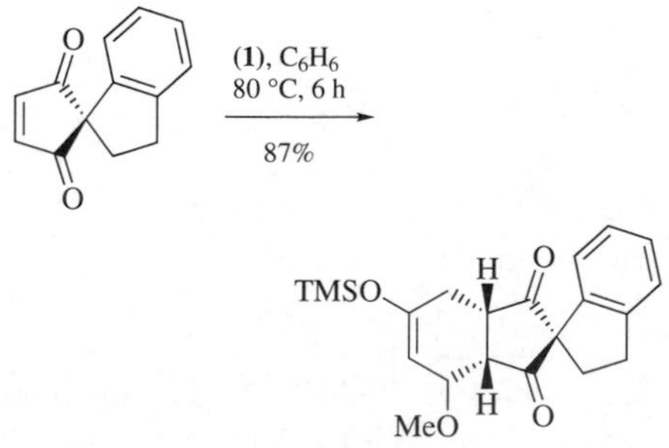

(14)

Cyclohexenones and Analogs. As described earlier, α,β-unsaturated aldehydes and ketones are good dienophiles and the resulting polyfunctional cyclohexenones have been used for the synthesis of ajugarin I (eq 15)[18] and damscones.[19] The diene can even react with (trifluoromethyl)ethylenes to give the adducts regioselectively.[20]

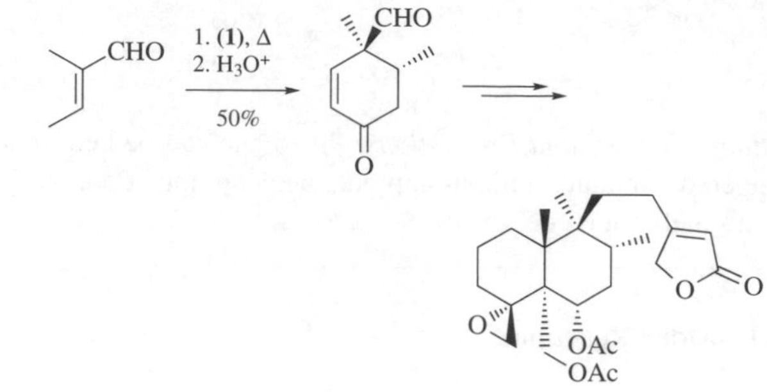

(15)

Ajugarin I

Doubly activated dienophiles such as methylenemalonates,[21] alkylidenemalononitriles[22] and 2-trifluoromethylacrylates[23] easily react with the diene to give enones and dienones (eq 16), from which the cannabinoid skeleton[22] and trifluororetinal have been synthesized.[23]

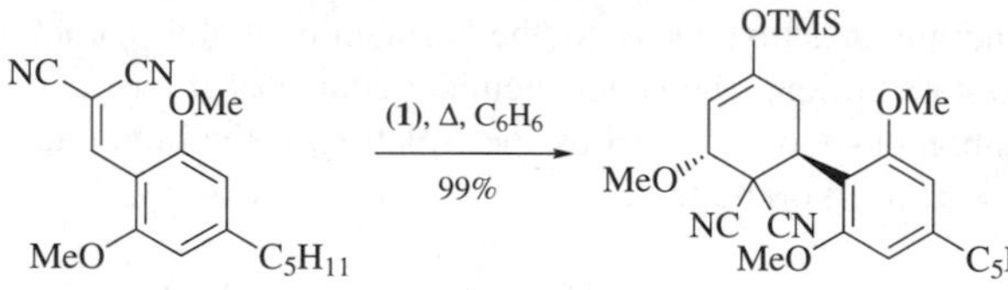

(16)

Dienones are directly obtained using dienophiles with additional functionality such as sulfoxide (eq 17)[24–26] or acetate.[27] The process has been applied to the synthesis of prephenate,[24,27] pretyrosines,[25] and tazettine.[26]

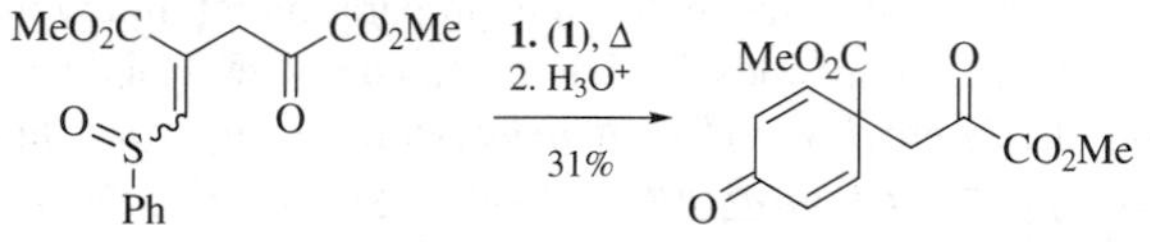

(17)

Vinyl sulfones are useful dienophiles for the preparation of monocyclic cyclohexenones[28,29] or phenols.[30] Reactions with nitro alkenes give precursors for cyclic nitro alkenes[31] and amino alcohols.[32]

Fused Alicyclic Ring Systems. As discussed earlier, even cyclohexenones or cyclohexene esters are useful dienophiles and various types of adducts with fused rings have been synthesized from these reaction partners. For example, methyl 1,4-cyclohexadienecarboxylate gives the *cis*-hexalin carboxylate, which is employed for the synthesis of vernolepin (eq 18).[33]

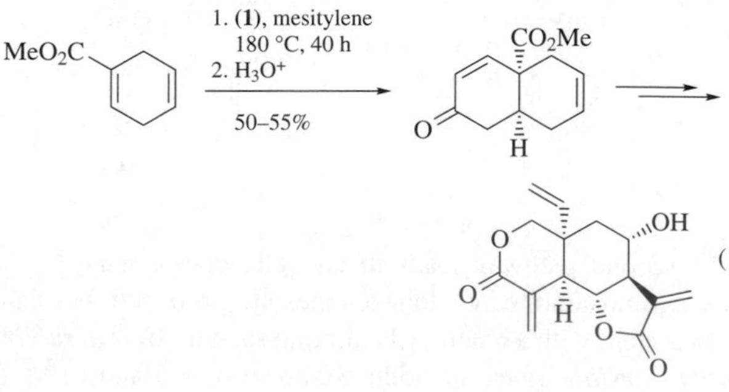

(18)

Vernolepin

This general procedure for the construction of *cis*-fused decalin or extended systems has been applied to the synthesis of pentalenolactone,[34] both racemic and natural quassinoids, picrasin B, quassin (eq 19),[35] the compactin skeleton,[36] and (±)-halipanicine.[37]

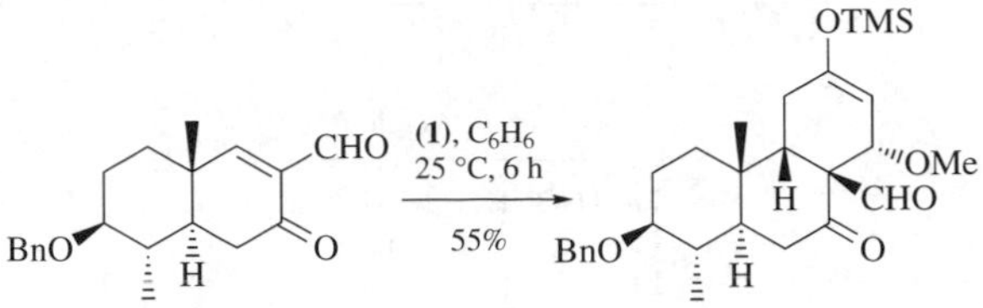

(19)

Reactions with cyclopentene dienophiles give *cis*-fused hydrindenones.[38,39] In the case of 3-alkoxycarbonylcyclopentadienone, the alkoxycarbonyl group is the principal substituent for controlling the orientation of the cycloaddition (eq 20).[40] With the ene-dione of fused cyclopentanes, single regioisomers are produced because the carbonyl group (asterisk) is oriented out of the plane of the other conjugated enone (eq 21).[41] Using an activated dioxopyrroline as the dienophile, the spirocyclic nucleus of the Erythrina alkaloid skeleton is constructed and the synthesis of Erythrina alkaloids has been achieved (eq 22).[42]

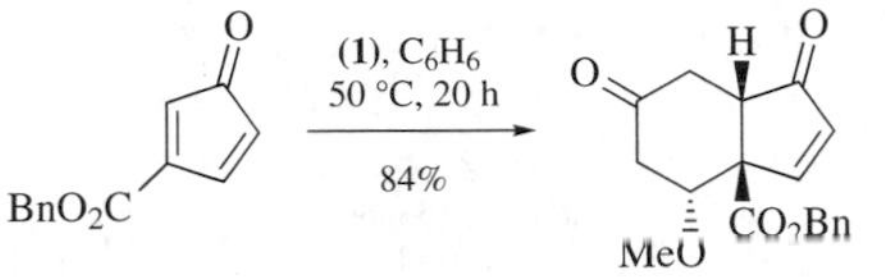

(20)

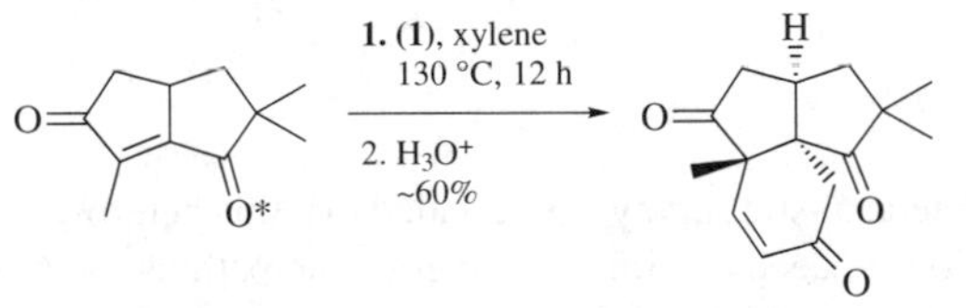

(21)

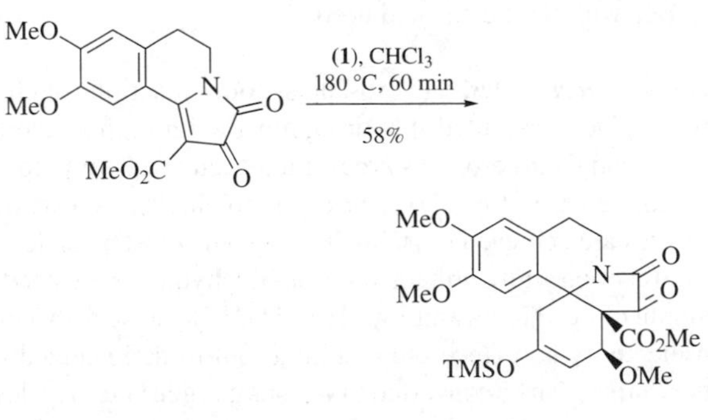

(22)

Cycloaddition with 1-acetylcyclobutene gives a bicyclic adduct, which is transformed to a cyclooctadienone (eq 23).[43] Exceptional stereoselectivity is also observed with bridged enones and a strained enone is a very reactive dienophile, giving the *exo* adduct exclusively in excellent yield (eq 24).[44]

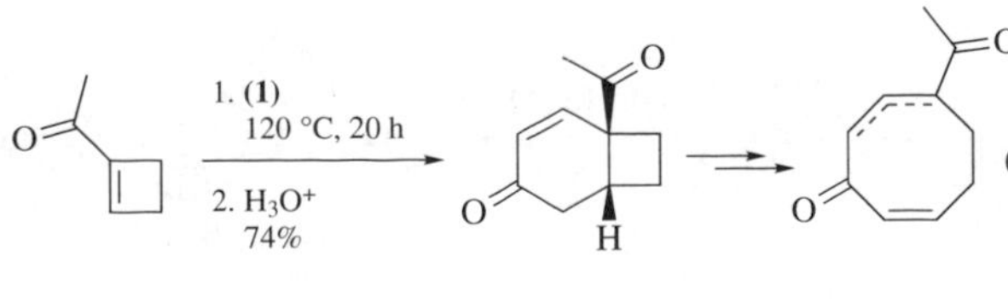

(23)

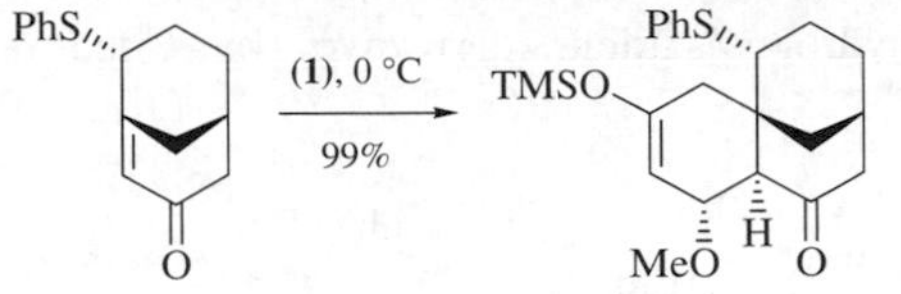

(24)

Lactones are poor dienophiles under normal conditions, but, under high pressure, they smoothly afford the corresponding *endo* adducts (eq 25).[45]

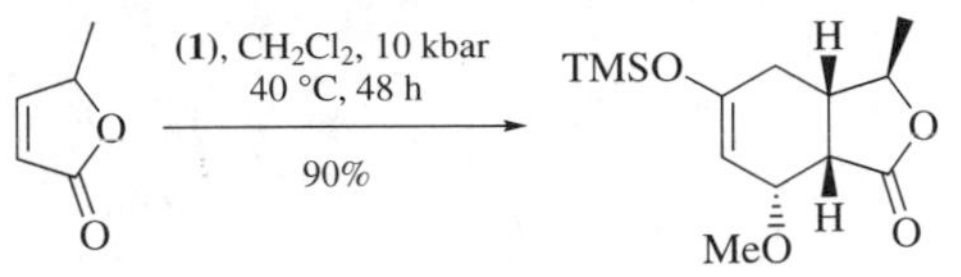

(25)

On the other hand, γ-lactams (eq 26) and δ-lactams (eq 27) are much better dienophiles and give hydroisoindolines[46] and hydroisoquinoline-type manzamine skeletons.[47] Even 2,4-bis(methoxycarbonyl)furan reacts with this diene to give fused heterocycles (eq 28).[48]

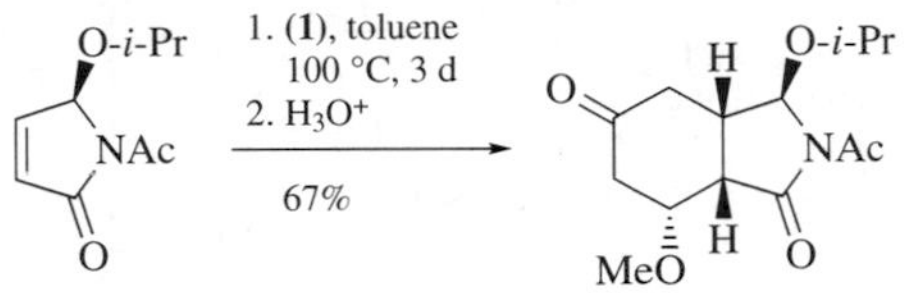

(26)

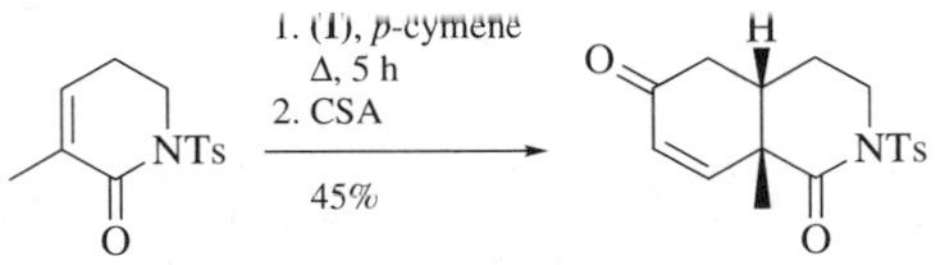

(27)

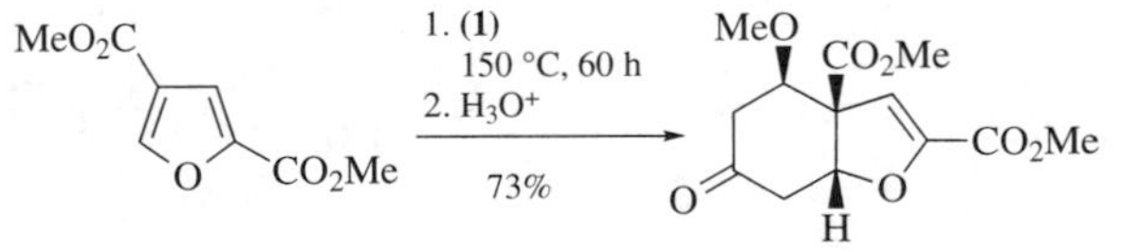

(28)

Miscellaneous. Danishefsky's diene reacts with strong electrophiles such as 2-methylene-1,3-cyclopentanedione to give Michael-type adducts at low temperature, but, on heating, it gives the usual cycloadduct (eq 29).[49]

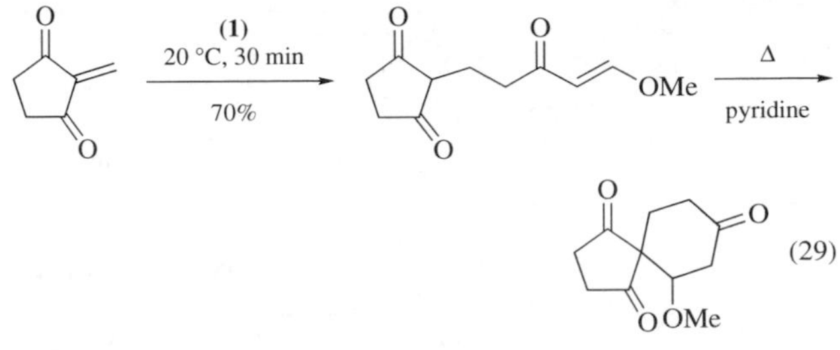

(29)

The electron-rich enol ether moiety of the diene reacts with carbenes preferentially to give cyclopropane analogs, which afford hydroguaiazulene derivatives by Cope rearrangement (eq 30).[50]

(1), $Rh_2(OAc)_4$, CH_2Cl_2, Δ, 30 min; 220 °C; 59% (30)

Strained alkenes are useful dienophiles and tetrachlorocyclopropene reacts with the diene at ambient temperature to give substituted tropones and tropolones via bicycloheptenes (eq 31).[51] Polyhalocyclobutenes give benzocyclobutenones in good yield on a preparative scale (eq 32).[52]

(1), CCl_4, rt, 4 d, 56% (31)

1. (1), 120 °C, 3.5 h; 2. H^+; 53%; NaOMe, MeOH, 96% (32)

Hetero-Diels–Alder Reactions.

General Aspects. Activated hetero-alkenes such as carbonyls and imines with electron-withdrawing substituents react with the diene at somewhat elevated temperatures to give dihydropyrones (eq 33)[53–55] and dihydropyridones, respectively (eq 34).[54,56,57] In these cases, however, reactions are limited only to special hetero-alkenes.

1. (1), THF, 55 °C, 24 h; 2. H_3O^+; ~30%? (33)

1. (1), C_6H_6, Δ, 15 h; 2. H_3O^+; 84% (34)

A major generalization of this heterocycle synthesis has been achieved by the reaction of hetero-alkenes with the diene in the presence of Lewis acid catalysts. In these cases, even hetero-alkenes without activating groups can react under mild conditions to give the corresponding heterocycles (eqs 35 and 36).[58,59]

(1), C_6H_6, $ZnCl_2$ (0.5 equiv), rt, 24–48 h (35)

R	Yield (%)
$PhCH_2OCH_2$	87
Ph	65
Et	48
$CbzNHCH_2$	80

(1), THF, $ZnCl_2$ (0.3 equiv), rt, 36–48 h, 68% (36)

The most common catalyst is ***Zinc Chloride***, but $ZnBr_2$, $MgCl_2$, $MgBr_2$, BF_3 etherate, $TiCl_4$, and alkylaluminum chlorides are also useful. In some cases, even the lanthanide reagent $Eu(fod)_3$ can catalyze the cycloaddition to give cyclic acetals,[60] and partial asymmetric induction is observed using $Eu(hfc)_3$ as a catalyst (eq 37).[61]

PhCHO; (1), $CDCl_3$, $Eu(hfc)_3$ (0.5 mol %), rt, ~24 h; 82%; 18% ee (37)

A mechanistic survey has revealed that this hetero-Diels–Alder reaction proceeds mainly via a pericyclic pathway with *endo* addition, but an aldol-type stepwise process is also followed and the ratio varies with the Lewis acid used.[62]

Oxygen Heterocycles. Hetero-Diels–Alder reactions of chiral aldehydes give substituted dihydropyrones with high diastereoselectivity and Cram products are formed exclusively or predominantly. In the case of α-alkoxy or α-amino aldehydes, excellent selectivities are achieved (eqs 38–40).[63,64] The diastereoselectivity of the reaction with a protected serine-aldehyde is influenced by the amount of Lewis acid employed (eq 41).[65] Because of this high stereoselectivity and since the resulting multifunctionalized pyrone derivatives can be converted to various oxygen heterocycles, a number of natural products, especially carbohydrates and analogs, have been synthesized by this method. The following are representative examples: spiroacetals (*Mus musculus* pheromone[64a] and avermectins[66]); hexoses (fucose and daunosamine[67]); unusual sugars (tunicaminyluracil,[68] octosyl acid A,[69] and galantinic acid[70]); *C*-glycosides (papulacandins[71]).

(1), THF, $MgBr_2$, 80%; only β-H (38)

(1), THF, $MgBr_2$ 67% (39)

β:α = 6.5:1

(1), THF, $ZnCl_2$ 68% (40)

β:α = 9:1

(1), CH_2Cl_2, $ZnCl_2$ rt (41)

Amount of catalyst	Ratio (β:α)	Yield (%)
2 mol %	1:1.3	94
7	18:1	89
70	60:1	57
5 (in THF)		0

Even remote chiral centers afford rather high diastereoselectivity, giving the desired isomers as major products; these are convertible to mevinolin and pravastatin (eq 42).[72]

(1), CH_2Cl_2, $TiCl_4$ –40 °C, 30 min 65% (42)

β:α = 90:10

Nitrogen and Sulfur Heterocycles. As mentioned earlier, dihydropyridones and arylquinolizinones are readily prepared (eq 43).[73] Reaction with triazolediones without a catalyst gives pyridazine derivatives.[74] The diene reacts with in situ formed thioaldehydes without a catalyst to give thiopyranones (eq 44).[75]

(1), THF, $ZnCl_2$ rt, 1 h 73% (43)

(1), CH_2Cl_2 *hν*, N_2 (44)

R = $PhCH_2CH_2$	84%
R = $PhSeCH_2CH_2$	73%
R = Ph	74%

Miscellaneous. Extremely polar hetero-alkenes react smoothly with the diene to give heterocycles. For example, reaction with nitrosobenzene at 0 °C gives substituted oxazinones (eq 45).[76] Phosphabenzenes are prepared by the reaction with chlorophosphorus ylides formed in situ (eq 46).[77] Arsabenzene is formed using the arsine ylide.[78] Reaction with singlet oxygen yields cyclic peroxides.[79]

(1), CH_2Cl_2, 0 °C no catalyst 95% (45)

DABCO Et_2O; (1) –50 °C, 5 h rt, 12 h 70% (46)

Asymmetric Synthesis. Asymmetric Diels–Alder reactions with dienophiles containing a chiral auxiliary usually give adducts with high diastereoselectivity. For example, reaction with a nitro alkene having a chiral sulfoxide gives the hydrindenone with 95% ee after elimination of the sulfoxide.[80] In the case of 5-(−)-menthyloxy-5*H*-furanone, the reaction gives two diastereomers via both *endo* and *exo* attack, but diastereofacial selectivity is exclusive (eq 47).[81] Reaction with *O*-β-D-glucosyljuglone gives a single adduct (eq 48); this is a versatile precursor for optically active anthracyclinones.[82]

1. (1), 100 °C, 20 h 2. TBAF 77% (47)

~100% de

(1), C_6H_6 Δ 92% (48)

~100% de

In the hetero-Diels–Alder reaction, the imine derived from D-galactosylamine gives the (*S*)-adduct preferentially, from which (*S*)-anabasin is obtained.[83] Imines of optically active amino acids give substituted pyridones with high diastereoselectivity (eq 49).[84]

(1), Lewis acid (49)

R^1	LA, Sol	Temp (°C)	Ratio	Yield
Pr	$ZnCl_2$, THF	0	15:85	11%
i-Pr	Et_2AlCl, CH_2Cl_2	−78 to −20	97: 3	45%
i-Pr	$MeAlCl_2$, CH_2Cl_2	−78 to −20	93: 7	50%
$EtOCO(CH_2)_2$	Et_2AlCl, CH_2Cl_2	−78 to −20	93: 7	50%

Catalytic asymmetric hetero-Diels–Alder reactions using chiral acyloxyboranes[85] or oxazaborolidines[86] give substituted hydropyranones and hydropyridones with ca. 70–95% ee (eqs 50 and 51).

PhCHO → (1), (20 mol %), −78 °C, 4–9 h (50)

R	ee	Yield
o-$MeOC_6H_4$	79%	80%
1,3,5-$Me_3C_6H_2$	95%	47%

RCHO → (1), (20 mol %), EtCN, −78 °C then TFA, 80–100%; 67–82% ee (51)

Related Reagents. 1-Acetoxy-1,3-butadiene; 1,1-Dimethoxy-3-trimethylsilyloxy-1,3-butadiene; 1-Methoxy-1,3-bis-(trimethylsilyloxy)-1,3-butadiene; 2-Methoxy-1,3-butadiene; 2-Methoxy-3-phenylthio-1,3-butadiene; 2-Methoxy-1-phenylthio-1,3-butadiene; 1-Trimethylsilyloxy-1,3-butadiene; 2-Trimethylsilyloxy-1,3-butadiene.

1. (a) Danishefsky, S. *ACR* **1981**, *14*, 400. (b) Petrzilka, M.; Grayson, J. I. *S* **1981**, 753. (c) Boger, D. L.; Weinreb, S. M. *Hetero Diels–Alder Methodology in Organic Synthesis*; Academic: London, **1987**.
2. (a) Danishefsky, S.; Kitahara, T. *JACS* **1974**, *96*, 7807. (b) Danishefsky, S.; Kitahara, T.; Yan, C. F.; Morris, J. *JACS* **1979**, *101*, 6996.
3. Danishefsky, S.; Kitahara, T.; Schuda, P. F. *OSC* **1990**, *7*, 312.
4. Cazeau, P.; Duboudin, F.; Moulines, F.; Babot, O.; Dunogues, J. *T* **1987**, *43*, 2089.
5. Hansson, L.; Carlson, R. *ACS* **1989**, *43*, 188.
6. Danishefsky, S.; Kitahara, T. *JOC* **1975**, *40*, 538.
7. Boeckman, R. K., Jr.; Dolak, T. M.; Culos, K. O. *JACS* **1978**, *100*, 7098.
8. Cameron, D. W.; Feutrill, G. I.; Keep, P. L. C. *TL* **1989**, *30*, 5173.
9. Krohn, K. *TL* **1980**, *21*, 3557.
10. Tanaka, H.; Yoshioka, T.; Shimauchi, Y.; Yoshimoto, A.; Ishikura, T.; Naganawa, H.; Takeuchi, T.; Umezawa, H. *TL* **1984**, *25*, 3355.
11. Cameron, D. W.; Feutrill, G. I.; Griffiths, P. G.; O'Brien, D. G. *TL* **1991**, *32*, 6179.
12. Krohn, K.; Tolkiehn, K. *CB* **1979**, *112*, 3453.
13. Farina, F.; Prados, P. *TL* **1979**, 477.
14. Tegmo-Larsson, I. M.; Rozeboom, M. D.; Houk, K. N. *TL* **1981**, *22*, 2043.
15. Gesson, J. P.; Jacquesy, J. C.; Renoux, B. *TL* **1983**, *24*, 2761.
16. Jackson, D. A.; Stoodley, R. J. *CC* **1981**, 478.
17. Evans, J. C.; Klix, R. C.; Bach, R. D. *JOC* **1988**, *53*, 5519.
18. Jones, P. S.; Ley, S. V.; Simpkins, N. S.; Whittle, A. J. *T* **1986**, *42*, 6519.
19. Kitahara, T.; Takagi, Y.; Matsui, M. *ABC* **1979**, *43*, 2359.
20. Ojima, I.; Yatabe, M.; Fuchikami, T. *JOC* **1982**, *47*, 2051.
21. Marx, J. N.; Bombach, E. J. *TL* **1977**, 2391.
22. ApSimon, J. W.; Holmes, A. M.; Johnson, I. *CJC* **1982**, *60*, 308.
23. Hanzawa, Y.; Suzuki, M.; Kobayashi, Y.; Taguchi, T.; Iitaka, Y. *JOC* **1991**, *56*, 1718.
24. (a) Danishefsky, S.; Harayama, T.; Singh, R. K. *JACS* **1979**, *101*, 7008. (b) Danishefsky, S.; Hirama, M. *JACS* **1977**, *99*, 7740.
25. Danishefsky, S.; Morris, J.; Clizbe, L. A. *JACS* **1981**, *103*, 1602.
26. Danishefsky, S.; Morris, J.; Mullen, G.; Gammill, R. *JACS* **1982**, *104*, 7591.
27. (a) Ramage, R.; McLeod, A. M. *CC* **1984**, 1008. (b) Ramage, R.; McLeod, A. M. *T* **1986**, *42*, 3251.
28. Kinney, W. A.; Crouse, G. D.; Paquette, L. A. *JOC* **1983**, *48*, 4986.
29. Chou, T. S.; Hung, S. C. *JOC* **1988**, *53*, 3020.
30. Hayakawa, K.; Nishiyama, H.; Kanematsu, K. *JOC* **1985**, *50*, 512.
31. Corey, E. J.; Estreicher, H. *JACS* **1978**, *100*, 6294.
32. Kraus, G. A.; Thurston, J.; Thomas, P. J.; Jacobson, R. A.; Su, Y. *TL* **1988**, *29*, 1879.
33. Danishefsky, S.; Kitahara, T.; Schuda, P. F.; Etheredge, S. J. *JACS* **1976**, *98*, 3028; **1977**, *99*, 6066.
34. Danishefsky, S.; Hirama, M.; Gombatz, K.; Harayama, T.; Berman, E.; Schuda, P. F. *JACS* **1978**, *100*, 6536; **1979**, *101*, 7020.
35. (a) Voyle, M.; Dunlap, N. K.; Watt, D. S.; Anderson, O. P. *JOC* **1983**, *48*, 3242. (b) Kim, M.; Kawada, K.; Gross, R. S.; Watt, D. S. *JOC* **1990**, *55*, 504.
36. Rosen, T.; Taschner, M. J.; Thomas, J. A.; Heathcock, C. H. *JOC* **1985**, *50*, 1190.
37. Nakamura, H.; Ye, B.; Murai, A. *TL* **1992**, *33*, 8113.
38. Tobe, Y.; Iseki, T.; Kakiuchi, K.; Odaira, Y. *TL* **1984**, *25*, 3895.
39. Lin, H. S.; Paquette, L. A. *OS* **1989**, *67*, 163.
40. Nantz, M. H.; Fuchs, P. L. *JOC* **1987**, *52*, 5298.
41. Danishefsky, S.; Kahn, M. *TL* **1981**, *22*, 489.
42. (a) Tsuda, Y.; Ohshima, T.; Sano, T.; Toda, J. *H* **1982**, *19*, 2027. (b) Sano, T.; Toda, J.; Kashiwaba, N.; Ohshima, T.; Tsuda, Y. *CPB* **1987**, *35*, 479.
43. Fujiwara, T.; Ohsaka, T.; Inoue, T.; Takeda, T. *TL* **1988**, *29*, 6283.
44. Kraus, G. A.; Hon, Y. S.; Sy, J.; Raggon, J. *JOC* **1988**, *53*, 1397.
45. Branchadell, V.; Sodupe, M.; Ortuno, R. M.; Oliva, A.; Gomez-Pardo, D.; Guingant, A.; D'Angelo, J. *JOC* **1991**, *56*, 4135.
46. Koot, W. J.; Hiemstra, H.; Speckamp, W. N. *JOC* **1992**, *57*, 1059.
47. Torisawa, Y.; Nakagawa, M.; Hosaka, T.; Tanabe, K.; Lai, Z.; Ogata, K.; Nakata, T.; Oishi, T.; Hino, T. *JOC* **1992**, *57*, 5741.
48. Wenkert, E.; Piettre, S. R. *JOC* **1988**, *53*, 5850.
49. Bunnelle, W. H.; Meyer, L. A. *JOC* **1988**, *53*, 4038.
50. Cantrell, W. R., Jr.; Davies, H. M. L. *JOC* **1991**, *56*, 723.
51. Banwell, M. G.; Knight, J. H. *CC* **1987**, 1082.
52. South, M. S.; Liebeskind, L. S. *JOC* **1982**, *47*, 3815.
53. Keana, J. F. W.; Eckler, P. E. *JOC* **1976**, *41*, 2850.

54. Jung, M. E.; Shishido, K.; Light, L.; Davis, L. *TL* **1981**, *22*, 4607.
55. Belanger, J.; Landry, N. L.; Pare, J. R. J.; Jankowski, K. *JOC* **1982**, *47*, 3649.
56. Kloek, J. A.; Leschinsky, K. L. *JOC* **1979**, *44*, 305.
57. Abramovitch, R. A.; Stowers, J. R. *H* **1984**, *22*, 671.
58. (a) Danishefsky, S.; Kerwin, J. F., Jr.; Kobayashi, S. *JACS* **1982**, *104*, 358. (b) Danishefsky, S.; Kerwin, J. F., Jr. *JOC* **1982**, *47*, 1597.
59. (a) Danishefsky, S.; Kerwin, J. F., Jr. *JOC* **1982**, *47*, 3183. (b) Danishefsky, S.; Kerwin, J. F., Jr. *TL* **1982**, *23*, 3739.
60. Bednarski, M.; Danishefsky, S. *JACS* **1983**, *105*, 3716.
61. Bednarski, M.; Maring, C.; Danishefsky, S. *TL* **1983**, *24*, 3451.
62. Danishefsky, S.; Larson, E.; Askin, D.; Kato, N. *JACS* **1985**, *107*, 1246.
63. Danishefsky, S.; Kobayashi, S.; Kerwin, J. F., Jr. *JOC* **1982**, *47*, 1981.
64. (a) Danishefsky, S. J.; Pearson, W. H.; Harvey, D. F. *JACS* **1984**, *106*, 2455, 2456. (b) Danishefsky, S. J.; Pearson, W. H.; Harvey, D. F.; Maring, C. J.; Springer, J. P. *JACS* **1985**, *107*, 1256.
65. Garner, P.; Ramakanth, S. *JOC* **1986**, *51*, 2609.
66. Danishefsky, S. J.; Armistead, D. M.; Wincott, F. E.; Selnick, H. G.; Hungate, R. *JACS* **1987**, *109*, 8117; **1989**, *111*, 2967.
67. Danishefsky, S.; Maring, C. *JACS* **1985**, *107*, 1269.
68. (a) Danishefsky, S.; Barbachyn, M. *JACS* **1985**, *107*, 7761. (b) Danishefsky, S. J.; DeNinno, S. L.; Chen, S. H.; Boisvert, L.; Barbachyn, M. *JACS* **1989**, *111*, 5810.
69. Danishefsky, S. J.; Hungate, R. *JACS* **1986**, *108*, 2486.
70. Golebiowski, A.; Kozak, J.; Jurczak, J. *TL* **1989**, *30*, 7103.
71. Danishefsky, S. J.; Phillips, G.; Ciufolini, M. *Carbohydr. Res.* **1987**, *171*, 317.
72. (a) Wovkulich, P. M.; Tang, P. C.; Chadha, N. K.; Batcho, A. D.; Barrish, J. C.; Uskokovic, M. R. *JACS* **1989**, *111*, 2596. (b) Daniewski, A. R.; Wovkulich, P. M.; Uskokovic, M. R. *JOC* **1992**, *57*, 7133.
73. Vacca, J. P. *TL* **1985**, *26*, 1277.
74. Johnson, M. P.; Moody, C. J. *JCS(P1)* **1985**, 71.
75. (a) Vedejs, E.; Eberlein, T. H.; Mazur, D. J.; McClure, C. K.; Perry, D. A.; Ruggeri, R.; Schwartz, E.; Stults, J. S.; Varie, D. L.; Wilde, R. G.; Wittenberger, S. *JOC* **1986**, *51*, 1556. (b) Vedejs, E.; Stults, J. S.; Wilde, R. G. *JACS* **1988**, *110*, 5452.
76. McClure, K. F.; Danishefsky, S. J. *JOC* **1991**, *56*, 850.
77. Pellon, P.; Hamelin, J. *TL* **1986**, *27*, 5611.
78. Himdi-Kabbab, S.; Pellon, P.; Hamelin, J. *TL* **1989**, *30*, 349.
79. Clennan, E. L.; L'Esperance, R. P. *TL* **1983**, *24*, 4291.
80. Fuji, K.; Tanaka, K.; Abe, H.; Itoh, A.; Node, M.; Taga, T.; Miwa, Y.; Shiro, M. *TA* **1991**, *2*, 1319.
81. De Jong, J. C.; Van Bolhuis, F.; Feringa, B. L. *TA* **1991**, *2*, 1247.
82. Beagley, B.; Curtis, A. D. M.; Pritchard, R. G.; Stoodley, R. J. *JCS(P1)* **1992**, 1981.
83. Pfrengle, W.; Kunz, H. *JOC* **1989**, *54*, 4261.
84. Waldmann, H.; Braun, M. *JOC* **1992**, *57*, 4444.
85. (a) Gao, Q.; Maruyama, T.; Mouri, M.; Yamamoto, H. *JOC* **1992**, *57*, 1951. (b) Hattori, K.; Yamamoto, H. *JOC* **1992**, *57*, 3264.
86. Corey, E. J.; Cywin, C. L.; Roper, T. D. *TL* **1992**, *33*, 6907.

Takeshi Kitahara
The University of Tokyo, Japan

1-Methoxy-1-(trimethylsilyloxy)propene[1]

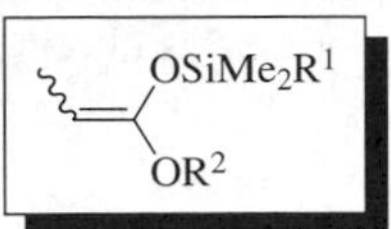

(**1**; R^1 = Me, R^2 = Me)		
[34880-70-1]	$C_7H_{16}O_2Si$	(MW 160.32)
(*E*)-(**1**)		
[72658-09-4]		
(*Z*)-(**1**)		
[72658-03-8]		
(**2**; R^1 = *t*-Bu, R^2 = Et)		
[83165-77-9]	$C_{11}H_{24}O_2Si$	(MW 216.44)
(*E*)-(**2**)		
[89043-55-0]		
(*Z*)-(**2**)		
[73967-98-3]		

(reactive nucleophile in Mukaiyama-type aldol reactions[2,3] with aldehydes,[15–29] imines,[30] acetals,[36] dialkoxycarbenium ions,[40] and orthoesters;[41] Michael-type additions to α,β-unsaturated carbonyl compounds;[42–46] substitution reactions with allylic alcohols and esters;[56] acylations;[53] aminations;[54] hydroxylations;[55] and pericyclic reactions;[59] silylating agent[67])

Alternate Name: methylketene methyl trimethylsilyl acetal.

Physical Data: (*Z*)-(**1**): bp 57 °C/18 mmHg; ^{1}H NMR data have been reported.[11] (*Z*)- and (*E*)-(**2**): ^{1}H and ^{13}C NMR data have been reported.[8b]

Solubility: highly sol organic solvents.

Preparative Methods: since the first preparation of a ketene trialkylsilyl acetal by Petrov in 1959 by the reaction of triethylsilane and methyl acrylate,[4] several alternative pathways have been explored.[5] Ainsworth and co-workers prepared 1-methoxy-1-(trimethylsilyloxy)propene in 90% yield by deprotonation of methyl propionate with ***Lithium Diisopropylamide*** at −78 °C followed by addition of ***Chlorotrimethylsilane***.[6] Reaction of the lithium enolate of methyl acetate with TMS-Cl provides 65% *O*-silylated and 35% *C*-silylated products (eq 1).[7] Substitution on the alcohol portion of the ester favors *C*-silylation while substitution on the α-carbon favors *O*-silylation. An increase in the steric bulk of the silylating agent (e.g. use of ***t-Butyldimethylchlorosilane***) results in significantly more *O*-silylation.

(eq 1): MeC(O)OMe —[1. LDA, −78 °C; 2. TMSCl, −78 to 25 °C]→ CH_2=C(OTMS)(OMe) + $TMSCH_2C(O)OMe$ (1)

Silyl ketene acetal geometry is controlled by the selective formation of the (*E*)- and (*Z*)-ester enolate. Deprotonation of methyl propionate with LDA in THF at −78 °C gives, upon silylation, a 6:94 ratio of the 'thermodynamic' (*Z*) and 'kinetic' (*E*) silyl ketene acetals (eq 2).[8] As noted by Ireland and co-workers, a change in the reaction solvent results in a reversal in selectivity.

In a THF–23% HMPA mixture, the silyl ketene acetals are isolated in a 84:16 ratio. In THF–45% DMPU ratios of up to 98:2 are obtained. With bulkier amide bases and in the absence of dipolar additives, a (*Z*):(*E*) ratio of 5:95 is observed.[9,10]

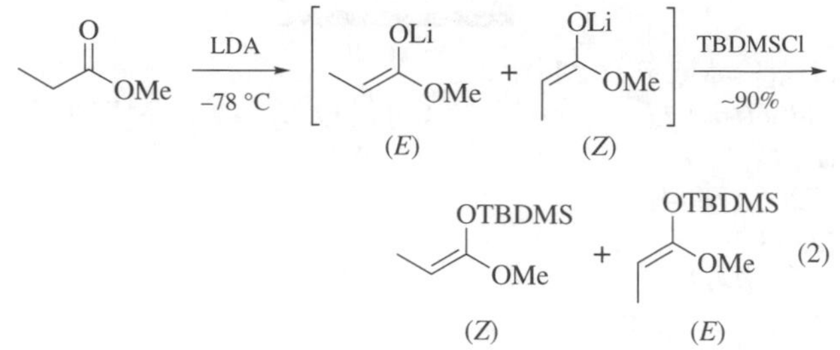

(2)

THF	(*Z*):(*E*) = 6:94
THF, 23% HMPA	(*Z*):(*E*) = 84:16
THF, 45% DMPU	(*Z*):(*E*) = >95: 5

Treatment of alkyl carboxylates with trialkylsilyl triflates in the presence of ***Triethylamine*** yields the thermodynamically more stable (*Z*)-ketene acetals at rt; however, a mixture of *O*- and *C*-silylated products is generally obtained.[11] The (*Z*)-isomers are also prepared by the reaction of triethylsilyl perchlorates with aliphatic esters.[12] The action of zinc powder on α-bromo esters in the presence of TMS-Cl and TMEDA–Et_3N leads mainly to (*E*)-ketene acetals, while the reaction of trialkylsilanes on alkyl acrylates catalyzed by Wilkinson's reagent (***Chlorotris(triphenylphosphine)rhodium(I)***) gives the corresponding (*Z*)-isomers (*Z*:*E* >98:2).[13]

Handling, Storage, and Precautions: easily hydrolyzed and oxidized by exposure to air. Storage is possible in sealed tubes at rt. Use in a fume hood.

Alkylation. The α-*t*-alkylation of silyl ketene acetals was reported by Reetz and Schwellnus.[14]

Additions to Aldehydes, Imines, and Oximes. Silyl ketene acetals add to aldehydes[2,3,19] in the presence of a Lewis acid ($TiCl_4$, BF_3, $SnCl_4$, $TrClO_4$, cationic iron complexes,[15] HgI_2,[16] lanthanides,[17] and clay montmorillonite).[18] High diastereofacial selectivity in Lewis acid-mediated additions of methylketene acetals to aldehydes was reported by Heathcock and Flippin.[20–24] The scope of this reaction was further extended by the introduction of camphor and *N*-methylephedrine derived chiral silyl ketene acetals.[25–28] The asymmetric variant of the Mukaiyama reaction also provides an efficient *anti*-selective chiral propionate equivalent (eqs 3 and 4). Reetz and Fox used a catalytic amount of ***Lithium Perchlorate*** suspended in CH_2Cl_2 for Mukaiyama aldol reactions of aldehydes with a silyl ketene acetal.[29]

High diastereoselectivities have been achieved in the Lewis acid-mediated addition of silyl ketene acetals to imines (eq 5).[30,31] In related applications, additions of *O*-silylated ketene acetals to β-lactams were used in the synthesis of carbapenem antibiotics.[32–34] Sekiya and co-workers have demonstrated that silyl ketene acetals add to *O*-benzyl oxime ethers in the presence of catalytic amounts of ***Trimethylsilyl Trifluoromethanesulfonate*** to afford β-benzyloxy amino esters.[35]

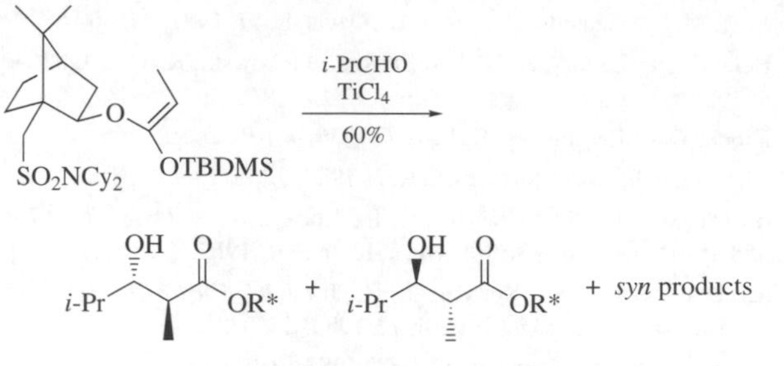

(3)

7.5:92.5

syn:*anti* = 1.8:98.2

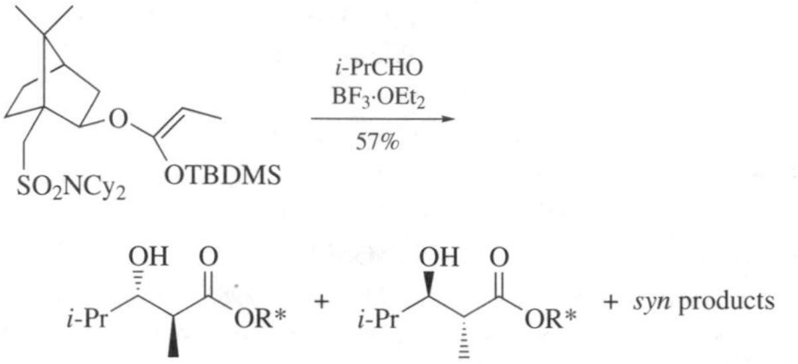

(4)

93.5:6.5

syn:*anti* = 6.5:93.5

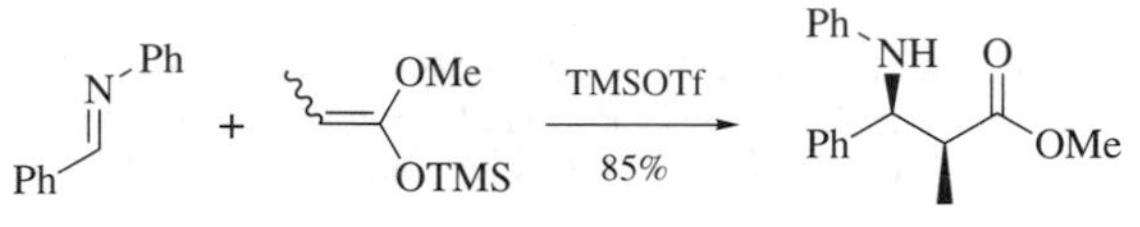

(5)

(*Z*):(*E*) = 25:75

syn only

Additions to Acetals, Dialkoxycarbenium Ions, and Orthoesters. Much attention has focused on the use of methylketene derivatives for the opening of chiral acetals.[36] Trimethylsilyl triflate-catalyzed addition of (*E*)-methylketene methyl trimethylsilyl acetal to benzaldehyde dimethyl acetal gives *syn*- and *anti*-esters in a 50:50 ratio, whereas the (*Z*)-silyl ketene acetal results in a 55:45 ratio (eq 6).[37]

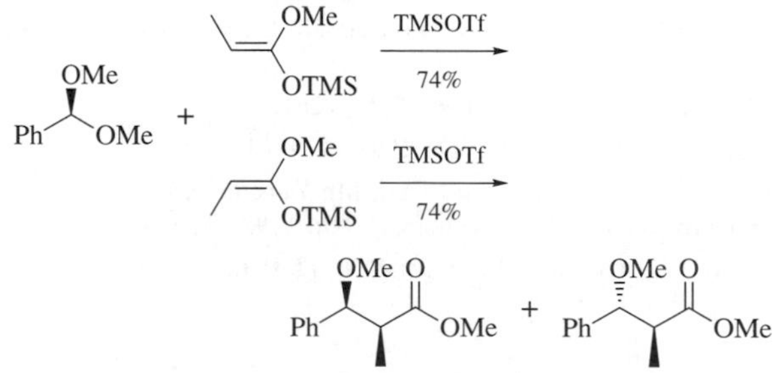

(6)

syn:*anti* = 50:50

syn:*anti* = 55:45

A chain-extended trimethylsilyl ketene acetal was used by Johnson et al. in combination with ***Titanium(IV) Chloride*** for a stereoselective acetal opening in the preparation of enantiomerically pure mevinolin analogs (eq 7).[38,39]

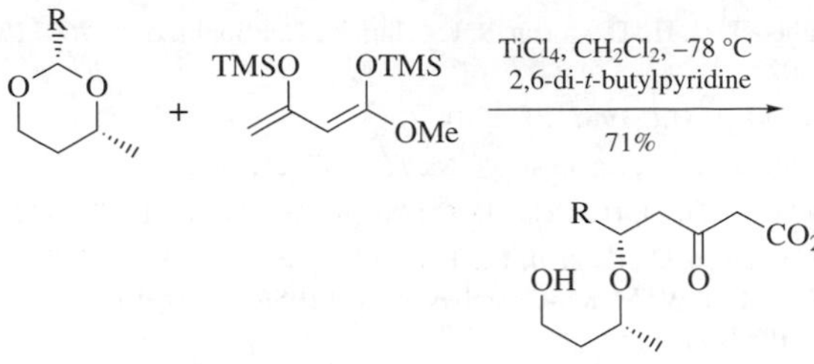

(7)

1,3-Dioxolan-2-ylium cations react with silyl ketene acetals to give β-keto esters, selectively monoprotected at the ketone carbonyl (eq 8).[40]

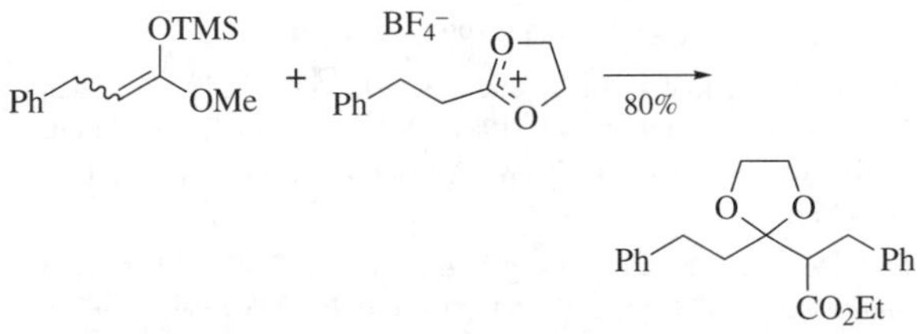

(8)

The reaction of an enolic orthoester with silyl ketene acetals results in the regiospecific formation of diketo ester monoacetals.[41]

Additions to α,β-Unsaturated Carbonyl Compounds. The $TiCl_4$-catalyzed addition of silyl ketene acetals to α,β-unsaturated enones is often superior to enolate addition protocols, especially for the preparation of β-quaternary centers.[42,43] Danishefsky and co-workers reported an interesting tandem Michael–aldol sequence (eq 9).[44] Acetonitrile or high pressure are often sufficient to promote additions in the absence of a Lewis acid.[45,46] In a related process, aromatic heterocycles can be prepared by addition of silyl ketene acetals to nitrobenzenes.[47]

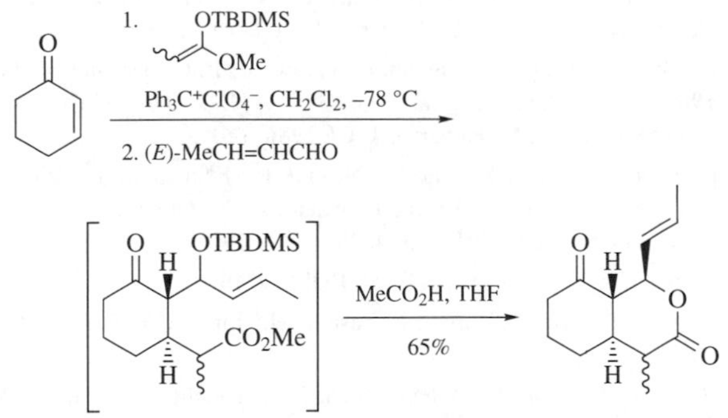

(9)

Alkylations with Allylic Alcohols and Esters. Pearson and Schkeryantz[48] used a lithium perchlorate-promoted substitution of a tertiary alcohol with ketene silyl acetals for the preparation of a key intermediate in the synthesis of (±)-γ-lycorane.[49,50] Grieco and co-workers have also used cyclopropyl alcohols in substitution reactions of this type.[51] Pd^0-catalyzed coupling of allyl acetates and ketene silyl acetals leads to cyclopropane derivatives via alkylation of the central carbon (C-2) of the allyl group.[52]

Acylation, Amination, and Hydroxylation. Acylation of silyl ketene acetals with acid chlorides occurs regioselectively and results in β-keto esters after acid hydrolysis.[53] α-Amino acid esters are prepared by amination of silyl ketene acetals with 3-acetoxyaminoquinazolin-4(3*H*)-ones followed by N–N bond cleavage.[54] Silyl ketene acetals may be epoxidized by peroxyacids and subsequently cleaved with fluoride ion to reveal the α-hydroxy esters.[55] Related sequences involve $^1O_2$,[56] and ***Lead(IV) Acetate***,[57] as well as camphor-derived auxiliaries.[58]

Claisen Rearrangements. Silyl ketene acetals of allyl esters undergo rapid Claisen rearrangements,[8b,59] at or above rt to give γ,δ-unsaturated carboxylic acid derivatives (eq 10).[60]

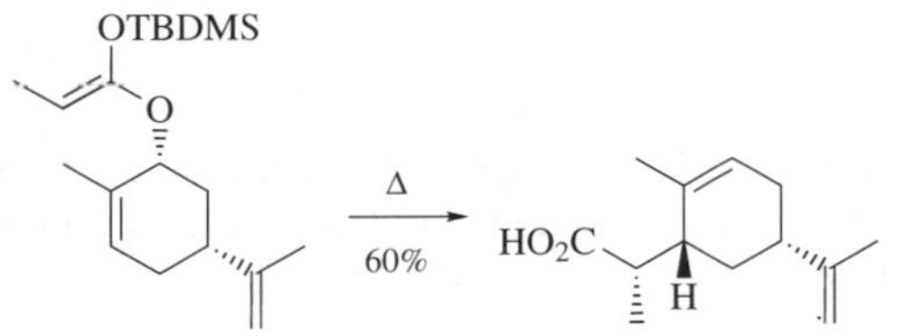

(10)

Reactions with Sulfur-Containing Electrophiles. Paterson has developed an alkylation of geminal phenylthio chlorides with silyl enol ethers as a general method to prepare 1,5 dicarbonyl compounds (eq 11).[61] Alkylation of the silyl enol ether gives a homoallylic sulfide. Ozonolysis of the alkene with concomitant sulfur oxidation produces an unstable sulfoxide which is thermally extruded to give a mixture of alkene isomers.

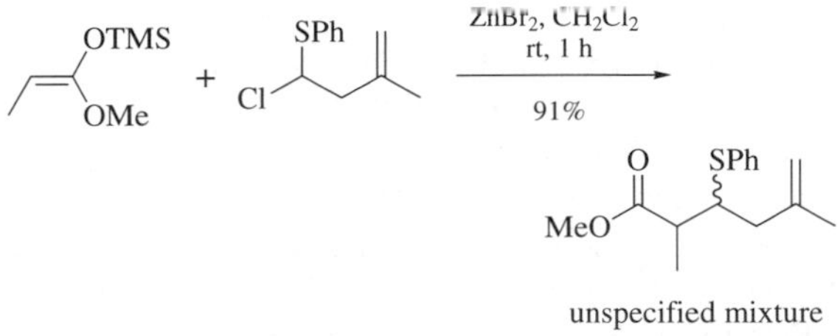

(11)

The ketene acetal has also been reported to react with dithianyl cations. Treatment of 1,3-dithian-2-ylium tetrafluoroborate with the silyl enol ether results in a β-dithio ester which can subsequently be hydrolyzed to a β-keto ester (eq 12).[62]

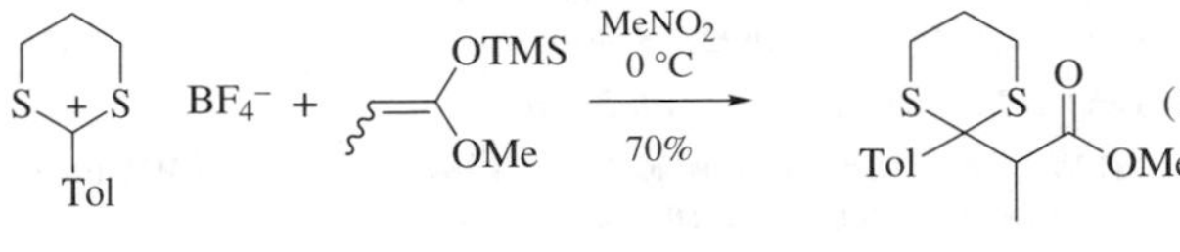

(12)

Cycloadditions and Photochemistry. Silyl ketene acetals undergo inverse electron demand hetero-Diels–Alder reactions with enones.[63] Photochemically induced Diels–Alder and [2 + 2] cycloadditions of silyl ketene acetals with dienes have been reported by Schuster and co-workers.[64] Cyclopropanes with vicinal donor and acceptor ligands will heterolytically cleave to form zwitterionic intermediates. In the presence of the ketene silyl acetal, a [3 + 2] cycloaddition reaction can occur with the zwitterionic intermediate to form highly substituted cyclopentenones (eq 13).[65]

$$\text{(MeO)}_2\text{-cyclopropane(Et)}\text{-}CO_2Et + \text{MeCH=C(OTMS)(OMe) (3 equiv)} \xrightarrow[70\%]{TiCl_4,\ CH_2Cl_2,\ -78\ °C\ \text{to rt}} \text{cyclopentenone (Et, OMe, } CO_2Et) + \text{cyclopentenone (Et, } CO_2Et) \quad 7:3 \tag{13}$$

Reactions with Nitrenes. *N*-Protected α-amino esters are produced by the reaction of (ethoxycarbonyl)nitrene with methylketene ethyl trimethylsilyl acetal.[66]

Use as a Silylating Agent. 1-Methoxy-1-(trimethylsilyloxy)propene has been used as a mild silylating agent for phenols, alcohols, carboxylic acids, thiols, and amides.[67]

Related Reagents. Ketene Bis(trimethylsilyl) Acetal; Ketene *t*-Butyldimethylsilyl Methyl Acetal; Ketene Diethyl Acetal; 1-Methoxy-3-methyl-1-trimethylsilyloxy-1,3-butadiene; 1-Methoxy-2-trimethylsilyl-1-(trimethylsilyloxy)ethylene; Methylketene Bis(trimethylsilyl) Acetal; Methylketene Dimethyl Acetal; Tris(trimethylsilyloxy)ethylene.

---

1. Brassard, P. In *The Chemistry of Ketenes, Allenes and Related Compounds*; Patai, S., Ed.; Wiley: New York, 1980; Part 2, p 487.
2. Mukaiyama, T. *OR* **1982**, *28*, 203.
3. Mukaiyama, T.; Murakami, M. *S* **1987**, 1043.
4. (a) Petrov, A. D.; Sadykh-Zade, S. I.; Filatova, I. *JGU* **1959**, *29*, 2896. (b) Yoshi, E.; Kobayashi, Y.; Koizumi, T.; Oribe, T. *CPB* **1974**, *22*, 2767.
5. (a) Krüger, C. R.; Rochow, E. G. *JOM* **1964**, *1*, 476. (b) Lutsenko, I. F.; Baukov, Y. I.; Burlachenko, G. S.; Khasapov, B. N. *JOM* **1966**, *5*, 20. (c) Kuo, N.-Y.; Chen, F.; Ainsworth, C.; Bloomfield, J. J. *CC* **1971**, 136.
6. Ainsworth, C.; Chen, F.; Kuo, Y.-N. *JOM* **1972**, *46*, 59.
7. Rathke, M. W.; Sullivan, D. F. *SC* **1973**, *3*, 67.
8. (a) Ireland, R. E.; Mueller, R. H.; Willard, A. K. *JACS* **1976**, *98*, 2868. (b) Ireland, R. E.; Wipf, P.; Armstrong, J. D., III. *JOC* **1991**, *56*, 650.
9. Corey, E. J.; Gross, A. W. *TL* **1984**, *25*, 495.
10. For the effects of hexane in enolization mixtures, see: Munchhof, M. J.; Heathcock, C. H. *TL* **1992**, *33*, 8005.
11. Emde, H.; Simchen, G. *LA* **1983**, 816.
12. (a) Wilcox, C. S.; Babston, R. E. *TL* **1984**, *25*, 699. (b) Wilcox, C. S.; Babston, R. E. *JOC* **1984**, *49*, 1451.
13. Slougui, N.; Rousseau, G. *SC* **1987**, *17*, 1.
14. Reetz, M. T.; Schwellnus, K. *TL* **1978**, 1455.
15. (a) Bach, T.; Fox, D. N. A.; Reetz, M. T. *CC* **1992**, 1634. (b) For a cationic ruthenium catalyst for the Mukaiyama crossed aldol reaction, see: Odenkirk, W.; Whelan, J.; Bosnich, B. *TL* **1992**, *33*, 5729. (c) For the use of TBSCl–$InCl_3$ as catalyst, see: Mukaiyama, T.; Ohno, T.; Han, J. S.; Kobayashi, S. *CL* **1991**, 949.
16. Dicker, I. B. *JOC* **1993**, *58*, 2324.
17. Mikami, K.; Terada, M.; Nakai, T. *TA* **1991**, *2*, 993.
18. Onaka, M.; Mimura, T.; Ohno, R.; Izumi, Y. *TL* **1989**, *30*, 6341.
19. For a recent review, see: Gennari, C. *COS* **1991**, *2*, 629.
20. Heathcock, C. H.; Flippin, L. A. *JACS* **1983**, *105*, 1667.
21. Heathcock, C. H.; Davidsen, S. K.; Hug, K. T.; Flippin, L. A. *JOC* **1986**, *51*, 3027.
22. Reetz, M. T. *PAC* **1985**, *57*, 1781.
23. Barrett, A. G. M.; Raynham, T. M. *TL* **1987**, *28*, 5615.
24. Terada, M.; Gu, J. H.; Deka, D. C.; Mikami, K.; Nakai, T. *CL* **1992**, 29.
25. (a) Helmchen, G.; Leikauf, U.; Taufer-Knöpfel, I. *AG(E)* **1985**, *24*, 874. (b) Oppolzer, W.; Marco-Contelles, J. *HCA* **1986**, *69*, 1699. (c) Oppolzer, W. *T* **1987**, *43*, 1969.
26. (a) Gennari, C.; Bernardi, A.; Colombo, L.; Scolastico, C. *JACS* **1985**, *107*, 5812. (b) Gennari, C.; Colombo, L.; Bertolini, G.; Schimperna, G. *JOC* **1987**, *52*, 2754.
27. Reetz, M. T.; Kunisch, F.; Heitmann, P. *TL* **1986**, *27*, 4721.
28. Reetz, M. T.; Kesseler, K. *JOC* **1985**, *50*, 5434.
29. Reetz, M. T.; Fox, D. N. A. *TL* **1993**, *34*, 1119.
30. (a) Ojima, I.; Inaba, S.; Yoshida, K. *TL* **1977**, 3643. (b) Guanti, G.; Narisano, E.; Banfi, L. *TL* **1987**, *28*, 4331. (c) For addition to *N*-silylimines, see: Colvin, E. W.; McGarry, D.; Nugent, M. J. *T* **1988**, *44*, 4157.
31. For reviews, see: (a) Evans, D. A.; Nelson, J. V.; Taber, T. R. *Top. Stereochem.* **1982**, *13*, 1. (b) Kleinman, E. F. *COS* **1991**, *2*, 893.
32. Kita, Y.; Shibata, N.; Tohjo, T.; Yoshida, N. *JCS(P1)* **1992**, *57*, 5054.
33. Murahashi, S.-I.; Saito, T.; Naota, T.; Kumobayashi, H.; Akutagawa, S. *TL* **1991**, *32*, 5991.
34. For the Lewis acid assisted condensation between a 5-methoxyisoxazolidine and silyl ketene acetals, see: Kozikowski, A. P.; Stein, P. D. *JACS* **1985**, *107*, 2569.
35. Ikeda, K.; Yoshinaga, Y.; Achiwa, K.; Sekiya, M. *CL* **1984**, 369.
36. Mukaiyama, T.; Murakami, M. *S* **1987**, 1043.
37. Murata, S.; Suzuki, M.; Noyori, R. *T* **1988**, *44*, 4259.
38. Johnson, W. S.; Kelson, A. B.; Elliott, J. D. *TL* **1988**, *29*, 3757.
39. Brownbridge, P.; Chan, T. H.; Brook, M. A.; Kang, G. J. *CJC* **1983**, *61*, 688.
40. Hayashi, Y.; Wariishi, K.; Mukaiyama, T. *CL* **1987**, 1243.
41. (a) Collins, D. J.; Dosen, M.; Jhingran, A. G. *TL* **1990**, *31*, 421. (b) Collins, D. J.; Choo, G. L. P.; Obrist, H. *AJC* **1990**, *43*, 617.
42. (a) Danishefsky, S. J.; Vaughan, K.; Gadwood, R. C.; Tsuzuki, K. *JACS* **1980**, *102*, 4262. (b) Schultz, A. G.; Godfrey, J. D. *JOC* **1976**, *41*, 3494. (c) Kobayashi, S.; Mukaiyama, T. *CL* **1986**, 1805.
43. (a) Otera, J.; Fujita, Y.; Sato, T.; Nozaki, H.; Fukuzumi, S.; Fujita, M. *JOC* **1992**, *57*, 5054; (b) Sato, T.; Wakahara, Y.; Otera, J.; Nozaki, H.; Fukuzumi, S. *JACS* **1991**, *113*, 4028.
44. Danishefsky, S. J.; Audia, J. E. *TL* **1988**, *29*, 1371.
45. Kita, Y.; Segawa, J.; Haruta, J.; Yasuda, H.; Tamura, Y. *JCS(P1)* **1982**, 1099.
46. Bunce, R. A.; Schlecht, M. F.; Dauben, W. G.; Heathcock, C. H. *TL* **1983**, *24*, 4943.
47. Rajanbabu, T. V.; Chenard, B. L.; Petti, M. A. *JOC* **1986**, *51*, 1704.
48. Grieco, P. A.; Clark, J. D.; Jagoe, C. T. *JACS* **1991**, *113*, 5488.
49. Pearson, W. H.; Schkeryantz, J. M. *JOC* **1992**, *57*, 6783.
50. Pearson, W. H.; Schkeryantz, J. M. *JOC* **1992**, *57*, 2986.
51. Grieco, P. A.; Collins, J. L.; Henry, K. J. *TL* **1992**, *33*, 4735.
52. (a) Carfagna, C.; Mariani, L.; Musco, A.; Sallese, G. *JOC* **1991**, *56*, 3924. (b) Carfagna, C.; Musco, A.; Sallese, G., Santi, R.; Fiorani, T. *JOC* **1991**, *56*, 261.
53. Rathke, M. W.; Sullivan, D. F. *TL* **1973**, 1297.
54. Atkinson, R. S.; Kelly, B. J.; Williams, J. *T* **1992**, *48*, 7713.
55. Rubottom, G. M.; Marrero, R. *SC* **1981**, *11*, 505.
56. Adam, W.; del Fierro, J. *JOC* **1978**, *43*, 1159.
57. Rubottom, G. M.; Gruber, J. M.; Marrero, R.; Juve, H. D.; Kim, C.-W. *JOC* **1983**, *48*, 4940.
58. Oppolzer, W.; Dudfield, P. *HCA* **1985**, *68*, 216.

59. Wipf, P. *COS* **1991**, *5*, 827.
60. Ireland, R. E.; Wipf, P.; Xiang, J.-N. *JOC* **1991**, *56*, 3572.
61. Khan, H. A.; Paterson, I. *TL* **1982**, *23*, 2399.
62. Stahl, I. *CB* **1985**, *118*, 3159.
63. Maier, M.; Schmidt, R. R. *LA* **1985**, 2261.
64. (a) Akbulut, N.; Schuster, G. B. *TL* **1988**, *29*, 5125. (b) Akbulut, N.; Hartsough, D.; Kim, J.-I.; Schuster, G. B. *JOC* **1989**, *54*, 2549.
65. Saigo, K.; Shimada, S.; Shibasaki, T.; Hasegawa, M. *CL* **1990**, 1093.
66. Chang, Y. H.; Chiu, F.-T.; Zon, G. *JOC* **1981**, *46*, 342.
67. (a) Kita, Y.; Tohma, H.; Inagaki, M.; Hatanaka, K.; Yakura, T. *JACS* **1992**, *114*, 2175. (b) Kita, Y.; Haruta, J.; Segawa, J.; Tamura, Y. *TL* **1979**, 4311.

Peter Wipf
University of Pittsburgh, PA, USA

Dennis Wright & Mark C. McMills
Ohio University, Athens, OH, USA

Methyl Acrylate

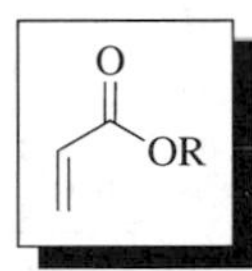

(**1**; R = Me)		
[96-33-3]	$C_4H_6O_2$	(MW 86.10)
(**2**; R = Et)		
[140-88-5]	$C_5H_8O_2$	(MW 100.13)
(**3**; R = *n*-Bu)		
[141-32-2]	$C_7H_{12}O_2$	(MW 128.19)
(**4**; R = *t*-Bu)		
[1663-39-4]	$C_7H_{12}O_2$	(MW 128.19)

(electrophile in conjugate addition reactions; dienophile or dipolarophile in cycloaddition reactions; acceptor in radical addition reactions; used in ene reactions)

Alternate Name: methyl propenoate.
Physical Data: (**1**) mp −75 °C; bp 80 °C. (**2**) mp −71 °C; bp 99 °C. (**3**) bp 145 °C. (**4**) bp 61–63 °C/60 mmHg.
Solubility: sol most organic solvents; slightly sol water.
Purification: wash repeatedly with aqueous NaOH to remove inhibitors, wash with H_2O, dry over $CaCl_2$, and distill under reduced pressure.
Handling, Storage, and Precautions: store at 0 °C in the dark (material will polymerize if exposed to light). Inhibited with up to 200 ppm hydroquinone monomethyl ether. Lachrymator and potential vesicant. Use in a fume hood. The ethyl ester (**2**) is a cancer suspect agent.

Conjugate Additions. Acrylic acid esters have been used as Michael acceptors for a variety of nucleophiles. A typical example using an amino alcohol is shown in eq 1.[1] This example illustrates chemoselectivity (amine over alcohol) and demonstrates that primary amines can undergo multiple additions.[2] This reaction has been performed with anilines,[3] imines,[4] guanidines,[5] hydrazones,[6] xanthines,[7] and pyrrolopyrimidines.[8] Other heteroatomic nucleophiles have also been used, including alcohols,[9] thiols,[10] halides,[11] and phosphorus reagents.[12]

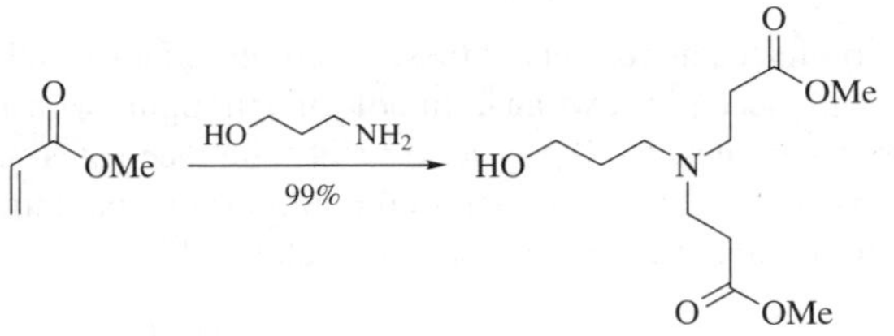

(1)

A variety of carbon nucleophiles have also been used with acrylates in conjugate additions. An example using an enamine is shown in eq 2.[13,14] This reaction has been extended to the synthesis of optically active esters through the use of chiral amines (eq 3).[15,16]

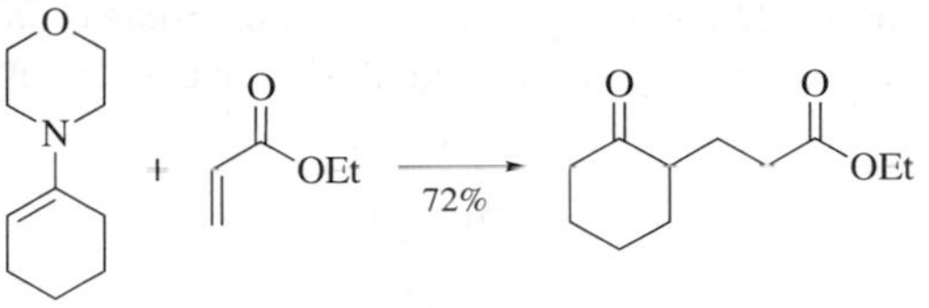

(2)

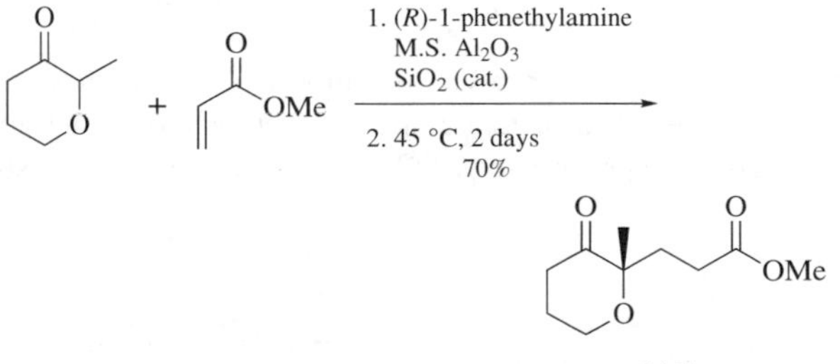

(3)

Stabilized azaallyl anions have also been used in asymmetric conjugate additions with methyl acrylate, as shown in eq 4.[17,18] Other stabilized anions used in this reaction include hydrazones,[19] malonates,[20] α-cyano anions,[21] ester enolates,[22] α-nitro anions,[23] α-sulfonyl anions,[24] phosphorus ylides,[25] and organozinc reagents.[26]

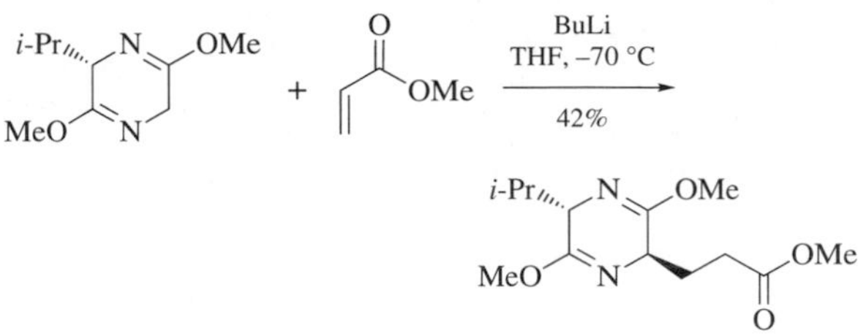

(4)

Enolates formed upon Michael addition of nucleophiles to acrylates have been trapped with a variety of electrophiles. An example of an intermolecular trapping with an aldehyde is shown in eq 5.[27,28] Other traps have included ketones,[29] *N*-tosyl imines,[30] and Michael acceptors.[31]

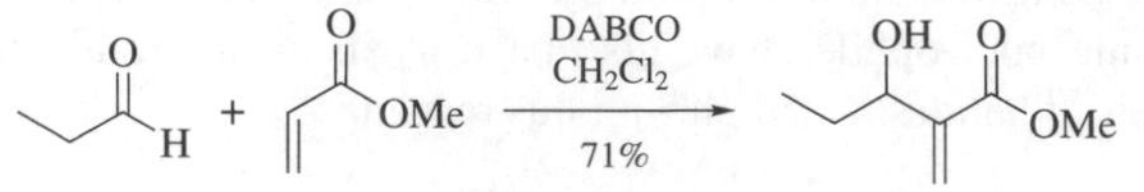

DABCO, CH_2Cl_2, 71% (5)

Intramolecular variants of these reactions are also well known. For example, eq 6 shows an intramolecular trapping with an enone to produce a bicyclic ring system.[32] Both chiral enolates[33] and chiral acrylates[34] have been used in this reaction. Other intramolecular traps have included esters[35] and ketones.[36]

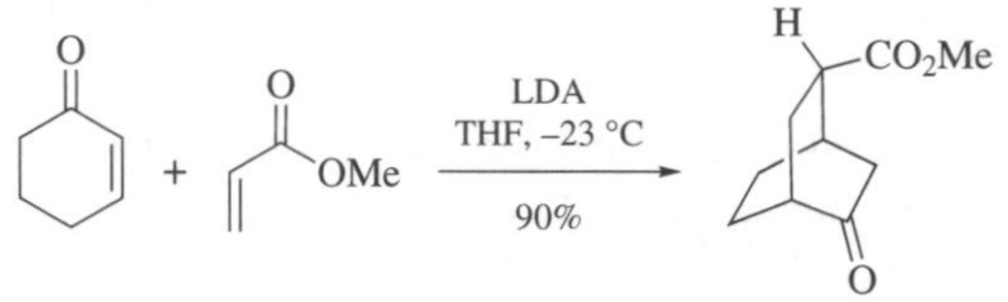

LDA, THF, −23 °C, 90% (6)

Cycloaddition Reactions. Acrylates are commonly used as dienophiles in Diels–Alder reactions. *Endo* products predominate when stereochemistry is involved. A simple example is shown in eq 7.[32b,37]

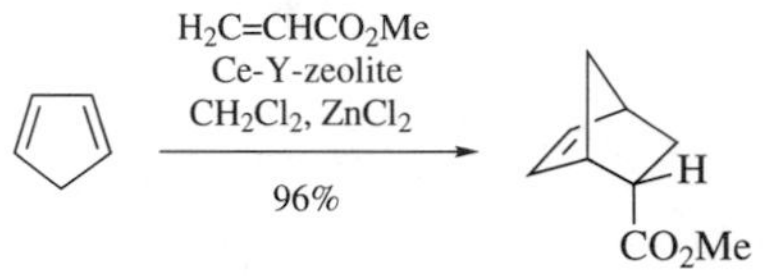

$H_2C{=}CHCO_2Me$, Ce-Y-zeolite, CH_2Cl_2, $ZnCl_2$, 96% (7)

A wide variety of attempts at asymmetric induction have been reported for this reaction. Chiral acrylate esters have been used in the cycloaddition process (eq 8),[38,39] as well as chiral catalysts.[40]

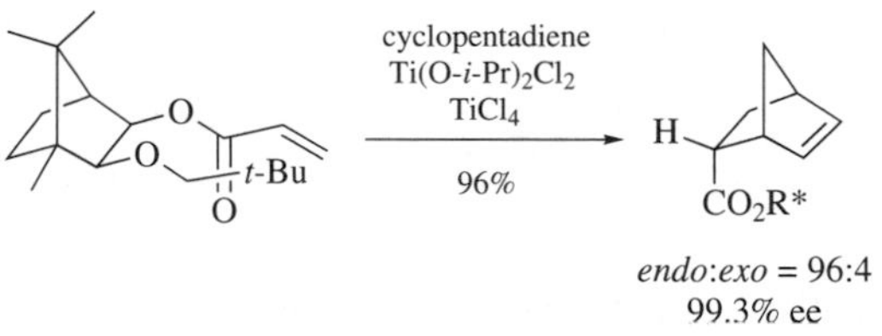

cyclopentadiene, Ti(O-*i*-Pr)$_2$Cl$_2$, $TiCl_4$, 96% (8)

endo:*exo* = 96:4
99.3% ee

Acrylates have further utility as dipolarophiles and have been used to trap nitrile oxides (eq 9),[41] nitrones,[42] azomethine ylides,[43] and azomethine imines.[44] They have also been used in hetero Diels–Alder reactions[45] and [2 + 2] cycloadditions.[46]

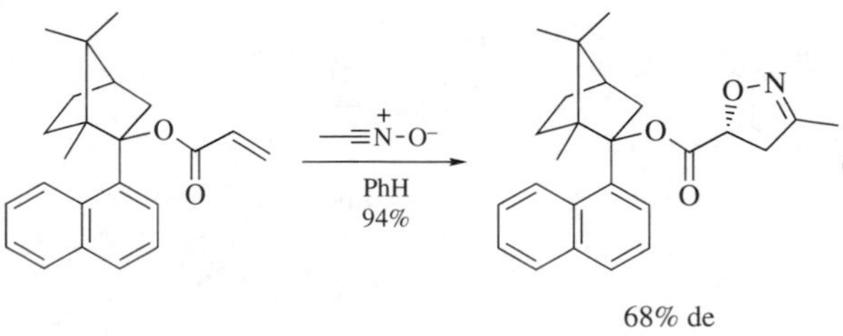

PhH, 94% (9)

68% de

Radical Trap. Acrylates have been used as traps for alkyl radicals in radical chain processes.[47] A variety of radical precursors may be used and cyclization often precedes intermolecular trapping. Frequently, it is necessary to use a large excess of the acrylate to enhance trapping. The example in eq 10 shows the reaction of cyclohexyl iodide with methyl acrylate in the presence of ***Tris(trimethylsilyl)silane***.[48]

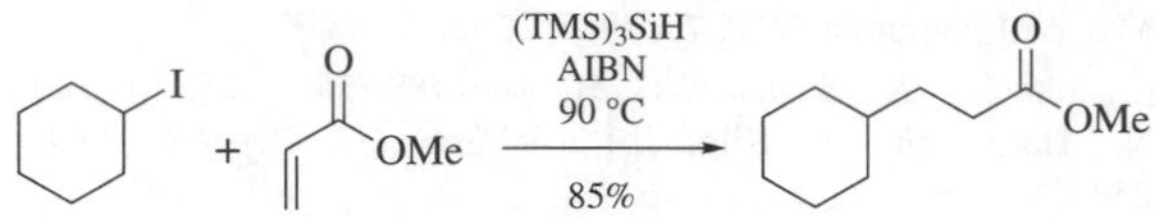

(TMS)$_3$SiH, AIBN, 90 °C, 85% (10)

Curran has developed atom transfer cycloaddition as a means of forming new rings and maintaining high levels of functionality in the final products (eq 11).[49] Acyl radicals have also been trapped by acrylates to produce 1,4-dicarbonyl compounds.[50] Ketones reduced with ***Samarium(II) Iodide*** have been trapped with ethyl acrylate to give the corresponding lactones (eq 12).[51]

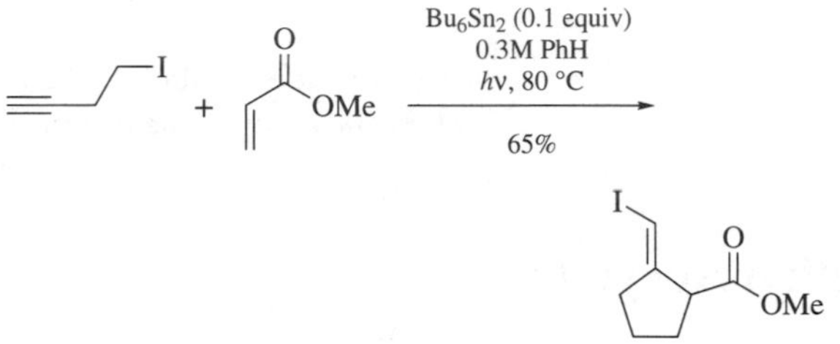

Bu_6Sn_2 (0.1 equiv), 0.3M PhH, *hν*, 80 °C, 65% (11)

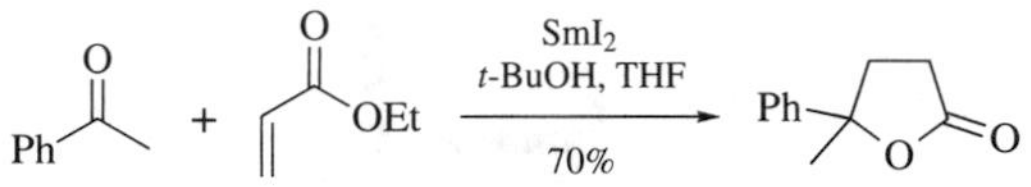

SmI_2, *t*-BuOH, THF, 70% (12)

Arylations and Vinylations. Aryl and vinyl halides and triflates have been coupled to acrylates in the presence of palladium catalysts to produce the corresponding unsaturated esters (Heck reaction). This method is very versatile and stereospecific when substituted vinyl halides are used. Coupling reactions with an aryl bromide,[52] a vinyl triflate,[53] and a vinyl halide[54] are shown in eqs 13–15.

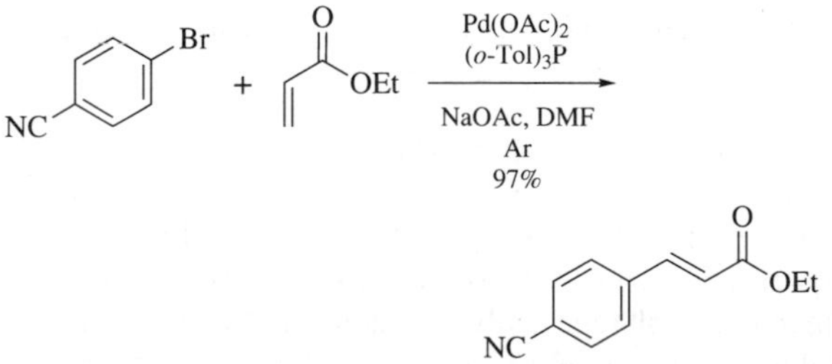

Pd(OAc)$_2$, (*o*-Tol)$_3$P, NaOAc, DMF, Ar, 97% (13)

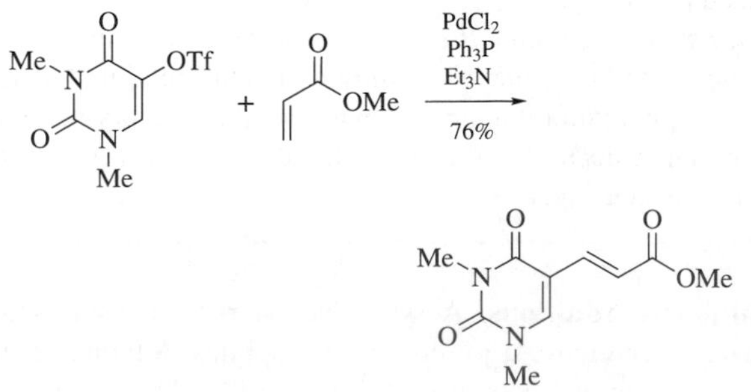

$PdCl_2$, Ph_3P, Et_3N, 76% (14)

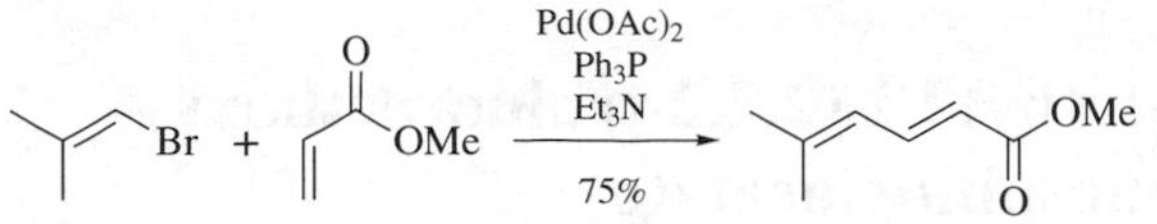

$Pd(OAc)_2$, Ph_3P, Et_3N; 75% (15)

Acrylates have also been oxidatively coupled to indoles and furans using palladium salts (eq 16). These reactions appear to proceed via π-complexation to the heterocyclic double bond, conversion to the σ-complex, addition to the acrylate, and subsequent reductive elimination of the palladium species.[55,56]

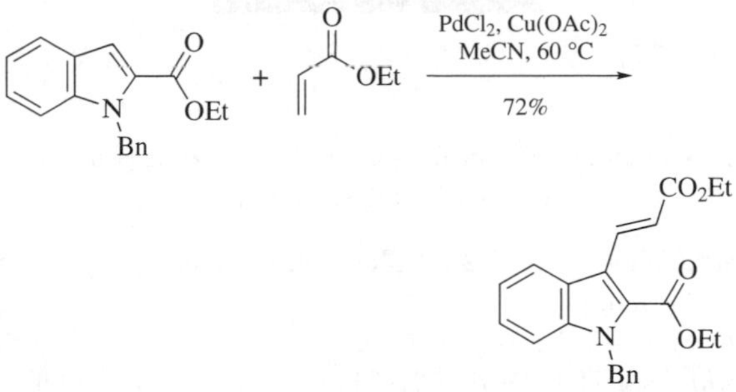

$PdCl_2$, $Cu(OAc)_2$, MeCN, 60 °C; 72% (16)

An interesting extension of this palladium chemistry, shown in eq 17, involves the vinylation of an imino iodide.[57] Under the same conditions, the corresponding imino chlorides are recovered unchanged.

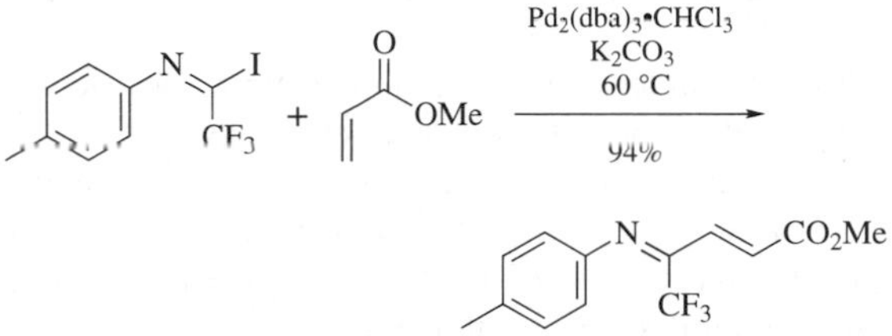

$Pd_2(dba)_3{\cdot}CHCl_3$, K_2CO_3, 60 °C; 94% (17)

Ene Reactions. Acrylates have been used in Lewis acid catalyzed ene reactions. Methyl acrylate reacts with (−)-β-pinene (eq 18) at rt with catalysis by ***Aluminum Chloride***.[58] This reaction can be stereoselective and may proceed better if a salt mixture is present.[59]

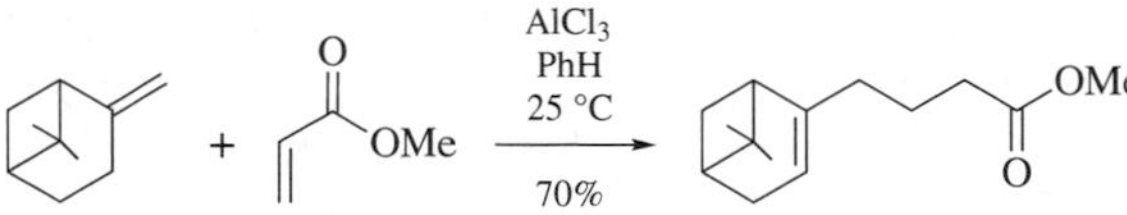

$AlCl_3$, PhH, 25 °C; 70% (18)

Transesterifications. The transesterification of acrylates is best carried out using ***p-Toluenesulfonic Acid*** in the presence of hydroquinone as an inhibitor of polymerization. A representative example is shown in eq 19.[60]

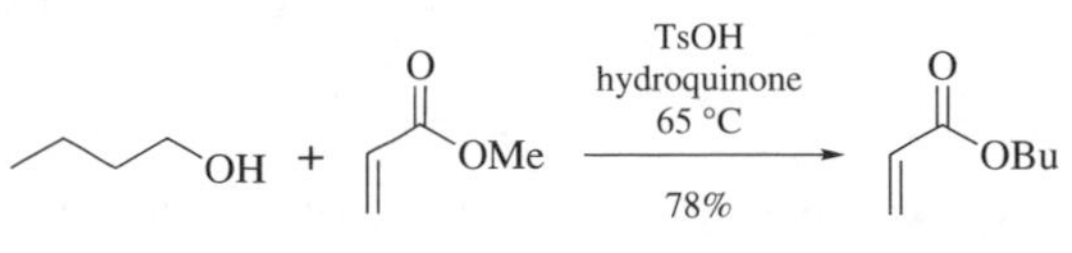

TsOH, hydroquinone, 65 °C; 78% (19)

Related Reagents. Acrylic Acid; Ethyl Acrylate; Ethyl 2-(Bromomethyl)acrylate; Ethyl 3,3-Diethoxyacrylate; Ethyl 3-Ethoxyacrylate; Ethyl α-(Hydroxymethyl)acrylate; Ethyl β-(1-Pyrrolidinyl)acrylate; Methyl 3-Nitroacrylate; Methyl 2-Trimethylsilylacrylate; 8-Phenylmenthyl Acrylate; Trimethyl 2-Phosphonoacrylate.

1. Wadsworth, D. H. *OSC* **1988**, *6*, 75.
2. (a) Baltzly, R.; Phillips, A. P. *JACS* **1949**, *71*, 3419. (b) Baldwin, J. E.; Harwood, L. M.; Lombard, M. J. *T* **1984**, *40*, 4363. (c) Mozingo, R.; McCracken, J. H. *OSC* **1955**, *3*, 258. (d) Jones, R. A. Y.; Katritzky, A. R.; Trepanier, D. L. *JCS(B)* **1971**, 1300.
3. (a) Braunholtz, J. T.; Mann, F. G. *JCS* **1957**, 4166. (b) Barluenga, J.; Villamaña, J.; Yus, M. *S* **1981**, 375.
4. Wessjohann, L.; McGaffin, G.; de Meijere, A. *S* **1989**, 359.
5. Kim, Y. H.; Lee, N. J. *H* **1983**, *20*, 1769.
6. Barluenga, J.; Palacios, F.; Viña, S.; Gotor, V. *JHC* **1986**, *23*, 447.
7. Kalcheva, V.; Stoyanova, D.; Simova, S. *LA* **1989**, 1251.
8. West, R. A. *JOC* **1963**, *28*, 1991.
9. Rehberg, C. E.; Dixon, M. B.; Fisher, C. H. *JACS* **1946**, *68*, 544.
10. (a) Kharasch, M. S.; Fuchs, C. F. *JOC* **1948**, *13*, 97. (b) Bakuzis, P.; Bakuzis, M. L. F. *JOC* **1981**, *46*, 235. (c) Fehnel, E. A.; Carmack, M. *OSC* **1963**, *4*, 669. (d) Mukaiyama, T.; Izawa, T.; Saigo, K.; Takei, H. *CL* **1973**, 355.
11. Mozingo, R.; Patterson, L. A. *OSC* **1955**, *3*, 576.
12. (a) Boyd, E. A.; Corless, M.; James, K.; Regan, A. C. *TL* **1990**, *31*, 2933. (b) Green, K. *TL* **1989**, *30*, 4807. (c) Thottathil, J. K.; Ryono, D. E.; Przybyla, C. A.; Moniot, J. L.; Neubeck, R. *TL* **1984**, *25*, 4741. (d) Beer, P. D.; Edwards, R. C.; Hall, C. D.; Jennings, J. R.; Cozens, R. J. *CC* **1980**, 351.
13. Fritz, H.; Fischer, O. *T* **1964**, *20*, 1737.
14. (a) Kinney, W. A.; Coghlan, M. J.; Paquette, L. A. *JACS* **1985**, *107*, 7352. (b) Borne, R. F.; Fifer, E. K.; Waters, I. W. *JMC* **1984**, *27*, 1271. (c) Barluenga, J.; Jardón, J.; Gotor, V. *S* **1988**, 146.
15. Desmaële, D.; Pain, G.; D'Angelo, J. *TA* **1992**, *3*, 863.
16. (a) Matsuyama, H.; Fujii, S.; Kamigata, N. *H* **1991**, *32*, 1875. (b) Stetin, C.; De Jeso, B.; Pommier, J.-C. *JOC* **1985**, *50*, 3863. (c) Ito, Y.; Sawamura, M.; Kominami, K.; Saegusa, T. *TL* **1985**, *26*, 5303.
17. Schollkopf, U.; Pettig, D.; Busse, U.; Egert, E.; Dyrbusch, M. *S* **1986**, 737.
18. (a) Minowa, N.; Hirayama, M.; Fukatsu, S. *TL* **1984**, *25*, 1147. (b) Belokon, Y. N.; Bulychev, A. G.; Ryzhov, M. G.; Vitt, S. V.; Batsanov, A. S.; Struchkov, Y. T.; Bakhmutov, V. I.; Belikov, V. M. *JCS(P1)* **1986**, 1865. (c) Kanemasa, S.; Tatsukawa, A.; Wada, E. *JOC* **1991**, *56*, 2875. (d) achiral 1-azaallyl anion: Hua, D. H., Bharathi, S. N.; Takusagawa, F.; Tsujimoto, A.; Panangadan, J. A. K.; Hung, M.-H.; Bravo, A. A.; Erpelding, A. M. *JOC* **1989**, *54*, 5659.
19. Baldwin, J. E.; Adlington, R. M.; Jain, A. U.; Kolhe, J. N.; Perry, M. W. D. *T* **1986**, *42*, 4247.
20. Floyd, D. E.; Miller, S. E. *JOC* **1951**, *16*, 882.
21. (a) Kubota, Y.; Nemoto, H.; Yamamoto, Y. *JOC* **1991**, *56*, 7195. (b) Cheng, A.; Uyeno, E.; Polgar, W.; Toll, L.; Lawson, J. A.; DeGraw, J. I.; Loew, G.; Camerman, A.; Camerman, N. *JMC* **1986**, *29*, 531.
22. (a) Kraus, G. A.; Roth, B. *TL* **1977**, 3129. (b) asymmetric synthesis: Aoki, S.; Sasaki, S.; Koga, K. *TL* **1989**, *30*, 7229. (c) Luthman, K.; Orbe, M.; Waglund, T.; Claesson, A. *JOC* **1987**, *52*, 3777.
23. (a) Moffett, R. B. *OSC* **1963**, *4*, 652. (b) Chasar, D. W. *S* **1982**, *10*, 841. (c) White, D. A.; Baizer, M. M. *TL* **1973**, 3597.
24. Trost, B. M.; Schmuff, N. R. *JACS* **1985**, *107*, 396.
25. Wanner, M. J.; Koomen, G. J. *S* **1988**, 325.
26. (a) Caronna, T.; Citterio, A.; Clerici, A. *OPP* **1974**, *6*, 299. (b) Sustmann, R.; Hopp, P.; Holl, P. *TL* **1989**, *30*, 689.
27. Brown, J. M.; Evans, P. L.; James, A. P. *OS* **1989**, *68*, 64.

28. For attempted asymmetric induction in this process, see (a) Basavaiah, D.; Gowriswari, V. V. L.; Sarma, P. K. S.; Rao, P. D. *TL* **1990**, *31*, 1621. (b) Drewes, S. E.; Emslie, N. D.; Karodia, N.; Khan, A. A. *CB* **1990**, *123*, 1447.
29. Basavaiah, D.; Gowriswari, V. V. L. *SC* **1989**, *19*, 2461.
30. (a) Bertenshaw, S.; Kahn, M. *TL* **1989**, *30*, 2731. (b) Perlmutter, P.; Teo, C. C. *TL* **1984**, *25*, 5951.
31. (a) Barco, A.; Benetti, S.; Casolari, A.; Pollini, G. P.; Spalluto, G. *TL* **1990**, *31*, 4917. (b) Posner, G. H.; Shulman-Roskes, E. M. *T* **1992**, *23*, 4677.
32. (a) White, K. B.; Reusch, W. *T* **1978**, *34*, 2439. (b) Lee, R. A. *TL* **1973**, 3333.
33. Zhao, R.-B.; Zhao, Y.-F.; Song, G.-Q.; Wu, Y.-L. *TL* **1990**, *31*, 3559.
34. Spitzner, D.; Wagner, P.; Simon, A.; Peters, K. *TL* **1989**, *30*, 547.
35. Wada, A.; Yamamoto, H.; Ohki, K.; Nagai, S.; Kanatomo, S. *JHC* **1992**, *29*, 911.
36. Marino, J. P.; Katterman, L. C. *CC* **1979**, 946.
37. (a) Narayama, Y. V. S.; Pillai, C. N. *SC* **1991**, *21*, 783. (b) Hashimoto, Y.; Saigo, K.; Machida, S.; Hasegawa, M. *TL* **1990**, *31*, 5625. (c) Cativiela, C.; Fraile, J. M.; Garcia, J. I.; Mayoral, J. A. Pires, E.; Figueras, F.; de Mènorval, L. C. *T* **1992**, *48*, 6467.
38. (a) Oppolzer, W.; Chapuis, C.; Dao, G. M.; Reichlin, D.; Godel, T. *TL* **1982**, *23*, 4781. (b) Oppolzer, W.; Chapuis, C.; Bernardinelli, G. *TL* **1984**, *25*, 5885.
39. (a) Corey, E. J.; Cheng, X.-M.; Cimprich, K. A. *TL* **1991**, *32*, 6839. (b) Stähle, W.; Kunz, H. *SL* **1991**, 260. (c) Poll, T.; Metter, J. O.; Helmchen, G. *AG(E)* **1985**, *24*, 112.
40. (a) Hawkins, J. M.; Loren, S. *JACS* **1991**, *113*, 7794. (b) Ketter, A.; Glahsl, G.; Herrmann, R. *JCR(M)* **1990**, 2118.
41. (a) Olsson, T.; Stern, K.; Westman, G.; Sundell, S. *T* **1990**, *46*, 2473. (b) Zhang, R.; Chen, J. *S* **1990**, 817.
42. Padwa, A.; Fisera, L.; Koehler, K. F.; Rodriguez, A.; Wong, G. S. K. *JOC* **1984**, *49*, 276.
43. (a) Allway, P.; Grigg, R. *TL* **1991**, *32*, 5817. (b) Padwa. A.; Haffmanns, G.; Tomas, M. *JOC* **1984**, *49*, 3314.
44. Zlicar, M.; Stanovnik, B.; Tisler, M. *T* **1992**, *48*, 7965.
45. (a) Chehna, M.; Pradere, J. P.; Guingant, A. *SC* **1987**, *17*, 1971. (b) Sainte, F.; Serckx-Poncin, B.; Hesbain-Frisque, A.-M.; Ghosez, L. *JACS* **1982**, *104*, 1428.
46. Guerry, P.; Neier, R. *CC* **1989**, 1727.
47. (a) Giese, B. *AG(E)* **1983**, *22*, 753. (b) Curran, D. P. *S* **1988**, *417*, 489.
48. Giese, B.; Kopping, B; Chatgilialoglu, C. *TL* **1989**, *30*, 681.
49. Curran, D. P.; Chen, M.-H. *JACS* **1987**, *109*, 6558.
50. Schwartz, C. E.; Curran, D. P. *JACS* **1990**, *112*, 9272.
51. Fukuzawa, S.-I.; Nakanishi, A.; Fujinami, T.; Sakai, S. *CC* **1986**, 624.
52. Spencer, A. *JOM* **1983**, *258*, 101.
53. Hirota, K.; Kitade, Y.; Isobe, Y.; Maki, Y. *H* **1987**, *26*, 355.
54. Dieck, H. A.; Heck, R. F. *JOC* **1975**, *40*, 1083.
55. (a) Murakami, Y.; Yokoyama, Y.; Aoki, T. *H* **1984**, *22*, 1493. (b) Itahara, T.; Ikeda, M.; Sakakibara, T. *JCS(P1)* **1983**, 1361.
56. Itahara, T.; Ouseto, F. *S* **1984**, 488.
57. Uneyama, K.; Watanabe, H. *TL* **1991**, *32*, 1459.
58. Snider, B. B. *JOC* **1974**, *39*, 255.
59. Åkermark, B.; Ljungqvist, A. *JOC* **1978**, *43*, 4387.
60. Rehberg, C. E. *OSC* **1955**, *3*, 146.

Duane A. Burnett & Margaret E. Browne
Schering-Plough Research Institute, Kenilworth, NJ, USA

Methyl Bis(2,2,2-trifluoroethoxy)-phosphinylacetate

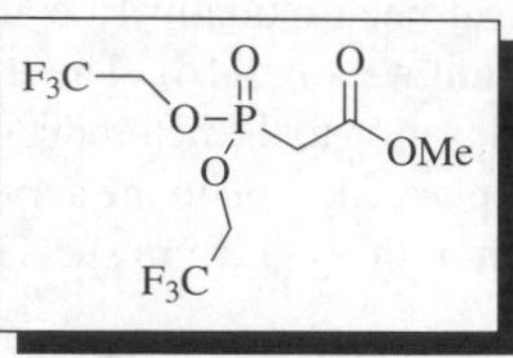

[88738-78-7] $C_7H_9F_6O_5P$ (MW 318.13)

(Horner–Emmons reagent for the production of α,β-unsaturated esters with (*Z*) selectivity)

Alternate Name: bis(2,2,2-trifluoroethyl) (methyoxycarbonyl-methyl)phosphonate.
Physical Data: *d* 1.504 g cm^{-3}.
Solubility: sol THF, DME, CH_2Cl_2.
Form Supplied in: oil; widely available.
Preparative Methods: prepared by reaction of methyl dichlorophosphonoacetate with trifluoroethanol in benzene.
Purification: silica gel chromatography.
Handling, Storage, and Precautions: irritant.

Modified Wittig Reagent. The Horner–Emmons reaction carried out by reacting methyl bis(trifluoroethoxy)phosphinylacetate with an aldehyde in the presence of a strongly dissociated base yields α,β-unsaturated esters with a strong preference for the (*Z*) geometry. The reaction is not overly sensitive to the type of aldehyde used. Examples are shown in eqs 1–3.[1,2]

PhCHO + $(F_3CCH_2O)_2P(O)CH_2CO_2Me$ —[KN(TMS)$_2$, THF, 18-crown-6, 95%]→ (*Z*)-PhCH=CHCO$_2$Me (>50:1) (1)

CH$_3$CH=CHCH=CHCHO + $(F_3CCH_2O)_2P(O)CH_2CO_2Me$ —[KN(TMS)$_2$, THF, 18-crown-6]→ triene CO$_2$Me (2)

The strongly dissociated base is necessary to maximize the (*Z*) selectivity of the alkenation reaction (see eq 4).[1,3]

In general, the reagent is slow to react with ketones, especially aryl ketones, providing little or no reaction (eq 5).[4] The majority of references to this reagent involve aldehydes as the coupling partner.[5]

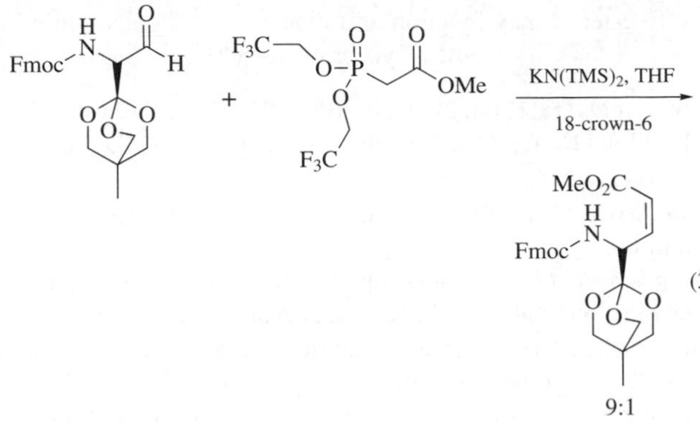

(3)

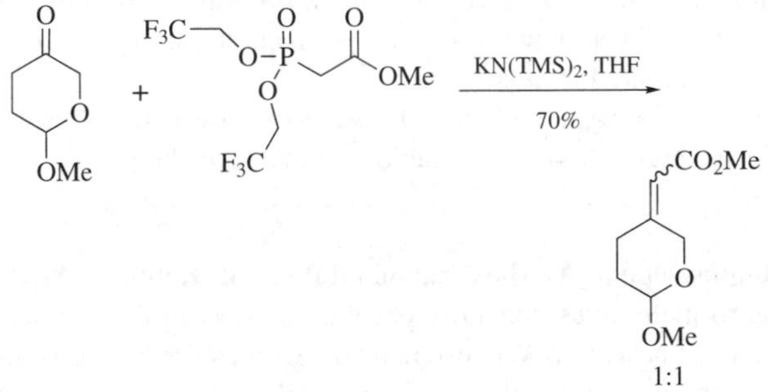

9:1

(4)

1:1

no reaction

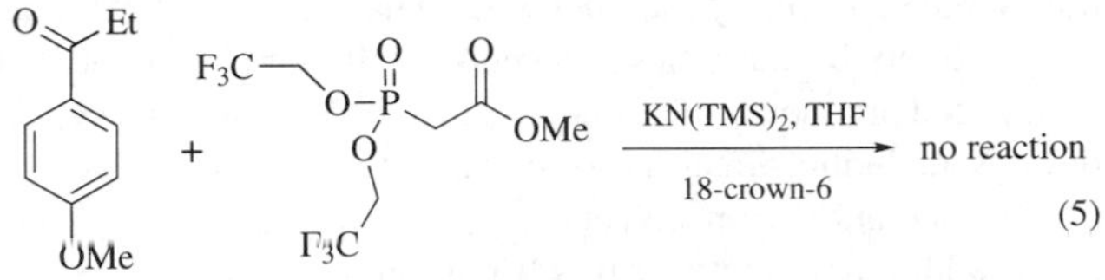

(5)

Other Uses. The anion of methyl bis(trifluoroethoxy)phosphinylacetate can undergo an S_N2 reaction with suitable electrophiles to make a more complex trisubstituted modified Wittig reagent, which can undergo alkenation reactions.[6]

Related Reagents. *t*-Butyl Trimethylsilylacetate; (Ethoxycarbonylmethylene)triphenylphosphorane; Ethyl (Methyldiphenylsilyl)acetate; Ethyl Trimethylsilylacetate; (Methoxycarbonylmethylene)triphenylphosphorane; Triethyl Phosphonoacetate; Trimethyl Phosphonoacetate.

1. Still, W. C.; Gennari, C. *TL* **1983**, *24*, 4405.
2. Blaskovich, M. A.; Lajoie, G. A. *JACS* **1993**, *115*, 5021.
3. López Tudanca, P. L.; Jones, K.; Brownbridge, P. *JCS(P1)* **1992**, 533.
4. Eguchi, T.; Aoyama, T.; Kakinuma, K. *TL* **1992**, *33*, 5545.
5. For recent examples, see: (a) Horton, D.; Koh, D. *TL* **1993**, *34*, 2283. (b) Hatakeyama, S.; Sugawara, K.; Takano, S. *CC* **1993**, 125. (c) Nakata, M.; Osumi, T.; Ueno, A.; Kimura, T.; Tamai, T.; Tatsuta, K. *BCJ* **1992**, *65*, 2974. (d) Yokokawa, F.; Hamada, Y.; Shioiri, T. *SL* **1992**, 703. (e) Spada, M. R.; Ubukata, M.; Isono, K *H* **1992**, *34*, 1147.
6. Marshall, J. A.; DeHoff, B. S.; Cleary, D. G. *JOC* **1986**, *51*, 1735.

Brad DeHoff

Syntex Technology Center, Boulder, CO, USA

Methylcopper–Boron Trifluoride Etherate[1]

$MeCu/BF_3{\cdot}OEt_2$

(MeCu)		
[1184-53-8]	CH_3Cu	(MW 78.59)
($BF_3{\cdot}OEt_2$)		
[109-63-7]	$C_4H_{10}BF_3O$	(MW 141.95)

(selective nucleophilic reagent for conjugate additions[2] and substitution reactions[1])

Preparative Method: prepared by the addition of ***Boron Trifluoride Etherate*** to a suspension of ***Methylcopper*** in Et_2O or THF at −75 °C.[2]

Handling, Storage, and Precautions: see entries for ***Boron Trifluoride Etherate*** and ***Methylcopper***. Use in a fume hood.

Conjugate Additions. The combination of an alkylcopper with boron trifluoride produces a nucleophilic reagent which is more reactive than the alkylcopper alone and which can exhibit regioselectivity on par with organocuprate reagents. For example, in each of two syntheses of the sesquiterpene modhephene, RCu/BF_3 was used to establish a quaternary center by conjugate addition to a 3-alkyl-2-cyclopentenone moiety (eq 1).[3,4] Alkenoates and alkenoic acids are also suitable substrates.[2,5] In the development of methodology for asymmetric synthesis, the alkylcopper/BF_3 combination was found to be effective in preparing β-substituted carboxylic acids.[6] High enantiomeric excesses (94–98%) were obtained by the use of a camphorsulfonamide auxiliary (eq 2).[6a]

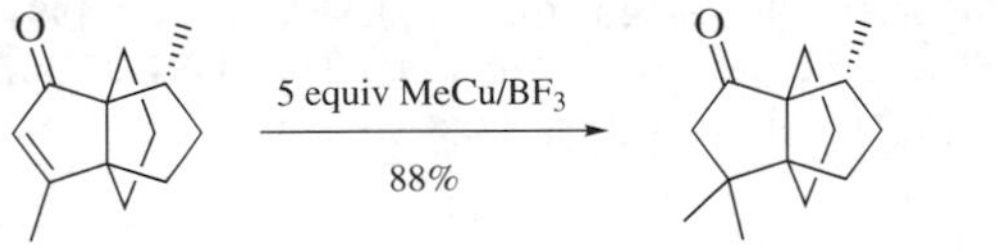

(1)

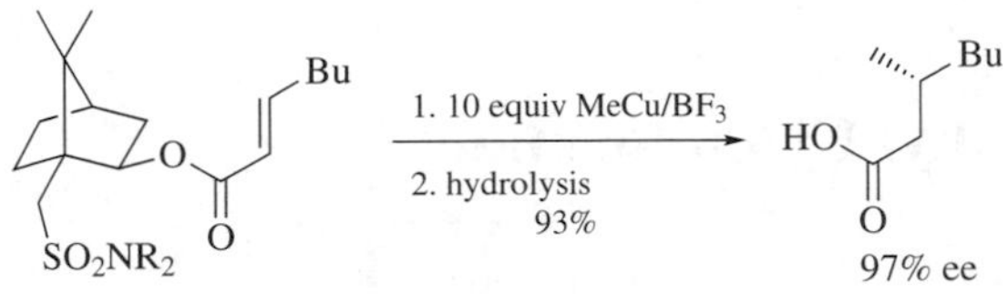

(2)

Enhancements of rate and regioselectivity in conjugate additions have also been observed with the combinations of alkylcopper/halosilane[7] and alkylcopper/***Aluminum Chloride***.[8]

Nucleophilic Substitutions. Acetals are converted to the corresponding monosubstitution products on treatment with alkylcopper/BF_3.[9,10] Similarly, orthoformates give acetals in high yield.[9] Chiral cyclic acetals undergo this monosubstitution reaction stereoselectively (eq 3), although cuprates with BF_3 were found to give better results.[10]

(3)

Allylic halides, ethers, esters, and alcohols suffer regioselective γ-attack by alkylcopper/BF_3.[11] Propargyl chloride and acetate afford 1,2-heptadiene. The stereoselectivity of these substitutions has been less rigorously examined, but appears to be less dramatic than the regioselectivity. The regioselectivity of attack on γ-acetoxy-α,β-enoates by alkylcopper/$AlCl_3$ has also been documented.[12]

Related Reagents. Boron Trifluoride Etherate; Lithium Cyano(methyl)cuprate; Lithium Dimethylcuprate; Lithium Dimethylcuprate–Boron Trifluoride; Methylcopper; Phenylcopper–Boron Trifluoride Etherate.

1. Yamamoto, Y. *AG(E)* **1986**, *25*, 947.
2. Yamamoto, Y.; Yamamoto, S.; Yatagai, H.; Ishihara, Y.; Maruyama, K. *JOC* **1982**, *47*, 119.
3. Karpf, M.; Dreiding, A. S. *TL* **1980**, *21*, 4569.
4. Schostarez, H.; Paquette, L. A. *JACS* **1981**, *103*, 722.
5. Yamamoto, Y.; Maruyama, K. *JACS* **1978**, *100*, 3240.
6. (a) Oppolzer, W.; Dudfield, P.; Stevenson, T.; Godel, T. *HCA* **1985**, *68*, 212. (b) Oppolzer, W.; Moretti, R.; Bernardinelli, G. *TL* **1986**, *27*, 4713.
7. (a) Bergdahl, M.; Lindstedt, E.-L.; Nilsson, M.; Olsson, T. *T* **1988**, *44*, 2055. (b) Matsuzawa, S.; Horiguchi, Y.; Nakamura, E.; Kuwajima, I. *T* **1989**, *45*, 349. (c) Bergdahl, M.; Lindstedt, E.-L.; Nilsson, M.; Olsson, T. *T* **1989**, *45*, 535. (d) Bergdahl, M.; Lindstedt, E.-L.; Olsson, T. *JOM* **1989**, *365*, C11.
8. Ibuka, T.; Tabushi, E.; Yasuda, M. *CPB* **1983**, *31*, 128.
9. Ghribi, A.; Alexakis, A.; Normant, J. F. *TL* **1984**, *25*, 3075.
10. Normant, J. F.; Alexakis, A.; Ghribi, A.; Mangeney, P. *T* **1989**, *45*, 507.
11. (a) Yamamoto, Y.; Yamamoto, S.; Yatagai, H.; Maruyama, K. *JACS* **1980**, *102*, 2318. (b) Kang, J.; Cho, W.; Lee, W. K. *JOC* **1984**, *49*, 1838. (c) Ghribi, A.; Alexakis, A.; Normant, J. F. *TL* **1984**, *25*, 3079.
12. Ibuka, T.; Minakata, H. *SC* **1980**, *10*, 119.

John N. Haseltine
Georgia Institute of Technology, Atlanta, GA, USA

Methyl Cyanoformate[1]

(**1**; R = Me)		
[17640-15-2]	$C_3H_3NO_2$	(MW 85.07)
(**2**; R = Et)		
[623-49-4]	$C_4H_5NO_2$	(MW 99.10)
(**3**; R = $PhCH_2$)		
[5532-86-5]	$C_9H_7NO_2$	(MW 161.17)

(agent for the regioselective methoxycarbonylation of carbanions;[1,2] reacts with organocadmium reagents to form α-keto esters;[3] may function as a dienophile,[4] dipolarophile,[5] or radical cyanating agent[6])

Physical Data: (**1**) mp 26 °C; bp 100–101 °C; *d* 1.072 g cm^{-3}. (**2**) bp 115–116 °C; *d* 1.003 g cm^{-3}. (**3**) bp 66–67 °C/0.6 mmHg; *d* 1.105 g cm^{-3}.

Solubility: sol all common organic solvents; dec by H_2O, alcohols, amines.

Form Supplied in: colorless liquid; methyl cyanoformate, as well as the ethyl and benzyl analogs, is available commercially.

Preparative Methods: small quantities of cyanoformate esters (up to 30 g) may be conveniently prepared from alkyl chloroformates by procedures employing phase-transfer catalysis with either ***18-Crown-6***[7] or ***Tetra-n-butylammonium Bromide***,[8] but several workers have found the products to be unsatisfactory when prepared on a larger scale.

Handling, Storage, and Precautions: store over 4Å molecular sieves; highly toxic; flammable; use in a fume hood.

Regioselective Methoxycarbonylation of Ketones. Methyl cyanoformate gives generally excellent results in the regiocontrolled synthesis of β-keto esters by the *C*-acylation of preformed lithium enolates (eq 1)[1,2] and is normally superior to the more traditional acylating agents such as acyl halides, anhydrides,[9] and CO_2,[10] partly because these reagents afford variable amounts of *O*-acylated products.[11] The enolates may be generated in a variety of ways, including direct enolization of ketones with suitable bases (eq 2),[2] liberation from silyl enol ethers and acetates (eq 3),[12] conjugate additions of cuprates to α,β-unsaturated ketones[13] (eq 4),[14] or by the reduction of enones by lithium in liquid ammonia (eq 5).[1,12]

(1)

(2)

(3)

(4)

(5)

Lithium enolates derived from sterically unencumbered cyclohexanones undergo preferential axial acylation (eq 6), whereas equatorial acylation is favored with $\Delta^{1(9)}$-2-octalones (eq 7),[12] even in the absence of an alkyl substituent at C-10.[15]

OTMS; t-Bu; 1. MeLi, THF, −20 °C; 2. $NCCO_2Me$, HMPA, −78 to 0 °C; O; CO_2Me; t-Bu; 7%; + O; CO_2Me; t-Bu; 75% (6)

R; O; 1. Li, NH_3, t-BuOH, Et_2O, −33 °C; 2. $NCCO_2Me$, −78 to 0 °C; R; O; H; CO_2Me; + R; O; H; CO_2Me (7)

R = H	53%	14%
R = Me	78%	0%

For compounds in which the β-carbon of the enolate is sterically hindered, treatment with methyl cyanoformate may result in variable degrees of *O*-acylation, although this problem may be ameliorated by the use of diethyl ether as the solvent. In several cases a switch from predominantly *O*-acylation in THF to predominant *C*-acylation in diethyl ether has been observed (eqs 8–10).[12]

O; 1. Me_2CuLi, −10 °C; 2. $NCCO_2Me$, −78 to 0 °C; OCO_2Me; + O; CO_2Me (8)

solvent: THF	87%	0%
solvent: Et_2O	6%	78%

OTMS; OCO_2Me; + O; CO_2Me; H (9)

MeLi, THF, −20 °C; $NCCO_2Me$, −78 to 0 °C	78%	0%
BuLi, Et_2O, −10 °C; $NCCO_2Me$, −78 to 0 °C	9%	68%

O; OMe; 1. Li, NH_3, t-BuOH, −33 °C; 2. $NCCO_2Me$, −78 to 0 °C; $MeOCO_2$; H; OMe + O; H; CO_2Me; OMe (10)

solvent: THF	67%	8%
solvent: Et_2O	0%	71%

A comparative study of lithium, sodium, and potassium enolates indicated that the lithium derivatives reacted most satisfactorily.[2] There may be substrates for which the thermodynamic enolates are required, however, and the sodium and potassium enolates may therefore be selected. Good results have been reported with these intermediates (eqs 11 and 12), although the latter afford significant amounts of *O*-acylated products.[16,17] Quite apart from the issue of regioselectivity, the cyanoformate-based procedure is exceptionally reliable and makes it possible to prepare β-keto esters from ketones under especially mild conditions. It is not only the method of choice with sensitive substrates,[16] but it will often ensure superior results with more robust compounds as well.

OTBDMS; H; O; 1. LDA, THF, −78 °C or NaHMDS, THF, 0 °C or KHMDS, THF, 0 °C; 2. $NCCO_2Me$, −78 °C; OTBDMS; H; CO_2Me; O; + OTBDMS; CO_2Me; H; O; + OTBDMS; H; OCO_2Me (11)

LDA	4:1:0
NaHMDS	1:7:0
KHMDS	0:1:1.5

O O; O; 1. KHMDS, THF, −78 °C; 2. $NCCO_2Me$, HMPA, −78 to 20 °C; O O; CO_2Me; O; 58%; + O O; $MeOCO_2$; 24% (12)

Methoxycarbonylation of Miscellaneous Carbon Acids. The title reagent has also been applied to the methoxycarbonylation of esters (eq 13),[18] lactones (eq 14),[19] phosphonates (eq 15),[20] imines (eq 16),[21] and the *N*-acylation of lactams (eq 17).[22]

CO_2Me; 1. LDA, THF, −78 °C; 2. $NCCO_2Me$, −78 °C; 98%; CO_2Me; MeO_2C (13)

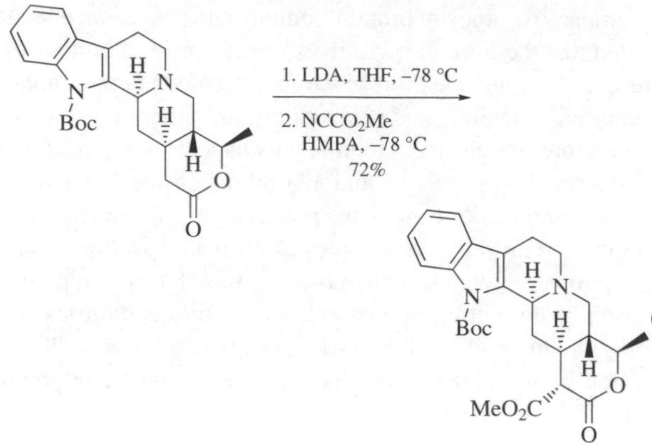

1. LDA, THF, −78 °C
2. $NCCO_2Me$, HMPA, −78 °C
72%
(14)

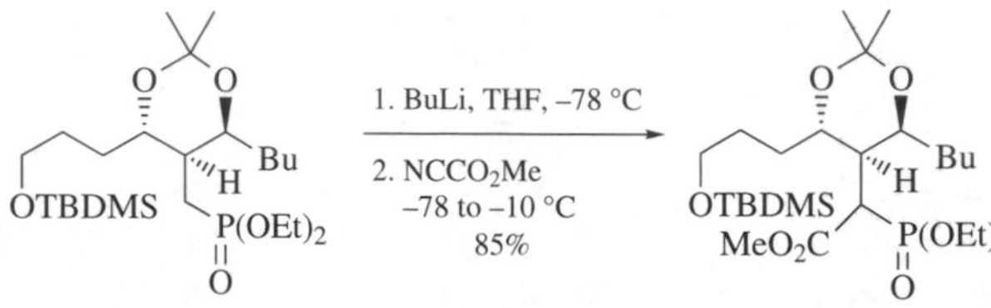

1. BuLi, THF, −78 °C
2. $NCCO_2Me$, −78 to −10 °C
85%
(15)

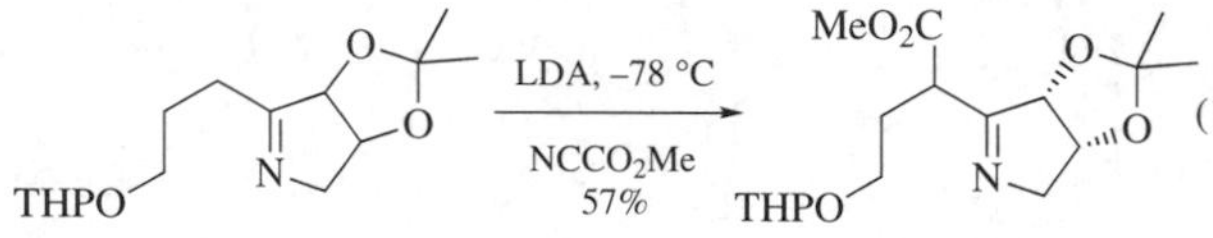

LDA, −78 °C
$NCCO_2Me$
57%
(16)

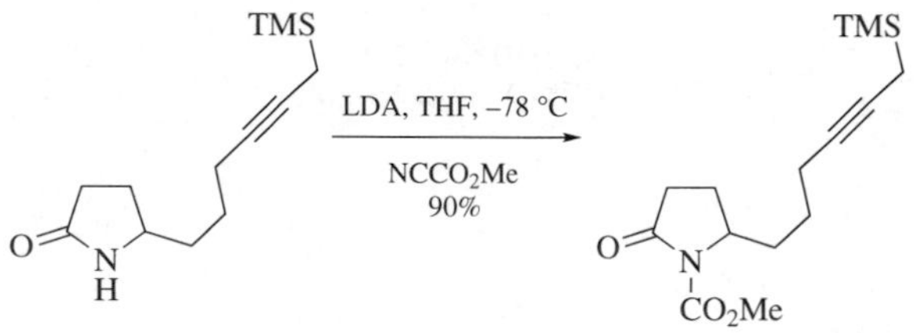

LDA, THF, −78 °C
$NCCO_2Me$
90%
(17)

Higher Alkyl Cyanoformates. A range of other alkyl cyanoformates has been successfully utilized for the acylation of enolate anions, including ethyl,[23] allyl,[24] benzyl,[25] and *p*-methoxybenzyl,[26] but not *t*-butyl cyanoformate, which appears to be insufficiently reactive. Enantiomerically enriched cyanoformates derived from (+)-menthol, (−)-borneol, and the Oppolzer alcohol were reported to furnish good chemical yields, but the level of enantioselectivity was disappointingly low (eq 18).[27]

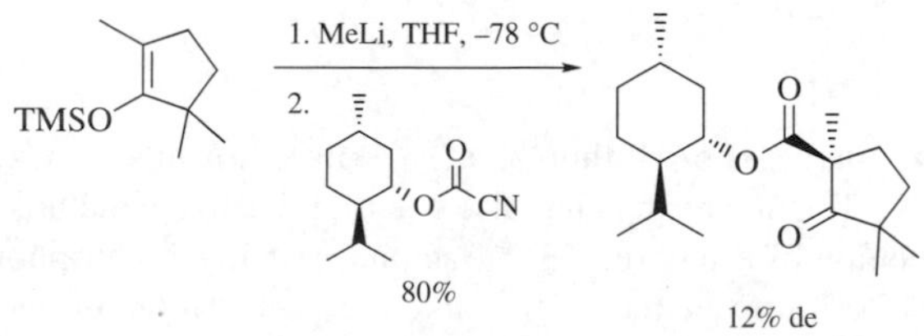

1. MeLi, THF, −78 °C
2. (menthyl cyanoformate)
80%
12% de
(18)

Additions to the Nitrile Group. Ethyl cyanoformate reacts with organocadmium reagents to afford α-keto esters (eq 19),[3] and with malonate esters, β-keto esters, and other active methylene compounds to give α-aminoacrylates (eq 20).[28] Both processes require catalysis with Lewis acids, of which ***Zinc Chloride*** has proven to be the most effective.

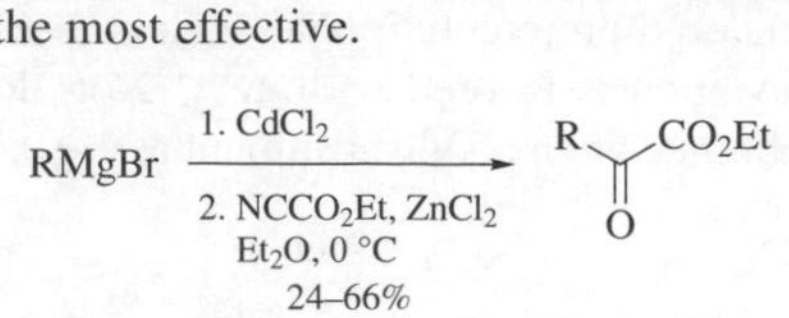

RMgBr → 1. $CdCl_2$; 2. $NCCO_2Et$, $ZnCl_2$, Et_2O, 0 °C; 24–66% → $RCOCO_2Et$ (19)

R = Ph, *i*-Pr, *s*-Bu, cyclohexyl, cyclopentyl

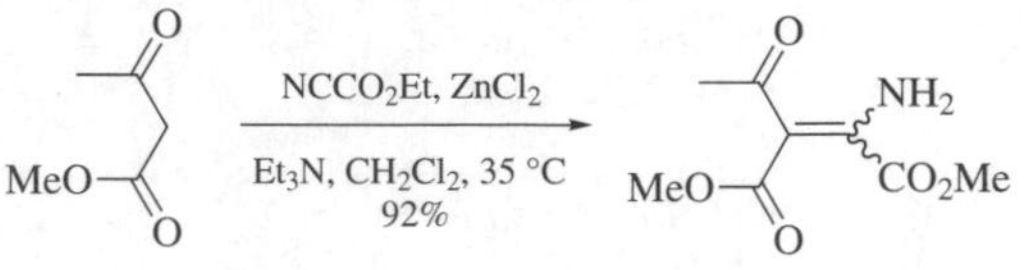

$NCCO_2Et$, $ZnCl_2$
Et_3N, CH_2Cl_2, 35 °C
92%
(20)

Cycloadditions. Methyl and ethyl cyanoformate have been reported to undergo [4+2] cycloadditions, e.g. with cyclopentadienones[4] and 2-alkyl-1-ethoxybuta-1,3-dienes to form pyridines (eq 21),[29] and with cyclobutadienes to form Dewar pyridines (eq 22).[30] Ethyl cyanoformate is also an effective dipolarophile, undergoing 1,3-dipolar addition to azides (eq 23)[31] and cyclic carbonyl ylides (eq 24).[5]

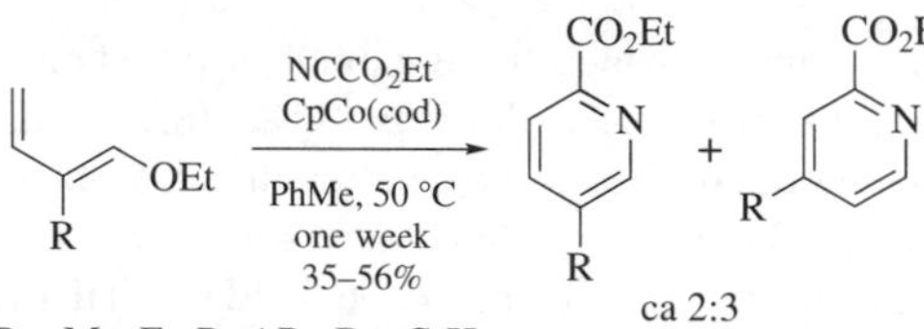

$NCCO_2Et$, CpCo(cod)
PhMe, 50 °C, one week
35–56%
ca 2:3
(21)

R = Me, Et, Pr, *i*-Pr, Bu, C_5H_{11}

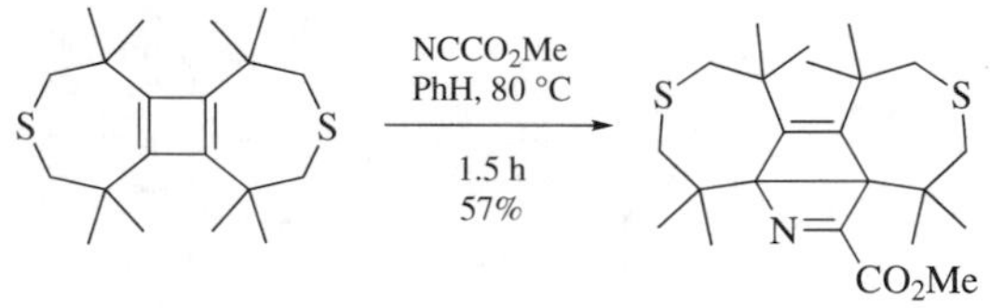

$NCCO_2Me$, PhH, 80 °C
1.5 h
57%
(22)

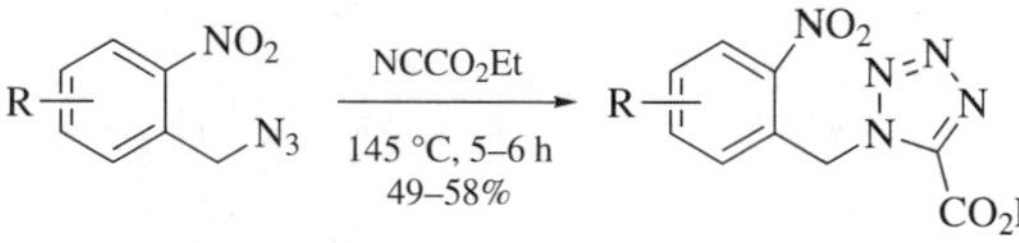

$NCCO_2Et$
145 °C, 5–6 h
49–58%
(23)

R = H, 5-Me. 4-Cl, 5-Cl

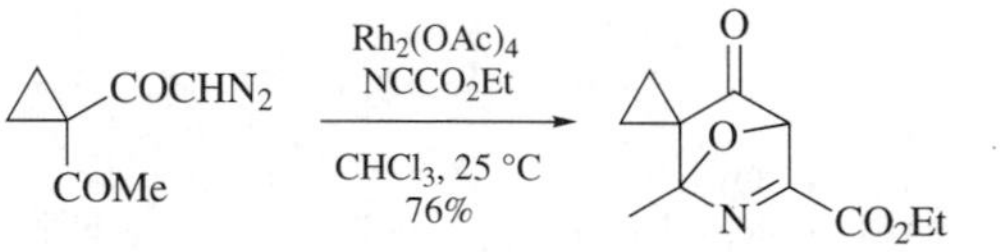

$Rh_2(OAc)_4$, $NCCO_2Et$
$CHCl_3$, 25 °C
76%
(24)

Radical Cyanation. The peroxide-initiated radical cyanation of cyclohexane and 2,3-dimethylbutane with methyl cyanoformate has been carried out in 72% and 77% yield, respectively.[6]

Related Reactions. β-Keto ester formation from ketones may be achieved directly with dialkyl carbonates[32] and dialkyl dicarbonates,[33] or indirectly with dialkyl oxalates,[34] methyl magnesium carbonate,[35] and ethyl diethoxyphosphinyl formate.[36] Regiocontrol is problematical, however, and is more reliably effected by trapping enolates with carbon dioxide,[10] carbon disulfide,[37] or carbon oxysulfide[38] followed by methylation. In the latter cases, the dithio and thiol esters are converted into the parent carboxy esters by mercury(II)-catalyzed hydrolysis. The chemistry of acyl cyanides, but excluding cyanoformates, has been the subject of several reviews.[39]

Related Reagents. Acetyl Cyanide; Carbon Dioxide; Carbon Oxysulfide; *N*,*N*′-Carbonyldiimidazole; Diethyl Carbonate; Methyl Chloroformate; Methyl Magnesium Carbonate.

---

1. Crabtree, S. R.; Mander, L. N.; Sethi, S. P. *OS* **1991**, *70*, 256.
2. Mander, L. N.; Sethi, S. P. *TL* **1983**, *24*, 5425.
3. Akiyama, Y.; Kawasaki, T.; Sakamoto, M. *CL* **1983**, 1231.
4. Padwa, A.; Akiba, M.; Cohen, L. A.; Gingrich, H. L.; Kamigata, N. *JACS* **1982**, *104*, 286.
5. Padwa, A.; Chinn, R. L.; Hornbuckle, S. F.; Zhang, Z. J. *JOC* **1991**, *56*, 3271.
6. Tanner, D. D.; Rahimi, P. M. *JOC* **1979**, *44*, 1674.
7. Childs, M. E.; Weber, W. P. *JOC* **1976**, *41*, 3486.
8. Nii, Y.; Okano, K.; Kobayashi, S.; Ohno, M. *TL* **1979**, 2517.
9. Caine, D. In *Carbon-Carbon Bond Formation*; Augustine, R. L., Ed.; Dekker: New York, 1979; Vol. 1, pp 250–258.
10. (a) Stork, G.; Rosen, P.; Goldman, N.; Coombs, R. V.; Tsuji, J. *JACS* **1965**, *87*, 275. (b) Caine, D. *OR* **1976**, *23*, 1.
11. (a) House, H. O. *Modern Synthetic Reactions*; Benjamin: Menlo Park, 1972; pp 760–763. (b) Black, T. H. *OPP* **1989**, *21*, 179. (c) Seebach, D.; Weller, T.; Protschuk, G.; Beck, A. K.; Hoekstra, M. S. *HCA* **1981**, *64*, 716. (d) cf. Ref. 5, p 258, footnote 69.
12. Crabtree, S. R.; Chu, W.-L. A.; Mander, L. N. *SL* **1990**, 169.
13. (a) Ihara, M.; Suzuki, T.; Katogi, M.; Taniguchi, N.; Fukumoto, K. *JCS(P1)* **1992**, 865. (b) Haynes, R. K.; Katsifis, A. G. *CC* **1987**, 340.
14. Hashimoto, S.; Kase, S.; Shinoda, T.; Ikegami, S. *CL* **1989**, 1063.
15. cf. Mathews, R. S.; Girgenti, S. J.; Folkers, E. A. *CC* **1970**, 708.
16. Ziegler, F. E.; Klein, S. I.; Pati, U. K.; Wang, T.-F. *JACS* **1985**, *107*, 2730.
17. Schuda, P. F.; Phillips, J. L.; Morgan T. M. *JOC* **1986**, *51*, 2742.
18. Ziegler, F. E.; Sobolov, S. B. *JACS* **1990**, *112*, 2749.
19. (a) Hanessian, S.; Faucher, A.-M. *JOC* **1991**, *56*, 2947. (b) Ziegler, F. E.; Cain, W. T.; Kneisly, A.; Stirchak, E. P.; Wester, R. T. *JACS* **1988**, *110*, 5442. (c) Leonard, J.; Ouali, D.; Rahman, S. K. *TL* **1990**, *31*, 739.
20. McLure, C. K.; Jung, K.-Y. *JOC* **1991**, *56*, 2326.
21. Bennet, R. B.; Cha, J. K. *TL* **1990**, *31*, 5437.
22. (a) Melching, K. H.; Hiemstra, H.; Klaver, W. J.; Speckamp, W. N. *TL* **1986**, *27*, 4799. (b) Esch, P. M.; Hiemstra, H.; Klaver, W. J.; Speckamp, W. N. *H* **1987**, *26*, 75. (c) Pirrung, F. O. H.; Rutjes, F. P. J. T.; Hiemstra, H.; Speckamp, W. N. *TL* **1990**, *31*, 5365.
23. Mori, K.; Ikunaka, M. *T* **1987**, *43*, 45.
24. Barton, D. H. R.; Donnelly, D. M. X.; Finet, J. P.; Guiry, P. J.; Kielty, J. M. *TL* **1990**, *31*, 6637.
25. Hashimoto, S.; Miyazaki, Y.; Shinoda, T.; Ikegami, S. *CC* **1990**, 1100.
26. (a) Winkler, J. D.; Henegar, K. E.; Williard, P. G. *JACS* **1987**, *109*, 2850. (b) Henegar, K. E.; Winkler, J. D. *TL* **1987**, *28*, 1051.
27. Kunisch, F.; Hobert, K.; Welzel, P. *TL* **1985**, *26*, 5433.
28. Iimori, T.; Nii, Y.; Izawa T.; Kobayashi, S.; Ohno, M. *TL* **1979**, 2525.
29. Potthoff, B.; Breitmaier, E. *S* **1986**, 584.
30. (a) Krebs, A.; Franken, E.; Müller, S. *TL* **1981**, *22*, 1675. (b) Fink, J.; Regitz, M. *BSF(2)* **1985**, 239.
31. Klaubert, D. H.; Bell, S. C.; Pattison, T. W. *JHC* **1985**, *22*, 333.
32. Deslongchamps, P.; Ruest, L. *OS* **1974**, *54*, 151.
33. Hellou, J.; Kingston, J. F.; Fallis, A. G. *S* **1984**, 1014.
34. Snyder, H. R.; Brooks, L. A.; Shapiro, S. H. *OSC* **1943**, *2*, 531.
35. (a) Stiles, M. *JACS* **1959**, *81*, 2598. (b) Pelletier, S. W.; Chappell, R. L.; Parthasarathy, P. C.; Lewin, N. *JOC* **1966**, 1747.
36. Shahak, I. *TL* **1966**, 2201.
37. Kende, A. S.; Becker, D. A. *SC* **1982**, *12*, 829.
38. Vedejs, E.; Nader, B. *JOC* **1982**, *47*, 3193.
39. (a) Thesing, J.; Witzel, D.; Brehm, A. *AE* **1956**, *68*, 425. (b) Bayer, O. *MOC* **1977**, *7/2c*, 2487. (c) Hunig, S.; Schaller, R. *AG(E)* **1982**, *21*, 36.

Lewis N. Mander
The Australian National University, Canberra, Australia

Methyl Dilithioacetoacetate

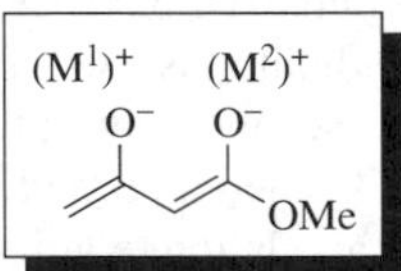

[30568-00-4]		
($M^1 = M^2 = Li^+$)		
[53437-05-1]	$C_5H_6Li_2O_3$	(MW 127.99)
($M^1 = Li^+$, $M^2 = Na^+$)		
[64670-05-9]	$C_5H_6LiNaO_3$	(MW 144.04)

(four-carbon chain, all four carbons of which can be reacted regioselectively; versatile starting material in synthesis of complex molecules; undergoes γ-alkylation and γ-acylation)

Solubility: sol THF, ether, dimethoxyethane, and HMPA; slightly sol hydrocarbon solvents.
Form Supplied in: generated in situ.
Analysis of Reagent Purity: NMR spectroscopy or mass spectrometry of D_2O quench.[1]
Preparative Methods: dianion (**1**) is prepared in a dry solvent, usually THF or diethyl ether, by adding the β-keto ester to 1.1 equiv of ***Sodium Hydride***, which has been washed free of mineral oil, and then adding 1.05 equiv of ***n-Butyllithium*** at 0 °C (eq 1).[1] It may also be prepared by treating the β-keto ester with 2.0 equiv of ***Lithium Diisopropylamide*** or ***Lithium 2,2,6,6-Tetramethylpiperidide*** at 0 °C.[1] These methods may be used to generated the dianions of a wide range of β-keto esters. The dianion (**1**) may be quenched with D_2O and the degree of dianion formation determined by integration of the ^1H NMR methyl peak at δ 2.2 relative to the ester signal or by mass spectrometry.[1] Occasionally it is necessary to add HMPA to dissolve the dianions of more substituted β-keto esters.

O=C(CH3)CH2CO2Me —(NaH, BuLi or 2 equiv LiN(*i*-Pr)2)→ (**1**) (1)

Handling, Storage, and Precautions: generated in solution; stable for hours at 0 °C; reacts with moisture and oxygen.

Introduction. Acetoacetate esters provide a four-carbon chain to homologate simpler molecules. The utility of this homologation is that all four carbons subsequently can be reacted regioselectively. Thus the acetoacetate unit is one of the most versatile starting materials available in the synthesis of complex natural and unnatural products.

Alkylation of Methyl Acetoacetate Dianion. The dianion (**1**) reacts with primary and secondary halides and sulfonates to give the γ-alkylated products (**2**) in good yield (eq 2 and Table 1). Halides are the most common alkylating agents. However, sulfonates will also react. The triflate is effective in the case of hindered alkylating agents.[7] Primary halides react without difficulty. Secondary halides are less reactive and some do not react under these conditions. Tertiary halides do not undergo alkylation. The reaction is compatible with a number of other functional groups and protecting groups. It is not necessary to protect alcohol groups, provided excess dianion is used. Following monoalkylation of (**1**), addition of a second equiv of base will regenerate a dianion which can be alkylated a second time (eq 3).[57] Highest yields in this one-pot bisalkylation required the addition of ***N,N′-Dimethylpropyleneurea***.[57] Other esters of acetoacetate, such as ethyl, *t*-butyl, and 2-trimethylsilylethyl, have also been used in these reactions.

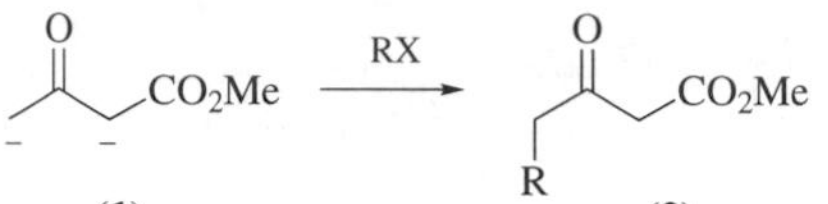

(2)

(3)

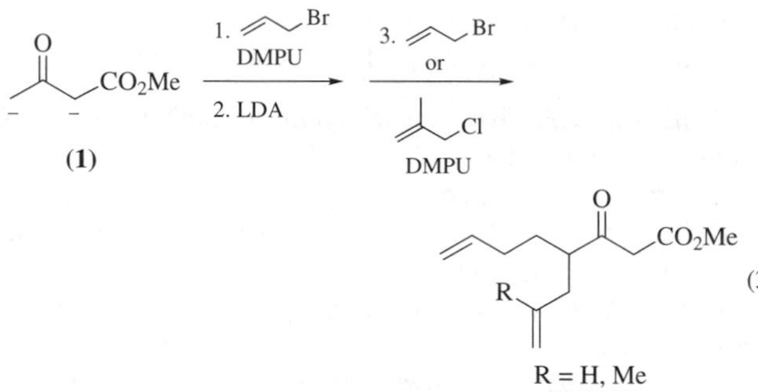

R = H, Me
57–58%

Table 1 Alkylation of Acetoacetate Dianions with Halides, Sulfonates, and Epoxides (eqs 2, 3, and 5)

Alkylating agent	Yield (%)	Ref.
Alkyl halides	50–90	1–26, 54, 56
Alkyl sulfonates	30–90	4b, 7, 53
Allylic halides	40–100	1, 5–7, 9, 14, 27–46
Alkynyl halides	85–90	25, 47, 48
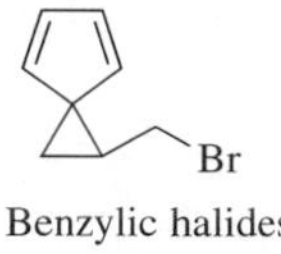 (spiro cyclopentadiene–cyclopropyl bromide)	43	49
Benzylic halides	40–85	1, 18, 50–52
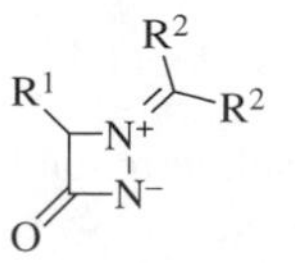 (diazetidinone azomethine imine)	30–50	55
Epoxides	50–90	16, 59–61

A recent example which exemplifies the complex carbon skeletons which can be assembled from acetoacetate dianions is taken from the synthesis of latrunculin A (eq 4).[58] The dianion of the β-keto ester (**3**) reacts with the phosphonium salt (**4**) which is generated in situ. The resulting phosphorane is treated with an aldehyde to give the diene β-keto ester (**5**), all in one pot. The next higher vinylog of (**4**) yields (*E,E,Z*)-trienes in the same reaction.[58c]

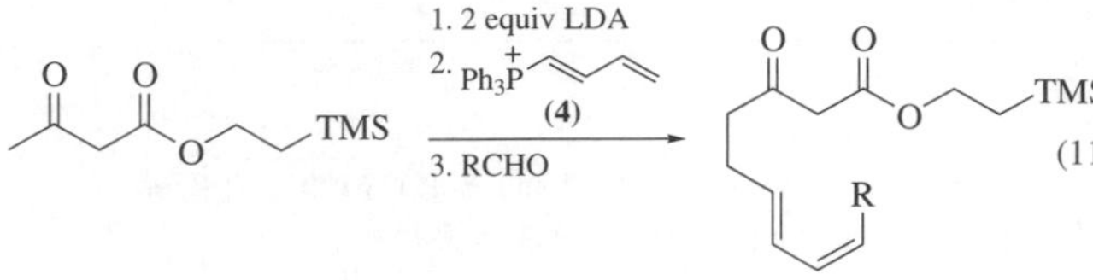

(11)

Epoxides react with (**1**) to give alcohols (Table 1). In the case of simple epoxides the initially formed alcohols often could not be isolated, because they underwent cyclization to tetrahydrofuranylidene-2-acetates (**6**) (eq 5).[16] In the case of more complex epoxides the alcohols usually could be isolated in good yield and in most cases the opening of the epoxide ring was regioselective.[68–70]

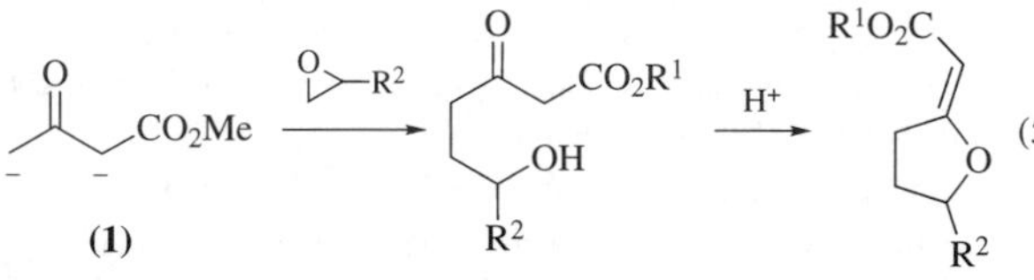

(5)

The dianion of methyl acetoacetate reacts with other simple electrophiles including benzenesulfenyl halides,[62] alkyl nitrates,[63] chlorophosphates,[64] and carbon dioxide.[65] Often the reaction is very clean, leading to the γ-substituted product.

The dianion (**1**) undergoes aldol condensations with a variety of aldehydes (Table 2). The resulting δ-hydroxy-β-keto esters (**7**) can be cyclized to the β-keto-δ-lactones, oxidized or dehydrated (eq 6). The stereoselectivity in the addition to chiral aldehydes is low, although it can be improved with the addition of Lewis acids.[77] In the case of α,β-unsaturated aldehydes and ketones, only 1,2-addition is observed.

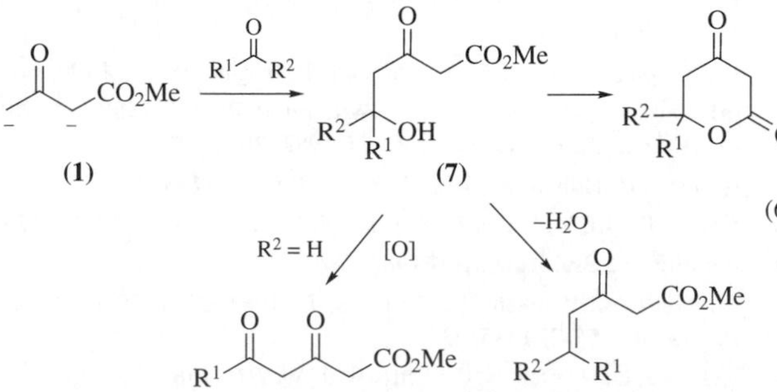

(6)

Dianion (**1**) can be acylated. A limited amount of work has been carried out using nitriles as the acylating agent and the product was found to be the enamino β-keto ester (**8**) (eq 7).[92]

Table 2 Aldol Condensations of Acetoacetate Dianions with Aldehydes and Ketones (eq 6)

R^1COR^2	Yield (%)	Ref.
Aliphatic aldehydes	25–90	66–69, 71–77
Aromatic aldehydes	70–95	67, 68, 70
Conjugated aldehydes[a]	50–100	50, 68, 78–85
Aliphatic ketones	25–70	67, 68, 86, 87
Conjugated ketones[a]	60–85	67, 88, 89
Aromatic ketones	60–95	67, 90
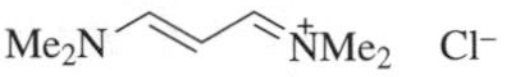 Me_2N–CH=CH–CH=$\overset{+}{N}Me_2$ Cl^-	17	91

[a] Yield of 1,2-addition product.

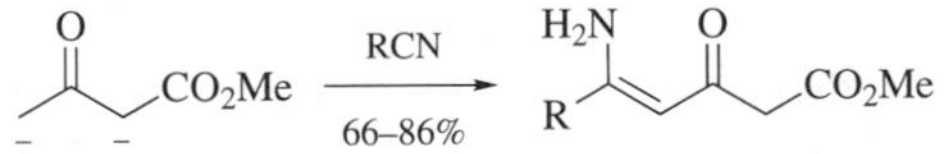

(1) —RCN, 66–86%→ (8) R = Me, Ph (7)

A much larger range of ester derivatives have been condensed with dianion (**1**) (Table 3). Because the monoanion of the initial product (**9**) is more acidic than the monoanion (**10**) of acetoacetate, the starting dianion (**1**) is quenched as the reaction proceeds. Initially this problem was overcome by adding an extra equiv of base during the reaction.[92] However, the problem does limit the yield of (**9**) (eq 8). Now an excess of dianion (**1**) is usually used to optimize the yield of acylated product (**9**). It was also found that addition of ***Boron Trifluoride Etherate*** improves the yield of (**9**).[93] Addition of TMEDA also leads to increased yields in these acylations.[94]

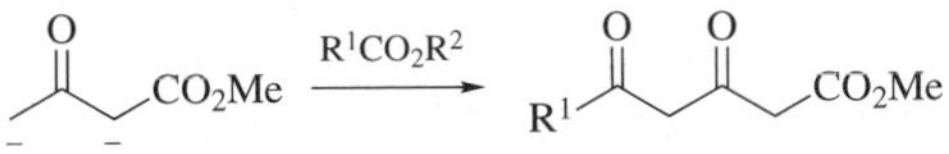

(1) —$R^1CO_2R^2$→ (9) (8)

Table 3 Acylation Reactions of Acetoacetate Dianions (eqs 8–10)

Acylating reagent	Yield (%)	Ref.
Aliphatic esters	45–80	92, 93, 96c
Aromatic esters	40–85	92–94
Lactones	40	95–96d
Diesters	30–95	100–102
Amides	60–85	93
Bisamides	45–70	93
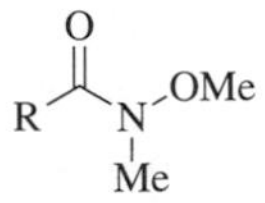 RC(O)N(OMe)Me	40–90	96b, 103

The acylated products have been used in the synthesis of many polyketide aromatic natural products.[96] For example, condensation of (**1**) with the monoanion (**10**) of methyl acetoacetate yields the triketo ester (**11**), which can be cyclized to the orsellinate (**12**) (eq 9).[92,97] The conversion of (**1**) to (**12**) can be carried out in one pot by refluxing the solution of the dianion (**1**) and the monoanion (**10**) (eq 9).[98,99] These condensations could be applied to dicarboxylic acid derivatives to yield polyketo diesters which readily cyclized (eq 10).[96,101] Acylations with amides in the presence of $BF_3 \cdot Et_2O$ led to improved yields of the triketo esters (**11**).[93] This was further refined with the use of *N*-methoxy-*N*-methylamides as acylating agents.[96b,103]

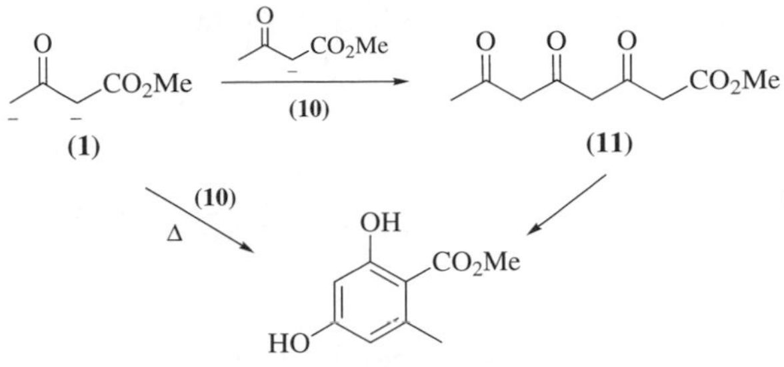

(1) —(10)→ (11); (1) —(10), Δ→ (12); (11) → (12) (9)

Table 4 Synthetic Equivalents to the Acetoacetate Dianion

Synthetic equivalent	Reaction	Ref.
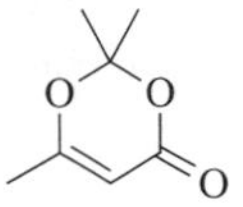 2,2,6-trimethyl-1,3-dioxin-4-one	Aldol	107
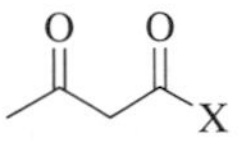 CH_3COCH_2COX, X = NMe_2, S-*t*-Bu	Alkylation, aldol, acylation	108, 109
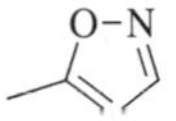 3-methylisoxazole	Alkylation, aldol, acylation	110
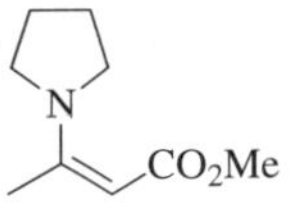 pyrrolidine enamine, CO_2Me	Alkylation	111
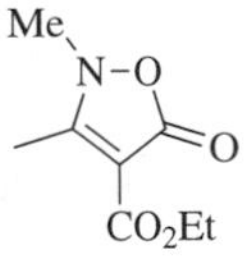 Me, N–O, =O, CO_2Et	Alkylation	112
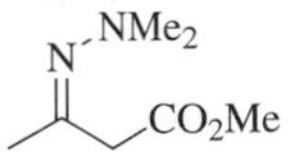 N–NMe_2 hydrazone, CO_2Me	Aldol	113
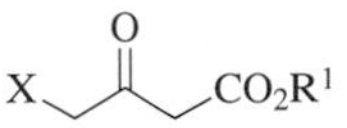 $XCH_2COCH_2CO_2R^1$, X = $PO(OR_2)_2$, $P(O)Ph_2$	Wittig	114–116
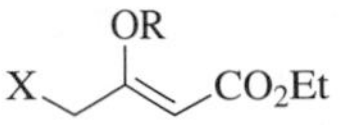 $XCH_2C(OR)=CHCO_2Et$, X = $PO(OMe)_2$, PPh_3^+	Wittig	117, 118
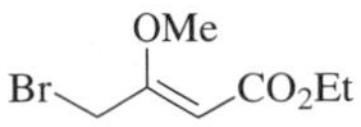 $BrCH_2C(OMe)=CHCO_2Et$	Reformatsky	119
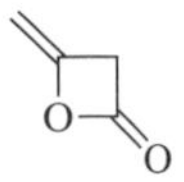 diketene	Alkylation	121
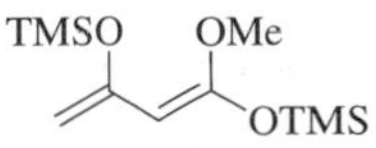 TMSO, OMe, OTMS (**13**)	Alkylation, aldol, acylation	120

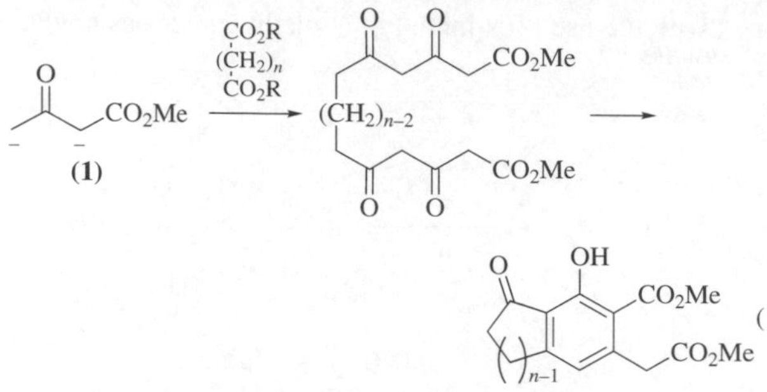

(10)

Dianion (**1**) undergoes 1,4-addition to certain Michael acceptors such as α,β-unsaturated thioamides,[104] α,β-unsaturated sulfoxides,[105] and α,β-unsaturated sulfones.[106]

Dianion (**1**) has proven to be so useful that several synthetic equivalents to it have been developed and these are listed in Table 4. Most of these reagents are generated under the strongly basic conditions that (**1**) is formed under. However, the bistrimethylsilyl ether (**13**) reacts with alkylating and acylating reagents in the presence of Lewis acids and provides a very useful complement to (**1**).

Related Reagents. Acetoacetic Acid; Acetone; Acetone Cyclohexylimine; Acetone Hydrazone; Diketene; Dimethyl 1,3-Acetonedicarboxylate; Ethyl Acetoacetate; Ethyl 4-Chloroacetoacetate; Ethyl 4-(Triphenylphosphorsyanylidene)acetoacetate; *N*-Methoxy-*N*-methylacetamide; 1-Methoxy-1,3-bis(trimethylsilyloxy)-1,3-butadiene; 2,4-Pentanedione; Trimethylsilylacetone; 2,2,6-Trimethyl-4H-1,3-dioxin-4-one.

---

1. (a) Weiler, L. *JACS* **1970**, *92*, 6702; (b) Huckin, S. N.; Weiler, L. *JACS* **1974**, *96*, 1082.
2. Utaka, M.; Watabu, H.; Higashi, H.; Sakai, T.; Tsuboi, S.; Torii, S. *JOC* **1990**, *55*, 3917.
3. Shono, T.; Kise, N.; Fujimoto, T.; Tominaga, N.; Morita, H. *JOC* **1992**, *57*, 7175.
4. (a) Sum, P. E.; Weiler, L. *CJC* **1979**, *57*, 1475; (b) Alderdice, M.; Spino, C.; Weiler, L. *CJC* **1993**, *71*, 1955.
5. Kates, S. A.; Dombroski, M. A.; Snider, B. B. *JOC* **1990**, *55*, 2427.
6. Snider, B. B.; Mohan, R.; Kates, S. A. *JOC* **1985**, *50*, 3659.
7. Armstrong, R. J.; Weiler, L. *CJC*, **1986**, *64*, 584.
8. Stoilov, I.; Back, T. G.; Thompson, J. E.; Djerassi, C. *T* **1986**, *42*, 4156.
9. (a) Holton, R. A.; Zoeller, J. R. *JACS* **1985**, *107*, 2124; (b) Lee, W. Y.; Jang, S. Y.; Kim, M.; Park, O. S. *SC* **1992**, *22*, 1283.
10. Welch, S. C.; Assercq, J. M.; Loh, J. P.; Glase, S. A. *JOC* **1987**, *52*, 1440.
11. Trost, B. M. *JOC* **1974**, *39*, 2648.
12. Sum, P. E.; Weiler, L. *CC* **1977**, 91.
13. (a) Taylor, E. C.; LaMattina, J. L. *JOC* **1978**, *43*, 1200; (b) Moyer, M. P.; Feldman, P. L.; Rapoport, H. *JOC* **1985**, *50*, 5223; (c) Brooks, D. W.; Kellog, R. P.; Cooper, C. S. *JOC* **1987**, *52*, 192.
14. (a) Sum, F. W.; Weiler, L. *CC* **1978**, 985; (b) Sum, F. W.; Weiler, L. *T* **1981**, *37* (Suppl. 1), 303.
15. Stumpp, M. C.; Schmidt, R. R. *T* **1986**, *42*, 5941.
16. Bryson, T. A. *JOC* **1973**, *38*, 3428.
17. Paulvannan, K.; Schwarz, J. B.; Stille, J. R. *TL* **1993**, *34*, 215.
18. (a) Heslin, J. C.; Moody, C. J.; Slawin, A. M. Z.; Williams, D. J. *TL* **1986**, *27*, 1403; (b) Hedtmann, U.; Hobert, K.; Klintz, R.; Welzel, P.; Frelek, J.; Strangmanndiekmann, M.; Klone, A.; Pongs, O. *AG(E)* **1989**, *28*, 1515.
19. Overman, L. E.; Rabinowitz, M. H. *JOC* **1993**, *58*, 3235.
20. Meyer, S. D.; Miwa, T.; Nakatsuka, M.; Schreiber, S. L. *JOC* **1992**, *57*, 5058.
21. Lee, B. H.; Biswas, A.; Miller, M. J. *JOC* **1986**, *51*, 106.
22. Kato, M.; Kamat, V. P.; Yoshikoshi, A. *S* **1988**, 699.
23. Zoretic, P. A.; Yu, B. C.; Biggers, M. S.; Casper, M. L. *JOC* **1990**, *55*, 3954.
24. Hoye, T. R.; Hanson, P. R.; Kovelsky, A. C.; Ocain, T. D.; Zhuang, Z. *JACS* **1991**, *113*, 9369.
25. Kotsuki, H.; Ushio, Y.; Kadota, I.; Ochi, M. *JOC* **1989**, *54*, 5153.
26. Sum, P. E.; Weiler, L. *CJC* **1977**, *55*, 996.
27. Hayes, T. K.; Villani, R.; Weinreb, S. M. *JACS* **1988**, *110*, 5533.
28. Holmes, A. B.; Swithenbank, C.; Williams, S. F. *CC* **1986**, 265.
29. Snider, B. B.; Patricia, J. J. *JOC* **1989**, *54*, 38.
30. Casey, C. P.; Marten, D. F. *SC* **1973**, *3*, 321.
31. (a) Sum, F. W.; Weiler, L. *TL* **1979**, 707; (b) Sum, F. W.; Weiler, L. *JACS* **1979**, *101*, 4401.
32. Schreiber, S. L.; Kelly, S. E.; Porco, J. A.; Sammakia, T.; Suh, E. M. *JACS* **1988**, *110*, 6210.
33. (a) Chen, K. M.; Joullié, M. M. *TL* **1984**, *25*, 3795; (b) Chen, K. M.; Semple, J. E.; Joullié, M. M. *JOC* **1985**, *50*, 3997.
34. Mori, K.; Mori, H. *T* **1987**, *43*, 4097.
35. Corey, E. J.; Ueda, Y.; Ruden, R. A. *TL* **1975**, 4347.
36. Clark, R. D.; Heathcock, C. H. *TL* **1975**, 529.
37. (a) Skeean, R. W.; Trammell, G. L.; White, J. D. *TL* **1976**, 525; (b) White, J. D.; Skeean, R. W.; Trammell, G. L. *JOC* **1985**, *50*, 1939.
38. Joulain, D.; Moreau, C.; Pfau, M. *T* **1973**, *29*, 143.
39. Borch, R. F.; Ho, B. C. *JOC* **1977**, *42*, 1225.
40. Corey, E. J.; Tius, M. A.; Das, J. *JACS* **1980**, *102*, 1742.
41. McMurry, J. E.; Erion, M. D. *JACS* **1985**, *107*, 2712.
42. (a) Corey, E. J.; Reid, J. G.; Myers, A. G.; Hahl, R. W. *JACS* **1987**, *109*, 918; (b) Corey, E. J.; Hahl, R. W. *TL* **1989**, *30*, 3023.
43. (a) Piers, E.; Ruediger, E. H. *JOC* **1980**, *45*, 1725; (b) Piers, E.; Jung, G.; Ruediger, E. H. *CJC* **1987**, *65*, 670.
44. Taber, D. F. *JACS* **1977**, *99*, 3513.
45. Kondo, K.; Umemoto, T.; Takahatake, Y.; Tunemoto, D. *TL* **1977**, 113.
46. Sharma, V. K.; Shahriari-Zavareh, H.; Garratt, P. J.; Sondheimer, F. *JOC* **1983**, *48*, 2379.
47. Crich, D.; Fortt, S. M. *T* **1989**, *45*, 6581.
48. Corey, E. J.; Seibel, W. L. *TL* **1986**, *27*, 905.
49. Gallacher, G.; Ng, A. S.; Attah-Poku, S. K.; Antczak, K.; Alward, S. J.; Kingston, J. F.; Fallis, A. G. *CJC* **1984**, *62*, 1709.
50. Hrytsak, M.; Durst, T. *H* **1987**, *26*, 2393.
51. Acheson, R. M.; Lee, G. C. M. *JCS(P1)* **1987**, 2321.
52. Scott, J. W.; Buchschacher, D.; Labler, L.; Meier, W.; Fürst, A. *HCA* **1974**, *57*, 1217.
53. Angle, S. R.; Louie, M. S. *JOC* **1991**, *56*, 2853.
54. Salzmann, T. N.; Ratcliffe, R. W.; Christensen, B. G. *TL* **1980**, *21*, 1193.
55. Taylor, E. C.; Davies, H. M. L. *JOC* **1984**, *49*, 113.
56. Ray, P. S.; Jaxa-Chamiec, A. A. *H* **1990**, *31*, 1777.
57. Dombroski, M. A.; Snider, B. B. *T* **1992**, *48*, 1417.
58. (a) White, J. D.; Kawaski, M. *JACS* **1990**, *112*, 4991; (b) White, J. D.; Kawaski, M. *JOC* **1992**, *57*, 5292; (c) White, J. D.; Jensen, M. S. *TL* **1992**, *33*, 577.

59. Yamaguchi, M.; Hirao, I. *CL* **1985**, 337.
60. Kieczykowski, G.; Roberts, M. R.; Schlessinger, R. H. *JOC* **1978**, *43*, 788.
61. Kieczykowski, G.; Schlessinger, R. H. *JACS* **1978**, *100*, 1938.
62. (a) Hiroi, K.; Matsuda, Y.; Sato, S. *S* **1979**, 621; (b) Hiroi, K.; Matsuda, Y.; Sato, S. *CPB* **1979**, *27*, 2338.
63. Duthaler, R. O. *HCA* **1983**, *66*, 1475.
64. Karanewsky, D. S.; Badia, M. C.; Ciosek, Jr., C. P.; Robl, J. A.; Sofia, M. J.; Simpkins, L. M.; DeLange, B.; Harrity, T. W.; Biller, S. A.; Gordon, E. M. *JMC* **1990**, *33*, 2952.
65. Cha, J. K.; Harris, T. M.; Ray, J. A.; Venkataraman, H. *TL* **1989**, *30*, 3505.
66. Davies, M. J.; Heslin, J. C.; Moody, C. J. *JCS(P1)* **1989**, 2473.
67. (a) Huckin, S. N.; Weiler, L. *TL* **1971**, 4835; (b) Huckin, S. N.; Weiler, L. *CJC* **1974**, *52*, 2157.
68. Peterson, J. R.; Winter, T. J.; Miller, C. P. *SC* **1988**, *18*, 949.
69. Bonini, C.; Racioppi, R.; Righi, G.; Viggiani, L. *JOC* **1993**, *58*, 802.
70. Kresze, G.; Morper, M.; Bijev, A. *TL* **1977**, 2259.
71. Seebach, D.; Meyer, H. *AG(E)* **1974**, *13*, 77.
72. Meyer, H. H. *LA* **1984**, 977.
73. Akita, H.; Yamada, H.; Matsukura, H.; Nakata, T.; Oishi, T. *TL* **1990**, *31*, 1731.
74. (a) Wang, N. Y.; Hsu, C. T.; Sih, C. J. *JACS* **1981**, *103*, 6538; (b) Hsu, C. T.; Wang, N. Y.; Latimer, L. H.; Sih, C. J. *JACS* **1983**, *105*, 593; (c) Girotra, N. N.; Wendler, N. L. *TL* **1982**, *23*, 5501.
75. Sliskovic, D. R.; Roth, B. D.; Wilson, M. W.; Hoefle, M. L.; Newton, R. S. *JMC* **1990**, *33*, 31; also see: Sliskovic, D. R.; Blankley, C. J.; Krause, B. R.; Newton, R. S.; Picard, J. A.; Roark, W. H.; Roth, B. D.; Sekerke, C.; Shaw, M. K.; Stanfield, R. L. *JMC* **1992**, *35*, 2095.
76. Roth, B. D.; Ortwine, D. F.; Hoefle, M. L.; Stratton, C. D.; Sliskovic, D. R.; Wilson, M. W.; Newton, R. S. *JMC* **1990**, *33*, 21.
77. Urabe, H.; Matsuka, T.; Sato, F. *TL* **1992**, *33*, 4179.
78. White, J. D.; Skeean, R. W. *JACS* **1978**, *100*, 6296.
79. Nagasawa, K.; Zako, Y.; Ishihara, H.; Shimizu, I. *TL* **1991**, *32*, 4937.
80. Henkel, B.; Kunath, A.; Schick, H. *LA* **1992**, 809.
81. (a) Batty, D.; Crich, D.; Fortt, S. M. *CC* **1989**, 1366; (b) Batty, D.; Crich, D.; Fortt, S. M. *JCS(P1)* **1990**, 2875.
82. Willard, P. G.; Grab, L. A.; de Laszlo, S. E. *JOC* **1983**, *48*, 1123.
83. Barton, D. H. R.; Dressaire, G.; Willis, B. J.; Barrett, A. G. M.; Pfeffer, M. *JCS(P1)* **1982**, 665.
84. Stokker, G. E.; Hoffman, W. F.; Alberts, A. W.; Cragoe, Jr., E. J.; Deana, A. A.; Gilfillan, J. L.; Huff, J. W.; Novello, F. C.; Prugh, J. D.; Smith, R. L.; Willard, A. K. *JMC* **1985**, *28*, 347.
85. Kondo, K.; Umemoto, T.; Yako, K.; Tunemoto, D. *TL* **1978**, 3927.
86. Tanabe, Y.; Ohno, N. *JOC* **1988**, *53*, 1560.
87. Monti, S. A.; Chen, S. C.; Yang, Y. L.; Yuan, S. S.; Bourgeois, O. P. *JOC* **1978**, *21*, 4062.
88. (a) Bérubé, G.; Fallis, A. G. *TL* **1989**, *30*, 4045; (b) Bérubé, G.; Fallis, A. G. *CJC* **1991**, *69*, 77; (c) Alward, S. J.; Fallis, A. G. *CJC* **1984**, *62*, 1709.
89. Matsumoto, M.; Watanabe, N. *H* **1986**, *24*, 3149.
90. Kingston, J. F.; Weiler, L. *CJC* **1977**, *55*, 785.
91. Nair, V.; Cooper, C. S. *JOC* **1981**, *46*, 4759.
92. (a) Huckin, S. N.; Weiler, L. *TL* **1972**, 2405; (b) Huckin, S. N.; Weiler, L. *CJC* **1974**, *52*, 1343.
93. (a) Yamaguchi, M.; Shibato, K.; Hirao, I. *CL* **1985**, 1145; (b) Yamaguchi, M.; Shibato, K.; Nakashima, H.; Minami, T. *T* **1988**, *44*, 4767.
94. Narasimhan, N. S.; Ammanamanchi, R. K. *JOC* **1983**, *48*, 3945.
95. Stockinger, H.; Schmidt, U. *LA* **1976**, 1617.
96. (a) Harris, T. M.; Harris, C. M. *T* **1977**, *33*, 2159; (b) Gilbreath, S. G.; Harris, C. M.; Harris, T. M. *JACS* **1988**, *110*, 6172; (c) Harris, T. M.; Harris, C. M.; Oster, T. A.; Brown, Jr., L. E.; Lee, J. Y. C. *JACS* **1988**, *110*, 6180; (d) Harris, T. M.; Harris, C. M.; Kuzma, P. C.; Lee, J. Y. C.; Mahalingam, S.; Gilbreath, G. S. *JACS* **1988**, *110*, 6186.
97. Harris, T. M.; Murray, T. P.; Harris, C. M.; Gumulka, M. *CC* **1974**, 362.
98. Chiarello, J.; Joullié, M. M. *T* **1988**, *44*, 41.
99. Barrett, A. G. M.; Morris, T. M.; Barton, D. H. R. *JCS(P1)* **1980**, 2272.
100. Parker, K. A.; Breault, G. A. *TL* **1986**, *27*, 3835.
101. (a) Yamaguchi, M.; Hasebe, K.; Minami, T. *TL* **1986**, *27*, 2401; (b) Yamaguchi, M.; Hasebe, K.; Higashi, H.; Uchida, M.; Irie, A.; Minami, T. *JOC* **1990**, *55*, 1611.
102. Minami, T.; Takahashi, K.; Hiyama, T. *TL* **1993**, *34*, 513.
103. Hanamoto, T.; Hiyama, T. *TL* **1988**, *29*, 6467.
104. Tamaru, Y.; Harada, T.; Yoshida, Z. *JACS* **1979**, *101*, 1316.
105. (a) Brown, P. J.; Jones, D. N.; Khan, M. A.; Meanwell, N. A. *TL* **1983**, *24*, 405; (b) Brown, P. J.; Jones, D. N.; Khan, M. A.; Meanwell, N. A. *JCS(P1)* **1984**, 2049.
106. Pyne, S. G.; Spellmeyer, D. C.; Chen, S.; Fuchs, P. L. *JACS* **1982**, *104*, 5728.
107. Lichenthaler, F. W.; Dinges, J.; Fukuda, Y. *AG(E)* **1991**, *30*, 1339.
108. Hubbard, J. S.; Harris, T. M. *TL* **1978**, 4601.
109. (a) Booth, P. M.; Fox, C. M. J.; Ley, S. V. *TL* **1983**, *24*, 5143; (b) Booth, P. M.; Fox, C. M. J.; Ley, S. V. *JCS(P1)* **1987**, 121; (c) Clarke, T.; Ley, S. V. *JCS(P1)* **1987**, 131.
110. Vinick, F. J.; Pan, Y.; Gschwend, H. W. *TL* **1978**, 4221.
111. Yoshimoto, M.; Ishida, N.; Hiraoka, T. *TL* **1973**, 39.
112. Doleschall, G. *TL* **1988**, *29*, 6339; see also Tischler, S. A.; Weiler, L. *TL* **1979**, 4903.
113. Corey, E. J.; Enders, D. *CB* **1978**, *111*, 1362.
114. (a) Bodalski, R.; Pietrusiewicz, K. M.; Monkiewicz, J.; Koszuk, J. *TL* **1980**, *21*, 2287; (b) Moorhoff, C. M.; Schneider, D. F. *TL* **1987**, *28*, 559.
115. (a) van der Gen, A.; van den Goorbergh, J. A. M. *TL* **1980**, *21*, 3621; (b) Rigby, J. H.; Qabar, M. *JOC* **1989**, *54*, 5852.
116. Ley, S. V.; Smith, S. C.; Woodward, P. R. *T* **1992**, *48*, 1145.
117. Bestmann, H. J.; Saalfrank, R. W. *AG(E)* **1970**, *9*, 367.
118. Hudlicky, T.; Radesca-Kwart, L.; Li, L.; Bryant, T. *TL* **1988**, *29*, 3283.
119. Israili, Z. H.; Smissman, E. E. *JOC* **1976**, *41*, 4070.
120. (a) Chan, T. H.; Brownbridge, P. *CC* **1979**, 578; (b) Brownbridge, P.; Chan, T. H. *TL* **1979**, 4437; (c) Brownbridge, P.; Chan, T. H.; Brook, M. A.; Kang, G. J. *CJC* **1983**, *61*, 688; (d) Molander, G. A.; Andrews, S. W. *TL* **1989**, *30*, 2351; (e) Evans, D. A.; Black, W. C. *JACS* **1992**, *114*, 2260.
121. (a) Izawa, T.; Mukaiyama, T. *CL* **1974**, 1189; (b) Izawa, T.; Mukaiyama, T. *CL* **1975**, 161; (c) Izawa, T.; Mukaiyama, T. *CL* **1978**, 409.

Larry Weiler
University of British Columbia, Vancouver, BC, Canada

Methylenetriphenylphosphorane[1]

$Ph_3P{=}CH_2$

[3487-44-3] $C_{19}H_{17}P$ (MW 276.33)

(methylenating agent for aldehydes and ketones;[1] can convert lactols into hydroxy alkenes;[2] can form a lithio derivative;[3] can be transformed into a β-ketophosphorus ylide[4–6])

Physical Data: mp 96 °C; yellow crystals turning white in air.

Solubility: sol ether, THF, DME, benzene, toluene, DMSO; water and protic solvents destroy the reagent completely.

Form Supplied in: not available commercially; prepared as a solution immediately prior to use.

Analysis of Reagent Purity: ylide concentration can be calibrated by color end-point titration of an aliquot with PhCHO in ether or $PhCO_2H$ in THF at −20 °C under N_2.

Preparative Methods: usually prepared by treatment of a suspension of ***Methyltriphenylphosphonium Bromide*** (or iodide) in an appropriate solvent with a strong base. Butyllithium, dimsylsodium, *t*-BuOK, and sodamide are commonly used. Typically, a suspension of dry methyltriphenylphosphonium bromide in dry ether or THF is treated with ***n-Butyllithium*** at 0 °C under N_2 and the resulting solution is stirred at rt for 1 h. A yellow to deep orange color indicates the ylide formation. The dimsylsodium (***Sodium Methylsulfinylmethylide***) procedure[7] requires heating a suspension of NaH in dry DMSO at 75–80 °C for 1 h followed by ice-cooling prior to the addition of phosphonium bromide. A 'salt-free' ylide is usually prepared using ***Sodium Amide*** in liquid ammonia[8] or in refluxing THF.[9]

Handling, Storage, and Precautions: should always be handled under N_2 or Ar. The ylide should be used as prepared for best results.

Methylenation of Aldehydes and Ketones. The most versatile utility of $Ph_3P{=}CH_2$ is Wittig methylenation[10] of aldehydes and ketones.[1,10] Under salt-free conditions, 1,2-oxaphosphetanes have been observed as the only intermediate in the methylenation of aldehydes and ketones.[8] Even in cases where a lithium base is used, the classically favored betaine intermediate could not be observed. Decomposition of 1,2-oxaphosphetanes upon warming then provides terminal alkenes from aldehydes and 1,1-disubstituted alkenes from ketones. Hydroxy aldehydes[11a] and ketones[11b] can be directly converted to hydroxy alkenes when excess reagent is employed (eqs 1 and 2). Often one-carbon homologation of an alcohol can be achieved through a sequence of oxidation, Wittig methylenation, and hydroboration (eq 3).[12]

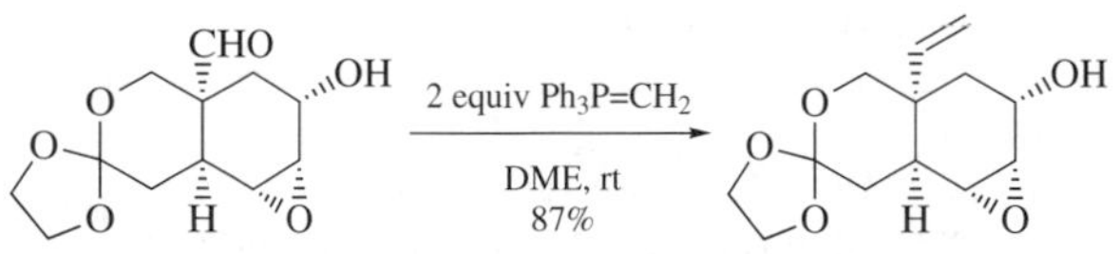

2 equiv $Ph_3P{=}CH_2$, DME, rt, 87% (1)

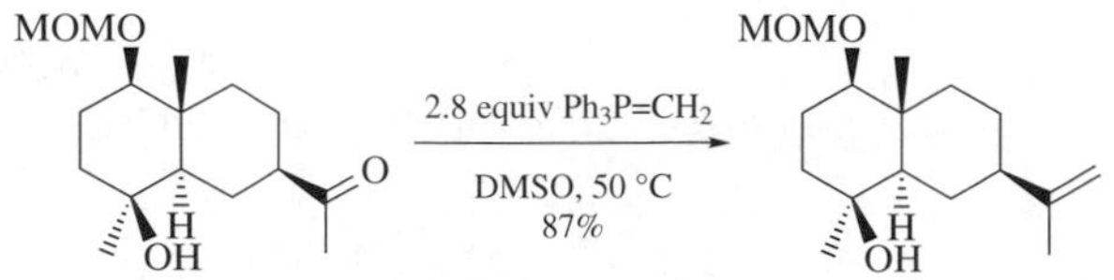

2.8 equiv $Ph_3P{=}CH_2$, DMSO, 50 °C, 87% (2)

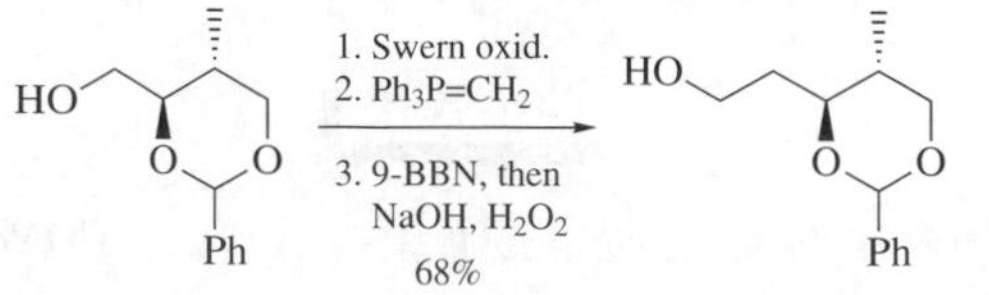

1. Swern oxid. 2. $Ph_3P{=}CH_2$ 3. 9-BBN, then NaOH, H_2O_2, 68% (3)

Vinyl cyclopropanes[13a] and cyclobutanes[13b] are conveniently prepared from corresponding cyclopropyl and cyclobutyl carbonyl compounds. Likewise, vinyl epoxides[14] and aziridines[15] can be prepared (eqs 4 and 5). α,β-Unsaturated aldehydes and ketones are readily converted to 1,3-dienes (eqs 6 and 7).[16]

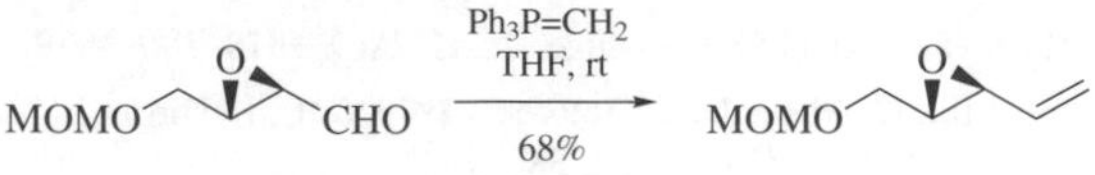

$Ph_3P{=}CH_2$, THF, rt, 68% (4)

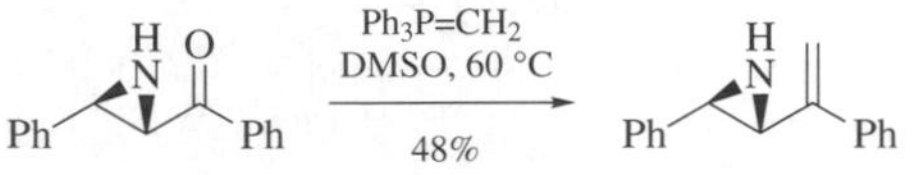

$Ph_3P{=}CH_2$, DMSO, 60 °C, 48% (5)

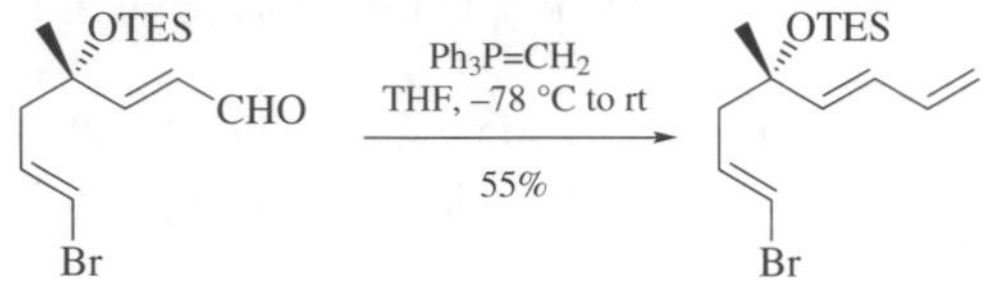

$Ph_3P{=}CH_2$, THF, −78 °C to rt, 55% (6)

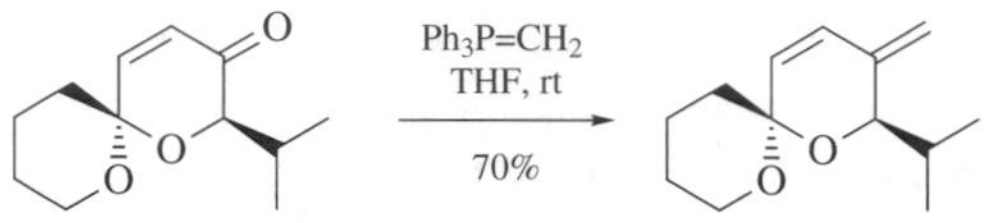

$Ph_3P{=}CH_2$, THF, rt, 70% (7)

Substituted 1,3-dienes can be prepared by the treatment of corresponding enals and enones with $Ph_3P{=}CH_2$. Some examples include dienyl iodide,[17a] silyldiene,[17b] and dienyltin[17c] (eqs 8–10). Under carefully controlled reaction conditions, dienol ethers and sulfides are obtained from vinylogous formates and thioesters, respectively (eqs 11 and 12).[18] The 1,3-dienes prepared this way are often used as Diels–Alder dienes.

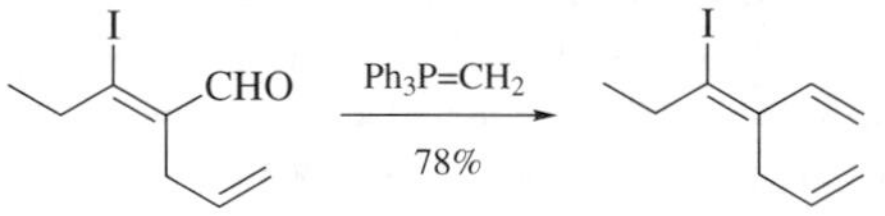

$Ph_3P{=}CH_2$, 78% (8)

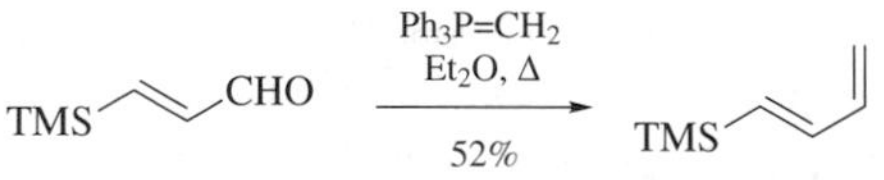

$Ph_3P{=}CH_2$, Et_2O, Δ, 52% (9)

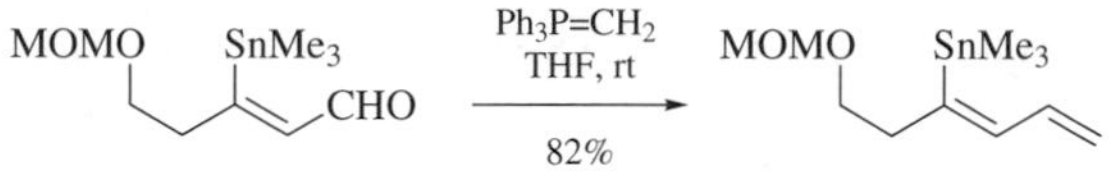

$Ph_3P{=}CH_2$, THF, rt, 82% (10)

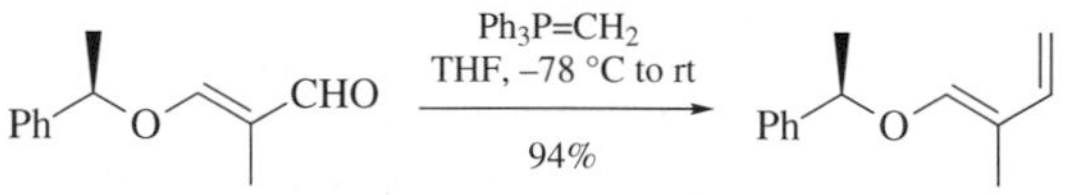

$Ph_3P{=}CH_2$, THF, −78 °C to rt, 94% (11)

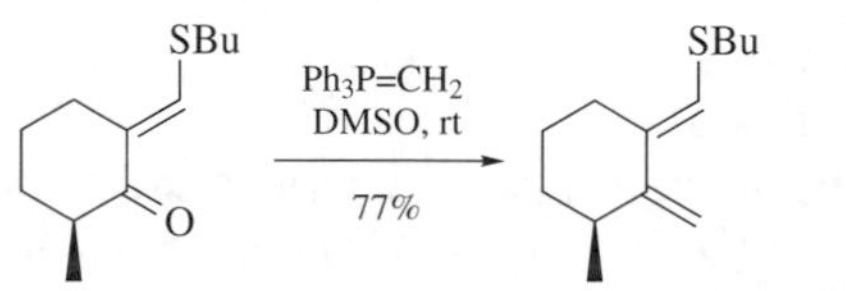

$Ph_3P{=}CH_2$, DMSO, rt, 77% (12)

α-Thioalkoxy aldehydes give rise to allylic sulfides in reaction with $Ph_3P{=}CH_2$.[19] Wittig methylenation can also be effected even in the presence of unusual functional groups such as a phosphonate ester,[20] *O*-silyl oxime,[21] peracetal,[22] and ketene dithioacetal[23] (eqs 13–16).

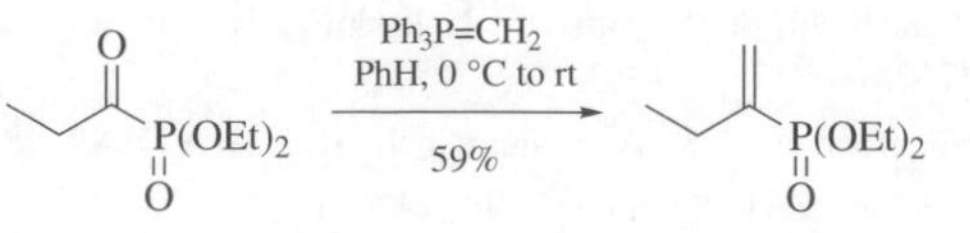

$Ph_3P{=}CH_2$, PhH, 0 °C to rt, 59% (13)

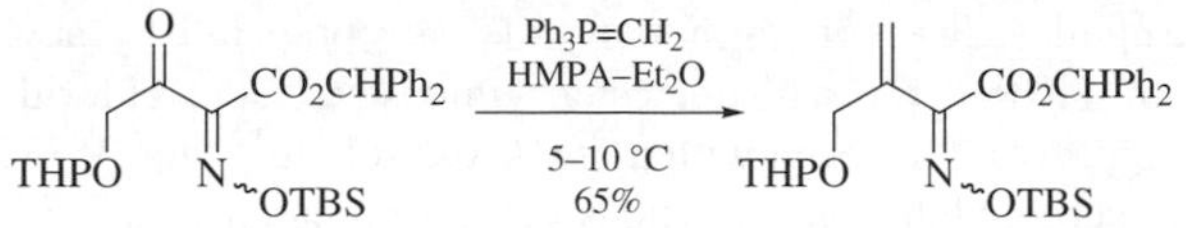

(14)

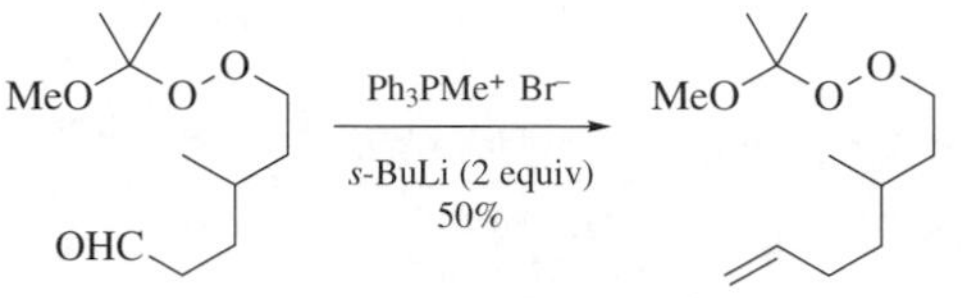

(15)

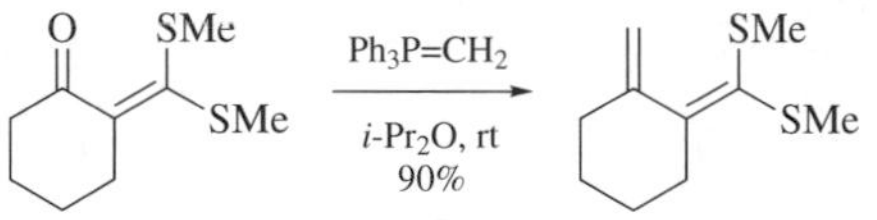

(16)

However, sterically hindered and enolizable ketones are not readily methylenated with $Ph_3P{=}CH_2$.[24] Methylenation of a sterically hindered ketone generally requires a substantial excess of the reagent at elevated temperature and a minimum amount of solvent, as demonstrated in a modhephene synthesis (eq 17).[24a] In this particular example, it was necessary to add the ketone to a preheated (92 °C) solution of a sevenfold excess of $Ph_3P{=}CH_2$ in toluene. One practical modification used to increase the yield of methylenation of enolizable ketones is the repeated additions of stoichiometric amounts of water and the reagent.[24b] Cyclooctanone was methylenated in ether in 89% yield (44%, otherwise) after five cycles of additions (eq 18).

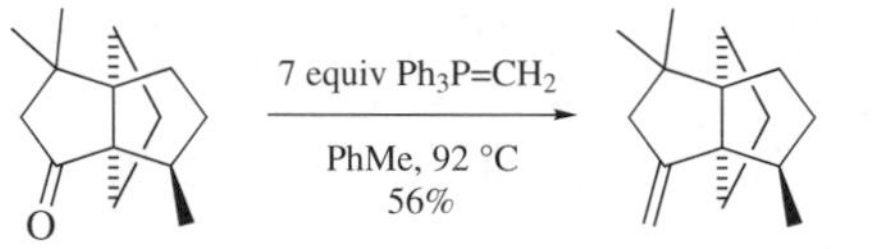

(17)

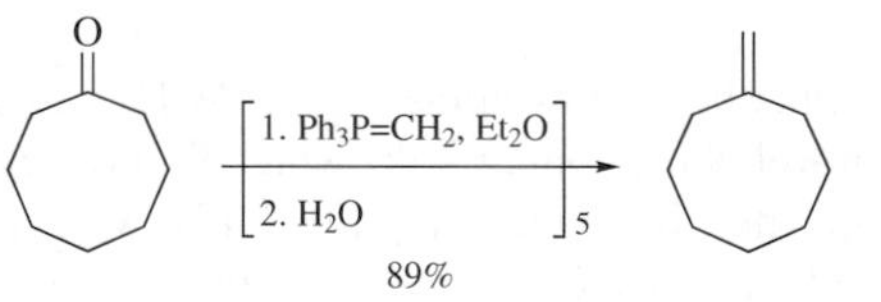

(18)

Other problems commonly observed in Wittig methylenation are β-elimination,[25a,b] retro-aldol,[25c,d] and α-ketol rearrangement,[25e] all of which are caused by the basic nature of $Ph_3P{=}CH_2$ (eqs 19–21).

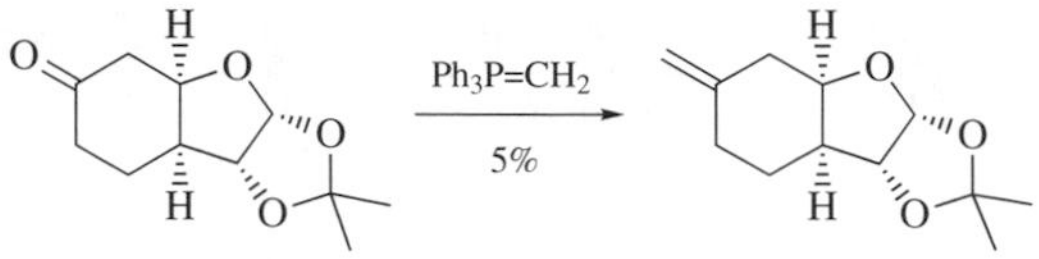

(19)

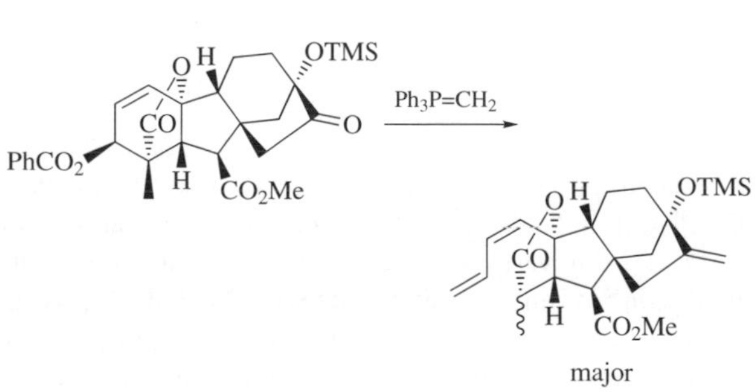

(20)

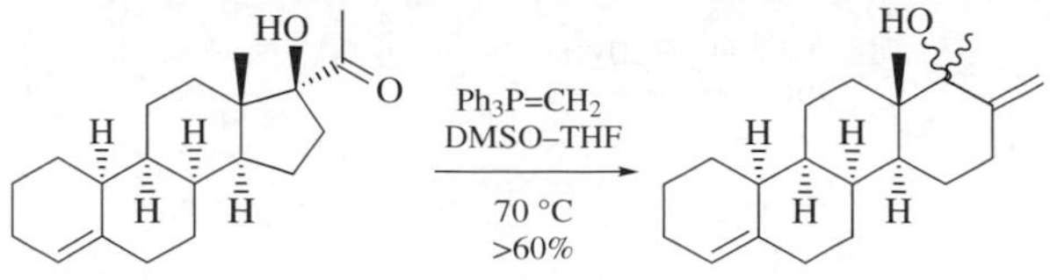

(21)

Methylenation of Lactols and Carbinollactams. Lactols readily undergo methylenation with excess $Ph_3P{=}CH_2$ to give hydroxy terminal alkenes (eqs 22 and 23).[2] γ-Lactones,[26] as well as δ-lactones,[27] can be converted to acyclic alkenic products via reductive methylenation (eq 24). α-Substituted lactols are sometimes disposed to a rapid equilibration with the more stable conformers through their acyclic tautomers, giving rise to the epimerized product (eq 25).[26e] ω-Carbinollactams also undergo methylenation to give corresponding alkenic amides (eq 26).[28]

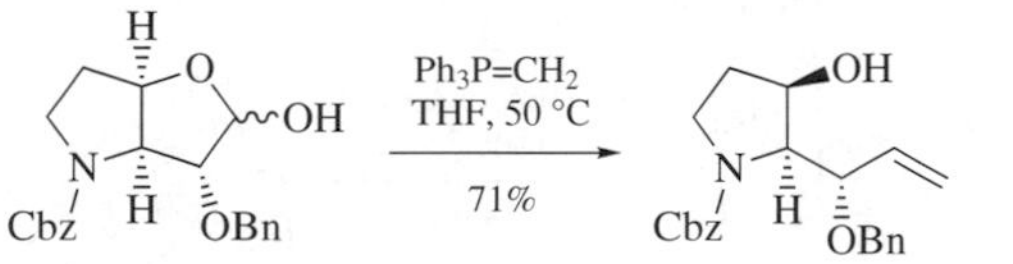

(22)

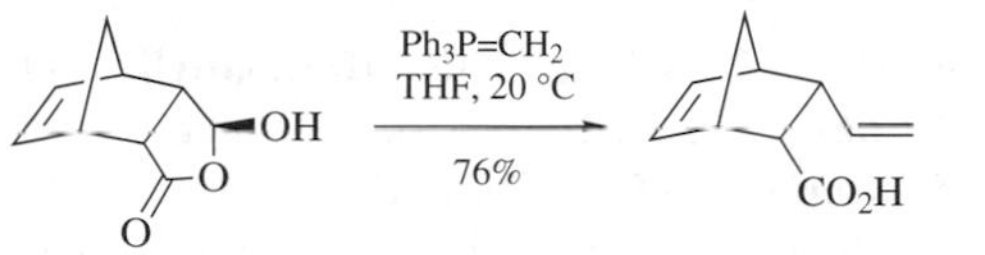

(23)

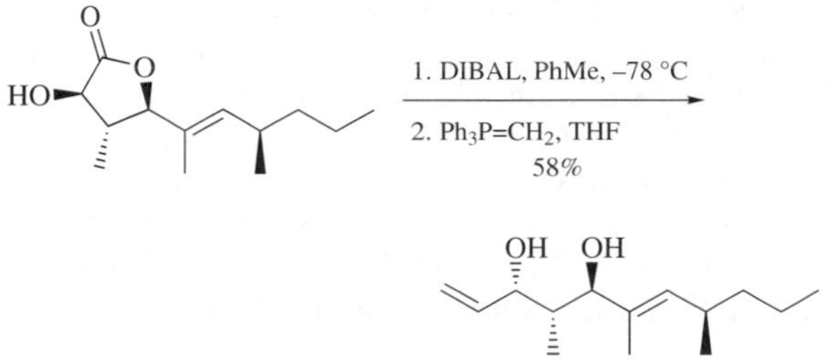

(24)

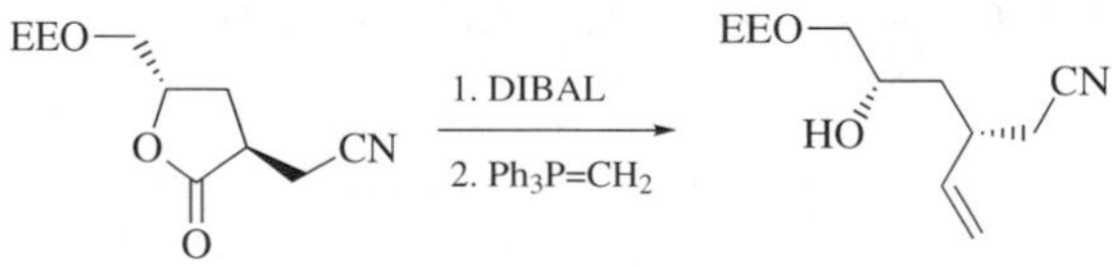

(25)

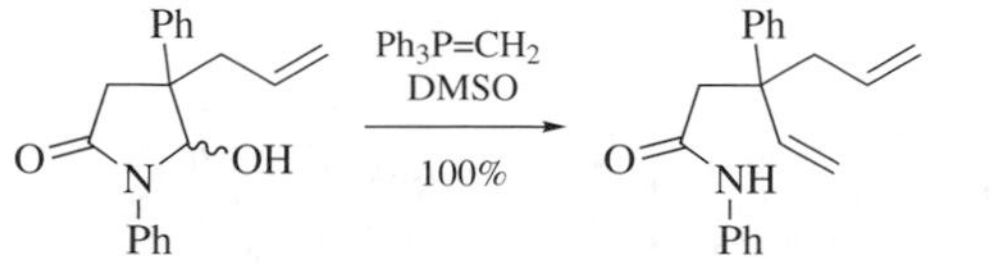

(26)

Reaction with Esters and Carbonyl Equivalents. $Ph_3P{=}CH_2$ can react with esters to give either acetylated or isopropenyl compounds, depending on the nature of the ester and the reaction conditions.[29] Aromatic esters usually give rise to isopropenyl compounds, while aliphatic esters result in a mixture of both (eqs 27 and 28).[29a] Polar aprotic solvents and salt-free conditions favor the formation of isopropenyl compounds.[29b] This method has been used recently in the synthesis of the unnatural enantiomer of rothrockene (eq 29).[30]

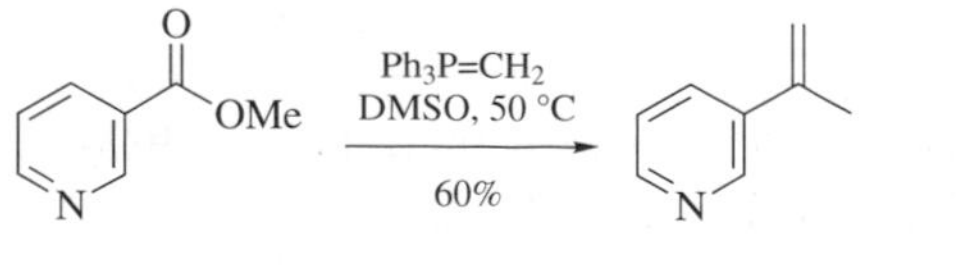

(27)

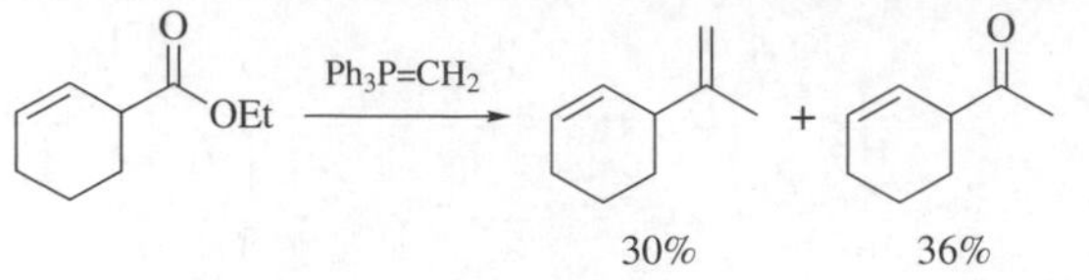

(28)

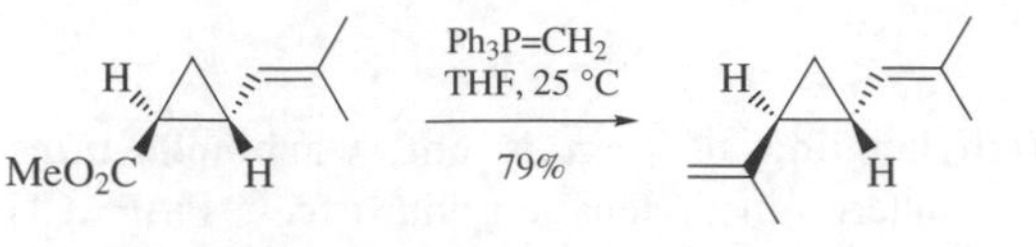

(29)

Carbonyl equivalents such as mixed acetals[31] and β-acyloxy enol esters[32] also react with $Ph_3P{=}CH_2$ to provide corresponding alkenic products (eqs 30 and 31).

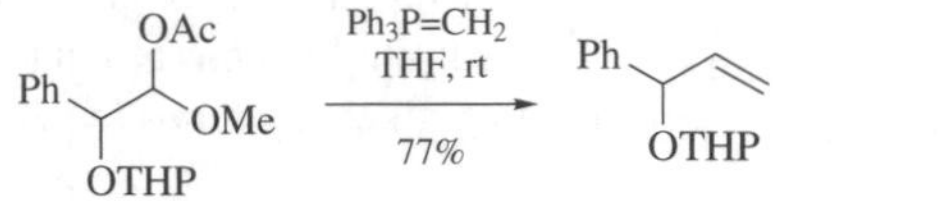

(30)

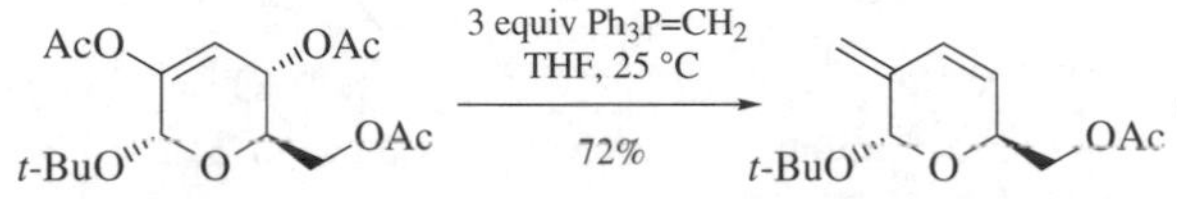

(31)

Formation of Allylic and Homoallylic Alcohols via Oxido Ylides. $Ph_3P{=}CH_2$ can be activated by lithiation with ***t-Butyllithium*** (or ***s-Butyllithium***) in ether.[3] α-Lithiomethylenetriphenylphosphorane ($Ph_3P{=}CHLi$) thus formed reacts with a hindered ketone, which is unaffected by $Ph_3P{=}CH_2$, to give a methylenated compound. Reaction of $Ph_3P{=}CHLi$ with 2 equiv of aldehyde results in formation of a *trans*-allylic alcohol via a β-oxido ylide (eq 32). $Ph_3P{=}CHLi$ also reacts with an epoxide to give a γ-oxido ylide which in turn couples with an aldehyde to provide a *trans*-homoallylic alcohol (eq 33). β-Oxido ylides are also formed in a SCOOPY reaction[33a] in which a mixture of *cis*- and *trans*-allylic alcohols is usually obtained (eq 34).[33b]

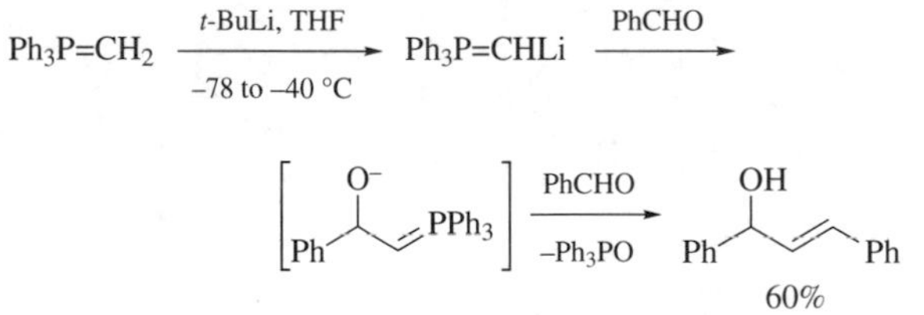

(32)

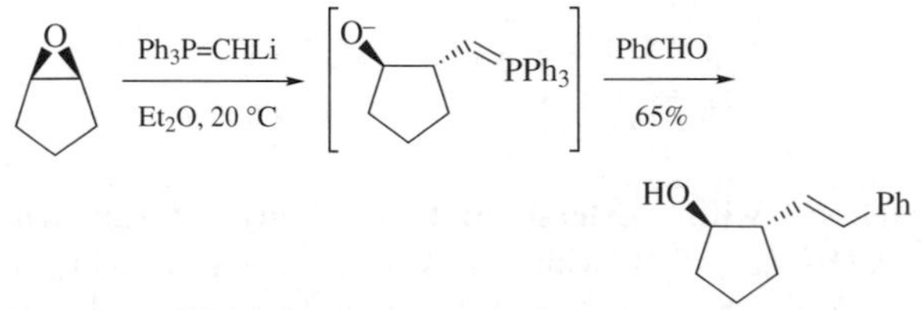

(33)

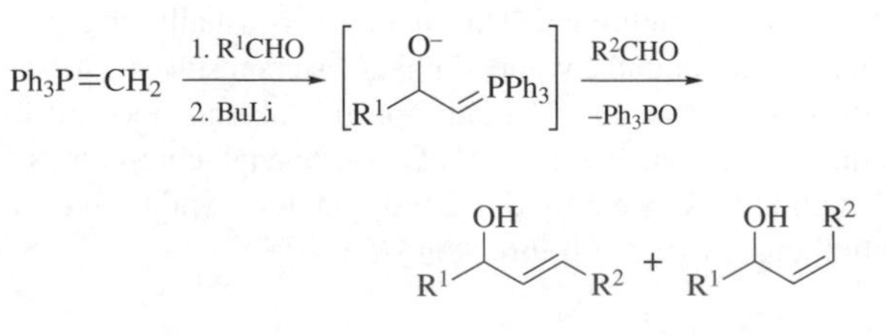

(34)

Formation of Extended Phosphorus Ylides. Various phosphorus ylides can be produced by reaction of $Ph_3P{=}CH_2$ with electrophiles. β-Ketophosphorus ylides are prepared by acylation of $Ph_3P{=}CH_2$ with an ester,[4] thioester,[5] acid chloride,[6] acylimidazole,[34] or even with a 5(4*H*)-oxazolone[35] (eqs 35–39). β-Thiocarbonylphosphorus ylides can be prepared in a similar manner.[36]

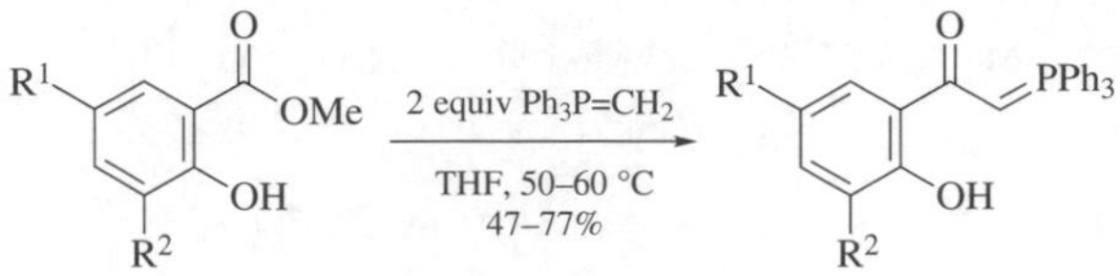

(35)

R^1 = H, OMe
R^2 = H, Me, Pr, allyl

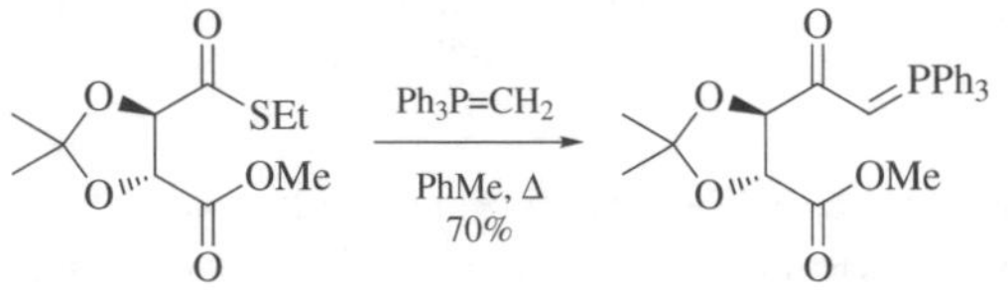

(36)

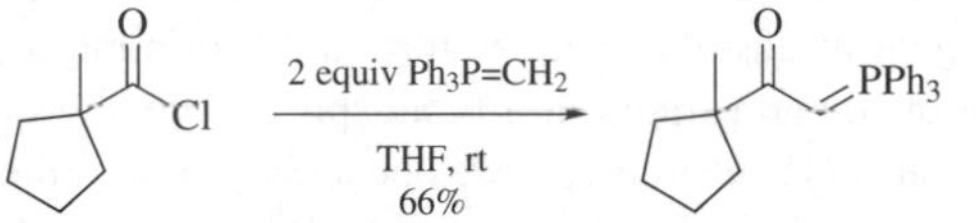

(37)

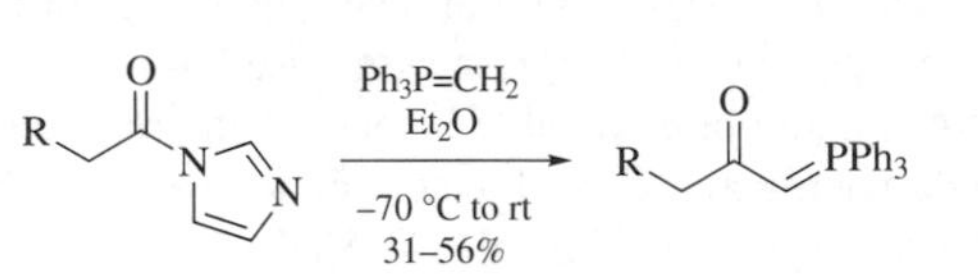

(38)

R = $CyCH_2CH_2$, C_7H_{15}, $MeOCH_2CH_2CH_2$

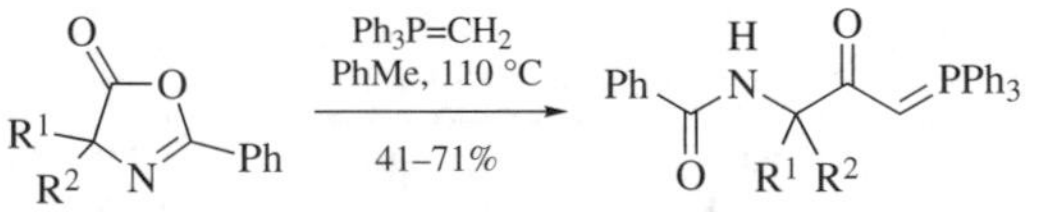

(39)

$R^1 = R^2$ = Ph
R^1 = Me, R^2 = Ph

On the other hand, alkylation of $Ph_3P{=}CH_2$ with alkyl halides provides novel phosphonium salts which in turn can yield ylides upon treatment with a base (eq 40).[37] Likewise, four- through seven-membered cyclic phosphorus ylides can be prepared by alkylation of corresponding alkyl dihalides with $Ph_3P{=}CH_2$ followed by deprotonation (eq 41).[38] Alkylation can also be effected with 1-alkylbenzotriazole,[39a] epoxide,[39b,c] or lactone.[39d] In the latter case, lactones may undergo either alkylation or acylation depending upon the reaction conditions used.[39d,e]

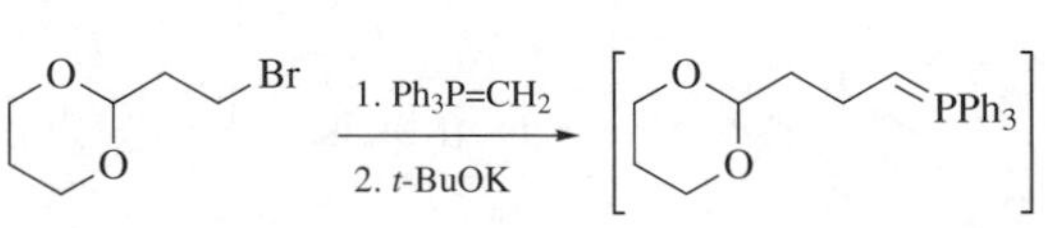

(40)

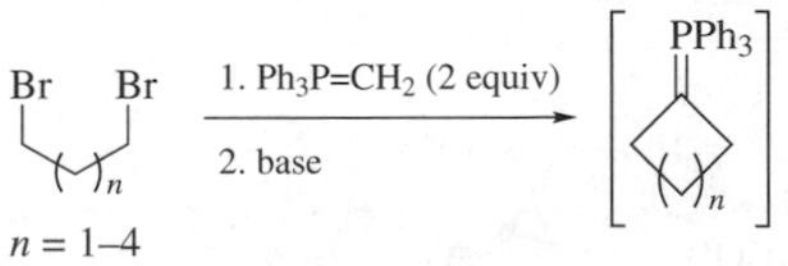

(41)

n = 1–4

Other electrophiles which react with $Ph_3P{=}CH_2$ include a disulfide,[40] sulfonyl halide,[41] halosilane,[42] selenenyl halide,[43] and cyanate[44] to yield various ylides after treatment with a base. Transylidation[45] of phosphonium salts with $Ph_3P{=}CH_2$ provides yet another entry to extended ylides (eq 42).[46] Phosphacumulene ylides have been prepared via double transylidation with

$Ph_3P{=}CH_2$ (eq 43).[47] Allylidenephosphoranes are also formed from the reaction of $Ph_3P{=}CH_2$ with dialkylaluminum alkylidenamides (eq 44).[48]

1. $(R_FCO)_2O$ 2. $Ph_3P{=}CH_2$ 3. PhLi, $-Ph_3PO$ (42)

$R_F = CF_3$ or C_3F_7

$Ph_3{=}CH_2$ (3 equiv) → $Ph_3P{=}\bullet{=}\bullet{=}X$ (43)

$X = CR^1R^2$, NR, S; Hal = Cl, Br

0.5 $[R^1, R^2, AlR^3_2]_2$ → $Ph_3{=}CH_2$ (44)

R^1, R^2 = H or alkyl; R^3 = alkyl

Related Reagents. μ-Chlorobis(cyclopentadienyl)(dimethylaluminum)-μ-methylenetitanium; Dibromomethane-Zinc-Titanium(IV) Chloride; Diiodomethane–Zinc–Titanium(IV) Chloride; Ethyltriphenylphosphonium Bromide; Methyltriphenylphosphonium Bromide; Phenylselenomethyllithium; Phenylthiomethyllithium; Trimethylaluminum-Dichlorobis(η^5-cyclopentadienyl)titanium; Trimethylsilylmethyllithium; Trimethylsilylmethylmagnesium Chloride.

1. (a) Maercker, A. *OR* **1965**, *14*, 270. (b) House, H. O. *Modern Synthetic Reactions*; Benjamin: Menlo Park, CA, 1972; pp 682–709. (c) Maryanoff, B. E.; Reitz, A. B. *CRV* **1989**, *89*, 863.
2. (a) Nishimura, Y.; Kondo, S.; Umezawa, H. *JOC* **1985**, *50*, 5210. (b) Dhavale, D. D.; Tagliavini, E.; Trombini, C.; Umani-Ronchi, A. *TL* **1988**, *29*, 6163. (c) Magnus, P.; Cairns, P. M. *JACS* **1986**, *108*, 217.
3. (a) Corey, E. J.; Kang, J. *JACS* **1982**, *104*, 4724. (b) Corey, E. J.; Kang, J.; Kyler, K. *TL* **1985**, *26*, 555. (c) Schaub, B.; Jenny, T.; Schlosser, M. *TL* **1984**, *25*, 4097.
4. (a) Zammattio, F.; Brion, J. D.; Ducrey, P.; Le Baut, G. *S* **1992**, 375. (b) Baldoli, C.; Licandro, E.; Maiorana, S.; Menta, E.; Papagni, A. *S* **1987**, 288. (c) Shen, Y.; Qiu, W.; Xin, Y.; Huang, Y. *S* **1984**, 924.
5. (a) Savage, I.; Thomas, E. J. *CC* **1989**, 717. (b) Bestmann, H. J.; Roth, D. *AG(E)* **1990**, *29*, 99. (c) Bestmann, H. J.; Stransky, W.; Vostrowsky, O. *CB* **1976**, *109*, 1694.
6. (a) Koller, M.; Karpf, M.; Dreiding, A. S. *HCA* **1983**, *66*, 2760. (b) Ronald, R. C.; Wheeler, C. J. *JOC* **1983**, *48*, 138. (c) Bestmann, H. J.; Arnason, B. *CB* **1962**, *95*, 1513.
7. Greenwald, R.; Chaykovsky, M.; Corey, E. J. *JOC* **1963**, *28*, 1128.
8. Vedejs, E.; Meier, G. P.; Snoble, K. A. J. *JACS* **1981**, *103*, 2823.
9. Koster, R.; Simic, D.; Grassberger, M. A. *LA* **1970**, *739*, 211.
10. (a) Wittig, G.; Geissler, G. *LA* **1953**, *580*, 44. (b) Wittig, G.; Schoellkopf, U. *CB* **1954**, *87*, 1318. (c) Wittig, G.; Schoellkopf, U. *OSC* **1973**, *5*, 751.
11. (a) Danishefsky, S.; Schuda, P. F.; Kitahara, T.; Etheredge, S. J. *JACS* **1977**, *99*, 6066. (b) Wijnberg, J. B. P. A.; Jenniskens, L. H. D.; Brunekreef, G. A.; De Groot, A. *JOC* **1990**, *55*, 941.
12. Toshima, H.; Yoshida, S.; Suzuki, T.; Nishiyama, S.; Yamamura, S. *TL* **1989**, *30*, 6721.
13. (a) Feldman, K. S.; Simpson, R. E. *JACS* **1989**, *111*, 4878. (b) Mori, K.; Miyake, M. *T* **1987**, *43*, 2229.
14. (a) Shibuya, H.; Kawashima, K.; Narita, N.; Kitagawa, I. *CPB* **1992**, *40*, 1166. (b) Marino, J. P.; Abe, H. *S* **1980**, 872.
15. Attia, M. E. M.; Gelas-Mialhe, Y.; Vessiere, R. *CJC* **1983**, *61*, 2126.
16. (a) Collins, P. W.; Shone, R. L.; Perkins, W. E.; Gasiecki, A. F.; Kalish, V. J.; Kramer, S. W.; Bianchi, R. G. *JMC* **1992**, *35*, 694. (b) DeShong, P.; Waltermire, R. E.; Ammon, H. L. *JACS* **1988**, *110*, 1901.
17. (a) Piers, E.; Skerlj, R. T. *JOC* **1987**, *52*, 4421. (b) Jung, M. E.; Gaede, B. *T.* **1979**, *35*, 621. (c) Piers, E.; Tillyer, R. D. *JCS(P1)* **1989**, 2124.
18. (a) Zadel, G.; Rieger, R.; Breitmaier, E. *LA* **1991**, 1343. (b) Lubineau, A.; Queneau, Y. *JOC* **1987**, *52*, 1001. (c) De Groot, A.; Jansen, B. J. M. *S* **1978**, 52.
19. Sato, T.; Hiramura, Y.; Otera, J.; Nozaki, H. *TL* **1989**, *30*, 2821.
20. Yamashita, M.; Kojima, M.; Yoshida, H.; Ogata, T.; Inokawa, S. *BCJ* **1980**, *53*, 1625.
21. Yamamoto, S.; Itani, H.; Takahashi, H.; Tsuji, T.; Nagata, W. *TL* **1984**, *25*, 4545.
22. Dussault, P.; Sahli, A. *TL* **1990**, *31*, 5117.
23. Masson, S.; Thuillier, A. *TL* **1980**, *21*, 4085.
24. (a) Smith, A. B., III; Jerris, P. J. *JOC* **1982**, *47*, 1845. (b) Adlercreutz, P.; Magnusson, G. *ACS* **1980**, *B34*, 647. (c) Olah, G. A.; Wu, A.-H.; Farooq, O. *JOC* **1989**, *54*, 1375. (d) Clawson, L.; Buchwald, S. L.; Grubbs, R. H. *TL* **1984**, *25*, 5733. (e) Pine, S. H.; Shen, G. S.; Hoang, H. *S* **1991**, 165. (f) Hibino, J.-i.; Okazoe, T.; Takai, K.; Nozaki, H. *TL* **1985**, *26*, 5579. (g) Stille, J. R.; Grubbs, R. H. *JACS* **1986**, *108*, 855. (h) Paquette, L. A.; Stevens, K. E. *CJC* **1984**, *62*, 2415.
25. Yeung, B. W. A.; Contelles, J. L. M.; Fraser-Reid, B. *CC* **1989**, 1160. (b) Tronchet, J. M. J.; Cottet, C.; Barbalat-Rey, F. *HCA* **1975**, *58*, 1501. (c) Mander, L. N.; Turner, J. V. *TL* **1981**, *22*, 4149. (d) Caine, D.; Crews, E.; Salvino, J. M. *TL* **1983**, *24*, 2083. (e) Keuss, H. A. C. M.; Lakeman, J. *T* **1976**, *32*, 1541.
26. (a) Balestra, M.; Wittman, M. D.; Kallmerten, J. *TL* **1988**, *29*, 6905. (b) Cha, J. K.; Cooke, R. J. *TL* **1987**, *28*, 5473. (c) Mulzer, J.; Steffen, U.; Zorn, L.; Schneider, C.; Weinhold, E.; Muench, W.; Rudert, R.; Luger, P.; Hartl, H. *JACS* **1988**, *110*, 4640. (d) Terlinden, R.; Boland, W.; Jaenicke, L. *HCA* **1983**, *66*, 466. (e) Kitahara, T.; Mori, K. *T* **1984**, *40*, 2935.
27. (a) Posner, G. H.; Crouch, R. D.; Kinter, C. M.; Carry, J.-C. *JOC* **1991**, *56*, 6981. (b) RajanBabu, T. V.; Fukunaga, T.; Reddy, G. S. *JACS* **1989**, *111*, 1759.
28. (a) Oostveen, A. R. C.; De Boer, J. J. J.; Speckamp, W. N. *H* **1977**, *7*, 171. (b) De Boer, J. J. J.; Speckamp, W. N. *TL* **1975**, 4039.
29. (a) Uijttewaal, A. P.; Jonkers, F. L.; Van der Gen, A. *TL* **1975**, 1439. (b) Uijttewaal, A. P.; Jonkers, F. L.; Van der Gen, A. *JOC* **1978**, *43*, 3306.
30. Epstein, W. W.; Gaudioso, L. A.; Brewster, G. B. *JOC* **1984**, *49*, 2748.
31. Mandai, T.; Irei, H.; Kawada, M.; Otera, J. *TL* **1984**, *25*, 2371.
32. Hanessian, S.; Demailly, G.; Chapleur, Y.; Leger, S. *CC* **1981**, 1125.
33. (a) Schlosser, M.; Christmann, F. K.; Piskala, A.; Coffinet, D. *S* **1971**, *29*, and refs therein. (b) Schlosser, M.; Coffinet, D. *S* **1971**, 380.
34. Miyano, M.; Stealey, M. A. *JOC* **1975**, *40*, 2840.
35. Erba, E.; Gelmi, M. L.; Pocar, D. *CB* **1988**, *121*, 1519.
36. (a) Bestmann, H. J.; Schaper, W. *TL* **1979**, 243. (b) Yoshida, H.; Matsuura, H.; Ogata, T.; Inokawa, S. *BCJ* **1975**, *48*, 2907.
37. (a) Bazureau, J. P.; Person, D.; Le Corre, M. *TL* **1989**, *30*, 3065. (b) Tanaka, A.; Suzuki, M.; Yamashita, K. *ABC* **1986**, *50*, 1069.
38. Bestmann, H. J.; Kranz, E. *AG(E)* **1967**, *6*, 81.
39. (a) Katritzky, A. R.; Jiang, J.; Greenhill, J. V. *JOC* **1993**, *58*, 1987. (b) Turcant, A.; Le Corre, M. *TL* **1976**, 1277. (c) Enholm, E. J.; Satici, H.; Prasad, G. *JOC* **1990**, *55*, 324. (d) Kise, H.; Arase, Y.; Shiraishi, S.; Seno, M.; Asahara, T. *CC* **1976**, 299. (e) Le Roux, J.; Le Corre, M. *CC* **1989**, 1464.
40. Flynn, G. A. *JOC* **1983**, *48*, 4125.
41. (a) Van Leusen, A. M.; Reith, B. A.; Iedema, A. J. W.; Strating, J. *RTC* **1972**, *91*, 37. (b) Stackhouse, J.; Westheimer, F. H. *JOC* **1981**, *46*, 1891.
42. Bestmann, H. J.; Bomhard, A.; Dostalek, R.; Pichl, R.; Riemer, R.; Zimmermann, R. *S* **1992**, 787.
43. Renard, M.; Hevesi, L. *T* **1985**, *41*, 5939.

44. Bestmann, H. J.; Pfohl, S. *LA* **1974**, 1688.

45. Bestmann, H. J. *CB* **1962**, *95*, 58.

46. Shen, Y.; Qiu, W. *TL* **1987**, *28*, 4283.

47. (a) Bestmann, H. J. *AG(E)* **1977**, *16*, 349. (b) Bestmann, H. J.; Schmid, G. *TL* **1975**, 4025. (c) Bestmann, H. J.; Schmid, G. *AG(E)* **1974**, *13*, 273.

48. (a) Bogdanovic, B.; Konstantinovic, S. *S* **1972**, 481. (b) Bogdanovic, B.; Koster, J. B. *LA* **1975**, 692.

Kevin C. Lee
DuPont Agricultural Products, Newark, DE, USA

(1*R*,2*S*)-*N*-Methylephedrine[1]

Ph, HO, NR$_2$

(**1**; R = Me)		
[552-79-4]	$C_{11}H_{17}NO$	(MW 179.29)
(**2**; R = *n*-Bu)		
[115651-77-9]	$C_{17}H_{29}NO$	(MW 263.47)
(**3**; R = CH_2=$CHCH_2$–)		
[150296-38-1]	$C_{15}H_{21}NO$	(MW 231.37)
(**4**; R = $Ph(CH_2)_4$)		
[132284-82-3]	$C_{29}H_{37}NO$	(MW 415.67)
(**5**; R = –$(CH_2)_5$–)		
[133576-76-8]	$C_{14}H_{21}NO$	(MW 219.36)

(chiral ligand for the enantioselective reduction of ketones with lithium aluminum hydride; chiral auxiliary for the diastereoselective aldol condensation; chiral catalyst for the enantioselective Darzens reaction; chiral catalyst for the enantioselective alkylation of aldehydes with dialkylzincs; chiral catalyst for the enantioselective conjugate addition of dialkylzincs to enones; chiral catalyst for the enantioselective alkylation of imines with dialkylzincs; chiral catalyst for the enantioselective Michael addition of nitromethane to α,β-unsaturated ketones)

Alternate Name: [*R*-(*R**,*S**)]-α-[1-(dimethylamino)ethyl]-benzenemethanol.
Physical Data: (**1**) mp 86–88 °C; $[\alpha]_D^{21}$ −29.2° (*c* 5, MeOH). (**2**) bp 170 °C/2 mmHg; $[\alpha]_D^{25}$ + 24.4° (*c* 2, hexane).
Solubility: sol many organic solvents.
Form Supplied in: (**1**) colorless crystals; (**2**) colorless oil; (**1**) and (**2**) are commercially available.

Enantioselective Reduction of Ketones with Lithium Aluminum Hydride–*N*-Methylephedrine. Aryl alkyl ketones and α-alkynic ketones are reduced enantioselectively by a chiral complex of ***Lithium Aluminum Hydride***, *N*-methylephedrine (**1**), and 3,5-dimethylphenol (molar ratio, 1:1:2) to afford optically active alcohols with 75–90% ee (eq 1).[2]

The optically active alkynyl alcohols are converted into the corresponding optically active 4-alkyl-γ-butyrolactones[3] and 4-alkylbutenolides.[4]

(1), $LiAlH_4$; 3,5-dimethylphenol — 75–90% ee (1)

A chiral complex of (**1**), $LiAlH_4$, and *N*-ethylaniline (molar ratio, 1:1:2) reduces aryl alkyl ketones to optically active alcohols in high ee.[5] α,β-Unsaturated ketones are reduced enantioselectively to afford optically active (*S*)-allylic alcohols with 80–98% ee. An intermediate in an anthracyclinone synthesis is prepared in 92% ee by the enantioselective reduction of a cyclic α,β-unsaturated ketone (eq 2).[6]

(1), $LiAlH_4$, PhNHEt — 92% ee (2)

A chiral complex of (**1**), $LiAlH_4$, and 2-ethylaminopyridine (molar ratio, 1:1:2), prepared in refluxing ether for 3 h, reduces cyclic ketones to (*R*)-alcohols in 75–96% ee.[7] Advantages of the enantioselective reduction of ketones with $LiAlH_4$ modified with (**1**) and additives are the ready availability of (**1**) in either enantiomeric form and easy removal of (**1**) from the reaction mixture by washing with dilute acid.

anti*-Selective Aldol Condensation and Related Reactions.** Silyl ketene acetals react with aldehydes in the presence of ***Titanium(IV) Chloride to give β-hydroxy esters.[8] The silyl ketene acetal derived from (1*R*,2*S*)-(**1**)-*O*-propionate reacts with benzaldehyde in the presence of $TiCl_4$ and ***Triphenylphosphine*** to afford the *anti*-α-methyl-β-hydroxy ester in 94% de (eq 3).[9]

PhCHO, $TiCl_4$, 90% — 94% de, *anti*:*syn* >30:1 (3)

When the same silyl ketene acetal is reacted with benzylideneaniline in the presence of $TiCl_4$, the *anti*-β-amino ester is obtained (*anti*/*syn* > 10/1). Cyclization of the β-amino ester affords the *trans*-β-lactam in 95% ee (eq 4).[10]

PhCH=NPh, $TiCl_4$; $LiN(TMS)_2$ — 79%, 95% ee (4)

The reaction of this silyl ketene acetal with ***Di-t-butyl Azodicarboxylate*** in the presence of $TiCl_4$ affords the adduct in 45–70% yield with ca. 90% de. The subsequent treatment of the adduct with ***Trifluoroacetic Acid*** and ***Lithium Hydroxide*** affords (*R*)-α-hydrazo acids (eq 5).[11]

$$\text{Ph-CH(O-C(OTMS)=CHR)-CH(Me)-NMe}_2 \xrightarrow[\text{2. TFA, LiOH}]{\text{1. BocN=NBoc, TiCl}_4} \text{HO}_2\text{C-CH(R)-NHNH}_2 \quad \text{ca. 90\% ee} \tag{5}$$

Enantioselective Butylation of Carbonyl Compounds with Lithium Tetra-*n*-butylaluminate Modified with (1). The reaction between lithium tetra-*n*-butylaluminate and (**1**) forms the chiral lithium alkoxytri-*n*-butylaluminate. This chiral ate complex reduces carbonyl compounds to form secondary and tertiary alcohols in 8–31% ee (eq 6).[12]

$$R^1C(O)R^2 \xrightarrow{(1),\ LiAlBu_4} R^1R^2C^*(Bu)(OH) \quad \text{8–31\% ee} \tag{6}$$

Enantioselective Darzens Reaction. An enantioselective Darzens reaction between ethyl methyl ketone and chloromethyl *p*-tolyl sulfone in the presence of a chiral ammonium salt derived from (**1**) and chloromethylpolystyrene affords an optically active α,β-epoxy sulfone in 23% ee.[13]

Catalytic Enantioselective Alkylation of Aldehydes with Dialkylzincs[1]. The chiral *N*,*N*-dialkylnorephedrines, analogs of (**1**), are highly efficient catalysts for the enantioselective addition of dialkylzincs to aliphatic and aromatic aldehydes.[14,15] Optically active aliphatic and aromatic secondary alcohols with high ee are obtained using *N*,*N*-dialkylnorephedrines (4–6 mol%) as chiral catalyst precursors. When (1*S*,2*R*)-*N*,*N*-dialkylnorephedrine is used as a chiral catalyst precursor, prochiral aldehydes are attacked at the *si* face to afford (*S*)-alcohols (when the priority order is $R^1 > R^2$) (eq 7).

$$R^1CHO \xrightarrow[R^2_2Zn]{(1S,2R)\text{-catalyst}} R^1\text{-}(S)\text{-CH(OH)-}R^2 \qquad R^1CHO \xrightarrow[R^2_2Zn]{(1R,2S)\text{-catalyst}} R^1\text{-}(R)\text{-CH(OH)-}R^2 \tag{7}$$

Table 1 Effect of *N*-Alkyl Substituents of (1*S*,2*R*)-*N*,*N*-Dialkylnorephedrine as Chiral Ligand for the Addition of Diethylzinc to 3-Methylbutanal to Yield (*S*)-5-Methylhexan-3-ol

Entry	*N*-Alkyl substituent	Yield (%)	ee (%)
1	Me (**1**)	53	53
2	Et	95	83
3	*n*-Pr	90	87
4	*n*-Bu (DBNE) (**2**)	92	93
5	n-C_5H_{11}	91	85
6	n-C_6H_{13}	85	83
7	n-C_7H_{15}	80	79
8	n-C_8H_{17}	53	76
9	$-(CH_2)_5-$ (**5**)	81	70

N-Alkyl substituents on the (1*S*,2*R*)-*N*,*N*-di-*n*-alkylnorephedrines have a significant effect on the enantioselectivity of the addition of diethylzinc to aldehydes (3-methylbutanal). As shown in Table 1, the optical purity of the product [(*S*)-5-methylhexane-3-ol] increases as the chain length of the *N*-*n*-alkyl substituent increases and reaches a peak of 93% ee at a chain length of four carbons (Table 1, entry 4). Thus, among *N*,*N*-di-*n*-alkylnorephedrines examined, (1*S*,2*R*)-*N*,*N*-di-*n*-butylnorephedrine (DBNE) (**2**) is the best chiral catalyst precursor.[14,15]

As shown in Table 2 (eq 7), the advantages of *N*,*N*-di-*n*-alkylnorephedrines (most typically DBNE) over other chiral catalysts for the enantioselective addition of dialkylzincs to aldehydes are as follows:

1) DBNE is highly enantioselective for the alkylation of *aliphatic* aldehydes (Table 2, entries 5–11) as well as for the alkylation of *aromatic* aldehydes (Table 2, entries 1–4). Most of the other types of chiral catalysts are effective only for the alkylation of *aromatic* aldehydes. Thus, various types of optically active *aliphatic* alcohols are first synthesized using DBNE (Table 2, entries 5–11). (It should be noted that the structures of aliphatic alcohols synthesized by asymmetric reduction of ketones or by asymmetric hydroboration of alkenes have been somewhat limited.)

2) The dialkylzinc additions catalyzed by *N*,*N*-di-*n*-alkylnorephedrines (most typically DBNE) are not limited to *primary* organometallic reagents. Diisopropylzinc (with a *secondary* alkyl substituent) adds to benzaldehyde in the presence of a catalytic amount of DBNE to afford the corresponding alcohol with high ee (entry 4).[15] The reaction of diisopropylzinc in the presence of other types of catalysts may result in the reduction of aldehydes.

3) *N*,*N*-Di-*n*-alkylnorephedrines are readily synthesized in a one-pot reaction between norephedrine and alkyl iodide in the presence of potassium carbonate.[14,15] (DBNE is commercially available.)

4) Either enantiomer of the *N*,*N*-di-*n*-alkylnorephedrines are available. Therefore by using the appropriate enantiomer of *N*,*N*-di-*n*-alkylnorephedrine as a chiral catalyst precursor, the optically active alcohol of the desired configuration with the same ee can be synthesized (entries 8 and 9).

Optically active fluorine-containing alcohols (91–93% ee) (entries 12 and 13)[16] and deuterio alcohols (84–94% ee) (entries 14 and 15)[17] are synthesized, respectively, by the enantioselective alkylation of fluorine-containing aldehyde and deuterio aldehyde using DBNE.

(1*S*,2*R*)-1-Phenyl-2-(1-pyrrolidinyl)propan-1-ol (**6**) (entry 10) and (1*S*,2*R*)-*N*,*N*-diallylnorephedrine (**3**) (entry 11) are also highly enantioselective catalyst precursors.[15]

Enantioselective Addition of Various Organozinc Reagents to Aldehydes. Catalytic enantioselective addition of dialkynylzinc reagents to aldehydes using (1*S*,2*R*)-DBNE (20 mol %) affords optically active (*R*)-alkynyl alcohols with 43% ee in 99% yield.[18] When an alkylalkynylzinc is used with (1*S*,2*R*)-DBNE (5 mol %), an alkynyl alcohol with 40% ee is obtained.[18] When

2-phenylzinc bromide is reacted with an aldehyde in the presence of 1 equiv of the lithium salt of (1*R*,2*S*)-(**1**), the corresponding alkynyl alcohol is obtained in 88% ee (eq 8).[19]

Table 2 Enantioselective Addition of Dialkylzincs to Aldehydes using Norephedrine-Derived Chiral Catalyst (eq 7)

Entry	R^1	R^2	Catalyst (6 mol %)	Yield (%)	ee (%)	Config.
1	Ph	Et	(1*S*,2*R*)-DBNE	100	90	(*S*)
2	2-$MeOC_6H_4$	Et	(1*S*,2*R*)-DBNE	100	94	(*S*)
3	Ph	Et	(1*S*,2*R*)-DBNE	94	95	(*S*)
4	Ph	*i*-Pr	(1*S*,2*R*)-DBNE	73	91	(*S*)
5	Me_2CHCH_2	Et	(1*S*,2*R*)-DBNE	92	93	–
6	*n*-C_6H_{13}	Me	(1*S*,2*R*)-DBNE	70	90	(*S*)
7	*n*-C_6H_{13}	*n*-Pr	(1*S*,2*R*)-DBNE	100	90	–
8	*n*-C_8H_{17}	Et	(1*S*,2*R*)-DBNE	95	87	(*S*)
9	*n*-C_8H_{17}	Et	(1*R*,2*S*)-DBNE	99	87	(*R*)
10	*n*-C_8H_{17}	Et	(1*S*,2*R*)-(**6**)	87	>95	(*S*)
11	*n*-C_8H_7	Et	(1*S*,2*R*)-(**3**)	61	88	(*S*)
12	4-$CF_3C_6H_4$	Et	(1*S*,2*R*)-DBNE	92	91	(*S*)
13	4-FC_6H_4	Et	(1*S*,2*R*)-DBNE	83	93	–
14	PhCDO	Et	(1*S*,2*R*)-DBNE	92	94	(*S*)
15	$Me(CH_2)_5CDO$	Et	(1*S*,2*R*)-DBNE	79	84	(*S*)

R^1CHO → [$(R^2C\equiv C)_2Zn$, (1*S*,2*R*)-DBNE] → $R^1CH(OH)C\equiv CR^2$; → [$PhC\equiv CZnR^2$, (1*S*,2*R*)-DBNE] → $R^1CH(OH)C\equiv CPh$, 39–40% ee; → [$PhC\equiv CZnBr$, (1*R*,2*S*)-Li(**1**)] → $R^1CH(OH)C\equiv CPh$, 88% ee (8)

Alkenylzinc bromides add to aldehydes to afford optically active allyl alcohols with 88% ee in 80% yield using a stoichiometric amount of the lithium salt of (1*S*,2*R*)-(**1**) (eq 9).[20]

RCHO + alkenyl-ZnBr → [(1*S*,2*R*)-Li(**1**), 80%] → R-CH(OH)-alkenyl, 88% ee (9)

A mixture of phenyl Grignard and zinc halide adds to aldehydes in the presence of a stoichiometric amount of (1*R*,2*S*)-DBNE to afford optically active phenyl alcohols with 82% ee in 90% yield (eq 10).[21]

$2\ PhMgBr + ZnCl_2$ → [PhM] (M = metal) → [RCHO, (1*R*,2*S*)-DBNE] → RCH(OH)Ph, 41–90%, 68–82% ee (10)

Difurylzinc adds to aldehydes in the presence of a stoichiometric amount of the lithium salt of (1*S*,2*R*)-*N*,*N*-bis(4-phenylbutyl)norephedrine (**4**) to afford optically active furylalcohols with 73% ee in 58% yield (eq 11).[22a]

RCHO + (2-furyl)$_2$Zn → [Li salt of **4** (Ph, OLi, N(CH$_2$CH$_2$CH$_2$CH$_2$Ph)$_2$), LiCl, 58%] → RCH(OH)(2-furyl), 73% ee (11)

Enantioselective addition of a Reformatsky reagent to aldehydes[22b] and ketones[22c] in the presence of DBNE or *N*,*N*-diallylnorephedrine (**3**) affords the corresponding β-hydroxy esters in up to 75% ee (eq 12).

RC(O)CH$_3$ + $BrZnCH_2CO_2$-*t*-Bu → [**3**, 38–65%] → RC(OH)(CH$_3$)CH$_2$C(O)O-*t*-Bu, 73–75% ee (12)

Enantioselective Addition of Dialkylzincs to Aldehydes with Functional Groups. Enantioselective and chemoselective addition of dialkylzincs to formyl esters using (1*S*,2*R*)-DBNE as a catalyst affords optically active hydroxy esters. The subsequent hydrolysis of the esters affords the corresponding optically active alkyl substituted lactones with up to 95% ee (eq 13).[23]

$HC(O)(CH_2)_nCO_2R^1 + R^2{}_2Zn$ → [cat. (1*S*,2*R*)-DBNE] → $R^2CH(OH)(CH_2)_nCO_2R^1$ → [1. NaOH, 2. H$^+$] → $n = 2$: γ-lactone, 95–97%, 85–92% ee; $n = 3$: δ-lactone, 97–98%, 88–95% ee (13)

Enantio- and chemoselective addition of diethylzinc to keto aldehydes using DBNE as a chiral ligand affords optically active hydroxy ketones with 91% ee in 84% yield (eq 14).[24] This reaction cannot be realized by Grignard reagents or alkyllithium reagents because of the strong reactivity towards both aldehydes and ketones.

4-PhC(O)C$_6$H$_4$CHO + Et_2Zn → [cat. (1*S*,2*R*)-DBNE, 84%] → 4-PhC(O)C$_6$H$_4$C*H(OH)Et, 91% ee (14)

Enantioselective addition of dialkylzinc to furyl aldehydes using DBNE as a chiral catalyst affords optically active furyl alcohols in up to 94% ee (eqs 15 and 16).[25]

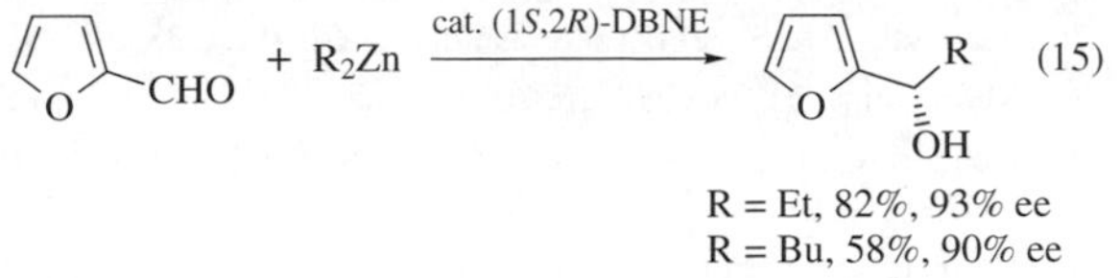

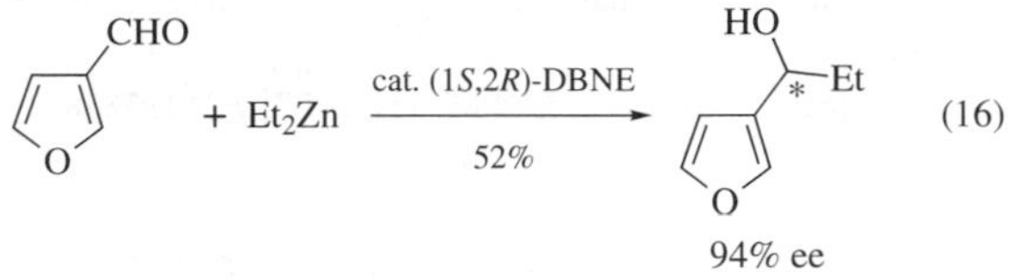

cat. (1S,2R)-DBNE (15)

R = Et, 82%, 93% ee
R = Bu, 58%, 90% ee

cat. (1S,2R)-DBNE, 52% (16)

94% ee

Enantioselective additions of dialkylzincs to 4-(diethoxymethyl)benzaldehyde,[26] 3-pyridinecarbaldehyde,[27] terephthalyl aldehyde,[28] and 2-bromobenzaldehyde[29] using DBNE as a chiral catalyst afford, after appropriate treatment, optically active hydroxy aldehydes,[26] pyridyl alcohols (eq 17),[27] diols (eq 18),[28] and 3-alkylphthalides (eq 19),[29] respectively, with high ee.

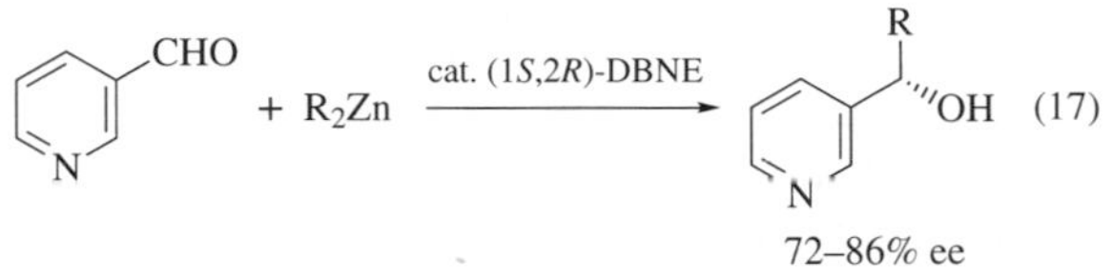

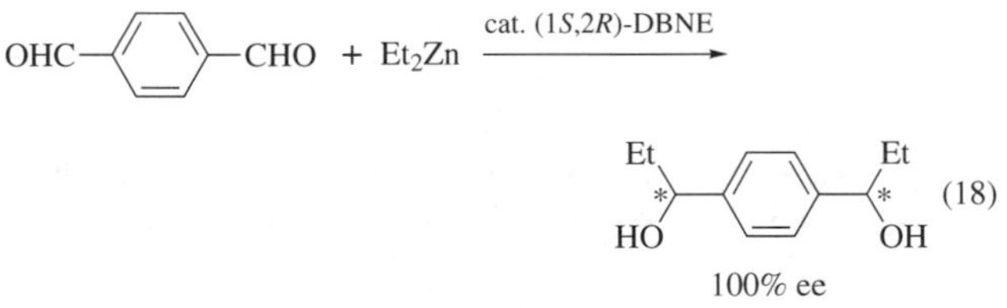

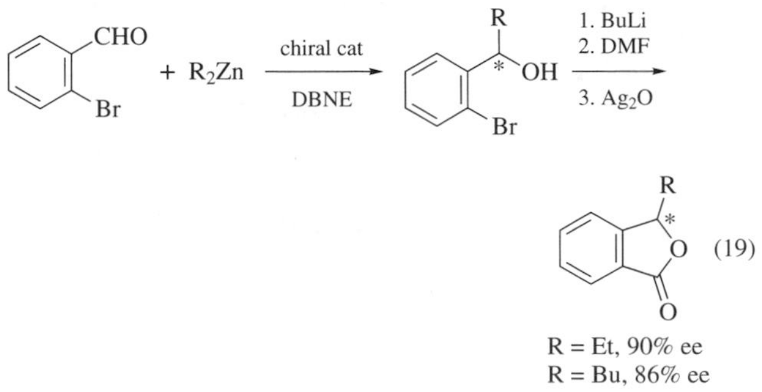

cat. (1S,2R)-DBNE (17)

72–86% ee

cat. (1S,2R)-DBNE (18)

100% ee

chiral cat DBNE; 1. BuLi 2. DMF 3. Ag_2O (19)

R = Et, 90% ee
R = Bu, 86% ee

A highly functionalized chiral aldehyde when treated with Et_2Zn using (1R,2S)-DBNE as a chiral catalyst affords the optically active alcohol with 82% de in 98% yield (eq 20).[30] The alcohol has been further elaborated into (+)-lepicidin.

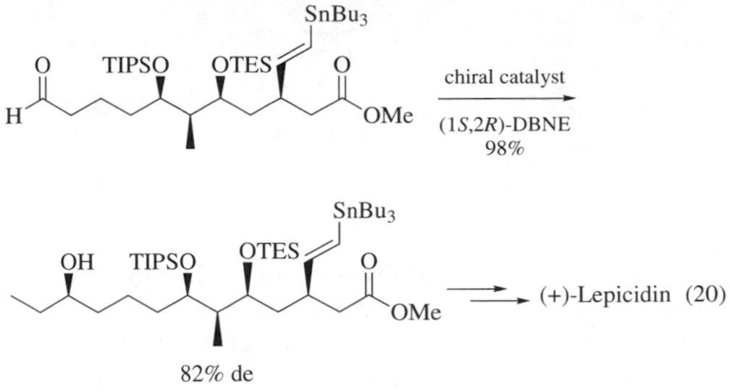

chiral catalyst (1S,2R)-DBNE 98%

→ (+)-Lepicidin (20)

82% de

Stereoselective Addition of Dialkylzincs to Chiral Aldehydes. Stereoselective addition of dibutylzinc to racemic 2-phenylpropanal using (1S,2R)-DBNE as a chiral catalyst affords optically active alcohols (84% ee, 92% ee) as a result of the *si* face attack of the aldehyde regardless of its configuration (eq 21).[31]

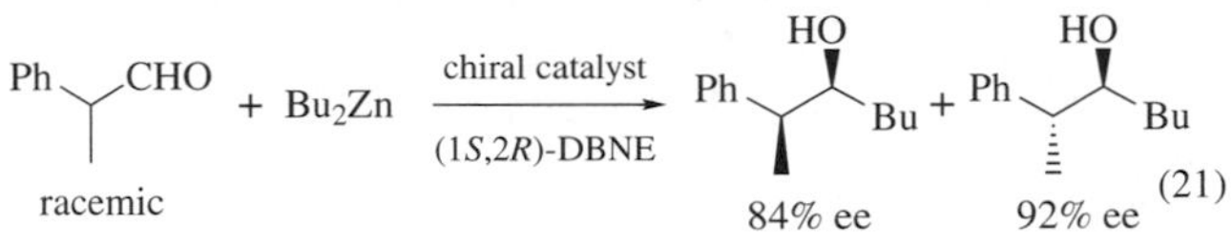

chiral catalyst (1S,2R)-DBNE (21)

racemic; 84% ee; 92% ee

By changing the configuration of the chiral catalyst precursor (DBNE), stereoselective synthesis of optically active *syn* (78% de) and *anti* (91% de) 1,3-diols has been reported in the addition of diethylzinc to optically active β-alkoxyaldehyde (eq 22).[32] The method has an advantage over the $R_2Zn–TiCl_4$ method,[33] which is only *anti* selective.

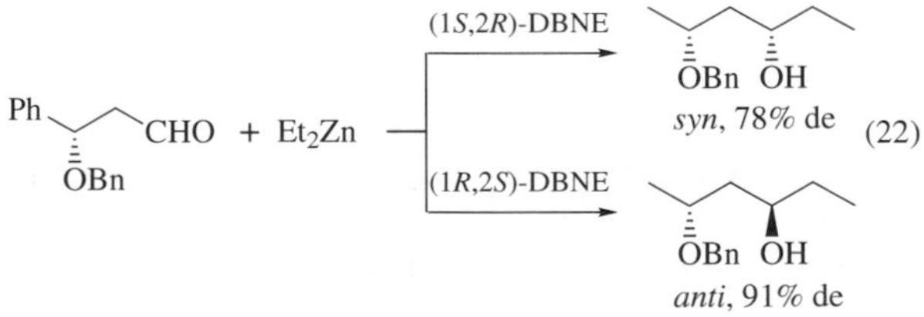

(1S,2R)-DBNE: *syn*, 78% de (22)
(1R,2S)-DBNE: *anti*, 91% de

Catalytic Enantioselective Conjugate Addition of Dialkylzincs to Enones. A chiral nickel complex modified with DBNE and an achiral ligand such as 2,2′-bipyridyl in acetonitrile/toluene is an highly enantioselective catalyst for the addition of dialkylzincs to enones.[34] β-Substituted ketones with up to 90% ee are obtained (eq 23).[34c] The method is the first highly enantioselective catalytic conjugate addition of an organometallic reagent to an enone.

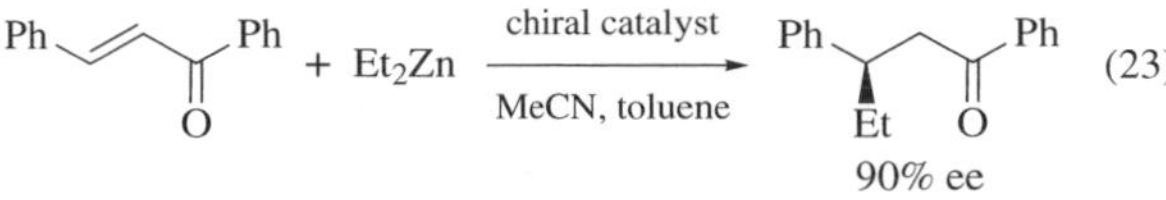

chiral catalyst, MeCN, toluene (23)

90% ee

chiral catalyst (1S,2R)-DBNE–$Ni(acac)_2$–2,2′-bipyridyl

In addition, a chiral amino alcohol [1-phenyl-2-(1-piperidinyl)propan-1-ol] mediates the reaction without using any nickel compound to afford the adduct in 94% ee (eq 24).[34d]

Ph–CH=CH–C(=O)–*t*-Bu + Et_2Zn → (96%) Ph–CH(Et)–CH$_2$–C(=O)–*t*-Bu, 94% ee (24)

Enantioselective Addition of Dialkylzincs to Imines. Enantioselective addition of dialkylzincs to *N*-diphenylphosphinoylimines in the presence of DBNE or its analog affords optically active phosphoramides. Subsequent hydrolysis affords optically active amines in up to 91% ee (eq 25).[35] When the amount of DBNE is catalytic (10 mol %), the enantioselectivity is 75% ee. One of the advantages of this method over the alkyllithium method[36] is the use of a lesser amount of chiral ligand.

Ar–CH=N–P(O)Ph$_2$ → (R_2Zn, (1*S*,2*R*)-DBNE or morpholino analog) Ar–CH(R)–NH–P(O)Ph$_2$ → (1. H^+; 2. OH^-) Ar–CH(R)–NH_2, 84%, 91% ee (25)

Diethylzinc also adds to *N*-(amidobenzyl)benzotriazoles (masked *N*-acylimines) in the presence of DBNE to afford an optically active amide with 76% ee (eq 26).[37]

PhCH=N–C(O)Me → (Et_2Zn, (1*S*,2*R*)-DBNE, 46%) Ph–*CH(Et)–NH–C(O)Me, 76% ee (26)

Asymmetric Michael Addition of Nitromethane to Enone. *N*-Methylephedrinium fluoride catalyzes the Michael addition of nitromethane to chalcone to afford the adduct with 23% ee in 50% yield (eq 27).[38]

Ph–CH=CH–C(O)–Ph → ($MeNO_2$, *N*-methylephedrinium fluoride, 50%) Ph–*CH(CH_2NO_2)–CH$_2$–C(O)–Ph, 23% ee (27)

Related Reagents. Diethylzinc; 2,2-Dimethyl-α,α,α′,α′-tetraphenyl-1,3-dioxolane-4,5-dimethanolatotitanium Diisopropoxide; (*S*)-Diphenyl(1-methylpyrrolidin-2-yl)methanol; (1*R*,2*S*)-Ephedrine; Lithium Aluminium Hydride.

1. Soai, K.; Niwa, S. *CRV* **1992**, *92*, 833.
2. (a) Jacquet, I.; Vigneron, J. P. *TL* **1974**, 2065. (b) Vigneron, J. P.; Bloy, V. *TL* **1979**, 2683.
3. Vigneron, J. P.; Bloy, V. *TL* **1980**, *21*, 1735.
4. Vigneron, J. P.; Méric, R.; Dhaenens, M. *TL* **1980**, *21*, 2057.
5. Terashima, S.; Tanno, N.; Koga, K. *CC* **1980**, 1026.
6. (a) Terashima, S.; Tanno, N.; Koga, K. *CL* **1980**, 981. (b) Terashima, S.; Hayashi, M.; Koga, K. *TL* **1980**, *21*, 2749, 2753.
7. Kawasaki, M.; Suzuki, Y.; Terashima, S. *CL* **1984**, 239.
8. Mukaiyama, T. *OR* **1982**, *28*, 203.
9. Gennari, C.; Bernardi, A.; Colombo, L.; Scolastico, C. *JACS* **1985**, *107*, 5812.
10. Gennari, C.; Venturini, I.; Gislon, G.; Schimperna, G. *TL* **1987**, *28*, 227.
11. Gennari, C.; Colombo, L.; Bertolini, G. *JACS* **1986**, *108*, 6394.
12. Boireau, G.; Abenhaïm, D.; Bourdais, J.; Henry-Basch, E. *TL* **1976**, 4781.
13. Colonna, S.; Fornasier, R.; Pfeiffer, U. *JCS(P1)* **1978**, 8.
14. Soai, K.; Yokoyama, S.; Ebihara, K.; Hayasaka, T. *CC* **1987**, 1690.
15. (a) Soai, K.; Yokoyama, S.; Hayasaka, T. *JOC* **1991**, *56*, 4264. (b) Soai, K.; Hayase, T.; Takai, K.; Sugiyama, T. *JOC* **1994**, *59*, 7908.
16. Soai, K.; Hirose, Y.; Niwa, S. *JFC* **1992**, *59*, 5.
17. Soai, K.; Hirose, Y.; Sakata, S. *BCJ* **1992**, *65*, 1734.
18. Niwa, S.; Soai, K. *JCS(P1)* **1990**, 937.
19. Tombo, G. M. R.; Didier, E.; Loubinoux, B. *SL* **1990**, 547.
20. Oppolzer, W.; Radinov, R. N. *TL* **1991**, *32*, 5777.
21. Soai, K.; Kawase, Y.; Oshio, A. *JCS(P1)* **1991**, 1613.
22. (a) Soai, K.; Kawase, Y. *JCS(P1)* **1990**, 3214. (b) Soai, K.; Kawase, Y. *TA* **1991**, *2*, 781. (c) Soai, K.; Oshio, A.; Saito, T. *CC* **1993**, 811.
23. Soai, K.; Yokoyama, S.; Hayasaka, T.; Ebihara, K. *CL* **1988**, 843.
24. Soai, K.; Watanabe, M.; Koyano, M. *CC* **1989**, 534.
25. (a) Soai, K.; Kawase, Y.; Niwa, S. *H* **1989**, *29*, 2219. (b) Van Oeveren, A.; Menge, W.; Feringa, B. L. *TL* **1989**, *30*, 6427.
26. Soai, K.; Hori, H.; Kawahara, M. *TA* **1990**, *1*, 769.
27. Soai, K. Hori, H.; Niwa, S. *H* **1989**, *29*, 2065.
28. Soai, K.; Hori, H.; Kawahara, M. *CC* **1992**, 106.
29. Soai, K.; Hori, H.; Kawahara, M. *TA* **1991**, *2*, 253.
30. Evans, D. A.; Black, W. C. *JACS* **1992**, *114*, 2260.
31. (a) Soai, K.; Niwa, S.; Hatanaka, T. *CC* **1990**, 709. (b) Niwa, S.; Hatanaka, T.; Soai, K. *JCS(P1)* **1991**, 2025.
32. Soai, K.; Hatanaka, T.; Yamashita, T. *CC* **1992**, 927.
33. Reetz, M. T.; Jung, A. *JACS* **1983**, *105*, 4833.
34. (a) Soai, K.; Yokoyama, S.; Hayasaka, T.; Ebihara, K. *JOC* **1988**, *53*, 4148. (b) Soai, K.; Hayasaka, T.; Ugajin, S.; Yokoyama, S. *CL* **1988**, 1571. (c) Soai, K.; Hayasaka, T.; Ugajin, S. *CC* **1989**, 516. (d) Soai, K.; Okudo, M.; Okamoto, M. *TL* **1991**, *32*, 95.
35. Soai, K.; Hatanaka, T.; Miyazawa, T. *CC* **1992**, 1097.
36. (a) Tomioka, K.; Inoue, I.; Shindo, M.; Koga, K. *TL* **1991**, *32*, 3095. (b) Itsuno, S.; Yanaka, H.; Hachisuka, C.; Ito, K. *JCS(P1)*, **1991**, 1341.
37. Katritzky, A. R.; Harris, P. A. *TA* **1992**, *3*, 437.
38. Colonna, S.; Hiemstra, H.; Wynberg, H. *CC* **1978**, 238.

Kenso Soai
Science University of Tokyo, Japan

Methyl Formate

$R^1C(=O)OR^2$

(**1**; R^1 = H, R^2 = Me)		
[107-31-3]	$C_2H_4O_2$	(MW 60.06)
(**2**; R^1 = H, R^2 = Et)		
[109-94-4]	$C_3H_6O_2$	(MW 74.09)
(**3**; R^1 = H, R^2 = Pr)		
[110-74-7]	$C_4H_8O_2$	(MW 88.12)
(**4**; R^1 = H, R^2 = Bu)		
[592-84-7]	$C_5H_{10}O_2$	(MW 102.15)
(**5**; R^1 = H, R^2 = *t*-Bu)		
[762-75-4]	$C_5H_{10}O_2$	(MW 102.15)
(**6**; R^1 = D, R^2 = Me)		
[23731-38-6]	$C_2H_3DO_2$	(MW 61.06)

(formylating agent for carbonyl compounds;[1–3] electrophile for organometallic agents[4])

Physical Data: (**1**) bp 34 °C; *d* 0.974 g cm^{-3}. (**2**) bp 52–54; *d* 0.917 g cm^{-3}. (**3**) bp 80–81 °C; *d* 0.904 g cm^{-3}. (**4**) bp 106–107 °C; *d* 0.892 g cm^{-3}. (**5**) bp 82–83 °C; *d* 0.872 g cm^{-3}. (**6**) bp 32.1 °C; *d* 0.990 g cm^{-3}.

Solubility: partially sol water; miscible with alcohol, ether.

Form Supplied in: liquid; widely available in 97–99% purity.

Handling, Storage, and Precautions: most formates are colorless, flammable, irritant, and moisture-sensitive liquids. Inhalation of methyl formate vapor produces nasal and conjunctional irritation, retching, narcosis, death from pulmonary effects. Ethyl formate is irritating to skin, mucous membranes, and, in high concentrations, narcotic. Use in a fume hood.

Among all the formates listed, methyl and ethyl formate are the most widely used agents in organic synthesis. These formates are mainly used as a formylating agent for carbonyl compounds. As shown in eq 1, cyclohexanone (**7**) is converted into the enolate (**8**) in the presence of either ***Sodium Ethoxide*** or ***Sodium Hydride*** and condensed with ethyl formate to give 2-hydroxymethylene cyclohexanone (**9**) or its tautomer, 2-formylcyclohexanone (**10**).[1i]

(**7**) —base→ (**8**) O^-M^+ —HCO_2R→ (**9**) =CHOH ⇌ (**10**) CHO (1)

To generate enolate (**8**) from the carbonyl compounds, sodium alkoxide[1] or sodium hydride (or ***Potassium Hydride***)[2] is generally used. Examples for using ***Potassium t-Butoxide*** are also reported.[3] The solvent can be either benzene, toluene, ether, dimethoxyethane, or THF. **Caution**: gas evolution has been cited in the formylation of cyclopentanone with ethyl formate in a solution of potassium *t*-butoxide in THF.[3] Among the carbonyl compounds, ketones are the most commonly used substrates for the formylation. A few examples for formylating lactones[5] and esters[6] are also reported.

A classic example of the synthetic utility is illustrated (eq 2) in the formylation of a cyclic ketone to block one α-position (**11** to **12** to **13a–c**) and alkylating the α′-position (**13a–c** to **14a–c**). The blocking group is then removed (**14a–c** to **15** to **16**).[1b,1c,1g]

(**11**) —base, HCO_2R→ (**12**) =CHOH —HNMePh; K_2CO_3, *i*-PrI or TsOH, BuSH→

(**13a**) R = NMePh, 71% from (**11**); (**13b**) R = O-*i*-Pr, 93% from (**11**); (**13c**) R = SBu, 64% from (**11**) —base, MeI→ (**14a**) R = NMePh; (**14b**) R = O-*i*-Pr; (**14c**) R = SBu —H^+→

(**15**) R = OH or SH —KOH, Δ→ (**16**) 68% from (**13a**), 43% from (**13b**), 66% from (**13c**) (2)

Organometallic agents are combined with the formates to give alcohols or aldehydes.[4] For example, treatment of an ethereal solution of 5-pentenylmagnesium bromide with methyl formate at 0 °C gives undeca-1,10-dien-6-ol in 86% yield (eq 3).[4c]

MgBr —HCO_2Me, Et_2O, 0 °C, 86%→ OH (3)

The carbanion generated from the deprotonation of *O,S*-acetal (**17**) reacts with ethyl formate at −80 °C to give the aldehyde (**18**) in 89% yield (eq 4).[4d]

(**17**) SEt —LDA, HCO_2Et, THF, −80 °C, 89%→ (**18**) EtS CHO (4)

The formates are also reported to react in the following unusual transformations. The reaction of Nitrobenzene with the formates, catalyzed by $PdCl_2(PPh_3)_2$, together with Bu_3PO/KBr under carbon monoxide atmosphere, to gives *N*-phenylcarbamates in 29–63% yield (eq 5).[7]

$$PhNO_2 \xrightarrow[Bu_3PO,\ KBr\ \ 29\text{–}63\%]{CO,\ HCO_2R\ \ PdCl_2(PPh_3)_2} PhNHCO_2R \quad (5)$$

Benzylic, aryl, and alkyl halides react with the formates under carbon monoxide atmosphere, in the presence of $(RhLCl)_2$ (L = 1,5-hexadiene), KI, and, optionally, $Pd(PPh_3)_4$, to give the corresponding carboxylate esters in 20–94% yield (eq 6).[8]

$$R^1X \xrightarrow[(L\,=\,1,5\text{-hexadiene})\ \ 20\text{–}94\%]{CO,\ HCO_2R\ \ KI,\ (RhLCl)_2} R^1CO_2R^2 \quad (6)$$

The formates add regioselectively to alkenes or alkynes, such as *p*-methylstyrene, catalyzed by ***Tetrakis(triphenylphosphine)palladium(0)*** and ***1,4-Bis(diphenylphosphino)butane*** under carbon monoxide atmosphere, to give the linear isomer and the branched isomer (6.5:1 ratio) in 67% yield (eq 7).[9]

$$p\text{-}MeC_6H_4CH{=}CH_2 \xrightarrow[(Ph_2P)_2(CH_2)_4\ \ toluene,\ 150\ ^\circ C\ \ 67\%]{CO,\ HCO_2Bu\ \ Pd(PPh_3)_4}$$

$$p\text{-}MeC_6H_4(CH_2)_2CO_2Bu + p\text{-}MeC_6H_4CH(CO_2Bu)Me \quad (7)$$
$$6.5:1$$

Related Reagents. Acetic Formic Anhydride; *t*-Butoxybis(dimethylamino)methane; Carbon Monoxide; Dimethylchloromethyleneammonium Chloride; *N*,*N*-Dimethylformamide; Formyl Chloride; *N*-Methyl-*N*-(2-pyridyl)formamide.

1. (a) Prelog, V.; Geyer, U. *HCA* **1945**, *28*, 1677. (b) Birch, A. J.; Robinson, R. *JCS* **1944**, 501. (c) Johnson, W. S.; Posvic, H. *JACS* **1947**, *69*, 1361. (d) Johnson, W. S.; Petersen, J. W.; Gutsche, C. D. *JACS* **1947**, *69*, 2942. (e) Wilds, A. L.; Shunk, C. H. *JACS* **1950**, *72*, 2388. (f) Frank, R. L.; Varland, R. H. *OSC* **1955**, *3*, 829. (g) Ireland, R. E.; Marshall, J. A. *CI(L)* **1960**, 1534; *JOC* **1962**, *27*, 1615, 1620. (h) Clinton, R. O.; Manson, A. J.; Stonner, F. W.; Neumann, H. C.; Christiansen, R. G.; Clarke, R. L.; Ackerman, J. H.; Page, D. F.; Dean, J. W.; Dickinson, W. B.; Carabateas, C. *JACS* **1961**, *83*, 1478. (i) Ainsworth, C. *OSC* **1963**, *4*, 536. (j) Schenone, P.; Bignardi, G.; Morasso, S. *JHC* **1972**, *9*, 1341. (k) Boatman, S.; Harris, T. M.; Hauser, C. R. *OSC* **1973**, *5*, 187.
2. (a) Weisenborn, F. L.; Remy, D. C.; Jacobs, T. L. *JACS* **1954**, *76*, 552. (b) Corey, E. J.; Cane, D. E. *JOC* **1971**, *36*, 3070. (c) Eaton, P. E.; Jobe, P. G. *S* **1983**, 796. (d) Denmark, S. E.; Habermas, K. L.; Hite, G. A. *HCA* **1988**, *71*, 168. (e) Peet, N. P.; LeTourneau, M. E. *H* **1991**, *32*, 41.
3. Myers, A. G.; Harrington, P. M.; Kuo, E. Y. *JACS* **1991**, *113*, 694.
4. (a) Coleman, G. H.; Craig, D. *OSC* **1943**, *2*, 179. (b) Barbot, F.; Miginiac, P. *JOM* **1977**, *132*, 445. (c) Cresp, T. M.; Probert, C. L.; Sondheimer, F. *TL* **1978**, 3955. (d) Boehme, H.; Sutoyo, P. N. *AP* **1983**, *316*, 505. (e) Katritzky, A. R.; Akutagawa, K.; Jones, R. A. *SC* **1988**, *18*, 1151.
5. (a) Rakhit, S.; Gut, M. *JOC* **1964**, *29*, 229, 859. (b) Harmon, A. D.; Hutchinson, C. R. *TL* **1973**, 1293. (c) Yamada, K.; Kato, M.; Hirata, Y. *TL* **1973**, 2745. (d) Murray, A. W.; Reid, R. G. *CC* **1984**, 132. (e) Lehmann, J.; Neugebauer, M.; Marquardt, N. *AP* **1990**, *323*, 117.
6. (a) Holmes, H. L.; Trevoy, L. W. *OSC* **1955**, *3*, 300. (b) Thenappan, A.; Burton, D. J. *JFC* **1990**, *48*, 153.
7. Lin, I. J. B.; Chang, C. S. *J. Mol. Catal.* **1992**, *73*, 167.
8. Buchan, C.; Hamel, N.; Woell, J. B.; Alper, H. *CC* **1986**, 167.
9. (a) Alper, H.; Saldana-Maldonado, M.; Lin, I. J. B. *J. Mol. Catal.* **1988**, *49*, L27. (b) Lin, I. J. B.; Alper, H. *CC* **1989**, 248.

Chiu-Hong Lin
The Upjohn Company, Kalamazoo, MI, USA

Methyllithium

[917-54-4] CH_3Li (MW 21.98)

(nucleophilic methylating agent for several functional groups; cleaves protecting groups; synthesis of other methyl organometallics, i.e. of Cu, Ga, Ti, Mg, Si, Ir, B; can function as a base; reduces transition metals)

Physical Data: unsolvated solid is pyrophoric; densities listed below.

Solubility: insol hydrocarbon solvents; moderately sol ethereal solvents; reacts exothermically with water and protic solvents to generate methane.

Form Supplied in: commonly available as >1.4 M in diethyl ether, d >0.732 g mL^{-1}; >1.5 M in diethyl ether (with ~1.0 equiv of LiBr), d >0.852 g mL^{-1}. New formulations are >1 M in THF/cumene with 0.08 M Me_2Mg,[1] d 0.86 g mL^{-1}; 70 wt % solid[2] complexed with 22 wt % diethyl ether, 6% LiCl, and ~1% LiOH.

Analysis of Reagent Purity: titration of 0.5 N 2-butanol in toluene (10 mL) with methyllithium via tared syringe at 15 °C using 2,2′-biquinoline as an indicator.

Handling, Storage, and Precautions: methyllithium in diethyl ether is pyrophoric in absence of LiBr, according to US Dept of Transportation regulations (49 CFR 173, Appendix E), while all other available formulations (solution and solid) are nonpyrophoric. Store in a cool, dry place in a tightly sealed container under an inert atmosphere.

Introduction. Most preparations involve use of diethyl ether[3] or THF[1] from which solvated solids[4,5] may be isolated. Unsolvated methyllithium[1,6] preparations would allow systematic investigation of solvent[7] and halide[8] effects. A procedure for preparing low-halide methyllithium is available.[8b] Aggregation and crystal studies have been reported.[9]

Addition Reactions. Methyllithium can have different stereoselectivity compared to methyl Grignard reagents in 1,2-additions to cyclic carbohydrates,[10] heterocycles,[11] and acyclic[12] substrates. The origin of the high diastereoselectivity in nucleophilic addition to ketones (eq 1) is based on chelation control.[13a] Stereoselectivity can also be affected by prior complexation of the carbonyl compound with aluminum and other Lewis acids.[13b]

–40 °C, MeLi, ether; 70–98%

R = Bu, Ph, allyl; >99:1 % de (1)

Conversion of organolithiums to organometallics of titanium,[14] cerium,[15] and zinc[16] has lead to improved chemo- and stereoselectivity. For instance, conversion of MeLi to a chiral titanate prior to 1,2-addition to benzaldehyde resulted in 91% ee (eq 2).[14a]

1. MeLi; 2. PhCHO; 87%; 91% ee (2)

Methyl ketones can be prepared from acids[17] and esters, usually with greater success by subsequent treatment with ***Chlorotrimethylsilane***.[18,19] Methyllithium is a convenient deacylating agent for alkoxycarboxyl- and benzoyl-protected secondary hydrazides.[20] Addition of methyllithium with ***Lithium Bromide*** to silyl ketone (**1**) gave silyl enol ethers (**2**) with high stereochemical purity when warmed to 0 °C (eq 3).[21]

1. MeLi, –78 °C; 2. –20 to 0 °C; 55–88%; (**1**) → (**2**); *(E)*:*(Z)* = 99:1 (3)

Similarly, almost quantitative yields of 2,2-difluoroenol silyl ethers were obtained from a perfluoroacylsilane.[22] Enolates, free of amine, can be prepared by cleavage of silyl enol ethers[23] and enol trifluoromethanesulfinates with methyllithium.[24]

Of 1,2-additions to carbon–nitrogen multiple bonds, nitriles, imines, and oxazolidine are the most useful substrates. For instance, addition of methyllithium in cumene/THF to protected cyanohydrins gave excellent yields of methyl steroidal ketones after hydrolysis.[25] Addition of aryl heterocycles, such as tetrazines[26] and pyrazines,[27] are problematic.

Conjugate addition is routinely achieved by prior transformation to the corresponding cuprate[28,29] or even aluminate.[30] In absence of any co-metals, methyllithium will undergo 1,4-addition to naphthalimines[31] in the presence of ***Hexamethylphosphoric Triamide*** (eq 4). Without HMPA, methyllithium undergoes 1,2-additions to naphthalimines[31] and oxazolidines.[32]

1. MeLi, 2.5 equiv HMPA, –10 °C; 2. MeI, –78 °C; 3. 10% HCl, 25 °C, 15 h; 4. $NaBH_4$, MeOH, 0 °C; 60%

1. MeLi, THF, –78 °C; 2. *i*-PrOH, aq HCl; 95% (4)

Secondary allylic alcohols, 3-substituted with SO_2Ph and $SiMe_3$, will undergo conjugate addition to give only the *syn*-adduct.[33] Asymmetric addition to sulfines in the presence of chiral ligands proceeded with only modest stereoselectivity.[34]

Miscellaneous Reactions. Methyllithium was the base of choice for the metalation of *o*-allylbenzamide, which cyclizes to a naphthol derivative (eq 5).[35]

2.2 equiv MeLi, THF; –78 °C to rt, 8 h; 77% (5)

Rearrangement of *gem*-dibromides of cyclopropyl derivatives with methyllithium leads to mixtures of allenes[36a] and monobromide derivatives.[36] Substituted epoxides,[37] silacyclopentane,[38] and cyclotriphosphazene[39] can be ring opened. The transmetalation of trimethylstannyl compounds to form vinyl-, allenyl-, and propargyllithium reagents is usually best done with methyllithium.[40] Zero-valence Pd for coupling alkenyl halides and organolithiums can be prepared from ***Palladium(II) Chloride*** and methyllithium.[41] Even preparation and utility of transition metal-activated organic compounds of molybdenum[42a] and tungsten[42b] are reported.

Related Reagents. Cerium(III) Chloride; Dichlorodimethyltitanium; Dilithium Pentamethyltricuprate; Dilithium Trimethylcuprate; Lithium Cyano(methyl)cuprate; Lithium Dimethylcuprate; Lithium Trimethylzincate; Methylcopper; Methylmagnesium Bromide; Methyltitanium Triisopropoxide; Methyltitanium Trichloride.

1. Morrison, R. C.; Rathman, T. L. (FMC Lithium Division) U.S. Patent 4 976 886, 1990.
2. Rittmeyer, P. *Brit. Assoc. for Chem. Spec., Chem. Spec. Europe*; Publisher: Location, 1993; p 17.
3. Houk, K. N.; Rondan, N. G.; Schleyer, P. v. R.; Kaufmann, E.; Clark, T. *JACS* **1985**, *107*, 2821.
4. Ogle, C. A.; Huckabee, B. K.; Johnson, H. C., IV; Sims, P. F.; Winslow, S. D.; Pinkerton, A. A. *OM* **1993**, *12*, 1960.
5. Deberitz, J.; Weiss, W. (Metallgesellschaft A.-G.) Eur. Patent 340 819, 1989 (*CA* **1990**, *112*, 77 529t).
6. Brown, T. L.; Rogers, M. T. *JACS* **1957**, *79*, 1859.
7. (a) Fujisawa, T.; Funabora, M.; Ukaji, Y.; Sato, T. *CL* **1988**, 59. (b) Ko, K.-Y.; Eliel, E. L. *JOC* **1986**, *51*, 5353.
8. (a) Savignac, P.; Teulade, M.-P.; Patois, C. *HC* **1990**, *1*, 211. (b) Lusch, M. J.; Phillips, W. V.; Sieloff, R. F.; Nomura, G. S.; House, H. O. *OSC* **1990**, *7*, 346.
9. (a) West, P.; Waack, R. *JACS* **1967**, *89*, 4395. (b) Köster, H.; Thoennes, D.; Weiss, E. *JOM* **1978**, *160*, 1. (c) Weiss, E.; Lucken, E. A. C. *JOM* **1964**, *2*, 197.
10. (a) Yoshimura, J.; Sato, K. *Carbohydr. Res.* **1983**, *123*, 341. (b) Mukaiyama, T.; Soai, K.; Sato, T.; Shimizu, H.; Suzuki, K. *JACS* **1979**, *101*, 1455.
11. (a) Wade, P. A.; Price, D. T.; Carroll, P. J.; Dailey, W. P. *JOC* **1990**, *55*, 3051. (b) Wade, P. A.; Price, D. T.; McCauley, J. P.; Carroll, P. J. *JOC* **1985**, *50*, 2804.
12. Hosokawa, T.; Yagi, T.; Ataka, Y.; Murahashi, S.-I. *BCJ* **1988**, *61*, 3380.
13. (a) Chikashita, H.; Nakamura, Y; Uemura, H. Itoh, K. *CL* **1992**, 439. (b) Maruoka, K.; Itoh, T.; Sakurai, M.; Nonoshita, K.; Yamamoto, H. *JACS* **1988**, *110*, 3588.

14. (a) Seebach, D.; Beck, A. K.; Imwinkelried, R.; Roggo, S.; Wonnocott, A. *HCA* **1987**, *70*, 954. (b) Reetz, M. T.; Kyung. S. H.; Huellmann, M. *T* **1986**, *42*, 2931. (c) Reetz, M. T.; Hugel, H.; Dresely, K. *T* **1987**, *43*, 109. (d) Schmidt, B; Seebach, D. *AG(E)* **1991**, *30*, 99.
15. (a) Imamoto, T.; Sugiura, Y. *JOM* **1985**, *285*, C21. (b) Paquette, L. A.; Learn, K. S.; *JACS* **1986**, *108*, 7873. (c) Denmark, S. E.; Weber, T.; Piotrowski, D. W. *JACS* **1987**, *109*, 2224. (d) Johnson, C. R.; Tait, B. D. *JOC* **1987**, *52*, 281.
16. (a) Kitamura, M.; Suga, S.; Kawai, K.; Noyori, R. *JACS* **1986**, *108*, 6071. (b) Morita, Y. Suzuki, M. Noyori, R. *JOC* **1989**, *54*, 1785. (c) Erdik, E. *T* **1987**, *43*, 2203.
17. Bare, T. M.; House, H. O. *OSC* **1973**, *5*, 775.
18. Overman, L. E.; Rishton, G. M. *OS* **1993**, *71*, 56 and 63.
19. (a) Rubottom, G. M.; Kim, C. *JOC* **1983**, *48*, 1550. (b) Jorgenson, M. J. *OR* **1970**, *18*, 1–98.
20. Tschamber, T.; Streith, J. *H* **1990**, *30*, 551.
21. Reich, H. J.; Holtan, R. C.; Bolm, C. *JACS* **1990**, *112*, 5609.
22. Jin, F.; Jiang, B.; Xu, Y. *TL* **1992**, *33*, 1221.
23. Bertz, S. H.; Jelinski, L. W.; Dabbagh, G. *CC* **1983**, 388.
24. Lee, S.-H.; Hulce, M. *SL* **1992**, 485.
25. Carruthers, N. I.; Garshasb, S. *JOC* **1992**, *57*, 961.
26. Wilkes, M. C. *JHC* **1991**, *28*, 1163.
27. Rizzi, G. P. *JOC* **1974**, *39*, 3598.
28. Lipshutz, B. H.; Sengupta, S. *OR* **1992**, *41*, 135.
29. Rossiter, B. E.; Swingle, N. M. *CRV* **1992**, *92*, 771.
30. Maruoka, K.; Nonoshita, K.; Yamamoto, H. *TL* **1987**, *28*, 5723.
31. Meyers, A. I.; Brown, J. D.; Laucher, D. *TL* **1987**, *28*, 5279.
32. Pridgen, L. N.; Mokhallalati, M. K.; Wu, M.-J. *JOC* **1992**, *57*, 1237.
33. Kitamura, M. Isobe, M; Ichikawa, Y.; Goto, T. *JACS* **1984**, *106*, 3252.
34. Rewinkel, J. B. M; Porskamp, P. A. T. W.; Zwanenburg, B. *RTC* **1988**, *107*, 563.
35. Sibi, M. P.; Dankwardt, J. W.; Snieckus, V. *JOC* **1986**, *51*, 271.
36. (a) Skatteboel, L.; Stenstroem, Y.; Stjerna, M.-B. *ACS* **1988**, *B42*, 475. (b) Lukin, K. A.; Zefirov, N. S.; Yufit, D. S.; Struchkov, Y. T. *T* **1992**, *48*, 9977. (c) Molchanov, A. P.; Kalyamin, S. A.; Kostikov, R. R. *JOU* **1992**, *28*, 102 (*CA* **1992**, *117*, 170 833a)
37. (a) Wender, P. A.; Erhardt, J. M.; Letendre, L. J. *JACS* **1981**, *103*, 2114. (b) Alcaide, B.; Areces, P.; Borredon, E.; Biurrun, C.; Castells, J. P.; Plumet, J. *H* **1990**, *31*, 1997.
38. Maercker, A.; Stoetzel, R. *JOM* **1984**, *269*, C40.
39. Harris, P. J. Fadeley, C. L. *IC* **1983**, *22*, 561.
40. Reich, H. J.; Reich, I. L.; Yelm, K. E.; Holladay, J. E.; Gschneidner, D. *JACS* **1993**, *115*, 6625. Reich, H. J.; Mason, J. D.; Holladay, J. E. *CC* **1993**, 1481.
41. Murahashi, S.-I.; Naota, T.; Tanigawa, Y. *OS* **1984**, *62*, 39.
42. (a) Kauffmann, T.; Jordan, J.; Voss, K.-U. *AG(E)* **1991**, *30*, 1138. (b) Kauffmann, T.; Kieper, G. *AG(E)* **1984**, *23*, 532.

Terry L. Rathman
FMC Corporation Lithium Division, Bessemer City, NC, USA

Methylmagnesium Bromide[1]

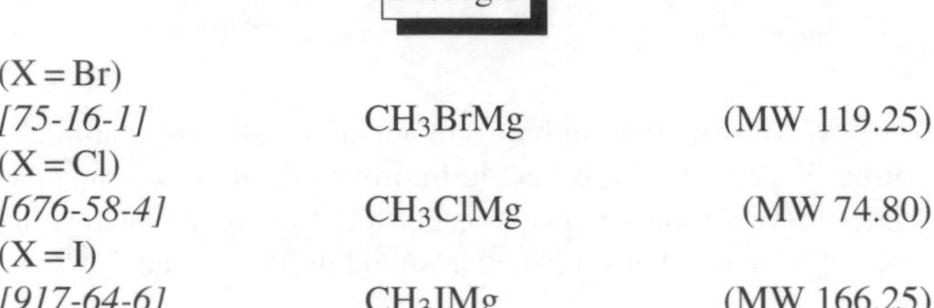

(X = Br)		
[75-16-1]	CH_3BrMg	(MW 119.25)
(X = Cl)		
[676-58-4]	CH_3ClMg	(MW 74.80)
(X = I)		
[917-64-6]	CH_3IMg	(MW 166.25)

(adds to many unsaturated functional groups; methyl can displace halide and like groups; can function as a strong base and a Lewis acid)[1b–f]

Physical Data: NMR, IR, kinetic, and calorimetric observations indicate that MeMgBr in Et_2O and THF, and MeMgCl in Et_2O, actually are mixtures of MeMgX, Me_2Mg, and MgX_2.[2] MeMgBr is essentially monomeric in THF, but it and MeMgI are associated significantly in Et_2O except at low concentrations.[3] Solids isolated include $MeMgBr(THF)_3$,[4] shown by X-ray diffraction to have five-coordinate Mg, and (low-melting) $MeMgX(Et_2O)_2$ (X = Br or I).[5] Complete removal of Et_2O leaves a mixture of Me_2Mg and MgX_2 (X = Cl or Br).[6]

Solubility: MeMgBr and MeMgI sol Et_2O; MeMgCl and MeMgBr sol THF. MgI_2 precipitates when MeMgI is prepared in THF, leaving mainly Me_2Mg in solution; some $MgCl_2$ generally precipitates when MeMgCl is prepared in Et_2O, particularly if the solution stands for several days. All are insoluble in hydrocarbons.

Form Supplied in: solutions of MeMgCl in THF, MeMgBr and MeMgI in Et_2O and Bu_2O, and MeMgBr in toluene/THF are commercially available.

Analysis of Reagent Purity: a small excess is used ordinarily to ensure that sufficient reagent is present, but concentration should be determined when exact stoichiometry is important.[7] R–Mg is conveniently determined by hydrolysis of an aliquot followed by addition of excess acid and back-titration with base.[8] Since this procedure does not distinguish R–Mg from HO–Mg and RO–Mg formed from reaction of the Grignard reagent with water and oxygen, respectively, a direct titration[9] more specific for R–Mg or a double titration procedure[10] should be used when such exposure is suspected. A qualitative color test is convenient for detecting the presence of Grignard reagent.[11]

Preparative Methods: [1c,e,12,13] in spite of their commercial availability, solutions of MeMgCl,[14,15] MeMgBr,[14,16] and MeMgI[17] are often prepared from reaction of a methyl halide and Mg, usually in Et_2O or THF.

Handling, Storage, and Precautions: Grignard reagents react readily with oxygen, water, and carbon dioxide. Reactants and apparatus used in their preparations and reactions should be dry throughout preparation, storage, and use; Grignard reagent solutions should be maintained under nitrogen or argon, though the vapor of a volatile solvent (especially Et_2O) provides some protection for short periods of time. Methyl Grignard reagents do not significantly attack Et_2O or THF at normal reaction temperatures.

Structure.[1a,c] The structure of a Grignard reagent, an important influence on reactions, is more elaborate than implied by the formula RMgX. In solution, a Grignard reagent is a mixture (Schlenk equilibrium) of RMgX, R_2Mg, and MgX_2, the composition varying with solvent and X. Mg is most commonly four-coordinate in solids but may have even higher coordination;[18] all evidence indicates similar coordination in solutions. The additional bonds to Mg result from some combination of association by bridging of the X atom (or R group) between two Mg atoms and coordination by donor molecules (usually solvent). Coordination by donor groups of substrates can play an important role in reactions.

Workup Procedures.[13] Workup of a Grignard reaction requires adding a proton source and removing magnesium salts. This is done conveniently by adding an aqueous HCl or H_2SO_4 solution and separating the organic and aqueous layers. When products cannot tolerate acids, however, a saturated aqueous ammonium chloride solution is generally used, either a large volume[19] sufficient to dissolve all magnesium salts or, alternatively, just enough[15] to precipitate the magnesium salts completely, leaving a nearly anhydrous solution of the product.

Representative Applications.[1h–f] A Grignard reagent is used in an extraordinary variety of reactions, the majority leading to attachment of its organic group to an electrophilic carbon atom of the substrate. Most common are additions to carbonyl groups (eq 1),[20] nitriles, and imines (eq 2),[21] and displacements of halide and like leaving groups. Although generally less reactive than organolithium compounds, Grignard reagents are often used for convenience, because of a different reactivity pattern, or to minimize competing reactions. MeMgX and MeLi, for example, sometimes exhibit quite different regioselectivities (eq 1) and stereoselectivities (eq 2). The key step in the synthesis of chiral $O{=}CHCH(Me)NH_2$ (eq 2) also illustrates the common practice in Grignard reagent additions of using a disposable group whose coordination or steric effects impart stereoselectivity.

(1)

MeMgBr, THF, –78 °C	60%	–
MeLi, TMEDA, THF, –107 °C	9%	87%

Grignard reagents are strong bases and Lewis acids, a problem in many reactions but of use in others. MeMgX, for example, extracts a proton from most N–H and O–H bonds and from particularly acidic C–H bonds. Volumetric measurement of the methane evolved from reaction with a compound of MeMgI was once a significant procedure (Zerewitinoff determination) for quantitative determination of its 'active hydrogens'.[22] MeMgX still is used as a base, for example in forming magnesium amides (eq 3)[23] and enolates. Cleavage of aryl–alkyl ethers (eq 4)[24] is a displacement that must require the Lewis acid properties of the Grignard reagent. Me_2Mg is sometimes used in place of MeMgX to minimize reactions, such as isomerization of oxiranes to aldehydes or ketones,[25] that involve Lewis acidity but not displacement.

(2)

MeMgBr, Et_2O, reflux, 24 h	80:20
MeMgBr, toluene, 20 °C, 1 h	94:6
MeLi, Et_2O, –78 °C to rt	2:98

MeMgI, Et_2O, THF (3)

MeMgI, 155–165 °C, 80% (4)

Strongly coordinating solvents tend to reduce reactivity in additions to unsaturated functions but to increase reactivity in reactions with alkyl halides and as bases. The greater rate and stereoselectivity of an addition (eq 2) in toluene than in Et_2O illustrate the importance of solvent.

Transition metal impurities, often introduced in the Mg used to prepare a Grignard reagent, can lead to unwanted products,[26] and where this is a problem particularly pure Mg should be used. Catalytic or stoichiometric amounts of transition metal compounds, of course, often are added deliberately to Grignard reagents.[27] Examples are copper compounds[28] to promote 1,4-additions to α,β-unsaturated carbonyl compounds (one variant of organocuprate chemistry) and nickel compounds[29] to promote substitution of aryl and vinyl halides.

Related Reagents. Chloromagnesium Dimethylcuprate; Copper(I) Bromide; Copper(I) Chloride; Copper(I) Iodide; Methylcopper; Methylcopper–Boron Trifluoride Etherate; Methylcopper–Tributylphosphine; Methyllithium; Methyltitanium Trichloride.

1. (a) Lindsell, W. E. In *Comprehensive Organometallic Chemistry*; Wilkinson G.; Stone, F. G. A.; Abel, E. W., Eds.; Pergamon: Oxford, 1982; Chapter 4. (b) Wakefield, B. J. In *Comprehensive Organometallic Chemistry*; Wilkinson G.; Stone, F. G. A.; Abel, E. W., Eds.; Pergamon: Oxford, 1982; Chapter 44. (c) Nützel, K. *MOC* **1973**, *13*/2a, 47. (d) Raston, C. L.; Salem, G. In *The Chemistry of the Metal–Carbon Bond*; Hartley, F. R.; Ed.; Wiley: Chichester, 1987; Vol. 4, Chapter 2. (e) Old but still extremely useful is Kharasch, M.; Reinmuth, O. *Grignard*

Reactions of Nonmetallic Substances; Prentice-Hall: New York, 1954. (f) Ioffe, S. T.; Nesmeyanov, A. N. *The Organic Compounds of Magnesium, Beryllium, Calcium, Strontium and Barium;* North-Holland: Amsterdam, 1967.

2. Ashby, E. C.; Laemmle, J.; Neumann, H. M. *ACR* **1974**, *7*, 272.
3. Walker, F. W.; Ashby, E. C. *JACS* **1969**, *91*, 3845.
4. Vallino, M. *JOM* **1969**, *20*, 1.
5. Kress, J.; Novak, A. *JOM* **1975**, *99*, 199.
6. Weiss, E. *CB* **1965**, *98*, 2805.
7. Crompton, T. R. *Comprehensive Organometallic Analysis*; Plenum: New York, 1987; Chapters 2, 3, and 6.
8. Gilman, H.; Zoellner, E. A.; Dickey, J. B. *JACS* **1929**, *51*, 1576.
9. Bergbreiter, D. E.; Pendergrass, E. *JOC* **1981**, *46*, 219.
10. Vlismas, T.; Parker, R. D. *JOM* **1967**, *10*, 193.
11. Gilman, H.; Schulze, F. *JACS* **1925**, *47*, 2002.
12. Bickelhaupt, F. In *Inorganic Reactions and Methods*; Hagen, A. P., Ed.; VCH: New York, 1989, Vol. 10, Section 5.4.2.2.1.
13. *FF* **1967**, *1*, 415.
14. Salinger, R. M.; Mosher, H. S. *JACS* **1964**, *86*, 1782.
15. Coburn, E. R. *OSC* **1955**, *3*, 696.
16. Colonge, J.; Marey, R. *OSC* **1963**, *4*, 601.
17. Callen, J. E.; Dornfeld, C. A.; Coleman, G. H. *OSC* **1955**, *3*, 26.
18. Markies, P. R.; Akkerman, O. S.; Bickelhaupt, F.; Smeets, W. J. J.; Spek, A. L. *Adv. Organomet. Chem.* **1991**, *32*, 147.
19. Skattebøl, L.; Jones, E. R. H.; Whiting, M. C. *OSC* **1963**, *4*, 792.
20. Liotta, D.; Saindane, M.; Barnum, C. *JOC* **1981**, *46*, 3369.
21. Alexakis, A.; Lensen, N.; Tranchier, J.-P.; Mangeney, P. *JOC* **1992**, *57*, 4563. Alexakis, A.; Lensen, N.; Mangeney, P. *TL* **1991**, *32*, 1171.
22. Siggia, S.; Hanna, J. G. *Quantitative Organic Analysis via Functional Groups*, 4th ed.; Wiley: New York, 1979; Chapter 8.
23. Kametani, T.; Huang, S.-P.; Yokohama, S.; Suzuki, Y.; Ihara, M. *JACS* **1980**, *102*, 2060.
24. Mechoulam, R.; Gaoni, Y. *JACS* **1965**, *87*, 3273.
25. Christensen, B. G.; Strachan, R. G.; Trenner, N. R.; Arison, B. H.; Hirschmann, R.; Chemerda, J. M. *JACS* **1960**, *82*, 3995.
26. Ashby, E. C.; Neumann, H. M.; Walker, F. W.; Laemmle, J.; Chao, L.-C. *JACS* **1973**, *95*, 3330; Ashby, E. C.; Wiesemann, T. L. *JACS* **1978**, *100*, 189.
27. Felkin, H.; Swierczewski, G. *T* **1975**, *31*, 2735.
28. Erdik., E. *T* **1984**, *40*, 641; Rossiter, B. E.; Swingle, N. M. *CRV* **1992**, *92*, 771.
29. Tamao, K. *COS* **1991**, *3*, 435; Jolly, P. W. In *Comprehensive Organometallic Chemistry*; Wilkinson G.; Stone, F. G. A.; Abel, E. W., Eds.; Pergamon: Oxford, 1982; Chapter 56.5.

Herman G. Richey, Jr.
The Pennsylvania State University, University Park, PA, USA

N-Methyl-*N*-(2-pyridyl)formamide

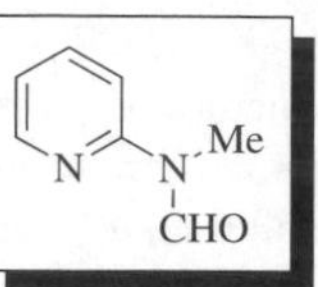

[67242-59-5] $C_7H_8N_2O$ (MW 136.17)

(formylating reagent for Grignard reagents[1] and certain organolithiums;[6] sequential addition of two Grignard reagents can provide unsymmetrical secondary alcohols[3])

Alternate Name: 2-(*N*-formyl-*N*-methyl)aminopyridine.
Physical Data: bp 71–72 °C/0.05 mmHg; *d* 1.137 g cm^{-3}.
Form Supplied in: colorless liquid; commercially available.
Analysis of Reagent Purity: IR, NMR.
Handling, Storage, and Precautions: moisture sensitive; irritant; should be kept under argon or nitrogen.

General Discussion. *N*-Methyl-*N*-(2-pyridyl)formamide (**1**) is a good formylating reagent for certain organometallics. Dropwise addition of Grignard reagents to (**1**) provides good yields of aldehydes (eq 1).[1]

$$\text{RM} \xrightarrow{(\mathbf{1})} \text{RCHO} \quad (1)$$

R = alkyl or aryl; M = MgBr

Various Grignard reagents have been studied and yields are good to excellent. The formylation is effective and facile because of a six-membered chelate complex formed in situ; this inhibits further reaction at low temperature (eq 2).[1,2] On workup with aqueous acid the complex is protonated and dissociates to give the aldehyde and amine salt. The aldehyde is isolated by ether extraction, and the byproduct, 2-(methylamino)pyridine, can be recovered by neutralization of the aqueous layer with sodium hydroxide and extraction with CH_2Cl_2. The amine can be economically recycled to (**1**) in 90% yield by formylation with a formic acid–acetic anhydride mixture.[3]

RM —(1), 0 °C→ [chelate complex: N, N–Me, M, O, R] —H_3O^+→ [2-pyridyl–$\overset{+}{N}$H$_2$Me] + RCHO (2)

R = alkyl or aryl; M = MgBr

Using a modification of this methodology, unsymmetrical secondary alcohols can be synthesized via a one-pot reaction using two different Grignard reagents (eq 3).[3] Similarly, the vinyl Grignard reagent (**2**) has been formylated using the reagent (**1**) (eq 4).[4]

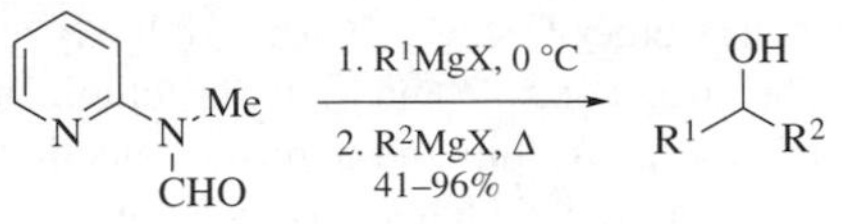

R^1 = alkyl or aryl; R^2 = alkyl or aryl (3)

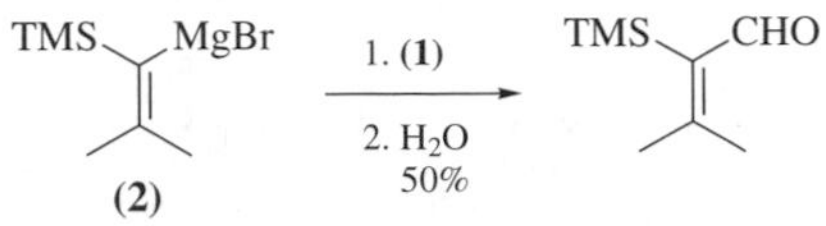

(4)

Using a one-pot reaction, a dithioester has been converted to a protected α-keto carbaldehyde using the reagent (**1**) (eq 5).[5] Heterocycles can be converted to their formyl derivatives by lithiation and reaction with (**1**) (eq 6).[6] The importance of the reagent (**1**) has been demonstrated by its application in natural product synthesis.[7]

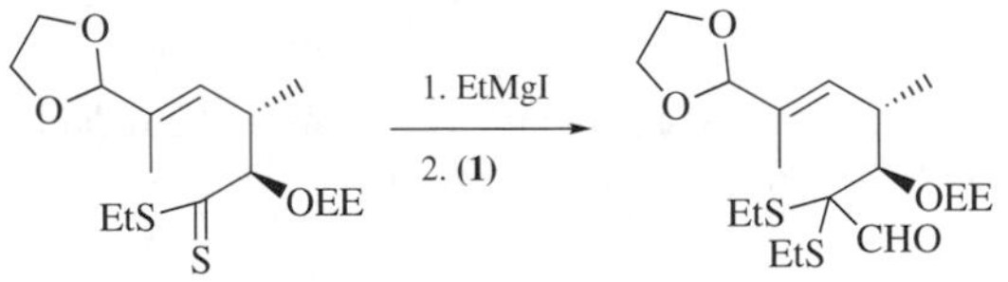

(5)

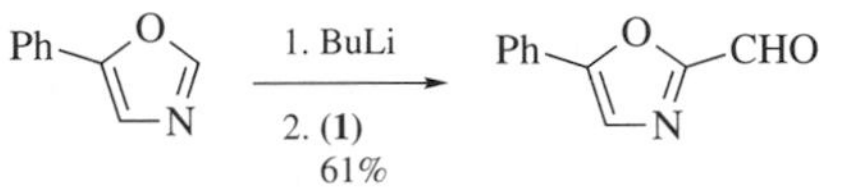

(6)

Related Reagents. Carbon Monoxide; Diethyl Phenyl Orthoformate; Dimethylchloromethyleneammonium Chloride; *N,N*-Dimethylformamide; *N*-Formylpiperidine; *N*-Methyl formanilide.

1. Comins, D. L.; Meyers, A. I. *S* **1978**, 403.
2. Amaratunga, W.; Frechet, J. M. J. *TL* **1983**, *24*, 1143.
3. Comins, D. L.; Dernell, W. *TL* **1981**, *22*, 1085.
4. Munstedt, R.; Wannagat, U. *LA* **1985**, 944.
5. Meyers, A. I.; Tait, T. A.; Comins, D. L. *TL* **1978**, 4657.
6. Pridgen, L. N.; Shilcrat, S. C. *S* **1984**, 1048.
7. (a) Meyers, A. I.; Comins, D. L.; Roland, D. M.; Henning, R.; Shimizu, K. *JACS* **1979**, *101*, 7104. (b) Meyers, A. I.; Roland, D. M.; Comins, D. L.; Henning, R.; Fleming, M. P.; Shimizu, K. *JACS* **1979**, *101*, 4732. (c) Guthrie, A. E.; Semple, J. E.; Joullie, M. M. *JOC* **1982**, *47*, 2369.

Daniel L. Comins & Sajan P. Joseph
North Carolina State University, Raleigh, NC, USA

Methyltitanium Trichloride[1]

$MeTiCl_3$

[2747-38-8] CH_3Cl_3Ti (MW 169.27)

(nonbasic nucleophilic reagent with high Lewis acidity;[1,2] adds to aldehydes, ketones, and acetals with high chemo- and stereoselectivity;[1,2] reagent for the S_N1 substitution of tertiary halides[3] and the geminal dimethylation of carbonyl compounds[4,5])

Alternate Names: trichloromethyltitanium; methyltitanium(IV) trichloride.
Physical Data: mp 29 °C; bp 37 °C/1 mmHg.
Solubility: sol most commonly used aprotic solvents, including pentane, Et_2O, THF, CH_2Cl_2. Forms octahedral complexes with bidentate ligands such as glycol ethers, ethylenediamines, and diphosphines as well as with solvents such as Et_2O and THF.
Form Supplied in: purple crystals; not available commercially.
Analysis of Reagent Purity: ^1H NMR: 2.9 ppm (s).
Preparative Methods: $MeTiCl_3$:[6] ether-free reagent, suitable for chelation-controlled carbonyl additions and S_N1 substitutions, can be prepared from a 2:1 mixture of ***Titanium(IV) Chloride*** and Me_2Zn[5] in CH_2Cl_2 at −78 °C. The resulting solution can be used as is or distilled under vacuum. A 1:1 mixture of $TiCl_4$ and Me_2Zn produces ***Dichlorodimethyltitanium***, which is more effective for S_N1 substitutions.[5] $MeLi–TiCl_4–Et_2O$:[2] solutions of $MeTiCl_3$ in Et_2O can be more conveniently prepared from ***Methyllithium*** and $TiCl_4$. Other organometallics, such as ***Methylmagnesium Bromide***, can also be used. In the workup, simple quenching with ice water is preferred. The use of ***Sodium Carbonate*** may lead to cumbersome emulsions of TiO_2, which can be avoided with NH_4F or ***Potassium Fluoride***.
Handling, Storage, and Precautions: moisture sensitive; stable at low temperature in the dark. At rt, it decomposes over several hours. Me_2Zn is very pyrophoric in a pure form.

Organotitanium Reagents. While many titanium compounds are used widely as key components of Ziegler–Natta polymerization catalysts,[1a,1c] a variety of organotitanium derivatives have found great utility in organic synthesis as well. Compounds of the general type R^1TiX_3 (X = Cl, OR^2, or NR^3_2), studied extensively by Reetz and others,[1] often add to carbonyl groups with a high degree of chemo-, regio-, diastereo-, and enantioselectivity as compared with organomagnesium (RMgX, Grignard) or organolithium reagents (RLi). The synthetic advantages of these organotitanium reagents arise from the modulation of their nucleophilicity and Lewis acidity and basicity, as well as their improved solubility in organic solvents, and their increased steric and electronic requirements. Among the various commonly used derivatives, $MeTiCl_3$ is the most Lewis acidic, while ***Methyltitanium Triisopropoxide*** is the least acidic. Other homologs of type $RTiCl_3$ are more difficult to prepare and have found only limited synthetic use. The similar reagent Me_2TiCl_2 is more nucleophilic than $MeTiCl_3$ and is more effective for S_N1 substitutions.[5]

Chemoselective Addition to Aldehydes and Ketones. $MeTiCl_3$ reacts quickly with aldehydes at low temperatures, while similar reactions with ketones require higher temperatures. Among other related organometallics, $MeTiCl_3$ exhibits one of the highest levels of aldehyde selectivities (eq 1).[2] In the presence of phosphines, however, an in situ blocking of the aldehyde group allows the reagent to react selectively with ketones in the presence of aldehydes.[7]

(1)

$MeTiCl_3$	99: 1
$MeTi(O\text{-}i\text{-}Pr)_3$	85:15
$MeZrCl_3$	74:26
MeLi	70:30

In general, $MeTiCl_3$ shows higher reactivity towards ketones than $MeTi(O\text{-}i\text{-}Pr)_3$. Discrimination between different ketones is also most effective with this reagent (eq 2).[2]

(2)

$MeTiCl_3$	99: 1
$MeTi(O\text{-}i\text{-}Pr)_3$	97: 3
MeLi	39:61

A wide variety of functional groups are tolerated under the reaction conditions, including ester, cyano, or nitro groups (eq 3).[2]

(3)

$MeTiCl_3$	76%
$MeTi(O\text{-}i\text{-}Pr)_3$	15%

Being the least basic, $MeTiCl_3$ is the reagent of choice for the addition of methyl groups to highly enolizable ketones (eq 4).[2]

(4)

$MeTiCl_3$	90%
MeLi	60%

Stereoselective Additions to Aldehydes and Ketones. The reagent, formed in situ from the combination of MeLi and $TiCl_4$ in Et_2O, adds to chiral aldehydes according to Cram's rule with higher stereoselectivity than ether-free preformed $MeTiCl_3$ or even $MeTi(O\text{-}i\text{-}Pr)_3$ (eq 5).[2,8]

(5)

syn *anti*

$MeTi(OPh)_3$	93: 7
$MeZr(O\text{-}i\text{-}Pr)_3$	90:10
$MeLi–TiCl_4–Et_2O$	90:10
$MeTi(O\text{-}i\text{-}Pr)_3$	88:12
$MeTiCl_3$	81:19
MeMgBr	66:34
MeLi	65:35

The addition of ether-free $MeTiCl_3$ to α-alkoxy aldehydes proceeds with chelation control[9] and high diastereofacial selectivity, while $MeTi(O\text{-}i\text{-}Pr)_3$ exhibits the opposite selectivity (eq 6).[10,11] The postulated titanium chelates have been observed directly by ^{13}C NMR spectroscopy.[11]

$MeTiX_3$, CH_2Cl_2, –78 °C

(6)

$MeTiCl_3$	92: 8
$MeTi(O\text{-}i\text{-}Pr)_3$	8:92

Similar chelation control with 1,3-asymmetric induction takes place with β-alkoxy aldehydes (eq 7).[12]

$MeTiCl_3$, CH_2Cl_2, –78 °C

(7)

90:10

Although preformed $MeTiCl_3$ and the combination of $Me_2Zn–TiCl_4$ often behave similarly, in some cases they exhibit significantly different diastereofacial selectivities (eq 8).[13]

(8)

$MeTiCl_3$	140: 1
Me_2CuLi	80:20
$Me_2Zn–TiCl_4$	68:32
$MeTi(O\text{-}i\text{-}Pr)_3$	58:42
MeLi	58:42
MeMgCl	37:63
$MeMgCl–ZnBr_2$	22:78
$Me_2CuLi–Et_3B$	15:85

Chelation of $MeTiCl_3$ with neighboring sulfur groups is also possible (eq 9).

$MeTiCl_3$, Et_2O, CH_2Cl_2, –30 °C to 0 °C, 4 h, 75%

(9)

syn:anti = 2:98

$MeTiCl_3$ adds efficiently and stereoselectively to the carbonyl group of β-keto sulfoxides, despite the facile enolization of these compounds. Presumably this process involves chelation with the sulfoxide oxygen (eq 10).[14]

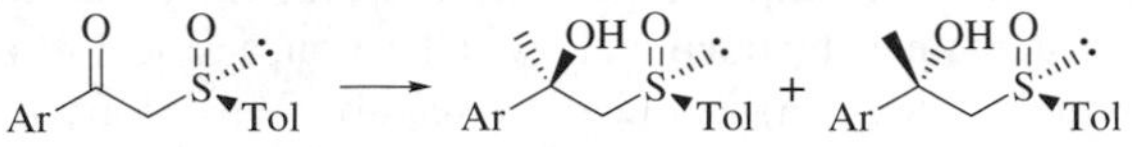

(10)

$MeTiCl_3$, Et_2O	96%	97: 3
Me_2Al, PhMe	71%	13:87

Additions to Acetals. $MeTiCl_3$ reacts readily with acetals at low temperature even in the presence of ketones. With chiral acetals, the reaction is highly stereoselective (eq 11).[15]

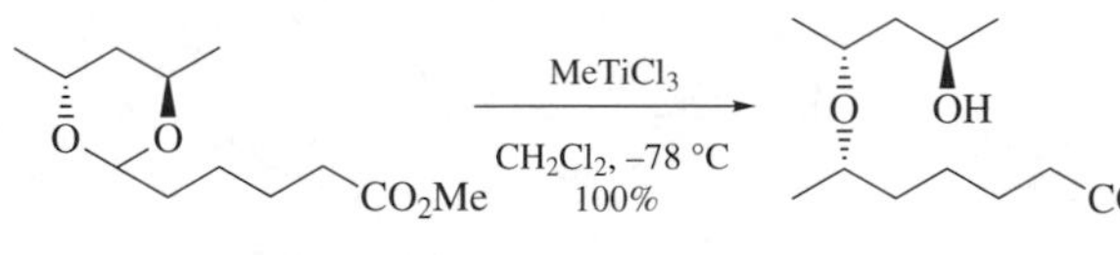

$MeTiCl_3$, CH_2Cl_2, −78 °C, 100% (11)

>95:5

S_N1 Methylation of Tertiary Alkyl Halides. The high Lewis acidity of $MeTiCl_3$ and the related derivative Me_2TiCl_2, obtained from $MeTiCl_3$ or Me_2Zn and 1 equiv of $TiCl_4$, makes them suitable for the S_N1 substitution of tertiary chlorides (eq 12)[3] or bromides (eq 13). Primary or secondary halides do not undergo this reaction, which presumably proceeds via a carbocation.

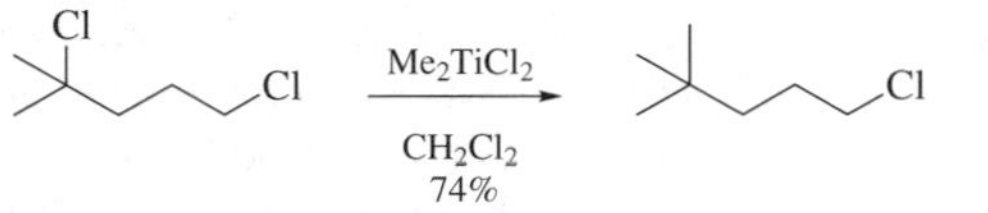

Me_2TiCl_2, CH_2Cl_2, 74% (12)

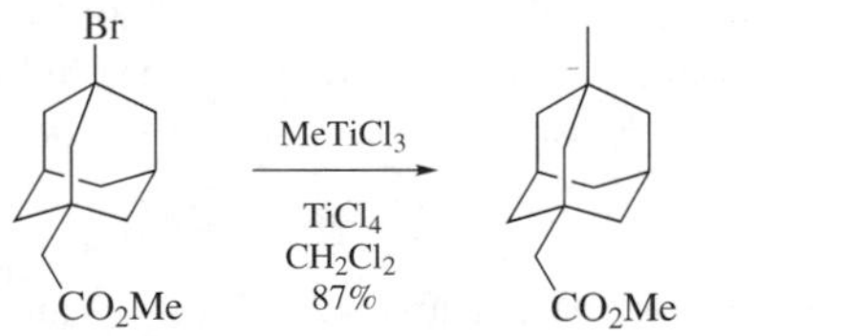

$MeTiCl_3$, $TiCl_4$, CH_2Cl_2, 87% (13)

Geminal Dimethylation of Carbonyl Compounds. Reaction of ketones with $MeTiCl_3$, or more effectively with Me_2TiCl_2[16,17] or $Me_2Zn–TiCl_4$,[5] gives the corresponding geminal dimethyl derivatives directly[4] (eq 14).[16] This process presumably involves an initial addition to the carbonyl followed by ionization and S_N1 substitution by a second methyl group.

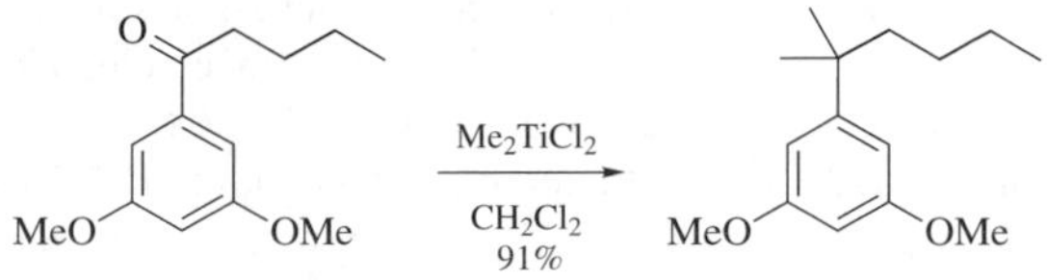

Me_2TiCl_2, CH_2Cl_2, 91% (14)

When a combination of Me_2Zn and $TiCl_4$ is used in this reaction, the geminal dimethylation products are maximized when at least 2 equiv of each is utilized.[5] Otherwise, the corresponding tertiary alcohol or tertiary chloride are obtained as major products (eq 15).[5]

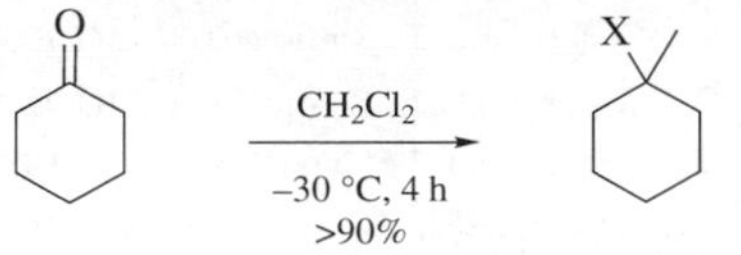

CH_2Cl_2, −30 °C, 4 h, >90% (15)

$TiCl_4$ (equiv)	Me_2Zn (equiv)	X
1	1	OH
3	1	Cl
2	2	Me

The reaction works well with a variety of aliphatic and aromatic ketones, with the exception of α,β-unsaturated derivatives, which give a mixture of products. It is even possible to generate adjacent quaternary carbons in this manner (eq 16).[5]

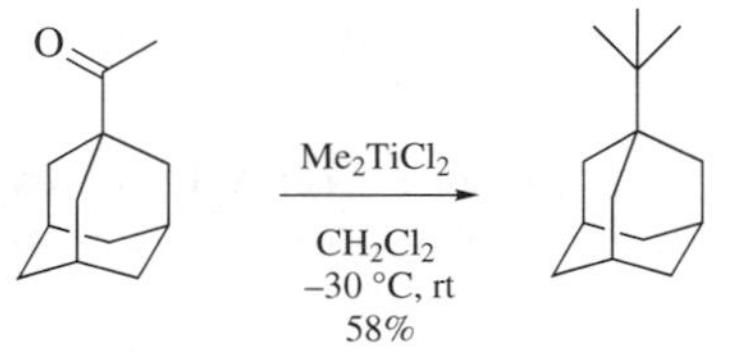

Me_2TiCl_2, CH_2Cl_2, −30 °C, rt, 58% (16)

Although aliphatic aldehydes give alcohols under these reaction conditions, aromatic aldehydes undergo geminal dimethylation to form isopropyl derivatives (eq 17).[17]

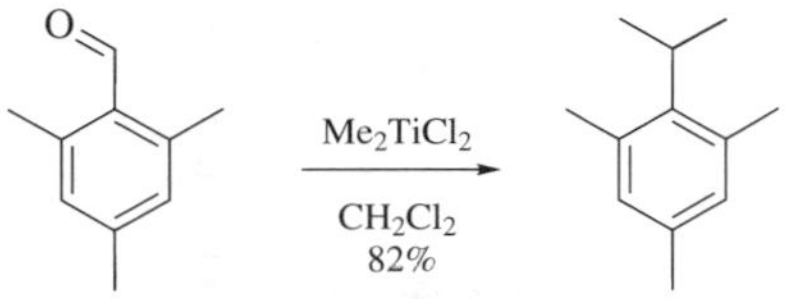

Me_2TiCl_2, CH_2Cl_2, 82% (17)

Similarly, acid chlorides undergo a direct geminal trimethylation (eq 18).[5]

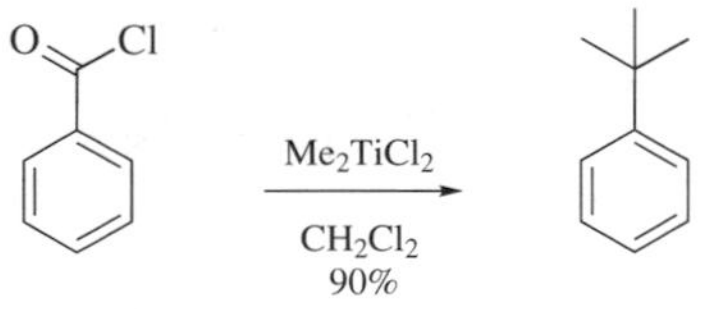

Me_2TiCl_2, CH_2Cl_2, 90% (18)

Related Reagents. Cadmium Chloride; Dilithium Trimethylcuprate; Methylmagnesium Bromide; Methyllithium; Methyltitanium Triisopropoxide; Titanium(IV) Chloride; Trimethylaluminum.

---

1. (a) Wailes, P. C.; Coutts, R. S. P.; Weigold, H. *Organometallic Chemistry of Titanium, Zirconium and Hafnium*; Academic: New York, 1974. (b) Reetz, M. T. *Top. Curr. Chem.* **1982**, *106*, 1. (c) Bottrill, M.; Gavens, P. D.; Kelland, J. W.; McMeeking, J. In *Comprehensive Organometallic Chemistry*; Wilkinson, G., Ed.; Pergamon: Oxford, 1982; Vol. 3, p 433. (d) Seebach, D.; Weidmann, B.; Widler, L. In *Transition Metals in Organic Synthesis*; Schaffold, R., Ed.; Wiley: New York, **1983**, Vol. 3, pp 217–353. (e) Weidmann, B.; Seebach, D. *AG(E)* **1983**, *22*, 31. (f) Reetz, M. T. *Organotitanium Reagents in Organic Synthesis*; Springer: Berlin, 1986. (g) Ferreri, C.; Palumbo, G.; Caputo, R. *COS* **1991**, *1*, 139. (h) Reetz, M. T. *ACR* **1993**, *26*, 462.
2. Reetz, M. T.; Kyung, S. H.; Hüllmann, M. *T* **1986**, *42*, 2931.

3. Reetz, M. T.; Westermann, J.; Steinbach, R. *AG(E)* **1980**, *19*, 901.
4. Reetz, M. T.; Westermann, J.; Steinbach, R. *AG(E)* **1980**, *19*, 900.
5. Reetz, M. T.; Westermann, J.; Kyung, S.-H. *CB* **1985**, *118*, 1050.
6. Clark, R. J. H.; Coles, M. A. *Inorg. Synth.* **1976**, *16*, 120.
7. Kauffmann, T.; Abel, T.; Schreer, M. *AG(E)* **1988**, *27*, 944.
8. Reetz, M. T.; Steinbach, R.; Westermann, J.; Peter, R. *AG(E)* **1980**, *19*, 1011.
9. Reetz, M. T. *AG(E)* **1984**, *23*, 556.
10. Reetz, M. T.; Kesseler, K.; Schmidtberger, S.; Wenderoth, B.; Steinbach, R. *AG(E)* **1983**, *22*, 989.
11. Reetz, M. T.; Raguse, B.; Seitz, T. *T* **1993**, *49*, 8561.
12. Reetz, M. T.; Jung, A. *JACS* **1983**, *105*, 4833.
13. Baldwin, S. W.; McIver, J. M. *TL* **1991**, *32*, 1937.
14. Fujisawa, T.; Fujimura, A.; Ukaji, Y. *CL* **1988**, 1541.
15. Mori, A.; Maruoka, K.; Yamamoto, H. *TL* **1984**, *25*, 4421.
16. Reetz, M. T.; Westermann, J.; Steinbach, R. *CC* **1981**, 237.
17. Reetz, M. T.; Kyung, S.-H. *CB* **1987**, *120*, 123.

Nicos A. Petasis & Irini Akritopoulou-Zanze
University of Southern California, Los Angeles, CA, USA

α-Methyltoluene-2,α-sultam[1]

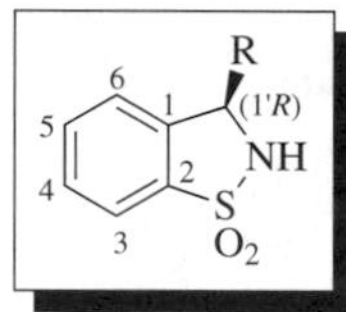

(1′*S*)-(**1a**; R = Me)
[130973-57-8] $C_8H_9NO_2S$ (MW 183.25)
(1′*R*)-(**1a**)
[130973-53-4]
(1′*S*)-(**1b**; R = *t*-Bu)
[137694-01-0] $C_{11}H_{15}NO_2S$ (MW 225.34)
(1′*R*)-(**1b**)
[137694-00-9]

(chiral auxiliary: *N*-enoyl derivatives undergo highly stereoselective Diels–Alder reactions with cyclopentadiene[2] and 1,3-dipolar cycloadditions with nitrile oxides;[3] enolates of *N*-acyl derivatives participate in highly stereoselective alkylations, acylations, and aldolizations[4])

Physical Data: (**1a**) mp 92 °C. (1′*S*)-(**1a**) $[\alpha]_D^{20}$ −30.0° (*c* 1.21, $CHCl_3$). (1′*R*)-(**1a**) $[\alpha]_D^{20}$ +31.0° (*c* 0.6, EtOH). (**1b**) mp 129–130 °C. (1′*S*)-(**1b**) $[\alpha]_D^{20}$ −53.9° (*c* 1.00, $CHCl_3$). (**1b**) has been incorrectly assigned.[3]

Preparative Methods: both enantiomers of the α-methyl sultam may be prepared on a multigram scale in optically pure form by asymmetric hydrogenation of imine (**2a**) followed by simple crystallization (eq 1).[5] The (*R*)-enantiomer of the α-*t*-butyl sultam may also be prepared in enantiomerically pure form by asymmetric reduction of imine (**2b**) followed by fractional crystallization.[3] However, multigram quantities of either enantiomer of the α-*t*-butyl sultam may be prepared by derivatization of the racemic auxiliary (obtained in 98% yield from reaction of (**2b**) with ***Sodium Borohydride*** in MeOH) with ***10-Camphorsulfonyl Chloride***[1], separation of the resulting diastereomers by fractional crystallization, and acidolysis.[3] Prochiral imines (**2a**) and (**2b**) are readily prepared from inexpensive Saccharine by treatment with ***Methyllithium*** (73%) and ***t-Butyllithium*** (66%), respectively.

Ru_2Cl_4[(+)-(*R*)-BINAP]$_2$, Et_3N
H_2 (4 atm)
CH_2Cl_2, EtOH
72% (cryst) [>99% ee (crude)]
Ru_2Cl_4[(−)-(*S*)-BINAP]$_2$, Et_3N (1)
(**2a**) R = Me
(**2b**) R = *t*-Bu
71% (cryst) [>99% ee (crude)]

Handling, Storage, and Precautions: these auxiliaries are white crystalline solids which are stable indefinitely at ambient temperature in sealed containers.

Introduction. The toluene-2,α-sultams are recently introduced relatives of the well established ***10,2-Camphorsultam*** chiral auxiliary and have been designed to provide similar high levels of face discrimination in reactions of pendent prochiral functionality. Features that distinguish them include high crystallinity and facile NMR and HPLC analysis of derivatives, favorable acylation and aldolization characteristics of derived *N*-acyl enolates, and improved cleavage characteristics.

Preparation of Derivatives. *N*-Enoyl[2,3] and *N*-acyl[4] sultam derivatives are readily prepared using either ***Sodium Hydride***–acid chloride or ***Triethylamine***–acid chloride single-step protocols. Various alternative derivatization procedures that work for the 10,2-camphorsultam auxiliary would also be expected to be effective.

Reactions of *N*-Enoyl and *N*-Acyl Derivatives.

[4 + 2] Diels–Alder Cycloadditions (Alkene → Six-Membered Cycloadduct)[2]. *N*-Acryloyl-α-methyltoluene-2,α-sultam (**3a**) participates in highly *endo* and C(α)-*re* π-face selective Lewis acid promoted Diels–Alder reactions with ***Cyclopentadiene***, ***1,3-Butadiene***, and ***Isoprene*** (eq 2 and Table 1). These levels of induction compare favorably with most alternative auxiliaries, including the 10,2-camphorsultam. However, *N*-crotonyl-α-methyltoluene-2,α-sultam (*ent*-**3b**) reacts with cyclopentadiene with only moderate π-face selectivity (cf. 93% de with 10,2-camphorsultam). Unusually high *endo* selectivity is observed for the non-Lewis acid-catalyzed reaction of sultam (**3a**) with cyclopentadiene, but again the π-face selectivity is only moderate. The corresponding reactions of both α-*t*-butyl- and α-benzyltoluene-2,α-sultams are less selective.

(3a) R^1 = H
(3b) R^1 = Me

Lewis acid CH_2Cl_2

(4) (5) (2)

1,3-Dipolar Cycloadditions with Nitrile Oxides (Alkene → Isoxazoline)[3]. 1,3-Dipolar cycloaddition reactions of *N*-acryloyl-α-*t*-butyltoluene-2,α-sultam (**6**) with various nitrile oxides give isoxazolines with extremely high C(α)-*re* π-facial control (eq 3). The levels of selectivity exceed those obtainable with the 10,2-camphorsultam auxiliary and are comparable to the highest levels reported for such cycloadditions.[6] The corresponding reactions of α-methyltoluene-2,α-sultams are less selective.

RCH=NOH
NCS, Et_3N, pyridine
CH_2Cl_2, toluene, rt
or
hexane, rt

(6) (7) (3)

R = Me*, Et*, *t*-Bu, Ph, CH_2O-*t*-Bu
77–88% (cryst) [90–96% de (crude)]
*using *ent*-(**6**) → *ent*-(**7**)

Acylation, Alkylation, and Aldolization (Acyl Species → α-, β-, or α/β-Functionalized Acyl Product).[3] Alkylation reactions of sodium enolates of various *N*-acyl-α-methyltoluene-2,α-sultams with selected (both "activated" and "nonactivated") alkyl iodides and bromides proceed with good C(α)-*re* stereocontrol (90–99% de). Analogous acylations with various acid chlorides can also be performed, giving β-keto products (97–99% de). Selective reduction of these latter products with ***Zinc Borohydride*** (chelate controlled, 82.6–98.2% de) or N-Selectride (nonchelate controlled, 95.8–99.6% de) can provide *syn*- and *anti*-aldol derivatives, respectively.[3]

Syn-aldol derivatives may also be obtained directly from boryl enolates of the same *N*-acyl-α-methyltoluene-2,α-sultams by condensation with aliphatic and aromatic aldehydes (eq 4).[3,7] The high C(α)-*si* topicity of these reactions parallels but exceeds that when using the 10,2-camphorsultam auxiliary and is the result of an analogous transition state.[3] It is noteworthy, however, that aldolizations of α-methyltoluene-2,α-sultam derivatives generally proceed to completion with just a small excess of aldehyde (1–1.2 equiv, cf. 2–3 equiv when 10,2-camphorsultam mediated). This may be ascribed to the lack of acidic protons α to the SO_2 group in the Saccharine-derived auxiliary.

1. Et_2BOTf, *i*-Pr_2NEt, CH_2Cl_2, –5 °C
2. RCHO, –78 °C
3. aq phosphate buffer

(4)

R = Me, *i*-Pr, *i*-Bu, Ph
71–84% (cryst); [R = *i*-Pr, 95% (FC)]
(>99% of crude)

Nondestructive Auxiliary Cleavage. The toluene-2,α-sultam auxiliaries are even more readily cleaved from derivatives than the 10,2-camphorsultam auxiliary. Following *N*-acyl bond cleavage, simple extraction and crystallization usually effect almost quantitative recovery of enantiomerically pure auxiliary which may be re-used if desired.

Enantiomerically pure carboxylic acids are routinely obtained from *N*-acylsultams by ***Hydrogen Peroxide*** assisted saponification with ***Lithium Hydroxide*** in aqueous THF.[2,4] Alternatively, transesterification can be effected under 'neutral' conditions in allyl alcohol containing ***Titanium Tetraisopropoxide***, giving the corresponding allyl esters which can be isomerized/hydrolyzed with Wilkinson's catalyst (***Chlorotris(triphenylphosphine)rhodium(I)***) in EtOH–H_2O. This provides a convenient route to carboxylic acids containing base-sensitive functionality.[8] Primary alcohols are obtained by treatment with L-Selectride (***Lithium Tri-s-butylborohydride***) in THF at ambient temperature.[3]

The α-methyltoluene-2,α-sultam auxiliary is also displaced by a variety of dilithiated alkyl phenyl sulfones.[7,9] This unique procedure provides direct access to synthetically useful β-oxo sulfones which may be further functionalized or simply subjected to reductive desulfonation to give alkyl ketones. A particularly striking use of this method is the preparation of β-oxo sulfone (**8**), a key intermediate in a concise synthesis of (−)-probably should be semicorrole (eq 5).[7] Remarkably, the $MeCLi_2SO_2Ph$ reagent attacks selectively the C(4)-imide C=O group in preference to the C(6)-ester C=O group and no epimerization occurs at C(3) or C(1′).

TBSO

$EtSO_2Ph$
BuLi (2 equiv)
THF, –78 °C

[(−)-semicorole numbering]

(**8**) 72% (FC) 89% (FC) (5)

Related Reagents. (*S*)-4-Benzyl-2-oxazolidinone; 10,2-Camphorsultam; 10-Dicyclohexylsulfonamidoisoborneol; (*R*,*R*)-

Table 1 Intermolecular Diels–Alder Reactions of *N*-Enoyl Sultams (**3a**) or (**3b**) → (**4**) and (**3a**) → (**5**) (eq 2)

Dienophile	Diene	Lewis acid[a]	Temp. (°C)	Time (h)	Adduct	Yield crude (cryst.) (%)	de crude (cryst.) (%)
(**3a**)	Cyclopentadiene	None	25		(**4**) (R^1 = H)	95[b]	62
(**3a**)	Cyclopentadiene	Me_2AlCl	−98	0.2	(**4**) (R^1 = H)	97[c] (83)	93 (>99)
(**3a**)	1,3-Butadiene	$EtAlCl_2$	−78	18	(**5**) (R^2 = H)	79	90
(**3a**)	Isoprene	Me_2AlCl	−78	7	(**5**) (R^2 = Me)	87	92
ent-(**3b**)	Cyclopentadiene	Me_2AlCl	−78	24	*ent*-(**4**) (R^1 = Me)	74[d] (58)	59 (>99)

[a] 1.6–2.0 equiv. [b] 96% *endo*. [c] >99% *endo*. [d] 97% *endo*.

1,2-Diphenyl-1,2-di-aminoethane- *N,N′*-bis[3,5-bis(trifluoromethyl)benzenesulfonamide]; (*S*)-2-Methoxymethylpyrrolidine; (4*S*,5*S*)-4-Methoxymethyl-2-methyl-5-phenyl-2-oxazoline; 1,1,2-Triphenyl-1,2-ethanediol.

---

1. Ganem, B. *Chemtracts–Org. Chem.* **1990**, 435.
2. (a) Oppolzer, W.; Wills, M.; Kelly, M. J.; Signer, M.; Blagg, J. *TL* **1990**, *31*, 5015. (b) Oppolzer, W.; Seletsky, B. M.; Bernardinelli, G. *TL* **1994**, *35*, 3509.
3. Oppolzer, W.; Kingma, A. J.; Pillai, S. K. *TL* **1991**, *32*, 4893.
4. Oppolzer, W.; Rodriguez, I.; Starkemann, C.; Walther, E. *TL* **1990**, *31*, 5019.
5. Oppolzer, W.; Wills, M.; Starkemann, C.; Bernardinelli, G. *TL* **1990**, *31*, 4117.
6. Curran, D. P.; Jeong, K. S.; Heffner, T. A.; Rebek, J., Jr. *JACS* **1989**, *111*, 9238.
7. Oppolzer, W.; Rodriguez, I. *HCA* **1993**, *76*, 1275.
8. Oppolzer, W. Lienard, P. *HCA* **1992**, *75*, 2572.
9. Oppolzer, W.; Rodriguez, I. *HCA* **1993**, *76*, 1282.

Alan C. Spivey
University of Cambridge, UK

1-Methyl-1-(trimethylsilyl)allene[1]

SiR_3 / R^2 / R^1

(**1a**; $R_3 = Me_3$, $R^1 = Me$, $R^2 = H$)[4b,9]
[74542-82-8] $C_7H_{14}Si$ (MW 126.30)
(**1b**; $R_3 = t\text{-BuMe}_2$, $R^1 = Me$, $R^2 = H$)[7]
[99035-24-2] $C_{10}H_{20}Si$ (MW 168.39)
(**1c**; $R_3 = (i\text{-Pr})_3$, $R^1 = Me$, $R^2 = H$)[7]
[120789-53-9] $C_{13}H_{26}Si$ (MW 210.48)
(**1d**; $R_3 = Me_3$, $R^1 = H$, $R^2 = Me$)[4b]
[74542-82-8] $C_7H_{14}Si$ (MW 126.30)

(propargylic anion equivalents;[2,3] three-carbon synthons for [3 + 2] annulations leading to five-membered compounds including cyclopentenes,[4] dihydrofurans,[5] pyrrolines,[5] isoxazoles,[6] furans,[7] and azulenes[8])

Physical Data: (**1a**) bp 111 °C; bp 54–56 °C/90 mmHg; (**1b**) bp 62–65 °C/30 mmHg; (**1d**) bp 107–110 °C.

Solubility: sol CH_2Cl_2, benzene, THF, Et_2O, and most organic solvents.

Form Supplied in: colorless liquid; not commercially available.

Analysis of Reagent Purity: (**1a**) IR (neat) 2955, 2910, 2900, 2860, 1935, 1440, 1400, 1250, 935, 880, 830, 805, 750, and 685 cm^{-1}; ^{1}H NMR (250 MHz, $CDCl_3$) δ 0.08 (s, 9H), 1.67 (t, 3H, $J = 3.3$), and 4.25 (q, 2H, $J = 3.3$); ^{13}C NMR (67.9 MHz, $CDCl_3$) δ −2.1, 15.1, 67.3, 89.1, and 209.1.[4b,9]

Preparative Methods: 1-methyl-1-(trialkylsilyl)allenes can be conveniently prepared by the method of Vermeer.[9,10] Silyl-substituted propargyl mesylates thus undergo S_N2' displacement by the organocopper reagent generated from methylmagnesium chloride, ***Copper(I) Bromide***, and ***Lithium Bromide***. 1-Methyl-1-(trimethylsilyl)allene is produced in 52% yield from commercially available (trimethylsilyl)propargyl alcohol in this fashion (eq 1).[4b,9] The *t*-butyldimethylsilyl and triisopropylsilyl analogs are synthesized by the same method in 90% and 58% yield, respectively.[7] Propargyl alcohols bearing these and other trialkylsilyl groups can be prepared by treatment of ***Propargyl Alcohol*** with ***n-Butyllithium*** and the appropriate trialkylsilyl chloride.[7] Allenylsilanes bearing other C-1 substituents can be prepared in an analogous manner by using the appropriate Grignard reagents.

HO—≡—TMS → [MeMgCl, THF; then $MeSO_2Cl$; −78 °C to rt, 2.5 h] → [MeMgCl, CuBr, LiBr, THF; −78 °C to rt, 2.5 h] → (**1a**) (TMS) (1)

1-Methyl-1-(trialkylsilyl)allenes can be alkylated at C-3 with a variety of alkyl halides by treatment of the allenylsilane with *n*-butyllithium and the desired alkyl halide (e.g. eq 2).[7] In addition to ethyl bromide, alkylating agents such as *n*-heptyl bromide,[6] 1,2-dibromobutane,[7] and 5-bromo-1-pentene[8] have been employed in this reaction.

(**1b**) (TBDMS) → [1.05 equiv BuLi, 0.99 equiv EtBr; THF, −78 °C; 84%] → Et, TBDMS (2)

3-Methyl-1-(trimethylsilyl)allene (**1d**) is prepared by the direct silylation of the lithium derivative of allene.[4b] Sequential treatment of a THF solution of allene with 1.0 equiv of ***Lithium 2,2,6,6-Tetramethylpiperidide*** (−78 °C, 3 h) and 1.05 equiv of

Chlorotrimethylsilane (−78 °C to 25 °C, 12 h) affords (**1d**) in 41% yield after distillation (eq 3).

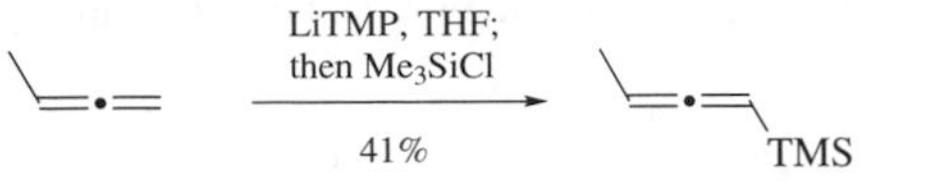

LiTMP, THF; then Me_3SiCl, 41% → TMS (**1d**) (3)

Purification: allenes (**1a**), (**1b**), and (**1c**) are purified by distillation at reduced pressure or by column chromatography. Allene (**1d**) is distilled at atmospheric pressure. The allenylsilanes obtained by the Vermeer method typically contain up to 7–8% of the trialkylsilyl-1-butyne isomer produced by S_N2 reaction. This mixture can be used directly in most subsequent reactions without further purification. If desired, however, the alkynyl contaminant can be selectively removed by treatment of the mixture with ***Silver(I) Nitrate*** in (10:1) methanol–water at room temperature for one hour.[9] 1-Methyl-1-(trimethylsilyl)allene is obtained in 79% yield after pentane extraction and distillation.

Handling, Storage, and Precautions: 1-methyl-1-(trimethylsilyl)allene is stable indefinitely when stored under nitrogen in the refrigerator.

Propargylic Anion Equivalents. Due to the silicon β-effect, allenylsilanes react with electrophiles at the 3-position in a fashion analogous to the behavior of allyl- and propargylsilanes.[11] Allenylsilanes can thus function as propargylic anion equivalents. Particularly important is the reaction of (trimethylsilyl)allenes with aldehydes and ketones to provide a regiocontrolled route to homopropargylic alcohols of a variety of substitution types. Allenylsilanes substituted at the 1-position undergo the addition to carbonyl compounds in the presence of ***Titanium(IV) Chloride*** to afford homopropargylic alcohols directly (eq 4).[2]

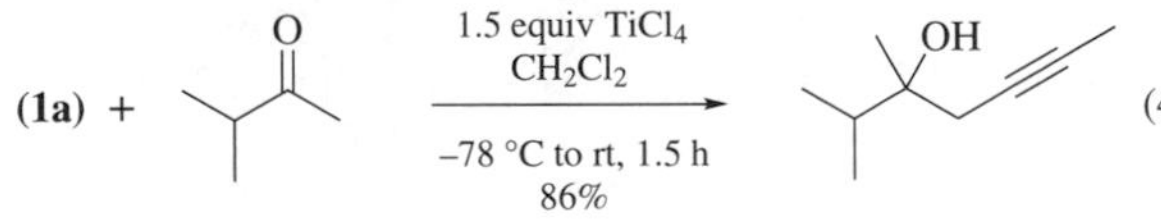

(**1a**) + [ketone] → 1.5 equiv $TiCl_4$, CH_2Cl_2; −78 °C to rt, 1.5 h; 86% → OH (4)

In contrast, allenylsilanes lacking a substituent at C-1 react with carbonyl compounds to produce a mixture of the desired homopropargylic alcohols and (trimethylsilyl)vinyl chlorides. This initial product can be converted to the desired alkyne using the method of Cunico and Dexheimer:[12] Exposure of the crude mixture of allenylsilane adducts to 2.5 equiv of ***Potassium Fluoride*** in DMSO furnishes the homopropargylic alcohols in good yield (eq 5).[2]

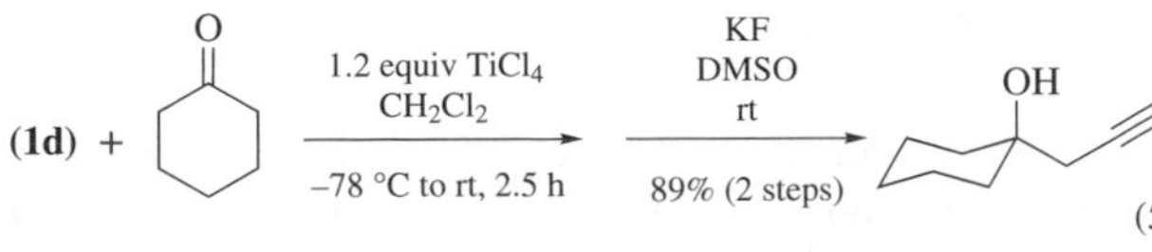

(**1d**) + [cyclohexanone] → 1.2 equiv $TiCl_4$, CH_2Cl_2; −78 °C to rt, 2.5 h → KF, DMSO, rt; 89% (2 steps) → OH (5)

A number of methods have been reported for the preparation of homopropargylic alcohols.[13] Alcohols having the substitution pattern represented in structure (**3**), for example, can be prepared using 3-alkyl-substituted allenyltitanium,[14] -alanate,[15] and -zinc compounds.[16] Unfortunately, these methods are not applicable to the synthesis of type (**4**) products. Zweifel has shown that allenyldialkylboranes (generated via the reaction of lithium chloropropargylide with trialkylboranes) combine with aldehydes (but not ketones) to produce type (**4**) homopropargylic alcohols.[17] The preparation of type (**2**) homopropargylic alcohols is discussed in the article on ***(Trimethylsilyl)allene***.

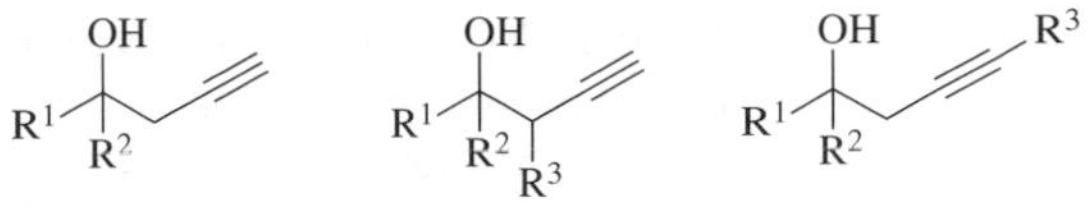

OH, R^1, R^2 (**2**) OH, R^1, R^2, R^3 (**3**) OH, R^1, R^2, R^3 (**4**)

Santelli has demonstrated that allenylsilanes without a C-1 substituent undergo conjugate addition to α,β-unsaturated acyl cyanides to give δ,ε-alkynic acyl cyanides.[3]

Three-Carbon Synthon for [3 + 2] Annulations. Danheiser and co-workers have exploited allenylsilanes as the three-carbon components in a [3 + 2] annulation strategy for the synthesis of a variety of five-membered carbocycles and heterocycles. The pathway by which a typical annulation proceeds is shown in eq 6. Reaction of the 2-carbon component (the 'allenophile') at C-3 of the allenylsilane is followed by rapid rearrangement of the silicon-stabilized vinyl cation. Ring closure then affords the five-membered product.

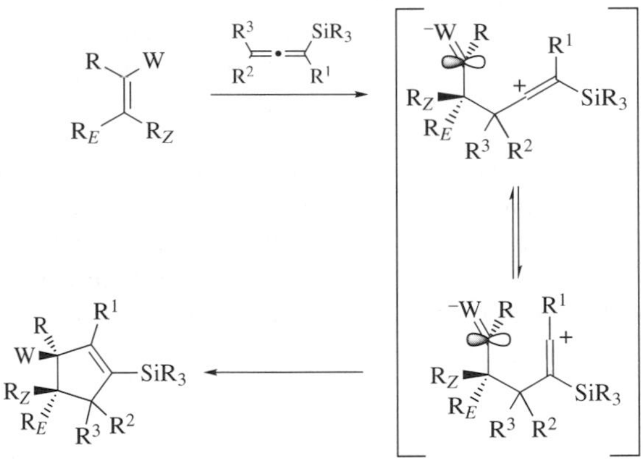

R, W, R_E, R_Z + R^3, SiR_3, R^2, R^1 → [$^-$W, R, R^1, +, SiR_3, R_Z, R_E, R^3, R^2 ⇌ $^-$W, R, R^1, +, SiR_3, R_Z, R_E, R^3, R^2] → R, R^1, W, SiR_3, R_Z, R_E, R^3, R^2 (6)

Synthesis of Five-Membered Carbocycles. 1-Substituted allenylsilanes react with α,β-unsaturated carbonyl compounds in the presence of titanium tetrachloride to produce cyclopentenes.[4] For example, carvone and 1-methyl-1-(trimethylsilyl)allene react smoothly to give a *cis*-fused hydrinenone (eq 7).[4d]

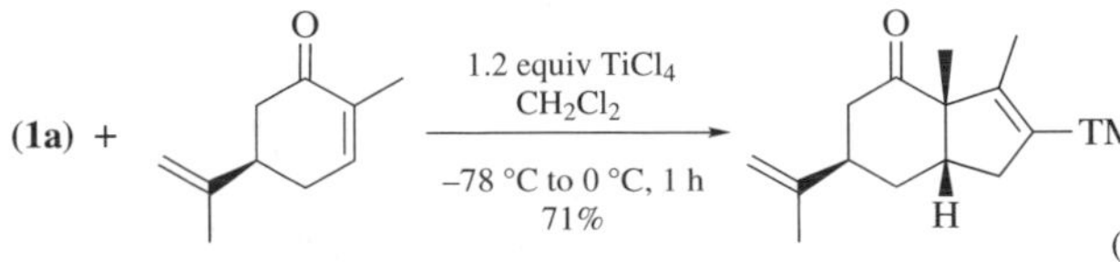

(**1a**) + [carvone] → 1.2 equiv $TiCl_4$, CH_2Cl_2; −78 °C to 0 °C, 1 h; 71% → O, TMS, H (7)

As illustrated above, the reaction proceeds with a strong preference for suprafacial addition of the allene to the allenophile, thus permitting the stereocontrolled synthesis of a variety of mono- and polycyclic systems. Both cyclic and acyclic enones participate in the reaction. Spiro-fused products are obtained from α-alkylidene ketone substrates (eq 8).[4b]

(1a) + [2-isopropylidenecyclohexanone] —1.5 equiv $TiCl_4$, CH_2Cl_2, rt, 1 h, 86%→ (8)

When α,β-unsaturated acyl silanes are employed, the type of product formed varies depending on the trialkylsilyl substituent of the acyl silane: five-membered carbocycles are produced from reaction with *t*-butyldimethylsilyl derivatives, whereas six-membered carbocycles are obtained from trimethylsilyl compounds (eq 9).[4c]

(1a) + R_3SiC(O)C(Me)=CH_2: R_3Si = *t*-$BuMe_2Si$ → TBDMS/TMS cyclopentene; R_3Si = Me_3Si → TMS cyclohexenone (9)

Allenylsilanes lacking a C-1 alkyl substituent do not function efficiently as three-carbon synthons in the [3 + 2] annulation. This phenomenon is attributable to the relative instability of the terminal vinyl cation intermediate required according to the proposed mechanism for the annulations (eq 6). Fully substituted five-membered rings result from annulations employing allenylsilanes substituted at both C-1 and C-3.[4]

Synthesis of 1,3-Dihydrofurans. (*t*-Butyldimethylsilyl)allenes combine with aldehydes to produce dihydrofurans (eq 10).[5] In a typical reaction, the aldehyde and 1.1 equiv of titanium tetrachloride are premixed at −78 °C in methylene chloride for 10 min. The allenylsilane (1.2 equiv) is then added, and the reaction mixture is stirred in the cold for 15–45 min.

(1b) + cyclohexyl-CHO —$TiCl_4$, CH_2Cl_2, 76%→ dihydrofuran-TBDMS (10)

In reactions of a C-3 substituted allenylsilane with achiral aldehydes, *cis*-substituted dihydrofurans are the predominant products (eq 11).[5]

allenyl-TBDMS —*t*-BuCHO, $TiCl_4$, CH_2Cl_2, 76%→ *t*-Bu dihydrofuran-TBDMS (11)

(Trimethylsilyl)allenes are unsuitable for this [3 + 2] annulation, as the intermediate carbocations undergo chloride-initiated desilylation to produce alkynic byproducts. This unwelcome reaction pathway is suppressed when the bulkier (*t*-butyldimethylsilyl)allenes are employed.

Synthesis of Pyrrolizinones. Cyclic *N*-acyl imine derivatives combine with (*t*-butyldimethylsilyl)allenes to afford nitrogen heterocycles (eq 12).[5] The *N*-acyliminium ions are generated from ethoxypyrrolidinones in the presence of titanium tetrachloride.

(1b) + [NH, OEt lactam] —$TiCl_4$, CH_2Cl_2, 67%→ [N-bicyclic, H]—TBDMS (12)

Synthesis of Furans and Isoxazoles. Electrophilic species of the general form Y≡X$^+$ serve as 'heteroallenophiles', combining with allenylsilanes in a regiocontrolled [3 + 2] annulation method. As illustrated in the mechanism shown in eq 13, addition of the heteroallenophile at C-3 of the allenylsilane produces a vinyl cation stabilized by hyperconjugative interaction with the adjacent carbon–silicon σ-bond. A 1,2-trialkylsilyl shift then occurs to generate an isomeric vinyl cation, which is intercepted by nucleophilic X. Elimination of H$^+$ furnishes the aromatic heterocycle.

$X^+\equiv Y$ + R^2-allene(SiR_3)(R^1) → ... → heterocycle with R^1, SiR_3, R^2 (13)

Isoxazoles are obtained when the heteroallenophile is nitrosonium ion. Thus, reaction of commercially available ***Nitrosonium Tetrafluoroborate*** with allenylsilanes in acetonitrile at −30 °C affords silyl-substituted isoxazoles in good yield (eq 14).[6]

(1b) —$NOBF_4$, MeCN, 87%→ TBDMS isoxazole (14)

In a variation of the above procedure, (trimethylsilyl)allenes are employed in a one-pot procedure that produces 5-substituted and 3,5-disubstituted isoxazoles lacking the 4-silyl substituent (eq 15). Desilylation is encouraged by addition of water and warming the reaction mixture to 65–70 °C after the initial annulation.[6] Alternatively, addition of electrophilic reagents to the reaction mixture leads to isoxazoles with C-4 substituents such as Br, COMe, etc.

TMS/Cy allene —$NOBF_4$, MeCN; then H_2O, Δ, 63%→ Cy isoxazole (15)

The heteroaromatic strategy is extended to the synthesis of furans when acylium ions are employed as the heteroallenophile (eq 16).[7] Acylium ions are generated in situ via the reaction of acyl chlorides and ***Aluminum Chloride***. Typically, the allenylsilane is added to a solution of 1.0 equiv each of $AlCl_3$ and the acyl chloride in methylene chloride at −20 °C. The reaction is complete in 1 h at −20 °C.

TIPS (1c) → PhCH$_2$COCl, AlCl$_3$, CH$_2$Cl$_2$, 60% → Ph, TIPS furan (16)

Intramolecular [3 + 2] annulation affords bicyclic furans (eq 17).[7]

TBDMS allene acid chloride → AlCl$_3$, CH$_2$Cl$_2$, 76% → TBDMS bicyclic furan (17)

(t-Butyldimethylsilyl)- and (triisopropylsilyl)allenes are superior to their trimethylsilyl counterparts for this annulation, presumably due to the ability of the larger trialkylsilyl groups to suppress undesirable desilylation reactions. In annulations involving allenylsilanes which lack C-3 substituents, the bulkier triisopropylsilyl derivatives are superior to t-butyldimethylsilyl analogs.

Synthesis of Azulenes. Reaction of tropylium cations with allenylsilanes produces substituted azulenes.[8] Typically, commercially available ***Tropylium Tetrafluoroborate*** (2 equiv) is employed. The second equivalent dehydrogenates the dihydroazulene intermediate to produce the aromatic product. ***Poly(4-vinylpyridine)*** (poly (4-VP)) or methyltrimethoxysilane is used to scavenge the HBF$_4$ produced in the reaction.

TBDMS allene → 2 equiv TpBF$_4$, poly(4-VP), MeCN, rt, 25 h, 52–55% → azulene–TBDMS (18)

The azulene synthesis proceeds best with 1,3-dialkyl (t-butyldimethylsilyl)allenes. (Trimethylsilyl)allenes desilylate to generate propargyl-substituted cycloheptatrienes as significant byproducts. As observed in the other [3 + 2] annulations discussed already, allenylsilanes lacking C-1 alkyl substituents do not participate in the reaction.

Synthesis of Silylalkynes via Ene Reactions. (Trimethylsilyl)allenes undergo ene reactions with ***4-Phenyl-1,2,4-triazoline-3,5-dione*** and other reactive enophiles to give silylalkenes.[18]

Related Reagents. Allene; Allenylboronic Acid; Allenyllithium; Allyltrimethylsilane; Propargylmagnesium Bromide; (Trimethylsilyl)allene.

1. Review: Panek, J. S. *COS* **1991**, *2*, 579.
2. (a) Danheiser, R. L.; Carini, D. J. *JOC* **1980**, *45*, 3925. (b) Danheiser, R. L.; Carini, D. J.; Kwasigroch, C. A. *JOC* **1986**, *51*, 3870.
3. (a) Jellal, A.; Santelli, M. *TL* **1980**, *21*, 4487. (b) Santelli, M.; Abed, D. E.; Jellal, A. *JOC* **1986**, *51*, 1199.
4. (a) Danheiser, R. L.; Carini, D. J.; Basak, A. *JACS* **1981**, *103*, 1604. (b) Danheiser, R. L.; Carini, D. J.; Fink, D. M.; Basak, A. *T* **1983**, *39*, 935. (c) Danheiser, R. L.; Fink, D. M. *TL* **1985**, *26*, 2513. (d) Danheiser, R. L.; Fink, D. M.; Tsai, Y.-M. *OS* **1988**, *66*, 8.
5. Danheiser, R. L.; Kwasigroch, C. A.; Tsai, Y.-M. *JACS* **1985**, *107*, 7233.
6. Danheiser, R. L.; Becker, D. A. *H* **1987**, *25*, 277.
7. Danheiser, R. L.; Stoner, E. J.; Koyama, H.; Yamashita, D. S.; Klade, C. A. *JACS* **1989**, *111*, 4407.
8. Becker, D. A.; Danheiser, R. L. *JACS* **1989**, *111*, 329.
9. Danheiser, R. L.; Tsai, Y.-M.; Fink, D. M. *OS* **1988**, *66*, 1.
10. Westmijze, H.; Vermeer, P. *S* **1979**, 390.
11. For reviews on allylsilanes and related systems, see Ref. 1 and: Fleming, I.; Dunogues, J.; Smithers, R. *OR* **1989**, *37*, 57; Fleming, I. *COS* **1991**, *2*, 563.
12. Cunico, R. F.; Dexheimer, E. M. *JACS* **1972**, *94*, 2868.
13. For reviews of the chemistry of propargylic anion equivalents, see: (a) Yamamoto, H. *COS* **1991**, *2*, 81. (b) Epsztein, R. In *Comprehensive Carbanion Chemistry*; Buncel, E.; Durst, T., Eds.; Elsevier: Amsterdam, 1984; Part B, pp 107–176. (c) Moreau, J.-L. In *The Chemistry of Ketenes, Allenes, and Related Compounds*; Patai, S., Ed.; Wiley: New York, 1978, pp 343–381.
14. (a) Furuta, K.; Ishiguro, M.; Haruta, R.; Ikeda, N.; Yamamoto, H. *BCJ* **1984**, *57*, 2768. (b) Ishiguro, M.; Ikeda, N.; Yamamoto, H. *JOC* **1982**, *47*, 2225.
15. Hahn, G.; Zweifel, G. *S* **1983**, 883.
16. Zweifel, G.; Hahn, G. *JOC* **1984**, *49*, 4565.
17. Zweifel, G.; Backlund, S. J.; Leung, T. *JACS* **1978**, *100*, 5561.
18. Laporterie, A.; Dubac, J.; Manuel, G.; Deleris, G.; Kowalski, J.; Dunogues, J.; Calas, R. *T* **1978**, *34*, 2669.

Katherine L. Lee & Rick L. Danheiser
Massachusetts Institute of Technology, Cambridge, MA, USA

Methyl Vinyl Ketone[1]

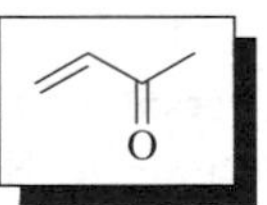

[78-94-4] C_4H_6O (MW 70.10)
[59120-04-6]

(reagent for attachment of 3-oxobutyl side chains, i.e. Michael additions;[2] reagent for the annulation of cyclohexenones[3])

Alternate Names: MVK; 3-buten-2-one.

Physical Data: fp −6 °C; bp 81.4 °C, 36.5–36.8 °C/145 mmHg, 33–34 °C/130 mmHg, 32–34 °C/60 mmHg; d (20 °C) 0.8636 g cm^{-3}, (25 °C) 0.8407 g cm^{-3}; n_D^{20} 1.4086; liquid with pungent odor.

Solubility: sol water, methanol, ethanol, ether, acetone, glacial acetic acid; slightly sol hydrocarbons; forms binary azeotrope with water, bp 75 °C (12% water).

Form Supplied in: clear liquid stabilized with 0.1% acetic acid and 0.05% or 1% hydroquinone.

Handling, Storage, and Precautions: should be kept cold. Polymerizes upon standing in the pure form. Readily absorbed through skin; lachrymator; highly toxic and flammable. For best results, dry over K_2CO_3 and freshly distill at reduced pressure. Use in a fume hood.

Introduction. As a synthetic reagent, methyl vinyl ketone is commonly viewed as a 3-oxobutyl synthon which undergoes predominantly 1,4 conjugate addition (Michael addition) resulting in the attachment of a 3-ketoalkyl side chain. The resulting products often undergo a subsequent cyclization step, thus resulting in a cyclohexenone derivative; this tandem process is the well-known Robinson annulation.[4] In most cases, MVK is the electrophilic partner, but there are a few important nucleophilic (umpolung) derivatives of MVK which also are useful for accomplishing ring construction (see below). MVK itself is particularly susceptible to anionic polymerization, which limited its early use under basic conditions; however, it should be noted that MVK will polymerize under certain acidic and radical conditions as well.[5] The many efforts to improve the Robinson ring synthesis have led to monumental synthetic advances with respect to both nucleophilic reactants and MVK synthetic equivalents (3-oxobutyl synthons).

Aside from inducing the anionic polymerization of MVK, enolates as nucleophilic partners in alkylations are limited by *O*- vs. *C*-alkylation, mono- vs. polyalkylation, rapid proton transfer, and the ability to control the regioselectivity of their generation. These problems have found solutions through the development of trimethylsilyl enol ethers,[6] in the utilization of enamine chemistry,[7] and by employing metalloenamines (imine anions).[8] With each of these developments in the nucleophilic partner, improvements in MVK alkylation have resulted in a greatly expanded use of the reagent. Nevertheless, a number of MVK synthetic equivalents have been successfully developed and are briefly reviewed in Table 1. Note that in this context, 'MVK equivalent' is defined as any reagent which contains all four carbon atoms; also the 3′-oxo functionality is in place or is one transformation away.

Methyl Vinyl Ketone versus its Synthetic Equivalents (as Alkylating Agents). Numerous MVK synthetic equivalents have been developed, primarily to circumvent the polymerization problem, and these reagents should also be explored as possible sources of the 3-oxobutyl synthon. However, each reagent is handicapped by the fact that additional synthetic steps are involved in its preparation, and the newly alkylated adduct often must be transformed into the 3-oxobutyl side chain. Furthermore the manipulation of the side chain to arrive at the keto side chain may require harsh conditions not compatible with other functional groups present in the molecule.

One of the first reagents developed as an MVK equivalent was ***1,3-Dichloro-2-butene*** (**1**) (the Wichterle reagent).[9] Relative to MVK, it is generally regarded as a superior alkylating agent, but the stringent hydrolytic conditions for conversion of the vinyl chloride to its corresponding ketone limits its application. An excellent alternative reagent of the same type (i.e. allylic halides) is the iodovinylsilane (**2**).[10] The use of this reagent leads to high yields of alkylated adducts, and the subsequent conversion of the alkylated vinylsilanes to their corresponding ketones is achieved by oxidative rearrangement with ***m-Chloroperbenzoic Acid***. These vinylsilanes can be used with enolates and enamines alike.

Robinson himself recognized that MVK polymerization was compromising the efficiency of the annulation protocol and he developed the first in a series of compounds which serve as precursors to MVK. The synthetic equivalents (**3**),[11] (**4**),[12] and (**5**)[13]undergo β-elimination to generate MVK in situ, but often the results are still not satisfactory (albeit improved) over the use of MVK itself. The principal problems involve poor yields and polyalkylated adducts and, if the synthetic sequence is extended further, unusual cyclization products.

The strategy of carbonyl protection in MVK in order to retard base-catalyzed polymerization fostered interest in a group of β-halo acetals, mixed acetals, and thioacetals of the type (**6**). After alkylation with such reagents, the 3-oxo functionality can be easily regenerated under mildly acidic conditions. The halo acetals (**6**; X = Br, I)[14,15] are prepared in two steps by the conjugate addition of the halo acid to MVK followed by acetalization of the β-halo ketone. The corresponding tosylate (**6**; X = OTs)[16] has also been prepared by a similar, albeit longer, route. At first glance, these reagents would seem ideal for appending the 3-oxobutyl side chain, by alkylation followed by mild acidic deprotection of the carbonyl. Unfortunately, in most cases these reagents are poor alkylating agents. A notable exception is the general annulation procedure (eq 1) which employs imine anions (metalloenamines) as the nucleophilic partner and the bromo acetal (**6**; X = Br) as the MVK equivalent. The sequence commences with the Wadsworth–Emmons alkenation of ketones to afford 2-azadienes, which can be isolated but are usually treated with ***n-Butyllithium*** to generate the highly reactive metalloenamine. The imine anions thus generated can be trapped with the bromo acetal (**6**; X = Br). Hydrolytic workup followed by aldol cyclization–dehydration completes the preparation of 4,4-disubstituted cyclohexenones in yields of 55–65% from the ketones.[17]

Two other MVK equivalents are the α-trialkylsilyl derivatives of MVK itself (**7**; R = Me, Et).[18,19] The trialkylsilyl group stabilizes the incipient negative charge after Michael addition and provides some steric bulk to impede polymerization. The silyl group can then be removed under basic conditions, usually those employed for a cyclization step. These MVK equivalents are especially important in that they allow effective Michael additions to vinyl ketones under aprotic conditions. The major disadvantage of these reagents is the multistep linear synthesis required for their preparation.

Ph; Li; N; P(O)(OEt)$_2$; O; R^1; R^2; Ph; N; R^1; R^2; 1. BuLi; 2. Br; O; O; 3. H_3O^+; 4. KOH, MeOH; O; R^1; R^2 (1)

Table 1 Methyl Vinyl Ketone Synthetic Equivalents (3'-Oxobutyl Synthons)

Alkylating agents and Michael acceptors	Conditions to regenerate ketone	Nucleophilic agents (umpolung type)	Conditions to regenerate ketone
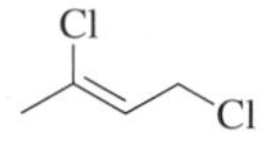 (1)	H_2SO_4 (conc.) or $Hg(OAc)_2$, AcOH	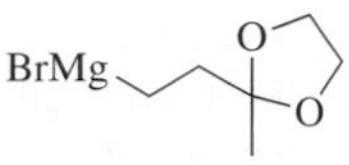 (8)	1 N HCl, rt
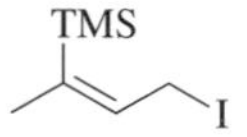 (2)	*m*-CPBA		
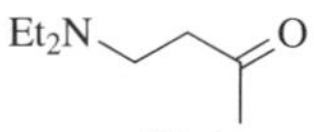 (3)	None	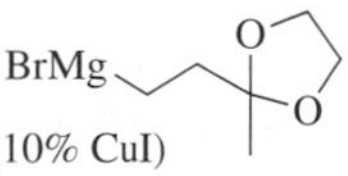 (10% CuI) (9)	1 N HCl, rt
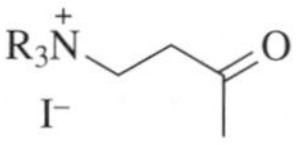 (4)	None		
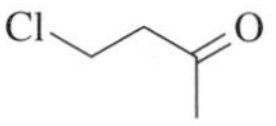 (5)	None		
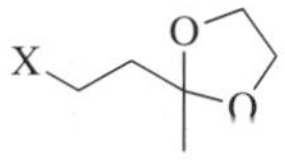 (6) X = Br, I, OTs	1 N HCl, rt	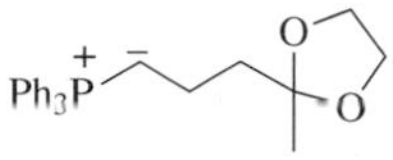 (10)	1 N HCl, rt
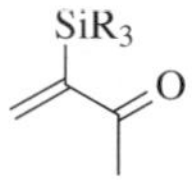 (7) R = Me, Et	NaOMe, MeOH	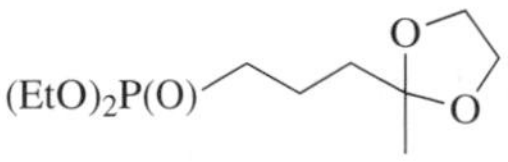 (11)	1 N HCl, rt

Methyl Vinyl Ketone versus its Synthetic Equivalents (Umpolung Reagents). In those instances in which the 3-oxobutyl side chain is to be connected to pre-existing carbonyl (or other electrophilic) centers, several umpolung-type MVK equivalents have been developed. Although somewhat capricious in its stability, the Grignard reagent (**8**)[20] derived from the bromo acetal (**6**; X = Br) has been successfully employed in a number of synthetic applications. Particularly useful is the fact that the acetal moiety may be unmasked in the usual hydrolytic workup or carried on as an acetal for removal at a later stage in the synthesis.[21] Special mention should be made of the Grignard reagent's use in the presence of catalytic copper salts (**9**).[22] This technique allows for the conjugate addition of the acetal moiety to α,β-unsaturated ketones as well as the direct attachment of the side chain to less complex systems. The related homologous Wittig reagent (**10**)[23] was developed in the late 1960s and the Horner–Emmons variant (**11**) appeared shortly thereafter.[24] Both compounds have been employed as reacting partners with lactones, ultimately leading to cyclohexanone products via an intramolecular Wittig alkenation process.

1,4-Conjugate Additions (Michael Addition Reactions, Achiral). There are numerous examples of the use of MVK in conjugate 1,4-addition reactions, i.e. the Michael reaction. Although most of these applications are framed in the context of an annulation procedure (e.g. the Robinson annulation), many of these reactions are worthy of discussion in their own right. The prototypical Michael addition of an enolate with MVK is, in fact, the first step of the Robinson annulation (eq 2).

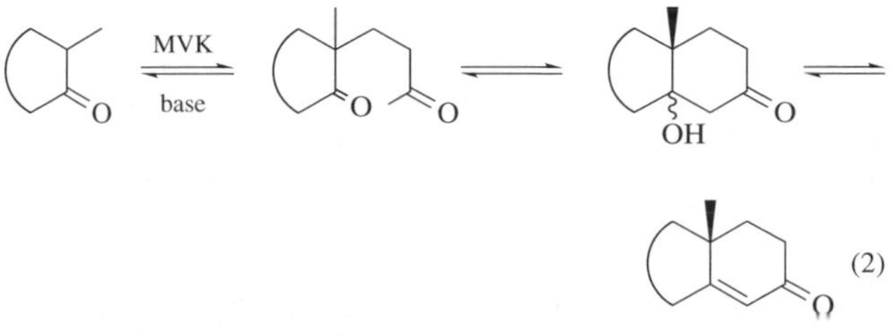

MVK, base (2)

Under basic conditions (e.g. alkylation reactions employing enolates), MVK suffers a high degree of polymerization, leading to the development of the aforementioned MVK equivalents. However, there are derivatives of ketones which are particularly suitable for Michael addition reactions, most notably the silyl enol ethers.[25] The development of silyl enol ethers addressed the most serious problems encountered in the alkylations of enolates. Firstly, the reactions are conducted in aprotic media under anhydrous conditions, thus avoiding proton transfer side reactions. Secondly, by judicious choice of the base used to generate the intermediate enolate, the regiochemistry of the alkylation can be controlled. However, the strongly basic conditions required for the regeneration of the enolates themselves preclude successful additions to MVK, resulting in the formation of polymer. Only recently have trimethylsilyl enol ethers been efficiently added to MVK with the assistance of Lewis acid catalysis (eqs 3 and 4).[26,27] The acidic conditions not only serve to activate MVK towards Michael addition, but also assist in unmasking the enolate as well. These procedures are usually conducted under relatively mild aprotic conditions which also retard MVK polymerization.

OTMS; MVK; ROH, $BF_3{\bullet}Et_2O$; $MeNO_2$, –20 °C; O; O (3)

OTMS; MVK; $Bu_2Sn(OTf)_2$; CH_2Cl_2; O; O (4)

β-Keto esters, α-nitro ketones, and α-keto sulfones are but a few of the stabilized enolate derivatives which have been used successfully in efficient Michael additions to MVK. One of the most troublesome examples involving β-keto esters was the condensation of methyl cyclohexanonecarboxylate with MVK under equilibrating conditions,[28] but this reaction has been developed into a viable synthetic procedure with the use of high pressure techniques under otherwise mild conditions (eq 5).[29]

O; CO_2Me; MVK; 15 kbar, Et_3N; MeCN, 20 °C; O; CO_2Me; O (5)

Another stabilized enolate family includes the α-nitro ketones; although the nitro group activates the formation of the enolate, it may be necessary to remove it in a subsequent procedure (eq 6).[30] Nevertheless, the overall yields of the products good.

Simple nitro alkanes have recently been used in an exceptionally mild procedure for the attachment of the 3-oxobutyl side chain using MVK (eq 7). This protocol highlights the use of ***Alumina*** at room temperature for the conditions of the Michael addition, and the yields are good to moderate.[31]

R^1; O; R^2; NO_2; MVK; PPh_3, THF, rt; 75–90%; R^1; O; R^2; O; NO_2

Bu_3SnH, AIBN; THF, 80 °C; R^1; O; O; R^2 (6)

R^1; NO_2; R^2; MVK; Al_2O_3, rt; R^1; NO_2; O; R^2 (7)

The Michael addition of α-cyano amines to MVK (eq 8) has also been used for the preparation of 1,4-diones,[32] which are historically difficult to prepare.[33] The reaction occurs under mild conditions; however, the use of ***Hexamethylphosphoric Triamide*** as the solvent limits the scale and the widespread use of this procedure. In this sequence, acidic hydrolysis of the alkylated α-cyano amine leads to the 1,4-diketone.

CN; R^1; N; Ph; R^2; Me; 1. HMPA; 2. MVK, LiBr; 3. H_3O^+; R^1; O; R^2; O (8)

γ-Methylthio allylic compounds have also served as the nucleophilic partner in reactions with MVK (eq 9).[34] The procedure employs the 'hyperbasic media' of the ***Lithium Diisopropylamide***/HMPA complex[35] and generates a quaternary carbon center in the product.

MeS; CO_2Me; MVK; LDA–HMPA; THF; O; MeS; CO_2Me (9)

Recently, dithioenamines have emerged as useful synthetic intermediates in general alkylation methods.[36] In their Michael additions with MVK in the presence of Lewis acid catalysts, an intermediate iminium dithiane is generated which is somewhat sluggishly hydrolyzed; however, the overall yields of alkylated adducts are impressive (eq 10).[37]

S; S; R^1; NR^2_2; 1. MVK; 1.0 equiv $ZnCl_2$; 2. H_3O^+, 0 °C, 2 days; 98%; S; S; R^1; O; O (10)

1,4-Conjugate Additions (Michael Addition Reactions, Chiral). In recent years, perhaps the most significant development in synthetic organic chemistry has been the explosion in the field of asymmetric induction. As the reactive intermediates of the enolate, enamine, and imine anions were developed into their chiral counterparts, not only were standard alkylation reactions studied but also Michael additions to α,β-unsaturated systems. It is not surprising that the prototypical model studies involved MVK as the Michael acceptor. While most of these methods are used in asymmetric ring synthesis (see below), some examples terminate at the alkylation step with MVK. All of these methods feature the chiral auxiliary in the nucleophilic reagent. One of the earliest examples of asymmetric induction using MVK was developed by S. Yamada and utilizes chiral pyrrolidino enamines (eq 11).[38] Although the yields of the Michael additions are moderate and the asymmetric induction itself is not particularly impressive by today's standards, this methodology laid the groundwork for many improvements which were to follow. A notable advance in chiral imine alkylations with MVK is the method of d'Angelo, in which the conjugate addition is catalyzed by ***Titanium(IV) Chloride*** (eq 12).[39] A particularly attractive feature of this method is that the chiral auxiliary, ***(S)-α-Methylbenzylamine***, is readily available, and the chemical yields are good with high asymmetric induction.

1. MVK
2. H_3O^+
3. pyrrolidine, HOAc (11)

1. MVK, $TiCl_4$
2. H_3O^+ (12)
91% ee

Another chiral enamine which has been used with MVK takes advantage of the high nucleophilicity of imine anions (i.e. metalated enamines). In this method the intermediate enolate from the Michael addition is trapped as its TMS enol ether, which is subsequently hydrolyzed in the workup (eq 13).[40] The yields are fair and the asymmetric induction is reasonable.

1. LDA
2. MVK, TMSCl
−100 °C
66% (13)
87% ee

A chiral reagent within the family of Horner–Emmons phosphonates has been used with MVK, but the results were disappointing in that the major product was the dialkylated adduct (eq 14). In addition, preparations of these reagents are often multistep and tedious.[41]

1. 2.0 equiv LDA
2. MVK, −80 °C

10% + 33% (14)

In a recent synthesis of (+)-*O*-methyljoubertiamine, Taber utilized very mild conditions for the addition of a chiral enolate to MVK in which the chiral auxiliary is the effective naphthyl camphor system (eq 15).[42]

MVK
K_2CO_3 (15)

Several polymer-based methods which induce some asymmetry to certain substrates have been reported. The degree of asymmetric induction is not particularly high (66% ee) and as yet does not appear to be very general. One such example is the cobalt based diamine polymer which provides modest yields of the Michael adduct (eq 16).[43]

MVK
Co(acac)
diamine (cat) (16)

In terms of high yields and efficiency of asymmetric induction, the Evans procedure using chiral acyl oxazolidones stands apart as a method of choice. In the MVK example the chiral reagent adds, with the assistance of titanium tetrachloride activation at 0 °C, with an impressive 88% chemical yield and 99% ee (eq 17). The oxazolidone can be recycled and the method is general for a variety of α,β-unsaturated systems.[44]

MVK
$TiCl_4$ (1.05 equiv)
0 °C
88% (17)
99:1

Alkylation/Cyclization Tandems: The Robinson Annulation. The most prevalent use of MVK in modern synthesis is the annulation of a six-membered ring onto a pre-existing ketone: the Robinson annulation (see above). In its original format,[45] the sequence involves the generation of an enolate under basic conditions followed by Michael addition to MVK, thus affording a 1,5-dione. The dione can be subjected to basic conditions which induce aldol cyclization–dehydration, resulting in a newly formed cyclohexenone fused to the starting ketone (eq 2). Since its inception, the sequence has been handicapped by a variety of problems: (1) proton transfer; (2) regiochemistry of Michael addition to MVK; (3) *O*- vs. *C*-alkylation; (4) polyalkylation; (5) poor yields in the conjugate addition step; and (6) base-promoted polymerization of MVK itself. In spite of these drawbacks, the transformation is so synthetically important that development and refinement of Robinson's protocol continues to this day. Most of the problems which are unique to MVK are often overcome by the use of a synthetic equivalent (see above). This section will deal with the annulation procedures which involve MVK itself and they can be grouped into linear vs. spiroannulation sequences. In addition, some of these methods involve asymmetric induction, resulting in chiral cyclohexenones. When chirality is involved, it is usually induced in the alkylation (Michael addition) step, but there are a few examples in which the aldol cyclization of the dione or keto aldehyde features the chiral auxiliary.

The Linear Robinson Annulation. This was the earliest class of Robinson annulation procedure and represents the majority of examples. When simple ketones are used in this sequence with MVK, the results are often unsatisfactory. However, when acidic dicarbonyl compounds are employed, the results are quite viable and practical. For example, 1,3-diones may be treated with a catalytic amount of base to generate a stable enolate which is subsequently trapped with MVK. The alkylated adduct is then cyclized–dehydrated under standard Knoevenagel conditions (eq 18). The acidic conditions of the cyclization help retard reversible β-elimination of MVK.[46]

MVK, KOH (cat), MeOH; pyrrolidine, $-H_2O$, 63–65% (18)

In a classic case of stereochemical control, chiral auxiliaries were not involved, but rather a simple reversal of order in alkylations of enolates resulting in the preparation of isomeric decalenones (eq 19).[47] Although this is a dialkylation procedure leading to a specific product, the overall strategy of stereochemical control in the Robinson annulation has been extensively studied and discussed.[48]

(19)

With the advent of kinetic, sterically hindered bases (such as LDA), which could be used under aprotic conditions, the regiochemistry of enolates could be controlled. In the following sequence, a slight excess of the ketone allows the Michael addition to occur at the most hindered site via the thermodynamic enolate, thus resulting in a quaternary carbon center with stereochemical control. Aldol cyclization–dehydration then affords the octalones in 60% yield (eq 20).[49]

1. 0.9 equiv LDA, THF, 20 °C; 2. MVK, −78 °C; 3. KOH, EtOH; 60% (20)

A highly successful approach to Robinson annulation involves the use of trimethylsilyl enol ethers with MVK under Lewis acid conditions, which thwarts MVK polymerization. The yields of the Michael addition itself are greatly improved (see above), and the resulting diones can be cyclized–dehydrated under standard aldol (basic) conditions. This particular example affords the octalones in 89% overall yield from MVK (eq 21).[27]

OTMS; MVK, $Bu_2Sn(OTf)_2$, CH_2Cl_2; NaOMe, MeOH, 89% (based on MVK) (21)

The Chiral Robinson Annulation. Perhaps the most important extensions involving the Robinson annulation protocol have concerned the production of chiral cyclohexenones. In general, the asymmetric induction takes place at the Michael addition step with a chiral reactive intermediate and MVK, although there are a few examples in which the aldol cyclization–dehydration step affords the stereochemical control. However, great care must be exercised in the cyclization step in order to avoid epimerization at the newly created chiral center. One of the earliest methods for chiral Robinson annulation is a general procedure in which a chiral ***Pyrrolidine*** enamine is condensed with MVK; the resulting keto imines are then hydrolyzed to give the chiral δ-keto aldehydes, which in turn are cyclized under (acidic) Knoevenagel conditions, finally affording chiral 4,4-disubstituted cyclohexenones (eq 22).

R^1, R^2; 1. MVK; 2. H_3O^+; 3. pyrrolidine, HOAc (22)

This methodology was highlighted in the chiral total synthesis of the *Sceletium* alkaloid (+)-mesembrine.[50] More recently, an even milder, shorter sequence was used in the key step for the synthesis of the *Sceletium* alkaloid (+)-*O*-methyljoubertiamine. In this case, the chiral auxiliary was the naphthylcamphor system (see above), which governed the condensation of the enolate with MVK using mild ***Potassium Carbonate*** as the base (eq 23).[42]

OMe; CHO; Naphth; MVK, K_2CO_3; RO, Ar (23)

The use of chiral imines has recently received attention as reactive intermediates for Michael additions to MVK. Their advantage is that they are derived from α-methylbenzylamine, which is affordable and can be recycled. The Michael addition itself is activated with titanium tetrachloride and gives good yields (61%) of the Michael adduct with high ee (91%). Hydrolysis concomitant with cyclization provides the octalone (eq 24).[51]

MVK, $TiCl_4$; 1. HOAc; 2. NaOMe, MeOH, 61%; 91% ee (24)

A very useful procedure for inducing chirality in the cyclization step was reported some years ago but has recently been optimized for scale-up. ***2-Methyl-1,3-cyclopentanedione*** serves as the reactive enolate system for the Michael addition with MVK, providing the trione in quantitative yield. The chiral cyclization is a modified Knoevenagel reaction using ***(S)-Proline*** as the asymmetric catalyst. This sequence provides the enedione in 93% optical purity (eq 25).[52]

DMF, 16 °C, 100%; (93% optical purity) (25)

The Spirocyclic Robinson Annulation. Procedures which attach two carbocyclic rings to one another by a single carbon atom (spiroannulation)[53] are not commonplace. The major synthetic problem to overcome in such an operation is the construction of the common, quaternary carbon center. Although there are a number of ingenious ways to construct quaternary carbon centers,[54] those involving spiroannulations with MVK are generally limited to the reactions of trisubstituted enamines with MVK as the conjugate addition step of the sequence (eq 26).[55] The starting carbonyl compounds are usually aldehydes, but there is also a highly efficient homologation procedure which commences from ketones.[56] In either case, the intermediate at the alkylation stage is a δ-keto imine which is subjected to mild acid hydrolysis, thus generating a δ-keto aldehyde, which is then cyclized–dehydrated with base. Enamines from ***Piperidine*** and ***Morpholine*** have been used successfully in this procedure. There is flexibility in the carbonyl starting materials as well: five-, six-, seven-, and eight-membered carbonyl systems have all been satisfactory substrates in this methodology. The spiroannulation procedure of MVK with enamines is summarized below (Table 2).

1. MVK, EtOH, Δ; 2. HOAc, NaOAc, Δ; 3. NaOH, Δ; $n = 1, 2, 4$ (26)

Table 2 Spiroannulation Procedures Involving Enamines and Methyl Vinyl Ketone

Ketone	Enamine	Yield of spirocycle (%)	Ref.
Cyclopentane-carbaldehyde	Piperidino	59	55c
Isopropylidene-cyclopentane-carbaldehyde	Piperidino	45–65	55a
Cyclohexane-carbaldehyde	Piperidino	45–65	55b,c
Cyclohexanone	Morpholino	41	56
2-Methylcyclo-hexanone	Morpholino	36	56
4-*t*-Butylcyclo-hexanone	Morpholino	41	56
Isophorone	Morpholino	30	56
2-Norbornanone	Morpholino	41	56
Cycloheptane-carbaldehyde	Piperidino	45–65	55b,c
Cyclooctane-carbaldehyde	Piperidino	45–65	55b,c

Related Reagents. 2-(2-Bromoethyl)-2-methyl-1,3-dioxolane; t-Butyl (*E*)-4-Iodo-2-methyl-2-butenoate; 4-Chloro-2-butanone; 1,4-Dichloro-2-butanone; 1,3-Dichloro-2-butene; 1-Methoxy-3-buten-2-one; Methyl 3-Oxo-4-pentenoate; 2-Methyl-6-vinylpyridine; 1,7-Octadien-3-one; *trans*-3-Penten-2-one; 4-Phenylsulfonyl-2-butanone Ethylene Acetal; 3-Triethylsilyl-3-buten-2-one; 3-Trimethylsilyl-3-buten-2-one; 2-Trimethylsilyloxy-1,3-butadiene.

1. *FF* **1967**, *1*, 697; **1969**, *2*, 283; **1975**, *5*, 464; **1977**, *6*, 407; **1979**, *7*, 247; **1982**, *10*, 272; **1986**, *12*, 329.
2. Review: Bergmann, E. D.; Ginsburg, D.; Pappo, R. *OR* **1959**, *10*, 179.
3. Review: Jung, M. E. *T* **1976**, *32*, 3.
4. Review: Gawley, R. E. *S* **1976**, 777.
5. Nicholson, J. W. *The Chemistry of Polymers*; Royal Society of Chemistry: Cambridge, 1991.
6. House, H. O.; Czuba, L. J.; Gall, M.; Olmstead, H. D. *JOC* **1969**, *34*, 2324.
7. (a) Stork, G.; Brizzolara, A.; Landesman, H.; Szmuszkovicz, J.; Terrell, R. *JACS* **1963**, *85*, 207. (b) Heathcock, C. H.; Ellis, J. E.; McMurry, J. E.; Coppolino, A. *TL* **1971**, 4995.
8. Martin, S. F.; Phillips, G. W.; Puckette, T. A.; Colapret, J. A. *JACS* **1980**, *102*, 5866 and references therein.
9. (a) Wichterle, O.; Procházka, J.; Hofman, J. *CCC* **1948**, *13*, 300. (b) Prelog, V.; Barman, P.; Zimmermann, M. *HCA* **1949**, *32*, 1284. (c) Julia, M. *BSB* **1954**, *21*, 780. (d) Marshall, J. A.; Schaeffer, D. J. *JOC* **1965**, *30*, 3642. (e) House, H. O. *Modern Synthetic Reactions*, 2nd ed.; Benjamin: Menlo Park, CA, 1972.
10. (a) Stork, G.; Jung, M. E. *JACS* **1974**, *96*, 3682. (b) Stork, G.; Jung, M. E.; Colvin, E.; Noel, Y. *JACS* **1974**, *96*, 3684.
11. Balasubramanian, K.; John, J. P.; Swaminathan, S. *S* **1974**, 51.
12. (a) Cornforth, J. W.; Robinson, R. *JCS* **1949**, 1855. (b) McQuillin, F. J.; Robinson, R. *JCS* **1938**, 1097.
13. (a) Taylor, D. A. H. *JCS* **1961**, 3319. (b) Pinder, A. R.; Williams, R. A. *JCS* **1963**, 2773. (c) Halsall, T. G.; Theobald, D. W.; Walshaw, K. B. *JCS* **1964**, 1029. (d) Theobald, D. W. *T* **1966**, *22*, 2869.

14. (a) Stork, G. *PAC* **1964**, *9*, 131. (b) Stork, G.; Borch, R. *JACS* **1964**, *86*, 935. (c) Brown, E.; Dahl, R. *BSF(2)* **1972**, 4292. (d) Sato, T.; Kawara, T.; Sakata, K.; Fujisawa, T. *BCJ* **1981**, *54*, 505. (e) Stowell, J. C.; Keith, D. R.; King, B. T. *OSC* **1990**, *7*, 59.
15. (a) Stowell, J. C.; King, B. T.; Hauck, H. F., Jr. *JOC* **1983**, *48*, 5381. (b) Murai, A.; Ono, M.; Masamune, T. *CC* **1977**, 573. (c) Crombie, L.; Tuchinda, P.; Powell, M. J. *JCS(P1)* **1982**, 1477. (d) Trost, B. M.; Kunz, R. A. *JOC* **1974**, *39*, 2475. (e) Trost, B. M.; Conway, W. P.; Strege, P. E.; Dietsche, T. J. *JACS* **1974**, *96*, 7165. (f) Trost, B. M.; Bridges, A. J. *JOC* **1975**, *40*, 2014. (g) Solas, D.; Wolinsky, J. *JOC* **1983**, *48*, 670. (h) Trost, B. M.; Kunz, R. A. *JACS* **1975**, *97*, 7152. (i) Kametani, T.; Suzuki, Y.; Furuyama, H.; Honda, T. *JOC* **1983**, *48*, 31.
16. (a) Yardley, J. P.; Rees, R. W.; Smith, H. *JMC* **1967**, *10*, 1088. (b) Danishefsky, S.; Cavanaugh, R. *JACS* **1968**, *90*, 520.
17. Martin, S. F.; Phillips, G. W.; Puckette, T. A.; Colapret, J. A. *JACS* **1980**, *102*, 5866.
18. (a) Stork, G.; Singh, J. *JACS* **1974**, *96*, 6181. (b) Boeckman, R. K., Jr. *JACS* **1973**, *95*, 6867. (c) Boeckman, R. K., Jr. *JACS* **1974**, *96*, 6179. (d) Boeckman, R. K., Jr.; Blum, D. M.; Ganem, B.; Halvey, N. *OSC* **1988**, *6*, 1033.
19. Stork, G.; Ganem, B. *JACS* **1973**, *95*, 6152.
20. (a) Ponaras, A. A. *TL* **1976**, 3105. (b) Büchi, G.; Wüest, H. *JOC* **1969**, *34*, 1122.
21. Martin, S. F.; Puckette, T. A.; Colapret, J. A. *JOC* **1979**, *44*, 3391.
22. (a) Snider, B. B.; Cartaya-Marin, C. P. *JOC* **1984**, *49*, 153. (b) Fujisawa, T.; Sato, T.; Kawara, T.; Noda, A. *TL* **1982**, *23*, 3193. (c) Gras, J.-L. *JOC* **1981**, *46*, 3738. (d) Paquette, L. A.; Galemmo, R. A., Jr.; Caille, J.-C.; Valpey, R. S. *JOC* **1986**, *51*, 686.
23. Henrick, C. A.; Böhme, E.; Edwards, J. A.; Fried, J. H. *JACS* **1968**, *90*, 5926.
24. Sturtz, G. *BSF(2)* **1964**, 2340.
25. Review: Kuwajima, I.; Nakamura, E. *ACR* **1985**, *18*, 181.
26. Duhamel, P.; Hennequin, L.; Poirier, N.; Poirier, J.-M. *TL* **1985**, *26*, 6201.
27. (a) Sato, T.; Wakahara, Y.; Otera, J.; Nozaki, H. *T* **1991**, *47*, 9773. (b) Sato, T.; Wakahara, Y.; Otera, J.; Nozaki, H. *JACS* **1992**, *113*, 4028.
28. (a) Metzger, J. D.; Baker, M. W.; Morris, R. J. *JOC* **1972**, *37*, 789. (b) Marshall, J. A.; Warne, T. M., Jr. *JOC* **1971**, *36*, 178.
29. Dauben, W. G.; Bunce, R. A. *JOC* **1983**, *48*, 4642.
30. Ono, N.; Miyake, H.; Kaji, A. *CC* **1983**, 875.
31. (a) Rosini, G.; Marotta, E.; Ballini, R.; Petrini, M. *S* **1986**, 237. (b) Ballini, R.; Petrini, M.; Rosini, G. *S* **1987**, 711. (c) Ballini, R.; Petrini, M.; Marcantoni, E.; Rosini, G. *S* **1988**, 231. (d) Bergbreiter, D. E.; LaLonde, J. J. *JOC* **1987**, *52*, 1601.
32. (a) Ahlbrecht, H.; Kompter, H.-M. *S* **1983**, 645. (b) Ivanov, I. C.; Sulay, P. B.; Dantchev, D. K. *LA* **1983**, 753.
33. For a review see: Miyakoshi, T. *OPP* **1989**, *21*, 661.
34. Kende, A. S.; Constantinides, D.; Lee, S. J.; Liebeskind, L. *TL* **1975**, 405.
35. Review: Mordini, A. In *Advances in Carbanion Chemistry*; Snieckus, V., Ed.; JAI: Greenwich, CT, 1992; Vol. 1.
36. Page, P. C. B.; Harkin, S. A.; Marchington, A. P.; van Niel, M. B. *T* **1989**, *45*, 3819.
37. Page, P. C. B.; Harkin, S. A.; Marchington, A. P. *SC* **1989**, *19*, 1655.
38. Sone, T.; Terashima, S.; Yamada, S. *CPB* **1976**, *24*, 1273.
39. (a) Pfau, M.; Revial, G.; Guingant, A.; d'Angelo, J. *JACS* **1985**, *107*, 273. (b) Hickmott, P. W.; Rae, B. *TL* **1985**, *26*, 2577. (c) Fourtinon, M.; De Jeso, B.; Pommier, J.-C. *JOM* **1985**, *289*, 239.
40. Tamioka, K.; Seo, W.; Ando, K.; Koga, K. *TL* **1987**, *28*, 6637.
41. El Achqar, A.; Boumzebra, M.; Roumestant, M.-L.; Viallefont, P. *T* **1988**, *44*, 5319.
42. Taber, D. F.; Mack, J. F.; Reingold, A. L.; Geib, S. J. *JOC* **1989**, *54*, 3831.
43. (a) Cram, D. J.; Sogah, G. D. Y. *CC* **1981**, 625. (b) Brunner, H.; Hammer, B. *AG(E)* **1984**, *23*, 312. (c) Li, T.-T.; Wu, Y.-L. *TL* **1988**, *29*, 4039. (d) Polymer based: Hodge, P.; Khoshdel, E.; Waterhouse, J. *JCS(P1)* **1983**, 2205.
44. Evans, D. A.; Bilodeau, M. T.; Somers, T. C.; Clardy, J.; Cherry, D.; Kato, Y. *JOC* **1991**, *56*, 5750.
45. (a) Rapson, W. S.; Robinson, R. *JCS* **1935**, 1285. (b) Du Feu, E. C.; McQuillin, F. J.; Robinson, R. *JCS* **1937**, 53.
46. (a) Ramachandran, S.; Newman, M. S. *OS* **1961**, *41*, 38. (b) Ramachandran, S.; Newman, M. S. *OSC* **1973**, *5*, 486. (c) Mekler, A. B.; Ramachandran, S.; Swaminathan, S.; Newman, M. S. *OS* **1961**, *41*, 56. (d) Meckler, A. B.; Ramachandran, S.; Swaminathan, S.; Newman, M. S. *OSC* **1973**, *5*, 743. (e) Newman, M. S.; Mekler, A. B. *JACS* **1960**, *82*, 4039.
47. Ireland, R. E.; Kierstead, R. C. *JOC* **1966**, *31*, 2543.
48. (a) See ref. 9(e). (b) Conia, J.-M. *Rec. Chem. Prog.* **1963**, *24*, 43. (c) House, H. O. *Rec. Chem. Prog.* **1967**, *28*, 99.
49. (a) Huffman, J. W.; Rowe, C. D.; Matthews, F. J. *JOC* **1982**, *47*, 1438. (b) Ziegler, F. E.; Wang, K. J. *JOC* **1983**, *48*, 3349. (c) Pariza, R. J.; Fuchs, P. L. *JOC* **1983**, *48*, 2306. (d) Chen, E. Y. *SC* **1983**, *13*, 927. (e) Dauben, W. G.; Bunce, R. A. *JOC* **1983**, *48*, 4642.
50. Yamada, S.; Otani, G. *TL* **1971**, 1133.
51. (a) Pfau, M.; Revial, G.; Guingant, A.; d'Angelo, J. *JACS* **1985**, *107*, 273. (b) Volpe, T.; Revial, G.; Pfau, M.; d'Angelo, J. *TL* **1987**, *28*, 2367. (c) Revial, G. *TL* **1989**, *30*, 4121.
52. Hajos, Z. G.; Parrish, D. R. *OSC* **1990**, *7*, 363.
53. Review: Krapcho, A. P. *S* **1974**, 383.
54. Review: Martin, S. F. *T* **1979**, *36*, 419.
55. (a) Hutchins, R. O.; Natale, N. R.; Taffer, I. M.; Zipkin, R. *SC* **1984**, *14*, 445. (b) Kane, V. V. *SC* **1976**, *6*, 237. (c) Kane, V. V.; Jones, M., Jr. *OS* **1983**, *61*, 129. (d) Kane, V. V.; Jones, M., Jr. *OSC* **1990**, *7*, 473.
56. (a) Martin, S. F. *JOC* **1976**, *41*, 3337. (b) Martin, S. F.; Gompper, R. *JOC* **1974**, *39*, 2814.

John A. Colapret & Paul T. Buonora
Lamar University, Beaumont, TX, USA

Nickel(II) Acetylacetonate[1]

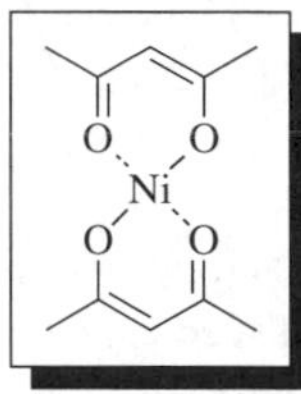

[3264-82-2] $C_{10}H_{14}O_4Ni$ (MW 256.93)

(catalyst for oligomerization, telomerization, hydrosilylation, hydrogenation, reduction, cross-coupling, oxidation, conjugate addition, addition to multiple bonds, and rearrangements[1])

Alternate Name: bis(acetylacetonato)nickel(II).
Physical Data: pale green solid, mp 240 °C (dec); see also ***2,4-Pentanedione***.
Solubility: sol ethers and aromatic and halogenated hydrocarbons.
Analysis of Reagent Purity: atomic absorption is the method most commonly used.
Preparative Methods: commercially available; can be prepared from ***Nickel(II) Chloride***.[1d]
Purification: recrystallize from benzene and sublime under vacuum (10^{-3} mmHg).[1c]
Handling, Storage, and Precautions: nickel is now recognized as a cancer suspect agent as well as a possible teratogen and due precautions should be taken when handling the reagent. The anhydrous solid is stable, but is an irritant and hygroscopic and should preferably be stored in a sealed container to preclude contact with air and moisture. Solutions are more susceptible to atmospheric oxidation.

Catalysts for Oligomerization, Cooligomerization, and Telomerization. Active catalytic systems used for these reactions have been prepared from many Ni^{II} salts and Ni^0 complexes. The most common are $Ni(acac)_2$, nickel halides, and nickel–alkene complexes. The reactive species are formed from the combination of Ni^0, a Lewis acid usually based on aluminum, and a suitable ligand. They are referred to as Ziegler catalysts. When Ni^{II} salts are used, a reducing agent is required to produce the active Ni^0. The Lewis acid present, commonly a trialkylaluminum, is generally sufficiently reactive to reduce the nickel salt. Other reducing agents such as ***n-Butyllithium***,[2] ***Sodium Borohydride***,[3] or an electric current[4] have also been used. The oligomerization reactions are usually done in hydrocarbon or halogenated solvents. The mechanism is believed to involve nickel hydride, formed in situ via β-hydride elimination. First, an alkene inserts into the Ni–H bond. Further insertion of a second alkene into the C–Ni bond and reductive elimination regenerates nickel hydride and produces the oligomer. While some very interesting carbocyclic systems can be accessed via these catalysts, the product distribution can vary depending on the nickel precatalyst/Lewis acid/ligand combination chosen. The better activity resulting from using one precatalyst over another for a specific transformation is not always well understood, and often results from careful tuning of the catalytic system. Below are those in which $Ni(acac)_2$ has been successful. Other articles and monographs on ***Nickel*** should also be consulted.[1a,1b]

Oligomerization and Cooligomerization of Alkenes. Cooligomerization of monoalkenes is generally of limited synthetic interest due to the formation of many isomers and oligomers and the difficulty in establishing conditions displaying suitable selectivity. It is, however, very important in the industrial production of lower alkenes. This topic is certainly well beyond the scope of this article. Some cyclic and bicyclic alkenes have proven more suitable (eq 1), the rigidity of which provides lesser opportunities for isomerizations by ways of insertion/migration.[5,6]

Ni[Li], Et_3Al (1)

Cooligomerization of 1,3-Dienes. The cyclotrimerization of butadiene is performed in the presence of a ligand-free nickel catalyst, giving cyclododecatrienes (eq 2),[7] with the all-*trans*-1,5,9-isomer as the major product. Lesser amounts of the other double-bond isomers are also found, the quantity of which are temperature and concentration dependent.[8] Substituted 1,3-dienes have not received as much attention due to the large number of isomers formed during the reaction, as well as the much lower reaction rates. The cyclooligomerization can be stopped at the stage of the dimer by introducing a phosphite or phosphine ligand.[9] In the reaction of butadiene the major product is 1,5-cyclooctadiene (eq 3),[9] which is accompanied by small amounts of divinylcyclopropane and 4-vinylcyclohexene. The three products are believed to originate from the same di-π-allylnickel intermediate (eq 4).[10] The proportion of these intermediates and consequently the product distribution is affected by the presence of a ligand, the effects of which have been related to their electronic and steric nature. The less basic and more bulky ligands favor a di-π-allyl intermediate which eventually closes by reductive elimination to the 1,5-cod via a terminal, rather than internal, di-σ-allyl complex.[10]

3 (butadiene), $Ni(acac)_2$, Et_2AlOEt (2)

2 (butadiene), $Ni(acac)_2$, Et_2AlOEt, ligand (3)

L–Ni, NiL, NiL, NiL (4)

Codimerizations of a variety of 1,3-dienes provide a convenient access to substituted 1,5-cyclooctadienes (eq 5)[11] and cyclohexenes (eq 6).[12] The presence of an activating group such as an ester can provide further stabilization of the σ-allyl/π-allyl complex, increasing the selectivity of the reaction (eq 7).[13]

Ni(acac)$_2$–ligand, Et$_2$Al(OEt) (5)

Ni(acac)$_2$–ligand, *i*-Bu$_3$Al (6)

Ni(acac)$_2$, PPh$_3$, AlCl$_3$; CO$_2$Me; CO$_2$Me (7)

The cooligomerization of 1,3-dienes and alkenes involves two molecules of the diene and one of the alkene to form substituted cyclodecadienes (eq 8)[14] and variable amounts of the linear cotrimers. In the case of symmetrical alkenes it is possible to obtain a single product, in contrast to unsymmetrical alkenes in which mixtures of isomers result, in addition to those resulting from the other modes of oligomerization (i.e. linear vs. cyclic). These reactions work best in the presence of the catalytic system Ni(acac)$_2$/ligand/diethylethoxyaluminum. This is among the few methods giving direct access to 10-membered rings.

Ni(acac)$_2$–ligand, Et$_2$Al(OEt); H; H (8)

Disubstituted alkynes as well as allenes have also served as substrates in cooligomerizations with butadienes. With the former, *cis*-4,5-divinylcyclohexenes and mixtures of cyclodecadienes result. Other nickel precatalysts, such as ***Bis(1,5-cyclooctadiene)nickel(0)***, have, however, proved to be more specific with regard to the type and number of products formed.[15]

Alkynes. The oligomerization of unsubstituted (eq 9)[16] and monosubstituted alkynes, also known as the Reppe reaction, produces variable mixtures of linear dimers, 1,2,4- and 1,3,5-trisubstituted benzenes, and cyclooctatetraene[17] isomers. Disubstituted alkynes are known not to undergo such a process although they can be used in cooligomerizations with mono- and unsubstituted alkynes. The product distribution depends on the nature of the ligands.[18] Weak ligands such as acac[19] or cod favor cyclotetramerization while stronger ones such as PPh$_3$ induce cyclotrimerization.[20] When coordinating solvents such as pyridine or DMF are used, only the linear dimer is formed.[21]

4 H—≡—H, Ni(acac)$_2$ (9)

Strained systems are also reactive and their synthetic utility has been elegantly demonstrated.[22] In the synthesis of 12-nor-13-acetoxymodhephene, advantage was taken of the facile Ni^0-induced transannulation that can take place in eight-membered rings, made possible by coordination of Ni^0 to the two double bonds (eq 10).[23]

20 mol% Ni(acac)$_2$, DIBAL, PPh$_3$, toluene, 91%; CO$_2$Et; EtO$_2$C; OAc (10)

Telomerization is also an important reaction catalyzed by Ni(acac)$_2$ and other nickel species in the presence of ligands and reducing agents.[24] In this process, a diene is inserted in a 1,2- or 1,4-fashion into an activated C–H bond (eq 11)[25] or into the X–H bond of an alcohol (eq 12),[26] phenol, amine, or silane.[27]

Ph, O; Ni(acac)$_2$, PhP(O-*i*-Pr)$_2$, NaBH$_4$; Ph, O (11)

OH; Ni(acac)$_2$, PhP(O-*i*-Pr)$_2$, NaBH$_4$; OEt (12)

1,7-Diynes also undergo cyclization with 1,8-insertion into a Si–H bond.[28] Finally, dimethylsilane adds efficiently and with excellent regioselectivity to functionalized electron-rich or -poor alkenes in a 1,2-fashion, providing access to a wide variety of silane-containing substrates.[29]

A catalytic asymmetric version of the codimerization of alkenes has also appeared.[30] Enantio-pure phosphine ligands are used to induce chirality and, in selected examples, have resulted in appreciable levels of asymmetric induction (eq 13).[31,32]

Ni(acac)$_2$ + homochiral phosphine ligand; H; * ; 80% ee (13)

Nickel hydride generated in situ has been postulated as the active catalyst in the transformations described above and, not surprisingly, a large number of precatalysts have been reported. Only those dealing specifically with Ni(acac)$_2$ have been described here. A more comprehensive account is provided in the excellent reviews by Jolly and Wilke.[1a,1b]

Catalyst for Oxidations. A very efficient protocol for the air oxidation of ketones, alkenes, and aldehydes, catalyzed by nickel (1,3-diketonates) has been developed by Mukaiyama and co-workers. Baeyer–Villiger oxidation of ketones to esters and lactones (eq 14)[33] is achieved in moderate to high yields with 1% Ni(acac)$_2$ or Ni(dpm)$_2$ (dpm = dipivaloylmethanato) and 2–3 molar equivalents of isovaleraldehyde or benzaldehyde under one atmosphere of air or oxygen, in 1,2-dichloroethane at ambient temperature. High regioselectivities were observed for unsymmetrical ketones (eq 15).[33] The aldehyde, which functions as a reducing agent, is converted to the acid in high yield.[34] This represents a mild, convenient, and ecologically sound oxidation protocol.

1% $Ni(dpm)_2$, *i*-PrCHO, O_2

n	%
1	84
2	91
3	33

(14)

$Ni(dpm)_2$, *i*-PrCHO, O_2, 99% (15)

The epoxidation of alkenes[35] can also be realized in moderate to high yields under the same conditions using an aldehyde, or under more forcing conditions using an alcohol,[36] as the coreducing agent. Very low stereospecificity is obtained for disubstituted alkenes, which provide nearly equal mixtures of *cis*- and *trans*-epoxides, irrespective of the initial alkene geometry. This is, however, a good protocol for some 1,1-symmetrical trisubstituted alkenes.[37] Little information is available on the chemoselectivity and compatibility of other functional groups under the reaction conditions.

Catalyst for Conjugate Additions. The conjugate addition of organozinc reagents in a 1,4-fashion to α,β-unsaturated ketones in the presence of 1–10 mol% of $Ni(acac)_2$ in ethereal or aromatic solvents proceeds in good to excellent yields (eq 16).[38] The reaction is fast and the conditions are very mild. This represents a nice alternative to organocopper reagents since diorganozinc reagents are much more stable than their organocopper analogs and can be used at ambient temperatures without decomposition. Addition occurs even with the most sterically demanding substrates, as demonstrated by the synthesis of (±)-β-cuparenone by Greene and co-workers (eq 17).[39] Aryl, *t*-butyl, cyclohexyl, and alkenylzinc reagents have also been used. It should be noted that only one of the two alkyl groups is transferred and monoalkylzinc halides do not add, which results in the loss of one equivalent of the nucleophile. Articles on copper reagents should be consulted for related transformations.

Me_2Zn, toluene–THF (10:1), 4–20 °C, 1.5 h, 90% (16)

$Ni(acac)_2$, Et_2O, 67% (17)

In the presence of $Ni(acac)_2$ and a coreducing agent such as ***Diisobutylaluminum Hydride*** (DIBAL), alkenylzirconium reagents[40] (but not alkylzirconium[41] undergo 1,4-additions to α,β-enones in 60–95% yield, which decrease with substitution at C-1 of the organometallic. Similarly, the complex formed by addition of $Ni(acac)_2$ and DIBAL (1:1) catalyzes the 1,4-addition of dialkylaluminum acetylides to α,β-enones (eq 18).[42] Again, the reaction works well even with highly hindered substrates. Trimethylsilylalkynes add efficiently but the yields are considerably lower with acetylene itself. Trialkylaluminums have been used directly on enones with in situ generation of the active nickel species.[43] Other nucleophiles such as lithium thiophenoxide[43] and 1,3-dicarbonyls (eq 19)[44,45] can be used but generally require more forcing conditions. Although the mechanism of these nickel-catalyzed additions has not been clearly established, it is believed to involve ketyl radicals resulting from single-electron transfer (SET) from a nickel(I) species.[46]

t-Bu—≡—$AlMe_2$, $Ni(acac)_2$, DIBAL, 49% (18)

$Ni(acac)_2$, DMF, $CH_2(CO_2Et)_2$, 18 h, 180 °C, 70% (19)

Catalytic Enantioselective Conjugate Additions. In recent years, much work has been done to achieve enantioselective addition of organozinc reagents to various enones using chiral ligands.[47,46b] Respectable levels of induction have been achieved with selected aromatic substrates, but attempts with simpler enones such as cyclohexenone have been unsuccessful. The accepted model to test the activity of a particular catalytic system is the addition of ***Diethylzinc*** to chalcone. Yields are generally good and enantiomeric excesses (ee) range from 25 to 95%. A variety of bi- and tetradentate ligands have been designed and the most successful are shown in eq 20. Nonlinear relationships between the ee of the ligand and the ee of the final product have been frequently observed.[48,46b]

$ZnEt_2$ 1–5 mol%, $Ni(acac)_2$, ligand (20)

R = Ph or Me

90% ee; R = *t*-Bu, $CH_2Si(OEt)_3$ 95% ee; 78% ee

70% ee; 90% ee; 60% ee

Catalyst for Cross-Coupling Reactions. While not as popular as other Ni^{II} precatalysts, Ni^0 complexes, or other metals such as Pd^0 complexes, $Ni(acac)_2$ has found many applications as a catalyst for cross-coupling reactions. It is fairly stable and sufficiently soluble to be used in a large variety of solvents. Aryl halides,[49] silyl enol ethers,[50] enol phosphates[51] and, more recently, aryl-*O*-carbamates and triflates[52] have been successfully coupled with alkyl, aryl, and alkenyl organometallics (eq 21).[53] The yields are generally good and chemoselectivity can be achieved in some cases (eq 22).[54,52]

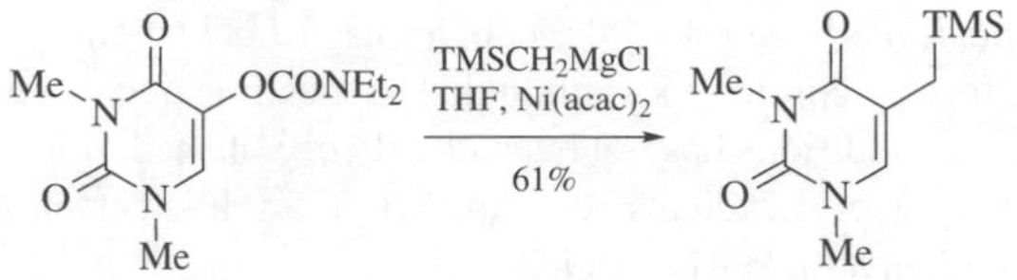

(21)

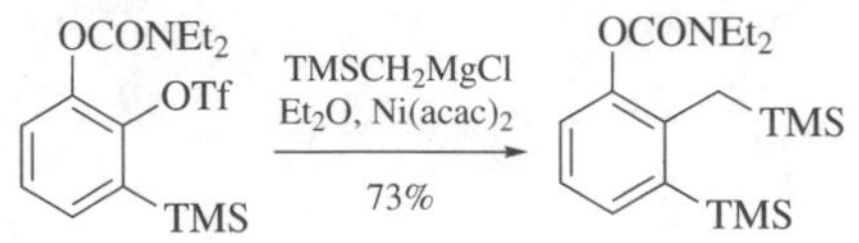

(22)

The addition of mono- and bidentate phosphine ligands in ethers or aromatic solvents often has a beneficial effect on the yield and rate of the reaction. The mechanism has been well documented and involves an oxidative addition of the aryl halide to Ni^0 followed by addition of the organometallic species to the Ni^{II} complex. Reductive elimination of the cross-coupling product and regeneration of Ni^0 completes the catalytic cycle. The coupling of stereodefined 1-alkenylalanes and zirconium with a variety of aryl halides proceeds in good yield and stereospecifically (eq 23).[55] Vinylic sulfones can serve as the electrophilic partner, coupling efficiently with ***Phenyl-*** or ***Methylmagnesium Bromide*** (eq 24).[56] An interesting example was reported in which both electrophilic and nucleophilic components are borne by the same carbon. In this example, an α-(bromomagnesium)sulfone dimerizes head-to-head to produce a mixture of *cis*- and *trans*-alkenes (eq 25).[57]

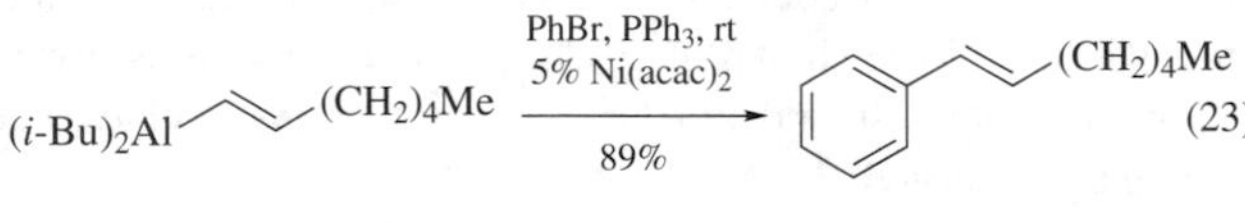

(23)

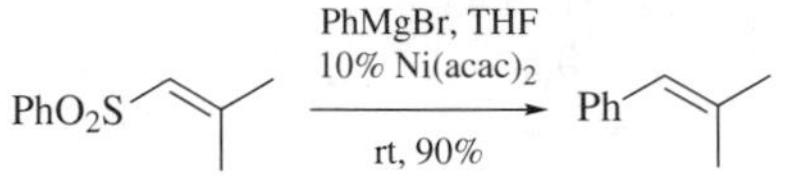

(24)

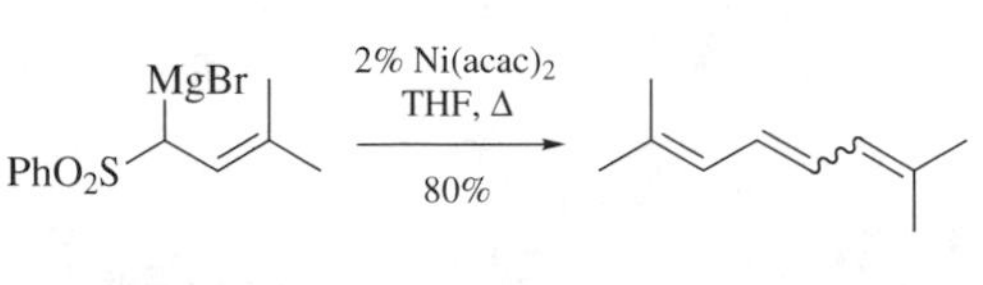

(25)

Again, $Ni(acac)_2$ is only one of the many precatalysts which have been used (often interchangeably) for these transformations. The choice of proper precatalyst has, in many instances, been the result of a more or less exhaustive screening of the five or six most common ligands. Other metal ions such as palladium, copper, or chromium are often similarly effective in promoting the transformation.

Other Additions. Grignard reagents add regioselectively and stereoselectively to substituted trimethysilylalkynes, in presence of $Ni(acac)_2$ and ***Trimethylaluminum***, to give a *syn* addition product (9:1). These additions occur exclusively at the carbon-substituted end of the alkyne.[58] If a heteroatom, such as an ether or an amine, is present on the carbon chain, isomerization to the *trans* product occurs (eq 26). The intermediate organometallic species can be subsequently trapped with a variety of electrophiles.

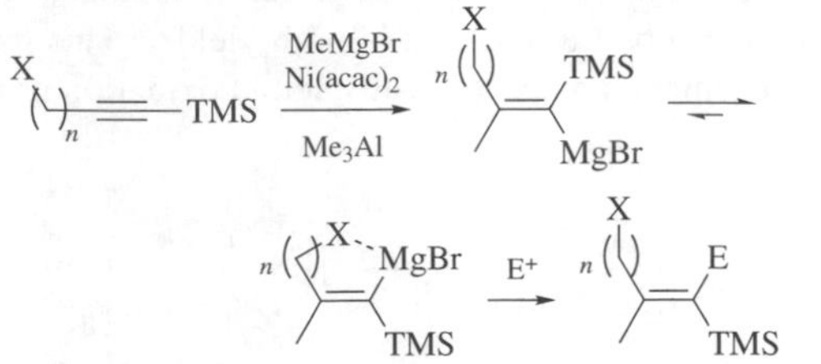

(26)

Functionalization of an unactivated terminal alkene is possible via remote chelation of a halomagnesium alkoxide (eq 27).[59] Addition of Grignard reagents to propargyl chlorides proceeds in a S_N2' fashion to provide good yields of the corresponding allenes. Methyl ketones can be prepared by $Ni(acac)_2$-catalyzed addition of trimethylaluminum to a nitrile followed by hydrolysis.[60a]

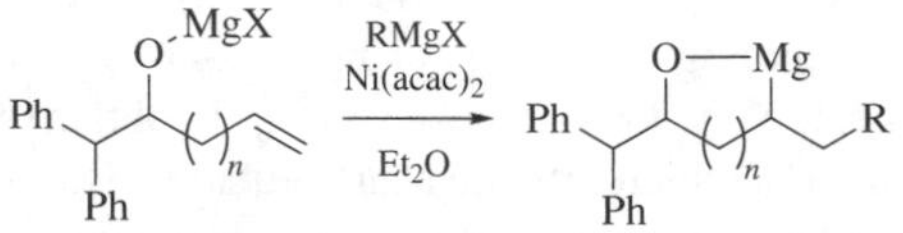

(27)

$Ni(acac)_2$ has been shown to promote the intramolecular coupling of a vinyl iodide with an aldehyde in the presence of ***Chromium(II) Chloride***, to form a 13-membered ring lactone. However, only a moderate stereoselectivity was observed, as both diastereomeric alcohols were produced (60% yield).[60b]

In the presence of a catalytic amount of $Ni(acac)_2$ (in addition to other metals), ***Cyanotrimethylsilane*** reacts smoothly with acetals or orthoesters derived from α,β-unsaturated carbonyls to give the corresponding *O*-methyl cyanohydrins under neutral conditions (eq 28).[61]

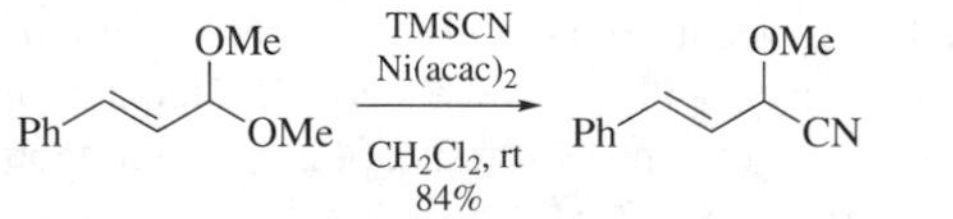

(28)

β-Diketonates add to ***Malononitrile*** in moderate yield to give β-amino nitriles (eq 29).[62] In contrast, β-keto esters, malonate diesters, and β-keto amides are all poor substrates. Other electrophiles have been used such as cyanogen, benzoyl cyanide,[63] and ***Trichloroacetonitrile***.[64] Reaction of acetylacetonate with cyanogen has been reported to produce highly substituted pyrimidines in moderate yields.[65]

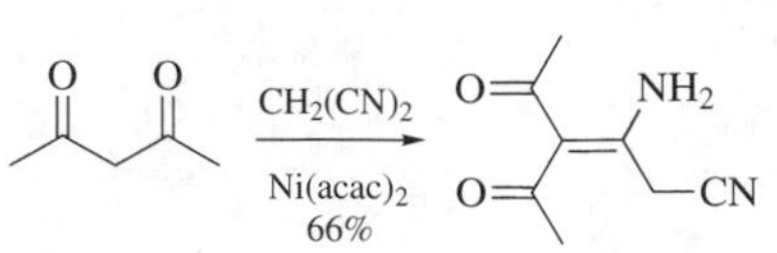

(29)

Optically active trialkylsilanes react with ***Vinylmagnesium Bromide*** via a pentacoordinate intermediate to produce, after loss of a hydride, tetrasubstituted silanes with almost complete retention of configuration.[66]

Reductions. $Ni(acac)_2$ has seldom been used as a reduction catalyst. Hydrogenation of alkenes has been realized under photochemical conditions in the presence of a ketone as a sensitizer.[67] Monohydrogenation of 1,4-cyclohexadiene was effected using a homogeneous nickel catalyst generated from $Ni(acac)_2$, $Et_3Al_2Cl_3$, and PPh_3. Substituted cyclohexadienes produce mixtures of cyclohexene isomers, isomerization of which was shown to be promoted by the catalyst itself. A zeolite-supported complex formed between $Ni(acac)_2$ and an optically active 2-(aminocarbonyl)pyrrolidine ligand catalyzes the asymmetric hydrogenation of ethyl (*Z*)-α-benzoylaminocinnamate with ee values up to 85%. Other metals that have been studied have resulted in even higher enantioselectivities.[68]

Vinylic sulfones are reduced in fair to good yield to the corresponding alkenes with retention of configuration upon treatment with *n*-butylmagnesium bromide and a catalytic amount of $Ni(acac)_2$.[69] 2-Arenesulfonyl-1,3-dienes are also reduced stereospecifically to conjugated (*Z,E*)-dienes under the same conditions (eq 30).[70]

SO_2Ph / Et / C_6H_{11} —[BuMgBr, $Ni(acac)_2$; THF, rt, 1 h; 51%]→ Et / C_6H_{11} (30)

Miscellaneous Uses of $Ni(acac)_2$. The isomerization of aldoximes to amides catalyzed by $Ni(acac)_2$ and ***Palladium(II) Acetylacetonate*** has been described.[71] The determination of the absolute configuration of vicinal glycols and amino alcohols complexed with Ni(acac) in protic or aprotic organic solvents has been claimed to be feasible by examination of the induced CD.[72] However, this method is not very general.

Related Reagents. Bis(1,5-cyclooctadiene)nickel(0); Chromium(II) Chloride–Nickel(II) Chloride; Nickel Boride; Nickel(II) Bromide; Nickel Catalysts (Heterogeneous); Nickel(II) Chloride; Tetrakis(triphenylphosphine)nickel(0).

---

1. (a) Jolly, P. W.; Wilke, G. In *The Organic Chemistry of Nickel*; Academic: New York, 1975; Vols. 1 and 2. (b) Jolly, P. W. In *Comprehensive Organometallic Chemistry*; Wilkinson, G., Ed.; Pergamon: New York, 1982; Vols. 7 and 8. (c) Schmidt, F. K.; Ratovskii, G. V.; Dmitrieva, T. V.; Ivleva, I. N.; Borodko, Y. G. *JOM* **1983**, *256*, 309. (d) Canoira, L.; Rodrigez, J. G. *JHC* **1985**, *22*, 1511.
2. (a) Beger, J.; Duschek, C.; Fullbier, H. *ZC* **1973**, *13*, 59. (b) Beger, J.; Duschek, C.; Fullbier, H.; Gaube, W. *JPR* **1974**, *316*, 26.
3. (a) Furukawa, J.; Kiji, J.; Mitani, S.; Yoshikawa, S.; Yamamoto, K.; Sasakawa, E. *CL* **1972**, 1211. (b) Baker, R.; Halliday, D. E.; Smith, T. N. *CC* **1971**, 1583.
4. Ohta, T.; Ebina, K.; Yamazaki, N. *BCJ* **1971**, *44*, 1321.
5. Bogdanovic, B.; Henc, B.; Meister, B.; Pauling, H.; Wilke, G. *AG* **1972**, *84*, 1070.
6. Bogdanovic, B.; Henc, B.; Karmann, H.-G.; Nussel, H.-G.; Walter, D.; Wilke, G. *Ind. Eng. Chem.* **1970**, *62*, 34.
7. Bogdanovic, B.; Heimbach, P.; Kroner, M.; Wilke, G.; Hoffmann, E. G.; Brandt, J. *LA* **1969**, *727*, 143.
8. Heimbach, P.; Jolly, P. W.; Wilke, G. *Adv. Organomet. Chem.* **1970**, *8*, 29.
9. Brenner, W.; Heimbach, P.; Hey, H.; Muller, E. W.; Wilke, G. *LA* **1969**, *727*, 161.
10. Jolly, P. W.; Wilke, G. In *The Organic Chemistry of Nickel*; Academic: New York, 1975; Vol. 2, p 147.
11. Heimbach, P.; Meyer, R. V.; Wilke, G. *LA* **1975**, 743.
12. Seidov, N. M.; Geidarov, M. A. *Dokl. Akad. Nauk. Azerb. SSR.* **1972**, *28*, 33 (*CA* **1973**, *79*, 32 623).
13. (a) Heimbach, P.; Jolly, P. W.; Wilke, G. *Adv. Organomet. Chem.* **1970**, *8*, 29. (b) Garratt, P. J.; Wyatt, M. *CC* **1974**, 251.
14. Lappert, M. F.; Takahashi, S. *CC* **1972**, 1272.
15. Brenner, W.; Heimbach, P.; Ploner, K.-J.; Thomel, F. *LA* **1973**, 1882.
16. (a) Reppe, W.; Schlichting, O.; Meister, H. *LA* **1948**, *560*, 93. (b) Heimbach, P.; Ploner, K.-J.; Thomel, F. *AG* **1971**, *83*, 285. (c) Fahey, D. R. *JOC* **1972**, *37*, 4471. (d) Benson, R. E.; Lindsey, R. V. Jr. *JACS* **1959**, *81*, 4247. (e) Benson, R. E.; Lindsey, R. V. Jr. *JACS* **1959**, *81*, 4250.
17. (a) Cope, A. C.; Rugen, D. F. *JACS* **1952**, *74*, 3215. (b) Cope, A. C.; Pike, R. M. *JACS* **1953**, *75*, 3220. (c) Cope, A. C.; Campbell, H. C. *JACS* **1951**, *73*, 3536. (d) Cope, A. C.; Campbell, H. C. *JACS* **1952**, *74*, 179.
18. (a) Reikhsfeld, V. O.; Lein, B. I.; Makovetskii, K. L. *Proc. Acad. Sci. USSR* **1970**, *190*, 31. (b) Schauzer, G. N.; Eichler, S. *CB* **1962**, *95*, 550.
19. (a) Hagihara, N. *J. Chem. Soc. Jpn.* **1952**, *73*, 323 (*CA* **1953**, *47*, 10 490). (b) Hagihara, N. *J. Chem. Soc. Jpn.* **1952**, *73*, 373 (*CA* **1953**, *47*, 10 491).
20. (a) Schauzer, G. N.; Eichler, S. *CB* **1962**, *95*, 550. (b) Reppe, W.; Kutepow, N. Von-Magin, A. *AG* **1969**, *81*, 717. (c) Wittig, G.; Fritze, P. *LA* **1968**, *712*, 79.
21. Chukhadzhyan, G. A.; Sarkisyan, E. L.; Elbakyan, T. S. *JOU* **1972**, *8*, 1133.
22. (a) Noyori, R.; Suzuki, T.; Kumagai, Y.; Takaya, H. *JACS* **1971**, *93*, 5894. (b) Noyori, R.; Suzuki, T.; Takaya, H. *JACS* **1971**, *93*, 5896. (c) Noyori, R.; Odagi, T.; Takaya, H. *JACS* **1970**, *92*, 5780. (d) Noyori, R.; Ishigami, T.; Hayashi, N.; Takaya, H. *JACS* **1973**, *95*, 1674.
23. Yamago, S.; Nakamura, E. *T* **1989**, *45*, 3081.
24. (a) Kiso, Y.; Kumada, M.; Tamao, K.; Umeno, M. *JOM* **1973**, *50*, 297. (b) Beger, J.; Duschek, C.; Fullbier, H. *ZC* **1973**, *13*, 59.
25. Baker, R.; Halliday, D. E.; Smith, T. N. *JOM* **1972**, *35*, C61.
26. Lappert, M. F.; Takahashi, S. *CC* **1972**, 1272.
27. (a) Ohta, T.; Ebina, K.; Yamazaki, N. *BCJ* **1971**, *44*, 1321. (b) Beger, J.; Duschek, C.; Fullbier, H. *ZC* **1973**, *13*, 59.
28. Tamao, K.; Kobayashi, K.; Ito, Y. *JACS* **1989**, *111*, 6478.
29. Salimgareeva, I. M.; Kaverin, V. V.; Yur'ev, V. P. *JOM* **1978**, *148*, 23.
30. Arbeiten, N.; Bogdanovic, B.; Henc, B.; Losler, A.; Meister, B.; Pauling, H.; Wilke, G. *AG* **1973**, *85*, 1013.
31. Bogdanovic, B.; Henc, B.; Meister, B.; Pauling, H.; Wilke, G. *AG* **1972**, *84*, 1070.
32. Bogdanovic, B.; Henc, B.; Karmann, H.-G.; Nussel, H.-G.; Walter, D.; Wilke, G. *Ind. Eng. Chem.* **1970**, *62*, 34.
33. Yamada, T.; Takahashi, K.; Kato, K.; Takai, T.; Inoki, S.; Mukaiyama, T. *CL* **1991**, 641.
34. Yamada, T.; Takai, T.; Rhode, O.; Mukaiyama, T. *CL* **1991**, 1.
35. (a) Yamada, T.; Takai, T.; Rhode, O.; Mukaiyama, T. *BCJ* **1991**, *64*, 2109. (b) Bouhlel, E.; Laszlo, P.; Levart, M.; Montaufier, M. T.; Singh, G. P. *TL* **1993**, *34*, 1123.
36. Mukaiyama, T.; Takai, T.; Yamada, T.; Rhode, O. *CL* **1990**, 1661.
37. (a) Yamada, T.; Rhode, O.; Takai, T.; Mukaiyama, T. *CL* **1991**, 5. (b) Nishida, Y.; Fujimoto, T.; Tanaka, N. *CL* **1992**, 1291.
38. Petrier, C.; De Souza Barbosa, J. C.; Dupuy, C.; Luche, J.-L. *JOC* **1985**, *50*, 5761.
39. (a) Greene, A. E.; Lansard, J.-Ph.; Luche, J.-L.; Petrier, C. *JOC* **1984**, *49*, 931. (b) Casares, A.; Maldonado, L. A. *SC* **1976**, *6*, 11.
40. Dayrit, F. M.; Schwartz, J. *JACS* **1981**, *103*, 4466.
41. Schwartz, J.; Loots, M. J.; Kosugi, H. *JACS* **1980**, *102*, 1333.
42. (a) Schwartz, J.; Carr, D. B.; Hansen, R. T.; Dayrit, F. M. *JOC* **1980**, *45*, 3053. (b) Hansen, R. T.; Carr, D. B.; Schwartz, J. *JACS* **1978**, *100*, 2244.
43. Fukamiya, N.; Oki, M.; Aratani, T. *CI(L)* **1981**, *17*, 606.
44. Basato, M.; Corain, B.; De Roni, P.; Favero, G.; Jaforte, R. *J. Mol. Catal.* **1987**, *42*, 115.
45. Shafizadeh, F.; Ward, D. D.; Pang, D. *Carbohydr. Res.* **1982**, *102*, 217.
46. (a) Dayrit, F. M.; Schwartz, J. *JACS* **1981**, *103*, 4466. (b) Bolm, C.; Ewald, M.; Felder, M. *CB* **1992**, *125*, 1205, 1781.

47. (a) Corma, A.; Iglesias, M.; Martin, M. V.; Rubio, J.; Sanchez, F. *TA* **1992**, *3*, 845. (b) Uemura, M.; Miyake, R.; Nakayama, K.; Hayashi, Y. *TA* **1992**, *3*, 713. (c) Bolm, C.; Felder, M.; Muller, J. *SL* **1992**, 439. (d) Botteghi, C.; Paganelli, S.; Schionato, A.; Boga, C.; Fava, A. *J. Mol. Catal.* **1991**, *66*, 7. (e) Bolm, C.; Ewald, M. *TL* **1990**, *31*, 5011. (f) Soai, K.; Hayasaka, T.; Ugajin, S. *CC* **1989**, 516.
48. Bolm, C. *TA* **1991**, *2*, 701.
49. (a) Ibuki, E.; Ozasa, S.; Fujioka, Y.; Okada, M.; Terada, K. *BCJ* **1980**, *53*, 821. (b) Rodrigez, J. G.; Canoira, L. *React. Kinet. Catal. Lett.* **1989**, *38*, 337.
50. Hayashi, T.; Katsuro, Y.; Kumada, M. *TL* **1980**, *21*, 3915.
51. Hayashi, T.; Katsuro, Y.; Okamoto, Y.; Kumada, M. *TL* **1981**, *22*, 4449.
52. Sengupta, S.; Leite, M.; Raslan, D. S.; Quesnelle, C.; Snieckus, V. *JOC* **1992**, *57*, 4066.
53. Negishi, E.; Takahashi, T.; Baba, S.; Van Horn, D. E.; Okukado, N. *JACS* **1987**, *109*, 2393.
54. Eapen, K. C.; Dua, S. S.; Tamborski, C. *JOC* **1984**, *49*, 478.
55. Negishi, E.; Takahashi, T.; Baba, S.; Van Horn, D. E.; Okukado, N. *JACS* **1987**, *109*, 2393.
56. Fabre, J. L.; Julia, M.; Verpeaux, J. N. *BSF* **1985**, 762.
57. Julia, M.; Verpeaux, J. N. *TL* **1982**, *23*, 2457.
58. (a) Snider, B. B.; Conn, R. S. E.; Karras, M. *TL* **1979**, 1679. (b) Conn, R. S. E.; Karras, M.; Snider, B. B. *Isr. J. Chem.* **1984**, *24*, 108.
59. Eisch, J. J.; Merkley, J. H. *JACS* **1979**, *101*, 1148.
60. (a) Bagnell, L.; Jeffrey, E. A.; Meisters, A.; Mole, T. *AJC* **1974**, *27*, 2577. (b) Shreiber, S. L.; Meyers, H. V. *JACS* **1988**, *110*, 5198.
61. Mukayiama, T.; Soga, T.; Takenoshit, H. *CL* **1989**, 997.
62. Cesare, V. A.; Gandolfi, V.; Corain, B.; Basato, M. *J. Mol. Catal.* **1986**, *36*, 339.
63. Basato, M.; Corain, B.; Cofler, M.; Veronese, A. C.; Zanotti, G. *CC* **1984**, 1593.
64. Veronese, A. C.; Talmelli, C.; Gandolfi, V.; Corain, B.; Basato, M. *J. Mol. Catal.* **1986**, *34*, 195.
65. Basato, M.; Corain, B.; Marcomini, A.; Valle, G.; Zanotti, G. *JCS(P2)* **1984**, 965.
66. Corriu, R. J. P.; Masse, J. P. R.; Meunier, B. *JOM* **1973**, *55*, 73.
67. Chow, Y. L.; Li, H.; Yang, M. S. *CJC* **1988**, *66*, 2920.
68. Corma, A.; Iglesias, M.; Del Pino, C.; Sanchez, F. *JOM* **1992**, *431*, 233.
69. Fabre, J. L.; Julia, M. *TL* **1983**, *24*, 4311.
70. Cuvigny, T.; Fabre, J. L.; Herve du Penhoat, C.; Julia, M. *TL* **1983**, *24*, 4319.
71. Leusink, A. J.; Meerbeek, T. G.; Noltes, J. G. *RTC* **1976**, *95*, 123.
72. Dillon, J.; Nakanishi, K. *JACS* **1975**, *97*, 5409.

Julien Doyon
The Ohio State University, Columbus, OH, USA

Nickel(II) Chloride[1]

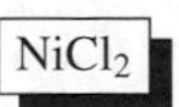

($NiCl_2$)		
[7718-54-9]	Cl_2Ni	(MW 129.59)
($NiCl_2 \cdot 6H_2O$)		
[7791-20-0]	$Cl_2H_{12}NiO_6$	(MW 237.71)

(mild Lewis acid;[2–5] catalyst for coupling reactions,[1a,7–12] and in combination with complex hydrides as a selective reducing agent[16,33,36–38])

Physical Data: mp 1001 °C; *d* 3.550 g cm^{-3}.
Solubility: sol H_2O, alcohol; insol most organic solvents.
Form Supplied in: yellow solid when anhydrous, green solid for the hydrate; widely available.
Drying: for anhydrous nickel chloride, standard procedure for drying metal chlorides can be used by refluxing with ***Thionyl Chloride*** followed by removal of excess $SOCl_2$.[45]
Handling, Storage, and Precautions: nickel(II) is reputed to be toxic and a cancer suspect agent. Use in a fume hood.

Mild Lewis Acid. Nickel chloride serves as a mild Lewis acid which promotes the regioselective rearrangement of dienols in aqueous *t*-BuOH at 60 °C in satisfactory yield (eq 1).[2] Brønsted acids give dehydration products, whereas other Lewis acids such as ***Nickel(II) Acetate***, ***Palladium(II) Chloride***, and ***Copper(II) Chloride*** proved less effective than nickel chloride and yield a mixture of rearranged and dehydration products. When anhydrous alcohol solvent is used, rearranged products bearing terminal alkoxy groups are obtained.

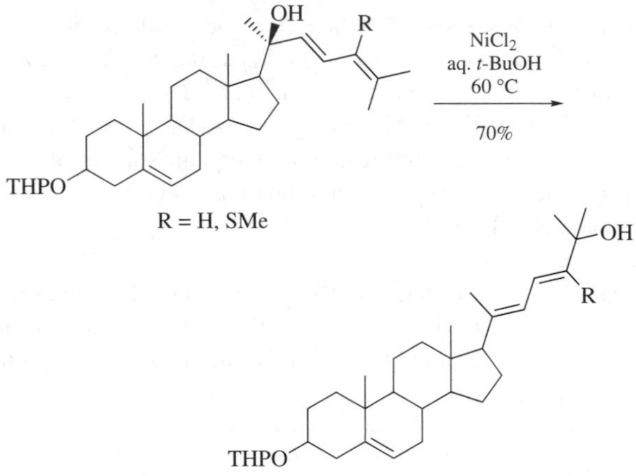

(1)

In the presence of a catalytic amount of $NiCl_2$, ***Cyanotrimethylsilane*** smoothly reacts with acetals or orthoesters derived from aromatic and α,β-unsaturated carbonyl compounds to give the corresponding α-cyano derivatives under neutral conditions (eq 2).[3] $NiCl_2$ can also accelerate the conversion of acrylamide to ethyl acrylate[4a] and catalyze the amination of 5,8-quinolinediones.[4b] The ring-opening reaction of epoxides with $LiAlR_4$ is catalyzed by $NiCl_2$ or ***Nickel(II) Bromide*** (eq 3).[5]

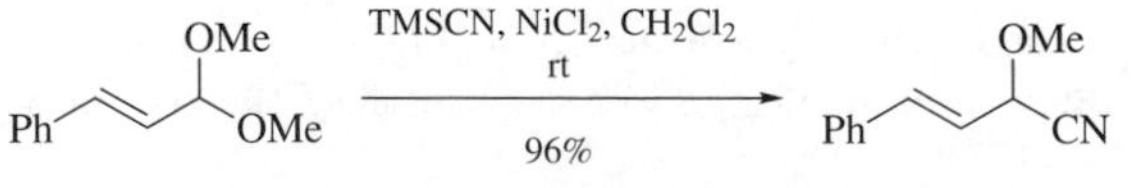

(2)

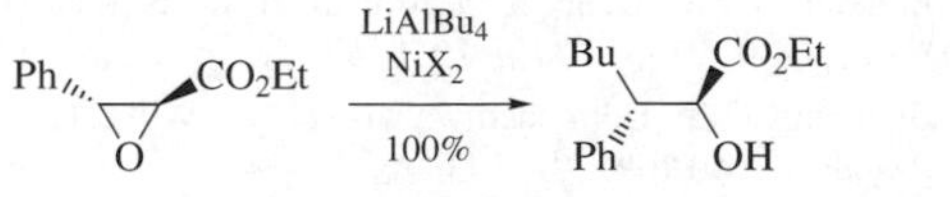

(3)

Nickel(II) Chloride–Chromium(II) Chloride. Although the ***Chromium(II) Chloride***-mediated reaction of an aldehyde with a vinylic iodide provides a useful entry for the preparation of allylic alcohol,[1a,6] the presence of a catalytic amount of $NiCl_2$ is essential to ensure the completion of the reaction.[7–12] Vinyl iodides (eq 4)[7] or triflates[8a] are commonly used. Alkynyl iodides behave similarly (eq 5).[9] Silyl enol ethers or enol phosphates are unreactive.

The stereochemistry of iodoalkenes is retained in the majority of cases with the exceptions of trisubstituted *cis*-iodoalkenes and *cis*-iodoenones, which afford exclusively the *trans*-alkenes instead of the expected *cis*-alkenes.[7a]

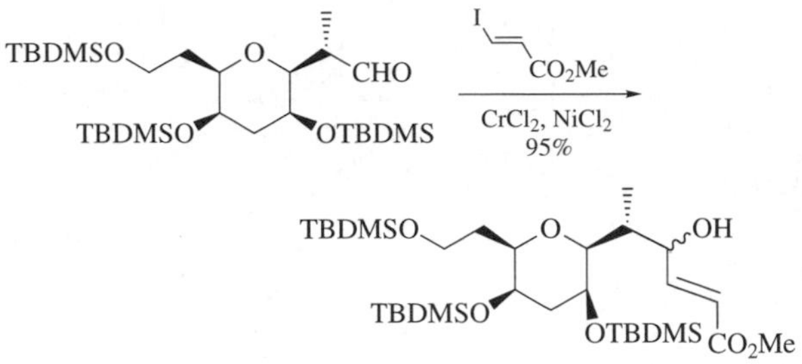

(4)

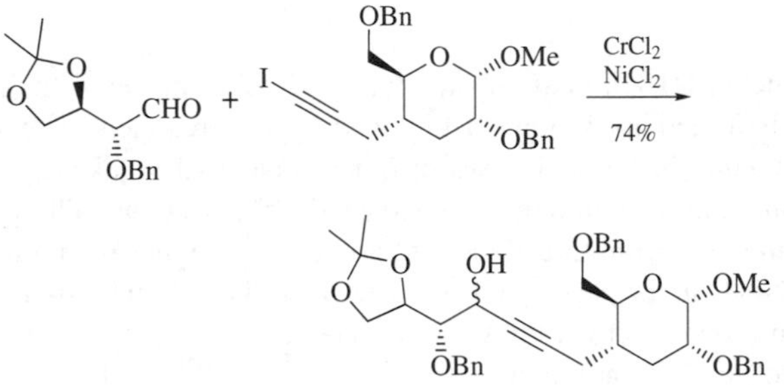

(5)

Functional groups such as esters, amides, nitriles, ketones, acetals, ethers, silyl ethers (TBDMS or TBDPS), alcohols, alkenes, and triple bonds are stable under the reaction conditions. Substrates containing structural complexity can be employed in this transformation. Thus, the reaction served as the key step for the formation of C(7)–C(8) and C(84)–C(85) bonds in the total synthesis of palytoxin,[10] as well as for the synthesis of other natural products and C-saccharides. The reagent has also been proved to be useful in the cyclization of the aldehydes (eqs 6 and 7).[11,12]

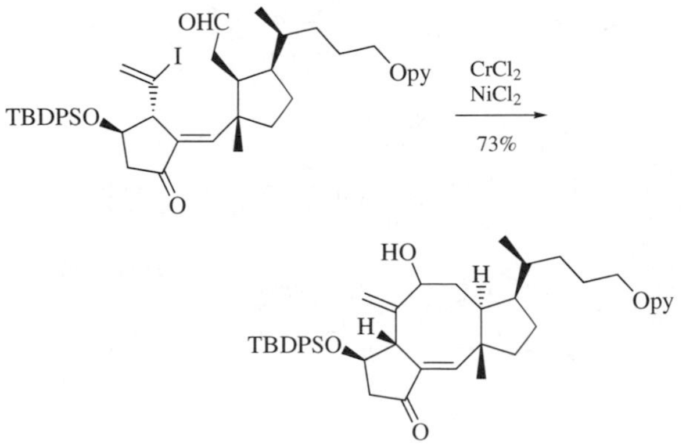

(6)

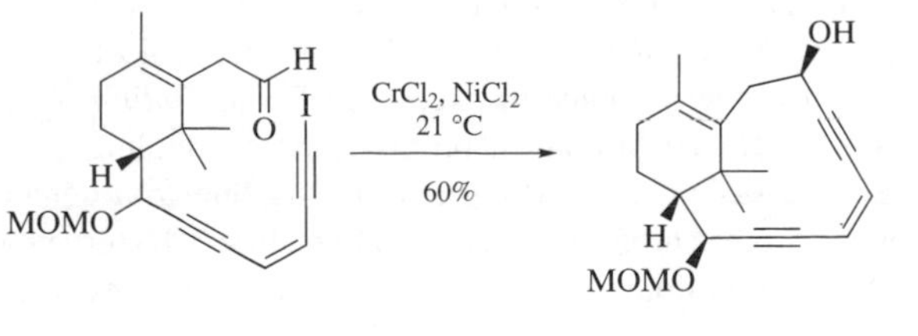

(7)

A simple and selective method for the conversion of an aldehyde to vinyl iodides, (*E*)-RCH=CHX, by means of ***Chromium(II) Chloride–Haloform*** has been developed (eq 8).[13]

$$\text{PhCHO} \xrightarrow[\text{THF, 91\%}]{\text{CHI}_3,\ \text{CrCl}_2} \text{PhCH=CHI} \quad (8)$$

DMF happens to be the most effective solvent for this coupling reaction. The reaction goes slowly but cleanly in the DMSO solvent.[8a] The presence of a phosphine ligand in the nickel catalyst gives a diene sideproduct.[4a] Nevertheless, the latter reagent has been used in the intramolecular cyclization of enynes (eq 9).[14]

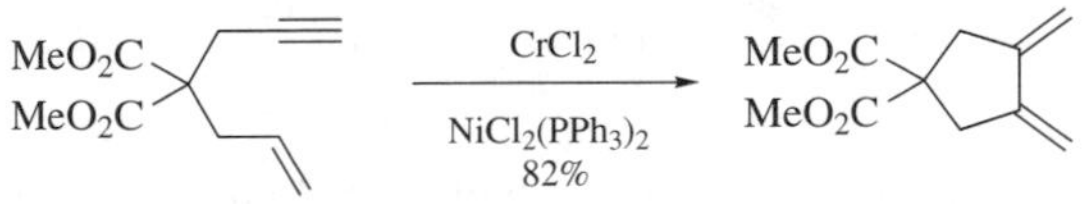

(9)

Monosubstituted α,β-unsaturated aldehydes are converted to cyclopropanols in the presence of $NiCl_2/CrCl_2$ in moderate yields (eq 10).[15]

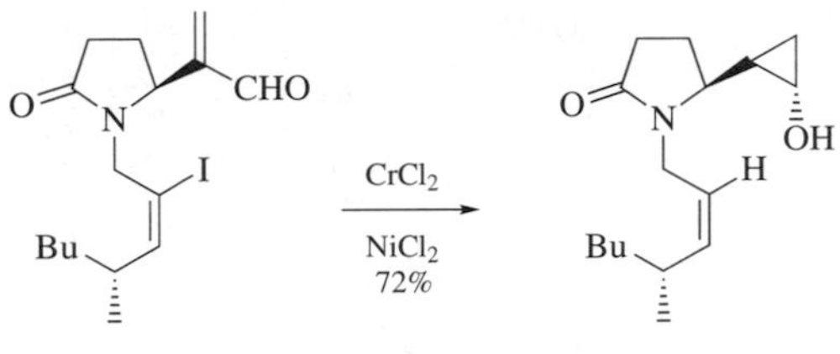

(10)

Selective Reductions. Low-valent transition metal complexes generated in situ from metal halides and reducing agents are particularly useful for the selective reduction of various functionalities.[16] Nickel chloride and nickel bromide have demonstrated a unique role in these reduction reactions. To illustrate this, in the presence of an equimolar quantity $NiCl_2$, ***Lithium Aluminum Hydride*** can reduce alkenes to alkanes in excellent yields.[17] Under similar conditions at −40 °C, alkynes are reduced to *cis*-alkenes in good yield.[17] Haloalkanes are also smoothly converted into the corresponding hydrocarbons under these conditions.[18a,b] Even chlorobenzene and 1-bromoadamantane can be reduced efficiently by this reagent. ***Sodium Hydride*** in the presence of $NiCl_2$ or $NiBr_2$ and a sodium alkoxide can also serve a similar purpose.[16c,18c]

The N–O bond in isoxazolidines is cleaved efficiently by $LiAlH_4/NiCl_2$ at −40 °C (eq 11).[19] Styrene oxide yields β-phenylethanol in 95% yield by this complex reagent, whereas $LiAlH_4$ alone gives α-phenylethanol.[19]

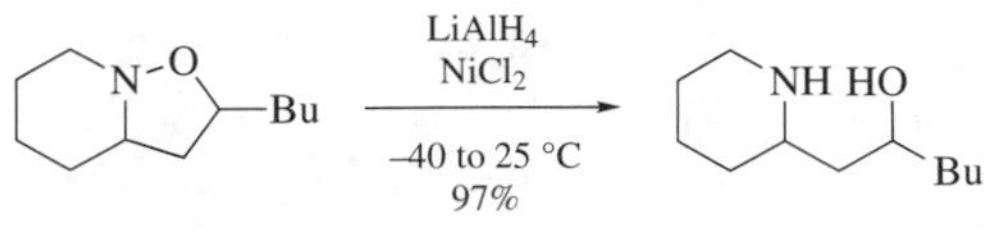

(11)

Nickel Boride, prepared in situ from the reaction of nickel chloride and ***Sodium Borohydride***, behaves like ***Raney Nickel***.[1b] In DMF, the dark brown/black solution comprises an efficient catalyst for alkene hydrogenation. The carbon–carbon double bonds of the α,β-unsaturated carbonyl compounds are reduced selectively (eq 12).[20] It is noted that carbon–sulfur bonds are selectively reduced under similar conditions (eq 13).[16b,21] Thiols, sulfides, disulfides, dithioacetals, as well as sulfoxides can all be hydrodesulfurized smoothly. Sulfones, on the other hand, remain intact under the reaction conditions.[21d,e]

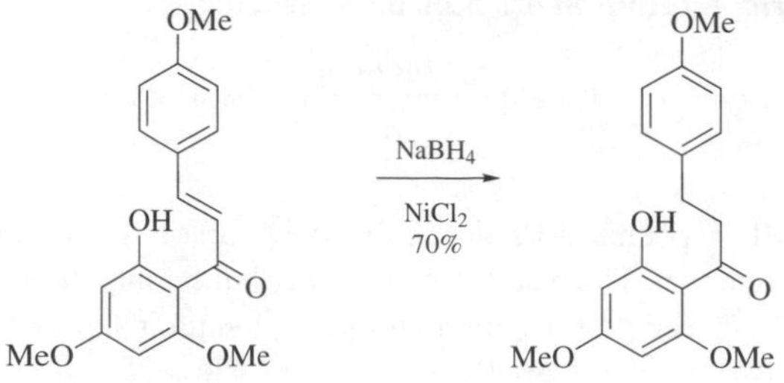

(12)

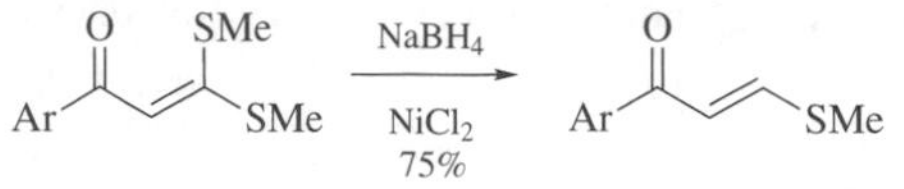

(13)

Reduction of α-halo ketones with nickel boride produces the corresponding ketones.[22] The carbon–oxygen bonds in allylic ethers,[23a] benzylic esters,[23b] as well as aryl tosylates[23c] are reduced to the corresponding C–H bonds (eq 14).

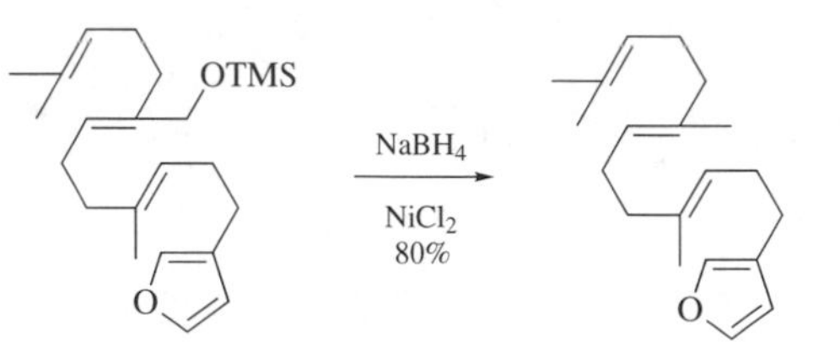

(14)

Upon treatment with $NiCl_2/NaBH_4$, nitro,[24a–c] azide,[24d,e] and oxime[24f] groups are smoothly transformed into amino groups in good yields. Carbon–carbon double bonds are occasionally reduced under these conditions.[24a,f] Nitro and cyano groups are also reduced to amines by the reagent mixture $NiCl_2/B_2H_6$.[25] Ketones, aldehydes, carboxylic acids, alkenes, esters, and amides, moieties are unaffected under these conditions.

Addition of TMSCN to an allene is catalyzed by nickel boride generated in situ, although the reaction is nonstereoselective (eq 15).[26]

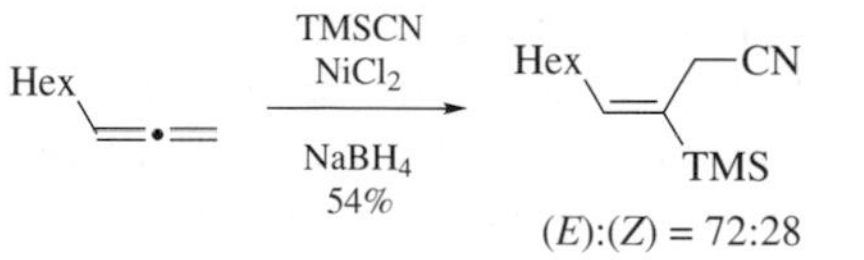

(15)

(*E*):(*Z*) = 72:28

Treatment of ***Diphenylacetylene*** with excess TMSCN in the presence of the $NiCl_2$/***Diisobutylaluminum Hydride*** or $NiCl_2$/***Triethylaluminum*** catalyst affords a substituted pyrrole in high yield (eq 16).[27]

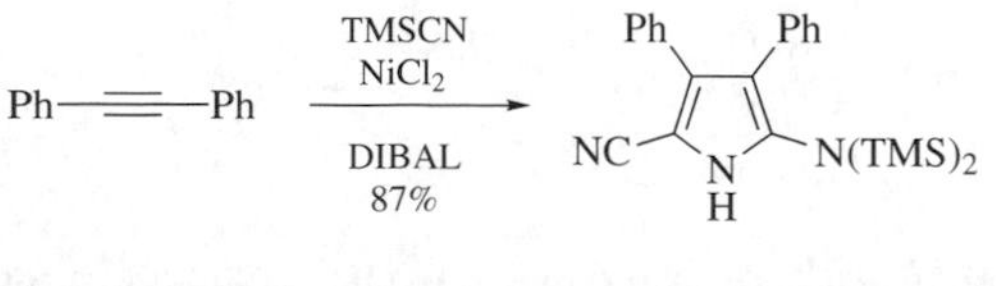

(16)

Hydrosilylation of conjugated dienes with $HSiR_3$ is catalyzed by $NiCl_2/Et_3Al$ in excellent yield; 1,4-addition is observed exclusively (eq 17).[28]

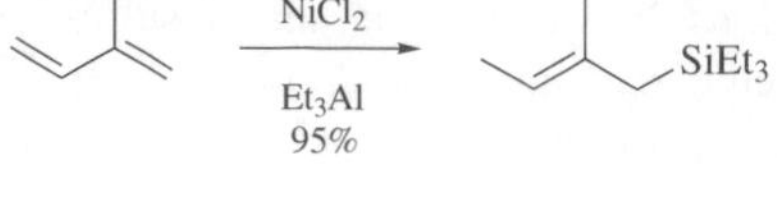

(17)

A combination of ***Aluminum*** and $NiCl_2$ promotes the selective reduction of α,β-enones to the corresponding saturated carbonyl compounds (eq 18).[29] Both nitro groups[30] and aryl ketones[29] are reduced to amines and benzylic alcohols, respectively.

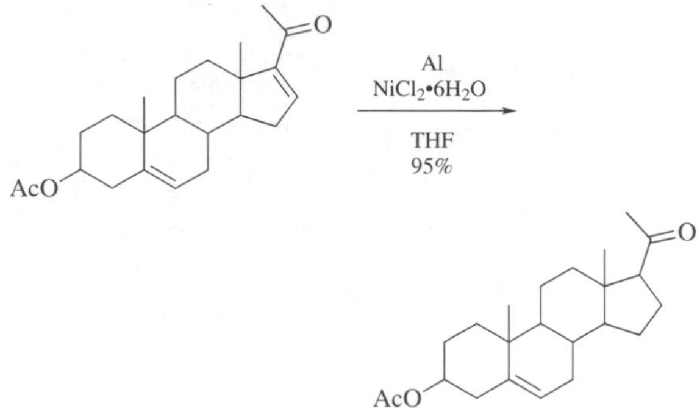

(18)

Nickel(II) Chloride–Zinc. Finely divided nickel with high catalytic activity is readily obtained by the treatment of $NiCl_2$ with ***Zinc*** dust.[31] This reagent reduces aldehydes, alkenes, and aromatic nitro compounds in good yields.[32a] Nitriles as well as aryl ketones give a mixture of reduced products under these conditions. $Zn/NiCl_2$ in the presence ***Ammonia***/NH_4^+ buffer (pH 6–10)[3] has been shown to effect the selective reduction of α,β-enones to the corresponding saturated carbonyl compounds.[32b] Aryl, allyl, and alkyl halides are reduced by water, zinc, and a catalytic amount of $NiCl_2$, ***Triphenylphosphine***, and iodide ion.[32c]

Reductive Heck-Like Reactions. Reductive Heck-like reactions (eq 19) can be achieved when alkyl, aryl, and vinyl bromides are treated with zinc/$NiCl_2 \cdot 6H_2O$ in the presence of an excess quantity of α,β-unsaturated esters.[33] A trace amount of water is essential for this conversion. Similar reactions are observed when alkenes are treated with iodofluoroacetate or iododifluoroacetate under the same conditions (eq 20).[34] Tandem reaction can also occur to give cyclic products (eq 21).[35]

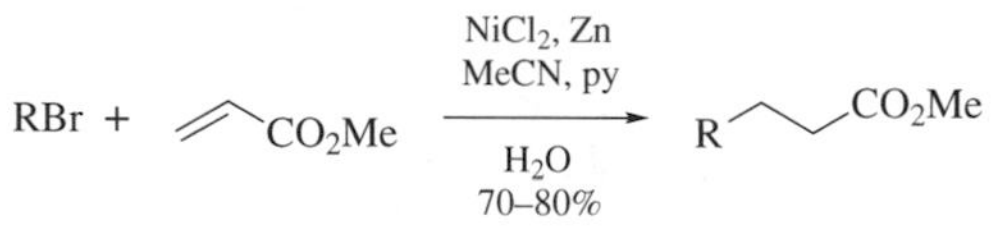

(19)

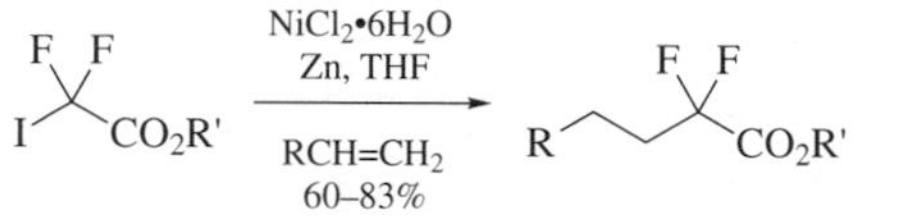

(20)

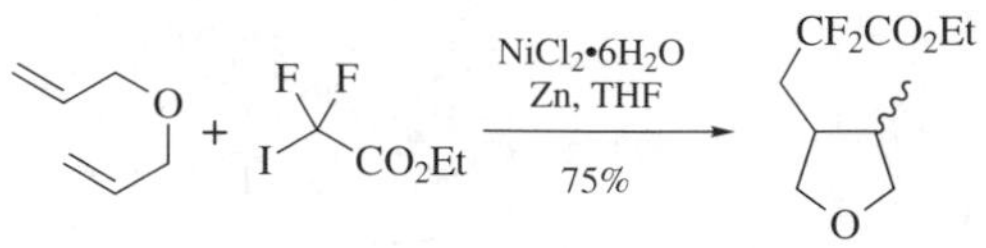

(21)

Homocoupling Reactions. In the absence of a Michael acceptor, aryl and vinyl halides undergo dimerization reaction upon treatment with the $NiCl_2/Zn$ reagent.[36–38] Under sonication conditions and in the presence of excess Ph_3P and ***Sodium Iodide*** in DMF, the $NiCl_2/Zn$ reagent promotes homocoupling of aryl triflates in good yields.[36] Bipyridyls having electron-donating groups, such as methoxy groups, are obtained in satisfactory yields under these conditions (eq 22).[37] Thiophene derivatives behave similarly.[38] Vinyl bromides dimerize to yield the corresponding butadienes.[39] It is interesting that the presence of iodide ion or thiourea can accelerate the reaction.

$NiCl_2$, Ph_3P, Zn, DMF, 50 °C; 88.5% (22)

Cross-Coupling Reactions. Most cross-coupling reactions using nickel catalysts require phosphine ligands and are therefore discussed in detail under ***Dichlorobis(triphenylphosphine)nickel(II)***. The reaction of aryl iodides or bromides with trialkyl phosphites in the presence of $NiCl_2$ is the premier method for preparing dialkyl arylphosphonates (eq 23).[40a,b] Thermolysis of allyl phosphite in the presence of $NiCl_2$ yields the corresponding allyl phosphonates (eq 24).[40c]

ArI + $P(OEt)_3$ —($NiCl_2$, 160 °C, 94%)→ $ArP(O)(OEt)_2$ (23)

$(EtO)_2P$–O–allyl —($NiCl_2$, 80 °C, 85%)→ $(EtO)_2P(O)$–allyl (24)

Miscellaneous Reactions. Symmetrical alkynes in the presence of $NiCl_2$ or $NiBr_2$ and ***Magnesium*** undergo trimerization to give the corresponding hexasubstituted aromatic compounds (eq 25). Terminal alkynes yield a mixture of regioisomers.[41]

Et—≡—Et —(NiX_2, Mg, 40–90%)→ hexaethylbenzene (25)

X = Cl, Br, I

Grignard reagents activated by a catalytic quantity of $NiCl_2$ can substitute the germanium–hydrogen bond with a germanium–carbon bond (eq 26).[42] It is noted that the stereochemistry of the original organogermane is retained.

i-PrPh(Naph)GeH —($H_2C{=}CHCH_2MgBr$, $NiCl_2$, 99%)→ *i*-PrPh(Naph)$GeCH_2CH{=}CH_2$ (26)

Hydromagnesiation of a styrene with ***Ethylmagnesium Bromide*** followed by treatment with ***Carbon Dioxide*** gives the 2-arylpropionic acids in good yield (eq 27).[43]

$NiCl_2$, EtMgBr, CO_2; 82% (CO_2H) (27)

Thermolysis of 1-phenyl-3,4-dimethylphosphole in the presence of $NiCl_2$ yields the corresponding nickel complex of the dimeric product. The ligand can be liberated upon treatment with ***Sodium Cyanide*** (eq 28).[44]

P–Ph —($NiCl_2$)→ —(NaCN)→ P(Ph)–P(Ph) dimer, 30% (28)

Related Reagents. Chromium(II) Chloride–Nickel(II) Chloride; Dichlorobis(triphenylphosphine)nickel(II); Lithium Aluminum Hydride–Nickel(II) Chloride; Nickel(II) Acetate; Nickel Boride; Nickel(II) Bromide.

1. (a) Cintas, P. *S* **1992**, 248. (b) Ganem, B.; Osby, J. O. *CR* **1986**, *86*, 763.
2. (a) Kyler, K. S.; Watt, D. S. *JACS* **1983**, *105*, 619. (b) Kyler, K. S.; Bashir-Hashemi, A.; Watt, D. S. *JOC* **1984**, *49*, 1084.
3. Mukaiyama, T.; Soga, T.; Takenoshita, H. *CL* **1989**, 997.
4. (a) Czarnik, A. W. *TL* **1984**, *25*, 4875. (b) Yoshida, K.; Yamamoto, M.; Ishiguro, M. *CL* **1986**, 1059.
5. Boireau, G.; Abenhaim, D.; Bernardon, C.; Henry-Basch, E.; Sabourault, B. *TL* **1975**, 2521. Boireau, G.; Abenhaim, D.; Henry-Basch, E. *T* **1980**, *36*, 3061.
6. Takai, K.; Kimura, K.; Kuroda, T.; Hiyama, T.; Nozaki, H. *TL* **1983**, *24*, 5281.
7. (a) Jin, H.; Uenishi, J.-i.; Christ, W. J.; Kishi, Y. *JACS* **1986**, *108*, 5644. (b) Aicher, T. D.; Buszek, K. R.; Fang, F. G.; Forsyth, C. J.; Jung, S. H.; Kishi, Y.; Matelich, M. C.; Scola, P. M.; Spero, D. M.; Yeon, S. K. *JACS* **1992**, *114*, 3162. (c) Aicher, T. D.; Buszek, K. R.; Fang, F. G.; Forsyth, C. J.; Jung, S. H.; Kishi, Y.; Scola, P. M. *TL* **1992**, *33*, 1549. (d) Dyer, U. C.; Kishi, Y. *JOC* **1988**, *53*, 3383. (e) Goekjian, P. G.; Wu, T.-C.; Kang, H.-Y.; Kishi, Y. *JOC* **1987**, *52*, 4823. (f) Chen, S. H.; Horvath, R. F.; Joglar, J.; Fisher, M. J.; Danishefsky, S. J. *JOC* **1991**, *56*, 5834.
8. (a) Takai, K.; Tagashira, M.; Kuroda, T.; Oshima, K.; Utimoto, K.; Nozaki, H. *JACS* **1986**, *108*, 6048. (b) Angell, R.; Parsons, P. J.; Naylor, A.; Tyrrell, E. *SL* **1992**, 599.
9. (a) Wang, Y.; Babirad, S. A.; Kishi, Y. *JOC* **1992**, *57*, 468. (b) Aicher, T. D.; Kishi, Y. *TL* **1987**, *28*, 3463.
10. Armstrong, R. W.; Beau, J. M.; Cheon, S. H.; Christ, W. J.; Fujioka, H.; Ham, W.-H.; Hawkins, L. D.; Jin, H.; Kang, S. H.; Kishi, Y.; Martinelli, M. J.; McWhorter, W. W., Jr.; Mizuno, M. Nakata, M.; Stutz, A. E.; Talamas, F. X.; Taniguchi, M.; Tino, J. A.; Ueda, K.; Uenishi, J. I.; White, J. B.; Yonaga, M. *JACS* **1989**, *111*, 7525. (b) Kishi, Y. *PAC* **1989**, *61*, 313.
11. (a) Rowley, M.; Tsukamoto, M.; Kishi, Y. *JACS* **1989**, *111*, 2735. (b) Rowley, M.; Kishi, Y. *TL* **1988**, *29*, 4909.
12. (a) Crévisy, C.; Beau, J. M. *TL* **1991**, *32*, 3171. (b) Lu, Y.-F.; Harwig, C. W.; Fallis, A. G. *JOC* **1993**, *58*, 4204.
13. Takai, K.; Nitta, K.; Utimoto, K. *JACS* **1986**, *108*, 7408.
14. Trost, B. M.; Tour, J. M. *JACS* **1987**, *109*, 5268.
15. Montgomery, D.; Reynolds, K.; Stevenson, P. *CC* **1993**, 363.
16. (a) Pons, J.-M.; Santelli, M. *T* **1988**, *44*, 4295. (b) Luh, T.-Y.; Ni, Z.-J. *S* **1990**, 89. (c) Caubère, P. *AG(E)* **1983**, *22*, 599.
17. Ashby, E. C.; Lin, J. J. *JOC* **1978**, *43*, 2567.
18. (a) Ashby, E. C.; Lin, J. J. *TL* **1977**, 4481. (b) Ashby E. C.; Lin, J. J. *JOC* **1978**, *43*, 1263. (c) Brunet, J. J.; Vanderesse, R.; Caubere, P. *JOM* **1978**, *157*, 125.
19. Tufariello, J. J.; Meckler, H.; Pushpananda, K.; Senaratne, A. *T* **1985**, *41*, 3447.

20. (a) Dhawan, D.; Grover, S. K. *SC* **1992**, *22*, 2405. (b) Abe, N.; Fujisaki, F.; Sumoto, K.; Miyano, S. *CPB* **1991**, *39*, 1167.
21. (a) Myrboh, B.; Singh, L. W.; Ila, H.; Junjappa, H. *S* **1982**, 307. (b) Euerby, M. R.; Waigh, R. D. *SC* **1986**, *16*, 779. (c) Nishio, T.; Omote, Y. *CL* **1979**, 1223. (d) Truce, W. E.; Perry, F. M. *JOC* **1965**, *30*, 1316. (e) Back, T. G. *CC* **1984**, 1417.
22. Sarma, J. C.; Borbaruah, M.; Sharma, R. P. *TL* **1985**, *26*, 4657.
23. (a) He, Y.; Pan, X.; Zhao, H.; Wang, S. *SC* **1989**, *19*, 3051. (b) Sharma, D. N.; Sarma, R. P. *TL* **1985**, *26*, 371. (c) Wang, F.; Chiba, K.; Tada, M. *JCS(P1)* **1992**, 1897.
24. (a) Nose, A.; Kudo, T. *CPB* **1988**, *36*, 1529. (b) Hanaya, K., Fujita, N.; Kudo, H. *CI(L)* **1973**, 794. (c) Osby, J. O.; Ganem, B. *TL* **1985**, *26*, 6413. (d) Sarma, J. C.; Sharma, R. P. *CI(L)* **1987**, 764. (e) Rao, H. S. P.; Reddy, K. S.; Turnbull, K.; Borchers, V. *SC* **1992**, *22*, 1339. (f) Ipaktschi, J. *CB* **1984**, *117*, 856.
25. (a) Nose, A., Kudo, T. *CPB* **1986**, *34*, 3905. (b) Satoh, T.; Suzuki, S.; Suzuki, Y.; Miyaji, Y.; Imai, Z. *TL* **1969**, 4555.
26. Chatani, N.; Takeyasu, T.; Hanafusa, T. *TL* **1986**, *27*, 1841.
27. Chatani, N.; Hanafusa, T. *TL* **1986**, *27*, 4201.
28. Lappert, M. F.; Nile, T. A.; Takahashi, S. *JOM* **1974**, *72*, 425.
29. Hazarika, M. J.; Barua, N. C. *TL* **1989**, *30*, 6567.
30. Sarmah, P.; Barua, N. C. *TL* **1990**, *31*, 4065.
31. (a) Sakai, K.; Watanabe, K. *BCJ* **1967**, *40*, 1548. (b) Rieke, R. D.; Kavaliunas, A. V.; Rhyne, L. D.; Fraser, D. J. J. *JACS* **1979**, *101*, 246.
32. (a) Nose, A.; Kudo, T. *CPB* **1990**, *38*, 2097. (b) Petrier, C.; Luche, J.-L. *TL* **1987**, *28*, 2351. (c) Colon, I. *JOC* **1982**, *47*, 2622.
33. Sustmann, R.; Hopp, P.; Holl, P. *TL* **1989**, *30*, 689.
34. (a) Wang, Y.; Yang, Z.-Y.; Burton, D. J. *TL* **1992**, *33*, 2137. (b) Yang, Z.-Y.; Burton, D. J. *JOC* **1992**, *57*, 5144.
35. Yang, Z.-Y.; Burton, D. J. *TL* **1991**, *32*, 1019.
36. Yamashita, J.; Inoue, Y.; Kondo, T.; Hashimoto, H. *CL* **1986**, 407.
37. (a) Tiecco, M.; Testaferri, L.; Tingoli, M.; Chianelli, D.; Montanucci, M. *S* **1984**, 736. (b) Tiecco, M.; Tingoli, M.; Testaferri, L.; Chianelli, D.; Wenkert, E. *T* **1986**, *42*, 1475. (c) Tiecco, M.; Tingoli, M.; Testaferri, L.; Bartoli, D.; Chianelli, D. *T* **1989**, *45*, 2857.
38. Sone, T.; Umetsu, Y.; Sato, K. *BCJ* **1991**, *64*, 864.
39. Takagi, K.; Hayama, N. *CL* **1983**, 637.
40. (a) Tavs, P. *CB* **1970**, *103*, 2428. (b) Balthazor, T. M.; Grabiak, R. C. *JOC* **1980**, *45*, 5425. (c) Lu, X.; Zhu, J. *JOM* **1986**, *304*, 239.
41. (a) Mauret, P.; Alphonse, P. *JOM* **1984**, *276*, 249. (b) Mauret, P.; Alphonse, P. *JOC* **1982**, *47*, 3322. (c) Alphonse, P.; Moyen, F.; Mazerolles, P. *JOM* **1988**, *345*, 209.
42. Carre, F. H.; Corriu, R. J. P. *JOM* **1974**, *74*, 49.
43. Amano, T.; Ota, T.; Yoshikawa, K.; Sano, T.; Ohuchi, Y.; Sato, F.; Shiono, M.; Fujita, Y. *BCJ* **1986**, *59*, 1656.
44. (a) Mercier, F.; Mathey, F.; Fischer, J.; Nelson, J. H. *JACS* **1984**, *106*, 425. (b) Mercier, F.; Mathey, F.; Fischer, J.; Nelson, J. H. *IC* **1985**, *24*, 4141.
45. Pray, A. R. *Inorg. Synth.* **1957**, *5*, 153.

Tien-Yau Luh & Yu-Tsai Hsieh
National Taiwan University, Taipei, Taiwan

Nitroethylene[1]

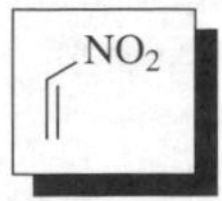

[3638-64-0] $C_2H_3NO_2$ (MW 73.06)

(Diels–Alder dienophile;[3] Michael acceptor[3])

Physical Data: bp 38–39 °C/80 mmHg.[2]

Analysis of Reagent Purity: titration against tetraphenylcyclopentadienone.[3]

Preparative Methods: best prepared by dehydration of 2-nitroethanol using phthalic anhydride (eq 1).[2,3] It can also be generated for reactions in situ from several precursors, especially 2-nitroethyl phenyl sulfoxide.[4]

HO–CH₂CH₂–NO_2 → (phthalic anhydride, 140–180 °C, 80 mmHg, 80%) → $CH_2=CH-NO_2$ (1)

Handling, Storage, and Precautions: lachrymatory oil. Polymerizes readily in the presence of water and violently with base. Purified compound darkens quickly, but 10% solutions in dry benzene can be kept for 6 months or longer when refrigerated.[3]

Cycloadditions. Nitroethylene is a very electron-deficient and highly reactive dienophile in Diels–Alder reactions with electron-rich or unactivated dienes. Its reactions with cyclopentadiene derivatives exhibit particularly high reactivity and selectivity, proceeding at temperatures as low as −100 °C to form only the *endo* product (eq 2).[3,5] Many other cyclic dienes show similar selectivity.[6] The adducts are readily converted to ketones, making nitroethylene a useful ketene equivalent.[5]

(eq 2: OMe-substituted cyclopentadiene + nitroethylene, −60 °C → MeO-bicyclic NO_2 adduct) (2)

Reactions with acyclic dienes proceed much more slowly, and are often carried out at 80–120 °C.[7] These reactions exhibit high regioselectivity with simple dienes.[8] Subsequent alkylation and denitration (***Tri-n-butylstannane***) of the adducts (eq 3) effects the overall synthetic equivalent of a regiocontrolled Diels–Alder reaction with 1-alkenes.[8]

Nitroethylene is also a reactive dipolarophile in 1,3-dipolar cycloadditions.[3,9] An interesting observation, rationalizable from FMO considerations, is that the regiochemistry of nitroethylene reactions with nitrones is often reversed from that observed with less electron-deficient alkenes (eq 4).[9a,c]

1. HCHO, NaOH
2. Ac_2O–py
3. Bu_3SnH, AIBN

80 °C, 5 h
86%

63%

95:5 isomer mixture

(3)

2:1

(4)

Radical Additions. Nitroethylene is a reactive and useful acceptor for nucleophilic radicals (eq 5).[10]

(5)

Nucleophilic Additions. A wide variety of nucleophiles have been used in conjugate additions to nitroethylene.[3,11] Avoiding the facile base-mediated polymerization of nitroethylene is crucial to the success of these additions. A broad study found that additions of amines with pK_a values between 2 and 8 are highly successful, while more basic amines lead to polymerization.[12] An exception is the pyrrolidine synthesis in eq 6, in which a second Michael addition traps the intermediate nitronate anion.[13]

25 °C
'good yield'

(6)

'Traditional' Michael additions of β-dicarbonyl compounds to nitroethylene in protic solvents are best carried out under the most mildly basic conditions possible (KF catalyzed,[14] for example). 'Kinetic' Michael additions of preformed enolates at low temperatures have also been reasonably successful.[15]

Related Reagents. 2-Acetoxyacrylonitrile; Acrylonitrile; 2-Chloroacrylonitrile; Ethyl Acrylate; Ethyl 3,3-Diethoxyacrylate; Methyl Acrylate; Methyl 3-Nitroacrylate; Nitromethane; 1-Nitro-1-propene; Phenylsulfinylethylene; Phenyl Vinyl Sulfide; Vinyl Acetate; 9-Vinyl-9-borabicyclo[3.3.1]nonane.

---

1. For a review of nitroalkene chemistry, see: Barrett, A. G. M.; Graboski, G. G. *CRV* **1986**, *86*, 751.
2. Buckley, G. D.; Scaife, C. W. *JCS* **1947**, 1471.
3. Ranganathan, D.; Rao, C. B.; Ranganathan, S.; Mehrotra, A, K.; Iyengar, R. *JOC* **1980**, *45*, 1185.
4. Ranganathan, S.; Ranganathan, D.; Singh, S. K. *TL* **1987**, *28*, 2893. See also ref 11.
5. Ranganathan, S.; Ranganathan, D.; Mehrotra, A. K. *JACS* **1974**, *96*, 5261.
6. (a) Posner, G. H.; Nelson, T. D.; Kinter, C. M.; Johnson, N. *JOC* **1992**, *57*, 4083. (b) Van Tamelen, E. E.; Zawacky, S. R. *TL* **1985**, *26*, 2833. (c) Corey, E. J.; Myers, A. G. *JACS* **1985**, *107*, 5574.
7. (a) Ono, N.; Kamimura, A.; Miyake, H.; Hamamoto, I.; Kaji, A. *JOC* **1985**, *50*, 3692. (b) Drake, N. L.; Kraekel, C. M. *JOC* **1961**, *26*, 41. (c) Kaplan, R. B.; Shechter, H. *JOC* **1961**, *26*, 982. (d) Zutterman, F.; Krief, A. *JOC* **1983**, *48*, 1135. (e) Ono, N.; Miyake, H.; Kamimura, A.; Tsukui, N.; Kaji, A. *TL* **1982**, *23*, 2957.
8. Ono, N.; Miyake, H.; Kamimura, A.; Kaji, A. *JCS(P1)* **1987**, 1929.
9. (a) Padwa, A.; Fisera, L.; Koehler, K. F.; Rodriguez, A.; Wong, G. S. K. *JOC* **1984**, *49*, 276. (b) Sasaki, T.; Eguchi, S.; Yamaguchi, M.; Esaki, T. *JOC* **1981**, *46*, 1800. (c) Sims, J.; Houk, K. N. *JACS* **1973**, *95*, 5798. (d) Baranski, A.; Cholewka, E. *Pol. J. Chem.* **1991**, *65*, 319. (e) Padwa, A.; Goldstein, S. I. *CJC* **1984**, *62*, 2506.
10. (a) Barton, D. H. R.; Crich, D.; Kretzschmar, G. *TL* **1984**, *25*, 1055. (b) Sumi, K.; Di Fabio, R.; Hanessian, S. *TL* **1992**, *33*, 749.
11. (a) Lambert, A.; Scaife, C. W.; Smith, A. E. W. *JCS* **1947**, 1474. (b) Heath, R. L.; Lambert, A. *JCS* **1947**, 1477. (c) Heath, R. L.; Piggott, H. A. *JCS* **1947**, 1481. (d) Heath, R. L.; Rose, J. D. *JCS* **1947**, 1486. (e) Pelter, A.; Hughes, L. *CC* **1977**, 913. (f) Confalone, P. N.; Lollar, E. D.; Pizzolato, G.; Uskokivic, M. R. *JACS* **1978**, *100*, 6291.
12. Ranganathan, D.; Ranganathan, S.; Bamezai, S. *TL* **1982**, *23*, 2789.
13. Barco, A.; Benetti, S.; Casolari, A.; Pollini, G. P.; Spalluto, G. *TL* **1990**, *31*, 3039.
14. Yanami, T.; Kato, M.; Yoshikoshi, A. *CC* **1975**, 726.
15. Curran, D. P.; Jacobs, P. B.; Elliott, R. L.; Kim, B. H. *JACS* **1987**, *109*, 5280. Chavdarian, C. G.; Seeman, J. I.; Wooten, J. B. *JOC* **1983**, *48*, 492.

Daniel A. Singleton
Texas A&M University, College Station, TX, USA

Nitromethane

$MeNO_2$

[75-52-5] CH_3NO_2 (MW 61.05)

(building block in synthesis; polar solvent)

Physical Data: mp −28.5 °C; bp 101 °C; *d* 1.13 g cm^{-3}; dipole moment 3.5 D.
Solubility: completely misc most organic solvents; slightly sol petroleum ether, water; sol alkaline solution.
Form Supplied in: colorless liquid, widely available. 3
Purification: purified by drying over $MgSO_4$ and distilling; a small acidic forerun is discarded.
Handling, Storage, and Precautions: stable compound. It is advised not to distill large quantities at reduced pressure. Flammable; toxic. Gives shock- and heat-sensitive alkali and heavy metal salts. Do not dry the sodium salt.

Introduction. Nitromethane is a common starting material for the synthesis of aliphatic nitro compounds, which serve as valuable building blocks, providing access to a variety of other functionalized products such as 2-nitro alcohols, nitroalkanes, nitroalkenes, hydrocarbons, amines, oximes, carbonyl compounds, and heterocycles.[1] Nitromethane reacts with alkali metal hydroxides and alkoxides to form the metal nitronates which are used as in situ generated reagents. The alkali metal nitronates are unstable and it is advisable not to isolate them. The sodium salt of nitromethane undergoes violent decomposition or detonation on heating and drying.

The Nitroaldol Reaction (Henry Reaction)[2].

Reactions with Aldehydes. In the presence of base, nitromethane reacts with aliphatic aldehydes in an aldol-type reaction with formation of 2-nitro alcohols (nitroaldols). Due to its reversibility, the reaction is normally carried out in the presence of only catalytic quantities of base, although in certain cases stoichiometric amounts are used to precipitate the product.[3] The reaction of nitromethane with aldehydes in the presence of one equivalent of base gives the salt of the *aci*-nitro tautomer of the product, which must be carefully acidified to avoid the Nef reaction. Alkali metal hydroxides, alkoxides or carbonates, and tertiary amines are effective catalysts.[2a] Elimination of water, the aldol reaction, and the Cannizzaro reaction are competing side reactions.[2a,b] 2-Nitro alcohols are unstable compounds and care has to be exercised during workup (decomposition during distillation). Fluoride ion catalysts often give higher yields.[4] The reaction has been carried out in the absence of solvent using powdered NaOH,[5] Al_2O_3,[6] and Al_2O_3-supported KF[7] as catalysts. The 2-nitro alcohols are readily converted into a variety of functionalities (eq 1).[1b,4,5,7–13,16–21]

(1)

Aromatic aldehydes react smoothly with nitromethane under the same conditions.[14] Elimination of water to give β-nitrostyrenes takes place on acidification.[2a] Synthesis of β-nitrostyrenes can be accomplished in one step by heating aromatic aldehydes in acetic acid and with NH_4OAc as catalyst.[15] Dehydration of 2-nitro alcohols to nitroalkenes has been accomplished with ***Methanesulfonyl Chloride***,[16] phthalic anhydride,[17] ***1,3-Dicyclohexylcarbodiimide***,[18] pivaloyl chloride,[19] ***Trifluoroacetic Anhydride***,[20] and ***Acetic Anhydride***.[20] Heating 2-nitro alcohols in dichloromethane in the presence of basic aluminium oxide is another mild method for the synthesis of nitro alkenes.[21] Nitroalkenes can be converted into a variety of functionalities (eq 2)[2,5,22,24–30] and have found utility as heterodienes in hetero Diels–Alder reactions and as reactants in Lewis acid promoted tandem [4 + 2]/[3 + 2] cycloadditions.[23]

A catalytic asymmetric nitroaldol reaction has been developed.[31] The catalyst is formed from ***Lanthanum(III) Chloride***, the dilithium salt of ***(R)-1,1′-Bi-2,2′-naphthol***, NaOH, and H_2O in a 1:1:1:10 molar ratio in THF. Nitromethane reacts with various aldehydes in the presence of 10% of the catalyst to give 2-nitro alcohols in high yields and with an ee as high as 93% (eq 3). Chiral binaphthol complexes of other rare earth metal trichlorides show similar catalytic effects. The optical purities of nitroaldols obtained using these rare earth metal complexes as asymmetric catalysts are highly dependent on the amount of water present and on the relation between substrate and radius of the rare earth metals.

(2)

MeNO₂, –50 °C; 10% BINOL–La complex; 76%; 93% ee (3)

Nitromethane can react with two mol equiv of an aldehyde.[2a] The reaction with the second aldehyde molecule proceeds slower than that with the first due to the lower acidity of the α-proton in 2-nitro alcohols and enhanced steric hindrance. With sterically hindered aldehydes, the reaction with the second aldehyde is difficult.[2b,d] Seebach et al.[32] synthesized the multiple coupling reagent 2-nitro-2-propenyl 2,2-dimethylpropanoate (***NNP 2-Nitro-3-pivaloyloxy propene***) from nitromethane. In the initial step, nitromethane reacts with two moles of formaldehyde to give 1,3-dihydroxy-2-nitropropane in 95% yield. Subsequent acylation with two mol equiv of pivaloyl chloride and elimination of pivalic acid gives NNP. The reaction may be run on a 40 to 200 g scale without problems (eq 4). NNP allows succesive introduction of two different nucleophiles Nu^1 and Nu^2 (eq 5).

$2\,CH_2O + CH_2{=}NO_2Na \xrightarrow[95\%]{MeOH}$ HO–CH₂–C(=NO₂Na)–CH₂–OH $\xrightarrow[75\%]{H^+}$

HO–CH₂–CH(NO₂)–CH₂–OH $\xrightarrow[95\%]{t\text{-BuCOCl, }CH_2Cl_2}$

t-BuCO₂–CH₂–CH(NO₂)–CH₂–O₂C-t-Bu $\xrightarrow[45\%]{NaOAc,\ Et_2O}$ CH₂=C(NO₂)–CH₂–O₂C-t-Bu (4)

NNP

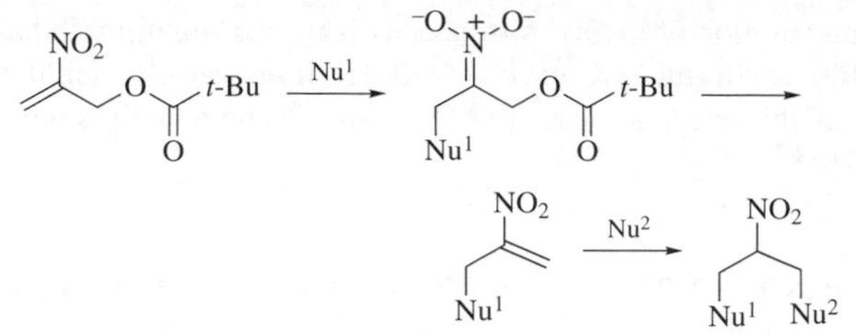

(5)

With dialdehydes, nitromethane forms cyclic nitro compounds.[33] Nitromethane reacts with glyoxal at pH 10 to give a mixture of isomeric inositol derivatives from which *neo*-inositol, one of 14 possible diastereomers, precipitates (72% yield) (eq 6).

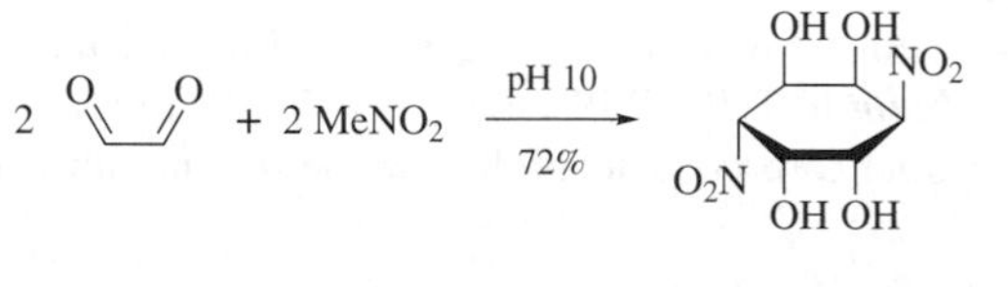

(6)

Tartaraldehyde gives a mixture of stereomeric nitro-2,3,4,5-cyclopentanetetraols.[33] With *o*-phthalaldehydes, nitromethane reacts in alcoholic alkali to give, after acidification, 2-nitro-3-hydroxyindenes.[33] Dialdehydes derived from periodate cleavage of sugars react with nitromethane in a one-pot cyclization reaction to give, frequently, one predominant stereoisomer (eq 7).[33,34]

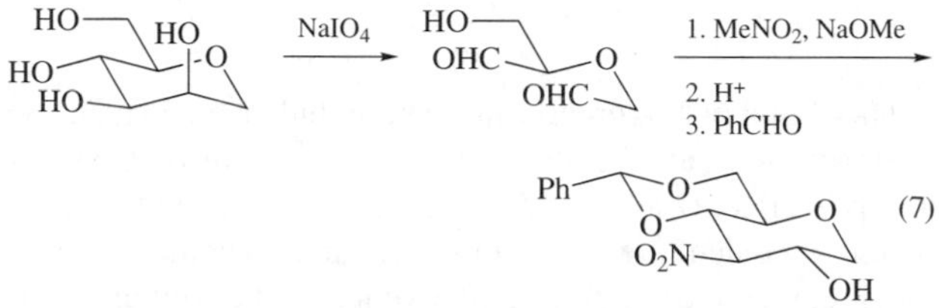

(7)

meso-1,5-Dialdehydes react with nitromethane in methanol and a catalytic amount of NaOH to give a crude mixture of 2,6-dihydroxynitrocyclohexanes from with the major *trans,trans*-isomer precipitates (eq 8).[35] The diacetates of carbocyclic and heterocyclic six-membered compounds formed by this reaction can be saponified enantioselectively with pig liver esterase (PLE) to give monoacetates of >95% ee.[35]

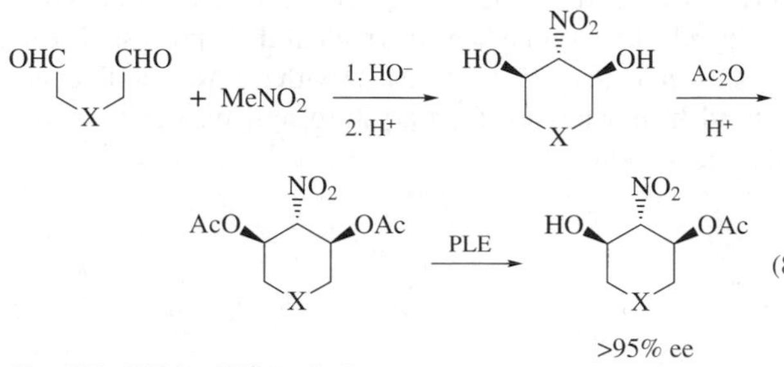

(8)

$X = CH_2$, CHMe, CHOEt, O, S

In the presence of primary or secondary amines, aldehydes react with nitromethane in a Mannich-type reaction.[2d]

Reactions with Ketones. Unlike other nitroalkanes, nitromethane often gives satisfactory yields in the reaction with ketones.[2b] The reaction is complex and depends on the ratio of reactants, base, temperature, and time (eq 9).[2a]

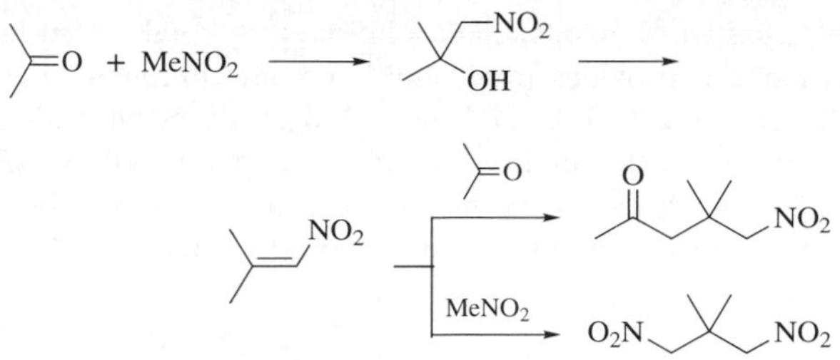

(9)

With alkali metal hydroxides or alkoxides, quaternary ammonium hydroxides, primary or tertiary amines, or ***Tetra-n-butylammonium Fluoride*** under pressure[36] it is usually possible to stop the reaction at the nitro alcohol stage.[2d] When the reaction is catalyzed by secondary amines, nitroalkenes are isolated.[37] From 3β-hydroxyandrost-5-en-17-one, nitromethane, and 1% ***1,2-Diaminoethane*** as catalyst, an exocyclic nitroalkene is obtained (eq 10),[38] although exocyclic nitroalkenes often rearrange to the endocyclic β,γ-nitroalkenes. With *N,N*-dimethylethylenediamine as base, it is possible to selectively synthesize allylic nitro compounds from both acyclic and alicyclic ketones (eq 11).[39]

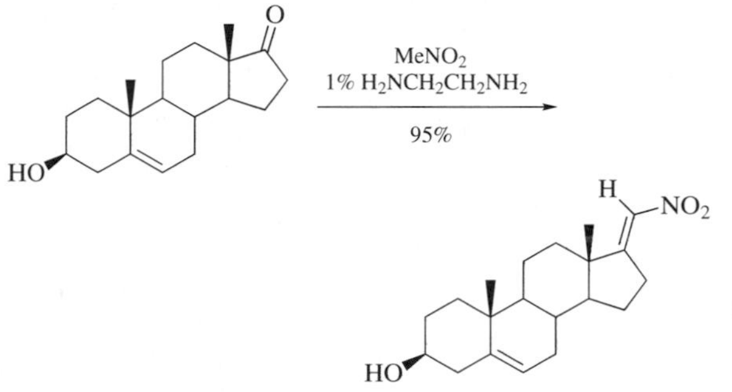

(10)

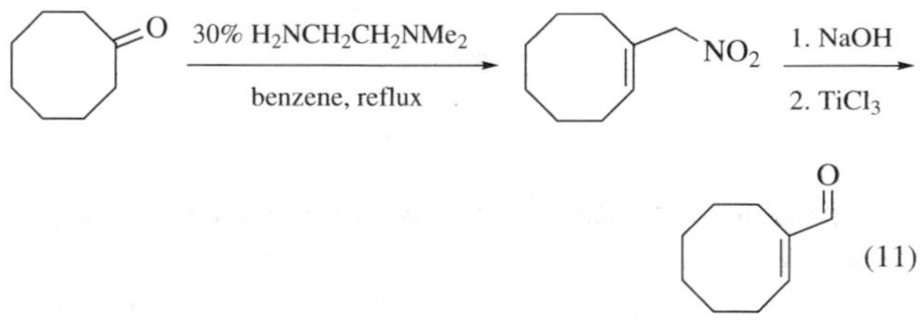

(11)

Nitromethane has been used in the Tiffeneau–Demjanov ring expansion of cyclic ketones by reduction of the nitro alcohol to the β-hydroxy amine and diazotization (eq 12).[40]

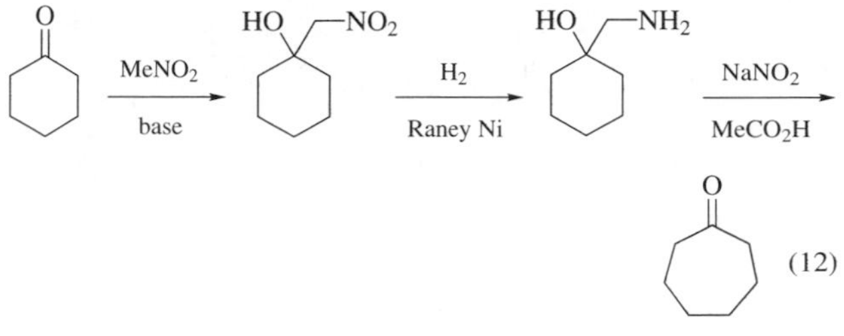

(12)

Michael Reactions. Conjugate addition of nitromethane to activated double bonds is another important C–C bond forming reaction (eq 13). Unsaturated aldehydes give rise to competing 1,2-addition (Henry reaction), which can be controlled to some extent by the choice of catalyst.[2a] Michael additions with nitromethane are catalyzed in homogeneous solution with catalysts such as alkali metal hydroxides in alcohol,[41] organic nitrogen bases,[42] and fluoride ions.[43] The reaction is also catalyzed in heterogeneous systems with alumina[44] or Al_2O_3-supported KF or CsF.[45] Nitromethane reacts smoothly with two moles of Michael acceptors to give a variety of coupling products (eq 13).[43,44]

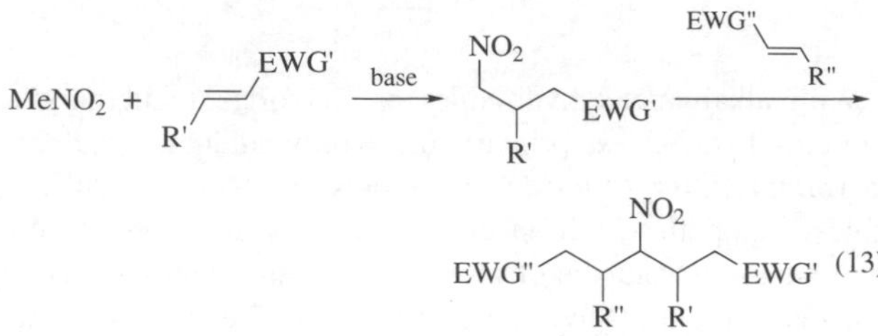

(13)

EWG = CHO, COR, CO_2R, CN, SOR, SO_2R, NO_2

Nitromethane has been employed in cyclopropanation.[46] The product is formed by Michael addition of nitromethane to a double bond with two geminal electron withdrawing groups followed by elimination of the nitro group (eq 14).

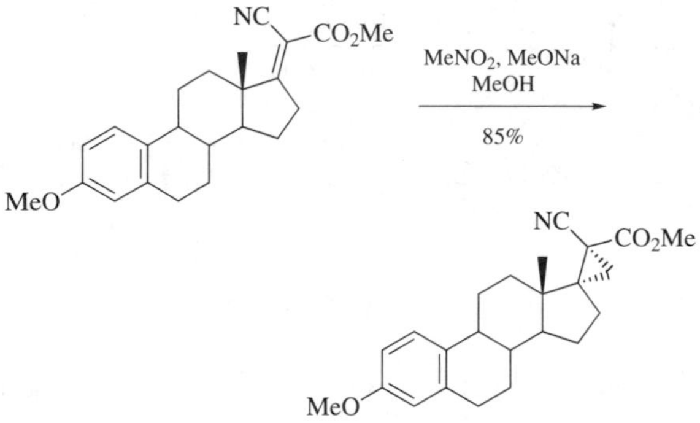

$MeNO_2$, MeONa, MeOH, 85% (14)

Acylation of Nitromethane. With few exceptions, acylation of sodium methanenitronate with acyl halides or anhydrides occurs on oxygen. The unstable products rearrange into hydroxamic acid derivatives. *C*-Acylation can be accomplished with acylimidazoles,[47] acyl cyanide,[48] and phenyl benzoates[49] and from benzoic acids by the action of ***Diethyl Phosphorocyanidate*** (eq 15).[50]

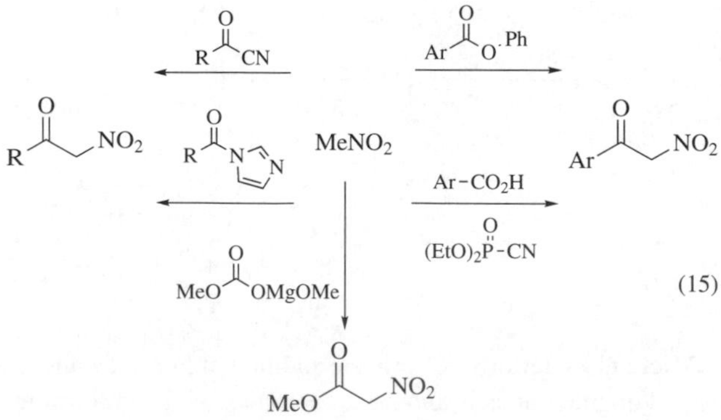

(15)

Carboxylation can be accomplished by reaction with ***Methyl Magnesium Carbonate*** (Stiles reagent).[51] Another method for the preparation of methyl nitroacetate is by heating nitromethane at 160 °C in the presence of KOH, followed by acidification of the nitronate salt at −15 °C in the methanolic solution (eq 16).[52]

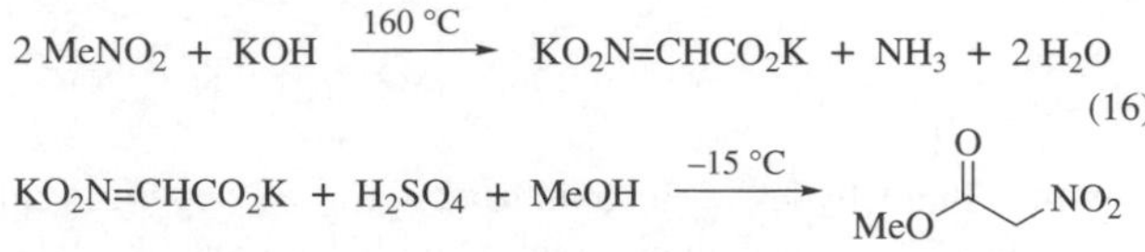

$$2\,MeNO_2 + KOH \xrightarrow{160\ °C} KO_2N{=}CHCO_2K + NH_3 + 2\,H_2O \quad (16)$$

$$KO_2N{=}CHCO_2K + H_2SO_4 + MeOH \xrightarrow{-15\ °C} MeO_2CCH_2NO_2$$

The dilithium salt of nitromethane reacts with carboxylic esters at −30 °C to form the β-oxo nitronate, which must be carefully acidified at −90 °C with acetic acid (eq 17).[53] The dilithium salt of nitromethane is formed by treatment of nitromethane with ***n-Butyllithium*** in THF/HMPA at −65 °C.[53] The dilithium salt is a much harder carbon nucleophile than the sodium nitronate.

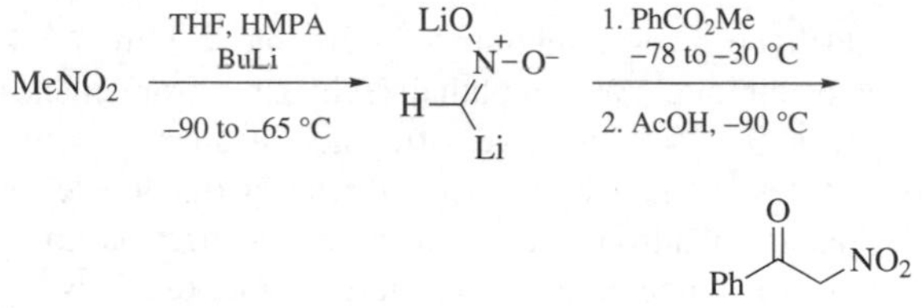

$MeNO_2$ → (THF, HMPA, BuLi, −90 to −65 °C) → 1. $PhCO_2Me$, −78 to −30 °C; 2. AcOH, −90 °C → $PhCOCH_2NO_2$ (17)

Alkylation of Nitromethane. The ambident methanenitronate ion reacts with alkyl iodides to give a mixture of *C*-alkylated and primarily *O*-alkylated compounds. The alkyl nitronates of nitromethane are very unstable and decompose to carbonyl compounds and formaldoxime.[54] The sodium salt of nitromethane can be selectively *C*-benzylated with 1-benzyl-2,4,6-triphenylpyridinium tetrafluoroborate in good to moderate yields.[55] The pyridinium cations are readily available from ***2,4,6-Triphenylpyrylium Tetrafluoroborate*** and the corresponding benzylamines.[55]

The dilithium salt of nitromethane is *C*-alkylated with 1-iodohexane in moderate yield.[56] Kornblum et al. has reported on a general high yielding method for the *C*-alkylation of nitromethane with tertiary nitro compounds (eq 18).[57] The tertiary nitro group is substituted with nitromethane in a radical chain process. The reaction is carried out at 25 °C in DMSO with exposure to fluorescent light and the molar ratio of tertiary nitro compound, nitromethane, and NaH is 1:4:8.

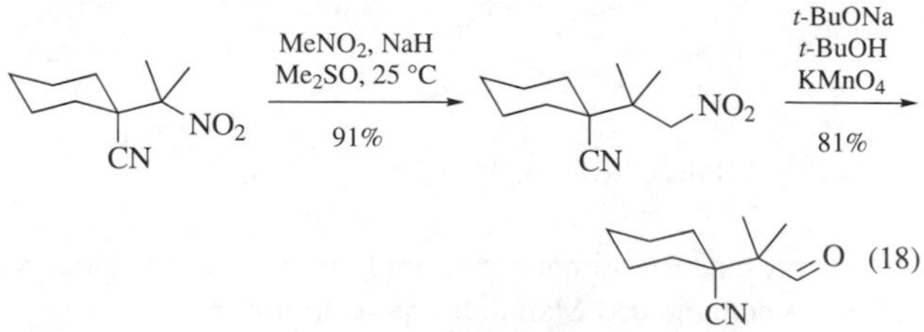

$MeNO_2$, NaH, Me_2SO, 25 °C, 91%; *t*-BuONa, *t*-BuOH, $KMnO_4$, 81% (18)

1,3-Dipolar Cycloadditions of Nitromethane. Nitromethane can be *O*-silylated with ***Chlorotrimethylsilane*** in the presence of ***Triethylamine***.[58a] The trimethylsilyl methanenitronate is unstable and dimerizes, but it can be trapped with an activated alkene in a 1,3-dipolar cycloaddition to give isoxazolidines which, upon acid treatment, give isoxazolines (eq 19).[58]

$$\mathrm{MeNO_2} \xrightarrow[\mathrm{Et_3N}]{\mathrm{TMSCl}} \left[\mathrm{H_2C{=}N^+(O^-)OTMS}\right] \longrightarrow \mathrm{TMSO{-}N{=}CH{-}CH{=}N^+(O^-)OTMS} \qquad (19)$$

(with butadiene) → 5-vinyl-N-OTMS-isoxazolidine $\xrightarrow{\mathrm{H^+}}$ 5-vinylisoxazoline $\xrightarrow{\mathrm{Et_3N}}$ CH₂=CH–CH(OH)–CH₂–CN

In contrast to other primary nitro compounds, the corresponding nitrile oxide (fulminic acid) of nitromethane is not formed in the Mukaiyama reaction.[59] When nitromethane is treated with ***Phenyl Isocyanate***, α-nitroacetanilide is formed first and is subsequently transformed to the nitrile oxide with a second mol equiv of phenyl isocyanate (eq 20). The nitrile oxide reacts with terminal double bonds in a regioselective manner to give 3,5-substituted isoxazolines.[59]

$$\mathrm{MeNO_2 + Ph{-}N{=}C{=}O \longrightarrow PhNHC(O)CH_2NO_2 \xrightarrow{Ph{-}N{=}C{=}O}}$$

$$\mathrm{PhNHC(O){-}C{\equiv}N^+{-}O^-} \xrightarrow{\text{CH}_2\text{=CHR}} \text{5-R-isoxazoline-3-C(O)NHPh} \qquad (20)$$

Related Reagents. *t*-Butyldimethylsilyl Ethylnitronate; *O*,*O*-Dilithio-1-nitropropene; Ethyl Nitroacetate; Lithium α-Lithiomethanenitronate; Nitroethane; Nitroethylene; 1-Nitro-1-propene; 1-Nitropropane; 2-Nitropropane; 2-Nitro-3-pivaloyloxypropene; Phenylsulfonylnitromethane; (Phenylthio)nitromethane.

---

1. (a) Seebach, D.; Colvin, E. W.; Lehr, F.; Weller, H. *C* **1979**, *33*, 1. (b) Rosini, G.; Ballini, R. *S* **1988**, 833.
2. (a) v. Schickh, O.; Apel, G.; Padeken, H. G.; Schwarz H. H.; Segnitz, A. *MOC* **1971**, 10/1 (b) Rosini, G. *COS* **1991**, *2*, 321. (c) Jones, G. *OR* **1967**, *15*, 204. (d) Baer, H. H.; Urbas, L. In *The Chemistry of the Nitro and Nitroso Groups*; Feuer, H., Ed.; Wiley: New York, 1970; Part 2, p 75. (e) Hass, H. B.; Riley, E. F. *CR* **1943**, *32*, 373.
3. (a) Dauben, H. J., Jr.; Ringold, H. J.; Wade, R. H.; Pearson, D. L.; Anderson, A. G., Jr. *OSC* **1963**, *4*, 221. (b) Noland, W. E. *OSC* **1973**, *5*, 833.
4. Wollenberg, R. H.; Miller, S. J. *TL* **1978**, 3219.
5. Bachman, G. B.; Maleski, R. J. *JOC* **1972**, *37*, 2810.
6. (a) Rosini, G.; Ballini, R.; Petrini, M.; Sorrenti, P. *S* **1985**, 515. (b) Rosini, G.; Ballini, R.; Sorrenti, P. *S* **1983**, 1014.
7. Melot, J.-M.; Texier-Boullet, F.; Foucaud, A. *TL* **1986**, *27*, 493.
8. Colvin, E. W.; Seebach, D. *CC* **1978**, 689.
9. (a) Noland, W. E. *CR* **1955**, *55*, 137. (b) Ballini, R.; Petrini, M. *TL* **1989**, *30*, 5329. (c) Clark, J. H.; Cork, D. G.; Gibbs, H. W. *JCS(P1)* **1983**, 2253.
10. Petrini, M.; Ballini, R.; Rosini, G. *S* **1987**, 713.
11. Wehrli, P. A.; Schaer, B. *JOC* **1977**, *42*, 3956.
12. Rosini, G.; Ballini, R.; Sorrenti, P.; Petrini, M. *S* **1984**, 607.
13. Ono, N.; Kaji, A. *S*, **1986**, 693.
14. (a) Worrall. D. E. *OSC* **1956**, *1*, 405. (b) Schales, O.; Graefe. H. A. *JACS* **1952**, *74*, 4486.
15. Raiford, L. C.; Fox, D. E. *JOC* **1944**, *9*, 170.
16. Melton, J.; McMurry, J. E. *JOC* **1975**, *40*, 2138.
17. Buckley, G. D.; Scaife, C. W. *JCS* **1947**, 1471.
18. Knochel, P.; Seebach, D. *S* **1982**, 1017.
19. Knochel, P.; Seebach, D. *TL* **1982**, *23*, 3897.
20. Denmark, S. E.; Moon, Y.-C.; Cramer, C. J.; Dappen, M. S.; Senanayake, C. B. W. *T* **1990**, *46*, 7373.
21. Ballini, R.; Castagnani, R.; Petrini, M. *JOC* **1992**, *57*, 2160.
22. (a) Barrett, A. G.; Graboski, G. G. *CR* **1986**, *86*, 751. (b) Posner. G. H.; Crouch, R. D. *T* **1990**, *46*, 7509.
23. (a) Denmark, S. E.; Senanayake, C. B. W.; Ho, G.-D. *T* **1990**, *46*, 4857. (b) Denmark, S. E.; Senanayake, C. B. W. *JOC* **1993**, *58*, 1853.
24. Denmark, S. E.; Marcin, L. R. *JOC* **1993**, *58*, 3850.
25. (a) Kabalka, G. W.; Guindi, L. H. M. *T* **1990**, *46*, 7443. (b) Rylander, P. *Catalytic Hydrogenation in Organic Synthesis*; Academic: New York, 1979. (c) Erne, M.; Ramirez, F. *HCA* **1950**, *33*, 912.
26. Aizpurua, J. M.; Oiarbide, M.; Palomo, C. *TL* **1987**, *28*, 5365.
27. Torii, S.; Tanaka, H.; Katoh, T. *CL* **1983**, 607.
28. Varma, R. S.; Varma, M.; Kabalka, G. W. *TL* **1985**, *26*, 3777.
29. Ono, N.; Kamimura, A.; Kaji, A. *TL* **1984**, *25*, 5319.
30. (a) Ono, N.; Miyake, H.; Kaji, A. *CC* **1982**, 33. (b) Enders, D.; Meyer, O.; Raabe, G. *S* **1992**, 1242.
31. (a) Sasai, H.; Suzuki, T.; Arai, S.; Arai, T.; Shibasaki, M. *JACS* **1992**, *114*, 4418. (b) Sasai, H.; Suzuki, T.; Itoh, N.; Shibasaki, M. *TL* **1993**, *34*, 851. (c) Sasai, H.; Suzuki, T.; Itoh, N.; Arai, S.; Shibasaki, M. *TL* **1993**, *34*, 2657.
32. Seebach, D.; Knochel, P. *HCA* **1984**, *67*, 261.
33. Lichtenthaler, F. W. *AG(E)* **1964**, *3*, 211.
34. (a) Wade, P. A.; Giuliano, R. M. In *Nitro Compounds*; Feuer, H.; Nielsen, A. T., Eds.; VCH: New York, 1990; p 137. (b) Sakakibara, T.; Nomura, Y.; Sudoh, R. *Carbohydr. Res.* **1983**, *124*, 53. (c) Fujimaki, I.; Kuzuhara, H. *J. Carbohydr. Chem.* **1982**, *1*, 145.
35. Eberle, M.; Egli, M.; Seebach, D. *HCA* **1988**, *71*, 1.
36. Matsumoto, K. *AG*, **1984** *96*, 599.
37. Ho, T.-L.; Wong, C. M. *S* **1974**, 196.
38. Barton, D. H. R.; Motherwell, W. B.; Zard, S. Z. *BSF(2)* **1983**, 61.
39. Tamura, R.; Sato, M.; Oda, D. *JOC* **1986**, *51*, 4368.
40. Dauben, H. J.; Ringold, H. J.; Wade, R. H.; Pearson, D. L.; Anderson, A. G. *OSC* **1963**, *4*, 221; Smith, P. A. S.; Baer, D. R. *OR* **1960**, *11*, 157.
41. Asaoka, M.; Mukuta, T.; Takei, H. *TL* **1981**, *22*, 725.
42. (a) Bäckvall, J.-E.; Ericsson, A. M.; Plobeck, N. A.; Juntunen, S. K. *TL* **1992**, *33*, 131. (b) Ono, N.; Kamimura, A.; Miyake, H.; Hamamoto, I.; Kaji, A. *JOC* **1985**, *59*, 3692. (c) Ono, N.; Kamimura, A.; Kaji, A. *S* **1984**, 226.
43. (a) Anderson, D. A.; Hwu, J. R. *JOC* **1990**, *55*, 511. (b) Anderson, D. A.; Hwu, J. R. *JCS(P1)* **1989**, 1694. (c) Clark, J. H.; Miller, J. M.; So, K. H. *JCS(P1)* **1978**, 941. (d) Colonna, S.; Hiemstra, H.; Wynberg, H. *JCS(P1)* **1978**, 238. (e) Belsky, I. *CC* **1977**, 237.
44. Rosini, G.; Ballini, R.; Petrini, M.; Marotta. E. *AG* **1986**, *98*, 935
45. (a) Bergbreiter, D. E.; Lalonde, J. J. *JOC* **1987**, *52*, 1601. (b) Clark, J. H.; Cork, D. G.; Robertson, M. S. *CL* **1983**, 1145.
46. Annen, K.; Hofmeister; H; Laurent, H.; Seeger, A.; Wiechert, R. *CB* **1978**, *111*, 3094.
47. (a) Baker, D. C.; Putt, S. R. *S* **1978**, 478. (b) Crumbie, R. L.; Nimitz, J. S.;, Mosher, H. S. *JOC* **1982**, *47*, 4040.

48. Bachman, G. B.; Hokama, T. *JACS* **1959**, *81*, 4882.
49. Field, G. F.; Zally, W. J. *S* **1979**, 295.
50. Hamada, Y.; Ando, K.; Shioiri, T. *CPB* **1981**, *29*, 259.
51. Stiles, M.; Finkbeiner, H. L. *JACS* **1959**, *81*, 505.
52. Zen, S.; Koyama; M.; Koto, S. *OSC* **1988**, *6*, 797.
53. Lehr, F.; Gonnermann, J.; Seebach, D. *HCA* **1979**, *62*, 2258.
54. (a) Kerber, R. C.; Urry, G. W.; Kornblum, N. *JACS* **1965**, *87*, 4520. (b) Kornblum, N.; Brown, R. A. *JACS* **1964**, *86*, 2681. (c) Kornblum, N.; Brown, R. A. *JACS* **1963**, *85*, 1359.
55. Katritzky, A. R.; De Ville, G.; Patel, R. C. *CC* **1979**, 602.
56. Seebach, D.; Henning, R.; Lehr, F.; Gonnermann, J. *TL* **1977**, 1161.
57. Kornblum, H.; Erickson, A. S. *JOC* **1981**, *46*, 1037.
58. (a) Torssell, K. B. G.; Zeuthen, O. *ACS* **1978**, *B32*, 118. (b) Das N. B.; Torssell, K. B. G. *T* **1983**, *39*, 2247.
59. (a) Mukaiyama, T.; Hoshino, T *JACS* **1960**, *82*, 5339. (b) Paul, R.; Tchelitcheff, S. *BSF* **1963**, 140.

Kurt B. G. Torssell & Kurt. V. Gothelf
Aarhus University, Denmark

Palladium(II) Acetate[1]

$Pd(OAc)_2$

[3375-31-3]	$C_4H_6O_4Pd$	(MW 224.52)
(trimer)		
[53189-26-7]		

(homogenous oxidation catalyst[3] that, in the presence of suitable co-reagents, will effect the activation of alkenic and aromatic compounds towards oxidative inter- and intramolecular nucleophilic attack by carbon, heteroatom, and hydride nucleophiles[1,3,4,5])

Alternate Name: bis(acetato)palladium; diacetatopalladium(II); palladium diacetate.

Physical Data: mp 205 °C (dec).

Solubility: sol organic solvents such as chloroform, methylene chloride, acetone, acetonitrile, diethyl ether. Dissolves with decomposition in aq HCl and aq KI solutions. Insol water and aqueous solutions of NaCl, NaOAc, $NaNO_3$ as well as in alcohols and petroleum ether. Decomposes when heated with alcohols.

Form Supplied in: orange-brown crystals; generally available.

Preparative Methods: preparation of palladium diacetate from palladium sponge was developed by Wilkinson et al.[2]

Purification: palladium nitrate impurities can be removed by recrystallization from glacial acetic acid in the presence of palladium sponge.

Handling, Storage, and Precautions: can be stored in air. Low toxicity.

General Considerations. Salts of palladium that are soluble in organic media, for example $Pd(OAc)_2$, ***Dilithium Tetrachloropalladate(II)***, and $PdCl_2(RCN)_2$, are among the most extensively used transition metal complexes in metal-mediated organic synthesis. Palladium acetate participates in several reaction types, the most important being: (i) Pd^{II}-mediated activation of alkenes towards nucleophilic attack by (reversible) formation of Pd^{II}–alkene complexes, (ii) activation of aromatic, benzylic, and allylic C–H bonds, and (iii) as a precursor for Pd^0 in Pd^0-mediated activation of aryl, vinyl, or allyl halides or acetates by oxidative addition to form palladium(II)–aryl, –vinyl and –(π)-allyl species, respectively.[1b] All reactions proceed via organopalladium(II) species which can undergo a number of synthetically useful transformations.

Alkenes complexed to Pd^{II} are readily attacked by nucleophiles such as water, alcohols, carboxylates, amines, and stabilized carbon nucleophiles (eq 1). Attack occurs predominantly from the face opposite to that of the metal (*trans* attack), thus forming a new carbon–nucleophile bond and a carbon–metal σ-bond.

The σ-complex obtained is usually quite reactive and unstable, and can undergo a number of synthetically useful transformations such as β-hydrogen elimination (eq 1) to give a vinyl substituted alkene and insertion of CO (eq 1) or alkenes (eq 1) into the carbon–palladium bond, which permit further functionalization of the original alkene. The same general chemistry is observed for complexes generated from Pd^0 (eq 2). Heck vinyl couplings and carbonylations together with allylic nucleophilic substitution reactions are among the synthetically most interesting reactions employing palladium acetate.[5]

$$R^1CH{=}CH_2 \xrightarrow{Pd^{II}} R^1CH{=}CH_2{\cdot}Pd^{II} \xrightarrow{Nu:} R^1CH(Nu)CH_2Pd^{II}$$

$$R^1CH(Nu)CH_2Pd^{II} \begin{cases} \xrightarrow{-\text{'PdH'}} R^1C(Nu){=}CH_2 \\ \xrightarrow[\text{2. } R^2OH]{\text{1. CO}} R^1CH(Nu)CH_2CO_2R^2 \\ \xrightarrow[\text{2. 'PdH'}]{\text{1. } CH_2{=}CHR^3} R^1CH(Nu)CH_2\text{∼}CH{=}CHR^3 \end{cases} \quad (1)$$

$$R^1X \xrightarrow[\text{or } Pd(PR_3)_4]{Pd(OAc)_2,\ PR_3} R^1Pd^{II}X \xrightarrow[-HX]{R^2H} R^1PdR^2 \xrightarrow[-Pd^0]{} R^1R^2 \quad (2)$$

The transformations in eqs 1 and 2 ultimately produce palladium(0), while palladium(II) is required to activate alkenes (eq 1). Thus, if such a process is to be run using catalytic amounts of the noble metal, a way to rapidly regenerate palladium(II) in the presence of both substrate and product is required. Often this reoxidation step is problematic in palladium(II)-catalyzed nucleophilic addition processes, and reaction conditions have to be tailored to fit a particular type of transformation. A number of very useful catalytic processes, supplementing the processes that employ stoichiometric amounts of the metal, have been developed.[1,3–5]

Oxidative Functionalization of Alkenes with Heteroatom Nucleophiles.

Oxidation of Terminal Alkenes to Methyl Ketones. The oxidation of ethylene to acetaldehyde with water acting as the nucleophile using a $Pd^{II}Cl_2$–$Cu^{II}Cl_2$ catalyst (see ***Palladium(II) Chloride*** and ***Palladium(II) Chloride-Copper(II) Chloride***) under an oxygen atmosphere is known as the Wacker process. On a laboratory scale the reaction conveniently allows the transformation of a wide variety of terminal alkenes to methyl ketones.[6] Some synthetic procedures that employ $Pd(OAc)_2$ in chloride-free media have been developed (eq 3).

$$RCH_2CH{=}CH_2 \xrightarrow[\substack{H_2O,\ DMF \\ \text{cat. acid} \\ 70\text{–}90\%}]{\substack{\text{cat } Pd(OAc)_2 \\ \text{st. oxidant}}} RCH_2C(O)CH_3 \quad (3)$$

By this, both the use of the highly corrosive reagent combination $PdCl_2$–$CuCl_2$ and the occurrence of chlorinated byproducts are avoided. The stoichiometric oxidant used in these reactions can be a peroxide,[7] ***1,4-Benzoquinone***,[8] or molecular ***Oxygen***.[8a,9] An electrode-mediated process has also been described.[10]

Other Heteroatom Nucleophiles. Alcohols and carboxylic acids also add to metal-activated alkenes,[1a] and processes for the industrial conversion of ethylene to vinyl acetate and acetals are well established.[1c] However, these processes have not been extensively used with more complex alkenes. In contrast, a number of intramolecular versions of the processes have been developed, a few examples of which are given here. Allylphenols cyclize readily in the presence of palladium(II) to form benzofurans (eq 4). Catalytic amounts of palladium acetate can be used if the reaction is carried out under 1 atm of molecular oxygen with copper diacetate as cooxidant, or in the presence of ***t-Butyl Hydroperoxide***. If instead of palladium acetate a chiral π-allylpalladium acetate complex is used, the cyclization proceeds to yield 2-vinyl-2,3-dihydrobenzofuran with up to 26% ee.[11]

Pd^{II}, oxidant; MeOH, 35 °C, 12 h

22–26% ee (4)

Methyl Glyoxylate adducts of *N*-Boc-protected allylic amines cyclize in the presence of a catalytic amount of palladium acetate and excess ***Copper(II) Acetate*** to 5-(1-alkenyl)-2-(methoxycarbonyl)oxazolidines (eq 5).[12] These heterocycles are easily converted to unsaturated *N*-Boc protected β-amino alcohols through anodic oxidation and mild hydrolysis.

$Pd(OAc)_2$, $Cu(OAc)_2$; DMSO, 70 °C, 2 h; 76% (5)

Nitrogen nucleophiles such as amines, and in intramolecular reactions amides and tosylamides, readily add to alkenes complexed to Pd^{II} derived from $PdCl_2(RCN)_2$ (see ***Palladium(II) Chloride***) with reactivity and regiochemical features paralleling those observed for oxygen nucleophiles.[3,4] Intramolecular nucleophilic attack by heteroatom nucleophiles also occurs in conjunction with other palladium-catalyzed processes presented in the following sections.

Allylic C–H Bond Activation. Internal alkenes, in particular cyclic ones, can be transformed into allylic acetates in a palladium-catalyzed oxidation (eq 6).[13] With benzoquinone (BQ) as stoichiometric oxidant or electron transfer mediator,[9a] the allylic acetoxylation proceeds with high selectivity for the allylic product and usually in excellent yield.

cat $Pd(OAc)_2$, BQ or ox/cat BQ; HOAc, 60 °C; 77–90% (6)

This one-step transformation of an alkene to an allylic acetate compares well with other methods of preparation such as hydride reduction of α,β-unsaturated carbonyl compounds followed by esterification. The scope and limitations of the reaction have been investigated.[14] The allylic acetoxylation proceeds via a π-allylpalladium intermediate,[15] and as a result, substituted and linear alkenes generally give several isomeric allylic acetates. With oxygen nucleophiles the reaction is quite general, and reactants and products are stable towards the reaction conditions. This is normally not yet the case with nitrogen nucleophiles, although one intramolecular palladium-catalyzed allylic amination mechanistically related to allylic acetoxylation has been reported.[16]

Functionalization of Conjugated Dienes. Electrophilic transition metals, particularly palladium(II) salts which do not form stable complexes with 1,3-dienes, do activate these substrates to undergo a variety of synthetically useful reactions with heteroatom nucleophiles.[17] Some examples are presented below.

Telomerization. Conjugated dienes combine with nucleophiles such as water, amines, alcohols, enamines and stabilized carbanions in the presence of palladium acetate and ***Triphenylphosphine*** to produce dimers with incorporation of one equivalent of the nucleophile.[1,18] Telomerization of butadiene (eq 7) yields linear 1,6- and 1,7-dienes and has been used for the synthesis of a variety of naturally occurring materials.[19]

2 (butadiene) + Nu-H → (cat $Pd(OAc)_2$, cat PPh_3) major + minor (7)

Oxidative 1,4-Functionalization. The regio- and stereoselective palladium-catalyzed oxidative 1,4-functionalization of 1,3-dienes (eq 8) constitutes a synthetically useful process.[20–23]

+ X^- + Y^- → (cat $Pd(OAc)_2$, BQ or ox/cat BQ; HOAc, rt) (8)

X = OAc, O_2CR, OR
Y = OAc, O_2CR, OR, Cl

$Pd(OAc)_2$, BQ, HOAc; (a) LiOAc (L = OAc): 91–95% *trans*; (b) LiOAc, cat LiCl (L = Cl): 98–100% *cis* (9)

A selective catalytic reaction that gives high yields of 1,4-diacetoxy-2-alkenes occurs in acetic acid in the presence of a lithium carboxylate and benzoquinone. The latter reagents act as the activating ligand and reoxidant for palladium(0).[24] The reaction can be made catalytic also in benzoquinone by the use of ***Manganese Dioxide***,[20] electrochemistry,[25] or metal-activated molecular oxygen[9a] as stoichiometric oxidant. If the reaction is carried out in alcoholic solvent in the presence of a catalytic amount of a nonnucleophilic acid, *cis*-1,4-dialkoxides can be obtained.[23] An important feature of the 1,4-diacetoxylation reaction is the ease by which the relative sterochemistry of the two acetoxy substituents can be controlled (eq 9).

The first step in the reaction sequence is a regioselective and stereoselective *trans*-acetoxypalladation of one of the double bonds, thus forming a π-allylpalladium(II) intermediate, which is then attacked by a second nucleophile. By variation of the concentration of chloride ions, reactions selective for either the *trans*-diacetate or the *cis*-diacetate (eq 9) can be accomplished. The use of other chloride salts resulted in poor selectivity. The selectivity for the *trans* product under chloride-free conditions is further enhanced if the reaction is carried out in the presence of a sulfoxide co-catalyst.[26] Enzymatic hydrolysis of the *cis-meso*-diacetate yields *cis*-1-acetoxy-4-hydroxy-2-cyclohexene in more than 98% ee,[27] thus giving access to a useful starting material for enantioselective synthesis.[28]

In a related catalytic procedure, run in the presence of a stoichiometric amount of ***Lithium Chloride*** (eq 10), it is possible to obtain *cis*-1-acetoxy-4-chloro-2-alkenes with high 1,4-selectivity and in high chemical yield.[21] A selective nucleophilic substitution of the chloro group in the chloro acetate, either by palladium catalysis or by classical methods (eq 10), and subsequent elaboration of the acetoxy group, offer a number of useful transformations.[22] The methodology has been applied to, for example, a synthesis of a naturally occurring 2,5-disubstituted pyrrolidine, some tropane alkaloids, and perhydrohistrionicotoxin.[29]

cat $Pd(OAc)_2$, LiCl, LiOAc, BQ; HOAc–pentane, rt, 8 h, 75% → Cl–C₆H₈–OAc (>98% *cis*)

(a) cat $Pd(OAc)_2$, PPh_3, Nu, THF, 25 °C → Nu–C₆H₈–OAc

(b) Nu, MeCN, 80 °C → Nu···C₆H₈–OAc (10)

The use of two different nucleophiles can lead to unsymmetrical dicarboxylates.[30] Palladium-catalyzed oxidation of 1,3-cyclohexadiene in acetic acid in the presence of CF_3CO_2H/LiO_2CCF_3, with MnO_2 and catalytic benzoquinone, yielded 70% of *trans*-1-acetoxy-4-trifluoroacetoxy-2-cyclohexene (more than 92% *trans*), with a selectivity for the unsymmetrical product of more than 92%. 1,3-Cycloheptadiene afforded the *cis* addition product in 58% yield with a selectivity for the unsymmetrical product of more than 95%. Since the two carboxylato groups have different reactivity, for example toward hydrolysis, further transformations can be carried out at one allylic position without affecting the other.

Intramolecular versions of the 1,4-oxidations have been developed.[31] In these reactions the internal nucleophile can be a carboxylate, an alkoxide, or nitrogen functionality, and the result of the first nucleophilic attack is the regioselective and stereoselective formation of a *cis*-fused heterocycle (eq 11).

X = O, NR

(a) cat $Pd(OAc)_2$, LiOAc, BQ, acetone, 20 °C → AcO···(cis-fused heterocycle)

(b) + LiCl (cat) → AcO–(cis-fused heterocycle) (11)

(c) + LiCl (1 equiv) → Cl–(cis-fused heterocycle)

The second attack can be directed as described above to yield either an overall *trans* or *cis* product in >70% yield. With internal nucleophiles linked to the 1-position of the 1,3-diene, spirocyclization occurs. The synthetic power of the method has been demonstrated in the total syntheses of heterocyclic natural products,[32] and further developed into a tandem cyclization of linear diene amides (eq 12) to yield bicyclic compounds with trisubstituted nitrogen centers.[33]

dienyl amide ($C(O)NH_2$) → cat $Pd(OAc)_2$, $CuCl_2$, O_2; THF, 60 °C, 24 h, 85% → bicyclic lactam (12)

Functionalization of Alkenes with Palladium-Activated Carbon Nucleophiles.

Heck Coupling[5]. The 'Heck reaction' is the common name for the coupling of an organopalladium species with an alkene and includes both inter- and intramolecular reaction types. However, no general reaction conditions exist and the multitude of variations can sometimes seem confusing.

The original version of the Heck reaction involved the coupling of an alkene with an organomercury(II) salt in the presence of stoichiometric amounts of palladium(II),[34] a method still used in nucleoside chemistry.[35] The finding that the organomercury reagent can be replaced by an organic halide, however, greatly increased the versatility of the process.[36] The modified process is catalyzed by zerovalent palladium, either in the form of preformed tertiary phosphine complexes or, preferentially, formed in situ from palladium acetate (eq 13).

$$R^1X \xrightarrow[\text{or cat Pd}^0(\text{PAr}_3)_4]{\text{cat. Pd(OAc)}_2,\ \text{PR}_3} R^1Pd^{II}X \xrightarrow[-HX]{+HR^2} R^1Pd^{II}R^2 \xrightarrow[-Pd^0PR^1_3]{} R^1R^2 \quad (13)$$

R^1 = Ar, vinyl; X = hal, OTf; R^2 = vinyl

The active catalyst may be kept in solution, by carrying out reactions in the presence of tertiary phosphines such as ***Triphenylphosphine***,[37] or rather tri(*o*-tolyl)phosphine,[38] which is now the phosphine most widely employed in Heck coupling reactions.[5] Other ligands successfully employed include tris(2,6-dimethoxyphenyl)phosphine and the bidentate ligands ***1,2-Bis(diphenylphosphino)ethane*** (dppe), ***1,3-Bis(diphenylphosphino)propane*** (dppp), ***1,4-Bis-(diphenylphosphino)butane*** (dppb), and ***1,1'-Bis(diphenylphos-***

phino)ferrocene (dppf). Coupling reactions can occur in homogenous aqueous media if a water-soluble palladium ligand, trisodium 3,3′,3′-(phosphinetriyl)tribenzenesulfonate, is employed. This greatly facilitates workup procedures, and good yields of coupled products were obtained from reactions of aryl and alkyl iodides with alkenes, alkynes, and allylic acetates.[39] In all cases, an inert atmosphere and the presence of a base, normally ***Triethylamine***, is required.

Phase-Transfer Conditions. The Heck conditions described above are not useful, however, for a large number of alkenic substrates.[40] A sometimes serious drawback is the high temperature (ca. 100 °C) often required. Upon addition of tetrabutylammonium chloride ('phase-transfer conditions' or 'Jeffery conditions'), aromatic halides or enol triflates react under mild conditions with vinylic substrates or allylic alcohols.[5,41] Variations of these conditions include the optional or additional presence of silver or thallium salts. The effect of using different salts, bases, catalysts, solvents, and protecting groups in the coupling of aminoacrylates with iodobenzene has been studied.[42]

Cross Coupling. In cross-coupling reactions, an aryl, vinyl, or acyl halide or triflate undergoes a palladium-catalyzed Heck-type coupling to an aryl-, vinyl-, or alkyl-metal reagent (eq 14) to give a new carbon–carbon bond.[5]

$$R^1X \xrightarrow{+Pd^0} R^1Pd^{II}X \xrightarrow[-MX]{+R^2M} R^1PdR^2 \xrightarrow{-Pd^0} R^1R^2 \quad (14)$$

Mg, Zn, and Zr are examples of metals used in cross-coupling reactions,[43] but, in particular, organostannanes have been employed in mild and selective palladium acetate-catalyzed couplings with organic halides and triflates (Stille reaction).[44] Aryl arenesulfonates undergo a cross-coupling reaction with various organostannanes in the presence of palladium diacetate, dppp, and LiCl in DMF.[45] An advantage of the arylsulfonates over triflates is that the former are solids whereas the latter are liquids. Also, arylboranes and boronic acids also undergo a palladium-catalyzed cross-coupling with alkyl halides, although the catalysts of choice are ***Tetrakis(triphenylphosphine)palladium(0)***, ***Dichloro[1,4-bis(diphenylphosphino)butane]palladium(II)***, or ***Dichloro[1,1′-bis(diphenylphosphino)ferrocene]palladium(II)*** .[46]

Arylation of Alkenes by Coupling and Cross Coupling. Alkenes can be functionalized with palladium-activated arenes, yielding styrene derivatives in a process applicable to a wide range of substrate combinations. An early demonstration of the possibilities of the Heck arylation was the coupling of 3-bromopyridine with *N*-3-butenylphthalimide (eq 15), the first step of four in a total synthesis of nornicotine.[47]

cat $Pd(OAc)_2$, $P(o\text{-Tol})_3$ (1:4); Et_3N, N_2, 100 °C, 10 h, 37%

NPhth

(15)

N-Vinylimides readily undergo palladium-catalyzed vinylic substitution with aryl bromides to yield 2-styryl- and 2-phenylethylimines. With aryl iodides (eq 16), the reaction proceeds even in the absence of added phosphine,[48] which opens the possibility of a sequential disubstitution of bromoiodoarenes.

Br, I + NPhth → cat $Pd(OAc)_2$, Et_3N, 100 °C, 20 h, 75% → NPhth, Br (16)

Vicinal dibromides undergo a twofold coupling reaction with monosubstituted alkenes to yield 1,3,5-trienes (eq 17). The reaction, catalyzed by palladium acetate in the presence of triphenylphosphine and triethylamine, can also be applied to aromatic tri-and tetrabromides.[49]

Br, Br + 2 CO_2Me → cat $Pd(OAc)_2$, cat PPh_3; Et_3N, DMF, 90 °C, 40 h, 55%

CO_2Me, CO_2Me → inert atm., xylene; 140 °C, 5 h, 89% → CO_2Me, CO_2Me (17)

A double coupling of 2-amidoacrylates with 3,3′-diiodobiphenyl constitutes a key step in a short preparation of a biphenomycin B analog.[50] Palladium acetate-catalyzed double coupling reactions of 1,8-diiodonaphthalene with substituted alkenes and alkynes under phase-transfer conditions are useful also for the synthesis of various acenaphthene and acenaphthylene derivatives.[51]

1,2-Disubstituted alkenes are generally less reactive towards coupling than are monosubstituted alkenes. However, the use of the more reactive aryl iodides can result in reasonable yields of the coupled product, usually as a mixture of (*E*) and (*Z*) isomers.[52] The reaction has been applied to a coupling of 2-iodoaniline derivatives with ***Dimethyl Maleate*** (eq 18), the product of which spontaneously cyclizes to form quinolone derivatives in 30–70% yield. If, instead, the 2-iodoaniline is coupled with ***Isoprene*** or cyclohexadiene in the presence of palladium acetate, triphenylphosphine, and triethylamine, indole and carbazole derivatives are obtained by a coupling followed by intramolecular nucleophilic attack by the heteroatom.[53]

X, I, NH_2 + CO_2Me, CO_2Me → cat $Pd(OAc)_2$, Et_3N, 100 °C → X, CO_2Me, N, H, O (18)

X = H (71%), OH (55%), Br (30%)

2-Alkylidenetetrahydrofurans can be prepared via intramolecular oxypalladation and subsequent coupling by treatment of aryl or alkyl alkynic alcohols with ***n-Butyllithium*** followed by palladium acetate and triphenylphosphine. The reaction proceeds to yield furans in moderate yields.[54]

Formation of Dienes and Enynes by Coupling and Cross Coupling. The vinylation of **Methyl Acrylate, Methyl Vinyl Ketone,**

or acrolein with (*E*) or (*Z*) vinylic halides under phase-transfer conditions gives high yields of (*E,E*) (eq 19) or (*E,Z*) (eq 20) conjugated dienoates, dienones, and dienals, respectively.[55] Coupling of vinyl halides or triflates with α,β- or β,γ-unsaturated acids under phase-transfer conditions yields vinyl lactones.[56]

$$(E)\text{-BuCH=CHI} + \text{CH}_2\text{=CHCO}_2\text{Me} \xrightarrow[\text{DMF, rt, 4 h; 96\%}]{\text{cat Pd(OAc)}_2,\ \text{K}_2\text{CO}_3,\ \text{NBu}_4\text{Cl}} (E,E)\text{-BuCH=CHCH=CHCO}_2\text{Me} \quad 99\%\ (E,E) \qquad (19)$$

$$(Z)\text{-BuCH=CHI} + \text{CH}_2\text{=CHCO}_2\text{Me} \xrightarrow[\text{DMF, rt, 1 h; 90\%}]{\text{cat Pd(OAc)}_2,\ \text{K}_2\text{CO}_3,\ \text{NBu}_4\text{Cl}} (E,Z)\text{-BuCH=CHCH=CHCO}_2\text{Me} \quad 95\%\ (E,Z) \qquad (20)$$

Commercially available trimethylvinylsilanes can be vinylated using either vinyl triflates or vinyl iodides in the presence of silver salts, in a reaction catalyzed by palladium acetate in the presence of triethylamine. The resulting 3-substituted 1-trimethylsilyl-1,3-dienes are obtained in reasonable to good yields.[57]

Alkenylpentafluorosilicates derived from terminal alkynes react readily with allylic substrates in a palladium-catalyzed cross-coupling reaction to yield (*E*)-1,4-dienes (eq 21).[58] Treatment of 1-alkenylstannanes with t-BuOOH in the presence of 10% of palladium acetate gives 1,3-dienes (eq 22), whereas coupling between 1- and 2-alkenylstannanes provides 1,4-dienes in good yields (eq 23).[59]

$$\text{Bu-CH=CH-SiF}_5\text{K}_2 + \text{CH}_2\text{=CHCH}_2\text{Cl} \xrightarrow[\text{THF, rt, 24 h; 71\%}]{\text{cat Pd(OAc)}_2} \text{Bu-CH=CH-CH}_2\text{CH=CH}_2 \qquad (21)$$

$$2\ \text{R-CH=CH-SnEt}_3 \xrightarrow[t\text{-BuOOH, PhH}]{\text{cat Pd(OAc)}_2} \text{R-CH=CH-CH=CH-R} \qquad (22)$$

R = Ph, 80%, (*E*):(*Z*) = 4:1
R = C_6H_{13}, 76%, only (*E*)

$$\text{Ph-CH=CH-SnEt}_3 + \text{CH}_2\text{=CHCH}_2\text{SnEt}_3 \xrightarrow[68\%]{\text{as eq 22}} \text{Ph-CH=CH-CH}_2\text{CH=CH}_2 \ \text{only}\ (E) \qquad (23)$$

Cross coupling of enol triflates under neutral conditions with allyl-, vinyl-, or alkynylstannanes in the presence of palladium diacetate and triphenylphosphine proceeds to give high yields of 1,4- and 1,3-dienes and 1,3-enynes, respectively (eq 24).[60]

$$\text{TfO-cyclohexenyl-CO}_2\text{Et} + \text{RSnBu}_3 \xrightarrow[\text{THF, 55 °C}]{\text{cat Pd(OAc)}_2,\ \text{cat PPh}_3\ (1:2)} \text{R-cyclohexenyl-CO}_2\text{Et} \qquad (24)$$

$\text{CH}_2\text{=CHCH}_2\text{SnBu}_3$ 81%
Ph-C≡C-SnBu_3 78%

Terminal alkynes react to form 1-en-3-ynes in a process catalyzed by palladium acetate and tris(2,6-dimethoxyphenyl)phosphine. A number of functional groups such as internal alkenes, esters, and alcohols are tolerated, and good yields of homo- (eq 25) as well as hetero-coupled enynes (eq 26) are obtained.[61]

$$2\ \text{C}_7\text{H}_{15}\text{-C≡CH} \xrightarrow[\text{PhH, rt; 63\%}]{\text{cat Pd(OAc)}_2,\ \text{cat P(2,6-(MeO)}_2\text{C}_6\text{H}_3)_3} \text{CH}_2\text{=C(C}_7\text{H}_{15})\text{-C≡C-C}_7\text{H}_{15} \qquad (25)$$

$$\text{Ph-C≡CH} + \text{CH}_3\text{C≡C-SO}_2\text{Ph} \xrightarrow[\text{PhH, rt; 91\%}]{\text{Pd(OAc)}_2,\ \text{Ar}_3\text{P}} \text{PhO}_2\text{S-CH=C(CH}_3)\text{-C≡C-Ph} \qquad (26)$$

An interesting approach to 1-en-5-ynes is the palladium-catalyzed tandem coupling of a *cis*-alkenyl iodide, a cyclic alkene, and a terminal alkyne (eq 27). With norbornene as the alkene, the coupling occurs in a stereodefined manner, and the enyne products are obtained in good yields.[62] ***Potassium Cyanide*** can be used instead of an alkyne to yield the corresponding cyanoalkene.[63]

$$\text{norbornene} + \text{HC≡C-R} + \text{I-CH=CH-CH(OTBDMS)C}_5\text{H}_{11} \xrightarrow[\text{Et}_2\text{NH, DMF, 80 °C, 12 h}]{\text{cat Pd(OAc)}_2,\ \text{PPh}_3\ (1:4),\ \text{CuI, Bu}_4\text{NCl}} \text{exo-2-(R-C≡C)-3-(CH=CH-CH(OTBDMS)C}_5\text{H}_{11})\text{norbornane} \qquad (27)$$

Formation of Aldehydes, Ketones, and Allylic Dienols by Coupling to Allylic Alcohols. Allylic alcohols can be coupled with aryl or vinyl halides or triflates. The outcome of the reaction depends on the coupling agent and the reaction conditions. Thus arylation of allylic alcohols under Heck conditions constitutes a convenient route to 3-aryl aldehydes and 3-aryl ketones (eq 28).[64]

$$\text{R}^2\text{-CH=CH-CH(OH)R}^1 + \text{PhI} \xrightarrow[\text{MeCN, reflux; 50–95\%}]{\text{cat Pd}^{\text{II}},\ \text{Et}_3\text{N}} \text{Ph-CH(R}^2)\text{-CH}_2\text{-C(=O)R}^1 \qquad (28)$$

Coupling of primary allylic alcohols with vinyl halides carried out under phase-transfer conditions (cat $Pd(OAc)_2$ in the presence of Ag_2CO_3 and n-Bu_4NHSO_4 in acetonitrile) gave 4-enals,[65] whereas secondary allylic alcohols, when treated with a vinyl halide or enol triflate, afforded conjugated dienols with good chemoselectivity, regiochemistry, and stereoselectivity.[66] Since the coupling reaction under these conditions proceeds without affecting the carbon bearing the alcohol functional group, it was possible to prepare optically active dienols from vinyl iodides and optically active allylic alcohols (eq 29).[67]

$$\text{I-CH=CH-(CH}_2)_3\text{CO}_2\text{Me} + \text{CH}_2\text{=CH-CH(OH)C}_5\text{H}_{11} \xrightarrow[\text{DMF, 45 °C; 75\%}]{\text{cat Pd(OAc)}_2,\ \text{Ag}^+\ \text{or Tl}^+} \text{MeO}_2\text{C(CH}_2)_3\text{CH=CH-CH=CH-CH(OH)C}_5\text{H}_{11} \qquad (29)$$

Formation of Allyl and Aryl Primary Allylic and Homoallylic Alcohols from Vinyl Epoxides and Oxetanes. Vinylic epoxides can be coupled with aryl (eq 30) or vinyl (eq 31) iodides or triflates to form allylic alcohols in 40–90% yield.[68] When employing palladium acetate as the catalyst, a reducing agent such as sodium formate is required in addition to the salts normally present under phase transfer conditions.

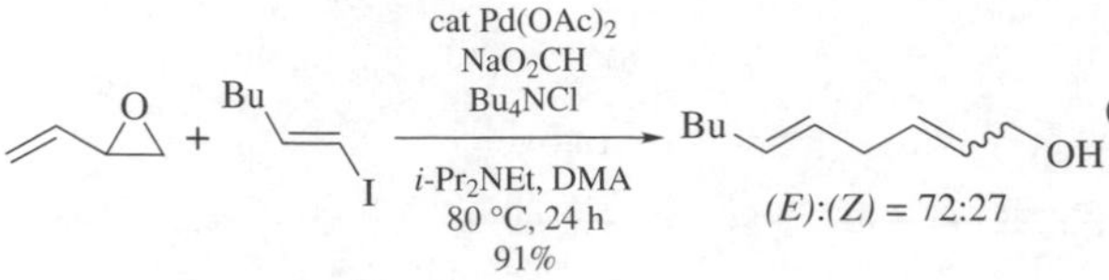

cat $Pd(OAc)_2$, NaO_2CH, Bu_4NCl, *i*-Pr_2NEt, DMA, 80 °C, 24 h, 91% (30)

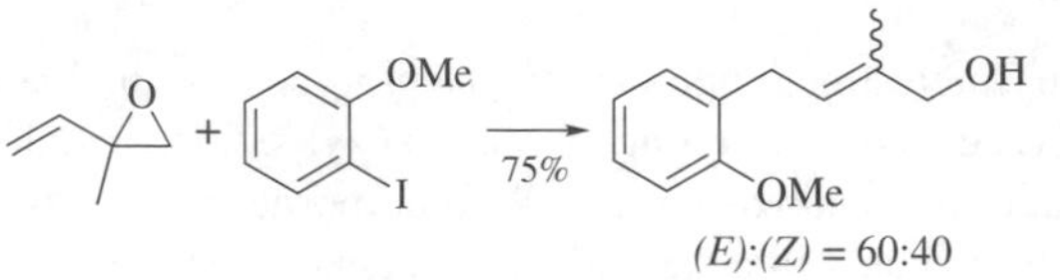

(*E*):(*Z*) = 72:27

75% (31)

(*E*):(*Z*) = 60:40

Vinyloxetane couples with aryl or vinyl iodides or triflates to form homoallylic alcohols under essentially the same reaction conditions (eq 32).[69] The process has also been applied to the preparation of aryl-substituted 3-alkenamides from 4-alkenyl-2-azetidinones (eq 33).[70]

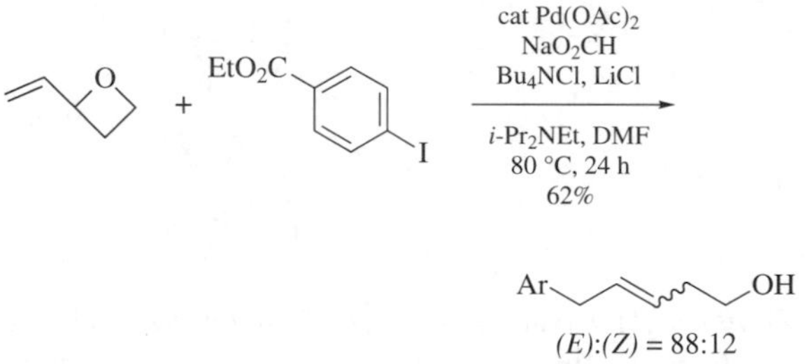

cat $Pd(OAc)_2$, NaO_2CH, Bu_4NCl, LiCl, *i*-Pr_2NEt, DMF, 80 °C, 24 h, 62% (32)

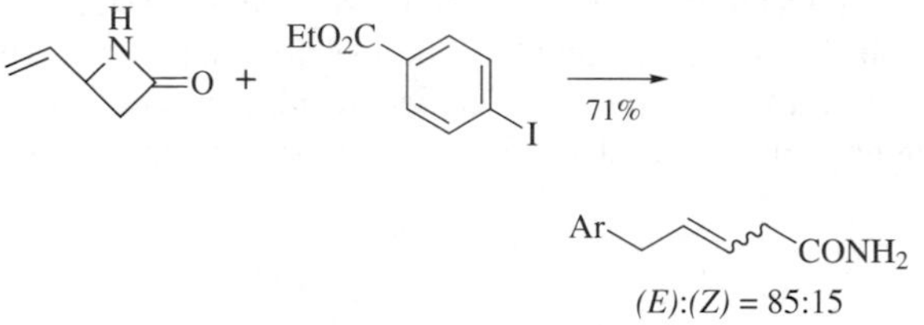

(*E*):(*Z*) = 88:12

71% (33)

(*E*):(*Z*) = 85:15

Homoallylic alcohols can also be prepared using a one-pot transformation of homopropargyl alcohols. Intramolecular hydrosilylation followed by a palladium-catalyzed coupling of the in situ generated alkenoxysilane with an aryl or alkenyl halide, in the presence of fluoride ions, affords the alcohol product.[71] This process has also been applied to the preparation of 1,3-dienes.

Carbonylation. ***Carbon Monoxide*** readily inserts into Pd–C σ-bonds. The resulting acylpalladium intermediate can react intermolecularly or intramolecularly with amines or alcohols to form ketones, amides, or esters, respectively, or with alkenes to yield unsaturated ketones.[1a,5] Thus treatment of vinyl triflates with $Pd(OAc)_2$, PPh_3, and MeOH in DMF results in one-carbon homologation of the original ketone to α,β-unsaturated esters.[72] Benzopyrans with a *cis*-fused γ-lactone can be prepared in high yield from *o*-disubstituted arenes by carbonylation of the intermediate formed upon intramolecular attack of the phenol on the terminal alkene (eq 34). The sequence affords the *cis*-fused lactone, regardless of the relative stereochemistry of the hydroxide and the methylenepalladium in the intermediate.[73]

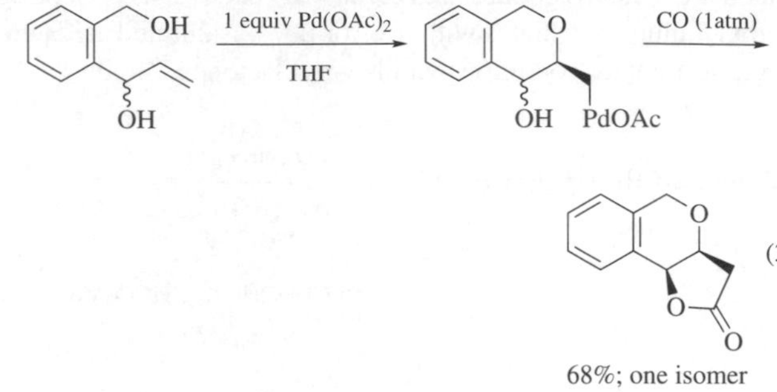

1 equiv $Pd(OAc)_2$, THF; CO (1atm) (34)

68%; one isomer

Vinyl triflates undergo carbonylative coupling with terminal alkynes to yield alkenyl alkynyl ketones in a reaction catalyzed by palladium acetate and dppp in the presence of triethylamine.[74] When applied to 2-hydroxyaryl iodides (eq 35), subsequent attack by the hydroxyl group on the alkyne yielded flavones and aurones. The cyclization result depends on the reaction conditions. ***1,8-Diazabicyclo[5.4.0]undec-7-ene*** as base in DMF yields mainly the six-membered ring flavone, whereas the only product observed when employing potassium acetate in anisole was the five-membered ring aurone.[75]

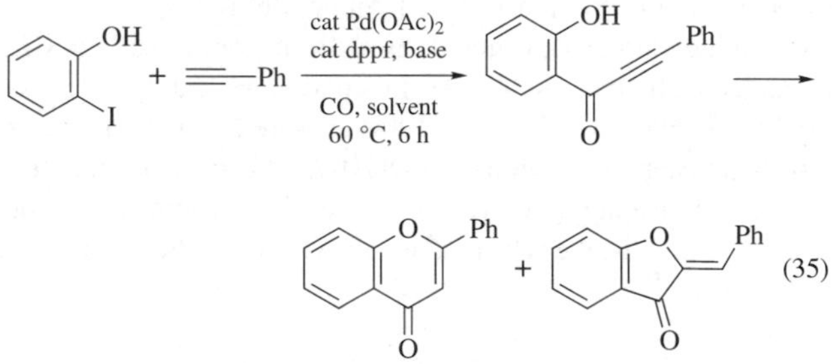

cat $Pd(OAc)_2$, cat dppf, base, CO, solvent, 60 °C, 6 h (35)

base	sol	
AcOK	anisole	0:100
DBU	DMF	92:8

Chiral α,β-unsaturated oxazolines can be obtained by a carbonylation–amidation of enol triflates or aryl halides with chiral amino alcohols (eq 36).[76] The palladium catalyst can be either $Pd(PPh_3)_4$, ***Bis(dibenzylideneacetone)palladium(0)*** and PPh_3, or $Pd(OAc)_2$ and dppp in the presence of triethylamine.

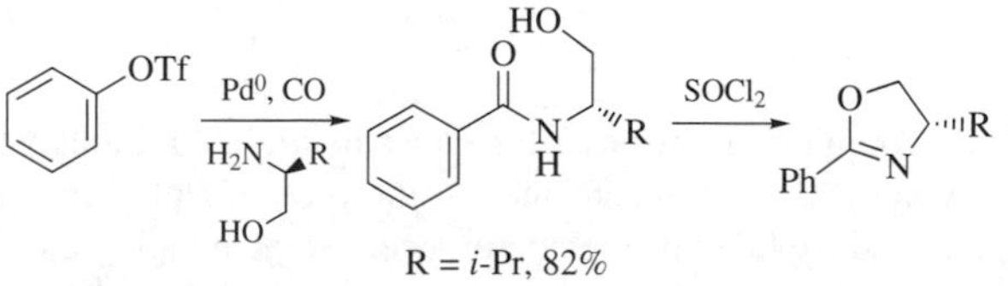

Pd^0, CO; $SOCl_2$ (36)

R = *i*-Pr, 82%

N-Substituted phthalimides are obtained from coupling *o*-dihalo aromatics with carbon monoxide and primary amines. The best catalysts for this reaction, however, were $PdCl_2L_2$ species.[77]

Formation of Heterocyclic Compounds. Coupling reactions of 2-halophenols or anilines with molecules containing functionalities that allow the heteroatom nucleophile to form a heterocycle either by intramolecular oxy- or amino-palladation of an alkene, or by lactone or lactam formation, has already been mentioned in the preceding sections.[78] In addition to these powerful techniques, carbon–heteroatom bonds can be constructed in steps prior to the cyclization. For example, the enamine 3-((2-bromoaryl)amino)cyclohex-2-en-1-one undergoes a palladium-catalyzed intramolecular coupling to yield 1,2-dihydrocarbazoles in moderate yields.[79] Intramolecular coupling of 2-iodoaryl allyl amines gave high yields of indoles under phase-transfer conditions (eq 37).[80] The corresponding aryl allyl ethers require the additional presence of sodium formate in order to give benzofurans in good yields (eq 38).

cat $Pd(OAc)_2$, base, Bu_4NCl, DMF (37)

R	Base	Time	Temp.	Yield
H	Na_2CO_3	24 h	25 °C	97%
Me	Et_3N	48 h	25 °C	81%
MeCO	NaOAc	24 h	80 °C	90%

cat $Pd(OAc)_2$, Na_2CO_3, NaO_2CH, Bu_4NCl, DMF, 80 °C, 48 h (38)

R = H (47%), Me (83%), C_5H_{11} (83%), Ph (81%)

The principle has been applied to the preparation of pharmaceutically interesting heterocyclic compounds,[81] and to the assembly of fused or bridged polycyclic systems containing quaternary centers.[82]

Formation of Carbocycles.

By Intramolecular Heck Coupling. 1-Bromo-1,5-dienes and 2-bromo-1,6-dienes cyclize in the presence of ***Piperidine*** and a palladium acetate–tri-*o*-tolylphosphine catalyst to produce cyclopentene derivatives (eq 39).[83] 2-Bromo-1,7-octadiene, when subjected to the same reaction conditions, cyclized to yield a mixture of six and five-membered ring products, whereas competing dimerization and polymerization was observed for the more reactive 2-bromo-1,5 dienes.

cat $Pd(OAc)_2$, $P(o\text{-Tol})_3$, piperidine, 100 °C, 66 h, 71% (39)

The influence of phosphine ligands, added salts, and the type of metal catalyst on the selectivity of the cyclization have been studied.[84] With K_2CO_3 as base, Wilkinson's catalyst (***Chlorotris(triphenylphosphine)rhodium(I)***) showed higher selectivity for the formation of 1,2-dimethylenecyclopentanes over 1-methylene-2-cyclohexenes than the palladium acetate–triphenylphosphine catalyst.

The palladium-catalyzed cyclization of acyclic polyenes to form polycyclic systems (eq 40) constitutes a very powerful further development of the above method. σ-Alkylpalladium intermediates, produced in an intramolecular Heck reaction, can be efficiently trapped by neighboring alkenes to give bis-cyclization products of either spiro or fused geometry. The second cyclization also produces a σ-alkylpalladium intermediate which can also be trapped.

cat $Pd(OAc)_2$, PPh_3, Ag_2CO_3, MeCN, rt (40)

R	n	Ratio
Me	1	1:1.5
Me	2	100:0 (1:1 H_α:H_β)
H	1	0:100

1-Iodo-1,4- and -1,5-dienes can be transformed into α-methylenecyclopentenones and -hexenones, respectively, by palladium-catalyzed carbonylation and subsequent intramolecular coupling.[85] Better results, however, were obtained using ***Tetrakis(triphenylphosphine)palladium(0)***.

Via (π-Allyl)palladium Intermediates. Allylic substitution, by nucleophilic attack on (π-allyl)palladium complexes generated from allylic substrates, are most often catalyzed by Pd^0–phosphine complexes.[86,87] There are, however, a few examples of intramolecular reactions where the active catalyst is generated in situ from palladium acetate. For example, ethyl 3-oxo-8-phenoxy-6-octenoate reacts to yield cyclic ketones in the presence of catalytic amounts of palladium diacetate and a phosphine or phosphite ligand (eq 41).[88] The product distribution between five- or seven-membered rings depends on the ligand employed and the solvent used. With a chiral phosphine, (*E*)-methyl 3-oxo-9-methoxycarbonyloxy-7-nonenoate was cyclized to give (*R*)-3-vinylcyclohexane with 41–48% ee.[89]

cat $Pd(OAc)_2$, phosphine, MeCN; major; cat $Pd(OAc)_2$, phosphine; cat $Pd(OAc)_2$, phosphite (41)

Another example is based on the palladium-catalyzed 1,4-chloroacetoxylation methodology,[21,22,29] where a common intermediate, by proper choice of reaction conditions, can be transformed into *cis*- or *trans*-annulated products.[89]

By Cyclization of Alkenyl Silyl Enol Ethers. Treatment of alkenyl silyl enol ethers with stoichiometric amounts of palladium acetate induces an intramolecular attack to form carbacycles (eqs 42 and 43). Good to high yields of α,β-unsaturated ketones were obtained.[90]

1 equiv $Pd(OAc)_2$, MeCN, rt, 10 h, 87% (42)

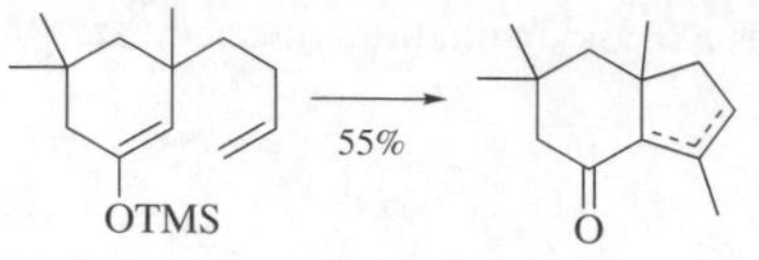

(43)

two isomers 1:1

With slightly different substrates, the observed products were not α,β-unsaturated ketones but nonconjugated bicycloalkenones.[91] The method, which affords bridged (eq 44) as well as spirocyclic (eq 45) bicycloalkenones in acceptable to good yields, has been applied to the preparation of bicyclo[3.3.1]nonadienones[92] and to a total synthesis of quadrone.[93]

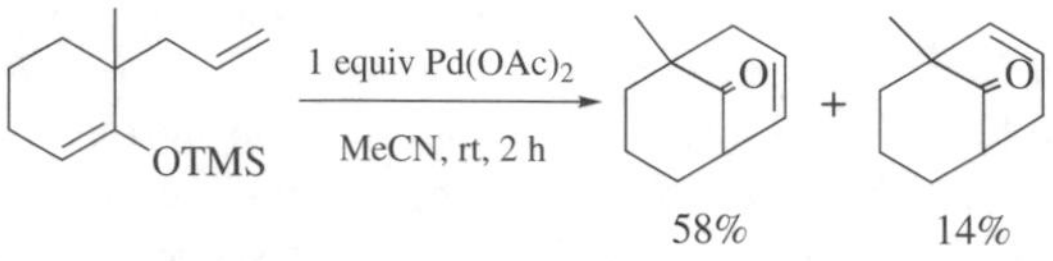

(44)

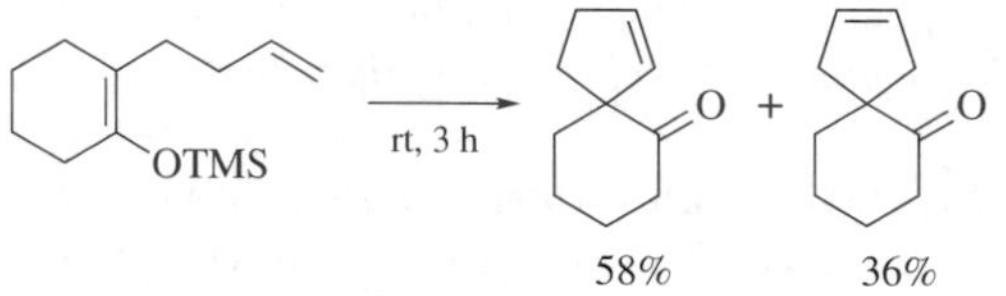

(45)

By Cyclization of Simple Dienes. Treatment of 1,5-dienes with catalytic amounts of $Pd(OAc)_2$ and benzoquinone with MnO_2 as stoichiometric oxidant in acetic acid leads to an oxidative cyclization reaction (eqs 46, 47).[94] The reaction normally yield cyclopentanes with acetate and exomethylene groups in a 1,3-configurational relationship.[95]

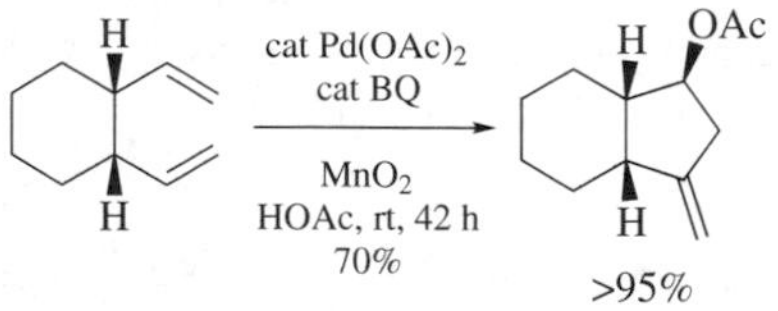

(46)

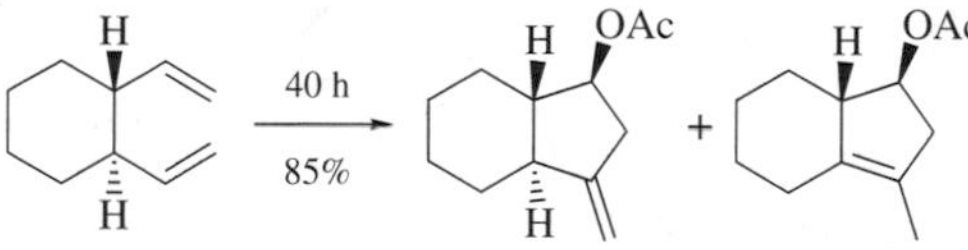

(47)

The selectivity of the reaction depends strongly upon the structure of the starting alkene. Substituents in the 1,3- and/or 4-positions of the diene are tolerated, but not in the 2- and 5-positions; thus the reaction most likely proceeds via an acetoxypalladation of the 1,2-double bond followed by insertion of the 5,6-alkene into the palladium–carbon σ-bond and subsequent reductive elimination.[96] The cyclization is compatible with the presence of several types of functional groups such as alcohols, acetate (even in the allylic position), ethers, nitriles, and carboxylic acids. An improved diastereoselectivity was observed in reactions carried out with chiral nucleophiles in the presence of water-containing molecular sieves.[97] The synthetic utility of the reaction was demonstrated by a synthesis of diquinanes.[98]

By Cycloisomerization of Enynes. When 1,6-enynes, prepared by a $Pd(PPh_3)_4$-catalyzed coupling of an allylic carboxylate with dimethyl propargylmalonate anion, is treated with a catalytic amount of a palladium(II) species, a carbocyclization leading to cyclopentanes carrying an exocyclic double bond occurs (eq 48).[99] Yields of 1,4-dienes ranging from 50% to 85% are observed. If the enyne has oxygen substituents in the allylic positions, the reaction instead yields a 1,3-diene (eq 49).[100] Cycloisomerization could also be induced for internal enynes carrying alkynic electron-withdrawing substituents.[101]

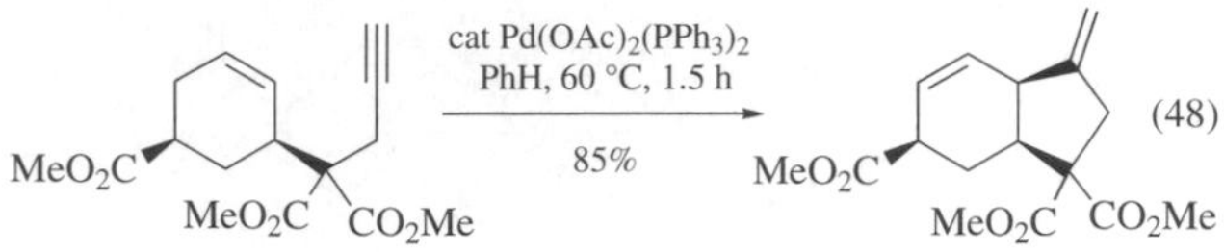

(48)

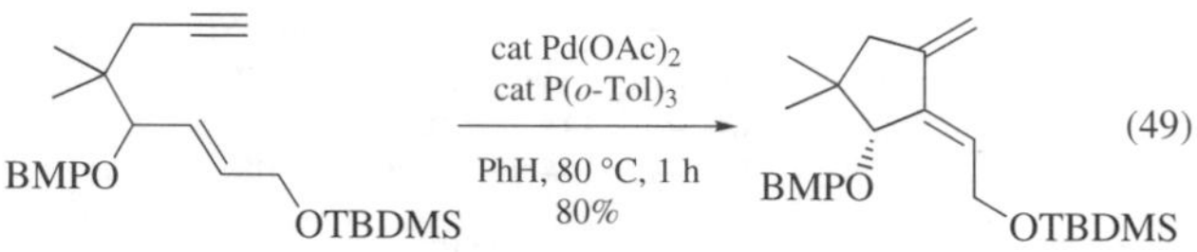

(49)

By Cycloaddition. Palladium acetate, combined with $(i\text{-PrO})_3P$, catalyzes the [2 + 3] cycloaddition of trimethylenemethane to alkenes carrying electron-withdrawing substituents (eq 50). The yields of five-membered carbocycle varied from 35–89%.[102] With 1,3-dienes, a [4 + 3] cycloaddition gave seven-membered ring products in good yield (eq 51), and in some cases excellent diastereomeric ratios were observed.[102]

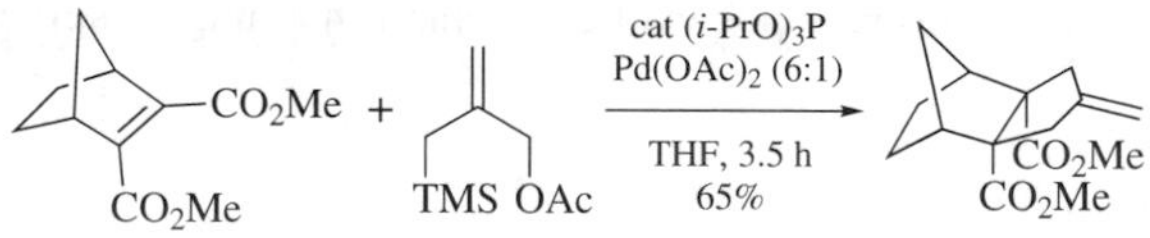

(50)

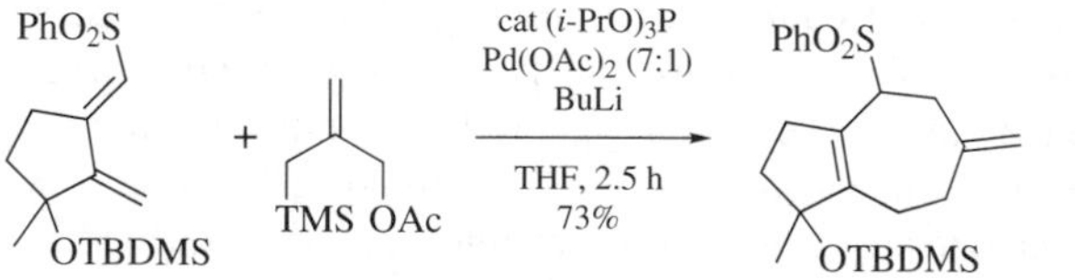

(51)

By Cyclopropanation. Alkenes undergo a cyclopropanation reaction with diazo compounds (caution)[103] such as ***Diazomethane*** or ***Ethyl Diazoacetate*** in the presence of a catalytic amount of palladium acetate.[104] With diazomethane, a selective cyclopropanation of terminal double bonds can be obtained (eq 52).[105]

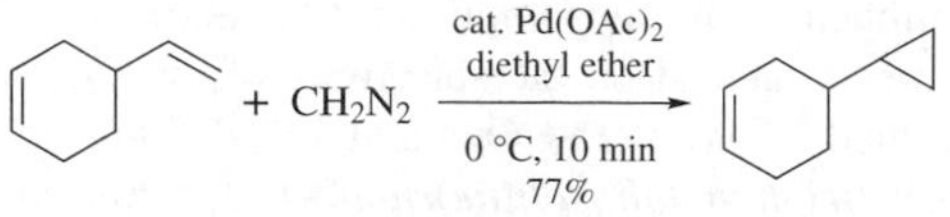

(52)

With diazo esters, the regioselectivity in transition metal-catalyzed cyclopropanation of dienes and trienes was generally not as good with palladium acetate as with a rhodium carboxylate catalyst,[106] although both palladium and rhodium carboxylates were better catalysts for the reaction than ***Copper(II) Trifluoromethanesulfonate***. α,β-Unsaturated carbonyl compounds also undergo palladium-catalyzed cyclopropanation, yielding the corresponding cyclopropyl ketones (eq 53) and esters (eq 54).[107]

$$\text{PhCH=CHCOR} \xrightarrow[85\text{–}98\%]{\text{CH}_2\text{N}_2,\ \text{cat Pd(OAc)}_2} \text{Ph, H-cyclopropyl-COR} \quad (53)$$

$$\text{PhCH=CHCOR} \xrightarrow[50\%]{\text{N}_2\text{CHCO}_2\text{Et},\ \text{cat Pd(OAc)}_2} \text{Ph-cyclopropyl(COR)(CO}_2\text{Et)} \quad (54)$$

Asymmetric cyclopropanations of α,β-unsaturated carboxylic acid derivatives with CH_2N_2 proceeds in greater than 97.6% diastereomeric excess when Oppolzer's sultam (***10,2-Camphorsultam***) is used as a chiral handle.[108] The stereoselectivity of the reaction was found to be temperature dependent, with the best results obtained at higher temperatures. A coupling of norbornene and a *cis*-alkenyl iodide in the presence of a hydride donor resulted in a cyclopropanation of the norbornene (eq 55).[65]

$$\text{norbornene} + \text{I-CH=CH-CH(OR)C}_5\text{H}_{11} \xrightarrow[84\%]{\text{Pd(OAc)}_2,\ \text{PPh}_3,\ \text{HO}_2\text{CH, Et}_3\text{N}} \text{product} \quad (55)$$

Other examples of palladium-catalyzed cyclopropanation are intramolecular processes catalyzed by, for example, ***Dichloro[1, 2-bis(diphenylphosphino)ethane]palladium(II)***,[109] ***Tetrakis(triphenylphosphine)palladium(0)***,[110] or ***Bis(allyl)di-μ-chlorodipalladium***.[111]

Oxidations.

Carbonyl Compounds by Oxidation of Alcohols and Aldehydes. Salts of palladium, in particular $PdCl_2$ in the presence of a base, catalyze the CCl_4 oxidation of alcohols to aldehydes and ketones. Allylic alcohols carrying a terminal double bond are transformed to 4,4,4-trichloro ketones at 110 °C, but yield halohydrins at 40 °C. These can be transformed to the corresponding trichloro ketones under catalysis of palladium acetate (eq 56).[112] The latter transformation is useful for the formation of ketones from internal alkenes provided the halohydrin formation is regioselective.

$$\text{BuCH(OH)CH=CH}_2 \xrightarrow{\text{CCl}_4} \text{BuCH(OH)CHClCH}_2\text{CCl}_3 \xrightarrow[\text{K}_2\text{CO}_3,\ \text{PhH, 110 °C},\ 57\%]{\text{Pd(OAc)}_2,\ \text{P(}o\text{-Tol)}_3} \text{BuCOCH}_2\text{CH}_2\text{CCl}_3 \quad (56)$$

Secondary alcohols can be oxidized in high yield to the corresponding ketones by bromobenzene in a reaction catalyzed by palladium acetate in the presence of a base and a phosphine ligand. These reaction conditions, when applied to Δ^2-, Δ^3-, and Δ^4-unsaturated secondary alcohols, yielded product mixtures. When the stoichiometric oxidant was bromomesitylene and a $Pd(OAc)_2$:PPh_3 ratio of 1:2 was used, the oxidation proceeded smoothly for a wide variety of alcohols (eqs 57 and 58).[113]

$$\text{alcohol} \xrightarrow[\text{NaH, Ox.}]{\text{Pd(OAc)}_2,\ \text{PPh}_3\ (1:2)} \text{ketone} \quad (57)$$

Ox = PhBr, 48%, MesBr, 77%

$$\text{OH} \xrightarrow[\text{NaH, Ox.}]{\text{Pd(OAc)}_2,\ \text{PPh}_3\ (1:2)} \text{CHO} \quad (58)$$

Ox = PhBr, 100%

Oxidation of aldehydes in the presence of ***Morpholine*** proceeded effectively to yield 50–100% of the corresponding morpholine amides.[114]

α,β-Unsaturated Ketones and Aldehydes by Oxidation of Enolates. Palladium diacetate-mediated dehydrosilylation of silyl enol ethers proceeds to yield unsaturated ketones in high chemical yield and with good selectivity for the formation of (*E*)-alkenes (eqs 59 and 60).[115] Although stoichiometric amounts of $Pd(OAc)_2$ are employed, this method for dehydrogenation has been employed in key steps in the total synthesis of some polycyclic natural products.[116]

$$\text{OTMS} \xrightarrow[\text{MeCN, rt, 30 h}]{0.5\ \text{equiv Pd(OAc)}_2,\ 0.5\ \text{equiv BQ}} \text{enone (94\%)} + \text{ketone (5\%)} \quad (59)$$

$$\text{OTMS} \xrightarrow[\text{rt, 5 h}]{0.5\ \text{equiv Pd(OAc)}_2,\ 0.5\ \text{equiv BQ, MeCN}} \text{enone (85\%)} + \text{ketone (8\%)} \quad (60)$$

Oxidation of primary vinyl methyl ethers yields α,β-unsaturated aldehydes. The method has been applied to a transformation of saturated aldehydes to one-carbon homologated unsaturated aldehydes (eq 61) by a Wittig reaction (e.g. ***Methoxymethylene Triphenylphosphorane***) and subsequent palladium acetate-mediated oxidation.[117] The oxidations, which were carried out in $NaHCO_3$-containing aqueous acetonitrile, yielded 50–96% of the unsaturated aldehydes.

$$\text{Ph(CH}_2)_3\text{CHO} \xrightarrow{\text{Wittig}} \text{Ph(CH}_2)_3\text{CH=CHOMe} \xrightarrow[\text{aq NaHCO}_3,\ \text{MeCN},\ 0\ \text{°C, 1 h, rt, 1 h}]{0.5\ \text{equiv Pd(OAc)}_2,\ \text{Cu(OAc)}_2} \text{Ph(CH}_2)_2\text{CH=CHCHO} \quad (61)$$

92% (*E*)

Allyl β-keto carboxylates and allyl enol carbonates undergo a palladium-catalyzed decarboxylation–dehydrogenation to yield α,β-unsaturated ketones in usually high chemical yield and with good selectivity.[118] Following this approach, it was possible to obtain 2-methyl-2-cyclopentenone in two steps from diallyl adipate in a procedure that could be convenient for large-scale preparations (eq 62).[119]

1. NaH, toluene 95 °C; 2. MeI, Et_4NCl 55 °C, 4 h 87%; 10% $Pd(OAc)_2$ MeCN, 80 °C 35 min 79% (62)

Activation of Phenyl and Benzyl C–H bonds: Oxidation of Aromatics. If palladium diacetate is heated in an aromatic solvent, oxidation of the solvent by cleavage–substitution of a C–H bond occurs, resulting in a mixture of products.[120] Depending on the reaction conditions, biaryls and phenyl or benzyl acetates are isolated. Seemingly small changes can result in large changes in product distribution (eq 63). For example, the oxidation of toluene by a palladium(II) salt yields benzyl acetate in reactions mediated by palladium acetate, whereas bitolyls are the major products in reactions carried out in the presence of chloride ions (eq 63).[121]

1 equiv $Pd(OAc)_2$ HOAc–AcO^- >98% (OAc); $PdCl_2$ HOAc–AcO^- 66% (63)

Oxygen Nucleophiles. A reagent such as permanganate oxidizes toluene to benzoic acid,[122] whereas benzylic oxidation by palladium acetate results in benzyl alcohol derivatives. The oxidation is favored by electron-releasing substituents in the phenyl ring.[123] Catalytic amounts of palladium acetate and tin diacetate, in combination with air, effects an efficient palladium-catalyzed benzylic oxidation of toluene and xylenes. For the latter substrates, the α,α'-diacetate is the main product.[124] A mixed palladium diacetate–copper diacetate catalyst has also been found to selectively catalyze the benzylic acyloxylation of toluene (eq 64).[125]

cat $Pd(OAc)_2$ cat $Cu(OAc)_2$ lauric acid; O_2 flow 165 °C, 5 h 50% (OCOR) (64)

Benzene can be oxidized to phenol by molecular oxygen in the presence of catalytic amounts of palladium diacetate and 1,10-phenanthroline (eq 65).[126] If potassium peroxydisulfate is used as a stoichiometric oxidant with 2,2′-bipyridyl as a ligand, a process yielding mainly *m*-acetoxylated aromatics results (eq 66).[127]

cat $Pd(OAc)_2$ cat 1,10-phenanthroline; 30 atm O_2 + CO (1:1) HOAc, 180 °C, 1 h (OH) (65)

12–13 turnovers/Pd

cat $Pd(OAc)_2$ cat 2.2′-bipy oxidant; reflux, 4 h 65% (OAc) (66)

90% ring oxidation
o:*m*:*p* = 6:59:36

Palladium diacetate in ***Trifluoroacetic Acid*** ($Pd(O_2CCF_3)_2$) gives a mixture of *o*- and *p*-trifluoroacetoxylated products.[128] The reagent is also capable of oxidizing saturated hydrocarbons such as adamantane and methane. In the presence of carbon monoxide and with sodium acetate as co-catalyst, carbonylation of aromatic C–H bonds occurs, eventually yielding acid anhydrides.[129]

Naphthalenes and methylbenzenes can be oxidized to *p*-quinones by aqueous H_2O_2 in acetic acid catalyzed by a Pd^{II}–DOWEX polystyrene resin. Yields and selectivities are generally higher for the methylnaphthalenes (50–65% *p*-quinone) than for methylbenzenes (3–8%).[130]

Carbon Nucleophiles. Palladium-mediated homocoupling of substituted arenes generally yields mixtures of all possible coupling products. If the reaction is carried out with a catalytic amount of palladium diacetate and with ***Thallium(III) Trifluoroacetate*** as stoichiometric oxidant (eq 67), aryls carrying substituents such as alkyl or halide afford mainly the 4,4′-biaryls in yields ranging from 60% (R = ethyl) to 98% (R = H).[131] Biaryls can also be formed without the palladium catalyst.[132]

cat $Pd(OAc)_2$ $Tl^{III}(CF_3CO_2)_3$; CF_3CO_2H (67)

R = Me, 40 h, 95% (74% 4,4′)

Oxidative substitution of aromatics with a heteroatom substituent in a benzylic position generally yields *o*-substituted products.[1b,5] The reaction probably proceeds via a cyclopalladated phenylpalladium species (eq 68), which decomposes to form substituted products. For example, the alkylation of a number of acetanilides proceeds with high selectivity for the *o*-alkylated product.[133]

NHCOMe; 1.5 equiv $Pd(OAc)_2$; MeI, MeCN 8 h, 60 °C; NHCOMe (68)

81%

With *t*-butyl perbenzoate as hydrogen acceptor, it is possible to couple benzene or furans with alkenes. In the absence of alkene, benzoxylation of the aromatic compound is observed.[134]

When heated in palladium acetate-containing acetic acid, diphenyl ether, diphenylamine, benzophenone, and benzanilide gave high yields of cyclized products (eq 69). A large number of ring substituents were tolerated in the cyclization.[135]

1–2 equiv $Pd(OAc)_2$; HOAc, reflux 40–90% (69)

X = O, NH, CO

Oxidation of benzoquinones and naphthoquinones by palladium diacetate in arene-containing acetic acid gave the corresponding aryl-substituted quinones (eq 70).[136] Treatment of 1,4-naphthoquinone with aromatic heterocycles, for example furfural, 2-acetylfuran, 2-acetylthiophene, and 4-pyrone, yielded the corresponding 2-heteroaryl-substituted 1,4-naphthoquinones.

$Pd(OAc)_2$, arene, HOAc, reflux, 14 h (70)

arene = C_6H_6 (85%), 2,5-$Me_2C_6H_4$ (78%), 2,5-$Cl_2C_6H_4$ (70%)

Palladium-Catalyzed Reductions.

Reduction of Alkynes. Alkynes are selectively reduced to (*Z*)-alkenes by a reduction catalyst prepared from NaH, *t*-$C_5H_{11}OH$, and $Pd(OAc)_2$ (6:2:1) in THF. The reactions, carried out in the presence of quinoline under near atmospheric pressure of H_2, are self-terminating at the semihydrogenated stage, and are more selective than the corresponding reductions catalyzed by Lindlar's catalyst. Omitting the *t*-$C_5H_{11}OH$ gave a catalyst that effected complete reduction.[137]

Alkenyldialkylboranes from internal alkynes undergo palladium acetate-catalyzed protonolysis to yield (*Z*)-alkenes under neutral conditions and (*E*)-alkenes in the presence of Et_3N.[138]

Hydrogenolysis of Allylic Heterosubstituents. Chemoselective removal of an allylic heterosubstituent in the presence of sensitive functional groups is a sometimes difficult transformation since nucleophilic displacement with hydride donors is efficient only if the heterosubstituent is a good leaving group or the hydride donor is powerful. However, removal of an allylic heterosubstituent is a reaction readily performed by Pd^0.[87] The resulting (π-allyl)palladium complexes are readily attacked by hydride nucleophiles (eq 71). Thus, mild hydride donors such as ***Sodium Borohydride*** or ***Sodium Cyanoborohydride*** can be employed.[139] Treatment of allylic oxygen, sulfur, and selenium functional groups with a combination of $Pd(PPh_3)_4$ and ***Lithium Triethylborohydride*** yielded the corresponding hydride-substituted compounds with good regio- and stereoselectivity, with the more highly substituted (*E*)-alkene as the predominant product (eq 71).[140] Similar results are observed for all hydride donor reagents but one: that derived from formic acid yields predominantly or exclusively the less substituted alkene (eq 71).[142]

R–CH=CH–CH$_2$OAc or R–CH(OAc)–CH=CH$_2$ —Pd^0, THF or dioxane, reflux→ [(π-allyl)Pd^{II}] —(a) H^-→ >80% (*E*); —(b) HCO_2^-→ (71)

The regio- and stereoselective hydride attack on the more substituted terminus of (π-allyl)palladium complexes derived from allylic formates has been applied to the palladium acetate–*n*-Bu_3P-catalyzed formation of ring junctions in hydrindane, decalin, and steroid systems, and to stereospecific generation of steroidal side-chain epimers.[141]

Deoxygenation of Carbonyls. Carbonyl compounds can be deoxygenated to form alkenes in a palladium-catalyzed reduction of enol triflates (eq 72). The reaction is quite general, and has been applied to aryl as well as alkyl enol triflates.[142]

OTBDMS → TfO ... OTBDMS —1. cat $Pd(OAc)_2$, PPh_3, Bu_3N, HCO_2H, DMF; 2. deprot. TBDMS, 85% two steps→ OH (72)

Related Reagents. Bis(benzonitrile)dichloropalladium(II); Bis(dibenzylideneacetone)palladium(0); Bis[1,2-bis(diphenylphosphino)ethane]palladium(0); Bis(triphenylphosphine)palladium(II) Acetate; Dilithium Tetrachloropalladate(II); Palladium(II) Acetylacetonate; Palladium(II) Chloride; Palladium(II) Chloride–Copper(I) Chloride; Palladium(II) Chloride–Copper(II) Chloride; Palladium(II) Trifluoroacetate; Sodium Hydride–Palladium(II) Acetate–Sodium *t*-Pentoxide; Tetrakis(triphenylphosphine)palladium(0); Thallium(III) Trifluoroacetate–Palladium(II) Acetate.

1. (a) Tsuji, J. *Organic Synthesis with Palladium Compounds*; Springer: Berlin, 1980. (b) Collman, J. P.; Hegedus, L. S.; Norton, J. R.; Finke, R. G. *Principles and Applications of Organotransition Metal Chemistry*; University Science Books: Mill Valley, CA, 1987. (c) Tsuji, J. *S* **1990**, 739.
2. Stevenson, T. A.; Morehouse, S. M.; Powell, A. R.; Heffer, J. P.; Wilkinson, G. *JCS* **1965**, 3632.
3. (a) Bäckvall, J. E. *ACR* **1983**, *16*, 335. (b) Henry, P. M. In *Catalysis by Metal Complexes*; Reidel: Dordrecht, 1980; Vol. 2. (c) Davison, S. F.; Maitlis, P. M. In *Organic Synthesis by Oxidation with Metal Compounds*; Plenum: New York, 1986.
4. Hegedus, L. S. *COS* **1991**, *4*, 551.
5. Heck, R. F. *Palladium Reagents in Organic Synthesis*; Academic: London, 1985.
6. For example, see: Tsuji, J. *S* **1984**, 369.
7. (a) Roussel, M.; Mimoun, H. *JOC* **1980**, *45*, 5387. (b) Mimoun, H.; Charpentier, R.; Mitschler, A.; Fischer, J.; Weiss, R. *JACS* **1980**, *102*, 1047.
8. (a) Bäckvall, J. E.; Hopkins, R. B. *TL* **1988**, *29*, 2885. (b) Miller, D. G.; Wayner, D. D. M. *JOC* **1990**, *55*, 2924.
9. (a) Bäckvall, J. E.; Hopkins, R. B.; Grennberg, H.; Mader, M. M.; Awasthi, A. K. *JACS* **1990**, *112*, 5160. (b) Srinivasan, S.; Ford, W. T. *J. Mol. Catal.* **1991**, *64*, 291.
10. Miller, D. G.; Wayner, D. D. M. *CJC* **1992**, *70*, 2485.
11. (a) Hosokawa, T.; Miyagi, S.; Murahashi, S. I.; Sonoda, A. *JOC* **1978**, *43*, 2752. (b) Hosokawa, T.; Okuda, C.; Murahashi, S. I. *JOC* **1985**, *50*, 1282.
12. van Benthem, R. A. T. M.; Hiemstra, H.; Speckamp, W. N. *JOC* **1992**, *57*, 6083.
13. Heumann, A.; Åkermark, B.; Hansson, S.; Rein, T. *OS* **1991**, *68*, 109.
14. Hansson, S.; Heumann, A.; Rein, T.; Åkermark, B. *JOC* **1990**, *55*, 975.
15. (a) Grennberg, H.; Simon, V.; Bäckvall, J. E. *CC* **1994**, 265. (b) Wolfe, S.; Campbell, P. C. G. *JACS* **1971**, *93*, 1497.
16. Heathcock, C. H.; Stafford, J. A., Clark, D. L. *JOC* **1992**, *57*, 2575.

17. Bäckvall, J. E. In *Advances in Metal-Organic Chemistry*; JAI: Greenwich, CT, 1989; Vol. 1, pp 135–175.
18. Hegedus, L. S. In *Comprehensive Carbanion Chemistry*; Buncel, E.; Durst, T., Eds.; Elsevier: Amsterdam, 1984; pp 1–64.
19. (a) Takahashi, T.; Minami, I.; Tsuji, J. *TL* **1981**, *22*, 2651. (b) Tsuji, J. *PAC* **1981**, *53*, 2371.
20. Bäckvall, J. E.; Byström, S. E.; Nordberg, R. E. *JOC* **1984**, *49*, 4619.
21. Bäckvall, J. E.; Nyström, J. E.; Nordberg, R. E. *JACS* **1985**, *107*, 3676.
22. (a) Nyström, J. E.; Rein, T.; Bäckvall, J. E. *OS* **1989**, *67*, 105. (b) Bäckvall, J. E.; Vågberg, J. O. *OS* **1992**, *69*, 38.
23. Bäckvall, J. E.; Vågberg, J. O. *JOC* **1988**, *53*, 5695.
24. The mechanism has been investigated: (a) Bäckvall, J. E.; Gogoll, A. *TL* **1988**, *29*, 2243. (b) Grennberg, H.; Gogoll, A.; Bäckvall, J. E. *OM* **1993**, *12*, 1790.
25. Bäckvall, J. E.; Gogoll, A. *CC* **1987**, 1236.
26. Grennberg, H.; Gogoll, A.; Bäckvall, J. E. *JOC* **1991**, *56*, 5808.
27. Kazlaukas, R. J.; Weissfloch, A. N. E.; Rappaport, A. T.; Cuccia, L. A. *JOC* **1991**, *56*, 2656.
28. (a) Schink, H. E.; Bäckvall, J. E. *JOC* **1992**, *57*, 1588. (b) Bäckvall, J. E.; Gatti, R.; Schink, H. E. *S* **1993**, 343.
29. (a) Bäckvall, J. E.; Schink, H. E.; Renko, Z. D. *JOC* **1990**, *55*, 826. (b) Schink, H. E.; Pettersson, H.; Bäckvall, J. E. *JOC* **1991**, *56*, 2769. (c) Tanner, D.; Sellén, M.; Bäckvall, J. E. *JOC* **1989**, *54*, 3374.
30. Bäckvall, J. E.; Vågberg, J.; Nordberg, R. E. *TL* **1984**, *25*, 2717.
31. (a) Bäckvall, J. E. *PAC* **1992**, *64*, 429. (b) Bäckvall, J. E.; Andersson, P. G. *JACS* **1992**, *114*, 6374. (c) Bäckvall, J. E.; Granberg, K. L.; Andersson, P. G.; Gatti, R.; Gogoll, A. *JOC* **1993**, *58*, 5445.
32. Bäckvall, J. E.; Andersson, P. G.; Stone, G. B.; Gogoll, A. *JOC* **1991**, *56*, 2988.
33. Andersson, P. G.; Bäckvall, J. E. *JACS* **1992**, *114*, 8696.
34. Heck, R. F. *JACS* **1968**, *90*, 5518 and 5526.
35. For example; Hacksell, U.; Daves, G. D., Jr. *JOC* **1983**, *48*, 2870.
36. Heck, R. F. *OR* **1982**, *27*, 345.
37. Heck, R. F.; Nolley, J. P., Jr. *JOC* **1972**, *37*, 2320. (b) Dieck, H. A.; Heck, R. F. *JACS* **1974**, *96*, 1133.
38. Ziegler, C. B., Jr.; Heck, R. F. *JOC* **1978**, *43*, 2941.
39. Genet, J-P.; Blart, E.; Savignac, M. *SL* **1992**, 715.
40. For examples of more systematic investigations, see (a) Spencer, A. J. *JOM* **1983**, *258*, 101. (b) Andersson, C. M.; Hallberg, A.; Daves, G. D., Jr. *JOC* **1987**, *52*, 3529.
41. Jeffery, T. *CC* **1984**, 1287.
42. Carlström, A-S.; Frejd, T. *ACS* **1992**, *46*, 163.
43. (a) Mg: Tamao, K.; Sumitani, K.; Kiso, Y.; Zembayashi, M.; Fijioka, A.; Komada, S.; Nakajima, I.; Minato, A.; Kumada, M. *BCJ* **1976**, *49*, 1958. (b) Zn, Zr: Negishi, E. *ACR* **1982**, *15*, 340.
44. Stille, J. K. *AG(E)* **1986**, *25*, 508.
45. Badone, D.; Cecchi, R.; Guzzi, U. *JOC* **1992**, *57*, 6321.
46. For example: (a) Miyaura, N.; Yamada, K.; Suginome, H.; Suzuki, A. *JACS* **1985**, *107*, 972. (b) Mitchell, M. B.; Wallbank, P. J. *TL* **1991**, *32*, 2273. (c) Miyaura, N.; Ishiyama, T.; Sasaki, H.; Ishikawa, M.; Satoh, M.; Suzuki, A. *JACS* **1989**, *111*, 314.
47. Frank, W. C.; Kim, Y. C.; Heck, R. F. *JOC* **1978**, *43*, 2947.
48. Ziegler, C. B., Jr.; Heck, R. F. *JOC* **1978**, *43*, 2949.
49. Lansky, A.; Reiser, O.; de Meijere, A. *SL* **1990**, 405.
50. Carlström, A-S.; Frejd, T. *CC* **1991**, 1216.
51. Dyker, G. *JOC* **1993**, *58*, 234.
52. Cortese, N. A.; Ziegler, C. B., Jr.; Hrnjes, B. J.; Heck, R. F. *JOC* **1978**, *43*, 2952.
53. O'Connor, J. M.; Stallman, B. J.; Clark, W. G.; Shu, A. Y. L.; Spada, R. E.; Stevenson, T. M.; Dieck, H. A. *JOC* **1983**, *48*, 807.
54. Luo, F-T.; Schreuder, I.; Wang, R-T. *JOC* **1992**, *57*, 2213.
55. Jeffery, T. *TL* **1985**, *26*, 2667.
56. Larock, R. C.; Leuck, D. J.; Harrison, L. W. *TL* **1988**, *29*, 6399.
57. Karabelas, K.; Hallberg, A. *JOC* **1988**, *53*, 4909.
58. Yoshida, J.; Tamao, K.; Takahashi, M.; Kumada, M. *TL* **1978**, 2161.
59. Kanemoto, S.; Matsubara, S.; Oshima, K.; Utimoto, K.; Nozaki, H. *CL* **1987**, 5.
60. Houpis, I. N. *TL* **1991**, *32*, 46.
61. Trost, B. M.; Chan, C.; Ruhter, G. *JACS* **1987**, *109*, 3486.
62. Torii, S.; Okumoto, H.; Kotani, T.; Nakayasu, S.; Ozaki, H. *TL* **1992**, *33*, 3503.
63. Torii, S.; Okumoto, H.; Ozaki, H.; Nakayasu, S.; Tadokoro, T.; Kotani, T. *TL* **1992**, *33*, 3499.
64. (a) Melpolder, J. B.; Heck, R. F. *JOC* **1976**, *41*, 265. (b) Chalk, A. J.; Magennis, S. A. *JOC* **1976**, *41*, 273. (c) Buntin, S. A.; Heck, R. F. *OSC* **1990**, *7*, 361.
65. Jeffery, T. *TL* **1990**, *31*, 6641.
66. (a) Jeffery, T. *CC* **1991**, 324. (b) Bernocchi, E.; Cacchi, S.; Ciattini, S. G.; Morera, E.; Ortar, G. *TL* **1992**, *33*, 3073.
67. Jeffery, T. *TL* **1993**, *34*, 1133.
68. (a) Larock, R. C.; Leung, W-Y. *JOC* **1990**, *55*, 6244. (b) Larock, R. C.; Ding, S. *JOC* **1993**, *58*, 804.
69. Larock, R. C.; Ding, S.; Tu, C. *SL* **1993**, 145.
70. Larock, R. C.; Ding, S. *TL* **1993**, *34*, 979.
71. Tamao, K.; Kobayashi, K.; Ito, Y. *TL* **1989**, *30*, 6051.
72. Cacchi, S.; Morera, E.; Ortar, G. *TL* **1985**, *26*, 1109.
73. Semmelhack, M. F.; Bodurow, C.; Baum, M. *TL* **1984**, *25*, 3171.
74. Ciattini, P. G.; Morera, E.; Ortar, G. *TL* **1991**, *32*, 6449.
75. Ciattini, P. G.; Morera, E.; Ortar, G.; Rossi, S. S. *T* **1991**, *47*, 6449.
76. Meyers, A. I.; Robichaud, A. J.; McKennon, M. J. *TL* **1992**, *33*, 1181.
77. Perry, R. J.; Turner, S. R. *JOC* **1991**, *56*, 6573.
78. See Refs. 3, 4, 11, 12, 31, 52–54, 73, and 75–76 cited above.
79. Iida, H.; Yuasa, Y.; Kibayashi, C. *JOC* **1980**, *45*, 2938.
80. (a) Larock, R. C.; Babu, S. *TL* **1987**, *28*, 5291. (b) Larock, R. C.; Stinn, D. E. *TL* **1988**, *29*, 4687.
81. For example: Macor, J. E.; Blank, D. H.; Post, R. J.; Ryan, K. *TL* **1992**, *33*, 8011.
82. Abelman, M. M.; Oh. T.; Overman, L. E. *JOC* **1987**, *52*, 4133.
83. Narula, C. K.; Mak, K. T.; Heck, R. F. *JOC* **1983**, *48*, 2792.
84. (a) Grigg, R.; Stevenson, P.; Worakun, T. *T* **1988**, *44*, 2033. (b) Grigg, R.; Stevenson, P.; Worakun, T. *CC* **1985**, 971.
85. Tour, J. M.; Negishi, E. I. *JACS* **1985**, *107*, 8289.
86. Godleski, S. A. *COS* **1991**, *4*, 585.
87. Trost, B. M. *AG(E)* **1989**, *28*, 1173.
88. For example: (a) Tsuji, J.; Kobayashi, Y.; Kataoka, H.; Takahashi, T. *TL* **1980**, *21*, 1475. (b) Yamamoto, K.; Tsuji, J. *TL* **1982**, *23*, 3089.
89. Bäckvall, J. E.; Vågberg, J. O.; Granberg, K. L. *TL* **1989**, *30*, 617.
90. Ito, Y.; Aoyama, H.; Hirao, T.; Mochizuki, A.; Saegusa, T. *JACS* **1979**, *101*, 494.
91. Kende, A. S.; Roth, B.; Sanfilippo, P. J. *JACS* **1982**, *104*, 1784.
92. Kende, A. S.; Battista, R. A.; Sandoval, S. B. *TL* **1984**, *25*, 1341.
93. Kende, A. S.; Roth, B.; Sanfilippo, P. J.; Blacklock, T. J. *JACS* **1982**, *104*, 5808.
94. (a) Antonsson, T.; Heumann, A.; Moberg, C. *CC* **1986**, 518. (b) Antonsson, T.; Moberg, C.; Tottie, L.; Heumann, A. *JOC* **1989**, *54*, 4914.
95. Antonsson, T.; Moberg, C.; Tottie, L.; Heumann, A. *JOC* **1989**, *54*, 4914.
96. Moberg, C.; Sutin, L.; Heumann, A. *ACS* **1991**, *45*, 77.
97. Tottie, L.; Baeckström, P.; Moberg, C.; Tegenfeldt, J.; Heumann, A, *JOC* **1992**, *57*, 6579.

98. Moberg, C.; Nordström, K.; Helquist, P. *S* **1992**, 685.
99. Trost, B. M.; Lautens, M. *JACS* **1985**, *107*, 1781.
100. Trost, B. M.; Chung, J. Y. L. *JACS* **1985**, *107*, 4586.
101. Trost, B. M.; MacPherson, D. T. *JACS* **1987**, *109*, 3483.
102. Trost, B. M. *AG(E)* **1986**, *25*, 1.
103. Black, H. T. *Aldrichim. Acta* **1983**, *16*, 3.
104. (a) Pulissen, R.; Hubert, A. J.; Teyssie, P. *TL* **1972**, 1465. (b) Kottwitz, J.; Vorbrüggen, H. *S* **1975**, 636. (c) Radüchel, B.; Mende, U.; Cleve, G.; Hoyer, G. A.; Vorbrüggen, H. *TL* **1975**, 633.
105. Suda, M. *S* **1981**, 714.
106. (a) Anciaux, A. J.; Demonceau, A.; Noels, A. F.; Warin, R.; Hubert, A. J.; Teyssié, P. *T* **1983**, *39*, 2169. (b) Anciaux, A. J.; Hubert, A. J.; Noels, A. F.; Petiniot, N.; Teyssié, P. *JOC* **1980**, *45*, 695.
107. Mende, U.; Radüchel, B.; Skuballa, W.; Vorbrüggen, H. *TL* **1975**, 629.
108. Vallgårda, J.; Hacksell, U. *TL* **1991**, *32*, 5625.
109. Genet, J. P.; Balabane, M.; Charbonnier, F. *TL* **1982**, *23*, 5027.
110. Genet, J. P.; Piau, F. *JOC* **1981**, *46*, 2414.
111. Hegedus, L. S.; Darlington, W. H.; Russell, C. E. *JOC* **1980**, *45*, 5193.
112. (a) Nagashima, H.; Sato, K.; Tsuji, J. *T* **1985**, *23*, 5645. (b) Tsuji, J.; Nagashima, H.; Sato, K. *TL* **1982**, *23*, 3085.
113. (a) Tamaru, Y.; Yamamoto, Y.; Yamada, Y.; Yoshida, Z. *TL* **1979**, 1401. (b) Tamaru, Y.; Inoue, K.; Yamada, Y.; Yoshida, Z. *TL* **1981**, *22*, 1801. (c) Tamaru, Y.; Yamada, Y.; Inoue, K.; Yamamoto, Y.; Yoshida, Z. *JOC* **1983**, *48*, 1286.
114. Tamaru, Y.; Yamada, Y.; Yoshida, Z. *S* **1983**, 474.
115. Ito, Y.; Hirao, T.; Saegusa, T. *JOC* **1978**, *43*, 1011.
116. For example: (a) Aphidicolin: Trost, B. M.; Nishimura, Y.; Yamamoto, K. *JACS* **1979**, *101*, 1328. (b) Isabelin: Wender, P. A.; Lechleiter, J. C. *JACS* **1980**, *102*, 6340. (c) Helenalin: Roberts, M. R.; Schlessinger, R. H. *JACS* **1979**, *101*, 7626.
117. Takayama, H.; Koike, T.; Aimi, N.; Sakai, S. *JOC* **1992**, *57*, 2173.
118. Shimizu, I.; Tsuji, J. *JACS* **1982**, *104*, 5844.
119. Tsuji, J.; Nisar, M.; Shimizu, I.; Minami, I. *S* **1984**, 1009.
120. Henry, P. M. *JOC* **1971**, *36*, 1886.
121. Bryant, D. R.; McKeon, J. E.; Ream, B. C. *TL* **1968**, 3371.
122. For example: Solomons, T. W. G. *Fundamentals of Organic Chemistry*, Wiley: New York, 1986.
123. Bushweller, C. H. *TL* **1968**, 6123.
124. (a) Bryant, D. R.; McKeon, J. E.; Ream, B. C. *JOC* **1968**, *33*, 4123. (b) Bryant, D. R.; McKeon, J. E.; Ream, B. C. *JOC* **1969**, *34*, 1107.
125. Goel, A. B. *ICA* **1986**, *121*, L11.
126. Jintuko, T.; Takaki, K.; Fujiwara, Y.; Fuchita, Y.; Hiraki, K. *BCJ* **1990**, *63*, 438.
127. Eberson, L.; Jönsson, L. *ACS* **1976**, *B30*, 361.
128. Sen, A.; Gretz, E.; Oliver, T. F.; Jiang, Z. *NJC* **1989**, *13*, 755.
129. Ugo, R.; Chiesa, A. *JCS(P1)* **1987**, 2625.
130. Yamaguchi, S.; Inoue, M.; Enomoto, S. *BCJ* **1986**, *59*, 2881.
131. Yatsimirsky, A. K.; Deiko, S. A.; Ryabov, A. D. *T* **1983**, *39*, 2381.
132. McKillop, A.; Turrell, A. G.; Young, D. W.; Taylor, E. C. *JACS* **1980**, *102*, 6504.
133. Tremont, S. J.; Rahman, H. U. *JACS* **1984**, *106*, 5759.
134. Tsuji, J.; Nagashima, H. *T* **1984**, *40*, 2699.
135. Åkermark, B.; Eberson, L.; Jonsson, E.; Pettersson, E. *JOC* **1975**, *40*, 1365.
136. Itahara, T. *JOC* **1985**, *50*, 5546.
137. Brunet, J-J.; Caubere, P. *JOC* **1984**, *49*, 4058.
138. (a) Yatagai, H.; Yamamoto, Y.; Maruyama, K. *CC* **1978**, 702. (b) Yatagai, H.; Yamamoto, Y.; Maruyama, K. *CC* **1977**, 852.
139. (a) Hutchins, R. O.; Learn, K.; Fulton, R. P. *TL* **1980**, *21*, 27. (b) Keinan, E.; Greenspoon, N. *TL* **1982**, *23*, 241.
140. Hutchins, R. O.; Learn, K. *JOC* **1982**, *47*, 4380.
141. (a) Mandai, T.; Matsumoto, T.; Kawada, M.; Tsuji, J. *JOC* **1992**, *57*, 1326. (b) Mandai, T.; Matsumoto, T.; Kawada, M.; Tsuji, J. *JOC* **1992**, *57*, 6090.
142. (a) Cacchi, S.; Morera, E.; Ortar, G. *OS* **1991**, *68*, 138. (b) Peterson, G. A.; Kunng, F-A.; McKallum, J. S.; Wulff, W. D. *TL* **1987**, *28*, 1381. (c) Paquette, L. A.; Meister, P. G.; Friedrich, D.; Sauer, D. R. *JACS* **1993**, *115*, 49.

Helena Grennberg
University of Uppsala, Sweden

Palladium(II) Chloride[1]

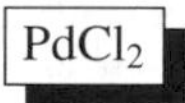

[7647-10-1] Cl_2Pd (MW 177.32)

(used as an oxidizing agent and to a lesser extent as a source of Pd^0 complexes)

Physical Data: mp 678 °C (dec).
Solubility: slightly sol H_2O; sol H_2O in the presence of chloride ion; sol aqueous HCl; sol PhCN, forming $Pd(PhCN)_2Cl_2$; insol organic solvents.
Form Supplied in: commercially available as a rust-colored stable powder or crystalline solid.
Handling, Storage, and Precautions: air stable; not hygroscopic.

General Considerations. Many of the reactions described below can be accomplished using derivatives of palladium chloride such as ***Potassium Tetrachloropalladate(II)***, ***Disodium Tetrachloropalladate(II)***, ***Bis(benzonitrile)dichloropalladium(II)***, dichlorobis(acetonitrile)palladium, and dichlorobis(triphenylphosphine)palladium. The physical properties of these alternative reagents are described under their separate entries, but their chemistry is included in this article.

Synthetic applications of $PdCl_2$ and its derivatives can be classified into three types: use as oxidizing agents, use as Pd^{II} catalysts, and use as a source of Pd^0 catalysts. Characteristic features of these applications are briefly summarized below.

Use as Oxidizing Agents. $PdCl_2$ and ***Palladium(II) Acetate*** are representative Pd^{II} salts used for various oxidation reactions, but their uses are different. For example, oxidative reactions of aromatic compounds are possible only with $Pd(OAc)_2$; $PdCl_2$ and its derivatives cannot be used. Oxidation reactions of various substrates with $PdCl_2$ are stoichiometric and Pd^0 is formed after the oxidation. Sometimes, but not always, Pd^0 can be reoxidized in situ to Pd^{II} with proper reoxidizing agents. In such a case, the oxidation reaction can be carried out with a catalytic amount of $PdCl_2$. Examples of reoxidants include $CuCl_2$, CuCl, $Cu(OAc)_2$, MnO_2, HNO_3, benzoquinone, alkyl nitrites, H_2O_2, and organic peroxides. Since solubility of $PdCl_2$ in water and organic solvents is small, the more soluble ***Dilithium Tetrachloropalladate(II)***, Na_2PdCl_4, K_2PdCl_4, and $Pd(PhCN)_2Cl_2$ are sometimes used for similar purposes.

Use as Pd^{II} Catalyst. $Pd(PhCN)_2Cl_2$ is used as a homogeneous Pd^{II} catalyst for some non-oxidative reactions such as rearrangement reactions.

Use as Source of Pd^0 Catalyst. Pd^{II} salts are reduced to Pd^0 catalysts with various reducing agents. Although $Pd(OAc)_2$ is more convenient for this purpose than $PdCl_2$ and its derivatives, $PdCl_2$ derivatives are used in many cases. Typically, $Pd(Ph_3P)_2Cl_2$ is reduced to form a Pd^0 phosphine complex.

Oxidations.

Oxidative Reactions of Alkenes.[2] Oxidative reactions of alkenes can be classified into two types: oxidative substitution and oxidative addition, as shown in eq 1. Here X^- and Y^- represent nucleophiles such as HO^-, RO^-, RCO_2^-, R_2N^- and CO, as well as soft carbon nucleophiles such as active methylene compounds.

R–CH=CH$_2$ + $PdCl_2$ —(X^-)→ R–CH(X)–CH$_2$–PdCl → R–C(X)=CH$_2$ + Pd^0 + HCl; or —(HY)→ R–CH(X)–CH$_2$–Y + Pd^0 + HCl (1)

X^- and Y^- = nucleophiles

Reaction with Water.[2a,b] Oxidation of ethylene to acetaldehyde under oxygen atmosphere is an industrial process called the Wacker process. $PdCl_2$ and ***Copper(II) Chloride*** in aqueous HCl are used as the catalysts. As shown by eq 2, the Wacker process comprises three unit reactions; $CuCl_2$ is a unique reoxidant of Pd^0.

$$\begin{aligned} CH_2{=}CH_2 + H_2O + PdCl_2 &\longrightarrow MeCHO + 2\,HCl + Pd^0 \\ Pd^0 + 2\,CuCl_2 &\longrightarrow PdCl_2 + 2\,CuCl \\ 2\,CuCl + 2\,HCl + 0.5\,O_2 &\longrightarrow 2\,CuCl_2 + 2\,H_2O \\ \hline CH_2{=}CH_2 + 0.5\,O_2 &\longrightarrow MeCHO \end{aligned} \quad (2)$$

Higher terminal alkenes are also oxidized in organic solvents containing water; DMF is most widely used as the solvent.[3] On a laboratory scale the oxidation can be carried out easily in a way similar to the hydrogenation of alkenes under atmospheric pressure. Instead of Pd black and hydrogen, the oxidation is carried out with $PdCl_2$ and the copper salt under an oxygen atmosphere at room temperature using a similar apparatus. Since the reaction proceeds under mild neutral conditions, many functional groups such as esters, acetals, THP ethers, alcohols, halogens, and amines are tolerated. The ketones obtained by the oxidation are sometimes chlorinated with $CuCl_2$ to give chloro ketones as byproducts. For this reason, nonchlorinating ***Copper(I) Chloride*** is recommended as the reoxidizing agent. This is easily preoxidized to the Cu^{II} state with oxygen.[4] In a laboratory synthesis, a stoichiometric amount of ***1,4-Benzoquinone*** is conveniently used as the reoxidant.

The reaction is a unique method for the one-step synthesis of ketones from alkenes, and allows alkenes to be regarded as masked ketones which are stable to acids, bases, and nucleophiles. Particularly useful is the oxidation of terminal alkenes, which provides methyl ketones (eq 3).[5] As a typical application, the allylation of a ketone, followed by the oxidation, affords a 1,4-diketone. A cyclopentenone can then be prepared by an aldol condensation (eq 4).[5] The annulation method has widespread uses in the synthesis of natural products such as pentalenene,[6] muscone,[7] and coriolin.[8] 1,5-Diketones are prepared by 3-butenylation of a ketone followed by the oxidation. This process has been used to prepare cyclohexenones (eq 5).[5]

R–CH=CH$_2$ —($PdCl_2$, CuCl; O_2, DMF)→ RCOMe (3)

cyclohexanone → 2-allylcyclohexanone —($PdCl_2$, CuCl; O_2, DMF, 68%)→ 2-(2-oxopropyl)cyclohexanone → bicyclic cyclopentenone (4)

2-(3-butenyl)-2-(methoxycarbonyl)cyclohexanone —($PdCl_2$, CuCl; O_2, DMF, 58%)→ 2-(3-oxobutyl)-2-(methoxycarbonyl)cyclohexanone → bicyclic cyclohexenone (5)

Simple internal alkenes are difficult to oxidize. However, the regioselective oxidation of internal alkenes takes place in the presence of suitably disposed oxygen functional groups by neighboring group participation. For example, α,β-unsaturated esters are oxidized to β-keto esters using Na_2PdCl_4 as catalyst and ***t-Butyl Hydroperoxide*** as the reoxidant (eq 6).[9] Allylic ethers are oxidized to β-alkoxy ketones which can be converted to α,β-unsaturated ketones for use in annulation reactions (eq 7).[10] Cyclohexene and cyclopentene can not be oxidized under the usual conditions, but are oxidized to cyclohexanone and cyclopentanone under different conditions. For example, chloride-free Pd^{II} salts, prepared from $Pd(OAc)_2$ and $HClO_4$, H_2SO_4, or HBF_4, are active catalysts (eq 8).[11] For additional examples of the Wacker process, see ***Palladium(II) Chloride-Copper(I) Chloride*** and ***Palladium(II) Chloride-Copper(II) Chloride***.

CH$_3$CH=CH–CO$_2$Me —(Na_2PdCl_4; *t*-BuOOH, 83%)→ CH$_3$COCH$_2$CO$_2$Me (6)

CH$_3$CH=CH–CH$_2$OBn —($PdCl_2$, CuCl; DMF, 67%)→ CH$_3$COCH$_2$CH$_2$OBn —(2-methylcyclohexanone derivative, MeONa)→ bicyclic enone (7)

cyclohexene —($Pd(OAc)_2$, $HClO_4$; benzoquinone)→ cyclohexanone (8)

Reaction with Alcohols and Phenols.[2c] The reaction of alcohols with terminal alkenes affords acetals of ketones (eq 9).[12] An elegant application of the reaction was a brevicomin synthesis (eq 10).[13]

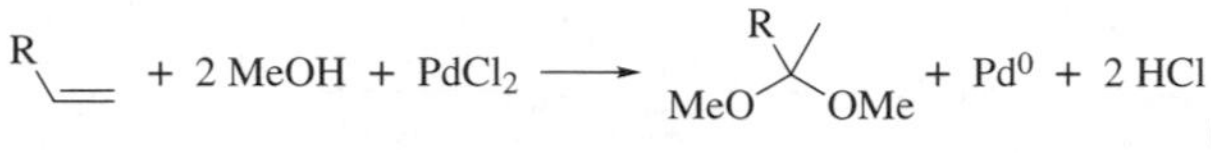

$$\text{RCH=CH}_2 + 2\,\text{MeOH} + \text{PdCl}_2 \longrightarrow \text{R(Me)C(OMe)}_2 + \text{Pd}^0 + 2\,\text{HCl} \quad (9)$$

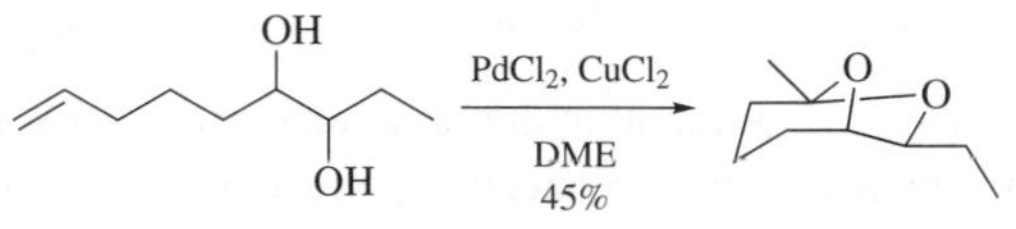

(10)

Alkenes with an electron-withdrawing group such as styrene, ***Acrylonitrile***, and acrylate are converted to acetals of the aldehydes rather than the ketones. The reaction of styrene with ethylene glycol affords the cyclic acetal (eq 11).[12a] 3,3-Dimethoxypropionitrile is produced commercially using methyl nitrite as the reoxidant. The nitrite can be regenerated easily by the oxidation of NO with oxygen (eq 12).[14]

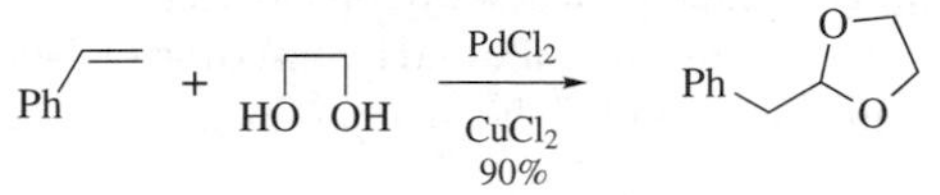

(11)

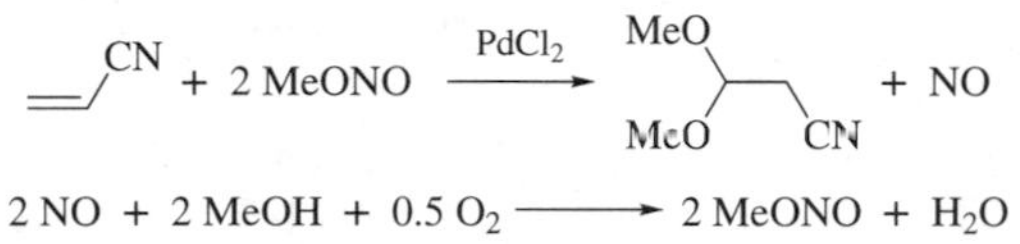

$$\text{CH}_2\text{=CHCN} + 2\,\text{MeONO} \xrightarrow{\text{PdCl}_2} \text{(MeO)}_2\text{CHCH}_2\text{CN} + \text{NO} \quad (12)$$

$$2\,\text{NO} + 2\,\text{MeOH} + 0.5\,\text{O}_2 \longrightarrow 2\,\text{MeONO} + \text{H}_2\text{O}$$

The intramolecular reaction of phenols or enols affords furans and pyrans (eq 13).[15]

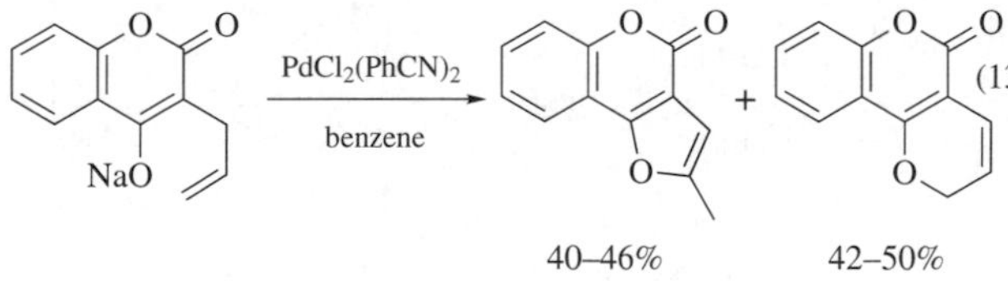

(13)

Reaction with Carboxylic Acids.[2c] The intramolecular reaction of carboxylic acids with alkenes affords unsaturated lactones (eq 14).[16]

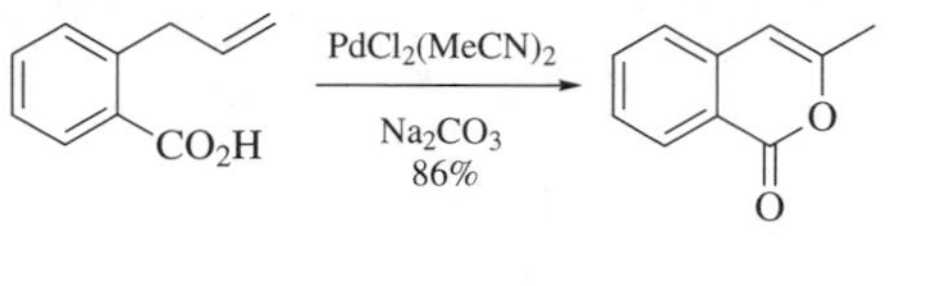

(14)

Reaction with Amines and Amides.[2c] Reaction of amines with alkenes proceeds most smoothly as an intramolecular version. Amides can be used in the intramolecular reaction to afford various heterocyclic compounds. In the example shown in eq 15, it should be noticed that the Pd^{II} species is regenerated by the β-elimination of OH, rather than the β-hydrogen. For this reason the reaction proceeds catalytically without a Pd^0 reoxidant.[17]

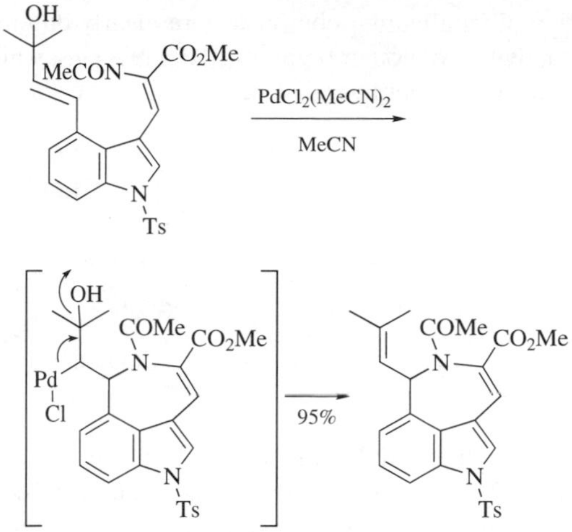

(15)

Reaction with Carbon Nucleophiles. The cyclooctadiene (cod) complex of $PdCl_2$ [**Dichloro (1,5-cyclooctadiene) palladium(II)**], which is insoluble in organic solvents, reacts in ether with malonate or acetoacetate under mild heterogeneous conditions; facile carbon–carbon bond formation takes place to give a new complex in a quantitative yield. Further intramolecular reaction of the complex with a base affords the cyclopropane derivative. Attack of a second malonate on the complex yields the [3.3.0] system (eq 16).[18] Carbopalladation of the double bond of *N*-vinylcarbamate with acetoacetate at −78 °C, and subsequent carbonylation of the Pd–carbon bond, proceeds smoothly to yield the carbocarbonylation product in 92% yield (eq 17).[19]

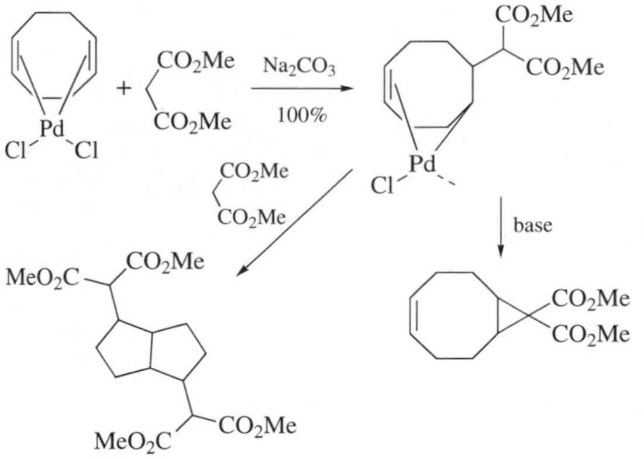

(16)

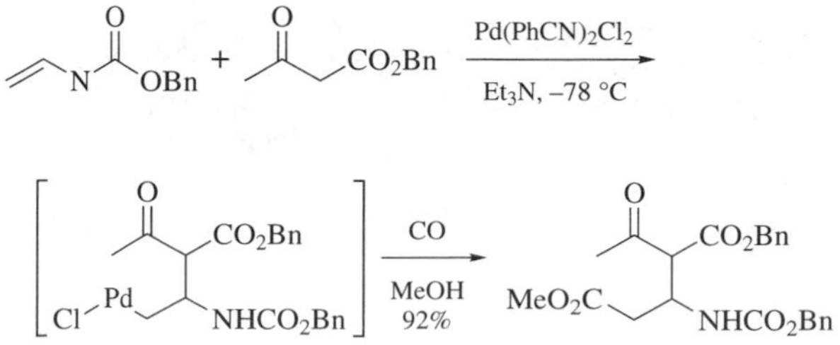

(17)

π-Allylpalladium Complex Formation[20]. π-Allylpalladium complexes are prepared by the reaction of alkenes with $PdCl_2$ or its soluble forms under various conditions (eq 18).[21] These π-allylpalladium chloride complexes react with carbon nucleophiles in DMSO as a coordinating solvent to form carbon–carbon bonds.[22] Thus π-allylpalladium complexes are clearly different in chemical reactivity from other organometallic reagents, which normally react with electrophiles (eq 19).

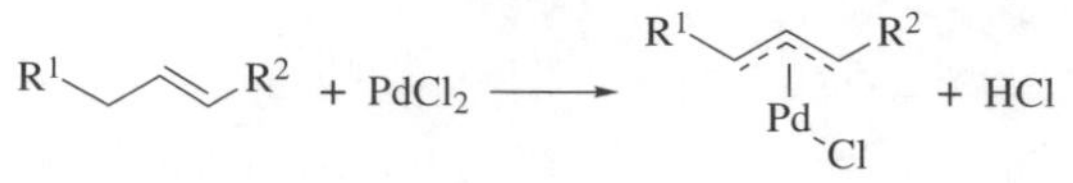

(18)

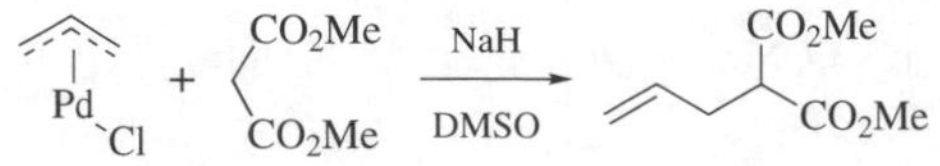

(19)

Based on this reaction, allylic alkylation of alkenes is possible. Active methylene compounds, such as malonates and β-keto esters, can be introduced to a steroid skeleton by the reaction of the steroidal π-allylpalladium complex in DMSO (eq 20).[23] The reaction of carbon nucleophiles also proceeds in the presence of an excess of ***Triphenylphosphine*** (eq 21).[24]

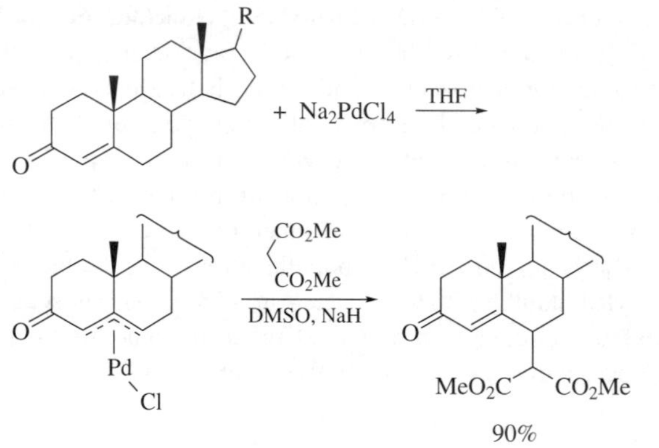

(20)

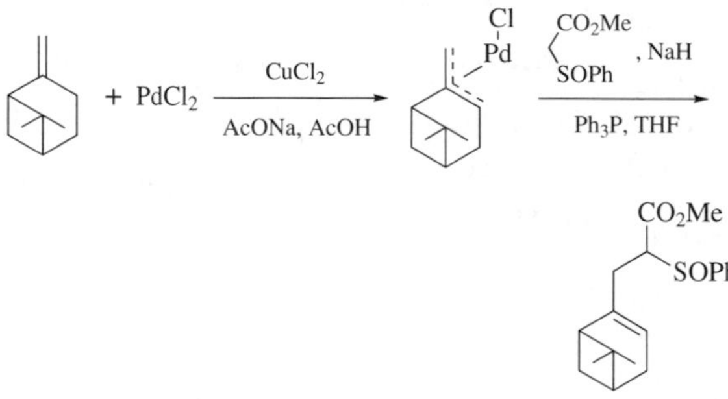

(21)

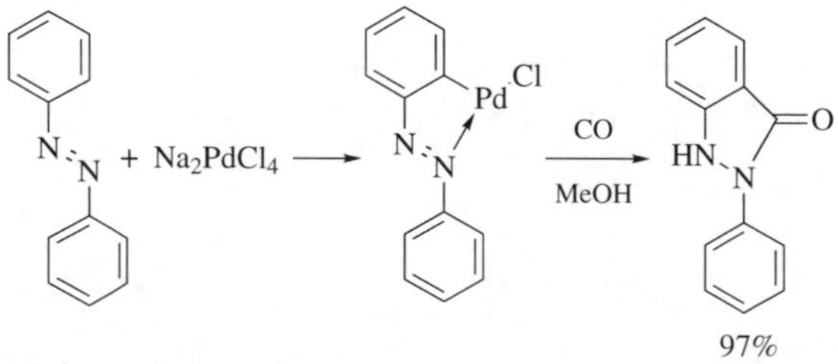

(22)

***ortho*-Palladation of Aromatic Compounds and Cyclopalladation of Allyl and Homoallyl Compounds**[25]**.** Azobenzene,[26] *N,N*-dimethylbenzylamine,[27] and related aromatic compounds react with Na_2PdCl_4 in ethanol to form stable *ortho*-palladation complexes. These carbon–palladium σ-bonded complexes are useful for the preparation of *ortho*-substituted aromatic compounds by the facile insertion of alkenes, alkynes, and CO. For example, insertion of CO to the azobenzene complex affords 2-aryl-3-indazolone (eq 22),[28] and facile insertion of styrene to the benzylamine complex yields a stilbene derivative (eq 23).[1a,29]

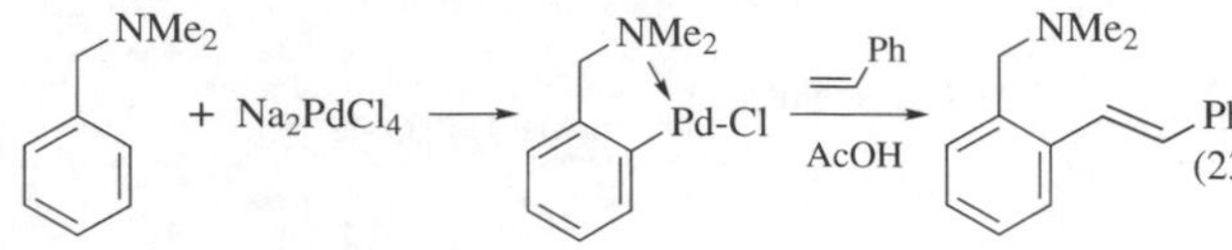

(23)

The cyclopalladation of allylic or homoallylic amines and sulfides proceeds due to the chelating effect of N and S atoms, and has been used for functionalization of alkenes. For example, isopropyl 3-butenyl sulfide is carbopalladated with methyl cyclopentanecarboxylate and Li_2PdCl_4. Reduction of the chelated complex with ***Sodium Cyanoborohydride*** affords the alkylated keto ester in 96% yield (eq 24).[30] Functionalization of 3-*N,N*-dimethylaminocyclopentene for the synthesis of a prostaglandin skeleton has been carried out via a *N*-chelated palladium complex as an intermediate. In the first step, malonate was introduced regio- and stereoselectively by carbopalladation (eq 25).[31] Elimination of a β-hydrogen generated a new cyclopentene, and its oxypalladation with 2-chloroethanol, followed by insertion of 1-octen-3-one and β-elimination, afforded the final product.

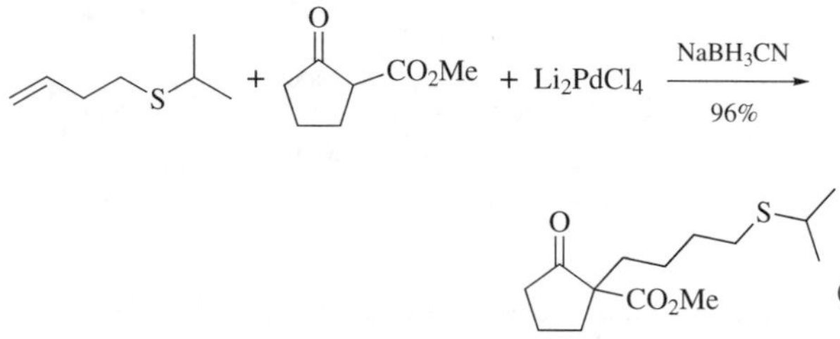

(24)

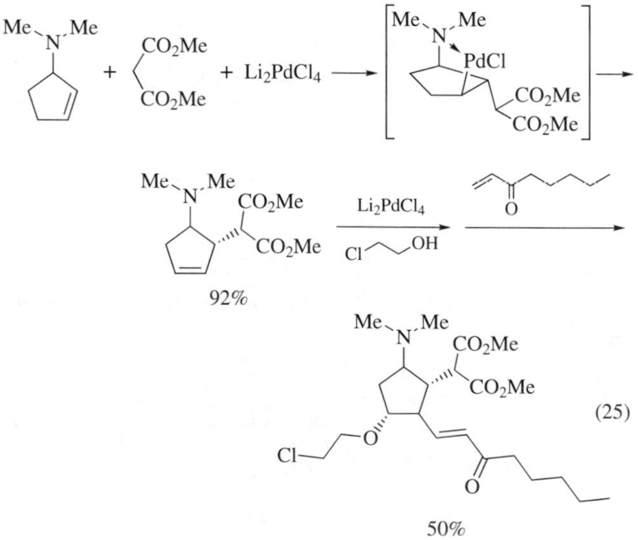

(25)

Oxidative Carbonylation.[32]

Oxidative Carbonylation of Alkenes. Oxidative carbonylation of alkenes with $PdCl_2$ in benzene affords β-chloroacyl chlorides (eq 26).[33] Oxidative carbonylation of alkenes in alcohol affords α,β-unsaturated esters and β-alkoxy esters by monocarbonylation and succinate derivatives by dicarbonylation (eq 27).[34]

(26)

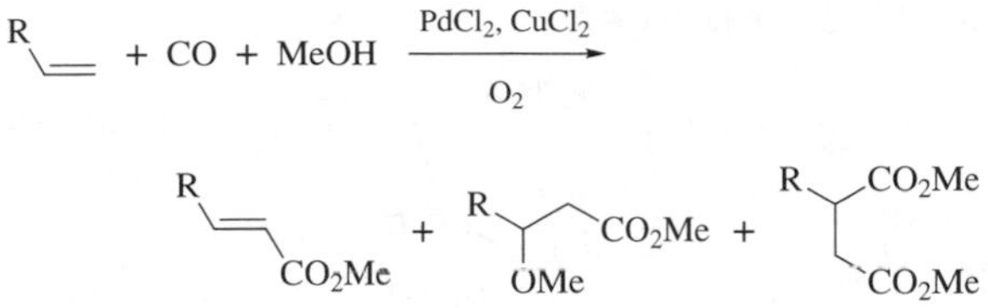

(27)

Intramolecular oxycarbonylation and aminocarbonylation are also known. As an example, frenolicin has been synthesized using oxycarbonylation at 1.1 atm of ***Carbon Monoxide*** as a key step (eq 28).[35] The intramolecular aminopalladation of a carbamate group and subsequent carbonylation of the substituted 3-hydroxy-4-pentenylamine proceeds smoothly in AcOH (eq 29).[36]

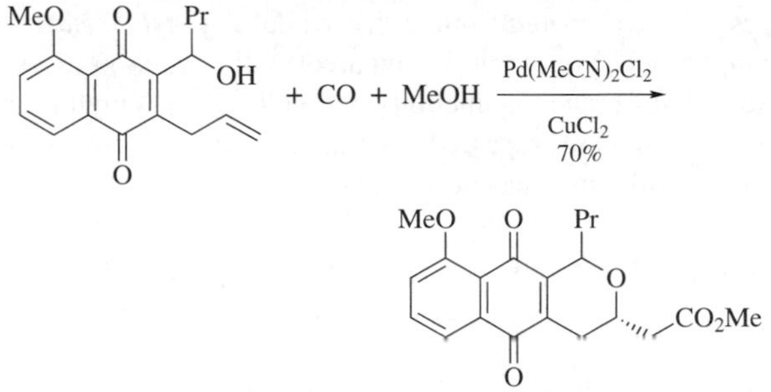

(28)

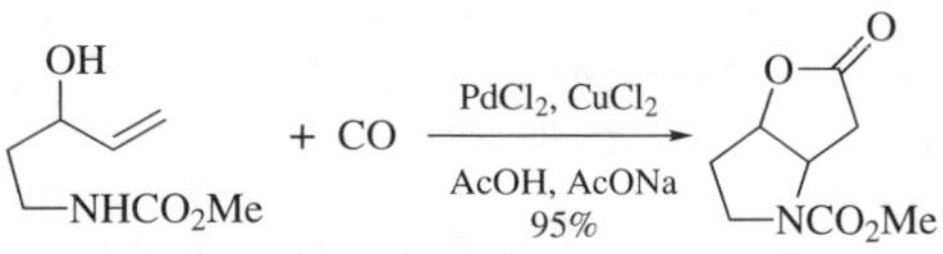

(29)

Oxidative Carbonylation of Alkynes. Terminal alkynes are carbonylated to give acetylenecarboxylates using $PdCl_2$ and $CuCl_2$ as catalysts (eq 30).[37] The acetylenecarboxylate in a β-lactam has been prepared by this procedure and then converted to a β-keto ester (eq 31).[38]

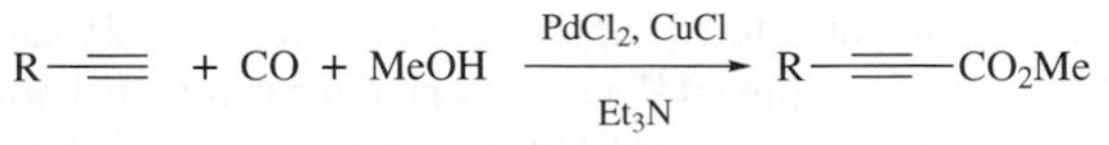

(30)

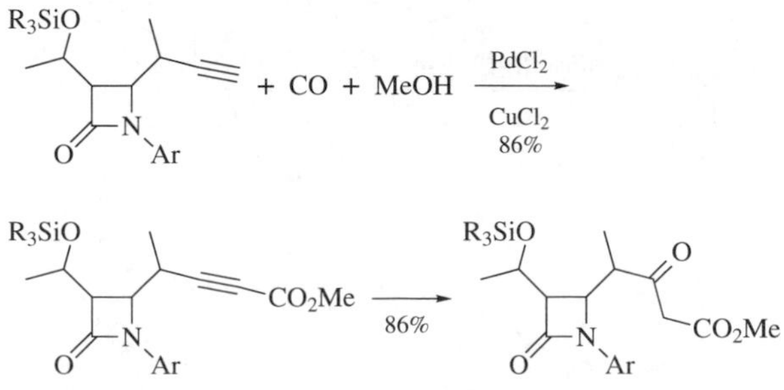

(31)

Oxidative dicarbonylation of acetylene with $Pd(PhCN)_2Cl_2$ in benzene affords the chlorides of maleic, fumaric, and muconic acids (eq 32).[39] Methyl muconate is obtained by passing acetylene and oxygen through MeOH containing thiourea and a catalytic amount of $PdCl_2$.[40] The oxidative dicarbonylation of alkynes produces maleate derivatives as a main product using $PdCl_2$ and $CuCl_2$ as catalysts under oxygen in alcohol.[41]

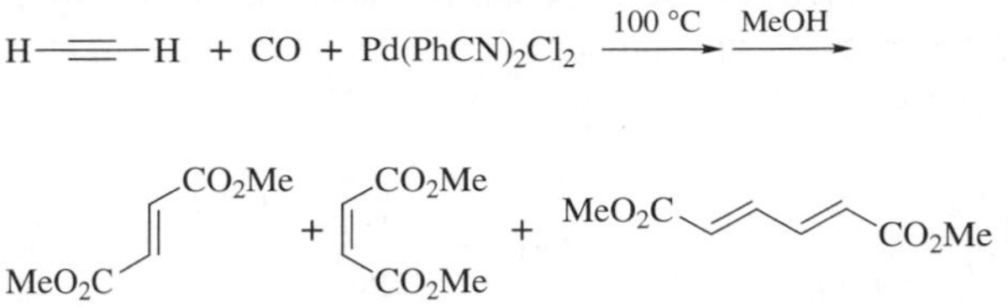

(32)

Oxidative Carbonylation of Alcohols. Oxalates and carbonates are formed by the oxidative carbonylation of alcohols. The reaction can be made catalytic by using $PdCl_2$ and $CuCl_2$ under oxygen in the alcohol.[42] Either oxalate or carbonate is obtained chemoselectively under different conditions (eq 33). Alkyl oxalates are produced commercially using alkyl nitrites as reoxidants (eq 34).[43]

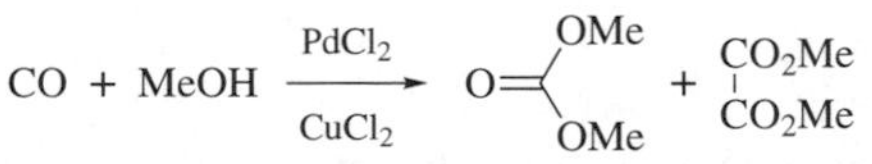

$$CO + MeOH \xrightarrow[CuCl_2]{PdCl_2} O{=}C(OMe)_2 + \begin{matrix}CO_2Me\\ |\\ CO_2Me\end{matrix} \quad (33)$$

$$2\ BuONO + 2\ CO \xrightarrow{PdCl_2} \begin{matrix}CO_2Bu\\ |\\ CO_2Bu\end{matrix} + 2\ NO$$

$$2\ NO + 2\ BuOH + 0.5\ O_2 \longrightarrow 2\ BuONO + H_2O \quad (34)$$

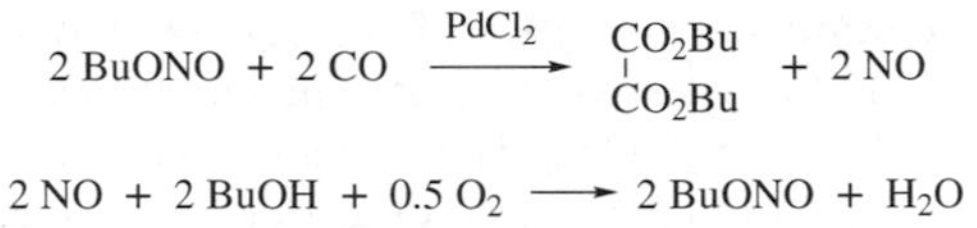

Reactions via Transmetallation of Organometallic Reagents. Transmetalation of organometallic compounds of Hg, B, Sn, Si, Tl, etc., with $PdCl_2$ produces the reactive organopalladium species, which undergoes insertion and coupling reactions. Aryl- or alkenylpalladium complexes, generated in situ from aryl- or alkenylmercury compounds, undergo insertion reactions with alkenes;[44,45] an example is shown in eq 35.[46] The arylmercury compound with 1,3-cyclohexadiene and Li_2PdCl_4 generates a π-allylpalladium intermediate, which then attacks the amide group intramolecularly to yield the cyclized product (eq 35).[46] CO insertion produces ketones and esters.[47] The *ortho*-thallation of benzoic acid and subsequent transmetalation with Pd^{II} generates a reactive arylpalladium complex, which reacts with butadiene to give an isocoumarin (eq 36).[48]

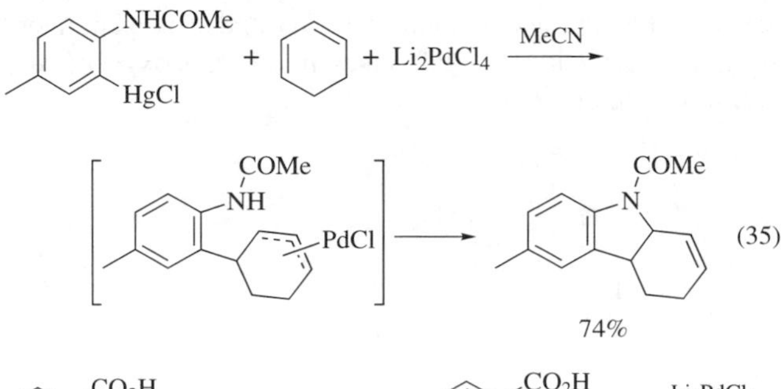

(35)

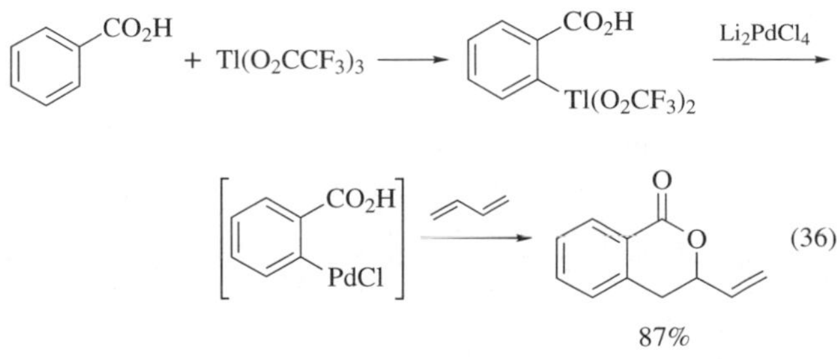

(36)

α,β-Unsaturated esters are obtained by the carbonylation of alkenylboranes[49] and alkenyl- or arylpentafluorosilicates (eq 37).[50] Conjugated dienes and diaryls are formed by the coupling of alkenyl- and arylstannanes. The homocoupling of the vinylstannane of benzoquinone is catalyzed by $PdCl_2(PhCN)_2$ with benzoquinone as the reoxidant (eq 38).[51]

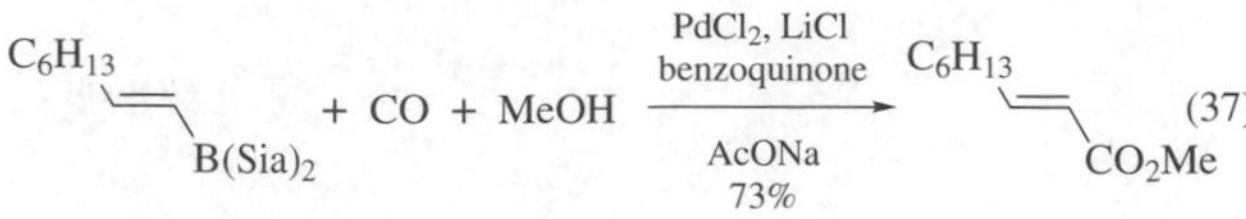

(37)

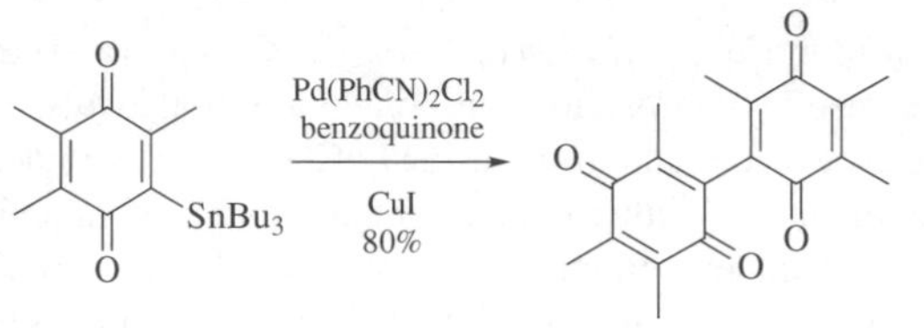

(38)

Miscellaneous Oxidation Reactions. Some oxidative reactions can be carried out only with $Pd(OAc)_2$, but not with $PdCl_2$. However, $Pd(OAc)_2$ can be generated in situ by the reaction of $PdCl_2$ with AcOK or AcONa. The oxidative coupling of aromatic rings is a typical example of a $Pd(OAc)_2$-promoted reaction. The following coupling reaction proceeds by $Pd(OAc)_2$ generated in situ from $PdCl_2$ (eq 39).[52]

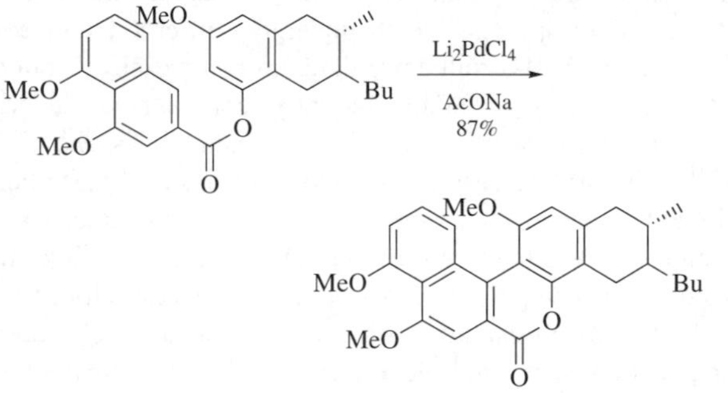

(39)

The oxidative rearrangement of the propargylic ester below proceeds with a catalytic amount of $PdBr_2$ under oxygen. Interestingly, the reoxidation of Pd^0 takes place with oxygen without addition of other reoxidants (eq 40).[53]

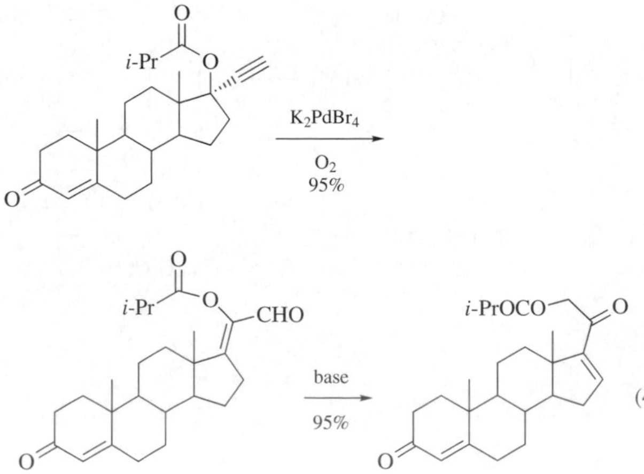

(40)

Catalytic Reactions with Pd^{II}.

Exchange Reactions of Vinyl Ethers and Esters[54]. Vinyl ethers are activated by Pd^{II}. Exchange with other alcohols to give mixtures of acetals and vinyl ethers is catalyzed by $PdCl_2$ (eq 41).[55] This reaction was used as the key step in the total synthesis of rhizobitoxine (eq 42).[56]

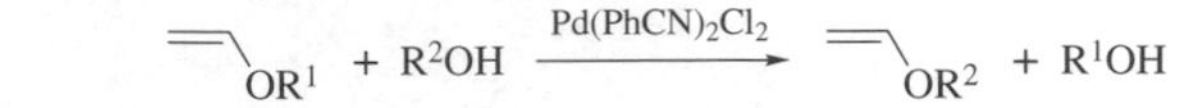

(41)

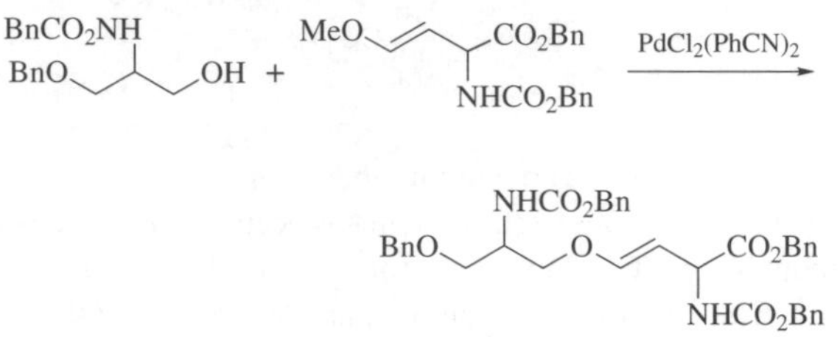

(42)

The exchange reaction of the acid component of vinyl esters with other acids is catalyzed by $PdCl_2$ (eq 43).[54] Thus various vinyl esters are prepared from easily available ***Vinyl Acetate***. As an example, vinyl itaconate is prepared by the reaction of vinyl acetate with itaconic monomethyl ester (eq 44).[57] *N*-Vinyllactams and cyclic imides are prepared by the exchange reaction of lactams and imides with vinyl acetate (eq 45).[58]

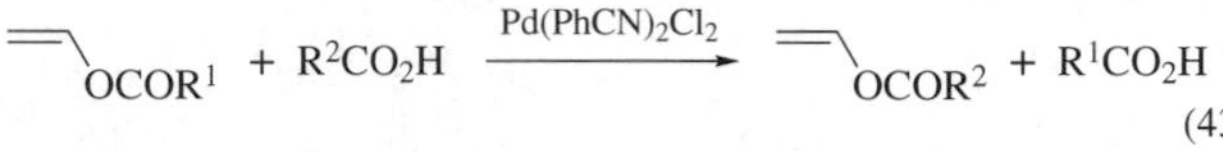

(43)

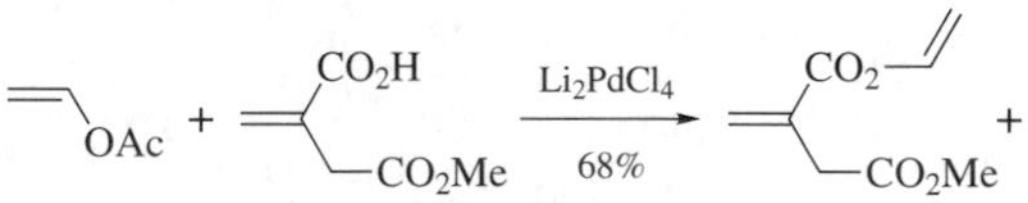

(44)

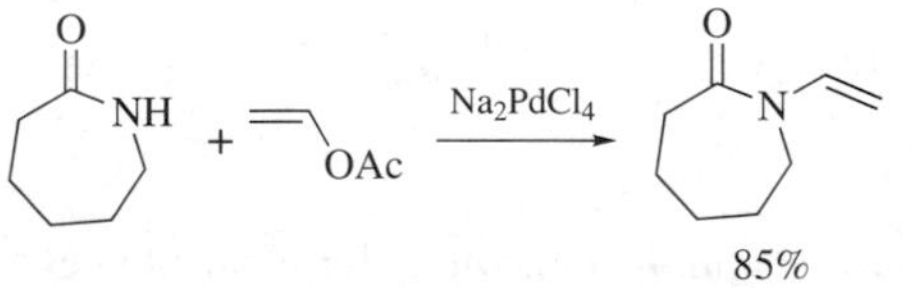

(45)

Pd^{II}-Catalyzed Rearrangement Reactions. Cope rearrangements are accelerated by catalytic amounts of $Pd(PhCN)_2Cl_2$, such that they proceed at room temperature in benzene or CH_2Cl_2 (eq 46).[59] Successful Pd^{II} catalysis appears to require that atoms 2 and 5 of the substituted 1,5-hexadienes have one H and one 'nonhydrogen' substituent.[60] Oxy–Cope rearrangements proceed at room temperature using $Pd(PhCN)_2Cl_2$ catalysis (eq 47).[61]

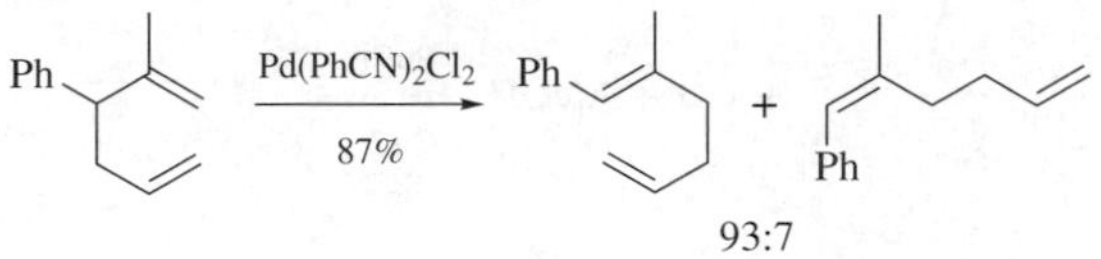

(46)

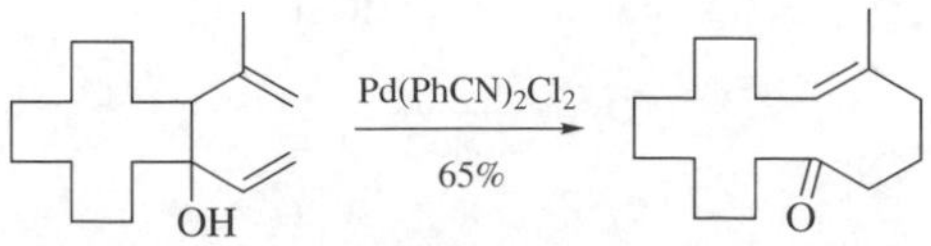

(47)

The $Pd(PhCN)_2Cl_2$-catalyzed Claisen rearrangement of allyl vinyl ethers has been studied to a lesser extent. The Claisen rearrangement shown in eq 48 proceeds smoothly even at room temperature to give the *syn* product with high diastereoselectivity.[62]

The Claisen rearrangement of 2-(allylthio)pyrimidin-4-(3*H*)-one affords the *N*-1 allylation product as a main product rather than the *N*-3 allylation product (eq 49).[63]

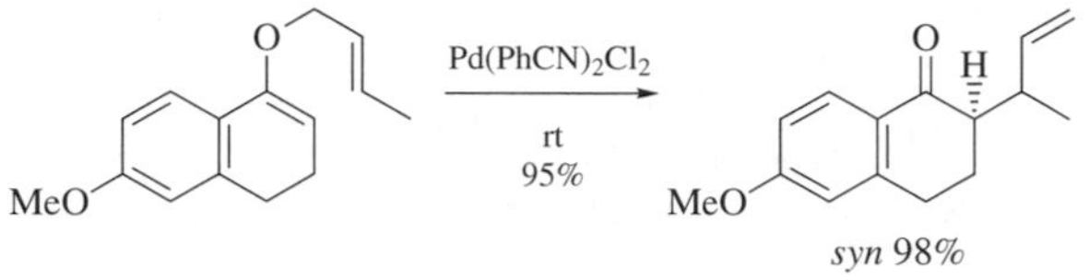

$Pd(PhCN)_2Cl_2$, rt, 95% — *syn* 98% (48)

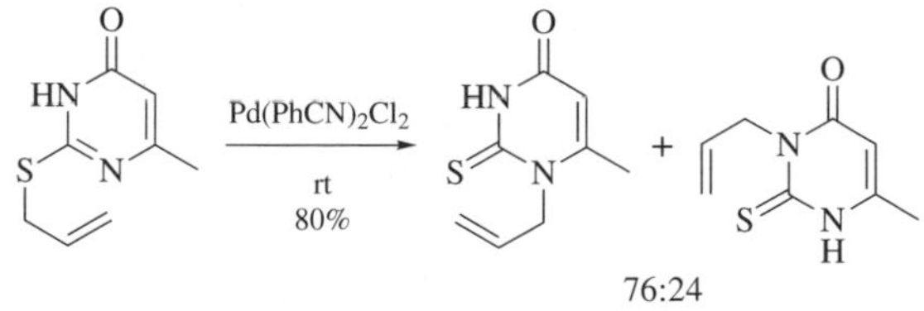

$Pd(PhCN)_2Cl_2$, rt, 80% — 76:24 (49)

The rearrangement of allylic esters, a useful reaction, is catalyzed efficiently by Pd^{II}.[64] The allylic rearrangement shown in eq 50, used in a prostaglandin synthesis, proceeds in one direction irreversibly, yielding the thermodynamically more stable product possibly due to steric reasons.[65] The diacetate of a 1,5-diene-3,4-diol is isomerized to the more stable conjugated diene with complete transfer of chirality (eq 51).[66] The Pd^{II}-catalyzed allylic rearrangement has been explained by an oxypalladation or cyclization-induced rearrangement. It is mechanistically different from rearrangements catalyzed by Pd^0 complexes, which proceed by formation of π-allylpalladium intermediates.

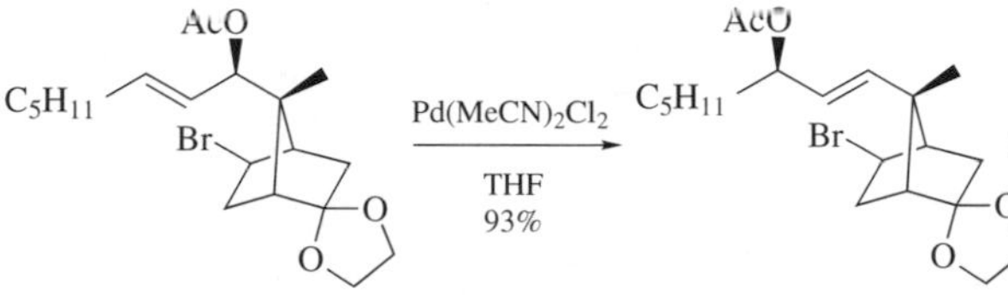

$Pd(MeCN)_2Cl_2$, THF, 93% (50)

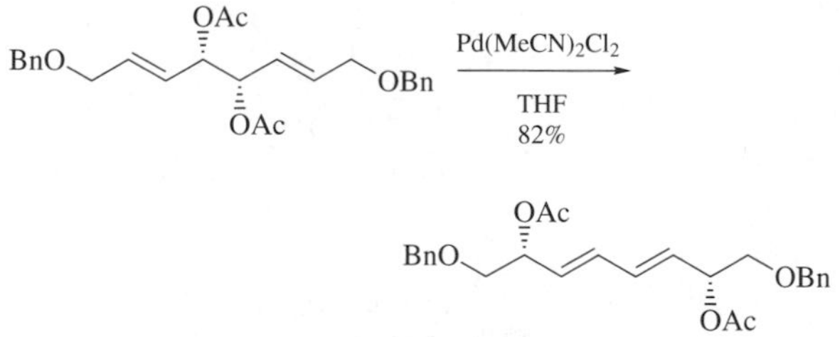

$Pd(MeCN)_2Cl_2$, THF, 82% (51)

Skeletal rearrangements of some strained compounds, such as bullvalene to bicyclo[4.2.2]deca-2,4,7,9-tetraene,[67] cubane to cuneane,[68] hexamethyl Dewar benzene to hexamethylbenzene (eq 52),[69] and quadricyclane to norbornadiene (eq 53),[70] are catalyzed by derivatives of $PdCl_2$.

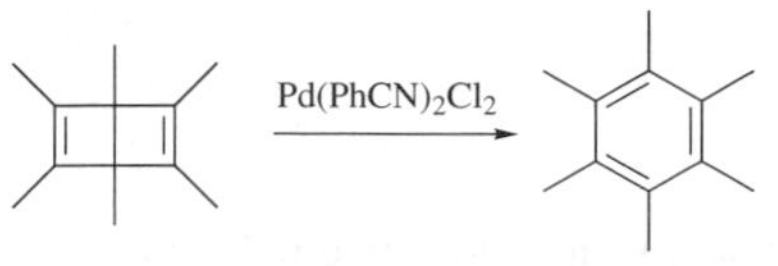

$Pd(PhCN)_2Cl_2$ (52)

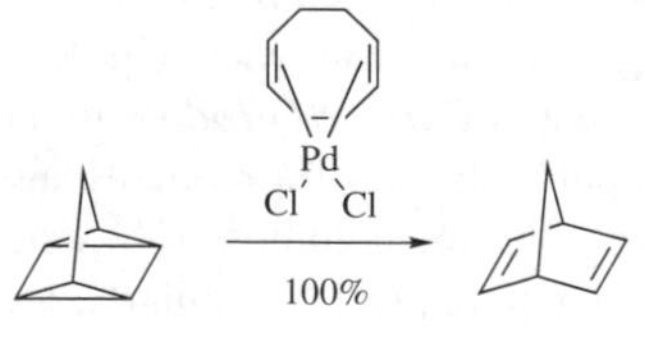

100% (53)

Intramolecular Reactions of Alkynes with Carboxylic Acids, Alcohols, and Amines. Addition of carboxylic acids, alcohols, and amines to alkynes via oxypalladation and aminopalladation proceeds with catalysis by Pd^{II} salts. Intramolecular additions are particularly facile.[71] Unsaturated γ-lactones are obtained by the treatment of 3-alkynoic acid and 4-alkynoic acid with $Pd(PhCN)_2Cl_2$ in THF in the presence of Et_3N (eq 54), and δ-lactones are obtained from 5-alkynoic acids.[72] 5-Hydroxyalkynes are converted to the cyclic enol ethers (eq 55).[71] The oxypalladation is a *trans* addition. Thus stereoselective enol ether formation by reaction of the alkynoic alcohol with $Pd(PhCN)_2Cl_2$, followed by reduction with ***Ammonium Formate***, has been applied to the synthesis of prostacyclin (eq 56).[73] Intramolecular addition of amines affords cyclic imines. 3-Alkynylamines are cyclized to 1-pyrrolines while 5-alkynylamines are converted to 2,3,4,5-tetrahydropyridines (eq 57).[74]

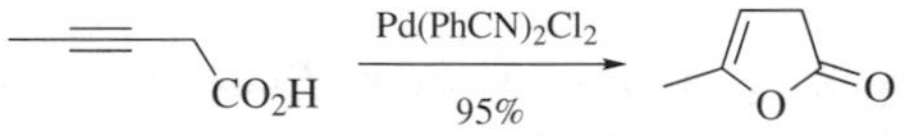

$Pd(PhCN)_2Cl_2$, 95% (54)

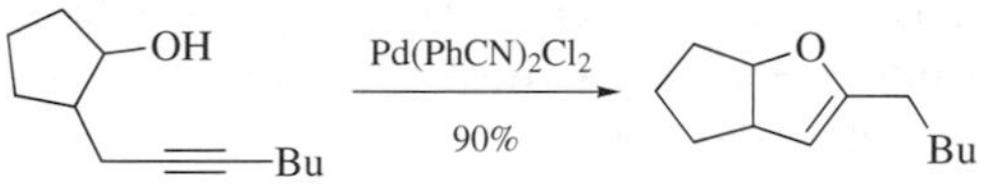

$Pd(PhCN)_2Cl_2$, 90% (55)

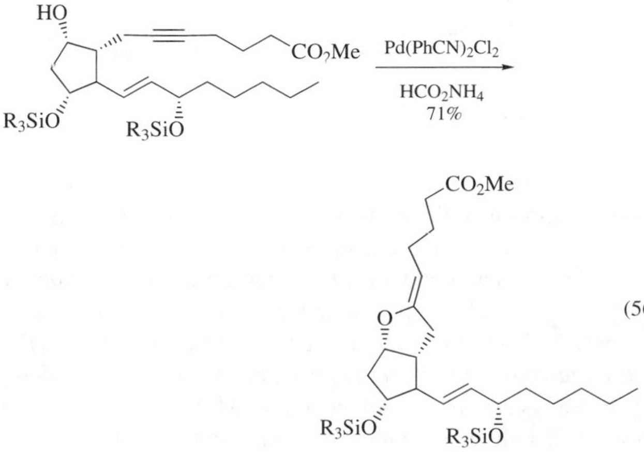

$Pd(PhCN)_2Cl_2$, HCO_2NH_4, 71% (56)

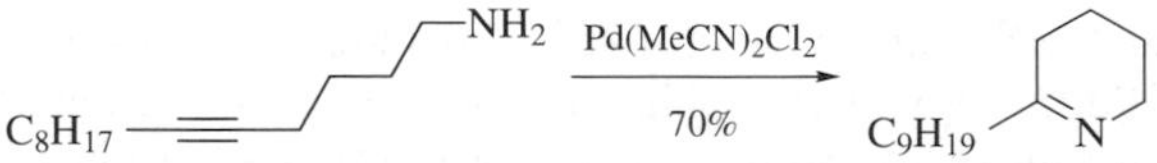

$Pd(MeCN)_2Cl_2$, 70% (57)

Simple alkynes cannot be hydrated with a palladium catalyst, but triple bonds are hydrated regioselectively to yield ketones with participation of suitably located carbonyl or hydroxy groups. 1,5-Diketones are prepared by the participation of a 5-keto group (eq 58).[75] 4-Hydroxyalkynes are converted to 4-hydroxy ketones and then oxidized to 1,4-diketones (eq 59).[71]

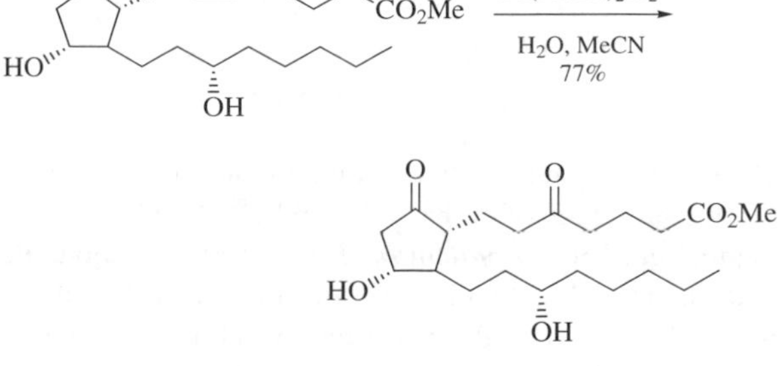

$Pd(MeCN)_2Cl_2$, H_2O, MeCN, 77% (58)

$$\text{(59)}$$

Cyclopentenone formation by the isomerization of 3-acetoxy-1,4-enynes is catalyzed by $Pd(PhCN)_2Cl_2$ (eq 60).[76]

$$\text{(60)}$$

Generation of Carbenes from Diazo Compounds. Both $PdCl_2$ and $Pd(OAc)_2$ are used for carbene generation from azo compounds.[77] The cyclopentenone carboxylates have been prepared by intramolecular insertions of the carbenes generated from α-diazo-β-keto esters (eq 61).[78]

$$\text{(61)}$$

Generation of Pd^0 catalysts. Pd^0 catalysts can be generated in situ from Pd^{II} in the presence or absence of phosphine ligands. ***Tetrakis(triphenylphosphine)palladium(0)*** is a commercially available Pd^0 complex used frequently as a catalyst, but it is air unstable. Therefore in situ generation of $Pd^0(Ph_3P)_n$ catalysts by the reduction of Pd^{II} in the presence of Ph_3P is convenient to use. In many cases the in situ reduction to Pd^0 takes place without addition of reducing agents. Alkenes, alcohols, CO, and phosphines, present in the reaction medium, behave as the reducing agent and react with Pd^{II} to give Pd^0. Generation of Pd^0 by reduction of $Pd(OAc)_2$ with phosphines has been reported.[79] Similarly, $PdCl_2$ and its derivatives have been converted to Pd^0 species with phosphines and bases.

$$\text{(62)}$$

$PdCl_2$ itself is used for the carbonylation of an aryl iodide in the presence of a base (eq 62).[80] More frequently, ***Bis(benzonitrile)dichloropalladium(II)*** is used for various Pd^0-catalyzed reactions. The coupling reaction of an acyl chloride with a disilane is catalyzed by Pd^0, generated from $Pd(PhCN)_2Cl_2$ and Ph_3P (eq 63).[81] The intermolecular coupling of a vinylenedistannane with two alkenyl iodides has been carried out using $Pd(PhCN)_2Cl_2$ without addition of Ph_3P in a total synthesis of rapamycin (eq 64).[82]

$$\text{(63)}$$

$$\text{(64)}$$

30% recovery of starting material

$$\text{(65)}$$

Dichlorobis(triphenylphosphine)palladium is used for Pd^0-catalyzed reactions without adding a reducing agent. For example, the coupling of terminal alkynes with halides is carried out with $Pd(Ph_3P)_2Cl_2$ and ***Copper(I) Iodide*** in the presence of ***Triethylamine*** without addition of a reducing agent. Hexaethynylbenzene is prepared by the coupling of hexabromobenzene with trimethylsilylacetylene (eq 65).[83] Similarly, the carbonylation of cinnamyl acetate, to give naphthyl acetate, is carried out in the presence of Et_3N (eq 66).[84]

$$\text{PhCH=CHCH}_2\text{OAc} + \text{CO} + \text{Ac}_2\text{O} \xrightarrow[\text{Et}_3\text{N, 76\%}]{\text{Pd(Ph}_3\text{P)}_2\text{Cl}_2} \text{1-naphthyl acetate} \quad (66)$$

In some cases, $Pd(Ph_3P)_2Cl_2$ is reduced to Pd^0 in situ with reducing agents such as metal hydrides, and used for Pd^0 catalyzed reactions. For example, $Pd(Ph_3P)_2Cl_2$ is reduced with ***Diisobutylaluminum Hydride*** and used for coupling reactions (eq 67).[85]

$$\text{3-bromocyclohexenone} + \text{Cp}_2\text{ZrCl-CH=CH-Bu} \xrightarrow[i\text{-Bu}_2\text{AlH}]{\text{Pd(Ph}_3\text{P)}_2\text{Cl}_2} \text{3-(CH=CH-Bu)cyclohexenone} \quad (67)$$

The carbonylation of alkenes in alcohols to give saturated esters proceeds smoothly with $PdCl_2$ or $Pd(Ph_3P)_2Cl_2$ as a catalyst (eq 68).[86] Alkynes are carbonylated efficiently to give α,β-unsaturated esters with the same catalyst in the presence of ***Iodomethane*** (eq 69).[87] In some reactions the Pd^0 species generated from $PdCl_2–Ph_3P$ and $Pd(OAc)_2–Ph_3P$ show different reactivities. For example, in the carbonylation of ***1,3-Butadiene***, 3-pentenoate is obtained with $PdCl_2–Ph_3P$, while 3,8-nonadienoate is obtained with $Pd(OAc)_2–Ph_3P$. The presence of chloride anion in the coordination sphere of palladium gives different catalytic activity (eq 70).[88]

$$\text{RCH=CH}_2 + \text{CO} + \text{MeOH} \xrightarrow{\text{Pd(Ph}_3\text{P)}_2\text{Cl}_2} \text{RCH}_2\text{CH}_2\text{CO}_2\text{Me} + \text{RCH(CH}_3\text{)CO}_2\text{Me} \quad (68)$$

$$\text{3,4-methylenedioxyphenylacetylene} + \text{CO} + \text{Et}_2\text{NH} \xrightarrow[\text{MeI, 92\%}]{\text{Pd(Ph}_3\text{P)}_2\text{Cl}_2} \text{ArC(=CH}_2\text{)CONEt}_2 \quad (69)$$

$$\text{CH}_2\text{=CHCH=CH}_2 + \text{CO} + \text{MeOH} \longrightarrow \begin{cases} \xrightarrow[\text{96\%}]{\text{Pd(OAc)}_2\text{, Ph}_3\text{P}} \text{CH}_2\text{=CH(CH}_2\text{)}_3\text{CH=CHCH}_2\text{CO}_2\text{Me} \\ \xrightarrow{\text{PdCl}_2\text{, Ph}_3\text{P}} \text{CH}_3\text{CH=CHCH}_2\text{CO}_2\text{Me} \end{cases} \quad (70)$$

Related Reagents. Bis(benzonitrile)dichloropalladium(II); (*R*)- & (*S*)-2,2′-Bis(diphenylphosphino)-1,1′-binaphthyl; Dichloro[1,1′-bis(diphenylphosphino)ferrocene]palladium(II); Dichloro[1,2-bis(diphenylphosphino)ethane]palladium(II); Dichloro[1,4-bis(diphenylphosphino)butane]palladium(II); Dichlorobis(tri-o-tolylphosphine)palladium(II); Dichloro(1,5-cyclooctadiene)palladium(II); Palladium(II) Acetate; Palladium(II) Chloride–Silver(I) Acetate; Palladium(II) Chloride–Copper(I) Chloride; Palladium(II) Chloride–Copper(II) Chloride.

---

1. (a) Tsuji, J. *ACR* **1969**, *2*, 144. (b) Tsuji, J. *Organic Synthesis with Palladium Compounds*; Springer: Berlin, 1980. (c) Henry, P. M. *Palladium Catalyzed Oxidation of Hydrocarbons*; Reidel: Dordrecht, 1980. (d) Trost, B. M.; Verhoeven, T. R. In *Comprehensive Organometallic Chemistry*; Wilkinson, G., Ed.; Pergamon: Oxford, 1982; Vol 8, pp 799–938. (e) Heck, R. F. *Palladium Reagents in Organic Syntheses*, Academic: New York, 1985.
2. (a) Tsuji, J. *S* **1984**, 369. (b) Tsuji, J. *COS* **1991**, *7*, 449. (c) Hegedus, L. S. *COS* **1991**, *4*, 551 and 571.
3. Clement, W. H., Selwitz, C. M. *JOC* **1964**, *29*, 241.
4. Tsuji, J.; Nagashima, H.; Nemoto, H. *OS* **1984**, *62*, 9.
5. Tsuji J., Shimizu, I., Yamamoto, K. *TL* **1976**, 2975.
6. Mehta, G.; Rao, K. S. *JACS* **1986**, *108*, 8015.
7. Tsuji, J.; Yamada, T.; Shimizu, I. *JOC* **1980**, *45*, 5209.
8. Iseki, K.; Yamazaki, M.; Shibasaki, M.; Ikegami, S. *T* **1981**, *37*, 4411.
9. Tsuji, J.; Nagashima, H.; Hori, K. *CL* **1980**, 257.
10. Tsuji, J.; Nagashima, H.; Hori, K. *TL* **1982**, *23*, 2679.
11. Miller, D. G.; Wayner, D. D. M. *JOC* **1990**, *55*, 2924.
12. (a) Lloyd, W. G.; Luberoff, B. J. *JOC* **1969**, *34*, 3949. (b) Hosokawa, T.; Nakajima, F.; Iwasa, S.; Murahashi, S. *CL* **1990**, 1387.
13. (a) Byrom, N. T.; Grigg, R.; Kongkathip, B. *CC* **1976**, 216. (b) Byrom, N. T.; Grigg, R.; Kongkathip, B.; Reimer, G.; Wade, A. R. *JCS(P1)* **1984**, 1643.
14. Matsui, K.; Uchiumi, S.; Iwayama, A.; Umezu, T. Eur. Pat. Appl. 55 108, 1976 (*CA* **1976**, *85*, 192 173).
15. Kimar, R. J.; Krupadanam, G. L. D.; Srimanarayana, G. *S* **1977**, 122.
16. (a) Kasahara, A.; Izumi, T.; Sato, K.; Maemura, M.; Hayasaka, T. *BCJ* **1977**, *50*, 1899, (b) Korte, D. E.; Hegedus, L. S.; Wirth, R. K. *JOC* **1977**, *42*, 1329.
17. Harrington, P. J.; Hegedus, L. S.; McDaniel, K. F. *JACS* **1987**, *109*, 4335.
18. Tsuji, J.; Takahashi, H. *JACS* **1965**, *87*, 3275. (b) Tsuji, J.; Takahashi, H. *JACS* **1968**, *90*, 2387.
19. (a) Wieber, G. M.; Hegedus, L. S.; Akermark, B.; Michalson, E. T. *JOC* **1989**, *54*, 4649. (b) Montgomery, J.; Wieber, G. M.; Hegedus, L. S. *JACS* **1990**, *112*, 6255.
20. (a) Tsuji, J. In *The Chemistry of the Metal-Carbon Bond*; Patai, S., Ed.; Wiley: New York, 1985; Vol 3, pp 163–199. (b) Godleski, S. A. *COS* **1991**, *4*, 585. (c) Trost B. M. *ACR* **1980**, *13*, 385.
21. (a) Huttel, R.; Christ, H. *CB* **1963**, *96*, 3101; **1965**, *98*, 1753. (b) Huttel, R.; McNiff, M. *CB* **1973**, *106*, 1789. (c) Volger, H. C. *RTC* **1969**, *88*, 225. (d) Morelli, D.; Ugo, R.; Conti, F.; Donati, M. *CC* **1967**, 801. (e) Trost, B. M.; Strege, P. E.; Weber, L.; Fullerton, T. J.; Dietsche, T. J. *JACS* **1978**, *100*, 3407.
22. Tsuji, J.; Takahashi, H.; Morikawa, M. *TL* **1965**, 4387.
23. (a) Jackson, W. R.; Strauss, J. U. *TL* **1975**, 2591. (b) Collins, D. J.; Jackson, W. R.; Timms, R. N. *AJC* **1977**, *30*, 2167. (c) Collins, D. J.; Jackson, W. R.; Timms, R. N. *TL* **1976**, 495.
24. Trost, B. M.; Fullerton, T. J. *JACS* **1973**, *95*, 292.
25. For reviews see (a) Bruce, M. I. *AG(E)* **1977**, *16*, 73. (b) Omae, I. *CRV* **1987**, *87*, 287. (c) Newkome, G. R.; Puckett, W. E.; Gupta, V. K.; Kiefer, G, E. *CRV* **1986**, *86*, 451. (d) Ryabov, A. D. *S* **1985**, 233.
26. Cope, A. C.; Siekman, R. W. *JACS* **1965**, *87*, 3272.
27. Cope, A. C.; Friedrich, E. C. *JACS* **1968**, *90*, 909.
28. Takahashi, H.; Tsuji, J. *JOM* **1967**, *10*, 511.
29. (a) Julia, M.; Duteil, M.; Lallemand, J. Y. *JOM* **1975**, *102*, 239. (b) Holton, R. A. *TL* **1977**, 355.
30. Holton, R. A.; Kjonaas, R. A. *JOM* **1977**, *142*, C15.

31. Holton, R. A. *JACS* **1977**, *99*, 8083.
32. (a) Colquhoun, H. M; Thompson, D. J.; Twigg, M. V. *Carbonylation*; Plenum: New York, 1991. (b) Thompson, D. J. *COS* **1991**, *3*, 1015.
33. (a) Tsuji, J.; Morikawa, M.; Kiji, J. *TL* **1963**, 1061. (b) Tsuji, J.; Morikawa, M.; Kiji, J. *JACS* **1964**, *86*, 8451.
34. Fenton, D. M.; Steinwand, P. J. *JOC* **1972**, *37*, 2034.
35. (a) Semmelhack, M. F.; Bozell, J. J.; Sato, T.; Wulff, W.; Spiess, E.; Zask, A. *JACS* **1982**, *104*, 5850. (b) Semmelhack, M. F.; Zask, A. *JACS* **1983**, *105*, 2034.
36. (a) Tamaru, Y.; Hojo, M.; Yoshida, Z. *JOC* **1988**, *53*, 5731. (b) Tamuru, Y.; Hojo, M.; Higashimura, H.; Yoshida, Z. *JACS* **1988**, *110*, 3994.
37. Tsuji, J.; Takahashi, M.; Takahashi, T. *TL* **1980**, *21*, 849.
38. Prasad, J. S.; Liebeskind, L. S. *TL* **1987**, *28*, 1857.
39. Tsuji, J.; Morikawa, M; Iwamoto, N. *JACS* **1964**, *86*, 2095.
40. Chiusoli, G. P.; Venturello, C.; Merzoni, S. *CI(L)* **1968**, 977.
41. Alper, H.; Despeyroux, B.; Woell, J. B. *TL* **1983**, *24*, 5691.
42. Fenton, D. M.; Steinwand, P. J. *JOC* **1974**, *39*, 701.
43. Ube Industries, Ltd. Belg. Patent 870 268, 1979 (*CA* **1979**, *91*, 4958).
44. For a review, see: Larock, R. C. *AG* **1978**, *90*, 28.
45. Heck, R. F. *JACS* **1968**, *90*, 5518, 5526, 5531, 5535, 5538, 5542.
46. Larock, R. C.; Harrison, L. W.; Hsu, M. H. *JOC* **1984**, *49*, 3664.
47. Heck, R. F. *JACS* **1968**, *90*, 5546.
48. (a) Larock, R. C.; Varaprath, S.; Lau, H. H.; Fellows, C. A. *JACS* **1984**, *106*, 5274. (b) Larock, R. C.; Liu, C.-L.; Lau, H. H.; Varaprath, S. *TL* **1984**, *25*, 4459.
49. Miyaura, N.; Suzuki, A. *CL* **1981**, 879.
50. (a) Tamao, K.; Kakui, T.; Kumada, M. *TL* **1979**, 619. (b) Yoshida, J. I.; Kohei, T.; Yamamoto, H.; Kakui, T.; Uchida, T.; Kumada, M. *OM* **1982**, *1*, 542.
51. Liebeskind, L. S.; Riesinger, S. W. *TL* **1991**, *32*, 5681.
52. Bringmann, G.; Reuscher, H. *TL* **1989**, *30*, 5249.
53. (a) Kataoka, H.; Watanabe, K.; Miyazaki, K.; Tahara, S.; Ogu, K.; Matsuoka, R.; Goto, K. *CL* **1990**, 1705. (b) Kataoka, H.; Watanabe, K.; Goto, K. *TL* **1990**, *31*, 4181.
54. Henry, P. M. *ACR* **1973**, *6*, 16.
55. (a) McKeon, J. E.; Fitton, P.; Griswold, A. A. *T* **1972**, *28*, 227. (b) McKeon, J. E.; Fitton, P. *T* **1972**, *28*, 233.
56. Keith, D. D.; Tortora, J. A.; Ineichen, K.; Leimgruber, W. *T* **1975**, *31*, 2633.
57. Bjorkquist, D. W.; Bush, R. D.; Ezra, F. S.; Keough, T. *JOC* **1986**, *51*, 3192.
58. Bayer, E.; Geckeler, K. *AG(E)* **1979**, *18*, 533.
59. For reviews, see: (a) Lutz, R. P. *CRV* **1984**, *84*, 205. (b) Overman, L. E. *AG(E)* **1984**, *23*, 579.
60. (a) Overman, L. E.; Knoll, F. M. *JACS* **1980**, *102*, 865. (b) Overman, L. E.; Jacobsen, J. *JACS* **1982**, *104*, 7225.
61. Bluthe, N.; Malacria, M.; Gore, J. *TL* **1983**, *24*, 1157.
62. Mikami, K.; Takahashi, K.; Nakai, T. *TL* **1987**, *28*, 5879.
63. Mizutani, M.; Sanemitsu, Y.; Tamaru, Y.; Yoshida, Z. *JOC* **1985**, *50*, 764.
64. Overman, L. E.; Knoll, F. M. *TL* **1979**, 321.
65. (a) Grieco, P. A.; Takigawa, T.; Bongers, S. L.; Tanaka, H. *JACS* **1980**, *102*, 7587. (b) Danishefsky, S. J.; Cabal, M. P.; Chow, K. *JACS* **1989**, *111*, 3456.
66. Saito, S.; Hamano, S.; Moriyama, H.; Okada, K.; Moriwake, T. *TL* **1988**, *29*, 1157.
67. Vedejs, E. *JACS* **1968**, *90*, 4751.
68. Cassar, L.; Eaton, P. E.; Halpern, J. *JACS* **1970**, *92*, 6366.
69. Dietl, H.; Maitlis, P. M. *CC* **1967**, 759.
70. Hogeveen, H.; Volger, H. C. *JACS* **1967**, *89*, 2486.
71. Utimoto, K. *PAC* **1983**, *55*, 1845.
72. Lambert, C.; Utimoto, K.; Nozaki, H. *TL* **1984**, *25*, 5323.
73. Suzuki, M.; Yanagisawa, A.; Noyori, R. *JACS* **1988**, *110*, 4718.
74. Fukuda, Y.; Matsubara, S.; Utimoto, K. *JOC* **1991**, *56*, 5812.
75. Imi, K.; Imai, K.; Utimoto, K. *TL* **1987**, *28*, 3127.
76. Rautenstrauch, V. *JOC* **1984**, *49*, 950.
77. (a) Paulissen, R.; Hubert, A. J.; Teyssie, P. *TL* **1972**, 1465. (b) Anciaux, A. J.; Hubert, A. J.; Noels, A. F.; Petiniot, N.; Teyssie, P. *JOC* **1980**, *45*, 695.
78. Taber, D. F.; Amedio, J. C.; Sherrill, R. G. *JOC* **1986**, *51*, 3382.
79. (a) Amatore, C.; Jutand, A.; M'Barki, M. A. *OM* **1992**, *11*, 3009. (b) Ozawa, F.; Kubo, A.; Hayashi, T. *CL* **1992**, 2177. (c) Mandai, T.; Matsumoto, T.; Tsuji, J.; Saito, S. *TL* **1993**, *34*, 2513.
80. Takahashi, T.; Nagashima, T.; Tsuji, J. *CL* **1980**, 369.
81. (a) Rich, J. D. *JACS* **1989**, *111*, 5886. (b) Kraft, T. E.; Rich, J. D.; McDermott, P. J. *JOC* **1990**, *55*, 5430.
82. Nicolaou, K. C.; Chakraborty, T. K.; Piscopio, A. D.; Minowa, N.; Bertinato, P. *JACS* **1993**, *115*, 4419.
83. (a) Diercks, R.; Vollhardt, K. P. C. *JACS* **1986**, *108*, 3150. (b) Schwager, H.; Spyroudis, S.; Vollhardt, K. P. C. *JOM* **1990**, *382*, 191.
84. (a) Matsuzaka, H.; Hiroe, Y.; Iwasaki, M.; Ishii, Y.; Koyasu, Y.; Hidai, M. *JOC* **1988**, *53*, 3832. (b) Kurosawa, H.; Ikeda, I. *JOM* **1992**, *428*, 289.
85. Okukado, N.; Van Horn, D. E.; Klima, W. L.; Negishi, E. *TL* **1978**, 1027.
86. (a) Tsuji, J.; Morikawa, M.; Kiji, J. *TL* **1963**, 1437. (b) Bittler, K.; Kutepow, N. V.; Neubauer, O.; Reis, H. *AG* **1968**, *80*, 352.
87. Torii, S.; Okumoto, H.; Sadakane, M.; Xu, L. H. *CL* **1991**, 1673.
88. (a) Hosaka, S.; Tsuji, J. *T* **1971**, *27*, 3821. (b) Tsuji, J.; Mori, Y.; Hara, M. *T* **1972**, *28*, 3721. (c) Billups, W. E.; Walker, W. E.; Shields, T. C. *CC* **1971**, 1067.

Jiro Tsuji
Okayama University of Science, Japan

(*R*)-Pantolactone[1]

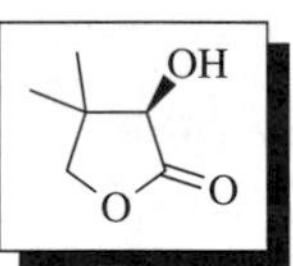

[599-04-2] $C_6H_{10}O$ (MW 130.16)

(effective chiral auxiliary in diastereoselective Diels–Alder reactions,[1] and for diastereoselective addition to ketenes;[2] used as a chiral pool reagent; also used as a covalently bound resolving agent[3])

Alternate Name: (*R*)-dihydro-3-hydroxy-4,4-dimethyl-2(3*H*)-furanone.
Physical Data: mp 92 °C; bp 120–122 °C/15 mmHg; $[\alpha]_D^{25}$ −50.7° (*c* 2.05, H_2O).
Solubility: sol water, alcohols, benzene, ether, chlorocarbons, THF.
Form Supplied in: crystalline white solid; commercially available.
Handling, Storage, and Precautions: hygroscopic.

Availability. Although commercially available via the degradation of pantothenic acid, (*R*)-pantolactone is also conveniently prepared by enantioselective reduction of its corresponding keto lactone employing homogeneous catalysis,[4a–g] or by microbial methods.[5] The (*S*)-enantiomer has been prepared by inversion of the natural product in 90% yield and 97% ee via triflate activation, acetate displacement, and ***Lithium Hydroxide*** hydrolysis.[6] The enantiomers were also prepared by resolution of the racemate with (*R*)- and (*S*)-phenethylamine.[7] A gas chromatographic method exists for ee determination.[8]

Diels–Alder Reactions. (*R*)-Pantolactone is one of the most effective chiral auxiliaries for preparative scale Diels–Alder additions of simple enoate esters in the presence of Lewis acids (eq 1).[9]

1.0 equiv $TiCl_4$; CH_2Cl_2, pet ether, −10 °C; 97:3; LiOH; THF, H_2O; >99.9% ee (1)

CH_2Cl_2, pet ether; 0.1–0.3 equiv $TiCl_4$, 0 °C; 93–97% de; LiOH; THF, H_2O; 71–74%; >99.5% ee

R* = ; R = H, Me

Endo–exo selectivity typically ranges from 20:1 to 45:1 with a maximum of 97.5:2.5 diastereoselection. Preparatively convenient reaction conditions are employed (CH_2Cl_2, CH_2Cl_2/cyclohexane; temp. approx. 0 °C; ca. 0.3 M concentration; and 0.1–1.0 molar equiv of Lewis acid). Products are typically crystalline and brought to high optical purity by recrystallization. Epimerization-free hydrolysis is effected with LiOH in THF/water. This procedure has been successfully applied in a nine-step synthesis of cyclosarkomycin in 17% overall yield (eq 2),[10] and to syntheses of the sandalwood fragrances.[11]

0.75 equiv $TiCl_4$; CH_2Cl_2, −24 °C; + *exo* isomers (2); 98:2

8 steps; 78%, >99% ee; Cyclosarkomycin 17% overall yield

The cyclohexane unit of the C(30) stereocenter of the C(18)–C(35) segment of FK-506 was established in excellent yield and de employing the same concept (eq 3).[6]

$TiCl_4$; CH_2Cl_2, −32 °C; 91% de; 1. LiOH, H_2O–THF; 2. *m*-CPBA, Et_3N; 61% overall; 91% recovered (3)

(*E*)-2-Cyanocinnamates have been similarly used as dienophiles. An *endo–exo* selectivity of 85:15 at a diastereoselectivity of 99:1 was obtained (eqs 4 and 5).[12,13]

0.75 equiv $TiCl_4$; CH_2Cl_2, −40 °C; 79%; 99:1 de (4)

0.75 equiv $TiCl_4$; CH_2Cl_2, −40 °C; 94:6 (5)

Further variations in dienophile have been equally successful (eqs 6 and 7),[14] including applications to the Michael reaction (eq 7)[15] and in the synthesis of a prostaglandin intermediate (eq 8).[16]

0.13 equiv $TiCl_4$; CH_2Cl_2, −23 °C (6)

R = H, 86% 98% de
R = Me, 80% 96% de

Nuc^-; DMF; 90–100% de (7)

Nuc = *t*-Bu–C$_6$H$_4$–S$^-$; $Me_2C^-NO_2$; $EtCH^-NO_2$; acetylacetonate anion

CH_2OBn ; COR* + ; $TiCl_4$, CH_2Cl_2–pet ether, –60 °C, 74%

BnO ; COR* ; 98:2 de → Corey lactone (8)

Ketene Additions. Reaction of the ketene derived from ibuprofen (Ar = *p*-isobutylphenyl) with (*R*)-pantolactone in the presence of simple tertiary amine bases in apolar solvents yielded >99% de favoring the (*R*,*R*)-ester (eq 9).[3] The reaction is first order in each component and possesses a pronounced deuterium isotope effect (k_H/k_D 4). The ketene from naproxen (Ar = 2-(6-methoxynaphthyl)) affords a de of 80% under similar conditions.

Ar = *p*-isobutylphenyl ; Me_3N, heptane, –78 °C ; OR* ; 99% de (9)

Extension of this work to a series of bromo- and iodoketenes proceeds with good to excellent de (eq 10).[17] Reaction of the products with azide ion affords a ready entry into amino acid synthesis (eq 11). However, with R = aryl, no selectivity was noted, possibly due to base-mediated epimerization under the reaction conditions.

Me_2NEt, THF, –78 °C ; 75–95% de (10)

R = Et, *t*-Bu, *i*-Pr, $PhCH_2$, Ph_2CH
X = Br or I, X^1 = Cl or Br

NaN_3, DMF, R = *t*-Bu, X = Br ; LiOH, THF, H_2O ; *t*-Bu, CO_2H, N_3 (11)

LAH, THF, 80 °C ; , H^+ ; 1. $COCl_2$, DMSO, Et_3N 2. Wittig–Horner 3. DIBAL ; Sharpless asymmetric epoxidation ; OR → Elfamycins (12)

Chiral Pool Reagent. (*R*)-Pantolactone has been used as a source of chiral fragments for synthesis. Applications include use in the syntheses of the elfamycins (eq 12)[18] and the bryostatins (eq 13).[19a,b] It has also been used to prepare potentially useful chiral epoxide synthons possessing a quaternary *gem*-dimethyl carbon.[20]

LAH, THF, 80 °C ; , H^+ ; OMPM ; S, BnO, OMPM → Bryostatins (13)

Miscellaneous Applications. Only one attempt to use (*R*)-pantolactone as an enantioselective protonating agent for enolates has been reported.[21] A series of structurally diverse chiral alcohols afforded modest ee's with (*R*)-pantolactone affording the largest ee noted for the series. The complexities of attempting a protonation of this sort in the presence of base and under exchanging conditions are discussed. Finally, the lactone has been used to resolve chiral acids by crystallization and chromatographic techniques applied to the (*R*)-pantolactone-derived esters.[3,22,23]

Related Reagents. (*S*)-4-Benzyl-2-oxazolidinone; (*R*,*R*)-2-*t*-Butyl-5-methyl-1,3-dioxolan-4-one; (*R*)-2-*t*-Butyl-6-methyl-4*H*-1,3-dioxin-4-one; 10,2-Camphorsultam; Dihydro-5-(hydroxymethyl)-2(3*H*)-furanone; (*S*)-Ethyl Lactate; Ethyl Mandelate; 3-Hydroxyisoborneol; Methyl (4*R*,5*R*)-(*E*)-3-(1,3-Dimethyl-4,5-diphenyl-2-imidazolidinyl)propenoate; α-Methyltoluene-2,α-sultam; (−)-8-Phenylmenthol; 8-Phenylmenthyl Acrylate; 1,1,2-Triphenyl-1,2-ethanediol.

1. Helmchen, G.; Hady, A. F. A.; Hartmann, H.; Karge, R.; Krotz, A.; Sartor, K.; Urmann, M. *PAC* **1989**, *61*, 409.
2. Larsen, R. D.; Corley, E. G.; Davis, P.; Reider, P. J.; Grabowski, E. J. J. *JACS* **1989**, *111*, 7650.
3. Duke, C. C.; Wells, R. J. *AJC* **1987**, *40*, 1641.
4. (a) Ojima, I.; Kogure, T.; Yoda, Y. *OS* **1985**, *63*, 18. (b) Ojima, I.; Kogure, T.; Terasaki, T.; Achiwa, K. *JOC* **1978**, *43*, 3444. (c) Morimoto, T.; Takahashi, H.; Fujii, K.; Chiba, M.; Achiwa, K. *CL* **1986**, 2061. (d) Hatat, C.; Karim, A.; Kokel, N.; Mortreux, A.; Petit, F. *TL* **1988**, *29*, 3675. (e) Genet, J. P.; Pinel, C.; Mallart, S.; Juge, S.; Cailhol, N.; Laffitte, J. A. *TL* **1992**, *33*, 5343. (f) Takahashi, H.; Hattori, M.; Chiba, M.; Morimoto, T.; Achiwa, K. *TL* **1986**, *27*, 4477. (g) Hatat, C.; Karim, A.; Kokel, N.; Mortreux, A.; Petit, F. *NJC* **1990**, *14*, 141.
5. Shimizu, S.; Yamada, H.; Hata, H.; Morishita, T.; Akutsu, S.; Kawamura, M. *ABC* **1987**, *51*, 289.
6. Corey, E. J.; Huang, H. C. *TL* **1989**, *30*, 5235.
7. Nohira, H.; Nohira, M.; Yoshida, S.; Osada, A.; Terunuma, D. *BCJ* **1988**, *61*, 1395.
8. Brunner, H.; Forster, St. *M* **1992**, *123*, 659.
9. Poll, T.; Sobczak, A.; Hartmann, H.; Helmchen, G. *TL* **1985**, *26*, 3095.
10. Linz, G.; Weetman, J.; Hadey, A. A. F.; Helmchen, G. *TL* **1989**, *30*, 5599.
11. Krotz, A.; Helmchen, G. *TA* **1990**, *1*, 537.
12. Cativiela, C.; Mayoral, J. A.; Avenoza, A.; Peregrina, J. M.; Lahoz, F. J.; Gimeno, S. *JOC* **1992**, *57*, 4664.

13. Avenoza, A.; Cativiela, C.; Mayoral, J. A.; Peregrina, J. M. *TA* **1992**, *3*, 913.
14. Hanzawa, Y.; Suzuki, M.; Kobayashi, Y.; Taguchi, T.; Iitaka, Y. *JOC* **1991**, *56*, 1718.
15. Knol, J.; Jansen, J. F. G. A.; Van Bolhuis, F.; Feringa, B. L. *TL* **1991**, *32*, 7465.
16. Miyaji, K.; Arai, K.; Ohara, Y.; Takahashi, Y. U.S. Patent 4 837 344, 1989.
17. Durst, T.; Koh, K. *TL* **1992**, *33*, 6799.
18. Dolle, R. E.; Nicolaou, K. C. *JACS* **1985**, *107*, 1691.
19. (a) DeBrabander, J.; Vanhessche, K.; Vandewalle, M. *TL* **1991**, *32*, 2821. (b) Roy, R.; Rey, A. W.; Charon, M.; Molino, R. *CC* **1989**, 1308.
20. Lavallée, P.; Ruel, R.; Grenier, L.; Bissonnette, M. *TL* **1986**, *27*, 679.
21. Gerlach, U.; Hünig, S. *AG(E)* **1987**, *26*, 1283.
22. Allan, R. D.; Bates, M. C.; Drew, C. A.; Duke, R. K.; Hambley, T. W.; Johnston, G. A. R.; Mewett, K. N.; Spence, I. *T* **1990**, *46*, 2511.
23. Mash, E. A.; Arterburn, J. B.; Fryling, J. A.; Mitchell, S. H. *JOC* **1991**, *56*, 1088.

Edward J. J. Grabowski
Merck Research Laboratories, Rahway, NJ, USA

Paraformaldehyde

$(CH_2O)_n$

[30525-89-4] CH_2O (MW 30.03)

(convenient source of anhydrous monomeric formaldehyde for aldol condensation;[1] electrophilic source of halomethyl[2] and hydroxymethyl groups;[3] convenient source of Mannich reaction intermediates;[4] homologation of alkynes;[5] synthesis of macrocyclic ligands[6])

Physical Data: mp 160–165 °C.
Solubility: sol organic solvents.
Form Supplied in: solid; widely available.
Handling, Storage, and Precautions: generally stable; irritant.

Introduction. The need to carry out one-carbon homologations during the course of organic syntheses is widespread. Since ***Formaldehyde*** itself is a hygroscopic gas which forms the hydrate easily, the ability to utilize functional equivalents of formaldehyde is quite useful. The proper choice for the synthetic equivalent of a tenuous species like formaldehyde is dictated by the context. Unlike the two other most common precursors, formalin and trioxane, paraformaldehyde is rather easy to handle. It is compatible with most organic media and displays useful reaction patterns. The polymeric acetal cleaves easily under acidic conditions and, to a lesser extent, under basic and neutral conditions, to afford one-carbon homologs. On those occasions when a solution of the monomeric formaldehyde is needed, the best approach is the thermal cracking of paraformaldehyde, with the evolved gaseous monomer bubbled into a solution and used immediately.

Reactions with Enolates and Enols. As the functional equivalent of formaldehyde, paraformaldehyde can afford aldol products, especially under acidic conditions wherein the enol intermediate is generated and conditions lead to unsaturated carbonyl compounds. Methyl ketones undergo reaction (eq 1) with paraformaldehyde to afford the vinyl ketones under acidic conditions.[7] Similar results are obtained with other active hydrogen reactants such as the treatment of methyloxazolines with paraformaldehyde and acid.[8] Under neutral aqueous conditions, multiple aldol reactions with cyclohexanone have been reported (eq 2).[9]

$+ (CH_2O)_n$ — H_2N^+Me (anilinium) $CF_3CO_2^-$, 90% → (1)

$+ (CH_2O)_n + H_2O$ — 35 °C → HOH_2C, HOH_2C, OH, CH_2OH, CH_2OH (2)

Noteworthy observations in this particular reaction are the ultimate reduction of the ketone, presumably via a crossed Cannizzaro reaction, and the ability to utilize large excesses of paraformaldehyde without major difficulties. Of course, both phenomena are related to the absence of α-hydrogens and resultant inability of formaldehyde to enolize.

While paraformaldehyde can be effective in enolate chemistry,[10] the more common practice, when condensation onto a preformed enolate is required, is to generate a solution of monomeric formaldehyde through pyrolysis.[1,11] One can isolate the resultant hydroxy ketone (eq 3),[12] or force dehydration through the use of ***Methanesulfonyl Chloride***,[12] or a variety of organometallic agents.[13] The monomer generated in this way also reacts readily with the dianions of carboxylic acids.[14]

1. LDA, –78 °C; 2. CH_2O (g), –20 °C; >95% (3)

Halomethyl and Hydroxymethyl Electrophiles. The chloromethylation of aromatic[2] and heteroaromatic compounds[15] can be accomplished using paraformaldehyde, ***Zinc Chloride***, and ***Aluminum Chloride***. Similar chemistry allows bromomethylation of arenes.[16] Halomethyl esters can be obtained from carboxylic acids as well.[17] The hydroxymethylation of anthraquinones has been reviewed[3] and paraformaldehyde can be used for the protection of alcohols as the methoxymethyl group.[18–20]

The Mannich Reaction and other Iminium Ion Reactions. While the trapping of an iminium ion (generated by the acid-catalyzed reaction of an amine and paraformaldehyde) by enols[4] has been well investigated (eq 4),[21,22] a variety of other nucleophiles are available, such as allylstannanes.[23] Of particular note are those reactions which bring about cyclization. In this context a vinylsilane (eq 5)[24] is effective, as is an alkene (eq 6).[25]

$+ (CH_2O)_n + Et_2NH_2Cl$ — 70% → NEt_2 (4)

Homologation of Alkynes[5]. While the terminal alkyne group is a relatively weak nucleophile, strong Lewis acids can bring about homologation, albeit with some isomerization (eq 7).

Macrocyclic Ligands[6]. By making use of template control, aza macrocycles can be formed which act as organometallic ligands and sequestering agents (eq 8).

(as nickel complex)

Related Reagents. Dimethyl(methylene)ammonium Iodide; (Dimethylamino)methyl Methyl Ether; Formaldehyde; Formaldehyde–Dimethylamine; Formaldehyde–Hydrogen Chloride; Formaldehyde–Hydrogen Bromide.

1. Stork, G.; d'Angelo, J. *JACS* **1974**, *96*, 7114.
2. Fieser, L.; Seligman, A. M. *JACS* **1935**, *57*, 942.
3. Krohn, K. *T* **1990**, *46*, 291.
4. Blicke, F. F. *OR* **1942**, *1*, 303.
5. Rodini, D. J.; Snider, B. B. *TL* **1980**, *21*, 3857.
6. Kang, S.-G.; Jung, S.-J.; Kweon, J. K.; Kim, M.-S. *Polyhedron* **1993**, *12*, 353.
7. Gras, J.-L. *TL* **1978**, 2955.
8. Meyers, A. I.; Kovelesky, A. C. *TL* **1969**, 4809.
9. Wittcoff, H. *OSC* **1963**, *4*, 907.
10. Ueno, Y.; Setoi, H.; Okawara, M. *TL* **1978**, 3753.
11. Stork, G.; Isobe, M. *JACS* **1975**, *97*, 6260.
12. Grieco, P. A.; Hiroi, K. *CC* **1972**, 1317.
13. Tsuji, J.; Nisar, M.; Minami, I. *TL* **1986**, *27*, 2483.
14. Pfeffer, P. E.; Sibert, L. S. *JOC* **1970**, *35*, 262.
15. Kochetkov, N. K.; Khomutova, E. D.; Bazilevskii, M. V. *ZOB* **1958**, *28*, 2736 (*CA* **1959**, *53*, 9187).
16. van der Made, A. W.; van der Made, R. H. *JOC* **1993**, *58*, 1262.
17. Knochel, P.; Chou, T.-S.; Joubert, C.; Rajagopal, D. *JOC* **1993**, *58*, 588.
18. Edwards, J. A.; Calzada, M. C.; Bowers, A. *JMC* **1964**, *7*, 528.
19. Kapnang, H.; Charles, G.; Sondengam, B. L.; Hemo, J. H. *TL* **1977**, 3469.
20. Barluenga, J.; Bayon, A. M.; Asensio, G. *CC* **1984**, 1334.
21. Wilds, A. L.; Nowak, R. M.; McCaleb, K. E. *OSC* **1963**, *4*, 281.
22. Hagemeyer, H. J., Jr. *JACS* **1949**, *71*, 1119.
23. Grieco, P. A.; Bahsos, A. *JOC* **1987**, *52*, 1378.
24. Overman, L. E.; Bell, K. E. *JACS* **1981**, *103*, 1851.
25. Shiotani, S.; Kometani, T. *TL* **1976**, 767.

Richard T. Taylor
Miami University, Oxford, OH, USA

(2*R*,4*R*)-2,4-Pentanediol[1]

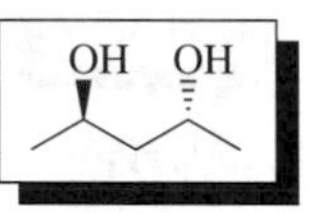

[42075-32-1] $C_5H_{12}O_2$ (MW 104.17)

(diol used for the preparation of chiral acetals[1])

Physical Data: mp 48–50 °C; bp 111–113 °C/19 mmHg.
Form Supplied in: white solid; widely available.
Preparative Methods: via asymmetric hydrogenation of ***2,4-Pentanedione***[2] Its enantiomer *[72345-23-4]* is also available by the same method.
Purification: recrystallization from ether.
Handling, Storage, and Precautions: should be stored in a tightly closed container since it is hygroscopic.

Cleavage of Acetals. Acetals of 2,4-pentanediol are easily prepared from aldehydes via standard procedures (e.g. cat. PPTS, PhH, Dean–Stark removal of H_2O). These acetals have been cleaved with a variety of nucleophiles in the presence of Lewis acids to yield hydroxy ethers with high (typically 90–95% de) diastereoselectivities. Oxidation and β-elimination then provides enantiomerically enriched alcohols (eq 1). Nucleophiles have included allylsilanes to produce homoallylic alcohols,[3] alkynylsilanes to give propargylic alcohols,[4] Me_3SiCN to provide (after hydrolysis) α-hydroxy acids,[5] and enol silyl ethers, α-silyl ketones, or silyl ketene acetals to yield aldol-type products.[6] The same strategy using organometallic reagents/Lewis acid combinations (e.g. $RCu/BF_3\cdot OEt_2$,[7] $RMgX/TiCl_4$,[8] $RLi/TiCl_4$,[8,9] $R_2Zn/TiCl_4$[9]) is a general route to secondary alcohols. Other nucleophile/Lewis acid combinations that have been used include alkynylstannanes/$TiCl_4$[10] and zinc enolates/$TiCl_4$.[11] These acetals have also been used in polyene cyclizations.[3a,12]

Nu-TMS, $TiCl_4$; 1. [O], 2. acid/base (1)

2,4-Pentanediol is often superior to other diols such as 2,3-butanediol for these reactions because of higher distereoselectivities in reactions with nucleophiles and the more facile cleavage of the resulting hydroxy ether by oxidation–β-elimination.[3] Removal of the chiral auxiliary is usually carried out with ***Pyridinium Chlorochromate*** oxidation followed by β-elimination using KOH,[3] K_2CO_3,[13] piperidinium acetate,[6] dibenzylammonium trifluoroacetate,[14] or DBU.[4c] In some cases, 1,3-butanediol is preferred because the final β-elimination may be effected under milder conditions.[14]

A detailed study of the mechanism and origin of stereoselectivity in reactions of allyltrimethylsilane with dioxane acetals has been published.[15]

Reduction of Acetals. Reductions of acetals of 2,4-pentanediol can provide (after removal of the chiral auxiliary by oxidation and β elimination) secondary alcohols with good enantioselectivity. The choice of reagents dictates the configuration of the final product. Use of ***Dibromoalane*** gives products from selective *syn* cleavage of the acetal while ***Triethylsilane/Titanium(IV) Chloride*** gives the more usual *anti* cleavage products (eq 2).[13]

Br_2AlH → 95% ee; Et_3SiH, $TiCl_4$ → 95% ee (2)

Elimination of Acetals. Treatment of 2,4-pentanediol acetals of *meso* ketones with ***Triisobutylaluminum*** gives enol ethers with high diastereoselectivities (eq 3).[16]

i-Bu_3Al → >80% de (3)

Acetals as Chiral Auxiliaries. There have been many applications of acetals of 2,4-pentanediol as chiral auxiliaries to control the diastereoselectivity of reactions on another functional group.[1] Examples include cyclopropanation of alkenyl dioxanes,[17] lithium amide-mediated isomerization of epoxides to allylic alcohols,[18] and addition of dioxane-substituted Grignard reagents[19] or organolithiums[20] to aldehydes.

Other Uses. Acetals of 2,4-pentanediol have also been prepared in order to determine the enantiomeric purity of aldehydes and ketones by analysis of diastereomers by GC or NMR.[21] ***2,3-Butanediol***[22] is more commonly used for this purpose but has been shown to be less effective in some cases.

Related Reagents. 2,3-Butanediol; Dimethyl L-Tartrate; 3-Hydroxyisoborneol; 1,1,2-Triphenyl-1,2-ethanediol.

1. Alexakis, A.; Mangeney, P. *TA* **1990**, *1*, 477.
2. (a) Ito, K.; Harada, T.; Tai, A. *BCJ* **1980**, *53*, 3367. (b) Tai, A.; Kikukawa, T.; Sugimura, T.; Inoue, Y.; Osawa, T.; Fujii, S. *CC* **1991**, 795, 1324.
3. (a) Bartlett, P. A.; Johnson, W. S.; Elliott, J. D. *JACS* **1983**, *105*, 2088. (b) Johnson, W. S.; Crackett, P. H.; Elliott, J. D.; Jagodzinski, J. J.; Lindell, S. D.; Natarajan, S. *TL* **1984**, *25*, 3951.
4. (a) Johnson, W. S.; Elliott, R.; Elliott, J. D. *JACS* **1983**, *105*, 2904. (b) Tabor, A. B.; Holmes, A. B.; Baker, R. *CC* **1989**, 1025. (c) Holmes, A. B.; Tabor, A. B.; Baker, R. *JCS(P1)* **1991**, 3301, 3307.
5. (a) Elliott, J. D.; Choi, V. M. F.; Johnson, W. S. *JOC* **1983**, *48*, 2295. (b) Choi, V. M. F.; Elliott, J. D.; Johnson, W. S. *TL* **1984**, *25*, 591. (c) Solladié-Cavallo, A.; Suffert, J.; Gordon, M. *TL* **1988**, *29*, 2955.
6. (a) Johnson, W. S.; Edington, C.; Elliott, J. D.; Silverman, I. R. *JACS* **1984**, *106*, 7588. (b) Elliott, J. D.; Steele, J.; Johnson, W. S. *TL* **1985**, *26*, 2535.
7. (a) Alexakis, A.; Mangeney, P.; Ghribi, A.; Marek, I.; Sedrani, R.; Guir, C.; Normant, J. F. *PAC* **1988**, *60*, 49. (b) Normant, J. F.; Alexakis, A.; Ghribi, A.; Mangeney, P. *T* **1989**, *45*, 507.
8. Lindell, S. D.; Elliott, J. D.; Johnson, W. S. *TL* **1984**, *25*, 3947.
9. Mori, A.; Marvoka, K.; Yamamoto, H. *TL* **1984**, *25*, 4421.
10. Yamamoto, Y.; Abe, H.; Nishii, S.; Yamada, J. *JCS(P1)* **1991**, 3253.
11. (a) Basile, T.; Tagliavini, E.; Trombini, C.; Umani-Ronchi, A. *CC* **1989**, 596. (b) Basile, T.; Tagiavini, E.; Trombini, C.; Umani-Ronchi, A. *S* **1990**, 305.
12. Johnson, W. S.; Elliott, J. D.; Hanson, G. J. *JACS* **1984**, *106*, 1138.
13. (a) Ishihara, K.; Mori, A.; Arai, I.; Yamamoto, H. *TL* **1986**, *27*, 983. (b) Ishihara, K.; Mori, A.; Yamamoto, H. *TL* **1986**, *27*, 987. (c) Ishihara, K.; Mori, A.; Arai, I.; Yamamoto, H. *T* **1987**, *43*, 755.
14. Silverman, I. R.; Edington, C.; Elliott, J. D.; Johnson, W. S. *JOC* **1987**, *52*, 180.
15. Denmark, S. E.; Almstead, N. G. *JACS* **1991**, *113*, 8089.
16. (a) Naruse, Y.; Yamamoto, H. *TL* **1986**, *27*, 1363. (b) Mori, A.; Yamamoto, H. *JOC* **1985**, *50*, 5446. (c) Naruse, Y.; Yamamoto, H. *T* **1988**, *44*, 6021. (d) Kaino, M.; Naruse, Y.; Ishihara, K.; Yamamoto, H. *JOC* **1990**, *55*, 5814. (e) Underiner, T. L.; Paquette, L. A. *JOC* **1992**, *57*, 5438.
17. (a) Arai, I.; Mori, A.; Yamamoto, H. *JACS* **1985**, *107*, 8254. (b) Mori, A.; Arai, I.; Yamamoto, H. *T* **1986**, *42*, 6447.
18. Yoshikawa, M.; Sugimura, T.; Tai, A. *CL* **1990**, 1003.
19. Kaino, M.; Ishihara, K.; Yamamoto, H. *BCJ* **1989**, *62*, 3736.
20. Chikashita, H.; Yuasa, T.; Itoh, K. *CL* **1992**, 1457.
21. (a) Fukutani, Y.; Maruoka, K.; Yamamoto, H. *TL* **1984**, *25*, 5911. (b) Furuta, K.; Kanematsu, A.; Yamamoto, H.; Takaoka, S. *TL* **1989**, *30*, 7231. (c) Nonoshita, K.; Banno, H.; Maruoka, K.; Yamamoto, H. *JACS* **1990**, *112*, 316.
22. Lemière, G. L.; Dommisse, R. A.; Lepoivre, J. A.; Alderweireldt, F. C.; Hiemstra, H.; Wynberg, H.; Jones, J. B.; Toone, E. J. *JACS* **1987**, *109*, 1363.

J. Michael Chong
University of Waterloo, Ontario, Canada

Phenyllithium[1]

PhLi

[591-51-5] C_6H_5Li (MW 84.05)

(organometallic agent useful as nucleophile for addition and substitution reactions,[1,3–20] and in the preparation of organolithium building blocks via metalation[32] and lithium–metalloid exchange reaction[1])

Physical Data: X-ray structures of the PMDTA-complexed monomer,[2i] the TMEDA-complexed dimer,[2a] the diethyl ether-complexed tetramer,[2b] and the PhLi:LiBr (3:1) mixed tetramer[2b] have been reported. Cryoscopic[2h] and NMR[2c–e] studies have shown that a monomer–dimer equilibrium exists in THF solution and that monomers exist in THF/HMPA[2f] and THF/PMDTA.[2c] PhLi is tetrameric in ether.

Solubility: sol ether solvents; sol hydrocarbon solvents only by the addition of donor solvents/additives such as ether or TMEDA.[2a]

Form Supplied in: commercially available in cyclohexane/ether solution.

Analysis of Reagent Purity: standard titration methods should be used.

Preparative Methods: prepared from phenyl halide (PhCl,[2j] PhBr,[2k,l] PhI[2m]) and lithium metal in ether. Several procedures have been used to prepare solid, salt-free PhLi.[2n–r] The synthesis of ^{13}C-labeled PhLi (*ipso* carbon only) has been described.[2h]

Handling, Storage, and Precautions: moisture and oxygen sensitive. Concentrated solutions are pyrophoric. Ether solutions slowly undergo proton abstraction to generate benzene and should be stored in the freezer.

Addition and Substitution Reactions. The greatest number of synthetic methodologies using PhLi involve carbon–carbon single bond formation through nucleophilic addition and substitution reactions. PhLi undergoes nucleophilic addition in a 1,2-fashion to a variety of carbonyl containing compounds including ***1,4-Benzoquinone***,[3a,b] cyclobutenediones,[3c] γ-amino enals,[3d] *N*-(alkyl)phthalimides,[3e] optically-active dienone–iron tricarbonyl complexes,[3f] α-silyl ketones[4] (eq 1),[4a] and β-lactams (eq 2).[5] PhLi adds similarly to cumulated carbonyl groups: isocyanates afford amides,[1] ketenes[6] yield enolates (eq 3),[6b] and ***Carbon Dioxide***[7] produces phenyl ketones (eq 4).[7b]

TMS 1. PhLi 2. H+ (1)

N— —OMe + PhLi THF, ether TMEDA, –78 °C 75%

Ph N H OMe (2)

=•=O 1. PhLi 2. TMSCl OTMS Ph (3)

t-BuLi + —Li 1. CO_2, pentane–ether, –40 °C 2. NH_4Cl, Δ 92% t-Bu (4)

In general, organolithium compounds preferentially undergo addition to the carbonyl of α,β-unsaturated carbonyl compounds rather than Michael addition. However, conjugate addition is promoted by increased solvent polarity, delocalized organolithium compounds, and lower temperatures which allow for thermodynamic control.[1,9] Conjugate addition of PhLi to vinyl sulfones[8] and α,β-unsaturated esters,[10a] ketones,[10b] and imines[11] has been reported. Thiophilic attack of thiocarbonyl compounds by PhLi has been observed for thioketones,[12] thioketenes (eq 5),[13] and thiocarbonates.[14] Addition of PhLi to carbon–nitrogen double and triple bonds, including the formal carbon–nitrogen double bond of azaaromatic heterocycles such as pyridine[15a] and quinoline,[15b] is general.[15] Arylation of 3,4-dihydro-β-carboline was accomplished by activation with ***Boron Trifluoride Etherate*** (eq 6).[15c] Similarily, the addition of PhLi to 2-isoxazolines[15d] and oxime ethers[15e] in the presence of boron trifluoride has been demonstrated. Addition reactions to nitriles containing no α-hydrogens have been reported to afford ketimines and ketones effectively.[16] Lithium aldimines have been prepared through the addition of PhLi to isocyanides.[17]

t-Bu S=• + —Li ether –78 °C 81% t-Bu S (5)

N + —Li BF_3•OEt_2 THF, –23 °C 75% N H N N H (6)

PhLi undergoes alkylation with primary halides, allylic halides (eq 7),[18a] and benzyl halides.[1a–c,18] Although not nearly as prevalent, displacement of sulfonate has also been reported.[1,19] Nucleophilic ring opening of epoxides with PhLi is general.[20] The reaction of PhLi with α,β-epoxy silanes yields β-hydroxy silanes which subsequently undergo Peterson reaction to afford alkenes (eq 8).[20d] Addition reactions to oxetanes[21a] and substituted aziridines[21b] are known, but have comparatively little utility.

H H Br + —Li ether cyclohexane 62% H H (7)

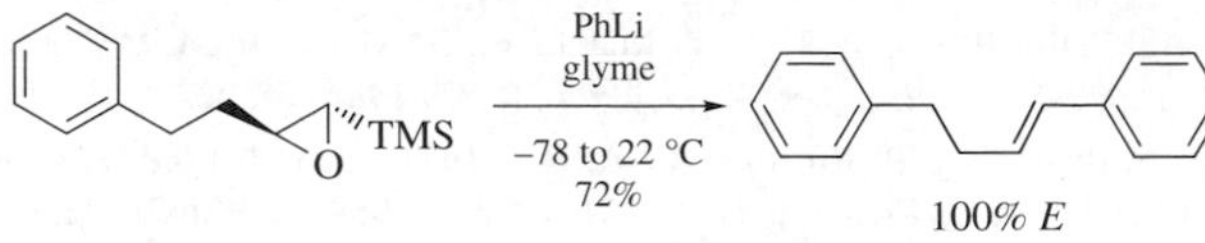

Nucleophilic addition and substitution reactions with PhLi often may be accompanied or followed by other reactions such as lithium–metalloid exchange, enolization, metalation, or carbanionic rearrangements.

Lithium–Metalloid Exchange Reaction. Phenyllithium undergoes lithium metalloid exchange[1] which is a reversible process with an equilibrium constant that reflects the difference in stability of PhLi and the new organolithium species generated. Mechanistic studies of this reaction abound.[1d,e] Lithium–halogen exchange is most prevalent, although related transmetalations involving selenium,[22] tellurium,[23] tin,[24] and mercury[25] are also common. In general, PhLi/M exchange is characteristically fast (I > Te, Sn > Br > Se > Cl)[1d] even at low temperatures. Exchange rates increase with increased solvent polarity, but are slowed by the presence of lithium salts.[2f,26] PhLi/M exchange reactions have been utilized for the preparation of functionalized aryllithiums (e.g. 4-methoxyphenyllithium,[27a] lithio-nitrobenzenes),[27b,c] substituted benzyllithiums (eq 9),[28] 3-thienyllithium derivatives (eq 10),[29] ***Vinyllithium***,[24c] 2-lithiophosphinines,[30] and other stabilized organolithium reagents which are not readily available by metalation or the reduction of halides. ***Allyllithium*** generated through reaction of PhLi and ***Tetraallylstannane***,[24d,e] can be converted with ***Potassium t-Butoxide*** to the more reactive allylpotassium which subsequently undergoes conjugate addition to vinyl sulfones (eq 11).[31]

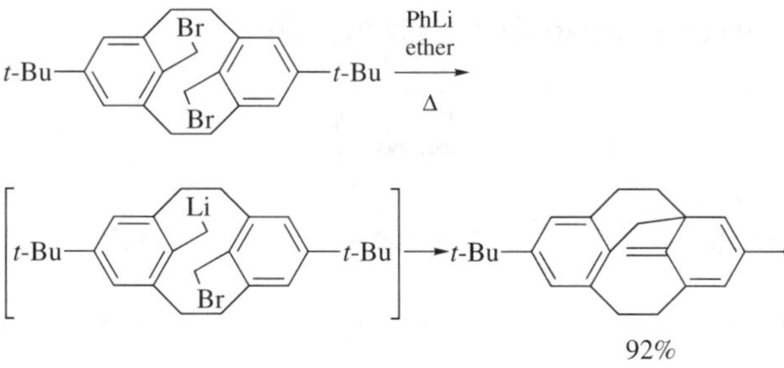

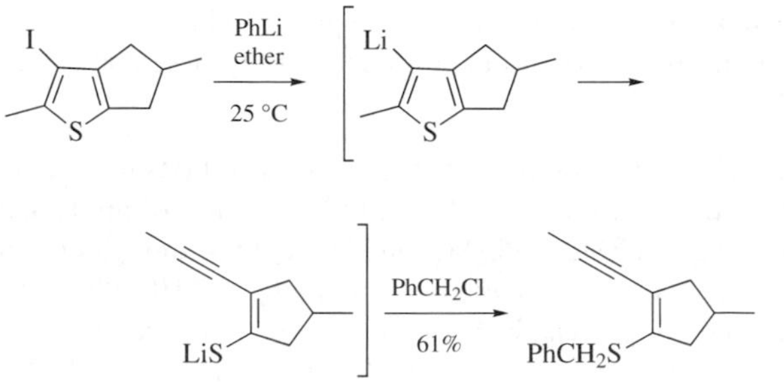

Lithiation Reactions[32–36]**.** Replacement of hydrogen by lithium in an organic compound is perhaps the most versatile method for preparing organolithium compounds. Although not nearly as prevalent as ***n-Butyllithium***, PhLi has been used as a base in heteroatom-facilitated lithiations, lithiations of relatively strong hydrocarbon acids (pK_a of benzene = 43),[37] and in the formation of stabilized anions where a milder reagent is warranted. Thiophenes,[32] sulfones,[8] nitriles,[33] 1,3-dimethoxybenzenes,[34] and α-methylpyridines[35] (eq 12)[35a] have been metalated with PhLi. Reaction of 2 equiv of PhLi with a β-epoxy sulfone resulted in base-catalyzed β-elimination and subsequent conjugate addition to produce a dianion which was alkylated with ***Iodomethane*** (eq 13).[8] PhLi has been used to deprotonate secondary amines to form lithium amides.[36]

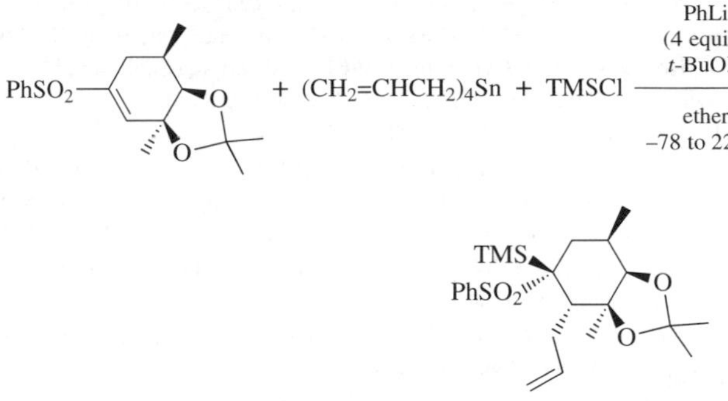

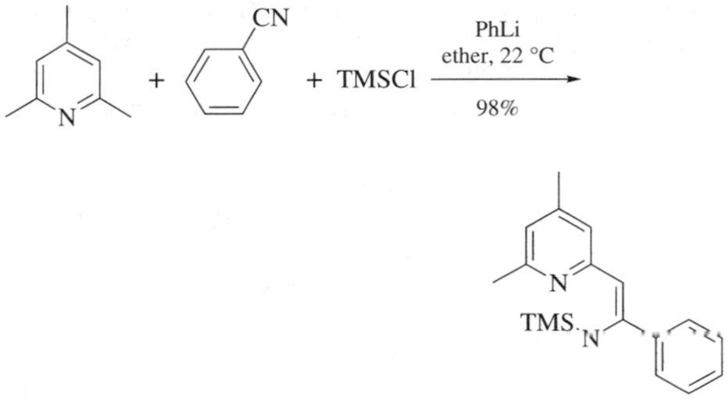

Related Reagents. *n*-Butyllithium; *s*-Butyllithium; *t*-Butyllithium; Hexamethylphosphoric Triamide; Lithium Diphenylcuprate; Lithium Di-*p*-tolylcuprate; Phenylmagnesium Bromide; Phenylsodium; Phenylzinc Chloride; *N,N,N′,N′*-Tetramethylethylenediamine.

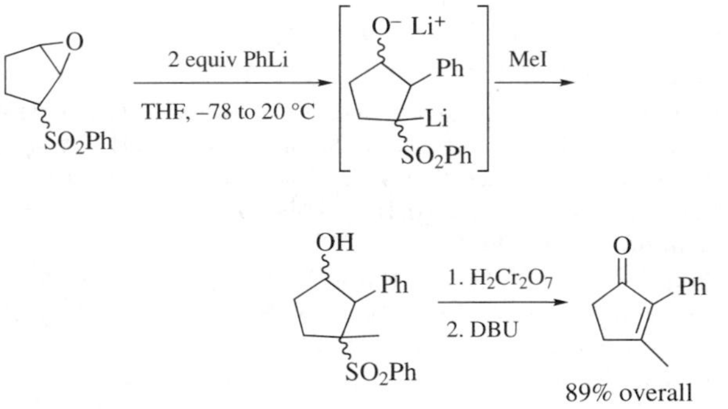

1. (a) Wakefield, B. J. *The Chemistry of Organolithium Compounds*; Pergamon: Oxford, 1974. (b) Wakefield, B. J. In *Comprehensive Organometallic Chemistry*; Wilkinson, G.; Stone, F. G. A.; Abel E. W., Eds.; Pergamon: Oxford, 1982; Vol. 7, Chapter 44. (c) Wardell, J. L. In *The Chemistry of the Metal–Carbon Bond*; Wiley: New York, 1987; Vol. 4, Chapter 1. (d) Bailey, W. F.; Patricia, J. J. *JOM* **1988**, *352*, 1. (e) Reich, H. J.; Green, D. P.; Phillips, N. H.; Borst, J. P.; Reich, I. L. *PS* **1992**, *67*, 83.

2. (a) Weiss, E.; Thoennes, D. *CB* **1978**, *111*, 3157. (b) Hope, H.; Power, P. P. *JACS* **1983**, *105*, 5320. (c) Bauer, W.; Winchester, W. R.; Schleyer,

P. v. R. *OM* **1987**, *6*, 2371. (d) Reich, H. J.; Green, D. P.; Phillips, N. H. *JACS* **1991**, *113*, 1414. (e) Jackman, L. M.; Scarmoutzos, L. M. *JACS* **1984**, *106*, 4627. (f) Reich, H. J.; Green, D. P. *JACS* **1989**, *111*, 8729. (g) Bauer, W.; Seebach, D. *HCA* **1984**, *67*, 1972. (h) Seebach, D.; Hässig, R.; Gabriel, J. *HCA* **1983**, *66*, 308. (i) Weiss, E.; Schümann, U.; Kopf, J. *AG(E)* **1985**, *24*, 215. (j) Esmay, D. I., *Adv. Chem. Ser.* **1959**, *23*, 47. (k) Gilman, H.; Morton, J. W. *OR* **1954**, *8*, 286. (l) Gilman, H.; Caj, B. J. *JOC* **1957**, *22*, 1165. (m) Müller, E.; Ludsteck, D. *CB* **1954**, *87*, 1887. (n) Wittig, G.; Meyer, F. J.; Lange, G. *LA* **1951**, *571*, 167. (o) Waack, R.; Doran, M. A. *JACS* **1963**, *85*, 1651. (p) Fraenkel, W. E.; Dagagi, S.; Kobayashi, S. *J. Phys. Chem.* **1961**, *83*, 3585. (q) Schlosser, M. *JOM* **1967**, *8*, 193. (r) Wehman, E.; Jastrzebski, J. T. B. H.; Ernsting, J.; Grove, D. M.; van Koten, G. *JOM* **1988**, *353*, 133.

3. (a) Alonso, F.; Yus, M. *T* **1991**, *47*, 7471. (b) Alonso, F.; Yus, M. *TL* **1992**, *48*, 2709. (c) Liebeskind, L. S.; Wirtz, K. R. *JOC* **1990**, *55*, 5350. (d) Reetz, M. T.; Wang, F.; Harms, K. *CC* **1991**, 1309. (e) Braun, L. L.; Torian, B. E. *JHC* **1984**, *21*, 293. (f) Franck-Neumann, M.; Chemla, P.; Martina, D. *SL* **1990**, 641.
4. (a) Ruden, R. A.; Gaffney, B. L. *SC* **1975**, *5*, 15. (b) Barrett, A. G. M.; Flygare, J. A. *JOC* **1991**, *56*, 638.
5. Kano, S.; Ebata, T.; Shibuya, S. *CPB* **1979**, *27*, 2450.
6. (a) Beel, J. A.; Vejvoda, E. *JACS* **1954**, *76*, 905. (b) Baigrie, L. M.; Seiklay, H. R.; Tidwell, T. *JACS* **1985**, *107*, 5391. (c) Seebach, D.; Laube, T.; Häner, R. *JACS* **1985**, *107*, 5396.
7. (a) Levine, R.; Karten, M. J.; Kadunce, W. M. *JOC* **1975**, *40*, 1770. (b) Breitmaier, E.; Zadel, G. *AG(E)* **1992**, *31*, 1035.
8. Fuchs, P. L.; Conrad, P. C. *JACS* **1978**, *100*, 346.
9. Cohen, T.; Abraham, W. D.; Myers, M. *JACS* **1987**, *109*, 7923.
10. (a) Tanaka, J.; Kanemasa, S.; Ninomiya, Y.; Tsuge, O. *BCJ* **1990**, *63*, 476. (b) Stern, A. J.; Swenton, J. S. *CC* **1988**, 1255.
11. (a) Meyers, A. I.; Barner, B. A. *JOC* **1986**, *51*, 120. (b) Kundig, E. P.; Liu, R. G.; Ripa, A. *HCA* **1992**, *75*, 2657.
12. (a) Beak, P.; Worley, J. W. *JACS* **1970**, *92*, 4142. (b) Ohno, A.; Nakamura, K.; Uohama, M.; Oka, S. *CL* **1975**, 983.
13. Schaumann, E.; Walter, W. *CB* **1974**, *107*, 3562.
14. Beak, P. Worley, J. W. *JACS* **1972**, *94*, 597.
15. (a) Overberger, C. G.; Lombardino, J. G.; Hiskey, R. G. *JACS* **1957**, *79*, 6430. (b) Uno, H.; Okada, S.; Suzuki, H. *JHC* **1991**, *28*, 346. (c) Nakagawa, M.; Kawate, T.; Yamazaki, H.; Hino, T. *CC* **1990**, 991. (d) Uno, H.; Terakawa, T.; Suzuki, H. *CL* **1989**, 1079. (e) Uno, H.; Terakawa, T.; Suzuki, H. *SL* **1991**, 559.
16. (a) Itsuno, S.; Hachisuka, C.; Kitano, K.; Ito, K. *TL* **1992**, *33*, 627. (b) Zimmerman, H. E.; Wright, C. W. *JACS* **1992**, *114*, 363. (c) Walborsky, H. M.; Niznik, G. E. *JOC* **1972**, *37*, 187.
17. Walborsky, H. M.; Periasamy, M. P. *JOC* **1974**, *39*, 611.
18. (a) Peterson, P. E.; Grant, G. *JOC* **1991**, *56*, 16. (b) Brown, H. C.; Rangaishenvi, M. V. *TL* **1990**, *31*, 7115. (c) Merkel, D.; Köbrich, G. *CB* **1973**, *106*, 2040.
19. (a) Tomalia, D. A.; Falk, J. C. *JHC* **1972**, *9*, 891. (b) Kasatkin, A. N. *IZV* **1988**, 2159.
20. (a) Cristol, S. J.; Douglass, J. R.; Meek, J. S. *JACS* **1951**, *73*, 816. (b) Aithie, G. C. M.; Miller, J. A. *TL* **1975**, 4419. (c) Rychnovsky, S. D.; Griesgraber, G.; Zeller, S.; Skalitzky, D. J. *JOC* **1991**, *56*, 5161. (d) Ukaji, Y.; Yoshida, A.; Fujisawa, T. *CL* **1990**, 157.
21. (a) Searles, S. *JACS* **1951**, *73*, 125. (b) Eis, M. J.; Ganem, B. *TL* **1985**, *26*, 1153.
22. (a) Reich, H. J. In *Organoselenium Chemistry*; Liotta, D., Ed.; Wiley: New York, 1987; p 243. (b) Dumont, W.; Bayet, P.; Krief, A. *AG(E)* **1974**, *13*, 243.
23. Tomoki, H.; Kambe, N.; Ogawa, A.; Miyoshi, N.; Murai, S.; Sonoda, N. *AG(E)* **1987**, *26*, 1187.
24. (a) Reich, H. J.; Phillips, N. H. *PAC* **1987**, *59*, 1021. (b) Seyferth, D.; Weiner, M. A. *OSC* **1973**, *5*, 452. (c) Seyferth, D.; Weiner, M. A. *JACS* **1961**, *83*, 3583. (d) Bristow, G. S. *Aldrichim. Acta* **1984**, *17*, 75. (e) Eisch, J. J. *OS* **1981**, *2*, 92.
25. (a) Eaton, P. E.; Cunkle, G. T.; Marchioro, G.; Martin, R. M. *JACS* **1987**, *109*, 948. (b) Maercker, A.; Dujardin, R. *AG* **1984**, *96*, 222.
26. (a) Batalov, A. P.; Rostokin, G. A. *JGU* **1971**, *41*, 154, 1740, 1743. (b) Batalov, A. P.; Rostokin, G. A. *JGU* **1973**, *43*, 959. (c) Winkler, H. J. S.; Winkler, H. *JACS* **1966**, *88*, 964, 969.
27. (a) Gilman, H.; Towle, J. L.; Spatz, S. M. *JACS* **1946**, *68*, 2017. (b) Köbrich, G.; Buck, P. *CB* **1970**, *103*, 1412. (c) Lucchesini, F. *T* **1992**, *48*, 9951.
28. Tashiro, M.; Yamato, T. *CL* **1982**, 61.
29. Frejd, T.; Karlsson, J. O.; Gronowitz, S. *JOC* **1981**, *46*, 3132.
30. (a) Lefloch, P.; Carmichael, D.; Mathey, F. *OM* **1991**, *10*, 2432. (b) Lefloch, P.; Carmichael, D.; Ricard, L.; Mathey, F.; Jutand, A.; Amatore, C. *OM* **1992**, *11*, 2475.
31. Fuchs, P. L.; Anderson, M. B. *JOC* **1990**, *55*, 337.
32. (a) Gilman, H.; Morton, J. W. *OR* **1954**, *8*, 286. (b) Gschwend, H. W.; Rodriguez, H. R. *OR* **1976**, *26*, 1. (c) Beak, P.; Snieckus, V. *ACR* **1982**, *15*, 306. (d) Beak, P.; Zajdel, W. J.; Reitz, D. B. *CRV* **1984**, *84*, 471.
33. (a) Weinstock, J.; Boekelheide, V. *OSC* **1963**, *4*, 641. (b) Cason, J.; Sumrell, G.; Mitchell, R. S. *JOC* **1950**, *15*, 850.
34. (a) Catlin, E. R.; Hassell, C. H. *JCS(C)* **1971**, 460. (b) Lambooy, J. P. *JACS* **1956**, *78*, 771.
35. (a) Compagnon, P.-L.; Gasquez, F.; Kimny, T. *S* **1986**, 948. (b) Lee, J. W.; Anderson, W. K. *SC* **1992**, *22*, 369.
36. (a) Huisgen, R.; Konig, H. *CB* **1959**, *92*, 203. (b) Iida, H.; Yuasa, Y.; Kibayashi, C. *S* **1977**, 879.
37. Streitwieser, A., Jr.; Scannon, P. J.; Niemeyer, H. M. *JACS* **1972**, *94*, 7936.

D. Patrick Green
The Dow Chemical Co., Midland, MI, USA

Phenylmagnesium Bromide[1]

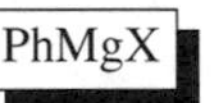

(X = Br)		
[100-58-3]	C_6H_5BrMg	(MW 181.32)
(X = Cl)		
[100-59-4]	C_6H_5ClMg	(MW 136.87)

(adds to many unsaturated functional groups; phenyl can displace halide and like groups; functions as a strong base and a Lewis acid[1b–f])

Physical Data: NMR and calorimetric observations indicate that PhMgBr in Et_2O and THF and PhMgCl in THF are actually mixtures of PhMgX, Ph_2Mg, and MgX_2.[2,3] Although PhMgBr and PhMgCl are essentially monomeric in THF, PhMgBr is associated significantly in Et_2O except at low concentrations.[3,4] Solids isolated include $PhMgBr(Et_2O)_2$[5] and $PhMgBr(THF)_2$,[6] both shown by X-ray diffraction studies to have four-coordinate Mg.

Solubility: both sol Et_2O, THF; insol hydrocarbons.

Form Supplied in: solutions of PhMgBr in Et_2O and THF and of PhMgCl in THF and THF/toluene are commercially available.

Analysis of Reagent Purity: see ***Methylmagnesium Bromide***.

Preparative Methods: in spite of their commercial availability, solutions of PhMgBr[8] and PhMgCl[9] are frequently prepared from reaction of a phenyl halide and Mg, usually in Et_2O or

THF.[1c,e,7] Preparation of PhMgBr in Et_2O is routine, but the formation of PhMgCl from PhCl and magnesium is difficult to initiate and sustain. By contrast, preparation is generally successful in THF.[9] Use of activated[10] ***Magnesium*** makes possible otherwise difficult preparations, allows the use of lower temperatures which may minimize undesired reactions with other functional groups in the reagent, and allows the use of Et_2O, other less favorable ethers, and hydrocarbons. Stirring Mg under nitrogen is one activation procedure;[11] other effective procedures (chemical treatment of the Mg,[12] preparation of Mg by reduction of salts,[13] use of ethylene bromide as a coreagent during the preparation[14]) can introduce other species.

Handling, Storage, and Precautions: see ***Methylmagnesium Bromide***. Phenyl Grignard reagents do not significantly attack Et_2O or THF at normal reaction temperatures.

Representative Applications[1b–f]. See also ***Methylmagnesium Bromide***. Phenyl Grignard reagents are used most frequently in reactions that result in attachment of the organic group to an electrophilic carbon atom of the substrate, most often of a carbonyl group, nitrile, or other unsaturated function. The oxazoline function, easily transformed into other functions, facilitates displacement of a methoxyl group (eq 1), probably both by its electronic effects and by coordinating with the Grignard reagent.[15]

PhMgBr; THF, −78 °C to rt; 100% (1)

Related Reagents. Copper(I) Bromide; Copper(I) Chloride; Copper(I) Iodide; Lithium Diphenylcuprate; Magnesium; Phenyllithium; Phenylzinc Chloride.

---

1. (a) Lindsell, W. E. In *Comprehensive Organometallic Chemistry*; Wilkinson, G.; Stone, F. G. A.; Abel, E. W., Eds.; Pergamon: Oxford, 1982; Chapter 4. (b) Wakefield, B. J. In *Comprehensive Organometallic Chemistry*; Wilkinson G.; Stone, F. G. A.; Abel, E. W., Eds.; Pergamon: Oxford, 1982; Chapter 44. (c) Nützel, K. *MOC* **1973**, *13/2a*, 47. (d) Raston, C. L.; Salem, G. In *The Chemistry of the Metal–Carbon Bond*; Hartley, F. R., Ed.; Wiley: Chichester, 1987; Vol. 4, Chapter 2. (e) Old but still extremely useful is: Kharasch, M.; Reinmuth, O. *Grignard Reactions of Nonmetallic Substances*; Prentice–Hall: New York, 1954. (f) Ioffe, S. T.; Nesmeyanov, A. N. *The Organic Compounds of Magnesium, Beryllium, Calcium, Strontium and Barium*; North-Holland: Amsterdam, 1967.
2. Evans, D. F.; Fazakerley, G. V. *JCS(A)* **1971**, 184.
3. Smith, M. B.; Becker, W. E. *T* **1966**, *22*, 3027; Smith, M. B.; Becker, W. E. *T* **1967**, *23*, 4215.
4. Walker, F. W.; Ashby, E. C. *JACS* **1969**, *91*, 3845.
5. Stucky, G.; Rundle, R. E. *JACS* **1964**, *86*, 4825.
6. Schröder, F. A. *CB* **1969**, *102*, 2035.
7. Bickelhaupt, F. In *Inorganic Reactions and Methods*; Hagen, A. P., Ed.; VCH: New York, 1989; Vol. 10, Section 5.4.2.2.1; *FF* **1967**, *1*, 415.
8. Allen, C. F. H.; Converse, S. *OSC* **1941**, *1*, 226; Hiers, G. S. *OSC* **1941**, *1*, 550.
9. Ramsden, H. E.; Balint, A. E.; Whitford, W. R.; Walburn, J. J.; Cserr, R. *JOC* **1957**, *22*, 1202.
10. Reviews: Ref. 7; Lai, Y.-H. *S* **1981**, 585.
11. Baker, K. V.; Brown, J. M.; Hughes, N.; Skarnulis, A. J.; Sexton, A. *JOC* **1991**, *56*, 698.
12. Oppolzer, W.; Schneider, P. *TL* **1984**, *25*, 3305; Oppolzer, W.; Cunningham, A. F. *TL* **1986**, *27*, 5467; Bönnemann, H.; Bogdanović, B.; Brinkmann, R.; He, D.-W.; Spliethoff, B. *AG(E)* **1983**, *22*, 728. Also see Bartmann, E.; Bogdanović, B.; Janke, N.; Liao, S.; Schlichte, K.; Spliethoff, B.; Treber, J.; Westeppe, U.; Wilczok, U. *CB* **1990**, *123*, 1517.
13. Rieke, R. D.; Bales, S. E. *JACS* **1974**, *96*, 1775. Rieke, R. D.; Bales, S. E.; Hudnall, P. M.; Burns, T. P.; Poindexter, G. S. *OSC* **1988**, *6*, 845 (note particularly footnote 19).
14. Pearson, D. E.; Cowan, D.; Beckler, J. D. *JOC* **1959**, *24*, 504.
15. Gant, T. G.; Meyers, A. I. *JACS* **1992**, *114*, 1010.

Herman G. Richey, Jr.
The Pennsylvania State University, University Park, PA, USA

(−)-8-Phenylmenthol

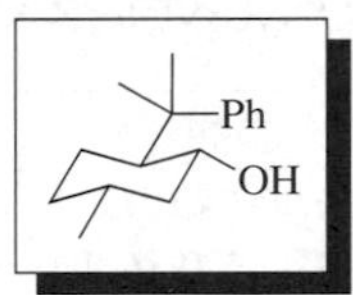

(1*R*,2*S*,5*R*)
[65253-04-5] $C_{16}H_{24}O$ (MW 232.37)
(1*S*,2*R*,5*R*)
[100101-42-6]
(1*S*,2*R*,5*S*)
[57707-91-2]

(chiral auxiliary for asymmetric induction)

Alternate Name: (1*R*,2*S*,5*R*)-5-methyl-2-(1-methyl-1-phenylethyl)cyclohexanol.
Physical Data: [α_D −26° (*c* 2, EtOH); *d* 0.999 g cm^{-3}.
Solubility: sol organic solvents.
Form Supplied in: commercially available as a colorless oil (98%) and as the chloroacetate ester.
Analysis of Reagent Purity: NMR and $[\alpha]_D$.
Preparative Methods: (1*R*,2*S*,5*R*)-(−)-8-phenylmenthol is prepared by the reaction of ***Phenylmagnesium Bromide*** (cat. ***Copper(I) Iodide***) with (+)-pulegone, equilibration of the resulting conjugate addition product, and reduction of the ketone (***Sodium***, 2-propanol).[1] The (1*R*,2*S*,5*R*)-isomer is accompanied by the (1*S*,2*R*,5*R*)-isomer and is conveniently separated by recrystallization of the chloroacetate ester.[2] This separation is essential to obtaining high optical yields since it has been shown that the two diastereomers have opposite chiral directing ability.[3] The preparation of the enantiomeric (1*S*,2*R*,5*S*)-(+)-8-phenylmenthol from (+)-pulegone has been reported.[4]
Handling, Storage, and Precautions: no special precautions are necessary other than those used for combustible organic compounds.

Chiral Auxiliary for Asymmetric Induction. Numerous derivatives of (−)-8-phenylmenthol have been utilized for asymmetric induction studies. These include inter-[5] and intramolecular[6] Diels–Alder reactions, dihydroxylations,[7] and intramolecular ene reactions[8] of α,β-unsaturated 8-phenylmenthol esters. These reactions usually proceed in moderate to good yield with high diastereofacial selectivity. α-Keto esters of 8-phenylmenthol (see ***8-Phenylmenthyl Pyruvate***) have been used for asymmetric addition to the keto group,[9] as well as for asymmetric [2 + 2] photoadditions[10] and nucleophilic alkylation.[11] Ene reactions of α-imino esters of 8-phenylmenthol with alkenes provide a direct route to α-amino acids of high optical purity.[12]

Vinyl and butadienyl ethers of 8-phenylmenthol have been prepared and the diastereofacial selectivity of nitrone[13] and Diels–Alder[14] cycloadditions, respectively, have been evaluated. α-Anions of 8-phenylmenthol esters also show significant diastereofacial selectivity in aldol condensations[15] and enantioselective alkene formation by reaction of achiral ketones with 8-phenylmenthyl phosphonoacetate gives de up to 90%.[16]

Related Reagents. (*S*)-Ethyl Lactate; Ethyl Mandelate; (*R*)-Pantolactone; 8-Phenylmenthyl Acrylate; 8-Phenylmenthyl Crotonate; 8-Phenylmenthyl Glyoxylate; 8-Phenylmenthyl Pyruvate.

1. Corey, E. J.; Ensley, H. E. *JACS* **1975**, *97*, 6908.
2. (a) Herzog, H.; Scharf, H. D. *S* **1986**, 420. (b) Ort, O. *OS* **1987**, *65*, 203. (c) See also Cervinka, O.; Svatos, A.; Masojidkova, M. *CCC* **1990**, *55*, 491.
3. Whitesell, J. K.; Liu, C. L.; Buchanan, C. M.; Chen, H. H.; Minton, M. A. *JOC* **1986**, *51*, 551. Whitesell, J. K. *CRV* **1992**, *92*, 953.
4. Ensley, H. E.; Parnell, C. A.; Corey, E. J. *JOC* **1978**, *43*, 1610.
5. Oppolzer, W.; Kurth, M.; Reichlin, D.; Chapuis, C.; Mohnhaupt, M.; Moffat, F. *HCA* **1981**, *64*, 2802.
6. Roush, W. R.; Gillis, H. R.; Ko, A. I. *JACS* **1982**, *104*, 2269.
7. Hatakeyama, S.; Matsui, Y.; Suzuki, M.; Sakurai, K.; Takano, S. *TL* **1985**, *26*, 6485.
8. (a) Oppolzer, W.; Robbiani, C.; Bättig, K. *HCA* **1980**, *63*, 2015. (b) Oppolzer, W.; Robbiani, C.; Bättig, K. *T* **1984**, *40*, 1391.
9. (a) Grossen, P.; Herold, P.; Mohr, P.; Tamm, C. *HCA* **1984**, *67*, 1625. (b) Sugimura, H.; Yoshida, K. *JOC* **1993**, *58*, 4484. (c) Solladie-Cavallo, A.; Bencheqroun, M. *TA* **1991**, *2*, 1165. (d) Comins, D. L.; Baevsky, M. F.; Hong, H. *JACS* **1992**, *114*, 10 971.
10. (a) Koch, H.; Scharf, H. D.; Runsink, J.; Leismann, H. *CB* **1985**, *118*, 1485. (b) Nehrings, A.; Scharf, H.-D.; Runsink, J. *AG(E)* **1985**, *24*, 877.
11. Hamon, D. P. G.; Massy-Westropp, R. A.; Razzino, P. *T* **1992**, *48*, 5163.
12. Mikami, K.; Kaneko, M.; Yajima, T. *TL* **1993**, *34*, 4841.
13. Carruthers, W.; Coggins, P.; Weston, J. B. *CC* **1991**, 117.
14. (a) Thiem, R.; Rotscheidt, K.; Breitmaier, E. *S* **1989**, 836. (b) Danishefsky, S.; Bednarski, M.; Izawa, T.; Maring, C. *JOC* **1984**, *49*, 2290.
15. Corey, E. J.; Peterson, R. T. *TL* **1985**, *26*, 5025.
16. (a) Gais, H. J.; Schmiedl, G.; Ball, W. A. *TL* **1988**, *29*, 1773. (b) Rehwinkel, H.; Skupsch, J.; Vorbrüggen, H. *TL* **1988**, *29*, 1775.

Harry E. Ensley, Matthew Beggs & Yinghong Gao
Tulane University, New Orleans, LA, USA

α-Phenylsulfonylethyllithium

R–CH(SO$_2$Ph)Li

(R = H)
[69291-71-0] $C_8H_9LiO_2S$ (MW 176.18)

(versatile stabilized carbon nucleophile for forming C–C, C=C, and C≡C bonds with a wide range of carbon and heteroatom electrophiles[1–3])

Physical Data: pK_a in DMSO 31.0.
Solubility: sol THF, Et_2O.
Preparative Methods: prepared in situ as needed by metalation of ethyl phenyl sulfone with ***n-Butyllithium*** or ***Lithium Diisopropylamide***.
Handling, Storage, and Precautions: moisture sensitive; handle under nitrogen or argon.

General Considerations. The sulfone group is perhaps second only to the carbonyl group in its versatility and utility. It serves the dual role of C–H activator and leaving group under a wide range of conditions,[1] thereby enabling creation of up to three C–C bonds from a single functional group. Alkyl aryl sulfones have a pK_a of ca. 31 and are therefore quantitatively deprotonated by strong bases such as ***n-Butyllithium***, ***Lithium Diisopropylamide***, or Grignard reagents. The resultant colored carbanions (yellow or red–orange)[4] are usually stable at rt. It is also possible to form an α,α-dianion which reacts with even relatively poor electrophiles.[5] In the following discussion, α-phenylsulfonylethyllithium serves as a paradigm for the class of 'a^{1}d^1' reagents represented by ArSO$_2$–CH(M)R.

C_8H_{17}–CH(Me)–CH$_2$–CH(SO$_2$Ph)Li + BnO–CH(Me)–CH(Me)–CH$_2$Br → (THF–HMPA, –52 °C, 4 h, 70%) C_8H_{17}–CH(Me)–CH$_2$–CH(SO$_2$Ph)–CH$_2$–CH(Me)–CH(Me)OBn (1)

α-Phenylsulfonylethyllithiums as Donors. α-Arylsulfonylalkyllithiums react with all the typical carbon electrophiles to form a C–C bond in generally good yield; some indication of the scope of the procedure can be gleaned from eqs 1–9. For example, alkylation with primary alkyl bromides (or iodides; eq 1),[6] tosylates (eq 2),[7] or triflates[8] occurs even when there is an α-branch in the chain. In the case of sluggish reactions, additives such as ***Hexamethylphosphoric Triamide*** or ***N,N,N′,N′-Tetramethylethylenediamine*** can be used to accelerate alkylation.

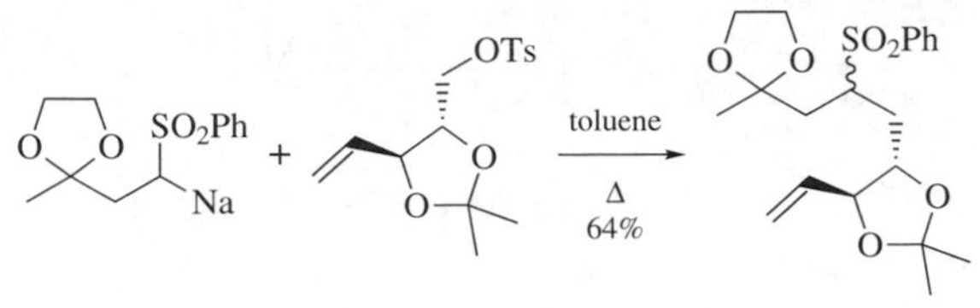

(2)

The reaction of sulfonyl carbanions with halocarbenoids (see ***Tribromomethyllithium***) gives a 1,1-dibromoalkene or a 1-bromoalkene (eq 3).[9] The reaction probably does not involve a carbene intermediate.

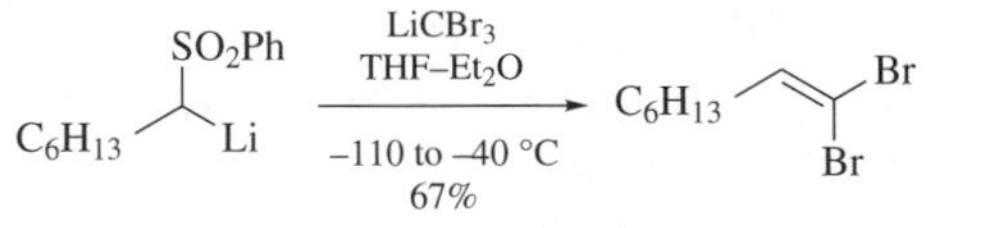

(3)

Terminal epoxides react slowly with sulfonyl carbanions such as the homoenolate equivalent (**1**) (eq 4).[10] With disubstituted epoxides and cyclic epoxides the reactions are slower still. For example, reaction of the lithio derivative of ethyl phenyl sulfone with cyclopentene oxide occurs in excellent yield (98%) after 10 h reflux in toluene.[11] It has been reported that, in some cases, the addition of a Lewis acid (***Magnesium Bromide***,[12] ***Boron Trifluoride Etherate***,[13,14] ***Titanium Tetraisopropoxide***,[15] MeOAl(*i*-Bu)$_2$[16]) or HMPA[17] improves the yield dramatically.

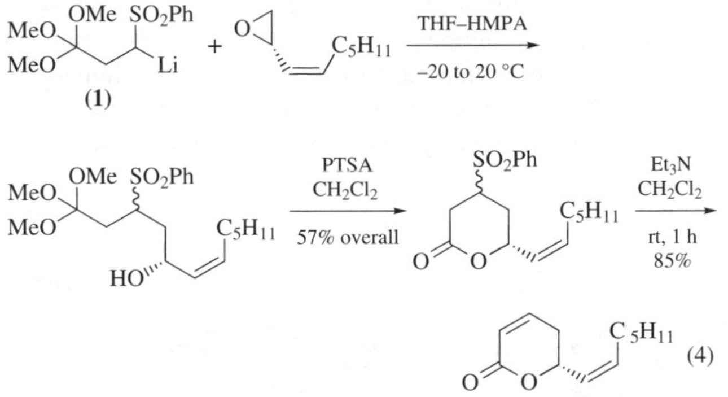

(4)

In the presence of HMPA, the homoenolate equivalent (**1**) underwent conjugate addition to cyclohexenone. In the absence of HMPA, a mixture of 1,2- and 1,4-adducts was obtained (eq 5).[18,19]

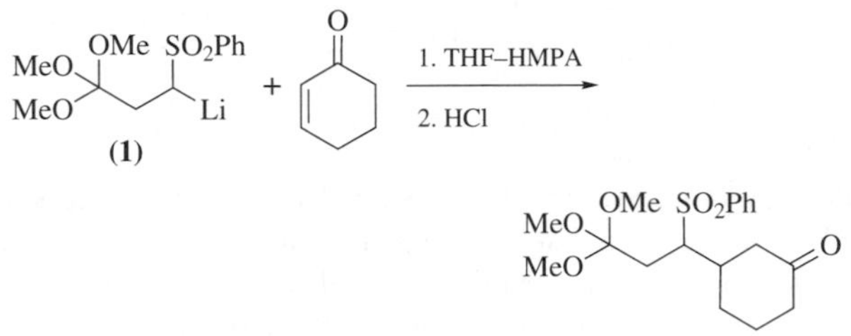

(5)

The addition of metalated sulfones to aldehydes is reversible and in simple cases the reaction displays modest selectivity for the *erythro* isomer (eq 6).[20] The reverse reaction is favored when the adducts are sterically compressed (e.g. ketone adducts) or when the sulfone anion is stabilized by conjugation (i.e. allylic or benzylic sulfones) or proximate heteroatoms. However, in unfavorable cases the position of the equilibrium can be tuned by varying the metal. For example, the lithio sulfone (**2**) did not give a stable adduct with aldehyde (**4**) but the 'ate' complex derived from the lithio derivative and BF_3 gave the desired adduct (**5**) (eq 7).[21]

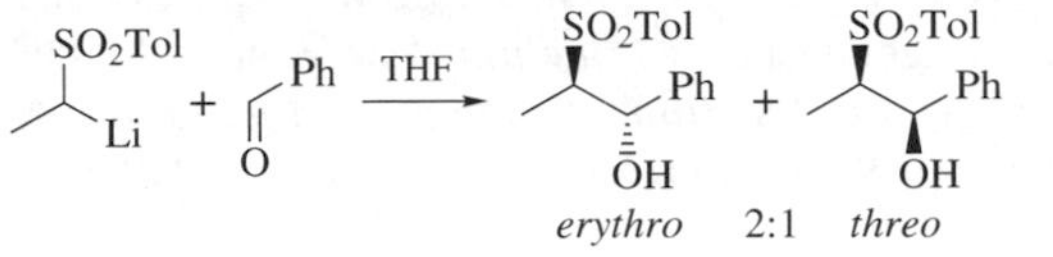

(6)

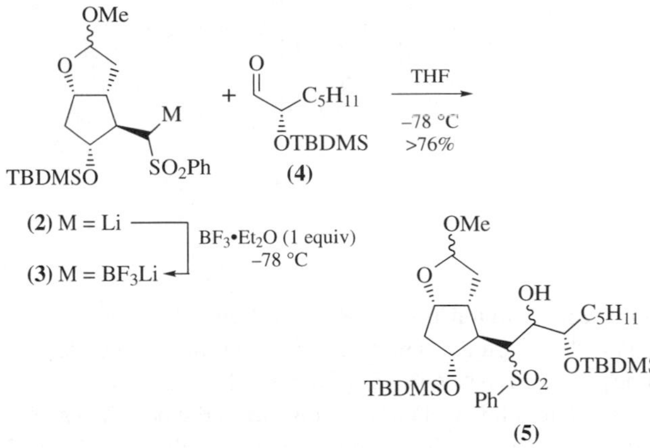

(7)

α-Phenylsulfonylethyllithium adds to acylsilanes to give an adduct which undergoes a Brook rearrangement with subsequent loss of benzenesulfinate anion. The product of the reaction is an enol silane (eq 8).[22]

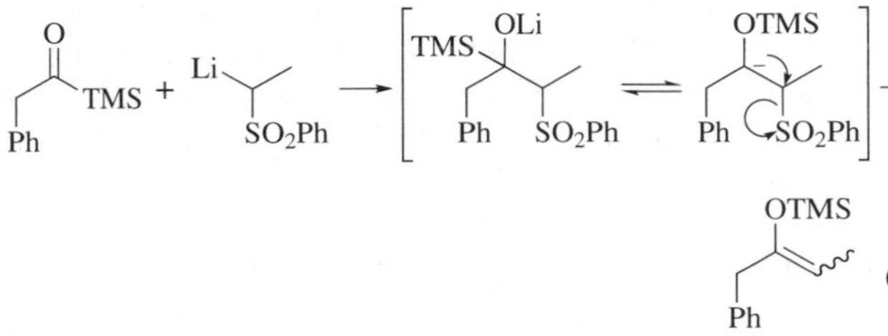

(8)

Reactions of the metalated sulfones with esters,[23,24] lactones,[25,26] amides and carbonates[27] lead to the corresponding β-keto sulfone (eq 9).[23] The β-keto sulfones thus formed display chemistry reminiscent of β-keto esters in their enhanced acidity and tendency to undergo *C*- and *O*-alkylation and acylation.

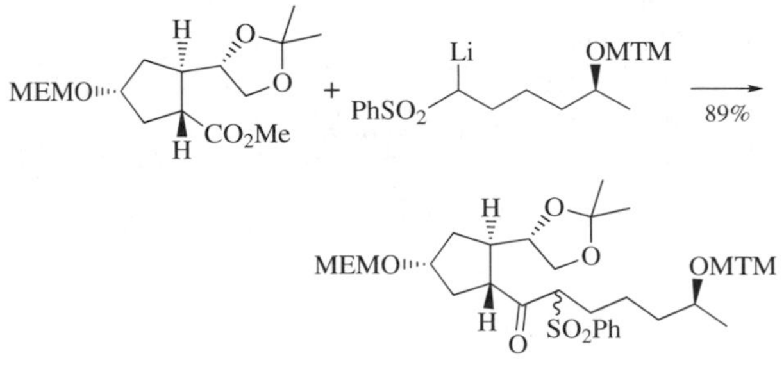

(9)

Aryl Sulfones as Acceptors: Desulfonylation Reactions. The sulfone group is a powerful electron acceptor which can be cleaved under a wide range of conditions. We have already seen that mild base will cause elimination of benzenesulfinate from β-arylsulfonyl-substituted carbonyl derivatives (eq 5). α-Arylsulfonyl-substituted carbonyl derivatives undergo reductive desulfonylation[28] under very mild conditions using ***Aluminum Amalgam*** (eq 10),[29] but in the absence of carbonyl activation, reductive cleavage of the sulfone group requires much stronger reducing agents such as ***Sodium Amalgam*** in MeOH buffered with Na_2HPO_4,[30] ***Sodium–Ammonia***,[31] ***Magnesium*** in refluxing MeOH,[32] or ***Samarium(II) Iodide*** in THF–HMPA.[33]

X = SO_2Ph → X = H: Al(Hg), THF–H_2O (10)

The Julia Alkenation and Related Reactions. In 1973, Julia and Paris[34] reported a new connective and regioselective alkene synthesis (eq 10) based on the reductive elimination of β-acyloxy sulfones. The Julia alkenation is now one of the principal methods for fragment linkage in complex natural product synthesis.[35,36] Mono-, di-, tri-, and tetrasubstituted alkenes can be prepared in moderate to good yield, depending on the substrate. The three-step sequence, illustrated in eq 11, entails (a) condensation of a metalated sulfone with an aldehyde or ketone, (b) *O*-functionalization of the adduct as the acetate, benzoate, or mesylate (to prevent retroaldolization), and (c) reductive elimination using 6% Na(Hg) in THF–MeOH (3:1) at −20 °C.[37] In favorable cases, step (b) can be omitted and the reductive elimination performed on a β-hydroxy sulfone intermediate. Potential problems attending each step have been summarized.[35]

THF, −90 °C; Ac_2O, DMAP, py; Na(Hg), Na_2HPO_4, MeOH, −20 °C, 30% overall (100% *E*) (11)

A detailed investigation of the scope and stereochemistry of the reductive elimination leading to 1,2-disubstituted alkenes revealed high *trans* stereoselectivity which is *independent* of the stereochemistry of the β-acyloxy sulfone adducts.[37,38] Furthermore, the stereoselectivity increases with increasing steric congestion about the nascent alkene and maximum yields and rate are observed for the formation of conjugated dienes and trienes. The Julia procedure has also been adapted to the synthesis of alkynes (eq 12).[39,40]

76%; 1. NaH, 2. $(EtO)_2POCl$; Na(Hg), THF–DMSO, 0 °C, 53% (3 steps) (12)

There are two further variants of the Julia alkenation which deserve wider recognition. Both methods surmount the inherent limitation in scale of the reductive elimination step imposed by the use of Na(Hg). The first method involves a radical-induced elimination of thiocarbonyl derivatives of β-hydroxy sulfones,[41] as illustrated in eq 13.[42]

1.5 equiv Bu_3SnH, AIBN (cat.), toluene, 95 °C, 86% (13)

The second recent variant, developed by Julia and co-workers, avoids reductive elimination altogether and provides a remarkable one-pot connective synthesis of alkenes.[43] The procedure, illustrated in eq 14, involves condensation of an aldehyde or ketone with a lithiated benzothiazolyl alkyl sulfone to give an adduct which first cyclizes and then fragments with extrusion of sulfur dioxide, benzothiazolone (which then tautomerizes to 2-hydroxybenzothiazole), and the alkene. Generally a mixture of (*E*)- and (*Z*)-alkenes is obtained, but in sterically hindered substrates the (*E*) isomer can be obtained selectively. The same reaction has been observed with the pyridinyl sulfone analogs, in which case the separable β-hydroxy sulfone intermediates undergo stereospecific *anti* elimination to the corresponding alkene.

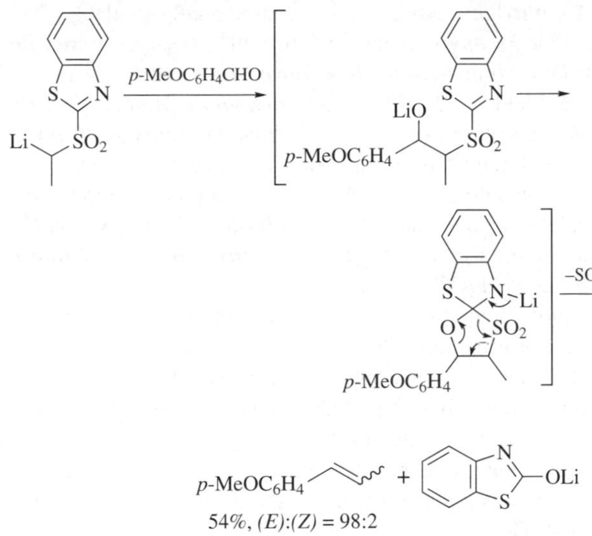

p-MeOC6H4CHO —SO2 (14)

p-MeOC$_6$H$_4$ + OLi (14)

54%, (E):(Z) = 98:2

Alternative Desulfonylation/C–C Bond-Forming Procedures. The foregoing discussion has focussed on reductive methods for removing sulfones and, in the case of the Julia alkenation, desulfonylation is accompanied by the formation of a new C–C bond. Another method which accomplishes desulfonylation with the concomitant construction of a C=C bond entails fluoride-induced elimination of β-silyl sulfones (see ***(2-Phenylsulfonylethyl)trimethylsilane***).

Allylic sulfones undergo S_N2' displacement by cyanocuprates with high *syn* stereoselectivity (eq 15).[44] Stabilized enolates also displace allylic arylsulfonyl groups in the presence of Pd^0 or Ni^0 catalysts.[45]

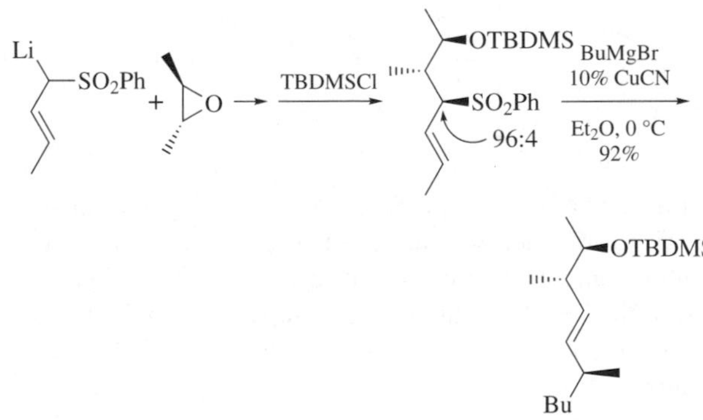

TBDMSCl; BuMgBr, 10% CuCN, Et_2O, 0 °C, 92%; 96:4 (15)

Allylic and tertiary alkyl sulfones can also participate in electrophilic cyclizations in the presence of ***Aluminum Chloride*** (Friedel–Crafts reaction; eq 16).[46]

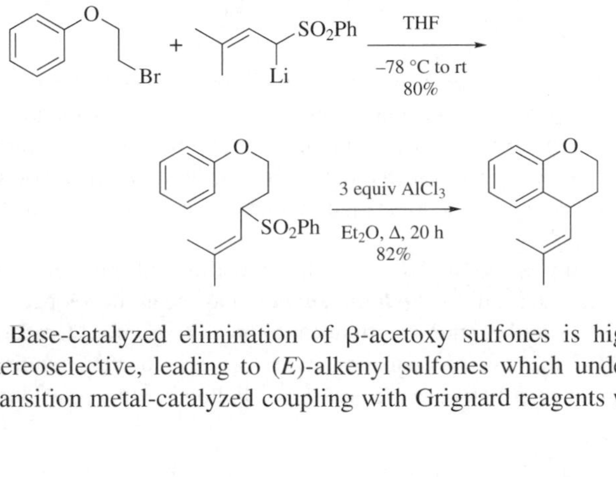

THF, −78 °C to rt, 80%; 3 equiv $AlCl_3$, Et_2O, Δ, 20 h, 82% (16)

Base-catalyzed elimination of β-acetoxy sulfones is highly stereoselective, leading to (*E*)-alkenyl sulfones which undergo transition metal-catalyzed coupling with Grignard reagents with retention of configuration to provide a stereoselective synthesis of trisubstituted alkenes.[47] Either ***Nickel(II) Acetylacetonate***, ***Tris(acetylacetonato)iron(III)***, or ***Iron(III) Chloride*** can be used as the catalyst (eq 17).[48]

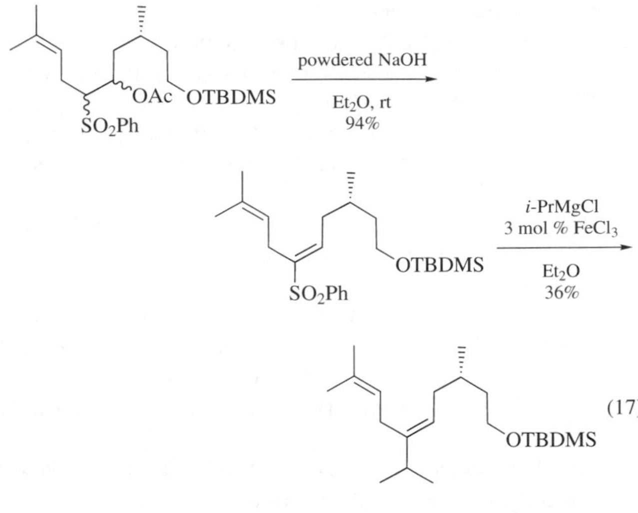

powdered NaOH, Et_2O, rt, 94%; i-PrMgCl, 3 mol % $FeCl_3$, Et_2O, 36% (17)

Oxidative Desulfonylation of Arylsulfonylalkyllithiums. Aryloxysulfonylalkyllithiums can be converted to ketones in one pot by reaction with ***Oxodiperoxymolybdenum(pyridine)(hexamethylphosphoric triamide)***,[49] ***Bis(trimethylsilyl) Peroxide***,[50] or ***Chlorodimethoxyborane/m-Chloroperbenzoic Acid***.[51]

Related Reagents. Allyl Phenyl Sulfone; Chloromethyl Phenyl Sulfone; Dimethyl Sulfone; Methyl Phenyl Sulfone; Phenylsulfonylacetylene; Phenylsulfonylallene; 4-Phenylsulfonyl-2-butanone Ethylene Acetal; Phenylsulfonylethylene; (2-Phenylsulfonylethyl)trimethylsilane; 3-(Phenylsulfonyl)propanal Ethylene Acetal; Phenylsulfonyl(trimethylsilyl)methane; p-Toluenesulfinylmethyllithium.

---

1. *The Chemistry of Sulphoxides and Sulphones*; Patai, S.; Rappoport, Z.; Stirling, C.; Eds.; Wiley: New York, 1988.
2. Solladié, G. *COS* **1991**, *6*, 133.
3. Simpkins, N. S. *Sulphones in Organic Synthesis*; Pergamon: Oxford, 1993.
4. Bongini, A.; Savoia, D.; Umani-Ronchi, A. *JOM* **1976**, *112*, 1.
5. Heathcock, C. H.; Finkelstein, B. L.; Jarvi, E. T.; Radel, P. A.; Hadley, C. R. *JOC* **1988**, *53*, 1922.
6. Larchevêque, M.; Sanner, C.; Azerad, R.; Buisson, D. *T* **1988**, *44*, 6407.
7. Wershofen, S.; Scharf, H.-D. *S* **1988**, 854.
8. Kennedy, R. M.; Abiko, A.; Takemasa, T.; Okumoto, H.; Masamune, S. *TL* **1988**, *29*, 451.
9. Charreau, P.; Julia, M.; Verpeaux, J.-N. *BSF(2)* **1990**, *127*, 275.
10. Carretero, J. C.; Ghosez, L. *TL* **1988**, *29*, 2059.
11. Nwaukwa, S. O.; Lee, S.; Keehn, P. M. *SC* **1986**, *16*, 309.
12. Marshall, J. A.; Andrews, R. C. *JOC* **1985**, *50*, 1602.
13. Nakata, T.; Saito, K.; Oishi, T. *TL* **1986**, *27*, 6345.
14. Achmatowicz, B.; Marzak, S.; Wicha, J. *CC* **1987**, 1226.
15. Greck, C.; Grice, P.; Ley, S. V.; Wonnacott, A. *TL* **1986**, *27*, 5277.
16. Spaltenstein, A.; Carpino, P. A.; Miyake, F.; Hopkins, P. B. *JOC* **1987**, *52*, 3759.

17. Scherkenbeck, J.; Barth, M.; Thiel, U.; Metten, K.-H.; Heinemann, F.; Welzel, P. *T* **1988**, *44*, 6325.
18. De Lombaert, S.; Nemery, I.; Roekens, B.; Carretero, J. C.; Kimmel, T.; Ghosez, L. *TL* **1986**, *27*, 5099.
19. Binns, M. R.; Haynes, R. K.; Katsifis, A. G.; Schober, P. A.; Vonwiller, S. C. *JOC* **1989**, *54*, 1960.
20. Truce, W. E.; Klingler, T. C. *JOC* **1970**, *35*, 1834.
21. Achmatowicz, B.; Baranowska, E.; Daniewski, A. R.; Pankowski, J.; Wicha, J. *TL* **1985**, *26*, 5597.
22. Reich, H. J.; Rusek, J. J.; Olson, R. E. *JACS* **1979**, *101*, 2225.
23. Trost, B. M.; Lynch, J.; Renaut, P.; Steinman, D. H. *JACS* **1986**, *108*, 284.
24. Julia, M.; Launay, M.; Stacino, J.-P.; Verpeaux, J.-N. *TL* **1982**, *23*, 2465.
25. Brimble, M. A.; Rush, C. J.; Williams, G. M.; Baker, E. N. *JCS(P1)* **1990**, 414.
26. Brimble, M. A.; Officer, D. L.; Williams, G. M. *TL* **1988**, *29*, 3609.
27. Babudri, F.; Florio, S.; Vitrani, A. M.; Di Nunno, L. *JCS(P1)* **1984**, 1899.
28. Corey, E. J.; Chaykovsky, M. *JACS* **1965**, *87*, 1345.
29. Isobe, M.; Ichikawa, Y.; Bai, D.-L.; Masaki, H.; Goto, T. *T* **1987**, *43*, 4767.
30. Trost, B. M.; Arndt, H. C.; Strege, P. E.; Verhoeven, T. R. *TL* **1976**, 3477.
31. Marshall, J. A.; Cleary, D. G. *JOC* **1986**, *51*, 858.
32. Brown, A. C.; Carpino, L. A. *JOC* **1985**, *50*, 1749.
33. Künzer, H.; Stahnke, M.; Sauer, G.; Wiechert, R. *TL* **1991**, *32*, 1949.
34. Julia, M.; Paris, J.-M. *TL* **1973**, 4833.
35. Kocienski, P. J. *COS* **1991**, *6*, 975.
36. Kelly, S. E. *COS* **1991**, *1*, 729.
37. Kocienski, P. J.; Lythgoe, B.; Ruston, S. *JCS(P1)* **1978**, 829.
38. Kocienski, P. J.; Lythgoe, B.; Roberts, D. A. *JCS(P1)* **1978**, 834.
39. Lythgoe, B.; Waterhouse, I. *JCS(P1)* **1980**, 1405.
40. Bartlett, P.; Green, F. R.; Rose, E. H. *JACS* **1978**, *100*, 4852.
41. Lythgoe, B.; Waterhouse, I. *TL* **1977**, 4223.
42. Williams, D. R.; Moore, J. L.; Yamada, M. *JOC* **1986**, *51*, 3916.
43. Baudin, J. B.; Hareau, G.; Julia, S. A.; Ruel, O. *BSF(2)* **1993**, *130*, 336.
44. Trost, B. M.; Merlic, C. A. *JACS* **1988**, *110*, 5216.
45. Trost, B. M.; Schmuff, N. R.; Miller, M. J. *JACS* **1980**, *102*, 5979.
46. Trost, B. M.; Ghadiri, M. R. *JACS* **1984**, *106*, 7260.
47. Fabre, J.-L.; Julia, M.; Verpeaux, J.-N. *BSF* **1985**, 772.
48. Alvarez, E.; Cuvigny, T.; Hervé du Penhoat, C.; Julia, M. *T* **1988**, *44*, 111.
49. Little, R. D.; Myong, S. O. *TL* **1980**, *21*, 3339.
50. Hwu, J. R. *JOC* **1983**, *48*, 4432.
51. Baudin, J.-B.; Julia, M.; Rolando, C. *TL* **1985**, *26*, 2333.

Georges Hareau & Philip Kocienski
Southampton University, UK

Phenyl(trichloromethyl)mercury[1]

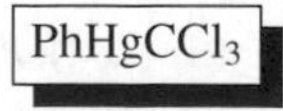

[3294-57-3] $C_7H_5Cl_3Hg$ (MW 396.06)

(transfers $:CCl_2$ to many unsaturated systems, to heteroatoms, and to several σ-bonds)

Physical Data: mp 117–118 °C.

Solubility: sol ether, benzene, chloroform.

Form Supplied in: colorless solid; commercially available.

Preparative Methods: prepared from PhHgBr either by reaction with ***Potassium t-Butoxide–Chloroform*** in benzene at 10 °C in 66% yield,[2] or by treatment with ***Sodium Methoxide–Ethyl Trichloroacetate*** in 62% yield.[3] ***Phenylmercury(II) Chloride*** furnishes $PhHgCCl_3$ by reaction with ***Sodium Trichloroacetate*** in refluxing DME (65%).[4] A two-phase approach from PhHgCl and $CHCl_3$ using aqueous ***Sodium Hydroxide–Potassium Fluoride*** and small amounts of ***Benzyltriethylammonium Chloride*** furnishes $PhHgCCl_3$ in 72% yield.[5]

Purification: recrystallization from chloroform.

Handling, Storage, and Precautions: flame-dried glassware and an atmosphere of dry nitrogen are used for the reactions. Organomercury(II) compounds should be carefully handled and skin contact and inhalation should be particularly avoided. All reactions must be conducted in a well-ventilated fume hood. The reagent is light sensitive and should be stored under nitrogen at 4 °C.

Cyclopropanation of Alkenes. A wide and representative set of alkenes react with $PhHgCCl_3$ to afford *gem*-dihalocyclopropanes in good yield (eq 1).[6] Reactions are carried out under nitrogen, using dry benzene as solvent, and can be easily monitored by TLC analysis. When all the starting mercurial has been consumed, PhHgCl is filtered off, the solvent and the excess alkene are removed in vacuo, and the resulting cyclopropanes are generally purified by distillation. The simplicity of product isolation represents an important synthetic advantage of this procedure.

cyclohexene → ($PhHgCCl_3$, benzene, 80 °C, 36 h) → 7,7-dichlorobicyclo[4.1.0]heptane (89%) + PhHgCl (99%) (1)

A distinctive feature of $PhHgCCl_3$ as a cyclopropanating reagent is that it reacts with alkenes that are electron-poor (eq 2).[6] In contrast, attempts to cyclopropanate tetrachloroethylene using *t*-BuOK–$CHCl_3$ or thermal decomposition of Cl_3CCO_2Na give only modest (less than 10%) yields of hexachlorocyclopropane.[7]

$Cl_2C{=}CCl_2$ → ($PhHgCCl_3$, 120 °C, 36 h) → hexachlorocyclopropane (63%) + PhHgCl (2)

$PhHgCCl_3$ competes well in several aspects with $PhHgCCl_2Br$ as a $:CCl_2$ transfer reagent for *gem*-dichlorocyclopropanation of alkenes; in particular, the former is a cheaper reagent and is more easily prepared. Conversely, the latter behaves as a milder source of $:CCl_2$, requiring shorter reaction times that enhance its synthetic potential.[6]

A simple modification of the experimental protocol, involving addition of ***Sodium Iodide***, has been developed. It decreases either the reaction time or the reaction temperature of the cyclopropanation with respect to that required when using α-haloorganomercurials.[8] Following this procedure, *trans*-stilbene smoothly undergoes cyclopropanation in high yield (eq 3).[9]

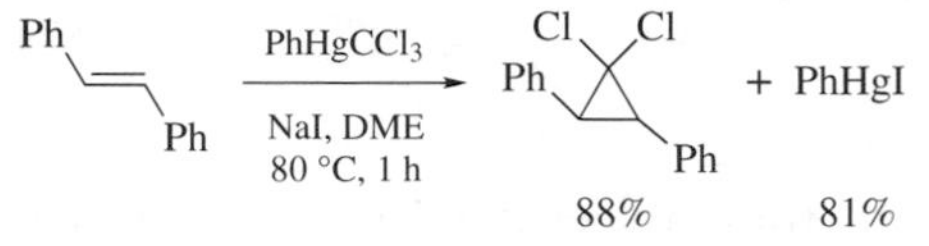

(3)

Cyclopropanation of cyclohexene by the above-mentioned conditions, (3 h, 85 °C) proceeded in 91% yield, while the cyclopropane derivative is formed in only 15% yield in the absence of NaI. Interestingly, the NaI method allows cyclopropanation by $PhHgCCl_3$ to take place at rt (eq 4).[9]

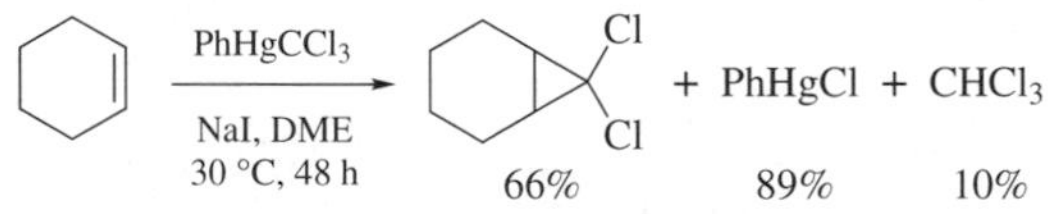

(4)

The activating effect of added NaI has been interpreted in terms of iodide attack at mercury and concomitant displacement of $^-CCl_3$, which generates in turn the dihalocarbene species ($^-CX_3 \rightarrow :CX_2 + X^-$). Consistent with that assumption, the reaction of NaI with equimolecular amounts of $PhHgCCl_3$ in dry acetone at 25 °C yields products that arise from ketone trapping by CCl_3^-.[9] The NaI-based procedure has some limitations. The reaction is sluggish with alkenes like tetrachloroethylene, reputed to be quite unreactive towards $:CCl_2$. Also, in good agreement with the electrophilic nature of the intermediate, the observed trend of alkene reactivity in this process ($R_2C{=}CR_2 > R_2C{=}CHR > R_2C{=}CH_2 > RCH{=}CHR > RCH{=}CH_2$)[9] matches well with that observed for the reactivity of carbenes generated by thermal treatment of Cl_3CCO_2Na in DME.

A major advantage of $PhHgCCl_3$ in *gem*-dichlorocyclopropanation reactions is realized when dealing with base-sensitive alkenes. In these cases, conventional carbene generation by treatment of chloroform with strong bases does not afford significant yields of cyclopropanes.[6] Many examples are known, and several synthetic applications have been reported. For example, 2-vinyl-4,4,6,6-tetramethyl-1,3,2-dioxaborinane is cleanly cyclopropanated with $PhHgCCl_3$ (eq 5);[10] cyclopropanation of this compound through thermal decomposition of Cl_3CCO_2Na as a source of $:CCl_2$ gives the desired cyclopropane in only 4% yield.[10]

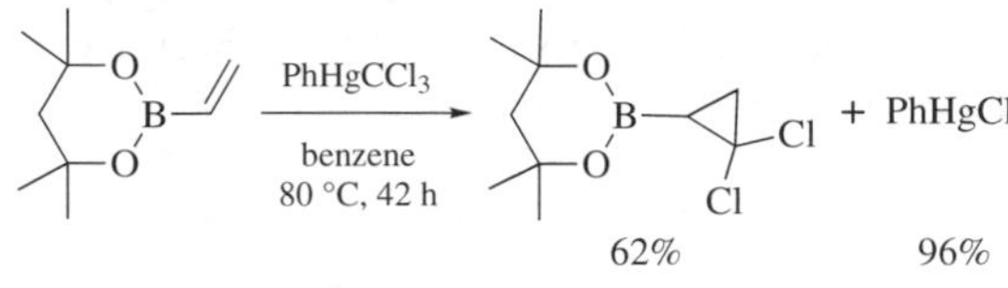

(5)

Diketene is also smoothly *gem*-dichlorocyclopropanated by $PhHgCCl_3$ (eq 6).[11] Interestingly, further reactions on the resulting dichlorocyclopropanespirolactone do not induce opening of the three-membered ring, in contrast with the behavior of the cyclopropanation product obtained by reaction of diketene with α-diazo ketones.[11]

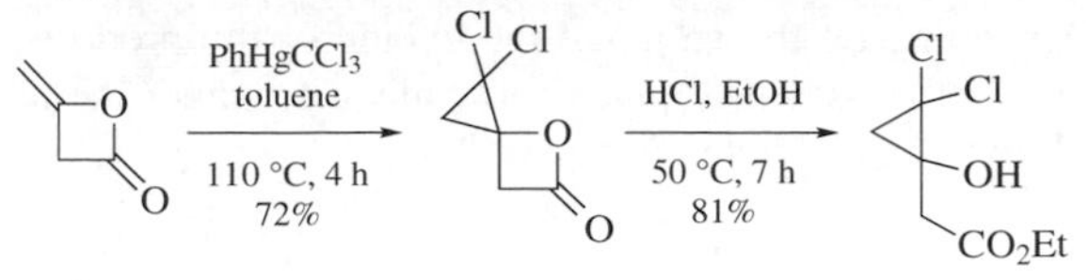

(6)

Vinyl Acetate, a base-sensitive compound that cannot be cyclopropanated by traditional methods, reacts readily with trihalomercurial reagents,[6] forming *gem*-dichlorocyclopropanes which are useful synthons leading to pyrazole rings (eq 7).[12]

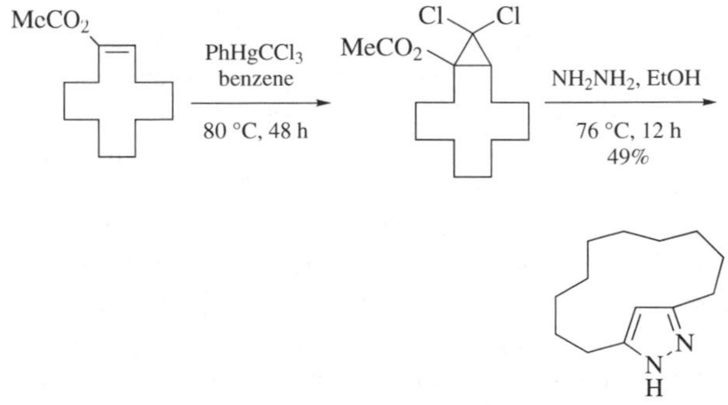

(7)

In this respect, if the size of the cycloalkenyl acetate is reduced then the yield of metacyclophane goes down, and the 1,2-function becomes significant. Treatment of cyclohexenyl acetate with $PhHgCCl_3$ and further reaction of the resulting *gem*-dichlorocyclopropane with ***Hydrazine*** yields exclusively 4,5,6,7-tetrahydroindazole (1,2-ring fusion) in 40% yield.[13] 2-Aminopyrimidines are also accessible by treating the α,α-dichlorocyclopropyl acetates with ***Guanidine*** (eq 8).[14]

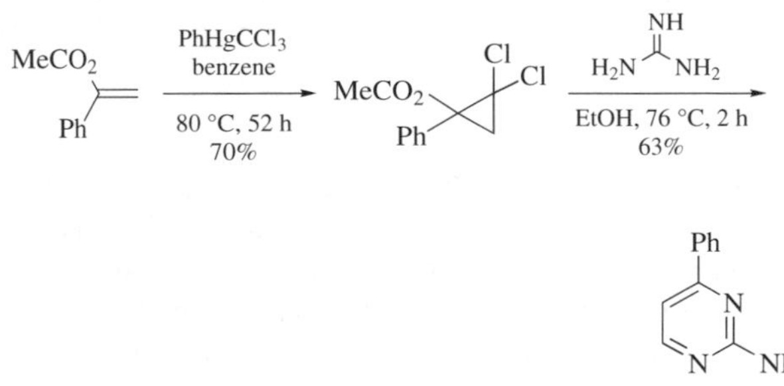

(8)

Similarly, the enol acetate function of steroids reacts with $PhHgCCl_3$ to give the corresponding cyclopropane derivatives; these are useful compounds for furnishing new skeletal rearrangements through ring-expansion reactions in basic media (eq 9).[15]

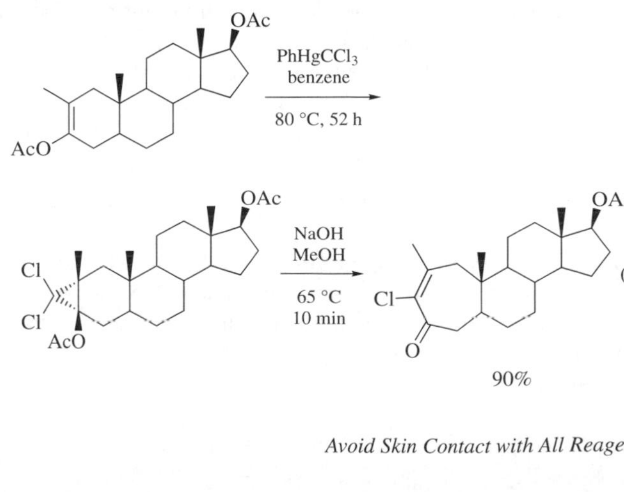

(9)

Depending on the strain of the cyclopropane products, ring-opening processes take place spontaneously under the reaction conditions, as illustrated in eq 10.[16]

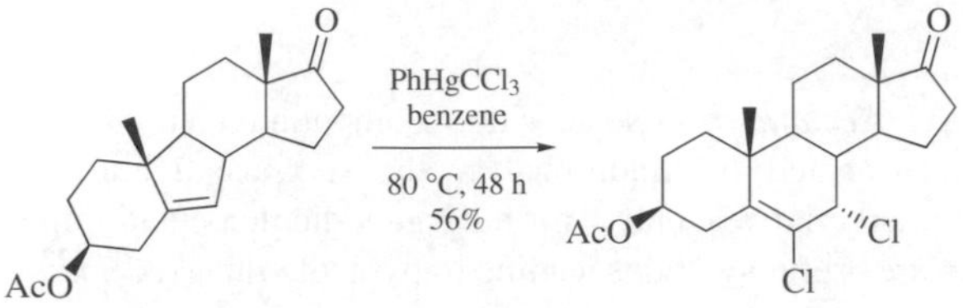

(10)

Carbene transfer from $PhHgCCl_3$ to steroid 5-enyl acetates occurs from the less-hindered side, the so-called α-face.[17] Thus the stereochemical outcome is generally opposite to that observed when using the Simmons–Smith procedure, suggesting that pre-coordination of the reagent does not occur.[17]

The cyclopropane derived from norbornene and $PhHgCCl_3$ also undergoes spontaneous ring expansion under the reaction conditions, affording 3,4-dichlorobicyclo[3.2.1]oct-2-ene.[18] This sequence, cyclopropanation/thermal ring expansion, has been successfully applied to cyclophane synthesis by starting from substituted indenes[19] and indoles.[20] Recently, cyclopropanation of alkenes has been used for the preparation of 1*H*-cyclopropa[*a*]naphthalene in a multistep sequence (eq 11).[21]

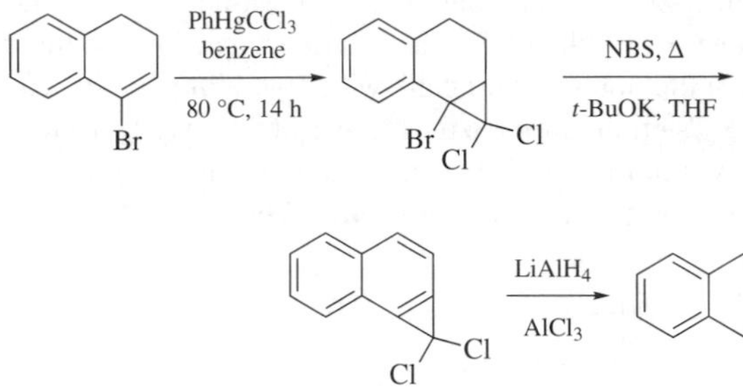

(11)

In short, the merit of $PhHgCCl_3$ as a **:**CCl_2 transfer reagent to cyclopropanate alkenes is derived from experimental facts; namely, the carbene is generated under neutral conditions at moderate reaction temperatures. The absence of side reactions clearly underlines the aforementioned comments. The accepted mechanism for the thermally induced cyclopropanation of alkenes with $PhHgCCl_3$ involves carbene generation from the α-halo organomercurial in the rate limiting step, followed by fast trapping of **:**CCl_2 by the alkene. The body of knowledge derived from the above described reactions supports this assumption. Of special relevance in this context is the easy cyclopropanation of vinyl acetates that clearly rules out participation of the trichloromethyl anion; in fact, all attempts to cyclopropanate this base-sensitive alkene using *t*-BuOK–$CHCl_3$ as carbene source were unsuccessful. Indirect evidence is provided by kinetic studies of related processes using $PhHgCCl_2Br$ as the carbene precursor.[22] Further support for the proposed mechanism is obtained from gas-phase pyrolysis studies on $PhHgCCl_3$,[23] and subsequent characterization of the thermally generated carbene by matrix isolation techniques.[24]

Reaction with Methoxynaphthalene. $PhHgCCl_3$ adds to the C-3–C-4 double bond of 1-methoxynaphthalene to give, after ring expansion under the reaction conditions, 5-chloro-2,3-benzotropone (eq 12),[25] a structure which was initially misassigned.[25a]

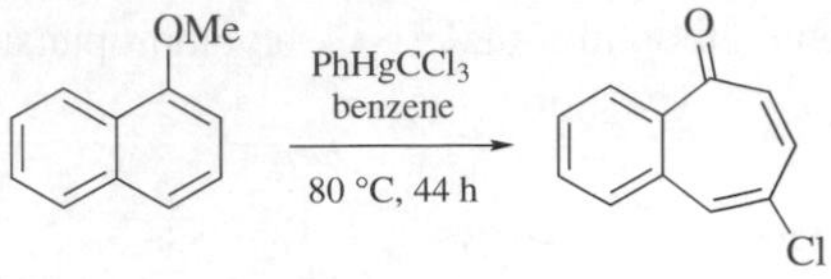

(12)

Reaction with Diiodoacetylene. The addition of *gem*-dichlorocarbene (generated from $PhHgCCl_2Br$) to different alkynes has been described.[26] On this ground, some of the products of the reaction mixture obtained from $PhHgCCl_3$ and IC≡CI were mechanistically interpreted to arise from initial addition of dichlorocarbene to the diiodoacetylene. When a 2:1 ratio of $PhHgCCl_3$/IC≡CI was used, 1-iodo-3,3,3-trichloropropyne and tetrachlorocyclopropene were formed as the major components of the reaction mixture (eq 13).[27]

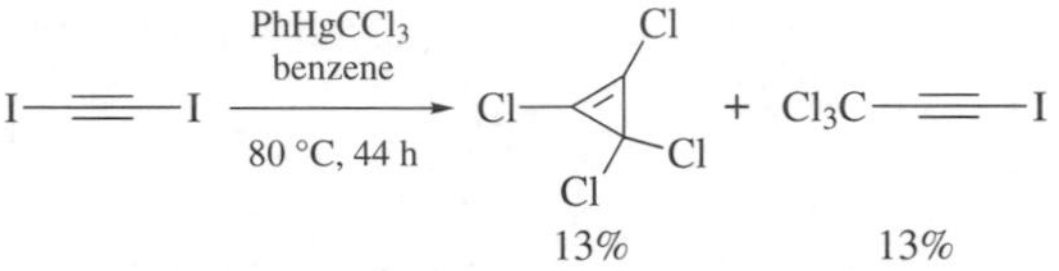

(13)

Reaction with 1,3-Dienes. 1,1,2,2,3,3-Hexamethyl-4,5-bis(methylene)cyclopentane is an interesting 1,3-diene to check whether 1,4-addition is a feasible reaction path for a given transformation. Its reaction with $PhHgCCl_3$ shows clearly that this possibility is a true alternative for a singlet carbene (eq 14).[28]

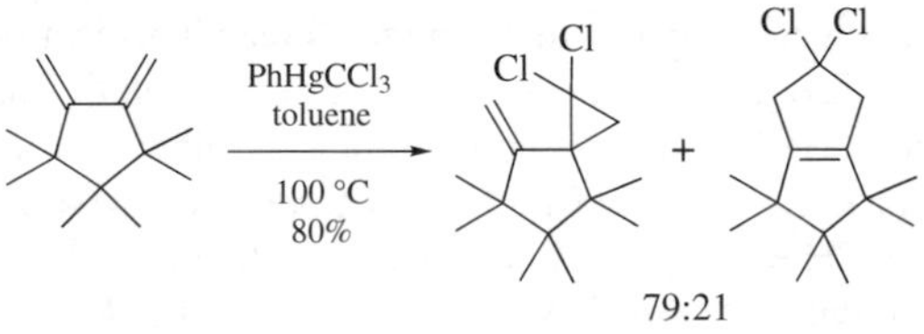

(14)

Reaction with Nitrogen- and Other Heteroatom-Substituted Compounds. Tertiary amines react with $PhHgCCl_3$ in boiling benzene; for instance, Et_3N gives EtCl, PhHgCl, and *N,N*-diethyltrichlorovinylamine ($Et_2NCCl{=}CCl_2$) in low yield.[29] Short reaction times have been noticed, suggesting an initial nucleophilic substitution reaction on mercury by the amine and $^-CCl_3$ displacement. Subsequent decomposition reactions would eventually lead to the nitrogen ylide ($Et_3N^+–CCl_2–^-CCl_2$) from which reaction products will be formed.[29] These findings are consistent with the interesting behavior shown by several allylamines towards $PhHgCCl_3$ (eq 15).[30] The yield of cyclopropane increases as the nucleophilic character of the nitrogen decreases. For more nucleophilic nitrogen atoms, competitive enamine formation could be the major pathway.

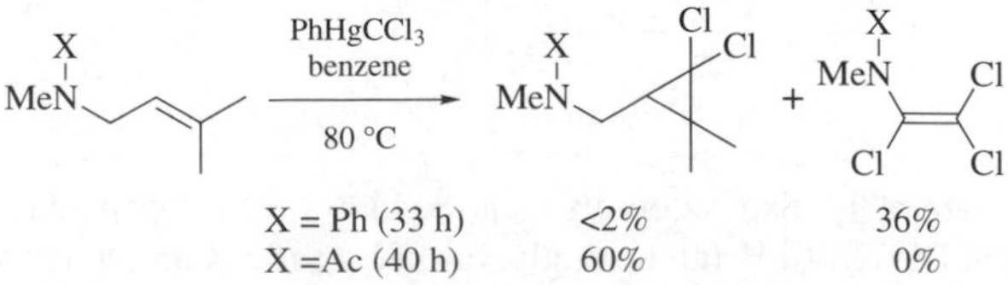

(15)

X = Ph (33 h)	<2%	36%
X =Ac (40 h)	60%	0%

The reaction of *N,N*-dimethylaniline with $PhHgCCl_3$ has also been described.[31] In this case, a novel class of compound is formed via electrophilic substitution on the starting amine, along with the expected products arising from an attack on nitrogen. Thermal reaction of $PhHgCCl_3$ with the C=N double bond should offer a simple entry to aziridines. However, its synthetic utility is severely limited due to the presence of PhHgCl, formed in the reaction.[32] Cyclopropanation was attempted in the presence of NaI[9] to overcome this problem. Although remarkable examples have been recorded (eq 16), the procedure lacks generality since the PhHgI formed catalyzes aziridine ring opening at rt.[32]

(16) [Scheme: $Ph_2C{=}NCH_2Ph$ → ($PhHgCCl_3$, NaI, DME, rt, 8 h, 98%) → 2,2-dichloro-3,3-diphenyl-1-benzylaziridine]

The C=N function of azirines reacts with $PhHgCCl_3$, giving ring-opened products (eq 17).[33] A minor byproduct, identified as the corresponding *N*-vinylaziridine, results from further attack of $:CCl_2$ on the C=N rather than the C=C bond of the azadiene.[33]

(17) [Scheme: 3-phenyl-2-methyl-2H-azirine → ($PhHgCCl_3$, benzene, 80 °C, 45 h, 41%) → Cl, Cl, Ph azadiene]

The reaction of oxygen-containing compounds with $:CCl_2$ generated from α-halo organomercurials ($PhHgXCCl_2$) is known, but most of the work deals with $PhHgCCl_2Br$ (reactions with ROH,[34] RCO_2H,[35] RCHO,[36] and RCOR[37] have been studied). $PhHgCCl_3$ attack at oxygen of the carbonyl group of several unsaturated steroidal ketones triggers skeletal rearrangements, and subsequent aromatization of the A ring.[38] $PhHgCCl_3$ is reported to deoxygenate pyridine *N*-oxides by carbene attack onto the oxygen atom to afford deoxygenated products.[39]

With respect to sulfur-containing substrates, sulfoximines react with $PhHgCCl_3$ to furnish sulfur ylides in modest yields via nitrogen displacement.[40] Of synthetic interest is the reaction of penams with $PhHgCCl_3$, offering a nonbasic approach to the corresponding azetidinones via thiazolidine ring opening (eq 18).[41] For these substrates, addition of NaI not only permits the use of milder reaction conditions, but results in a different type of product formation,[41b] probably due to $^{-}CCl_3$ acting as a base.

(18) [Scheme: penam methyl ester (CO_2Me) → (1. $PhHgCCl_3$, toluene, 110 °C, 3 h; 2. Et_3N, rt, 1 h; 94%) → azetidinone with $SCHCl_2$, MeO_2C; or → (1. $PhHgCCl_3$, NaI, toluene, 0 °C, 1.5 h; 2. AcCl, 0 °C, 0.5 h; 67%) → azetidinone with SAc, MeO_2C]

Insertion Reactions. $PhHgCCl_3$ easily inserts $:CCl_2$ into the Si–H bond.[42] In this respect, good yields of insertion product are obtained with silanes having bulky groups (eq 19).[42a] In contrast, carbenes generated following the *t*-BuOK–$CHCl_3$ protocol fail to react with these silicon compounds. Optically active silanes have been used to study the mechanism.[43] Since retention of configuration is mainly observed,[43a] the insertion process is seen to take place through a three-centered transition state.[43b]

$$t\text{-}Bu_3SiH \xrightarrow[120\ °C]{PhHgCCl_3} \underset{55\%}{t\text{-}Bu_3SiCHCl_2} + PhHgCl \qquad (19)$$

Insertion of $:CCl_2$, generated by thermal treatment of $PhHgCCl_3$, into C–H bonds usually does not compete with the addition to C=C bonds in typical alkenes, although some examples have been described in steroidal systems.[17] Several examples of insertion into Si–Br and Ge–Cl bonds are also known for $PhHgCCl_3$.[44]

Related Reagents. Bis(trichloromethyl)mercury; Chloroform; Ethyl Trichloroacetate; Iodo(iodomethyl)mercury; Phenyl(tribromomethyl)mercury; Potassium Hydroxide–Carbon Tetrachloride; Sodium Trichloroacetate; Trichloroacetic Acid; Trichloromethyllithium.

---

1. (a) Seyferth, D. *PAC* **1970**, *23*, 391. (b) Seyferth, D. *ACR* **1972**, *5*, 65. (c) Zeller, K.-P.; Straub, H. *MOC* **1974**, *13/2b*, 351. (d) Larock, R. C. *Organomercury Compounds in Organic Synthesis*; Springer: Berlin, 1985; Chapter X, pp 327–413.
2. (a) Reutov, O. A.; Lovtsova, A. N. *DOK* **1961**, *139*, 622. (b) Seyferth, D.; Burlitch, J. M. *JACS* **1962**, *84*, 1757. (c) Seyferth, D.; Burlitch, J. M. *JOM* **1965**, *4*, 127.
3. Schweizer, E. E.; O'Neill, G. J. *JOC* **1963**, *28*, 851.
4. Logan, T. J. *OSC* **1973**, *5*, 969.
5. Fedorynski, D.; Makosza, M. *JOM* **1973**, *51*, 89.
6. Seyferth, D.; Burlitch, J. M.; Minasz, R. J.; Mui, J. Y.-P.; Simmons, H. D., Jr.; Treiber, A. J. H.; Dowd, S. R. *JACS* **1965**, *87*, 4259.
7. Moore, W. R.; Krikorian, S. E.; La Prade, J. E. *JOC* **1963**, *28*, 1404.
8. Seyferth, D.; Mui, J. Y.-P.; Gordon, M. E.; Burlitch, J. M. *JACS* **1965**, *87*, 681.
9. Seyferth, D.; Gordon, M. E.; Mui, J. Y.-P.; Burlitch, J. M. *JACS* **1967**, *89*, 959.
10. Woods, W. G.; Bengelsdorf, I. S. *JOC* **1966**, *31*, 2769.
11. Kato, T.; Chiba, T.; Sato, R.; Yashima, T. *JOC* **1980**, *45*, 2020.
12. Parham, W. E.; Dooley, J. F. *JACS* **1967**, *89*, 985.
13. Parham, W. E.; Dooley, J. F. *JOC* **1968**, *33*, 1476.
14. Parham, W. E.; Dooley, J. F.; Meilahn, M. K.; Greidanus, J. W. *JOC* **1969**, *34*, 1474.
15. Crabbé, P.; Luche, J.-L.; Damiano, J.-C.; Luche, M.-J.; Cruz, A. *JOC* **1979**, *44*, 2929.
16. Rosen, P.; Karasiewicz, R. *JOC* **1973**, *38*, 289.
17. Bond, F. T.; Cornelia, R. H. *CC* **1968**, 1189.
18. Jefford, C. W.; Hill, D. T.; Gore, J.; Waegell, B. *HCA* **1972**, *55*, 790.
19. Parham, W. E.; Egberg, D. C.; Montgomery, W. C. *JOC* **1973**, *38*, 1207.
20. (a) Parham, W. E.; Davenport, R. W.; Biasotti, J. B. *TL* **1969**, 557. (b) Parham, W. E.; Davenport, R. W.; Biasotti, J. B. *JOC* **1970**, *35*, 3775.
21. (a) Müller, P.; Nguyen-Thi, H.-C. *HCA* **1984**, *67*, 467. (b) Müller, P.; Bernardinelli, G.; Gadoy-Nguyen Thi, H. C. *HCA* **1989**, *72*, 1627.
22. (a) Seyferth, D.; Mui, J. Y.-P.; Burlitch, J. M. *JACS* **1967**, *89*, 4953. (b) Seyferth, D.; Mui, J. Y.-P.; Damrauer, R. *JACS* **1968**, *90*, 6182.
23. Mal'tsev, A. K.; Mikaelyan, R. G.; Nefedov, O. M. *IZV* **1971**, 199 (*CA* **1971**, *75*, 19 592q).
24. Mal'tsev, A. K.; Mikaelyan, R. G.; Nefedov, O. M.; Hauge, R. H.; Margrave, J. L. *PNA* **1971**, *68*, 3238.
25. (a) Saraf, S. D. *S* **1971**, 264. (b) Ebine, S.; Hoshino, M.; Machiguchi, T. *BCJ* **1971**, *44*, 3480.
26. Seyferth, D.; Damrauer, R. *JOC* **1966**, *31*, 1660.
27. Cohen, H. M.; Keough, A. H. *JOC* **1966**, *31*, 3428.

28. Mayr, H.; Heigl, U. W. *AG(E)* **1985**, *24*, 579.
29. Seyferth, D.; Gordon, M. E.; Damrauer, R. *JOC* **1967**, *32*, 469.
30. Parham, W. E.; Potoski, J. R. *JOC* **1967**, *32*, 278.
31. Saraf, S. D. *CJC* **1969**, *47*, 1173.
32. Meilahn, M. K.; Olsen, D. K.; Brittain, W. J.; Anders, R. T. *JOC* **1978**, *43*, 1346.
33. Hassner, A.; Currie, J. O., Jr.; Steinfeld, A. S.; Atkinson, R. F. *JACS* **1973**, *95*, 2982.
34. Seyferth, D.; Mai, V. A.; Mui, J. Y.-P.; Darragh, K. V. *JOC* **1966**, *31*, 4079.
35. Seyferth, D.; Mui, J. Y.-P. *JACS* **1966**, *88*, 4672.
36. Martin, C. W.; Lund, P. R.; Rapp, E.; Landgrebe, J. A. *JOC* **1978**, *43*, 1071.
37. Seyferth, D.; Tronich, W.; Smith, W. E.; Hopper, S. P. *JOM* **1974**, *67*, 341.
38. Berkoz, B.; Lewis, G. S.; Edwards, J. A. *JOC* **1970**, *35*, 1060.
39. Schweizer, E. E.; O'Neill, G. J. *JOC* **1963**, *28*, 2460.
40. Iqbal, J.; Rahman, W. *JOM* **1979**, *169*, 141.
41. (a) Kang, J.; Im, W. B.; Choi, S.-g.; Lim, D.; Choi, Y. R.; Cho, H. G.; Lee, J. H. *H* **1989**, *29*, 209. (b) Kang, J.; Lim, D. S. *SL* **1990**, 611.
42. (a) Weidenbruch, M.; Peter, W.; Pierrard, C. *AG(E)* **1976**, *15*, 43. (b) Lukevics, E.; Sturkovich, R.; Goldberg, Yu.; Gaukham, A. *JOM* **1988**, *345*, 19.
43. (a) Sommer, L. H.; Ulland, L. A.; Ritter, A. *JACS* **1968**, *90*, 4486. (b) Sommer, L. H.; Ulland, L. A.; Parker, G. A. *JACS* **1972**, *94*, 3469.
44. (a) Weidenbruch, M.; Pierrard, C. *JOM* **1974**, *71*, C29. (b) Weidenbruch, M.; Pierrard, C. *CB* **1977**, *110*, 1545.

José Barluenga, Miguel Tomás & José M. González
Universidad de Oviedo, Spain

Potassium Cyanide[1]

KCN

[151-50-8] CKN (MW 65.12)

(reagent for the synthesis of nitriles,[2] cyanohydrins,[3] and α-amino nitriles;[4] catalyst for benzoin condensation[5] and transesterification[6])

Physical Data: mp 634 °C; *d* 1.520 g cm^{-3}.

Solubility: sol 2 parts cold water, 1 part boiling water, 2 parts glycerol, 100 parts ethanol, 25 parts methanol. The aq soln is strongly alkaline (pH = 11.0) and rapidly dec.

Form Supplied in: white granular powder or fused pieces; widely available.

Purification: see Perrin[7] for purification and recrystallization procedures.

Handling, Storage, and Precautions: highly toxic; may be fatal if inhaled, swallowed, or absorbed through the skin. Exposure may cause nausea, dizziness, headache, lung irritation, and cyanosis (toxicity data: orl-hmn LDLo 2857 μg kg^{-1}, orl-rat LD_{50} 5 mg kg^{-1}). Air and moisture sensitive; keep in a tightly closed container and store in dry area. Use only in a chemical fume hood. Do not breath the dust or allow it to get in eyes, on skin, or on clothing. In case of fire, water spray must be used with caution since direct contact of KCN with water or steam will cause decomposition, liberating highly toxic HCN gas as well as generating a highly hazardous solution of dissolved cyanide which must be kept out of sewers and watercourses. KCN also forms explosive mixtures, sometimes spontaneously, with chlorates, nitrates, and nitrogen trichloride plus ammonia. Incompatible with acids, strong oxidizing agents, alkaloids, chloral hydrate, iodine, and metallic salts. Thermal decomposition produces toxic fumes of hydrogen cyanide, carbon monoxide, carbon dioxide, and nitrogen oxides.

Nitrile Synthesis. The transformation of alcohols to nitriles is an important method for the elongation of carbon chains or the synthesis of useful intermediates in organic synthesis.[8] The usual procedure involves activation of alcohols by conversion into halides or sulfonates, followed by displacement with a cyano group. Among the methods for direct one-pot conversion of alcohols to nitriles,[9,10] the most general and convenient is treatment with KCN in the presence of ***Tri-n-butylphosphine***, CCl_4, and ***18-Crown-6*** (eq 1).[10] Displacement of primary, benzylic, or allylic halides with KCN produces nitriles in very good yields; the same reaction proceeds in moderate yield with secondary halides,[8,11] while it fails completely for tertiary halides, giving elimination under these conditions. The use of KCN in MeCN containing 18-crown-6 produces cyanide anions ('naked' anions)[12] which are strong nucleophiles and displace halides smoothly and quantitatively (eq 2).[13] Mesylates are converted into nitriles in about 70% yield by reaction with KCN in H_2O/C_6H_6 at 5–6 h reflux with phase-transfer catalysis.[14] 2,3′-*O*-Isopropylidene-5′-*O*-tosylates of nucleosides react with KCN only when the reagent has been complexed with 18-crown-6 (eq 3),[15] while reaction of substituted benzylamines with cyanide ions furnish the corresponding benzonitriles.[16] Ketones are converted into nitriles by way of tosylhydrazones[17a,b] or methoxycarbonyl hydrazones.[17c] Thus the tosylhydrazone derivative of a ketone upon treatment with KCN in MeOH/AcOH is converted into a cyano hydrazide, which is heated at 180 °C for 2 h to give the nitrile (eq 4).[18] Hydrocyanation of conjugated double[19] or isolated double[20] and triple[21] bonds proceed smoothly with KCN. Similarly, KCN in combination with NH_4Cl results in 1,4-addition of HCN to cholestanone derivatives, affording the 5α- and 5β-cyano ketones in 1:1 ratio.[22] The use of organoaluminum cyanides increases the selectivity of the reaction.[19]

$$C_8H_{17}OH \xrightarrow[\text{Bu}_3\text{P, 18-crown-6, CCl}_4\text{, MeCN, 76\%}]{\text{KCN}} C_8H_{17}CN \quad (1)$$

$$C_6H_{13}Br \xrightarrow[\text{18-crown-6, MeCN, 100\%}]{\text{KCN}} C_6H_{13}CN \quad (2)$$

TsO–(2′,3′-O-isopropylidene nucleoside)–Adenine → NC–(2′,3′-O-isopropylidene nucleoside)–Adenine (3)
KCN; 18-crown-6, MeCN, 55%

$$\text{C}_6\text{H}_{10}\text{=NNHTs} \xrightarrow[\text{MeOH, AcOH}\ 85\%]{\text{KCN}} \text{C}_6\text{H}_{10}(\text{NHNHTs})(\text{CN}) \xrightarrow[-\text{N}_2,\ -\text{TsH}\ 60\%]{180\ °\text{C}} \text{C}_6\text{H}_{11}\text{–C}\equiv\text{N} \quad (4)$$

Vinyl bromides are converted into α,β-unsaturated nitriles by reaction with KCN in the presence of ***Tetrakis(triphenylphosphine)palladium(0)*** and 18-crown-6. The reaction is highly stereospecific and proceeds in almost quantitative yield (eq 5).[23] Vinyl sulfones on exposure to KCN and ***Dicyclohexano-18-crown-6*** in refluxing *t*-butyl alcohol are smoothly converted to α,β-unsaturated nitriles, providing a particularly useful route for the construction of trisubstituted alkenes.[24]

$$\text{Bu–CH=CH–Br} \xrightarrow[\text{Pd(PPh}_3)_4,\ 18\text{-crown-6}\ \ \text{C}_6\text{H}_6\ \ 98\%]{\text{KCN}} \text{Bu–CH=CH–C}\equiv\text{N} \quad (5)$$

isom. purity 99%

Treatment of benzaldoximes with KCN in the presence of a phase-transfer catalyst (18-crown-6 or quaternary ammonium salt) produces a 3:1 mixture of a benzonitrile and benzamide (eq 6),[25] thus providing a general, one-pot conversion of aromatic aldehydes to aryl nitriles. Photochemically induced cyanation of arenes[26] furnishes only moderate to low yields of aryl nitriles, but the use of KCN in conjunction with 18-crown-6 ('naked' cyanide ion) doubles the yield.[27] Photolysis of arylthallium ditrifluoroacetates in the presence of excess KCN proceeds via in situ formation of the complex ions $[\text{ArTl(CN)}_3]^-\text{K}^+$ and produces aryl nitriles in good yields.[28] Cyanation of aromatic heterocycles by means of KCN and ***Benzyltriethylammonium Chloride*** followed by reaction with an acyl chloride furnishes the corresponding Reissert products (eq 7)[29] in better yields than the conventional conditions.[30]

$$\text{PhCH=NOH} \xrightarrow[\text{MeCN}]{\text{KCN, 18-crown-6}} \underset{76\%}{\text{PhCN}} + \underset{24\%}{\text{PhCONH}_2} \quad (6)$$

$$\text{quinoline} \xrightarrow[\text{H}_2\text{O, CH}_2\text{Cl}_2\ \ 57\%]{\text{KCN, R}_4\text{N}^+\text{Cl}^-,\ \text{EtCOCl}} \text{1-(EtOC(O))-2-CN-1,2-dihydroquinoline} \quad (7)$$

The classical method for the synthesis of acyl cyanides is the reaction of an acyl halide with heavy metal cyanides at high temperatures.[31] Recent modifications using aryl iodide[32] or a phase-transfer catalyst[33] have led to some improvements. On the contrary, aroyl cyanides are prepared by reaction of aroyl chlorides with KCN in MeCN. The presence of trace amounts of water markedly accelerates the rate and improves the yield (eq 8).[34] However, the use of too much water lowers the yield of the product due to hydrolysis and dimer formation.[34] Aromatic or heteroaromatic acyl cyanides may be prepared at lower reaction temperatures by ultrasonic treatment of the heterogeneous mixture of an aroyl chloride and KCN in MeCN (eq 9).[35] Several phenylacetonitriles are obtained in good yield by reaction of the appropriate benzyl halide with the cyanide form of an anion exchange resin.[36] The direct conversion of aromatic iodides into aroyl nitriles in good yields is accomplished by their carbonylation catalyzed with ***Iodo(phenyl)bis(triphenylphosphine)palladium(II)***[37] and in situ reaction with KCN (eq 10).[38] Cyanoformates are obtained from the corresponding chloroformates by reaction with KCN in the presence of 18-crown-6 as catalyst (eq 11).[39]

$$\text{4-MeOC}_6\text{H}_4\text{COCl} \xrightarrow[\text{MeCN, H}_2\text{O}\ \ 80\ °\text{C, 90 min}\ \ 79\%\ \text{(by GLC)}]{\text{KCN}} \text{4-MeOC}_6\text{H}_4\text{COCN} \quad (8)$$

$$\text{4-MeC}_6\text{H}_4\text{COCl} \xrightarrow[\text{MeCN, ultrasound}\ \ 50\ °\text{C}\ \ 83\%]{\text{KCN}} \text{4-MeC}_6\text{H}_4\text{COCN} \quad (9)$$

$$\text{4-MeOC}_6\text{H}_4\text{I} \xrightarrow[\text{THF}\ \ 90\%]{\text{KCN, CO}\ \ \text{PhPdI(PPh}_3)_2} \text{4-MeOC}_6\text{H}_4\text{COCN} + \text{KI} \quad (10)$$

$$\text{BuOCOCl} \xrightarrow[18\text{-crown-6}\ \ 90\%]{\text{KCN}} \text{BuOCOCN} \quad (11)$$

α-Cyanoenamines are obtained either by conversion of amides to α-chloroenamines and subsequent treatment with KCN in MeCN (eq 12),[40] or starting with aldehydes and proceeding via α-chloroaldimines and reaction with KCN in MeOH (eq 13).[41] These compounds are recognized as reactive intermediates since they can easily be converted to the corresponding α-diketones or dihydropyrazine derivatives.[40]

$$\text{Me}_2\text{CHCONMe}_2 \xrightarrow[\text{Et}_3\text{N}]{\text{HCl, COCl}_2} \text{Me}_2\text{C=C(Cl)NMe}_2 \xrightarrow[\text{MeCN, }\Delta\ \ 80\%]{\text{KCN}} \text{Me}_2\text{C=C(CN)NMe}_2 \quad (12)$$

$$\text{Me}_2\text{CHCHO} \xrightarrow{\text{1. MeNH}_2\ \ \text{2. NCS, CCl}_4\ \ \text{3. KCN, MeCN}} \text{Me}_2\text{C=C(CN)NHMe} \xrightarrow[82\%]{\text{EtNH}_2} \text{Me}_2\text{C=C(CN)N(Me)Et} \quad (13)$$

Cyanotributylstannane, which is an efficient cyanation agent for acyl chlorides, is obtained in very good yield by treatment of ***Tri-n-butylchlorostannane*** with KCN in MeCN (eq 14).[42] Similarly, ***Cyanotrimethylsilane*** is prepared by treatment of ***Chlorotrimethylsilane*** with H_2SO_4 and KCN (eq 15).[43]

$$\text{Bu}_3\text{SnCl} \xrightarrow[18\text{-crown-6}\ \ 80\ °\text{C, 7 h}\ \ 83\%]{\text{KCN, MeCN}} \text{Bu}_3\text{SnCN} + \text{KCl} \quad (14)$$

$$2\ \text{TMSCl} + \text{H}_2\text{SO}_4 \xrightarrow{-2\text{HCl}} [\text{TMSO}]_2\text{SO}_2 \xrightarrow[-\text{K}_2\text{SO}_4\ \ 65\text{–}83\%]{2\ \text{KCN}} 2\ \text{TMSCN} \quad (15)$$

Cyanohydrin Synthesis. Cyanohydrins are versatile starting materials for the synthesis of several classes of compounds.[44] They are prepared by treatment of aldehydes or ketones with KCN and AcOH in various organic solvents.[45] This is an equilibrium reaction which proceeds smoothly except with sterically hindered ketones or aromatic aldehydes, in which case the benzoin condensation competes. Cyanohydrin synthesis was first applied to carbohydrates by Kiliani[46] in 1885 as the key step of a route (modified later by Fischer[47]) which allows the construction of two epimeric aldoses having one carbon atom more than the parent aldose. More recent studies have confirmed that the ratio of epimers (eq 16) is pH dependent.[48] Thus reaction of D-arabinose with KCN at acidic pH furnishes D-gluconназвания nitrile as the predominant epimer, while under alkaline conditions D-mannonitrile is the predominant epimer. Similarly treatment of formaldehyde with KCN gave glyconitrile (eq 17).[49] Cyanohydrin synthesis is important in the construction of steroid sidechains.[50] Thus 17-keto steroids are transformed to corticoids via their 17β-cyanohydrins (eq 18).[51] Treatment of β-hydroxy ketones with KCN/TMSCN in the presence of ***Zinc Iodide*** produces *syn*-β-hydroxy cyanohydrins in high distereomeric excess (eq 19).[52]

(16)

$$CH_2O + KCN + H_2O \xrightarrow{80\%} HOCH_2CN + KOH \quad (17)$$

KCN, MeOH, AcOH, 98% (18)

KCN, TMSCN, ZnI_2, 75%; de >95% (19)

Optically active cyanohydrins are obtained using KCN and enzymatic catalysis.[53] Thus treatment of benzaldehyde with oxynitrilase (extracts of ground defatted almonds) in 1N KCN/AcOH buffer (pH 5.4) furnishes exclusively (*R*)-(+)-α-hydroxyphenylacetonitrile (eq 20).[54] The other enantiomer is obtained by kinetic resolution of racemic cyanohydrins either by using a lipoprotein lipase from *Pseudomonas* species[55] or by incubation with *Pichia miso* (IAM 4682), which hydrolyzes selectively the acetate of the (*R*)-enantiomer, leaving the (*S*)-enantiomer intact.[56] Silylated cyanohydrins are obtained in high yields by a one-pot reaction using KCN in combination with TMSCl (eq 21).[57,58] The presence of ZnI_2 enhances the rate of cyanosilylation, while the presence of 18-crown-6 has very little effect on the yield or the rate.[57]

oxynitrilase, KCN, AcOH, pH = 5.4; EtOH, 0 °C, 95%; ee >98% (20)

KCN, TMSCl; ZnI_2, MeCN, Δ, 96% (21)

α-Amino Nitriles. The preparation of α-amino nitriles is the key reaction of the Strecker synthesis,[59] which is considered to be the most convenient experimental protocol for the preparation of α-amino acids[4] and 1,2-diamines[60] on preparative scale. The classical procedure in which an aldehyde or ketone is treated with KCN, NH_4Cl,[59] is sensitive to the nature of the ketone substitution, involves lengthy reaction times and tedious work-up procedures, and leads to considerable amounts of cyanohydrin byproduct.[61] Modifications using Al_2O_3 and ultrasound in MeCN have improved the yield slightly.[62] On the contrary, stepwise synthesis either by displacement of the hydroxy group of α-amino alcohols with cyanide[63] using KCN, AcOH, or via cyanohydrins[64] by treatment with ammonia or NaN_3 have greatly improved the yields. Recently, however, two simple one-pot procedures have been reported. These produce α-amino nitriles efficiently, in almost quantitative yields, by treatment of carbonyl compounds with KCN and either alkyl- or benzylamines in an aqueous solution of $NaHSO_3$ (eq 22)[65] or benzylamine in methanolic AcOH (eq 23).[66] It is noteworthy that the α-amino nitrile synthesis (Strecker method) with substituted cyclic ketones furnishes preferentially α-amino acids of one geometrical isomer (eq 24),[67] while the opposite isomer is obtained via the spirohydantoin synthesis (Bucherer method) (eq 25).[67] Optically active α-amino nitriles are obtained when a chiral auxiliary is used. Thus reaction of *R*-(−)-2-phenylglycinol with isobutylaldehyde furnishes the two diastereomers in an 84:16 ratio (eq 26).[68]

$NaHSO_3$, $MeNH_2$, KCN; H_2O, rt, 88% (22)

KCN, $BnNH_2$, AcOH; MeOH, 60 °C, 92% (23)

KCN, NH_4Cl; H_2O, EtOH, 6 days, 60%; HCl, Δ, 12 h, 71% (24)

KCN, $(NH_4)_2CO_3$; H_2O, EtOH, 24 h, 82%; 1. H_2SO_4 2. $Ba(OH)_2$, 150 °C, 35% (25)

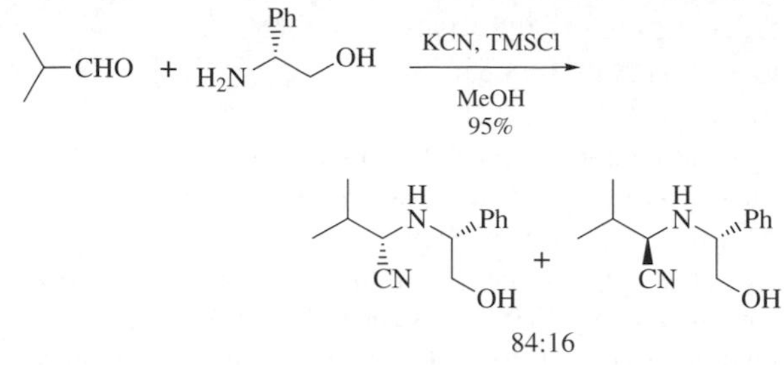

(26)

84:16

Benzoin Condensation. Aromatic[69] or heteroaromatic aldehydes[70] are condensed in a KCN-catalyzed process to form benzoins (α-hydroxy ketones) (eq 27). The reaction mechanism involves the formation of a cyano-stabilized carbanion,[71] facilitated by the use of an aprotic solvent. KCN is a highly specific catalyst, since it performs three functions:[72] (i) acts as a nucleophile, (ii) permits loss of the aldehyde proton with its electron-withdrawing ability, and (iii) at the end, acts as a leaving group (See also ***3-Benzyl-4-methyl-1,3-thiazolium Chloride***). The benzoin-type reaction has been extended to include a catalyzed addition of aldehydes to α,β-unsaturated nitriles.[73] Thus 4-keto nitriles are obtained by addition of aromatic aldehydes to α,β-unsaturated nitriles under KCN catalysis (eq 28). Similarly, benzoin-type condensation of tetraphthalaldehyde bisbisulfite with benzaldehyde under KCN catalysis furnishes the bisbenzoin product (eq 29).[74]

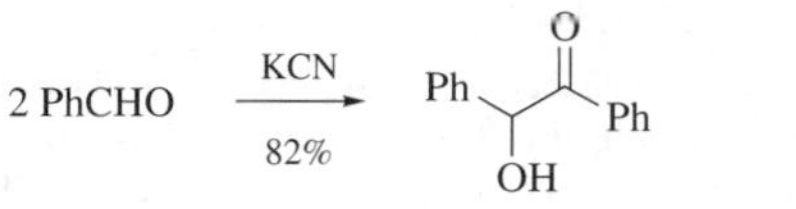

(27)

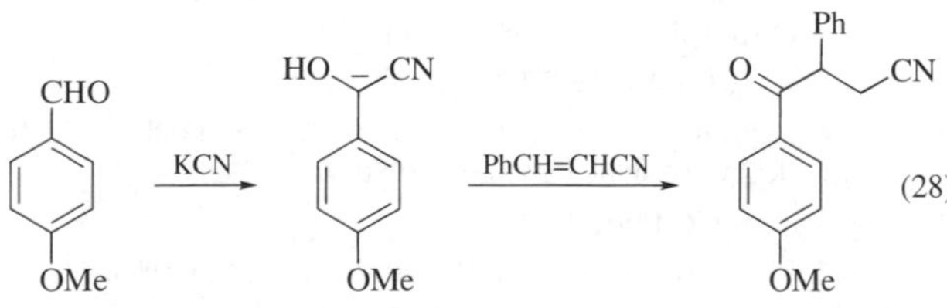

(28)

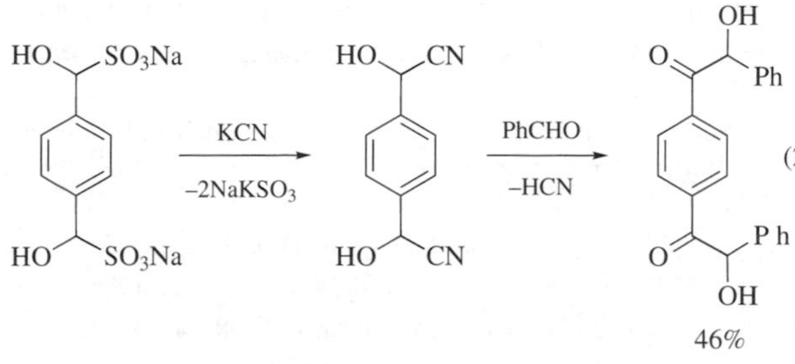

(29)

46%

Transesterification. KCN is a mild and effective catalyst for the transesterification of α,β-unsaturated esters without isomerization of the conjugated double bond. Thus methyl *trans,trans*-farnesoate is readily converted into the ethyl ester with very slight *cis*/*trans* isomerization (eq 30).[75] Similarly, KCN in combination with 18-crown-6 is an effective catalyst for cyanoesterification.[76] On the other hand, treatment of δ-lactones with methanol in the presence of KCN results in lactone ring opening and methyl ester formation (eq 31).[77]

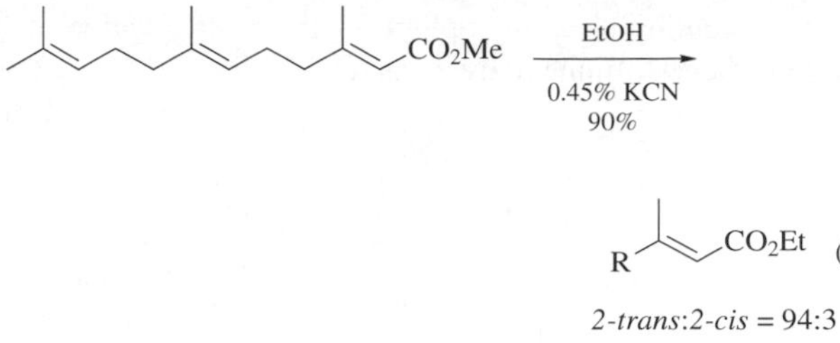

(30)

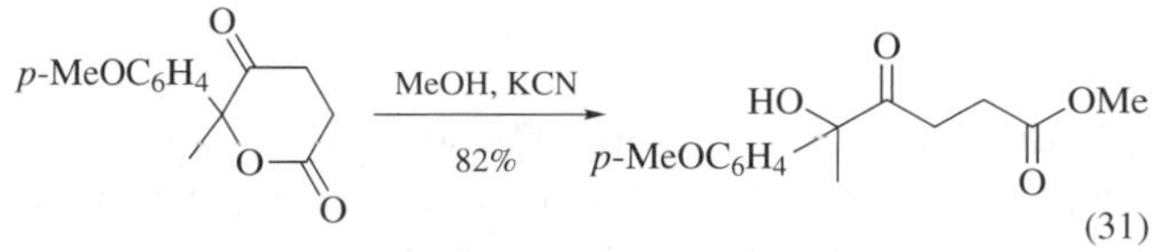

2-*trans*:2-*cis* = 94:3

(31)

Ring Formation and Ring Opening. Reaction of α-chloro ketimines with KCN in MeOH results in nucleophilic addition and subsequent intramolecular nucleophilic substitution, yielding the corresponding α-cyano aziridines (eq 32).[78] Treatment of 2-bromodeoxybenzoin with KCN in the presence of solid adsorbent (silica gel or alumina) gives stereoselectively *cis*-2,3-diphenyl-2-cyanooxirane in good yield (eq 33).[79] Cold aqueous ethanolic KCN converts *N*-methyl-*C*-phenyl nitrone into 1-methyl-4,5-diphenylimidazole via the intermediate cyanoimine (eq 34).[80] *N*-Allyl-*N*-3-butenyl-*N*-Cbz-amine on treatment with thexylborane, followed by cyanidation, affords an azacyclone, which can be transformed into δ-coniceine by catalytic reduction (eq 35).[81] On the other hand, alkaline solutions of KCN in the presence of ***Tetra-n-butylammonium Bromide*** open 1,2-epoxides, yielding β-hydroxy nitriles (eq 36).[82] Similarly activated cyclopropane rings are cleaved by reaction with KCN.[83]

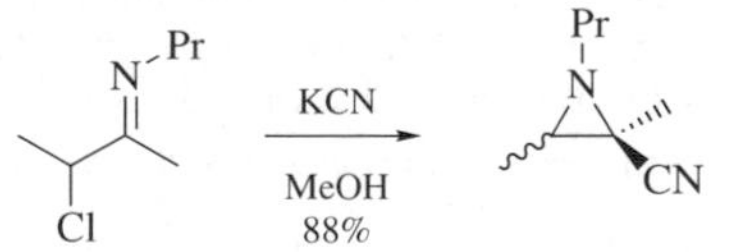

(32)

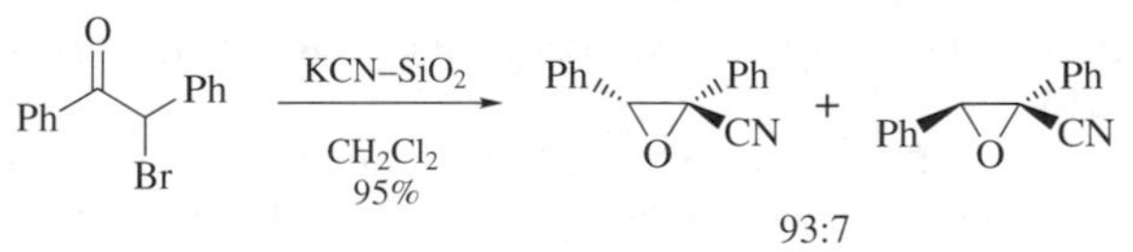

(33)

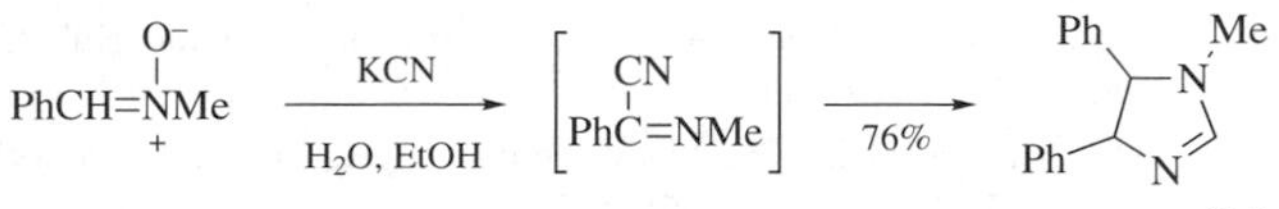

93:7

(34)

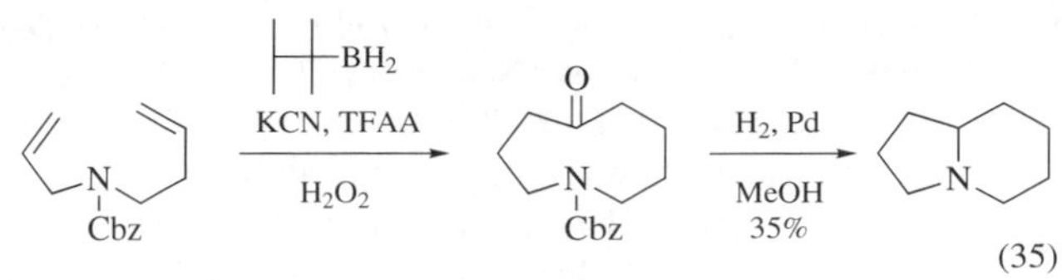

(35)

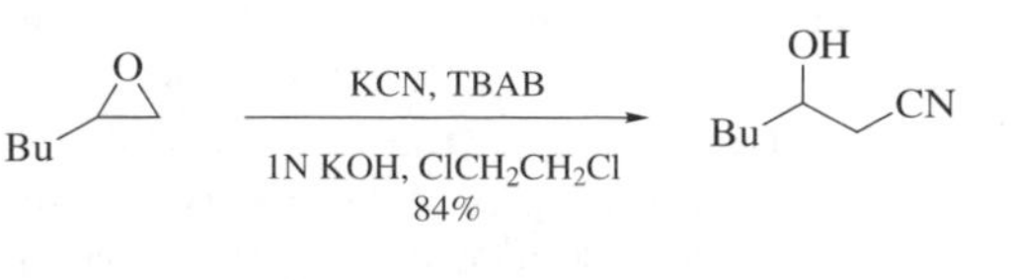

(36)

Aromatization. KCN in DMF is an effective reagent for the conversion of 2,4-cyclohexadien-1-ols to benzene derivatives.[84]

The same conditions were applied for the aromatization of 2,4-cyclohexadiene-6-imine derivatives (eq 37).[85]

$$\text{(2,4-cyclohexadiene-6-imine benzoate)} \xrightarrow[\text{DMF, 82\%}]{\text{KCN}} \text{PhC(O)NH-aryl} \quad (37)$$

Reduction. 0.1 N aqueous KCN solutions of aromatic nitro compounds upon UV irradiation undergo deoxygenation to the corresponding nitroso compounds (eq 38).[86]

$$PhNO_2 \xrightarrow[H_2O]{h\nu,\ KCN} PhNO \quad (38)$$

Related Reagents. Acetone Cyanohydrin; Acetyl Cyanide; 3-Benzyl-4-methyl-1,3-thiazolium Chloride; *t*-Butyldimethylsilyl Cyanide; Cyanotributylstannane; Cyanotrimethylsilane; Diethylaluminum Cyanide; Hydrogen Cyanide; Tetraethylammonium Cyanide; Zinc Cyanide.

---

1. Fatiadi, A. J. In *The Chemistry of Functional Groups, Suppl. C*; Patai, S.; Rappoport, Z. Eds.; Wiley: New York, 1983; Part 2, Chapter 26, pp 1057–1303.
2. (a) Buehller, C. A.; Pearson, D. E. *Survey of Organic Syntheses*; Wiley: New York, 1977; Vol. 2, Chapter 19. (b) Ellis, G. P.; Thomas, I. L. In *Progress in Medicinal Chemistry*; Ellis, G. P.; West, G. B., Eds.; North-Holland: Amsterdam, 1974; Vol. 10, Chapter 6.
3. (a) Friedrich, K. In *The Chemistry of Functional Groups, Suppl. C*; Patai, S.; Rappoport, Z., Eds.; Wiley: New York, 1983; Part 2, Chapter 28, pp 1345–1390. (b) Albright, J. D. *T* **1983**, *39*, 3207.
4. Williams, R. M. *Synthesis of Optically Active α-Amino Acids*; Pergamon: Oxford, 1989, pp 208–229.
5. Ide, W. S.; Buck, J. S. *OR* **1948**, *4*, 269.
6. Birch, A. J.; Corrie, J. E. T.; MacDonald, P. L.; Rao, G. S. *JCS(P1)* **1972**, 1186.
7. (a) Perrin, D. D.; Armarego, W. L. F.; Perrin, D. R. *Purification of Laboratory Chemicals*, 2nd edn.; Pergamon: Oxford, 1980. (b) Adisesh, A.; Taylor, M. *J. Phys. Chem.* **1962**, *66*, 2426.
8. March, J. *Advanced Organic Chemistry*, 3rd edn.; McGraw-Hill: New York, 1985; pp 429–430.
9. (a) Schwartz, M. A.; Zoda, M.; Vishnuvajjala, B.; Miami, I. *JOC* **1976**, *41*, 2502. (b) Kurtz, P. *LA* **1951**, *572*, 23.
10. Mizuno, A.; Hamada, Y.; Shioiri, T. *S* **1980**, 1007.
11. Friedrich, K.; Wallenfels, K. In *The Chemistry of the Cyano Group*; Rappoport, Z. Ed. Interscience: New York, 1970; pp 77–86.
12. Liotta, C. L.; Grisdale, E. E. *TL* **1975**, 4205.
13. Cook, F. L.; Bowers, C. W.; Liotta, C. L. *JOC* **1974**, *39*, 3416.
14. Newman, M. S.; Barbee, T. G.; Blakesley, C. N.; Din, Z.; Gromelski, S.; Khanna, V. K.; Lee, L.-F.; Radhakrishnan, J.; Robey, R. L.; Sankaran, V.; Sankarappa, S. K.; Springer, J. M. *JOC* **1975**, *40*, 2863.
15. Meyer, W.; Bohnke, E.; Follman, H. *AG(E)* **1976**, *15*, 49.
16. Short, J. H.; Dunnigan, D. A.; Ours, C. W. *T* **1973**, *29*, 1931.
17. (a) Cacchi, S.; Caglioti, L.; Paolucci, G. *S* **1975**, 120. (b) Orere, D. M.; Reese, C. B. *CC* **1977**, 280. (c) Wender, P. A.; Eissenstat, M. A.; Sapuppo, N.; Ziegler, F. E. *OSC* **1988**, *6*, 334.
18. Cacchi, S.; Caglioti, L.; Paolucci, G. *CI(L)* **1972**, 213.
19. Nagata, W.; Yoshioka, M. *OR* **1977**, *25*, 255.
20. Davey, W.; Tivey, D. J. *JCS* **1958**, 1230.
21. Cobb, T. *JACS* **1923**, *45*, 604.
22. Nagata, W.; Yoshioka, M.; Hirai, S. *TL* **1962**, 461.
23. Yamamura, K.; Murahashi, S.-I. *TL* **1977**, 4429.
24. Taber, D. F.; Saleh, S. A. *JOC* **1981**, *46*, 4817.
25. Rasmussen, J. K. *CL* **1977**, 1295.
26. (a) Vink, J. A. J.; Verheijdt, P. L.; Cornelisse, J.; Havinga, E. *T* **1972**, *28*, 5081. (b) Mizuno, K.; Pac, C.; Sakurai, H. *CC* **1975**, 553.
27. (a) Beugelmans, R.; LeGoff, M.-T.; Pusset, J.; Roussi, G. *CC* **1976**, 377. (b) Beugelmans, R.; LeGoff, M.-T.; Pusset, J.; Roussi, G. *TL* **1976**, 2305.
28. Taylor, E. C.; Altland, H. W.; Danforth, R. H.; McGillivray, G.; McKillop, A. *JACS* **1970**, *92*, 3520.
29. Koizumi, T.; Takeda, K.; Yoshida, K.; Yoshi, E. *S* **1977**, 497.
30. Reuss, R. H.; Smith, N. G.; Winters, L. J. *JOC* **1974**, *39*, 2027.
31. Oakwood, T. S.; Weisgerber, C. A. *OSC* **1955**, *3*, 112.
32. Haase, K.; Hoffman, H. M. R. *AG(E)* **1982**, *21*, 83.
33. Koenig, K. E.; Weber, W. P. *TL* **1974**, 2275.
34. Tanaka, M.; Koyanagi, M. *S* **1981**, 973.
35. Ando, T.; Kawate, T.; Yamawaki, J.; Hanafusa, T. *S* **1983**, 637.
36. Gordon, M.; DePamphilis, M. L.; Griffin, C. E. *JOC* **1963**, *28*, 698.
37. Fitton, P.; Rick, E. A. *JOM* **1971**, *28*, 287.
38. Tanaka, M. *BCJ* **1981**, *54*, 637.
39. Childs, M. E.; Weber, W. P. *JOC* **1976**, *41*, 3486.
40. Toye, J.; Ghosez, L. *JACS* **1975**, *97*, 2276.
41. DeKimpe, N.; Verhé, R.; DeBuyck, L.; Schamp, N. *S* **1979**, 741.
42. Tanaka, M. *TL* **1980**, *21*, 2959.
43. *FF* **1988**, *14*, 107.
44. Brussee, J.; Loos, W. T.; Kruse, C. G.; Van der Gen, A. *T* **1990**, *46*, 979.
45. (a) Haroutounian, S. A.; Georgiadis, M. P.; Delitheos, A. K.; Bailar, J. C. *Eur. J. Med. Chem.* **1987**, *22*, 325. (b) Gasc, J. C.; Nédélec, L. *TL* **1971**, 2005; (c) Nitta, I.; Fujimori, S.; Ueno, H. *BCJ* **1985**, *58*, 978.
46. (a) Kiliani, H. *CB* **1885**, *18*, 3066. (b) Kiliani, H. *CB* **1886**, *19*, 221, 767, 3029. (c) Kiliani, H. *CB* **1888**, *21*, 915.
47. Fischer, E. *CB* **1889**, *22*, 2204.
48. (a) Serianni, A. S.; Nunez, H. A.; Barker, R. *JOC* **1980**, *45*, 3329. (b) Blazer, R. M.; Whaley, T. W. *JACS* **1980**, *102*, 5082.
49. Gaudry, R. *OSC* **1955**, *3*, 436.
50. (a) Livingston, D. A.; Petre, J. E.; Bergh, C. L. *JACS* **1990**, *112*, 6449. (b) Reid, J. G.; Debiak-Krook, T. *TL* **1990**, *31*, 3669.
51. Carruthers, N. I.; Garshasb, S.; McPhail, A. T. *JOC* **1992**, *57*, 961.
52. Brunet, E.; Batra, M. S.; Aguilar, F. J.; Ruano, J. L. G. *TL* **1991**, *32*, 5423.
53. (a) Becker, W.; Pfeil, E. *JACS* **1966**, *88*, 4299. (b) Oku, J.-I.; Inoue, S. *CC* **1981**, 229. (c) Effenberger, F.; Ziegler, T.; Förster *AG(E)* **1987**, *26*, 458.
54. Brussee, J.; Roos, E. C.; Van der Gen, A. *TL* **1988**, *29*, 4485.
55. Wang, Y.-F.; Chen, S.-T.; Liu, K. K.-C.; Wong, C.-H. *TL* **1989**, *30*, 1917.
56. Ohta, H.; Kimura, Y.; Sugano, Y.; Sugai, T. *T* **1989**, *45*, 5469.
57. Rasmussen, J. K.; Heilmann, S. M. *S* **1978**, 219.
58. (a) Takahashi, T.; Yokoyama, H.; Haino, T.; Yamada, H. *JOC* **1992**, *57*, 3521. (b) Takahashi, T.; Kitamuta, K.; Nemoto, H.; Tsuji, J. *TL* **1983**, *24*, 3489.
59. (a) Strecker, A. *LA* **1850**, *75*, 27. (b) Stein, G. A.; Bronner, H. A.; Pfister, K. *JACS* **1955**, *77*, 700.
60. (a) Georgiadis, M. P.; Haroutounian, S. A.; Bailar, J. C. *JHC* **1988**, *25*, 995. (b) Freifelder, M.; Hasbrouck, R. B. *JACS* **1960**, *82*, 696.
61. (a) Edward, J. T.; Jitrangsri, C. *CJC* **1975**, *53*, 3339. (b) Taillades, J.; Commeyras, A. *T* **1974**, *30*, 127, 2493, 3407.
62. Dappen, M. S.; Pellicciari, R.; Natalini, B.; Monahan, J. B.; Chiorri, C.; Cordi, A. A. *JMC* **1991**, *34*, 161.

63. Fadel, A. *T* **1991**, *47*, 6265.
64. Davis, J. W. *JOC* **1978**, *43*, 3980.
65. (a) DeKimpe, N.; Sulmon, P.; Stevens, C. *T* **1991**, *47*, 4723. (b) Inaba, T.; Fujita, M.; Ogura, K. *JOC* **1991**, *56*, 1274.
66. Georgiadis, M. P.; Haroutounian, S. A. *S* **1989**, 616.
67. (a) Munday, L. *Nature* **1961**, *190*, 1103. (b) Haroutounian, S. A.; Georgiadis, M. P.; Polissiou, M. G. *JHC* **1989**, *26*, 1283.
68. Chakraborty, T. K.; Reddy, G. V.; Azhar-Hussain, K. *TL* **1991**, *32*, 7597.
69. (a) Shinkai, S.; Yamashita, T.; Kusano, Y.; Ide, T.; Manabe, O. *JACS* **1980**, *102*, 2335. (b) Reardon, W. C.; Wilson, J. E.; Trisler, J. C. *JOC* **1974**, *39*, 1596.
70. Hayashi, E.; Higashino, T. *H* **1979**, *12*, 837.
71. Kuebrich, J. P.; Schowen, R. L.; Wang, M.-S.; Lupes, M. E. *JACS* **1971**, *93*, 1214.
72. (a) March, J. *Advanced Organic Chemistry*, 3rd edn.; McGraw-Hill: New York, 1985; pp. 859–860. (b) Ide, W. S.; Buck, J. S. *OR* **1948**, *4*, 269.
73. (a) Stetter, H.; Schreckenberg, M. *CB* **1974**, *107*, 210. (b) Stetter, H. *AG(E)* **1976**, *15*, 639. (c) Stetter, H.; Kuhlmann, H.; Lorenz, G. *OSC* **1988**, *8*, 866.
74. Kratzer, R. H.; Paciorek, K. L.; Karle, D. W. *JOC* **1976**, *41*, 2230.
75. Mori, K.; Tominga, M.; Matsui, M.; Takigawa, T. *S* **1973**, 790.
76. Nishizawa, M.; Adachi, K.; Hayashi, Y. *CC* **1984**, 1637.
77. Georgiadis, M. P.; Apostolopoulos, C. D.; Haroutounian, S. A. *JHC* **1991**, *28*, 599.
78. (a) DeKimpe, N.; Moens, L.; Verhe, R.; DeBuyck, L.; Schamp, N. *CC* **1982**, 19. (b) DeKimpe, N.; DeCock, W.; Stevens, C. *T* **1992**, *48*, 2739.
79. Takahashi, K.; Nishizuka, T.; Iida, H. *TL* **1981**, *22*, 2389.
80. Cawkill, E.; Clark, N. G. *JCS(P1)* **1980**, 244.
81. Garst, M. E.; Bonfiglio, J. N. *TL* **1981**, *22*, 2075.
82. Mitchell, D.; Koenig, T. M. *TL* **1992**, *33*, 3281.
83. Gilbert, J. C.; Luo, T.; Davis, R. E. *TL* **1975**, 2545.
84. (a) Barton, D. H. R.; Magnus, P. D.; Pearson, M. J. *JCS(C)* **1971**, 2231. (b) Adams, R.; Brower, K. R. *JACS* **1956**, *78*, 4770. (c) Adams, R.; Dunbar, J. E. *JACS* **1956**, *78*, 4774.
85. Stahl, M. A.; Kenesky, B. F.; Berbee, R. P. M.; Richards, M.; Heine, H. W. *JOC* **1980**, *45*, 1197.
86. Petersen, W. C.; Letsinger, R. L. *TL* **1971**, 2197.

Serkos A. Haroutounian
Agricultural University of Athens, Greece

Propargyl Chloride

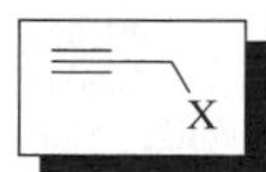

(X = Cl)		
[624-65-7]	C_3H_3Cl	(MW 74.51)
(X = Br)		
[106-96-7]	C_3H_3Br	(MW 118.96)
(X = OTf)		
[41029-46-3]	$C_4H_3F_3O_3S$	(MW 188.14)

(three-carbon alkylating agent;[1] acetone carbenium ion equivalent and annulating agent;[2] propargylic or allenic nucleophile after conversion to an organometallic[3])

Alternate Name: 3-chloro-1-propyne.

Physical Data: X = Cl: bp 58 °C; *d* 1.030 g cm^{-3}. X = Br: bp 88–90 °C; *d* 1.579 g cm^{-3}; *d* (80% toluene solution) 1.335 g cm^{-3}.

Form Supplied in: propargyl chloride is supplied as a liquid in 98% purity. Because of a reported shock-sensitivity of the neat material, propargyl bromide is supplied as an 80% (by weight) solution in toluene.

Preparative Method: propargyl triflate is not commercially available but is prepared by treating ***Propargyl Alcohol*** with ***Pyridine/Trifluoromethanesulfonic Anhydride***.[4]

Handling, Storage, and Precautions: in the presence of copper(I) halides, propargyl halides can isomerize to a mixture of the parent compound and a halopropadiene.[5] Because of an instability at ambient temperature, the triflate is typically prepared in solution and used directly. Purification of the triflate by distillation is possible, albeit in low yield.[6]

Electrophilic Alkylation. As an electrophile, propargyl chloride has been employed in the propynylation of a variety of substrates, including Grignard reagents,[1,7] carbanions,[8] enolates,[9] carbonyl compounds,[10] alcohols,[4,11] phenols,[12] carboxylates,[13] amines,[14] amides,[15] sulfonamides,[16] sulfones,[17] thiols,[18] thioamides,[19] and alkyl phosphonates.[20] An unusual *N*-alkylation of a phenothiazine has been reported whereby the 1-propynyl derivative is isolated.[21] Because of the basic nature of the above alkylation reactions, isomerization to the allene can take place.[22] The extent to which isomerization occurs is dependent upon the reaction time, temperature, and the specific base employed.[15a,23] Trialkyl borates[24] and alkylaluminum species,[25] in the presence of a copper(I) halide, couple with propargyl bromide (preferred over the chloride) to give exclusively the terminal allene. By contrast, the reaction of bromopropadiene (derived from propargyl bromide) with an alkylaluminum affords solely the terminal alkyne.[26] Alkylation of a trithiocarbonate, followed by cyclization, leads to the formation of a thione.[27] In a similar manner, thioamides can be alkylated and further elaborated to thiazoles[28] or thiophenes.[29]

Annulation Reactions. When the alkylation with propargyl bromide is followed by hydration with ***Mercury(II) Oxide*** the net result is equivalent to an alkylation with ***Aceton***. More often, when alkylation occurs α to a carbonyl, the product is further treated with base to give a fused cyclopentenone (eq 1) (see also ***Isopropenyl Acetate***, ***Methallyl Chloride***, ***2-Methoxyallyl Bromide***).[2,30] A radical-based annulation is also possible. By generating a radical center δ to the triple bond, cyclization to a five-membered ring system with an exocyclic methylene occurs.[16,31]

CO₂Et — 1. HC≡CCH₂Br, NaOEt; 2. BF₃•Et₂O, HgO, H⁺ (62%) — CO₂Et — KOH (73%) — (1)

Sigmatropic Rearrangements. Because of the triple bond, alkylation with propargyl chloride can provide intermediates further capable of intramolecular rearrangement. Crotyl propargyl ethers undergo [2,3]-Wittig rearrangement with high *erythro* or *threo* selectivity, dependent upon the starting alkene geometry.[32] The corresponding chiral analog allows for an asymmetric [2,3]-Wittig rearrangement with a high degree of enantioselectivity.[33] Under thermal conditions, propargylated phenols undergo a [3,3]-sigmatropic rearrangement to give an allenyl enol. A [1,5]-hydrogen shift, followed by electrocyclic ring closure results in the formation of a benzopyran (eq 2).[34,35]

OH; $HC{\equiv}CCH_2Br$, K_2CO_3, acetone, 55%; PhCl, 60%

(2)

Cycloadditions. When appropriately appended, the propargyl group can serve as a dienophile for the intramolecular Diels–Alder reaction with ***Furan***,[36] styrene,[37] and triazine[38] derivatives. Under the typically basic reaction conditions, cycloaddition is believed to proceed via the allenyl ether intermediates. Propargyl bromide can also function as a 1,3-dipolarophile in the [3 + 2] cycloaddition reaction with nitrile oxides to form isoxazoles.[39]

Nucleophilic Additions. As a halide, propargyl bromide can be converted to a variety of organometallic species, which then add to aldehydes and ketones. The Grignard reagent[40] is proposed to exist in the allenic form,[41] and reacts to form predominantly homopropargylic alcohols. Whereas the organozinc reagent reacts to give both α-allenic and β-alkynic alcohols,[42] the organoaluminum reagent can be used to generate products which are free of allenic isomers.[43] The reaction between carbonyl compounds and the tin reagent favors α-allenic alcohol formation.[44] However, the mixed metal system of tin and aluminum allows for selective formation of the homopropargylic alcohol.[45] A chiral tin complex can be formed with propargyl bromide and (+)-diethyl tartrate and made to react with aldehydes to give enantiomerically enriched α-allenic alcohols.[46] Other organometallic species prepared from propargyl halides include those of chromium,[3] lead,[47] and mercury.[48]

MgBr; 1. $B(OMe)_3$, 2. H_3O^+; $B(OH)_2$; 1. D(–)-DET, 2. RCHO, 74%; 1. L(+)-DET, 2. RCHO, 56%; OH, 87% ee; OH, 92% ee

(3)

The Grignard reagent reacts with to give, after hydrolysis, allenylboronic acid (eq 3). Further reaction with diethyl tartrate (DET) generates a chiral allenylboronic ester which enantioselectively adds to aldehydes to give chiral homopropargylic alcohols.[49]

Because of the acidic C–H bond, deprotonation generates the propargylic anion which may also serve as a nucleophile. In addition to simple acylation,[50] lithium chloropropargylide adds to trialkylboranes to give a lithium trialkylalkynylborate. Dependent upon the subsequent reaction conditions, substituted allenes,[51] alkynes,[52] or homopropargylic and α-allenic alcohols[53] may be formed.

Related Reagents. Allenylboronic Acid; Allenyllithium; Bromoacetone; 3-Chloro-1-propynyllithium; 3-Iodo-1-trimethylsilylpropyne; 2-Methoxyallyl Bromide; Propargyl Alcohol; Propargylmagnesium Bromide; (Trimethylsilyl)allene.

---

1. Taniguchi, H.; Mathai, I. M.; Miller, S. I. *T* **1966**, *22*, 867.
2. Islam, A. M.; Raphael, R. A. *JCS* **1952**, 4086.
3. Place, P.; Verniere, C.; Gore, J. *T* **1981**, *37*, 1359.
4. Beard, C. D.; Baum, K.; Grakauskas, V. *JOC* **1973**, *38*, 3673.
5. Jacobs, T. L.; Brill, W. F. *JACS* **1953**, *75*, 1314.
6. Vedejs, E.; Engler, D. A.; Mullins, M. J. *JOC* **1977**, *42*, 3109.
7. Ward, J. P.; van Dorp, D. A. *RTC* **1966**, *85*, 117.
8. (a) Corey, E. J.; Boger, D. L. *TL* **1978**, 13. (b) Barluenga, J.; Tomas, M.; Suarez-Sabrino, A. *SL* **1990**, 351. (c) Bartoli, G.; Bosco, M.; Cimarelli, C.; Dalpozzo, R.; Palmieri, G. *SL* **1991**, 229.
9. (a) Herrmann, J. L.; Kieczykowski, G. R.; Schelessinger, R. H. *TL* **1973**, 2433. (b) Crimmins, M. T.; Mascarella, S. W.; Deloach, J. A. *JOC* **1984**, *49*, 3033. (c) Ikegami, S.; Uchiyama, H.; Hayama, T.; Yamaguchi, M.; Katsuki, T. *T* **1988**, *44*, 5333. (d) Oppolzer, W.; Moretti, R.; Thomi, S. *TL* **1989**, *30*, 5603. (e) Pettig, D.; Horwell, D. C. *S* **1990**, 465. (f) Negrete, G. R.; Konopelski, J. P. *TA* **1991**, *2*, 105.
10. (a) Queignec, R.; Lambert, F.; Aboutaj, M.; Kirschleger, B. *SC* **1988**, *18*, 1213. (b) Sambasivarao, K.; Kubiak, G.; Lannoye, G.; Cook, J. M. *JOC* **1988**, *53*, 5173. (c) Valli, V. L. K.; Sarma, G. V. M.; Choudary, B. M. *IJC(B)* **1990**, *29*, 481.
11. (a) Forsyth, C. J.; Clardy, J. *JACS* **1988**, *110*, 5911. (b) Kanematsu, K.; Nagashima, S. *CC* **1989**, 1028. (c) Neeson, S. J.; Stevenson, P. J. *T* **1989**, *45*, 6239. (d) Sharma, G. V. M.; Vepachedu, S. R. *T* **1991**, *47*, 519.
12. (a) Sebok, P.; Timar, T.; Jaszberenyi, J. C.; Batta, G. *H* **1988**, *27*, 2595. (b) Sumathi, T.; Balasubramanian, K. K. *TL* **1990**, *31*, 3775.
13. Dekeyser, J. L.; DeCock, C. J. C.; Poupaert, J. H.; Dumont, P. *JOC* **1988**, *53*, 4859.
14. (a) Hillard, R. L.; Parnell, C. A.; Vollhardt, K. P. C. *T* **1983**, *39*, 905. (b) Corriu, R. J. P.; Huynh, V.; Moreau, J. J. E. *TL* **1984**, *25*, 1887. (c) Filali, A.; Yaouanc, J. J.; Handel, H. *AG(E)* **1991**, *30*, 560.
15. (a) Amstutz, R.; Ringdahl, B.; Karlen, B.; Roch, M.; Jenden, D. J. *JMC* **1985**, *28*, 1760. (b) Bram, G.; Galons, H.; Labidalle, S.; Loupy, A.; Miocque, M.; Petit, A.; Pigeon, P.; Sansoulet, J. *BSF(2)* **1989**, 247.
16. Boger, D. L.; Coleman, R. S. *JACS* **1988**, *110*, 4796.
17. (a) Regis, R. R.; Doweyko, A. M. *TL* **1982**, *23*, 2539. (b) Zhang, Z.; Liu, G. J.; Wang, Y. L.; Wang, Y. *SC* **1989**, *19*, 1167.
18. (a) Arai, Y.; Takadoi, M.; Kontani, T.; Shiro, M.; Koizumi, T. *CL* **1990**, 1581. (b) Petrosyan, V. A.; Niyazymbetov, M. E.; Konyushkin, L. D.; Litvinov, V. P. *S* **1990**, 841.
19. Majumdar, K. C.; Chattopadhyay, S. K.; Gupta, A. K. *IJC(B)* **1990**, *29*, 1138.
20. Genet, J. P.; Uziel, J.; Touzin, A. M.; Juge, S. *S* **1990**, 41.
21. Zaugg, H. E.; Swett, L. R.; Stone, G. R. *JOC* **1958**, *23*, 1389.
22. Ringdahl, B.; Muhi-Elden, Z.; Ljunggren, C.; Karlen, B.; Resul, B.; Dahlbom, R.; Jenden, D. J. *Acta Pharm. Suec.* **1979**, *16*, 89.

23. Diez-Barra, E.; de la Hoz, A.; Sanchez-Migallon, A.; Tejeda, J. *SC* **1990**, *20*, 2849.
24. Miyaura, N.; Itoh, M.; Suzuki, A. *BCJ* **1977**, *50*, 2199.
25. Sato, F.; Oguro, K.; Sato, M. *CL* **1978**, 805.
26. Sato, F.; Kodama, H.; Sato, M. *CL* **1978**, 789.
27. (a) Haley, N. F. *TL* **1978**, 5161. (b) Haley, N. F.; Fichtner, M. W. *JOC* **1980**, *45*, 175.
28. Bhattacharjee, S. S.; Asokan, C. V.; Ila, H.; Junjappa, H. *S* **1982**, 1062.
29. Bhattacharjee, S. S.; Ila, H.; Junjappa, H. *S* **1983**, 410.
30. (a) Llyod, D.; Rowe, F. *JCS* **1953**, 3718. (b) Kloster-Jensen, E.; Kovats, E.; Eschenmoser, A.; Heilbronner, E. *HCA* **1956**, *39*, 1051.
31. (a) Boger, D. L.; Ishizaki, T.; Wysocki, R. J.; Munk, S. A.; Kitos, P. A.; Suntornwat, O. *JACS* **1989**, *111*, 6461. (b) Pak, H.; Dickson, J. K.; Fraser-Reid, B. *JOC* **1989**, *54*, 5357. (c) Pak, H.; Canalda, I. I.; Fraser-Reid, B. *JOC* **1990**, *55*, 3009. (d) Sharma, G. V. M.; Vepachedu, S. R. *TL* **1990**, *31*, 4931. (e) Knapp, S.; Gibson, F. S.; Choe, Y. H. *TL* **1990**, *31*, 5397.
32. Mikami, K.; Azuma, K.-I.; Nakai, T. *T* **1984**, *40*, 2303.
33. Sayo, N.; Azuma, K.; Mikami, K.; Nakai, T. *TL* **1984**, *25*, 565.
34. (a) Rodighiero, P.; Manzini, P.; Pastorini, G.; Bordin, F. *H* **1987**, *24*, 485. (b) Majumdar, K. C.; Khan, A. T.; De, R. N. *SC* **1988**, *18*, 1589.
35. See also: Zsindely, J.; Schmid, H. *HCA* **1968**, *51*, 1510.
36. (a) Hayakawa, K.; Yodo, M.; Ohsuki, S.; Kanematsu, K. *JACS* **1984**, *106*, 6735. (b) Yamaguchi, Y.; Tatsuta, N.; Soejima, S.; Hayakawa, K.; Kanematsu, K. *H* **1990**, *30*, 223.
37. Kanematsu, K.; Tsuruoka, M.; Takaoka, Y.; Sasaki, T. *H* **1991**, *32*, 859.
38. Sagi, M.; Sato, O.; Konno, S.; Yamanaka, H. *H* **1989**, *29*, 2253.
39. Chiarino, D.; Sala, A.; Napoletano, M. *SC* **1988**, *18*, 1171.
40. Sondheimer, F.; Amiel, Y.; Gaoni, Y. *JACS* **1962**, *84*, 270.
41. Bogdanovic, B.; Janke, N.; Kinzelmann, H. G. *CB* **1990**, *123*, 1507.
42. (a) Henbest, H. B.; Jones, E. R. H.; Walls, I. M. S. *JCS* **1949**, 2696. (b) Golse, R.; Gavarret, J.; Demange, M. G. *BSF(2)* **1950**, 285. (c) Brown, J. B.; Henbest, H. B.; Jones, E. R. H. *JCS* **1950**, 3634. (d) Karrer, P.; Eugster, C. H. *HCA* **1951**, *34*, 28. (e) Gaudemar, M. *BSF(2)* **1962**, 974. (f) Friedrich, L. E.; de Vera, N.; Hamilton, M. *SC* **1980**, *10*, 637. (g) Fuganti, C.; Servi, S.; Zirotti, C. *TL* **1983**, *24*, 5285. (h) Papadopoulou, M. V. *CB* **1989**, *122*, 2017.
43. (a) Lauger, P.; Prost, M.; Charlier, R. *HCA* **1959**, *42*, 2379. (b) Schneider, D. F.; Weedon, B. C. L. *JCS(C)* **1967**, 1686.
44. (a) Mukaiyama, T.; Harada, T. *CL* **1981**, 621. (b) Iyoda, M.; Kanao, Y.; Nishizaki, M.; Oda, M. *BCJ* **1989**, *62*, 3380. (c) Wu, S.; Huang, B.; Gao, X. *SC* **1990**, *20*, 1279.
45. Nokami, J.; Tamaoka, T.; Koguchi, T.; Okawara, R. *CL* **1984**, 1939.
46. Boldrini, G. P.; Tagliavini, E.; Trombini, C.; Umani-Ronchi, A. *CC* **1986**, 685.
47. Tanaka, H.; Hamatani, T.; Yamashita, S.; Torii, S. *CL* **1986**, 1461.
48. Larock, R. C.; Chow, M.-S. *TL* **1984**, *25*, 2727.
49. Haruta, R.; Ishiguro, M.; Ikeda, N.; Yamamoto, H. *JACS* **1982**, *104*, 7667.
50. Olomucki, M.; Le Gall, J.-Y.; Barrand, I. *CC* **1982**, *22*, 1290.
51. Leung, T.; Zweifel, G. *JACS* **1974**, *96*, 5620.
52. Hara, S.; Satoh, Y.; Suzuki, A. *CL* **1982**, 1289.
53. Zweifel, G.; Backlund, S. J.; Leung, T. *JACS* **1978**, *100*, 5561.

Mark A. Krook
The Upjohn Company, Kalamazoo, MI, USA

Tetrakis(triphenylphosphine)nickel(0)[1]

$Ni(PPh_3)_4$

[15133-82-1] $C_{72}H_{60}P_4Ni$ (MW 1107.89)

(a source of nickel(0) useful for coupling reactions of organic halides,[2] and the cyclooligomerization of cumulenes[3])

Physical Data: mp 123–128 °C (N_2).

Solubility: sol DMF, DMA, THF, acetonitrile, benzene; slightly sol Et_2O; very slightly sol *n*-heptane, EtOH.

Form Supplied in: widely available as a red powder of greater than 98% purity.

Preparative Methods: the standard preparation involves the reduction of ***Nickel(II) Acetylacetonate*** with ***Triethylaluminum*** in the presence of ***Triphenylphosphine***. To 21.3 g of anhydrous $Ni(acac)_2$ and 125 g PPh_3 in 800 mL of Et_2O under N_2 at 0 °C is slowly added 28.0 g Et_3Al. The reddish-brown precipitate is collected, washed with Et_2O, and twice dissolved in benzene and reprecipitated by the addition of *n*-heptane to give about 50 g (55%) of $Ni(PPh_3)_4$.[4]

Handling, Storage, and Precautions: highly oxygen sensitive. Special inert-atmosphere techniques must be used.[5] Should be stored at 0 °C. Cancer suspect agent.

Coupling Reactions of Organic Halides. The zerovalent nickel complex $Ni(PPh_3)_4$ reacts with organic halides by oxidative addition into the carbon–halogen bonds to give organonickel(II) intermediates.[6] These intermediates, which are typically not isolated, can react with a variety of nucleophilic reagents to replace the original halide with another group.[2] With a catalytic amount of $Ni(PPh_3)_4$, aryl halides react with halide salts to give halogen exchange,[7] and with MgH_2 to give hydrogenation.[8] Aryl halides and triflates react with cyanide salts to give nitriles.[9] Aryl bromides and iodides both work well as substrates, and many functional groups are tolerated. However, *o*-substituents tend to slow the reactions down and give lower yields. Oxidative addition to benzylic halides occurs with racemization.[6a] Insertion of an alkene into the organonickel intermediate leads to overall coupling via a β-hydride elimination pathway.[10] This type of reaction has been utilized intramolecularly in the synthesis of indole and oxindole derivatives (eq 1).[11] The organonickel intermediate derived from MeI and $Ni(PPh_3)_4$ can be used for the regioselective alkylation of epoxides.[12]

Homocoupling of aryl and alkenyl halides can occur with stoichiometric amounts of nickel(0) (see ***Bis(1,5-cyclooctadiene)-nickel(0)***),[13] or catalytically in the presence of an added reducing agent such as zinc (see ***Tris(triphenylphosphine)-nickel(0)***).[14] $Ni(PPh_3)_4$ has been used for the homocoupling of 2-halopyridines,[15] and α-halo ketones.[16] Cross coupling of distinct aryl halides by this method is generally not efficient due to extensive symmetrical coupling. Intramolecular couplings of bis(aryl halides), however, do not suffer from this limitation and give cyclized products.[17] These couplings work well for many different tether lengths (eq 2) and this strategy was utilized in the synthesis of the macrocyclic ketone alnusone.[18] The mild conditions of this reaction permit a wide variety of functional groups to be present without concern for the formation of byproducts or decomposition. This is in contrast to the copper catalyzed Ullmann reaction, which often requires harsh thermal conditions (>200 °C).[19]

1. $Ni(PPh_3)_4$, 2. NH_4Cl (aq); products 73% + 26% (1)

$Ni(PPh_3)_4$, DMF, 55 °C (2)

n = 2, 81%; 3, 83%; 4, 76%; 5, 85%; 6, 38%

Coupling Reactions with Organometallic Compounds. The cross-coupling of aryl and alkenyl halides may be effected by the reaction of organometallic reagents with organic halides in the presence of $Ni(PPh_3)_4$. As before, oxidative addition of Ni^0 to the organic halide is the initial step in the mechanism. The transiently formed organonickel intermediate can then accept an organic group from the added metal reagent through ligand substitution, which leads, upon reductive elimination, to replacement of the original halide with this new group and regeneration of a catalytic nickel(0) species. While initial studies in this area were confined to highly reactive organomagnesium and -lithium reagents,[20] which limited their applicability in synthesis, more recent studies have shown that organic complexes of aluminum, zinc, and zirconium, among others, are all active for the coupling reaction.[21]

Recent uses of Grignard reagents in the presence of $Ni(PPh_3)_4$ include the reaction with ***Trichloroethylene*** to give 1,1-dichloroalkenes,[22] with 1,2-dichloroethylenes to produce vinyl chlorides stereoselectively,[23] and with phenolic ethers to give arenes,[24] and the reaction of $TMSCH_2MgCl$ with vinyl iodides to give allyl silanes.[25] In addition, lithium enolates have been coupled with aryl halides inter- and intramolecularly.[26]

Alkenylalanes readily couple with aryl and alkenyl halides in the presence of $Ni(PPh_3)_4$ to give styrene and butadiene derivatives in good yields (eq 3).[21] *Trans*-alkenylalanes, which are readily obtainable by the carboalumination of alkynes, couple with retention of configuration at the double bond to provide an efficient route to *trans*-alkenes.[27] (*E*)-1,2-Dichloroethene was coupled with an alkenylalane to give a 1-chloro-(*E,E*)-1,3-diene in 80% yield.[28]

Alkenylzirconium reagents are similar to alkenylalanes in that they can be easily prepared by carbometalation of alkynes. They may be coupled with aryl, alkenyl, and alkynyl halides (eq 3).[21] The coupling is facilitated by the presence of added salts containing Zn or Cd such as ***Zinc Chloride***.[29]

C_5H_{11}–CH=CH–M → (Ni(PPh$_3$)$_4$, THF, rt, PhI) → C_5H_{11}–CH=CH–Ph (3)

M = Al(i-Bu)$_2$, 91%; $ZrCp_2Cl$, 96%

Arylzinc compounds also react with aryl halides in good yields with very little homocoupling.[30] This mild coupling method was used with great success in the synthesis of a bisbenzocyclooctadiene lignan, steganone (eq 4).[31] Alkylzincs formed from the Reformatsky reagent also react with aryl halides to give arylacetic acid esters.[32] Benzylzinc compounds react with alkenyl halides in a similar fashion; however, palladium catalysts were found to be more suitable for this reaction due to extensive isomerization to the conjugated isomers in the presence of nickel.[33] Knochel's (dialkoxyboryl)methylzinc reagents have also been coupled with greater success using palladium catalysts.[34]

1. Ni(PPh$_3$)$_4$, THF, –20 °C; 2. dil HCl, reflux, CH_2Cl_2, 80% (4)

In general, the coupling reactions of aryl halides may be catalyzed with similar efficiency by palladium catalysts such as ***Tetrakis(triphenylphosphine)palladium(0)*** or $Cl_2Pd(PPh_3)_2$–***Diisobutylaluminum Hydride***. The nickel catalysts tend to be slightly more reactive, however, easily entering into reactions with both organic iodides and bromides, while the palladium catalysts sometimes require activated bromides to react.[30a] In the case of alkenyl halides, the nickel catalysts have shown slight ($\leq$10%) stereochemical scrambling when coupled with alkenylalanes, whereas with the palladium catalysts the stereochemical integrity is $\geq$97%.[27b] Palladium catalysts also have the added advantage that they are compatible with the nitro group, which nullifies the catalytic activity of the nickel(0) complexes,[30a] and coupling with boranes is possible.[35]

Cyclooligomerization of Cumulenes. Ni(PPh$_3$)$_4$ reacts catalytically with cumulenes via a [2 + 2] cycloaddition to give [4]radialenes[36] and in a [2 + 2 + 2] fashion to give [6]radialenes. The choice of solvent affects the selectivity of some reactions, with benzene providing dimers and DMF leading primarily to the trimeric products.[37] Several extremely strained cyclobutanes have been prepared by this method.[38] Interestingly, the cumulene starting materials (**3**) can be prepared by the Ni^0 mediated coupling of 1,1-dihaloalkenes (**1**) and the subsequent Ni^0 catalyzed elimination of the 2,3-dihalo-1,3-butadiene products (**2**). Thus [4]radialenes (**4**) may be prepared in one step from 1,1-dihaloalkenes (**1**) (eq 5).[39]

(**1**) → (Ni(PPh$_3$)$_4$, Et_4Ni, toluene, –78 to 50 °C) → [(**2**)] → [(**3**)] → 17% (**4**) (5)

Other Uses. Terminal alkenes may be prepared by the oxidative addition of primary halides to Ni(PPh$_3$)$_4$ followed by β-hydride elimination.[40] Cyclopropanation of electron-deficient alkenes is possible by the reaction of the complex with *gem*-dibromides or diazoalkanes.[41] Methylenecyclopropane is linearly trimerized selectively in the presence of this catalyst.[42] Tethered diynes react inter- and intramolecularly with alkynes in [2 + 2 + 2] cycloadditions to give tetralins (eq 6).[43]

Ni(PPh$_3$)$_4$, THF, rt (6)

n = 1, 70%; 2, 46%

Related Reagents. Bis(1,5-cyclooctadiene)nickel(0); Bis(triphenylphosphine)nickel(0); Nickel(II) Acetylacetonate; Tetrakis(triphenylphosphine)palladium(0); Tris(triphenylphosphine)nickel(0).

1. Jolly, P. W.; Wilke, G. *The Organic Chemistry of Nickel*; Academic: New York, 1974; Vols. 1 and 2.
2. Jolly, P. W. In *Comprehensive Organometallic Chemistry*; Wilkinson, G.; Stone, F. G. A.; Abel, E. W., Eds.; Pergamon: New York 1982; Chapter 56.5, p 713.
3. Iyoda, M.; Kuwatani, Y.; Oda, M. *JACS* **1989**, *111*, 3761.
4. Schunn, R. A. *Inorg. Synth.* **1972**, *13*, 124.
5. Shriver, D. F. *The Manipulation of Air-Sensitive Compounds*, McGraw-Hill: New York, 1969.
6. (a) Stille, J. K.; Cowell, A. B. *JOM* **1977**, *124*, 253. (b) Tsou, T. T.; Kochi, J. K. *JACS* **1979**, *101*, 6319.
7. Tsou, T. T.; Kochi, J. K. *JOC* **1980**, *45*, 1930.
8. Carfagna, C.; Musco, A.; Pontellini, R. *J. Mol. Catal.* **1989**, *54*, L23.
9. (a) Cassar, L. *JOM* **1973**, *54*, C57. (b) Chambers, M. R. I.; Widdowson, D. A. *JCS(P1)* **1989**, 1365.
10. Kron, T. E.; Lopatina, V. S.; Morozova, L. N.; Lebedev, S. A.; Isaeva, L. S.; Kravtsov, D. N.; Petrov, É. S. *BAU* **1989**, 703.
11. (a) Mori, M.; Ban, Y. *TL* **1976**, 1803. (b) Mori, M.; Ban, Y. *TL* **1976**, 1807. (c) Canoira, L.; Rodriguez, J. G. *JHC* **1985**, *22*, 1511. (d) Canoira, L.; Rodriguez, J. G. *JCR(S)* **1988**, 68.
12. Hase, T.; Miyashita, A.; Nohira, H. *CL* **1988**, 219.
13. Semmelhack, M. F.; Helquist, P.; Jones, J. D.; Keller, L.; Mendelson, L.; Ryono, L. S.; Smith, J. G.; Stauffer, R. D. *JACS* **1981**, *103*, 6460.

14. Zembayashi, M.; Tamao, K.; Yoshida, J-I.; Kumada, M. *TL* **1977**, 4089.
15. Tiecco, M.; Testaferri, L.; Tingoli, M.; Chianelli, D; Montanucci, M. *S* **1984**, *736*.
16. Iyoda, M.; Sakaitani, M.; Kojima, A.; Oda, M. *TL* **1985**, *26*, 3719.
17. (a) Whiting, D. A.; Wood, A. F. *JCS(P1)* **1980**, *623*. (b) Colquhoun, H. M.; Dudman, C. C.; Thomas, M.; O'Mahoney, C. A.; Williams, D. J. *JCS(C)* **1990**, *336*.
18. Semmelhack, M. F.; Ryono, L. S. *JACS* **1975**, *97*, 3873. and reference 13.
19. (a) Ullmann, F.; Bielecki, J. *CB* **1901**, *34*, 2147. (b) Normant, J. F. *S* **1972**, 63. (c) Fanta, P. E. *S* **1974**, 9. (d) Sainsbury, M. *T* **1980**, *36*, 3327.
20. (a) Corriu, R. J. P.; Masse, J. P. *JCS(C)* **1972**, *144*. (b) Tamao, K.; Sumitani, K.; Kumada, M. *JACS* **1972**, *94*, 4374. (c) Reference 2 and references therein.
21. Negishi, E-i.; Takahashi, T.; Baba, S.; Van Horn, D. E.; Okukado, N. *JACS* **1987**, *109*, 2393.
22. Ratovelomanana, V.; Linstrumelle, G.; Normant, J-F. *TL* **1985**, *26*, 2575.
23. Ratovelomanana, V.; Linstrumelle, G. *SC* **1984**, *14*, 179.
24. Johnstone, R. A. W.; McLean, W. N. *TL* **1988**, *29*, 5553.
25. Negishi, E-i.; Luo, F-T.; Rand, C. L. *TL* **1982**, *23*, 27.
26. Semmelhack, M. F.; Stauffer, R. D.; Rogerson, T. D. *TL* **1973**, 4519.
27. (a) Negishi, E-i.; Baba, S. *JCS(C)* **1976**, *596*. (b) Baba, S.; Negishi, E-i. *JACS* **1976**, *98*, 6729.
28. Ratovelomanana, V.; Linstrumelle, G. *TL* **1984**, *25*, 6001.
29. (a) Okukado, N.; Van Horn, D. E.; Klima, W. L.; Negishi, E-i. *TL* **1978**, 1027. (b) Negishi, E-i.; Okukado, N.; King, A. O.; Van Horn, D. E.; Spiegel, B. I. *JACS* **1978**, *100*, 2254.
30. (a) Negishi, E-i.; King, A. O.; Okukado, N. *JOC* **1977**, *42*, 1821. (b) Negishi, E-i.; Takahashi, T.; King, A. O. *OS* **1988**, *66*, 67.
31. (a) Larson, E. R.; Raphael, R. A. *TL* **1979**, 5041. (b) Ziegler, F. E.; Schwartz, J. A. *JOC* **1978**, *43*, 985.
32. (a) Fauvarque, J. F.; Jutand, A. *JOM* **1977**, *132*, C17. (b) Fauvarque, J. F.; Jutand, A. *JOM* **1979**, *177*, 273.
33. Negishi, E-i.; Matsushita, H.; Okukado, N. *TL* **1981**, *22*, 2715.
34. Watanabe, T.; Miyaura, N.; Suzuki, A. *JOM* **1993**, *444*, C1.
35. Miyaura, N.; Yamada, K.; Suginome, H.; Suzuki, A. *JACS* **1985**, *107*, 972.
36. (a) Hagelee, L.; West, R.; Calabrese, J.; Norman, J. *JACS* **1979**, *101*, 4888. (b) Iyoda, M.; Kuwatani, Y.; Oda, M. *JACS* **1989**, *111*, 3761.
37. Iyoda, M.; Tanaka, S.; Nose, M.; Oda, M. *JCS(C)* **1983**, 1058.
38. (a) Pasto, D. J.; Mitra, D. K. *JOC* **1982**, *47*, 1382. (b) Hashmi, S.; Polborn, K.; Szeimies, G. *CB* **1989**, 2399.
39. (a) Iyoda, M.; Sakaitani, M.; Miyazaki, T.; Oda, M. *CL* **1984**, *2005*. (b) Iyoda, M.; Tanaka, S.; Otani, H.; Nose, M.; Oda, M. *JACS* **1988**, *110*, 8494.
40. Henningsen, M. C.; Jeropoulos, S.; Smith, E. H. *JOC* **1989**, *54*, 3015.
41. (a) Kanai, H.; Hiraki, N. *CL* **1979**, 761. (b) Kanai, H.; Hiraki, N.; Iida, S. *BCJ* **1983**, *56*, 1025. (c) Nakamura, A.; Yoshida, T.; Cowie, M.; Otsuka, S.; Ibers, J. A. *JACS* **1977**, *99*, 2108.
42. Binger, P.; Brinkmann, A.; McMeeking, J. *LA* **1977**, 1065.
43. Bhatarah, P.; Smith, E. H. *JCS(P1)* **1992**, 2163.

Paul A. Wender & Thomas E. Smith
Stanford University, CA, USA

Tetrakis(triphenylphosphine)palladium(0)[1]

[14221-01-3] $C_{72}H_{60}P_4Pd$ (MW 1155.62)

(catalyzes carbon–carbon bond formation of organometallics with a wide variety of electrophiles;[2] in combination with other reagents, catalyzes the reduction of a variety of functional groups;[3] catalyzes carbon–metal (Sn, Si) bond formation;[4] catalyzes deprotection of the allyloxycarbonyl group[5])

Physical Data: mp has been reported to vary between 100–116 °C (dec) and is not a good indication of purity.
Solubility: insol saturated hydrocarbons; moderately sol many other organic solvents including $CHCl_3$, DME, THF, DMF, PhMe, benzene.
Form Supplied in: yellow, crystalline solid from various sources. The quality of batches from the same source have been noted to be highly variable and can dramatically alter the expected reactivity.
Preparative Methods: readily prepared by the reduction of $PdCl_2(Ph_3P)_2$[6] with ***Hydrazine*** or by the reaction of ***Tris(dibenzylideneacetone)dipalladium*** with ***Triphenylphosphine***.[7]
Handling, Storage, and Precautions: is air and light sensitive and should be stored in an inert atmosphere in the absence of light. It can be handled for short periods quickly in the air but best results are achieved by handling in a glove box or glove bag under argon or nitrogen.

Direct Carbon–Carbon Bond Formation. One of the most attractive features of $Pd(Ph_3P)_4$ is its ability to catalyze carbon–carbon bond formation under mild conditions by the cross-coupling of organometallic (typically organoaluminum,[2e] -boron,[2f] -copper, -magnesium, -tin,[2b–d] or -zinc[2e] reagents) and unsaturated electrophilic partners (halides or sulfonates such as trifluoromethanesulfonates[2d] (triflates)). Although $Pd(Ph_3P)_4$ is the catalyst of choice in many of these reactions, numerous other Pd^0 and Pd^{II} catalysts have been used successfully.

Symmetrical or unsymmetrical biaryls are efficiently produced by the $Pd(Ph_3P)_4$ catalyzed cross-coupling of aryl halides or aryl triflates (I, Br > OTf in terms of rate of reaction[8a]) with a variety of metalated aromatics such as arylboronic acids,[8] arylstannanes,[9] aryl Grignards[10] and arylzincs[11] (eqs 1 and 2). Reactions employing $ArB(OH)_2$ are carried out in aqueous base (2M Na_2CO_3 or K_3PO_4) while the remainder are conducted under anhydrous conditions. A recent report documents the reaction of $ArB(OR)_2$ with ArBr under nonaqueous conditions in the presence of ***Thallium(I) Carbonate***.[8c] Reactions employing ArOTf and $ArSnR_3$ as coupling partners require greater than stoichiometric amounts of ***Lithium Chloride***.[9a,b] Acceleration of the reaction rate has been noted in the coupling of $ArSnMe_3$ and either ArOTf or ArBr/I by the addition of catalytic ***Copper(I) Bromide***[9c] or stoichiometric ***Silver(I) Oxide***,[9f] respectively. In many cases, one or both of the aromatic species can be heteroaromatic such as pyridine, furan, thiophene, quinoline, oxazole, thiazole, or indole.[8e,9f,g,11]

Symmetrical biaryls have been prepared in excellent yields by the $Pd(Ph_3P)_4$-catalyzed homocoupling of ArBr/I under phase transfer conditions.[12]

$$\text{PhX} + \text{M–Ph} \xrightarrow{Pd(Ph_3P)_4} \text{Ph–Ph} \quad (1)$$

M = MgBr, X = I 70%
M = ZnCl, X = OSO_2F 95%

$$2\text{-}(CONEt_2)C_6H_4B(OH)_2 + \text{Br–}(2\text{-methylphenyl}) \xrightarrow[\text{2. LDA, THF}]{\text{1. } Pd(Ph_3P)_4\text{, aq } Na_2CO_3} \text{9-phenanthrol} \quad (2)$$

92%

In a similar manner, $Pd(Ph_3P)_4$ catalyzes the cross-coupling of metalated alkenes, such as vinylaluminum reagents,[13] vinylboronates or -boronic acids,[8d,14] vinylstannanes,[15] vinylsilanes,[16] vinylzincs or -zirconates,[17] vinylcuprates[18] or -copper reagents[19] and vinyl Grignards[20] with vinyl halides (I, Br), triflates, or phosphates to form 1,3-dienes (eq 3). Vinylallenes,[17f] dienyl sulfides,[14h] and dienyl ethers[17d] have also been prepared using this strategy. In most instances the dienes are formed with retention of double bond geometry in both reacting partners.[14g] However, the formation of (*E,Z*)- and (*Z,Z*)-diene combinations has been documented to suffer from poor reaction yields or scrambling of alkene geometry in some cases (eq 4).[14a,b,18–20] A dramatic rate enhancement has been noted in the coupling of vinyl iodides with vinylboronic acids by replacing the aqueous bases that are normally used (NaOEt or NaOH) with ***Thallium(I) Hydroxide***[14f,i] or ***Thallium(I) Carbonate***.[21] LiCl[15a,c] is required when vinyl triflates are used in coupling reactions with vinylstannanes. Intramolecular versions of the vinylstannane–vinyl triflate coupling have been reported[22] and the necessity of adding LiCl in these reactions has been debated.[22a] 1,3-Dienes have been prepared via the $Pd(Ph_3P)_4$-catalyzed reaction between allylic alcohols and aldehydes in the presence of ***Phenyl Isocyanate*** and ***Tri-n-butylphosphine*** (eq 5).[23] (*E/E*):(*E/Z*) ratios range from 1:1 to 4:1.

$$R^1\text{CH=CHX} + \text{M–CH=CH}R^2 \xrightarrow{Pd(Ph_3P)_4} R^1\text{CH=CH–CH=CH}R^2 \quad (3)$$

R^1 = Hex, X = Br; M = B(DOB), R^2 = Bu 86%
R^1 = Bu, X = I; M = $ZrCp_2Cl$, R^2 = Hex 93%

$$(Z)\text{-}R^1\text{CH=CHX} + (Z)\text{-M–CH=CH}R^2 \xrightarrow{Pd(Ph_3P)_4} (Z,Z)\text{-}R^1\text{CH=CH–CH=CH}R^2 \quad (4)$$

R^1 = Bu, X = Br; M = $B(sia)_2$, R^2 = Hex 49%
R^1 = *t*-Bu, X = I; M = $Cu{\bullet}MgCl_2$, R^2 = Pent 53%

$$\text{PhCH=CHCH}_2\text{OH} + C_7H_{15}CHO \xrightarrow[\text{PhNCO, }Bu_3P]{Pd(Ph_3P)_4} \text{PhCH=CH–CH=CH}C_7H_{15} \quad (5)$$

82%, 4:1

Similar cross-coupling procedures have been used to prepare styrenes by the reaction of metalated aromatics with vinyl halides/triflates[11d,24] or, conversely, metalated alkenes with aromatic halides/triflates[9b,16,25] in the presence of $Pd(Ph_3P)_4$ (eq 6).

Typically, ArCl are poor substrates in $Pd(PPh_3)_4$-catalyzed coupling reactions. However, by forming the chromium tricarbonyl complex of the aryl chloride, a facile coupling reaction with vinylstannanes can be achieved (eq 7).[26]

$$4\text{-MeOC}_6H_4Br \xrightarrow[Pd(Ph_3P)_4\text{, PhMe}]{\text{CH}_2\text{=CHSnBu}_3} 4\text{-MeOC}_6H_4\text{CH=CH}_2 \quad (6)$$

76%

$$4\text{-MeOC}_6H_4Cl\cdot Cr(CO)_3 \xrightarrow[\text{2. } I_2]{\text{1. CH}_2\text{=CHSnBu}_3\text{, } Pd(Ph_3P)_4\text{, THF}} 4\text{-MeOC}_6H_4\text{CH=CH}_2 \quad (7)$$

68%

Enynes and arenynes are available from the $Pd(Ph_3P)_4$-catalyzed coupling of metalated alkynes (Mg,[20] Al,[13,27] Zn,[17f,28] Sn[15a,b,29]) with vinyl or aryl halides, triflates or phosphates (eq 8). Alternatively, 1-haloalkynes and metalated alkenes (B[14c] or Zn[11a,17a]) can be utilized in similar procedures.

$$\text{BuCH=CHI} + \text{M–C≡CH} \xrightarrow{Pd(Ph_3P)_4} \text{BuCH=CH–C≡CH} \quad (8)$$

M = $SnMe_3$ 50%
M = ZnCl 71%

Enynes and arenynes can also be prepared by the $Pd(Ph_3P)_4$-catalyzed reaction between vinyl halides (I, Br, or Cl) and 1-alkynes in the presence of ***Copper(I) Iodide*** and an amine base such as RNH_2 (R = Bu, Pr), Et_2NH, or Et_3N.[30] A modified procedure employing aqueous base under phase-transfer conditions has also been described (eq 9).[31] The arenyne products derived from such coupling reactions provide ready access to substituted indoles (eq 10).[29c,30d]

$$\text{BuCH=CHI} + C_5H_{11}\text{C≡CH} \xrightarrow[R_4NCl\text{, PhH}]{Pd(Ph_3P)_4\text{, CuI, aq NaOH}} \text{BuCH=CH–C≡C}C_5H_{11} \quad (9)$$

50%

$$\text{MeO}_2\text{C-aryl(OTf)(NHAc)} \xrightarrow[\text{2. }(MeCN)_2PdCl_2]{\text{1. } Bu_3Sn\text{–C≡C–}i\text{-Pr, } Pd(Ph_3P)_4} \text{MeO}_2\text{C-indole(N-Ac, 2-}i\text{-Pr)} \quad (10)$$

62%

$Pd(Ph_3P)_4$ catalyzes the coupling of simple alkyl metals and vinyl halides (Br, I) or triflates to form substituted alkenes (eq 11). Alkylboron,[32] alkyl Grignard,[20,33] alkylzinc,[34] and alkylaluminum[13] reagents have been particularly useful in this regard. A variety of functional groups on either reacting partner are tolerated and the reaction proceeds with retention of alkene geometry (usually >98%), providing stereochemically pure, highly substituted alkenes. For example, allylsilanes (eq 12)[35] and vinylcyclopropanes (eq 13)[36] have been prepared employing ***Trimethylsilylmethylmagnesium Chloride*** and cyclopropylzinc chloride, respectively, as the organometallic partner.

$$\text{Hex}\!\sim\!\sim\text{I} \xrightarrow[\text{Pd(Ph}_3\text{P)}_4]{\text{EtMgBr}} \text{Hex}\!\sim\!\sim\text{Et} \quad 87\%\ (E),\ 85\%\ (Z) \tag{11}$$

$$\text{Pr--CH=CH--I} \xrightarrow[\text{TMSCH}_2\text{MgCl, THF; 84\%}]{\text{Pd(Ph}_3\text{P)}_4\text{, rt}} \text{Pr--CH=CH--TMS} \tag{12}$$

$$\text{PO}\sim\triangle\text{--ZnCl} \xrightarrow[\text{I--CH=CH--CH}_2\text{OP}]{\text{Pd(Ph}_3\text{P)}_4\text{, THF, rt}} \text{PO}\sim\triangle\text{--CH=CH--CH}_2\text{OP} \quad \text{P = TBDMS, 82\%} \tag{13}$$

Ketones are obtained by the $Pd(Ph_3P)_4$-catalyzed coupling of acid chlorides with organometallic reagents (eq 14). Organozinc[37] and organocopper[38] reagents have been used most successfully, while other reports document the utility of R_4Pb (R = Bu, Et),[39] R_4Sn (R = Me, Bu, Ph),[40] and R_2Hg (R = Et, Ph)[41] reagents in this reaction. For those cases in which the organometallic reagent is alkenic, complete retention of alkene geometry is observed (see eq 14). In addition, the formation of tertiary alcohols is not observed under the conditions employed. A modified procedure that substitutes alkyl chloroformates for acid chlorides leads to an efficient preparation of esters (eq 15).[37b,38] In a related reaction, $Pd(Ph_3P)_4$ catalyzes the coupling reaction between substituted aryl- or alkylsulfonyl chlorides and vinyl- or allylstannanes, providing a general route to sulfones (eq 16).[42]

$$\text{3-furyl-COCl} \xrightarrow[i\text{-Pr--CH=CH--Cu, MgCl; 85\%}]{\text{Pd(Ph}_3\text{P)}_4\text{, THF}} \text{3-furyl-CO--CH=CH--}i\text{-Pr} \tag{14}$$

$$o\text{-tolyl-ZnCl} \xrightarrow[\text{THF; 73\%}]{\text{MeOCOCl, Pd(Ph}_3\text{P)}_4} o\text{-tolyl-CO}_2\text{Me} \tag{15}$$

$$\text{Ph--CH=CH--SnBu}_3 \xrightarrow[\text{THF, 70 °C; 90\%}]{\text{MeSO}_2\text{Cl, Pd(Ph}_3\text{P)}_4} \text{Ph--CH=CH--SO}_2\text{Me} \tag{16}$$

The $Pd(Ph_3P)_4$-catalyzed reaction of various allylic electrophiles with carbon-based nucleophiles is a very useful method for the formation of C–C bonds under relatively mild conditions (eq 17) and has been extensively reviewed elsewhere.[2a,43] The most commonly used electrophilic substrates for $Pd(Ph_3P)_4$-catalyzed allylic substitution reactions are allylic esters, carbonates, phosphates, carbamates, halides, sulfones, and epoxides. More recently, allylic alcohols themselves have been demonstrated to be useful substrates.[44] Commonly employed nucleophiles include soft stabilized carbanions such as malonate and, to a lesser extent, a variety of organometallic reagents (Al, Grignards, Sn, Zn, Zr). A few selected examples begin to illustrate the scope of this reaction in terms of the general patterns of reactivity with respect to regioselectivity (eqs 18 and 19) and stereoselectivity (eqs 17 and 18). It should be noted that a variety of other Pd catalysts (including ***Palladium(II) Acetate–1,2-Bis(diphenylphosphino)ethane*** and ***Bis(dibenzylideneacetone)palladium(0)***) have been shown to be useful in these alkylations. In addition, certain nitrogen-, sulfur-, and oxygen-based reagents are suitable nucleophilic substrates. The utility of some of these latter reagents are covered in subsequent sections.

$$\text{cis-5-(CO}_2\text{Me)cyclohex-2-enyl OAc} \xrightarrow[\text{Pd(Ph}_3\text{P)}_4\text{; 90\%}]{\text{NaCH(CO}_2\text{Me)}_2\text{, rt, THF}} \text{cis-5-(CO}_2\text{Me)cyclohex-2-enyl-CH(CO}_2\text{Me)}_2 \tag{17}$$

$$\text{Oct-epoxide-CH=CH}_2 \xrightarrow[\text{Pd(Ph}_3\text{P)}_4\text{; 84\%}]{\text{CH}_2\text{(CO}_2\text{Me)}_2\text{, rt, THF}} \text{Oct-CH(OH)-CH=CH-CH}_2\text{-CH(CO}_2\text{Me)}_2 \quad (E):(Z) = 98:2 \tag{18}$$

$$\text{PhO}_2\text{S-CH}_2\text{-CO-O-CH}_2\text{CH}_2\text{-CH=CH-CH}_2\text{OAc} \xrightarrow[\text{THF, rt}]{\text{NaH, Pd(Ph}_3\text{P)}_4} \text{8-membered lactone (94\%)} + \text{6-membered lactone (6\%)} \tag{19}$$

Aldehydes and α-bromo ketones or esters are efficiently coupled in an aldol reaction in the presence of ***Diethylaluminum Chloride–Tri-n-butylstannyllithium*** (or ***Tin(II) Chloride***) and catalytic $Pd(Ph_3P)_4$, providing β-hydroxy ketones or esters (eq 20).[45]

$$\text{R-CO-CH}_2\text{Br} \xrightarrow[\text{Bu}_3\text{SnLi–Et}_2\text{AlCl, Pd(Ph}_3\text{P)}_4\text{, THF, 0 °C}]{\text{Ph-CH=CH-CHO}} \text{R-CO-CH}_2\text{-CH(OH)-CH=CH-Ph} \quad \text{R = Ph, 70\%; EtO, 75\%} \tag{20}$$

Carbonylative Carbon–Carbon Bond Formation. A general, mild (50 °C), and high yielding conversion of halides and triflates into aldehydes via $Pd(Ph_3P)_4$-catalyzed carbonylation (1–3 atm CO) in the presence of ***Tri-n-butylstannane*** has been described (eq 21).[46] The range of usable substrates is extensive and includes ArI, benzyl and allyl halides, and vinyl iodides and triflates. The reaction has been extended to include ArBr by carrying out the carbonylation at 80 °C under pressure (50 atm CO), using poly(methylhydrosiloxane) (PMHS) instead of tin hydride.[47]

$$p\text{-MeO}_2\text{C-C}_6\text{H}_4\text{-I} \xrightarrow[\text{Bu}_3\text{SnH, THF, 50 °C; 90\%}]{\text{Pd(Ph}_3\text{P)}_4\text{, 3 atm CO}} p\text{-MeO}_2\text{C-C}_6\text{H}_4\text{-CHO} \tag{21}$$

$Pd(Ph_3P)_4$ catalyzes the carbonylation of benzyl[48a] and vinyl[48b] bromides under phase-transfer conditions in the presence of hydroxide to form the corresponding carboxylic acids. A wide variety of substitution is tolerated and the products are formed in moderate to excellent yield at room temperature and at normal pressure (1 atm CO). Extension of the reaction to the formation of esters from aryl, alkyl, and vinyl bromides has been described.[49] These transformations usually require a co-catalyst system of

$Pd(Ph_3P)_4$ and [(1,5-cyclohexadiene)$RhCl]_2$ in the presence of either $M(OR)_4$ (M = Ti, Zr) or $M(OR)_3$ (M = B, Al) (eq 22).

Ph–CH=CH–Br → [$Pd(Ph_3P)_4$, [(1,5-hd)$RhCl]_2$; CO, $M(OR)_n$] → Ph–CH=CH–CO_2R (22)

$Ti(OBu)_4$ (no Rh cat) 85%
$Al(OEt)_3$ 66%

Vinyl triflates serve as substrates for $Pd(Ph_3P)_4$-catalyzed carbonylation and have been converted into the corresponding esters[50] or ketones[15c,51] (eq 23).

4-*t*-Bu-cyclohexenyl OTf → [$Pd(Ph_3P)_4$, Me_3SnR; CO (15–50 psi), THF] → 4-*t*-Bu-cyclohexenyl C(O)R (23)

R = $CHCH_2$, 76%; Ph, 93%

Aromatic and Vinyl Nitriles. Aromatic halides (Br, I) have been converted into nitriles in excellent yields by $Pd(Ph_3P)_4$ catalysis in the presence of ***Sodium Cyanide/Alumina***,[52] ***Potassium Cyanide***,[53] or ***Cyanotrimethylsilane***[54] (eq 24). While the latter two procedures require the use of ArI as substrates, a more extensive range of substituents are tolerated than the alternative method employing ArBr. A $Pd(Ph_3P)_4$-catalyzed extrusion of CO from aromatic and heteroaromatic acyl cyanides (readily available from cyanohydrins) at 120 °C provides aryl nitriles in excellent yields (eq 25).[55]

4-Cl-C_6H_4–I → [$Pd(Ph_3P)_4$, TMSCN; Et_3N, reflux; 70%] → 4-Cl-C_6H_4–CN (24)

4-Cl-C_6H_4–COCN → [$Pd(Ph_3P)_4$, PhH; 98%] → 4-Cl-C_6H_4–CN (25)

Similarly, vinyl halides (Br, Cl) provide vinyl nitriles upon treatment with $Pd(Ph_3P)_4$/***Potassium Cyanide/18-Crown-6***.[56]

Carbon–Heteroatom (N, S, O, Sn, Si, Se, P) Bond Formation. Primary and secondary amines (but not ammonia) undergo reaction with allylic acetates,[57] halides,[58] phosphates,[59] and nitro compounds[60] in the presence of $Pd(Ph_3P)_4$ to provide the corresponding allylic amines (eq 26). A variety of ammonia equivalents have been demonstrated to be useful in this $Pd(Ph_3P)_4$-catalyzed alkylation, including 4,4′-dimethoxybenzhydrylamine,[57c] NaNHTs,[58] and NaN_3[61] (eq 26). Both allylic phosphates and chlorides react faster than the corresponding acetates[58,59a] and (*Z*)-alkenes are isomerized to the (*E*)-isomers.[57a,59a] The use of primary amines as nucleophiles in the synthesis of secondary allyl amines is sometimes problematic since the amine that is formed undergoes further alkylation to form the tertiary amine. Thus hydroxylamines have been shown to be useful primary amine equivalents (eq 27) since the reaction products are easily reduced to secondary amines.[59b]

Ph–CH=CH–CH_2OX → [$Pd(Ph_3P)_4$, THF, rt; NaN_3 or $HNEt_2$] → Ph–CH=CH–CH_2Nu (26)

X = $PO(OEt)_2$, Nu = NEt_2 68%
X = Ac, Nu = N_3 88%

PhCH(OAc)CH=CHCH$_3$ → [MeNHOH•HCl, $Pd(Ph_3P)_4$; NaOH, THF, rt; 99%] → PhCH=CHCH(N(OH)Me)CH$_3$ (27)

Allyl sulfones can be obtained by the $Pd(Ph_3P)_4$-catalyzed reaction of allylic acetates[62] and allylic nitro compounds[63,64] with $NaSO_2Ar$ (eq 28). $Pd(Ph_3P)_4$ also catalyzes the addition of HOAc to vinyl epoxides, providing a facile entry into 1,4-hydroxy acetates.[65]

Ph–CH=CH–CH_2OAc → [$NaTs•4H_2O$, $Pd(Ph_3P)_4$; THF–MeOH, rt; 98%] → Ph–CH=CH–CH_2SO_2Tol (28)

Aryl halides (Br, I) have been converted in good yields into the corresponding arylstannanes or -silanes by treatment with R_6Sn_2 (R = Bu, Me)[4,66] or ***Hexamethyldisilane***,[67] respectively, in the presence of catalytic $Pd(Ph_3P)_4$ (eq 29). Aryl[9a,b] and vinyl[15b,68] triflates produce aryl- and vinylstannanes under similar conditions, provided ***Hexamethyldistannane*** and LiCl are used as coreactants. In some cases, the presence of additional Ph_3P has been observed to improve yields.[68] By using the Vinyl halides can also be converted into vinylstannanes with $(Ph_3Sn)_2$Zn–TMEDA complex.[69]

4-Ac-C_6H_4–I → [$Pd(Ph_3P)_4$, Bu_6Sn_2; THF, 80 °C; 83%] → 4-Ac-C_6H_4–$SnBu_3$ (29)

Acyltrimethylstannanes can be prepared in moderate yields by the treatment of acid chlorides with Me_6Sn_2 and catalytic $Pd(Ph_3P)_4$ in refluxing THF.[70c] However, $Pd(Ph_3P)_2Cl_2$ is a superior catalyst for this transformation when using sterically bulky or electron-poor acyl halides.

$Pd(Ph_3P)_4$ catalyzes the addition of R_6Sn_2,[70] R_6Si_2,[71] ***Diphenyl Disulfide***, and ***Diphenyl Diselenide***[72] across the triple bond of 1-alkynes to provide the respective (*Z*)-1,2-addition products (eq 30). The (*Z*)-distannanes can be partially isomerized to the (*E*)-isomers by photolysis.[70] The reaction cannot be extended to include internal alkynes containing alkyl substituents, but allenes do undergo 1,2-addition of disilane[73] and ditin.[74] In a similar fashion, $Pd(PPh_3)_4$ catalyzes the regio- and stereospecific addition of $R_3Sn–SiR_3$ to 1-alkynes, the (*Z*)-1-silyl-2-stannylalkene isomers being the sole products (eq 30).[75]

HO–CH_2CH_2–C≡CH → [$Pd(Ph_3P)_4$; MN] → HO–CH_2CH_2–C(M)=CH–N (30)

M = N = SPh 79%
M = N = $SnBu_3$ 59%
M = $SnMe_3$, N = TMS 51%

α,β-Acetylenic esters react with R_6Sn_2 in the presence of $Pd(Ph_3P)_4$ at room temperature to provide only the (*Z*)-2,3-distannylalkenoates (eq 31).[76] When heated to 75–95 °C, clean isomerization to the (E)-isomers is observed. The corresponding amides are also useful substrates but provide either the (E)-isomers directly or (E/Z)-distannane mixtures.[76] Surprisingly, under similar reaction conditions, α,β-alkynic aldehydes and ketones form (Z)-β-stannyl enals and enones in excellent yields (eq 32).[77]

$$R{-}{\equiv}{-}CO_2Me \xrightarrow[\text{THF, rt; 66–90\%}]{Me_6Sn_2,\ Pd(Ph_3P)_4} (Me_3Sn)(R)C{=}C(CO_2Me)(SnMe_3) \quad (31)$$

R = Me, TBDMSOCH$_2$

$$PO(CH_2)_2{-}C{\equiv}C{-}COX \xrightarrow[\text{THF, 80 °C}]{Me_6Sn_2,\ Pd(Ph_3P)_4} PO(CH_2)_2C(SnMe_3){=}CH{-}COX \quad (32)$$

P = TBDMS; X = H, 87%; Me, 90%

The $Pd(Ph_3P)_4$-catalyzed hydrostannylation of 1-alkynes with R_3SnH provides mixtures of vinylstannane regio- and stereoisomers, the ratios depending upon the nature of the alkyne substituents and the R group of the tin reagent.[78] In general, when using alkyl-substituted 1-alkynes, ***Triphenylstannane*** provides (E)-1-stannylalkenes[78a] as the major products, while 2-stannylalkenes are obtained predominantly with $(Bu_3Sn)_2Zn$[78c] (eq 33). The $Pd(Ph_3P)_4$ mediated *cis* addition of R_3Sn–H across symmetrical internal alkynes has also been demonstrated to be generally high yielding.[78b] The hydrostannylation of α,β-unsaturated nitriles with $Bu_3SnH/Pd(Ph_3P)_4$ is regioselective, providing α-stannyl nitriles as the sole products.[79]

$$C_{10}H_{21}{-}C{\equiv}CH \xrightarrow{Pd(Ph_3P)_4} C_{10}H_{21}CH{=}CHSnR_3 + C_{10}H_{21}C(SnR_3){=}CH_2 \quad (33)$$

Ph_3SnH, 77% 89:11
$(Bu_3Sn)_2Zn$, 70% <5:>95

The $Pd(Ph_3P)_4$-mediated preparation of allyl- and benzylstannanes has been achieved by treatment of allylic acetates with $Et_2AlSnBu_3$[80] or by reaction of benzyl halides (Br, Cl) with R_6Sn_2.[66c] In a similar fashion, allyl- and benzylsilanes are prepared by the $Pd(Ph_3P)_4$-catalyzed reaction of R_6Si_2 with allylic halides[81] and benzyl halides,[4] respectively. Allylstannanes have also been prepared by the addition of Bu_3SnH to 1,3-dienes in the presence of $Pd(Ph_3P)_4$ (eq 34).[82]

$$\text{1,3-cyclohexadiene} \xrightarrow[\text{PhH, rt; 78\%}]{Pd(Ph_3P)_4,\ Bu_3SnH} \text{3-(tributylstannyl)cyclohexene } ({-}SnBu_3) \quad (34)$$

Dialkyl aryl- and vinylphosphonates ($RPO(OR')_2$) are readily prepared in excellent yields by the reaction of aryl or vinyl bromides, respectively, with dialkyl phosphite ($HPO(OR')_2$) in the presence of catalytic $Pd(Ph_3P)_4$ and Et_3N.[83] In a related process, unsymmetrical alkyl diarylphosphinates (ArPhPO(OR)) are obtained in good yields, regardless of aromatic substitution, by the $Pd(Ph_3P)_4$-catalyzed coupling reaction between aryl bromides and alkyl benzenephosphonites (HPhPO(OR)).[84]

Oxidation Reactions. α-Bromo ketones are dehydrobrominated to produce enones in low to good yields, especially when the products are phenolic, by treatment with stoichiometric $Pd(Ph_3P)_4$ in hot benzene.[85] Primary and secondary alcohols are oxidized in the presence of PhBr, base (NaH or K_2CO_3) and $Pd(Ph_3P)_4$ as catalyst to the corresponding aldehydes or ketones.[86] The practical advantages of these methods to alternative strategies have yet to be demonstrated.

Reduction Reactions. At elevated temperatures (100–110 °C), ArBr and ArI are reduced to ArH in the presence of catalytic $Pd(Ph_3P)_4$ and reducing agents such as HCO_2Na (eq 35),[3,87] NaOMe,[88] and poly(methylhydrosiloxane) (PMHS)/Bu_3N.[3] Aldehydes, ketones, esters, acids, and nitro substituents are unaffected. ArCl are poor substrates unless the aromatic nucleus is substituted with NO_2.[88] ArOTf are reduced in poorer yield under similar conditions; $Pd(OAc)_2$ and $Pd(Ph_3P)_2Cl_2$ are superior catalysts with these substrates.[89] A limited number of examples of the $Pd(PPh_3)_4$-catalyzed reduction of vinyl bromides and triflates to alkenes in the presence of HCO_2Na[3] and Bu_3SnH,[15a,b] respectively, have been described.

$$X{-}C_6H_4{-}Br \xrightarrow[\text{DMF, 100 °C}]{Pd(Ph_3P)_4,\ HCO_2Na} X{-}C_6H_5 \quad (35)$$

X = CHO, 80%; COMe, 84%

$Pd(Ph_3P)_4$ catalyzes the reductive displacement of a variety of allylic substituents with hydride transfer reagents. Allylic acetates are reduced to simple alkenes in the presence of ***Sodium Cyanoborohydride***,[90] PMHS,[91] ***Samarium(II) Iodide***/*i*-PrOH,[92] or Bu_3SnH[93] (eq 36). $Bu_3SnH/Pd(Ph_3P)_4$ also reduces allylic amines[93] and thiocarbamates.[94] All of these reductive procedures are accompanied by positional and/or geometrical isomerization of the alkene to an extent dependent upon the substrate structure. Interestingly, the $Pd(Ph_3P)_4$-catalyzed reduction of allylic sulfones with ***Sodium Borohydride*** is high yielding with no double bond positional isomerization observed.[95] With the more bulky reducing agent, ***Lithium Triethylborohydride***, a range of allylic groups, such as methyl, phenyl, and silyl ethers, sulfides, sulfones, selenides, and chlorides, are reduced to alkenes in the presence of $Pd(Ph_3P)_4$ and excess Ph_3P with little or no loss of alkene regio- or stereochemistry (eq 37).[96] An interesting $Pd(Ph_3P)_4$-catalyzed reduction of allylic acetates, tosylates, and chlorides to the corresponding alkenes has been described that uses n-BuZnCl as the hydride source.[97] This procedure proceeds with high levels of regio- and stereoselectivity, but the scope has yet to be explored.

$$\text{geranyl-type allylic acetate }({-}OAc) \xrightarrow[\text{[H]}]{Pd(Ph_3P)_4} R{-}CH(CH_3){=}CH{-}CH_3 + R{-}CH(CH_3){-}CH{=}CH_2 \quad (36)$$

[H] = $NaBH_3CN$, 68% 42:58
[H] = SmI_2, i-PrOH, 92% 93:7

$$Hept{-}CH{=}CH{-}CH_2{-}X \xrightarrow[Ph_3P,\ THF]{LiBHEt_3,\ Pd(Ph_3P)_4} Hept{-}CH{=}CH{-}CH_3 \quad (37)$$

X = SPh, 90% (E):(Z) = 94:5
X = OTBDMS, 80% (E):(Z) = 98:1

Acid chlorides[98] and acyl selenides[99] are efficiently reduced in the presence of Bu_3SnH under $Pd(Ph_3P)_4$ catalysis to provide the corresponding aldehydes in good to excellent yields without the formation of ester or alcohol byproducts (eq 38). A wide variety of substrate substituents, such as alkenes, nitriles, bromides, and nitro groups, are tolerated.

$$\mathrm{RC(O)X} \xrightarrow[\text{PhH, rt}]{\mathrm{Bu_3SnH,\ Pd(Ph_3P)_4}} \mathrm{RC(O)H} \quad (38)$$

R = Hex, X = Cl 77%
R = Oct, X = SePh 69%

The conjugate reduction of α,β-unsaturated ketones and aldehydes to the saturated analogs can be accomplished in the presence of $Pd(Ph_3P)_4$ and hydride transfer reagents such as Bu_3SnH[100] or mixed systems of Bu_3SnH/HOAc or $Bu_3SnH/ZnCl_2$.[79] A potentially more versatile, general, and selective reduction procedure involves a three-component system of $Pd(Ph_3P)_4/Ph_2SiH_2/ZnCl_2$ (eq 39).[101] α,β-Unsaturated esters and nitriles are untouched using this latter method.

$$\xrightarrow[\text{CHCl}_3\text{, rt; 96\%}]{\mathrm{Ph_2SiH_2,\ ZnCl_2,\ Pd(Ph_3P)_4}} \quad (39)$$

α-Bromo ketones are reductively debrominated to the parent ketones in the presence of $Pd(Ph_3P)_4$ and Ph_2SiH_2/K_2CO_3 at room temperature.[102] However, ***Hexacarbonylmolybdenum*** appears to be a superior catalyst for this conversion. Reductive debromination of α-bromo ketones, acids, and nitriles can also be accomplished by $Pd(Ph_3P)_4$ catalysis using PMHS/Bn_3N,[87b] HCO_2Na,[87b] or Me_6Si_2[103] as hydrogen donors but under much more drastic conditions (110–170 °C).

Removal of Allyloxycarbonyl (Aloc) and Allyl Protecting Groups. The allyloxycarbonyl protecting group[5] has been used extensively for the protection of alcohols[104] and amines,[105] including the amines of nucleotide bases,[106] *N*-terminal amines of amino acids and peptides,[107] and amino sugars.[108] $Pd(Ph_3P)_4$ mediates the high yielding removal of the Aloc group in the presence of nucleophilic allyl scavengers (eq 40) such as Bu_3SnH, 2-ethylhexanoic acid, dimedone, malonate, ***Ammonium Formate***, ***N-Hydroxysuccinimide***, and various amines including ***Morpholine*** and ***Pyrrolidine***. The deprotections are effected at or below room temperature, often in the presence of excess Ph_3P. The mild conditions used in these procedures leave most of the common N or O protecting groups intact, including Boc, TBDMS, MMT, DMT, and carbonates.

$$\xrightarrow[\mathrm{BuNH_2,\ Ph_3P,\ THF,\ rt;\ 100\%}]{\mathrm{Pd(Ph_3P)_4,\ HCO_2H}} \quad (40)$$

The $Pd(Ph_3P)_4$-catalyzed removal of the *O*-allyl protecting group has been described for a number of systems.[5] For example, allyl esters are efficiently cleaved to the parent acid in chemically sensitive systems such as penicillins[105,109] and glycopeptides.[110] The internucleotide phosphate linkage, protected as the allyl phospho(III)triester, remains intact upon deprotection under $Pd(Ph_3P)_4$ catalysis.[111] Allyl ethers that protect the anomeric hydroxy in carbohydrates (mono- and disaccharides) are efficiently removed under $Pd(Ph_3P)_4$ catalysis in hot (80 °C) HOAc.[112]

Rearrangements, Isomerizations, and Eliminations. $Pd(Ph_3P)_4$ catalyzes several [3,3]-sigmatropic rearrangements including those of *O*-allyl phosphoro- and phosphonothionates to the corresponding *S*-allyl thiolates (eq 41)[113] and allylic *N*-phenylformimidates to *N*-allyl-*N*-phenylformamides (eq 42).[114] The 3-aza-Cope rearrangement of *N*-allylenamines to γ,δ-unsaturated imines (eq 43)[115] and a Claisen rearrangement but with no allyl inversion[116] have also been described. In general, these $Pd(Ph_3P)_4$-catalyzed rearrangements provide compounds that are either not accessible via thermal reactions or are produced only under much more forcing reaction conditions. For those cases in which the allyl moiety contains substituents that may lead to regioisomers upon rearrangement, the less substituted isomer is usually favored[113,115] (see eq 41), although exceptions are known.[114]

$$\mathrm{(EtO)_2P(S)O{-}allyl} \xrightarrow[\text{DME, 80 °C; 93\%}]{\mathrm{Pd(Ph_3P)_4}} \mathrm{(EtO)_2P(O)S{-}allyl} \quad (41)$$

$$\xrightarrow[\text{THF, refux; 100\%}]{\mathrm{Pd(Ph_3P)_4}} \quad (42)$$

$$\xrightarrow[\text{PhH, 50 °C; 82\%}]{\mathrm{Pd(Ph_3P)_4,\ TFAA}} \quad (43)$$

$Pd(Ph_3P)_4$ catalyzes stereoselective, intramolecular metalloene reactions of acetoxydienes in HOAc, efficiently generating a range of cyclic 1,4-dienes (eq 44).[117] The reactions proceed in good to excellent yields and have been extended to include the preparation of pyrrolidines, piperidines, and tetrahydrofurans by incorporating N and O atoms into the bridge that tethers the reactive alkenes.

$$\xrightarrow[\text{HOAc, 75 °C}]{\mathrm{Pd(Ph_3P)_4}} \quad (44)$$

X = $C(CO_2Me)_2$, 52% 10:90
X = $NCOCF_3$, 67% 28:72

The isomerization of allylic acetates is a useful method for allylic oxygen interconversion. Although these reactions are typically carried out in the presence of Pd^{II} catalysts such as $Pd(OAc)_2$, $Pd(Ph_3P)_4$ has proven to be useful in the 1,3-rearrangement of α-cyanoallylic acetates to γ-acetoxy-α,β-unsatu rated nitriles (eq 45).[118] These compounds are conveniently transformed into furans.

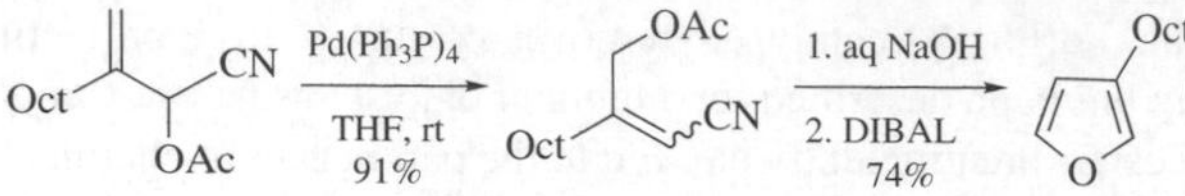

(45)

Several 1,3-diene syntheses involving elimination reactions that are catalyzed by $Pd(Ph_3P)_4$ have been reported. The first involves the Et_3N-mediated elimination of HOAc from allylic acetates in refluxing THF.[119] A complementary procedure involves the $Pd(Ph_3P)_4$-catalyzed decarboxylative elimination of β-acetoxycarboxylic acids (eq 46).[120] The substrates are easily prepared by the condensation of enals and carboxylate enolates; irrespective of the diastereomeric mixture, (*E*)-alkenes are formed in a highly stereocontrolled manner. The geometry of the double bond present in the enal precursor remains unaffected in the elimination and the reaction is applicable to the formation of 1,3-cyclohexadienes.

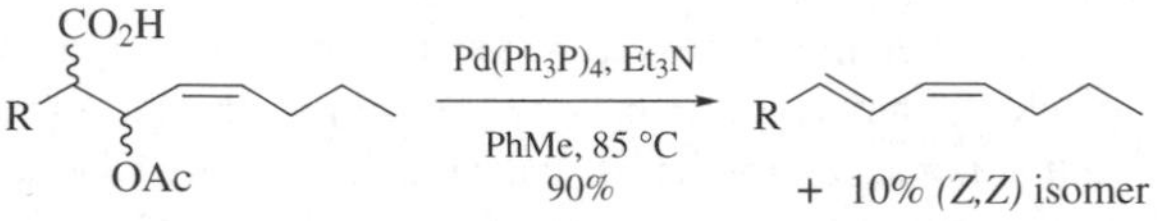

(46)

Monoepoxides of simple cyclic 1,3-dienes are smoothly converted in good yield to β,γ-unsaturated ketones in the presence of $Pd(Ph_3P)_4$ catalyst (eq 47).[121] Other vinyl epoxides, such as those in open chains or in cyclic compounds in which the double bond is not in the ring, are converted under similar conditions into dienols.

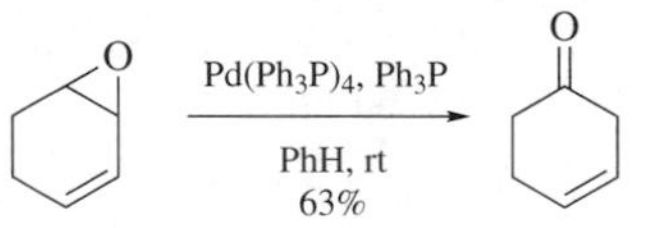

(47)

A mixture of $Pd(Ph_3P)_4$/dppe catalyzes the transformation of α,β-epoxy ketones, readily available via several methods, into β-diketones (eq 48).[122] The reaction is applicable to both cyclic and acyclic compounds, although epoxy ketones bearing an α-alkyl group are poor substrates.

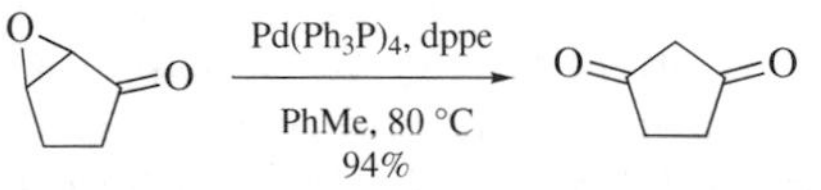

(48)

Related Reagents. Bis(dibenzylidene acetone)palladium(0); Bis[1,2-bis(diphenylphosphino)ethane]palladium(0); Bis(triphenylphosphine)[1,2-bis(diphenylphosphino)ethane]palladium(0); Diphenylsilane–Tetrakis(triphenylphosphine)palladium(0)–Zinc Chloride; Iodo(phenyl)bis(triphenylphosphine)palladium(II); Palladium(II) Acetate; Palladium(II) Chloride.

1. (a) Heck, R. F. *Palladium Reagents in Organic Syntheses*; Academic: London, 1985. (b) Tsuji, J. *Organic Synthesis with Palladium Compounds*; Springer: Berlin, 1980.
2. (a) Frost, C. G.; Howarth, J.; Williams, J. M. J. *TA* **1991**, *3*, 1089. (b) Stille, J. K. *PAC* **1985**, *57*, 1771. (c) Stille, J. K. *AG(E)* **1986**, *25*, 508. (d) Scott, W. J.; Stille, J. K. *ACR* **1988**, *21*, 47. (e) Negishi, E. *ACR* **1982**, *15*, 340. (f) Suzuki, A. *PAC* **1985**, *57*, 1749. (g) Beletskaya, I. P. *JOM* **1983**, *250*, 551. (h) Mitchell, T. N. *JOM* **1986**, *304*, 1.
3. Pri-Bar, I.; Buchman, O. *JOC* **1986**, *51*, 734.
4. Azarian, D.; Dua, S. S.; Eaborn, C.; Walton, D. R. M. *JOM* **1976**, *117*, C55.
5. Greene, T. W.; Wuts, P. G. M. *Protective Groups in Organic Synthesis*, 2nd ed.; Wiley: New York, 1991.
6. Coulson, D. *Inorg. Synth* **1972**, *13*, 121.
7. Ito, T.; Hasegawa, S.; Takahashi, Y.; Ishii, Y. *JOM* **1974**, *73*, 401.
8. (a) Fu, J.; Snieckus, V. *TL* **1990**, *31*, 1665. (b) Miyaura, N.; Yanagi, T.; Suzuki, A. *SC* **1981**, *11*, 513. (c) Sato, M.; Miyaura, N.; Suzuki, A. *CL* **1989**, 1405. (d) Takayuki, O.; Miyaura, N.; Suzuki, A. *JOC* **1993**, *58*, 2201. (e) Sharp, M. J.; Snieckus, V. *TL* **1985**, *26*, 5997. (f) Sharp, M. J.; Cheng, W.; Snieckus, V. *TL* **1987**, *28*, 5093. (g) Miller, R. B.; Dugar, S. *OM* **1984**, *3*, 1261. (h) Thompson, W. J.; Gaudino, J. *JOC* **1984**, *49*, 5237. (i) Huth, A.; Beetz, I.; Schumann, I. *T* **1989**, *45*, 6679.
9. (a) Echavarren, A. M.; Stille, J. K. *JACS* **1987**, *109*, 5478. (b) Echavarren, A. M.; Stille, J. K. *JACS* **1988**, *110*, 4051. (c) Gómez-Bengoa, E.; Echavarren, A. M. *JOC* **1991**, *56*, 3497. (d) Dondoni, A.; Fantin, G.; Fogagnolo, M.; Medici, A.; Pedrini, P. *S* **1987**, 693. (e) Clough, J. M.; Mann, I. S.; Widdowson, D. A. *TL* **1987**, *28*, 2645. (f) Malm, J.; Björk, P.; Gronowitz, S.; Hörnfeldt, A.-B. *TL* **1992**, *33*, 2199. (g) Achab, S.; Guyot, M.; Potier, P. *TL* **1993**, *34*, 2127.
10. Widdowson, D. A.; Zhang, Y.-Z. *T* **1986**, *42*, 2111.
11. (a) Negishi, E.; Luo, F.-T.; Frisbee, R.; Matsushita, H. *H* **1982**, *18*, 117. (b) Negishi, E.; Takahashi, T.; King, A. O. *OS* **1988**, *66*, 67. (c) Roth, G. P.; Fuller, C. E. *JOC* **1991**, *56*, 3493. (d) Arcadi, A.; Burini, A.; Cacchi, S.; Delmastro, M.; Marinelli, F.; Pietroni, B. *SL* **1990**, 47. (e) Pelter, A.; Rowlands, M.; Clements, G. *S* **1987**, 51. (f) Pelter, A.; Rowlands, M.; Jenkins, I. H. *TL* **1987**, *28*, 5213.
12. Torii, S.; Tanaka, H.; Morisaki, K. *TL* **1985**, *26*, 1655.
13. Takai, K.; Sato, M.; Oshima, K.; Nozaki, H. *BCJ* **1984**, *57*, 108.
14. (a) Miyaura, N.; Yamada, K.; Suzuki, A. *TL* **1979**, 3437. (b) Miyaura, N.; Suginome, H.; Suzuki, A. *TL* **1981**, *22*, 127. (c) Miyaura, N.; Yamada, K.; Suginome, H.; Suzuki, A. *JACS* **1985**, *107*, 972. (d) Miyaura, N.; Satoh, M.; Suzuki, A. *TL* **1986**, *27*, 3745. (e) Satoh, M.; Miyaura, N.; Suzuki, A. *CL* **1986**, 1329. (f) Uenishi, J.; Beau, J.-M.; Armstrong, R. W.; Kishi, Y. *JACS* **1987**, *109*, 4756. (g) Miyaura, N.; Suginome, H.; Suzuki, A. *T* **1983**, *39*, 3271. (h) Ishiyama, T.; Miyaura, N.; Suzuki, A. *CL* **1987**, 25. (i) Roush, W. R.; Moriarty, K. J.; Brown, B. B. *TL* **1990**, *31*, 6509.
15. (a) Scott, W. J.; Crisp, G. T.; Stille, J. K. *JACS* **1984**, *106*, 4630. (b) Scott, W. J.; Stille, J. K. *JACS* **1986**, *108*, 3033. (c) Scott, W. J.; Crisp, G. T.; Stille, J. K. *OS* **1990**, *68*, 116. (d) Stille, J. K.; Groh, B. L. *JACS* **1987**, *109*, 813.
16. Hatanaka, Y.; Hiyama, T. *TL* **1990**, *31*, 2719.
17. (a) Negishi, E.; Okukado, N.; King, A. O.; Van Horn, D. E.; Spiegel, B. I. *JACS* **1978**, *100*, 2254. (b) Negishi, E.; Takahashi, T.; Baba, S. *OS* **1988**, *66*, 60. (c) Negishi, E.; Luo, F.-T. *JOC* **1983**, *48*, 1562. (d) Negishi, E.; Takahashi, T.; Baba, S.; Van Horn, D. E.; Okukado, N. *JACS* **1987**, *109*, 2393. (e) Okukado, N.; Van Horn, D. E.; Klima, W. L.; Negishi, E. *TL* **1978**, 1027. (f) Ruitenberg, K.; Kleijn, H.; Elsevier, C. J.; Meijer, J.; Vermeer, P. *TL* **1981**, *22*, 1451.
18. (a) Jabri, N.; Alexakis, A.; Normant, J. F. *TL* **1981**, *22*, 959. (b) Jabri, N.; Alexakis, A.; Normant, J. F. *BSF(2)* **1983**, 321.
19. (a) Jabri, N.; Alexakis, A.; Normant, J. F. *TL* **1982**, *23*, 1589. (b) Jabri, N.; Alexakis, A.; Normant, J. F. *BSF(2)* **1983**, 332.
20. Dang, H. P.; Linstrumelle, G. *TL* **1978**, 191.
21. Hoshino, Y.; Miyaura, N.; Suzuki, A. *BCJ* **1988**, *61*, 3008.
22. (a) Piers, E.; Friesen, R. W.; Keay, B. A. *CC* **1985**, 809. (b) Stille, J. K.; Tanaka, M. *JACS* **1987**, *109*, 3785.
23. Okukado, N.; Uchikawa, O.; Nakamura, Y. *CL* **1988**, 1449.
24. (a) Miller, R. B.; Al-Hassan, M. *JOC* **1985**, *50*, 2121. (b) McCague, R. *TL* **1987**, *28*, 701.
25. (a) McKean, D. R.; Parrinello, G.; Renaldo, A. F.; Stille, J. K. *JOC* **1987**, *52*, 422. (b) Miyaura, N.; Suzuki, A. *CC* **1979**, 866. (c) Miyaura, N.; Maeda, K.; Suginome, H.; Suzuki, A. *JOC* **1982**, *47*, 2117.
26. Scott, W. J. *CC* **1987**, 1755.
27. (a) Takai, K.; Oshima, K.; Nozaki, H. *TL* **1980**, *21*, 2531. (b) Sato, M.; Takai, K.; Oshima, K.; Nozaki, H. *TL* **1981**, *22*, 1609.

28. (a) King, A. O.; Okukado, N.; Negishi, E. *CC* **1977**, 683. (b) King, A. O.; Negishi, E. *JOC* **1978**, *43*, 358. (c) Negishi, E.; Okukado, N.; Lovich, S. F.; Luo, F.-T. *JOC* **1984**, *49*, 2629. (d) Carpita, A.; Rossi, R. *TL* **1986**, *27*, 4351. (e) Chen, Q.-Y.; He, Y.-B. *TL* **1987**, *28*, 2387.
29. (a) Castedo, L.; Mouriño, A.; Sarandeses, L. A. *TL* **1986**, *27*, 1523. (b) Stille, J. K.; Simpson, J. H. *JACS* **1987**, *109*, 2138. (c) Rudisill, D. E.; Stille, J. K. *JOC* **1989**, *54*, 5856.
30. (a) Ratovelomanana, V.; Linstrumelle, G. *TL* **1981**, *22*, 315. (b) Ratovelomanana, V.; Linstrumelle, G. *SC* **1981**, *11*, 917. (c) Jeffery-Luong, T.; Linstrumelle, G. *S* **1983**, 32. (d) Arcadi, A.; Cacchi, S.; Marinelli, F. *TL* **1989**, *30*, 2581. (e) Scott, W. J.; Peña, M. R.; Swärd, K.; Stoessel, S. J.; Stille, J. K. *JOC* **1985**, *50*, 2302. (f) Mandai, T.; Nakata, T.; Murayama, H.; Yamaoki, H.; Ogawa, M.; Kawada, M.; Tsuji, J. *TL* **1990**, *31*, 7179.
31. Rossi, R.; Carpita, A.; Quirici, M. G.; Gaudenzi, M. L. *T* **1982**, *38*, 631.
32. Hoshino, Y.; Ishiyama, T.; Miyaura, N.; Suzuki, A. *TL* **1988**, *29*, 3983.
33. (a) Murahashi, S.-I.; Yamamura, M.; Yanagisawa, K.; Mita, N.; Kondo, K. *JOC* **1979**, *44*, 2408. (b) Huynh, C.; Linstrumelle, G. *TL* **1979**, 1073.
34. (a) Negishi, E.; Valente, L. F.; Kobayashi, M. *JACS* **1980**, *102*, 3298. (b) Negishi, E.; Zhang, Y.; Cederbaum, F. E.; Webb, M. B. *JOC* **1986**, *51*, 4080. (c) Tamaru, Y.; Ochiai, H.; Nakamura, T.; Yoshida, Z. *TL* **1986**, *27*, 955.
35. Negishi, E.; Luo, F.-T.; Rand, C. L. *TL* **1982**, *23*, 27.
36. Piers, E.; Jean, M.; Marrs, P. S. *TL* **1987**, *28*, 5075.
37. (a) Sato, T.; Itoh, T.; Fujisawa, T. *CL* **1982**, 1559. (b) Negishi, E.; Bagheri, V.; Chatterjee, S.; Luo, F.-T., Miller, J. A.; Stoll, A. T. *TL* **1983**, *24*, 5181. (c) Tamaru, Y.; Ochiai, H.; Nakamura, T.; Yoshida, Z. *OS* **1989**, *67*, 98.
38. Jabri, N.; Alexakis, A.; Normant, J. F. *T* **1986**, *42*, 1369.
39. Yamada, J.; Yamamoto, Y. *CC* **1987**, 1302.
40. Kosugi, M.; Shimizu, Y.; Migita, T. *CL* **1977**, 1423.
41. Takagi, K.; Okamoto, T.; Sakakibara, Y.; Ohno, A.; Oka, S.; Hayama, N. *CL* **1975**, 951.
42. Labadie, S. S. *JOC* **1989**, *54*, 2496.
43. (a) Trost, B. M. *AG(E)* **1989**, *28*, 1173. (b) Tsuji, J.; Minami, I. *ACR* **1987**, *20*, 140. (c) Trost, B. M. *ACR* **1980**, *13*, 385. (d) Trost, B. M. *T* **1977**, *33*, 2615.
44. Starý, I.; Stará, I. G.; Kocovský, P. *TL* **1993**, *34*, 179.
45. Matsubara, S.; Tsuboniwa, N.; Morizawa, Y.; Oshima, K.; Nozaki, H. *BCJ* **1984**, *57*, 3242.
46. (a) Baillargeon, V. P.; Stille, J. K. *JACS* **1983**, *105*, 7175. (b) Baillargeon, V. P.; Stille, J. K. *JACS* **1986**, *108*, 452.
47. Pri-Bar, I.; Buchman, O. *JOC* **1984**, *49*, 4009.
48. (a) Alper, H.; Hashem, K.; Heveling, J. *OM* **1982**, *1*, 775. (b) Galamb, V.; Alper, H. *TL* **1983**, *24*, 2965.
49. (a) Woell, J. B.; Fergusson, S. B.; Alper, H. *JOC* **1985**, *50*, 2134. (b) Hashem, K. E.; Woell, J. B.; Alper, H. *TL* **1984**, *25*, 4879. (c) Alper, H.; Antebi, S.; Woell, J. B. *AG(E)* **1984**, *23*, 732.
50. Hashimoto, H.; Furuichi, K.; Miwa, T. *CC* **1987**, 1002.
51. Crisp, G. T.; Scott, W. J.; Stille, J. K. *JACS* **1984**, *106*, 7500.
52. Dalton, J. R.; Regen, S. L. *JOC* **1979**, *44*, 4443.
53. Sekiya, A.; Ishikawa, N. *CL* **1975**, 277.
54. Chatani, N.; Hanafusa, T. *JOC* **1986**, *51*, 4714.
55. Murahashi, S.-I.; Naota, T.; Nakajima, N. *JOC* **1986**, *51*, 898.
56. Yamamura, K.; Murahashi, S.-I. *TL* **1977**, 4429.
57. (a) Genêt, J. P.; Balabane, M.; Bäckvall, J. E.; Nyström, J. E. *TL* **1983**, *24*, 2745. (b) Nyström, J. E.; Rein, T.; Bäckvall, J. E. *OS* **1989**, *67*, 105. (c) Trost, B. M.; Keinan, E. *JOC* **1979**, *44*, 3451.
58. Byström, S. E.; Aslanian, R.; Bäckvall, J. E. *TL* **1985**, *26*, 1749.
59. (a) Tanigawa, Y.; Nishimura, K.; Kawasaki, A.; Murahashi, S.-I. *TL* **1982**, *23*, 5549. (b) Murahashi, S.-I.; Imada, Y.; Taniguchi, Y.; Kodera, Y. *TL* **1988**, *29*, 2973.
60. Tamura, R.; Hegedus, L. S. *JACS* **1982**, *104*, 3727.
61. Murahashi, S.-I.; Tanigawa, Y.; Imada, Y.; Taniguchi, Y. *TL* **1986**, *27*, 227.
62. Inomata, K.; Yamamoto, T.; Kotake, H. *CL* **1981**, 1357.
63. (a) Tamura, R.; Hayashi, K.; Kakihana, M.; Tsuji, M.; Oda, D. *TL* **1985**, *26*, 851. (b) Ono, N.; Hamamoto, I.; Yanai, T.; Kaji, A. *CC* **1985**, 523.
64. Tamura, R.; Hayashi, K.; Kakihana, M.; Tsuji, M.; Oda, D. *CL* **1985**, 229.
65. Deardorff, D. R.; Myles, D. C. *OS* **1989**, *67*, 114.
66. (a) Kosugi, M.; Shimizu, K.; Ohtani, A.; Migita, T. *CL* **1981**, 829. (b) Kosugi, M.; Ohya, T.; Migita, T. *BCJ* **1983**, *56*, 3855. (c) Azizian, H.; Eaborn, C.; Pidcock, A. *JOM* **1981**, *215*, 49.
67. (a) Matsumoto, H.; Nagashima, S.; Yoshihiro, K.; Nagai, Y. *JOM* **1975**, *85*, C1. (b) Matsumoto, H.; Yoshihiro, K.; Nagashima, S.; Watanabe, H.; Nagai, Y. *JOM* **1977**, *128*, 409.
68. Wulff, W. D.; Peterson, G. A.; Bauta, W. E.; Chan, K.-S.; Faron, K. L.; Gilbertson, S. R.; Kaesler, R. W.; Yang, D. C.; Murray, C. K. *JOC* **1986**, *51*, 277.
69. Nonaka, T.; Okuda, Y.; Matsubara, S.; Oshima, K.; Utimoto, K.; Nozaki, H. *JOC* **1986**, *51*, 4716.
70. (a) Mitchell, T. N.; Amamria, A.; Killing, H.; Rutschow, D. *JOM* **1983**, *241*, C45. (b) Mitchell, T. N.; Amamria, A.; Killing, H.; Rutschow, D. *JOM* **1986**, *304*, 257. (c) Mitchell, T. N.; Kwetkat, K. *JOM* **1992**, *439*, 127.
71. Watanabe, H.; Kobayashi, M.; Saito, M.; Nagai, Y. *JOM* **1981**, *216*, 149.
72. Kuniyasu, H.; Ogawa, A.; Miyazaki, S.-I.; Ryu, I.; Kambe, N.; Sonoda, N. *JACS* **1991**, *113*, 9796.
73. Watanabe, H.; Saito, M.; Sutou, N.; Kishimoto, K.; Inose, J.; Nagai, Y. *JOM* **1982**, *225*, 343.
74. Killing, H.; Mitchell, T. N. *OM* **1984**, *3*, 1318.
75. (a) Chenard, B. L.; Laganis, E. D.; Davidson, F.; RajanBabu, T. V. *JOC* **1985**, *50*, 3666. (b) Chenard, B. L.; Van Zyl, C. M. *JOC* **1986**, *51*, 3561. (c) Mitchell, T. N.; Killing, H.; Dicke, R.; Wickenkamp, R. *CC* **1985**, 354. (d) Mitchell, T. N.; Wickenkamp, R.; Amamria, A.; Dicke, R.; Schneider, U. *JOC* **1987**, *52*, 4868.
76. Piers, E.; Skerlj, R. T. *CC* **1986**, 626.
77. Piers, E.; Tillyer, R. D. *JCS(P1)* **1989**, 2124.
78. (a) Ichinose, Y.; Oda, H.; Oshima, K.; Utimoto, K. *BCJ* **1987**, *60*, 3468. (b) Miyake, H.; Yamamura, K. *CL* **1989**, 981. (c) Matsubara, S.; Hibino, J.-I.; Morizawa, Y.; Oshima, K.; Nozaki, H. *JOM* **1985**, *285*, 163.
79. Four, P.; Guibe, F. *TL* **1982**, *23*, 1825.
80. Trost, B. M.; Herndon, J. W. *JACS* **1984**, *106*, 6835.
81. Matsumoto, H.; Yako, T.; Nagashima, S.; Motegi, T.; Nagai, Y. *JOM* **1978**, *148*, 97.
82. Miyake, H.; Yamamura, K. *CL* **1992**, 507.
83. (a) Hirao, T.; Masunaga, T.; Ohshiro, Y.; Agawa, T. *TL* **1980**, *21*, 3595. (b) Hirao, T.; Masunaga, T.; Yamada, N.; Ohshiro, Y.; Agawa, T. *BCJ* **1982**, *55*, 909.
84. Xu, Y.; Li, Z.; Xia, J.; Guo, H.; Huang, Y. *S* **1983**, 377.
85. Townsend, J. M.; Reingold, I. D.; Kendall, M. C. R.; Spencer, T. A. *JOC* **1975**, *40*, 2976.
86. Tamaru, Y.; Yamada, Y.; Inoue, K.; Yamamoto, Y.; Yoshida, Z. *JOC* **1983**, *48*, 1286.
87. Helquist, P. *TL* **1978**, 1913.
88. Zask, A.; Helquist, P. *JOC* **1978**, *43*, 1619.
89. (a) Chen, Q.-Y.; He, Y.-B.; Yang, Z.-Y. *CC* **1986**, 1452. (b) Peterson, G. A.; Kunng, F.-A.; McCallum, J. S.; Wulff, W. D. *TL* **1987**, *28*, 1381.
90. Hutchins, R. O.; Learn, K.; Fulton, R. P. *TL* **1980**, *21*, 27.
91. Keinan, E.; Greenspoon, N. *JOC* **1983**, *48*, 3545.
92. Tabuchi, T.; Inanaga, J.; Yamaguchi, M. *TL* **1986**, *27*, 601.
93. Keinan, E.; Greenspoon, N. *TL* **1982**, *23*, 241.
94. Yamamoto, Y.; Hori, A.; Hutchinson, C. R. *JACS* **1985**, *107*, 2471.

95. Kotake, H.; Yamamoto, T.; Kinoshita, H. *CL* **1982**, 1331.
96. Hutchins, R. O.; Learn, K. *JOC* **1982**, *47*, 4380.
97. Matsushita, H.; Negishi, E. *JOC* **1982**, *47*, 4161.
98. (a) Guibe, F.; Four, P.; Riviere, H. *CC* **1980**, 432. (b) Four, P.; Guibe, F. *JOC* **1981**, *46*, 4439.
99. Kuniyasu, H.; Ogawa, A.; Higaki, K.; Sonoda, N. *OM* **1992**, *11*, 3937.
100. Keinan, E.; Gleize, P. A. *TL* **1982**, *23*, 477.
101. (a) Keinan, E.; Greenspoon, N. *TL* **1985**, *26*, 1353. (b) Keinan, E.; Greenspoon, N. *JACS* **1986**, *108*, 7314.
102. Perez, D.; Greenspoon, N.; Keinan, E. *JOC* **1987**, *52*, 5570.
103. Urata, H.; Suzuki, H.; Moro-Oka, Y.; Ikawa, T. *JOM* **1982**, *234*, 367.
104. Guibe, F.; Saint M'Leux, Y. *TL* **1981**, *22*, 3591.
105. Jeffrey, P. D.; McCombie, S. W. *JOC* **1982**, *47*, 587.
106. Hayakawa, Y.; Kato, H.; Uchiyama, M.; Kajino, H.; Noyori, R. *JOC* **1986**, *51*, 2400.
107. (a) Kunz, H.; Unverzagt, C. *AG(E)* **1984**, *23*, 436. (b) Kinoshita, H.; Inomata, K.; Kameda, T.; Kotake, H. *CL* **1985**, 515.
108. Boullanger, P.; Descotes, G. *TL* **1986**, *27*, 2599.
109. Deziel, R. *TL* **1987**, *28*, 4371.
110. (a) Kunz, H.; Waldmann, H. *AG(E)* **1984**, *23*, 71. (b) Kunz, H.; Waldmann, H. *HCA* **1985**, *68*, 618. (c) Friedrich-Bochnitschek, S.; Waldmann, H.; Kunz, H. *JOC* **1989**, *54*, 751.
111. Hayakawa, Y.; Uchiyama, M.; Kato, H.; Noyori, R. *TL* **1985**, *26*, 6505.
112. Nakayama, K.; Uoto, K.; Higashi, K.; Soga, T.; Kusama, T. *CPB* **1992**, *40*, 1718.
113. (a) Tamaru, Y.; Yoshida, Z.; Yamada, Y.; Mukai, K.; Yoshioka, H. *JOC* **1983**, *48*, 1293. (b) Yamada, Y.; Mukai, K.; Yoshioka, H.; Tamaru, Y.; Yoshida, Z. *TL* **1979**, 5015.
114. Ikariya, T.; Ishikawa, Y.; Hirai, K.; Yoshikawa, S. *CL* **1982**, 1815.
115. Murahashi, S.-I.; Makabe, Y.; Kunita, K. *JOC* **1988**, *53*, 4489.
116. Trost, B. M.; Runge, T. A.; Jungheim, L. N. *JACS* **1980**, *102*, 2840.
117. Oppolzer, W. *AG(E)* **1989**, *28*, 38.
118. Mandai, T.; Hashio, S.; Goto, J.; Kawada, M. *TL* **1981**, *22*, 2187.
119. Trost, B. M.; Verhoeven, T. R.; Fortunak, J. M. *TL* **1979**, 2301.
120. Trost, B. M.; Fortunak, J. M. *JACS* **1980**, *102*, 2843.
121. Suzuki, M.; Oda, Y.; Noyori, R. *JACS* **1979**, *101*, 1623.
122. Suzuki, M.; Watanabe, A.; Noyori, R. *JACS* **1980**, *102*, 2095.

Richard W. Friesen
Merck Frosst Centre for Therapeutic Research, Quebec, Canada

Thexylborane[1]

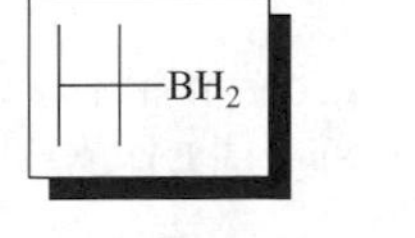

[3688-24-2] $C_6H_{15}B$ (MW 98.02)

(readily available monoalkylborane useful for regioselective hydroboration of alkenes and dienes; thexylalkylboranes and thexyldialkylboranes are useful intermediates for the synthesis of unsymmetrical ketones, cyclic ketones, *trans* disubstituted alkenes, conjugated dienes, and diols[1])

Alternate Name: (1,1,2-trimethylpropyl)borane.

Physical Data: generally prepared in situ; dimeric in THF. It can be isolated as a liquid, mp −34.7 to −32.3 °C.[2]

Solubility: sol ether, hydrocarbon, and halocarbon solvents; THF is generally the solvent of choice; reacts rapidly with protic solvents.[1a]

Form Supplied in: not commercially available.

Analysis of Reagent Purity: analyzed by NMR and IR spectroscopy and by hydrogen evolution upon reaction with methanol.[1a,3]

Preparative Method: most conveniently prepared from ***Borane–Tetrahydrofuran*** and 2,3-dimethyl-2-butene in THF (eq 1).[1a,2]

BH_3; THF, 0 °C; BH_2 (1)

Handling, Storage, and Precautions: very reactive with oxygen and moisture; must be handled using standard techniques for handling air-sensitive materials.[3] The reagent is reported to be stable for at least a week when stored at 0 °C in THF solution under N_2.[1a] However, at rt the boron atom slowly migrates from the tertiary position to the primary (3% in 8 days).[4] Use in a fume hood.

Hydroboration of Alkenes. Reactions of alkenes with thexylborane have been extensively studied and several reviews have appeared.[1] Thexylborane ($ThxBH_2$) in THF reacts with 2 equiv of relatively unhindered alkenes to form thexyldialkylboranes. The regioselectivity in the hydroboration of alkenes with thexylborane is similar to that of borane in THF. For example, hydroboration of 2 equiv of 1-hexene with thexylborane followed by oxidation produces a 95:5 mixture of 1-hexanol and 2-hexanol. Hydroboration of a terminal monosubstituted alkene with thexylborane in a 1:1 ratio gives a mixture of both the thexylmonoalkylborane and thexyldialkylborane. (Preparation of thexylmonoalkylboranes from monosubstituted alkenes can be accomplished using ***Chloro(thexyl)borane-Dimethyl Sulfide***). With most disubstituted and some trisubstituted alkenes, it is possible to prepare the corresponding thexylmonoalkylborane by treating 1 equiv of the alkene with thexylborane at −20 to −25 °C.[5] Thexylmonoalkylboranes can hydroborate relatively unhindered alkenes to form thexyldialkylboranes containing two different alkyl groups (eq 2).[6]

BH_2; THF, −25 °C; B, H; B (2)

Hydroboration of a hindered alkene with either thexylborane or thexylmonoalkylborane is slow and is accompanied by dehydroboration of the thexylmonoalkylborane, producing a monoalkylborane and 2,3-dimethyl-2-butene. Lower reaction temperatures and the presence of excess 2,3-dimethyl-2-butene in the reaction may reduce the amount of dehydroboration.[6] Sterically hindered alkenes can be hydroborated under high pressure (6000 atm) to produce highly hindered trialkylboranes, such as trithexylborane,[7] but this procedure does not appear to be practical for synthetic purposes.

Mixed thexyldialkylboranes have also been prepared by treating thexylborane consecutively with different halomagnesium or lithium dialkylcuprates.[8] This procedure offers the advantage of being able to introduce methyl or aryl groups onto the boron atom.

The hydroboration of either terminal or internal alkynes with thexylborane in a 2:1 ratio gives good yields of the expected thexyldialkenylboranes, but the reaction of thexylborane with 1 equiv of a terminal alkyne is reported to give at most 20% of the thexylalkenylborane.[1a,4]

Complete dehydroboration of the thexyl group in thexylmonoalkylboranes can be achieved by treating thexylmonoalkylboranes with a fourfold excess of ***Triethylamine***.[5] This reaction provides a general method for the synthesis of monoalkylboranes as the triethylamine complexes. More recently, it has been found that thexylborane–triethylamine or thexylborane–***N,N,N′,N′-Tetramethylethylenediamine*** adducts can be used to hydroborate hindered alkenes directly (with concomitant loss of the thexyl group as 2,3-dimethyl-2-butene) to give the monoalkylborane–amine adducts.[9] ***Monoisopinocampheylborane***, a useful chiral hydroborating reagent, can be prepared by this reaction.[10]

Thexylborane is the reagent of choice for the hydroboration of dienes to form *B*-thexylboracyclanes (eq 3)[11] since the reaction of borane with dienes tends to give polymeric products.[12] Stereoselective cyclic hydroboration of dienes by thexylborane has been employed to prepare acyclic diols with 1,3-, 1,4-, and 1,5-asymmetric induction (eq 4).[13] In the cyclic hydroboration of appropriately substituted 1,5-dienes to yield 1,5-diols, 1,2-asymmetric induction was employed as a key step (eq 5) in the synthesis of the Prelog–Djerassi lactone.[14]

BH_2 72% (3)

1. BH_2 2. [O]

OH HO + OH HO (4)

6:1

TBDMSO BH_2 [O]

TBDMSO OH OH + TBDMSO OH OH (5)

3.8:1

Cyclic hydroboration of allyl vinyl ethers provides a highly stereoselective synthesis of 1,3-diols with *syn* stereochemistry (eq 6).[15] The *syn* stereoselectivity observed in the cyclic hydroboration of allyl vinyl ethers is opposite to the stereoselectivity observed in the acyclic hydroboration of allylic alcohols (see eq 9).[18] Remote stereocontrol in the hydroboration–reduction of enones with thexylborane has also been reported (eq 7).[16] This reaction is proposed to occur by a rapid hydroboration of the carbon–carbon double bond followed by an intramolecular reduction of the carbonyl group by the intermediate dialkylborane.

Bu O 1. BH_2 2. [O] Bu OH OH (6)

syn:*anti* = >200:1

O 1. BH_2 2. [O] 84%

OH OH + OH OH (7)

15:1

Acyclic diastereoselection in the hydroboration of alkenes has been reported by several groups. Evans and co-workers[17] observed high levels of 1,3-asymmetric induction in the hydroboration of a number of terminal alkenes bearing substituents at the 2-position of the alkene and a proximal chiral center, as illustrated in eq 8. These workers concluded that the diastereoselection is directed primarily by the nearest chiral center in each of the substrates, and they proposed a transition state model to account for the observed diastereoselectivity. The hydroboration of acyclic secondary allylic alcohols with thexylborane, yielding 1,3-diols with high *anti* (or *threo*) diastereoselection, has also been reported (eq 9).[18] The diastereoselection observed in this reaction does not require the use of a protecting group on the allylic alcohol. Thexylborane proves to be the hydroboration reagent of choice for reaction with trisubstituted alkenes, but terminal alkenes give higher diastereoselection with the more sterically demanding reagents ***9-Borabicyclo[3.3.1]nonane*** or ***Dicyclohexylborane***.

TMSO O N Ph O O (X_N) BH_2 [O]

OH TMSO O X_N + OH TMSO O X_N (8)

85:15

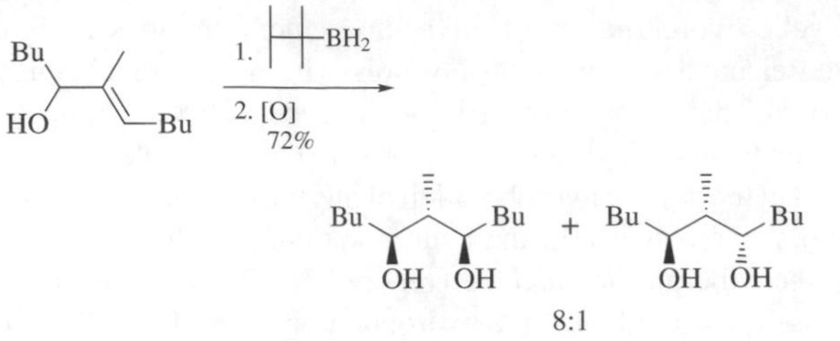

(9)

8:1

Selective Functional Group Reductions. The selective reduction of functional groups by thexylborane has been reported,[19a] and the relative reactivity of thexylborane with common functional groups has been compared to that of other borane reagents (***Diborane***, ***Disiamylborane***, and ***Chloro(thexyl)borane–Dimethyl Sulfide***).[19] Acidic hydrogens in –OH, $-CO_2H$, and $-SO_3H$ groups react at moderate to rapid rates with thexylborane with the evolution of hydrogen. Aldehydes generally react rapidly with thexylborane yielding alcohols after hydrolysis. However, ketones and most other carbonyl groups react only slowly with thexylborane. Carboxylic acids can be reduced to aldehydes by 2.5 equiv of thexylborane, but Chloro(thexyl)borane–Dimethyl Sulfide is the borane reagent of choice for this transformation. Nitriles, oximes, epoxides, and aromatic nitro compounds are reduced only slowly, and alkyl nitro compounds, disulfides, sulfones, and tosylates do not react with thexylborane.

Use of Thexyldialkylboranes in Synthesis. Thexyldialkylboranes are particularly useful in synthetic reaction sequences due to the availability of a variety of these compounds and the low migratory aptitude of the tertiary thexyl group in most rearrangement reactions. Several reviews of the role of boron in synthesis have appeared.[1,2,20]

Synthesis of Ketones. Reaction of thexyldialkylboranes with ***Carbon Monoxide*** in the presence of water followed by oxidation provides a novel route to the synthesis of ketones (eq 10).[21] Yields in this reaction are generally in the 50–80% range. The observed migratory aptitude for the alkyl groups is: primary > secondary ≫ tertiary. Cyclic ketones can be prepared from dienes by cyclic hydroboration with thexylborane followed by carbonylation and oxidation (eq 11).[22] This annulation procedure stereoselectively provides the *trans*-ring fusion in both the indanone and decalone systems. This methodology has been successfully employed in the stereospecific annulation of 1,5-diene substrates to form the *trans*-hydroazulene nucleus.[23]

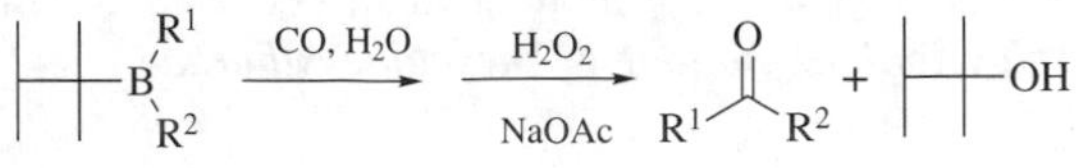

(10)

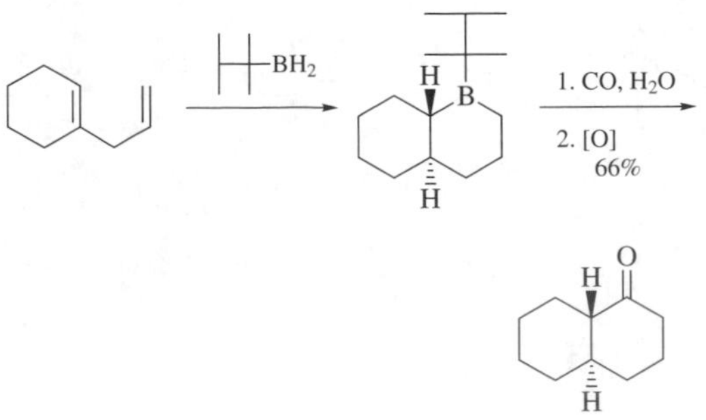

(11)

An alternate procedure to convert thexyldialkylboranes into ketones via an intermediate cyanoborate has been reported (eq 12)[24] which avoids the use of carbon monoxide under high pressure. Yields are quite good, generally >75%, and no special equipment is required for the reaction. This procedure has been employed in the synthesis of the heterocyclic ring system of δ-coniceine (**1**) (eq 13).[25]

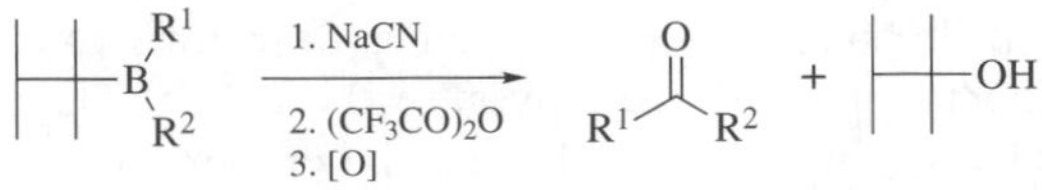

(12)

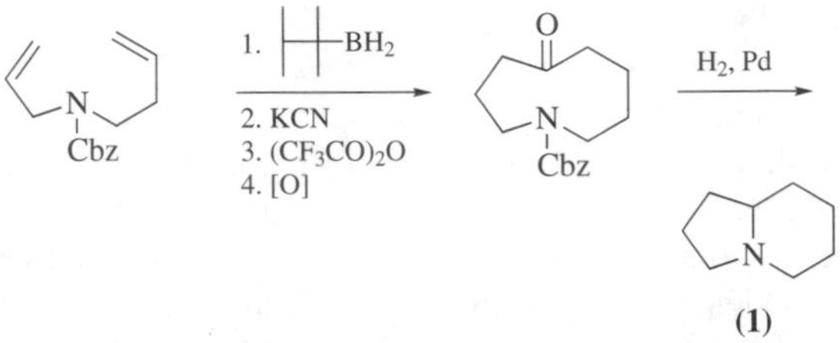

(13)

(1)

Synthesis of Alkenes. The stereoselective synthesis of (*E*)-disubstituted alkenes[26] can be carried out by treating a thexylmonoalkylborane (**2**) with a 1-bromo-1-alkyne to yield the intermediate thexylalkyl(1-bromo-1-alkenyl)borane (**3**), which is then treated with ***Sodium Methoxide*** to induce a stereospecific rearrangement of the alkyl group (R^1) from boron to carbon. Stereospecific protonolysis of the resulting intermediate (**4**) provides the (*E*)-alkene in high isomeric purity and good to excellent yield (eq 14).

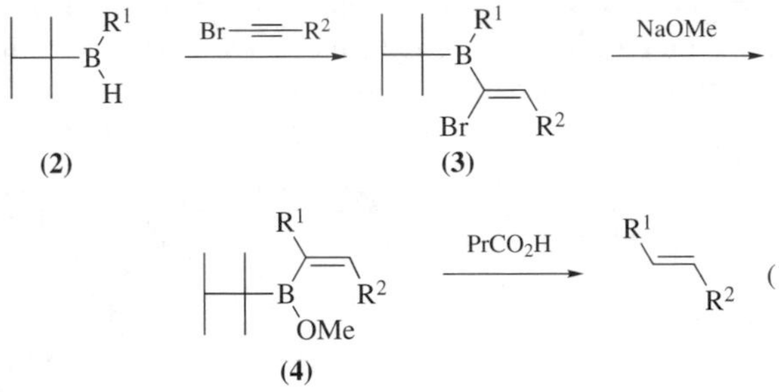

(2) **(3)**

(4) (14)

Trisubstituted alkenes can be synthesized[27] by first treating a thexyldialkylborane (**5**) with an alkynyllithium to yield a thexyldialkylalkynylborate (**6**), which is then alkylated with migration of an alkyl group from boron to carbon to give the thexylalkylvinylboranes (**7** and **8**) (eq 15). After acid hydrolysis, the trisubstituted alkenes (**9** and **10**) are obtained in >70% yield in a ratio of about 9:1. The major product (**9**) in all cases is the isomer in which the migrating group (R^1) and the group introduced by alkylation (R^3) are *cis* to each other. Various alkylating agents are effective in this reaction including primary alkyl iodides, benzyl and allyl bromides, ***Dimethyl Sulfate***, and ***Triethyloxonium Tetrafluoroborate***.

Synthesis of Dienes. Thexylborane reacts with 2 equiv of 1-alkyne to give the corresponding thexyldialkenylborane (**11**) in good yield (eq 16).[4] However, when (**11**) is treated with ***Sodium Hydroxide*** and ***Iodine*** to induce rearrangement, an almost equal mixture of the desired (*E,Z*)-diene (**12**) and alkene (**13**) results where either the alkenyl group or the thexyl group migrates with approximately equal facility.[28] This problem can be surmounted by selectively oxidizing the thexyl group in the thexyldialkenylborane (**11**) with ***Trimethylamine N-Oxide*** before the reaction with iodine and sodium methoxide (eq 17).[28] This procedure appears to be useful only for the synthesis of symmetrical (*E,Z*)-dienes (**12**) due to the difficulty in preparing unsymmetrical thexyldialkenylboranes in a stepwise hydroboration sequence from two different 1-alkynes.

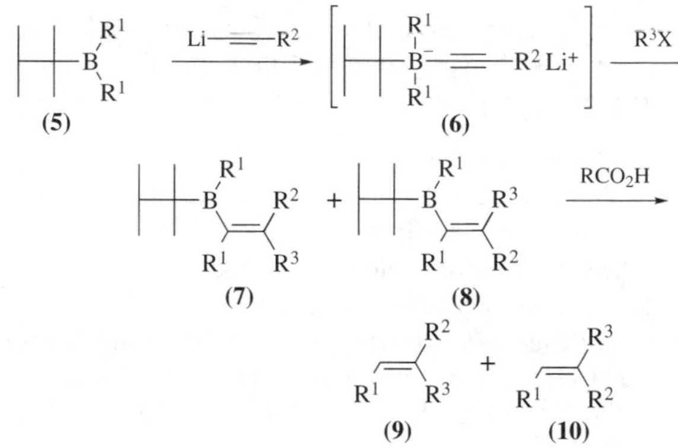

(15)

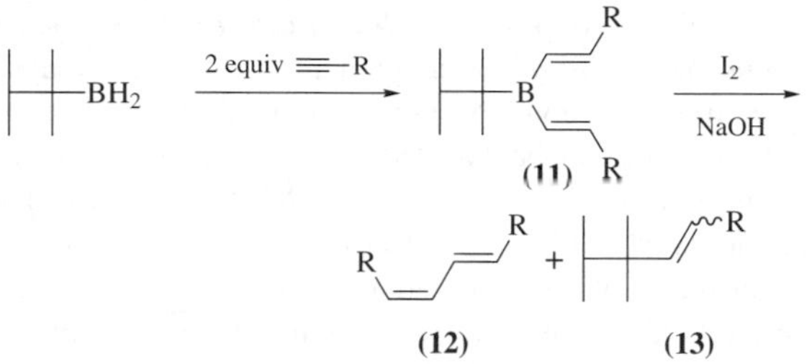

(16)

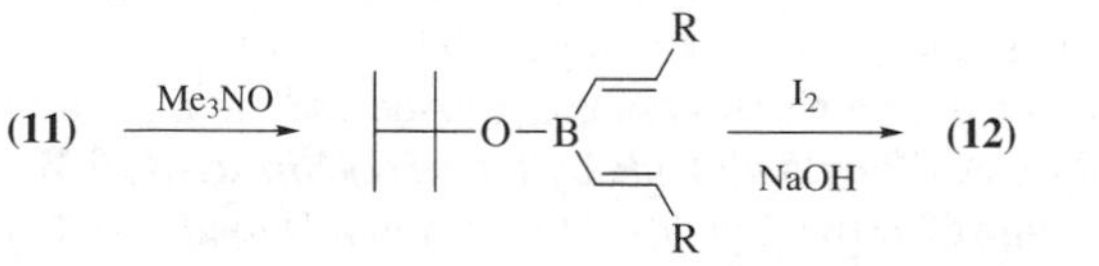

(17)

However, thexylborane will react with 1-chloro-1-alkynes and 1-bromo-1-alkynes (but not with 1-iodo-1-alkynes) to give thexyl(1-haloalkenyl)boranes (**14**) in high yield (eq 18).[29] This observation enabled the development of a general and highly stereospecific synthesis of conjugated (*E,E*)-dienes. The borane (**14**) reacts stereoselectively with a 1-alkyne to give the thexyldialkenylborane (**15**), which is then treated with sodium methoxide to produce (**16**). Compound (**16**) is protonolyzed with refluxing isobutyric acid to give the (*E,E*)-diene (**17**) in 53–63% yield and >98% isomeric purity.[29] The intermediate (**16**) is also transformed into the (*E*)-enone (**18**) in 50% yield by oxidation with ***Hydrogen Peroxide*** and NaOAc (eq 19).[29]

A stereoselective synthesis of 1,2,3-butatrienes was also developed based on the reaction of thexylborane with 1-halo-1-alkynes. In this synthesis, 2 equiv of a 1-iodo-1-alkyne react with thexylborane to give the thexyldialkenylborane (**19**) that rearranges to the 1,2,3-butatriene (**21**) upon treatment with 2 equiv of sodium methoxide (eq 20).[29b,30] Although the yields of (**21**) are only moderate (47%, R = butyl; 29%, R = cyclohexyl), the products are of high isomeric purity.

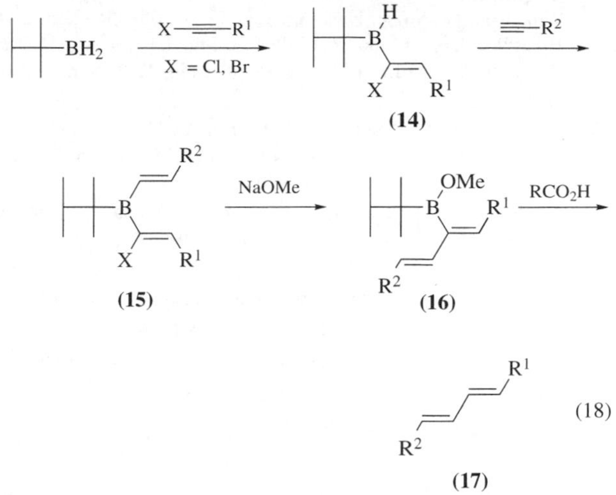

(18)

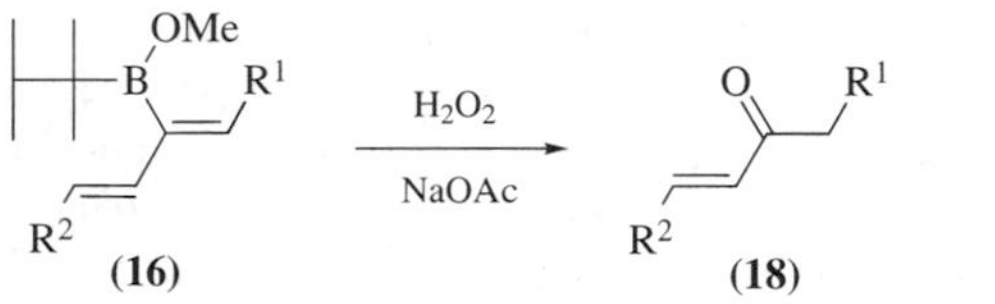

(19)

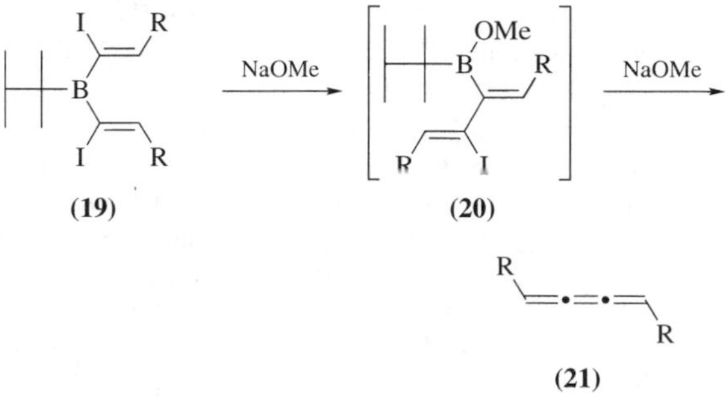

(20)

Synthesis of Carboxylic Acids. A convenient procedure for the direct oxidation of organoboranes from terminal alkenes to carboxylic acids has been reported.[31] The oxidation gives high yields with a number of different organoboranes derived from a variety of borane reagents including thexylborane. Several different oxidizing agents (***Pyridinium Dichromate***, ***Sodium Dichromate*** in aqueous H_2SO_4, and ***Chromium(VI) Oxide*** in 90% aqueous acetic acid) are effective for the reaction. For example, 2-methyl-1-pentene is converted to 2-methylpentanoic acid in 86% yield (eq 21).

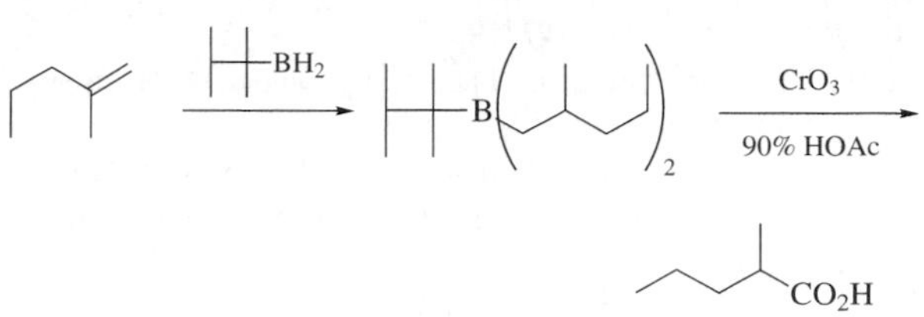

(21)

Related Reagents. 9-Borabicyclo[3.3.1]nonane; Borane–Tetrahydrofuran; Chloro(thexyl)borane–Dimethyl Sulfide; Diborane; Dicyclohexylborane; Disiamylborane; Monoisopinocampheylborane.

---

1. (a) Negishi, E.; Brown, H. C. *S* **1974**, 77. (b) Brown, H. C.; Negishi, E.; Zaidlewicz, M. In *Comprehensive Organometallic Chemistry*;

Wilkinson, G.; Stone, F. G. A.; Abel, E. W., Eds.; Pergamon: Oxford, 1982; Vol. 7, pp 111–363. (c) Pelter, A.; Smith, K.; Brown, H. C. *Borane Reagents*; Academic: London, 1988. (d) Smith, K.; Pelter, A. *COS* **1991**, *8*, 709.

2. (a) Brown, H. C.; Kramer, G. W.; Levy, A.; Midland, M. M. *Organic Syntheses via Boranes*; Wiley: New York, 1975; p 31. (b) Brown, H. C.; Mandal, A. K. *JOC* **1992**, *57*, 4970.
3. (a) Brown, H. C.; Kramer, G. W.; Levy, A.; Midland, M. M. *Organic Synthesis via Boranes*; Wiley: New York, 1975; Chapter 9. (b) Schwier, J. R.; Brown, H. C. *JOC* **1993**, *58*, 1546.
4. Zweifel, G.; Brown, H. C. *JACS* **1963**, *85*, 2066.
5. Brown, H. C.; Negishi, E.; Katz, J.-J. *JACS* **1975**, *97*, 2791.
6. Brown, H. C.; Katz, J.-J.; Lane, C. F.; Negishi, E. *JACS* **1975**, *97*, 2799.
7. Rice, J. E.; Okamoto, Y. *JOC* **1982**, *47*, 4189.
8. Whitely, C. G. *TL* **1984**, *25*, 5563.
9. Brown, H. C.; Yoon, N. M.; Mandal, A. K. *JOM* **1977**, *135*, C10.
10. Brown, H. C.; Mandal, A. K.; Yoon, N. M.; Singaram, B.; Schwier, J. R.; Jadhav, P. K. *JOC* **1982**, *47*, 5069.
11. Brown, H. C.; Negishi, E. *JACS* **1972**, *94*, 3567.
12. Brown, H. C.; Negishi, E.; Burk, P. L. *JACS* **1972**, *94*, 3561.
13. (a) Still, W. C.; Darst, K. P. *JACS* **1980**, *102*, 7385. (b) Harada, T.; Matsuda, Y.; Wada, I.; Uchimura, J.; Oku, A. *CC* **1990**, 21.
14. Morgans, D. J., Jr. *TL* **1981**, *22*, 3721.
15. Harada, T.; Matsuda, Y.; Uchimura, J.; Oku, A. *CC* **1989**, 1429.
16. Harada, T.; Matsuda, Y.; Imanaka, S.; Oku, A. *CC* **1990**, 1641.
17. (a) Evans, D. A.; Barttoli, J.; Godel, T. *TL* **1982**, *23*, 4577. (b) Evans, D. A.; Barttoli, J. *TL* **1982**, *23*, 807.
18. Still, W. C.; Barrish, J. C. *JACS* **1983**, *105*, 2487.
19. (a) Brown, H. C.; Heim, P.; Yoon, N. M. *JOC* **1972**, *37*, 2942. (b) Brown, H. C.; Nazer, B.; Cha, J. S.; Sikorski, J. A. *JOC* **1986**, *51*, 5264.
20. (a) Coveney, D. J. *COS* **1991**, *3*, 793. (b) Carruthers, W. *Some Modern Methods of Organic Synthesis*, 3rd ed.; Cambridge University Press: Cambridge, 1986; pp 294–317. (c) Thomas, S. E. *Organic Synthesis: The Roles of Boron and Silicon*; Oxford University Press: Oxford, 1991; pp 1–46.
21. (a) Brown, H. C.; Negishi, E. *JACS* **1967**, *89*, 5285. (b) Negishi, E.; Brown, H. C. *S* **1972**, 196. (c) Brown, H. C. *ACR* **1969**, *2*, 65.
22. (a) Brown, H. C.; Negishi, E. *JACS* **1967**, *89*, 5477. (b) Brown, H. C.; Negishi, E. *CC* **1968**, 594.
23. Stevenson, J. W. S.; Bryson, T. A. *CL* **1984**, 5.
24. (a) Pelter, A.; Hutchings, M. G.; Smith, K. *CC* **1970**, 1529. (b) Pelter, A.; Hutchings, M. G.; Smith, K. *CC* **1971**, 1048. (c) Pelter, A.; Hutchings, M. G.; Smith, K. *CC* **1973**, 186. (d) Pelter, A.; Smith, K.; Hutchings, M. G.; Rowe, K. *JCS(P1)* **1975**, 129. (e) Pelter, A.; Hutchings, M. G.; Smith, K.; Williams, D. J. *JCS(P1)* **1975**, 145.
25. Garst, M. E.; Bonfiglio, J. N. *TL* **1981**, *22*, 2075.
26. (a) Negishi, E.; Katz, J.-J., Brown, H. C. *S* **1972**, 555. (b) Corey, E. J.; Ravindranathan, T. *JACS* **1972**, *94*, 4013.
27. Pelter, A.; Subrahmanyam, C.; Laub, R. J.; Gould, K. J.; Harrison, C. R. *TL* **1975**, *19*, 1633.
28. Zwiefel, G.; Polston, N. L.; Whitney, C. C. *JACS* **1968**, *90*, 6243.
29. (a) Negishi, E.; Yoshida, T. *CC* **1973**, 606. (b) Negishi, E.; Yoshida, T.; Abramovitch, A.; Lew, G.; Williams, R. M. *T* **1991**, *47*, 343.
30. Yoshida, T.; Williams, R. M.; Negishi, E. *JACS* **1974**, *96*, 3688.
31. Brown, H. C.; Kulkarni, S. V.; Khanna, V. V.; Patil, V. D.; Racherla, U. S. *JOC* **1992**, *57*, 6173.

William S. Mungall
Hope College, Holland, MI, USA

p-Toluenesulfonyl Azide[1]

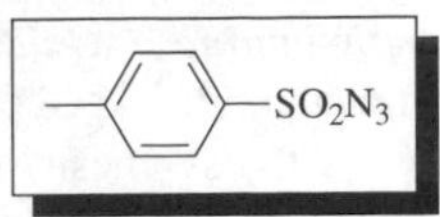

[941-55-9] $C_7H_7N_3O_2S$ (MW 197.24)

(introduction of azide and diazo groups into organic compounds;[2] serves as a source of nitrene and 1,3-azide dipoles for [3 + 2] cycloadditions;[3] used for the synthesis of *N*-tosylphosphinimines,[4] -sulfimines, and -sulfoximines[5])

Alternate Names: tosyl azide.
Physical Data: mp 21–22 °C; bp 110–115 °C/0.001 mmHg;[6] d^{25} 1.286 g cm^{-3}; n^{25}_{589} 1.55010.
Solubility: sol chloroform, diethyl ether, acetone.
Form Supplied in: oily colorless liquid.
Analysis of Reagent Purity: IR (neat) $\nu = 2130$ (strong, N=N=N), 1380 and 1180 (strong, SO_2) cm^{-1}; ^1H NMR ($CDCl_3$) $\delta = 2.47$ (s, CH_3), 7.40 (d, $J = 8$ Hz, *m*-$SO_2C_6H_4$), 7.84 (d, $J = 8$ Hz, *o*-$SO_2C_6H_4$).[7]
Preparative Methods: the simplest synthetic method for tosyl azide is the reaction of ***p*-Toluenesulfonyl Chloride** with ***Sodium Azide***. Tosyl azide prepared in ethanol may contain 7–20% of ethyl *p*-toluenesulfonate;[8] the formation of this byproduct can be avoided by working in aqueous acetone instead of ethanol.[9,10] The ^1H NMR spectrum of tosyl azide prepared (99% yield) according to Curphey exhibits signals for traces of dichloromethane (<1%);[10] the compound is pure by TLC, while HPLC reveals traces (>1.5%) of an impurity. Polymer-bound tosyl chloride can be transformed to polymer-bound tosyl azide.[11] A polymer-bound phase-transfer catalyst may also be employed (94% yield).[12] Further methods for the preparation of tosyl azide are the oxidation of ***p*-Toluenesulfonylhydrazide** by ***Iron(III) Nitrate-K10 Montmorillonite Clay*** (83% yield),[13] by ***Nitrogen Dioxide*** in CCl_4 (95% yield),[14] or by ***Nitrosonium Tetrafluoroborate*** (85% yield).[15]
Handling, Storage, and Precautions: crystallizes to a white solid at −20 °C and can be stored indefinitely at this temperature. Particular care is required for all reactions in which tosyl azide is heated at or above 100 °C. The initial temperature of the explosive decomposition is about 120 °C.[16] Severe explosions during the attempted distillation of tosyl azide have been reported.[6,17] Polymer-bound tosyl azide can be stored indefinitely at room temperature and, in contrast to tosyl azide itself, is not sensitive to mechanical shock. Use in a fume hood.

Synthesis of Azides. Tosyl azide acts as a transfer reagent to introduce the azide group to anions of C–H acidic compounds such as malonate derivatives (eq 1);[18] aryl azides are obtained from the reactions of aryl Grignard reagents with tosyl azide after cleavage of the primarily formed triazene salt intermediates with sodium pyrophosphate (eq 2);[19] azidothiophenes and -bithiophenes[20] or azidotetrazolium salts[21] can be prepared similarly (using organolithium compounds). The anions of primary amines, generated from the latter by treatment with Grignard reagents[22] or preferably with ***Sodium Hydride*** in THF,[23] also react with tosyl azide to furnish azides (diazo group transfer to primary amines, eq 3). α-Azido sulfones are formed in one step by reactions of aliphatic nitro compounds with ***Potassium Hydride*** in THF and subsequent treatment with tosyl azide (eq 4).[24]

$$\text{EtO}_2\text{C-CH(Ph)-CO}_2\text{Et} \xrightarrow[\text{2. TsN}_3\ \ 77\%]{\text{1. NaH, ether, 20 °C}} \text{EtO}_2\text{C-C(Ph)(N}_3\text{)-CO}_2\text{Et} \quad (1)$$

$$\text{Ar-Br} \xrightarrow[\text{2. TsN}_3\text{, 0 °C; 3. Na}_4\text{P}_2\text{O}_7\cdot 10\text{H}_2\text{O}]{\text{1. Mg, ether}} \text{Ar-N}_3 \quad (2)$$

Ar = phenyl, 73%; mesityl, 63%

$$\text{Ar-NH}_2 \xrightarrow[\text{2. TsN}_3\ \ 44\text{–}49\%]{\text{1. MeMgBr, ether}} \text{Ar-N}_3 \quad (3)$$

Ar = Ph, p-tolyl, p-ClC$_6$H$_4$

$$\text{C}_6\text{H}_{11}\text{-NO}_2 \xrightarrow[\text{2. TsN}_3\text{, –10 °C; 3. NaOH, H}_2\text{O}\ \ 56\%]{\text{1. KH, THF, 20 °C}} \text{1-azido-1-(p-tolylsulfonyl)cyclohexane} \quad (4)$$

Synthesis of Diazo Compounds[2]. The synthesis of diazo compounds from a carbon nucleophile and an arenesulfonyl azide, preferably tosyl azide (eq 5), is known as diazo group transfer.[25] The actual reaction in a somewhat different form has been known for a long time;[26] it was first employed for the synthesis of diazocyclopentadiene from the cyclopentadienide anion and tosyl azide[27] and has since been developed intensively.

$$\text{XYCH}_2 \xrightarrow[-\text{TsNH}_2]{\text{N}_2\text{=N-Ts (base)}} \text{XYC=N}_2$$

X, Y = CO_2R, CHO, COR, $CONR_2$, CN, NO_2, SOR, POR_2, $PO(OR)_2$, Ar

$$\text{MeCOCH}_2\text{CO}_2\text{-}t\text{-Bu} \xrightarrow[\text{2. TsN}_3\ \ -\text{TsNH}_2\ \ 95\%]{\text{1. NEt}_3\text{, MeCN, 20 °C}} \text{MeCOC(=N}_2\text{)CO}_2\text{-}t\text{-Bu} \xrightarrow[70\%]{\text{MeONa, MeOH, 0 °C}} \text{N}_2\text{=CHCO}_2\text{-}t\text{-Bu} \quad (5)$$

When X and Y are suitable electron-attracting groups such as CO_2R, COR, NO_2, SO_2R, Ar, etc., the required carbanion can be generated by treatment of the corresponding methylene compound with an appropriate base (NEt_3, pyridine/piperidine, NaOEt, *t*-BuOK). The synthesis of *t*-butyl diazoacetate is an explicit example.[8] The use of a phase-transfer catalyst is sometimes advantageous for the generation of the carbanion.[28] When only one electron-attracting, activating group is present in the substrate, diazo group transfer can be effected by an indirect route in which, for example, a ketone is transformed to an α-formyl ketone which is, in turn, converted by treatment with tosyl azide in the presence of a base to the desired diazo compound by cleavage of *N*-formyltosyl amide (deformylating diazo group transfer, eq 6).[29]

$$\text{cyclohexanone} \xrightarrow[\text{EtONa, EtOH}]{\text{HCO}_2\text{Et}} \text{2-formylcyclohexanone (CHO)} \xrightarrow[\text{2. TsN}_3\text{, –15 °C}\ \ -\text{TsNHCHO}\ \ 90\%]{\text{1. NEt}_3\text{, CH}_2\text{Cl}_2} \text{2-diazocyclohexanone (N}_2\text{)} \quad (6)$$

Tosyl azide is the reagent employed most frequently. However, ***p-Dodecylbenzenesulfonyl Azide***,[16] *p*-carboxybenzenesulfonyl azide,[30] polymer-bound tosyl azide,[11] trifluoromethanesulfonyl azide (triflyl azide),[31] (azidochloromethylene)dimethylammonium chloride,[32] ***2,4,6-Triisopropylbenzenesulfonyl Azide*** (trisyl azide),[33] and others[34] have also been used. 2-Azido-3-ethyl-1,3-benzothiazolium tetrafluoroborate and 2-azido-1-ethylpyridinium tetrafluoroborate[35] have proven to be useful as diazo group transfer reagents since they can also be employed in weakly acidic media with base-sensitive methylene compounds. In addition to the requirement for a less hazardous reagent than tosyl azide, the separation of the inevitably formed amine from the desired diazo compound represents the main stimulant for the continued search for novel diazo group transfer reagents.

Supplementary to the classic diazo group transfer to activated methylene compounds described briefly above, the reactions of tosyl azide with enamines, enol ethers, ynamines, ynethers, activated alkenes and alkynes, cyclopropenes, and methylenephosphoranes may also be included in this reaction type in its widest sense, when the respective reactions give rise to diazo compounds (see the examples in eqs 7 and 8);[36] see also cycloaddition reactions (below) as an alternative.

$$\text{1-Me-3,3-dimethyl-2-methyleneindoline} \xrightarrow{\text{TsN}_3} \text{2-(=N-Ts)-indoline} + \text{H}_2\text{C=N}_2\ \ 87\% \quad (7)$$

$$\text{Me(Ph)N-C}\equiv\text{CH} \xrightarrow{\text{TsN}_3} \text{5-(N(Me)Ph)-1-Ts-1,2,3-triazole} \xrightarrow[\text{opening}]{\text{ring}} \text{Me(Ph)N-C(=N-Ts)-CH=N}_2\ \ 83\% \quad (8)$$

Cycloaddition Reactions. The 1,3-dipolar cycloadditions of tosyl azide to alkenes primarily furnish the 4,5-dihydrotriazoles which can seldom be isolated.[37] Generally, subsequent reactions such as the [3 + 2] cycloreversion to *N*-tosylimines and diazo compounds (diazo group transfer), the cleavage of nitrogen in *N*-tosylaziridines, or the concomitant 1,2-shift of a substituent in *N*-tosylimines with rearranged molecular skeletons (eq 9) predominate.[38] All processes can occur in parallel; however, there is a pronounced reaction selectivity in dependence on the group X (the electron-donating substituent).

From the plethora of synthetic applications of tosyl azide, its examples of the reactions with nonstrained cycloalkenes leading to *N*-tosylimines (eq 10),[39] with cyclic enamines which give amidines (eq 11),[40] with indoles to furnish 2-tosyliminoindoles (eq 12),[41] and with enol ethers giving rise to imidates in quantitative yield (eq 13)[42] are mentioned as examples. When the alkene carries a further substituent capable of elimination, the cycloadditions give rise to 1,2,3-triazoles directly (eq 14).[43] The *N*-tosyl group can be readily removed by hydrolysis to liberate the corresponding NH compound. Thus in this sense, tosyl azide can be considered as a synthetic equivalent of the difficult to handle hydrogen azide (***Hydrazoic Acid***, HN_3).

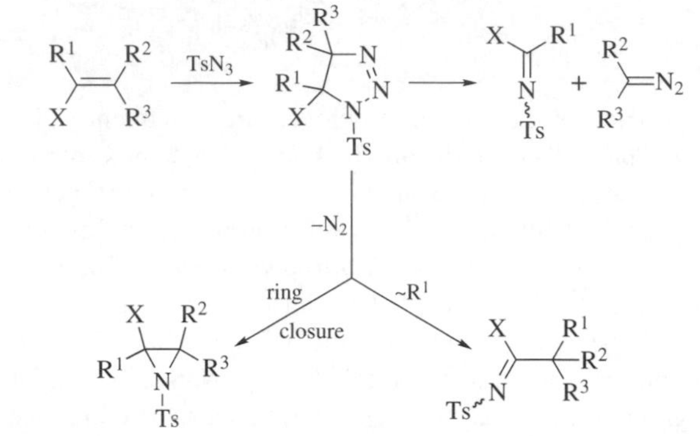

(9)

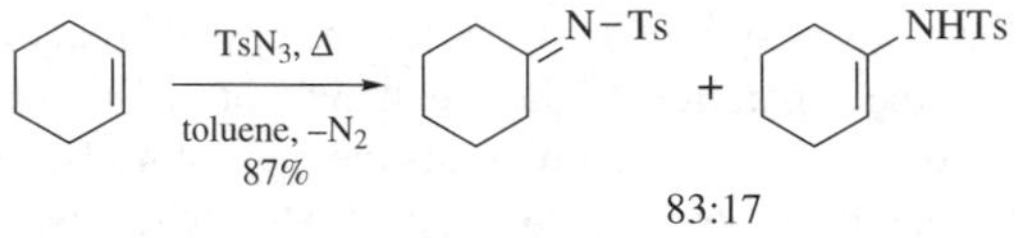

(10)

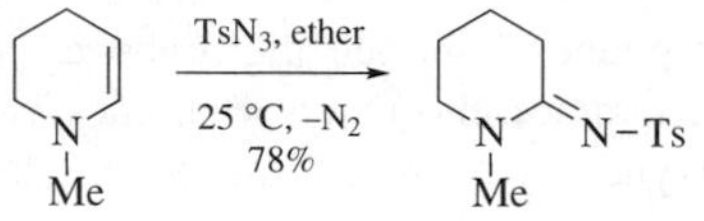

83:17

(11)

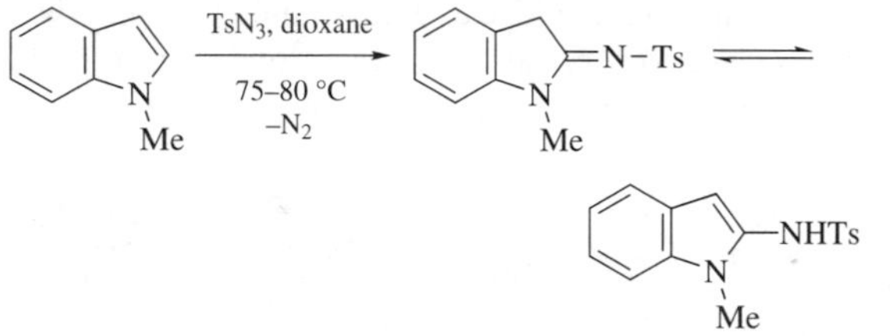

(12)

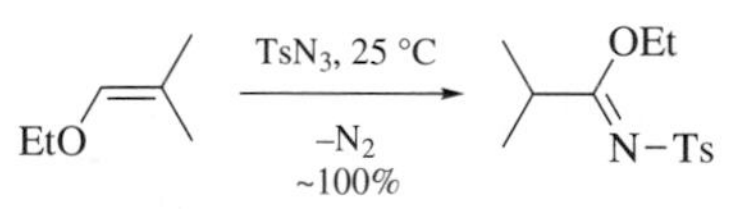

(13)

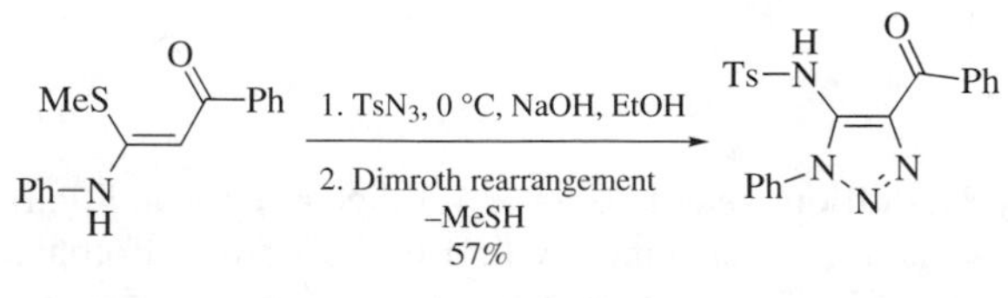

(14)

Miscellaneous. The reactions of aryl and heteroaryl hydrazides with tosyl azide in the presence of NaOH under phase-transfer conditions give rise to the corresponding aromatic and heteroaromatic compounds (eq 15).[44] After oxidative workup, reactions with trialkylboranes yield alkyl aryl sulfides (eq 16).[45]

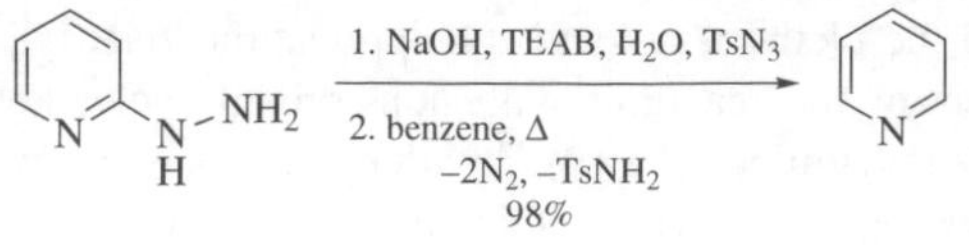

(15)

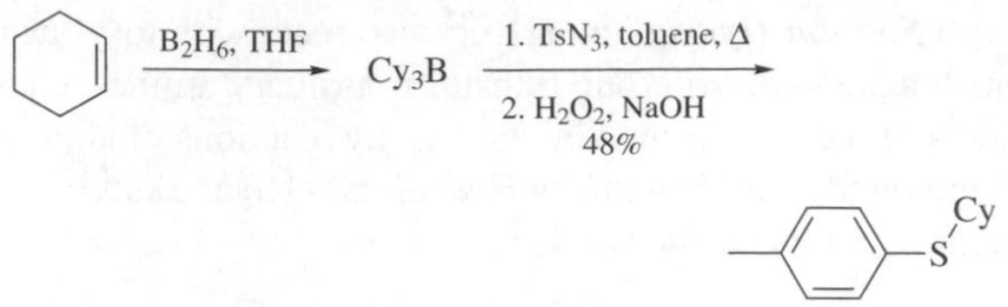

(16)

Related Reagents. *p*-Acetamidobenzenesulfonyl Azide; Benzenesulfonyl Azide; 4-Bromobenzenesulfonyl Azide; Diphenyl Phosphorazidate; *p*-Dodecylbenzenesulfonyl Azide; Methanesulfonyl Azide; *o*-Nitrobenzenesulfonyl Azide; *p*-Toluenesulfonylhydrazide; 2,4,6-Triisopropylbenzenesulfonyl Azide.

1. (a) Grundmann, C. *MOC* **1965**, *10/4*, 777. (b) *The Chemistry of the Azido Group*; Patai, S., Ed.; Wiley: London, 1971. (c) Scriven, E. F. V.; Turnbull, K. *CRV* **1988**, *88*, 351.
2. (a) Regitz, M.; Maas, G. *Diazo Compounds. Properties and Synthesis;* Academic: Orlando, 1986; pp 326–435. (b) Böhshar, M.; Fink, J.; Heydt, H.; Wagner, O.; Regitz, M. *MOC* **1990**, *E14b*, 961.
3. Lwowski, W. *1,3-Dipolar Cycloaddition Chemistry*; Padwa, A., Ed.; Wiley: New York, 1984; Vol. 1, p 559.
4. (a) Heydt, H.; Regitz, M. *MOC* **1982**, *E2*, 96. (b) Laszlo, P.; Polla, E. *TL* **1984**, *25*, 4651. (c) For the synthesis of arsinimines by the same method, see Cadogan, J. I. G.; Gosney, I. *JCS(P1)* **1974**, 460.
5. (a) Haake, M. *MOC* **1985**, *E11*, 901. (b) Haake, M. *MOC* **1985**, *E11*, 1304.
6. Caution: the distillation of *p*-toluenesulfonyl azide should be avoided if at all possible since severe explosions have been reported: Spencer, H. *Chem. Br.* **1981**, *17*, 106.
7. Hua, D. H.; Peacock, N. J.; Meyers, C. Y. *JOC* **1980**, *45*, 1717.
8. Regitz, M.; Hocker, J.; Liedhegener, A. *OSC* **1973**, *5*, 179.
9. Breslow, D. S.; Sloan, M. F.; Newberg, N. R.; Renfrow, W. B. *JACS* **1969**, *91*, 2273.
10. Curphey, T. J. *OPP* **1981**, *13*, 112.
11. Roush, W. R.; Feitler, D.; Rebĕk, J. *TL* **1974**, 1391.
12. Kumar, S. M. *SC* **1987**, *17*, 1015.
13. Laszlo, P.; Polla, E. *TL* **1984**, *25*, 3701.
14. Kim, Y. H.; Kim, K.; Shim, S. B. *TL* **1986**, *27*, 4749.
15. Pozsgay, V.; Jennings, H. J. *TL* **1987**, *28*, 5091.
16. Hazen, G. G.; Weinstock, L. M.; Connell, R.; Bollinger, F. W. *SC* **1981**, *11*, 947.
17. Rewicki, D.; Tuchscherer, C. *AG(E)* **1972**, *11*, 44.
18. (a) Weininger, S. J.; Kohen, S.; Mataka, S.; Koga, G.; Anselme, J.-P. *JOC* **1974**, *39*, 1591. (b) Kozikowski, A. P.; Greco, M. N. *JOC* **1984**, *49*, 2310.
19. Smith, P. A. S.; Rowe, C. D.; Bruner, L. B. *JOC* **1969**, *34*, 3430.
20. (a) Spagnolo, P.; Zanirato, P. *JOC* **1978**, *43*, 3539. (b) Spagnolo, P.; Zanirato, P.; Gronowitz, S. *JOC* **1982**, *47*, 3177.
21. Weiss, R.; Lowack, R. H. *AG(E)* **1991**, *30*, 1162.
22. (a) Fischer, W.; Anselme, J.-P. *JACS* **1967**, *89*, 5284. (b) Anselme, J.-P.; Fischer, W. *T* **1969**, *25*, 855.
23. Quast, H.; Eckert, P. *LA* **1974**, 1727.
24. Koft, E. R. *JOC* **1987**, *52*, 3466.
25. (a) Regitz, M. *AG(E)* **1967**, *6*, 733. (b) Regitz, M. *S* **1972**, 351.
26. (a) Dimroth, O. *LA* **1910**, *373*, 336. (b) Curtius, T.; Klavehn, W. *JPR(2)* **1926**, *112*, 65.
27. von Doering, W.; De Puy, C. H. *JACS* **1953**, *75*, 5955.

28. (a) Ledon, H. *S* **1974**, 347. (b) Ledon, H. J. *OS* **1980**, *59*, 66. (c) Nakajima, M.; Anselme, J.-P. *TL* **1976**, 4421. (d) Gonzáles, A.; Gálvez, C. *S* **1981**, 741.
29. Regitz, M.; Rüter, J.; Liedhegener, A. *OS* **1971**, *51*, 86.
30. Hendrickson, J. B.; Wolf, W. A. *JOC* **1968**, *33*, 3610.
31. Cavender, C. J.; Shiner, V. J., Jr. *JOC* **1972**, *37*, 3567.
32. Kokel, B.; Viehe, H. G. *AG(E)* **1980**, *19*, 716.
33. Lombardo, L.; Mander, L. N. *S* **1980**, 368.
34. Döpp, D.; Döpp, H. *MOC* **1990**, *E14B*, 1052.
35. (a) Balli, H.; Kersting, F. *LA* **1961**, *647*, 1. (b) Balli, H.; Low, R.; Müller, V.; Rempfler, H.; Sezen-Gezgin, G. *HCA* **1978**, *61*, 97.
36. Regitz, M.; Maas, G. *Diazo Compounds. Properties and Synthesis*; Academic: Orlando, 1986; p 384.
37. Example: $X = NR_2, R^1, R^2$ = cycloalkenyl, R^3 = H: Pocar, D.; Ripamonti, M. C.; Stradi, R.; Trimarco, P. *JHC* **1977**, *14*, 173.
38. (a) Croce, P. D.; Stradi, R. *T* **1977**, *33*, 865. (b) Bourgois, J.; Mathieu, A.; Texier, F. *JHC* **1984**, *21*, 513. (c) Quast, H.; Regnat, D.; Balthasar, J.; Banert, K.; Peters, E.-M.; Peters, K.; von Schnering, H. G. *LA* **1991**, 409.
39. Abramovitch, R. A.; Knaus, G. N.; Pavlin, M.; Holcomb, W. D. *JCS(P1)* **1974**, 2169.
40. (a) Ritchie, A. C.; Rosenberger, M. *JCS(C)* **1968**, 227. (b) Warren, B. K.; Knaus, E. E. *JHC* **1987**, *24*, 1413.
41. (a) Bailey, A. S.; Scattergood, R.; Warr, W. A. *JCS(C)* **1971**, 2479. (b) Harmon, R. E.; Wellman, G.; Gupta, S. K. *JHC* **1972**, *9*, 1191.
42. Gerlach, O.; Reiter, P. L.; Effenberger, F. *LA* **1974**, 1895.
43. Chakrasali, R. T.; Ila, H.; Junjappa, H. *S* **1988**, 851.
44. Stanovnik, B.; Tišler, M.; Kunaver, M.; Gabrijelčič, D.; Kočevar, M. *TL* **1978**, 3059.
45. Ortiz, M.; Larson, G. L. *SC* **1982**, *12*, 43.

Heinrich Heydt & Manfred Regitz
University of Kaiserslautern, Germany

p-Toluenesulfonylhydrazide[1]

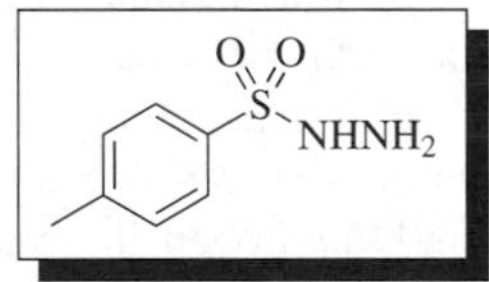

[1576-35-8] $C_7H_{10}N_2O_2S$ (MW 186.26)

(used as a source of diazene;[2] condensed with ketones and aldehydes to form hydrazones that can be converted into reactive intermediates such as diazoalkanes, carbenes, carbenium ions, alkyllithiums, or umpolung synthons;[7] used in 1,3-dipolar cycloaddition reactions to form N-heterocycles; used to make propargyl aldehydes and ketones via the Eschenmoser fragmentation;[44] ketone hydrazones are deoxygenated with mild reagents in a modified Wolff–Kishner reduction[50])

Alternate Name: tosyl hydrazide.
Physical Data: mp 108–110 °C (dec).
Solubility: sol virtually all organic solvents except hydrocarbons; insol water.
Form Supplied in: solid; widely available.
Handling, Storage, and Precautions: is thermally labile at elevated temperatures, but can be stored at or below room temperature. It is a toxic, potentially flammable solid which should be handled with gloves under inert atmosphere.

Alkene Reduction/Allylic Diazene Rearrangement. *p*-Toluenesulfonylhydrazide may be thermolyzed in solution to generate diazene (diimide) for alkene reduction.[2] It is the least reactive member of the common arenesulfonylhydrazides, with ***Mesitylenesulfonylhydrazide*** being 24 times more reactive and ***2,4,6-Triisopropylbenzenesulfonylhydrazide*** being 380 times more reactive under base-catalyzed decomposition conditions. The latter has largely supplanted tosyl hydrazide as a source of diazene because of the low-temperature convenience.[3,4]

Sigmatropic rearrangement of allylic and propargylic diazene intermediates is well established and is a reliable means of alkene synthesis.[5] Several options are available to generate these intermediates, such as hydride reduction of tosylhydrazones (refer to eqs 14 and 15), elimination of *p*-toluenesulfinic acid from an allylic hydrazine (eqs 1 and 2), Wolff–Kishner reduction of an α,β-unsaturated ketone, or oxidation of alkylhydrazines.[5]

p-Tolyl sulfones have been prepared in excellent yield by halide displacement with toluenesulfinate ion.[6]

Nucleophilic Addition to Hydrazones. The bulk of tosyl hydrazide's synthetic utility comes from its condensation products with ketones and aldehydes.[7,8] Synthetically useful addition reactions typically only succeed with nonenolizable ketone hydrazones.[9,10] The treatment of enolizable ketone tosyl hydrazones with alkyllithium nucleophiles invariably results in the formation of a dianion rather than addition to the azomethine bond. In contrast, aldehyde tosylhydrazones do not form dianions but instead undergo addition with both alkyllithiums and cuprate reagents to give reductive alkylation products.[11,12] A versatile extension of this methodology involves either vinyllithium addition to *N*-*t*-butyldimethylsilyl tosylhydrazones or the complementary 1,2-addition of alkyllithium to α,β-unsaturated *N*-*t*-butyldimethylsilyl tosylhydrazones. Reductive addition in either case produces an *N*-TBDMS allylic diazene intermediate which undergoes protodesilylation and subsequent 1,5-sigmatropic rearrangement. (*E*)-Alkenes are prepared with high stereoselectivity due to $A_{1,3}$-strain in the transition state (eqs 1 and 2), and epimerization of adjacent stereocenters is not observed.[5]

Other nucleophiles, such as cyanide ion, similarly add to aldehyde and ketone tosylhydrazones, which constitutes a useful one-carbon homologation.[13] ***Diphenylphosphine Oxide*** or dimethyl phosphite addition followed by decomposition of the intermediate with base or ***Sodium Borohydride*** gives alkyldiphenylphosphine oxides and *s*-alkanephosphonates, respectively.[14]

Formation of tosylhydrazones from α-halo ketones, followed by vinylogous halide elimination using mild base, gives *p*-toluenesulfonylazoalkenes, which undergo S_N2'-type addition of strong nucleophiles (eq 3).[15] This sequence constitutes a synthetically useful carbonyl umpolung synthon for introducing nucleophiles in the α-position of ketones. Examples of simple alkyllithium and 1,3-diketone enolate nucleophilic addition are also known.[16]

In another example, thermolysis of the tosylhydrazone of an α-hydroxy steroid in ethylene glycol without base resulted in reductive dehydroxylation, presumably through rearrangement of an intermediate epoxy diazene.[17]

Ph / N–NH(Ts) ; TBDMSOTf, NEt_3 ; THF, −78 °C ; N–N(Ts)TBDMS

1. BuLi, −78 °C ; 2. AcOH, −20 °C, CF_3CH_2OH ; −$TolSO_2H$; −TBDMS ; $-N_2$

(1)

(E):(Z) = 12:1, 88%

1. Li–CH=CH–Pr, −78 °C ; 2. AcOH, −20 °C, CF_3CH_2OH ; 86%

(2)

(E):(Z) = >20:1

Na_2CO_3 ; 1. Nu^- ; 2. H^+ (3)

X = Cl, Br

Nu^- =	
Ph_2CuLi	75%
Me_2CuLi	~50%
PhS^-	78%
$PhSe^-$	68%

Electrophilic Addition to Hydrazone Anions. α-Deprotonation of tosylhydrazones with 2 equiv of strong base give highly nucleophilic azaenolates that react with a variety of electrophiles.[18] Trisyl hydrazones have also received considerable attention for this transformation[19] (see ***2,4,6-Triisopropylbenzenesulfonylhydrazide***. However, some cyclohexanones, such as 4-*t*-butylcyclohexanone, require tosyl hydrazones for successful alkylation since the corresponding trisyl hydrazones undergo significant Shapiro elimination (see below) even at −78 °C.[18c]

The regioselectivity of deprotonation for unsymmetrical tosylhydrazones often is high, and is dictated by three factors. In many cases the configuration of the C=N double bond plays an important role since the sulfonamide anion induces *syn* deprotonation in ethereal solvents.[20] Thus the ratio of α vs. α′ anions reflects the C=N isomeric ratio, which is typically 85:15 (*E*):(*Z*) or greater in unsymmetrical ketones.[21,22] The '*syn*-directing effect' is inoperative in TMEDA solvent, allowing excellent regioselectivity of deprotonation, independent of the hydrazone geometrical isomer population.[23] Electron-stabilizing groups in the α-position direct deprotonation, and finally, the acidity of α-hydrogens follows the generalization that methyl > methylene > methine with only a few exceptions.

Bamford–Stevens and Shapiro Reactions. Tosylhydrazones may be converted into a variety of reactive intermediates depending upon the specific reaction conditions employed. The protic Bamford–Stevens reaction conditions (ethylene glycol; NaOR; heat) give diazoalkane intermediates that, on occasion, can be isolated but more often undergo in situ protonation followed by loss of dinitrogen to give cationic intermediates. These intermediates tend to give the more substituted double bond, but predicting regioselectivity is dubious, and cationic rearrangements are common (eq 4).[24]

TsNHN ; Na^0 ; ethylene glycol

38% + 17% (4)

Triisopropylbenzenesulfonyl and mesitylenesulfonyl hydrazones are far superior to tosyl hydrazones if isolation of the diazo compound is desired, primarily because much lower temperatures are required to induce the sulfinate elimination.[25]

The aprotic Bamford–Stevens reaction involves deprotonation of the acidic sulfonamide hydrogen in an aprotic solvent (e.g. glyme) to form a salt (Li, Na, and K are common) which is pyrolyzed or photolyzed to generate carbenes (eq 5).[26]

$TsNHNH_2$, AcOH, Δ ; toluene 82% ; NBS, DBU ; DMF 81% ; NaH, *hν*, dimethyl fumarate ; 55% ; CO_2Me (5)

Five- and six-membered ring ketones usually undergo 1,2-hydride migration to give preferentially the more substituted alkene.[27,12,4] Regioselectivity is usually modest but high selectivity is found in some instances, such as the synthesis of furanose glycals,[28] Δ^2 unsaturation in *trans*-A/B steroids, and Δ^3 unsaturation in *cis*-A/B steroids.[29] Nevertheless, examples of competitive cyclopropanation[30] and C–H bond insertion in medium-sized rings[31] abound.

The Shapiro reaction is extremely useful for the regioselective conversion of ketones into alkenes.[1] The anions of tosylhydrazones eliminate *p*-toluenesulfinate and dinitrogen between 0 °C and room temperature over the course of several hours to give the corresponding least-substituted alkene. Acyclic azaenolates have a strong preference for E_{CC} geometry and react equally well to yield (*Z*)-alkenes stereoselectively unless α-branching causes too much allylic strain in the transition state.[1a,3,8,32] This method has been used extensively in the synthesis of natural products such as alkaloids,[33] prostaglandins,[34] and especially terpenes (eq 6).[24a,35]

1. $TsNHNH_2$ 2. LDA −78 to 0 °C 76% (6)

The intermediate in this reaction is an alkenyllithium, which can be trapped under certain conditions with electrophiles,[21,36] providing general access to allylic alcohols, α,β-unsaturated aldehydes and acids,[37] vinyl halides, vinyl sulfides,[38] and vinylsilanes.[39] Use of trisylhydrazones typically gives superior results because the intermediate alkenyllithium is formed at lower temperature and does not require excess base to circumvent quenching of the intermediate by orthometalation.[40,21] Tandem one-pot electrophilic addition/Shapiro elimination sequences reliably provide alkenes with highly variable substitution patterns (eq 7), including tetrasubstitution.[18c]

Tosylhydrazones of α,β-unsaturated ketones likewise participate in the Shapiro elimination to give 2-lithioalkadiene intermediates which may be protonated to give the corresponding dienes, or be trapped with electrophiles.[41] Regioselectivity cannot be predicted with certainty, but in general an α′-proton is removed (giving the cross-conjugated azaenolate) in preference to a γ-proton, once again analogous to the kinetic deprotonation of α,β-unsaturated ketones.[42] Tosylhydrazones of β-keto esters are deprotonated to give trianions that decompose to give β,γ-unsaturated esters in fair yield.[22,8] Cyclic 1,3-diketone monotosylhydrazones analogously give enones when treated with ***Potassium Carbonate***.[43]

NHTs; *t*-Bu; 1. *s*-BuLi 2. BuI, −78 °C; NTs, Bu, *t*-Bu; *s*-BuLi −78 to 0 °C; NTS, Bu, 2 Li^+, *t*-Bu; −LiTs $-N_2$; Li, Bu, *t*-Bu; DMF; CHO, Bu, *t*-Bu, 43% (7)

Tanabe-Eschenmoser Fragmentation and Related Eliminations. Preparation of macrocyclic cycloalkynones by fragmentation of α,β-epoxy ketones is also possible with this reagent (eq 8).[44]

NHTs; base; N–Ts; O (8)

These fragmentation reaction conditions are too harsh to give the corresponding propargylic aldehydes in good yield.[45] More recently, other reagents have proved more efficacious for this transformation (see ***2,4-Dinitrobenzenesulfonylhydrazide*** and ***Mesitylenesulfonylhydrazide***).

Arenesulfonylhydrazones with leaving groups in the α-position proceed through well-documented intermediates or related fragmentation reactions. For instance, treatment of α-bromo tosylhydrazones with mild base triggers vinylogous halide elimination to give tosylazoalkenes (eq 9, and also refer to eq 3 and related text). Addition of strong base to these compounds provides 2-lithioalkadienes[46] and is a complementary route to the enone method described above.

NHTs, Br, *t*-Bu; NEt_3 −78 °C; NTs, *t*-Bu; MeLi; Li, *t*-Bu (9)

In acyclic cases the same intermediates seem to eliminate differently to give alkynes (eq 10).[47] eq 11 illustrates a Shapiro elimination in which the resultant vinyllithium intermediate fragments by displacing alkoxide to give the unanticipated hydroxyalkyne in high yield.[48] If deprotonation occurs on the α-carbon not occupied by the leaving group, the alkenyllithium intermediate fragments to a hydroxyallene in fair yield (eq 12).[49]

NHTs, X; base; N–Ts, H, H; base; H (10)

X = F, OSO_2Me

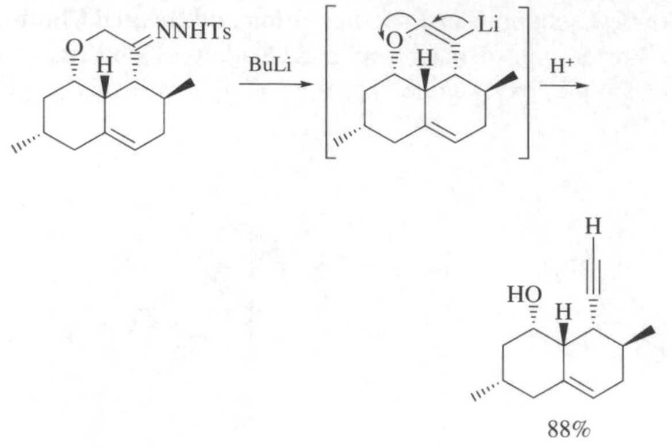

(11)

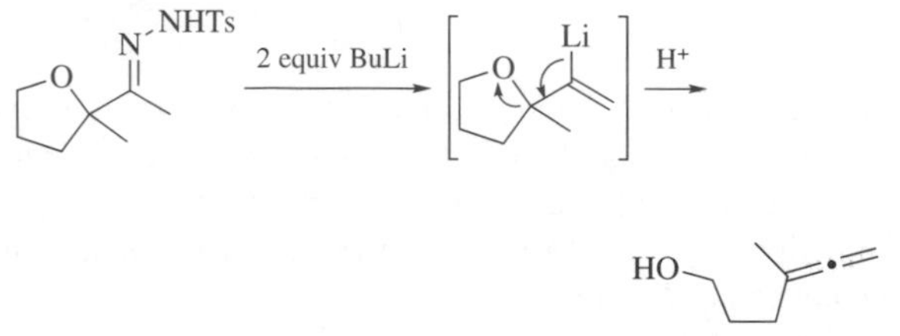

88%

(12)

Reductions of Tosylhydrazones[50]. Reduction of aldehyde and ketone tosylhydrazones in a Wolff–Kishner-type reaction was first described using ***Lithium Aluminium Hydride*** which was effective in several cases, but proved to be basic enough to produce Shapiro elimination side products.[51,52] More recently, milder reductions such as ***Sodium Borohydride***, ***Sodium Trimethoxyborohydride***, ***Sodium Triacetoxyborohydride***, ***Sodium Cyanoborohydride***, and ***Catecholborane*** have greatly expanded the scope and utility of these reductions well beyond the classic Wolff–Kishner and Clemmenson reductions, to become one of the mildest deoxygenation procedures known.

$NaBH_4$ is an effective tosylhydrazone reducing agent when used in methanolic dioxane ($NaBH(OMe)_3$ is most likely the active reducing species), acetic acid ($NaBH(OAc)_3$ is most likely the active reducing species), and less commonly in 2-propanol and DMF (eq 13).[53]

Selective formation of hydrazones (using a stoichiometric amount of tosyl hydrazide) in the presence of relatively hindered or deactivated ketones is easily achieved in high yield and, in the case of eq 13, the ketones are not reduced under the reaction conditions.[54] Reductions occur between room temperature and 70 °C for most tosylhydrazones without epimerization of α-stereocenters.[55] The reaction proceeds by reductive elimination of *p*-toluenesulfinate and extrusion of dinitrogen to form an alkyllithium, which is quenched by solvent. Use of $NaBD_4$ in AcOH or AcOD allows for mono- and dideuterio incorporation.[53c] Catecholborane is an equally effective reducing agent.[56]

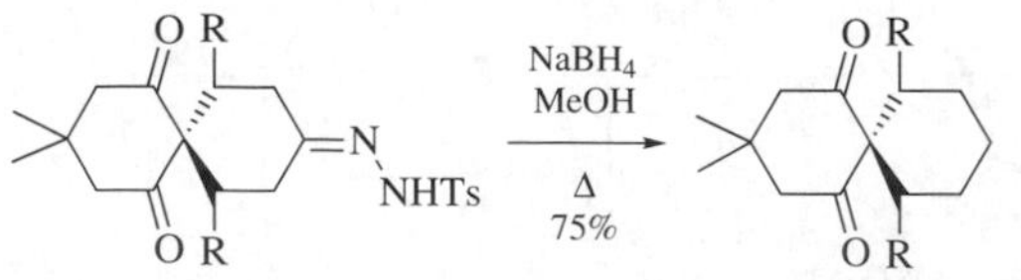

(13)

$NaBH_3CN$ has proved particularly serviceable for reducing tosylhydrazones that can be generated in situ, since iminium ions reduce much faster than carbonyl groups with this reagent.[57] The reactions are usually carried out in acidic 1:1 DMF—sulfolane at 100 °C, but even milder conditions have been used,[50b,58] providing products in fair to good yield.[59]

Reduction of α,β-unsaturated tosylhydrazones with the boron reagents gives 1,2-hydride delivery followed by extrusion of toluenesulfinic acid to give an allylic diazene. Note that this process gives a similar intermediate to that obtained by alkyl- and alkenyllithium addition to *N*-TBDMS tosylhydrazones (eqs 1 and 2). Thus, reduction of these systems gives allylic migration of the double bond independent of thermodynamic preference, and is an excellent means of deconjugating double bonds with an intervening methylene (eqs 14 and 15).[60]

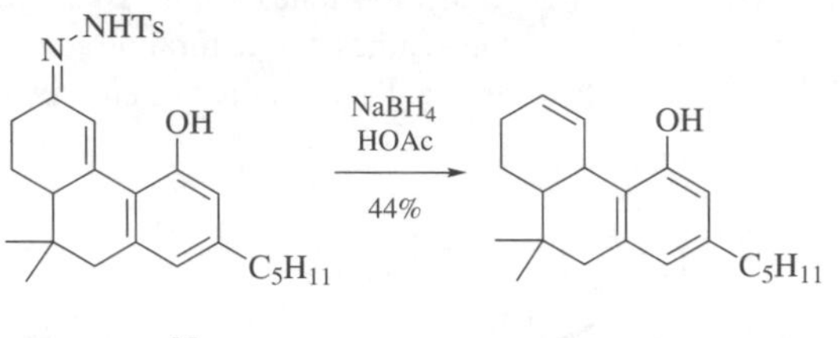

(14)

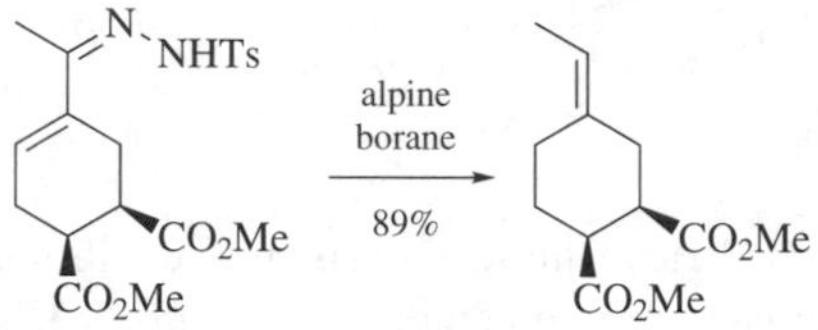

(15)

The $NaBH_4$/MeOH conditions do not give good yields in this process due to predominating Bamford–Stevens-type reactions.[61] $NaBH_4$ and $NaBH_3CN$ work equally well in AcOH but catecholborane is the best reagent for this transformation because 1,4-addition of hydride is minimal. Catecholborane adds axially to bridgehead enones and undergoes 1,5-sigmatropic reduction on the convex face to give *cis*-decalin systems with good selectivity.[62]

Miscellaneous. Intramolecular 1,3-dipolar cycloadditions of intermediate diazoalkanes derived from tosylhydrazones have been conducted under ***Boron Trifluoride Etherate*** or aprotic Bamford–Stevens reaction conditions to give pyrazolines and related compounds.[63,30a] [4 + 2] Cycloadditions of α,β-unsaturated aldehyde tosylhydrazones gave low yields of dihydropyridines.[64] 1,2-Enone transpositions have been described using vinylsilanes and vinyl sulfides.[65] Tosylhydrazones may also be converted into alkyl hydroperoxides via a two-step procedure.[66]

Related Reagents. 2,4-Dinitrobenzenesulfonylhydrazide; Hydrazine; Mesitylenesulfonylhydrazide; 2,4,6-Triisopropylbenzenesulfonylhydrazide.

---

1. (a) Shapiro, R. H. *OR* **1976**, *23*, 1976. (b) Chamberlin, A. R.; Bloom, S. H. *OR* **1990**, *39*, 1.
2. van Tamelin, E. E.; Dewey, R. S.; Timmons, R. J. *JACS* **1961**, *83*, 3725.
3. Cusack, N. J.; Reese, C. B.; Risius, A. C.; Roozpeikar, B. *T* **1976**, *32*, 2157.
4. Cusack, N. J.; Reese, C. B.; Roozpeikar, B. *CC* **1972**, 1132.
5. See Myers, A. G.; Kukkola, P. J. *JACS* **1990**, *112*, 8208 and references cited therein.
6. Ballini, R.; Marcantoni, E.; Petrini, M. *T* **1989**, *45*, 6791.
7. Bertz, S. H.; Dabbagh, G. *JOC* **1983**, *48*, 116.
8. Vinczer, P.; Novak, L.; Szantay, C. *SC* **1984**, *14*, 281.
9. Meinwald, J.; Uno, F. *JACS* **1968**, *90*, 800.

10. Herz, J. E.; Gonzales, E. *CC* **1969**, 1395.
11. Vedejs, E.; Stolle, W. T. *TL* **1977**, 135.
12. Bertz, S. H. *TL* **1980**, *21*, 3151.
13. (a) Cacchi, S.; Caglioti, L.; Paolucci, G. *S* **1975**, 120. (b) Jiricny, J.; Orere, D. M.; Reese, C. B. *JCS(P1)* **1980**, 1487.
14. (a) Bertz, S. H. *JACS* **1981**, *103*, 5932. (b) Mislanker, D. G.; Mugrage, B.; Darling, S. D. *TL* **1981**, *22*, 4619. (c) Inokawa, S.; Nakatsukasa, Y.; Horisaki, M.; Yamashita, M.; Yoshida, H.; Ogata, T. *S* **1977**, 179.
15. (a) Sacks, C. E.; Fuchs, P. L. *JACS* **1975**, *97*, 7372. (b) Reese, C. B.; Sanders, H. P. *JCS(P1)* **1982**, 2719. (c) Cacchi, S.; Felici, M.; Rosini, G. *JCS(P1)* **1977**, 1260.
16. Bernardi, L.; Masi, P.; Rosini, G. *AC(R)* **1973**, *63*, 601.
17. Paryzek, Z.; Martynow, J. *JCS(P1)* **1991**, 243.
18. (a) Shapiro, R. H.; Lipton, M. F.; Kolonko, K. J.; Buswell, R. L.; Capuano, L. A. *TL* **1975**, 1811. (b) Lipton, M. F.; Shapiro, R. H. *JOC* **1978**, *43*, 1409. (c) Bond, F. T.; DiPietro, R. A. *JOC* **1981**, *46*, 1315.
19. (a) Adlington, R. M.; Barrett, A. G. M. *CC* **1978**, 1071. (b) Adlington, R. M.; Barrett, A. G. M. *CC* **1979**, 1122. (c) Adlington, R. M.; Barrett, A. G. M. *JCS(P1)* **1981**, 2848.
20. Bergbreiter, D. E.; Newcomb, M. *Asymmetric Syntheses*; Academic: San Diego, 1983; Vol. 2.
21. Chamberlin, A. R.; Bond, F. T. *JOC* **1978**, *43*, 147.
22. Bunnell, C. A.; Fuchs, P. L. *JOC* **1977**, *42*, 2614.
23. (a) Chamberlin, A. R.; Bond, F. T. *S* **1979**, 44. (b) Stemke, J. E.; Bond, F. T. *TL* **1975**, 1815.
24. (a) Kutney, J. P.; Singhi, A. K. *CJC* **1983**, *61*, 1111. (b) Shapiro, R. H.; Heath, M. J. *JACS* **1967**, *89*, 5734. (c) Fetizon, M.; Jaudon, P. *T* **1977**, *33*, 2079. (d) Satyanarayana, N.; Nayak, U. R. *SC* **1985**, *15*, 1107.
25. Dudman, C. C.; Reese, C. B. *S* **1982**, 419.
26. (a) Mangholz, S. E.; Vasella, A. *HCA* **1991**, *74*, 2100. (b) Stierman, T. J.; Johnson, R. P. *JACS* **1985**, *107*, 3971. (c) Nickon, A.; Zurer, P. S. J. *JOC* **1981**, *46*, 4685.
27. Coxon, J. M.; Hartshorn, M. P.; Kirk, D. N.; Wilson, M. A. *T* **1969**, *25*, 3107.
28. Nair, V.; Sinhababu, A. K. *JOC* **1978**, *43*, 5013.
29. Caglioti, L.; Grasselli, P.; Selva, A. *G* **1964**, *94*, 537.
30. (a) Schultz, A. G.; Eng, K. K.; Kullnig, R. K. *TL* **1986**, *27*, 2331. (b) Remy, D. C.; King, S. W.; Cochran, D.; Springer, J. P.; Hirshfield, J. *JOC* **1985**, *50*, 4120.
31. Luyten, M. A.; Keese, R. *T* **1986**, *42*, 1687.
32. Kolonko, K. J.; Shapiro, R. H. *JOC* **1978**, *43*, 1404.
33. Ihara, M.; Ishida, Y.; Abe, M.; Toyota, M.; Fukumoto, K.; Kametani, T. *CL* **1985**, 1127.
34. Yamazaki, M.; Shibasaki, M.; Ikegami, S. *CL* **1981**, 1245.
35. (a) Fujita, T.; Ohtsuka, T.; Shirahama, H.; Matsumoto, T. *TL* **1982**, *23*, 4091. (b) Carda, M.; Arno, M.; Marco, J. A. *T* **1986**, *42*, 3655. (c) Scott, W. L.; Evans, D. A. *JACS* **1972**, *94*, 4779. (d) Grieco, P. A.; Nishizawa, M. *JOC* **1977**, *42*, 1717. (e) Girotra, N. N.; Reamer, R. A.; Wendler, N. L. *TL* **1984**, *25*, 5371.
36. Stemke, J. E.; Chamberlin, A. R.; Bond, F. T. *TL* **1976**, 2947.
37. Traas, P. C.; Boelens, H.; Takken, H. J. *TL* **1976**, 2287.
38. Nakai, T.; Mimura, T. *TL* **1979**, 531.
39. Paquette, L. A.; Fristad, W. E.; Dime, D. S.; Bailey, T. R. *JOC* **1980**, *45*, 3017.
40. Adlington, R. M.; Barrett, A. G. M. *ACR* **1983**, *16*, 55.
41. (a) van Tamelen, E. E.; Zawacky, S.; Russell, R. K.; Carlson, J. G. *JACS* **1983**, *105*, 142. (b) Dauben, W. G.; Lorber, M. E.; Vietmeyer, N. D.; Shapiro, R. H.; Duncan, J. H.; Tomer, K. *JACS* **1968**, *90*, 4762.
42. (a) Dauben, W. G.; Rivers, G. T.; Zimmerman, W. T. *JACS* **1977**, *99*, 3414. (b) Caille, J. C.; Farnier, M.; Guilard, R. *CJC* **1986**, *64*, 824. (c) Bütikofer, P.-A.; Eugster, C. H. *HCA* **1983**, *66*, 1148.
43. Hiegel, G. A.; Burk, P. *JOC* **1973**, *38*, 3637.
44. (a) Eschenmoser, A.; Felix, D.; Ohloff, G. *HCA* **1967**, *50*, 708. (b) Tanabe, M.; Crowe, D. F.; Dehn, R. L. **1967**, 3739, 3943.
45. (a) Corey, E. J. *JOC* **1975**, *40*, 579. (b) Felix, D.; Schreiber, J.; Piers, K.; Horn, U.; Eschenmoser, A. *HCA* **1968**, *51*, 1461.
46. Lightner, A.; Bouman, T. D.; Gawronski, J. K.; Gawronski, K.; Chappuis, J. L.; Crist, B. V.; Hansen, A. E. *JACS* **1981**, *103*, 5314.
47. Wieland, P. *HCA* **1970**, *53*, 171.
48. Hecker, S. J.; Heathcock, C. H. *JACS* **1986**, *108*, 4586.
49. Foster, A. M.; Agosta, W. C. *JOC* **1972**, *37*, 61.
50. For a recent, comprehensive review, see (a) Hutchins, R. O.; Hutchins, M. K. *COS* **1991**, *8*, 327. (b) Hutchins, R. O.; Natale, N. R. *OPP* **1979**, *11*, 201.
51. Caglioti, L.; Magi, M. *T* **1963**, *19*, 1127.
52. Fischer, M.; Pelah, Z.; Williams, D. H.; Djerassi, C. *CB* **1965**, *98*, 3236.
53. (a) Caglioti, L.; Grasselli, P. *CI(L)* **1964**, 153. (b) Caglioti, L. *T* **1966**, *22*, 487. (c) Hutchins, R. O.; Natale, N. R. *JOC* **1978**, *43*, 2299. (d) Gribble, G. W.; Lord, P. D.; Skotnicki, J.; Dietz, S. E.; Eaton, J. T.; Johnson, J. L. *JACS* **1974**, *96*, 7812.
54. Hoeve, W. T.; Wynberg, H. *JOC* **1980**, *45*, 2925.
55. (a) Caporusso, A. M.; Giacomelli, G.; Lardicci, L. *JCS(P1)* **1979**, 3139. (b) Tavernier, D.; Hosten, N.; Anteunis, M. *S* **1979**, 613.
56. (a) Kalbalka, G. W. *OPP* **1977**, *9*, 275. (b) Kalbalka, G. W.; Summers, S. T. *JOC* **1981**, *46*, 1217. (c) Kalbalka, G. W.; Chandler, J. H. *SC* **1979**, *9*, 275.
57. Borch, R. F.; Bernstein, M. D.; Durst, H. D. *JACS* **1971**, *93*, 2897.
58. (a) Nair, V.; Sinhababu, A. K. *JOC* **1978**, *43*, 5013. (b) Edwards, O. E.; Paryzek, Z. *CJC* **1983**, *61*, 1973. (c) Tochtermann, W.; Pahl, A.; Peters, E.-M.; Peters, K.; von Schnering, H. G. *CB* **1988**, *121*, 493.
59. (a) Yamamoto, Y.; Maruyama, K. *CC* **1984**, 904. (b) Bohlmann, F.; Lonitz, M. *CB* **1980**, *113*, 2410.
60. (a) Srebnik, M.; Lander, N.; Breuer, A.; Mechoulam, R. *JCS(P1)* **1984**, 2881. (b) Penco, S.; Angelucci, F.; Ballabio, M.; Barchielli, G.; Suarato, A.; Vanotti, E.; Vigevani, A.; Arcamone, F. *T* **1984**, *40*, 4677.
61. Grandi, R.; Messerotti, W.; Pagnoni, U. M.; Trave, R. *JOC* **1977**, *42*, 1352.
62. Kuo, C. H.; Patchett, A. A.; Wendler, N. L. *JOC* **1983**, *48*, 1991.
63. Miyashi, T.; Nishizawa, Y.; Fujii, Y.; Yamakawa, K.; Kamata, M.; Akao, S.; Mukai, T. *JACS* **1986**, *108*, 1617.
64. Allcock, S. J.; Gilchrist, T. L.; Shuttleworth, S. J. *T* **1991**, *47*, 10 053.
65. (a) Fristad, W. E.; Bailey, T. R.; Paquette, L. A. *JOC* **1978**, *43*, 1623. (b) Mimura, T.; Nakai, T. *CL* **1981**, 1579.
66. (a) Bloodworth, A. J.; Korkodilos, D. *TL* **1991**, *47*, 6953. (b) Caglioti, L.; Gasparrini, F.; Misiti, D.; Palmieri, G. *T* **1978**, *34*, 135.

A. Richard Chamberlin & James E. Sheppeck II
University of California, Irvine, CA, USA

p-Tolylsulfinylmethyllithium[1]

[1519-39-7] C_8H_9LiOS (MW 160.18)

(asymmetric synthesis; adds diastereoselectively to carbonyl compounds,[1] esters,[1] imines,[1b,f] nitrones,[2] and nitrile oxides[3])

Physical Data: stable in THF even above 0 °C.
Solubility: sol THF.
Form Supplied in: generated in situ from precursors, ***(R)-(+)-Methyl p-Tolyl Sulfoxide***, mp 73–75 °C, $[\alpha]^{20}$ +145° (c = 2, acetone), and (*S*)-(−)-methyl *p*-tolyl sulfoxide, mp 75–77 °C, $[\alpha]^{20}$ −147° (c = 2, acetone), both of which are commercially available and can also be prepared by a number of methods.[1,4]
Preparative Method: prepared in situ by the reaction of enantiomers of methyl *p*-tolyl sulfoxide with ***Lithium Diethylamide*** or ***Lithium Diisopropylamide*** in THF at 0 to −78 °C.[1]
Handling, Storage, and Precautions: highly sensitive to air and moisture. Its generation and handling must be carried out under an inert atmosphere of nitrogen or argon. Normally, preparation is carried out immediately prior to use.

Introduction. Most of the synthetic applications are carried out with (*R*)-(+)-*p*-tolylsulfinylmethyllithium. The carbanion undergoes copper-promoted oxidative dimerization to the optically pure 1,2-bis-sulfoxide (eq 1).[5]

p-Tol–S(O)–CH2Li (R)-(+) → [anhydrous $CuCl_2$, −78 °C, 45%] → p-Tol–S(O)–CH2CH2–S(O)–p-Tol (R,R) (1)

Addition to Carbonyl Compounds. The addition of *p*-tolylsulfinylmethyllithium to carbonyl compounds gives the β-hydroxy-α-*p*-tolylsulfinyl adducts in poor diastereoselectivity; normally, an approximately 1:1 diastereomeric mixture is obtained.[6] The diastereoselection with benzaldehyde is improved to about 80% with the addition of ***Zinc Chloride***.[7] Although proceeding with a poor diastereoselectivity, the reaction can be applied to the asymmetric synthesis of optically active alcohols. The two diastereomeric adducts can be separated by chromatography. Desulfinylation of each diastereomer gives the corresponding optically active alcohol (eq 2).[6] Optically active alcohols can also be prepared from epoxides (eq 3).[6]

p-Tol–S(O)–Me → [Et_2NLi, THF, −20 °C] → p-Tol–S(O)–CH2Li → [PhCHO, THF, −20 °C, 84%] → two diastereomeric β-hydroxy sulfoxides (1:1) → [Raney Ni, 60%] → each enantiomer of 1-phenylethanol (2)

p-Tol–S(O)–CH2Li → [cyclohexene oxide, DME, reflux, 64%] → 65% + 35% diastereomeric 2-(p-tolylsulfinylmethyl)cyclohexanols → [Raney Ni, 97%] → 65% + 35% 2-methylcyclohexanols (3)

(*R*)-(+)-*trans*-β-Styryl *p*-tolyl sulfoxide, produced by the reaction of the benzaldehyde adduct with excess ***Sodium Hydride*** and ***Iodomethane***, undergoes selective Michael reaction with diethyl sodiomalonate to give the (R_C,R_S) product as the major isomer, which can be obtained in pure form by fractional crystallization. The product can be transformed into optically active (−)-3-phenylbutyric acid (eq 4).[8]

p-Tol–S(O)–CH2CH(OH)Ph → [NaH, MeI, 75%] → p-Tol–S(O)–CH=CHPh (R)-(+) → [$CH_2(CO_2Et)_2$, 82%] → p-Tol–S(O)–CH2CH(Ph)CH(CO_2Et)2, (R_CR_S):(S_CR_S) = 8:2 → on (R_CR_S) → → PhCH(Me)CH2CO_2H (R)-(−) (4)

The dianion of ***(R)-(+)-3-(p-Tolylsulfinyl)propionic Acid***, prepared from the alkylation product of *p*-tolylsulfinylmethyllithium with lithium bromoacetate, adds to aldehydes to give diastereomeric lactones which can be separated by chromatography; their pyrolyses give optically pure 5-substituted furan-2(5*H*)-ones (eq 5).[9]

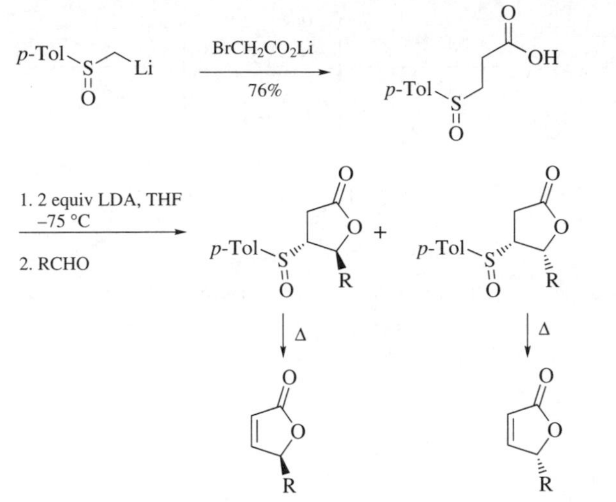

(5)

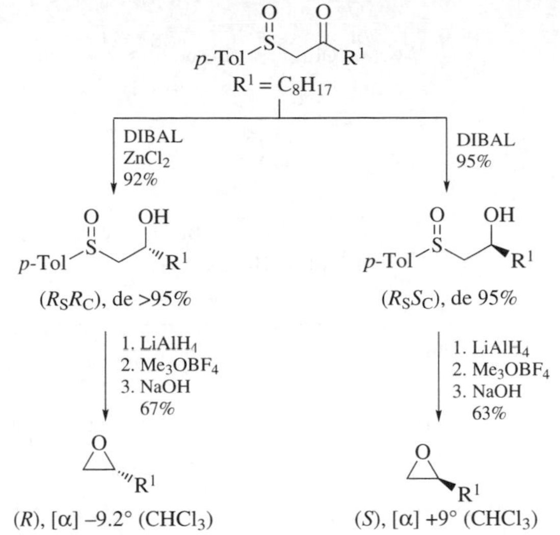

(8)

Addition to Esters. Racemic *p*-tolylsulfinylmethyllithium reacts with (*R*)-(−)-menthyl benzoate in enantiodifferentiating fashion to give the corresponding optically active β-keto sulfoxide together with optically active methyl *p*-tolyl sulfoxide which has the opposite configuration.[10] A similar result is obtained in the reaction of ethyl carboxylates in the presence of (−)-sparteine (eq 6).[11] The optically active β-keto sulfoxides, which are very important chiral synthons for the synthesis of optically active materials, are normally prepared by the reaction of (*R*)-*p*-tolylsulfinylmethyllithium with carboxylic acid esters.[1a] *p*-Tolylsulfinylmethyl phenyl ketone reacts with ***Diethylaluminum Cyanide*** in toluene diastereoselectively to give one diastereomer of the corresponding cyanohydrin which can be transformed into the optically active alcohol (eq 7).[12]

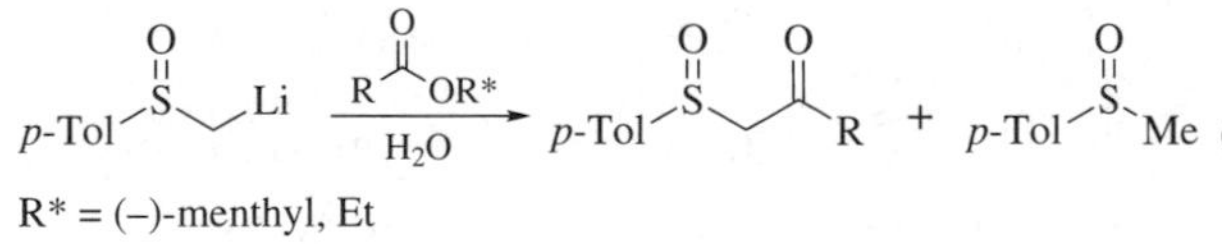

(6)

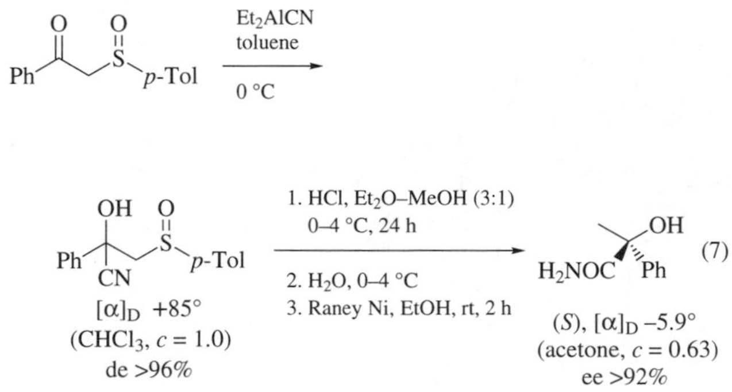

R* = (−)-menthyl, Et

(7)

β-Hydroxy sulfoxides of opposite stereochemistry can be prepared in very high diastereomeric excesses (90–95%) by the reduction of β-keto sulfoxides with ***Lithium Aluminium Hydride***, ***Diisobutylaluminum Hydride***, or DIBAL/$ZnCl_2$.[13] The optically pure diastereomers of the β-hydroxy sulfoxides can be transformed into optically active epoxides (eq 8),[14] alcohols (eq 9),[15] and 4-substituted butenolides (eq 10).[16]

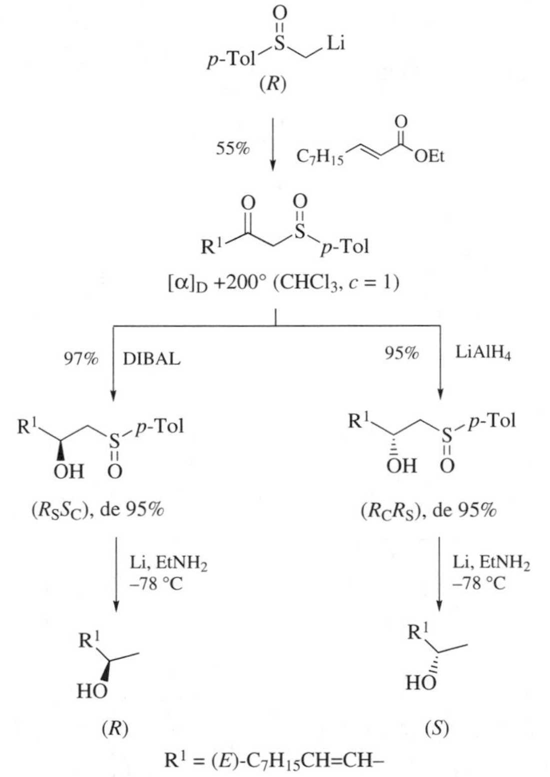

(9)

R^1 = (*E*)-C_7H_{15}CH=CH–

Addition to Arenesulfinic Esters. The reaction of (*R*)-(+)-*p*-tolylsulfinylmethyllithium with ***(−)-(1R,2S,5R)-Menthyl (S)-p-Toluenesulfinate*** gives (*S,S*)-bis(*p*-tolylsulfinyl)methane. (*S*)-(−)-*p*-Tolylsulfinylmethyllithium reacts similarly with (−)-menthyl (*S*)-*p*-toluenesulfinate, producing (*R,S*)-bis(*p*-tolylsulfinyl)methane (eq 11).[4c] In contrast to the poor diastereoselection in the reaction of (*R*)-(+)-*p*-tolylsulfinylmethyllithium with carbonyl compounds, (*S,S*)-(+)-bis(*p*-tolylsulfinyl)methyllithium reacts with aromatic aldehydes with high diastereoselectivity. Its reaction with α,β-unsaturated aldehydes yields 4-substituted (*E,E*)-(*S,S*)-1,1-bis(*p*-tolylsulfinyl)-1,3-butadiene (eq 12).[17]

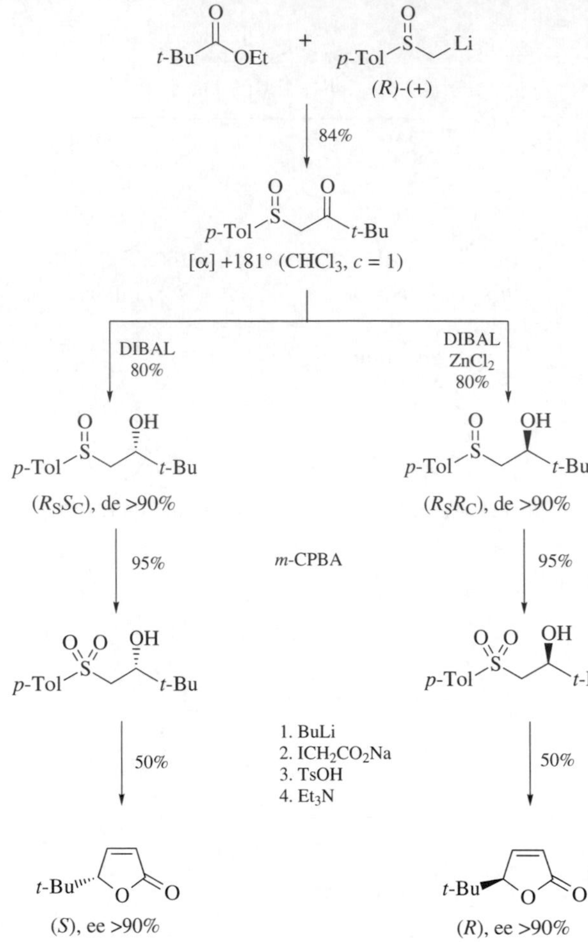

t-Bu–CO–OEt + p-Tol–S(O)–CH$_2$Li (R)-(+) → 84% → p-Tol–S(O)–CH$_2$–CO–t-Bu, [α] +181° ($CHCl_3$, c = 1)

DIBAL 80% → (R_SS_C), de >90%; DIBAL, $ZnCl_2$ 80% → (R_SR_C), de >90%

m-CPBA 95%

1. BuLi 2. ICH_2CO_2Na 3. TsOH 4. Et_3N — 50%

(S), ee >90%; (R), ee >90% (10)

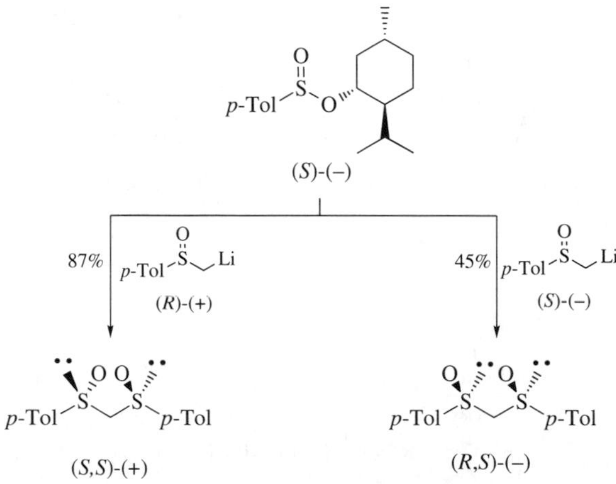

(S)-(–)

87% p-Tol–S(O)–CH$_2$Li (R)-(+) → (S,S)-(+); 45% p-Tol–S(O)–CH$_2$Li (S)-(–) → (R,S)-(–) (11)

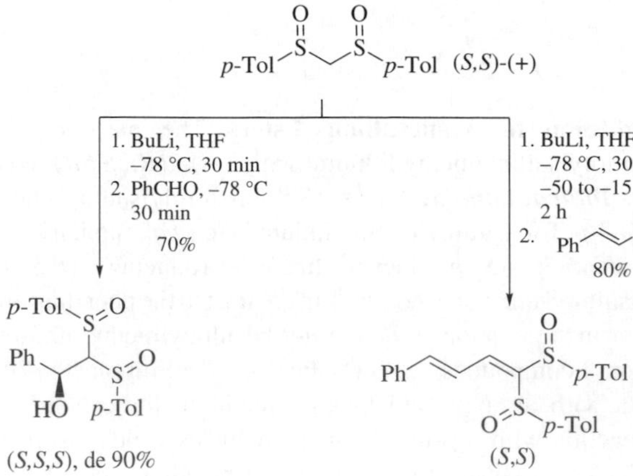

p-Tol–S(O)–CH$_2$–S(O)–p-Tol (S,S)-(+)

1. BuLi, THF –78 °C, 30 min 2. PhCHO, –78 °C 30 min 70% → (S,S,S), de 90%

1. BuLi, THF –78 °C, 30 min –50 to –15 °C 2 h 2. Ph–CH=CH–CHO 80% → (S,S) (12)

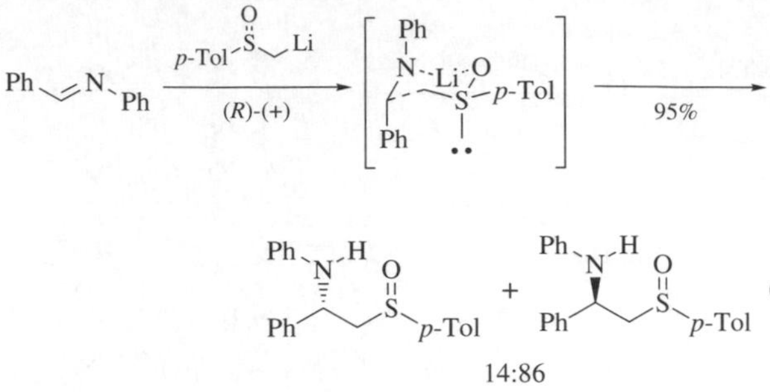

Ph–CH=N–Ph + p-Tol–S(O)–CH$_2$Li (R)-(+) → [intermediate] → 95% → 14:86 (13)

MeO, MeO dihydroisoquinoline + p-Tol–S(O)–CH$_2$Li 92% → de 92:8

2,3-dimethoxybenzaldehyde (CHO, MeO, MeO), $NaBH_3CN$ 87%

1. TFAA 2. Δ 82%

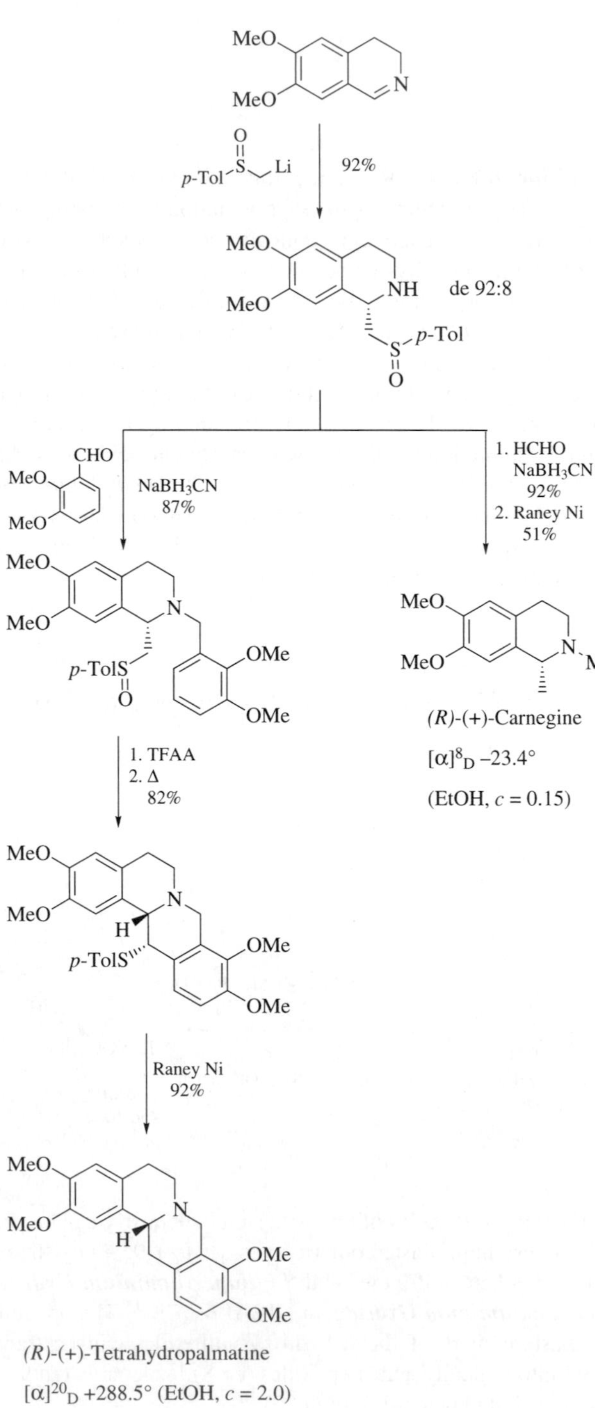

Raney Ni 92%

(R)-(+)-Tetrahydropalmatine

$[\alpha]^{20}_D$ +288.5° (EtOH, c = 2.0) (14)

1. HCHO $NaBH_3CN$ 92% 2. Raney Ni 51%

(R)-(+)-Carnegine

$[\alpha]^{8}_D$ –23.4°

(EtOH, c = 0.15)

Addition to Imines. The addition of (*R*)-(+)-*p*-tolylsulfinylmethyllithium to benzylideneaniline produces (R_SR_S)-(+)-*N*-phenyl-2-amino-2-phenylethyl *p*-tolyl sulfoxide in high diastereoselectivity (eq 13).[18] Acyclic and cyclic imines react similarly. The chair transition state can account for the diastereoselective process. Detailed investigations reveal that diastereoselectivity depends on several factors, i.e. the temperature used for deprotonation of methyl *p*-tolyl sulfoxide (optimum conditions: LDA, THF, 0 °C)[18c] and kinetically and thermodynamically controlled reaction conditions. Equilibrium between the diastereomeric adducts is thought to occur over the time scale of the addition reaction. Diastereoselectivity is better under kinetic control than under the thermodynamic control.[18b] This methodology has been applied to the syntheses of (*R*)-(+)-carnegine and (*R*)-(+)-tetrahydropalmatine (eq 14).[19]

(*R*)-(+)-p-Tolylsulfinylmethylmagnesium bromide reacts with 4-bromobutanenitrile to provide (*R*)-(+)-4,5-dihydro-2-(*p*-tolylsulfinylmethyl)-3*H*-pyrrole in low yield (eq 15).[20] This pyrrole derivative, which can be prepared in 92% yield by reaction of α-lithiated 3,4-dihydro-5-methyl-2*H*-pyrrole with (*S*)-(−)-menthyl *p*-toluenesulfinate, is utilized in the syntheses of (+)-elaeokanine A and (−)-elaeokanine B (eq 16).[20b] A similar approach has been utilized in the synthesis of yohimban alkaloids (eq 17).[20c]

p-Tol–S(O)–Me (*R*)-(+); *i*-Pr_2NMgBr, $Br(CH_2)_3CN$, 18%; LDA 92%

$[\alpha]^{20}_D$ +146° (CH_2Cl_2, c = 0.645) (15)

LDA, I(CH₂)₃I 61%; $NaBH_4$, 81%; 4:1; 76% 90%; 1. LDA, 2. Δ; 75% 90%; PCC 89%; PCC 93% (16)

(−)-Elaeokanine B $[\alpha]^{22}_D$ −76° ($CHCl_3$, c = 0.4)

(−)-Elaeokanine B $[\alpha]^{22}_D$ −49° ($CHCl_3$, c = 0.4)

(+)-Elaeokanine A $[\alpha]^{22}_D$ +49° ($CHCl_3$, c = 0.5)

LDA; (*S*)-(−)-menthyl *p*-toluenesulfinate 95%; LDA, CO_2Me 42%; 1:7; 1. Raney Ni, 82%; 2. $LiAlH_4$, 92% (17)

(−)-Alloyohimban $[\alpha]^{22}_D$ −166° (py, c = 0.4)

Addition to Nitrones and Nitrile Oxides. The reaction with 6,7-dimethoxy-3,4-dihydroisoquinoline *N*-oxide in the presence of quinidine is a highly diastereoselective process. The *N*-hydroxytetrahydroisoquinoline adduct can be desulfurized to (*R*)-(+)-salsolidine (eq 18).[2b] The addition of (*R*)-(+)-*p*-

tolylsulfinylmethyllithium to arylnitrile oxides gives β-oximino sulfoxides with high optical purity (eq 19).[3]

THF, –78 °C
68%

(18)

8:92

Raney Ni

(R)-(+)-Sasolidine
$[\alpha]_D$ +55.7° (EtOH, c = 1.40)

THF, –78 °C
78%

(19)

$[\alpha]_D^{25}$ +53.8° ($CHCl_3$, c = 1)

Related Reagents. (R)-(+)-t-Butyl 2-(p-Tolylsulfinyl)-acetate; Lithium Methylsulfinylmethylide; Methyl Phenyl Sulfone; (S)- or (R)-Menthyl p-Toluenesulfinate; (R)-(+)-Methyl p-Tolyl Sulfoxide; Phenylthiomethyllithium; (R)-(+)-p-Tolylsulfinylacetic Acid.

1. (a) Solladié, G. *S* **1981**, 185. (b) Walker, A. J. *TA* **1992**, *3*, 961. (c) Durst, T. In *Comprehensive Organic Chemistry*; Barton, D. H. R., Ed.; Pergamon: Oxford, 1979; Vol. 3, Chapter 11.6. (d) *Asymmetric Synthesis*; Morrison, J. D., Ed.; Academic: New York, 1983; Vol. 2. (e) Burbachyn, M. R.; Johnson, C. R. In *Asymmetric Synthesis*; Morrison, J. D., Ed.; Academic: New York, 1984; Vol. 3, Chapter 2. (f) Ogura, K. *COS* **1991**, *1*, Chapter 2.3. (g) Krief, A. *COS* **1991**, *3*, Chapter 1.3.
2. (a) Pyne, S. G.; Hajipour, A. R. *T* **1992**, *48*, 9385. (b) Murahashi, S.-I.; Sun, J.; Tsuda, T. *TL* **1993**, *34*, 2645.
3. Annunziata, R.; Cinquini, M. *S* **1982**, 929.
4. (a) Andersen, K. K. *TL* **1962**, 93. (b) Mislow, K.; Green, M. M.; Laur, P.; Melillo, J. T.; Simmons, T.; Ternay, A. L., Jr. *JACS* **1965**, *87*, 1958. (c) Kunieda, N.; Nokami, J.; Kinoshita, M. *BCJ* **1976**, *49*, 256. (d) Solladié, G.; Hutt, J.; Girardin, A. *S* **1987**, 173.
5. Maryanoff, C. A.; Maryanoff, B. E.; Tang, R.; Mislow, K. *JACS* **1973**, *95*, 5839.
6. (a) Tsuchihashi, G.-I.; Iriuchijima, S.; Ishibashi, M. *TL* **1972**, 4605. (b) Kunieda, N.; Kinoshita, M.; Nokami, J. *CL* **1977**, 289.
7. Braun, M.; Hild, W. *CB* **1984**, *117*, 413.
8. (a) Tsuchihashi, G.-I.; Mitamura, S.; Inoue, S.; Ogura, K. *TL* **1973**, 323. (b) Iwasaki, F.; Mitamura, S.; Tsuchihashi, G.-I. *BCJ* **1978**, *51*, 2530.
9. Albinati, A.; Bravo, P.; Ganazzoli, F.; Resnati, G.; Viani, F. *JCS(P1)* **1986**, 1405.
10. Kunieda, N.; Suzuki, A.; Kinoshita, M. *BCJ* **1981**, *54*, 1143.
11. Kunieda, N.; Kinoshita, M. *PS* **1981**, *10*, 383.
12. Ruano, J. L. G.; Castro, A. M. M.; Rodriguez, J. H. *TL* **1991**, *32*, 3195.
13. (a) Solladié, G.; Greck, C.; Demailly, G.; Solladié-Cavallo, A. *TL* **1982**, *23*, 5047. (b) Carreño, M. C.; Ruano, J. L. G.; Martín, A. M.; Pedregal, C.; Rodriguez, J. H.; Rubio, A.; Sanchez, J.; Solladié, G. *JOC* **1990**, *55*, 2120.
14. Solladié, G.; Demailly, G.; Greck, C. *TL* **1985**, *26*, 435.
15. (a) Solladié, G.; Demailly, G.; Greck, C. *JOC* **1985**, *50*, 1552. (b) Bravo, P.; Frigerio, M.; Resnati, G. *JOC* **1990**, *55*, 4216. (c) Solladie, G.; Fernandez, I.; Maestro, C. *TL* **1991**, *32*, 509.
16. Solladié, G.; Fréchou, C.; Demailly, G.; Greck, C. *JOC* **1986**, *51*, 1912.
17. Solladié, G.; Colobert, F.; Ruiz, P.; Hamdouchi, C.; Carreño, M. C.; Ruano, J. L. G. *TL* **1991**, *32*, 3695.
18. (a) Tsuchihashi, G.-I.; Iriuchijima, S.; Maniwa, K. *TL* **1973**, 3389. (b) Pyne, S. G.; Dikic, B. *CC* **1989**, 826. (c) Ronan, B.; Marchalin, S.; Samuel, O.; Kagan, H. B. *TL* **1988**, *47*, 6101.
19. (a) Pyne, S. G.; Chapman, S. L. *CC* **1986**, 1688. (b) Pyne, S. G.; Dikic, B. *JOC* **1990**, *55*, 1932.
20. (a) Hua, D. H.; Bharathi, S. N.; Takusagawa, F.; Tsujimoto, A.; Panangadan, J. A. K.; Hung, M.-H.; Bravo, A. A.; Erpelding, A. M. *JOC* **1989**, *54*, 5659. (b) Hua, D. H.; Bharathi, S. N.; Robinson, P. D.; Tsujimoto, A. *JOC* **1990**, *55*, 2128. (c) Hua, D. H.; Bharathi, S. N.; Panangadan, J. A. K.; Tsujimoto, A. *JOC* **1991**, *56*, 6998.

Vichai Reutrakul & Manat Pohmakotr
Mahidol University, Bangkok, Thailand

p-Tolylsulfonylmethyl Isocyanide[1]

[36635-61-7] $C_9H_9NO_2S$ (MW 195.26)

(reductive cyanation of ketones[2a] and aldehydes;[2b] synthesis of azoies (pyrroles, oxazoles, imidazoles, thiazoles, etc.) by delivering a C–N–C fragment to polarized double bonds;[3] connective reagent for coupling of alkyl halides (or carbonyl compounds) by a CO[4a] or a CH_2 bridge;[4b] preparation of (formal) Knoevenagel condensation products from aldehydes and ketones[5])

Alternate Name: tosylmethyl isocyanide; TosMIC.
Physical Data: mp 116–117 °C (dec).
Solubility: sol THF, CH_2Cl_2, $CHCl_3$, DME, AcOEt, benzene; slightly sol Et_2O, EtOH, MeOH.
Form Supplied in: white to near-white, odorless, commercially available solid.
Analysis of Reagent Purity: IR (Nujol) 2150 (N=C, 1320, 1155 cm^{-1} (SO_2); ^1H NMR ($CDCl_3$) δ 2.5 (s, CH_3), 4.6 (s, CH_2); ^{13}C NMR ($CDCl_3$) δ 61.1 (CH_2), 165.7 (N=C).

Preparative Method: by dehydration of *N*-(*p*-tolylsulfonylmethyl)formamide.[6]
Handling, Storage, and Precautions: shelf stable.

Introduction. *p*-Tolylsulfonylmethyl isocyanide is the best known compound of a series of (hetero-)substituted derivatives of ***Methyl Isocyanide***, which includes ***Diethyl Isocyanomethylphosphonate***, ***p-Tolylsulfonylmethyl Isocyanide***, ***Methyl N-(p-Tolylsulfonylmethyl)thiobenzimidate*** (a TosMIC derivative), and ***Ethyl Isocyanoacetate***. TosMIC is a multipurpose synthesis reagent, and by far the most versatile and most widely applicable reagent of the above series.

Typical Appllications. One characteristic example of each category of TosMIC applications will be given first (to be followed by further illustrative examples below).

Reductive Cyanations. Most ketones are converted with TosMIC in one operation into cyanides (introduction of a one-carbon unit) using ***Potassium t-Butoxide*** (1–7 equiv) in nonprotic solvents (e.g. DME, DMSO) (eq 1).[2a] The reductive cyanation of aldehydes is carried out at lower temperatures and needs addition of MeOH in the final stage of the process (eq 2).[2b,7] In eqs 1 and 2, the carbonyl oxygen is removed (unlike the well known addition of HCN); hence the name of this process: reductive cyanation.

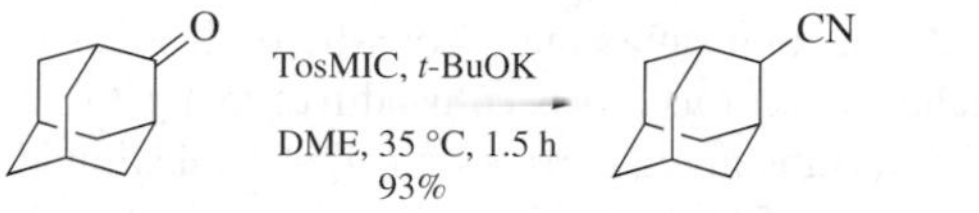

TosMIC, *t*-BuOK; DME, 35 °C, 1.5 h; 93% (1)

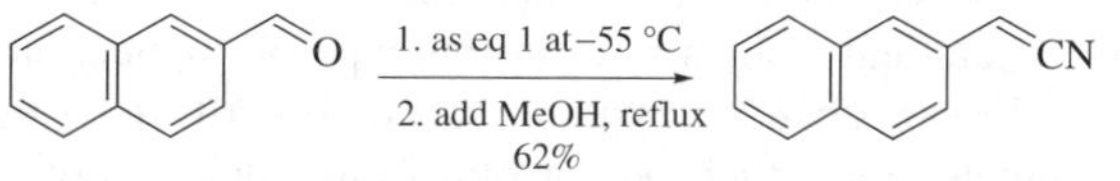

1. as eq 1 at –55 °C; 2. add MeOH, reflux; 62% (2)

Synthesis of Azoles. Reaction of TosMIC with aldehydes in protic solvents, e.g. MeOH at 20 °C, leads to oxazolines (eq),[8a] whereas oxazoles are formed in refluxing MeOH by elimination of *p*-toluenesulfinic acid (salt) (eq 4).[5c,8a] Reaction of TosMIC with acid chlorides, anhydrides, or esters leads to oxazoles in which the tosyl group is retained.[8]

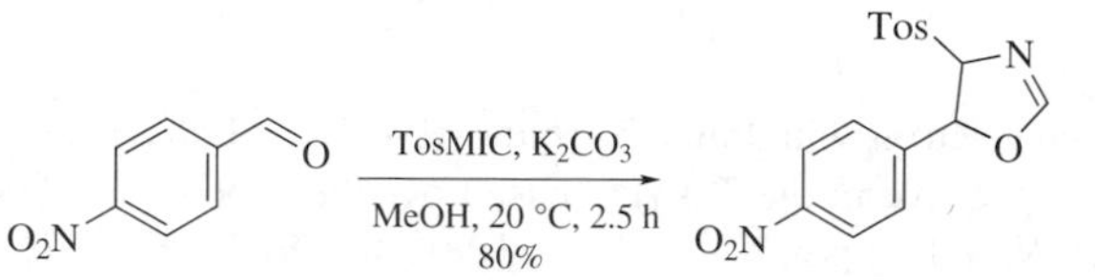

O_2N–C$_6$H$_4$–CHO → TosMIC, K_2CO_3; MeOH, 20 °C, 2.5 h; 80% → Tos-oxazoline (3)

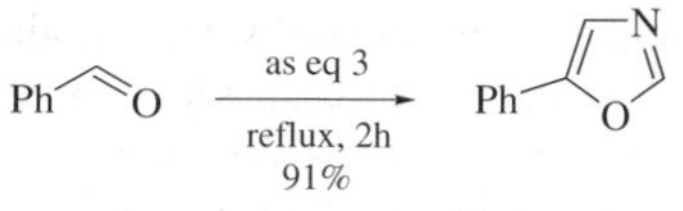

Ph–CHO → as eq 3; reflux, 2h; 91% → Ph-oxazole (4)

Imidazoles are obtained analogously from imines,[3b] thiazoles from dithioesters or CS_2,[9] and 1,2,4-triazoles from diazonium salts.[10] More widely applied, however, is the synthesis of pyrroles from TosMIC and Michael acceptors (eq 5).[11] The pyrroles are formed in one operation; dihydropyrrole derivatives (compare eqs 3 and 4) are not usually observed. The pyrrole ring positions 1, 2, and 5 remain intrinsically unsubstituted, which is one of the virtues of this method, since such pyrroles otherwise have to be prepared by temporarily using protective groups at these positions.[11]

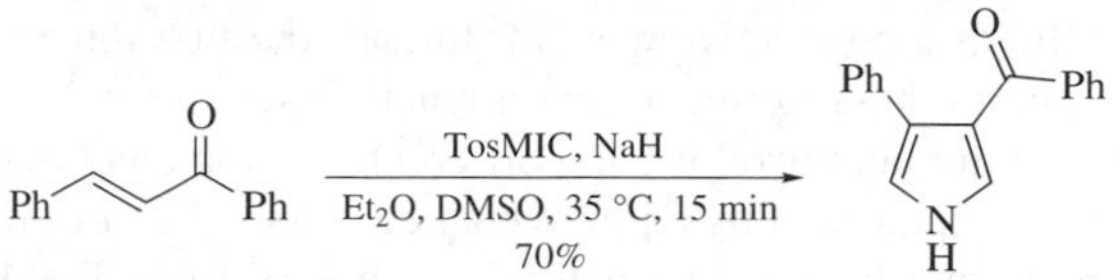

Ph–CH=CH–CO–Ph → TosMIC, NaH; Et_2O, DMSO, 35 °C, 15 min; 70% → 3-benzoyl-4-phenylpyrrole (N–H) (5)

Connective Reagent. The two methylene hydrogens of TosMIC have been replaced consecutively by base-induced alkylations (eqs 6 and 7).[12] Reactions as in eq 6 are carried out under phase-transfer catalysis (PTC) conditions to prevent twofold alkylation. In the product of eq 7, two alkyl halides (MeI and octyliodine) are connected by the TosMIC methylene group; hence the name: connective reagent. This application of TosMIC is complementary to other reagents with activated methylenes, such as ***1,3-Dithiane***[13] and ***Methylthiomethyl p-Tolyl Sulfone***. Acid hydrolysis of the geminal Tos and N=C groups produces ketones (i.e. a CO bridge, eq 7), whereas reduction with ***Lithium*** in liquid ammonia provides a methylene bridge (eq 8).[4b] Monosubstituted TosMIC derivatives, as obtained by eq 6, may replace TosMIC in eqs 3–5 to provide azoles with an additional substituent, e.g. an additional Me at position 2 of the pyrrole ring in eq 5.[3c]

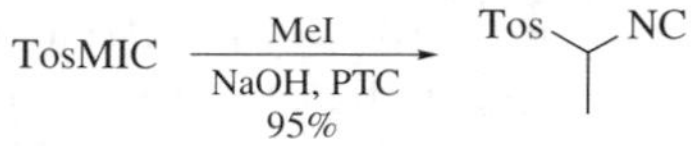

TosMIC → MeI; NaOH, PTC; 95% → Tos–CH(Me)–NC (6)

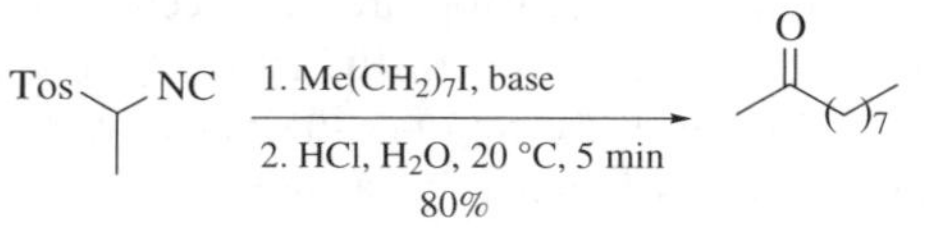

Tos–CH(Me)–NC → 1. $Me(CH_2)_7I$, base; 2. HCl, H_2O, 20 °C, 5 min; 80% → $CH_3CO(CH_2)_7CH_3$ (7)

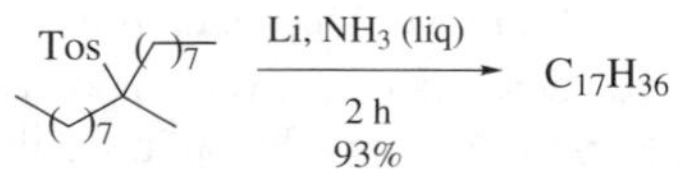

→ Li, NH_3 (liq); 2 h; 93% → $C_{17}H_{36}$ (8)

Knoevenagel Condensation Products. Reactions of TosMIC with aldehydes or ketones have been directed such that formal Knoevenagel condensation products are formed by overall elimination of H_2O (as contrasted to the reductive cyanation in eqs 1 and 2). In fact, this process requires two steps (eqs 9 and 10).[5b,d] Knoevenagel products have also been obtained by applying Peterson alkenation conditions to TosMIC (eq 11).[14]

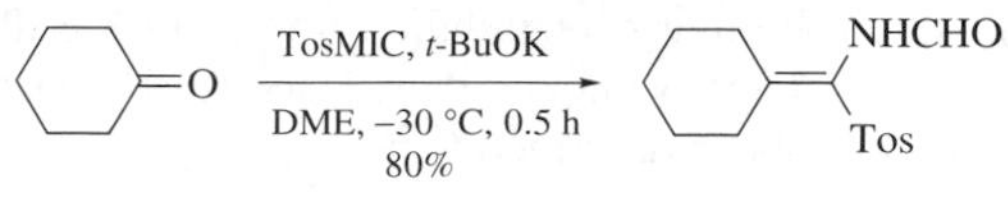

cyclohexanone → TosMIC, *t*-BuOK; DME, –30 °C, 0.5 h; 80% → cyclohexylidene–C(NHCHO)(Tos) (9)

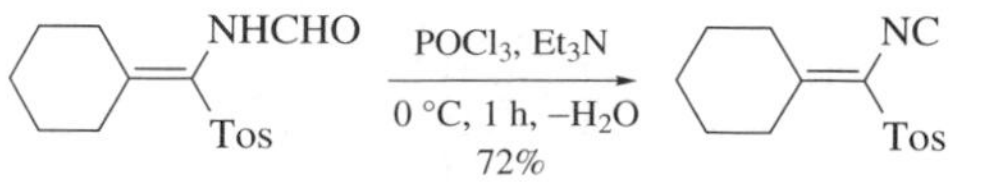

cyclohexylidene–C(NHCHO)(Tos) → $POCl_3$, Et_3N; 0 °C, 1 h, $-H_2O$; 72% → cyclohexylidene–C(NC)(Tos) (10)

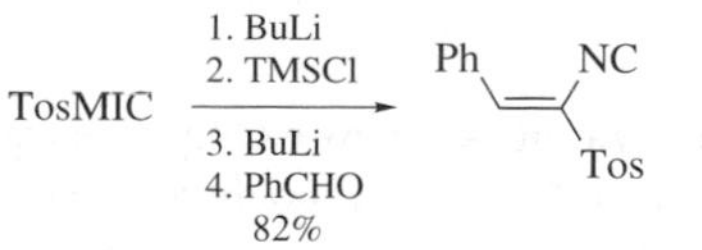

TosMIC → 1. BuLi; 2. TMSCl; 3. BuLi; 4. PhCHO; 82% → PhCH=C(NC)(Tos) (11)

The isocyano carbon of TosMIC is directly involved in the initial stages of the reductive cyanations, the azole syntheses, and the (formal) Knoevenagel condensations, but not in the application of TosMIC as a connective reagent. In fact, the first three types of reaction are based on a common reaction scheme;[2a,15] different products are obtained by using different reaction conditions. This point is emphasized by eq 12 which describes the three different products that have been obtained separately from TosMIC and benzaldehyde.[2b,5d,8a]

PhCHO + TosMIC: (a) 1. t-BuOK, DME, −50 °C, 10-60 min; 2. MeOH, reflux; cf. eqs 1 and 2 → $PhCH_2CN$; (b) K_2CO_3, MeOH; see eq 4; 91% → 5-phenyloxazole; (c) 1. t-BuOK, DME, −35 °C, 30 min; 2. $POCl_3$, Et_3N, 0 °C; cf. eqs 9-11; 47% → PhCH=C(NC)Tos (12)

Fundamental Aspects of TosMIC Chemistry. In this section, the interrelation of TosMIC reactions is discussed briefly. TosMIC is prepared in two steps: a Mannich reaction of ***p-Toluenesulfinic Acid*** (TosH), formaldehyde, and formamide gives *N*(*p*-tolylsulfonylmethyl)formamide (TosMIC precursor), which is dehydrated ($POCl_3/Et_3N$ or *i*-Pr_2NH) to TosMIC.[6] The Mannich reaction is reversible, but after dehydration to TosMIC the reverse reaction (of the TosMIC precursor) is blocked. Both TosMIC and the TosMIC precursor are *N,S*-acetals of formaldehyde. The methylene group of TosMIC is highly activated (estimated $pK_a = 14$) by the two electron-withdrawing substituents. Deprotonation of TosMIC has been achieved with an array of bases, ranging from ***Potassium Carbonate*** in MeOH to ***n-Butyllithium*** in THF. Even dilithio-MIC has been reported.[8b] Thus TosMIC is a formaldehyde derivative of reversed polarity (umpolung);[13] eqs 6–8 are based on this principle. Acid-catalyzed hydration of dialkylated TosMIC derivatives provides formamides, which upon hydrolysis (reversed Mannich) lead to the product ketones (eqs 7 and 35–38). In other applications of TosMIC, attack of the TosMIC anion on a carbonyl carbon is followed (or accompanied) by ring closure of the carbonyl oxygen to the electrophilic isocyano carbon to form an oxazoline (eq 3). Base-induced ring opening of these oxazolines gives α,β-unsaturated tosylformamides (eq 9), which have been dehydrated subsequently (eqs 10 and 12). Base-induced elimination of TosH from these unsaturated tosyl formamides gives *N*-formylketenimines ($R_2C{=}C{=}C{=}N\ CH{=}O$; not identified), which through nucleophilic removal of the formyl group eventually produce cyanides (eqs 1, 2, and 13–16). Details of these reactions have been described.[2a,15]

1-Bn-indole-3-carbaldehyde → TosMIC; 1. *t*-BuOK, DME, -50 °C; 2. MeOH, reflux, 15 min; 61% → 1-Bn-indole-3-acetonitrile (13)

Reductive Cyanation. Few examples have been described of the reaction of TosMIC with aldehydes. Examples are given in eqs 2 and 13.[2b,16] The reaction with ketones, however, is very general (eqs 1 and 14–16).[2a,15,16] Only severely sterically hindered ketones (e.g. di-*t*-butyl ketone) and readily enolizable ketones (e.g. benzyl phenyl ketone) will not undergo the reductive cyanation reaction.[2a]

t-BuC(O)Me → TosMIC; cf. eqs 3 and 4; 99% → oxazoline (Tos, *t*-Bu) → *t*-BuOK, THF; 20 °C, 10 min; 85% → *t*-BuCH(Me)CN (14)

Steroid 17-ketone (3-ethylene ketal) → TosMIC, 10 equiv *t*-BuOK; t-BuOH, DME, 20 °C, 3h; 88% → 17-CN; 17α:17β = 3:7 (15)

1,4-cyclohexanedione monoethylene ketal → TosMIC, *t*-BuOK; DME, 0-20 °C, 3h; 75% → 4-cyano ketal (16)

Synthesis of Azoles (Pyrroles in Particular). The reaction of TosMIC, or monosubstituted TosMIC derivatives (see eq 6), with Michael acceptors has been used frequently for the synthesis of 1,2,5-unsubstituted pyrroles and 1,5-unsubstituted pyrroles, respectively (eqs 5 and 21–24; compare eqs 25–29, 31, and 32). Over 50 papers and patents refer to this type of application of TosMIC.[17] Fewer papers deal with the use of TosMIC in the synthesis of other azoles, such as oxazoles from aldehydes (eq 4), imidazoles from imines (eq 17),[3b] thiazoles from dithioesters or CS_2 (eq 18),[9] and 1,2,4-triazoles from diazonium salts (eq 19).[10] Eqs 20 and 21 show that monosubstituted TosMIC molecules react similarly to give azoles with an additional substituent.[3c]

O_2N-C6H4-CH=N−Ph → TosMIC, K_2CO_3; MeOH, DME, 20 °C, 16h; 82% → 1-Ph-5-(4-O_2N-C6H4)imidazole (17)

The usual electron-withdrawing groups (EWG) in Michael acceptors are operative in the TosMiC based pyrrole synthesis (COR, CO_2R, CN, NO_2; eqs 5, 21–25, 27, and 28). Only for EWG = CHO, will the TosMIC anion react preferentially with the aldehyde group to form oxazoles,[5c,11] as in also the case with aromatic aldehydes (cf. eqs 4 and 20). Several base/solvent systems have been used in the TosMIC pyrrole syntheses; the trend is to use an excess of base (up to 3 equiv) with less reactive Michael acceptors are used, whereas lower temperatures (to −70 °C) are recommended for the more reactive Michael acceptors. TosMIC has been used for the synthesis of antibiotically active pyrroles such as verrucarine E (eq 22)[18a] and 3-cyano-4-(2,3-dichlorophenyl)pyrrole (Fenpiclonil (see eq 31). Three separate reactions have been reported between TosMIC and methyl sorbate (a dienyl Michael acceptor): depending on the conditions of the reaction, one the two monopyrroles or an oxazole is formed (eq 23).[8b] A bipyrrole has been synthesized from ethyl sorbate using 2 equiv of TosMIC. However, this process requires three steps. After the introduction of the first pyrrole ring, the remaining α,β-double bond needs to be activated (by *N*-sulfonylation of the first-formed pyrrole ring) before the second pyrrole ring can be realized (eq 24).[19a] Recently, a heptapyrrole has been developed along related lines.[19b] Other pyrrole syntheses based on TosMIC will be discussed in the next section.

CS_2 → TosMIC, 1. Bu_4NBr, $CHCl_3$, NaOH, 20 °C, 1.5 h; 2. MeI, 3 h; 90% (18)

Me_2N–C6H4–$\overset{+}{N}\equiv N\ BF_4^-$ → TosMIC, K_2CO_3, DMSO, MeOH, H_2O; 94% (19)

Tos–CH(CH3)–NC + furfural → t-BuOK, MeOH, 40 °C; 64% (20)

Ph–CH=CH–CN + Tos–CH(Bn)–NC → NaH, DMSO, Et_2O, 20 °C; 80% (21)

BnO–CH2–CH=CH–CO–CH3 → TosMIC, 1. NaH, Et_2O, DMSO 40 °C, 20 min; 2. H_2, Pd/C, AcOH 20 °C; 69% → Verrucarine E (22)

Methyl sorbate (CO_2Me) + TosMIC → NaH, DMSO, Et_2O, 20 °C, 1h, 60%; BuLi, THF, −70 to 0 °C, 2h, 34%; 2 equiv BuLi, THF, −70 to 0 °C, 2h, 25% (23)

Ethyl sorbate (CO_2Et) → TosMIC, NaH, Et_2O, DMSO, 20 °C, 3 h, 80% → TosMIC, 1. NaH, THF, $PhSO_2Cl$; 2. NaH, $HN(TMS)_2$, THF, 0 °C, 15 min; 74% → PhO_2S-bipyrrole, CO_2Et (24)

Knoevenagel-Type Condensation Products; Synthesis of (Di)Vinylpyrroles, Indoles, 3-Nitropyrroles, 3-Cyanopyrroles, and 3,4-Dialkylpyrroles. Two different types of reactions are possible with the (formal) Knoevenagel condensation products (eqs 9–12) of TosMIC. They can act as Michael acceptors (eqs 29–32) or as monosubstituted TosMIC derivatives. Eqs 25–28 provide examples of the latter type of reaction. γ-Deprotonation of the condensation product of TosMIC and cyclopentanone (eq 25)[5b] produces an allylic anion, which reacts exclusively via its α-carbon with Michael acceptors to form pyrroles in much the same way as in eq 21. The difference, clearly, is that in eq 25 a 2-(alk-1-enyl)pyrrole is formed. These 2-vinyl-pyrroles are efficient precursors for an alternate synthesis of indoles provided that they bear a second "vinyl' substituent at position 3. Thermal or photochemical electrocyclic ring closure of the 6π-electron system, followed by dehydrogenation, leads to indoles (eqs 26–28). When the "3-vinyl" substituent is part of an aromatic ring, electrocyclization is achieved photochemically (eq 26) but not thermally.[5b] Thermal ring closure is effective with normal 3-vinyl substituents, obtained via dienic and trienic Michael acceptors (eqs 27 and 28, respectively).[5b,20] The main product of eq 28 is formed by an intramolecular Diels–Alder reaction of the initially formed dihydroindole (not shown).[20]

In yet another synthesis of pyrroles, the Knoevenagel condensation products of TosMIC and aldehydes (eqs 11 and 12) are used as Michael acceptors (eqs 29–32). Eq 29 shows a highly efficient, one-step synthesis of 3-nitropyrroles (difficult to access otherwise),[21b] in which the ***Nitromethane*** carbon links the isocyano- and β-carbons of the Michael acceptor into a pyrrole ring.[21a] The same product of eq 29 has been obtained in the reaction of TosMIC and β-nitrostyrene in only 27% yield.[22] No reaction takes place when the nitromethane in eq 29 is replaced by acetonitrile. 3-Cyanopyrroles, however, are obtained effectively when cyanoacetate is used instead. Eq 31[3a] gives an example in the form of an alternate synthesis of a commercially employed seed-protecting agent, which is produced from TosMIC and ethyl α-cyano-2,3-dichlorocinnamate.[21c] This approach even applies to the synthesis of 3,4-dialkylpyrroles (eq 32),[3a] which obviously cannot be prepared from TosMIC and alkenes (compare eq 5).

Ph–CH=CH–CO–Ph + cyclopentylidene(NC)(Tos) → t-BuOK, THF, −50 to −20 °C, 1 h; 95% (25)

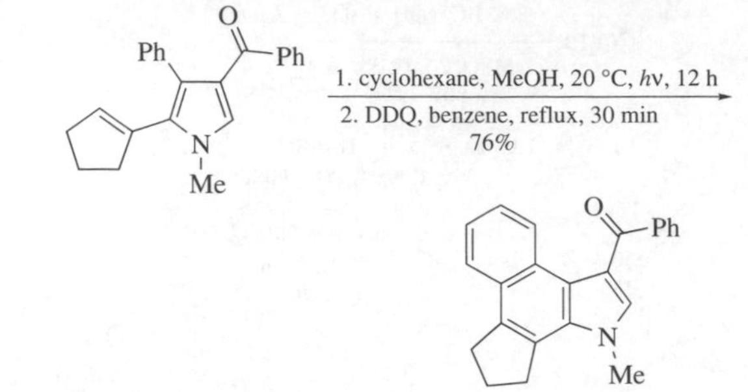

1. cyclohexane, MeOH, 20 °C, $h\nu$, 12 h
2. DDQ, benzene, reflux, 30 min
76%

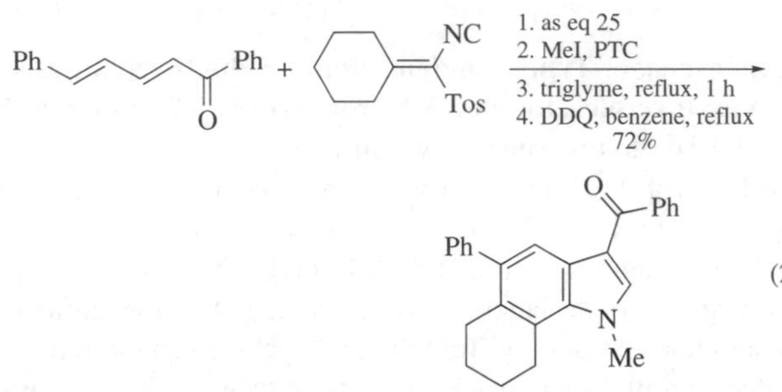

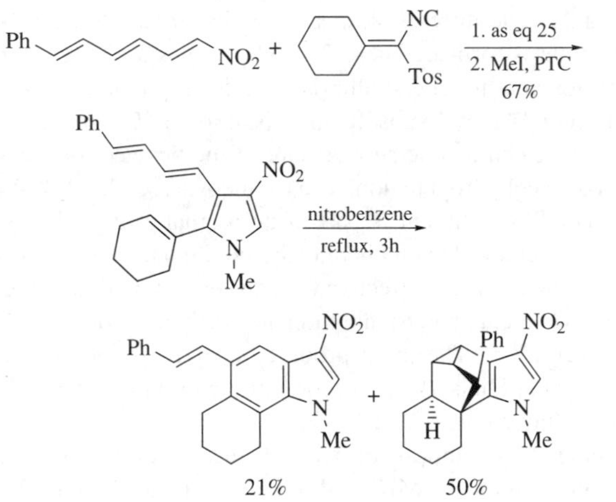

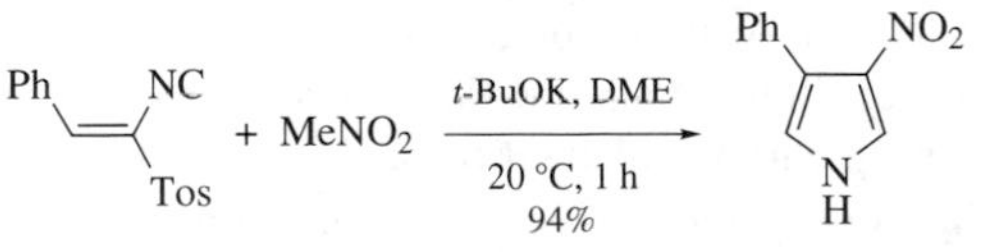

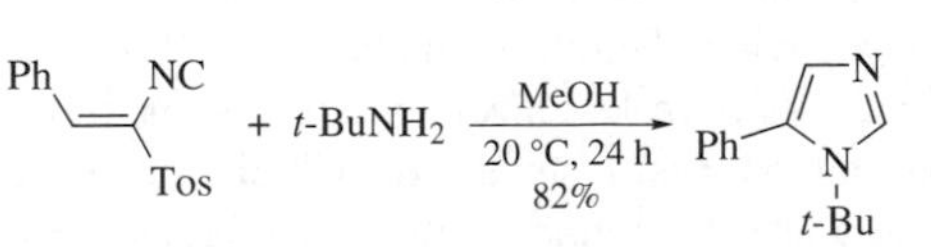

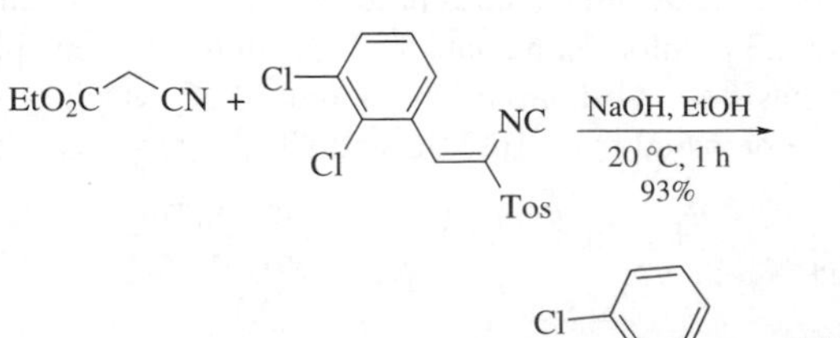

(26)

1. as eq 25
2. MeI, PTC
3. triglyme, reflux, 1 h
4. DDQ, benzene, reflux
72%

(27)

1. as eq 25
2. MeI, PTC
67%

nitrobenzene
reflux, 3h

21% 50%

(28)

t-BuOK, DME
20 °C, 1 h
94%

(29)

MeOH
20 °C, 24 h
82%

(30)

NaOH, EtOH
20 °C, 1 h
93%

Fenpiclonil®

(31)

t-BuOK, DME
20 °C, 1h
33%

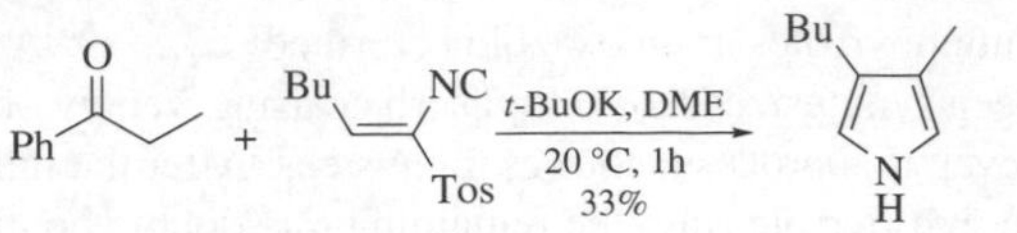

(32)

TosMIC as a Connective Reagent. The principle of using TosMIC as a connective reagent to form CO or CH_2 bridges has been exploited in several different ways (eqs 6–8 and 33–38). Eq 33 describes the synthesis of muscalure, a pheromone of the common house fly.[4b] The isocyano group is retained at the bridging carbon when electroreduction is applied (eq 34).[23] A practical synthesis of cyclobutanone is based on the use of TosMIC as an intramolecular connective reagent (eq 35),[24a] a method which has been extended to the first synthesis of (*R*)- and (*S*)-2-methylcyclobutanone from *rac*-1,3-dibromobutane and a chiral analog of TosMIC, (+)-neomenthylsulfonylmethyl isocyanide.[24b] Eq 36 gives another recent example of the synthesis of cyclic ketones.[24c] Several symmetrical and unsymmetrical acyclic diketones have been prepared by using TosMIC twice (eq 37).[4a] TosMIC has been used extensively in the construction of hydroxyacetyl side chains of corticosteroids and acetyl side chains as in progesterone (eq 38),[5e] starting from 17-oxo steroids.

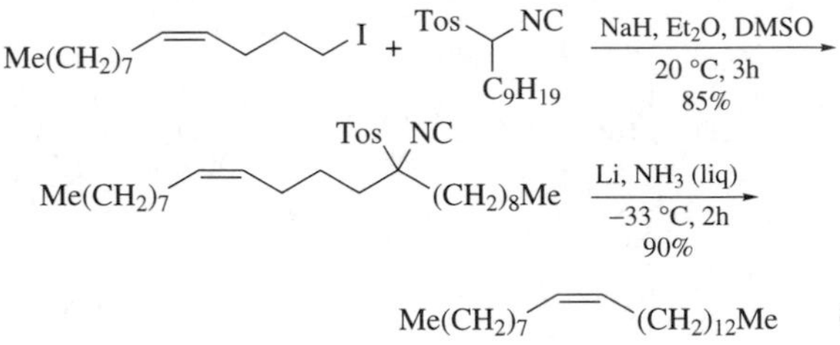

NaH, Et_2O, DMSO
20 °C, 3h
85%

Li, NH_3 (liq)
−33 °C, 2h
90%

Muscalure

(33)

TosMIC Analogs and Derivatives. Three labeled TosMIC compounds have been reported: $Tos^{14}CH_2N{=}C$,[2a] $TosCH_2N{=}^{13}C$,[18a] and $TosCH_2{}^{15}N{=}C$.[18b] Derivatives of TosMIC, of which one or both methylene hydrogens are replaced by alkyl, aryl, silyl, or alkylidene groups, have been included in the above discussion. Furthermore, the *p*-tolyl group of TosMIC has been replaced by various other aryl or alkyl groups,[1a,b] among which are several chiral groups.[24b,25a] The TosMIC sulfur has been made a stereogenic center by replacing one oxygen for a TosN= group.[25b] Also, both oxygens of the sulfonyl group have been formally removed, providing the reagent ***p-Tolylsulfonylmethyl Isocyanide***. Finally, the isocyano carbon of TosMIC has been equipped with two substituents, for example Ph and MeS groups,[26a] two MeO groups,[26a] or a $Ph_3C{-}N{=}$ group.[26b] All analogs and derivatives of TosMIC show TosMIC-like chemistry, but they have applied much less frequently than TosMIC itself.

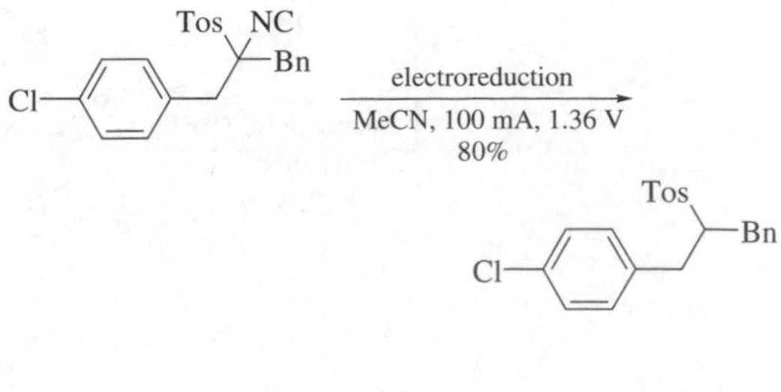

electroreduction
MeCN, 100 mA, 1.36 V
80%

(34)

Br Br TosMIC, NaH, DMSO Et_2O, 35 °C, 2h 71%

Tos NC H_2O, H_2SO_4, sulfolane 110 °C, 2.5 h 84% O (35)

Br Br OMe TosMIC 1. NaH, DMF, 20 °C, 5h 2. HCl, H_2O

O OMe OMe O + O O OMe OMe OMe O (36)

11% 14%

2 Tos NC + Br Br 1. NaH, DMSO, Et_2O, reflux, 45 min 2. HCl, H_2O, ET_2O, CH_2Cl_2, 20 °C, 15 min 73%

O O (37)

CN Tos MeO 1. $NaBH_4$, EtOH, THF, 20 °C, 1.5 h 2. MeI, PTC 3. HCl, H2O, CH2Cl2, 20 °C, 45 min 93%

CN O MeO Progestrone

Related Reagents. *N*,*N*-Diethylaminoacetonitrile; Diethyl Isocyanomethylphosphonate; Ethyl Isocyanoacetate; Isocyanomethyllithium; 2-Lithio-1,3-dithiane; Methoxyacetonitrile; Methyl Isocyanide; Methylthiomethyl *p*-Tolyl Sulfone; Methyl *N*-(*p*-Tolylsulfonylmethyl)thiobenzimidate; 1,1,3,3-Tetramethylbutyl Isocyanide; *p*-Tolylsulfonylmethyl Isocyanide.

---

1. (a) van Leusen, A. M. In *Perspectives in the Organic Chemistry of sulfur*; Zwanenburg, B.; Klunder, A. J. H., Eds.; Elsevier; Amsterdam, 1987; pp 119–144. (b) Leusen, A. M. *Lect. Heterocycl. Chem.* **1980**, *5*, S111. (c) *FF* **1974**, *4*, 514; **1975**, *5*, 684; **1977**, *6*, 600; **1979**, *7*, 377; **1980**, *8*, 493; **1982**, *10*, 409; **1984**, *11*, 539; **1986**, *12*, 511; **1988**, *13*, 313.
2. (a) Oldenziel, O. H.; van Leusen, D.; van Leusen, A. M. *JOC* **1977**, *42*, 3114. (b) van Leusen, A. M.; Oomkes, P. G. *SC* **1980**, *10*, 399.
3. (a) van Leusen, D.; van Echten, E.; Leusen, A. M. *JOC* **1992**, *57*, 2245, and footnote 3 therein. (b) van Leusen, A. M.; Wildeman, J.; Oldenziel, O. H. *JOC* **1977**, *42*, 1153. (c) Possel, O.; van Leusen, A. M. *H* **1977**, *7*, 77.
4. (a) van Leusen, A. M.; Oosterwijk, R.; van Echten, E.; van Leusen, D. *RTC* **1985**, *105*, 50. (b) Yadav, J. S.; Reddy, P. S.; Joshi, B. V. *T* **1988**, *44*, 7243.
5. (a) van Leusen, D.; van Echten, E.; van Leusen, A. M. *RTC* **1992**, *111*, 469. (b) Moskal, J.; van Leusen, A. M. *JOC* **1986**, *51*, 4131. (c) Moskal, J.; van Stralen, R.; Postma, D.; van Leusen, A. M. *TL* **1986**, *27*, 2173. (d) van Leusen, A. M.; Schaart, F. J.; van Leusen, D. *RTC* **1979**, *98*, 258. (e) van Leusen, D.; van Leusen, A. M. *S* **1991**, 531.
6. (a) Hoogenboom, B. E.; Oldenziel, O. H.; van Leusen, A. M. *OSC* **1988**, *6*, 987; meanwhile the yield of the first reaction step has been improved from 42–47% to 86–90%, see: Tezaki, K.; Nakayama, S.; Miyazaki, Y., Sugita, Y. Jpn. Patent 61 186 359 (*CA* **1987**, *106*, 13 138t); Barendse, N. C. M. Eur, Patent 242 001 (*CA* **1988**, *109*, 24 508s). (b) Obrecht, R.; Herrmann, R.; Ugi, I. *S* **1985**, 400.
7. Oldenziel, O. H.; Wildeman, J.; van Leusen, A. M. *OSC* **1988**, *6*, 41.
8. (a) van Leusen, A. M.; Hoogenboom, B. E.; Siderius, H. *TL* **1972**, 2369. (b) van Nispen, S. P. J. M.; Mensink, C.; van Leusen, A. M. *TL* **1980**, *21*, 3723.
9. van Leusen, A. M.; Wildeman, J. *S* **1977**, 501.
10. van Leusen, A. M.; Hoogenboom, B. E.; Houwing, H. A. *JOC* **1976**, *41*, 711.
11. van Leusen, A. M.; Siderius, H.; Hoogenboom, B. E.; van Leusen, D. *TL* **1972**, 5337.
12. (a) van Leusen, A. M.; Bouma, R. J.; Possel, O. *TL* **1975**, 3487. (b) van Leusen, A. M.; Possel, O. *TL* **1977**, 4229.
13. Gröbel, B.-T.; Seebach, D. *S* **1977**, 357.
14. van Leusen, A. M.; Wildeman, J. *RTC* **1982**, *101*, 202.
15. van Leusen, D.; van Leusen, A. M. *RTC* **1991**, *110*, 402.
16. (a) Merour, J. Y.; Buzar, A. *SC* **1988**, *18*, 2331. (b) Bull, J. R.; Tuinman, A. *T* **1975**, *31*, 2151. (c) Becker, D. P.; Flynn, D. L. *S* **1992**, 1080.
17. See footnote 3 in Ref. 3a.
18. (a) Gossauer, A.; Suhl, K. *HCA* **1976**, *59*, 1698. (b) Cappon, J. J.; Witters, K. D.; Verdegem, P. J. E.; Hoek, A. C.; Luiten, R. J. H.; Raap, J.; Lugtenburg, J. *RTC* **1994**, *113*, 318.
19. (a) Magnus, P.; Gallagher, T.; Schultz, J.; Or, Y.-S.; Ananthanarayan, T. P. *JACS* **1987**, *109*, 2706. (b) Magnus, P.; Danikiewicz, W.; Katoh, T.; Huffman, J. C.; Folting, K. *JACS* **1990**, *112*, 2465.
20. Leusink, F. R.; ten Have, R.; van den Berg, K. J.; van Leusen, A. M. *CC* **1992**, 1401.
21. (a) van Leusen, D.; Flentge, E.; van Leusen, A. M. *T* **1991**, *47*, 4639. (b) Barton, D. H. R.; Kervagoret, J.; Zard, S. Z. *T* **1990**, *46*, 7587. (c) See footnotes 8 and 9 in Ref. 3a.
22. Ref. 21a; meanwhile the yield of that reaction has been improved to 70% by using NaH in DME at −40 °C at lower concentrations (ten Have, R., unpublished results).
23. Hesz, U.; Brosig, H.; Fehlhammer, W. P. *TL*, **1991**, *32*, 5539.
24. (a) van Leusen, D.; van Leusen, A. M. *S* **1980**, 325. (b) van Leusen, D.; Rouwette, P. H. F. M.; van Leusen, A. M. *JOC* **1981**. *46*, 5159 ((*R*)- and (*S*)-2-methylcyclobutanone were not obtained enantiomerically pure). (c) Breitenbach, J.; Vögtle, F. *S S* **1992**, 41.
25. (a) Hundscheid, F. J. A.; Tandon, V. K.; Rouwette, P. H. F. M.; van Leusen, A. M. *T* **1987**, *43*, 5073. (b) van Leusen, D.; van Leusen, A. M. *RTC* **1984**, *103*, 41.
26. (a) Houwing, H. A.; Wildeman, J.; van Leusen, A. M. *JHC* **1981**, *18*, 1133. (b) van Leusen, A. M.; Jeuring, H. J.; Wildeman, J.; van Nispen, S. P. J. M. *JOC* **1981**, *46*, 2069.

Albert M. van Leusen & Daan van Leusen
Groningen University, The Netherlands

Tri-n-butylchlorostannane

n-Bu_3SnCl

[1461-22-9] $C_{12}H_{27}ClSn$ (MW 325.55)

(precursor of other organotributyltin reagents;[1] allylic tributyltins for the diastereoselective synthesis of homoallyl alcohols;[2] tributyltin enolates for alkylation[3] and regioselective C–C bond formation[4])

Alternate Name: tributyltin chloride.
Physical Data: bp 171–173 °C/25 mmHg; n_D^{20} 1.400; d 1.200 g cm^{-3}; fp >110 °C.
Form Supplied in: colorless liquid; widely available.
Solubility: sol ether, THF, hexane, CH_2Cl_2, MeOH.
Handling, Storage, and Precautions: in general, tributyltin compounds are highly toxic. The LD_{50} value for oral administration of Bu_3SnCl to rats is 122–349 μg g^{-1}. This reagent is absorbed through the skin to induce temporary skin burns. It must be handled with gloves in a well-ventilated hood. The reagent is air-stable, but slowly decomposes in presence of moisture.

Precursor to other Tributyltin Reagents. Since the chlorine atom can be readily replaced by a variety of nucleophiles, tributyltin chloride is widely used for the preparation of other synthetically useful organotin compounds[1] such as ***Tri-n-butylstannane***,[5] ***Tri-n-butyl(methoxy)stannane***,[6] diethylaminotributyltin,[7] ***Cyanotributylstannane***,[8] ***Tri-n-butyltin Azide***,[9] ***Hexabutyldistannane***,[10] and ***Bis(tri-n-butyltin) Oxide*** (eq 1).[11]

Bu_3SnCl → (NaOMe) Bu_3SnOMe; (LiAlH$_4$) Bu_3SnH; (NaOH) $Bu_3SnOSnBu_3$; (Na) $Bu_3SnSnBu_3$; (Et$_2$NLi) Bu_3SnNEt_2; (KCN) Bu_3SnCN; (NaN$_3$) Bu_3SnN_3 (1)

Bu_3SnCl →(Li) Bu_3SnLi →(1. $C_6H_{13}CHO$, −78 °C; 2. ClCH$_2$OCH(CH$_3$)CH$_3$-type chloroether, PhNMe$_2$, 0°C; 97%)

C_6H_{13}CH(Bu_3Sn)–O–CH(CH$_3$)–O–Et →(1. BuLi, −78 °C; 2. cyclohexanone; 81%) 1-hydroxycyclohexyl–CH(C_6H_{13})–O–CH(CH$_3$)–O–Et (2)

OTBDMS-substituted 4-chlorocyclopentenone →($(Bu_3Sn)_2CuLi$, −25 °C, 84%) OTBDMS-substituted 4-($SnBu_3$)cyclopentenone (3)

Tri-n-butylstannyllithium is readily prepared by the treatment of tributyltin chloride with ***Lithium*** metal[12] and is a valuable reagent for the preparation of organotin compounds.[1a]

The stannyllithium is also prepared from n-Bu_3SnH and ***Lithium Diisopropylamide***, or from $Bu_3SnSnBu_3$ and ***n-Butyllithium***. Among these preparative methods, the procedure using n-Bu_3SnH provides the highest yields of n-Bu_3SnLi.[10a] The reaction of the stannyllithium reagent with aldehydes followed by alcohol protection affords α-alkoxystannanes, which are treated with n-butyllithium to yield α-alkoxyllithium derivatives (eq 2).[14] ***Tri-n-butylstannylcopper*** (or cuprate) reagents are prepared from the stannyllithium and copper salts.[15] These reagents are employed for the preparation of β-stannyl-α-β-unsaturated carbonyl compounds (eq 3).[16]

Allyl-, Alkenyl-, and Alkynyltin Reagents. Simple allyltins are prepared by the reaction of Grignard reagents with tributyltin chloride.[17] γ-Alkoxyallyltins are obtained by trapping the allyl anions generated from allyl ethers.[2a] In this case, tributyltin chloride attacks the γ-position of the allyl anions to produce the (Z)-allyltins, owing to the strong coordination ability of the oxygen atom toward the lithium atom. γ-Methoxyallyltin reacts with aldehydes in the presence of Lewis acids to give *syn*-1,2-diol derivatives with high diastereoselectivity (eq 4).[2] An intramolecular allylic tin-aldehyde condensation has been applied to the synthesis of medium-ring cyclic ethers (eq 5).[18]

MeO–CH$_2$CH=CH$_2$ →(s-BuLi, TMEDA, −78 °C) [Li-chelated allyl anion] →(Bu_3SnCl, −78 to 0°C, 90%) (Z)-MeOCH=CHCH$_2SnBu_3$ →(PhCHO, $BF_3·OEt_2$, −78 °C, 86%) Ph–CH(OH)–CH(OMe)–CH=CH$_2$ (syn) + (anti) 10:1 (4)

Tetrahydropyranyl aldehyde with (Z)-alkoxyallyl $SnBu_3$ side chain →($BF_3·OEt_2$, −78 °C, 99%) fused oxepane with OH and vinyl groups (5)

HC≡C–TMS →(PrCu, (EtO)$_2$P, −20 °C) [Pr–CH=C(TMS)Cu] →(Bu_3SnCl, HMPA, −50 °C, 56%) Pr–CH=C(TMS)$SnBu_3$ (6)

C_6H_{13}C≡CH →(DIBAL, 0 °C) [C_6H_{13}CH=CH–Al(i-Bu)$_2$] →(Bu_3SnCl, LiCl, −25 °C, 80%) C_6H_{13}CH=CH–$SnBu_3$ (7)

C_8H_{17}C≡CH →($Bu_3SnAlEt_2$, CuCN, −30 °C) [C_8H_{17}C($SnBu_3$)=CH–AlEt$_2$] →(allyl bromide, Pd(PPh$_3$)$_2$Cl$_2$, DIBAL, 25 °C, 89%) C_8H_{17}C($SnBu_3$)=CH–CH$_2$CH=CH$_2$ (8)

Tributyltin chloride reacts with alkenylcopper reagents, generated by the reaction of alkylcoppers and terminal alkynes, to give (*Z*)-alkenyltin derivatives (eq 6).[19] In contrast, the reaction of the alkenylaluminum obtained by hydroalumination affords the (*E*)-isomer (eq 7).[20] Dienyltin derivatives are prepared via the hydrozirconation of 1-en-3-yne.[21] Terminal alkynes react with n-$Bu_2SnAlEt_2$, prepared in situ from n-Bu_3SnLi and ***Diethylaluminum Chloride***, in the presence of Cu^I or Pd^0 catalysts, to give, 1,2-dimetallo-1-alkenes with high regio- and stereoselectivity. These intermediates can be selectively functionalized at the vinyl-aluminum bond to provide alkenyltins (eq 8).[13]

Alkynyltins are easily made by the reaction of lithium or sodium acetylides with tributyltin chloride.[22] These compounds undergo coupling with vinyl iodides in the presence of a palladium catalyst to give conjugated enynes (eq 9).

NaC≡CH —($Bu_3SnAlEt_2$, 75%)→ $Bu_3SnC≡CH$ —(THPO~~~I, $(MeCN)_2PdCl_2$, 25 °C, 68%)→ THPO~~~≡ (9)

Tin Enolates. Tin enolates prepared by transmetalation of lithium enolates with tributyltin chloride undergo rapid aldol condensations with aldehydes (eq 10).[23] This approach has been used to enhance the yield of the alkylation of the enolate generated by the conjugate addition of a lithium enolate to cyclopentenone (eq 11).[3] A complicated mixture is obtained in the absence of the tin reagent.

OLi, Et —(1. Bu_3SnCl; 2. PhCHO, −78 °C; 92%)→ Ph, OH, O, Et + Ph, OH, O, Et (10)

46:54

O (cyclopentenone) —(1. TMS, OLi, H, OMe, −78 °C; 2. Bu_3SnCl, HMPA; 3. $BrCH_2C≡CEt$, −50 °C; 54%)→ O, CO_2Me, TMS (11)

CO_2Et —(1. LDA, −78 °C; 2. Bu_3SnCl, −20 °C; 83%)→ α, CO_2Et, γ, Bu_3Sn

(*E*):(*Z*) = 13:87

PhCHO, $BF_3{\cdot}OEt_2$, −78 °C, 100% → CO_2Et, Ph, OH + CO_2Et, Ph, OH

>99:1

PhCOCl, $PhCH_2Pd(PPh_3)_2Cl$, 60 °C, 70% → Ph, O, CO_2Et (12)

(*E*):(*Z*) = 1:8

The reaction of lithium dienolates with tributyltin chloride gives γ-stannylated α,β-unsaturated esters, which are stable enough to be isolated by distillation (eq 12).[4] The aldol-type condensation of the tin-masked dienolates proceeds at the α-position of the dienolates with high regio- and stereoselectivity in the presence of Lewis acid.[4a,b] On the other hand, the palladium-catalyzed coupling reaction with organic halides takes place at the position directly bonded to tin.[4a,c]

Related Reagents. Tributyltin chloride is used more frequently than ***Chlorotrimethylstannane***, perhaps owing to its lower toxicity. The latter reagent is useful in reactions which require a sterically compact trialkylstannyl substituent. See also Chlorotriphenylstannane; Lithium Phenylthio-(trimethylstannyl)cuprate; Methyl Tributylstannyl Sulfide; Tri-*n*-butylfluorostannane; Tri-*n*-butylstannylcopper; Tri-*n*-butylstannyllithium; Tri-*n*-butyl(methoxy)stannane; Tri-*n*-butyltin Trifluoromethanesulfonate; Trimethylstannylcopper–Dimethyl Sulfide; Trimethylstannyllithium.

1. (a) Pereyre, M.; Quintard, J. P.; Rahm, A. *Tin in Organic Synthesis*; Butterworth: London, 1987. (b) Davies, A. G.; Smith, P. J. In *Comprehensive Organometallic Chemistry*; Wilkinson, G., Ed.; Pergamon: Oxford, 1982; Vol. 2, p. 519.
2. (a) Yamamoto, Y. *ACR* **1987**, *20*, 243. (b) Yamamoto, Y.; Saito, Y.; Maruyama, K. *JOM* **1985**, *292*, 311. (c) Koreeda, M.; Tanaka, Y. *TL* **1987**, *28*, 143. (d) Keck, G. E.; Abbott, D. E.; Wiley, M. R. *TL* **1987**, *28*, 139. (e) Yamamoto, Y.; Kobayashi, K.; Okano, H.; Kadota, I. *JOC* **1992**, *57*, 7003.
3. Nishiyama, H.; Sakuta, K.; Itoh, K. *TL* **1984**, *25*, 2487.
4. (a) Yamamoto, Y.; Hatsuya, S.; Yamada, J. *JOC* **1990**, *55*, 3118. (b) Yamamoto, Y.; Hatsuya, S.; Yamada, J. *CC* **1987**, 561. (c) Yamamoto, Y.; Hatsuya, S.; Yamada, J. *CC* **1988**, 86.
5. Kuivila, H. G. *S* **1970**, 499.
6. Alleston, D. L.; Davies, A. G. *JCS* **1962**, 2050.
7. Jones, K.; Lappert, M. F. *JCS* **1965**, 1944.
8. Tanaka, M. *TL* **1980**, *21*, 2959.
9. Kricheldort, H. R.; Leppert, E. *S* **1976**, 329.
10. Kocheshkov, K, A.; Nesmeyanov, A. N.; Puzyreva, V. P. *CB* **1936**, *69*, 1639.
11. Van der Kerk, G. J. M.; Luijten, J. G. A. *J. Appl. Chem.* **1956**, *6*, 49.
12. Tamborski, C.; Ford, F. E.; Soloski, E. J. *JOC* **1963**, *28*, 237.
13. (a) Sharma, S.; Oehlschlager, A. C. *JOC* **1989**, *54*, 5064. (b) Hibino, J.; Matsubara, S.; Morizawa, Y.; Oshima, K.; Nozaki, H. *TL* **1984**, *25*, 2151.
14. Still, W. C. *JACS* **1978**, *100*, 1481.
15. Piers, E.; Morton, H. E.; Chong, J. M. *CJC* **1987**, *65*, 78.
16. (a) Gill, M.; Bainton, H. P.; Rickards, R. W. *TL* **1981**, *22*, 1437. (b) Seitz, D. E.; Lee, S.-H. *TL* **1981**, *22*, 4909.
17. (a) Seyferth, D.; Weiner, M. A. *JOC* **1961**, *26*, 4797. (b) Grignon, J.; Servens, C.; Pereyre, M. *JOM* **1975**, *96*, 225.
18. (a) Yamada, J.; Asano, T.; Kadota, I.; Yamamoto, Y. *JOC* **1990**, *55*, 6066. (b) Kadota, I.; Gevorgyan, V.; Yamada, J.; Yamamoto, Y. *SL* **1991**, 823. (c) Yamamoto, Y.; Yamada, J.; Kadota, I. *TL* **1991**, *32*, 7069. (d) Gevorgyan, V.; Kadota, I.; Yamamoto, Y. *TL* **1993**, *34*, 1313.
19. Obayashi, M.; Utimoto, K. Nozaki, H. *JOM* **1979**, *177*, 145.
20. Groh, B. L. *TL* **1991**, *32*, 7647.
21. Fryzuk, M. D.; Bates, G. S.; Stone, C. *TL* **1986**, *27*, 1537.
22. Stille, J. K.; Simpson, J. H. *JACS* **1987**, *109*, 2138.
23. Yamamoto, Y.; Yatagai, H.; Maruyama, K. *CC* **1981**, 162.

Isao Kadota & Yoshinori Yamamoto
Tohoku University, Sendai, Japan

Tri-*n*-butylstannyllithium

n-Bu_3SnLi

[4226-01-1] $C_{12}H_{27}LiSn$ (MW 297.04)

(reagent for the synthesis of α-alkoxystannanes,[1] β-stannyl ketones,[1] allenylstannanes,[2] allylstannanes,[3] acylstannanes,[4] and vinylstannanes[5])

Alternate Name: tributyltinlithium.

Preparative Methods: bney the reaction of ***Hexabutyldistannane*** and ***n-Butyllithium***, or by the deprotonation of ***Tri-n-butylstannane***.

Handling, Storage, and Precautions: should be used soon after preparation; however, storage of a THF solution for short periods of time (<12 h) at 0 °C is possible. For some applications the purity of the starting material used for generation of the tributyltinlithium reagent is critical. Tributyltin hydride should be distilled prior to deprotonation. Sensitive to air and moisture. Volatile tetraalkylstannanes are toxic and should be handled in a well-ventilated fume hood. Contact with the eyes and skin should be avoided.

General Discussion. Tributyltinlithium can be prepared from hexabutylditin by reaction with butyllithium[6] or with ***Lithium*** metal,[7] or from ***Tri-n-butylchlorostannane*** and lithium metal.[7b] A more convenient method which eliminates the need to separate the tetraalkyltin byproducts involves the deprotonation of tributyltin hydride by ***Lithium Diisopropylamide***.[1] One of the most extensive uses of tributyltinlithium is in the synthesis of α-alkoxystannanes. Condensation of tributyltinlithium and aldehydes or ketones leads to unstable α-hydroxystannanes which can be protected by a variety of α-chloro ethers (eq 1).[1,8]

$C_6H_{13}CHO$ → (Bu_3SnLi) → $C_6H_{13}CH(OH)SnBu_3$ → (Cl–CH(Me)–O–Et, *i*-Pr_2NEt) → $C_6H_{13}CH(OEE)SnBu_3$ 97% (1)

In reactions with α-substituted aldehydes, the addition of tributyltinlithium resulted in the same selectivity as unhindered Grignard reagents, while β-oxygenated aldehydes gave a higher degree of stereocontrol due to a cyclic chelate mechanism. The α-alkoxystannane can be transmetalated by reaction with butyllithium to provide an α-alkoxy organolithium reagent (eq 2).[6] The relative order of stabilities of α-alkoxyorganolithio reagents obtained from transmetalation of the corresponding α-alkoxystannanes has been defined.[8b] The lithio reagent appears to be stable at low temperatures and can be employed in a variety of subsequent reactions, including alkylation, addition to carbonyls, and transmetalation to organocuprate derivatives.[9] A synthesis of dendrolasin was accomplished using the α-alkoxystannane methodology,[1] offering several advantages over alternative approaches which involved a nonregioselective allyl anion addition (eq 3).

Ph–CH₂–CH(Me)–CHO → (1. Bu_3SnLi; 2. MOMCl) → Ph–CH₂–CH(Me)–CH(OMOM)SnBu₃ (two diastereomers) → (1. BuLi; 2. MeCOMe) → Ph–CH₂–CH(Me)–CH(OMOM)–C(OH)Me₂ (two diastereomers) (2)

3-furyl–CH(OEE)–SnBu₃ → (1. BuLi; 2. Cl-allyl geranyl chloride; 3. Li, NH_3; 66% overall) → dendrolasin (3)

Conjugate addition of α-alkoxyorganocuprates to cyclic enones followed by an intramolecular Mukaiyama aldol reaction provided a ring annulated tetrahydrofuran (eq 4).[10] The overall procedure is equivalent to regiocontrolled cycloaddition of an unstabilized carbonyl ylide. α-Alkoxyorganostannanes are also precursors for the stereoselective synthesis of enol ethers by a retro Diels–Alder reaction[11] or by transmetalation induced β-elimination.[12]

$Me_2C(OMOM)SnBu_3$ → (BuLi, 0.5 equiv CuCN) → $[Me_2C(OMOM)]_2Cu(CN)Li_2$ + cyclopentenone → (TMSCl, THF) → OTMS silyl enol ether → ($TiCl_4$) → bicyclic tetrahydrofuranone, 73% overall (4)

Addition of tributyltinlithium to 4-*t*-butylcyclohexanone was shown to be a reversible process.[8] Under thermodynamic conditions, axial addition of the tin anion predominated (93:7), while under kinetic conditions (addition at −100 °C) the equatorial stannane predominated (25:75). Both the axial and equatorial stannane underwent transmetalation and alkylation with retention of configuration. The tin anion addition/transmetalation/alkylation sequence provided access to axially substituted cyclohexanols, in contrast to the equatorially substituted products obtained by simple organometallic addition reactions (eq 5).

4-*t*-Bu-cyclohexanone → (1. Bu_3SnLi; 2. MOMCl; 60%) → 1-$SnBu_3$-1-OMOM-4-*t*-Bu-cyclohexane, axial:equatorial = 93:7 → (1. BuLi; 2. MeI, Me_2SO_4) → 1-Me-1-OMOM-4-*t*-Bu-cyclohexane (5) α-Alkoxyallylstannanes are also obtained by the direct addition of tributyltinlithium to unsaturated aldehydes.[3] The α-alkoxyallylstannanes undergo Lewis acid-catalyzed rearrangement to provide γ-alkoxyallylstannanes.[13] Lewis acid-catalyzed alkylation of ketene silyl acetals with acetal-protected α-alkoxystannanes followed by transmetalation and ring closure of the derived amide resulted in a novel synthesis of furanones (eq 6).[14]

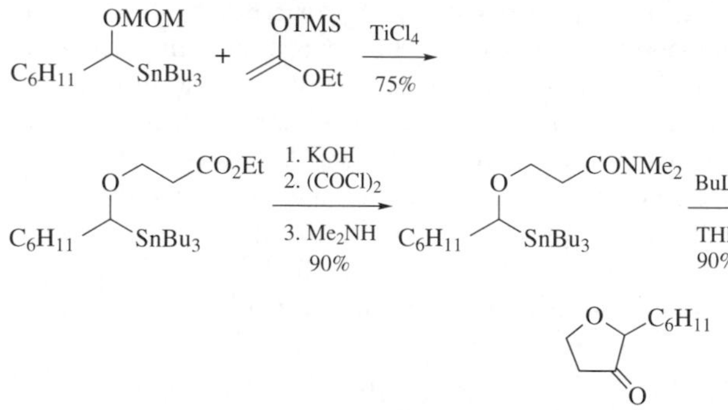

(6)

Enantiopure α-alkoxystannanes obtained by the asymmetric reduction of acylstannanes[15] have been employed in the synthesis of optically pure α-alkoxy carboxylic acids,[16] as well as transformed into the corresponding α-alkoxyorganocuprate reagents via the intermediate α-alkoxy lithio species.[17] Transmetalation of enantiopure α-alkoxyorganostannanes and the subsequent alkylation reaction was shown to occur with complete retention of configuration.[18] Access to enantiopure α-alkoxyallylstannanes has also been realized by the asymmetric reduction of unsaturated acylstannanes.[19] An alternative route to enantiopure α-alkoxyorganostannanes involves the direct displacement of α-chloro boronate esters.[20] Mitsunobu inversion of enantiopure α-alkoxyorganostannanes with ***Phthalimide*** provides a route to optically active α-aminoalkyl stannanes.[21] An alternative route to nonracemic (α-aminoalkyl)tributylstannanes has been realized by the direct displacement of sulfones (eq 7).[22]

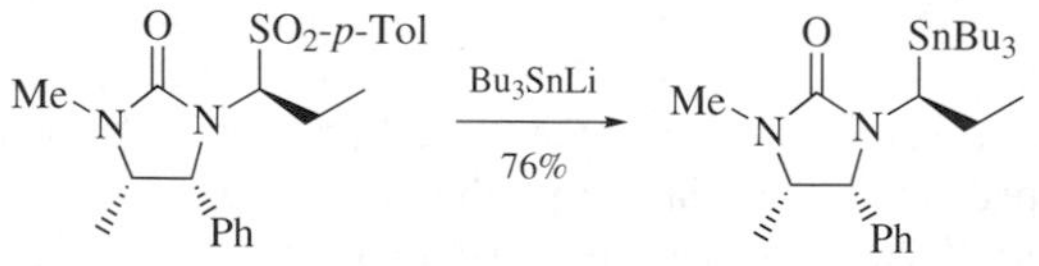

(7)

The (α-aminoalkyl)organolithio anions can also be obtained by transmetalation using butyllithium; however, the α-amino lithio species are not as configurationally stable as the α-alkoxy lithio derivatives. Racemic α-aminostannanes are available by the reaction of tributyltinlithium and iminium ions.[23]

Tributyltinlithium also undergoes direct displacement or nucleophilic addition reactions with a variety of other electrophiles. The synthesis of allenylstannanes is accomplished by tributyltinlithium addition to propargylic halides.[2] The direct synthesis of acylstannanes can be accomplished by the reaction of tributyltinlithium and ethyl esters in the presence of ***Boron Trifluoride Etherate***.[4] Addition of tributyltinlithium to thionolactones results in a functionalized stannane that may be transmetalated and alkylated with allyl halides.[24] Addition of tributyltinlithium to epoxides followed by stereospecific elimination of the derived β-acetoxystannane provides a method for the stereospecific transformation of epoxides to alkenes with retention of stereochemistry.[25]

A different elimination process results in the conversion of β-stannylvinyl sulfones to γ-hydroxyvinylstannanes.[26] 1,4-Addition of tributyltinlithium to the unsaturated sulfone provides an intermediate enolate which is trapped by reaction with an aldehyde. The alkoxide formed then undergoes an elimination reaction in which the alkoxide assists in the 1,2-elimination of the stannyl and sulfone groups (eq 8).

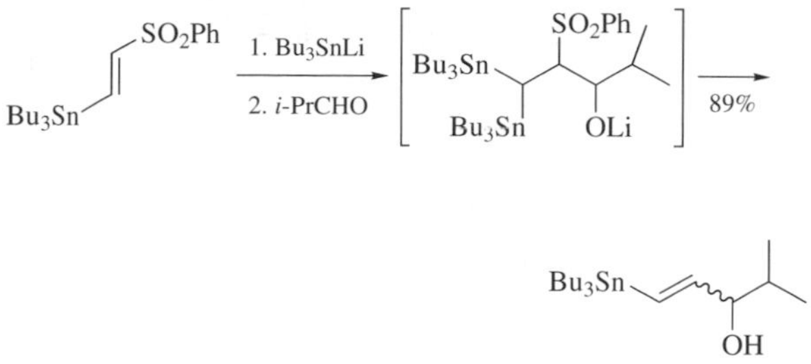

(8)

(*E*):(*Z*) = 60:40

Tributyltinlithium does not always undergo clean displacement or addition reactions. In a synthesis of 1-methoxy-1,3-dienes, tributyltinlithium promotes metal–halogen exchange rather than displacement of an allyl chloride (eq 9).[27] A similar competing metal halogen exchange process was observed in the stannylation of chlorosilanes.[28]

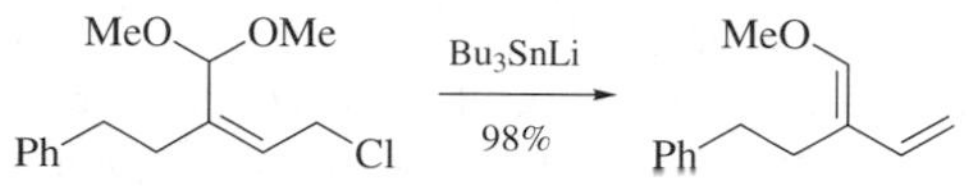

(9)

1,4-Addition of tributyltinlithium to enones occurs readily in THF or THF/NH_3. In contrast, ethereal tributyltinlithium adds only in a 1,2-fashion to cyclohexenone.[6] The β-trialkylstannyl ketone can be further transformed by alkyllithium addition to the carbonyl followed by oxidation with excess ***Dipyridine Chromium(VI) Oxide***, resulting in an overall enone transposition. Tributyltinlithium will even add to β,β-disubstituted unsaturated esters (eq 10).[6,29] This relatively unusual reactivity is attributed to the long Sn–C bond (2.2 Å).

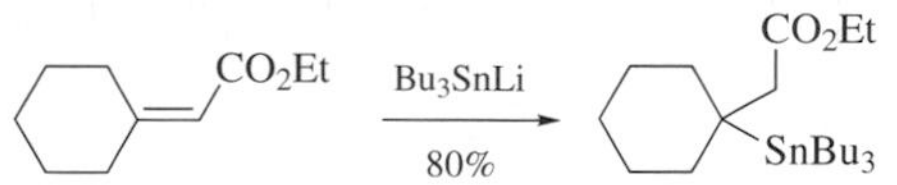

(10)

Although the β-stannyl carbonyl moiety is a reactive functional group, many synthetic reactions can be accomplished on the molecule without loss of the tin.[29] The 1,4-addition product obtained from cycloalkenones can also undergo transmetalation and alkylation if the ketone is first protected as an enol ether (eq 11).[30]

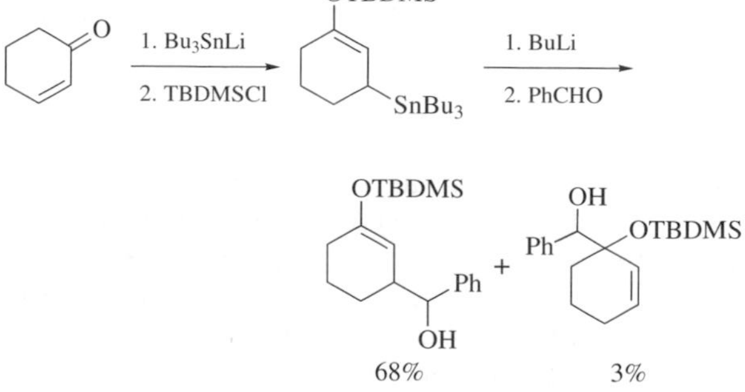

(11)

A similar sequence has been carried out on the morpholine enamine of β-tributylstannylcyclohexanone.[31] 1,4-Addition of tributyltinlithium to 2-phenylselenocyclopentenone with in situ alkylation of the enolate results in a 2-alkyl-2-phenylseleno-3-tributylstannylcyclopentanone. Destannyl-selenylation can be readily accomplished using a variety of Lewis acids or fluoride sources, ultimately to provide a 2-substituted cyclopentenone derivative.[32] The synthesis of a prostaglandin precursor was achieved using this methodology (eq 12).

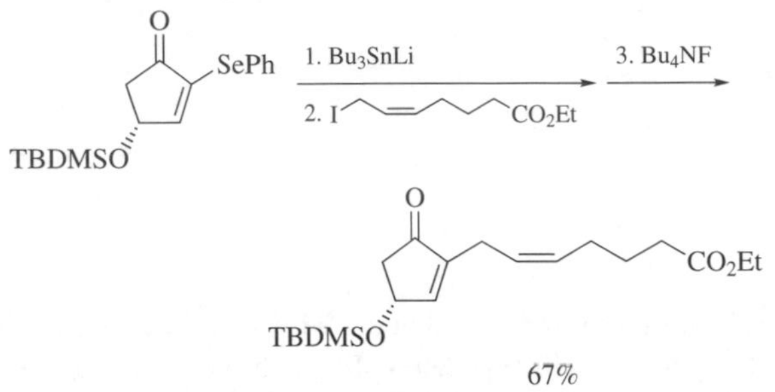

1. Bu_3SnLi; 2. I–CH₂CH=CH(CH₂)₃CO_2Et; 3. Bu_4NF — 67% (12)

1,4-Addition products obtained from tributyltinlithium addition to cycloalkenones are also useful precursors for a variety of rearrangement and fragmentation reactions. Lewis acid-catalyzed rearrangement of the alcohol derived from reduction or alkyllithium addition to the carbonyl results in the formation of cyclopropanes.[33] If the ketone is not reduced prior to the addition of a Lewis acid, the cyclopropanol intermediate undergoes fragmentation in situ to provide either the destannylated ketone or the ring contracted product.[34] The cyclopropanol product could be isolated only when the tributyltin bearing carbon was a primary carbon, or when the carbonyl was an aldehyde. The direction of cleavage is consistent with protonation at the less substituted carbon of the cyclopropanol. Treatment of the β-tributylstannyl ketones obtained by 1,4-addition with ***m*-Chloroperbenzoic Acid** resulted in a tin-directed Baeyer–Villager reaction, ultimately providing the alkene ester fragmentation product (eq 13).[35]

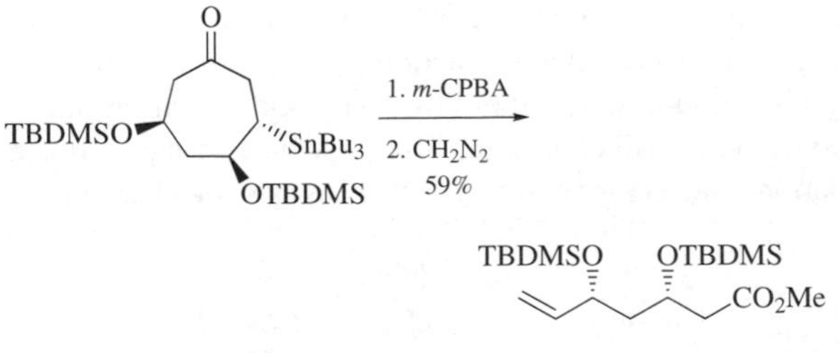

1. *m*-CPBA; 2. CH_2N_2 — 59% (13)

Migration of the β-tributylalkyl substituent occurred preferentially over *t*-butyl or naphthyl groups; therefore the reaction resulted in a complete reversal of the normal migratory aptitude realized in the Baeyer–Villager reaction. The directing effects of the tin group were also extended to the Beckmann fragmentation of the derived β-tributylstannyl oximes (eq 14).[35]

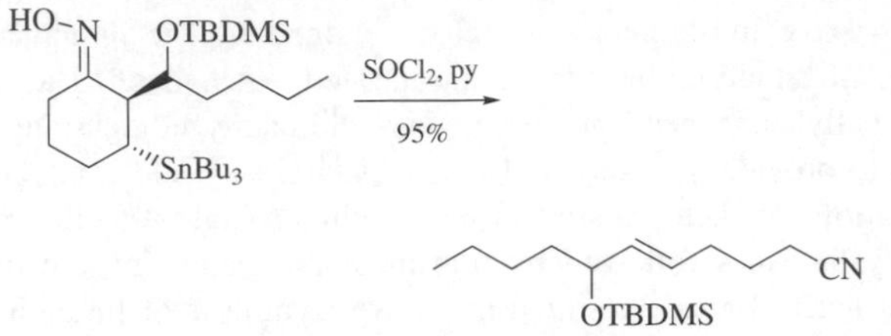

$SOCl_2$, py — 95% (14)

Oxidative fragmentation of β-tributylstannyl cycloalkanols with iodosobenzene results in the generation of ring-opened keto alkenes.[36] The fragmentation reaction is stereoelectronically controlled and requires an antiperiplanar arrangement of breaking bonds. β-Tributylstannylcyclohexanol derivatives which cannot adopt the correct orientation do not fragment, but rather undergo reaction at a butyl group on tin.[36c] A similar oxidative fragmentation can also be achieved on a cyclic hemiacetal, resulting in formation of a macrocyclic lactone (eq 15).[37]

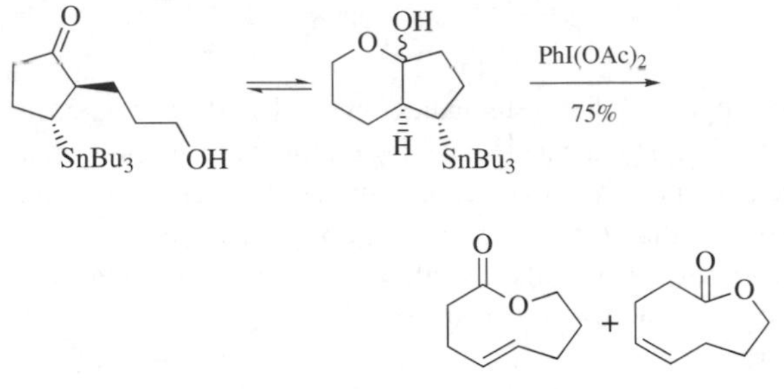

$PhI(OAc)_2$ — 75% (15)

Oxidative fragmentation can also be induced using ***Lead(IV) Acetate***. An intramolecular dipolar cycloaddition reaction has been accomplished by in situ generation of a nitrile oxide and an alkene by the fragmentation reaction (eq 16).[38]

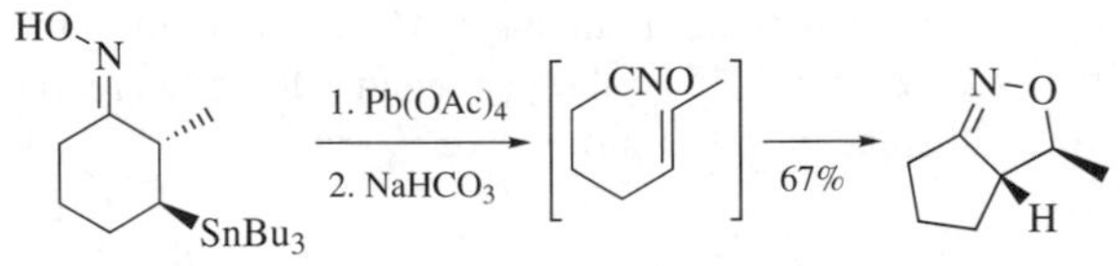

1. $Pb(OAc)_4$; 2. $NaHCO_3$ — 67% (16)

A one-pot, four-component annulation procedure has also been developed which is initiated by tributyltinlithium 1,4-addition to an enone.[39] The first intermediate enolate then acts as the nucleophile for a Michael addition reaction to a second enone, followed by a subsequent Michael reaction to an enoate. This sequence results in a cyclic hemiacetal containing a β-tributylstannyl moiety. Lead tetraacetate oxidative fragmentation then results in the generation of a macrocyclic lactone (eq 17).

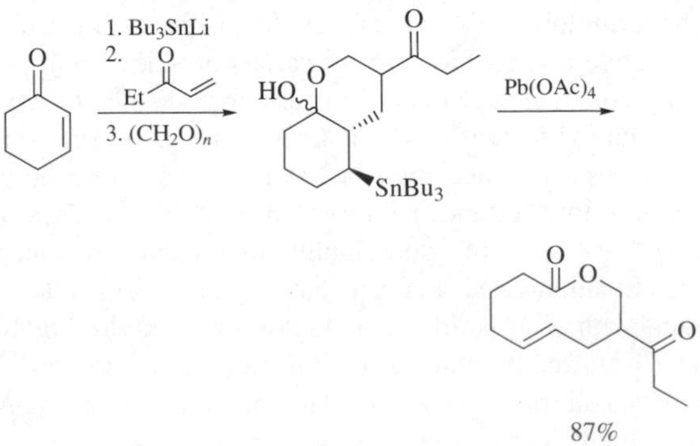

1. Bu_3SnLi; 2. Et-CO-CH=CH₂; 3. $(CH_2O)_n$; $Pb(OAc)_4$ — 87% (17)

Transmetalation of tributyltinlithium to a variety of other organometallic reagents has also been accomplished. Organotincuprate reagents undergo 1,4-addition to enones,[40] addition to alkynes,[41] enynes,[42] and addition–elimination reactions to provide β-tributylstannyl unsaturated esters[5] and amides.[43] Tributyltin boron and aluminum reagents have also been prepared.[44]

Related Reagents. Diethyl(tributylstannyl)aluminum; Dilithium (Trimethylstannyl)(2-thienyl)cyanocuprate; Lithium Phenylthio(trimethylstannyl)cuprate; Tri-*n*-butylchlorostannane; Tri-*n*-butylstannylcopper; Tri-*n*-butyl[(methoxymethoxy)-methyl]stannane; Trimethylstannylcopper-Dimethyl Sulfide; Trimethylstannyllithium.

---

1. Still, W. C. *JACS* **1978**, *100*, 1481.
2. Marshall, J. A.; Wang, X.-J. *JOC* **1990**, *55*, 6246.
3. Pratt, A. J.; Thomas, E. J. *CC* **1982**, 1115.
4. Capperucci, A.; Degl'Innocenti, A.; Faggi, C.; Reginato, G.; Ricci, A. *JOC* **1989**, *54*, 2966.
5. Seitz, D. E.; Lee, S.-H. *TL* **1981**, *21*, 4909.
6. Still, W. C. *JACS* **1977**, *99*, 4836.
7. (a) Quintard, J.-P.; Hauvette-Frey, S.; Pereyre, M. *JOM* **1978**, *159*, 147. (b) Newcomb, M.; Smith, M. G. *JOM* **1982**, *228*, 61.
8. (a) Sawyer, J. S.; Macdonald, T. L.; McGarvey, G. J. *JACS* **1984**, *106*, 3376. (b) Sawyer, J. S.; Kucerovy, A.; Macdonald, T. L.; McGarvey, G. J. *JACS* **1988**, *110*, 842.
9. (a) Linderman, R. J.; Godfrey, A.; Horne, K. *T* **1989**, *45*, 495. (b) Linderman, R. J.; McKenzie, J. R. *TL* **1988**, *29*, 3911.
10. Linderman, R. J.; Godfrey, A. *JACS* **1988**, *110*, 6249.
11. McGarvey, G. J.; Bajwa, J. S. *JOC* **1984**, *49*, 4091.
12. McGarvey, G. J.; Kimura, M.; Kucerovy, A. *TL* **1985**, *26*, 1419.
13. Marshall, J. A.; Welmaker, G. S. *JOC* **1992**, *57*, 7158.
14. (a) Linderman, R. J.; Graves, D. M.; Kwochka, W. R.; Ghannam, A. F.; Anklekar, T. V. *JACS* **1990**, *112*, 7438. (b) Linderman, R. J.; Viviani, F. G.; Kwochka, W. R. *TL* **1992**, *33*, 3571.
15. (a) Chan, P. C.-M.; Chong, J. M. *JOC* **1988**, *53*, 5584. (b) Marshall, J. A.; Gung, W. Y. *T* **1989**, *45*, 1043.
16. (a) Chong, J. M.; Mar, E. K. *T* **1989**, *45*, 7709. (b) Chong, J. M.; Mar, E. K. *TL* **1990**, *31*, 1981. (c) Chan, P. C.-M.; Chong, J. M. *TL* **1990**, *31*, 1985.
17. (a) Hutchinson, D. K.; Fuchs, P. L. *JACS* **1987**, *109*, 4930. (b) Linderman, R. J.; Griedel, B. D. *JOC* **1990**, *55*, 5428. (c) Linderman, R. J.; Griedel, B. D. *JOC* **1991**, *56*, 5491. (d) Prandi, J.; Audin, C.; Beau, J.-M. *TL* **1991**, *32*, 769.
18. Still, W. C.; Sreekumar, C. *JACS* **1980**, *102*, 1201.
19. Marshall, J. A.; Yashunsky, D. V. *JOC* **1991**, *56*, 5493.
20. Matteson, D. S.; Tripathy, P. B.; Sarkar, A.; Sadhu, K. M. *JACS* **1989**, *111*, 4399.
21. Chong, J. M.; Park, S. B. *JOC* **1992**, *57*, 2220.
22. (a) Pearson, W. H.; Lindbeck, A. C. *JACS* **1991**, *113*, 8546. (b) Pearson, W. H.; Lindbeck, A. C.; Kampf, J. W. *JACS* **1993**, *115*, 2622.
23. Elissondo, B.; Verlhac, J. B.; Quintard, J.-P.; Pereyre, M. *JOM* **1988**, *339*, 267.
24. Nicolaou, K. C.; McGarry, D. G.; Somers, P. K.; Veale, C. A.; Furst, G. T. *JACS* **1987**, *109*, 2504.
25. Jousseaume, B.; Noiret, N.; Pereyre, M.; Francès, J.-M.; Pétroud, M. *OM* **1992**, *11*, 3910.
26. Ochiai, M.; Ukita, T.; Fujita, E. *TL* **1983**, *24*, 4025.
27. Fujiwara, S.; Katsumura, S.; Isoe, S. *TL* **1990**, *31*, 691.
28. Quintard, J.-P.; Dumartin, G.; Guerin, C.; Dubac, J.; Laporterie, A. *JOM* **1984**, *266*, 123.
29. Jephcote, V. J.; Thomas, E. J. *JCS(P1)* **1991**, 429.
30. Chenard, B. L. *TL* **1986**, *27*, 2805.
31. Ahlbrecht, H.; Weber, P. *S* **1992**, 1018.
32. Kusuda, S.; Watanabe, Y.; Ueno, Y.; Toru, T. *JOC* **1992**, *57*, 3145.
33. (a) Kadow, J. F.; Johnson, C. R. *TL* **1984**, *25*, 5255. (b) Fleming, I.; Urch, C. J. *JOM* **1985**, *285*, 173.
34. Sato, T.; Watanabe, M.; Watanabe, T.; Onoda, Y.; Murayama, E. *JOC* **1988**, *53*, 1894.
35. Bakale, R. P.; Scialdone, M. A.; Johnson, C. R. *JACS* **1990**, *112*, 6729.
36. (a) Ochiai, M.; Ukita, T.; Nagao, Y.; Fujita, E. *CC* **1984**, 1007. (b) Ochiai, M.; Ukita, T.; Nagao, Y.; Fujita, E. *CC* **1985**, 637. (c) Ochiai, M.; Ukita, T.; Iwaki, S.; Nagao, Y.; Fujita, E. *JOC* **1989**, *54*, 4832.
37. Ochiai, M.; Iwaki, S. Ukita, T.; Nagao, Y. *CL* **1987**, 133.
38. Nishiyama, H.; Arai, H.; Ohki, T.; Itoh, K. *JACS* **1985**, *107*, 5310.
39. (a) Posner, G. H.; Asirvatham, E. *TL* **1986**, *21*, 663. (b) Posner, G. H.; Webb, K. S.; Asirvatham, E.; Jew, S.-S.; Degl'Innocenti, A. *JACS* **1988**, *110*, 4754.
40. Piers, E.; Morton, H. E.; Chong, J. M. *CJC* **1987**, *65*, 78.
41. Matsubara, S.; Hibino, J.-I.; Morizawa, Y.; Oshima, K.; Nozaki, H. *JOM* **1985**, *285*, 163.
42. Aksela, R.; Oehlschlager, A. C. *T* **1991**, *47*, 1163,
43. Imanieh, H.; MacLeod, D.; Quayle, P.; Zhao, Y. *TL* **1992**, *33*, 405.
44. Singh, S. M.; Oehlschlager, A. C. *CJC* **1991**, *69*, 1872.

Russell J. Linderman
North Carolina State University, Raleigh, NC, USA

Trichloroethylene

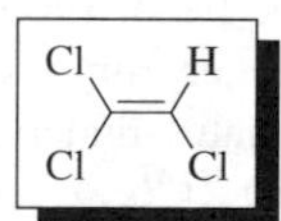

[79-01-6] C_2HCl_3 (MW 131.38)

(starting material for the preparation of dichloroacetylene;[1–4] reagent for the preparation of alkynic ethers;[12] reagent for ethynylation and vinylation)

Alternate Name: trichloroethene.
Physical Data: bp 86.9 °C; *d* 1.464 g cm^{-3}.
Solubility: practically insol water; miscible with ether, alcohol, chloroform.
Drying: can be dried by distillation from phosphorus pentoxide.
Handling, Storage, and Precautions: is slowly decomposed (with formation of HCl) by light in the presence of moisture. Avoid prolonged exposure to excessive heat. Moderate exposures to humans can cause symptoms similar to alcohol inebriation. Higher concentrations can have a narcotic effect. Deaths occurring after heavy exposure have been attributed to ventricular fibrillation. Found to induce hepatocellular carcinomas in mice. Dichloroacetylene (see below) is known to be toxic and, in its pure form, explosive. Use in a fume hood.

Preparation of Dichloroacetylene. A dichloroacetylene–ether complex is readily prepared from trichloroethylene with

Lithium Hexamethyldisilazide in hexanes at −78 °C[1] or with ***Potassium Hydride*** and catalytic methanol in THF at room temperature.[2] Alternatively, the dichloroacetylene–ether complex can be obtained from a mixture of trichloroethylene and diethyl ether in an aqueous solution of ***Sodium Hydroxide*** in the presence of either a phase transfer catalyst[3] or a catalytic amount of ***Dimethyl Sulfoxide***.[4]

Dichloroacetylene reacts with tertiary ketone or ester enolates to give chloroethynyl adducts in 64–90% yields.[5] The chloroethynyl group can be converted to the ethynyl derivative using ***Copper*** powder in HOAc/THF, or can be directly reduced (H_2/Lindlar catalyst) to the vinyl derivative (eq 1).[5] This method is limited to tertiary enolates.

1. LDA, HMPA, −78 °C
2. Cl—≡—Cl

64%

77% 96% (1)

Additional enolate addition/elimination reactions with chloroacetylenes that yield alkynyl derivatives have been reported for ***1-Chloro-2-phenylacetylene***[5] and 1-chloro-2-phenylthioacetylene.[5] Similar alkynylations have been reported for tertiary β-dicarbonyl enolates using alkynyllead triacetates[6] or alkynyliodonium tetrafluoroborates.[7] Direct vinylations have been reported using alkenyllead triacetates,[8] vinyl halides and nickel catalysis,[9] or enol ether iron complexes.[10] Dichloroacetylene reacts with alkylthiolates under mild conditions to yield bis(alkylthio)acetylenes (eq 2).[11]

Cl—≡—Cl → (t-BuSH, KH, THF, 0 °C; 98%) t-BuS—≡—S-t-Bu (2)

Preparation of Alkynic Ethers. Secondary alcohols, on treatment in THF with 2 equiv of potassium hydride and trichloroethylene, followed by ***n-Butyllithium*** and a primary iodide (or water), are converted in 66–87% yield to alkynic ethers (eq 3).[12] The potassium alkoxides react with trichloroethylene first to generate and then to add to dichloroacetylene. The resulting dichloro enol ethers on treatment with *n*-butyllithium are converted to the corresponding acetylides, which are then either alkylated or protonated.

1. Cl_2C=CHCl, KH
2. BuLi, MeI

87% (3)

Reaction with Other Nucleophiles. Trichloroethylene reacts with the enolate of 2,6-dimethylcyclohexenone (***Lithium Diisopropylamide***, HMPA, in THF, −78 °C) to yield the dichlorovinyl adduct (**1**) in 60% yield.[5,13] While the dichlorovinyl adduct may serve as a masked ethynyl group,[13] the reagent of choice is the independently generated dichloroacetylene described above. Under more vigorous conditions (***Sodium Hydride***, HMPA, refluxing THF) the tertiary enolate of diethyl ethylmalonate reacts to give dichlorovinyl adduct (**2**) in 64% yield.[5] These reactions have been demonstrated to proceed through the in situ generation of dichloroacetylene.[14] Comparable reactions have been reported for CClF=CHCl,[5] CF_2=CHF,[15,16] CF_2=CH_2,[16] and ***Chlorotrifluoroethylene***.[16]

(**1**) (**2**)

Trichloroethylene reacts with ***Sodium Amide*** and excess dimethylamine to yield bis(dimethylamino)acetylene (eq 4).[17] With lithium dialkylamides in ether as the base in the presence of excess dialkylamines, the intermediate 1,2-dichloro dialkyl enamines or chloro ketene aminals can be isolated (eq 5).[18] Trichloroethylene also reacts with secondary aliphatic amines in the presence of an aqueous solution of NaOH and a catalytic quantity of ***Benzyltriethylammonium Chloride*** to give glycinamides (eq 6).[19]

Cl_2C=CHCl → ($NaNH_2$, Me_2NH; 60%) Me_2N—≡—NMe_2 (4)

1. $LiNR_2$, ether, −70 °C
2. HNR_2, −70 °C to rt

70–75%

1. $LiNR_2$, ether, −70 °C
2. HNR_2, −70 °C to rt

80–85% (5)

R_2NH, NaOH, phase transfer catalyst; 71–83% (6)

Trichloroethylene will undergo palladium-catalyzed coupling with arylmagnesium halides to yield (*Z*)-1,2-dichloroethylenes (eq 7).[20] Under alternate conditions with alkylmagnesium halides, 1,1-dichloroalkenes are obtained (eq 8).[21]

PhMgBr, ether, $PdCl_2$(dppb); 55–76% (7)

CyMgBr, ether, benzene, $Pd(PPh_3)_4$; 81% (8)

Related Reagents. 1-Chloro-2-phenylacetylene; Chlorotrifluoroethylene; 1,2-Dichloroethylene.

1. Kende, A. S.; Fludzinski, P. *S* **1982**, 455.
2. Denis, J.-N.; Moyano, A.; Greene, A. E. *JOC* **1987**, *52*, 3461.

3. Pielichowski, J.; Popielarz, R. *S* **1984**, 433.
4. Pielichowski, J.; Bogdal, D. *JPR* **1989**, *331*, 145.
5. (a) Kende, A. S.; Fludzinski, P. *TL* **1982**, *23*, 2373. (b) Kende, A. S.; Fludzinski, P.; Hill, J. H.; Swenson, W.; Clardy, J. *JACS* **1984**, *106*, 3551.
6. (a) Moloney, M. G.; Pinhey, J. T.; Roche, E. G. *JCS(P1)* **1989**, 333. (b) Hashimoto, S.; Miyazaki, Y.; Shinoda, T.; Ikegami, S. *CC* **1990**, 1100.
7. Ochiai, M.; Ito, T.; Takaoka, Y.; Masaki, Y.; Kunishima, M.; Tani, S.; Nagao, Y. *CC* **1990**, 118.
8. Moloney, M. G.; Pinhey, J. T. *JCS(P1)* **1988**, 2847.
9. Millard, A. A.; Rathke, M. W. *JACS* **1977**, *99*, 4833.
10. (a) Chang, T. C.; Rosenblum, M.; Samuels, S. B. *JACS* **1980**, *102*, 5930. (b) Chang, T. C.; Rosenblum, M.; Simms, N. *OS* **1988**, *66*, 95.
11. Riera, A.; Cabre, F.; Moyano, A.; Pericas, M. A.; Santamaria, J. *TL* **1990**, *31*, 2169.
12. (a) Moyano, A.; Charbonnier, F.; Greene, A. E. *JOC* **1987**, *52*, 2919. (b) Denmark, S. E.; Senanayake, C. B. W.; Ho, G.-D. *T* **1990**, *46*, 4857. (c) Almansa, C.; Moyano, A.; Pericas, M. A.; Serratosa, F. *S* **1988**, 707. (d) Loffler, A.; Himbert, G. *S* **1992**, 495.
13. (a) Kende, A. S.; Benechie, M.; Curran, D. P.; Fludzinski, P.; Swenson, W.; Clardy, J. *TL* **1979**, 4513. (b) Kende, A. S.; Fludzinski, P. *OS* **1986**, *64*, 73.
14. Kende, A. S.; Fludzinski, P. *TL* **1982**, *23*, 2369.
15. Kende, A. S.; Fludzinski, P. *JOC* **1983**, *48*, 1384.
16. Crouse, G. D.; Webster, J. D. *JOC* **1992**, *57*, 6643.
17. Rene, L.; Janousek, Z.; Viehe, H.-G. *S* **1982**, 645.
18. van der Heiden, R.; Brandsma, L. *S* **1987**, 76.
19. Pielichowski, J.; Popielarz, R. *T* **1984**, *40*, 2671.
20. Minato, A.; Suzuki, K.; Tamao, K. *JACS* **1987**, *109*, 1257.
21. Ratovelomanana, V.; Linstrumelle, G.; Normant, J.-F. *TL* **1985**, *26*, 2575.

Pawel Fludzinski
Lilly Research Centre, Windlesham, UK

Triethylaluminum[1]

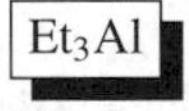

[97-93-8] $C_6H_{15}Al$ (MW 114.19)

(Lewis acid and source of nucleophilic ethyl groups;[2–5] couples with alkenyl halides in the presence of a transition metal catalyst;[8–11] selective hydrocyanation agent in combination with HCN[12])

Physical Data: mp −58 °C; bp 62 °C/0.8 mmHg; *d* 0.835 g cm^{-3} (25 °C).

Solubility: freely miscible with saturated and aromatic hydrocarbons; reacts violently with H_2O and protic solvents.

Form Supplied in: as a neat liquid in a stainless container or as a solution in hydrocarbon solvents (hexane, heptane, toluene).

Analysis of Reagent Purity: brochures from manufacturers, describe an apparatus and method for assay.

Handling, Storage, and Precautions: indefinitely stable under an inert atmosphere. The neat liquid or dense solutions are highly pyrophoric. Solutions more dilute than a certain concentration are not pyrophoric and are safer to handle. The nonpyrophoric limits are 13 wt % in isopentane, 12 wt % in hexane, and 12 wt % in heptane, respectively. Use of halogenated hydrocarbons as solvents should be avoided because of possible explosive reactions sometimes observed for mixtures of CCl_4 and organoaluminums.

Ethylation. Et_3Al, like other organoaluminums, can act as a Lewis acid to activate Lewis basic functionalities and also as a captor of electrophilic species by ethylation, as illustrated in eq 1.[2] Substitution reactions of glycosyl fluorides (eq 2)[3] and bromides[4] can be effected with Et_3Al. γ-Lactols react with Et_3Al in the presence of ***Boron Trifluoride Etherate*** to deliver 2,5-disubstituted tetrahydrofurans stereoselectively (eq 3).[5]

Et_3Al, hexane, 0 °C, 75% (1)

70:30

Et_3Al, toluene, 0 °C, 93% (2)

α:β = 6:1

BF_3•OEt_2, Et_3Al, CH_2Cl_2, −78 °C to rt, 61% (3)

trans:*cis* = 13:1

Conjugate addition of Et_3Al to 2-nitrofurans provides, after hydrolysis, dihydro-2(3*H*)-furanones (eq 4).[6] Regioselective addition of Et_3Al to unsymmetrical 1,1′-azodicarbonyl compounds has been reported.[7]

Et_3Al, hexane–benzene, −95 to 0 °C, 84% (4)

Cross coupling of Et_3Al with aryl phosphates[8] or alkynyl bromides[9] proceeds with a Ni catalyst to provide alkylated arenes or alkynes (eqs 5 and 6). While Pd or Cu complexes are used as the catalyst for the coupling of Et_3Al with carboxylic acid chlorides or thioesters (eq 7),[10] ***Iron(III) Chloride*** is used for reactions of propargyl acetates to give substituted allenes (eq 8).[11]

Et_3Al, cat $NiCl_2$(dppp), Et_2O, rt, 95% (5)

s-Bu—≡—Br → *s*-Bu—≡—Et; Et_3Al, cat Ni(mesal)$_2$, pentane, 25 °C, 84% (6)

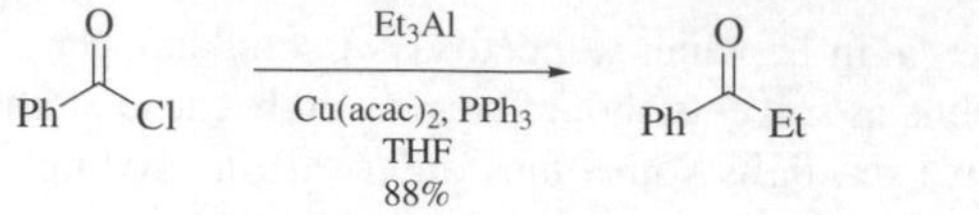

(7)

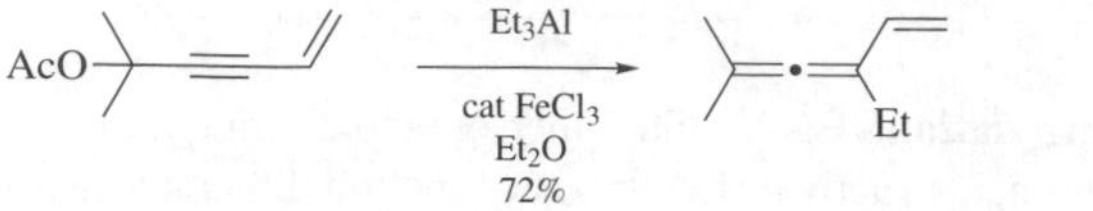

(8)

Lewis Acid. Et_3Al–***Hydrogen Cyanide*** and ***Diethylaluminum Cyanide*** are two optional reagents for conjugate hydrocyanation of α,β-unsaturated ketones.[12,13] The results from these two reagents often differ. Due to a rather slow reaction rate between Et_3Al and HCN, the former reagent contains a proton source (HCN) which can quench aluminum enolate intermediates. An impressive example is shown in eq 9. Preformed Et_2AlCN gives *cis*-isomer, whereas HCN–Et_3Al leads to *trans*-isomer.[14] A variant involves trapping of the enolates as TMS ethers by use of ***Cyanotrimethylsilane***–Et_3Al (eq 10).[14]

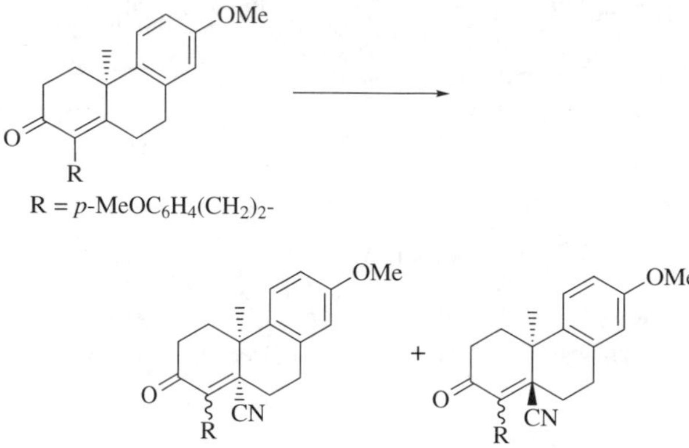

R = p-MeOC₆H₄(CH₂)₂-

(9)

Et_2AlCN, Ph, 99% 15:1
Et_3Al–HCN, THF, 89% 1:5

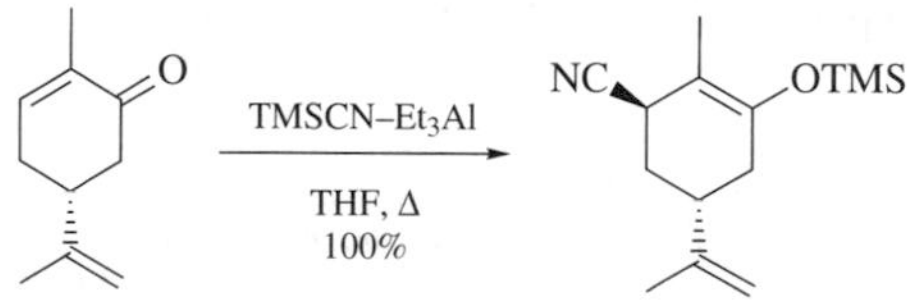

(10)

The HCN–Et_3Al system has been used to cleave oxiranes in steroids[15a] or carbohydrates (eq 11) to give β-cyano alcohols.[15b]

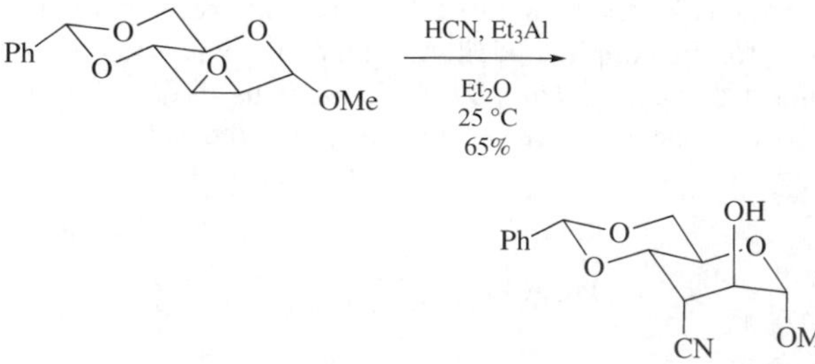

(11)

The ate complex, formed on metalation of allyl *i*-propyl ether with ***s-Butyllithium*** followed by addition of Et_3Al, reacts with carbonyl compounds at the α-position in *syn*-selective manner (eq 12).[16]

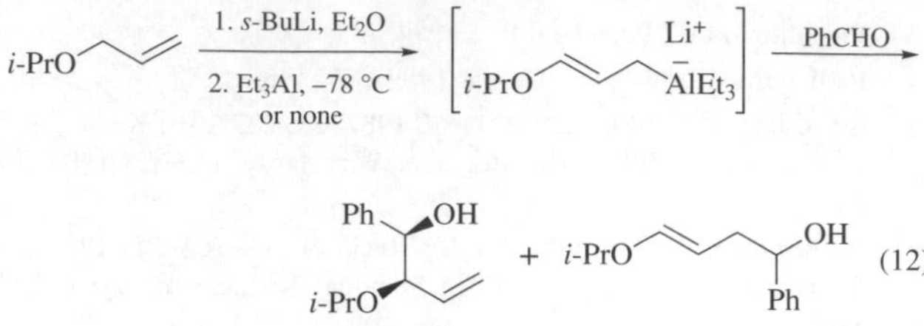

(12)

syn:*anti* = 92:8

none, 95% 28:72
Et_3Al, 81% >99:<1

Rearrangements. Allyl vinyl ethers undergo [3,3]-sigmatropic rearrangements promoted by Et_3Al, which also effects subsequent ethylation of the resulting aldehydes (eq 13). Use of ***Triisobutylaluminum*** leads to primary alcohols by β-hydride reduction.[17]

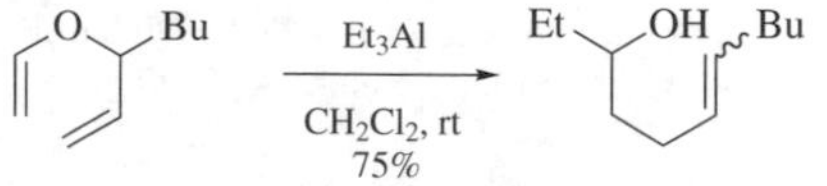

(13)

Alkylative Beckmann rearrangements of oxime sulfonates are promoted by trialkylaluminums. The rearrangements give the imines, which are reduced with ***Diisobutylaluminum Hydride*** to the corresponding amines (eq 14).[18] Related alkylative Beckmann fragmentations have also been reported.[19]

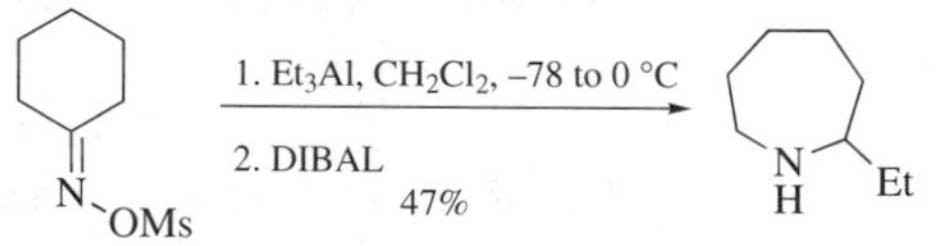

(14)

Et_3Al promotes stereospecific pinacol-type rearrangements of chiral β-methanesulfonyloxy alcohols. Aryl[20a] or alkenyl groups[20b] cleanly take part in the 1,2-migration to provide a range of α-chiral ketones (eq 15). The 1,2-migration of alkyl groups is effected by the more Lewis acidic ***Diethylaluminum Chloride***.[20c] The reagent combination of DIBAL and Et_3Al effects the reductive 1,2-rearrangement of α-mesyloxy ketones (eq 16).[20d,e]

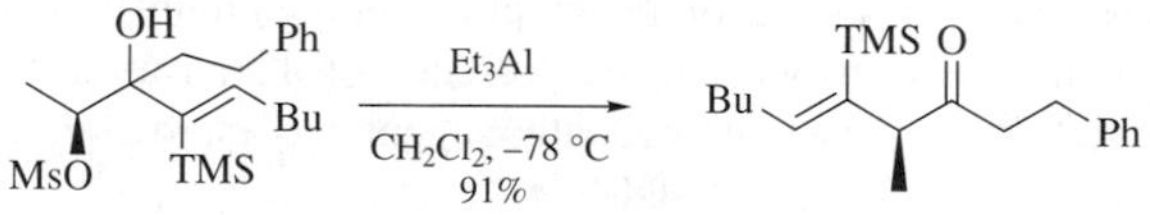

(15)

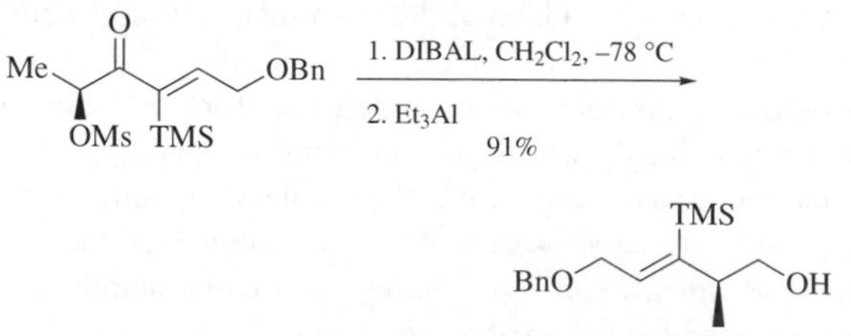

(16)

Cyclopropanation. The reagent combination of ***Diiodomethane*** and Et_3Al (or other organoaluminums) leads to cyclopropanation of alkenes (eq 17).[21]

$$\text{PhC(OTMS)=CH}_2 \xrightarrow[\text{CH}_2\text{Cl}_2,\ \text{rt}\ 86\%]{\text{Et}_3\text{Al, CH}_2\text{I}_2} \text{1-phenyl-1-(OTMS)cyclopropane} \quad (17)$$

Related Reagents. Cyanotrimethylsilane; Diethylaluminum Chloride; Diethylaluminum Cyanide; Hydrogen Cyanide; Titanium(IV) Chloride–Triethylaluminum; Triisobutylaluminum; Trimethylaluminum.

1. (a) Mole, T.; Jeffery, E. A. *Organoaluminum Compounds*; Elsevier: Amsterdam, 1972. (b) Reinheckel, H.; Haage, K.; Jahnke, D. *Organomet. Chem. Rev. A* **1969**, *4*, 47. (c) Lehmkuhl, H.; Ziegler, K.; Gellert, H. G. *MOC* **1970**, *8/4*. (d) Negishi, E. *JOM Libr.* **1976**, *1*, 93. (e) Yamamoto, H.; Nozaki, H. *AG(E)* **1985**, *17*, 169. (f) Negishi, E. *Organometallics in Organic Synthesis*; Wiley: New York, 1980; Vol. 1, pp 286–393. (g) Eisch, J. J. In *Comprehensive Organometallic Chemistry*, Wilkinson, G.; Stone, F. G. A.; Abel, E. W., Eds.; Pergamon: Oxford, 1982; Vol. 1, pp 555–682. (h) Zietz, J. R. Jr.; Robinson, G. C.; Lindsay, K. L. In *Comprehensive Organometallic Chemistry*, Wilkinson, G.; Stone, F. G. A.; Abel, E. W., Eds.; Pergamon: Oxford, 1982; Vol. 7, pp 365–464. (i) Maruoka, K.; Yamamoto, H. *AG(E)* **1985**, *24*, 668. (j) Maruoka, K. Yamamoto, H. *T* **1988**, *44*, 5001.
2. Hashimoto, S.; Kitagawa, Y.; Iemura, S.; Yamamoto, H.; Nozaki, H. *TL* **1976**, 2615.
3. (a) Posner, G. H.; Haines, S. R. *TL* **1985**, *26*, 1823. (b) Nicolaou, K. C.; Dolle, R. E.; Chucholowski, A.; Randall, J. L. *CC* **1984**, 1153.
4. Tolstikov, G. A.; Prokhorova, N. A.; Spivak, A. Yu.; Khalilov, L. M.; Sultanmuratova, V. R. *ZOK* **1991**, *27*, 2101.
5. Tomooka, K.; Matsuzawa, K.; Suzuki, K.; Tsuchihashi, G. *TL* **1987**, *28*, 6339.
6. Pecunioso, A.; Menicagli, R. *JCR(S)* **1988**, 228.
7. Yamamoto, Y.; Yumoto, M.; Yamada, J. *TL* **1991**, *32*, 3079.
8. Hayashi, T.; Katsuro, Y.; Okamoto, Y.; Kumada, M. *TL* **1981**, *22*, 4449.
9. Giacomelli, G.; Lardicci, L. *TL* **1978**, 2831.
10. (a) Takai, K.; Oshima, K.; Nozaki, H. *BCJ* **1981**, *54*, 1281. (b) Wakamatsu, K.; Okuda, Y.; Oshima, K.; Nozaki, H. *BCJ* **1985**, *58*, 2425.
11. Tolstikov, G. A.; Romanova, T. Yu.; Kuchin, A. V. *JOM* **1985**, *285*, 71.
12. (a) Nagata, W.; Yoshioka, M. *OS* **1972**, *52*, 100. (b) Nagata, W.; Yoshioka, M. *OR* **1977**, *25*, 255.
13. Ireland, R. E.; Dawson, M. I.; Welch, S. C.; Hagenbach, A.; Bordner, J.; Trus, B. *JACS* **1973**, *95*, 7829.
14. (a) Utimoto, K.; Obayashi, M.; Shishiyama, Y.; Inoue, M.; Nozaki, H. *TL* **1980**, *21*, 3389. (b) Utimoto, K.; Wakabayashi, Y.; Horiie, T.; Inoue, M.; Shishiyama, Y.; Obayashi, M.; Nozaki, H. *T* **1983**, *39*, 967.
15. (a) Nagata, W.; Yoshioka, M.; Okumura, T. *TL* **1966**, 847. (b) Davidson, B. E.; Guthrie, R. D.; McPhail, A. T. *CC* **1968**, 1273.
16. (a) Yamamoto, Y.; Yatagai, H.; Maruyama, K. *JOC* **1980**, *45*, 195. (b) Yamamoto, Y.; Yatagai, H.; Saito, Y.; Maruyama, K. *JOC* **1984**, *49*, 1096. (c) Yamamoto, Y.; Saito, Y.; Maruyama, K. *JOM* **1985**, *292*, 311.
17. (a) Takai, K.; Mori, I.; Oshima, K.; Nozaki, H. *TL* **1981**, *22*, 3985. (b) Takai, K.; Mori, I.; Oshima, K.; Nozaki, H. *BCJ* **1984**, *57*, 446.
18. (a) Hattori, K.; Matsumura, Y.; Miyazaki, T.; Maruoka, K.; Yamamoto, H. *JACS* **1981**, *103*, 7368. (b) Sakane, S.; Matsumura, Y.; Yamamura, Y.; Ishida, Y.; Maruoka, K.; Yamamoto, H. *JACS* **1983**, *105*, 672. (c) Maruoka, K.; Miyazaki, T.; Ando, M.; Matsumura, Y.; Sakane, S.; Hattori, K.; Yamamoto, H. *JACS* **1983**, *105*, 2831.
19. Fujioka, H.; Yamanaka, T.; Takuma, K.; Miyazaki, M.; Kita, Y. *CC* **1991**, 533.
20. (a) Suzuki, K.; Katayama, E.; Tsuchihashi, G. *TL* **1983**, *24*, 4997. (b) Suzuki, K.; Katayama, E.; Tsuchihashi, G. *TL* **1984**, *25*, 1817. (c) Suzuki, K.; Tomooka, K.; Tsuchihashi, G. *TL* **1984**, *25*, 4253. (d) Suzuki, K.; Tomooka, K.; Katayama, E.; Matsumoto, T.; Tsuchihashi, G. *JACS* **1986**, *108*, 5221. (e) Suzuki, K.; Katayama, E.; Matsumoto, T.; Tsuchihashi, G. *TL* **1984**, *25*, 3715.
21. (a) Maruoka, K.; Fukutani, Y.; Yamamoto, H. *JOC* **1985**, *50*, 4412. (b) Maruoka, K.; Sakane, S.; Yamamoto, H. *OS* **1988**, *67*, 176.

Keisuke Suzuki & Tetsuya Nagasawa
Keio University, Yokohama, Japan

Triethylborane

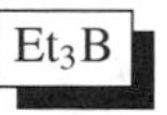

[97-94-9] $C_6H_{15}B$ (MW 98.02)

(precursor of triethylborohydride reducing agents;[1] enoxytriethylborates and -diethylboranes for aldol[19,22] and alkylation[17] reactions; regio- and stereoselective reactions of allyltriethylborates;[26,27] alkylating reagents;[1b] stereocontrol of carbanion reactions;[39,41] radical initiator[44])

Physical Data: mp −93 °C; bp 95 °C; *d* 0.677 g cm^{-3}.
Solubility: sol ethanol, acetone, THF, ether, hexane, benzene, $CHCl_3$.
Form Supplied in: colorless liquid; widely available. Since it is highly flammable, triethylborane diluted with hexane or THF (1.0 M solution) is also available.
Handling, Storage, and Precautions: neat triethylborane ignites instantaneously upon contact with air. The reagent should be handled with a syringe under Ar or N_2 atmosphere. It is stable toward moisture and water, and reputed to be toxic. Use in a fume hood.

Reducing Reagents. ***Lithium Triethylborohydride***, known as 'Super-Hydride', is prepared in THF by the reaction of lithium hydride with triethylborane (eq 1).[1] Alkali metal trialkylborohydrides are exceptionally powerful nucleophilic reducing agents capable of cleaving cyclic ethers,[2] reducing hindered halides,[3] *p*-toluenesulfonate esters of hindered and cyclic alcohols,[4] epoxides,[5] and activated alkenes[6] rapidly and quantitatively to the desired products. The advantage of $LiEt_3BH$ is especially evident in the reduction of labile bicyclic epoxides (eq 2). Thus benzonorbornadiene oxide, which invariably gives rearranged products with conventional reducing agents, undergoes facile reduction with $LiEt_3BH$ yielding 93% of *exo*-benzonorborneol in >99.9% isomeric purity. The reactivity of the trialkylborohydrides and the stereochemical course of their reactions are strongly influenced by the steric bulk of the alkyl group on boron.[7] The lithium hydride route (eq 1) provides a convenient entry only to the relatively unhindered lithium trialkylborohydrides. However, ***Potassium Hydride*** reacts rapidly and quantitatively with the hindered trialkylboranes, such as tri-*s*-butylborane, yielding the corresponding sterically hindered trialkylborohydride ***Potassium Tri-s-butylborohydride***.[8] A general synthesis of lithium trialkylborohydrides has been developed using lithium trimethoxyaluminohydride (eq 3).[9]

$$\text{LiH} + \text{BEt}_3 \xrightarrow[25\,^\circ\text{C}]{\text{THF}} \text{LiEt}_3\text{BH} \quad (1)$$

Super-Hydride

(2)

$LiEt_3BH$, THF, 65 °C, 24 h	93%	<0.1%
$LiAlH_4$, Et_2O, reflux, 24 h	15%	85%
BH_3, THF, reflux, 4 h	54%	23%
Li, ethylenediamine, 50 °C, 24 h	31%	15%

$$\text{R}_3\text{B} + \text{LiAl(OMe)}_3\text{H} \xrightarrow[\text{0.25 h, 100\%}]{\text{THF, 25 °C}} \text{LiR}_3\text{BH} + \text{Al(OMe)}_3\downarrow \quad (3)$$

R = Et, Bu, *s*-Bu, *i*-Bu

The combination of ***Lithium Tri-t-butoxyaluminum Hydride*** and triethylborane induces a rapid, essentially quantitative, reductive ring opening of THF to produce 1-butanol upon hydrolysis (eq 4).[10] Cyclohexene oxide and 1-methylcyclohexene oxide are instantaneously and quantitatively cleaved to their corresponding carbinols. Oxetane is readily cleaved to give 1-propanol in 98% yield. The reductive cleavage of both tetrahydropyran and oxepan is very sluggish and incomplete.

$$\text{THF} + \text{LiAl(O-}t\text{-Bu)}_3\text{H} \xrightarrow[\text{exothermic (~45 °C), 5 min}]{\text{Et}_3\text{B, THF}} \xrightarrow[100\%]{\text{H}_2\text{O}} \text{BuOH} \quad (4)$$

Triethyl- or triisopropylborane/***Trifluoromethanesulfonic Acid*** (triflic acid) is a convenient reagent for the selective reduction of hydroxy substituted carboxylic acids, ketones, and aldehydes to yield the corresponding carbonyl compounds (eq 5).[11] Not only tertiary hydroxy but also primary, secondary, and benzylic hydroxy groups are reduced in good to high yields. In general, the triisopropylborane/triflic acid system gives better results than triethylborane/triflic acid.

$$+ \text{Et}_3\text{B} + \text{CF}_3\text{SO}_3\text{H} \xrightarrow[\text{−30 °C, 35 min, then rt, 6 h; 71\%}]{\text{ClF}_2\text{CCCl}_2\text{F}} \quad (5)$$

OH

Preparation of R_2BOR' is generally carried out by treating alcohols R′OH with trialkylboranes R_3B in the presence of activating reagents like pivalic acid[13] or air.[14] However, Et_2BOMe can be prepared simply by mixing Et_3B with MeOH in THF at room temperature.[12] With the Et_2BOMe as a chelating agent, *syn*-1,3-diols are prepared in >98% stereochemical purity by reducing β-hydroxy ketones with ***Sodium Borohydride***.[12]

Enoxytriethylborates and Enoxydiethylboranes. Potassium enolates of ketones react with an unhindered trialkylborane such as triethylborane to form a potassium enoxytriethylborate, which undergoes selective α-monoalkylation with alkyl halides in high yields (eq 6).[15] In the absence of Et_3B, the potassium enolate itself gives a mixture of 43% mono- and 31% diallylated cyclohexanone along with 28% of recovered cyclohexanone. Monomethylation, -benzylation, and -propargylation of acetophenone also proceed in high yield in the presence of Et_3B. Lithium enolates, such as those obtained from acetophenone and cyclohexanone, do not form the corresponding enoxytriethylborates. Use of ***Potassium Hexamethyldisilazide*** as a base at −78 °C generates the less stable enolate (**1**) with high regioselectivity, while use of potassium hydride at 25 °C generates the most stable enolate (**2**) with ≥90% regioselectivity (eq 7).[16] The alkylations of these enolates proceed without complication in the presence of Et_3B (eq 7). Comparable regioselectivities are observed in the alkylations of 2-heptanone.

1. KH, THF, rt; 2. BEt_3, rt → [$O\bar{B}Et_3$ K^+] → 1. allyl Br, rt; 2. NaOH, H_2O; 90% (6)

1. $KH(TMS)_2$, THF, −78 °C; 2. BEt_3 → $OBEt_3K$ (**1**) → MeI, 86% overall → 93%; 2,6- (7)

1. KH, THF, 25 °C; 2. Et_3B → $OBEt_3K$ (**2**) → MeI, 79% overall → 90%; 2,2-

Allylation of potassium enoxyborates can be catalyzed by ***Tetrakis(triphenylphosphine)palladium(0)***.[17] Zinc enolates, readily obtained by treating lithium enolates with dry ***Zinc Chloride***, also undergo the Pd-catalyzed allylation with high regio- and stereoselectivities. Overall retention is observed with respect to the allylic cation center (eq 8).[17] In the presence of $Pd(PPh_3)_4$ catalyst and 2 equiv of BEt_3, lithium enolates of cyclopentanone and cyclohexanone derivatives react with (*E*)- or (*Z*)-allylic acetate (**3**) to provide (*E*)- or (*Z*)-allylation products with high stereospecificity (eq 9).[18] Both the Pd catalyst and BEt_3 are essential for the stereospecific allylation.

OM, Ph → cat. $Pd(PPh_3)_4$, THF, rt → CO_2H, O, Ph (8)

M = BEt_3K or ZnCl

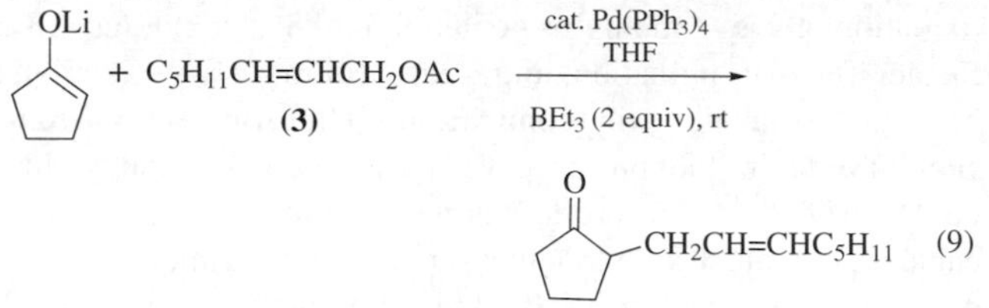

OLi + $C_5H_{11}CH{=}CHCH_2OAc$ (3) — cat. $Pd(PPh_3)_4$, THF, BEt_3 (2 equiv), rt → $CH_2CH{=}CHC_5H_{11}$ (9)

(*E*)-(**3**)	77%	(*E*):(*Z*) = 98:2
(*Z*)-(**3**)	73%	(*Z*):(*E*) = >97:3

The aldol reaction of preformed lithium enolates with aldehydes in the presence of trialkylboranes, such as BEt_3 and $B(n\text{-}Bu)_3$, leads to product mixtures rich in the more stable *anti*-aldol (eq 10).[19] Use of 3 equiv of BEt_3 gives high *anti* selectivity, while the stereoselectivity is low when 1 equiv of BEt_3 is used (eq 10). When lithium enolates are generated from silyl enol ethers and ***n-Butyllithium*** in THF, use of 1 equiv of BEt_3 is enough to produce high *anti* selectivity. The condensation of the lithio dianion of ethyl 3-hydroxybutyrate with *N*-anisyl cinnamylideneimine in the presence of Et_3B produces excellent 1′,3-*syn*/3,4-*cis* stereoselectivity (eq 11), whereas 1′,3-*syn*/3,4-*trans* selectivity is obtained in the presence of *t*-BuMgCl.[20] Aldol condensation of acetaldehyde and benzaldehyde with the lithium enolate of ethyl *N,N*-dimethylglycine in the presence of 1 equiv of Et_3B results in the formation of the corresponding *syn* 3-hydroxy-2-amino acid esters with excellent stereocontrol (>95% de).[21] The stereochemical outcome of these reactions is rationalized via the selective formation of the (*Z*)-enolate of ethyl *N,N*-dimethylglycine in the presence of triethylborane.

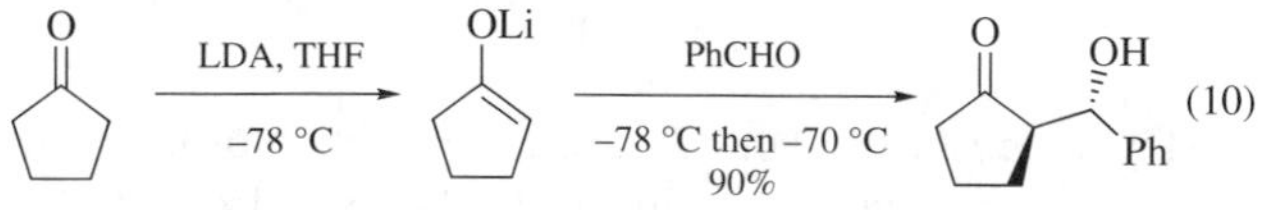

LDA, THF, −78 °C; PhCHO, −78 °C then −70 °C, 90% (10)

1 equiv BEt_3, *anti*:*syn* = 1:1
3 equiv BEt_3, *anti*:*syn* = 97:3

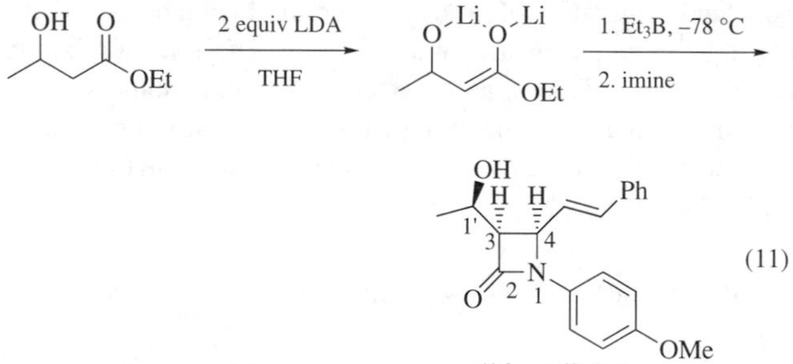

2 equiv LDA, THF; 1. Et_3B, −78 °C; 2. imine (11)

1′,3-*syn*/3,4-*cis*

There are several methods for generation of enoxyboranes (boron enolates or enol borirates).[22] Ketenes react with dialkylthioalkylboranes, $R_2BS(t\text{-}Bu)$, to yield alkenyloxyboranes formally derived from thioesters.[22a] A variety of ketones and carboxylic acid derivatives are converted to boron enolates upon treatment with dialkylboryl triflates in the presence of a tertiary amine, and the subsequent aldol condensation of these boron enolates has been studied.[22b,c] Trialkylboranes readily react with diazoacetaldehyde to give alkenyloxyboranes.[22d] Trialkylboranes spontaneously transfer an alkyl group to the β-position of β-unsubstituted α,β-unsaturated aldehydes and ketones to give alkenyloxyboranes, which are produced regio- but not stereospecifically.[22e] In the presence of 1–10 mol % of diethylboryl pivalate, Et_3B and ketones $RCOCH_2R'$ react at 85–110 °C to give diethyl(vinyloxy)boranes, $Et_2BOCR{=}CHR'$, in 70–90% yield.[22f] Reaction of α-bromo ketones with ***Triphenylsilane*** in the presence of Et_3B provides boron enolates which react with carbonyl compounds to give β-hydroxy ketones in good yields (eq 12).[23] The Et_3B-induced Reformatsky type reaction of α-iodo ketones with aldehydes or ketones proceeds without Ph_3SnH.[23,24] α-Bromocyclopentanone and -cyclohexanone provide *anti*-adducts with high diastereoselectivity (78–100%), whereas the reaction of 7-bromo-6-dodecanone with benzaldehyde gives a 65:35 mixture of the *syn*- and *anti*-adduct. It is proposed that vinyloxy(diethyl)boranes are involved as intermediates.

+ PhCHO — 1. benzene, 25 °C, Et_3B; 2. Ph_3SnH, benzene, 25 °C; 74% (12)

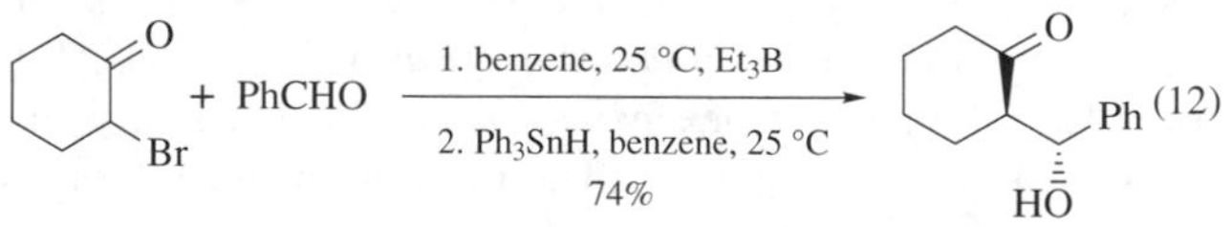

anti:syn = 78:22

Allylborates. 2-Butenyllithium reacts with aldehydes to afford the *anti*- and *syn*-β-methylhomoallyl alcohols in nearly equal amounts. However, if trialkylboranes such as Et_3B are present, the *anti*-product predominates (eq 13).[25] The corresponding allylic borate complexes are presumably involved as intermediates. Lithium allylic boronates, prepared by the addition of trialkylboranes (Et_3B, ***Tri-n-butylborane***, or *n*-Bu-9-BBN) to an ether solution of allylic lithium compounds, regioselectively react with allylic halides to produce head-to-tail 1,5-dienes (eq 14).[26] Regio- and stereocontrol via boron ate complexes is applicable to not only simple allylic but also heteroatom-substituted allylic anions.[27] The allyloxy carbanions (**4**) generally react with alkyl halides at the α-position, but react with carbonyl compounds at the γ-position. The Et_3B (or ***Triethylaluminum***) ate complexes of (**4**) react with aldehydes, ketones, and reactive halides at the α-position. The (alkylthio)allyl carbanion (**5**) reacts with alkyl halides at the γ-position, but with carbonyl compounds at the α-position. The Et_3B (or Et_3Al) ate complexes of (**5**) react with aldehydes, ketones, and allylic halides at the α-position. In general, the aluminum ate complex gives higher regioselectivity than the boron ate complex. The regioselectivity of Me_3Si- or pyrrolidine (*N*-atom)-substituted allylic anions is also controlled by the addition of Et_3B (or Et_3Al).[27] Either branched or linear homoallyl alcohols may be prepared by the reaction of (phenylselenyl)allyl carbanion with aldehydes and triethylborane under appropriate reaction conditions (eq 15).[28] The ethyl group of Et_3B in the initially formed ate complex $PhSeCH(BEt_3)CH{=}CH_2\ Li^+$ undergoes a facile migration from boron to the α-carbon to give (**6**), which reacts with benzaldehyde to give the linear adduct. The prolonged reaction period at higher temperatures induces the allylic rearrangement of (**6**) to (**7**), resulting in the formation of the branched adduct.

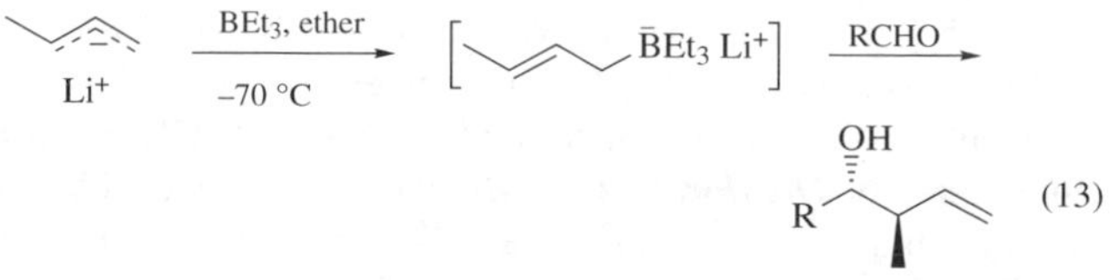

Li^+ — BEt_3, ether, −70 °C → [$\bar{B}Et_3\ Li^+$] — RCHO → (13)

	anti:*syn*
R = Ph, 90%	82:18
R = Me, 78%	85:15

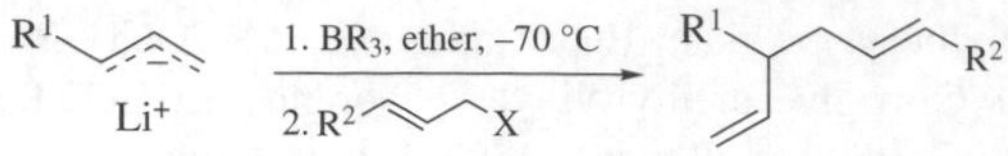

Alkylating Reagents. Monoalkylation of ketones is accomplished by reaction of trialkylboranes with α-bromo ketones under the influence of ***Potassium t-Butoxide*** in THF.[29] For example, α-bromocyclohexanone reacts with Et_3B to give α-ethylcyclohexanone (eq 16).[29] The reaction involves formation of the anion of the α-bromo ketone, formation of the boron ate complex, and rearrangement of Et from boron to the α-carbon. The use of potassium 2,6-di-*t*-butylphenoxide as a base, instead of *t*-BuOK, provides better results.[30] α-Bromoacetone, chloroacetonitrile, ethyl bromoacetate, and ethyl dibromoacetate are alkylated using this hindered base and R_3B.[30] The reaction of Et_3B with ethyl 4-bromocrotonate in the presence of one equiv of the new base affords ethyl 3-hexenoate (79% *trans*).[31] Monoalkylation of dichloroacetonitrile with Et_3B is achieved in 89% yield, and the dialkylation is carried out by using 2 equiv of base and 2 equiv of Et_3B (97% yield).[32]

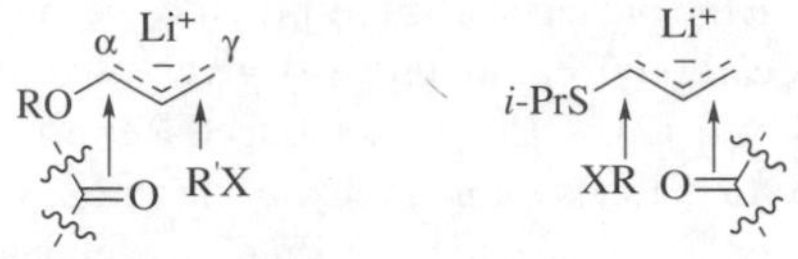

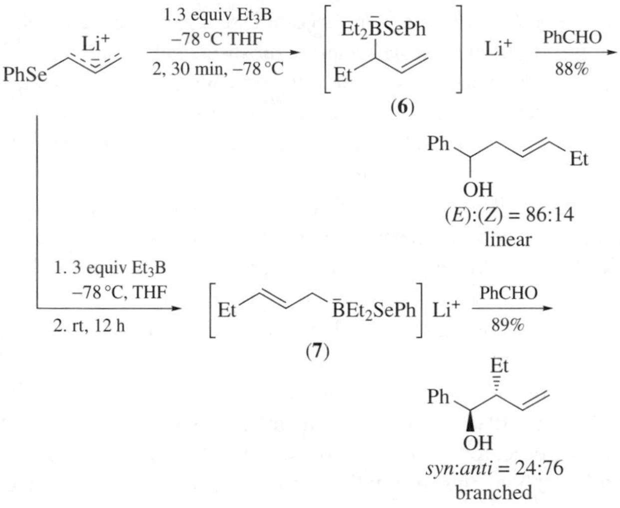

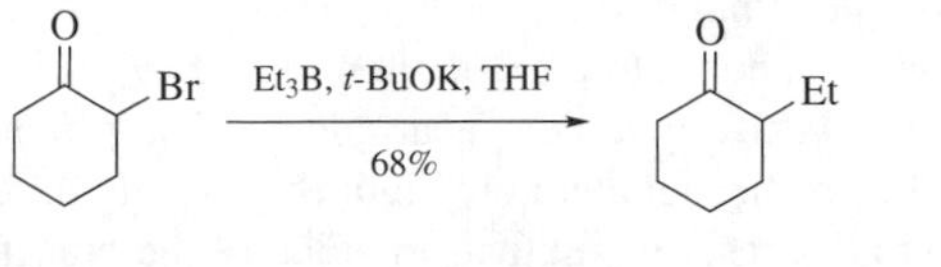

Trialkylcarbinols are prepared by the reaction of trialkylboranes with carbon monoxide in diglyme followed by oxidation with ***Hydrogen Peroxide*** (eq 17).[1b] Alternatively, trialkylcarbinols are obtained by the reaction of trialkylboranes with ***Chlorodifluoromethane*** (or ***Dichloromethyl Methyl Ether***) under the influence of lithium triethylmethoxide,[1b,33] or by the cyanidation reaction of trialkylboranes with ***Sodium Cyanide–Trifluoroacetic Anhydride*** followed by oxidation.[1b,34] Bromination of triethylborane under irradiation in the presence of water followed by oxidation gives 3-methyl-3-pentanol in 88% yield (eq 18).[35] Successful α-bromination–migration requires slow addition of bromine to avoid polybromination. The use of ***N-Bromosuccinimide*** in the presence of water increases the yield in eq 18 to 97%.[36] The bromination–migration reaction is applicable to simple trialkylboranes and dialkylborinic acids. The cross-coupling reaction of *B*-alkyl-9-borabicyclo[3.3.1]nonanes (*B*-R-9-BBN) with 1-halo-1-alkenes or haloarenes (R′X) in the presence of a catalytic amount of ***Dichloro[1,1′-bis(diphenylphosphino)ferrocene]palladium(II)*** and bases, such as NaOH and K_2CO_3, gives the corresponding alkenes or arenes (R–R′).[37] The use of catalytic amounts of $Cl_2Pd[PPh_3]_2$ in combination with ***Bis(acetylacetonato)zinc(II)*** effects carbonylative coupling of trialkylboranes with aryl iodides to give unsymmetrical ketones in 60–80% yields (eq 19).[38]

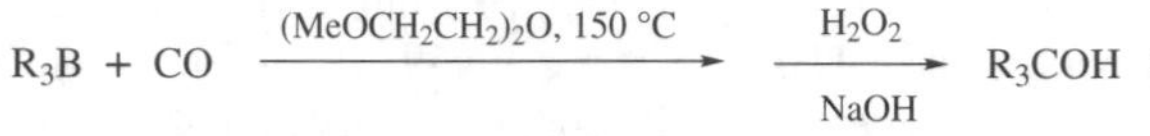

Alkynes are easily synthesized by the reaction of iodine with alkyne 'ate' complexes, readily formed in situ from R_3B and lithium acetylides (eq 20).[1b] Treatment of the alkyne 'ate' complexes with mild electrophiles E^+ results in β-attack on the triple bond and a migration of the organic group R from boron to carbon (eq 20).[1b] The protonation reaction with HX yields a mixture of *cis*- and *trans* alkenes, and mixtures of alkene isomers are also obtained in reactions involving MeI, MeOTs, allyl bromide, and oxirane. However, a single stereoisomer results from the reactions with other electrophiles.

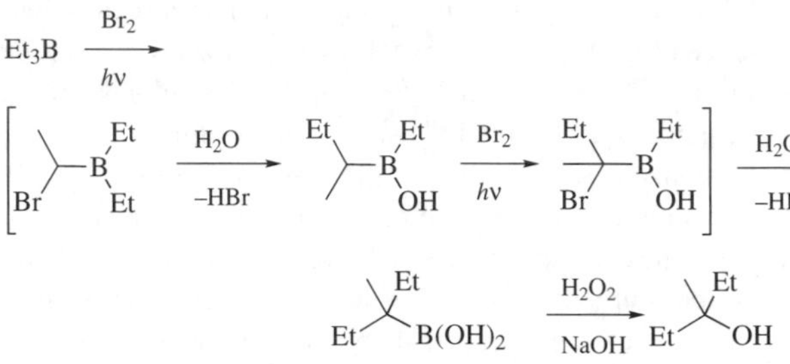

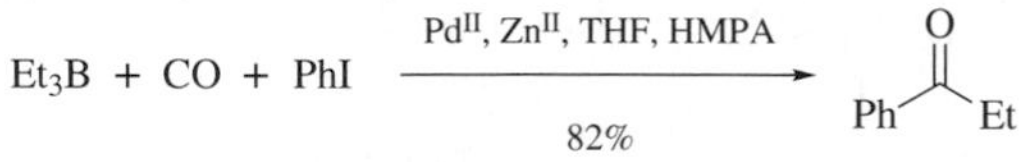

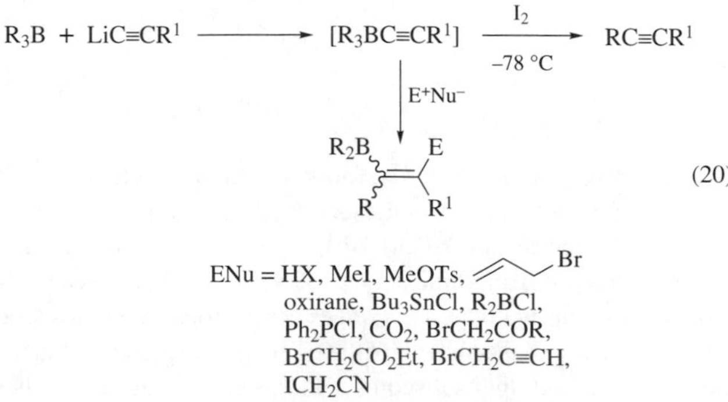

ENu = HX, MeI, MeOTs, allyl bromide, oxirane, Bu_3SnCl, R_2BCl, Ph_2PCl, CO_2, $BrCH_2COR$, $BrCH_2CO_2Et$, $BrCH_2C{\equiv}CH$, ICH_2CN

Stereochemical Control Element. Triethylborane acts as a stereo- and regiocontrol element in certain carbanionic reactions; several examples have been demonstrated in eqs 10, 11, and 13–15. Triethylborane-mediated epimerization of a 1αmethylcarbapenem intermediate proceeds with high stereoselectivity to give the 1β-methyl diastereomer (eq 21).[39] The 1β-methyl derivative is also obtained via alkylation of an 2-azetidinon-4-ylacetic acid derivative by using LDA–Et_3Al–MeI.[39] The deuteration of α-lithiobenzyl methyl sulfoxide in the presence of Et_3Al occurs with inversion, while the reaction in the absence of the additive occurs with retention; the use of Et_3B gives a mixture of the retention and inversion product.[40] The reagent RCu·BEt_3, prepared in situ from RCu and BEt_3 in ether at −70 °C, adds to α,β-alkynic carbonyl compounds with high stereospecificity, which cannot be achieved with conventional reagents such as R_2CuLi (eq 22).[41]

Lewis Acids and Radical Reactions. Methylenecyclopropanes react with 2-cyclopentenone in the presence of a Ni^0 catalyst such as ***Bis(1,5-cyclooctadiene)nickel(0)***, ***Triphenylphosphine***, and triethylborane to afford 6-methylenebicyclo[3.3.0]octan-2-ones (eq 23).[42] Treatment of tantalum–alkyne complexes with dimethylhydrazones and ***Trimethylaluminum*** in a DME, benzene, and THF solvent system at 45 °C gives (*E*)-allylic hydrazines stereoselectively, although the use of Et_3B results in formation of the product in very low yield.[43]

TBDMSO ... CO_2Me (azetidinone) → 1. LDA, THF, −80 °C; 2. 2 equiv Et_3B, hexane, −75 °C; 3. HOAc in THF, −75 °C → TBDMSO ... CO_2Me (21)

β:α = 93:7

$MeO_2CC{\equiv}CCO_2Me$ → 1. BuCu•BEt_3, −70 °C, ether; 2. MeOH, rt → MeO_2C / CO_2Me / Bu (22)

BuCu•BEt_3	85%	(*Z*):(*E*) = >99:1
BuCu	95%	= 85:15
Bu_2CuLi	98%	= 60:40

R (methylenecyclopropane) + 2-cyclopentenone → cat. Ni^0, PPh_3, BEt_3; 68% → bicyclic ketone, *trans*, R (23)

$R = C_5H_{11}$ *trans*:*cis* = 85:15

Trialkylboranes do not undergo facile addition reactions to carbonyl groups. However, rapid conjugate addition reactions occur with α,β-unsaturated carbonyl compounds, such as ***Acrolein*** and ***Methyl Vinyl Ketone*** (eq 24).[1b] The reaction proceeds through a radical mechanism. Trialkylboranes also participate in facile radical chain reactions with disulfides (e.g. ***Diphenyl Disulfide***), producing the corresponding thioethers (RSPh).[1b] Triphenylgermane adds easily to alkynes (RC≡CH) in the presence of Et_3B to give (*E*)- or (*Z*)-alkenyltriphenylgermanes ($RCH{=}CHGePh_3$) in good yields.[44] The (*Z*)-isomers predominate at −78 °C, whereas the hydrogermylation at 60 °C favors the (*E*)-isomer. Similarly, Et_3B is as efficient as ***Azobisisobutyronitrile*** for initiation of the hydrostannylation of alkynes, resulting in vinyltins.[45] The reaction is sluggish in the absence of oxygen. Triethylborane can also initiate radical cyclization of unsaturated alkynes to vinylstannanes (eq 25).[45] The 1,4-reduction of α,β-unsaturated ketones and aldehydes with ***Triphenylstannane*** or ***Tri-n-butylstannane*** proceeds in the presence of Et_3B to give the corresponding saturated ketones and aldehydes in good yields, whereas the same reaction of α,β-unsaturated esters with Ph_3SnH affords the tin hydride conjugate adduct.[46] Thiols[47a] and perfluoroalkyl iodides[47b] undergo similar addition reactions to alkynes in the presence of catalytic amounts of Et_3B. Treatment of 1-allyloxy-1-phenyl-2-bromo-1-silacyclopentanes with Bu_3SnH in the presence of catalytic amounts of Et_3B provides the cyclization products, which can be converted to 1,4,6-triol derivatives (eq 26).[48] Alkoxymethyl radicals (2-oxahex-5-enyl or 2-oxahept-6-enyl radicals), generated conveniently from phenylseleno precursors upon treatment with AIBN or Et_3B, cyclize to afford substituted tetrahydrofurans and tetrahydropyrans.[49]

R_3B + acrolein → R–CH=CH–$O{-}BR_2$ → H_2O → R–CH$_2$CH$_2$CHO (24)

Enyne ether → Bu_3SnH, toluene, 25 °C; BEt_3 (0.2 equiv); 96% → tetrahydrofuran vinylstannane ($SnBu_3$) (25)

Silacyclopentane (X, Si, Ph, O, R^1, R^2) → Bu_3SnH, benzene, 25 °C; BEt_3 (0.2 equiv) →

X = Br, I

Bicyclic (Ph, Si, O, H, R^1, R^2, X) → 1. H_2O_2, KF, $KHCO_3$; 2. Ac_2O, py → CHR^1R^2, H, H, OAc, OAc, OAc (26)

Related Reagents. *B*-Allyl-9-borabicyclo[3.3.1]nonane; Chlorodiethylborane; Crotyldimethoxyborane; Di-*n*-butylboryl Trifluoromethanesulfonate; Lithium Triethylborohydride; Potassium Tri-*s*-butylborohydride; Tri-*n*-butylborane; Triethylaluminum; Trimethylaluminum.

1. (a) Krishnamurthy, S. *Aldrichim. Acta* **1974**, *7*, 55. (b) Pelter, A.; Smith, K.; Brown, H. C. *Borane Reagents*, Academic: London, 1988.
2. Brown, H. C.; Krishnamurthy, S.; Coleman, R. A. *JACS* **1972**, *94*, 1750. Brown, H. C.; Krishnamurthy, S. *CC* **1972**, 868.
3. Brown, H. C.; Krishnamurthy, S. *JACS* **1973**, *95*, 1669.
4. Krishnamurthy, S.; Brown, H. C. *JOC* **1976**, *41*, 3064; Krishnamurthy, S. *JOM* **1978**, *156*, 171.
5. Krishnamurthy, S.; Schubert, R. M.; Brown, H. C. *JACS* **1973**, *95*, 8486.
6. Brown, H. C.; Kim, S. C. *JOC* **1984**, *49*, 1064.
7. Brown, H. C.; Krishnamurthy, S.; Hubbard, J. L. *JACS* **1978**, *100*, 3343.

8. Brown, C. A. *JACS* **1973**, *95*, 4100.
9. Brown, H. C.; Krishnamurthy, S.; Hubbard, J. L. *JOM* **1979**, *166*, 271. Brown, H. C.; Hubbard, J. L.; Singaram, B. *T* **1981**, *37*, 2359.
10. Krishnamurthy, S.; Brown, H. C. *JOC* **1979**, *44*, 3678.
11. Olah, G. A.; Wu, A. *S* **1991**, 407.
12. Chen, K.-M.; Gunderson, K. G.; Hardtmann, G. E.; Prasad, K.; Repic, O.; Shapiro, M. J. *CL* **1987**, 1923.
13. Köster, R.; Fenzl, W.; Seidel, G. *LA* **1975**, 352.
14. Narasaka, K.; Pai, F.-C. *T* **1984**, *40*, 2233.
15. Negishi, E.; Idacavage, M. J.; DiPasquale, F.; Silveira, A., Jr. *TL* **1979**, 845. Rathke, M. W.; Lindert, A. *SC* **1978**, *8*, 9.
16. Negishi, E.; Chatterjee, S. *TL* **1983**, *24*, 1341.
17. Negishi, E.; Matsushita, H.; Chatterjee, S.; John, R. A. *JOC* **1982**, *47*, 3188. Negishi, E.; John, R. A. *JOC* **1983**, *48*, 4098.
18. Luo, F.-T.; Negishi, E. *TL* **1985**, *26*, 2177.
19. Yamamoto, Y.; Yatagai, H.; Maruyama, K. *TL* **1982**, *23*, 2387.
20. Georg, G. I.; Akgün, E. *TL* **1990**, *31*, 3267.
21. Georg, G. I.; Akgün, E. *TL* **1991**, *32*, 5521.
22. (a) Inomata, K.; Muraki, M.; Mukaiyama, T. *BCJ* **1973**, *46*, 1807. (b) Mukaiyama, T.; Inone, T. *CL* **1976**, 559. (c) Evans, D. A.; Nelson, J. V.; Vogel, E.; Taber, T. R. *JACS* **1981**, *103*, 3099. (d) Hooz, J.; Morrison, G. F. *CJC* **1970**, *48*, 868. (e) Fenzl, W.; Köster, R.; Zimmerman, H.-J. *LA* **1975**, 2201. (f) Fenzl, W.; Köster, R. *LA* **1975**, 1322.
23. Nozaki, K.; Oshima, K.; Utimoto, K. *TL* **1988**, *29*, 1041.
24. Maruoka, K.; Hirayama, N.; Yamamoto, H. *Polyhedron* **1990**, *9*, 223.
25. Yamamoto, Y.; Yatagai, H.; Maruyama, K. *CC* **1980**, 1072.
26. Yamamoto, Y.; Yatagai, H.; Maruyama, K. *JACS* **1981**, *103*, 1969.
27. (a) Yamamoto, Y.; Yatagai, H.; Saito, Y.; Maruyama, K. *JOC* **1984**, *49*, 1096. (b) Yamamoto, Y.; Yatagai, H.; Maruyama, K. *CL* **1979**, *385*; *CC* **1979**, 157.
28. Yamamoto, Y.; Saito, Y.; Maruyama, K. *JOC* **1983**, *48*, 5408.
29. Brown, H. C.; Rogić, M. M.; Rathke, M. W. *JACS* **1968**, *90*, 6218.
30. Brown, H. C.; Nambu, H.; Rogić, M. M. *JACS* **1969**, *91*, 6852; *JACS* **1969**, *91*, 6854; *JACS* **1969**, *91*, 6855.
31. Brown, H. C.; Nambu, H. *JACS* **1970**, *92*, 1761.
32. Nambu, H.; Brown, H. C. *JACS* **1970**, *92*, 5790.
33. Brown, H. C.; Carlson, B. A.; Prager, R. H. *JACS* **1971**, *93*, 2070.
34. Pelter, A.; Hutchings, M. G.; Rowe, K.; Smith, K. *JCS(P1)* **1975**, 138.
35. Lane, C. F.; Brown, H. C. *JACS* **1971**, *93*, 1025.
36. Brown, H. C.; Yamamoto, Y. *S* **1972**, 699.
37. Miyaura, N.; Ishiyama, T.; Sasaki, H.; Ishikawa, M.; Satoh, M.; Suzuki, A. *JACS* **1989**, *111*, 314.
38. Wakita, Y.; Yasunaga, T.; Akita, M.; Kojima, M. *JOM* **1986**, *301*, C17.
39. Bender, D. R.; DeMarco, A. M.; Melillo, D. G.; Riseman, S. M.; Shinkai, I. *JOC* **1992**, *57*, 2411.
40. Yamamoto, Y.; Maruyama, K. *CC* **1980**, 239.
41. Yamamoto, Y.; Yatagai, H.; Maruyama, K. *JOC* **1979**, *44*, 1744.
42. Binger, P.; Schäfer, B. *TL* **1988**, *29*, 4539.
43. Takai, K.; Miwatashi, S.; Kataoka, Y.; Utimoto, K. *CL* **1992**, 99.
44. Ichinose, Y.; Nozaki, K.; Wakamatsu, K.; Oshima, K.; Utimoto, K. *TL* **1987**, *28*, 3709.
45. Nozaki, K.; Oshima, K.; Utimoto, K. *T* **1989**, *45*, 923; *BCJ* **1987**, *60*, 3465.
46. Nozaki, K.; Oshima, K.; Utimoto, K. *BCJ* **1991**, *64*, 2585.
47. (a) Ichinose, Y.; Wakamatsu, K.; Nozaki, K.; Birbaum, J.-L.; Oshima, K.; Utimoto, K. *CL* **1987**, 1647. (b) Takeyama, Y.; Ichinose, Y.; Oshima, K.; Utimoto, K. *TL* **1989**, *30*, 3159.
48. Matsumoto, K.; Miura, K.; Oshima, K.; Utimoto, K. *TL* **1992**, *33*, 7031.
49. Rawal, V. H.; Singh, S. P.; Dufour, C.; Michoud, C. *JOC* **1991**, *56*, 5245.

Yoshinori Yamamoto
Tohoku University, Sendai, Japan

Triethyl Orthoacetate[1]

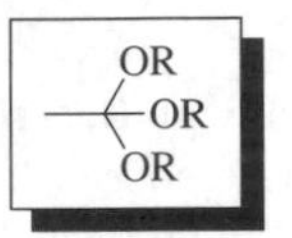

(**1**; R = Et)
[78-39-7] $C_8H_{18}O_3$ (MW 162.26)
(**2**; R = Me)
[1445-45-0] $C_5H_{12}O_3$ (MW 120.17)

(diethyl acetal of ethyl acetate; provides γ,δ-unsaturated ethyl esters from α,β-unsaturated alcohols through the Claisen rearrangement; converts diols to epoxides of the same relative stereochemistry; gives 2-methyl-1,3-heteroatom ring systems)

Physical Data: (**1**) bp 144–146 °C; $d = 0.8847$ g cm^{-3}; (**2**) bp 107–109 °C; $d = 0.9438$ g cm^{-3}.

Orthoester Exchange with Alcohols and Water. Orthoesters (1,1,1-trialkoxyalkanes, 'ester acetals') are acid labile, base stable, masked forms of carboxylic acid esters. Mild acid-catalyzed hydrolysis of an orthoester produces an ester and an alcohol, in this instance, ethyl acetate and ethanol. Thus ethyl orthoacetate, and orthoesters in general, can serve to drive the equilibrium in Fischer esterifications of carboxylic acids with ethanol by forming low-boiling ethyl acetate and ethanol from the water generated during the esterification.

Higher molecular weight orthoacetates (orthoesters) of nonallylic alcohols are prepared by acid-catalyzed exchange, with removal of the lower boiling alcohol to shift the equilibrium. When this reaction is conducted with a 1,2-diol such as L(+)-diethyl tartrate in the presence of excess ethyl orthoacetate, two of the ethoxy groups are exchanged to form a 2-ethoxy-2-methyl-1,3-dioxolane (eq 1).

HO CO2Et / HO CO2Et —[MeC(OEt)3, H2SO4]→ EtO (2-methyl-1,3-dioxolane) CO2Et / CO2Et (1)

Interconversion of Orthoesters and Haloesters. This exchange is an important element in a reaction sequence designed to convert a diol to the corresponding epoxide without loss of stereochemistry. Thus the dioxolane of D(−)-**2,3-Butanediol**, which is formed from methyl orthoacetate, is cleaved with trityl chloride to produce the β-acetoxy chloride, which, upon treatment with KOH, affords D(+)-2,3-epoxybutane (eq 2). The cleavage procedure, which is also effective with 2-ethoxy-2-methyl-1,3-dioxanes (eq 3), occurs through an S_N2 mechanism.[2] Because the formation of methyl trityl ether can complicate isolation, ***Chlorotrimethylsilane***, which gives volatile byproducts, is an effective replacement for trityl chloride in the cleavage of dioxolanes.[3] ***Phosphorus(V) Chloride*** has been found to be effective in the cleavage of a dioxolane derived from ethyl orthoformate (eq 4).[4] These reactions proceed through the formation of a cyclic dioxolenium (eqs 2 and 4) or dioxenium (eq 3) ion, an intermediate that can be accessed in the reverse direction to produce an orthoester. This procedure is typified by the transformation of the tetra-*O*-acetyl glucosyl bromide of eq 5.[5] The remaining ethoxy group of the cyclic orthoester can be exchanged intramolecularly by removing ethanol from the equilibrium even if an unfavorable conformation is required (eq 6).[6]

Ph3CCl, CH2Cl2; KOH (2)

Ph3CCl, CH2Cl2, 83% (3)

PCl5, CH2Cl2, 86% (4)

s-collidine, Bu4N+ Br−, 1 equiv EtOH, 85% (5)

2,6-lutidine hydrobromide, 4Å mol. sieves, CH2Cl2, 65% (6)

Orthoester Claisen Rearrangement. The principal use of triethyl and trimethyl orthoacetate and their more substituted analogs[7] is in the orthoester Claisen rearrangement.[1c] The reagent, which is often used as a solvent, is heated in the presence of an allylic alcohol and a weak acid catalyst with removal of ethanol (or methanol) (eq 7).

MeC(OEt)3, weak H+ (cat.); −EtOH (7)

The reaction was first applied in the synthesis of the triterpene squalene (**1**), using a symmetrical bis-allylic alcohol (eq 8).[8] The trisubstituted (*E*)-alkene is formed in >98% purity in excellent yield. The ester group of the product is converted into a secondary isopropenyl carbinol and it is then subjected to a second orthoacetate rearrangement.

(**1**)

MeC(OEt)3, propionic acid (cat.) (8)

The (*E*)-alkene selectivity is typical of the rearrangement of secondary allylic carbinols. Enantiomerically pure allylic alcohols (**2**) and (**3**), both of which are accessible from a common source, undergo rearrangement to give the same enantiomerically pure ester (**4**) (eq 9).[9] The results of eqs 8 and 9 demonstrate that propionic acid is capable of forming the dioxenium carbocation from the orthoacetate without isomerizing or racemizing the allylic alcohol and that the rearrangement is suprafacial.

(**2**) (*S*)-(*E*); (**3**) (*R*)-(*Z*); MeC(OEt)3, propionic acid (cat.); (**4**) (*S*)-(*E*) (9)

The ***Propionic Acid*** catalyst is often consumed in the form of the propionate ester of the allylic alcohol, thereby stopping the rearrangement process (eq 10).

$$EtCO_2H + MeC(OR)_3 \longrightarrow EtCO_2C(Me)(OR)_2 + ROH$$
$$EtCO_2C(Me)(OR)_2 + R'OH \xrightarrow{H^+} EtCO_2R' + MeCO_2R + ROH \quad (10)$$

Esterification can occur in reactions that are slow because of constraints imposed by the transition state. The second step of eq 10 is repressed by using a hindered carboxylic acid, thereby making the rearrangement competitive with esterification. Thus ***Pivalic Acid*** (5 mol %) as a catalyst leads to a 74% yield of ester (**6**); propionic acid affords lower yields and appreciable quantities of the propionate ester of alcohol (**5**) (eq 11).[10a] Alternatively, 2,4-dinitrophenol can be employed as a catalyst (eq 12).[10b] Rearrangement of allylic alcohol (**7**) at 100 °C is slow because of the low boiling point of methyl orthoacetate (bp 107–9 °C). When propionic acid is employed as the catalyst, the methyl ester is obtained in 50% yield along with esterified allylic alcohol. On occasion, the intermediate ketene acetal can undergo elimination to produce dienes if the rate of rearrangement is slow (eq 13).[11]

(**5**) MeC(OEt)3, pivalic acid (cat.), 140 °C, 74% (**6**) (11)

(**7**) MeC(OMe)3, 2,4-dinitrophenol (cat.), 100 °C, 17 h, 73% (12)

MeC(OMe)3, propionic acid (cat.), 137 °C, 24 h (13)

Carbon–Carbon Bond Formation. Orthoacetates undergo carbon–carbon bond formation with carbon nucleophiles in the presence of acid catalysts to provide acetals (eqs 14 and 15),[12a,b] enol ethers (eq 16),[13] and azulenes (eq 17).[14] The azulene formation is believed to be an [8 + 2] electrocyclic reaction wherein the ketene acetal from ethyl orthoacetate is functioning as the two-carbon component.

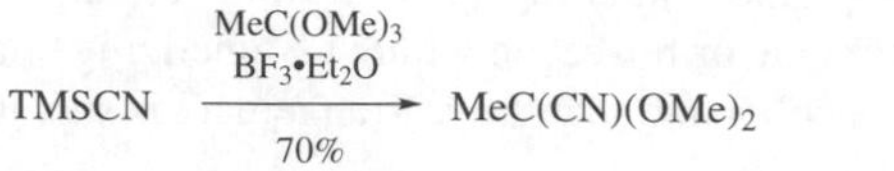

TMSCN —[$MeC(OMe)_3$, BF_3•Et_2O; 70%]→ $MeC(CN)(OMe)_2$ (14)

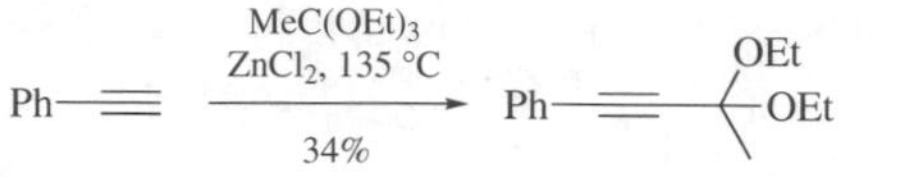

Ph—≡ —[$MeC(OEt)_3$, $ZnCl_2$, 135 °C; 34%]→ Ph—≡—C(OEt)(OEt) (15)

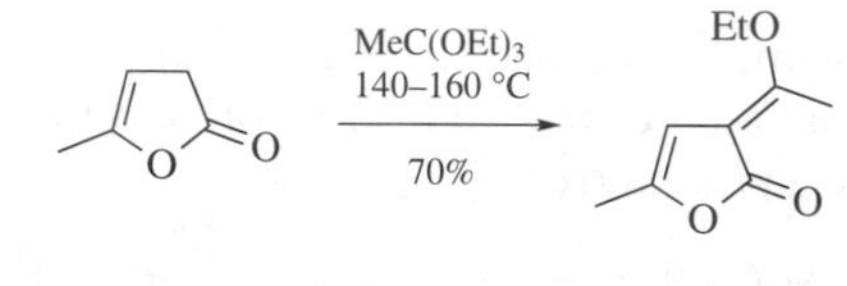

—[$MeC(OEt)_3$, 140–160 °C; 70%]→ EtO (16)

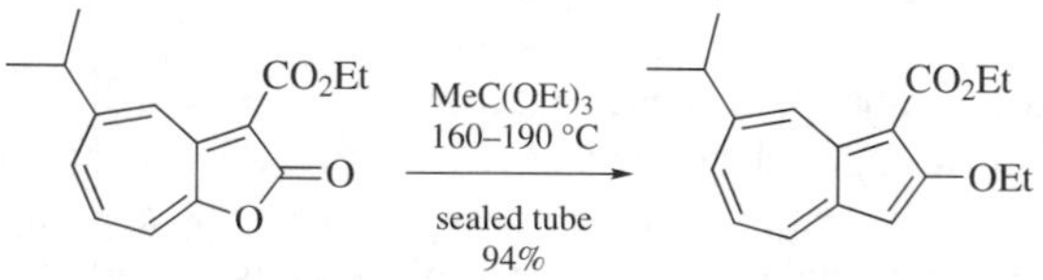

CO_2Et —[$MeC(OEt)_3$, 160–190 °C, sealed tube; 94%]→ CO_2Et, OEt (17)

Exchange with Heteroatoms. Alkenes are converted into epoxides by 90% ***Hydrogen Peroxide*** in the presence of ethyl orthoacetate. Trisubstituted and 1,2-disubstituted alkenes give yields in excess of 80%. The peroxide is thought to exchange with an ethoxy group to give the diethyl acetal of a peroxy acid as the reactive intermediate.[15] Orthoacetates exchange with one equivalent of a primary amine to give an acetimidate. A second equivalent produces an acetamidine (eq 18).[16]

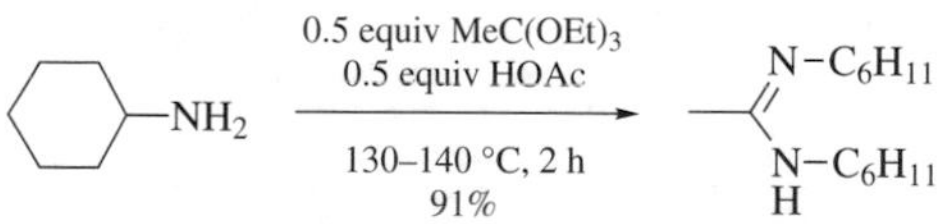

$-NH_2$ —[0.5 equiv $MeC(OEt)_3$, 0.5 equiv HOAc, 130–140 °C, 2 h; 91%]→ $N-C_6H_{11}$, $N(H)-C_6H_{11}$ (18)

Formation of Heterocycles. Intramolecular versions of eq 18 lead to heterocyclic systems through the intermediate acetimidate. By this method, imidazoles (eq 19),[17] oxazoles,[18] benzotriazepinones (eq 20),[19] quinazolinones,[20] dihydroimidazolium,[21] and tetrahydropyrimidinium[21] salts are prepared. In eq 19, the C-2 carbon and the C-2 methyl group of the product are derived from the acetic acid portion of methyl orthoacetate; the *N*-methyl group comes from the methoxy group of the orthoacetate by alkylation of the dioxenium ion. The oxazoline *N*-oxide (eq 21) is a reactive species that can participate in [3 + 2] cycloadditions.[22]

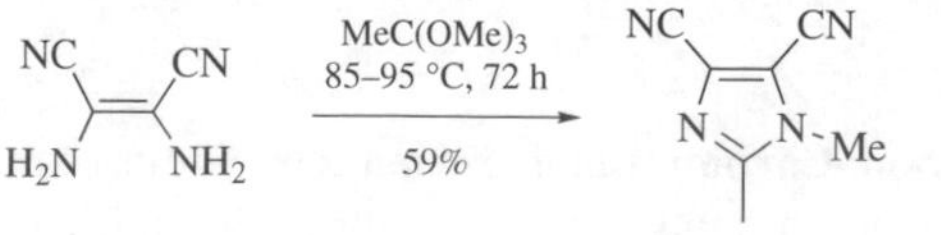

NC, CN, H_2N, NH_2 —[$MeC(OMe)_3$, 85–95 °C, 72 h; 59%]→ NC, CN, N, N-Me (19)

(8)

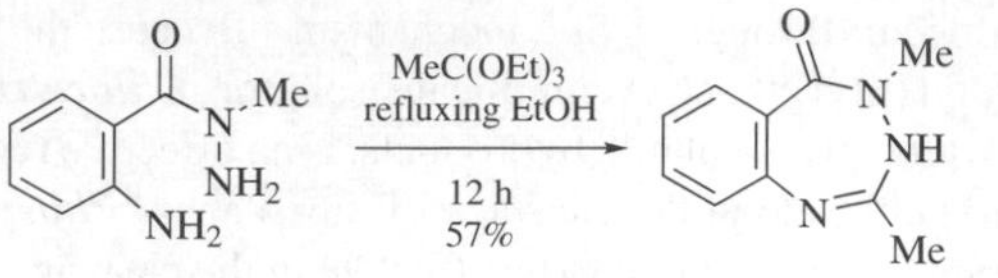

O, N-Me, NH_2, NH_2 —[$MeC(OEt)_3$, refluxing EtOH, 12 h; 57%]→ O, N-Me, NH, N=, Me (20)

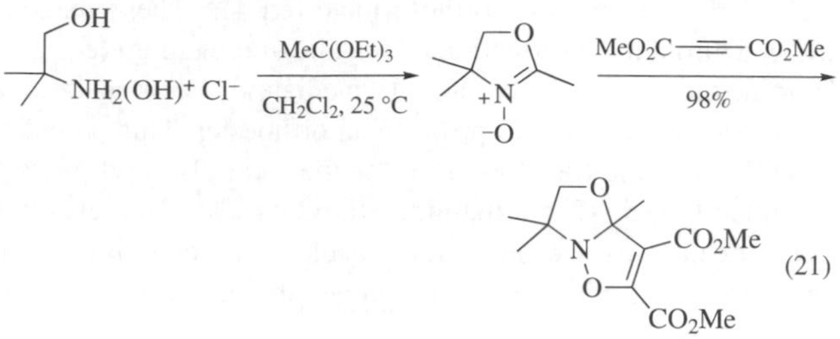

OH, $NH_2(OH)^+ Cl^-$ —[$MeC(OEt)_3$, CH_2Cl_2, 25 °C]→ O, +N, -O —[MeO_2C—≡—CO_2Me; 98%]→ O, N, O, CO_2Me, CO_2Me (21)

Reaction of Acetimidate Intermediates with Carbon Nucleophiles. Compounds with active methylene groups react with acetimidate-like intermediates to form carbon–carbon bonds in addition to heterocyclic rings. The reaction can be conducted in both the inter-[23] (eq 22) and intramolecular sense (eq 23).[24]

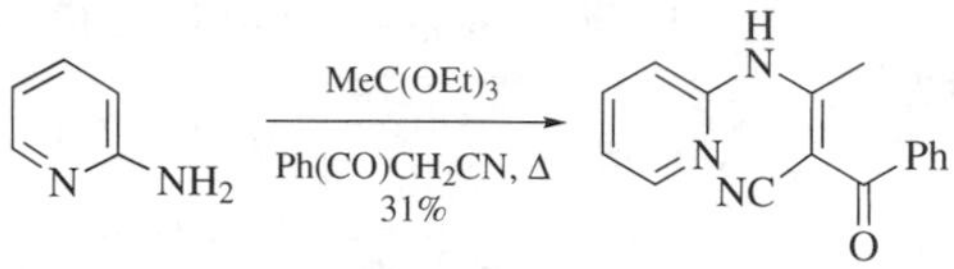

N, NH_2 —[$MeC(OEt)_3$, $Ph(CO)CH_2CN$, Δ; 31%]→ H, N, N, NC, Ph, O (22)

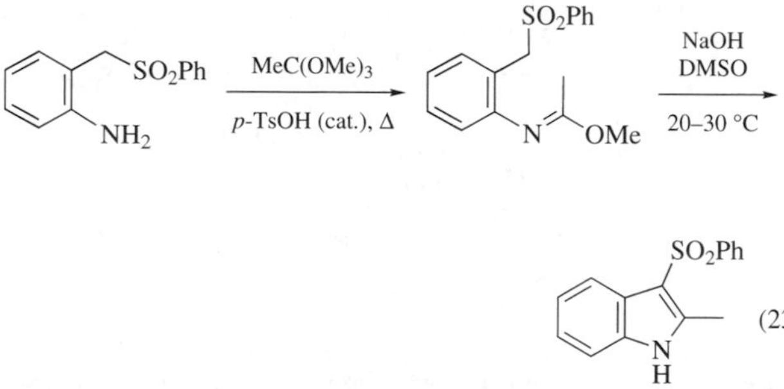

SO_2Ph, NH_2 —[$MeC(OMe)_3$, *p*-TsOH (cat.), Δ]→ SO_2Ph, N, OMe —[NaOH, DMSO, 20–30 °C]→ SO_2Ph, N, H (23)

Related Reagents. Diethyl Phenyl Orthoformate; *N*,*N*-Dimethylacetamide Dimethyl Acetal; *N*,*N*-Dimethylformamide Diethyl Acetal; Ethoxyacetylene; Ethyl Vinyl Ether; Ketene Bis(trimethylsilyl) Acetal; Ketene *t*-Butyldimethylsilyl Methyl Acetal; Methylketene Dimethyl Acetal; 1-Methoxy-1-(trimethylsilyloxy)propene; Triethyl Orthoformate.

---

1. (a) Pindur, U. in *The Chemistry of Acid Derivatives. Suppl. B*; Patai, S., Ed.; Wiley: Chichester, 1992; Vol. 2, Pt. 2, Chapter 17. (b) Cordes, E. H. In *The Chemistry of Carboxylic Acids and Esters*; Patai, S., Ed.; Interscience: London, 1969; Chapter 13. (c) DeWolfe, R. H. *Carboxylic Ortho Acid Derivatives*; Academic: New York, 1970. (d) Ziegler, F. E. *CRV* **1988**, *88*, 1423. (e) Ogliaruso, M. A.; Wolfe, J. F. *Synthesis of Carboxylic Acids, Esters and Their Derivatives*, Patai, S.; Rappoport, Z., Eds.; Wiley: New York, 1991.
2. Newman, M. S.; Chen, C. H. *JACS* **1973**, *95*, 278.
3. Newman, M. S.; Olson, D. R. *JOC* **1973**, *38*, 4203.
4. Nicolaou, K. C.; Papahatjis, D. P.; Claremon, D. A.; Magolda, R. L.; Dolle, R. E. *JOC* **1985**, *50*, 1440.
5. Lemieux, R. U.; Morgan, A. R. *CJC* **1965**, *43*, 2199.
6. Katano, K.; Chang, P-I.; Millar, A.; Pozsgay, V.; Minster, D. K.; Ohgi, T.; Hecht, S. M. *JOC* **1985**, *50*, 5807.

7. (a) Daub, G. W.; Teramura, D. H.; Bryant, K. E.; Burch, M. T. *JOC* **1981**, *46*, 1485. (b) Posner, G. H.; Kinter, C. M. *JOC* **1990**, *55*, 3967.
8. Johnson, W. S.; Werthemann, L.; Bartlett, W. R.; Brocksom, T. J.; Li, T-T.; Faulkner, D. J.; Petersen, M. R. *JACS* **1970**, *92*, 741.
9. (a) Chan, K.-K.; Cohen, N.; De Noble, J. P.; Specian, A. C., Jr.; Saucy, G. *JOC* **1976**, *91*, 3497. (b) Cohen, N.; Eichel, W. F.; Lopresti, R. J.; Neukom, C.; Saucy, G. *JOC* **1976**, *76*, 3505, 3512.
10. (a) Ziegler, F. E.; Bennett, G. B. *JACS* **1973**, *95*, 7458. (b) Miles, D. H.; Loew, P.; Johnson, W. S.; Kluge, A. F.; Meinwald, J. *TL* **1972**, 3019.
11. Dauben, W. G.; Dietsche, T. J. *JOC* **1972**, *37*, 1212.
12. (a) Utimoto, K.; Wakabayashi, Y.; Shishiyama, Y.; Inoue, M.; Nozaki, H. *TL* **1981**, *22*, 4279. (b) Howk, B. W.; Sauer, J. C. *JACS* **1958**, *80*, 4607.
13. Wolfers, H.; Kraatz, U.; Korte, F. *CB* **1976**, *109*, 1061.
14. Nozoe, T.; Wakabayashi, H.; Shindo, K.; Ishikawa, S.; Wu, C-P.; Yang, P-W. *H* **1991**, *32*, 213.
15. Rebek, J., Jr.; McCready, R. *TL* **1979**, 1001.
16. Taylor, E. C.; Ehrhart, W. A. *JOC* **1963**, *28*, 1108. For reactions with ureas, see: Whitehead, C. W.; Traverso, J. J. *JACS* **1955**, *77*, 5872.
17. Johnson, S. J. *S* **1991**, 75.
18. LaMattina, J. L. *JOC* **1980**, *45*, 2261.
19. (a) Sunder, S.; Peet, N. P.; Trepanier, D. L. *JOC* **1976**, *41*, 2732. (b) Leiby, R. W.; Heindel, N. D. *JOC* **1976**, *41*, 2736. (c) Peet, N. P.; Sunder, S.; Barbuch, R. J. *JHC* **1983**, *20*, 511.
20. Peet, N. P.; Sunder, S.; Cregge, R. J. *JOC* **1976**, *41*, 2733.
21. Saba, S.; Brescia, A.; Kaloustian, M. K. *TL* **1991**, *32*, 5031.
22. Coates, R. M.; Ashburn, S. P. *JOC* **1984**, *49*, 3127.
23. Junek, H.; Schmidt, H.-W.; Gfrerer, G. *S* **1982**, 791.
24. Wojciechowski, K.; Mākosza, M. *S* **1986**, 651.

Frederick E. Ziegler, Makonen Belema,
Patrick G. Harran, & Renata X. Kover
Yale University, New Haven, CT, USA

Triethyl Orthoformate[1]

(**1**; R = Et)		
[122-51-0]	$C_7H_{16}O_3$	(MW 148.23)
(**2**; R = Me)		
[149-73-5]	$C_4H_{10}O_3$	(MW 106.14)

(precursor for higher analogs by reaction with alcohols,[1c] including cyclic orthoesters from polyols;[2] for deoxygenation of 1,2-diols, affording alkenes;[3] acetalization of carbonyl compounds;[4] a dehydrating agent[5] for enol ether formation;[6] esterification of acids;[7] formylation of active methylene compounds,[8] heteroatom nucleophiles[9] and organometallic reagents;[10] formylation of electron-rich species; dialkoxycarbenium ion precursor;[11] solvent for thallium trinitrate reactions[12])

Alternate Name: triethoxymethane.
Physical Data: (**1**) bp 146 °C; $d = 0.891$ g cm^{-3}; (**2**) bp 102 °C; $d = 0.970$ g cm^{-3}.
Solubility: sol most organic solvents.
Form Supplied in: clear liquid; widely available.
Purification: distillation.
Handling, Storage, and Precautions: highly moisture sensitive; flammable; irritant with high volatility. Use in a fume hood.

Introduction. The orthoformates are a remarkably useful group of reagents. They are shelf-stable, yet highly reactive. As alkylating agents, they readily transfer the associated alkyl group, a large variety of which are easily available. As formylation reagents, they are reactive under both acidic and basic conditions. The choice of ester is often arbitrary in this context.

Transesterification.

Higher Orthoformates. While there are many ways to obtain esters of orthoformic acid,[1c] an easy method takes advantage of the rapid equilibrium among orthoesters (eq 1). By starting with the lowest analog, trimethyl orthoformate, essentially complete conversion to the higher esters is possible by carrying out the reaction at such a temperature as to distill away the evolving methanol. It is important that reactions be carried out under anhydrous conditions to avoid formation of the formate through hydrolysis.

$$HC(OR)_3 + 3\,R'OH \rightleftharpoons HC(OR')_3 + 3\,ROH \qquad (1)$$

Cyclic Orthoformates[2]. When the above strategy is applied to polyols, cyclic orthoformates can be isolated. Most common are the cyclization of 1,3-diols (eq 2)[2,13] and 1,2-diols,[14] as well as the formation of caged structures from the use of polyols (eq 3).[15,16]

(MeO)$_3$CH, *p*-TsOH; 12 h, reflux; 43% (2)

(MeO)$_3$CH; *p*-TsOH, DMF (3)

Those orthoformates obtained from 1,2-diols (eq 4)[3] can undergo cycloelimination upon pyrolysis to afford alkenes in high yield.[17] There are a variety of methods for carrying out this overall process,[18] but the orthoester route is competitive if the alkene is thermally stable.

(EtO)$_3$CH, 95.3%; 175 °C, 70% (4)

Acetals and Enol Ethers. The conversion of the orthoformate to formate is energetically favored. As a result, the acetalization of ketones by orthoformate is a highly favored process and allows the formation of acetals under exceedingly mild conditions. A wide variety of ketones can be converted into dimethyl acetals by the action of trimethyl orthoformate and ***p-Toluenesulfonic Acid***. The methyl formate thus evolved is distilled away.[4]

The process is general and allows isolation of quite sensitive acetals (eq 5).[19] The technique accommodates protection of α,β-unsaturated carbonyl compounds.[20,21,22] While some form of acid catalysis is usually needed, there is great flexibility in the choice of acid, including Amberlyst 15[23] and, in particular, ***Montmorillonite K10***.[24] Cyclic acetals are also accessible (eq 6).[25]

1. O_3, MeOH, –70 °C; 2. Me_2S, $(MeO)_3CH$, MeOH, HCl; 90% (5)

1. CH_2=$CHCH_2MgBr$, $CeCl_3$; 2. $(MeO)_3CH$ (6)

Upon distillation of some acetals,[26] or upon attempted acetalization of highly conjugated species,[26,27] the enol ether can also be observed. The choice of acid often determines whether the acetal or the enol ether is isolated.

Orthoformates are also useful in promoting the formation of other acetals[5] by functioning as dehydrating agents. This function is useful for β-lactone formation as well (eq 7).[28,29]

$(MeO)_3CH$, benzene, reflux; Δ, 83% overall (7)

Esterification. In a related process, orthoformates are good esterification agents; they operate on carboxylic acids (eq 8),[7] sulfonic acids,[30] and carboxyboranes,[31] often without the need for acid catalysis.

CO_2H + $(EtO)_3CH$ —Δ, no acid→ CO_2Et (8)

Formylation. The orthoformate carbon is highly reactive in a number of bond-forming reactions. It is capable of reaction under both electrophilic and nucleophilic conditions and serves as a formylation reagent.

Active Methylene Compounds. Triethyl orthoformate can formylate diethyl malonate under slightly acidic conditions.[8] With less activated compounds it can be induced to undergo a Mannich reaction,[32] and can also formylate a cyclohexanone enolate anion (eq 9).[33]

OLi —$(MeO)_3CH$, BF_3•Et_2O, 79%→ O, OMe, OMe (9)

This reaction is noteworthy in its propensity for *C*-alkylation and the fact that the protected acetal raises the pK_a of the product relative to the unprotected β-dicarbonyl compound.

Heteroatom Nucleophiles. The formylation of anilines is well known[34] and provides entry to a large array of functional groups.[35] Of greater interest is the ability of the product to be trapped in a subsequent reaction to afford heterocycles (eq 10).[9,36–38] Orthoformates also react readily with phosphorus nucleophiles.[39]

$(EtO)_2P$(O), O, N, NH_2, NH_2 —HCl, $(EtO)_3CH$, DMF, Δ, 63%→ $(EtO)_2P$(O), O, N, NH, O (10)

Organometallic Reactions. In addition to the enolate reactions described above, orthoformates can also carry out the formal formylation of Grignard reagents.[10]

Electrophilic Formylation. It has been shown that the dialkoxycarbenium ion can be readily formed by acid treatment of orthoformates. While the reactive ion can be isolated and used directly,[11] the typical practice is to generate it in situ.

While not as universal as the Gatterman–Koch reaction, the cation works well for the formylation of activated aromatic compounds (eq 11).[40,41] Of greater synthetic utility is the effective formylation of alkynes (eq 12)[42] and alkenes (see below).

HO, OH —$(EtO)_3CH$, CH_2Cl_2, $AlCl_3$→ HO, OH, OEt, OEt —H_2O, 64% overall→ HO, OH, CHO (11)

Ph—≡ —$(EtO)_3CH$, ZnI_2, 75%→ Ph—≡—CH(OEt)(OEt) (12)

Reactions of silyl enol ethers with dialkoxycarbenium ions result in α-formyl ketones (eq 13),[43] much like those achieved above through the use of enolate anions. With a dienol silane (eq 14), regioselective γ-formylation is achieved.[44] In extended alkenic systems, cationic cyclization (eq 15) can be realized.[45]

OTMS + $(MeO)_3CH$ —(H^+, 72%)→ O OMe OMe (13)

OTMS + $(RO)_3CH$ —(85%)→ O OR OR (14)

$R = CH_2{=}CHCH_2$

—($SnCl_4$, $(MeO)_3CH$, –20 °C)→ OH

O H MeO 24% + O H MeO Cl 25% + O Cl MeO 11% (15)

Solvent for Thallium Trinitrate Oxidations. While ***Thallium(III) Nitrate*** oxidations of aromatic ketones and chalcones often gives rise to mixtures of products, the use of trimethyl orthoformate as solvent gives substantially cleaner reactions and higher yields (eq 16).[12]

Ar O —(TTN, MeOH, $(MeO)_3CH$, 70–95%)→ Ar OMe CO_2Me (16)

Related Reagents. Diethyl Phenyl Orthoformate; Dimethoxycarbenium Tetrafluoroborate; Dimethylchloromethyleneammonium Chloride; *N,N*-Dimethylformamide; *N,N*-Dimethylformamide Diethyl Acetal; Methyl Formate.

---

1. (a) DeWolfe, R. H. *Carboxylic Ortho Acid Derivatives*; Academic: New York, 1970. (b) Ghosh, S.; Ghatak, U. R. *Proc. Ind. Acad. Sci.* **1988**, *100*, 235. (c) DeWolfe, R. H. *S* **1974**, 153.
2. Denmark, S. E.; Almstead, N. G. *JOC* **1991**, *56*, 6458.
3. Camps, P.; Cardellach, J.; Font, J.; Ortuno, R. M.; Ponsati, O. *T* **1982**, *38*, 2395.
4. Napolitano, E.; Fiaschi, R.; Mastrorilli, E. *S* **1986**, 122.
5. Marquet, A.; Dvolaitzky, M.; Kagan, H. B.; Mamlok, L.; Ouannes, C.; Jacques, J. *BSF* **1961**, 1822.
6. Wohl, R. A. *S* **1974**, 38.
7. Cohen, H.; Mier, J. D. *CI(L)* **1965**, 349.
8. Parham, W. E.; Reed, L. J. *OSC* **1955**, *3*, 395.
9. Harden, M. R.; Jarvest, R. L.; Parratt, M. J. *JCS(P1)* **1992**, 2259.
10. Bachman, G. B. *OSC* **1943**, *2*, 323.
11. Pindur, U.; Flo, C. *SC* **1989**, *19*, 2307.
12. Taylor, E. C.; Robey, R. L.; Liu, K.-T.; Favre, B.; Bozimo, H. T.; Conley, R. A.; Chiang, C.-S.; McKillop, A.; Ford, M. E. *JACS* **1976**, *98*, 3037.
13. Gardi, R.; Vitali, R.; Ercoli, A. *TL* **1961**, 448.
14. Takasu, M.; Naruse, Y.; Yamamoto, H. *TL* **1988**, *29*, 1947.
15. Stetter, H.; Steinacker, K. H. *CB* **1953**, *86*, 790.
16. Yu, K.-L.; Fraser-Reid, B. *TL* **1988**, *29*, 979.
17. Burgstahler, A. W.; Boger, D. L.; Naik, N. C. *T* **1976**, *32*, 309.
18. Block, E. *OR* **1984**, *30*, 457.
19. Frickel, F. *S* **1974**, 507.
20. Taylor, E. C.; Conley, R. A.; Johnson, D. K.; McKillop, A. *JOC* **1977**, *42*, 4167.
21. van Allen, J. A. *OSC* **1963**, *4*, 21.
22. Wengel, J.; Lau, J.; Pedersen, E. B.; Nielson, C. M. *JOC* **1991**, *56*, 3591.
23. Patwardhan, S. A.; Dev, S. *S* **1974**, 348.
24. Taylor, E. C.; Chiang, C.-S. *S* **1977**, 467.
25. Rychnovsky, S. D.; Griesgraber, G. *CC* **1993**, 291.
26. Meek, E. G.; Turnbull, J. H.; Wilson, W. *JCS* **1953**, 811.
27. van Hulle, F.; Sipido, V.; Vandewalle, M. *TL* **1973**, 2213.
28. Rogic, M. M.; van Peppe, J. F.; Klein, K. P.; Demmin, T. R. *JOC* **1974**, *39*, 3424.
29. Blume, R. C. *TL* **1969**, 1047.
30. Padmapriya, A. A.; Just, G.; Lewis, N. G. *SC* **1985**, *15*, 1057.
31. Mittakanti, M.; Feakes, D. A.; Morse, K. W. *S* **1992**, 380.
32. El Cherif, S.; Rene, L. *S* **1988**, 138.
33. Suzuki, S.; Yanagisawa, A.; Noyori, R. *TL* **1982**, *23*, 3595.
34. Roberts, R. M.; Vogt, P. J. *JACS* **1956**, *78*, 4778.
35. Crochet, R. A.; Blanton, C. D., Jr. *S* **1974**, 55.
36. Jenkins, G. L.; Knevel, A. M.; Davis, C. S. *JOC* **1961**, *26*, 274.
37. Patridge, M. W.; Slorach, S. A.; Vipond, H. J. *JCS* **1964**, 3670.
38. Lee, K.-J.; Kim, S. H.; Kim, S.; Cho, Y. R. *S* **1992**, 929.
39. Baille, A. C.; Cornell, C. L.; Wright, B. J.; Wright, K. *TL* **1992**, *33*, 5133.
40. Treibs, W. *TL* **1967**, 4707.
41. Gross, H.; Rieche, A.; Matthey, G. *CB* **1963**, *96*, 308.
42. Howk, B. W.; Sauer, J. C. *OSC* **1963**, *4*, 801.
43. Mukaiyama, T.; Iwakiri, H. *CL* **1985**, 1363.
44. Pirrung, M. C.; Thomson, S. A. *TL* **1986**, *27*, 2703.
45. Perron-Sierra, F.; Promo, M. A.; Martin, V. A.; Albizati, K. F. *JOC* **1991**, *56*, 6188.

Richard T. Taylor
Miami University, Oxford, OH, USA

Triethyl Phosphonoacetate[1]

$(EtO)_2P(O)CH_2CO_2Et$

[867-13-0] $C_8H_{17}O_5P$ (MW 224.22)

(treatment with a base gives a phosphoryl-stabilized carbanion that reacts with aldehydes,[1] ketones,[1] nitrones,[2] epoxides,[3] nitroso compounds,[4] ketenes,[5] alkyl halides,[6c] and halogenating agents[6a,b])

Alternate Names: ethyl (diethoxyphosphinyl)acetate; diethyl (ethoxycarbonyl)methylphosphonate.

Physical Data: bp 152–153 °C/20 mmHg; *d* 1.130 g cm^{-3}.
Solubility: sol common organic solvents.
Form Supplied in: colorless liquid; widely available.
Handling, Storage, and Precautions: irritant; store under N_2.

Reactions with Aldehydes and Ketones. $(EtO)_2P(O)CH_2CO_2Et$ in either DME, THF, DMF, MeCN, dioxane, benzene, or toluene is deprotonated by ***Sodium Hydride***,[7] ***Sodium Amide***,[8] metal alkoxides,[9] and other bases to give a phosphoryl-stabilized carbanion, $[(EtO)_2P(O)\overline{C}HCO_2Et]\ M^+$. Subsequent reaction of the carbanion with an aldehyde or ketone gives an (*E*)-α,β-unsaturated ester as the major product in a Horner–Wadsworth–Emmons (HWE) reaction (eqs 1–4)[1] (see also ***Trimethyl Phosphonoacetate*** and ***Methyl Bis(2,2,2-trifluoroethoxy)phosphinylacetate***). The phosphate ester byproduct is water soluble, unlike the Wittig byproduct (Ph_3PO), and is easily removed at the completion of the reaction. Ketones and aliphatic aldehydes often give the (*Z*)-α,β-unsaturated ester in addition to the (*E*)-isomer,[1c,d] the isomer ratio being influenced by the reaction solvent (eq 1).[7a] Some ketones and aldehydes that are unreactive to Wittig alkenation conditions (e.g. 3-keto steriods[9,10] and glucose derivatives[11]) will react with the carbanion of $(EtO)_2P(O)CH_2CO_2Et$. The Peterson alkenation reaction of lithio **Ethyl Trimethylsilylacetate** with aldehydes and ketones offers a mild alternative to both the HWE and Wittig approaches to α,β-unsaturated esters.[12] The Peterson reaction often gives the (*Z*)-isomer as the major product.[12b] (*Z*)-Alkenes are also obtained from HWE reactions using bis(2,2,2-trifluoroethyl) phosphonates (see also ***Trimethyl Phosphonoacetate Methyl Bis(trifluoroethoxy)phosphonylacetate***).[13]

[(EtO)₂P(O)CHCO₂Et]Na — PhH or EtOH (1)

EtOH (*E*):(*Z*) = 3:1
PhH (*E*):(*Z*) = 1:19

Milder reaction conditions have been developed to increase yields, accommodate sensitive substrates, and minimize undesired side reactions such as double bond migrations, Knoevenagel condensation, Cannizzaro reaction, and Michael addition.[1] HWE reactions of aldehydes and $(EtO)_2P(O)CH_2CO_2Et$ have been carried out in two-phase liquid–liquid conditions (aqueous $NaOH/CH_2Cl_2$) in the presence of a phase transfer catalyst, thus eliminating the need for anhydrous solvents.[14] Superior yields and reduced reaction times were obtained using aromatic aldehydes and ketones under solid–liquid two-phase conditions (powdered NaOH or KOH/THF) (eq 2).[14b,c]

Cl–C₆H₄–CHO; KOH powder, THF, rt, 15 min 96%; or K_2CO_3, H_2O, 20 °C, 10 h 97% (2)

K_2CO_3, toluene, 20 °C, 48 h; 87%

Aliphatic and particularly aromatic aldehydes give high yields of (*E*)-α,β-unsaturated esters using $(EtO)_2P(O)CH_2CO_2Et$ and K_2CO_3 (eq 2)[15] or Cs_2CO_3[16] under solid–liquid conditions. K_2CO_3 has also been used in highly concentrated (6–10 molar) aqueous solutions under heterogeneous liquid–liquid conditions in the absence of organic solvent (eq 2).[15a,c] The combination of aqueous K_2CO_3 and $(EtO)_2P(O)CH_2CO_2Et$ has been employed elsewhere, e.g. under these conditions glutaraldehyde and succinaldehyde give 1-ethoxycarbonyl-6-cyclohexenol and 1-ethoxycarbonyl-5-cyclopentenol, respectively,[17] and aqueous formaldehyde gives ***Ethyl α-(Hydroxymethyl)acrylate***[15a] rather than the normal HWE product. The use of D_2O, rather than water, gives a general preparation of α-deuterated α,β-unsaturated esters[18] (%D > 90).

Furfural reacts with $(EtO)_2P(O)CH_2CO_2Et$ and activated ***Barium Hydroxide*** catalyst at 70 °C in dioxane containing a trace of water to give (*E*)-ethyl 3-(2′-furyl)acrylate in quantitative yield.[19] Reaction times are reduced in comparison with other solid–liquid procedures.[19] No secondary reactions were observed, in contrast to aqueous base catalysis, and the method is applicable to hindered aldehydes.[19] A reduction in both the amount of $Ba(OH)_2$ catalyst required and the reaction time is obtained using sonochemical, rather than thermal activation.[20]

$(EtO)_2P(O)CH_2CO_2Et$, DIPEA, LiCl, MeCN; rt, 24 h, 85% (3)

(*E*):(*Z*) >50:1

The reaction of an aldehyde with $(EtO)_2P(O)CH_2CO_2Et$ and ***Diisopropylethylamine*** (DIPEA)[21,22] or ***1,8-Diazabicyclo[5.4.0]undec-7-ene*** (DBU)[21,23] in a stirred suspension of ***Lithium Chloride*** in dry MeCN gives excellent yields of (*E*)-α,β-unsaturated esters (eq 3).[21] The method is widely used with base-sensitive aldehydes.[21–23] Few reports have appeared using a ketone rather than an aldehyde under these conditions.[21,24] The use of DIPEA, rather than DBU, has been reported to give less epimerization of the aldehyde component (eq 4).[22b] An interesting extension of this reaction proceeds by an in situ oxidation and HWE coupling (eq 5).[22c] The combination of $(EtO)_2P(O)CH_2CO_2Et$, TEA, LiCl and an aldehyde also gives high yields of an α,β-unsaturated ester.[25] The use of $(CF_3CH_2O)_2P(O)CH_2CO_2Et$ has extended the scope of this reaction to include ketones.[25,26] Other catalysts for the HWE reaction include ***Alumina***, and ***Potassium Fluoride-***

Alumina.[27] Polymer-bound phosphonates have also received some attention.[28]

		(4)
$(EtO)_2P(O)CH_2CO_2Et$, DIPEA, LiCl, rt	95:5	65%
$(EtO)_2P(O)CH_2CO_2Et$, DBU, LiCl, rt	22:78	85%
$Ph_3PCH_2CO_2Et$, toluene, reflux	58:42	76%

1.DMSO, $(COCl)_2$ DIPEA, −78 °C 2. $(EtO)_2P(O)CH_2CO_2Et$ (5)

Diisobutylaluminum Hydride (DIBAL) reduction of saturated esters[29] and lactones[29a] at −78 °C in the presence of the NaH-derived carbanion of $(EtO)_2P(O)CH_2CO_2Et$ gives good yields of the homologous esters with >90% (*E*) stereochemistry and with little or no overreduction of either the starting material or product ester (eq 6).[29a] The reaction of $(EtO)_2P(O)CH_2CO_2Et$ with either glucopyranoses[11b] or hydroxy phthalides in the presence of 1 mol % of tetra-*n*-hexylammonium bromide[30] gives ring-opened alkenes in good yields.

1.0 equiv DIBAL $[(EtO)_2P(O)CHCO_2Et]Na$ THF, −78 °C 62% (6)

The addition of an acyl phosphonate [e.g. $(EtO)_2P(O)COPh$] to the NaH-derived phosphoryl carbanion of $(EtO)_2P(O)CH_2CO_2Et$, gives the (*Z*)-alkene as the major product.[31] The analogous Wittig alkenation reaction gives (*E*)-alkenes.[31]

Other Reactions. Nitrones react with $(EtO)_2P(O)CH_2CO_2Et$ and NaH in DME or alkali metal alkoxides in alcoholic solvents to give aziridines and cyclic enamines (eq 7).[2] Cyclopropanes are prepared by refluxing an epoxide with $(EtO)_2P(O)CH_2CO_2Et$,[3a] or the corresponding sodium phosphoryl carbanion in dioxane or DME[3b] (eq 8).[3c,d]

$(EtO)_2P(O)CH_2CO_2Et$ (7)

NaH, DME, reflux	35%	–
t-BuONa, *t*-BuOH, reflux	75%	–
MeONa, MeOH, reflux	6%	36%

$(EtO)_2P(O)CH_2CO_2Et$ 1. NaH, dioxane 2. reflux, 90 h 54% (8)

7-Pteridinones are prepared by treating the sodium phosphoryl carbanion of $(EtO)_2P(O)CH_2CO_2Et$ with 4,6-diamino-5-nitrosopyrimidines (eq 9).[4] Ketenes react with the NaH-derived carbanion of $(EtO)_2P(O)CH_2CO_2Et$ in DME at 50 °C to give allenes (eq 10).[5,6a] α-Alkylated and arylated phosphonates are prepared by reacting the carbanion of $(EtO)_2P(O)CH_2CO_2Et$ with an allyl[6c] and benzyl[32] halide, respectively. Triethyl phosphonoacetate is also readily halogenated at the α-position.[6a,b]

$[(EtO)_2P(O)CHCO_2Et]Na$ THF, rt 90% (9)

$(EtO)_2P(O)CH_2CO_2Et$ 1. NaH, DME 2. rt to 50 °C 32% (10)

Related Reagents. *t*-Butyl α-Lithiobis(trimethylsilyl)-acetate; Diethyl Acetonylphosphonate; (Ethoxycarbonylmethyl)triphenylphosphonium Bromide; Ethyl (Methyldiphenylsilyl)acetate; Ethyl Trimethylsilylacetate; (Methoxycarbonylmethylene)triphenylphosphorane; Methyl Bis(2,2,2-trifluoroethoxy)phosphinylacetate; Trimethyl Phosphonoacetate; Trimethyl 2-Phosphonoacrylate.

1. (a) Boutagy, J.; Thomas, R. *CRV* **1974**, *74*, 87. (b) Wadsworth, W. S., Jr. *OR* **1977**, *25*, 73. (c) Larsen, R. O.; Aksnes, G. *PS* **1983**, *16*, 339. (d) Maryanoff, B. E.; Reitz, A. B. *CRV* **1989**, *89*, 863.
2. (a) Zbaida, S.; Breuer, E. *JOC* **1982**, *47*, 1073. (b) Breuer, E.; Zbaida, S. *JOC* **1977**, *42*, 1904.
3. (a) Denney, D. B.; Vill, J. J.; Boskin, M. J. *JACS* **1962**, *84*, 3944. (b) Smith, D. J. H. *Organophosphorus Reagents in Organic Synthesis*; Cadogan, J. I. G., Ed.; Academic: London, 1979; pp 214–217 and references therein. (c) Fraser-Reid, B.; Carthy, B. J. *CJC* **1972**, *50*, 2928. (d) Fitzsimmons, B. J.; Fraser-Reid, B. *T* **1984**, *40*, 1279.
4. (a) Youssefyeh, R. D.; Kalmus, A. *CC* **1970**, 1371. (b) Taylor, E. C.; Evans, B. E. *CC* **1971**, 189.
5. Kresze, G.; Runge, W.; Ruch, E. *LA* **1972**, *756*, 112.
6. (a) Wadsworth, W. S., Jr.; Emmons, W. D. *JACS* **1961**, *83*, 1733. (b) McKenna, C. H.; Khawli, L. A. *JOC* **1986**, *51*, 5467. (c) Kirschleger, B.; Queignec, R. *S* **1986**, 926.
7. (a) Wender, P. A.; Eissenstat, M. A.; Filosa, M. P. *JACS* **1979**, *101*, 2196. (b) Baggiolini, E. G.; Iacobelli, J. A.; Hennessy, B. M.; Batcho, A. D.; Sereno, J. F.; Uskokovic, M. R. *JOC* **1986**, *51*, 3098.
8. Takahashi, H.; Fujiwara, K.; Ohta, M. *BCJ* **1962**, *35*, 1498.
9. Bose, A. K.; Dahill, R. T., Jr. *JOC* **1965**, *30*, 505.
10. Bose, A. K.; Dahill, R. T., Jr. *TL* **1963**, 959.
11. (a) Tadano, K.; Idogaki, Y.; Yamada, H.; Suami, T. *JOC* **1987**, *52*, 1201. (b) Monti, D.; Gramatica, P.; Speranza, G.; Manitto, P. *TL* **1987**, *28*, 5047.

12. (a) Shimoji, K.; Taguchi, H.; Oshima, K.; Yamamoto, H.; Nozaki, H. *JACS* **1974**, *96*, 1620. (b) Strekowski, L.; Visnick, M.; Battiste, M. A. *TL* **1984**, *25*, 5603.
13. Still, W. C.; Gennari, C. *TL* **1983**, *24*, 4405.
14. (a) Piechucki, C. *S* **1974**, 869. (b) Texier-Boullet, F.; Foucaud, A. *S* **1979**, 884. (c) Texier-Boullet, F.; Foucaud, A. *TL* **1980**, *21*, 2161.
15. (a) Villieras, J.; Rambaud, M. *S* **1983**, 300. (b) Mouloungui, Z.; Delmas, M.; Gaset, A. *SC* **1984**, *14*, 701. (c) Villieras, J.; Rambaud, M.; Graff, M. *TL* **1985**, *26*, 53. (d) Mouloungui, Z.; Elmestour, R.; Delmas, M.; Gaset, A. *T* **1992**, *48*, 1219.
16. Mouloungui, Z.; Murengezi, I.; Delmas, M.; Gaset, A. *SC* **1988**, *18*, 1241.
17. Graff, M.; Al Dilaimi, A.; Seguineau, P.; Rambaud, M.; Villieras, J. *TL* **1986**, *27*, 1577.
18. Seguineau, P.; Villieras, J. *TL* **1988**, *29*, 477.
19. Sinisterra, J. V.; Mouloungui, Z.; Delmas, M.; Gaset, A. *S* **1985**, 1097.
20. (a) Fuentes, A.; Marinas, J. M.; Sinisterra, J. V. *TL* **1987**, *28*, 2951. (b) Sinisterra, J. V.; Fuentes, A.; Marinas, J. M. *JOC* **1987**, *52*, 3875.
21. Blanchette, M. A.; Choy, W.; Davis, J. T.; Essenfeld, A. P.; Masamune, S.; Roush, W. R.; Sakai, T. *TL* **1984**, *25*, 2183.
22. (a) Dauben, W. G.; Greenfield, L. J. *JOC* **1992**, *57*, 1597. (b) Guanti, G.; Banfi, L.; Narisano, E.; Riva, R. *TL* **1992**, *33*, 2221. (c) Blackwell, C. M.; Davidson, A. H.; Launchbury, S. B.; Lewis, C. N.; Morrice, E. M.; Reeve, M. M.; Roffey, J. A. R.; Tipping, A. S.; Todd, R. S. *JOC* **1992**, *57*, 5596.
23. (a) Takacs, J. M.; Myoung, Y. C. *TL* **1992**, *33*, 317. (b) Gravier-Pelletier, C.; Dumas, J.; Le Merrer, Y.; Depezay, J.-C. *T* **1992**, *48*, 2441.
24. Courtneidge, J. L.; Bush, M.; Loh, L.-S. *JCS(P1)* **1992**, 1539.
25. (a) Rathke, M. W.; Nowak, M. *JOC* **1985**, *50*, 2624. (b) Tsukamoto, T.; Kitazume, T. *CC* **1992**, 540. (c) Tietze, L. F.; Wünsch, J. R. *AG(E)* **1991**, *30*, 1697.
26. Rathke, M. W.; Bouhlel, E. *SC* **1990**, *20*, 869.
27. Texier-Boullet, F.; Villemin, D.; Ricard, M.; Moison, H.; Foucaud, A. *T* **1985**, *41*, 1259.
28. Qureshi, A. E.; Ford, W. T. *Brit. Polym. J.* **1984**, *16*, 231.
29. (a) Takacs, J. M.; Helle, M. A.; Seely, F. L. *TL* **1986**, *27*, 1257. (b) Johnson, S. J.; Kesten, S. R.; Wise, L. D. *JOC* **1992**, *57*, 4746. (c) Ikemoto, N.; Schreiber, S. L. *JACS* **1992**, *114*, 2524.
30. Trost, B. M.; Rivers, G. T.; Gold, J. M. *JOC* **1980**, *45*, 1835.
31. Harris, R. L. N.; McFadden, H. G. *AJC* **1984**, *37*, 417.
32. Rodriguez, M.; Heitz, A.; Martinez, J. *TL* **1990**, *31*, 7319.

Andrew Abell
University of Canterbury, Christchurch, New Zealand

Trifluoromethyltrimethylsilane

Me_3SiCF_3

[81290-20-2] $C_4H_9F_3Si$ (MW 142.22)

(nucleophilic trifluoromethylating agent;[1,2] difluorocarbene precursor[1])

Physical Data: bp 54–55 °C; d^{20} 0.963 g cm^{-3}.
Solubility: sol THF, ether, CH_2Cl_2.
Form Supplied in: colorless liquid; commercially available.
Preparative Methods: is prepared[2,6] based on the original procedure of Ruppert,[7] by the reaction of ***Chlorotrimethylsilane*** with the complex of trifluoromethyl bromide and hexaethylphosphorous triamide in benzonitrile (eq 1). Other less convenient procedures for its preparation are also reported.[8,9]

$$\text{TMSCl} + \text{CF}_3\text{Br} \xrightarrow[75\%]{(\text{Et}_2\text{N})_3\text{P, benzonitrile}} \text{TMSCF}_3 \quad (1)$$

Handling, Storage, and Precautions: is an acid-, base-, and moisture-sensitive compound and should be stored under anhydrous conditions in a refrigerator. Use in a fume hood.

Introduction. $TMSCF_3$ is a valuable reagent for trifluoromethylation of electrophilic substrates under nucleophilic catalysis or initiation.[1,2] Homologous trifluoromethyltrialkylsilanes have also been used for the same purpose.[3–5]

$TMSCF_3$ reacts as a trifluoromethide equivalent with a wide variety of electrophilic substrates such as carbonyl compounds, sulfonyl fluorides, sulfoxides, deactivated aromatics, sulfur dioxide, and alkenes, etc.

Reactions with Carbonyl Compounds. $TMSCF_3$ reacts with aldehydes in the presence of a catalytic amount of ***Tetra-n-butylammonium Fluoride*** (TBAF) in THF to form the corresponding trifluoromethylated carbinols in good to excellent yields following aqueous hydrolysis of the silyl ethers (eq 2).[2,3,6,10] The reaction also works very well for ketones under the same conditions, with the exception of extremely hindered ones such as 1,7,7-trimethylbicyclo[2.2.1]heptan-2-one, di-1-adamantyl ketone, and fenchone. The reaction has been characterized as a fluoride-induced autocatalytic reaction.[3,10] Other initiators such as ***Tris(dimethylamino)sulfonium Difluorotrimethylsilicate*** (TASF), ***Potassium Fluoride***, $Ph_3SnF_2^-$, and RO^- can also be used for these reactions. For the reactions of $TMSCF_3$ and perfluorinated ketones and pentafluorobenzaldehyde, excess of KF is needed.[16,17]

$$\text{RC(O)R'} \xrightarrow[\text{cat. TBAF, THF}]{\text{TMSCF}_3} \text{RR'C(OTMS)CF}_3 \xrightarrow{\text{aq. HCl}} \text{RR'C(OH)CF}_3 \;\; (60\text{–}90\%) \quad (2)$$

$TMSCF_3$ has been used in the preparation of polycyclic aromatic carcinogens, the key step being the addition of $TMSCF_3$ to the carbonyl group (eq 3).[11]

(3)

A series of tripeptides containing the trifluoromethyl group has been prepared as potent inhibitors of human leukocyte elastase (HLE) by using $TMSCF_3$.[12] $TMSCF_3$ has also been used to prepare trifluoromethyl analogs of L-fucose and 6-deoxy-D-ribose.[13] Gassman et al. have prepared 1-trifluoromethylindene starting from 1-indanone.[14]

1,2-Diketones such as benzil give only the monoadduct. On the other hand, the highly enolizable cyclohexane-1,3-dione did not give the addition product with $TMSCF_3$.[15]

A series of α,β-conjugated enones and ynones react with $TMSCF_3$ to give predominant 1,2-addition products (eq 4).[3,15]

1. $TMSCF_3$, F^-, THF; 2. aq. HCl; 50–90% (4)

Simple unactivated esters do not react with $TMSCF_3$.[3] However, activated esters such as trifluoroacetic acid esters do react.[3] Cyclic esters, i.e. lactones, react with $TMSCF_3$ to give the corresponding adducts.[3] An efficient and simple synthesis of trifluoropyruvic acid monohydrate has been developed starting from di-*t*-butyl oxalate.[18]

Direct trifluoromethylation of α-keto esters gave Mosher's acid derivatives (eq 5).[19]

$TMSCF_3$, cat. TBAF, THF; aq. HCl (5)

Acyl halides such as benzoyl chloride react with $TMSCF_3$ to give a mixture of the trifluoroacetophenone and hexafluorocumyl alcohol in the presence of equimolar TBAF.[3] Cyclic anhydrides react readily with $TMSCF_3$; however, a stochiometric amount of TBAF is required. Acyclic anhydrides react less cleanly.[3]

Simple amides, such as benzamide and acetamide, do not react with $TMSCF_3$ even with molar quantities of TBAF.[3] However, an activated amide carbonyl such as in *N*-trifluoroacetylpiperidine reacts to give an adduct which upon subsequent hydrolysis gives hexafluoroacetone trihydrate.[15] Imidazolidinetriones react with $TMSCF_3$ to give 5-trifluoromethyl-5-hydroxyimidazolidine-2,4-diones upon aqueous acid workup.[20] Imides such as *N*-methylsuccinimide react smoothly to afford the hemiaminal adducts (eq 6).[15]

1. $TMSCF_3$, F^-, THF; 2. aq. HCl (6)

Sulfur Derivatives. Kirchmer and Patel have reported[17] the preparation of trifluoromethylsulfinyl fluorides, sulfonyl fluorides, sulfoxides, and sulfuranes using $TMSCF_3$ in the presence of a catalytic amount of KF.

$TMSCF_3$ reacts with ***Dimethyl Sulfoxide*** in the presence of catalytic amount of TBAF to give Me_2CF_3SOTMS.[17] Arylsulfonyl fluorides react with $TMSCF_3$ to give the corresponding trifluoromethyl sulfones.[1,21]

Sulfur Dioxide reacts with $TMSCF_3$ in the presence of TMSONa to give the sodium trifluoromethyl sulfinate. The sulfinate has been further oxidized to trifluoromethanesulfonic acid in 30% overall yield.[1]

Nitroso Group. Nitrosobenzene reacts with $TMSCF_3$ to afford the *O*-silylated trifluoromethylated hydroxylamine quantitatively.[1]

Aromatic Compounds. Nucleophilic trifluoromethylation of aromatic compounds containing nitro, fluoro, and trifluoromethyl groups as substituents using $TMSCF_3$ has been investigated.[22,15] Yagupolskii et al. have reported that $TMSCF_3$–TASF (1:1) reacts with 1,2,4,5-tetrakis(trifluoromethyl)benzene at −30 °C to give the stable carbanion salt (eq 7).[23]

$TMSCF_3$, TASF; $(Me_2N)_3S^+$ (7)

Phosphorus Compounds. Treatment of $(BuO)_2P(O)F$ with $TMSCF_3$ and a catalytic amount of KF gave $(BuO)_2P(O)CF_3$ in 93% isolated yield.[1]

Generation of Difluorocarbene. Treatment of $TMSCF_3$ with an anhydrous fluoride source such as TASF in THF results in the generation of singlet difluorocarbene. In the presence of an acceptor such as tetramethylethylene, the corresponding adduct can be isolated.[1]

Related Reagents. Chlorodifluoromethane; Dibromodifluoromethane; Sodium Chlorodifluoroacetate; 2,2,2-Trifluoroethyl *p*-Toluenesulfonate; Trifluoroiodomethane; Trifluoromethylcopper(I); *S*-(Trifluoromethyl)dibenzothiophenium Triflate; Trifluoromethyl Hypofluorite.

---

1. Prakash, G. K. S. In *Synthetic Fluorine Chemistry*; Olah, G. A.; Chambers, R. D.; Prakash, G. K. S., Eds.; Wiley: Chichester, 1992, Chapter 10.
2. Bosmans, J. P. *Janssen Chim. Acta* **1992**, *10*, 22 (*CA* **1992**, *117*, 7213q).
3. Krishnamurti, R.; Bellew, D. R.; Prakash, G. K. S. *JOC* **1991**, *56*, 984.
4. Stahly, G. P.; Bell, D. R. *JOC* **1989**, *54*, 2873.
5. Urata, H.; Fuchikami, T. *TL* **1991**, *32*, 91.
6. Ramaiah, P.; Krishnamurti, R.; Prakash, G. K. S. *OS* **1995**, *72*, 232.
7. Ruppert, I.; Schlich, K.; Volbach, W. *TL* **1984**, *25*, 2195.
8. Pawelke, G. *JFC* **1989**, *42*, 429.
9. Eaborn, C.; Griffiths, R. W.; Pidcock, A. *JOM* **1982**, *225*, 331.
10. Prakash, G. K. S.; Krishnamurti, R.; Olah, G. A. *JACS* **1989**, *111*, 393.
11. Coombs, M. M.; Zepik, H. H. *CC* **1992**, 1376.
12. Skiles, J. W.; Fuchs, V.; Miao, C.; Sorcek, R.; Grozinger, K. G.; Mauldin, S. C.; Vitous, J.; Mui, P. W.; Jacober, S.; Chow, G.; Matteo, M.; Skoog,

M.; Weldon, S. M.; Possanza, G.; Keirns, J.; Letts, G.; Rosenthal, A. S. *JMC* **1992**, *35*, 641.
13. Bansal, R. C.; Dean, B.; Hakomori, S-I.; Toyokuni, T. *CC* **1991**, 796.
14. Gassman, P. G.; Ray, J. A.; Wenthold, P. G.; Mickelson, J. W. *JOC* **1991**, *56*, 5143.
15. Kantamneni, S. Ph.D Dissertation, University of Southern California, July 1993.
16. Kotun, S. P.; Anderson, J. D. O.; DesMarteau, D. D. *JOC* **1992**, *57*, 1124.
17. Patel, N. R.; Kirchmeier, R. L. *IC* **1992**, *31*, 2537.
18. Broicher, V.; Geffken, D. *TL* **1989**, *30*, 5243.
19. Ramaiah, P.; Prakash, G. K. S. *SL* **1991**, 643.
20. Broicher, V.; Geffken, D. *AP* **1990**, *323*, 929.
21. Kolomeitsev, A. A.; Movchun, V. N.; Kondratenko, N. V.; Yagupolskii, Yu. L. *S* **1990**, 1151.
22. Bardin, V. V.; Kolomeitsev, A. A.; Furin, G. G.; Yagupolskii, Yu. L. *IZV* **1990**, 1693 (*CA* **1991**, *115*, 279 503c).
23. Kolomeitsev, A. A.; Movchun, V. N.; Yagupolskii, Yu. L. *TL* **1992**, *33*, 6191.

George. A. Olah, G. K. Surya Prakash,
Qi Wang & Xing-Ya Li
University of Southern California, Los Angeles, CA, USA

Trimethylaluminum[1]

Me_3Al

[75-24-1] C_3H_9Al (MW 72.10)

(Lewis acid with methylation ability;[2–8] can effect useful C–C bond forming reactions including carboalumination,[14] methylenation,[15,16] and cyclopropanation;[17] can serve as a precursor to various sophisticated Lewis acids[27] or chiral catalysts[28,29])

Physical Data: mp 15 °C; bp 127 °C; *d* 0.743 g cm^{-3} (30 °C).
Solubility: freely miscible with saturated and aromatic hydrocarbons; reacts violently with H_2O and protic solvents.
Form Supplied in: clear colorless liquid; widely available as a neat liquid in a stainless container or 1–2 M solution in hydrocarbon solvents (hexane, heptane, toluene).
Analysis of Reagent Purity: brochures from manufacturers describe an apparatus and method for assay.
Handling, Storage, and Precautions: indefinitely stable under an inert atmosphere. Neat trimethylaluminum or concentrated solutions are highly pyrophoric. Solutions that are more dilute are not pyrophoric and are safer to handle. The nonpyrophoric limits are 11 wt % in hexane and 14 wt % in heptane. Halogenated hydrocarbons should be avoided because of possible explosive reactions sometimes observed for mixtures of CCl_4 and organoaluminums.

Methylation. The Lewis acidity of Me_3Al has been used to activate electronegative atoms, such as oxygens, halogens, and other functional groups. It also methylates the resulting electrophilic species. Bridgehead methylation of bicyclo[2.2.2]octyl bromide (eq 1)[2a] and regioselective methylation of terpene derivatives[3a] are early examples that show these features. More recent examples include methylation of bromododecahedrane[2b] and unusual aromatic substitutions of *N*-hydroxyaniline derivatives (eq 2).[3b] Substitution reactions of glycosyl fluorides,[4] benzyl or glycal acetates,[5a,b] and sulfonyl groups[6] at activated positions can be effected with Me_3Al (eqs 3–6).

Br (bicyclo[2.2.2]octyl bromide) —(Me_3Al, CH_2Cl_2; −70 to 0 °C; 98%)→ (1)

N(OTMS)(CH$_2$Ph)-phenyl —(Me_3Al; 0 °C; 71%)→ ortho-methyl NH–CH$_2$Ph + para-methyl NH–CH$_2$Ph, 3:2 (2)

BnO / BnO / OBn / OBn glycosyl F —(Me_3Al; toluene, 0 °C; 95%)→ C-methyl glycoside (3)

$(OC)_3Cr$ OAc —(Me_3Al, CH_2Cl_2; −78 to 0 °C; 99%)→ $(OC)_3Cr$ methyl (4)

OAc, Ph, O —(Me_3Al, CH_2Cl_2; −20 to 0 °C; 96%)→ Ph, O; *trans*:*cis* = 96:4 (5)

N, SO_2Ph, O, O —(Me_3Al; CH_2Cl_2, rt; 75%)→ N, O, O (6)

Boron Trifluoride Etherate alters the reactivity of γ- or δ-lactols toward Me_3Al (eq 7).[7] Me_3Al traps isocyanates generated from certain sigmatropic rearrangements (eq 8).[8]

TBDPSO, O, OH —(Me_3Al; CH_2Cl_2)→ TBDPSO, OH, OH + TBDPSO, O (7)

without BF_3•OEt_2	67%	0%
with BF_3•OEt_2	0%	75%
		trans:*cis* = 10:1

O, O, NH_2 —(Tf_2O, *i*-Pr_2NEt; CH_2Cl_2, −78 °C)→ [R, N=•=O] —(Me_3Al; −78 °C)→ HN, O; 91% (8)

In connection with the asymmetric Sharpless epoxidation, the regioselective cleavage of 2,3-epoxy alcohols and their derivatives has been well studied. In contrast to organocuprates, which provide 1,3-diols via C-2 attack, trialkylaluminiums give 1,2-diols via C-3 attack (eq 9).[9a,b] It should be noted that substrates with a phenyl group at C-3 undergo the methylation largely in retentive manner.[9c,d] ***n*-Butyllithium** catalyzes the regioselective β-addition to α- and β-alkoxy epoxides (eq 10).[10a,b] This regioselectivity also holds for epoxy acids (eq 11).[9d] A mixed reagent, Me_3Al and water (5:3), effects the regioselective and stereospecific methylation of γ,δ-epoxy acrylates (eq 12).[10c,d]

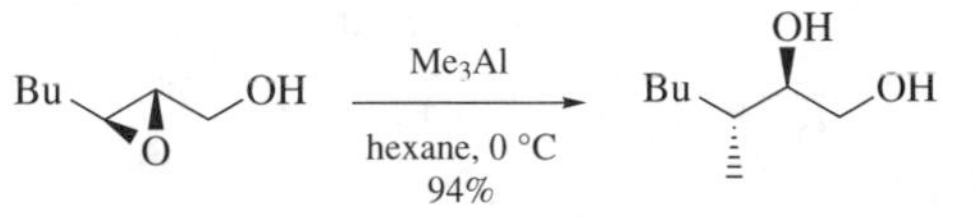

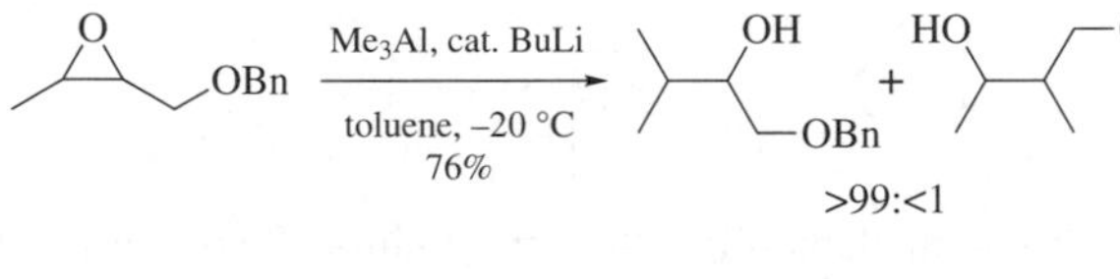

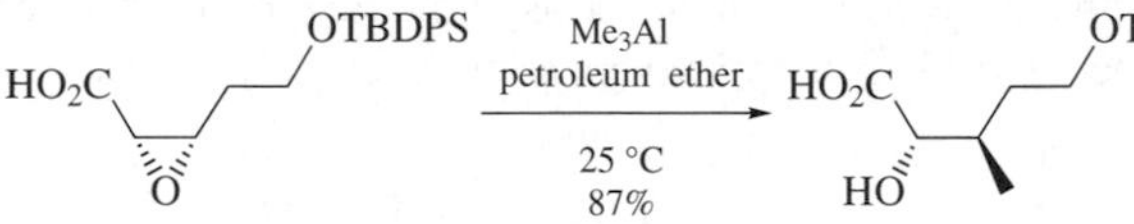

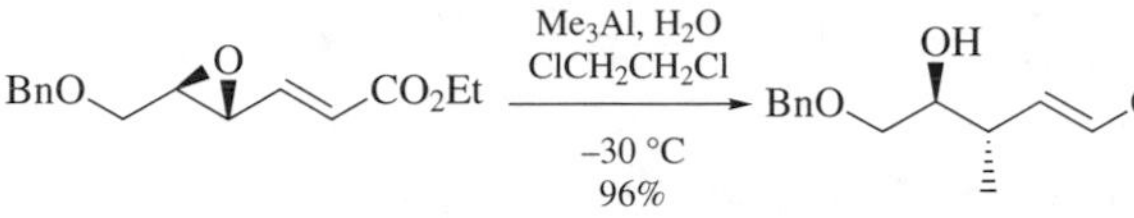

Asymmetric reactions via regioselective ring cleavage of chiral unsaturated acetals have been reported. The regioselectivity is sensitive to the solvent employed (eq 13).[11a,b] Recent developments have been described by Ishihara et al.[11c,d]

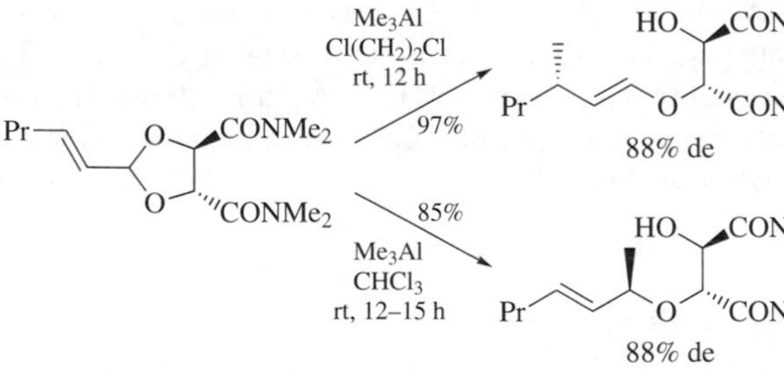

Cross couplings of organoaluminums with vinyl halides,[13b] enol phosphates,[12a] carboxylic acid chlorides, and thioesters[12b,c] are achieved by employing Pd or Cu catalysts (eqs 14 and 15). Unsaturated *N*-acylsulfoximines[13a] or *C*-alkylated purine nucleosides are thus prepared (eq 16).[13b]

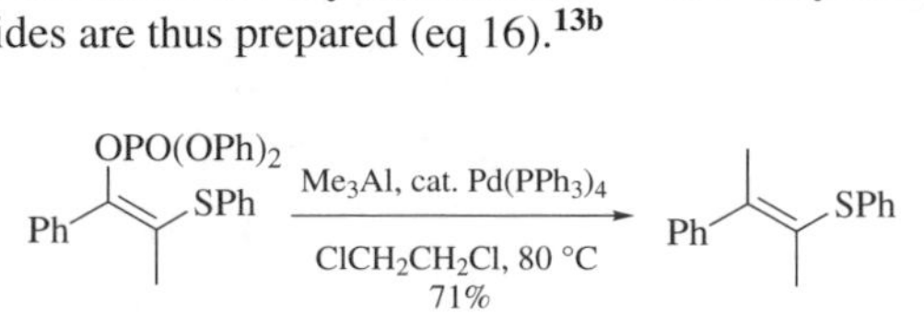

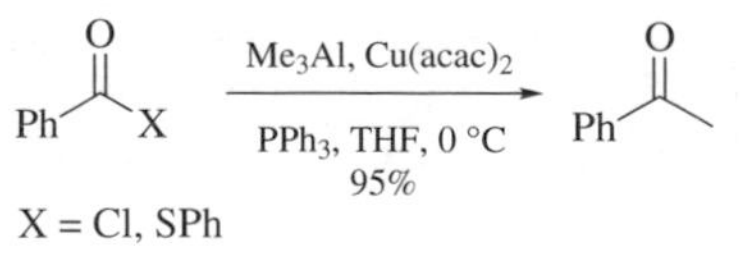

Other useful C–C bond formations using Me_3Al include the carboalumination of 1-alkynes. The procedure with ***Trimethylaluminum–Dichlorobis(η^5-cyclopentadienyl)zirconium*** (eq 17) has broad applicability,[14a–c] and has been used in many natural product syntheses (eq 18).[14d] Direct manipulation of the alkenylalanes provides a stereoselective route to trisubstituted alkenes (eq 19).[14e,f]

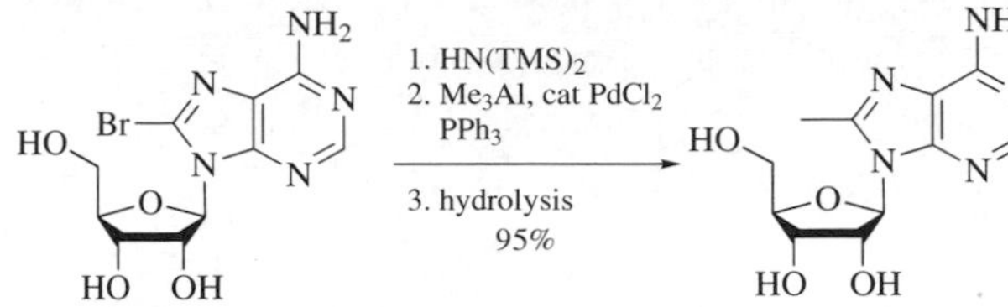

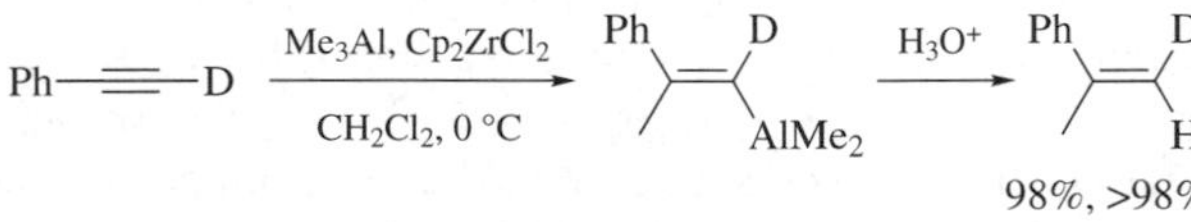

The Tebbe reagent (see ***μ-Chlorobis(cyclopentadienyl)-(dimethylaluminum)-μ-methylenetitanium***) is conveniently generated in situ by the reaction of ***Dichlorobis(cyclopentadienyl)titanium*** and Me_3Al, which, in contrast to the Wittig reaction, methylenates ester carbonyls to provide enol ethers (eq 20).[15] See also ***Trimethylaluminum-Dichlorobis(η^5-cyclopentadienyl)titanium***. A triad reagent, ***Diiodomethane–Zinc***–Me_3Al, effects chemoselective methylenation of aldehydes (eq 21),[16] while the combination of CH_2I_2 and Me_3Al (or other organoaluminums) leads to cyclopropanation of alkenes (eq 22).[17] The regiochemical course of the latter reaction is markedly different from the Simmons–Smith reaction. Regioselective addition of polyhalomethanes to alkenes is induced by Me_3Al (eq 23).[18]

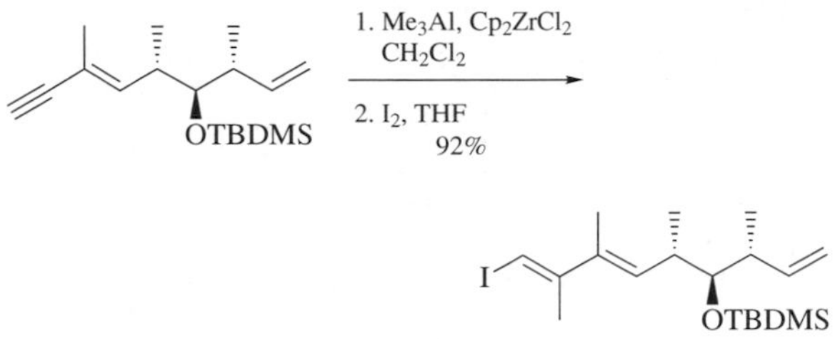

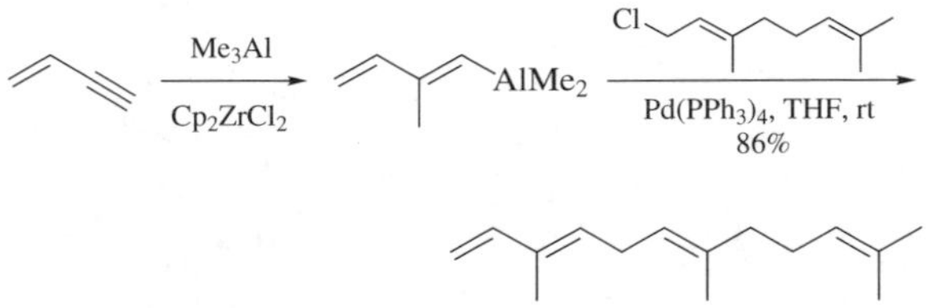

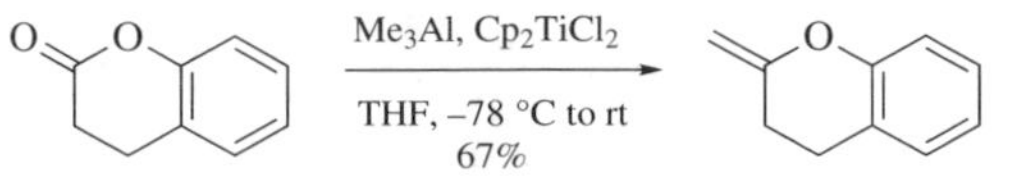

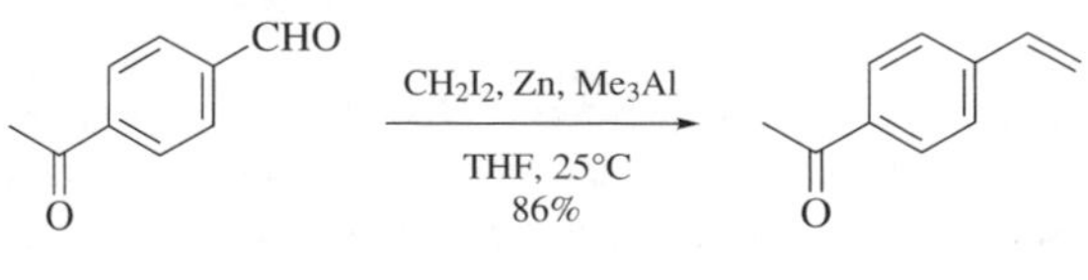

CH_2I_2, Me_3Al / hexane, rt / 96% (22)

Ph + CF_3I → Me_3Al / CH_2Cl_2, –25 °C / 76% → Ph, I, CF_3 (23)

Carbon–heteroatom bond formation is often facilitated by the co-addition of Me_3Al. Aminolysis of esters is achieved by the action of amines or their hydrochlorides in the presence of Me_3Al (eq 24),[19a–c] the latter procedure being particularly useful when volatile amines are concerned (see ***Dimethylaluminum Amide***). The technique is also applicable to the synthesis of acid hydrazides.[19d] Selenoformamides are available from the corresponding formamides by reaction with $(Me_2Al)_2Se$, generated in situ from $(Bu_3Sn)_2Se$ (eq 25).[20] The corresponding tellurium chemistry works also. Me_3Al–LiSPh offers a tandem Michael–aldol approach to the oxahydrindene subunit of avermectins (eq 26).[21a]

OMe → $Me_2NH{\cdot}HCl$, Me_3Al / benzene, 50 °C / 100% → NMe_2 (24)

(H, O, N, Ph, Me) → $(Bu_3Sn)_2Se$, 2 equiv Me_3Al / toluene, THF, 40 °C / 91% → (H, Se, N, Ph, Me) (25)

CO_2Me, O, O, H, OMe → Me_3Al, LiSPh / THF, 0 °C / 89% → HO, CO_2Me, SPh, O, H, OMe (26)

The steric course of reduction of cyclic imines is impressively reversed by a mixed reagent, ***Lithium Aluminum Hydride***–Me_3Al (eq 27).[21b]

$C_{11}H_{23}$, N → H^- → $C_{11}H_{23}$, N, H + $C_{11}H_{23}$, N, H (27)

DIBAL, CH_2Cl_2	>99:<1	
$LiAlH_4$–Me_3Al, THF	5:95	

Rearrangements. Allyl vinyl ethers undergo [3,3]-sigmatropic rearrangements in the presence of Me_3Al, which also captures the resulting aldehyde (eq 28).[22] In contrast, use of ***Triisobutylaluminum*** results in a primary alcohol.

O, Bu → Me_3Al, $ClCH_2CH_2Cl$ / 25 °C, 15 min / 91% → OH, Bu; (*E*):(*Z*) = 47:53 (28)

Me_3Al effects sequential Beckmann rearrangement–methylation of oxime sulfonates. The rearranged imines can be stereoselectively reduced with ***Diisobutylaluminum Hydride*** to give amines (eq 29, cf. eq 27).[23a–c] Related alkylative ring enlargement[24] or alkylative Beckmann fragmentation[23d] reactions are also known (eqs 30 and 31).

H, H, N-OTs → 1. Me_3Al / 2. DIBAL / 57% → H, N, H H (29)

Bn, N, OCO_2Et → Me_3Al / CH_2Cl_2, 0–25 °C / 74% → N, Bn (30)

BnO, NOAc → Me_3Al / CH_2Cl_2, 0 °C / 79% → BnO, CN (31)

β-Hydroxy methanesulfonates or *N*-nitrosulfoximines have been activated by Me_3Al to induce asymmetric pinacol-type rearrangement (eq 32)[25] or carbocyclization reactions (eq 33).[26]

TMS, HO, $MeSO_2O$ → Me_3Al / CH_2Cl_2, –78 °C / 88% → TMS, O (32)

TMS, C_7H_{15}, O=S, Ph, NNO_2 → Me_3Al / CH_2Cl_2, Δ / 96% → C_7H_{15} (33)

Lewis Acid. Me_3Al serves as the precursor to various sophisticated Lewis acids with different acidities, such as $MeAl(OTf)_2$[26] and $MeAl(OC_6F_5)_2$,[11c,d] or other acids with unique spatial environments (eq 34).[27] Chiral Lewis acids, such as (**1**)[28] and (**2**),[29] serve as catalyst for asymmetric Diels–Alder reactions. The catalyst generated from (**3**) and Me_3Al has been used for asymmetric cyanohydrin synthesis.[30]

Me_3Al + 2 HO–(2,6-di-*t*-Bu-C$_6$H$_2$)–R → (R–Ar–O)$_2$Al–Me (*t*-Bu groups) (34)

MAD	R = Me
MAT	R = *t*-Bu
MABR	R = Br

(**1**): $SiAr_3$, O, Al–Me, O, $SiAr_3$

(**2**): Ph, Ph, CF_3SO_2N, NSO_2CF_3, Al, Me

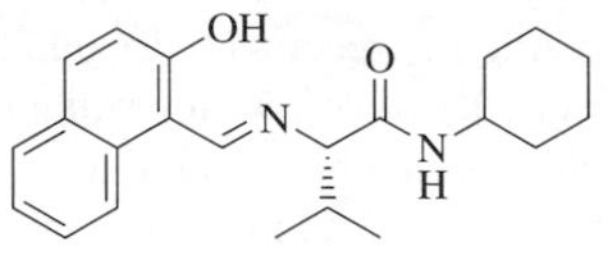

(3)

Related Reagents. μ-Chlorobis(cyclopentadienyl)(dimethylaluminum)-μ-methylenetitanium; Dimethylaluminum Amide; Dimethylaluminum Chloride; Dimethylaluminum Iodide; Dimethylaluminum Methylselenolate; Dimethylaluminum Trifluoromethanesulfonate; Methyltitanium Trichloride; Triethylaluminum; Trimethylaluminum–Dichlorobis(η^5-cyclopentadienyl)zirconium.

---

1. (a) Mole, T.; Jeffery, E. A. *Organoaluminum Compounds*; Elsevier: Amsterdam, 1972. (b) Reinheckel, H.; Haage, K.; Jahnke, D. *Organomet. Chem. Rev. A* **1969**, *4*, 47. (c) Lehmkuhl, H.; Ziegler, K.; Gellert, H. G. *MOC* **1970**, *XIII/4*, 1. (d) Negishi, E. *JOM Libr.* **1976**, *1*, 93. (e) Yamamoto, H.; Nozaki, H. *AG(E)* **1985**, *17*, 169. (f) Negishi, E. *Organometallics in Organic Synthesis*; Wiley: New York, 1980; Vol 1, pp 286–393. (g) Eisch, J. J. *Comprehensive Organometallic Chemistry*; Wilkinson, G.; Stone, F. G. A.; Abel, E. W., Eds.; Pergamon: Oxford, 1982; Vol. 1, pp 555–682. (h) Zietz, J. R., Jr.; Robinson, G. C.; Lindsay, K. L. *Comprehensive Organometallic Chemistry*; Wilkinson, G.; Stone, F. G. A.; Abel, E. W., Eds.; Pergamon: Oxford, 1982; Vol. 7, pp 365–464. (i) Maruoka, K.; Yamamoto, H. *AG(E)* **1985**, *24*, 668. (j) Maruoka, K.; Yamamoto, H. *T* **1988**, *44*, 5001.
2. (a) Della, E. W.; Bradshaw, T. K. *JOC* **1975**, *40*, 1638. (b) Paquette, L. A.; Weber, J. C.; Kobayashi, T.; Miyahara, Y. *JACS* **1988**, *110*, 8591.
3. Hashimoto, S.; Kitagawa, Y.; Iemura, S.; Yamamoto, H.; Nozaki, H. *TL* **1976**, 2615.
4. (a) Nicolaou, K. C.; Dolle, R. E.; Chucholowski, A.; Randall, J. L. *CC* **1984**, 1153. (b) Posner, G. H.; Haines, S. R. *TL* **1985**, *26*, 1823.
5. (a) Maruoka, K.; Nonoshita, K.; Itoh, T.; Yamamoto, H. *CL* **1987**, 2215. (b) Uemura, M.; Isobe, K.; Hayashi, Y. *TL* **1985**, *26*, 767.
6. Brown, D. S.; Charreau, P.; Hannson, T.; Ley, S. V. *T* **1991**, *47*, 1311.
7. Tomooka, K.; Matsuzawa, K.; Suzuki, K.; Tsuchihashi, G. *TL* **1987**, *28*, 6339.
8. Ichikawa, Y.; Yamazaki, M.; Isobe, M. *JCS(P1)* **1993**, 2429.
9. (a) Suzuki, T.; Saimoto, H.; Tomioka, H.; Oshima, K.; Nozaki, H. *TL* **1982**, *23*, 3597. (b) Roush, W. R.; Adam, M. A.; Peseckis, S. M. *TL* **1983**, *24*, 1377. (c) Takano, S.; Yanase, M.; Ogasawara, K. *H* **1989**, *29*, 249. (d) Still, W. C.; Ohmizu, H. *JOC* **1981**, *46*, 5242.
10. (a) Pfaltz, A.; Mattenberger, A. *AG(E)* **1982**, *21*, 71. (b) Flippin, L. A.; Brown, P. A.; Jalali-Araghi, K. *JOC* **1989**, *54*, 3588. (c) Miyashita, M.; Hoshino, M.; Yoshikoshi, A. *JOC* **1991**, *56*, 6483. (d) Miyashita, M.; Hoshino, M.; Yoshikoshi, A.; Kawamine, K.; Yoshihara, K.; Irie, H. *CL* **1992**, 1101.
11. (a) Fujiwara, J.; Fukutani, Y.; Hasegawa, K.; Maruoka, K.; Yamamoto, H. *JACS* **1984**, *106*, 5004. (b) Maruoka, K.; Nakai, S.; Sakurai, M.; Yamamoto, H. *S* **1986**, 130. (c) Ishihara, K.; Hanaki, N.; Yamamoto, H. *JACS* **1991**, *113*, 7074. (d) Ishihara, K.; Hanaki, N.; Yamamoto, H. *JACS* **1993**, *115*, 10 695.
12. (a) Wakamatsu, K.; Okuda, Y.; Oshima, K.; Nozaki, H. *BCJ* **1985**, *58*, 2425. (b) Takai, K.; Oshima, K.; Nozaki, H. *BCJ* **1981**, *54*, 1281. (c) Takai, K.; Sato, M.; Oshima, K.; Nozaki, H. *BCJ* **1984**, *57*, 108.
13. (a) Paley, R. S.; Snow, S. R. *TL* **1990**, *31*, 5853. (b) Hirota, K.; Kitade, Y.; Kanbe, Y.; Maki, Y. *JOC* **1992**, *57*, 5268.
14. (a) Van Horn, D. E.; Negishi, E. *JACS* **1978**, *100*, 2252. (b) Negishi, E.; Van Horn, D. E.; Yoshida, T. *JACS* **1985**, *107*, 6639. (c) Reviews: Negishi, E. *PAC* **1981**, *53*, 2333; Negishi, E. *ACR* **1987**, *20*, 65. (d) Barrett, A. G. M.; Edmunds, J. J.; Malecha, J. W.; Parkinson, C. J. *CC* **1992**, *57*, 1240. (e) Matsushita, H.; Negishi, E. *JACS* **1981**, *103*, 2882. (f) Negishi, E.; Matsushita, H. *OS* **1984**, *62*, 31.
15. (a) Pine, S. H.; Kim, G.; Lee, V. *OS* **1990**, *69*, 72. (b) Cannizzo, L. F.; Grubbs, R. H. *JOC* **1985**, *50*, 2386. (c) Pine, S. H.; Zahler, R.; Evans, D. A.; Grubbs, R. H. *JACS* **1980**, *102*, 3270.
16. Okazoe, T.; Hibino, J.; Takai, K.; Nozaki, H. *TL* **1985**, *26*, 5581.
17. (a) Maruoka, K.; Fukutani, Y.; Yamamoto, H. *JOC* **1985**, *50*, 4412. (b) Maruoka, K.; Sakane, S.; Yamamoto, H. *OS* **1988**, *67*, 176.
18. Maruoka, K.; Sano, H.; Fukutani, Y.; Yamamoto, H. *CL* **1985**, 1689.
19. (a) Basha, A.; Lipton, M.; Weinreb, S. M. *TL* **1977**, 4171. (b) Lipton, M. F.; Basha, A.; Weinreb, S. M. *OS* **1979**, *59*, 49. (c) Levin, J. I.; Turos, E.; Weinreb, S. M. *SC* **1982**, *12*, 989. (d) Benderly, A.; Stavchansky, S. *TL* **1988**, *29*, 739.
20. Segi, M.; Kojima, A.; Nakajima, T.; Suga, S. *SL* **1991**, 105.
21. (a) Armistead, D. M.; Danishefsky, S. J. *TL* **1987**, *28*, 4959. (b) Matsumura, Y.; Maruoka, K.; Yamamoto, H. *TL* **1982**, *23*, 1929.
22. (a) Takai, K.; Mori, I.; Oshima, K.; Nozaki, H. *TL* **1981**, *22*, 3985. (b) Takai, K.; Mori, I.; Oshima, K.; Nozaki, H. *BCJ* **1984**, *57*, 446.
23. (a) Hattori, K.; Matsumura, Y.; Miyazaki, T.; Maruoka, K.; Yamamoto, H. *JACS* **1981**, *103*, 7368. (b) Sakane, S.; Matsumura, Y.; Yamamura, Y.; Ishida, Y.; Maruoka, K.; Yamamoto, H. *JACS* **1983**, *105*, 672. (c) Maruoka, K.; Miyazaki, T.; Ando, M.; Matsumura, Y.; Sakane, S.; Hattori, K.; Yamamoto, H. *JACS* **1983**, *105*, 2831. (d) Fujioka, H.; Yamanaka, T.; Takuma, K.; Miyazaki, M.; Kita, Y. *CC* **1991**, 533.
24. Fujiwara, J.; Sano, H.; Maruoka, K.; Yamamoto, H. *TL* **1984**, *25*, 2367.
25. (a) Suzuki, K.; Ohkuma, T.; Tsuchihashi, G. *TL* **1985**, *26*, 861. (b) Suzuki, K.; Ohkuma, T.; Miyazawa, M.; Tsuchihashi, G. *TL* **1986**, *27*, 373. (c) Honda, Y.; Morita, E.; Tsuchihashi, G. *CL* **1986**, 277.
26. Trost, B. M.; Matsuoka, R. T. *SL* **1992**, 27.
27. (a) Maruoka, K.; Itoh, T.; Yamamoto, H. *JACS* **1985**, *107*, 4573. (b) Maruoka, K.; Itoh, T.; Sakurai, M.; Nonoshita, K.; Yamamoto, H. *JACS* **1988**, *110*, 3588. (c) Maruoka, K.; Ooi, T.; Yamamoto, H. *JACS* **1990**, *112*, 9011.
28. Maruoka, K.; Itoh, T.; Shirasaka, T.; Yamamoto, H. *JACS* **1988**, *110*, 310.
29. (a) Corey, E. J.; Sarshar, S.; Bordner, J. *JACS* **1992**, *114*, 7938. (b) Corey, E. J.; Imwinkelried, R.; Pikul, S.; Xiang, Y. B. *JACS* **1989**, *111*, 5493. (c) Corey, E. J.; Imai, N.; Pikul, S. *TL* **1991**, *32*, 7517. (d) Corey, E. J. *PAC* **1990**, *62*, 1209.
30. (a) Mori, A.; Ohno, H.; Nitta, H.; Tanaka, K.; Inoue, S. *SL* **1991**, 563. (b) Ohno, H.; Nitta, H.; Tanaka, K.; Mori, A.; Inoue, S. *JOC* **1992**, *57*, 6778.

Keisuke Suzuki & Tetsuya Nagasawa
Keio University, Yokohama, Japan

Trimethylaluminum–Dichlorobis(η^5-cyclopentadienyl)zirconium

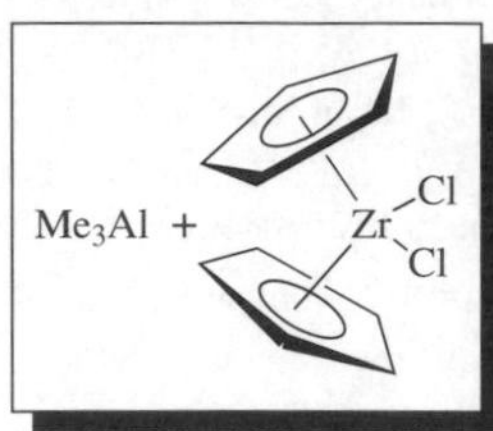

(Me_3Al)		
[75-24-1]	C_3H_9Al	(MW 72.10)
(Cl_2ZrCp_2)		
[1291-32-2]	$C_{10}H_{10}Cl_2Zr$	(MW 292.32)

(regioselective and stereoselective methylalumination of alkynes; regioselective carbometallation of metallated alkynes)

Physical Data: Me_3Al: bp 125 °C; *d* 0.752 g cm^{-3}. Cl_2ZrCp_2: mp 245 °C.

Form Supplied in: Me_3Al: liquid, available neat or as solution in hexanes or toluene. Cl_2ZrCp_2: colorless crystals. Both are commercially available.

Preparative Method: obtained simply by mixing the two components.

Handling, Storage, and Precautions: Me_3Al is extremely pyrophoric and must be used and stored under an inert atmosphere of N_2 or Ar. Cp_2ZrCl_2 may be handled in air, but it appears advisable to avoid extensive exposure to air, moisture, and light. Use in a fume hood.

NMR Behavior. The ^1H NMR spectrum of a mixture of ***Trimethylaluminum*** and ***Dichlorobis(cyclopentadienyl)-zirconium*** (2:1 ratio) in 1,2-dichloroethane at 30 °C, the usual conditions under which its reactions are carried out, exhibits two broad signals (half width = 4 Hz) at δ −0.31 and 6.40 ppm. When this mixture is quenched with THF (2 equiv), the high-field Me signal splits into three, at δ 0.25, −0.75, and −0.95, while the downfield Cp signal splits into two, at δ 6.45 and 6.22. Analysis of these signals indicates that the THF-quenched mixture consists of Cl_2ZrCp_2, $Cl(Me)ZrCp_2$, $Me_3Al{\cdot}THF$, and $Me_2AlCl{\cdot}THF$.[1–3] Thus, under the reaction conditions, Me_3Al and Cl_2ZrCp_2 undergo Me–Cl exchange, which is rapid on the NMR timescale, to form $Cl(Me)ZrCp_2$ (eq 1).

$$Me_3Al + Cl_2ZrCp_2 \rightleftharpoons Me_2Al(\mu\text{-}Me)(\mu\text{-}Cl)ZrCp_2Cl \rightleftharpoons Me_2Al(\mu\text{-}Cl)_2ZrCp_2Me \rightleftharpoons Me_2AlCl + Cl(Me)ZrCp_2 \quad (1)$$

Controlled Methylalumination of Alkynes. Controlled carbometallation of alkynes provides a very selective and efficient route to trisubstituted alkenes, many of which are important building blocks for a wide variety of natural products, especially those of terpenoid origin.[4] Alkynes react with organoalanes to give carbometallation products. However, with terminal alkynes, metallation of the terminal carbon atom is the major pathway, and the regioselectivity of the carbometallation product, obtained in generally poor yields, is low.[5,6] With internal alkynes the reaction requires higher temperatures, at which the product of initial carbometallation competes with the trialkylalane for the alkyne.[6,7] The use of transition metal complexes in conjunction with the organoalane reagent leads to a dramatic improvement of the chemo- and regioselectivity of the carboalumination reaction. Negishi and co-workers have reported that alkynes react with an organometallic reagent obtained by mixing Me_3Al and Cl_2ZrCp_2 to give alkenyl metals in high yields.[1] The reaction is not complicated by the known hydrogen abstraction of terminal alkynes with organozirconium[8] or organoaluminum[5,6] compounds. Studies by NMR spectroscopy have indicated that the product of carbometallation is largely an organoalane. Therefore the reaction is catalytic with respect to zirconium. Indeed, the reaction of 1 equiv of phenylacetylene with 2 equiv of Me_3Al in the presence of 0.1 equiv of Cl_2ZrCp_2 for 12 h at 25 °C produces, after hydrolysis, α-methylstyrene in high yield.[1] The stereoselectivity of the reaction is excellent (>98 % *cis* addition) (eq 2), as is the regioselectivity in the cases of terminal alkynes (eqs 2 and 3).[1] A recent modification of the reaction conditions by using a stoichiometric amount of H_2O leads to a remarkable acceleration of the reaction.[9] Internal alkynes can also be stereoselectively methylaluminated in high yields (eq 4).[1]

$$Ph{-}{\equiv}{-}D \xrightarrow[Cl_2ZrCp_2]{Me_3Al} Ph(Me)C{=}C(D)AlMe_2 + Ph(Me_2Al)C{=}C(D)Me \xrightarrow[98\%]{H_2O} Ph(Me)C{=}CHD + PhCH{=}C(D)Me \quad (2)$$

>98% *(Z)*; 95:5

$$Hex{-}{\equiv}{-}H \xrightarrow[\text{2. } H_2O \text{ quantitative}]{\text{1. } Me_3Al,\ Cl_2ZrCp_2} Hex(Me)C{=}CH_2 + HexCH{=}CHMe \quad (3)$$

95:5

$$Bu{-}{\equiv}{-}Bu \xrightarrow[\text{2. } H_2O \text{ 89\%}]{\text{1. } Me_3Al,\ Cl_2ZrCp_2} Bu(Me)C{=}CHBu \quad (4)$$

>98% (Z)

The Zr-catalyzed methylalumination can accommodate various heterofunctional groups such as OH, OTBDMS, SPh, I, and TMS.[10] Of particular synthetic utility are those alkynes containing heterofunctional groups in the propargylic or homopropargylic position. Interestingly, and in marked contrast with some other known carbometallation reactions of propargyl and homopropargyl derivatives,[11] this reaction is very highly stereoselective (>98%) and regioselective (92–100%, except for the case of homopropargyl phenyl sulfide with which the reaction is 83% regioselective), the regioselectivity being essentially the same as that observed with simple terminal alkynes (eq 5). The yields of methylalumination products are generally in the 50–90%

range. Table 1 illustrates the scope of this methylalumination reaction.

$$\text{Z}\!-\!(\,)_n\!-\!\equiv \xrightarrow[\text{Cl}_2\text{ZrCp}_2]{\text{Me}_3\text{Al}} \text{Z}\!-\!(\,)_n\!-\!\text{C(Me)=CH–AlMe}_2 \quad (5)$$

Z = OH, OTBDMS, SPh, I, TMS, etc.; n = 1 or 2

Table 1 Carboalumination of Terminal Alkynes Catalyzed by Zirconocene Dichloride

	Major product	
Alkyne	Yield (%)	Regioselectivity[a] (%)
$HC{\equiv}CC_6H_{13}$	≥98	≥95[1]
$HC{\equiv}CPh$	≥98	≥95[1]
$HC{\equiv}CCH{=}CH_2$	≥86	≥95[12]
$HC{\equiv}CC(Me){=}CH_2$	70	≥95[13a]
$HC{\equiv}C(CH_2)_2CH{=}CMe_2$	72	≥95[13a]
$HC{\equiv}CCH_2OH$	41	94[10a]
$HC{\equiv}CCH(OH)C_5H_{11}$	60	>98[10a]
$HC{\equiv}CCH_2SPh$	78	>98[10a]
$HC{\equiv}CCH_2SiMe_3$	63	≥95[10b]
$HC{\equiv}C(CH_2)_2OH$	87	94[10a]
$HC{\equiv}C(CH_2)_2I$	74	>98[10a]
$HC{\equiv}C(CH_2)_2SPh$	78	83[10a]

[a] The major regioisomer is (E)-$R^1(Me)C{=}CHAlR^2_2$, and the stereoisomeric purity in each case is ≥98%.

From the viewpoint of synthetic applications, alkenylalanes are very useful reagents. They react with a wide variety of electrophiles, and the C–Al bond can be readily converted into C–X (X = H, D, halogen, Hg, B, Zr, Cu, C, etc.) with essentially complete retention of stereochemistry via protonolysis,[6] deuterolysis,[6] halogenolysis,[13] transmetallation,[14] reactions with various carbon electrophiles such as $ClCO_2Et$, CO_2, $(CH_2O)_n$, $ClCH_2OMe$,[15] and epoxides[16] (preferably after ate complexation), and Pd- or Ni-catalyzed cross-coupling reactions.[12,17] Applications include syntheses of allylic or homoallylic alcohols,[15,16] conjugated dienes or enynes,[14c] 1,4-dienes,[12] and 1,5-dienes or enynes.[18] This methodology has been applied to the synthesis of various natural products such as geraniol,[15] farnesol,[18] monocyclofarnesol,[19] α-farnesene,[12] dendrolasin,[20] mokupalide,[20] vitamin A,[21] brassinolide,[22] milbemycin,[23] verrucarin,[24] udoteatrial,[25] and zoapatanol.[26]

The methylalumination of alkynes with $Me_3Al–Cl_2ZrCp_2$ appears to involve direct addition of a Me–Al bond, rather than a Me–Zr bond, to alkynes via a four-centered process which is facilitated by a $ZrCp_2$ species, as depicted in eq 6.[2] However, the reaction of 1-pentynyldimethylalane with either this reagent system or preformed $Cl(Me)ZrCp_2$ must involve direct addition of the Me–Zr bond, as depicted in eq 7.[27]

$$\text{R–}\equiv \xrightarrow[\text{Cl}_2\text{ZrCp}_2]{\text{Me}_3\text{Al}} [\text{R–}\equiv \cdots \text{Me–}\overset{\delta+}{\text{Al}}(\text{Me})\text{--X--}\overset{\delta-}{\text{ZrCp}_2\text{ClY}}] \longrightarrow \text{R(Me)C=CH–AlMeX} + \text{ZrCp}_2\text{ClY} \quad (6)$$

X, Y = Me or Cl

$$\text{Pr–}\equiv\text{–AlMe}_2 \xrightarrow[\text{Cl}_2\text{ZrCp}_2]{\text{Me}_3\text{Al}} [\text{Pr–}\equiv\text{–}\overset{\delta-}{\text{AlMe}_2}\cdots\text{Cl}\cdots\overset{\delta+}{\text{ZrCp}_2}\text{–Me}] \longrightarrow \text{Pr(Me)C=C(AlMe}_2\text{)(ZrCp}_2\text{Cl)} \; (\mathbf{1}) \quad (7)$$

Methylalumination of 1-Metalloalkynes. The methylmetallation reaction of 1-alkynes that are terminally metallated with $AlMe_2$ or $SiMe_3$ with $Me_3Al–Cl_2ZrCp_2$ is a highly regioselective process, leading to the formation of 1,1-dimetalloalkenes (eqs 7 and 8).[27,28] Whereas the reaction of alkynylsilanes gives the methylaluminated products (the reaction can proceed with a catalytic amount of Cl_2ZrCp_2),[28] that of alkynyldimethylalanes is a methylzirconation reaction which can also be achieved by using preformed $Cl(Me)ZrCp_2$.[27] It is noteworthy that the methylalumination of 1,4-bis(trimethylsilyl)-1,3-butadiyne gives exclusively the product of *trans* addition.[28]

$$\text{TMS–}\equiv\text{–}\equiv\text{–TMS} \xrightarrow[\text{Cl}_2\text{ZrCp}_2]{\text{Me}_3\text{Al}} \text{Me(TMS–C}\equiv\text{C)C=C(TMS)(AlMe}_2) \quad (8)$$

1,1-Dimetalloalkenes are of potential interest for the development of selective routes to tri- and tetrasubstituted alkenes through differentiation of the two metal groups. Some promising results have been obtained along this line. Thus, for example, treatment of (**1**) with ***Acetyl Chloride*** in the presence of ***Aluminum Chloride*** gives a 92:8 mixture of (Z)- and (E)-4-methyl-3-hepten-2-one in 61% yield.[27]

However, the most extensive synthetic application of these 1,1-dimetalloalkenes to date is the synthesis of cycloalkenylmetals. Treatment of 1-(trimethylsilyl)-4-bromo-1-butyne with Me_3Al (2 equiv) in the presence of Cl_2ZrCp_2 (1 equiv) in $(CH_2Cl)_2$ at 25 °C gives a cyclobutene (**2**) in 92% yield after 6 h, rather than the expected methylmetallation product (eq 9).[29]

$$\text{Br–CH}_2\text{CH}_2\text{–}\equiv\text{–TMS} \xrightarrow[\text{Cl}_2\text{ZrCp}_2]{\text{Me}_3\text{Al}} [\text{BrCH}_2\text{CH}_2\text{(Me)C=C(TMS)(AlMe}_2)] \longrightarrow \text{1-TMS-2-methylcyclobutene} \; (\mathbf{2}) \quad (9)$$

The reaction evidently involves the intermediacy of the methylalumination product shown in eq 9. A π-type cyclization involving the intermediacy of a cyclopropylcarbinyl species must be operating,[29] since (1) alkylation of alkenylalanes with alkyl halides does not occur under comparable conditions, (2) the Zr-catalyzed methylalumination is known to give exclusively or predominantly *cis* addition products which are the 'wrong' isomers for a σ-type cyclization, and (3) the reaction is regioselective but nonregiospecific (eq 10).[30] These results are consistent with the mechanism shown in eq 11.

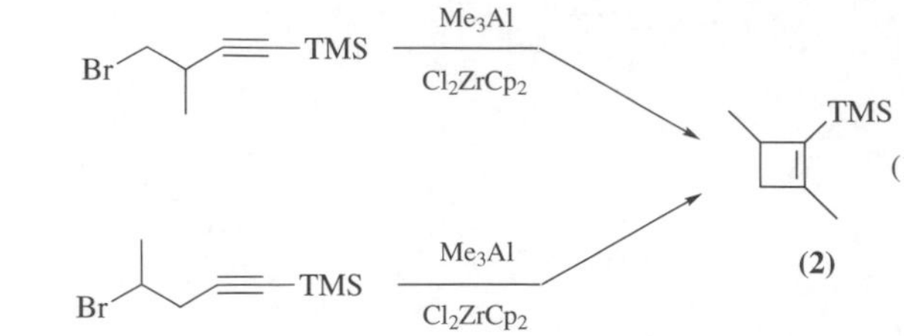

(10)

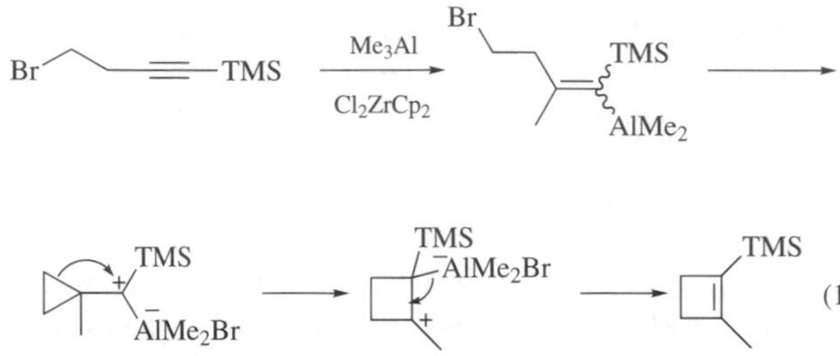

(2)

(11)

The reaction is not limited to the formation of four-membered rings. The methylalumination reaction of 1-(trimethylsilyl)-6-bromo-1-hexyne gives 1-(trimethylsilyl)-1-cyclohexene as the major product.[31] On the other hand, the corresponding reaction of 1-(trimethylsilyl)-5-bromo-1-pentyne gives only the noncyclized methylalumination product.[31]

Synthetically more attractive is the reaction of ω-halo-1-alkynylalanes (**3**) with $Me_3Al–Cl_2ZrCp_2$, which gives the corresponding cycloalkenylmetals (**4**), readily convertible to a wide variety of cyclobutenyl derivatives (eq 12).[30] The same method can be applied to the synthesis of cyclopentenyl or cyclohexenyl analogs of (**4**), such as (**5**) and (**6**), in good yields.[32]

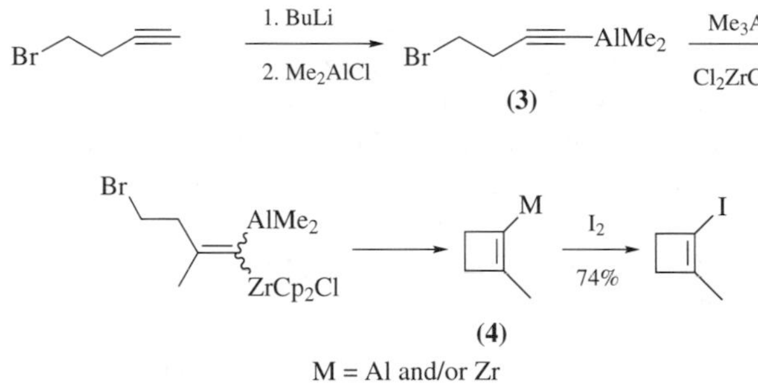

(3)

(12)

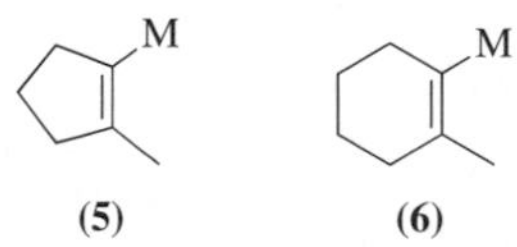

(4)

M = Al and/or Zr

(5) (6)

Ziegler–Natta Polymerization of Alkenes. Some reagent systems consisting of methylaluminum and zirconocene derivatives have been used as homogeneous Ziegler–Natta polymerization catalysts. Satisfactory activity levels have been attained through the use of an additional component. For example, the use of Al_2O_3, $MgCl_2$, or SiO_2-supported Cl_2ZrCp_2 with Me_3Al promotes the polymerization of propene with a fairly good activity.[33] The most satisfactory Cl_2ZrCp_2-based system known to date makes use of methylaluminoxanes which are obtained by controlled hydrolysis of Me_3Al.[34]

Related Reagents. Chlorobis(cyclopentadienyl)methylzirconium; Dichlorobis(cyclopentadienyl)zirconium; Lithium Dimethylcuprate; Methylcopper; Trimethylaluminum.

---

1. (a) Van Horn, D. E.; Negishi, E. *JACS* **1978**, *100*, 2252. (b) Negishi, E.; Van Horn, D. E. *Organomet. Synth.* **1986**, *3*, 467. (c) For a review see: Negishi, E. *PAC* **1981**, *53*, 2333.
2. Yoshida, T.; Negishi, E. *JACS* **1981**, *103*, 4985.
3. Negishi, E.; Van Horn, D. E.; Yoshida, T. *JACS* **1985**, *107*, 6639.
4. Devon, T. K.; Scott, A. I. *Handbook of Naturally Occurring Compounds*; Academic: New York, 1972 and 1975; Vols. 1 and 2.
5. Mole, T.; Surtees, J. R. *AJC* **1964**, *17*, 1229.
6. Mole, T.; Jeffrey, E. A. *Organoaluminum Compounds*; Elsevier: Amsterdam, 1972; Chapter 11.
7. Wilke, G.; Muller, H. *LA* **1960**, *629*, 222.
8. Wailes, P. C.; Weigold, H.; Bell, A. P. *JOM* **1971**, *33*, 181.
9. Wipf, P.; Lim, S. *AG(E)* **1993**, *32*, 1068.
10. (a) Rand, C. L.; Van Horn, D. E.; Moore, M. W.; Negishi, E. *JOC* **1981**, *46*, 4093. (b) Negishi, E.; Luo, F.-T.; Rand, C. L. *TL* **1982**, *23*, 27.
11. (a) Normant, J. F.; Alexakis, A. *S* **1981**, 841. (b) Normant, J. F. *JOM. Libr.* **1976**, *1*, 219. (c) Brown, D. C.; Nichols, S. A.; Gilpin, A. B.; Thompson, D. W. *JOC* **1979**, *44*, 3457.
12. (a) Matsushita, H.; Negishi, E. *JACS* **1981**, *103*, 2882. (b) Negishi, E.; Matsushita, H. *OS* **1984**, *62*, 31.
13. (a) Negishi, E.; Van Horn, D. E.; King, A. O.; Okukado, N. *S* **1979**, 501. (b) Zweifel, G.; Whitney, C. C. *JACS* **1967**, *89*, 2753.
14. (a) With $HgCl_2$: Negishi, E.; Jadhav, K. P.; Daotien, N. *TL* **1982**, *23*, 2085. (b) With *B*-methoxy-9-borabicyclo[3.3.1]nonane or X_2ZrCp_2: Negishi, E.; Boardman, L. D. *TL* **1982**, *23*, 3327. (c) With $ZnCl_2$: Negishi, E.; Okukado, N.; King, A. O.; Van Horn, D. E.; Spiegel, B. I. *JACS* **1978**, *100*, 2254. (d) To alkenylcopper: Wipf, P.; Smitrovich, J. H.; Moon, C.-W. *JOC* **1992**, *57*, 3178. (e) Ireland, R. E.; Wipf, P. *JOC* **1990**, *55*, 1425.
15. Okukado, N.; Negishi, E. *TL* **1978**, 2357.
16. Kobayashi, M.; Valente, L. F.; Negishi, E.; Patterson, W.; Silveira, A., Jr. *S* **1980**, 1034.
17. (a) Negishi, E.; Matsushita, H.; Okukado, N. *TL* **1981**, *22*, 2715. (b) Matsushita, H.; Negishi, E. *JCS(C)* **1982**, 160. (c) Chatterjee, S.; Negishi, E. *JOC* **1985**, *50*, 3406.
18. Negishi, E.; Valente, L. F.; Kobayashi, M. *JACS* **1980**, *102*, 3298.
19. Negishi, E.; King, A. O.; Klima, W. L.; Patterson, W.; Silveira, A., Jr. *JOC* **1980**, *45*, 2526.
20. Kobayashi, M.; Negishi, E. *JOC* **1980**, *45*, 5223.
21. Negishi, E.; Owczarczyk, Z. *TL* **1991**, *32*, 6683.
22. (a) Fung, S.; Siddall, J. B. *JACS* **1980**, *102*, 6580. (b) Mori, K.; Sakakibara, M.; Okada, K. *T* **1984**, *40*, 1767.
23. Williams, D. R.; Barner, B. A.; Nishitani, K.; Phillips, J. G. *JACS* **1982**, *104*, 4708.
24. Roush, W. R.; Blizzard, T. A. *JOC* **1983**, *48*, 758; **1984**, *49*, 1772, 4332.
25. Whitesell, J. K.; Fisher, M.; Jardine, P. D. S. *JOC* **1983**, *48*, 1556.
26. Cookson, R. C.; Liverton, N. J. *JCS(P1)* **1985**, 1589.
27. Yoshida, T.; Negishi, E. *JACS* **1981**, *103*, 1276.
28. (a) Kusumoto, T.; Nishide, K.; Hiyama, T. *CL* **1985**, 1409. (b) Kusumoto, T.; Nishide, K.; Hiyama, T. *BCJ* **1990**, *63*, 1947.
29. Negishi, E.; Boardman, L. D.; Tour, J. M.; Sawada, H.; Rand, C. L. *JACS* **1983**, *105*, 6344.
30. Boardman, L. D.; Bagheri, V.; Sawada, H.; Negishi, E. *JACS* **1984**, *106*, 6105.

31. Negishi, E.; Boardman, L. D.; Sawada, H.; Bagheri, V.; Stoll, A. T.; Tour, J. M.; Rand, C. L. *JACS* **1988**, *110*, 5383.
32. Negishi, E.; Sawada, H.; Tour, J. M.; Wei, Y. *JOC* **1988**, *53*, 913.
33. Soga, K.; Kaminaka, M. *Makromol. Chem.* **1993**, *194*, 1745.
34. (a) Kaminsky, W.; Steiger, R. *Polyhedron* **1988**, *7*, 2375. (b) Kaminsky, W.; Miri, M.; Sinn, H.; Woldt, R. *Makromol. Chem., Rapid Commun.* **1983**, *4*, 417.

Ei-ichi Negishi & Danièle Choueiry
Purdue University, West Lafayette, IN, USA

2,2,6-Trimethyl-4*H*-1,3-dioxin-4-one[1]

[5394-63-8] $C_7H_{10}O_3$ (MW 142.17)

(acetylketene equivalent, used for acetoacetylation and cycloaddition reactions; can be functionalized to provide substituted acylketenes; diketene equivalent)

Alternate Name: diketene–acetone adduct.
Physical Data: mp 12–13 °C; bp 65–67 °C/0.2 mmHg; *d* 1.088 g cm^{-3}.
Solubility: miscible with most organic solvents.
Form Supplied in: neat liquid; 95% (major impurity acetone).
Analysis of Reagent Purity: GC (keep injector port 200–225 °C); NMR.
Preparative Methods: the diketene–acetone adduct (**1**) is one of a series of 1,3-dioxin-4-ones which can be prepared by the reaction of diketene with an aldehyde or ketone in the presence of an acidic catalyst[2] (eq 1) or a quaternary ammonium salt.[3] An alternate preparatory method is the reaction of a β-keto acid with acetone.[4]

cat. H_2SO_4, 90 °C, 90% → (**1**) (1)

Purification: high-vacuum distillation at temperatures below 80 °C.
Handling, Storage, and Precautions: generally nonhazardous; anticipate release of acetone upon heating over 80 °C; will also produce CO_2 if heated in presence of water; extremely good solvent: will dissolve some plastic tubing. Use in a fume hood.

Chemical Transformations: Acetylketene Generation. Dioxinone (**1**) generates acetylketene (**2**) and acetone upon heating at temperatures around 120 °C,[5] via a retro Diels–Alder reaction. The acetylketene thus produced reacts in situ with nucleophiles, including alcohols, amines, and thiols, to provide acetoacetic acid derivatives, e.g. (**3**) (eq 2).[6] The diketene–acetone adduct is thus a convenient, safe, and nonlachrymatory alternative to diketene. In practice, dioxinone (**1**) is often added to a solution of the nucleophile in xylene at 110 °C and the acetone is removed by evaporation or distillation; no catalyst is required. Most of these acetoacetylation reactions are complete within 20 min at 120 °C if the acetone is removed efficiently, although polymeric nucleophiles may require higher reaction temperatures or longer times. Changing the substituents at C-2 will alter the temperature at which the acetylketene is liberated (see below). Irradiation at 254 nm will also provide acetylketene from (**1**).[7] Nucleophilic ring opening can also be effected with base (***Potassium Carbonate***, MeOH).[8]

(**1**) ~120 °C ⇌ [(**2**)] + acetone; 1-methylcyclohexanol, 91% → (**3**) (2)

Virtually every synthetic transformation which utilizes dioxinone (**1**) ultimately involves the generation of an acylketene or a β-keto ester, often following elaboration of the dioxinone nucleus.

Preparation of Heterocycles. Dioxinone (**1**), as an acetylketene equivalent, reacts with a wide variety of functional groups in a manner analogous to diketene[9] to provide heterocycles. For example, heating dioxinone (**1**) in the presence of electron-rich alkenes, such as enamines or enol ethers, will provide the corresponding pyrone derivative (**4**) via a [4+2] cycloaddition reaction, as illustrated in the preparation of 2,6-dideoxy sugar (**5**) (eq 3).[10,11]

(**1**) + OBu (vinyl ether); toluene, 2.5 h, 84% → (**4**); 1. DIBAL 2. MeI, NaH 3. BH_3•SMe_2 4. $NaBO_3$ → (**5**) (3)

DeMayo Reactions. The [2+2] photocycloaddition of alkenes to dioxinone (**1**) (eq 4), followed by reduction and then cyclization of the resultant retro-aldol product, affords cyclohexenones (**6**).[12]

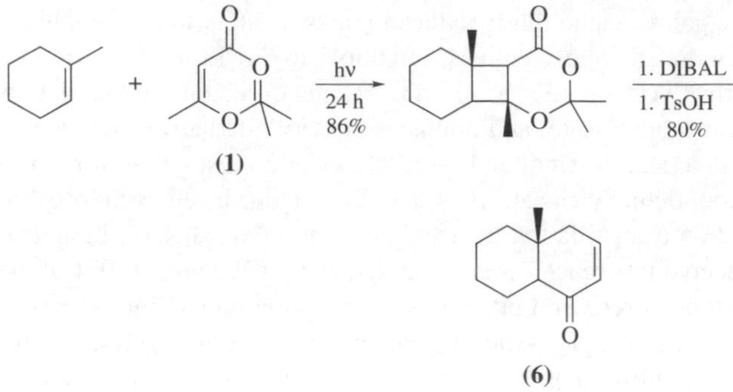

(4)

(6)

Synthesis of Modified Dioxinones. Many dioxinones are modified prior to acylketene generation, either at C-5 or on the C-6 methyl group, to enhance the synthetic utility.[1] Thus dioxinone (**1**) can be halogenated[13] at C-5 or on the 6-methyl group[14] prior to further synthetic elaboration (eqs 5 and 6). The latter bromomethyl dioxinone (**7**) is readily converted into a Wittig[15] or Horner–Emmons reagent.[14]

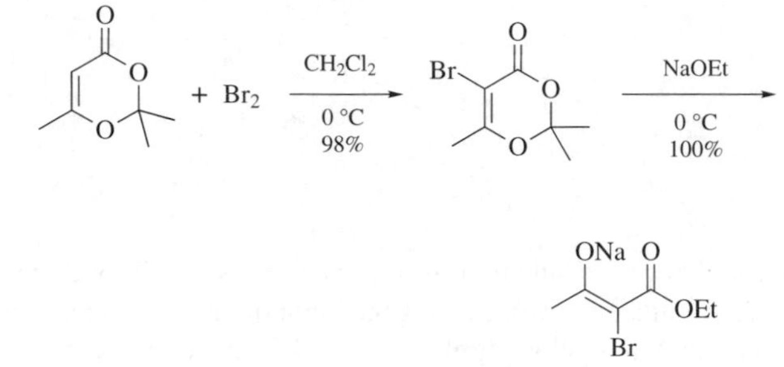

(5)

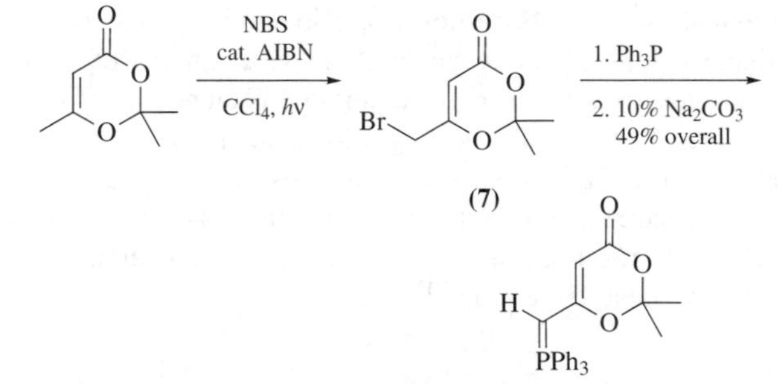

(6)

An alternate approach to modification of dioxinone (**1**) on the C-6 methyl group is via deprotonation (eq 7),[16] sometimes accompanied by silyl enol ether formation (eq 8).

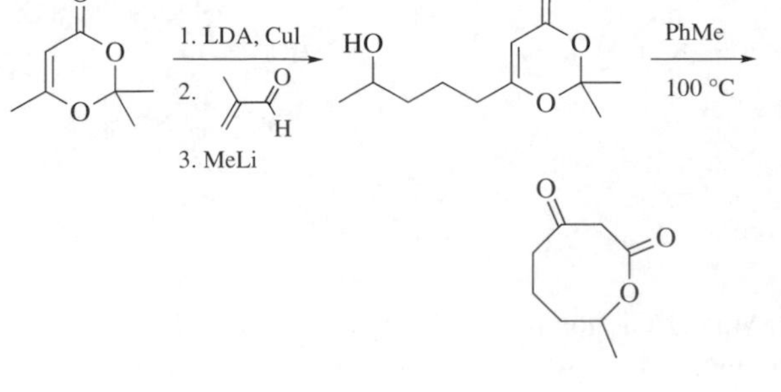

(7)

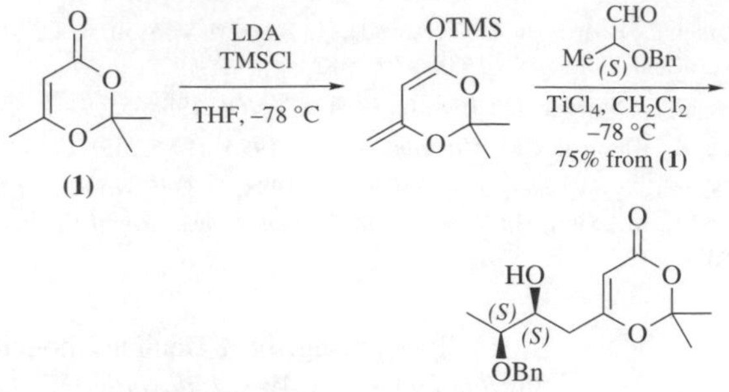

(8)

The resulting elaborated dioxinones have been used to effect ring closure of large rings via intramolecular ketene trappings,[16] methoxide-mediated ring openings to the β-keto esters,[16c,17] and in intramolecular photocycloaddition reactions (eq 9),[18] thereby demonstrating the synthetic utility and versatility of the diketene–acetone adduct (**1**). Both (*R*)- and (*S*)-2-*t*-butyl dioxinones are enantiopure acetoacetate derivatives that are useful for asymmetric synthesis.

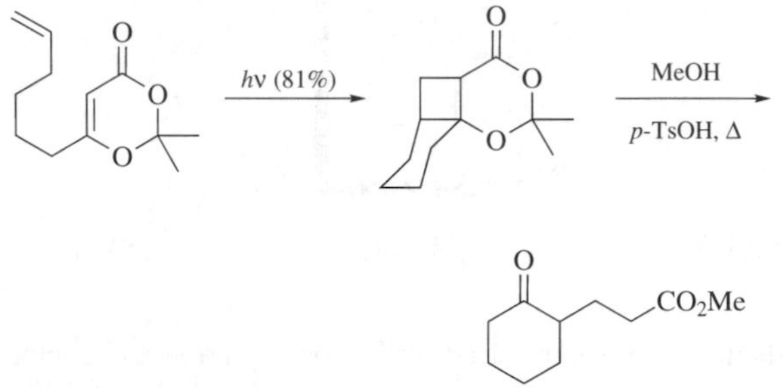

(9)

Related Reagents. Acetoacetic Acid; (*R*)-2-*t*-Butyl-6-methyl-4*H*-1,3-dioxin-4-one; Diketene; Ethyl Acetoacetate; Ethyl 4-Chloroacetoacetate; Ethyl 3-Hydroxybutanoate; Ethyl 4-(Triphenylphosphoranylidene)acetoacetate; Methyl Dilithioacetoacetate.

1. (a) *FF* **1967**, *1*, 256; **1982**, *10*, 424. (b) Kaneko, C.; Sato, M.; Sakaki, J.; Abe, Y. *JHC* **1990**, *27*, 25.
2. Carroll, M. F.; Bader, A. R. *JACS* **1952**, *74*, 6305; **1953**, *75*, 5400.
3. Dehmlow, E. V.; Shamout, A. R. *LA* **1982**, 1753.
4. Sato, M.; Ogasawanra, H.; Oi, K.; Kato, T. *CPB* **1983**, *31*, 1896.
5. Clemens, R. J.; Witzeman, J. S. *JACS* **1989**, *111*, 2186.
6. Clemens, R. J.; Hyatt, J. A. *JOC* **1985**, *50*, 2431.
7. Sato, M.; Ogasawara, H.; Takayama, K.; Kaneko, C. *H* **1987**, *26*, 2611.
8. Sato, M.; Sakaki, J.; Sugita, J.; Yasuda, S.; Sakoda, H.; Kaneko, C. *T* **1991**, *47*, 5689.
9. Clemens, R. J. *CRV* **1986**, *86*, 241.
10. Sato, M.; Ogasawara, H.; Kato, T. *CPB* **1984**, *32*, 2602.
11. Coleman, R. S.; Fraser, J. R. *JOC* **1993**, *58*, 385.
12. Baldwin, S. W.; Wilkinson, J. M. *JACS* **1980**, *102*, 3634.
13. Clemens, R. J. U.S. Patents 4 582 913 and 4 633 013, 1986.
14. Boeckman, R. K., Jr.; Thomas, A. J. *JOC* **1982**, *47*, 2823.
15. Bodurow, C.; Carr, M. A.; Moore, L. L. *OPP* **1990**, *22*, 109.
16. (a) Petasis, N. A.; Patane, M. A. *CC* **1990**, 836. (b) Sugita, Y.; Sakaki, J.; Sato, M.; Kaneko, C. *JCSPI* **1992**, 2855. (c) Lichtenthaler, F. W.; Dinges, J.; Fukuda, Y. *AG(E)* **1991**, *30*, 1339.

17. Sato, M.; Sugita, Y.; Abiko, Y.; Kaneko, C. *TA* **1992**, *3*, 1157.
18. Winkler, J. D.; Hey, J. P.; Hannon, F. J. *H* **1987**, *25*, 55.

Robert J. Clemens
Eastman Chemical Company, Kingsport, TN, USA

Trimethylsilylacetylene

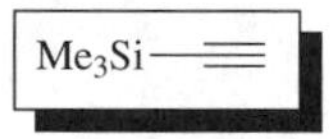

[1066-54-2] $C_5H_{10}Si$ (MW 98.24)

(ethynylation by palladium(0)-catalyzed coupling/condensation with aryl and vinyl halides[1] and triflates,[2] or by nucleophilic attack of the corresponding acetylide on electrophilic centers;[3,4] reacts with alkyl iodides,[5] tin hydrides,[6] and dichloroketene[7] in a regioselective and stereoselective manner)

Alternate Names: trimethylsilylethyne; TMSA.
Physical Data: bp 53 °C; *d* 0.695 g cm^{-3}.
Solubility: sol all organic solvents.
Form Supplied in: colorless transparent liquid; supplied in ampules.
Handling, Storage, and Precautions: once transferred from the ampule to a sample bottle, it can be stored for long periods without loss of purity and material if stored cold. It is a flammable liquid classified as an irritant. Use in a fume hood.

Ethynylations. All ethynylation processes involve two steps: (a) coupling of the trimethylsilylethynyl group to the substrate either by a palladium(0)-catalyzed reaction or by nucleophilic attack of the derived acetylide on an electrophilic center; and (b) replacement of the trimethylsilyl group with a proton. Although ***2-Methylbut-3-yn-2-ol*** can be used for ethynylation and is a much cheaper reagent than TMSA, the advantage of TMSA is in the mild conditions needed for removal of the trimethylsilyl group. Deprotection of the 2-methyl-2-hydroxybutynyl group requires heating in toluene at >70 °C with NaH[8] or NaOH,[9] whereas replacement of the trimethylsilyl group occurs at room temperature and can be effected with dilute aqueous methanol solutions of NaOH or KOH,[10] with LiOH in aqueous THF,[11] or with mild bases such as K_2CO_3[12] or Na_2CO_3[13] in MeOH and KF in aqueous DMF.[14] The yields usually range from good to high.

Palladium(0)-Catalyzed Coupling Reactions. Treatment of the title reagent with vinyl (eq 1)[1b] and aryl (eq 2)[1c] halides and triflates (eqs 3 and 4)[2c,d] and an appropriate Pd catalyst affords vinylalkynes in good yield. For halides, the order of reactivity is I > Br ≫ Cl.[15] Vinyl chlorides undergo this reaction, and one-step diethynylation of dichloroethylenes can be achieved in good yield (eq 5).[16] Aromatic chlorides, however, undergo ethynylation with terminal alkynes only if there is a strong electron-withdrawing group, such as nitro, on the ring.[15]

≡—TMS; $(Ph_3P)_4Pd$, 80 °C; *N*-methylpyrrolidine; 66% (1)

≡—TMS; $(Ph_3P)_2PdCl_2$; Et_3N; 80%. K_2CO_3, MeOH; 25 °C; 84% (2)

≡—TMS; $(Ph_3P)_2PdCl_2$; THF, *i*-Pr_2NH; 63% (3)

≡—TMS; $(Ph_3P)_2PdCl_2$; Et_3N, DMF; 87% (4)

≡—TMS; $(Ph_3P)_4Pd$, CuI; C_6H_6, $BuNH_2$; 79% (5)

Heteroaromatic and heterocyclic vinyl halides and triflates can also be effectively ethynylated using TMSA (eqs 6 and 7).[17] The product heterocycles can be elaborated into polycyclic compounds (eq 8).[18]

≡—TMS; $(Ph_3P)_2PdCl_2$; CuI, Et_3N; 74%. NaOH, MeOH; 94% (6)

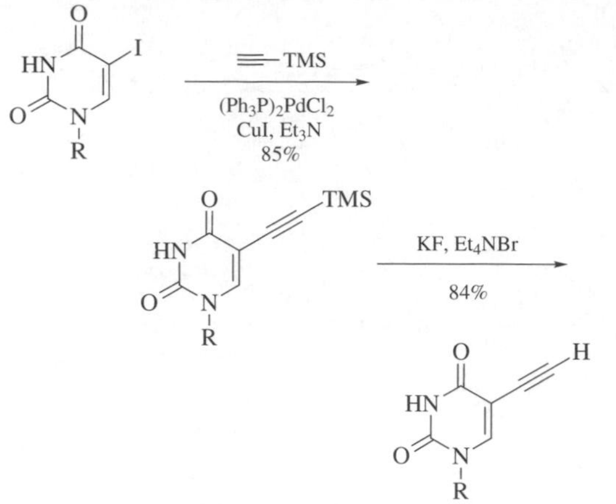

(7)

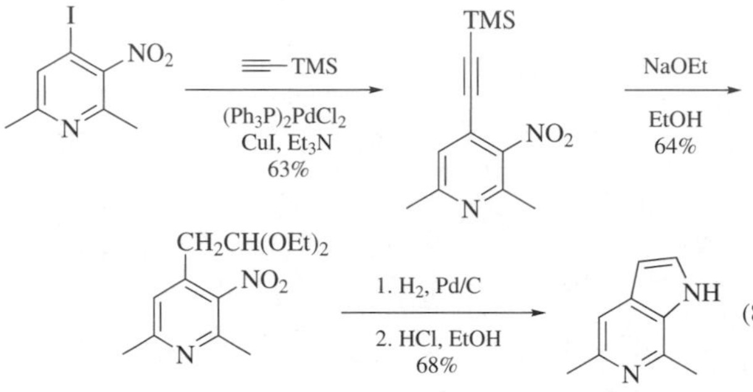

(8)

Aromatic compounds with nitro groups *ortho* to the alkyne have been converted into indoles (eq 9).[19] Alkynic ketones have been prepared from either the corresponding acyl halides[20] or by carbonylative ethynylation[21] of vinyl halides and triflates with TMSA.

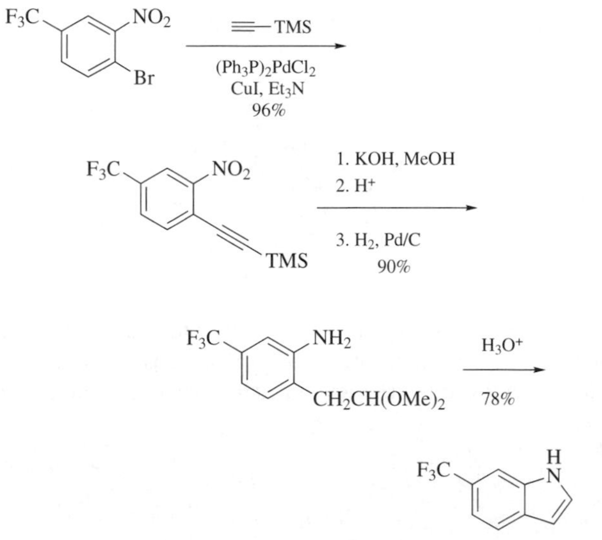

(9)

Generally, the palladium(0)-catalyzed reaction requires a base to deprotonate the terminal alkyne. Often, the solvent for the reaction is an amine which also serves as base. Alternatively, the amine can be used in slightly more than a stoichiometric amount in nonbasic solvents such as THF,[2c] benzene,[16b] or DMF.[1b,22] Sodium alkoxides[22] and acetate[2a] have also been used as bases. Various Pd^{II} and Pd^{0} complexes are effective for the coupling reaction. In one case, Pd^{0}-catalyzed coupling of TMSA with a vinyl halide gave appreciable amounts (40–45%) of a fulvene instead of the expected enyne when the reaction was conducted with $(MeCN)_2PdCl_2$ in the absence of CuI (eq 10).[23] With appropriate choice of reaction conditions, however, it is possible to reduce or totally eliminate this fulvene formation (eq 10).

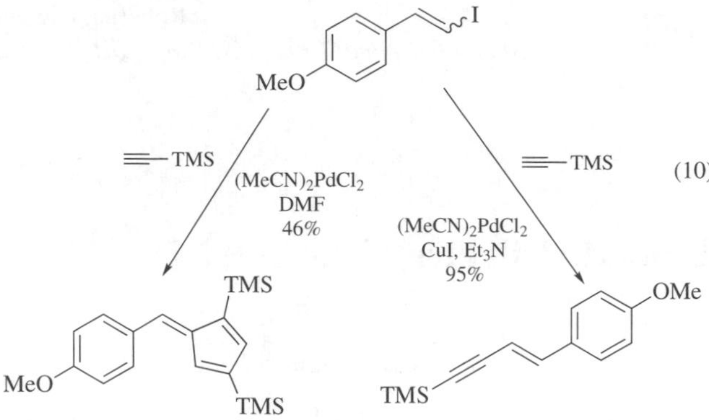

(10)

Reaction of Trimethylsilylacetylides with Electrophiles. ***Lithium (Trimethylsilyl)acetylide*** is easily generated from TMSA and ***n-Butyllithium***[4c,3b] or ***Methyllithium***[24] at low temperatures. The corresponding Grignard reagent can be prepared from TMSA and ***Ethylmagnesium Bromide***. The zinc chloride derivative can be generated by transmetalation of the lithium acetylide.[4a] A cerate derivative has also been described.[25] All of these acetylides add to the carbonyl group of ketones[26] and aldehydes[3,24] to generate alcohols. The tertiary sulfide shown in eq 11 reacts smoothly with zinc acetylides without elimination, a complication that occurs with the more basic Li and Mg acetylides.[4a]

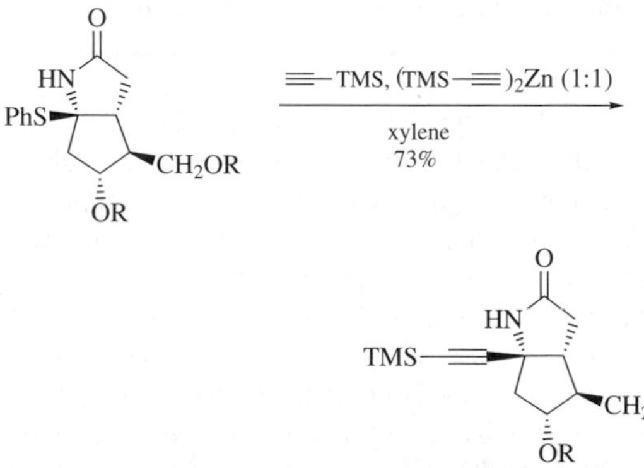

(11)

Radical-Initiated and Transition Metal-Catalyzed Additions. Some radical and transition metal-catalyzed additions to TMSA are unique when compared with additions to other terminals alkynes, because they show remarkable regioselectivity and/or stereoselectivity. The regioselectivity of a metal-catalyzed addition may be complementary to that of a radical-initiated process (eq 12). For example, rhodium[27] and molybdenum[28] complex-catalyzed additions of trialkyltin or triaryltin hydrides to TMSA give mainly the 1,1-disubstituted ethylenes, whereas radical hydrostannylation through sonication[29] or ***Triethylborane***[6] initiation gives the 1,2-adducts with the (*E*)-isomers predominating. Other terminal alkynes undergo radical or metal-catalyzed hydrostannylation with either poorer or reverse selectivity.

$$\text{TMS}-\equiv + R_3SnH \xrightarrow[76\text{–}86\%]{} $$

R = Ph, Bu

$$\text{(E)-1,2-}\ (\text{TMS})(H)C{=}C(H)(SnR_3) + \text{(Z)-1,2-}\ (\text{TMS})(H)C{=}C(H)(SnR_3) + \text{1,1-}\ (\text{TMS})(SnR_3)C{=}CH_2 \quad (12)$$

Et_3B-initiated radical addition of various alkyl iodides to TMSA occurs with high regioselectivity, giving predominantly or exclusively (*Z*)-1-iodo-1-trimethylsilyl-2-alkylethylenes (eq 13).[5]

$$R^1-\equiv + RI \xrightarrow{Et_3B} (E)\ R^1(I)C{=}CH(R) + (Z)\ R^1(I)C{=}CH(R) \quad (13)$$

R^1 = TMS, R = *i*-Pr, 79% 0:100

R^1 = Ph, R = *i*-Pr, 81% 21:79

Cycloaddition Reactions. Unlike [2+2] cyclo additions involving other alkynes, TMSA adds to dichloroketene[7,30] and keteniminium salts[31] in a highly regioselective manner to give cyclobutanones (eqs 14 and 15). The regiochemistry in this cycloaddition is opposite to that predicted from the electronic effects of the trimethylsilyl group and has been explained using MO considerations.[30] The cycloaddition of TMSA to Fischer carbene complexes to provide naphthoquinones has also been reported (eq 16).[32]

$$Cl_3C(C{=}O)Cl \xrightarrow[60\%]{Zn(Cu),\ Et_2O,\ POCl_3,\ TMSC{\equiv}CMe} \text{cyclobutenones (7:5)}$$

$$Cl_3C(C{=}O)Cl \xrightarrow[67\%]{Zn(Cu),\ Et_2O,\ TMSC{\equiv}CH} \text{cyclobutenones (99:1)} \quad (14)$$

$$Me_2C{=}C{=}\overset{+}{N}Me_2\ TiCl_5^- \xrightarrow[\text{2. 0.1 N NaOH, 88\%}]{\text{1. TMSC≡CH, −40 °C}} \text{3-TMS-4,4-dimethylcyclobutenone} \quad (15)$$

$$\text{MeO(}p\text{-MeC}_6H_4\text{)C=Mo(CO)}_5 \xrightarrow[\text{2. oxidation, 69\%}]{\text{1. TMSiC≡CH, toluene, 90 °C}} \text{2-TMS-6-methyl-1,4-naphthoquinone} \quad (16)$$

Related Reagents. Acetylene; Bis(trimethylsilyl)acetylene; Ethoxy(trimethylsilyl)acetylene; Lithium (Trimethylsilyl)acetylide; 2-Methylbut-3-yn-2-ol; Tri-*n*-butylstannylacetylene; (Trimethylsilyl)ethynylcopper(I).

---

1. (a) Jeffery-Luong, T.; Linstrumelle, G. *S* **1983**, *32*. (b) Nwokogu, G. *JOC* **1985**, *50*, 3900. (c) Austin, W. B.; Bilow, N.; Kellaghan, W. J.; Lau, K. S. Y. *JOC* **1981**, *46*, 2280. (d) Myers, A. G.; Alauddin, M. M.; Fuhry, M. A. M.; Dragovich, P. S.; Finney, N. S.; Harrington, P. M. *TL* **1989**, *50*, 6997. (e) Konno, S.; Fujihara, S.; Yamanaka, S. *H* **1984**, *22*, 2245. (f) Feldman, K. S. *TL* **1982**, *23*; 3031.
2. (a) Cacchi, S. *S* **1986**, 320. (b) Suffert, J.; Bruckner, R. *TL* **1991**, *32*, 1453. (c) Bruckner, R.; Scheuplein, S. W.; Suffert, J. *TL* **1991**, *32*, 1449. (d) Chen, Q.-Y., Yang, Z.-Y. *TL* **1986**, *27*, 1171.
3. (a) Kuroda, S.; Katsuki, T.; Yamaguchi, M. *TL* **1987**, *28*, 803. (b) Kitano, Y.; Matsumoto, T.; Sato, F. *T* **1988**, *44*, 4073.
4. (a) Mori, S.; Iwakura, H.; Takechi, S. *TL* **1988**, *29*, 5391. (b) Baumeler, A.; Brade, W.; Eugster, C. H. *HCA* **1990**, *73*, 700. (c) White, J. D.; Somers, T. C.; Reddy, G. N. *JACS* **1986**, *108*, 5352.
5. Ichinose, Y.; Matsunaga, S.; Fugami, K.; Oshima, K.; Utimoto, K. *TL* **1989**, *30*, 3155.
6. Nozaki, K.; Oshima, K.; Utimoto, K. *JACS* **1987**, *109*, 2547.
7. Hassner, A.; Dillon, J. L., Jr. *JOC* **1983**, *48*, 3382.
8. Havens, S. J.; Hergenrother, P. M. *JOC* **1985**, *50*, 1763.
9. Sabourin, E. T.; Onopchenko, A. *JOC* **1983**, *48*, 5135.
10. (a) Bakthavachalam, V.; d'Alarco, M.; Leonard, N. J. *JOC* **1984**, *49*, 289. (b) Jensen, B. J.; Hergenrother, P. M. *J. Polym. Sci., Polym. Chem. Ed* **1985**, *23*, 2233.
11. Magnus, P.; Annoura, H.; Harling, J. *JOC* **1990**, *55*, 1709.
12. Havens, S. J.; Hergenrother, P. M. *J. Polym. Sci., Polym. Chem. Ed.* **1984**, *22*, 3011.
13. Jensen, B. J.; Hergenrother, P. M.; Nwokogu, G. *J. Macromol. Sci.–Pure Appl. Chem.* **1993**, *A30*, 449.
14. Semmelhack, M. F.; Neu, T.; Foubelo, F. *TL* **1992**, *33*, 3277.
15. Fitton, P.; Rick, A. E. *JOM* **1971**, *28*, 287.
16. (a) Ratovelomanana, V.; Hammoud, A.; Linstrumelle, G. *TL* **1987**, *28*, 1649. (b) Vollhardt, K. P. C.; Winn, L. S. *TL* **1985**, *26*, 709.
17. (a) Tilley, J. W.; Zawoiski, S. *JOC* **1988**, *53*, 386. (b) Robins, M. J., Barr, P. J. *JOC* **1983**, *48*, 1854. (c) Konno, S.; Fujimura, S.; Yamanaka, H. *H* **1984**, *22*, 2245.
18. Sakamoto, T.; Kondo, Y.; Yamanaka, H. *H* **1984**, *22*, 1347.
19. (a) Tischler, A. N.; Lanza, T. J. *TL* **1986**, *27*, 1653. (b) Sakamoto, T.; Kondo, Y.; Yamanaka, H. *H* **1986**, *24*, 31.
20. Logue, M. W.; Teng, K. *JOC* **1982**, *47*, 2549.
21. Ciattini, P. G.; Morera, E.; Ortar, G. *TL* **1991**, *32*, 6449.
22. Cassar, L. *JOM* **1975**, *93*, 253.
23. Lee, G. C. M.; Tobias, B.; Holmes, J. M.; Harcourt, D. A.; Garst, M. E. *JACS* **1990**, *112*, 9330.
24. Wenkert, E.; Leftin, M. H.; Michelotti, E. L. *JOC* **1985**, *50*, 1122.
25. Tamura, Y.; Sasho, M.; Ohe, H.; Akai, S.; Kita, Y. *TL* **1985**, *26*, 1549.
26. (a) Thies, R. W., Daruwala, K. P. *JOC* **1987**, *52*, 3798. (b) Kiesewetter, D. O.; Katzenellenbogen, J. A.; Kilbourn, M. R.; Welch, M. J. *JOC* **1984**, *49*, 4900.
27. Kikukawa, K.; Umekawa, H.; Wada, F.; Matsuda, T. *CL* **1988**, *5*, 881.
28. Zhang, X. H.; Guibe, F.; Balavoine, G. *JOC* **1990**, *55*, 1857.
29. Nakamura, E.; Machii, D.; Inubushi, T. *JACS* **1989**, *111*, 6849.

30. Danheiser, R. L.; Sard, H. *TL* **1982**, *24*, 23.
31. Schmidt, C.; Sahraoui-Taleb, S.; Differding, E.; Dehasse-De Lombaert; C. G.; Ghosez, L. *TL* **1984**, *25*, 5043.
32. Dotz, H. K.; Larbig, H. *JOM* **1991**, *405*, C38.

Godson C. Nwokogu
Hampton University, VA, USA

Trimethylsilyldiazomethane[1]

Me_3Si–$CH=N_2$

[18107-18-1] $C_4H_{10}N_2Si$ (MW 114.25)

(one-carbon homologation reagent; stable, safe substitute for diazomethane; [C–N–N] 1,3-dipole for the preparation of azoles[1])

Physical Data: bp 96 °C/775 mmHg; n_D^{25} 1.4362.[2]
Solubility: sol most organic solvents; insol H_2O.
Form Supplied in: commercially available as 2 M and 10 w/w% solutions in hexane, and 10 w/w% solution in CH_2Cl_2.
Analysis of Reagent Purity: concentration in hexane is determined by ^1H NMR analysis.[3]
Preparative Method: prepared by the diazo-transfer reaction of ***Trimethylsilylmethylmagnesium Chloride*** with ***Diphenyl Phosphorazidate*** (DPPA) (eq 1).[3]

TMS⁀Cl —(Mg)→ TMS⁀MgCl —($(PhO)_2P(O)N_3$)→ TMS⁀N_2 (1)

Handling, Storage, and Precautions: should be protected from light.

One-Carbon Homologation. Along with its lithium salt, which is easily prepared by lithiation of trimethylsilyldiazomethane ($TMSCHN_2$) with ***n-Butyllithium***, $TMSCHN_2$ behaves in a similar way to ***Diazomethane*** as a one-carbon homologation reagent. $TMSCHN_2$ is acylated with aromatic acid chlorides in the presence of ***Triethylamine*** to give α-trimethylsilyl diazo ketones. In the acylation with aliphatic acid chlorides, the use of 2 equiv of $TMSCHN_2$ without triethylamine is recommended. The crude diazo ketones undergo thermal Wolff rearrangement to give the homologated carboxylic acid derivatives (eqs 2 and 3).[4]

1-Naphthoyl chloride (COCl) —(1. $TMSCHN_2$, Et_3N; 2. $PhNH_2$, 180 °C, 2,4,6-trimethylpyridine; 80%)→ 1-naphthyl–CH₂CONHPh (2)

N-CO₂Bn-prolinoyl chloride (COCl) —(1. 2 equiv $TMSCHN_2$; 2. $PhCH_2OH$, 180 °C, 2,4,6-trimethylpyridine; 77%)→ N-CO₂Bn-pyrrolidine–CH_2CO_2Bn (3)

Various ketones react with $TMSCHN_2$ in the presence of ***Boron Trifluoride Etherate*** to give the chain or ring homologated ketones (eqs 4–6).[5] The bulky trimethylsilyl group of $TMSCHN_2$ allows for regioselective methylene insertion (eq 5). Homologation of aliphatic and alicyclic aldehydes with $TMSCHN_2$ in the presence of ***Magnesium Bromide*** smoothly gives methyl ketones after acidic hydrolysis of the initially formed β-keto silanes (eq 7).[6]

$PhCOCH_2Ph$ —($TMSCHN_2$, BF_3•Et_2O; CH_2Cl_2, –15 °C, 1 h; 74%)→ $PhCOCH_2CH_2Ph$ (4)

2-Methylcyclohexanone —($TMSCHN_2$, BF_3•Et_2O; CH_2Cl_2, –15 °C, 4 h; 69%)→ 2-methylcycloheptanone (5)

Fluorenone —($TMSCHN_2$, BF_3•Et_2O; CH_2Cl_2, –15 to –10 °C, 3 h; 80%)→ 9-phenanthrol (OH) (6)

t-BuCHO —(1. $TMSCHN_2$, $MgBr_2$; 2. 10% aq HCl; 89%)→ *t*-BuCOMe (7)

O-Methylation of carboxylic acids, phenols, enols, and alcohols can be accomplished with $TMSCHN_2$ under different reaction conditions. $TMSCHN_2$ instantaneously reacts with carboxylic acids in benzene in the presence of methanol at room temperature to give methyl esters in nearly quantitative yields (eq 8).[7] This method is useful for quantitative gas chromatographic analysis of fatty acids. Similarly, *O*-methylation of phenols and enols with $TMSCHN_2$ can be accomplished, but requires the use of ***Diisopropylethylamine*** (eqs 9 and 10).[8] Although methanol is recommended in these *O*-methylation reactions, methanol is not the methylating agent. Various alcohols also undergo *O*-methylation with $TMSCHN_2$ in the presence of 42% aq. ***Tetrafluoroboric Acid***, smoothly giving methyl ethers (eq 11).[9]

HO–C_6H_4–CO_2H —($TMSCHN_2$, MeOH–benzene; rt, 30 min)→ HO–C_6H_4–CO_2Me, quantitative (8)

2-HO–C_6H_4–CO_2Me —($TMSCHN_2$, *i*-Pr_2NEt, MeOH–MeCN; rt, 15 h; 78%)→ 2-MeO–C_6H_4–CO_2Me (9)

PhCOCH₂CO₂Et —($TMSCHN_2$, *i*-Pr_2NEt, MeOH–MeCN; rt, 15 h; 89%)→ Ph(MeO)C=CHCO₂Et (10)

tertiary alcohol (OH, $(CH_2)_8$ chain) —($TMSCHN_2$, 42% aq HBF_4; CH_2Cl_2, 0 °C, 2 h; 74%)→ methyl ether (OMe) (11)

Alkylation of the lithium salt of $TMSCHN_2$ (TMSC(Li)N_2) gives α-trimethylsilyl diazoalkanes which are useful for the preparation of vinylsilanes and acylsilanes. Decomposition of α-trimethylsilyl diazoalkanes in the presence of a catalytic amount of ***Copper(I) Chloride*** gives mainly (*E*)-vinylsilanes (eq 12),[10] while replacement of CuCl with rhodium(II) pivalate affords (*Z*)-vinylsilanes as the major products (eq 12).[11] Oxidation of α-trimethylsilyl diazoalkanes with ***m*-Chloroperbenzoic Acid** in a two-phase system of benzene and phosphate buffer (pH 7.6) affords acylsilanes (α-keto silanes) (eq 12).[12]

(12) CuCl, benzene, reflux, 0.5 h: 89%, (*E*):(*Z*) = 95:5; $Rh_2(OPiv)_4$, benzene, reflux, 0.5 h: 84%, (*E*):(*Z*) = 22:78; *m*-CPBA, benzene, phosphate buffer (pH 7.6), 0 °C, 5 min; rt, 1 h: 65%

(*E*)-β-Trimethylsilylstyrenes are formed by reaction of alkanesulfonyl chlorides with $TMSCHN_2$ in the presence of triethylamine (eq 13).[13] $TMSC(Li)N_2$ reacts with carbonyl compounds to give α-diazo-β-hydroxy silanes which readily decompose to give α,β-epoxy silanes (eq 14).[14] However, benzophenone gives diphenylacetylene under similar reaction conditions (eq 15).[15]

(13) $TMSCHN_2$, Et_3N, THF, 0 °C, 1 h; 70%

(14) PhCOMe, $TMSC(Li)N_2$, THF, −70 °C, 30 min; Δ, $-N_2$

(15) PhCOPh → Ph—≡—Ph; $TMSC(Li)N_2$, Et_2O, rt, 2 h; 80%

Silylcyclopropanes are formed by reaction of alkenes with $TMSCHN_2$ in the presence of either ***Palladium(II) Chloride*** or CuCl depending upon the substrate (eqs 16 and 17).[16] Silylcyclopropanones are also formed by reaction with trialkylsilyl and germyl ketenes (eq 18).[17]

(16) $TMSCHN_2$, $PdCl_2$, benzene, 60 °C, 1 h; 79%; (*E*):(*Z*) = 1:1.4

(17) PhCH=CH$_2$; $TMSCHN_2$, CuCl, benzene, rt, 1 h; 46%; (*E*):(*Z*) = 1:4.8

(18) $TMSCHN_2$, Et_2O, −10 °C to rt; 90%

[C–N–N] Azole Synthon. $TMSCHN_2$, mainly as its lithium salt, $TMSC(Li)N_2$, behaves like a 1,3-dipole for the preparation of [C–N–N] azoles. The reaction mode is similar to that of diazomethane but not in the same fashion. $TMSC(Li)N_2$ (2 equiv) reacts with carboxylic esters to give 2-substituted 5-trimethylsilyltetrazoles (eq 19).[18] Treatment of thiono and dithio esters with $TMSC(Li)N_2$ followed by direct workup with aqueous methanol gives 5-substituted 1,2,3-thiadiazoles (eq 20).[19] While reaction of di-*t*-butyl thioketone with $TMSCHN_2$ produces the episulfide with evolution of nitrogen (eq 21),[20] its reaction with $TMSC(Li)N_2$ leads to removal of one *t*-butyl group to give the 1,2,3-thiadiazole (eq 21).[20]

(19) 2 equiv $TMSC(Li)N_2$, Et_2O, 0 °C, 3 h; 88%

(20) 1. $TMSC(Li)N_2$; 2. H_2O–MeOH; X = OMe or SMe; X = OMe, 79%; SMe, 84%

(21) *t*-Bu$_2$C=S; $TMSCHN_2$, Et_2O, 0 °C, 2 h; rt, 1 h; 89%; $TMSC(Li)N_2$, Et_2O, 0 °C, 1 h; 82%

$TMSCHN_2$ reacts with activated nitriles only, such as cyanogen halides, to give 1,2,3-triazoles.[21] In contrast with this, $TMSC(Li)N_2$ smoothly reacts with nitriles including aromatic, heteroaromatic, and aliphatic nitriles, giving 4-substituted 5-trimethylsilyl-1,2,3-triazoles (eq 22).[22] However, reaction of α,β-unsaturated nitriles with $TMSC(Li)N_2$ in Et_2O affords 3(or 5)-trimethylsilylpyrazoles, in which the nitrile group acts as a leaving group (eq 23).[23] Although α,β-unsaturated nitriles bearing bulky substituents at the α- and/or β-positions of the nitrile group undergo reaction with $TMSC(Li)N_2$ to give pyrazoles, significant amounts of 1,2,3-triazoles are also formed. Changing the reaction solvent from Et_2O to THF allows for predominant formation of pyrazoles (eq 24).[23] Complete exclusion of the formation of 1,2,3-triazoles can be achieved when the nitrile group is replaced by a phenylsulfonyl species.[24] Thus reaction of α,β-unsaturated sulfones with $TMSC(Li)N_2$ affords pyrazoles in excellent yields (eq 25). The geometry of the double bond of α,β-unsaturated sulfones is not critical in the reaction. When both a cyano and a sulfonyl group are present as a leaving group, elimination of the sulfonyl group occurs preferentially (eq 26).[24] The trimethylsilyl group attached to the heteroaromatic products is easily removed with 10% aq. KOH in EtOH or HCl–KF.

(22) $TMSC(Li)N_2$, Et_2O, 0 °C, 2.5 h; 92%

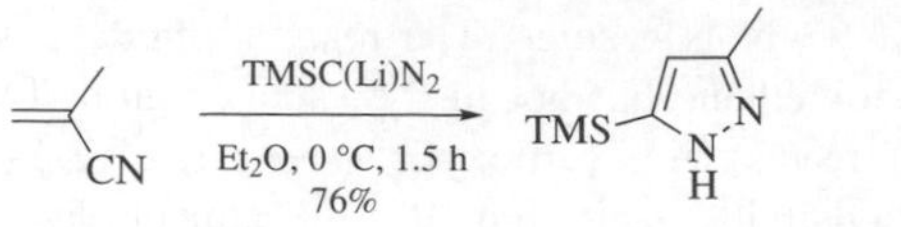

(23)

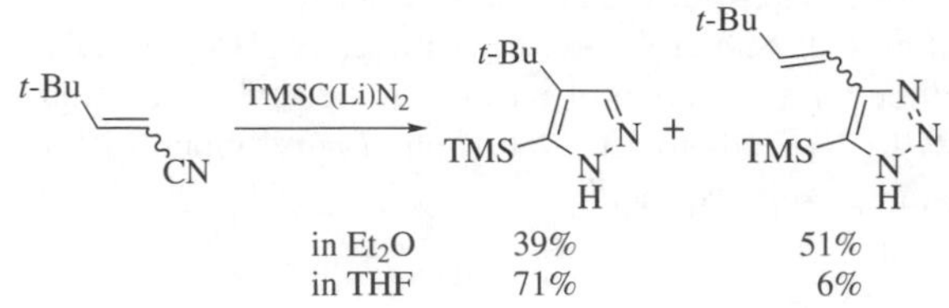

(24)

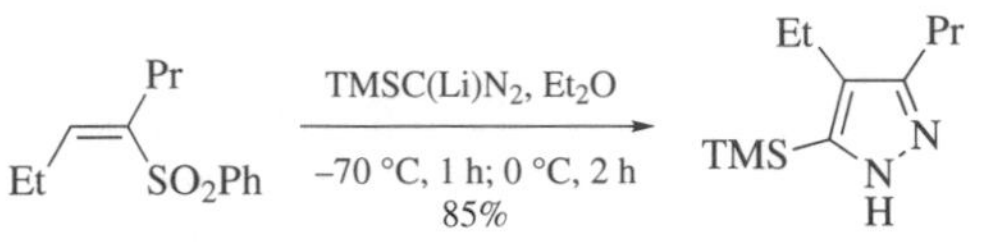

(25)

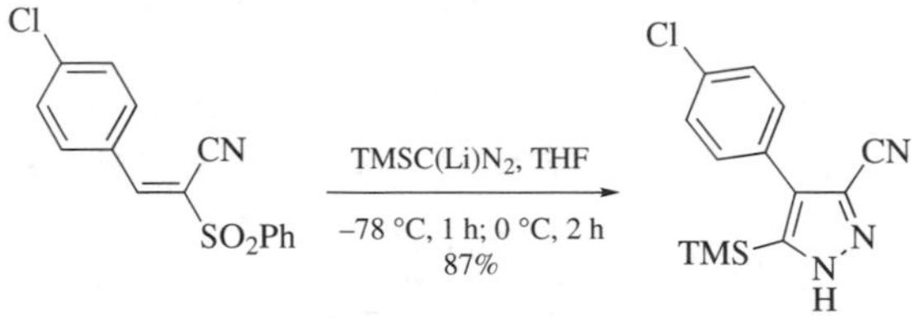

(26)

Various 1,2,3-triazoles can be prepared by reaction of $TMSC(Li)N_2$ with heterocumulenes. Reaction of isocyanates with $TMSC(Li)N_2$ gives 5-hydroxy-1,2,3-triazoles (eq 27).[25] It has been clearly demonstrated that the reaction proceeds by a stepwise process and not by a concerted 1,3-dipolar cycloaddition mechanism. Isothiocyanates also react with $TMSC(Li)N_2$ in THF to give lithium 1,2,3-triazole-5-thiolates which are treated in situ with alkyl halides to furnish 1-substituted 4-trimethylsilyl-5-alkylthio-1,2,3-triazoles in excellent yields (eq 28).[26] However, changing the reaction solvent from THF to Et_2O causes a dramatic solvent effect. Thus treatment of isothiocyanates with $TMSC(Li)N_2$ in Et_2O affords 2-amino-1,3,4-thiadiazoles in good yields (eq 28).[27] Reaction of ketenimines with $TMSC(Li)N_2$ smoothly proceeds to give 1,5-disubstituted 4-trimethylsilyl-1,2,3-triazoles in high yields (eq 29).[28] Ketenimines bearing an electron-withdrawing group at one position of the carbon–carbon double bond react with $TMSC(Li)N_2$ to give 4-aminopyrazoles as the major products (eq 30).[29]

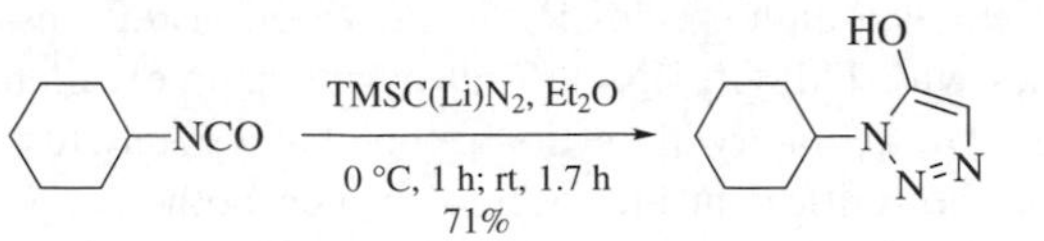

(27)

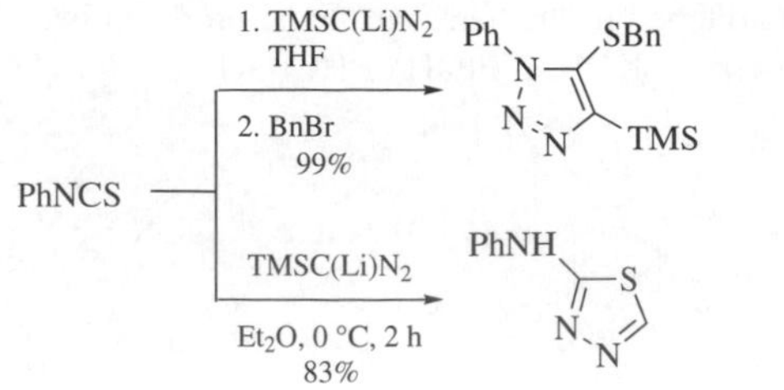

(28)

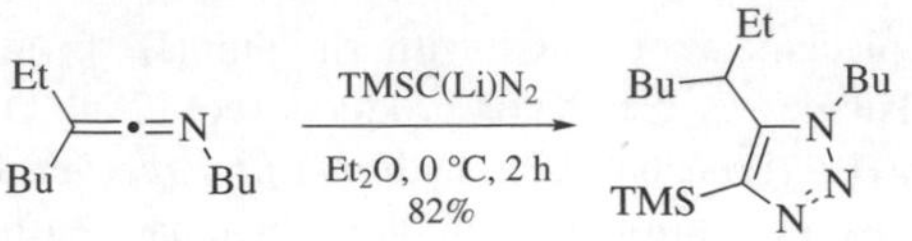

(29)

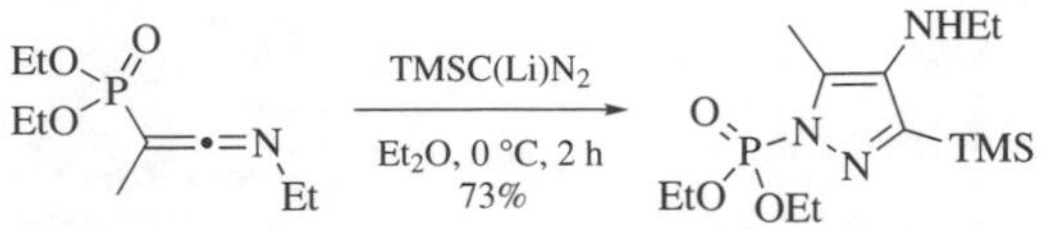

(30)

Pyrazoles are formed by reaction of $TMSCHN_2$ or $TMSC(Li)N_2$ with some alkynes (eqs 31 and 32)[24,30] and quinones (eq 33).[31] Some miscellaneous examples of the reactivity of $TMSCHN_2$ or its lithium salt are shown in eqs 34–36.[20,31,32]

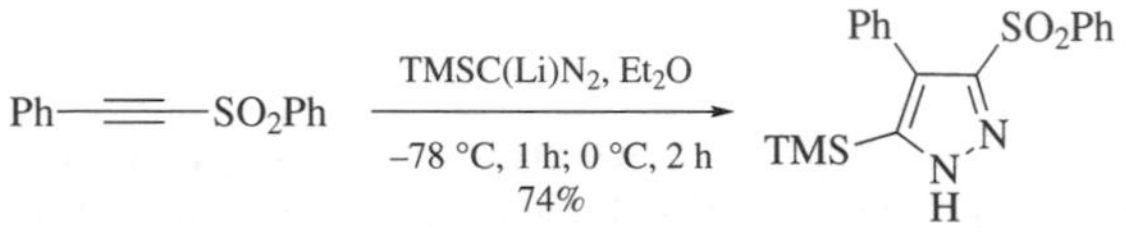

(31)

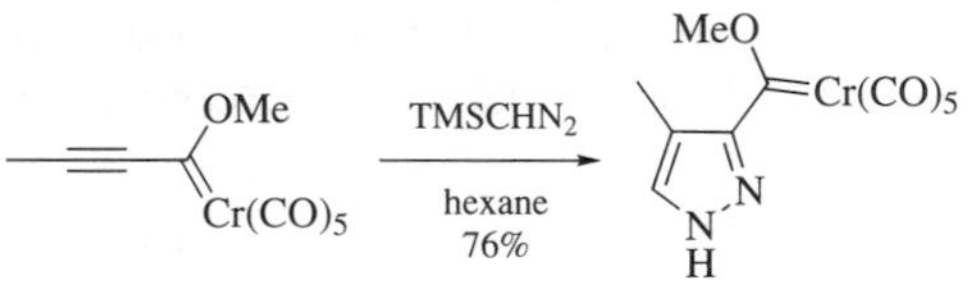

(32)

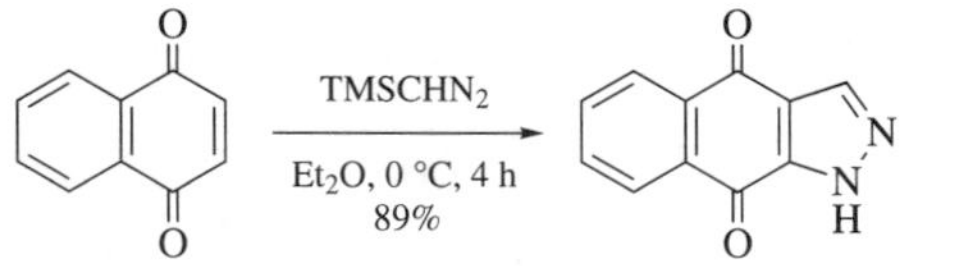

(33)

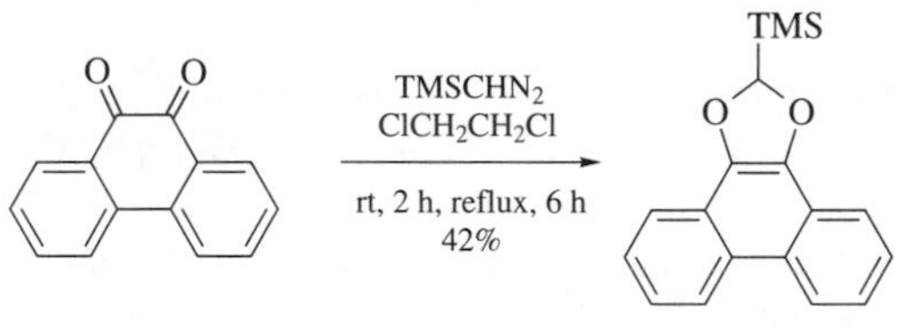

(34)

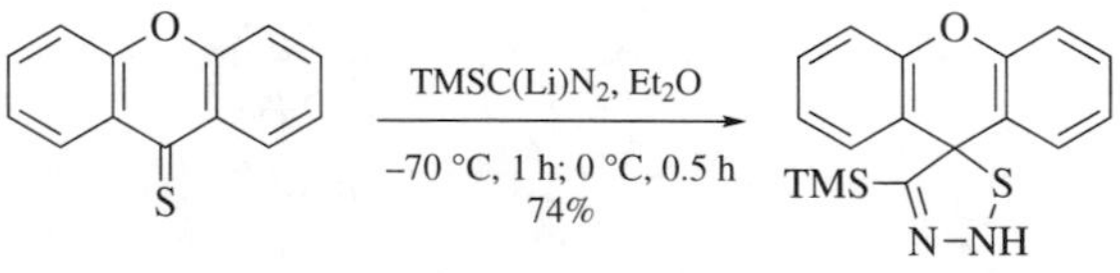

(35)

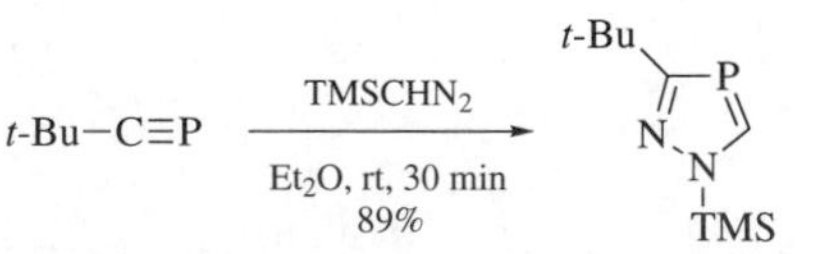

(36)

Related Reagents. Benzylsulfonyldiazomethane; Diazomethane; Diazo(trimethylsilyl)methyllithium; Diethyl Diazomethylphosphonate; 3-Methyl-1-*p*-tolyltriazene.

1. (a) Shioiri, T.; Aoyama, T. *J. Synth. Org. Chem. Jpn* **1986**, *44*, 149 (*CA* **1986**, *104*, 168525q). (b) Aoyama, T. *YZ* **1991**, *111*, 570 (*CA* **1992**, *116*, 58332q). (c) Anderson, R.; Anderson, S. B. In *Advances in Silicon Chemistry*; Larson, G. L., Ed.; JAI: Greenwich, CT, 1991; Vol. 1, pp 303–325. (d) Shioiri, T.; Aoyama, T. In *Advances in the Use of Synthons in Organic Chemistry*; Dondoni, A., Ed.; JAI: London, 1993; Vol. 1, pp 51–101.
2. Seyferth, D.; Menzel, H.; Dow, A. W.; Flood, T. C. *JOM* **1972**, *44*, 279.
3. Shioiri, T.; Aoyama, T.; Mori, S. *OS* **1990**, *68*, 1.
4. Aoyama, T.; Shioiri, T. *CPB* **1981**, *29*, 3249.
5. Hashimoto, N.; Aoyama, T.; Shioiri, T. *CPB* **1982**, *30*, 119.
6. Aoyama, T.; Shioiri, T. *S* **1988**, 228.
7. Hashimoto, N.; Aoyama, T.; Shioiri, T. *CPB* **1981**, *29*, 1475.
8. Aoyama, T.; Terasawa, S.; Sudo, K.; Shioiri, T. *CPB* **1984**, *32*, 3759.
9. Aoyama, T.; Shioiri, T. *TL* **1990**, *31*, 5507.
10. Aoyama, T.; Shioiri, T. *TL* **1988**, *29*, 6295.
11. Aoyama, T.; Shioiri, T. *CPB* **1989**, *37*, 2261.
12. Aoyama, T.; Shioiri, T. *TL* **1986**, *27*, 2005.
13. Aoyama, T.; Toyama, S.; Tamaki, N.; Shioiri, T. *CPB* **1983**, *31*, 2957.
14. Schöllkopf, U.; Scholz, H.-U. *S* **1976**, 271.
15. Colvin, E. W.; Hamill, B. J. *JCS(P1)* **1977**, 869.
16. Aoyama, T.; Iwamoto, Y.; Nishigaki, S.; Shioiri, T. *CPB* **1989**, *37*, 253.
17. Zaitseva, G. S.; Lutsenko, I. F.; Kisin, A. V.; Baukov, Y. I.; Lorberth, J. *JOM* **1988**, *345*, 253.
18. Aoyama, T.; Shioiri, T. *CPB* **1982**, *30*, 3450.
19. Aoyama, T.; Iwamoto, Y.; Shioiri, T. *H* **1986**, *24*, 589.
20. Shioiri, T.; Iwamoto, Y.; Aoyama, T. *H* **1987**, *26*, 1467.
21. Crossman, J. M.; Haszeldine, R. N.; Tipping, A. E. *JCS(D)* **1973**, 483.
22. Aoyama, T.; Sudo, K.; Shioiri, T. *CPB* **1982**, *30*, 3849.
23. Aoyama, T,; Inoue, S.; Shioiri, T. *TL* **1984**, *25*, 433.
24. Asaki, T.; Aoyama, T.; Shioiri, T. *H* **1988**, *27*, 343.
25. Aoyama, T.; Kabeya, M.; Fukushima, A.; Shioiri, T. *H* **1985**, *23*, 2363.
26. Aoyama, T.; Kabeya, M.; Shioiri, T. *H* **1985**, *23*, 2371.
27. Aoyama, T.; Kabeya, M.; Fukushima, A.; Shioiri, T. *H* **1985**, *23*, 2367.
28. Aoyama, T.; Katsuta, S.; Shioiri, T. *H* **1989**, *28*, 133.
29. Aoyama, T.; Nakano, T.; Marumo, K.; Uno, Y.; Shioiri, T. *S* **1991**, 1163.
30. Chan, K. S.; Wulff, W. D. *JACS* **1986**, *108*, 5229.
31. Aoyama, T.; Nakano, T.; Nishigaki, S.; Shioiri, T. *H* **1990**, *30*, 375.
32. Rösch, W.; Hees, U.; Regitz, M. *CB* **1987**, *120*, 1645.

Takayuki Shioiri & Toyohiko Aoyama
Nagoya City University, Japan

Trimethylsilylmethylmagnesium Chloride[1]

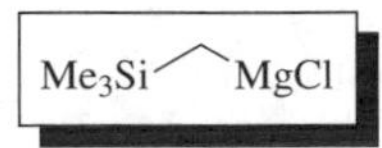

[13170-43-9] $C_4H_{11}ClMgSi$ (MW 147.00)

(methylenation of carbonyl compounds;[1b,2] provides a variety of methods to prepare allylsilanes[1k])

Solubility: sol ethereal solvents; reacts with protic solvents.

Preparative Method: from ***(Chloromethyl)trimethylsilane*** and ***Magnesium*** in an ethereal solvent.[3,4]

Handling, Storage, and Precautions: this Grignard reagent reacts with protic solvents.

Peterson Alkenation. Trimethylsilylmethylmagnesium chloride (**1**) reacts with carbonyl compounds to give β-hydroxysilanes (**2**).[2,4,5] These silanes can then be eliminated to provide an alkene under acidic or basic conditions, such as with ***Sodium Hydride*** or ***Potassium Hydride*** (eq 1).[1a,1b,5,6] The elimination can also be accomplished by ***Acetyl Chloride*** or ***Thionyl Chloride***.[7] For the introduction of *exo*-methylene groups, reagent (**1**) has been found to be superior to a Wittig approach;[8] the silicon reagent reacts rapidly and the byproduct is simple to remove.[1b]

TMS⌒Cl —(Mg, Et_2O)→ TMS⌒MgCl (**1**) —(R^1R^2CO)→ TMS–CH_2–C(R^1)(R^2)OH (**2**) —(H_3O^+ or base)→ CH_2=C(R^1)(R^2) (1)

This methodology has found application in the carbohydrate field for homologation of a saccharide,[9] as other functional groups can be tolerated.[1b,10] The resultant alkene can be functionalized in a wide variety of ways.[11] The use of ***Paraformaldehyde*** as electrophile provides a simple method to ***2-(Trimethylsilyl)ethanol***.[12]

Many of the uses of the Grignard reagent (**1**) are complementary to those of ***Trimethylsilylmethyllithium***, although the cerium reagent derived from the lithium analog provides higher yields with enolizable aldehydes and ketones.[13]

With α,β-unsaturated carbonyl compounds the Grignard reagent reacts in a 1,2-manner,[14] although 1,4-addition can be observed in certain cases.[15] The resultant β-hydroxysilane from a 1,2-addition can be isomerized to the β-ketosilane with a rhodium catalyst.[16] In the presence of copper(I), the Grignard reagent (**1**) reacts in a 1,4-manner with α,β-unsaturated carbonyl compounds.[15,17]

The use of substituted carbonyl compounds allows for the formation of functionalized alkenes; for example, α,β-epoxy ketones afford the monoepoxide of a diene.[18] The successive treatment of α-chlorocarbonyl compounds with (**1**) and ***Lithium*** powder provides a regioselective entry to allylsilanes (eq 2).[19]

R^1C(=O)CH(Cl)CH_2R^2 —(1. (**1**), THF, Et_2O; 2. Li, –78 °C)→ TMSCH_2C(R^1)=CHR^2 (2)

Reagent (**1**) does also react with imines that, in turn, can be generated in situ.[20]

Reaction with Carboxylic Acid Derivatives. In addition to carbonyl compounds, reagent (**1**) also reacts with carboxylic acid derivatives.[4] Thus lactones provide hydroxy allylsilanes (**3**) (eq 3).[21]

Reaction of excess (**1**) with esters provides the tertiary alcohol in an analogous manner,[1d] and subsequent elimination provides the allylsilane.[22] The addition of the second equivalent of (**1**), however, is dependent on the steric requirements of the intermediate β-silyl ketone.[23] The addition of ***Chlorotrimethylsilane*** to the reaction mixture has been advocated as higher yields of the resultant allylsilane are obtained.[24] The use of ***Cerium(III) Chloride*** with Grignard reagent (**1**) promotes nucleophilic attack on esters, and the allylsilanes can be obtained in high yield (eq 4).[25] The yields seem to be higher than those for the analogous reaction between an acid chloride and the cerium reagent prepared from trimethylsilylmethyllithium.[26] 1,2-Addition is observed with α,β-unsaturated esters.[25] Other functional groups, such as acetals, thioacetals, halogens, hydroxy, acetates, and sulfides, can be incorporated at the α-position of the ester group without detrimental effects.[27]

Reaction with diketene in the presence of a nickel catalyst gives 3-(trimethylsilylmethyl)but-3-enoic acid (**4**) (eq 5).[28]

The Grignard reagent (**1**) does react with acid chlorides to provide ketones after hydrolytic workup.[2] The use of a copper(I) catalyst allows isolation of β-silyl ketones (eq 6).[29]

β-Silyl ketones are hydrolytically unstable and can be converted to the desilylated ketone by simple acid or base treatment,[30] or used in a Peterson alkenation reaction to provide enones.[1b,29a] They are also precursors to silyl enol ethers by rearrangement.[29c,31] Reaction of the β-silyl ketone with a vinyl Grignard reagent provides a rapid entry to 2-substituted 1,3-dienes by a Peterson protocol.[32]

Reaction of (**1**) with ***Carbon Dioxide*** provides ***Trimethylsilylacetic Acid***,[33] while treatment of (**1**) with ***Ethyl Chloroformate*** gives ***Ethyl Trimethylsilylacetate***.[34] The use of ethyl oxalyl chloride as substrate for (**1**) provides a simple preparation of ethyl 2-(trimethylsilylmethyl)propenoate.[35] Condensation of (**1**) with benzonitrile resulted in isolation of acetophenone (64%) and desoxybenzoin (45%).[4]

Reaction with Alkyl Halides. The Grignard reagent (**1**) can be alkylated by allyl halides to afford the homoallylsilane.[4,36] The use of a nickel(II) catalyst facilitates coupling of the Grignard reagent (**1**) to vinyl halides[37] and aryl halides, triflates and *O*-carbamates[38] and provides a useful method for the preparation of allyl- and benzylsilanes (eq 7). Palladium catalysis is also effective to couple (**1**) with vinyl halides.[37b,39]

Reaction with Sulfur Compounds. In a similar coupling reaction to those of alkyl halides, (**1**) reacts in the presence of a nickel catalyst with allylic dithioacetals to yield 1-(trimethylsilyl)butadienes (eq 8).[40]

Dithioacetals derived from alkyl aryl ketones react with the Grignard reagent (**1**) in the presence of a nickel catalyst.[41,42] The use of an orthothioester as substrate in place of a thioacetal provides 1,3-bis(trimethylsilyl)propenes as a mixture of the (*E*)- and (*Z*)-isomers.[43] With α-oxoketene dithioacetals (**5**), the Grignard reagent (**1**) reacts, in the presence of ***Copper(I) Iodide***, to provide 1-trimethylsilyloxy-3-methylthio-1,3-dienes(eq 9).[44]

Reaction with Other Electrophiles. Reaction occurs between (**1**) and epoxides to yield γ-hydroxysilanes.[45] Michael addition is observed with nitroalkenes; the silyl group can then promote a Nef reaction to afford β-silyl ketones (eq 10).[46]

With aromatic nitro compounds, nucleophilic addition of (**1**) occurs on the aromatic nucleus (eq 11).[47]

$$\text{p-X-C}_6\text{H}_4\text{NO}_2 \xrightarrow{\text{(1), THF}} \text{2-(TMS-CH}_2\text{)-4-X-C}_6\text{H}_3\text{NO}_2 \quad (11)$$

A nickel catalyst allows reaction between (**1**) and an enol phosphate,[48] silyl enol ether,[49] or substituted dihydrofurans[50] and dihydropyrans[51] to afford allylsilanes. Additional functionality can be tolerated in the substrate.[52]

Coupling of (**1**) with propargylic tosylates, mesylates, or acetates in the presence of copper(I) leads to α-trimethylsilylallenes,[53] while the use of a propargyl alcohol substrate leads to the substituted allylsilane (eq 12).[54]

$$\text{propargyl alcohol (OH)} \xrightarrow[\text{2. pH 9; 64\%}]{\text{1. (1), } s\text{-BuMgCl, CuCl, Et}_2\text{O}} \text{TMS-CH}_2\text{-C(=CH}_2\text{)-CH}_2\text{OH} \quad (12)$$

With alkynes, copper-catalyzed addition of the Grignard reagent (**1**) provides the allylsilane.[55] The intermediate vinylcopper reagent can also be trapped with electrophiles.[56]

The reagent (**1**) also reacts with wide variety of electrophiles, including carbon dioxide,[33] cyanogen,[57] silyl chlorides,[3,58] metal halides,[59] and ***Phosphorus(III) Chloride***.[59a] With methyl phenyl sulfinate, reaction of (**1**) leads to the formation of (phenylthio)(trimethylsiloxy)methane, a protected form of formaldehyde, by a sila-Pummerer rearrangement.[60]

The Grignard reagent (**1**) is the precursor of numerous organometallic compounds that contain the trimethylsilylmethyl ligand.[61]

Related Reagents. Allyltrimethylsilane; Bis(trimethylsilyl)-methane; (Diisopropoxymethylsilyl)methylmagnesium Chloride; Methylenetriphenylphosphorane; Tri-*n*-butyl(trimethylsilylmethyl)stannane; (Trimethylsilyl)allene; Trimethylsilylmethylpotassium; (Trimethylstannylmethyl)lithium; Trimethylsilylmethyllithium; Trimethylsilylmethylcopper; Trimethylsilylmethyl Trifluoromethanesulfonate; Tris(trimethylsilyl)methane.

---

1. (a) Chan, T.-H. *ACR* **1977**, *10*, 442. (b) Ager, D. J. *OR* **1990**, *38*, 1. (c) Ager, D. J. *S* **1984**, 384. (d) Fleming, I. In *Comprehensive Organic Chemistry*; Barton, D. H. R.; Ollis, W. D., Eds.; Pergamon: Oxford, 1979; Vol. 3; p 541. (e) Weber, W. P. *Silicon Reagents for Organic Synthesis–Concepts in Organic Chemistry*; Springer: New York, 1983; Vol. 14. (f) Colvin, E. W. *Silicon in Organic Synthesis*; Butterworths: London, 1981. (g) Colvin, E. W. *CSR* **1978**, *7*, 15. (h) Magnus, P. *Aldrichim. Acta* **1980**, *13*, 43. (i) Magnus, P. D.; Sarkar, T.; Djuric, S. In *Comprehensive Organometallic Chemistry*; Wikinson, G.; Stone, F. G. A.; Abel, E. W., Eds.; Pergamon: Oxford, 1982; Vol. 7, p 515. (j) Birkofer, L.; Stuhl, O. *Top. Curr. Chem.* **1980**, *88*, 33. (k) Chan, T. H.; Fleming, I. *S* **1979**, 761.
2. Chan, T. H.; Chang, E.; Vinokur, E. *TL* **1970**, 1137.
3. Sommer, L. H.; Goldberg, G. M.; Gold, J.; Whitmore, F. C. *JACS* **1947**, *69*, 980.
4. Hauser, C. R.; Hance, C. R. *JACS* **1952**, *74*, 5091.
5. Peterson, D. J. *JOC* **1968**, *33*, 780.
6. (a) Hudrlik, P. F.; Peterson, D. *JACS* **1975**, *97*, 1464. (b) Hudrlik, P. F.; Peterson, D. *TL* **1974**, 1133.
7. Chan, T. H.; Chang, E. *JOC* **1974**, *39*, 3264.
8. (a) Boeckman, R. K. Jr.; Silver, S. M. *TL* **1973**, 3497. (b) Akiyama, T.; Ohnari, M.; Shima, H.; Ozaki, S. *SL* **1991**, 831.
9. (a) Jones, K.; Wood, W. W. *JCS(P1)* **1987**, 537. (b) Ferrier, R. J.; Stütz, A. E. *Carbohydr. Res.* **1990**, *205*, 283. (c) Udodong, U. E.; Fraser-Reid, B. *JOC* **1988**, *53*, 2131. (d) Udodong, U. E.; Fraser-Reid, B. *JOC* **1989**, *54*, 2103.
10. Lin, J.; Nikaido, M. M.; Clark, G. *JOC* **1987**, *52*, 3745.
11. (a) Glänzer, B. I.; Györgydeák, Z.; Bernet, B.; Vasella, A. *HCA* **1991**, *74*, 343. (b) Ager, D. J.; East, M. B. *T* **1993**, *49*, 5683.
12. Mancini, M.; Honek, J. F. *TL* **1982**, *23*, 3249.
13. Johnson, C. R.; Tait, B. D. *JOC* **1987**, *52*, 281.
14. (a) Carter, M. J.; Fleming, I. *CC* **1976**, 679. (b) Pillot, J.-P.; Dunoguès, J.; Calas, R. *JCR(S)* **1977**, 268.
15. Taylor, R. T.; Galloway, J. G. *JOM* **1981**, *220*, 295.
16. (a) Sato, S.; Okada, H.; Matsuda, I.; Izumi, Y. *TL* **1984**, *25*, 769. (b) Matsuda, I.; Okada, H.; Sato, S.; Izumi, Y. *TL* **1984**, *25*, 3879.
17. (a) Hatanaka, Y.; Kuwajima, I. *JOC* **1986**, *51*, 1932. (b) Horiguchi, Y.; Kataoka, Y.; Kuwajima, I. *TL* **1989**, *30*, 3327. (c) Fujiwara, T.; Suda, A.; Takeda, T. *CL* **1991**, 1619.
18. Kitahara, T.; Kurata, H.; Mori, K. *T* **1988**, *44*, 4339.
19. Barluenga, J.; Fernández-Simón, J. L.; Concellón, J. M.; Yus, M. *TL* **1989**, *30*, 5927.
20. Sisko, J.; Weinreb, S. M. *JOC* **1990**, *55*, 393.
21. Ochiai, M.; Fujita, E.; Arimoto, M.; Yamaguchi, H. *CPB* **1985**, *33*, 989.
22. de Raadt, A.; Stütz, A. E. *Carbohydr. Res.* **1991**, *220*, 101.
23. Fleming, I.; Pearce, A. *JCS(P1)* **1981**, 251.
24. Box, V. G. S.; Brown, D. P. *H* **1991**, *32*, 1273.
25. Narayanan, B. A.; Bunelle, W. H. *TL* **1987**, *28*, 6261.
26. (a) Anderson, M. B.; Fuchs, P. L. *SC* **1987**, *17*, 621. (b) Hojo, M.; Ohsumi, K.; Hosomi, A. *TL* **1992**, *33*, 5981.
27. (a) Lee, T. V.; Channon, J. A.; Cregg, C.; Porter, J. R.; Roden, F. S.; Yeoh, H. T.-L. *T* **1989**, *45*, 5877. (b) Calò, V.; Lopez, L.; Pesce, G. *JOM* **1988**, *353*, 405.
28. (a) Itoh, K.; Fukui, M.; Kurachi, Y. *CC* **1977**, 500. (b) Itoh, K.; Yogo, T.; Ishii, Y. *CL* **1977**, 103. (c) Lee, T. V.; Boucher, R. J.; Rockell, C. J. M. *TL* **1988**, *29*, 689.
29. (a) Fürstner, A.; Kollegger, G.; Weidmann, H. *JOM* **1991**, *414*, 295. (b) Pearlman, B. A.; McNamara, J. M.; Hasan, I.; Hatakeyama, S.; Sekizaki, H.; Kishi, Y. *JACS* **1981**, *103*, 4248. (c) Yamamoto, Y.; Ohdoi, K.; Nakatani, M.; Akiba, K.-y. *CL* **1984**, 1967. (d) Koerwitz, F. L.; Hammond, G. B.; Wiemer, D. F. *JOC* **1989**, *54*, 738.
30. Whitmore, F. C.; Sommer, L. H.; Gold, J.; Van Strien, R. E. *JACS* **1947**, *69*, 1551.
31. Larson, G. L.; Montes de López-Cepero, I.; Torres, L. E. *TL* **1984**, *25*, 1673.
32. Brown, P. A.; Bonnert, R. V.; Jenkins, P. R.; Lawrence, N. J.; Selim, M. R. *JCS(P1)* **1991**, 1893.
33. Sommer, L. H.; Gold, J. R.; Goldberg, G. M.; Marans, N. S. *JACS* **1949**, *71*, 1509.
34. Gold, J. R.; Sommer, L. H.; Whitmore, F. C. *JACS* **1948**, *70*, 2874.
35. Haider, A. *S* **1985**, 271.
36. Hosomi, A.; Masunari, T.; Tominaga, Y.; Yanagi, T.; Hojo, M. *TL* **1990**, *31*, 6201.
37. (a) Hayashi, T.; Kabeta, K.; Hamachi, I.; Kumada, M. *TL* **1983**, *24*, 2865. (b) Negishi, E.-i.; Luo, F.-T.; Rand, C. L. *TL* **1982**, *23*, 27. (c) Kitano, Y.; Matsumoto, T.; Wakasa, T.; Okamoto, S.; Shimazaki, T.; Kobayashi, Y.; Sato, F.; Miyaji, K.; Arai, K. *TL* **1987**, *28*, 6351.
38. (a) Tamao, K.; Sumitani, K.; Kiso, Y.; Zembayashi, M.; Fujioka, A.; Kodama, S.-i.; Nakajima, I.; Minato, A.; Kumada, M. *BCJ* **1976**, *49*, 1958. (b) Sengupta, S.; Leite, M.; Raslan, D. S.; Quesnelle, C.; Snieckus, V. *JOC* **1992**, *57*, 4066.
39. Andreini, B. P.; Carpita, A.; Rossi, R.; Scamuzzi, B. *T* **1989**, *45*, 5621.
40. Ni, Z.-J.; Luh, T.-Y. *JOC* **1988**, *53*, 5582.

41. (a) Ni, Z.-J.; Luh, T.-Y. *CC* **1988**, 1011. (b) Wong, K.-T.; Ni, Z.-J.; Luh, T.-Y. *JCS(P1)* **1991**, 3113. (c) Cheng, W.-L.; Luh, T.-Y. *JOC* **1992**, *57*, 3516. (d) Wong, K.-T.; Luh, T.-Y. *CC* **1992**, 564. (e) Wong, K.-T.; Luh, T.-Y. *JACS* **1992**, *114*, 7308.
42. (a) Ni, Z.-J.; Luh, T.-Y. *JOC* **1988**, *53*, 2129. (b) Ni, Z.-J.; Yang, P.-F.; Ng, D. K. P.; Tzeng, Y.-L.; Luh, T.-Y. *JACS* **1990**, *112*, 9356.
43. Tzeng, Y.-L.; Luh, T.-Y.; Fang, J.-M. *CC* **1990**, 399.
44. Tominga, Y.; Kamio, C.; Hosomi, A. *CL* **1989**, 1761.
45. (a) Sommer, L. H.; Van Strien, R. E.; Whitmore, F. C. *JACS* **1949**, *71*, 3056. (b) Fleming, I.; Loreto, M. A.; Wallace, I. H. M.; Michael, J. P. *JCS(P1)* **1986**, 349. (c) Camp Schuda, A. D.; Mazzocchi, P. H.; Fritz, G.; Morgan, T. *S* **1986**, 309.
46. Hwu, J. R.; Gilbert, B. A. *JACS* **1991**, *113*, 5917.
47. (a) Bartoli, G.; Bosco, M.; Dalpozzo, R.; Todesco, P. E. *JOC* **1986**, *51*, 3694. (b) Bartoli, G.; Bosco, M.; Dalpozzo, R.; Todesco, P. E. *CC* **1988**, 807. (c) Bartoli, G.; Bosco, M.; Dal Pozzo, R.; Petrini, M. *T* **1987**, *43*, 4221.
48. (a) Hayashi, T.; Fujiwa, T.; Okamata, Y.; Katsuro, Y.; Kumada, M. *S* **1981**, 1001. (b) Danishefsky, S. J.; Mantlo, N. *JACS* **1988**, *110*, 8129.
49. Hayashi, T.; Katsuro, Y.; Kumada, M. *TL* **1980**, *21*, 3915.
50. Wadman, S.; Whitby, R.; Yeates, C.; Kocieński, P.; Cooper, K. *CC* **1987**, 241.
51. Kocieński, P.; Dixon, N. J.; Wadman, S. *TL* **1988**, *29*, 2353.
52. Pettersson, L.; Frejd, T.; Magnusson, G. *TL* **1987**, *28*, 2753.
53. (a) Westmijze, H.; Vermeer, P. *S* **1979**, 390. (b) Montury, M.; Psaume, B.; Goré, J. *TL* **1980**, *21*, 163. (c) Trost, B. M.; Urabe, H. *JACS* **1990**, *112*, 4982.
54. (a) Kleijn, H.; Vermeer, P. *JOC* **1985**, *50*, 5143. (b) Trost, B. M.; Matelich, M. C. *S* **1992**, 151. (c) Trost, B. M.; Matelich, M. C. *JACS* **1991**, *113*, 9007.
55. Foulon, J. P.; Bourgain-Commerçon, M.; Normant, J. F. *T* **1986**, *42*, 1389.
56. Foulon, J. P.; Bourgain-Commerçon, M.; Normant, J. F. *T* **1986**, *42*, 1399.
57. Prober, M. *JACS* **1955**, *77*, 3224.
58. Sommer, L. H.; Mitch, F. A.; Goldberg, G. M. *JACS* **1949**, *71*, 2746.
59. (a) Seyferth, D.; Freyer, W. *JOC* **1961**, *26*, 2604. (b) Westerhausen, M.; Rademacher, B.; Poll, W. *JOM* **1991**, *421*, 175.
60. (a) Brook, A. G.; Anderson, D. G. *CJC* **1968**, *46*, 2115. (b) Ager, D. J. *CSR* **1982**, *11*, 493.
61. Armitage, D. A. In *Comprehensive Organometallic Chemistry*; Wilkinson, G.; Stone, F. G. A.; Abel, E. W., Eds.; Pergamon: Oxford, 1982; Vol. 2, p 1.

David J. Ager
The NutraSweet Company, Mount Prospect, IL, USA

1-Trimethylsilyloxy-1,3-butadiene

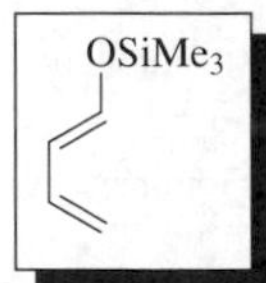

[6651-43-0] $C_7H_{14}OSi$ (MW 142.30)

(easily prepared[1] reactive diene for Diels–Alder reactions and other cycloadditions;[2–21] reactive silyl enol ether for aldol and Michael reactions,[22–28] and electrophilic additions[29–32])

Physical Data: bp 131 -b°C, bp 49.5 °C/25 mmHg; *d* 0.811 g cm^{-3}.

Solubility: sol most standard organic solvents.

Form Supplied in: liquid commercially available (98% purity) as an approximately 85:15 mixture of (*E*)- and (*Z*)-isomers.

Preparative Methods: can be prepared easily.[1]

Handling, Storage, and Precautions: is a flammable liquid and is moisture sensitive.

Diels–Alder Reactions and Other Cycloadditions. Although commercially available, 1-trimethylsilyloxy-1,3-butadiene (**1**) can be easily prepared by silylation of ***Crotonaldehyde***.[1] It has often been used as a reactive diene in Diels–Alder reactions. For example, reaction with ***Dimethyl Acetylenedicarboxylate*** (**2**) affords the cyclohexadiene diester (**3**) in 68% yield. This initial Diels–Alder adduct can be converted into two different aromatic products by the proper choice of conditions: namely, thermal elimination affords the phthalate (**4**), while oxidation produces the phenol (**5**), both in good yield (eq 1).[2] Reaction with ***Methyl 3-Nitroacrylate*** (**6**) followed by hydrolysis of the initial adduct (**7**) and elimination of the β-nitro group leads to the cyclohexadienol (**8**) (eq 2).[3] Many other Diels–Alder reactions of this type have been carried out using (**1**), as shown in Table 1.[4] In general, the *endo* adduct is favored, especially at lower temperatures.

(1) + (2) → (3): PhH, 50 °C, 2 days, 68%; (3) → (5): CrO_3•2py, 60%; (3) → (4): PhMe, Δ, 85% from (1) (1)

(1) + (6) → (7): PhH, 25 °C, 100%; (7) → (8): 1. H_3O^+, 78%; 2. K_2CO_3, MeOH, 68% (2)

Cyclic enones, lactones, and lactams have also been used often as dienophiles in Diels–Alder reactions with (**1**), again giving mainly the *endo* adduct,[5] e.g. (**12**) reacted with (**1**) to give (**13**) as the major product in 74% yield (eq 3).[5a] As mentioned earlier, the initial adducts are often oxidized (either directly or after hydrolysis of the silyl ether) with Jones reagent to the corresponding enone,[6] e.g. (**14**) to give (**15**) (eq 4).[6a] This corresponds to the annulation of a cyclohexenone unit onto an existing enone or unsaturated lactone unit and has been used often in synthesis, e.g. in the preparation of aureolic acid derivatives such as (**17**) from (**16**) and (**1**) (eq 5).[6c]

Table 1 Diels–Alder Reactions of 1-Trimethylsilyl-1,3-butadiene (**1**)

Dienophile	Conditions	Yield	Ratio (**10**):(**11**)
Z = CHO, R = H	50 °C, 48 h	88%	11:1[4a]
Z = CHO, R = H	130 °C, 7 h	53%	86:14[4b]
Z = CHO, R = Me	150 °C, 7 h	69%	65:35[4b]
Z = CHO, R = CO_2Me	40 °C, 48 h	>90%	83:17[4c]
Z = R = COPh	80 °C, 5 h	93%	no stereochem. given[4d]
Z = NO_2, R = TMS	100 °C, 37 h	83%	100:0[4e]
Z = NO_2, R = OCOPh	25 °C	90%	6:1[4f]

(1) + (12) → 1. PhMe, 100–110 °C, 3 d; 2. TBAF, KF, THF, 25 °C; 74% → (13) (3)

(1) + (14) → 1. 12 kbar, CH_2Cl_2, 50 °C, 65 h, 85%; 2. H_2CrO_4, 0 °C, 60% → (15) (4)

(1) + (16) → 1. 125 °C, 3 d; 2. aq. NH_4Cl, 83%; 3. H_2CrO_4, 0 °C, 80% → (17) (5)

There are some exceptions to the preference for *endo* stereochemistry, especially with unsaturated sulfones.[7] A curious and useful reversal of the stereochemical preference has been reported, namely Diels–Alder cycloaddition of the pyrazolecarboxylate (**18**) with (**1**) followed by photochemical elimination of nitrogen gave mainly the *endo* adduct (**19n**) (eq 6), whereas cycloaddition of (**1**) with the cyclopropenecarboxylate (**20**) gave nearly exclusively the *exo* adduct (**19x**) (eq 7).[8]

(1) + (18) → 1. 12 kbar, CH_2Cl_2, 18 h; 2. *hν*; 89% → (19x) + (19n), 1:5 (6)

(1) + (20) → 10 kbar, CH_2Cl_2, 18 h, 78% → (19x) + (19n), 50:1 (7)

Stereocontrol with acyclic dienophiles can also be high,[9] e.g. (**21**) giving only (**22**) (eq 8),[9a] although the relative stereochemistry of the adjacent allylic center can cause nearly stereorandom addition as well.[9a] The versatility of the initial adducts has been evidenced most clearly in the synthesis of anthraquinones and their derivatives. The adducts of (**1**) with various quinones and substituted quinones have been transformed into simple aromatics,[10] phenols,[11] e.g. (**23**) gave (**24**) (eq 9),[11a] cyclohexenones,[12] e.g. (**25**) gave (**26**) (eq 10),[12a] and allylic alcohols,[13] all in excellent yields.

(1) + (21) → 12 kbar, CH_2Cl_2, 40 °C, 64 h, 100% → (22) (8)

(23) → 1. (**1**), CH_2Cl_2, 78 h, Δ, 79%; 2. H_2CrO_4, 90%; 3. $AlCl_3$, 90% → (24) (9)

(25) → 1. (**1**), CH_2Cl_2, 78 h, Δ, 75%; 2. H_2CrO_4, 90% → (26) (10)

The reaction can be carried out with excellent enantiocontrol (generally 80% or better) using juglone (**27**) and a catalyst prepared from ***Trimethyl Borate*** and a tartaramide to give (**28**) (eq 11).[13a] Finally, with metal complexes of tropene and tropone systems, one can obtain either normal [4 + 2] cycloadditions[14a] or novel [6 + 4] cycloadditions, e.g. (**29**) giving (**30**) (eq 12).[14b,c] Other [4 + 2] cycloadditions have also been reported, using as dienophiles aldehydes[15] to give pyran derivatives such as (**31**), vinyl chlorides,[16] singlet oxygen,[17] nitrosoalkanes[18] to give compounds such as (**32**), and phosphaalkenes (eq 13).[19] Several [2 + 2] cyclizations are known, namely carbene and ketene additions to (**1**), all of which occur at the unsubstituted double bond.[20] One clever use of (**1**) in synthesis is the tandem [2 + 2]–Cope process which converts (**33**) into (**34**) (eq 14).[20d] Finally, a nickel(0)-catalyzed [4 + 4] cyclization of (**1**) gives the *trans*-3,4-bis(silyloxy)-1,5-cyclooctadiene (**35**) in excellent yield (eq 15).[21]

(**27**) → (**28**) ~80% ee: (**1**), $B(OMe)_3$, HO(CONHR)CH–CH(CONHR)OH; 73–94% (11)

(**29**) [$Cr(CO)_3$ complex] + (**1**) → (**30**): 1. *hν*, 0 °C; 2. $P(OMe)_3$; 56% (12)

(**31**) *trans*/*cis* mixture ← RO_2CCHO, 14–98% — (**1**) — PhN=O, CH_2Cl_2, 0 °C, 60% → (**32**) (13)

(**33**) + (**1**) → (**34**): 0.01 equiv $Rh_2(OAc)_4$; CH_2Cl_2, reflux; 86% (14)

(**1**) → (**35**): $Ni(acac)_2$, Et_2AlOEt; PPh_3, PhH, 60 °C, 18 d; 90% (15)

Aldol Condensations. The nucleophilicity of the γ-carbon atom in 1-trimethylsilyloxy-1,3-butadiene (**1**) allows Lewis acid-catalyzed aldol condensations to be carried out, with (**1**) acting as the equivalent of the enolate of ***crotonaldehyde***.[22–26] The products usually have the (*E*) stereochemistry about the newly formed double bond. Orthoesters are good electrophiles;[22,1d] when orthoformate is used as the electrophile, the (*E*)-monoacetal of glutacondialdehyde (**36**) is formed in good yield.[22e] This was then used in a very short synthesis of the antiviral agent AZT (**37**) (eq 16).[22e] 1,3-Dithienium salts and 2-alkoxydithiolanes have also been used to produce the dithioacetals corresponding to (**36**).[23] Simple acetals and α-chloro ethers can be used as the electrophiles with (**1**) to generate initially δ-alkoxy-α,β-unsaturated aldehydes and then the doubly unsaturated aldehydes after treatment with base.[24,1c]

A clever approach to the synthesis of indole from pyrrole involves the condensation of (**1**) with the endoperoxides derived from *N*-alkoxycarbonylpyrrole as shown in eq 17.[25a] An approach to the piperidine alkaloids uses similar chemistry.[25b,c]

(**1**) → (**36**): $HC(OMe)_3$, $ZnCl_2$, 65%; (**36**) → (**37**): 7 steps (16)

N-CO_2Me pyrrole → 1. $^1O_2$; 2. (**1**), $SnCl_2$; 62% → 3-(N-CO_2Me-pyrrol-2-yl)propenyl CHO → 1. $SnCl_4$; 2. KOH → indole (17)

Finally, a conceptually similar reaction involves the addition of (**1**) to an acyl imminium salt to give an intermediate (**38**) (eq 18) which was then used in an intramolecular Diels–Alder approach to the heteroyohimboid alkaloids.[26] A Michael adduct is formed when (**1**) is allowed to react with 2,2-bis(phenylsulfonyl)styrene, probably as the result of a Diels–Alder reaction in which the adduct reverses to a zwitterion and internally deprotonates.[27] Also the lithium enolate derived from (**1**) has been added in a Michael fashion to α,β-unsaturated ketones to give, after silylation, the same adducts as the direct Diels–Alder reaction but at much lower temperatures.[28]

Dihydro-β-carboline → crotonyl COCl; (**1**), THF, –78 °C; 78% → (**38**) (18)

Electrophilic Additions. Finally, various electrophiles (other than the carbon species in aldol and other condensations) have been added to (**1**) in good yield, namely: ***Benzenesulfenyl Chloride***,[29] ***Bromine*** or ***N-Bromosuccinimide***,[30] and ***t-Butyl Hypochlorite***.[31] There are also several reports of the reaction of (**1**) with various transition metal electrophiles which generate either a silyloxydiene bound to the metal[32] or a crotonaldehyde unit bound to the metal.[33]

Related Reagents. 1-Acetoxy-1,3-butadiene; Benzyl 1,3-Butadiene-1-carbamate; Crotonaldehyde; 1,1-Dimethoxy-3-trimethylsilyloxy-1,3-butadiene; 3-Hydroxy-2-pyrone; 2-Methoxy 1,3-butadiene; 2-Methoxy-3-phenylthio-1,3-butadiene;

1-Methoxy-3-trimethylsilyloxy-1,3-butadiene; 1-Methoxy-1,3-bis(trimethylsilyloxy)-1,3-butadiene; 2-Methoxy-1-phenylthio-1,3-butadiene; 2-Trimethylsilyloxy-1,3-butadiene.

---

1. (a) Belg. Patent 670 769, 1966 (*CA* **1966**, *65*, 5487d). (b) Cazeau, P.; Frainnet, E. *BSF(2)* **1972**, 1658. (c) Ishida, A.; Mukaiyama, T. *BCJ* **1977**, *50*, 1161. (d) Makin, S. M.; Kruglikova, R. I.; Popova, T. P.; Chernyshev, A. I. *JOU* **1982**, *18*, 834. (e) Makin, S. M.; Kruglikova, R. I.; Shavrygina, O. A.; Chernyshev, A. I.; Popova, T. P.; Tung, N. F. *JOU* **1982**, *18*, 250. (f) Cazeau, P.; Duboudin, F.; Moulines, F.; Babot, O.; Dunoguès, J. *T* **1987**, *43*, 2089. (g) Iqbal, J.; Khan, M. A. *SC* **1989**, *19*, 515.
2. Yamamoto, K.; Suzuki, S.; Tsuji, J. *CL* **1978**, 649.
3. Danishefsky, S. J.; Prisbylla, M. P.; Hiner, S. *JACS* **1978**, *100*, 2918.
4. (a) Kurth, M. J.; Brown, E. G.; Hendra, E.; Hope, H. *JOC* **1985**, *50*, 1115. (b) Makin, S. M.; Tung, N. F.; Shavrygina, O. A.; Arshava, B. M.; Romanova, I. A. *JOU* **1983**, *19*, 640. (c) Kakushima, M.; Scott, D. E. *CJC* **1979**, *57*, 1399. (d) Oida, T.; Tanimoto, S.; Sugimoto, T.; Okano, M. *S* **1980**, 131. (e) Padwa, A.; MacDonald, J. G. *JOC* **1983**, *48*, 3189. (f) Kraus, G. A.; Thurston, J.; Thomas, P. J.; Jacobson, R. A.; Su, Y. *TL* **1988**, *29*, 1879.
5. (a) Koot, W.-J.; Hiemstra, H.; Speckamp, W. N. *JOC* **1992**, *57*, 1059. (b) Alonso, D.; Font, J.; Ortuño, R. M.; d'Angelo, J.; Guingant, A.; Bois, C. *T* **1991**, *47*, 5895. (c) Tsuda, Y.; Horiguchi, Y.; Sano, T. *H* **1976**, *4*, 1355.
6. (a) Ortuño, R. M.; Guingant, A.; d'Angelo, J. *TL* **1988**, *29*, 6989. (b) Moriarty, K. J.; Shen, C.-C.; Paquette, L. A. *SL* **1990**, *5*, 263. (c) Franck, R. W.; Subramanian, C. S.; John, T. V.; Blount, J. F. *TL* **1984**, *25*, 2439. (d) Kallmerten, J. *TL* **1984**, *25*, 2843. (e) Card, P. J. *JOC* **1982**, *47*, 2169.
7. (a) Cossu, S.; Delogu, G.; De Lucchi, O.; Fabbri, D.; Licini, G. *AG(E)* **1989**, *28*, 766. (b) Hayakawa, K.; Nishiyama, H.; Kanematsu, K. *JOC* **1985**, *50*, 512.
8. Rigby, J. H.; Kierkus, P. C. *JACS* **1989**, *111*, 4125.
9. (a) Casas, R.; Parella, T.; Branchadell, V.; Oliva, A.; Ortuño, R. M. *T* **1992**, *48*, 2659. (b) Serrano, J. A.; Cáceres, L. E.; Román, E. *JCS(P1)* **1992**, 941.
10. (a) Echavarren, A.; Prados, P.; Fariña, F. *T* **1984**, *40*, 4561. (b) Kraus, G. A.; Molina, M. T.; Walling, J. A. *CC* **1986**, 1568. (c) Lee, J.; Snyder, J. K. *JOC* **1990**, *55*, 4995. (d) Lee, J.; Tang, J.; Snyder, J. K. *TL* **1987**, *28*, 3427.
11. (a) Laugraud, S.; Guingant, A.; Chassagrand, C.; d'Angelo, J. *JOC* **1988**, *53*, 1557. (b) Kraus, G. A.; Chen, L. *JOC* **1991**, *56*, 5098. (c) McKenzie, T. C.; Hassen, W.; Macdonald, J. F. *TL* **1987**, *28*, 5435. (d) Cameron, D. W.; Feutrill, G. I.; Gibson, C. L.; Read, R. W. *TL* **1985**, *26*, 3887.
12. (a) Kraus, G. A.; Walling, J. A. *TL* **1986**, *27*, 1873. (b) Laugraud, S.; Guingant, A.; d'Angelo, J. *TL* **1989**, *30*, 83.
13. (a) Maruoka, K.; Sakurai, M.; Fujiwara, J.; Yamamoto, H. *TL* **1986**, *27*, 4895. (b) Kraus, G. A.; Taschner, M. J. *JOC* **1980**, *45*, 1174. (c) Carretero, J. C.; Cuevas, J. C.; Echavarren, A.; Fariña, F.; Prados, P. *JCR(S)* **1984**, 6.
14. (a) Rigby, J. H.; Ogbu, C. O. *TL* **1990**, *31*, 3385. (b) Rigby, J. H.; Ateeq, H. S.; Charles, N. R.; Cuisiat, S. V.; Ferguson, M. D.; Henshilwood, J. A.; Krueger, A. C.; Ogbu, C. O.; Short, K. M.; Heeg, M. J. *JACS* **1993**, *115*, 1382. (c) Rigby, J. H.; Ateeq, H. S. *JACS* **1990**, *112*, 6442.
15. (a) Cervinka, O.; Svatos, A.; Trska, P.; Pech, P. *CCC* **1990**, *55*, 230. (b) Achmatowicz, O., Jr.; Bialecka-Florjanczyk, E. *T* **1990**, *46*, 5317. (c) Bélanger, J.; Landry, N. L.; Paré, J. R. J.; Jankowski, K. *JOC* **1982**, *47*, 3649.
16. (a) South, M. S.; Liebeskind, L. S. *JOC* **1982**, *47*, 3815. (b) Seitz, G.; v. Gemmern, R. *S* **1987**, 953.
17. Clennan, E. L.; L'Esperance, R. P. *TL* **1983**, *24*, 4291.
18. McClure, K. F.; Danishefsky, S. J. *JOC* **1991**, *56*, 850.
19. Märkl, G.; Kallmünzer, A. *TL* **1989**, *30*, 5245.
20. (a) Wenkert, E.; Goodwin, T. E.; Ranu, B. C. *JOC* **1977**, *42*, 2137. (b) Tomilov, Yu. V.; Kostitsyn, A. B.; Shulishov, E. V.; Nefedov, O. M. *S* **1990**, 246. (c) Brady, W. T.; Lloyd, R. M. *JOC* **1981**, *46*, 1322. (d) Cantrell, W. R., Jr.; Davies, H. M. L. *JOC* **1991**, *56*, 723.
21. (a) Tenaglia, A.; Brun, P.; Waegele, B. *JOM* **1985**, *285*, 343. (b) Brun, P.; Tenaglia, A.; Waegele, B. *TL* **1983**, *24*, 385.
22. (a) Makin, S. M.; Raifel'd, Yu. E.; Zil'berg, L. L.; Arshava, B. M. *JOU* **1984**, *20*, 189. (b) Akgün, E.; Pindur, U. *S* **1984**, 227. (c) Akgün, E.; Pindur, U. *M* **1984**, *115*, 587. (d) Duhamel, L.; Ple, G.; Ramondenc, Y. *TL* **1989**, *30*, 7377. (e) Jung, M. E.; Gardiner, J. M. *JOC* **1991**, *56*, 2614.
23. (a) Paterson, I.; Price, L. G. *TL* **1981**, *22*, 2833. (b) Hatanaka, K.; Tanimoto, S.; Sugimoto, T.; Okano, M. *TL* **1981**, *22*, 3243.
24. (a) Makin, S. M.; Kruglikova, R. I.; Popova, T. P. *JOU* **1982**, *18*, 1001. (b) Makin, S. M.; Shavrygina, O. A.; Kruglikova, R. I.; Mikerin, I. E.; Tung, N. F.; Ermakova, G. A. *JOU* **1983**, *19*, 1994. (c) Makin, S. M.; Dymshakova, G. M.; Granenkina, L. S. *JOU* **1986**, *22*, 256. (d) Ishida, A.; Mukaiyama, T. *BCJ* **1978**, *51*, 2077. (e) Ishida, A.; Mukaiyama, T. *CL* **1977**, 467. (f) Ibragimov, M. A.; Lazareva, M. I.; Smit, W. A. *S* **1985**, 880.
25. (a) Natsume, M.; Muratake, H. *TL* **1979**, 3477. (b) Natsume, M.; Muratake, H. *H* **1980**, *14*, 615. (c) Natsume, M.; Muratake, H. *H* **1981**, *16*, 973.
26. Martin, S. A.; Benage, B.; Geraci, L. S.; Hunter, J. E.; Mortimore, M. *JACS* **1991**, *113*, 6161.
27. De Lucchi, O.; Fabbri, D.; Lucchini, V. *T* **1992**, *48*, 1485.
28. Kraus, G. A.; Sugimoto, H. *TL* **1977**, 3929.
29. Fleming, I.; Goldhill, J.; Paterson, I. *TL* **1979**, 3205.
30. Duhamel, L.; Guillemont, J.; Le Gallic, Y.; Plé, G.; Poirier, J.-M.; Ramondenc, Y.; Chabardes, P. *TL* **1990**, *31*, 3129.
31. Decor, J. P. Ger. Patent 2 708 281, 1977 (*CA* **1977**, *87*, 200 795w).
32. Tolstikov, G. A.; Miftakhov, M. S.; Menakov, Yu. B.; Lomakina, S. I. *ZOB* **1976**, *46*, 2630.
33. (a) Benyunes, S. A.; Day, J. P.; Green, M.; Al-Saadoon, A. W.; Waring, T. L. *AG(E)* **1990**, *29*, 1416. (b) Benyunes, S. A.; Green, M.; Grimshire, M. J. *OM* **1989**, *8*, 2268.

Michael E. Jung
University of California, Los Angeles, CA, USA

2-Trimethylsilyloxy-1,3-butadiene

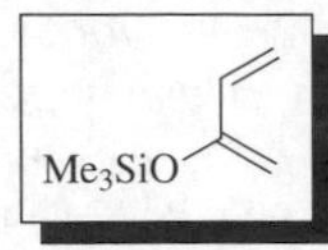

[38053-91-7] $C_7H_{14}OSi$ (MW 142.30)

(easily prepared[1] reactive diene for Diels–Alder reactions[2–40] and other cycloadditions;[41–44] reactive silyl enol ether for aldol, Michael reactions, and electrophilic additions[45–46])

Physical Data: bp 50–55°C/50 mmHg; *d* 0.811 g cm^{-3}.
Solubility: sol most standard organic solvents.
Form Supplied in: liquid available commercially.
Preparative Methods: can be prepared easily from methyl vinyl ketone.[1]
Handling, Storage, and Precautions: is a flammable liquid and is moisture sensitive.

Diels–Alder Reactions. Although commercially available, 2-trimethylsilyloxy-1,3-butadiene (**1**) can be easily prepared by silylation of ***Methyl Vinyl Ketone***.[1] It has been often used as a reactive diene in Diels–Alder reactions. In nearly all cases with a strongly activated dienophile, the trimethylsilyloxy group ends up 1,4 to the activating group in the adduct. Several representative examples are shown in Table 1.[1–14] The initial Diels–Alder adduct is a silyl enol ether and as such can be converted into several different products by simple silyl enol ether chemistry, e.g. the adduct (**3**) from the reaction of (**1**) with ***Dimethyl Fumarate*** (**2**) (eq 1) can be hydrolyzed to the cyclohexanone (**4**) in either acid or base, converted into the α-bromo or α-hydroxy ketone (**5**) or (**6**) on treatment with ***Bromine*** or ***N-Bromosuccinimide*** and ***m-Chloroperbenzoic Acid***, respectively, and finally via a Mukaiyama-type aldol (PhCHO/***Titanium(IV) Chloride***) to the enone (**7**).[1b,2]

TMSO (**1**) + MeO_2C–CH=CH–CO_2Me (**2**) → (PhMe, 110 °C, 18 h, 77%) → (**3**) (1)

(**4**) 95% (**5**) X = Br, 70% (**6**) X = OH, 61% (**7**) 80%

Alkynic dienophiles also work well, e.g. ethyl propiolate (see ***Methyl Propiolate***) and (**1**) produce the expected cyclohexadiene ester in 77% yield.[15] In addition to simple dienophiles, a number of allenic dienophiles have also been utilized, as shown in Table 2.[16–19] Cyclic dienophiles, enones, lactams, etc., often react with (**1**) to give fused bicyclic systems, but generally the yields are only in the 25–45% range,[20–24] e.g. reaction of (**1**) with (**13**) gives (**14**) in 27% yield (eq 2).[20b]

(**1**) + (**13**) → (1. 0.5 equiv. $SnCl_4$, Et_2O or $SnCl_2$, Et_2O, 0 °C; 2. aq. workup; 27%) → (**14**) (2)

More reactive dienophiles, e.g. β-nitroenones,[21] enediones,[22] 1,2-disulfonylethylene,[23] etc., give higher yields, as do Lewis acids,[24] e.g. ***Zinc Bromide*** (eq 3),[24a] and the use of high pressure.[25] Stereofacial differentiation of (**1**) with cyclic enones,[26] lactones,[27] and lactams[28] is generally excellent, e.g. (**1**) reacts with (**17a,b**) to give (**18a,b**) with high selectivity in good yield (eq 4).[26] Reaction of (**1**) with quinones has often been used[29] in syntheses of anthraquinones and their derivatives, e.g. to prepare the A ring of the anthracyclines.[29b–g] Double Michael reactions[30] have been used quite often instead of Diels–Alder reactions to provide the same products, often in higher yields. The preparation of the azabicyclic ketone (**20**) is an interesting example of a tandem Diels–Alder reaction–aldol-type condensation (eq 5).[31] An example of an aldol reaction followed by a Diels–Alder is also known.[32]

(**1**) + (**15**) → (1. $ZnBr_2$, CH_2Cl_2, 0 °C, 1 h; 2. aq. workup; 92%) → (**16**) (3)

(**1**) + (**17**) → (for **a**: $SnCl_4$, CH_2Cl_2, –78 °C, 1 h; for **b**: $EtAlCl_2$, toluene, 25 °C, 2–4 h) → (**18**) (4)

(**17a**) R = TMS, R^1 = H
(**17b**) R = C(Me)=CH_2, R^1 = Me

(**18a**) 66% 100:0
(**18b**) 73% 95:5

Table 1 Diels–Alder Reactions of 2-Trimethylsilyl-1,3-butadiene (**1**)

(**1**) + (**8**) → (**9**) + (**10**)

Z	R^1	R^2	R^3	Conditions	Yield (%)	Ratio	Ref.
CN	CN	CN	CN	PhH, 25 °C	90	100:0	1f
CO_2Me	H	H	H	toluene, 110 °C, 18 h	35	100:0	1b, 2
COMe	H	H	H	toluene, 110 °C, 18 h	60	100:0	1b, 2
CO_2Et	H	CO_2Et	H	toluene, 110 °C, 18 h	77	100:0	1b, 2
CO_2Et	H	H	CO_2Et	toluene, 110 °C, 18 h	39	100:0	1b, 2
CO_2Et	H	H	CO_2Et	toluene, 140 °C, 7 h	95	100:0	1c
CHO	H	H	H	toluene, 110 °C, 24 h	81[b]	100:0	3
CO_2Me	Me	H	H	xylene, 140 °C, 48 h	68	100:0	4
CHO	Me	H	H	toluene, 110 °C, 45 h	54	100:0	5
NO_2	H	CO_2Me	H	PhH, 25 °C, 42 h	100 (72)[a]	100:0	6
NO_2	H	OCOPh	H	–	95	100:0	7
CO_2Me	H	CO_2-*t*-Bu	H	CH_2Cl_2, –20 °C, MAD, 50 h	48	99:1	8
SO_2Ph	SO_2Ph	H	H	CH_2Cl_2, 20 °C, 24 h	100 (79)[a]	100:0	9
SO_2Ph	SO_2Ph	Me	H	$LiClO_4$, Et_2O, 50 °C, 24 h	95	100:0	10
SO_2Ph	SO_2Ph	Ph	H	$LiClO_4$, Et_2O, 50 °C, 24 h	90	100:0	10
SO_2Ph	H	CF_3	H	toluene, 110 °C, 4 h	89 (68)[a]	66:23	11
CF_3	H	H	H	neat, 150 °C, 72 h	17	76:24	12
$CO_2CMe_2O_2C$		$(CH_2)_2CX_2(CH_2)_2$		PhH, 80 °C, 22 h	97	100:0	13
COCR=CRO		H	H	toluene, 110 °C, 12 h	(53)[a]	100:0	14

[a] Yield of cyclohexanone after hydrolysis. [b] Product is 93% pure.

(**1**) + (**19**) → (**20**) (4 equiv $TiCl_4$, CH_2Cl_2, 70 to 25 °C, 82%) (5)

Table 2 Allenic Dienophiles in Diels–Alder Reactions with (**1**)

(**1**) + (**11**) → (**12**)

Z	R	Conditions	Yield (%)	Ref.
CO_2TMS	CO_2TMS	PhH, 82 °C, 20 h	80	16
CO_2Et	CO_2Et	PhH, 82 °C, 14 h	97	17
SO_2Ph	H	160 °C, 3 h	79	18
CO_2TMS	$CH_2OTBDMS$	PhMe, 110 °C, 24 h	62	19

Finally, there is one example of (**1**) acting as the dienophile, namely in the Diels–Alder reaction with α-nitrosostyrene, where 6-vinyl-6-silyloxy-4,5-dihydrooxazine is formed in 65% yield.[33]

Heterodienophiles. Many heterodienophiles react with (**1**).[34–37] Mesoxalate, glyoxalate, and hexafluoroacetone all give the 4-pyranone silyl ether derivatives in fair yield.[34] The corresponding thiomesoxalate affords the 3-thiopyranone-4,4-diester in good yield,[35a,b] as does an α-oxosulfine,[35c] while alkyl thioformates yield the 4-thiopyranones.[35d] By far the largest group of heterodienophiles used are imine derivatives.[36–37] Activated imines, e.g. diacyl imines,[36] afford the 4-piperidones, while aryl-fused dihydropyridines, e.g. (**21**), furnish fused 4-piperidones, e.g. (**22**), in good yield from reactions with (**1**) in the presence of a Lewis acid (eq 6).[37]

(**21**) → (**22**) ((**1**), $BF_3•OEt_2$, 91%) (6)

Finally, several imines give products of an aldol-type condensation rather than Diels–Alder adducts in reactions with (**1**) in the presence of Lewis acids.[38] Phosphaalkenes react with (**1**) to give ultimately the aromatic phosphinine[39] and dimethyl azodicarboxylate affords the expected [4+2] adduct in 83% yield.[40]

[2+2] and Other Cycloadditions, Carbene Additions. Several [2+2] cyclizations of (**1**) with various alkenes are known.[41] Quite often these vinyl cyclobutyl silyl ethers can be thermally rearranged to the normal Diels–Alder product,[16,19,41] e.g. (**23**) giving (**25**) via (**24**) (eq 7).[41a] These [2+2] adducts can also be transformed via palladium catalysis into α-methylenecyclopentanones[41d,e] in good yield, e.g. (**26**) gives (**28**) via (**27**) (eq 8).[41d] Metal complexes of ***1,3,5-Cycloheptatriene*** react with (**1**) to give a novel [6+4] cycloaddition, e.g. (**29**) giving (**30**) (eq 9).[42]

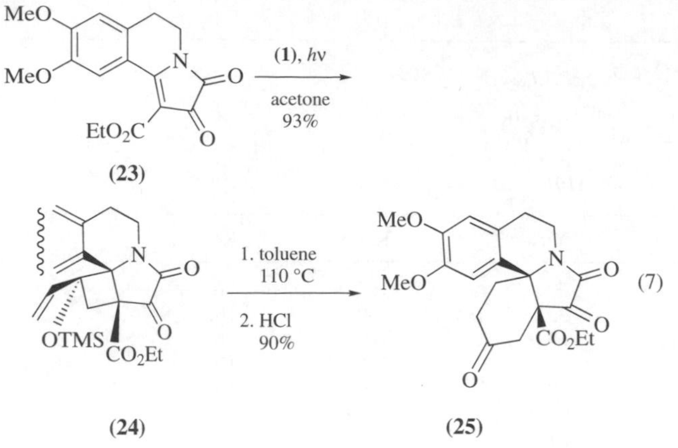

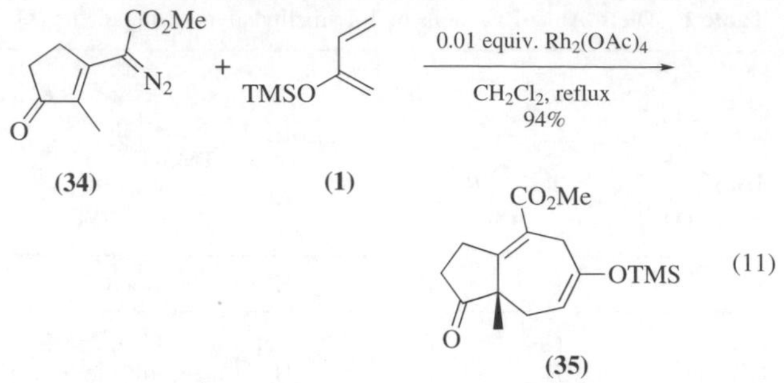

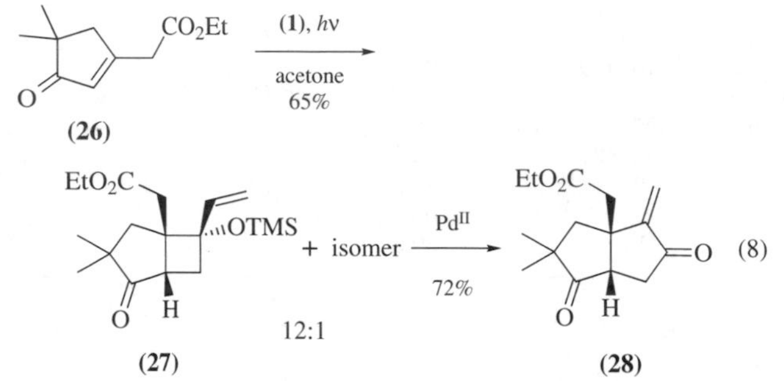

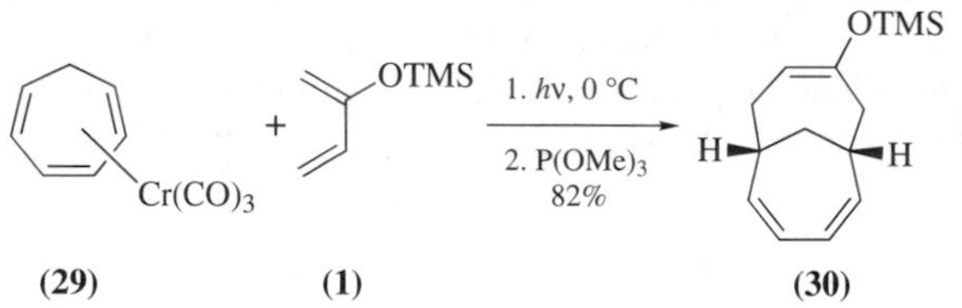

(7)

(8)

(9)

Even [3 + 2] cycloadditions occur when cyclopropyl ketones are treated with (**1**) under photolytic or Lewis acid conditions to produce cyclopentane systems.[43] Cyclopropanations of (**1**) are well known, in all cases adding to the more electron-rich silyloxy alkene.[1a,44] The vinylcyclopropanol silyl ethers can be converted into a number of different products, e.g. cyclopentanones, 2-methylcyclobutanones, vinyl ketones, and β-alkoxy ketones (eq 10). One clever use of (**1**) in synthesis is the tandem [2 + 2]–Cope process which converts (**34**) into (**35**) (eq 11).[44i]

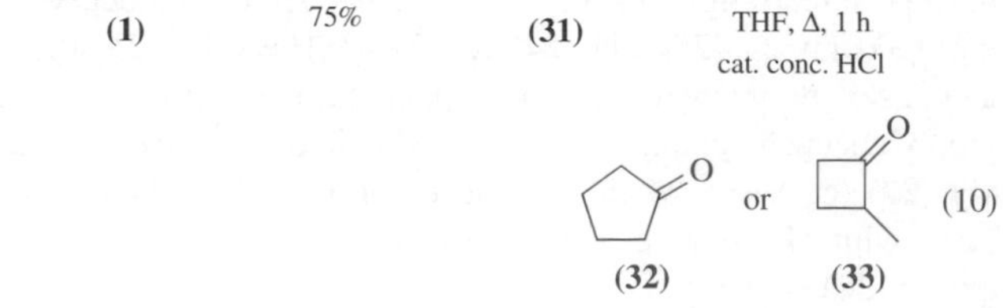

(10)

(11)

Aldol Condensations and Electrophilic Additions. 2-Trimethylsilyloxybutadiene (**1**) can act as the enolate of methyl vinyl ketone (MVK) towards reactive carbon atoms in an aldol-like process to give the products of overall addition of MVK to reactive centers.[45] Various strong electrophiles have been added to (**1**), e.g. bromine, NBS, *m*-CPBA, alkylating agents, and acylating agents, usually reacting at the carbon of the silyl enol ether to give α-bromo, α-silyloxy, or α-alkyl ketones in good yields.[46] With acylating agents, however, the *O*-acylated products are obtained.[46e]

Related Reagents. 1-Acetoxy-1,3-butadiene; 1,3-Butadiene; 2-Methoxy-1,3-butadiene; 2-Methoxy-1-phenylthio-1,3-butadiene; 2-Methoxy-3-phenylthio-1,3-butadiene; 1-Methoxy-3-trimethylsilyloxy-1,3-butadiene; 3-Methoxyisoprene; Methyl Vinyl Ketone; 1-Trimethylsilyloxy-1,3-butadiene.

---

1. (a) Girard, C.; Amice, P.; Barnier, J. P.; Conia, J. M. *TL* **1974**, 3329. (b) Jung, M. E.; McCombs, C. A. *TL* **1976**, 2935. (c) Jung, M. E.; McCombs, C. A. *OSC* **1988**, *6*, 445. (d) Cazeau, P.; Moulines, F.; LaPorte, O.; Duboudin, F. *JOM* **1980**, *201*, C9. (e) Cazeau, P.; Duboudin, F.; Moulines, F.; Babot, O.; Dunoguès, J. *T* **1987**, *43*, 2089. (f) Cazeau, P.; Frainnet, E. *BSF(2)* **1972**, 1658.
2. Jung, M. E.; McCombs, C. A.; Takeda, Y.; Pan, Y.-G. *JACS* **1981**, *103*, 6677.
3. Yin, T.-K.; Lee, J. G.; Borden, W. T. *JOC* **1985**, *50*, 531.
4. Baldwin, J. E.; Broline, B. M. *JOC* **1982**, *47*, 1385.
5. Rigby, J. H.; Kotnis, A. S. *TL* **1987**, *28*, 4943.
6. Danishefsky, S. J.; Prisbylla, M. P.; Hiner, S. *JACS* **1978**, *100*, 2918.
7. Kraus, G. A.; Thurston, J.; Thomas, P. J.; Jacobson, R. A.; Su, Y. *TL* **1988**, *29*, 1879.
8. Maruoka, K.; Saito, S.; Yamamoto, H. *JACS* **1992**, *114*, 1089.
9. Rao, Y. K.; Nagarajan, M. *S* **1984**, 757.
10. De Lucchi, O.; Fabbri, D.; Lucchini, V. *T* **1992**, *48*, 1485.
11. Taguchi, T.; Hosoda, A.; Tomizawa, G.; Kawara, A.; Masuo, T.; Suda, Y.; Nakajima, M.; Kobayashi, Y. *CPB* **1987**, *35*, 909.
12. Ojima, I.; Yatabe, M.; Fuchikami, T. *JOC* **1982**, *47*, 2051.
13. Bell, V. L.; Holmes, A. B.; Hsu, S.-Y.; Mock, G. A.; Raphael, R. A. *JCS(P1)* **1986**, 1502.
14. Ko, B.-S.; Oritani, T.; Yamashita, K. *ABC* **1990**, *54*, 2199.
15. Ackland, D. A.; Pinhey, J. T. *JSC(P1)* **1987**, 2689.
16. Jung, M. E.; Node, M.; Pfluger, R.; Lyster, M. A.; Lowe, J. A., III *JOC* **1982**, *47*, 1150.
17. Kozikowski, A. P.; Schmiesing, R. *SC* **1978**, *8*, 363.
18. Hayakawa, K.; Nishiyama, H.; Kanematsu, K. *JOC* **1985**, *50*, 512.
19. Jung, M. E.; Lowe, J. A., III; Lyster, M. A.; Node, M.; Pfluger, R.; Brown, R. W. *T* **1984**, *40*, 4751.

20. (a) Liu, H.-J.; Ngooi, T. K. *SC* **1982**, *12*, 715. (b) Liu, H.-J.; Ngooi, T. K.; Browne, E. N. C. *CJC* **1988**, *66*, 3143. (c) Kraus, G. A.; Gottschalk, P. *JOC* **1984**, *49*, 1153. (d) Ghosh, S.; Saha, S. *TL* **1985**, *26*, 5325. (e) Ghosh, S.; Saha Roy, S.; Saha, G. *T* **1988**, *44*, 6235. (f) Sano, T.; Toda, J.; Kashiwaba, N.; Ohshima, T.; Tsuda, Y. *CPB* **1987**, *35*, 479. (g) Rigby, J. H.; Ogbu, C. O. *TL* **1990**, *31*, 3385. (h) Seitz, G.; v. Gemmern, R. *S* **1987**, 953. (i) Wulff, W. D.; Yang, D. C. *JACS* **1984**, *106*, 7565.
21. Corey, E. J.; Estreicher, H. *TL* **1981**, *22*, 603.
22. Danishefsky, S.; Kahn, M. *TL* **1981**, *22*, 489.
23. De Lucchi, O.; Fabbri, D.; Cossu, S.; Valle, G. *JOC* **1991**, *56*, 1888.
24. (a) de Oliveira Imbroisi, D.; Simpkins, N. S. *JCS(P1)* **1991**, 1815. (b) de Oliveira Imbroisi, D.; Simpkins, N. S. *TL* **1989**, *30*, 4309. (c) Fujiwara, T.; Ohsaka, T.; Inoue, T.; Takeda, T. *TL* **1988**, *29*, 6283.
25. (a) Laugraud, S.; Guingant, A.; d'Angelo, J. *TL* **1989**, *30*, 83. (b) Branchadell, V.; Sodupe, M.; Ortuño, R. M.; Oliva, A.; Gomez-Pardo, D.; Guingant, A.; d'Angelo, J. *JOC* **1991**, *56*, 4135.
26. (a) Asaoka, M.; Nishimura, K.; Takei, H. *BCJ* **1990**, *63*, 407. (b) Haaksma, A. A.; Jansen, B. J. M.; de Groot, A. *T* **1992**, *48*, 3121.
27. de Jong, J. C.; van Bolhuis, F.; Feringa, B. L. *TA* **1991**, *2*, 1247.
28. (a) Koot, W.-J.; Hiemstra, H.; Speckamp, W. N. *JOC* **1992**, *57*, 1059. (b) Baldwin, S. W.; Greenspan, P.; Alaimo, C.; McPhail, A. T. *TL* **1991**, *32*, 5877.
29. (a) Boisvert, L.; Brassard, P. *JOC* **1988**, *53*, 4052. (b) Tamura, Y.; Wada, A.; Sasho, M.; Fukunaga, K.; Maeda, H.; Kita, Y. *JOC* **1982**, *47*, 4376. (c) Tamura, Y.; Sasho, M.; Akai, S.; Wada, A.; Kita, Y. *T* **1984**, *40*, 4539. (d) Carretero, J. C.; Cuevas, J. C.; Echavarren, A.; Fariña, F.; Prados, P. *JCR(S)* **1984**, 6. (e) Cameron, D. W.; Feutrill, G. I.; Griffiths, P. G.; Merrett, B. K. *TL* **1986**, *27*, 2421. (f) Preston, P. N.; Winwick, T.; Morley, J. O. *JCS(P1)* **1985**, 39. (g) Laugraud, S.; Guingant, A.; d'Angelo, J. *TL* **1992**, *33*, 1289.
30. (a) Mukaiyama, T.; Sagawa, Y.; Kobayashi, S. *CL* **1986**, 1821. (b) Inokuchi, T.; Kurokawa, Y.; Kusumoto, M.; Tanigawa, S.; Takagishi, S.; Torii, S. *BCJ* **1989**, *62*, 3739. (c) see also ref 20c.
31. Yang, T.-K.; Hung, S.-M.; Lee, D.-S.; Hong, A.-W.; Cheng, C.-C. *TL* **1989**, *30*, 4973.
32. Simpkins, N. S. *T* **1991**, *47*, 323.
33. Hippeli, C.; Reissig, H.-U. *S* **1987**, 77.
34. (a) Bélanger, J.; Landry, N. L.; Paré, J. R. J.; Jankowski, K. *JOC* **1982**, *47*, 3649. (b) Kirmse, W.; Mrotzeck, U. *CB* **1988**, *121*, 485. (c) Ishihara, T.; Shinjo, H.; Inoue, Y.; Ando, T. *JFC* **1983**, *22*, 1.
35. (a) Larsen, S. D. *JACS* **1988**, *110*, 5932. (b) Kirby, G. W.; McGregor, W. M. *JCS(P1)* **1990**, 3175. (c) Rewunkel, J. G. M.; Zwanenberg, B. *RTC* **1990**, *101*, 190. (d) Herczegh, P.; Zsély, M.; Bognár, R. *TL* **1986**, *27*, 1509.
36. (a) Jung, M. E.; Shishido, K.; Light, L.; Davis, L. *TL* **1981**, *22*, 4607. (b) Hamley, P.; Holmes, A. B.; Kee, A.; Ladduwahetty, T.; Smith, D. F. *SL* **1991**, 29.
37. (a) Vacca, J. P. *TL* **1985**, *26*, 1277. (b) Ryan, K. M.; Reamer, R. A.; Volante, R. P.; Shinkai, I. *TL* **1987**, *28*, 2103. (c) Huff, J. R.; Baldwin, J. J.; deSolms, S. J.; Guare, J. P.; Hunt, C. A.; Randall, W. C.; Sanders, W. S.; Smith, S. J.; Vacca, J. P.; Zrada, M. M. *JMC* **1988**, *31*, 641.
38. (a) Ueda, Y.; Maynard, S. C. *TL* **1985**, *26*, 6309. (b) Schrader, T.; Steglich, W. *S* **1990**, 1153. (c) Hartmann, P.; Obrecht, J.-P. *SC* **1988**, *18*, 553.
39. (a) Märkl, G.; Kallmünzer, A. *TL* **1989**, *30*, 5245. (b) Markovskiii, L. N.; Romanenko, V. D.; Kachkovskaya, L. S. *JGU* **1985**, *55*, 2488.
40. Vartanyan, R. S.; Gyul'budagyan, A.; Khanamiryan, A. Kh.; Mkrtumyan, E. N.; Kazaryan, Zh. V. *KGS* **1991**, 783.
41. (a) Sano, T.; Toda, J.; Ohshima, T.; Tsuda, Y. *CPB* **1992**, *40*, 873. (b) Sano, T.; Toda, J.; Horiguchi, Y.; Imafuku, K.; Tsuda, Y. *H* **1981**, *16*, 1463. (c) Sano, T.; Toda, J.; Tsuda, Y.; Yamaguchi, K.; Sakai, S.-I. *CPB* **1984**, *32*, 3255. (d) Demuth, M.; Pandey, B.; Wietfeld, B.; Said, H.; Viader, J. *HCA* **1988**, *71*, 1392. (e) de Almeida-Barbosa, L.-C.; Mann, J. *JCS(P1)* **1992**, 337. (f) Dolbier, W. R., Jr.; Piedrahita, C.; Houk, K. N.; Strozier, R. W.; Gandour, R. W. *TL* **1978**, 2231. (g) Hassner, A.; Naisdorf, S. *TL* **1986**, *27*, 6389.
42. (a) Rigby, J. H.; Ateeq, H. S. *JACS* **1990**, *112*, 6442. (b) Rigby, J. H.; Ateeq, H. S.; Charles, N. R.; Cuisiat, S. V.; Ferguson, M. D.; Henshilwood, J. A.; Krueger, A. C.; Ogbu, C. O.; Short, K. M.; Heeg, M. J. *JACS* **1993**, *115*, 1382.
43. (a) Demuth, M.; Wietfeld, B.; Pandey, B.; Schaffner, K. *AG(E)* **1985**, *24*, 763. (b) Komatsu, M.; Suehiro, I.; Horiguchi, Y.; Kuwajima, I. *SL* **1991**, 771.
44. (a) Conia, J. M. *JCR(S)* **1978**, 182. (b) Girard, C.; Conia, J. M. *TL* **1974**, 3333. (c) Olofson, R. A.; Lotts, K. D.; Barber, G. N. *TL* **1976**, 3779. (d) Kunkel, E.; Reichelt, I.; Reissig, H.-U. *LA* **1984**, 512. (e) Zschiesche, R.; Reissig, H.-U. *LA* **1987**, 387. (f) Kuehne, M.; Pitner, J. B. *JOC* **1989**, *54*, 4553. (g) Shi, G.; Xu, Y. *JOC* **1990**, *55*, 3383. (h) de Meijere, A.; Schulz, T.-J.; Kostikov, R. R.; Graupner, F.; Murr, T.; Beifeldt, T. *S* **1991**, 547. (i) Cantrell, W. R., Jr.; Davies, H. M. L. *JOC* **1991**, *56*, 723.
45. (a) Natsume, M.; Muratake, H. *TL* **1979**, 3477. (b) Conde-Frieboes, K.; Hoppe, D. *SL* **1990**, 99.
46. (a) Lüönd, R. M.; Cuomo, J.; Neier, R. W. *JOC* **1992**, *57*, 5005. (b) Herman, T.; Carlson, R. *TL* **1989**, *30*, 3657. (c) Pennanen, S. I. *SC* **1985**, *15*, 865, 1063. (d) Takeda, K.; Ayabe, A.; Kawashima, H.; Harigaya, Y. *TL* **1992**, *33*, 951. (e) Olofson, R. A.; Cuomo, J. *TL* **1980**, *21*, 819.

Michael E. Jung
University of California, Los Angeles, CA, USA

1,1,2-Triphenyl-1,2-ethanediol[1,2]

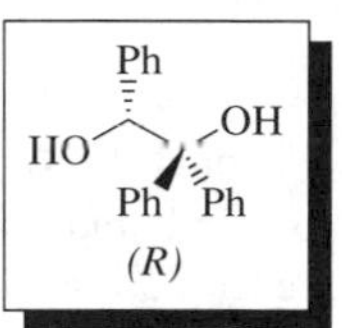

[95061-46-4] $C_{20}H_{18}O_2$ (MW 290.38)

(derived chiral monoesters undergo stereoselective aldol additions; formation of *O*-silyl orthoesters and cyclic phosphonates)

Physical Data: mp 126 °C. (*R*): $[\alpha]_D^{25}$ +214° (*c* = 1RM, ethanol), +220° (*c* = 1, 95% ethanol); (*S*): $[\alpha]_D^{25}$ −217° (*c* = 1, ethanol).
Solubility: sol dichloromethane, chloroform, THF, ethanol; insol hexane.
Form Supplied in: white solid; the (*R*)-form is commercially available.
Preparative Methods: (*R*)-1,1,2-triphenylethane-1,2-diol [(*R*)-(**1**)] is easily available from commercial (*R*)-***Mandelic Acid***, which is first esterified to give methyl mandelate and then treated with ***Phenylmagnesium Bromide*** (3.5 equiv). In an analogous way, (*S*)-(**1**) is accessible from (*S*)-mandelic acid, which is also commercially available (eq 1).[2]

HO–CH(Ph)–CO_2H —(MeOH, H_2SO_4, 92%)→ HO–CH(Ph)–CO_2Me —(PhMgBr, Et_2O, 66%)→ HO–CH(Ph)–C(OH)Ph Ph (1)

anti*-Selective and Diastereofacially Selective Aldol Additions.** 2-Trimethylsilyloxy-1,2,2-triphenylethyl propionate, which is prepared from (*R*)-(**1**) by esterification with propionyl chloride and subsequent silylation of the tertiary hydroxy group, reacts in a highly stereoselective manner upon deprotonation, transmetalation with ***Dichlorobis(cyclopentadienyl)zirconium, and addition to 2-methylpropanal. The diastereoselectivity is 96:4, which is the ratio of the major product to the sum of all other diastereomers. Subsequent reduction with ***Lithium Aluminium Hydride*** affords (2*S*,3*R*)-2,4-dimethyl-1,3-pentanediol in 95% ee (eq 2).[3] *anti*-Selective aldol additions which deliver chiral nonracemic products have been a longstanding problem of asymmetric synthesis.[4] Doubly deprotonated 2-hydroxy-1,2,2-triphenylethyl propionate has been applied in a total synthesis of dolastatin.[5]

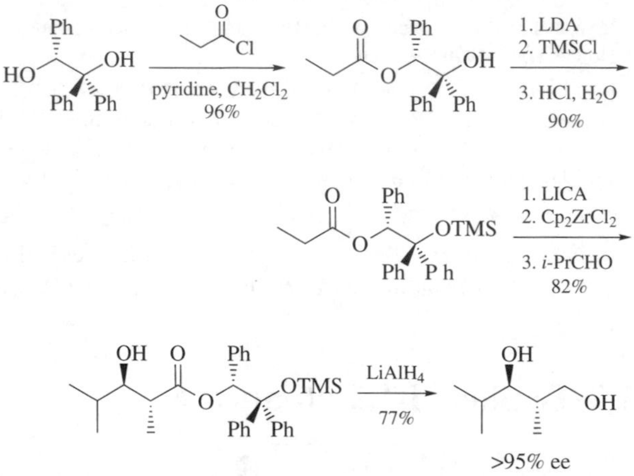

pyridine, CH_2Cl_2 96%; 1. LDA 2. TMSCl 3. HCl, H_2O 90%; 1. LICA 2. Cp_2ZrCl_2 3. *i*-PrCHO 82%; $LiAlH_4$ 77%; >95% ee (2)

When 1,1,2-triphenylethane-1,2-diol-derived esters are submitted to a monodeprotonation and subsequently treated with ***Chlorotrimethylsilane***, the formation of 2-trimethylsilyloxy-1,3-dioxolanes results. The orthoester moiety thus obtained serves as a protecting group for carboxylic acids (eq 3); it is stable towards alkyllithium reagents and can be cleaved under nonacidic conditions by alkaline hydrolysis.[6]

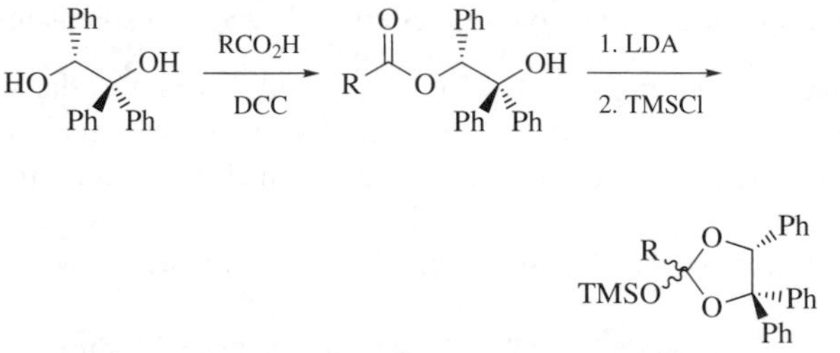

RCO_2H, DCC; 1. LDA 2. TMSCl (3)

Methanephosphonyl dichloride reacts with (*R*)-(**1**) to give 2-methyl-4,4,5-triphenyl-2-oxo-1,3,2-dioxaphospholane (eq 4); the (R_P,R_C) diastereomer forms predominantly (9:1).[7]

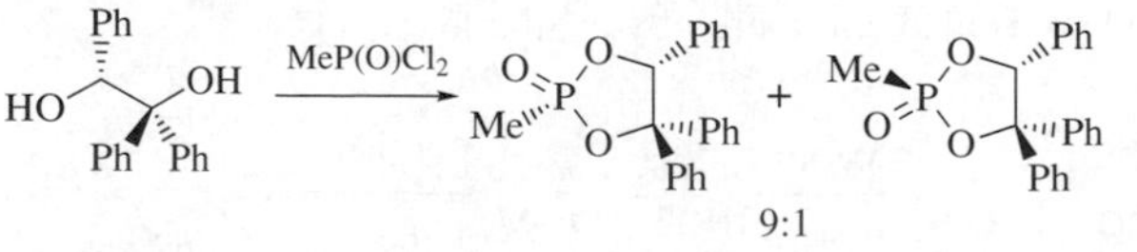

$MeP(O)Cl_2$; 9:1 (4)

A series of enantiomerically pure 1,1-diaryl-2-phenylethane-1,2-diols is available from methyl mandelate by addition of the corresponding substituted arylmagnesium bromides or aryllithium reagents.[2b,8]

Related Reagents. *N*,*N*′-Bis[3,5-bis(trifluoromethyl)benzenesulfonamide]; *trans*-2,5-Bis(methoxymethyl)pyrrolidine; (*S*)-4-Benzyl-2-oxazolidinone; (*R*)-2-*t*-Butyl-6-methyl-4*H*-1,3-dioxin-4-one; (*R*)-(+)-*t*-Butyl 2-(*p*-Tolylsulfinyl)propionate; 10,2-Camphorsultam; Chloro(cyclopentadienyl)bis[3-*O*-(1,2:5,6-di-*O*-isopropylidene-α-D-glucofuranosyl)]titanium; 10-Dicyclohexylsulfonamidoisoborneol; Diisopinocampheylboron Trifluoromethanesulfonate; (*R*,*R*)-2,5-Dimethylborolane; (*R*,*R*)-1,2-Diphenyl-1,2-diaminoethane; 2-Hydroxy-1,2,2-triphenylethyl Acetate; α-Methyltoluene-2,α-sultam; 3-Propionylthiazolidine-2-thione.

1. (a) McKenzie, A.; Wren, H. *JCS* **1910**, *97*, 473. (b) Roger, R.; McKay, W. B. *JCS* **1931**, 2229.
2. (a) Braun, M.; Devant, R. *TL* **1984**, *25*, 5031. (b) Devant, R.; Mahler, U.; Braun, M. *CB* **1988**, *121*, 397. (c) Braun, M.; Gräf, S.; Herzog, S. *OS* **1993**, *72*, 32.
3. (a) Braun, M.; Sacha, H. *AG* **1991**, *103*, 1369; *AG(E)* **1991**, *30*, 1318. (b) Sacha, H.; Waldmüller, D.; Braun, M. *CB* **1994**, *127*, 1959.
4. For reviews, see: (a) Braun, M. In *Advances in Carbanion Chemistry*; Snieckus, V., Ed.; JAI: Greenwich, CT, 1992; Vol. 1, pp 177–247; (b) Braun, M.; Sacha, H. *JPR* **1993**, 653.
5. (a) Pettit, G. R.; Singh, S. B. U.S. Patent 4 978 744, 1990, (*CA* **1991**, *114*, 164 824v). (b) Pettit, G. R.; Singh, S. B.; Herald, D. L.; Lloyd-Williams, P.; Kantoci, D.; Burkett, D. D.; Barkóczy, J.; Hogan, F.; Wardlaw, T. R. *JOC* **1994**, *59*, 6287.
6. Waldmüller, D.; Braun, M.; Steigel, A. *SL* **1991**, 160.
7. Brodesser, B.; Braun, M. *PS* **1989**, *44*, 217.
8. Prasad, K.; Chen, K.-M.; Repic, O.; Hardtmann, G. E. *TA* **1990**, *1*, 703.

Manfred Braun
Heinrich-Heine-Universität, Düsseldorf, Germany

Triphenylphosphine–Carbon Tetrabromide[1]

$Ph_3P–CBr_4$

(Ph_3P)		
[603-35-0]	$C_{18}H_{15}P$	(MW 262.30)
(CBr_4)		
[558-13-4]	CBr_4	(MW 331.61)

(reagent combination for the conversion of alcohols to bromides, aldehydes and ketones to dibromoalkenes, and terminal alkynes to 1-bromoalkynes; carboxyl activation)

Physical Data: Ph_3P: mp 79–81 °C; bp 377 °C; *d* 1.0749 g cm^{-3}. CBr_4: mp 90–91 °C; bp 190 °C; *d* 3.273 g cm^{-3}.
Solubility: sol MeCN, CH_2Cl_2, pyridine, DMF.
Preparative Method: the reactive species is generated in situ by reaction of Ph_3P with CBr_4.
Handling, Storage, and Precautions: Ph_3P is an irritant; CBr_4 is toxic and a cancer suspect agent. All solvents used must be carefully dried because the intermediates are all susceptible to hydrolysis. This reagent should be used in a fume hood.

Conversion of Alcohols to Alkyl Bromides. The reaction of alcohols with ***Triphenylphosphine*** and ***Carbon Tetrabromide*** results in the formation of alkyl bromides. The conditions are sufficiently mild to allow for the efficient conversion of alcohols into the corresponding bromides. The uridine derivative (eq 1) is transformed into its bromide with Ph_3P and CBr_4.[2]

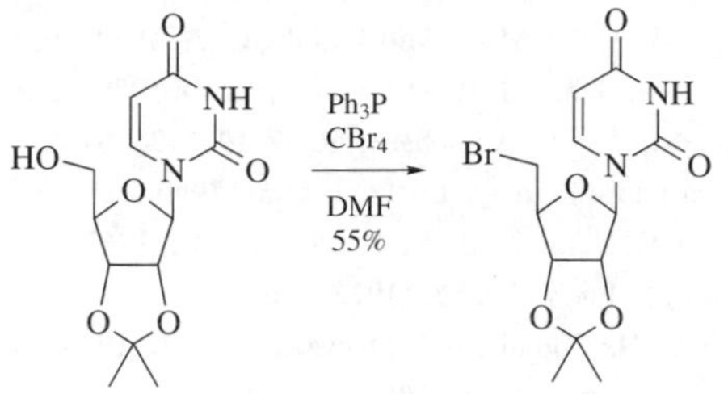

(1)

The geometry about the double bond of an allylic alcohol is usually not compromised (eq 2).[3] Allylic rearrangement is also not commonly observed. Although ketones are known to react with this reagent combination, the reaction of an allylic alcohol has been selectively achieved in the presence of a ketone (eq 3).[4]

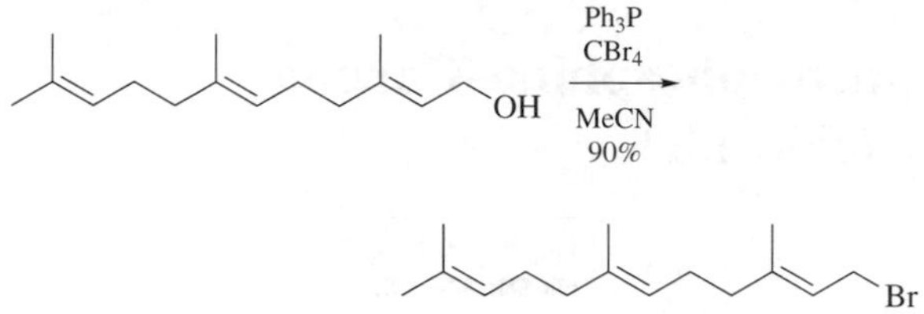

(2)

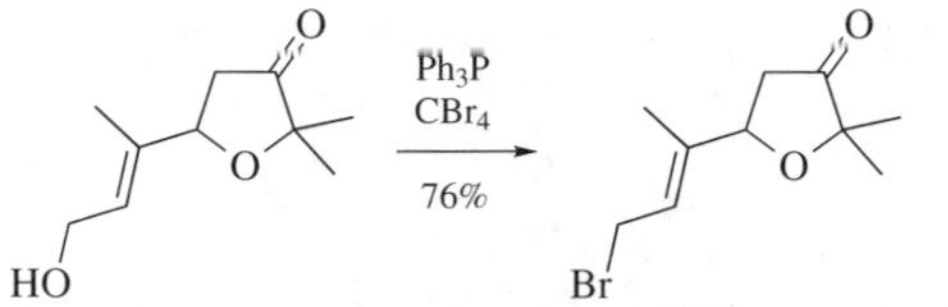

(3)

Regioselective bromination of primary alcohols in the presence of secondary alcohols is possible. These reactions are usually performed in pyridine as solvent. The reaction of the methyl glucopyranoside (eq 4) results in the selective formation of the primary bromide in 98% yield.[5] An investigation of this reaction with a chiral deuterated neopentyl alcohol yielded a partially racemized bromide.[6]

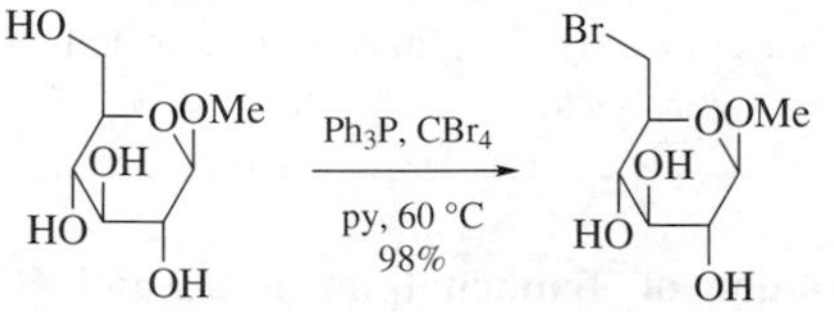

(4)

Silyl ether protected alcohols (eq 5) have been converted directly into the corresponding bromides with Ph_3P and CBr_4. The reaction works best if 1.5 equiv of acetone are added.[7] Tetrahydropyranyl ether protected alcohols have also been directly transformed into the bromides using this reagent combination. The reaction has been reported to proceed with inversion of configuration (eq 6).[8] If unsaturation is appropriately placed within a tetrahydropyranyl (eq 7) or a methoxymethyl (eq 8) protected alcohol, cyclization occurs to afford tetrahydropyrans.[9] The conversion of an alcohol to the bromide without complications with a methoxymethyl protected alcohol in the molecule is possible (eq 9).[10]

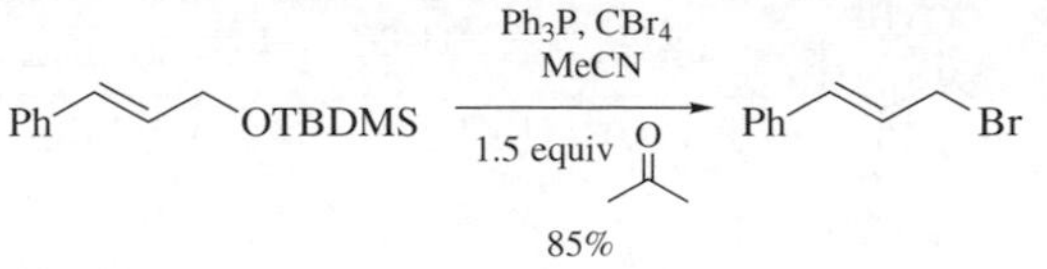

(5)

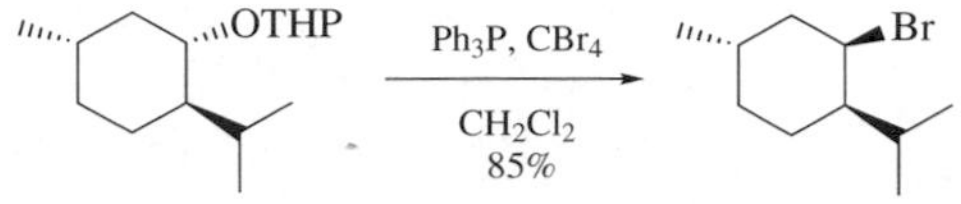

(6)

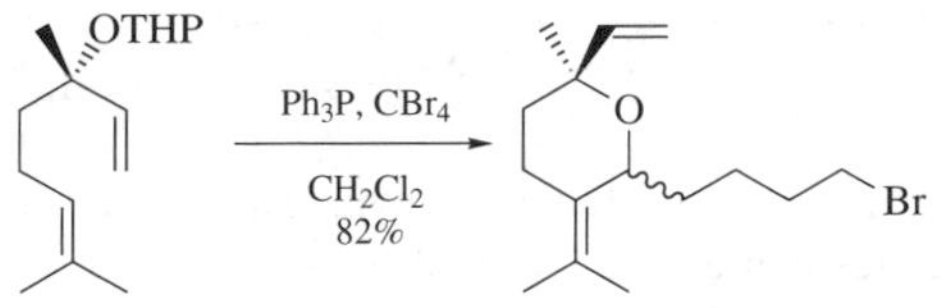

(7)

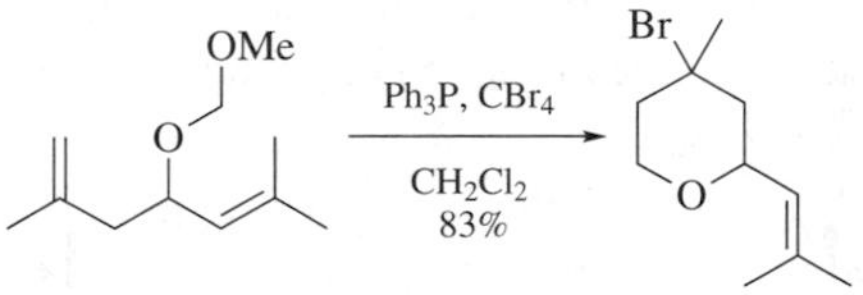

(8)

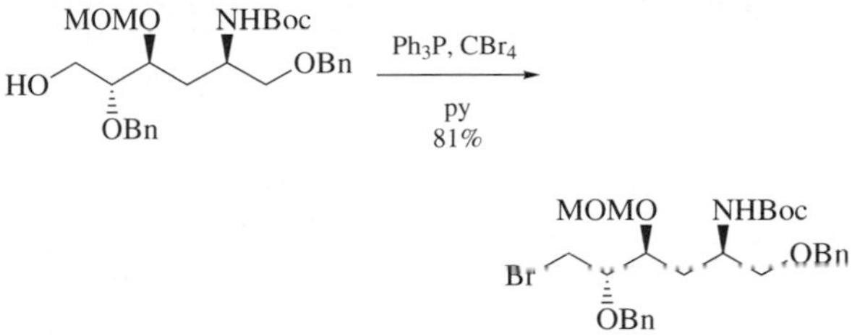

(9)

Amides from Carboxylic Acids. *N*-Methoxy-*N*-methyl amides can be prepared from carboxylic acids and the amine hydrochlorides. In the case of an α-phenyl carboxylic acid (eq 10), the amide is formed in 71% yield and no racemization is detected.[11]

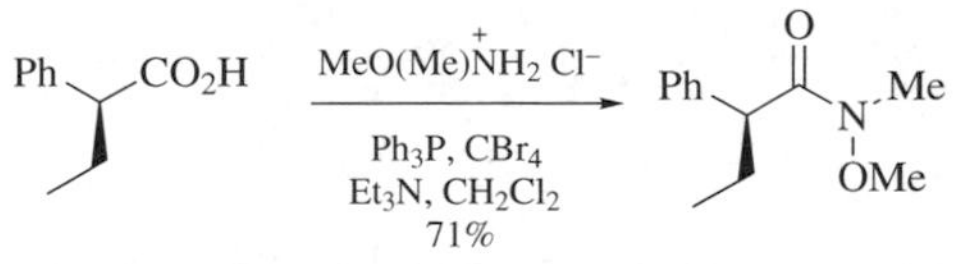

(10)

Dibromoalkenes from Aldehydes and Ketones. Benzaldehyde is transformed into the dibromoalkene in 84% yield when treated with Ph_3P and CBr_4 (eq 11).[12] An alternative procedure for the conversion of an aldehyde to the dibromoalkene uses ***Zinc*** dust in place of an excess of the phosphine. This allows the amount of Ph_3P and CBr_4 to be reduced to 2 equiv each as opposed to 4 equiv. This procedure gives comparable results to the original procedure. The dibromoalkenes can be reacted with ***n-Butyllithium*** to form the intermediate lithium acetylide. The acetylides can then be reacted with electrophiles such as H_2O (eq 12) and CO_2 (eq 13). This offers a convenient method for the formyl to ethynyl conversion.[13]

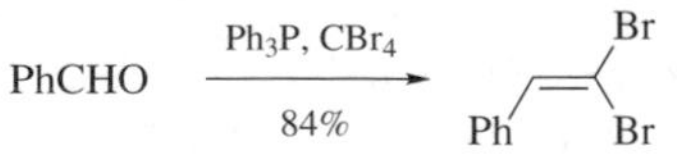

(11)

OTHP, C_5H_{11}, H, O — Ph_3P, CBr_4 / Zn (dust) → THPO, Br, C_5H_{11}, Br — 1. BuLi / 2. H_2O → OTHP, C_5H_{11}, H (12)

62% overall

$C_7H_{15}CHO$ — Ph_3P, CBr_4 / Zn (dust) → 1. BuLi / 2. CO_2 → C_7H_{15}–≡–CO_2H (13)

~78% overall

Ketones are converted to dibromomethylene derivatives. These intermediates can be transformed to isopropylidene compounds by reaction with ***Lithium Dimethylcuprate*** and ***Iodomethane*** (eq 14).[14] No racemization was reported for the chain extension of the aldehyde derived from (*S*)-ethyl lactate under the reaction conditions (eq 15).[15]

Ph_3P, CBr_4 → Me_2CuLi / MeI → (14)

~80% overall

MEMO, H, O — Ph_3P, CBr_4 → MEMO, Br, Br (15)

~70% from (*S*)-ethyl lactate

β-Bromo Enones from 1,3-Diketones. The reaction of Ph_3P and CBr_4 with a 1,3-diketone efficiently converts it to the β-bromo enone (eq 16).[16]

O, OH — Ph_3P, CBr_4 / 85% → O, Br (16)

1-Bromoalkynes from Terminal Alkynes. Terminal alkynes on reaction with Ph_3P and CBr_4 afford 1-bromoalkynes in high yield (eq 17).[17]

OAc, H — Ph_3P, CBr_4 / 92% → OAc, Br (17)

Related Reagents. 1,2-Bis(diphenylphosphino)ethane Tetrabromide; Bromine-Triphenyl Phosphite; Phosphorous(III) Bromide; Phosphorous(V) Bromide; Thionyl Bromide; Tribromomethyllithium; Triphenylphosphine; Triphenylphosphine *N*-Bromosuccinimide; Triphenylphosphine Carbon Tetrachloride; Triphenylphosphine Dibromide.

1. Castro, B. R. *OR* **1983**, *29*, 1.
2. Verheyden, J. P. H.; Moffatt, J. G. *JOC* **1972**, *37*, 2289.
3. Axelrod, E. H.; Milne, G. M.; van Tamelen, E. E. *JACS* **1970**, *92*, 2139.
4. Kang, S. H.; Hong, C. Y. *TL* **1987**, *28*, 675.
5. Kashem, A.; Anisuzzaman, M.; Whistler, R. L. *CR* **1978**, *61*, 511.
6. Weiss, R. G.; Snyder, E. I. *JOC* **1971**, *36*, 403.
7. Mattes, H.; Benezra, C. *TL* **1987**, *28*, 1697.
8. Wagner, A.; Heitz, M.-P.; Mioskowski, C. *TL* **1989**, *30*, 557.
9. Wagner, A.; Heitz, M.-P.; Mioskowski, C. *TL* **1989**, *30*, 1971.
10. Clinch, K.; Vasella, A.; Schauer, R. *TL* **1987**, *28*, 6425.
11. Einhorn, J.; Einhorn, C.; Luche, J.-L. *SC* **1990**, *20*, 1105.
12. Ramirez, F.; Desai, N. B.; McKelvie, N. *JACS* **1962**, *84*, 1745.
13. Corey, E. J.; Fuchs, P. L. *TL* **1972**, 3769.
14. Posner, G. H.; Loomis, G. L.; Sawaya, H. S. *TL* **1975**, 1373.
15. Mahler, H.; Braun, M. *TL* **1987**, *28*, 5145.
16. Gruber, L.; Tömösközi, I.; Radics, L. *S* **1975**, 708.
17. Wagner, A.; Heitz, M.-P.; Mioskowski, C. *TL* **1990**, *31*, 3141.

Michael J. Taschner

The University of Akron, OH, USA

Triphenylphosphine–Carbon Tetrachloride[1]

$Ph_3P–CCl_4$

(Ph_3P)		
[605-35-0]	$C_{18}H_{15}P$	(MW 262.30)
(CCl_4)		
[56-23-5]	CCl_4	(MW 153.81)

(reagent combination for the conversion of a number of functional groups into their corresponding chlorides and for dehydrations)

Physical Data: Ph_3P: mp 79–81 °C; bp 377 °C; *d* 1.0749 g cm^{-3}. CCl_4: mp −23 °C; bp 77 °C; *d* 1.594 g cm^{-3}.
Solubility: sol CCl_4, MeCN, CH_2Cl_2, 1,2-dichloroethane.
Preparative Method: reactive intermediates are generated in situ by reaction of Ph_3P and CCl_4.
Handling, Storage, and Precautions: Ph_3P is an irritant; CCl_4 is toxic and a cancer suspect agent; use in a fume hood. Solvents must be carefully dried because the intermediates are all susceptible to hydrolysis.

Combination of Triphenylphosphine and Carbon Tetrachloride. This reagent combination is capable of performing a range of chlorinations and dehydrations. The reactions are typically run using the so-called two-component or three-component systems. Carbon tetrachloride can function as both the reagent and solvent. However, the rates of the reactions are highly solvent-dependent, with MeCN providing the fastest rates.[1]

Conversion of Alcohols to Alkyl Chlorides. The reaction of alcohols with ***Triphenylphosphine*** and carbon tetrachloride results in the formation of alkyl chlorides.[2] The mild, neutral conditions allow for the efficient conversion of even sensitive alcohols into the corresponding chlorides (eqs 1 and 2).[3,4] The reaction typically proceeds with inversion of configuration.[5] In eq 3, it is interesting to note that the reaction not only proceeds with inversion of configuration but also no acyloxy migration is observed.[6]

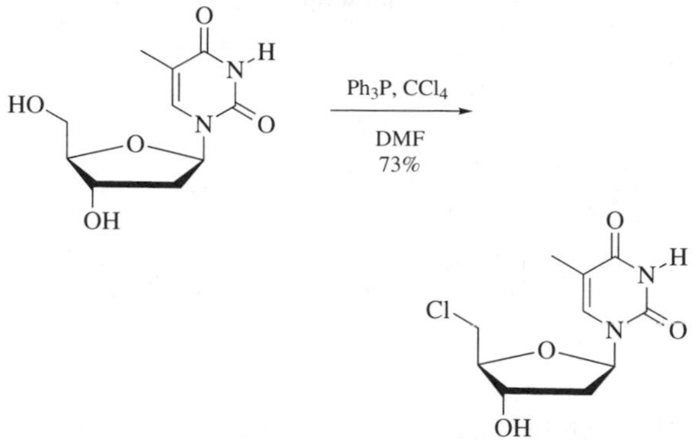

(1)

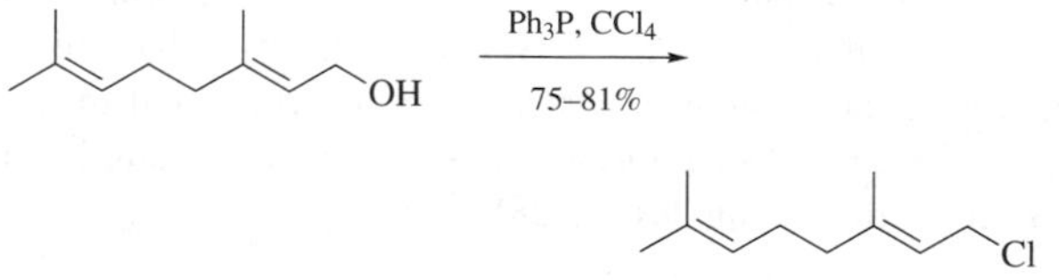

(2)

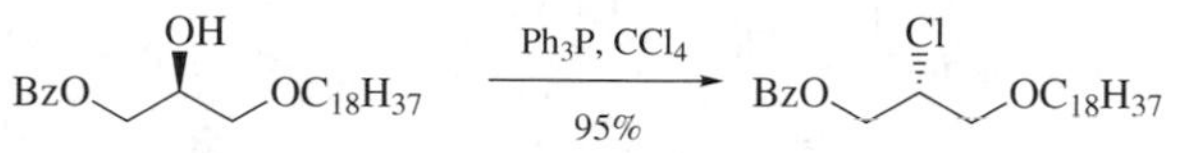

(3)

The conversion of an allylic alcohol to an allylic chloride occurs with no or minimal allylic rearrangement (eqs 4 and 5).[7] For the synthesis of low boiling allylic alcohols, it is advantageous to substitute hexachloroacetone (HCA) for CCl_4 (eq 6). The stereochemical integrity of the double bond also remains intact under these conditions.[8]

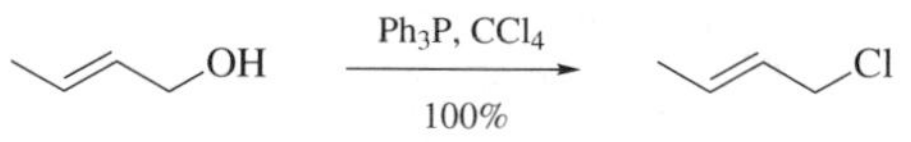

(4)

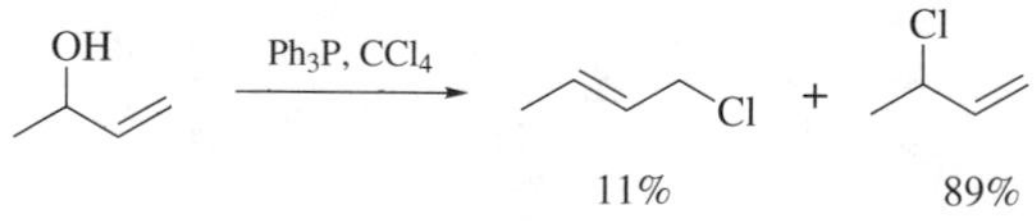

(5)

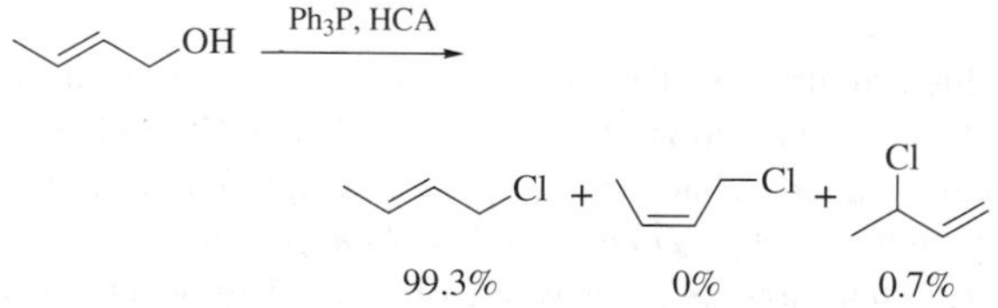

(6)

If the conversion of the alcohol to the chloride is attempted in refluxing MeCN, dehydration to form the alkene occurs (eqs 7 and 8).[9] Occasionally the separation of the product from the triphenylphosphine oxide produced in the reaction can be problematic. This can be overcome by using a polymer-supported phosphine.[10] Simple filtration and evaporation of the solvent are all that is required under these conditions. Not only is the workup facilitated, but the rate of the reaction is also increased by employing the supported reagent.[10c]

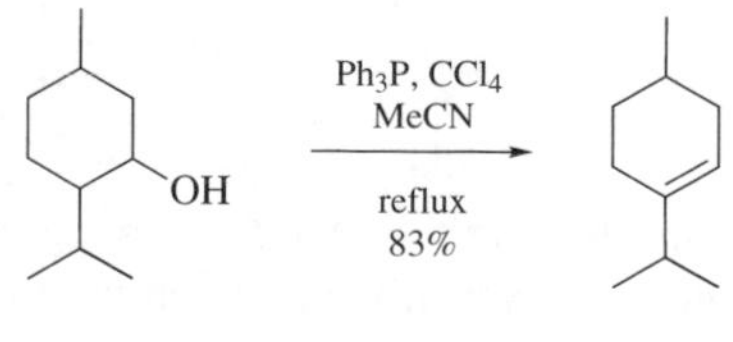

(7)

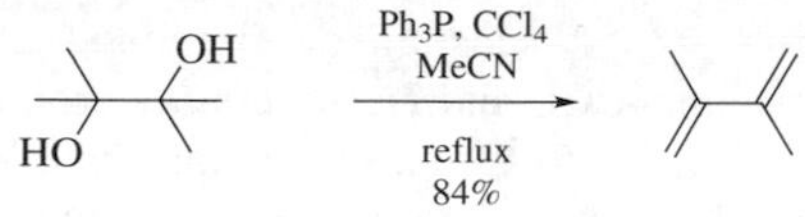

(8)

Conversion of Acids to Acid Chlorides. The reaction of carboxylic acids with triphenylphosphine–CCl_4 reportedly produces acid chlorides in good yield under mild conditions (eq 9).[11] These conditions will allow acid sensitive functional groups to survive. Phosphoric mono- and diesters are successfully converted into the phosphoric monoester dichlorides and diester chlorides, respectively. The reaction of the diethyl ester does not produce the acid chloride. Instead, the anhydride is formed. Phosphinic acid chlorides can also be prepared from the corresponding phosphinic acid under these conditions (eq 10).[12]

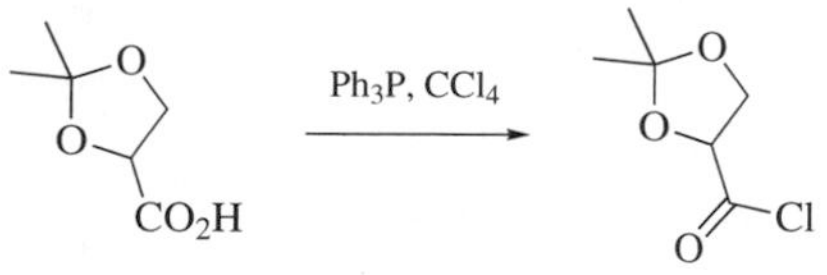

(9)

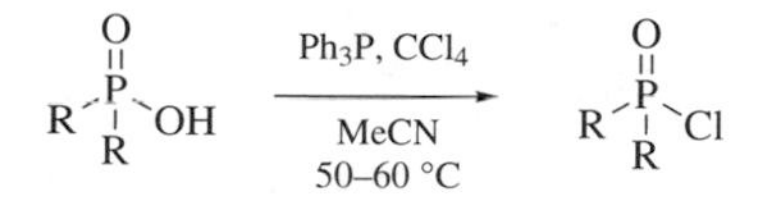

(10)

Epoxides to *cis*-1,2-Dichloroalkanes. Epoxides are converted into *cis*-1,2-dichlorides by the action of triphenylphosphine in refluxing CCl_4.[13] Cyclohexene oxide forms *cis*-1,2-dichlorocyclohexane (eq 11) contaminated by a trace of the *trans*-isomer. Cyclopentene oxide gives only the *cis*-isomer.

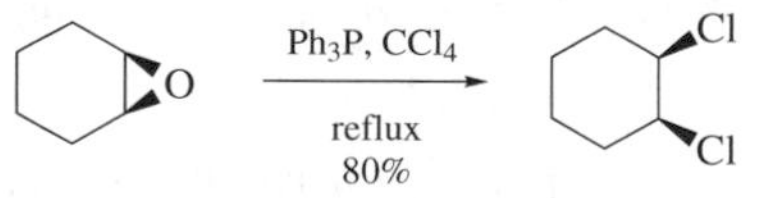

(11)

Dehydrations. Diols may be cyclodehydrated to the corresponding cyclic ethers. The reaction is most effective for 1,4-diols (eq 12). Dehydration of 1,3- and 1,5-diols is not as successful, except in the case of the configurationally constrained 1,5-diol shown in eq 13.[14] The reaction of *trans*-1,2-cyclohexanediol with the reagent affords *trans*-2-chlorocyclohexanol with none of the *cis*-isomer or the *trans*-dichloride being detected. If the reaction is performed in the presence of K_2CO_3 as an HCl scavenger, the epoxide is formed (eq 14).[14] The yields are not as good with substituted acyclic diols.[15]

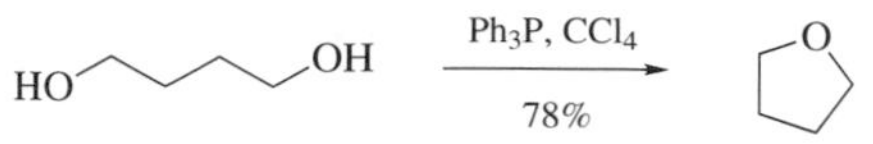

(12)

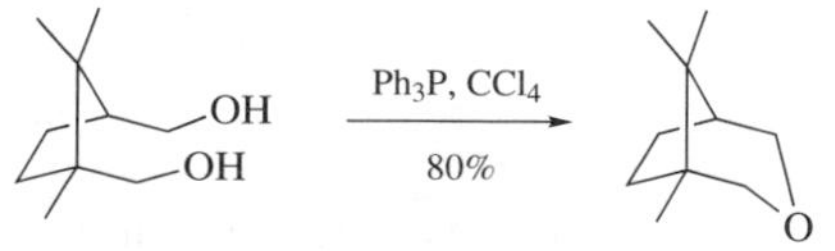

(13)

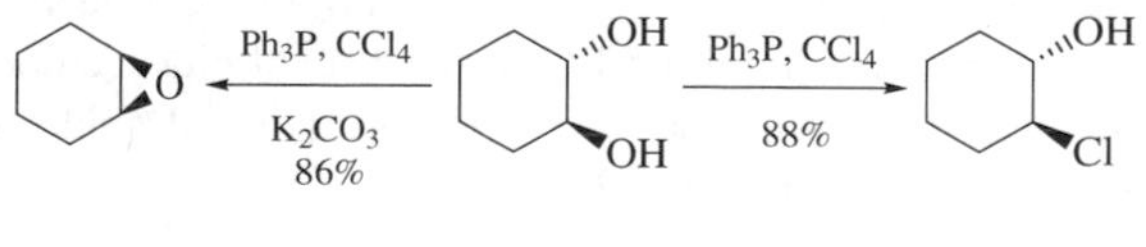

(14)

Dehydration of *N*-substituted β-amino alcohols with triphenylphosphine–CCl_4 and ***Triethylamine*** produces aziridines in good yield (eq 15).[16] This reaction has been successfully employed in the preparation of stable arene imines (eq 16).[17] Azetidines can be obtained from the corresponding 3-aminoalkanols (eq 17).[18] Additionally, reaction of 2-(3-hydroxypropyl)piperidine under these conditions yields octahydroindolizine (eq 18).

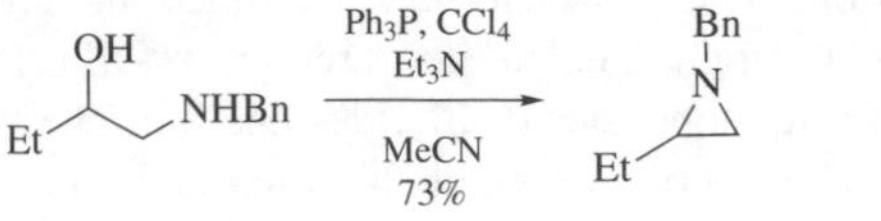

(15)

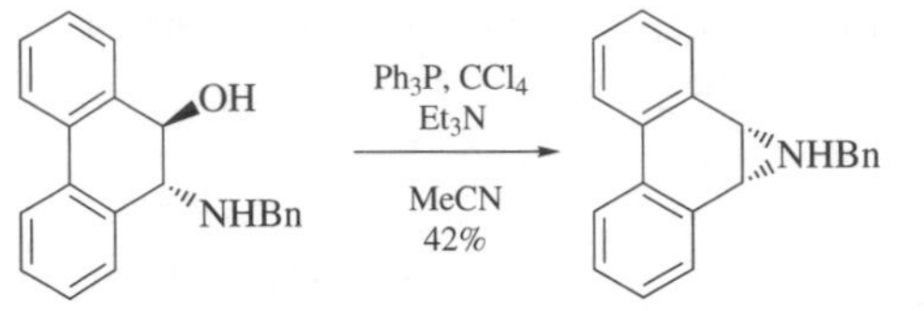

(16)

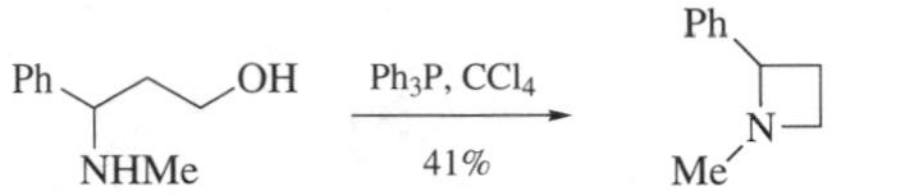

(17)

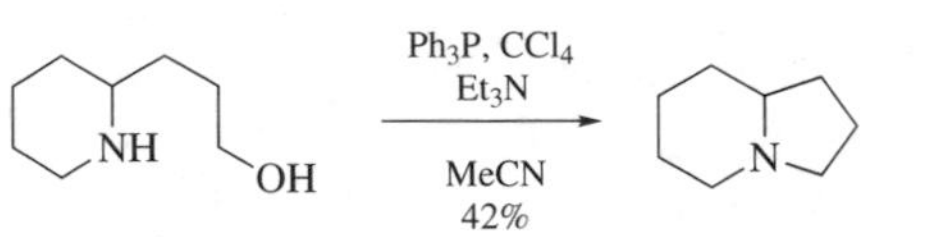

(18)

Substituted hydroxamic acids successfully cyclize to form β-lactams as long as Et_3N is present (eq 19). In the absence of the base, complex mixtures are formed.[19] Unsubstituted amides can be converted into nitriles via dehydration (eq 20).[20] This is the reagent of choice for the transformation of the amide to the nitrile in eq 21.[21]

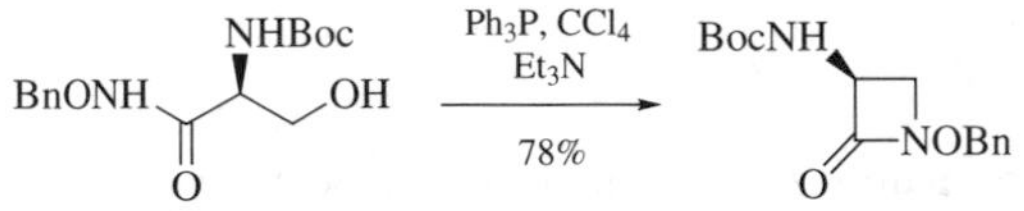

(19)

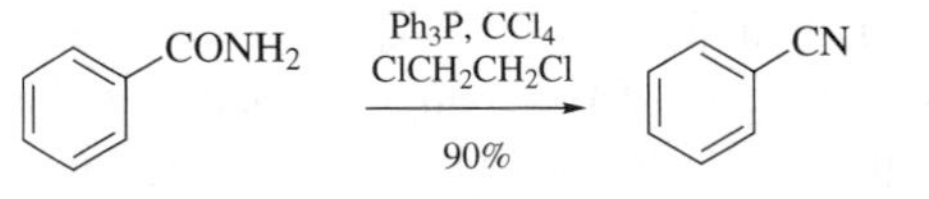

(20)

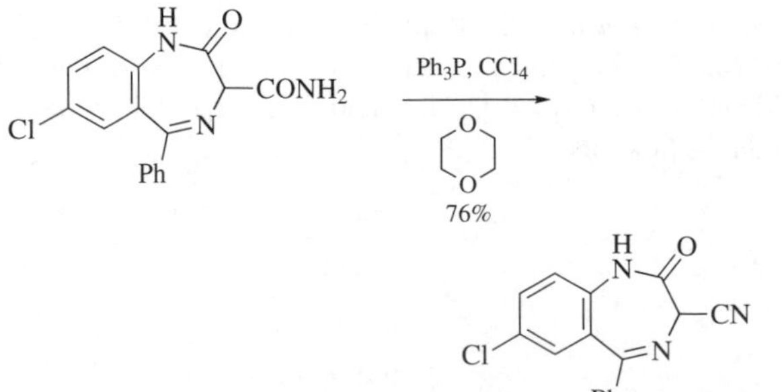

(21)

Nitriles can also be obtained from aldoximes using this reagent (eq 22).[22] Ketoximes produce imidoyl chlorides via a Beckmann rearrangement under these conditions (eq 23).[23] Imidoyl chlorides are also available by the reaction of monosubstituted amides with $Ph_3P–CCl_4$ in acetonitrile (eq 24).[24]

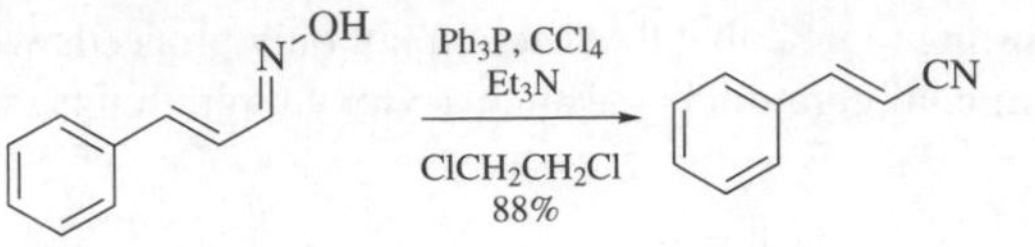

(22)

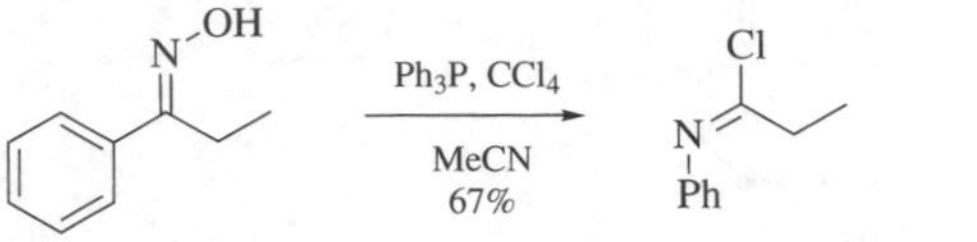

(23)

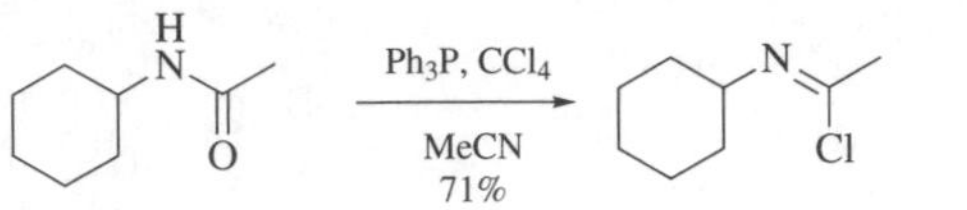

(24)

N,N,N′-Trisubstituted ureas afford chloroformamidine derivatives (eq 25),[25] while *N,N′*-disubstituted ureas and thioureas produce carbodiimides (eq 26).[26] When carbamoyl chlorides are treated with the reagent in MeCN, they are converted into isocyanates (eq 27).[27] Dehydration of *N*-substituted formamides provides access to isocyanides (eq 28).[28]

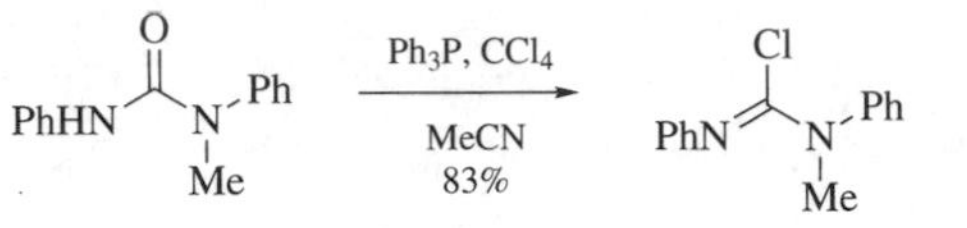

(25)

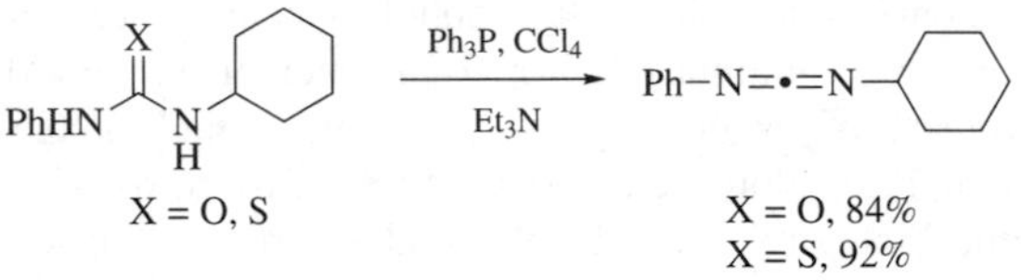

(26)

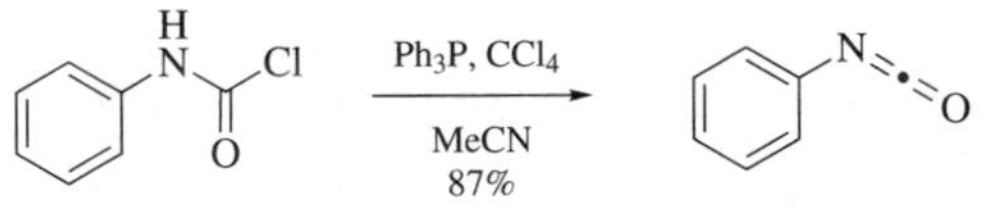

(27)

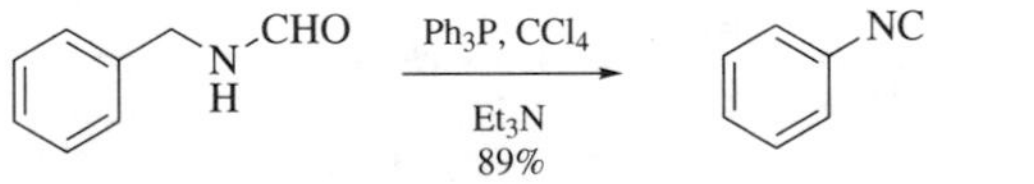

(28)

Amide Formation. The synthesis of an amide can be accomplished by initial reaction of an acid with $Ph_3P–CCl_4$ and then reaction of the intermediate with 2 equiv of the appropriate amine. A tertiary amine, such as ***Diisopropylethylamine***, can be employed as the HCl scavenger in cases where one would not want to waste any of a potentially valuable amine.[29] This method has been used in the construction of an amide in the synthesis of the skeleton of the lycorine alkaloids (eq 29).[30]

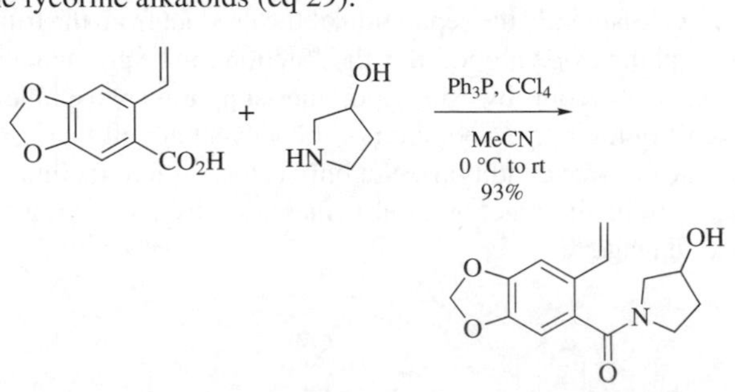

(29)

The method is effective for peptide coupling in that the yields are typically good; however, the reaction is often accompanied by racemization of the product.[31] The racemization problem can be suppressed, but this involves a change in the phosphine employed[32] or slightly modified reaction conditions. These modified conditions have been used successfully in the construction of a hexapeptide without racemization (eq 30).[33]

$$\text{Cbz–Leu–Ala–Val–Phe–Gly–OH} + \text{Pro–OBn} \xrightarrow[\text{Et}_3\text{N}\ 68\%]{\text{Ph}_3\text{P, CCl}_4} \text{Cbz–Leu–Ala–Val–Phe–Gly–Pro–OBn} \quad (30)$$

If amino alcohols, amino thiols, or diamines are used, the intermediate amides cyclodehydrate under the reaction conditions to form Δ^2-oxazolines (eq 31), Δ^2-oxazines, Δ^2-thiazolines, or Δ^2-imidazolines.[33] The reaction requires the use of 3 equiv of the $Ph_3P–CCl_4$ reagent. The reaction is reported to fail if a commercial sample of the polymer-supported phosphine is used instead of triphenylphosphine.

(31) [scheme: PhCH₂CO₂H + H₂N–C(CH₃)₂–CH₂OH → (Ph₃P, CCl₄; Et₃N; 67%) 2-benzyl-4,4-dimethyl-Δ²-oxazoline]

1,1-Dichloroalkenes and Vinyl Chlorides. The reaction of aldehydes produces a mixture of the 1,1-dichloroalkene and the dichloromethylene derivatives. These are the result of reaction of the in situ generated ***Dichloromethylenetriphenylphosphorane*** and ***Triphenylphosphine Dichloride***, respectively. Benzaldehyde (eq 32) produces a 1:1 mixture of the dichloroalkene and benzal chloride in 72% yield.[34]

$$\text{PhCHO} \xrightarrow[72\%]{\text{Ph}_3\text{P, CCl}_4} \text{PhCH=CCl}_2 + \text{PhCHCl}_2 \quad (32)$$

1:1

Ketones can also be used in this transformation, but sometimes enolizable ketones lead to the formation of the vinyl chloride derived from the enol. This is usually more of a problem for six-membered ring ketones than for five-membered ring ketones. Cyclopentanone yields predominantly the 1,1-dichloroalkene (eq 33), while cyclohexanone provides mainly the vinyl chloride (eq 34).[35]

(33) [scheme: cyclopentanone → (Ph₃P, CCl₄; 65%) (dichloromethylene)cyclopentane + 1-chlorocyclopentene, 95:5]

(34) [scheme: cyclohexanone → (Ph₃P, CCl₄; 65%) (dichloromethylene)cyclohexane + 1-chlorocyclohexene, 7:93]

Related Reagents. Catechylphosphorus Trichloride; *N*-Chlorosuccinimide-Dimethyl Sulfide; Dichloromethyl Methyl Ether; Dichloromethylene triphenylphosphorane; Diethyl Dichloromethylphosphonate; Oxalyl Chloride; Phosphorous(III) Chloride; Phosphorous(V) Chloride; Phosphorousoxy Chloride; Thionyl Chloride; Triphenylphosphine Dichloride; Triphenylphosphine-Hexachloroacetone.

---

1. (a) Appel, R. *AG(E)* **1975**, *14*, 801. (b) Appel, R.; Halstenberg, M. *Organophosphorus Reagents in Organic Synthesis*; Cadogan, J. I. G., Ed.; Academic: New York, 1979; pp 387–431.
2. (a) Lee, J. B.; Nolan, T. J. *CJC* **1966**, *44*, 1331. (b) Lee, J. B.; Downie, I. W. *T* **1967**, *23*, 359.
3. Verheyden, J. P. H.; Moffat, J. G. *JOC* **1972**, *37*, 2289.
4. Calzada, J. G.; Hooz, J. *OS* **1974**, *54*, 63.
5. Weiss, R. G.; Snyder, E. I. *JOC* **1970**, *35*, 1627.
6. Aneja, R.; Davies, A. P.; Knaggs, J. A. *CC* **1973**, 110.
7. Snyder, E. I. *JOC* **1972**, *37*, 1466.
8. Majid, R. M.; Fruchey, O. S.; Johnson, W. L. *TL* **1977**, 2999.
9. Appel, R.; Wihler, H.-D. *CB* **1976**, *109*, 3446.
10. (a) Harrison, C. R.; Hodge, P. *CC* **1975**, 622. (b) Regen, S. L.; Lee, D. P. *JOC* **1975**, *40*, 1669. (c) Harrison, C. R.; Hodge, P. *CC* **1978**, 813.
11. Lee, J. B. *JACS* **1966**, *88*, 3440.
12. Appel, R.; Einig, H. *Z. Anorg. Allg. Chem.* **1975**, *414*, 236.
13. Isaacs, N. S.; Kirkpatrick, D. *TL* **1972**, 3869.
14. Barry, C. N.; Evans, S. A. *JOC* **1981**, *46*, 3361.
15. Barry, C. N.; Evans, S. A. *TL* **1983**, *24*, 661.
16. Appel, R.; Kleinstück, R. *CB* **1974**, *107*, 5.
17. Ittah, Y.; Shahak, I.; Blum, J. *JOC* **1978**, *43*, 397.
18. Stoilova, V.; Trifonov, L. S.; Orahovats, A. S. *S* **1979**, 105.
19. Miller, M. J.; Mattingly, P. G.; Morrison, M. A.; Kerwin, J. F. *JACS* **1980**, *102*, 7026.
20. (a) Yamato, E.; Sugasawa, S. *TL* **1970**, 4383. (b) Appel, R.; Kleinstück, R.; Ziehn, K.-D. *CB* **1971**, *104*, 1030.
21. Juanin, R.; Arnold, W. *HCA* **1973**, *56*, 2569.
22. Appel, R.; Kohnke, J. *CB* **1971**, *104*, 2023.
23. Appel, R.; Warning, K. *CB* **1975**, *108*, 1437.
24. Appel, R.; Warning, K.; Ziehn, K.-D. *CB* **1973**, *106*, 3450.
25. (a) Appel, R.; Warning, K.; Ziehn, K.-D. *CB* **1974**, *107*, 698. (b) Appel, R.; Warning, K.; Ziehn, K.-D. *CB* **1973**, *106*, 2093.
26. Appel, R.; Warning, K.; Ziehn, K.-D. *CB* **1971**, *104*, 1335.
27. Appel, R.; Warning, K.; Ziehn, K.-D.; Gilak, A. *CB* **1974**, *107*, 2671.
28. Appel, R.; Kleinstück, R.; Ziehn, K.-D. *AG(E)* **1971**, *10*, 132.
29. Barstow, L. E.; Hruby, V. J. *JOC* **1971**, *36*, 1305.
30. Stork, G.; Morgans, D. J. *JACS* **1979**, *101*, 7110.
31. Wieland, T.; Seeliger, A. *CB* **1971**, *104*, 3992.
32. Takeuchi, Y.; Yamada, S. *CPB* **1974**, *22*, 832.
33. Appel, R.; Bäumer, G.; Strüver, W. *CB* **1975**, *108*, 2680.
34. Rabinowitz, R.; Marcus, R. *JACS* **1962**, *84*, 1312.
35. Isaacs, N.; Kirkpatrick, D. *CC* **1972**, 443.

Michael J. Taschner
The University of Akron, OH, USA

Vinyllithium[1]

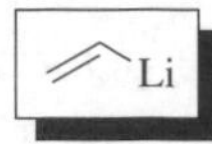

[917-57-7] C_2H_3Li (MW 33.99)

(introduction of vinyl groups[2,3])

Physical Data: white solid; no melting point; decomposes upon heating. The ^{1}H,[4] ^{13}C,[5] and ^{7}Li NMR[6] and UV–vis[7] spectra of vinyllithium have all been recorded.

Solubility: sol THF, ether; slightly sol pentane, hexane;[8] insol benzene.[9]

Analysis of Reagent Purity: determined by simple acid titration,[8] by reaction with vanadium pentoxide followed by titration with standard permanganate solution,[10] by hydrolysis followed by measurement of ethylene evolution,[11] or by the Gilman procedure.[12]

Preparative Methods: prepared by a variety of methods. Direct metalation of ***Ethylene*** with ***Lithium*** metal in dimethoxymethane yields vinyllithium contaminated with lithium hydride, and other organolithio species, such as butyllithium and 1,4-dilithiobutane.[13] Direct formation from ethylene can also be accomplished using ***Potassium t-Butoxide***, ***n-Butyllithium***, and TMEDA (***N,N,N′,N′-Tetramethylethylenediamine***) in hexane at −40 °C, followed by treatment of the vinylpotassium with ***Lithium Bromide***.[14] The lithium–halogen exchange reaction has been used, starting from vinyl chloride and lithium/sodium dispersions,[10,12] or from vinyl bromide with 2 equiv of ***t-Butyllithium*** in either a Trapp solvent mixture at −120 °C[15,16] or ether at −78 °C.[17] This is the most convenient procedure, but lithium halide is formed during the reaction and is present as a contaminant. Halide-free vinyllithium can be prepared by transmetalation of ***Tetravinylstannane*** with phenyl- or butyllithium,[8] of tetravinyllead with either lithium metal or ***Phenyllithium***,[9] or of divinylmercury with lithium dispersion in pentane.[11]

Purification: ether solutions can be evaporated under an inert atmosphere and the resulting solid washed with anhydrous pentane or hexane.

Handling, Storage, and Precautions: as with most alkyllithium reagents, vinyllithium reacts readily with moisture. Solid vinyllithium should be maintained at −25 °C or below, under an inert atmosphere. The solid suffers a loss in activity with time and should be used immediately upon preparation.[9] Solutions of the reagent in ether or THF, under an inert atmosphere, are stable for up to 1 week at rt.[8] Solid vinyllithium is violently pyrophoric, yielding a brilliant red flash on contact with air.[9]

Vinyl Reagents. Vinylcuprates are easily assembled from the reaction of vinyllithium and copper species. ***Lithium Divinylcuprate*** (eq 1) has been prepared from vinyllithium using the dimethyl sulfide complex of a copper(I) halide,[18] or from ***Copper(I) Cyanide***.[19] Mixed cuprates are available by the addition of vinyllithium to (2-thienyl)Cu(CN)Li,[20] $(MeCu)_n$,[18] or ***1-Pentynylcopper(I)***.[21] In all of the mixed cuprates the vinyl ligand is preferentially transferred.

$$(Me_2S)CuBr \xrightarrow[Et_2O,\ -40\ ^\circ C]{\text{vinyl-Li (2 equiv)}} (CH_2{=}CH)_2CuLi \quad (1)$$

Vinylsilanes can be prepared by reaction of a silyl halide with vinyllithium.[22] Alternatively, the reaction of ***Bis(trimethylsilyl) Peroxide*** with vinyllithium yields an 80:20 mixture of ***Vinyltrimethylsilane*** and the trimethylsilyl enol ether (eq 2).[23] Use of the corresponding vinyl Grignard reagent produces exclusively the enol ether.

$$\text{TMSO–OTMS} \xrightarrow[45\%]{\text{vinyl-Li}} CH_2{=}CH\text{–TMS} + CH_2{=}CH\text{–OTMS} \quad (80{:}20) \quad (2)$$

Trialkylvinylborates can be prepared by the treatment of trialkylboranes with vinyllithium in ether at 0 °C.[24] Use of a vinyl Grignard reagent to form the vinyl borate leads to lower yields in subsequent reactions. Dimesitylvinylborane is synthesized by lithium–halogen exchange of vinyllithium with fluorodimesitylborane.[25] Dialkyl disulfides[15] or diphenylphosphinodithioates (**1**)[26] react with vinyllithium to produce vinyl alkyl, or aryl, sulfides in good yields (eq 3).

$$\underset{(\mathbf{1})}{Ph_2P(=S)\text{–SPh}} \xrightarrow[\text{THF, HMPA},\ -78\text{ to }-40\ ^\circ C,\ 71\%]{\text{vinyl-Li}} CH_2{=}CH\text{–SPh} \quad (3)$$

Ring Formation Reactions. Vinyllithium reacts with carbonyl compounds, providing a convenient method for the introduction of a double bond for subsequent ring formation. Addition to a vinyl-substituted cyclobutanone, followed by treatment with ***Potassium Hydride***, gives the *cis-* and *trans*-cyclooctenones in 62% yield after oxy-Cope rearrangement (eq 4).[2] The vinyllithium addition yields only one diastereomer according to NMR and HPLC analysis. Similar sigmatropic reactions have been used in bicyclic compounds. Vinyllithium adds to the carbonyl in 5-phenylbicyclo[3.2.0]heptan-6-ones (eq 5) which, when subsequently treated with potassium hydride, undergo a [1,3]-sigmatropic rearrangement of the 1-vinylcyclobutanols to yield the ring expanded product (**2**).[27] An oxy-Cope rearrangement has been used to construct bicyclo[5.3.1]undecenones after addition of vinyllithium to bicyclo[3.1.1]heptan-6-ones.[28] An intramolecular Diels–Alder reaction has been used to construct Isomer G, a component of thujopsene. The addition of vinyllithium to the acid (**3**) produces the precursor for the Diels–Alder cyclization (eq 6).[29] Further elaboration leads to Isomer G.

Other cyclizations have, also utilized vinyl groups introduced by reaction of vinyllithium with carbonyl groups, as in the base-catalyzed cyclization used in the synthesis of aphidicolin,[30] the formation of substituted quinones from dimethyl squarate,[31] and the cyclization using sodium in THF in the synthesis of (±)-patchoulol.[32]

Nucleophilic Additions. Reaction of vinyllithium with 1-iodooctane in THF gives 1-decene in 92% yield. The use of 1,4-dibromobutane gives 1,7-octadiene in 67% yield.[33] Alkenes are also available from the dimerization of vinyllithium with anhydrous ***Cerium(III) Chloride*** in THF.[34] Aryl or alkyl vinyl ketones are readily formed by treating carboxylic acids with 2 equiv of vinyllithium in DME at 5–10 °C. Propionic acid produces 1-penten-3-one in 92% yield.[3]

Substituted enol silyl ethers are available from vinyllithium addition to silyl ketones (eq 7). Addition of the vinyllithium to the silyl ketone results in a silyloxyallyllithium which can be trapped with various electrophiles to yield the substituted enol silyl ethers.[17] Vinyllithium can also be added to epoxides to give the ring-opened product,[35] added to aldimines in toluene at −42 °C to produce allylamines,[36] and inserted *ortho* to either a 2,6-di-*t*-butyl-4-methoxyphenyl ester substituent[37] or an oxazoline group on a naphthalene ring.[38]

Electrophile	E	Yield
MeSSMe	MeS	89%
MeI	Me	83%
Pr—≡	H	79%

Related Vinylmetal Reagents. Perfluorinated vinyllithium has been prepared by transmetalation of phenyltris(perfluorovinyl)tin with 3 equiv of phenyllithium in ether at −35 to −40 °C,[39] or by lithium–bromine exchange of perfluorovinyl bromide with 2 equiv of ***t-Butyllithium*** in THF at −110 °C.[16] Vinylsodium is prepared by metalation of ethylene using pentylsodium with sodium isopropoxide.[40] Vinylpotassium is made from vinyl chloride by metal–halogen exchange using a 90% potassium/sodium alloy.[41]

Related Reagents. 1,3-Butadienyl-1-lithium; Isopropenyllithium; Lithium Divinylcuprate; Lithium Divinylcuprate–Tributylphosphine; 1-Propenyllithium; 1-(Trimethylsilyl)vinyllithium; (*E*)-2-(Trimethylsilyl)vinyllithium; Vinylcopper; Vinylmagnesium Bromide; Vinylmagnesium Bromide–Methylcopper; Vinylmagnesium Chloride–Copper(I) Chloride; Vinyltributylstannane; Vinyltrimethylsilane.

1. (a) Seyferth, D. *Prog. Inorg. Chem.* **1962**, *3*, 129. (b) Wakefield, B. J. *The Chemistry of Organolithium Compounds*; Pergamon: New York, 1974.
2. Gadwood, R. C.; Lett, R. M. *JOC* **1982**, *47*, 2268.
3. Floyd, J. C. *TL* **1974**, 2877.
4. Johnson, C. S., Jr.; Weiner, M. A.; Waugh, J. S.; Seyferth, D. *JACS* **1961**, *83*, 1306.
5. van Dongen, J. P. C. M.; van Dijkman, H. W. D.; de Bie, M. J. A. *RTC* **1974**, *93*, 29.
6. Scherr, P. A.; Hogan, R. J.; Oliver, J. P. *JACS* **1974**, *96*, 6055.
7. Waack, R.; Doran, M. A. *JACS* **1963**, *85*, 1651.
8. Seyferth, D.; Weiner, M. A. *JACS* **1961**, *83*, 3583.
9. Juenge, E. C.; Seyferth, D. *JOC* **1961**, *26*, 563.
10. West, R.; Glaze, W. H. *JOC* **1961**, *26*, 2096.
11. Bartocha, B.; Douglas, C. M.; Gray, M. Y. *ZN(B)* **1959**, *14*, 809.
12. Smith, W. N., Jr. *JOM* **1974**, *82*, 7.
13. Rautenstrauch, V. *AG(E)* **1975**, *14*, 259.
14. Brandsma, L.; Verkruijsse, H. D.; Schade, C.; von Rague Schleyer, P. *CC* **1986**, 260.
15. Neumann, H.; Seebach, D. *TL* **1976**, 4839.
16. Neumann, H.; Seebach, D. *CB* **1978**, *111*, 2785.
17. (a) Reich, H. J.; Holtan, R. C.; Bolm, C. *JACS* **1990**, *112*, 5609. (b) Reich, H. J.; Olson, R. E.; Clark, M. C. *JACS* **1980**, *102*, 1423.
18. House, H. O.; Chu, C.-Y.; Wilkins, J. M.; Umen, M. J. *JOC* **1975**, *40*, 1460.
19. (a) Lautens, M.; Di Felice, C.; Huboux, A. *TL* **1989**, *30*, 6817. (b) Boring, D. L.; Sindelar, R. D. *JOC* **1988**, *53*, 3617.
20. Lipshutz, B. H.; Kozlowski, J. A.; Parker, D. A.; Nguyen, S. L.; McCarthy, K. E. *JOM* **1985**, *285*, 437.
21. Corey, E. J.; Beames, D. J. *JACS* **1972**, *94*, 7210.
22. (a) Auner, N. *Z. Anorg. Allg. Chem.* **1988**, *558*, 87. (b) Soderquist, J. A.; Rivera, I.; Negron, A. *JOC* **1989**, *54*, 4051.
23. Camici, L.; Dembech, P.; Ricci, A.; Seconi, G.; Taddei, M. *T* **1988**, *44*, 4197.
24. Utimoto, K.; Uchida, K.; Nozaki, H. *T* **1977**, *33*, 1949.
25. Brown, N. M. D.; Davidson, F.; Wilson, J. W. *JOM* **1981**, *209*, 1.
26. Yoshifuji, M.; Hanafusa, F.; Inamoto, N. *CL* **1979**, *6*, 723.
27. Snider, B. B.; Niwa, M. *TL* **1988**, *29*, 3175.
28. Snider, B. B.; Allentoff, A. J. *JOC* **1991**, *56*, 321.
29. Tavares, R. F.; Katten, E. *TL* **1977**, 1713.
30. Krafft, M. E.; Kennedy, R. M.; Holton, R. A. *TL* **1986**, *27*, 2087.

31. Gayo, L. M.; Winters, M. P.; Moore, H. W. *JOC* **1992**, *57*, 6896.
32. Bertrand, M.; Teisseire, P.; Pelerin, G. *TL* **1980**, *21*, 2055.
33. Millon, J.; Lorne, R.; Linstrumelle, G. *S* **1975**, 434.
34. Imamoto, T.; Hatajima, T.; Ogata, K. *TL* **1991**, *32*, 2787.
35. Rychnovsky, S. D.; Griesgraber, G.; Zeller, S.; Skalitzky, D. J. *JOC* **1991**, *56*, 5161.
36. Tomioka, K.; Inoue, I.; Shindo, M.; Koga, K. *TL* **1991**, *32*, 3095.
37. Tomioka, K.; Shindo, M.; Koga, K. *JOC* **1990**, *55*, 2276.
38. (a) Meyers, A. I.; Lutomski, K. A.; Laucher, D. *T* **1988**, *44*, 3107. (b) Meyers, A. I.; Higashiyama, K. *JOC* **1987**, *52*, 4592. (c) Rawson, D. J.; Meyers, A. I. *JOC* **1991**, *56*, 2292.
39. Seyferth, D.; Wada, T.; Raab, G. *TL* **1960**, 20.
40. Morton, A. A.; Marsh, F. D.; Coombs, R. D.; Lyons, A. L.; Penner, S. E.; Ramsden, H. E.; Baker, V. B.; Little, E. L.; Letsinger, R. L. *JACS* **1950**, *72*, 3785.
41. Anderson, R. G.; Silverman, M. B.; Ritter, D. M. *JOC* **1958**, *23*, 750.

Eric K. Eisenhart

Rohm and Haas Company, Spring House, PA, USA

Vinylmagnesium Bromide[1]

MgBr

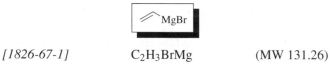

[1826-67-1] C_2H_3BrMg (MW 131.26)

(introduction of a vinyl unit)

Form Supplied in: widely available in solution in THF (1.0 M).

Preparative Methods: prepared in situ from vinyl bromide and ***Magnesium*** in THF.[2]

Handling, Storage, and Precautions: the solution in THF must be handled with care since it is flammable and may ignite on contact with moisture and air. It should be stored and handled in an inert atmosphere (argon or nitrogen) and measured or transferred by means of a hypodermic syringe or cannula. The solution may deteriorate with time and the concentration of the reagent should be titrated before use.[3]

General Discussion. Vinylmagnesium bromide reacts with various electrophiles (e.g. aldehydes, ketones, esters, lactones, anhydrides, acid chlorides, carboxylic acids, nitriles, amides, epoxides, allyl halides) (eq 1).[1,2,4]

Br —(Mg, THF, 35–40 °C)→ MgBr —(E^+)→ E (1)

Complexation of vinylmagnesium bromide with TDA-1 [$N(CH_2CH_2OCH_2CH_2OMe)_3$] gives a powder which can react as the conventional Grignard reagent in hydrocarbon solvents (toluene, cyclohexane) (eq 2).[5]

$C_7H_{15}CHO$ —(MgBr, TDA-1, toluene, 20 °C, 89%)→ C_7H_{15}(OH) (2)

Vinylmagnesium bromide reacts with a silyl ketone to give a tertiary allylic alcohol, which is a useful intermediate for the synthesis of 2-substituted 1,3-dienes (eq 3).[6] A fluorinated 1,3-butadiene bearing a siloxy group is prepared by treatment of a trifluoroacetylsilane with vinylmagnesium bromide (eq 4).[7]

C_6H_{13} (O) TMS —(MgBr, THF, 94%)→ C_6H_{13} HO TMS —(AcONa, AcOH, 70%)→ C_6H_{13} (3)

F_3C (O) $SiPh_3$ —(MgBr, THF, 96%)→ F F $OSiPh_3$ (4)

An organocerium reagent, prepared from vinylmagnesium bromide and anhydrous ***Cerium(III) Chloride***, reacts with easily enolizable ketones to afford addition products in good yield (eq 5).[8] Under copper catalysis, vinylmagnesium bromide undergoes conjugate addition to α,β-unsaturated ketones very efficiently (eqs 6 and 7).[9–11]

Ph Ph =O —(MgBr, $CeCl_3$, THF, 100%)→ Ph Ph OH (5)

O —(MgBr, CuI, THF, 80%)→ O + O 1:1 (6)

O —(MgBr, CuBr, TMSCl, HMPA, THF, 97%)→ TMSO (7)

In the presence of copper(I) salts (2–10%) in THF, vinylmagnesium bromide reacts with alkyl halides (eq 8)[12] and epoxides. In the latter case, pure ***Copper(I) Iodide*** (or ***Copper(I) Bromide***–dimethyl sulfide) is needed for the ring opening of cyclohexene oxide with high stereospecificity (eq 9).[13]

MgBr + $C_8H_{17}I$ —(CuI (5%), THF, 82%)→ C_8H_{17} (8)

O —(MgBr, THF, catalyst)→ OH + OH (9)

Catalyst (10%)	Yield (%)	
CuI (99.999% purity)	70%	100:0
CuI (98% purity)	70%	55:45
CuBr•Me_2S	65%	100:0

In the presence of nickel or palladium complexes, vinylmagnesium bromide couples efficiently with aryl halides (eq 10),[14] vinyl halides (eqs 11 and 12),[15,16] and vinyl carbamates (eq 13).[17] The transmetalation of vinylmagnesium bromide occurs easily to give useful vinyl metals (eqs 14 and 15).[18,19]

MeO–(2-naphthyl)–Br + CH_2=CHMgBr —[Ni(dmpe)Cl_2, THF; 86%]→ MeO–(2-naphthyl)–CH=CH_2 (10)

C_5H_{11}–CH=CH–CH=CH–Cl + CH_2=CHMgBr —[Ni(PPh_3)$_4$, C_6H_6, THF; 91%]→ C_5H_{11}–CH=CH–CH=CH–CH=CH_2 (11)

Ph–CH=CH–Br + CH_2=CHMgBr —[Pd(PPh_3)$_4$, C_6H_6, THF; 91%]→ Ph–CH=CH–CH=CH_2 (12)

Ph–CH(OTIPS)–CH_2–C(=CH_2)–OCONEt_2 + CH_2=CHMgBr —[Ni(acac)$_2$, THF, Et_2O; 61%]→ Ph–CH(OTIPS)–CH_2–C(=CH_2)–CH=CH_2 (13)

CH_2=CHMgBr + Bu_2SnCl_2 —[THF; 74–91%]→ (CH_2=CH)$_2$SnBu_2 (14)

CH_2=CHMgBr + *i*-PrO–B(pinacolato) —[1. THF, –78 °C; 2. HCl, 25 °C; 68%]→ CH_2=CH–B(pinacolato) (15)

Related Reagents. Isopropenylmagnesium Bromide; Lithium Divinylcuprate; Vinylcopper; Vinyllithium; Vinylmagnesium Bromide–Copper(I) Iodide; Vinylmagnesium Bromide–Methylcopper; Vinylmagnesium Chloride–Copper(I) Chloride.

1. (a) Normant, H. *Adv. Org. Chem.* **1960**, *2*, 1. (b) Ioffe, S. T.; Nesmeyanov, A. N. *Methods of Elemento-Organic Chemistry*; North-Holland: Amsterdam, 1967; Vol. 2, p 1. (c) Negishi, E. I. *Organometallics in Organic Synthesis*; Wiley: New York, 1980. (d) Brandsma, L.; Verkruijsse, H. D. *Preparative Polar Organometallic Chemistry*; Springer: 1987; Vol. 1, p 46. (e) Erdik, E. *T* **1984**, *40*, 641.
2. Normant, H. *BSF(2)* **1957**, 728.
3. Watson, S. C.; Eastham, J. F. *JOM* **1967**, *9*, 165.
4. (a) Normant, H.; Ficini, J. *BSF(2)* **1956**, 1441. (b) Cuvigny, T.; Normant, H. *BSF(2)* **1961**, 2423. (c) Ficini, J.; Normant, H. *BSF(2)* **1964**, 1294. (d) Normant, J. *BSF(2)* **1963**, 1888. (e) Seyferth, D.; Vaughan, L. G. *JOM* **1963**, *1*, 138. (f) Kuwajima, I.; Kato, M. *CC* **1979**, 708. (g) Holt, D. A. *TL* **1981**, *22*, 2243. (h) Gadwood, R. C.; Lett, R. M. *JOC* **1982**, *47*, 2268.
5. Boudin, A.; Cerveau, G.; Chuit, C.; Corriu, J. P.; Reye, C. *T* **1989**, *45*, 171.
6. Brown, P. A.; Bonnert, R. V.; Jenkins, P. R.; Lawrence, N. J.; Selim, M. R. *JCS(P1)* **1991**, 1893.
7. (a) Jin, F.; Xu, Y.; Huang, W. *JCS(P1)* **1993**, 795. (b) Jin, F.; Xu, Y.; Huang, W. *CC* **1993**, 814.
8. Imamoto, T.; Takiyama, N.; Nakamura, K.; Hatajima, T.; Kamiya, Y. *JACS* **1989**, *111*, 4392.
9. (a) Posner, G. H. *OR* **1972**, *19*, 1. (b) Posner, G. H. *OR* **1975**, *22*, 253. (c) Normant, J. F. *S* **1972**, 63. (d) Lipshutz, B. H. *S* **1987**, 325. (e) Lipshutz, B. H. *COS* **1991**, *1*, 107. (f) Lipshutz, B. H.; Sengupta, S. *OR* **1992**, *41*, 135.
10. (a) Harayama, T.; Cho, H.; Inubushi, Y. *TL* **1977**, 3273. (b) House, H. O.; Chu, C.-Y. Phillipps, W. V.; Sayer, T. S. B.; Yau, C. C. *JOC* **1977**, *42*, 1709.
11. Matsuzawa, S.; Horiguchi, Y.; Nakamura, E.; Kuwajima, I. *T* **1989**, *45*, 349.
12. Derguini-Boumechal, F.; Linstrumelle, G. *TL* **1976**, 3225.
13. Henin, F.; Muzart, J. *SC* **1984**, *14*, 1355.
14. Nugent, W. A.; McKinney, R. J. *JOC* **1985**, *50*, 5370.
15. (a) Ratovelomanana, V.; Linstrumelle, G. *BSF(2)* **1987**, 174 (*CA* **1988**, *108*, 5773n). (b) Dang, H. P.; Linstrumelle, G. *TL* **1978**, 191.
16. Murahashi, S. I.; Yamamura, M.; Yanagisawa, K. I.; Mita, N.; Kondo, K. *JOC* **1979**, *44*, 2408.
17. Tsukazaki, M.; Snieckus, V. *TL* **1993**, *34*, 411.
18. Seyferth, D. *OSC* **1963**, *4*, 258.
19. Wallace, R. H.; Zong, K. K. *TL* **1992**, *33*, 6941.

Gérard Linstrumelle & Mouâd Alami
Ecole Normale Supérieure, Paris, France

Vinyltributylstannane[1]

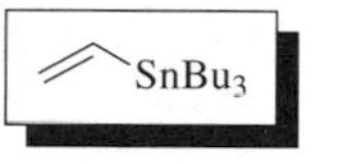

[7486-35-3] $C_{14}H_{30}Sn$ (MW 317.15)

(vinyl nucleophile in palladium-mediated cross coupling reactions;[2] occasionally used as a vinyllithium precursor[3])

Alternate Name: vinyltributyltin.
Physical Data: bp 104–106 °C/3.5 mmHg, 95 °C/1.5 mmHg; *d* 1.085 g cm^{-3}.
Solubility: sol common organic solvents; insol H_2O.
Form Supplied in: neat oil; widely available. Occasionally contaminated with small amounts of Bu_3SnCl.
Analysis of Reagent Purity: the pure compound shows one nonpolar spot by TLC (hexane) and one peak in the GC. ^{1}H NMR can also be useful.
Preparative Methods: conveniently prepared by the reaction of ***Vinylmagnesium Bromide*** with ***Tri-n-butylchlorostannane***.[4]
Purification: large quantities of material are purified by fractional distillation. Small amounts may be filtered through a plug of silica.
Handling, Storage, and Precautions: is air- and water-stable; requires no special handling or storage. Organostannanes are toxic and should only be used in a well ventilated hood. All glassware should be rinsed in a KOH/EtOH bath during cleaning.

Palladium-Mediated Cross Coupling (the Stille Reaction). Vinyltributyltin is commonly used as a nucleophile in palladium-mediated cross-coupling reactions with organic electrophiles.[2] The electrophile may be a halide or a pseudohalide which contains an sp or sp^2 carbon at or immediately adjacent to the electrophilic center. Halide reactivity varies in the order: I > Br ≫ Cl. Either Pd^0 or Pd^{II} catalysts may be used to mediate the reaction, the latter being reduced in situ by excess stannane. Acid chlorides react with vinyltributyltin in the presence of

Benzylchlorobis(triphenylphosphine)palladium(II) in HMPA or chloroform at 65 °C (eq 1).[5] In the latter solvent, yields are lower but the workup is significantly easier. Often the introduction of air into the reaction accelerates cross coupling of acid chlorides. Imidoyl chlorides (eq 2)[6] and vinyl halides also undergo Pd-mediated cross coupling (eqs 3 and 4).[7,8] Coupling of aryl halides typically requires electron-poor aromatics or aryl iodides (eq 5).[9]

0.4 mol% $PdBn(Cl)(PPh_3)_2$; $CHCl_3$, 0.33 h, 65 °C; 71% (1)

0.8 mol% $PdCl_2(PPh_3)_2$; xylene, 120 °C, 10 h; 67% (2)

$PdCl_2(PPh_3)_2$; DMF, rt, 18 h; 72% (3)

5 mol% $PdCl_2(PPh_3)_2$; DMF, rt, 12 h (4)

18 mol% $PdCl_2(PPh_3)_2$; toluene, 100 °C; 75–95% (5)

Hegedus et al. report a case in which an iodide attached to an electron-rich alkene fails to couple under normal conditions.[10] Treatment of the vinyl halide under Heck alkenation conditions in the absence of solvent gives a moderate yield of the desired product (eq 6).[10] Use of weakly donating stabilizing ligands, such as tri(2-furyl)phosphine and triphenylarsine, accelerates transmetalation, allowing coupling under conditions more gentle than those required with normal palladium catalysts [$Pd(PPh_3)_4$, $PdCl_2(PPh_3)_2$, etc.] (eq 7).[11]

10 mol% $Pd(OAc)_2$, $P(o\text{-}Tol)_3$, Bu_3N; 100 °C; 36% (6)

5 mol% $Pd(dba)_2$, THF, 65 °C; 10 mol% tri(2-furyl)phosphine (7)

A number of pseudohalides act as electrophiles in palladium-catalyzed cross couplings. Vinyl and aryl triflates couple with a wide variety of organostannanes in the presence of excess of ***Lithium Chloride*** (eq 8).[12–14] Cross coupling with aryl triflates is more difficult than that with vinyl triflates, typically requiring either a more nucleophilic catalyst [***Dichloro[1,1′-bis(diphenylphosphino)ferrocene]palladium(II)***, (dppf)$PdCl_2$)] or higher temperatures (dioxane at 98 °C) (eq 9).[15] Use of vinyl[16] or phenyl[17] fluorosulfonates allows lower cost at the expense of lower stability with respect to the analogous triflate (eq 10). Arenesulfonates have also been coupled with vinyltributyltin.[18] However, the generality of this variation is not yet clear.

4 mol% $Pd(PPh_3)_4$; 3 equiv LiCl, THF (8)

$Pd(PPh_3)_4$; dioxane, 98 °C, 4 h (9)

$Pd(OAc)_2$; CH_2Cl_2, 2 h, rt (10)

Other electrophiles which undergo cross coupling include hypervalent iodine compounds (eq 11)[19] and aryl diazonium salts (eq 12).[20] Coupling of aryl diazonium salts requires excess vinyltributyltin and 1 equiv of the catalyst to avoid significant byproduct formation.[20] Carbonylative couplings can be achieved if cross coupling is conducted under 1–3 atm of ***Carbon Monoxide*** (eq 13).[14] Vinyltributyltin has also been used in tandem Heck alkenation/Stille couplings, in which the stannane reacts with an alkyne and an electrophile in the presence of ***Tetrakis(triphenylphosphine)palladium(0)*** (eq 14).[21,22]

(11) Ph–CH=CH–I^+–Ph BF_4^- + CH₂=CH–$SnBu_3$ → [5 mol% $Pd(MeCN)_2Cl_2$, DMF, rt, 5 min, 79%]

(12) PhN_2BF_4 + CH₂=CH–$SnBu_3$ → [$Pd(dba)_2$, MeCN, rt, 5 min] styrene + biphenyl + stilbene

(13) aryl triflate + CH₂=CH–$SnBu_3$ → [4 mol% $PdCl_2(dppf)$, 1 atm CO, 3 equiv LiCl, DMF, 75 °C, 21 h, 67%]

(14) Ph–≡ + CH₂=CH–$SnBu_3$ → [$Pd(PPh_3)_4$, TMSI, dioxane, 8 h, 80%]

Transmetalation. Transmetalation from tin offers a facile synthesis of vinyl nucleophiles. Treatment of vinyltributyltin with ***Methyllithium*** affords methyltributyltin and ***Vinyllithium*** (eq 15).[3] Similarly, treatment with higher order cuprates causes quantitative in situ formation of the vinylcuprate (eq 16).[23] Vinyltributyltin will transfer a vinyl group in reactions with both trimethylchlorogermane and tetrachlorogermane.[24] Transmetalation with trimethylchlorogermane requires the presence of a catalytic amount of ***Aluminum Chloride***. Treatment of an excess of vinyltributyltin with ***9-Bromo-9-borabicyclo[3.3.1]nonane*** affords ***9-Vinyl-9-borabicyclo[3.3.1]nonane*** while avoiding the manipulation of air-sensitive compounds.[25] Excess vinyltributyltin is necessary to remove adventitious protic acid, and a greater than usual amount of H_2O_2/NaOH is used during oxidative hydrolysis to allow for reaction with bromotributyltin. Iodosulfonylation of vinyltributyltin followed by elimination of HI (DBU) affords the vinyl sulfone (eq 17),[26] which can subsequently be used in Diels–Alder reactions or can be lithiated.[27]

(15) CH₂=CH–$SnBu_3$ → [MeLi, –45 °C, THF] CH₂=CH–Li → [1. oxazoline; 2. iodide; 99%]

(16) CH₂=CH–$SnBu_3$ → [1. $Me_2Cu(CN)Li_2$, THF, rt, 1 h; 2. cyclohexenone, –78 °C] 3-vinylcyclohexanone

(17) CH₂=CH–$SnBu_3$ → [1. p-Me$C_6H_4SO_2I$; 2. DBU, CH_2Cl_2; 35%] $ArSO_2$CH=CH$SnBu_3$

Related Reagents. *trans*-1,2-Bis(tributylstannyl)ethylene; 1-Ethoxy-2-tributylstannylethylene; Lithium (1-Pentynyl)(2-tri-*n*-butylstannylvinyl)cuprate; Tetravinylstannane; Tri-*n*-butylstannylacetylene; (*E*)-1-Tri-*n*-butylstannyl-2-trimethylsilylethylene; (*E*)-1-Trimethylsilyl-2-trimethylstannylethylene; 9-Vinyl-9-borabicyclo[3.3.1]nonane; Vinylcopper; Vinyllithium; Vinylmagnesium Bromide; Vinylmagnesium Bromide–Copper(I) Iodide; Vinylmagnesium Bromide–Methylcopper; Vinylmagnesium Chloride–Copper(I) Chloride; Vinyltrimethylsilane.

1. (a) Pereyre, M.; Quintard, J.-P.; Rahm, A. *Tin in Organic Synthesis*; Butterworths: London, 1987. (b) Davies, A. G.; Smith, P. J. In *Comprehensive Organometallic Chemistry*; Wilkinson, G., Ed.; Pergamon: Oxford, 1982; Chapter 11. (c) Ingham, R. K.; Rosenberg, S. D.; Gilman, H. *CRV* **1960**, *60*, 459.
2. (a) Stille, J. K. *AG(E)* **1986**, *25*, 508. (b) Scott, W. J.; McMurry, J. E. *ACR* **1988**, *21*, 47.
3. Robichaud, A. J.; Meyers, A. I. *JOC* **1991**, *56*, 2607.
4. Seyferth, D.; Stone, F. G. A. *JACS* **1957**, *79*, 515.
5. Stille, J. K.; Labadie, J. W.; Tueting, D. *JOC* **1983**, *48*, 4634.
6. Kosugi, M; Koshiba, M.; Atoh, A.; Sano, H.; Migita, T. *BCJ* **1986**, *59*, 677.
7. Forsyth, C. J.; Clardy, J. *JACS* **1988**, *110*, 5911.
8. Wender, P. A.; Tebbe, M. J. *S* **1991**, 1089.
9. Marsais, F.; Pineau, P.; Nivolliers, F.; Mallet, M.; Turck, A.; Godard, A.; Queguiner, G. *JOC* **1992**, *57*, 565.
10. Hegedus, L. S.; Holden, M. S. *JOC* **1986**, *51*, 1171.
11. Farina, V.; Baker, S. R.; Benigni, D. A.; Sapino, C. *TL* **1988**, *29*, 5739.
12. Scott, W. J.; Stille, J. K. *JACS* **1986**, *108*, 3033.
13. Peña, M. R.; Stille, J. K. *JACS* **1989**, *111*, 5417.
14. Echavarren, A. M.; Stille, J. K. *JACS* **1988**, *110*, 1557.
15. Echavarren, A. M.; Stille, J. K. *JACS* **1987**, *109*, 5478.
16. Roth, G. P.; Sapino, C. *TL* **1991**, *32*, 4073.
17. Roth, G. P.; Fuller, C. E. *JOC* **1991**, *56*, 3493.

18. Badone, D.; Cecchi, R.; Guzzi, U. *JOC* **1992**, *57*, 6321.
19. Moriarty, R. M.; Epa, W. R. *TL* **1992**, *33*, 4095.
20. Kikukawa, K.; Kono, K.; Wada, F.; Matsuda, T. *JOC* **1983**, *48*, 1333.
21. Chatani, N.; Amishiro, N.; Murai, S. *JACS* **1991**, *113*, 7778.
22. Kosugi, M.; Tamura, H.; Sano, H.; Migita, T. *TL* **1989**, *45*, 961.
23. Behling, J. R.; Babiak, K. A.; Ng, J. S.; Campbell, A. L. *JACS* **1988**, *110*, 2641.
24. Mironov, V. F.; Nuridzhanyan, A. K. *Metalloorg. Khim.* **1992**, *5*, 705 (*CA* **1992**, *117*, 234 145n).
25. Singleton, D. A.; Martinez, J. P.; Ndip, G. M. *JOC* **1992**, *57*, 5768.
26. Rasset-Deloge, C.; Martinez-Fresneda, P.; Vaultier, M. *BSF* **1992**, *129*, 285.
27. Ochiai, M.; Ukita, T.; Fujita, E. *TL* **1983**, *24*, 4025.

William J. Scott & Alessandro F. Moretto

Bayer Pharmaceuticals Division, West Haven, CT, USA

Zinc/Copper Couple[1]

Zn(Cu)

[12019-27-1] Zn (MW 65.39)

(an activated form of zinc metal that can be used for cyclopropanation;[1] conjugate addition of alkyl iodides to enones;[2] preparation of dichloroketene;[3] preparation of 2-oxyallyl cations for cycloaddition reactions[4])

Physical Data: see ***Zinc***.
Solubility: insol organic solvents and H_2O.
Form Supplied in: reddish-brown or dark gray powder; often prepared directly before use.
Preparative Methods: although zinc–copper couple is commercially available, freshly prepared reagents are often more active. Of the many preparations described in the literature,[1] LeGoff's is both simple and results in a very active zinc–copper couple.[5] Zinc dust (35 g) was added to a rapidly stirred solution of 2.0 g $Cu(OAc)_2 \cdot H_2O$ (see ***Copper(II) Acetate***) in 50 mL of hot acetic acid. After 30 s the couple was allowed to settle, decanted, and washed once with acetic acid and three times with ether. A less reactive couple can be prepared from granular zinc using the same procedure. Modified procedures for preparing zinc–copper couples are given in many of the references.
Handling, Storage, and Precautions: zinc–copper couple deteriorates in moist air and should be stored under nitrogen. Very active zinc–copper couple is oxygen sensitive. It evolves hydrogen on contact with strong aqueous acids.

Introduction. Zinc–copper couple is an active form of zinc metal, and for many reactions the presence of copper does not appear to exert any special influence beyond activating the surface of the zinc. Many of the reactions of zinc–copper couple could probably be effected using zinc metal activated in situ with ***Chlorotrimethylsilane*** or ***1,2-Dibromoethane***,[6] but the advantage of zinc–copper couple is that it is a storable form of activated zinc that is easily prepared and handled.

Simmons–Smith Cyclopropanation.[1,7] Zinc–copper couple reduces ***Diiodomethane*** to generate ***Iodomethylzinc Iodide***, which is in equilibrium with bis(iodomethyl)zinc and zinc iodide.[8] This Simmons-Smith reagent is widely used to cyclopropanate alkenes. As shown in eqs 1 and 2, the cyclopropanation is stereospecific, with methylene adding cis to the starting alkene.[7]

CH_2I_2, Zn(Cu) (1)

CH_2I_2, Zn(Cu) (2)

Hydroxyl groups or sterically accessible ethers will direct the Simmons–Smith reagent to the proximate face of an alkene,[9] and this stereoselectivity has been used extensively in synthesis (eq 3).[10]

CH_2I_2, Zn(Cu), 85% (3)

One important application of directed cyclopropanation is the stereoselective introduction of methyl groups in terpene synthesis. In eq 4 the allylic alcohol is again used to direct the cyclopropanation to the desired face of the ketone.[11] Oxidation to the ketone and lithium–ammonia reduction gave the β-methyl ketone. This stepwise approach to β-methyl substitution of an enone complements a ***Lithium Dimethylcuprate*** addition because it can be carried out on β-disubstituted enones and the stereochemistry is reliably predictable based on the alcohol stereochemistry.

Simmons–Smith, 51%; Jones, 63%; 1. $LiNH_3$, 2. Jones, 52% (4)

Wenkert developed an alternative to the classic alkylation of ketones by cyclopropanating the derived enol ether, followed by acid catalyzed hydrolysis of the cyclopropyl ether. This sequence was used to introduce the angular methyl group of (−)-valeranone with complete control of stereochemistry (eq 5).[12]

CH_2I_2, Zn(Cu), 91%; 1. Jones oxidation, 2. Wolff–Kishner, 43%; HCl, H_2O, MeOH, Δ, 100%; (−)-Valeranone (5)

Mash has found that optically pure acetals prepared from chiral diols and cycloalkanones will cyclopropanate with good diastereoselectivity (eq 6), presumably through selective coordination of the Simmons-Smith reagent to the more accessible acetal oxygen. Hydrolysis of the acetals leads to bicyclo[*n*.1.0]alkanones in good to excellent optical purity.

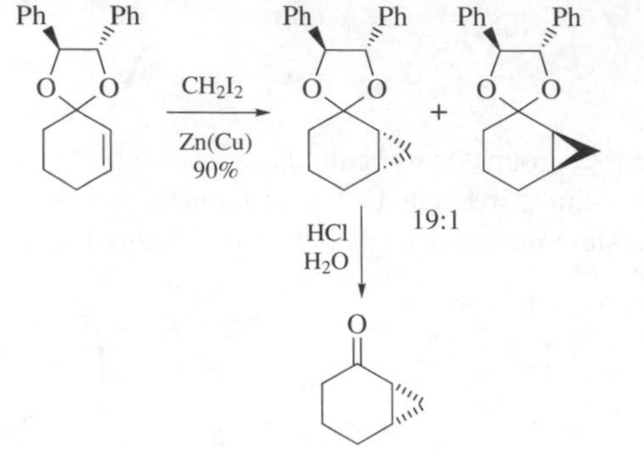

(6)

(10)

Alkyl Zinc Preparations and Reactions. Dialkylzinc compounds have been prepared from the corresponding alkyl iodides and zinc–copper couple,[13] but are more commonly prepared using activated zinc or from the appropriate Grignard reagent and ***Zinc Chloride***. Zinc–copper couple is useful for preparing a variety of unusual alkyl zinc compounds such as (iodomethyl)zinc iodide. Apart from cyclopropanations, (iodomethyl)zinc iodide reacts with ***Tri-n-butylchlorostannane*** to give ***(Iodomethyl)tri-n-butylstannane***, an important reagent in the preparation of α-alkoxylithium reagents (eq 7).[14]

$$CH_2I_2 \xrightarrow[\text{2. Bu}_3\text{SnCl}]{\text{1. Zn(Cu)}} ICH_2SnBu_3 \xrightarrow[\text{NaH}]{\text{PhCH}_2\text{OH}} PhCH_2OCH_2SnBu_3 \quad (7)$$

Propargyl bromide (see ***Propargyl Chloride***) reacts with zinc–copper couple in the presence of an aromatic aldehyde to produce the alkyne adduct in good yield (eq 8). Presumably the allenylzinc bromide is formed in situ and reacts with 1,3-rearrangement; none of the allene adduct is reported.[15]

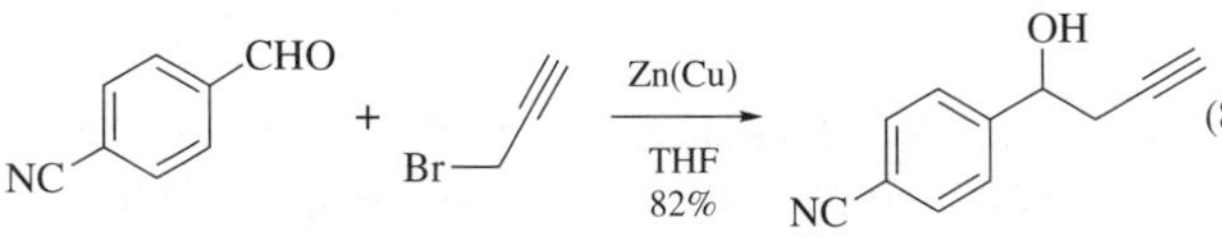

(8)

Yoshida's group found that zinc–copper couple was very effective in the preparation of zinc homoenolates and bishomoenolates. Reaction of ethyl 4-iodobutyrate with zinc–copper couple in benzene and DMA gave the homoenolate (**1**), which coupled with acyl chlorides, vinyl iodides, or vinyl triflates in the presence of ***Tetrakis(triphenylphosphine)palladium(0)*** (eq 9).[16]

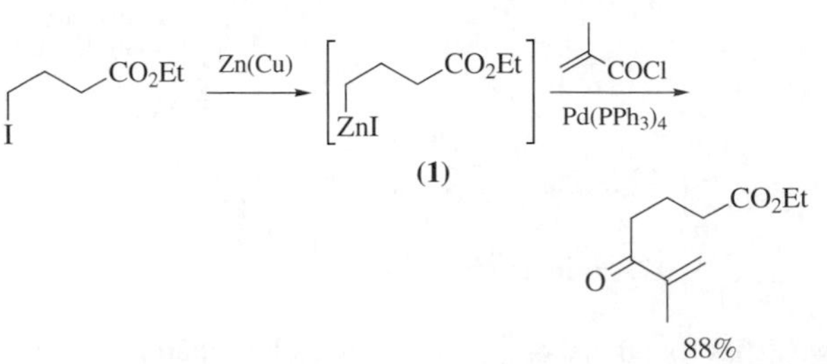

(9)

Zinc homoenolates such as ***(3-Ethoxy-3-oxopropyl)iodozinc*** prepared in this manner can be transmetalated to titanium homoenolates and coupled with aldehydes (eq 10),[17] or reacted with ***t-Butyldimethylchlorosilane*** to give the cyclopropyl ethers.[18]

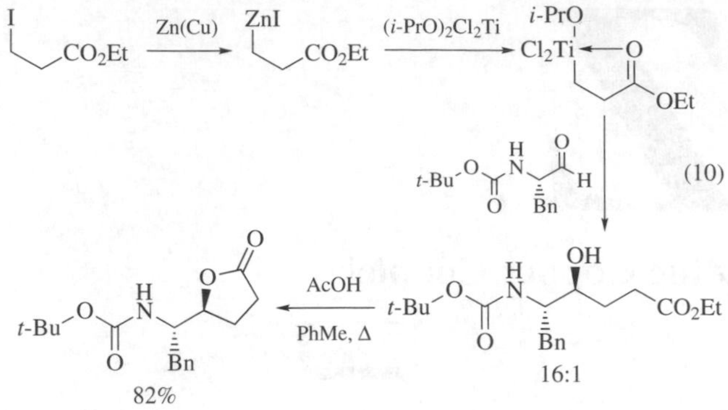

Zinc–copper couple leads to higher yield and faster reactions in the Reformatsky reaction see ***Ethyl Bromozincacetate*** than the more commonly used zinc metal (eq 11).[19]

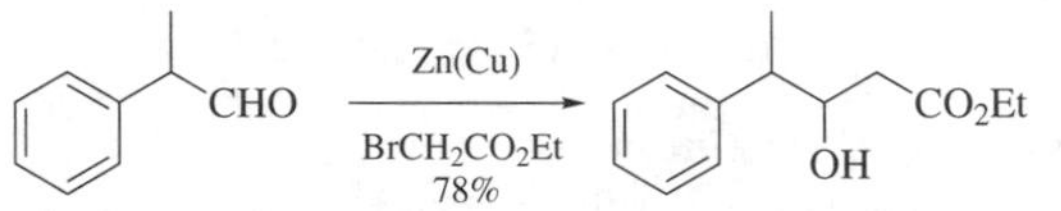

(11)

Semmelhack reported an unusual cyclization reaction that probably proceeds through an allylic zinc intermediate (eq 12).[20] Model cyclizations were carried out using allylic bromides, but the actual cyclization was performed on the allylic sulfonium salt. In contrast to the result with zinc–copper couple, cyclization of the sulfonium salt intermediate with ***Bis(1,5-cyclooctadiene)nickel(0)*** gave the other *cis*-fused lactone in 43% overall yield.

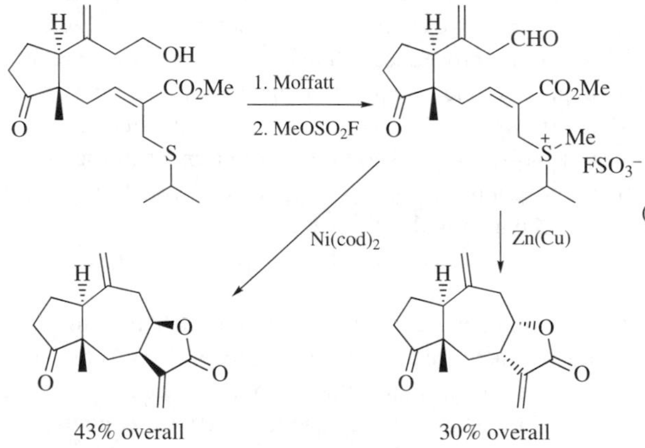

(12)

Conjugate Additions of Alkyl Iodides to Alkenes. An intriguing new carbon–carbon bond forming reaction has been developed that promises to be very useful in synthesis. Luche's group found that alkyl iodides will add to alkenes bearing a strong electron-withdrawing group when sonicated with zinc–copper couple in an aqueous solvent.[21] This reaction is notable because the conditions are very mild, it is compatible with many reactive functional groups, and moderately hindered carbon–carbon bonds can be formed in good yield. Luche proposed that the reaction proceeds through a radical intermediate generated in the reduction of the alkyl iodide with zinc–copper couple, but the evidence is not completely clear. When the reaction is conducted in a solvent system with a good deuterium atom donor and a good proton source, the proton is incorporated into the product, whereas when the role of the hydrogen and deuterium are reversed, deuterium is incorporated into the product (eq 13).[2]

Zn(Cu), solvent; solvent = Me_2CDOH, H_2O, X = H; solvent = EtOD, D_2O, X = D; Zn(Cu); H^+ or D^+; Zn(Cu) or I^- (13)

A radical intermediate should give the opposite incorporation pattern, but Luche invokes a rapid reduction of the α-keto radical to give the enolate, which would give the observed incorporation. Thus the addition can still proceed through a radical even though the quenching is ionic. The best direct evidence of radical involvement is cyclization of the radical intermediate in eq 14, but the yield is quite low.

Zn(Cu), sonicate; 10% (14)

Several other features of this reaction are noteworthy: both water and copper are required, and (*S*)-octyl iodide couples well but leads to a racemic product.

The conjugate addition reactions of alkyl iodides work well in complex and highly functionalized molecules. The side chain of a vitamin D analog was prepared using the reaction shown in eq 15.[22] Unlike most transition metal-catalyzed reactions, this procedure did not affect the vinyl triflate.

CO_2Me; Zn, CuI,))); EtOH, H_2O; 65% (15)

Surprisingly, the coupling reaction will also work in carbohydrate systems. The alkyl iodide derived from ribose was coupled in good yield with 2-butenenitrile (eq 16), without undergoing reductive elimination to form an alkene.[23] This new coupling procedure promises to be very useful in synthesis.

CN; Zn, CuI; Et_2O, THF, H_2O; 53% (16)

Reductive Coupling of Carbonyl Compounds. Carbonyl compounds can be reductively dimerized to diols or alkenes. A pinacol cross-coupling reaction between α,β-unsaturated ketones and acetone has been reported. When carvone is sonicated with zinc dust and ***Copper(II) Chloride*** in acetone/water, the cross-coupling product is isolated in 92% yield (eq 17).[24] Acetone itself does not dimerize under the reaction conditions, and β,β-disubstituted enones fail to react, presumably due to a higher reduction potential.

Zn, $CuCl_2$; acetone, H_2O; sonicate (17)

The McMurry coupling uses low valent titanium to reductively dimerize carbonyl compounds to alkenes.[25] The reagent is usually prepared from ***Titanium(III) Chloride*** and an added reducing agent like ***Potassium***, ***Lithium***, or ***Lithium Aluminium Hydride***. One of the best reagents is prepared from $TiCl_3(DME)_{1.5}$ and zinc–copper couple (eq 18).[26] The same dimerization gives only 12% yield with $TiCl_3/LiAlH_4$.

$TiCl_3(DME)_{1.5}$; Zn(Cu); 87% (18)

Preparation of Dichloroketene. ***Dichloroketene*** can be prepared by reducing ***Trichloroacetyl Chloride*** with zinc–copper couple.[3] Dichloroketene undergoes facile [2 + 2] cycloadditions with alkenes[27] and alkynes[28] to give cyclobutanones and cyclobutenones.[29] Recently the use of a chiral auxiliary has been reported to control the facial selectivity of dichloroketene additions (eq 19).[30] Ring expansion with ***Diazomethane*** gave the dichlorocyclopentanone, and subsequent reduction removed the chiral auxiliary and gave the optically enriched cyclopentenone.

CCl_3COCl; Zn(Cu); 92%; 95:5; CH_2N_2; $Cr(ClO_4)_2$; 63% (19)

Dichloroketene will react with vinyl sulfoxides to give γ-lactones with efficient stereochemical transfer from the sulfoxide stereogenic center (eq 20).[31] Optically pure vinyl sulfoxides give lactones with high enantiomeric excesses. The chlorine atoms can be removed by reduction with zinc–copper couple or ***Aluminum Amalgam*** to give the lactone.

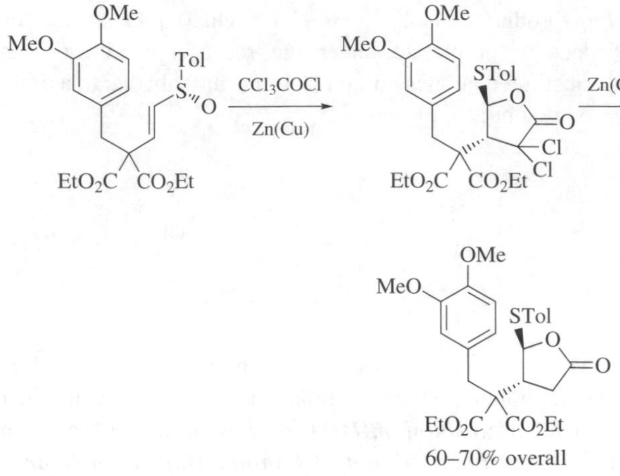

(20)

60–70% overall

Reduction of α-halo acyl chlorides normally gives ketenes, but Hoffmann has reported an unusual reductive trimerization of 1-bromocyclopropanecarbonyl chloride that does not proceed through the corresponding ketene.[32] Reduction with zinc–copper couple in MeCN gave the unusual trimer in 20% yield (eq 21), whereas the corresponding ketene is known to dimerize. Hoffmann suggests that a Reformatsky-type intermediate may react with a second molecule of acyl chloride rather than eliminating to generate the ketene.

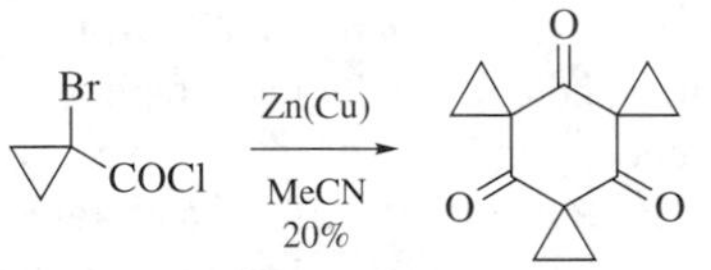

(21)

Preparation of 2-Oxyallyl Cations. Zinc–copper couple reduces α,α′-dibromo ketones to zinc oxyallyl cations, and these allyl cations will undergo [3 + 4] cycloadditions with dienes (eq 22).[33]

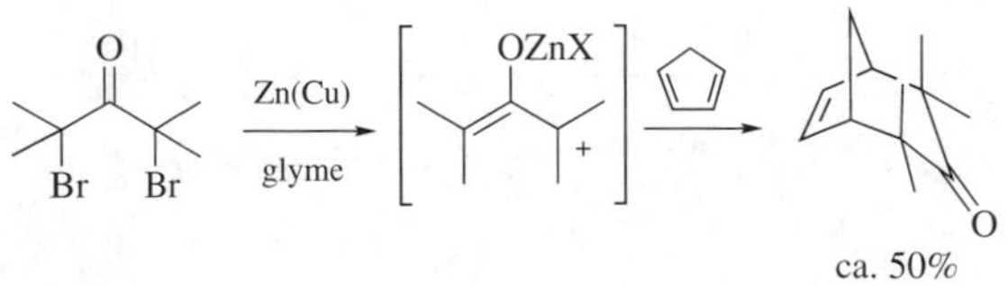

(22)

ca. 50%

The original procedure used zinc–copper couple directly, but more recently sonicating the reactions has been reported to improve the yields significantly: the coupling in eq 23 gives 60% yield without sonication and 91% yield with sonication.[4]

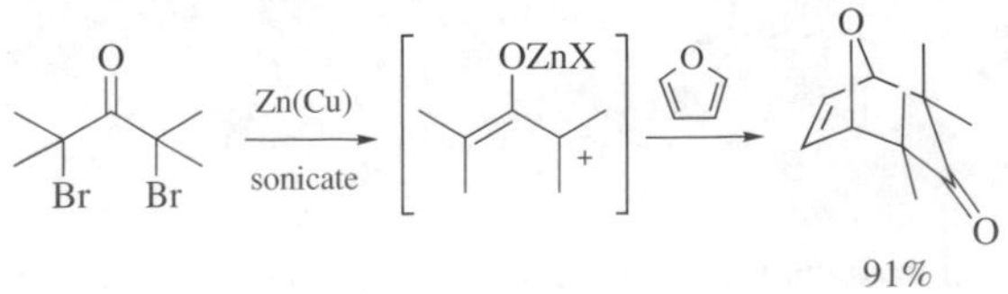

(23)

91%

These reactions can be carried out using ***Nonacarbonyldiiron*** as the reducing agent in comparable yield, but the iron reagent is quite toxic and does not offer any advantage in simple cases.[34]

Reduction of α,α′-dibromo ketones under slightly different conditions, using *N*-methylformamide as the solvent without diene or sonication, leads to 1,4-diketones (eq 24).[35] Unsymmetrically substituted ketones give statistical mixtures of products.

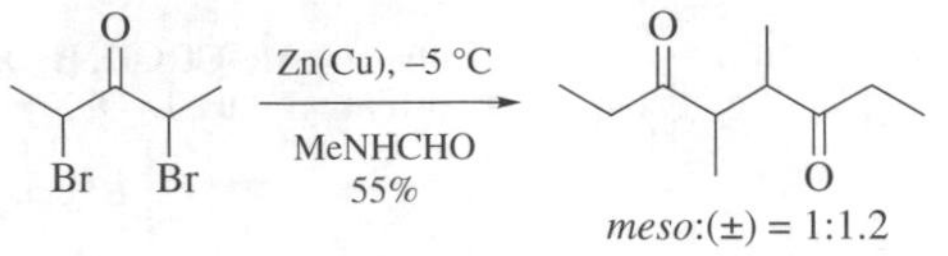

(24)

meso:(±) = 1:1.2

Reduction of 1,2- and 1,3-Dihalides and Halo Ethers. ***Zinc–Acetic Acid*** is commonly used to reduce 1,2-dihalides to alkenes. Several modifications of this procedure use zinc–copper couple as the reducing agent.[36] Bromooxiranes prepared from allyl alcohols are reduced with zinc–copper couple and sonication to give the 1,3-transposed allylic alcohol (eq 25).[37] The oxirane ring strain facilitates the alkene formation; notice that the tetrahydrofuran in eq 16 does not form an alkene.

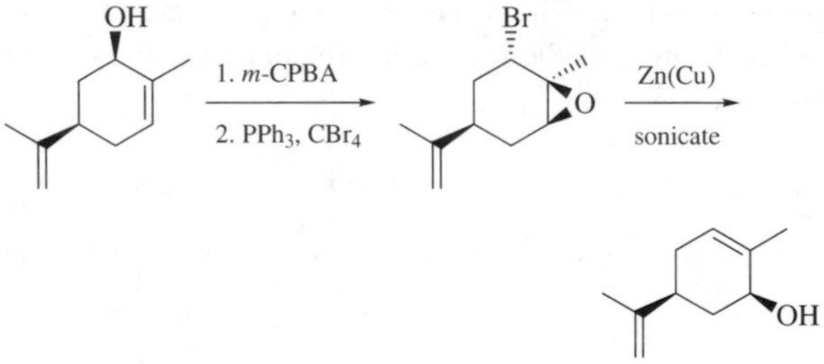

(25)

The 2,2,2-trichloroethyl group is used as a protecting group for many functional groups such as carboxylic acids and carbamic acids (amines), and is usually removed by reduction with zinc metal. In some cases it is advantageous to remove these protecting groups with zinc–copper couple.[38]

Under forcing conditions, oxiranes can be reduced to the corresponding alkene. Reaction of eupachloroxin with zinc–copper couple in refluxing ethanol for 3 d gave eupachlorin without reducing the many other functional groups in the molecule (eq 26).[39] The reduction is stereoselective: *cis*-oxiranes give *cis*-alkenes and *trans*-oxiranes give *trans*-alkenes.

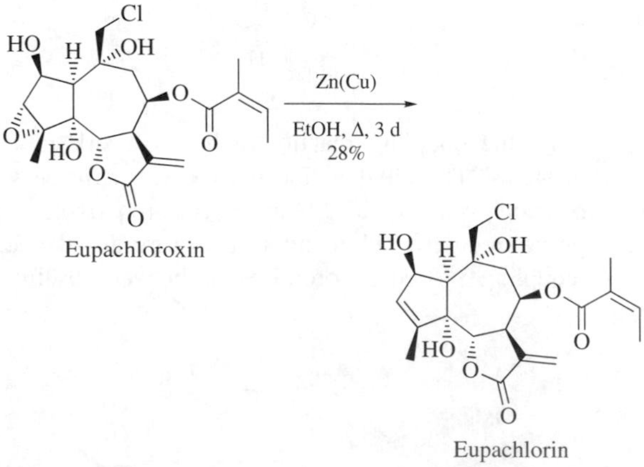

Eupachloroxin

(26)

Eupachlorin

Reduction of 1,3-dibromides with zinc–copper couple gives cyclopropanes (eq 27).[40] Cyclopropanes are also formed when

2-bromoethyloxiranes are reduced using zinc–copper couple with sonication.[41]

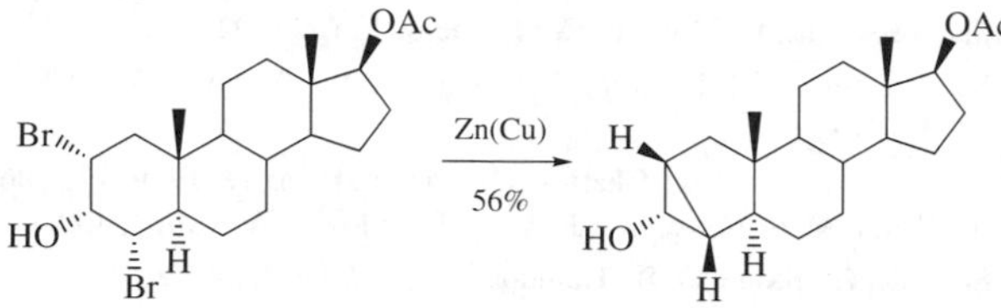

(27)

Reduction of C–C Multiple Bonds. Alkenes with powerful electron-withdrawing groups like esters, ketones, or nitriles are reduced with zinc–copper couple in refluxing methanol (eq 28).[42] Isolated alkenes are not reduced.

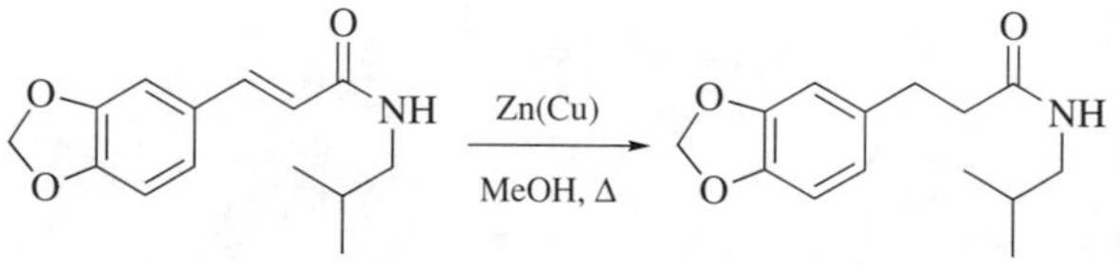

(28)

Unactivated alkynes are reduced to (*Z*)-alkenes under the same conditions.[43] Terminal alkynes are reduced to vinyl groups without the over-reduction that can sometimes accompany Birch-type reductions. This reduction was recently applied to the synthesis of leukotriene B_4 (eq 29).[44]

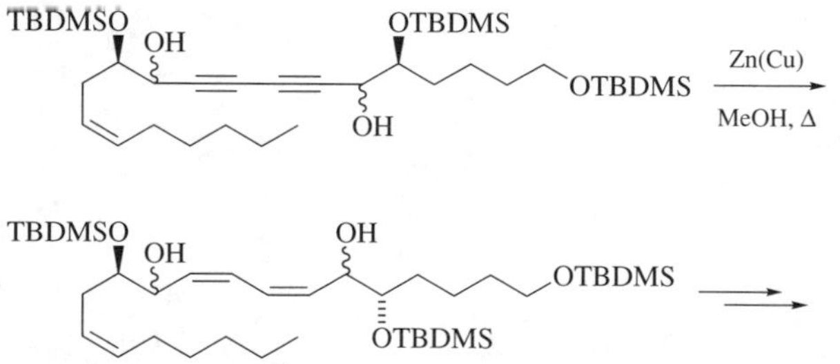

Leukotriene B_4 (29)

Reduction of Other Functional Groups. Zinc metal has been used to reduce many functional groups and zinc–copper couple would be effective in most of these reactions. Zinc–copper couple reduces alkyl halide to the corresponding hydrocarbon, and in the presence of D_2O this leads to an expedient synthesis of specifically deuterated materials (eq 30).[45]

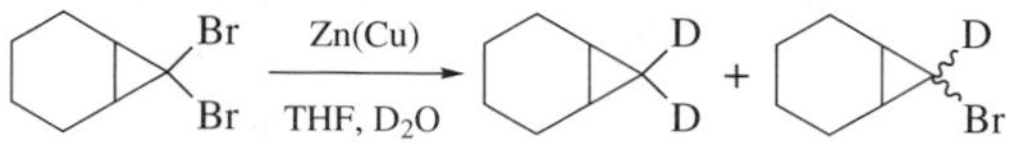

(30)

21%, 96% d_2 62%, 94% d_2

Aromatic nitro compounds are reduced to the protected hydroxylamines, *N*-acetoxy-*N*-acetylarylamines, with zinc–copper couple in the presence of acetic acid and acetic anhydride. In contrast, zinc metal reduces 4-nitrostilbene to azoxystilbene.[46] Oximes can be reduced to amines with zinc–copper couple, and this procedure is reported to be much more effective than an ***Aluminum Amalgam*** reduction (eq 31).[47]

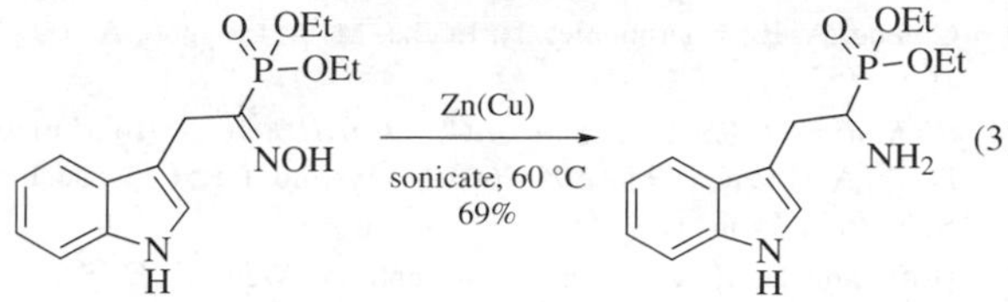

(31)

Related Reagents. Chloroiodomethane–Zinc/Copper Couple; Dibromomethane–Zinc/Copper Couple; Diethylzinc; Diethylzinc-Iodoform; Diiodomethane; Iodomethylzinc Iodide; Potassium Iodide–Zinc/Copper Couple; Titanium(III) Chloride–Zinc/ Copper Couple.

---

1. Simmons, H. E.; Cairns, T. L.; Vladuchick, S. A.; Hoiness, C. M. *OR* **1973**, *20*, 1.
2. (a) Luche, J. L.; Allavena, C.; Petrier, C.; Dupuy, C. *TL* **1988**, *29*, 5373. (b) Dupuy, C.; Petrier, C.; Sarandeses, L. A.; Luche, J. L. *SC* **1991**, *21*, 643.
3. Hassner, A.; Krepski, L. R. *JOC* **1978**, *43*, 3173.
4. Joshi, N. L.; Hoffmann, H. M. R. *TL* **1986**, *27*, 687.
5. LeGoff, E. *JOC* **1964**, *29*, 2048.
6. Sidduri, A.; Rozema, M. J.; Knochel, P. *JOC* **1993**, *58*, 2694.
7. Simmons, H. E.; Smith, R. D. *JACS* **1959**, *81*, 4256.
8. Blanchard, E. P.; Simmons, H. E. *JACS* **1964**, *86*, 1337.
9. (a) Chan, J. H. H.; Rickborn, B. *JACS* **1968**, *90*, 6406. (b) Dauben, W. G.; Berezin, G. H. *JACS* **1963**, *85*, 468.
10. Ando, M.; Sayama, S.; Takase, K. *JOC* **1985**, *50*, 251.
11. Packer, R. A.; Whitehurst, J. S. *JCS(P1)* **1978**, 110.
12. (a) Wenkert, E.; Mueller, R. A.; Reardon, E. J.; Sathe, S. S.; Scharf, D. J.; Tosi, G. *JACS* **1970**, *92*, 7428. (b) Wenkert, E.; Berges, D. A.; Golob, N. F. *JACS* **1978**, *100*, 1263.
13. Moorhouse, S.; Wilkinson, G. *JCS(D)* **1974**, 2187.
14. (a) Still, W. C. *JACS* **1978**, *100*, 1481. (b) Seyferth, D.; Andrews, S. B. *JOM* **1971**, *30*, 151.
15. Papadopoulou, M. V. *CB* **1989**, *122*, 2017.
16. Tamara, Y.; Ochiai, H.; Nakamura, T.; Tsubaki, K.; Yoshida, Z. *TL* **1985**, *26*, 5559. (b) Tamara, Y.; Ochiai, H.; Nakamura, T.; Yoshida, Z. *TL* **1986**, *27*, 955. (c) Tamara, Y.; Ochiai, H.; Nakamura, T.; Yoshida, Z. *OR* **1988**, *67*, 98.
17. DeCamp, A. E.; Kawaguchi, A. T.; Volante, R. P.; Shinkai, I. *TL* **1991**, *32*, 1867.
18. Yasui, K.; Fugami, K.; Tanaka, S.; Tamaru, Y.; Ii, A.; Yoshida, Z.; Saidi, M. R. *TL* **1992**, *33*, 785.
19. Santaniello, E.; Manzocchi, A. *S* **1977**, 698.
20. (a) Semmelhack, M. F.; Yamashita, A.; Tomesch, J. C.; Hirotsu, K. *JACS* **1978**, *100*, 5565. (b) Semmelhack, M. F.; Wu, E. S. C. *JACS* **1976**, *98*, 3384.
21. (a) Petrier, C.; Dupuy, C.; Luche, J. L. *TL* **1986**, *27*, 3149. (b) Luche, J. L.; Allavena, C. *TL* **1988**, *29*, 5369.
22. Sestelo, J. P.; Mascarenas, J. L.; Castedo, L.; Mourino, A. *JOC* **1993**, *58*, 118.
23. Blanchard, P.; Kortbi, M. S. E.; Fourrey, J.-L.; Robert-Gero, M. *TL* **1992**, *33*, 3319.
24. Delair, P.; Luche, J,-L. *CC* **1989**, 398.
25. McMurry, J. E.; Flemming, M. P. *JACS* **1974**, *96*, 4708.
26. McMurry, J. E.; Lectka, T.; Rico, J. G. *JOC* **1989**, *54*, 3748.
27. Brady, W. T. *T* **1981**, *37*, 2949.
28. Hassner, A.; Dillon, J. L., Jr. *JOC* **1983**, *48*, 3382.
29. (a) Danheiser, R. L.; Sard, H. *TL* **1983**, *24*, 23. (b) Ammann, A. A.; Rey, M.; Dreiding, A. S. *HCA* **1987**, *70*, 321.

30. Greene, A. E.; Charbonnier, F.; Luche, M.-J.; Moyano, A. *JACS* **1987**, *109*, 4752.
31. (a) Marino, J. P.; Neisser, M. *JACS* **1981**, *103*, 7687. (b) Marino, J. P.; Perez, A. D. *JACS* **1984**, *106*, 7643. (c) Marino, J. P.; Laborde, E.; Paley, R. S. *JACS* **1988**, *110*, 966.
32. Hoffmann, H. M. R.; Eggert, U.; Walenta, A.; Weineck, E.; Schhomburg, D.; Wartchow, R.; Allen, F. H. *JOC* **1989**, *54*, 6096.
33. Hoffmann, H. M. R.; Clemens, K. E.; Smithers, R. H. *JACS* **1972**, *94*, 3940.
34. Noyori, R.; Hayakawa, Y. *OR* **1983**, *29*, 163.
35. Chassin, C.; Schmidt, E. A.; Hoffmann, H. M. R. *JACS* **1974**, *96*, 606.
36. Santaniello, E.; Hadd, H. H.; Caspi, E. *J. Steroid Biochem.* **1975**, *6*, 1505.
37. (a) Sarandeses, L. A.; Mourino, A.; Luche, J.-L. *CC* **1991**, 818. (b) Sarandeses, L. A.; Luche, J.-L. *JOC* **1992**, *57*, 2757.
38. Imai, J.; Torrence, P. F. *JOC* **1981**, *46*, 4061.
39. (a) Kupchan, S. M.; Maruyama, M. *JOC* **1971**, *36*, 1187. (b) Ekong, D. E. U.; Okogun, J. I.; Sondengam, B. L. *JCS(P1)* **1975**, 2118.
40. Templeton, J. F.; Wie, C. W. *CJC* **1975**, *53*, 1693.
41. Sarandeses, L. A.; Mourino, A.; Luche, J.-L. *CC* **1992**, 798.
42. Sondengam, B. L.; Fomum, Z. T.; Charles, G.; Akam, T. M. *JCS(P1)* **1983**, 1219.
43. (a) Sondengam, B. L.; Charles, G.; Akam, T. M. *TL* **1980**, *21*, 1069. (b) Veliev, M. G.; Guseinov, M. M.; Mamedov, S. A. *S* **1981**, 400.
44. Solladie, G.; Stone, G. B.; Hamdouchi, C. *TL* **1993**, *34*, 1807.
45. (a) Blakenship, R. B.; Burdett, K. A.; Swenton, J. S. *JOC* **1974**, *39*, 2300. (b) Stephenson, L. M.; Gemmer, R. V.; Currect, S. P. *JOC* **1977**, *42*, 212.
46. Franz, R.; Neumann, H.-G. *Carcinogenesis* **1986**, *7*, 183.
47. Rogers, R. S.; Stern, M. K. *SL* **1992**, 708.

Scott D. Rychnovsky & Jay P. Powers
University of Minnesota, Minneapolis, MN, USA

LIST OF CONTRIBUTORS

Andrew Abell — *University of Canterbury, Christchurch, New Zealand*
- Triethyl Phosphonoacetate 645

David J. Ager — *The NutraSweet Company, Mount Prospect, IL, USA*
- Trimethylsilylmethylmagnesium Chloride 665

Irini Akritopoulou-Zanze — *University of Southern California, Los Angeles, CA, USA*
- Methyltitanium Trichloride 515

Henry A. Alegria — *University of South Carolina, Columbia, SC, USA*
- 1,3-Butadiene 144

Toyohiko Aoyama — *Nagoya City University, Japan*
- Trimethylsilyldiazomethane 662

José Barluenga — *Universidad de Oviedo, Spain*
- Phenyl(trichloromethyl)mercury 582

Albert K. Beck — *Eidgenössische Technische Hochschule, Zürich, Switzerland*
- (*R*)-2-*t*-Butyl-6-methyl-4*H*-1,3-dioxin-4-one 158

Matthew Beggs — *Tulane University, New Orleans, LA, USA*
- (−)-8-Phenylmenthol 577

Makonen Belema — *Yale University, New Haven, CT, USA*
- Triethyl Orthoacetate 640

Steven H. Bertz — *LONZA, Annandale, NJ, USA*
- Copper(I) Bromide 214
- Copper(I) Chloride 220
- Copper(I) Cyanide 227
- Copper(I) Iodide 230

Samit K. Bhattacharya — *University of Pennsylvania, Philadelphia, PA, USA*
- Acetylene 56

Robert K. Boeckman, Jr. — *University of Rochester, NY, USA*
- Lithium Bis(1-ethoxyvinyl)cuprate 450

Roger Bolton — *University of Surrey, Guildford, UK*
- Allyl Bromide 64

Manfred Braun — *Heinrich-Heine-Universität, Düsseldorf, Germany*
- 1,1,2-Triphenyl-1,2-ethanediol 675

Margaret E. Browne — *Schering-Plough Research Institute, Kenilworth, NJ, USA*
- Methyl Acrylate 487

Jarmila Brunckova — *University of Illinois at Chicago, IL, USA*
- Iodoform 421

Paul T. Buonora
Lamar University, Beaumont, TX, USA
- Methyl Vinyl Ketone 523

Carmen E. Burgos-Lepley
Cortech, Denver, CO, USA
- Glycidol 406

Duane A. Burnett
Schering-Plough Research Institute, Kenilworth, NJ, USA
- Methyl Acrylate 487

Vito Capriati
University of Bari, Italy
- Allylmagnesium Bromide 69

Kenneth C. Caster
Union Carbide Corporation, South Charleston, WV, USA
- Ethoxyacetylene 352

A. Richard Chamberlin
University of California, Irvine, CA, USA
- *p*-Toluenesulfonylhydrazide 611

Naoto Chatani
Osaka University, Japan
- Carbon Monoxide 169

J. Michael Chong
University of Waterloo, Ontario, Canada
- (2*R*,4*R*)-2,4-Pentanediol 572

Danièle Choueiry
Purdue University, West Lafayette, IN, USA
- Trimethylaluminum–Dichlorobis(η^5-cyclopentadienyl)zirconium 654

Robert J. Clemens
Eastman Chemical Company, Kingsport, TN, USA
- 2,2,6-Trimethyl-4*H*-1,3-dioxin-4-one 657
- Diketene 287

John A. Colapret
Lamar University, Beaumont, TX, USA
- Methyl Vinyl Ketone 523

Robert S. Coleman
University of South Carolina, Columbia, SC, USA
- 1,3-Butadiene 144

Françoise Colobert
University Louis Pasteur, Strasbourg, France
- (−)-(1*R*,2*S*,5*R*)-Menthyl (*S*)-*p*-Toluenesulfinate 468
- (*R*)-(+)-*t*-Butyl 2-(*p*-Tolylsulfinyl)acetate 162

Daniel L. Comins
North Carolina State University, Raleigh, NC, USA
- *N*,*N*-Dimethylformamide 312
- *N*-Methyl-*N*-(2-pyridyl)formamide 514

Arnel B. Concepcion
Nagoya University, Japan
- Formaldehyde 394

James M. Cook
University of Wisconsin-Milwaukee, WI, USA
- Dimethyl 1,3-Acetonedicarboxylate 301

Nicholas D. P. Cosford
SIBIA, La Jolla, CA, USA
- Copper(II) Bromide 217
- Copper(II) Chloride 223

Paul L. Creger
Ann Arbor, MI, USA
- Dilithioacetate 289

Michael T. Crimmins
University of North Carolina at Chapel Hill, NC, USA
- Ketene Diethyl Acetal 438

Saverio Florio — *University of Bari, Italy*
- Allylmagnesium Bromide 69

Pawel Fludzinski — *Lilly Research Centre, Windlesham, UK*
- Trichloroethylene 631

David C. Forbes — *University of Illinois, Urbana-Champaign, IL, USA*
- Dimethoxycarbenium Tetrafluoroborate 295

Georg Fráter — *Givaudan-Roure Research Ltd, Duebendorf, Switzerland*
- Ethyl 3-Hydroxybutanoate 385

Fillmore Freeman — *University of California, Irvine, CA, USA*
- Ethyl Cyanoacetate 379

Richard W. Friesen — *Merck Frosst Centre for Therapeutic Research, Quebec, Canada*
- Tetrakis(triphenylphosphine)palladium(0) 596

Alois Fürstner — *Max-Planck-Institut für Kohlenforschung, Mülheim, Germany*
- Ethyl Bromozincacetate 372

James R. Gage — *The Upjohn Company, Kalamazoo, MI, USA*
- *(R,R)*-1,2-Diphenyl-1,2-diaminoethane*N,N'*-Bis[3,5-bis(trifluoromethyl)-benzenesulfonamide] 336

Yinghong Gao — *Tulane University, New Orleans, LA, USA*
- (−)-8-Phenylmenthol 577

David S. Garvey — *Abbott Laboratories, Abbott Park, IL, USA*
- Di-*n*-butylboryl Trifluoromethane-sulfonate 252

Cesare Gennari — *Università di Milano, Italy*
- Ketene *t*-Butyldimethylsilyl Methyl Acetal 434

Manuka Ghosh — *University of Notre Dame, IN, USA*
- Chlorosulfonyl Isocyanate 196

Paul R. Giles — *University of Sheffield, UK*
- Dimethylchloromethyleneammonium Chloride 309

José M. González — *Universidad de Oviedo, Spain*
- Phenyl(trichloromethyl)mercury 582

Kurt. V. Gothelf — *Aarhus University, Denmark*
- Nitromethane 541

Mark T. Goulet — *Merck Research Laboratories, Rahway, NJ, USA*
- *B*-Allyldiisopinocampheylborane 67

Edward J. J. Grabowski — *Merck Research Laboratories, Rahway, NJ, USA*
- (*R*)-Pantolactone 568
- (*S*)-Ethyl Lactate 389

Helena Grennberg — *University of Uppsala, Sweden*
- Palladium(II) Acetate 547

D. Patrick Green — *The Dow Chemical Co., Midland, MI, USA*
- Phenyllithium 574

Ellen M. Leahy — *Affymax Research Institute, Palo Alto, CA, USA*
- Chlorotrimethylsilane 202

James W. Leahy — *University of California, Berkeley, CA, USA*
- Dichloroketene 256

Katherine L. Lee — *Massachusetts Institute of Technology, Cambridge, MA, USA*
- 1-Methyl-1-(trimethylsilyl)allene 520

Kevin C. Lee — *DuPont Agricultural Products, Newark, DE, USA*
- Methylenetriphenylphosphorane 499

Albert M. van Leusen — *Groningen University, The Netherlands*
- *p*-Tolylsulfonylmethyl Isocyanide 620

Daan van Leusen — *Groningen University, The Netherlands*
- *p*-Tolylsulfonylmethyl Isocyanide 620

Russell J. Linderman — *North Carolina State University, Raleigh, NC, USA*
- Tri-*n*-butylstannyllithium 628

Gérard Linstrumelle — *Ecole Normale Supérieure, Paris, France*
- Vinylmagnesium Bromide 684

Chiu-Hong Lin — *The Upjohn Company, Kalamazoo, MI, USA*
- Methyl Formate 509

Shou-Yuan Lin — *Clemson University, SC, USA*
- Lithium Divinylcuprate 459

Chin Liu — *Searle Research & Development, Skokie, IL, USA*
- Dimethylsulfoxonium Methylide 330

Shengrong Li — *Auburn University, AL, USA*
- Copper(II) Acetylacetonate 213

Xing-Ya Li — *University of Southern California, Los Angeles, CA, USA*
- Trifluoromethyltrimethylsilane 648

Tien-Yau Luh — *National Taiwan University, Taipei, Taiwan*
- Nickel(II) Chloride 536

Gerhard Maas — *University of Kaiserslautern, Germany*
- Benzenediazonium Tetrafluoroborate 83

David J. Madar — *Indiana University, Bloomington, IN, USA*
- Diisopropyl 2-Crotyl-1,3,2-dioxa-borolane-4,5-dicarboxylate 285

Lewis N. Mander — *The Australian National University, Canberra, Australia*
- Methyl Cyanoformate 492

Lawrence R. Marcin — *University of Illinois at Urbana-Champaign, IL, USA*
- *trans*-2,5-Dimethylpyrrolidine 325

Charles M. Marson — *University of Sheffield, UK*
- Dimethylchloromethyleneammonium Chloride 309

Bruce E. Maryanoff — *R. W. Johnson Pharmaceutical Research Institute, Spring House, PA, USA*
- Formylmethylenetriphenylphosphorane 404

Kazuo Matsumoto — *Tanabe Seiyaku Co., Osaka, Japan*
- Ethyl Isocyanoacetate 387

David F. McComsey — *R. W. Johnson Pharmaceutical Research Institute, Spring House, PA, USA*
- Formylmethylenetriphenylphosphorane 404

Mark E. McDonnell — *The R. W. Johnson Pharmaceutical Research Institute, Spring House, PA, USA*
- (Ethoxycarbonylmethylene)triphenyl-phosphorane 355

Patrick G. McDougal — *Reed College, Portland, OR, USA*
- 2-Chloroacrylonitrile 178

Jeffrey A. McKinney — *Zeneca Pharmaceuticals, Wilmington, DE, USA*
- Formaldehyde Dimethyl Thioacetal Monoxide 401

Mark C. McMills — *Ohio University, Athens, OH, USA*
- 1-Methoxy-1-(trimethylsilyloxy)propene 483

Albert I. Meyers — *Colorado State University, Fort Collins, CO, USA*
- (*S*)-*N*, *N*-Dimethyl-*N*′-(1-*t*-butoxy-3-methyl-2-butyl)formamidine 308
- *N*′-*t*-Butyl-*N*,*N*-dimethylformamidine 153

M. Mark Midland — *University of California, Riverside, CA, USA*
- Ethynylmagnesium Bromide 392
- Lithium Acetylide 446

Marvin J. Miller — *University of Notre Dame, IN, USA*
- Chlorosulfonyl Isocyanate 196

Terence N. Mitchell — *University of Dortmund, Germany*
- Hexabutyldistannane 411
- Hexamethyldistannane 413

J. Mittendorf — *Bayer, Wuppertal, Germany*
- (2*S*)-(+)-2,5-Dihydro-2-isopropyl-3,6-dimethoxypyrazine 275

Thomas M. Mitzel — *The Ohio State University, Columbus, OH, USA*
- Ketene 431

Christopher J. Moody — *Loughborough University of Technology, UK*
- Diethyl Diazomethylphosphonate 265

Alessandro F. Moretto — *Bayer Pharmaceuticals Division, West Haven, CT, USA*
- Vinyltributylstannane 685

William S. Mungall — *Hope College, Holland, MI, USA*
- Thexylborane 604

Shinji Murai — *Osaka University, Japan*
- Carbon Monoxide 169

Tetsuya Nagasawa — *Keio University, Yokohama, Japan*
- Triethylaluminum 633
- Trimethylaluminum 650

Takeshi Nakai — *Tokyo Institute of Technology, Japan*
- Diethylaluminum Cyanide 262

Eiichi Nakamura — *Tokyo Institute of Technology, Japan*
- 1-Ethoxy-1-(trimethylsilyloxy)-cyclopropane 362

G. K. Surya Prakash — *University of Southern California, Los Angeles, CA, USA*
- Trifluoromethyltrimethylsilane 648

P. Sivaramakrishnan Ramamoorthy — *The University of Michigan, Ann Arbor, MI, USA*
- Cyclopropyldiphenylsulfonium Tetrafluoroborate 241

Jorge Ramos-Veguilla — *University of Puerto Rico, Rio Piedras, Puerto Rico*
- Methoxymethylenetriphenylphosphorane 471

Terry L. Rathman — *FMC Corporation Lithium Division, Bessemer City, NC, USA*
- Methyllithium 510

Mundla S. Reddy — *University of Wisconsin-Milwaukee, WI, USA*
- Dimethyl 1,3-Acetonedicarboxylate 301

Manfred Regitz — *University of Kaiserslautern, Germany*
- *p*-Toluenesulfonyl Azide 608

Allen B. Reitz — *The R. W. Johnson Pharmaceutical Research Institute, Spring House, PA, USA*
- (Ethoxycarbonylmethylene)triphenyl-phosphorane 355

Lynn Resnick — *Brown University, Providence, RI, USA*
- *N,N*-Dimethylacetamide Dimethyl Acetal 299

Vichai Reutrakul — *Mahidol University, Bangkok, Thailand*
- *p*-Tolylsulfinylmethyllithium 616

Herman G. Richey, Jr. — *The Pennsylvania State University, University Park, PA, USA*
- Phenylmagnesium Bromide 576
- Methylmagnesium Bromide 512

Gérard Romeder — *Lonza, Basel, Switzerland*
- Diethyl Malonate 267
- Hydrogen Cyanide 415

Scott D. Rychnovsky — *University of Minnesota, Minneapolis, MN, USA*
- Zinc/Copper Couple 689

Toshiyasu Sakakura — *National Institute of Materials & Chemical Research, Tsukuba, Japan*
- Benzenediazonium Tetrafluoroborate 83

Tarek Sammakia — *University of Colorado, Boulder, CO, USA*
- Diazomethane 244

Mark E. Schnute — *Stanford University, CA, USA*
- 3-Hydroxyisoborneol 417

William J. Scott — *Bayer Pharmaceuticals Division, West Haven, CT, USA*
- Vinyltributylstannane 685

John F. Sebastian — *Miami University, Oxford, OH, USA*
- 1-Ethoxyvinyllithium 365

Dieter Seebach — *Eidgenössische Technische Hochschule Zürich, Switzerland*
- (*R,R*)-2-*t*-Butyl-5-methyl-1,3-dioxolan-4-one 160
- (*R*)-2-*t*-Butyl-6-methyl-4*H*-1,3-dioxin-4-one 158

Chi-Yung Shen — *National Chung-Hsing University, Taichung, Republic of China*
- 1,4-Benzoquinone 88

Reagent Formula Index

Subject Index

Reference Abbreviations

ABC	Agric. Biol. Chem.
AC(R)	Ann. Chim. (Rome)
ACR	Acc. Chem. Res.
ACS	Acta Chem. Scand.
AF	Arzneim.-Forsch.
AG	Angew. Chem.
AG(E)	Angew. Chem., Int. Ed. Engl.
AJC	Aust. J. Chem.
AK	Ark. Kemi
ANY	Ann. N. Y. Acad. Sci.
AP	Arch. Pharm. (Weinheim, Ger.)
B	Biochemistry
BAU	Bull. Acad. Sci. USSR, Div. Chem. Sci.
BBA	Biochim. Biophys. Acta
BCJ	Bull. Chem. Soc. Jpn.
BJ	Biochem. J.
BML	Bioorg. Med. Chem. Lett.
BSB	Bull. Soc. Chim. Belg.
BSF(2)	Bull. Soc. Chem. Fr. Part 2
C	Chimia
CA	Chem. Abstr.
CB	Ber. Dtsch. Chem. Ges./Chem. Ber.
CC	Chem. Commun./J. Chem. Soc., Chem. Commun.
CCC	Collect. Czech. Chem. Commun.
CED	J. Chem. Eng. Data
CI(L)	Chem. Ind. (London)
CJC	Can. J. Chem.
CL	Chem. Lett.
COS	Comprehensive Organic Synthesis
CPB	Chem. Pharm. Bull.
CR(C)	C. R. Hebd. Seances Acad. Sci., Ser. C
CRV	Chem. Rev.
CS	Chem. Ser.
CSR	Chem. Soc. Rev.
CZ	Chem.-Ztg.
DOK	Dokl. Akad. Nauk SSSR
E	Experientia
FES	Farmaco Ed. Sci.
FF	Fieser & Fieser
G	Gazz. Chim. Ital.
H	Heterocycles
HC	Heteroatom Chem.
HCA	Helv. Chim. Acta
IC	Inorg. Chem.
ICA	Inorg. Chim. Acta
IJC(B)	Indian J. Chem., Sect. B
IJS(B)	Int. J. Sulfur Chem., Part B
IZV	Izv. Akad. Nauk SSSR, Ser. Khim.
JACS	J. Am. Chem. Soc.
JBC	J. Biol. Chem.
JCP	J. Chem. Phys.
JCR(M)	J. Chem. Res. (M)
JCR(S)	J. Chem. Res. (S)
JCS	J. Chem. Soc.
JCS(C)	J. Chem. Soc. (C)
JCS(D)	J. Chem. Soc., Dalton Trans.
JCS(F)	J. Chem. Soc., Faraday Trans.
JCS(P1)	J. Chem. Soc., Perkin Trans. 1
JCS(P2)	J. Chem. Soc., Perkin Trans. 2
JFC	J. Fluorine Chem.
JGU	J. Gen. Chem. USSR (Engl. Transl.)
JHC	J. Heterocycl. Chem.
JIC	J. Indian Chem. Soc.
JMC	J. Med. Chem.
JMR	J. Magn. Reson.
JOC	J. Org. Chem.
JOM	J. Organomet. Chem.
JOU	J. Org. Chem. USSR (Engl. Transl.)
JPOC	J. Phys. Org. Chem.
JPP	J. Photochem. Photobiol.
JPR	J. Prakt. Chem.
JPS	J. Pharm. Sci.
KGS	Khim. Geterotsikl. Soedin.
LA	Justus Liebigs Ann. Chem./Liebigs Ann. Chem.
M	Monatsh. Chem.
MOC	Methoden Org. Chem. (Houben-Weyl)
MRC	Magn. Reson. Chem.
N	Naturwissenschaften
NJC	Nouv. J. Chim.
NKK	Nippon Kagaku Kaishi
OM	Organometallics
OMR	Org. Magn. Reson.
OPP	Org. Prep. Proced. Int.
OR	Org. React.
OS	Org. Synth.
OSC	Org. Synth., Coll. Vol.
P	Phytochemistry
PAC	Pure Appl. Chem.
PIA(A)	Proc. Indian Acad. Sci., Sect. A